| DIMENSION | METRIC | METRIC/ENGLISH |
|---|---|---|
| Power, heat transfer rate | $1\ \text{W} = 1\ \text{J/s}$<br>$1\ \text{kW} = 1000\ \text{W} = 1.341\ \text{hp}$<br>$1\ \text{hp}^{\ddagger} = 745.7\ \text{W}$ | $1\ \text{kW} = 3412.14\ \text{Btu/h}$<br>$\quad = 737.56\ \text{lbf} \cdot \text{ft/s}$<br>$1\ \text{hp} = 550\ \text{lbf} \cdot \text{ft/s} = 0.7068\ \text{Btu/s}$<br>$\quad = 42.41\ \text{Btu/min} = 2544.5\ \text{Btu/h}$<br>$\quad = 0.74570\ \text{kW}$<br>$1\ \text{boiler hp} = 33{,}475\ \text{Btu/h}$<br>$1\ \text{Btu/h} = 1.055056\ \text{kJ/h}$<br>$1\ \text{ton of refrigeration} = 200\ \text{Btu/min}$ |
| Pressure | $1\ \text{Pa} = 1\ \text{N/m}^2$<br>$1\ \text{kPa} = 10^3\ \text{Pa} = 10^{-3}\ \text{MPa}$<br>$1\ \text{atm} = 101.325\ \text{kPa} = 1.01325\ \text{bars}$<br>$\quad = 760\ \text{mmHg at } 0°\text{C}$<br>$\quad = 1.03323\ \text{kgf/cm}^2$<br>$1\ \text{mmHg} = 0.1333\ \text{kPa}$ | $1\ \text{Pa} = 1.4504 \times 10^{-4}\ \text{psia}$<br>$\quad = 0.020886\ \text{lbf/ft}^2$<br>$1\ \text{psia} = 144\ \text{lbf/ft}^2 = 6.894757\ \text{kPa}$<br>$1\ \text{atm} = 14.696\ \text{psia} = 29.92\ \text{inHg at } 30°\text{F}$<br>$1\ \text{inHg} = 3.387\ \text{kPa}$ |
| Specific heat | $1\ \text{kJ/kg} \cdot °\text{C} = 1\ \text{kJ/kg} \cdot \text{K}$<br>$\quad = 1\ \text{J/g} \cdot °\text{C}$ | $1\ \text{Btu/lbm} \cdot °\text{F} = 4.1868\ \text{kJ/kg} \cdot °\text{C}$<br>$1\ \text{Btu/lbmol} \cdot \text{R} = 4.1868\ \text{kJ/kmol} \cdot \text{K}$<br>$1\ \text{kJ/kg} \cdot °\text{C} = 0.23885\ \text{Btu/lbm} \cdot °\text{F}$<br>$\quad = 0.23885\ \text{Btu/lbm} \cdot \text{R}$ |
| Specific volume | $1\ \text{m}^3\text{/kg} = 1000\ \text{L/kg}$<br>$\quad = 1000\ \text{cm}^3\text{/g}$ | $1\ \text{m}^3\text{/kg} = 16.02\ \text{ft}^3\text{/lbm}$<br>$1\ \text{ft}^3\text{/lbm} = 0.062428\ \text{m}^3\text{/kg}$ |
| Temperature | $T(\text{K}) = T(°\text{C}) + 273.15$<br>$\Delta T(\text{K}) = \Delta T(°\text{C})$ | $T(\text{R}) = T(°\text{F}) + 459.67 = 1.8T(\text{K})$<br>$T(°\text{F}) = 1.8\ T(°\text{C}) + 32$<br>$\Delta T(°\text{F}) = \Delta T(\text{R}) = 1.8^{*}\ \Delta T(\text{K})$ |
| Thermal conductivity | $1\ \text{W/m} \cdot °\text{C} = 1\ \text{W/m} \cdot \text{K}$ | $1\ \text{W/m} \cdot °\text{C} = 0.57782\ \text{Btu/h} \cdot \text{ft} \cdot °\text{F}$ |
| Thermal resistance | $1°\text{C/W} = 1\ \text{K/W}$ | $1\ \text{K/W} = 0.52750°\text{F/h} \cdot \text{Btu}$ |
| Velocity | $1\ \text{m/s} = 3.60\ \text{km/h}$ | $1\ \text{m/s} = 3.2808\ \text{ft/s} = 2.237\ \text{mi/h}$<br>$1\ \text{mi/h} = 1.46667\ \text{ft/s}$<br>$1\ \text{mi/h} = 1.609\ \text{km/h}$ |
| Viscosity, dynamic | $1\ \text{kg/m} \cdot \text{s} = 1\ \text{N} \cdot \text{s/m}^2 = 1\ \text{Pa} \cdot \text{s} = 10\ \text{poise}$ | $1\ \text{kg/m} \cdot \text{s} = 2419.1\ \text{lbf/ft} \cdot \text{h}$<br>$\quad = 0.020886\ \text{lbf} \cdot \text{s/ft}^2$<br>$\quad = 5.8016 \times 10^{-6}\ \text{lbf} \cdot \text{h/ft}^2$ |
| Viscosity, kinematic | $1\ \text{m}^2\text{/s} = 10^4\ \text{cm}^2\text{/s}$<br>$1\ \text{stoke} = 1\ \text{cm}^2\text{/s} = 10^{-4}\ \text{m}^2\text{/s}$ | $1\ \text{m}^2\text{/s} = 10.764\ \text{ft}^2\text{/s} = 3.875 \times 10^4\ \text{ft}^2\text{/h}$<br>$1\ \text{m}^2\text{/s} = 10.764\ \text{ft}^2\text{/s}$ |
| Volume | $1\ \text{m}^3 = 1000\ \text{L} = 10^6\ \text{cm}^3\ (\text{cc})$ | $1\ \text{m}^3 = 6.1024 \times 10^4\ \text{in}^3 = 35.315\ \text{ft}^3$<br>$\quad = 264.17\ \text{gal (U.S.)}$<br>$1\ \text{U.S. gallon} = 231\ \text{in}^3 = 3.7854\ \text{L}$<br>$1\ \text{fl ounce} = 29.5735\ \text{cm}^3 = 0.0295735\ \text{L}$<br>$1\ \text{U.S. gallon} = 128\ \text{fl ounces}$ |

*Exact conversion factor between metric and English units.

‡Mechanical horsepower. The electrical horsepower is taken to be exactly 746 W.

D0221356

# Fundamentals of
# THERMAL-FLUID
# SCIENCES

# Fundamentals of
# THERMAL-FLUID
# SCIENCES

FOURTH
EDITION

**YUNUS A.
ÇENGEL**
*Department of
Mechanical
Engineering
University of Nevada,
Reno*

**JOHN M.
CIMBALA**
*Department of
Mechanical and
Nuclear Engineering
The Pennsylvania
State University*

**ROBERT H.
TURNER**
*Department of
Mechanical
Engineering
University of Nevada,
Reno*

Mc
Graw
Hill

*Connect
Learn
Succeed*™

The McGraw-Hill Companies

Connect
Learn
Succeed™

FUNDAMENTALS OF THERMAL-FLUID SCIENCES, FOURTH EDITION

This book is printed on acid-free paper.

1 2 3 4 5 6 7 8 9 0 RJE/RJE 1 0 9 8 7 6 5 4 3 2 1

ISBN 978-0-07-338020-9
MHID 0-07-338020-2

Vice President & Editor-in-Chief: *Marty Lange*
Vice President EDP/Central Publishing Services: *Kimberly Meriwether David*
Global Publisher: *Raghothaman Srinivasan*
Executive Editor: *Bill Stenquist*
Senior Marketing Manager: *Curt Reynolds*
Developmental Editor: *Lorraine K. Buczek*
Project Manager: *Robin A. Reed*
Design Coordinator: *Brenda A. Rolwes*
Cover Designer: *Studio Montage, St. Louis, Missouri*
Senior Photo Research Coordinator: *Lori Hancock*
Cover Image: *© Getty Images/RF*
Buyer: *Sherry L. Kane*
Media Project Manager: *Balaji Sundararaman*
Compositor: *RPK Editorial Services, Inc.*
Typeface: *10/12 Times Roman*
Printer: *R.R. Donnelley*

**Library of Congress Cataloging-in-Publication Data**

Çengel, Yunus A.
   Fundamentals of thermal-fluidsciences / Yunus A. Çengel, John M. Cimbala, Robert H. Turner. —
4th ed.
      p.   cm.
   Includes bibliographical references and index.
   ISBN 0-07-338020-2 (alk. paper)
   1.  Thermodynamics.  2.  Heat--Transmission.  3.  Fluid mechanics.  I.  Cimbala, John M. II. Title.
TJ265.C42 2011
621.402—dc22                                                              2010051756

www.mhhe.com

He who will not economize will have to agonize.
**—Confucius**

We must think as people of action and act as people of thought.
**—Henri Bergson**

The world is a dangerous place to live; not because of the people who are evil,
but because of the people who don't do anything about it.
**—Albert Einstein**

Since there is unanimity in what is good but conflict in what is better, sometimes good is better than the better.
**—Said Nursi**

Men die of fright and live of confidence.
**—Henry David Thoreau**

They may forget what you said, but they will never forget how you make them feel.
**—Carol Buchner**

Nothing will ever be attempted if all possible objections must first be overcome.
**—Samuel Johnson**

He who is devoid of the power to forgive is devoid of the power to love. There is some good in the worst of us
and some evil in the best of us. When we discover this, we are less prone to hate our enemies.
**—Martin Luther King, Jr.**

The art of being wise is knowing what to overlook.
**—William James**

A leader is one who knows the way, goes the way, and shows the way.
**—John C. Maxwell**

Obstacles are those frightful things you see when you take your eyes off your goal.
**—Henry Ford**

Never worry about numbers. Help one person at a time, and always start with the person nearest you.
**—Mother Teresa**

The highest reward for a man's toil is not what he gets for it but what he becomes by it.
**—John Ruskin**

And in the end it's not the years in your life that count. It's the life in your years.
**—Abraham Lincoln**

# McGRAW-HILL DIGITAL OFFERINGS INCLUDE

## McGRAW-HILL CREATE™

Craft your teaching resources to match the way you teach! With McGraw-Hill Create™, www.mcgrawhillcreate.com, you can easily rearrange chapters, combine material from other content sources, and quickly upload content you have written like your course syllabus or teaching notes. Find the content you need in Create by searching through thousands of leading McGraw-Hill textbooks. Arrange your book to fit your teaching style. Create even allows you to personalize your book's appearance by selecting the cover and adding your name, school, and course information. Order a Create book and you'll receive a complimentary print review copy in 3–5 business days or a complimentary electronic review copy (eComp) via email in minutes. Go to www.mcgrawhillcreate.com today and register to experience how McGraw-Hill Create™ empowers you to teach *your* students *your* way.

## McGRAW-HILL HIGHER EDUCATION AND BLACKBOARD HAVE TEAMED UP

Blackboard, the Web-based course-management system, has partnered with McGraw-Hill to better allow students and faculty to use online materials and activities to complement face-to-face teaching. Blackboard features exciting social learning and teaching tools that foster more logical, visually impactful and active learning opportunities for students. You'll transform your closed-door classrooms into communities where students remain connected to their educational experience 24 hours a day.

This partnership allows you and your students access to McGraw-Hill's Create™ right from within your Blackboard course—all with one single sign-on. McGraw-Hill and Blackboard can now offer you easy access to industry leading technology and content, whether your campus hosts it, or we do. Be sure to ask your local McGraw-Hill representative for details.

## ELECTRONIC TEXTBOOK OPTIONS

This text is offered through CourseSmart for both instructors and students. CourseSmart is an online resource where students can purchase the complete text online at almost half the cost of a traditional text. Purchasing the eTextbook allows students to take advantage of CourseSmart's web tools for learning, which include full text search, notes and highlighting, and email tools for sharing notes between classmates. To learn more about CourseSmart options, contact your sales representative or visit www.CourseSmart.com.

**Yunus A. Çengel** is Professor Emeritus of Mechanical Engineering at the University of Nevada, Reno. He received his B.S. in mechanical engineering from Istanbul Technical University and his M.S. and Ph.D. in mechanical engineering from North Carolina State University. His areas of interest are renewable energy, energy efficiency, energy policies, heat transfer enhancement, and engineering education. He served as the director of the Industrial Assessment Center (IAC) at the University of Nevada, Reno, from 1996 to 2000. He has led teams of engineering students to numerous manufacturing facilities in Northern Nevada and California to perform industrial assessments, and has prepared energy conservation, waste minimization, and productivity enhancement reports for them. He has also served as an advisor for various government organizations and corporations.

Dr. Çengel is also the author or coauthor of the widely adopted textbooks *Thermodynamics: An Engineering Approach* (7th ed., 2011), *Heat and Mass Transfer: Fundamentals and Applications* (4th ed., 2011), *Introduction to Thermodynamics and Heat Transfer* (2nd ed., 2008), *Fluid Mechanics: Fundamentals and Applications* (2nd ed., 2010), and *Essentials of Fluid Mechanics: Fundamentals and Applications* (1st ed., 2008), all published by McGraw-Hill. Some of his textbooks have been translated into Chinese, Japanese, Korean, Thai, Spanish, Portuguese, Turkish, Italian, Greek, and French.

Dr. Çengel is the recipient of several outstanding teacher awards, and he has received the ASEE Meriam/Wiley Distinguished Author Award for excellence in authorship in 1992 and again in 2000. Dr. Çengel is a registered Professional Engineer in the State of Nevada, and is a member of the American Society of Mechanical Engineers (ASME) and the American Society for Engineering Education (ASEE).

**John M. Cimbala** is Professor of Mechanical Engineering at The Pennsylvania State University, University Park. He received his B.S. in Aerospace Engineering from Penn State and his M.S. in Aeronautics from the California Institute of Technology (CalTech). He received his Ph.D. in Aeronautics from CalTech in 1984 under the supervision of Professor Anatol Roshko, to whom he will be forever grateful. His research areas include experimental and computational fluid mechanics and heat transfer, turbulence, turbulence modeling, turbomachinery, indoor air quality, and air pollution control. During the academic year 1993–94, Professor Cimbala took a sabbatical leave from the University and worked at NASA Langley Research Center, where he advanced his knowledge of computational fluid dynamics (CFD) and turbulence modeling.

Dr. Cimbala is the co-author of three other textbooks: *Fluid Mechanics: Fundamentals and Applications* (2010), *Essentials of Fluid Mechanics: Fundamentals and Applications* (2008), both published by McGraw-Hill, and *Indoor Air Quality Engineering: Environmental Health and Control of Indoor Pollutants* (2003), published by Marcel-Dekker, Inc. He has also contributed to parts of other books and is the author or co-author of dozens of journal and conference papers. More information can be found at www.mne.psu.edu/cimbala.

Professor Cimbala is the recipient of several outstanding teaching awards and views his book writing as an extension of his love of teaching. He is a member of the American Institute of Aeronautics and Astronautics (AIAA), the American Society of Mechanical Engineers (ASME), the American Society for Engineering Education (ASEE), and the American Physical Society (APS). In 2009 he was chosen to be an ASME Fellow.

**Robert H. Turner** is Professor Emeritus of Mechanical Engineering at the University of Nevada, Reno (UNR). He earned a B.S. and M.S. from the University of California at Berkeley, and his Ph.D. from UCLA, all in mechanical engineering. He worked in industry for 18 years, including nine years at Cal Tech's Jet Propulsion Laboratory (JPL). Dr. Turner then joined the University of Nevada in 1983. His research interests include solar and renewable energy applications, thermal sciences, and energy conservation. He established and was the first director of the Industrial Assessment Center at the University of Nevada.

For 20 years Dr. Turner has designed the solar components of many houses. In 1994–95, in a cooperative effort between UNR and Erciyes University in Kayseri, Turkey, he designed and oversaw construction of the fully instrumented Solar Research Laboratory at Erciyes University, featuring 130 square meters of site-integrated solar collectors. His interest in applications has led Dr. Turner to maintain an active consulting practice.

Dr. Turner is a registered Professional Engineer and is a member of the American Society of Mechanical Engineers (ASME) and the American Society of Heating, Refrigeration, and Air Conditioning Engineers (ASHRAE).

# CONTENTS

# CHAPTER TWENTY

## NATURAL CONVECTION 831

# CHAPTER TWENTY-ONE

## RADIATION HEAT TRANSFER 869

# CHAPTER TWENTY-TWO

## HEAT EXCHANGERS 927

# APPENDIX 1

## PROPERTY TABLES AND CHARTS (SI UNITS) 973

# APPENDIX 2

## PROPERTY TABLES AND CHARTS (ENGLISH UNITS) 1017

# P R E F A C E

## BACKGROUND

This text is an abbreviated version of standard thermodynamics, fluid mechanics, and heat transfer texts, covering topics that the engineering students are most likely to need in their professional lives. The thermodynamics portion of this text is based on the text *Thermodynamics: An Engineering Approach* by Y. A. Çengel and M. A. Boles, the fluid mechanics portion is based on *Fluid Mechanics: Fundamentals and Applications* by Y. A. Çengel and J. M. Cimbala, and the heat transfer portion is based on *Heat and Mass Transfer: Fundamentals and Applications* by Y. A. Çengel and A. J. Ghajar, all published by McGraw-Hill. Most chapters are practically independent of each other and can be covered in any order. The text is well-suited for curriculums that have a common introductory course or a two-course sequence on thermal-fluid sciences or on thermodynamics and heat transfer.

It is recognized that all topics of thermodynamics, fluid mechanics, and heat transfer cannot be covered adequately in a typical three-semester-hour course, and, therefore, sacrifices must be made from the depth if not from the breadth. Selecting the right topics and finding the proper level of depth and breadth are no small challenge for the instructors, and this text is intended to serve as the ground for such selection. Students in a combined thermal-fluids course can gain a basic understanding of energy and energy interactions, various mechanisms of heat transfer, and fundamentals of fluid flow. Such a course can also instill in students the confidence and the background to do further reading of their own and to be able to communicate effectively with specialists in thermal-fluid sciences.

## OBJECTIVES

This book is intended for use as a textbook in a first course in thermal-fluid sciences for undergraduate engineering students in their junior or senior year, and as a reference book for practicing engineers. Students are assumed to have an adequate background in calculus, physics, and engineering mechanics. The objectives of this text are

- To cover the *basic principles* of thermodynamics, fluid mechanics, and heat transfer.
- To present numerous and diverse real-world *engineering examples* to give students a feel for how thermal-fluid sciences are applied in engineering practice.
- To develop an *intuitive understanding* of thermal-fluid sciences by emphasizing the physics and physical arguments.

The text contains sufficient material to give instructors flexibility and to accommodate their preferences on the right blend of thermodynamics, fluid mechanics, and heat transfer for their students. By careful selection of topics, an instructor can spend one-third, one-half, or two-thirds of the course on thermodynamics and the rest on selected topics of fluid mechanics and heat transfer.

## PHILOSOPHY AND GOAL

The philosophy that contributed to the warm reception of the first edition of this book has remained unchanged. Namely, our goal is to offer an engineering textbook that

- Communicates directly to the minds of tomorrow's engineers in a *simple yet precise* manner.
- Leads students towards a clear understanding and firm grasp of the *basic principles* of thermal-fluid sciences.
- Encourages *creative thinking* and development of a *deeper understanding* and *intuitive feel* for thermal-fluid sciences.
- Is *read* by students with *interest* and *enthusiasm* rather than being used as an aid to solve problems.

Special effort has been made to appeal to readers' natural curiosity and to help students explore the various facets of the exciting subject area of thermal-fluid sciences. The enthusiastic response we received from the users of the previous editions—from small colleges to large universities all over the world—indicates that our objectives have largely been achieved. It is our philosophy that the best way to learn is by practice. Therefore, special effort is made throughout the book to reinforce material that was presented earlier.

Yesterday's engineers spent a major portion of their time substituting values into the formulas and obtaining numerical results. However, now formula manipulations and number crunching are being left to computers. Tomorrow's engineer will need to have a clear understanding and a firm grasp of the *basic principles* so that he or she can understand even the most complex problems, formulate them, and interpret the results. A conscious effort is made to emphasize these basic principles while also providing students with a look at how modern tools are used in engineering practice.

## NEW IN THIS EDITION

All the popular features of the previous editions are retained while new ones are added. The main body of the text remains largely unchanged except that an existing chapter is removed and a new chapter is added. The most significant changes in this edition are highlighted below.

### A NEW CHAPTER ON POWER AND REFRIGERATION CYCLES

The new Chapter 9 on Power and Refrigeration Cycles, which was removed in the previous edition and was made available as a web chapter, is put back in this edition by the request of a large number of users. This chapter exposes students to the foundations of power generation and refrigeration in a well-ordered and compact manner. To make room for this change, the Chapter 11 on Fluid Kinematics, which the user surveys indicated to be the least used chapter in the previous edition, is removed.

### UPGRADED ARTWORK

To enhance real-life apperance, a large number of line artwork in the entire text is upgaded to realistic three-dimensional figures.

### NEW EXAMPLES AND END-OF-CHAPTER PROBLEMS

A large number of the problems in the text are modified and many problems are replaced by new ones. Also, some of the solved examples are replaced by new ones.

### A THOROUGH UPDATE OF ALL CHAPTERS

The entire text is thoroughly reviewed, and a large number of modifications are made in content and organization to make the text more relevant and current.

## FEATURED IN THIS EDITION

### EARLY INTRODUCTION OF THE FIRST LAW OF THERMODYNAMICS

The first law of thermodynamics is now introduced early in Chapter 3, "Energy, Energy Transfer, and General Energy Analysis." This introductory chapter sets the framework of establishing a general understanding of various forms of energy, mechanisms of energy transfer, the concept of energy balance, thermo-economics, energy conversion, and conversion efficiency using familiar settings that involve mostly electrical and mechanical forms of energy. It also exposes students to some exciting real-world applications of thermodynamics early in the course, and helps them establish a sense of the monetary value of energy.

### UPDATED STEAM AND REFRIGERANT-134A TABLES

Updated steam and refrigerant-134a tables based on the most current property data from EES are used throughout the book. An added advantage of this update is that students will get the same result when solving problems whether they use steam or refrigerant properties from EES or property tables in the Appendices.

### LEARNING OBJECTIVES

Each chapter begins with an *overview* of the material to be covered and chapter-specific *learning objectives* to introduce the material and to set goals, and ends with a summary of key concepts and equations.

## LEARNING TOOLS

### EMPHASIS ON PHYSICS

A distinctive feature of this book is its emphasis on the physical aspects of subject matter in addition to mathematical representations and manipulations. The authors believe that the emphasis in undergraduate education should remain on *developing a sense of underlying physical mechanisms* and a *mastery of solving practical problems* that an engineer is likely to face in the real world. Developing an intuitive understanding should also make the course a more motivating and worthwhile experience for the students.

## EFFECTIVE USE OF ASSOCIATION

An observant mind should have no difficulty understanding engineering sciences. After all, the principles of engineering sciences are based on our *everyday experiences* and *experimental observations.* A more physical, intuitive approach is used throughout this text. Frequently, *parallels are drawn* between the subject matter and students' everyday experiences so that they can relate the subject matter to what they already know.

## SELF-INSTRUCTING

The material in the text is introduced at a level that an average student can follow comfortably. It speaks to students, not over students. In fact, it is *self-instructive.* Noting that the principles of science are based on experimental observations, most of the derivations in this text are largely based on physical arguments, and thus they are easy to follow and understand.

## EXTENSIVE USE OF ARTWORK

Figures are important learning tools that help the students "get the picture." The text makes effective use of graphics, and it contains a great number of figures and illustrations. Figures attract attention and stimulate curiosity and interest. Some of the figures in this text are intended to serve as a means of emphasizing some key concepts that would otherwise go unnoticed; some serve as page summaries.

## CHAPTER OPENERS AND SUMMARIES

Each chapter begins with an overview of the material to be covered and chapter objectives. A *summary* is included at the end of each chapter for a quick review of basic concepts and important relations.

## NUMEROUS WORKED-OUT EXAMPLES

Each chapter contains several worked-out *examples* that clarify the material and illustrate the use of the basic principles. An *intuitive* and *systematic* approach is used in the solution of the example problems, with particular attention to the proper use of units.

## A WEALTH OF REAL-WORLD END-OF-CHAPTER PROBLEMS

The end-of-chapter problems are grouped under specific topics in the order they are covered to make problem selection easier for both instructors and students. Within each group of problems are *Concept Questions*, indicated by "C" to check the students' level of understanding of basic concepts. The problems under *Review Problems* are more comprehensive in nature and are not directly tied to any specific section of a chapter—in some cases they require review of material learned in previous chapters. The problems under the *Design and Essay Problems* title are intended to encourage students to make engineering judgments, to conduct independent exploration of topics of interest, and to communicate their findings in a professional manner. Several economics- and safety-related problems are incorporated throughout to enhance cost and safety awareness among engineering students. Answers to selected problems are listed immediately following the problem for convenience to students.

## A SYSTEMATIC SOLUTION PROCEDURE

A well-structured approach is used in problem solving while maintaining an informal conversational style. The problem is first stated and the objectives are identified, and the assumptions made are stated together with their justifications.

The properties needed to solve the problem are listed separately. Numerical values are used together with their units to emphasize that numbers without units are meaningless, and unit manipulations are as important as manipulating the numerical values with a calculator. The significance of the findings is discussed following the solutions. This approach is also used consistently in the solutions presented in the Instructor's Solutions Manual.

## RELAXED SIGN CONVENTION

The use of a formal sign convention for heat and work is abandoned as it often becomes counterproductive. A physically meaningful and engaging approach is adopted for interactions instead of a mechanical approach. Subscripts "in" and "out," rather than the plus and minus signs, are used to indicate the directions of interactions.

## A CHOICE OF SI ALONE OR SI / ENGLISH UNITS

In recognition of the fact that English units are still widely used in some industries, both SI and English units are used in this text, with an emphasis on SI. The material in this text can be covered using combined SI/English units or SI units alone, depending on the preference of the instructor. The property tables and charts in the appendices are presented in both units, except the ones that involve dimensionless quantities. Problems, tables, and charts in English units are designated by "E" after the number for easy recognition, and they can be ignored easily by the SI users.

## CONVERSION FACTORS

Frequently used conversion factors and physical constants are listed on the inner cover pages of the text for easy reference.

# SUPPLEMENTS

The following supplements are available to the adopters of the book.

# STUDENT RESOURCE DVD
## ENGINEERING EQUATION SOLVER (EES)

Packaged free with every new text, the Student Resource DVD contains the Limited Academic Version of EES (Engineering Equation Solver) software with scripted solutions to selected text problems.

Developed by Sanford Klein and William Beckman from the University of Wisconsin—Madison, this software combines equation-solving capability and engineering property data. EES can do optimization, parametric analysis, and linear and nonlinear regression, and provides publication-quality plotting capabilities. Thermodynamics and transport properties for air, water, and many other fluids are built in, and EES allows the user to enter property data or functional relationships. It requires less time than a calculator for data entry and allows more time for thinking critically about modeling and solving engineering problems. Look for the EES icons in the homework problems sections of the text.

## TEXT WEBSITE (www.mhhe.com/cengel)

Web support is provided for the text on the text specific website. Visit this web site for general text information, errata, and author information. The site also includes resources for students including a list of helpful web links. The instructor side of the site includes the solutions manual, the text's images in PowerPoint form, and more!

## COSMOS (Available to Instructors Only)

McGraw-Hill's COSMOS (Complete Online Solutions Manual Organization System) allows instructors to streamline the creation of assignments, quizzes, and tests by using problems and solutions from the textbook, as well as their own custom material. COSMOS is now available online at http://cosmos.mhhe.com/

# ACKNOWLEDGMENTS

We would like to acknowledge with appreciation the numerous and valuable comments, suggestions, criticisms, and praise of these academic evaluators:

Alexis Abramson
*Case Western Reserve University*

David Novo
*McMaster University*

James Patrick Abulencia
*Manhattan College*

Mahadevan Padmanabhan
*Worcester Polytechnic Institute*

Suresh Chandra
*North Carolina A&T State University*

Narender P. Reddy
*University of Akron*

Paul Gannon
*Montana State University*

Brian Sangeorzan
*Oakland University*

Bill Hutzel
*Purdue University*

Udyavar Satyanand
*University of Texas at Arlington*

Thomas M. Kiehne
*University of Texas at Austin*

Mark Snyder
*Illinois Institute of Technology*

James Mayhew
*Rose-Hulman Inst of Tech*

Kambiz Vafai
*University of California, Riverside*

Their suggestions have greatly helped to improve the quality of this text. Special thanks are due to Mehmet Kanoglu of the University of Gaziantep, Turkey, for his valuable contributions and his critical review of the manuscript and for his special attention to accuracy and detail. We also would like to thank our students who provided plenty of feedback from their perspectives. Finally, we would like to express our appreciation to our wives and our children for their continued patience, understanding, and support throughout the preparation of this text.

Yunus A. Çengel
John M. Cimbala
Robert H. Turner

# INTRODUCTION AND OVERVIEW

**M**any engineering systems involve the transfer, transport, and conversion of energy, and the sciences that deal with these subjects are broadly referred to as *thermal-fluid sciences*. Thermal-fluid sciences are usually studied under the subcategories of *thermodynamics, heat transfer, and fluid mechanics*. We start this chapter with an overview of these sciences, and give some historical background. Then we review the unit systems that will be used, and discuss dimensional homogeneity. We then present an intuitive systematic *problem solving technique* that can be used as a model in solving engineering problems, followed by a discussion of the proper place of software packages in engineering. Finally, we discuss accuracy and significant digits in engineering measurements and calculations.

## Objectives

Objectives of this chapter are to:

- Be acquainted with the engineering sciences thermodynamics, heat transfer, and fluid mechanics, and understand the basic concepts of thermal-fluid sciences.

- Be comfortable with the metric SI and English units commonly used in engineering.

- Develop an intuitive systematic problem-solving technique.

- Learn the proper use of software packages in engineering.

- Develop an understanding of accuracy and significant digits in calculations.

# 1–1 · INTRODUCTION TO THERMAL-FLUID SCIENCES

The word *thermal* stems from the Greek word *therme,* which means *heat.* Therefore, thermal sciences can loosely be defined as the sciences that deal with heat. The recognition of different forms of energy and its transformations has forced this definition to be broadened. Today, the physical sciences that deal with energy and the transfer, transport, and conversion of energy are usually referred to as **thermal-fluid sciences** or just **thermal sciences**. Traditionally, the thermal-fluid sciences are studied under the subcategories of thermodynamics, heat transfer, and fluid mechanics. In this book we present the basic principles of these sciences, and apply them to situations that engineers are likely to encounter in their practice.

The design and analysis of most thermal systems such as power plants, automotive engines, and refrigerators involve all categories of thermal-fluid sciences as well as other sciences (Fig. 1–1). For example, designing the radiator of a car involves the determination of the amount of energy transfer from a knowledge of the properties of the coolant using *thermodynamics,* the determination of the size and shape of the inner tubes and the outer fins using *heat transfer,* and the determination of the size and type of the water pump using *fluid mechanics.* Of course the determination of the materials and the thickness of the tubes requires the use of material science as well as strength of materials. The reason for studying different sciences separately is simply to facilitate learning without being overwhelmed. Once the basic principles are mastered, they can then be synthesized by solving comprehensive real-world practical problems. But first we present an overview of thermal-fluid sciences.

**FIGURE 1–1**

The design of many engineering systems, such as this solar hot water system, involves thermal-fluid sciences.

## Application Areas of Thermal-Fluid Sciences

All activities in nature involve some interaction between energy and matter; thus it is hard to imagine an area that does not relate to thermal-fluid sciences in some manner. Therefore, developing a good understanding of basic principles of thermal-fluid sciences has long been an essential part of engineering education.

Thermal-fluid sciences are commonly encountered in many engineering systems and other aspects of life, and one does not need to go very far to see some application areas of them. In fact, one does not need to go anywhere. The heart is constantly pumping blood to all parts of the human body, various energy conversions occur in trillions of body cells, and the body heat generated is constantly rejected to the environment. The human comfort is closely tied to the rate of this metabolic heat rejection. We try to control this heat transfer rate by adjusting our clothing to the environmental conditions. Also, any defect in the heart and the circulatory system is a major cause for alarm.

Other applications of thermal-fluid sciences are right where one lives. An ordinary house is, in some respects, an exhibition hall filled with wonders of thermal-fluid sciences. Many ordinary household utensils and appliances are designed, in whole or in part, by using the principles of thermal-fluid sciences. Some examples include the electric or gas range, the heating and air-conditioning systems, the refrigerator, the humidifier, the pressure cooker, the

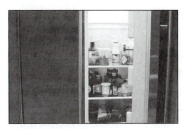

Refrigeration systems
© The McGraw-Hill Companies,
Inc/Jill Braaten, photographer.

Boats
© Vol. 5/Photo Disc/Getty RF.

Aircraft and spacecraft
© Vol. 1/Photo Disc/Getty RF.

Power plants
© Vol. 57/Photo Disc/Getty RF.

Human body
© Vol. 110/Photo Disc/Getty RF.

Cars
Photo by John M. Cimbala.

Wind turbines
© Vol. 17/Photo Disc/Getty RF.

Air conditioning systems
© The McGraw-Hill Companies,
Inc/Jill Braaten, photographer.

Industrial applications
Courtesy UMDE Engineering, Contracting,
and Trading. Used by permission.

**FIGURE 1–2**
Some application areas of thermal-fluid sciences.

water heater, the shower, the iron, the plumbing and sprinkling systems, and even the computer, the TV, and the DVD player. On a larger scale, thermal-fluid sciences play a major part in the design and analysis of automotive engines, rockets, jet engines, and conventional or nuclear power plants, solar collectors, the transportation of water, crude oil, and natural gas, the water distribution systems in cities, and the design of vehicles from ordinary cars to airplanes (Fig. 1–2). The energy-efficient home that you may be living in, for example, is designed on the basis of minimizing heat loss in winter and heat gain in summer. The size, location, and the power input of the fan of your computer is also selected after a thermodynamic, heat transfer, and fluid flow analysis of the computer.

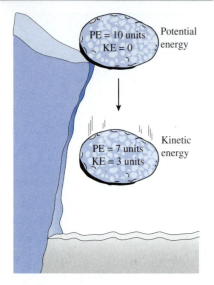

**FIGURE 1–3**
Energy cannot be created or destroyed; it can only change forms (the first law).

**FIGURE 1–4**
Conservation of energy principle for the human body.

**FIGURE 1–5**
Heat flows in the direction of decreasing temperature.

# 1–2 · THERMODYNAMICS

Thermodynamics can be defined as the science of *energy*. Although everybody has a feeling of what energy is, it is difficult to give a precise definition for it. Energy can be viewed as the ability to cause changes.

The name *thermodynamics* stems from the Greek words *therme* (heat) and *dynamis* (power), which is most descriptive of the early efforts to convert heat into power. Today the same name is broadly interpreted to include all aspects of energy and energy transformations including power generation, refrigeration, and relationships among the properties of matter.

One of the most fundamental laws of nature is the **conservation of energy principle**. It simply states that during an interaction, energy can change from one form to another but the total amount of energy remains constant. That is, energy cannot be created or destroyed. A rock falling off a cliff, for example, picks up speed as a result of its potential energy being converted to kinetic energy (Fig. 1–3). The conservation of energy principle also forms the backbone of the diet industry: A person who has a greater energy input (food) than energy output (exercise) will gain weight (store energy in the form of fat), and a person who has a smaller energy input than output will lose weight (Fig. 1–4). The change in the energy content of a body or any other system is equal to the difference between the energy input and the energy output, and the energy balance is expressed as $E_{in} - E_{out} = \Delta E$.

The **first law of thermodynamics** is simply an expression of the conservation of energy principle, and it asserts that *energy* is a thermodynamic property. The **second law of thermodynamics** asserts that energy has *quality* as well as *quantity,* and actual processes occur in the direction of decreasing quality of energy. For example, a cup of hot coffee left on a table eventually cools, but a cup of cool coffee in the same room never gets hot by itself (Fig. 1–5). The high-temperature energy of the coffee is degraded (transformed into a less useful form at a lower temperature) once it is transferred to the surrounding air.

Although the principles of thermodynamics have been in existence since the creation of the universe, thermodynamics did not emerge as a science until the construction of the first successful atmospheric steam engines in England by Thomas Savery in 1697 and Thomas Newcomen in 1712. These engines were very slow and inefficient, but they opened the way for the development of a new science.

The first and second laws of thermodynamics emerged simultaneously in the 1850s, primarily out of the works of William Rankine, Rudolph Clausius, and Lord Kelvin (formerly William Thomson). The term *thermodynamics* was first used in a publication by Lord Kelvin in 1849. The first thermodynamic textbook was written in 1859 by William Rankine, a professor at the University of Glasgow.

It is well-known that a substance consists of a large number of particles called *molecules*. The properties of the substance naturally depend on the behavior of these particles. For example, the pressure of a gas in a container is the result of momentum transfer between the molecules and the walls of the container. However, one does not need to know the behavior of the gas particles to determine the pressure in the container. It would be sufficient to attach a pressure gage to the container. This macroscopic approach to the study of thermodynamics that does not require a knowledge of the behavior of individual particles is called **classical thermodynamics**. It provides a direct and easy way to the solution of engineering problems. A more elaborate approach, based on the average behavior of large groups of individual particles, is called **statistical thermodynamics**. This microscopic approach is rather involved and is used in this text only in the supporting role.

## 1–3 ■ HEAT TRANSFER

We all know from experience that a cold canned drink left in a room warms up and a warm canned drink left in a refrigerator cools down. This is accomplished by the transfer of *energy* from the warm medium to the cold one. The energy transfer is always from the higher temperature medium to the lower temperature one, and the energy transfer stops when the two mediums reach the same temperature.

Energy exists in various forms. In heat transfer we are primarily interested in **heat**, which is *the form of energy that can be transferred from one system to another as a result of temperature difference.* The science that deals with the determination of the *rates* of such energy transfers is **heat transfer**.

You may be wondering why we need to undertake a detailed study on heat transfer. After all, we can determine the amount of heat transfer for any system undergoing any process using a thermodynamic analysis alone. The reason is that thermodynamics is concerned with the *amount* of heat transfer as a system undergoes a process from one equilibrium state to another, and it gives no indication about *how long* the process will take. A thermodynamic analysis simply tells us how much heat must be transferred to realize a specified change of state to satisfy the conservation of energy principle.

In practice we are more concerned about the rate of heat transfer (heat transfer per unit time) than we are with the amount of it. For example, we can determine the amount of heat transferred from a thermos bottle as the hot coffee inside cools from 90°C to 80°C by a thermodynamic analysis alone. But a typical user or designer of a thermos bottle is primarily interested in *how long* it will be before the hot coffee inside cools to 80°C, and a thermodynamic analysis cannot answer this question. Determining the rates of heat transfer to or from a system and thus the times of heating or cooling, as well as the variation of the temperature, is the subject of *heat transfer* (Fig. 1–6).

Thermodynamics deals with equilibrium states and changes from one equilibrium state to another. Heat transfer, on the other hand, deals with systems that lack thermal equilibrium, and thus it is a *nonequilibrium* phenomenon. Therefore, the study of heat transfer cannot be based on the principles of thermodynamics alone. However, the laws of thermodynamics lay the framework for the science of heat transfer. The *first law* requires that the rate of energy transfer into a system be equal to the rate of increase of the energy of that system. The *second law* requires that heat be transferred in the direction of decreasing temperature (Fig. 1–7). This is like a car parked on an inclined road must go downhill in the direction of decreasing elevation when its brakes are released. It is also analogous to the electric current flowing in the direction of decreasing voltage or the fluid flowing in the direction of decreasing total pressure.

**FIGURE 1–6**

We are normally interested in how long it takes for the hot coffee in a thermos bottle to cool to a certain temperature, which cannot be determined from a thermodynamic analysis alone.

**FIGURE 1–7**

Heat flows in the direction of decreasing temperature.

The basic requirement for heat transfer is the presence of a *temperature difference*. There can be no net heat transfer between two bodies that are at the same temperature. The temperature difference is the *driving force* for heat transfer, just as the *voltage difference* is the driving force for electric current flow and *pressure difference* is the driving force for fluid flow. The rate of heat transfer in a certain direction depends on the magnitude of the *temperature gradient* (the temperature difference per unit length or the rate of change of temperature) in that direction. The larger the temperature gradient, the higher the rate of heat transfer.

## 1–4 ▪ FLUID MECHANICS

**Mechanics** is the oldest physical science that deals with both stationary and moving bodies under the influence of forces. The branch of mechanics that deals with bodies at rest is called **statics**, while the branch that deals with bodies in motion is called **dynamics**. The subcategory **fluid mechanics** is defined as the science that deals with the behavior of fluids at rest (*fluid statics*) or in motion (*fluid dynamics*), and the interaction of fluids with solids or other fluids at the boundaries. Fluid mechanics is also referred to as **fluid dynamics** by considering fluids at rest as a special case of motion with zero velocity (Fig. 1–8).

Fluid mechanics itself is also divided into several categories. The study of the motion of fluids that can be approximated as incompressible (such as liquids, especially water, and gases at low speeds) is usually referred to as **hydrodynamics**. A subcategory of hydrodynamics is **hydraulics**, which deals with liquid flows in pipes and open channels. **Gas dynamics** deals with the flow of fluids that undergo significant density changes, such as the flow of gases through nozzles at high speeds. The category **aerodynamics** deals with the flow of gases (especially air) over bodies such as aircraft, rockets, and automobiles at high or low speeds. Some other specialized categories such as **meteorology**, **oceanography**, and **hydrology** deal with naturally occurring flows.

You will recall from physics that a substance exists in three primary phases: solid, liquid, and gas. (At very high temperatures, it also exists as plasma.) A substance in the liquid or gas phase is referred to as a **fluid**. Distinction between a solid and a fluid is made on the basis of the substance's ability to resist an applied shear (or tangential) stress that tends to change its shape. A solid can resist an applied shear stress by deforming, whereas a *fluid deforms continuously under the influence of a shear stress*, no matter how small. In solids, stress is proportional to *strain*, but in fluids, stress is proportional to *strain rate*. When a constant shear force is applied, a solid eventually stops deforming at some fixed strain angle, whereas a fluid never stops deforming and approaches a constant *rate* of strain.

Consider a rectangular rubber block tightly placed between two plates. As the upper plate is pulled with a force $F$ while the lower plate is held fixed, the rubber block deforms, as shown in Fig. 1–9. The angle of deformation $\alpha$ (called the *shear strain* or *angular displacement*) increases in proportion to the applied force $F$. Assuming there is no slip between the rubber and the plates, the upper surface of the rubber is displaced by an amount equal to the displacement of the upper plate while the lower surface remains stationary.

**FIGURE 1–8**
Fluid mechanics deals with liquids and gases in motion or at rest.
© *Vol. 16/Photo Disc.*

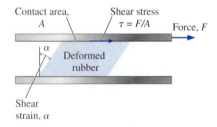

**FIGURE 1–9**
Deformation of a rubber block placed between two parallel plates under the influence of a shear force. The shear stress shown is that on the rubber—an equal but opposite shear stress acts on the upper plate.

In equilibrium, the net force acting on the upper plate in the horizontal direction must be zero, and thus a force equal and opposite to $F$ must be acting on the plate. This opposing force that develops at the plate–rubber interface due to friction is expressed as $F = \tau A$, where $\tau$ is the shear stress and A is the contact area between the upper plate and the rubber. When the force is removed, the rubber returns to its original position. This phenomenon would also be observed with other solids such as a steel block provided that the applied force does not exceed the elastic range. If this experiment were repeated with a fluid (with two large parallel plates placed in a large body of water, for example), the fluid layer in contact with the upper plate would move with the plate continuously at the velocity of the plate no matter how small the force $F$. The fluid velocity would decrease with depth because of friction between fluid layers, reaching zero at the lower plate.

You will recall from statics that stress is defined as force per unit area and is determined by dividing the force by the area upon which it acts. The normal component of a force acting on a surface per unit area is called the normal stress, and the tangential component of a force acting on a surface per unit area is called **shear stress** (Fig. 1–10). In a fluid at rest, the normal stress is called **pressure**. A fluid at rest is at a state of zero shear stress. When the walls are removed or a liquid container is tilted, a shear develops as the liquid moves to re-establish a horizontal free surface.

In a liquid, groups of molecules can move relative to each other, but the volume remains relatively constant because of the strong cohesive forces between the molecules. As a result, a liquid takes the shape of the container it is in, and it forms a free surface in a larger container in a gravitational field. A gas, on the other hand, expands until it encounters the walls of the container and fills the entire available space. This is because the gas molecules are widely spaced, and the cohesive forces between them are very small. Unlike liquids, a gas in an open container cannot form a free surface (Fig. 1–11).

Although solids and fluids are easily distinguished in most cases, this distinction is not so clear in some borderline cases. For example, *asphalt* appears and behaves as a solid since it resists shear stress for short periods of time. When these forces are exerted over extended periods of time, however, the asphalt deforms slowly, behaving as a fluid. Some plastics, lead, and slurry mixtures exhibit similar behavior. Such borderline cases are beyond the scope of this text. The fluids we deal with in this text will be clearly recognizable as fluids.

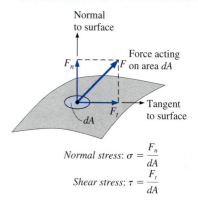

Normal stress: $\sigma = \dfrac{F_n}{dA}$

Shear stress: $\tau = \dfrac{F_t}{dA}$

**FIGURE 1–10**
The normal stress and shear stress at the surface of a fluid element. For fluids at rest, the shear stress is zero and pressure is the only normal stress.

**FIGURE 1–11**
Unlike a liquid, a gas does not form a free surface, and it expands to fill the entire available space.

# 1–5 ▪ IMPORTANCE OF DIMENSIONS AND UNITS

Any physical quantity can be characterized by **dimensions**. The magnitudes assigned to the dimensions are called **units**. Some basic dimensions such as mass $m$, length $L$, time $t$, and temperature $T$ are selected as **primary** or **fundamental dimensions**, while others such as velocity $V$, energy $E$, and volume $V$ are expressed in terms of the primary dimensions and are called **secondary dimensions**, or **derived dimensions**.

A number of unit systems have been developed over the years. Despite strong efforts in the scientific and engineering community to unify the world with a single unit system, two sets of units are still in common use today: the **English system**, which is also known as the *United States Customary System*

(USCS), and the metric **SI** (from *Le Système International d' Unités*), which is also known as the *International System.* The SI is a simple and logical system based on a decimal relationship between the various units, and it is being used for scientific and engineering work in most of the industrialized nations, including England. The English system, however, has no apparent systematic numerical base, and various units in this system are related to each other rather arbitrarily (12 in = 1 ft, 1 mile = 5280 ft, 4 qt = gal, etc.), which makes it confusing and difficult to learn. The United States is the only industrialized country that has not yet fully converted to the metric system.

The systematic efforts to develop a universally acceptable system of units dates back to 1790 when the French National Assembly charged the French Academy of Sciences to come up with such a unit system. An early version of the metric system was soon developed in France, but it did not find universal acceptance until 1875 when *The Metric Convention Treaty* was prepared and signed by 17 nations, including the United States. In this international treaty, meter and gram were established as the metric units for length and mass, respectively, and a *General Conference of Weights and Measures* (CGPM) was established that was to meet every six years. In 1960, the CGPM produced the SI, which was based on six fundamental quantities, and their units were adopted in 1954 at the Tenth General Conference of Weights and Measures: *meter* (m) for length, *kilogram* (kg) for mass, *second* (s) for time, *ampere* (A) for electric current, *degree Kelvin* (°K) for temperature, and *candela* (cd) for luminous intensity (amount of light). In 1971, the CGPM added a seventh fundamental quantity and unit: *mole* (mol) for the amount of matter.

Based on the notational scheme introduced in 1967, the degree symbol was officially dropped from the absolute temperature unit, and all unit names were to be written without capitalization even if they were derived from proper names (Table 1–1). However, the abbreviation of a unit was to be capitalized if the unit was derived from a proper name. For example, the SI unit of force, which is named after Sir Isaac Newton (1647–1723), is *newton* (not Newton), and it is abbreviated as N. Also, the full name of a unit may be pluralized, but its abbreviation cannot. For example, the length of an object can be 5 m or 5 meters, *not* 5 ms or 5 meter. Finally, no period is to be used in unit abbreviations unless they appear at the end of a sentence. For example, the proper abbreviation of meter is m (not m.).

The recent move toward the metric system in the United States seems to have started in 1968 when Congress, in response to what was happening in the rest of the world, passed a Metric Study Act. Congress continued to promote a voluntary switch to the metric system by passing the Metric Conversion Act in 1975. A trade bill passed by Congress in 1988 set a September 1992 deadline for all federal agencies to convert to the metric system. However, the deadlines were relaxed later with no clear plans for the future.

The industries that are heavily involved in international trade (such as the automotive, soft drink, and liquor industries) have been quick in converting to the metric system for economic reasons (having a single worldwide design, fewer sizes, smaller inventories, etc.). Today, nearly all the cars manufactured in the United States are metric. Most car owners probably do not realize this until they try an English socket wrench on a metric bolt. Most industries, however, resisted the change, thus slowing down the conversion process.

**TABLE 1–1**

The seven fundamental (or primary) dimensions and their units in SI

| Dimension | Unit |
|---|---|
| Length | meter (m) |
| Mass | kilogram (kg) |
| Time | second (s) |
| Temperature | kelvin (K) |
| Electric current | ampere (A) |
| Amount of light | candela (cd) |
| Amount of matter | mole (mol) |

**FIGURE 1–12**
The SI unit prefixes are used in all branches of engineering.

**TABLE 1–2**

Standard prefixes in SI units

| Multiple | Prefix |
|----------|--------|
| $10^{24}$ | yotta, Y |
| $10^{21}$ | zetta, Z |
| $10^{18}$ | exa, E |
| $10^{15}$ | peta, P |
| $10^{12}$ | tera, T |
| $10^{9}$ | giga, G |
| $10^{6}$ | mega, M |
| $10^{3}$ | kilo, k |
| $10^{2}$ | hecto, h |
| $10^{1}$ | deka, da |
| $10^{-1}$ | deci, d |
| $10^{-2}$ | centi, c |
| $10^{-3}$ | milli, m |
| $10^{-6}$ | micro, $\mu$ |
| $10^{-9}$ | nano, n |
| $10^{-12}$ | pico, p |
| $10^{-15}$ | femto, f |
| $10^{-18}$ | atto, a |
| $10^{-21}$ | zepto, z |
| $10^{-24}$ | yocto, y |

Presently the United States is a dual-system society, and it will stay that way until the transition to the metric system is completed. This puts an extra burden on today's engineering students, since they are expected to retain their understanding of the English system while learning, thinking, and working in terms of the SI. Given the position of the engineers in the transition period, both unit systems are used in this text, with particular emphasis on SI units.

As pointed out, the SI is based on a decimal relationship between units. The prefixes used to express the multiples of the various units are listed in Table 1–2. They are standard for all units, and the student is encouraged to memorize them because of their widespread use (Fig. 1–12).

## Some SI and English Units

In SI, the units of mass, length, and time are the kilogram (kg), meter (m), and second (s), respectively. The respective units in the English system are the pound-mass (lbm), foot (ft), and second (s). The pound symbol *lb* is actually the abbreviation of *libra*, which was the ancient Roman unit of weight. The English retained this symbol even after the end of the Roman occupation of Britain in 410. The mass and length units in the two systems are related to each other by

$$1 \text{ lbm} = 0.45359 \text{ kg}$$

$$1 \text{ ft} = 0.3048 \text{ m}$$

In the English system, force is usually considered to be one of the primary dimensions and is assigned a nonderived unit. This is a source of confusion and error that necessitates the use of a dimensional constant ($g_c$) in many formulas. To avoid this nuisance, we consider force to be a secondary dimension whose unit is derived from Newton's second law, that is,

$$\text{Force} = (\text{Mass})(\text{Acceleration})$$

or

$$F = ma \qquad (1\text{--}1)$$

In SI, the force unit is the newton (N), and it is defined as the *force required to accelerate a mass of 1 kg at a rate of 1 m/s²*. In the English system, the force unit is the **pound-force** (lbf) and is defined as the *force required to accelerate a mass of 32.174 lbm (1 slug) at a rate of 1 ft/s²* (Fig. 1–13). That is,

$$1 \text{ N} = 1 \text{ kg·m/s}^2$$

$$1 \text{ lbf} = 32.174 \text{ lbm·ft/s}^2$$

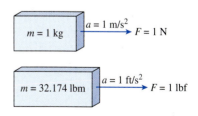

**FIGURE 1–13**
The definition of the force units.

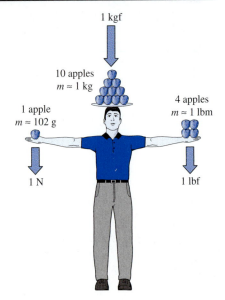

**FIGURE 1–14**

The relative magnitudes of the force units newton (N), kilogram-force (kgf), and pound-force (lbf).

**FIGURE 1–15**

A body weighing 150 lbf on earth will weigh only 25 lbf on the moon.

A force of 1 N is roughly equivalent to the weight of a small apple ($m = 102$ g), whereas a force of 1 lbf is roughly equivalent to the weight of four medium apples ($m_{total} = 454$ g), as shown in Fig. 1–14. Another force unit in common use in many European countries is the *kilogram-force* (kgf), which is the weight of 1 kg mass at sea level (1 kgf = 9.807 N).

The term **weight** is often incorrectly used to express mass, particularly by the "weight watchers." Unlike mass, weight $W$ is a *force*. It is the gravitational force applied to a body, and its magnitude is determined from Newton's second law,

$$W = mg \quad \text{(N)} \qquad \text{(1–2)}$$

where $m$ is the mass of the body, and $g$ is the local gravitational acceleration ($g$ is 9.807 m/s$^2$ or 32.174 ft/s$^2$ at sea level and 45° latitude). An ordinary bathroom scale measures the gravitational force acting on a body. The weight of a unit volume of a substance is called the **specific weight** $\gamma$ and is determined from $\gamma = \rho g$, where $\rho$ is density.

The mass of a body remains the same regardless of its location in the universe. Its weight, however, changes with a change in gravitational acceleration. A body weighs less on top of a mountain since $g$ decreases with altitude. On the surface of the moon, an astronaut weighs about one-sixth of what she or he normally weighs on earth (Fig. 1–15).

At sea level a mass of 1 kg weighs 9.807 N, as illustrated in Fig. 1–16. A mass of 1 lbm, however, weighs 1 lbf, which misleads people to believe that pound-mass and pound-force can be used interchangeably as pound (lb), which is a major source of error in the English system.

It should be noted that the *gravity force* acting on a mass is due to the *attraction* between the masses, and thus it is proportional to the magnitudes of the masses and inversely proportional to the square of the distance between them. Therefore, the gravitational acceleration $g$ at a location depends on the *local density* of the earth's crust, the *distance* to the center of the earth, and to a lesser extent, the positions of the moon and the sun. The value of $g$ varies with location from 9.832 m/s$^2$ at the poles (9.789 at the equator) to 7.322 m/s$^2$ at 1000 km above sea level. However, at altitudes up to 30 km, the variation of $g$ from the sea-level value of 9.807 m/s$^2$ is less than 1 percent. Therefore, for most practical purposes, the gravitational acceleration can be assumed to be *constant* at 9.807 m/s$^2$, often rounded to 9.81 m/s$^2$. It is interesting to note that at locations below sea level, the value of $g$ increases with distance from the sea level, reaches a maximum at about 4500 m, and then starts decreasing. (What do you think the value of $g$ is at the center of the earth?)

The primary cause of confusion between mass and weight is that mass is usually measured *indirectly* by measuring the *gravity force* it exerts. This approach also assumes that the forces exerted by other effects such as air buoyancy and fluid motion are negligible. This is like measuring the distance to a star by measuring its red shift, or measuring the altitude of an airplane by measuring barometric pressure. Both of these are also indirect measurements. The correct *direct* way of measuring mass is to compare it to a known mass. This is cumbersome, however, and it is mostly used for calibration and measuring precious metals.

*Work*, which is a form of energy, can simply be defined as force times distance; therefore, it has the unit "newton-meter (N·m)," which is called a **joule** (J). That is,

$$1 \text{ J} = 1 \text{ N·m} \qquad \text{(1–3)}$$

A more common unit for energy in SI is the kilojoule (1 kJ = $10^3$ J). In the English system, the energy unit is the **Btu** (British thermal unit), which is defined as the energy required to raise the temperature of 1 lbm of water at 68°F by 1°F. In the metric system, the amount of energy needed to raise the temperature of 1 g of water at 14.5°C by 1°C is defined as 1 **calorie** (cal), and 1 cal = 4.1868 J. The magnitudes of the kilojoule and Btu are almost identical (1 Btu = 1.0551 kJ). Here is a good way to get a feel for these units: If you light a typical match and let it burn itself out, it yields approximately one Btu (or one kJ) of energy (Fig. 1–17).

The unit for time rate of energy is joule per second (J/s), which is called a **watt** (W). In the case of work, the time rate of energy is called *power*. A commonly used unit of power is horsepower (hp), which is equivalent to 746 W. Electrical energy typically is expressed in the unit kilowatt-hour (kWh), which is equivalent to 3600 kJ. An electric appliance with a rated power of 1 kW consumes 1 kWh of electricity when running continuously for one hour. When dealing with electric power generation, the units kW and kWh are often confused. Note that kW or kJ/s is a unit of power, whereas kWh is a unit of energy. Therefore, statements like "the new wind turbine will generate 50 kW of electricity per year" are meaningless and incorrect. A correct statement should be something like "the new wind turbine with a rated power of 50 kW will generate 120,000 kWh of electricity per year."

## Dimensional Homogeneity

We all know that apples and oranges do not add. But we somehow manage to do it (by mistake, of course). In engineering, all equations must be *dimensionally homogeneous*. That is, every term in an equation must have the same unit (Fig. 1–18). If, at some stage of an analysis, we find ourselves in a position to add two quantities that have different units, it is a clear indication that we have made an error at an earlier stage. So checking dimensions can serve as a valuable tool to spot errors.

**FIGURE 1–16**
The weight of a unit mass at sea level.

**FIGURE 1–17**
A typical match yields about one Btu (or one kJ) of energy if completely burned.
*Photo by John M. Cimbala.*

**FIGURE 1–18**
To be dimensionally homogeneous, all the terms in an equation must have the same dimensions.
*BLONDIE © KING FEATURES SYNDICATE.*

**EXAMPLE 1–1**    **Electric Power Generation by a Wind Turbine**

A school is paying $0.09/kWh for electric power. To reduce its power bill, the school installs a wind turbine (Fig. 1–19) with a rated power of 30 kW. If the turbine operates 2200 hours per year at the rated power, determine the amount of electric power generated by the wind turbine and the money saved by the school per year.

**SOLUTION**    A wind turbine is installed to generate electricity. The amount of electric energy generated and the money saved per year are to be determined.
*Analysis*    The wind turbine generates electric energy at a rate of 30 kW or 30 kJ/s. Then the total amount of electric energy generated per year becomes

Total energy = (Energy per unit time)(Time interval)
= (30 kW)(2200 h)
= **66,000 kWh**

**FIGURE 1–19**
A wind turbine, as discussed
in Example 1–1.

*Courtesy of Steve Stadler, Oklahoma
Wind Power Initiative.*

The money saved per year is the monetary value of this energy determined as

$$\text{Money saved} = (\text{Total energy})(\text{Unit cost of energy})$$
$$= (66,000 \text{ kWh})(\$0.09/\text{kWh})$$
$$= \mathbf{\$5940}$$

*Discussion* The annual electric energy production also could be determined
in kJ by unit manipulations as

$$\text{Total energy} = (30 \text{ kW})(2200 \text{ h})\left(\frac{3600 \text{ s}}{1 \text{ h}}\right)\left(\frac{1 \text{ kJ/s}}{1 \text{ kW}}\right) = 2.38 \times 10^8 \text{ kJ}$$

which is equivalent to 66,000 kWh (1 kWh = 3600 kJ).

We all know from experience that units can give terrible headaches if they
are not used carefully in solving a problem. However, with some attention
and skill, units can be used to our advantage. They can be used to check formulas; sometimes they can even be used to *derive* formulas, as explained in
the following example.

**FIGURE 1–20**
Schematic for Example 1–2.

*EXAMPLE 1–2*     **Obtaining Formulas from Unit Considerations**

A tank is filled with oil whose density is $\rho = 850 \text{ kg/m}^3$. If the volume of the
tank is $V = 2 \text{ m}^3$, determine the amount of mass $m$ in the tank.

**SOLUTION**   The volume of an oil tank is given. The mass of oil is to be
determined.
*Assumptions*   Oil is a nearly incompressible substance and thus its density is
constant.
*Analysis*   A sketch of the system just described is given in Fig. 1–20. Suppose we forgot the formula that relates mass to density and volume. However, we know that mass has the unit of kilograms. That is, whatever
calculations we do, we should end up with the unit of kilograms. Putting the
given information into perspective, we have

$$\rho = 850 \text{ kg/m}^3 \qquad \text{and} \qquad V = 2 \text{ m}^3$$

It is obvious that we can eliminate $m^3$ and end up with kg by multiplying
these two quantities. Therefore, the formula we are looking for should be

$$m = \rho V$$

Thus,

$$m = (850 \text{ kg/m}^3)(2 \text{ m}^3) = \mathbf{1700 \text{ kg}}$$

*Discussion*   Note that this approach may not work for more complicated formulas. Nondimensional constants also may be present in the formulas, and
these cannot be derived from unit considerations alone.

You should keep in mind that a formula that is not dimensionally homogeneous is definitely wrong (Fig. 1–21), but a dimensionally homogeneous formula is not necessarily right.

## Unity Conversion Ratios

Just as all nonprimary dimensions can be formed by suitable combinations of primary dimensions, *all nonprimary units (secondary units) can be formed by combinations of primary units.* Force units, for example, can be expressed as

$$N = kg \frac{m}{s^2} \quad and \quad lbf = 32.174 \, lbm \frac{ft}{s^2}$$

They can also be expressed more conveniently as **unity conversion ratios** as

$$\frac{N}{kg \cdot m/s^2} = 1 \quad and \quad \frac{lbf}{32.174 \, lbm \cdot ft/s^2} = 1$$

Unity conversion ratios are identically equal to 1 and are unitless, and thus such ratios (or their inverses) can be inserted conveniently into any calculation to properly convert units (Fig. 1–22). You are encouraged to always use unity conversion ratios such as those given here when converting units. Some textbooks insert the archaic gravitational constant $g_c$ defined as $g_c = 32.174 \, lbm \cdot ft/lbf \cdot s^2 = kg \cdot m/N \cdot s^2 = 1$ into equations in order to force units to match. This practice leads to unnecessary confusion and is strongly discouraged by the present authors. We recommend that you instead use unity conversion ratios.

**FIGURE 1–21**
Always check the units in your calculations.

**FIGURE 1–22**
Every unity conversion ratio (as well as its inverse) is exactly equal to one. Shown here are a few commonly used unity conversion ratios.

*Courtesy of Steve Stadler, Oklahoma Wind Power Initiative. Used by permission.*

---

**EXAMPLE 1–3    The Weight of One Pound-Mass**

Using unity conversion ratios, show that 1.00 lbm weighs 1.00 lbf on earth (Fig. 1–23).

**SOLUTION**  A mass of 1.00 lbm is subjected to standard earth gravity. Its weight in lbf is to be determined.
**Assumptions**  Standard sea-level conditions are assumed.
**Properties**  The gravitational constant is $g = 32.174 \, ft/s^2$.
**Analysis**  We apply Newton's second law to calculate the weight (force) that corresponds to the known mass and acceleration. The weight of any object is equal to its mass times the local value of gravitational acceleration. Thus,

$$W = mg = (1.00 \, lbm)(32.174 \, ft/s^2)\left(\frac{1 \, lbf}{32.174 \, lbm \cdot ft/s^2}\right) = \textbf{1.00 lbf}$$

**Discussion**  The quantity in large parentheses in this equation is a unity conversion ratio. Mass is the same regardless of its location. However, on some other planet with a different value of gravitational acceleration, the weight of 1 lbm would differ from that calculated here.

**FIGURE 1–23**
A mass of 1 lbm weighs 1 lbf on earth.

**FIGURE 1–24**
A quirk in the metric system of units.

When you buy a box of breakfast cereal, the printing may say "Net weight: One pound (454 grams)." (See Fig. 1–24.) Technically, this means that the cereal inside the box weighs 1.00 lbf on earth and has a *mass* of 453.6 g (0.4536 kg). Using Newton's second law, the actual weight of the cereal on earth is

$$W = mg = (453.6\text{ g})(9.81\text{ m/s}^2)\left(\frac{1\text{ N}}{1\text{ kg·m/s}^2}\right)\left(\frac{1\text{ kg}}{1000\text{ g}}\right) = 4.49\text{ N}$$

## 1–6 ▪ PROBLEM-SOLVING TECHNIQUE

The first step in learning any science is to grasp the fundamentals and to gain a sound knowledge of it. The next step is to master the fundamentals by testing this knowledge. This is done by solving significant real-world problems. Solving such problems, especially complicated ones, requires a systematic approach. By using a step-by-step approach, an engineer can reduce the solution of a complicated problem into the solution of a series of simple problems (Fig. 1–25). When you are solving a problem, we recommend that you use the following steps zealously as applicable. This will help you avoid some of the common pitfalls associated with problem solving.

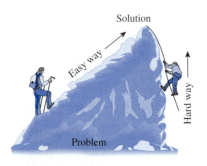

**FIGURE 1–25**
A step-by-step approach can greatly simplify problem solving.

### Step 1: Problem Statement
In your own words, briefly state the problem, the key information given, and the quantities to be found. This is to make sure that you understand the problem and the objectives before you attempt to solve the problem.

### Step 2: Schematic
Draw a realistic sketch of the physical system involved, and list the relevant information on the figure. The sketch does not have to be something elaborate, but it should resemble the actual system and show the key features. Indicate any energy and mass interactions with the surroundings. Listing the given information on the sketch helps one to see the entire problem at once. Also, check for properties that remain constant during a process (such as temperature during an isothermal process), and indicate them on the sketch.

### Step 3: Assumptions and Approximations
State any appropriate assumptions and approximations made to simplify the problem to make it possible to obtain a solution. Justify the questionable assumptions. Assume reasonable values for missing quantities that are necessary. For example, in the absence of specific data for atmospheric pressure, it can be taken to be 1 atm. However, it should be noted in the analysis that the atmospheric pressure decreases with increasing elevation. For example, it drops to 0.83 atm in Denver (elevation 1610 m) (Fig. 1–26).

### Step 4: Physical Laws
Apply all the relevant basic physical laws and principles (such as the conservation of mass), and reduce them to their simplest form by utilizing the assumptions made. However, the region to which a physical law is applied

must be clearly identified first. For example, the increase in speed of water flowing through a nozzle is analyzed by applying conservation of mass between the inlet and outlet of the nozzle.

## Step 5: Properties

Determine the unknown properties at known states necessary to solve the problem from property relations or tables. List the properties separately, and indicate their source, if applicable.

## Step 6: Calculations

Substitute the known quantities into the simplified relations and perform the calculations to determine the unknowns. Pay particular attention to the units and unit cancellations, and remember that a dimensional quantity without a unit is meaningless. Also, don't give a false implication of high precision by copying all the digits from the screen of the calculator—round the results to an appropriate number of significant digits (see p. 18).

## Step 7: Reasoning, Verification, and Discussion

Check to make sure that the results obtained are reasonable and intuitive, and verify the validity of the questionable assumptions. Repeat the calculations that resulted in unreasonable values. For example, insulating a water heater that uses $80 worth of natural gas a year cannot result in savings of $200 a year (Fig. 1–27).

Also, point out the significance of the results, and discuss their implications. State the conclusions that can be drawn from the results, and any recommendations that can be made from them. Emphasize the limitations under which the results are applicable, and caution against any possible misunderstandings and using the results in situations where the underlying assumptions do not apply. For example, if you determined that wrapping a water heater with a $20 insulation jacket will reduce the energy cost by $30 a year, indicate that the insulation will pay for itself from the energy it saves in less than a year. However, also indicate that the analysis does not consider labor costs, and that this will be the case if you install the insulation yourself.

Keep in mind that the solutions you present to your instructors, and any engineering analysis presented to others, is a form of communication. Therefore neatness, organization, completeness, and visual appearance are of utmost importance for maximum effectiveness (Fig. 1–28). Besides, neatness also serves as a great checking tool since it is very easy to spot errors and inconsistencies in neat work. Carelessness and skipping steps to save time often end up costing more time and unnecessary anxiety.

The approach described here is used in the solved example problems without explicitly stating each step, as well as in the Solutions Manual of this text. For some problems, some of the steps may not be applicable or necessary. For example, often it is not practical to list the properties separately. However, we cannot overemphasize the importance of a logical and orderly approach to problem solving. Most difficulties encountered while solving a problem are not due to a lack of knowledge; rather, they are due to a lack of organization. You are strongly encouraged to follow these steps in problem solving until you develop your own approach that works best for you.

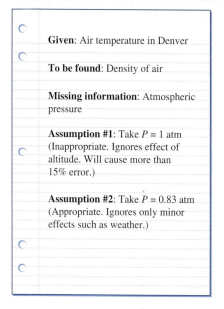

**Given**: Air temperature in Denver

**To be found**: Density of air

**Missing information**: Atmospheric pressure

**Assumption #1**: Take $P = 1$ atm (Inappropriate. Ignores effect of altitude. Will cause more than 15% error.)

**Assumption #2**: Take $P = 0.83$ atm (Appropriate. Ignores only minor effects such as weather.)

**FIGURE 1–26**
The assumptions made while solving an engineering problem must be reasonable and justifiable.

Energy use:                    $80/yr

Energy saved
by insulation:              $200/yr

IMPOSSIBLE!

**FIGURE 1–27**
The results obtained from an engineering analysis must be checked for reasonableness.

**FIGURE 1–28**
Neatness and organization are highly valued by employers.

# Engineering Software Packages

You may be wondering why we are about to undertake an in-depth study of the fundamentals of another engineering science. After all, almost all such problems we are likely to encounter in practice can be solved using one of several sophisticated software packages readily available in the market today. These software packages not only give the desired numerical results, but also supply the outputs in colorful graphical form for impressive presentations. It is unthinkable to practice engineering today without using some of these packages. This tremendous computing power available to us at the touch of a button is both a blessing and a curse. It certainly enables engineers to solve problems easily and quickly, but it also opens the door for abuses and misinformation. In the hands of poorly educated people, these software packages are as dangerous as sophisticated powerful weapons in the hands of poorly trained soldiers.

Thinking that a person who can use the engineering software packages without proper training on fundamentals can practice engineering is like thinking that a person who can use a wrench can work as a car mechanic. If it were true that the engineering students do not need all these fundamental courses they are taking because practically everything can be done by computers quickly and easily, then it would also be true that the employers would no longer need high-salaried engineers since any person who knows how to use a word-processing program can also learn how to use those software packages. However, the statistics show that the need for engineers is on the rise, not on the decline, despite the availability of these powerful packages.

We should always remember that all the computing power and the engineering software packages available today are just *tools*, and tools have meaning only in the hands of masters. Having the best word-processing program does not make a person a good writer, but it certainly makes the job of a good writer much easier and makes the writer more productive (Fig. 1–29). Hand calculators did not eliminate the need to teach our children how to add or subtract, and the sophisticated medical software packages did not take the place of medical school training. Neither will engineering software packages replace the traditional engineering education. They will simply cause a shift in emphasis in the courses from mathematics to physics. That is, more time will be spent in the classroom discussing the physical aspects of the problems in greater detail, and less time on the mechanics of solution procedures.

All these marvelous and powerful tools available today put an extra burden on today's engineers. They must still have a thorough understanding of the fundamentals, develop a "feel" of the physical phenomena, be able to put the data into proper perspective, and make sound engineering judgments, just like their predecessors. However, they must do it much better, and much faster, using more realistic models because of the powerful tools available today. The engineers in the past had to rely on hand calculations, slide rules, and later hand calculators and computers. Today they rely on software packages. The easy access to such power and the possibility of a simple misunderstanding or misinterpretation causing great damage make it more important today than ever to have solid training in

**FIGURE 1–29**

An excellent word-processing program does not make a person a good writer; it simply makes a good writer a more efficient writer.

*© Vol. 80/PhotoDisc/Getty RF.*

the fundamentals of engineering. In this text we make an extra effort to put the emphasis on developing an intuitive and physical understanding of natural phenomena instead of on the mathematical details of solution procedures.

## Engineering Equation Solver (EES)

EES is a program that solves systems of linear or nonlinear algebraic or differential equations numerically. It has a large library of built-in thermodynamic property functions as well as mathematical functions, and allows the user to supply additional property data. Unlike some software packages, EES does not solve engineering problems; it only solves the equations supplied by the user. Therefore, the user must understand the problem and formulate it by applying any relevant physical laws and relations. EES saves the user considerable time and effort by simply solving the resulting mathematical equations. This makes it possible to attempt significant engineering problems not suitable for hand calculations, and to conduct parametric studies quickly and conveniently. EES is a very powerful yet intuitive program that is very easy to use, as shown in Example 1–4.

---

**EXAMPLE 1–4    Solving a System of Equations with EES**

The difference of two numbers is 4, and the sum of the squares of these two numbers is equal to the sum of the numbers plus 20. Determine these two numbers.

**SOLUTION**   Relations are given for the difference and the sum of the squares of two numbers. They are to be determined.

*Analysis*   We start the EES program by double-clicking on its icon, open a new file, and type the following on the blank screen that appears:

$$x - y = 4$$

$$x^\wedge 2 + y^\wedge 2 = x + y + 20$$

which is an exact mathematical expression of the problem statement with x and y denoting the unknown numbers. The solution to this system of two nonlinear equations with two unknowns is obtained by a single click on the "calculator" icon on the taskbar. It gives

$$\mathbf{x = 5} \quad \text{and} \quad \mathbf{y = 1}$$

*Discussion*   Note that all we did is formulate the problem as we would on paper; EES took care of all the mathematical details of solution. Also note that equations can be linear or nonlinear, and they can be entered in any order with unknowns on either side. Friendly equation solvers such as EES allow the user to concentrate on the physics of the problem without worrying about the mathematical complexities associated with the solution of the resulting system of equations.

**Given:**

| | |
|---|---|
| Volume: | $V = 3.75$ L |
| Density: | $\rho = 0.845$ kg/L |

(3 significant digits)

**Also,** $3.75 \times 0.845 = 3.16875$

**Find:**

Mass: $m = \rho V = 3.16875$ kg

*Rounding to 3 significant digits:*

$m = 3.17$ kg

**FIGURE 1–30**

A result with more significant digits than that of given data falsely implies more accuracy.

## A Remark on Significant Digits

In engineering calculations, the information given is not known to more than a certain number of significant digits, usually three digits. Consequently, the results obtained cannot possibly be accurate to more significant digits. Reporting results in more significant digits implies greater accuracy than exists, and it should be avoided.

For example, consider a 3.75-L container filled with gasoline whose density is 0.845 kg/L, and try to determine its mass. Probably the first thought that comes to your mind is to multiply the volume and density to obtain 3.16875 kg for the mass, which falsely implies that the mass determined is accurate to six significant digits. In reality, however, the mass cannot be more accurate than three significant digits since both the volume and the density are accurate to three significant digits only. Therefore, the result should be rounded to three significant digits, and the mass should be reported to be 3.17 kg instead of what appears in the screen of the calculator. The result 3.16875 kg would be correct only if the volume and density were given to be 3.75000 L and 0.845000 kg/L, respectively. The value 3.75 L implies that we are fairly confident that the volume is accurate within $\pm 0.01$ L, and it cannot be 3.74 or 3.76 L. However, the volume can be 3.746, 3.750, 3.753, etc., since they all round to 3.75 L (Fig. 1–30). It is more appropriate to retain all the digits during intermediate calculations, and to do the rounding in the final step since this is what a computer will normally do.

When solving problems, we will assume the given information to be accurate to at least three significant digits. Therefore, if the length of a pipe is given to be 40 m, we will assume it to be 40.0 m in order to justify using three significant digits in the final results. You should also keep in mind that all experimentally determined values are subject to measurement errors and such errors will reflect in the results obtained. For example, if the density of a substance has an uncertainty of 2 percent, then the mass determined using this density value will also have an uncertainty of 2 percent.

You should also be aware that we sometimes knowingly introduce small errors in order to avoid the trouble of searching for more accurate data. For example, when dealing with liquid water, we just use the value of 1000 kg/m³ for density, which is the density value of pure water at 0°C. Using this value at 75°C will result in an error of 2.5 percent since the density at this temperature is 975 kg/m³. The minerals and impurities in the water will introduce additional error. This being the case, you should have no reservation in rounding the final results to a reasonable number of significant digits. Besides, having a few percent uncertainty in the results of engineering analysis is usually the norm, not the exception.

## SUMMARY

In this chapter, some basic concepts of thermal-fluid sciences are introduced and discussed. The physical sciences that deal with energy and the transfer, transport, and conversion of energy are referred to as *thermal-fluid sciences,* and they are studied under the subcategories of thermodynamics, heat transfer, and fluid mechanics.

*Thermodynamics* is the science that primarily deals with energy. The *first law of thermodynamics* is simply an expression of the conservation of energy principle, and it asserts that *energy* is a thermodynamic property. The *second law of thermodynamics* asserts that energy has *quality* as well as *quantity,* and actual processes occur in the direction of decreasing quality of energy. Determining the rates of heat transfer to or from a system and thus the times of cooling or heating, as well as the variation of the temperature, is the subject of *heat transfer.* The basic requirement for heat transfer is the presence of a *temperature difference.* A substance in the liquid or gas phase is referred to as a *fluid. Fluid mechanics* is the science that deals with the behavior of fluids at rest (*fluid statics*) or in motion (*fluid dynamics*), and the interaction of fluids with solids or other fluids at the boundaries.

In engineering calculations, it is important to pay particular attention to the units of the quantities to avoid errors caused by inconsistent units, and to follow a systematic approach. It is also important to recognize that the information given is not known to more than a certain number of significant digits, and the results obtained cannot possibly be accurate to more significant digits.

When solving a problem, it is recommended that a step-by-step approach be used. Such an approach involves stating the problem, drawing a schematic, making appropriate assumptions, applying the physical laws, listing the relevant properties, making the necessary calculations, and making sure that the results are reasonable.

## REFERENCES AND SUGGESTED READINGS

**1.** American Society for Testing and Materials. *Standards for Metric Practice.* ASTM E 380-79, January 1980.

**2.** Y. A. Çengel and A. J. Ghajar. *Heat and Mass Transfer: Fundamentals and Applications.* 4th ed. New York: McGraw-Hill, 2011.

**3.** Y. A. Çengel and M. A. Boles. *Thermodynamics. An Engineering Approach.* 7th ed. New York: McGraw-Hill, 2011.

**4.** Y. A. Çengel and John M. Cimbala. *Fluid Mechanics: Fundamentals and Applications.* 2nd ed. New York: McGraw-Hill, 2010.

## PROBLEMS*

### Thermodynamics, Heat Transfer, and Fluid Mechanics

**1–1C** Why does a bicyclist pick up speed on a downhill road even when he is not pedaling? Does this violate the conservation of energy principle?

**1–2C** An office worker claims that a cup of cold coffee on his table warmed up to 80°C by picking up energy from the surrounding air, which is at 25°C. Is there any truth to his claim? Does this process violate any thermodynamic laws?

**1–3C** One of the most amusing things a person can experience is that in certain parts of the world a still car on neutral going uphill when its brakes are released. Such occurrences are even broadcast on TV. Can this really happen or is it a bad eyesight? How can you verify if a road is really uphill or downhill?

**1–4C** How does the science of heat transfer differ from the science of thermodynamics?

**1–5C** What is the driving force for (*a*) heat transfer, (*b*) electric current, and (*c*) fluid flow?

**1–6C** Why is heat transfer a nonequilibrium phenomenon?

**1–7C** Can there be any heat transfer between two bodies that are at the same temperature but at different pressures?

**1–8C** Define stress, normal stress, shear stress, and pressure.

---

* Problems designated by a "C" are concept questions, and students are encouraged to answer them all. Problems designated by an "E" are in English units, and the SI users can ignore them. Problems with the 🖥 icon are solved using EES, and complete solutions together with parametric studies are included on the enclosed DVD. Problems with the 📷 icon are comprehensive in nature, and are intended to be solved with a computer, preferably using the EES software that accompanies this text.

### Mass, Force, and Units

**1–9C** In a news article, it is stated that a recently developed geared turbofan engine produces 15,000 pounds of thrust to propel the aircraft forward. Is "pound" mentioned here lbm or lbf? Explain.

**1–10C** Explain why the light-year has the dimension of length.

**1–11C**  What is the net force acting on a car cruising at a constant velocity of 70 km/h (*a*) on a level road and (*b*) on an uphill road?

**1–12E**  A man weighs 210 lbf at a location where $g = 32.10$ ft/s$^2$. Determine his weight on the moon, where $g = 5.47$ ft/s$^2$. *Answer: 35.8 lbf*

**1–13**  Determine the mass and the weight of the air contained in a room whose dimensions are 6 m $\times$ 6 m $\times$ 8 m. Assume the density of the air is 1.16 kg/m$^3$.  *Answers: 334.1 kg, 3277 N*

**1–14**  At 45° latitude, the gravitational acceleration as a function of elevation $z$ above sea level is given by $g = a - bz$, where $a = 9.807$ m/s$^2$ and $b = 3.32 \times 10^{-6}$ s$^{-2}$. Determine the height above sea level where the weight of an object will decrease by 0.5 percent.  *Answer: 14.770 m*

**1–15**  What is the weight, in N, of an object with a mass of 200 kg at a location where $g = 9.6$ m/s$^2$?

**1–16E**  The constant-pressure specific heat of air at 25°C is 1.005 kJ/kg·°C. Express this value in kJ/kg·K, J/g·°C, kcal/kg·°C, and Btu/lbm·°F.

**1–17**  A 3-kg rock is thrown upward with a force of 200 N at a location where the local gravitational acceleration is 9.79 m/s$^2$. Determine the acceleration of the rock, in m/s$^2$.

**1–18**  Solve Prob. 1–17 using EES (or other) software. Print out the entire solution, including the numerical results with proper units.

**1–19**  While solving a problem, a person ends up with the equation $E = 25$ kJ $+ 7$ kJ/kg at some stage. Here $E$ is the total energy and has the unit of kilojoules. Determine how to correct the error and discuss what may have caused it.

**1–20**  A 4-kW resistance heater in a water heater runs for 2 hours to raise the water temperature to the desired level. Determine the amount of electric energy used in both kWh and kJ.

**1–21**  The gas tank of a car is filled with a nozzle that discharges gasoline at a constant flow rate. Based on unit considerations of quantities, obtain a relation for the filling time in terms of the volume $V$ of the tank (in L) and the discharge rate of gasoline $\dot{V}$ (in L/s).

**1–22**  A pool of volume $V$ (in m$^3$) is to be filled with water using a hose of diameter $D$ (in m). If the average discharge velocity is $V$ (in m/s) and the filling time is $t$ (in s), obtain a relation for the volume of the pool based on until considerations of quantities involved.

## Solving Engineering Problems and EES

**1–23C**  What is the value of the engineering software packages in (*a*) engineering education and (*b*) engineering practice?

**1–24**  Determine a positive real root of this equation using EES:

$$2x^3 - 10x^{0.5} - 3x = -3$$

**1–25**  Solve this system of two equations with two unknowns using EES:

$$x^3 - y^2 = 7.75$$
$$3xy + y = 3.5$$

**1–26**  Solve this system of three equations with three unknowns using EES:

$$2x - y + z = 5$$
$$3x^2 + 2y = z + 2$$
$$xy + 2z = 8$$

**1–27**  Solve this system of three equations with three unknowns using EES:

$$x^2y - z = 1$$
$$x - 3y^{0.5} + xz = -2$$
$$x + y - z = 2$$

**1–28E**  Specific heat is defined as the amount of energy needed to increase the temperature of a unit mass of a substance by one degree. The specific heat of water at room temperature is 4.18 kJ/kg·°C in SI unit system. Using the unit conversion function capability of EES, express the specific heat of water in (*a*) kJ/kg·K, (*b*) Btu/lbm·°F, (*c*) Btu/lbm·R, and (*d*) kcal/kg·°C units. *Answers: (a) 4.18, (b) (c) (d) 0.9984*

## Review Problems

**1–29**  A lunar exploration module weighs 4000 N at a location where $g = 9.8$ m/s$^2$. Determine the weight of this module in newtons when it is on the moon where $g = 1.64$ m/s$^2$.

**1–30**  The weight of bodies may change somewhat from one location to another as a result of the variation of the gravitational acceleration $g$ with elevation. Accounting for this variation using the relation in Prob. 1–14, determine the weight of an 80-kg person at sea level ($z = 0$), in Denver ($z = 1610$ m), and on the top of Mount Everest ($z = 8848$ m).

**1–31E**  A man goes to a traditional market to buy a steak for dinner. He finds a 12-oz steak (1 lbm = 16 oz) for $3.15. He then goes to the adjacent international market and finds a 300-g steak of identical quality for $2.95. Which steak is the better buy?

**1–32E** What is the weight of a 1-kg substance in N, kN, kg·m/s², kgf, lbm·ft/s², and lbf?

**1–33** Consider the flow of air through a wind turbine whose blades sweep an area of diameter $D$ (in m). The average air velocity through the swept area is $V$ (in m/s). On the bases of the units of the quantities involved, show that the mass flow rate of air (in kg/s) through the swept area is proportional to air density, the wind velocity, and the square of the diameter of the swept area.

**1–34** The drag force exerted on a car by air depends on a dimensionless drag coefficient, the density of air, the car velocity, and the frontal area of the car. That is, $F_D =$ function $(C_{\text{Drag}}, A_{\text{front}}, \rho, V)$. Based on unit considerations alone, obtain a relation for the drag force.

**FIGURE P1–34**

### Design and Essay Problems

**1–35** Write an essay on the various mass- and volume-measurement devices used throughout history. Also, explain the development of the modern units for mass and volume.

# THERMODYNAMICS

# BASIC CONCEPTS OF THERMODYNAMICS

**E**very science has a unique vocabulary associated with it, and thermodynamics is no exception. Precise definition of basic concepts forms a sound foundation for the development of a science and prevents possible misunderstandings. We start this chapter with a discussion of some basic concepts such as *system, state, state postulate, equilibrium, and process.* We also discuss *temperature* and *temperature scales* with particular emphasis on the International Temperature Scale of 1990. We then present *pressure,* which is the normal force exerted by a fluid per unit area and discuss *absolute* and *gage* pressures, the variation of pressure with depth, and pressure measurement devices, such as manometers and barometers. Careful study of these concepts is essential for a good understanding of the topics in the following chapters.

## Objectives

Objectives of this chapter are to:

- Identify the unique vocabulary associated with thermodynamics through the precise definition of basic concepts to form a sound foundation for the development of the principles of thermodynamics.

- Explain the basic concepts of thermodynamics such as system, state, state postulate, equilibrium, process, and cycle.

- Review concepts of temperature, temperature scales, pressure, and absolute and gage pressure.

# 2–1 ■ SYSTEMS AND CONTROL VOLUMES

A system is defined as a *quantity of matter or a region in space chosen for study*. The mass or region outside the system is called the **surroundings**. The real or imaginary surface that separates the system from its surroundings is called the **boundary** (Fig. 2–1). The boundary of a system can be *fixed* or *movable*. Note that the boundary is the contact surface shared by both the system and the surroundings. Mathematically speaking, the boundary has zero thickness, and thus it can neither contain any mass nor occupy any volume in space.

Systems may be considered to be *closed* or *open,* depending on whether a fixed mass or a fixed volume in space is chosen for study. A **closed system** (also known as a **control mass** or just *system* when the context makes it clear) consists of a fixed amount of mass, and no mass can cross its boundary. That is, no mass can enter or leave a closed system, as shown in Fig. 2–2. But energy, in the form of heat or work, can cross the boundary; and the volume of a closed system does not have to be fixed. If, as a special case, even energy is not allowed to cross the boundary, that system is called an **isolated system**.

Consider the piston-cylinder device shown in Fig. 2–3. Let us say that we would like to find out what happens to the enclosed gas when it is heated. Since we are focusing our attention on the gas, it is our system. The inner surfaces of the piston and the cylinder form the boundary, and since no mass is crossing this boundary, it is a closed system. Notice that energy may cross the boundary, and part of the boundary (the inner surface of the piston, in this case) may move. Everything outside the gas, including the piston and the cylinder, is the surroundings.

An **open system**, or a **control volume**, as it is often called, is a properly selected region in space. It usually encloses a device that involves mass flow such as a compressor, turbine, or nozzle. Flow through these devices is best studied by selecting the region within the device as the control volume. Both mass and energy can cross the boundary of a control volume.

A large number of engineering problems involve mass flow in and out of a system and, therefore, are modeled as *control volumes*. A water heater, a car radiator, a turbine, and a compressor all involve mass flow and should be analyzed as control volumes (open systems) instead of as control masses (closed systems). In general, *any arbitrary region in space* can be selected as a control volume. There are no concrete rules for the selection of control volumes, but the proper choice certainly makes the analysis much easier. If we were to analyze the flow of air through a nozzle, for example, a good choice for the control volume would be the region within the nozzle.

The boundaries of a control volume are called a *control surface,* and they can be real or imaginary. In the case of a nozzle, the inner surface of the nozzle forms the real part of the boundary, and the entrance and exit areas form the imaginary part, since there are no physical surfaces there (Fig. 2–4a).

A control volume can be fixed in size and shape, as in the case of a nozzle, or it may involve a moving boundary, as shown in Fig. 2–4b. Most control volumes, however, have fixed boundaries and thus do not involve any moving boundaries. A control volume can also involve heat and work interactions just as a closed system, in addition to mass interaction.

**FIGURE 2–1**

System, surroundings, and boundary.

**FIGURE 2–2**

Mass cannot cross the boundaries of a closed system, but energy can.

**FIGURE 2–3**

A closed system with a moving boundary.

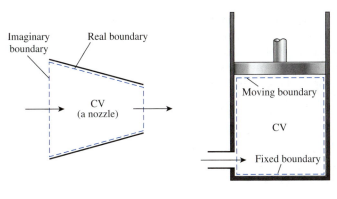

(a) A control volume with real and imaginary boundaries

(b) A control volume with fixed and moving boundaries

**FIGURE 2–4**

A control volume can involve fixed, moving, real, and imaginary boundaries.

As an example of an open system, consider the water heater shown in Fig. 2–5. Let us say that we would like to determine how much heat we must transfer to the water in the tank in order to supply a steady stream of hot water. Since hot water will leave the tank and be replaced by cold water, it is not convenient to choose a fixed mass as our system for the analysis. Instead, we can concentrate our attention on the volume formed by the interior surfaces of the tank and consider the hot and cold water streams as mass leaving and entering the control volume. The interior surfaces of the tank form the control surface for this case, and mass is crossing the control surface at two locations.

In an engineering analysis, the system under study *must* be defined carefully. In most cases, the system investigated is quite simple and obvious, and defining the system may seem like a tedious and unnecessary task. In other cases, however, the system under study may be rather involved, and a proper choice of the system may greatly simplify the analysis.

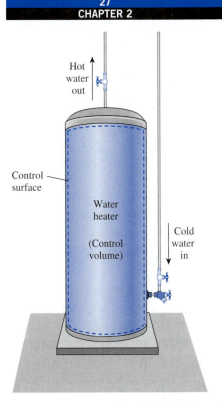

**FIGURE 2–5**

An open system (a control volume) with one inlet and one exit.

## 2–2 ▪ PROPERTIES OF A SYSTEM

Any characteristic of a system is called a **property**. Some familiar properties are pressure $P$, temperature $T$, volume $V$, and mass $m$. The list can be extended to include less familiar ones such as viscosity, thermal conductivity, modulus of elasticity, thermal expansion coefficient, electric resistivity, and even velocity and elevation.

Properties are considered to be either *intensive* or *extensive*. **Intensive properties** are those that are independent of the mass of a system, such as temperature, pressure, and density. **Extensive properties** are those whose values depend on the size—or extent—of the system. Total mass, total volume, and total momentum are some examples of extensive properties. An easy way to determine whether a property is intensive or extensive is to divide the system into two equal parts with an imaginary partition, as shown in Fig. 2–6. Each part will have the same value of intensive properties as the original system, but half the value of the extensive properties.

**FIGURE 2–6**

Criterion to differentiate intensive and extensive properties.

Generally, uppercase letters are used to denote extensive properties (with mass $m$ being a major exception), and lowercase letters are used for intensive properties (with pressure $P$ and temperature $T$ being the obvious exceptions).

Extensive properties per unit mass are called **specific properties**. Some examples of specific properties are specific volume ($v = V/m$) and specific total energy ($e = E/m$).

## Continuum

Matter is made up of atoms that are widely spaced in the gas phase. Yet it is very convenient to disregard the atomic nature of a substance and view it as a continuous, homogeneous matter with no holes, that is, a **continuum**. The continuum idealization allows us to treat properties as point functions and to assume the properties vary continually in space with no jump discontinuities. This idealization is valid as long as the size of the system we deal with is large relative to the space between the molecules. This is the case in practically all problems, except some specialized ones. The continuum idealization is implicit in many statements we make, such as "the density of water in a glass is the same at any point."

To have a sense of the distance involved at the molecular level, consider a container filled with oxygen at atmospheric conditions. The diameter of the oxygen molecule is about $3 \times 10^{-10}$ m and its mass is $5.3 \times 10^{-26}$ kg. Also, the *mean free path* of oxygen at 1 atm pressure and 20°C is $6.3 \times 10^{-8}$ m. That is, an oxygen molecule travels, on average, a distance of $6.3 \times 10^{-8}$ m (about 200 times of its diameter) before it collides with another molecule.

Also, there are about $3 \times 10^{16}$ molecules of oxygen in the tiny volume of 1 mm$^3$ at 1 atm pressure and 20°C (Fig. 2–7). The continuum model is applicable as long as the characteristic length of the system (such as its diameter) is much larger than the mean free path of the molecules. At very high vacuums or very high elevations, the mean free path may become large (for example, it is about 0.1 m for atmospheric air at an elevation of 100 km). For such cases the **rarefied gas flow theory** should be used, and the impact of individual molecules should be considered. In this text we will limit our consideration to substances that can be modeled as a continuum.

**FIGURE 2–7**

Despite the large gaps between molecules, a substance can be treated as a continuum because of the very large number of molecules even in an extremely small volume.

## 2–3 · DENSITY AND SPECIFIC GRAVITY

**Density** is defined as *mass per unit volume* (Fig. 2–8).

*Density:*
$$\rho = \frac{m}{V} \quad (\text{kg/m}^3) \tag{2–1}$$

The reciprocal of density is the **specific volume** $v$, which is defined as *volume per unit mass*. That is,

$$v = \frac{V}{m} = \frac{1}{\rho} \tag{2–2}$$

For a differential volume element of mass $\delta m$ and volume $\delta V$, density can be expressed as $\rho = \delta m/\delta V$.

The density of a substance, in general, depends on temperature and pressure. The density of most gases is proportional to pressure and inversely proportional to temperature. Liquids and solids, on the other hand, are essentially

**FIGURE 2–8**

Density is mass per unit volume; specific volume is volume per unit mass.

incompressible substances, and the variation of their density with pressure is usually negligible. At 20°C, for example, the density of water changes from 998 kg/m³ at 1 atm to 1003 kg/m³ at 100 atm, a change of just 0.5 percent. The density of liquids and solids depends more strongly on temperature than it does on pressure. At 1 atm, for example, the density of water changes from 998 kg/m³ at 20°C to 975 kg/m³ at 75°C, a change of 2.3 percent, which can still be neglected in many engineering analyses.

Sometimes the density of a substance is given relative to the density of a well-known substance. Then it is called **specific gravity**, or **relative density**, and is defined as *the ratio of the density of a substance to the density of some standard substance at a specified temperature* (usually water at 4°C, for which $\rho_{H_2O} = 1000$ kg/m³). That is,

*Specific gravity:* $$SG = \frac{\rho}{\rho_{H_2O}} \qquad (2\text{--}3)$$

Note that the specific gravity of a substance is a dimensionless quantity. However, in SI units, the numerical value of the specific gravity of a substance is exactly equal to its density in g/cm³ or kg/L (or 0.001 times the density in kg/m³) since the density of water at 4°C is 1 g/cm³ = 1 kg/L = 1000 kg/m³. The specific gravity of mercury at 0°C, for example, is 13.6. Therefore, its density at 0°C is 13.6 g/cm³ = 13.6 kg/L = 13,600 kg/m³. The specific gravities of some substances at 0°C are given in Table 2–1. Note that substances with specific gravities less than 1 are lighter than water, and thus they would float on water.

The weight of a unit volume of a substance is called **specific weight** and is expressed as

*Specific weight:* $$\gamma_s = \rho g \qquad (N/m^3) \qquad (2\text{--}4)$$

where $g$ is the gravitational acceleration.

The densities of liquids are essentially constant, and thus they can often be approximated as being incompressible substances during most processes without sacrificing much in accuracy.

## 2–4 · STATE AND EQUILIBRIUM

Consider a system not undergoing any change. At this point, all the properties can be measured or calculated throughout the entire system, which gives us a set of properties that completely describes the condition, or the **state**, of the system. At a given state, all the properties of a system have fixed values. If the value of even one property changes, the state will change to a different one. In Fig. 2–9 a system is shown at two different states.

Thermodynamics deals with *equilibrium* states. The word **equilibrium** implies a state of balance. In an equilibrium state there are no unbalanced potentials (or driving forces) within the system. A system in equilibrium experiences no changes when it is isolated from its surroundings.

There are many types of equilibrium, and a system is not in thermodynamic equilibrium unless the conditions of all the relevant types of equilibrium are satisfied. For example, a system is in **thermal equilibrium** if the temperature is the same throughout the entire system, as shown in Fig. 2–10. That is, the system involves no temperature differential, which is the driving force for

**TABLE 2–1**

Specific gravities of some substances at 0°C

| Substance | SG |
|---|---|
| Water | 1.0 |
| Blood | 1.05 |
| Seawater | 1.025 |
| Gasoline | 0.7 |
| Ethyl alcohol | 0.79 |
| Mercury | 13.6 |
| Wood | 0.3–0.9 |
| Gold | 19.2 |
| Bones | 1.7–2.0 |
| Ice | 0.92 |
| Air (at 1 atm) | 0.0013 |

(a) State 1          (b) State 2

**FIGURE 2–9**

A system at two different states.

(a) Before          (b) After

**FIGURE 2–10**

A closed system reaching thermal equilibrium.

heat flow. **Mechanical equilibrium** is related to pressure, and a system is in mechanical equilibrium if there is no change in pressure at any point of the system with time. However, the pressure may vary within the system with elevation as a result of gravitational effects. For example, the higher pressure at a bottom layer is balanced by the extra weight it must carry, and, therefore, there is no imbalance of forces. The variation of pressure as a result of gravity in most thermodynamic systems is relatively small and usually disregarded. If a system involves two phases, it is in **phase equilibrium** when the mass of each phase reaches an equilibrium level and stays there. Finally, a system is in **chemical equilibrium** if its chemical composition does not change with time, that is, no chemical reactions occur. A system will not be in equilibrium unless all the relevant equilibrium criteria are satisfied.

## The State Postulate

As noted earlier, the state of a system is described by its properties. But we know from experience that we do not need to specify all the properties in order to fix a state. Once a sufficient number of properties are specified, the rest of the properties assume certain values automatically. That is, specifying a certain number of properties is sufficient to fix a state. The number of properties required to fix the state of a system is given by the **state postulate**:

> The state of a simple compressible system is completely specified by two independent, intensive properties.

A system is called a **simple compressible system** in the absence of electrical, magnetic, gravitational, motion, and surface tension effects. These effects are due to external force fields and are negligible for most engineering problems. Otherwise, an additional property needs to be specified for each effect that is significant. If the gravitational effects are to be considered, for example, the elevation $z$ needs to be specified in addition to the two properties necessary to fix the state.

The state postulate requires that the two properties specified be independent to fix the state. Two properties are **independent** if one property can be varied while the other one is held constant. Temperature and specific volume, for example, are always independent properties, and together they can fix the state of a simple compressible system (Fig. 2–11). Temperature and pressure, however, are independent properties for single-phase systems, but are dependent properties for multiphase systems. At sea level ($P = 1$ atm), water boils at 100°C, but on a mountaintop where the pressure is lower, water boils at a lower temperature. That is, $T = f(P)$ during a phase-change process; thus, temperature and pressure are not sufficient to fix the state of a two-phase system. Phase-change processes are discussed in detail in Chap. 4.

## 2–5 ▪ PROCESSES AND CYCLES

Any change that a system undergoes from one equilibrium state to another is called a **process**, and the series of states through which a system passes during a process is called the **path** of the process (Fig. 2–12). To describe a process completely, one should specify the initial and final states of the process, as well as the path it follows, and the interactions with the surroundings.

**FIGURE 2–11**

The state of nitrogen is fixed by two independent, intensive properties.

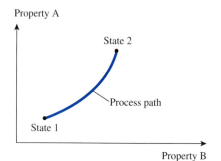

**FIGURE 2–12**

A process between states 1 and 2 and the process path.

When a process proceeds in such a manner that the system remains infinitesimally close to an equilibrium state at all times, it is called a **quasi-static**, or **quasi-equilibrium**, **process**. A quasi-equilibrium process can be viewed as a sufficiently slow process that allows the system to adjust itself internally so that properties in one part of the system do not change any faster than those at other parts.

This is illustrated in Fig. 2–13. When a gas in a piston-cylinder device is compressed suddenly, the molecules near the face of the piston will not have enough time to escape and they will have to pile up in a small region in front of the piston, thus creating a high-pressure region there. Because of this pressure difference, the system can no longer be said to be in equilibrium, and this makes the entire process nonquasi-equilibrium. However, if the piston is moved slowly, the molecules will have sufficient time to redistribute and there will not be a molecule pileup in front of the piston. As a result, the pressure inside the cylinder will always be nearly uniform and will rise at the same rate at all locations. Since equilibrium is maintained at all times, this is a quasi-equilibrium process.

It should be pointed out that a quasi-equilibrium process is an idealized process and is not a true representation of an actual process. But many actual processes closely approximate it, and they can be modeled as quasi-equilibrium with negligible error. Engineers are interested in quasi-equilibrium processes for two reasons. First, they are easy to analyze; second, work-producing devices deliver the most work when they operate on quasi-equilibrium processes. Therefore, quasi-equilibrium processes serve as standards to which actual processes can be compared.

Process diagrams plotted by employing thermodynamic properties as coordinates are very useful in visualizing the processes. Some common properties that are used as coordinates are temperature $T$, pressure $P$, and volume $V$ (or specific volume $v$). Figure 2–14 shows the $P$-$V$ diagram of a compression process of a gas.

Note that the process path indicates a series of equilibrium states through which the system passes during a process and has significance for quasi-equilibrium processes only. For nonquasi-equilibrium processes, we are not able to characterize the entire system by a single state, and thus we cannot speak of a process path for a system as a whole. A nonquasi-equilibrium process is denoted by a dashed line between the initial and final states instead of a solid line.

The prefix *iso-* is often used to designate a process for which a particular property remains constant. An **isothermal process**, for example, is a process during which the temperature $T$ remains constant; an **isobaric process** is a process during which the pressure $P$ remains constant; and an **isochoric** (or **isometric**) **process** is a process during which the specific volume $v$ remains constant.

A system is said to have undergone a **cycle** if it returns to its initial state at the end of the process. That is, for a cycle the initial and final states are identical.

## The Steady-Flow Process

The terms *steady* and *uniform* are used frequently in engineering, and thus it is important to have a clear understanding of their meanings. The term *steady* implies *no change with time*. The opposite of steady is *unsteady,* or *transient.*

(a) Slow compression
(quasi-equilibrium)

(b) Very fast compression
(nonquasi-equilibrium)

**FIGURE 2–13**

Quasi-equilibrium and nonquasi-equilibrium compression processes.

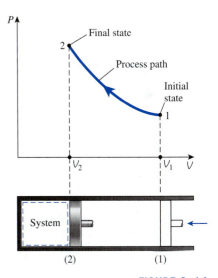

**FIGURE 2–14**

The $P$-$V$ diagram of a compression process.

**FIGURE 2–15**

During a steady-flow process, fluid properties within the control volume may change with position but not with time.

**FIGURE 2–16**

Under steady-flow conditions, the mass and energy contents of a control volume remain constant.

**FIGURE 2–17**

Two bodies reaching thermal equilibrium after being brought into contact in an isolated enclosure.

The term *uniform,* however, implies *no change with location* over a specified region. These meanings are consistent with their everyday use (steady girl-friend, uniform properties, etc.).

A large number of engineering devices operate for long periods of time under the same conditions, and they are classified as *steady-flow devices.* Processes involving such devices can be represented reasonably well by a somewhat idealized process, called the **steady-flow process,** which can be defined as a *process during which a fluid flows through a control volume steadily* (Fig. 2–15). That is, the fluid properties can change from point to point within the control volume, but at any fixed point they remain the same during the entire process. Therefore, the volume $V$, the mass $m$, and the total energy content $E$ of the control volume remain constant during a steady-flow process (Fig. 2–16).

Steady-flow conditions can be closely approximated by devices that are intended for continuous operation such as turbines, pumps, boilers, condensers, and heat exchangers or power plants or refrigeration systems. Some cyclic devices, such as reciprocating engines or compressors, do not satisfy any of the conditions stated above since the flow at the inlets and the exits will be pulsating and not steady. However, the fluid properties vary with time in a periodic manner, and the flow through these devices can still be analyzed as a steady-flow process by using time-averaged values for the properties.

## 2–6 ▪ TEMPERATURE AND THE ZEROTH LAW OF THERMODYNAMICS

Although we are familiar with temperature as a measure of "hotness" or "coldness," it is not easy to give an exact definition for it. Based on our physiological sensations, we express the level of temperature qualitatively with words like *freezing cold, cold, warm, hot,* and *red-hot.* However, we cannot assign numerical values to temperatures based on our sensations alone. Furthermore, our senses may be misleading. A metal chair, for example, will feel much colder than a wooden one even when both are at the same temperature.

Fortunately, several properties of materials change with temperature in a *repeatable* and *predictable* way, and this forms the basis for accurate temperature measurement. The commonly used mercury-in-glass thermometer, for example, is based on the expansion of mercury with temperature. Temperature is also measured by using several other temperature-dependent properties.

It is a common experience that a cup of hot coffee left on the table eventually cools off and a cold drink eventually warms up. That is, when a body is brought into contact with another body that is at a different temperature, heat is transferred from the body at higher temperature to the one at lower temperature until both bodies attain the same temperature (Fig. 2–17). At that point, the heat transfer stops, and the two bodies are said to have reached **thermal equilibrium.** The equality of temperature is the only requirement for thermal equilibrium.

The **zeroth law of thermodynamics** states that if two bodies are in thermal equilibrium with a third body, they are also in thermal equilibrium with each other. It may seem silly that such an obvious fact is called one of the basic laws of thermodynamics. However, it cannot be concluded from the other laws of thermodynamics, and it serves as a basis for the validity of temperature measurement. By replacing the third body with a thermometer, the zeroth law can be restated as *two bodies are in thermal equilibrium if both have the same temperature reading even if they are not in contact.*

The zeroth law was first formulated and labeled by R. H. Fowler in 1931. As the name suggests, its value as a fundamental physical principle was recognized more than half a century after the formulation of the first and the second laws of thermodynamics. It was named the zeroth law since it should have preceded the first and the second laws of thermodynamics.

## Temperature Scales

Temperature scales enable us to use a common basis for temperature measurements, and several have been introduced throughout history. All temperature scales are based on some easily reproducible states such as the freezing and boiling points of water, which are also called the *ice point* and the *steam point,* respectively. A mixture of ice and water that is in equilibrium with air saturated with vapor at 1 atm pressure is said to be at the ice point, and a mixture of liquid water and water vapor (with no air) in equilibrium at 1 atm pressure is said to be at the steam point.

The temperature scales used in the SI and in the English system today are the **Celsius scale** (formerly called the *centigrade scale;* in 1948 it was renamed after the Swedish astronomer A. Celsius, 1702–1744, who devised it) and the **Fahrenheit scale** (named after the German instrument maker G. Fahrenheit, 1686–1736), respectively. On the Celsius scale, the ice and steam points were originally assigned the values of 0 and 100°C, respectively. The corresponding values on the Fahrenheit scale are 32 and 212°F. These are often referred to as *two-point scales* since temperature values are assigned at two different points.

In thermodynamics, it is very desirable to have a temperature scale that is independent of the properties of any substance or substances. Such a temperature scale is called a **thermodynamic temperature scale**, which is developed later in conjunction with the second law of thermodynamics. The thermodynamic temperature scale in the SI is the **Kelvin scale**, named after Lord Kelvin (1824–1907). The temperature unit on this scale is the **kelvin**, which is designated by K (not °K; the degree symbol was officially dropped from kelvin in 1967). The lowest temperature on the Kelvin scale is absolute zero, or 0 K. Then it follows that only one nonzero reference point needs to be assigned to establish the slope of this linear scale. Using nonconventional refrigeration techniques, scientists have approached absolute zero kelvin (they achieved 0.000000002 K in 1989).

The thermodynamic temperature scale in the English system is the **Rankine scale**, named after William Rankine (1820–1872). The temperature unit on this scale is the **rankine**, which is designated by R.

A temperature scale that turns out to be nearly identical to the Kelvin scale is the **ideal-gas temperature scale**. The temperatures on this scale are measured using a **constant-volume gas thermometer**, which is basically a rigid vessel filled with a gas, usually hydrogen or helium, at low pressure.

This thermometer is based on the principle that *at low pressures, the temperature of a gas is proportional to its pressure at constant volume.* That is, the temperature of a gas of fixed volume varies *linearly* with pressure at sufficiently low pressures. Then the relationship between the temperature and the pressure of the gas in the vessel can be expressed as

$$T = a + bP \qquad (2\text{–}5)$$

where the values of the constants $a$ and $b$ for a gas thermometer are determined experimentally. Once $a$ and $b$ are known, the temperature of a medium can be calculated from this relation by immersing the rigid vessel of the gas thermometer into the medium and measuring the gas pressure when thermal equilibrium is established between the medium and the gas in the vessel whose volume is held constant.

An ideal-gas temperature scale can be developed by measuring the pressures of the gas in the vessel at two reproducible points (such as the ice and the steam points) and assigning suitable values to temperatures at those two points. Considering that only one straight line passes through two fixed points on a plane, these two measurements are sufficient to determine the constants $a$ and $b$ in Eq. 2–5. Then the unknown temperature $T$ of a medium corresponding to a pressure reading $P$ can be determined from that equation by a simple calculation. The values of the constants will be different for each thermometer, depending on the type and the amount of the gas in the vessel, and the temperature values assigned at the two reference points. If the ice and steam points are assigned the values 0°C and 100°C, respectively, then the gas temperature scale will be identical to the Celsius scale. In this case the value of the constant $a$ (which corresponds to an absolute pressure of zero) is determined to be −273.15°C regardless of the type and the amount of the gas in the vessel of the gas thermometer. That is, on a *P-T* diagram, all the straight lines passing through the data points in this case will intersect the temperature axis at −273.15°C when extrapolated, as shown in Fig. 2–18. This is the lowest temperature that can be obtained by a gas thermometer, and thus we can obtain an *absolute gas temperature scale* by assigning a value of zero to the constant $a$ in Eq. 2–5. In that case, Eq. 2–5 reduces to $T = bP$, and thus we need to specify the temperature at only *one* point to define an absolute gas temperature scale.

It should be noted that the absolute gas temperature scale is not a thermodynamic temperature scale, since it cannot be used at very low temperatures (due to condensation) and at very high temperatures (due to dissociation and ionization). However, absolute gas temperature is identical to the thermodynamic temperature in the temperature range in which the gas thermometer can be used. Thus, we can view the thermodynamic temperature scale at this point as an absolute gas temperature scale that utilizes an "ideal" or "imaginary" gas that always acts as a low-pressure gas regardless of the temperature. If such a gas thermometer existed, it would read zero kelvin at absolute zero pressure, which corresponds to −273.15°C on the Celsius scale (Fig. 2–19).

The Kelvin scale is related to the Celsius scale by

$$T(\text{K}) = T(°\text{C}) + 273.15 \qquad (2\text{–}6)$$

The Rankine scale is related to the Fahrenheit scale by

$$T(\text{R}) = T(°\text{F}) + 459.67 \qquad (2\text{–}7)$$

**FIGURE 2–18**

*P* versus *T* plots of the experimental data obtained from a constant-volume gas thermometer using four different gases at different (but low) pressures.

**FIGURE 2–19**

A constant-volume gas thermometer would read −273.15°C at absolute zero pressure.

It is common practice to round the constant in Eq. 2–6 to 273 and that in Eq. 2–7 to 460.

The temperature scales in the two unit systems are related by

$$T(\text{R}) = 1.8T(\text{K}) \tag{2–8}$$

$$T(°\text{F}) = 1.8T(°\text{C}) + 32 \tag{2–9}$$

A comparison of various temperature scales is given in Fig. 2–20.

The reference temperature chosen in the original Kelvin scale was 273.15 K (or 0°C), which is the temperature at which water freezes (or ice melts) and water exists as a solid–liquid mixture in equilibrium under standard atmospheric pressure (the *ice point*). At the Tenth General Conference on Weights and Measures in 1954, the reference point was changed to a much more precisely reproducible point, the *triple point* of water (the state at which all three phases of water coexist in equilibrium), which is assigned the value 273.16 K. The Celsius scale was also redefined at this conference in terms of the ideal-gas temperature scale and a single fixed point, which is again the triple point of water with an assigned value of 0.01°C. The boiling temperature of water (the *steam point*) was experimentally determined to be again 100.00°C, and thus the new and old Celsius scales were in good agreement.

## The International Temperature Scale of 1990 (ITS-90)

The *International Temperature Scale of 1990*, which supersedes the International Practical Temperature Scale of 1968 (IPTS-68), 1948 (ITPS-48), and 1927 (ITS-27), was adopted by the International Committee of Weights and Measures at its meeting in 1989 at the request of the Eighteenth General Conference on Weights and Measures. The ITS-90 is similar to its predecessors except that it is more refined with updated values of fixed temperatures, has an extended range, and conforms more closely to the thermodynamic temperature scale. On this scale, the unit of thermodynamic temperature $T$ is again the kelvin (K), defined as the fraction 1/273.16 of the thermodynamic temperature of the triple point of water, which is sole defining fixed point of both the ITS-90 and the Kelvin scale and is the most important thermometric fixed point used in the calibration of thermometers to ITS-90.

The unit of Celsius temperature is the degree Celsius (°C), which is by definition equal in magnitude to the kelvin (K). A temperature difference may be expressed in kelvins or degrees Celsius. The ice point remains the same at 0°C (273.15 K) in both ITS-90 and ITPS-68, but the steam point is 99.975°C in ITS-90 (with an uncertainty of ±0.005°C) whereas it was 100.000°C in IPTS-68. The change is due to precise measurements made by gas thermometry by paying particular attention to the effect of sorption (the impurities in a gas absorbed by the walls of the bulb at the reference temperature being desorbed at higher temperatures, causing the measured gas pressure to increase).

The ITS-90 extends upward from 0.65 K to the highest temperature practically measurable in terms of the Planck radiation law using monochromatic radiation. It is based on specifying definite temperature values on a

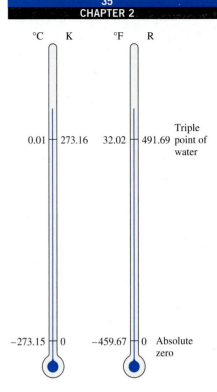

**FIGURE 2–20**
Comparison of temperature scales.

**FIGURE 2–21**

Comparison of magnitudes of various temperature units.

number of fixed and easily reproducible points to serve as benchmarks and expressing the variation of temperature in a number of ranges and subranges in functional form.

In ITS-90, the temperature scale is considered in four ranges. In the range of 0.65 to 5 K, the temperature scale is defined in terms of the vapor pressure—temperature relations for $^3$He and $^4$He. Between 3 and 24.5561 K (the triple point of neon), it is defined by means of a properly calibrated helium gas thermometer. From 13.8033 K (the triple point of hydrogen) to 1234.93 K (the freezing point of silver), it is defined by means of platinum resistance thermometers calibrated at specified sets of defining fixed points. Above 1234.93 K, it is defined in terms of the Planck radiation law and a suitable defining fixed point such as the freezing point of gold (1337.33 K).

We emphasize that the magnitudes of each division of 1 K and 1°C are identical (Fig. 2–21). Therefore, when we are dealing with temperature differences $\Delta T$, the temperature interval on both scales is the same. Raising the temperature of a substance by 10°C is the same as raising it by 10 K. That is

$$\Delta T(K) = \Delta T(°C) \qquad (2\text{–}10)$$

$$\Delta T(R) = \Delta T(°F) \qquad (2\text{–}11)$$

Some thermodynamic relations involve the temperature $T$ and often the question arises of whether it is in K or °C. If the relation involves temperature differences (such as $a = b\Delta T$), it makes no difference and either can be used. However, if the relation involves temperatures only instead of temperature differences (such as $a = bT$) then K must be used. When in doubt, it is always safe to use K because there are virtually no situations in which the use of K is incorrect, but there are many thermodynamic relations that will yield an erroneous result if °C is used.

---

**EXAMPLE 2–1**     **Expressing Temperature Rise in Different Units**

During a heating process, the temperature of a system rises by 10°C. Express this rise in temperature in K, °F, and R.

**SOLUTION**   The temperature rise of a system is to be expressed in different units.
**Analysis**   This problem deals with temperature changes, which are identical in Kelvin and Celsius scales. Then,

$$\Delta T(K) = \Delta T(°C) = \mathbf{10\ K}$$

The temperature changes in Fahrenheit and Rankine scales are also identical and are related to the changes in Celsius and Kelvin scales through Eqs. 2–8 and 2–11:

$$\Delta T(R) = 1.8\ \Delta T(K) = (1.8)(10) = \mathbf{18\ R}$$

and

$$\Delta T(°F) = \Delta T(R) = \mathbf{18°F}$$

**Discussion**   Note that the units °C and K are interchangeable when dealing with temperature differences.

# 2–7 ▪ PRESSURE

**Pressure** is defined as *a normal force exerted by a fluid per unit area.* We speak of pressure only when we deal with a gas or a liquid. The counterpart of pressure in solids is *normal stress*. Since pressure is defined as force per unit area, it has the unit of newtons per square meter ($N/m^2$), which is called a **pascal** (Pa). That is,

$$1\ Pa = 1\ N/m^2$$

The pressure unit pascal is too small for pressures encountered in practice. Therefore, its multiples *kilopascal* ($1\ kPa = 10^3\ Pa$) and *megapascal* ($1\ MPa = 10^6\ Pa$) are commonly used. Three other pressure units commonly used in practice, especially in Europe, are *bar, standard atmosphere,* and *kilogram-force per square centimeter:*

$$1\ bar = 10^5\ Pa = 0.1\ MPa = 100\ kPa$$

$$1\ atm = 101{,}325\ Pa = 101.325\ kPa = 1.01325\ bars$$

$$1\ kgf/cm^2 = 9.807\ N/cm^2 = 9.807 \times 10^4\ N/m^2 = 9.807 \times 10^4\ Pa$$

$$= 0.9807\ bar$$

$$= 0.9679\ atm$$

Note that the pressure units bar, atm, and $kgf/cm^2$ are almost equivalent to each other. In the English system, the pressure unit is *pound-force per square inch* ($lbf/in^2$, or psi), and $1\ atm = 14.696\ psi$. The pressure units $kgf/cm^2$ and $lbf/in^2$ are also denoted by $kg/cm^2$ and $lb/in^2$, respectively, and they are commonly used in tire gages. It can be shown that $1\ kgf/cm^2 = 14.223\ psi$.

Pressure is also used for solids as synonymous to *normal stress,* which is force acting perpendicular to the surface per unit area. For example, a 150-pound person with a total foot imprint area of 50 in$^2$ exerts a pressure of 150 lbf/50 in$^2$ = 3.0 psi on the floor (Fig. 2–22). If the person stands on one foot, the pressure doubles. If the person gains excessive weight, he or she is likely to encounter foot discomfort because of the increased pressure on the foot (the size of the foot does not change with weight gain). This also explains how a person can walk on fresh snow without sinking by wearing large snowshoes, and how a person cuts with little effort when using a sharp knife.

The actual pressure at a given position is called the **absolute pressure**, and it is measured relative to absolute vacuum (i.e., absolute zero pressure). Most pressure-measuring devices, however, are calibrated to read zero in the atmosphere (Fig. 2–23), and so they indicate the difference between the absolute pressure and the local atmospheric pressure. This difference is called the **gage pressure**. Pressures below atmospheric pressure are called **vacuum pressures** and are measured by vacuum gages that indicate the difference between the atmospheric pressure and the absolute pressure. Absolute, gage, and vacuum pressures are all positive quantities and are related to each other by

$$P_{gage} = P_{abs} - P_{atm} \tag{2–12}$$

$$P_{vac} = P_{atm} - P_{abs} \tag{2–13}$$

This is illustrated in Fig. 2–24.

150 pounds    300 pounds

$A_{feet} = 50\ in^2$

$P = 3\ psi$    $P = 6\ psi$

$$P = \sigma_n = \frac{W}{A_{feet}} = \frac{150\ lbf}{50\ in^2} = 3\ psi$$

**FIGURE 2–22**

The normal stress (or "pressure") on the feet of a chubby person is much greater than on the feet of a slim person.

**FIGURE 2–23**

Some basic pressure gages.

*Dresser Instruments, Dresser, Inc.*
*Used by permission.*

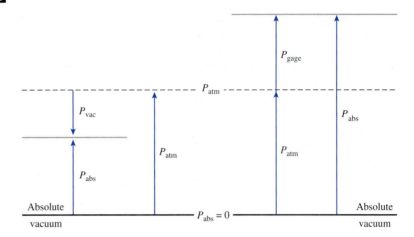

**FIGURE 2–24**

Absolute, gage, and vacuum pressures.

Like other pressure gages, the gage used to measure the air pressure in an automobile tire reads the gage pressure. Therefore, the common reading of 32 psi (2.25 kgf/cm$^2$) indicates a pressure of 32 psi above the atmospheric pressure. At a location where the atmospheric pressure is 14.3 psi, for example, the absolute pressure in the tire is $32 + 14.3 = 46.3$ psi.

In thermodynamic relations and tables, absolute pressure is almost always used. Throughout this text, the pressure $P$ will denote *absolute pressure* unless specified otherwise. Often the letters "a" (for absolute pressure) and "g" (for gage pressure) are added to pressure units (such as psia and psig) to clarify what is meant.

---

*EXAMPLE 2–2*    **Absolute Pressure of a Vacuum Chamber**

A vacuum gage connected to a chamber reads 5.8 psi at a location where the atmospheric pressure is 14.5 psi. Determine the absolute pressure in the chamber.

**SOLUTION**   The gage pressure of a vacuum chamber is given. The absolute pressure in the chamber is to be determined.
*Analysis*   The absolute pressure is easily determined from Eq. 2–13 to be

$$P_{abs} = P_{atm} - P_{vac} = 14.5 - 5.8 = \textbf{8.7 psi}$$

*Discussion*   Note that the local value of the atmospheric pressure is used when determining the absolute pressure.

---

Pressure is the *compressive force* per unit area, and it gives the impression of being a vector. However, pressure at any point in a fluid is the same in all directions. That is, it has magnitude but not a specific direction, and thus it is a scalar quantity.

## Variation of Pressure with Depth

It will come as no surprise to you that pressure in a fluid at rest does not change in the horizontal direction. This can be shown easily by considering with each other. It may seem silly that such an obvious fact is called one of the basic laws of thermodynamics. However, it cannot be concluded from the other laws of thermodynamics, and it serves as a basis for the validity of temperature measurement. By replacing the third body with a thermometer, the zeroth law can be restated as *two bodies are in thermal equilibrium if both have the same temperature reading even if they are not in contact.*

The zeroth law was first formulated and labeled by R. H. Fowler in 1931. As the name suggests, its value as a fundamental physical principle was recognized more than half a century after the formulation of the first and the second laws of thermodynamics. It was named the zeroth law since it should have preceded the first and the second laws of thermodynamics.

## Temperature Scales

Temperature scales enable us to use a common basis for temperature measurements, and several have been introduced throughout history. All temperature scales are based on some easily reproducible states such as the freezing and boiling points of water, which are also called the *ice point* and the *steam point,* respectively. A mixture of ice and water that is in equilibrium with air saturated with vapor at 1 atm pressure is said to be at the ice point, and a mixture of liquid water and water vapor (with no air) in equilibrium at 1 atm pressure is said to be at the steam point.

The temperature scales used in the SI and in the English system today are the **Celsius scale** (formerly called the *centigrade scale;* in 1948 it was renamed after the Swedish astronomer A. Celsius, 1702–1744, who devised it) and the **Fahrenheit scale** (named after the German instrument maker G. Fahrenheit, 1686–1736), respectively. On the Celsius scale, the ice and steam points were originally assigned the values of 0 and 100°C, respectively. The corresponding values on the Fahrenheit scale are 32 and 212°F. These are often referred to as *two-point scales* since temperature values are assigned at two different points.

In thermodynamics, it is very desirable to have a temperature scale that is independent of the properties of any substance or substances. Such a temperature scale is called a **thermodynamic temperature scale**, which is developed later in conjunction with the second law of thermodynamics. The thermodynamic temperature scale in the SI is the **Kelvin scale**, named after Lord Kelvin (1824–1907). The temperature unit on this scale is the **kelvin**, which is designated by K (not °K; the degree symbol was officially dropped from kelvin in 1967). The lowest temperature on the Kelvin scale is absolute zero, or 0 K. Then it follows that only one nonzero reference point needs to be assigned to establish the slope of this linear scale. Using nonconventional refrigeration techniques, scientists have approached absolute zero kelvin (they achieved 0.000000002 K in 1989).

The thermodynamic temperature scale in the English system is the **Rankine scale**, named after William Rankine (1820–1872). The temperature unit on this scale is the **rankine**, which is designated by R.

A temperature scale that turns out to be nearly identical to the Kelvin scale is the **ideal-gas temperature scale**. The temperatures on this scale are measured using a **constant-volume gas thermometer**, which is basically a rigid vessel filled with a gas, usually hydrogen or helium, at low pressure.

This thermometer is based on the principle that *at low pressures, the temperature of a gas is proportional to its pressure at constant volume*. That is, the temperature of a gas of fixed volume varies *linearly* with pressure at sufficiently low pressures. Then the relationship between the temperature and the pressure of the gas in the vessel can be expressed as

$$T = a + bP \tag{2–5}$$

where the values of the constants $a$ and $b$ for a gas thermometer are determined experimentally. Once $a$ and $b$ are known, the temperature of a medium can be calculated from this relation by immersing the rigid vessel of the gas thermometer into the medium and measuring the gas pressure when thermal equilibrium is established between the medium and the gas in the vessel whose volume is held constant.

An ideal-gas temperature scale can be developed by measuring the pressures of the gas in the vessel at two reproducible points (such as the ice and the steam points) and assigning suitable values to temperatures at those two points. Considering that only one straight line passes through two fixed points on a plane, these two measurements are sufficient to determine the constants $a$ and $b$ in Eq. 2–5. Then the unknown temperature $T$ of a medium corresponding to a pressure reading $P$ can be determined from that equation by a simple calculation. The values of the constants will be different for each thermometer, depending on the type and the amount of the gas in the vessel, and the temperature values assigned at the two reference points. If the ice and steam points are assigned the values 0°C and 100°C, respectively, then the gas temperature scale will be identical to the Celsius scale. In this case the value of the constant $a$ (which corresponds to an absolute pressure of zero) is determined to be −273.15°C regardless of the type and the amount of the gas in the vessel of the gas thermometer. That is, on a *P-T* diagram, all the straight lines passing through the data points in this case will intersect the temperature axis at −273.15°C when extrapolated, as shown in Fig. 2–18. This is the lowest temperature that can be obtained by a gas thermometer, and thus we can obtain an *absolute gas temperature scale* by assigning a value of zero to the constant $a$ in Eq. 2–5. In that case, Eq. 2–5 reduces to $T = bP$, and thus we need to specify the temperature at only *one* point to define an absolute gas temperature scale.

It should be noted that the absolute gas temperature scale is not a thermodynamic temperature scale, since it cannot be used at very low temperatures (due to condensation) and at very high temperatures (due to dissociation and ionization). However, absolute gas temperature is identical to the thermodynamic temperature in the temperature range in which the gas thermometer can be used. Thus, we can view the thermodynamic temperature scale at this point as an absolute gas temperature scale that utilizes an "ideal" or "imaginary" gas that always acts as a low-pressure gas regardless of the temperature. If such a gas thermometer existed, it would read zero kelvin at absolute zero pressure, which corresponds to −273.15°C on the Celsius scale (Fig. 2–19).

The Kelvin scale is related to the Celsius scale by

$$T(\text{K}) = T(°\text{C}) + 273.15 \tag{2–6}$$

The Rankine scale is related to the Fahrenheit scale by

$$T(\text{R}) = T(°\text{F}) + 459.67 \tag{2–7}$$

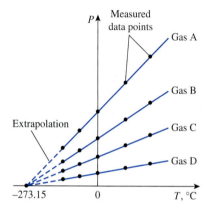

**FIGURE 2–18**

*P* versus *T* plots of the experimental data obtained from a constant-volume gas thermometer using four different gases at different (but low) pressures.

**FIGURE 2–19**

A constant-volume gas thermometer would read −273.15°C at absolute zero pressure.

a thin horizontal layer of fluid and doing a force balance in any horizontal direction. However, this is not the case in the vertical direction in a gravity field. Pressure in a fluid increases with depth because more fluid rests on deeper layers, and the effect of this "extra weight" on a deeper layer is balanced by an increase in pressure (Fig. 2–25).

To obtain a relation for the variation of pressure with depth, consider a rectangular fluid element of height $\Delta z$, length $\Delta x$, and unit depth (into the page) in equilibrium, as shown in Fig. 2–26. Assuming the density of the fluid $\rho$ to be constant, a force balance in the vertical $z$-direction gives

$$\sum F_z = ma_z = 0: \qquad P_2 \Delta x - P_1 \Delta x - \rho g \, \Delta x \, \Delta z = 0 \qquad \text{(2–14)}$$

where $W = mg = \rho g \, \Delta x \, \Delta z$ is the weight of the fluid element. Dividing by $\Delta x$ and rearranging gives

$$\Delta P = P_2 - P_1 = \rho g \, \Delta z = \gamma_s \, \Delta z \qquad \text{(2–15)}$$

where $\gamma_s = \rho g$ is the *specific weight* of the fluid. Thus, we conclude that the pressure difference between two points in a constant density fluid is proportional to the vertical distance $\Delta z$ between the points and the density $\rho$ of the fluid. In other words, pressure in a fluid increases linearly with depth. This is what a diver experiences when diving deeper in a lake. For a given fluid, the vertical distance $\Delta z$ is sometimes used as a measure of pressure, and it is called the *pressure head*.

We also conclude from Eq. 2–15 that for small to moderate distances, the variation of pressure with height is negligible for gases because of their low density. The pressure in a tank containing a gas, for example, can be considered to be uniform since the weight of the gas is too small to make a significant difference. Also, the pressure in a room filled with air can be assumed to be constant (Fig. 2–27).

If we take point 1 to be at the free surface of a liquid open to the atmosphere (Fig. 2–28), where the pressure is the atmospheric pressure $P_{\text{atm}}$, then the pressure at a depth $h$ from the free surface becomes

$$P = P_{\text{atm}} + \rho g h \quad \text{or} \quad P_{\text{gage}} = \rho g h \qquad \text{(2–16)}$$

Liquids are essentially incompressible substances, and thus the variation of density with depth is negligible. This is also the case for gases when the elevation change is not very large. The variation of density of liquids or gases with temperature can be significant, however, and may need to be considered when high accuracy is desired. Also, at great depths such as those encountered in oceans, the change in the density of a liquid can be significant because of the compression by the tremendous amount of liquid weight above.

The gravitational acceleration $g$ varies from 9.807 m/s² at sea level to 9.764 m/s² at an elevation of 14,000 m where large passenger planes cruise. This is a change of just 0.4 percent in this extreme case. Therefore, $g$ can be assumed to be constant with negligible error.

For fluids whose density changes significantly with elevation, a relation for the variation of pressure with elevation can be obtained by dividing Eq. 2–14 by $\Delta x \, \Delta z$, and taking the limit as $\Delta z \to 0$. It gives

$$\frac{dP}{dz} = -\rho g \qquad \text{(2–17)}$$

**FIGURE 2–25**

The pressure of a fluid at rest increases with depth (as a result of added weight).

**FIGURE 2–26**

Free-body diagram of a rectangular fluid element in equilibrium.

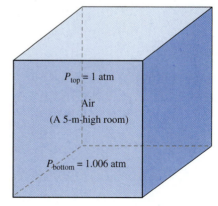

**FIGURE 2–27**

In a room filled with a gas, the variation of pressure with height is negligible.

**FIGURE 2–28**

Pressure in a liquid at rest increases linearly with distance from the free surface.

The negative sign is due to our taking the positive $z$ direction to be upward so that $dP$ is negative when $dz$ is positive since pressure decreases in an upward direction. When the variation of density with elevation is known, the pressure difference between points 1 and 2 can be determined by integration to be

$$\Delta P = P_2 - P_1 = -\int_1^2 \rho g \, dz \qquad (2\text{–}18)$$

For constant density and constant gravitational acceleration, this relation reduces to Eq. 2–15, as expected.

Pressure in a fluid at rest is independent of the shape or cross section of the container. It changes with the vertical distance, but remains constant in other directions. Therefore, the pressure is the same at all points on a horizontal plane in a given fluid. The Dutch mathematician Simon Stevin (1548–1620) published in 1586 the principle illustrated in Fig. 2–29. Note that the pressures at points $A$, $B$, $C$, $D$, $E$, $F$, and $G$ are the same since they are at the same depth, and they are interconnected by the same static fluid. However, the pressures at points $H$ and $I$ are not the same since these two points cannot be interconnected by the same fluid (i.e., we cannot draw a curve from point $I$ to point $H$ while remaining in the same fluid at all times), although they are at the same depth. (Can you tell at which point the pressure is higher?) Also, the pressure force exerted by the fluid is always normal to the surface at the specified points.

A consequence of the pressure in a fluid remaining constant in the horizontal direction is that *the pressure applied to a confined fluid increases the pressure throughout by the same amount.* This is called **Pascal's law**, after Blaise Pascal (1623–1662). Pascal also knew that the force applied by a fluid is proportional

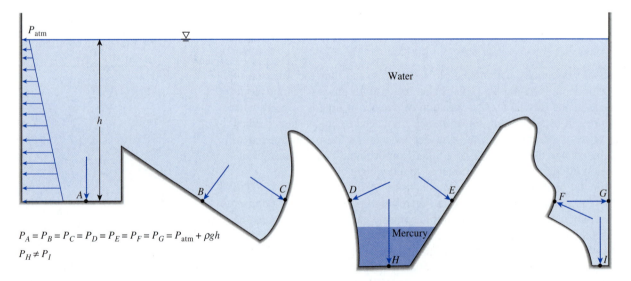

$P_A = P_B = P_C = P_D = P_E = P_F = P_G = P_{atm} + \rho g h$

$P_H \neq P_I$

**FIGURE 2–29**

The pressure is the same at all points on a horizontal plane in a given fluid regardless of geometry, provided that the points are interconnected by the same fluid.

to the surface area. He realized that two hydraulic cylinders of different areas could be connected, and the larger could be used to exert a proportionally greater force than that applied to the smaller. "Pascal's machine" has been the source of many inventions that are a part of our daily lives such as hydraulic brakes and lifts. This is what enables us to lift a car easily by one arm, as shown in Fig. 2–30. Noting that $P_1 = P_2$ since both pistons are at the same level (the effect of small height differences is negligible, especially at high pressures), the ratio of output force to input force is determined to be

$$P_1 = P_2 \rightarrow \frac{F_1}{A_1} = \frac{F_2}{A_2} \rightarrow \frac{F_2}{F_1} = \frac{A_2}{A_1} \qquad \text{(2–19)}$$

The area ratio $A_2/A_1$ is called the *ideal mechanical advantage* of the hydraulic lift. Using a hydraulic car jack with a piston area ratio of $A_2/A_1 = 10$, for example, a person can lift a 1000-kg car by applying a force of just 100 kgf (=981 N).

**FIGURE 2–30**

Lifting of a large weight by a small force by the application of Pascal's law.

# 2–8 ▪ THE MANOMETER

We notice from Eq. 2–15 that an elevation change of $\Delta z$ in a fluid at rest corresponds to $\Delta P/\rho g$, which suggests that a fluid column can be used to measure pressure differences. A device based on this principle is called a **manometer**, and it is commonly used to measure small and moderate pressure differences. A manometer mainly consists of a glass or plastic U-tube containing one or more fluids such as mercury, water, alcohol, or oil. To keep the size of the manometer to a manageable level, heavy fluids such as mercury are used if large pressure differences are anticipated.

Consider the manometer shown in Fig. 2–31 that is used to measure the pressure in the tank. Since the gravitational effects of gases are negligible, the pressure anywhere in the tank and at position 1 has the same value. Furthermore, since pressure in a fluid does not vary in the horizontal direction within a fluid, the pressure at point 2 is the same as the pressure at point 1, $P_2 = P_1$.

The differential fluid column of height $h$ is in static equilibrium, and it is open to the atmosphere. Then the pressure at point 2 is determined directly from Eq. 2–16 to be

$$P_2 = P_{atm} + \rho g h \qquad \text{(2–20)}$$

where $\rho$ is the density of the fluid in the tube. Note that the cross-sectional area of the tube has no effect on the differential height $h$, and thus the pressure exerted by the fluid. However, the diameter of the tube should be large enough (more than a few millimeters) to ensure that the surface tension effect and thus the capillary rise is negligible.

**FIGURE 2–31**

The basic manometer.

---

**EXAMPLE 2–3**     **Measuring Pressure with a Manometer**

A manometer is used to measure the pressure in a tank. The fluid used has a specific gravity of 0.85, and the manometer column height is 55 cm, as shown in Fig. 2–32. If the local atmospheric pressure is 96 kPa, determine the absolute pressure within the tank.

**FIGURE 2–32**

Schematic for Example 2–3.

**FIGURE 2–33**

In stacked-up fluid layers, the pressure change across a fluid layer of density $\rho$ and height $h$ is $\rho g h$.

Many engineering problems and some manometers involve multiple immiscible fluids of different densities stacked on top of each other. Such systems can be analyzed easily by remembering that (1) the pressure change across a fluid column of height $h$ is $\Delta P = \rho g h$, (2) pressure increases downward in a given fluid and decreases upward (i.e., $P_{bottom} > P_{top}$), and (3) two points at the same elevation in a continuous fluid at rest are at the same pressure.

The last principle, which is a result of *Pascal's law,* allows us to "jump" from one fluid column to the next in manometers without worrying about pressure change as long as we don't jump over a different fluid, and the fluid is at rest. Then the pressure at any point can be determined by starting with a point of known pressure and adding or subtracting $\rho g h$ terms as we advance toward the point of interest. For example, the pressure at the bottom of the tank in Fig. 2–33 can be determined by starting at the free surface where the pressure is $P_{atm}$, moving downward until we reach point 1 at the bottom, and setting the result equal to $P_1$. It gives

$$P_{atm} + \rho_1 g h_1 + \rho_2 g h_2 + \rho_3 g h_3 = P_1$$

In the special case of all fluids having the same density, this relation reduces to Eq. 2–20, as expected.

Manometers are particularly well-suited to measure pressure drops across a horizontal flow section between two specified points due to the presence of a device such as a valve or heat exchanger or any resistance to flow. This is done by connecting the two legs of the manometer to these two points, as shown in Fig. 2–34. The working fluid can be either a gas or a liquid whose density is $\rho_1$. The density of the manometer fluid is $\rho_2$, and the differential fluid height is $h$.

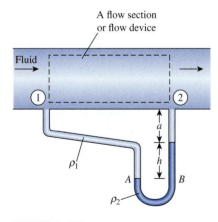

**FIGURE 2–34**

Measuring the pressure drop across a flow section or a flow device by a differential manometer.

A relation for the pressure difference $P_1 - P_2$ can be obtained by starting at point 1 with $P_1$, moving along the tube by adding or subtracting the $\rho g h$ terms until we reach point 2, and setting the result equal to $P_2$:

$$P_1 + \rho_1 g(a + h) - \rho_2 g h - \rho_1 g a = P_2 \qquad \text{(2–21)}$$

Note that we jumped from point $A$ horizontally to point $B$ and ignored the part underneath since the pressure at both points is the same. Simplifying,

$$P_1 - P_2 = (\rho_2 - \rho_1)gh \qquad \text{(2–22)}$$

Note that the distance $a$ has no effect on the result, but must be included in the analysis. Also, when the fluid flowing in the pipe is a gas, then $\rho_1 \ll \rho_2$ and the relation in Eq. 2–22 simplifies to $P_1 - P_2 \cong \rho_2 g h$.

---

### EXAMPLE 2–4    Measuring Pressure with a Multifluid Manometer

The water in a tank is pressurized by air, and the pressure is measured by a multifluid manometer as shown in Fig. 2–35. The tank is located on a mountain at an altitude of 1400 m where the atmospheric pressure is 85.6 kPa. Determine the air pressure in the tank if $h_1 = 0.1$ m, $h_2 = 0.2$ m, and $h_3 = 0.35$ m. Take the densities of water, oil, and mercury to be 1000 kg/m³, 850 kg/m³, and 13,600 kg/m³, respectively.

**SOLUTION**   The pressure in a pressurized water tank is measured by a multifluid manometer. The air pressure in the tank is to be determined.

**Assumption**   The air pressure in the tank is uniform (i.e., its variation with elevation is negligible due to its low density), and thus we can determine the pressure at the air–water interface.

**Properties**   The densities of water, oil, and mercury are given to be 1000 kg/m³, 850 kg/m³, and 13,600 kg/m³, respectively.

**Analysis**   Starting with the pressure at point 1 at the air–water interface, moving along the tube by adding or subtracting the $\rho g h$ terms until we reach point 2, and setting the result equal to $P_{atm}$ since the tube is open to the atmosphere gives

$$P_1 + \rho_{water} g h_1 + \rho_{oil} g h_2 - \rho_{mercury} g h_3 = P_{atm}$$

Solving for $P_1$ and substituting,

$$P_1 = P_{atm} - \rho_{water} g h_1 - \rho_{oil} g h_2 + \rho_{mercury} g h_3$$

$$= P_{atm} + g(\rho_{mercury} h_3 - \rho_{water} h_1 - \rho_{oil} h_2)$$

$$= 85.6 \text{ kPa} + (9.81 \text{ m/s}^2)[(13{,}600 \text{ kg/m}^3)(0.35\text{m}) - 1000 \text{ kg/m}^3)(0.1 \text{ m})$$

$$- (850 \text{ kg/m}^3)(0.2 \text{ m})]\left(\frac{1 \text{ N}}{1 \text{ kg·m/s}^2}\right)\left(\frac{1 \text{ kPa}}{1000 \text{ N/m}^2}\right)$$

$$= \textbf{130 kPa}$$

**Discussion**   Note that jumping horizontally from one tube to the next and realizing that pressure remains the same in the same fluid simplifies the analysis considerably. Also note that mercury is a toxic fluid, and mercury manometers and thermometers are being replaced by ones with safer fluids because of the risk of exposure to mercury vapor during an accident.

**FIGURE 2–35**

Schematic for Example 2–4. (Drawing not to scale.)

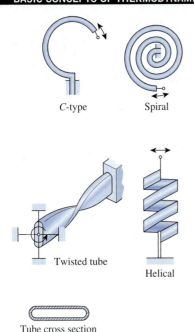

*C*-type      Spiral

Twisted tube

Helical

Tube cross section

**FIGURE 2–36**

Various types of Bourdon tubes used to measure pressure.

# Other Pressure Measurement Devices

Another type of commonly used mechanical pressure measurement device is the **Bourdon tube**, named after the French engineer and inventor Eugene Bourdon (1808–1884), which consists of a hollow metal tube bent like a hook whose end is closed and connected to a dial indicator needle (Fig. 2–36). When the tube is open to the atmosphere, the tube is undeflected, and the needle on the dial at this state is calibrated to read zero (gage pressure). When the fluid inside the tube is pressurized, the tube stretches and moves the needle in proportion to the pressure applied.

Electronics have made their way into every aspect of life, including pressure measurement devices. Modern pressure sensors, called **pressure transducers**, use various techniques to convert the pressure effect to an electrical effect such as a change in voltage, resistance, or capacitance. Pressure transducers are smaller and faster, and they can be more sensitive, reliable, and precise than their mechanical counterparts. They can measure pressures from less than a millionth of 1 atm to several thousands of atm.

A wide variety of pressure transducers is available to measure gage, absolute, and differential pressures in a wide range of applications. *Gage pressure transducers* use the atmospheric pressure as a reference by venting the back side of the pressure-sensing diaphragm to the atmosphere, and they give a zero signal output at atmospheric pressure regardless of altitude. The *absolute pressure transducers* are calibrated to have a zero signal output at full vacuum. *Differential pressure transducers* measure the pressure difference between two locations directly instead of using two pressure transducers and taking their difference.

**Strain-gage pressure transducers** work by having a diaphragm deflect between two chambers open to the pressure inputs. As the diaphragm stretches in response to a change in pressure difference across it, the strain gage stretches and a Wheatstone bridge circuit amplifies the output. A capacitance transducer works similarly, but capacitance change is measured instead of resistance change as the diaphragm stretches.

**Piezoelectric transducers**, also called solid-state pressure transducers, work on the principle that an electric potential is generated in a crystalline substance when it is subjected to mechanical pressure. This phenomenon, first discovered by brothers Pierre and Jacques Curie in 1880, is called the piezoelectric (or press-electric) effect. Piezoelectric pressure transducers have a much faster frequency response compared to the diaphragm units and are very suitable for high-pressure applications, but they are generally not as sensitive as the diaphragm-type transducers.

# 2–9 ▪ THE BAROMETER AND ATMOSPHERIC PRESSURE

Atmospheric pressure is measured by a device called a **barometer**; thus, the atmospheric pressure is often referred to as the *barometric pressure*.

The Italian Evangelista Torricelli (1608–1647) was the first to conclusively prove that the atmospheric pressure can be measured by inverting a mercury-filled tube into a mercury container that is open to the atmosphere, as shown in Fig. 2–37. The pressure at point *B* is equal to the atmospheric

$W = \rho g h A$

$P_{\text{atm}}$

**FIGURE 2–37**

The basic barometer.

pressure, and the pressure at $C$ can be taken to be zero since there is only mercury vapor above point $C$ and the pressure is very low relative to $P_{atm}$ and can be neglected to an excellent approximation. Writing a force balance in the vertical direction gives

$$P_{atm} = \rho g h \qquad (2\text{–}23)$$

where $\rho$ is the density of mercury, $g$ is the local gravitational acceleration, and $h$ is the height of the mercury column above the free surface. Note that the length and the cross-sectional area of the tube have no effect on the height of the fluid column of a barometer (Fig. 2–38).

A frequently used pressure unit is the *standard atmosphere*, which is defined as the pressure produced by a column of mercury 760 mm in height at 0°C ($\rho_{Hg}$ = 13,595 kg/m³) under standard gravitational acceleration ($g$ = 9.807 m/s²). If water instead of mercury were used to measure the standard atmospheric pressure, a water column of about 10.3 m would be needed. Pressure is sometimes expressed (especially by weather forecasters) in terms of the height of the mercury column. The standard atmospheric pressure, for example, is 760 mmHg (29.92 inHg) at 0°C. The unit mmHg is also called the **torr** in honor of Torricelli. Therefore, 1 atm = 760 torr and 1 torr = 133.3 Pa.

The standard atmospheric pressure $P_{atm}$ changes from 101.325 kPa at sea level to 89.88, 79.50, 54.05, 26.5, and 5.53 kPa at altitudes of 1000, 2000, 5000, 10,000, and 20,000 meters, respectively. The standard atmospheric pressure in Denver (elevation = 1610 m), for example, is 83.4 kPa.

Remember that the atmospheric pressure at a location is simply the weight of the air above that location per unit surface area. Therefore, it changes not only with elevation but also with weather conditions.

The decline of atmospheric pressure with elevation has far-reaching ramifications in daily life. For example, cooking takes longer at high altitudes since water boils at a lower temperature at lower atmospheric pressures. Nose bleeding is a common experience at high altitudes since the difference between the blood pressure and the atmospheric pressure is larger in this case, and the delicate walls of veins in the nose are often unable to withstand this extra stress.

For a given temperature, the density of air is lower at high altitudes, and thus a given volume contains less air and less oxygen. So it is no surprise that we tire more easily and experience breathing problems at high altitudes. To compensate for this effect, people living at higher altitudes develop more efficient lungs. Similarly, a 2.0-L car engine will act like a 1.7-L car engine at 1500 m altitude (unless it is turbocharged) because of the 15 percent drop in pressure and thus 15 percent drop in the density of air (Fig. 2–39). A fan or compressor will displace 15 percent less air at that altitude for the same volume displacement rate. Therefore, larger cooling fans may need to be selected for operation at high altitudes to ensure the specified mass flow rate. The lower pressure and thus lower density also affects lift and drag: airplanes need a longer runway at high altitudes to develop the required lift, and they climb to very high altitudes for cruising for reduced drag and thus better fuel efficiency.

**FIGURE 2–38**
The length or the cross-sectional area of the tube has no effect on the height of the fluid column of a barometer, provided that the tube diameter is large enough to avoid surface tension (capillary) effects.

**FIGURE 2–39**
At high altitudes, a car engine generates less power and a person gets less oxygen because of the lower density of air.

### EXAMPLE 2–5    Measuring Atmospheric Pressure with a Barometer

Determine the atmospheric pressure at a location where the barometric reading is 740 mm Hg and the gravitational acceleration is $g = 9.81$ m/s$^2$. Assume the temperature of mercury to be 10°C, at which its density is 13,570 kg/m$^3$.

**SOLUTION** The barometric reading at a location in height of mercury column is given. The atmospheric pressure is to be determined.
**Assumptions** The temperature of mercury is 10°C.
**Properties** The density of mercury is given to be 13,570 kg/m$^3$.
**Analysis** From Eq. 2–23, the atmospheric pressure is determined to be

$$P_{atm} = \rho g h$$

$$= (13{,}570 \text{ kg/m}^3)(9.81 \text{ m/s}^2)(0.74 \text{ m})\left(\frac{1 \text{ N}}{1 \text{ kg·m/s}^2}\right)\left(\frac{1 \text{ kPa}}{1000 \text{ N/m}^2}\right)$$

$$= \mathbf{98.5 \text{ kPa}}$$

**Discussion** Note that density changes with temperature, and thus this effect should be considered in calculations.

**FIGURE 2–40**

Schematic for Example 2–6; and the free-body diagram of the piston.

### EXAMPLE 2–6    Effect of Piston Weight on Pressure in a Cylinder

The piston of a vertical piston–cylinder device containing a gas has a mass of 60 kg and a cross-sectional area of 0.04 m$^2$, as shown in Fig. 2–40. The local atmospheric pressure is 0.97 bar, and the gravitational acceleration is 9.81 m/s$^2$. (*a*) Determine the pressure inside the cylinder. (*b*) If some heat is transferred to the gas and its volume is doubled, do you expect the pressure inside the cylinder to change?

**SOLUTION** A gas is contained in a vertical cylinder with a heavy piston. The pressure inside the cylinder and the effect of volume change on pressure are to be determined.
**Assumptions** Friction between the piston and the cylinder is negligible.
**Analysis** (*a*) The gas pressure in the piston–cylinder device depends on the atmospheric pressure and the weight of the piston. Drawing the free-body diagram of the piston as shown in Fig. 2–40 and balancing the vertical forces yield

$$PA = P_{atm}A + W$$

Solving for $P$ and substituting,

$$P = P_{atm} + \frac{mg}{A}$$

$$= 0.97 \text{ bar} + \frac{(60 \text{ kg})(9.81 \text{ m/s}^2)}{(0.04 \text{ m}^2)}\left(\frac{1 \text{ N}}{1 \text{ kg·m/s}^2}\right)\left(\frac{1 \text{ bar}}{10^5 \text{ N/m}^2}\right)$$

$$= \mathbf{1.12 \text{ bar}}$$

(*b*) The volume change will have no effect on the free-body diagram drawn in part (*a*), and therefore the pressure inside the cylinder will remain the same.
**Discussion** If the gas behaves as an ideal gas, the absolute temperature doubles when the volume is doubled at constant pressure.

**EXAMPLE 2–7**  **Hydrostatic Pressure in a Solar Pond with Variable Density**

Solar ponds are small artificial lakes of a few meters deep that are used to store solar energy. The rise of heated (and thus less dense) water to the surface is prevented by adding salt at the pond bottom. In a typical salt gradient solar pond, the density of water increases in the gradient zone, as shown in Fig. 2–41, and the density can be expressed as

$$\rho = \rho_0 \sqrt{1 + \tan^2\left(\frac{\pi}{4}\frac{z}{H}\right)}$$

where $\rho_0$ is the density on the water surface, $z$ is the vertical distance measured downward from the top of the gradient zone, and $H$ is the thickness of the gradient zone. For $H = 4$ m, $\rho_0 = 1040$ kg/m$^3$, and a thickness of 0.8 m for the surface zone, calculate the gage pressure at the bottom of the gradient zone.

**SOLUTION**  The variation of density of saline water in the gradient zone of a solar pond with depth is given. The gage pressure at the bottom of the gradient zone is to be determined.

**Assumptions**  The density in the surface zone of the pond is constant.

**Properties**  The density of brine on the surface is given to be 1040 kg/m$^3$.

**Analysis**  We label the top and the bottom of the gradient zone as 1 and 2, respectively. Noting that the density of the surface zone is constant, the gage pressure at the bottom of the surface zone (which is the top of the gradient zone) is

$$P_1 = \rho g h_1 = (1040\ \text{kg/m}^3)(9.81\ \text{m/s}^2)(0.8\ \text{m})\left(\frac{1\ \text{kN}}{1000\ \text{kg·m/s}^2}\right) = 8.16\ \text{kPa}$$

since 1 kN/m$^2$ = 1 kPa. The differential change in hydrostatic pressure across a vertical distance of $dz$ is given by

$$dP = \rho g\, dz$$

Integrating from the top of the gradient zone (point 1 where $z = 0$) to any location $z$ in the gradient zone (no subscript) gives

$$P - P_1 = \int_0^z \rho g\, dz \qquad \rightarrow \qquad P = P_1 + \int_0^z \rho_0 \sqrt{1 + \tan^2\left(\frac{\pi}{4}\frac{z}{H}\right)}\, g\, dz$$

Sun

Increasing salinity and density

$\rho_0 = 1040$ kg/m$^3$

Surface zone  ①

$H = 4$ m  Gradient zone

Storage zone  ②

**FIGURE 2–41**

Schematic for Example 2–7.

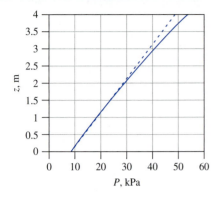

**FIGURE 2–42**

The variation of gage pressure with depth in the gradient zone of the solar pond.

Performing the integration gives the variation of gage pressure in the gradient zone to be

$$P = P_1 + \rho_0 g \frac{4H}{\pi} \sinh^{-1}\left(\tan \frac{\pi}{4} \frac{z}{H}\right)$$

Then the pressure at the bottom of the gradient zone ($z = H = 4$ m) becomes

$$P_2 = 8.16 \text{ kPa} + (1040 \text{ kg/m}^3)(9.81 \text{ m/s}^2)\frac{4(4 \text{ m})}{\pi}$$

$$\times \sinh^{-1}\left(\tan \frac{\pi}{4} \frac{4}{4}\right)\left(\frac{1 \text{ kN}}{1000 \text{ kg}\cdot\text{m/s}^2}\right)$$

$$= 54.0 \text{ kPa (gage)}$$

**Discussion** The variation of gage pressure in the gradient zone with depth is plotted in Fig. 2–42. The dashed line indicates the hydrostatic pressure for the case of constant density at 1040 kg/m³ and is given for reference. Note that the variation of pressure with depth is not linear when density varies with depth.

## SUMMARY

In this chapter, the basic concepts of thermodynamics are introduced and discussed. A system of fixed mass is called a *closed system,* or *control mass,* and a system that involves mass transfer across its boundaries is called an *open system,* or *control volume.* The mass-dependent properties of a system are called *extensive properties* and the others *intensive properties. Density* is mass per unit volume, and *specific volume* is volume per unit mass.

A system is said to be in *thermodynamic equilibrium* if it maintains thermal, mechanical, phase, and chemical equilibrium. Any change from one state to another is called a *process.* A process with identical end states is called a *cycle.* During a *quasi-static* or *quasi-equilibrium process,* the system remains practically in equilibrium at all times. The state of a simple, compressible system is completely specified by two independent, intensive properties.

The *zeroth law of thermodynamics* states that two bodies are in thermal equilibrium if both have the same temperature reading even if they are not in contact.

The temperature scales used in the SI and the English system today are the *Celsius scale* and the *Fahrenheit scale,* respectively. They are related to absolute temperature scales by

$$T(\text{K}) = T(°\text{C}) + 273.15$$

$$T(\text{R}) = T(°\text{F}) + 459.67$$

The magnitudes of each division of 1 K and 1°C are identical, and so are the magnitudes of each division of 1 R and 1°F. Therefore,

$$\Delta T(\text{K}) = \Delta T(°\text{C}) \quad \text{and} \quad \Delta T(\text{R}) = \Delta T(°\text{F})$$

The normal force exerted by a fluid per unit area is called *pressure,* and its unit is the *pascal,* 1 Pa = 1 N/m². The pressure relative to absolute vacuum is called the *absolute pressure,* and the difference between the absolute pressure and the local atmospheric pressure is called the *gage pressure.* Pressures below atmospheric pressure are called *vacuum pressures.* The absolute, gage, and vacuum pressures are related by

$$P_{\text{gage}} = P_{\text{abs}} - P_{\text{atm}} \quad \text{(for pressures above } P_{\text{atm}})$$

$$P_{\text{vac}} = P_{\text{atm}} - P_{\text{abs}} \quad \text{(for pressures below } P_{\text{atm}})$$

The pressure at a point in a fluid has the same magnitude in all directions. The variation of pressure with elevation is given by

$$\frac{dP}{dz} = -\rho g$$

where the positive $z$ direction is taken to be upward. When the density of the fluid is constant, the pressure difference across a fluid layer of thickness $\Delta z$ is

$$\Delta P = P_2 - P_1 = \rho g \, \Delta z$$

The absolute and gage pressures in a liquid open to the atmosphere at a depth $h$ from the free surface are

$$P = P_{atm} + \rho g h \quad \text{or} \quad P_{gage} = \rho g h$$

Small to moderate pressure differences are measured by a *manometer*. The pressure in a stationary fluid remains constant in the horizontal direction. *Pascal's principle* states that the pressure applied to a confined fluid increases the pressure throughout by the same amount. The atmospheric pressure is measured by a *barometer* and is given by

$$P_{atm} = \rho g h$$

where $h$ is the height of the liquid column.

## REFERENCES AND SUGGESTED READINGS

**1.** A. Bejan. *Advanced Engineering Thermodynamics*. 3rd ed. New York: Wiley, 2006.

**2.** J. A. Schooley. *Thermometry*. Boca Raton, FL: CRC Press, 1986.

## PROBLEMS*

### Systems, Properties, State, and Processes

**2–1C** You have been asked to do a metabolism (energy) analysis of a person. How would you define the system for this purpose? What type of system is this?

**2–2C** You are trying to understand how a reciprocating air compressor (a piston-cylinder device) works. What system would you use? What type of system is this?

**2–3C** How might you define a system to study the depletion of ozone in the upper layers of the earth's atmosphere?

**2–4C** What is the difference between intensive and extensive properties?

**2–5C** Is the weight of a system an extensive or intensive property?

**2–6C** The molar specific volume of a system $\bar{v}$ is defined as the ratio of the volume of the system to the number of moles of substance contained in the system. Is this an extensive or intensive property?

**2–7C** For a system to be in thermodynamic equilibrium, do the temperature and the pressure have to be the same everywhere?

**2–8C** What is a quasi-equilibrium process? What is its importance in engineering?

**2–9C** Define the isothermal, isobaric, and isochoric processes.

**2–10C** What is the state postulate?

**2–11C** How would you describe the state of the water in a bathtub? How would you describe the process that this water experiences as it cools?

**2–12C** When analyzing the acceleration of gases as they flow through a nozzle, what would you choose as your system? What type of system is this?

**2–13C** When is a steady-flow process?

**2–14** [EES] The density of atmospheric air varies with elevation, decreasing with increasing altitude. (*a*) Using the data given in the table, obtain a relation for the variation of density with elevation, and calculate the density at an elevation of 7000 m. (*b*) Calculate the mass of the atmosphere using the correlation you obtained. Assume the earth to be a perfect sphere with a radius of 6377 km, and take the thickness of the atmosphere to be 25 km.

---

* Problems designated by a "C" are concept questions, and students are encouraged to answer them all. Problems designated by an "E" are in English units, and the SI users can ignore them. Problems with the 🖭 icon are solved using EES, and complete solutions together with parametric studies are included on the enclosed DVD. Problems with the 🖥 icon are comprehensive in nature, and are intended to be solved with a computer, preferably using the EES software that accompanies this text.

| $z$, km | $\rho$, kg/m$^3$ |
|---------|------------------|
| 6377 | 1.225 |
| 6378 | 1.112 |
| 6379 | 1.007 |
| 6380 | 0.9093 |
| 6381 | 0.8194 |
| 6382 | 0.7364 |
| 6383 | 0.6601 |
| 6385 | 0.5258 |
| 6387 | 0.4135 |
| 6392 | 0.1948 |
| 6397 | 0.08891 |
| 6402 | 0.04008 |

## Temperature

**2–15C**  What is the zeroth law of thermodynamics?

**2–16C**  What are the ordinary and absolute temperature scales in the SI and the English system?

**2–17C**  Consider an alcohol and a mercury thermometer that read exactly 0°C at the ice point and 100°C at the steam point. The distance between the two points is divided into 100 equal parts in both thermometers. Do you think these thermometers will give exactly the same reading at a temperature of, say, 60°C? Explain.

**2–18**  The deep body temperature of a healthy person is 37°C. What is it in kelvins?

**2–19E**  What is the temperature of the heated air at 150°C in °F and R?

**2–20**  The temperature of a system rises by 45°C during a heating process. Express this rise in temperature in kelvins.

**2–21E**  The flash point of an engine oil is 363°F. What is the absolute flash-point temperature in K and R?

**2–22E**  The temperature of ambient air in a certain location is measured to be −40°C. Express this temperature in Fahrenheit (°F), Kelvin (K), and Rankine (R) units.

**2–23E**  The temperature of water changes by 10°F during a process. Express this temperature change in Celcius (°C), Kelvin (K), and Rankine (R) units.

**2–24E**  Humans are most comfortable when the temperature is between 65°F and 75°F. Express these temperature limits in °C. Convert the size of this temperature range (10°F) to K, °C, and R. Is there any difference in the size of this range as measured in relative or absolute units?

## Pressure, Manometer, and Barometer

**2–25C**  What is the difference between gage pressure and absolute pressure?

**2–26C**  A health magazine reported that physicians measured 100 adults' blood pressure using two different arm positions: parallel to the body (along the side) and perpendicular to the body (straight out). Readings in the parallel position were up to 10 percent higher than those in the perpendicular position, regardless of whether the patient was standing, sitting, or lying down. Explain the possible cause for the difference.

**2–27C**  Someone claims that the absolute pressure in a liquid of constant density doubles when the depth is doubled. Do you agree? Explain.

**2–28C**  A tiny steel cube is suspended in water by a string. If the lengths of the sides of the cube are very small, how would you compare the magnitudes of the pressures on the top, bottom, and side surfaces of the cube?

**2–29C**  Express Pascal's law, and give a real-world example of it.

**2–30E**  The pressure at the exit of an air compressor is 150 psia. What is this pressure in kPa?

**2–31**  The pressure in a compressed air storage tank is 1500 kPa. What is the tank's pressure in (a) kN and m units; (b) kg, m, and s units; and (c) kg, km, and s units?

**2–32E**  The pressure in a water line is 1500 kPa. What is the line pressure in (a) lb/ft$^2$ units and (b) lbf/in$^2$ (psi) units?

**2–33E**  If the pressure inside a rubber balloon is 1500 mm Hg, what is this pressure in pounds-force per square inch (psi)?  *Answer:* 29.0 psi

**2–34**  The pressure of the helium inside a toy balloon is 1250 mm Hg. What is this pressure in kPa?

**2–35**  The water in a tank is pressurized by air, and the pressure is measured by a multifluid manometer as shown in Fig. P2–35. Determine the gage pressure of air in the tank if $h_1 = 0.2$ m, $h_2 = 0.3$ m, and $h_3 = 0.46$ m. Take the densities of water, oil, and mercury to be 1000 kg/m$^3$, 850 kg/m$^3$, and 13,600 kg/m$^3$, respectively.

**FIGURE P2–35**

**2–36** Determine the atmospheric pressure at a location where the barometric reading is 750 mm Hg. Take the density of mercury to be 13,600 kg/m³.

**2–37** The gage pressure in a liquid at a depth of 3 m is read to be 42 kPa. Determine the gage pressure in the same liquid at a depth of 9 m.

**2–38** The absolute pressure in water at a depth of 5 m is read to be 145 kPa. Determine (a) the local atmospheric pressure, and (b) the absolute pressure at a depth of 5 m in a liquid whose specific gravity is 0.85 at the same location.

**2–39E** Show that 1 kgf/cm² = 14.223 psi.

**2–40E** The diameters of the pistons shown in Fig. P2–40E are $D_1$ = 3 in and $D_2$ = 1.5 in. Determine the pressure in chamber 3, in psia, when the other pressures are $P_1$ = 150 psia and $P_2$ = 250 psia.

**FIGURE P2–40E**

**2–41** The piston diameters in Fig. P2–40E are $D_1$ = 10 cm and $D_2$ = 4 cm. When the pressure in chamber 2 is 2000 kPa and the pressure in chamber 3 is 700 kPa, what is the pressure in chamber 1, in kPa? *Answer: 908 kPa*

**2–42** Consider a 70-kg woman who has a total foot imprint area of 400 cm². She wishes to walk on the snow, but the snow cannot withstand pressures greater than 0.5 kPa. Determine the minimum size of the snowshoes needed (imprint area per shoe) to enable her to walk on the snow without sinking.

**2–43** A vacuum gage connected to a tank reads 30 kPa at a location where the barometric reading is 750 mm Hg. Determine the absolute pressure in the tank. Take $\rho_{Hg}$ = 13,590 kg/m³. *Answer: 70.0 kPa*

**2–44E** The vacuum pressure of a condenser is given to be 80 kPa. If the atmospheric pressure is 98 kPa, what is the gage pressure and absolute pressure in kPa, kN/m², lbf/in², psi, and mm Hg.

**2–45** The barometer of a mountain hiker reads 740 mbars at the beginning of a hiking trip and 630 mbars at the end. Neglecting the effect of altitude on local gravitational acceleration, determine the vertical distance climbed. Assume an average air density of 1.20 kg/m³. *Answer: 934 m*

**2–46** Consider a 1.75-m-tall man standing vertically in water and completely submerged in a pool. Determine the difference between the pressures acting at the head and at the toes of the man, in kPa.

**2–47** A gas is contained in a vertical, frictionless piston–cylinder device. The piston has a mass of 3.2 kg and a cross-sectional area of 35 cm². A compressed spring above the piston exerts a force of 150 N on the piston. If the atmospheric pressure is 95 kPa, determine the pressure inside the cylinder. *Answer: 147 kPa*

**FIGURE P2–47**

**2–48** Reconsider Prob. 2–47. Using EES (or other) software, investigate the effect of the spring force in the range of 0 to 500 N on the pressure inside the cylinder. Plot the pressure against the spring force, and discuss the results.

**2–49** Both a gage and a manometer are attached to a gas tank to measure its pressure. If the reading on the pressure gage is 80 kPa, determine the distance between the two fluid levels of the manometer if the fluid is (a) mercury ($\rho$ = 13,600 kg/m³) or (b) water ($\rho$ = 1000 kg/m³).

**FIGURE P2–49**

**2–50** Reconsider Prob. 2–49. Using EES (or other) software, investigate the effect of the manometer fluid density in the range of 800 to 13,000 kg/m³ on the differential fluid height of the manometer. Plot the differential fluid height against the density, and discuss the results.

**2–51** A manometer containing oil ($\rho = 850$ kg/m³) is attached to a tank filled with air. If the oil-level difference between the two columns is 36 cm and the atmospheric pressure is 98 kPa, determine the absolute pressure of the air in the tank. *Answer:* 101 kPa

**2–52** A mercury manometer ($\rho = 13,600$ kg/m³) is connected to an air duct to measure the pressure inside. The difference in the manometer levels is 15 mm, and the atmospheric pressure is 100 kPa. (*a*) Judging from Fig. P2–52, determine if the pressure in the duct is above or below the atmospheric pressure. (*b*) Determine the absolute pressure in the duct.

**FIGURE P2–52**

**2–53** Repeat Prob. 2–52 for a differential mercury height of 45 mm.

**2–54E** Blood pressure is usually measured by wrapping a closed air-filled jacket equipped with a pressure gage around the upper arm of a person at the level of the heart. Using a mercury manometer and a stethoscope, the systolic pressure (the maximum pressure when the heart is pumping) and the diastolic pressure (the minimum pressure when the heart is resting) are measured in mm Hg. The systolic and diastolic pressures of a healthy person are about 120 mm Hg and 80 mm Hg, respectively, and are indicated as 120/80. Express both of these gage pressures in kPa, psi, and meter water column.

**2–55** The maximum blood pressure in the upper arm of a healthy person is about 120 mm Hg. If a vertical tube open to the atmosphere is connected to the vein in the arm of the person, determine how high the blood will rise in the tube. Take the density of the blood to be 1050 kg/m³.

**FIGURE P2–55**

**2–56** Determine the pressure exerted on a diver at 30 m below the free surface of the sea. Assume a barometric pressure of 101 kPa and a specific gravity of 1.03 for seawater. *Answer:* 404.0 kPa

**2–57** Consider a U-tube whose arms are open to the atmosphere. Now water is poured into the U-tube from one arm, and light oil ($\rho = 790$ kg/m³) from the other. One arm contains 70-cm-high water, while the other arm contains both fluids with an oil-to-water height ratio of 4. Determine the height of each fluid in that arm.

**FIGURE P2–57**

**2–58** Freshwater and seawater flowing in parallel horizontal pipelines are connected to each other by a double U-tube manometer, as shown in Fig. P2–58. Determine the pressure difference between the two pipelines. Take the density of seawater at that location to be $\rho = 1035$ kg/m³. Can the air column be ignored in the analysis?

**FIGURE P2–58**

**2–59** Repeat Prob. 2–58 by replacing the air with oil whose specific gravity is 0.72.

**2–60** Calculate the absolute pressure, $P_1$, of the manometer shown in Fig. P2–60 in kPa. The local atmospheric pressure is 758 mm Hg.

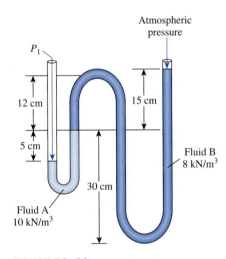

**FIGURE P2–60**

**2–61** Consider the manometer in Fig. P2–60. If the specific weight of fluid A is 100 kN/m³, what is the absolute pressure, in kPa, indicated by the manometer when the local atmospheric pressure is 90 kPa?

**2–62** Consider the manometer in Fig. P2–60. If the specific weight of fluid B is 12 kN/m³, what is the absolute pressure, in kPa, indicated by the manometer when the local atmospheric pressure is 720 mm Hg?

**2–63** The gage pressure of the air in the tank shown in Fig. P2–63 is measured to be 80 kPa. Determine the differential height $h$ of the mercury column.

**FIGURE P2–63**

**2–64** Repeat Prob. 2–63 for a gage pressure of 40 kPa.

**2–65** The top part of a water tank is divided into two compartments, as shown in Fig. P2–65. Now a fluid with an unknown density is poured into one side, and the water level rises a certain amount on the other side to compensate for this effect. Based on the final fluid heights shown on the figure, determine the density of the fluid added. Assume the liquid does not mix with water.

**FIGURE P2–65**

**2–66** Consider the system shown in Fig. P2–66. If a change of 0.7 kPa in the pressure of air causes the brine–mercury interface in the right column to drop by 5 mm in the brine level in the right column while the pressure in the brine pipe remains constant, determine the ratio of $A_2/A_1$.

**FIGURE P2–66**

**2–67** A multifluid container is connected to a U-tube, as shown in Fig. P2–67. For the given specific gravities and fluid column heights, determine the gage pressure at $A$. Also determine the height of a mercury column that would create the same pressure at $A$. *Answers:* 0.471 kPa, 0.353 cm

**FIGURE P2–67**

**Review Problems**

**2–68** The pilot of an airplane reads the altitude 9000 m and the absolute pressure 25 kPa when flying over a city. Calculate the local atmospheric pressure in that city in kPa and in mm Hg. Take the densities of air and mercury to be 1.15 kg/m³ and 13,600 kg/m³, respectively.

**FIGURE P2–68**

**2–69E** The efficiency of a refrigerator increases by 3 percent for each °C rise in the minimum temperature in the device. What is the increase in the efficiency for each (a) K, (b) °F, and (c) R rise in temperature?

**2–70E** The boiling temperature of water decreases by about 3°C for each 1000-m rise in altitude. What is the decrease in the boiling temperature in (a) K, (b) °F, and (c) R for each 1000-m rise in altitude?

**2–71E** Hyperthermia of 5°C (i.e., 5°C rise above the normal body temperature) is considered fatal. Express this fatal level of hyperthermia in (a) K, (b) °F, and (c) R.

**2–72E** A house is losing heat at a rate of 2700 kJ/h per °C temperature difference between the indoor and the outdoor temperatures. Express the rate of heat loss from this house per (a) K, (b) °F, and (c) R difference between the indoor and the outdoor temperature.

**2–73** The average temperature of the atmosphere in the world is approximated as a function of altitude by the relation

$$T_{atm} = 288.15 - 6.5z$$

where $T_{atm}$ is the temperature of the atmosphere in K and $z$ is the altitude in km with $z = 0$ at sea level. Determine the average temperature of the atmosphere outside an airplane that is cruising at an altitude of 12,000 m.

**2–74** Joe Smith, an old-fashioned engineering student, believes that the boiling point of water is best suited for use as the reference point on temperature scales. Unhappy that the boiling point corresponds to some odd number in the current absolute temperature scales, he has proposed a new absolute temperature scale that he calls the Smith scale. The temperature unit on this scale is *smith*, denoted by S, and the boiling point of water on this scale is assigned to be 1000 S. From a thermodynamic point of view, discuss if it is an acceptable temperature scale. Also, determine the ice point of water on the Smith scale and obtain a relation between the Smith and Celsius scales.

**2–75E** It is well-known that cold air feels much colder in windy weather than what the thermometer reading indicates because of the "chilling effect" of the wind. This effect is due to the increase in the convection heat transfer coefficient with increasing air velocities. The *equivalent wind chill temperature* in °F is given by [ASHRAE, *Handbook of Fundamentals* (Atlanta, GA, 1993), p. 8.15]

$$T_{equiv} = 91.4 - (91.4 - T_{ambient})$$
$$\times (0.475 - 0.0203V + 0.304\sqrt{V})$$

where $V$ is the wind velocity in mi/h and $T_{ambient}$ is the ambient air temperature in °F in calm air, which is taken to be air with light winds at speeds up to 4 mi/h. The constant 91.4°F in the given equation is the mean skin temperature of a resting person in a comfortable environment. Windy air at temperature

$T_{ambient}$ and velocity $V$ will feel as cold as the calm air at temperature $T_{equiv}$. Using proper conversion factors, obtain an equivalent relation in SI units where $V$ is the wind velocity in km/h and $T_{ambient}$ is the ambient air temperature in °C.

*Answer:*     $T_{equiv} = 33.0 - (33.0 - T_{ambient})$
              $\times (0.475 - 0.0126V + 0.240\sqrt{V})$

**2–76E** Reconsider Prob. 2–75E. Using EES (or other) software, plot the equivalent wind chill temperatures in °F as a function of wind velocity in the range of 4 to 40 mph for the ambient temperatures of 20, 40, and 60°F. Discuss the results.

**2–77** An air-conditioning system requires a 20-m-long section of 15-cm diameter duct work to be laid underwater. Determine the upward force the water will exert on the duct. Take the densities of air and water to be 1.3 kg/m³ and 1000 kg/m³, respectively.

**2–78** Balloons are often filled with helium gas because it weighs only about one-seventh of what air weighs under identical conditions. The buoyancy force, which can be expressed as $F_b = \rho_{air}gV_{balloon}$, will push the balloon upward. If the balloon has a diameter of 12 m and carries two people, 85 kg each, determine the acceleration of the balloon when it is first released. Assume the density of air is $\rho = 1.16$ kg/m³, and neglect the weight of the ropes and the cage. *Answer: 22.4 m/s²*

Helium
$D = 12$ m
$p_{He} = \frac{1}{7}\rho_{air}$

$m = 170$ kg

**FIGURE P2–78**

**2–79** Reconsider Prob. 2–78. Using EES (or other) software, investigate the effect of the number of people carried in the balloon on acceleration. Plot the acceleration against the number of people, and discuss the results.

**2–80** Determine the maximum amount of load, in kg, the balloon described in Prob. 2–78 can carry. *Answer: 900 kg*

**2–81** The lower half of a 10-m-high cylindrical container is filled with water ($\rho = 1000$ kg/m³) and the upper half with oil that has a specific gravity of 0.85. Determine the pressure difference between the top and bottom of the cylinder. *Answer: 90.7 kPa*

Oil
SG = 0.85

$h = 10$ m

Water
$\rho = 1000$ kg/m³

**FIGURE P2–81**

**2–82** A vertical, frictionless piston–cylinder device contains a gas at 180 kPa absolute pressure. The atmospheric pressure outside is 100 kPa, and the piston area is 25 cm². Determine the mass of the piston.

**2–83** A pressure cooker cooks a lot faster than an ordinary pan by maintaining a higher pressure and temperature inside. The lid of a pressure cooker is well sealed, and steam can escape only through an opening in the middle of the lid. A separate metal piece, the petcock, sits on top of this opening and prevents steam from escaping until the pressure force overcomes the weight of the petcock. The periodic escape of the steam in this manner prevents any potentially dangerous pressure buildup and keeps the pressure inside at a constant value. Determine the mass of the petcock of a pressure cooker whose operation pressure is 100 kPa gage and has an opening cross-sectional area of 4 mm². Assume an atmospheric pressure of 101 kPa, and draw the free-body diagram of the petcock. *Answer: 40.8 g*

$P_{atm} = 101$ kPa

Petcock

$A = 4$ mm²

Pressure cooker

**FIGURE P2–83**

**2–84** A glass tube is attached to a water pipe, as shown in Fig. P2–84. If the water pressure at the bottom of the tube is 120 kPa and the local atmospheric pressure is 99 kPa, determine how high the water will rise in the tube, in m. Take the density of water to be 1000 kg/m³.

**FIGURE P2–84**

**2–85** When measuring small pressure differences with a manometer, often one arm of the manometer is inclined to improve the accuracy of reading. (The pressure difference is still proportional to the *vertical* distance and not the actual length of the fluid along the tube.) The air pressure in a circular duct is to be measured using a manometer whose open arm is inclined 45° from the horizontal, as shown in Fig. P2–85  The density of the liquid in the manometer is 0.81 kg/L, and the vertical distance between the fluid levels in the two arms of the manometer is 12 cm. Determine the gage pressure of air in the duct and the length of the fluid column in the inclined arm above the fluid level in the vertical arm.

**FIGURE P2–85**

**2–86E** Consider a U-tube whose arms are open to the atmosphere. Now equal volumes of water and light oil ($\rho$ = 49.3 lbm/ft$^3$) are poured from different arms. A person blows from the oil side of the U-tube until the contact surface of the two fluids moves to the bottom of the U-tube, and thus the liquid levels in the two arms are the same. If the fluid height in each arm is 30 in, determine the gage pressure the person exerts on the oil by blowing.

**FIGURE P2–86E**

**2–87** Intravenous infusions are usually driven by gravity by hanging the fluid bottle at sufficient height to counteract the blood pressure in the vein and to force the fluid into the body. The higher the bottle is raised, the higher the flow rate of the fluid will be. (*a*) If it is observed that the fluid and the blood pressures balance each other when the bottle is 80 cm above the arm level, determine the gage pressure of the blood. (*b*) If the gage pressure of the fluid at the arm level needs to be 15 kPa for sufficient flow rate, determine how high the bottle must be placed. Take the density of the fluid to be 1020 kg/m$^3$.

**FIGURE P2–87**

**2–88E** A water pipe is connected to a double-U manometer as shown in Fig. P2–88E at a location where the local atmospheric pressure is 14.2 psia. Determine the absolute pressure at the center of the pipe.

**FIGURE P2–88E**

**2–89** The average atmospheric pressure on earth is approximated as a function of altitude by the relation $P_{atm} = 101.325 \, (1 - 0.02256z)^{5.256}$, where $P_{atm}$ is the atmospheric pressure in kPa and $z$ is the altitude in km with $z = 0$ at sea level. Determine the approximate atmospheric pressures at Atlanta ($z = 306$ m), Denver ($z = 1610$ m), Mexico City ($z = 2309$ m), and the top of Mount Everest ($z = 8848$ m).

## Design and Essay Problems

**2–90** Write an essay on different temperature measurement devices. Explain the operational principle of each device, its advantages and disadvantages, its cost, and its range of applicability. Which device would you recommend for use in the following cases: taking the temperatures of patients in a doctor's office, monitoring the variations of temperature of a car engine block at several locations, and monitoring the temperatures in the furnace of a power plant?

# ENERGY, ENERGY TRANSFER, AND GENERAL ENERGY ANALYSIS

**W**hether we realize it or not, energy is an important part of most aspects of daily life. The quality of life, and even its sustenance, depends on the availability of energy. Therefore, it is important to have a good understanding of the sources of energy, the conversion of energy from one form to another, and the ramifications of these conversions.

Energy exists in numerous forms such as thermal, mechanical, electric, chemical, and nuclear. Even mass can be considered a form of energy. Energy can be transferred to or from a closed system (a fixed mass) in two distinct forms: *heat* and *work*. For control volumes, energy can also be transferred by mass flow. An energy transfer to or from a closed system is *heat* if it is caused by a temperature difference. Otherwise it is *work*, and it is caused by a force acting through a distance.

We start this chapter with a discussion of various forms of energy and energy transfer by heat. We then introduce various forms of work and discuss energy transfer by work. We continue with developing a general intuitive expression for the *first law of thermodynamics*, also known as the *conservation of energy principle*, which is one of the most fundamental principles in nature, and we then demonstrate its use. Finally, we discuss the efficiencies of some familiar energy conversion processes. Detailed treatments of the first law of thermodynamics for closed systems and control volumes are given in Chaps. 5 and 6, respectively.

## Objectives

Objectives of this chapter are to:

- Introduce the concept of energy and define its various forms.
- Discuss the nature of internal energy.
- Define the concept of heat and the terminology associated with energy transfer by heat.
- Discuss the three mechanisms of heat transfer: conduction, convection, and radiation.
- Define the concept of work, including electrical work and several forms of mechanical work.
- Introduce the first law of thermodynamics, energy balances, and mechanisms of energy transfer to or from a system.
- Determine that a fluid flowing across a control surface of a control volume carries energy across the control surface in addition to any energy transfer across the control surface that may be in the form of heat and/or work.
- Define energy conversion efficiencies.

**FIGURE 3–1**

A refrigerator operating with its door open in a well-sealed and well-insulated room.

**FIGURE 3–2**

A fan running in a well-sealed and well-insulated room will raise the temperature of air in the room.

# 3–1 · INTRODUCTION

We are familiar with the conservation of energy principle, which is an expression of the first law of thermodynamics, back from our high school years. We are told repeatedly that energy cannot be created or destroyed during a process; it can only change from one form to another. This seems simple enough, but let's test ourselves to see how well we understand and truly believe in this principle.

Consider a room whose door and windows are tightly closed, and whose walls are well-insulated so that heat loss or gain through the walls is negligible. Now let's place a refrigerator in the middle of the room with its door open, and plug it into a wall outlet (Fig. 3–1). You may even use a small fan to circulate the air in order to maintain temperature uniformity in the room. Now, what do you think will happen to the average temperature of air in the room? Will it be increasing or decreasing? Or will it remain constant?

Probably the first thought that comes to mind is that the average air temperature in the room will decrease as the warmer room air mixes with the air cooled by the refrigerator. Some may draw our attention to the heat generated by the motor of the refrigerator, and may argue that the average air temperature may rise if this heating effect is greater than the cooling effect. But they will get confused if it is stated that the motor is made of superconducting materials, and thus there is hardly any heat generation in the motor.

Heated discussion may continue with no end in sight until we remember the conservation of energy principle that we take for granted: If we take the entire room—including the air and the refrigerator—as the system, which is an adiabatic closed system since the room is well-sealed and well-insulated, the only energy interaction involved is the electrical energy crossing the system boundary and entering the room. The conservation of energy requires the energy content of the room to increase by an amount equal to the amount of the electrical energy drawn by the refrigerator, which can be measured by an ordinary electric meter. The refrigerator or its motor does not store this energy. Therefore, this energy must now be in the room air, and it will manifest itself as a rise in the air temperature. The temperature rise of air can be calculated on the basis of the conservation of energy principle using the properties of air and the amount of electrical energy consumed. What do you think would happen if we had a window air conditioning unit instead of a refrigerator placed in the middle of this room? What if we operated a fan in this room instead (Fig. 3–2)?

Note that energy is conserved during the process of operating the refrigerator placed in a room—the electrical energy is converted into an equivalent amount of thermal energy stored in the room air. If energy is already conserved, then what are all those speeches on energy conservation and the measures taken to conserve energy? Actually, by "energy conservation" what is meant is the conservation of the *quality* of energy, not the quantity. Electricity, which is of the highest quality of energy, for example, can always be converted to an equal amount of thermal energy (also called *heat*). But only a small fraction of thermal energy, which is the lowest quality of energy, can be converted back to electricity, as we discuss in Chap. 7. Think about the things that you can do with the electrical energy that the refrigerator has consumed, and the air in the room that is now at a higher temperature.

Now if asked to name the energy transformations associated with the operation of a refrigerator, we may still have a hard time answering because all we see is electrical energy entering the refrigerator and heat dissipated from the refrigerator to the room air. Obviously there is need to study the various forms of energy first, and this is exactly what we do next, followed by a study of the mechanisms of energy transfer.

## 3–2 ▪ FORMS OF ENERGY

Energy can exist in numerous forms such as thermal, mechanical, kinetic, potential, electric, magnetic, chemical, and nuclear, and their sum consti- tutes the **total energy** $E$ of a system. The total energy of a system on a *unit mass* basis is denoted by $e$ and is expressed as

$$e = \frac{E}{m} \qquad \text{(kJ/kg)} \qquad \textbf{(3–1)}$$

Thermodynamics provides no information about the absolute value of the total energy. It deals only with the *change* of the total energy, which is what matters in engineering problems. Thus the total energy of a system can be assigned a value of zero ($E = 0$) at some convenient reference point. The change in total energy of a system is independent of the reference point selected. The decrease in the potential energy of a falling rock, for example, depends on only the elevation difference and not the reference level selected.

In thermodynamic analysis, it is often helpful to consider the various forms of energy that make up the total energy of a system in two groups: *macroscopic* and *microscopic*. The **macroscopic** forms of energy are those a system possesses as a whole with respect to some outside reference frame, such as kinetic and potential energies (Fig. 3–3). The **microscopic** forms of energy are those related to the molecular structure of a system and the degree of the molecular activity, and they are independent of out- side reference frames. The sum of all the microscopic forms of energy is called the **internal energy** of a system and is denoted by $U$.

The term *energy* was coined in 1807 by Thomas Young, and its use in thermodynamics was proposed in 1852 by Lord Kelvin. The term *internal energy* and its symbol $U$ first appeared in the works of Rudolph Clausius and William Rankine in the second half of the nineteenth century, and it eventually replaced the alternative terms *inner work, internal work,* and *intrinsic energy* commonly used at the time.

The macroscopic energy of a system is related to motion and the influ- ence of some external effects such as gravity, magnetism, electricity, and surface tension. The energy that a system possesses as a result of its motion relative to some reference frame is called **kinetic energy** (KE). When all parts of a system move with the same velocity, the kinetic energy is expressed as

$$\text{KE} = m\frac{V^2}{2} \qquad \text{(kJ)} \qquad \textbf{(3–2)}$$

**FIGURE 3–3**

The macroscopic energy of an object changes with velocity and elevation.

or, on a unit mass basis,

$$\text{ke} = \frac{V^2}{2} \qquad \text{(kJ/kg)} \qquad \textbf{(3–3)}$$

where $V$ denotes the velocity of the system relative to some fixed reference frame. The kinetic energy of a rotating solid body is given by $\frac{1}{2}I\omega^2$ where $I$ is the moment of inertia of the body and $\omega$ is the angular velocity.

The energy that a system possesses as a result of its elevation in a gravitational field is called **potential energy** (PE) and is expressed as

$$\text{PE} = mgz \qquad \text{(kJ)} \qquad \textbf{(3–4)}$$

or, on a unit mass basis,

$$\text{pe} = gz \qquad \text{(kJ/kg)} \qquad \textbf{(3–5)}$$

where $g$ is the gravitational acceleration and $z$ is the elevation of the center of gravity of a system relative to some arbitrarily selected reference level.

The magnetic, electric, and surface tension effects are significant in some specialized cases only and are usually ignored. In the absence of such effects, the total energy of a system consists of the kinetic, potential, and internal energies and is expressed as

$$E = U + \text{KE} + \text{PE} = U + m\frac{V^2}{2} + mgz \qquad \text{(kJ)} \qquad \textbf{(3–6)}$$

or, on a unit mass basis,

$$e = u + \text{ke} + \text{pe} = u + \frac{V^2}{2} + gz \qquad \text{(kJ/kg)} \qquad \textbf{(3–7)}$$

Most closed systems remain stationary during a process and thus experience no change in their kinetic and potential energies. Closed systems whose velocity and elevation of the center of gravity remain constant during a process are frequently referred to as **stationary systems**. The change in the total energy $\Delta E$ of a stationary system is identical to the change in its internal energy $\Delta U$. In this text, a closed system is assumed to be stationary unless stated otherwise.

Control volumes typically involve fluid flow for long periods of time, and it is convenient to express the energy flow associated with a fluid stream in the rate form. This is done by incorporating the **mass flow rate** $\dot{m}$, which is *the amount of mass flowing through a cross section per unit time*. It is related to the **volume flow rate** $\dot{V}$, which is the volume of a fluid flowing through a cross section per unit time, by

*Mass flow rate:* $\qquad \dot{m} = \rho\dot{V} = \rho A_c V_{\text{avg}} \qquad \text{(kg/s)} \qquad \textbf{(3–8)}$

which is analogous to $m = \rho V$. Here $\rho$ is the fluid density, $A_c$ is the cross-sectional area of flow, and $V_{\text{avg}}$ is the average flow velocity normal to $A_c$. The dot over a symbol is used to indicate *time rate* throughout the book. Then the energy flow rate associated with a fluid flowing at a rate of $\dot{m}$. is (Fig. 3–4)

*Energy flow rate:* $\qquad \dot{E} = \dot{m}e \qquad \text{(kJ/s or kW)} \qquad \textbf{(3–9)}$

which is analogous to $E = me$.

**FIGURE 3–4**

Mass and energy flow rates associated with the flow of steam in a pipe of inner diameter $D$ with an average velocity of $V_{\text{avg}}$.

## Some Physical Insight to Internal Energy

Internal energy is defined earlier as the sum of all the *microscopic* forms of energy of a system. It is related to the *molecular structure* and the degree of *molecular activity,* and can be viewed as the sum of the *kinetic* and *potential* energies of the molecules.

To have a better understanding of internal energy, let us examine a system at the molecular level. The molecules of a gas move through space with some velocity, and thus possess some kinetic energy. This is known as the *translational energy.* The atoms of polyatomic molecules rotate about an axis, and the energy associated with this rotation is the *rotational kinetic energy.* The atoms of a polyatomic molecule may also vibrate about their common center of mass, and the energy associated with this back-and-forth motion is the *vibrational kinetic energy.* For gases, the kinetic energy is mostly due to translational and rotational motions, with vibrational motion becoming significant at higher temperatures. The electrons in an atom rotate about the nucleus, and thus possess *rotational kinetic energy.* Electrons at outer orbits have larger kinetic energies. Electrons also spin about their axes, and the energy associated with this motion is the *spin energy.* Other particles in the nucleus of an atom also possess spin energy. The portion of the internal energy of a system associated with the kinetic energies of the molecules is called the **sensible energy** (Fig. 3–5). The average velocity and the degree of activity of the molecules are proportional to the temperature of the gas. Therefore, at higher temperatures, the molecules possess higher kinetic energies, and as a result the system has a higher internal energy.

The internal energy is also associated with various *binding forces* between the molecules of a substance, between the atoms within a molecule, and between the particles within an atom and its nucleus. The forces that bind the *molecules* to each other are, as one would expect, strongest in solids and weakest in gases. If sufficient energy is added to the molecules of a solid or liquid, the molecules overcome these molecular forces and break away, turning the substance into a gas. This is a phase-change process. Because of this added energy, a system in the gas phase is at a higher internal energy level than it is in the solid or the liquid phase. The internal energy associated with the phase of a system is called the **latent energy.** The phase-change process can occur without a change in the chemical composition of a system. Most practical problems fall into this category, and one does not need to pay any attention to the forces binding the atoms in a molecule to each other.

An atom consists of neutrons and positively charged protons bound together by very strong nuclear forces in the nucleus, and negatively charged electrons orbiting around it. The internal energy associated with the atomic bonds in a molecule is called **chemical energy.** During a chemical reaction, such as a combustion process, some chemical bonds are destroyed while others are formed. As a result, the internal energy changes. The nuclear forces are much larger than the forces that bind the electrons to the nucleus. The tremendous amount of energy associated with the strong bonds within the nucleus of the atom itself is called **nuclear energy** (Fig. 3–6). Obviously, we need not be concerned with nuclear energy in thermodynamics unless, of course, we deal with fusion or fission reactions. A chemical reaction involves changes in the structure of the electrons of the atoms, but a nuclear reaction involves changes in the core or nucleus. Therefore, an

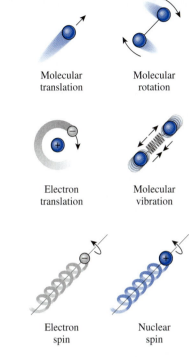

| Molecular translation | Molecular rotation |
| Electron translation | Molecular vibration |
| Electron spin | Nuclear spin |

**FIGURE 3–5**

The various forms of microscopic energies that make up *sensible* energy.

Sensible and latent energy

Chemical energy

Nuclear energy

**FIGURE 3–6**

The internal energy of a system is the sum of all forms of the microscopic energies.

atom preserves its identity during a chemical reaction but loses it during a nuclear reaction. Atoms may also possess *electric* and *magnetic dipole-moment energies* when subjected to external electric and magnetic fields due to the twisting of the magnetic dipoles produced by the small electric currents associated with the orbiting electrons.

The forms of energy already discussed, which constitute the total energy of a system, can be *contained* or *stored* in a system, and thus can be viewed as the *static* forms of energy. The forms of energy not stored in a system can be viewed as the *dynamic* forms of energy or as *energy interactions*. The dynamic forms of energy are recognized at the system boundary as they cross it, and they represent the energy gained or lost by a system during a process. The only two forms of energy interactions associated with a closed system are **heat transfer** and **work**. An energy interaction is heat transfer if its driving force is a temperature difference. Otherwise it is work, as explained in the next section. A control volume can also exchange energy via mass transfer since any time mass is transferred into or out of a system, the energy content of the mass is also transferred with it.

In daily life, we frequently refer to the sensible and latent forms of internal energy as *heat*, and we talk about heat content of bodies. In thermodynamics, however, we usually refer to those forms of energy as **thermal energy** to prevent any confusion with *heat transfer*.

Distinction should be made between the macroscopic kinetic energy of an object as a whole and the microscopic kinetic energies of its molecules that constitute the sensible internal energy of the object (Fig. 3–7). The kinetic energy of an object is an *organized* form of energy associated with the orderly motion of all molecules in one direction in a straight path or around an axis. In contrast, the kinetic energies of the molecules are completely *random* and highly *disorganized*. As you will see in later chapters, the organized energy is much more valuable than the disorganized energy, and a major application area of thermodynamics is the conversion of disorganized energy (heat) into organized energy (work). You will also see that the organized energy can be converted to disorganized energy completely, but only a fraction of disorganized energy can be converted to organized energy by specially built devices called *heat engines* (like car engines and power plants). A similar argument can be given for the macroscopic potential energy of an object as a whole and the microscopic potential energies of the molecules.

**FIGURE 3–7**
The *macroscopic* kinetic energy is an organized form of energy and is much more useful than the disorganized *microscopic* kinetic energies of the molecules.

## More on Nuclear Energy

The best known fission reaction involves the split of the uranium atom (the U-235 isotope) into other elements and is commonly used to generate electricity in nuclear power plants (440 of them in 2004, generating 363,000 MW worldwide), to power nuclear submarines and aircraft carriers, and even to power spacecraft as well as building nuclear bombs.

The percentage of electricity produced by nuclear power is 78 percent in France, 25 percent in Japan, 28 percent in Germany, and 20 percent in the United States. The first nuclear chain reaction was achieved by Enrico Fermi in 1942, and the first large-scale nuclear reactors were built in 1944 for the purpose of producing material for nuclear weapons. When a uranium-235 atom absorbs a neutron and splits during a fission process, it produces a cesium-140 atom, a rubidium-93 atom, 3 neutrons, and $3.2 \times 10^{-11}$ J of energy. In practical terms, the complete fission of 1 kg of uranium-235 releases $6.73 \times 10^{10}$ kJ of heat, which is more than the heat released when 3000 tons of coal are burned. Therefore, for the same amount of fuel, a nuclear fission reaction releases several million times more energy than a chemical reaction. The safe disposal of used nuclear fuel, however, remains a concern.

Nuclear energy by fusion is released when two small nuclei combine into a larger one. The huge amount of energy radiated by the sun and the other stars originates from such a fusion process that involves the combination of two hydrogen atoms into a helium atom. When two heavy hydrogen (deuterium) nuclei combine during a fusion process, they produce a helium-3 atom, a free neutron, and $5.1 \times 10^{-13}$ J of energy (Fig. 3–8).

Fusion reactions are much more difficult to achieve in practice because of the strong repulsion between the positively charged nuclei, called the *Coulomb repulsion*. To overcome this repulsive force and to enable the two nuclei to fuse together, the energy level of the nuclei must be raised by heating them to about 100 million °C. But such high temperatures are found only in the stars or in exploding atomic bombs (the A-bomb). In fact, the uncontrolled fusion reaction in a hydrogen bomb (the H-bomb) is initiated by a small atomic bomb. The uncontrolled fusion reaction was achieved in the early 1950s, but all the efforts since then to achieve controlled fusion by massive lasers, powerful magnetic fields, and electric currents to generate power have failed.

(*a*) Fission of uranium

(*b*) Fusion of hydrogen

**FIGURE 3–8**

The fission of uranium and the fusion of hydrogen during nuclear reactions, and the release of nuclear energy.

---

**EXAMPLE 3–1    A Car Powered by Nuclear Fuel**

An average car consumes about 5 L of gasoline a day, and the capacity of the fuel tank of a car is about 50 L. Therefore, a car needs to be refueled once every 10 days. Also, the density of gasoline ranges from 0.68 to 0.78 kg/L, and its lower heating value is about 44,000 kJ/kg (that is, 44,000 kJ of heat is released when 1 kg of gasoline is completely burned). Suppose all the problems associated with the radioactivity and waste disposal of nuclear fuels are resolved, and a car is to be powered by U-235. If a new car comes equipped with 0.1-kg of the nuclear fuel U-235, determine if this car will ever need refueling under average driving conditions (Fig. 3–9).

**Solution**   A car powered by nuclear energy comes equipped with nuclear fuel. It is to be determined if this car will ever need refueling.
**Assumptions**   1 Gasoline is an incompressible substance with an average density of 0.75 kg/L. 2 Nuclear fuel is completely converted to thermal energy.
**Analysis**   The mass of gasoline used per day by the car is

$$m_{gasoline} = (\rho V)_{gasoline} = (0.75 \text{ kg/L})(5 \text{ L/day}) = 3.75 \text{ kg/day}$$

Noting that the heating value of gasoline is 44,000 kJ/kg, the energy supplied to the car per day is

$$E = (m_{gasoline})(\text{Heating value})$$

$$= (3.75 \text{ kg/day})(44,000 \text{ kJ/kg}) = 165,000 \text{ kJ/day}$$

**FIGURE 3–9**

Schematic for Example 3–1.

The complete fission of 0.1 kg of uranium-235 releases

$$(6.73 \times 10^{10}\,\text{kJ/kg})(0.1\,\text{kg}) = 6.73 \times 10^9\,\text{kJ}$$

of heat, which is sufficient to meet the energy needs of the car for

$$\text{No. of days} = \frac{\text{Energy content of fuel}}{\text{Daily energy use}} = \frac{6.73 \times 10^9\,\text{kJ}}{165{,}000\,\text{kJ/day}} = \textbf{40,790 days}$$

which is equivalent to about 112 years. Considering that no car will last more than 100 years, this car will never need refueling. It appears that nuclear fuel of the size of a cherry is sufficient to power a car during its lifetime.

**Discussion** Note that this problem is not quite realistic since the necessary critical mass cannot be achieved with such a small amount of fuel. Further, all of the uranium cannot be converted in fission, again because of the critical mass problems after partial conversion.

**FIGURE 3–10**

Mechanical energy is a useful concept for flows that do not involve significant heat transfer or energy conversion, such as the flow of gasoline from an underground tank into a car.

© *Corbis RF.*

## Mechanical Energy

Many engineering systems are designed to transport a fluid from one location to another at a specified flow rate, velocity, and elevation difference, and the system may generate mechanical work in a turbine or it may consume mechanical work in a pump or fan during this process (Fig. 3–10). These systems do not involve the conversion of nuclear, chemical, or thermal energy to mechanical energy. Also, they do not involve any heat transfer in any significant amount, and they operate essentially at constant temperature. Such systems can be analyzed conveniently by considering the *mechanical forms of energy* only and the frictional effects that cause the mechanical energy to be lost (i.e., to be converted to thermal energy that usually cannot be used for any useful purpose).

The **mechanical energy** can be defined as *the form of energy that can be converted to mechanical work completely and directly by an ideal mechanical device such as an ideal turbine.* Kinetic and potential energies are the familiar forms of mechanical energy. Thermal energy is not mechanical energy, however, since it cannot be converted to work directly and completely (the second law of thermodynamics).

A pump transfers mechanical energy to a fluid by raising its pressure, and a turbine extracts mechanical energy from a fluid by dropping its pressure. Therefore, the pressure of a flowing fluid is also associated with its mechanical energy. In fact, the pressure unit Pa is equivalent to Pa = N/m$^2$ = N·m/m$^3$ = J/m$^3$, which is energy per unit volume, and the product $Pv$ or its equivalent $P/\rho$ has the unit J/kg, which is energy per unit mass. Note that pressure itself is not a form of energy but a pressure force acting on a fluid through a distance produces work, called *flow work*, in the amount of $P/\rho$ per unit mass. Flow work is expressed in terms of fluid properties, and it is convenient to view it as part of the energy of a flowing fluid and call it *flow energy*. Therefore, the mechanical energy of a flowing fluid can be expressed on a unit mass basis as

$$e_{\text{mech}} = \frac{P}{\rho} + \frac{V^2}{2} + gz \tag{3–10}$$

where $P/\rho$ is the *flow energy*, $V^2/2$ is the *kinetic energy*, and $gz$ is the *potential energy* of the fluid, all per unit mass. It can also be expressed in rate form as

$$\dot{E}_{mech} = \dot{m}e_{mech} = \dot{m}\left(\frac{P}{\rho} + \frac{V^2}{2} + gz\right) \qquad (3\text{--}11)$$

where $\dot{m}$ is the mass flow rate of the fluid. Then the mechanical energy change of a fluid during incompressible ($\rho$ = constant) flow becomes

$$\Delta e_{mech} = \frac{P_2 - P_1}{\rho} + \frac{V_2^2 - V_1^2}{2} + g(z_2 - z_1) \qquad \text{(kJ/kg)} \qquad (3\text{--}12)$$

and

$$\Delta\dot{E}_{mech} = \dot{m}\Delta e_{mech} = \dot{m}\left(\frac{P_2 - P_1}{\rho} + \frac{V_2^2 - V_1^2}{2} + g(z_2 - z_1)\right) \qquad \text{(kW)} \qquad (3\text{--}13)$$

Therefore, the mechanical energy of a fluid does not change during flow if its pressure, density, velocity, and elevation remain constant. In the absence of any irreversible losses, the mechanical energy change represents the mechanical work supplied to the fluid (if $\Delta e_{mech} > 0$) or extracted from the fluid (if $\Delta e_{mech} < 0$). The maximum (ideal) power generated by a turbine, for example, is $\dot{W}_{max} = \dot{m}\Delta e_{mech}$, as shown in Fig. 3–11.

$\dot{W}_{max} = \dot{m}\Delta e_{mech} = \dot{m}g(z_1 - z_4) = \dot{m}gh$
since $P_1 \approx P_4 = P_{atm}$ and $V_1 = V_4 \approx 0$
(*a*)

$\dot{W}_{max} = \dot{m}\Delta e_{mech} = \dot{m}\dfrac{(P_2 - P_3)}{\rho} = \dot{m}\dfrac{\Delta P}{\rho}$
since $V_2 \approx V_3$ and $z_2 = z_3$
(*b*)

**FIGURE 3–11**

Mechanical energy is illustrated by an ideal hydraulic turbine coupled with an ideal generator. In the absence of irreversible losses, the maximum produced power is proportional to (*a*) the change in water surface elevation from the upstream to the downstream reservoir or (*b*) (close-up view) the drop in water pressure from just upstream to just downstream of the turbine.

---

**EXAMPLE 3–2    Wind Energy**

A site evaluated for a wind farm is observed to have steady winds at a speed of 8.5 m/s (Fig. 3–12). Determine the wind energy (*a*) per unit mass, (*b*) for a mass of 10 kg, and (*c*) for a flow rate of 1154 kg/s for air.

**Solution**    A site with a specified wind speed is considered. Wind energy per unit mass, for a specified mass, and for a given mass flow rate of air are to be determined.

**Assumptions**    Wind flows steadily at the specified speed.

**Analysis**    The only harvestable form of energy of atmospheric air is the kinetic energy, which is captured by a wind turbine.

(*a*) Wind energy per unit mass of air is

$$e = ke = \frac{V^2}{2} = \frac{(8.5 \text{ m/s})^2}{2}\left(\frac{1 \text{ J/kg}}{1 \text{ m}^2/\text{s}^2}\right) = \mathbf{36.1 \text{ J/kg}}$$

(*b*) Wind energy for an air mass of 10 kg is

$$E = me = (10 \text{ kg})(36.1 \text{ J/kg}) = \mathbf{361 \text{ J}}$$

(*c*) Wind energy for a mass flow rate of 1154 kg/s is

$$\dot{E} = \dot{m}e = (1154 \text{ kg/s})(36.1 \text{ J/kg})\left(\frac{1 \text{ kW}}{1000 \text{ J/s}}\right) = \mathbf{41.7 \text{ kW}}$$

**Discussion**    It can be shown that the specified mass flow rate corresponds to a 12-m diameter flow section when the air density is 1.2 kg/m³. Therefore, a wind turbine with a wind span diameter of 12 m has a power generation potential of 41.7 kW. Real wind turbines convert about one-third of this potential to electric power.

8.5 m/s

**FIGURE 3–12**

Potential site for a wind farm as discussed in Example 3–2.

© Vol. 36/PhotoDisc/Getty RF.

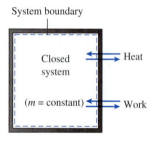

**FIGURE 3–13**

Energy can cross the boundaries of a closed system in the form of heat and work.

**FIGURE 3–14**

Temperature difference is the driving force for heat transfer. The larger the temperature difference, the higher is the rate of heat transfer.

# 3–3 ▪ ENERGY TRANSFER BY HEAT

Energy can cross the boundary of a closed system in two distinct forms: *heat* and *work* (Fig. 3–13). It is important to distinguish between these two forms of energy. Therefore, they will be discussed first, to form a sound basis for the development of the laws of thermodynamics.

We know from experience that a can of cold soda left on a table eventually warms up and that a hot baked potato on the same table cools down. When a body is left in a medium that is at a different temperature, energy transfer takes place between the body and the surrounding medium until thermal equilibrium is established, that is, the body and the medium reach the same temperature. The direction of energy transfer is always from the higher temperature body to the lower temperature one. Once the temperature equality is established, energy transfer stops. In the processes described above, energy is said to be transferred in the form of heat.

**Heat** is defined as *the form of energy that is transferred between two systems (or a system and its surroundings) by virtue of a temperature difference* (Fig. 3–14). That is, an energy interaction is heat only if it takes place because of a temperature difference. Then it follows that there cannot be any heat transfer between two systems that are at the same temperature.

Several phrases in common use today—such as heat flow, heat addition, heat rejection, heat absorption, heat removal, heat gain, heat loss, heat storage, heat generation, electrical heating, resistance heating, frictional heating, gas heating, heat of reaction, liberation of heat, specific heat, sensible heat, latent heat, waste heat, body heat, process heat, heat sink, and heat source—are not consistent with the strict thermodynamic meaning of the term *heat*, which limits its use to the *transfer* of thermal energy during a process. However, these phrases are deeply rooted in our vocabulary, and they are used by both ordinary people and scientists without causing any misunderstanding since they are usually interpreted properly instead of being taken literally. (Besides, no acceptable alternatives exist for some of these phrases.) For example, the phrase *body heat* is understood to mean *the thermal energy content* of a body. Likewise, *heat flow* is understood to mean *the transfer of thermal energy,* not the flow of a fluidlike substance called heat, although the latter incorrect interpretation, which is based on the caloric theory, is the origin of this phrase. Also, the transfer of heat into a system is frequently referred to as *heat addition* and the transfer of heat out of a system as *heat rejection*. Perhaps there are thermodynamic reasons for being so reluctant to replace *heat* by *thermal energy*: It takes less time and energy to say, write, and comprehend *heat* than it does *thermal energy*.

Heat is energy in transition. It is recognized only as it crosses the boundary of a system. Consider the hot baked potato one more time. The potato contains energy, but this energy is heat transfer only as it passes through the skin of the potato (the system boundary) to reach the air, as shown in Fig. 3–15. Once in the surroundings, the transferred heat becomes part of the internal energy of the surroundings. Thus, in thermodynamics, the term *heat* simply means *heat transfer.*

A process during which there is no heat transfer is called an **adiabatic process** (Fig. 3–16). The word *adiabatic* comes from the Greek word *adiabatos,* which means *not to be passed.* There are two ways a process can be adiabatic: Either the system is well insulated so that only a negligible amount of heat can pass through the boundary, or both the system and the surroundings are at the same temperature and therefore there is no driving force (temperature difference) for heat transfer. An adiabatic process should not be confused with an isothermal process. Even though there is no heat transfer during an adiabatic process, the energy content and thus the temperature of a system can still be changed by other means such as work.

As a form of energy, heat has energy units, kJ (or Btu) being the most common one. The amount of heat transferred during the process between two states (states 1 and 2) is denoted by $Q_{12}$, or just $Q$. Heat transfer *per unit mass* of a system is denoted $q$ and is determined from

$$q = \frac{Q}{m} \qquad \text{(kJ/kg)} \qquad \text{(3–14)}$$

Sometimes it is desirable to know the *rate of heat transfer* (the amount of heat transferred per unit time) instead of the total heat transferred over some time interval (Fig. 3–17). The heat transfer rate is denoted $\dot{Q}$, where the overdot stands for the time derivative, or "per unit time." The heat transfer rate $\dot{Q}$ has the unit kJ/s, which is equivalent to kW. When $\dot{Q}$ varies with time, the amount of heat transfer during a process is determined by integrating $\dot{Q}$ over the time interval of the process:

$$Q = \int_{t_1}^{t_2} \dot{Q} \, dt \qquad \text{(kJ)} \qquad \text{(3–15)}$$

When $\dot{Q}$ remains constant during a process, this relation reduces to

$$Q = \dot{Q} \, \Delta t \qquad \text{(kJ)} \qquad \text{(3–16)}$$

where $\Delta t = t_2 - t_1$ is the time interval during which the process takes place.

## Historical Background on Heat

Heat has always been perceived to be something that produces in us a sensation of warmth, and one would think that the nature of heat is one of the first things understood by mankind. However, it was only in the middle of the nineteenth century that we had a true physical understanding of the nature of heat, thanks to the development at that time of the **kinetic theory**, which treats molecules as tiny balls that are in motion and thus possess kinetic energy. Heat is then defined as the energy associated with the random motion of atoms and molecules. Although it was suggested in the eighteenth and early nineteenth centuries that heat is the manifestation of motion at the molecular level (called the *live force*), the prevailing view of heat until the middle of the nineteenth century was based on the caloric theory proposed by the French chemist Antoine Lavoisier (1744–1794) in 1789. The caloric theory asserts that heat is a fluidlike substance called the **caloric** that is a massless, colorless, odorless, and tasteless substance that can be poured from one body into another (Fig. 3–18). When caloric was added to

**FIGURE 3–15**

Energy is recognized as heat transfer only as it crosses the system boundary.

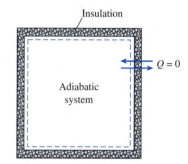

**FIGURE 3–16**

During an adiabatic process, a system exchanges no heat with its surroundings.

**FIGURE 3–17**

The relationships among $q$, $Q$, and $\dot{Q}$.

**FIGURE 3–18**

In the early nineteenth century, heat was thought to be an invisible fluid called the *caloric* that flowed from warmer bodies to the cooler ones.

a body, its temperature increased; and when caloric was removed from a body, its temperature decreased. When a body could not contain any more caloric, much the same way as when a glass of water could not dissolve any more salt or sugar, the body was said to be saturated with caloric. This interpretation gave rise to the terms *saturated liquid* and *saturated vapor* that are still in use today.

The caloric theory came under attack soon after its introduction. It maintained that heat is a substance that could not be created or destroyed. Yet it was known that heat can be generated indefinitely by rubbing one's hands together or rubbing two pieces of wood together. In 1798, the American Benjamin Thompson (Count Rumford) (1754–1814) showed in his papers that heat can be generated continuously through friction. The validity of the caloric theory was also challenged by several others. But it was the careful experiments of the Englishman James P. Joule (1818–1889) published in 1843 that finally convinced the skeptics that heat was not a substance after all, and thus put the caloric theory to rest. Although the caloric theory was totally abandoned in the middle of the nineteenth century, it contributed greatly to the development of thermodynamics and heat transfer.

Heat is transferred by three mechanisms: conduction, convection, and radiation. **Conduction** is the transfer of energy from the more energetic particles of a substance to the adjacent less energetic ones as a result of interaction between particles. **Convection** is the transfer of energy between a solid surface and the adjacent fluid that is in motion, and it involves the combined effects of conduction and fluid motion. **Radiation** is the transfer of energy due to the emission of electromagnetic waves (or photons). An overview of the three mechanisms of heat transfer is given at the end of this chapter as a Topic of Special Interest.

## 3–4 ▪ ENERGY TRANSFER BY WORK

Work, like heat, is an energy interaction between a system and its surroundings. As mentioned earlier, energy can cross the boundary of a closed system in the form of heat or work. Therefore, *if the energy crossing the boundary of a closed system is not heat, it must be work.* Heat is easy to recognize: Its driving force is a temperature difference between the system and its surroundings. Then we can simply say that an energy interaction that is not caused by a temperature difference between a system and its surroundings is work. More specifically, *work is the energy transfer associated with a force acting through a distance.* A rising piston, a rotating shaft, and an electric wire crossing the system boundaries are all associated with work interactions.

Work is also a form of energy transferred like heat and, therefore, has energy units such as kJ. The work done during a process between states 1 and 2 is denoted by $W_{12}$, or simply $W$. The work done *per unit mass* of a system is denoted by $w$ and is expressed as

$$w = \frac{W}{m} \qquad (\text{kJ/kg}) \qquad (3\text{–}17)$$

**FIGURE 3–19**

The relationships among $w$, $W$, and $\dot{W}$.

The work done *per unit time* is called **power** and is denoted $\dot{W}$. (Fig. 3–19). The unit of power is kJ/s, or kW.

Heat and work are *directional quantities*, and thus the complete description of a heat or work interaction requires the specification of both the *magnitude* and *direction*. One way of doing that is to adopt a sign convention. The generally accepted **formal sign convention** for heat and work interactions is as follows: *heat transfer to a system and work done by a system are positive; heat transfer from a system and work done on a system are negative*. Another way is to use the subscripts *in* and *out* to indicate direction (Fig. 3–20). For example, a work input of 5 kJ can be expressed as $W_{in} = 5$ kJ, while a heat loss of 3 kJ can be expressed as $Q_{out} = 3$ kJ. When the direction of a heat or work interaction is not known, we can simply *assume* a direction for the interaction (using the subscript *in* or *out*) and solve for it. A positive result indicates the assumed direction is right. A negative result, on the other hand, indicates that the direction of the interaction is the opposite of the assumed direction. This is just like assuming a direction for an unknown force when solving a statics problem, and reversing the direction when a negative result is obtained for the force. We will use this *intuitive approach* in this book as it eliminates the need to adopt a formal sign convention and the need to carefully assign negative values to some interactions.

Note that a quantity that is transferred to or from a system during an interaction is not a property since the amount of such a quantity depends on more than just the state of the system. Heat and work are *energy transfer mechanisms* between a system and its surroundings, and there are many similarities between them:

1. Both are recognized at the boundaries of a system as they cross the boundaries. That is, both heat and work are *boundary* phenomena.
2. Systems possess energy, but not heat or work.
3. Both are associated with a *process,* not a state. Unlike properties, heat or work has no meaning at a state.
4. Both are *path functions* (i.e., their magnitudes depend on the path followed during a process as well as the end states).

**Path functions** have **inexact differentials** designated by the symbol $\delta$. Therefore, a differential amount of heat or work is represented by $\delta Q$ or $\delta W$, respectively, instead of $dQ$ or $dW$. Properties, however, are **point functions** (i.e., they depend on the state only, and not on how a system reaches that state), and they have **exact differentials** designated by the symbol $d$. A small change in volume, for example, is represented by $dV$, and the total volume change during a process between states 1 and 2 is

$$\int_1^2 dV = V_2 - V_1 = \Delta V$$

That is, the volume change during process 1–2 is always the volume at state 2 minus the volume at state 1, regardless of the path followed (Fig. 3–21). The total work done during process 1–2, however, is

$$\int_1^2 \delta W = W_{12} \qquad (not\ \Delta W)$$

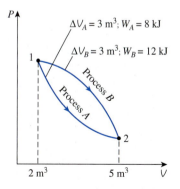

Surroundings

$Q_{in}$
$Q_{out}$

System

$W_{in}$
$W_{out}$

**FIGURE 3–20**

Specifying the directions of heat and work.

$\Delta V_A = 3$ m³; $W_A = 8$ kJ

$\Delta V_B = 3$ m³; $W_B = 12$ kJ

Process B

Process A

1

2

2 m³        5 m³      $V$

**FIGURE 3–21**

Properties are point functions; but heat and work are path functions (their magnitudes depend on the path followed).

That is, the total work is obtained by following the process path and adding the differential amounts of work ($\delta W$) done along the way. The integral of $\delta W$ is not $W_2 - W_1$ (i.e., the work at state 2 minus work at state 1), which is meaningless since work is not a property and systems do not possess work at a state.

**FIGURE 3–22**

Schematic for Example 3–3.

---

### EXAMPLE 3–3    Burning of a Candle in an Insulated Room

A candle is burning in a well-insulated room. Taking the room (the air plus the candle) as the system, determine (*a*) if there is any heat transfer during this burning process and (*b*) if there is any change in the internal energy of the system.

**Solution**   A candle burning in a well-insulated room is considered. It is to be determined whether there is any heat transfer and any change in internal energy.

*Analysis*   (*a*) The interior surfaces of the room form the system boundary, as indicated by the dashed lines in Fig. 3–22. As pointed out earlier, heat is recognized as it crosses the boundaries. Since the room is well insulated, we have an adiabatic system and no heat will pass through the boundaries. Therefore, $Q = 0$ for this process.

(*b*) The internal energy involves energies that exist in various forms (sensible, latent, chemical, nuclear). During the process just described, part of the chemical energy is converted to sensible energy. Since there is no increase or decrease in the total internal energy of the system, $\Delta U = 0$ for this process.

---

**FIGURE 3–23**

Schematic for Example 3–4.

---

### EXAMPLE 3–4    Heating of a Potato in an Oven

A potato initially at room temperature (25°C) is being baked in an oven that is maintained at 200°C, as shown in Fig. 3–23. Is there any heat transfer during this baking process?

**Solution**   A potato is being baked in an oven. It is to be determined whether there is any heat transfer during this process.

*Analysis*   This is not a well-defined problem since the system is not specified. Let us assume that we are observing the potato, which will be our system. Then the outer surface of the skin of the potato can be viewed as the system boundary. Part of the energy in the oven will pass through the skin to the potato. Since the driving force for this energy transfer is a temperature difference, this is a heat transfer process.

---

### EXAMPLE 3–5    Heating of an Oven by Work Transfer

A well-insulated electric oven is being heated through its heating element. If the entire oven, including the heating element, is taken to be the system, determine whether this is a heat or work interaction.

**Solution** A well-insulated electric oven is being heated by its heating element. It is to be determined whether this is a heat or work interaction.

**Analysis** For this problem, the interior surfaces of the oven form the system boundary, as shown in Fig. 3–24. The energy content of the oven obviously increases during this process, as evidenced by a rise in temperature. This energy transfer to the oven is not caused by a temperature difference between the oven and the surrounding air. Instead, it is caused by *electrons* crossing the system boundary and thus doing work. Therefore, this is a work interaction.

**FIGURE 3–24**

Schematic for Example 3–5.

**EXAMPLE 3–6**     **Heating of an Oven by Heat Transfer**

Answer the question in Example 3–5 if the system is taken as only the air in the oven without the heating element.

**Solution** The question in Example 3–5 is to be reconsidered by taking the system to be only the air in the oven.

**Analysis** This time, the system boundary will include the outer surface of the heating element and will not cut through it, as shown in Fig. 3–25. Therefore, no electrons will be crossing the system boundary at any point. Instead, the energy generated in the interior of the heating element will be transferred to the air around it as a result of the temperature difference between the heating element and the air in the oven. Therefore, this is a heat transfer process.

**Discussion** For both cases, the amount of energy transfer to the air is the same. These two examples show that an energy transfer can be heat or work, depending on how the system is selected.

**FIGURE 3–25**

Schematic for Example 3–6.

## Electrical Work

It was pointed out in Example 3–5 that electrons crossing the system boundary do electrical work on the system. In an electric field, electrons in a wire move under the effect of electromotive forces, doing work. When $N$ coulombs of electrical charge move through a potential difference $\mathbf{V}$, the electrical work done is

$$W_e = \mathbf{V}N$$

which can also be expressed in the rate form as

$$\dot{W}_e = \mathbf{V}I \qquad \text{(W)} \tag{3–18}$$

where $\dot{W}_e$ is the **electrical power** and $I$ is the number of electrical charges flowing per unit time, that is, the *current* (Fig. 3–26). In general, both $\mathbf{V}$ and $I$ vary with time, and the electrical work done during a time interval $\Delta t$ is expressed as

$$W_e = \int_1^2 \mathbf{V}I \, dt \qquad \text{(kJ)} \tag{3–19}$$

When both $\mathbf{V}$ and $I$ remain constant during the time interval $\Delta t$, it reduces to

$$W_e = \mathbf{V}I \, \Delta t \qquad \text{(kJ)} \tag{3–20}$$

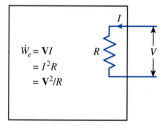

**FIGURE 3–26**

Electrical power in terms of resistance $R$, current $I$, and potential difference $\mathbf{V}$.

**FIGURE 3–27**

The work done is proportional to the force applied ($F$) and the distance traveled ($s$).

**FIGURE 3–28**

If there is no movement, no work is done.

*BLONDIE © KING FEATURES SYNDICATE.*

**FIGURE 3–29**

Energy transmission through rotating shafts is commonly encountered in practice.

# 3–5 ▪ MECHANICAL FORMS OF WORK

There are several different ways of doing work, each in some way related to a force acting through a distance (Fig. 3–27). In elementary mechanics, the work done by a constant force $F$ on a body displaced a distance $s$ in the direction of the force is given by

$$W = Fs \qquad \text{(kJ)} \qquad (3\text{–}21)$$

If the force $F$ is not constant, the work done is obtained by adding (i.e., integrating) the differential amounts of work,

$$W = \int_1^2 F \, ds \qquad \text{(kJ)} \qquad (3\text{–}22)$$

Obviously, one needs to know how the force varies with displacement to perform this integration. Equations 3–21 and 3–22 give only the magnitude of the work. The sign is easily determined from physical considerations: The work done on a system by an external force acting in the direction of motion is negative, and work done by a system against an external force acting in the opposite direction to motion is positive.

There are two requirements for a work interaction between a system and its surroundings to exist: (1) there must be a *force* acting on the boundary, and (2) the boundary must *move*. Therefore, the presence of forces on the boundary without any displacement of the boundary does not constitute a work interaction. Likewise, the displacement of the boundary without any force to oppose or drive this motion (such as the expansion of a gas into an evacuated space) is not a work interaction since no energy is transferred.

In many thermodynamic problems, mechanical work is the only form of work involved. It is associated with the movement of the boundary of a system or with the movement of the entire system as a whole (Fig. 3–28). Some common forms of mechanical work are discussed next.

## Shaft Work

Energy transmission with a rotating shaft is very common in engineering practice (Fig. 3–29). Often the torque T applied to the shaft is constant, which means that the force $F$ applied is also constant. For a specified constant torque, the work done during $n$ revolutions is determined as follows: A force $F$ acting through a moment arm $r$ generates a torque T of (Fig. 3–30)

$$\text{T} = Fr \quad \rightarrow \quad F = \frac{\text{T}}{r} \qquad (3\text{–}23)$$

This force acts through a distance $s$, which is related to the radius $r$ by

$$s = (2\pi r)n \qquad (3\text{–}24)$$

Then the shaft work is determined from

$$W_{\text{sh}} = Fs = \left(\frac{\text{T}}{r}\right)(2\pi rn) = 2\pi n\text{T} \qquad \text{(kJ)} \qquad (3\text{–}25)$$

The power transmitted through the shaft is the shaft work done per unit time, which can be expressed as

$$\dot{W}_{\text{sh}} = 2\pi \dot{n}\text{T} \qquad \text{(kW)} \qquad (3\text{–}26)$$

where $\dot{n}$ is the number of revolutions per unit time.

### EXAMPLE 3–7    Power Transmission by the Shaft of a Car

Determine the power transmitted through the shaft of a car when the torque applied is 200 N·m and the shaft rotates at a rate of 4000 revolutions per minute (rpm).

**Solution** The torque and the rpm for a car engine are given. The power transmitted is to be determined.

**Analysis** A sketch of the car is given in Fig. 3–31. The shaft power is determined directly from

$$\dot{W}_{sh} = 2\pi\dot{n}T = (2\pi)\left(4000 \, \frac{1}{min}\right)(200 \, \text{N·m})\left(\frac{1 \, min}{60 \, s}\right)\left(\frac{1 \, kJ}{1000 \, \text{N·m}}\right)$$

$$= \mathbf{83.8 \, kW} \qquad \text{(or 112 hp)}$$

**Discussion** Note that power transmitted by a shaft is proportional to torque and the rotational speed.

$$\dot{W}_{sh} = 2\pi\dot{n}T$$

$$\text{Torque} = Fr$$

**FIGURE 3–30**

Shaft work is proportional to the torque applied and the number of revolutions of the shaft.

$$\dot{n} = 4000 \, \text{rpm}$$
$$T = 200 \, \text{N·m}$$

**FIGURE 3–31**

Schematic for Example 3–7.

## Spring Work

It is common knowledge that when a force is applied on a spring, the length of the spring changes (Fig. 3–32). When the length of the spring changes by a differential amount $dx$ under the influence of a force $F$, the work done is

$$\delta W_{spring} = F \, dx \qquad \text{(3–27)}$$

To determine the total spring work, we need to know a functional relationship between $F$ and $x$. For linear elastic springs, the displacement $x$ is proportional to the force applied (Fig. 3–33). That is,

$$F = kx \qquad \text{(kN)} \qquad \text{(3–28)}$$

where $k$ is the spring constant and has the unit kN/m. The displacement $x$ is measured from the undisturbed position of the spring (that is, $x = 0$ when $F = 0$). Substituting Eq. 3–28 into Eq. 3–27 and integrating yield

$$W_{spring} = \tfrac{1}{2}k\left(x_2^2 - x_1^2\right) \qquad \text{(kJ)} \qquad \text{(3–29)}$$

where $x_1$ and $x_2$ are the initial and the final displacements of the spring, respectively, measured from the undisturbed position of the spring.

There are many other forms of mechanical work. Next we introduce some of them briefly.

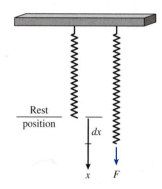

Rest position

$dx$

$x$    $F$

**FIGURE 3–32**

Elongation of a spring under the influence of a force.

## Work Done on Elastic Solid Bars

Solids are often modeled as linear springs because under the action of a force they contract or elongate, as shown in Fig. 3–34, and when the force is lifted, they return to their original lengths, like a spring. This is true as long as the force is in the elastic range, that is, not large enough to cause permanent (plastic) deformations. Therefore, the equations given for a linear spring can also be used for elastic solid bars. Alternately, we can determine

**FIGURE 3–33**

The displacement of a linear spring doubles when the force is doubled.

**FIGURE 3–34**

Solid bars behave as springs under the influence of a force.

**FIGURE 3–35**

Stretching a liquid film with a movable wire.

the work associated with the expansion or contraction of an elastic solid bar by replacing pressure $P$ by its counterpart in solids, *normal stress* $\sigma_n = F/A$, in the work expression:

$$W_{elastic} = \int_1^2 F \, dx = \int_1^2 \sigma_n A \, dx \qquad \text{(kJ)} \qquad \textbf{(3–30)}$$

where $A$ is the cross-sectional area of the bar. Note that the normal stress has pressure units.

## Work Associated with the Stretching of a Liquid Film

Consider a liquid film such as soap film suspended on a wire frame (Fig. 3–35). We know from experience that it will take some force to stretch this film by the movable portion of the wire frame. This force is used to overcome the microscopic forces between molecules at the liquid–air interfaces. These microscopic forces are perpendicular to any line in the surface, and the force generated by these forces per unit length is called the **surface tension** $\sigma_s$, whose unit is N/m. Therefore, the work associated with the stretching of a film is also called *surface tension work*. It is determined from

$$W_{surface} = \int_1^2 \sigma_s \, dA \qquad \text{(kJ)} \qquad \textbf{(3–31)}$$

where $dA = 2b \, dx$ is the change in the surface area of the film. The factor 2 is due to the fact that the film has two surfaces in contact with air. The force acting on the movable wire as a result of surface tension effects is $F = 2b\sigma_s$ where $\sigma_s$ is the surface tension force per unit length.

## Work Done to Raise or to Accelerate a Body

When a body is raised in a gravitational field, its potential energy increases. Likewise, when a body is accelerated, its kinetic energy increases. The conservation of energy principle requires that an equivalent amount of energy must be transferred to the body being raised or accelerated. Remember that energy can be transferred to a given mass by heat and work, and the energy transferred in this case obviously is not heat since it is not driven by a temperature difference. Therefore, it must be work. Then we conclude that (1) the work transfer needed to raise a body is equal to the change in the potential energy of the body, and (2) the work transfer needed to accelerate a body is equal to the change in the kinetic energy of the body (Fig. 3–36). Similarly, the potential or kinetic energy of a body represents the work that can be obtained from the body as it is lowered to the reference level or decelerated to zero velocity.

This discussion together with the consideration for friction and other losses form the basis for determining the required power rating of motors used to drive devices such as elevators, escalators, conveyor belts, and ski lifts. It also plays a primary role in the design of automotive and aircraft engines, and in the determination of the amount of hydroelectric power that can be produced from a given water reservoir, which is simply the potential energy of the water relative to the location of the hydraulic turbine.

## EXAMPLE 3–8    Power Needs of a Car to Climb a Hill

Consider a 1200-kg car cruising steadily on a level road at 90 km/h. Now the car starts climbing a hill that is sloped 30° from the horizontal (Fig. 3–37). If the velocity of the car is to remain constant during climbing, determine the additional power that must be delivered by the engine.

**Solution**    A car is to climb a hill while maintaining a constant velocity. The additional power needed is to be determined.

**Analysis**    The additional power required is simply the work that needs to be done per unit time to raise the elevation of the car, which is equal to the change in the potential energy of the car per unit time:

$$\dot{W}_g = mg \ \Delta z/\Delta t = mgV_{\text{vertical}}$$

$$= (1200 \text{ kg})(9.81 \text{ m/s}^2)(90 \text{ km/h})(\sin 30°)\left(\frac{1 \text{ m/s}}{3.6 \text{ km/h}}\right)\left(\frac{1 \text{ kJ/kg}}{1000 \text{ m}^2/\text{s}^2}\right)$$

$$= 147 \text{ kJ/s} = \mathbf{147 \ kW} \qquad \text{(or 197 hp)}$$

**Discussion**    Note that the car engine will have to produce almost 200 hp of additional power while climbing the hill if the car is to maintain its velocity.

**FIGURE 3–36**

The energy transferred to a body while being raised is equal to the change in its potential energy.

## EXAMPLE 3–9    Power Needs of a Car to Accelerate

Determine the power required to accelerate a 900-kg car shown in Fig. 3–38 from rest to a velocity of 80 km/h in 20 s on a level road.

**Solution**    The power required to accelerate a car to a specified velocity is to be determined.

**Analysis**    The work needed to accelerate a body is simply the change in the kinetic energy of the body,

$$W_a = \tfrac{1}{2}m(V_2^2 - V_1^2) = \tfrac{1}{2}(900 \text{ kg})\left[\left(\frac{80{,}000 \text{ m}}{3600 \text{ s}}\right)^2 - 0^2\right]\left(\frac{1 \text{ kJ/kg}}{1000 \text{ m}^2/\text{s}^2}\right)$$

$$= 222 \text{ kJ}$$

The average power is determined from

$$\dot{W}_a = \frac{W_a}{\Delta t} = \frac{222 \text{ kJ}}{20 \text{ s}} = \mathbf{11.1 \ kW} \qquad \text{(or 14.9 hp)}$$

**Discussion**    This is in addition to the power required to overcome friction, rolling resistance, and other imperfections.

$m = 1200$ kg

90 km/h

30°

**FIGURE 3–37**

Schematic for Example 3–8.

# Nonmechanical Forms of Work

The treatment in Section 3–5 represents a fairly comprehensive coverage of mechanical forms of work except the *moving boundary work* that is covered in Chap. 5. Some work modes encountered in practice are not mechanical in nature. However, these nonmechanical work modes can be treated in a similar manner by identifying a *generalized force F* acting in

0 ——→ 80 km/h

$m = 900$ kg

**FIGURE 3–38**

Schematic for Example 3–9.

the direction of a *generalized displacement x*. Then the work associated with the differential displacement under the influence of this force is determined from $\delta W = F dx$.

Some examples of nonmechanical work modes are **electrical work**, where the generalized force is the *voltage* (the electrical potential) and the generalized displacement is the *electrical charge,* as discussed earlier; **magnetic work**, where the generalized force is the *magnetic field strength* and the generalized displacement is the total *magnetic dipole moment;* and **electrical polarization work**, where the generalized force is the *electric field strength* and the generalized displacement is the *polarization of the medium* (the sum of the electric dipole rotation moments of the molecules). Detailed consideration of these and other nonmechanical work modes can be found in specialized books on these topics.

## 3–6 ▪ THE FIRST LAW OF THERMODYNAMICS

So far, we have considered various forms of energy such as heat $Q$, work $W$, and total energy $E$ individually, and no attempt is made to relate them to each other during a process. The *first law of thermodynamics,* also known as *the conservation of energy principle,* provides a sound basis for studying the relationships among the various forms of energy and energy interactions. Based on experimental observations, the first law of thermodynamics states that *energy can be neither created nor destroyed during a process; it can only change forms.* Therefore, every bit of energy should be accounted for during a process.

We all know that a rock at some elevation possesses some potential energy, and part of this potential energy is converted to kinetic energy as the rock falls (Fig. 3–39). Experimental data show that the decrease in potential energy ($mg\,\Delta z$) exactly equals the increase in kinetic energy $[m(V_2^2 - V_1^2)/2]$ when the air resistance is negligible, thus confirming the conservation of energy principle for mechanical energy.

Consider a system undergoing a series of *adiabatic* processes from a specified state 1 to another specified state 2. Being adiabatic, these processes obviously cannot involve any heat transfer, but they may involve several kinds of work interactions. Careful measurements during these experiments indicate the following: *For all adiabatic processes between two specified states of a closed system, the net work done is the same regardless of the nature of the closed system and the details of the process.* Considering that there are an infinite number of ways to perform work interactions under adiabatic conditions, this statement appears to be very powerful, with a potential for far-reaching implications. This statement, which is largely based on the experiments of Joule in the first half of the nineteenth century, cannot be drawn from any other known physical principle and is recognized as a fundamental principle. This principle is called the **first law of thermodynamics** or just the **first law**.

A major consequence of the first law is the existence and the definition of the property *total energy E.* Considering that the net work is the same for all adiabatic processes of a closed system between two specified states, the value of the net work must depend on the end states of the system only, and thus it must correspond to a change in a property of the system. This property

**FIGURE 3–39**

Energy cannot be created or destroyed; it can only change forms.

is the *total energy*. Note that the first law makes no reference to the value of the total energy of a closed system at a state. It simply states that the *change* in the total energy during an adiabatic process must be equal to the net work done. Therefore, any convenient arbitrary value can be assigned to total energy at a specified state to serve as a reference point.

Implicit in the first law statement is the conservation of energy. Although the essence of the first law is the existence of the property *total energy,* the first law is often viewed as a statement of the *conservation of energy* principle. Next, we develop the first law or the conservation of energy relation with the help of some familiar examples using intuitive arguments.

First, we consider some processes that involve heat transfer but no work interactions. The potato baked in the oven is a good example for this case (Fig. 3–40). As a result of heat transfer to the potato, the energy of the potato will increase. If we disregard any mass transfer (moisture loss from the potato), the increase in the total energy of the potato becomes equal to the amount of heat transfer. That is, if 5 kJ of heat is transferred to the potato, the energy increase of the potato will also be 5 kJ.

As another example, consider the heating of water in a pan on top of a range (Fig. 3–41). If 15 kJ of heat is transferred to the water from the heating element and 3 kJ of it is lost from the water to the surrounding air, the increase in energy of the water will be equal to the net heat transfer to water, which is 12 kJ.

Now consider a well-insulated (i.e., adiabatic) room heated by an electric heater as our system (Fig. 3–42). As a result of electrical work done, the energy of the system will increase. Since the system is adiabatic and cannot have any heat transfer to or from the surroundings ($Q = 0$), the conservation of energy principle dictates that the electrical work done on the system must equal the increase in energy of the system.

Next, let us replace the electric heater with a paddle wheel (Fig. 3–43). As a result of the stirring process, the energy of the system will increase. Again, since there is no heat interaction between the system and its surroundings ($Q = 0$), the shaft work done on the system must show up as an increase in the energy of the system.

Many of you have probably noticed that the temperature of air rises when it is compressed (Fig. 3–44). This is because energy is transferred to the air in the form of boundary work. In the absence of any heat transfer ($Q = 0$), the entire boundary work will be stored in the air as part of its total energy. The conservation of energy principle again requires that the increase in the energy of the system be equal to the boundary work done on the system.

We can extend these discussions to systems that involve various heat and work interactions simultaneously. For example, if a system gains 12 kJ of heat during a process while 6 kJ of work is done on it, the increase in the energy of the system during that process is 18 kJ (Fig. 3–45). That is, the change in the energy of a system during a process is simply equal to the net energy transfer to (or from) the system.

## Energy Balance

In the light of the preceding discussions, the conservation of energy principle can be expressed as follows: *The net change (increase or decrease) in the total energy of the system during a process is equal to the difference*

**FIGURE 3–40**

The increase in the energy of a potato in an oven is equal to the amount of heat transferred to it.

**FIGURE 3–41**

In the absence of any work interactions, the energy change of a system is equal to the net heat transfer.

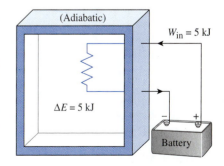

**FIGURE 3–42**

The work (electrical) done on an adiabatic system is equal to the increase in the energy of the system.

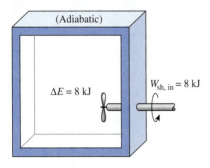

**FIGURE 3–43**

The work (shaft) done on an adiabatic system is equal to the increase in the energy of the system.

**FIGURE 3–44**

The work (boundary) done on an adiabatic system is equal to the increase in the energy of the system.

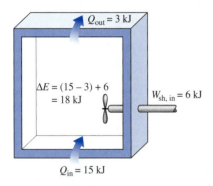

**FIGURE 3–45**

The energy change of a system during a process is equal to the *net* work and heat transfer between the system and its surroundings.

*between the total energy entering and the total energy leaving the system during that process.* That is,

$$\left(\begin{array}{c}\text{Total energy}\\\text{entering the system}\end{array}\right) - \left(\begin{array}{c}\text{Total energy}\\\text{leaving the system}\end{array}\right) = \left(\begin{array}{c}\text{Change in the total}\\\text{energy of the system}\end{array}\right)$$

or

$$E_{\text{in}} - E_{\text{out}} = \Delta E_{\text{system}}$$

This relation is often referred to as the **energy balance** and is applicable to any kind of system undergoing any kind of process. The successful use of this relation to solve engineering problems depends on understanding the various forms of energy and recognizing the forms of energy transfer.

## Energy Change of a System, $\Delta E_{\text{system}}$

The determination of the energy change of a system during a process involves the evaluation of the energy of the system at the beginning and at the end of the process, and taking their difference. That is,

Energy change = Energy at final state − Energy at initial state

or

$$\Delta E_{\text{system}} = E_{\text{final}} - E_{\text{initial}} = E_2 - E_1 \quad \text{(3–32)}$$

Note that energy is a property, and the value of a property does not change unless the state of the system changes. Therefore, the energy change of a system is zero if the state of the system does not change during the process. Also, energy can exist in numerous forms such as internal (sensible, latent, chemical, and nuclear), kinetic, potential, electric, and magnetic, and their sum constitutes the *total energy E* of a system. In the absence of electric, magnetic, and surface tension effects (i.e., for simple compressible systems), the change in the total energy of a system during a process is the sum of the changes in its internal, kinetic, and potential energies and can be expressed as

$$\Delta E = \Delta U + \Delta \text{KE} + \Delta \text{PE} \quad \text{(3–33)}$$

where

$$\Delta U = m(u_2 - u_1)$$
$$\Delta \text{KE} = \tfrac{1}{2}m(V_2^2 - V_1^2)$$
$$\Delta \text{PE} = mg(z_2 - z_1)$$

When the initial and final states are specified, the values of the specific internal energies $u_1$ and $u_2$ can be determined directly from the property tables or thermodynamic property relations.

Most systems encountered in practice are stationary, that is, they do not involve any changes in their velocity or elevation during a process (Fig. 3–46). Thus, for **stationary systems**, the changes in kinetic and potential energies are zero (that is, $\Delta \text{KE} = \Delta \text{PE} = 0$), and the total energy change relation in Eq. 3–33 reduces to $\Delta E = \Delta U$ for such systems. Also, the energy of a system during a process will change even if only one form of its energy changes while the other forms of energy remain unchanged.

# Mechanisms of Energy Transfer, $E_{in}$ and $E_{out}$

Energy can be transferred to or from a system in three forms: *heat, work,* and *mass flow.* Energy interactions are recognized at the system boundary as they cross it, and they represent the energy gained or lost by a system during a process. The only two forms of energy interactions associated with a fixed mass or closed system are *heat transfer* and *work.*

1. **Heat Transfer**, **$Q$** Heat transfer to a system (heat gain) increases the energy of the molecules and thus the internal energy of the system, and heat transfer from a system (heat loss) decreases it since the energy transferred out as heat comes from the energy of the molecules of the system.

2. **Work Transfer**, **$W$** An energy interaction that is not caused by a temperature difference between a system and its surroundings is work. A rising piston, a rotating shaft, and an electrical wire crossing the system boundaries are all associated with work interactions. Work transfer to a system (i.e., work done on a system) increases the energy of the system, and work transfer from a system (i.e., work done by the system) decreases it since the energy transferred out as work comes from the energy contained in the system. Car engines and hydraulic, steam, or gas turbines produce work while compressors, pumps, and mixers consume work.

3. **Mass Flow**, **$m$** Mass flow in and out of the system serves as an additional mechanism of energy transfer. When mass enters a system, the energy of the system increases because mass carries energy with it (in fact, mass is energy). Likewise, when some mass leaves the system, the energy contained within the system decreases because the leaving mass takes out some energy with it. For example, when some hot water is taken out of a water heater and is replaced by the same amount of cold water, the energy content of the hot-water tank (the control volume) decreases as a result of this mass interaction (Fig. 3–47).

**FIGURE 3–46**

For stationary systems,
$\Delta KE = \Delta PE = 0$; thus $\Delta E = \Delta U$.

**FIGURE 3–47**

The energy content of a control volume can be changed by mass flow as well as heat and work interactions.

Noting that energy can be transferred in the forms of heat, work, and mass, and that the net transfer of a quantity is equal to the difference between the amounts transferred in and out, the energy balance can be written more explicitly as

$$E_{in} - E_{out} = (Q_{in} - Q_{out}) + (W_{in} - W_{out}) + (E_{mass,in} - E_{mass,out}) = \Delta E_{system} \quad \text{(3–34)}$$

where the subscripts "in" and "out" denote quantities that enter and leave the system, respectively. All six quantities on the right side of the equation represent "amounts," and thus they are *positive* quantities. The direction of any energy transfer is described by the subscripts "in" and "out."

The heat transfer $Q$ is zero for adiabatic systems, the work transfer $W$ is zero for systems that involve no work interactions, and the energy transport with mass $E_{mass}$ is zero for systems that involve no mass flow across their boundaries (i.e., closed systems).

Energy balance for any system undergoing any kind of process can be expressed more compactly as

$$\underbrace{E_{\text{in}} - E_{\text{out}}}_{\substack{\text{Net energy transfer} \\ \text{by heat, work, and mass}}} = \underbrace{\Delta E_{\text{system}}}_{\substack{\text{Change in internal, kinetic,} \\ \text{potential, etc., energies}}} \quad \text{(kJ)} \qquad \text{(3–35)}$$

or, in the **rate form**, as

$$\underbrace{\dot{E}_{\text{in}} - \dot{E}_{\text{out}}}_{\substack{\text{Rate of net energy transfer} \\ \text{by heat, work, and mass}}} = \underbrace{dE_{\text{system}}/dt}_{\substack{\text{Rate of change in internal,} \\ \text{kinetic, potential, etc., energies}}} \quad \text{(kW)} \qquad \text{(3–36)}$$

For constant rates, the total quantities during a time interval $\Delta t$ are related to the quantities per unit time as

$$Q = \dot{Q}\,\Delta t, \quad W = \dot{W}\,\Delta t, \quad \text{and} \quad \Delta E = (dE/dt)\,\Delta t \qquad \text{(kJ)} \qquad \text{(3–37)}$$

The energy balance can be expressed on a **per unit mass** basis as

$$e_{\text{in}} - e_{\text{out}} = \Delta e_{\text{system}} \qquad \text{(kJ/kg)} \qquad \text{(3–38)}$$

which is obtained by dividing all the quantities in Eq. 3–35 by the mass $m$ of the system. Energy balance can also be expressed in the differential form as

$$\delta E_{\text{in}} - \delta E_{\text{out}} = dE_{\text{system}} \quad \text{or} \quad \delta e_{\text{in}} - \delta e_{\text{out}} = de_{\text{system}} \qquad \text{(3–39)}$$

For a closed system undergoing a **cycle**, the initial and final states are identical, and thus $\Delta E_{\text{system}} = E_2 - E_1 = 0$. Then the energy balance for a cycle simplifies to $E_{\text{in}} - E_{\text{out}} = 0$ or $E_{\text{in}} = E_{\text{out}}$. Noting that a closed system does not involve any mass flow across its boundaries, the energy balance for a cycle can be expressed in terms of heat and work interactions as

$$W_{\text{net,out}} = Q_{\text{net,in}} \quad \text{or} \quad \dot{W}_{\text{net,out}} = \dot{Q}_{\text{net,in}} \qquad \text{(for a cycle)} \qquad \text{(3–40)}$$

That is, the net work output during a cycle is equal to net heat input (Fig. 3–48).

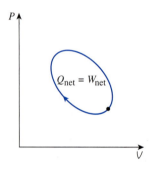

**FIGURE 3–48**

For a cycle $\Delta E = 0$, thus $Q = W$.

---

**EXAMPLE 3–10**    **Cooling of a Hot Fluid in a Tank**

A rigid tank contains a hot fluid that is cooled while being stirred by a paddle wheel. Initially, the internal energy of the fluid is 800 kJ. During the cooling process, the fluid loses 500 kJ of heat, and the paddle wheel does 100 kJ of work on the fluid. Determine the final internal energy of the fluid. Neglect the energy stored in the paddle wheel.

**Solution**  A fluid in a rigid tank looses heat while being stirred. The final internal energy of the fluid is to be determined.

*Assumptions*  **1** The tank is stationary and thus the kinetic and potential energy changes are zero, $\Delta$KE = $\Delta$PE = 0. Therefore, $\Delta E = \Delta U$ and internal energy is the only form of the system's energy that may change during this process. **2** Energy stored in the paddle wheel is negligible.

*Analysis*  Take the contents of the tank as the *system* (Fig. 3–49). This is a *closed system* since no mass crosses the boundary during the process. We observe that the volume of a rigid tank is constant, and thus there is no moving boundary work. Also, heat is lost from the system and shaft work is done on the system. Applying the energy balance on the system gives

$Q_{\text{out}} = 500 \text{ kJ}$

$U_1 = 800 \text{ kJ}$
$U_2 = ?$

$W_{\text{sh, in}} = 100 \text{ kJ}$

Fluid

**FIGURE 3–49**

Schematic for Example 3–10.

$$\underbrace{E_{in} - E_{out}}_{\substack{\text{Net energy transfer} \\ \text{by heat, work, and mass}}} = \underbrace{\Delta E_{system}}_{\substack{\text{Change in internal, kinetic,} \\ \text{potential, etc., energies}}}$$

$$W_{sh,in} - Q_{out} = \Delta U = U_2 - U_1$$

$$100 \text{ kJ} - 500 \text{ kJ} = U_2 - 800 \text{ kJ}$$

$$U_2 = \textbf{400 kJ}$$

Therefore, the final internal energy of the system is 400 kJ.

---

### EXAMPLE 3–11    Acceleration of Air by a Fan

A fan that consumes 20 W of electric power when operating is claimed to discharge air from a ventilated room at a rate of 1.0 kg/s at a discharge velocity of 8 m/s (Fig. 3–50). Determine if this claim is reasonable.

**Solution**  A fan is claimed to increase the velocity of air to a specified value while consuming electric power at a specified rate. The validity of this claim is to be investigated.

**Assumptions**  The ventilating room is relatively calm, and air velocity in it is negligible.

**Analysis**  First, let's examine the energy conversions involved: The motor of the fan converts part of the electrical power it consumes to mechanical (shaft) power, which is used to rotate the fan blades in air. The blades are shaped such that they impart a large fraction of the mechanical power of the shaft to air by mobilizing it. In the limiting ideal case of no losses (no conversion of electrical and mechanical energy to thermal energy) in steady operation, the electric power input will be equal to the rate of increase of the kinetic energy of air. Therefore, for a control volume that encloses the fan-motor unit, the energy balance can be written as

$$\underbrace{\dot{E}_{in} - \dot{E}_{out}}_{\substack{\text{Rate of net energy transfer} \\ \text{by heat, work, and mass}}} = \underbrace{dE_{system}/dt}_{\substack{\text{Rate of change in internal, kinetic,} \\ \text{potential, etc., energies}}}^{\nearrow 0 \text{ (steady)}} = 0 \quad \rightarrow \quad \dot{E}_{in} = \dot{E}_{out}$$

$$\dot{W}_{elect,\,in} = \dot{m}_{air}\, ke_{out} = \dot{m}_{air} \frac{V_{out}^2}{2}$$

Solving for $V_{out}$ and substituting gives the maximum air outlet velocity to be

$$V_{out} = \sqrt{\frac{2\dot{W}_{elect,in}}{\dot{m}_{air}}} = \sqrt{\frac{2(20 \text{ J/s})}{1.0 \text{ kg/s}} \left(\frac{1 \text{ m}^2/\text{s}^2}{1 \text{ J/kg}}\right)} = 6.3 \text{ m/s}$$

which is less than 8 m/s. Therefore, the claim is **false**.

**Discussion**  The conservation of energy principle requires the energy to be preserved as it is converted from one form to another, and it does not allow any energy to be created or destroyed during a process. From the first law point of view, there is nothing wrong with the conversion of the entire electrical energy into kinetic energy. Therefore, the first law has no objection to air velocity reaching 6.3 m/s—but this is the upper limit. Any claim of higher velocity is in violation of the first law, and thus impossible. In reality, the air velocity will be considerably lower than 6.3 m/s because of the losses associated with the conversion of electrical energy to mechanical shaft energy, and the conversion of mechanical shaft energy to kinetic energy or air.

8 m/s        Fan

Air

**FIGURE 3–50**
Schematic for Example 3–11.

$\dot{Q}_{out}$

Room

$\dot{W}_{elect.\,in}$

Fan

**FIGURE 3–51**

Schematic for Example 3–12.

## EXAMPLE 3–12    Heating Effect of a Fan

A room is initially at the outdoor temperature of 25°C. Now a large fan that consumes 200 W of electricity when running is turned on (Fig. 3–51). The heat transfer rate between the room and the outdoor air is given as $\dot{Q} = UA(T_i - T_o)$ where $U = 6$ W/m²·°C is the overall heat transfer coefficient, $A = 30$ m² is the exposed surface area of the room, and $T_i$ and $T_o$ are the indoor and outdoor air temperatures, respectively. Determine the indoor air temperature when steady operating conditions are established.

**Solution**    A large fan is turned on and kept on in a room that looses heat to the outdoors. The indoor air temperature is to be determined when steady operation is reached.

**Assumptions**    **1** Heat transfer through the floor is negligible. **2** There are no other energy interactions involved.

**Analysis**    The electricity consumed by the fan is energy input for the room, and thus the room gains energy at a rate of 200 W. As a result, the room air temperature tends to rise. But as the room air temperature rises, the rate of heat loss from the room increases until the rate of heat loss equals the electric power consumption. At that point, the temperature of the room air, and thus the energy content of the room, remains constant, and the conservation of energy for the room becomes

$$\underbrace{\dot{E}_{in} - \dot{E}_{out}}_{\substack{\text{Rate of net energy transfer} \\ \text{by heat, work, and mass}}} = \underbrace{dE_{system}/dt}_{\substack{\text{Rate of change in internal, kinetic,} \\ \text{potential, etc., energies}}} {}^{\nearrow 0\,(\text{steady})} = 0 \quad \rightarrow \quad \dot{E}_{in} = \dot{E}_{out}$$

$$\dot{W}_{elect,in} = \dot{Q}_{out} = UA(T_i - T_o)$$

Substituting,

$$200\ \text{W} = (6\ \text{W/m}^2\cdot\text{°C})(30\ \text{m}^2)(T_i - 25\text{°C})$$

It gives

$$T_i = \mathbf{26.1°C}$$

Therefore, the room air temperature will remain constant after it reaches 26.1°C.

**Discussion**    Note that a 200-W fan heats a room just like a 200-W resistance heater. In the case of a fan, the motor converts part of the electric energy it draws to mechanical energy in the form of a rotating shaft while the remaining part is dissipated as heat to the room air because of the motor inefficiency (no motor converts 100 percent of the electric energy it receives to mechanical energy, although some large motors come close with a conversion efficiency of over 97 percent). Part of the mechanical energy of the shaft is converted to kinetic energy of air through the blades, which is then converted to thermal energy as air molecules slow down because of friction. At the end, the entire electric energy drawn by the fan motor is converted to thermal energy of air, which manifests itself as a rise in temperature.

**FIGURE 3–52**

Fluorescent lamps lighting a classroom as discussed in Example 3–13.

© Vol. 24/PhotoDisc/Getty RF.

## EXAMPLE 3–13    Annual Lighting Cost of a Classroom

The lighting needs of a classroom are met by 30 fluorescent lamps, each consuming 80 W of electricity (Fig. 3–52). The lights in the classroom are kept on for 12 hours a day and 250 days a year. For a unit electricity cost of

7 cents per kWh, determine annual energy cost of lighting for this class-room. Also, discuss the effect of lighting on the heating and air-conditioning requirements of the room.

**Solution**  The lighting of a classroom by fluorescent lamps is considered. The annual electricity cost of lighting for this classroom is to be determined, and the lighting's effect on the heating and air-conditioning requirements is to be discussed.

**Assumptions**  The effect of voltage fluctuations is negligible so that each fluorescent lamp consumes its rated power.

**Analysis**  The electric power consumed by the lamps when all are on and the number of hours they are kept on per year are

$$\text{Lighting power} = (\text{Power consumed per lamp}) \times (\text{No. of lamps})$$

$$= (80 \text{ W/lamp})(30 \text{ lamps})$$

$$= 2400 \text{ W} = 2.4 \text{ kW}$$

$$\text{Operating hours} = (12 \text{ h/day})(250 \text{ days/year}) = 3000 \text{ h/year}$$

Then the amount and cost of electricity used per year become

$$\text{Lighting energy} = (\text{Lighting power})(\text{Operating hours})$$

$$= (2.4 \text{ kW})(3000 \text{ h/year}) = 7200 \text{ kWh/year}$$

$$\text{Lighting cost} = (\text{Lighting energy})(\text{Unit cost})$$

$$= (7200 \text{ kWh/year})(\$0.07/\text{kWh}) = \mathbf{\$504/year}$$

Light is absorbed by the surfaces it strikes and is converted to thermal energy. Disregarding the light that escapes through the windows, the entire 2.4 kW of electric power consumed by the lamps eventually becomes part of thermal energy of the classroom. Therefore, the lighting system in this room reduces the heating requirements by 2.4 kW, but increases the air-conditioning load by 2.4 kW.

**Discussion**  Note that the annual lighting cost of this classroom alone is over $500. This shows the importance of energy conservation measures. If incandescent light bulbs were used instead of fluorescent tubes, the lighting costs would be four times as much since incandescent lamps use four times as much power for the same amount of light produced.

---

*EXAMPLE 3–14*    **Conservation of Energy for an Oscillating Steel Ball**

The motion of a steel ball in a hemispherical bowl of radius $h$ shown in Fig. 3–53 is to be analyzed. The ball is initially held at the highest location at point $A$, and then it is released. Obtain relations for the conservation of energy of the ball for the cases of frictionless and actual motions.

**Solution**  A steel ball is released in a bowl. Relations for the energy balance are to be obtained.

**Assumptions**  The motion is frictionless, and thus friction between the ball, the bowl, and the air is negligible.

**Analysis**  When the ball is released, it accelerates under the influence of gravity, reaches a maximum velocity (and minimum elevation) at point $B$ at

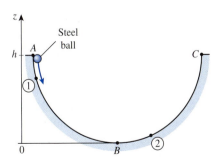

**FIGURE 3–53**

Schematic for Example 3–14.

**FIGURE 3–54**

The definition of performance is not limited to thermodynamics only.

*BLONDIE © KING FEATURES SYNDICATE.*

Water heater

| Type | Efficiency |
|------|------------|
| Gas, conventional | 55% |
| Gas, high-efficiency | 62% |
| Electric, conventional | 90% |
| Electric, high-efficiency | 94% |

**FIGURE 3–55**

Typical efficiencies of conventional and high-efficiency electric and natural gas water heaters.

*© The McGraw-Hill Companies, Inc./Jill Braaten, photographer.*

the bottom of the bowl, and moves up toward point $C$ on the opposite side. In the ideal case of frictionless motion, the ball will oscillate between points $A$ and $C$. The actual motion involves the conversion of the kinetic and potential energies of the ball to each other, together with overcoming resistance to motion due to friction (doing frictional work). The general energy balance for any system undergoing any process is

$$\underbrace{E_{in} - E_{out}}_{\substack{\text{Net energy transfer} \\ \text{by heat, work, and mass}}} = \underbrace{\Delta E_{system}}_{\substack{\text{Change in internal, kinetic,} \\ \text{potential, etc., energies}}}$$

Then the energy balance for the ball for a process from point 1 to point 2 becomes

$$-w_{friction} = (ke_2 + pe_2) - (ke_1 + pe_1)$$

or

$$\frac{V_1^2}{2} + gz_1 = \frac{V_2^2}{2} + gz_2 + w_{friction}$$

since there is no energy transfer by heat or mass and no change in the internal energy of the ball (the heat generated by frictional heating is dissipated to the surrounding air). The frictional work term $w_{friction}$ is often expressed as $e_{loss}$ to represent the loss (conversion) of mechanical energy into thermal energy.

For the idealized case of frictionless motion, the last relation reduces to

$$\frac{V_1^2}{2} + gz_1 = \frac{V_2^2}{2} + gz_2 \quad \text{or} \quad \frac{V^2}{2} + gz = C = \text{constant}$$

where the value of the constant is $C = gh$. That is, *when the frictional effects are negligible, the sum of the kinetic and potential energies of the ball remains constant.*

**Discussion** This is certainly a more intuitive and convenient form of the conservation of energy equation for this and other similar processes such as the swinging motion of the pendulum of a wall clock.

## 3–7 ■ ENERGY CONVERSION EFFICIENCIES

*Efficiency* is one of the most frequently used terms in thermodynamics, and it indicates how well an energy conversion or transfer process is accomplished. Efficiency is also one of the most frequently misused terms in thermodynamics and a source of misunderstandings. This is because efficiency is often used without being properly defined first. Next, we will clarify this further and define some efficiencies commonly used in practice.

Performance or efficiency, in general, can be expressed in terms of the desired output and the required input as (Fig. 3–54)

$$\text{Performance} = \frac{\text{Desired output}}{\text{Required input}} \tag{3–41}$$

If you are shopping for a water heater, a knowledgeable salesperson will tell you that the efficiency of a conventional electric water heater is about 90 percent (Fig. 3–55). You may find this confusing, since the heating elements of electric water heaters are resistance heaters, and the efficiency of all resistance heaters is 100 percent as they convert all the electrical energy they consume into thermal energy. A knowledgeable salesperson will clarify this by explaining that the heat losses from the hot-water tank to the surrounding air amount to 10 percent of the electrical energy consumed, and the **efficiency of a water heater** is defined as the ratio of the *energy delivered to the house by hot water* to the *energy supplied to the water heater.* A clever salesperson may even talk you into buying a more expensive water heater with thicker insulation that has an efficiency of 94 percent. If you are a knowledgeable consumer and have access to natural gas, you will probably purchase a gas water heater whose efficiency is only 55 percent since a gas unit costs about the same as an electric unit to purchase and install, but the annual energy cost of a gas unit will be much less than that of an electric unit.

Perhaps you are wondering how the efficiency for a gas water heater is defined, and why it is much lower than the efficiency of an electric heater. As a general rule, the efficiency of equipment that involves the combustion of a fuel is based on the **heating value of the fuel**, which is *the amount of heat released when a unit amount of fuel at room temperature is completely burned and the combustion products are cooled to the room temperature* (Fig. 3–56). Then the performance of combustion equipment can be characterized by **combustion efficiency**, defined as

**FIGURE 3–56**

The definition of the heating value of gasoline.

$$\eta_{\text{combustion}} = \frac{Q}{\text{HV}} = \frac{\text{Amount of heat released during combustion}}{\text{Heating value of the fuel burned}} \qquad \textbf{(3–42)}$$

A combustion efficiency of 100 percent indicates that the fuel is burned completely and the stack gases leave the combustion chamber at room temperature, and thus the amount of heat released during a combustion process is equal to the heating value of the fuel.

Most fuels contain hydrogen, which forms water when burned, and the heating value of a fuel will be different, depending on whether the water in combustion products is in the liquid or vapor form. The heating value is called the *lower heating value*, or LHV, when the water leaves as a vapor, and the *higher heating value*, or HHV, when the water in the combustion gases is completely condensed and thus the heat of vaporization is also recovered. The difference between these two heating values is equal to the product of the amount of water and the enthalpy of vaporization of water at room temperature. For example, the lower and higher heating values of gasoline are 44,000 kJ/kg and 47,300 kJ/kg, respectively. An efficiency definition should make it clear whether it is based on the higher or lower heating value of the fuel. Efficiencies of cars and jet engines are normally based on *lower heating values* since water normally leaves as a vapor in the exhaust gases, and it is not practical to try to recuperate the heat of vaporization. Efficiencies of furnaces, on the other hand, are based on *higher heating values*.

The efficiency of space heating systems of residential and commercial buildings is usually expressed in terms of the **annual fuel utilization efficiency**, or **AFUE**, which accounts for the combustion efficiency as well as other losses such as heat losses to unheated areas and start-up and cool-down losses. The AFUE of most new heating systems is about 85 percent,

**TABLE 3–1**

The efficacy of different lighting systems

| Type of lighting | Efficacy, lumens/W |
|---|---|
| *Combustion* | |
| Candle | 0.3 |
| Kerosene lamp | 1–2 |
| *Incandescent* | |
| Ordinary | 6–20 |
| Halogen | 15–35 |
| *Fluorescent* | |
| Compact | 40–87 |
| Tube | 60–120 |
| *High-intensity discharge* | |
| Mercury vapor | 40–60 |
| Metal halide | 65–118 |
| High-pressure sodium | 85–140 |
| Low-pressure sodium | 70–200 |
| *Solid-State* | |
| LED | 20–160 |
| OLED | 15–60 |
| Theoretical limit | 300* |

*This value depends on the spectral distribution of the assumed ideal light source. For white light sources, the upper limit is about 300 lm/W for metal halide, 350 lm/W for fluorescents, and 400 lm/W for LEDs. Spectral maximum occurs at a wavelength of 555 nm (green) with a light output of 683 lm/W.

15 W          60 W

**FIGURE 3–57**

A 15-W compact fluorescent lamp provides as much light as a 60-W incandescent lamp.

although the AFUE of some old heating systems is under 60 percent. The AFUE of some new high-efficiency furnaces exceeds 96 percent, but the high cost of such furnaces cannot be justified for locations with mild to moderate winters. Such high efficiencies are achieved by reclaiming most of the heat in the flue gases, condensing the water vapor, and discharging the flue gases at temperatures as low as 38°C (or 100°F) instead of about 200°C (or 400°F) for the conventional models.

For *car engines,* the work output is understood to be the power delivered by the crankshaft. But for power plants, the work output can be the mechanical power at the turbine exit, or the electrical power output of the generator.

A generator is a device that converts mechanical energy to electrical energy, and the effectiveness of a generator is characterized by the **generator efficiency**, which is the ratio of the *electrical power output* to the *mechanical power input.* The *thermal efficiency* of a power plant, which is of primary interest in thermodynamics, is usually defined as the ratio of the net shaft work output of the turbine to the heat input to the working fluid. The effects of other factors are incorporated by defining an **overall efficiency** for the power plant as the ratio of the *net electrical power output* to the *rate of fuel energy input.* That is,

$$\eta_{overall} = \eta_{combustion}\,\eta_{thermal}\,\eta_{generator} = \frac{\dot{W}_{net,electric}}{HHV \times \dot{m}_{net}} \tag{3–43}$$

The overall efficiencies are about 26–30 percent for gasoline automotive engines, 34–40 percent for diesel engines, and up to 60 percent for large power plants.

We are all familiar with the conversion of electrical energy to *light* by incandescent lightbulbs, fluorescent tubes, and high-intensity discharge lamps. The efficiency for the conversion of electricity to light can be defined as the ratio of the energy converted to light to the electrical energy consumed. For example, common incandescent lightbulbs convert about 5 percent of the electrical energy they consume to light; the rest of the energy consumed is dissipated as heat, which adds to the cooling load of the air conditioner in summer. However, it is more common to express the effectiveness of this conversion process by **lighting efficacy**, which is defined as the *amount of light output in lumens per W of electricity consumed.*

The efficacy of different lighting systems is given in Table 3–1. Note that a compact fluorescent lightbulb produces about four times as much light as an incandescent lightbulb per W, and thus a 15-W fluorescent bulb can replace a 60-W incandescent lightbulb (Fig. 3–57). Also, a compact fluorescent bulb lasts about 10,000 h, which is 10 times as long as an incandescent bulb, and it plugs directly into the socket of an incandescent lamp. Therefore, despite their higher initial cost, compact fluorescents reduce the lighting costs considerably through reduced electricity consumption. Sodium-filled high-intensity discharge lamps provide the most efficient lighting, but their use is limited to outdoor use because of their yellowish light.

We can also define efficiency for cooking appliances since they convert electrical or chemical energy to heat for cooking. The **efficiency of a cooking appliance** can be defined as the ratio of the *useful energy transferred to the* *food to the energy consumed by the appliance* (Fig. 3–58). Electric ranges are more efficient than gas ranges, but it is much cheaper to cook with natural gas than with electricity because of the lower unit cost of natural gas (Table 3–2).

The cooking efficiency depends on user habits as well as the individual appliances. Convection and microwave ovens are inherently more efficient than conventional ovens. On average, convection ovens save about *one-third* and microwave ovens save about *two-thirds* of the energy used by conventional ovens. The cooking efficiency can be increased by using the smallest oven for baking, using a pressure cooker, using an electric slow cooker for stews and soups, using the smallest pan that will do the job, using the smaller heating element for small pans on electric ranges, using flat-bottomed pans on electric burners to assure good contact, keeping burner drip pans clean and shiny, defrosting frozen foods in the refrigerator before cooking, avoiding preheating unless it is necessary, keeping the pans covered during cooking, using timers and thermometers to avoid overcooking, using the self-cleaning feature of ovens right after cooking, and keeping inside surfaces of microwave ovens clean.

Using energy-efficient appliances and practicing energy conservation measures help our pocketbooks by reducing our utility bills. It also helps the **environment** by reducing the amount of pollutants emitted to the atmosphere during the combustion of fuel at home or at the power plants where electricity is generated. The combustion of *each therm of natural gas* produces 6.4 kg of carbon dioxide, which causes global climate change; 4.7 g of nitrogen oxides and 0.54 g of hydrocarbons, which cause smog; 2.0 g of carbon monoxide, which is toxic; and 0.030 g of sulfur dioxide, which causes acid rain. Each therm of natural gas saved eliminates the emission of these pollutants while saving $0.60 for the average consumer in the United States. Each kWh of electricity conserved saves 0.4 kg of coal and 1.0 kg of $CO_2$ and 15 g of $SO_2$ from a coal power plant.

$$\text{Efficiency} = \frac{\text{Energy utilized}}{\text{Energy supplied to appliance}}$$

$$= \frac{3 \text{ kWh}}{5 \text{ kWh}} = 0.60$$

**FIGURE 3–58**

The efficiency of a cooking appliance represents the fraction of the energy supplied to the appliance that is transferred to the food.

**TABLE 3–2**

Energy costs of cooking a casserole with different appliances*

[From J. T. Amann, A. Wilson, and K. Ackerly, *Consumer Guide to Home Energy Savings*, 9th ed., American Council for an Energy-Efficient Economy, Washington, D.C., 2007, p. 163.]

| Cooking appliance | Cooking temperature | Cooking time | Energy used | Cost of energy |
|---|---|---|---|---|
| Electric oven | 350°F (177°C) | 1 h | 2.0 kWh | $0.19 |
| Convection oven (elect.) | 325°F (163°C) | 45 min | 1.39 kWh | $0.13 |
| Gas oven | 350°F (177°C) | 1 h | 0.112 therm | $0.13 |
| Frying pan | 420°F (216°C) | 1 h | 0.9 kWh | $0.09 |
| Toaster oven | 425°F (218°C) | 50 min | 0.95 kWh | $0.09 |
| Crockpot | 200°F (93°C) | 7 h | 0.7 kWh | $0.07 |
| Microwave oven | "High" | 15 min | 0.36 kWh | $0.03 |

*Assumes a unit cost of $0.095/kWh for electricity and $1.20/therm for gas.

38%    Gas Range

73%    Electric Range

**FIGURE 3–59**

Schematic of the 73 percent efficient electric heating unit and 38 percent efficient gas burner discussed in Example 3–15.

**EXAMPLE 3–15    Cost of Cooking with Electric and Gas Ranges**

The efficiency of cooking appliances affects the internal heat gain from them since an inefficient appliance consumes a greater amount of energy for the same task, and the excess energy consumed shows up as heat in the living space. The efficiency of open burners is determined to be 73 percent for electric units and 38 percent for gas units (Fig. 3–59). Consider a 2-kW electric burner at a location where the unit costs of electricity and natural gas are $0.09/kWh and $1.20/therm, respectively. Determine the rate of energy consumption by the burner and the unit cost of utilized energy for both electric and gas burners.

**Solution**   The operation of electric and gas ranges is considered. The rate of energy consumption and the unit cost of utilized energy are to be determined.

**Analysis**   The efficiency of the electric heater is given to be 73 percent. Therefore, a burner that consumes 2 kW of electrical energy will supply

$$\dot{Q}_{\text{utilized}} = (\text{Energy input}) \times (\text{Efficiency}) = (2\ \text{kW})(0.73) = \mathbf{1.46\ kW}$$

of useful energy. The unit cost of utilized energy is inversely proportional to the efficiency, and is determined from

$$\text{Cost of utilized energy} = \frac{\text{Cost of energy input}}{\text{Efficiency}} = \frac{\$0.09/\text{kWh}}{0.73} = \mathbf{\$0.123/kWh}$$

Noting that the efficiency of a gas burner is 38 percent, the energy input to a gas burner that supplies utilized energy at the same rate (1.46 kW) is

$$\dot{Q}_{\text{input, gas}} = \frac{\dot{Q}_{\text{utilized}}}{\text{Efficiency}} = \frac{1.46\ \text{kW}}{0.38} = \mathbf{3.84\ kW} \qquad (=13{,}100\ \text{Btu/h})$$

since 1 kW = 3412 Btu/h. Therefore, a gas burner should have a rating of at least 13,100 Btu/h to perform as well as the electric unit.

Noting that 1 therm = 29.3 kWh, the unit cost of utilized energy in the case of a gas burner is determined to be

$$\text{Cost of utilized energy} = \frac{\text{Cost of energy input}}{\text{Efficiency}} = \frac{\$1.20/29.3\ \text{kWh}}{0.38}$$

$$= \mathbf{\$0.108/kWh}$$

**Discussion**   The cost of utilized gas is less than that of utilized electricity. Therefore, despite its higher efficiency, cooking with an electric burner will cost about 14 percent more compared to a gas burner in this case. This explains why cost-conscious consumers always ask for gas appliances, and it is not wise to use electricity for heating purposes.

## Efficiencies of Mechanical and Electrical Devices

The transfer of mechanical energy is usually accomplished by a rotating shaft, and thus mechanical work is often referred to as *shaft work*. A pump or a fan receives shaft work (usually from an electric motor) and transfers it to the fluid as mechanical energy (less frictional losses). A turbine, on the other hand, converts the mechanical energy of a fluid to shaft work. In the absence of any irreversibilities such as friction, mechanical energy can be converted entirely from one mechanical form to another, and the **mechanical efficiency** of a device or process can be defined as (Fig. 3–60)

$$\eta_{mech} = \frac{\text{Mechanical energy output}}{\text{Mechanical energy input}} = \frac{E_{mech,out}}{E_{mech,in}} = 1 - \frac{E_{mech,loss}}{E_{mech,in}} \qquad (3\text{–}44)$$

A conversion efficiency of less than 100 percent indicates that conversion is less than perfect and some losses have occurred during conversion. A mechanical efficiency of 97 percent indicates that 3 percent of the mechanical energy input is converted to thermal energy as a result of frictional heating, and this will manifest itself as a slight rise in the temperature of the fluid.

In fluid systems, we are usually interested in increasing the pressure, velocity, and/or elevation of a fluid. This is done by *supplying mechanical energy* to the fluid by a pump, a fan, or a compressor (we will refer to all of them as pumps). Or we are interested in the reverse process of *extracting mechanical energy* from a fluid by a turbine and producing mechanical power in the form of a rotating shaft that can drive a generator or any other rotary device. The degree of perfection of the conversion process between the mechanical work supplied or extracted and the mechanical energy of the fluid is expressed by the **pump efficiency** and **turbine efficiency**, defined as

$$\eta_{pump} = \frac{\text{Mechanical energy increase of the fluid}}{\text{Mechanical energy input}} = \frac{\Delta\dot{E}_{mech,fluid}}{\dot{W}_{shaft,in}} = \frac{\dot{W}_{pump,u}}{\dot{W}_{pump}} \qquad (3\text{–}45)$$

where $\Delta\dot{E}_{mech,fluid} = \dot{E}_{mech,out} - \dot{E}_{mech,in}$ is the rate of increase in the mechanical energy of the fluid, which is equivalent to the **useful pumping power** $\dot{W}_{pump,u}$ supplied to the fluid, and

$$\eta_{turbine} = \frac{\text{Mechanical energy output}}{\text{Mechanical energy decrease of the fluid}} = \frac{\dot{W}_{shaft,out}}{|\Delta\dot{E}_{mech,fluid}|} = \frac{\dot{W}_{turbine}}{\dot{W}_{turbine,e}} \qquad (3\text{–}46)$$

where $|\Delta\dot{E}_{mech,fluid}| = \dot{E}_{mech,in} - \dot{E}_{mech,out}$ is the rate of decrease in the mechanical energy of the fluid, which is equivalent to the mechanical power extracted from the fluid by the turbine $\dot{W}_{turbine,e}$, and we use the absolute value sign to avoid negative values for efficiencies. A pump or turbine efficiency of 100 percent indicates perfect conversion between the shaft work and the mechanical energy of the fluid, and this value can be approached (but never attained) as the frictional effects are minimized.

Electrical energy is commonly converted to *rotating mechanical energy* by electric motors to drive fans, compressors, robot arms, car starters, and so forth. The effectiveness of this conversion process is characterized by the *motor efficiency* $\eta_{motor}$, which is the ratio of the *mechanical energy output* of the motor to the *electrical energy input*. The full-load motor efficiencies range from about 35 percent for small motors to over 97 percent for large high-efficiency motors. The difference between the electrical energy consumed and the mechanical energy delivered is dissipated as waste heat.

The mechanical efficiency should not be confused with the **motor efficiency** and the **generator efficiency**, which are defined as

*Motor:* $\qquad \eta_{motor} = \dfrac{\text{Mechanical power output}}{\text{Electric power input}} = \dfrac{\dot{W}_{shaft,out}}{\dot{W}_{elect,in}} \qquad (3\text{–}47)$

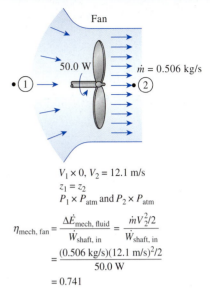

Fan

50.0 W $\qquad \dot{m} = 0.506$ kg/s

$V_1 \times 0$, $V_2 = 12.1$ m/s
$z_1 = z_2$
$P_1 \times P_{atm}$ and $P_2 \times P_{atm}$

$$\eta_{mech,\ fan} = \frac{\Delta\dot{E}_{mech,\ fluid}}{\dot{W}_{shaft,\ in}} = \frac{\dot{m}V_2^2/2}{\dot{W}_{shaft,\ in}}$$

$$= \frac{(0.506\ \text{kg/s})(12.1\ \text{m/s})^2/2}{50.0\ \text{W}}$$

$$= 0.741$$

**FIGURE 3–60**

The mechanical efficiency of a fan is the ratio of the rate of increase of the mechanical energy of air to the mechanical power input.

$\eta_{\text{turbine}} = 0.75$  $\eta_{\text{generator}} = 0.97$

$\dot{W}_{\text{elect, out}}$

Turbine

Generator

$\eta_{\text{turbine-gen}} = \eta_{\text{turbine}} \eta_{\text{generator}}$
$= 0.75 \times 0.97$
$= 0.73$

**FIGURE 3–61**

The overall efficiency of a turbine–generator is the product of the efficiency of the turbine and the efficiency of the generator, and represents the fraction of the mechanical power of the fluid converted to electrical power.

and

*Generator:* $\qquad \eta_{\text{generator}} = \dfrac{\text{Electric power output}}{\text{Mechanical power input}} = \dfrac{\dot{W}_{\text{elect,out}}}{\dot{W}_{\text{shaft,in}}}$ $\qquad$ **(3–48)**

A pump is usually packaged together with its motor, and a turbine with its generator. Therefore, we are usually interested in the **combined** or **overall efficiency** of pump–motor and turbine–generator combinations (Fig. 3–61), which are defined as

$$\eta_{\text{pump}-\text{motor}} = \eta_{\text{pump}} \eta_{\text{motor}} = \frac{\dot{W}_{\text{pump},u}}{\dot{W}_{\text{elect,in}}} = \frac{\Delta \dot{E}_{\text{mech,fluid}}}{\dot{W}_{\text{elect,in}}} \qquad \textbf{(3–49)}$$

and

$$\eta_{\text{turbine}-\text{gen}} = \eta_{\text{turbine}} \eta_{\text{generator}} = \frac{\dot{W}_{\text{elect,out}}}{\dot{W}_{\text{turbine},e}} = \frac{\dot{W}_{\text{elect,out}}}{|\Delta \dot{E}_{\text{mech,fluid}}|} \qquad \textbf{(3–50)}$$

All the efficiencies just defined range between 0 and 100 percent. The lower limit of 0 percent corresponds to the conversion of the entire mechanical or electric energy input to thermal energy, and the device in this case functions like a resistance heater. The upper limit of 100 percent corresponds to the case of perfect conversion with no friction or other irreversibilities, and thus no conversion of mechanical or electric energy to thermal energy.

①

$h = 70$ m

②

$\dot{m} = 1500$ kg/s

Generator $\qquad$ Turbine

**FIGURE 3–62**

Schematic for Example 3–16.

**EXAMPLE 3–16** $\qquad$ **Power Generation from a Hydroelectric Plant**

Electric power is to be generated by installing a hydraulic turbine–generator at a site 70 m below the free surface of a large water reservoir that can supply water at a rate of 1500 kg/s steadily (Fig. 3–62). If the mechanical power output of the turbine is 800 kW and the electric power generation is 750 kW, determine the turbine efficiency and the combined turbine–generator efficiency of this plant. Neglect losses in the pipes.

**Solution** A hydraulic turbine-generator installed at a large reservoir is to generate electricity. The combined turbine–generator efficiency and the turbine efficiency are to be determined.

*Assumptions* **1** The water elevation in the reservoir remains constant. **2** The mechanical energy of water at the turbine exit is negligible.

*Analysis* We take the free surface of water in the reservoir to be point 1 and the turbine exit to be point 2. We also take the turbine exit as the reference level ($z_2 = 0$) so that the potential energies at 1 and 2 are $pe_1 = gz_1$ and $pe_2 = 0$. The flow energy $P/\rho$ at both points is zero since both 1 and 2 are open to the atmosphere ($P_1 = P_2 = P_{\text{atm}}$). Further, the kinetic energy at both points is zero ($ke_1 = ke_2 = 0$) since the water at point 1 is essentially motionless, and the kinetic energy of water at turbine exit is assumed to be negligible. The potential energy of water at point 1 is

$$pe_1 = gz_1 = (9.81 \text{ m/s}^2)(70 \text{ m})\left(\frac{1 \text{ kJ/kg}}{1000 \text{ m}^2/\text{s}^2}\right) = 0.687 \text{ kJ/kg}$$

Then the rate at which the mechanical energy of water is supplied to the turbine becomes

$$|\Delta \dot{E}_{\text{mech,fluid}}| = \dot{m}(e_{\text{mech,in}} - e_{\text{mech,out}}) = \dot{m}(\text{pe}_1 - 0) = \dot{m}\text{pe}_1$$

$$= (1500 \text{ kg/s})(0.687 \text{ kJ/kg})$$

$$= 1031 \text{ kW}$$

The combined turbine–generator and the turbine efficiency are determined from their definitions to be

$$\eta_{\text{turbine}-\text{gen}} = \frac{\dot{W}_{\text{elect,out}}}{|\Delta \dot{E}_{\text{mech,fluid}}|} = \frac{750 \text{ kW}}{1031 \text{ kW}} = 0.727 \quad \text{or} \quad \textbf{72.7\%}$$

$$\eta_{\text{turbine}} = \frac{\dot{W}_{\text{elect,out}}}{|\Delta \dot{E}_{\text{mech,fluid}}|} = \frac{800 \text{ kW}}{1031 \text{ kW}} = 0.776 \quad \text{or} \quad \textbf{77.6\%}$$

Therefore, the reservoir supplies 1031 kW of mechanical energy to the turbine, which converts 800 kW of it to shaft work that drives the generator, which then generates 750 kW of electric power.

*Discussion* This problem can also be solved by taking point 1 to be at the turbine inlet, and using flow energy instead of potential energy. It would give the same result since the flow energy at the turbine inlet is equal to the potential energy at the free surface of the reservoir.

**EXAMPLE 3–17    Cost Savings Associated with High-Efficiency Motors**

A 60-hp electric motor (a motor that delivers 60 hp of shaft power at full load) that has an efficiency of 89.0 percent is worn out and is to be replaced by a 93.2 percent efficient high-efficiency motor (Fig. 3–63). The motor operates 3500 hours a year at full load. Taking the unit cost of electricity to be $0.08/kWh, determine the amount of energy and money saved as a result of installing the high-efficiency motor instead of the standard motor. Also, determine the simple payback period if the purchase prices of the standard and high-efficiency motors are $4520 and $5160, respectively.

**Solution** A worn-out standard motor is to be replaced by a high-efficiency one. The amount of electrical energy and money saved as well as the simple payback period are to be determined.

*Assumptions* The load factor of the motor remains constant at 1 (full load) when operating.

*Analysis* The electric power drawn by each motor and their difference can be expressed as

$$\dot{W}_{\text{electric in,standard}} = \dot{W}_{\text{shaft}}/\eta_{\text{st}} = (\text{Rated power})(\text{Load factor})/\eta_{\text{st}}$$

$$\dot{W}_{\text{electric in,efficient}} = \dot{W}_{\text{shaft}}/\eta_{\text{eff}} = (\text{Rated power})(\text{Load factor})/\eta_{\text{eff}}$$

$$\text{Power savings} = \dot{W}_{\text{electric in,standard}} - \dot{W}_{\text{electric in,efficient}}$$

$$= (\text{Rated power})(\text{Load factor})(1/\eta_{\text{st}} - 1/\eta_{\text{eff}})$$

60 hp

$\eta = 89.0\%$

Standard motor

60 hp

$\eta = 93.2\%$

High-efficiency motor

**FIGURE 3–63**

Schematic for Example 3–17.

where $\eta_{st}$ is the efficiency of the standard motor, and $\eta_{eff}$ is the efficiency of the comparable high-efficiency motor. Then the annual energy and cost savings associated with the installation of the high-efficiency motor become

$$\text{Energy savings} = (\text{Power savings})(\text{Operating hours})$$

$$= (\text{Rated power})(\text{Operating hours})(\text{Load factor})(1/\eta_{st} - 1/\eta_{eff})$$

$$= (60 \text{ hp})(0.7457 \text{ kW/hp})(3500 \text{ h/year})(1)(1/0.89 - 1/0.93.2)$$

$$= \textbf{7929 kWh/year}$$

$$\text{Cost savings} = (\text{Energy savings})(\text{Unit cost of energy})$$

$$= (7929 \text{ kWh/year})(\$0.08/\text{ kWh})$$

$$= \textbf{\$634/year}$$

Also,

$$\text{Excess initial cost} = \text{Purchase price differential} = \$5160 - \$4520 = \$640$$

This gives a simple payback period of

$$\text{Simple payback period} = \frac{\text{Excess initial cost}}{\text{Annual cost savings}} = \frac{\$640}{\$634/\text{year}} = \textbf{1.01 year}$$

***Discussion*** Note that the high-efficiency motor pays for its price differential within about one year from the electrical energy it saves. Considering that the service life of electric motors is several years, the purchase of the higher efficiency motor is definitely indicated in this case.

## SUMMARY

The sum of all forms of energy of a system is called *total energy*, which consists of internal, kinetic, and potential energy for simple compressible systems. *Internal energy* represents the molecular energy of a system and may exist in sensible, latent, chemical, and nuclear forms.

*Mass flow rate* $\dot{m}$ is defined as the amount of mass flowing through a cross section per unit time. It is related to the *volume flow rate* $\dot{V}$, which is the volume of a fluid flowing through a cross section per unit time, by

$$\dot{m} = \rho\dot{V} = \rho A_c V_{avg}$$

The energy flow rate associated with a fluid flowing at a rate of $\dot{m}$ is

$$\dot{E} = \dot{m}e$$

which is analogous to $E = me$.

The *mechanical energy* is defined as *the form of energy that can be converted to mechanical work completely and directly by a mechanical device such as an ideal turbine*. It is expressed on a unit mass basis and rate form as

$$e_{mech} = \frac{P}{\rho} + \frac{V^2}{2} + gz$$

and

$$\dot{E}_{mech} = \dot{m}e_{mech} = \dot{m}\left(\frac{P}{\rho} + \frac{V^2}{2} + gz\right)$$

where $P/\rho$ is the *flow energy*, $V^2/2$ is the *kinetic energy*, and $gz$ is the *potential energy* of the fluid per unit mass. Energy can cross the boundaries of a closed system in the form of heat or work. For control volumes, energy can also be transported by mass. If the energy transfer is due to a temperature difference between a closed system and its surroundings, it is *heat*; otherwise, it is *work*.

Work is the energy transferred as a force acts on a system through a distance. Various forms of work are expressed as follows:

*Electrical work:* $W_e = \mathbf{V}I\,\Delta t$

*Shaft work:* $W_{sh} = 2\pi n\mathrm{T}$

*Spring work:* $W_{spring} = \dfrac{1}{2}k\left(x_2^2 - x_1^2\right)$

The *first law of thermodynamics* is essentially an expression of the conservation of energy principle, also called the *energy balance*. The general mass and energy balances for *any system* undergoing *any process* can be expressed as

$$\underbrace{E_{in} - E_{out}}_{\substack{\text{Net energy transfer} \\ \text{by heat, work, and mass}}} = \underbrace{\Delta E_{system}}_{\substack{\text{Change in internal, kinetic,} \\ \text{potential, etc., energies}}} \quad \text{(kJ)}$$

It can also be expressed in the *rate form* as

$$\underbrace{\dot{E}_{in} - \dot{E}_{out}}_{\substack{\text{Rate of net energy transfer} \\ \text{by heat, work, and mass}}} = \underbrace{dE_{system}/dt}_{\substack{\text{Rate of change in internal,} \\ \text{kinetic, potential, etc., energies}}} \quad \text{(kW)}$$

The efficiencies of various devices are defined as

$$\eta_{pump} = \frac{\Delta\dot{E}_{mech,fluid}}{\dot{W}_{shaft,in}} = \frac{\dot{W}_{pump,u}}{\dot{W}_{pump}}$$

$$\eta_{turbine} = \frac{\dot{W}_{shaft,out}}{|\Delta\dot{E}_{mech,fluid}|} = \frac{\dot{W}_{turbine}}{\dot{W}_{turbine,e}}$$

$$\eta_{motor} = \frac{\text{Mechanical power output}}{\text{Electric power input}} = \frac{\dot{W}_{shaft,out}}{\dot{W}_{elect,in}}$$

$$\eta_{generator} = \frac{\text{Electric power output}}{\text{Mechanical power input}} = \frac{\dot{W}_{elect,out}}{\dot{W}_{shaft,in}}$$

$$\eta_{pump-motor} = \eta_{pump}\eta_{motor} = \frac{\Delta\dot{E}_{mech,fluid}}{\dot{W}_{elect,in}}$$

$$\eta_{turbine-gen} = \eta_{turbine}\eta_{generator} = \frac{\dot{W}_{elect,out}}{|\Delta\dot{E}_{mech,fluid}|}$$

## REFERENCES AND SUGGESTED READINGS

**1.** ASHRAE *Handbook of Fundamentals.* SI version. Atlanta, GA: American Society of Heating, Refrigerating, and Air-Conditioning Engineers, Inc., 1993.

**2.** Y. A. Çengel. "An Intuitive and Unified Approach to Teaching Thermodynamics." ASME International Mechanical Engineering Congress and Exposition, Atlanta, Georgia, AES-Vol. 36, pp. 251–260, November 17–22, 1996.

## PROBLEMS*

### Forms of Energy

**3–1C** What is the difference between the macroscopic and microscopic forms of energy?

* Problems designated by a "C" are concept questions, and students are encouraged to answer them all. Problems designated by an "E" are in English units, and the SI users can ignore them. Problems with the ✹ icon are solved using EES, and complete solutions together with parametric studies are included on the enclosed DVD. Problems with the 💻 icon are comprehensive in nature, and are intended to be solved with a computer, preferably using the EES software that accompanies this text.

**3–2C** What is total energy? Identify the different forms of energy that constitute the total energy.

**3–3C** How are heat, internal energy, and thermal energy related to each other?

**3–4C** What is mechanical energy? How does it differ from thermal energy? What are the forms of mechanical energy of a fluid stream?

**3–5C** Natural gas, which is mostly methane $CH_4$, is a fuel and a major energy source. Can we say the same about hydrogen gas, $H_2$?

**3–6E** Calculate the total kinetic energy, in Btu, of an object with a mass of 15 lbm when its velocity is 100 ft/s. *Answer:* 3.0 Btu

**3–7** Calculate the total kinetic energy, in kJ, of an object whose mass is 100 kg and whose velocity is 20 m/s.

**3–8E** The specific potential energy of an object with respect to some datum level is given by $gz$ where $g$ is the local gravitational acceleration and $z$ is the elevation of the object above the datum. Determine the specific potential energy, in Btu/lbm, of an object elevated 100 ft above a datum at a location where $g = 32.1$ ft/s$^2$.

**3–9E** Calculate the total potential energy, in Btu, of an object with a mass of 200 lbm when it is 10 ft above a datum level at a location where standard gravitational acceleration exists.

**3–10** Calculate the total potential energy, in kJ, of an object whose mass is 20 kg when it is located 20 m below a datum level in a location where $g = 9.5$ m/s$^2$.

**3–11** A person gets into an elevator at the lobby level of a hotel together with his 30-kg suitcase, and gets out at the 10th floor 35 m above. Determine the amount of energy consumed by the motor of the elevator that is now stored in the suitcase.

**3–12** Electric power is to be generated by installing a hydraulic turbine–generator at a site 160 m below the free surface of a large water reservoir that can supply water at a rate of 3500 kg/s steadily. Determine the power generation potential.

**3–13** At a certain location, wind is blowing steadily at 10 m/s. Determine the mechanical energy of air per unit mass and the power generation potential of a wind turbine with 60-m-diameter blades at that location. Take the air density to be 1.25 kg/m$^3$.

**3–14** A water jet that leaves a nozzle at 60 m/s at a flow rate of 120 kg/s is to be used to generate power by striking the buckets located on the perimeter of a wheel. Determine the power generation potential of this water jet.

**3–15** Two sites are being considered for wind power generation. In the first site, the wind blows steadily at 7 m/s for 3000 hours per year, whereas in the second site the wind blows at 10 m/s for 2000 hours per year. Assuming the wind velocity is negligible at other times for simplicity, determine which is a better site for wind power generation. *Hint:* Note that the mass flow rate of air is proportional to wind velocity.

**3–16** A river flowing steadily at a rate of 175 m$^3$/s is considered for hydroelectric power generation. It is determined that a dam can be built to collect water and release it from an elevation difference of 80 m to generate power. Determine how much power can be generated from this river water after the dam is filled.

**3–17** Consider a river flowing toward a lake at an average velocity of 3 m/s at a rate of 500 m$^3$/s at a location 90 m above the lake surface. Determine the total mechanical energy of the river water per unit mass and the power generation potential of the entire river at that location.

**FIGURE P3–17**

### Energy Transfer by Heat and Work

**3–18C** In what forms can energy cross the boundaries of a closed system?

**3–19C** When is the energy crossing the boundaries of a closed system heat and when is it work?

**3–20C** What is an adiabatic process? What is an adiabatic system?

**3–21C** What are point and path functions? Give some examples.

**3–22C** Consider an automobile traveling at a constant speed along a road. Determine the direction of the heat and work interactions, taking the following as the system: (*a*) the car radiator, (*b*) the car engine, (*c*) the car wheels, (*d*) the road, and (*e*) the air surrounding the car.

**3–23C** The length of a spring can be changed by (*a*) applying a force to it or (*b*) changing its temperature (i.e., thermal expansion). What type of energy interaction between the system (spring) and surroundings is required to change the length of the spring in these two ways?

**3–24C** A personal computer is to be examined from a thermodynamic perspective. Determine the direction of the work and heat transfers (in or out) when the (*a*) keyboard, (*b*) monitor, (*c*) processing unit, and (*d*) all of these are taken as the system.

**3–25** A small electrical motor produces 5 W of mechanical power. What is this power in (*a*) N, m, and s units; and (*b*) kg, m, and s units? *Answers:* (*a*) 5 N·m/s, (*b*) 5 kg·m$^2$/s$^3$

**3–26E** A model aircraft internal-combustion engine produces 10 W of power. How much power is this in (*a*) lbf·ft/s and (*b*) hp?

### Mechanical Forms of Work

**3–27C** A car is accelerated from rest to 85 km/h in 10 s. Would the energy transferred to the car be different if it were accelerated to the same speed in 5 s?

**3–28** Determine the energy required to accelerate an 800-kg car from rest to 100 km/h on a level road. *Answer:* 309 kJ

**3–29E** A construction crane lifts a prestressed concrete beam weighing 3 tons from the ground to the top of piers that are 24 ft above the ground. Determine the amount of work done considering (*a*) the beam and (*b*) the crane as the system. Express your answers in both lbf·ft and Btu.

**3–30E** A man weighing 180 lbf is pushing a cart that weighs 100 lbf with its contents up a ramp that is inclined at an angle of 10° from the horizontal. Determine the work needed to move along this ramp a distance of 100 ft considering (*a*) the man and (*b*) the cart and its contents as the system. Express your answers in both lbf·ft and Btu.

**FIGURE P3–30E**

*© The McGraw-Hill Companies, Inc./Lars A. Niki, photographer.*

**3–31E** The force $F$ required to compress a spring a distance $x$ is given by $F - F_0 = kx$ where $k$ is the spring constant and $F_0$ is the preload. Determine the work required to compress a spring whose spring constant is $k = 200$ lbf/in a distance of one inch starting from its free length where $F_0 = 0$ lbf. Express your answer in both lbf·ft and Btu.

**FIGURE P3–31E**

**3–32E** A spherical soap bubble with a surface-tension of 0.005 lbf/ft is expanded from a diameter of 0.5 in to 2.0 in. How much work, in Btu, is required to expand this bubble? *Answer:* $2.11 \times 10^{-6}$ Btu

**3–33E** A 0.5-in diameter, 12-in-long steel rod with a Young's modulus of 30,000 lbf/in² is stretched 0.125 in. How much work does this require, in Btu? The strain work is given by $W = \frac{V_0 E}{2}(\varepsilon_2^2 - \varepsilon_1^2)$ where $V_0$ is the original volume of the solid, $E$ is Young's modulus and $\varepsilon$ is the strain at the beginning and ending of the process.

**3–34E** A spring whose spring constant is 200 lbf/in has an initial force of 100 lbf acting on it. Determine the work, in Btu, required to compress it another 1 inch.

**3–35** How much work, in kJ, can a spring whose spring constant is 3 kN/cm produce after it has been compressed 3 cm from its unloaded length?

**3–36** A ski lift has a one-way length of 1 km and a vertical rise of 200 m. The chairs are spaced 20 m apart, and each chair can seat three people. The lift is operating at a steady speed of 10 km/h. Neglecting friction and air drag and assuming that the average mass of each loaded chair is 250 kg, determine the power required to operate this ski lift. Also estimate the power required to accelerate this ski lift in 5 s to its operating speed when it is first turned on.

**3–37** Determine the power required for a 1150-kg car to climb a 100-m-long uphill road with a slope of 30° (from horizontal) in 12 s (*a*) at a constant velocity, (*b*) from rest to a final velocity of 30 m/s, and (*c*) from 35 m/s to a final velocity of 5 m/s. Disregard friction, air drag, and rolling resistance. *Answers:* (*a*) 47.0 kW, (*b*) 90.1 kW, (*c*) −10.5 kW

**FIGURE P3–37**

**3–38** A damaged 1200-kg car is being towed by a truck. Neglecting the friction, air drag, and rolling resistance, determine the extra power required (*a*) for constant velocity on a level road, (*b*) for constant velocity of 50 km/h on a 30° (from horizontal) uphill road, and (*c*) to accelerate on a level road from stop to 90 km/h in 12 s. *Answers:* (*a*) 0, (*b*) 81.7 kW, (*c*) 31.3 kW

### The First Law of Thermodynamics

**3–39C** For a cycle, is the net work necessarily zero? For what kind of systems will this be the case?

**3–40C** What are the different mechanisms for transferring energy to or from a control volume?

**3–41C** On a hot summer day, a student turns his fan on when he leaves his room in the morning. When he returns in the evening, will the room be warmer or cooler than the neighboring rooms? Why? Assume all the doors and windows are kept closed.

**3–42E** One way to improve the fuel efficiency of a car is to use tires that have a lower rolling resistance—tires that roll with less resistance, and highway tests at 65 mph showed that tires with the lowest rolling resistance can improve the fuel efficiency by nearly 2 mpg (miles per gallon). Consider a car that gets 35 mpg on high rolling resistance tires and is driven 15,000 miles per year. For a fuel cost of $2.20/gal, determine how much money will be saved per year by switching to low rolling resistance tires.

**3–43** An adiabatic closed system is accelerated from 0 m/s to 30 m/s. Determine the specific energy change of this system, in kJ/kg.

**3–44** An adiabatic closed system is raised 100 m at a location where the gravitational acceleration is 9.8 m/s². Determine the energy change of this system, in kJ/kg.

**3–45E** A water pump increases the water pressure from 10 psia to 50 psia. Determine the power input required, in hp, to pump 1.2 ft³/s of water. Does the water temperature at the inlet have any significant effect on the required flow power? *Answer:* 12.6 hp

**3–46** A classroom that normally contains 40 people is to be air-conditioned with window air-conditioning units of 5-kW cooling capacity. A person at rest may be assumed to dissipate heat at a rate of about 360 kJ/h. There are 10 light-bulbs in the room, each with a rating of 100 W. The rate of heat transfer to the classroom through the walls and the windows is estimated to be 15,000 kJ/h. If the room air is to be maintained at a constant temperature of 21°C, determine the number of window air-conditioning units required. *Answer:* 2 units

**3–47** The lighting needs of a storage room are being met by 6 fluorescent light fixtures, each fixture containing four lamps rated at 60 W each. All the lamps are on during operating hours of the facility, which are 6 AM to 6 PM 365 days a year. The storage room is actually used for an average of 3 h a day. If the price of electricity is $0.08/kWh, determine the amount of energy and money that will be saved as a result of installing motion sensors. Also, determine the simple payback period if the purchase price of the sensor is $32 and it takes 1 hour to install it at a cost of $40.

**3–48** Consider a room that is initially at the outdoor temperature of 20°C. The room contains a 100-W lightbulb, a 110-W TV set, a 200-W refrigerator, and a 1000-W iron. Assuming no heat transfer through the walls, determine the rate of increase of the energy content of the room when all of these electric devices are on.

**3–49** A fan is to accelerate quiescent air to a velocity of 8 m/s at a rate of 9 m³/s. Determine the minimum power that must be supplied to the fan. Take the density of air to be 1.18 kg/m³. *Answer:* 340 W

**3–50E** Consider a fan located in a 3 ft × 3 ft square duct. Velocities at various points at the outlet are measured, and the average flow velocity is determined to be 22 ft/s. Taking the air density to 0.075 lbm/ft³, estimate the minimum electric power consumption of the fan motor.

**3–51**  The driving force for fluid flow is the pressure difference, and a pump operates by raising the pressure of a fluid (by converting the mechanical shaft work to

flow energy). A gasoline pump is measured to consume 3.8 kW of electric power when operating. If the pressure differential between the outlet and inlet of the pump is measured to be 7 kPa and the changes in velocity and elevation are negligible, determine the maximum possible volume flow rate of gasoline.

$\Delta P = 7$ kPa

Pump

**FIGURE P3–51**

**3–52** An escalator in a shopping center is designed to move 30 people, 75 kg each, at a constant speed of 0.8 m/s at 45° slope. Determine the minimum power input needed to drive this escalator. What would your answer be if the escalator velocity were to be doubled?

**3–53** An automobile moving through the air causes the air velocity (measured with respect to the car) to decrease and fill a larger flow channel. An automobile has an effective flow channel area of 3 m². The car is traveling at 90 km/h on a day when the barometric pressure is 70 cm of mercury and the temperature is 20°C. Behind the car, the air velocity (with respect to the car) is measured to be 82 km/h, and the temperature is 20°C. Determine the power required to move this car through the air and the area of the effective flow channel behind the car.

Flow channel

**FIGURE P3–53**

### Energy Conversion Efficiencies

**3–54C** What is mechanical efficiency? What does a mechanical efficiency of 100 percent mean for a hydraulic turbine?

**3–55C** How is the combined pump–motor efficiency of a pump and motor system defined? Can the combined pump–motor efficiency be greater than either the pump or the motor efficiency?

**3–56C** Define turbine efficiency, generator efficiency, and combined turbine–generator efficiency.

**3–57C** Can the combined turbine-generator efficiency be greater than either the turbine efficiency or the generator efficiency? Explain.

**3–58** Consider a 24-kW hooded electric open burner in an area where the unit costs of electricity and natural gas are $0.10/kWh and $1.20/therm (1 therm = 105,500 kJ), respectively. The efficiency of open burners can be taken to be 73 percent for electric burners and 38 percent for gas burners. Determine the rate of energy consumption and the unit cost of utilized energy for both electric and gas burners.

**3–59** A 75-hp (shaft output) motor that has an efficiency of 91.0 percent is worn out and is replaced by a high-efficiency 75-hp motor that has an efficiency of 95.4 percent. Determine the reduction in the heat gain of the room due to higher efficiency under full-load conditions.

**3–60** A 90-hp (shaft output) electric car is powered by an electric motor mounted in the engine compartment. If the motor has an average efficiency of 91 percent, determine the rate of heat supply by the motor to the engine compartment at full load.

**3–61** A 75-hp (shaft output) motor that has an efficiency of 91.0 percent is worn out and is to be replaced by a high-efficiency motor that has an efficiency of 95.4 percent. The motor operates 4368 hours a year at a load factor of 0.75. Taking the cost of electricity to be $0.08/kWh, determine the amount of energy and money saved as a result of installing the high-efficiency motor instead of the standard motor. Also, determine the simple payback period if the purchase prices of the standard and high-efficiency motors are $5449 and $5520, respectively.

**3–62E** The steam requirements of a manufacturing facility are being met by a boiler whose rated heat input is 5.5 × $10^6$ Btu/h. The combustion efficiency of the boiler is measured to be 0.7 by a hand-held flue gas analyzer. After tuning up the boiler, the combustion efficiency rises to 0.8. The boiler operates 4200 hours a year intermittently. Taking the unit cost of energy to be $4.35/$10^6$ Btu, determine the annual energy and cost savings as a result of tuning up the boiler.

**3–63E** [EES] Reconsider Prob. 3–62E. Using EES (or other) software, study the effects of the unit cost of energy, the new combustion efficiency on the annual energy, and cost savings. Let the efficiency vary from 0.7 to 0.9, and the unit cost to vary from $4 to $6 per million Btu. Plot the annual energy and cost savings against the efficiency for unit costs of $4, $5, and $6 per million Btu, and discuss the results.

**3–64** An exercise room has eight weight-lifting machines that have no motors and four treadmills each equipped with a 2.5-hp (shaft output) motor. The motors operate at an average load factor of 0.7, at which their efficiency is 0.77. During peak evening hours, all 12 pieces of exercising equipment are used continuously, and there are also two people doing light exercises while waiting in line for one piece of the equipment. Assuming the average rate of heat dissipation from people in an exercise room is 525 W, determine the rate of heat gain of the exercise room from people and the equipment at peak load conditions.

**3–65** A room is cooled by circulating chilled water through a heat exchanger located in a room. The air is circulated through the heat exchanger by a 0.25-hp (shaft output) fan. Typical efficiency of small electric motors driving 0.25-hp equipment is 54 percent. Determine the rate of heat supply by the fan–motor assembly to the room.

**3–66** The water in a large lake is to be used to generate electricity by the installation of a hydraulic turbine-generator at a location where the depth of the water is 50 m (Fig. 3–62). Water is to be supplied at a rate of 5000 kg/s. If the electric power generated is measured to be 1862 kW and the generator efficiency is 95 percent, determine (a) the overall efficiency of the turbine—generator, (b) the mechanical efficiency of the turbine, and (c) the shaft power supplied by he turbine to the generator.

**3–67** At a certain location, wind is blowing steadily at 7 m/s. Determine the mechanical energy of air per unit mass and the power generation potential of a wind turbine with 80-m-diameter blades at that location. Also determine the actual electric power generation assuming an overall efficiency of 30 percent. Take the air density to be 1.25 kg/m$^3$.

**3–68** [EES] Reconsider Prob. 3–67. Using EES (or other) software, investigate the effect of wind velocity and the blade span diameter on wind power generation. Let the velocity vary from 5 to 20 m/s in increments of 5 m/s, and the diameter vary from 20 to 120 m in increments of 20 m. Tabulate the results, and discuss their significance.

**3–69** Water is pumped from a lake to a storage tank 20 m above at a rate of 70 L/s while consuming 20.4 kW of electric power. Disregarding any frictional losses in the pipes and any changes in kinetic energy, determine (a) the overall efficiency of the pump–motor unit and (b) the pressure difference between the inlet and the exit of the pump.

**FIGURE P3–69**

**3–70** Large wind turbines with blade span diameters of over 100 m are available for electric power generation. Consider a wind turbine with a blade span diameter of 100 m installed at a site subjected to steady winds at 8 m/s. Taking the overall efficiency of the wind turbine to be 32 percent and the air density to be 1.25 kg/m³, determine the electric power generated by this wind turbine. Also, assuming steady winds of 8 m/s during a 24-hour period, determine the amount of electric energy and the revenue generated per day for a unit price of $0.06/kWh for electricity.

**3–71E** A water pump delivers 6 hp of shaft power when operating. If the pressure differential between the outlet and the inlet of the pump is measured to be 1.2 psi when the flow rate is 15 ft³/s and the changes in velocity and elevation are negligible, determine the mechanical efficiency of this pump.

**3–72** Water is pumped from a lower reservoir to a higher reservoir by a pump that provides 20 kW of shaft power. The free surface of the upper reservoir is 45 m higher than that of the lower reservoir. If the flow rate of water is measured to be 0.03 m³/s, determine mechanical power that is converted to thermal energy during this process due to frictional effects.

**FIGURE P3–72**

**3–73** The water behind Hoover Dam in Nevada is 206 m higher than the Colorado River below it. At what rate must water pass through the hydraulic turbines of this dam to produce 100 MW of power if the turbines are 100 percent efficient?

**FIGURE P3–73**

*Photo by Lynn Betts, USDA Natural Resources Conservation Service.*

**3–74** An oil pump is drawing 35 kW of electric power while pumping oil with $\rho = 860$ kg/m³ at a rate of 0.1 m³/s. The inlet and outlet diameters of the pipe are 8 cm and 12 cm, respectively. If the pressure rise of oil in the pump is measured to be 400 kPa and the motor efficiency is 90 percent, determine the mechanical efficiency of the pump.

**FIGURE P3–74**

**3–75E** An 80-percent efficient pump with a power input of 20 hp is pumping water from a lake to a nearby pool at a rate of 1.5 ft³/s through a constant-diameter pipe. The free surface of the pool is 80 ft above that of the lake. Determine the mechanical power used to overcome frictional effects in piping. *Answer:* 2.37 hp

**3–76** A wind turbine is rotating at 15 rpm under steady winds flowing through the turbine at a rate of 42,000 kg/s. The tip velocity of the turbine blade is measured to be 250 km/h. If 180 kW power is produced by the turbine, determine (*a*) the average velocity of the air and (*b*) the conversion efficiency of the turbine. Take the density of air to be 1.31 kg/m³.

### Review Problems

**3–77** Consider a classroom for 55 students and one instructor, each generating heat at a rate of 100 W. Lighting is provided by 18 fluorescent lightbulbs, 40 W each, and the ballasts consume an additional 10 percent. Determine the rate of internal heat generation in this classroom when it is fully occupied.

**3–78** Consider a homeowner who is replacing his 25-year-old natural gas furnace that has an efficiency of 55 percent. The homeowner is considering a conventional furnace that has an efficiency of 82 percent and costs $1600 and a high-efficiency furnace that has an efficiency of 95 percent and costs $2700. The homeowner would like to buy the high-efficiency furnace if the savings from the natural gas pay for the additional cost in less than 8 years. If the homeowner presently pays $1200 a year for heating, determine if he should buy the conventional or high-efficiency model.

**3–79E** The energy contents, unit costs, and typical conversion efficiencies of various energy sources for use in water heaters are given as follows: 1025 Btu/ft³, $0.012/ft³, and 55 percent for natural gas; 138,700 Btu/gal, $1.15/gal, and 55 percent for heating oil; and 1 kWh/kWh, $0.084/kWh, and 90 percent for electric heaters, respectively. Determine the lowest-cost energy source for water heaters.

**3–80** A homeowner is considering these heating systems for heating his house: Electric resistance heating with $0.09/kWh and 1 kWh = 3600 kJ, gas heating with $1.24/therm and 1 therm = 105,500 kJ, and oil heating with $1.25/gal and 1 gal of oil = 138,500 kJ. Assuming efficiencies of 100 percent for the electric furnace and 87 percent for the gas and oil furnaces, determine the heating system with the lowest energy cost.

**3–81** A typical household pays about $1200 a year on energy bills, and the U.S. Department of Energy estimates that 46 percent of this energy is used for heating and cooling, 15 percent for heating water, 15 percent for refrigerating and freezing, and the remaining 24 percent for lighting, cooking, and running other appliances. The heating and cooling costs of a poorly insulated house can be reduced by up to 30 percent by adding adequate insulation. If the cost of insulation is $200, determine how long it will take for the insulation to pay for itself from the energy it saves.

**3–82** The U.S. Department of Energy estimates that up to 10 percent of the energy use of a house can be saved by caulking and weatherstripping doors and windows to reduce air leaks at a cost of about $60 for materials for an average home with 12 windows and 2 doors. Caulking and weatherstripping every gas-heated home properly would save enough energy to heat about 4 million homes. The savings can be increased by installing storm windows. Determine how long it will take for the caulking and weatherstripping to pay for itself from the energy they save for a house whose annual energy use is $1500.

**3–83E** The energy stored in the spring of a railroad car is 5000 lbf·ft. What is this energy in (a) lbm, ft, and s units; (b) lbf and yd units; and (c) lbm, mile, and hour units?

**3–84E** The force required to expand the gas in a gas spring a distance $x$ is given by

$$F = \frac{\text{Constant}}{x^k}$$

where the constant is determined by the geometry of this device and $k$ is determined by the gas used in the device. One such device has a constant of 200 lbf·in$^{1.4}$ and $k = 1.4$. Determine the work, in Btu, required to compress this device from 1 in to 4 in. *Answer: 0.0228 Btu*

**3–85E** A man weighing 180 lbf pushes a block weighing 100 lbf along a horizontal plane. The dynamic coefficient of friction between the block and plane is 0.2. Assuming that the block is moving at constant speed, calculate the work required to move the block a distance of 100 ft considering (a) the man and (b) the block as the system. Express your answers in both lbf·ft and Btu.

**3–86E** Water is pumped from a 200-ft-deep well into a 100-ft-high storage tank. Determine the power, in kW, that would be required to pump 200 gallons per minute.

**3–87** A grist mill of the 1800s employed a water wheel that was 14 m high; 320 liters per minute of water flowed on to the wheel near the top. How much power, in kW, could this water wheel have produced? *Answer: 0.732 kW*

**3–88** Windmills slow the air and cause it to fill a larger channel as it passes through the blades. Consider a circular windmill with a 7-m-diameter rotor in a 10 m/s wind on a day when the atmospheric pressure is 100 kPa and the temperature is 20°C. The wind speed behind the windmill is measured at 9 m/s. Determine the diameter of the wind channel downstream from the rotor and the power produced by this windmill, presuming that the air is incompressible.

Flow channel

W

**FIGURE P3–88**

**3–89** In a hydroelectric power plant, 65 m³/s of water flows from an elevation of 90 m to a turbine, where electric power is generated. The overall efficiency of the turbine–generator is 84 percent. Disregarding frictional losses in piping, estimate the electric power output of this plant. *Answer: 48.2 MW*

① 

65 m³/s

90 m

② 

Generator   Turbine

$\eta_{\text{turbine–gen}} = 84\%$

**FIGURE P3–89**

**3–90** The demand for electric power is usually much higher during the day than it is at night, and utility companies often sell power at night at much lower prices to encourage consumers to use the available power generation capacity and to avoid building new expensive power plants that will be used only a short time during peak periods. Utilities are also willing to purchase power produced during the day from private parties at a high price.

Suppose a utility company is selling electric power for $0.03/kWh at night and is willing to pay $0.08/kWh for power produced during the day. To take advantage of this opportunity, an entrepreneur is considering building a large reservoir 40 m above the lake level, pumping water from the lake to the reservoir at night using cheap power, and letting the water flow from the reservoir back to the lake during the day, producing power as the pump–motor operates as a turbine–generator during reverse flow. Preliminary analysis shows that a water flow rate of 2 m³/s can be used in either direction. The combined pump–motor and turbine–generator efficiencies are expected to be 75 percent each. Disregarding the frictional losses in piping and assuming the system operates for 10 h each in the pump and turbine modes during a typical day, determine the potential revenue this pump–turbine system can generate per year.

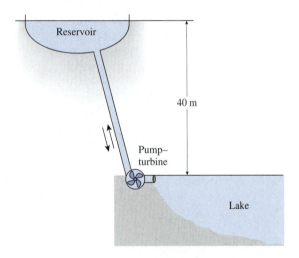

**FIGURE P3–90**

## Design and Essay Problems

**3–91** Solar energy reaching the earth is about 1350 W/m² outside the earth's atmosphere, and 950 W/m² on earth's surface normal to the sun on a clear day. Someone is marketing 2 m × 3 m photovoltaic cell panels with the claim that a single panel can meet the electricity needs of a house. How do you evaluate this claim? Photovoltaic cells have a conversion efficiency of about 15 percent.

**3–92** Find out the prices of heating oil, natural gas, and electricity in your area, and determine the cost of each per kWh of energy supplied to the house as heat. Go through your utility bills and determine how much money you spent for heating last January. Also determine how much your January heating bill would be for each of the heating systems if you had the latest and most efficient system installed.

**3–93** Prepare a report on the heating systems available in your area for residential buildings. Discuss the advantages and disadvantages of each system and compare their initial and operating costs. What are the important factors in the selection of a heating system? Give some guidelines. Identify the conditions under which each heating system would be the best choice in your area.

**3–94** An electrical-generation utility sometimes pumps liquid water into an elevated reservoir during periods of low electrical consumption. This water is used to generate electricity during periods when the demand for electricity exceeds the utility's ability to produce electricity. Discuss this energy-storage scheme from a conversion efficiency perspective as compared to storing a compressed phase-changing substance.

**3–95** Some engineers have suggested that air compressed into tanks can be used to propel personal transportation vehicles. Current compressed-air tank technology permits us to compress and safely hold air at up to 4000 psia. Tanks made of composite materials require about 10 lbm of construction materials for each 1 ft³ of stored gas. Approximately 0.01 hp is required per pound of vehicle weight to move a vehicle at a speed of 30 miles per hour. What is the maximum range that this vehicle can have? Account for the weight of the tanks only and assume perfect conversion of the energy in the compressed air.

# PROPERTIES OF PURE SUBSTANCES

**W**e start this chapter with the introduction of the concept of a *pure substance* and a discussion of the physics of phase-change processes. We then illustrate the various property diagrams and $P$-$v$-$T$ surfaces of pure substances. After demonstrating the use of the property tables, the hypothetical substance *ideal gas* and the *ideal-gas equation of state* are discussed. The *compressibility factor*, which accounts for the deviation of real gases from ideal-gas behavior, is introduced.

## Objectives

Objectives of this chapter are to:

- Introduce the concept of a pure substance.

- Discuss the physics of phase-change processes.

- Illustrate the $P$-$v$, $T$-$v$, and $P$-$T$ property diagrams and $P$-$v$-$T$ surfaces of pure substances.

- Demonstrate the procedures for determining thermodynamic properties of pure substances from tables of property data.

- Describe the hypothetical substance "ideal gas" and the ideal-gas equation of state.

- Apply the ideal-gas equation of state in the solution of typical problems.

- Introduce the compressibility factor, which accounts for the deviation of real gases from ideal-gas behavior.

**FIGURE 4–1**

Nitrogen and gaseous air are
pure substances.

(a) $H_2O$          (b) Air

**FIGURE 4–2**

A mixture of liquid and gaseous water
is a pure substance, but a mixture of
liquid and gaseous air is not.

**FIGURE 4–3**

The molecules in a solid are kept at
their positions by the large springlike
intermolecular forces.

# 4–1 ▪ PURE SUBSTANCE

A substance that has a fixed chemical composition throughout is called a **pure substance**. Water, nitrogen, helium, and carbon dioxide, for example, are all pure substances.

A pure substance does not have to be of a single chemical element or compound, however. A mixture of various chemical elements or compounds also qualifies as a pure substance as long as the mixture is homogeneous. Air, for example, is a mixture of several gases, but it is often considered to be a pure substance because it has a uniform chemical composition (Fig. 4–1). However, a mixture of oil and water is not a pure substance. Since oil is not soluble in water, it will collect on top of the water, forming two chemically dissimilar regions.

A mixture of two or more phases of a pure substance is still a pure substance as long as the chemical composition of all phases is the same (Fig. 4–2). A mixture of ice and liquid water, for example, is a pure substance because both phases have the same chemical composition. A mixture of liquid air and gaseous air, however, is not a pure substance since the composition of liquid air is different from the composition of gaseous air, and thus the mixture is no longer chemically homogeneous. This is due to different components in air condensing at different temperatures at a specified pressure.

# 4–2 ▪ PHASES OF A PURE SUBSTANCE

We all know from experience that substances exist in different phases. At room temperature and pressure, copper is a solid, mercury is a liquid, and nitrogen is a gas. Under different conditions, each may appear in a different phase. Even though there are three principal phases—solid, liquid, and gas—a substance may have several phases within a principal phase, each with a different molecular structure. Carbon, for example, may exist as graphite or diamond in the solid phase. Helium has two liquid phases; iron has three solid phases. Ice may exist at seven different phases at high pressures. A phase is identified as having a distinct molecular arrangement that is homogeneous throughout and separated from the others by easily identifiable boundary surfaces. The two phases of $H_2O$ in iced water represent a good example of this.

When studying phases or phase changes in thermodynamics, one does not need to be concerned with the molecular structure and behavior of different phases. However, it is very helpful to have some understanding of the molecular phenomena involved in each phase, and a brief discussion of phase transformations follows.

Intermolecular bonds are strongest in solids and weakest in gases. One reason is that molecules in solids are closely packed together, whereas in gases they are separated by relatively large distances.

The molecules in a **solid** are arranged in a three-dimensional pattern (lattice) that is repeated throughout (Fig. 4–3). Because of the small distances between molecules in a solid, the attractive forces of molecules on each other are large and keep the molecules at fixed positions (Fig. 4–4). Note that the attractive forces between molecules turn to repulsive forces as the distance between the molecules approaches zero, thus preventing the molecules from piling up on top of each other. Even though the molecules in a solid cannot move relative to each other, they continually oscillate about their equilibrium positions. The velocity of the molecules during these oscillations depends on the temperature. At sufficiently high temperatures, the velocity (and thus the momentum) of the molecules may reach a point where the intermolecular forces are partially overcome and groups of molecules break away (Fig. 4–5). This is the beginning of the melting process.

The molecular spacing in the **liquid** phase is not much different from that of the solid phase, except the molecules are no longer at fixed positions relative to each other and they can rotate and translate freely. In a liquid, the intermolecular forces are weaker relative to solids, but still relatively strong compared with gases. The distances between molecules generally experience a slight increase as a solid turns liquid, with water being a notable exception.

In the **gas** phase, the molecules are far apart from each other, and a molecular order is nonexistent. Gas molecules move about at random, continually colliding with each other and the walls of the container they are in. Particularly at low densities, the intermolecular forces are very small, and collisions are the only mode of interaction between the molecules. Molecules in the gas phase are at a considerably higher energy level than they are in the liquid or solid phases. Therefore, the gas must release a large amount of its energy before it can condense or freeze.

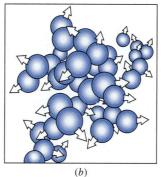

**FIGURE 4–4**

In a solid, the attractive and repulsive forces between the molecules tend to maintain them at relatively constant distances from each other.
*BLONDIE © KING FEATURES SYNDICATE.*

# 4–3 ▪ PHASE-CHANGE PROCESSES OF PURE SUBSTANCES

There are many practical situations where two phases of a pure substance coexist in equilibrium. Water exists as a mixture of liquid and vapor in the boiler and the condenser of a steam power plant. The refrigerant turns from liquid to vapor in the freezer of a refrigerator. Even though many home

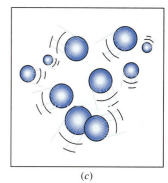

(a)      (b)      (c)

**FIGURE 4–5**

The arrangement of atoms in different phases: (*a*) molecules are at relatively fixed positions in a solid, (*b*) groups of molecules move about each other in the liquid phase, and (*c*) molecules move about at random in the gas phase.

**FIGURE 4–6**

At 1 atm and 20°C, water exists in the liquid phase (*compressed liquid*).

**FIGURE 4–7**

At 1 atm pressure and 100°C, water exists as a liquid that is ready to vaporize (*saturated liquid*).

**FIGURE 4–8**

As more heat is transferred, part of the saturated liquid vaporizes (*saturated liquid–vapor mixture*).

owners consider the freezing of water in underground pipes as the most important phase-change process, attention in this section is focused on the liquid and vapor phases and their mixture. As a familiar substance, water is used to demonstrate the basic principles involved. Remember, however, that all pure substances exhibit the same general behavior.

## Compressed Liquid and Saturated Liquid

Consider a piston–cylinder device containing liquid water at 20°C and 1 atm pressure (state 1, Fig. 4–6). Under these conditions, water exists in the liquid phase, and it is called a **compressed liquid**, or a **subcooled liquid**, meaning that it is *not about to vaporize*. Heat is now transferred to the water until its temperature rises to, say, 40°C. As the temperature rises, the liquid water expands slightly, and so its specific volume increases. To accommodate this expansion, the piston moves up slightly. The pressure in the cylinder remains constant at 1 atm during this process since it depends on the outside barometric pressure and the weight of the piston, both of which are constant. Water is still a compressed liquid at this state since it has not started to vaporize.

As more heat is transferred, the temperature keeps rising until it reaches 100°C (state 2, Fig. 4–7). At this point water is still a liquid, but any heat addition will cause some of the liquid to vaporize. That is, a phase-change process from liquid to vapor is about to take place. A liquid that is *about to vaporize* is called a **saturated liquid**. Therefore, state 2 is a saturated liquid state.

## Saturated Vapor and Superheated Vapor

Once boiling starts, the temperature stops rising until the liquid is completely vaporized. That is, the temperature will remain constant during the entire phase-change process if the pressure is held constant. This can easily be verified by placing a thermometer into boiling pure water on top of a stove. At sea level (*P* = 1 atm), the thermometer will always read 100°C if the pan is uncovered or covered with a light lid. During a boiling process, the only change we will observe is a large increase in the volume and a steady decline in the liquid level as a result of more liquid turning to vapor.

Midway about the vaporization line (state 3, Fig. 4–8), the cylinder contains equal amounts of liquid and vapor. As we continue transferring heat, the vaporization process continues until the last drop of liquid is vaporized (state 4, Fig. 4–9). At this point, the entire cylinder is filled with vapor that is on the borderline of the liquid phase. Any heat loss from this vapor will cause some of the vapor to condense (phase change from vapor to liquid). A vapor that is *about to condense* is called a **saturated vapor**. Therefore, state 4 is a saturated vapor state. A substance at states between 2 and 4 is referred to as a **saturated liquid–vapor mixture** since the *liquid and vapor phases coexist* in equilibrium at these states.

Once the phase-change process is completed, we are back to a single-phase region again (this time vapor), and further transfer of heat results in an increase in both the temperature and the specific volume (Fig. 4–10). At state 5, the temperature of the vapor is, let us say, 300°C; and if we transfer

some heat from the vapor, the temperature may drop somewhat but no condensation will take place as long as the temperature remains above 100°C
(for $P = 1$ atm). A vapor that is *not about to condense* (i.e., not a saturated
vapor) is called a **superheated vapor**. Therefore, water at state 5 is a superheated vapor. This constant-pressure phase-change process is illustrated on a
$T$-$v$ diagram in Fig. 4–11.

If the entire process described here is reversed by cooling the water while
maintaining the pressure at the same value, the water will go back to state 1,
retracing the same path, and in so doing, the amount of heat released will
exactly match the amount of heat added during the heating process.

In our daily life, water implies liquid water and steam implies water
vapor. In thermodynamics, however, both water and steam usually mean
only one thing: $H_2O$.

# Saturation Temperature and Saturation Pressure

It probably came as no surprise to you that water started to boil at 100°C.
Strictly speaking, the statement "water boils at 100°C" is incorrect. The correct statement is "water boils at 100°C at 1 atm pressure." The only reason
water started boiling at 100°C was because we held the pressure constant at
1 atm (101.325 kPa). If the pressure inside the cylinder were raised to 500 kPa
by adding weights on top of the piston, water would start boiling at 151.8°C.
That is, *the temperature at which water starts boiling depends on the pressure;
therefore, if the pressure is fixed, so is the boiling temperature.*

At a given pressure, the temperature at which a pure substance changes
phase is called the **saturation temperature** $T_{sat}$. Likewise, at a given temperature, the pressure at which a pure substance changes phase is called the
**saturation pressure** $P_{sat}$. At a pressure of 101.325 kPa, $T_{sat}$ is 99.97°C.
Conversely, at a temperature of 99.97°C, $P_{sat}$ is 101.325 kPa. (At 100.00°C,
$P_{sat}$ is 101.42 kPa in the ITS-90 discussed in Chap. 2.)

Saturation tables that list the saturation pressure against the temperature (or the saturation temperature against the pressure) are available for

**FIGURE 4–9**

At 1 atm pressure, the temperature
remains constant at 100°C until the
last drop of liquid is vaporized
(*saturated vapor*).

**FIGURE 4–10**

As more heat is transferred, the
temperature of the vapor starts to rise
(*superheated vapor*).

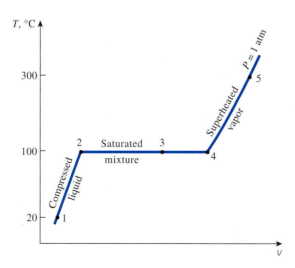

**FIGURE 4–11**

$T$-$v$ diagram for the heating process of
water at constant pressure.

### TABLE 4–1

Saturation (boiling) pressure of water at various temperatures

| Temperature, $T$, °C | Saturation pressure, $P_{sat}$, kPa |
|---|---|
| −10 | 0.26 |
| −5 | 0.40 |
| 0 | 0.61 |
| 5 | 0.87 |
| 10 | 1.23 |
| 15 | 1.71 |
| 20 | 2.34 |
| 25 | 3.17 |
| 30 | 4.25 |
| 40 | 7.39 |
| 50 | 12.35 |
| 100 | 101.4 |
| 150 | 476.2 |
| 200 | 1555 |
| 250 | 3976 |
| 300 | 8588 |

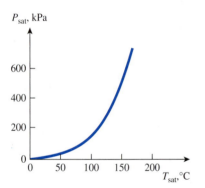

**FIGURE 4–12**

The liquid–vapor saturation curve of a pure substance (numerical values are for water).

practically all substances. A partial listing of such a table is given in Table 4–1 for water. This table indicates that the pressure of water changing phase (boiling or condensing) at 25°C must be 3.17 kPa, and the pressure of water must be maintained at 3976 kPa (about 40 atm) to have it boil at 250°C. Also, water can be frozen by dropping its pressure below 0.61 kPa.

It takes a large amount of energy to melt a solid or vaporize a liquid. The amount of energy absorbed or released during a phase-change process is called the **latent heat**. More specifically, the amount of energy absorbed during melting is called the **latent heat of fusion** and is equivalent to the amount of energy released during freezing. Similarly, the amount of energy absorbed during vaporization is called the **latent heat of vaporization** and is equivalent to the energy released during condensation. The magnitudes of the latent heats depend on the temperature or pressure at which the phase change occurs. At 1 atm pressure, the latent heat of fusion of water is 333.7 kJ/kg and the latent heat of vaporization is 2256.5 kJ/kg.

During a phase-change process, pressure and temperature are obviously dependent properties, and there is a definite relation between them, that is, $T_{sat} = f(P_{sat})$. A plot of $T_{sat}$ versus $P_{sat}$, such as the one given for water in Fig. 4–12, is called a **liquid–vapor saturation curve**. A curve of this kind is characteristic of all pure substances.

It is clear from Fig. 4–12 that $T_{sat}$ increases with $P_{sat}$. Thus, a substance at higher pressures boils at higher temperatures. In the kitchen, higher boiling temperatures mean shorter cooking times and energy savings. A beef stew, for example, may take 1 to 2 h to cook in a regular pan that operates at 1 atm pressure, but only 20 min in a pressure cooker operating at 3 atm absolute pressure (corresponding boiling temperature: 134°C).

The atmospheric pressure, and thus the boiling temperature of water, decreases with elevation. Therefore, it takes longer to cook at higher altitudes than it does at sea level (unless a pressure cooker is used). For example, the standard atmospheric pressure at an elevation of 2000 m is 79.50 kPa, which corresponds to a boiling temperature of 93.3°C as opposed to 100°C at sea level (zero elevation). The variation of the boiling temperature of water with altitude at standard atmospheric conditions is given in Table 4–2. For each 1000 m increase in elevation, the boiling temperature drops by a little over 3°C. Note that the atmospheric pressure at a location, and thus the boiling temperature, changes slightly with the weather conditions. But the corresponding change in the boiling temperature is no more than about 1°C.

## Some Consequences of $T_{sat}$ and $P_{sat}$ Dependence

We mentioned earlier that a substance at a specified pressure boils at the saturation temperature corresponding to that pressure. This phenomenon allows us to control the boiling temperature of a substance by simply controlling the pressure, and it has numerous applications in practice. Below we give some examples. The natural drive to achieve phase equilibrium by allowing some liquid to evaporate is at work behind the scenes.

Consider a sealed can of *liquid refrigerant-134a* in a room at 25°C. If the can has been in the room long enough, the temperature of the refrigerant in the can is also 25°C. Now, if the lid is opened slowly and some refrigerant is allowed to escape, the pressure in the can will start dropping until it reaches the atmospheric pressure. If you are holding the can, you will notice its temperature dropping rapidly, and even ice forming outside the can if the air is humid. A thermometer inserted in the can will register −26°C when the pressure drops to 1 atm, which is the saturation temperature of refrigerant-134a at that pressure. The temperature of the liquid refrigerant will remain at −26°C until the last drop of it vaporizes.

Another aspect of this interesting physical phenomenon is that a liquid cannot vaporize unless it absorbs energy in the amount of the latent heat of vaporization, which is 217 kJ/kg for refrigerant-134a at 1 atm. Therefore, the rate of vaporization of the refrigerant depends on the rate of heat transfer to the can: the larger the rate of heat transfer, the higher the rate of vaporization. The rate of heat transfer to the can and thus the rate of vaporization of the refrigerant can be minimized by insulating the can heavily. In the limiting case of no heat transfer, the refrigerant will remain in the can as a liquid at −26°C indefinitely.

The boiling temperature of *nitrogen* at atmospheric pressure is −196°C (see Table A–3a). This means the temperature of liquid nitrogen exposed to the atmosphere must be −196°C since some nitrogen will be evaporating. The temperature of liquid nitrogen remains constant at −196°C until it is depleted. For this reason, nitrogen is commonly used in low-temperature scientific studies (such as superconductivity) and cryogenic applications to maintain a test chamber at a constant temperature of −196°C. This is done by placing the test chamber into a liquid nitrogen bath that is open to the atmosphere. Any heat transfer from the environment to the test section is absorbed by the nitrogen, which evaporates isothermally and keeps the test chamber temperature constant at −196°C (Fig. 4–13). The entire test section must be insulated heavily to minimize heat transfer and thus liquid nitrogen consumption. Liquid nitrogen is also used for medical purposes to burn off unsightly spots on the skin. This is done by soaking a cotton swab in liquid nitrogen and wetting the target area with it. As the nitrogen evaporates, it freezes the affected skin by rapidly absorbing heat from it.

A practical way of cooling leafy vegetables is **vacuum cooling**, which is based on *reducing the pressure* of the sealed cooling chamber to the saturation pressure at the desired low temperature, and evaporating some water from the products to be cooled. The heat of vaporization during evaporation is absorbed from the products, which lowers the product temperature. The saturation pressure of water at 0°C is 0.61 kPa, and the products can be cooled to 0°C by lowering the pressure to this level. The cooling rate can be increased by lowering the pressure below 0.61 kPa, but this is not desirable because of the danger of freezing and the added cost.

In vacuum cooling, there are two distinct stages. In the first stage, the products at ambient temperature, say at 25°C, are loaded into the chamber, and the operation begins. The temperature in the chamber remains constant until the *saturation pressure* is reached, which is 3.17 kPa at 25°C. In the second stage that follows, saturation conditions are maintained inside at progressively *lower pressures* and the corresponding *lower temperatures* until the desired temperature is reached (Fig. 4–14).

### TABLE 4–2

Variation of the standard atmospheric pressure and the boiling (saturation) temperature of water with altitude

| Elevation m | Atmospheric pressure, kPa | Boiling temperature, °C |
|---|---|---|
| 0 | 101.33 | 100.0 |
| 1,000 | 89.55 | 96.5 |
| 2,000 | 79.50 | 93.3 |
| 5,000 | 54.05 | 83.3 |
| 10,000 | 26.50 | 66.3 |
| 20,000 | 5.53 | 34.7 |

**FIGURE 4–13**

The temperature of liquid nitrogen exposed to the atmosphere remains constant at −196°C, and thus it maintains the test chamber at −196°C.

**FIGURE 4–14**

The variation of the temperature of fruits and vegetables with pressure during vacuum cooling from 25°C to 0°C.

**FIGURE 4–15**

In 1775, ice was made by evacuating the air space in a water tank.

Vacuum cooling is usually more expensive than the conventional refrigerated cooling, and its use is limited to applications that result in much faster cooling. Products with large surface area per unit mass and a high tendency to release moisture such as lettuce and spinach are well-suited for vacuum cooling. Products with low surface area to mass ratio are not suitable, especially those that have relatively impervious peels such as tomatoes and cucumbers. Some products such as mushrooms and green peas can be vacuum cooled successfully by wetting them first.

The vacuum cooling just described becomes **vacuum freezing** if the vapor pressure in the vacuum chamber is dropped below 0.61 kPa, the saturation pressure of water at 0°C. The idea of making ice by using a vacuum pump is nothing new. Dr. William Cullen actually made ice in Scotland in 1775 by evacuating the air in a water tank (Fig. 4–15).

**Package icing** is commonly used in small-scale cooling applications to remove heat and keep the products cool during transit by taking advantage of the large latent heat of fusion of water, but its use is limited to products that are not harmed by contact with ice. Also, ice provides *moisture* as well as *refrigeration*.

## 4–4 · PROPERTY DIAGRAMS FOR PHASE-CHANGE PROCESSES

The variations of properties during phase-change processes are best studied and understood with the help of property diagrams. Next, we develop and discuss the $T$-$v$, $P$-$v$, and $P$-$T$ diagrams for pure substances.

### 1  The $T$-$v$ Diagram

The phase-change process of water at 1 atm pressure was described in detail in the last section and plotted on a $T$-$v$ diagram in Fig. 4–11. Now we repeat this process at different pressures to develop the $T$-$v$ diagram.

Let us add weights on top of the piston until the pressure inside the cylinder reaches 1 MPa. At this pressure, water has a somewhat smaller specific volume than it does at 1 atm pressure. As heat is transferred to the water at this new pressure, the process follows a path that looks very much like the process path at 1 atm pressure, as shown in Fig. 4–16, but there are some noticeable differences. First, water starts boiling at a much higher temperature (179.9°C) at this pressure. Second, the specific volume of the saturated liquid is larger and the specific volume of the saturated vapor is smaller than the corresponding values at 1 atm pressure. That is, the horizontal line that connects the saturated liquid and saturated vapor states is much shorter.

As the pressure is increased further, this saturation line continues to shrink, as shown in Fig. 4–16, and it becomes a point when the pressure reaches 22.06 MPa for the case of water. This point is called the **critical point**, and it is defined as *the point at which the saturated liquid and saturated vapor states are identical.*

The temperature, pressure, and specific volume of a substance at the critical point are called, respectively, the *critical temperature* $T_{cr}$, *critical pressure* $P_{cr}$, and *critical specific volume* $v_{cr}$. The critical-point properties of

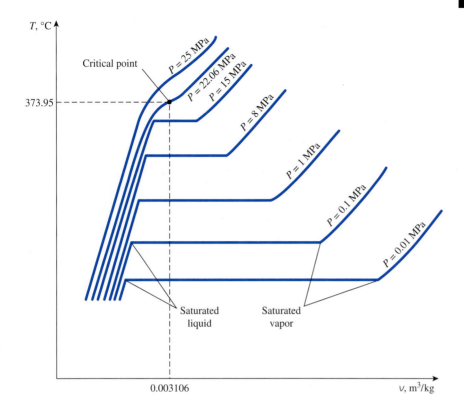

**FIGURE 4–16**

$T$-$v$ diagram of constant-pressure phase-change processes of a pure substance at various pressures (numerical values are for water).

water are $P_{cr}$ = 22.06 MPa, $T_{cr}$ = 373.95°C, and $v_{cr}$ = 0.003106 m³/kg. For helium, they are 0.23 MPa, −267.85°C, and 0.01444 m³/kg. The critical properties for various substances are given in Table A–1 in the appendix.

At pressures above the critical pressure, there is not a distinct phase-change process (Fig. 4–17). Instead, the specific volume of the substance continually increases, and at all times there is only one phase present. Eventually, it resembles a vapor, but we can never tell when the change has occurred. Above the critical state, there is no line that separates the compressed liquid region and the superheated vapor region. However, it is customary to refer to the substance as superheated vapor at temperatures above the critical temperature and as compressed liquid at temperatures below the critical temperature.

The saturated liquid states in Fig. 4–16 can be connected by a line called the **saturated liquid line**, and saturated vapor states in the same figure can be connected by another line, called the **saturated vapor line**. These two lines meet at the critical point, forming a dome as shown in Fig. 4–18a. All the compressed liquid states are located in the region to the left of the saturated liquid line, called the **compressed liquid region**. All the superheated vapor states are located to the right of the saturated vapor line, called the **superheated vapor region**. In these two regions, the substance exists in a single phase, a liquid or a vapor. All the states that involve both phases in equilibrium are located under the dome, called the **saturated liquid–vapor mixture region**, or the **wet region**.

**FIGURE 4–17**

At supercritical pressures ($P > P_{cr}$), there is no distinct phase-change (boiling) process.

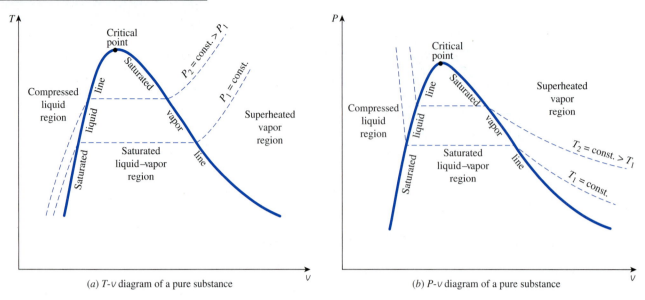

(a) $T$-$v$ diagram of a pure substance

(b) $P$-$v$ diagram of a pure substance

**FIGURE 4–18**

Property diagrams of a pure substance.

## 2 The *P-v* Diagram

The general shape of the $P$-$v$ diagram of a pure substance is very much like the $T$-$v$ diagram, but the $T =$ constant lines on this diagram have a downward trend, as shown in Fig. 4–18b.

Consider again a piston–cylinder device that contains liquid water at 1 MPa and 150°C. Water at this state exists as a compressed liquid. Now the weights on top of the piston are removed one by one so that the pressure inside the cylinder decreases gradually (Fig. 4–19). The water is allowed to exchange heat with the surroundings so its temperature remains constant. As the pressure decreases, the volume of the water increases slightly. When the pressure reaches the saturation-pressure value at the specified temperature (0.4762 MPa), the water starts to boil. During this vaporization process, both the temperature and the pressure remain constant, but the specific volume increases. Once the last drop of liquid is vaporized, further reduction in pressure results in a further increase in specific volume. Notice that during the phase-change process, we did not remove any weights. Doing so would cause the pressure and therefore the temperature to drop [since $T_{sat} = f(P_{sat})$], and the process would no longer be isothermal.

When the process is repeated for other temperatures, similar paths are obtained for the phase-change processes. Connecting the saturated liquid and the saturated vapor states by a curve, we obtain the $P$-$v$ diagram of a pure substance, as shown in Fig. 4–18b.

## Extending the Diagrams to Include the Solid Phase

The two equilibrium diagrams developed so far represent the equilibrium states involving the liquid and the vapor phases only. However, these diagrams can easily be extended to include the solid phase as well as the

**FIGURE 4–19**

The pressure in a piston–cylinder device can be reduced by reducing the weight of the piston.

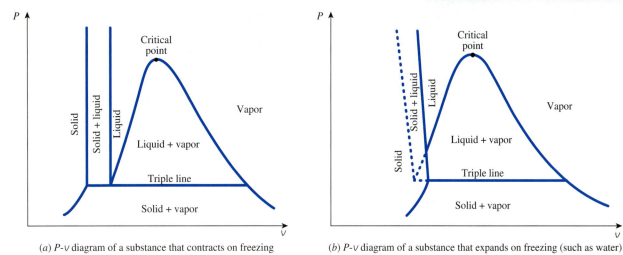

(a) P-v diagram of a substance that contracts on freezing

(b) P-v diagram of a substance that expands on freezing (such as water)

**FIGURE 4–20**

P-v diagrams of different substances.

solid–liquid and the solid–vapor saturation regions. The basic principles discussed in conjunction with the liquid–vapor phase-change process apply equally to the solid–liquid and solid–vapor phase-change processes. Most substances contract during a solidification (i.e., freezing) process. Others, like water, expand as they freeze. The P-v diagrams for both groups of substances are given in Figs. 4–20a and 4–20b. These two diagrams differ only in the solid–liquid saturation region. The T-v diagrams look very much like the P-v diagrams, especially for substances that contract on freezing.

The fact that water expands upon freezing has vital consequences in nature. If water contracted on freezing as most other substances do, the ice formed would be heavier than the liquid water, and it would settle to the bottom of rivers, lakes, and oceans instead of floating at the top. The sun's rays would never reach these ice layers, and the bottoms of many rivers, lakes, and oceans would be covered with ice at times, seriously disrupting marine life.

We are all familiar with two phases being in equilibrium, but under some conditions all three phases of a pure substance coexist in equilibrium (Fig. 4–21). On P-v or T-v diagrams, these triple-phase states form a line called the **triple line**. The states on the triple line of a substance have the same pressure and temperature but different specific volumes. The triple line appears as a point on the P-T diagrams and, therefore, is often called the **triple point**. The triple-point temperatures and pressures of various substances are given in Table 4–3. For water, the triple-point temperature and pressure are 0.01°C and 0.6117 kPa, respectively. That is, all three phases of water coexist in equilibrium only if the temperature and pressure have precisely these values. No substance can exist in the liquid phase in stable equilibrium at pressures below the triple-point pressure. The same can be said for temperature for substances that contract on freezing. However, substances at high pressures can exist in the liquid phase at temperatures below the triple-point temperature. For example,

**FIGURE 4–21**

At triple-point pressure and temperature, a substance exists in three phases in equilibrium.

## TABLE 4–3

### Triple-point temperatures and pressures of various substances

| Substance | Formula | $T_{tp}$, K | $P_{tp}$, kPa |
|---|---|---|---|
| Acetylene | $C_2H_2$ | 192.4 | 120 |
| Ammonia | $NH_3$ | 195.40 | 6.076 |
| Argon | A | 83.81 | 68.9 |
| Carbon (graphite) | C | 3900 | 10,100 |
| Carbon dioxide | $CO_2$ | 216.55 | 517 |
| Carbon monoxide | CO | 68.10 | 15.37 |
| Deuterium | $D_2$ | 18.63 | 17.1 |
| Ethane | $C_2H_6$ | 89.89 | $8 \times 10^{-4}$ |
| Ethylene | $C_2H_4$ | 104.0 | 0.12 |
| Helium 4 ($\lambda$ point) | He | 2.19 | 5.1 |
| Hydrogen | $H_2$ | 13.84 | 7.04 |
| Hydrogen chloride | HCl | 158.96 | 13.9 |
| Mercury | Hg | 234.2 | $1.65 \times 10^{-7}$ |
| Methane | $CH_4$ | 90.68 | 11.7 |
| Neon | Ne | 24.57 | 43.2 |
| Nitric oxide | NO | 109.50 | 21.92 |
| Nitrogen | $N_2$ | 63.18 | 12.6 |
| Nitrous oxide | $N_2O$ | 182.34 | 87.85 |
| Oxygen | $O_2$ | 54.36 | 0.152 |
| Palladium | Pd | 1825 | $3.5 \times 10^{-3}$ |
| Platinum | Pt | 2045 | $2.0 \times 10^{-4}$ |
| Sulfur dioxide | $SO_2$ | 197.69 | 1.67 |
| Titanium | Ti | 1941 | $5.3 \times 10^{-3}$ |
| Uranium hexafluoride | $UF_6$ | 337.17 | 151.7 |
| Water | $H_2O$ | 273.16 | 0.61 |
| Xenon | Xe | 161.3 | 81.5 |
| Zinc | Zn | 692.65 | 0.065 |

*Source:* Data from National Bureau of Standards (U.S.) Circ., 500 (1952).

water cannot exist in liquid form in equilibrium at atmospheric pressure at temperatures below 0°C, but it can exist as a liquid at −20°C at 200 MPa pressure. Also, ice exists at seven different solid phases at pressures above 100 MPa.

There are two ways a substance can pass from the solid to vapor phase: either it melts first into a liquid and subsequently evaporates, or it evaporates directly without melting first. The latter occurs at pressures below the triple-point value, since a pure substance cannot exist in the liquid phase at those pressures (Fig. 4–22). Passing from the solid phase directly into the vapor phase is called **sublimation**. For substances that have a triple-point pressure above the atmospheric pressure such as solid $CO_2$ (dry ice), sublimation is the only way to change from the solid to vapor phase at atmospheric conditions.

**FIGURE 4–22**

At low pressures (below the triple-point value), solids evaporate without melting first (*sublimation*).

## 3   The *P-T* Diagram

Figure 4–23 shows the *P-T* diagram of a pure substance. This diagram is often called the **phase diagram** since all three phases are separated from each other by three lines. The sublimation line separates the solid and vapor

$P$

Substances
that expand
on freezing

Substances
that contract
on freezing

Critical
point

Melting

Melting

Liquid

Vaporization

Solid

Triple point

Vapor

Sublimation

$T$

**FIGURE 4–23**

$P$-$T$ diagram of pure substances.

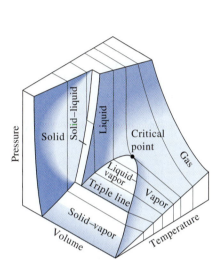

Pressure

Solid

Solid–liquid

Liquid

Critical
point

Gas

Liquid–
vapor

Triple line

Vapor

Solid–vapor

Volume

Temperature

regions, the vaporization line separates the liquid and vapor regions, and the melting (or fusion) line separates the solid and liquid regions. These three lines meet at the triple point, where all three phases coexist in equilibrium. The vaporization line ends at the critical point because no distinction can be made between liquid and vapor phases above the critical point. Substances that expand and contract on freezing differ only in the melting line on the $P$-$T$ diagram.

**FIGURE 4–24**

$P$-$v$-$T$ surface of a substance that *contracts* on freezing.

## The $P$-$v$-$T$ Surface

The state of a simple compressible substance is fixed by any two independent, intensive properties. Once the two appropriate properties are fixed, all the other properties become dependent properties. Remembering that any equation with two independent variables in the form $z = z(x, y)$ represents a surface in space, we can represent the $P$-$v$-$T$ behavior of a substance as a surface in space, as shown in Figs. 4–24 and 4–25. Here $T$ and $v$ may be viewed as the independent variables (the base) and $P$ as the dependent variable (the height).

All the points on the surface represent equilibrium states. All states along the path of a quasi-equilibrium process lie on the $P$-$v$-$T$ surface since such a process must pass through equilibrium states. The single-phase regions appear as curved surfaces on the $P$-$v$-$T$ surface, and the two-phase regions as surfaces perpendicular to the $P$-$T$ plane. This is expected since the projections of two-phase regions on the $P$-$T$ plane are lines.

All the two-dimensional diagrams we have discussed so far are merely projections of this three-dimensional surface onto the appropriate planes. A $P$-$v$ diagram is just a projection of the $P$-$v$-$T$ surface on the $P$-$v$ plane, and a $T$-$v$ diagram is nothing more than the bird's-eye view of this surface.

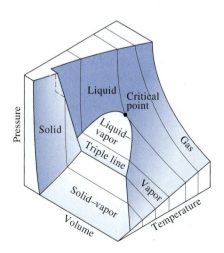

Pressure

Solid

Liquid

Critical
point

Liquid–
vapor

Triple line

Gas

Vapor

Solid–vapor

Volume

Temperature

**FIGURE 4–25**

$P$-$v$-$T$ surface of a substance that *expands* on freezing (like water).

The *P-v-T* surfaces present a great deal of information at once, but in a thermodynamic analysis it is more convenient to work with two-dimensional diagrams, such as the *P-v* and *T-v* diagrams.

## 4–5 ▪ PROPERTY TABLES

For most substances, the relationships among thermodynamic properties are too complex to be expressed by simple equations. Therefore, properties are frequently presented in the form of tables. Some thermodynamic properties can be measured easily, but others cannot and are calculated by using the relations between them and measurable properties. The results of these measurements and calculations are presented in tables in a convenient format. In the following discussion, the steam tables are used to demonstrate the use of thermodynamic property tables. Property tables of other substances are used in the same manner.

For each substance, the thermodynamic properties are listed in more than one table. In fact, a separate table is prepared for each region of interest such as the superheated vapor, compressed liquid, and saturated (mixture) regions. Property tables are given in the appendices in both SI and English units. The tables in English units carry the same number as the corresponding tables in SI, followed by an identifier E. Tables A–6 and A–6E, for example, list properties of superheated water vapor, the former in SI and the latter in English units. Before we get into the discussion of property tables, we define a new property called *enthalpy*.

### Enthalpy—A Combination Property

A person looking at the tables will notice two new properties: enthalpy *h* and entropy *s*. Entropy is a property associated with the second law of thermodynamics, and we will not use it until it is properly defined in Chap. 8. However, it is appropriate to introduce enthalpy at this point.

In the analysis of certain types of processes, particularly in power generation and refrigeration (Fig. 4–26), we frequently encounter the combination of properties $u + Pv$. For the sake of simplicity and convenience, this combination is defined as a new property, **enthalpy**, and given the symbol *h*:

$$h = u + Pv \qquad \text{(kJ/kg)} \qquad \text{(4–1)}$$

or,

$$H = U + PV \qquad \text{(kJ)} \qquad \text{(4–2)}$$

Both the total enthalpy *H* and specific enthalpy *h* are simply referred to as enthalpy since the context clarifies which one is meant. Notice that the equations given above are dimensionally homogeneous. That is, the unit of the pressure–volume product may differ from the unit of the internal energy by only a factor (Fig. 4–27). For example, it can be easily shown that 1 kPa·m$^3$ = 1 kJ. In some tables encountered in practice, the internal energy *u* is frequently not listed, but it can always be determined from $u = h - Pv$.

The widespread use of the property enthalpy is due to Professor Richard Mollier, who recognized the importance of the group $u + Pv$ in the analysis

**FIGURE 4–26**

The combination $u + Pv$ is frequently encountered in the analysis of control volumes.

kPa·m$^3$ ≡ kJ
kPa·m$^3$/kg ≡ kJ/kg
bar·m$^3$ ≡ 100 kJ
MPa·m$^3$ ≡ 1000 kJ
psi·ft$^3$ ≡ 0.18505 Btu

**FIGURE 4–27**

The product *pressure* × *volume* has energy units.

of steam turbines and in the representation of the properties of steam in tabular and graphical form (as in the famous Mollier chart). Mollier referred to the group $u + Pv$ as *heat content* and *total heat*. These terms were not quite consistent with the modern thermodynamic terminology and were replaced in the 1930s by the term *enthalpy* (from the Greek word *enthalpien*, which means *to heat*).

## 1a Saturated Liquid and Saturated Vapor States

The properties of saturated liquid and saturated vapor for water are listed in Tables A–4 and A–5. Both tables give the same information. The only difference is that in Table A–4 properties are listed under temperature and in Table A–5 under pressure. Therefore, it is more convenient to use Table A–4 when *temperature* is given and Table A–5 when *pressure* is given. The use of Table A–4 is illustrated in Fig. 4–28.

The subscript $f$ is used to denote properties of a saturated liquid, and the subscript $g$ to denote the properties of saturated vapor. These symbols are commonly used in thermodynamics and originated from German. Another subscript commonly used is $fg$, which denotes the difference between the saturated vapor and saturated liquid values of the same property. For example,

$$v_f = \text{specific volume of saturated liquid}$$

$$v_g = \text{specific volume of saturated vapor}$$

$$v_{fg} = \text{difference between } v_g \text{ and } v_f \text{ (that is, } v_{fg} = v_g - v_f)$$

The quantity $h_{fg}$ is called the **enthalpy of vaporization** (or latent heat of vaporization). It represents the amount of energy needed to vaporize a unit mass of saturated liquid at a given temperature or pressure. It decreases as the temperature or pressure increases and becomes zero at the critical point.

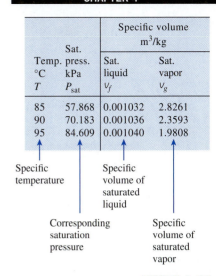

| | | Specific volume m³/kg | |
|---|---|---|---|
| Sat. Temp. press. °C kPa $T$ $P_{sat}$ | | Sat. liquid $v_f$ | Sat. vapor $v_g$ |
| 85 | 57.868 | 0.001032 | 2.8261 |
| 90 | 70.183 | 0.001036 | 2.3593 |
| 95 | 84.609 | 0.001040 | 1.9808 |

Specific temperature — Corresponding saturation pressure — Specific volume of saturated liquid — Specific volume of saturated vapor

**FIGURE 4–28**
A partial list of Table A–4.

---

**EXAMPLE 4–1    Pressure of Saturated Liquid in a Tank**

A rigid tank contains 50 kg of saturated liquid water at 90°C. Determine the pressure in the tank and the volume of the tank.

**SOLUTION** A rigid tank contains saturated liquid water. The pressure and volume of the tank are to be determined.
**Analysis** The state of the saturated liquid water is shown on a $T$-$v$ diagram in Fig. 4–29. Since saturation conditions exist in the tank, the pressure must be the saturation pressure at 90°C:

$$P = P_{\text{sat @ 90°C}} = \mathbf{70.183\ kPa} \qquad \text{(Table A–4)}$$

The specific volume of the saturated liquid at 90°C is

$$v = v_{f\text{ @ 90°C}} = 0.001036\ \text{m}^3/\text{kg} \qquad \text{(Table A–4)}$$

Then the total volume of the tank becomes

$$V = mv = (50\ \text{kg})(0.001036\ \text{m}^3/\text{kg}) = \mathbf{0.0518\ m^3}$$

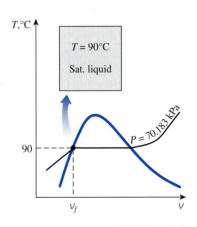

**FIGURE 4–29**
Schematic and $T$-$v$ diagram for Example 4–1.

$P$, psia

50

$T = 280.99°F$

$v_g$

$v$

**FIGURE 4–30**

Schematic and $P$-$v$ diagram for
Example 4–2.

$P$, kPa

100

$v_f$   $v_g$   $v$

**FIGURE 4–31**

Schematic and $P$-$v$ diagram for
Example 4–3.

**EXAMPLE 4–2    Temperature of Saturated Vapor in a Cylinder**

A piston–cylinder device contains 2 ft³ of saturated water vapor at 50-psia pressure. Determine the temperature and the mass of the vapor inside the cylinder.

**SOLUTION**  A cylinder contains saturated water vapor. The temperature and the mass of vapor are to be determined.
**Analysis**  The state of the saturated water vapor is shown on a $P$-$v$ diagram in Fig. 4–30. Since the cylinder contains saturated vapor at 50 psia, the temperature inside must be the saturation temperature at this pressure:

$$T = T_{\text{sat @ 50 psia}} = \mathbf{280.99°F} \qquad \text{(Table A–5E)}$$

The specific volume of the saturated vapor at 50 psia is

$$v = v_{g \text{ @ 50 psia}} = 8.5175 \text{ ft}^3/\text{lbm} \qquad \text{(Table A–5E)}$$

Then the mass of water vapor inside the cylinder becomes

$$m = \frac{V}{v} = \frac{2 \text{ ft}^3}{8.5175 \text{ ft}^3/\text{lbm}} = \mathbf{0.235 \text{ lbm}}$$

**EXAMPLE 4–3    Volume and Energy Change during Evaporation**

A mass of 200 g of saturated liquid water is completely vaporized at a constant pressure of 100 kPa. Determine (a) the volume change and (b) the amount of energy transferred to the water.

**SOLUTION**  Saturated liquid water is vaporized at constant pressure. The volume change and the energy transferred are to be determined.
**Analysis**  (a) The process described is illustrated on a $P$-$v$ diagram in Fig. 4–31. The volume change per unit mass during a vaporization process is $v_{fg}$, which is the difference between $v_g$ and $v_f$. Reading these values from Table A–5 at 100 kPa and substituting yield

$$v_{fg} = v_g - v_f = 1.6941 - 0.001043 = 1.6931 \text{ m}^3/\text{kg}$$

Thus,

$$\Delta V = m v_{fg} = (0.2 \text{ kg})(1.6931 \text{ m}^3/\text{kg}) = \mathbf{0.3386 \text{ m}^3}$$

(b) The amount of energy needed to vaporize a unit mass of a substance at a given pressure is the enthalpy of vaporization at that pressure, which is $h_{fg} = 2257.5$ kJ/kg for water at 100 kPa. Thus, the amount of energy transferred is

$$m h_{fg} = (0.2 \text{ kg})(2257.5 \text{ kJ/kg}) = \mathbf{451.5 \text{ kJ}}$$

**Discussion**  Note that we have considered the first four decimal digits of $v_{fg}$ and disregarded the rest. This is because $v_g$ has significant numbers to the first four decimal places only, and we do not know the numbers in the other decimal places. Copying all the digits from the calculator would mean that we are assuming $v_g = 1.694100$, which is not necessarily the case. It could very well be that $v_g = 1.694138$ since this number, too, would truncate to 1.6941. All the digits in our result (1.6931) are significant. But if we did not truncate the result, we would obtain $v_{fg} = 1.693057$, which falsely implies that our result is accurate to the sixth decimal place.

## 1b Saturated Liquid–Vapor Mixture

During a vaporization process, a substance exists as part liquid and part vapor. That is, it is a mixture of saturated liquid and saturated vapor (Fig. 4–32). To analyze this mixture properly, we need to know the proportions of the liquid and vapor phases in the mixture. This is done by defining a new property called the **quality** $x$ as the ratio of the mass of vapor to the total mass of the mixture:

$$x = \frac{m_{vapor}}{m_{total}} \qquad (4\text{--}3)$$

where

$$m_{total} = m_{liquid} + m_{vapor} = m_f + m_g$$

Quality has significance for *saturated mixtures* only. It has no meaning in the compressed liquid or superheated vapor regions. Its value is between 0 and 1. The quality of a system that consists of *saturated liquid* is 0 (or 0 percent), and the quality of a system consisting of *saturated vapor* is 1 (or 100 percent). In saturated mixtures, quality can serve as one of the two independent intensive properties needed to describe a state. Note that *the properties of the saturated liquid are the same whether it exists alone or in a mixture with saturated vapor*. During the vaporization process, only the amount of saturated liquid changes, not its properties. The same can be said about a saturated vapor.

A saturated mixture can be treated as a combination of two subsystems: the saturated liquid and the saturated vapor. However, the amount of mass for each phase is usually not known. Therefore, it is often more convenient to imagine that the two phases are mixed well, forming a homogeneous mixture (Fig. 4–33). Then the properties of this "mixture" will simply be the average properties of the saturated liquid–vapor mixture under consideration. Here is how it is done.

Consider a tank that contains a saturated liquid–vapor mixture. The volume occupied by saturated liquid is $V_f$, and the volume occupied by saturated vapor is $V_g$. The total volume $V$ is the sum of the two:

$$V = V_f + V_g$$

$$V = mv \longrightarrow m_t v_{avg} = m_f v_f + m_g v_g$$

$$m_f = m_t - m_g \longrightarrow m_t v_{avg} = (m_t - m_g)v_f + m_g v_g$$

Dividing by $m_t$ yields

$$v_{avg} = (1 - x)v_f + x v_g$$

since $x = m_g/m_t$. This relation can also be expressed as

$$v_{avg} = v_f + x v_{fg} \qquad (\text{m}^3/\text{kg}) \qquad (4\text{--}4)$$

where $v_{fg} = v_g - v_f$. Solving for quality, we obtain

$$x = \frac{v_{avg} - v_f}{v_{fg}} \qquad (4\text{--}5)$$

Based on this equation, quality can be related to the horizontal distances on a $P$-$v$ or $T$-$v$ diagram (Fig. 4–34). At a given temperature or pressure, the numerator of Eq. 4–5 is the distance between the actual state and the saturated liquid state, and the denominator is the length of the entire horizontal line that connects the saturated liquid and saturated vapor states. A state of 50 percent quality lies in the middle of this horizontal line.

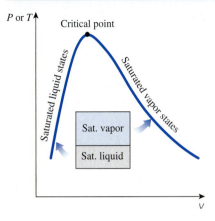

**FIGURE 4–32**

The relative amounts of liquid and vapor phases in a saturated mixture are specified by the *quality x*.

**FIGURE 4–33**

A two-phase system can be treated as a homogeneous mixture for convenience.

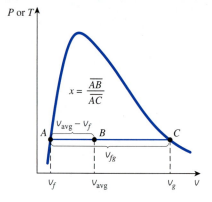

**FIGURE 4–34**

Quality is related to the horizontal distances on $P$-$v$ and $T$-$v$ diagrams.

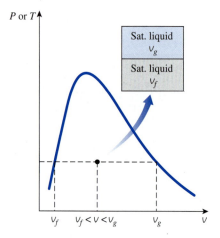

**FIGURE 4–35**

The $v$ value of a saturated liquid–vapor mixture lies between the $v_f$ and $v_g$ values at the specified $T$ or $P$.

The analysis given above can be repeated for internal energy and enthalpy with the following results:

$$u_{avg} = u_f + x u_{fg} \qquad (kJ/kg) \tag{4–6}$$

$$h_{avg} = h_f + x h_{fg} \qquad (kJ/kg) \tag{4–7}$$

All the results are of the same format, and they can be summarized in a single equation as

$$y_{avg} = y_f + x y_{fg}$$

where $y$ is $v$, $u$, or $h$. The subscript "avg" (for "average") is usually dropped for simplicity. The values of the average properties of the mixtures are always *between* the values of the saturated liquid and the saturated vapor properties (Fig. 4–35). That is,

$$y_f \leq y_{avg} \leq y_g$$

Finally, all the saturated-mixture states are located under the saturation curve, and to analyze saturated mixtures, all we need are saturated liquid and saturated vapor data (Tables A–4 and A–5 in the case of water).

---

**EXAMPLE 4–4**   **Pressure and Volume of a Saturated Mixture**

A rigid tank contains 10 kg of water at 90°C. If 8 kg of the water is in the liquid form and the rest is in the vapor form, determine (*a*) the pressure in the tank and (*b*) the volume of the tank.

**SOLUTION**   A rigid tank contains saturated mixture. The pressure and the volume of the tank are to be determined.

**Analysis**   (*a*) The state of the saturated liquid–vapor mixture is shown in Fig. 4–36. Since the two phases coexist in equilibrium, we have a saturated mixture, and the pressure must be the saturation pressure at the given temperature:

$$P = P_{sat\ @\ 90°C} = \mathbf{70.183\ kPa} \qquad (\text{Table A–4})$$

(*b*) At 90°C, we have $v_f = 0.001036$ m³/kg and $v_g = 2.3593$ m³/kg (Table A–4). One way of finding the volume of the tank is to determine the volume occupied by each phase and then add them:

$$V = V_f + V_g = m_f v_f + m_g v_g$$
$$= (8\ kg)(0.001036\ m^3/kg) + (2\ kg)(2.3593\ m^3/kg)$$
$$= \mathbf{4.73\ m^3}$$

Another way is to first determine the quality $x$, then the average specific volume $v$, and finally the total volume:

$$x = \frac{m_g}{m_t} = \frac{2\ kg}{10\ kg} = 0.2$$
$$v = v_f + x v_{fg}$$
$$= 0.001036\ m^3/kg + (0.2)[(2.3593 - 0.001036)\ m^3/kg]$$
$$= 0.473\ m^3/kg$$

and

$$V = mv = (10 \text{ kg})(0.473 \text{ m}^3/\text{kg}) = 4.73 \text{ m}^3$$

***Discussion*** The first method appears to be easier in this case since the masses of each phase are given. In most cases, however, the masses of each phase are not available, and the second method becomes more convenient.

$$T = 90°C$$
$$m_g = 2 \text{ kg}$$
$$m_f = 8 \text{ kg}$$

$$P = 70.183 \text{ kPa}$$

90

$v_f = 0.001036$   $v_g = 2.3593$   $v$, m³/kg

**FIGURE 4–36**

Schematic and $T$-$v$ diagram for Example 4–4.

---

### EXAMPLE 4–5    Properties of Saturated Liquid–Vapor Mixture

An 80-L vessel contains 4 kg of refrigerant-134a at a pressure of 160 kPa. Determine (*a*) the temperature, (*b*) the quality, (*c*) the enthalpy of the refrigerant, and (*d*) the volume occupied by the vapor phase.

**SOLUTION**   A vessel is filled with refrigerant-134a. Some properties of the refrigerant are to be determined.

***Analysis*** (*a*) The state of the saturated liquid–vapor mixture is shown in Fig. 4–37. At this point we do not know whether the refrigerant is in the compressed liquid, superheated vapor, or saturated mixture region. This can be determined by comparing a suitable property to the saturated liquid and saturated vapor values. From the information given, we can determine the specific volume:

$$v = \frac{V}{m} = \frac{0.080 \text{ m}^3}{4 \text{ kg}} = 0.02 \text{ m}^3/\text{kg}$$

At 160 kPa, we read

$$v_f = 0.0007437 \text{ m}^3/\text{kg}$$
$$v_g = 0.12348 \text{ m}^3/\text{kg}$$
(Table A–12)

R-134a
$$P = 160 \text{ kPa}$$
$$m = 4 \text{ kg}$$

$$T = -15.60°C$$

160

$v_f = 0.0007437$   $v_g = 0.12348$   $v$, m³/kg
$h_f = 31.21$   $h_g = 241.11$   $h$, kJ/kg

**FIGURE 4–37**

Schematic and $P$-$v$ diagram for Example 4–5.

Obviously, $v_f < v < v_g$, and, the refrigerant is in the saturated mixture region. Thus, the temperature must be the saturation temperature at the specified pressure:

$$T = T_{\text{sat @ 160 kPa}} = \mathbf{-15.60°C}$$

(*b*) Quality can be determined from

$$x = \frac{v - v_f}{v_{fg}} = \frac{0.02 - 0.0007437}{0.12348 - 0.0007437} = \mathbf{0.157}$$

(*c*) At 160 kPa, we also read from Table A–12 that $h_f = 31.21$ kJ/kg and $h_{fg} = 209.90$ kJ/kg. Then,

$$h = h_f + xh_{fg}$$
$$= 31.21 \text{ kJ/kg} + (0.157)(209.90 \text{ kJ/kg})$$
$$= \mathbf{64.2 \text{ kJ/kg}}$$

(*d*) The mass of the vapor is

$$m_g = xm_t = (0.157)(4 \text{ kg}) = 0.628 \text{ kg}$$

and the volume occupied by the vapor phase is

$$V_g = m_g v_g = (0.628 \text{ kg})(0.12348 \text{ m}^3/\text{kg}) = \mathbf{0.0775 \text{ m}^3} \text{ (or 77.5 L)}$$

The rest of the volume (2.5 L) is occupied by the liquid.

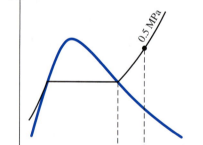

| $T,°C$ | $v$ m³/kg | $u$ kJ/kg | $h$ kJ/kg |
|---|---|---|---|
| | $P = 0.1$ MPa (99.61°C) | | |
| Sat. | 1.6941 | 2505.6 | 2675.0 |
| 100 | 1.6959 | 2506.2 | 2675.8 |
| 150 | 1.9367 | 2582.9 | 2776.6 |
| ⋮ | ⋮ | ⋮ | ⋮ |
| 1300 | 7.2605 | 4687.2 | 5413.3 |
| | $P = 0.5$ MPa (151.83°C) | | |
| Sat. | 0.37483 | 2560.7 | 2748.1 |
| 200 | 0.42503 | 2643.3 | 2855.8 |
| 250 | 0.47443 | 2723.8 | 2961.0 |

**FIGURE 4–38**

A partial listing of Table A–6.

## 2 Superheated Vapor

Property tables are also available for saturated solid–vapor mixtures. Properties of saturated ice–water vapor mixtures, for example, are listed in Table A–8. Saturated solid–vapor mixtures can be handled just as saturated liquid–vapor mixtures.

In the region to the right of the saturated vapor line and at temperatures above the critical point temperature, a substance exists as superheated vapor. Since the superheated region is a single-phase region (vapor phase only), temperature and pressure are no longer dependent properties and they can conveniently be used as the two independent properties in the tables. The format of the superheated vapor tables is illustrated in Fig. 4–38.

In these tables, the properties are listed against temperature for selected pressures starting with the saturated vapor data. The saturation temperature is given in parentheses following the pressure value.

Compared to saturated vapor, superheated vapor is characterized by

Lower pressures ($P < P_{sat}$ at a given $T$)
Higher tempreatures ($T > T_{sat}$ at a given $P$)
Higher specific volumes ($v > v_g$ at a given $P$ or $T$)
Higher internal energies ($u > u_g$ at a given $P$ or $T$)
Higher enthalpies ($h > h_g$ at a given $P$ or $T$)

**FIGURE 4–39**

At a specified $P$, superheated vapor exists at a higher $h$ than the saturated vapor (Example 4–7).

**EXAMPLE 4–6    Internal Energy of Superheated Vapor**

Determine the internal energy of water at 20 psia and 400°F.

**SOLUTION**  The internal energy of water at a specified state is to be determined.
**Analysis**  At 20 psia, the saturation temperature is 227.92°F. Since $T > T_{sat}$, the water is in the superheated vapor region. Then the internal energy at the given temperature and pressure is determined from the superheated vapor table (Table A–6E) to be

$$u = \mathbf{1145.1 \ Btu/lbm}$$

**EXAMPLE 4–7    Temperature of Superheated Vapor**

Determine the temperature of water at a state of $P = 0.5$ MPa and $h = 2890$ kJ/kg.

**SOLUTION**  The temperature of water at a specified state is to be determined.
**Analysis**  At 0.5 MPa, the enthalpy of saturated water vapor is $h_g = 2748.1$ kJ/kg. Since $h > h_g$, as shown in Fig. 4–39, we again have superheated vapor. Under 0.5 MPa in Table A–6 we read

| $T$, °C | $h$, kJ/kg |
|---|---|
| 200 | 2855.8 |
| 250 | 2961.0 |

Obviously, the temperature is between 200 and 250°C. By linear interpolation it is determined to be

$$T = 216.3°C$$

## 3  Compressed Liquid

Compressed liquid tables are not as commonly available, and Table A–7 is the only compressed liquid table in this text. The format of Table A–7 is very much like the format of the superheated vapor tables. One reason for the lack of compressed liquid data is the relative independence of compressed liquid properties from pressure. Variation of properties of compressed liquid with pressure is very mild. Increasing the pressure 100 times often causes properties to change less than 1 percent.

In the absence of compressed liquid data, a general approximation is *to treat compressed liquid as saturated liquid at the given temperature* (Fig. 4–40). This is because the compressed liquid properties depend on temperature much more strongly than they do on pressure. Thus,

$$y \cong y_{f @ T} \tag{4–8}$$

for compressed liquids, where $y$ is $v$, $u$, or $h$. Of these three properties, the property whose value is most sensitive to variations in the pressure is the enthalpy $h$. Although the above approximation results in negligible error in $v$ and $u$, the error in $h$ may reach undesirable levels. However, the error in $h$ at low to moderate pressures and temperatures can be reduced significantly by evaluating it from

$$h \cong h_{f @ T} + v_{f @ T}(P - P_{sat @ T}) \tag{4–9}$$

instead of taking it to be just $h_f$. Note, however, that the approximation in Eq. 4–9 does not yield any significant improvement at moderate to high temperatures and pressures, and it may even backfire and result in greater error due to overcorrection at very high temperatures and pressures (*see* Kostic, Ref. 4).

In general, a compressed liquid is characterized by

  Higher pressures ($P > P_{sat}$ at a given $T$)

  Lower tempreatures ($T < T_{sat}$ at a given $P$)

  Lower specific volumes ($v < v_f$ at a given $P$ or $T$)

  Lower internal energies ($u < u_f$ at a given $P$ or $T$)

  Lower enthalpies ($h < h_f$ at a given $P$ or $T$)

But unlike superheated vapor, the compressed liquid properties are not much different from the corresponding saturated liquid values.

**FIGURE 4–40**

A compressed liquid may be approximated as a saturated liquid at the given temperature.

---

**EXAMPLE 4–8**    **Approximating Compressed Liquid as Saturated Liquid**

Determine the internal energy of compressed liquid water at 80°C and 5 MPa, using (a) data from the compressed liquid table and (b) saturated liquid data. What is the error involved in the second case?

$T$, °C

$T = 80°C$
$P = 5$ MPa

5 MPa

80

$u \cong u_{f\,@\,80°C}$

$u$

**FIGURE 4–41**

Schematic and $T$-$u$ diagram for Example 4–8.

**SOLUTION** The exact and approximate values of the internal energy of liquid water are to be determined.

**Analysis** At 80°C, the saturation pressure of water is 47.416 kPa, and since 5 MPa $> P_{sat}$, we obviously have compressed liquid, as shown in Fig. 4–41.

(a) From the compressed liquid table (Table A–7)

$$\left. \begin{array}{l} P = 5 \text{ MPa} \\ T = 80°C \end{array} \right\} \quad u = \textbf{333.82 kJ/kg}$$

(b) From the saturation table (Table A–4), we read

$$u \cong u_{f\,@\,80°C} = \textbf{334.97 kJ/kg}$$

The error involved is

$$\frac{334.97 - 333.82}{333.82} \times 100 = \textbf{0.34\%}$$

which is less than 1 percent.

# Reference State and Reference Values

The values of $u$, $h$, and $s$ cannot be measured directly, and they are calculated from measurable properties using the relations between thermodynamic properties. However, those relations give the *changes* in properties, not the values of properties at specified states. Therefore, we need to choose a convenient *reference state* and assign a value of *zero* for a convenient property or properties at that state. For water, the state of saturated liquid at 0.01°C is taken as the reference state, and the internal energy and entropy are assigned zero values at that state. For refrigerant-134a, the state of saturated liquid at −40°C is taken as the reference state, and the enthalpy and entropy are assigned zero values at that state. Note that some properties may have negative values as a result of the reference state chosen.

It should be mentioned that sometimes different tables list different values for some properties at the same state as a result of using a different reference state. However, in thermodynamics we are concerned with the *changes* in properties, and the reference state chosen is of no consequence in calculations as long as we use values from a single consistent set of tables or charts.

**EXAMPLE 4–9** **The Use of Steam Tables to Determine Properties**

Determine the missing properties and the phase descriptions in the following table for water:

|  | $T$, °C | $P$, kPa | $u$, kJ/kg | $x$ | Phase description |
|---|---|---|---|---|---|
| (a) |  | 200 |  | 0.6 |  |
| (b) | 125 |  | 1600 |  |  |
| (c) |  | 1000 | 2950 |  |  |
| (d) | 75 | 500 |  |  |  |
| (e) |  | 850 |  | 0.0 |  |

**SOLUTION** Properties and phase descriptions of water are to be determined at various states.

*Analysis* (*a*) The quality is given to be $x = 0.6$, which implies that 60 percent of the mass is in the vapor phase and the remaining 40 percent is in the liquid phase. Therefore, we have saturated liquid–vapor mixture at a pressure of 200 kPa. Then the temperature must be the saturation temperature at the given pressure:

$$T = T_{\text{sat @ 200 kPa}} = \mathbf{120.21°C} \qquad \text{(Table A–5)}$$

At 200 kPa, we also read from Table A–5 that $u_f = 504.50$ kJ/kg and $u_{fg} = 2024.6$ kJ/kg. Then the average internal energy of the mixture is

$$
\begin{aligned}
u &= u_f + x u_{fg} \\
&= 504.50 \text{ kJ/kg} + (0.6)(2024.6 \text{ kJ/kg}) \\
&= \mathbf{1719.26 \text{ kJ/kg}}
\end{aligned}
$$

(*b*) This time the temperature and the internal energy are given, but we do not know which table to use to determine the missing properties because we have no clue as to whether we have saturated mixture, compressed liquid, or superheated vapor. To determine the region we are in, we first go to the saturation table (Table A–4) and determine the $u_f$ and $u_g$ values at the given temperature. At 125°C, we read $u_f = 524.83$ kJ/kg and $u_g = 2534.3$ kJ/kg. Next we compare the given $u$ value to these $u_f$ and $u_g$ values, keeping in mind that

$$
\begin{aligned}
&\text{if} \quad u < u_f &&\text{we have } \textit{compressed liquid} \\
&\text{if} \quad u_f \le u \le u_g &&\text{we have } \textit{saturated mixture} \\
&\text{if} \quad u > u_g &&\text{we have } \textit{superheated vapor}
\end{aligned}
$$

In our case the given $u$ value is 1600, which falls between the $u_f$ and $u_g$ values at 125°C. Therefore, we have saturated liquid–vapor mixture. Then the pressure must be the saturation pressure at the given temperature:

$$P = P_{\text{sat @ 125°C}} = \mathbf{232.23 \text{ kPa}} \qquad \text{(Table A–4)}$$

The quality is determined from

$$x = \frac{u - u_f}{u_{fg}} = \frac{1600 - 524.83}{2009.5} = \mathbf{0.535}$$

The criteria above for determining whether we have compressed liquid, saturated mixture, or superheated vapor can also be used when enthalpy $h$ or specific volume $v$ is given instead of internal energy $u$, or when pressure is given instead of temperature.

(*c*) This is similar to case (*b*), except pressure is given instead of temperature. Following the argument given above, we read the $u_f$ and $u_g$ values at the specified pressure. At 1 MPa, we have $u_f = 761.39$ kJ/kg and $u_g = 2582.8$ kJ/kg. The specified $u$ value is 2950 kJ/kg, which is greater than the $u_g$ value at 1 MPa. Therefore, we have superheated vapor, and the temperature at this state is determined from the superheated vapor table by interpolation to be

$$T = \mathbf{395.2°C} \qquad \text{(Table A–6)}$$

We would leave the quality column blank in this case since quality has no meaning for a superheated vapor.

(*d*) In this case the temperature and pressure are given, but again we cannot tell which table to use to determine the missing properties because we do

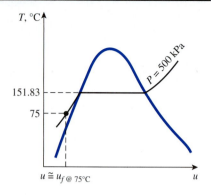

**FIGURE 4–42**

At a given $P$ and $T$, a pure substance will exist as a compressed liquid if $T < T_{sat @ P}$.

not know whether we have saturated mixture, compressed liquid, or super-heated vapor. To determine the region we are in, we go to the saturation table (Table A–5) and determine the saturation temperature value at the given pressure. At 500 kPa, we have $T_{sat} = 151.83°C$. We then compare the given $T$ value to this $T_{sat}$ value, keeping in mind that

if $\qquad T < T_{sat @ given P} \qquad$ we have *compressed liquid*

if $\qquad T = T_{sat @ given P} \qquad$ we have *saturated mixture*

if $\qquad T > T_{sat @ given P} \qquad$ we have *superheated vapor*

In our case, the given $T$ value is 75°C, which is less than the $T_{sat}$ value at the specified pressure. Therefore, we have compressed liquid (Fig. 4–42), and normally we would determine the internal energy value from the compressed liquid table. But in this case the given pressure is much lower than the lowest pressure value in the compressed liquid table (which is 5 MPa), and therefore we are justified to treat the compressed liquid as saturated liquid at the given temperature (*not* pressure):

$$u \cong u_{f @ 75°C} = \mathbf{313.99 \ kJ/kg} \qquad \text{(Table A–4)}$$

We would leave the quality column blank in this case since quality has no meaning in the compressed liquid region.

(*e*) The quality is given to be $x = 0$, and thus we have saturated liquid at the specified pressure of 850 kPa. Then the temperature must be the saturation temperature at the given pressure, and the internal energy must have the saturated liquid value:

$$T = T_{sat @ 850 kPa} = \mathbf{172.94°C}$$

$$u = u_{f @ 850 kPa} = \mathbf{731.00 \ kJ/kg} \qquad \text{(Table A–5)}$$

# 4–6 ▪ THE IDEAL-GAS EQUATION OF STATE

Property tables provide very accurate information about the properties, but they are bulky and vulnerable to typographical errors. A more practical and desirable approach would be to have some simple relations among the properties that are sufficiently general and accurate.

Any equation that relates the pressure, temperature, and specific volume of a substance is called an **equation of state**. Property relations that involve other properties of a substance at equilibrium states are also referred to as equations of state. There are several equations of state, some simple and others very complex. The simplest and best-known equation of state for substances in the gas phase is the ideal-gas equation of state. This equation predicts the $P$-$v$-$T$ behavior of a gas quite accurately within some properly selected region.

*Gas* and *vapor* are often used as synonymous words. The vapor phase of a substance is customarily called a *gas* when it is above the critical temperature. *Vapor* usually implies a gas that is not far from a state of condensation.

In 1662, Robert Boyle, an Englishman, observed during his experiments with a vacuum chamber that the pressure of gases is inversely proportional to their volume. In 1802, J. Charles and J. Gay-Lussac,

Frenchmen, experimentally determined that at low pressures the volume of a gas is proportional to its temperature. That is,

$$P = R\left(\frac{T}{v}\right)$$

or

$$Pv = RT \tag{4–10}$$

where the constant of proportionality $R$ is called the **gas constant**. Equation 4–10 is called the **ideal-gas equation of state**, or simply the **ideal-gas relation**, and a gas that obeys this relation is called an **ideal gas**. In this equation, $P$ is the absolute pressure, $T$ is the absolute temperature, and $v$ is the specific volume.

The gas constant $R$ is different for each gas (Fig. 4–43) and is determined from

$$R = \frac{R_u}{M} \quad \text{(kJ/kg·K or kPa·m}^3\text{/kg·K)}$$

where $R_u$ is the **universal gas constant** and $M$ is the molar mass (also called *molecular weight*) of the gas. The constant $R_u$ is the same for all substances, and its value is

$$R_u = \begin{cases} 8.31447 \text{ kJ/kmol·K} \\ 8.31447 \text{ kPa·m}^3\text{/kmol·K} \\ 0.0831447 \text{ bar·m}^3\text{/kmol·K} \\ 1.98588 \text{ Btu/lbmol·R} \\ 10.7316 \text{ psia·ft}^3\text{/lbmol·R} \\ 1545.37 \text{ ft·lbf/lbmol·R} \end{cases} \tag{4–11}$$

The **molar mass** $M$ can simply be defined as *the mass of one mole* (also called a *gram-mole*, abbreviated gmol) *of a substance in grams*, or *the mass of one kmol* (also called a *kilogram-mole*, abbreviated kgmol) *in kilograms*. In English units, it is the mass of 1 lbmol in lbm. Notice that the molar mass of a substance has the same numerical value in both unit systems because of the way it is defined. When we say the molar mass of nitrogen is 28, it simply means the mass of 1 kmol of nitrogen is 28 kg, or the mass of 1 lbmol of nitrogen is 28 lbm. That is, $M = 28$ kg/kmol $= 28$ lbm/lbmol. The mass of a system is equal to the product of its molar mass $M$ and the mole number $N$:

$$m = MN \quad \text{(kg)} \tag{4–12}$$

The values of $R$ and $M$ for several substances are given in Table A–1.

The ideal-gas equation of state can be written in several different forms:

$$V = mv \longrightarrow PV = mRT \tag{4–13}$$

$$mR = (MN)R = NR_u \longrightarrow PV = NR_uT \tag{4–14}$$

$$V = N\bar{v} \longrightarrow P\bar{v} = R_uT \tag{4–15}$$

where $\bar{v}$ is the molar specific volume, that is, the volume per unit mole (in m$^3$/kmol or ft$^3$/lbmol). A bar above a property denotes values on a *unit-mole basis* throughout this text (Fig. 4–44).

| Substance | $R$, kJ/kg·K |
|-----------|--------------|
| Air | 0.2870 |
| Helium | 2.0769 |
| Argon | 0.2081 |
| Nitrogen | 0.2968 |

**FIGURE 4–43**

Different substances have different gas constants.

| Per unit mass | Per unit mole |
|---------------|---------------|
| $v$, m$^3$/kg | $\bar{v}$, m$^3$/kmol |
| $u$, kJ/kg | $\bar{u}$, kJ/kmol |
| $h$, kJ/kg | $\bar{h}$, kJ/kmol |

**FIGURE 4–44**

Properties per unit mole are denoted with a bar on the top.

**FIGURE 4–45**

The ideal-gas relation often is not applicable to real gases; thus, care should be exercised when using it.

*BLONDIE © KING FEATURES SYNDICATE.*

$T = 25°C$
$P_g = 210$ kPa

**FIGURE 4–46**

Schematic for Example 4–10.

By writing Eq. 4–13 twice for a fixed mass and simplifying, the properties of an ideal gas at two different states are related to each other by

$$\frac{P_1 V_1}{T_1} = \frac{P_2 V_2}{T_2} \tag{4–16}$$

An ideal gas is an *imaginary* substance that obeys the relation $Pv = RT$ (Fig. 4–45). It has been experimentally observed that the ideal-gas relation given closely approximates the *P-v-T* behavior of real gases at low densities. At low pressures and high temperatures, the density of a gas decreases, and the gas behaves as an ideal gas under these conditions. What constitutes low pressure and high temperature is explained later.

In the range of practical interest, many familiar gases such as air, nitrogen, oxygen, hydrogen, helium, argon, neon, krypton, and even heavier gases such as carbon dioxide can be treated as ideal gases with negligible error (often less than 1 percent). Dense gases such as water vapor in steam power plants and refrigerant vapor in refrigerators, however, should not be treated as ideal gases. Instead, the property tables should be used for these substances.

---

**EXAMPLE 4–10**    **Temperature Rise of Air in a Tire During a Trip**

The gage pressure of an automobile tire is measured to be 210 kPa before a trip and 220 kPa after the trip at a location where the atmospheric pressure is 95 kPa (Fig. 4–46). Assuming the volume of the tire remains constant and the air temperature before the trip is 25°C, determine air temperature in the tire after the trip.

**SOLUTION**   The pressure in an automobile tire is measured before and after a trip. The temperature of air in the tire after the trip is to be determined.

*Assumptions*   **1** The volume of the tire remains constant. **2** Air is an ideal gas.

*Properties*   The local atmospheric pressure is 95 kPa.

*Analysis*   The absolute pressures in the tire before and after the trip are

$$P_1 = P_{\text{gage.1}} + P_{\text{atm}} = 210 + 95 = 305 \text{ kPa}$$
$$P_2 = P_{\text{gage.2}} + P_{\text{atm}} = 220 + 90 = 315 \text{ kPa}$$

Note that air is an ideal gas and the volume is constant, the air temperatures after the trip is determined to be

$$\frac{P_1 V_1}{T_1} = \frac{P_2 V_2}{T_2} \longrightarrow T_2 = \frac{P_2}{P_1} T_1 = \frac{315 \text{ kPa}}{305 \text{ kPa}} (25 + 273 \text{ K}) = 307.8 \text{ K} = 34.8°C$$

Therefore, the absolute temperature of air in the tire will increase by 6.9% during this trip.

*Discussion*   Note that the air temperature has risen nearly 10°C during this trip. This shows the importance of measuring the tire pressures before long trips to avoid errors due to temperature rise of air in tire. Also note that the unit Kelvin is used for temperature in the ideal gas relation.

## Is Water Vapor an Ideal Gas?

This question cannot be answered with a simple yes or no. The error involved in treating water vapor as an ideal gas is calculated and plotted in Fig. 4–47. It is clear from this figure that at pressures below 10 kPa, water vapor can be treated as an ideal gas, regardless of its temperature, with negligible error (less than 0.1 percent). At higher pressures, however, the ideal-gas assumption yields unacceptable errors, particularly in the vicinity of the critical point and the saturated vapor line (over 100 percent). Therefore, in air-conditioning applications, the water vapor in the air can be treated as an ideal gas with essentially no error since the pressure of the water vapor is very low. In steam power plant applications, however, the pressures involved are usually very high; therefore, ideal-gas relations should not be used.

## 4–7 ▪ COMPRESSIBILITY FACTOR—A MEASURE OF DEVIATION FROM IDEAL-GAS BEHAVIOR

The ideal-gas equation is very simple and thus very convenient to use. However, as illustrated in Fig. 4–47, gases deviate from ideal-gas behavior significantly at states near the saturation region and the critical point.

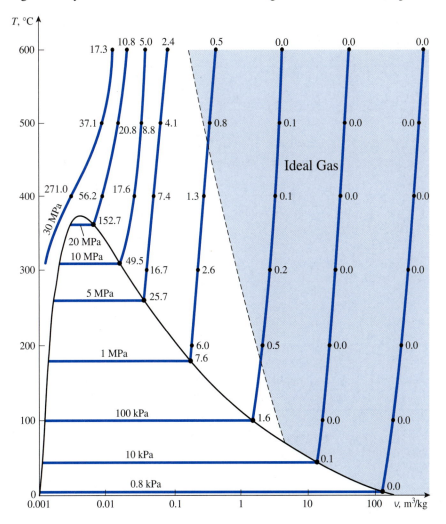

**FIGURE 4–47**

Percentage of error ($[|v_{table} - v_{ideal}|/v_{table}] \times 100$) involved in assuming steam to be an ideal gas, and the region where steam can be treated as an ideal gas with less than 1 percent error.

**FIGURE 4–48**

The compressibility factor is unity for ideal gases.

This deviation from ideal-gas behavior at a given temperature and pressure can accurately be accounted for by the introduction of a correction factor called the **compressibility factor** $Z$ defined as

$$Z = \frac{Pv}{RT} \tag{4–17}$$

or

$$Pv = ZRT \tag{4–18}$$

It can also be expressed as

$$Z = \frac{v_{actual}}{v_{ideal}} \tag{4–19}$$

where $v_{ideal} = RT/P$. Obviously, $Z = 1$ for ideal gases. For real gases $Z$ can be greater than or less than unity (Fig. 4–48). The farther away $Z$ is from unity, the more the gas deviates from ideal-gas behavior.

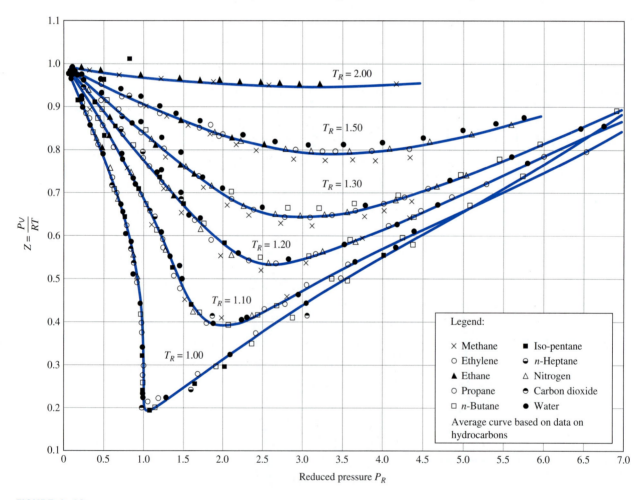

**FIGURE 4–49**

Comparison of $Z$ factors for various gases.

Source: *Gour-Jen Su, "Modified Law of Corresponding States,"* Ind. Eng. Chem. *(international ed.) 38 (1946), p. 803.*

We have said that gases follow the ideal-gas equation closely at low pressures and high temperatures. But what exactly constitutes low pressure or high temperature? Is $-100°C$ a low temperature? It definitely is for most substances but not for air. Air (or nitrogen) can be treated as an ideal gas at this temperature and atmospheric pressure with an error under 1 percent. This is because nitrogen is well over its critical temperature ($-147°C$) and away from the saturation region. At this temperature and pressure, however, most substances would exist in the solid phase. Therefore, the pressure or temperature of a substance is high or low relative to its critical temperature or pressure.

Gases behave differently at a given temperature and pressure, but they behave very much the same at temperatures and pressures normalized with respect to their critical temperatures and pressures. The normalization is done as

$$P_R = \frac{P}{P_{cr}} \quad \text{and} \quad T_R = \frac{T}{T_{cr}} \tag{4–20}$$

Here $P_R$ is called the **reduced pressure** and $T_R$ the **reduced temperature**. The $Z$ factor for all gases is approximately the same at the same reduced pressure and temperature. This is called the **principle of corresponding states**. In Fig. 4–49, the experimentally determined $Z$ values are plotted against $P_R$ and $T_R$ for several gases. The gases seem to obey the principle of corresponding states reasonably well. By curve-fitting all the data, we obtain the **generalized compressibility chart** that can be used for all gases (Fig. A–28).

The following observations can be made from the generalized compressibility chart:

1. At very low pressures ($P_R \ll 1$), gases behave as an ideal gas regardless of temperature (Fig. 4–50),
2. At high temperatures ($T_R > 2$), ideal-gas behavior can be assumed with good accuracy regardless of pressure (except when $P_R \gg 1$).
3. The deviation of a gas from ideal-gas behavior is greatest in the vicinity of the critical point (Fig. 4–51).

**FIGURE 4–50**

At very low pressures, all gases approach ideal-gas behavior (regardless of their temperature).

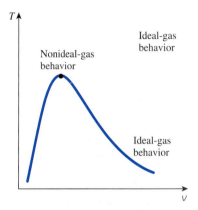

**FIGURE 4–51**

Gases deviate from the ideal-gas behavior the most in the neighborhood of the critical point.

---

**EXAMPLE 4–11     The Use of Generalized Charts**

Determine the specific volume of refrigerant-134a at 1 MPa and 50°C, using (*a*) the ideal-gas equation of state and (*b*) the generalized compressibility chart. Compare the values obtained to the actual value of 0.021796 m³/kg and determine the error involved in each case.

**SOLUTION**  The specific volume of refrigerant-134a is to be determined assuming ideal- and nonideal-gas behavior.
*Analysis*  The gas constant, the critical pressure, and the critical temperature of refrigerant-134a are determined from Table A–1 to be

$$R = 0.0815 \text{ kPa·m}^3/\text{kg·K}$$
$$P_{cr} = 4.059 \text{ MPa}$$
$$T_{cr} = 374.2 \text{ K}$$

(a) The specific volume of refrigerant-134a under the ideal-gas assumption is

$$v = \frac{RT}{P} = \frac{(0.0815 \text{ kPa} \cdot \text{m}^3/\text{kg} \cdot \text{K})(323 \text{ K})}{1000 \text{ kPa}} = \mathbf{0.026325 \text{ m}^3/\text{kg}}$$

Therefore, treating the refrigerant-134a vapor as an ideal gas would result in an error of (0.026325 − 0.021796)/0.021796 = **0.208**, or 20.8 percent in this case.

(b) To determine the correction factor $Z$ from the compressibility chart, we first need to calculate the reduced pressure and temperature:

$$\left. \begin{array}{l} P_R = \dfrac{P}{P_{cr}} = \dfrac{1 \text{ MPa}}{4.059 \text{ MPa}} = 0.246 \\[2mm] T_R = \dfrac{T}{T_{cr}} = \dfrac{323 \text{ K}}{374.2 \text{ K}} = 0.863 \end{array} \right\} \quad Z = 0.84$$

Thus

$$v = Z v_{\text{ideal}} = (0.84)(0.026325 \text{ m}^3/\text{kg}) = \mathbf{0.022113 \text{ m}^3/\text{kg}}$$

**Discussion** The error in this result is less than **2 percent**. Therefore, in the absence of tabulated data, the generalized compressibility chart can be used with confidence.

When $P$ and $v$, or $T$ and $v$, are given instead of $P$ and $T$, the generalized compressibility chart can still be used to determine the third property, but it would involve tedious trial and error. Therefore, it is necessary to define one more reduced property called the **pseudo-reduced specific volume** $v_R$ as

$$v_R = \frac{v_{\text{actual}}}{RT_{cr}/P_{cr}} \tag{4-21}$$

Note that $v_R$ is defined differently from $P_R$ and $T_R$. It is related to $T_{cr}$ and $P_{cr}$ instead of $v_{cr}$. Lines of constant $v_R$ are also added to the compressibility charts, and this enables one to determine $T$ or $P$ without having to resort to time-consuming iterations (Fig. 4–52).

**FIGURE 4–52**

The compressibility factor can also be determined from a knowledge of $P_R$ and $v_R$.

$$\left. \begin{array}{l} P_R = \dfrac{P}{P_{cr}} \\[2mm] v_R = \dfrac{v}{RT_{cr}/P_{cr}} \end{array} \right\} \begin{array}{l} Z = \ldots \\ (\text{Fig. A–15}) \end{array}$$

---

**EXAMPLE 4–12     Using Generalized Charts to Determine Pressure**

Determine the pressure of water vapor at 600°F and 0.51431 ft³/lbm, using (a) the steam tables, (b) the ideal-gas equation, and (c) the generalized compressibility chart.

**SOLUTION** The pressure of water vapor is to be determined in three different ways.

**Analysis** A sketch of the system is given in Fig. 4–53. The gas constant, the critical pressure, and the critical temperature of steam are determined from Table A–1E to be

$$R = 0.5956 \text{ psia} \cdot \text{ft}^3/\text{lbm} \cdot \text{R}$$

$$P_{cr} = 3200 \text{ psia}$$

$$T_{cr} = 1164.8 \text{ R}$$

**FIGURE 4–53**

Schematic for Example 4–12.

In figure 4–53:
H₂O
$T = 600°\text{F}$
$v = 0.51431 \text{ ft}^3/\text{lbm}$
$P = ?$

(a) The pressure at the specified state is determined from Table A–6E to be

$$\left. \begin{array}{l} v = 0.51431 \text{ ft}^3/\text{lbm} \\ T = 600°F \end{array} \right\} \quad P = \textbf{1000 psia}$$

This is the experimentally determined value, and thus it is the most accurate.
(b) The pressure of steam under the ideal-gas assumption is determined from the ideal-gas relation to be

$$P = \frac{RT}{v} = \frac{(0.5956 \text{ psia·ft}^3/\text{lbm·R})(1060 \text{ R})}{0.51431 \text{ ft}^3/\text{lbm}} = \textbf{1228 psia}$$

Therefore, treating the steam as an ideal gas would result in an error of $(1228 - 1000)/1000 = 0.228$, or 22.8 percent in this case.
(c) To determine the correction factor $Z$ from the compressibility chart (Fig. A–28), we first need to calculate the pseudo-reduced specific volume and the reduced temperature:

$$\left. \begin{array}{l} v_R = \dfrac{v_{\text{actual}}}{RT_{\text{cr}}/P_{\text{cr}}} = \dfrac{(0.51431 \text{ ft}^3/\text{lbm})(3200 \text{ psia})}{(0.5956 \text{ psia·ft}^3/\text{lbm·R})(1164.8 \text{ R})} = 2.372 \\[2ex] T_R = \dfrac{T}{T_{\text{cr}}} = \dfrac{1060 \text{ R}}{1164.8 \text{ R}} = 0.91 \end{array} \right\} \quad P_R = 0.33$$

Thus,

$$P = P_R P_{\text{cr}} = (0.33)(3200 \text{ psia}) = \textbf{1056 psia}$$

**Discussion** Using the compressibility chart reduced the error from 22.8 to 5.6 percent, which is acceptable for most engineering purposes (Fig. 4–54). A bigger chart, of course, would give better resolution and reduce the reading errors. Notice that we did not have to determine $Z$ in this problem since we could read $P_R$ directly from the chart.

| | P, psia |
|---|---|
| Exact | 1000 |
| Z chart | 1056 |
| Ideal gas | 1228 |
| (from Example 4-12) | |

**FIGURE 4–54**
Results obtained by using the compressibility chart are usually within a few percent of actual values.

The ideal-gas equation of state is very simple, but its range of applicability is limited. It is desirable to have equations of state that represent the $P\text{-}v\text{-}T$ behavior of substances accurately over a larger region with no limitations. Such equations are naturally more complicated. Several equations have been proposed for this purpose including the *van der Waals* equation which is one of the earliest, the *Beattie-Bridgeman* equation of state because which is one of the best known and is reasonably accurate, and the *Benedict-Webb-Rubin* equation because which is one of the more recent and is very accurate.

## SUMMARY

A substance that has a fixed chemical composition throughout is called a *pure substance*. A pure substance exists in different phases depending on its energy level. In the liquid phase, a substance that is not about to vaporize is called a *compressed* or *subcooled liquid*. In the gas phase, a substance that is not about to condense is called a *superheated vapor*.

During a phase-change process, the temperature and pressure of a pure substance are dependent properties. At a given pressure, a substance changes phase at a fixed temperature, called the *saturation temperature*. Likewise, at a given temperature, the pressure at which a substance changes phase is called the *saturation pressure*. During a boiling process, both the liquid and the vapor phases coexist in equilibrium, and under this condition the liquid is called *saturated liquid* and the vapor *saturated vapor*.

In a saturated liquid–vapor mixture, the mass fraction of vapor is called the *quality* and is expressed as

$$x = \frac{m_{vapor}}{m_{total}}$$

Quality may have values between 0 (saturated liquid) and 1 (saturated vapor). It has no meaning in the compressed liquid or superheated vapor regions. In the saturated mixture region, the average value of any intensive property $y$ is determined from

$$y = y_f + xy_{fg}$$

where $f$ stands for saturated liquid and $g$ for saturated vapor.

In the absence of compressed liquid data, a general approximation is to treat a compressed liquid as a saturated liquid at the given *temperature,*

$$y \cong y_{f @ T}$$

where $y$ stands for $v$, $u$, or $h$.

The state beyond which there is no distinct vaporization process is called the *critical point*. At supercritical pressures, a substance gradually and uniformly expands from the liquid to vapor phase. All three phases of a substance coexist in equilibrium at states along the *triple line* characterized by triple-line temperature and pressure. The compressed liquid has lower $v$, $u$, and $h$ values than the saturated liquid at the same $T$ or $P$. Likewise, superheated vapor has higher $v$, $u$, and $h$ values than the saturated vapor at the same $T$ or $P$.

Any relation among the pressure, temperature, and specific volume of a substance is called an *equation of state*. The simplest and best-known equation of state is the *ideal-gas equation of state,* given as

$$Pv = RT$$

where $R$ is the gas constant. Caution should be exercised in using this relation since an ideal gas is a fictitious substance. Real gases exhibit ideal-gas behavior at relatively low pressures and high temperatures.

The deviation from ideal-gas behavior can be properly accounted for by using the *compressibility factor $Z$*, defined as

$$Z = \frac{Pv}{RT} \quad \text{or} \quad Z = \frac{v_{actual}}{v_{ideal}}$$

The $Z$ factor is approximately the same for all gases at the same *reduced temperature* and *reduced pressure,* which are defined as

$$T_R = \frac{T}{T_{cr}} \quad \text{and} \quad P_R = \frac{P}{P_{cr}}$$

where $P_{cr}$ and $T_{cr}$ are the critical pressure and temperature, respectively. This is known as the *principle of corresponding states*. When either $P$ or $T$ is unknown, it can be determined from the compressibility chart with the help of the *pseudo-reduced specific volume,* defined as

$$v_R = \frac{v_{actual}}{RT_{cr}/P_{cr}}$$

The $P$-$v$-$T$ behavior of substances can be represented more accurately by more complex equations of state.

## REFERENCES AND SUGGESTED READINGS

1. ASHRAE *Handbook of Fundamentals*. SI version. Atlanta, GA: American Society of Heating, Refrigerating, and Air-Conditioning Engineers, Inc., 1993.

2. ASHRAE *Handbook of Refrigeration*. SI version. Atlanta, GA: American Society of Heating, Refrigerating, and Air-Conditioning Engineers, Inc., 1994.

3. A. Bejan. *Advanced Engineering Thermodynamics*. 3rd ed. New York: Wiley, 2006.

4. M. Kostic. *Analysis of Enthalpy Approximation for Compressed Liquid Water*. Journal of Heat Transfer, Vol. 128, p. 421–426, May 2006.

# PROBLEMS*

## Pure Substances, Phase-Change Processes, Property Diagrams

**4–1C**  What is the difference between saturated liquid and compressed liquid?

**4–2C**  What is the difference between saturated vapor and superheated vapor?

**4–3C**  Is there any difference between the intensive properties of saturated vapor at a given temperature and the vapor of a saturated mixture at the same temperature?

**4–4C**  If the pressure of a substance is increased during a boiling process, will the temperature also increase or will it remain constant? Why?

**4–5C**  Why are the temperature and pressure dependent properties in the saturated mixture region?

**4–6C**  What is the difference between the critical point and the triple point?

**4–7C**  Is it possible to have water vapor at $-10°C$?

**4–8C**  A househusband is cooking beef stew for his family in a pan that is (a) uncovered, (b) covered with a light lid, and (c) covered with a heavy lid. For which case will the cooking time be the shortest? Why?

**4–9C**  How does the boiling process at supercritical pressures differ from the boiling process at subcritical pressures?

## Property Tables

**4–10C**  A perfectly fitting pot and its lid often stick after cooking, and it becomes very difficult to open the lid when the pot cools down. Explain why this happens and what you would do to open the lid.

**4–11C**  It is well known that warm air in a cooler environment rises. Now consider a warm mixture of air and gasoline on top of an open gasoline can. Do you think this gas mixture will rise in a cooler environment?

**4–12C**  Does the amount of heat absorbed as 1 kg of saturated liquid water boils at 100°C have to be equal to the amount of heat released as 1 kg of saturated water vapor condenses at 100°C?

---

* Problems designated by a "C" are concept questions, and students are encouraged to answer them all. Problems designated by an "E" are in English units, and the SI users can ignore them. Problems with the ● icon are solved using EES, and complete solutions together with parametric studies are included on the enclosed DVD. Problems with the ▣ icon are comprehensive in nature, and are intended to be solved with a computer, preferably using the EES software that accompanies this text.

**4–13C**  Does the reference point selected for the properties of a substance have any effect on thermodynamic analysis? Why?

**4–14C**  What is the physical significance of $h_{fg}$? Can it be obtained from a knowledge of $h_f$ and $h_g$? How?

**4–15C**  Does $h_{fg}$ change with pressure? How?

**4–16C**  Is it true that it takes more energy to vaporize 1 kg of saturated liquid water at 100°C than it would at 120°C?

**4–17C**  What is quality? Does it have any meaning in the superheated vapor region?

**4–18C**  Which process requires more energy: completely vaporizing 1 kg of saturated liquid water at 1 atm pressure or completely vaporizing 1 kg of saturated liquid water at 8 atm pressure?

**4–19C**  Can quality be expressed as the ratio of the volume occupied by the vapor phase to the total volume? Explain.

**4–20C**  In the absence of compressed liquid tables, how is the specific volume of a compressed liquid at a given $P$ and $T$ determined?

**4–21C**  In 1775, Dr. William Cullen made ice in Scotland by evacuating the air in a water tank. Explain how that device works, and discuss how the process can be made more efficient.

**4–22**  Complete this table for $H_2O$:

| $T$, °C | $P$, kPa | $v$, m³/kg | Phase description |
|---|---|---|---|
| 50 | | 7.72 | |
| | 400 | | Saturated vapor |
| 250 | 500 | | |
| 110 | 350 | | |

**4–23**  Reconsider Prob. 4–22. Using EES (or other) software, determine the missing properties of water. Repeat the solution for refrigerant-134a, refrigerant-22, and ammonia.

**4–24E**  Complete this table for $H_2O$:

| $T$, °F | $P$, psia | $u$, Btu/lbm | Phase description |
|---|---|---|---|
| 300 | | 782 | |
| | 40 | | Saturated liquid |
| 500 | 120 | | |
| 400 | 400 | | |

**4–25E**  Reconsider Prob. 4–24E. Using EES (or other) software, determine the missing properties of water. Repeat the solution for refrigerant-134a, refrigerant-22, and ammonia.

**4–26** Complete this table for $H_2O$:

| T, °C | P, kPa | h, kJ/kg | x | Phase description |
|---|---|---|---|---|
| | 200 | | 0.7 | |
| 140 | | 1800 | | |
| | 950 | | 0.0 | |
| 80 | 500 | | | |
| | 800 | 3162.2 | | |

**4–27** Complete this table for refrigerant-134a:

| T, °C | P, kPa | v, m³/kg | Phase description |
|---|---|---|---|
| −12 | 320 | | |
| 30 | | 0.0065 | |
| | 550 | | Saturated vapor |
| 60 | 600 | | |

**4–28** If sufficient data are provided, complete the blank cells in the following table of properties of water. In the last column describe the condition of water as compressed liquid, saturated mixture, superheated vapor, or insufficient information; and, if applicable, give the quality.

| P, kPa | T, °C | v, m³/kg | h, kJ/kg | Condition description and quality (if applicable) |
|---|---|---|---|---|
| 200 | | | 2706.3 | |
| | 130 | | | 0.650 |
| | 400 | | 3277.0 | |
| 800 | 30 | | | |
| 450 | 147.90 | | | |

**4–29E** Complete this table for refrigerant-134a:

| T, °F | P, psia | h, Btu/lbm | x | Phase description |
|---|---|---|---|---|
| | 80 | 78 | | |
| 15 | | | 0.6 | |
| 10 | 70 | | | |
| | 180 | 129.46 | | |
| 110 | | | 1.0 | |

**4–30** A piston–cylinder device contains 0.85 kg of refrigerant-134a at −10°C. The piston that is free to move has a mass of 12 kg and a diameter of 25 cm. The local atmospheric pressure is 88 kPa. Now, heat is transferred to refrigerant-134a until the temperature is 15°C. Determine (a) the final pressure, (b) the change in the volume of the cylinder, and (c) the change in the enthalpy of the refrigerant-134a.

R-134a
0.85 kg
−10°C

Q

**FIGURE P4–30**

**4–31E** R-134a. whose specific volume is 0.4618 ft³/lbm, flows through a tube at 120 psia. What is the temperature in the tube?

**4–32** One kilogram of water fills a 150-L rigid container at an initial pressure of 2 MPa. The container is then cooled to 40°C. Determine the initial temperature and the final pressure of the water.

$H_2O$
1 kg
150 L
2 MPa

Q

**FIGURE P4–32**

**4–33** 10-kg of R-134a fill a 1.348-m³ rigid container at an initial temperature of −40°C. The container is then heated until the pressure is 200 kPa. Determine the final temperature and the initial pressure. *Answer:* 66.3°C, 51.25 kPa

**4–34E** 2-lbm of water at 500 psia intially fill the 1.5-ft³ left chamber of a partitioned system. The right chamber's volume is also 1.5 ft³, and it is initially evacuated. The partition is now ruptured, and heat is transferred to the water until its temperature is 300°F. Determine the final pressure of water, in psia, and the total internal energy, in Btu, at the final state.

Water
2 lbm
1.5 ft³
500 psia

Evacuated
1.5 ft³

**FIGURE P4–34E**

**4–35** A 9-m³ container is filled with 300 kg of R-134a at 10°C. What is the specific enthalpy of the R-134a in the container?

**4–36** Refrigerant-134a at 200 kPa and 25°C flows through a refrigeration line. Determine its specific volume.

**4–37E** A spring-loaded piston-cylinder device is initially filled with 0.2 lbm of an R-134a liquid-vapor mixture whose temperature is −30°F and whose quality is 80 percent. The spring constant in the spring force relation $F = kx$ is 37 lbf/in, and the piston diameter is 12 in. The R-134a undergoes a process that increases its volume by 40 percent. Calculate the final temperature and enthalpy of the R-134a. *Answers:* 81.5°F, 120 Btu/lbm

**FIGURE P4–37E**

**4–38E** One pound-mass of water fills a 2.4264-ft³ weighted piston-cylinder device at a temperature of 600°F. The piston-cylinder device is now cooled until its temperature is 200°F. Determine the final pressure of water, in psia, and the volume, in ft³. *Answers:* 250 psia, 0.01663 ft³

**4–39** Ten kilograms of R-134a fill a 1.595-m³ weighted piston-cylinder device at a temperature of −26.4°C. The container is now heated until the temperature is 100°C. Determine the final volume of the R-134a.

**FIGURE P4–39**

**4–40E** One pound-mass of water fills a container whose volume is 2 ft³. The pressure in the container is 100 psia. Calculate the total internal energy and enthalpy in the container. *Answers:* 661 Btu, 698 Btu

**4–41** Three kilograms of water in a container have a pressure of 100 kPa and temperature of 250°C. What is the volume of this container?

**4–42** 10-kg of R-134a at 300 kPa fills a rigid container whose volume is 14 L. Determine the temperature and total enthalpy in the container. The container is now heated until the pressure is 600 kPa. Determine the temperature and total enthalpy when the heating is completed.

**FIGURE P4–42**

**4–43** 100-kg of R-134a at 200 kPa are contained in a piston-cylinder device whose volume is 12.322 m³. The piston is now moved until the volume is one-half its original size. This is done such that the pressure of the R-134a does not change. Determine the final temperature and the change in the total internal energy of the R-134a.

**4–44** Water initially at 300 kPa and 250°C is contained in a piston-cylinder device fitted with stops. The water is allowed to cool at constant pressure until it exists as a saturated vapor and the piston rests on the stops. Then the water continues to cool until the pressure is 100 kPa. On the *T-v* diagrams sketch, with respect to the saturation lines, the process curves passing through both the initial, intermediate, and final states of the water. Label the *T*, *P*, and *v* values for end states on the process curves. Find the overall change in internal energy between the initial and final states per unit mass of water.

**FIGURE P4–44**

**4–45E** The atmospheric pressure at a location is usually specified at standard conditions, but it changes with the weather conditions. As the weather forecasters frequently state, the atmospheric pressure drops during stormy weather and it rises during clear and sunny days. If the pressure difference between the two extreme conditions is given to be 0.2 in of mercury, determine how much the boiling temperatures of water will vary as the weather changes from one extreme to the other.

**4–46** A person cooks a meal in a 30-cm-diameter pot that is covered with a well-fitting lid and lets the food cool to the room temperature of 20°C. The total mass of the food and the pot is 8 kg. Now the person tries to open the pan by lifting the lid up. Assuming no air has leaked into the pan during cooling, determine if the lid will open or the pan will move up together with the lid.

**4–47** Water is boiled at 1 atm pressure in a 25-cm-internal-diameter stainless steel pan on an electric range. If it is observed that the water level in the pan drops by 10 cm in 45 min, determine the rate of heat transfer to the pan.

**4–48** Repeat Prob. 4–47 for a location at 2000-m elevation where the standard atmospheric pressure is 79.5 kPa.

**4–49** Saturated steam coming off the turbine of a steam power plant at 40°C condenses on the outside of a 3-cm-outer-diameter, 35-m-long tube at a rate of 130 kg/h. Determine the rate of heat transfer from the steam to the cooling water flowing through the pipe.

**4–50** Water in a 5-cm-deep pan is observed to boil at 98°C. At what temperature will the water in a 40-cm-deep pan boil? Assume both pans are full of water.

**4–51** Water is being heated in a vertical piston–cylinder device. The piston has a mass of 20 kg and a cross-sectional area of 100 cm². If the local atmospheric pressure is 100 kPa, determine the temperature at which the water starts boiling.

**4–52** A rigid tank with a volume of 1.8 m³ contains 15 kg of saturated liquid–vapor mixture of water at 90°C. Now the water is slowly heated. Determine the temperature at which the liquid in the tank is completely vaporized. Also, show the process on a *T-v* diagram with respect to saturation lines. *Answer:* 202.9°C

**4–53** A piston–cylinder device contains 0.005 m³ of liquid water and 0.9 m³ of water vapor in equilibrium at 600 kPa. Heat is transferred at constant pressure until the temperature reaches 200°C.

(a) What is the initial temperature of the water?
(b) Determine the total mass of the water.
(c) Calculate the final volume.
(d) Show the process on a *P-v* diagram with respect to saturation lines.

**FIGURE P4–53**

**4–54** Reconsider Prob. 4–53. Using EES (or other) software, investigate the effect of pressure on the total mass of water in the tank. Let the pressure vary from 0.1 MPa to 1 MPa. Plot the total mass of water against pressure, and discuss the results. Also, show the process in Prob. 4–53 on a *P-v* diagram using the property plot feature of EES.

**4–55E** Superheated water vapor at 180 psia and 500°F is allowed to cool at constant volume until the temperature drops to 250°F. At the final state, determine (a) the pressure, (b) the quality, and (c) the enthalpy. Also, show the process on a *T-v* diagram with respect to saturation lines. *Answers:* (a) 29.84 psia, (b) 0.219, (c) 426.0 Btu/lbm

**4–56E** Reconsider Prob. 4–55E. Using EES (or other) software, investigate the effect of initial pressure on the quality of water at the final state. Let the pressure vary from 100 psi to 300 psi. Plot the quality against initial pressure, and discuss the results. Also, show the process in Prob. 4–55E on a *T-v* diagram using the property plot feature of EES.

**4–57** Determine the specific volume, internal energy, and enthalpy of compressed liquid water at 80°C and 20 MPa using the saturated liquid approximation. Compare these values to the ones obtained from the compressed liquid tables.

**4–58** Reconsider Prob. 4–57. Using EES (or other) software, determine the indicated properties of compressed liquid, and compare them to those obtained using the saturated liquid approximation.

**4–59** A piston–cylinder device contains 0.8 kg of steam at 300°C and 1 MPa. Steam is cooled at constant pressure until one-half of the mass condenses.

(a) Show the process on a *T-v* diagram.
(b) Find the final temperature.
(c) Determine the volume change.

**4–60** A rigid tank contains water vapor at 250°C and an unknown pressure. When the tank is cooled to 124°C, the vapor starts condensing. Estimate the initial pressure in the tank. *Answer:* 0.30 MPa

**4–61** A piston-cylinder device initially contains 1.4-kg saturated liquid water at 200°C. Now heat is transferred to the water until the volume quadruples and the cylinder contains saturated vapor only. Determine (a) the volume of the tank, (b) the final temperature and pressure, and (c) the internal energy change of the water.

**FIGURE P4–61**

**4–62E** How much error would one expect in determining the specific enthalpy by applying the incompressible-liquid approximation to water at 1500 psia and 400°F?

**4–63** How much error would result in calculating the specific volume and enthalpy of water at 20 MPa and 140°C by using the incompressible-liquid approximation?

**4–64** 100 grams of R-134a initially fill a weighted piston-cylinder device at 60 kPa and −20°C. The device is then heated until the temperature is 100°C. Determine the change in the device's volume as a result of the heating. *Answer:* 0.0168 m³

**FIGURE P4–64**

**4–65** A rigid vessel contains 2 kg of refrigerant-134a at 800 kPa and 120°C. Determine the volume of the vessel and the total internal energy. *Answers:* 0.0753 m³, 655.7 kJ

**4–66** A 0.5-m³ vessel contains 10 kg of refrigerant-134a at −20°C. Determine (a) the pressure, (b) the total internal energy, and (c) the volume occupied by the liquid phase. *Answers:* (a) 132.82 kPa, (b) 904.2 kJ, (c) 0.00489 m³

**4–67** Pressure-enthalpy state diagrams are useful for studying refrigeration and like systems. Using EES (or other) software and actual property data, plot a pressure-enthalpy diagram for R-134a that includes the saturation lines. Also, sketch isothermal and constant-entropy processes on this diagram.

**4–68** A 0.3-m³ rigid vessel initially contains saturated liquid-vapor mixture of water at 150°C. The water is now heated until it reaches the critical state. Determine the mass of the liquid water and the volume occupied by the liquid at the initial state. *Answers:* 96.1 kg, 0.105 m³

## Ideal Gas

**4–69C** What is the difference between mass and molar mass? How are these two related?

**4–70C** Under what conditions is the ideal-gas assumption suitable for real gases?

**4–71C** What is the difference between $R$ and $R_u$? How are these two related?

**4–72C** Propane and methane are commonly used for heating in winter, and the leakage of these fuels, even for short periods, poses a fire danger for homes. Which gas leakage do you think poses a greater risk for fire? Explain.

**4–73** What is the specific volume of nitrogen at 300 kPa and 227°C?

**4–74E** A 3-ft³ container is filled with 2-lbm of oxygen at a pressure of 80 psia. What is the temperature of the oxygen?

**4–75** A 2-kg mass of helium is maintained at 300 kPa and 27°C in a rigid container. How large is the container, in m³?

**4–76** A spherical balloon with a diameter of 9 m is filled with helium at 27°C and 200 kPa. Determine the mole number and the mass of the helium in the balloon. *Answers:* 30.6 kmol, 123 kg

**4–77** Reconsider Prob. 4–76. Using EES (or other) software, investigate the effect of the balloon diameter on the mass of helium contained in the balloon for the pressures of (a) 100 kPa and (b) 200 kPa. Let the diameter vary from 5 m to 15 m. Plot the mass of helium against the diameter for both cases.

**4–78** A 1-m³ tank containing air at 25°C and 500 kPa is connected through a valve to another tank containing 5 kg of air at 35°C and 200 kPa. Now the valve is opened, and the entire system is allowed to reach thermal equilibrium with the surroundings, which are at 20°C. Determine the volume of the second tank and the final equilibrium pressure of air. *Answers:* 2.21 m³, 284.1 kPa

**4–79E** 2.3 lbmol of air at 65°F and 32 psia is contained in an elastic tank. What is the volume of the tank? If the volume is doubled at the same pressure, determine the final temperature.

**4–80** A rigid tank contains an ideal gas at 1227°C and 200 kPa gage. The gas is cooled until the gage pressure reads 50 kPa. If the atmospheric pressure is 100 kPa, determine the final temperature of the gas.

**4–81** A rigid tank whose volume is unknown is divided into two parts by a partition. One side of the tank contains an ideal gas at 927°C. The other side is evacuated and has a volume twice the size of the part containing the gas. The partition is now removed and the gas expands to fill the entire tank. Heat is now applied to the gas until the pressure equals the initial pressure. Determine the final temperature of the gas.   *Answer:* 3327°C

**FIGURE P4–81**

**4–82** Argon in the amount of 0.6 kg fills a 0.05-m³ piston-cylinder device at 550 kPa. The piston is now moved by changing the weights until the volume is twice its original size. During this process, argon's temperature is maintained constant. Determine the final pressure in the device.

**4–83** The pressure in an automobile tire depends on the temperature of the air in the tire. When the air temperature is 25°C, the pressure gage reads 210 kPa. If the volume of the tire is 0.025 m³, determine the pressure rise in the tire when the air temperature in the tire rises to 50°C. Also, determine the amount of air that must be bled off to restore pressure to its original value at this temperature. Assume the atmospheric pressure is 100 kPa.

$V = 0.025$ m³
$T = 25$°C
$P_g = 210$ kPa

AIR

**FIGURE P4–83**

## Compressibility Factor

**4–84C** What is the physical significance of the compressibility factor $Z$?

**4–85C** What is the principle of corresponding states?

**4–86C** How are the reduced pressure and reduced temperature defined?

**4–87** Determine the specific volume of superheated water vapor at 15 MPa and 350°C, using (*a*) the ideal-gas equation, (*b*) the generalized compressibility chart, and (*c*) the steam tables. Also determine the error involved in the first two cases. *Answers:* (*a*) 0.01917 m³/kg, 67.0 percent, (*b*) 0.01246 m³/kg, 8.5 percent, (*c*) 0.01148 m³/kg

**4–88** Reconsider Prob. 4–87. Solve the problem using the generalized compressibility factor feature of the EES software. Again using EES, compare the specific volume of water for the three cases at 15 MPa over the temperature range of 350 to 600°C in 25°C intervals. Plot the percent error involved in the ideal-gas approximation against temperature, and discuss the results.

**4–89** Determine the specific volume of refrigerant-134a vapor at 0.9 MPa and 70°C based on (*a*) the ideal-gas equation, (*b*) the generalized compressibility chart, and (*c*) data from tables. Also, determine the error involved in the first two cases.

**4–90** Determine the specific volume of superheated water vapor at 3.5 MPa and 450°C based on (*a*) the ideal-gas equation, (*b*) the generalized compressibility chart, and (*c*) the steam tables. Determine the error involved in the first two cases.

**4–91E** Ethane in a rigid vessel is to be heated from 80 psia and 100°F until its temperature is 540°F. What is the final pressure of the ethane as predicted by the compressibility chart?

**4–92** Ethylene is heated at constant pressure from 5 MPa and 20°C to 200°C. Using the compressibility chart, determine the change in the ethylene's specific volume as a result of this heating.   *Answer:* 0.0172 m³/kg

**4–93** Saturated water vapor at 350°C is heated at constant pressure until its volume has doubled. Determine the final temperature using the ideal gas equation of state, the compressibility charts, and the steam tables.

**4–94E** Saturated water vapor at 400°F is heated at constant pressure until its volume has doubled. Determine the final temperature using the ideal gas equation of state, the compressibility charts, and the steam tables.

**4–95** Methane at 8 MPa and 300 K is heated at constant pressure until its volume has increased by 50 percent. Determine the final temperature using the ideal gas equation of state and the compressibility factor. Which of these two results is more accurate?

**4–96** What is the percentage of error involved in treating carbon dioxide at 5 MPa and 25°C as an ideal gas?   *Answer:* 45 percent

**4–97** Carbon dioxide gas enters a pipe at 3 MPa and 500 K at a rate of 2 kg/s. $CO_2$ is cooled at constant pressure as it flows in the pipe and the temperature of $CO_2$ drops to 450 K at the exit. Determine the volume flow rate and the density of carbon dioxide at the inlet and the volume flow rate at the exit of the pipe using (a) the ideal-gas equation and (b) the generalized compressibility chart. Also, determine (c) the error involved in the first case.

3 MPa
500 K
2 kg/s

$CO_2$ ⟶

450 K

**FIGURE P4–97**

**4–98** Determine the specific volume of nitrogen gas at 10 MPa and 150 K based on (a) the idea-gas equation and (b) the generalized compressibility chart. Compare these results with the experimental value of 0.002388 m³/kg, and determine the error involved in each case. *Answers:* (a) 0.004452 m³/kg, 86.4 percent; (b) 0.002404 m³/kg, 0.7 percent

## Review Problems

**4–99** A rigid tank contains nitrogen gas at 227°C and 100 kPa gage. The gas is heated until the gage pressure reads 250 kPa. If the atmospheric pressure is 100 kPa, determine the final temperature of the gas, in °C.

$P_{atm} = 100$ kPa

Nitrogen gas
227°C
100 kPa (gage)

Q

**FIGURE P4–99**

**4–100** Carbon-dioxide gas at 3 MPa and 500 K flows steadily in a pipe at a rate of 0.4 kmol/s. Determine (a) the volume and mass flow rates and the density of carbon dioxide at this state. If $CO_2$ is cooled at constant pressure as it flows in the pipe so that the temperature of $CO_2$ drops to 450 K at the exit of the pipe, determine (b) the volume flow rate at the exit of the pipe.

3 MPa
500 K
0.4 kmol/s

$CO_2$ ⟶

450 K

**FIGURE P4–100**

**4–101** The combustion in a gasoline engine may be approximated by a constant volume heat addition process. There exists the air–fuel mixture in the cylinder before the combustion and the combustion gases after it, and both may be approximated as air, an ideal gas. In a gasoline engine, the cylinder conditions are 1.4 MPa and 450°C before the combustion and 1600°C after it. Determine the pressure at the end of the combustion process. *Answer:* 3.63 MPa

Combustion
chamber
1.4 MPa
450°C

**FIGURE P4–101**

**4–102** Combustion in a diesel engine may be modeled as a constant-pressure heat addition process with air in the cylinder before and after combustion. Consider a diesel engine with cylinder conditions of 950 K and 75 cm³ before combustion, and 150 cm³ after it. The engine operates with an air–fuel ratio of 22 kg air/kg fuel (the mass of the air divided by the mass of the fuel). Determine the temperature after the combustion process.

Combustion
chamber
950 K
75 cm³

**FIGURE P4–102**

**4–103** One kilogram of R-134a fills a 0.090 m³ rigid container at an initial temperature of −40°C. The container is then heated until the pressure is 280 kPa. Determine the initial pressure and final temperature. *Answers:* 51.25 kPa, 50°C

**4–104E** One pound-mass of water fills a 2.649 ft³ weighted piston-cylinder device at a temperature of 400°F. The piston-cylinder device is now cooled until its temperature is 100°F. Determine the final pressure and volume of the water.

**FIGURE P4–104E**

**4–105** Ethane at 10 MPa and 100°C is heated at constant pressure until its volume has increased by 60 percent. Determine the final temperature using (a) the ideal gas equation of state and (b) the compressibility factor. Which of these two results is the more accurate?

**4–106** A 20-m³ tank contains nitrogen at 23°C and 600 kPa. Some nitrogen is allowed to escape until the pressure in the tank drops to 400 kPa. If the temperature at this point is 20°C, determine the amount of nitrogen that has escaped. *Answer: 44.6 kg*

**4–107** A 4-L rigid tank contains 2 kg of saturated liquid–vapor mixture of water at 50°C. The water is now slowly heated until it exists in a single phase. At the final state, will the water be in the liquid phase or the vapor phase? What would your answer be if the volume of the tank were 400 L instead of 4 L?

**FIGURE P4–107**

**4–108** The gage pressure of an automobile tire is measured to be 200 kPa before a trip and 220 kPa after the trip at a location where the atmospheric pressure is 90 kPa. Assuming the volume of the tire remains constant at 0.035 m³, determine the percent increase in the absolute temperature of the air in the tire.

**4–109** Steam at 400°C has a specific volume of 0.02 m³/kg. Determine the pressure of the steam based on (a) the ideal-gas equation, (b) the generalized compressibility chart, and (c) the steam tables. *Answers: (a) 15,529 kPa, (b) 12,574 kPa, (c) 12,515 kPa*

**4–110** A tank whose volume is unknown is divided into two parts by a partition. One side of the tank contains 0.03 m³

of refrigerant-134a that is a saturated liquid at 1.2 MPa, while the other side is evacuated. The partition is now removed, and the refrigerant fills the entire tank. If the final state of the refrigerant is 30°C and 400 kPa, determine the volume of the tank.

**FIGURE P4–110**

**4–111** Reconsider Prob. 4–110. Using EES (or other) software, investigate the effect of the initial pressure of refrigerant-134a on the volume of the tank. Let the initial pressure vary from 0.5 to 1.5 MPa. Plot the volume of the tank versus the initial pressure, and discuss the results.

**4–112** Liquid propane is commonly used as a fuel for heating homes, powering vehicles such as forklifts, and filling portable picnic tanks. Consider a propane tank that initially contains 5 L of liquid propane at the environment temperature of 20°C. If a hole develops in the connecting tube of a propane tank and the propane starts to leak out, determine the temperature of propane when the pressure in the tank drops to 1 atm. Also, determine the total amount of heat transfer from the environment to the tank to vaporize the entire propane in the tank.

**FIGURE P4–112**

**4–113** Repeat Prob. 4–112 for isobutane.

**4–114** A tank contains helium at 77°C and 110 kPa gage. The helium is heated in a process by heat transfer from the surroundings such that the helium reaches a final equilibrium state at 300°C. Determine the final gage pressure of the helium. Assume atmospheric pressure is 100 kPa.

**4–115** If sufficient data are provided, complete the blank cells in the following table of properties of water. In the last column describe the condition of water as compressed liquid, saturated mixture, superheated vapor, or insufficient information; and, if applicable, give the quality.

| P, kPa | T, °C | v, m³/kg | u, kJ/kg | Condition description and quality (if applicable) |
|--------|-------|----------|----------|-------------------------------------------------|
|        | 250   |          | 2728.9   |                                                 |
| 300    |       |          | 1560.0   |                                                 |
| 101.42 | 100   |          |          |                                                 |
| 3000   | 180   |          |          |                                                 |

**4–116** If sufficient data are provided, complete the blank cells in the following table of properties of water. In the last column describe the condition of water as compressed liquid, saturated mixture, superheated vapor, or insufficient information; and, if applicable, give the quality.

| P, kPa | T, °C | v, m³/kg | u, kJ/kg | Condition description and quality (if applicable) |
|--------|-------|----------|----------|-------------------------------------------------|
| 200    |       |          | 2529.1   |                                                 |
|        | 125   |          |          | x = 0.650                                       |
|        | 400   |          | 2967.2   |                                                 |
| 1000   | 30    |          |          |                                                 |
| 120.90 | 150   |          |          |                                                 |

**4–117** Water initially at 300 kPa and 250°C is contained in a constant volume tank. The water is allowed to cool until its pressure is 150 kPa. On the P-v and T-v diagrams sketch, with respect to the saturation lines, the process curve passing through both the initial and final states of the water. Label the end states on the process curve. Also, on both the P-v and T-v diagrams, sketch the isotherms passing through both states and show their values, in °C, on the isotherms.

### Design and Essay Problems

**4–118** In an article on tire maintenance, it is stated that tires lose air over time, and pressure losses as high as 90 kPa (13 psi) per year are measured. The article recommends checking tire pressure at least once a month to avoid low tire pressure that hurts fuel efficiency and causes uneven thread wear on tires. Taking the beginning tire pressure to be 220 kPa (gage) and the atmospheric pressure to be 100 kPa, determine the fraction of air that can be lost from a tire per year.

**4–119** It is well known that water freezes at 0°C at atmospheric pressure. The mixture of liquid water and ice at 0°C is said to be at stable equilibrium since it cannot undergo any changes when it is isolated from its surroundings. However, when water is free of impurities and the inner surfaces of the container are smooth, the temperature of water can be lowered to −2°C or even lower without any formation of ice at atmospheric pressure. But at that state even a small disturbance can initiate the formation of ice abruptly, and the water temperature stabilizes at 0°C following this sudden change. The water at −2°C is said to be in a *metastable state*. Write an essay on metastable states and discuss how they differ from stable equilibrium states.

# ENERGY ANALYSIS OF CLOSED SYSTEMS

I n Chap. 3, we considered various forms of energy and energy transfer, and we developed a general relation for the conservation of energy principle or energy balance. Then in Chap. 4, we learned how to determine the thermodynamics properties of substances. In this chapter, we apply the energy balance relation to systems that do not involve any mass flow across their boundaries; that is, closed systems.

We start this chapter with a discussion of the *moving boundary work* or *P dV work* commonly encountered in reciprocating devices such as automotive engines and compressors. We continue by applying the *general energy balance* relation, which is simply expressed as $E_{in} - E_{out} = \Delta E_{system}$, to systems that involve pure substance. Then we define *specific heats*, obtain relations for the internal energy and enthalpy of *ideal gases* in terms of specific heats and temperature changes, and perform energy balances on various systems that involve ideal gases. We repeat this for systems that involve solids and liquids, which are approximated as *incompressible substances*.

## Objectives

Objectives of this chapter are to:

- Examine the moving boundary work or *P dV* work commonly encountered in reciprocating devices such as automotive engines and compressors.

- Identify the first law of thermodynamics as simply a statement of the conservation of energy principle for closed (fixed mass) systems.

- Develop the general energy balance applied to closed systems.

- Define the specific heat at constant volume and the specific heat at constant pressure.

- Relate the specific heats to the calculation of the changes in internal energy and enthalpy of ideal gases.

- Describe incompressible substances and determine the changes in their internal energy and enthalpy.

- Solve energy balance problems for closed (fixed mass) systems that involve heat and work interactions for general pure substances, ideal gases, and incompressible substances.

**FIGURE 5–1**

The work associated with a moving boundary is called *boundary work.*

**FIGURE 5–2**

A gas does a differential amount of work $\delta W_b$ as it forces the piston to move by a differential amount $ds$.

# 5–1 ▪ MOVING BOUNDARY WORK

One form of mechanical work frequently encountered in practice is associated with the expansion or compression of a gas in a piston–cylinder device. During this process, part of the boundary (the inner face of the piston) moves back and forth. Therefore, the expansion and compression work is often called **moving boundary work**, or simply **boundary work** (Fig. 5–1). Some call it the $P\,dV$ work for reasons explained later. Moving boundary work is the primary form of work involved in *automobile engines*. During their expansion, the combustion gases force the piston to move, which in turn forces the crankshaft to rotate.

The moving boundary work associated with real engines or compressors cannot be determined exactly from a thermodynamic analysis alone because the piston usually moves at very high speeds, making it difficult for the gas inside to maintain equilibrium. Then the states through which the system passes during the process cannot be specified, and no process path can be drawn. Work, being a path function, cannot be determined analytically without a knowledge of the path. Therefore, the boundary work in real engines or compressors is determined by direct measurements.

In this section, we analyze the moving boundary work for a *quasi-equilibrium process*, a process during which the system remains nearly in equilibrium at all times. A quasi-equilibrium process, also called a *quasi-static process*, is closely approximated by real engines, especially when the piston moves at low velocities. Under identical conditions, the work output of the engines is found to be a maximum, and the work input to the compressors to be a minimum when quasi-equilibrium processes are used in place of nonquasi-equilibrium processes. Below, the work associated with a moving boundary is evaluated for a quasi-equilibrium process.

Consider the gas enclosed in the piston–cylinder device shown in Fig. 5–2. The initial pressure of the gas is $P$, the total volume is $V$, and the cross-sectional area of the piston is $A$. If the piston is allowed to move a distance $ds$ in a quasi-equilibrium manner, the differential work done during this process is

$$\delta W_b = F\,ds = PA\,ds = P\,dV \tag{5–1}$$

That is, the boundary work in the differential form is equal to the product of the absolute pressure $P$ and the differential change in the volume $dV$ of the system. This expression also explains why the moving boundary work is sometimes called the $P\,dV$ work.

Note in Eq. 5–1 that $P$ is the absolute pressure, which is always positive. However, the volume change $dV$ is positive during an expansion process (volume increasing) and negative during a compression process (volume decreasing). Thus, the boundary work is positive during an expansion process and negative during a compression process. Therefore, Eq. 5–1 can be viewed as an expression for boundary work output, $W_{b,\text{out}}$. A negative result indicates boundary work input (compression).

The total boundary work done during the entire process as the piston moves is obtained by adding all the differential works from the initial state to the final state:

$$W_b = \int_1^2 P\,dV \qquad (\text{kJ}) \tag{5–2}$$

This integral can be evaluated only if we know the functional relationship between $P$ and $V$ during the process. That is, $P = f(V)$ should be available. Note that $P = f(V)$ is simply the equation of the process path on a $P$-$V$ diagram.

The quasi-equilibrium expansion process described is shown on a $P$-$V$ diagram in Fig. 5–3. On this diagram, the differential area $dA$ is equal to $P\,dV$, which is the differential work. The total area $A$ under the process curve 1–2 is obtained by adding these differential areas:

$$\text{Area} = A = \int_1^2 dA = \int_1^2 P\,dV \tag{5–3}$$

A comparison of this equation with Eq. 5–2 reveals that *the area under the process curve on a $P$-$V$ diagram is equal, in magnitude, to the work done during a quasi-equilibrium expansion or compression process of a closed system.* (On the $P$-$v$ diagram, it represents the boundary work done per unit mass.)

A gas can follow several different paths as it expands from state 1 to state 2. In general, each path will have a different area underneath it, and since this area represents the magnitude of the work, the work done will be different for each process (Fig. 5–4). This is expected, since work is a path function (i.e., it depends on the path followed as well as the end states). If work were not a path function, no cyclic devices (car engines, power plants) could operate as work-producing devices. The work produced by these devices during one part of the cycle would have to be consumed during another part, and there would be no net work output. The cycle shown in Fig. 5–5 produces a net work output because the work done by the system during the expansion process (area under path $A$) is greater than the work done on the system during the compression part of the cycle (area under path $B$), and the difference between these two is the net work done during the cycle (the colored area).

If the relationship between $P$ and $V$ during an expansion or a compression process is given in terms of experimental data instead of in a functional form, obviously we cannot perform the integration analytically. We can, however, plot the $P$-$V$ diagram of the process using these data points and calculate the area underneath graphically to determine the work done.

Strictly speaking, the pressure $P$ in Eq. 5–2 is the pressure at the inner surface of the piston. It becomes equal to the pressure of the gas in the cylinder only if the process is quasi-equilibrium and thus the entire gas in the cylinder is at the same pressure at any given time. Equation 5–2 can also be used for nonquasi-equilibrium processes provided that the pressure *at the inner face of the piston* is used for $P$. (Besides, we cannot speak of the pressure of a *system* during a nonquasi-equilibrium process since properties are defined for equilibrium states only.) Therefore, we can generalize the boundary work relation by expressing it as

$$W_b = \int_1^2 P_i\,dV \tag{5–4}$$

where $P_i$ is the pressure at the inner face of the piston.

Note that work is a mechanism for energy interaction between a system and its surroundings, and $W_b$ represents the amount of energy transferred

**FIGURE 5–3**

The area under the process curve on a $P$-$V$ diagram represents the boundary work.

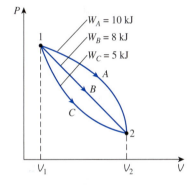

**FIGURE 5–4**

The boundary work done during a process depends on the path followed as well as the end states.

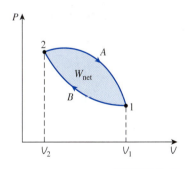

**FIGURE 5–5**

The net work done during a cycle is the difference between the work done by the system and the work done on the system.

from the system during an expansion process (or to the system during a compression process). Therefore, it has to appear somewhere else and we must be able to account for it since energy is conserved. In a car engine, for example, the boundary work done by the expanding hot gases is used to overcome friction between the piston and the cylinder, to push atmospheric air out of the way, and to rotate the crankshaft. Therefore,

$$W_b = W_{\text{friction}} + W_{\text{atm}} + W_{\text{crank}} = \int_1^2 (F_{\text{friction}} + P_{\text{atm}}A + F_{\text{crank}})\,dx \qquad \textbf{(5–5)}$$

Of course the work used to overcome friction appears as frictional heat and the energy transmitted through the crankshaft is transmitted to other components (such as the wheels) to perform certain functions. But note that the energy transferred by the system as work must equal the energy received by the crankshaft, the atmosphere, and the energy used to overcome friction. The use of the boundary work relation is not limited to the quasi-equilibrium processes of gases only. It can also be used for solids and liquids.

### EXAMPLE 5–1    Boundary Work for a Constant-Volume Process

A rigid tank contains air at 500 kPa and 150°C. As a result of heat transfer to the surroundings, the temperature and pressure inside the tank drop to 65°C and 400 kPa, respectively. Determine the boundary work done during this process.

**SOLUTION**   Air in a rigid tank is cooled, and both the pressure and temperature drop. The boundary work done is to be determined.
*Analysis*   A sketch of the system and the *P-V* diagram of the process are shown in Fig. 5–6. The boundary work can be determined from Eq. 5–2 to be

$$W_b = \int_1^2 P\,d\cancelto{0}{V} = 0$$

*Discussion*   This is expected since a rigid tank has a constant volume and $dV = 0$ in this equation. Therefore, there is no boundary work done during this process. That is, the boundary work done during a constant-volume process is always zero. This is also evident from the *P-V* diagram of the process (the area under the process curve is zero).

### EXAMPLE 5–2    Boundary Work for a Constant-Pressure Process

A frictionless piston–cylinder device contains 10 lbm of steam at 60 psia and 320°F. Heat is now transferred to the steam until the temperature reaches 400°F. If the piston is not attached to a shaft and its mass is constant, determine the work done by the steam during this process.

**SOLUTION**   Steam in a piston cylinder device is heated and the temperature rises at constant pressure. The boundary work done is to be determined.
*Analysis*   A sketch of the system and the *P-v* diagram of the process are shown in Fig. 5–7.

**FIGURE 5–6**
Schematic and *P-V* diagram for Example 5–1.

*Assumption* The expansion process is quasi-equilibrium.

*Analysis* Even though it is not explicitly stated, the pressure of the steam within the cylinder remains constant during this process since both the atmospheric pressure and the weight of the piston remain constant. Therefore, this is a constant-pressure process, and, from Eq. 5–2

$$W_b = \int_1^2 P \, dV = P_0 \int_1^2 dV = P_0(V_2 - V_1) \tag{5-6}$$

or

$$W_b = mP_0(v_2 - v_1)$$

since $V = mv$. From the superheated vapor table (Table A–6E), the specific volumes are determined to be $v_1 = 7.4863$ ft³/lbm at state 1 (60 psia, 320°F) and $v_2 = 8.3548$ ft³/lbm at state 2 (60 psia, 400°F). Substituting these values yields

$$W_b = (10 \text{ lbm})(60 \text{ psia})[(8.3548 - 7.4863) \text{ ft}^3/\text{lbm}] \left( \frac{1 \text{ Btu}}{5.404 \text{ psia·ft}^3} \right)$$

$$= \textbf{96.4 Btu}$$

*Discussion* The positive sign indicates that the work is done by the system. That is, the steam used 96.4 Btu of its energy to do this work. The magnitude of this work could also be determined by calculating the area under the process curve on the *P-V* diagram, which is simply $P_0 \Delta V$ for this case.

**FIGURE 5–7**

Schematic and *P-v* diagram for Example 5–2.

---

**EXAMPLE 5–3    Isothermal Compression of an Ideal Gas**

A piston–cylinder device initially contains 0.4 m³ of air at 100 kPa and 80°C. The air is now compressed to 0.1 m³ in such a way that the temperature inside the cylinder remains constant. Determine the work done during this process.

**SOLUTION** Air in a piston–cylinder device is compressed isothermally. The boundary work done is to be determined.

*Analysis* A sketch of the system and the *P-V* diagram of the process are shown in Fig. 5–8.

*Assumptions* **1** The compression process is quasi-equilibrium. **2** At specified conditions, air can be considered to be an ideal gas since it is at a high temperature and low pressure relative to its critical-point values.

*Analysis* For an ideal gas at constant temperature $T_0$,

$$PV = mRT_0 = C \quad \text{or} \quad P = \frac{C}{V}$$

where $C$ is a constant. Substituting this into Eq. 5–2, we have

$$W_b = \int_1^2 P \, dV = \int_1^2 \frac{C}{V} dV = C \int_1^2 \frac{dV}{V} = C \ln \frac{V_2}{V_1} = P_1 V_1 \ln \frac{V_2}{V_1} \tag{5-7}$$

In Eq. 5–7, $P_1 V_1$ can be replaced by $P_2 V_2$ or $mRT_0$. Also, $V_2/V_1$ can be replaced by $P_1/P_2$ for this case since $P_1 V_1 = P_2 V_2$.

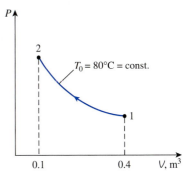

**FIGURE 5–8**

Schematic and *P-V* diagram for Example 5–3.

Substituting the numerical values into Eq. 5–7 yields

$$W_b = (100 \text{ kPa})(0.4 \text{ m}^3)\left(\ln\frac{0.1}{0.4}\right)\left(\frac{1 \text{ kJ}}{1 \text{ kPa·m}^3}\right)$$

$$= -55.5 \text{ kJ}$$

**Discussion** The negative sign indicates that this work is done on the system (a work input), which is always the case for compression processes.

## Polytropic Process

During actual expansion and compression processes of gases, pressure and volume are often related by $PV^n = C$, where $n$ and $C$ are constants. A process of this kind is called a **polytropic process** (Fig. 5–9). Below we develop a general expression for the work done during a polytropic process. The pressure for a polytropic process can be expressed as

$$P = CV^{-n} \tag{5–8}$$

Substituting this relation into Eq. 5–2, we obtain

$$W_b = \int_1^2 P \, dV = \int_1^2 CV^{-n} \, dV = C\frac{V_2^{-n+1} - V_1^{-n+1}}{-n + 1} = \frac{P_2 V_2 - P_1 V_1}{1 - n} \tag{5–9}$$

since $C = P_1 V_1^n = P_2 V_2^n$. For an ideal gas ($PV = mRT$), this equation can also be written as

$$W_b = \frac{mR(T_2 - T_1)}{1 - n} \qquad n \neq 1 \qquad \text{(kJ)} \tag{5–10}$$

For the special case of $n = 1$ the boundary work becomes

$$W_b = \int_1^2 P \, dV = \int_1^2 CV^{-1} \, dV = PV \ln\left(\frac{V_2}{V_1}\right)$$

For an ideal gas this result is equivalent to the isothermal process discussed in the previous example.

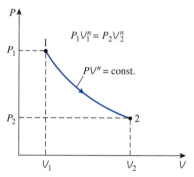

**FIGURE 5–9**

Schematic and $P$-$V$ diagram for a polytropic process.

---

### EXAMPLE 5–4    Expansion of a Gas against a Spring

A piston–cylinder device contains 0.05 m³ of a gas initially at 200 kPa. At this state, a linear spring that has a spring constant of 150 kN/m is touching the piston but exerting no force on it. Now heat is transferred to the gas, causing the piston to rise and to compress the spring until the volume inside the cylinder doubles. If the cross-sectional area of the piston is 0.25 m², determine (a) the final pressure inside the cylinder, (b) the total work done by the gas, and (c) the fraction of this work done against the spring to compress it.

**SOLUTION** A gas in a piston–cylinder device equipped with a linear spring expands as a result of heating. The final gas pressure, the total work done, and the fraction of the work done to compress the spring are to be determined.

*Assumptions* **1** The expansion process is quasi-equilibrium. **2** The spring is linear in the range of interest.

*Analysis* A sketch of the system and the *P-V* diagram of the process are shown in Fig. 5–10.

(*a*) The enclosed volume at the final state is

$$V_2 = 2V_1 = (2)(0.05 \text{ m}^3) = 0.1 \text{ m}^3$$

Then the displacement of the piston (and of the spring) becomes

$$x = \frac{\Delta V}{A} = \frac{(0.1 - 0.05) \text{ m}^3}{0.25 \text{ m}^2} = 0.2 \text{ m}$$

The force applied by the linear spring at the final state is

$$F = kx = (150 \text{ kN/m})(0.2 \text{ m}) = 30 \text{ kN}$$

The additional pressure applied by the spring on the gas at this state is

$$P = \frac{F}{A} = \frac{30 \text{ kN}}{0.25 \text{ m}^2} = 120 \text{ kPa}$$

Without the spring, the pressure of the gas would remain constant at 200 kPa while the piston is rising. But under the effect of the spring, the pressure rises linearly from 200 kPa to

$$200 + 120 = \textbf{320 kPa}$$

at the final state.

(*b*) An easy way of finding the work done is to plot the process on a *P-V* diagram and find the area under the process curve. From Fig. 5–10 the area under the process curve (a trapezoid) is determined to be

$$W = \text{area} = \frac{(200 + 320) \text{ kPa}}{2} \left[(0.1 - 0.05) \text{ m}^3\right] \left(\frac{1 \text{ kJ}}{1 \text{ kPa} \cdot \text{m}^3}\right) = \textbf{13 kJ}$$

Note that the work is done by the system.

(*c*) The work represented by the rectangular area (region I) is done against the piston and the atmosphere, and the work represented by the triangular area (region II) is done against the spring. Thus,

$$W_{\text{spring}} = \tfrac{1}{2}\left[(320 - 200) \text{ kPa}\right](0.05 \text{ m}^3)\left(\frac{1 \text{ kJ}}{1 \text{ kPa} \cdot \text{m}^3}\right) = \textbf{3 kJ}$$

*Discussion* This result could also be obtained from

$$W_{\text{spring}} = \tfrac{1}{2}k(x_2^2 - x_1^2) = \tfrac{1}{2}(150 \text{ kN/m})\left[(0.2 \text{ m})^2 - 0^2\right]\left(\frac{1 \text{ kJ}}{1 \text{ kN} \cdot \text{m}}\right) = 3 \text{ kJ}$$

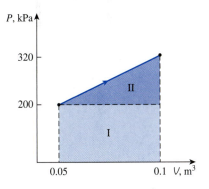

**FIGURE 5–10**
Schematic and *P-V* diagram for Example 5–4.

# 5–2 · ENERGY BALANCE FOR CLOSED SYSTEMS

Energy balance for any system undergoing any kind of process was expressed as (see Chap. 3)

$$\underbrace{E_{\text{in}} - E_{\text{out}}}_{\substack{\text{Net energy transfer} \\ \text{by heat, work, and mass}}} = \underbrace{\Delta E_{\text{system}}}_{\substack{\text{Change in internal, kinetic,} \\ \text{potential, etc., energies}}} \quad \text{(kJ)} \qquad \textbf{(5–11)}$$

or, in the **rate form**, as

$$\underbrace{\dot{E}_{in} - \dot{E}_{out}}_{\substack{\text{Rate of net energy transfer} \\ \text{by heat, work, and mass}}} = \underbrace{dE_{system}/dt}_{\substack{\text{Rate of change in internal,} \\ \text{kinetic, potential, etc., energies}}} \quad \text{(kW)} \qquad \textbf{(5–12)}$$

For constant rates, the total quantities during a time interval $\Delta t$ are related to the quantities per unit time as

$$Q = \dot{Q}\,\Delta t, \quad W = \dot{W}\,\Delta t, \quad \text{and} \quad \Delta E = (dE/dt)\,\Delta t \qquad \text{(kJ)} \qquad \textbf{(5–13)}$$

The energy balance can be expressed on a **per unit mass** basis as

$$e_{in} - e_{out} = \Delta e_{system} \qquad \text{(kJ/kg)} \qquad \textbf{(5–14)}$$

which is obtained by dividing all the quantities in Eq. 5–11 by the mass $m$ of the system. Energy balance can also be expressed in the differential form as

$$\delta E_{in} - \delta E_{out} = dE_{system} \quad \text{or} \quad \delta e_{in} - \delta e_{out} = de_{system} \qquad \textbf{(5–15)}$$

For a closed system undergoing a **cycle**, the initial and final states are identical, and thus $\Delta E_{system} = E_2 - E_1 = 0$. Then, the energy balance for a cycle simplifies to $E_{in} - E_{out} = 0$ or $E_{in} = E_{out}$. Noting that a closed system does not involve any mass flow across its boundaries, the energy balance for a cycle can be expressed in terms of heat and work interactions as

$$W_{net,out} = Q_{net,in} \quad \text{or} \quad \dot{W}_{net,out} = \dot{Q}_{net,in} \qquad \text{(for a cycle)} \qquad \textbf{(5–16)}$$

That is, the net work output during a cycle is equal to net heat input (Fig. 5–11).

The energy balance (or the first-law) relations already given are intuitive in nature and are easy to use when the magnitudes and directions of heat and work transfers are known. However, when performing a general analytical study or solving a problem that involves an unknown heat or work interaction, we need to assume a direction for the heat or work interactions. In such cases, it is common practice to use the classical thermodynamics sign convention and to assume heat to be transferred *into the system* (heat input) in the amount of $Q$ and work to be done *by the system* (work output) in the amount of $W$, and then to solve the problem. The energy balance relation in that case for a closed system becomes

$$Q_{net,in} - W_{net,out} = \Delta E_{system} \quad \text{or} \quad Q - W = \Delta E \qquad \textbf{(5–17)}$$

where $Q = Q_{net,in} = Q_{in} - Q_{out}$ is the *net heat input* and $W = W_{net,out} = W_{out} - W_{in}$ is the *net work output*. Obtaining a negative quantity for $Q$ or $W$ simply means that the assumed direction for that quantity is wrong and should be reversed. Various forms of this "traditional" first-law relation for closed systems are given in Fig. 5–12.

The first law cannot be proven mathematically, but no process in nature is known to have violated the first law, and this should be taken as sufficient proof. Note that if it were possible to prove the first law on the basis of other physical principles, the first law then would be a consequence of those principles instead of being a fundamental physical law itself.

As energy quantities, heat and work are not that different, and you probably wonder why we keep distinguishing them. After all, the change in the

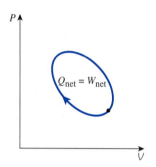

**FIGURE 5–11**

For a cycle $\Delta E = 0$, thus $Q = W$.

**FIGURE 5–12**

Various forms of the first-law relation for closed systems.

General   $Q - W = \Delta E$

Stationary systems   $Q - W = \Delta U$

Per unit mass   $q - w = \Delta e$

Differential form   $\delta q - \delta w = de$

energy content of a system is equal to the amount of energy that crosses the system boundaries, and it makes no difference whether the energy crosses the boundary as heat or work. It seems as if the first-law relations would be much simpler if we had just one quantity that we could call *energy interaction* to represent both heat and work. Well, from the first-law point of view, heat and work are not different at all. From the second-law point of view, however, heat and work are very different, as is discussed in later chapters.

---

### EXAMPLE 5–5  Electric Heating of a Gas at Constant Pressure

A piston–cylinder device contains 25 g of saturated water vapor that is maintained at a constant pressure of 300 kPa. A resistance heater within the cylinder is turned on and passes a current of 0.2 A for 5 min from a 120-V source. At the same time, a heat loss of 3.7 kJ occurs. (*a*) Show that for a closed system the boundary work $W_b$ and the change in internal energy $\Delta U$ in the first-law relation can be combined into one term, $\Delta H$, for a constant-pressure process. (*b*) Determine the final temperature of the steam.

**SOLUTION**  Saturated water vapor in a piston–cylinder device expands at constant pressure as a result of heating. It is to be shown that $\Delta U + W_b = \Delta H$, and the final temperature is to be determined.

*Assumptions*  **1** The tank is stationary and thus the kinetic and potential energy changes are zero, $\Delta KE = \Delta PE = 0$. Therefore, $\Delta E = \Delta U$ and internal energy is the only form of energy of the system that may change during this process. **2** Electrical wires constitute a very small part of the system, and thus the energy change of the wires can be neglected.

*Analysis*  We take the contents of the cylinder, including the resistance wires, as the *system* (Fig. 5–13). This is a *closed system* since no mass crosses the system boundary during the process. We observe that a piston–cylinder device typically involves a moving boundary and thus boundary work $W_b$. The pressure remains constant during the process and thus $P_2 = P_1$. Also, heat is lost from the system and electrical work $W_e$ is done on the system.

(*a*) This part of the solution involves a general analysis for a closed system undergoing a quasi-equilibrium constant-pressure process, and thus we consider a general closed system. We take the direction of heat transfer $Q$ to be to the system and the work $W$ to be done by the system. We also express the work as the sum of boundary and other forms of work (such as electrical and shaft). Then, the energy balance can be expressed as

$$\underbrace{E_{in} - E_{out}}_{\substack{\text{Net energy transfer} \\ \text{by heat, work, and mass}}} = \underbrace{\Delta E_{system}}_{\substack{\text{Change in internal, kinetic,} \\ \text{potential, etc., energies}}}$$

$$Q - W = \Delta U + \cancelto{0}{\Delta KE} + \cancelto{0}{\Delta PE}$$

$$Q - W_{other} - W_b = U_2 - U_1$$

For a constant-pressure process, the boundary work is given as $W_b = P_0(V_2 - V_1)$. Substituting this into the preceding relation gives

$$Q - W_{other} - P_0(V_2 - V_1) = U_2 - U_1$$

However,

$$P_0 = P_2 = P_1 \;\rightarrow\; Q - W_{other} = (U_2 + P_2V_2) - (U_1 + P_1V_1)$$

**FIGURE 5–13**

Schematic and $P$-$v$ diagram for Example 5–5.

**FIGURE 5–14**

For a closed system undergoing a quasi-equilibrium, $P$ = constant process, $\Delta U + W_b = \Delta H$. Note that this relation is NOT valid for closed systems processes during which pressure DOES NOT remain constant.

Also $H = U + PV$, and thus

$$Q - W_{other} = H_2 - H_1 \qquad \text{(kJ)} \qquad \text{(5–18)}$$

which is the desired relation (Fig. 5–14). This equation is very convenient to use in the analysis of closed systems undergoing a **constant-pressure quasi-equilibrium process** since the boundary work is automatically taken care of by the enthalpy terms, and one no longer needs to determine it separately.

(b) The only other form of work in this case is the electrical work, which can be determined from

$$W_e = \mathbf{V}I\,\Delta t = (120\text{ V})(0.2\text{ A})(300\text{ s})\left(\frac{1\text{ kJ/s}}{1000\text{ VA}}\right) = 7.2\text{ kJ}$$

$$\text{State 1:} \quad \left.\begin{array}{l} P_1 = 300\text{ kPa} \\ \text{sat. vapor} \end{array}\right\} \quad h_1 = h_{g\,@\,300\text{ kPa}} = 2724.9\text{ kJ/kg} \qquad \text{(Table A–5)}$$

The enthalpy at the final state can be determined directly from Eq. 5–18 by expressing heat transfer from the system and work done on the system as negative quantities (since their directions are opposite to the assumed directions). Alternately, we can use the general energy balance relation with the simplification that the boundary work is considered automatically by replacing $\Delta U$ by $\Delta H$ for a constant-pressure expansion or compression process:

$$\underbrace{E_{in} - E_{out}}_{\substack{\text{Net energy transfer} \\ \text{by heat, work, and mass}}} = \underbrace{\Delta E_{system}}_{\substack{\text{Change in internal, kinetic,} \\ \text{potential, etc., energies}}}$$

$$W_{e,in} - Q_{out} - W_b = \Delta U$$

$$W_{e,in} - Q_{out} = \Delta H = m(h_2 - h_1) \qquad (\text{since } P = \text{constant})$$

$$7.2\text{ kJ} - 3.7\text{ kJ} = (0.025\text{ kg})(h_2 - 2724.9)\text{ kJ/kg}$$

$$h_2 = 2864.9\text{ kJ/kg}$$

Now the final state is completely specified since we know both the pressure and the enthalpy. The temperature at this state is

$$\text{State 2:} \quad \left.\begin{array}{l} P_2 = 300\text{ kPa} \\ h_2 = 2864.9\text{ kJ/kg} \end{array}\right\} \quad T_2 = \mathbf{200°C} \qquad \text{(Table A–6)}$$

Therefore, the steam will be at 200°C at the end of this process.

**Discussion** Strictly speaking, the potential energy change of the steam is not zero for this process since the center of gravity of the steam rose somewhat. Assuming an elevation change of 1 m (which is rather unlikely), the change in the potential energy of the steam would be 0.0002 kJ, which is very small compared to the other terms in the first-law relation. Therefore, in problems of this kind, the potential energy term is always neglected.

---

**EXAMPLE 5–6    Unrestrained Expansion of Water**

A rigid tank is divided into two equal parts by a partition. Initially, one side of the tank contains 5 kg of water at 200 kPa and 25°C, and the other side is evacuated. The partition is then removed, and the water expands into the

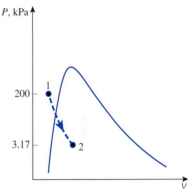

**FIGURE 5–15**

Schematic and P-v diagram for Example 5–6.

entire tank. The water is allowed to exchange heat with its surroundings until the temperature in the tank returns to the initial value of 25°C. Determine (a) the volume of the tank, (b) the final pressure, and (c) the heat transfer for this process.

**SOLUTION** One half of a rigid tank is filled with liquid water while the other side is evacuated. The partition between the two parts is removed and water is allowed to expand and fill the entire tank while the temperature is maintained constant. The volume of the tank, the final pressure, and the heat transfer are to be to determined.

**Assumptions** **1** The system is stationary and thus the kinetic and potential energy changes are zero, $\Delta KE = \Delta PE = 0$ and $\Delta E = \Delta U$. **2** The direction of heat transfer is to the system (heat gain, $Q_{in}$). A negative result for $Q_{in}$ indicates the assumed direction is wrong and thus it is a heat loss. **3** The volume of the rigid tank is constant, and thus there is no energy transfer as boundary work. **4** There is no electrical, shaft, or any other kind of work involved.

**Analysis** We take the contents of the tank, including the evacuated space, as the *system* (Fig. 5–15). This is a *closed system* since no mass crosses the system boundary during the process. We observe that the water fills the entire tank when the partition is removed (possibly as a liquid–vapor mixture).

(a) Initially the water in the tank exists as a compressed liquid since its pressure (200 kPa) is greater than the saturation pressure at 25°C (3.1698 kPa). Approximating the compressed liquid as a saturated liquid at the given temperature, we find

$$v_1 \cong v_{f@25°C} = 0.001003 \text{ m}^3/\text{kg} \cong 0.001 \text{ m}^3/\text{kg} \quad \text{(Table A–4)}$$

Then the initial volume of the water is

$$V_1 = mv_1 = (5 \text{ kg})(0.001 \text{ m}^3/\text{kg}) = 0.005 \text{ m}^3$$

The total volume of the tank is twice this amount:

$$V_{tank} = (2)(0.005 \text{ m}^3) = \mathbf{0.01 \text{ m}^3}$$

(b) At the final state, the specific volume of the water is

$$v_2 = \frac{V_2}{m} = \frac{0.01 \text{ m}^3}{5 \text{ kg}} = 0.002 \text{ m}^3/\text{kg}$$

which is twice the initial value of the specific volume. This result is expected since the volume doubles while the amount of mass remains constant.

At 25°C: $v_f = 0.001003 \text{ m}^3/\text{kg}$ and $v_g = 43.340 \text{ m}^3/\text{kg}$ (Table A–4)

Since $v_f < v_2 < v_g$, the water is a saturated liquid–vapor mixture at the final state, and thus the pressure is the saturation pressure at 25°C:

$$P_2 = P_{sat@25°C} = \mathbf{3.1698 \text{ kPa}} \quad \text{(Table A–4)}$$

(c) Under stated assumptions and observations, the energy balance on the system can be expressed as

$$\underbrace{E_{in} - E_{out}}_{\substack{\text{Net energy transfer} \\ \text{by heat, work, and mass}}} = \underbrace{\Delta E_{system}}_{\substack{\text{Change in internal, kinetic,} \\ \text{potential, etc., energies}}}$$

$$Q_{in} = \Delta U = m(u_2 - u_1)$$

**FIGURE 5–16**

Expansion against a vacuum involves no work and thus no energy transfer.

Notice that even though the water is expanding during this process, the system chosen involves fixed boundaries only (the dashed lines) and therefore the moving boundary work is zero (Fig. 5–16). Then $W = 0$ since the system does not involve any other forms of work. (Can you reach the same conclusion by choosing the water as our system?) Initially,

$$u_1 \cong u_{f\,@\,25°C} = 104.83 \text{ kJ/kg}$$

The quality at the final state is determined from the specific volume information:

$$x_2 = \frac{v_2 - v_f}{v_{fg}} = \frac{0.002 - 0.001}{43.34 - 0.001} = 2.3 \times 10^{-5}$$

Then

$$u_2 = u_f + x_2 u_{fg}$$
$$= 104.83 \text{ kJ/kg} + (2.3 \times 10^{-5})(2304.3 \text{ kJ/kg})$$
$$= 104.88 \text{ kJ/kg}$$

Substituting yields

$$Q_{in} = (5 \text{ kg})[(104.88 - 104.83) \text{ kJkg}] = \mathbf{0.25\ kJ}$$

***Discussion*** The positive sign indicates that the assumed direction is correct, and heat is transferred to the water.

**FIGURE 5–17**

It takes different amounts of energy to raise the temperature of different substances by the same amount.

# 5–3 ▪ SPECIFIC HEATS

We know from experience that it takes different amounts of energy to raise the temperature of identical masses of different substances by one degree. For example, we need about 4.5 kJ of energy to raise the temperature of 1 kg of iron from 20 to 30°C, whereas it takes about 9 times this energy (41.8 kJ to be exact) to raise the temperature of 1 kg of liquid water by the same amount (Fig. 5–17). Therefore, it is desirable to have a property that will enable us to compare the energy storage capabilities of various substances. This property is the specific heat.

The **specific heat** is defined as *the energy required to raise the temperature of a unit mass of a substance by one degree* (Fig. 5–18). In general, this energy depends on how the process is executed. In thermodynamics, we are interested in two kinds of specific heats: **specific heat at constant volume** $c_v$ and **specific heat at constant pressure** $c_p$.

Physically, the specific heat at constant volume $c_v$ can be viewed as *the energy required to raise the temperature of the unit mass of a substance by one degree as the volume is maintained constant.* The energy required to do the same as the pressure is maintained constant is the specific heat at constant pressure $c_p$. This is illustrated in Fig. 5–19. The specific heat at constant pressure $c_p$ is always greater than $c_v$ because at constant pressure the system is allowed to expand and the energy for this expansion work must also be supplied to the system.

Now we attempt to express the specific heats in terms of other thermodynamic properties. First, consider a fixed mass in a stationary closed system

**FIGURE 5–18**

Specific heat is the energy required to raise the temperature of a unit mass of a substance by one degree in a specified way.

undergoing a constant-volume process (and thus no expansion or compression work is involved). The conservation of energy principle $e_{in} - e_{out} = \Delta e_{system}$ for this process can be expressed in the differential form as

$$\delta e_{in} - \delta e_{out} = du$$

The left-hand side of this equation represents the net amount of energy transferred to the system. From the definition of $c_v$, this energy must be equal to $c_v\,dT$, where $dT$ is the differential change in temperature. Thus,

$$c_v\,dT = du \qquad \text{at constant volume}$$

or

$$c_v = \left(\frac{\partial u}{\partial T}\right)_v \tag{5-19}$$

Similarly, an expression for the specific heat at constant pressure $c_p$ can be obtained by considering a constant-pressure expansion or compression process. It yields

$$c_p = \left(\frac{\partial h}{\partial T}\right)_p \tag{5-20}$$

Equations 5–19 and 5–20 are the defining equations for $c_v$ and $c_p$, and their interpretation is given in Fig. 5–20.

Note that $c_v$ and $c_p$ are expressed in terms of other properties; thus, they must be properties themselves. Like any other property, the specific heats of a substance depend on the state that, in general, is specified by two independent, intensive properties. That is, the energy required to raise the temperature of a substance by one degree is different at different temperatures and pressures (Fig. 5–21). But this difference is usually not very large.

A few observations can be made from Eqs. 5–19 and 5–20. First, these equations are *property relations* and as such *are independent of the type of processes*. They are valid for *any* substance undergoing *any* process. The only relevance $c_v$ has to a constant-volume process is that $c_v$ happens to be the energy transferred to a system during a constant-volume process per unit mass, per unit degree rise in temperature. This is how the values of $c_v$ are determined. This is also how the name *specific heat at constant volume* originated. Likewise, the energy transferred to a system per unit mass per unit temperature rise during a constant-pressure process happens to be equal to $c_p$. This is how the values of $c_p$ can be determined and also explains the origin of the name *specific heat at constant pressure*.

Another observation that can be made from Eqs. 5–19 and 5–20 is that $c_v$ is related to the changes in *internal energy* and $c_p$ to the changes in *enthalpy*. In fact, it would be more proper to define $c_v$ as *the change in the internal energy of a substance per unit change in temperature at constant volume*. Likewise, $c_p$ can be defined as *the change in the enthalpy of a substance per unit change in temperature at constant pressure*. In other words, $c_v$ is a measure of the variation of internal energy of a substance with temperature, and $c_p$ is a measure of the variation of enthalpy of a substance with temperature.

Both the internal energy and enthalpy of a substance can be changed by the transfer of *energy* in any form, with heat being only one of them. Therefore,

**FIGURE 5–19**

Constant-volume and constant-pressure specific heats $c_v$ and $c_p$ (values given are for helium gas).

**FIGURE 5–20**

Formal definitions of $c_v$ and $c_p$.

**FIGURE 5–21**

The specific heat of a substance changes with temperature.

the term *specific energy* is probably more appropriate than the term *specific heat*, which implies that energy is transferred (and stored) in the form of heat.

A common unit for specific heats is kJ/kg·°C or kJ/kg·K. Notice that these two units are *identical* since $\Delta T(°C) = \Delta T(K)$, and 1°C change in temperature is equivalent to a change of 1 K. The specific heats are sometimes given on a *molar basis*. They are then denoted by $\overline{c}_v$ and $\overline{c}_p$ and have the unit kJ/kmol·°C or kJ/kmol·K.

## 5–4 ▪ INTERNAL ENERGY, ENTHALPY, AND SPECIFIC HEATS OF IDEAL GASES

We defined an ideal gas as a gas whose temperature, pressure, and specific volume are related by

$$Pv = RT$$

It has been demonstrated mathematically and experimentally (Joule, 1843) that for an ideal gas the internal energy is a function of the temperature only. That is,

$$u = u(T) \tag{5–21}$$

Thermometer

Water

Air
(high pressure)      Evacuated

**FIGURE 5–22**

Schematic of the experimental apparatus used by Joule.

In his classical experiment, Joule submerged two tanks connected with a pipe and a valve in a water bath, as shown in Fig. 5–22. Initially, one tank contained air at a high pressure and the other tank was evacuated. When thermal equilibrium was attained, he opened the valve to let air pass from one tank to the other until the pressures equalized. Joule observed no change in the temperature of the water bath and assumed that no heat was transferred to or from the air. Since there was also no work done, he concluded that the internal energy of the air did not change even though the volume and the pressure changed. Therefore, he reasoned, the internal energy is a function of temperature only and not a function of pressure or specific volume. (Joule later showed that for gases that deviate significantly from ideal-gas behavior, the internal energy is not a function of temperature alone.)

Using the definition of enthalpy and the equation of state of an ideal gas, we have

$$\left.\begin{array}{r} h = u + Pv \\ Pv = RT \end{array}\right\} \quad h = u + RT$$

Since $R$ is constant and $u = u(T)$, it follows that the enthalpy of an ideal gas is also a function of temperature only:

$$h = h(T) \tag{5–22}$$

$$u = u(T)$$
$$h = h(T)$$
$$c_v = c_v(T)$$
$$c_p = c_p(T)$$

**FIGURE 5–23**

For ideal gases, $u$, $h$, $c_v$, and $c_p$ vary with temperature only.

Since $u$ and $h$ depend only on temperature for an ideal gas, the specific heats $c_v$ and $c_p$ also depend, at most, on temperature only. Therefore, at a given temperature, $u$, $h$, $c_v$, and $c_p$ of an ideal gas have fixed values regardless of the specific volume or pressure (Fig. 5–23). Thus, for ideal gases, the partial derivatives in Eqs. 5–19 and 5–20 can be replaced by ordinary derivatives. Then, the differential changes in the internal energy and enthalpy of an ideal gas can be expressed as

$$du = c_v(T)\, dT \tag{5–23}$$

and

$$dh = c_p(T)\, dT \tag{5–24}$$

The change in internal energy or enthalpy for an ideal gas during a process from state 1 to state 2 is determined by integrating these equations:

$$\Delta u = u_2 - u_1 = \int_1^2 c_v(T)\, dT \qquad (\text{kJ/kg}) \tag{5–25}$$

and

$$\Delta h = h_2 - h_1 = \int_1^2 c_p(T)\, dT \qquad (\text{kJ/kg}) \tag{5–26}$$

To carry out these integrations, we need to have relations for $c_v$ and $c_p$ as functions of temperature.

At low pressures, all real gases approach ideal-gas behavior, and therefore their specific heats depend on temperature only. The specific heats of real gases at low pressures are called *ideal-gas specific heats*, or *zero-pressure specific heats*, and are often denoted $c_{p0}$ and $c_{v0}$. Accurate analytical expressions for ideal-gas specific heats, based on direct measurements or calculations from statistical behavior of molecules, are available and are given as third-degree polynomials in the appendix (Table A–2c) for several gases. A plot of $\overline{c}_{p0}(T)$ data for some common gases is given in Fig. 5–24.

The use of ideal-gas specific heat data is limited to low pressures, but these data can also be used at moderately high pressures with reasonable accuracy as long as the gas does not deviate from ideal-gas behavior significantly.

The integrations in Eqs. 5–25 and 5–26 are straightforward but rather time-consuming and thus impractical. To avoid these laborious calculations, $u$ and $h$ data for a number of gases have been tabulated over small temperature intervals. These tables are obtained by choosing an arbitrary reference point and performing the integrations in Eqs. 5–25 and 5–26 by treating state 1 as the reference state. In the ideal-gas tables given in the appendix, zero kelvin is chosen as the reference state, and both the enthalpy and the internal energy are assigned zero values at that state (Fig. 5–25). The choice of the reference state has no effect on $\Delta u$ or $\Delta h$ calculations. The $u$ and $h$ data are given in kJ/kg for air (Table A–21) and usually in kJ/kmol for other gases. The unit kJ/kmol is very convenient in the thermodynamic analysis of chemical reactions.

Some observations can be made from Fig. 5–24. First, the specific heats of gases with complex molecules (molecules with two or more atoms) are higher and increase with temperature. Also, the variation of specific heats with temperature is smooth and may be approximated as linear over small temperature intervals (a few hundred degrees or less). Therefore, the specific heat functions in Eqs. 5–25 and 5–26 can be replaced by the constant average specific heat values. Then, the integrations in these equations can be performed, yielding

$$u_2 - u_1 = c_{v,\text{avg}}(T_2 - T_1) \qquad (\text{kJ/kg}) \tag{5–27}$$

and

$$h_2 - h_1 = c_{p,\text{avg}}(T_2 - T_1) \qquad (\text{kJ/kg}) \tag{5–28}$$

**FIGURE 5–24**

Ideal-gas constant-pressure specific heats for some gases (see Table A–2c for $c_p$ equations).

**FIGURE 5–25**

In the preparation of ideal-gas tables, 0 K is chosen as the reference temperature.

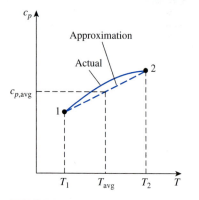

**FIGURE 5–26**

For small temperature intervals, the specific heats may be assumed to vary linearly with temperature.

**FIGURE 5–27**

The relation $\Delta u = c_v \, \Delta T$ is valid for *any* kind of process, constant-volume or not.

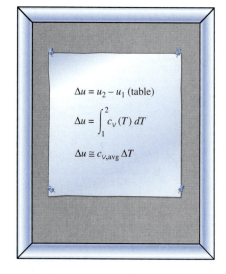

**FIGURE 5–28**

Three ways of calculating $\Delta u$.

The specific heat values for some common gases are listed as a function of temperature in Table A–2b. The average specific heats $c_{p,\text{avg}}$ and $c_{v,\text{avg}}$ are evaluated from this table at the average temperature $(T_1 + T_2)/2$, as shown in Fig. 5–26. If the final temperature $T_2$ is not known, the specific heats may be evaluated at $T_1$ or at the anticipated average temperature. Then $T_2$ can be determined by using these specific heat values. The value of $T_2$ can be refined, if necessary, by evaluating the specific heats at the new average temperature.

Another way of determining the average specific heats is to evaluate them at $T_1$ and $T_2$ and then take their average. Usually both methods give reasonably good results, and one is not necessarily better than the other.

Another observation that can be made from Fig. 5–24 is that the ideal-gas specific heats of *monatomic gases* such as argon, neon, and helium remain constant over the entire temperature range. Thus, $\Delta u$ and $\Delta h$ of monatomic gases can easily be evaluated from Eqs. 5–27 and 5–28.

Note that the $\Delta u$ and $\Delta h$ relations given previously are not restricted to any kind of process. They are valid for all processes. The presence of the constant-volume specific heat $c_v$ in an equation should not lead one to believe that this equation is valid for a constant-volume process only. On the contrary, the relation $\Delta u = c_{v,\text{avg}} \, \Delta T$ is valid for *any* ideal gas undergoing *any* process (Fig. 5–27). A similar argument can be given for $c_p$ and $\Delta h$.

To summarize, there are three ways to determine the internal energy and enthalpy changes of ideal gases (Fig. 5–28):

1. By using the tabulated $u$ and $h$ data. This is the easiest and most accurate way when tables are readily available.
2. By using the $c_v$ or $c_p$ relations as a function of temperature and performing the integrations. This is very inconvenient for hand calculations but quite desirable for computerized calculations. The results obtained are very accurate.
3. By using average specific heats. This is very simple and certainly very convenient when property tables are not available. The results obtained are reasonably accurate if the temperature interval is not very large.

## Specific Heat Relations of Ideal Gases

A special relationship between $c_p$ and $c_v$ for ideal gases can be obtained by differentiating the relation $h = u + RT$, which yields

$$dh = du + R \, dT$$

Replacing $dh$ by $c_p \, dT$ and $du$ by $c_v \, dT$ and dividing the resulting expression by $dT$, we obtain

$$c_p = c_v + R \qquad (\text{kJ/kg·K}) \qquad (5\text{–}29)$$

This is an important relationship for ideal gases since it enables us to determine $c_v$ from a knowledge of $c_p$ and the gas constant $R$.

When the specific heats are given on a molar basis, $R$ in the above equation should be replaced by the universal gas constant $R_u$ (Fig. 5–29).

$$\bar{c}_p = \bar{c}_v + R_u \qquad (\text{kJ/kmol·K}) \qquad (5\text{–}30)$$

At this point, we introduce another ideal-gas property called the **specific heat ratio** $k$, defined as

$$k = \frac{c_p}{c_v} \tag{5-31}$$

The specific ratio also varies with temperature, but this variation is very mild. For monatomic gases, its value is essentially constant at 1.667. Many diatomic gases, including air, have a specific heat ratio of about 1.4 at room temperature.

Air at 300 K

$\left. \begin{array}{l} c_v = 0.718 \text{ kJ/kg·K} \\ R = 0.287 \text{ kJ/kg·K} \end{array} \right\}$ $c_p = 1.005$ kJ/kg·K

or

$\left. \begin{array}{l} \bar{c}_v = 20.80 \text{ kJ/kmol·K} \\ R_u = 8.314 \text{ kJ/kmol·K} \end{array} \right\}$ $\bar{c}_p = 29.114$ kJ/kmol·K

**FIGURE 5–29**

The $c_p$ of an ideal gas can be determined from a knowledge of $c_v$ and $R$.

---

**EXAMPLE 5–7**   **Evaluation of the $\Delta u$ of an Ideal Gas**

Air at 300 K and 200 kPa is heated at constant pressure to 600 K. Determine the change in internal energy of air per unit mass, using (a) data from the air table (Table A–21), (b) the functional form of the specific heat (Table A–2c), and (c) the average specific heat value (Table A–2b).

**SOLUTION**   The internal energy change of air is to be determined in three different ways.

**Assumptions**   At specified conditions, air can be considered to be an ideal gas since it is at a high temperature and low pressure relative to its critical-point values.

**Analysis**   The internal energy change $\Delta u$ of ideal gases depends on the initial and final temperatures only, and not on the type of process. Thus, the following solution is valid for any kind of process.

(a) One way of determining the change in internal energy of air is to read the $u$ values at $T_1$ and $T_2$ from Table A–21 and take the difference:

$$u_1 = u_{@ \, 300 \, K} = 214.07 \text{ kJ/kg}$$

$$u_2 = u_{@ \, 600 \, K} = 434.78 \text{ kJ/kg}$$

Thus,

$$\Delta u = u_2 - u_1 = (434.78 - 214.07) \text{ kJ/kg} = \textbf{220.71 kJ/kg}$$

(b) The $\bar{c}_p(T)$ of air is given in Table A–2c in the form of a third-degree polynomial expressed as

$$\bar{c}_p(T) = a + bT + cT^2 + dT^3$$

where $a = 28.11$, $b = 0.1967 \times 10^{-2}$, $c = 0.4802 \times 10^{-5}$, and $d = -1.966 \times 10^{-9}$. From Eq. 5–30,

$$\bar{c}_v(T) = \bar{c}_p - R_u = (a - R_u) + bT + cT^2 + dT^3$$

From Eq. 5–25,

$$\Delta \bar{u} = \int_1^2 \bar{c}_v(T) \, dT = \int_{T_1}^{T_2} \left[ (a - R_u) + bT + cT^2 + dT^3 \right] dT$$

Performing the integration and substituting the values, we obtain

$$\Delta \bar{u} = 6447 \text{ kJ/kmol}$$

The change in the internal energy on a unit-mass basis is determined by dividing this value by the molar mass of air (Table A–1):

$$\Delta u = \frac{\Delta \bar{u}}{M} = \frac{6447 \text{ kJ/kmol}}{28.97 \text{ kg/kmol}} = 222.5 \text{ kJ/kg}$$

which differs from the tabulated value by 0.8 percent.

(c) The average value of the constant-volume specific heat $c_{v,\text{avg}}$ is determined from Table A–2b at the average temperature of $(T_1 + T_2)/2 = 450$ K to be

$$c_{v,\text{avg}} = c_{v \text{ @ } 450 \text{ K}} = 0.733 \text{ kJ/kg·K}$$

Thus,

$$\Delta u = c_{v,\text{avg}}(T_2 - T_1) = (0.733 \text{ kJ/kg·K})[(600 - 300)\text{K}]$$
$$= 220 \text{ kJ/kg}$$

**Discussion**  This answer differs from the tabulated value (220.71 kJ/kg) by only 0.4 percent. This close agreement is not surprising since the assumption that $c_v$ varies linearly with temperature is a reasonable one at temperature intervals of only a few hundred degrees. If we had used the $c_v$ value at $T_1 = 300$ K instead of at $T_{\text{avg}}$, the result would be 215.4 kJ/kg, which is in error by about 2 percent. Errors of this magnitude are acceptable for most engineering purposes.

---

### EXAMPLE 5–8    Heating of a Gas in a Tank by Stirring

An insulated rigid tank initially contains 1.5 lbm of helium at 80°F and 50 psia. A paddle wheel with a power rating of 0.02 hp is operated within the tank for 30 min. Determine (a) the final temperature and (b) the final pressure of the helium gas.

**SOLUTION**  Helium gas in an insulated rigid tank is stirred by a paddle wheel. The final temperature and pressure of helium are to be determined.
**Assumptions**  1 Helium is an ideal gas since it is at a very high temperature relative to its critical-point value of −451°F. 2 Constant specific heats can be used for helium. 3 The system is stationary and thus the kinetic and potential energy changes are zero, $\Delta \text{KE} = \Delta \text{PE} = 0$ and $\Delta E = \Delta U$. 4 The volume of the tank is constant, and thus there is no boundary work. 5 The system is adiabatic and thus there is no heat transfer.
**Analysis**  We take the contents of the tank as the *system* (Fig. 5–30). This is a *closed system* since no mass crosses the system boundary during the process. We observe that there is shaft work done on the system.

(a) The amount of paddle-wheel work done on the system is

$$W_{\text{sh}} = \dot{W}_{\text{sh}} \Delta t = (0.02 \text{ hp})(0.5 \text{ h})\left(\frac{2545 \text{ Btu/h}}{1 \text{ hp}}\right) = 25.45 \text{ Btu}$$

Under the stated assumptions and observations, the energy balance on the system can be expressed as

$$\underbrace{E_{\text{in}} - E_{\text{out}}}_{\substack{\text{Net energy transfer} \\ \text{by heat, work, and mass}}} = \underbrace{\Delta E_{\text{system}}}_{\substack{\text{Change in internal, kinetic,} \\ \text{potential, etc., energies}}}$$

$$W_{\text{sh,in}} = \Delta U = m(u_2 - u_1) = mc_{v,\text{avg}}(T_2 - T_1)$$

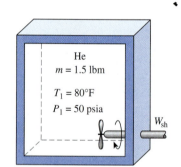

He
$m = 1.5$ lbm
$T_1 = 80°F$
$P_1 = 50$ psia
$W_{\text{sh}}$

$P$, psia
$P_2$ —————● 2
$50$ —————● 1
$V_2 = V_1$    $V$

**FIGURE 5–30**

Schematic and $P$-$V$ diagram for Example 5–8.

As we pointed out earlier, the ideal-gas specific heats of monatomic gases (helium being one of them) are constant. The $c_v$ value of helium is determined from Table A–2E$a$ to be $c_v$ = 0.753 Btu/lbm·°F. Substituting this and other known quantities into the above equation, we obtain

$$25.45 \text{ Btu} = (1.5 \text{ lbm})(0.753 \text{ Btu/lbm·°F})(T_2 - 80°F)$$

$$T_2 = \textbf{102.5°F}$$

(b) The final pressure is determined from the ideal-gas relation

$$\frac{P_1 V_1}{T_1} = \frac{P_2 V_2}{T_2}$$

where $V_1$ and $V_2$ are identical and cancel out. Then the final pressure becomes

$$\frac{50 \text{ psia}}{(80 + 460) \text{ R}} = \frac{P_2}{(102.5 + 460)\text{R}}$$

$$P_2 = \textbf{52.1 psia}$$

**Discussion** Note that the pressure in the ideal-gas relation is always the absolute pressure.

---

**EXAMPLE 5–9    Heating of a Gas by a Resistance Heater**

A piston–cylinder device initially contains 0.5 m³ of nitrogen gas at 400 kPa and 27°C. An electric heater within the device is turned on and is allowed to pass a current of 2 A for 5 min from a 120-V source. Nitrogen expands at constant pressure, and a heat loss of 2800 J occurs during the process. Determine the final temperature of nitrogen.

**SOLUTION** Nitrogen gas in a piston–cylinder device is heated by an electric resistance heater. Nitrogen expands at constant pressure while some heat is lost. The final temperature of nitrogen is to be determined.

**Assumptions** **1** Nitrogen is an ideal gas since it is at a high temperature and low pressure relative to its critical-point values of −147°C, and 3.39 MPa. **2** The system is stationary and thus the kinetic and potential energy changes are zero, $\Delta KE = \Delta PE = 0$ and $\Delta E = \Delta U$. **3** The pressure remains constant during the process and thus $P_2 = P_1$. **4** Nitrogen has constant specific heats at room temperature.

**Analysis** We take the contents of the cylinder as the *system* (Fig. 5–31). This is a *closed system* since no mass crosses the system boundary during the process. We observe that a piston–cylinder device typically involves a moving boundary and thus boundary work, $W_b$. Also, heat is lost from the system and electrical work $W_e$ is done on the system.

First, let us determine the electrical work done on the nitrogen:

$$W_e = \textbf{V}I \, \Delta t = (120 \text{ V})(2 \text{ A})(5 \times 60 \text{ s})\left(\frac{1 \text{ kJ/s}}{1000 \text{ VA}}\right) = 72 \text{ kJ}$$

The mass of nitrogen is determined from the ideal-gas relation:

$$m = \frac{P_1 V_1}{RT_1} = \frac{(400 \text{ kPa})(0.5 \text{ m}^3)}{(0.297 \text{ kPa·m}^3/\text{kg·K})(300 \text{ K})} = 2.245 \text{ kg}$$

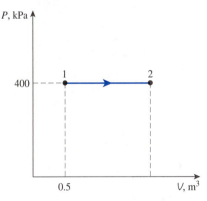

**FIGURE 5–31**

Schematic and $P$-$V$ diagram for Example 5–9.

Under the stated assumptions and observations, the energy balance on the system can be expressed as

$$\underbrace{E_{\text{in}} - E_{\text{out}}}_{\substack{\text{Net energy transfer} \\ \text{by heat, work, and mass}}} = \underbrace{\Delta E_{\text{system}}}_{\substack{\text{Change in internal, kinetic,} \\ \text{potential, etc., energies}}}$$

$$W_{e,\text{in}} - Q_{\text{out}} - W_{b,\text{out}} = \Delta U$$

$$W_{e,\text{in}} - Q_{\text{out}} = \Delta H = m(h_2 - h_1) = mc_p(T_2 - T_1)$$

since $\Delta U + W_b \equiv \Delta H$ for a closed system undergoing a quasi-equilibrium expansion or compression process at constant pressure. From Table A–2a, $c_p = 1.039$ kJ/kg·K for nitrogen at room temperature. The only unknown quantity in the previous equation is $T_2$, and it is found to be

$$72 \text{ kJ} - 2.8 \text{ kJ} = (2.245 \text{ kg})(1.039 \text{ kJ/kg·K})(T_2 - 27°C)$$

$$T_2 = \mathbf{56.7°C}$$

**Discussion** Note that we could also solve this problem by determining the boundary work and the internal energy change rather than the enthalpy change.

---

**EXAMPLE 5–10    Heating of a Gas at Constant Pressure**

A piston–cylinder device initially contains air at 150 kPa and 27°C. At this state, the piston is resting on a pair of stops, as shown in Fig. 5–32, and the enclosed volume is 400 L. The mass of the piston is such that a 350-kPa pressure is required to move it. The air is now heated until its volume has doubled. Determine (a) the final temperature, (b) the work done by the air, and (c) the total heat transferred to the air.

**SOLUTION** Air in a piston–cylinder device with a set of stops is heated until its volume is doubled. The final temperature, work done, and the total heat transfer are to be determined.

**Assumptions** 1 Air is an ideal gas since it is at a high temperature and low pressure relative to its critical-point values. 2 The system is stationary and thus the kinetic and potential energy changes are zero, $\Delta KE = \Delta PE = 0$ and $\Delta E = \Delta U$. 3 The volume remains constant until the piston starts moving, and the pressure remains constant afterwards. 4 There are no electrical, shaft, or other forms of work involved.

**Analysis** We take the contents of the cylinder as the *system* (Fig. 5–32). This is a *closed system* since no mass crosses the system boundary during the process. We observe that a piston-cylinder device typically involves a moving boundary and thus boundary work, $W_b$. Also, the boundary work is done by the system, and heat is transferred to the system.

(a) The final temperature can be determined easily by using the ideal-gas relation between states 1 and 3 in the following form:

$$\frac{P_1 V_1}{T_1} = \frac{P_3 V_3}{T_3} \longrightarrow \frac{(150 \text{ kPa})(V_1)}{300 \text{ K}} = \frac{(350 \text{ kPa})(2V_1)}{T_3}$$

$$T_3 = \mathbf{1400 \text{ K}}$$

Air
$V_1 = 400$ L
$P_1 = 150$ kPa
$T_1 = 27°C$

$Q$

**FIGURE 5–32**

Schematic and *P-V* diagram for Example 5–10.

(b) The work done could be determined by integration, but for this case it is much easier to find it from the area under the process curve on a P-V diagram, shown in Fig. 5–32:

$$A = (V_2 - V_1)P_2 = (0.4 \text{ m}^3)(350 \text{ kPa}) = 140 \text{ m}^3\cdot\text{kPa}$$

Therefore,

$$W_{13} = \textbf{140 kJ}$$

The work is done by the system (to raise the piston and to push the atmospheric air out of the way), and thus it is work output.

(c) Under the stated assumptions and observations, the energy balance on the system between the initial and final states (process 1–3) can be expressed as

$$\underbrace{E_{\text{in}} - E_{\text{out}}}_{\substack{\text{Net energy transfer} \\ \text{by heat, work, and mass}}} = \underbrace{\Delta E_{\text{system}}}_{\substack{\text{Change in internal, kinetic,} \\ \text{potential, etc., energies}}}$$

$$Q_{\text{in}} - W_{b,\text{out}} = \Delta U = m(u_3 - u_1)$$

The mass of the system can be determined from the ideal-gas relation:

$$m = \frac{P_1 V_1}{RT_1} = \frac{(150 \text{ kPa})(0.4 \text{ m}^3)}{(0.287 \text{ kPa}\cdot\text{m}^3/\text{kg}\cdot\text{K})(300 \text{ K})} = 0.697 \text{ kg}$$

The internal energies are determined from the air table (Table A–21) to be

$$u_1 = u_{@ \text{ 300 K}} = 214.07 \text{ kJ/kg}$$

$$u_3 = u_{@ \text{ 1400 K}} = 1113.52 \text{ kJ/kg}$$

Thus,

$$Q_{\text{in}} - 140 \text{ kJ} = (0.697 \text{ kg})[(1113.52 - 214.07) \text{ kJ/kg}]$$

$$Q_{\text{in}} = \textbf{767 kJ}$$

**Discussion**  The positive sign verifies that heat is transferred to the system.

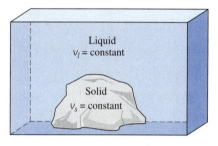

**FIGURE 5–33**

The specific volumes of incompressible substances remain constant during a process.

## 5–5 ▪ INTERNAL ENERGY, ENTHALPY, AND SPECIFIC HEATS OF SOLIDS AND LIQUIDS

A substance whose specific volume (or density) is constant is called an **incompressible substance**. The specific volumes of solids and liquids essentially remain constant during a process (Fig. 5–33). Therefore, liquids and solids can be approximated as incompressible substances without sacrificing much in accuracy. The constant-volume assumption should be taken to imply that the energy associated with the volume change is negligible compared with other forms of energy. Otherwise, this assumption would be ridiculous for studying the thermal stresses in solids (caused by volume change with temperature) or analyzing liquid-in-glass thermometers.

It can be mathematically shown that the constant-volume and constant-pressure specific heats are identical for incompressible substances (Fig. 5–34). Therefore, for solids and liquids, the subscripts on

**FIGURE 5–34**

The $c_v$ and $c_p$ values of incompressible substances are identical and are denoted by $c$.

$c_p$ and $c_v$ can be dropped, and both specific heats can be represented by a single symbol $c$. That is,

$$c_p = c_v = c \qquad \text{(5–32)}$$

This result could also be deduced from the physical definitions of constant-volume and constant-pressure specific heats. Specific heat values for several common liquids and solids are given in Table A–3.

## Internal Energy Changes

Like those of ideal gases, the specific heats of incompressible substances depend on temperature only. Thus, the partial differentials in the defining equation of $c_v$ can be replaced by ordinary differentials, which yield

$$du = c_v \, dT = c(T) \, dT \qquad \text{(5–33)}$$

The change in internal energy between states 1 and 2 is then obtained by integration:

$$\Delta u = u_2 - u_1 = \int_1^2 c(T) \, dT \qquad \text{(kJ/kg)} \qquad \text{(5–34)}$$

The variation of specific heat $c$ with temperature should be known before this integration can be carried out. For small temperature intervals, a $c$ value at the average temperature can be used and treated as a constant, yielding

$$\Delta u \cong c_{\text{avg}}(T_2 - T_1) \qquad \text{(kJ/kg)} \qquad \text{(5–35)}$$

## Enthalpy Changes

Using the definition of enthalpy $h = u + Pv$ and noting that $v = $ constant, the differential form of the enthalpy change of incompressible substances can be determined by differentiation to be

$$dh = du + v \, dP + P \, dv\nearrow^{0} = du + v \, dP \qquad \text{(5–36)}$$

Integrating,

$$\Delta h = \Delta u + v \, \Delta P \cong c_{\text{avg}} \, \Delta T + v \, \Delta P \qquad \text{(kJ/kg)} \qquad \text{(5–37)}$$

For *solids*, the term $v \, \Delta P$ is insignificant and thus $\Delta h = \Delta u \cong c_{\text{avg}}\Delta T$. For *liquids*, two special cases are commonly encountered:

1. *Constant-pressure processes,* as in heaters ($\Delta P = 0$): $\Delta h = \Delta u \cong c_{\text{avg}}\Delta T$
2. *Constant-temperature processes,* as in pumps ($\Delta T = 0$): $\Delta h = v \, \Delta P$

For a process between states 1 and 2, the last relation can be expressed as $h_2 - h_1 = v(P_2 - P_1)$. By taking state 2 to be the compressed liquid state at a given $T$ and $P$ and state 1 to be the saturated liquid state at the same temperature, the enthalpy of the compressed liquid can be expressed as

$$h_{@P,T} \cong h_{f @ T} + v_{f @ T}(P - P_{\text{sat} @ T}) \qquad \text{(5–38)}$$

as discussed in Chap. 4. This is an improvement over the assumption that the enthalpy of the compressed liquid could be taken as $h_f$ at the given temperature (that is, $h_{@ P,T} \cong h_{f @ T}$). However, the contribution of the last term is often very small, and is neglected. (Note that at high temperature and pressures, Eq. 5–38 may overcorrect the enthalpy and result in a larger error than the approximation $h \cong h_{f \,@\, T}$.)

---

**EXAMPLE 5–11    Enthalpy of Compressed Liquid**

Determine the enthalpy of liquid water at 100°C and 15 MPa (*a*) by using compressed liquid tables, (*b*) by approximating it as a saturated liquid, and (*c*) by using the correction given by Eq. 5–38.

**SOLUTION** The enthalpy of liquid water is to be determined exactly and approximately.

**Analysis** At 100°C, the saturation pressure of water is 101.42 kPa, and since $P > P_{\text{sat}}$, the water exists as a compressed liquid at the specified state.

(*a*) From compressed liquid tables, we read

$$\left. \begin{array}{l} P = 15 \text{ MPa} \\ T = 100°C \end{array} \right\} \quad h = \textbf{430.39 kJ/kg} \qquad \text{(Table A–7)}$$

This is the exact value.

(*b*) Approximating the compressed liquid as a saturated liquid at 100°C, as is commonly done, we obtain

$$h \cong h_{f \,@\, 100°C} = \textbf{419.17 kJ/kg}$$

This value is in error by about 2.6 percent.

(*c*) From Eq. 5–38,

$$h_{@P,T} \cong h_{f \,@\, T} + v_{f \,@\, T}(P - P_{\text{sat} \,@\, T})$$

$$= (419.17 \text{ kJ/kg}) + (0.001 \text{ m}^3 \text{ kg})[(15,\!000 - 101.42) \text{ kPa}]\left( \frac{1 \text{ kJ}}{1 \text{ kPa} \cdot \text{m}^3} \right)$$

$$= \textbf{434.07 kJ/kg}$$

**Discussion** Note that the correction term reduced the error from 2.6 to about 1 percent in this case. However, this improvement in accuracy is often not worth the extra effort involved.

---

**EXAMPLE 5–12    Cooling of an Iron Block by Water**

A 50-kg iron block at 80°C is dropped into an insulated tank that contains 0.5 m³ of liquid water at 25°C. Determine the temperature when thermal equilibrium is reached.

**SOLUTION** An iron block is dropped into water in an insulated tank. The final temperature when thermal equilibrium is reached is to be determined.

**Assumptions** **1** Both water and the iron block are incompressible substances. **2** Constant specific heats at room temperature can be used for water and the iron. **3** The system is stationary and thus the kinetic and potential energy changes are zero, $\Delta KE = \Delta PE = 0$ and $\Delta E = \Delta U$. **4** There are no electrical, shaft, or other forms of work involved. **5** The system is well-insulated and thus there is no heat transfer.

**FIGURE 5–35**
Schematic for Example 5–12.

*Analysis* We take the entire contents of the tank as the *system* (Fig. 5–35). This is a *closed system* since no mass crosses the system boundary during the process. We observe that the volume of a rigid tank is constant, and thus there is no boundary work. The energy balance on the system can be expressed as

$$\underbrace{E_{in} - E_{out}}_{\substack{\text{Net energy transfer} \\ \text{by heat, work, and mass}}} = \underbrace{\Delta E_{system}}_{\substack{\text{Change in internal, kinetic,} \\ \text{potential, etc., energies}}}$$

$$0 = \Delta U$$

The total internal energy $U$ is an extensive property, and therefore it can be expressed as the sum of the internal energies of the parts of the system. Then the total internal energy change of the system becomes

$$\Delta U_{sys} = \Delta U_{iron} + \Delta U_{water} = 0$$

$$[mc(T_2 - T_1)]_{iron} + [mc(T_2 - T_1)]_{water} = 0$$

The specific volume of liquid water at or about room temperature can be taken to be 0.001 m³/kg. Then the mass of the water is

$$m_{water} = \frac{V}{v} = \frac{0.5 \text{ m}^3}{0.001 \text{ m}^3/\text{kg}} = 500 \text{ kg}$$

The specific heats of iron and liquid water are determined from Table A–3 to be $c_{iron}$ = 0.45 kJ/kg·°C and $c_{water}$ = 4.18 kJ/kg·°C. Substituting these values into the energy equation, we obtain

$$(50 \text{ kg})(0.45 \text{ kJ/kg·°C})(T_2 - 80°C) + (500 \text{ kg})(4.18 \text{ kJ/kg·°C})(T_2 - 25°C) = 0$$

$$T_2 = \mathbf{25.6°C}$$

Therefore, when thermal equilibrium is established, both the water and iron will be at 25.6°C.

*Discussion* The small rise in water temperature is due to its large mass and large specific heat.

**FIGURE 5–36**

Schematic for Example 5–13

### EXAMPLE 5–13　Heating of Aluminum Rods in a Furnace

Long cylindrical aluminum rods ($\rho$ = 2700 kg/m³ and $c_p$ = 0.973 kJ/kg·K) of 5-cm diameter are heat treated from 20°C to an average temperature of 400°C by drawing them at a velocity of 8 m/min through a long oven. Determine the rate of heat transfer to the rods in the oven.

**SOLUTION** Aluminum rods are to be heated in an oven to a specified average temperature. The rate of heat transfer to the rods is to be determined.

*Assumptions* **1** The thermal properties of the rods are constant. **2** There are no changes in kinetic and potential energies. **3** The balls are at a uniform temperature when they leave the oven.

*Analysis* Aluminum rods pass through the oven at a constant speed of 8 m/min. That is, an external observer will see that an 8-m long section of cold rods enters and an 8-m long section of hot rods leaves the oven every minute. We take the 8-m long section of the rod as the system. The energy balance for this closed system can be expressed as

$$\underbrace{E_{in} - E_{out}}_{\substack{\text{Net energy transfer} \\ \text{by heat, work, and mass}}} = \underbrace{\Delta E_{system}}_{\substack{\text{Change in internal, kinetic,} \\ \text{potential, etc., energies}}}$$

$$Q_{in} = \Delta U_{rod} = m(u_2 - u_1)$$
$$Q_{in} = mc(T_2 - T_1)$$

The density and specific heat of the rods are given to be $\rho = 2700$ kg/m³ and $c = 0.973$ kJ/kg·K = 0.973 kJ/kg·°C. The amount of heat transfer to an 8-m long section of the rod as it is heated to the specified temperature is determined to be

$$m = \rho V = \rho \frac{\pi D^2}{4} L = (2700 \text{ kg/m}^3) \frac{\pi (0.05 \text{ m})^2}{4} (8 \text{ m}) = 42.41 \text{ kg}$$

$$Q_{in} = mc\,(T_2 - T_1) = (42.41 \text{ kg})\,(0.973 \text{ kJ/kg·°C})\,(400 - 20)°C$$
$$= 15{,}680 \text{ kJ (per 8-m section)}$$

Considering that an 8-m long section of the rods is heated every minute, the rate of heat transfer to the rods in the oven becomes

$$\dot{Q}_{in} = Q_{in}/\Delta t = 15{,}680 \text{ kJ/min} = \textbf{261 kJ/s}$$

**Discussion** This problem can also be solved by working with the *rate* form of the equations as

$$\dot{m} = \rho \dot{V} = \rho \frac{\pi D^2}{4} L/\Delta t = \rho \frac{\pi D^2}{4} V = (2700 \text{ kg/m}^3) \frac{\pi (0.05 \text{ m})^2}{4} (8 \text{ m/min})$$

$$= 42.41 \text{ kg/min}$$

$$\dot{Q}_{in} = \dot{m} c (T_2 - T_1) = (42.41 \text{ kg/min})\,(0.973 \text{ kJ/kg·°C})\,(400 - 20)°C$$
$$= 15{,}680 \text{ kJ/min}$$

which is identical to the result obtained before.

## SUMMARY

Work is the energy transferred as a force acts on a system through a distance. The most common form of mechanical work is the *boundary work*, which is the work associated with the expansion and compression of substances. On a *P-V* diagram, the area under the process curve represents the boundary work for a quasi-equilibrium process. Various forms of boundary work are expressed as follows:

(1) General

$$W_b = \int_1^2 P\,dV$$

(2) Isobaric process

$$W_b = P_0(V_2 - V_1) \qquad (P_1 = P_2 = P_0 = \text{constant})$$

(3) Polytropic process

$$W_b = \frac{P_2 V_2 - P_1 V_1}{1 - n} \quad (n \neq 1) \qquad (PV^n = \text{constant})$$

(4) Isothernal process of an ideal gas

$$W_b = P_1 V_1 \ln \frac{V_2}{V_1}$$

$$= mRT_0 \ln \frac{V_2}{V_1} \qquad (PV = mRT_0 = \text{constant})$$

The first law of thermodynamics is essentially an expression of the conservation of energy principle, also called the energy balance. The general energy balances for *any system* undergoing *any process* can be expressed as

$$\underbrace{E_{in} - E_{out}}_{\substack{\text{Net energy transfer} \\ \text{by heat, work, and mass}}} = \underbrace{\Delta E_{system}}_{\substack{\text{Change in internal, kinetic,} \\ \text{potential, etc., energies}}}$$

It can also be expressed in the *rate form* as

$$\underbrace{\dot{E}_{in} - \dot{E}_{out}}_{\substack{\text{Rate of net energy transfer} \\ \text{by heat, work, and mass}}} = \underbrace{dE_{system}/dt}_{\substack{\text{Rate of change in internal,} \\ \text{kinetic, potential, etc., energies}}}$$

Taking heat transfer *to* the system and work done *by* the system to be positive quantities, the energy balance for a closed system can also be expressed as

$$Q - W = \Delta U + \Delta KE + \Delta PE$$

where

$$W = W_{other} + W_b$$
$$\Delta U = m(u_2 - u_1)$$
$$\Delta KE = \tfrac{1}{2}m(V_2^2 - V_1^2)$$
$$\Delta PE = mg(z_2 - z_1)$$

For a *constant-pressure process*, $W_b + \Delta U = \Delta H$. Thus,

$$Q - W_{other} = \Delta H + \Delta KE + \Delta PE$$

Note that the relation above is limited to constant pressure processes of closed system, and is NOT valid for processes during which pressure varies.

The amount of energy needed to raise the temperature of a unit mass of a substance by one degree is called the *specific heat at constant volume* $c_v$ for a constant-volume process and the *specific heat at constant pressure* $c_p$ for a constant-pressure process. They are defined as

$$c_v = \left(\frac{\partial u}{\partial T}\right)_v \quad \text{and} \quad c_p = \left(\frac{\partial h}{\partial T}\right)_p$$

For ideal gases $u$, $h$, $c_v$, and $c_p$ are functions of temperature alone. The $\Delta u$ and $\Delta h$ of ideal gases are expressed as

$$\Delta u = u_2 - u_1 = \int_1^2 c_v(T)\, dT \cong c_{v,avg}(T_2 - T_1)$$

$$\Delta h = h_2 - h_1 = \int_1^2 c_p(T)\, dT \cong c_{p,avg}(T_2 - T_1)$$

For ideal gases, $c_v$ and $c_p$ are related by

$$c_p = c_v + R$$

where $R$ is the gas constant. The *specific heat ratio* $k$ is defined as

$$k = \frac{c_p}{c_v}$$

For *incompressible substances* (liquids and solids), both the constant-pressure and constant-volume specific heats are identical and denoted by $c$:

$$c_p = c_v = c$$

The $\Delta u$ and $\Delta h$ of incompressible substances are given by

$$\Delta u = \int_1^2 c(T)\, dT \cong c_{avg}(T_2 - T_1)$$

$$\Delta h = \Delta u + v\Delta P$$

## REFERENCES AND SUGGESTED READINGS

**1.** ASHRAE *Handbook of Fundamentals.* SI version. Atlanta, GA: American Society of Heating, Refrigerating, and Air-Conditioning Engineers, Inc., 1993.

**2.** ASHRAE *Handbook of Refrigeration.* SI version. Atlanta, GA: American Society of Heating, Refrigerating, and Air-Conditioning Engineers, Inc., 1994.

# PROBLEMS*

## Moving Boundary Work

**5–1C** Is the boundary work associated with constant-volume systems always zero?

**5–2C** An ideal gas at a given state expands to a fixed final volume first at constant pressure and then at constant temperature. For which case is the work done greater?

**5–3** Show that $1 \text{ kPa}\cdot\text{m}^3 = 1 \text{ kJ}$.

**5–4** The volume of 1 kg of helium in a piston-cylinder device is initially 7 m³. Now helium is compressed to 3 m³ while its pressure is maintained constant at 150 kPa. Determine the initial and final temperatures of helium as well as the work required to compress it, in kJ.

**5–5** A piston–cylinder device initially contains 0.07 m³ of nitrogen gas at 130 kPa and 120°C. The nitrogen is now expanded polytropically to a state of 100 kPa and 100°C. Determine the boundary work done during this process.

**5–6** A piston–cylinder device with a set of stops initially contains 0.3 kg of steam at 1.0 MPa and 400°C. The location of the stops corresponds to 60 percent of the initial volume. Now the steam is cooled. Determine the compression work if the final state is (*a*) 1.0 MPa and 250°C and (*b*) 500 kPa. (*c*) Also determine the temperature at the final state in part (*b*).

**FIGURE P5–6**

**5–7** A piston–cylinder device initially contains 0.07 m³ of nitrogen gas at 130 kPa and 180°C. The nitrogen is now expanded to a pressure of 80 kPa polytropically with a polytropic exponent whose value is equal to the specific heat ratio

(called *isentropic expansion*). Determine the final temperature and the boundary work done during this process.

**5–8** A mass of 5 kg of saturated water vapor at 300 kPa is heated at constant pressure until the temperature reaches 200°C. Calculate the work done by the steam during this process. *Answer:* 165.9 kJ

**5–9** A frictionless piston–cylinder device initially contains 50 L of saturated liquid refrigerant-134a. The piston is free to move, and its mass is such that it maintains a pressure of 500 kPa on the refrigerant. The refrigerant is now heated until its temperature rises to 70°C. Calculate the work done during this process. *Answer:* 1600 kJ

**FIGURE P5–9**

**5–10** Reconsider Prob. 5–9. Using EES (or other) software, investigate the effect of pressure on the work done. Let the pressure vary from 200 to 1200 kPa. Plot the work done versus the pressure, and discuss the results. Explain why the plot is not linear. Also plot the process described in Prob. 5–9 on the *P-v* diagram.

**5–11** 1-m³ of saturated liquid water at 200°C is expanded isothermally in a closed system until its quality is 80 percent. Determine the total work produced by this expansion, in kJ.

**5–12** A mass of 2.4 kg of air at 150 kPa and 12°C is contained in a gas-tight, frictionless piston–cylinder device. The air is now compressed to a final pressure of 600 kPa. During the process, heat is transferred from the air such that the temperature inside the cylinder remains constant. Calculate the work input during this process. *Answer:* 272 kJ

**5–13** Determine the boundary work done by a gas during an expansion process if the pressure and volume values at various states are measured to be 300 kPa, 1 L; 290 kPa, 1.1 L; 270 kPa, 1.2 L; 250 kPa, 1.4 L; 220 kPa, 1.7 L; and 200 kPa, 2 L.

**5–14** During some actual expansion and compression processes in piston–cylinder devices, the gases have been observed to satisfy the relationship $PV^n = C$, where *n* and *C* are constants. Calculate the work done when a gas expands from 350 kPa and 0.03 m³ to a final volume of 0.2 m³ for the case of $n = 1.5$.

---

* Problems designated by a "C" are concept questions, and students are encouraged to answer them all. Problems designated by an "E" are in English units, and the SI users can ignore them. Problems with the ☀ icon are solved using EES, and complete solutions together with parametric studies are included on the enclosed DVD. Problems with the 💻 icon are comprehensive in nature, and are intended to be solved with a computer, preferably using the EES software that accompanies this text.

**5–15**  Reconsider Prob. 5–14. Using the EES (or other) software, plot the process described in the problem on a $P$-$V$ diagram, and investigate the effect of the polytropic exponent $n$ on the boundary work. Let the polytropic exponent vary from 1.1 to 1.6. Plot the boundary work versus the polytropic exponent, and discuss the results.

**5–16** A frictionless piston–cylinder device contains 2 kg of nitrogen at 100 kPa and 300 K. Nitrogen is now compressed slowly according to the relation $PV^{1.4} = $ constant until it reaches a final temperature of 360 K. Calculate the work input during this process. *Answer: 89 kJ*

**FIGURE P5–16**

**5–17**  The equation of state of a gas is given as $\overline{v}(P + 10/\overline{v}^2) = R_uT$, where the units of $\overline{v}$ and $P$ are m$^3$/kmol and kPa, respectively. Now 0.2 kmol of this gas is expanded in a quasi-equilibrium manner from 2 to 4 m$^3$ at a constant temperature of 350 K. Determine (*a*) the unit of the quantity 10 in the equation and (*b*) the work done during this isothermal expansion process.

**5–18**  Reconsider Prob. 5–17. Using the integration feature of the EES software, calculate the work done, and compare your result with the "hand-calculated" result obtained in Prob. 5–17. Plot the process described in the problem on a $P$-$v$ diagram.

**5–19** Carbon dioxide contained in a piston–cylinder device is compressed from 0.3 to 0.1 m$^3$. During the process, the pressure and volume are related by $P = aV^{-2}$, where $a = 8$ kPa·m$^6$. Calculate the work done on the carbon dioxide during this process. *Answer: 53.3 kJ*

**5–20E** During an expansion process, the pressure of a gas changes from 15 to 100 psia according to the relation $P = aV + b$, where $a = 5$ psia/ft$^3$ and $b$ is a constant. If the initial volume of the gas is 7 ft$^3$, calculate the work done during the process. *Answer: 181 Btu*

**5–21** A piston–cylinder device initially contains 0.25 kg of nitrogen gas at 130 kPa and 180°C. The nitrogen is now expanded isothermally to a pressure of 80 kPa. Determine the boundary work done during this process. *Answer: 16.3 kJ*

**FIGURE P5–21**

**5–22** A piston–cylinder device contains 0.15 kg of air initially at 2 MPa and 350°C. The air is first expanded isothermally to 500 kPa, then compressed polytropically with a polytropic exponent of 1.2 to the initial pressure, and finally compressed at the constant pressure to the initial state. Determine the boundary work for each process and the net work of the cycle.

**5–23** 1-kg of water that is initially at 90°C with a quality of 10 percent occupies a spring-loaded piston–cylinder device, such as that in Fig. P5–23. This device is now heated until the pressure rises to 800 kPa and the temperature is 250°C. Determine the total work produced during this process, in kJ. *Answer: 24.5 kJ*

**FIGURE P5–23**

**5–24** 1.5-kg water that is initially at 1 MPa and 30 percent quality occupies a spring-loaded piston–cylinder device. This device is now cooled until the water is a saturated liquid at 100°C. Calculate the total work produced during this process, in kJ.

**5–25** An ideal gas undergoes two processes in a piston-cylinder device as follows:

1-2 Polytropic compression from $T_1$ and $P_1$ with a polytropic exponent $n$ and a compression ratio of $r = V_1/V_2$.

2-3 Constant pressure expansion at $P_3 = P_2$ until $V_3 = V_1$.

(*a*) Sketch the processes on a single $P$-$V$ diagram.

(*b*) Obtain an expression for the ratio of the compression-to-expansion work as a function of $n$ and $r$.

(*c*) Find the value of this ratio for values of $n = 1.4$ and $r = 6$. *Answers: (b)* $\dfrac{1}{n-1}\left(\dfrac{1-r^{1-n}}{r-1}\right)$, *(c)* 0.256

## Closed System Energy Analysis

**5–26E** Complete the table below on the basis of the conservation of energy principle for a closed system.

| $Q_{in}$ Btu | $W_{out}$ Btu | $E_1$ Btu | $E_2$ Btu | $m$ lbm | $e_2 - e_1$ Btu/lbm |
|------|------|------|------|------|------|
| 350 | — | 1020 | 860 | 3 | — |
| 350 | 130 | 550 | — | 5 | — |
| — | 260 | 600 | — | 2 | 150 |
| −500 | — | 1400 | 900 | 7 | — |
| — | −50 | 1000 | — | 3 | −200 |

**5–27E** A closed system like that shown in Fig. P5–27E is operated in an adiabatic manner. First, 15,000 lbf·ft of work are done by this system. Then, work is applied to the stirring device to raise the internal energy of the fluid by 10.28 Btu. Calculate the net increase in the internal energy of this system.

**FIGURE P5–27E**

**5–28E** A closed system undergoes a process in which there is no internal energy change. During this process, the system produces $1.1 \times 10^6$ lbf·ft of work. Calculate the heat transfer for this process, in Btu.

**5–29** A rigid container equipped with a stirring device contains 1.5 kg of motor oil. Determine the rate of specific energy increase when heat is transferred to the oil at a rate of 1 W, and 1.5 W of power is applied to the stirring device.

**5–30** A well-insulated rigid tank contains 2 kg of a saturated liquid–vapor mixture of water at 150 kPa. Initially, three-quarters of the mass is in the liquid phase. An electric resistor placed in the tank is connected to a 110-V source, and a current of 8 A flows through the resistor when the switch is turned on. Determine how long it will take to vaporize all the liquid in the tank. Also, show the process on a $T$-$v$ diagram with respect to saturation lines.

**FIGURE P5–30**

**5–31** Reconsider Prob. 5–30. Using EES (or other) software, investigate the effect of the initial mass of water on the length of time required to completely vaporize the liquid. Let the initial mass vary from 1 to 10 kg. Plot the vaporization time against the initial mass, and discuss the results.

**5–32** Saturated water vapor at 200°C is isothermally condensed to a saturated liquid in a piston–cylinder device. Calculate the heat transfer and the work done during this process, in kJ/kg. *Answers:* 1940 kJ/kg, 196 kJ/kg

**FIGURE P5–32**

**5–33** A rigid 10-L vessel initially contains a mixture of liquid water and vapor at 100°C with 12.3 percent quality. The mixture is then heated until its temperature is 150°C. Calculate the heat transfer required for this process. *Answer:* 46.9 kJ

**FIGURE P5–33**

**5–34** A fixed mass of saturated water vapor at 300 kPa is isothermally cooled until it is a saturated liquid. Calculate the amount of heat rejected during this process, in kJ/kg.

**5–35** Saturated water vapor in a closed system is condensed by cooling it at constant pressure to a saturated liquid at 40 kPa. Determine the heat transferred and the work done during this process, in kJ/kg.

**5–36** An insulated piston–cylinder device contains 5 L of saturated liquid water at a constant pressure of 175 kPa. Water is stirred by a paddle wheel while a current of 8 A flows for 45 min through a resistor placed in the water. If one-half of the liquid is evaporated during this constant-pressure process and the paddle-wheel work amounts to 400 kJ, determine the voltage of the source. Also, show the process on a *P-v* diagram with respect to saturation lines. *Answer: 224 V*

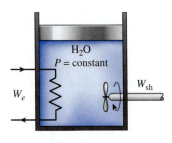

**FIGURE P5–36**

**5–37** A piston–cylinder device initially contains steam at 200 kPa, 200°C, and 0.4 m³. At this state, a linear spring ($F \propto x$) is touching the piston but exerts no force on it. Heat is now slowly transferred to the steam, causing the pressure and the volume to rise to 250 kPa and 0.6 m³, respectively. Show the process on a *P-v* diagram with respect to saturation lines and determine (*a*) the final temperature, (*b*) the work done by the steam, and (*c*) the total heat transferred. *Answers: (a) 606°C, (b) 45 kJ, (c) 288 kJ*

**5–38** Reconsider Prob. 5–37. Using EES (or other) software, investigate the effect of the initial temperature of steam on the final temperature, the work done, and the total heat transfer. Let the initial temperature vary from 150 to 250°C. Plot the final results against the initial temperature, and discuss the results.

**5–39** A 30-L electrical radiator containing heating oil is placed in a 50-m³ room. Both the room and the oil in the radiator are initially at 10°C. The radiator with a rating of 1.8 kW is now turned on. At the same time, heat is lost from the room at an average rate of 0.35 kJ/s. After some time, the average temperature is measured to be 20°C for the air in the room, and 50°C for the oil in the radiator. Taking the density and the specific heat of the oil to be 950 kg/m³ and 2.2 kJ/kg·°C, respectively, determine how long the heater is kept on. Assume the room is well-sealed so that there are no air leaks.

**FIGURE P5–39**

**5–40** Steam at 75 kPa and 8 percent quality is contained in a spring-loaded piston–cylinder device, as shown in Fig. P5–40, with an initial volume of 2 m³. Steam is now heated until its volume is 5 m³ and its pressure is 225 kPa. Determine the heat transferred to and the work produced by the steam during this process.

**FIGURE P5–37**

**FIGURE P5–40**

**5–41** Refrigerant-134a at 600 kPa and 150°C is contained in a spring-loaded piston–cylinder device with an initial volume of 0.3 m³. The refrigerant is now cooled until its temperature is −30°C and its volume is 0.1 m³. Determine the heat transferred to and the work produced by the refrigerant during this process. *Answers: 1849 kJ heat out, 68.4 kJ work in*

**5–42E** Saturated R-134a vapor at 100°F is condensed at constant pressure to a saturated liquid in a closed piston–cylinder system. Calculate the heat transfer and work done during this process, in Btu/lbm.

**5–43** 0.8-kg of saturated liquid R-134a with an initial temperature of −5°C is contained in a well-insulated, weighted piston–cylinder device. This device contains an electrical resistor, as shown in Fig. P5–43, to which 10 volts are applied causing a current of 2 amperes to flow through the resistor. Determine the time required for the refrigerant to be converted to a saturated vapor, and the final temperature.

*I*
$\Delta V$

**FIGURE P5–43**

**5–44** Two tanks (Tank A and Tank B) are separated by a partition. Initially Tank A contains 2-kg steam at 1 MPa and 300°C while Tank B contains 3-kg saturated liquid–vapor mixture at 150°C with a vapor mass fraction of 50 percent. The partition is now removed and the two sides are allowed to mix until the mechanical and thermal equilibrium are established. If the pressure at the final state is 300 kPa, determine (*a*) the temperature and quality of the steam (if mixture) at the final state and (*b*) the amount of heat lost from the tanks.

| Tank A | Tank B |
| --- | --- |
| 2 kg | 3 kg |
| 1 MPa | 150°C |
| 300°C | $x = 0.5$ |

*Q*

**FIGURE P5–44**

## Specific Heats, $\Delta u$, and $\Delta h$ of Ideal Gases

**5–45C** In the relation $\Delta u = mc_v\Delta T$, what is the correct unit of $c_v$: kJ/kg·°C or kJ/kg·K?

**5–46C** Is the relation $\Delta u = mc_{v,avg}\Delta T$ restricted to constant-volume processes only, or can it be used for any kind of process of an ideal gas?

**5–47C** Is the relation $\Delta h = mc_{p,avg}\Delta T$ restricted to constant-pressure processes only, or can it be used for any kind of process of an ideal gas?

**5–48C** Is the energy required to heat air from 295 to 305 K the same as the energy required to heat it from 345 to 355 K? Assume the pressure remains constant in both cases.

**5–49C** A fixed mass of an ideal gas is heated from 50 to 80°C at a constant pressure of (*a*) 1 atm and (*b*) 3 atm. For which case do you think the energy required will be greater? Why?

**5–50C** A fixed mass of an ideal gas is heated from 50 to 80°C at a constant volume of (*a*) 1 m³ and (*b*) 3 m³. For which case do you think the energy required will be greater? Why?

**5–51C** When undergoing a given temperature change, which of the two gases—air or oxygen—experiences the largest change in its (*a*) enthalpy, $h$ and (*b*) internal energy, $u$?

**5–52** Show that for an ideal gas $\bar{c}_p = \bar{c}_v + R_u$.

**5–53** What is the change in the enthalpy, in kJ/kg, of oxygen as its temperature changes from 150 to 200°C? Is there any difference if the temperature change were from 0 to 50°C? Does the pressure at the beginning and end of this process have any effect on the enthalpy change?

**5–54E** Air is compressed from 20 psia and 70°F to 150 psia in a compressor. The compressor is operated such that the air temperature remains constant. Calculate the change in the specific volume of air as it passes through this compressor.

**5–55** The temperature of 2 kg of neon is increased from 20 to 180°C. Calculate the change in the total internal energy of the neon, in kJ. Would the internal energy change be any different if the neon were replaced with argon?

**5–56** Calculate the change in the enthalpy of argon, in kJ/kg, when it is cooled from 100 to 25°C. If neon had undergone this same change of temperature, would its enthalpy change have been any different?

**5–57E** Determine the enthalpy change $\Delta h$ of oxygen, in Btu/lbm, as it is heated from 800 to 1500 R, using (*a*) the empirical specific heat equation as a function of temperature (Table A–2E*c*), (*b*) the $c_p$ value at the average temperature (Table A–2E*b*), and (*c*) the $c_p$ value at room temperature (Table A–2E*a*). *Answers: (a) 170.1 Btu/lbm, (b) 178.5 Btu/lbm, (c) 153.3 Btu/lbm*

**5–58**  Determine the internal energy change $\Delta u$ of hydrogen, in kJ/kg, as it is heated from 200 to 800 K, using (a) the empirical specific heat equation as a function of temperature (Table A–2c), (b) the $c_v$ value at the average temperature (Table A–2b), and (c) the $c_v$ value at room temperature (Table A–2a).

### Closed-System Energy Analysis: Ideal Gases

**5–59C**  Is it possible to compress an ideal gas isothermally in an adiabatic piston–cylinder device? Explain.

**5–60**  1-kg of oxygen is heated from 20 to 120°C. Determine the amount of heat transfer required when this is done during a (a) constant-volume process and (b) isobaric process.

**FIGURE P5–60**

**5–61E**  A rigid tank contains 20 lbm of air at 50 psia and 80°F. The air is now heated until its pressure doubles. Determine (a) the volume of the tank and (b) the amount of heat transfer.  *Answers: (a)* 80 ft³, *(b)* 1898 Btu

**5–62E**  Determine the internal energy change of air in a rigid container as its temperature is increased by transferring 50 Btu/lbm of heat.

**5–63E**  Nitrogen gas to 20 psia and 100°F initially occupies a volume of 1 ft³ in a rigid container equipped with a stirring paddle wheel. After 5000 lbf·ft of paddle wheel work is done on nitrogen, what is its final temperature?  *Answer:* 489°F

**5–64**  Consider a piston-cylinder device that contains nitrogen gas as the system. Initially, the system is at 1 MPa and 427°C. It now undergoes an isobaric process until its temperature is 27°C. Determine the final pressure and the heat transfer, in kJ/kg, associated with this process.  *Answer:* 416 kJ/kg

**5–65**  A 4-m × 5-m × 6-m room is to be heated by a baseboard resistance heater. It is desired that the resistance heater be able to raise the air temperature in the room from 5 to 25°C within 11 min. Assuming no heat losses from the room and an atmospheric pressure of 100 kPa, determine the required power of the resistance heater. Assume constant specific heats at room temperature.  *Answer:* 3.28 kW

**5–66**  A student living in a 4-m × 6-m × 6-m dormitory room turns on her 150-W fan before she leaves the room on a summer day, hoping that the room will be cooler when she comes back in the evening. Assuming all the doors and windows are tightly closed and disregarding any heat transfer through the walls and the windows, determine the temperature in the room when she comes back 10 h later. Use specific heat values at room temperature, and assume the room to be at 100 kPa and 15°C in the morning when she leaves.  *Answer:* 58.2°C

**FIGURE P5–66**

**5–67E**  A 3-ft³ adiabatic rigid container is divided into two equal volumes by a thin membrane, as shown in Fig. P5–67E. Initially, one of these chambers is filled with air at 100 psia and 100°F while the other chamber is evacuated. Determine the internal energy change of the air when the membrane is ruptured. Also determine the final air pressure in the container.

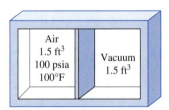

**FIGURE P5–67E**

**5–68**  A closed system containing 2-kg of air undergoes an isothermal process from 600 kPa and 200°C to 80 kPa. Determine the initial volume of this system, the work done, and the heat transfer during this process.  *Answers:* 0.453 m³, 547 kJ, 547 kJ

**5–69** A piston–cylinder device containing argon gas as the system undergoes an isothermal process from 200 kPa and 100°C to 50 kPa. During the process, 1500 kJ of heat is transferred to the system. Determine the mass of this system and the amount of work produced.

**5–70** Argon is compressed in a polytropic process with $n = 1.2$ from 120 kPa and 10°C to 800 kPa in a piston–cylinder device. Determine the work produced and heat transferred during this compression process, in kJ/kg.

**FIGURE P5–70**

**5–71** A spring-loaded piston-cylinder device contains 1 kg of carbon dioxide. This system is heated from 100 kPa and 25°C to 1000 kPa and 300°C. Determine the total heat transfer to and work produced by this system.

**FIGURE P5–71**

**5–72** Air is contained in a variable-load piston-cylinder device equipped with a paddle wheel. Initially, air is at 500 kPa and 27°C. The paddle wheel is now turned by an external electric motor until 50 kJ/kg of work has been transferred to air. During this process, heat is transferred to maintain a constant air temperature while allowing the gas volume to triple. Calculate the required amount of heat transfer, in kJ/kg. *Answer: 44.6 kJ/kg*

**FIGURE P5–72**

**5–73** A mass of 15 kg of air in a piston–cylinder device is heated from 25 to 77°C by passing current through a resistance heater inside the cylinder. The pressure inside the cylinder is held constant at 300 kPa during the process, and a heat loss of 60 kJ occurs. Determine the electric energy supplied, in kWh. *Answer: 0.235 kWh*

**FIGURE P5–73**

**5–74** A piston–cylinder device contains 1.5 kg of nitrogen initially at 100 kPa and 17°C. The nitrogen is now compressed slowly in a polytropic process during which $PV^{1.3} =$ constant until the volume is reduced by one-half. Determine the work done and the heat transfer for this process.

**5–75** Reconsider Prob. 5–74. Using EES (or other) software, plot the process described in the problem on a $P$-$V$ diagram, and investigate the effect of the polytropic exponent $n$ on the boundary work and heat transfer. Let the polytropic exponent vary from 1.1 to 1.6. Plot the boundary work and the heat transfer versus the polytropic exponent, and discuss the results.

**5–76**  A piston–cylinder device, with a set of stops on the top, initially contains 3 kg of air at 200 kPa and 27°C. Heat is now transferred to the air, and the piston rises until it hits the stops, at which point the volume is twice the initial volume. More heat is transferred until the pressure inside the cylinder also doubles. Determine the work done and the amount of heat transfer for this process. Also, show the process on a P-v diagram.

**5–77** Air is contained in a cylinder device fitted with a piston-cylinder. The piston initially rests on a set of stops, and a pressure of 300 kPa is required to move the piston. Initially, the air is at 100 kPa and 27°C and occupies a volume of 0.4 m³. Determine the amount of heat transferred to the air, in kJ, while increasing the temperature to 1200 K. Assume air has constant specific heats evaluated at 300 K.   *Answer: 340 kJ*

Air
100 kPa
27°C
0.4 m³

**FIGURE P5–77**

**5–78** Air is contained in a piston-cylinder device at 600 kPa and 927°C, and occupies a volume of 0.8 m³. The air undergoes and isothermal (constant temperature) process until the pressure in reduced to 300 kPa. The piston is now fixed in place and not allowed to move while a heat transfer process takes place until the air reaches 27°C.

(a) Sketch the system showing the energies crossing the boundary and the P-V diagram for the combined processes.

(b) For the combined processes determine the net amount of heat transfer, in kJ, and its direction.

Assume air has constant specific heats evaluated at 300 K.

**5–79** Argon is contained in a cylinder device fitted with a piston. Initially, the argon is at 100 kPa and 27°C and occupies a volume of 0.4 m³. The argon is first compressed while the temperature is held constant until the volume is 0.2 m³. Then the argon expands while the pressure is held constant until the volume is 0.6 m³.

(a) Sketch both processes on a single P-V diagram showing the direction of the processes and label the end states as 1, 2, and 3.

(b) Determine the total amount of net heat transferred to the argon, in kJ, during the combined processes.
*Answer: (b) 172 kJ*

**5–80** An ideal gas contained in a piston-cylinder device undergoes an isothermal compression process which begins with an initial pressure and volume of 100 kPa and 0.6 m³, respectively. During the process there is a heat transfer of 60 kJ from the ideal gas to the surroundings. Determine the volume and pressure at the end of the process.   *Answers: 0.221 m³, 272 kPa*

### Closed-System Energy Analysis: Solids and Liquids

**5–81** A 1-kg block of iron is heated from 25 to 75°C. What is the change in the iron's total internal energy and enthalpy?

**5–82E** The state of liquid water is changed from 50 psia and 50°F to 2000 psia and 100°F. Determine the change in the internal energy and enthalpy of water on the basis of the (a) compressed liquid tables, (b) incompressible substance approximation and property tables, and (c) specific-heat model.

**5–83E** During a picnic on a hot summer day, all the cold drinks disappeared quickly, and the only available drinks were those at the ambient temperature of 75°F. In an effort to cool a 12-fluid-oz drink in a can, a person grabs the can and starts shaking it in the iced water of the chest at 32°F. Using the properties of water for the drink, determine the mass of ice that will melt by the time the canned drink cools to 45°F.

**5–84** Consider a 1000-W iron whose base plate is made of 0.5-cm-thick aluminum alloy 2024-T6 ($\rho = 2770$ kg/m³ and $c_p = 875$ J/kg·°C). The base plate has a surface area of 0.03 m². Initially, the iron is in thermal equilibrium with the ambient air at 22°C. Assuming 90 percent of the heat generated in the resistance wires is transferred to the plate, determine the minimum time needed for the plate temperature to reach 200°C.

**FIGURE P5–84**

**5–85** Stainless steel ball bearings ($\rho$ = 8085 kg/m³ and $c_p$ = 0.480 kJ/kg·°C) having a diameter of 1.2 cm are to be quenched in water at a rate of 800 per minute. The balls leave the oven at a uniform temperature of 900°C and are exposed to air at 25°C for a while before they are dropped into the water. If the temperature of the balls drops to 850°C prior to quenching, determine the rate of heat transfer from the balls to the air.

**5–86** Carbon steel balls ($\rho$ = 7833 kg/m³ and $c_p$ = 0.465 kJ/kg·°C) 8 mm in diameter are annealed by heating them first to 900°C in a furnace, and then allowing them to cool slowly to 100°C in ambient air at 35°C. If 2500 balls are to be annealed per hour, determine the total rate of heat transfer from the balls to the ambient air.   *Answer: 542 W*

**FIGURE P5–86**

**5–87** An ordinary egg can be approximated as a 5.5-cm-diameter sphere. The egg is initially at a uniform temperature of 8°C and is dropped into boiling water at 97°C. Taking the properties of the egg to be $\rho$ = 1020 kg/m³ and $c_p$ = 3.32 kJ/kg·°C, determine how much heat is transferred to the egg by the time the average temperature of the egg rises to 80°C.

**5–88E** In a production facility, 1.2-in-thick 2-ft × 2-ft square brass plates ($\rho$ = 532.5 lbm/ft³ and $c_p$ = 0.091 Btu/lbm·°F) that are initially at a uniform temperature of 75°F are heated by passing them through an oven at 1300°F at a rate of 300 per minute. If the plates remain in the oven until their average temperature rises to 1000°F, determine the rate of heat transfer to the plates in the furnace.

**FIGURE P5–88E**

**5–89** Long cylindrical steel rods ($\rho$ = 7833 kg/m³ and $c_p$ = 0.465 kJ/kg·°C) of 8-cm diameter are heat-treated by drawing them at a velocity of 2 m/min through an oven maintained at 900°C. If the rods enter the oven at 30°C and leave at a mean temperature of 700°C, determine the rate of heat transfer to the rods in the oven.

**5–90** An electronic device dissipating 25 W has a mass of 20 g and a specific heat of 850 J/kg·°C. The device is lightly used, and it is on for 5 min and then off for several hours, during which it cools to the ambient temperature of 25°C. Determine the highest possible temperature of the device at the end of the 5-min operating period. What would your answer be if the device were attached to a 0.5-kg aluminum heat sink? Assume the device and the heat sink to be nearly isothermal.

**5–91** Reconsider Prob. 5–90. Using EES (or other) software, investigate the effect of the mass of the heat sink on the maximum device temperature. Let the mass of heat sink vary from 0 to 1 kg. Plot the maximum temperature against the mass of heat sink, and discuss the results.

**5–92** If you ever slapped someone or got slapped yourself, you probably remember the burning sensation. Imagine you had the unfortunate occasion of being slapped by an angry person, which caused the temperature of the affected area of your face to rise by 1.8°C (ouch!). Assuming the slapping hand has a mass of 1.2 kg and about 0.150 kg of the tissue on the face and the hand is affected by the incident, estimate the velocity of the hand just before impact. Take the specific heat of the tissue to be 3.8 kJ/kg·K.

### Review Problems

**5–93** Does the concept of a specific heat property (either at constant volume or constant pressure) have a meaning for substances that are undergoing a phase change? Why or why not?

**5–94** 10-kg of nitrogen is heated from 20°C to 250°C. Determine the total amount of heat required when this is done in (a) a constant-volume process and (b) an isobaric process.

**5–95** Consider a classroom that is losing heat to the outdoors at a rate of 12,000 kJ/h. If there are 40 students in class, each dissipating sensible heat at a rate of 84 W, determine if it is necessary to turn the heater in the classroom on to prevent the room temperature from dropping.

**5–96E** Air in the amount of 2 lbm is contained in a well-insulated, rigid vessel equipped with a stirring paddle wheel. The initial state of this air is 30 psia and 60°F. How much work, in Btu, must be transferred to the air with the paddle wheel to raise the air pressure to 40 psia? Also, what is the final temperature of air?

**FIGURE P5–96E**

**5–97** Air is expanded in a polytropic process with $n = 1.5$ from 2 MPa and 300°C to 80 kPa in a piston-cylinder device. Determine the final temperature of the air.

**5–98** Nitrogen at 100 kPa and 25°C in a rigid vessel is heated until its pressure is 300 kPa. Calculate the work done and the heat transferred during this process, in kJ/kg.

**5–99** A well-insulated rigid vessel contains 3 kg of saturated liquid water at 40°C. The vessel also contains an electrical resistor that draws 10 amperes when 50 volts are applied. Determine the final temperature in the vessel after the resistor has been operating for 30 minutes. *Answer:* 119°C

**5–100** A frictionless piston–cylinder device initially contains air at 100 kPa and 0.15 m³. At this state, a linear spring ($F \propto x$) is touching the piston but exerts no force on it. The air is now heated to a final state of 0.45 m³ and 800 kPa. Determine (a) the total work done by the air and (b) the work done against the spring. Also, show the process on a P-v diagram. *Answers:* (a) 135 kJ, (b) 105 kJ

**FIGURE P5–100**

**5–101** A mass of 5 kg of saturated liquid–vapor mixture of water is contained in a piston–cylinder device at 125 kPa. Initially, 2 kg of the water is in the liquid phase and the rest is in the vapor phase. Heat is now transferred to the water, and the piston, which is resting on a set of stops, starts moving when

the pressure inside reaches 300 kPa. Heat transfer continues until the total volume increases by 20 percent. Determine (a) the initial and final temperatures, (b) the mass of liquid water when the piston first starts moving, and (c) the work done during this process. Also, show the process on a P-v diagram.

**FIGURE P5–101**

**5–102E** A spherical balloon contains 7 lbm of air at 30 psia and 600 R. The balloon material is such that the pressure inside is always proportional to the square of the diameter. Determine the work done when the volume of the balloon doubles as a result of heat transfer. *Answer:* 376 Btu

**5–103E** Reconsider Prob. 5–102E. Using the integration feature of the EES software, determine the work done. Compare the result with your "hand-calculated" result.

**5–104** A mass of 12 kg of saturated refrigerant-134a vapor is contained in a piston–cylinder device at 240 kPa. Now 300 kJ of heat is transferred to the refrigerant at constant pressure while a 110-V source supplies current to a resistor within the cylinder for 6 min. Determine the current supplied if the final temperature is 70°C. Also, show the process on a T-v diagram with respect to the saturation lines. *Answer:* 12.8 A

**FIGURE P5–104**

**5–105** Saturated water vapor at 200°C is condensed to a saturated liquid at 50°C in a spring-loaded piston-cylinder device. Determine the heat transfer for this process, in kJ/kg.

**5–106** A piston–cylinder device contains helium gas initially at 150 kPa, 20°C, and 0.5 m³. The helium is now compressed in a polytropic process ($PV^n$ = constant) to 400 kPa and 140°C. Determine the heat loss or gain during this process.  *Answer: 11.2 kJ loss*

**FIGURE P5–106**

**5–107** Nitrogen gas is expanded in a polytropic process with $n = 1.45$ from 2 MPa and 1200 K to 200 kPa in a piston–cylinder device. How much work is produced and heat is transferred during this expansion process, in kJ/kg?

**5–108** A frictionless piston–cylinder device and a rigid tank initially contain 12 kg of an ideal gas each at the same temperature, pressure, and volume. It is desired to raise the temperatures of both systems by 15°C. Determine the amount of extra heat that must be supplied to the gas in the cylinder which is maintained at constant pressure to achieve this result. Assume the molar mass of the gas is 25.

**5–109** A passive solar house that is losing heat to the outdoors at an average rate of 50,000 kJ/h is maintained at 22°C at all times during a winter night for 10 h. The house is to be heated by 50 glass containers each containing 20 L of water that is heated to 80°C during the day by absorbing solar energy. A thermostat-controlled 15-kW back-up electric resistance heater turns on whenever necessary to keep the house at 22°C. (a) How long did the electric heating system run that night? (b) How long would the electric heater run that night if the house incorporated no solar heating?  *Answers: (a) 4.77 h, (b) 9.26 h*

**FIGURE P5–109**

**5–110** An 1800-W electric resistance heating element is immersed in 40 kg of water initially at 20°C. Determine how long it will take for this heater to raise the water temperature to 80°C.

**5–111** One ton (1000 kg) of liquid water at 50°C is brought into a well-insulated and well-sealed 4-m × 5-m × 6-m room initially at 15°C and 95 kPa. Assuming constant specific heats for both air and water at room temperature, determine the final equilibrium temperature in the room.  *Answer: 49.2°C*

**5–112** A 4-m × 5-m × 6-m room is to be heated by one ton (1000 kg) of liquid water contained in a tank that is placed in the room. The room is losing heat to the outside at an average rate of 8000 kJ/h. The room is initially at 20°C and 100 kPa and is maintained at an average temperature of 20°C at all times. If the hot water is to meet the heating requirements of this room for a 24-h period, determine the minimum temperature of the water when it is first brought into the room. Assume constant specific heats for both air and water at room temperature.

**5–113** The energy content of a certain food is to be determined in a bomb calorimeter that contains 3 kg of water by burning a 2-g sample of it in the presence of 100 g of air in the reaction chamber. If the water temperature rises by 3.2°C when equilibrium is established, determine the energy content of the food, in kJ/kg, by neglecting the thermal energy stored in the reaction chamber and the energy supplied by the mixer. What is a rough estimate of the error involved in neglecting the thermal energy stored in the reaction chamber?  *Answer: 20,060 kJ/kg*

**FIGURE P5–113**

**5–114** A 68-kg man whose average body temperature is 39°C drinks 1 L of cold water at 3°C in an effort to cool down. Taking the average specific heat of the human body to be 3.6 kJ/kg·°C, determine the drop in the average body temperature of this person under the influence of this cold water.

**5–115** A 0.3-L glass of water at 20°C is to be cooled with ice to 5°C. Determine how much ice needs to be added to the water, in grams, if the ice is at (a) 0°C and (b) −20°C. Also determine how much water would be needed if the cooling is to be done with cold water at 0°C. The melting temperature and the heat of fusion of ice at atmospheric pressure are 0°C and 333.7 kJ/kg, respectively, and the density of water is 1 kg/L.

**5–116** ![EES] Reconsider Prob. 5–115. Using EES (or other) software, investigate the effect of the initial temperature of the ice on the final mass required. Let the ice temperature vary from –26 to 0°C. Plot the mass of ice against the initial temperature of ice, and discuss the results.

**5–117** A rigid tank containing 0.4 m³ of air at 400 kPa and 30°C is connected by a valve to a piston–cylinder device with zero clearance. The mass of the piston is such that a pressure of 200 kPa is required to raise the piston. The valve is now opened slightly, and air is allowed to flow into the cylinder until the pressure in the tank drops to 200 kPa. During this process, heat is exchanged with the surroundings such that the entire air remains at 30°C at all times. Determine the heat transfer for this process.

**FIGURE P5–117**

**5–118** A well-insulated 4-m × 4-m × 5-m room initially at 10°C is heated by the radiator of a steam heating system. The radiator has a volume of 15 L and is filled with superheated vapor at 200 kPa and 200°C. At this moment both the inlet and the exit valves to the radiator are closed. A 120-W fan is used to distribute the air in the room. The pressure of the steam is observed to drop to 100 kPa after 30 min as a result of heat transfer to the room. Assuming constant specific heats for air at room temperature, determine the average temperature of air in 30 min. Assume the air pressure in the room remains constant at 100 kPa.

**FIGURE P5–118**

**5–119** Consider a well-insulated horizontal rigid cylinder that is divided into two compartments by a piston that is free to move but does not allow either gas to leak into the other side. Initially, one side of the piston contains 1 m³ of $N_2$ gas at 500 kPa and 120°C while the other side contains 1 m³ of He gas at 500 kPa and 40°C. Now thermal equilibrium is established in the cylinder as a result of heat transfer through the piston. Using constant specific heats at room temperature, determine the final equilibrium temperature in the cylinder. What would your answer be if the piston were not free to move?

**FIGURE P5–119**

**5–120** Repeat Prob. 5–119 by assuming the piston is made of 8 kg of copper initially at the average temperature of the two gases on both sides. *Answer:* 83.7°C

**5–121** ![EES] Reconsider Prob. 5–120. Using EES (or other) software, investigate the effect of the mass of the copper piston on the final equilibrium temperature. Let the mass of piston vary from 1 to 10 kg. Plot the final temperature against the mass of piston, and discuss the results.

**5–122** An insulated piston-cylinder device initially contains 1.4-kg saturated liquid water at 200°C. Now an electric resistor placed in the tank is turned on for 20 min until the volume quadruples. Determine (a) the volume of the tank, (b) the final temperature, and (c) the electrical power rating of the resistor. *Answers:* (a) 0.00648 m³, (b) 200°C, (c) 0.0623 kW

**FIGURE P5–122**

**5–123** A vertical 12-cm diameter piston–cylinder device contains an ideal gas at the ambient conditons of 1 bar and 24°C. Initially, the inner face of the piston is 20 cm from the base of the cylinder. Now an external shaft connected to the piston exerts a force corresponding to a boundary work input of 0.1 kJ. The temperature of the gas remains constant during the process. Determine (*a*) the amount of heat transfer, (*b*) the final pressure in the cylinder, and (*c*) the distance that the piston is displaced.

**5–124** A piston–cylinder device initially contains 0.35-kg steam at 3.5 MPa, superheated by 7.4°C. Now the steam loses heat to the surroundings and the piston moves down, hitting a set of stops at which point the cylinder contains saturated liquid water. The cooling continues until the cylinder contains water at 200°C. Determine (*a*) the final pressure and the quality (if mixture), (*b*) the boundary work, (*c*) the amount of heat transfer when the piston first hits the stops, (*d*) and the total heat transfer.

**FIGURE P5–124**

## Design and Essay Problems

**5–125** Compressed gases and phase-changing liquids are used to store energy in rigid containers. What are the advantages and disadvantages of each substance as a means of storing energy?

**5–126** Design an experiment complete with instrumentation to determine the specific heats of a liquid using a resistance heater. Discuss how the experiment will be conducted, what measurements need to be taken, and how the specific heats will be determined. What are the sources of error in your system? How can you minimize the experimental error? How would you modify this system to determine the specific heat of a solid?

**5–127** You are asked to design a heating system for a swimming pool that is 2 m deep, 25 m long, and 25 m wide. Your client desires that the heating system be large enough to raise the water temperature from 20 to 30°C in 3 h. The rate of heat loss from the water to the air at the outdoor design conditions is determined to be 960 W/m², and the heater must also be able to maintain the pool at 30°C at those conditions. Heat losses to the ground are expected to be small and can be disregarded. The heater considered is a natural gas furnace whose efficiency is 80 percent. What heater size (in kW input) would you recommend to your client?

**5–128** It is claimed that fruits and vegetables are cooled by 6°C for each percentage point of weight loss as moisture during vacuum cooling. Using calculations, demonstrate if this claim is reasonable.

**5–129** A 1982 U.S. Department of Energy article (FS #204) states that a leak of one drip of hot water per second can cost $1.00 per month. Making reasonable assumptions about the drop size and the unit cost of energy, determine if this claim is reasonable.

# MASS AND ENERGY ANALYSIS OF CONTROL VOLUMES

In Chap. 5, we applied the general energy balance relation expressed as $E_{in} - E_{out} = \Delta E_{system}$ to closed systems. In this chapter, we extend the energy analysis to systems that involve mass flow across their boundaries i.e., control volumes, with particular emphasis to steady-flow systems.

We start this chapter with the development of the general *conservation of mass* relation for control volumes, and we continue with a discussion of flow work and the energy of fluid streams. We then apply the energy balance to systems that involve *steady-flow processes* and analyze the common steady-flow devices such as nozzles, diffusers, compressors, turbines, throttling devices, mixing chambers, and heat exchangers. Finally, we apply the energy balance to general *unsteady-flow processes* such as the charging and discharging of vessels.

## Objectives

The objectives of this chapter are to:

- Develop the conservation of mass principle.

- Apply the conservation of mass principle to various systems including steady- and unsteady-flow control volumes.

- Apply the first law of thermodynamics as the statement of the conservation of energy principle to control volumes.

- Identify the energy carried by a fluid stream crossing a control surface as the sum of internal energy, flow work, kinetic energy, and potential energy of the fluid and to relate the combination of the internal energy and the flow work to the property enthalpy.

- Solve energy balance problems for common steady-flow devices such as nozzles, compressors, turbines, throttling valves, mixers, heaters, and heat exchangers.

- Apply the energy balance to general unsteady-flow processes with particular emphasis on the uniform-flow process as the model for commonly encountered charging and discharging processes.

**FIGURE 6–1**

Mass is conserved even during chemical reactions.

# 6–1 ■ CONSERVATION OF MASS

Conservation of mass is one of the most fundamental principles in nature. We are all familiar with this principle, and it is not difficult to understand. A person does not have to be a rocket scientist to figure out how much vinegar-and-oil dressing is obtained by mixing 100 g of oil with 25 g of vinegar. Even chemical equations are balanced on the basis of the conservation of mass principle. When 16 kg of oxygen reacts with 2 kg of hydrogen, 18 kg of water is formed (Fig. 6–1). In an electrolysis process, the water separates back to 2 kg of hydrogen and 16 kg of oxygen.

Mass, like energy, is a conserved property, and it cannot be created or destroyed during a process. However, mass $m$ and energy $E$ can be converted to each other according to the well-known formula proposed by Albert Einstein (1879–1955):

$$E = mc^2 \tag{6–1}$$

where $c$ is the speed of light in a vacuum, which is $c = 2.9979 \times 10^8$ m/s. The equation suggests that there is equivalence between mass and energy. All physical and chemical systems exhibit energy interactions with their surroundings. But the amount of energy interacted is equivalent to an extremely small mass compared to the system's total mass. For example, when 1 kg of liquid water is formed from oxygen and hydrogen at normal atmospheric conditions, the amount of energy released is 15.8 MJ, which corresponds to a mass of $1.76 \times 10^{-10}$ kg. However, in nuclear reactions, the mass equivalence of the amount of energy interacted is a significant fraction of the total mass involved. Therefore, in most engineering analysis, we consider both mass and energy as conserved quantities.

For *closed systems*, the conservation of mass principle is implicitly used by requiring that the mass of the system remain constant during a process. For *control volumes*, however, mass can cross the boundaries, and so we must keep track of the amount of mass entering and leaving the control volume.

## Mass and Volume Flow Rates

The amount of mass flowing through a cross section per unit time is called the **mass flow rate** and is denoted by $\dot{m}$. The dot over a symbol is used to indicate *time rate of change*, as explained in Chap. 3.

A fluid usually flows into or out of a control volume through pipes or ducts. The differential mass flow rate of fluid flowing across a small area element $dA_c$ on a flow cross section is proportional to $dA_c$ itself, the fluid density $\rho$, and the component of the flow velocity normal to $dA_c$, which we denote as $V_n$, and is expressed as (Fig. 6–2)

$$\delta\dot{m} = \rho V_n \, dA_c \tag{6–2}$$

Note that both $\delta$ and $d$ are used to indicate differential quantities, but $\delta$ is typically used for quantities (such as heat, work, and mass transfer) that are *path functions* and have *inexact differentials*, while $d$ is used for quantities (such as properties) that are *point functions* and have *exact differentials*. For flow through an annulus of inner radius $r_1$ and outer radius $r_2$, for example,

$$\int_1^2 dA_c = A_{c2} - A_{c1} = \pi\left(r_2^2 - r_1^2\right) \text{ but } \int_1^2 \delta\dot{m} = \dot{m}_{\text{total}} \text{ (total mass flow rate through}$$

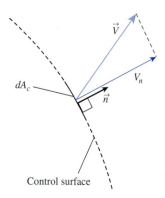

**FIGURE 6–2**

The normal velocity $V_n$ for a surface is the component of velocity perpendicular to the surface.

the annulus), not $\dot{m}_2 - \dot{m}_1$. For specified values of $r_1$ and $r_2$, the value of the integral of $dA_c$ is fixed (thus the names point function and exact differential), but this is not the case for the integral of $d\dot{m}$, (thus the names path function and inexact differential).

The mass flow rate through the entire cross-sectional area of a pipe or duct is obtained by integration:

$$\dot{m} = \int_{A_c} \delta\dot{m} = \int_{A_c} \rho V_n \, dA_c \qquad (\text{kg/s}) \qquad \textbf{(6–3)}$$

While Eq. 6–3 is always valid (in fact it is *exact*), it is not always practical for engineering analyses because of the integral. We would like instead to express mass flow rate in terms of average values over a cross section of the pipe. In a general compressible flow, both $\rho$ and $V_n$ vary across the pipe. In many practical applications, however, the density is essentially uniform over the pipe cross section, and we can take $\rho$ outside the integral of Eq. 6–3. Velocity, however, is *never* uniform over a cross section of a pipe because of the fluid sticking to the surface and thus having zero velocity at the wall (the no-slip condition). Rather, the velocity varies from zero at the walls to some maximum value at or near the centerline of the pipe. We define the **average velocity** $V_{avg}$ as the average value of $V_n$ across the entire cross section (Fig. 6–3),

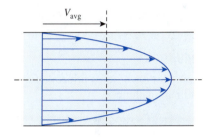

*Average velocity:*
$$V_{avg} = \frac{1}{A_c} \int_{A_c} V_n \, dA_c \qquad \textbf{(6–4)}$$

where $A_c$ is the area of the cross section normal to the flow direction. Note that if the velocity were $V_{avg}$ all through the cross section, the mass flow rate would be identical to that obtained by integrating the actual velocity profile. Thus for incompressible flow or even for compressible flow where $\rho$ is uniform across $A_c$, Eq. 6–3 becomes

$$\dot{m} = \rho V_{avg} A_c \qquad (\text{kg/s}) \qquad \textbf{(6–5)}$$

For compressible flow, we can think of $\rho$ as the bulk average density over the cross section, and then Eq. 6–5 can still be used as a reasonable approximation.

For simplicity, we drop the subscript on the average velocity. Unless otherwise stated, $V$ denotes the average velocity in the flow direction. Also, $A_c$ denotes the cross-sectional area normal to the flow direction.

The volume of the fluid flowing through a cross section per unit time is called the **volume flow rate** $\dot{V}$ (Fig. 6–4) and is given by

$$\dot{V} = \int_{A_c} V_n \, dA_c = V_{avg} A_c = V A_c \qquad (\text{m}^3/\text{s}) \qquad \textbf{(6–6)}$$

An early form of Eq. 6–6 was published in 1628 by the Italian monk Benedetto Castelli (circa 1577–1644). Note that most fluid mechanics textbooks use $Q$ instead of $\dot{V}$ for volume flow rate. We use $\dot{V}$ to avoid confusion with heat transfer.

The mass and volume flow rates are related by

$$\dot{m} = \rho \dot{V} = \frac{\dot{V}}{v} \qquad \textbf{(6–7)}$$

**FIGURE 6–5**

Conservation of mass principle for an ordinary bathtub.

where $v$ is the specific volume. This relation is analogous to $m = \rho V = V/v$, which is the relation between the mass and the volume of a fluid in a container.

## Conservation of Mass Principle

The **conservation of mass principle** for a control volume can be expressed as: *The net mass transfer to or from a control volume during a time interval $\Delta t$ is equal to the net change (increase or decrease) in the total mass within the control volume during $\Delta t$.* That is,

$$\begin{pmatrix} \text{Total mass entering} \\ \text{the CV during } \Delta t \end{pmatrix} - \begin{pmatrix} \text{Total mass leaving} \\ \text{the CV during } \Delta t \end{pmatrix} = \begin{pmatrix} \text{Net change of mass} \\ \text{within the CV during } \Delta t \end{pmatrix}$$

or

$$m_{in} - m_{out} = \Delta m_{CV} \qquad (\text{kg}) \qquad (6\text{–}8)$$

where $\Delta m_{CV} = m_{final} - m_{initial}$ is the change in the mass of the control volume during the process (Fig. 6–5). It can also be expressed in *rate form* as

$$\dot{m}_{in} - \dot{m}_{out} = dm_{CV}/dt \qquad (\text{kg/s}) \qquad (6\text{–}9)$$

where $\dot{m}_{in}$ and $\dot{m}_{out}$ are the total rates of mass flow into and out of the control volume, and $dm_{CV}/dt$ is the time rate of change of mass within the control volume boundaries. Equations 6–8 and 6–9 are often referred to as the **mass balance** and are applicable to any control volume undergoing any kind of process.

Consider a control volume of arbitrary shape, as shown in Fig. 6–6. The mass of a differential volume $dV$ within the control volume is $dm = \rho \, dV$. The total mass within the control volume at any instant in time $t$ is determined by integration to be

*Total mass within the CV:* $\qquad m_{CV} = \displaystyle\int_{CV} \rho \, dV \qquad (6\text{–}10)$

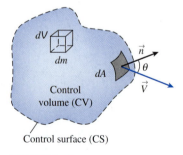

**FIGURE 6–6**

The differential control volume $dV$ and the differential control surface $dA$ used in the derivation of the conservation of mass relation.

Then the time rate of change of the amount of mass within the control volume can be expressed as

*Rate of change of mass within the CV:* $\qquad \dfrac{dm_{CV}}{dt} = \dfrac{d}{dt}\displaystyle\int_{CV} \rho \, dV \qquad (6\text{–}11)$

For the special case of no mass crossing the control surface (i.e., the control volume is a closed system), the conservation of mass principle reduces to $dm_{CV}/dt = 0$. This relation is valid whether the control volume is fixed, moving, or deforming.

Now consider mass flow into or out of the control volume through a differential area $dA$ on the control surface of a fixed control volume. Let $\vec{n}$ be the outward unit vector of $dA$ normal to $dA$ and $\vec{V}$ be the flow velocity at $dA$ relative to a fixed coordinate system, as shown in Fig. 6–6. In general, the velocity may cross $dA$ at an angle $\theta$ off the normal of $dA$, and the mass flow rate is proportional to the normal component of velocity $\vec{V}_n = \vec{V} \cos \theta$ ranging from a maximum outflow of $\vec{V}$ for $\theta = 0$ (flow is normal to $dA$) to a minimum of zero for $\theta = 90°$ (flow is tangent to $dA$) to a maximum *inflow* of $V$ for $\theta = 180°$ (flow is normal to $dA$ but in the opposite direction). Making use of the concept of dot product of two vectors, the magnitude of the normal component of velocity can be expressed as

*Normal component of velocity:*   $V_n = V \cos \theta = \vec{V} \cdot \vec{n}$   **(6–12)**

The mass flow rate through $dA$ is proportional to the fluid density $\rho$, normal velocity $V_n$, and the flow area $dA$, and can be expressed as

*Differential mass flow rate:*   $\delta\dot{m} = \rho V_n \, dA = \rho(V \cos \theta) \, dA = \rho(\vec{V} \cdot \vec{n}) \, dA$   **(6–13)**

The net flow rate into or out of the control volume through the entire control surface is obtained by integrating $\delta\dot{m}$ over the entire control surface,

*Net mass flow rate:*   $\dot{m}_{net} = \int_{CS} \delta\dot{m} = \int_{CS} \rho V_n \, dA = \int_{CS} \rho(\vec{V} \cdot \vec{n}) \, dA$   **(6–14)**

Note that $\vec{V} \cdot \vec{n} = V \cos \theta$ is positive for $\theta < 90°$ (outflow) and negative for $\theta > 90°$ (inflow). Therefore, the direction of flow is automatically accounted for, and the surface integral in Eq. 6–14 directly gives the *net* mass flow rate. A positive value for $\dot{m}_{net}$ indicates net outflow, and a negative value indicates a net inflow of mass.

Rearranging Eq. 6–9 as $dm_{CV}/dt + \dot{m}_{out} - \dot{m}_{in} = 0$, the conservation of mass relation for a fixed control volume can then be expressed as

*General conservation of mass:*   $\dfrac{d}{dt}\int_{CV} \rho \, dV + \int_{CS} \rho(\vec{V} \cdot \vec{n}) \, dA = 0$   **(6–15)**

It states that *the time rate of change of mass within the control volume plus the net mass flow rate through the control surface is equal to zero.*

Splitting the surface integral in Eq. 6–15 into two parts—one for the outgoing flow streams (positive) and one for the incoming streams (negative)—the general conservation of mass relation can also be expressed as

$$\dfrac{d}{dt}\int_{CV} \rho \, dV + \sum_{out} \rho |V_n| \, dA - +\sum_{in} \rho |V_n| \, dA = 0 \qquad \textbf{(6–16)}$$

where $A$ represents the area for an inlet or outlet, and the summation signs are used to emphasize that *all* the inlets and outlets are to be considered. Using the definition of mass flow rate, Eq. 6–16 can also be expressed as

$$\dfrac{d}{dt}\int_{CV} \rho \, dV = \sum_{in} \dot{m} - \sum_{out} \dot{m} \quad \text{or} \quad \dfrac{dm_{CV}}{dt} = \sum_{in} \dot{m} - \sum_{out} \dot{m} \qquad \textbf{(6–17)}$$

Equations 6–15 and 6–16 are also valid for moving or deforming control volumes provided that the *absolute velocity* $\vec{V}$ is replaced by the *relative velocity* $\vec{V}_r$, which is the fluid velocity relative to the control surface.

## Mass Balance for Steady-Flow Processes

During a steady-flow process, the total amount of mass contained within a control volume does not change with time ($m_{CV}$ = constant). Then the conservation of mass principle requires that the total amount of mass

$\dot{m}_1 = 2$ kg/s    $\dot{m}_2 = 3$ kg/s

CV

$\dot{m}_3 = \dot{m}_1 + \dot{m}_2 = 5$ kg/s

**FIGURE 6–7**

Conservation of mass principle
for a two-inlet–one-outlet
steady-flow system.

entering a control volume equal the total amount of mass leaving it. For a garden hose nozzle in steady operation, for example, the amount of water entering the nozzle per unit time is equal to the amount of water leaving it per unit time.

When dealing with steady-flow processes, we are not interested in the amount of mass that flows in or out of a device over time; instead, we are interested in the amount of mass flowing per unit time, that is, *the mass flow rate* $\dot{m}$. The conservation of mass principle for a general steady-flow system with multiple inlets and outlets can be expressed in rate form as (Fig. 6–7)

*Steady flow:*
$$\sum_{in} \dot{m} = \sum_{out} \dot{m} \quad \text{(kg/s)} \tag{6–18}$$

It states that *the total rate of mass entering a control volume is equal to the total rate of mass leaving it.*

Many engineering devices such as nozzles, diffusers, turbines, compressors, and pumps involve a single stream (only one inlet and one outlet). For these cases, we denote the inlet state by the subscript 1 and the outlet state by the subscript 2, and drop the summation signs. Then Eq. 6–18 reduces, for *single-stream steady-flow systems*, to

*Steady flow (single stream):* $\quad \dot{m}_1 = \dot{m}_2 \quad \rightarrow \quad \rho_1 V_1 A_1 = \rho_2 V_2 A_2 \tag{6–19}$

## Special Case: Incompressible Flow

The conservation of mass relations is simplified even further when the fluid is incompressible, which is usually the case for liquids. Canceling the density from both sides of the general steady-flow relation gives

*Steady, incompressible flow:*
$$\sum_{in} \dot{V} = \sum_{out} \dot{V} \quad \text{(m}^3/\text{s)} \tag{6–20}$$

For single-stream steady-flow systems it becomes

*Steady, incompressible flow (single stream):* $\quad \dot{V}_1 = \dot{V}_2 \longrightarrow V_1 A_1 = V_2 A_2 \tag{6–21}$

$\dot{m}_2 = 2$ kg/s
$\dot{V}_2 = 0.8$ m³/s

Air
compressor

$\dot{m}_1 = 2$ kg/s
$\dot{V}_1 = 1.4$ m³/s

**FIGURE 6–8**

During a steady-flow process,
volume flow rates are not necessarily
conserved although mass flow
rates are.

It should always be kept in mind that there is no such thing as a "conservation of volume" principle. Therefore, the volume flow rates into and out of a steady-flow device may be different. The volume flow rate at the outlet of an air compressor is much less than that at the inlet even though the mass flow rate of air through the compressor is constant (Fig. 6–8). This is due to the higher density of air at the compressor exit. For steady flow of liquids, however, the volume flow rates, as well as the mass flow rates, remain nearly constant since liquids are essentially incompressible (constant-density) substances. Water flow through the nozzle of a garden hose is an example of the latter case.

The conservation of mass principle requires every bit of mass to be accounted for during a process. If you can balance your checkbook (by keeping track of deposits and withdrawals, or by simply observing the "conservation of money" principle), you should have no difficulty applying the conservation of mass principle to engineering systems.

## EXAMPLE 6–1    Water Flow through a Garden Hose Nozzle

A garden hose attached with a nozzle is used to fill a 10-gal bucket. The inner diameter of the hose is 2 cm, and it reduces to 0.8 cm at the nozzle exit (Fig. 6–9). If it takes 50 s to fill the bucket with water, determine (a) the volume and mass flow rates of water through the hose, and (b) the average velocity of water at the nozzle exit.

**SOLUTION**   A garden hose is used to fill a water bucket. The volume and mass flow rates of water and the exit velocity are to be determined.
**Assumptions**   **1** Water is an incompressible substance. **2** Flow through the hose is steady. **3** There is no waste of water by splashing.
**Properties**   We take the density of water to be 1000 kg/m³ = 1 kg/L.
**Analysis**   (a) Noting that 10 gal of water are discharged in 50 s, the volume and mass flow rates of water are

$$\dot{V} = \frac{V}{\Delta t} = \frac{10 \text{ gal}}{50 \text{ s}} \left( \frac{3.7854 \text{ L}}{1 \text{ gal}} \right) = \textbf{0.757 L/s}$$

$$\dot{m} = \rho \dot{V} = (1 \text{ kg/L})(0.757 \text{ L/s}) = \textbf{0.757 kg/s}$$

(b) The cross-sectional area of the nozzle exit is

$$A_e = \pi r_e^2 = \pi (0.4 \text{ cm})^2 = 0.5027 \text{ cm}^2 = 0.5027 \times 10^{-4} \text{ m}^2$$

The volume flow rate through the hose and the nozzle is constant. Then the average velocity of water at the nozzle exit becomes

$$V_e = \frac{\dot{V}}{A_e} = \frac{0.757 \text{ L/s}}{0.5027 \times 10^{-4} \text{ m}^2} \left( \frac{1 \text{ m}^3}{1000 \text{ L}} \right) = \textbf{15.1 m/s}$$

**Discussion**   It can be shown that the average velocity in the hose is 2.4 m/s. Therefore, the nozzle increases the water velocity by over six times.

**FIGURE 6–9**
Schematic for Example 6–1.

## EXAMPLE 6–2    Discharge of Water from a Tank

A 4-ft-high, 3-ft-diameter cylindrical water tank whose top is open to the atmosphere is initially filled with water. Now the discharge plug near the bottom of the tank is pulled out, and a water jet whose diameter is 0.5 in streams out (Fig. 6–10). The average velocity of the jet is approximated as $V = \sqrt{2gh}$, where $h$ is the height of water in the tank measured from the center of the hole (a variable) and $g$ is the gravitational acceleration. Determine how long it takes for the water level in the tank to drop to 2 ft from the bottom.

**SOLUTION**   The plug near the bottom of a water tank is pulled out. The time it takes for half of the water in the tank to empty is to be determined.
**Assumptions**   **1** Water is an incompressible substance. **2** The distance between the bottom of the tank and the center of the hole is negligible compared to the total water height. **3** The gravitational acceleration is 32.2 ft/s².
**Analysis**   We take the volume occupied by water as the control volume. The size of the control volume decreases in this case as the water level drops,

**FIGURE 6–10**
Schematic for Example 6–2.

and thus this is a variable control volume. (We could also treat this as a fixed control volume that consists of the interior volume of the tank by disregarding the air that replaces the space vacated by the water.) This is obviously an unsteady-flow problem since the properties (such as the amount of mass) within the control volume change with time.

The conservation of mass relation for a control volume undergoing any process is given in rate form as

$$\dot{m}_{\text{in}} - \dot{m}_{\text{out}} = \frac{dm_{\text{CV}}}{dt} \tag{1}$$

During this process no mass enters the control volume ($\dot{m}_{\text{in}} = 0$), and the mass flow rate of discharged water can be expressed as

$$\dot{m}_{\text{out}} = (\rho VA)_{\text{out}} = \rho \sqrt{2gh} A_{\text{jet}} \tag{2}$$

where $A_{\text{jet}} = \pi D^2_{\text{jet}}/4$ is the cross-sectional area of the jet, which is constant. Noting that the density of water is constant, the mass of water in the tank at any time is

$$m_{\text{CV}} = \rho \mathcal{V} = \rho A_{\text{tank}} h \tag{3}$$

where $A_{\text{tank}} = \pi D^2_{\text{tank}}/4$ is the base area of the cylindrical tank. Substituting Eqs. 2 and 3 into the mass balance relation (Eq. 1) gives

$$-\rho \sqrt{2gh} A_{\text{jet}} = \frac{d(\rho A_{\text{tank}} h)}{dt} \rightarrow -\rho \sqrt{2gh} (\pi D^2_{\text{jet}}/4) = \frac{\rho (\pi D^2_{\text{tank}}/4)\, dh}{dt}$$

Canceling the densities and other common terms and separating the variables give

$$dt = \frac{D^2_{\text{tank}}}{D^2_{\text{jet}}} \frac{dh}{\sqrt{2gh}}$$

Integrating from $t = 0$ at which $h = h_0$ to $t = t$ at which $h = h_2$ gives

$$\int_0^t dt = -\frac{D^2_{\text{tank}}}{D^2_{\text{jet}} \sqrt{2g}} \int_{h_0}^{h_2} \frac{dh}{\sqrt{h}} \rightarrow t = \frac{\sqrt{h_0} - \sqrt{h_2}}{\sqrt{g/2}} \left( \frac{D_{\text{tank}}}{D_{\text{jet}}} \right)^2$$

Substituting, the time of discharge is

$$t = \frac{\sqrt{4\,\text{ft}} - \sqrt{2\,\text{ft}}}{\sqrt{32.2/2\,\text{ft/s}^2}} \left( \frac{3 \times 12\,\text{in}}{0.5\,\text{in}} \right)^2 = 757\,\text{s} = \textbf{12.6 min}$$

Therefore, half of the tank is emptied in 12.6 min after the discharge hole is unplugged.

*Discussion* Using the same relation with $h_2 = 0$ gives $t = 43.1$ min for the discharge of the entire amount of water in the tank. Therefore, emptying the bottom half of the tank takes much longer than emptying the top half. This is due to the decrease in the average discharge velocity of water with decreasing $h$.

## 6–2 · FLOW WORK AND THE ENERGY OF A FLOWING FLUID

Unlike closed systems, control volumes involve mass flow across their boundaries, and some work is required to push the mass into or out of the control volume. This work is known as the **flow work**, or **flow energy**, and is necessary for maintaining a continuous flow through a control volume.

To obtain a relation for flow work, consider a fluid element of volume $V$ as shown in Fig. 6–11. The fluid immediately upstream forces this fluid element to enter the control volume; thus, it can be regarded as an imaginary piston. The fluid element can be chosen to be sufficiently small so that it has uniform properties throughout.

If the fluid pressure is $P$ and the cross-sectional area of the fluid element is $A$ (Fig. 6–12), the force applied on the fluid element by the imaginary piston is

$$F = PA \tag{6–22}$$

To push the entire fluid element into the control volume, this force must act through a distance $L$. Thus, the work done in pushing the fluid element across the boundary (i.e., the flow work) is

$$W_{\text{flow}} = FL = PAL = PV \quad \text{(kJ)} \tag{6–23}$$

The flow work per unit mass is obtained by dividing both sides of this equation by the mass of the fluid element:

$$w_{\text{flow}} = Pv \quad \text{(kJ/kg)} \tag{6–24}$$

The flow work relation is the same whether the fluid is pushed into or out of the control volume (Fig. 6–13).

It is interesting that unlike other work quantities, flow work is expressed in terms of properties. In fact, it is the product of two properties of the fluid. For that reason, some people view it as a *combination property* (like enthalpy) and refer to it as *flow energy, convected energy,* or *transport energy* instead of flow work. Others, however, argue rightfully that the product $PV$ represents energy for flowing fluids only and does not represent any form of energy for nonflow (closed) systems. Therefore, it should be treated as work. This controversy is not likely to end, but it is comforting to know that both arguments yield the same result for the energy balance equation. In the discussions that follow, we consider the flow energy to be part of the energy of a flowing fluid, since this greatly simplifies the energy analysis of control volumes.

### Total Energy of a Flowing Fluid

As we discussed in Chap. 3, the total energy of a simple compressible system consists of three parts: internal, kinetic, and potential energies (Fig. 6–14). On a unit-mass basis, it is expressed as

$$e = u + \text{ke} + \text{pe} = u + \frac{V^2}{2} + gz \quad \text{(kJ/kg)} \tag{6–25}$$

where $V$ is the velocity and $z$ is the elevation of the system relative to some external reference point.

**FIGURE 6–11**

Schematic for flow work.

**FIGURE 6–12**

In the absence of acceleration, the force applied on a fluid by a piston is equal to the force applied on the piston by the fluid.

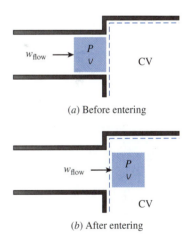

(a) Before entering

(b) After entering

**FIGURE 6–13**

Flow work is the energy needed to push a fluid into or out of a control volume, and it is equal to $Pv$.

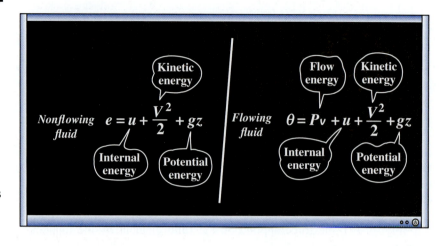

**FIGURE 6–14**

The total energy consists of three parts for a nonflowing fluid and four parts for a flowing fluid.

The fluid entering or leaving a control volume possesses an additional form of energy—the *flow energy* $Pv$, as already discussed. Then the total energy of a **flowing fluid** on a unit-mass basis (denoted by $\theta$) becomes

$$\theta = Pv + e = Pv + (u + \text{ke} + \text{pe}) \qquad \text{(6–26)}$$

But the combination $Pv + u$ has been previously defined as the enthalpy $h$. So the relation in Eq. 6–26 reduces to

$$\theta = h + \text{ke} + \text{pe} = h + \frac{V^2}{2} + gz \qquad (\text{kJ/kg}) \qquad \text{(6–27)}$$

By using the enthalpy instead of the internal energy to represent the energy of a flowing fluid, one does not need to be concerned about the flow work. The energy associated with pushing the fluid into or out of the control volume is automatically taken care of by enthalpy. In fact, this is the main reason for defining the property enthalpy. From now on, the energy of a fluid stream flowing into or out of a control volume is represented by Eq. 6–27, and no reference will be made to flow work or flow energy.

## Energy Transport by Mass

Noting that $\theta$ is total energy per unit mass, the total energy of a flowing fluid of mass $m$ is simply $m\theta$, provided that the properties of the mass $m$ are uniform. Also, when a fluid stream with uniform properties is flowing at a mass flow rate of $\dot{m}$, the rate of energy flow with that stream is $\dot{m}\theta$ (Fig. 6–15). That is,

**FIGURE 6–15**

The product $\dot{m}_i\theta_i$ is the energy transported into control volume by mass per unit time.

*Amount of energy transport:* $\quad E_{\text{mass}} = m\theta = m\left(h + \dfrac{V^2}{2} + gz\right) \qquad (\text{kJ}) \qquad \text{(6–28)}$

*Rate of energy transport:* $\quad \dot{E}_{\text{mass}} = \dot{m}\theta = \dot{m}\left(h + \dfrac{V^2}{2} + gz\right) \qquad (\text{kW}) \qquad \text{(6–29)}$

When the kinetic and potential energies of a fluid stream are negligible, as is often the case, these relations simplify to $E_{\text{mass}} = mh$ and $\dot{E}_{\text{mass}} = \dot{m}h$.

In general, the total energy transported by mass into or out of the control volume is not easy to determine since the properties of the mass at each inlet or exit may be changing with time as well as over the cross section.

Thus, the only way to determine the energy transport through an opening as a result of mass flow is to consider sufficiently small differential masses $\delta m$ that have uniform properties and to add their total energies during flow.

Again noting that $\theta$ is total energy per unit mass, the total energy of a flowing fluid of mass $\delta m$ is $\theta \, \delta m$. Then the total energy transported by mass through an inlet or exit ($m_i \theta_i$ and $m_e \theta_e$) is obtained by integration. At an inlet, for example, it becomes

$$E_{\text{in,mass}} = \int_{m_i} \theta_i \, \delta m_i = \int_{m_i} \left( h_i + \frac{V_i^2}{2} + gz_i \right) \delta m_i \qquad \text{(6–30)}$$

Most flows encountered in practice can be approximated as being steady and one-dimensional, and thus the simple relations in Eqs. 6–28 and 6–29 can be used to represent the energy transported by a fluid stream.

### EXAMPLE 6–3    Energy Transport by Mass

Steam is leaving a 4-L pressure cooker whose operating pressure is 150 kPa (Fig. 6–16). It is observed that the amount of liquid in the cooker has decreased by 0.6 L in 40 min after the steady operating conditions are established, and the cross-sectional area of the exit opening is 8 mm². Determine (a) the mass flow rate of the steam and the exit velocity, (b) the total and flow energies of the steam per unit mass, and (c) the rate at which energy leaves the cooker by steam.

**SOLUTION** Steam leaves a pressure cooker at a specified pressure. The velocity, flow rate, the total and flow energies, and the rate of energy transfer by mass are to be determined.
*Assumptions* **1** The flow is steady, and the initial start-up period is disregarded. **2** The kinetic and potential energies are negligible, and thus they are not considered. **3** Saturation conditions exist within the cooker at all times so that steam leaves the cooker as a saturated vapor at the cooker pressure.
*Properties* The properties of saturated liquid water and water vapor at 150 kPa are $v_f = 0.001053$ m³/kg, $v_g = 1.1594$ m³/kg, $u_g = 2519.2$ kJ/kg, and $h_g = 2693.1$ kJ/kg (Table A–5).
*Analysis* (a) Saturation conditions exist in a pressure cooker at all times after the steady operating conditions are established. Therefore, the liquid has the properties of saturated liquid and the exiting steam has the properties of saturated vapor at the operating pressure. The amount of liquid that has evaporated, the mass flow rate of the exiting steam, and the exit velocity are

$$m = \frac{\Delta V_{\text{liquid}}}{v_f} = \frac{0.6 \text{ L}}{0.001053 \text{ m}^3/\text{kg}} \left( \frac{1 \text{ m}^3}{1000 \text{ L}} \right) = 0.570 \text{ kg}$$

$$\dot{m} = \frac{m}{\Delta t} = \frac{0.570 \text{ kg}}{40 \text{ min}} = 0.0142 \text{ kg/min} = \textbf{2.37} \times \textbf{10}^{-4} \textbf{ kg/s}$$

$$V = \frac{\dot{m}}{\rho_g A_c} = \frac{\dot{m} v_g}{A_c} = \frac{(2.37 \times 10^{-4} \text{ kg/s})(1.1594 \text{ m}^3/\text{kg})}{8 \times 10^{-6} \text{ m}^2} = \textbf{34.3 m/s}$$

Steam

150 kPa

Pressure cooker

**FIGURE 6–16**
Schematic for Example 6–3.

**FIGURE 6–17**

Many engineering systems such as power plants operate under steady conditions.

© Vol. 57/PhotoDisc/Getty RF.

**FIGURE 6–18**

Under steady-flow conditions, the mass and energy contents of a control volume remain constant.

**FIGURE 6–19**

Under steady-flow conditions, the fluid properties at an inlet or exit remain constant (do not change with time).

(b) Noting that $h = u + Pv$ and that the kinetic and potential energies are disregarded, the flow and total energies of the exiting steam are

$$e_{flow} = Pv = h - u = 2693.1 - 2519.2 = \mathbf{173.9\ kJ/kg}$$

$$\theta = h + ke + pe \cong h = \mathbf{2693.1\ kJ/kg}$$

Note that the kinetic energy in this case is $ke = V^2/2 = (34.3\ m/s)^2/2 = 588\ m^2/s^2 = 0.588\ kJ/kg$, which is small compared to enthalpy.

(c) The rate at which energy is leaving the cooker by mass is simply the product of the mass flow rate and the total energy of the exiting steam per unit mass,

$$\dot{E}_{mass} = \dot{m}\theta = (2.37 \times 10^{-4}\ kg/s)(2693.1\ kJ/kg) = 0.638\ kJ/s = \mathbf{0.638\ kW}$$

**Discussion** The numerical value of the energy leaving the cooker with steam alone does not mean much since this value depends on the reference point selected for enthalpy (it could even be negative). The significant quantity is the difference between the enthalpies of the exiting vapor and the liquid inside (which is $h_{fg}$) since it relates directly to the amount of energy supplied to the cooker.

## 6–3 ▪ ENERGY ANALYSIS OF STEADY-FLOW SYSTEMS

A large number of engineering devices such as turbines, compressors, and nozzles operate for long periods of time under the same conditions once the transient start-up period is completed and steady operation is established, and they are classified as *steady-flow devices* (Fig. 6–17). Processes involving such devices can be represented reasonably well by a somewhat idealized process, called the **steady-flow process**, which was defined in Chap. 2 as *a process during which a fluid flows through a control volume steadily*. That is, the fluid properties can change from point to point within the control volume, but at any point, they remain constant during the entire process. (Remember, *steady* means *no change with time*.)

During a steady-flow process, no intensive or extensive properties *within the control volume* change with time. Thus, the volume $V$, the mass $m$, and the total energy content $E$ of the control volume remain constant (Fig. 6–18). As a result, the boundary work is zero for steady-flow systems (since $V_{CV} =$ constant), and the total mass or energy entering the control volume must be equal to the total mass or energy leaving it (since $m_{CV} =$ constant and $E_{CV} =$ constant). These observations greatly simplify the analysis.

The fluid properties at an inlet or exit remain constant during a steady-flow process. The properties may, however, be different at different inlets and exits. They may even vary over the cross section of an inlet or an exit. However, all properties, including the velocity and elevation, must remain constant with time at a fixed point at an inlet or exit. It follows that the mass flow rate of the fluid at an opening must remain constant during a steady-flow process (Fig. 6–19). As an added simplification, the fluid properties at an opening are usually considered to be uniform (at some average value) over the cross section. Thus, the fluid properties at an inlet or exit may be specified by the average single values. Also, the *heat* and *work* interactions between a steady-flow system and its surroundings do not change with time. Thus, the power delivered by a system and the rate of heat transfer to or from a system remain constant during a steady-flow process.

The *mass balance* for a general steady-flow system was given in Sec. 6–1 as

$$\sum_{\text{in}} \dot{m} = \sum_{\text{out}} \dot{m} \qquad (\text{kg/s}) \qquad \text{(6–31)}$$

The mass balance for a single-stream (one-inlet and one-outlet) steady-flow system was given as

$$\dot{m}_1 = \dot{m}_2 \quad \longrightarrow \quad \rho_1 V_1 A_1 = \rho_2 V_2 A_2 \qquad \text{(6–32)}$$

where the subscripts 1 and 2 denote the inlet and the exit states, respectively, $\rho$ is density, $V$ is the average flow velocity in the flow direction, and $A$ is the cross-sectional area normal to flow direction.

During a steady-flow process, the total energy content of a control volume remains constant ($E_{\text{CV}} = \text{constant}$), and thus the change in the total energy of the control volume is zero ($\Delta E_{\text{CV}} = 0$). Therefore, the amount of energy entering a control volume in all forms (by heat, work, and mass) must be equal to the amount of energy leaving it. Then the rate form of the general energy balance reduces for a steady-flow process to

$$\underbrace{\dot{E}_{\text{in}} - \dot{E}_{\text{out}}}_{\substack{\text{Rate of net energy transfer} \\ \text{by heat, work, and mass}}} = \underbrace{dE_{\text{system}}/dt}_{\substack{\text{Rate of change in internal, kinetic,} \\ \text{potential, etc., energies}}} \nearrow^{0 \text{ (steady)}} = 0 \qquad \text{(6–33)}$$

or

*Energy balance:* $\qquad \underbrace{\dot{E}_{\text{in}}}_{\substack{\text{Rate of net energy transfer in} \\ \text{by heat, work, and mass}}} = \underbrace{\dot{E}_{\text{out}}}_{\substack{\text{Rate of net energy transfer out} \\ \text{by heat, work, and mass}}} \qquad (\text{kW}) \qquad \text{(6–34)}$

Noting that energy can be transferred by heat, work, and mass only, the energy balance in Eq. 6–34 for a general steady-flow system can also be written more explicitly as

$$\dot{Q}_{\text{in}} + \dot{W}_{\text{in}} + \sum_{\text{in}} \dot{m}\theta = \dot{Q}_{\text{out}} + \dot{W}_{\text{out}} + \sum_{\text{out}} \dot{m}\theta \qquad \text{(6–35)}$$

or

$$\dot{Q}_{\text{in}} + \dot{W}_{\text{in}} + \underbrace{\sum_{\text{in}} \dot{m}\left(h + \frac{V^2}{2} + gz\right)}_{\text{for each inlet}} = \dot{Q}_{\text{out}} + \dot{W}_{\text{out}} + \underbrace{\sum_{\text{out}} \dot{m}\left(h + \frac{V^2}{2} + gz\right)}_{\text{for each exit}} \qquad \text{(6–36)}$$

since the energy of a flowing fluid per unit mass is $\theta = h + \text{ke} + \text{pe} = h + V^2/2 + gz$. The energy balance relation for steady-flow systems first appeared in 1859 in a German thermodynamics book written by Gustav Zeuner.

Consider, for example, an ordinary electric hot-water heater under steady operation, as shown in Fig. 6–20. A cold-water stream with a mass flow rate $\dot{m}$ is continuously flowing into the water heater, and a hot-water stream of the same mass flow rate is continuously flowing out of it. The water heater (the control volume) is losing heat to the surrounding air at a rate of $\dot{Q}_{\text{out}}$,

**FIGURE 6–20**

A water heater in steady operation.

and the electric heating element is supplying electrical work (heating) to the water at a rate of $\dot{W}_{in}$. On the basis of the conservation of energy principle, we can say that the water stream experiences an increase in its total energy as it flows through the water heater that is equal to the electric energy supplied to the water minus the heat losses.

The energy balance relation just given is intuitive in nature and is easy to use when the magnitudes and directions of heat and work transfers are known. When performing a general analytical study or solving a problem that involves an unknown heat or work interaction, however, we need to assume a direction for the heat or work interactions. In such cases, it is common practice to assume heat to be transferred *into the system* (heat input) at a rate of $\dot{Q}$, and work produced *by the system* (work output) at a rate of $\dot{W}$, and then solve the problem. The first-law or energy balance relation in that case for a general steady-flow system becomes

$$\dot{Q} - \dot{W} = \sum_{out} \underbrace{\dot{m}\left(h + \frac{V^2}{2} + gz\right)}_{\text{for each exit}} - \sum_{in} \underbrace{\dot{m}\left(h + \frac{V^2}{2} + gz\right)}_{\text{for each inlet}} \tag{6-37}$$

Obtaining a negative quantity for $\dot{Q}$ or $\dot{W}$ simply means that the assumed direction is wrong and should be reversed. For single-stream devices, the steady-flow energy balance equation becomes

$$\dot{Q} - \dot{W} = \dot{m}\left[h_2 - h_1 + \frac{V_2^2 - V_1^2}{2} + g(z_2 - z_1)\right] \tag{6-38}$$

Dividing Eq. 6–38 by $\dot{m}$ gives the energy balance on a unit-mass basis as

$$q - w = h_2 - h_1 + \frac{V_2^2 - V_1^2}{2} + g(z_2 - z_1) \tag{6-39}$$

where $q = \dot{Q}/\dot{m}$ and $w = \dot{W}/\dot{m}$ are the heat transfer and work done per unit mass of the working fluid, respectively. When the fluid experiences negligible changes in its kinetic and potential energies (that is, $\Delta ke \cong 0$, $\Delta pe \cong 0$), the energy balance equation is reduced further to

$$q - w = h_2 - h_1 \tag{6-40}$$

The various terms appearing in the above equations are as follows:

$\dot{Q}$ = **rate of heat transfer between the control volume and its surroundings**. When the control volume is losing heat (as in the case of the water heater), $\dot{Q}$ is negative. If the control volume is well insulated (i.e., adiabatic), then $\dot{Q} = 0$.

$\dot{W}$ = **power**. For steady-flow devices, the control volume is constant; thus, there is no boundary work involved. The work required to push mass into and out of the control volume is also taken care of by using enthalpies for the energy of fluid streams instead of internal energies. Then $\dot{W}$ represents the remaining forms of work done per unit time (Fig. 6–21). Many steady-flow devices, such as turbines, compressors, and pumps, transmit power through a shaft, and $\dot{W}$ simply becomes the shaft power for those devices. If the control surface is crossed by electric wires (as in the case of an electric water heater), $\dot{W}$ represents the electrical work done per unit time. If neither is present, then $\dot{W} = 0$.

**FIGURE 6–21**

Under steady operation, shaft work and electrical work are the only forms of work a simple compressible system may involve.

$\Delta h = h_2 - h_1$. The enthalpy change of a fluid can easily be determined by reading the enthalpy values at the exit and inlet states from the tables. For ideal gases, it can be approximated by $\Delta h = c_{p,avg}(T_2 - T_1)$. Note that $(kg/s)(kJ/kg) \equiv kW$.

$\Delta ke = (V_2^2 - V_1^2)/2$. The unit of kinetic energy is $m^2/s^2$, which is equivalent to J/kg (Fig. 6–22). The enthalpy is usually given in kJ/kg. To add these two quantities, the kinetic energy should be expressed in kJ/kg. This is easily accomplished by dividing it by 1000. A velocity of 45 m/s corresponds to a kinetic energy of only 1 kJ/kg, which is a very small value compared with the enthalpy values encountered in practice. Thus, the kinetic energy term at low velocities can be neglected. When a fluid stream enters and leaves a steady-flow device at about the same velocity ($V_1 \cong V_2$), the change in the kinetic energy is close to zero regardless of the velocity. Caution should be exercised at high velocities, however, since small changes in velocities may cause significant changes in kinetic energy (Fig. 6–23).

$\Delta pe = g(z_2 - z_1)$. A similar argument can be given for the potential energy term. A potential energy change of 1 kJ/kg corresponds to an elevation difference of 102 m. The elevation difference between the inlet and exit of most industrial devices such as turbines and compressors is well below this value, and the potential energy term is always neglected for these devices. The only time the potential energy term is significant is when a process involves pumping a fluid to high elevations and we are interested in the required pumping power.

# 6–4 · SOME STEADY-FLOW ENGINEERING DEVICES

Many engineering devices operate essentially under the same conditions for long periods of time. The components of a steam power plant (turbines, compressors, heat exchangers, and pumps), for example, operate nonstop for months before the system is shut down for maintenance (Fig. 6–24). Therefore, these devices can be conveniently analyzed as steady-flow devices.

**FIGURE 6–22**

The units $m^2/s^2$ and J/kg are equivalent.

| $V_1$ m/s | $V_2$ m/s | $\Delta ke$ kJ/kg |
|---|---|---|
| 0 | 45 | 1 |
| 50 | 67 | 1 |
| 100 | 110 | 1 |
| 200 | 205 | 1 |
| 500 | 502 | 1 |

**FIGURE 6–23**

At very high velocities, even small changes in velocities can cause significant changes in the kinetic energy of the fluid.

5-Stage low pressure compressor (LPC)

Cold end drive flange

LPC bleed air collector

14-Stage high pressure compressor

Combustor

Fuel system manifolds

2-Stage high pressure turbine

5-Stage low pressure turbine

Hot end drive flange

**FIGURE 6–24**

A land-based gas turbine used for electric power production. This is a General Electric LM5000 turbine. It has a length of 6.2 m, it weighs 12.5 tons, and produces 55.2 MW at 3600 rpm with steam injection.

*Courtesy of GE Energy.*

In this section, some common steady-flow devices are described, and the thermodynamic aspects of the flow through them are analyzed. The conservation of mass and the conservation of energy principles for these devices are illustrated with examples.

# 1    Nozzles and Diffusers

Nozzles and diffusers are commonly utilized in jet engines, rockets, spacecraft, and even garden hoses. A **nozzle** is a device that *increases the velocity of a fluid* at the expense of pressure. A **diffuser** is a device that *increases the pressure of a fluid* by slowing it down. That is, nozzles and diffusers perform opposite tasks. The cross-sectional area of a nozzle decreases in the flow direction for subsonic flows and increases for supersonic flows. The reverse is true for diffusers.

The rate of heat transfer between the fluid flowing through a nozzle or a diffuser and the surroundings is usually very small ($\dot{Q} \approx 0$) since the fluid has high velocities, and thus it does not spend enough time in the device for any significant heat transfer to take place. Nozzles and diffusers typically involve no work ($\dot{W} = 0$) and any change in potential energy is negligible ($\Delta pe \cong 0$). But nozzles and diffusers usually involve very high velocities, and as a fluid passes through a nozzle or diffuser, it experiences large changes in its velocity (Fig. 6–25). Therefore, the kinetic energy changes must be accounted for in analyzing the flow through these devices ($\Delta ke \neq 0$).

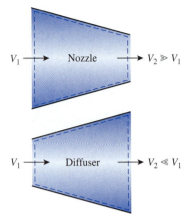

**FIGURE 6–25**

Nozzles and diffusers are shaped so that they cause large changes in fluid velocities and thus kinetic energies.

**FIGURE 6–26**

Schematic for Example 6–4.

---

**EXAMPLE 6–4    Deceleration of Air in a Diffuser**

Air at 10°C and 80 kPa enters the diffuser of a jet engine steadily with a velocity of 200 m/s. The inlet area of the diffuser is 0.4 m². The air leaves the diffuser with a velocity that is very small compared with the inlet velocity. Determine (*a*) the mass flow rate of the air and (*b*) the temperature of the air leaving the diffuser.

**SOLUTION**    Air enters the diffuser of a jet engine steadily at a specified velocity. The mass flow rate of air and the temperature at the diffuser exit are to be determined.

*Assumptions*   **1** This is a steady-flow process since there is no change with time at any point and thus $\Delta m_{CV} = 0$ and $\Delta E_{CV} = 0$. **2** Air is an ideal gas since it is at a high temperature and low pressure relative to its critical-point values. **3** The potential energy change is zero, $\Delta pe = 0$. **4** Heat transfer is negligible. **5** Kinetic energy at the diffuser exit is negligible. **6** There are no work interactions.

*Analysis*   We take the *diffuser* as the system (Fig. 6–26). This is a *control volume* since mass crosses the system boundary during the process. We observe that there is only one inlet and one exit and thus $\dot{m}_1 = \dot{m}_2 = \dot{m}$.

(*a*) To determine the mass flow rate, we need to find the specific volume of the air first. This is determined from the ideal-gas relation at the inlet conditions:

$$v_1 = \frac{RT_1}{P_1} = \frac{(0.287 \text{ kPa·m}^3/\text{kg·K})(283 \text{ K})}{80 \text{ kPa}} = 1.015 \text{ m}^3/\text{kg}$$

Then,

$$\dot{m} = \frac{1}{v_1} V_1 A_1 = \frac{1}{1.015 \text{ m}^3/\text{kg}} (200 \text{ m/s})(0.4 \text{ m}^2) = \textbf{78.8 kg/s}$$

Since the flow is steady, the mass flow rate through the entire diffuser remains constant at this value.

(b) Under stated assumptions and observations, the energy balance for this steady-flow system can be expressed in the rate form as

$$\underbrace{\dot{E}_{in} - \dot{E}_{out}}_{\substack{\text{Rate of net energy transfer} \\ \text{by heat, work, and mass}}} = \underbrace{dE_{system}/dt}_{\substack{\text{Rate of change in internal, kinetic,} \\ \text{potential, etc., energies}}}^{\nearrow 0 \text{ (steady)}} = 0$$

$$\dot{E}_{in} = \dot{E}_{out}$$

$$\dot{m}\left( h_1 + \frac{V_1^2}{2} \right) = \dot{m}\left( h_2 + \frac{V_2^2}{2} \right) \qquad (\text{since } \dot{Q} \cong 0, \dot{W} = 0, \text{ and } \Delta pe \cong 0)$$

$$h_2 = h_1 - \frac{V_2^2 - V_1^2}{2}$$

The exit velocity of a diffuser is usually small compared with the inlet velocity ($V_2 \ll V_1$); thus, the kinetic energy at the exit can be neglected. The enthalpy of air at the diffuser inlet is determined from the air table (Table A–21) to be

$$h_1 = h_{@ \, 283 \, K} = 283.14 \text{ kJ/kg}$$

Substituting, we get

$$h_2 = 283.14 \text{ kJ/kg} - \frac{0 - (200 \text{ m/s})^2}{2}\left( \frac{1 \text{ kJ/kg}}{1000 \text{ m}^2/\text{s}^2} \right)$$

$$= 303.14 \text{ kJ/kg}$$

From Table A–21, the temperature corresponding to this enthalpy value is

$$T_2 = \textbf{303 K}$$

*Discussion* This result shows that the temperature of the air increases by about 20°C as it is slowed down in the diffuser. The temperature rise of the air is mainly due to the conversion of kinetic energy to internal energy.

---

## EXAMPLE 6–5     Acceleration of Steam in a Nozzle

Steam at 250 psia and 700°F steadily enters a nozzle whose inlet area is 0.2 ft². The mass flow rate of steam through the nozzle is 10 lbm/s. Steam leaves the nozzle at 200 psia with a velocity of 900 ft/s. Heat losses from the nozzle per unit mass of the steam are estimated to be 1.2 Btu/lbm. Determine (a) the inlet velocity and (b) the exit temperature of the steam.

**SOLUTION**    Steam enters a nozzle steadily at a specified flow rate and velocity. The inlet velocity of steam and the exit temperature are to be determined.

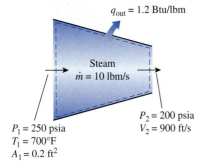

$q_{out} = 1.2$ Btu/lbm

Steam
$\dot{m} = 10$ lbm/s

$P_1 = 250$ psia
$T_1 = 700°F$
$A_1 = 0.2$ ft$^2$

$P_2 = 200$ psia
$V_2 = 900$ ft/s

**FIGURE 6–27**

Schematic for Example 6–5.

**Assumptions** **1** This is a steady-flow process since there is no change with time at any point and thus $\Delta m_{CV} = 0$ and $\Delta E_{CV} = 0$. **2** There are no work interactions. **3** The potential energy change is zero, $\Delta pe = 0$.

**Analysis** We take the *nozzle* as the system (Fig. 6–27). This is a *control volume* since mass crosses the system boundary during the process. We observe that there is only one inlet and one exit and thus $\dot{m}_1 = \dot{m}_2 = \dot{m}$.

(*a*) The specific volume and enthalpy of steam at the nozzle inlet are

$$\left. \begin{array}{l} P_1 = 250 \text{ psia} \\ T_1 = 700°F \end{array} \right\} \quad \begin{array}{l} v_1 = 2.6883 \text{ ft}^3/\text{lbm} \\ h_1 = 1371.4 \text{ Btu/lbm} \end{array} \quad \text{(Table A–6E)}$$

Then,

$$\dot{m} = \frac{1}{v_1} V_1 A_1$$

$$10 \text{ lbm/s} = \frac{1}{2.6883 \text{ ft}^3/\text{lbm}} (V_1)(0.2 \text{ ft}^2)$$

$$V_1 = \textbf{134.4 ft/s}$$

(*b*) Under stated assumptions and observations, the energy balance for this steady-flow system can be expressed in the rate form as

$$\underbrace{\dot{E}_{in} - \dot{E}_{out}}_{\substack{\text{Rate of net energy transfer} \\ \text{by heat, work, and mass}}} = \underbrace{dE_{system}/dt}_{\substack{\text{Rate of change in internal, kinetic,} \\ \text{potential, etc., energies}}} \overset{\nearrow 0 \text{ (steady)}}{=} 0$$

$$\dot{E}_{in} = \dot{E}_{out}$$

$$\dot{m}\left( h_1 + \frac{V_1^2}{2} \right) = \dot{Q}_{out} + \dot{m}\left( h_2 + \frac{V_2^2}{2} \right) \quad \text{(since } \dot{W} = 0, \text{ and } \Delta pe \cong 0)$$

Dividing by the mass flow rate $\dot{m}$ and substituting, $h_2$ is determined to be

$$h_2 = h_1 - q_{out} - \frac{V_2^2 - V_1^2}{2}$$

$$= (1371.4 - 1.2) \text{ Btu/lbm} - \frac{(900 \text{ ft/s})^2 - (134.4 \text{ ft/s})^2}{2}\left( \frac{1 \text{ Btu/lbm}}{25{,}037 \text{ ft}^2/\text{s}^2} \right)$$

$$= 1354.4 \text{ Btu/lbm}$$

Then,

$$\left. \begin{array}{l} P_2 = 200 \text{ psia} \\ h_2 = 1354.4 \text{ Btu/lbm} \end{array} \right\} \quad T_2 = \textbf{662.0°F} \quad \text{(Table A–6E)}$$

**Discussion** Note that the temperature of steam drops by 38.0°F as it flows through the nozzle. This drop in temperature is mainly due to the conversion of internal energy to kinetic energy. (The heat loss is too small to cause any significant effect in this case.)

## 2   Turbines and Compressors

In steam, gas, or hydroelectric power plants, the device that drives the electric generator is the turbine. As the fluid passes through the turbine, work is done against the blades, which are attached to the shaft. As a result, the shaft rotates, and the turbine produces work.

Compressors, as well as pumps and fans, are devices used to increase the pressure of a fluid. Work is supplied to these devices from an external source through a rotating shaft. Therefore, compressors involve work inputs. Even though these three devices function similarly, they do differ in the tasks they perform. A *fan* increases the pressure of a gas slightly and is mainly used to mobilize a gas. A *compressor* is capable of compressing the gas to very high pressures. *Pumps* work very much like compressors except that they handle liquids instead of gases.

Note that turbines produce power output whereas compressors, pumps, and fans require power input. Heat transfer from turbines is usually negligible ($\dot{Q} \approx 0$) since they are typically well insulated. Heat transfer is also negligible for compressors unless there is intentional cooling. Potential energy changes are negligible for all of these devices ($\Delta pe \cong 0$). The velocities involved in these devices, with the exception of turbines and fans, are usually too low to cause any significant change in the kinetic energy ($\Delta ke \cong 0$). The fluid velocities encountered in most turbines are very high, and the fluid experiences a significant change in its kinetic energy. However, this change is usually very small relative to the change in enthalpy, and thus it is often disregarded.

---

### EXAMPLE 6–6    Compressing Air by a Compressor

Air at 100 kPa and 280 K is compressed steadily to 600 kPa and 400 K. The mass flow rate of the air is 0.02 kg/s, and a heat loss of 16 kJ/kg occurs during the process. Assuming the changes in kinetic and potential energies are negligible, determine the necessary power input to the compressor.

**SOLUTION**   Air is compressed steadily by a compressor to a specified temperature and pressure. The power input to the compressor is to be determined.
**Assumptions**   **1** This is a steady-flow process since there is no change with time at any point and thus $\Delta m_{CV} = 0$ and $\Delta E_{CV} = 0$. **2** Air is an ideal gas since it is at a high temperature and low pressure relative to its critical-point values. **3** The kinetic and potential energy changes are zero, $\Delta ke = \Delta pe = 0$.
**Analysis**   We take the *compressor* as the system (Fig. 6–28). This is a *control volume* since mass crosses the system boundary during the process. We observe that there is only one inlet and one exit and thus $\dot{m}_1 = \dot{m}_2 = \dot{m}$. Also, heat is lost from the system and work is supplied to the system.

Under stated assumptions and observations, the energy balance for this steady-flow system can be expressed in the rate form as

$$\underbrace{\dot{E}_{in} - \dot{E}_{out}}_{\substack{\text{Rate of net energy transfer} \\ \text{by heat, work, and mass}}} = \underbrace{dE_{system}/dt}_{\substack{\text{Rate of change in internal, kinetic,} \\ \text{potential, etc., energies}}} \nearrow^{0 \text{ (steady)}} = 0$$

$$\dot{E}_{in} = \dot{E}_{out}$$
$$\dot{W}_{in} + \dot{m}h_1 = \dot{Q}_{out} + \dot{m}h_2 \qquad (\text{since } \Delta ke = \Delta pe \cong 0)$$
$$\dot{W}_{in} = \dot{m}q_{out} + \dot{m}(h_2 - h_1)$$

$q_{out} = 16$ kJ/kg
$P_2 = 600$ kPa
$T_2 = 400$ K
Air
$\dot{m} = 0.02$ kg/s
$\dot{W}_{in} = ?$
$P_1 = 100$ kPa
$T_1 = 280$ K

**FIGURE 6–28**
Schematic for Example 6–6.

The enthalpy of an ideal gas depends on temperature only, and the enthalpies of the air at the specified temperatures are determined from the air table (Table A–21) to be

$$h_1 = h_{@\ 280\ K} = 280.13\ \text{kJ/kg}$$
$$h_2 = h_{@\ 400\ K} = 400.98\ \text{kJ/kg}$$

Substituting, the power input to the compressor is determined to be

$$\dot{W}_{in} = (0.02\ \text{kg/s})(16\ \text{kJ/kg}) + (0.02\ \text{kg/s})(400.98 - 280.13)\ \text{kJ/kg}$$
$$= \mathbf{2.74\ kW}$$

**Discussion**  Note that the mechanical energy input to the compressor manifests itself as a rise in enthalpy of air and heat loss from the compressor.

---

$P_1 = 2$ MPa
$T_1 = 400°C$
$V_1 = 50$ m/s
$z_1 = 10$ m

Steam turbine

$\dot{W}_{out} = 5$ MW

$P_2 = 15$ kPa
$x_2 = 90\%$
$V_2 = 180$ m/s
$z_2 = 6$ m

**FIGURE 6–29**

Schematic for Example 6–7.

---

**EXAMPLE 6–7**    **Power Generation by a Steam Turbine**

The power output of an adiabatic steam turbine is 5 MW, and the inlet and the exit conditions of the steam are as indicated in Fig. 6–29.

(a) Compare the magnitudes of $\Delta h$, $\Delta ke$, and $\Delta pe$.
(b) Determine the work done per unit mass of the steam flowing through the turbine.
(c) Calculate the mass flow rate of the steam.

**SOLUTION**  The inlet and exit conditions of a steam turbine and its power output are given. The changes in kinetic energy, potential energy, and enthalpy of steam, as well as the work done per unit mass and the mass flow rate of steam are to be determined.

**Assumptions**  **1** This is a steady-flow process since there is no change with time at any point and thus $\Delta m_{CV} = 0$ and $\Delta E_{CV} = 0$. **2** The system is adiabatic and thus there is no heat transfer.

**Analysis**  We take the *turbine* as the system. This is a *control volume* since mass crosses the system boundary during the process. We observe that there is only one inlet and one exit and thus $\dot{m}_1 = \dot{m}_2 = \dot{m}$. Also, work is done by the system. The inlet and exit velocities and elevations are given, and thus the kinetic and potential energies are to be considered.

(a) At the inlet, steam is in a superheated vapor state, and its enthalpy is

$$\left.\begin{array}{l} P_1 = 2\ \text{MPa} \\ T_1 = 400°C \end{array}\right\}\quad h_1 = 3248.4\ \text{kJ/kg}\qquad \text{(Table A–6)}$$

At the turbine exit, we obviously have a saturated liquid–vapor mixture at 15-kPa pressure. The enthalpy at this state is

$$h_2 = h_f + x_2 h_{fg} = [225.94 + (0.9)(2372.3)]\ \text{kJ/kg} = 2361.01\ \text{kJ/kg}$$

Then

$$\Delta h = h_2 - h_1 = (2361.01 - 3248.4)\ \text{kJ/kg} = \mathbf{-887.39\ kJ/kg}$$

$$\Delta ke = \frac{V_2^2 - V_1^2}{2} = \frac{(180\ \text{m/s})^2 - (50\ \text{m/s})^2}{2}\left(\frac{1\ \text{kJ/kg}}{1000\ \text{m}^2/\text{s}^2}\right) = \mathbf{14.95\ kJ/kg}$$

$$\Delta pe = g(z_2 - z_1) = (9.81\ \text{m/s}^2)[(6 - 10)\ \text{m}]\left(\frac{1\ \text{kJ/kg}}{1000\ \text{m}^2/\text{s}^2}\right) = \mathbf{-0.04\ kJ/kg}$$

(b) The energy balance for this steady-flow system can be expressed in the rate form as

$$\underbrace{\dot{E}_{\text{in}} - \dot{E}_{\text{out}}}_{\substack{\text{Rate of net energy transfer} \\ \text{by heat, work, and mass}}} = \underbrace{dE_{\text{system}}/dt}_{\substack{\text{Rate of change in internal, kinetic,} \\ \text{potential, etc., energies}}} \nearrow^{0 \text{ (steady)}} = 0$$

$$\dot{E}_{\text{in}} = \dot{E}_{\text{out}}$$

$$\dot{m}\left(h_1 + \frac{V_1^2}{2} + gz_1\right) = \dot{W}_{\text{out}} + \dot{m}\left(h_2 + \frac{V_2^2}{2} + gz_2\right) \qquad \text{(since } \dot{Q} = 0\text{)}$$

Dividing by the mass flow rate $\dot{m}$ and substituting, the work done by the turbine per unit mass of the steam is determined to be

$$w_{\text{out}} = -\left[(h_2 - h_1) + \frac{V_2^2 - V_1^2}{2} + g(z_2 - z_1)\right] = -(\Delta h + \Delta\text{ke} + \Delta\text{pe})$$

$$= -[-887.39 + 14.95 - 0.04]\text{ kJ/kg} = \textbf{872.48 kJ/kg}$$

(c) The required mass flow rate for a 5-MW power output is

$$\dot{m} = \frac{\dot{W}_{\text{out}}}{w_{\text{out}}} = \frac{5000\text{ kJ/s}}{872.48\text{ kJ/kg}} = \textbf{5.73 kg/s}$$

**Discussion** Two observations can be made from these results. First, the change in potential energy is insignificant in comparison to the changes in enthalpy and kinetic energy. This is typical for most engineering devices. Second, as a result of low pressure and thus high specific volume, the steam velocity at the turbine exit can be very high. Yet the change in kinetic energy is a small fraction of the change in enthalpy (less than 2 percent in our case) and is therefore often neglected.

## 3 Throttling Valves

Throttling valves are *any kind of flow-restricting devices* that cause a significant pressure drop in the fluid. Some familiar examples are ordinary adjustable valves, capillary tubes, and porous plugs (Fig. 6–30). Unlike turbines, they produce a pressure drop without involving any work. The pressure drop in the fluid is often accompanied by a *large drop in temperature*, and for that reason throttling devices are commonly used in refrigeration and air-conditioning applications. The magnitude of the temperature drop (or, sometimes, the temperature rise) during a throttling process is governed by a property called the *Joule-Thomson coefficient*.

Throttling valves are usually small devices, and the flow through them may be assumed to be adiabatic ($q \cong 0$) since there is neither sufficient time nor large enough area for any effective heat transfer to take place. Also, there is no work done ($w = 0$), and the change in potential energy, if any, is very small ($\Delta\text{pe} \cong 0$). Even though the exit velocity is often considerably higher than the inlet velocity, in many cases, the increase in kinetic energy is insignificant ($\Delta\text{ke} \cong 0$). Then the conservation of energy equation for this single-stream steady-flow device reduces to

$$h_2 \cong h_1 \qquad \text{(kJ/kg)} \tag{6-41}$$

(a) An adjustable valve

(b) A porous plug

(c) A capillary tube

**FIGURE 6–30**

Throttling valves are devices that cause large pressure drops in the fluid.

That is, enthalpy values at the inlet and exit of a throttling valve are the same. For this reason, a throttling valve is sometimes called an *isenthalpic device*. Note, however, that for throttling devices with large exposed surface areas such as capillary tubes, heat transfer may be significant.

To gain some insight into how throttling affects fluid properties, let us express Eq. 6–41 as follows:

$$u_1 + P_1 v_1 = u_2 + P_2 v_2$$

or

$$\text{Internal energy} + \text{Flow energy} = \text{Constant}$$

Thus the final outcome of a throttling process depends on which of the two quantities increases during the process. If the flow energy increases during the process ($P_2 v_2 > P_1 v_1$), it can do so at the expense of the internal energy. As a result, internal energy decreases, which is usually accompanied by a drop in temperature. If the product $Pv$ decreases, the internal energy and the temperature of a fluid will increase during a throttling process. In the case of an ideal gas, $h = h(T)$, and thus the temperature has to remain constant during a throttling process (Fig. 6–31).

Throttling
valve

Ideal gas $\xrightarrow{\quad T_1 \quad}$ $\quad T_2 = T_1$
$h_1$ $\qquad h_2 = h_1$

**FIGURE 6–31**

The temperature of an ideal gas does not change during a throttling ($h =$ constant) process since $h = h(T)$.

Throttling
valve

$u_1 = 94.79$ kJ/kg $\qquad u_2 = 88.79$ kJ/kg
$P_1 v_1 = 0.68$ kJ/kg $\qquad P_2 v_2 = 6.68$ kJ/kg
$(h_1 = 95.47$ kJ/kg$)$ $\quad (h_2 = 95.47$ kJ/kg$)$

**FIGURE 6–32**

During a throttling process, the enthalpy (flow energy + internal energy) of a fluid remains constant. But internal and flow energies may be converted to each other.

---

**EXAMPLE 6–8**  **Expansion of Refrigerant-134a in a Refrigerator**

Refrigerant-134a enters the capillary tube of a refrigerator as saturated liquid at 0.8 MPa and is throttled to a pressure of 0.12 MPa. Determine the quality of the refrigerant at the final state and the temperature drop during this process.

**SOLUTION**  Refrigerant-134a that enters a capillary tube as saturated liquid is throttled to a specified pressure. The exit quality of the refrigerant and the temperature drop are to be determined.

**Assumptions**  **1** Heat transfer from the tube is negligible. **2** Kinetic energy change of the refrigerant is negligible.

**Analysis**  A capillary tube is a simple flow-restricting device that is commonly used in refrigeration applications to cause a large pressure drop in the refrigerant. Flow through a capillary tube is a throttling process; thus, the enthalpy of the refrigerant remains constant (Fig. 6–32).

At inlet: $\left.\begin{array}{l} P_1 = 0.8 \text{ MPa} \\ \text{sat. liquid} \end{array}\right\}$ $\quad\begin{array}{l} T_1 = T_{\text{sat @ 0.8 MPa}} = 31.31°C \\ h_1 = h_{f\text{ @ 0.8 MPa}} = 95.47 \text{ kJ/kg} \end{array}$  (Table A–12)

At exit: $\begin{array}{l} P_2 = 0.12 \text{ MPa} \\ (h_2 = h_1) \end{array} \longrightarrow \begin{array}{l} h_f = 22.49 \text{ kJ/kg} \qquad T_{\text{sat}} = -22.32°C \\ h_g = 236.97 \text{ kJ/kg} \end{array}$

Obviously $h_f < h_2 < h_g$; thus, the refrigerant exists as a saturated mixture at the exit state. The quality at this state is

$$x_2 = \frac{h_2 - h_f}{h_{fg}} = \frac{95.47 - 22.49}{236.97 - 22.49} = \mathbf{0.340}$$

Since the exit state is a saturated mixture at 0.12 MPa, the exit temperature must be the saturation temperature at this pressure, which is −22.32°C. Then the temperature change for this process becomes

$$\Delta T = T_2 - T_1 = (-22.32 - 31.31)°C = \mathbf{-53.63°C}$$

**Discussion** Note that the temperature of the refrigerant drops by 53.63°C during this throttling process. Also note that 34.0 percent of the refrigerant vaporizes during this throttling process, and the energy needed to vaporize this refrigerant is absorbed from the refrigerant itself.

# 4a   Mixing Chambers

In engineering applications, mixing two streams of fluids is not a rare occurrence. The section where the mixing process takes place is commonly referred to as a **mixing chamber**. The mixing chamber does not have to be a distinct "chamber." An ordinary T-elbow or a Y-elbow in a shower, for example, serves as the mixing chamber for the cold- and hot-water streams (Fig. 6–33).

The conservation of mass principle for a mixing chamber requires that the sum of the incoming mass flow rates equal the mass flow rate of the outgoing mixture.

Mixing chambers are usually well insulated ($q \cong 0$) and usually do not involve any kind of work ($w = 0$). Also, the kinetic and potential energies of the fluid streams are usually negligible (kc $\cong$ 0, pe $\cong$ 0). Then all there is left in the energy equation is the total energies of the incoming streams and the outgoing mixture. The conservation of energy principle requires that these two equal each other. Therefore, the conservation of energy equation becomes analogous to the conservation of mass equation for this case.

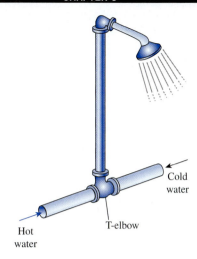

**FIGURE 6–33**

The T-elbow of an ordinary shower serves as the mixing chamber for the hot- and the cold-water streams.

---

**EXAMPLE 6–9    Mixing of Hot and Cold Waters in a Shower**

Consider an ordinary shower where hot water at 140°F is mixed with cold water at 50°F. If it is desired that a steady stream of warm water at 110°F be supplied, determine the ratio of the mass flow rates of the hot to cold water. Assume the heat losses from the mixing chamber to be negligible and the mixing to take place at a pressure of 20 psia.

**SOLUTION**   In a shower, cold water is mixed with hot water at a specified temperature. For a specified mixture temperature, the ratio of the mass flow rates of the hot to cold water is to be determined.

**Assumptions   1** This is a steady-flow process since there is no change with time at any point and thus $\Delta m_{CV} = 0$ and $\Delta E_{CV} = 0$. **2** The kinetic and potential energies are negligible, ke $\cong$ pe $\cong$ 0. **3** Heat losses from the system are negligible and thus $\dot{Q} \cong 0$. **4** There is no work interaction involved.

**Analysis**   We take the *mixing chamber* as the system (Fig. 6–34). This is a *control volume* since mass crosses the system boundary during the process. We observe that there are two inlets and one exit.

Under the stated assumptions and observations, the mass and energy balances for this steady-flow system can be expressed in the rate form as follows:

*Mass balance:*   $\dot{m}_{in} - \dot{m}_{out} = dm_{system}/dt \nearrow^{0 \text{ (steady)}} = 0$

$$\dot{m}_{in} = \dot{m}_{out} \rightarrow \dot{m}_1 + \dot{m}_2 = \dot{m}_3$$

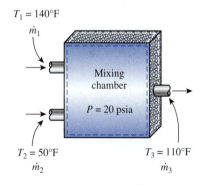

$T_1 = 140°F$

$\dot{m}_1$

Mixing chamber

$P = 20$ psia

$T_2 = 50°F$

$\dot{m}_2$

$T_3 = 110°F$

$\dot{m}_3$

**FIGURE 6–34**

Schematic for Example 6–9.

$$\textit{Energy balance:} \quad \underbrace{\dot{E}_{\text{in}} - \dot{E}_{\text{out}}}_{\substack{\text{Rate of net energy transfer} \\ \text{by heat, work, and mass}}} = \underbrace{dE_{\text{system}}/dt}_{\substack{\text{Rate of change in internal, kinetic,} \\ \text{potential, etc., energies}}} \overset{\nearrow\,0\text{ (steady)}}{} = 0$$

$$\dot{E}_{\text{in}} = \dot{E}_{\text{out}}$$

$$\dot{m}_1 h_1 + \dot{m}_2 h_2 = \dot{m}_3 h_3 \quad (\text{since } \dot{Q} \cong 0, \dot{W} = 0, \text{ke} \cong \text{pe} \cong 0)$$

Combining the mass and energy balances,

$$\dot{m}_1 h_1 + \dot{m}_2 h_2 = (\dot{m}_1 + \dot{m}_2) h_3$$

Dividing this equation by $\dot{m}_2$ yields

$$y h_1 + h_2 = (y + 1) h_3$$

where $y = \dot{m}_1 / \dot{m}_2$ is the desired mass flow rate ratio.

The saturation temperature of water at 20 psia is 227.92°F. Since the temperatures of all three streams are below this value ($T < T_{\text{sat}}$), the water in all three streams exists as a compressed liquid (Fig. 6–35). A compressed liquid can be approximated as a saturated liquid at the given temperature. Thus,

$$h_1 \cong h_{f\,@\,140°F} = 107.99 \text{ Btu/lbm}$$

$$h_2 \cong h_{f\,@\,50°F} = 18.07 \text{ Btu/lbm}$$

$$h_3 \cong h_{f\,@\,110°F} = 78.02 \text{ Btu/lbm}$$

Solving for $y$ and substituting yields

$$y = \frac{h_3 - h_2}{h_1 - h_3} = \frac{78.02 - 18.07}{107.99 - 78.02} = \mathbf{2.0}$$

**Discussion** Note that the mass flow rate of the hot water must be twice the mass flow rate of the cold water for the mixture to leave at 110°F.

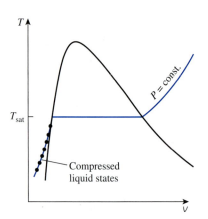

**FIGURE 6–35**

A substance exists as a compressed liquid at temperatures below the saturation temperatures at the given pressure.

## 4b Heat Exchangers

As the name implies, **heat exchangers** are devices where two moving fluid streams exchange heat without mixing. Heat exchangers are widely used in various industries, and they come in various designs.

The simplest form of a heat exchanger is a *double-tube* (also called *tube-and-shell*) *heat exchanger*, shown in Fig. 6–36. It is composed of two concentric pipes of different diameters. One fluid flows in the inner pipe, and the other in the annular space between the two pipes. Heat is transferred from the hot fluid to the cold one through the wall separating them. Sometimes the inner tube makes a couple of turns inside the shell to increase the heat transfer area, and thus the rate of heat transfer. The mixing chambers discussed earlier are sometimes classified as *direct-contact heat exchangers*.

The conservation of mass principle for a heat exchanger in steady operation requires that the sum of the inbound mass flow rates equal the sum of the outbound mass flow rates. This principle can also be expressed as follows: *Under steady operation, the mass flow rate of each fluid stream flowing through a heat exchanger remains constant.*

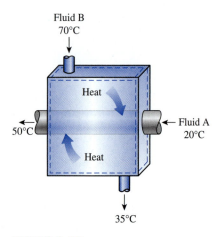

**FIGURE 6–36**

A heat exchanger can be as simple as two concentric pipes.

FIGURE 6–37
The heat transfer associated with a heat exchanger may be zero or nonzero depending on how the control volume is selected.

(a) System: Entire heat exchanger ($Q_{CV} = 0$)

(b) System: Fluid A ($Q_{CV} \neq 0$)

Heat exchangers typically involve no work interactions ($w = 0$) and negligible kinetic and potential energy changes ($\Delta ke \cong 0$, $\Delta pe \cong 0$) for each fluid stream. The heat transfer rate associated with heat exchangers depends on how the control volume is selected. Heat exchangers are intended for heat transfer between two fluids *within* the device, and the outer shell is usually well insulated to prevent any heat loss to the surrounding medium.

When the entire heat exchanger is selected as the control volume, $\dot{Q}$ becomes zero, since the boundary for this case lies just beneath the insulation and little or no heat crosses the boundary (Fig. 6–37). If, however, only one of the fluids is selected as the control volume, then heat will cross this boundary as it flows from one fluid to the other and $\dot{Q}$ will not be zero. In fact, $\dot{Q}$ in this case will be the rate of heat transfer between the two fluids.

---

### EXAMPLE 6–10    Cooling of Refrigerant-134a by Water

Refrigerant-134a is to be cooled by water in a condenser. The refrigerant enters the condenser with a mass flow rate of 6 kg/min at 1 MPa and 70°C and leaves at 35°C. The cooling water enters at 300 kPa and 15°C and leaves at 25°C. Neglecting any pressure drops, determine (*a*) the mass flow rate of the cooling water required and (*b*) the heat transfer rate from the refrigerant to water.

**SOLUTION** Refrigerant-134a is cooled by water in a condenser. The mass flow rate of the cooling water and the rate of heat transfer from the refrigerant to the water are to be determined.

*Assumptions* **1** This is a steady-flow process since there is no change with time at any point and thus $\Delta m_{CV} = 0$ and $\Delta E_{CV} = 0$. **2** The kinetic and potential energies are negligible, ke $\cong$ pe $\cong$ 0. **3** Heat losses from the system are negligible and thus $\dot{Q} \cong 0$. **4** There is no work interaction.

*Analysis* We take the *entire heat exchanger* as the system (Fig. 6–38). This is a *control volume* since mass crosses the system boundary during the process. In general, there are several possibilities for selecting the control volume for multiple-stream steady-flow devices, and the proper choice depends on the situation at hand. We observe that there are two fluid streams (and thus two inlets and two exits) but no mixing.

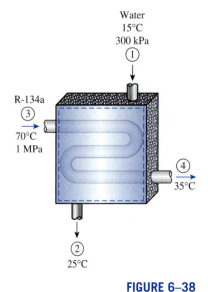

FIGURE 6–38
Schematic for Example 6–10.

(*a*) Under the stated assumptions and observations, the mass and energy balances for this steady-flow system can be expressed in the rate form as follows:

*Mass balance:* $\qquad\qquad\qquad \dot{m}_{in} = \dot{m}_{out}$

for each fluid stream since there is no mixing. Thus,

$$\dot{m}_1 = \dot{m}_2 = \dot{m}_w$$

$$\dot{m}_3 = \dot{m}_4 = \dot{m}_R$$

*Energy balance:* $\quad \underbrace{\dot{E}_{in} - \dot{E}_{out}}_{\substack{\text{Rate of net energy transfer} \\ \text{by heat, work, and mass}}} = \underbrace{dE_{system}/dt}_{\substack{\text{Rate of change in internal, kinetic,} \\ \text{potential, etc., energies}}}\nearrow^{0 \text{ (steady)}} = 0$

$$\dot{E}_{in} = \dot{E}_{out}$$

$$\dot{m}_1 h_1 + \dot{m}_3 h_3 = \dot{m}_2 h_2 + \dot{m}_4 h_4 \qquad (\text{since } \dot{Q} \cong 0, \dot{W} = 0, \text{ke} \cong \text{pe} \cong 0)$$

Combining the mass and energy balances and rearranging give

$$\dot{m}_w(h_1 - h_2) = \dot{m}_R(h_4 - h_3)$$

Now we need to determine the enthalpies at all four states. Water exists as a compressed liquid at both the inlet and the exit since the temperatures at both locations are below the saturation temperature of water at 300 kPa (133.52°C). Approximating the compressed liquid as a saturated liquid at the given temperatures, we have

$$h_1 \cong h_{f@15°C} = 62.982 \text{ kJ/kg}$$
$$h_2 \cong h_{f@25°C} = 104.83 \text{ kJ/kg}$$
$$\qquad\qquad\qquad\qquad\qquad\qquad\qquad \text{(Table A–4)}$$

The refrigerant enters the condenser as a superheated vapor and leaves as a compressed liquid at 35°C. From refrigerant-134a tables,

$$\left.\begin{array}{l} P_3 = 1 \text{ MPa} \\ T_3 = 70°C \end{array}\right\} \quad h_3 = 303.85 \text{ kJ/kg} \qquad \text{(Table A–13)}$$

$$\left.\begin{array}{l} P_4 = 1 \text{ MPa} \\ T_4 = 35°C \end{array}\right\} \quad h_4 \cong h_{f@35°C} = 100.87 \text{ kJ/kg} \qquad \text{(Table A–11)}$$

Substituting, we find

$$\dot{m}_w(62.982 - 104.83) \text{ kJ/kg} = (6 \text{ kg/min})[(100.87 - 303.85) \text{ kJ/kg}]$$

$$\dot{m}_w = \textbf{29.1 kg/min}$$

(*b*) To determine the heat transfer from the refrigerant to the water, we have to choose a control volume whose boundary lies on the path of heat transfer. We can choose the volume occupied by either fluid as our control volume. For no particular reason, we choose the volume occupied by the water. All the assumptions stated earlier apply, except that the heat transfer is no longer zero. Then assuming heat to be transferred to water, the energy balance for this single-stream steady-flow system reduces to

$$\underbrace{\dot{E}_{in} - \dot{E}_{out}}_{\substack{\text{Rate of net energy transfer} \\ \text{by heat, work, and mass}}} = \underbrace{dE_{system}/dt}_{\substack{\text{Rate of change in internal, kinetic,} \\ \text{potential, etc., energies}}}\nearrow^{0 \text{ (steady)}} = 0$$

$$\dot{E}_{in} = \dot{E}_{out}$$

$$\dot{Q}_{w,in} + \dot{m}_w h_1 = \dot{m}_w h_2$$

Rearranging and substituting,

$$\dot{Q}_{w,\text{in}} = \dot{m}_w(h_2 - h_1) = (29.1 \text{ kg/min})[(104.83 - 62.982) \text{ kJ/kg}]$$

$$= \textbf{1218 kJ/min}$$

**Discussion** Had we chosen the volume occupied by the refrigerant as the control volume (Fig. 6–39), we would have obtained the same result for $\dot{Q}_{R,\text{out}}$ since the heat gained by the water is equal to the heat lost by the refrigerant.

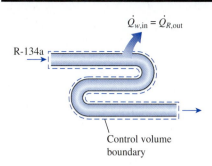

**FIGURE 6–39**

In a heat exchanger, the heat transfer depends on the choice of the control volume.

# 5 Pipe and Duct Flow

The transport of liquids or gases in pipes and ducts is of great importance in many engineering applications. Flow through a pipe or a duct usually satisfies the steady-flow conditions and thus can be analyzed as a steady-flow process. This, of course, excludes the transient start-up and shut-down periods. The control volume can be selected to coincide with the interior surfaces of the portion of the pipe or the duct that we are interested in analyzing.

Under normal operating conditions, the amount of heat gained or lost by the fluid may be very significant, particularly if the pipe or duct is long (Fig. 6–40). Sometimes heat transfer is desirable and is the sole purpose of the flow. Water flow through the pipes in the furnace of a power plant, the flow of refrigerant in a freezer, and the flow in heat exchangers are some examples of this case. At other times, heat transfer is undesirable, and the pipes or ducts are insulated to prevent any heat loss or gain, particularly when the temperature difference between the flowing fluid and the surroundings is large. Heat transfer in this case is negligible.

If the control volume involves a heating section (electric wires), a fan, or a pump (shaft), the work interactions should be considered (Fig. 6–41). Of these, fan work is usually small and often neglected in energy analysis.

The velocities involved in pipe and duct flow are relatively low, and the kinetic energy changes are usually insignificant. This is particularly true when the pipe or duct diameter is constant and the heating effects are negligible. Kinetic energy changes may be significant, however, for gas flow in ducts with variable cross-sectional areas especially when the compressibility effects are significant. The potential energy term may also be significant when the fluid undergoes a considerable elevation change as it flows in a pipe or duct.

**FIGURE 6–40**

Heat losses from a hot fluid flowing through an uninsulated pipe or duct to the cooler environment may be very significant.

**FIGURE 6–41**

Pipe or duct flow may involve more than one form of work at the same time.

---

### EXAMPLE 6–11    Electric Heating of Air in a House

The electric heating systems used in many houses consist of a simple duct with resistance heaters. Air is heated as it flows over resistance wires. Consider a 15-kW electric heating system. Air enters the heating section at 100 kPa and 17°C with a volume flow rate of 150 m³/min. If heat is lost from the air in the duct to the surroundings at a rate of 200 W, determine the exit temperature of air.

**SOLUTION** The electric heating system of a house is considered. For specified electric power consumption and air flow rate, the air exit temperature is to be determined.

**FIGURE 6–42**

Schematic for Example 6–11.

$\Delta h = 1.005 \, \Delta T$ (kJ/kg)

**FIGURE 6–43**

The error involved in $\Delta h = c_p \, \Delta T$, where $c_p = 1.005$ kJ/kg·°C, is less than 0.5 percent for air in the temperature range −20 to 70°C.

**Assumptions** **1** This is a steady-flow process since there is no change with time at any point and thus $\Delta m_{CV} = 0$ and $\Delta E_{CV} = 0$. **2** Air is an ideal gas since it is at a high temperature and low pressure relative to its critical-point values. **3** The kinetic and potential energy changes are negligible, $\Delta ke \cong \Delta pe \cong 0$. **4** Constant specific heats at room temperature can be used for air.

**Analysis** We take the *heating section portion of the duct* as the system (Fig. 6–42). This is a *control volume* since mass crosses the system boundary during the process. We observe that there is only one inlet and one exit and thus $\dot{m}_1 = \dot{m}_2 = \dot{m}$. Also, heat is lost from the system and electrical work is supplied to the system.

At temperatures encountered in heating and air-conditioning applications, $\Delta h$ can be replaced by $c_p \, \Delta T$ where $c_p = 1.005$ kJ/kg·°C—the value at room temperature—with negligible error (Fig. 6–43). Then the energy balance for this steady-flow system can be expressed in the rate form as

$$\underbrace{\dot{E}_{in} - \dot{E}_{out}}_{\substack{\text{Rate of net energy transfer} \\ \text{by heat, work, and mass}}} = \underbrace{dE_{system}/dt}_{\substack{\text{Rate of change in internal, kinetic,} \\ \text{potential, etc., energies}}} \nearrow^{0 \text{ (steady)}} = 0$$

$$\dot{E}_{in} = \dot{E}_{out}$$

$$\dot{W}_{e,in} + \dot{m}h_1 = \dot{Q}_{out} + \dot{m}h_2 \quad (\text{since } \Delta ke \cong \Delta pe \cong 0)$$

$$\dot{W}_{e,in} - \dot{Q}_{out} = \dot{m}c_p(T_2 - T_1)$$

From the ideal-gas relation, the specific volume of air at the inlet of the duct is

$$v_1 = \frac{RT_1}{P_1} = \frac{(0.287 \text{ kPa·m}^3/\text{kg·K})(290 \text{ K})}{100 \text{ kPa}} = 0.832 \text{ m}^3/\text{kg}$$

The mass flow rate of the air through the duct is determined from

$$\dot{m} = \frac{\dot{V}_1}{v_1} = \frac{150 \text{ m}^3/\text{min}}{0.832 \text{ m}^3/\text{kg}} \left(\frac{1 \text{ min}}{60 \text{ s}}\right) = 3.0 \text{ kg/s}$$

Substituting the known quantities, the exit temperature of the air is determined to be

$$(15 \text{ kJ/s}) - (0.2 \text{ kJ/s}) = (3 \text{ kg/s})(1.005 \text{ kJ/kg·°C})(T_2 - 17)\text{°C}$$

$$T_2 = \mathbf{21.9°C}$$

**Discussion** Note that heat loss from the duct reduces the exit temperature of air.

# 6–5 · ENERGY ANALYSIS OF UNSTEADY-FLOW PROCESSES

During a steady-flow process, no changes occur within the control volume; thus, one does not need to be concerned about what is going on within the boundaries. Not having to worry about any changes within the control volume with time greatly simplifies the analysis.

Many processes of interest, however, involve *changes* within the control volume with time. Such processes are called *unsteady-flow*, or *transient-flow*, processes. The steady-flow relations developed earlier are obviously not applicable to these processes. When an unsteady-flow process is analyzed, it is important to keep track of the mass and energy contents of the control volume as well as the energy interactions across the boundary.

Some familiar unsteady-flow processes are the charging of rigid vessels from supply lines (Fig. 6–44), discharging a fluid from a pressurized vessel, driving a gas turbine with pressurized air stored in a large container, inflating tires or balloons, and even cooking with an ordinary pressure cooker.

Unlike steady-flow processes, unsteady-flow processes start and end over some finite time period instead of continuing indefinitely. Therefore in this section, we deal with changes that occur over some time interval $\Delta t$ instead of with the rate of changes (changes per unit time). An unsteady-flow system, in some respects, is similar to a closed system, except that the mass within the system boundaries does not remain constant during a process.

Another difference between steady- and unsteady-flow systems is that steady-flow systems are fixed in space, size, and shape. Unsteady-flow systems, however, are not (Fig. 6–45). They are usually stationary; that is, they are fixed in space, but they may involve moving boundaries and thus boundary work.

The *mass balance* for any system undergoing any process can be expressed as (*see* Sec. 6–1)

$$m_{in} - m_{out} = \Delta m_{system} \quad \text{(kg)} \tag{6–42}$$

where $\Delta m_{system} = m_{final} - m_{initial}$ is the change in the mass of the system. For control volumes, it can also be expressed more explicitly as

$$m_i - m_e = (m_2 - m_1)_{CV} \tag{6–43}$$

where $i$ = inlet, $e$ = exit, 1 = initial state, and 2 = final state of the control volume. Often one or more terms in the equation above are zero. For example, $m_i = 0$ if no mass enters the control volume during the process, $m_e = 0$ if no mass leaves, and $m_1 = 0$ if the control volume is initially evacuated.

The energy content of a control volume changes with time during an unsteady-flow process. The magnitude of change depends on the amount of energy transfer across the system boundaries as heat and work as well as on the amount of energy transported into and out of the control volume by mass during the process. When analyzing an unsteady-flow process, we must keep track of the energy content of the control volume as well as the energies of the incoming and outgoing flow streams.

The general energy balance was given earlier as

*Energy balance:*
$$\underbrace{E_{in} - E_{out}}_{\substack{\text{Net energy transfer} \\ \text{by heat, work, and mass}}} = \underbrace{\Delta E_{system}}_{\substack{\text{Change in internal, kinetic,} \\ \text{potential, etc., energies}}} \quad \text{(kJ)} \tag{6–44}$$

The general unsteady-flow process, in general, is difficult to analyze because the properties of the mass at the inlets and exits may change during a process. Most unsteady-flow processes, however, can be represented reasonably well by the **uniform-flow process**, which involves the following idealization: *The fluid flow at any inlet or exit is uniform and steady, and thus the fluid properties do not change with time or position over the cross*

**FIGURE 6–44**

Charging of a rigid tank from a supply line is an unsteady-flow process since it involves changes within the control volume.

**FIGURE 6–45**

The shape and size of a control volume may change during an unsteady-flow process.

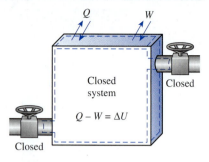

**FIGURE 6–46**

The energy equation of a uniform-flow system reduces to that of a closed system when all the inlets and exits are closed.

**FIGURE 6–47**

A uniform-flow system may involve electrical, shaft, and boundary work all at once.

*section of an inlet or exit. If they do, they are averaged and treated as constants for the entire process.*

Note that unlike the steady-flow systems, the state of an unsteady-flow system may change with time, and that the state of the mass leaving the control volume at any instant is the same as the state of the mass in the control volume at that instant. The initial and final properties of the control volume can be determined from the knowledge of the initial and final states, which are completely specified by two independent intensive properties for simple compressible systems.

Then the energy balance for a uniform-flow system can be expressed explicitly as

$$\left( Q_{in} + W_{in} + \sum_{in} m\theta \right) - \left( Q_{out} + W_{out} + \sum_{out} m\theta \right) = (m_2 e_2 - m_1 e_1)_{system} \quad \textbf{(6–45)}$$

where $\theta = h + ke + pe$ is the energy of a fluid stream at any inlet or exit per unit mass, and $e = u + ke + pe$ is the energy of the nonflowing fluid within the control volume per unit mass. When the kinetic and potential energy changes associated with the control volume and fluid streams are negligible, as is usually the case, the energy balance above simplifies to

$$Q - W = \sum_{out} mh - \sum_{in} mh + (m_2 u_2 - m_1 u_1)_{system} \quad \textbf{(6–46)}$$

where $Q = Q_{net,in} = Q_{in} - Q_{out}$ is the net heat input and $W = W_{net,out} = W_{out} - W_{in}$ is the net work output. Note that if no mass enters or leaves the control volume during a process ($m_i = m_e = 0$, and $m_1 = m_2 = m$), this equation reduces to the energy balance relation for closed systems (Fig. 6–46). Also note that an unsteady-flow system may involve boundary work as well as electrical and shaft work (Fig. 6–47).

Although both the steady-flow and uniform-flow processes are somewhat idealized, many actual processes can be approximated reasonably well by one of these with satisfactory results. The degree of satisfaction depends on the desired accuracy and the degree of validity of the assumptions made.

---

**EXAMPLE 6–12    Charging of a Rigid Tank by Steam**

A rigid, insulated tank that is initially evacuated is connected through a valve to a supply line that carries steam at 1 MPa and 300°C. Now the valve is opened, and steam is allowed to flow slowly into the tank until the pressure reaches 1 MPa, at which point the valve is closed. Determine the final temperature of the steam in the tank.

**SOLUTION** A valve connecting an initially evacuated tank to a steam line is opened, and steam flows in until the pressure inside rises to the line level. The final temperature in the tank is to be determined.

*Assumptions* **1** This process can be analyzed as a *uniform-flow process* since the properties of the steam entering the control volume remain constant during the entire process. **2** The kinetic and potential energies of the streams are negligible, ke ≅ pe ≅ 0. **3** The tank is stationary and thus its kinetic and potential energy changes are zero; that is, $\Delta KE = \Delta PE = 0$ and $\Delta E_{system} = \Delta U_{system}$.

**4** There are no boundary, electrical, or shaft work interactions involved. **5** The tank is well insulated and thus there is no heat transfer.

*Analysis* We take the *tank* as the system (Fig. 6–48). This is a *control volume* since mass crosses the system boundary during the process. We observe that this is an unsteady-flow process since changes occur within the control volume. The control volume is initially evacuated and thus $m_1 = 0$ and $m_1 u_1 = 0$. Also, there is one inlet and no exits for mass flow.

Noting that microscopic energies of flowing and nonflowing fluids are represented by enthalpy $h$ and internal energy $u$, respectively, the mass and energy balances for this uniform-flow system can be expressed as

*Mass balance:* $\quad m_{\text{in}} - m_{\text{out}} = \Delta m_{\text{system}} \quad \rightarrow \quad m_i = m_2 - \overset{0}{\cancel{m_1}} = m_2$

*Energy balance:* $\quad \underbrace{E_{\text{in}} - E_{\text{out}}}_{\substack{\text{Net energy transfer} \\ \text{by heat, work, and mass}}} = \underbrace{\Delta E_{\text{system}}}_{\substack{\text{Change in internal, kinetic,} \\ \text{potential, etc., energies}}}$

$$m_i h_i = m_2 u_2 \qquad (\text{since } W = Q = 0, \text{ ke} \cong \text{ pe} \cong 0, m_1 = 0)$$

Combining the mass and energy balances gives

$$u_2 = h_i$$

That is, the final internal energy of the steam in the tank is equal to the enthalpy of the steam entering the tank. The enthalpy of the steam at the inlet state is

$$\left. \begin{array}{l} P_i = 1 \text{ MPa} \\ T_i = 300°C \end{array} \right\} \quad h_i = 3051.6 \text{ kJ/kg} \qquad (\text{Table A–6})$$

which is equal to $u_2$. Since we now know two properties at the final state, it is fixed and the temperature at this state is determined from the same table to be

$$\left. \begin{array}{l} P_2 = 1 \text{ MPa} \\ u_2 = 3051.6 \text{ kJ/kg} \end{array} \right\} \quad T_2 = \mathbf{456.1°C}$$

(*a*) Flow of steam into an evacuated tank

(*b*) The closed-system equivalence

**FIGURE 6–48**

Schematic for Example 6–12.

Steam
$T_i = 300°C$

$T_2 = 456.1°C$

**FIGURE 6–49**

The temperature of steam rises from 300 to 456.1°C as it enters a tank as a result of flow energy being converted to internal energy.

*Discussion* Note that the temperature of the steam in the tank has increased by 156.1°C. This result may be surprising at first, and you may be wondering where the energy to raise the temperature of the steam came from. The answer lies in the enthalpy term $h = u + Pv$. Part of the energy represented by enthalpy is the flow energy $Pv$, and this flow energy is converted to sensible internal energy once the flow ceases to exist in the control volume, and it shows up as an increase in temperature (Fig. 6–49).

*Alternative solution* This problem can also be solved by considering the region within the tank and the mass that is destined to enter the tank as a closed system, as shown in Fig. 6–48b. Since no mass crosses the boundaries, viewing this as a closed system is appropriate.

During the process, the steam upstream (the imaginary piston) will push the enclosed steam in the supply line into the tank at a constant pressure of 1 MPa. Then the boundary work done during this process is

$$W_{b,in} = -\int_1^2 P_i \, dV = -P_i(V_2 - V_1) = -P_i[V_{tank} - (V_{tank} + V_i)] = P_i V_i$$

where $V_i$ is the volume occupied by the steam before it enters the tank and $P_i$ is the pressure at the moving boundary (the imaginary piston face). The energy balance for the closed system gives

$$\underbrace{E_{in} - E_{out}}_{\substack{\text{Net energy transfer} \\ \text{by heat, work, and mass}}} = \underbrace{\Delta E_{system}}_{\substack{\text{Change in internal, kinetic,} \\ \text{potential, etc., energies}}}$$

$$W_{b,in} = \Delta U$$
$$m_i P_i v_i = m_2 u_2 - m_i u_i$$
$$u_2 = u_i + P_i v_i = h_i$$

since the initial state of the system is simply the line conditions of the steam. This result is identical to the one obtained with the uniform-flow analysis. Once again, the temperature rise is caused by the so-called flow energy or flow work, which is the energy required to move the fluid during flow.

---

**EXAMPLE 6–13    Discharge of Heated Air at Constant Temperature**

An insulated 8-m³ rigid tank contains air at 600 kPa and 400 K. A valve connected to the tank is now opened, and air is allowed to escape until the pressure inside drops to 200 kPa. The air temperature during the process is maintained constant by an electric resistance heater placed in the tank. Determine the electrical energy supplied to air during this process.

**SOLUTION** Pressurized air in an insulated rigid tank equipped with an electric heater is allowed to escape at constant temperature until the pressure inside drops to a specified value. The amount of electrical energy supplied to air is to be determined.

*Assumptions* **1** This is an unsteady process since the conditions within the device are changing during the process, but it can be analyzed as a uniform-flow process since the exit conditions remain constant. **2** Kinetic and potential energies are negligible. **3** The tank is insulated and thus heat transfer is negligible. **4** Air is an ideal gas with variable specific heats.

**FIGURE 6–50**

Schematic for Example 6–13.

*Analysis* We take the contents of the tank as the system, which is a control volume since mass crosses the boundary (Fig. 6–50). Noting that the microscopic energies of flowing and nonflowing fluids are represented by enthalpy $h$ and internal energy $u$, respectively, the mass and energy balances for this uniform-flow system can be expressed as

*Mass balance:* $\qquad m_{in} - m_{out} = \Delta m_{system} \quad \rightarrow \quad m_e = m_1 - m_2$

*Energy balance:* $\qquad \underbrace{E_{in} - E_{out}}_{\substack{\text{Net energy transfer} \\ \text{by heat, work, and mass}}} = \underbrace{\Delta E_{system}}_{\substack{\text{Change in internal, kinetic,} \\ \text{potential, etc., energies}}}$

$$W_{e,in} - m_e h_e = m_2 u_2 - m_1 u_1 \quad (\text{since } Q \cong \text{ke} \cong \text{pe} \cong 0)$$

The gas constant of air is $R = 0.287 \text{ kPa·m}^3/\text{kg·K}$ (Table A–1). The initial and final masses of air in the tank and the discharged amount are determined from the ideal gas relation to be

$$m_1 = \frac{P_1 V_1}{R T_1} = \frac{(600 \text{ kPa})(8 \text{ m}^3)}{(0.287 \text{ kPa·m}^3/\text{kg·K})(400 \text{ K})} = 41.81 \text{ kg}$$

$$m_2 = \frac{P_2 V_2}{R T_2} = \frac{(200 \text{ kPa})(8 \text{ m}^3)}{(0.287 \text{ kPa·m}^3/\text{kg·K})(400 \text{ K})} = 13.94 \text{ kg}$$

$$m_e = m_1 - m_2 = 41.81 - 13.94 = 27.87 \text{ kg}$$

The enthalpy and internal energy of air at 400 K are $h_e = 400.98 \text{ kJ/kg}$ and $u_1 = u_2 = 286.16 \text{ kJ/kg}$ (Table A–21). The the electrical energy supplied to air is determined from the energy balance to be

$$W_{e,in} = m_e h_e + m_2 u_2 - m_1 u_1$$

$$= (27.87 \text{ kg})(400.98 \text{ kJ/kg}) + (13.94 \text{ kg})(286.16 \text{ kJ/kg})$$

$$-(41.81 \text{ kg})(286.16 \text{ kJ/kg})$$

$$= 3200 \text{ kJ} = \mathbf{0.889 \text{ kWh}}$$

since 1 kWh = 3600 kJ.

*Discussion* If the temperature of discharged air changes during the process, the problem can be solved with reasonable accuracy by evaluating $h_e$ at the average discharge temperature $T_e = (T_2 + T_1)/2$, and treating it as constant.

# SUMMARY

The *conservation of mass principle* states that the net mass transfer to or from a system during a process is equal to the net change (increase or decrease) in the total mass of the system during that process, and is expressed as

$$m_{in} - m_{out} = \Delta m_{system} \quad \text{and} \quad \dot{m}_{in} - \dot{m}_{out} = dm_{system}/dt$$

where $\Delta m_{system} = m_{final} - m_{initial}$ is the change in the mass of the system during the process, $\dot{m}_{in}$ and $\dot{m}_{out}$ are the total rates of mass flow into and out of the system, and $dm_{system}/dt$ is the rate of change of mass within the system boundaries. The relations above are also referred to as the mass balance and are applicable to any system undergoing any kind of process.

The amount of mass flowing through a cross section per unit time is called the *mass flow rate*, and is expressed as

$$\dot{m} = \rho V A$$

where $\rho$ = density of fluid, $V$ = average fluid velocity normal to $A$, and $A$ = cross-sectional area normal to flow direction. The volume of the fluid flowing through a cross section per unit time is called the *volume flow rate* and is expressed as

$$\dot{V} = VA = \dot{m}/\rho$$

The work required to push a unit mass of fluid into or out of a control volume is called *flow work* or *flow energy*, and is expressed as $w_{flow} = Pv$. In the analysis of control volumes, it is convenient to combine the flow energy and internal energy into *enthalpy*. Then the total energy of a flowing fluid is expressed as

$$\theta = h + \text{ke} + \text{pe} = h + \frac{V^2}{2} + gz$$

The total energy transported by a flowing fluid of mass $m$ with uniform properties is $m\theta$. The rate of energy transport by a fluid with a mass flow rate of $\dot{m}$ is $\dot{m}\theta$. When the kinetic and potential energies of a fluid stream are negligible, the amount and rate of energy transport become $E_{mass} = mh$ and $\dot{E}_{mass} = \dot{m}h$, respectively.

The *first law of thermodynamics* is essentially an expression of the conservation of energy principle, also called the *energy balance*. The general mass and energy balances for *any system* undergoing *any process* can be expressed as

$$\underbrace{E_{in} - E_{out}}_{\substack{\text{Net energy transfer} \\ \text{by heat, work, and mass}}} = \underbrace{\Delta E_{system}}_{\substack{\text{Changes in internal, kinetic,} \\ \text{potential, etc., energies}}}$$

It can also be expressed in the *rate form* as

$$\underbrace{\dot{E}_{in} - \dot{E}_{out}}_{\substack{\text{Rate of net energy transfer} \\ \text{by heat, work, and mass}}} = \underbrace{dE_{system}/dt}_{\substack{\text{Rate of change in internal, kinetic,} \\ \text{potential, etc., energies}}}$$

Thermodynamic processes involving control volumes can be considered in two groups: steady-flow processes and unsteady-flow processes. During a *steady-flow process*, the fluid flows through the control volume steadily, experiencing no change with time at a fixed position. The mass and energy content of the control volume remain constant during a steady-flow process. Taking heat transfer *to* the system and work done *by* the system to be positive quantities, the

conservation of mass and energy equations for steady-flow processes are expressed as

$$\sum_{in} \dot{m} = \sum_{out} \dot{m}$$

$$\dot{Q} - \dot{W} = \sum_{out} \dot{m}\underbrace{\left( h + \frac{V^2}{2} + gz \right)}_{\text{for each exit}} - \sum_{in} \dot{m}\underbrace{\left( h + \frac{V^2}{2} + gz \right)}_{\text{for each inlet}}$$

These are the most general forms of the equations for steady-flow processes. For single-stream (one-inlet–one-exit) systems such as nozzles, diffusers, turbines, compressors, and pumps, they simplify to

$$\dot{m}_1 = \dot{m}_2 \longrightarrow \frac{1}{v_1} V_1 A_1 = \frac{1}{v_2} V_2 A_2$$

$$\dot{Q} - \dot{W} = \dot{m}\left[ h_2 - h_1 + \frac{V_2^2 - V_1^2}{2} + g(z_2 - z_1) \right]$$

In these relations, subscripts 1 and 2 denote the inlet and exit states, respectively.

Most unsteady-flow processes can be modeled as a *uniform-flow process*, which requires that the fluid flow at any inlet or exit is uniform and steady, and thus the fluid properties do not change with time or position over the cross section of an inlet or exit. If they do, they are averaged and treated as constants for the entire process. When kinetic and potential energy changes associated with the control volume and the fluid streams are negligible, the mass and energy balance relations for a uniform-flow system are expressed as

$$m_{in} - m_{out} = \Delta m_{system}$$

$$Q - W = \sum_{out} mh - \sum_{in} mh + (m_2 u_2 - m_1 u_1)_{system}$$

where $Q = Q_{net,in} = Q_{in} - Q_{out}$ is the net heat input and $W = W_{net,out} = W_{out} - W_{in}$ is the net work output.

When solving thermodynamic problems, it is recommended that the general form of the energy balance $E_{in} - E_{out} = \Delta E_{system}$ be used for all problems, and simplify it for the particular problem instead of using the specific relations given here for different processes.

## REFERENCES AND SUGGESTED READINGS

**1.** ASHRAE Handbook of Fundamentals. SI version. Atlanta, GA: American Society of Heating, Refrigerating, and Air-Conditioning Engineers, Inc., 1993.

**2.** ASHRAE Handbook of Refrigeration. SI version. Atlanta, GA: American Society of Heating, Refrigerating, and Air-Conditioning Engineers, Inc., 1994.

**3.** Y. A. Çengel and J. M. Cimbala, *Fluid Mechanics: Fundamentals and Applications, Second Edition.* New York: McGraw-Hill, 2009.

# PROBLEMS*

## Conservation of Mass

**6–1C** Define mass and volume flow rates. How are they related to each other?

**6–2C** When is the flow through a control volume steady?

**6–3C** Does the amount of mass entering a control volume have to be equal to the amount of mass leaving during an unsteady-flow process?

**6–4C** Consider a device with one inlet and one outlet. If the volume flow rates at the inlet and at the outlet are the same, is the flow through this device necessarily steady? Why?

**6–5E** A pneumatic accumulator arranged to maintain a constant pressure as air enters or leaves it is set for 200 psia. Initially, the volume is 0.2 ft³ and the temperature is 80°F. Air is now added to the accumulator until its volume is 1 ft³ and its temperature is 80°F. How much air has been added to the accumulator?

**6–6E** A steady-flow compressor is used to compress helium from 15 psia and 70°F at the inlet to 200 psia and 600°F at the outlet. The outlet area and velocity are 0.01 ft² and 100 ft/s, respectively, and the inlet velocity is 50 ft/s. Determine the mass flow rate and the inlet area. *Answers: 0.0704 lbm/s, 0.133 ft²*

**6–7** Air enters a nozzle steadily at 2.21 kg/m³ and 40 m/s and leaves at 0.762 kg/m³ and 180 m/s. If the inlet area of the nozzle is 90 cm², determine (*a*) the mass flow rate through the nozzle, and (*b*) the exit area of the nozzle. *Answers: (a) 0.796 kg/s, (b) 58 cm²*

**6–8** Water enters the constant 130-mm inside-diameter tubes of a boiler at 7 MPa and 65°C, and leaves the tubes at 6 MPa and 450°C with a velocity of 80 m/s. Calculate the velocity of the water at the tube inlet and the inlet volume flow rate.

**6–9** A hair dryer is basically a duct of constant diameter in which a few layers of electric resistors are placed. A small fan pulls the air in and forces it through the resistors where it is heated. If the density of air is 1.20 kg/m³ at the inlet and 0.95 kg/m³ at the exit, determine the percent increase in the velocity of air as it flows through the dryer.

**FIGURE P6–9**

**6–10** A 1-m³ rigid tank initially contains air whose density is 1.18 kg/m³. The tank is connected to a high-pressure supply line through a valve. The valve is opened, and air is allowed to enter the tank until the density in the tank rises to 7.20 kg/m³. Determine the mass of air that has entered the tank. *Answer: 6.02 kg*

**6–11** Air enters the 1-m² inlet of an aircraft engine at 100 kPa and 20°C with a velocity of 180 m/s. Determine the volume flow rate, in m³/s, at the engine's inlet and the mass flow rate, in kg/s, at the engine's exit.

**6–12** A pump increases the water pressure from 70 kPa at the inlet to 700 kPa at the outlet. Water enters this pump at 15°C through a 1-cm-diameter opening and exits through a 1.5-cm-diameter opening. Determine the velocity of the water at the inlet and outlet when the mass flow rate through the pump is 0.5 kg/s. Will these velocities change significantly if the inlet temperature is raised to 40°C?

**FIGURE P6–12**

**6–13** Refrigerant-134a enters a 28-cm diameter pipe steadily at 200 kPa and 20°C with a velocity of 5 m/s. The refrigerant gains heat as it flows and leaves the pipe at 180 kPa and 40°C. Determine (*a*) the volume flow rate of the refrigerant at the inlet, (*b*) the mass flow rate of the refrigerant, and (*c*) the velocity and volume flow rate at the exit.

**6–14** A smoking lounge is to accommodate 15 heavy smokers. The minimum fresh air requirement for smoking lounges is specified to be 30 L/s per person (ASHRAE, Standard 62, 1989). Determine the minimum required flow rate of fresh air that needs to be supplied to the lounge, and the diameter of the duct if the air velocity is not to exceed 8 m/s.

**FIGURE P6–14**

**6–15** The minimum fresh air requirement of a residential building is specified to be 0.35 air change per hour (ASHRAE, Standard 62, 1989). That is, 35 percent of the entire air contained in a residence should be replaced by fresh outdoor air every hour. If the ventilation requirement of a 3.0-m-high, 200-m² residence is to be met entirely by a fan, determine the flow capacity in L/min of the fan that needs to be installed. Also determine the diameter of the duct if the air velocity is not to exceed 4 m/s.

### Flow Work and Energy Transfer by Mass

**6–16C** What are the different mechanisms for transferring energy to or from a control volume?

**6–17C** What is flow energy? Do fluids at rest possess any flow energy?

**6–18C** How do the energies of a flowing fluid and a fluid at rest compare? Name the specific forms of energy associated with each case.

**6–19E** A water pump increases the water pressure from 10 psia to 50 psia. Determine the flow work, in Btu/lbm, required by the pump.

**6–20** An air compressor compresses 6 L of air at 120 kPa and 20°C to 1000 kPa and 400°C. Determine the flow work, in kJ/kg, required by the compressor. *Answer:* 109 kJ/kg

**6–21E** Steam is leaving a pressure cooker whose operating pressure is 20 psia. It is observed that the amount of liquid in the cooker has decreased by 0.6 gal in 45 minutes after the steady operating conditions are established, and the cross-sectional area of the exit opening is 0.15 in². Determine (a) the mass flow rate of the steam and the exit velocity, (b) the total and flow energies of the steam per unit mass, and (c) the rate at which energy is leaving the cooker by steam.

**6–22** Air flows steadily in a pipe at 300 kPa, 77°C, and 25 m/s at a rate of 18 kg/min. Determine (a) the diameter of the pipe, (b) the rate of flow energy, (c) the rate of energy

transport by mass, and (d) the error involved in part (c) if the kinetic energy is neglected.

### Steady-Flow Energy Balance: Nozzles and Diffusers

**6–23C** Can a steady-flow system involve boundary work?

**6–24C** A diffuser is an adiabatic device that decreases the kinetic energy of the fluid by slowing it down. What happens to this *lost* kinetic energy?

**6–25C** The kinetic energy of a fluid increases as it is accelerated in an adiabatic nozzle. Where does this energy come from?

**6–26C** Is heat transfer to or from the fluid desirable as it flows through a nozzle? How will heat transfer affect the fluid velocity at the nozzle exit?

**6–27** Air at 80 kPa and 127°C enters an adiabatic diffuser steadily at a rate of 6000 kg/h and leaves at 100 kPa. The velocity of the air stream is decreased from 230 to 30 m/s as it passes through the diffuser. Find (a) the exit temperature of the air and (b) the exit area of the diffuser.

**6–28** Air enters an adiabatic nozzle steadily at 300 kPa, 200°C, and 45 m/s and leaves at 100 kPa and 180 m/s. The inlet area of the nozzle is 110 cm². Determine (a) the mass flow rate through the nozzle, (b) the exit temperature of the air, and (c) the exit area of the nozzle. *Answers:* (a) 1.09 kg/s, (b) 185°C, (c) 79.9 cm²

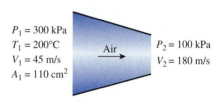

$P_1 = 300$ kPa
$T_1 = 200$°C
$V_1 = 45$ m/s
$A_1 = 110$ cm²

Air

$P_2 = 100$ kPa
$V_2 = 180$ m/s

**FIGURE P6–28**

**6–29** Reconsider Prob. 6–28. Using EES (or other) software, investigate the effect of the inlet area on the mass flow rate, exit temperature, and the exit area. Let the inlet area vary from 50 cm² to 150 cm². Plot the final results against the inlet area, and discuss the results.

**6–30E** The stators in a gas turbine are designed to increase the kinetic energy of the gas passing through them adiabatically. Air enters a set of these nozzles at 300 psia and 700°F with a velocity of 80 ft/s and exits at 250 psia and 645°F. Calculate the velocity at the exit of the nozzles.

**6–31** The diffuser in a jet engine is designed to decrease the kinetic energy of the air entering the engine compressor without any work or heat interactions. Calculate the velocity at the exit of a diffuser when air at 100 kPa and 20°C enters it with a velocity of 500 m/s and the exit state is 200 kPa and 90°C.

**FIGURE P6–31**

© *Stockbyte/Punchstock.*

**6–32** Steam enters a nozzle at 400°C and 800 kPa with a velocity of 10 m/s, and leaves at 300°C and 200 kPa while losing heat at a rate of 25 kW. For an inlet area of 800 cm$^2$, determine the velocity and the volume flow rate of the steam at the nozzle exit. *Answers: 606 m/s, 2.74 m$^3$/s*

400°C
800 kPa   Steam →   300°C
10 m/s                      200 kPa

$\dot{Q}$

**FIGURE P6–32**

**6–33** Steam at 3 MPa and 400°C enters an adiabatic nozzle steadily with a velocity of 40 m/s and leaves at 2.5 MPa and 300 m/s. Determine (*a*) the exit temperature and (*b*) the ratio of the inlet to exit area $A_1/A_2$.

**6–34E** Air at 13 psia and 50°F enters an adiabatic diffuser steadily with a velocity of 600 ft/s and leaves with a low velocity at a pressure of 14.5 psia. The exit area of the diffuser is 4 times the inlet area. Determine (*a*) the exit temperature and (*b*) the exit velocity of the air.

$P_1$ = 13 psia
$T_1$ = 50°F     Air →     $P_2$ = 14.5 psia
$V_1$ = 600 ft/s           $V_2 \ll V_1$
                           $A_2 = 4A_1$

**FIGURE P6–34E**

**6–35** Carbon dioxide enters an adiabatic nozzle steadily at 1 MPa and 500°C with a mass flow rate of 6000 kg/h and leaves at 100 kPa and 450 m/s. The inlet area of the nozzle is 40 cm$^2$. Determine (*a*) the inlet velocity and (*b*) the exit temperature. *Answers: (a) 60.8 m/s, (b) 685.8 K*

**6–36** Refrigerant-134a at 700 kPa and 120°C enters an adiabatic nozzle steadily with a velocity of 20 m/s and leaves at 400 kPa and 30°C. Determine (*a*) the exit velocity and (*b*) the ratio of the inlet to exit area $A_1/A_2$.

**6–37** Refrigerant-134a enters a diffuser steadily as saturated vapor at 800 kPa with a velocity of 120 m/s, and it leaves at 900 kPa and 40°C. The refrigerant is gaining heat at a rate of 2 kJ/s as it passes through the diffuser. If the exit area is 80 percent greater than the inlet area, determine (*a*) the exit velocity and (*b*) the mass flow rate of the refrigerant. *Answers: (a) 60.8 m/s, (b) 1.308 kg/s*

**6–38** Steam at 4 MPa and 400°C enters a nozzle steadily with a velocity of 60 m/s, and it leaves at 2 MPa and 300°C. The inlet area of the nozzle is 50 cm$^2$, and heat is being lost at a rate of 75 kJ/s. Determine (*a*) the mass flow rate of the steam, (*b*) the exit velocity of the steam, and (*c*) the exit area of the nozzle.

## Turbines and Compressors

**6–39C** Consider an adiabatic turbine operating steadily. Does the work output of the turbine have to be equal to the decrease in the energy of the steam flowing through it?

**6–40C** Consider an air compressor operating steadily. How would you compare the volume flow rates of the air at the compressor inlet and exit?

**6–41C** Will the temperature of air rise as it is compressed by an adiabatic compressor? Why?

**6–42C** Somebody proposes the following system to cool a house in the summer: Compress the regular outdoor air, let it cool back to the outdoor temperature, pass it through a turbine, and discharge the cold air leaving the turbine into the house. From a thermodynamic point of view, is the proposed system sound?

**6–43** Refrigerant-134a enters a compressor at 100 kPa and –24°C with a flow rate of 1.35 m$^3$/min and leaves at 800 kPa and 60°C. Determine the mass flow rate of R-134a and the power input to the compressor.

**6–44** Refrigerant-134a enters a compressor at 180 kPa as a saturated vapor with a flow rate of 0.35 m$^3$/min and leaves at 700 kPa. The power supplied to the refrigerant during compression process is 2.35 kW. What is the temperature of R-134a at the exit of the compressor? *Answer: 48.8°C*

**6–45** Steam enters an adiabatic turbine at 10 MPa and 500°C and leaves at 10 kPa with a quality of 90 percent. Neglecting the changes in kinetic and potential energies, determine the mass flow rate required for a power output of 5 MW. *Answer: 4.852 kg/s*

**6–46E** Steam flows steadily through a turbine at a rate of 45,000 lbm/h, entering at 1000 psia and 900°F and leaving at 5 psia as saturated vapor. If the power generated by the turbine is 4 MW, determine the rate of heat loss from the steam.

**6–47** An adiabatic air compressor compresses 10 L/s of air at 120 kPa and 20°C to 1000 kPa and 300°C. Determine (a) the work required by the compressor, in kJ/kg, and (b) the power required to drive the air compressor, in kW.

1 MPa
300°C

Compressor

120 kPa
20°C
10 L/s

**FIGURE P6–47**

**6–48** Argon gas enters an adiabatic turbine steadily at 1600 kPa and 450°C with a velocity of 55 m/s and leaves at 150 kPa with a velocity of 150 m/s. The inlet area of the turbine is 60 cm². If the power output of the turbine is 190 kW, determine the exit temperature of the argon.

**6–49** Helium is to be compressed from 120 kPa and 310 K to 700 kPa and 430 K. A heat loss of 20 kJ/kg occurs during the compression process. Neglecting kinetic energy changes, determine the power input required for a mass flow rate of 90 kg/min.

$P_2 = 700$ kPa
$T_2 = 430$ K

20 kJ/kg

He
$\dot{m} = 90$ kg/min

$\dot{W}_{in}$

$P_1 = 120$ kPa
$T_1 = 310$ K

**FIGURE P6–49**

**6–50** Carbon dioxide enters an adiabatic compressor at 100 kPa and 300 K at a rate of 0.5 kg/s and leaves at 600 kPa and 450 K. Neglecting kinetic energy changes, determine (a) the volume flow rate of the carbon dioxide at the compressor

inlet and (b) the power input to the compressor. *Answers: (a)* 0.28 m³/s, (b) 68.8 kW

**6–51** An adiabatic gas turbine expands air at 1300 kPa and 500°C to 100 kPa and 127°C. Air enters the turbine through a 0.2-m² opening with an average velocity of 40 m/s, and exhausts through a 1-m² opening. Determine (a) the mass flow rate of air through the turbine and (b) the power produced by the turbine. *Answers: (a)* 46.9 kg/s, (b) 18.3 MW

**6–52E** Air is expanded by an adiabatic gas turbine from 500 psia and 800°F to 60 psia and 250°F. If the volume flow rate at the exit is 50 ft³/s; the inlet area is 0.6 ft²; and the outlet area is 1.2 ft²; determine the power produced by this turbine.

**6–53** Steam flows steadily into a turbine with a mass flow rate of 26 kg/s and a negligible velocity at 6 MPa and 600°C. The steam leaves the turbine at 0.5 MPa and 200°C with a velocity of 180 m/s. The rate of work done by the steam in the turbine is measured to be 20 MW. If the elevation change between the turbine inlet and exit is negligible, determine the rate of heat transfer associated with this process. *Answer:* 455 kW

5 MPa
600°C
20 kg/s

Steam
20 kg/s

10 kPa
x = 0.85

0.1 MPa
2 kg/s

**FIGURE P6–53**

**6–54** Determine the power input for a compressor that compresses helium from 150 kPa and 20°C to 400 kPa and 200°C. Helium enters this compressor through a 0.1-m² pipe at a velocity of 15 m/s. What is the increase in the flow power during this process?

## Throttling Valves

**6–55C** Someone claims, based on temperature measurements, that the temperature of a fluid rises during a throttling process in a well-insulated valve with negligible friction. How do you evaluate this claim? Does this process violate any thermodynamic laws?

**6–56C** Would you expect the temperature of air to drop as it undergoes a steady-flow throttling process? Explain.

**6–57C** Would you expect the temperature of a liquid to change as it is throttled? Explain.

**6–58C** During a throttling process, the temperature of a fluid drops from 30 to −20°C. Can this process occur adiabatically?

**6–59** An adiabatic capillary tube is used in some refrigeration systems to drop the pressure of the refrigerant from the condenser level to the evaporator level. The R-134a enters the capillary tube as a saturated liquid at 50°C, and leaves at −20°C. Determine the quality of the refrigerant at the inlet of the evaporator.

**6–60** Saturated liquid-vapor mixture of water, called wet steam, in a steam line at 2000 kPa is throttled to 100 kPa and 120°C. What is the quality in the steam line? *Answer: 0.957*

Throttling valve

Steam
2 MPa

100 kPa
120°C

**FIGURE P6–60**

**6–61** Refrigerant-134a at 800 kPa and 25°C is throttled to a temperature of −20°C. Determine the pressure and the internal energy of the refrigerant at the final state. *Answers: 133 kPa, 80.7 kJ/kg*

**6–62** A well-insulated valve is used to throttle steam from 8 MPa and 350°C to 2 MPa. Determine the final temperature of the steam. *Answer: 285°C*

**6–63** Reconsider Prob. 6–62. Using EES (or other) software, investigate the effect of the exit pressure of steam on the exit temperature after throttling. Let the exit pressure vary from 6 to 1 MPa. Plot the exit temperature of steam against the exit pressure, and discuss the results.

**6–64E** Refrigerant-134a enters the expansion valve of a refrigeration system at 160 psia as a saturated liquid and leaves at 30 psia. Determine the temperature and internal energy changes across the valve.

### Mixing Chambers and Heat Exchangers

**6–65C** Consider a steady-flow mixing process. Under what conditions will the energy transported into the control volume by the incoming streams be equal to the energy transported out of it by the outgoing stream?

**6–66C** Consider a steady-flow heat exchanger involving two different fluid streams. Under what conditions will the amount of heat lost by one fluid be equal to the amount of heat gained by the other?

**6–67C** When two fluid streams are mixed in a mixing chamber, can the mixture temperature be lower than the temperature of both streams? Explain.

**6–68** Hot and cold streams of a fluid are mixed in a rigid mixing chamber. The hot fluid flows into the chamber at a mass flow rate of 5 kg/s with an energy in the amount of 150 kJ/kg. The cold fluid flows into the chamber with a mass flow rate of 15 kg/s and carries energy in the amount of 50 kJ/kg. There is heat transfer to the surroundings from the mixing chamber in the amount of 5.5 kW. The mixing chamber operates in a steady-flow manner and does not gain or loose energy or mass with time. Determine the energy carried from the mixing chamber by the fluid mixture per unit mass of fluid, in kJ/kg.

**6–69** A hot-water stream at 80°C enters a mixing chamber with a mass flow rate of 0.5 kg/s where it is mixed with a stream of cold water at 20°C. If it is desired that the mixture leave the chamber at 42°C, determine the mass flow rate of the cold-water stream. Assume all the streams are at a pressure of 250 kPa. *Answer: 0.865 kg/s*

$T_1 = 80°C$
$\dot{m}_1 = 0.5$ kg/s

$H_2O$
$(P = 250$ kPa$)$   $T_3 = 42°C$

$T_2 = 20°C$
$\dot{m}_2$

**FIGURE P6–69**

**6–70E** Water at 65°F and 20 psia is heated in a chamber by mixing it with saturated water vapor at 20 psia. If both streams enter the mixing chamber at the same mass flow rate, determine the temperature and the quality of the exiting stream. *Answers: 228°F, 0.415*

**6–71** A stream of refrigerant-134a at 1 MPa and 20°C is mixed with another stream at 1 MPa and 80°C. If the mass flow rate of the cold stream is twice that of the hot one, determine the temperature and the quality of the exit stream.

**6–72** Reconsider Prob. 6–71. Using EES (or other) software, investigate the effect of the mass flow rate of the cold stream of R-134a on the temperature and the quality of the exit stream. Let the ratio of the mass flow rate of the cold stream to that of the hot stream vary from 1 to 4. Plot the mixture temperature and quality against the cold-to-hot mass flow rate ratio, and discuss the results.

**6–73E** Steam is to be condensed on the shell side of a heat exchanger at 85°F. Cooling water enters the tubes at 60°F at a rate of 138 lbm/s and leaves at 73°F. Assuming the heat exchanger to be well-insulated, determine the rate of heat transfer in the heat exchanger and the rate of condensation of the steam.

**6–74** Steam enters the condenser of a steam power plant at 20 kPa and a quality of 95 percent with a mass flow rate of 20,000 kg/h. It is to be cooled by water from a nearby river by circulating the water through the tubes within the condenser. To prevent thermal pollution, the river water is not allowed to experience a temperature rise above 10°C. If the steam is to leave the condenser as saturated liquid at 20 kPa, determine the mass flow rate of the cooling water required. *Answer: 297.7 kg/s*

$\dot{m}_3 = 20{,}000 \text{ kg/h}$
$P_3 = 20 \text{ kPa}$
$x_3 = 0.95$

Steam

Water
$T_1$

$T_1 + 10°C$

$P_4 = 20 \text{ kPa}$
Sat. liquid

**FIGURE P6–74**

**6–75** A heat exchanger is to cool ethylene glycol ($c_p = 2.56$ kJ/kg·°C) flowing at a rate of 3.2 kg/s from 80°C to 40°C by water ($c_p = 4.18$ kJ/kg·°C) that enters at 20°C and leaves at 70°C. Determine (*a*) the rate of heat transfer and (*b*) the mass flow rate of water.

**6–76** Reconsider Prob. 6–75. Using EES (or other) software, investigate the effect of the inlet temperature of cooling water on the mass flow rate of water. Let the inlet temperature vary from 10 to 40°C, and assume the exit temperature to remain constant. Plot the mass flow rate of water against the inlet temperature, and discuss the results.

**6–77** Cold water ($c_p = 4.18$ kJ/kg·°C) leading to a shower enters a thin-walled double-pipe counter-flow heat exchanger at 15°C at a rate of 0.60 kg/s and is heated to 45°C by hot water ($c_p = 4.19$ kJ/kg·°C) that enters at 100°C at a rate of 3 kg/s. Determine the rate of heat transfer in the heat exchanger and the exit temperature of the hot water.

**6–78** Air ($c_p = 1.005$ kJ/kg·°C) is to be preheated by hot exhaust gases in a cross-flow heat exchanger before it enters the furnace. Air enters the heat exchanger at 95 kPa and 20°C at a rate of 0.6 m³/s. The combustion gases ($c_p = 1.10$ kJ/kg·°C) enter at 160°C at a rate of 0.95 kg/s and leave at 95°C. Determine the rate of heat transfer to the air and its outlet temperature.

Air
95 kPa
20°C
0.6 m³/s

Exhaust gases
0.95 kg/s
95°C

**FIGURE P6–78**

**6–79E** An open feedwater heater heats the feedwater by mixing it with hot steam. Consider an electrical power plant with an open feedwater heater that mixes 0.1 lbm/s of steam at 10 psia and 200°F with 2.0 lbm/s of feedwater at 10 psia and 100°F to produce 10 psia and 120°F feedwater at the outlet. The diameter of the outlet pipe is 0.5 ft. Determine the mass flow rate and feedwater velocity at the outlet. Would the outlet flow rate and velocity be significantly different if the temperature at the outlet were 180°F?

**6–80E** In a steam heating system, air is heated by being passed over some tubes through which steam flows steadily. Steam enters the heat exchanger at 30 psia and 400°F at a rate of 15 lbm/min and leaves at 25 psia and 212°F. Air enters at 14.7 psia and 80°F and leaves at 130°F. Determine the volume flow rate of air at the inlet.

**6–81** Hot exhaust gases of an internal combustion engine are to be used to produce saturated water vapor at 2 MPa pressure. The exhaust gases enter the heat exchanger at 400°C at a rate of 32 kg/min while water enters at 15°C. The heat exchanger is not well insulated, and it is estimated that 10 percent of heat given up by the exhaust gases is lost to the surroundings. If the mass flow rate of the exhaust gases is 15 times that of the water, determine (*a*) the temperature of the exhaust gases at the heat exchanger exit and (*b*) the rate of heat transfer to the water. Use the constant specific heat properties of air for the exhaust gases.

**FIGURE P6–81**

**6–82** A chilled-water heat-exchange unit is designed to cool 5 m³/s of air at 100 kPa and 30°C to 100 kPa and 18°C by using water at 8°C. Determine the maximum water outlet temperature when the mass flow rate of the water is 2 kg/s. *Answer:* 16.3°C

**FIGURE P6–82**

**6–83** The condenser of a refrigeration cycle is basically a heat exchanger in which a refrigerant is condensed by rejecting heat to a cooling fluid. Refrigerant-134a enters a condenser at 1200 kPa and 85°C with a flow rate of 0.042 kg/s and leaves at the same pressure subcooled by 6.3°C. The condensation is done by cooling water which experiences a temperature rise of 12°C in the condenser. Determine (*a*) the rate of heat transferred to the water in the condenser, in kJ/min and (*b*) the mass flow rate of water, in kg/min. *Answers:* (*a*) 525 kJ/min, (*b*) 10.5 kg/min

**6–84** The evaporator of a refrigeration cycle is basically a heat exchanger in which a refrigerant is evaporated by absorbing heat from a fluid. Refrigerant-22 enters an evaporator at 200 kPa with a quality of 22 percent and a flow rate of 2.25 L/h. R-22 leaves the evaporator at the same pressure superheated by 5°C. The refrigerant is evaporated by absorbing heat from air whose flow rate is 0.5 kg/s. Determine (*a*) the rate of heat absorbed from the air and (*b*) the temperature change of air. The properties of R-22 at the inlet and exit of the condenser are $h_1$ = 220.2 kJ/kg, $v_1$ = 0.0253 m³/kg, and $h_2$ = 398.0 kJ/kg.

### Pipe and Duct Flow

**6–85** Argon steadily flows into a constant pressure heater at 300 K and 100 kPa with a mass flow rate of 6.24 kg/s. Heat transfer in the rate of 150 kW is supplied to the argon as it

flows through the heater. (*a*) Determine the argon temperature at the heater exit, in °C. (*b*) Determine the argon volume flow rate at the heater exit, in m³/s.

**6–86** Saturated liquid water is heated in a steady-flow steam boiler at a constant pressure of 2 MPa at a rate of 4 kg/s to an outlet temperature of 250°C. Determine the rate of heat transfer in the boiler.

**6–87E** Water enters a boiler at 500 psia as a saturated liquid and leaves at 600°F at the same pressure. Calculate the heat transfer per unit mass of water.

**6–88** A 110-volt electrical heater is used to warm 0.3 m³/s of air at 100 kPa and 15°C to 100 kPa and 30°C. How much current in amperes must be supplied to this heater?

**6–89E** The fan on a personal computer draws 0.5 ft³/s of air at 14.7 psia and 70°F through the box containing the CPU and other components. Air leaves at 14.7 psia and 80°F. Calculate the electrical power, in kW, dissipated by the PC components. *Answer:* 0.0948 kW

**FIGURE P6–89E**
© *Vol. 55/PhotoDisc/Getty Images.*

**6–90** A desktop computer is to be cooled by a fan. The electronic components of the computer consume 60 W of power under full-load conditions. The computer is to operate in environments at temperatures up to 45°C and at elevations up to 3400 m where the average atmospheric pressure is 66.63 kPa. The exit temperature of air is not to exceed 60°C to meet the reliability requirements. Also, the average velocity of air is not to exceed 110 m/min at the exit of the computer case where the fan is installed to keep the noise level down. Determine the flow rate of the fan that needs to be installed and the diameter of the casing of the fan.

**6–91** Repeat Prob. 6–90 for a computer that consumes 100 W of power.

**6–92E** Water enters the tubes of a cold plate at 70°F with an average velocity of 40 ft/min and leaves at 105°F. The diameter of the tubes is 0.25 in. Assuming 15 percent of the heat generated is dissipated from the components to the surroundings by convection and radiation, and the remaining 85 percent is removed by the cooling water, determine the amount of heat generated by the electronic devices mounted on the cold plate. *Answer:* 614 W

**6–93**  The components of an electronic system dissipating 180 W are located in a 1.4-m-long horizontal duct whose cross section is 20 cm × 20 cm. The components in the duct are cooled by forced air that enters the duct at 30°C and 1 atm at a rate of 0.6 m³/min and leaves at 40°C. Determine the rate of heat transfer from the outer surfaces of the duct to the ambient. *Answer: 63 W*

Natural
convection

40°C

180 W

1.4 m

30°C
0.6 m³/min

**FIGURE P6–93**

**6–94** Repeat Prob. 6–93 for a circular horizontal duct of diameter 10 cm.

**6–95** Consider a hollow-core printed circuit board 9 cm high and 18 cm long, dissipating a total of 15 W. The width of the air gap in the middle of the PCB is 0.25 cm. If the cooling air enters the 12-cm-wide core at 25°C and 1 atm at a rate of 0.8 L/s, determine the average temperature at which the air leaves the hollow core. *Answer: 46.0°C*

**6–96** A computer cooled by a fan contains eight PCBs, each dissipating 10 W power. The height of the PCBs is 12 cm and the length is 18 cm. The cooling air is supplied by a 25-W fan mounted at the inlet. If the temperature rise of air as it flows through the case of the computer is not to exceed 10°C, determine (a) the flow rate of the air that the fan needs to deliver and (b) the fraction of the temperature rise of air that is due to the heat generated by the fan and its motor. *Answers: (a) 0.0104 kg/s, (b) 24 percent*

Air
outlet

PCB, 10 W

Air
inlet

**FIGURE P6–96**

**6–97** Hot water at 90°C enters a 22-m section of a cast iron pipe whose inner diameter is 2.5 cm at an average velocity of 0.6 m/s. The outer surface of the pipe is exposed to the cold air at 10°C in a basement. If water leaves the basement at 88°C, determine the rate of heat loss from the water.

**6–98** Reconsider Prob. 6–97. Using EES (or other) software, investigate the effect of the inner pipe diameter on the rate of heat loss. Let the pipe diameter vary from 1.5 to 7.5 cm. Plot the rate of heat loss against the diameter, and discuss the results.

**6–99** A 5-m × 6-m × 8-m room is to be heated by an electric resistance heater placed in a short duct in the room. Initially, the room is at 15°C, and the local atmospheric pressure is 98 kPa. The room is losing heat steadily to the outside at a rate of 200 kJ/min. A 200-W fan circulates the air steadily through the duct and the electric heater at an average mass flow rate of 50 kg/min. The duct can be assumed to be adiabatic, and there is no air leaking in or out of the room. If it takes 15 min for the room air to reach an average temperature of 25°C, find (a) the power rating of the electric heater and (b) the temperature rise that the air experiences each time it passes through the heater.

**6–100E** The hot-water needs of a household are to be met by heating water at 55°F to 180°F by a parabolic solar collector at a rate of 4 lbm/s. Water flows through a 1.25-in-diameter thin aluminum tube whose outer surface is black-anodized in order to maximize its solar absorption ability. The centerline of the tube coincides with the focal line of the collector, and a glass sleeve is placed outside the tube to minimize the heat losses. If solar energy is transferred to water at a net rate of 400 Btu/h per ft length of the tube, determine the required length of the parabolic collector to meet the hot-water requirements of this house.

**6–101** A house has an electric heating system that consists of a 300-W fan and an electric resistance heating element placed in a duct. Air flows steadily through the duct at a rate of 0.6 kg/s and experiences a temperature rise of 7°C. The rate of heat loss from the air in the duct is estimated to be 300 W. Determine the power rating of the electric resistance heating element. *Answer: 4.22 kW*

**6–102** Steam enters a long, horizontal pipe with an inlet diameter of $D_1 = 12$ cm at 2 MPa and 300°C with a velocity of 3 m/s. Farther downstream, the conditions are 800 kPa and 250°C, and the diameter is $D_2 = 10$ cm. Determine (a) the mass flow rate of the steam and (b) the rate of heat transfer. *Answers: (a) 0.270 kg/s, (b) 19.9 kJ/s*

**6–103** Refrigerant-134a enters the condenser of a refrigerator at 900 kPa and 60°C, and leaves as a saturated liquid at the same pressure. Determine the heat transfer from the refrigerant per unit mass.

**FIGURE P6–103**

**6–104** Saturated liquid water is heated at constant pressure in a steady-flow device until it is a saturated vapor. Calculate the heat transfer, in kJ/kg, when the vaporization is done at a pressure of 800 kPa.

**6–105** Air at 300 K and 100 kPa steadily flows into a hair dryer having electrical work input of 1500 W. Because of the size of the air intake, the inlet velocity of the air is negligible. The air temperature and velocity at the hair dryer exit are 80°C and 21 m/s, respectively. The flow process is both constant pressure and adiabatic. Assume air has constant specific heats evaluated at 300 K. (*a*) Determine the air mass flow rate into the hair dryer, in kg/s. (*b*) Determine the air volume flow rate at the hair dryer exit, in m³/s. *Answers:* (*a*) 0.0280 kg/s, (*b*) 0.0284 m³/s

**FIGURE P6–105**

**6–106** ![EES] Reconsider Prob. 6–105. Using EES (or other) software, investigate the effect of the exit velocity on the mass flow rate and the exit volume flow rate. Let the exit velocity vary from 5 to 25 m/s. Plot the mass flow rate and exit volume flow rate against the exit velocity, and discuss the results.

## Charging and Discharging Processes

**6–107** An insulated rigid tank is initially evacuated. A valve is opened, and atmospheric air at 95 kPa and 17°C enters the tank until the pressure in the tank reaches 95 kPa, at which point the valve is closed. Determine the final temperature of the air in the tank. Assume constant specific heats. *Answer:* 406 K

**6–108** A rigid, insulated tank that is initially evacuated is connected through a valve to a supply line that carries helium at 200 kPa and 120°C. Now the valve is opened, and helium is allowed to flow into the tank until the pressure reaches 200 kPa, at which point the valve is closed. Determine the flow work of the helium in the supply line and the final temperature of the helium in the tank. *Answers:* 816 kJ/kg, 655 K

**FIGURE P6–108**

**6–109** A 2-m³ rigid tank initially contains air at 100 kPa and 22°C. The tank is connected to a supply line through a valve. Air is flowing in the supply line at 600 kPa and 22°C. The valve is opened, and air is allowed to enter the tank until the pressure in the tank reaches the line pressure, at which point the valve is closed. A thermometer placed in the tank indicates that the air temperature at the final state is 77°C. Determine (*a*) the mass of air that has entered the tank and (*b*) the amount of heat transfer. *Answers:* (*a*) 9.58 kg, (*b*) 339 kJ

**FIGURE P6–109**

**6–110** A 0.2-m³ rigid tank equipped with a pressure regulator contains steam at 2 MPa and 300°C. The steam in the tank is now heated. The regulator keeps the steam pressure constant by letting out some steam, but the temperature inside rises. Determine the amount of heat transferred when the steam temperature reaches 500°C.

**6–111** An insulated, vertical piston–cylinder device initially contains 10 kg of water, 6 kg of which is in the vapor phase. The mass of the piston is such that it maintains a constant pressure of 200 kPa inside the cylinder. Now steam at 0.5 MPa and 350°C is allowed to enter the cylinder from a supply line until all the liquid in the cylinder has vaporized. Determine (*a*) the final temperature in the cylinder and (*b*) the mass of the steam that has entered. *Answers:* (*a*) 120.2°C, (*b*) 19.07 kg

**FIGURE P6–111**

**6–112E** A scuba diver's 2-ft³ air tank is to be filled with air from a compressed air line at 120 psia and 100°F. Initially, the air in this tank is at 20 psia and 70°F. Presuming that the tank is well insulated, determine the temperature and mass in the tank when it is filled to 120 psia.

**6–113** An air-conditioning system is to be filled from a rigid container that initially contains 5 kg of liquid R-134a at 24°C. The valve connecting this container to the air-conditioning system is now opened until the mass in the container is 0.25 kg, at which time the valve is closed. During this time, only liquid R-134a flows from the container. Presuming that the process is isothermal while the valve is open, determine the final quality of the R-134a in the container and the total heat transfer. *Answers:* 0.506, 22.6 kJ

**FIGURE P6–113**

**6–114E** Oxygen is supplied to a medical facility from ten 1.5-ft³ compressed oxygen tanks. Initially, these tanks are at 1500 psia and 80°F. The oxygen is removed from these tanks slowly enough that the temperature in the tanks remains at 80°F. After two weeks, the pressure in the tanks is 300 psia. Determine the mass of oxygen used and the total heat transfer to the tanks.

**6–115** A 0.03-m³ rigid tank initially contains refrigerant-134a at 1.4 MPa and 100 percent quality. The tank is connected by a valve to a supply line that carries refrigerant-134a at 1.6 MPa and 36°C. Now the valve is opened, and the refrigerant is allowed to enter the tank. The valve is closed when it is observed that the tank contains saturated liquid at 1.6 MPa. Determine (*a*) the mass of the refrigerant that has entered the tank and (*b*) the amount of heat transfer. *Answers:* (*a*) 29.8 kg, (*b*) 697 kJ

**6–116** A 0.3-m³ rigid tank is filled with saturated liquid water at 200°C. A valve at the bottom of the tank is opened, and liquid is withdrawn from the tank. Heat is transferred to the water such that the temperature in the tank remains constant. Determine the amount of heat that must be transferred by the time one-half of the total mass has been withdrawn.

**FIGURE P6–116**

**6–117** A 0.12-m³ rigid tank contains saturated refrigerant-134a at 800 kPa. Initially, 25 percent of the volume is occupied by liquid and the rest by vapor. A valve at the bottom of the tank is now opened, and liquid is withdrawn from the tank. Heat is transferred to the refrigerant such that the pressure inside the tank remains constant. The valve is closed when no liquid is left in the tank and vapor starts to come out. Determine the total heat transfer for this process. *Answer:* 201.2 kJ

**6–118E** A 2-ft³ rigid tank contains saturated refrigerant-134a at 160 psia. Initially, 5 percent of the volume is occupied by liquid and the rest by vapor. A valve at the top of the tank is now opened, and vapor is allowed to escape slowly from the tank. Heat is transferred to the refrigerant such that the pressure inside the tank remains constant. The valve is closed when the last drop of liquid in the tank is vaporized. Determine the total heat transfer for this process.

**FIGURE P6–118E**

**6–119** A 0.4-m³ rigid tank initially contains refrigerant-134a at 14°C. At this state, 70 percent of the mass is in the vapor phase, and the rest is in the liquid phase. The tank is connected by a valve to a supply line where refrigerant at 1 MPa and 100°C flows steadily. Now the valve is opened slightly, and the refrigerant is allowed to enter the tank. When the pressure in the tank reaches 700 kPa, the entire refrigerant in the tank exists in the vapor phase only. At this point the valve is closed. Determine (a) the final temperature in the tank, (b) the mass of refrigerant that has entered the tank, and (c) the heat transfer between the system and the surroundings.

**6–120** The air-release flap on a hot-air balloon is used to release hot air from the balloon when appropriate. On one hot-air balloon, the air release opening has an area of 0.5 m², and the filling opening has an area of 1 m². During a two minute adiabatic flight maneuver, hot air enters the balloon at 100 kPa and 35°C with a velocity of 2 m/s; the air in the balloon remains at 100 kPa and 35°C; and air leaves the balloon through the air-release flap at velocity 1 m/s. At the start of this maneuver, the volume of the balloon is 75 m³. Determine the final volume of the balloon and work produced by the air inside the balloon as it expands the balloon skin.

**FIGURE P6–120**

© Vol. 55/PhotoDisc/Getty Images.

**6–121** An insulated 0.15-m³ tank contains helium at 3 MPa and 130°C. A valve is now opened, allowing some helium to escape. The valve is closed when one-half of the initial mass has escaped. Determine the final temperature and pressure in the tank. *Answers: 257 K, 956 kPa*

**6–122E** An insulated 60-ft³ rigid tank contains air at 75 psia and 120°F. A valve connected to the tank is now opened, and air is allowed to escape until the pressure inside drops to 30 psia. The air temperature during this process is maintained constant by an electric resistance heater placed in the tank. Determine the electrical work done during this process.

Air
$V = 60$ ft³
$P = 75$ psia        $W_{e,in}$
$T = 120°F$

**FIGURE P6–122E**

**6–123** A vertical piston–cylinder device initially contains 0.2 m³ of air at 20°C. The mass of the piston is such that it maintains a constant pressure of 300 kPa inside. Now a valve connected to the cylinder is opened, and air is allowed to escape until the volume inside the cylinder is decreased by one-half. Heat transfer takes place during the process so that the temperature of the air in the cylinder remains constant. Determine (a) the amount of air that has left the cylinder and (b) the amount of heat transfer. *Answers: (a) 0.357 kg, (b) 0*

**6–124** An insulated vertical piston–cylinder device initially contains 0.8 m³ of refrigerant-134a at 1.2 MPa and 120°C. A linear spring at this point applies full force to the piston. A valve connected to the cylinder is now opened, and refrigerant is allowed to escape. The spring unwinds as the piston moves down, and the pressure and volume drop to 0.6 MPa and 0.5 m³ at the end of the process. Determine (a) the amount of refrigerant that has escaped and (b) the final temperature of the refrigerant.

R-134a
0.8 m³
1.2 MPa
120°C

**FIGURE P6–124**

**6–125** A piston–cylinder device initially contains 0.6 kg of steam with a volume of 0.1 m³. The mass of the piston is such that it maintains a constant pressure of 800 kPa. The cylinder is connected through a valve to a supply line that carries steam at 5 MPa and 500°C. Now the valve is opened and steam is allowed to flow slowly into the cylinder until the volume of the cylinder doubles and the temperature in the cylinder reaches 250°C, at which point the valve is closed. Determine (a) the mass of steam that has entered and (b) the amount of heat transfer.

**FIGURE P6–125**

**6–126** An adiabatic piston-cylinder device equipped with a spring maintains the pressure inside at 300 kPa when the volume is 0, and 3000 kPa when the volume is 5 m³. The device is connected to a steam line maintained at 1500 kPa and 200°C, and initially the volume is 0. Determine the final temperature (and quality if appropriate) when valve is opened and steam in the line is allowed to enter the cylinder until the pressure inside matches that in the line. Also determine the total work produced during this adiabatic filling process.

**FIGURE P6–126**

**6–127** Repeat Prob. 6–126 when the line is filled with air that is maintained at 2000 kPa and 327°C.

## Review Problems

**6–128** Carbon dioxide gas enters a throttling valve at 5 MPa and 100°C and leaves at 100 kPa. Determine the temperature change during this process if $CO_2$ is assumed to be (a) an ideal gas and (b) a real gas. Real gas properties of $CO_2$ may be obtained from EES.

**FIGURE P6–128**

**6–129** Helium steadily enters a pipe with a mass flow rate of 8 kg/s at 427°C and 100 kPa and leaves the pipe at 27°C. The pressure during the process is constant at 100 kPa. (a) Determine the heat transfer for the process, in kW. (b) Determine the volume flow rate of the helium at the pipe exit, in m³/s.

**6–130** Underground water is being pumped into a pool whose cross section is 6 m × 9 m while water is discharged through a 7-cm-diameter orifice at a constant average velocity of 4 m/s. If the water level in the pool rises at a rate of 2.5 cm/min, determine the rate at which water is supplied to the pool, in m³/s.

**6–131** Air at 4.18 kg/m³ enters a nozzle that has an inlet-to-exit area ratio of 2:1 with a velocity of 120 m/s and leaves with a velocity of 380 m/s. Determine the density of air at the exit. *Answer:* 2.64 kg/m³

**6–132** An air compressor compresses 15 L/s of air at 120 kPa and 20°C to 800 kPa and 300°C while consuming 6.2 kW of power. How much of this power is being used to increase the pressure of the air versus the power needed to move the fluid through the compressor? *Answers:* 4.48 kW, 1.72 kW

**6–133** A steam turbine operates with 1.6 MPa and 350°C steam at its inlet and saturated vapor at 30°C at its exit. The mass flow rate of the steam is 16 kg/s, and the turbine produces 9000 kW of power. Determine the rate at which heat is lost through the casing of this turbine.

**FIGURE P6–133**

**6–134E** Nitrogen gas flows through a long, constant-diameter adiabatic pipe. It enters at 100 psia and 120°F and leaves at 50 psia and 70°F. Calculate the velocity of the nitrogen at the pipe's inlet and outlet.

**6–135** A 110-V electric hot-water heater warms 0.1 L/s of water from 18 to 30°C. Calculate the current in amperes that must be supplied to this heater. *Answer: 45.6 A*

**6–136** Water is boiled at 100°C electrically by a 3-kW resistance wire. Determine the rate of evaporation of water.

**FIGURE P6–136**

**6–137** Steam enters a long, insulated pipe at 1400 kPa, 350°C, and 10 m/s, and exits at 1000 kPa. The diameter of the pipe is 0.15 m at the inlet, and 0.1 m at the exit. Calculate the mass flow rate of the steam and its speed at the pipe outlet.

**6–138** The air flow in a compressed air line is divided into two equal streams by a T-fitting in the line. The compressed air enters this 2.5-cm diameter fitting at 1.6 MPa and 40°C with a velocity of 50 m/s. Each outlet has the same diameter as the inlet, and the air at these outlets has a pressure of 1.4 MPa and a temperature of 36°C. Determine the velocity of the air at the outlets and the rate of change of flow energy (flow power) across the T-fitting.

**FIGURE P6–138**

**6–139** Air enters a pipe at 65°C and 200 kPa and leaves at 60°C and 175 kPa. It is estimated that heat is lost from the pipe in the amount of 3.3 kJ per kg of air flowing in the pipe. The diameter ratio for the pipe is $D_1/D_2 = 1.4$. Using constant specific heats for air, determine the inlet and exit velocities of the air. *Answers: 29.9 m/s, 66.1 m/s*

**6–140** Steam enters a nozzle with a low velocity at 150°C and 200 kPa, and leaves as a saturated vapor at 75 kPa. There is a heat transfer from the nozzle to the surroundings in the amount of 26 kJ for every kilogram of steam flowing through the nozzle. Determine (a) the exit velocity of the steam and (b) the mass flow rate of the steam at the nozzle entrance if the nozzle exit area is 0.001 m².

**6–141** In a gas-fired boiler, water is boiled at 150°C by hot gases flowing through a stainless steel pipe submerged in water. If the rate of heat transfer from the hot gases to water is 74 kJ/s, determine the rate of evaporation of water.

**6–142** Cold water enters a steam generator at 20°C and leaves as saturated vapor at 200°C. Determine the fraction of heat used in the steam generator to preheat the liquid water from 20°C to the saturation temperature of 200°C.

**6–143** Cold water enters a steam generator at 20°C and leaves as saturated vapor at the boiler pressure. At what pressure will the amount of heat needed to preheat the water to saturation temperature be equal to the heat needed to vaporize the liquid at the boiler pressure?

**6–144** An ideal gas expands in an adiabatic turbine from 1200 K and 900 kPa to 700 K. Determine the turbine inlet volume flow rate of the gas, in m³/s, required to produce turbine work output at the rate of 350 kW. The average values of the specific heats for this gas over the temperature range and the gas constant are $c_p = 1.13$ kJ/kg·K, $c_v = 0.83$ kJ/kg·K, and $R = 0.30$ kJ/kg·K.

**6–145** Chickens with an average mass of 2.2 kg and average specific heat of 3.54 kJ/kg·°C are to be cooled by chilled water that enters a continuous-flow-type immersion chiller at 0.5°C. Chickens are dropped into the chiller at a uniform temperature of 15°C at a rate of 500 chickens per hour and are cooled to an average temperature of 3°C before they are taken out. The chiller gains heat from the surroundings at a rate of 200 kJ/h. Determine (a) the rate of heat removal from the chickens, in kW, and (b) the mass flow rate of water, in kg/s, if the temperature rise of water is not to exceed 2°C.

**6–146** Repeat Prob. 6–145 assuming heat gain of the chiller is negligible.

**6–147E** A refrigeration system is being designed to cool eggs ($\rho = 67.4$ lbm/ft³ and $c_p = 0.80$ Btu/lbm·°F) with an average mass of 0.14 lbm from an initial temperature of 90°F to a final average temperature of 50°F by air at 34°F at a rate of 10,000 eggs per hour. Determine (a) the rate of heat removal from the eggs, in Btu/h and (b) the required volume flow rate of air, in ft³/h, if the temperature rise of air is not to exceed 10°F.

**6–148** A glass bottle washing facility uses a well-agitated hot-water bath at 55°C that is placed on the ground. The bottles enter at a rate of 800 per minute at an ambient temperature of 20°C and leave at the water temperature. Each bottle has a mass of 150 g and removes 0.2 g of water as it leaves the bath wet. Make-up water is supplied at 15°C. Disregarding any heat losses from the outer surfaces of the bath, determine the rate at which (*a*) water and (*b*) heat must be supplied to maintain steady operation.

**6–149** Long aluminum wires of diameter 5 mm ($\rho = 2702$ kg/m$^3$ and $c_p = 0.896$ kJ/kg·°C) are extruded at a temperature of 350°C and are cooled to 50°C in atmospheric air at 25°C. If the wire is extruded at a velocity of 8 m/min, determine the rate of heat transfer from the wire to the extrusion room.

350°C

$T_{air} = 25$°C

8 m/min

Aluminum
wire

**FIGURE P6–149**

**6–150** Repeat Prob. 6–149 for a copper wire ($\rho = 8950$ kg/m$^3$ and $c_p = 0.383$ kJ/kg·°C).

**6–151** Steam at 40°C condenses on the outside of a 5-m-long, 3-cm-diameter thin horizontal copper tube by cooling water that enters the tube at 25°C at an average velocity of 2 m/s and leaves at 35°C. Determine the rate of condensation of steam. *Answer:* 0.0245 kg/s

Steam
40°C

Cooling
water

25°C

35°C

**FIGURE P6–151**

**6–152E** Steam at 60 psia and 350°F is mixed with water at 40°F and 60 psia steadily in an adiabatic device. Steam enters the device at a rate of 0.05 lbm/s, while the water enters at 1 lbm/s. Determine the temperature of the mixture leaving this device when the outlet pressure is 60 psia. *Answer:* 97°F

**6–153E** The average atmospheric pressure in Spokane, Washington (elevation = 2350 ft), is 13.5 psia, and the average winter temperature is 36.5°F. The pressurization test of a 9-ft-high, 4500-ft$^2$ older home revealed that the seasonal average infiltration rate of the house is 2.2 air changes per hour (ACH). That is, the entire air in the house is replaced completely 2.2 times per hour by the outdoor air. It is suggested that the infiltration rate of the house can be reduced by half to 1.1 ACH by winterizing the doors and the windows. If the house is heated by natural gas whose unit cost is $1.24/therm and the heating season can be taken to be six months, determine how much the home owner will save from the heating costs per year by this winterization project. Assume the house is maintained at 72°F at all times and the efficiency of the furnace is 0.92. Also assume the latent heat load during the heating season to be negligible.

**6–154** Determine the rate of sensible heat loss from a building due to infiltration if the outdoor air at −5°C and 90 kPa enters the building at a rate of 35 L/s when the indoors is maintained at 20°C.

**6–155** Consider a large classroom on a hot summer day with 150 students, each dissipating 60 W of sensible heat. All the lights, with 6.0 kW of rated power, are kept on. The room has no external walls, and thus heat gain through the walls and the roof is negligible. Chilled air is available at 15°C, and the temperature of the return air is not to exceed 25°C. Determine the required flow rate of air, in kg/s, that needs to be supplied to the room to keep the average temperature of the room constant. *Answer:* 1.49 kg/s

**6–156** A fan is powered by a 0.5-hp motor and delivers air at a rate of 85 m$^3$/min. Determine the highest value for the average velocity of air mobilized by the fan. Take the density of air to be 1.18 kg/m$^3$.

**6–157** An air-conditioning system requires airflow at the main supply duct at a rate of 130 m$^3$/min. The average velocity of air in the circular duct is not to exceed 8 m/s to avoid excessive vibration and pressure drops. Assuming the fan converts 80 percent of the electrical energy it consumes into kinetic energy of air, determine the size of the electric motor needed to drive the fan and the diameter of the main duct. Take the density of air to be 1.20 kg/m$^3$.

130 m$^3$/min

8 m/s

**FIGURE P6–157**

**6–158** An adiabatic air compressor is to be powered by a direct-coupled adiabatic steam turbine that is also driving a generator. Steam enters the turbine at 12.5 MPa and 500°C at a rate of 25 kg/s and exits at 10 kPa and a quality of 0.92. Air enters the compressor at 98 kPa and 295 K at a rate of 10 kg/s and exits at 1 MPa and 620 K. Determine the net power delivered to the generator by the turbine.

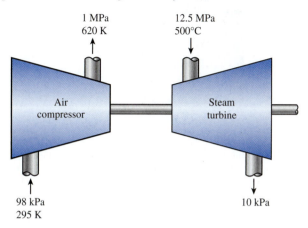

1 MPa
620 K

12.5 MPa
500°C

Air compressor

Steam turbine

98 kPa
295 K

10 kPa

**FIGURE P6–158**

**6–159** Determine the power input for a compressor that compresses helium from 150 kPa and 20°C to 400 kPa and 200°C. Helium enters this compressor through a 0.1-m² pipe at a velocity of 15 m/s.

**6–160** Refrigerant 134a enters a compressor with a mass flow rate of 5 kg/s and a negligible velocity. The refrigerant enters the compressor as a saturated vapor at 10°C and leaves the compressor at 1400 kPa with an enthalpy of 281.39 kJ/kg and a velocity of 50 m/s. The rate of work done on the refrigerant is measured to be 132.4 kW. If the elevation change between the compressor inlet and exit is negligible, determine the rate of heat transfer associated with this process, in kW.

1400 kPa
50 m/s

R-134a

5 kg/s
10°C
sat. vap.

**FIGURE P6–160**

**6–161** Submarines change their depth by adding or removing air from rigid ballast tanks, thereby displacing seawater in the tanks. Consider a submarine that has a 700 m³ air-ballast

tank originally partially filled with 100 m³ of air at 1500 kPa and 15°C. For the submarine to surface, air at 1500 kPa and 20°C is pumped into the ballast tank, until it is entirely filled with air. The tank is filled so quickly that the process is adiabatic and the seawater leaves the tank at 15°C. Determine the final temperature and mass of the air in the ballast tank.

**6–162** In the preceding problem, presume that air is added to the tank in such a way that the temperature and pressure of the air in the tank remain constant. Determine the final mass of the air in the ballast tank under this condition. Also determine the total heat transfer while the tank is being filled in this manner.

**6–163** A steel cylinder initially contains nitrogen gas at 200 kPa and 25°C. The cylinder is connected to a supply line containing nitrogen at 800 kPa and 25°C. A valve is opened, allowing nitrogen to flow into the cylinder until the pressure reaches 800 kPa. The cylinder internal volume is 0.1 m³, its mass (cylinder only) is 50 kg, and its specific heat is 0.43 kJ/kg·K. Find the final mass of nitrogen in the tank and the final temperature of the nitrogen assuming constant specific heats at room temperature for nitrogen, and assuming (a) no heat transfer from the nitrogen to the tank, and (b) rapid heat transfer between the nitrogen and tank such that the cylinder and nitrogen remain in thermal equilibrium during the process (with negligible heat transfer from the cylinder to its external surroundings.) *Answers:* (a) 379 K, 0.711 kg, (b) 301 K, 0.896 kg

**6–164** A tank with an internal volume of 1 m³ contains air at 800 kPa and 25°C. A valve on the tank is opened allowing air to escape and the pressure inside quickly drops to 150 kPa, at which point the valve is closed. Assume there is negligible heat transfer from the tank to the air left in the tank.
(a) Using the approximation $h_e \approx$ constant $= h_{e,avg} = 0.5(h_1 + h_2)$, calculate the mass withdrawn during the process.
(b) Consider the same process but broken into two parts. That is, consider an intermediate state at $P_2 = 400$ kPa, calculate the mass removed during the process from $P_1 = 800$ kPa to $P_2$ and then the mass removed during the process from $P_2$ to $P_3 = 150$ kPa, using the type of approximation used in part (a), and add the two to get the total mass removed.
(c) Calculate the mass removed if the variation of $h_e$ is accounted for.

Air
800 kPa
25°C, 1 m³

**FIGURE P6–164**

**6–165** A liquid R-134a bottle has an internal volume of 0.001 m³. Initially it contains 0.4 kg of R-134a (saturated mixture) at 26°C. A valve is opened and R-134a vapor only (no liquid) is allowed to escape slowly such that temperature remains constant until the mass of R-134a remaining is 0.1 kg. Find the heat transfer necessary with the surroundings to maintain the temperature and pressure of the R-134a constant.

**6–166** Steam enters a turbine steadily at 7 MPa and 600°C with a velocity of 60 m/s and leaves at 25 kPa with a quality of 95 percent. A heat loss of 20 kJ/kg occurs during the process. The inlet area of the turbine is 150 cm², and the exit area is 1400 cm². Determine (a) the mass flow rate of the steam, (b) the exit velocity, and (c) the power output.

**6–167** Reconsider Prob. 6–166. Using EES (or other) software, investigate the effects of turbine exit area and turbine exit pressure on the exit velocity and power output of the turbine. Let the exit pressure vary from 10 to 50 kPa (with the same quality), and the exit area to vary from 1000 to 3000 cm². Plot the exit velocity and the power outlet against the exit pressure for the exit areas of 1000, 2000, and 3000 cm², and discuss the results.

**6–168** In large gas-turbine power plants, air is preheated by the exhaust gases in a heat exchanger called the *regenerator* before it enters the combustion chamber. Air enters the regenerator at 1 MPa and 550 K at a mass flow rate of 800 kg/min. Heat is transferred to the air at a rate of 3200 kJ/s. Exhaust gases enter the regenerator at 140 kPa and 800 K and leave at 130 kPa and 600 K. Treating the exhaust gases as air, determine (a) the exit temperature of the air and (b) the mass flow rate of exhaust gases. *Answers:* (a) 775 K, (b) 14.9 kg/s

**6–169** It is proposed to have a water heater that consists of an insulated pipe of 5-cm diameter and an electric resistor inside. Cold water at 20°C enters the heating section steadily at a rate of 30 L/min. If water is to be heated to 55°C, determine (a) the power rating of the resistance heater and (b) the average velocity of the water in the pipe.

**6–170** An insulated vertical piston–cylinder device initially contains 0.11 m³ of air at 150 kPa and 22°C. At this state, a linear spring touches the piston but exerts no force on it. The cylinder is connected by a valve to a line that supplies air at 700 kPa and 22°C. The valve is opened, and air from the high-pressure line is allowed to enter the cylinder. The valve is turned off when the pressure inside the cylinder reaches 600 kPa. If the enclosed volume inside the cylinder doubles during this process, determine (a) the mass of air that entered the cylinder, and (b) the final temperature of the air inside the cylinder.

Air
$V_1 = 0.11 \text{ m}^3$
$P_1 = 150 \text{ kPa}$
$T_1 = 22°C$

$P_i = 700 \text{ kPa}$
$T_i = 22°C$

**FIGURE P6–170**

**6–171** A piston–cylinder device initially contains 2 kg of refrigerant-134a at 800 kPa and 80°C. At this state, the piston is touching on a pair of stops at the top. The mass of the piston is such that a 500-kPa pressure is required to move it. A valve at the bottom of the tank is opened, and R-134a is withdrawn from the cylinder. After a while, the piston is observed to move and the valve is closed when half of the refrigerant is withdrawn from the tank and the temperature in the tank drops to 20°C. Determine (a) the work done and (b) the heat transfer. *Answers:* (a) 11.6 kJ, (b) 60.7 kJ

**6–172** The pump of a water distribution system is powered by a 6-kW electric motor whose efficiency is 95 percent. The water flow rate through the pump is 18 L/s. The diameters of the inlet and outlet pipes are the same, and the elevation difference across the pump is negligible. If the pressures at the inlet and outlet of the pump are measured to be 100 kPa and 300 kPa (absolute), respectively, determine (a) the mechanical efficiency of the pump and (b) the temperature rise of water as it flows through the pump due to the mechanical inefficiency. *Answers:* (a) 63.2 percent, (b) 0.028°C

Water
18 L/s
300 kPa ②
$\eta_{\text{motor}} = 95\%$
Motor 6 kW
100 kPa ①
$\dot{W}_{\text{pump}}$

**FIGURE P6–172**

**6–173** A pressure cooker is a pot that cooks food much faster than ordinary pots by maintaining a higher pressure and temperature during cooking. The pressure inside the pot is controlled by a pressure regulator (the petcock) that keeps the pressure at a constant level by periodically allowing some steam to escape, thus preventing any excess pressure buildup. Pressure cookers, in general, maintain a gage pressure of 2 atm (or 3 atm absolute) inside. Therefore, pressure cookers cook at a temperature of about 133°C instead of 100°C, cutting the cooking time by as much as 70 percent while minimizing the loss of nutrients. The newer pressure cookers use a spring valve with several pressure settings rather than a weight on the cover.

A certain pressure cooker has a volume of 6 L and an operating pressure of 75 kPa gage. Initially, it contains 1 kg of water. Heat is supplied to the pressure cooker at a rate of 500 W for 30 min after the operating pressure is reached. Assuming an atmospheric pressure of 100 kPa, determine (*a*) the temperature at which cooking takes place and (*b*) the amount of water left in the pressure cooker at the end of the process. *Answers:* (*a*) 116.04°C, (*b*) 0.6 kg

**6–174** A $D_0$ = 10-m-diameter tank is initially filled with water 2 m above the center of a $D$ = 10-cm-diameter valve near the bottom. The tank surface is open to the atmosphere, and the tank drains through a $L$ = 100-m-long pipe connected to the valve. The friction factor of the pipe is given to be $f$ = 0.015, and the discharge velocity is expressed as

$$V = \sqrt{\frac{2gz}{1.5 + fL/D}}$$ where $z$ is the water height above the

center of the valve. Determine (*a*) the initial discharge velocity from the tank and (*b*) the time required to empty the tank. The tank can be considered to be empty when the water level drops to the center of the valve.

## Design and Essay Problems

**6–175E** Pneumatic nail drivers used in construction require 0.02 ft$^3$ of air at 100 psia and 1 Btu of energy to drive a single nail. You have been assigned the task of designing a compressed-air storage tank with enough capacity to drive 500 nails. The pressure in this tank cannot exceed 500 psia, and the temperature cannot exceed that normally found at a construction site. What is the maximum pressure to be used in the tank and what is the tank's volume?

**6–176** You have been given the responsibility of picking a steam turbine for an electrical-generation station that is to produce 300 MW of electrical power that will sell for $0.05 per kilowatt-hour. The boiler will produce steam at 700 psia and 700°F, and the condenser is planned to operate at 80°F. The cost of generating and condensing the steam is $0.01 per kilowatt-hour of electricity produced. You have narrowed your selection to the three turbines in the table below. Your criterion for selection is to pay for the equipment as quickly as possible. Which turbine should you choose?

| Turbine | Capacity (MW) | $\eta$ | Cost ($Million) | Operating Cost ($/kWh) |
|---|---|---|---|---|
| A | 50 | 0.9 | 5 | 0.01 |
| B | 100 | 0.92 | 11 | 0.01 |
| C | 100 | 0.93 | 10.5 | 0.015 |

**6–177E** You are to design a small, directional control rocket to operate in space by providing as many as 100 bursts of 5 seconds each with a mass flow rate of 0.5 lbm/s at a velocity of 400 ft/s. Storage tanks that will contain up to 3000 psia are available, and the tanks will be located in an environment whose temperature is 40°F. Your design criterion is to minimize the volume of the storage tank. Should you use a compressed-air or an R-134a system?

**6–178** An air cannon uses compressed air to propel a projectile from rest to a final velocity. Consider an air cannon that is to accelerate a 10-gram projectile to a speed of 300 m/s using compressed air, whose temperature cannot exceed 20°C. The volume of the storage tank is not to exceed 0.1 m$^3$. Select the storage volume size and maximum storage pressure that requires the minimum amount of energy to fill the tank.

# THE SECOND LAW OF THERMODYNAMICS

To this point, we have focused our attention on the first law of thermodynamics, which requires that energy be conserved during a process. In this chapter, we introduce the second law of thermodynamics, which asserts that processes occur in a certain direction and that energy has quality as well as quantity. A process cannot take place unless it satisfies both the first and second laws of thermodynamics. In this chapter, the thermal energy reservoirs, reversible and irreversible processes, heat engines, refrigerators, and heat pumps are introduced first. Various statements of the second law are followed by a discussion of the thermodynamic temperature scale. The Carnot cycle is introduced next, and the Carnot principles are discussed. Finally, the idealized Carnot heat engines, refrigerators, and heat pumps are examined.

## Objectives

The objectives of this chapter are to:

- Introduce the second law of thermodynamics.

- Identify valid processes as those that satisfy both the first and second laws of thermodynamics.

- Discuss thermal energy reservoirs, reversible and irreversible processes, heat engines, refrigerators, and heat pumps.

- Describe the Kelvin–Planck and Clausius statements of the second law of thermodynamics.

- Apply the second law of thermodynamics to cycles and cyclic devices.

- Apply the second law to develop the absolute thermodynamic temperature scale.

- Describe the Carnot cycle.

- Examine the Carnot principles, idealized Carnot heat engines, refrigerators, and heat pumps.

- Determine the expressions for the thermal efficiencies and coefficients of performance for reversible heat engines, heat pumps, and refrigerators.

**FIGURE 7–1**

A cup of hot coffee does not get hotter in a cooler room.

**FIGURE 7–2**

Transferring heat to a wire will not generate electricity.

**FIGURE 7–3**

Transferring heat to a paddle wheel will not cause it to rotate.

**FIGURE 7–4**

Processes occur in a certain direction, and not in the reverse direction.

# 7–1 ▪ INTRODUCTION TO THE SECOND LAW

In Chaps. 5 and 6, we applied the *first law of thermodynamics*, or the *conservation of energy principle*, to processes involving closed and open systems. As pointed out repeatedly in those chapters, energy is a conserved property, and no process is known to have taken place in violation of the first law of thermodynamics. Therefore, it is reasonable to conclude that a process must satisfy the first law to occur. However, as explained here, satisfying the first law alone does not ensure that the process will actually take place.

It is common experience that a cup of hot coffee left in a cooler room eventually cools off (Fig. 7–1). This process satisfies the first law of thermodynamics since the amount of energy lost by the coffee is equal to the amount gained by the surrounding air. Now let us consider the reverse process—the hot coffee getting even hotter in a cooler room as a result of heat transfer from the room air. We all know that this process never takes place. Yet, doing so would not violate the first law as long as the amount of energy lost by the air is equal to the amount gained by the coffee.

As another familiar example, consider the heating of a room by the passage of electric current through a resistor (Fig. 7–2). Again, the first law dictates that the amount of electric energy supplied to the resistance wires be equal to the amount of energy transferred to the room air as heat. Now let us attempt to reverse this process. It will come as no surprise that transferring some heat to the wires does not cause an equivalent amount of electric energy to be generated in the wires.

Finally, consider a paddle-wheel mechanism that is operated by the fall of a mass (Fig. 7–3). The paddle wheel rotates as the mass falls and stirs a fluid within an insulated container. As a result, the potential energy of the mass decreases, and the internal energy of the fluid increases in accordance with the conservation of energy principle. However, the reverse process, raising the mass by transferring heat from the fluid to the paddle wheel, does not occur in nature, although doing so would not violate the first law of thermodynamics.

It is clear from these arguments that processes proceed in a *certain direction* and not in the reverse direction (Fig. 7–4). The first law places no restriction on the direction of a process, but satisfying the first law does not ensure that the process can actually occur. This inadequacy of the first law to identify whether a process can take place is remedied by introducing another general principle, the *second law of thermodynamics*. We show later in this chapter that the reverse processes discussed above violate the second law of thermodynamics. This violation is easily detected with the help of a property, called *entropy*, defined in Chap. 8. A process cannot occur unless it satisfies both the first and the second laws of thermodynamics (Fig. 7–5).

There are numerous valid statements of the second law of thermodynamics. Two such statements are presented and discussed later in this chapter in relation to some engineering devices that operate on cycles.

The use of the second law of thermodynamics is not limited to identifying the direction of processes. The second law also asserts that energy has *quality* as well as quantity. The first law is concerned with the quantity of energy and the transformations of energy from one form to another with no regard to its quality. Preserving the quality of energy is a major concern to engineers, and the second law provides the necessary means to determine the quality as well as the degree of degradation of energy during a process. As discussed later in this chapter, more of high-temperature energy can be converted to work, and thus it has a higher quality than the same amount of energy at a lower temperature.

The second law of thermodynamics is also used in determining the *theoretical limits* for the performance of commonly used engineering systems, such as heat engines and refrigerators, as well as predicting the *degree of completion* of chemical reactions. The second law is also closely associated with the concept of *perfection*. In fact, the second law *defines* perfection for thermodynamic processes. It can be used to quantify the level of perfection of a process, and point the direction to eliminate imperfections effectively.

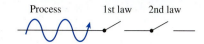

**FIGURE 7–5**

A process must satisfy both the first and second laws of thermodynamics to proceed.

# 7–2 ▪ THERMAL ENERGY RESERVOIRS

In the development of the second law of thermodynamics, it is very convenient to have a hypothetical body with a relatively large *thermal energy capacity* (mass × specific heat) that can supply or absorb finite amounts of heat without undergoing any change in temperature. Such a body is called a **thermal energy reservoir**, or just a reservoir. In practice, large bodies of water such as oceans, lakes, and rivers as well as the atmospheric air can be modeled accurately as thermal energy reservoirs because of their large thermal energy storage capabilities or thermal masses (Fig. 7–6). The *atmosphere*, for example, does not warm up as a result of heat losses from residential buildings in winter. Likewise, megajoules of waste energy dumped in large rivers by power plants do not cause any significant change in water temperature.

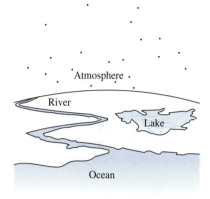

**FIGURE 7–6**

Bodies with relatively large thermal masses can be modeled as thermal energy reservoirs.

A *two-phase system* can also be modeled as a reservoir since it can absorb and release large quantities of heat while remaining at constant temperature. Another familiar example of a thermal energy reservoir is the *industrial furnace*. The temperatures of most furnaces are carefully controlled, and they are capable of supplying large quantities of thermal energy as heat in an essentially isothermal manner. Therefore, they can be modeled as reservoirs.

A body does not actually have to be very large to be considered a reservoir. Any physical body whose thermal energy capacity is large relative to the amount of energy it supplies or absorbs can be modeled as one. The air in a room, for example, can be treated as a reservoir in the analysis of the heat dissipation from a TV set in the room, since the amount of heat transfer from the TV set to the room air is not large enough to have a noticeable effect on the room air temperature.

A reservoir that supplies energy in the form of heat is called a **source**, and one that absorbs energy in the form of heat is called a **sink** (Fig. 7–7). Thermal energy reservoirs are often referred to as **heat reservoirs** since they supply or absorb energy in the form of heat.

Heat transfer from industrial sources to the environment is of major concern to environmentalists as well as to engineers. Irresponsible management of waste energy can significantly increase the temperature of portions of the environment, causing what is called *thermal pollution*. If it is not carefully controlled, thermal pollution can seriously disrupt marine life

**FIGURE 7–7**

A source supplies energy in the form of heat, and a sink absorbs it.

Work    No work

Heat

Heat

Water    Water

**FIGURE 7–8**

Work can always be converted to heat directly and completely, but the reverse is not true.

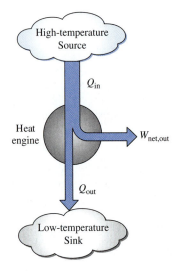

High-temperature
Source

$Q_{in}$

Heat
engine

$W_{net,out}$

$Q_{out}$

Low-temperature
Sink

**FIGURE 7–9**

Part of the heat received by a heat engine is converted to work, while the rest is rejected to a sink.

in lakes and rivers. However, by careful design and management, the waste energy dumped into large bodies of water can be used to improve the quality of marine life by keeping the local temperature increases within safe and desirable levels.

# 7–3 ▪ HEAT ENGINES

As pointed out earlier, work can easily be converted to other forms of energy, but converting other forms of energy to work is not that easy. The mechanical work done by the shaft shown in Fig. 7–8, for example, is first converted to the internal energy of the water. This energy may then leave the water as heat. We know from experience that any attempt to reverse this process will fail. That is, transferring heat to the water does not cause the shaft to rotate. From this and other observations, we conclude that work can be converted to heat directly and completely, but converting heat to work requires the use of some special devices. These devices are called **heat engines**.

Heat engines differ considerably from one another, but all can be characterized by the following (Fig. 7–9):

1. They receive heat from a high-temperature source (solar energy, oil furnace, nuclear reactor, etc.).
2. They convert part of this heat to work (usually in the form of a rotating shaft).
3. They reject the remaining waste heat to a low-temperature sink (the atmosphere, rivers, etc.).
4. They operate on a cycle.

Heat engines and other cyclic devices usually involve a fluid to and from which heat is transferred while undergoing a cycle. This fluid is called the **working fluid**.

The term *heat engine* is often used in a broader sense to include work-producing devices that do not operate in a thermodynamic cycle. Engines that involve internal combustion such as gas turbines and car engines fall into this category. These devices operate in a mechanical cycle but not in a thermodynamic cycle since the working fluid (the combustion gases) does not undergo a complete cycle. Instead of being cooled to the initial temperature, the exhaust gases are purged and replaced by fresh air-and-fuel mixture at the end of the cycle.

The work-producing device that best fits into the definition of a heat engine is the *steam power plant*, which is an external-combustion engine. That is, combustion takes place outside the engine, and the thermal energy released during this process is transferred to the steam as heat. The schematic of a basic steam power plant is shown in Fig. 7–10. This is a rather simplified diagram, and the discussion of actual steam power plants is given in later chapters. The various quantities shown on this figure are as follows:

$Q_{in}$ = amount of heat supplied to steam in boiler from a high-temperature source (furnace)

$Q_{out}$ = amount of heat rejected from steam in condenser to a low-temperature sink (the atmosphere, a river, etc.)

$W_{out}$ = amount of work delivered by steam as it expands in turbine

$W_{in}$ = amount of work required to compress water to boiler pressure

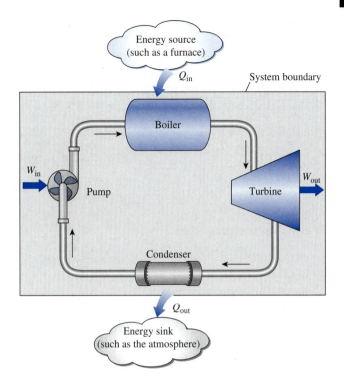

**FIGURE 7–10**
Schematic of a steam power plant.

Notice that the directions of the heat and work interactions are indicated by the subscripts *in* and *out*. Therefore, all four of the described quantities are always *positive*.

The net work output of this power plant is simply the difference between the total work output of the plant and the total work input (Fig. 7–11):

**FIGURE 7–11**
A portion of the work output of a heat engine is consumed internally to maintain continuous operation.

$$W_{net,out} = W_{out} - W_{in} \quad (kJ) \qquad (7–1)$$

The net work can also be determined from the heat transfer data alone. The four components of the steam power plant involve mass flow in and out, and therefore should be treated as open systems. These components, together with the connecting pipes, however, always contain the same fluid (not counting the steam that may leak out, of course). No mass enters or leaves this combination system, which is indicated by the shaded area on Fig. 7–10; thus, it can be analyzed as a closed system. Recall that for a closed system undergoing a cycle, the change in internal energy $\Delta U$ is zero, and therefore the net work output of the system is also equal to the net heat transfer to the system:

$$W_{net,out} = Q_{in} - Q_{out} \quad (kJ) \qquad (7–2)$$

## Thermal Efficiency

In Eq. 7–2, $Q_{out}$ represents the magnitude of the energy wasted in order to complete the cycle. But $Q_{out}$ is never zero; thus, the net work output of a heat engine is always less than the amount of heat input. That is, only part of the

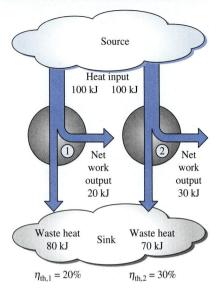

$\eta_{th,1} = 20\%$     $\eta_{th,2} = 30\%$

**FIGURE 7–12**

Some heat engines perform better than others (convert more of the heat they receive to work).

heat transferred to the heat engine is converted to work. The fraction of the heat input that is converted to net work output is a measure of the performance of a heat engine and is called the **thermal efficiency** $\eta_{th}$ (Fig. 7–12).

For heat engines, the desired output is the net work output, and the required input is the amount of heat supplied to the working fluid. Then the thermal efficiency of a heat engine can be expressed as

$$\text{Thermal efficiency} = \frac{\text{Net work output}}{\text{Total heat input}} \qquad (7\text{–}3)$$

or

$$\eta_{th} = \frac{W_{net,out}}{Q_{in}} \qquad (7\text{–}4)$$

It can also be expressed as

$$\eta_{th} = 1 - \frac{Q_{out}}{Q_{in}} \qquad (7\text{–}5)$$

since $W_{net,out} = Q_{in} - Q_{out}$.

Cyclic devices of practical interest such as heat engines, refrigerators, and heat pumps operate between a high-temperature medium (or reservoir) at temperature $T_H$ and a low-temperature medium (or reservoir) at temperature $T_L$. To bring uniformity to the treatment of heat engines, refrigerators, and heat pumps, we define these two quantities:

$Q_H$ = magnitude of heat transfer between the cyclic device and the high-temperature medium at temperature $T_H$

$Q_L$ = magnitude of heat transfer between the cyclic device and the low-temperature medium at temperature $T_L$

Notice that both $Q_L$ and $Q_H$ are defined as *magnitudes* and therefore are positive quantities. The direction of $Q_H$ and $Q_L$ is easily determined by inspection. Then, the net work output and thermal efficiency relations for any heat engine (shown in Fig. 7–13) can also be expressed as

$$W_{net,out} = Q_H - Q_L$$

and

$$\eta_{th} = \frac{W_{net,out}}{Q_H} \qquad \text{or} \qquad \eta_{th} = 1 - \frac{Q_L}{Q_H} \qquad (7\text{–}6)$$

The thermal efficiency of a heat engine is always less than unity since both $Q_L$ and $Q_H$ are defined as positive quantities.

Thermal efficiency is a measure of how efficiently a heat engine converts the heat that it receives to work. Heat engines are built for the purpose of converting heat to work, and engineers are constantly trying to improve the efficiencies of these devices since increased efficiency means less fuel consumption and thus lower fuel bills and less pollution.

The thermal efficiencies of work-producing devices are relatively low. Ordinary spark-ignition automobile engines have a thermal efficiency of about 25 percent. That is, an automobile engine converts about 25 percent of the chemical energy of the gasoline to mechanical work. This number is as high as 40 percent for diesel engines and large gas-turbine plants and as

**FIGURE 7–13**

Schematic of a heat engine.

high as 60 percent for large combined gas-steam power plants. Thus, even with the most efficient heat engines available today, almost one-half of the energy supplied ends up in the rivers, lakes, or the atmosphere as waste or useless energy (Fig. 7–14).

## Can We Save $Q_{out}$?

In a steam power plant, the condenser is the device where large quantities of waste heat is rejected to rivers, lakes, or the atmosphere. Then one may ask, can we not just take the condenser out of the plant and save all that waste energy? The answer to this question is, unfortunately, a firm *no* for the simple reason that without a heat rejection process in a condenser, the cycle cannot be completed. (Cyclic devices such as steam power plants cannot run continuously unless the cycle is completed.) This is demonstrated next with the help of a simple heat engine.

Consider the simple heat engine shown in Fig. 7–15 that is used to lift weights. It consists of a piston–cylinder device with two sets of stops. The working fluid is the gas contained within the cylinder. Initially, the gas temperature is 30°C. The piston, which is loaded with the weights, is resting on top of the lower stops. Now 100 kJ of heat is transferred to the gas in the cylinder from a source at 100°C, causing it to expand and to raise the loaded piston until the piston reaches the upper stops, as shown in the figure. At this point, the load is removed, and the gas temperature is observed to be 90°C.

The work done on the load during this expansion process is equal to the increase in its potential energy, say 15 kJ. Even under ideal conditions (weightless piston, no friction, no heat losses, and quasi-equilibrium expansion), the amount of heat supplied to the gas is greater than the work done since part of the heat supplied is used to raise the temperature of the gas.

Now let us try to answer this question: *Is it possible to transfer the 85 kJ of excess heat at 90°C back to the reservoir at 100°C for later use?* If it is, then we will have a heat engine that can have a thermal efficiency of 100 percent under ideal conditions. The answer to this question is again *no*, for the very

**FIGURE 7–14**

Even the most efficient heat engines reject almost one-half of the energy they receive as waste heat.

**FIGURE 7–15**

A heat-engine cycle cannot be completed without rejecting some heat to a low-temperature sink.

simple reason that heat is always transferred from a high-temperature medium to a low-temperature one, and never the other way around. Therefore, we cannot cool this gas from 90 to 30°C by transferring heat to a reservoir at 100°C. Instead, we have to bring the system into contact with a low-temperature reservoir, say at 20°C, so that the gas can return to its initial state by rejecting its 85 kJ of excess energy as heat to this reservoir. This energy cannot be recycled, and it is properly called *waste energy*.

We conclude from this discussion that every heat engine must *waste* some energy by transferring it to a low-temperature reservoir in order to complete the cycle, even under idealized conditions. The requirement that a heat engine exchange heat with at least two reservoirs for continuous operation forms the basis for the Kelvin–Planck expression of the second law of thermodynamics discussed later in this section.

**FIGURE 7–16**

Schematic for Example 7–1.

### EXAMPLE 7–1    Net Power Production of a Heat Engine

Heat is transferred to a heat engine from a furnace at a rate of 80 MW. If the rate of waste heat rejection to a nearby river is 50 MW, determine the net power output and the thermal efficiency for this heat engine.

**SOLUTION**  The rates of heat transfer to and from a heat engine are given. The net power output and the thermal efficiency are to be determined.
**Assumptions**  Heat losses through the pipes and other components are negligible.
**Analysis**  A schematic of the heat engine is given in Fig. 7–16. The furnace serves as the high-temperature reservoir for this heat engine and the river as the low-temperature reservoir. The given quantities can be expressed as

$$\dot{Q}_H = 80 \text{ MW} \quad \text{and} \quad \dot{Q}_L = 50 \text{ MW}$$

The net power output of this heat engine is

$$\dot{W}_{net,out} = \dot{Q}_H - \dot{Q}_L = (80 - 50) \text{ MW} = \mathbf{30 \text{ MW}}$$

Then the thermal efficiency is easily determined to be

$$\eta_{th} = \frac{\dot{W}_{net,out}}{\dot{Q}_H} = \frac{30 \text{ MW}}{80 \text{ MW}} = \mathbf{0.375} \ (\text{or } 37.5\%)$$

**Discussion**  Note that the heat engine converts 37.5 percent of the heat it receives to work.

### EXAMPLE 7–2    Fuel Consumption Rate of a Car

A car engine with a power output of 65 hp has a thermal efficiency of 24 percent. Determine the fuel consumption rate of this car if the fuel has a heating value of 19,000 Btu/lbm (that is, 19,000 Btu of energy is released for each lbm of fuel burned).

**SOLUTION**  The power output and the efficiency of a car engine are given. The rate of fuel consumption of the car is to be determined.
**Assumptions**  The power output of the car is constant.

*Analysis* A schematic of the car engine is given in Fig. 7–17. The car engine is powered by converting 24 percent of the chemical energy released during the combustion process to work. The amount of energy input required to produce a power output of 65 hp is determined from the definition of thermal efficiency to be

$$\dot{Q}_H = \frac{\dot{W}_{net,out}}{\eta_{th}} = \frac{65 \text{ hp}}{0.24} \left( \frac{2545 \text{ Btu/h}}{1 \text{ hp}} \right) = 689{,}270 \text{ Btu/h}$$

To supply energy at this rate, the engine must burn fuel at a rate of

$$\dot{m} = \frac{689{,}270 \text{ Btu/h}}{19{,}000 \text{ Btu/lbm}} = \textbf{36.3 lbm/h}$$

since 19,000 Btu of thermal energy is released for each lbm of fuel burned.
*Discussion* Note that if the thermal efficiency of the car could be doubled, the rate of fuel consumption would be reduced by half.

**FIGURE 7–17**
Schematic for Example 7–2.

# The Second Law of Thermodynamics: Kelvin–Planck Statement

We have demonstrated earlier with reference to the heat engine shown in Fig. 7–15 that, even under ideal conditions, a heat engine must reject some heat to a low-temperature reservoir in order to complete the cycle. That is, no heat engine can convert all the heat it receives to useful work. This limitation on the thermal efficiency of heat engines forms the basis for the Kelvin–Planck statement of the second law of thermodynamics, which is expressed as follows:

> It is impossible for any device that operates on a cycle to receive heat from a single reservoir and produce a net amount of work.

That is, a heat engine must exchange heat with a low-temperature sink as well as a high-temperature source to keep operating. The Kelvin–Planck statement can also be expressed as *no heat engine can have a thermal efficiency of 100 percent* (Fig. 7–18), or as *for a power plant to operate, the working fluid must exchange heat with the environment as well as the furnace.*

Note that the impossibility of having a 100 percent efficient heat engine is not due to friction or other dissipative effects. It is a limitation that applies to both the idealized and the actual heat engines. Later in this chapter, we develop a relation for the maximum thermal efficiency of a heat engine. We also demonstrate that this maximum value depends on the reservoir temperatures only.

**FIGURE 7–18**
A heat engine that violates the Kelvin–Planck statement of the second law.

# 7–4 ▪ REFRIGERATORS AND HEAT PUMPS

We all know from experience that heat is transferred in the direction of decreasing temperature, that is, from high-temperature mediums to low-temperature ones. This heat transfer process occurs in nature without requiring any devices. The reverse process, however, cannot occur by itself.

**FIGURE 7–19**
Basic components of a refrigeration system and typical operating conditions.

**FIGURE 7–20**
The objective of a refrigerator is to remove $Q_L$ from the cooled space.

The transfer of heat from a low-temperature medium to a high-temperature one requires special devices called **refrigerators**.

Refrigerators, like heat engines, are cyclic devices. The working fluid used in the refrigeration cycle is called a **refrigerant**. The most frequently used refrigeration cycle is the *vapor-compression refrigeration cycle*, which involves four main components: a compressor, a condenser, an expansion valve, and an evaporator, as shown in Fig. 7–19.

The refrigerant enters the compressor as a vapor and is compressed to the condenser pressure. It leaves the compressor at a relatively high temperature and cools down and condenses as it flows through the coils of the condenser by rejecting heat to the surrounding medium. It then enters a capillary tube where its pressure and temperature drop drastically due to the throttling effect. The low-temperature refrigerant then enters the evaporator, where it evaporates by absorbing heat from the refrigerated space. The cycle is completed as the refrigerant leaves the evaporator and reenters the compressor.

In a household refrigerator, the freezer compartment where heat is absorbed by the refrigerant serves as the evaporator, and the coils usually behind the refrigerator where heat is dissipated to the kitchen air serve as the condenser.

A refrigerator is shown schematically in Fig. 7–20. Here $Q_L$ is the magnitude of the heat removed from the refrigerated space at temperature $T_L$, $Q_H$ is the magnitude of the heat rejected to the warm environment at temperature $T_H$, and $W_{net,in}$ is the net work input to the refrigerator. As discussed before, $Q_L$ and $Q_H$ represent magnitudes and thus are positive quantities.

## Coefficient of Performance

The *efficiency* of a refrigerator is expressed in terms of the **coefficient of performance** (COP), denoted by COP$_R$. The objective of a refrigerator is to remove heat ($Q_L$) from the refrigerated space. To accomplish this objective, it requires a work input of $W_{net,in}$. Then the COP of a refrigerator can be expressed as

$$\text{COP}_R = \frac{\text{Desired output}}{\text{Required input}} = \frac{Q_L}{W_{net,in}} \qquad (7\text{–}7)$$

This relation can also be expressed in rate form by replacing $Q_L$ by $\dot{Q}_L$ and $W_{net,in}$ by $\dot{W}_{net,in}$.

The conservation of energy principle for a cyclic device requires that

$$W_{net,in} = Q_H - Q_L \qquad \text{(kJ)} \qquad (7\text{–}8)$$

Then the COP relation becomes

$$\text{COP}_R = \frac{Q_L}{Q_H - Q_L} = \frac{1}{Q_H/Q_L - 1} \qquad (7\text{–}9)$$

Notice that the value of $\text{COP}_R$ can be *greater than unity*. That is, the amount of heat removed from the refrigerated space can be greater than the amount of work input. This is in contrast to the thermal efficiency, which can never be greater than 1. In fact, one reason for expressing the efficiency of a refrigerator by another term—the coefficient of performance—is the desire to avoid the oddity of having efficiencies greater than unity.

**FIGURE 7–21**

The objective of a heat pump is to supply heat $Q_H$ into the warmer space.

## Heat Pumps

Another device that transfers heat from a low-temperature medium to a high-temperature one is the **heat pump**, shown schematically in Fig. 7–21. Refrigerators and heat pumps operate on the same cycle but differ in their objectives. The objective of a refrigerator is to maintain the refrigerated space at a low temperature by removing heat from it. Discharging this heat to a higher-temperature medium is merely a necessary part of the operation, not the purpose. The objective of a heat pump, however, is to maintain a heated space at a high temperature. This is accomplished by absorbing heat from a low-temperature source, such as well water or cold outside air in winter, and supplying this heat to the high-temperature medium such as a house (Fig. 7–22).

An ordinary refrigerator that is placed in the window of a house with its door open to the cold outside air in winter will function as a heat pump since it will try to cool the outside by absorbing heat from it and rejecting this heat into the house through the coils behind it (Fig. 7–23).

The measure of performance of a heat pump is also expressed in terms of the **coefficient of performance** $\text{COP}_{HP}$, defined as

$$\text{COP}_{HP} = \frac{\text{Desired output}}{\text{Required input}} = \frac{Q_H}{W_{net,in}} \qquad (7\text{–}10)$$

which can also be expressed as

$$\text{COP}_{HP} = \frac{Q_H}{Q_H - Q_L} = \frac{1}{1 - Q_L/Q_H} \qquad (7\text{–}11)$$

A comparison of Eqs. 7–7 and 7–10 reveals that

$$\text{COP}_{HP} = \text{COP}_R + 1 \qquad (7\text{–}12)$$

**FIGURE 7–22**

The work supplied to a heat pump is used to extract energy from the cold outdoors and carry it into the warm indoors.

for fixed values of $Q_L$ and $Q_H$. This relation implies that the coefficient of performance of a heat pump is always greater than unity since $COP_R$ is a positive quantity. That is, a heat pump will function, at worst, as a resistance heater, supplying as much energy to the house as it consumes. In reality, however, part of $Q_H$ is lost to the outside air through piping and other devices, and $COP_{HP}$ may drop below unity when the outside air temperature is too low. When this happens, the system usually switches to a resistance heating mode. Most heat pumps in operation today have a seasonally averaged COP of 2 to 3.

Most existing heat pumps use the cold outside air as the heat source in winter, and they are referred to as *air-source heat pumps*. The COP of such heat pumps is about 3.0 at design conditions. Air-source heat pumps are not appropriate for cold climates since their efficiency drops considerably when temperatures are below the freezing point. In such cases, geothermal (also called ground-source) heat pumps that use the ground as the heat source can be used. Geothermal heat pumps require the burial of pipes in the ground 1 to 2 m deep. Such heat pumps are more expensive to install, but they are also more efficient (up to 45 percent more efficient than air-source heat pumps). The COP of ground-source heat pumps can be as high as 6 in the cooling mode.

**Air conditioners** are basically refrigerators whose refrigerated space is a room or a building instead of the food compartment. A window air-conditioning unit cools a room by absorbing heat from the room air and discharging it to the outside. The same air-conditioning unit can be used as a heat pump in winter by installing it backwards as shown in Fig. 7–23. In this mode, the unit absorbs heat from the cold outside and delivers it to the room. Air-conditioning systems that are equipped with proper controls and a reversing valve operate as air conditioners in summer and as heat pumps in winter.

**FIGURE 7–23**
When installed backward, an air conditioner functions as a heat pump.
*BLONDIE © KING FEATURES SYNDICATE.*

## Performance of Refrigerators, Air-Conditioners, and Heat Pumps

The performance of air conditioners and heat pumps is often expressed in terms of the **energy efficiency ratio** (EER) or **seasonal energy efficiency ratio** (SEER) determined by following certain testing standards. SEER is the ratio the total amount of heat removed by an air conditioner or heat pump during a normal cooling season (in Btu) to the total amount of electricity consumed (in watt-hours, Wh), and it is a measure of seasonal performance of cooling equipment. EER, on the other hand, is a measure of the instantaneous energy efficiency, and is defined as the ratio of the rate of heat removal from the cooled space by the cooling equipment to the rate of electricity consumption in steady operation. Therefore, both EER and SEER have the unit Btu/Wh. Considering that 1 kWh = 3412 Btu and thus 1 Wh = 3.412 Btu, a devicet that removes 1 kWh of heat from the cooled space for each kWh of electricity it consumes (COP = 1) will have an EER of 3.412. Therefore, the relation between EER (or SEER) and COP is

$$EER = 3.412\,COP_R$$

To promote the efficient use of energy, governments worldwide have mandated minimum standards for the performance of energy consuming equipment. Most air conditioners or heat pumps in the market have SEER values from 13 to 21, which correspond to COP values of 3.8 to 6.2. Best performance is achieved using units equipped with variable-speed drives (also called inverters). Variable-speed compressors and fans allow the unit to operate at maximum efficiency for varying heating/cooling needs and weather conditions as determined by a microprocessor. In the air-conditioning mode, for example, they operate at higher speeds on hot days and at lower speeds on cooler days, enhancing both efficiency and comfort.

The EER or COP of a refrigerator decreases with decreasing refrigeration temperature. Therefore, it is not economical to refrigerate to a lower temperature than needed. The COPs of refrigerators are in the range of 2.7–3.0 for cutting and preparation rooms; 2.3–2.6 for meat, deli, dairy, and produce; 1.2–1.5 for frozen foods; and 1.0–1.2 for ice cream units. Note that the COP of freezers is about half of the COP of meat refrigerators, and thus it costs twice as much to cool the meat products with refrigerated air that is cold enough to cool frozen foods. It is good energy conservation practice to use separate refrigeration systems to meet different refrigeration needs.

---

### EXAMPLE 7–3    Heat Rejection by a Refrigerator

The food compartment of a refrigerator, shown in Fig. 7–24, is maintained at 4°C by removing heat from it at a rate of 360 kJ/min. If the required power input to the refrigerator is 2 kW, determine (a) the coefficient of performance of the refrigerator and (b) the rate of heat rejection to the room that houses the refrigerator.

**SOLUTION**   The power consumption of a refrigerator is given. The COP and the rate of heat rejection are to be determined.
**Assumptions**   Steady operating conditions exist.
**Analysis**   (a) The coefficient of performance of the refrigerator is

$$\text{COP}_R = \frac{\dot{Q}_L}{\dot{W}_{net,in}} = \frac{360 \text{ kJ/min}}{2 \text{ kW}} \left( \frac{1 \text{ kW}}{60 \text{ kJ/min}} \right) = 3$$

That is, 3 kJ of heat is removed from the refrigerated space for each kJ of work supplied.

(b) The rate at which heat is rejected to the room that houses the refrigerator is determined from the conservation of energy relation for cyclic devices,

$$\dot{Q}_H = \dot{Q}_L + \dot{W}_{net,in} = 360 \text{ kJ/min} + (2 \text{ kW}) \left( \frac{60 \text{ kJ/min}}{1 \text{ kW}} \right) = 480 \text{ kJ/min}$$

**Discussion**   Notice that both the energy removed from the refrigerated space as heat and the energy supplied to the refrigerator as electrical work eventually show up in the room air and become part of the internal energy of the air. This demonstrates that energy can change from one form to another, can move from one place to another, but is never destroyed during a process.

Food compartment
4°C

**FIGURE 7–24**

Schematic for Example 7–3.

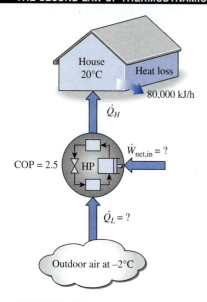

**FIGURE 7–25**

Schematic for Example 7–4.

EXAMPLE 7–4    Heating a House by a Heat Pump

A heat pump is used to meet the heating requirements of a house and maintain it at 20°C. On a day when the outdoor air temperature drops to −2°C, the house is estimated to lose heat at a rate of 80,000 kJ/h. If the heat pump under these conditions has a COP of 2.5, determine (a) the power consumed by the heat pump and (b) the rate at which heat is absorbed from the cold outdoor air.

**SOLUTION**   The COP of a heat pump is given. The power consumption and the rate of heat absorption are to be determined.

**Assumptions**   Steady operating conditions exist.

**Analysis**   (a) The power consumed by this heat pump, shown in Fig. 7–25, is determined from the definition of the coefficient of performance to be

$$\dot{W}_{net,in} = \frac{\dot{Q}_H}{COP_{HP}} = \frac{80{,}000 \text{ kJ/h}}{2.5} = \textbf{32{,}000 kJ/h} \text{ (or 8.9 kW)}$$

(b) The house is losing heat at a rate of 80,000 kJ/h. If the house is to be maintained at a constant temperature of 20°C, the heat pump must deliver heat to the house at the same rate, that is, at a rate of 80,000 kJ/h. Then the rate of heat transfer from the outdoor becomes

$$\dot{Q}_L = \dot{Q}_H - \dot{W}_{net,in} = (80{,}000 - 32{,}000) \text{ kJ/h} = \textbf{48{,}000 kJ/h}$$

**Discussion**   Note that 48,000 of the 80,000 kJ/h heat delivered to the house is actually extracted from the cold outdoor air. Therefore, we are paying only for the 32,000-kJ/h energy that is supplied as electrical work to the heat pump. If we were to use an electric resistance heater instead, we would have to supply the entire 80,000 kJ/h to the resistance heater as electric energy. This would mean a heating bill that is 2.5 times higher. This explains the popularity of heat pumps as heating systems and why they are preferred to simple electric resistance heaters despite their considerably higher initial cost.

# The Second Law of Thermodynamics: Clausius Statement

There are two classical statements of the second law—the Kelvin–Planck statement, which is related to heat engines and discussed in the preceding section, and the Clausius statement, which is related to refrigerators or heat pumps. The Clausius statement is expressed as follows:

> It is impossible to construct a device that operates in a cycle and produces no effect other than the transfer of heat from a lower-temperature body to a higher-temperature body.

It is common knowledge that heat does not, of its own volition, transfer from a cold medium to a warmer one. The Clausius statement does not imply that a cyclic device that transfers heat from a cold medium to a warmer one is impossible to construct. In fact, this is precisely what a common household refrigerator does. It simply states that a refrigerator cannot operate unless its compressor is driven by an external power source, such as an electric motor (Fig. 7–26). This way, the net effect on the surroundings involves the consumption of some energy in the form of work, in addition to the transfer of heat from a colder body to a warmer one. That is, it leaves a trace in the surroundings. Therefore, a household refrigerator is in complete compliance with the Clausius statement of the second law.

Both the Kelvin–Planck and the Clausius statements of the second law are negative statements, and a negative statement cannot be proved. Like any other physical law, the second law of thermodynamics is based on experimental observations. To date, no experiment has been conducted that contradicts the second law, and this should be taken as sufficient proof of its validity.

## Equivalence of the Two Statements

The Kelvin–Planck and the Clausius statements are equivalent in their consequences, and either statement can be used as the expression of the second law of thermodynamics. Any device that violates the Kelvin–Planck statement also violates the Clausius statement, and vice versa. This can be demonstrated as follows.

Consider the heat-engine-refrigerator combination shown in Fig. 7–27a, operating between the same two reservoirs. The heat engine is assumed to have, in violation of the Kelvin–Planck statement, a thermal efficiency of 100 percent, and therefore it converts all the heat $Q_H$ it receives to work $W$. This work is now supplied to a refrigerator that removes heat in the amount of $Q_L$ from the low-temperature reservoir and rejects heat in the amount of $Q_L + Q_H$ to the high-temperature reservoir. During this process, the high-temperature reservoir receives a net amount of heat $Q_L$ (the difference between $Q_L + Q_H$ and $Q_H$). Thus, the combination of these two devices can be viewed as a refrigerator, as shown in Fig. 7–27b, that transfers heat in an

**FIGURE 7–26**

A refrigerator that violates the Clausius statement of the second law.

(a) A refrigerator that is powered by a 100 percent efficient heat engine

(b) The equivalent refrigerator

**FIGURE 7–27**

Proof that the violation of the Kelvin–Planck statement leads to the violation of the Clausius statement.

amount of $Q_L$ from a cooler body to a warmer one without requiring any input from outside. This is clearly a violation of the Clausius statement. Therefore, a violation of the Kelvin–Planck statement results in the violation of the Clausius statement.

It can also be shown in a similar manner that a violation of the Clausius statement leads to the violation of the Kelvin–Planck statement. Therefore, the Clausius and the Kelvin–Planck statements are two equivalent expressions of the second law of thermodynamics.

## 7–5 ▪ REVERSIBLE AND IRREVERSIBLE PROCESSES

The second law of thermodynamics states that no heat engine can have an efficiency of 100 percent. Then one may ask, what is the highest efficiency that a heat engine can possibly have? Before we can answer this question, we need to define an idealized process first, which is called the *reversible process*.

The processes that were discussed at the beginning of this chapter occurred in a certain direction. Once having taken place, these processes cannot reverse themselves spontaneously and restore the system to its initial state. For this reason, they are classified as *irreversible processes*. Once a cup of hot coffee cools, it will not heat up by retrieving the heat it lost from the surroundings. If it could, the surroundings, as well as the system (coffee), would be restored to their original condition, and this would be a reversible process.

A **reversible process** is defined as a *process that can be reversed without leaving any trace on the surroundings* (Fig. 7–28). That is, both the system *and* the surroundings are returned to their initial states at the end of the reverse process. This is possible only if the net heat *and* net work exchange between the system and the surroundings is zero for the combined (original and reverse) process. Processes that are not reversible are called **irreversible processes**.

It should be pointed out that a system can be restored to its initial state following a process, regardless of whether the process is reversible or irreversible. But for reversible processes, this restoration is made without leaving any net change on the surroundings, whereas for irreversible processes, the surroundings usually do some work on the system and therefore does not return to their original state.

Reversible processes actually do not occur in nature. They are merely *idealizations* of actual processes. Reversible processes can be approximated by actual devices, but they can never be achieved. That is, all the processes occurring in nature are irreversible. You may be wondering, then, *why* we are bothering with such fictitious processes. There are two reasons. First, they are easy to analyze, since a system passes through a series of equilibrium states during a reversible process. Second, they serve as idealized models to which actual processes can be compared.

In daily life, the concepts of Mr. Right and Ms. Right are also idealizations, just like the concept of a reversible (perfect) process. People who insist on finding Mr. or Ms. Right to settle down are bound to remain Mr. or Ms. Single for the rest of their lives. The possibility of finding the perfect prospective mate is no higher than the possibility of finding a perfect (reversible)

(*a*) Frictionless pendulum

(*b*) Quasi-equilibrium expansion and compression of a gas

**FIGURE 7–28**

Two familiar reversible processes.

process. Likewise, a person who insists on perfection in friends is bound to have no friends.

Engineers are interested in reversible processes because work-producing devices such as car engines and gas or steam turbines *deliver the most work*, and work-consuming devices such as compressors, fans, and pumps *consume the least work* when reversible processes are used instead of irreversible ones (Fig. 7–29).

Reversible processes can be viewed as *theoretical limits* for the corresponding irreversible ones. Some processes are more irreversible than others. We may never be able to have a reversible process, but we can certainly approach it. The more closely we approximate a reversible process, the more work delivered by a work-producing device or the less work required by a work-consuming device.

The concept of reversible processes leads to the definition of the **second-law efficiency** for actual processes, which is the degree of approximation to the corresponding reversible processes. This enables us to compare the performance of different devices that are designed to do the same task on the basis of their efficiencies. The better the design, the lower the irreversibilities and the higher the second-law efficiency.

## Irreversibilities

The factors that cause a process to be irreversible are called **irreversibilities**. They include friction, unrestrained expansion, mixing of two fluids, heat transfer across a finite temperature difference, electric resistance, inelastic deformation of solids, and chemical reactions. The presence of any of these effects renders a process irreversible. A reversible process involves none of these. Some of the frequently encountered irreversibilities are discussed briefly below.

**Friction** is a familiar form of irreversibility associated with bodies in motion. When two bodies in contact are forced to move relative to each other (a piston in a cylinder, for example, as shown in Fig. 7–30), a friction force that opposes the motion develops at the interface of these two bodies, and some work is needed to overcome this friction force. The energy supplied as work is eventually converted to heat during the process and is transferred to the bodies in contact, as evidenced by a temperature rise at the interface. When the direction of the motion is reversed, the bodies are restored to their original position, but the interface does not cool, and heat is not converted back to work. Instead, more of the work is converted to heat while overcoming the friction forces that also oppose the reverse motion. Since the system (the moving bodies) and the surroundings cannot be returned to their original states, this process is irreversible. Therefore, any process that involves friction is irreversible. The larger the friction forces involved, the more irreversible the process is.

Friction does not always involve two solid bodies in contact. It is also encountered between a fluid and solid and even between the layers of a fluid moving at different velocities. A considerable fraction of the power produced by a car engine is used to overcome the friction (the drag force) between the air and the external surfaces of the car, and it eventually becomes part of the internal energy of the air. It is not possible to reverse

(a) Slow (reversible) process

(b) Fast (irreversible) process

**FIGURE 7–29**
Reversible processes deliver the most and consume the least work.

**FIGURE 7–30**
Friction renders a process irreversible.

(a) Fast compression

(b) Fast expansion

700 kPa    50 kPa

(c) Unrestrained expansion

**FIGURE 7–31**

Irreversible compression and expansion processes.

20°C
Heat
20°C
5°C

(a) An irreversible heat transfer process

20°C
Heat
5°C
2°C

(b) An impossible heat transfer process

**FIGURE 7–32**

(a) Heat transfer through a temperature difference is irreversible, and (b) the reverse process is impossible.

this process and recover that lost power, even though doing so would not violate the conservation of energy principle.

Another example of irreversibility is the **unrestrained expansion of a gas** separated from a vacuum by a membrane, as shown in Fig. 7–31. When the membrane is ruptured, the gas fills the entire tank. The only way to restore the system to its original state is to compress it to its initial volume, while transferring heat from the gas until it reaches its initial temperature. From the conservation of energy considerations, it can easily be shown that the amount of heat transferred from the gas equals the amount of work done on the gas by the surroundings. The restoration of the surroundings involves conversion of this heat completely to work, which would violate the second law. Therefore, unrestrained expansion of a gas is an irreversible process.

A third form of irreversibility familiar to us all is **heat transfer** through a finite temperature difference. Consider a can of cold soda left in a warm room (Fig. 7–32). Heat is transferred from the warmer room air to the cooler soda. The only way this process can be reversed and the soda restored to its original temperature is to provide refrigeration, which requires some work input. At the end of the reverse process, the soda will be restored to its initial state, but the surroundings will not be. The internal energy of the surroundings will increase by an amount equal in magnitude to the work supplied to the refrigerator. The restoration of the surroundings to the initial state can be done only by converting this excess internal energy completely to work, which is impossible to do without violating the second law. Since only the system, not both the system and the surroundings, can be restored to its initial condition, heat transfer through a finite temperature difference is an irreversible process.

Heat transfer can occur only when there is a temperature difference between a system and its surroundings. Therefore, it is physically impossible to have a reversible heat transfer process. But a heat transfer process becomes less and less irreversible as the temperature difference between the two bodies approaches zero. Then, heat transfer through a differential temperature difference $dT$ can be considered to be reversible. As $dT$ approaches zero, the process can be reversed in direction (at least theoretically) without requiring any refrigeration. Notice that reversible heat transfer is a conceptual process and cannot be duplicated in the real world.

The smaller the temperature difference between two bodies, the smaller the heat transfer rate will be. Any significant heat transfer through a small temperature difference requires a very large surface area and a very long time. Therefore, even though approaching reversible heat transfer is desirable from a thermodynamic point of view, it is impractical and not economically feasible.

## Internally and Externally Reversible Processes

A typical process involves interactions between a system and its surroundings, and a reversible process involves no irreversibilities associated with either of them.

A process is called **internally reversible** if no irreversibilities occur within the boundaries of the system during the process. During an internally reversible process, a system proceeds through a series of equilibrium states, and when the process is reversed, the system passes through exactly the same equilibrium states while returning to its initial state. That is, the paths of the forward and reverse processes coincide for an internally reversible process. The quasi-equilibrium process is an example of an internally reversible process.

A process is called **externally reversible** if no irreversibilities occur outside the system boundaries during the process. Heat transfer between a reservoir and a system is an externally reversible process if the outer surface of the system is at the temperature of the reservoir.

A process is called **totally reversible**, or simply **reversible**, if it involves no irreversibilities within the system or its surroundings (Fig. 7–33). A totally reversible process involves no heat transfer through a finite temperature difference, no nonquasi-equilibrium changes, and no friction or other dissipative effects.

As an example, consider the transfer of heat to two identical systems that are undergoing a constant-pressure (thus constant-temperature) phase-change process, as shown in Fig. 7–34. Both processes are internally reversible, since both take place isothermally and both pass through exactly the same equilibrium states. The first process shown is externally reversible also, since heat transfer for this process takes place through an infinitesimal temperature difference $dT$. The second process, however, is externally irreversible, since it involves heat transfer through a finite temperature difference $\Delta T$.

## 7–6 ▪ THE CARNOT CYCLE

We mentioned earlier that heat engines are cyclic devices and that the working fluid of a heat engine returns to its initial state at the end of each cycle. Work is done by the working fluid during one part of the cycle and on the working fluid during another part. The difference between these two is the net work delivered by the heat engine. The efficiency of a heat-engine cycle greatly depends on how the individual processes that make up the cycle are executed. The net work, thus the cycle efficiency, can be maximized by using processes that require the least amount of work and deliver the most, that is, by using *reversible processes*. Therefore, it is no surprise that the most efficient cycles are reversible cycles, that is, cycles that consist entirely of reversible processes.

Reversible cycles cannot be achieved in practice because the irreversibilities associated with each process cannot be eliminated. However, reversible cycles provide upper limits on the performance of real cycles. Heat engines and refrigerators that work on reversible cycles serve as models to which actual heat engines and refrigerators can be compared. Reversible cycles also serve as starting points in the development of actual cycles and are modified as needed to meet certain requirements.

Probably the best known reversible cycle is the **Carnot cycle**, first proposed in 1824 by French engineer Sadi Carnot. The theoretical heat engine that operates on the Carnot cycle is called the **Carnot heat engine**. The Carnot cycle is composed of four reversible processes—two isothermal and two adiabatic—and it can be executed either in a closed or a steady-flow system.

Consider a closed system that consists of a gas contained in an adiabatic piston–cylinder device, as shown in Fig. 7–35. The insulation of the cylinder

**FIGURE 7–33**

A reversible process involves no internal and external irreversibilities.

Heat

Thermal energy reservoir at 20.000...1°C

(*a*) Totally reversible

Heat

Thermal energy reservoir at 30°C

(*b*) Internally reversible

**FIGURE 7–34**

Totally and internally reversible heat transfer processes.

**FIGURE 7–35**

Execution of the Carnot cycle
in a closed system.

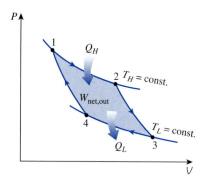

**FIGURE 7–36**

$P$-$V$ diagram of the Carnot cycle.

head is such that it may be removed to bring the cylinder into contact with
reservoirs to provide heat transfer. The four reversible processes that make
up the Carnot cycle are as follows:

**Reversible Isothermal Expansion** (process 1-2, $T_H$ = constant). Initially
(state 1), the temperature of the gas is $T_H$ and the cylinder head is in close
contact with a source at temperature $T_H$. The gas is allowed to expand
slowly, doing work on the surroundings. As the gas expands, the
temperature of the gas tends to decrease. But as soon as the temperature
drops by an infinitesimal amount $dT$, some heat is transferred from the
reservoir into the gas, raising the gas temperature to $T_H$. Thus, the gas
temperature is kept constant at $T_H$. Since the temperature difference
between the gas and the reservoir never exceeds a differential amount $dT$,
this is a reversible heat transfer process. It continues until the piston
reaches position 2. The amount of total heat transferred to the gas during
this process is $Q_H$.

**Reversible Adiabatic Expansion** (process 2-3, temperature drops from $T_H$
to $T_L$). At state 2, the reservoir that was in contact with the cylinder head
is removed and replaced by insulation so that the system becomes
adiabatic. The gas continues to expand slowly, doing work on the
surroundings until its temperature drops from $T_H$ to $T_L$ (state 3). The
piston is assumed to be frictionless and the process to be quasi-
equilibrium, so the process is reversible as well as adiabatic.

**Reversible Isothermal Compression** (process 3-4, $T_L$ = constant). At
state 3, the insulation at the cylinder head is removed, and the cylinder
is brought into contact with a sink at temperature $T_L$. Now the piston is
pushed inward by an external force, doing work on the gas. As the gas
is compressed, its temperature tends to rise. But as soon as it rises by
an infinitesimal amount $dT$, heat is transferred from the gas to the sink,
causing the gas temperature to drop to $T_L$. Thus, the gas temperature
remains constant at $T_L$. Since the temperature difference between the
gas and the sink never exceeds a differential amount $dT$, this is a
reversible heat transfer process. It continues until the piston reaches
state 4. The amount of heat rejected from the gas during this process
is $Q_L$.

**Reversible Adiabatic Compression** (process 4-1, temperature rises from
$T_L$ to $T_H$). State 4 is such that when the low-temperature reservoir is
removed, the insulation is put back on the cylinder head, and the gas is
compressed in a reversible manner, the gas returns to its initial state
(state 1). The temperature rises from $T_L$ to $T_H$ during this reversible
adiabatic compression process, which completes the cycle.

The $P$-$V$ diagram of this cycle is shown in Fig. 7–36. Remembering that
on a $P$-$V$ diagram the area under the process curve represents the boundary
work for quasi-equilibrium (internally reversible) processes, we see that the
area under curve 1-2-3 is the work done by the gas during the expansion
part of the cycle, and the area under curve 3-4-1 is the work done on the gas
during the compression part of the cycle. The area enclosed by the path of
the cycle (area 1-2-3-4-1) is the difference between these two and represents
the net work done during the cycle.

Notice that if we acted stingily and compressed the gas at state 3 adiabatically instead of isothermally in an effort *to save* $Q_L$, we would end up back at state 2, retracing the process path 3-2. By doing so we would save $Q_L$, but we would not be able to obtain any net work output from this engine. This illustrates once more the necessity of a heat engine exchanging heat with at least two reservoirs at different temperatures to operate in a cycle and produce a net amount of work.

The Carnot cycle can also be executed in a steady-flow system. It is discussed in later chapters in conjunction with other power cycles.

Being a reversible cycle, the Carnot cycle is the most efficient cycle operating between two specified temperature limits. Even though the Carnot cycle cannot be achieved in reality, the efficiency of actual cycles can be improved by attempting to approximate the Carnot cycle more closely.

## The Reversed Carnot Cycle

The Carnot heat-engine cycle just described is a totally reversible cycle. Therefore, all the processes that comprise it can be *reversed*, in which case it becomes the **Carnot refrigeration cycle**. This time, the cycle remains exactly the same, except that the directions of any heat and work interactions are reversed: Heat in the amount of $Q_L$ is absorbed from the low-temperature reservoir, heat in the amount of $Q_H$ is rejected to a high-temperature reservoir, and a work input of $W_{net,in}$ is required to accomplish all this.

The P-V diagram of the reversed Carnot cycle is the same as the one given for the Carnot cycle, except that the directions of the processes are reversed, as shown in Fig. 7–37.

**FIGURE 7–37**

P-V diagram of the reversed Carnot cycle.

## 7–7 ▪ THE CARNOT PRINCIPLES

The second law of thermodynamics puts limits on the operation of cyclic devices as expressed by the Kelvin–Planck and Clausius statements. A heat engine cannot operate by exchanging heat with a single reservoir, and a refrigerator cannot operate without a net energy input from an external source.

We can draw valuable conclusions from these statements. Two conclusions pertain to the thermal efficiency of reversible and irreversible (i.e., actual) heat engines, and they are known as the **Carnot principles** (Fig. 7–38), expressed as follows:

1. The efficiency of an irreversible heat engine is always less than the efficiency of a reversible one operating between the same two reservoirs.
2. The efficiencies of all reversible heat engines operating between the same two reservoirs are the same.

These two statements can be proved by demonstrating that the violation of either statement results in the violation of the second law of thermodynamics.

To prove the first statement, consider two heat engines operating between the same reservoirs, as shown in Fig. 7–39. One engine is reversible and the other is irreversible. Now each engine is supplied with the same amount of heat $Q_H$. The amount of work produced by the reversible heat engine is $W_{rev}$, and the amount produced by the irreversible one is $W_{irrev}$.

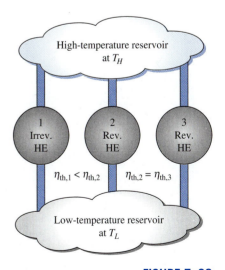

**FIGURE 7–38**

The Carnot principles.

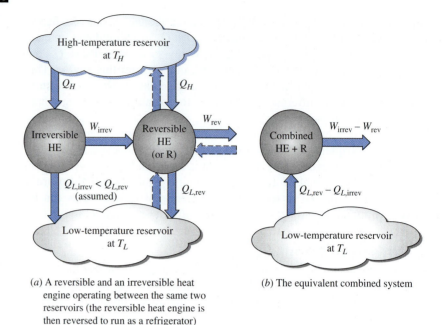

(a) A reversible and an irreversible heat engine operating between the same two reservoirs (the reversible heat engine is then reversed to run as a refrigerator)

(b) The equivalent combined system

**FIGURE 7–39**

Proof of the first Carnot principle.

In violation of the first Carnot principle, we assume that the irreversible heat engine is more efficient than the reversible one (that is, $\eta_{th,irrev} > \eta_{th,rev}$) and thus delivers more work than the reversible one. Now let the reversible heat engine be reversed and operate as a refrigerator. This refrigerator will receive a work input of $W_{rev}$ and reject heat to the high-temperature reservoir. Since the refrigerator is rejecting heat in the amount of $Q_H$ to the high-temperature reservoir and the irreversible heat engine is receiving the same amount of heat from this reservoir, the net heat exchange for this reservoir is zero. Thus, it could be eliminated by having the refrigerator discharge $Q_H$ directly into the irreversible heat engine.

Now considering the refrigerator and the irreversible engine together, we have an engine that produces a net work in the amount of $W_{irrev} - W_{rev}$ while exchanging heat with a single reservoir—a violation of the Kelvin–Planck statement of the second law. Therefore, our initial assumption that $\eta_{th,irrev} > \eta_{th,rev}$ is incorrect. Then we conclude that no heat engine can be more efficient than a reversible heat engine operating between the same reservoirs.

The second Carnot principle can also be proved in a similar manner. This time, let us replace the irreversible engine by another reversible engine that is more efficient and thus delivers more work than the first reversible engine. By following through the same reasoning, we end up having an engine that produces a net amount of work while exchanging heat with a single reservoir, which is a violation of the second law. Therefore, we conclude that no reversible heat engine can be more efficient than a reversible one operating between the same two reservoirs, regardless of how the cycle is completed or the kind of working fluid used.

# 7–8 · THE THERMODYNAMIC TEMPERATURE SCALE

A temperature scale that is independent of the properties of the substances that are used to measure temperature is called a **thermodynamic temperature scale**. Such a temperature scale offers great conveniences in thermodynamic calculations, and its derivation is given below using some reversible heat engines.

The second Carnot principle discussed in Section 7–7 states that all reversible heat engines have the same thermal efficiency when operating between the same two reservoirs (Fig. 7–40). That is, the efficiency of a reversible engine is independent of the working fluid employed and its properties, the way the cycle is executed, or the type of reversible engine used. Since energy reservoirs are characterized by their temperatures, the thermal efficiency of reversible heat engines is a function of the reservoir temperatures only. That is,

$$\eta_{th,rev} = g(T_H, T_L)$$

or

$$\frac{Q_H}{Q_L} = f(T_H, T_L) \qquad \text{(7–13)}$$

since $\eta_{th} = 1 - Q_L/Q_H$. In these relations $T_H$ and $T_L$ are the temperatures of the high- and low-temperature reservoirs, respectively.

The functional form of $f(T_H, T_L)$ can be developed with the help of the three reversible heat engines shown in Fig. 7–41. Engines A and C are supplied with the same amount of heat $Q_1$ from the high-temperature reservoir at $T_1$. Engine C rejects $Q_3$ to the low-temperature reservoir at $T_3$. Engine B receives the heat $Q_2$ rejected by engine A at temperature $T_2$ and rejects heat in the amount of $Q_3$ to a reservoir at $T_3$.

The amounts of heat rejected by engines B and C must be the same since engines A and B can be combined into one reversible engine operating between the same reservoirs as engine C and thus the combined engine will have the same efficiency as engine C. Since the heat input to engine C is the same as the heat input to the combined engines A and B, both systems must reject the same amount of heat.

Applying Eq. 7–13 to all three engines separately, we obtain

$$\frac{Q_1}{Q_2} = f(T_1, T_2), \quad \frac{Q_2}{Q_3} = f(T_2, T_3), \quad \text{and} \quad \frac{Q_1}{Q_3} = f(T_1, T_3)$$

Now consider the identity

$$\frac{Q_1}{Q_3} = \frac{Q_1}{Q_2} \frac{Q_2}{Q_3}$$

which corresponds to

$$f(T_1, T_3) = f(T_1, T_2) \cdot f(T_2, T_3)$$

A careful examination of this equation reveals that the left-hand side is a function of $T_1$ and $T_3$, and therefore the right-hand side must also be a function

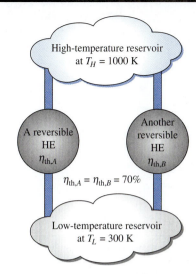

**FIGURE 7–40**

All reversible heat engines operating between the same two reservoirs have the same efficiency (the second Carnot principle).

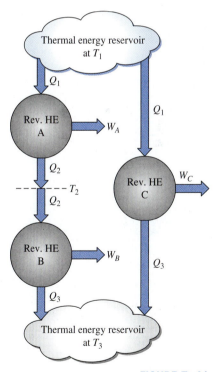

**FIGURE 7–41**

The arrangement of heat engines used to develop the thermodynamic temperature scale.

**FIGURE 7–42**

For reversible cycles, the heat transfer ratio $Q_H/Q_L$ can be replaced by the absolute temperature ratio $T_H/T_L$.

**FIGURE 7–43**

A conceptual experimental setup to determine thermodynamic temperatures on the Kelvin scale by measuring heat transfers $Q_H$ and $Q_L$.

of $T_1$ and $T_3$ only, and not $T_2$. That is, the value of the product on the right-hand side of this equation is independent of the value of $T_2$. This condition will be satisfied only if the function $f$ has the following form:

$$f(T_1, T_2) = \frac{\phi(T_1)}{\phi(T_2)} \quad \text{and} \quad f(T_2, T_3) = \frac{\phi(T_2)}{\phi(T_3)}$$

so that $\phi(T_2)$ will cancel from the product of $f(T_1, T_2)$ and $f(T_2, T_3)$, yielding

$$\frac{Q_1}{Q_3} = f(T_1, T_3) = \frac{\phi(T_1)}{\phi(T_3)} \tag{7–14}$$

This relation is much more specific than Eq. 7–13 for the functional form of $Q_1/Q_3$ in terms of $T_1$ and $T_3$.

For a reversible heat engine operating between two reservoirs at temperatures $T_H$ and $T_L$, Eq. 7–14 can be written as

$$\frac{Q_H}{Q_L} = \frac{\phi(T_H)}{\phi(T_L)} \tag{7–15}$$

This is the only requirement that the second law places on the ratio of heat transfers to and from the reversible heat engines. Several functions $\phi(T)$ satisfy this equation, and the choice is completely arbitrary. Lord Kelvin first proposed taking $\phi(T) = T$ to define a thermodynamic temperature scale as (Fig. 7–42)

$$\left(\frac{Q_H}{Q_L}\right)_{\text{rev}} = \frac{T_H}{T_L} \tag{7–16}$$

This temperature scale is called the **Kelvin scale**, and the temperatures on this scale are called **absolute temperatures**. On the Kelvin scale, the temperature ratios depend on the ratios of heat transfer between a reversible heat engine and the reservoirs and are independent of the physical properties of any substance. On this scale, temperatures vary between zero and infinity.

The thermodynamic temperature scale is not completely defined by Eq. 7–16 since it gives us only a ratio of absolute temperatures. We also need to know the magnitude of a kelvin. At the International Conference on Weights and Measures held in 1954, the triple point of water (the state at which all three phases of water exist in equilibrium) was assigned the value 273.16 K (Fig. 7–43). The *magnitude of a kelvin* is defined as 1/273.16 of the temperature interval between absolute zero and the triple-point temperature of water. The magnitudes of temperature units on the Kelvin and Celsius scales are identical (1 K ≡ 1°C). The temperatures on these two scales differ by a constant 273.15:

$$T(°C) = T(K) - 273.15 \tag{7–17}$$

Even though the thermodynamic temperature scale is defined with the help of the reversible heat engines, it is not possible, nor is it practical, to actually operate such an engine to determine numerical values on the absolute temperature scale. Absolute temperatures can be measured accurately by other means, such as the constant-volume ideal-gas thermometer together with extrapolation techniques as discussed in Chap. 2. The validity of Eq. 7–16 can be demonstrated from physical considerations for a reversible cycle using an ideal gas as the working fluid.

## 7–9 · THE CARNOT HEAT ENGINE

The hypothetical heat engine that operates on the reversible Carnot cycle is called the **Carnot heat engine**. The thermal efficiency of any heat engine, reversible or irreversible, is given by Eq. 7–6 as

$$\eta_{th} = 1 - \frac{Q_L}{Q_H}$$

where $Q_H$ is heat transferred to the heat engine from a high-temperature reservoir at $T_H$, and $Q_L$ is heat rejected to a low-temperature reservoir at $T_L$. For reversible heat engines, the heat transfer ratio in the above relation can be replaced by the ratio of the absolute temperatures of the two reservoirs, as given by Eq. 7–16. Then the efficiency of a Carnot engine, or any reversible heat engine, becomes

$$\eta_{th,rev} = 1 - \frac{T_L}{T_H} \qquad (7\text{–}18)$$

This relation is often referred to as the **Carnot efficiency**, since the Carnot heat engine is the best known reversible engine. *This is the highest efficiency a heat engine operating between the two thermal energy reservoirs at temperatures $T_L$ and $T_H$ can have* (Fig. 7–44). All irreversible (i.e., actual) heat engines operating between these temperature limits ($T_L$ and $T_H$) have lower efficiencies. An actual heat engine cannot reach this maximum theoretical efficiency value because it is impossible to completely eliminate all the irreversibilities associated with the actual cycle.

Note that $T_L$ and $T_H$ in Eq. 7–18 are *absolute temperatures*. Using °C or °F for temperatures in this relation gives results grossly in error.

The thermal efficiencies of actual and reversible heat engines operating between the same temperature limits compare as follows (Fig. 7–45):

$$\eta_{th} \begin{cases} < \eta_{th,rev} & \text{irreversible heat engine} \\ = \eta_{th,rev} & \text{reversible heat engine} \\ > \eta_{th,rev} & \text{impossible heat engine} \end{cases} \qquad (7\text{–}19)$$

Most work-producing devices (heat engines) in operation today have efficiencies under 40 percent, which appear low relative to 100 percent. However, when the performance of actual heat engines is assessed, the efficiencies should not be compared to 100 percent; instead, they should be compared to the efficiency of a reversible heat engine operating between the same temperature limits—because this is the true theoretical upper limit for the efficiency, not 100 percent.

The maximum efficiency of a steam power plant operating between $T_H = 1000$ K and $T_L = 300$ K is 70 percent, as determined from Eq. 7–18. Compared with this value, an actual efficiency of 40 percent does not seem so bad, even though there is still plenty of room for improvement.

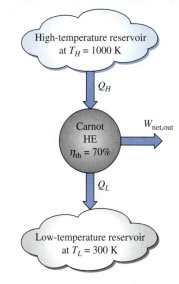

**FIGURE 7–44**

The Carnot heat engine is the most efficient of all heat engines operating between the same high- and low-temperature reservoirs.

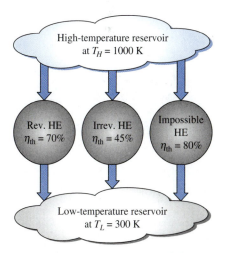

**FIGURE 7–45**

No heat engine can have a higher efficiency than a reversible heat engine operating between the same high- and low-temperature reservoirs.

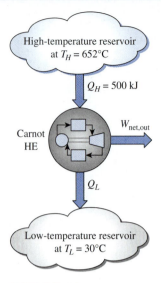

**FIGURE 7–46**

Schematic for Example 7–5.

**FIGURE 7–47**

The fraction of heat that can be converted to work as a function of source temperature (for $T_L = 303$ K).

It is obvious from Eq. 7–18 that the efficiency of a Carnot heat engine increases as $T_H$ is increased, or as $T_L$ is decreased. This is to be expected since as $T_L$ decreases, so does the amount of heat rejected, and as $T_L$ approaches zero, the Carnot efficiency approaches unity. This is also true for actual heat engines. *The thermal efficiency of actual heat engines can be maximized by supplying heat to the engine at the highest possible temperature* (limited by material strength) *and rejecting heat from the engine at the lowest possible temperature* (limited by the temperature of the cooling medium such as rivers, lakes, or the atmosphere).

---

**EXAMPLE 7–5**    **Analysis of a Carnot Heat Engine**

A Carnot heat engine, shown in Fig. 7–46, receives 500 kJ of heat per cycle from a high-temperature source at 652°C and rejects heat to a low-temperature sink at 30°C. Determine (a) the thermal efficiency of this Carnot engine and (b) the amount of heat rejected to the sink per cycle.

**SOLUTION**   The heat supplied to a Carnot heat engine is given. The thermal efficiency and the heat rejected are to be determined.
**Analysis**   (a) The Carnot heat engine is a reversible heat engine, and so its efficiency can be determined from Eq. 7–18 to be

$$\eta_{th,C} = \eta_{th,rev} = 1 - \frac{T_L}{T_H} = 1 - \frac{(30 + 273)\text{ K}}{(652 + 273)\text{ K}} = 0.672$$

That is, this Carnot heat engine converts 67.2 percent of the heat it receives to work.

(b) The amount of heat rejected $Q_L$ by this reversible heat engine is easily determined from Eq. 7–16 to be

$$Q_{L,rev} = \frac{T_L}{T_H} Q_{H,rev} = \frac{(30 + 273)\text{ K}}{(652 + 273)\text{ K}} (500\text{ kJ}) = 164\text{ kJ}$$

**Discussion**   Note that this Carnot heat engine rejects to a low-temperature sink 164 kJ of the 500 kJ of heat it receives during each cycle.

## The Quality of Energy

The Carnot heat engine in Example 7–5 receives heat from a source at 925 K and converts 67.2 percent of it to work while rejecting the rest (32.8 percent) to a sink at 303 K. Now let us examine how the thermal efficiency varies with the source temperature when the sink temperature is held constant.

The thermal efficiency of a Carnot heat engine that rejects heat to a sink at 303 K is evaluated at various source temperatures using Eq. 7–18 and is listed in Fig. 7–47. Clearly, the thermal efficiency decreases as the source temperature is lowered. When heat is supplied to the heat engine at 500 instead of 925 K, for example, the thermal efficiency drops from 67.2 to 39.4 percent.

That is, the fraction of heat that can be converted to work drops to 39.4 percent when the temperature of the source drops to 500 K. When the source temperature is 350 K, this fraction becomes a mere 13.4 percent.

These efficiency values show that energy has **quality** as well as quantity. It is clear from the thermal efficiency values in Fig. 7–47 that *more of the high-temperature thermal energy can be converted to work. Therefore, the higher the temperature, the higher the quality of the energy* (Fig. 7–48).

Large quantities of solar energy, for example, can be stored in large bodies of water called *solar ponds* at about 350 K. This stored energy can then be supplied to a heat engine to produce work (electricity). However, the efficiency of solar pond power plants is very low (under 5 percent) because of the low quality of the energy stored in the source, and the construction and maintenance costs are relatively high. Therefore, they are not competitive even though the energy supply of such plants is free. The temperature (and thus the quality) of the solar energy stored could be raised by utilizing concentrating collectors, but the equipment cost in that case becomes very high.

Work is a more valuable form of energy than heat since 100 percent of work can be converted to heat, but only a fraction of heat can be converted to work. When heat is transferred from a high-temperature body to a lower-temperature one, it is degraded since less of it now can be converted to work. For example, if 100 kJ of heat is transferred from a body at 1000 K to a body at 300 K, at the end we will have 100 kJ of thermal energy stored at 300 K, which has no practical value. But if this conversion is made through a heat engine, up to $1 - 300/1000 = 70$ percent of it could be converted to work, which is a more valuable form of energy. Thus 70 kJ of work potential is wasted as a result of this heat transfer, and energy is degraded.

**FIGURE 7–48**

The higher the temperature of the thermal energy, the higher its quality.

# 7–10 ▪ THE CARNOT REFRIGERATOR AND HEAT PUMP

A refrigerator or a heat pump that operates on the reversed Carnot cycle is called a **Carnot refrigerator**, or a **Carnot heat pump**. The coefficient of performance of any refrigerator or heat pump, reversible or irreversible, is given by Eqs. 7–9 and 7–11 as

$$COP_R = \frac{1}{Q_H/Q_L - 1} \quad \text{and} \quad COP_{HP} = \frac{1}{1 - Q_L/Q_H}$$

where $Q_L$ is the amount of heat absorbed from the low-temperature medium and $Q_H$ is the amount of heat rejected to the high-temperature medium. The COPs of all reversible refrigerators or heat pumps can be determined by replacing the heat transfer ratios in the above relations by the ratios of the absolute temperatures of the high- and low-temperature reservoirs, as expressed by Eq. 7–16. Then the COP relations for reversible refrigerators and heat pumps become

$$COP_{R,rev} = \frac{1}{T_H/T_L - 1} \quad \text{(7–20)}$$

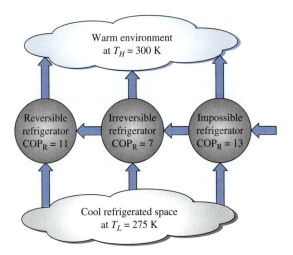

**FIGURE 7–49**

No refrigerator can have a higher COP than a reversible refrigerator operating between the same temperature limits.

and

$$\text{COP}_{\text{HP,rev}} = \frac{1}{1 - T_L/T_H} \tag{7–21}$$

*These are the highest coefficients of performance that a refrigerator or a heat pump operating between the temperature limits of $T_L$ and $T_H$ can have.* All actual refrigerators or heat pumps operating between these temperature limits ($T_L$ and $T_H$) have lower coefficients of performance (Fig. 7–49).

The coefficients of performance of actual and reversible refrigerators operating between the same temperature limits can be compared as follows:

$$\text{COP}_R \begin{cases} < \text{COP}_{R,\text{rev}} & \text{irreversible refrigerator} \\ = \text{COP}_{R,\text{rev}} & \text{reversible refrigerator} \\ > \text{COP}_{R,\text{rev}} & \text{impossible refrigerator} \end{cases} \tag{7–22}$$

A similar relation can be obtained for heat pumps by replacing all $\text{COP}_R$'s in Eq. 7–22 by $\text{COP}_{\text{HP}}$.

The COP of a reversible refrigerator or heat pump is the maximum theoretical value for the specified temperature limits. Actual refrigerators or heat pumps may approach these values as their designs are improved, but they can never reach them.

As a final note, the COPs of both the refrigerators and the heat pumps decrease as $T_L$ decreases. That is, it requires more work to absorb heat from lower-temperature media. As the temperature of the refrigerated space approaches zero, the amount of work required to produce a finite amount of refrigeration approaches infinity and $\text{COP}_R$ approaches zero.

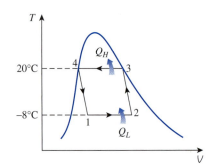

**FIGURE 7–50**

Schematic for Example 7–6.

**EXAMPLE 7–6**    **A Carnot Refrigeration Cycle Operating in the Saturation Dome**

A Carnot refrigeration cycle is executed in a closed system in the saturated liquid–vapor mixture region using 0.8 kg of refrigerant-134a as the working fluid (Fig. 7–50). The maximum and the minimum temperatures in the cycle are 20 and −8°C, respectively. It is known that the refrigerant is saturated liquid at the end of the heat rejection process, and the net work input to the cycle is 15 kJ. Determine the fraction of the mass of the refrigerant that vaporizes during the heat addition process, and the pressure at the end of the heat rejection process.

**SOLUTION** A Carnot refrigeration cycle is executed in a closed system. The mass fraction of the refrigerant that vaporizes during the heat addition process and the pressure at the end of the heat rejection process are to be determined.

*Assumptions* The refrigerator operates on the ideal Carnot cycle.

*Analysis* Knowing the high and low temperatures, the coefficient of performance of the cycle is

$$\text{COP}_R = \frac{1}{T_H/T_L - 1} = \frac{1}{(20 + 273\,\text{K})/(-8 + 273\,\text{K}) - 1} = 9.464$$

The amount of cooling is determined from the definition of the coefficient of performance to be

$$Q_L = \text{COP}_R \times W_{in} = (9.464)(15\,\text{kJ}) = 142\,\text{kJ}$$

The enthalpy of vaporization R-134a at −8°C is $h_{fg} = 204.52$ kJ/kg (Table A-11). Then the amount of refrigerant that vaporizes during heat absorption becomes

$$Q_L = m_{evap}h_{fg@-8°C} \rightarrow m_{evap} = \frac{142\,\text{kJ}}{204.52\,\text{kJ/kg}} = 0.694\,\text{kg}$$

Therefore, the fraction of mass that vaporized during heat addition process to the refrigerant is

$$\text{Mass fraction} = \frac{m_{evap}}{m_{total}} = \frac{0.694\,\text{kg}}{0.8\,\text{kg}} = \mathbf{0.868\ or\ 86.8\%}$$

The pressure at the end of heat rejection process is simply the saturation pressure at heat rejection temperature,

$$P_4 = P_{sat@20°C} = \mathbf{572.1\,kPa}$$

*Discussion* Carnot cycle is an idealized refrigeration cycle, thus it cannot be achieved in practice. Practical refrigeration cycles are analyzed in Chap. 9.

---

## EXAMPLE 7–7    Heating a House by a Carnot Heat Pump

A heat pump is to be used to heat a house during the winter, as shown in Fig. 7–51. The house is to be maintained at 21°C at all times. The house is estimated to be losing heat at a rate of 135,000 kJ/h when the outside temperature drops to −5°C. Determine the minimum power required to drive this heat pump.

**SOLUTION** A heat pump maintains a house at a constant temperature. The required minimum power input to the heat pump is to be determined.

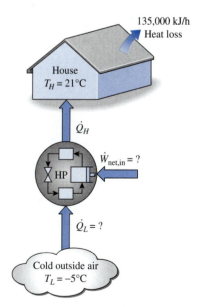

**FIGURE 7–51**

Schematic for Example 7–7.

**Assumptions** Steady operating conditions exist.

**Analysis** The heat pump must supply heat to the house at a rate of $Q_H =$ 135,000 kJ/h = 37.5 kW. The power requirements are minimum when a reversible heat pump is used to do the job. The COP of a reversible heat pump operating between the house and the outside air is

$$\text{COP}_{\text{HP,rev}} = \frac{1}{1 - T_L/T_H} = \frac{1}{1 - (-5 + 273 \text{ K})/(21 + 273 \text{ K})} = 11.3$$

Then, the required power input to this reversible heat pump becomes

$$\dot{W}_{\text{net,in}} = \frac{\dot{Q}_H}{\text{COP}_{\text{HP}}} = \frac{37.5 \text{ kW}}{11.3} = \textbf{3.32 kW}$$

**Discussion** This reversible heat pump can meet the heating requirements of this house by consuming electric power at a rate of 3.32 kW only. If this house were to be heated by electric resistance heaters instead, the power consumption would jump up 11.3 times to 37.5 kW. This is because in resistance heaters the electric energy is converted to heat at a one-to-one ratio. With a heat pump, however, energy is absorbed from the outside and carried to the inside using a refrigeration cycle that consumes only 3.32 kW. Notice that the heat pump does not create energy. It merely transports it from one medium (the cold outdoors) to another (the warm indoors).

# SUMMARY

The *second law of thermodynamics* states that processes occur in a certain direction, not in any direction. A process does not occur unless it satisfies both the first and the second laws of thermodynamics. Bodies that can absorb or reject finite amounts of heat isothermally are called *thermal energy reservoirs* or *heat reservoirs*.

Work can be converted to heat directly, but heat can be converted to work only by some devices called *heat engines*. The *thermal efficiency* of a heat engine is defined as

$$\eta_{\text{th}} = \frac{W_{\text{net,out}}}{Q_H} = 1 - \frac{Q_L}{Q_H}$$

where $W_{\text{net,out}}$ is the net work output of the heat engine, $Q_H$ is the amount of heat supplied to the engine, and $Q_L$ is the amount of heat rejected by the engine.

Refrigerators and heat pumps are devices that absorb heat from low-temperature media and reject it to higher-temperature ones. The performance of a refrigerator or a heat pump is expressed in terms of the *coefficient of performance*, which is defined as

$$\text{COP}_{\text{R}} = \frac{Q_L}{W_{\text{net,in}}} = \frac{1}{Q_H/Q_L - 1}$$

$$\text{COP}_{\text{HP}} = \frac{Q_H}{W_{\text{net,in}}} = \frac{1}{1 - Q_L/Q_H}$$

The *Kelvin–Planck statement* of the second law of thermodynamics states that no heat engine can produce a net amount of work while exchanging heat with a single reservoir only. The *Clausius statement* of the second law states that no device can transfer heat from a cooler body to a warmer one without leaving an effect on the surroundings.

Any device that violates the first or the second law of thermodynamics is called a *perpetual-motion machine*.

A process is said to be *reversible* if both the system and the surroundings can be restored to their original conditions. Any other process is *irreversible*. The effects such as friction, non-quasi-equilibrium expansion or compression, and heat transfer through a finite temperature difference render a process irreversible and are called *irreversibilities*.

The *Carnot cycle* is a reversible cycle that is composed of four reversible processes, two isothermal and two adiabatic. The *Carnot principles* state that the thermal efficiencies of all reversible heat engines operating between the same two reservoirs are the same, and that no heat engine is more efficient than a reversible one operating between the same two reservoirs. These statements form the basis for establishing a *thermodynamic temperature scale* related to the heat transfers between a reversible device and the high- and low-temperature reservoirs by

$$\left(\frac{Q_H}{Q_L}\right)_{\text{rev}} = \frac{T_H}{T_L}$$

Therefore, the $Q_H/Q_L$ ratio can be replaced by $T_H/T_L$ for reversible devices, where $T_H$ and $T_L$ are the absolute temperatures of the high- and low-temperature reservoirs, respectively.

A heat engine that operates on the reversible Carnot cycle is called a *Carnot heat engine*. The thermal efficiency of a Carnot heat engine, as well as all other reversible heat engines, is given by

$$\eta_{th,rev} = 1 - \frac{T_L}{T_H}$$

This is the maximum efficiency a heat engine operating between two reservoirs at temperatures $T_H$ and $T_L$ can have.

The COPs of reversible refrigerators and heat pumps are given in a similar manner as

$$COP_{R,rev} = \frac{1}{T_H/T_L - 1}$$

and

$$COP_{HP,rev} = \frac{1}{1' - T_L/T_H}$$

Again, these are the highest COPs a refrigerator or a heat pump operating between the temperature limits of $T_H$ and $T_L$ can have.

## REFERENCES AND SUGGESTED READINGS

**1.** *ASHRAE Handbook of Refrigeration,* SI version. Atlanta, GA: American Society of Heating, Refrigerating, and Air-Conditioning Engineers, Inc. 1994.

**2.** D. Stewart. "Wheels Go Round and Round, but Always Run Down." November 1986, *Smithsonian,* pp. 193–208.

## PROBLEMS*

### Second Law of Thermodynamics and Thermal Energy Reservoirs

**7–1C** Describe an imaginary process that satisfies the first law but violates the second law of thermodynamics.

**7–2C** Describe an imaginary process that satisfies the second law but violates the first law of thermodynamics.

**7–3C** Describe an imaginary process that violates both the first and the second laws of thermodynamics.

**7–4C** An experimentalist claims to have raised the temperature of a small amount of water to 150°C by transferring heat from high-pressure steam at 120°C. Is this a reasonable claim? Why? Assume no refrigerator or heat pump is used in the process.

**7–5C** What is a thermal energy reservoir? Give some examples.

**7–6C** Consider the process of baking potatoes in a conventional oven. Can the hot air in the oven be treated as a thermal energy reservoir? Explain.

* Problems designated by a "C" are concept questions, and students are encouraged to answer them all. Problems designated by an "E" are in English units, and the SI users can ignore them. Problems with the 🖭 icon are solved using EES, and complete solutions together with parametric studies are included on the enclosed DVD. Problems with the 🖳 icon are comprehensive in nature, and are intended to be solved with a computer, preferably using the EES software that accompanies this text.

**7–7C** Consider the energy generated by a TV set. What is a suitable choice for a thermal energy reservoir?

### Heat Engines and Thermal Efficiency

**7–8C** Is it possible for a heat engine to operate without rejecting any waste heat to a low-temperature reservoir? Explain.

**7–9C** What are the characteristics of all heat engines?

**7–10C** Baseboard heaters are basically electric resistance heaters and are frequently used in space heating. A home owner claims that her 5-year-old baseboard heaters have a conversion efficiency of 100 percent. Is this claim in violation of any thermodynamic laws? Explain.

**7–11C** What is the Kelvin–Planck expression of the second law of thermodynamics?

**7–12C** Does a heat engine that has a thermal efficiency of 100 percent necessarily violate (*a*) the first law and (*b*) the second law of thermodynamics? Explain.

**7–13C** In the absence of any friction and other irreversibilities, can a heat engine have an efficiency of 100 percent? Explain.

**7–14C** Are the efficiencies of all the work-producing devices, including the hydroelectric power plants, limited by the Kelvin–Planck statement of the second law? Explain.

**7–15C** Consider a pan of water being heated (*a*) by placing it on an electric range and (*b*) by placing a heating element in the water. Which method is a more efficient way of heating water? Explain.

**7–16E** A heat engine has a heat input of $3 \times 10^4$ Btu/h and a thermal efficiency of 40 percent. Calculate the power it will produce, in hp.

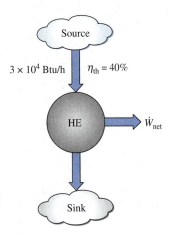

**FIGURE P7–16E**

**7–17** The thermal efficiency of a general heat engine is 40 percent, and it produces 30 hp. At what rate is heat transferred to this engine, in kJ/s?

**7–18** A 600-MW steam power plant, which is cooled by a nearby river, has a thermal efficiency of 40 percent. Determine the rate of heat transfer to the river water. Will the actual heat transfer rate be higher or lower than this value? Why?

**7–19** A heat engine that pumps water out of an underground mine accepts 700 kJ of heat and produces 250 kJ of work. How much heat does it reject, in kJ?

**7–20** A heat engine with a thermal efficiency of 40 percent rejects 1000 kJ/kg of heat. How much heat does it receive? *Answer:* 1667 kJ/kg

**7–21** A steam power plant with a power output of 150 MW consumes coal at a rate of 60 tons/h. If the heating value of the coal is 30,000 kJ/kg, determine the overall efficiency of this plant. *Answer:* 30.0 percent

**7–22** An automobile engine consumes fuel at a rate of 22 L/h and delivers 55 kW of power to the wheels. If the fuel has a heating value of 44,000 kJ/kg and a density of 0.8 g/cm³, determine the efficiency of this engine. *Answer:* 25.6 percent

**7–23** In 2001, the United States produced 51 percent of its electricity in the amount of $1.878 \times 10^{12}$ kWh from coal-fired power plants. Taking the average thermal efficiency to be 34 percent, determine the amount of thermal energy rejected by the coal-fired power plants in the United States that year.

**7–24** The Department of Energy projects that between the years 1995 and 2010, the United States will need to build new power plants to generate an additional 150,000 MW of electricity to meet the increasing demand for electric power. One possibility is to build coal-fired power plants, which cost $1300 per kW to construct and have an efficiency of 40 percent. Another possibility is to use the clean-burning Integrated Gasification Combined Cycle (IGCC) plants where the coal is subjected to heat and pressure to gasify it while removing sulfur and particulate matter from it. The gaseous coal is then burned in a gas turbine, and part of the waste heat from the exhaust gases is recovered to generate steam for the steam turbine. Currently the construction of IGCC plants costs about $1500 per kW, but their efficiency is about 48 percent. The average heating value of the coal is about 28,000,000 kJ per ton (that is, 28,000,000 kJ of heat is released when 1 ton of coal is burned). If the IGCC plant is to recover its cost difference from fuel savings in five years, determine what the price of coal should be in $ per ton.

**7–25** [EES] Reconsider Prob. 7–24. Using EES (or other) software, investigate the price of coal for varying simple payback periods, plant construction costs, and operating efficiency.

**7–26** Repeat Prob. 7–24 for a simple payback period of three years instead of five years.

**7–27E** Solar energy stored in large bodies of water, called solar pounds, is being used to generate electricity. If such a solar power plant has an efficiency of 4 percent and a net power output of 350 kW, determine the average value of the required solar energy collection rate, in Btu/h.

**7–28** A coal-burning steam power plant produces a net power of 300 MW with an overall thermal efficiency of 32 percent. The actual gravimetric air–fuel ratio in the furnace is calculated to be 12 kg air/kg fuel. The heating value of the coal is 28,000 kJ/kg. Determine (a) the amount of coal consumed during a 24-hour period and (b) the rate of air flowing through the furnace. *Answers:* (a) $2.89 \times 10^6$ kg, (b) 402 kg/s

## Refrigerators and Heat Pumps

**7–29C** What is the difference between a refrigerator and a heat pump?

**7–30C** What is the difference between a refrigerator and an air conditioner?

**7–31C** In a refrigerator, heat is transferred from a lower-temperature medium (the refrigerated space) to a higher-temperature one (the kitchen air). Is this a violation of the second law of thermodynamics? Explain.

**7–32C** A heat pump is a device that absorbs energy from the cold outdoor air and transfers it to the warmer indoors. Is this a violation of the second law of thermodynamics? Explain.

**7–33C** Define the coefficient of performance of a refrigerator in words. Can it be greater than unity?

**7–34C** Define the coefficient of performance of a heat pump in words. Can it be greater than unity?

**7–35C** A heat pump that is used to heat a house has a COP of 2.5. That is, the heat pump delivers 2.5 kWh of energy to the house for each 1 kWh of electricity it consumes. Is this a violation of the first law of thermodynamics? Explain.

**7–36C** A refrigerator has a COP of 1.5. That is, the refrigerator removes 1.5 kWh of energy from the refrigerated space for each 1 kWh of electricity it consumes. Is this a violation of the first law of thermodynamics? Explain.

**7–37C** What is the Clausius expression of the second law of thermodynamics?

**7–38C** Show that the Kelvin–Planck and the Clausius expressions of the second law are equivalent.

**7–39E** A residential heat pump has a coefficient of performance of 2.4. How much heating effect, in Btu/h, will result when 5 hp is supplied to this heat pump?

**FIGURE P7–39E**

**7–40** An air conditioner produces a 2-kW cooling effect while rejecting 2.5 kW of heat. What is its COP?

**7–41** A refrigerator used to cool a computer requires 3 kW of electrical power and has a COP of 1.4. Calculate the cooling effect of this refrigerator, in kW.

**7–42** A food department is kept at $-12°C$ by a refrigerator in an environment at 30°C. The total heat gain to the food department is estimated to be 3300 kJ/h and the heat rejection in the condenser is 4800 kJ/h. Determine the power input to the compressor, in kW and the COP of the refrigerator.

**FIGURE P7–42**

**7–43** A household refrigerator with a COP of 1.2 removes heat from the refrigerated space at a rate of 60 kJ/min. Determine (*a*) the electric power consumed by the refrigerator and (*b*) the rate of heat transfer to the kitchen air. *Answers: (a)* 0.83 kW, *(b)* 110 kJ/min

**7–44E** A commercial heat pump removes 10,000 Btu/h from the source, rejects 15,090 Btu/h to the sink, and requires 2 hp of power. What is this heat pump's coefficient of performance?

**7–45** A refrigerator used for cooling food in a grocery store is to produce a 25,000 kJ/h cooling effect, and it has a coefficient of performance of 1.60. How many kilowatts of power will this refrigerator require to operate? *Answer:* 4.34 kW

**7–46** A household refrigerator that has a power input of 450 W and a COP of 2.5 is to cool five large watermelons, 10 kg each, to 8°C. If the watermelons are initially at 20°C, determine how long it will take for the refrigerator to cool them. The watermelons can be treated as water whose specific heat is 4.2 kJ/kg·°C. Is your answer realistic or optimistic? Explain. *Answer:* 2240 s

**7–47** Bananas are to be cooled from 24 to 13°C at a rate of 215 kg/h by a refrigeration system. The power input to the refrigerator is 1.4 kW. Determine the rate of cooling, in kJ/min, and the COP of the refrigerator. The specific heat of banana above freezing is 3.35 kJ/kg·°C.

**7–48** A refrigerator is used to cool water from 23 to 5°C in a continuous manner. The heat rejected in the condenser is 570 kJ/min and the power is 2.65 kW. Determine the rate at which water is cooled, in L/min and the COP of the refrigerator. The specific heat of water is 4.18 kJ/kg·°C and its density is 1 kg/L. *Answers:* 5.46 L/min, 2.58

**7–49** A heat pump is used to maintain a house at a constant temperature of 23°C. The house is losing heat to the outside air through the walls and the windows at a rate of 60,000 kJ/h while the energy generated within the house from people, lights, and appliances amounts to 4000 kJ/h. For a COP of 2.5, determine the required power input to the heat pump.
*Answer: 6.22 kW*

FIGURE P7–49

**7–50E** Water enters an ice machine at 55°F and leaves as ice at 25°F. If the COP of the ice machine is 2.4 during this operation, determine the required power input for an ice production rate of 28 Ibm/h. (169 Btu of energy needs to be removed from each Ibm of water at 55°F to turn it into ice at 25°F.)

**7–51E** Consider an office room that is being cooled adequately by a 12,000 Btu/h window air conditioner. Now it is decided to convert this room into a computer room by installing several computers, terminals, and printers with a total rated power of 8.4 kW. The facility has several 7000 Btu/h air conditioners in storage that can be installed to meet the additional cooling requirements. Assuming a usage factor of 0.4 (i.e., only 40 percent of the rated power will be consumed at any given time) and additional occupancy of seven people, each generating heat at a rate of 100 W, determine how many of these air conditioners need to be installed to the room.

**7–52** Consider a building whose annual air-conditioning load is estimated to be 120,000 kWh in an area where the unit cost of electricity is $0.10/kWh. Two air conditioners are considered for the building. Air conditioner A has a seasonal average COP of 3.2 and costs $5500 to purchase and install. Air conditioner B has a seasonal average COP of 5.0 and costs $7000 to purchase and install. All else being equal, determine which air conditioner is a better buy.

FIGURE P7–52

**7–53** A house that was heated by electric resistance heaters consumed 1200 kWh of electric energy in a winter month. If this house were heated instead by a heat pump that has an average COP of 2.4, determine how much money the home owner would have saved that month. Assume a price of 8.5¢/kWh for electricity.

**7–54** Refrigerant-134a enters the condenser of a residential heat pump at 800 kPa and 35°C at a rate of 0.018 kg/s and leaves at 800 kPa as a saturated liquid. If the compressor consumes 1.2 kW of power, determine (*a*) the COP of the heat pump and (*b*) the rate of heat absorption from the outside air.

FIGURE P7–54

## Reversible and Irreversible Processes

**7–55C** Show that processes that use work for mixing are irreversible by considering an adiabatic system whose contents are stirred by turning a paddle wheel inside the system (e.g., stirring a cake mix with an electric mixer).

**7–56C** Show that processes involving rapid chemical reactions are irreversible by considering the combustion of a natural gas (e.g., methane) and air mixture in a rigid container.

**7–57C** A cold canned drink is left in a warmer room where its temperature rises as a result of heat transfer. Is this a reversible process? Explain.

**7–58C** A block slides down an inclined plane with friction and no restraining force. Is this process reversible or irreversible? Justify your answer.

**7–59C** Why are engineers interested in reversible processes even though they can never be achieved?

**7–60C** Why does a nonquasi-equilibrium compression process require a larger work input than the corresponding quasi-equilibrium one?

**7–61C** Why does a nonquasi-equilibrium expansion process deliver less work than the corresponding quasi-equilibrium one?

**7–62C** How do you distinguish between internal and external irreversibilities?

**7–63C** Is a reversible expansion or compression process necessarily quasi-equilibrium? Is a quasi-equilibrium expansion or compression process necessarily reversible? Explain.

## The Carnot Cycle and Carnot Principles

**7–64C** What are the four processes that make up the Carnot cycle?

**7–65C** What are the two statements known as the Carnot principles?

**7–66C** Somebody claims to have developed a new reversible heat-engine cycle that has a higher theoretical efficiency than the Carnot cycle operating between the same temperature limits. How do you evaluate this claim?

**7–67C** Somebody claims to have developed a new reversible heat-engine cycle that has the same theoretical efficiency as the Carnot cycle operating between the same temperature limits. Is this a reasonable claim?

**7–68C** Is it possible to develop (a) an actual and (b) a reversible heat-engine cycle that is more efficient than a Carnot cycle operating between the same temperature limits? Explain.

## Carnot Heat Engines

**7–69C** Is there any way to increase the efficiency of a Carnot heat engine other than by increasing $T_H$ or decreasing $T_L$?

**7–70C** Consider two actual power plants operating with solar energy. Energy is supplied to one plant from a solar pond at 80°C and to the other from concentrating collectors that raise the water temperature to 600°C. Which of these power plants will have a higher efficiency? Explain.

**7–71E** A heat engine operates between two thermal-energy reservoirs at 1260 R and 510 R. What is the maximum work the engine can produce per unit of heat that the engine receives from the source?

**7–72** From a work-production perspective, which is more valuable: (a) thermal energy reservoirs at 675 K and 325 K or (b) thermal energy reservoirs at 625 K and 275 K?

**7–73** A heat engine operates between a source at 477°C and a sink at 25°C. If heat is supplied to the heat engine at a steady rate of 65,000 kJ/min, determine the maximum power output of this heat engine.

**7–74** Reconsider Prob. 7–73. Using EES (or other) software, study the effects of the temperatures of the heat source and the heat sink on the power produced and the cycle thermal efficiency. Let the source temperature vary from 300 to 1000°C, and the sink temperature to vary from 0 to 50°C. Plot the power produced and the cycle efficiency against the source temperature for sink temperatures of 0°C, 25°C, and 50°C, and discuss the results.

**7–75E** A heat engine is operating on a Carnot cycle and has a thermal efficiency of 75 percent. The waste heat from this engine is rejected to a nearby lake at 60°F at a rate of 800 Btu/min. Determine (a) the power output of the engine and (b) the temperature of the source. *Answers:* (a) 56.6 hp, (b) 2080 R

**FIGURE P7–75E**

**7–76E** An inventor claims to have devised a cyclical engine for use in space vehicles that operates with a nuclear-fuel-generated energy source whose temperature is 1000 R and a sink at 550 R that radiates waste heat to deep space. He also claims that this engine produces 5 hp while rejecting heat at a rate of 15,000 Btu/h. Is this claim valid?

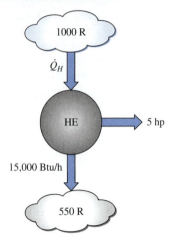

**FIGURE P7–76E**

**7–77** A heat engine receives heat from a heat source at 1200°C and has a thermal efficiency of 40 percent. The heat engine does maximum work equal to 500 kJ. Determine the heat supplied to the heat engine by the heat source, the heat rejected to the heat sink, and the temperature of the heat sink.

**7–78** A heat engine operating on the Carnot cycle has a measured work output of 900 kJ and heat rejection of 150 kJ to a heat reservoir at 27°C. Determine the heat supplied to the heat engine by the heat source, in kJ, and the temperature of the heat source, in °C.

**7–79** A heat engine is operating on a Carnot cycle and has a thermal efficiency of 75 percent. The waste heat from this engine is rejected to a nearby lake at 15°C at a rate of 14 kW. Determine the power output of the engine and the temperature of the source, in °C. *Answers: 42 kW, 879°C.*

**7–80** A geothermal power plant uses geothermal water extracted at 150°C at a rate of 210 kg/s as the heat source and produces 8000 kW of net power. The geothermal water leaves the plant at 90°C. If the environment temperature is 25°C, determine (*a*) the actual thermal efficiency, (*b*) the maximum possible thermal efficiency, and (*c*) the actual rate of heat rejection from this power plant.

**7–81** It is claimed that the efficiency of a completely reversible heat engine can be doubled by doubling the temperature of the energy source. Justify the validity of this claim.

## Carnot Refrigerators and Heat Pumps

**7–82C** How can we increase the COP of a Carnot refrigerator?

**7–83C** A homeowner buys a new refrigerator and a new air conditioner. Which one of these devices would you expect to have a higher COP? Why?

**7–84C** A homeowner buys a new refrigerator with no freezer compartment and a deep freezer for the new kitchen. Which of these devices would you expect to have a lower COP? Why?

**7–85C** In an effort to conserve energy in a heat-engine cycle, somebody suggests incorporating a refrigerator that will absorb some of the waste energy $Q_L$ and transfer it to the energy source of the heat engine. Is this a smart idea? Explain.

**7–86C** It is well established that the thermal efficiency of a heat engine increases as the temperature $T_L$ at which heat is rejected from the heat engine decreases. In an effort to increase the efficiency of a power plant, somebody suggests refrigerating the cooling water before it enters the condenser, where heat rejection takes place. Would you be in favor of this idea? Why?

**7–87C** It is well known that the thermal efficiency of heat engines increases as the temperature of the energy source increases. In an attempt to improve the efficiency of a power plant, somebody suggests transferring heat from the available energy source to a higher-temperature medium by a heat pump before energy is supplied to the power plant. What do you think of this suggestion? Explain.

**7–88** Determine the minimum work per unit of heat transfer from the source that is required to drive a refrigerator with thermal-energy reservoirs at 273 K and 303 K.

**7–89** An inventor claims to have developed a heat pump that produces a 200-kW heating effect for a 293 K heated zone while only using 75 kW of power and a heat source at 273 K. Justify the validity of this claim.

**7–90** A heat pump operates on a Carnot heat pump cycle with a COP of 8.7. It keeps a space at 26°C by consuming 4.25 kW of power. Determine the temperature of the reservoir from which the heat is absorbed and the heating load provided by the heat pump. *Answers: 265 K, 37.0 kW*

**7–91** A refrigerator is to remove heat from the cooled space at a rate of 300 kJ/min to maintain its temperature at −8°C. If the air surrounding the refrigerator is at 25°C, determine the minimum power input required for this refrigerator. *Answer: 0.623 kW*

**FIGURE P7–91**

**7–92** An inventor claims to have developed a refrigeration system that removes heat from the closed region at $-12°C$ and transfers it to the surrounding air at 25°C while maintaining a COP of 6.5. Is this claim reasonable? Why?

**7–93E** An air-conditioning system is used to maintain a house at 70°F when the temperature outside is 100°F. The house is gaining heat through the walls and the windows at a rate of 800 Btu/min, and the heat generation rate within the house from people, lights, and appliances amounts to 100 Btu/min. Determine the minimum power input required for this air-conditioning system. *Answer:* 1.20 hp

**7–94** A heat pump is used to maintain a house at 22°C by extracting heat from the outside air on a day when the outside air temperature is 2°C. The house is estimated to lose heat at a rate of 110,000 kJ/h, and the heat pump consumes 5 kW of electric power when running. Is this heat pump powerful enough to do the job?

**FIGURE P7–94**

**7–95E** A completely reversible refrigerator operates between thermal energy reservoirs at 450 R and 540 R. How many kilowatts of power are required for this device to produce a 15,000-Btu/h cooling effect?

**7–96** A refrigerator operating on the reversed Carnot cycle has a measured work input of 200 kW and heat rejection of 2000 kW to a heat reservoir at 27°C. Determine the cooling load supplied to the refrigerator, in kW, and the temperature of the heat source, in °C. *Answers:* 1800 kW, $-3°C$

**7–97** An 18,000 Btu/h split air conditioner running at full load is used to keep a room at 22°C in an environment at

33°C. The power input to the compressor is 3.4 kW. Determine (*a*) the rate of heat rejected in the condenser, in kJ/h, (*b*) the COP of the air conditioner, and (*c*) the rate of cooling, in Btu/h, if the air conditioner operated as a Carnot refrigerator for the same power input.

**7–98** A Carnot refrigerator absorbs heat from a space at 15°C at a rate of 16,000 kJ/h and rejects heat to a reservoir at 36°C. Determine the COP of the refrigerator, the power input, in kW, and the rate of heat rejected to high-temperature reservoir, in kJ/h.

**FIGURE P7–98**

**7–99** The performance of a heat pump degrades (i.e., its COP decreases) as the temperature of the heat source decreases. This makes using heat pumps at locations with severe weather conditions unattractive. Consider a house that is heated and maintained at 20°C by a heat pump during the winter. What is the maximum COP for this heat pump if heat is extracted from the outdoor air at (*a*) 10°C, (*b*) $-5°C$, and (*c*) $-30°C$?

**7–100E** A heat pump is to be used for heating a house in the winter. The house is to be maintained at 78°F at all times. When the temperature outdoors drops to 25°F, the heat losses from the house are estimated to be 55,000 Btu/h. Determine the minimum power required to run this heat pump if heat is extracted from (*a*) the outdoor air at 25°F and (*b*) the well water at 50°F.

**7–101** A Carnot heat pump is to be used to heat a house and maintain it at 25°C in winter. On a day when the average outdoor temperature remains at about 2°C, the house is estimated to lose heat at a rate of 55,000 kJ/h. If the heat pump consumes 6.6 kW of power while operating, determine (*a*) how long the heat pump ran on that day; (*b*) the total heating costs, assuming an average price of 8.5¢/kWh for electricity; and (*c*) the heating cost for the same day if resistance heating is used instead of a heat pump. *Answers:* (*a*) 4.29 h, (*b*) $2.41, (*c*) $31.2

**FIGURE P7–101**

**FIGURE P7–105**

**7–102** A Carnot heat engine receives heat from a reservoir at 900°C at a rate of 800 kJ/min and rejects the waste heat to the ambient air at 27°C. The entire work output of the heat engine is used to drive a refrigerator that removes heat from the refrigerated space at −5°C and transfers it to the same ambient air at 27°C. Determine (a) the maximum rate of heat removal from the refrigerated space and (b) the total rate of heat rejection to the ambient air. *Answers: (a) 4982 kJ/min, (b) 5782 kJ/min*

**7–103E** A Carnot heat engine receives heat from a reservoir at 1700°F at a rate of 700 Btu/min and rejects the waste heat to the ambient air at 80°F. The entire work output of the heat engine is used to drive a refrigerator that removes heat from the refrigerated space at 20°F and transfers it to the same ambient air at 80°F. Determine (a) the maximum rate of heat removal from the refrigerated space and (b) the total rate of heat rejection to the ambient air. *Answers: (a) 4200 Btu/min, (b) 4900 Btu/min*

**7–104** The structure of a house is such that it loses heat at a rate of 3800 kJ/h per °C difference between the indoors and outdoors. A heat pump that requires a power input of 4 kW is used to maintain this house at 24°C. Determine the lowest outdoor temperature for which the heat pump can meet the heating requirements of this house. *Answer: −9.5°C*

**7–105** An air-conditioner with refrigerant-134a as the working fluid is used to keep a room at 23°C by rejecting the waste heat to the outdoor air at 37°C. The room gains heat through the walls and the windows at a rate of 250 kJ/min while the heat generated by the computer, TV, and lights amounts to 900 W. The refrigerant enters the compressor at 400 kPa as a saturated vapor at a rate of 100 L/min and leaves at 1200 kPa and 70°C. Determine (a) the actual COP, (b) the maximum COP, and (c) the minimum volume flow rate of the refrigerant at the compressor inlet for the same compressor inlet and exit conditions. *Answers: (a) 3.45, (b) 21.1, (c) 16.3 L/min*

**7–106** Calculate and plot the COP of a completely reversible refrigerator as a function of the temperature of the sink up to 500 K with the temperature of the source fixed at 250 K.

**7–107** Derive an expression for the COP of a completely reversible refrigerator in terms of the thermal energy reservoir temperatures, $T_L$ and $T_H$.

## Review Problems

**7–108** A heat engine operates with a source at 1280 K and a sink at 290 K. What is the maximum work per unit of heat that the engine can remove from the source?

**7–109** A heat engine receives heat from a heat source at 1200°C and rejects heat to a heat sink at 50°C. The heat engine does maximum work equal to 500 kJ. Determine the heat supplied to the heat engine by the heat source and the heat rejected to the heat sink.

**7–110E** A heat pump creates a heating effect of 32,000 Btu/h for a space maintained at 530 R while using 1.8 kW of electrical power. What is the minimum temperature of the source that satisfies the second law of thermodynamics? *Answer: 428 R*

**7–111** A heat pump with a COP of 2.4 is used to heat a house. When running, the heat pump consumes 8 kW of electric power. If the house is losing heat to the outside at an average rate of 40,000 kJ/h and the temperature of the house is 3°C when the heat pump is turned on, determine how long it will take for the temperature in the house to rise to 22°C. Assume the house is well sealed (i.e., no air leaks) and take the entire mass within the house (air, furniture, etc.) to be equivalent to 2000 kg of air.

**7–112E** A refrigeration system uses water-cooled condenser for rejecting the waste heat. The system absorbs heat from a space at 25°F at a rate of 24,000 Btu/h. Water enters the condenser at 58°F at a rate of 1.45 lbm/s. The COP of the system is estimated to be 1.77. Determine (a) the power input to the system, in kW, (b) the temperature of the water at the exit of the condenser, in °F and (c) the maximum possible COP of the system. The specific heat of water is 1.0 Btu/bm·°F.

**7–113** Consider a Carnot heat-engine cycle executed in a closed system using 0.01 kg of refrigerant-134a as the working fluid. The cycle has a thermal efficiency of 15 percent, and the refrigerant-134a changes from saturated liquid to saturated vapor at 50°C during the heat addition process. Determine the net work output of this engine per cycle.

**7–114** A heat pump with a COP of 2.8 is used to heat an air-tight house. When running, the heat pump consumes 5 kW of power. If the temperature in the house is 7°C when the heat pump is turned on, how long will it take for the heat pump to raise the temperature of the house to 22°C? Is this answer realistic or optimistic? Explain. Assume the entire mass within the house (air, furniture, etc.) is equivalent to 1500 kg of air. *Answer:* 19.2 min

**7–115** Consider a Carnot heat-engine cycle executed in a closed system using 0.025 kg of steam as the working fluid. It is known that the maximum absolute temperature in the cycle is twice the minimum absolute temperature, and the net work output of the cycle is 60 kJ. If the steam changes from saturated vapor to saturated liquid during heat rejection, determine the temperature of the steam during the heat rejection process.

**7–116** Reconsider Prob. 7–115. Using EES (or other) software, investigate the effect of the net work output on the required temperature of the steam during the heat rejection process. Let the work output vary from 40 to 60 kJ.

**7–117** Consider a Carnot refrigeration cycle executed in a closed system in the saturated liquid–vapor mixture region using 0.96 kg of refrigerant-134a as the working fluid. It is known that the maximum absolute temperature in the cycle is 1.2 times the minimum absolute temperature, and the net work input to the cycle is 22 kJ. If the refrigerant changes from saturated vapor to saturated liquid during the heat rejection process, determine the minimum pressure in the cycle.

**7–118** Reconsider Prob. 7–117. Using EES (or other) software, investigate the effect of the net work input on the minimum pressure. Let the work input vary from 10 to 30 kJ. Plot the minimum pressure in the refrigeration cycle as a function of net work input, and discuss the results.

**7–119** Consider two Carnot heat engines operating in series. The first engine receives heat from the reservoir at 1800 K and rejects the waste heat to another reservoir at temperature $T$. The second engine receives this energy rejected by the first one, converts some of it to work, and rejects the rest to a reservoir at 300 K. If the thermal efficiencies of both engines are the same, determine the temperature $T$. *Answer:* 735 K

**7–120E** Calculate and plot the thermal efficiency of a completely reversible heat engine as a function of the source temperature up to 2000 R with the sink temperature fixed at 500 R.

**7–121** A Carnot heat engine receives heat at 900 K and rejects the waste heat to the environment at 300 K. The entire work output of the heat engine is used to drive a Carnot refrigerator that removes heat from the cooled space at −15°C at a rate of 250 kJ/min and rejects it to the same environment at 300 K. Determine (a) the rate of heat supplied to the heat engine and (b) the total rate of heat rejection to the environment.

**7–122** Reconsider Prob. 7–121. Using EES (or other) software, investigate the effects of the heat engine source temperature, the environment temperature, and the cooled space temperature on the required heat supply to the heat engine and the total rate of heat rejection to the environment. Let the source temperature vary from 500 to 1000 K, the environment temperature vary from 275 to 325 K, and the cooled space temperature vary from −20 to 0°C. Plot the required heat supply against the source temperature for the cooled space temperature of −15°C and environment temperatures of 275, 300, and 325 K, and discuss the results.

**7–123** A heat engine operates between two reservoirs at 800 and 20°C. One-half of the work output of the heat engine is used to drive a Carnot heat pump that removes heat from the cold surroundings at 2°C and transfers it to a house maintained at 22°C. If the house is losing heat at a rate of 62,000 kJ/h, determine the minimum rate of heat supply to the heat engine required to keep the house at 22°C.

**7–124E** An inventor claims to have developed a refrigerator that maintains the refrigerated space at 35°F while operating in a room where the temperature is 75°F and that has a COP of 13.5. Is this claim reasonable?

**7–125** Consider a Carnot heat-pump cycle executed in a steady-flow system in the saturated liquid–vapor mixture region using refrigerant-134a flowing at a rate of 0.22 kg/s as the working fluid. It is known that the maximum absolute temperature in the cycle is 1.2 times the minimum absolute temperature, and the net power input to the cycle is 5 kW. If the refrigerant changes from saturated vapor to saturated liquid during the heat rejection process, determine the ratio of the maximum to minimum pressures in the cycle.

**7–126** Replacing incandescent lights with energy-efficient fluorescent lights can reduce the lighting energy consumption to one-fourth of what it was before. The energy consumed by the lamps is eventually converted to heat, and thus switching to energy-efficient lighting also reduces the cooling load in summer but increases the heating load in winter. Consider a building that is heated by a natural gas furnace with an efficiency of 80 percent and cooled by an air conditioner with a COP of 3.5. If electricity costs

$0.08/kWh and natural gas costs $1.40/therm (1 therm = 105,500 kJ), determine if efficient lighting will increase or decrease the total energy cost of the building (a) in summer and (b) in winter.

**7–127** A heat pump supplies heat energy to a house at the rate of 140,000 kJ/h when the house is maintained at 25°C. Over a period of one month, the heat pump operates for 100 hours to transfer energy from a heat source outside the house to inside the house. Consider a heat pump receiving heat from two different outside energy sources. In one application the heat pump receives heat from the outside air at 0°C. In a second application the heat pump receives heat from a lake having a water temperature of 10°C. If electricity costs $0.085/kWh, determine the maximum money saved by using the lake water rather than the outside air as the outside energy source.

**7–128** The cargo space of a refrigerated truck whose inner dimensions are 12 m × 2.3 m × 3.5 m is to be precooled from 25°C to an average temperature of 5°C. The construction of the truck is such that a transmission heat gain occurs at a rate of 80 W/°C. If the ambient temperature is 25°C, determine how long it will take for a system with a refrigeration capacity of 8 kW to precool this truck.

80 W/°C          25°C

Refrigerated truck
12 m × 2.3 m × 3.5 m

25 to 5°C

**FIGURE P7–128**

**7–129** A refrigeration system is to cool bread loaves with an average mass of 350 g from 30 to −10°C at a rate of 1200 loaves per hour by refrigerated air at −30°C. Taking the average specific and latent heats of bread to be 2.93 kJ/kg·°C and 109.3 kJ/kg, respectively, determine (a) the rate of heat removal from the breads, in kJ/h; (b) the required volume

flow rate of air, in m³/h, if the temperature rise of air is not to exceed 8°C; and (c) the size of the compressor of the refrigeration system, in kW, for a COP of 1.2 for the refrigeration system.

**7–130** The drinking water needs of a production facility with 20 employees is to be met by a bubbler type water fountain. The refrigerated water fountain is to cool water from 22 to 8°C and supply cold water at a rate of 0.4 L per hour per person. Heat is transferred to the reservoir from the surroundings at 25°C at a rate of 45 W. If the COP of the refrigeration system is 2.9, determine the size of the compressor, in W, that will be suitable for the refrigeration system of this water cooler.

Cold water
8°C

Water inlet
22°C
0.4 L/h·person

Water reservoir

25°C

Refrigeration system

Water fountain

**FIGURE P7–130**

**7–131E** The "Energy Guide" label on a washing machine indicates that the washer will use $33 worth of hot water if the water is heated by a gas water heater at a natural gas rate of $1.21/therm. If the water is heated from 60 to 130°F, determine how many gallons of hot water an average family uses per week. Disregard the electricity consumed by the washer, and take the efficiency of the gas water heater to be 58 percent.

**7–132** A typical electric water heater has an efficiency of 95 percent and costs $250 a year to operate at a unit cost of electricity of $0.08/kWh. A typical heat pump–powered water heater has a COP of 3.3 but costs about $800 more to install. Determine how many years it will take for the heat pump water heater to pay for its cost differential from the energy it saves.

**FIGURE P7–135**

Water
heater

© *The McGraw-Hill Companies, Inc.*
*Jill Braaten, photographer.*

**FIGURE P7–132**

**7–133** Reconsider Prob. 7–132. Using EES (or other) software, investigate the effect of the heat pump COP on the yearly operation costs and the number of years required to break even. Let the COP vary from 2 to 5. Plot the payback period against the COP and discuss the results.

**7–134** A homeowner is trying to decide between a high-efficiency natural gas furnace with an efficiency of 97 percent and a ground-source heat pump with a COP of 3.5. The unit costs of electricity and natural gas are $0.092/kWh and $1.42/therm (1 therm = 105,500 kJ). Determine which system will have a lower energy cost.

**7–135** Cold water at 10°C enters a water heater at the rate of 0.02 m³/min and leaves the water heater at 50°C. The water heater receives heat from a heat pump that receives heat from a heat source at 0°C.

(*a*) Assuming the water to be an incompressible liquid that does not change phase during heat addition, determine the rate of heat supplied to the water, in kJ/s.

(*b*) Assuming the water heater acts as a heat sink having an average temperature of 30°C, determine the minimum power supplied to the heat pump, in kW.

**7–136** A heat pump receives heat from a lake that has an average winter time temperature of 6°C and supplies heat into a house having an average temperature of 27°C.

(*a*) If the house loses heat to the atmosphere at the rate of 64,000 kJ/h, determine the minimum power supplied to the heat pump, in kW.

(*b*) A heat exchanger is used to transfer the energy from the lake water to the heat pump. If the lake water temperature decreases by 5°C as it flows through the lake water-to-heat pump heat exchanger, determine the minimum mass flow rate of lake water, in kg/s. Neglect the effect of the lake water pump.

**FIGURE P7–136**

**7–137** Prove that a refrigerator's COP cannot exceed that of a completely reversible refrigerator that shares the same thermal energy reservoirs.

**7–138** Prove that the COP of all completely reversible refrigerators must be the same when the reservoir temperatures are the same.

**7–139** Derive an expression for the COP of a completely reversible heat pump in terms of the thermal energy reservoir temperatures, $T_L$ and $T_H$.

**7–140** A Carnot heat engine is operating between a source at $T_H$ and a sink at $T_L$. If it is desired to double the thermal efficiency of this engine, what should the new source temperature be? Assume the sink temperature is held constant.

## Design and Essay Problems

**7–141** The sun supplies electromagnetic energy to the earth. It appears to have an effective temperature of approximately 5800 K. On a clear summer day in North America, the energy incident on a surface facing the sun is approximately 0.95 kW/m2. The electromagnetic solar energy can be converted into thermal energy by being absorbed on a darkened surface. How might you characterize the work potential of the sun's energy when it is to be used to produce work?

**7–142** In the search to reduce thermal pollution and take advantage of renewable energy sources, some people have proposed that we take advantage of such sources as discharges from electrical power plants, geothermal energy, and ocean thermal energy. Although many of these sources contain an enormous amount of energy, the amount of work they are capable of producing is limited. How might you use the work potential to assign an "energy quality" to these proposed sources? Test your proposed "energy quality" measure by applying it to the ocean thermal source, where the temperature 30 m below the surface is perhaps 5°C lower than at the surface. Apply it also to the geothermal water source, where the temperature 2 to 3 km below the surface is perhaps 150°C hotter than at the surface.

**7–143** The maximum work that can be produced by a heat engine may be increased by lowering the temperature of its sink. Hence, it seems logical to use refrigerators to cool the sink and reduce its temperature. What is the fallacy in this logic?

**7–144** Using a timer (or watch) and a thermometer, conduct the following experiment to determine the rate of heat gain of your refrigerator. First make sure that the door of the refrigerator is not opened for at least a few hours so that steady operating conditions are established. Start the timer when the refrigerator stops running and measure the time $\Delta t_1$ it stays off before it kicks in. Then, measure the time $\Delta t_2$ it stays on. Noting that the heat removed during $\Delta t_2$ is equal to the heat gain of the refrigerator during $\Delta t_1 + \Delta t_2$ and using the power consumed by the refrigerator when it is running, determine the average rate of heat gain for your refrigerator, in W. Take the COP (coefficient of performance) of your refrigerator to be 1.3 if it is not available.

**7–145** Design a hydrocooling unit that can cool fruits and vegetables from 30 to 5°C at a rate of 20,000 kg/h under the following conditions:

The unit will be of flood type, which will cool the products as they are conveyed into the channel filled with water. The products will be dropped into the channel filled with water at one end and be picked up at the other end. The channel can be as wide as 3 m and as high as 90 cm. The water is to be circulated and cooled by the evaporator section of a refrigeration system. The refrigerant temperature inside the coils is to be −2°C, and the water temperature is not to drop below 1°C and not to exceed 6°C.

Assuming reasonable values for the average product density, specific heat, and porosity (the fraction of air volume in a box), recommend reasonable values for (a) the water velocity through the channel and (b) the refrigeration capacity of the refrigeration system.

# ENTROPY

I n Chap. 7, we introduced the second law of thermodynamics and applied it to cycles and cyclic devices. In this chapter, we apply the second law to processes. The first law of thermodynamics deals with the property *energy* and the conservation of it. The second law leads to the definition of a new property called *entropy*. Entropy is a somewhat abstract property, and it is difficult to give a physical description of it without considering the microscopic state of the system. Entropy is best understood and appreciated by studying its uses in commonly encountered engineering processes, and this is what we intend to do.

This chapter starts with a discussion of the Clausius inequality, which forms the basis for the definition of entropy, and continues with the increase of entropy principle. Unlike energy, entropy is a nonconserved property, and there is no such thing as *conservation of entropy*. Next, the entropy changes that take place during processes for pure substances, incompressible substances, and ideal gases are discussed, and a special class of idealized processes, called *isentropic processes*, is examined. Then, the reversible steady-flow work and the isentropic efficiencies of various engineering devices such as turbines and compressors are considered. Finally, entropy balance is introduced and applied to various systems.

## Objectives

The Objectives of this chapter are to:

- Apply the second law of thermodynamics to processes.
- Define a new property called *entropy* to quantify the second-law effects.
- Establish the *increase of entropy principle*.
- Calculate the entropy changes that take place during processes for pure substances, incompressible substances, and ideal gases.
- Examine a special class of idealized processes, called *isentropic processes*, and develop the property relations for these processes.
- Derive the reversible steady-flow work relations.
- Develop the isentropic efficiencies for various steady-flow devices.
- Introduce and apply the entropy balance to various systems.

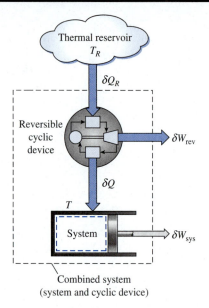

Thermal reservoir
$T_R$

$\delta Q_R$

Reversible
cyclic
device

$\delta W_{rev}$

$\delta Q$

$T$

System

$\delta W_{sys}$

Combined system
(system and cyclic device)

**FIGURE 8–1**

The system considered in the development of the Clausius inequality.

# 8–1 ■ ENTROPY

The second law of thermodynamics often leads to expressions that involve inequalities. An irreversible (i.e., actual) heat engine, for example, is less efficient than a reversible one operating between the same two thermal energy reservoirs. Likewise, an irreversible refrigerator or a heat pump has a lower coefficient of performance (COP) than a reversible one operating between the same temperature limits. Another important inequality that has major consequences in thermodynamics is the **Clausius inequality**. It was first stated by the German physicist R. J. E. Clausius (1822–1888), one of the founders of thermodynamics, and is expressed as

$$\oint \frac{\delta Q}{T} \leq 0$$

That is, *the cyclic integral of δQ/T is always less than or equal to zero.* This inequality is valid for all cycles, reversible or irreversible. The symbol $\oint$ (integral symbol with a circle in the middle) is used to indicate that the integration is to be performed over the entire cycle. Any heat transfer to or from a system can be considered to consist of differential amounts of heat transfer. Then the cyclic integral of δQ/T can be viewed as the sum of all these differential amounts of heat transfer divided by the temperature at the boundary.

To demonstrate the validity of the Clausius inequality, consider a system connected to a thermal energy reservoir at a constant thermodynamic (i.e., absolute) temperature of $T_R$ through a *reversible* cyclic device (Fig. 8–1). The cyclic device receives heat $\delta Q_R$ from the reservoir and supplies heat $\delta Q$ to the system whose temperature at that part of the boundary is $T$ (a variable) while producing work $\delta W_{rev}$. The system produces work $\delta W_{sys}$ as a result of this heat transfer. Applying the energy balance to the combined system identified by dashed lines yields

$$\delta W_C = \delta Q_R - dE_C$$

where $\delta W_C$ is the total work of the combined system ($\delta W_{rev} + \delta W_{sys}$) and $dE_C$ is the change in the total energy of the combined system. Considering that the cyclic device is a *reversible* one, we have

$$\frac{\delta Q_R}{T_R} = \frac{\delta Q}{T}$$

where the sign of $\delta Q$ is determined with respect to the system (positive if *to* the system and negative if *from* the system) and the sign of $\delta Q_R$ is determined with respect to the reversible cyclic device. Eliminating $\delta Q_R$ from the two relations above yields

$$\delta W_C = T_R \frac{\delta Q}{T} - dE_C$$

We now let the system undergo a cycle while the cyclic device undergoes an integral number of cycles. Then the preceding relation becomes

$$W_C = T_R \oint \frac{\delta Q}{T}$$

since the cyclic integral of energy (the net change in the energy, which is a property, during a cycle) is zero. Here $W_C$ is the cyclic integral of $\delta W_C$, and it represents the net work for the combined cycle.

It appears that the combined system is exchanging heat with a single thermal energy reservoir while involving (producing or consuming) work $W_C$ during a cycle. On the basis of the Kelvin–Planck statement of the second law, which states that *no system can produce a net amount of work while operating in a cycle and exchanging heat with a single thermal energy reservoir*, we reason that $W_C$ cannot be a work output, and thus it cannot be a positive quantity. Considering that $T_R$ is the thermodynamic temperature and thus a positive quantity, we must have

$$\oint \frac{\delta Q}{T} \leq 0 \qquad (8\text{–}1)$$

which is the *Clausius inequality*. This inequality is valid for all thermodynamic cycles, reversible or irreversible, including the refrigeration cycles.

If no irreversibilities occur within the system as well as the reversible cyclic device, then the cycle undergone by the combined system is internally reversible. As such, it can be reversed. In the reversed cycle case, all the quantities have the same magnitude but the opposite sign. Therefore, the work $W_C$, which could not be a positive quantity in the regular case, cannot be a negative quantity in the reversed case. Then it follows that $W_{C,\text{int rev}} = 0$ since it cannot be a positive or negative quantity, and therefore

$$\oint \left( \frac{\delta Q}{T} \right)_{\text{int rev}} = 0 \qquad (8\text{–}2)$$

for internally reversible cycles. Thus, we conclude that *the equality in the Clausius inequality holds for totally or just internally reversible cycles and the inequality for the irreversible ones.*

To develop a relation for the definition of entropy, let us examine Eq. 8–2 more closely. Here we have a quantity whose cyclic integral is zero. Let us think for a moment what kind of quantities can have this characteristic. We know that the cyclic integral of *work* is not zero. (It is a good thing that it is not. Otherwise, heat engines that work on a cycle such as steam power plants would produce zero net work.) Neither is the cyclic integral of heat.

Now consider the volume occupied by a gas in a piston–cylinder device undergoing a cycle, as shown in Fig. 8–2. When the piston returns to its initial position at the end of a cycle, the volume of the gas also returns to its initial value. Thus, the net change in volume during a cycle is zero. This is also expressed as

$$\oint dV = 0 \qquad (8\text{–}3)$$

That is, the cyclic integral of volume (or any other property) is zero. Conversely, a quantity whose cyclic integral is zero depends on the *state* only and not the process path, and thus it is a property. Therefore, the quantity $(\delta Q/T)_{\text{int rev}}$ must represent a property in the differential form.

$\oint dV = \Delta V_{\text{cycle}} = 0$

**FIGURE 8–2**

The net change in volume (a property) during a cycle is always zero.

Clausius realized in 1865 that he had discovered a new thermodynamic property, and he chose to name this property **entropy**. It is designated $S$ and is defined as

$$dS = \left(\frac{\delta Q}{T}\right)_{\text{int rev}} \qquad \text{(kJ/K)} \qquad \textbf{(8–4)}$$

Entropy is an extensive property of a system and sometimes is referred to as *total entropy*. Entropy per unit mass, designated $s$, is an intensive property and has the unit kJ/kg·K. The term *entropy* is generally used to refer to both total entropy and entropy per unit mass since the context usually clarifies which one is meant.

The entropy change of a system during a process can be determined by integrating Eq. 8–4 between the initial and the final states:

$$\Delta S = S_2 - S_1 = \int_1^2 \left(\frac{\delta Q}{T}\right)_{\text{int rev}} \qquad \text{(kJ/K)} \qquad \textbf{(8–5)}$$

Notice that we have actually defined the *change* in entropy instead of entropy itself, just as we defined the change in energy instead of the energy itself when we developed the first-law relation. Absolute values of entropy are determined on the basis of the third law of thermodynamics, which is discussed later in this chapter. Engineers are usually concerned with the *changes* in entropy. Therefore, the entropy of a substance can be assigned a zero value at some arbitrarily selected reference state, and the entropy values at other states can be determined from Eq. 8–5 by choosing state 1 to be the reference state ($S = 0$) and state 2 to be the state at which entropy is to be determined.

To perform the integration in Eq. 8–5, one needs to know the relation between $Q$ and $T$ during a process. This relation is often not available, and the integral in Eq. 8–5 can be performed for a few cases only. For the majority of cases we have to rely on tabulated data for entropy.

Note that entropy is a property, and like all other properties, it has fixed values at fixed states. Therefore, the entropy change $\Delta S$ between two specified states is the same no matter what path, reversible or irreversible, is followed during a process (Fig. 8–3).

Also note that the integral of $\delta Q/T$ gives us the value of entropy change *only if* the integration is carried out along an *internally reversible* path between the two states. The integral of $\delta Q/T$ along an irreversible path is not a property, and in general, different values will be obtained when the integration is carried out along different irreversible paths. Therefore, even for irreversible processes, the entropy change should be determined by carrying out this integration along some convenient *imaginary* internally reversible path between the specified states.

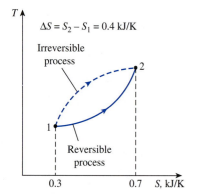

**FIGURE 8–3**

The entropy change between two specified states is the same whether the process is reversible or irreversible.

## A Special Case: Internally Reversible Isothermal Heat Transfer Processes

Recall that isothermal heat transfer processes are internally reversible. Therefore, the entropy change of a system during an internally reversible isothermal heat transfer process can be determined by performing the integration in Eq. 8–5:

$$\Delta S = \int_1^2 \left(\frac{\delta Q}{T}\right)_{\text{int rev}} = \int_1^2 \left(\frac{\delta Q}{T_0}\right)_{\text{int rev}} = \frac{1}{T_0} \int_1^2 (\delta Q)_{\text{int rev}}$$

which reduces to

$$\Delta S = \frac{Q}{T_0} \qquad \text{(kJ/K)} \qquad \text{(8–6)}$$

where $T_0$ is the constant temperature of the system and $Q$ is the heat transfer for the internally reversible process. Equation 8–6 is particularly useful for determining the entropy changes of thermal energy reservoirs that can absorb or supply heat indefinitely at a constant temperature.

Notice that the entropy change of a system during an internally reversible isothermal process can be positive or negative, depending on the direction of heat transfer. Heat transfer to a system increases the entropy of a system, whereas heat transfer from a system decreases it. In fact, losing heat is the only way the entropy of a system can be decreased.

---

**EXAMPLE 8–1**    **Entropy Change during an Isothermal Process**

A piston–cylinder device contains a liquid–vapor mixture of water at 300 K. During a constant-pressure process, 750 kJ of heat is transferred to the water. As a result, part of the liquid in the cylinder vaporizes. Determine the entropy change of the water during this process.

**SOLUTION**   Heat is transferred to a liquid–vapor mixture of water in a piston–cylinder device at constant pressure. The entropy change of water is to be determined.
**Assumptions**   No irreversibilities occur within the system boundaries during the process.
**Analysis**   We take the *entire water* (liquid + vapor) in the cylinder as the system (Fig. 8–4). This is a *closed system* since no mass crosses the system boundary during the process. We note that the temperature of the system remains constant at 300 K during this process since the temperature of a pure substance remains constant at the saturation value during a phase-change process at constant pressure.

The system undergoes an internally reversible, isothermal process, and thus its entropy change can be determined directly from Eq. 8–6 to be

$$\Delta S_{\text{sys,isothermal}} = \frac{Q}{T_{\text{sys}}} = \frac{750 \text{ kJ}}{300 \text{ K}} = \textbf{2.5 kJ/K}$$

**Discussion**   Note that the entropy change of the system is positive, as expected, since heat transfer is *to* the system.

$T = 300 \text{ K} = \text{const.}$

$\Delta S_{\text{sys}} = \dfrac{Q}{T} = 2.5 \dfrac{\text{kJ}}{\text{K}}$

$Q = 750 \text{ kJ}$

**FIGURE 8–4**
Schematic for Example 8–1.

---

# 8–2 ▪ THE INCREASE OF ENTROPY PRINCIPLE

Consider a cycle that is made up of two processes: process 1-2, which is arbitrary (reversible or irreversible), and process 2-1, which is internally reversible, as shown in Figure 8–5. From the Clausius inequality,

$$\oint \frac{\delta Q}{T} \leq 0$$

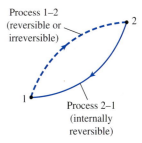

Process 1–2
(reversible or irreversible)

Process 2–1
(internally reversible)

**FIGURE 8–5**
A cycle composed of a reversible and an irreversible process.

or

$$\int_1^2 \frac{\delta Q}{T} + \int_2^1 \left(\frac{\delta Q}{T}\right)_{\text{int rev}} \leq 0$$

The second integral in the previous relation is recognized as the entropy change $S_1 - S_2$. Therefore,

$$\int_1^2 \frac{\delta Q}{T} + S_1 - S_2 \leq 0$$

which can be rearranged as

$$S_2 - S_1 \geq \int_1^2 \frac{\delta Q}{T} \qquad \text{(8–7)}$$

It can also be expressed in differential form as

$$dS \geq \frac{\delta Q}{T} \qquad \text{(8–8)}$$

where the equality holds for an internally reversible process and the inequality for an irreversible process. We may conclude from these equations that the entropy change of a closed system during an irreversible process is greater than the integral of $\delta Q/T$ evaluated for that process. In the limiting case of a reversible process, these two quantities become equal. We again emphasize that $T$ in these relations is the *thermodynamic temperature* at the *boundary* where the differential heat $\delta Q$ is transferred between the system and the surroundings.

The quantity $\Delta S = S_2 - S_1$ represents the *entropy change* of the system. For a reversible process, it becomes equal to $\int_1^2 \delta Q/T$, which represents the *entropy transfer* with heat.

The inequality sign in the preceding relations is a constant reminder that the entropy change of a closed system during an irreversible process is always greater than the entropy transfer. That is, some entropy is *generated* or *created* during an irreversible process, and this generation is due entirely to the presence of irreversibilities. The entropy generated during a process is called **entropy generation** and is denoted by $S_{\text{gen}}$. Noting that the difference between the entropy change of a closed system and the entropy transfer is equal to entropy generation, Eq. 8–7 can be rewritten as an equality as

$$\Delta S_{\text{sys}} = S_2 - S_1 = \int_1^2 \frac{\delta Q}{T} + S_{\text{gen}} \qquad \text{(8–9)}$$

Note that the entropy generation $S_{\text{gen}}$ is always a *positive* quantity or zero. Its value depends on the process, and thus it is *not* a property of the system. Also, in the absence of any entropy transfer, the entropy change of a system is equal to the entropy generation.

Equation 8–7 has far-reaching implications in thermodynamics. For an isolated system (or simply an adiabatic closed system), the heat transfer is zero, and Eq. 8–7 reduces to

$$\Delta S_{\text{isolated}} \geq 0 \qquad \text{(8–10)}$$

This equation can be expressed as *the entropy of an isolated system during a process always increases or, in the limiting case of a reversible process, remains constant.* In other words, it *never* decreases. This is known as the **increase of entropy principle**. Note that in the absence of any heat transfer, entropy change is due to irreversibilities only, and their effect is always to increase entropy.

Entropy is an extensive property, thus the total entropy of a system is equal to the sum of the entropies of the parts of the system. An isolated system may consist of any number of subsystems (Fig. 8–6). A system and its surroundings, for example, constitute an isolated system since both can be enclosed by a sufficiently large arbitrary boundary across which there is no heat, work, or mass transfer (Fig. 8–7). Therefore, a system and its surroundings can be viewed as the two subsystems of an isolated system, and the entropy change of this isolated system during a process is the sum of the entropy changes of the system and its surroundings, which is equal to the entropy generation since an isolated system involves no entropy transfer. That is,

$$S_{gen} = \Delta S_{total} = \Delta S_{sys} + \Delta S_{surr} \geq 0 \qquad (8\text{–}11)$$

where the equality holds for reversible processes and the inequality for irreversible ones. Note that $\Delta S_{surr}$ refers to the change in the entropy of the surroundings as a result of the occurrence of the process under consideration.

Since no actual process is truly reversible, we can conclude that some entropy is generated during a process, and therefore the entropy of the universe, which can be considered to be an isolated system, is continuously increasing. The more irreversible a process, the larger the entropy generated during that process. No entropy is generated during reversible processes ($S_{gen} = 0$).

Entropy increase of the universe is a major concern not only to engineers but also to philosophers, theologians, economists, and environmentalists since entropy is viewed as a measure of the disorder (or "mixed-up-ness") in the universe.

The increase of entropy principle does not imply that the entropy of a system cannot decrease. The entropy change of a system *can* be negative during a process (Fig. 8–8), but entropy generation cannot. The increase of entropy principle can be summarized as follows:

$$S_{gen} \begin{cases} > 0 & \text{Irreversible process} \\ = 0 & \text{Reversible process} \\ < 0 & \text{Impossible process} \end{cases}$$

This relation serves as a criterion in determining whether a process is reversible, irreversible, or impossible.

Things in nature have a tendency to change until they attain a state of equilibrium. The increase of entropy principle dictates that the entropy of an isolated system increases until the entropy of the system reaches a *maximum* value. At that point, the system is said to have reached an equilibrium state since the increase of entropy principle prohibits the system from undergoing any change of state that results in a decrease in entropy.

(Isolated)

$$\Delta S_{total} = \sum_{i=1}^{N} \Delta S_i > 0$$

**FIGURE 8–6**

The entropy change of an isolated system is the sum of the entropy changes of its components, and is never less than zero.

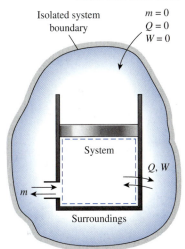

**FIGURE 8–7**

A system and its surroundings form an isolated system.

$$S_{gen} = \Delta S_{total} = \Delta S_{sys} + \Delta S_{surr} = 1 \text{ kJ/K}$$

**FIGURE 8–8**

The entropy change of a system can be negative, but the entropy generation cannot.

## Some Remarks about Entropy

In light of the preceding discussions, we draw the following conclusions:

1. Processes can occur in a *certain* direction only, not in *any* direction. A process must proceed in the direction that complies with the increase of entropy principle, that is, $S_{gen} \geq 0$. A process that violates this principle is impossible. This principle often forces chemical reactions to come to a halt before reaching completion.

2. Entropy is a *nonconserved property*, and there is *no* such thing as the *conservation of entropy principle*. Entropy is conserved during the idealized reversible processes only and increases during *all* actual processes.

3. The performance of engineering systems is degraded by the presence of irreversibilities, and *entropy generation* is a measure of the magnitudes of the irreversibilities present during that process. The greater the extent of irreversibilities, the greater the entropy generation. Therefore, entropy generation can be used as a quantitative measure of irreversibilities associated with a process. It is also used to establish criteria for the performance of engineering devices. This point is illustrated further in Example 8–2.

**FIGURE 8–9**

Schematic for Example 8–2.

---

**EXAMPLE 8–2**    **Entropy Generation during Heat Transfer Processes**

A heat source at 800 K loses 2000 kJ of heat to a sink at (*a*) 500 K and (*b*) 750 K. Determine which heat transfer process is more irreversible.

**SOLUTION**    Heat is transferred from a heat source to two heat sinks at different temperatures. The heat transfer process that is more irreversible is to be determined.

*Analysis*    A sketch of the reservoirs is shown in Fig. 8–9. Both cases involve heat transfer through a finite temperature difference, and therefore both are irreversible. The magnitude of the irreversibility associated with each process can be determined by calculating the total entropy change for each case. The total entropy change for a heat transfer process involving two reservoirs (a source and a sink) is the sum of the entropy changes of each reservoir since the two reservoirs form an adiabatic system.

Or do they? The problem statement gives the impression that the two reservoirs are in direct contact during the heat transfer process. But this cannot be the case since the temperature at a point can have only one value, and thus it cannot be 800 K on one side of the point of contact and 500 K on the other side. In other words, the temperature function cannot have a jump discontinuity. Therefore, it is reasonable to assume that the two reservoirs are separated by a partition through which the temperature drops from 800 K on one side to 500 K (or 750 K) on the other. In that case, the entropy change of the partition should also be considered when evaluating the total entropy change for this process. However, considering that entropy is a property and the values of properties depend on the state of a system, we can argue that the entropy change of the partition is zero since the partition appears to have undergone a *steady* process and thus experienced no change in its properties at any point. We base this argument on the fact that the temperature on both sides of the partition and thus through-out remains constant during this process. Therefore, we are justified to assume that $\Delta S_{partition} = 0$ since the entropy (as well as the energy) content of the partition remains constant during this process.

The entropy change for each reservoir can be determined from Eq. 8–6 since each reservoir undergoes an internally reversible, isothermal process.

(*a*) For the heat transfer process to a sink at 500 K:

$$\Delta S_{source} = \frac{Q_{source}}{T_{source}} = \frac{-2000 \text{ kJ}}{800 \text{ K}} = -2.5 \text{ kJ/K}$$

$$\Delta S_{sink} = \frac{Q_{sink}}{T_{sink}} = \frac{2000 \text{ kJ}}{500 \text{ K}} = +4.0 \text{ kJ/K}$$

and

$$S_{gen} = \Delta S_{total} = \Delta S_{source} + \Delta S_{sink} = (-2.5 + 4.0) \text{ kJ/K} = \textbf{1.5 kJ/K}$$

Therefore, 1.5 kJ/K of entropy is generated during this process. Noting that both reservoirs have undergone internally reversible processes, the entire entropy generation took place in the partition.

(*b*) Repeating the calculations in part (*a*) for a sink temperature of 750 K, we obtain

$$\Delta S_{source} = -2.5 \text{ kJ/k}$$

$$\Delta S_{sink} = +2.7 \text{ kJ/K}$$

and

$$S_{gen} = \Delta S_{total} = (-2.5 + 2.7) \text{ kJ/K} = \textbf{0.2 kJ/K}$$

The total entropy change for the process in part (*b*) is smaller, and therefore it is less irreversible. This is expected since the process in (*b*) involves a smaller temperature difference and thus a smaller irreversibility.

**Discussion** The irreversibilities associated with both processes could be eliminated by operating a Carnot heat engine between the source and the sink. For this case it can be shown that $\Delta S_{total} = 0$.

# 8–3 · ENTROPY CHANGE OF PURE SUBSTANCES

Entropy is a property, and thus the value of entropy of a system is fixed once the state of the system is fixed. Specifying two intensive independent properties fixes the state of a simple compressible system, and thus the value of entropy and of other properties at that state. Starting with its defining relation, the entropy change of a substance can be expressed in terms of other properties (see Sec. 8–7). But in general, these relations are too complicated and are not practical to use for hand calculations. Therefore, using a suitable reference state, the entropies of substances are evaluated from measurable property data following rather involved computations, and the results are tabulated in the same manner as the other properties such as $v$, $u$, and $h$ (Fig. 8–10).

The entropy values in the property tables are given relative to an arbitrary reference state. In steam tables the entropy of saturated liquid $s_f$ at 0.01°C is

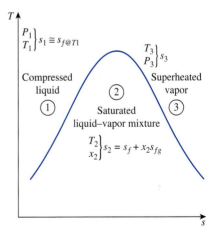

**FIGURE 8–10**

The entropy of a pure substance is determined from the tables (like other properties).

assigned the value of zero. For refrigerant-134a, the zero value is assigned to saturated liquid at $-40°C$. The entropy values become negative at temperatures below the reference value.

The value of entropy at a specified state is determined just like any other property. In the compressed liquid and superheated vapor regions, it can be obtained directly from the tables at the specified state. In the saturated mixture region, it is determined from

$$s = s_f + xs_{fg} \quad \text{(kJ/kg·K)}$$

where $x$ is the quality and $s_f$ and $s_{fg}$ values are listed in the saturation tables. In the absence of compressed liquid data, the entropy of the compressed liquid can be approximated by the entropy of the saturated liquid at the given temperature:

$$s_{@\,T,P} \cong s_{f\,@\,T} \quad \text{(kJ/kg·K)}$$

The entropy change of a specified mass $m$ (a closed system) during a process is simply

$$\Delta S = m\Delta s = m(s_2 - s_1) \quad \text{(kJ/K)} \tag{8–12}$$

which is the difference between the entropy values at the final and initial states.

When studying the second-law aspects of processes, entropy is commonly used as a coordinate on diagrams such as the $T\text{-}s$ and $h\text{-}s$ diagrams. The general characteristics of the $T\text{-}s$ diagram of pure substances are shown in Fig. 8–11 using data for water. Notice from this diagram that the constant-volume lines are steeper than the constant-pressure lines and the constant-pressure lines are parallel to the constant-temperature lines in the saturated liquid–vapor mixture region. Also, the constant-pressure lines almost coincide with the saturated liquid line in the compressed liquid region.

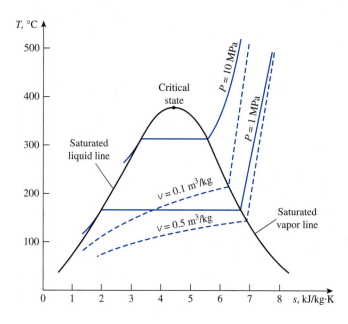

**FIGURE 8–11**

Schematic of the $T\text{-}s$ diagram for water.

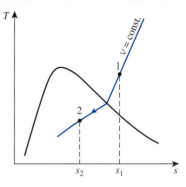

## EXAMPLE 8–3  Entropy Change of a Substance in a Tank

A rigid tank contains 5 kg of refrigerant-134a initially at 20°C and 140 kPa. The refrigerant is now cooled while being stirred until its pressure drops to 100 kPa. Determine the entropy change of the refrigerant during this process.

**SOLUTION**  The refrigerant in a rigid tank is cooled while being stirred. The entropy change of the refrigerant is to be determined.

**Assumptions**  The volume of the tank is constant and thus $v_2 = v_1$.

**Analysis**  We take the refrigerant in the tank as the *system* (Fig. 8–12). This is a *closed system* since no mass crosses the system boundary during the process. We note that the change in entropy of a substance during a process is simply the difference between the entropy values at the final and initial states. The initial state of the refrigerant is completely specified.

Recognizing that the specific volume remains constant during this process, the properties of the refrigerant at both states are

State 1:  $\left. \begin{array}{l} P_1 = 140 \text{ kPa} \\ T_1 = 20°C \end{array} \right\}$  $\begin{array}{l} s_1 = 1.0624 \text{ kJ/kg·K} \\ v_1 = 0.16544 \text{ m}^3/\text{kg} \end{array}$

State 2:  $\left. \begin{array}{l} P_2 = 100 \text{ kPa} \\ (v_2 = v_1) \end{array} \right\}$  $\begin{array}{l} v_f = 0.0007259 \text{ m}^3/\text{kg} \\ v_g = 0.19254 \text{ m}^3/\text{kg} \end{array}$

The refrigerant is a saturated liquid–vapor mixture at the final state since $v_f < v_2 < v_g$ at 100 kPa pressure. Therefore, we need to determine the quality first:

$$x_2 = \frac{v_2 - v_f}{v_{fg}} = \frac{0.16544 - 0.0007259}{0.19254 - 0.0007259} = 0.859$$

Thus,

$$s_2 = s_f + x_2 s_{fg} = 0.07188 + (0.859)(0.87995) = 0.8278 \text{ kJ/kg·K}$$

Then, the entropy change of the refrigerant during this process is

$$\Delta S = m(s_2 - s_1) = (5 \text{ kg})(0.8278 - 1.0624) \text{ kJ/kg·K}$$

$$= \mathbf{-1.173 \text{ kJ/K}}$$

**Discussion**  The negative sign indicates that the entropy of the system is decreasing during this process. This is not a violation of the second law, however, since it is the *entropy generation* $S_{gen}$ that cannot be negative.

**FIGURE 8–12**

Schematic and *T-s* diagram for Example 8–3.

## EXAMPLE 8–4  Entropy Change during a Constant-Pressure Process

A piston–cylinder device initially contains 3 lbm of liquid water at 20 psia and 70°F. The water is now heated at constant pressure by the addition of 3450 Btu of heat. Determine the entropy change of the water during this process.

**SOLUTION**  Liquid water in a piston–cylinder device is heated at constant pressure. The entropy change of water is to be determined.

**Assumptions**  1 The tank is stationary and thus the kinetic and potential energy changes are zero, $\Delta KE = \Delta PE = 0$. 2 The process is quasi-equilibrium. 3 The pressure remains constant during the process and thus $P_2 = P_1$.

**FIGURE 8–13**

Schematic and *T-s* diagram for Example 8–4.

*Analysis* We take the water in the cylinder as the *system* (Fig. 8–13). This is a *closed system* since no mass crosses the system boundary during the process. We note that a piston–cylinder device typically involves a moving boundary and thus boundary work $W_b$. Also, heat is transferred to the system.

Water exists as a compressed liquid at the initial state since its pressure is greater than the saturation pressure of 0.3632 psia at 70°F. By approximating the compressed liquid as a saturated liquid at the given temperature, the properties at the initial state are

*State 1:* $\left.\begin{array}{l} P_1 = 20 \text{ psia} \\ T_1 = 70°F \end{array}\right\}$ $\begin{array}{l} s_1 \cong s_{f@70°F} = 0.07459 \text{ Btu/lbm·R} \\ h_1 \cong h_{f@70°F} = 38.08 \text{ Btu/lbm} \end{array}$

At the final state, the pressure is still 20 psia, but we need one more property to fix the state. This property is determined from the energy balance,

$$\underbrace{E_{in} - E_{out}}_{\substack{\text{Net energy transfer} \\ \text{by heat, work, and mass}}} = \underbrace{\Delta E_{system}}_{\substack{\text{Change in internal, kinetic,} \\ \text{potential, etc., energies}}}$$

$$Q_{in} - W_b = \Delta U$$
$$Q_{in} = \Delta H = m(h_2 - h_1)$$
$$3450 \text{ Btu} = (3 \text{ lbm})(h_2 - 38.08 \text{ Btu/lbm})$$
$$h_2 = 1188.1 \text{ Btu/lbm}$$

since $\Delta U + W_b = \Delta H$ for a constant-pressure quasi-equilibrium process. Then,

*State 2:* $\left.\begin{array}{l} P_2 = 20 \text{ psia} \\ h_2 = 1188.1 \text{ Btu/lbm} \end{array}\right\}$ $\begin{array}{l} s_2 = 1.7761 \text{ Btu/lbm·R} \\ \text{(Table A-6E, interpolation)} \end{array}$

Therefore, the entropy change of water during this process is

$$\Delta S = m(s_2 - s_1) = (3 \text{ lbm})(1.7761 - 0.07459) \text{ Btu/lbm·R}$$
$$= \textbf{5.105 Btu/R}$$

## 8–4 · ISENTROPIC PROCESSES

We mentioned earlier that the entropy of a fixed mass can be changed by (1) heat transfer and (2) irreversibilities. Then it follows that the entropy of a fixed mass does not change during a process that is *internally reversible* and *adiabatic* (Fig. 8–14). A process during which the entropy remains constant is called an **isentropic process**. It is characterized by

*Isentropic process:* $\qquad \Delta s = 0 \quad \text{or} \quad s_2 = s_1 \qquad \text{(kJ/kg·K)} \qquad$ **(8–13)**

That is, a substance will have the same entropy value at the end of the process as it does at the beginning if the process is carried out in an isentropic manner.

Many engineering systems or devices such as pumps, turbines, nozzles, and diffusers are essentially adiabatic in their operation, and they perform best when the irreversibilities, such as the friction associated with the process, are minimized. Therefore, an isentropic process can serve as an appropriate model for actual processes. Also, isentropic processes enable us to define

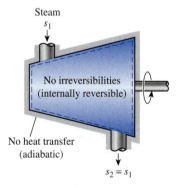

**FIGURE 8–14**

During an internally reversible, adiabatic (isentropic) process, the entropy remains constant.

efficiencies for processes to compare the actual performance of these devices to the performance under idealized conditions.

It should be recognized that a *reversible adiabatic* process is necessarily isentropic ($s_2 = s_1$), but an *isentropic* process is not necessarily a reversible adiabatic process. (The entropy increase of a substance during a process as a result of irreversibilities may be offset by a decrease in entropy as a result of heat losses, for example.) However, the term *isentropic process* is customarily used in thermodynamics to imply an *internally reversible, adiabatic process.*

**FIGURE 8–15**

Schematic and *T-s* diagram for Example 8–5.

---

### EXAMPLE 8–5    Isentropic Expansion of Steam in a Turbine

Steam enters an adiabatic turbine at 5 MPa and 450°C and leaves at a pressure of 1.4 MPa. Determine the work output of the turbine per unit mass of steam if the process is reversible.

**SOLUTION**    Steam is expanded in an adiabatic turbine to a specified pressure in a reversible manner. The work output of the turbine is to be determined.

**Assumptions**    **1** This is a steady-flow process since there is no change with time at any point and thus $\Delta m_{CV} = 0$, $\Delta E_{CV} = 0$, and $\Delta S_{CV} = 0$. **2** The process is reversible. **3** Kinetic and potential energies are negligible. **4** The turbine is adiabatic and thus there is no heat transfer.

**Analysis**    We take the *turbine* as the system (Fig. 8–15). This is a *control volume* since mass crosses the system boundary during the process. We note that there is only one inlet and one exit, and thus $\dot{m}_1 = \dot{m}_2 = \dot{m}$.

The power output of the turbine is determined from the rate form of the energy balance,

$$\underbrace{\dot{E}_{in} - \dot{E}_{out}}_{\substack{\text{Rate of net energy transfer} \\ \text{by heat, work, and mass}}} = \underbrace{dE_{system}/dt}_{\substack{\text{Rate of change in internal, kinetic,} \\ \text{potential, etc., energies}}}^{\nearrow 0 \text{ (steady)}} = 0$$

$$\dot{E}_{in} = \dot{E}_{out}$$

$$\dot{m}h_1 = \dot{W}_{out} + \dot{m}h_2 \qquad (\text{since } \dot{Q} = 0, \text{ke} \cong \text{pe} \cong 0)$$

$$\dot{W}_{out} = \dot{m}(h_1 - h_2)$$

The inlet state is completely specified since two properties are given. But only one property (pressure) is given at the final state, and we need one more property to fix it. The second property comes from the observation that the process is reversible and adiabatic, and thus isentropic. Therefore, $s_2 = s_1$, and

State 1:    $\left. \begin{array}{l} P_1 = 5 \text{ MPa} \\ T_1 = 450°C \end{array} \right\}$    $\begin{array}{l} h_1 = 3317.2 \text{ kJ/kg} \\ s_1 = 6.8210 \text{ kJ/kg·K} \end{array}$

State 2:    $\left. \begin{array}{l} P_2 = 1.4 \text{ MPa} \\ s_2 = s_1 \end{array} \right\}$    $h_2 = 2967.4 \text{ kJ/kg}$

Then, the work output of the turbine per unit mass of the steam becomes

$$w_{out} = h_1 - h_2 = 3317.2 - 2967.4 = \textbf{349.8 kJ/kg}$$

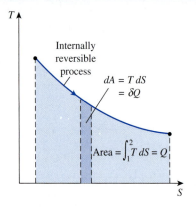

**FIGURE 8–16**

On a *T-S* diagram, the area under the process curve represents the heat transfer for internally reversible processes.

# 8–5 · PROPERTY DIAGRAMS INVOLVING ENTROPY

Property diagrams serve as great visual aids in the thermodynamic analysis of processes. We have used *P-v* and *T-v* diagrams extensively in previous chapters in conjunction with the first law of thermodynamics. In the second-law analysis, it is very helpful to plot the processes on diagrams for which one of the coordinates is entropy. The two diagrams commonly used in the second-law analysis are the *temperature-entropy* and the *enthalpy-entropy* diagrams.

Consider the defining equation of entropy (Eq. 8–4). It can be rearranged as

$$\delta Q_{\text{int rev}} = T\, dS \qquad (\text{kJ}) \qquad (8\text{–}14)$$

As shown in Fig. 8–16, $\delta Q_{\text{rev int}}$ corresponds to a differential area on a *T-S* diagram. The total heat transfer during an internally reversible process is determined by integration to be

$$Q_{\text{int rev}} = \int_1^2 T\, dS \qquad (\text{kJ}) \qquad (8\text{–}15)$$

which corresponds to the area under the process curve on a *T-S* diagram. Therefore, we conclude that *the area under the process curve on a T-S diagram represents heat transfer during an internally reversible process.* This is somewhat analogous to reversible boundary work being represented by the area under the process curve on a *P-V* diagram. Note that the area under the process curve represents heat transfer for processes that are internally (or totally) reversible. The area has no meaning for irreversible processes.

Equations 8–14 and 8–15 can also be expressed on a unit-mass basis as

$$\delta q_{\text{int rev}} = T\, ds \qquad (\text{kJ/kg}) \qquad (8\text{–}16)$$

and

$$q_{\text{int rev}} = \int_1^2 T\, ds \qquad (\text{kJ/kg}) \qquad (8\text{–}17)$$

To perform the integrations in Eqs. 8–15 and 8–17, one needs to know the relationship between *T* and *s* during a process. One special case for which these integrations can be performed easily is the *internally reversible isothermal process.* It yields

$$Q_{\text{int rev}} = T_0\, \Delta S \qquad (\text{kJ}) \qquad (8\text{–}18)$$

or

$$q_{\text{int rev}} = T_0\, \Delta s \qquad (\text{kJ/kg}) \qquad (8\text{–}19)$$

where $T_0$ is the constant temperature and $\Delta S$ is the entropy change of the system during the process.

An isentropic process on a *T-s* diagram is easily recognized as a *vertical-line segment.* This is expected since an isentropic process involves no heat transfer, and therefore the area under the process path must be zero (Fig. 8–17). The *T-s* diagrams serve as valuable tools for visualizing the second-law aspects of processes and cycles, and thus they are frequently used in thermodynamics. The *T-s* diagram of water is given in the appendix in Fig. A–9.

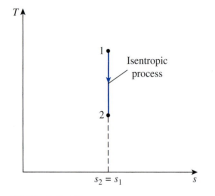

**FIGURE 8–17**

The isentropic process appears as a *vertical* line segment on a *T-s* diagram.

Another diagram commonly used in engineering is the enthalpy-entropy diagram, which is quite valuable in the analysis of steady-flow devices such as turbines, compressors, and nozzles. The coordinates of an h-s diagram represent two properties of major interest: enthalpy, which is a primary property in the first-law analysis of the steady-flow devices, and entropy, which is the property that accounts for irreversibilities during adiabatic processes. In analyzing the steady flow of steam through an adiabatic turbine, for example, the vertical distance between the inlet and the exit states $\Delta h$ is a measure of the work output of the turbine, and the horizontal distance $\Delta s$ is a measure of the irreversibilities associated with the process (Fig. 8–18).

The h-s diagram is also called a **Mollier diagram** after the German scientist R. Mollier (1863–1935). An h-s diagram is given in the appendix for steam in Fig. A–10.

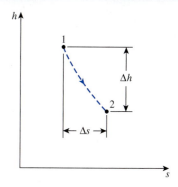

**FIGURE 8–18**

For adiabatic steady-flow devices, the vertical distance $\Delta h$ on an h-s diagram is a measure of work, and the horizontal distance $\Delta s$ is a measure of irreversibilities.

---

**EXAMPLE 8–6**    The T-S Diagram of the Carnot Cycle

Show the Carnot cycle on a T-S diagram and indicate the areas that represent the heat supplied $Q_H$, heat rejected $Q_L$, and the net work output $W_{net,out}$ on this diagram.

**SOLUTION**    The Carnot cycle is to be shown on a T-S diagram, and the areas that represent $Q_H$, $Q_L$, and $W_{net,out}$ are to be indicated.

*Analysis*    Recall that the Carnot cycle is made up of two reversible isothermal ($T$ = constant) processes and two isentropic ($s$ = constant) processes. These four processes form a rectangle on a T-S diagram, as shown in Fig. 8–19.

On a T-S diagram, the area under the process curve represents the heat transfer for that process. Thus the area A12B represents $Q_H$, the area A43B represents $Q_L$, and the difference between these two (the area in color) represents the net work since

$$W_{net,out} = Q_H - Q_L$$

Therefore, the area enclosed by the path of a cycle (area 1234) on a T-S diagram represents the net work. Recall that the area enclosed by the path of a cycle also represents the net work on a P-V diagram.

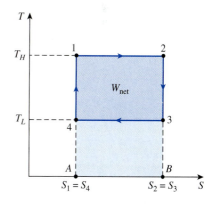

**FIGURE 8–19**

The T-S diagram of a Carnot cycle (Example 8–6).

---

# 8–6 ▪ WHAT IS ENTROPY?

It is clear from the previous discussion that entropy is a useful property and serves as a valuable tool in the second-law analysis of engineering devices. But this does not mean that we know and understand entropy well. Because we do not. In fact, we cannot even give an adequate answer to the question, what is entropy? Not being able to describe entropy fully, however, does not take anything away from its usefulness. We could not define *energy* either, but it did not interfere with our understanding of energy transformations and the conservation of energy principle. Granted, entropy is not a household word like energy. But with continued use, our understanding of entropy will deepen, and our appreciation of it will grow. The next discussion should shed some light on the physical meaning of entropy by considering the microscopic nature of matter.

Entropy,
kJ/kg·K

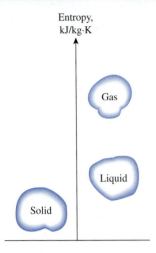

**FIGURE 8–20**

The level of molecular disorder (entropy) of a substance increases as it melts or evaporates.

Pure crystal
$T = 0$ K
Entropy = 0

**FIGURE 8–21**

A pure crystalline substance at absolute zero temperature is in perfect order, and its entropy is zero (the third law of thermodynamics).

Entropy can be viewed as a measure of *molecular disorder*, or *molecular randomness*. As a system becomes more disordered, the positions of the molecules become less predictable and the entropy increases. Thus, it is not surprising that the entropy of a substance is lowest in the solid phase and highest in the gas phase (Fig. 8–20). In the solid phase, the molecules of a substance continually oscillate about their equilibrium positions, but they cannot move relative to each other, and their position at any instant can be predicted with good certainty. In the gas phase, however, the molecules move about at random, collide with each other, and change direction, making it extremely difficult to predict accurately the microscopic state of a system at any instant. Associated with this molecular chaos is a high value of entropy.

When viewed microscopically (from a statistical thermodynamics point of view), an isolated system that appears to be at a state of equilibrium may exhibit a high level of activity because of the continual motion of the molecules. To each state of macroscopic equilibrium there corresponds a large number of possible microscopic states or molecular configurations. The entropy of a system is related to the total number of possible microscopic states of that system, called *thermodynamic probability p*, by the **Boltzmann relation**, expressed as

$$S = k \ln p \qquad\qquad \text{(8–20)}$$

where $k = 1.3806 \times 10^{-23}$ J/K is the **Boltzmann constant**. Therefore, from a microscopic point of view, the entropy of a system increases whenever the molecular randomness or uncertainty (i.e., molecular probability) of a system increases. Thus, entropy is a measure of molecular disorder, and the molecular disorder of an isolated system increases anytime it undergoes a process.

As mentioned earlier, the molecules of a substance in solid phase continually oscillate, creating an uncertainty about their position. These oscillations, however, fade as the temperature is decreased, and the molecules supposedly become motionless at absolute zero. This represents a state of ultimate molecular order (and minimum energy). Therefore, *the entropy of a pure crystalline substance at absolute zero temperature is zero* since there is no uncertainty about the state of the molecules at that instant (Fig. 8–21). This statement is known as the **third law of thermodynamics**. The third law of thermodynamics provides an absolute reference point for the determination of entropy. The entropy determined relative to this point is called **absolute entropy**, and it is extremely useful in the thermodynamic analysis of chemical reactions. Notice that the entropy of a substance that is not pure crystalline (such as a solid solution) is not zero at absolute zero temperature. This is because more than one molecular configuration exists for such substances, which introduces some uncertainty about the microscopic state of the substance.

Molecules in the gas phase possess a considerable amount of kinetic energy. However, we know that no matter how large their kinetic energies are, the gas molecules do not rotate a paddle wheel inserted into the container and produce work. This is because the gas molecules, and the energy they possess, are disorganized. Probably the number of molecules trying to rotate the wheel in one direction at any instant is equal to the number of molecules that are trying to rotate it in the opposite direction, causing the wheel to remain motionless. Therefore, we cannot extract any useful work directly from disorganized energy (Fig. 8–22).

Now consider a rotating shaft shown in Fig. 8–23. This time the energy of the molecules is completely organized since the molecules of the shaft are rotating in the same direction together. This organized energy can readily be used to perform useful tasks such as raising a weight or generating electricity. Being an organized form of energy, work is free of disorder or randomness and thus free of entropy. *There is no entropy transfer associated with energy transfer as work.* Therefore, in the absence of any friction, the process of raising a weight by a rotating shaft (or a flywheel) does not produce any entropy. Any process that does not produce a net entropy is reversible, and thus the process just described can be reversed by lowering the weight. Therefore, energy is not degraded during this process, and no potential to do work is lost.

Instead of raising a weight, let us operate the paddle wheel in a container filled with a gas, as shown in Fig. 8–24. The paddle-wheel work in this case is converted to the internal energy of the gas, as evidenced by a rise in gas temperature, creating a higher level of molecular disorder in the container. This process is quite different from raising a weight since the organized paddle-wheel energy is now converted to a highly disorganized form of energy, which cannot be converted back to the paddle wheel as the rotational kinetic energy. Only a portion of this energy can be converted to work by partially reorganizing it through the use of a heat engine. Therefore, energy is degraded during this process, the ability to do work is reduced, molecular disorder is produced, and associated with all this is an increase in entropy.

The *quantity* of energy is always preserved during an actual process (the first law), but the *quality* is bound to decrease (the second law). This decrease in quality is always accompanied by an increase in entropy. As an example, consider the transfer of 10 kJ of energy as heat from a hot medium to a cold one. At the end of the process, we still have the 10 kJ of energy, but at a lower temperature and thus at a lower quality.

Heat is, in essence, a form of *disorganized energy*, and some disorganization (entropy) flows with heat (Fig. 8–25). As a result, the entropy and the level of molecular disorder or randomness of the hot body decreases with the entropy and the level of molecular disorder of the cold body increases. The second law requires that the increase in entropy of the cold body be greater than the decrease in entropy of the hot body, and thus the net entropy of the combined system (the cold body and the hot body) increases. That is, the combined system is at a state of greater disorder at the final state. Thus we can conclude that processes can occur only in the direction of increased overall entropy or molecular disorder. That is, the entire universe is getting more and more chaotic everyday.

## Entropy and Entropy Generation in Daily Life

The concept of entropy can also be applied to other areas. Entropy can be viewed as a measure of disorder or disorganization in a system. Likewise, entropy generation can be viewed as a measure of disorder or disorganization generated during a process. The concept of entropy is not used in daily life nearly as extensively as the concept of energy, even though entropy is

**FIGURE 8–22**
Disorganized energy does not create much useful effect, no matter how large it is.

**FIGURE 8–23**
In the absence of friction, raising a weight by a rotating shaft does not create any disorder (entropy), and thus energy is not degraded during this process.

**FIGURE 8–24**
The paddle-wheel work done on a gas increases the level of disorder (entropy) of the gas, and thus energy is degraded during this process.

**FIGURE 8–25**

During a heat transfer process, the net entropy increases. (The increase in the entropy of the cold body more than offsets the decrease in the entropy of the hot body.)

**FIGURE 8–26**

The use of entropy (disorganization, uncertainty) is not limited to thermodynamics.

*BLONDIE © KING FEATURES SYNDICATE.*

readily applicable to various aspects of daily life. The extension of the entropy concept to nontechnical fields is not a novel idea. It has been the topic of several articles, and even some books. Next, we present several ordinary events and show their relevance to the concept of entropy and entropy generation.

Efficient people lead low-entropy (highly organized) lives. They have a place for everything (minimum uncertainty), and it takes minimum energy for them to locate something. Inefficient people, on the other hand, are disorganized and lead high-entropy lives. It takes them minutes (if not hours) to find something they need, and they are likely to create a bigger disorder as they are searching since they will probably conduct the search in a disorganized manner (Fig. 8–26). People leading high-entropy lifestyles are always on the run, and never seem to catch up.

You probably noticed (with frustration) that some people seem to *learn* fast and remember well what they learn. We can call this type of learning organized or low-entropy learning. These people make a conscientious effort to file the new information properly by relating it to their existing knowledge base and creating a solid information network in their minds. On the other hand, people who throw the information into their minds as they study, with no effort to secure it, may *think* they are learning. They are bound to discover otherwise when they need to locate the information, for example, during a test. It is not easy to retrieve information from a database that is, in a sense, in the gas phase. Students who have blackouts during tests should reexamine their study habits.

A *library* with a good shelving and indexing system can be viewed as a low-entropy library because of the high level of organization. Likewise, a library with a poor shelving and indexing system can be viewed as a high-entropy library because of the high level of disorganization. A library with no indexing system is like no library, since a book is of no value if it cannot be found.

Consider two identical buildings, each containing one million books. In the first building, the books are *piled* on top of each other, whereas in the second building they are *highly organized, shelved, and indexed* for easy reference. There is no doubt about which building a student will prefer to go to for checking out a certain book. Yet, some may argue from the first-law point of view that these two buildings are equivalent since the mass and knowledge content of the two buildings are identical, despite the high level of disorganization (entropy) in the first building. This example illustrates that any realistic comparisons should involve the second-law point of view.

Two *textbooks* that seem to be identical because both cover basically the same topics and present the same information may actually be *very* different depending on *how* they cover the topics. After all, two seemingly identical cars are not so identical if one goes only half as many miles as the other one on the same amount of fuel. Likewise, two seemingly identical books are not so identical if it takes twice as long to learn a topic from one of them as it does from the other. Thus, comparisons made on the basis of the first law only may be highly misleading.

Having a disorganized (high-entropy) *army* is like having no army at all. It is no coincidence that the command centers of any armed forces are among the primary targets during a war. One army that consists of

10 divisions is 10 times more powerful than 10 armies each consisting of a single division. Likewise, one country that consists of 10 states is more powerful than 10 countries, each consisting of a single state. The *United States* would not be such a powerful country if there were 50 independent countries in its place instead of a single country with 50 states. The European Union has the potential to be a new economic and political superpower. The old cliché "divide and conquer" can be rephrased as "increase the entropy and conquer."

We know that mechanical friction is always accompanied by entropy generation, and thus reduced performance. We can generalize this to daily life: *friction in the workplace* with fellow workers is bound to generate entropy, and thus adversely affect performance (Fig. 8–27). It results in reduced productivity.

We also know that *unrestrained expansion* (or explosion) and uncontrolled electron exchange (chemical reactions) generate entropy and are highly irreversible. Likewise, unrestrained opening of the mouth to scatter angry words is highly irreversible since this generates entropy, and it can cause considerable damage. A person who gets up in anger is bound to sit down at a loss. Hopefully, someday we will be able to come up with some procedures to quantify entropy generated during nontechnical activities, and maybe even pinpoint its primary sources and magnitude.

**FIGURE 8–27**

As in mechanical systems, friction in the workplace is bound to generate entropy and reduce performance.
© Vol. 26/PhotoDisc/Getty RF.

## 8–7 · THE *T ds* RELATIONS

Recall that the quantity $(\delta Q/T)_{\text{int rev}}$ corresponds to a differential change in the property *entropy*. The entropy change for a process, then, can be evaluated by integrating $\delta Q/T$ along some imaginary internally reversible path between the actual end states. For isothermal internally reversible processes, this integration is straightforward. But when the temperature varies during the process, we have to have a relation between $\delta Q$ and $T$ to perform this integration. Finding such relations is what we intend to do in this section.

The differential form of the conservation of energy equation for a closed stationary system (a fixed mass) containing a simple compressible substance can be expressed for an internally reversible process as

$$\delta Q_{\text{int rev}} - \delta W_{\text{int rev,out}} = dU \qquad \text{(8–21)}$$

But

$$\delta Q_{\text{int rev}} = T\,dS$$
$$\delta W_{\text{int rev,out}} = P\,dV$$

Thus,

$$T\,dS = dU + P\,dV \qquad \text{(kJ)} \qquad \text{(8–22)}$$

or

$$T\,ds = du + P\,dv \qquad \text{(kJ/kg)} \qquad \text{(8–23)}$$

This equation is known as the first *T ds*, or *Gibbs*, equation. Notice that the only type of work interaction a simple compressible system may involve as it undergoes an internally reversible process is the boundary work.

**FIGURE 8–28**

The $T\,ds$ relations are valid for both reversible and irreversible processes and for both closed and open systems.

The second $T\,ds$ equation is obtained by eliminating $du$ from Eq. 8–23 by using the definition of enthalpy ($h = u + Pv$):

$$
\left.
\begin{array}{l}
h = u + Pv \quad \longrightarrow \quad dh = du + P\,dv + v\,dP \\
\text{(Eq. 8–23)} \quad \longrightarrow \quad T\,ds = du + P\,dv
\end{array}
\right\} \; T\,ds = dh - v\,dP \qquad \textbf{(8–24)}
$$

Equations 8–23 and 8–24 are extremely valuable since they relate entropy changes of a system to the changes in other properties. Unlike Eq. 8–4, they are property relations and therefore are independent of the type of the processes.

These $T\,ds$ relations are developed with an internally reversible process in mind since the entropy change between two states must be evaluated along a reversible path. However, the results obtained are valid for both reversible and irreversible processes since entropy is a property and the change in a property between two states is independent of the type of process the system undergoes. Equations 8–23 and 8–24 are relations between the properties of a unit mass of a simple compressible system as it undergoes a change of state, and they are applicable whether the change occurs in a closed or an open system (Fig. 8–28).

Explicit relations for differential changes in entropy are obtained by solving for $ds$ in Eqs. 8–23 and 8–24:

$$
ds = \frac{du}{T} + \frac{P\,dv}{T} \qquad \textbf{(8–25)}
$$

and

$$
ds = \frac{dh}{T} - \frac{v\,dP}{T} \qquad \textbf{(8–26)}
$$

The entropy change during a process can be determined by integrating either of these equations between the initial and the final states. To perform these integrations, however, we must know the relationship between $du$ or $dh$ and the temperature (such as $du = c_v\,dT$ and $dh = c_p\,dT$ for ideal gases) as well as the equation of state for the substance (such as the ideal-gas equation of state $Pv = RT$). For substances for which such relations exist, the integration of Eq. 8–25 or 8–26 is straightforward. For other substances, we have to rely on tabulated data.

The $T\,ds$ relations for nonsimple systems, that is, systems that involve more than one mode of quasi-equilibrium work, can be obtained in a similar manner by including all the relevant quasi-equilibrium work modes.

## 8–8 ▪ ENTROPY CHANGE OF LIQUIDS AND SOLIDS

Recall that liquids and solids can be approximated as *incompressible substances* since their specific volumes remain nearly constant during a process. Thus, $dv \cong 0$ for liquids and solids, and Eq. 8–25 for this case reduces to

$$
ds = \frac{du}{T} = \frac{c\,dT}{T} \qquad \textbf{(8–27)}
$$

since $c_p = c_v = c$ and $du = c\,dT$ for incompressible substances. Then the entropy change during a process is determined by integration to be

*Liquids, solids:*
$$s_2 - s_1 = \int_1^2 c(T)\,\frac{dT}{T} \cong c_{avg}\ln\frac{T_2}{T_1} \qquad (kJ/kg\cdot K) \qquad \textbf{(8–28)}$$

where $c_{avg}$ is the *average* specific heat of the substance over the given temperature interval. Note that the entropy change of a truly incompressible substance depends on temperature only and is independent of pressure.

Equation 8–28 can be used to determine the entropy changes of solids and liquids with reasonable accuracy. However, for liquids that expand considerably with temperature, it may be necessary to consider the effects of volume change in calculations. This is especially the case when the temperature change is large.

A relation for isentropic processes of liquids and solids is obtained by setting the entropy change relation above equal to zero. It gives

*Isentropic:*
$$s_2 - s_1 = c_{avg}\ln\frac{T_2}{T_1} = 0 \quad \longrightarrow \quad T_2 = T_1 \qquad \textbf{(8–29)}$$

That is, the temperature of a truly incompressible substance remains constant during an isentropic process. Therefore, the isentropic process of an incompressible substance is also isothermal. This behavior is closely approximated by liquids and solids.

---

### EXAMPLE 8–7    Effect of Density of a Liquid on Entropy

Liquid methane is commonly used in various cryogenic applications. The critical temperature of methane is 191 K (or −82°C), and thus methane must be maintained below 191 K to keep it in liquid phase. The properties of liquid methane at various temperatures and pressures are given in Table 8–1. Determine the entropy change of liquid methane as it undergoes a process from 110 K and 1 MPa to 120 K and 5 MPa (a) using tabulated properties and (b) approximating liquid methane as an incompressible substance. What is the error involved in the latter case?

---

### TABLE 8–1

Properties of liquid methane

| Temp., $T$, K | Pressure, $P$, MPa | Density, $\rho$, kg/m³ | Enthalpy, $h$, kJ/kg | Entropy, $s$, kJ/kg·K | Specific heat, $c_p$, kJ/kg·K |
|---|---|---|---|---|---|
| 110 | 0.5 | 425.3 | 208.3 | 4.878 | 3.476 |
| | 1.0 | 425.8 | 209.0 | 4.875 | 3.471 |
| | 2.0 | 426.6 | 210.5 | 4.867 | 3.460 |
| | 5.0 | 429.1 | 215.0 | 4.844 | 3.432 |
| 120 | 0.5 | 410.4 | 243.4 | 5.185 | 3.551 |
| | 1.0 | 411.0 | 244.1 | 5.180 | 3.543 |
| | 2.0 | 412.0 | 245.4 | 5.171 | 3.528 |
| | 5.0 | 415.2 | 249.6 | 5.145 | 3.486 |

$P_2 = 5$ MPa
$T_2 = 120$ K

Heat

Methane
pump

$P_1 = 1$ MPa
$T_1 = 110$ K

**FIGURE 8–29**
Schematic for Example 8–7.

**SOLUTION** Liquid methane undergoes a process between two specified states. The entropy change of methane is to be determined by using actual data and by assuming methane to be incompressible.

**Analysis** (a) We consider a unit mass of liquid methane (Fig. 8–29). The properties of the methane at the initial and final states are

State 1: $\left.\begin{array}{l} P_1 = 1 \text{ MPa} \\ T_1 = 110 \text{ K} \end{array}\right\}$ $\begin{array}{l} s_1 = 4.875 \text{ kJ/kg·K} \\ c_{p1} = 3.471 \text{ kJ/kg·K} \end{array}$

State 2: $\left.\begin{array}{l} P_2 = 5 \text{ MPa} \\ T_2 = 120 \text{ K} \end{array}\right\}$ $\begin{array}{l} s_2 = 5.145 \text{ kJ/kg·K} \\ c_{p2} = 3.486 \text{ kJ/kg·K} \end{array}$

Therefore,

$$\Delta s = s_2 - s_1 = 5.145 - 4.875 = \mathbf{0.270 \text{ kJ/kg·K}}$$

(b) Approximating liquid methane as an incompressible substance, its entropy change is determined to be

$$\Delta s = c_{avg} \ln \frac{T_2}{T_1} = (3.4785 \text{ kJ/kg·K}) \ln \frac{120 \text{ K}}{110 \text{ K}} = \mathbf{0.303 \text{ kJ/kgK}}$$

since

$$c_{avg} = \frac{c_{p1} + c_{p2}}{2} = \frac{3.471 + 3.486}{2} = 3.4785 \text{ kJ/kg·K}$$

Therefore, the error involved in approximating liquid methane as an incompressible substance is

$$\text{Error} = \frac{|\Delta s_{actual} - \Delta s_{ideal}|}{\Delta s_{actual}} = \frac{|0.270 - 0.303|}{0.270} = \mathbf{0.122 \text{ (or 12.2\%)}}$$

**Discussion** This result is not surprising since the density of liquid methane changes during this process from 425.8 to 415.2 kg/m³ (about 3 percent), which makes us question the validity of the incompressible substance assumption. Still, this assumption enables us to obtain reasonably accurate results with less effort, which proves to be very convenient in the absence of compressed liquid data.

---

**EXAMPLE 8–8** **Economics of Replacing a Valve by a Turbine**

A cryogenic manufacturing facility handles liquid methane at 115 K and 5 MPa at a rate of 0.280 m³/s . A process requires dropping the pressure of liquid methane to 1 MPa, which is done by throttling the liquid methane by passing it through a flow resistance such as a valve. A recently hired engineer proposes to replace the throttling valve by a turbine in order to produce power while dropping the pressure to 1 MPa. Using data from Table 8–1, determine the maximum amount of power that can be produced by such a turbine. Also, determine how much this turbine will save the facility from electricity usage costs per year if the turbine operates continuously (8760 h/yr) and the facility pays $0.075/kWh for electricity.

**SOLUTION** Liquid methane is expanded in a turbine to a specified pressure at a specified rate. The maximum power that this turbine can produce and the amount of money it can save per year are to be determined.

*Assumptions* **1** This is a steady-flow process since there is no change with time at any point and thus $\Delta m_{CV} = 0$, $\Delta E_{CV} = 0$, and $\Delta S_{CV} = 0$. **2** The turbine is adiabatic and thus there is no heat transfer. **3** The process is reversible. **4** Kinetic and potential energies are negligible.

*Analysis* We take the *turbine* as the system (Fig. 8–30). This is a *control volume* since mass crosses the system boundary during the process. We note that there is only one inlet and one exit and thus $\dot{m}_1 = \dot{m}_2 = \dot{m}$.

The assumptions above are reasonable since a turbine is normally well insulated and it must involve no irreversibilities for best performance and thus *maximum* power production. Therefore, the process through the turbine must be *reversible adiabatic* or *isentropic*. Then, $s_2 = s_1$ and

State 1: $\left. \begin{array}{l} P_1 = 5 \text{ MPa} \\ T_1 = 115 \text{ K} \end{array} \right\}$  $\begin{array}{l} h_1 = 232.3 \text{ kJ/kg} \\ s_1 = 4.9945 \text{ kJ/kg·K} \\ \rho_1 = 422.15 \text{ kg/s} \end{array}$

State 2: $\left. \begin{array}{l} P_2 = 1 \text{ MPa} \\ s_2 = s_1 \end{array} \right\}$  $h_2 = 222.8 \text{ kJ/kg}$

Also, the mass flow rate of liquid methane is

$$\dot{m} = \rho_1 \dot{V}_1 = (422.15 \text{ kg/m}^3)(0.280 \text{ m}^3/\text{s}) = 118.2 \text{ kg/s}$$

Then the power output of the turbine is determined from the rate form of the energy balance to be

$$\underbrace{\dot{E}_{in} - \dot{E}_{out}}_{\substack{\text{Rate of net energy transfer} \\ \text{by heat, work, and mass}}} = \underbrace{dE_{\text{system}}/dt}_{\substack{\text{Rate of change in internal,} \\ \text{kinetic, potential, etc., energies}}}^{\nearrow^{0 \text{ (steady)}}} = 0$$

$$\dot{E}_{in} = \dot{E}_{out}$$

$$\dot{m}h_1 = \dot{W}_{out} + \dot{m}h_2 \qquad (\text{since } \dot{Q} = 0, \text{ ke} \cong \text{pe} \cong 0)$$

$$\dot{W}_{out} = \dot{m}(h_1 - h_2)$$

$$= (118.2 \text{ kg/s})(232.3 - 222.8) \text{ kJ/kg}$$

$$= \textbf{1123 kW}$$

For continuous operation ($365 \times 24 = 8760$ h), the amount of power produced per year is

$$\text{Annual power production} = \dot{W}_{out} \times \Delta t = (1123 \text{ kW})(8760 \text{ h/yr})$$

$$= 0.9837 \times 10^7 \text{ kWh/yr}$$

At \$0.075/kWh, the amount of money this turbine can save the facility is

$$\text{Annual power savings} = (\text{Annual power production})(\text{Unit cost of power})$$

$$= (0.9837 \times 10^7 \text{ kWh/yr})(\$0.075/\text{kWh})$$

$$= \textbf{\$737,800/yr}$$

That is, this turbine can save the facility \$737,800 a year by simply taking advantage of the potential that is currently being wasted by a throttling valve, and the engineer who made this observation should be rewarded.

**FIGURE 8–30**

A 1.0-MW liquified natural gas (LNG) turbine with 95-cm turbine runner diameter being installed in a cryogenic test facility.

*Courtesy of Ebara International Corporation, Cryodynamics Division, Sparks, Nevada.*

## 8–9 ▪ THE ENTROPY CHANGE OF IDEAL GASES

An expression for the entropy change of an ideal gas can be obtained from Eq. 8–25 or 8–26 by employing the property relations for ideal gases (Fig. 8–31). By substituting $du = c_v \, dT$ and $P = RT/v$ into Eq. 8–25, the differential entropy change of an ideal gas becomes

$$ds = c_v \frac{dT}{T} + R \frac{dv}{v} \tag{8–30}$$

The entropy change for a process is obtained by integrating this relation between the end states:

$$s_2 - s_1 = \int_1^2 c_v(T) \frac{dT}{T} + R \ln \frac{v_2}{v_1} \tag{8–31}$$

A second relation for the entropy change of an ideal gas is obtained in a similar manner by substituting $dh = c_p \, dT$ and $v = RT/P$ into Eq. 8–26 and integrating. The result is

$$s_2 - s_1 = \int_1^2 c_p(T) \frac{dT}{T} - R \ln \frac{P_2}{P_1} \tag{8–32}$$

The specific heats of ideal gases, with the exception of monatomic gases, depend on temperature, and the integrals in Eqs. 8–31 and 8–32 cannot be performed unless the dependence of $c_v$ and $c_p$ on temperature is known. Even when the $c_v(T)$ and $c_p(T)$ functions are available, performing long integrations every time entropy change is calculated is not practical. Then two reasonable choices are left: either perform these integrations by simply assuming constant specific heats or evaluate those integrals once and tabulate the results. Both approaches are presented next.

### Constant Specific Heats (Approximate Analysis)

Assuming constant specific heats for ideal gases is a common approximation, and we used this assumption before on several occasions. It usually simplifies the analysis greatly, and the price we pay for this convenience is some loss in accuracy. The magnitude of the error introduced by this assumption depends on the situation at hand. For example, for monatomic ideal gases such as helium, the specific heats are independent of temperature,

$Pv = RT$
$du = C_v \, dT$
$dh = C_p \, dT$

**FIGURE 8–31**

A broadcast from channel IG.

© Vol. 1/PhotoDisc/Getty RF.

and therefore the constant-specific-heat assumption introduces no error. For ideal gases whose specific heats vary almost linearly in the temperature range of interest, the possible error is minimized by using specific heat values evaluated at the average temperature (Fig. 8–32). The results obtained in this way usually are sufficiently accurate if the temperature range is not greater than a few hundred degrees.

The entropy-change relations for ideal gases under the constant-specific-heat assumption are easily obtained by replacing $c_v(T)$ and $c_p(T)$ in Eqs. 8–31 and 8–32 by $c_{v,avg}$ and $c_{p,avg}$, respectively, and performing the integrations. We obtain

$$s_2 - s_1 = c_{v,avg} \ln \frac{T_2}{T_1} + R \ln \frac{v_2}{v_1} \qquad \text{(kJ/kg·K)} \qquad \text{(8–33)}$$

and

$$s_2 - s_1 = c_{p,avg} \ln \frac{T_2}{T_1} - R \ln \frac{P_2}{P_1} \qquad \text{(kJ/kg·K)} \qquad \text{(8–34)}$$

Entropy changes can also be expressed on a unit-mole basis by multiplying these relations by molar mass:

$$\bar{s}_2 - \bar{s}_1 = \bar{c}_{v,avg} \ln \frac{T_2}{T_1} + R_u \ln \frac{v_2}{v_1} \qquad \text{(kJ/kmol·K)} \qquad \text{(8–35)}$$

and

$$\bar{s}_2 - \bar{s}_1 = \bar{c}_{p,avg} \ln \frac{T_2}{T_1} - R_u \ln \frac{P_2}{P_1} \qquad \text{(kJ/kmol·K)} \qquad \text{(8–36)}$$

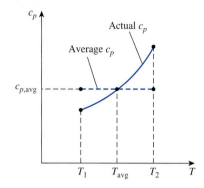

**FIGURE 8–32**

Under the constant-specific-heat assumption, the specific heat is assumed to be constant at some average value.

## Variable Specific Heats (Exact Analysis)

When the temperature change during a process is large and the specific heats of the ideal gas vary nonlinearly within the temperature range, the assumption of constant specific heats may lead to considerable errors in entropy-change calculations. For those cases, the variation of specific heats with temperature should be properly accounted for by utilizing accurate relations for the specific heats as a function of temperature. The entropy change during a process is then determined by substituting these $c_v(T)$ or $c_p(T)$ relations into Eq. 8–31 or 8–32 and performing the integrations.

Instead of performing these laborious integrals each time we have a new process, it is convenient to perform these integrals once and tabulate the results. For this purpose, we choose absolute zero as the reference temperature and define a function $s°$ as

$$s° = \int_0^T c_p(T) \frac{dT}{T} \qquad \text{(8–37)}$$

Obviously, $s°$ is a function of temperature alone, and its value is zero at absolute zero temperature. The values of $s°$ are calculated at various temperatures, and the results are tabulated in the appendix as a function of temperature for air. Given this definition, the integral in Eq. 8–32 becomes

$$\int_1^2 c_p(T) \frac{dT}{T} = s_2° - s_1° \qquad \text{(8–38)}$$

| T, K | s°, kJ/kg·K |
|------|-------------|
| . | . |
| . | . |
| . | . |
| 300 | 1.70203 |
| 310 | 1.73498 |
| 320 | 1.76690 |
| . | . |
| . | . |
| . | . |

(Table A–21)

**FIGURE 8–33**

The entropy of an ideal gas depends on both $T$ and $P$. The function $s°$ represents only the temperature-dependent part of entropy.

where $s_2^\circ$ is the value of $s°$ at $T_2$ and $s_1^\circ$ is the value at $T_1$. Thus,

$$s_2 - s_1 = s_2^\circ - s_1^\circ - R \ln \frac{P_2}{P_1} \qquad \text{(kJ/kg·K)} \qquad \text{(8–39)}$$

It can also be expressed on a unit-mole basis as

$$\bar{s}_2 - \bar{s}_1 = \bar{s}_2^\circ - \bar{s}_1^\circ - R_u \ln \frac{P_2}{P_1} \qquad \text{(kJ/kmol·K)} \qquad \text{(8–40)}$$

Note that unlike internal energy and enthalpy, the entropy of an ideal gas varies with specific volume or pressure as well as the temperature. Therefore, entropy cannot be tabulated as a function of temperature alone. The $s°$ values in the tables account for the temperature dependence of entropy (Fig. 8–33). The variation of entropy with pressure is accounted for by the last term in Eq. 8–39. Another relation for entropy change can be developed based on Eq. 8–31, but this would require the definition of another function and tabulation of its values, which is not practical.

---

**EXAMPLE 8–9    Entropy Change of an Ideal Gas**

Air is compressed from an initial state of 100 kPa and 17°C to a final state of 600 kPa and 57°C. Determine the entropy change of air during this compression process by using (a) property values from the air table and (b) average specific heats.

**SOLUTION**   Air is compressed between two specified states. The entropy change of air is to be determined by using tabulated property values and also by using average specific heats.
**Assumptions**   Air is an ideal gas since it is at a high temperature and low pressure relative to its critical-point values. Therefore, entropy change relations developed under the ideal-gas assumption are applicable.
**Analysis**   A sketch of the system and the T-s diagram for the process are given in Fig. 8–34. We note that both the initial and the final states of air are completely specified.

(a) The properties of air are given in the air table (Table A–21). Reading $s°$ values at given temperatures and substituting, we find

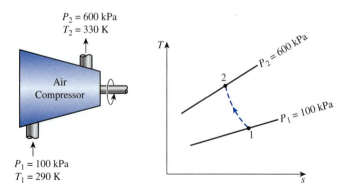

**FIGURE 8–34**

Schematic and T-s diagram for Example 8–9.

$$s_2 - s_1 = s_2^\circ - s_1^\circ - R \ln \frac{P_2}{P_1}$$

$$= [(1.79783 - 1.66802) \text{ kJ/kg·K}] - (0.287 \text{ kJ/kg·K}) \ln \frac{600 \text{ kPa}}{100 \text{ kPa}}$$

$$= -0.3844 \text{ kJ/kg·K}$$

(b) The entropy change of air during this process can also be determined approximately from Eq. 8–34 by using a $c_p$ value at the average temperature of 37°C (Table A–2b) and treating it as a constant:

$$s_2 - s_1 = c_{p,\text{avg}} \ln \frac{T_2}{T_1} - R \ln \frac{P_2}{P_1}$$

$$= (1.006 \text{ kJ/kg·K}) \ln \frac{330 \text{ K}}{290 \text{ K}} - (0.287 \text{ kJ/kg·K}) \ln \frac{600 \text{ kPa}}{100 \text{ kPa}}$$

$$= -0.3842 \text{ kJ/kg·K}$$

**Discussion** The two results above are almost identical since the change in temperature during this process is relatively small (Fig. 8–35). When the temperature change is large, however, they may differ significantly. For those cases, Eq. 8–39 should be used instead of Eq. 8–34 since it accounts for the variation of specific heats with temperature.

**FIGURE 8–35**

For small temperature differences, the exact and approximate relations for entropy changes of ideal gases give almost identical results.

## Isentropic Processes of Ideal Gases

Several relations for the isentropic processes of ideal gases can be obtained by setting the entropy-change relations developed previously equal to zero. Again, this is done first for the case of constant specific heats and then for the case of variable specific heats.

## Constant Specific Heats (Approximate Analysis)

When the constant-specific-heat assumption is valid, the isentropic relations for ideal gases are obtained by setting Eqs. 8–33 and 8–34 equal to zero. From Eq. 8–33,

$$\ln \frac{T_2}{T_1} = -\frac{R}{c_v} \ln \frac{v_2}{v_1}$$

which can be rearranged as

$$\ln \frac{T_2}{T_1} = \ln \left( \frac{v_1}{v_2} \right)^{R/c_v} \tag{8–41}$$

or

$$\left( \frac{T_2}{T_1} \right)_{s=\text{const.}} = \left( \frac{v_1}{v_2} \right)^{k-1} \qquad (\text{ideal gas}) \tag{8–42}$$

since $R = c_p - c_v$, $k = c_p/c_v$, and thus $R/c_v = k - 1$.

Equation 8–42 is the *first isentropic relation* for ideal gases under the constant-specific-heat assumption. The *second isentropic relation* is obtained in a similar manner from Eq. 8–34 with the following result:

$$\left( \frac{T_2}{T_1} \right)_{s=\text{const.}} = \left( \frac{P_2}{P_1} \right)^{(k-1)/k} \qquad (\text{ideal gas}) \tag{8–43}$$

**FIGURE 8–36**

The isentropic relations of ideal gases are valid for the isentropic processes of ideal gases only.

The *third isentropic relation* is obtained by substituting Eq. 8–43 into Eq. 8–42 and simplifying:

$$\left(\frac{P_2}{P_1}\right)_{s=\text{const.}} = \left(\frac{v_1}{v_2}\right)^k \qquad \text{(ideal gas)} \qquad \textbf{(8–44)}$$

Equations 8–42 through 8–44 can also be expressed in a compact form as

$$Tv^{k-1} = \text{constant} \qquad\qquad \textbf{(8–45)}$$

$$TP^{(1-k)/k} = \text{constant} \qquad \text{(ideal gas)} \qquad \textbf{(8–46)}$$

$$Pv^k = \text{constant} \qquad\qquad \textbf{(8–47)}$$

The specific heat ratio $k$, in general, varies with temperature, and thus an average $k$ value for the given temperature range should be used.

Note that the ideal-gas isentropic relations above, as the name implies, are strictly valid for isentropic processes only when the constant-specific-heat assumption is appropriate (Fig. 8–36).

## Variable Specific Heats (Exact Analysis)

When the constant-specific-heat assumption is not appropriate, the isentropic relations developed previously yields results that are not quite accurate. For such cases, we should use an isentropic relation obtained from Eq. 8–39 that accounts for the variation of specific heats with temperature. Setting this equation equal to zero gives

$$0 = s_2^\circ - s_1^\circ - R \ln \frac{P_2}{P_1}$$

or

$$s_2^\circ = s_1^\circ + R \ln \frac{P_2}{P_1} \qquad\qquad \textbf{(8–48)}$$

where $s_2^\circ$ is the $s^\circ$ value at the end of the isentropic process.

## Relative Pressure and Relative Specific Volume

Equation 8–48 provides an accurate way of evaluating property changes of ideal gases during isentropic processes since it accounts for the variation of specific heats with temperature. However, it involves tedious iterations when the volume ratio is given instead of the pressure ratio. This is quite an inconvenience in optimization studies, which usually require numerous repetitive calculations. To remedy this deficiency, we define two new dimensionless quantities associated with isentropic processes.

The definition of the first is based on Eq. 8–48, which can be rearranged as

$$\frac{P_2}{P_1} = \exp \frac{s_2^\circ - s_1^\circ}{R}$$

or

$$\frac{P_2}{P_1} = \frac{\exp(s_2^\circ / R)}{\exp(s_1^\circ / R)}$$

The quantity $\exp(s°/R)$ is defined as the **relative pressure** $P_r$. With this definition, the last relation becomes

$$\left(\frac{P_2}{P_1}\right)_{s=\text{const.}} = \frac{P_{r2}}{P_{r1}} \tag{8-49}$$

Note that the relative pressure $P_r$ is a *dimensionless* quantity that is a function of temperature only since $s°$ depends on temperature alone. Therefore, values of $P_r$ can be tabulated against temperature. This is done for air in Table A–21. The use of $P_r$ data is illustrated in Fig. 8–37.

Sometimes specific volume ratios are given instead of pressure ratios. This is particularly the case when automotive engines are analyzed. In such cases, one needs to work with volume ratios. Therefore, we define another quantity related to specific volume ratios for isentropic processes. This is done by utilizing the ideal-gas relation and Eq. 8–49:

$$\frac{P_1 v_1}{T_1} = \frac{P_2 v_2}{T_2} \rightarrow \frac{v_2}{v_1} = \frac{T_2}{T_1}\frac{P_1}{P_2} = \frac{T_2}{T_1}\frac{P_{r1}}{P_{r2}} = \frac{T_2/P_{r2}}{T_1/P_{r1}}$$

The quantity $T/P_r$ is a function of temperature only and is defined as **relative specific volume** $v_r$. Thus,

$$\left(\frac{v_2}{v_1}\right)_{s=\text{const.}} = \frac{v_{r2}}{v_{r1}} \tag{8-50}$$

Equations 8–49 and 8–50 are strictly valid for isentropic processes of ideal gases only. They account for the variation of specific heats with temperature and therefore give more accurate results than Eqs. 8–42 through 8–47. The values of $P_r$ and $v_r$ are listed for air in Table A–21.

**FIGURE 8–37**
The use of $P_r$ data for calculating the final temperature during an isentropic process.

---

### EXAMPLE 8–10    Isentropic Compression of Air in a Car Engine

Air is compressed in a car engine from 22°C and 95 kPa in a reversible and adiabatic manner. If the compression ratio $V_1/V_2$ of this engine is 8, determine the final temperature of the air.

**SOLUTION** Air is compressed in a car engine isentropically. For a given compression ratio, the final air temperature is to be determined.
**Assumptions** At specified conditions, air can be treated as an ideal gas. Therefore, the isentropic relations for ideal gases are applicable.
**Analysis** A sketch of the system and the $T$-$s$ diagram for the process are given in Fig. 8–38.

This process is easily recognized as being isentropic since it is both reversible and adiabatic. The final temperature for this isentropic process can be determined from Eq. 8–50 with the help of relative specific volume data (Table A–21), as illustrated in Fig. 8–39.

For closed systems: $$\frac{V_2}{V_1} = \frac{v_2}{v_1}$$

At $T_1 = 295$ K: $$v_{r1} = 647.9$$

From Eq. 8–50: $$v_{r2} = v_{r1}\left(\frac{v_2}{v_1}\right) = (647.9)\left(\frac{1}{8}\right) = 80.99 \rightarrow T_2 = \mathbf{662.7\ K}$$

**FIGURE 8–38**
Schematic and $T$-$s$ diagram for Example 8–10.

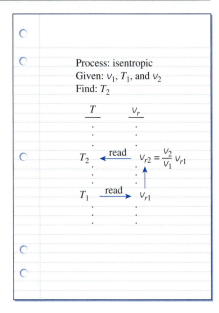

**FIGURE 8–39**

The use of $v_r$ data for calculating the final temperature during an isentropic process (Example 8–10).

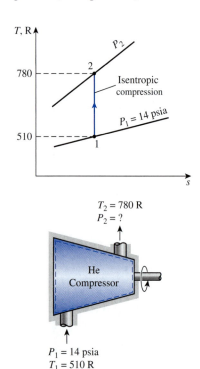

**FIGURE 8–40**

Schematic and $T$-$s$ diagram for Example 8–11.

Therefore, the temperature of air will increase by 367.7°C during this process.

**ALTERNATIVE SOLUTION** The final temperature could also be determined from Eq. 8–42 by assuming constant specific heats for air:

$$\left(\frac{T_2}{T_1}\right)_{s=\text{const.}} = \left(\frac{v_1}{v_2}\right)^{k-1}$$

The specific heat ratio $k$ also varies with temperature, and we need to use the value of $k$ corresponding to the average temperature. However, the final temperature is not given, and so we cannot determine the average temperature in advance. For such cases, calculations can be started with a $k$ value at the initial or the anticipated average temperature. This value could be refined later, if necessary, and the calculations can be repeated. We know that the temperature of the air will rise considerably during this adiabatic compression process, so we *guess* the average temperature to be about 450 K. The $k$ value at this anticipated average temperature is determined from Table A–2b to be 1.391. Then, the final temperature of air becomes

$$T_2 = (295\text{ K})(8)^{1.391-1} = 665.2\text{ K}$$

This gives an average temperature value of 480.1 K, which is sufficiently close to the assumed value of 450 K. Therefore, it is not necessary to repeat the calculations by using the $k$ value at this average temperature.

The result obtained by assuming constant specific heats for this case is in error by about 0.4 percent, which is rather small. This is not surprising since the temperature change of air is relatively small (only a few hundred degrees) and the specific heats of air vary almost linearly with temperature in this temperature range.

**EXAMPLE 8–11    Isentropic Compression of an Ideal Gas**

Helium gas is compressed by an adiabatic compressor from an initial state of 14 psia and 50°F to a final temperature of 320°F in a reversible manner. Determine the exit pressure of helium.

**SOLUTION**   Helium is compressed from a given state to a specified pressure isentropically. The exit pressure of helium is to be determined.
**Assumptions**   At specified conditions, helium can be treated as an ideal gas. Therefore, the isentropic relations developed earlier for ideal gases are applicable.
**Analysis**   A sketch of the system and the $T$-$s$ diagram for the process are given in Fig. 8–40.

The specific heat ratio $k$ of helium is 1.667 and is independent of temperature in the region where it behaves as an ideal gas. Thus the final pressure of helium can be determined from Eq. 8–43:

$$P_2 = P_1\left(\frac{T_2}{T_1}\right)^{k/(k-1)} = (14\text{ psia})\left(\frac{780\text{ R}}{510\text{ R}}\right)^{1.667/0.667} = \textbf{40.5 psia}$$

# 8–10 ▪ REVERSIBLE STEADY-FLOW WORK

The work done during a process depends on the path followed as well as on the properties at the end states. Recall that reversible (quasi-equilibrium) moving boundary work associated with closed systems is expressed in terms of the fluid properties as

$$W_b = \int_1^2 P \, dV$$

We mentioned that the quasi-equilibrium work interactions lead to the maximum work output for work-producing devices and the minimum work input for work-consuming devices.

It would also be very insightful to express the work associated with steady-flow devices in terms of fluid properties.

Taking the positive direction of work to be from the system (work output), the energy balance for a steady-flow device undergoing an internally reversible process can be expressed in differential form as

$$\delta q_{rev} - \delta w_{rev} = dh + dke + dpe$$

But

$$\left. \begin{array}{ll} \delta q_{rev} = T \, ds & \text{(Eq. 8–16)} \\ T \, ds = dh - v \, dP & \text{(Eq. 8–24)} \end{array} \right\} \quad \delta q_{rev} = dh - v \, dP$$

Substituting this into the relation above and canceling $dh$ yield

$$-\delta w_{rev} = v \, dP + dke + dpe$$

Integrating, we find

$$w_{rev} = -\int_1^2 v \, dP - \Delta ke - \Delta pe \qquad \text{(kJ/kg)} \qquad \textbf{(8–51)}$$

When the changes in kinetic and potential energies are negligible, this equation reduces to

$$w_{rev} = -\int_1^2 v \, dP \qquad \text{(kJ/kg)} \qquad \textbf{(8–52)}$$

Equations 8–51 and 8–52 are relations for the *reversible work output* associated with an internally reversible process in a steady-flow device. They will give a negative result when work is done on the system. To avoid the negative sign, Eq. 8–51 can be written for work input to steady-flow devices such as compressors and pumps as

$$w_{rev,in} = \int_1^2 v \, dP + \Delta ke + \Delta pe \qquad \textbf{(8–53)}$$

The resemblance between the $v \, dP$ in these relations and $P \, dv$ is striking. They should not be confused with each other, however, since $P \, dv$ is associated with reversible boundary work in closed systems (Fig. 8–41).

Obviously, one needs to know $v$ as a function of $P$ for the given process to perform the integration. When the working fluid is *incompressible*, the

$$w_{rev} = -\int_1^2 v \, dP$$

$w_{rev}$

(a) Steady-flow system

$w_{rev}$

$$w_{rev} = \int_1^2 P \, dv$$

(b) Closed system

**FIGURE 8–41**

Reversible work relations for steady-flow and closed systems.

specific volume $v$ remains constant during the process and can be taken out of the integration. Then Eq. 8–51 simplifies to

$$w_{rev} = -v(P_2 - P_1) - \Delta ke - \Delta pe \qquad (kJ/kg) \qquad \textbf{(8–54)}$$

For the steady flow of a liquid through a device that involves no work interactions (such as a nozzle or a pipe section), the work term is zero, and the equation above can be expressed as

$$v(P_2 - P_1) + \frac{V_2^2 - V_2^2}{2} + g(z_2 - z_1) = 0 \qquad \textbf{(8–55)}$$

which is known as the **Bernoulli equation** in fluid mechanics. It is developed for an internally reversible process and thus is applicable to incompressible fluids that involve no irreversibilities such as friction or shock waves. This equation can be modified, however, to incorporate these effects.

Equation 8–52 has far-reaching implications in engineering regarding devices that produce or consume work steadily such as turbines, compressors, and pumps. It is obvious from this equation that the reversible steady-flow work is closely associated with the specific volume of the fluid flowing through the device. *The larger the specific volume, the larger the reversible work produced or consumed by the steady-flow device* (Fig. 8–42). This conclusion is equally valid for actual steady-flow devices. Therefore, every effort should be made to keep the specific volume of a fluid as small as possible during a compression process to minimize the work input and as large as possible during an expansion process to maximize the work output.

In steam or gas power plants, the pressure rise in the pump or compressor is equal to the pressure drop in the turbine if we disregard the pressure losses in various other components. In steam power plants, the pump handles liquid, which has a very small specific volume, and the turbine handles vapor, whose specific volume is many times larger. Therefore, the work output of the turbine is much larger than the work input to the pump. This is one of the reasons for the wide-spread use of steam power plants in electric power generation.

If we were to compress the steam exiting the turbine back to the turbine inlet pressure before cooling it first in the condenser in order to "save" the heat rejected, we would have to supply all the work produced by the turbine back to the compressor. In reality, the required work input would be even greater than the work output of the turbine because of the irreversibilities present in both processes.

In gas power plants, the working fluid (typically air) is compressed in the gas phase, and a considerable portion of the work output of the turbine is consumed by the compressor. As a result, a gas power plant delivers less net work per unit mass of the working fluid.

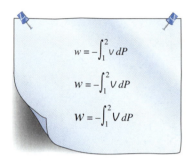

**FIGURE 8–42**

The larger the specific volume, the greater the work produced (or consumed) by a steady-flow device.

---

**EXAMPLE 8–12**    **Compressing a Substance in the Liquid versus Gas Phases**

Determine the compressor work input required to compress steam isentropically from 100 kPa to 1 MPa, assuming that the steam exists as (a) saturated liquid and (b) saturated vapor at the inlet state.

**SOLUTION** Steam is to be compressed from a given pressure to a specified pressure isentropically. The work input is to be determined for the cases of steam being a saturated liquid and saturated vapor at the inlet.

**Assumptions** **1** Steady operating conditions exist. **2** Kinetic and potential energy changes are negligible. **3** The process is given to be isentropic.

**Analysis** We take first the turbine and then the pump as the *system*. Both are *control volumes* since mass crosses the boundary. Sketches of the pump and the turbine together with the *T-s* diagram are given in Fig. 8–43.

(*a*) In this case, steam is a saturated liquid initially, and its specific volume is

$$v_1 = v_{f\,@\,100\,kPa} = 0.001043 \text{ m}^3/\text{kg} \qquad \text{(Table A–5)}$$

which remains essentially constant during the process. Thus,

$$
\begin{aligned}
w_{rev,in} &= \int_1^2 v\,dP \cong v_1(P_2 - P_1) \\
&= (0.001043 \text{ m}^3/\text{kg})[(1000 - 100)\text{ kPa}]\left(\frac{1 \text{ kJ}}{1 \text{ kPa·m}^3}\right) \\
&= \mathbf{0.94 \text{ kJ/kg}}
\end{aligned}
$$

(*b*) This time, steam is a saturated vapor initially and remains a vapor during the entire compression process. Since the specific volume of a gas changes considerably during a compression process, we need to know how *v* varies with *P* to perform the integration in Eq. 8–53. This relation, in general, is not readily available. But for an isentropic process, it is easily obtained from the second *T ds* relation by setting *ds* = 0:

$$
\left.
\begin{array}{l}
T\,ds = dh - v\,dP \quad \text{(Eq. 8-24)} \\
ds = 0 \quad \text{(isentropic process)}
\end{array}
\right\} \quad v\,dP = dh
$$

Thus,

$$w_{rev,in} = \int_1^2 v\,dP = \int_1^2 dh = h_2 - h_1$$

This result could also be obtained from the energy balance relation for an isentropic steady-flow process. Next we determine the enthalpies:

*State 1:*  
$$\left.\begin{array}{l} P_1 = 100 \text{ kPa} \\ \text{(sat. vapor)} \end{array}\right\} \quad \begin{array}{l} h_1 = 2675.0 \text{ kJ/kg} \\ s_1 = 7.3589 \text{ kJ/kg·K} \end{array} \qquad \text{(Table A–5)}$$

*State 2:*  
$$\left.\begin{array}{l} P_2 = 1 \text{ MPa} \\ s_2 = s_1 \end{array}\right\} \quad h_2 = 3194.5 \text{ kJ/kg} \qquad \text{(Table A–6)}$$

Thus,

$$w_{rev,in} = (3194.5 - 2675.0) \text{ kJ/kg} = \mathbf{519.5 \text{ kJ/kg}}$$

**Discussion** Note that compressing steam in the vapor form would require over 500 times more work than compressing it in the liquid form between the same pressure limits.

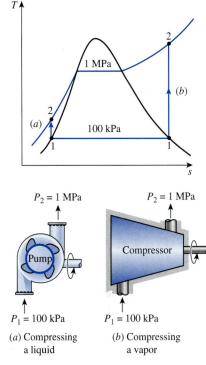

**FIGURE 8–43**

Schematic and *T-s* diagram for Example 8–12.

## Proof that Steady-Flow Devices Deliver the Most and Consume the Least Work When the Process Is Reversible

We have shown in Chap. 7 that cyclic devices (heat engines, refrigerators, and heat pumps) deliver the most work and consume the least when reversible processes are used. Now we demonstrate that this is also the case for individual devices such as turbines and compressors in steady operation.

Consider two steady-flow devices, one reversible and the other irreversible, operating between the same inlet and exit states. Again taking heat transfer to the system and work done by the system to be positive quantities, the energy balance for each of these devices can be expressed in the differential form as

Actual: $$\delta q_{act} - \delta w_{act} = dh + dke + dpe$$

Reversible: $$\delta q_{rev} - \delta w_{rev} = dh + dke + dpe$$

The right-hand sides of these two equations are identical since both devices are operating between the same end states. Thus,

$$\delta q_{act} - \delta w_{act} = \delta q_{rev} - \delta w_{rev}$$

or

$$\delta w_{rev} - \delta w_{act} = \delta q_{rev} - \delta q_{act}$$

However,

$$\delta q_{rev} = T\,ds$$

Substituting this relation into the preceding equation and dividing each term by $T$, we obtain

$$\frac{\delta w_{rev} - \delta w_{act}}{T} = ds - \frac{\delta q_{act}}{T} \geq 0$$

since

$$ds \geq \frac{\delta q_{act}}{T}$$

Also, $T$ is the absolute temperature, which is always positive. Thus,

$$\partial w_{rev} \geq \partial w_{act}$$

or

$$w_{rev} \geq w_{act}$$

Therefore, work-producing devices such as turbines ($w$ is positive) deliver more work, and work-consuming devices such as pumps and compressors ($w$ is negative) require less work when they operate reversibly (Fig. 8–44).

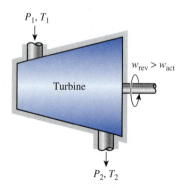

$P_1, T_1$

$w_{rev} > w_{act}$

Turbine

$P_2, T_2$

**FIGURE 8–44**

A reversible turbine delivers more work than an irreversible one if both operate between the same end states.

# 8–11 · ISENTROPIC EFFICIENCIES OF STEADY-FLOW DEVICES

We mentioned repeatedly that irreversibilities inherently accompany all actual processes and that their effect is always to downgrade the performance of devices. In engineering analysis, it would be very desirable to have some parameters that would enable us to quantify the degree of degradation of energy in these devices. In the last chapter we did this for cyclic devices, such as heat engines and refrigerators, by comparing the actual cycles to the idealized ones, such as the Carnot cycle. A cycle that was composed entirely of reversible processes served as the *model cycle* to which the actual cycles could be compared. This idealized model cycle enabled us to determine the theoretical limits of performance for cyclic devices under specified conditions and to examine how the performance of actual devices suffered as a result of irreversibilities.

Now we extend the analysis to discrete engineering devices working under steady-flow conditions, such as turbines, compressors, and nozzles, and we examine the degree of degradation of energy in these devices as a result of irreversibilities. However, first we need to define an ideal process that serves as a model for the actual processes.

Although some heat transfer between these devices and the surrounding medium is unavoidable, many steady-flow devices are intended to operate under adiabatic conditions. Therefore, the model process for these devices should be an adiabatic one. Furthermore, an ideal process should involve no irreversibilities since the effect of irreversibilities is always to downgrade the performance of engineering devices. Thus, the ideal process that can serve as a suitable model for adiabatic steady-flow devices is the *isentropic* process (Fig. 8–45).

The more closely the actual process approximates the idealized isentropic process, the better the device performs. Thus, it would be desirable to have a parameter that expresses quantitatively how efficiently an actual device approximates an idealized one. This parameter is the **isentropic** or **adiabatic efficiency**, which is a measure of the deviation of actual processes from the corresponding idealized ones.

Isentropic efficiencies are defined differently for different devices since each device is set up to perform different tasks. Next, we define the isentropic efficiencies of turbines, compressors, and nozzles by comparing the actual performance of these devices to their performance under isentropic conditions for the same inlet state and exit pressure.

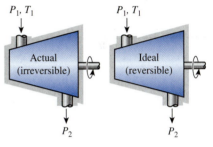

**FIGURE 8–45**

The isentropic process involves no irreversibilities and serves as the ideal process for adiabatic devices.

## Isentropic Efficiency of Turbines

For a turbine under steady operation, the inlet state of the working fluid and the exhaust pressure are fixed. Therefore, the ideal process for an adiabatic turbine is an isentropic process between the inlet state and the exhaust pressure. The desired output of a turbine is the work produced, and the **isentropic efficiency of a turbine** is defined as *the ratio of the actual work output of the turbine to the work output that would be*

**FIGURE 8–46**

The h-s diagram for the actual and isentropic processes of an adiabatic turbine.

*achieved if the process between the inlet state and the exit pressure were isentropic:*

$$\eta_T = \frac{\text{Actual turbine work}}{\text{Isentropic turbine work}} = \frac{w_a}{w_s} \qquad \text{(8–56)}$$

Usually the changes in kinetic and potential energies associated with a fluid stream flowing through a turbine are small relative to the change in enthalpy and can be neglected. Then, the work output of an adiabatic turbine simply becomes the change in enthalpy, and Eq. 8–56 becomes

$$\eta_T \cong \frac{h_1 - h_{2a}}{h_1 - h_{2s}} \qquad \text{(8–57)}$$

where $h_{2a}$ and $h_{2s}$ are the enthalpy values at the exit state for actual and isentropic processes, respectively (Fig. 8–46).

The value of $\varepsilon_T$ greatly depends on the design of the individual components that make up the turbine. Well-designed, large turbines have isentropic efficiencies above 90 percent. For small turbines, however, it may drop even below 70 percent. The value of the isentropic efficiency of a turbine is determined by measuring the actual work output of the turbine and by calculating the isentropic work output for the measured inlet conditions and the exit pressure. This value can then be used conveniently in the design of power plants.

**FIGURE 8–47**

Schematic and T-s diagram for Example 8–13.

**EXAMPLE 8–13    Isentropic Efficiency of a Steam Turbine**

Steam enters an adiabatic turbine steadily at 3 MPa and 400°C and leaves at 50 kPa and 100°C. If the power output of the turbine is 2 MW, determine (*a*) the isentropic efficiency of the turbine and (*b*) the mass flow rate of the steam flowing through the turbine.

**SOLUTION** Steam flows steadily in a turbine between inlet and exit states. For a specified power output, the isentropic efficiency and the mass flow rate are to be determined.

**Assumptions** **1** Steady operating conditions exist. **2** The changes in kinetic and potential energies are negligible.

**Analysis** A sketch of the system and the T-s diagram of the process are given in Fig. 8–47.

(*a*) The enthalpies at various states are

State 1:    $\left. \begin{array}{l} P_1 = 3 \text{ MPA} \\ T_1 = 400°C \end{array} \right\}$    $\begin{array}{l} h_1 = 3231.7 \text{ kJ/kg} \\ s_1 = 6.9235 \text{ kJ/kg·K} \end{array}$    (Table A–6)

State 2a:    $\left. \begin{array}{l} P_{2a} = 50 \text{ kPa} \\ T_{2a} = 100°C \end{array} \right\}$    $h_{2a} = 2682.4 \text{ kJ/kg}$    (Table A–6)

The exit enthalpy of the steam for the isentropic process $h_{2s}$ is determined from the requirement that the entropy of the steam remain constant $(s_{2s} = s_1)$:

State 2s: $\quad \begin{aligned} P_{2s} &= 50 \text{ kPa} \\ (s_{2s} &= s_1) \end{aligned} \quad \longrightarrow \quad \begin{aligned} s_f &= 1.0912 \text{ kJ/kg·K} \\ s_g &= 7.5931 \text{ kJ/kg·K} \end{aligned} \quad$ (Table A–5)

Obviously, at the end of the isentropic process steam exists as a saturated mixture since $s_f < s_{2s} < s_g$. Thus, we need to find the quality at state 2s first:

$$x_{2s} = \frac{s_{2s} - s_f}{s_{fg}} = \frac{6.9235 - 1.0912}{6.5019} = 0.897$$

and

$$h_{2s} = h_f + x_{2s}h_{fg} = 340.54 + 0.897(2304.7) = 2407.9 \text{ kJ/kg}$$

By substituting these enthalpy values into Eq. 8–57, the isentropic efficiency of this turbine is determined to be

$$\eta_T \cong \frac{h_1 - h_{2a}}{h_1 - h_{2s}} = \frac{3231.7 - 2682.4}{3231.7 - 2407.9} = \textbf{0.667, or 66.7\%}$$

(b) The mass flow rate of steam through this turbine is determined from the energy balance for steady-flow systems:

$$\dot{E}_{in} = \dot{E}_{out}$$

$$\dot{m}h_1 = \dot{W}_{a,out} + \dot{m}h_{2a}$$

$$\dot{W}_{a,out} = \dot{m}(h_1 - h_{2a})$$

$$2 \text{ MW}\left(\frac{1000 \text{ kJ/s}}{1 \text{ MW}}\right) = \dot{m}(3231.7 - 2682.4) \text{ kJ/kg}$$

$$\dot{m} = \textbf{3.64 kg/s}$$

# Isentropic Efficiencies of Compressors and Pumps

The **isentropic efficiency of a compressor** is defined as *the ratio of the work input required to raise the pressure of a gas to a specified value in an isentropic manner to the actual work input:*

$$\eta_C = \frac{\text{Isentropic compressor work}}{\text{Actual compressor work}} = \frac{w_s}{w_a} \qquad \textbf{(8–58)}$$

Notice that the isentropic compressor efficiency is defined with the *isentropic work input in the numerator* instead of in the denominator. This is because $w_s$ is a smaller quantity than $w_a$, and this definition prevents $\eta_C$ from becoming greater than 100 percent, which would falsely imply that the actual compressors performed better than the isentropic ones. Also notice that the inlet conditions and the exit pressure of the gas are the same for both the actual and the isentropic compressor.

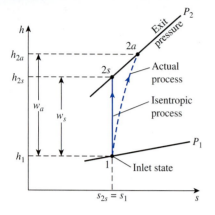

**FIGURE 8–48**

The *h-s* diagram of the actual and isentropic processes of an adiabatic compressor.

**FIGURE 8–49**

Compressors are sometimes intentionally cooled to minimize the work input.

When the changes in kinetic and potential energies of the gas being compressed are negligible, the work input to an adiabatic compressor becomes equal to the change in enthalpy, and Eq. 8–58 for this case becomes

$$\eta_C \cong \frac{h_{2s} - h_1}{h_{2a} - h_1} \tag{8–59}$$

where $h_{2a}$ and $h_{2s}$ are the enthalpy values at the exit state for actual and isentropic compression processes, respectively, as illustrated in Fig. 8–48. Again, the value of $\eta_C$ greatly depends on the design of the compressor. Well-designed compressors have isentropic efficiencies that range from 80 to 90 percent.

When the changes in potential and kinetic energies of a liquid are negligible, the isentropic efficiency of a pump is defined similarly as

$$\eta_P = \frac{w_s}{w_a} = \frac{v(P_2 - P_1)}{h_{2a} - h_1} \tag{8–60}$$

When no attempt is made to cool the gas as it is compressed, the actual compression process is nearly adiabatic and the reversible adiabatic (i.e., isentropic) process serves well as the ideal process. However, sometimes *compressors are cooled intentionally* by utilizing fins or a water jacket placed around the casing to reduce the work input requirements (Fig. 8–49). In this case, the isentropic process is not suitable as the model process since the device is no longer adiabatic and the isentropic compressor efficiency defined above is meaningless. A realistic model process for compressors that are intentionally cooled during the compression process is the *reversible isothermal process*. Then we can conveniently define an **isothermal efficiency** for such cases by comparing the actual process to a reversible isothermal one:

$$\eta_C = \frac{w_t}{w_a} \tag{8–61}$$

where $w_t$ and $w_a$ are the required work inputs to the compressor for the reversible isothermal and actual cases, respectively.

---

**EXAMPLE 8–14    Effect of Efficiency on Compressor Power Input**

Air is compressed by an adiabatic compressor from 100 kPa and 12°C to a pressure of 800 kPa at a steady rate of 0.2 kg/s. If the isentropic efficiency of the compressor is 80 percent, determine (*a*) the exit temperature of air and (*b*) the required power input to the compressor.

**SOLUTION**   Air is compressed to a specified pressure at a specified rate. For a given isentropic efficiency, the exit temperature and the power input are to be determined.

**Assumptions**   **1** Steady operating conditions exist. **2** Air is an ideal gas. **3** The changes in kinetic and potential energies are negligible.

*Analysis* A sketch of the system and the *T-s* diagram of the process are given in Fig. 8–50.

(*a*) We know only one property (pressure) at the exit state, and we need to know one more to fix the state and thus determine the exit temperature. The property that can be determined with minimal effort in this case is $h_{2a}$ since the isentropic efficiency of the compressor is given. At the compressor inlet,

$$T_1 = 285 \text{ K} \quad \rightarrow \quad h_1 = 285.14 \text{ kJ/kg} \qquad \text{(Table A–21)}$$

$$(P_{r1} = 1.1584)$$

The enthalpy of the air at the end of the isentropic compression process is determined by using one of the isentropic relations of ideal gases,

$$P_{r2} = P_{r1}\left(\frac{P_2}{P_1}\right) = 1.1584\left(\frac{800 \text{ kPa}}{100 \text{ kPa}}\right) = 9.2672$$

and

$$P_{r2} = 9.2672 \quad \rightarrow \quad h_{2s} = 517.05 \text{ kJ/kg}$$

Substituting the known quantities into the isentropic efficiency relation, we have

$$\eta_C \cong \frac{h_{2s} - h_1}{h_{2a} - h_1} \quad \rightarrow \quad 0.80 = \frac{(517.05 - 285.14) \text{ kJ/kg}}{(h_{2a} - 285.14) \text{ kJ/kg}}$$

Thus,

$$h_{2a} = 575.03 \text{ kJ/kg} \quad \rightarrow \quad T_{2a} = \mathbf{569.5 \text{ K}}$$

(*b*) The required power input to the compressor is determined from the energy balance for steady-flow devices,

$$\dot{E}_{in} = \dot{E}_{out}$$

$$\dot{m}h_1 + \dot{W}_{a,in} = \dot{m}h_{2a}$$

$$\dot{W}_{a,in} = \dot{m}(h_{2a} - h_1)$$

$$= (0.2 \text{ kg/s})[(575.03 - 285.14) \text{ kJ/kg}]$$

$$= \mathbf{58.0 \text{ kW}}$$

*Discussion* Notice that in determining the power input to the compressor, we used $h_{2a}$ instead of $h_{2s}$ since $h_{2a}$ is the actual enthalpy of the air as it exits the compressor. The quantity $h_{2s}$ is a hypothetical enthalpy value that the air would have if the process were isentropic.

**FIGURE 8–50**

Schematic and *T-s* diagram for Example 8–14.

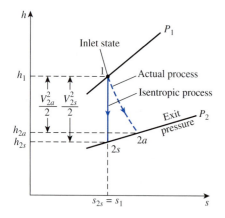

**FIGURE 8–51**

The h-s diagram of the actual and isentropic processes of an adiabatic nozzle.

**FIGURE 8–52**

Schematic and T-s diagram for Example 8–15.

## Isentropic Efficiency of Nozzles

Nozzles are essentially adiabatic devices and are used to accelerate a fluid. Therefore, the isentropic process serves as a suitable model for nozzles. The **isentropic efficiency of a nozzle** is defined as *the ratio of the actual kinetic energy of the fluid at the nozzle exit to the kinetic energy value at the exit of an isentropic nozzle for the same inlet state and exit pressure.* That is,

$$\eta_N = \frac{\text{Actual KE at nozzle exit}}{\text{Isentropic KE at nozzle exit}} = \frac{V_{2a}^2}{V_{2s}^2} \qquad (8\text{–}62)$$

Note that the exit pressure is the same for both the actual and isentropic processes, but the exit state is different.

Nozzles involve no work interactions, and the fluid experiences little or no change in its potential energy as it flows through the device. If, in addition, the inlet velocity of the fluid is small relative to the exit velocity, the energy balance for this steady-flow device reduces to

$$h_1 = h_{2a} + \frac{V_{2a}^2}{2}$$

Then the isentropic efficiency of the nozzle can be expressed in terms of enthalpies as

$$\eta_N \cong \frac{h_1 - h_{2a}}{h_1 - h_{2s}} \qquad (8\text{–}63)$$

where $h_{2a}$ and $h_{2s}$ are the enthalpy values at the nozzle exit for the actual and isentropic processes, respectively (Fig. 8–51). Isentropic efficiencies of nozzles are typically above 90 percent, and nozzle efficiencies above 95 percent are not uncommon.

---

**EXAMPLE 8–15**    **Effect of Efficiency on Nozzle Exit Velocity**

Air at 200 kPa and 950 K enters an adiabatic nozzle at low velocity and is discharged at a pressure of 80 kPa. If the isentropic efficiency of the nozzle is 92 percent, determine (a) the maximum possible exit velocity, (b) the exit temperature, and (c) the actual exit velocity of the air. Assume constant specific heats for air.

**SOLUTION**   The acceleration of air in a nozzle is considered. For specified exit pressure and isentropic efficiency, the maximum and actual exit velocities and the exit temperature are to be determined.

*Assumptions*  **1** Steady operating conditions exist. **2** Air is an ideal gas. **3** The inlet kinetic energy is negligible.

*Analysis*  A sketch of the system and the T-s diagram of the process are given in Fig. 8–52.

The temperature of air will drop during this acceleration process because some of its internal energy is converted to kinetic energy. This problem can be solved accurately by using property data from the air table. But we will assume constant specific heats (thus sacrifice some accuracy) to demonstrate their use. Let us guess the average temperature of the air to be about 800 K. Then, the average values of $c_p$ and $k$ at this anticipated average temperature are determined from Table A–2b to be $c_p = 1.099$ kJ/kg·K and $k = 1.354$.

(a) The exit velocity of the air will be a maximum when the process in the nozzle involves no irreversibilities. The exit velocity in this case is determined from the steady-flow energy equation. However, first we need to determine the exit temperature. For the isentropic process of an ideal gas we have:

$$\frac{T_{2s}}{T_1} = \left(\frac{P_{2s}}{P_1}\right)^{(k-1)/k}$$

or

$$T_{2s} = T_1 \left(\frac{P_{2s}}{P_1}\right)^{(k-1)/k} = (950 \text{ K})\left(\frac{80 \text{ kPa}}{200 \text{ kPa}}\right)^{0.354/1.354} = 748 \text{ K}$$

This gives an average temperature of 849 K, which is somewhat higher than the assumed average temperature (800 K). This result could be refined by reevaluating the $k$ value at 749 K and repeating the calculations, however, it is not warranted since the two average temperatures are sufficiently close (doing so would change the temperature by only 1.5 K, which is not significant).

Now we can determine the isentropic exit velocity of the air from the energy balance for this isentropic steady-flow process:

$$e_{\text{in}} = e_{\text{out}}$$

$$h_1 + \frac{V_1^2}{2} = h_{2s} + \frac{V_{2s}^2}{2}$$

or

$$V_{2s} = \sqrt{2(h_1 - h_{2s})} = \sqrt{2c_{p,\text{avg}}(T_1 - T_{2s})}$$

$$= \sqrt{2(1.099 \text{ kJ/kg·K})[(950 - 748) \text{ K}]\left(\frac{1000 \text{ m}^2/\text{s}^2}{1 \text{ kJ/kg}}\right)}$$

$$= \textbf{666 m/s}$$

(b) The actual exit temperature of the air is higher than the isentropic exit temperature evaluated above and is determined from

$$\eta_N \cong \frac{h_1 - h_{2a}}{h_1 - h_{2s}} = \frac{c_{p,\text{avg}}(T_1 - T_{2a})}{c_{p,\text{avg}}(T_1 - T_{2s})}$$

or

$$0.92 = \frac{950 - T_{2a}}{950 - 748} \quad \rightarrow \quad T_{2a} = \textbf{764 K}$$

That is, the temperature is 16 K higher at the exit of the actual nozzle as a result of irreversibilities such as friction. It represents a loss since this rise in temperature comes at the expense of kinetic energy (Fig. 8–53).

(c) The actual exit velocity of air can be determined from the definition of isentropic efficiency of a nozzle,

$$\eta_N = \frac{V_{2a}^2}{V_{2s}^2} \quad \rightarrow \quad V_{2a} = \sqrt{\eta_N V_{2s}^2} = \sqrt{0.92(666 \text{ m/s})^2} = \textbf{639 m/s}$$

**FIGURE 8–53**

A substance leaves actual nozzles at a higher temperature (thus a lower velocity) as a result of friction.

**FIGURE 8–54**

Energy and entropy balances for a system.

## 8–12 ▪ ENTROPY BALANCE

The property *entropy* is a measure of molecular disorder or randomness of a system, and the second law of thermodynamics states that entropy can be created but it cannot be destroyed. Therefore, the entropy change of a system during a process is greater than the entropy transfer by an amount equal to the entropy generated during the process within the system, and the *increase of entropy principle* for any system is expressed as (Fig. 8–54)

$$\begin{pmatrix} \text{Total} \\ \text{entropy} \\ \text{entering} \end{pmatrix} - \begin{pmatrix} \text{Total} \\ \text{entropy} \\ \text{leaving} \end{pmatrix} + \begin{pmatrix} \text{Total} \\ \text{entropy} \\ \text{generated} \end{pmatrix} = \begin{pmatrix} \text{Change in the} \\ \text{total entropy} \\ \text{of the system} \end{pmatrix}$$

or

$$S_{\text{in}} - S_{\text{out}} + S_{\text{gen}} = \Delta S_{\text{system}} \tag{8–64}$$

which is a verbal statement of Eq. 8–9. This relation is often referred to as the **entropy balance** and is applicable to any system undergoing any process. The entropy balance relation above can be stated as: *the entropy change of a system during a process is equal to the net entropy transfer through the system boundary and the entropy generated within the system.* Next, we discuss the various terms in that relation.

### Entropy Change of a System, $\Delta S_{\text{system}}$

Despite the reputation of entropy as being vague and abstract and the intimidation associated with it, entropy balance is actually easier to deal with than energy balance since, unlike energy, entropy does not exist in various forms. Therefore, the determination of entropy change of a system during a process involves evaluating entropy of the system at the beginning and at the end of the process and taking their difference. That is,

Entropy change = Entropy at final state − Entropy at initial state

or

$$\Delta S_{\text{system}} = S_{\text{final}} - S_{\text{initial}} = S_2 - S_1 \tag{8–65}$$

Note that entropy is a property, and the value of a property does not change unless the state of the system changes. Therefore, the entropy change of a system is zero if the state of the system does not change during the process. For example, the entropy change of steady-flow devices such as nozzles, compressors, turbines, pumps, and heat exchangers is zero during steady operation.

When the properties of the system are not uniform, the entropy of the system can be determined by integration from

$$S_{\text{system}} = \int s \, \delta m = \int_V s\rho \, dV \tag{8–66}$$

where $V$ is the volume of the system and $\rho$ is density.

# Mechanisms of Entropy Transfer, $S_{in}$ and $S_{out}$

Entropy can be transferred to or from a system by two mechanisms: *heat transfer* and *mass flow* (in contrast, energy is transferred by work also). Entropy transfer is recognized at the system boundary as it crosses the boundary, and it represents the entropy gained or lost by a system during a process. The only form of entropy interaction associated with a fixed mass or closed system is *heat transfer*, and thus the entropy transfer for an adiabatic closed system is zero.

## 1 Heat Transfer

Heat is, in essence, a form of disorganized energy, and some disorganization (entropy) will flow with heat. Heat transfer to a system increases the entropy of that system and thus the level of molecular disorder or randomness, and heat transfer from a system decreases it. In fact, heat rejection is the only way the entropy of a fixed mass can be decreased. The ratio of the heat transfer $Q$ at a location to the absolute temperature $T$ at that location is called the *entropy flow* or *entropy transfer* and is expressed as (Fig. 8–55)

Surroundings

$T_b = 400$ K

System

$Q = 500$ kJ

$S_{heat} = \dfrac{Q}{T_b}$

$= 1.25$ kJ/K

**FIGURE 8–55**

Heat transfer is always accompanied by entropy transfer in the amount of $Q/T$, where $T$ is the boundary temperature.

*Entropy transfer by heat transfer:* $\qquad S_{heat} = \dfrac{Q}{T} \qquad (T = \text{constant}) \qquad$ **(8–67)**

The quantity $Q/T$ represents the entropy transfer accompanied by heat transfer, and the direction of entropy transfer is the same as the direction of heat transfer since thermodynamic temperature $T$ is always a positive quantity.

When the temperature $T$ is not constant, the entropy transfer during a process 1-2 can be determined by integration (or by summation if appropriate) as

$$S_{heat} = \int_1^2 \frac{\delta Q}{T} \cong \sum \frac{Q_k}{T_k} \qquad \textbf{(8–68)}$$

where $Q_k$ is the heat transfer through the boundary at temperature $T_k$ at location $k$.

When two systems are in contact, the entropy transfer from the warmer system is equal to the entropy transfer into the cooler one at the point of contact. That is, no entropy can be created or destroyed at the boundary since the boundary has no thickness and occupies no volume.

Note that **work** is entropy-free, and no entropy is transferred by work. Energy is transferred by both heat and work, whereas entropy is transferred only by heat. That is,

*Entropy transfer by work:* $\qquad S_{work} = 0 \qquad$ **(8–69)**

The first law of thermodynamics makes no distinction between heat transfer and work; it considers them as *equals*. The distinction between heat transfer and work is brought out by the second law: *an energy interaction that is accompanied by entropy transfer is heat transfer, and an energy interaction that is not accompanied by entropy transfer is work*. That is, no entropy is exchanged during a work interaction between a system and its surroundings. Thus, only *energy* is exchanged during work interaction whereas both *energy* and *entropy* are exchanged during heat transfer (Fig. 8–56).

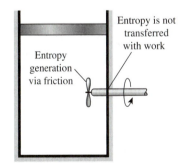

Entropy is not transferred with work

Entropy generation via friction

**FIGURE 8–56**

No entropy accompanies work as it crosses the system boundary. But entropy may be generated within the system as work is dissipated into a less useful form of energy.

**FIGURE 8–57**

Mass contains entropy as well as energy, and thus mass flow into or out of system is always accompanied by energy and entropy transfer.

## 2 Mass Flow

Mass contains entropy as well as energy, and the entropy and energy contents of a system are proportional to the mass. (When the mass of a system is doubled, so are the entropy and energy contents of the system.) Both entropy and energy are carried into or out of a system by streams of matter, and the rates of entropy and energy transport into or out of a system are proportional to the mass flow rate. Closed systems do not involve any mass flow and thus any entropy transfer by mass. When a mass in the amount of $m$ enters or leaves a system, entropy in the amount of $ms$, where $s$ is the specific entropy (entropy per unit mass entering or leaving), accompanies it (Fig. 8–57). That is,

*Entropy transfer by mass flow:* $\qquad S_{mass} = ms$ $\qquad$ **(8–70)**

Therefore, the entropy of a system increases by $ms$ when mass in the amount of $m$ enters and decreases by the same amount when the same amount of mass at the same state leaves the system. When the properties of the mass change during the process, the entropy transfer by mass flow can be determined by integration from

$$\dot{S}_{mass} = \int_{A_c} s\rho V_n \, dA_c \quad \text{and} \quad S_{mass} = \int s\,\delta m = \int_{\Delta t} \dot{S}_{mass}\, dt \qquad \text{(8–71)}$$

where $A_c$ is the cross-sectional area of the flow and $V_n$ is the local velocity normal to $dA_c$.

## Entropy Generation, $S_{gen}$

Irreversibilities such as friction, mixing, chemical reactions, heat transfer through a finite temperature difference, unrestrained expansion, nonquasi-equilibrium compression, or expansion always cause the entropy of a system to increase, and entropy generation is a measure of the entropy created by such effects during a process.

For a *reversible process* (a process that involves no irreversibilities), the entropy generation is zero and thus the *entropy change* of a system is equal to the *entropy transfer*. Therefore, the entropy balance relation in the reversible case becomes analogous to the energy balance relation, which states that *energy change* of a system during a process is equal to the *energy transfer* during that process. However, note that the energy change of a system equals the energy transfer for *any* process, but the entropy change of a system equals the entropy transfer only for a *reversible* process.

The entropy transfer by heat $Q/T$ is zero for adiabatic systems, and the entropy transfer by mass $ms$ is zero for systems that involve no mass flow across their boundary (i.e., closed systems).

Entropy balance for *any system* undergoing *any process* can be expressed more explicitly as

$$\underbrace{S_{in} - S_{out}}_{\substack{\text{Net entropy transfer} \\ \text{by heat and mass}}} + \underbrace{S_{gen}}_{\substack{\text{Entropy} \\ \text{generation}}} = \underbrace{\Delta S_{system}}_{\substack{\text{Change} \\ \text{in entropy}}} \qquad (kJ/K) \qquad \text{(8–72)}$$

or, in the **rate form**, as

$$\underbrace{\dot{S}_{in} - \dot{S}_{out}}_{\substack{\text{Rate of net entropy} \\ \text{transfer by heat} \\ \text{and mass}}} + \underbrace{\dot{S}_{gen}}_{\substack{\text{Rate of entropy} \\ \text{generation}}} = \underbrace{dS_{system}/dt}_{\substack{\text{Rate of change} \\ \text{in entropy}}} \quad \text{(kW/K)} \qquad \textbf{(8–73)}$$

where the rates of entropy transfer by heat transferred at a rate of $\dot{Q}$ and mass flowing at a rate of $\dot{m}$ are $\dot{S}_{heat} = \dot{Q}/T$ and $\dot{S}_{mass} = \dot{m}s$. The entropy balance can also be expressed on a **unit-mass basis** as

$$(s_{in} - s_{out}) + s_{gen} = \Delta s_{system} \qquad \text{(kJ/kg·K)} \qquad \textbf{(8–74)}$$

where all the quantities are expressed per unit mass of the system. Note that for a *reversible process*, the entropy generation term $S_{gen}$ drops out from all of the relations above.

The term $S_{gen}$ represents the entropy generation *within the system boundary* only (Fig. 8–58), and not the entropy generation that may occur outside the system boundary during the process as a result of external irreversibilities. Therefore, a process for which $S_{gen} = 0$ is *internally reversible*, but not necessarily *totally* reversible. The *total* entropy generated during a process can be determined by applying the entropy balance to an *extended system* that includes the system itself and its immediate surroundings where external irreversibilities might be occurring (Fig. 8–59). Also, the entropy change in this case is equal to the sum of the entropy change of the system and the entropy change of the immediate surroundings. Note that under steady conditions, the state and thus the entropy of the immediate surroundings (let us call it the "buffer zone") at any point does not change during the process, and the entropy change of the buffer zone is zero. The entropy change of the buffer zone, if any, is usually small relative to the entropy change of the system, and thus it is usually disregarded.

When evaluating the entropy transfer between an extended system and the surroundings, the boundary temperature of the extended system is simply taken to be the *environment temperature*.

## Closed Systems

A closed system involves *no mass flow* across its boundaries, and its entropy change is simply the difference between the initial and final entropies of the system. The *entropy change* of a closed system is due to the *entropy transfer* accompanying heat transfer and the *entropy generation* within the system boundaries. Taking the positive direction of heat transfer to be *to* the system, the general entropy balance relation (Eq. 8–72) can be expressed for a closed system as

*Closed system:* $\qquad \sum \dfrac{Q_k}{T_k} + S_{gen} = \Delta S_{system} = S_2 - S_1 \qquad \text{(kJ/K)} \qquad \textbf{(8–75)}$

The entropy balance relation above can be stated as:

The entropy change of a closed system during a process is equal to the sum of the net entropy transferred through the system boundary by heat transfer and the entropy generated within the system boundaries.

**FIGURE 8–58**

Mechanisms of entropy transfer for a general system.

**FIGURE 8–59**

Entropy generation outside system boundaries can be accounted for by writing an entropy balance on an extended system that includes the system and its immediate surroundings.

For an *adiabatic process* ($Q = 0$), the entropy transfer term in the above relation drops out and the entropy change of the closed system becomes equal to the entropy generation within the system boundaries. That is,

*Adiabatic closed system:*  $\qquad\qquad S_{gen} = \Delta S_{\text{adiabatic system}}$  (8–76)

Noting that any closed system and its surroundings can be treated as an adiabatic system and the total entropy change of a system is equal to the sum of the entropy changes of its parts, the entropy balance for a closed system and its surroundings can be written as

*System + Surroundings:*  $\qquad\qquad S_{gen} = \sum \Delta S = \Delta S_{\text{system}} + \Delta S_{\text{surroundings}}$  (8–77)

where $\Delta S_{\text{system}} = m(s_2 - s_1)$ and the entropy change of the surroundings can be determined from $\Delta S_{\text{surr}} = Q_{\text{surr}}/T_{\text{surr}}$ if its temperature is constant. At initial stages of studying entropy and entropy transfer, it is more instructive to start with the general form of the entropy balance (Eq. 8–72) and to simplify it for the problem under consideration. The specific relations above are convenient to use after a certain degree of intuitive understanding of the material is achieved.

## Control Volumes

The entropy balance relations for control volumes differ from those for closed systems in that they involve one more mechanism of entropy exchange: *mass flow across the boundaries*. As mentioned earlier, mass possesses entropy as well as energy, and the amounts of these two extensive properties are proportional to the amount of mass (Fig. 8–60).

Taking the positive direction of heat transfer to be *to* the system, the general entropy balance relations (Eqs. 8–72 and 8–73) can be expressed for control volumes as

$$\sum \frac{Q_k}{T_k} + \sum m_i s_i - \sum m_e s_e + S_{gen} = (S_2 - S_1)_{CV} \qquad \text{(kJ/K)} \qquad \text{(8–78)}$$

or, in the rate form, as

$$\sum \frac{\dot{Q}_k}{T_k} + \sum \dot{m}_i s_i - \sum \dot{m}_e s_e + \dot{S}_{gen} = dS_{CV}/dt \qquad \text{(kW/K)} \qquad \text{(8–79)}$$

This entropy balance relation can be stated as:

The rate of entropy change within the control volume during a process is equal to the sum of the rate of entropy transfer through the control volume boundary by heat transfer, the net rate of entropy transfer into the control volume by mass flow, and the rate of entropy generation within the boundaries of the control volume as a result of irreversibilities.

Most control volumes encountered in practice such as turbines, compressors, nozzles, diffusers, heat exchangers, pipes, and ducts operate steadily, and thus they experience no change in their entropy. Therefore, the entropy balance relation for a general **steady-flow process** can be obtained from Eq. 8–79 by setting $dS_{CV}/dt = 0$ and rearranging to give

*Steady-flow:*  $\qquad\qquad \dot{S}_{gen} = \sum \dot{m}_e s_e - \sum \dot{m}_i s_i - \sum \frac{\dot{Q}_k}{T_k}$  (8–80)

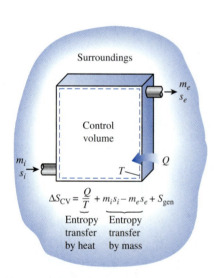

Surroundings

Control volume

$m_e$
$s_e$

$m_i$
$s_i$

$Q$

$T$

$$\Delta S_{CV} = \frac{Q}{T} + m_i s_i - m_e s_e + S_{gen}$$

Entropy transfer by heat   Entropy transfer by mass

**FIGURE 8–60**

The entropy of a control volume changes as a result of mass flow as well as heat transfer.

For *single-stream* (one inlet and one exit) steady-flow devices, the entropy balance relation simplifies to

*Steady-flow, single-stream:*
$$\dot{S}_{gen} = \dot{m}(s_e - s_i) - \sum \frac{\dot{Q}_k}{T_k} \qquad \text{(8-81)}$$

For the case of an *adiabatic* single-stream device, the entropy balance relation further simplifies to

*Steady-flow, single-stream, adiabatic:*
$$\dot{S}_{gen} = \dot{m}(s_e - s_i) \qquad \text{(8-82)}$$

which indicates that the specific entropy of the fluid must increase as it flows through an adiabatic device since $\dot{S}_{gen} \geq 0$ (Fig. 8–61). If the flow through the device is *reversible* and *adiabatic*, then the entropy remains constant, $s_e = s_i$, regardless of the changes in other properties.

$s_e \geq s_i$

$s_i$

**FIGURE 8–61**

The entropy of a substance always increases (or remains constant in the case of a reversible process) as it flows through a single-stream, adiabatic, steady-flow device.

---

### EXAMPLE 8–16    Entropy Generation in a Wall

Consider steady heat transfer through a 5-m × 7-m brick wall of a house of thickness 30 cm. On a day when the temperature of the outdoors is 0°C, the house is maintained at 27°C. The temperatures of the inner and outer surfaces of the brick wall are measured to be 20°C and 5°C, respectively, and the rate of heat transfer through the wall is 1035 W. Determine the rate of entropy generation in the wall, and the rate of total entropy generation associated with this heat transfer process.

**SOLUTION** Steady heat transfer through a wall is considered. For specified heat transfer rate, wall temperatures, and environment temperatures, the entropy generation rate within the wall and the total entropy generation rate are to be determined.

**Assumptions** **1** The process is steady, and thus the rate of heat transfer through the wall is constant. **2** Heat transfer through the wall is one-dimensional.

**Analysis** We first take the *wall* as the system (Fig. 8–62). This is a *closed system* since no mass crosses the system boundary during the process. We note that the entropy change of the wall is zero during this process since the state and thus the entropy of the wall do not change anywhere in the wall. Heat and entropy are entering from one side of the wall and leaving from the other side.

The rate form of the entropy balance for the wall simplifies to

$$\underbrace{\dot{S}_{in} - \dot{S}_{out}}_{\substack{\text{Rate of net entropy} \\ \text{transfer by heat} \\ \text{and mass}}} + \underbrace{\dot{S}_{gen}}_{\substack{\text{Rate of entropy} \\ \text{generation}}} = \underbrace{dS_{system}/dt}_{\substack{\text{Rate of change} \\ \text{in entropy}}} \nearrow^{0 \text{ (steady)}}$$

$$\left(\frac{\dot{Q}}{T}\right)_{in} - \left(\frac{\dot{Q}}{T}\right)_{out} + \dot{S}_{gen} = 0$$

$$\frac{1035 \text{ W}}{293 \text{ K}} - \frac{1035 \text{ W}}{278 \text{ K}} + \dot{S}_{gen} = 0$$

Therefore, the rate of entropy generation in the wall is

$$\dot{S}_{gen,wall} = \textbf{0.191 W/K}$$

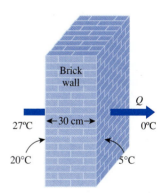

Brick wall

27°C    ←30 cm→    0°C

20°C    5°C

$Q$

**FIGURE 8–62**

Schematic for Example 8–16.

Note that entropy transfer by heat at any location is $Q/T$ at that location, and the direction of entropy transfer is the same as the direction of heat transfer.

To determine the rate of total entropy generation during this heat transfer process, we extend the system to include the regions on both sides of the wall that experience a temperature change. Then, one side of the system boundary becomes room temperature while the other side becomes the temperature of the outdoors. The entropy balance for this *extended system* (system + immediate surroundings) is the same as that given above, except the two boundary temperatures are now 300 and 273 K instead of 293 and 278 K, respectively. Then the rate of total entropy generation becomes

$$\frac{1035 \text{ W}}{300 \text{ K}} - \frac{1035 \text{ W}}{273 \text{ K}} + \dot{S}_{\text{gen,total}} = 0 \quad \rightarrow \quad \dot{S}_{\text{gen,total}} = \textbf{0.341 W/K}$$

**Discussion**  Note that the entropy change of this extended system is also zero since the state of air does not change at any point during the process. The differences between the two entropy generations is 0.150 W/K, and it represents the entropy generated in the air layers on both sides of the wall. The entropy generation in this case is entirely due to irreversible heat transfer through a finite temperature difference.

$P_1 = 7$ MPa
$T_1 = 450°C$

$P_2 = 3$ MPa

**FIGURE 8–63**

Schematic and *T-s* diagram for Example 8–17.

---

### EXAMPLE 8–17    Entropy Generation During a Throttling Process

Steam at 7 MPa and 450°C is throttled in a valve to a pressure of 3 MPa during a steady-flow process. Determine the entropy generated during this process and check if the increase of entropy principle is satisfied.

**SOLUTION**  Steam is throttled to a specified pressure. The entropy generated during this process is to be determined, and the validity of the increase of entropy principle is to be verified.

**Assumptions**  **1** This is a steady-flow process since there is no change with time at any point and thus $\Delta m_{\text{CV}} = 0$, $\Delta E_{\text{CV}} = 0$, and $\Delta S_{\text{CV}} = 0$. **2** Heat transfer to or from the valve is negligible. **3** The kinetic and potential energy changes are negligible, $\Delta \text{ke} = \Delta \text{pe} = 0$.

**Analysis**  We take the throttling valve as the *system* (Fig. 8–63). This is a *control volume* since mass crosses the system boundary during the process. We note that there is only one inlet and one exit and thus $\dot{m}_1 = \dot{m}_2 = \dot{m}$. Also, the enthalpy of a fluid remains nearly constant during a throttling process and thus $h_2\Sigma \cong h_1$.

The entropy of the steam at the inlet and the exit states is determined from the steam tables to be

$$\text{State 1:} \quad \begin{matrix} P_1 = 7 \text{ MPa} \\ T_1 = 450°C \end{matrix} \Big\} \quad \begin{matrix} h_1 = 3288.3 \text{ kJ/kg} \\ s_1 = 6.6353 \text{ kJ/kg·K} \end{matrix}$$

$$\text{State 2:} \quad \begin{matrix} P_2 = 3 \text{ MPa} \\ h_2 = h_1 \end{matrix} \Big\} \quad s_2 = 7.0046 \text{ kJ/kg·K}$$

Then, the entropy generation per unit mass of the steam is determined from the entropy balance applied to the throttling valve,

$$\underbrace{\dot{S}_{in} - \dot{S}_{out}}_{\substack{\text{Rate of net entropy} \\ \text{transfer by heat} \\ \text{and mass}}} + \underbrace{\dot{S}_{gen}}_{\substack{\text{Rate of entropy} \\ \text{generation}}} = \underbrace{dS_{system}/dt}_{\substack{\text{Rate of change} \\ \text{in entropy}}} \nearrow^{0\ (steady)}$$

$$\dot{m}s_1 - \dot{m}s_2 + \dot{S}_{gen} = 0$$

$$\dot{S}_{gen} = \dot{m}(s_2 - s_1)$$

Dividing by mass flow rate and substituting gives

$$s_{gen} = s_2 - s_1 = 7.0046 - 6.6353 = \mathbf{0.3693\ kJ/kg \cdot K}$$

This is the amount of entropy generated per unit mass of steam as it is throttled from the inlet state to the final pressure, and it is caused by unrestrained expansion. The increase of entropy principle is obviously satisfied during this process since the entropy generation is positive.

---

**EXAMPLE 8–18**     **Entropy Generated when a Hot Block Is Dropped in a Lake**

A 50-kg block of iron casting at 500 K is thrown into a large lake that is at a temperature of 285 K. The iron block eventually reaches thermal equilibrium with the lake water. Assuming an average specific heat of 0.45 kJ/kg·K for the iron, determine (a) the entropy change of the iron block, (b) the entropy change of the lake water, and (c) the entropy generated during this process.

**SOLUTION**  A hot iron block is thrown into a lake, and cools to the lake temperature. The entropy changes of the iron and of the lake as well as the entropy generated during this process are to be determined.
**Assumptions**  1 Both the water and the iron block are incompressible substances. 2 Constant specific heats can be used for the water and the iron. 3 The kinetic and potential energy changes of the iron are negligible, $\Delta KE = \Delta PE = 0$ and thus $\Delta E = \Delta U$.
**Properties**  The specific heat of the iron is 0.45 kJ/kg·K (Table A–3).
**Analysis**  We take the *iron casting* as the system (Fig. 8–64). This is a *closed system* since no mass crosses the system boundary during the process.
To determine the entropy change for the iron block and for the lake, first we need to know the final equilibrium temperature. Given that the thermal energy capacity of the lake is very large relative to that of the iron block, the lake will absorb all the heat rejected by the iron block without experiencing any change in its temperature. Therefore, the iron block will cool to 285 K during this process while the lake temperature remains constant at 285 K.

(a) The entropy change of the iron block can be determined from

$$\Delta S_{iron} = m(s_2 - s_1) = mc_{avg} \ln \frac{T_2}{T_1}$$

$$= (50\ kg)(0.45\ kJ/kg \cdot K) \ln \frac{285\ K}{500\ K}$$

$$= \mathbf{-12.65\ kJ/K}$$

**FIGURE 8–64**
Schematic for Example 8–18.

(b) The temperature of the lake water remains constant during this process at 285 K. Also, the amount of heat transfer from the iron block to the lake is determined from an energy balance on the iron block to be

$$\underbrace{E_{in} - E_{out}}_{\substack{\text{Net energy transfer} \\ \text{by heat, work, and mass}}} = \underbrace{\Delta E_{system}}_{\substack{\text{Change in internal, kinetic,} \\ \text{potential, etc., energies}}}$$

$$-Q_{out} = \Delta U = mc_{avg}(T_2 - T_1)$$

or

$$Q_{out} = mc_{avg}(T_1 - T_2) = (50 \text{ kg})(0.45 \text{ kJ/kg·K})(500 - 285) \text{ K} = 4838 \text{ kJ}$$

Then, the entropy change of the lake becomes

$$\Delta S_{lake} = \frac{Q_{lake}}{T_{lake}} = \frac{+4838 \text{ kJ}}{285 \text{ K}} = \textbf{16.97 kJ/K}$$

(c) The entropy generated during this process can be determined by applying an entropy balance on an *extended system* that includes the iron block and its immediate surroundings so that the boundary temperature of the extended system is at 285 K at all times:

$$\underbrace{S_{in} - S_{out}}_{\substack{\text{Net entropy transfer} \\ \text{by heat and mass}}} + \underbrace{S_{gen}}_{\substack{\text{Entropy} \\ \text{generation}}} = \underbrace{\Delta S_{system}}_{\substack{\text{Change} \\ \text{in entropy}}}$$

$$-\frac{Q_{out}}{T_b} + S_{gen} = \Delta S_{system}$$

or

$$S_{gen} = \frac{Q_{out}}{T_b} + \Delta S_{system} = \frac{4838 \text{ kJ}}{285 \text{ K}} - (12.65 \text{ kJ/K}) = \textbf{4.32 kJ/K}$$

**Discussion**   The entropy generated can also be determined by taking the iron block and the entire lake as the system, which is an isolated system, and applying an entropy balance. An isolated system involves no heat or entropy transfer, and thus the entropy generation in this case becomes equal to the total entropy change,

$$S_{gen} = \Delta S_{total} = \Delta S_{system} + \Delta S_{lake} = -12.65 + 16.97 = 4.32 \text{ kJ/K}$$

which is the same result obtained above.

**FIGURE 8–65**

Schematic for Example 8–19

### EXAMPLE 8–19   Entropy Generation in a Heat Exchanger

Air in a large building is kept warm by heating it with steam in a heat exchanger (Fig. 8-65). Saturated water vapor enters this unit at 35°C at a rate of 10,000 kg/h and leaves as saturated liquid at 32°C. Air at 1-atm pressure enters the unit at 20°C and leaves at 30°C at about the same pressure. Determine the rate of entropy generation associated with this process.

**SOLUTION**   Air is heated by steam in a heat exchanger. The rate of entropy generation associated with this process is to be determined.

*Assumptions* **1** Steady operating conditions exist. **2** The heat exchanger is well-insulated so that heat loss to the surroundings is negligible and thus heat transfer from the hot fluid is equal to the heat transfer to the cold fluid. **3** Changes in the kinetic and potential energies of fluid streams are negligible. **4** Air is an ideal gas with constant specific heats at room temperature. **5** The pressure of air remains constant.

*Analysis* The rate of entropy generation within the heat exchanger is determined by applying the rate form of the entropy balance on the entire heat exchanger:

$$\underbrace{\dot{S}_{\text{in}} - \dot{S}_{\text{out}}}_{\substack{\text{Rate of net entropy} \\ \text{transfer by heat} \\ \text{and mass}}} + \underbrace{\dot{S}_{\text{gen}}}_{\substack{\text{Rate of entropy} \\ \text{generation}}} = \underbrace{\Delta \dot{S}_{\text{system}}}_{\substack{\text{Rate of change} \\ \text{in entropy}}} \nearrow^{0 \text{ (steady)}}$$

$$\dot{m}_{\text{steam}} s_1 + \dot{m}_{\text{air}} s_3 - \dot{m}_{\text{steam}} s_2 - \dot{m}_{\text{air}} s_4 + \dot{S}_{\text{gen}} = 0$$

$$\dot{S}_{\text{gen}} = \dot{m}_{\text{steam}}(s_2 - s_1) + \dot{m}_{\text{air}}(s_4 - s_3)$$

The specific heat of air at room temperature is $c_p = 1.005$ kJ/kg·°C (Table A-2*a*). The properties of the steam at the inlet and exit states are

$$\left. \begin{array}{l} T_1 = 35°C \\ x_1 = 1 \end{array} \right\} \quad \begin{array}{l} h_1 = 2564.6 \text{ kJ/kg} \\ s_1 = 8.3517 \text{ kJ/kg·K} \end{array} \quad \text{(Table A–4)}$$

$$\left. \begin{array}{l} T_2 = 32°C \\ x_2 = 0 \end{array} \right\} \quad \begin{array}{l} h_2 = 134.10 \text{ kJ/kg} \\ s_2 = 0.4641 \text{ kJ/kg·K} \end{array} \quad \text{(Table A–4)}$$

From an energy balance the heat transferred from steam is equal to the heat transferred to the air. Then, the mass flow rate of air is determined to be

$$\dot{Q} = \dot{m}_{\text{steam}}(h_1 - h_2) = (10{,}000/3600 \text{ kg/s})(2564.6 - 134.10) \text{ kJ/kg} = 6751 \text{ kW}$$

$$\dot{m}_{\text{air}} = \frac{\dot{Q}}{c_p(T_4 - T_3)} = \frac{6751 \text{ kW}}{(1.005 \text{ kJ/kg·°C})(30 - 20)°C} = 671.7 \text{ kg/s}$$

Substituting into the entropy balance relation, the rate of entropy generation becomes

$$\dot{S}_{\text{gen}} = \dot{m}_{\text{steam}}(s_2 - s_1) + \dot{m}_{\text{air}}(s_4 - s_3)$$

$$= \dot{m}_{\text{steam}}(s_2 - s_1) + \dot{m}_{\text{air}} c_p \ln \frac{T_4}{T_3}$$

$$= (10{,}000/3600 \text{ kg/s})(0.4641 - 8.3517) \text{ kJ/kg·K}$$

$$+ (671.7 \text{ kg/s})(1.005 \text{ kJ/kg·K}) \ln \frac{303 \text{ K}}{293 \text{ K}}$$

$$= \textbf{0.745 kW/K}$$

*Discussion* Note that the pressure of air remains nearly constant as it flows through the heat exchanger, and thus the pressure term is not included in the entropy change expression for air.

## SUMMARY

The second law of thermodynamics leads to the definition of a new property called *entropy*, which is a quantitative measure of microscopic disorder for a system. Any quantity whose cyclic integral is zero is a property, and entropy is defined as

$$dS = \left(\frac{dQ}{T}\right)_{int\ rev}$$

For the special case of an internally reversible, isothermal process, it gives

$$\Delta S = \frac{Q}{T_0}$$

The inequality part of the Clausius inequality combined with the definition of entropy yields an inequality known as the *increase of entropy principle*, expressed as

$$S_{gen} \geq 0$$

where $S_{gen}$ is the *entropy generated* during the process. Entropy change is caused by heat transfer, mass flow, and irreversibilities. Heat transfer to a system increases the entropy, and heat transfer from a system decreases it. The effect of irreversibilities is always to increase the entropy.

The *entropy-change* and *isentropic relations* for a process can be summarized as follows:

**1.** *Pure substances:*

Any process: $\Delta s = s_2 - s_1$

Isentropic process: $s_2 = s_1$

**2.** *Incompressible substances:*

Any process: $s_2 - s_1 = c_{avg} \ln \frac{T_2}{T_1}$

Isentropic process: $T_2 = T_1$

**3.** *Ideal gases:*
*a.* Constant specific heats (approximate treatment):
Any process:

$$s_2 - s_1 = c_{v,avg} \ln \frac{T_2}{T_1} + R \ln \frac{v_2}{v_1}$$

$$s_2 - s_1 = c_{p,avg} \ln \frac{T_2}{T_1} - R \ln \frac{P_2}{P_1}$$

Isentropic process:

$$\left(\frac{T_2}{T_1}\right)_{s=const.} = \left(\frac{v_1}{v_2}\right)^{k-1}$$

$$\left(\frac{T_2}{T_1}\right)_{s=const.} = \left(\frac{P_2}{P_1}\right)^{(k-1)/k}$$

$$\left(\frac{P_2}{P_1}\right)_{s=const.} = \left(\frac{v_1}{v_2}\right)^{k}$$

*b.* Variable specific heats (exact treatment):
Any process:

$$s_2 - s_1 = s_2^\circ - s_1^\circ - R \ln \frac{P_2}{P_1}$$

Isentropic process:

$$s_2^\circ = s_1^\circ + R \ln \frac{P_2}{P_1}$$

$$\left(\frac{P_2}{P_1}\right)_{s=const.} = \frac{P_{r2}}{P_{r1}}$$

$$\left(\frac{v_2}{v_1}\right)_{s=const.} = \frac{v_{r2}}{v_{r1}}$$

where $P_r$ is the *relative pressure* and $v_r$ is the *relative specific volume*. The function $s^\circ$ depends on temperature only.

The *steady-flow work* for a reversible process can be expressed in terms of the fluid properties as

$$w_{rev} = -\int_1^2 v\,dP - \Delta ke - \Delta pe$$

For incompressible substances ($v$ = constant) it simplifies to

$$w_{rev} = -v(P_2 - P_1) - \Delta ke - \Delta pe$$

The work done during a steady-flow process is proportional to the specific volume. Therefore, $v$ should be kept as small as possible during a compression process to minimize the work input and as large as possible during an expansion process to maximize the work output.

The work input to a compressor can be reduced by using multistage compression with intercooling. For maximum savings from the work input, the pressure ratio across each stage of the compressor must be the same.

Most steady-flow devices operate under adiabatic conditions, and the ideal process for these devices is the isentropic process. The parameter that describes how efficiently a device approximates a corresponding isentropic device is called *isentropic* or *adiabatic efficiency*. It is expressed for turbines, compressors, and nozzles as follows:

$$\eta_T = \frac{\text{Actual turbine work}}{\text{Isentropic turbine work}} = \frac{w_a}{w_s} \cong \frac{h_1 - h_{2a}}{h_1 - h_{2s}}$$

$$\eta_C = \frac{\text{Isentropic compressor work}}{\text{Actual compressor work}} = \frac{w_s}{w_a} \cong \frac{h_{2s} - h_1}{h_{2a} - h_1}$$

$$\eta_N = \frac{\text{Actual KE at nozzle exit}}{\text{Isentropic KE at nozzle exit}} = \frac{V_{2a}^2}{V_{2s}^2} \cong \frac{h_1 - h_{2a}}{h_1 - h_{2s}}$$

In the relations above, $h_{2a}$ and $h_{2s}$ are the enthalpy values at the exit state for actual and isentropic *processes*, respectively.

The entropy balance for any system undergoing any process can be expressed in the general form as

$$\underbrace{S_{\text{in}} - S_{\text{out}}}_{\substack{\text{Net entropy transfer} \\ \text{by heat and mass}}} + \underbrace{S_{\text{gen}}}_{\substack{\text{Entropy} \\ \text{generation}}} = \underbrace{\Delta S_{\text{system}}}_{\substack{\text{Change} \\ \text{in entropy}}}$$

or, in the *rate form*, as

$$\underbrace{\dot{S}_{\text{in}} - \dot{S}_{\text{out}}}_{\substack{\text{Rate of net entropy} \\ \text{transfer by} \\ \text{heat and mass}}} + \underbrace{\dot{S}_{\text{gen}}}_{\substack{\text{Rate of entropy} \\ \text{generation}}} = \underbrace{dS_{\text{system}}/dt}_{\substack{\text{Rate of change} \\ \text{in entropy}}}$$

For a general *steady-flow process* it simplifies to

$$\dot{S}_{\text{gen}} = \sum \dot{m}_e s_e - \sum \dot{m}_i s_i - \sum \frac{\dot{Q}_k}{T_k}$$

## REFERENCES AND SUGGESTED READINGS

**1.** A. Bejan. *Advanced Engineering Thermodynamics*. 3rd ed. New York: Wiley Interscience, 2006.

**2.** A. Bejan. *Entropy Generation through Heat and Fluid Flow*. New York: Wiley Interscience, 1982.

**3.** Y. A. Çengel and H. Kimmel. "Optimization of Expansion in Natural Gas Liquefaction Processes." *LNG Journal*, U.K., May–June, 1998.

**4.** J. Rifkin. *Entropy*. New York: The Viking Press, 1980.

## PROBLEMS*

### Entropy and the Increase of Entropy Principle

**8–1C** Does the cyclic integral of work have to be zero (i.e., does a system have to produce as much work as it consumes to complete a cycle)? Explain.

**8–2C** A system undergoes a process between two fixed states first in a reversible manner and then in an irreversible manner. For which case is the entropy change greater? Why?

**8–3C** Is the value of the integral $\int_1^2 \delta Q/T$ the same for all processes between states 1 and 2? Explain.

**8–4C** To determine the entropy change for an irreversible process between states 1 and 2, should the integral $\int_1^2 \delta Q/T$ be performed along the actual process path or an imaginary reversible path? Explain.

**8–5C** Is an isothermal process necessarily internally reversible? Explain your answer with an example.

---

* Problems designated by a "C" are concept questions, and students are encouraged to answer them all. Problems designated by an "E" are in English units, and the SI users can ignore them. Problems with the ⊚ icon are solved using EES, and complete solutions together with parametric studies are included on the enclosed DVD. Problems with the ▨ icon are comprehensive in nature, and are intended to be solved with a computer, preferably using the EES software that accompanies this text.

**8–6C** How do the values of the integral $\int_1^2 \delta Q/T$ compare for a reversible and irreversible process between the same end states?

**8–7C** The entropy of a hot baked potato decreases as it cools. Is this a violation of the increase of entropy principle? Explain.

**8–8C** Is it possible to create entropy? Is it possible to destroy it?

**8–9C** When a system is adiabatic, what can be said about the entropy change of the substance in the system?

**8–10C** Work is entropy free, and sometimes the claim is made that work will not change the entropy of a fluid passing through an adiabatic steady-flow system with a single inlet and outlet. Is this a valid claim?

**8–11C** A piston–cylinder device contains helium gas. During a reversible, isothermal process, the entropy of the helium will (*never, sometimes, always*) increase.

**8–12C** A piston–cylinder device contains nitrogen gas. During a reversible, adiabatic process, the entropy of the nitrogen will (*never, sometimes, always*) increase.

**8–13C** A piston–cylinder device contains superheated steam. During an actual adiabatic process, the entropy of the steam will (*never, sometimes, always*) increase.

**8–14C** The entropy of steam will (*increase, decrease, remain the same*) as it flows through an actual adiabatic turbine.

**8–15C** The entropy of the working fluid of the ideal Carnot cycle (*increases, decreases, remains the same*) during the isothermal heat addition process.

**8–16C** The entropy of the working fluid of the ideal Carnot cycle (*increases, decreases, remains the same*) during the isothermal heat rejection process.

**8–17C** During a heat transfer process, the entropy of a system (*always, sometimes, never*) increases.

**8–18C** Steam is accelerated as it flows through an actual adiabatic nozzle. The entropy of the steam at the nozzle exit will be (*greater than, equal to, less than*) the entropy at the nozzle inlet.

**8–19C** Is it possible for the entropy change of a closed system to be zero during an irreversible process? Explain.

**8–20C** What three different mechanisms can cause the entropy of a control volume to change?

**8–21E** A completely reversible heat engine operates with a source at 1500 R and a sink at 500 R. If the entropy of the sink increases by 10 Btu/R, how much will the entropy of the source decrease? How much heat, in Btu, is transferred from the source?

**8–22** An energy-source at 1000 K transfers heat to a completely reversible heat engine. This engine transfers heat to a sink at 300 K. How much heat must be transferred from the energy-source to increase the entropy of the energy-sink by 20 kJ/K?

**8–23E** A heat engine accepts 200,000 Btu of heat from a source at 1500 R and rejects 100,000 Btu of heat to a sink at 600 R. Calculate the entropy change of all the components of this engine and determine if it is completely reversible. How much total work does it produce?

**8–24** Air is compressed by a 30-kW compressor from $P_1$ to $P_2$. The air temperature is maintained constant at 25°C during this process as a result of heat transfer to the surrounding medium at 17°C. Determine the rate of entropy change of the air. State the assumptions made in solving this problem. *Answer:* −0.101 kW/K

**8–25** Heat in the amount of 100 kJ is transferred directly from a hot reservoir at 1200 K to a cold reservoir at 600 K. Calculate the entropy change of the two reservoirs and determine if the increase of entropy principle is satisfied.

**FIGURE P8–25**

**8–26** In the previous problem, assume that the heat is transferred from the cold reservoir to the hot reservoir contrary to the Clausius statement of the second law. Prove that this violates the increase of entropy principle—as it must according to Clausius.

**8–27** A completely reversible heat pump produces heat at a rate of 300 kW to warm a house maintained at 24°C. The exterior air, which is at 7°C, serves as the source. Calculate the rate of entropy change of the two reservoirs and determine if this heat pump satisfies the second law according to the increase of entropy principle.

**FIGURE P8–27**

**8–28E** During the isothermal heat rejection process of a Carnot cycle, the working fluid experiences an entropy change of −0.7 Btu/R. If the temperature of the heat sink is 95°F, determine (*a*) the amount of heat transfer, (*b*) the entropy change of the sink, and (*c*) the total entropy change for this process. *Answers:* (*a*) 388.5 Btu, (*b*) 0.7 Btu/R, (*c*) 0

**FIGURE P8–28E**

**8–29** Refrigerant-134a enters the coils of the evaporator of a refrigeration system as a saturated liquid–vapor mixture at a pressure of 160 kPa. The refrigerant absorbs 180 kJ of heat from the cooled space, which is maintained at −5°C, and leaves as saturated vapor at the same pressure. Determine (*a*) the entropy change of the refrigerant, (*b*) the entropy change of the cooled space, and (*c*) the total entropy change for this process.

## Entropy Changes of Pure Substances

**8–30C**  Is a process that is internally reversible and adiabatic necessarily isentropic? Explain.

**8–31E**  2-lbm of water at 300 psia fill a weighted piston-cylinder device whose volume is 2.5 ft³. The water is then heated at constant pressure until the temperature reaches 500°F. Determine the resulting change in the water's total entropy.  *Answer: 0.474 Btu/R*

**8–32**  A well-insulated rigid tank contains 5 kg of a saturated liquid–vapor mixture of water at 150 kPa. Initially, three-quarters of the mass is in the liquid phase. An electric resistance heater placed in the tank is now turned on and kept on until all the liquid in the tank is vaporized. Determine the entropy change of the steam during this process.  *Answer: 19.2 kJ/K*

**FIGURE P8–32**

**8–33**  A rigid tank is divided into two equal parts by a partition. One part of the tank contains 2.5 kg of compressed liquid water at 400 kPa and 60°C while the other part is evacuated. The partition is now removed, and the water expands to fill the entire tank. Determine the entropy change of water during this process, if the final pressure in the tank is 40 kPa.  *Answer: 0.492 kJ/K*

**FIGURE P8–33**

**8–34**  Reconsider Prob. 8–33. Using EES (or other) software, evaluate and plot the entropy generated as a function of surrounding temperature, and determine the values of the surrounding temperatures that are valid for this problem. Let the surrounding temperature vary from 0 to 100°C. Discuss your results.

**8–35E**  A piston–cylinder device contains 2 lbm of refrigerant-134a at 120 psia and 100°F. The refrigerant is now cooled at constant pressure until it exists as a liquid at 50°F. Determine the entropy change of the refrigerant during this process.

**8–36**  An insulated piston–cylinder device contains 5 L of saturated liquid water at a constant pressure of 150 kPa. An electric resistance heater inside the cylinder is now turned on, and 2200 kJ of energy is transferred to the steam. Determine the entropy change of the water during this process.  *Answer: 5.72 kJ/K*

**8–37**  Calculate the change in the specific entropy of water as it is cooled at a constant pressure of 300 kPa from a saturated vapor to a saturated liquid by using a Gibbs equation $(T\,ds = dh - v\,dP)$. Use the steam tables to verify your result.

**8–38E**  Saturated R-134a vapor enters a compressor at 0°F. At compressor exit, the specific entropy is the same as that at the inlet, and the pressure is 60 psia. Determine the R-134a exit temperature and the change in the enthalpy of R-134a.

**FIGURE P8–38E**

**8–39**  Water vapor enters a turbine at 6 MPa and 400°C, and leaves the turbine at 100 kPa with the same specific entropy as that at the inlet. Calculate the difference between the specific enthalpy of the water at the turbine inlet and exit.

**8–40**  1-kg of R-134a initially at 600 kPa and 25°C undergoes a process during which the entropy is kept constant until the pressure drops to 100 kPa. Determine the final temperature of the R-134a and the final specific internal energy.

**8–41**  Refrigerant-134a is expanded isentropically from 800 kPa and 60°C at the inlet of a steady-flow turbine to 100 kPa at the outlet. The outlet area is 1 m², and the inlet area is 0.5 m². Calculate the inlet and outlet velocities when the mass flow rate is 0.5 kg/s.  *Answers: 0.030 m/s, 0.105 m/s*

**8–42** A heavily insulated piston–cylinder device contains 0.02 m³ of steam at 300 kPa and 200°C. Steam is now compressed in a reversible manner to a pressure of 1.2 MPa. Determine the work done on the steam during this process.

**8–43** Reconsider Prob. 8–42. Using EES (or other) software, evaluate and plot the work done on the steam as a function of final pressure as the pressure varies from 300 kPa to 1.2 MPa.

**8–44** A piston–cylinder device contains 1.2 kg of saturated water vapor at 200°C. Heat is now transferred to steam, and steam expands reversibly and isothermally to a final pressure of 800 kPa. Determine the heat transferred and the work done during this process.

**8–45** Reconsider Prob. 8–44. Using EES (or other) software, evaluate and plot the heat transferred to the steam and the work done as a function of final pressure as the pressure varies from the initial value to the final value of 800 kPa.

**8–46** Refrigerant-134a at 240 kPa and 20°C undergoes an isothermal process in a closed system until its quality is 20 percent. On per unit mass basis, determine how much work and heat transfer are required. *Answers: 37.0 kJ/kg, 172 kJ/kg*

R-134a
240 kPa
20°C

**FIGURE P8–46**

**8–47** Determine the heat transfer, in kJ/kg, for the reversible process 1-3 shown in Fig. P8–47.

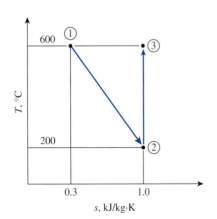

**FIGURE P8–47**

**8–48E** Calculate the heat transfer, in Btu/lbm, for the reversible process 1-3 shown in Fig. P8–48E. *Answer: 515 Btu/lbm*

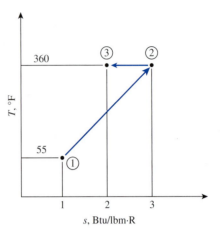

**FIGURE P8–48E**

**8–49** Calculate the heat transfer, in kJ/kg, for the reversible steady-flow process 1-3 shown in Fig. P8–49

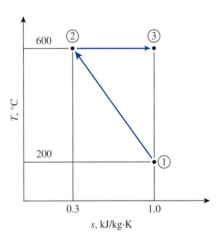

**FIGURE P8–49**

**8–50** Steam is expanded in an isentropic turbine with a single inlet and outlet. At the inlet, the steam is at 2 MPa and 360°C. The steam pressure at the outlet is 100 kPa. Calculate the work produced by this turbine, in kJ/kg.

**8–51** An isentropic steam turbine processes 5 kg/s of steam at 4 MPa, which is exhausted at 50 kPa and 100°C. Five percent of this flow is diverted for feedwater heating at 700 kPa. Determine the power produced by this turbine, in kW. *Answer: 6328 kW*

4 MPa
5 kg/s

4 MW

Steam
turbine

700 kPa

50 kPa
100°C

**FIGURE P8–51**

**8–52** Water at 70 kPa and 100°C is compressed isentropically in a closed system to 4 MPa. Determine the final temperature of the water and the work required, in kJ/kg, for this compression.

H₂O
70 kPa
100°C

**FIGURE P8–52**

**8–53** 0.5-kg of R-134a is expanded isentropically from 600 kPa and 30°C to 140 kPa. Determine the total heat transfer and work production for this expansion.

**8–54** Refrigerant-134a enters a steady-flow, adiabatic turbine as a saturated vapor at 1200 kPa and expands to 100 kPa. The power produced by the turbine is found to be 100 kW when the process is also reversible.

(a) Sketch the *T-s* diagram with respect to the saturation lines for this process.

(b) Determine the volume flow rate of the refrigerant-134a at the turbine exit, in m³/s.

*Answer: 0.376 m³/s*

**8–55** Two kilograms of saturated water vapor at 600 kPa are contained in a piston-cylinder device. The water expands adiabatically until the pressure is 100 kPa and is said to produce 700 kJ of work output.

(a) Determine the entropy change of the water, in kJ/kg·K.

(b) Is this process realistic? Using the *T-s* diagram for the process and the concepts of second law, support your answer.

**8–56** Refrigerant-134a enters a steady-flow adiabatic compressor as a saturated vapor at 320 kPa and is compressed to 1200 kPa. The minimum power supplied to the compressor is found to be 100 kW.

(a) Sketch the *T-s* diagram with respect to the saturation lines for this process.

(b) Determine the volume flow rate of the refrigerant-134a at the compressor inlet, in m³/s.

**8–57** Steam enters a steady-flow adiabatic nozzle with a low inlet velocity as a saturated vapor at 6 MPa and expands to 1.2 MPa.

(a) Under the conditions that the exit velocity is to be the maximum possible value, sketch the *T-s* diagram with respect to the saturation lines for this process.

(b) Determine the maximum exit velocity of the steam, in m/s.

*Answer: 764 m/s*

**8–58** A rigid, 20-L steam cooker is arranged with a pressure relief valve set to release vapor and maintain the pressure once the pressure inside the cooker reaches 150 kPa. Initially, this cooker is filled with water at 175 kPa with a quality of 10 percent. Heat is now added until the quality inside the cooker is 40 percent. Determine the minimum entropy change of the thermal energy reservoir supplying this heat.

**8–59** In the preceding problem, the water is stirred at the same time that it is being heated. Determine the minimum entropy change of the heat-supplying source if 100 kJ of work is done on the water as it is being heated.

**8–60** A piston–cylinder device contains 5 kg of steam at 100°C with a quality of 50 percent. This steam undergoes two processes as follows:

1-2 Heat is transferred to the steam in a reversible manner while the temperature is held constant until the steam exists as a saturated vapor.

2-3 The steam expands in an adiabatic, reversible process until the pressure is 15 kPa.

(a) Sketch these processes with respect to the saturation lines on a single *T-s* diagram.

(b) Determine the heat transferred to the steam in process 1-2, in kJ.

(c) Determine the work done by the steam in process 2-3, in kJ.

**8–61E** A 0.8-ft³ well-insulated rigid can initially contains refrigerant-134a at 140 psia and 50°F. Now a crack develops in the can, and the refrigerant starts to leak out slowly, Assuming the refrigerant remaining in the can has undergone a reversible, adiabatic process, determine the final mass in the can when the pressure drops to 30 psia.

**FIGURE P8–61E**

**8–62E** An electric windshield defroster is used to remove 0.25-in of ice from a windshield. The properties of the ice are $T_{sat} = 32°F$, $u_{if} = h_{if} = 144$ Btu/lbm, and $v = 0.01602$ ft$^3$/lbm. Determine the electrical energy required per square foot of windshield surface area to melt this ice and remove it as liquid water at 32°F. What is the minimum temperature at which the defroster may be operated? Assume that no heat is transferred from the defroster or ice to the surroundings.

## Entropy Change of Incompressible Substances

**8–63C** Consider two solid blocks, one hot and the other cold, brought into contact in an adiabatic container. After a while, thermal equilibrium is established in the container as a result of heat transfer. The first law requires that the amount of energy lost by the hot solid be equal to the amount of energy gained by the cold one. Does the second law require that the decrease in entropy of the hot solid be equal to the increase in entropy of the cold one?

**8–64** A 75-kg copper block initially at 110°C is dropped into an insulated tank that contains 160 L of water at 15°C. Determine the final equilibrium temperature and the total entropy change for this process.

**FIGURE P8–64**

**8–65** Ten grams of computer chips with a specific heat of 0.3 kJ/kg·K are initially at 20°C. These chips are cooled by placement in 5 grams of saturated liquid R-134a at −40°C. Presuming that the pressure remains constant while the chips are being cooled, determine the entropy change of (a) the chips, (b) the R-134a, and (c) the entire system. Is this process possible? Why?

**8–66** A 25-kg iron block initially at 350°C is quenched in an insulated tank that contains 100 kg of water at 18°C. Assuming the water that vaporizes during the process condenses back in the tank, determine the total entropy change during this process.

**8–67** A 30-kg aluminum block initially at 140°C is brought into contact with a 40-kg block of iron at 60°C in an insulated enclosure. Determine the final equilibrium temperature and the total entropy change for this process. *Answers: 109°C, 0.251 kJ/K*

**8–68** Reconsider Prob. 8–67. Using EES (or other) software, study the effect of the mass of the iron block on the final equilibrium temperature and the total entropy change for the process. Let the mass of the iron vary from 10 to 100 kg. Plot the equilibrium temperature and the total entropy change as a function of iron mass, and discuss the results.

**8–69** A 50-kg iron block and a 20-kg copper block, both initially at 80°C, are dropped into a large lake at 15°C. Thermal equilibrium is established after a while as a result of heat transfer between the blocks and the lake water. Determine the total entropy change for this process.

**FIGURE P8–69**

**8–70** An adiabatic pump is to be used to compress saturated liquid water at 10 kPa to a pressure to 15 MPa in a reversible manner. Determine the work input using (a) entropy data from the compressed liquid table, (b) inlet specific volume and pressure values, (c) average specific volume and pressure values. Also, determine the errors involved in parts (b) and (c).

↑ 15 MPa

10 kPa → Pump

**FIGURE P8–70**

## Entropy Change of Ideal Gases

**8–71C** Some properties of ideal gases such as internal energy and enthalpy vary with temperature only [that is, $u = u(T)$ and $h = h(T)$]. Is this also the case for entropy?

**8–72C** Can the entropy of an ideal gas change during an isothermal process?

**8–73C** An ideal gas undergoes a process between two specified temperatures, first at constant pressure and then at constant volume. For which case will the ideal gas experience a larger entropy change? Explain.

**8–74** Prove that the two relations for entropy change of ideal gases under the constant-specific-heat assumption (Eqs. 8–33 and 8–34) are equivalent.

**8–75** Starting with the second $T\,ds$ relation (Eq. 8–26), obtain Eq. 8–34 for the entropy change of ideal gases under the constant-specific-heat assumption.

**8–76** Starting with Eq. 8–34, obtain Eq. 8–43.

**8–77** Which of the two gases—helium or nitrogen—experiences the greatest entropy change as its state is changed from 2000 kPa and 427°C to 200 kPa and 27°C?

**8–78** Air is expanded from 2000 kPa and 500°C to 100 kPa and 50°C. Assuming constant specific heats, determine the change in the specific entropy of air.

**8–79E** What is the difference between the entropies of air at 15 psia and 70°F and air at 40 psia and 250°F per unit mass basis.

**8–80** Nitrogen at 900 kPa and 300°C is expanded adiabatically in a closed system to 100 kPa. Determine the minimum nitrogen temperature after the expansion.

**8–81E** Air at 15 psia and 70°F is compressed adiabatically in a closed system to 200 psia. What is the minimum temperature of the air following this compression?

**8–82** An insulated piston–cylinder device initially contains 300 L of air at 120 kPa and 17°C. Air is now heated for 15 min by a 200-W resistance heater placed inside the cylinder. The pressure of air is maintained constant during this process. Determine the entropy change of air, assuming (*a*) constant specific heats and (*b*) variable specific heats.

**8–83** A piston–cylinder device contains 0.75 kg of nitrogen gas at 140 kPa and 37°C. The gas is now compressed slowly in a polytropic process during which $PV^{1.3}$ = constant. The process ends when the volume is reduced by one-half. Determine the entropy change of nitrogen during this process.
*Answer:* −0.0385 kJ/K

**8–84** [EES] Reconsider Prob. 8–83. Using EES (or other) software, investigate the effect of varying the polytropic exponent from 1 to 1.4 on the entropy change of the nitrogen. Show the processes on a common $P$-$v$ diagram.

**8–85E** A mass of 15 lbm of helium undergoes a process from an initial state of 50 ft³/lbm and 80°F to a final state of 10 ft³/lbm and 200°F. Determine the entropy change of helium during this process, assuming (*a*) the process is reversible and (*b*) the process is irreversible.

**8–86** 1-kg of air at 200 kPa and 127°C is contained in a piston-cylinder device. Air is now allowed to expand in a reversible, isothermal process until its pressure is 100 kPa. Determine the amount of heat transferred to the air during this expansion.

**8–87** Argon is expanded in an isentropic turbine from 2 MPa and 500°C to 200 kPa. Determine the outlet temperature and the work produced by this turbine per unit mass of argon.

**8–88E** Air is compressed in an isentropic compressor from 15 psia and 70°F to 200 psia. Determine the outlet temperature and the work consumed by this compressor per unit mass of air. *Answers:* 1095 R, 138 Btu/lbm

200 psia

Air compressor

15 psia
70°F

**FIGURE P8–88E**

**8–89** An insulated rigid tank is divided into two equal parts by a partition. Initially, one part contains 5 kmol of an ideal gas at 250 kPa and 40°C, and the other side is evacuated. The partition is now removed, and the gas fills the entire tank. Determine the total entropy change during this process.
*Answer:* 28.81 kJ/K

**8–90** Air is compressed in a piston–cylinder device from 90 kPa and 22°C to 900 kPa in a reversible, adiabatic process. Determine the final temperature and the work done during this process, assuming (a) constant specific heats and (b) variable specific heats for air. *Answers: (a) 565 K, 197 kJ/kg*

**8–91**  Reconsider Prob. 8–90. Using EES (or other) software, evaluate and plot the work done and final temperature during the compression process as functions of the final pressure for the two cases as the final pressure varies from 100 to 1200 kPa.

**8–92** An insulated rigid tank contains 4 kg of argon gas at 450 kPa and 30°C. A valve is now opened, and argon is allowed to escape until the pressure inside drops to 200 kPa. Assuming the argon remaining inside the tank has undergone a reversible, adiabatic process, determine the final mass in the tank. *Answer: 2.46 kg*

Argon
4 kg
30°C
450 kPa

**FIGURE P8–92**

**8–93**  Reconsider Prob. 8–92. Using EES (or other) software, investigate the effect of the final pressure on the final mass in the tank as the pressure varies from 450 to 150 kPa, and plot the results.

**8–94E** Air enters an adiabatic nozzle at 60 psia, 540°F, and 200 ft/s and exits at 12 psia. Assuming air to be an ideal gas with variable specific heats and disregarding any irreversibilities, determine the exit velocity of the air.

**8–95** Air is expanded in an adiabatic nozzle during a polytropic process with $n = 1.3$. It enters the nozzle at 700 kPa and 100°C with a velocity of 30 m/s and exits at a pressure of 200 kPa. Calculate the air temperature and velocity at the nozzle exit.

700 kPa
100°C
30 m/s

Air

200 kPa

**FIGURE P8–95**

**8–96** Repeat Prob. 8–95 for the polytropic exponent of $n = 1.1$.

**8–97** Air at 427°C and 600 kPa is contained in a piston-cylinder device. The air expands adiabatically until the pressure is 100 kPa. Determine the mass of air needed to produce maximum work of 1000 kJ. Assume air has constant specific heats evaluated at 300 K. *Answer: 4.97 kg*

**8–98** Helium enters a steady-flow adiabatic compressor at 0.6 kg/s, 100 kPa, and 27°C with a low inlet velocity and is compressed to 600 kPa.

(a) Determine the exit temperature to make the work input and kinetic energy at the compressor exit have the minimum values.

(b) If the rate of work input into the compressor is measured to be a minimum with a value of 1000 kW, determine the compressor exit velocity, in m/s.

**8–99** 5-kg of air at 427°C and 600 kPa are contained in a piston-cylinder device. The air expands adiabatically until the pressure is 100 kPa and produces 600 kJ of work output. Assume air has constant specific heats evaluated at 300 K.

(a) Determine the entropy change of the air, in kJ/kg·K

(b) Since the process is adiabatic, is the process realistic? Using concepts of the second law, support your answer.

**8–100** A constant-volume tank contains 5 kg of air at 100 kPa and 327°C. The air is cooled to the surroundings temperature of 27°C. Assume constant specific heats at 300 K. (a) Determine the entropy change of the air in the tank during the process, in kJ/K, (b) determine the net entropy change of the universe due to this process, in kJ/K, and (c) sketch the processes for the air in the tank and the surroundings on a single T-s diagram. Be sure to label the initial and final states for both processes.

**8–101** A container filled with 45 kg of liquid water at 95°C is placed in a 90-m³ room that is initially at 12°C. Thermal equilibrium is established after a while as a result of heat transfer between the water and the air in the room. Using constant specific heats, determine (a) the final equilibrium temperature, (b) the amount of heat transfer between the water and the air in the room, and (c) the entropy generation. Assume the room is well sealed and heavily insulated.

Room
90 m³
12°C

Water
45 kg
95°C

**FIGURE P8–101**

**8–102** An ideal gas at 100 kPa and 15°C enters a steady-flow compressor. The gas is compressed to 600 kPa, and 10 percent of the mass that entered the compressor is removed for some other use. The remaining 90 percent of the inlet gas is compressed to 800 kPa before leaving the compressor. The entire compression process is assumed to be reversible and adiabatic. The power supplied to the compressor is measured to be 32 kW. If the ideal gas has constant specific heats such that $c_v = 0.8$ kJ/kg·K and $c_p = 1.1$ kJ/kg·K, (a) sketch the compression process on a $T$-$s$ diagram, (b) determine the temperature of the gas at the two compressor exits, in K, and (c) determine the mass flow rate of the gas into the compressor, in kg/s.

**8–103E** The well-insulated container shown in Fig. P8–103E is initially evacuated. The supply line contains air that is maintained at 200 psia and 100°F. The valve is opened until the pressure in the container is the same as the pressure in the supply line. Determine the minimum temperature in the container when the valve is closed.

Vessel

Valve

Supply line

**FIGURE P8–103E**

### Reversible Steady-Flow Work

**8–104C** In large compressors, the gas is frequently cooled while being compressed to reduce the power consumed by the compressor. Explain how cooling the gas during a compression process reduces the power consumption.

**8–105C** The turbines in steam power plants operate essentially under adiabatic conditions. A plant engineer suggests to end this practice. She proposes to run cooling water through the outer surface of the casing to cool the steam as it flows through the turbine. This way, she reasons, the entropy of the steam will decrease, the performance of the turbine will improve, and as a result the work output of the turbine will increase. How would you evaluate this proposal?

**8–106C** It is well known that the power consumed by a compressor can be reduced by cooling the gas during compression.

Inspired by this, somebody proposes to cool the liquid as it flows through a pump, in order to reduce the power consumption of the pump. Would you support this proposal? Explain.

**8–107E** Air is compressed isothermally from 13 psia and 90°F to 80 psia in a reversible steady-flow device. Calculate the work required, in Btu/lbm, for this compression. *Answer:* 68.5 Btu/lbm

**8–108** Saturated water vapor at 150°C is compressed in a reversible steady-flow device to 1000 kPa while its specific volume remains constant. Determine the work required, in kJ/kg.

**8–109E** Calculate the work produced, in Btu/lbm, for the reversible steady-flow process 1-3 shown in Fig. P8–109E.

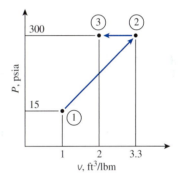

**FIGURE P8–109E**

**8–110** Calculate the work produced, in kJ/kg, for the reversible isothermal, steady-flow process 1-3 shown in Fig. P8–110 when the working fluid is an ideal gas.

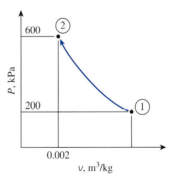

**FIGURE P8–110**

**8–111** Liquid water enters a 25-kW pump at 100-kPa pressure at a rate of 5 kg/s. Determine the highest pressure the liquid water can have at the exit of the pump. Neglect the kinetic and potential energy changes of water, and take the specific volume of water to be 0.001 m³/kg. *Answer:* 5100 kPa

**FIGURE P8–111**

**8–112** Consider a steam power plant that operates between the pressure limits of 5 MPa and 10 kPa. Steam enters the pump as saturated liquid and leaves the turbine as saturated vapor. Determine the ratio of the work delivered by the turbine to the work consumed by the pump. Assume the entire cycle to be reversible and the heat losses from the pump and the turbine to be negligible.

**8–113** [EES] Reconsider Prob. 8–112. Using EES (or other) software, investigate the effect of the quality of the steam at the turbine exit on the net work output. Vary the quality from 0.5 to 1.0, and plot the net work output as a function of this quality.

**8–114** Liquid water at 120 kPa enters a 7-kW pump where its pressure is raised to 5 MPa. If the elevation difference between the exit and the inlet levels is 10 m, determine the highest mass flow rate of liquid water this pump can handle. Neglect the kinetic energy change of water, and take the specific volume of water to be 0.001 m³/kg.

## Isentropic Efficiencies of Steady-Flow Devices

**8–115C** Describe the ideal process for an (*a*) adiabatic turbine, (*b*) adiabatic compressor, and (*c*) adiabatic nozzle, and define the isentropic efficiency for each device.

**8–116C** Is the isentropic process a suitable model for compressors that are cooled intentionally? Explain.

**8–117C** On a *T-s* diagram, does the actual exit state (state 2) of an adiabatic turbine have to be on the right-hand side of the isentropic exit state (state 2*s*)? Why?

**8–118E** Steam at 100 psia and 650°F is expanded adiabatically in a closed system to 10 psia. Determine the work produced, in Btu/lbm, and the final temperature of steam for an isentropic expansion efficiency of 80 percent. *Answers:* 132 Btu/lbm, 275°F

**8–119** Steam at 3 MPa and 400°C is expanded to 30 kPa in an adiabatic turbine with an isentropic efficiency of 92 percent. Determine the power produced by this turbine, in kW, when the mass flow rate is 2 kg/s.

**8–120** Repeat Prob. 8–119 for a turbine efficiency of 85 percent.

**8–121** Steam enters an adiabatic turbine at 7 MPa, 600°C, and 80 m/s and leaves at 50 kPa, 150°C, and 140 m/s. If the power output of the turbine is 6 MW, determine (*a*) the mass flow rate of the steam flowing through the turbine and (*b*) the isentropic efficiency of the turbine. *Answers:* (*a*) 6.95 kg/s, (*b*) 73.4 percent

**8–122E** Combustion gases enter an adiabatic gas turbine at 1540°F and 120 psia and leave at 60 psia with a low velocity. Treating the combustion gases as air and assuming an isentropic efficiency of 82 percent, determine the work output of the turbine. *Answer:* 71.7 Btu/lbm

**8–123** Air at 100 kPa and 20°C is compressed to 700 kPa steadily and adiabatically at a rate of 2 kg/s. Determine the power required to compress this air if the isentropic compression efficiency is 95 percent.

**8–124** Steam at 4 MPa and 350°C is expanded in an adiabatic turbine to 120 kPa. What is the isentropic efficiency of this turbine if the steam is exhausted as a saturated vapor?

**FIGURE P8–124**

**8–125** Air is expanded from 2 MPa and 327°C to 100 kPa in an adiabatic turbine. Determine the isentropic efficiency of this turbine if the air is exhausted at 0°C.

**8–126** A refrigeration unit compresses saturated R-134a vapor at 20°C to 1000 kPa. How much power is required to compress 0.5 kg/s of R-134a with a compressor efficiency of 85 percent? *Answer:* 6.78 kW

**8–127**  Refrigerant-134a enters an adiabatic compressor as saturated vapor at 100 kPa at a rate of 0.7 m³/min and exits at 1-MPa pressure. If the isentropic efficiency of the compressor is 87 percent, determine (a) the temperature of the refrigerant at the exit of the compressor and (b) the power input, in kW. Also, show the process on a T-s diagram with respect to saturation lines.

1 MPa

R-134a
Compressor

100 kPa
sat. vapor

**FIGURE P8–127**

**8–128** Reconsider Prob. 8–127. Using EES (or other) software, redo the problem by including the effects of the kinetic energy of the flow by assuming an inlet-to-exit area ratio of 1.5 for the compressor when the compressor exit pipe inside diameter is 2 cm.

**8–129** Air enters an adiabatic compressor at 100 kPa and 17°C at a rate of 2.4 m³/s, and it exits at 257°C. The compressor has an isentropic efficiency of 84 percent. Neglecting the changes in kinetic and potential energies, determine (a) the exit pressure of air and (b) the power required to drive the compressor.

**8–130** Air is compressed by an adiabatic compressor from 95 kPa and 27°C to 600 kPa and 277°C. Assuming variable specific heats and neglecting the changes in kinetic and potential energies, determine (a) the isentropic efficiency of the compressor and (b) the exit temperature of air if the process were reversible. *Answers: (a) 81.9 percent, (b) 505.5 K*

**8–131E** Argon gas enters an adiabatic compressor at 20 psia and 90°F with a velocity of 60 ft/s, and it exits at 200 psia and 240 ft/s. If the isentropic efficiency of the compressor is 80 percent, determine (a) the exit temperature of the argon and (b) the work input to the compressor.

**8–132E** Air enters an adiabatic nozzle at 45 psia and 940°F with low velocity and exits at 650 ft/s. If the isentropic efficiency of the nozzle is 85 percent, determine the exit temperature and pressure of the air.

**8–133E** Reconsider Prob. 8–132E. Using EES (or other) software, study the effect of varying the nozzle isentropic efficiency from 0.8 to 1.0 on both the exit temperature and pressure of the air, and plot the results.

**8–134** The exhaust nozzle of a jet engine expands air at 300 kPa and 180°C adiabatically to 100 kPa. Determine the air velocity at the exit when the inlet velocity is low and the nozzle isentropic efficiency is 96 percent.

**8–135E** An adiabatic diffuser at the inlet of a jet engine increases the pressure of the air that enters the diffuser at 13 psia and 30°F to 20 psia. What will the air velocity at the diffuser exit be if the diffuser isentropic efficiency defined as the ratio of the actual kinetic energy change to the isentropic kinetic energy change is 82 percent and the diffuser inlet velocity is 1000 ft/s? *Answer: 606 ft/s*

13 psia
30°F
1000 ft/s

Air

20 psia

**FIGURE P8–135E**

### Entropy Balance

**8–136E** Refrigerant-134a is expanded adiabatically from 100 psia and 100°F to a saturated vapor at 10 psia. Determine the entropy generation for this process, in Btu/lbm·R.

R-134a
100 psia
100°F

10 psia

**FIGURE P8–136E**

**8–137** Oxygen enters an insulated 12-cm-diameter pipe with a velocity of 70 m/s. At the pipe entrance, the oxygen is at 240 kPa and 20°C; and, at the exit, it is at 200 kPa and 18°C. Calculate the rate at which entropy is generated in the pipe.

**8–138** Nitrogen is compressed by an adiabatic compressor from 100 kPa and 25°C to 800 kPa and 307°C. Calculate the entropy generation for this process, in kJ/kg·K.

**8–139** Consider a family of four, with each person taking a 5-min shower every morning. The average flow rate through the shower head is 12 L/min. City water at 15°C is heated to 55°C in an electric water heater and tempered to 42°C by cold water at the T-elbow of the shower before being routed to the shower head. Determine the amount of entropy generated by this family per year as a result of taking daily showers.

**8–140** Cold water ($c_p$ = 4.18 kJ/kg·°C) leading to a shower enters a well-insulated, thin-walled, double-pipe, counterflow heat exchanger at 10°C at a rate of 0.95 kg/s and is heated to 70°C by hot water ($c_p$ = 4.19 kJ/kg·°C) that enters at 85°C at a rate of 1.6 kg/s. Determine (a) the rate of heat transfer and (b) the rate of entropy generation in the heat exchanger.

FIGURE P8–140

**8–141** Air ($c_p$ = 1.005 kJ/kg·°C) is to be preheated by hot exhaust gases in a cross-flow heat exchanger before it enters the furnace. Air enters the heat exchanger at 95 kPa and 20°C at a rate of 1.6 m³/s. The combustion gases ($c_p$ = 1.10 kJ/kg·°C) enter at 180°C at a rate of 2.2 kg/s and leave at 95°C. Determine (a) the rate of heat transfer to the air, (b) the outlet temperature of the air, and (c) the rate of entropy generation.

**8–142** A well-insulated, shell-and-tube heat exchanger is used to heat water ($c_p$ = 4.18 kJ/kg·°C) in the tubes from 20 to 70°C at a rate of 4.5 kg/s. Heat is supplied by hot oil ($c_p$ = 2.30 kJ/kg·°C) that enters the shell side at 170°C at a rate of 10 kg/s. Disregarding any heat loss from the heat exchanger, determine (a) the exit temperature of the oil and (b) the rate of entropy generation in the heat exchanger.

FIGURE P8–142

**8–143** Refrigerant-134a is throttled from 1200 kPa and 40°C to 200 kPa. Heat is lost from the refrigerant in the amount of 0.5 kJ/kg to the surroundings at 25°C. Determine (a) the exit temperature of the refrigerant and (b) the entropy generation during this process.

FIGURE P8–143

**8–144** In an ice-making plant, water at 0°C is frozen at atmospheric pressure by evaporating saturated R-134a liquid at −16°C. The refrigerant leaves this evaporator as a saturated vapor, and the plant is sized to produce ice at 0°C at a rate of 2500 kg/h. Determine the rate of entropy generation in this plant.   *Answer:* 0.0528 kW/K

FIGURE P8–144

**8–145E** Water at 20 psia and 50°F enters a mixing chamber at a rate of 300 lbm/min where it is mixed steadily with steam entering at 20 psia and 240°F. The mixture leaves the chamber at 20 psia and 130°F, and heat is lost to the surrounding air at 70°F at a rate of 180 Btu/min. Neglecting the changes in kinetic and potential energies, determine the rate of entropy generation during this process?

FIGURE P8–145E

**8–146E** Steam is to be condensed on the shell side of a heat exchanger at 120°F. Cooling water enters the tubes at 60°F at a rate of 92 lbm/s and leaves at 73°F. Assuming the heat exchanger to be well-insulated, determine (a) the rate of heat transfer in the heat exchanger and (b) the rate of entropy generation in the heat exchanger.

**8–147** In a dairy plant, milk at 4°C is pasteurized continuously at 72°C at a rate of 12 L/s for 24 hours a day and 365 days a year. The milk is heated to the pasteurizing temperature by hot water heated in a natural-gas-fired boiler that has an efficiency of 82 percent. The pasteurized milk is then cooled by cold water at 18°C before it is finally refrigerated back to 4°C. To save energy and money, the plant installs a regenerator that has an effectiveness of 82 percent. If the cost of natural gas is $1.04/therm (1 therm = 105,500 kJ), determine how much energy and money the regenerator will save this company per year and the annual reduction in entropy generation.

**FIGURE P8–147**

**8–148** Stainless-steel ball bearings ($\rho = 8085$ kg/m³ and $c_p = 0.480$ kJ/kg·°C) having a diameter of 1.8 cm are to be quenched in water at a rate of 1100 per minute. The balls leave the oven at a uniform temperature of 900°C and are exposed to air at 20°C for a while before they are dropped into the water. If the temperature of the balls drops to 850°C prior to quenching, determine (a) the rate of heat transfer from the balls to the air and (b) the rate of entropy generation due to heat loss from the balls to the air.

**8–149** An ordinary egg can be approximated as a 5.5-cm-diameter sphere. The egg is initially at a uniform temperature of 8°C and is dropped into boiling water at 97°C. Taking the properties of the egg to be $\rho = 1020$ kg/m³ and $c_p = 3.32$ kJ/kg·°C, determine (a) how much heat is transferred to the egg by the time the average temperature of the egg rises to 70°C and (b) the amount of entropy generation associated with this heat transfer process.

**FIGURE P8–149**

**8–150** Long cylindrical steel rods ($\rho = 7833$ kg/m³ and $c_p = 0.465$ kJ/kg·°C) of 10-cm diameter are heat treated by drawing them at a velocity of 3 m/min through a 7-m-long oven maintained at 900°C. If the rods enter the oven at 30°C and leave at 700°C, determine (a) the rate of heat transfer to the rods in the oven and (b) the rate of entropy generation associated with this heat transfer process.

**FIGURE P8–150**

**8–151** The inner and outer surfaces of a 3-m × 8-m brick wall of thickness 20 cm are maintained at temperatures of 20°C and 2°C, respectively. If the rate of heat transfer through the wall is 1550 W, determine the rate of entropy generation within the wall.

**8–152** For heat transfer purposes, a standing man can be modeled as a 30-cm-diameter, 170-cm-long vertical cylinder with both the top and bottom surfaces insulated and with the side surface at an average temperature of 34°C. If the rate of heat loss from this man to the environment at 20°C is 336 W, determine the rate of entropy transfer from the body of this person accompanying heat transfer, in W/K.

**8–153** A 1000-W iron is left on the ironing board with its base exposed to the air at 20°C. If the surface temperature is 400°C, determine the rate of entropy generation during this process in steady operation. How much of this entropy generation occurs within the iron?

**8–154E** A frictionless piston–cylinder device contains saturated liquid water at 40-psia pressure. Now 600 Btu of heat is transferred to water from a source at 1000°F, and part of the liquid vaporizes at constant pressure. Determine the total entropy generated during this process, in Btu/R.

**8–155E** Steam enters a diffuser at 20 psia and 240°F with a velocity of 900 ft/s and exits as saturated vapor at 240°F and 100 ft/s. The exit area of the diffuser is 1 ft². Determine (a) the mass flow rate of the steam and (b) the rate of entropy generation during this process. Assume an ambient temperature of 77°F.

**8–156** Steam enters an adiabatic nozzle at 2.5 MPa and 450°C with a velocity of 55 m/s and exits at 1 MPa and 390 m/s. If the nozzle has an inlet area of 6 cm², determine (a) the exit temperature and (b) the rate of entropy generation for this process. *Answers:* (a) 406°C, (b) 0.0783 kW/K

**8–157** Steam expands in a turbine steadily at a rate of 40,000 kg/h, entering at 8 MPa and 500°C and leaving at 40 kPa as saturated vapor. If the power generated by the turbine is 8.2 MW, determine the rate of entropy generation for this process. Assume the surrounding medium is at 25°C. *Answer:* 11.4 kW/K

**FIGURE P8–160**

8 MPa
500°C

Steam turbine

8.2 MW

40 kPa
sat. vapor

**FIGURE P8–157**

**8–158E** Air enters a compressor at ambient conditions of 15 psia and 60°F with a low velocity and exits at 150 psia, 620°F, and 350 ft/s. The compressor is cooled by the ambient air at 60°F at a rate of 1500 Btu/min. The power input to the compressor is 400 hp. Determine (a) the mass flow rate of air and (b) the rate of entropy generation.

**8–159** A hot-water stream at 70°C enters an adiabatic mixing chamber with a mass flow rate of 3.6 kg/s, where it is mixed with a stream of cold water at 20°C. If the mixture leaves the chamber at 42°C, determine (a) the mass flow rate of the cold water and (b) the rate of entropy generation during this adiabatic mixing process. Assume all the streams are at a pressure of 200 kPa.

**8–160** Liquid water at 200 kPa and 20°C is heated in a chamber by mixing it with superheated steam at 200 kPa and 150°C. Liquid water enters the mixing chamber at a rate of 2.5 kg/s, and the chamber is estimated to lose heat to the surrounding air at 25°C at a rate of 1200 kJ/min. If the mixture leaves the mixing chamber at 200 kPa and 60°C, determine (a) the mass flow rate of the superheated steam and (b) the rate of entropy generation during this mixing process. *Answers:* (a) 0.166 kg/s, (b) 0.333 kW/K

**8–161** A 0.18-m³ rigid tank is filled with saturated liquid water at 120°C. A valve at the bottom of the tank is now opened, and one-half of the total mass is withdrawn from the tank in the liquid form. Heat is transferred to water from a source at 230°C so that the temperature in the tank remains constant. Determine (a) the amount of heat transfer and (b) the total entropy generation for this process.

**8–162E** An iron block of unknown mass at 185°F is dropped into an insulated tank that contains 0.8 ft³ of water at 70°F. At the same time, a paddle wheel driven by a 200-W motor is activated to stir the water. Thermal equilibrium is established after 10 min with a final temperature of 75°F. Determine (a) the mass of the iron block and (b) the entropy generated during this process.

**8–163** A rigid tank contains 7.5 kg of saturated water mixture at 400 kPa. A valve at the bottom of the tank is now opened, and liquid is withdrawn from the tank. Heat is transferred to the steam such that the pressure inside the tank remains constant. The valve is closed when no liquid is left in the tank. If it is estimated that a total of 5 kJ of heat is transferred to the tank, determine (a) the quality of steam in the tank at the initial state, (b) the amount of mass that has escaped, and (c) the entropy generation during this process if heat is supplied to the tank from a source at 500°C.

### Review Problems

**8–164E** A heat engine whose thermal efficiency is 40 percent uses a hot reservoir at 1300 R and a cold reservoir at 500 R. Calculate the entropy change of the two reservoirs when 1 Btu of heat is transferred from the hot reservoir to the engine. Does this engine satisfy the increase of entropy principle? If the thermal efficiency of the heat engine is increased to 70 percent, will the increase of entropy principle still be satisfied?

**8–165** A refrigerator with a coefficient of performance of 4 transfers heat from a cold region at $-20°C$ to a hot region at $30°C$. Calculate the total entropy change of the regions when 1 kJ of heat is transferred from the cold region. Is the second law satisfied? Will this refrigerator still satisfy the second law if its coefficient of performance is 6?

**FIGURE P8–165**

**8–166** Calculate the rate of entropy change of all the components of a refrigerator that uses 10 kW of power, rejects 14 kW of heat, and has a high-temperature energy reservoir at 400 K and a low-temperature energy reservoir at 200 K. What is the rate of cooling produced by this refrigerator? Is this refrigerator completely reversible?

**8–167** What is the minimum internal energy that R-134a can achieve as it is compressed adiabatically from 200 kPa and 85 percent quality to 800 kPa in a closed system?

**8–168** Is it possible to cool and condense R-134a to a saturated liquid from 1000 kPa and 180°C in a closed system undergoing an isobaric, reversible process while exchanging heat with an isothermal energy reservoir at 100°C?

**8–169** It has been suggested that air at 100 kPa and 25°C can be cooled by first compressing it adiabatically in a closed system to 1000 kPa and then expanding it adiabatically back to 100 kPa. Is this possible?

**8–170E** Can 1 lbm of air in a closed system at 20 psia and 100°F be compressed adiabatically to 120 psia and a volume of 3 ft³?

**8–171E** A 100-lbm block of a solid material whose specific heat is 0.5 Btu/lbm·R is at 70°F. It is heated with 10 lbm of saturated water vapor that has a constant pressure of 14.7 psia. Determine the final temperature of the block and water, and the entropy change of (a) the block, (b) the water, and (c) the entire system. Is this process possible? Why?

**8–172** One kilogram of air is in a piston–cylinder apparatus that can only exchange heat with a reservoir at 300 K. Initially this air is at 100 kPa and 27°C. Someone claims that the air can be compressed to 250 kPa and 27°C. Determine if this claim is valid by performing a second-law analysis of the claimed process.

**FIGURE P8–172**

**8–173** A rigid tank contains 3.2 kg of water at 140°C and 400 kPa. Now 48 kJ of shaft work is done on the system and the final temperature in the tank is 80°C. If the entropy change of water is zero and the surroundings are at 15°C, determine (a) the final pressure in the tank, (b) the amount of heat transfer between the tank and the surroundings, and (c) the entropy generation during this process. *Answers:* (a) 47.4 kPa, (b) 163 kJ, (c) 0.565 kJ/K

**8–174** A horizontal cylinder is separated into two compartments by an adiabatic, frictionless piston. One side contains 0.2 m³ of nitrogen and the other side contains 0.1 kg of helium, both initially at 20°C and 95 kPa. The sides of the cylinder and the helium end are insulated. Now heat is added to the nitrogen side from a reservoir at 500°C until the pressure of the helium rises to 120 kPa. Determine (a) the final temperature of the helium, (b) the final volume of the nitrogen, (c) the heat transferred to the nitrogen, and (d) the entropy generation during this process.

**FIGURE P8–174**

**8–175** A 0.8-m³ rigid tank contains carbon dioxide ($CO_2$) gas at 250 K and 100 kPa. A 500-W electric resistance heater placed in the tank is now turned on and kept on for 40 min after which the pressure of $CO_2$ is measured to be 175 kPa. Assuming the surroundings to be at 300 K and using constant specific heats, determine (a) the final temperature of $CO_2$, (b) the net amount of heat transfer from the tank, and (c) the entropy generation during this process.

**FIGURE P8–175**

**FIGURE P8–179**

**8–176** Helium gas is throttled steadily from 300 kPa and 50°C. Heat is lost from the helium in the amount of 1.75 kJ/kg to the surroundings at 25°C and 100 kPa. If the entropy of the helium increases by 0.25 kJ/kg·K in the valve, determine (a) the exit pressure and temperature and (b) the entropy generation during this process. *Answers:* (a) 265 kPa, 49.7°C, (b) 0.256 kJ/kg·K

**8–177** Refrigerant-134a enters a compressor as a saturated vapor at 200 kPa at a rate of 0.03 m³/s and leaves at 700 kPa. The power input to the compressor is 10 kW. If the surroundings at 20°C experience an entropy increase of 0.008 kW/K, determine (a) the rate of heat loss from the compressor, (b) the exit temperature of the refrigerant, and (c) the rate of entropy generation.

**8–178** Air at 500 kPa and 400 K enters an adiabatic nozzle at a velocity of 30 m/s and leaves at 300 kPa and 350 K. Using variable specific heats, determine (a) the isentropic efficiency, (b) the exit velocity, and (c) the entropy generation.

Air
500 kPa
400 K
30 m/s

300 kPa
350 K

**FIGURE P8–178**

**8–179** An insulated tank containing 0.2 m³ of saturated water vapor at 350 kPa is connected to an initially evacuated, insulated piston–cylinder device. The mass of the piston is such that a pressure of 200 kPa is required to raise it. Now the valve is opened slightly, and part of the steam flows to the cylinder, raising the piston. This process continues until the pressure in the tank drops to 200 kPa. Assuming the steam that remains in the tank to have undergone a reversible adiabatic process, determine the final temperature (a) in the rigid tank and (b) in the cylinder.

**8–180** 3-kg of helium gas at 100 kPa and 27°C are adiabatically compressed to 900 kPa. If the isentropic compression efficiency is 80 percent, determine the required work input and the final temperature of helium.

**8–181** Refrigerant-134a at 600 kPa and 100°C undergoes a reversible, isothermal expansion to 200 kPa in a steady-flow device with one inlet and one outlet. Determine the power produced by this device and the rate of heat transfer when the flow rate through the device is 1 kg/s.

**8–182** An inventor claims to have invented an adiabatic steady-flow device with a single inlet-outlet that produces 100 kW when expanding 1 kg/s of air from 900 kPa and 300°C to 100 kPa. Is this claim valid?

**8–183** You are to expand a gas adiabatically from 3 MPa and 300°C to 80 kPa in a piston-cylinder device. Which of the two choices – air with an isentropic expansion efficiency of 90 percent or neon with an isentropic expansion efficiency of 80 percent – will produce the most work?

**8–184** An adiabatic capillary tube is used in some refrigeration systems to drop the pressure of the refrigerant from the condenser level to the evaporator level. R-134a enters the capillary tube as a saturated liquid at 50°C, and leaves at −12°C. Determine the rate of entropy generation in the capillary tube for a mass flow rate of 0.2 kg/s. *Answer:* 0.0077 kW/K

**FIGURE P8–184**

**8–185** Air is compressed steadily by a compressor from 100 kPa and 17°C to 700 kPa at a rate of 5 kg/min. Determine the minimum power input required if the process is (a) adiabatic and (b) isothermal. Assume air to be an ideal gas with variable specific heats, and neglect the changes in kinetic and potential energies. *Answers:* (a) 18.0 kW, (b) 13.5 kW

**8–186** Refrigerant-134a at 100 kPa and −20°C is compressed by an adiabatic 1.3-kW compressor to an exit state of 800 kPa and 60°C. Neglecting the changes in kinetic and potential energies, determine (a) the isentropic efficiency of the compressor, (b) the volume flow rate of the refrigerant at the compressor inlet, in L/min, and (c) the maximum volume flow rate at the inlet conditions that this adiabatic 0.7-kW compressor can handle without violating the second law.

**8–187** An adiabatic air compressor is to be powered by a direct-coupled adiabatic steam turbine that is also driving a generator. Steam enters the turbine at 12.5 MPa and 500°C at a rate of 25 kg/s and exits at 10 kPa and a quality of 0.92. Air enters the compressor at 98 kPa and 295 K at a rate of 10 kg/s and exits at 1 MPa and 620 K. Determine (a) the net power delivered to the generator by the turbine and (b) the rate of entropy generation within the turbine and the compressor during this process.

**FIGURE P8–187**

**8–188** Reconsider Prob. 8–187. Using EES (or other) software, determine the isentropic efficiencies for the compressor and turbine. Then use EES to study how varying the compressor efficiency over the range 0.6 to 0.8 and the turbine efficiency over the range 0.7 to 0.95 affect the net work for the cycle and the entropy generated for the process. Plot the net work as a function of the compressor efficiency for turbine efficiencies of 0.7, 0.8, and 0.9, and discuss your results.

**8–189** The explosion of a hot water tank in a school in Spencer, Oklahoma, in 1982 killed 7 people while injuring 33 others. Although the number of such explosions has decreased dramatically since the development of the ASME Pressure Vessel Code, which requires the tanks to be designed to withstand four times the normal operating pressures, they still occur as a result of the failure of the pressure relief valves and thermostats. When a tank filled with a high-pressure and high-temperature liquid ruptures, the sudden drop of the pressure of the liquid to the atmospheric level causes part of the liquid to flash into vapor, and thus to experience a huge rise in its volume. The resulting pressure wave that propagates rapidly can cause considerable damage.

Considering that the pressurized liquid in the tank eventually reaches equilibrium with its surroundings shortly after the explosion, the work that a pressurized liquid would do if allowed to expand reversibly and adiabatically to the pressure of the surroundings can be viewed as the *explosive energy* of the pressurized liquid. Because of the very short time period of the explosion and the apparent calm afterward, the explosion process can be considered to be adiabatic with no changes in kinetic and potential energies and no mixing with the air.

Consider a 80-L hot-water tank that has a working pressure of 0.5 MPa. As a result of some malfunction, the pressure in the tank rises to 2 MPa, at which point the tank explodes. Taking the atmospheric pressure to be 100 kPa and assuming the liquid in the tank to be saturated at the time of explosion, determine the total explosion energy of the tank in terms of the TNT equivalence. (The explosion energy of TNT is about 3250 kJ/kg, and 5 kg of TNT can cause total destruction of unreinforced structures within about a 7-m radius.) *Answer:* 1.972 kg TNT

**FIGURE P8–189**

**8–190** Using the arguments in the Prob. 8–189, determine the total explosion energy of a 0.35-L canned drink that explodes at a pressure of 1.2 MPa. To how many kg of TNT is this explosion energy equivalent?

**8–191** Air is expanded in an adiabatic turbine of 85 percent isentropic efficiency from an inlet state of 2200 kPa and 300°C to an outlet pressure of 200 kPa. Calculate the outlet temperature of air and the work produced by this turbine per unit mass of air.

**8–192** Air is expanded in an adiabatic turbine of 90 percent isentropic efficiency from an inlet state of 2200 kPa and 300°C to an outlet pressure of 200 kPa. Calculate the outlet temperature of air, the work produced by this turbine, and the entropy generation. *Answers:* 319 K, 258 kJ/kg, 0.0944 kJ/kg·K

**8–193** To control the power output of an isentropic steam turbine, a throttle valve is placed in the steam line supplying the turbine inlet, as shown in the figure. Steam at 6 MPa and

400°C is supplied to the throttle inlet, and the turbine exhaust pressure is set at 70 kPa. Compare the work produced by this steam turbine, in kJ/kg, when the throttle valve is completely open (so that there is no pressure loss) and when it is partially closed so that the pressure at the turbine inlet is 3 MPa.

**FIGURE P8–193**

**8–194** Oxygen tanks are filled by compressing the oxygen gas, as shown in the figure. An initially evacuated 1-m³ tank is to be filled to 13,000 kPa while the temperature of the oxygen in the tank remains constant at 20°C with this system. An isentropic compressor is used. The oxygen enters this compressor with a constant pressure and temperature of 150 kPa and 20°C. Determine the total work required by the compressor and the total heat transfer from the oxygen tank. *Answers:* 80,710 kJ, 93,720 kJ

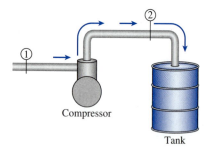

**FIGURE P8–194**

**8–195** Two rigid tanks are connected by a valve. Tank A is insulated and contains 0.3 m³ of steam at 400 kPa and 60 percent quality. Tank B is uninsulated and contains 2 kg of steam at 200 kPa and 250°C. The valve is now opened, and steam flows from tank A to tank B until the pressure in tank A drops to 200 kPa. During this process 300 kJ of heat is transferred from tank B to the surroundings at 17°C. Assuming the steam remaining inside tank A to have undergone a reversible adiabatic process, determine (*a*) the final temperature in each tank and (*b*) the entropy generated during this process. *Answers:* (*a*) 120.2°C, 116.1°C, (*b*) 0.498 kJ/K

**FIGURE P8–195**

**8–196** Heat is transferred steadily to boiling water in the pan through its flat bottom at a rate of 500 W. If the temperatures of the inner and outer surfaces of the bottom of the tank are 104 and 105°C, respectively, determine the rate of entropy generation within bottom of the pan, in W/K.

**FIGURE P8–196**

**8–197** A 1200-W electric resistance heating element whose diameter is 0.5 cm is immersed in 40 kg of water initially at 20°C. Assuming the water container is well-insulated, determine how long it will take for this heater to raise the water temperature to 50°C. Also, determine the entropy generated during this process, in kJ/K.

**8–198** In large steam power plants, the feedwater is frequently heated in closed feedwater heaters, which are basically heat exchangers, by steam extracted from the turbine at some stage. Steam enters the feedwater heater at 1 MPa and 200°C and leaves as saturated liquid at the same pressure. Feedwater enters the heater at 2.5 MPa and 50°C and leaves 10°C below the exit temperature of the steam. Neglecting any heat losses from the outer surfaces of the heater, determine (*a*) the ratio of the mass flow rates of the extracted steam and the feedwater heater and (*b*) the total entropy change for this process per unit mass of the feedwater.

**8–199** [EES] Reconsider Prob. 8–198. Using EES (or other) software, investigate the effect of the state of the steam at the inlet of the feedwater heater. Assume the entropy of the extraction steam is constant at the value for 1 MPa, 200°C and decrease the extraction steam pressure from 1 MPa to 100 kPa. Plot both the ratio of the mass flow rates of the extracted steam and the feedwater heater and the total entropy change for this process per unit mass of the feedwater as functions of the extraction pressure.

**8–200E**  A 5-ft³ rigid tank initially contains refrigerant-134a at 80 psia and 100 percent quality. The tank is connected by a valve to a supply line that carries refrigerant-134a at 160 psia and 80°F. The valve is now opened, allowing the refrigerant to enter the tank, and is closed when it is observed that the tank contains only saturated liquid at 120 psia. Determine (a) the mass of the refrigerant that entered the tank, (b) the amount of heat transfer with the surroundings at 100°F, and (c) the entropy generated during this process.

**8–201**  A passive solar house that is losing heat to the outdoors at 3°C at an average rate of 50,000 kJ/h is maintained at 22°C at all times during a winter night for 10 h. The house is to be heated by 50 glass containers, each containing 20 L of water that is heated to 80°C during the day by absorbing solar energy. A thermostat controlled 15 kW backup electric resistance heater turns on whenever necessary to keep the house at 22°C. Determine how long the electric heating system was on that night and the amount of entropy generated during the night.

## Design and Essay Problems

**8–202**  Compare the pumping of water to a higher elevation to isentropic compression of air into a compressed-air tank as means of storing energy for later use.

**8–203**  It is well-known that the temperature of a gas rises while it is compressed as a result of the energy input in the form of compression work. At high compression ratios, the air temperature may rise above the autoignition temperature of some hydrocarbons, including some lubricating oil. Therefore, the presence of some lubricating oil vapor in high-pressure air raises the possibility of an explosion, creating a fire hazard. The concentration of the oil within the compressor is usually too low to create a real danger. However, the oil that collects on the inner walls of exhaust piping of the compressor may cause an explosion. Such explosions have largely been eliminated by using the proper lubricating oils, carefully designing the equipment, intercooling between compressor stages, and keeping the system clean.

A compressor is to be designed for an industrial application in Los Angeles. If the compressor exit temperature is not to exceed 250°C for safety consideration, determine the maximum allowable compression ratio that is safe for all possible weather conditions for that area.

**8–204**  Obtain the following information about a power plant that is closest to your town: the net power output; the type and amount of fuel; the power consumed by the pumps, fans, and other auxiliary equipment; stack gas losses; temperatures at several locations; and the rate of heat rejection at the condenser. Using these and other relevant data, determine the rate of entropy generation in that power plant.

**8–205**  Enthalpy-entropy state diagrams (also known as *Mollier diagrams*) are more useful than temperature-entropy state diagrams for analyzing some systems. Using EES (or other) software and actual property data, plot an enthalpy-entropy diagram for water that includes the saturation lines. Also, sketch isothermal, isobaric, and constant volume processes on this diagram.

# POWER AND REFRIGERATION CYCLES

Two important areas of application for thermodynamics are power generation and refrigeration. Both power generation and refrigeration are usually accomplished by systems that operate on a thermodynamic cycle. Thermodynamic cycles can be divided into two general categories: *power cycles* and *refrigeration cycles.*

The devices or systems used to produce a net power output are often called *engines*, and the thermodynamic cycles they operate on are called *power cycles*. The devices or systems used to produce refrigeration are called *refrigerators, air conditioners,* or *heat pumps*, and the cycles they operate on are called *refrigeration cycles.*

Thermodynamic cycles can also be categorized as *gas cycles* or *vapor cycles*, depending on the *phase* of the working fluid—the substance that circulates through the cyclic device. In gas cycles, the working fluid remains in the gaseous phase throughout the entire cycle, whereas in vapor cycles the working fluid exists in the vapor phase during one part of the cycle and in the liquid phase during another part.

Thermodynamic cycles can be categorized yet another way: *closed* and *open cycles.* In closed cycles, the working fluid is returned to the initial state at the end of the cycle and is recirculated. In open cycles, the working fluid is renewed at the end of each cycle instead of being recirculated. In automobile engines, for example, the combustion gases are exhausted and replaced by fresh air–fuel mixture at the end of each cycle. The engine operates on a mechanical cycle, but the working fluid in this type of device does not go through a complete thermodynamic cycle.

Heat engines are categorized as *internal combustion* or *external combustion engines*, depending on how the heat is supplied to the working fluid. In external combustion engines (such as steam power plants), energy is supplied to the working fluid from an external source such as a furnace, a geo thermal well, a nuclear reactor, or even the sun. In internal combustion engines (such as automobile engines), this is done by burning the fuel within the system boundary. In this chapter, various gas power cycles are analyzed under some simplifying assumptions.

Steam is the most common working fluid used in vapor power cycles because of its many desirable characteristics, such as low cost, availability,

## Objectives
The objectives of this chapter are to:

- Evaluate the performance of gas power cycles.
- Develop simplifying assumptions applicable to gas power cycles.
- Review the operation of reciprocating engines.
- Solve problems based on the Otto and Diesel cycles.
- Solve problems based on the Brayton cycle and the Brayton cycle with regeneration.
- Analyze vapor power cycles in which the working fluid is alternately vaporized and condensed.
- Investigate ways to modify the basic Rankine vapor power cycle to increase the cycle thermal efficiency.
- Analyze the reheat vapor power cyles.
- Analyze the ideal vapor-compression refrigeration cycle.
- Discuss the operation of refrigeration and heat pump systems

**FIGURE 9–1**
Modeling is a powerful engineering tool that provides great insight and simplicity at the expense of some loss in accuracy.

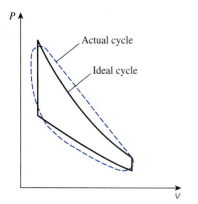

**FIGURE 9–2**
The analysis of many complex processes can be reduced to a manageable level by utilizing some idealizations.

**FIGURE 9–3**
Care should be exercised in the interpretation of the results from ideal cycles.

*BLONDIE © KING FEATURES SYNDICATE.*

and high enthalpy of vaporization. Other working fluids used include sodium, potassium, and mercury for high-temperature applications and some organic fluids such as benzene and the freons for low-temperature applications.

Steam power plants are commonly referred to as *coal plants*, *nuclear plants*, or *natural gas plants*, depending on the type of fuel used to supply heat to the steam. But the steam goes through the same basic cycle in all of them. Therefore, all can be analyzed in the same manner.

The most frequently used refrigeration cycle is the *vapor-compression refrigeration cycle* in which the refrigerant is vaporized and condensed alternately and is compressed in the vapor phase.

## 9–1 ■ BASIC CONSIDERATIONS IN THE ANALYSIS OF POWER CYCLES

Most power-producing devices operate on cycles, and the study of power cycles is an exciting and important part of thermodynamics. The cycles encountered in actual devices are difficult to analyze because of the presence of complicating effects, such as friction, and the absence of sufficient time for establishment of the equilibrium conditions during the cycle. To make an analytical study of a cycle feasible, we have to keep the complexities at a manageable level and utilize some idealizations (Fig. 9–1). When the actual cycle is stripped of all the internal irreversibilities and complexities, we end up with a cycle that resembles the actual cycle closely but is made up totally of internally reversible processes. Such a cycle is called an **ideal cycle** (Fig. 9–2).

A simple idealized model enables engineers to study the effects of the major parameters that dominate the cycle without getting bogged down in the details. The cycles discussed in this chapter are somewhat idealized, but they still retain the general characteristics of the actual cycles they represent. The conclusions reached from the analysis of ideal cycles are also applicable to actual cycles. The thermal efficiency of the Otto cycle, the ideal cycle for spark-ignition automobile engines, for example, increases with the compression ratio. This is also the case for actual automobile engines. The numerical values obtained from the analysis of an ideal cycle, however, are not necessarily representative of the actual cycles, and care should be exercised in their interpretation (Fig. 9–3). The simplified analysis presented in this chapter for various power cycles of practical interest may also serve as the starting point for a more in-depth study.

Heat engines are designed for the purpose of converting thermal energy to work, and their performance is expressed in terms of the **thermal efficiency** $\eta_{th}$, which is the ratio of the net work produced by the engine to the total heat input:

$$\eta_{th} = \frac{W_{net}}{Q_{in}} \quad \text{or} \quad \eta_{th} = \frac{w_{net}}{q_{in}} \tag{9–1}$$

Recall that heat engines that operate on a totally reversible cycle, such as the Carnot cycle, have the highest thermal efficiency of all heat engines

**FIGURE 9–4**

An automotive engine with the combustion chamber exposed.

*Courtesy of General Motors.*

operating between the same temperature levels. That is, nobody can develop a cycle more efficient than the *Carnot cycle.* Then the following question arises naturally: If the Carnot cycle is the best possible cycle, why do we not use it as the model cycle for all the heat engines instead of bothering with several so-called *ideal* cycles? The answer to this question is hardware-related. Most cycles encountered in practice differ significantly from the Carnot cycle, which makes it unsuitable as a realistic model. Each ideal cycle discussed in this chapter is related to a specific work-producing device and is an *idealized* version of the actual cycle.

The ideal cycles are *internally reversible,* but, unlike the Carnot cycle, they are not necessarily externally reversible. That is, they may involve irreversibilities external to the system such as heat transfer through a finite temperature difference. Therefore, the thermal efficiency of an ideal cycle, in general, is less than that of a totally reversible cycle operating between the same temperature limits. However, it is still considerably higher than the thermal efficiency of an actual cycle because of the idealizations utilized (Fig. 9–4).

The idealizations and simplifications commonly employed in the analysis of power cycles can be summarized as follows:

1. The cycle does not involve any *friction.* Therefore, the working fluid docs not experience any pressure drop as it flows in pipes or devices such as heat exchangers.
2. All expansion and compression processes take place in a *quasi-equilibrium* manner.
3. The pipes connecting the various components of a system are well insulated, and *heat transfer* through them is negligible.

Neglecting the changes in *kinetic* and *potential energies* of the working fluid is another commonly utilized simplification in the analysis of power cycles. This is a reasonable assumption since in devices that involve shaft work, such as turbines, compressors, and pumps, the kinetic and potential

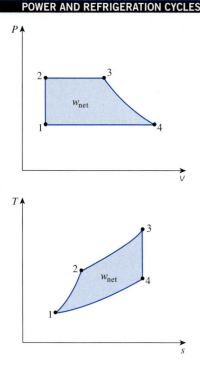

**FIGURE 9–5**

On both $P$-$v$ and $T$-$s$ diagrams, the area enclosed by the process curve represents the net work of the cycle.

energy terms are usually very small relative to the other terms in the energy equation. Fluid velocities encountered in devices such as condensers, boilers, and mixing chambers are typically low, and the fluid streams experience little change in their velocities, again making kinetic energy changes negligible. The only devices where the changes in kinetic energy are significant are the nozzles and diffusers, which are specifically designed to create large changes in velocity.

In the preceding chapters, *property diagrams* such as the $P$-$v$ and $T$-$s$ diagrams have served as valuable aids in the analysis of thermodynamic processes. On both the $P$-$v$ and $T$-$s$ diagrams, the area enclosed by the process curves of a cycle represents the net work produced during the cycle (Fig. 9–5), which is also equivalent to the net heat transfer for that cycle. The $T$-$s$ diagram is particularly useful as a visual aid in the analysis of ideal power cycles. An ideal power cycle does not involve any internal irreversibilities, and so the only effect that can change the entropy of the working fluid during a process is heat transfer.

On a $T$-$s$ diagram, a *heat-addition* process proceeds in the direction of increasing entropy, a *heat-rejection* process proceeds in the direction of decreasing entropy, and an *isentropic* (internally reversible, adiabatic) process proceeds at constant entropy. The area under the process curve on a $T$-$s$ diagram represents the heat transfer for that process. The area under the heat addition process on a $T$-$s$ diagram is a geometric measure of the total heat supplied during the cycle $q_{in}$, and the area under the heat rejection process is a measure of the total heat rejected $q_{out}$. The difference between these two (the area enclosed by the cyclic curve) is the net heat transfer, which is also the net work produced during the cycle. Therefore, on a $T$-$s$ diagram, the ratio of the area enclosed by the cyclic curve to the area under the heat-addition process curve represents the thermal efficiency of the cycle. *Any modification that increases the ratio of these two areas will also increase the thermal efficiency of the cycle.*

Although the working fluid in an ideal power cycle operates on a closed loop, the type of individual processes that comprises the cycle depends on the individual devices used to execute the cycle. In the Rankine cycle, which is the ideal cycle for steam power plants, the working fluid flows through a series of steady-flow devices such as the turbine and condenser, whereas in the Otto cycle, which is the ideal cycle for the spark-ignition automobile engine, the working fluid is alternately expanded and compressed in a piston–cylinder device. Therefore, equations pertaining to steady-flow systems should be used in the analysis of the Rankine cycle, and equations pertaining to closed systems should be used in the analysis of the Otto cycle.

## 9–2 ▪ THE CARNOT CYCLE AND ITS VALUE IN ENGINEERING

The Carnot cycle is composed of four totally reversible processes: isothermal heat addition, isentropic expansion, isothermal heat rejection, and isentropic compression. The $P$-$v$ and $T$-$s$ diagrams of a Carnot cycle are replotted in Fig. 9–6. The Carnot cycle can be executed in a closed system (a piston–cylinder device) or a steady-flow system (utilizing two turbines and two compressors, as shown in Fig. 9–7), and either a gas or a vapor can be utilized as the working fluid. The Carnot cycle is the most efficient cycle that can be executed between a heat source at temperature $T_H$ and a sink at temperature $T_L$, and its thermal efficiency is expressed as

$$\eta_{th,Carnot} = 1 - \frac{T_L}{T_H} \tag{9–2}$$

Reversible isothermal heat transfer is very difficult to achieve in reality because it would require very large heat exchangers and it would take a very long time (a power cycle in a typical engine is completed in a fraction of a second). Therefore, it is not practical to build an engine that would operate on a cycle that closely approximates the Carnot cycle.

The real value of the Carnot cycle comes from its being a standard against which the actual or the ideal cycles can be compared. The thermal efficiency of the Carnot cycle is a function of the sink and source temperatures only, and the thermal efficiency relation for the Carnot cycle (Eq. 9–2) conveys an important message that is equally applicable to both ideal and actual cycles: *Thermal efficiency increases with an increase in the average temperature at which heat is supplied to the system or with a decrease in the average temperature at which heat is rejected from the system.*

The source and sink temperatures that can be used in practice are not without limits, however. The highest temperature in the cycle is limited by the maximum temperature that the components of the heat engine, such as the piston or the turbine blades, can withstand. The lowest temperature is limited by the temperature of the cooling medium utilized in the cycle such as a lake, a river, or the atmospheric air.

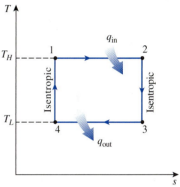

**FIGURE 9–6**
$P$-$v$ and $T$-$s$ diagrams of a Carnot cycle.

**FIGURE 9–7**
A steady-flow Carnot engine.

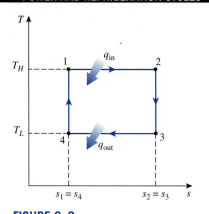

**FIGURE 9–8**
$T$-$s$ diagram for Example 9–1.

---

**EXAMPLE 9–1**    **Derivation of the Efficiency of the Carnot Cycle**

Show that the thermal efficiency of a Carnot cycle operating between the temperature limits of $T_H$ and $T_L$ is solely a function of these two temperatures and is given by Eq. 9–2.

**Solution**    It is to be shown that the efficiency of a Carnot cycle depends on the source and sink temperatures alone.

**Analysis**    The $T$-$s$ diagram of a Carnot cycle is redrawn in Fig. 9–8. All four processes that comprise the Carnot cycle are reversible, and thus the area under each process curve represents the heat transfer for that process. Heat is transferred to the system during process 1-2 and rejected during process 3-4. Therefore, the amount of heat input and heat output for the cycle can be expressed as

$$q_{in} = T_H(s_2 - s_1) \quad \text{and} \quad q_{out} = T_L(s_3 - s_4) = T_L(s_2 - s_1)$$

since processes 2-3 and 4-1 are isentropic, and thus $s_2 = s_3$ and $s_4 = s_1$. Substituting these into Eq. 9–1, we see that the thermal efficiency of a Carnot cycle is

$$\eta_{th} = \frac{w_{net}}{q_{in}} = 1 - \frac{q_{out}}{q_{in}} = 1 - \frac{T_L(s_2 - s_1)}{T_H(s_2 - s_1)} = 1 - \frac{T_L}{T_H}$$

**Discussion**    Notice that the thermal efficiency of a Carnot cycle is independent of the type of the working fluid used (an ideal gas, steam, etc.) or whether the cycle is executed in a closed or steady-flow system.

## 9–3 · AIR-STANDARD ASSUMPTIONS

In gas power cycles, the working fluid remains a gas throughout the entire cycle. Spark-ignition engines, diesel engines, and conventional gas turbines are familiar examples of devices that operate on gas cycles. In all these engines, energy is provided by burning a fuel within the system boundaries. That is, they are *internal combustion engines*. Because of this combustion process, the composition of the working fluid changes from air and fuel to combustion products during the course of the cycle. However, considering that air is predominantly nitrogen that undergoes hardly any chemical reactions in the combustion chamber, the working fluid closely resembles air at all times.

Even though internal combustion engines operate on a mechanical cycle (the piston returns to its starting position at the end of each revolution), the working fluid does not undergo a complete thermodynamic cycle. It is thrown out of the engine at some point in the cycle (as exhaust gases) instead of being returned to the initial state. Working on an open cycle is the characteristic of all internal combustion engines.

The actual gas power cycles are rather complex. To reduce the analysis to a manageable level, we utilize the following approximations, commonly known as the **air-standard assumptions**:

1. The working fluid is air, which continuously circulates in a closed loop and always behaves as an ideal gas.
2. All the processes that make up the cycle are internally reversible.

3. The combustion process is replaced by a heat-addition process from an external source (Fig. 9–9).
4. The exhaust process is replaced by a heat-rejection process that restores the working fluid to its initial state.

Another assumption that is often utilized to simplify the analysis even more is that air has constant specific heats whose values are determined at *room temperature* (25°C, or 77°F). When this assumption is utilized, the air-standard assumptions are called the **cold-air-standard assumptions**. A cycle for which the air-standard assumptions are applicable is frequently referred to as an **air-standard cycle**.

The air-standard assumptions previously stated provide considerable simplification in the analysis without significantly deviating from the actual cycles. This simplified model enables us to study qualitatively the influence of major parameters on the performance of the actual engines.

(a) Actual

(b) Ideal

**FIGURE 9–9**
The combustion process is replaced by a heat-addition process in ideal cycles.

## 9–4 · AN OVERVIEW OF RECIPROCATING ENGINES

Despite its simplicity, the reciprocating engine (basically a piston–cylinder device) is one of the rare inventions that has proved to be very versatile and to have a wide range of applications. It is the powerhouse of the vast majority of automobiles, trucks, light aircraft, ships, and electric power generators, as well as many other devices.

The basic components of a reciprocating engine are shown in Fig. 9–10. The piston reciprocates in the cylinder between two fixed positions called the **top dead center** (TDC)—the position of the piston when it forms the smallest volume in the cylinder—and the **bottom dead center** (BDC)—the position of the piston when it forms the largest volume in the cylinder. The distance between the TDC and the BDC is the largest distance that the piston can travel in one direction, and it is called the **stroke** of the engine. The diameter of the piston is called the **bore**. The air or air–fuel mixture is drawn into the cylinder through the **intake valve**, and the combustion products are expelled from the cylinder through the **exhaust valve**.

The minimum volume formed in the cylinder when the piston is at TDC is called the **clearance volume** (Fig. 9–11). The volume displaced by the piston as it moves between TDC and BDC is called the **displacement volume**. The ratio of the maximum volume formed in the cylinder to the minimum (clearance) volume is called the **compression ratio** $r$ of the engine:

$$r = \frac{V_{max}}{V_{min}} = \frac{V_{BDC}}{V_{TDC}} \tag{9–3}$$

Notice that the compression ratio is a *volume ratio* and should not be confused with the pressure ratio.

Another term frequently used in conjunction with reciprocating engines is the **mean effective pressure** (MEP). It is a fictitious pressure that, if it acted on the piston during the entire power stroke, would produce the same

**FIGURE 9–10**
Nomenclature for reciprocating engines.

(a) Displacement volume

(b) Clearance volume

**FIGURE 9–11**
Displacement and clearance volumes of a reciprocating engine.

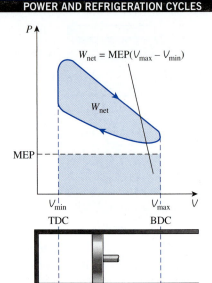

**FIGURE 9–12**
The net work output of a cycle is equivalent to the product of the mean effective pressure and the displacement volume.

amount of net work as that produced during the actual cycle (Fig. 9–12). That is,

$$W_{net} = MEP \times \text{Piston area} \times \text{Stroke} = MEP \times \text{Displacement volume}$$

or

$$MEP = \frac{W_{net}}{V_{max} - V_{min}} = \frac{w_{net}}{v_{max} - v_{min}} \quad (kPa) \qquad (9\text{–}4)$$

The mean effective pressure can be used as a parameter to compare the performances of reciprocating engines of equal size. The engine with a larger value of MEP delivers more net work per cycle and thus performs better.

Reciprocating engines are classified as **spark-ignition (SI) engines** or **compression-ignition (CI) engines**, depending on how the combustion process in the cylinder is initiated. In SI engines, the combustion of the air–fuel mixture is initiated by a spark plug. In CI engines, the air–fuel mixture is self-ignited as a result of compressing the mixture above its self-ignition temperature. In the next two sections, we discuss the *Otto* and *Diesel cycles,* which are the ideal cycles for the SI and CI reciprocating engines, respectively.

## 9–5 ▪ OTTO CYCLE: THE IDEAL CYCLE FOR SPARK-IGNITION ENGINES

The Otto cycle is the ideal cycle for spark-ignition reciprocating engines. It is named after Nikolaus A. Otto, who built a successful four-stroke engine in 1876 in Germany using the cycle proposed by Frenchman Beau de Rochas in 1862. In most spark-ignition engines, the piston executes four complete strokes (two mechanical cycles) within the cylinder, and the crankshaft completes two revolutions for each thermodynamic cycle. These engines are called **four-stroke** internal combustion engines. A schematic of each stroke as well as a *P-v* diagram for an actual four-stroke spark-ignition engine is given in Fig. 9–13*a*.

Initially, both the intake and the exhaust valves are closed, and the piston is at its lowest position (BDC). During the *compression stroke,* the piston moves upward, compressing the air–fuel mixture. Shortly before the piston reaches its highest position (TDC), the spark plug fires and the mixture ignites, increasing the pressure and temperature of the system. The high-pressure gases force the piston down, which in turn forces the crankshaft to rotate, producing a useful work output during the *expansion* or *power stroke.* At the end of this stroke, the piston is at its lowest position (the completion of the first mechanical cycle), and the cylinder is filled with combustion products. Now the piston moves upward one more time, purging the exhaust gases through the exhaust valve (the *exhaust stroke*), and down a second time, drawing in fresh air–fuel mixture through the intake valve (the *intake stroke*). Notice that the pressure in the cylinder is slightly above the atmospheric value during the exhaust stroke and slightly below during the intake stroke.

**FIGURE 9–13**

Actual and ideal cycles in spark-ignition engines and their $P$-$v$ diagrams.

In **two-stroke engines**, all four functions described above are executed in just two strokes: the power stroke and the compression stroke. In these engines, the crankcase is sealed, and the outward motion of the piston is used to slightly pressurize the air–fuel mixture in the crankcase, as shown in Fig. 9–14. Also, the intake and exhaust valves are replaced by openings in the lower portion of the cylinder wall. During the latter part of the power stroke, the piston uncovers first the exhaust port, allowing the exhaust gases to be partially expelled, and then the intake port, allowing the fresh air–fuel mixture to rush in and drive most of the remaining exhaust gases out of the cylinder. This mixture is then compressed as the piston moves upward during the compression stroke and is subsequently ignited by a spark plug.

The two-stroke engines are generally less efficient than their four-stroke counterparts because of the incomplete expulsion of the exhaust gases and the partial expulsion of the fresh air–fuel mixture with the exhaust gases. However, they are relatively simple and inexpensive, and they have high power-to-weight and power-to-volume ratios, which make them suitable for applications requiring small size and weight such as for motorcycles, chain saws, and lawn mowers (Fig. 9–15).

Advances in several technologies—such as direct fuel injection, stratified charge combustion, and electronic controls—brought about a renewed interest

**FIGURE 9–14**

Schematic of a two-stroke reciprocating engine.

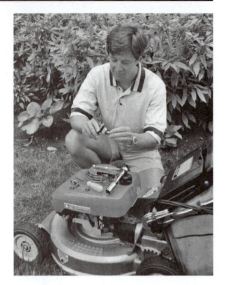

**FIGURE 9–15**
Two-stroke engines are commonly used in motorcycles and lawn mowers.

© Vol. 26/PhotoDisc/Getty RF.

in two-stroke engines that can offer high performance and fuel economy while satisfying the stringent emission requirements. For a given weight and displacement, a well-designed two-stroke engine can provide significantly more power than its four-stroke counterpart because two-stroke engines produce power on every engine revolution instead of every other one. In the new two-stroke engines, the highly atomized fuel spray that is injected into the combustion chamber toward the end of the compression stroke burns much more completely. The fuel is sprayed after the exhaust valve is closed, which prevents unburned fuel from being ejected into the atmosphere. With stratified combustion, the flame that is initiated by igniting a small amount of the rich fuel–air mixture near the spark plug propagates through the combustion chamber filled with a much leaner mixture, and this results in much cleaner combustion. Also, the advances in electronics have made it possible to ensure the optimum operation under varying engine load and speed conditions. Major car companies have research programs underway on two-stroke engines which are expected to make a comeback in the future.

The thermodynamic analysis of the actual four-stroke or two-stroke cycles described is not a simple task. However, the analysis can be simplified significantly if the air-standard assumptions are utilized. The resulting cycle, which closely resembles the actual operating conditions, is the ideal **Otto cycle**. It consists of four internally reversible processes:

1-2   Isentropic compression
2-3   Constant-volume heat addition
3-4   Isentropic expansion
4-1   Constant-volume heat rejection

The execution of the Otto cycle in a piston–cylinder device together with a $P$-$v$ diagram is illustrated in Fig. 9–13b. The $T$-$s$ diagram of the Otto cycle is given in Fig. 9–16.

The Otto cycle is executed in a closed system, and disregarding the changes in kinetic and potential energies, the energy balance for any of the processes is expressed, on a unit-mass basis, as

$$(q_{in} - q_{out}) + (w_{in} - w_{out}) = \Delta u \qquad (kJ/kg) \qquad \textbf{(9–5)}$$

No work is involved during the two heat transfer processes since both take place at constant volume. Therefore, heat transfer to and from the working fluid can be expressed as

$$q_{in} = u_3 - u_2 = c_v(T_3 - T_2) \qquad \textbf{(9–6a)}$$

and

$$q_{out} = u_4 - u_1 = c_v(T_4 - T_1) \qquad \textbf{(9–6b)}$$

Then the thermal efficiency of the ideal Otto cycle under the cold air standard assumptions becomes

$$\eta_{th,Otto} = \frac{w_{net}}{q_{in}} = 1 - \frac{q_{out}}{q_{in}} = 1 - \frac{T_4 - T_1}{T_3 - T_2} = 1 - \frac{T_1(T_4/T_1 - 1)}{T_2(T_3/T_2 - 1)}$$

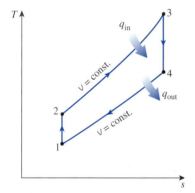

**FIGURE 9–16**
$T$-$s$ diagram of the ideal Otto cycle.

Processes 1-2 and 3-4 are isentropic, and $v_2 = v_3$ and $v_4 = v_1$. Thus,

$$\frac{T_1}{T_2} = \left(\frac{v_2}{v_1}\right)^{k-1} = \left(\frac{v_3}{v_4}\right)^{k-1} = \frac{T_4}{T_3} \tag{9–7}$$

Substituting these equations into the thermal efficiency relation and simplifying give

$$\eta_{\text{th,Otto}} = 1 - \frac{1}{r^{k-1}} \tag{9–8}$$

where

$$r = \frac{V_{\text{max}}}{V_{\text{min}}} = \frac{V_1}{V_2} = \frac{v_1}{v_2} \tag{9–9}$$

is the **compression ratio** and $k$ is the specific heat ratio $c_p/c_v$.

Equation 9–8 shows that under the cold-air-standard assumptions, the thermal efficiency of an ideal Otto cycle depends on the compression ratio of the engine and the specific heat ratio of the working fluid. The thermal efficiency of the ideal Otto cycle increases with both the compression ratio and the specific heat ratio. This is also true for actual spark-ignition internal combustion engines. A plot of thermal efficiency versus the compression ratio is given in Fig. 9–17 for $k = 1.4$, which is the specific heat ratio value of air at room temperature. For a given compression ratio, the thermal efficiency of an actual spark-ignition engine is less than that of an ideal Otto cycle because of the irreversibilities, such as friction, and other factors such as incomplete combustion.

We can observe from Fig. 9–17 that the thermal efficiency curve is rather steep at low compression ratios but flattens out starting with a compression ratio value of about 8. Therefore, the increase in thermal efficiency with the compression ratio is not as pronounced at high compression ratios. Also, when high compression ratios are used, the temperature of the air–fuel mixture rises above the autoignition temperature of the fuel (the temperature at which the fuel ignites without the help of a spark) during the combustion process, causing an early and rapid burn of the fuel at some point or points ahead of the flame front, followed by almost instantaneous inflammation of the end gas. This premature ignition of the fuel, called **autoignition**, produces an audible noise, which is called **engine knock**. Autoignition in spark-ignition engines cannot be tolerated because it hurts performance and can cause engine damage. The requirement that autoignition not be allowed places an upper limit on the compression ratios that can be used in spark-ignition internal combustion engines.

Improvement of the thermal efficiency of gasoline engines by utilizing higher compression ratios (up to about 12) without facing the autoignition problem has been made possible by using gasoline blends that have good antiknock characteristics, such as gasoline mixed with tetraethyl lead. Tetraethyl lead had been added to gasoline since the 1920s because it is an inexpensive method of raising the *octane rating*, which is a measure of the engine knock resistance of a fuel. Leaded gasoline, however, has a very undesirable side effect: it forms compounds during the combustion process that are hazardous to health and pollute the environment. In an effort to combat air pollution, the government adopted a policy in the mid-1970s that

**FIGURE 9–17**
Thermal efficiency of the ideal Otto cycle as a function of compression ratio $(k = 1.4)$.

resulted in the eventual phase-out of leaded gasoline. Unable to use lead, the refiners developed other techniques to improve the antiknock characteristics of gasoline. Most cars made since 1975 have been designed to use unleaded gasoline, and the compression ratios had to be lowered to avoid engine knock. The ready availability of high octane fuels made it possible to raise the compression ratios again in recent years. Also, owing to the improvements in other areas (reduction in overall automobile weight, improved aerodynamic design, etc.), today's cars have better fuel economy and consequently get more miles per gallon of fuel. This is an example of how engineering decisions involve compromises, and efficiency is only one of the considerations in final design.

The second parameter affecting the thermal efficiency of an ideal Otto cycle is the specific heat ratio $k$. For a given compression ratio, an ideal Otto cycle using a monatomic gas (such as argon or helium, $k = 1.667$) as the working fluid will have the highest thermal efficiency. The specific heat ratio $k$, and thus the thermal efficiency of the ideal Otto cycle, decreases as the molecules of the working fluid get larger (Fig. 9–18). At room temperature it is 1.4 for air, 1.3 for carbon dioxide, and 1.2 for ethane. The working fluid in actual engines contains larger molecules such as carbon dioxide, and the specific heat ratio decreases with temperature, which is one of the reasons that the actual cycles have lower thermal efficiencies than the ideal Otto cycle. The thermal efficiencies of actual spark-ignition engines range from about 25 to 30 percent.

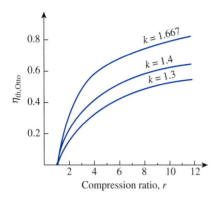

**FIGURE 9–18**

The thermal efficiency of the Otto cycle increases with the specific heat ratio $k$ of the working fluid.

---

**EXAMPLE 9–2**    **The Ideal Otto Cycle**

An ideal Otto cycle has a compression ratio of 8. At the beginning of the compression process, air is at 100 kPa and 17°C, and 800 kJ/kg of heat is transferred to air during the constant-volume heat-addition process. Accounting for the variation of specific heats of air with temperature, determine (a) the maximum temperature and pressure that occur during the cycle, (b) the net work output, (c) the thermal efficiency, and (d) the mean effective pressure for the cycle.

**Solution**   An ideal Otto cycle is considered. The maximum temperature and pressure, the net work output, the thermal efficiency, and the mean effective pressure are to be determined.

**Assumptions**   **1** The air-standard assumptions are applicable. **2** Kinetic and potential energy changes are negligible. **3** The variation of specific heats with temperature is to be accounted for.

**Analysis**   The $P$-$v$ diagram of the ideal Otto cycle described is shown in Fig. 9–19. We note that the air contained in the cylinder forms a closed system.

(a) The maximum temperature and pressure in an Otto cycle occur at the end of the constant-volume heat-addition process (state 3). But first we need to determine the temperature and pressure of air at the end of the isentropic compression process (state 2), using data from Table A–21:

$$T_1 = 290 \text{ K} \quad \rightarrow \quad u_1 = 206.91 \text{ kJ/kg}$$
$$v_{r1} = 676.1$$

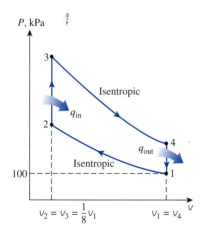

**FIGURE 9–19**

$P$-$v$ diagram for the Otto cycle discussed in Example 9–2.

Process 1-2 (isentropic compression of an ideal gas):

$$\frac{v_{r2}}{v_{r1}} = \frac{v_2}{v_1} = \frac{1}{r} \rightarrow v_{r2} = \frac{v_{r1}}{r} = \frac{676.1}{8} = 84.51 \rightarrow T_2 = 652.4 \text{ K}$$

$$u_2 = 475.11 \text{ kJ/kg}$$

$$\frac{P_2 v_2}{T_2} = \frac{P_1 v_1}{T_1} \rightarrow P_2 = P_1 \left(\frac{T_2}{T_1}\right)\left(\frac{v_1}{v_2}\right)$$

$$= (100 \text{ kPa})\left(\frac{652.4 \text{ K}}{290 \text{ K}}\right)(8) = 1799.7 \text{ kPa}$$

Process 2-3 (constant-volume heat addition):

$$q_{in} = u_3 - u_2$$

$$800 \text{ kJ/kg} = u_3 - 475.11 \text{ kJ/kg}$$

$$u_3 = 1275.11 \text{ kJ/kg} \rightarrow T_3 = \mathbf{1575.1 \text{ K}}$$

$$v_{r3} = 6.108$$

$$\frac{P_3 v_3}{T_3} = \frac{P_2 v_2}{T_2} \rightarrow P_3 = P_2 \left(\frac{T_3}{T_2}\right)\left(\frac{v_2}{v_3}\right)$$

$$= (1.7997 \text{ MPa})\left(\frac{1575.1 \text{ K}}{652.4 \text{ K}}\right)(1) = \mathbf{4.345 \text{ MPa}}$$

(b) The net work output for the cycle is determined either by finding the boundary ($P \, dV$) work involved in each process by integration and adding them or by finding the net heat transfer that is equivalent to the net work done during the cycle. We take the latter approach. However, first we need to find the internal energy of the air at state 4:

Process 3-4 (isentropic expansion of an ideal gas):

$$\frac{v_{r4}}{v_{r3}} = \frac{v_4}{v_3} = r \rightarrow v_{r4} = rv_{r3} = (8)(6.108) = 48.864 \rightarrow T_4 = 795.6 \text{ K}$$

$$u_4 = 588.74 \text{ kJ/kg}$$

Process 4-1 (constant-volume heat rejection):

$$-q_{out} = u_1 - u_4 \rightarrow q_{out} = u_4 - u_1$$

$$q_{out} = 588.74 - 206.91 = 381.83 \text{ kJ/kg}$$

Thus,

$$w_{net} = q_{net} = q_{in} - q_{out} = 800 - 381.83 = \mathbf{418.17 \text{ kJ/kg}}$$

(c) The thermal efficiency of the cycle is determined from its definition:

$$\eta_{th} = \frac{w_{net}}{q_{in}} = \frac{418.17 \text{ kJ/kg}}{800 \text{ kJ/kg}} = \mathbf{0.523 \text{ or } 52.3\%}$$

Under the cold-air-standard assumptions (constant specific heat values at room temperature), the thermal efficiency would be (Eq. 9–8)

$$\eta_{th,Otto} = 1 - \frac{1}{r^{k-1}} = 1 - r^{1-k} = 1 - (8)^{1-1.4} = 0.565 \text{ or } 56.5\%$$

which is considerably different from the value obtained above. Therefore, care should be exercised in utilizing the cold-air-standard assumptions.

(*d*) The mean effective pressure is determined from its definition, Eq. 9–4:

$$\text{MEP} = \frac{w_{net}}{v_1 - v_2} = \frac{w_{net}}{v_1 - v_1/r} = \frac{w_{net}}{v_1(1 - 1/r)}$$

where

$$v_1 = \frac{RT_1}{P_1} = \frac{(0.287 \text{ kPa·m}^3/\text{kg·K})(290 \text{ K})}{100 \text{ kPa}} = 0.832 \text{ m}^3/\text{kg}$$

Thus,

$$\text{MEP} = \frac{418.17 \text{ kJ/kg}}{(0.832 \text{ m}^3/\text{kg})(1 - \frac{1}{8})} \left(\frac{1 \text{ kPa·m}^3}{1 \text{ kJ}}\right) = \textbf{574 kPa}$$

***Discussion*** Note that a constant pressure of 574 kPa during the power stroke would produce the same net work output as the entire cycle.

## 9–6 ▪ DIESEL CYCLE: THE IDEAL CYCLE FOR COMPRESSION-IGNITION ENGINES

The Diesel cycle is the ideal cycle for CI reciprocating engines. The CI engine, first proposed by Rudolph Diesel in the 1890s, is very similar to the SI engine discussed in the last section, differing mainly in the method of initiating combustion. In spark-ignition engines (also known as *gasoline engines*), the air–fuel mixture is compressed to a temperature that is below the autoignition temperature of the fuel, and the combustion process is initiated by firing a spark plug. In CI engines (also known as *diesel engines*), the air is compressed to a temperature that is above the autoignition temperature of the fuel, and combustion starts on contact as the fuel is injected into this hot air. Therefore, the spark plug and carburetor are replaced by a fuel injector in diesel engines (Fig. 9–20).

In gasoline engines, a mixture of air and fuel is compressed during the compression stroke, and the compression ratios are limited by the onset of autoignition or engine knock. In diesel engines, only air is compressed during the compression stroke, eliminating the possibility of autoignition. Therefore, diesel engines can be designed to operate at much higher compression ratios, typically between 12 and 24. Not having to deal with the problem of autoignition has another benefit: many of the stringent requirements placed

**FIGURE 9–20**

In diesel engines, the spark plug is replaced by a fuel injector, and only air is compressed during the compression process.

on the gasoline can now be removed, and fuels that are less refined (thus less expensive) can be used in diesel engines.

The fuel injection process in diesel engines starts when the piston approaches TDC and continues during the first part of the power stroke. Therefore, the combustion process in these engines takes place over a longer interval. Because of this longer duration, the combustion process in the ideal Diesel cycle is approximated as a constant-pressure heat-addition process. In fact, this is the only process where the Otto and the Diesel cycles differ. The remaining three processes are the same for both ideal cycles. That is, process 1-2 is isentropic compression, 2-3 is constant-pressure heat addition, 3-4 is isentropic expansion, and 4-1 is constant-volume heat rejection. The similarity between the two cycles is also apparent from the $P$-$v$ and $T$-$s$ diagrams of the Diesel cycle, shown in Fig. 9–21.

Noting that the Diesel cycle is executed in a piston–cylinder device, which forms a closed system, the amount of heat transferred to the working fluid at constant pressure and rejected from it at constant volume can be expressed as

$$q_{in} - w_{b,out} = u_3 - u_2 \rightarrow q_{in} = P_2(v_3 - v_2) + (u_3 - u_2)$$

$$= h_3 - h_2 = c_p(T_3 - T_2) \quad \text{(9–10a)}$$

and

$$-q_{out} = u_1 - u_4 \rightarrow q_{out} = u_4 - u_1 = c_v(T_4 - T_1) \quad \text{(9–10b)}$$

Then the thermal efficiency of the ideal Diesel cycle under the cold-air-standard assumptions becomes

$$\eta_{th,Diesel} = \frac{w_{net}}{q_{in}} = 1 - \frac{q_{out}}{q_{in}} = 1 - \frac{T_4 - T_1}{k(T_3 - T_2)} = 1 - \frac{T_1(T_4/T_1 - 1)}{kT_2(T_3/T_2 - 1)}$$

We now define a new quantity, the **cutoff ratio** $r_c$, as the ratio of the cylinder volumes after and before the combustion process:

$$r_c = \frac{V_3}{V_2} = \frac{v_3}{v_2} \quad \text{(9–11)}$$

Utilizing this definition and the isentropic ideal-gas relations for processes 1-2 and 3-4, we see that the thermal efficiency relation reduces to

$$\eta_{th,Diesel} = 1 - \frac{1}{r^{k-1}} \left[ \frac{r_c^k - 1}{k(r_c - 1)} \right] \quad \text{(9–12)}$$

where $r$ is the compression ratio defined by Eq. 9–9. Looking at Eq. 9–12 carefully, one would notice that under the cold-air-standard assumptions, the efficiency of a Diesel cycle differs from the efficiency of an Otto cycle by the quantity in the brackets. This quantity is always greater than 1. Therefore,

$$\eta_{th,Otto} > \eta_{th,Diesel} \quad \text{(9–13)}$$

when both cycles operate on the same compression ratio. Also, as the cut-off ratio decreases, the efficiency of the Diesel cycle increases (Fig. 9–22). For the limiting case of $r_c = 1$, the quantity in the brackets becomes unity

(a) $P$-$v$ diagram

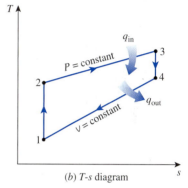

(b) $T$-$s$ diagram

**FIGURE 9–21**

$T$-$s$ and $P$-$v$ diagrams for the ideal Diesel cycle.

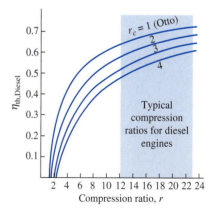

**FIGURE 9–22**

Thermal efficiency of the ideal Diesel cycle as a function of compression and cutoff ratios ($k = 1.4$).

(can you prove it?), and the efficiencies of the Otto and Diesel cycles become identical. Remember, though, that diesel engines operate at much higher compression ratios and thus are usually more efficient than the spark-ignition (gasoline) engines. The diesel engines also burn the fuel more completely since they usually operate at lower revolutions per minute and the air–fuel mass ratio is much higher than spark-ignition engines. Thermal efficiencies of large diesel engines range from about 35 to 40 percent.

The higher efficiency and lower fuel costs of diesel engines make them attractive in applications requiring relatively large amounts of power, such as in locomotive engines, emergency power generation units, large ships, and heavy trucks. As an example of how large a diesel engine can be, a 12-cylinder diesel engine built in 1964 by the Fiat Corporation of Italy had a normal power output of 25,200 hp (18.8 MW) at 122 rpm, a cylinder bore of 90 cm, and a stroke of 91 cm.

In modern high-speed compression ignition engines, fuel is injected into the combustion chamber much sooner compared to the early diesel engines. Fuel starts to ignite late in the compression stroke, and consequently part of the combustion occurs almost at constant volume. Fuel injection continues until the piston reaches the top dead center, and combustion of the fuel keeps the pressure high well into the expansion stroke. Thus, the entire combustion process can better be modeled as the combination of constant-volume and constant-pressure processes. The ideal cycle based on this concept is called the **dual cycle** and $P$-$v$ diagram for it is given in Fig. 9–23. The relative amounts of heat transferred during each process can be adjusted to approximate the actual cycle more closely. Note that both the Otto and the Diesel cycles can be obtained as special cases of the dual cycle. Dual cycle is a more realistic model than diesel cycle for representing modern, high-speed compression ignition engines.

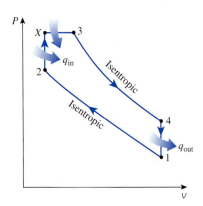

**FIGURE 9–23**

$P$-$v$ diagram of an ideal dual cycle.

---

**EXAMPLE 9–3     The Ideal Diesel Cycle**

An ideal Diesel cycle with air as the working fluid has a compression ratio of 18 and a cutoff ratio of 2. At the beginning of the compression process, the working fluid is at 14.7 psia, 80°F, and 117 in³. Utilizing the cold-air-standard assumptions, determine (a) the temperature and pressure of air at the end of each process, (b) the net work output and the thermal efficiency, and (c) the mean effective pressure.

**Solution**   An ideal Diesel cycle is considered. The temperature and pressure at the end of each process, the net work output, the thermal efficiency, and the mean effective pressure are to be determined.

**Assumptions**   **1** The cold-air-standard assumptions are applicable and thus air can be assumed to have constant specific heats at room temperature. **2** Kinetic and potential energy changes are negligible.

**Properties**   The gas constant of air is $R = 0.3704$ psia·ft³/lbm·R and its other properties at room temperature are $c_p = 0.240$ Btu/lbm·R, $c_v = 0.171$ Btu/lbm·R, and $k = 1.4$ (Table A–2E$a$).

*Analysis* The $P$-$V$ diagram of the ideal Diesel cycle described is shown in Fig. 9–24. We note that the air contained in the cylinder forms a closed system.

(*a*) The temperature and pressure values at the end of each process can be determined by utilizing the ideal-gas isentropic relations for processes 1-2 and 3-4. But first we determine the volumes at the end of each process from the definitions of the compression ratio and the cutoff ratio:

$$V_2 = \frac{V_1}{r} = \frac{117 \text{ in}^3}{18} = 6.5 \text{ in}^3$$

$$V_3 = r_c V_2 = (2)(6.5 \text{ in}^3) = 13 \text{ in}^3$$

$$V_4 = V_1 = 117 \text{ in}^3$$

Process 1-2 (isentropic compression of an ideal gas, constant specific heats):

$$T_2 = T_1 \left(\frac{V_1}{V_2}\right)^{k-1} = (540 \text{ R})(18)^{1.4-1} = \textbf{1716 R}$$

$$P_2 = P_1 \left(\frac{V_1}{V_2}\right)^{k} = (14.7 \text{ psia})(18)^{1.4} = \textbf{841 psia}$$

Process 2-3 (constant-pressure heat addition to an ideal gas):

$$P_3 = P_2 = \textbf{841 psia}$$

$$\frac{P_2 V_2}{T_2} = \frac{P_3 V_3}{T_3} \rightarrow T_3 = T_2 \left(\frac{V_3}{V_2}\right) = (1716 \text{ R})(2) = \textbf{3432 R}$$

Process 3-4 (isentropic expansion of an ideal gas, constant specific heats):

$$T_4 = T_3 \left(\frac{V_3}{V_4}\right)^{k-1} = (3432 \text{ R})\left(\frac{13 \text{ in}^3}{117 \text{ in}^3}\right)^{1.4-1} = \textbf{1425 R}$$

$$P_4 = P_3 \left(\frac{V_3}{V_4}\right)^{k} = (841 \text{ psia})\left(\frac{13 \text{ in}^3}{117 \text{ in}^3}\right)^{1.4} = \textbf{38.8 psia}$$

(*b*) The net work for a cycle is equivalent to the net heat transfer. But first we find the mass of air:

$$m = \frac{P_1 V_1}{R T_1} = \frac{(14.7 \text{ psia})(117 \text{ in}^3)}{(0.3704 \text{ psia·ft}^3/\text{lbm·R})(540 \text{ R})}\left(\frac{1 \text{ ft}^3}{1728 \text{ in}^3}\right) = 0.00498 \text{ lbm}$$

Process 2-3 is a constant-pressure heat-addition process, for which the boundary work and $\Delta u$ terms can be combined into $\Delta h$. Thus,

$$Q_{in} = m(h_3 - h_2) = m c_p (T_3 - T_2)$$

$$= (0.00498 \text{ lbm})(0.240 \text{ Btu/lbm·R})[(3432 - 1716) \text{ R}]$$

$$= 2.051 \text{ Btu}$$

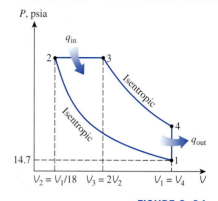

$P$, psia

$q_{in}$

$q_{out}$

$V_2 = V_1/18 \quad V_3 = 2V_2 \quad V_1 = V_4$

**FIGURE 9–24**
$P$-$V$ diagram for the ideal Diesel cycle discussed in Example 9–3.

Process 4-1 is a constant-volume heat-rejection process (it involves no work interactions), and the amount of heat rejected is

$$Q_{out} = m(u_4 - u_1) = mc_v(T_4 - T_1)$$
$$= (0.00498 \text{ lbm})(0.171 \text{ Btu/lbm·R})[(1425 - 540) \text{ R}]$$
$$= 0.754 \text{ Btu}$$

Thus,

$$W_{net} = Q_{in} - Q_{out} = 2.051 - 0.754 = \textbf{1.297 Btu}$$

Then the thermal efficiency becomes

$$\eta_{th} = \frac{W_{net}}{Q_{in}} = \frac{1.297 \text{ Btu}}{2.051 \text{ Btu}} = \textbf{0.632 or 63.2\%}$$

The thermal efficiency of this Diesel cycle under the cold-air-standard assumptions could also be determined from Eq. 9–12.

(c) The mean effective pressure is determined from its definition, Eq. 9–4:

$$\text{MEP} = \frac{W_{net}}{V_{max} - V_{min}} = \frac{W_{net}}{V_1 - V_2} = \frac{1.297 \text{ Btu}}{(117 - 6.5) \text{ in}^3}\left(\frac{778.17 \text{ lbf·ft}}{1 \text{ Btu}}\right)\left(\frac{12 \text{ in.}}{1 \text{ ft}}\right)$$
$$= \textbf{110 psia}$$

**Discussion** Note that a constant pressure of 110 psia during the power stroke would produce the same net work output as the entire Diesel cycle.

# 9–7 ▪ BRAYTON CYCLE: THE IDEAL CYCLE FOR GAS-TURBINE ENGINES

The Brayton cycle was first proposed by George Brayton for use in the reciprocating oil-burning engine that he developed around 1870. Today, it is used for gas turbines only where both the compression and expansion processes take place in rotating machinery. Gas turbines usually operate on an *open cycle,* as shown in Fig. 9–25. Fresh air at ambient conditions is drawn into the compressor, where its temperature and pressure are raised. The high-pressure air proceeds into the combustion chamber, where the fuel is burned at constant pressure. The resulting high-temperature gases then enter the turbine, where they expand to the atmospheric pressure while producing power. The exhaust gases leaving the turbine are thrown out (not recirculated), causing the cycle to be classified as an open cycle.

The open gas-turbine cycle described above can be modeled as a *closed cycle,* as shown in Fig. 9–26, by utilizing the air-standard assumptions. Here the compression and expansion processes remain the same, but the combustion process is replaced by a constant-pressure heat-addition process from an external source, and the exhaust process is replaced by a constant-pressure heat-rejection process to the ambient air. The ideal cycle that the working

**FIGURE 9–25**
An open-cycle gas-turbine engine.

fluid undergoes in this closed loop is the **Brayton cycle**, which is made up of four internally reversible processes:

1-2   Isentropic compression (in a compressor)

2-3   Constant-pressure heat addition

3-4   Isentropic expansion (in a turbine)

4-1   Constant-pressure heat rejection

The $T$-$s$ and $P$-$v$ diagrams of an ideal Brayton cycle are shown in Fig. 9–27. Notice that all four processes of the Brayton cycle are executed in steady-flow devices; thus, they should be analyzed as steady-flow processes. When the changes in kinetic and potential energies are neglected, the energy balance for a steady-flow process can be expressed, on a unit–mass basis, as

$$(q_{in} - q_{out}) + (w_{in} - w_{out}) = h_{exit} - h_{inlet} \tag{9–14}$$

Therefore, heat transfers to and from the working fluid are

$$q_{in} = h_3 - h_2 = c_p(T_3 - T_2) \tag{9–15a}$$

and

$$q_{out} = h_4 - h_1 = c_p(T_4 - T_1) \tag{9–15b}$$

Then the thermal efficiency of the ideal Brayton cycle under the cold-air-standard assumptions becomes

$$\eta_{th,Brayton} = \frac{w_{net}}{q_{in}} = 1 - \frac{q_{out}}{q_{in}} = 1 - \frac{c_p(T_4 - T_1)}{c_p(T_3 - T_2)} = 1 - \frac{T_1(T_4/T_1 - 1)}{T_2(T_3/T_2 - 1)}$$

Processes 1-2 and 3-4 are isentropic, and $P_2 = P_3$ and $P_4 = P_1$. Thus,

$$\frac{T_2}{T_1} = \left(\frac{P_2}{P_1}\right)^{(k-1)/k} = \left(\frac{P_3}{P_4}\right)^{(k-1)/k} = \frac{T_3}{T_4}$$

Substituting these equations into the thermal efficiency relation and simplifying give

$$\eta_{th,Brayton} = 1 - \frac{1}{r_p^{(k-1)/k}} \tag{9–16}$$

where

$$r_p = \frac{P_2}{P_1} \tag{9–17}$$

is the **pressure ratio** and $k$ is the specific heat ratio. Equation 9–16 shows that under the cold-air-standard assumptions, the thermal efficiency of an ideal Brayton cycle depends on the pressure ratio of the gas turbine and the specific heat ratio of the working fluid. The thermal efficiency increases with both of these parameters, which is also the case for actual gas turbines. A plot of thermal efficiency versus the pressure ratio is given in Fig. 9–28 for $k = 1.4$, which is the specific-heat-ratio value of air at room temperature.

The highest temperature in the cycle occurs at the end of the combustion process (state 3), and it is limited by the maximum temperature that the turbine

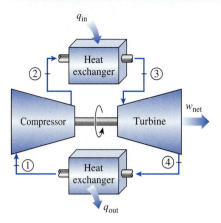

**FIGURE 9–26**
A closed-cycle gas-turbine engine.

(a) $T$-$s$ diagram

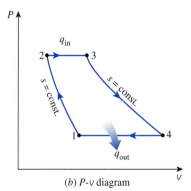

(b) $P$-$v$ diagram

**FIGURE 9–27**
$T$-$s$ and $P$-$v$ diagrams for the ideal Brayton cycle.

**FIGURE 9–28**
Thermal efficiency of the ideal Brayton cycle as a function of the pressure ratio.

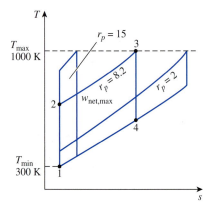

**FIGURE 9–29**
For fixed values of $T_{min}$ and $T_{max}$, the net work of the Brayton cycle first increases with the pressure ratio, then reaches a maximum at $r_p = (T_{max}/T_{min})^{k/[2(k-1)]}$, and finally decreases.

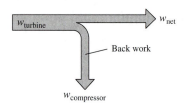

**FIGURE 9–30**
The fraction of the turbine work used to drive the compressor is called the back work ratio.

blades can withstand. This also limits the pressure ratios that can be used in the cycle. For a fixed turbine inlet temperature $T_3$, the net work output per cycle increases with the pressure ratio, reaches a maximum, and then starts to decrease, as shown in Fig. 9–29. Therefore, there should be a compromise between the pressure ratio (thus the thermal efficiency) and the net work output. With less work output per cycle, a larger mass flow rate (thus a larger system) is needed to maintain the same power output, which may not be economical. In most common designs, the pressure ratio of gas turbines ranges from about 11 to 16.

The air in gas turbines performs two important functions: It supplies the necessary oxidant for the combustion of the fuel, and it serves as a coolant to keep the temperature of various components within safe limits. The second function is accomplished by drawing in more air than is needed for the complete combustion of the fuel. In gas turbines, an air–fuel mass ratio of 50 or above is not uncommon. Therefore, in a cycle analysis, treating the combustion gases as air does not cause any appreciable error. Also, the mass flow rate through the turbine is greater than that through the compressor, the difference being equal to the mass flow rate of the fuel. Thus, assuming a constant mass flow rate throughout the cycle yields conservative results for open-loop gas-turbine engines.

The two major application areas of gas-turbine engines are *aircraft propulsion* and *electric power generation*. When it is used for aircraft propulsion, the gas turbine produces just enough power to drive the compressor and a small generator to power the auxiliary equipment. The high-velocity exhaust gases are responsible for producing the necessary thrust to propel the aircraft. Gas turbines are also used as stationary power plants to generate electricity as stand-alone units or in conjunction with steam power plants on the high-temperature side. In these plants, the exhaust gases of the gas turbine serve as the heat source for the steam. The gas-turbine cycle can also be executed as a closed cycle for use in nuclear power plants. This time the working fluid is not limited to air, and a gas with more desirable characteristics (such as helium) can be used.

The majority of the Western world's naval fleets already use gas-turbine engines for propulsion and electric power generation. The General Electric LM2500 gas turbines used to power ships have a simple-cycle thermal efficiency of 37 percent. The General Electric WR-21 gas turbines equipped with intercooling and regeneration have a thermal efficiency of 43 percent and produce 21.6 MW (29,040 hp). The regeneration also reduces the exhaust temperature from 600°C (1100°F) to 350°C (650°F). Air is compressed to 3 atm before it enters the intercooler. Compared to steam-turbine and diesel-propulsion systems, the gas turbine offers greater power for a given size and weight, high reliability, long life, and more convenient operation. The engine start-up time has been reduced from 4 h required for a typical steam-propulsion system to less than 2 min for a gas turbine. Many modern marine propulsion systems use gas turbines together with diesel engines because of the high fuel consumption of simple-cycle gas-turbine engines. In combined diesel and gas-turbine systems, diesel is used to provide for efficient low-power and cruise operation, and gas turbine is used when high speeds are needed.

In gas-turbine power plants, the ratio of the compressor work to the turbine work, called the **back work ratio**, is very high (Fig. 9–30). Usually more than one-half of the turbine work output is used to drive the compressor. The situation is even worse when the isentropic efficiencies of the compressor and the turbine are low. This is quite in contrast to steam power plants, where the back work ratio is only a few percent. This is not surprising, however, since a liquid is compressed in steam power plants instead of a gas, and the steady-flow work is proportional to the specific volume of the working fluid.

A power plant with a high back work ratio requires a larger turbine to provide the additional power requirements of the compressor. Therefore, the turbines used in gas-turbine power plants are larger than those used in steam power plants of the same net power output.

## Development of Gas Turbines

The gas turbine has experienced phenomenal progress and growth since its first successful development in the 1930s. The early gas turbines built in the 1940s and even 1950s had simple-cycle efficiencies of about 17 percent because of the low compressor and turbine efficiencies and low turbine inlet temperatures due to metallurgical limitations of those times. Therefore, gas turbines found only limited use despite their versatility and their ability to burn a variety of fuels. The efforts to improve the cycle efficiency concentrated in three areas:

**1. Increasing the turbine inlet (or firing) temperatures**  This has been the primary approach taken to improve gas-turbine efficiency. The turbine inlet temperatures have increased steadily from about 540°C (1000°F) in the 1940s to 1425°C (2600°F) and even higher today. These increases were made possible by the development of new materials and the innovative cooling techniques for the critical components such as coating the turbine blades with ceramic layers and cooling the blades with the discharge air from the compressor. Maintaining high turbine inlet temperatures with an air-cooling technique requires the combustion temperature to be higher to compensate for the cooling effect of the cooling air. However, higher combustion temperatures increase the amount of nitrogen oxides ($NO_x$), which are responsible for the formation of ozone at ground level and smog. Using steam as the coolant allowed an increase in the turbine inlet temperatures by 200°F without an increase in the combustion temperature. Steam is also a much more effective heat transfer medium than air.

**2. Increasing the efficiencies of turbomachinery components**  The performance of early turbines suffered greatly from the inefficiencies of turbines and compressors. However, the advent of computers and advanced techniques for computer-aided design made it possible to design these components aerodynamically with minimal losses. The increased efficiencies of the turbines and compressors resulted in a significant increase in the cycle efficiency.

**3. Adding modifications to the basic cycle**  The simple-cycle efficiencies of early gas turbines were practically doubled by incorporating intercooling, regeneration (or recuperation), and reheating, discussed in the next two sections. These improvements, of course, come at the expense of increased initial and operation costs, and they cannot be justified unless the decrease in fuel costs offsets the increase in other costs. The relatively low fuel prices, the general desire in the industry to minimize installation costs, and the tremendous increase in the simple-cycle efficiency to about 40 percent left little desire for opting for these modifications.

The first gas turbine for an electric utility was installed in 1949 in Oklahoma as part of a combined-cycle power plant. It was built by General Electric and produced 3.5 MW of power. Gas turbines installed until the mid-1970s suffered from low efficiency and poor reliability. In the past, the base-load electric power generation was dominated by large coal and nuclear power plants. However, there has been an historic shift toward natural gas–fired gas turbines because of their higher efficiencies, lower capital costs, shorter installation times, and better emission characteristics, and the abundance of natural gas supplies, and more and more electric utilities are using gas turbines for base-load power production as well as for peaking. The construction costs for gas-turbine power plants are roughly half that of comparable conventional fossil-fuel steam power plants, which were the primary base-load power plants until the early 1980s. More than half of all power plants to be installed in the foreseeable future are forecast to be gas-turbine or combined gas–steam turbine types.

A gas turbine manufactured by General Electric in the early 1990s had a pressure ratio of 13.5 and generated 135.7 MW of net power at a thermal efficiency of 33 percent in simple-cycle operation. A more recent gas turbine manufactured by General Electric uses a turbine inlet temperature of 1425°C (2600°F) and produces up to 282 MW while achieving a thermal efficiency of 39.5 percent in the simple-cycle mode. A 1.3-ton small-scale gas turbine labeled OP-16, built by the Dutch firm Opra Optimal Radial Turbine, can run on gas or liquid fuel and can replace a 16-ton diesel engine. It has a pressure ratio of 6.5 and produces up to 2 MW of power. Its efficiency is 26 percent in the simple-cycle operation, which rises to 37 percent when equipped with a regenerator.

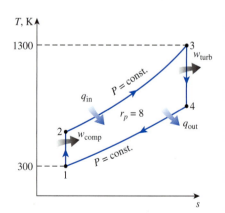

**FIGURE 9–31**

$T$-$s$ diagram for the Brayton cycle discussed in Example 9–4.

**EXAMPLE 9–4    The Simple Ideal Brayton Cycle**

A gas-turbine power plant operating on an ideal Brayton cycle has a pressure ratio of 8. The gas temperature is 300 K at the compressor inlet and 1300 K at the turbine inlet. Utilizing the air-standard assumptions, determine (a) the gas temperature at the exits of the compressor and the turbine, (b) the back work ratio, and (c) the thermal efficiency.

**Solution**   A power plant operating on the ideal Brayton cycle is considered. The compressor and turbine exit temperatures, back work ratio, and the thermal efficiency are to be determined.

*Assumptions*   1 Steady operating conditions exist. 2 The air-standard assumptions are applicable. 3 Kinetic and potential energy changes are negligible. 4 The variation of specific heats with temperature is to be considered.

*Analysis*   The $T$-$s$ diagram of the ideal Brayton cycle described is shown in Fig. 9–31. We note that the components involved in the Brayton cycle are steady-flow devices.

(a) The air temperatures at the compressor and turbine exits are determined from isentropic relations:

Process 1-2 (isentropic compression of an ideal gas):

$$T_1 = 300 \text{ K} \rightarrow h_1 = 300.19 \text{ kJ/kg}$$

$$P_{r1} = 1.386$$

$$P_{r2} = \frac{P_2}{P_1} P_{r1} = (8)(1.386) = 11.09 \rightarrow T_2 = \mathbf{540 \text{ K}} \qquad \text{(at compressor exit)}$$

$$h_2 = 544.35 \text{ kJ/kg}$$

Process 3-4 (isentropic expansion of an ideal gas):

$$T_3 = 1300 \text{ K} \rightarrow h_3 = 1395.97 \text{ kJ/kg}$$

$$P_{r3} = 330.9$$

$$P_{r4} = \frac{P_4}{P_3} P_{r3} = \left(\frac{1}{8}\right)(330.9) = 41.36 \rightarrow T_4 = \mathbf{770 \text{ K}} \qquad \text{(at turbine exit)}$$

$$h_4 = 789.37 \text{ kJ/kg}$$

(b) To find the back work ratio, we need to find the work input to the compressor and the work output of the turbine:

$$w_{comp,in} = h_2 - h_1 = 544.35 - 300.19 = 244.16 \text{ kJ/kg}$$

$$w_{turb,out} = h_3 - h_4 = 1395.97 - 789.37 = 606.60 \text{ kJ/kg}$$

Thus,

$$r_{bw} = \frac{w_{comp,in}}{w_{turb,out}} = \frac{244.16 \text{ kJ/kg}}{606.60 \text{ kJ/kg}} = \mathbf{0.403}$$

That is, 40.3 percent of the turbine work output is used just to drive the compressor.

(c) The thermal efficiency of the cycle is the ratio of the net power output to the total heat input:

$$q_{in} = h_3 - h_2 = 1395.97 - 544.35 = 851.62 \text{ kJ/kg}$$

$$w_{net} = w_{out} - w_{in} = 606.60 - 244.16 = 362.4 \text{ kJ/kg}$$

Thus,

$$\eta_{th} = \frac{w_{net}}{q_{in}} = \frac{362.4 \text{ kJ/kg}}{851.62 \text{ kJ/kg}} = \mathbf{0.426 \text{ or } 42.6\%}$$

The thermal efficiency could also be determined from

$$\eta_{th} = 1 - \frac{q_{out}}{q_{in}}$$

where

$$q_{out} = h_4 - h_1 = 789.37 - 300.19 = 489.2 \text{ kJ/kg}$$

*Discussion* Under the cold-air-standard assumptions (constant specific heat values at room temperature), the thermal efficiency would be, from Eq. 9–16,

$$\eta_{\text{th,Brayton}} = 1 - \frac{1}{r_p^{(k-1)/k}} = 1 - \frac{1}{8^{(1.4-1)/1.4}} = 0.448$$

which is sufficiently close to the value obtained by accounting for the variation of specific heats with temperature.

## Deviation of Actual Gas-Turbine Cycles from Idealized Ones

The actual gas-turbine cycle differs from the ideal Brayton cycle on several accounts. For one thing, some pressure drop during the heat-addition and heat-rejection processes is inevitable. More importantly, the actual work input to the compressor is more, and the actual work output from the turbine is less because of irreversibilities. The deviation of actual compressor and turbine behavior from the idealized isentropic behavior can be accurately accounted for by utilizing the isentropic efficiencies of the turbine and compressor as

$$\eta_C = \frac{w_s}{w_a} \cong \frac{h_{2s} - h_1}{h_{2a} - h_1} \tag{9–18}$$

and

$$\eta_T = \frac{w_a}{w_s} \cong \frac{h_3 - h_{4a}}{h_3 - h_{4s}} \tag{9–19}$$

where states $2a$ and $4a$ are the actual exit states of the compressor and the turbine, respectively, and $2s$ and $4s$ are the corresponding states for the isentropic case, as illustrated in Fig. 9–32. The effect of the turbine and compressor efficiencies on the thermal efficiency of the gas-turbine engines is illustrated below with an example.

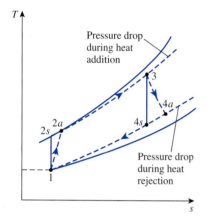

**FIGURE 9–32**

The deviation of an actual gas-turbine cycle from the ideal Brayton cycle as a result of irreversibilities.

---

**EXAMPLE 9–5     An Actual Gas-Turbine Cycle**

Assuming a compressor efficiency of 80 percent and a turbine efficiency of 85 percent, determine (*a*) the back work ratio, (*b*) the thermal efficiency, and (*c*) the turbine exit temperature of the gas-turbine cycle discussed in Example 9–4.

**Solution** The Brayton cycle discussed in Example 9–4 is reconsidered. For specified turbine and compressor efficiencies, the back work ratio, the thermal efficiency, and the turbine exit temperature are to be determined.

*Analysis* (*a*) The *T-s* diagram of the cycle is shown in Fig. 9–33. The actual compressor work and turbine work are determined by using the definitions of compressor and turbine efficiencies, Eqs. 9–18 and 9–19:

*Compressor:*
$$w_{\text{comp,in}} = \frac{w_s}{\eta_C} = \frac{244.16 \text{ kJ/kg}}{0.80} = 305.20 \text{ kJ/kg}$$

*Turbine:*
$$w_{\text{turb,out}} = \eta_T w_s = (0.85)(606.60 \text{ kJ/kg}) = 515.61 \text{ kJ/kg}$$

Thus,

$$r_{bw} = \frac{w_{comp,in}}{w_{turb,out}} = \frac{305.20\ kJ/kg}{515.61\ kJ/kg} = 0.592$$

That is, the compressor is now consuming 59.2 percent of the work produced by the turbine (up from 40.3 percent). This increase is due to the irreversibilities that occur within the compressor and the turbine.

(b) In this case, air leaves the compressor at a higher temperature and enthalpy, which are determined to be

$$w_{comp,in} = h_{2a} - h_1 \rightarrow h_{2a} = h_1 + w_{comp,in}$$

$$= 300.19 + 305.20$$

$$= 605.39\ kJ/kg \quad (and\ T_{2a} = 598\ K)$$

Thus,

$$q_{in} = h_3 - h_{2a} = 1395.97 - 605.39 = 790.58\ kJ/kg$$

$$w_{net} = w_{out} - w_{in} = 515.61 - 305.20 = 210.41\ kJ/kg$$

and

$$\eta_{th} = \frac{w_{net}}{q_{in}} = \frac{210.41\ kJ/kg}{790.58\ kJ/kg} = 0.266\ or\ 26.6\%$$

That is, the irreversibilities occurring within the turbine and compressor caused the thermal efficiency of the gas turbine cycle to drop from 42.6 to 26.6 percent. This example shows how sensitive the performance of a gas-turbine power plant is to the efficiencies of the compressor and the turbine. In fact, gas-turbine efficiencies did not reach competitive values until significant improvements were made in the design of gas turbines and compressors.

(c) The air temperature at the turbine exit is determined from an energy balance on the turbine:

$$w_{turb,out} = h_3 - h_{4a} \rightarrow h_{4a} = h_3 - w_{turb,out}$$

$$= 1395.97 - 515.61$$

$$= 880.36\ kJ/kg$$

Then, from Table A–21,

$$T_{4a} = 853\ K$$

**Discussion** The temperature at turbine exit is considerably higher than that at the compressor exit ($T_{2a} = 598$ K), which suggests the use of regeneration to reduce fuel cost.

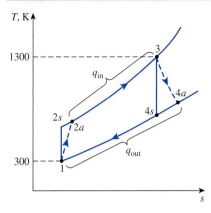

**FIGURE 9–33**
T-s diagram of the gas-turbine cycle discussed in Example 9–5.

# 9–8 · THE BRAYTON CYCLE WITH REGENERATION

In gas-turbine engines, the temperature of the exhaust gas leaving the turbine is often considerably higher than the temperature of the air leaving the compressor. Therefore, the high-pressure air leaving the compressor can be heated by transferring heat to it from the hot exhaust gases in a counter-flow

**FIGURE 9–34**
A gas-turbine engine with regenerator.

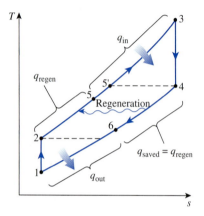

**FIGURE 9–35**
*T-s* diagram of a Brayton cycle with regeneration.

heat exchanger, which is also known as a *regenerator* or a *recuperator.* A sketch of the gas-turbine engine utilizing a regenerator and the *T-s* diagram of the new cycle are shown in Figs. 9–34 and 9–35, respectively.

The thermal efficiency of the Brayton cycle increases as a result of regeneration since the portion of energy of the exhaust gases that is normally rejected to the surroundings is now used to preheat the air entering the combustion chamber. This, in turn, decreases the heat input (thus fuel) requirements for the same net work output. Note, however, that the use of a regenerator is recommended only when the turbine exhaust temperature is higher than the compressor exit temperature. Otherwise, heat will flow in the reverse direction (*to* the exhaust gases), decreasing the efficiency. This situation is encountered in gas-turbine engines operating at very high pressure ratios.

The highest temperature occurring within the regenerator is $T_4$, the temperature of the exhaust gases leaving the turbine and entering the regenerator. Under no conditions can the air be preheated in the regenerator to a temperature above this value. Air normally leaves the regenerator at a lower temperature, $T_5$. In the limiting (ideal) case, the air exits the regenerator at the inlet temperature of the exhaust gases $T_4$. Assuming the regenerator to be well insulated and any changes in kinetic and potential energies to be negligible, the actual and maximum heat transfers from the exhaust gases to the air can be expressed as

$$q_{regen,act} = h_5 - h_2 \qquad (9\text{–}20)$$

and

$$q_{regen,max} = h_{5'} - h_2 = h_4 - h_2 \qquad (9\text{–}21)$$

The extent to which a regenerator approaches an ideal regenerator is called the **effectiveness $\epsilon$** and is defined as

$$\epsilon = \frac{q_{regen,act}}{q_{regen,max}} = \frac{h_5 - h_2}{h_4 - h_2} \qquad (9\text{–}22)$$

When the cold-air-standard assumptions are utilized, it reduces to

$$\epsilon \cong \frac{T_5 - T_2}{T_4 - T_2} \qquad (9\text{–}23)$$

A regenerator with a higher effectiveness obviously saves a greater amount of fuel since it preheats the air to a higher temperature prior to combustion.

However, achieving a higher effectiveness requires the use of a larger regenerator, which carries a higher price tag and causes a larger pressure drop. Therefore, the use of a regenerator with a very high effectiveness cannot be justified economically unless the savings from the fuel costs exceed the additional expenses involved. The effectiveness of most regenerators used in practice is below 0.85.

Under the cold-air-standard assumptions, the thermal efficiency of an ideal Brayton cycle with regeneration is

$$\eta_{th,regen} = 1 - \left(\frac{T_1}{T_3}\right)(r_p)^{(k-1)/k} \qquad (9\text{–}24)$$

Therefore, the thermal efficiency of an ideal Brayton cycle with regeneration depends on the ratio of the minimum to maximum temperatures as well as the pressure ratio. The thermal efficiency is plotted in Fig. 9–36 for various pressure ratios and minimum-to-maximum temperature ratios. This figure shows that regeneration is most effective at lower pressure ratios and low minimum-to-maximum temperature ratios.

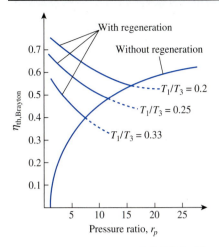

**FIGURE 9–36**

Thermal efficiency of the ideal Brayton cycle with and without regeneration.

---

### EXAMPLE 9–6    Actual Gas-Turbine Cycle with Regeneration

Determine the thermal efficiency of the gas-turbine described in Example 9–5 if a regenerator having an effectiveness of 80 percent is installed.

**Solution** The gas-turbine discussed in Example 9–5 is equipped with a regenerator. For a specified effectiveness, the thermal efficiency is to be determined.

**Analysis** The T-s diagram of the cycle is shown in Fig. 9–37. We first determine the enthalpy of the air at the exit of the regenerator, using the definition of effectiveness:

$$\epsilon = \frac{h_5 - h_{2a}}{h_{4a} - h_{2a}}$$

$$0.80 = \frac{(h_5 - 605.39) \text{ kJ/kg}}{(880.36 - 605.39) \text{ kJ/kg}} \rightarrow h_5 = 825.37 \text{ kJ/kg}$$

Thus,

$$q_{in} = h_3 - h_5 = (1395.97 - 825.37) \text{ kJ/kg} = 570.60 \text{ kJ/kg}$$

This represents a savings of 220.0 kJ/kg from the heat input requirements. The addition of a regenerator (assumed to be frictionless) does not affect the net work output. Thus,

$$\eta_{th} = \frac{w_{net}}{q_{in}} = \frac{210.41 \text{ kJ/kg}}{570.60 \text{ kJ/kg}} = \textbf{0.369 or 36.9\%}$$

**Discussion** Note that the thermal efficiency of the gas turbine has gone up from 26.6 to 36.9 percent as a result of installing a regenerator that helps to recuperate some of the thermal energy of the exhaust gases.

**FIGURE 9–37**

T-s diagram of the regenerative Brayton cycle described in Example 9–6.

# 9–9 · THE CARNOT VAPOR CYCLE

We have mentioned repeatedly that the Carnot cycle is the most efficient cycle operating between two specified temperature limits. Thus it is natural to look at the Carnot cycle first as a prospective ideal cycle for vapor power plants. If we could, we would certainly adopt it as the ideal cycle. As explained below, however, the Carnot cycle is not a suitable model for power cycles. Throughout the discussions, we assume *steam* to be the working fluid since it is the working fluid predominantly used in vapor power cycles.

Consider a steady-flow *Carnot cycle* executed within the saturation dome of a pure substance, as shown in Fig. 9–38a. The fluid is heated reversibly and isothermally in a boiler (process 1-2), expanded isentropically in a turbine (process 2-3), condensed reversibly and isothermally in a condenser (process 3-4), and compressed isentropically by a compressor to the initial state (process 4-1).

Several impracticalities are associated with this cycle:

**1.** Isothermal heat transfer to or from a two-phase system is not difficult to achieve in practice since maintaining a constant pressure in the device automatically fixes the temperature at the saturation value. Therefore, processes 1-2 and 3-4 can be approached closely in actual boilers and condensers. Limiting the heat transfer processes to two-phase systems, however, severely limits the maximum temperature that can be used in the cycle (it has to remain under the critical-point value, which is 374°C for water). Limiting the maximum temperature in the cycle also limits the thermal efficiency. Any attempt to raise the maximum temperature in the cycle involves heat transfer to the working fluid in a single phase, which is not easy to accomplish isothermally.

**2.** The isentropic expansion process (process 2-3) can be approximated closely by a well-designed turbine. However, the quality of the steam decreases during this process, as shown on the *T-s* diagram in Fig. 9–38a. Thus the turbine has to handle steam with low quality, that is, steam with a high moisture content. The impingement of liquid droplets on the turbine blades causes erosion and is a major source of wear. Thus steam with qualities less than about 90 percent cannot be tolerated in the operation of power plants. This problem could be eliminated by using a working fluid with a very steep saturated vapor line.

**3.** The isentropic compression process (process 4-1) involves the compression of a liquid–vapor mixture to a saturated liquid. There are two difficulties associated with this process. First, it is not easy to control the condensation process so precisely as to end up with the desired quality at state 4. Second, it is not practical to design a compressor that handles two phases.

Some of these problems could be eliminated by executing the Carnot cycle in a different way, as shown in Fig. 9–38b. This cycle, however, presents other problems such as isentropic compression to extremely high pressures and isothermal heat transfer at variable pressures. Thus we conclude that the Carnot cycle cannot be approximated in actual devices and is not a realistic model for vapor power cycles.

(a)

(b)

**FIGURE 9–38**

*T-s* diagram of two Carnot vapor cycles.

# 9–10 · RANKINE CYCLE: THE IDEAL CYCLE FOR VAPOR POWER CYCLES

Many of the impracticalities associated with the Carnot cycle can be eliminated by superheating the steam in the boiler and condensing it completely in the condenser, as shown schematically on a *T-s* diagram in Fig. 9–39. The cycle that results is the **Rankine cycle**, which is the ideal cycle for vapor power plants. The ideal Rankine cycle does not involve any internal irreversibilities and consists of the following four processes:

1-2   Isentropic compression in a pump
2-3   Constant pressure heat addition in a boiler
3-4   Isentropic expansion in a turbine
4-1   Constant pressure heat rejection in a condenser

Water enters the *pump* at state 1 as saturated liquid and is compressed isentropically to the operating pressure of the boiler. The water temperature increases somewhat during this isentropic compression process due to a slight decrease in the specific volume of water. The vertical distance between states 1 and 2 on the *T-s* diagram is greatly exaggerated for clarity. (If water were truly incompressible, would there be a temperature change at all during this process?)

Water enters the *boiler* as a compressed liquid at state 2 and leaves as a superheated vapor at state 3. The boiler is basically a large heat exchanger where the heat originating from combustion gases, nuclear reactors, or other sources is transferred to the water essentially at constant pressure. The boiler, together with the section where the steam is superheated (the superheater), is often called the *steam generator*.

The superheated vapor at state 3 enters the *turbine,* where it expands isentropically and produces work by rotating the shaft connected to an electric generator. The pressure and the temperature of steam drop during this process to the values at state 4, where steam enters the *condenser.* At this state, steam is usually a saturated liquid–vapor mixture with a high quality. Steam is condensed at constant pressure in the condenser, which is basically a large heat exchanger, by rejecting heat to a cooling medium such as a lake, a river, or the atmosphere. Steam leaves the condenser as saturated liquid and enters the pump, completing the cycle. In areas where water is precious, the power plants are cooled by air instead of water. This method of cooling, which is also used in car engines, is called *dry cooling.* Several power plants in the world, including some in the United States, use dry cooling to conserve water.

Remembering that the area under the process curve on a *T-s* diagram represents the heat transfer for internally reversible processes, we see that the area under process curve 2-3 represents the heat transferred to the water in the boiler and the area under the process curve 4-1 represents the heat rejected in the condenser. The difference between these two (the area enclosed by the cycle curve) is the net work produced during the cycle.

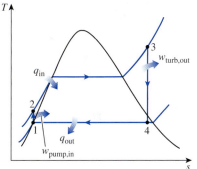

**FIGURE 9–39**
The simple ideal Rankine cycle.

## Energy Analysis of the Ideal Rankine Cycle

All four components associated with the Rankine cycle (the pump, boiler, turbine, and condenser) are steady-flow devices, and thus all four processes that make up the Rankine cycle can be analyzed as steady-flow processes. The kinetic and potential energy changes of the steam are usually small relative to the work and heat transfer terms and are therefore usually neglected. Then the *steady-flow energy equation* per unit mass of steam reduces to

$$(q_{in} - q_{out}) + (w_{in} - w_{out}) = h_e - h_i \qquad \text{(kJ/kg)} \tag{9–25}$$

The boiler and the condenser do not involve any work, and the pump and the turbine are assumed to be isentropic. Then the conservation of energy relation for each device can be expressed as follows:

*Pump* $(q = 0)$: 
$$w_{pump,in} = h_2 - h_1 \tag{9–26}$$

or,

$$w_{pump,in} = v(P_2 - P_1) \tag{9–27}$$

where

$$h_1 = h_{f @ P_1} \quad \text{and} \quad v \cong v_1 = v_{f @ P_1} \tag{9–28}$$

*Boiler* $(w = 0)$: 
$$q_{in} = h_3 - h_2 \tag{9–29}$$

*Turbine* $(q = 0)$: 
$$w_{turb,out} = h_3 - h_4 \tag{9–30}$$

*Condenser* $(w = 0)$: 
$$q_{out} = h_4 - h_1 \tag{9–31}$$

The *thermal efficiency* of the Rankine cycle is determined from

$$\eta_{th} = \frac{w_{net}}{q_{in}} = 1 - \frac{q_{out}}{q_{in}} \tag{9–32}$$

where

$$w_{net} = q_{in} - q_{out} = w_{turb,out} - w_{pump,in}$$

The conversion efficiency of power plants in the United States is often expressed in terms of **heat rate**, which is the amount of heat supplied, in Btu's, to generate 1 kWh of electricity. The smaller the heat rate, the greater the efficiency. Considering that 1 kWh = 3412 Btu and disregarding the losses associated with the conversion of shaft power to electric power, the relation between the heat rate and the thermal efficiency can be expressed as

$$\eta_{th} = \frac{3412 \ \text{(Btu/kWh)}}{\text{Heat rate (Btu/kWh)}} \tag{9–33}$$

For example, a heat rate of 11,363 Btu/kWh is equivalent to 30 percent efficiency.

The thermal efficiency can also be interpreted as the ratio of the area enclosed by the cycle on a *T-s* diagram to the area under the heat-addition process. The use of these relations is illustrated in the following example.

## EXAMPLE 9–7    The Simple Ideal Rankine Cycle

Consider a steam power plant operating on the simple ideal Rankine cycle. Steam enters the turbine at 3 MPa and 350°C and is condensed in the condenser at a pressure of 75 kPa. Determine the thermal efficiency of this cycle.

**SOLUTION** A steam power plant operating on the simple ideal Rankine cycle is considered. The thermal efficiency of the cycle is to be determined.
**Assumptions** 1 Steady operating conditions exist. 2 Kinetic and potential energy changes are negligible.
**Analysis** The schematic of the power plant and the *T-s* diagram of the cycle are shown in Fig. 9–40. We note that the power plant operates on the ideal Rankine cycle. Therefore, the pump and the turbine are isentropic, there are no pressure drops in the boiler and condenser, and steam leaves the condenser and enters the pump as saturated liquid at the condenser pressure.

First we determine the enthalpies at various points in the cycle, using data from steam tables (Tables A–4, A–5, and A–6):

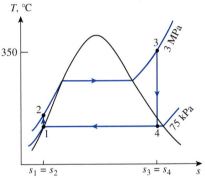

**FIGURE 9–40**
Schematic and *T-s* diagram for Example 9–7.

State 1:     $P_1 = 75 \text{ kPa}$ ⎫     $h_1 = h_{f @ 75 \text{ kPa}} = 384.44 \text{ kJ/kg}$
             Sat. liquid ⎭     $v_1 = v_{f @ 75 \text{ kPa}} = 0.001037 \text{ m}^3/\text{kg}$

State 2: $P_2 = 3 \text{ MPa}$

$\quad s_2 = s_1$

$$w_{pump,in} = v_1(P_2 - P_1) = (0.001037 \text{ m}^3/\text{kg})[(3000 - 75) \text{ kPa}]\left(\frac{1 \text{ kJ}}{1 \text{ kPa·m}^3}\right)$$

$$= 3.03 \text{ kJ/kg}$$

$$h_2 = h_1 + w_{pump,in} = (384.44 + 3.03) \text{ kJ/kg} = 387.47 \text{ kJ/kg}$$

State 3:     $P_3 = 3 \text{ MPa}$ ⎫     $h_3 = 3116.1 \text{ kJ/kg}$
             $T_3 = 350°C$ ⎭     $s_3 = 6.7450 \text{ kJ/kg·K}$

State 4:     $P_4 = 75 \text{ kPa}$    (sat. mixture)

$\quad s_4 = s_3$

$$x_4 = \frac{s_4 - s_f}{s_{fg}} = \frac{6.7450 - 1.2132}{6.2426} = 0.8861$$

$$h_4 = h_f + x_4 h_{fg} = 384.44 + 0.8861(2278.0) = 2403.0 \text{ kJ/kg}$$

Thus,

$$q_{in} = h_3 - h_2 = (3116.1 - 387.47) \text{ kJ/kg} = 2728.6 \text{ kJ/kg}$$

$$q_{out} = h_4 - h_1 = (2403.0 - 384.44) \text{ kJ/kg} = 2018.6 \text{ kJ/kg}$$

and

$$\eta_{th} = 1 - \frac{q_{out}}{q_{in}} = 1 - \frac{2018.6 \text{ kJ/kg}}{2728.6 \text{ kJ/kg}} = \mathbf{0.260 \text{ or } 26.0\%}$$

The thermal efficiency could also be determined from

$$w_{\text{turb,out}} = h_3 - h_4 = (3116.1 - 2403.0) \text{ kJ/kg} = 713.1 \text{ kJ/kg}$$

$$w_{\text{net}} = w_{\text{turb,out}} - w_{\text{pump,in}} = (713.1 - 3.03) \text{ kJ/kg} = 710.1 \text{ kJ/kg}$$

or

$$w_{\text{net}} = q_{\text{in}} - q_{\text{out}} = (2728.6 - 2018.6) \text{ kJ/kg} = 710.0 \text{ kJ/kg}$$

and

$$\eta_{\text{th}} = \frac{w_{\text{net}}}{q_{\text{in}}} = \frac{710.0 \text{ kJ/kg}}{2728.6 \text{ kJ/kg}} = \textbf{0.260 or 26.0\%}$$

That is, this power plant converts 26 percent of the heat it receives in the boiler to net work. An actual power plant operating between the same temperature and pressure limits will have a lower efficiency because of the irreversibilities such as friction.

**Discussion** Notice that the back work ratio ($r_{\text{bw}} = w_{\text{in}}/w_{\text{out}}$) of this power plant is 0.004, and thus only 0.4 percent of the turbine work output is required to operate the pump. Having such low back work ratios is characteristic of vapor power cycles. This is in contrast to the gas power cycles, which typically involve very high back work ratios (about 40 to 80 percent).

It is also interesting to note the thermal efficiency of a Carnot cycle operating between the same temperature limits

$$\eta_{\text{th,Carnot}} = 1 - \frac{T_{\text{min}}}{T_{\text{max}}} = 1 - \frac{(91.76 + 273) \text{ K}}{(350 + 273) \text{ K}} = 0.415$$

The difference between the two efficiencies is due to the large external irreversibility in the Rankine cycle caused by the large temperature difference between steam and combustion gases in the furnace.

## 9–11 ▪ DEVIATION OF ACTUAL VAPOR POWER CYCLES FROM IDEALIZED ONES

The actual vapor power cycle differs from the ideal Rankine cycle, as illustrated in Fig. 9–41a, as a result of irreversibilities in various components. Fluid friction and heat loss to the surroundings are the two common sources of irreversibilities.

*Fluid friction* causes pressure drops in the boiler, the condenser, and the piping between various components. As a result, steam leaves the boiler at a somewhat lower pressure. Also, the pressure at the turbine inlet is somewhat lower than that at the boiler exit due to the pressure drop in the connecting pipes. The pressure drop in the condenser is usually very small. To compensate for these pressure drops, the water must be pumped to a sufficiently higher pressure than the ideal cycle calls for. This requires a larger pump and larger work input to the pump.

The other major source of irreversibility is the *heat loss* from the steam to the surroundings as the steam flows through various components. To maintain the same level of net work output, more heat needs to be transferred to

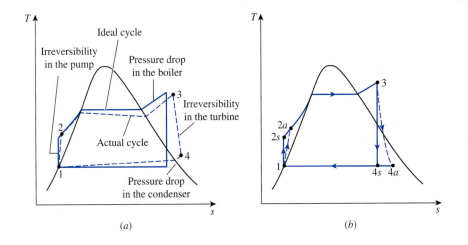

**FIGURE 9–41**
(a) Deviation of actual vapor power cycle from the ideal Rankine cycle. (b) The effect of pump and turbine irreversibilities on the ideal Rankine cycle.

the steam in the boiler to compensate for these undesired heat losses. As a result, cycle efficiency decreases.

Of particular importance are the irreversibilities occurring within the pump and the turbine. A pump requires a greater work input, and a turbine produces a smaller work output as a result of irreversibilities. Under ideal conditions, the flow through these devices is isentropic. The deviation of actual pumps and turbines from the isentropic ones can be accounted for by utilizing *isentropic efficiencies,* defined as

$$\eta_P = \frac{w_s}{w_a} = \frac{h_{2s} - h_1}{h_{2a} - h_1} \qquad (9\text{–}34)$$

and

$$\eta_T = \frac{w_a}{w_s} = \frac{h_3 - h_{4a}}{h_3 - h_{4s}} \qquad (9\text{–}35)$$

where states 2a and 4a are the actual exit states of the pump and the turbine, respectively, and 2s and 4s are the corresponding states for the isentropic case (Fig. 9–41b).

Other factors also need to be considered in the analysis of actual vapor power cycles. In actual condensers, for example, the liquid is usually sub-cooled to prevent the onset of *cavitation,* the rapid vaporization and conden-sation of the fluid at the low-pressure side of the pump impeller, which may damage it. Additional losses occur at the bearings between the moving parts as a result of friction. Steam that leaks out during the cycle and air that leaks into the condenser represent two other sources of loss. Finally, the power consumed by the auxiliary equipment such as fans that supply air to the furnace should also be considered in evaluating the overall performance of power plants.

The effect of irreversibilities on the thermal efficiency of a steam power cycle is illustrated below with an example.

**EXAMPLE 9–8    An Actual Steam Power Cycle**

A steam power plant operates on the cycle shown in Fig. 9–42. If the isentropic efficiency of the turbine is 87 percent and the isentropic efficiency of the pump is 85 percent, determine (*a*) the thermal efficiency of the cycle and (*b*) the net power output of the plant for a mass flow rate of 15 kg/s.

**SOLUTION**    A steam power cycle with specified turbine and pump efficiencies is considered. The thermal efficiency and the net power output are to be determined.

**Assumptions**    **1** Steady operating conditions exist. **2** Kinetic and potential energy changes are negligible.

**Analysis**    The schematic of the power plant and the *T-s* diagram of the cycle are shown in Fig. 9–42. The temperatures and pressures of steam at various points are also indicated on the figure. We note that the power plant involves steady-flow components and operates on the Rankine cycle, but the imperfections at various components are accounted for.

(*a*) The thermal efficiency of a cycle is the ratio of the net work output to the heat input, and it is determined as follows:

*Pump work input:*

$$w_{pump,in} = \frac{w_{s,pump,in}}{\eta_P} = \frac{v_1(P_2 - P_1)}{\eta_P}$$

$$= \frac{(0.001009 \text{ m}^3/\text{kg})[(16,000 - 9) \text{ kPa}]}{0.85}\left(\frac{1 \text{ kJ}}{1 \text{ kPa·m}^3}\right)$$

$$= 19.0 \text{ kJ/kg}$$

**FIGURE 9–42**
Schematic and *T-s* diagram for Example 9–8.

*Turbine work output:*

$$w_{\text{turb,out}} = \eta_T w_{s,\text{turb,out}}$$

$$= \eta_T(h_5 - h_{6s}) = 0.87(3583.1 - 2115.3)\ \text{kJ/kg}$$

$$= 1277.0\ \text{kJ/kg}$$

*Boiler heat input:* $\quad q_{\text{in}} = h_4 - h_3 = (3647.6 - 160.1)\ \text{kJ/kg} = 3487.5\ \text{kJ/kg}$

Thus,

$$w_{\text{net}} = w_{\text{turb,out}} - w_{\text{pump,in}} = (1277.0 - 19.0)\ \text{kJ/kg} = 1258.0\ \text{kJ/kg}$$

$$\eta_{\text{th}} = \frac{w_{\text{net}}}{q_{\text{in}}} = \frac{1258.0\ \text{kJ/kg}}{3487.5\ \text{kJ/kg}} = \mathbf{0.361\ or\ 36.1\%}$$

(*b*) The power produced by this power plant is

$$\dot{W}_{\text{net}} = \dot{m}(w_{\text{net}}) = (15\ \text{kg/s})(1258.0\ \text{kJ/kg}) = \mathbf{18.9\ MW}$$

**Discussion** Without the irreversibilities, the thermal efficiency of this cycle would be 43.0 percent (see Example 9–9c).

# 9–12 · HOW CAN WE INCREASE THE EFFICIENCY OF THE RANKINE CYCLE?

Steam power plants are responsible for the production of most electric power in the world, and even small increases in thermal efficiency can mean large savings from the fuel requirements. Therefore, every effort is made to improve the efficiency of the cycle on which steam power plants operate.

The basic idea behind all the modifications to increase the thermal efficiency of a power cycle is the same: *Increase the average temperature at which heat is transferred to the working fluid in the boiler, or decrease the average temperature at which heat is rejected from the working fluid in the condenser.* That is, the average fluid temperature should be as high as possible during heat addition and as low as possible during heat rejection. Next we discuss three ways of accomplishing this for the simple ideal Rankine cycle.

## Lowering the Condenser Pressure (*Lowers* $T_{\text{low,avg}}$)

Steam exists as a saturated mixture in the condenser at the saturation temperature corresponding to the pressure inside the condenser. Therefore, lowering the operating pressure of the condenser automatically lowers the temperature of the steam, and thus the temperature at which heat is rejected.

The effect of lowering the condenser pressure on the Rankine cycle efficiency is illustrated on a *T-s* diagram in Fig. 9–43. For comparison purposes, the turbine inlet state is maintained the same. The colored area on this diagram represents the increase in net work output as a result of lowering the condenser pressure from $P_4$ to $P_4'$. The heat input requirements also

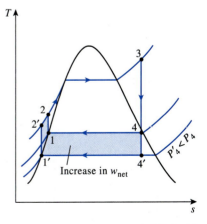

**FIGURE 9–43**

The effect of lowering the condenser pressure on the ideal Rankine cycle.

increase (represented by the area under curve 2′-2), but this increase is very small. Thus the overall effect of lowering the condenser pressure is an increase in the thermal efficiency of the cycle.

To take advantage of the increased efficiencies at low pressures, the condensers of steam power plants usually operate well below the atmospheric pressure. This does not present a major problem since the vapor power cycles operate in a closed loop. However, there is a lower limit on the condenser pressure that can be used. It cannot be lower than the saturation pressure corresponding to the temperature of the cooling medium. Consider, for example, a condenser that is to be cooled by a nearby river at 15°C. Allowing a temperature difference of 10°C for effective heat transfer, the steam temperature in the condenser must be above 25°C; thus the condenser pressure must be above 3.2 kPa, which is the saturation pressure at 25°C.

Lowering the condenser pressure is not without any side effects, however. For one thing, it creates the possibility of air leakage into the condenser. More importantly, it increases the moisture content of the steam at the final stages of the turbine, as can be seen from Fig. 9–43. The presence of large quantities of moisture is highly undesirable in turbines because it decreases the turbine efficiency and erodes the turbine blades. Fortunately, this problem can be corrected, as discussed next.

## Superheating the Steam to High Temperatures (*Increases T*$_{high,avg}$)

The average temperature at which heat is transferred to steam can be increased without increasing the boiler pressure by superheating the steam to high temperatures. The effect of superheating on the performance of vapor power cycles is illustrated on a *T-s* diagram in Fig. 9–44. The colored area on this diagram represents the increase in the net work. The total area under the process curve 3-3′ represents the increase in the heat input. Thus both the net work and heat input increase as a result of superheating the steam to a higher temperature. The overall effect is an increase in thermal efficiency, however, since the average temperature at which heat is added increases.

Superheating the steam to higher temperatures has another very desirable effect: It decreases the moisture content of the steam at the turbine exit, as can be seen from the *T-s* diagram (the quality at state 4′ is higher than that at state 4).

The temperature to which steam can be superheated is limited, however, by metallurgical considerations. Presently the highest steam temperature allowed at the turbine inlet is about 620°C (1150°F). Any increase in this value depends on improving the present materials or finding new ones that can withstand higher temperatures. Ceramics are very promising in this regard.

## Increasing the Boiler Pressure (*Increases T*$_{high,avg}$)

Another way of increasing the average temperature during the heat-addition process is to increase the operating pressure of the boiler, which automatically raises the temperature at which boiling takes place. This, in turn, raises the average temperature at which heat is transferred to the steam and thus raises the thermal efficiency of the cycle.

The effect of increasing the boiler pressure on the performance of vapor power cycles is illustrated on a *T-s* diagram in Fig. 9–45. Notice that for a

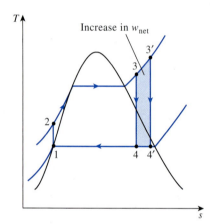

**FIGURE 9–44**
The effect of superheating the steam to higher temperatures on the ideal Rankine cycle.

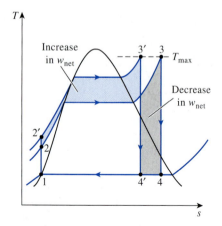

**FIGURE 9–45**
The effect of increasing the boiler pressure on the ideal Rankine cycle.

fixed turbine inlet temperature, the cycle shifts to the left and the moisture content of steam at the turbine exit increases. This undesirable side effect can be corrected, however, by reheating the steam, as discussed in the next section.

Operating pressures of boilers have gradually increased over the years from about 2.7 MPa (400 psia) in 1922 to over 30 MPa (4500 psia) today, generating enough steam to produce a net power output of 1000 MW or more in a large power plant. Today many modern steam power plants operate at supercritical pressures ($P > 22.06$ MPa) and have thermal efficiencies of about 40 percent for fossil-fuel plants and 34 percent for nuclear plants. There are over 150 supercritical-pressure steam power plants in operation in the United States. The lower efficiencies of nuclear power plants are due to the lower maximum temperatures used in those plants for safety reasons. The *T-s* diagram of a supercritical Rankine cycle is shown in Fig. 9–46.

The effects of lowering the condenser pressure, superheating to a higher temperature, and increasing the boiler pressure on the thermal efficiency of the Rankine cycle are illustrated below with an example.

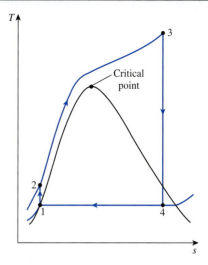

**FIGURE 9–46**
A supercritical Rankine cycle.

---

**EXAMPLE 9–9**    **Effect of Boiler Pressure and Temperature on Efficiency**

Consider a steam power plant operating on the ideal Rankine cycle. Steam enters the turbine at 3 MPa and 350°C and is condensed in the condenser at a pressure of 10 kPa. Determine (a) the thermal efficiency of this power plant, (b) the thermal efficiency if steam is superheated to 600°C instead of 350°C, and (c) the thermal efficiency if the boiler pressure is raised to 15 MPa while the turbine inlet temperature is maintained at 600°C.

**SOLUTION**    A steam power plant operating on the ideal Rankine cycle is considered. The effects of superheating the steam to a higher temperature and raising the boiler pressure on thermal efficiency are to be investigated.
**Analysis**    The *T-s* diagrams of the cycle for all three cases are given in Fig. 9–47.

(a) This is the steam power plant discussed in Example 9–7, except that the condenser pressure is lowered to 10 kPa. The thermal efficiency is determined in a similar manner:

State 1: $\left. \begin{array}{l} P_1 = 10 \text{ kPa} \\ \text{Sat. liquid} \end{array} \right\}$  $\begin{array}{l} h_1 = h_{f\ @\ 10\ kPa} = 191.81 \text{ kJ/kg} \\ v_1 = v_{f\ @\ 10\ kPa} = 0.00101 \text{ m}^3/\text{kg} \end{array}$

State 2: $P_2 = 3$ MPa
$s_2 = s_1$

$$w_{\text{pump,in}} = v_1(P_2 - P_1) = (0.00101 \text{ m}^3/\text{kg})[(3000 - 10) \text{ kPa}]\left(\frac{1 \text{ kJ}}{1 \text{ kPa·m}^3}\right)$$

$$= 3.02 \text{ kJ/kg}$$

$$h_2 = h_1 + w_{\text{pump,in}} = (191.81 + 3.02) \text{ kJ/kg} = 194.83 \text{ kJ/kg}$$

State 3: $\left. \begin{array}{l} P_3 = 3 \text{ MPa} \\ T_3 = 350°C \end{array} \right\}$  $\begin{array}{l} h_3 = 3116.1 \text{ kJ/kg} \\ s_3 = 6.7450 \text{ kJ/kg·K} \end{array}$

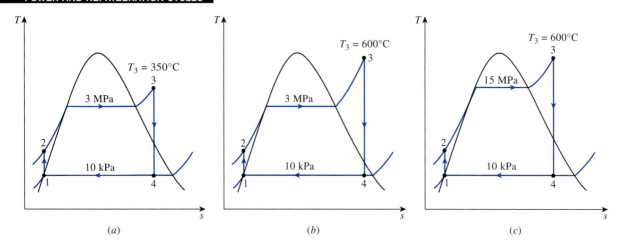

**FIGURE 9–47**
*T-s* diagrams of the three cycles discussed in Example 9–9.

*State 4:* $P_4 = 10$ kPa  (sat. mixture)

$$s_4 = s_3$$

$$x_4 = \frac{s_4 - s_f}{s_{fg}} = \frac{6.7450 - 0.6492}{7.4996} = 0.8128$$

Thus,

$$h_4 = h_f + x_4 h_{fg} = 191.81 + 0.8128(2392.1) = 2136.1 \text{ kJ/kg}$$

$$q_{in} = h_3 - h_2 = (3116.1 - 194.83) \text{ kJ/kg} = 2921.3 \text{ kJ/kg}$$

$$q_{out} = h_4 - h_1 = (2136.1 - 191.81) \text{ kJ/kg} = 1944.3 \text{ kJ/kg}$$

and

$$\eta_{th} = 1 - \frac{q_{out}}{q_{in}} = 1 - \frac{1944.3 \text{ kJ/kg}}{2921.3 \text{ kJ/kg}} = \textbf{0.334 or 33.4\%}$$

Therefore, the thermal efficiency increases from 26.0 to 33.4 percent as a result of lowering the condenser pressure from 75 to 10 kPa. At the same time, however, the quality of the steam decreases from 88.6 to 81.3 percent (in other words, the moisture content increases from 11.4 to 18.7 percent).

(*b*) States 1 and 2 remain the same in this case, and the enthalpies at state 3 (3 MPa and 600°C) and state 4 (10 kPa and $s_4 = s_3$) are determined to be

$$h_3 = 3682.8 \text{ kJ/kg}$$

$$h_4 = 2380.3 \text{ kJ/kg} \quad (x_4 = 0.915)$$

Thus,

$$q_{in} = h_3 - h_2 = 3682.8 - 194.83 = 3488.0 \text{ kJ/kg}$$

$$q_{out} = h_4 - h_1 = 2380.3 - 191.81 = 2188.5 \text{ kJ/kg}$$

and

$$\eta_{th} = 1 - \frac{q_{out}}{q_{in}} = 1 - \frac{2188.5 \text{ kJ/kg}}{3488.0 \text{ kJ/kg}} = \textbf{0.373 or 37.3\%}$$

Therefore, the thermal efficiency increases from 33.4 to 37.3 percent as a result of superheating the steam from 350 to 600°C. At the same time, the quality of the steam increases from 81.3 to 91.5 percent (in other words, the moisture content decreases from 18.7 to 8.5 percent).

(c) State 1 remains the same in this case, but the other states change. The enthalpies at state 2 (15 MPa and $s_2 = s_1$), state 3 (15 MPa and 600°C), and state 4 (10 kPa and $s_4 = s_3$) are determined in a similar manner to be

$$h_2 = 206.95 \text{ kJ/kg}$$

$$h_3 = 3583.1 \text{ kJ/kg}$$

$$h_4 = 2115.3 \text{ kJ/kg} \qquad (x_4 = 0.804)$$

Thus,

$$q_{in} = h_3 - h_2 = 3583.1 - 206.95 = 3376.2 \text{ kJ/kg}$$

$$q_{out} = h_4 - h_1 = 2115.3 - 191.81 = 1923.5 \text{ kJ/kg}$$

and

$$\eta_{th} = 1 - \frac{q_{out}}{q_{in}} = 1 - \frac{1923.5 \text{ kJ/kg}}{3376.2 \text{ kJ/kg}} = \textbf{0.430 or 43.0\%}$$

**Discussion** The thermal efficiency increases from 37.3 to 43.0 percent as a result of raising the boiler pressure from 3 to 15 MPa while maintaining the turbine inlet temperature at 600°C. At the same time, however, the quality of the steam decreases from 91.5 to 80.4 percent (in other words, the moisture content increases from 8.5 to 19.6 percent).

# 9–13 · THE IDEAL REHEAT RANKINE CYCLE

We noted in the last section that increasing the boiler pressure increases the thermal efficiency of the Rankine cycle, but it also increases the moisture content of the steam to unacceptable levels. Then it is natural to ask the following question:

*How can we take advantage of the increased efficiencies at higher boiler pressures without facing the problem of excessive moisture at the final stages of the turbine?*

Two possibilities come to mind:

**1.** Superheat the steam to very high temperatures before it enters the turbine. This would be the desirable solution since the average temperature at which heat is added would also increase, thus increasing the cycle efficiency. This is not a viable solution, however, since it requires raising the steam temperature to metallurgically unsafe levels.

**2.** Expand the steam in the turbine in two stages, and reheat it in between. In other words, modify the simple ideal Rankine cycle with a **reheat** process. Reheating is a practical solution to the excessive moisture problem in turbines, and it is commonly used in modern steam power plants.

**FIGURE 9–48**

The ideal reheat Rankine cycle.

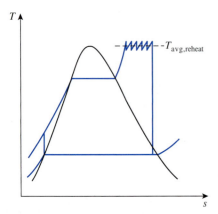

**FIGURE 9–49**

The average temperature at which heat is transferred during reheating increases as the number of reheat stages is increased.

The $T$-$s$ diagram of the ideal reheat Rankine cycle and the schematic of the power plant operating on this cycle are shown in Fig. 9–48. The ideal reheat Rankine cycle differs from the simple ideal Rankine cycle in that the expansion process takes place in two stages. In the first stage (the high-pressure turbine), steam is expanded isentropically to an intermediate pressure and sent back to the boiler where it is reheated at constant pressure, usually to the inlet temperature of the first turbine stage. Steam then expands isentropically in the second stage (low-pressure turbine) to the condenser pressure. Thus the total heat input and the total turbine work output for a reheat cycle become

$$q_{in} = q_{primary} + q_{reheat} = (h_3 - h_2) + (h_5 - h_4) \qquad (9\text{–}36)$$

and

$$w_{turb,out} = w_{turb,I} + w_{turb,II} = (h_3 - h_4) + (h_5 - h_6) \qquad (9\text{–}37)$$

The incorporation of the single reheat in a modern power plant improves the cycle efficiency by 4 to 5 percent by increasing the average temperature at which heat is transferred to the steam.

The average temperature during the reheat process can be increased by increasing the number of expansion and reheat stages. As the number of stages is increased, the expansion and reheat processes approach an isothermal process at the maximum temperature, as shown in Fig. 9–49. The use of more than two reheat stages, however, is not practical. The theoretical improvement in efficiency from the second reheat is about half of that which results from a single reheat. If the turbine inlet pressure is not high enough, double reheat would result in superheated exhaust. This is undesirable as it would cause the average temperature for heat rejection to increase and thus the cycle efficiency to decrease. Therefore, double reheat is used only on supercritical-pressure ($P > 22.06$ MPa) power plants. A third reheat stage would increase the cycle efficiency by about half of the improvement attained by the second reheat. This gain is too small to justify the added cost and complexity.

The reheat cycle was introduced in the mid-1920s, but it was abandoned in the 1930s because of the operational difficulties. The steady increase in boiler pressures over the years made it necessary to reintroduce single reheat in the late 1940s and double reheat in the early 1950s.

The reheat temperatures are very close or equal to the turbine inlet temperature. The optimum reheat pressure is about one-fourth of the maximum cycle pressure. For example, the optimum reheat pressure for a cycle with a boiler pressure of 12 MPa is about 3 MPa.

Remember that the sole purpose of the reheat cycle is to reduce the moisture content of the steam at the final stages of the expansion process. If we had materials that could withstand sufficiently high temperatures, there would be no need for the reheat cycle.

---

**EXAMPLE 9–10    The Ideal Reheat Rankine Cycle**

Consider a steam power plant operating on the ideal reheat Rankine cycle. Steam enters the high-pressure turbine at 15 MPa and 600°C and is condensed in the condenser at a pressure of 10 kPa. If the moisture content of the steam at the exit of the low-pressure turbine is not to exceed 10.4 percent, determine (a) the pressure at which the steam should be reheated and (b) the thermal efficiency of the cycle. Assume the steam is reheated to the inlet temperature of the high-pressure turbine.

**SOLUTION**   A steam power plant operating on the ideal reheat Rankine cycle is considered. For a specified moisture content at the turbine exit, the reheat pressure and the thermal efficiency are to be determined.

**Assumptions**  1 Steady operating conditions exist. 2 Kinetic and potential energy changes are negligible.

**Analysis**  The schematic of the power plant and the $T$-$s$ diagram of the cycle are shown in Fig. 9–50. We note that the power plant operates on the ideal reheat Rankine cycle. Therefore, the pump and the turbines are isentropic, there are no pressure drops in the boiler and condenser, and steam leaves the condenser and enters the pump as saturated liquid at the condenser pressure.

(a) The reheat pressure is determined from the requirement that the entropies at states 5 and 6 be the same:

*State 6:*     $P_6 = 10$ kPa

$x_6 = 0.896$   (sat. mixture)

$s_6 = s_f + x_6 s_{fg} = 0.6492 + 0.896(7.4996) = 7.3688$ kJ/kg·K

Also,

$h_6 = h_f + x_6 h_{fg} = 191.81 + 0.896(2392.1) = 2335.1$ kJ/kg

Thus,

*State 5:*     $T_5 = 600°C$ $\Big\}$  $P_5 = \mathbf{4.0\ MPa}$

$s_5 = s_6$        $h_5 = 3674.9$ kJ/kg

Therefore, steam should be reheated at a pressure of 4 MPa or lower to prevent a moisture content above 10.4 percent.

(b) To determine the thermal efficiency, we need to know the enthalpies at all other states:

*State 1:*     $P_1 = 10$ kPa $\Big\}$  $h_1 = h_{f@\ 10\ kPa} = 191.81$ kJ/kg

Sat. liquid        $v_1 = v_{f@\ 10\ kPa} = 0.00101$ m³/kg

**FIGURE 9–50**
Schematic and *T-s* diagram
for Example 9–10.

*State 2:*    $P_2 = 15$ MPa
$s_2 = s_1$

$$w_{\text{pump,in}} = v_1(P_2 - P_1) = (0.00101 \text{ m}^3/\text{kg})$$

$$\times [(15{,}000 - 10)\text{kPa}]\left(\frac{1 \text{ kJ}}{1 \text{ kPa}\cdot\text{m}^3}\right)$$

$$= 15.14 \text{ kJ/kg}$$

$$h_2 = h_1 + w_{\text{pump,in}} = (191.81 + 15.14) \text{ kJ/kg} = 206.95 \text{ kJ/kg}$$

*State 3:*    $\left.\begin{array}{l} P_3 = 15 \text{ MPa} \\ T_3 = 600°\text{C} \end{array}\right\}$    $\begin{array}{l} h_3 = 3583.1 \text{ kJ/kg} \\ s_3 = 6.6796 \text{ kJ/kg}\cdot\text{K} \end{array}$

*State 4:*    $\left.\begin{array}{l} P_4 = 4 \text{ MPa} \\ s_4 = s_3 \end{array}\right\}$    $\begin{array}{l} h_4 = 3155.0 \text{ kJ/kg} \\ (T_4 = 375.5°\text{C}) \end{array}$

Thus

$$q_{\text{in}} = (h_3 - h_2) + (h_5 - h_4)$$

$$= (3583.1 - 206.95) \text{ kJ/kg} + (3674.9 - 3155.0) \text{ kJ/kg}$$

$$= 3896.1 \text{ kJ/kg}$$

$$q_{\text{out}} = h_6 - h_1 = (2335.1 - 191.81) \text{ kJ/kg}$$

$$= 2143.3 \text{ kJ/kg}$$

and

$$\eta_{\text{th}} = 1 - \frac{q_{\text{out}}}{q_{\text{in}}} = 1 - \frac{2143.3 \text{ kJ/kg}}{3896.1 \text{ kJ/kg}} = \textbf{0.450 or 45.0\%}$$

*Discussion* This problem was solved in Example 9–9c for the same pressure and temperature limits but without the reheat process. A comparison of the two results reveals that reheating reduces the moisture content from 19.6 to 10.4 percent while increasing the thermal efficiency from 43.0 to 45.0 percent.

# 9–14 ▪ REFRIGERATORS AND HEAT PUMPS

We all know from experience that heat flows in the direction of decreasing temperature, that is, from high-temperature regions to low-temperature ones. This heat-transfer process occurs in nature without requiring any devices. The reverse process, however, cannot occur by itself. The transfer of heat from a low-temperature region to a high-temperature one requires special devices called **refrigerators**.

Refrigerators are cyclic devices, and the working fluids used in the refrigeration cycles are called **refrigerants**. A refrigerator is shown schematically in Fig. 9–51a. Here $Q_L$ is the magnitude of the heat removed from the refrigerated space at temperature $T_L$, $Q_H$ is the magnitude of the heat rejected to the warm space at temperature $T_H$, and $W_{net,in}$ is the net work input to the refrigerator. As discussed in Chap. 7, $Q_L$ and $Q_H$ represent magnitudes and thus are positive quantities.

Another device that transfers heat from a low-temperature medium to a high-temperature one is the **heat pump**. Refrigerators and heat pumps are essentially the same devices; they differ in their objectives only. The objective of a refrigerator is to maintain the refrigerated space at a low temperature by removing heat from it. Discharging this heat to a higher-temperature medium is merely a necessary part of the operation, not the purpose. The objective of a heat pump, however, is to maintain a heated space at a high temperature. This is accomplished by absorbing heat from a low-temperature source, such as well water or cold outside air in winter, and supplying this heat to a warmer medium such as a house (Fig. 9–51b).

The performance of refrigerators and heat pumps is expressed in terms of the **coefficient of performance** (COP), defined as

$$COP_R = \frac{\text{Desired output}}{\text{Required input}} = \frac{\text{Cooling effect}}{\text{Work input}} = \frac{Q_L}{W_{net,in}} \tag{9–38}$$

$$COP_{HP} = \frac{\text{Desired output}}{\text{Required input}} = \frac{\text{Heating effect}}{\text{Work input}} = \frac{Q_H}{W_{net,in}} \tag{9–38}$$

These relations can also be expressed in the rate form by replacing the quantities $Q_L$, $Q_H$, and $W_{net,in}$ by $\dot{Q}_L$, $\dot{Q}_H$, and $\dot{W}_{net,in}$, respectively. Notice that both $COP_R$ and $COP_{HP}$ can be greater than 1. A comparison of Eqs. 9–38 and 9–39 reveals that

$$COP_{HP} = COP_R + 1 \tag{9–40}$$

for fixed values of $Q_L$ and $Q_H$. This relation implies that $COP_{HP} > 1$ since $COP_R$ is a positive quantity. That is, a heat pump functions, at worst, as a resistance heater, supplying as much energy to the house as it consumes. In reality, however, part of $Q_H$ is lost to the outside air through piping and other devices, and $COP_{HP}$ may drop below unity when the outside air temperature is too low. When this happens, the system normally switches to the fuel (natural gas, propane, oil, etc.) or resistance-heating mode.

The *cooling capacity* of a refrigeration system—that is, the rate of heat removal from the refrigerated space—is often expressed in terms of **tons of refrigeration**. The capacity of a refrigeration system that can freeze 1 ton

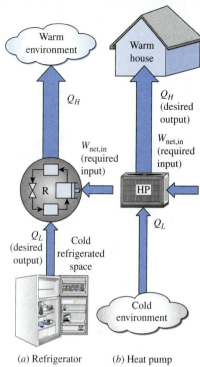

(a) Refrigerator    (b) Heat pump

**FIGURE 9–51**

The objective of a refrigerator is to remove heat ($Q_L$) from the cold medium; the objective of a heat pump is to supply heat ($Q_H$) to a warm medium.

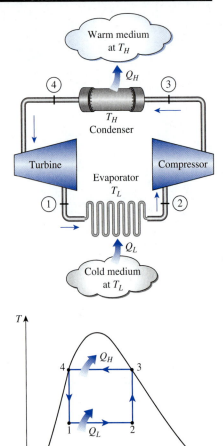

**FIGURE 9–52**

Schematic of a Carnot refrigerator and *T-s* diagram of the reversed Carnot cycle.

(2000 lbm) of liquid water at 0°C (32°F) into ice at 0°C in 24 h is said to be 1 ton. One ton of refrigeration is equivalent to 211 kJ/min or 200 Btu/min. The cooling load of a typical 200-m² residence is in the 3-ton (10-kW) range.

## 9–15 · THE REVERSED CARNOT CYCLE

Recall from Chap. 7 that the Carnot cycle is a totally reversible cycle that consists of two reversible isothermal and two isentropic processes. It has the maximum thermal efficiency for given temperature limits, and it serves as a standard against which actual power cycles can be compared.

Since it is a reversible cycle, all four processes that comprise the Carnot cycle can be reversed. Reversing the cycle does also reverse the directions of any heat and work interactions. The result is a cycle that operates in the counterclockwise direction on a *T-s* diagram, which is called the **reversed Carnot cycle**. A refrigerator or heat pump that operates on the reversed Carnot cycle is called a **Carnot refrigerator** or a **Carnot heat pump**.

Consider a reversed Carnot cycle executed within the saturation dome of a refrigerant, as shown in Fig. 9–52. The refrigerant absorbs heat isothermally from a low-temperature source at $T_L$ in the amount of $Q_L$ (process 1-2), is compressed isentropically to state 3 (temperature rises to $T_H$), rejects heat isothermally to a high-temperature sink at $T_H$ in the amount of $Q_H$ (process 3-4), and expands isentropically to state 1 (temperature drops to $T_L$). The refrigerant changes from a saturated vapor state to a saturated liquid state in the condenser during process 3-4.

The coefficients of performance of Carnot refrigerators and heat pumps are expressed in terms of temperatures as

$$\text{COP}_{R,\text{Carnot}} = \frac{1}{T_H/T_L - 1} \qquad (9\text{–}41)$$

and

$$\text{COP}_{HP,\text{Carnot}} = \frac{1}{1 - T_L/T_H} \qquad (9\text{–}42)$$

Notice that both COPs increase as the difference between the two temperatures decreases, that is, as $T_L$ rises or $T_H$ falls.

The reversed Carnot cycle is the *most efficient* refrigeration cycle operating between two specified temperature levels. Therefore, it is natural to look at it first as a prospective ideal cycle for refrigerators and heat pumps. If we could, we certainly would adapt it as the ideal cycle. As explained below, however, the reversed Carnot cycle is not a suitable model for refrigeration cycles.

The two isothermal heat transfer processes are not difficult to achieve in practice since maintaining a constant pressure automatically fixes the temperature of a two-phase mixture at the saturation value. Therefore, processes 1-2 and 3-4 can be approached closely in actual evaporators and condensers. However, processes 2-3 and 4-1 cannot be approximated closely in practice. This is because process 2-3 involves the compression of a liquid–vapor mixture, which requires a compressor that will handle two phases, and process 4-1 involves the expansion of high-moisture-content refrigerant in a turbine.

It seems as if these problems could be eliminated by executing the reversed Carnot cycle outside the saturation region. But in this case we have difficulty in maintaining isothermal conditions during the heat-absorption and heat-rejection processes. Therefore, we conclude that the reversed Carnot cycle cannot be approximated in actual devices and is not a realistic model for refrigeration cycles. However, the reversed Carnot cycle can serve as a standard against which actual refrigeration cycles are compared.

# 9–16 • THE IDEAL VAPOR-COMPRESSION REFRIGERATION CYCLE

Many of the impracticalities associated with the reversed Carnot cycle can be eliminated by vaporizing the refrigerant completely before it is compressed and by replacing the turbine with a throttling device, such as an expansion valve or capillary tube. The cycle that results is called the **ideal vapor-compression refrigeration cycle**, and it is shown schematically and on a *T-s* diagram in Fig. 9–53. The vapor-compression refrigeration cycle is the most widely used cycle for refrigerators, air-conditioning systems, and heat pumps. It consists of four processes:

1-2   Isentropic compression in a compressor
2-3   Constant-pressure heat rejection in a condenser
3-4   Throttling in an expansion device
4-1   Constant-pressure heat absorption in an evaporator

In an ideal vapor-compression refrigeration cycle, the refrigerant enters the compressor at state 1 as saturated vapor and is compressed isentropically to the condenser pressure. The temperature of the refrigerant increases during this isentropic compression process to well above the temperature of the surrounding medium. The refrigerant then enters the condenser as superheated vapor at state 2 and leaves as saturated liquid at state 3 as a result of heat rejection to the surroundings. The temperature of the refrigerant at this state is still above the temperature of the surroundings.

The saturated liquid refrigerant at state 3 is throttled to the evaporator pressure by passing it through an expansion valve or capillary tube. The temperature of the refrigerant drops below the temperature of the refrigerated space during this process. The refrigerant enters the evaporator at state 4 as a low-quality saturated mixture, and it completely evaporates by absorbing heat from the refrigerated space. The refrigerant leaves the evaporator as saturated vapor and reenters the compressor, completing the cycle.

In a household refrigerator, the tubes in the freezer compartment where heat is absorbed by the refrigerant serves as the evaporator. The coils behind the refrigerator, where heat is dissipated to the kitchen air, serve as the condenser (Fig. 9–54).

Remember that the area under the process curve on a *T-s* diagram represents the heat transfer for internally reversible processes. The area under the process curve 4-1 represents the heat absorbed by the refrigerant in the evaporator, and the area under the process curve 2-3 represents the heat rejected in the condenser. A rule of thumb is that the *COP improves by 2 to 4 percent for each °C the evaporating temperature is raised or the condensing temperature is lowered.*

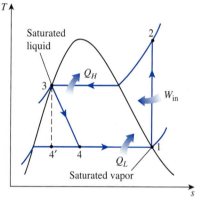

**FIGURE 9–53**

Schematic and *T-s* diagram for the ideal vapor-compression refrigeration cycle.

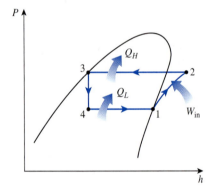

**FIGURE 9–54**

An ordinary household refrigerator.

**FIGURE 9–55**

The *P-h* diagram of an ideal vapor-compression refrigeration cycle.

Another diagram frequently used in the analysis of vapor-compression refrigeration cycles is the *P-h* diagram, as shown in Fig. 9–55. On this diagram, three of the four processes appear as straight lines, and the heat transfer in the condenser and the evaporator is proportional to the lengths of the corresponding process curves.

Notice that unlike the ideal cycles discussed before, the ideal vapor-compression refrigeration cycle is not an internally reversible cycle since it involves an irreversible (throttling) process. This process is maintained in the cycle to make it a more realistic model for the actual vapor-compression refrigeration cycle. If the throttling device were replaced by an isentropic turbine, the refrigerant would enter the evaporator at state 4′ instead of state 4. As a result, the refrigeration capacity would increase (by the area under process curve 4′-4 in Fig. 9–53) and the net work input would decrease (by the amount of work output of the turbine). Replacing the expansion valve by a turbine is not practical, however, since the added benefits cannot justify the added cost and complexity.

All four components associated with the vapor-compression refrigeration cycle are steady-flow devices, and thus all four processes that make up the cycle can be analyzed as steady-flow processes. The kinetic and potential energy changes of the refrigerant are usually small relative to the work and heat transfer terms, and therefore they can be neglected. Then the steady-flow energy equation on a unit–mass basis reduces to

$$(q_{\text{in}} - q_{\text{out}}) + (w_{\text{in}} - w_{\text{out}}) = h_e - h_i \tag{9–43}$$

The condenser and the evaporator do not involve any work, and the compressor can be approximated as adiabatic. Then the COPs of refrigerators and heat pumps operating on the vapor-compression refrigeration cycle can be expressed as

$$\text{COP}_R = \frac{q_L}{w_{\text{net,in}}} = \frac{h_1 - h_4}{h_2 - h_1} \tag{9–44}$$

and

$$\text{COP}_{HP} = \frac{q_H}{w_{\text{net,in}}} = \frac{h_2 - h_3}{h_2 - h_1} \tag{9–45}$$

where $h_1 = h_{g \, @ \, P_1}$ and $h_3 = h_{f \, @ \, P_3}$ for the ideal case.

Vapor-compression refrigeration dates back to 1834 when the Englishman Jacob Perkins received a patent for a closed-cycle ice machine using ether or other volatile fluids as refrigerants. A working model of this machine was built, but it was never produced commercially. In 1850, Alexander Twining began to design and build vapor-compression ice machines using ethyl ether, which is a commercially used refrigerant in vapor-compression systems. Initially, vapor-compression refrigeration systems were large and were mainly used for ice making, brewing, and cold storage. They lacked automatic controls and were steam-engine driven. In the 1890s, electric motor-driven smaller machines equipped with automatic controls started to replace the older units, and refrigeration systems began to appear in butcher shops and households. By 1930, the continued improvements made it possible to have vapor-compression refrigeration systems that were relatively efficient, reliable, small, and inexpensive.

## EXAMPLE 9–11　　The Ideal Vapor-Compression Refrigeration Cycle

A refrigerator uses refrigerant-134a as the working fluid and operates on an ideal vapor-compression refrigeration cycle between 0.14 and 0.8 MPa. If the mass flow rate of the refrigerant is 0.05 kg/s, determine (a) the rate of heat removal from the refrigerated space and the power input to the compressor, (b) the rate of heat rejection to the environment, and (c) the COP of the refrigerator.

**SOLUTION** A refrigerator operates on an ideal vapor-compression refrigeration cycle between two specified pressure limits. The rate of refrigeration, the power input, the rate of heat rejection, and the COP are to be determined.

**Assumptions** 1 Steady operating conditions exist. 2 Kinetic and potential energy changes are negligible.

**Analysis** The $T$-$s$ diagram of the refrigeration cycle is shown in Fig. 9–56. We note that this is an ideal vapor-compression refrigeration cycle, and thus the compressor is isentropic and the refrigerant leaves the condenser as a saturated liquid and enters the compressor as saturated vapor. From the refrigerant-134a tables, the enthalpies of the refrigerant at all four states are determined as follows:

$$P_1 = 0.14 \text{ MPa} \longrightarrow h_1 = h_{g @ 0.14 \text{ MPa}} = 239.16 \text{ kJ/kg}$$

$$s_1 = s_{g @ 0.14 \text{ MPa}} = 0.94456 \text{ kJ/kg·K}$$

$$\left.\begin{array}{l} P_2 = 0.8 \text{ MPa} \\ s_2 = s_1 \end{array}\right\} \quad h_2 = 275.39 \text{ kJ/kg}$$

$$P_3 = 0.8 \text{ MPa} \longrightarrow h_3 = h_{f @ 0.8 \text{ MPa}} = 95.47 \text{ kJ/kg}$$

$$h_4 \cong h_3 \text{ (throttling)} \longrightarrow h_4 = 95.47 \text{ kJ/kg}$$

(a) The rate of heat removal from the refrigerated space and the power input to the compressor are determined from their definitions:

$$\dot{Q}_L = \dot{m}(h_1 - h_4) = (0.05 \text{ kg/s})[(239.16 - 95.47) \text{ kJ/kg}] = \textbf{7.18 kW}$$

and

$$\dot{W}_{in} = \dot{m}(h_2 - h_1) = (0.05 \text{ kg/s})[(275.39 - 239.16) \text{ kJ/kg}] = \textbf{1.81 kW}$$

(b) The rate of heat rejection from the refrigerant to the environment is

$$\dot{Q}_H = \dot{m}(h_2 - h_3) = (0.05 \text{ kg/s})[(275.39 - 95.47) \text{ kJ/kg}] = \textbf{9.0 kW}$$

It could also be determined from

$$\dot{Q}_H = \dot{Q}_L + \dot{W}_{in} = 7.18 + 1.81 = 8.99 \text{ kW}$$

(c) The coefficient of performance of the refrigerator is

$$\text{COP}_R = \frac{\dot{Q}_L}{\dot{W}_{in}} = \frac{7.18 \text{ kW}}{1.81 \text{ kW}} = \textbf{3.97}$$

That is, this refrigerator removes about 4 units of thermal energy from the refrigerated space for each unit of electric energy it consumes.

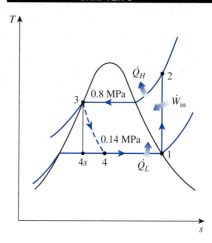

**FIGURE 9–56**

$T$-$s$ diagram of the ideal vapor-compression refrigeration cycle described in Example 9–11.

*Discussion* It would be interesting to see what happens if the throttling valve were replaced by an isentropic turbine. The enthalpy at state 4s (the turbine exit with $P_{4s} = 0.14$ MPa, and $s_{4s} = s_3 = 0.35404$ kJ/kg·K) is 88.94 kJ/kg, and the turbine would produce 0.33 kW of power. This would decrease the power input to the refrigerator from 1.81 to 1.48 kW and increase the rate of heat removal from the refrigerated space from 7.18 to 7.51 kW. As a result, the COP of the refrigerator would increase from 3.97 to 5.07, an increase of 28 percent.

## 9–17 • ACTUAL VAPOR-COMPRESSION REFRIGERATION CYCLE

An actual vapor-compression refrigeration cycle differs from the ideal one in several ways, owing mostly to the irreversibilities that occur in various components. Two common sources of irreversibilities are fluid friction (causes pressure drops) and heat transfer to or from the surroundings. The *T-s* diagram of an actual vapor-compression refrigeration cycle is shown in Fig. 9–57.

In the ideal cycle, the refrigerant leaves the evaporator and enters the compressor as *saturated vapor*. In practice, however, it may not be possible to control the state of the refrigerant so precisely. Instead, it is easier to design the system so that the refrigerant is slightly superheated at the compressor inlet. This slight overdesign ensures that the refrigerant is completely vaporized when it enters the compressor. Also, the line connecting the evaporator to the compressor is usually very long; thus the pressure drop caused by fluid friction and heat transfer from the surroundings to the refrigerant can be very significant. The result of superheating, heat gain in the connecting line, and pressure drops in the evaporator and the connecting line is an increase in the specific volume, thus an increase in the power input requirements to the compressor since steady-flow work is proportional to the specific volume.

The *compression process* in the ideal cycle is internally reversible and adiabatic, and thus isentropic. The actual compression process, however, involves frictional effects, which increase the entropy, and heat transfer, which may increase or decrease the entropy, depending on the direction. Therefore, the entropy of the refrigerant may increase (process 1-2) or decrease (process 1-2′) during an actual compression process, depending on which effects dominate. The compression process 1-2′ may be even more desirable than the isentropic compression process since the specific volume of the refrigerant and thus the work input requirement are smaller in this case. Therefore, the refrigerant should be cooled during the compression process whenever it is practical and economical to do so.

In the ideal case, the refrigerant is assumed to leave the condenser as *saturated liquid* at the compressor exit pressure. In reality, however, it is unavoidable to have some pressure drop in the condenser as well as in the lines connecting the condenser to the compressor and to the throttling valve. Also, it is not easy to execute the condensation process with such precision

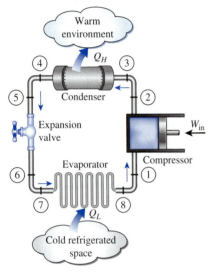

**FIGURE 9–57**

Schematic and *T-s* diagram for the actual vapor-compression refrigeration cycle.

that the refrigerant is a saturated liquid at the end, and it is undesirable to route the refrigerant to the throttling valve before the refrigerant is completely condensed. Therefore, the refrigerant is subcooled somewhat before it enters the throttling valve. We do not mind this at all, however, since the refrigerant in this case enters the evaporator with a lower enthalpy and thus can absorb more heat from the refrigerated space. The throttling valve and the evaporator are usually located very close to each other, so the pressure drop in the connecting line is small.

---

**EXAMPLE 9–12** **The Actual Vapor-Compression Refrigeration Cycle**

Refrigerant-134a enters the compressor of a refrigerator as superheated vapor at 0.14 MPa and $-10°C$ at a rate of 0.05 kg/s and leaves at 0.8 MPa and 50°C. The refrigerant is cooled in the condenser to 26°C and 0.72 MPa and is throttled to 0.15 MPa. Disregarding any heat transfer and pressure drops in the connecting lines between the components, determine (*a*) the rate of heat removal from the refrigerated space and the power input to the compressor, (*b*) the isentropic efficiency of the compressor, and (*c*) the coefficient of performance of the refrigerator.

**SOLUTION** A refrigerator operating on a vapor-compression cycle is considered. The rate of refrigeration, the power input, the compressor efficiency, and the COP are to be determined.
*Assumptions* **1** Steady operating conditions exist. **2** Kinetic and potential energy changes are negligible.
*Analysis* The *T-s* diagram of the refrigeration cycle is shown in Fig. 9–58. We note that the refrigerant leaves the condenser as a compressed liquid and enters the compressor as superheated vapor. The enthalpies of the refrigerant at various states are determined from the refrigerant tables to be

$$P_1 = 0.14 \text{ MPa} \atop T_1 = -10°C \Bigg\} \quad h_1 = 246.36 \text{ kJ/kg}$$

$$P_2 = 0.8 \text{ MPa} \atop T_2 = 50°C \Bigg\} \quad h_2 = 286.69 \text{ kJ/kg}$$

$$P_3 = 0.72 \text{ MPa} \atop T_3 = 26°C \Bigg\} \quad h_3 \cong h_{f @ 26°C} = 87.83 \text{ kJ/kg}$$

$$h_4 \cong h_3 \text{ (throttling)} \quad \longrightarrow \quad h_4 = 87.83 \text{ kJ/kg}$$

(*a*) The rate of heat removal from the refrigerated space and the power input to the compressor are determined from their definitions:

$$\dot{Q}_L = \dot{m}(h_1 - h_4) = (0.05 \text{ kg/s})[(246.36 - 87.83) \text{ kJ/kg}] = \textbf{7.93 kW}$$

and

$$\dot{W}_{in} = \dot{m}(h_2 - h_1) = (0.05 \text{ kg/s})[(286.69 - 246.36) \text{ kJ/kg}] = \textbf{2.02 kW}$$

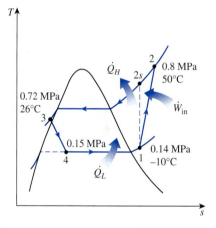

**FIGURE 9–58**
*T-s* diagram for Example 9–12.

(b) The isentropic efficiency of the compressor is determined from

$$\eta_C \cong \frac{h_{2s} - h_1}{h_2 - h_1}$$

where the enthalpy at state 2s ($P_{2s} = 0.8$ MPa and $s_{2s} = s_1 = 0.9724$ kJ/kg·K) is 284.21 kJ/kg. Thus,

$$\eta_C = \frac{284.21 - 246.36}{286.69 - 246.36} = \mathbf{0.939 \text{ or } 93.9\%}$$

(c) The coefficient of performance of the refrigerator is

$$COP_R = \frac{\dot{Q}_L}{\dot{W}_{in}} = \frac{7.93 \text{ kW}}{2.02 \text{ kW}} = \mathbf{3.93}$$

*Discussion* This problem is identical to the one worked out in Example 9–11, except that the refrigerant is slightly superheated at the compressor inlet and subcooled at the condenser exit. Also, the compressor is not isentropic. As a result, the heat removal rate from the refrigerated space increases (by 10.4 percent), but the power input to the compressor increases even more (by 11.6 percent). Consequently, the COP of the refrigerator decreases from 3.97 to 3.93.

# SUMMARY

The most efficient cycle operating between a heat source at temperature $T_H$ and a sink at temperature $T_L$ is the Carnot cycle, and its thermal efficiency is given by

$$\eta_{th, Carnot} = 1 - \frac{T_L}{T_H}$$

The actual gas cycles are rather complex. The approximations used to simplify the analysis are known as the *air-standard assumptions*. Under these assumptions, all the processes are assumed to be internally reversible; the working fluid is assumed to be air, which behaves as an ideal gas; and the combustion and exhaust processes are replaced by heat-addition and heat-rejection processes, respectively. The air-standard assumptions are called *cold-air-standard assumptions* if, in addition, air is assumed to have constant specific heats at room temperature.

In reciprocating engines, the *compression ratio r* and the *mean effective pressure* MEP are defined as

$$r = \frac{V_{max}}{V_{min}} = \frac{V_{BDC}}{V_{TDC}}$$

$$MEP = \frac{w_{net}}{V_{max} - V_{min}}$$

The *Otto cycle* is the ideal cycle for the spark-ignition reciprocating engines, and it consists of four internally reversible processes: isentropic compression, constant-volume

heat addition, isentropic expansion, and constant-volume rejection. Under cold-air-standard assumptions, the thermal efficiency of the ideal Otto cycle is

$$\eta_{th, Otto} = 1 - \frac{1}{r^{k-1}}$$

where $r$ is the compression ratio and $k$ is the specific heat ratio $c_p/c_v$.

The *Diesel cycle* is the ideal cycle for the compression-ignition reciprocating engines. It is very similar to the Otto cycle, except that the constant-volume heat-addition process is replaced by a constant-pressure heat-addition process. Its thermal efficiency under cold-air-standard assumptions is

$$\eta_{th, Diesel} = 1 - \frac{1}{r^{k-1}} \left[ \frac{r_c^k - 1}{k(r_c - 1)} \right]$$

where $r_c$ is the *cutoff ratio,* defined as the ratio of the cylinder volumes after and before the combustion process.

The ideal cycle for modern gas-turbine engines is the *Brayton cycle,* which is made up of four internally reversible processes: isentropic compression, constant-pressure heat addition, isentropic expansion, and constant-pressure heat rejection. Under cold-air-standard assumptions, its thermal efficiency is

$$\eta_{th, Brayton} = 1 - \frac{1}{r_p^{(k-1)/k}}$$

where $r_p = P_{max}/P_{min}$ is the pressure ratio and $k$ is the specific heat ratio. The thermal efficiency of the simple Brayton cycle increases with the pressure ratio.

The deviation of the actual compressor and the turbine from the idealized isentropic ones can be accurately accounted for by utilizing their isentropic efficiencies, defined as

$$\eta_C = \frac{w_s}{w_a} \cong \frac{h_{2s} - h_1}{h_{2a} - h_1}$$

and

$$\eta_T = \frac{w_a}{w_s} \cong \frac{h_3 - h_{4a}}{h_3 - h_{4s}}$$

where states 1 and 3 are the inlet states, $2a$ and $4a$ are the actual exit states, and $2s$ and $4s$ are the isentropic exit states.

In gas-turbine engines, the temperature of the exhaust gas leaving the turbine is often considerably higher than the temperature of the air leaving the compressor. Therefore, the high-pressure air leaving the compressor can be heated by transferring heat to it from the hot exhaust gases in a counter-flow heat exchanger, which is also known as a *regenerator.* The extent to which a regenerator approaches an ideal regenerator is called the *effectiveness* $\varepsilon$ and is defined as

$$\varepsilon = \frac{q_{regen,\ act}}{q_{regen,\ max}}$$

Under cold-air-standard assumptions, the thermal efficiency of an ideal Brayton cycle with regeneration becomes

$$\eta_{th,\ regen} = 1 - \left(\frac{T_1}{T_3}\right)(r_p)^{(k-1)/k}$$

where $T_1$ and $T_3$ are the minimum and maximum temperatures, respectively, in the cycle.

The *Carnot cycle* is not a suitable model for vapor power cycles because it cannot be approximated in practice. The model cycle for vapor power cycles is the *Rankine cycle,* which is composed of four internally reversible processes: constant-pressure heat addition in a boiler, isentropic expansion in a turbine, constant-pressure heat rejection in a condenser, and isentropic compression in a pump. Steam leaves the condenser as a saturated liquid at the condenser pressure.

The thermal efficiency of the Rankine cycle can be increased by increasing the average temperature at which heat is added to the working fluid and/or by decreasing the average temperature at which heat is rejected to the cooling medium. The average temperature during heat rejection can be decreased by lowering the turbine exit pressure. Consequently, the condenser pressure of most vapor power plants is well below the atmospheric pressure. The average temperature during heat addition can be increased by raising the boiler pressure or by superheating the fluid to high temperatures. There is a limit to the degree of superheating, however, since the fluid temperature is not allowed to exceed a metallurgically safe value.

Superheating has the added advantage of decreasing the moisture content of the steam at the turbine exit. Lowering the exhaust pressure or raising the boiler pressure, however, increases the moisture content. To take advantage of the improved efficiencies at higher boiler pressures and lower condenser pressures, steam is usually *reheated* after expanding partially in the high-pressure turbine. This is done by extracting the steam after partial expansion in the high-pressure turbine, sending it back to the boiler where it is reheated at constant pressure, and returning it to the low-pressure turbine for complete expansion to the condenser pressure. The average temperature during the reheat process, and thus the thermal efficiency of the cycle, can be increased by increasing the number of expansion and reheat stages. As the number of stages is increased, the expansion and reheat processes approach an isothermal process of expansion at maximum temperature. Reheating also decreases the moisture content at the turbine exit.

The transfer of heat from lower-temperature regions to higher-temperature ones is called *refrigeration.* Devices that produce refrigeration are called *refrigerators,* and the cycles on which they operate are called *refrigeration cycles.* The working fluids used in refrigerators are called *refrigerants.* Refrigerators used for the purpose of heating a space by transferring heat from a cooler medium are called *heat pumps.*

The performance of refrigerators and heat pumps is expressed in terms of *coefficient of performance* (COP), defined as

$$COP_R = \frac{\text{Desired output}}{\text{Required output}} = \frac{\text{Cooling effect}}{\text{Work input}} = \frac{Q_L}{W_{net,\ in}}$$

$$COP_{HP} = \frac{\text{Desired output}}{\text{Required output}} = \frac{\text{Heating effect}}{\text{Work input}} = \frac{Q_H}{W_{net,\ in}}$$

The standard of comparison for refrigeration cycles is the *reversed Carnot cycle.* A refrigerator or heat pump that operates on the reversed Carnot cycle is called a *Carnot refrigerator* or a *Carnot heat pump,* and their COPs are

$$COP_{R,\ Carnot} = \frac{1}{T_H/T_L - 1}$$

$$COP_{HP,\ Carnot} = \frac{1}{1 - T_L/T_H}$$

The most widely used refrigeration cycle is the *vapor-compression refrigeration cycle.* In an ideal vapor-compression refrigeration cycle, the refrigerant enters the compressor as a saturated vapor and is cooled to the saturated liquid state in the condenser. It is then throttled to the evaporator pressure and vaporizes as it absorbs heat from the refrigerated space.

## REFERENCES AND SUGGESTED READINGS

1. *ASHRAE, Handbook of Fundamentals.* Atlanta: American Society of Heating, Refrigerating, and Air-Conditioning Engineers, 1985.

2. R. L. Bannister and G. J. Silvestri. "The Evolution of Central Station Steam Turbines." *Mechanical Engineering,* February 1989, pp. 70–78.

3. R. L. Bannister, G. J. Silvestri, A. Hizume, and T. Fujikawa. "High Temperature Supercritical Steam Turbines." *Mechanical Engineering,* February 1987, pp. 60–65.

4. Y. A. Çengel and M. A. Boles. *Thermodynamics: An Engineering Approach.* 7th ed. New York: McGraw-Hill, 2011.

5. M. M. El-Wakil. *Powerplant Technology.* New York: McGraw-Hill, 1984.

6. *Heat Pump Systems—A Technology Review.* OECD Report, Paris, 1982.

7. L. C. Lichty. *Combustion Engine Processes.* New York: McGraw-Hill, 1967.

8. W. Siuru. "Two-stroke Engines: Cleaner and Meaner." *Mechanical Engineering,* June 1990, pp. 66–69.

9. H. Sorensen. *Energy Conversion Systems.* New York: John Wiley & Sons, 1983.

10. *Steam, Its Generation and Use.* 39th ed. New York: Babcock and Wilcox Co., 1978.

11. W. F. Stoecker and J. W. Jones. *Refrigeration and Air Conditioning.* 2nd ed. New York: McGraw-Hill, 1982.

12. C. F. Taylor. *The Internal Combustion Engine in Theory and Practice.* Cambridge, MA: M.I.T. Press, 1968.

13. J. Weisman and R. Eckart. *Modern Power Plant Engineering.* Englewood Cliffs, NJ: Prentice Hall, 1985.

## PROBLEMS*

### Actual and Ideal Cycles, Carnot Cycle, Air-Standard Assumptions, Reciprocating Engines

**9–1C** How does the thermal efficiency of an ideal cycle, in general, compare to that of a Carnot cycle operating between the same temperature limits?

**9–2C** What does the area enclosed by the cycle represent on a $P$-$v$ diagram? How about on a $T$-$s$ diagram?

**9–3C** What are the air-standard assumptions?

**9–4C** What is the difference between air-standard assumptions and the cold-air-standard assumptions?

**9–5C** What is the difference between the clearance volume and the displacement volume of reciprocating engines?

**9–6C** Define the compression ratio for reciprocating engines.

**9–7C** How is the mean effective pressure for reciprocating engines defined?

**9–8C** Can the mean effective pressure of an automobile engine in operation be less than the atmospheric pressure?

**9–9C** As a car gets older, will its compression ratio change? How about the mean effective pressure?

**9–10C** What is the difference between spark-ignition and compression-ignition engines?

**9–11C** Define the following terms related to reciprocating engines: stroke, bore, top dead center, and clearance volume.

**9–12E** What is the maximum possible thermal efficiency of a gas power cycle when using thermal energy reservoirs at 940°F and 40°F?

**9–13** An air-standard cycle with constant specific heats is executed in a closed piston-cylinder system and is composed of the following three processes:

1.2 Isentropic compression with a compression ratio $r = 6$, $T_1 = 27°C$, and $P_1 = 100$ kPa

2-3 Constant pressure heat addition

3-1 Constant volume heat rejection

Assume air has constant properties with $c_v = 0.718$ kJ/kg·K, $c_p = 1.005$ kJ/kg·K, $R = 0.287$ kJ/kg·K, and $k = 1.4$.

(*a*) Sketch the $P$-$v$ and $T$-$s$ diagrams for this cycle.

(*b*) Determine the back work ratio for this cycle.

**9–14** An air-standard cycle is executed within a closed piston-cylinder system and consists of three processes as follows:

1-2 $V = constant$ heat addition from 100 kPa and 27°C to 700 kPa

2-3 Isothermal expansion until $V_3 = 7V_2$

3-1 $P = constant$ heat rejection to the initial state

Assume air has constant properties with $c_v = 0.718$ kJ/kg·K, $c_p = 1.005$ kJ/kg·K, $R = 0.287$ kJ/kg·K, and $k = 1.4$.

---

* Problems designated by a "C" are concept questions, and students are encouraged to answer them all. Problems designated by an "E" are in English units, and the SI users can ignore them. Problems with the 🖲 icon are solved using EES, and complete solutions together with parametric studies are included on the enclosed DVD. Problems with the 🖳 icon are comprehensive in nature, and are intended to be solved with a computer, preferably using the EES software that accompanies this text.

(a) Sketch the P-$v$ and T-s diagrams for the cycle.

(b) Determine the ratio of the compression work to the expansion work (the back work ratio).

(c) Determine the cycle thermal efficiency.

*Answers: (b) 0.440, (c) 26.6 percent*

**9–15** An air-standard cycle with variable specific heats is executed in a closed system and is composed of the following four processes:

1-2 Isentropic compression from 100 kPa and 22°C to 600 kPa

2-3 $v$ = *constant* heat addition to 1500 K

3-4 Isentropic expansion to 100 kPa

4-1 $P$ = *constant* heat rejection to initial state

(a) Show the cycle on P-$v$ and T-s diagrams.

(b) Calculate the net work output per unit mass.

(c) Determine the thermal efficiency.

**9–16** [EES] Reconsider Prob. 9–15. Using EES (or other) software, study the effect of varying the temperature after the constant-volume heat addition from 1500 K to 2500 K. Plot the net work output and thermal efficiency as a function of the maximum temperature of the cycle. Plot the T-s and P-$v$ diagrams for the cycle when the maximum temperature of the cycle is 1500 K.

**9–17** An air-standard cycle is executed in a closed system with 0.004 kg of air and consists of the following three processes:

1-2 Isentropic compression from 100 kPa and 27°C to 1 MPa

2-3 $P$ = *constant* heat addition in the amount of 2.76 kJ

3-1 $P$ = $c_1 v$ + $c_2$ heat rejection to initial state ($c_1$ and $c_2$ are constants)

(a) Show the cycle on P-$v$ and T-s diagrams.

(b) Calculate the heat rejected.

(c) Determine the thermal efficiency.

Assume constant specific heats at room temperature.
*Answers: (b) 1.679 kJ, (c) 39.2 percent*

**9–18E** An air-standard cycle with variable specific heats is executed in a closed system and is composed of the following four processes:

1-2 $v$ = *constant* heat addition from 14.7 psia and 80°F in the amount of 300 Btu/lbm

2-3 $P$ = *constant* heat addition to 3200 R

3-4 Isentropic expansion to 14.7 psia

4-1 $P$ = *constant* heat rejection to initial state

(a) Show the cycle on P-$v$ and T-s diagrams.

(b) Calculate the total heat input per unit mass.

(c) Determine the thermal efficiency.

*Answers: (b) 612.4 Btu/lbm, (c) 24.2 percent*

**9–19E** Repeat Prob. 9–18E using constant specific heats at room temperature.

**9–20** Consider a Carnot cycle executed in a closed system with 0.6 kg of air. The temperature limits of the cycle are 300 and 1100 K, and the minimum and maximum pressures that occur during the cycle are 20 and 3000 kPa. Assuming constant specific heats, determine the net work output per cycle.

**9–21** Consider a Carnot cycle executed in a closed system with air as the working fluid. The maximum pressure in the cycle is 800 kPa while the maximum temperature is 750 K. If the entropy increase during the isothermal heat rejection process is 0.25 kJ/kg·K and the net work output is 100 kJ/kg, determine (a) the minimum pressure in the cycle, (b) the heat rejection from the cycle, and (c) the thermal efficiency of the cycle. (d) If an actual heat engine cycle operates between the same temperature limits and produces 5200 kW of power for an air flow rate of 90 kg/s, determine the second law efficiency of this cycle.

**9–22** An ideal gas Carnot cycle uses air as the working fluid, receives heat from a heat reservoir at 1027°C, is repeated 1500 times per minute, and has a compression ratio of 12. The compression ratio is defined as the volume ratio during the compression process. Determine the maximum temperature of the low-temperature heat reservoir, the cycle's thermal efficiency, and the amount of heat that must be supplied per cycle if this device is to produce 500 kW of power.
*Answers: 481 K, 63.0 percent, 31.8 kJ*

## Otto Cycle

**9–23C** How is the rpm (revolutions per minute) of an actual four-stroke gasoline engine related to the number of thermodynamic cycles? What would your answer be for a two-stroke engine?

**9–24C** How do the efficiencies of the ideal Otto cycle and the Carnot cycle compare for the same temperature limits? Explain.

**9–25C** What four processes make up the ideal Otto cycle?

**9–26C** Are the processes that make up the Otto cycle analyzed as closed-system or steady-flow processes? Why?

**9–27C** How does the thermal efficiency of an ideal Otto cycle change with the compression ratio of the engine and the specific heat ratio of the working fluid?

**9–28C** Why are high compression ratios not used in spark-ignition engines?

**9–29C** An ideal Otto cycle with a specified compression ratio is executed using (a) air, (b) argon, and (c) ethane as the working fluid. For which case will the thermal efficiency be the highest? Why?

**9–30C** What is the difference between fuel-injected gasoline engines and diesel engines?

**9–31** An ideal Otto cycle has a compression ratio of 10.5, takes in air at 90 kPa and 40°C, and is repeated 2500 times per minute. Using constant specific heats at room temperature, determine the thermal efficiency of this cycle and the rate of heat input if the cycle is to produce 90 kW of power.

**9–32** Repeat Prob. 9–31 for a compression ratio of 8.5.

**9–33** Someone has suggested that the air-standard Otto cycle is more accurate if the two isentropic processes are replaced with polytropic processes with a polytropic exponent $n = 1.3$. Consider such a cycle when the compression ratio is 8, $P_1 = 95$ kPa, $T_1 = 15°C$, and the maximum cycle temperature is 1200°C. Determine the heat transferred to and rejected from this cycle, as well as the cycle's thermal efficiency. Use constant specific heats at room temperature. *Answers: 835 kJ/kg, 420 kJ/kg, 49.8 percent*

**9–34** How do the results of Prob. 9-33 change when isentropic processes are used in place of the polytropic processes?

**9–35E** A six-cylinder, four-stroke, spark-ignition engine operating on the ideal Otto cycle takes in air at 14 psia and 65°F, and is limited to a maximum cycle temperature of 1600°F. Each cylinder has a bore of 3.5 in, and each piston has a stroke of 3.9 in. The minimum enclosed volume is 14 percent of the maximum enclosed volume. How much power will this engine produce when operated at 2500 rpm? Use constant specific heats at room temperature.

**9–36E** A spark-ignition engine has a compression ratio of 8, an isentropic compression efficiency of 85 percent, and an isentropic expansion efficiency of 95 percent. At the beginning of the compression, the air in the cylinder is at 13 psia and 60°F. The maximum gas temperature is found to be 2300°F by measurement. Determine the heat supplied per unit mass, the thermal efficiency, and the mean effective pressure of this engine when modeled with the Otto cycle. Use constant specific heats at room temperature. *Answers: 247 Btu/lbm, 47.5 percent, 49.0 psia*

**9–37** The compression ratio of an air-standard Otto cycle is 9.5. Prior to the isentropic compression process, the air is at 100 kPa, 35°C, and 600 cm³. The temperature at the end of the isentropic expansion process is 800 K. Using specific heat values at room temperature, determine (a) the highest temperature and pressure in the cycle; (b) the amount of heat transferred in, in kJ; (c) the thermal efficiency; and (d) the mean effective pressure. *Answers: (a) 1969 K, 6072 kPa, (b) 0.59 kJ, (c) 59.4 percent, (d) 652 kPa*

**9–38** Repeat Prob. 9–37, but replace the isentropic expansion process by a polytropic expansion process with the polytropic exponent $n = 1.35$.

**9–39E** An ideal Otto cycle with air as the working fluid has a compression ratio of 8. The minimum and maximum temperatures in the cycle are 540 and 2400 R. Accounting for the variation of specific heats with temperature, determine (a) the amount of heat transferred to the air during the heat-addition process, (b) the thermal efficiency, and (c) the thermal efficiency of a Carnot cycle operating between the same temperature limits.

**9–40E** Repeat Prob. 9–39E using argon as the working fluid.

**9–41** A four-cylinder, four-stroke, 1.6-L gasoline engine operates on the Otto cycle with a compression ratio of 11. The air is at 100 kPa and 37°C at the beginning of the compression process, and the maximum pressure in the cycle is 8 MPa. The compression and expansion processes may be modeled as polytropic with a polytropic constant of 1.3. Using constant specific heats at 850 K, determine (a) the temperature at the end of the expansion process, (b) the net work output and the thermal efficiency, (c) the mean effective pressure, (d) the engine speed for a net power output of 50 kW, and (e) the specific fuel consumption, in g/kWh, defined as the ratio of the mass of the fuel consumed to the net work produced. The air–fuel ratio, defined as the amount of air divided by the amount of fuel intake, is 16.

**9–42** When we double the compression ratio of an ideal Otto cycle, what happens to the maximum gas temperature and pressure when the state of the air at the beginning of the compression and the amount of heat addition remain the same? Use constant specific heats at room temperature.

**9–43** In a spark-ignition engine, some cooling occurs as the gas is expanded. This may be modeled by using a polytropic process in lieu of the isentropic process. Determine if the polytropic exponent used in this model will be greater than or less than the isentropic exponent.

## Diesel Cycle

**9–44C** How does a diesel engine differ from a gasoline engine?

**9–45C** How does the ideal Diesel cycle differ from the ideal Otto cycle?

**9–46C** For a specified compression ratio, is a diesel or gasoline engine more efficient?

**9–47C** Do diesel or gasoline engines operate at higher compression ratios? Why?

**9–48C** What is the cutoff ratio? How does it affect the thermal efficiency of a Diesel cycle?

**9–49** An ideal Diesel cycle has a compression ratio of 20 and a cutoff ratio of 1.3. Determine the maximum temperature of the air and the rate of heat addition to this cycle when it produces 250 kW of power and the state of the air at the beginning of the compression is 90 kPa and 15°C. Use constant specific heats at room temperature.

**9–50E** An ideal Diesel cycle has a maximum cycle temperature of 2300°F and a cutoff ratio of 1.4. The state of the air at the beginning of the compression is $P_1 = 14.4$ psia and $T_1 = 50°F$. This cycle is executed in a four-stroke, eight-cylinder engine with a cylinder bore of 4 in and a piston stroke of 4 in. The minimum volume enclosed in the cylinder is 4.5 percent of the maximum cylinder volume. Determine the power produced by this engine when it is operated at 1800 rpm. Use constant specific heats at room temperature.

**9–51** An air-standard dual cycle has a compression ratio of 14 and a cutoff ratio of 1.2. The pressure ratio during the constant-volume heat addition process is 1.5. Determine the thermal efficiency, amount of heat added, the maximum gas pressure and temperature when this cycle is operated at 80 kPa and 20°C at the beginning of the compression. Use constant specific heats at room temperature.

**9–52** Repeat Prob. 9–51 when the state of the air at the beginning of the compression is 80 kPa and −20°C.

**9–53E** An air-standard Diesel cycle has a compression ratio of 18.2. Air is at 80°F and 14.7 psia at the beginning of the compression process and at 3000 R at the end of the heat-addition process. Accounting for the variation of specific heats with temperature, determine (a) the cutoff ratio, (b) the heat rejection per unit mass, and (c) the thermal efficiency.

**9–54E** Repeat Prob. 9–53E using constant specific heats at room temperature.

**9–55** An ideal diesel engine has a compression ratio of 20 and uses air as the working fluid. The state of air at the beginning of the compression process is 95 kPa and 20°C. If the maximum temperature in the cycle is not to exceed 2200 K, determine (a) the thermal efficiency and (b) the mean effective pressure. Assume constant specific heats for air at room temperature. *Answers:* (a) 63.5 percent, (b) 933 kPa

**9–56** Repeat Prob. 9–55, but replace the isentropic expansion process by polytropic expansion process with the polytropic exponent $n = 1.35$.

**9–57** ⬛EES Reconsider Prob. 9–56. Using EES (or other) software, study the effect of varying the compression ratio from 14 to 24. Plot the net work output, mean effective pressure, and thermal efficiency as a function of the compression ratio. Plot the T-s and P-v diagrams for the cycle when the compression ratio is 20.

**9–58** A four-cylinder two-stroke 2.0-L diesel engine that operates on an ideal Diesel cycle has a compression ratio of 22 and a cutoff ratio of 1.8. Air is at 70°C and 97 kPa at the beginning of the compression process. Using the cold-air-standard assumptions, determine how much power the engine will deliver at 2300 rpm.

**9–59** Repeat Prob. 9–58 using nitrogen as the working fluid.

**9–60** An air-standard dual cycle has a compression ratio of 18 and a cutoff ratio of 1.1. The pressure ratio during constant-volume heat addition process is 1.1. At the beginning of the compression, $P_1 = 90$ kPa, $T_1 = 18°C$, and $V_1 = 0.003$ m³. How much power will this cycle produce when it is executed 4000 times per minute? Use constant specific heats at room temperature.

**9–61** Repeat Prob. 9–60 if the isentropic compression efficiency is 85 percent and the isentropic expansion efficiency is 90 percent. *Answer:* 9.26 kW

**9–62E** An ideal dual cycle has a compression ratio of 15 and a cutoff ratio of 1.4. The pressure ratio during constant-volume heat addition process is 1.1. The state of the air at the beginning of the compression is $P_1 = 14.2$ psia and $T_1 = 75°F$. Calculate the cycle's net specific work, specific heat addition, and thermal efficiency. Use constant specific heats at room temperature.

**9–63** A six-cylinder, four-stroke, 3.2-L compression-ignition engine operates on the ideal diesel cycle with a compression ratio of 19. The air is at 95 kPa and 67°C at the beginning of the compression process and the engine speed is 1750 rpm. The engine uses light diesel fuel with a heating value of 42,500 kJ/kg, an air–fuel ratio of 28, and a combustion efficiency of 98 percent. Using constant specific heats at 850 K, determine (a) the maximum temperature in the cycle and the cutoff ratio (b) the net work output per cycle and the thermal efficiency, (c) the mean effective pressure, (d) the net power output, and (e) the specific fuel consumption, in g/kWh, defined as the ratio of the mass of the fuel consumed to the net work produced. *Answers:* (a) 2244 K, 2.36, (b) 2.71 kJ, 57.4 percent, (c) 847 kPa, (d) 39.5 kW, (e) 151 g/kWh

## Ideal and Actual Gas-Turbine (Brayton) Cycles

**9–64C** What four processes make up the simple ideal Brayton cycle?

**9–65C** For fixed maximum and minimum temperatures, what is the effect of the pressure ratio on (a) the thermal efficiency and (b) the net work output of a simple ideal Brayton cycle?

**9–66C** What is the back work ratio? What are typical back work ratio values for gas-turbine engines?

**9–67C** Why are the back work ratios relatively high in gas-turbine engines?

**9–68C** How do the inefficiencies of the turbine and the compressor affect (a) the back work ratio and (b) the thermal efficiency of a gas-turbine engine?

**9–69E** A simple ideal Brayton cycle with air as the working fluid has a pressure ratio of 10. The air enters the compressor at 520 R and the turbine at 2000 R. Accounting for the variation of specific heats with temperature, determine (a) the air temperature at the compressor exit, (b) the back work ratio, and (c) the thermal efficiency.

**9–70** ⬛EES A simple Brayton cycle using air as the working fluid has a pressure ratio of 10. The minimum and maximum temperatures in the cycle are 295 and 1240 K. Assuming an isentropic efficiency of 83 percent for the compressor and 87 percent for the turbine, determine (a) the air temperature at the turbine exit, (b) the net work output, and (c) the thermal efficiency.

**9–71** ⬛EES Reconsider Prob. 9–70. Using EES (or other) software, allow the mass flow rate, pressure ratio, turbine inlet temperature, and the isentropic efficiencies of the turbine and compressor to vary. Assume the compressor inlet pressure is 100 kPa. Develop a general solution for the problem by taking advantage of the diagram window method for supplying data to EES software.

**9–72** Repeat Prob. 9–70 using constant specific heats at room temperature.

**9–73E** A simple ideal Brayton cycle uses helium as the working fluid; operates with 12 psia and 60°F at the compressor inlet; has a pressure ratio of 14; and a maximum cycle temperature of 1300°F. How much power will this cycle produce when the rate at which the helium is circulated about the cycle is 100 lbm/min? Use constant specific heats at room temperature.

**9–74E** Repeat Prob. 9-73E when the isentropic efficiency of the compressor is 95 percent.

**9–75** Consider a simple Brayton cycle using air as the working fluid; has a pressure ratio of 12; has a maximum cycle temperature of 600°C; and operates the compressor inlet at 90 kPa and 15°C. Which will have the greatest impact on the back-work ratio: a compressor isentropic efficiency of 90 percent or a turbine isentropic efficiency of 90 percent? Use constant specific heats at room temperature.

determine (a) the fraction of the turbine work output used to drive the compressor and (b) the thermal efficiency. Assume variable specific heats for air.

**9–78** Repeat Prob. 9–77 using constant specific heats at room temperature.

**9–79** An aircraft engine operates on a simple ideal Brayton cycle with a pressure ratio of 10. Heat is added to the cycle at a rate of 500 kW; air passes through the engine at a rate of 1 kg/s; and the air at the beginning of the compression is at 70 kPa and 0°C. Determine the power produced by this engine and its thermal efficiency. Use constant specific heats at room temperature.

**9–80** Repeat Prob. 9–79 for a pressure ratio of 15.

**9–81** A gas-turbine power plant operates on the simple Brayton cycle between the pressure limits of 100 and 2000 kPa. The working fluid is air, which enters the compressor at 40°C at a rate of 700 m³/min and leaves the turbine at 650°C. Using variable specific heats for air and assuming a compressor isentropic efficiency of 85 percent and a turbine isentropic efficiency of 88 percent, determine (a) the net power output, (b) the back work ratio, and (c) the thermal efficiency. *Answers: (a) 5404 kW, (b) 0.545, (c) 39.2 percent*

$Q_{in}$

$Q_{out}$

**FIGURE P9–75**

Combustion chamber

2 MPa

100 kPa
40°C

650°C

$\dot{W}_{net}$

**FIGURE P9–81**

**9–76** Air is used as the working fluid in a simple ideal Brayton cycle that has a pressure ratio of 12, a compressor inlet temperature of 300 K, and a turbine inlet temperature of 1000 K. Determine the required mass flow rate of air for a net power output of 70 MW, assuming both the compressor and the turbine have an isentropic efficiency of (a) 100 percent and (b) 85 percent. Assume constant specific heats at room temperature. *Answers: (a) 352 kg/s, (b) 1037 kg/s*

**9–77** Air enters the compressor of a gas-turbine engine at 300 K and 100 kPa, where it is compressed to 700 kPa and 580 K. Heat is transferred to air in the amount of 950 kJ/kg before it enters the turbine. For a turbine efficiency of 86 percent,

**9–82** A gas-turbine power plant operates on the simple Brayton cycle between the pressure limits of 100 and 800 kPa. Air enters the compressor at 30°C and leaves at 330°C at a mass flow rate of 200 kg/s. The maximum cycle temperature is 1400 K. During operation of the cycle, the net power output is measured experimentally to be 60 MW. Assume constant properties for air at 300 K with $c_v = 0.718$ kJ/kg·K, $c_p = 1.005$ kJ/kg·K, $R = 0.287$ kJ/kg·K, $k = 1.4$.

(a) Sketch the T-s diagram for the cycle.

(b) Determine the isentropic efficiency of the turbine for these operating conditions.

(c) Determine the cycle thermal efficiency.

**9–83** A gas-turbine power plant operates on a modified Brayton cycle shown in the figure with an overall pressure ratio of 8. Air enters the compressor at 0°C and 100 kPa. The maximum cycle temperature is 1500 K. The compressor and the turbines are isentropic. The high pressure turbine develops just enough power to run the compressor. Assume constant properties for air at 300 K with $c_v = 0.718$ kJ/kg·K, $c_p = 1.005$ kJ/kg·K, $R = 0.287$ kJ/kg·K, $k = 1.4$.

(a) Sketch the T-s diagram for the cycle. Label the data states.

(b) Determine the temperature and pressure at state 4, the exit of the high pressure turbine.

(c) If the net power output is 200 MW, determine mass flow rate of the air into the compressor, in kg/s.

*Answers:* (b) 1279 K, 457 kPa, (c) 442 kg/s

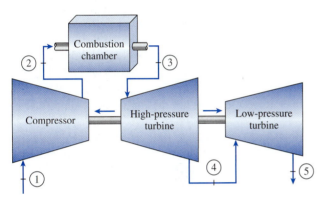

**FIGURE P9–83**

**9–84** A gas-turbine power plant operating on the simple Brayton has a pressure ratio of 7. Air enters the compressor at 0°C and 100 kPa. The maximum cycle temperature is 1500 K. The compressor has an isentropic efficiency of 80 percent, and the turbine has an isentropic efficiency of 90 percent. Assume constant properties for air at 300 K with $c_v = 0.718$ kJ/kg·K, $c_p = 1.005$ kJ/kg·K, $R = 0.287$ kJ/kg·K, $k = 1.4$.

(a) Sketch the T-s diagram for the cycle.

(b) If the net power output is 150 MW, determine the volume flow rate of the air into the compressor, in m³/s.

(c) For a fixed compressor inlet velocity and flow area, explain the effect of increasing compressor inlet temperature (i.e., summertime operation versus wintertime operation) on the inlet mass flow rate and the net power output with all other parameters of the problem being the same.

## Brayton Cycle with Regeneration

**9–85C** How does regeneration affect the efficiency of a Brayton cycle, and how does it accomplish it?

**9–86C** Somebody claims that at very high pressure ratios, the use of regeneration actually decreases the thermal efficiency of a gas-turbine engine. Is there any truth in this claim? Explain.

**9–87C** Define the effectiveness of a regenerator used in gas-turbine cycles.

**9–88C** In an ideal regenerator, is the air leaving the compressor heated to the temperature at (a) turbine inlet, (b) turbine exit, (c) slightly above turbine exit?

**9–89C** In 1903, Aegidius Elling of Norway designed and built an 11-hp gas turbine that used steam injection between the combustion chamber and the turbine to cool the combustion gases to a safe temperature for the materials available at the time. Currently there are several gas-turbine power plants that use steam injection to augment power and improve thermal efficiency. For example, the thermal efficiency of the General Electric LM5000 gas turbine is reported to increase from 35.8 percent in simple-cycle operation to 43 percent when steam injection is used. Explain why steam injection increases the power output and the efficiency of gas turbines. Also, explain how you would obtain the steam.

**9–90** A gas turbine for an automobile is designed with a regenerator. Air enters the compressor of this engine at 100 kPa and 20°C. The compressor pressure ratio is 8; the maximum cycle temperature is 800°C; and the cold air stream leaves the regenerator 10°C cooler than the hot air stream at the inlet of the regenerator. Assuming both the compressor and the turbine to be isentropic, determine the rates of heat addition and rejection for this cycle when it produces 150 kW. Use constant specific heats at room temperature. *Answers:* 303 kW, 153 kW

**FIGURE P9–90**

**9–91** Rework Prob. 9–90 when the compressor isentropic efficiency is 87 percent and the turbine isentropic efficiency is 93 percent.

**9–92** A gas turbine engine operates on the ideal Brayton cycle with regeneration, as shown in Fig. P9–90. Now the regenerator is rearranged so that the air streams of states 2 and 5 enter at one end of the regenerator and streams 3 and 6 exit at the other end (i.e., parallel flow arrangement of a heat exchanger). Consider such a system when air enters the compressor at 100 kPa and 20°C; the compressor pressure ratio is 7; the maximum cycle temperature is 727°C; and the difference between the hot and cold air stream temperatures is 6°C at the end of the regenerator where the cold stream leaves the regenerator. Is the cycle arrangement shown in the figure more or less efficient than this arrangement? Assume both the compressor and the turbine are isentropic, and use constant specific heats at room temperature.

**9–93** The 7FA gas turbine manufactured by General Electric is reported to have an efficiency of 35.9 percent in the simple-cycle mode and to produce 159 MW of net power. The pressure ratio is 14.7 and the turbine inlet temperature is 1288°C. The mass flow rate through the turbine is 1,536,000 kg/h. Taking the ambient conditions to be 30°C and 100 kPa, determine the isentropic efficiency of the turbine and the compressor. Also, determine the thermal efficiency of this gas turbine if a regenerator with an effectiveness of 65 percent is added.

**9–94** Reconsider Prob. 9–93. Using EES (or other) software, develop a solution that allows different isentropic efficiencies for the compressor and turbine and study the effect of the isentropic efficiencies on net work done and the heat supplied to the cycle. Plot the T-s diagram for the cycle.

**9–95** A Brayton cycle with regeneration using air as the working fluid has a pressure ratio of 7. The minimum and maximum temperatures in the cycle are 310 and 1150 K. Assuming an isentropic efficiency of 75 percent for the compressor and 82 percent for the turbine and an effectiveness of 65 percent for the regenerator, determine (a) the air temperature at the turbine exit, (b) the net work output, and (c) the thermal efficiency. *Answers: (a) 783 K, (b) 108.1 kJ/kg, (c) 22.5 percent*

**9–96** A stationary gas-turbine power plant operates on an ideal regenerative Brayton cycle ($\epsilon = 100$ percent) with air as the working fluid. Air enters the compressor at 95 kPa and 290 K and the turbine at 760 kPa and 1100 K. Heat is transferred to air from an external source at a rate of 75,000 kJ/s. Determine the power delivered by this plant (a) assuming constant specific heats for air at room temperature and (b) accounting for the variation of specific heats with temperature.

**9–97** Air enters the compressor of a regenerative gas-turbine engine at 310 K and 100 kPa, where it is compressed to 900 kPa and 650 K. The regenerator has an effectiveness of 80 percent, and the air enters the turbine at 1400 K. For a turbine efficiency of 90 percent, determine (a) the amount of heat transfer in the regenerator and (b) the thermal efficiency. Assume variable specific heats for air. *Answers: (a) 193 kJ/kg, (b) 40.0 percent*

**9–98** Repeat Prob. 9–97 using constant specific heats at room temperature.

**9–99** Repeat Prob. 9–97 for a regenerator effectiveness of 70 percent.

**9–100** Develop an expression for the thermal efficiency of an ideal Brayton cycle with an ideal regenerator of effectiveness 100 percent. Use constant specific heats at room temperature.

## Carnot Vapor Cycle

**9–101C** Why is the Carnot cycle not a realistic model for steam power plants?

**9–102E** Water enters the boiler of a steady-flow Carnot engine as a saturated liquid at 250 psia and leaves with a quality of 0.95. Steam leaves the turbine at a pressure of 40 psia. Show the cycle on a T-s diagram relative to the saturation lines, and determine (a) the thermal efficiency, (b) the quality at the end of the isothermal heat-rejection process, and (c) the net work output. *Answers: (a) 15.5 percent, (b) 0.137, (c) 122 Btu/lbm*

**9–103** A steady-flow Carnot cycle uses water as the working fluid. Water changes from saturated liquid to saturated vapor as heat is transferred to it from a source at 250°C. Heat rejection takes place at a pressure of 20 kPa. Show the cycle on a T-s diagram relative to the saturation lines, and determine (a) the thermal efficiency, (b) the amount of heat rejected, and (c) the net work output.

**9–104** Repeat Prob. 9–103 for a heat rejection pressure of 10 kPa.

**9–105** Consider a steady-flow Carnot cycle with water as the working fluid. The maximum and minimum temperatures in the cycle are 350 and 60°C. The quality of water is 0.891 at the beginning of the heat-rejection process and 0.1 at the end. Show the cycle on a T-s diagram relative to the saturation lines, and determine (a) the thermal efficiency, (b) the pressure at the turbine inlet, and (c) the net work output. *Answers: (a) 0.465, (b) 1.40 MPa, (c) 1623 kJ/kg*

## The Simple Rankine Cycle

**9–106C** What four processes make up the simple ideal Rankine cycle?

**9–107C** Consider a simple ideal Rankine cycle with fixed turbine inlet conditions. What is the effect of lowering the condenser pressure on

| | |
|---|---|
| Pump work input: | (a) increases, (b) decreases, (c) remains the same |
| Turbine work output: | (a) increases, (b) decreases, (c) remains the same |
| Heat supplied: | (a) increases, (b) decreases, (c) remains the same |
| Heat rejected: | (a) increases, (b) decreases, (c) remains the same |
| Cycle efficiency: | (a) increases, (b) decreases, (c) remains the same |
| Moisture content at turbine exit: | (a) increases, (b) decreases, (c) remains the same |

**9–108C** Consider a simple ideal Rankine cycle with fixed turbine inlet temperature and condenser pressure. What is the effect of increasing the boiler pressure on

| | | |
|---|---|---|
| Pump work input: | (a) increases, | (b) decreases, (c) remains the same |
| Turbine work output: | (a) increases, | (b) decreases, (c) remains the same |
| Heat supplied: | (a) increases, | (b) decreases, (c) remains the same |
| Heat rejected: | (a) increases, | (b) decreases, (c) remains the same |
| Cycle efficiency: | (a) increases, | (b) decreases, (c) remains the same |
| Moisture content at turbine exit: | (a) increases, | (b) decreases, (c) remains the same |

**9–109C** Consider a simple ideal Rankine cycle with fixed boiler and condenser pressures. What is the effect of superheating the steam to a higher temperature on

| | | |
|---|---|---|
| Pump work input: | (a) increases, | (b) decreases, (c) remains the same |
| Turbine work output: | (a) increases, | (b) decreases, (c) remains the same |
| Heat supplied: | (a) increases, | (b) decreases, (c) remains the same |
| Heat rejected: | (a) increases, | (b) decreases, (c) remains the same |
| Cycle efficiency: | (a) increases, | (b) decreases, (c) remains the same |
| Moisture content at turbine exit: | (a) increases, | (b) decreases, (c) remains the same |

**9–110C** How do actual vapor power cycles differ from idealized ones?

**9–111C** Compare the pressures at the inlet and the exit of the boiler for (a) actual and (b) ideal cycles.

**9–112C** The entropy of steam increases in actual steam turbines as a result of irreversibilities. In an effort to control entropy increase, it is proposed to cool the steam in the turbine by running cooling water around the turbine casing. It is argued that this will reduce the entropy and the enthalpy of the steam at the turbine exit and thus increase the work output. How would you evaluate this proposal?

**9–113C** Is it possible to maintain a pressure of 10 kPa in a condenser that is being cooled by river water entering at 20°C?

**9–114** An engineer has proposed that a simple ideal Rankine cycle that uses refrigerant-134a be used to produce work with heat from a low-temperature thermal energy reservoir. The boiler operates at 1.6 MPa, the condenser at 0.4 MPa, and the turbine inlet at 80°C. Determine the R-134a mass flow rate needed for this cycle to produce 750 kW of power and the thermal efficiency of the cycle.

**9–115** Refrigerant-134a is used as the working fluid in a simple ideal Rankine cycle which operates the boiler at 1400 kPa and the condenser at 10°C. The mixture at the exit of the turbine has a quality of 98 percent. Determine the turbine inlet temperature, the cycle thermal efficiency, and the back-work ratio of this cycle.

**9–116** A simple ideal Rankine cycle which uses water as the working fluid operates its condenser at 40°C and its boiler at 300°C. Calculate the work produced by the turbine, the heat supplied in the boiler, and the thermal efficiency of this cycle when the steam enters the turbine without any superheating.

**FIGURE P9–116**

**9–117E** The turbine of a steam power plant operating on a simple ideal Rankine cycle produces 1750 kW of power when the boiler is operated at 800 psia, the condenser at 3 psia, and the temperature at the turbine entrance is 900°F. Determine the rate of heat supply in the boiler, the rate of heat rejection in the condenser, and the thermal efficiency of the cycle.

**9–118E** A simple ideal Rankine cycle with water as the working fluid operates between the pressure limits of 2500 psia in the boiler and 5 psia in the condenser. What is the minimum temperature required at the turbine inlet such that the quality of the steam leaving the turbine is not below 80 percent. When operated at this temperature, what is the thermal efficiency of this cycle?

**9–119E** A steam Rankine cycle operates between the pressure limits of 2500 psia in the boiler and 1 psia in the condenser. The turbine inlet temperature is 800°F. The turbine isentropic efficiency is 90 percent, the pump losses are negligible, and the cycle is sized to produce 1000 kW of power. Calculate the mass flow rate through the boiler, the power produced by the turbine, the rate of heat supply in the boiler, and the thermal efficiency.

**9–120E** Reconsider Prob. 10–119E. How much error is caused in the thermal efficiency if the power required by the pump were completely neglected?

**9–121** Consider a steam power plant that operates on a simple ideal Rankine cycle and has a net power output of 45 MW. Steam enters the turbine at 7 MPa and 500°C and is cooled in the condenser at a pressure of 10 kPa by running cooling water from a lake through the tubes of the condenser at a rate of 2000 kg/s. Show the cycle on a *T-s* diagram with respect to saturation lines, and determine (*a*) the thermal efficiency of the cycle, (*b*) the mass flow rate of the steam, and (*c*) the temperature rise of the cooling water.  *Answers:* (*a*) 38.9 percent, (*b*) 36 kg/s, (*c*) 8.4°C

**9–122** Repeat Prob. 9–121 assuming an isentropic efficiency of 87 percent for both the turbine and the pump. *Answers:* (*a*) 33.8 percent, (*b*) 41.4 kg/s, (*c*) 10.5°C

**9–123** A simple Rankine cycle uses water as the working fluid. The boiler operates at 6000 kPa and the condenser at 50 kPa. At the entrance to the turbine, the temperature is 450°C. The isentropic efficiency of the turbine is 94 percent, pressure and pump losses are negligible, and the water leaving the condenser is subcooled by 6.3°C. The boiler is sized for a mass flow rate of 20 kg/s. Determine the rate at which heat is added in the boiler, the power required to operate the pumps, the net power produced by the cycle, and the thermal efficiency.  *Answers:* 59,660 kW, 122 kW, 18,050 kW, 30.3 percent.

**9–124** ⟨EES⟩ Using EES (or other) software, determine how much the thermal efficiency of the cycle in Prob. 9–123 would change if there were a 50 kPa pressure drop across the boiler.

**9–125** The net work output and the thermal efficiency for the Carnot and the simple ideal Rankine cycles with steam as the working fluid are to be calculated and compared. Steam enters the turbine in both cases at 5 MPa as a saturated vapor, and the condenser pressure is 50 kPa. In the Rankine cycle, the condenser exit state is saturated liquid and in the Carnot cycle, the boiler inlet state is saturated liquid. Draw the *T-s* diagrams for both cycles.

**9–126** Consider a coal-fired steam power plant that produces 120 MW of electric power. The power plant operates on a simple ideal Rankine cycle with turbine inlet conditions of 9 MPa and 550°C and a condenser pressure of 15 kPa. The coal has a heating value (energy released when the fuel is burned) of 29,300 kJ/kg. Assuming that 75 percent of this energy is transferred to the steam in the boiler and that the electric generator has an efficiency of 96 percent, determine (*a*) the overall plant efficiency (the ratio of net electric power output to the energy input as fuel) and (*b*) the required rate of coal supply.  *Answers:* (*a*) 28.4 percent, (*b*) 51.9 t/h

**9–127** The schematic of a single-flash geothermal power plant with state numbers is given in Fig. P9–127. Geothermal

resource exists as saturated liquid at 230°C. The geothermal liquid is withdrawn from the production well at a rate of 230 kg/s, and is flashed to a pressure of 500 kPa by an essentially isenthalpic flashing process where the resulting vapor is separated from the liquid in a separator and directed to the turbine. The steam leaves the turbine at 10 kPa with a moisture content of 10 percent and enters the condenser where it is condensed and routed to a reinjection well along with the liquid coming off the separator. Determine (*a*) the mass flow rate of steam through the turbine, (*b*) the isentropic efficiency of the turbine, (*c*) the power output of the turbine, and (*d*) the thermal efficiency of the plant (the ratio of the turbine work output to the energy of the geothermal fluid relative to standard ambient conditions).  *Answers:* (*a*) 38.2 kg/s, (*b*) 0.686, (*c*) 15.4 MW, (*d*) 7.6 percent

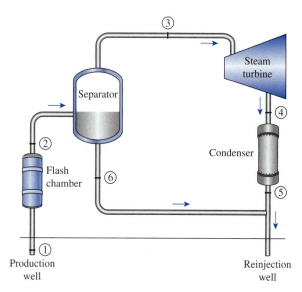

**FIGURE P9–127**

## The Reheat Rankine Cycle

**9–128C** How do the following quantities change when a simple ideal Rankine cycle is modified with reheating? Assume the mass flow rate is maintained the same.

| Pump work input: | (*a*) increases, (*b*) decreases, (*c*) remains the same |
|---|---|
| Turbine work output: | (*a*) increases, (*b*) decreases, (*c*) remains the same |
| Heat supplied: | (*a*) increases, (*b*) decreases, (*c*) remains the same |
| Heat rejected: | (*a*) increases, (*b*) decreases, (*c*) remains the same |
| Moisture content at turbine exit: | (*a*) increases, (*b*) decreases, (*c*) remains the same |

**9–129C**  Is there an optimal pressure for reheating the steam of a Rankine cycle? Explain.

**9–130C**  Consider a simple ideal Rankine cycle and an ideal Rankine cycle with three reheat stages. Both cycles operate between the same pressure limits. The maximum temperature is 700°C in the simple cycle and 450°C in the reheat cycle. Which cycle do you think will have a higher thermal efficiency?

**9–131**  An ideal reheat Rankine cycle with water as the working fluid operates the inlet of the high-pressure turbine at 8000 kPa and 450°C; the inlet of the low-pressure turbine at 500 kPa and 500°C; and the condenser at 10 kPa. Determine the mass flow rate through the boiler needed for this system to produce a net 5000 kW of power and the thermal efficiency of the cycle.

**9–132**  An ideal reheat Rankine cycle with water as the working fluid operates the boiler at 15,000 kPa, the reheater at 2000 kPa, and the condenser at 100 kPa. The temperature is 450°C at the entrance of the high-pressure and low-pressure turbines. The mass flow rate through the cycle is 1.74 kg/s. Determine the power used by pumps, the power produced by the cycle, the rate of heat transfer in the reheater, and the thermal efficiency of this system.

**9–133**  A steam power plant operates on the ideal reheat Rankine cycle. Steam enters the high-pressure turbine at 6 MPa and 400°C and leaves at 2 MPa. Steam is then reheated at constant pressure to 400°C before it expands to 20 kPa in the low-pressure turbine. Determine the turbine work output, in kJ/kg, and the thermal efficiency of the cycle. Also, show the cycle on a $T$-$s$ diagram with respect to saturation lines.

**9–134**  Reconsider Prob. 9–133. Using EES (or other) software, solve this problem by the diagram window data entry feature of EES. Include the effects of the turbine and pump efficiencies and also show the effects of reheat on the steam quality at the low-pressure turbine exit. Plot the cycle on a $T$-$s$ diagram with respect to the saturation lines. Discuss the results of your parametric studies.

**9–135E**  Consider a steam power plant that operates on the ideal reheat Rankine cycle. The plant maintains the inlet of the high-pressure turbine at 600 psia and 600°F, the inlet of the low-pressure turbine at 200 psia and 600°F, and the condenser at 10 psia. The net power produced by this plant is 5000 kW. Determine the rate of heat addition and rejection and the thermal efficiency of the cycle.

**9–136E**  In Prob. 9–135E, is there any advantage to operating the reheat section of the boiler at 100 psia rather than 200 psia while maintaining the same low-pressure turbine inlet temperature?

**9–137**  Consider a steam power plant that operates on the ideal reheat Rankine cycle. The plant maintains the boiler at 7000 kPa, the reheat section at 800 kPa, and the condenser at 10 kPa. The mixture quality at the exit of both turbines is 93 percent. Determine the temperature at the inlet of each turbine and the cycle's thermal efficiency. *Answers:* 373°C, 416°C, 37.6 percent

**FIGURE P9–137**

**9–138**  A steam power plant operates on an ideal reheat Rankine cycle between the pressure limits of 15 MPa and 10 kPa. The mass flow rate of steam through the cycle is 12 kg/s. Steam enters both stages of the turbine at 500°C. If the moisture content of the steam at the exit of the low-pressure turbine is not to exceed 10 percent, determine (*a*) the pressure at which reheating takes place, (*b*) the total rate of heat input in the boiler, and (*c*) the thermal efficiency of the cycle. Also, show the cycle on a $T$-$s$ diagram with respect to saturation lines.

**9–139**  A steam power plant operates on the reheat Rankine cycle. Steam enters the high-pressure turbine at 12.5 MPa and 550°C at a rate of 7.7 kg/s and leaves at 2 MPa. Steam is then reheated at constant pressure to 450°C before it expands in the low-pressure turbine. The isentropic efficiencies of the turbine and the pump are 85 percent and 90 percent, respectively. Steam leaves the condenser as a saturated liquid. If the moisture content of the steam at the exit of the turbine is not to exceed 5 percent, determine (*a*) the condenser pressure, (*b*) the net power output, and (*c*) the thermal efficiency.  *Answers:* (*a*) 9.73 kPa, (*b*) 10.2 MW, (*c*) 36.9 percent

**FIGURE P9–139**

## The Reversed Carnot Cycle

**9–140C**  Why do we study the reversed Carnot cycle even though it is not a realistic model for refrigeration cycles?

**9–141C**  Why is the reversed Carnot cycle executed within the saturation dome not a realistic model for refrigeration cycles?

**9–142**  A steady-flow Carnot refrigeration cycle uses refrigerant- 134a as the working fluid. The refrigerant changes from saturated vapor to saturated liquid at 40°C in the condenser as it rejects heat. The evaporator pressure is 100 kPa. Show the cycle on a T-s diagram relative to saturation lines, and determine (a) the coefficient of performance, (b) the amount of heat absorbed from the refrigerated space, and (c) the net work input.  *Answers: (a) 3.72, (b) 128 kJ/kg, (c) 34.6 kJ/kg*

**9–143E**  Refrigerant-134a enters the condenser of a steady-flow Carnot refrigerator as a saturated vapor at 90 psia, and it leaves with a quality of 0.05. The heat absorption from the refrigerated space takes place at a pressure of 30 psia. Show the cycle on a T-s diagram relative to saturation lines, and determine (a) the coefficient of performance, (b) the quality at the beginning of the heat-absorption process, and (c) the net work input.

## Ideal and Actual Vapor-Compression Refrigeration Cycles

**9–144C**  Does the ideal vapor-compression refrigeration cycle involve any internal irreversibilities?

**9–145C**  Why is the throttling valve not replaced by an isentropic turbine in the ideal vapor-compression refrigeration cycle?

**9–146C**  It is proposed to use water instead of refrigerant-134a as the working fluid in air-conditioning applications where the minimum temperature never falls below the freezing point. Would you support this proposal? Explain.

**9–147C**  In a refrigeration system, would you recommend condensing the refrigerant-134a at a pressure of 0.7 or 1.0 MPa if heat is to be rejected to a cooling medium at 15°C? Why?

**9–148C**  Does the area enclosed by the cycle on a T-s diagram represent the net work input for the reversed Carnot cycle? How about for the ideal vapor-compression refrigeration cycle?

**9–149C**  Consider two vapor-compression refrigeration cycles. The refrigerant enters the throttling valve as a saturated liquid at 30°C in one cycle and as subcooled liquid at 30°C in the other one. The evaporator pressure for both cycles is the same. Which cycle do you think will have a higher COP?

**9–150C**  The COP of vapor-compression refrigeration cycles improves when the refrigerant is subcooled before it enters the throttling valve. Can the refrigerant be subcooled indefinitely to maximize this effect, or is there a lower limit? Explain.

**9–151E**  A refrigerator operates on the ideal vapor-compression refrigeration cycle and uses refrigerant-134a as the working fluid. The condenser operates at 300 psia and the evaporator at 20°F. If an adiabatic, reversible expansion device were available and used to expand the liquid leaving the condenser, how much would the COP improve by using this device instead of the throttle device?  *Answer: 16.5 percent*

**9–152**  An ideal vapor-compression refrigeration cycle that uses refrigerant-134a as its working fluid maintains a condenser at 1000 kPa and the evaporator at 4°C. Determine this system's COP and the amount of power required to service a 400 kW cooling load.  *Answers: 6.46, 61.9 kW*

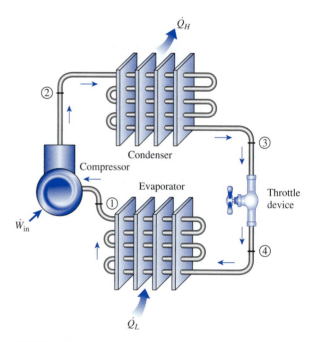

**FIGURE P9–152**

**9–153**  A refrigerator uses refrigerant-134a as the working fluid and operates on an ideal vapor-compression refrigeration cycle between 0.12 and 0.7 MPa. The mass flow rate of the refrigerant is 0.05 kg/s. Show the cycle on a T-s diagram with respect to saturation lines. Determine (a) the rate of heat removal from the refrigerated space and the power input to the compressor, (b) the rate of heat rejection to the environment, and (c) the coefficient of performance.  *Answers: (a) 7.41 kW, 1.83 kW, (b) 9.23 kW, (c) 4.06*

**9–154**  Repeat Prob. 9–153 for a condenser pressure of 0.9 MPa.

**9–155**  If the throttling valve in Prob. 9–153 is replaced by an isentropic turbine, determine the percentage increase in the COP and in the rate of heat removal from the refrigerated space.  *Answers: 4.2 percent, 4.2 percent*

**9–156** Refrigerant-134a enters the compressor of a refrigerator as superheated vapor at 0.20 MPa and −5°C at a rate of 0.07 kg/s, and it leaves at 1.2 MPa and 70°C. The refrigerant is cooled in the condenser to 44°C and 1.15 MPa, and it is throttled to 0.21 MPa. Disregarding any heat transfer and pressure drops in the connecting lines between the components, show the cycle on a *T-s* diagram with respect to saturation lines, and determine (*a*) the rate of heat removal from the refrigerated space and the power input to the compressor, (*b*) the isentropic efficiency of the compressor, and (*c*) the COP of the refrigerator.  *Answers:* (a) 9.42 kW, 3.63 kW, (b) 74.1 percent, (c) 2.60

**9–157E** A refrigerator uses refrigerant-134a as its working fluid and operates on the ideal vapor-compression refrigeration cycle. The refrigerant evaporates at 5°F and condenses at 180 psia. This unit serves a 45,000 Btu/h cooling load. Determine the mass flow rate of the refrigerant and the power that this unit will require.

**9–158E** Using EES (or other) software, repeat Prob. 9–157E if ammonia is used in place of refrigerant-134a.

**9–159** A commercial refrigerator with refrigerant-134a as the working fluid is used to keep the refrigerated space at −30°C by rejecting its waste heat to cooling water that enters the condenser at 18°C at a rate of 0.25 kg/s and leaves at 26°C. The refrigerant enters the condenser at 1.2 MPa and 65°C and leaves at 42°C. The inlet state of the compressor is 60 kPa and −34°C and the compressor is estimated to gain a net heat of 450 W from the surroundings. Determine (*a*) the quality of the refrigerant at the evaporator inlet, (*b*) the refrigeration load, (*c*) the COP of the refrigerator, and (*d*) the theoretical maximum refrigeration load for the same power input to the compressor.

**FIGURE P9–159**

**9–160** Refrigerant-134a enters the compressor of a refrigerator at 100 kPa and −20°C at a rate of 0.5 m³/min and leaves at 0.8 MPa. The isentropic efficiency of the compressor is 78 percent. The refrigerant enters the throttling valve at 0.75 MPa and 26°C and leaves the evaporator as saturated vapor at −26°C. Show the cycle on a *T-s* diagram with respect to saturation lines, and determine (*a*) the power input to the compressor, (*b*) the rate of heat removal from the refrigerated space, and (*c*) the pressure drop and rate of heat gain in the line between the evaporator and the compressor.  *Answers:* (a) 2.40 kW, (b) 6.17 kW, (c) 1.73 kPa, 0.203 kW

**9–161** Reconsider Prob. 9–160. Using EES (or other) software, investigate the effects of varying the compressor isentropic efficiency over the range 60 to 100 percent and the compressor inlet volume flow rate from 0.1 to 1.0 m³/min on the power input and the rate of refrigeration. Plot the rate of refrigeration and the power input to the compressor as functions of compressor efficiency for compressor inlet volume flow rates of 0.1, 0.5, and 1.0 m³/min, and discuss the results.

**9–162** A refrigerator uses refrigerant-134a as the working fluid and operates on the ideal vapor-compression refrigeration cycle except for the compression process. The refrigerant enters the evaporator at 120 kPa with a quality of 30 percent and leaves the compressor at 60°C. If the compressor consumes 450 W of power, determine (*a*) the mass flow rate of the refrigerant, (*b*) the condenser pressure, and (*c*) the COP of the refrigerator.  *Answers:* (a) 0.00727 kg/s, (b) 672 kPa, (c) 2.43

**FIGURE P9–162**

**9–163** The manufacturer of an air conditioner claims a seasonal energy efficiency ratio (SEER) of 16 (Btu/h)/W for one of its units. This unit operates on the normal vapor compression refrigeration cycle and uses refrigerant-22 as the working fluid. This SEER is for the operating conditions when the evaporator saturation temperature is $-5°C$ and the condenser saturation temperature is $45°C$. Selected data for refrigerant-22 are provided in the table below.

| $T$, °C | $P_{sat}$, kPa | $h_f$, kJ/kg | $h_g$, kJ/kg | $s_g$, kJ/kg·K |
|---|---|---|---|---|
| −5 | 421.2 | 38.76 | 248.1 | 0.9344 |
| 45 | 1728 | 101 | 261.9 | 0.8682 |

(a) Sketch the hardware and the T-s diagram for this air conditioner.

(b) Determine the heat absorbed by the refrigerant in the evaporator per unit mass of refrigerant-22, in kJ/kg.

(c) Determine the work input to the compressor and the heat rejected in the condenser per unit mass of refrigerant-22, in kJ/kg.

**9–164** An actual refrigerator operates on the vapor-compression refrigeration cycle with refrigerant-22 as the working fluid. The refrigerant evaporates at $-15°C$ and condenses at $40°C$. The isentropic efficiency of the compressor is 83 percent. The refrigerant is superheated by $5°C$ at the compressor inlet and subcooled by $5°C$ at the exit of the condenser. Determine (a) the heat removed from the cooled space and the work input, in kJ/kg and the COP of the cycle. Determine (b) the same parameters if the cycle operated on the ideal vapor-compression refrigeration cycle between the same evaporating and condensing temperatures.

The properties of R-22 in the case of actual operation are: $h_1 = 402.49$ kJ/kg, $h_2 = 454.00$ kJ/kg, $h_3 = 243.19$ kJ/kg

The properties of R-22 in the case of ideal operation are: $h_1 = 399.04$ kJ/kg, $h_2 = 440.71$ kJ/kg, $h_3 = 249.80$ kJ/kg

Note: state 1: compressor inlet, state 2: compressor exit, state 3: condenser exit, state 4: evaporator inlet.

**9–165** A refrigerator uses refrigerant-134a as the working fluid and operates on the vapor-compression refrigeration cycle. The evaporator and condenser pressures are 200 kPa and 1400 kPa, respectively. The isentropic efficiency of the compressor is 88 percent. The refrigerant enters the compressor at a rate of 0.025 kg/s superheated by $10.1°C$ and leaves the condenser subcooled by $4.4°C$. Determine (a) the rate of cooling provided by the evaporator, the power input, and the COP. Determine (b) the same parameters if the cycle operated on the ideal vapor-compression refrigeration cycle between the same pressure limits.

## Heat Pump Systems

**9–166C** Do you think a heat pump system will be more cost-effective in New York or in Miami? Why?

**9–167C** What is a water-source heat pump? How does the COP of a water-source heat pump system compare to that of an air-source system?

**9–168E** A heat pump uses refrigerant-134a as the working fluid and operates on the ideal vapor-compression refrigeration cycle. The condenser pressure is 100 psia while the evaporator temperature is $40°F$. What is the COP of this heat pump?

**9–169** A heat pump operates on the ideal vapor-compression refrigeration cycle and uses refrigerant-134a as the working fluid. The condenser operates at 1000 kPa and the evaporator at 200 kPa. Determine this system's COP and the rate of heat supplied to the evaporator when the compressor consumes 6 kW.

**9–170** A heat pump operates on the ideal vapor-compression cycle with refrigerant-134a as the working fluid. This heat pump is used to keep a space at $25°C$ by absorbing heat at the rate of 2.7 kW from geothermal water flowing through the evaporator. The evaporator operates at $20°C$, and the condenser operates at 1400 kPa. The compressor receives work equal to 20 kJ for each kilogram of refrigerant flowing into it.

(a) Sketch the hardware and the T-s diagram for this heat pump.
(b) Determine the rate of heat transfer to the heated space at $25°C$.
(c) Determine the COP of the heat pump.

Refrigerant-134a data: $T = 20°C$: $h_f = 79.3$ kJ/kg, $h_g = 261.6$ kJ/kg; $P = 1400$ kPa: $h_f = 127.2$ kJ/kg, $h_g = 276.2$ kJ/kg

**9–171** A building requires a 2-ton heat pump for maintaining the interior space at $27°C$ when the outside temperature is $5°C$. The heat pump operates on the normal vapor compression refrigeration cycle and uses refrigerant-134a as the working fluid. The heat pump operating conditions require an evaporator pressure of 240 kPa and a condenser pressure of 1600 kPa while the compressor has an isentropic efficiency of 85 percent. Selected data for refrigerant-134a are provided in the table below.

| $P$, kPa | $T_{sat}$, °C | $h_f$, kJ/kg | $h_g$, kJ/kg | $s_g$, kJ/kg·K |
|---|---|---|---|---|
| 240 | −5.37 | 43 | 244 | 0.9222 |
| 1600 | 57.92 | 134 | 275 | 0.8982 |

For R-134a at $P = 1600$ kPa and $s = 0.9222$ kJ/kg·K, $h = 285$ kJ/kg. Also, 1 ton = 211 kJ/min.

(a) Sketch the hardware and the T-s diagram for this heat pump system.
(b) Determine the power required to drive the heat pump, in kW and the COP. *Answers: (b) 2.14 kW, 3.29*

**9–172** A heat pump with refrigerant-134a as the working fluid is used to keep a space at $25°C$ by absorbing heat from geothermal water that enters the evaporator at $50°C$ at a rate of 0.065 kg/s and leaves at $40°C$. The refrigerant enters the evaporator at $20°C$ with a quality of 23 percent and leaves at the inlet pressure as saturated vapor. The refrigerant loses 300 W of heat to the surroundings as it flows through the compressor and the refrigerant leaves the compressor at 1.4 MPa at the same entropy as the inlet. Determine (a) the degrees of subcooling of the refrigerant in the condenser, (b) the mass flow rate of the refrigerant, (c) the heating load and the COP of the heat pump, and (d) the theoretical minimum power input to the compressor for the same heating load. *Answers: (a) 3.8°C, (b) 0.0194 kg/s, (c) 3.07 kW, 4.68, (d) 0.238 kW*

**FIGURE P9–172**

**9–173** Refrigerant-134a enters the condenser of a residential heat pump at 800 kPa and 55°C at a rate of 0.018 kg/s and leaves at 750 kPa subcooled by 3°C. The refrigerant enters the compressor at 200 kPa superheated by 4°C. Determine (a) the isentropic efficiency of the compressor, (b) the rate of heat supplied to the heated room, and (c) the COP of the heat pump. Also, determine (d) the COP and the rate of heat supplied to the heated room if this heat pump operated on the ideal vapor-compression cycle between the pressure limits of 200 and 800 kPa.

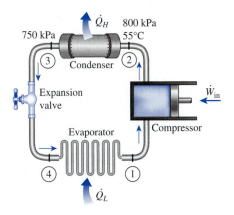

**FIGURE P9–173**

**Review Problems**

**9–174** An Otto cycle with a compression ratio of 7 begins its compression at 90 kPa and 15°C. The maximum cycle temperature is 1000°C. Utilizing air-standard assumptions, determine the thermal efficiency of this cycle using (a) constant specific heats at room temperature and (b) variable specific heats. *Answers:* (a) 54.1 percent, (b) 51.0 percent

**9–175E** A Diesel cycle has a compression ratio of 20 and begins its compression at 13 psia and 45°F. The maximum cycle temperature is 1800°F. Utilizing air-standard assumptions, determine the thermal efficiency of this cycle using (a) constant specific heats at room temperature and (b) variable specific heats.

**9–176E** A Brayton cycle with a pressure ratio of 12 operates with air entering the compressor at 13 psia and 20°F, and the turbine at 1000°F. Calculate the net specific work produced by this cycle treating the air as an ideal gas with (a) constant specific heats at room temperature and (b) variable specific heats.

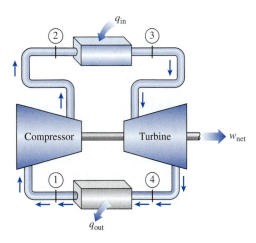

**FIGURE P9–176E**

**9–177** A four-stroke turbocharged V-16 diesel engine built by GE Transportation Systems to power fast trains produces 3500 hp at 1200 rpm. Determine the amount of power produced per cylinder per (a) mechanical cycle and (b) thermodynamic cycle.

**9–178** Consider a simple ideal Brayton cycle operating between the temperature limits of 300 and 1500 K. Using constant specific heats at room temperature, determine the pressure ratio for which the compressor and the turbine exit temperatures of air are equal.

**9–179** A four-cylinder spark-ignition engine has a compression ratio of 10.5, and each cylinder has a maximum volume of 0.4 L. At the beginning of the compression process, the air is at 98 kPa and 37°C, and the maximum temperature in the cycle is 2100 K. Assuming the engine to operate on the ideal Otto cycle, determine (a) the amount of heat supplied per cylinder, (b) the thermal efficiency, and (c) the number of revolutions per minute required for a net power output of 45 kW. Assume variable specific heats for air.

**9–180**  Reconsider Prob. 9–179. Using EES (or other) software, study the effect of varying the compression ratio from 5 to 11 on the net work done and the efficiency of the cycle. Plot the *P-v* and *T-s* diagrams for the cycle, and discuss the results.

**9–181** An ideal gas Carnot cycle uses helium as the working fluid and rejects heat to a lake at 15°C. Determine the pressure ratio, compression ratio, and minimum temperature of the heat source for this cycle to have a thermal efficiency of 50 percent. *Answers: 5.65, 2.83, 576 K*

**9–182E** Repeat Prob. 9–181 when the lake is at 60°F and the Carnot cycle's thermal efficiency is to be 60 percent.

**9–183** A typical hydrocarbon fuel produces 43,000 kJ/kg of heat when used in a spark-ignition engine. Determine the compression ratio required for an ideal Otto cycle to use 0.035 grams of fuel to produce 1 kJ of work. Use constant specific heats at room temperature. *Answer: 15.3*

**9–184** An ideal Otto cycle has a compression ratio of 9.2 and uses air as the working fluid. At the beginning of the compression process, air is at 98 kPa and 27°C. The pressure is doubled during the constant-volume heat-addition process. Accounting for the variation of specific heats with temperature, determine (*a*) the amount of heat transferred to the air, (*b*) the net work output, (*c*) the thermal efficiency, and (*d*) the mean effective pressure for the cycle.

**9–185** Repeat Prob. 9–184 using constant specific heats at room temperature.

**9–186E** An ideal dual cycle has a compression ratio of 14 and uses air as the working fluid. At the beginning of the compression process, air is at 14.7 psia and 120°F, and occupies a volume of 98 in³. During the heat-addition process, 0.6 Btu of heat is transferred to air at constant volume and 1.1 Btu at constant pressure. Using constant specific heats evaluated at room temperature, determine the thermal efficiency of the cycle.

**9–187** Consider a simple ideal Brayton cycle with air as the working fluid. The pressure ratio of the cycle is 6, and the minimum and maximum temperatures are 300 and 1300 K, respectively. Now the pressure ratio is doubled without changing the minimum and maximum temperatures in the cycle. Determine the change in (*a*) the net work output per unit mass and (*b*) the thermal efficiency of the cycle as a result of this modification. Assume variable specific heats for air. *Answers: (a) 41.5 kJ/kg, (b) 10.6 percent*

**9–188** Repeat Prob. 9–187 using constant specific heats at room temperature.

**9–189** A steam power plant operating on a simple ideal Rankine cycle maintains the boiler at 6000 kPa, the turbine inlet at 600°C, and the condenser at 50 kPa. Compare the thermal efficiency of this cycle when it is operated so that the liquid enters the pump as a saturated liquid against that when the liquid enters the pump 11.3°C cooler than a saturated liquid at the condenser pressure.

**9–190E** The Stillwater geothermal power plant in Nevada, which started full commercial operation in 1986, is designed to operate with seven identical units. Each of these seven units consists of a pair of power cycles, labeled Level I and Level II, operating on the simple Rankine cycle using an organic fluid as the working fluid.

The heat source for the plant is geothermal water (brine) entering the vaporizer (boiler) of Level I of each unit at 325°F at a rate of 384,286 lbm/h and delivering 22.79 MBtu/h ("M" stands for "million"). The organic fluid that enters the vaporizer at 202.2°F at a rate of 157,895 lbm/h leaves it at 282.4°F and 225.8 psia as saturated vapor. This saturated vapor expands in the turbine to 95.8°F and 19.0 psia and produces 1271 kW of electric power. About 200 kW of this power is

**FIGURE P9–190**
Schematic of a binary geothermal power plant.

*Courtesy of ORMAT Energy Systems, Inc.*

used by the pumps, the auxiliaries, and the six fans of the condenser. Subsequently, the organic working fluid is condensed in an air-cooled condenser by air that enters the condenser at 55°F at a rate of 4,195,100 lbm/h and leaves at 84.5°F. The working fluid is pumped and then preheated in a preheater to 202.2°F by absorbing 11.14 MBtu/h of heat from the geothermal water (coming from the vaporizer of Level II) entering the preheater at 211.8°F and leaving at 154.0°F.

Taking the average specific heat of the geothermal water to be 1.03 Btu/lbm·°F, determine (a) the exit temperature of the geothermal water from the vaporizer, (b) the rate of heat rejection from the working fluid to the air in the condenser, (c) the mass flow rate of the geothermal water at the preheater, and (d) the thermal efficiency of the Level I cycle of this geothermal power plant. *Answers:* (a) 267.4°F, (b) 29.7 MBtu/h, (c) 187,120 lbm/h, (d) 10.8 percent

**9–191** A steam power plant operates on an ideal Rankine cycle with two stages of reheat and has a net power output of 75 MW. Steam enters all three stages of the turbine at 550°C. The maximum pressure in the cycle is 10 MPa, and the minimum pressure is 30 kPa. Steam is reheated at 4 MPa the first time and at 2 MPa the second time. Show the cycle on a *T-s* diagram with respect to saturation lines, and determine (a) the thermal efficiency of the cycle, and (b) the mass flow rate of the steam. *Answers:* (a) 40.5 percent, (b) 48.5 kg/s

**9–192** Consider a steam power plant operating on the ideal Rankine cycle with reheat between the pressure limits of 25 MPa and 10 kPa with a maximum cycle temperature of 600°C and a moisture content of 8 percent at the turbine exit. For a reheat temperature of 600°C, determine the reheat pressures of the cycle for the cases of (a) single and (b) double reheat.

**9–193** Using EES (or other) software, investigate the effect of the condenser pressure on the performance of a simple ideal Rankine cycle. Turbine inlet conditions of steam are maintained constant at 10 MPa and 550°C while the condenser pressure is varied from 5 to 100 kPa. Determine the thermal efficiency of the cycle and plot it against the condenser pressure, and discuss the results.

**9–194** Using EES (or other) software, investigate the effect of superheating the steam on the performance of a simple ideal Rankine cycle. Steam enters the turbine at 3 MPa and exits at 10 kPa. The turbine inlet temperature is varied from 250 to 1100°C. Determine the thermal efficiency of the cycle and plot it against the turbine inlet temperature, and discuss the results.

**9–195** Using EES (or other) software, investigate the effect of number of reheat stages on the performance of an ideal Rankine cycle. The maximum and minimum pressures in the cycle are 15 MPa and 10 kPa, respectively, and steam enters all stages of the turbine at 500°C. For each case, maintain roughly the same pressure ratio across each turbine stage. Determine the thermal efficiency of the cycle and plot it against the number of reheat stages 1, 2, 4, and 8, and discuss the results.

**9–196** Consider a steady-flow Carnot refrigeration cycle that uses refrigerant-134a as the working fluid. The maximum and minimum temperatures in the cycle are 30 and −20°C, respectively. The quality of the refrigerant is 0.15 at the beginning of the heat absorption process and 0.80 at the end. Show the cycle on a *T-s* diagram relative to saturation lines, and determine (a) the coefficient of performance, (b) the condenser and evaporator pressures, and (c) the net work input.

**9–197** Rooms with floor areas of up to 15-m² are cooled adequately by window air conditioners whose cooling capacity is 5000 Btu/h. Assuming the COP of the air conditioner to be 3.5, determine the rate of heat gain of the room, in Btu/h, when the air conditioner is running continuously to maintain a constant room temperature.

**9–198** A heat pump water heater (HPWH) heats water by absorbing heat from the ambient air and transferring it to water. The heat pump has a COP of 3.4 and consumes 6 kW of electricity when running. Determine if this heat pump can be used to meet the cooling needs of a room most of the time for "free" by absorbing heat from the air in the room. The rate of heat gain of a room is usually less than 45,000 kJ/h.

**FIGURE P9–198**

**9–199** A heat pump that operates on the ideal vapor-compression cycle with refrigerant-134a is used to heat a house. The mass flow rate of the refrigerant is 0.32 kg/s. The condenser and evaporator pressures are 900 and 200 kPa, respectively. Show the cycle on a *T-s* diagram with respect to saturation lines, and determine (a) the rate of heat supply to the house, (b) the volume flow rate of the refrigerant at the compressor inlet, and (c) the COP of this heat pump.

**9–200** A ground-coupled heat pump operates on the ideal vapor-compression cycle with refrigerant-134a as the working fluid. This heat pump has a COP equal to 6 in this cooling mode and maintains a space at 25°C when absorbing heat from the space at the rate of 18 kW in the evaporator. The evaporator operates at 20°C, and the condenser operates at 1400 kPa. Ground water taken from a well at the rate of 0.32 kg/s absorbs the heat transferred from the refrigerant in the condenser.

(a) Sketch the hardware and the T-s diagram for this heat pump.
(b) If the ground water enters the condenser at 10°C, what will be the temperature of the cooling water as it is returned to the ground? Take $c_p$ = 4.18 kJ/kg·K for cooling water.
(c) Determine the COP of the heat pump.

Refrigerant-134a data: $T$ = 20°C: $h_f$ = 79.3 kJ/kg, $h_g$ = 261.6 kJ/kg; $P$ = 1400 kPa: $h_f$ = 127.2 kJ/kg, $h_g$ = 276.2 kJ/kg

**9–201** A refrigeration unit operates on the ideal vapor compression refrigeration cycle and uses refrigerant-22 as the working fluid. The operating conditions for this unit are evaporator saturation temperature of −5°C and the condenser saturation temperature of 45°C. Selected data for refrigerant-22 are provided in the table below.

| T, °C | $P_{sat}$, kPa | $h_f$, kJ/kg | $h_g$, kJ/kg | $s_g$, kJ/kg·K |
|---|---|---|---|---|
| −5 | 421.2 | 38.76 | 248.1 | 0.9344 |
| 45 | 1728 | 101 | 261.9 | 0.8682 |

For R-22 at $P$ = 1728 kPa and $s$ = 0.9344 kJ/kg·K, $T$ = 68.15°C and $h$ = 283.7 kJ/kg. Also, take $c_{p,air}$ = 1.005 kJ/kg·K.

(a) Sketch the hardware and the T-s diagram for this heat pump application.
(b) Determine the COP for this refrigeration unit.
(c) The evaporator of this unit is located inside the air handler of the building. The air flowing through the air handler enters the air handler at 27°C and is limited to a 20°C temperature drop. Determine the ratio of volume flow rate of air entering the air handler ($m_{air}^3$/min) to mass flow rate of R-22 ($kg_{R-22}$/s) through the air handler, in ($m_{air}^3$/min)/($kg_{R-22}$/s). Assume the air pressure is 100 kPa.

**9–202** Consider an ice producing plant that operates on the ideal vapor-compression refrigeration cycle and uses refrigerant-134a as the working fluid. The refrigeration cycle operating conditions require an evaporator pressure of 140 kPa and the condenser pressure of 1200 kPa. Cooling water flows through the water jacket surrounding the condenser and is supplied at the rate of 200 kg/s. The cooling water has a 10°C temperature rise as it flows through the water jacket. To produce ice, potable water is supplied to the chiller section of

the refrigeration cycle. For each kg of ice produced, 333 kJ of energy must be removed from the potable water supply.

(a) Sketch the hardware for all three working fluids of this refrigerant-ice making system and the T-s diagram for refrigeration cycle.
(b) Determine the mass flow rate of the refrigerant, in kg/s.
(c) Determine the mass flow rate of the potable water supply, in kg/s.

## Design and Essay Problems

**9–203** The amount of fuel introduced into a spark-ignition engine is used in part to control the power produced by the engine. Gasoline produces approximately 42,000 kJ/kg when burned with air in a spark-ignition engine. Develop a schedule for gasoline consumption and maximum cycle temperature versus power production for an Otto cycle with a compression ratio of 8.

**9–204** The weight of a diesel engine is directly proportional to the compression ratio ($W = kr$) because extra metal must be used to strengthen the engine for the higher pressures. Examine the net specific work produced by a diesel engine per unit of weight as the pressure ratio is varied and the specific heat input remains fixed. Do this for several heat inputs and proportionality constants $k$. Are there any optimal combinations of $k$ and specific heat inputs.

**9–205** Design a closed-system air-standard gas power cycle composed of three processes and having a minimum thermal efficiency of 20 percent. The processes may be isothermal, isobaric, isochoric, isentropic, polytropic, or pressure as a linear function of volume; however, the Otto, Diesel, Ericsson, and Stirling cycles may not be used. Prepare an engineering report describing your design, showing the system, P-v and T-s diagrams, and sample calculations.

**9–206** Write an essay on the most recent developments on the two-stroke engines, and find out when we might be seeing cars powered by two-stroke engines in the market. Why do the major car manufacturers have a renewed interest in two-stroke engines?

**9–207** Exhaust gases from the turbine of a simple Brayton cycle are quite hot and may be used for other thermal purposes. One proposed use is generating saturated steam at 110°C from water at 30°C in a boiler. This steam will be distributed to several buildings on a college campus for space heating. A Brayton cycle with a pressure ratio of 6 is to be used for this purpose. Plot the power produced, the flow rate of produced steam, and the maximum cycle temperature as functions of the rate at which heat is added to the cycle. The temperature at the turbine inlet is not to exceed 2000°C.

**9–208** Design a steam power cycle that can achieve a cycle thermal efficiency of at least 40 percent under the conditions that all turbines have isentropic efficiencies of 85 percent and all pumps have isentropic efficiencies of 60 percent. Prepare an engineering report describing your design. Your design report must include, but is not limited to, the following:

(a) Discussion of various cycles attempted to meet the goal as well as the positive and negative aspects of your design.
(b) System figures and *T-s* diagrams with labeled states and temperature, pressure, enthalpy, and entropy information for your design.
(c) Sample calculations.

**9–209** A 10-MW geothermal power plant is being considered at a site where geothermal water at 230°C is available. Geothermal water is to be flashed into a chamber to a lower pressure where part of the water evaporates. The liquid is returned to the ground while the vapor is used to drive the steam turbine. The pressures at the turbine inlet and the turbine exit are to remain above 200 kPa and 8 kPa, respectively. High-pressure flash chambers yield a small amount of steam with high exergy whereas lower-pressure flash chambers yield considerably more steam but at a lower exergy. By trying several pressures, determine the optimum pressure of the flash chamber to maximize the power production per unit mass of geothermal water withdrawn. Also, determine the thermal efficiency for each case assuming 10 percent of the power produced is used to drive the pumps and other auxiliary equipment.

**FIGURE P9–209**

**9–210** Design the condenser of a steam power plant that has a thermal efficiency of 40 percent and generates 10 MW of net electric power. Steam enters the condenser as saturated vapor at 10 kPa, and it is to be condensed outside horizontal tubes through which cooling water from a nearby river flows. The temperature rise of the cooling water is limited to 8°C, and the velocity of the cooling water in the pipes is limited to 6 m/s to keep the pressure drop at an acceptable level. From prior experience, the average heat flux based on the outer surface of the tubes can be taken to be 12,000 W/m². Specify the pipe diameter, total pipe length, and the arrangement of the pipes to minimize the condenser volume.

**9–211** Solar or photovoltaic (PV) cells convert sunlight to electricity and are commonly used to power calculators, satellites, remote communication systems, and even pumps. The conversion of light to electricity is called the *photoelectric effect*. It was first discovered in 1839 by Frenchman Edmond Becquerel, and the first PV module, which consisted of several cells connected to each other, was built in 1954 by Bell Laboratories. The PV modules today have conversion efficiencies of about 12 to 15 percent. Noting that the solar energy incident on a normal surface on earth at noontime is about 1000 W/m² during a clear day, PV modules on a 1-m² surface can provide as much as 150 W of electricity. The annual average daily solar energy incident on a horizontal surface in the United States ranges from about 2 to 6 kWh/m².

A PV-powered pump is to be used in Arizona to pump water for wildlife from a depth of 180 m at an average rate of 400 L/day. Assuming a reasonable efficiency for the pumping system, which can be defined as the ratio of the increase in the potential energy of the water to the electrical energy consumed by the pump, and taking the conversion efficiency of the PV cells to be 0.13 to be on the conservative side, determine the size of the PV module that needs to be installed, in m².

**FIGURE P9–211**

**9–212** A company owns a refrigeration system whose refrigeration capacity is 200 tons (1 ton of refrigeration = 211 kJ/min), and you are to design a forced-air cooling system for fruits whose diameters do not exceed 7 cm under the following conditions: The fruits are to be cooled from 28°C to an average temperature of 8°C. The air temperature is to remain above −2°C and below 10°C at all times, and the velocity of air approaching the fruits must remain under 2 m/s. The cooling section can be as wide as 3.5 m and as high as 2 m.

Assuming reasonable values for the average fruit density, specific heat, and porosity (the fraction of air volume in a box), recommend reasonable values for (a) the air velocity approaching the cooling section, (b) the product-cooling capacity of the system, in kg·fruit/h, and (c) the volume flow rate of air.

# FLUID MECHANICS

# INTRODUCTION AND PROPERTIES OF FLUIDS

I n the second part of the text we present the fundamentals of fluid mechanics. In this introductory chapter, we introduce the basic concepts commonly used in the analysis of fluid flow to avoid any misunderstandings. We start with a discussion of the no-slip condition, which is responsible for the development of boundary layers adjacent to the solid surfaces, and the no-temperature-jump condition. We continue with a discussion of the numerous ways of classification of fluid flow, such as *viscous versus inviscid regions of flow, internal versus external flow, compressible versus incompressible flow, laminar versus turbulent flow, natural versus forced flow,* and *steady versus unsteady flow*. Then we discuss the properties *vapor pressure* and *viscosity*, which play a dominant role in most aspects of fluid flow. Finally, we present the property *surface tension*, and determine the *capillary rise* from static equilibrium conditions.

## OBJECTIVES

The objectives of this chapter are to:

- Understand the basic concepts of Fluid Mechanics

- Recognize the various types of fluid flow problems encountered in practice.

- Understand the vapor pressure and its role in the occurrence of cavitation.

- Have a working knowledge of viscosity and the consequences of the frictional effects it causes in fluid flow.

- Calculate the capillary rises and drops due to the surface tension effects.

**FIGURE 10–1**

The development of a velocity profile due to the no-slip condition as a fluid flows over a blunt nose.

*"Hunter Rouse: Laminar and Turbulent Flow Film." Copyright IIHR-Hydroscience & Engineering, The University of Iowa. Used by permission.*

**FIGURE 10–2**

A fluid flowing over a stationary surface comes to a complete stop at the surface because of the no-slip condition.

# 10–1 ■ THE NO-SLIP CONDITION

Fluid flow is often confined by solid surfaces, and it is important to understand how the presence of solid surfaces affects fluid flow. We know that water in a river cannot flow through large rocks, and must go around them. That is, the water velocity normal to the rock surface must be zero, and water approaching the surface normally comes to a complete stop at the surface. What is not as obvious is that water approaching the rock at any angle also comes to a complete stop at the rock surface, and thus the tangential velocity of water at the surface is also zero.

Consider the flow of a fluid in a stationary pipe or over a solid surface that is nonporous (i.e., impermeable to the fluid). All experimental observations indicate that a fluid in motion comes to a complete stop at the surface and assumes a zero velocity relative to the surface. That is, a fluid in direct contact with a solid "sticks" to the surface, and there is no slip. This is known as the **no-slip condition**. The fluid property responsible for the no-slip condition and the development of the boundary layer is *viscosity* and is discussed in Sec. 10–4.

The photograph in Fig. 10–1 clearly shows the evolution of a velocity gradient as a result of the fluid sticking to the surface of a blunt nose. The layer that sticks to the surface slows the adjacent fluid layer because of viscous forces between the fluid layers, which slows the next layer, and so on. A consequence of the no-slip condition is that all velocity profiles must have zero values with respect to the surface at the points of contact between a fluid and a solid surface (Fig. 10–2). Therefore, the no-slip condition is responsible for the development of the velocity profile. The flow region adjacent to the wall in which the viscous effects (and thus the velocity gradients) are significant is called the **boundary layer**. Another consequence of the no-slip condition is the *surface drag*, or *skin friction drag*, which is the force a fluid exerts on a surface in the flow direction.

When a fluid is forced to flow over a curved surface, such as the back side of a cylinder, the boundary layer may no longer remain attached to the surface and separates from the surface—a process called **flow separation** (Fig. 10–3). We emphasize that the no-slip condition applies *everywhere* along the surface, even downstream of the separation point.

A phenomenon similar to the no-slip condition occurs in heat transfer. When two bodies at different temperatures are brought into contact, heat transfer occurs such that both bodies assume the same temperature at the points of contact. Therefore, a fluid and a solid surface have the same temperature at the points of contact. This is known as **no-temperature-jump condition**.

**FIGURE 10–3**

Flow separation during flow over a curved surface.

*From G. M. Homsy et al, "Multi-Media Fluid Mechanics," Cambridge Univ. Press (2001). ISBN 0-521-78748-3. Reprinted by permission.*

# 10–2 ▪ CLASSIFICATION OF FLUID FLOWS

Earlier we defined *fluid mechanics* as the science that deals with the behavior of fluids at rest or in motion, and the interaction of fluids with solids or other fluids at the boundaries. There is a wide variety of fluid flow problems encountered in practice, and it is usually convenient to classify them on the basis of some common characteristics to make it feasible to study them in groups. There are many ways to classify fluid flow problems, and here we present some general categories.

## Viscous versus Inviscid Regions of Flow

When two fluid layers move relative to each other, a friction force develops between them and the slower layer tries to slow down the faster layer. This internal resistance to flow is quantified by the fluid property *viscosity*, which is a measure of internal stickiness of the fluid. Viscosity is caused by cohesive forces between the molecules in liquids and by molecular collisions in gases. There is no fluid with zero viscosity, and thus all fluid flows involve viscous effects to some degree. Flows in which the frictional effects are significant are called **viscous flows**. However, in many flows of practical interest, there are *regions* (typically regions not close to solid surfaces) where viscous forces are negligibly small compared to inertial or pressure forces. Neglecting the viscous terms in such **inviscid flow regions** greatly simplifies the analysis without much loss in accuracy.

The development of viscous and inviscid regions of flow as a result of inserting a flat plate parallel into a fluid stream of uniform velocity is shown in Fig. 10–4. The fluid sticks to the plate on both sides because of the no-slip condition, and the thin boundary layer in which the viscous effects are significant near the plate surface is the *viscous flow region*. The region of flow on both sides away from the plate and largely unaffected by the presence of the plate is the *inviscid flow region*.

**FIGURE 10–4**

The flow of an originally uniform fluid stream over a flat plate, and the regions of viscous flow (next to the plate on both sides) and inviscid flow (away from the plate).

*Fundamentals of Boundary Layers,*
*National Committee from Fluid Mechanics Films,*
*© Education Development Center.*

## Internal versus External Flow

A fluid flow is classified as being internal or external, depending on whether the fluid flows in a confined space or over a surface. The flow of an unbounded fluid over a surface such as a plate, a wire, or a pipe is **external flow**. The flow in a pipe or duct is **internal flow** if the fluid is completely bounded by solid surfaces. Water flow in a pipe, for example, is internal flow, and airflow over a ball or over an exposed pipe during a windy day is external flow (Fig. 10–5). The flow of liquids in a duct is called *open-channel flow* if the duct is only partially filled with the liquid and there is a free surface. The flows of water in rivers and irrigation ditches are examples of such flows.

Internal flows are dominated by the influence of viscosity throughout the flow field. In external flows the viscous effects are limited to boundary layers near solid surfaces and to wake regions downstream of bodies.

**FIGURE 10–5**

External flow over a tennis ball, and the turbulent wake region behind.

*Courtesy NASA and Cislunar Aerospace, Inc.*

## Compressible versus Incompressible Flow

A flow is classified as being *compressible* or *incompressible,* depending on the level of variation of density during flow. Incompressibility is an approximation, in which the flow is said to be **incompressible** if the density remains nearly

**FIGURE 10–6**

Schlieren image of a small model of the space shuttle orbiter being tested at Mach 3 in the supersonic wind tunnel of the Penn State Gas Dynamics Lab. Several *oblique shocks* are seen in the air surrounding the spacecraft.

*Photo by G. S. Settles, Penn State University. Used by permission.*

Laminar

Transitional

Turbulent

**FIGURE 10–7**

Laminar, transitional, and turbulent flows over a flat plate.

*Courtesy ONERA, photograph by Werlé.*

constant throughout. Therefore, the volume of every portion of fluid remains unchanged over the course of its motion when the flow is approximated as incompressible.

The densities of liquids are essentially constant, and thus the flow of liquids is typically incompressible. Therefore, liquids are usually referred to as *incompressible substances.* A pressure of 210 atm, for example, causes the density of liquid water at 1 atm to change by just 1 percent. Gases, on the other hand, are highly compressible. A pressure change of just 0.01 atm, for example, causes a change of 1 percent in the density of atmospheric air.

When analyzing rockets, spacecraft, and other systems that involve high-speed gas flows (Fig. 10–6), the flow speed is often expressed in terms of the dimensionless **Mach number** defined as

$$\text{Ma} = \frac{V}{c} = \frac{\text{Speed of flow}}{\text{Speed of sound}}$$

where $c$ is the **speed of sound** whose value is 346 m/s in air at room temperature at sea level. A flow is called **sonic** when $\text{Ma} = 1$, **subsonic** when $\text{Ma} < 1$, **supersonic** when $\text{Ma} > 1$, and **hypersonic** when $\text{Ma} \gg 1$.

Liquid flows are incompressible to a high level of accuracy, but the level of variation of density in gas flows and the consequent level of approximation made when modeling gas flows as incompressible depends on the Mach number. Gas flows can often be approximated as incompressible if the density changes are under about 5 percent, which is usually the case when $\text{Ma} < 0.3$. Therefore, the compressibility effects of air at room temperature can be neglected at speeds under about 100 m/s.

Small density changes of liquids corresponding to large pressure changes can still have important consequences. The irritating "water hammer" in a water pipe, for example, is caused by the vibrations of the pipe generated by the reflection of pressure waves following the sudden closing of the valves.

## Laminar versus Turbulent Flow

Some flows are smooth and orderly while others are rather chaotic. The highly ordered fluid motion characterized by smooth layers of fluid is called **laminar**. The word *laminar* comes from the movement of adjacent fluid particles together in "laminae." The flow of high-viscosity fluids such as oils at low velocities is typically laminar. The highly disordered fluid motion that typically occurs at high velocities and is characterized by velocity fluctuations is called **turbulent** (Fig. 10–7). The flow of low-viscosity fluids such as air at high velocities is typically turbulent. A flow that alternates between being laminar and turbulent is called **transitional**. The experiments conducted by Osborne Reynolds in the 1880s resulted in the establishment of the dimensionless **Reynolds number, Re**, as the key parameter for the determination of the flow regime in pipes.

## Natural (or Unforced) versus Forced Flow

A fluid flow is said to be natural or forced, depending on how the fluid motion is initiated. In **forced flow**, a fluid is forced to flow over a surface or in a pipe by external means such as a pump or a fan. In **natural flows**, fluid motion is due to natural means such as the buoyancy effect, which manifests itself as the rise of warmer (and thus lighter) fluid and the fall of cooler (and thus denser) fluid (Fig. 10–8). In solar hot-water systems, for example, the thermosiphoning effect is commonly used to replace pumps by placing the water tank sufficiently above the solar collectors.

## Steady versus Unsteady Flow

The terms *steady* and *uniform* are used frequently in engineering, and thus it is important to have a clear understanding of their meanings. The term **steady** implies *no change of properties, velocity, temperature, etc., at a point with time.* The opposite of steady is **unsteady**. The term **uniform** implies *no change with location* over a specified region. These meanings are consistent with their everyday use (steady girlfriend, uniform distribution, etc.).

The terms *unsteady* and *transient* are often used interchangeably, but these terms are not synonyms. In fluid mechanics, *unsteady* is the most general term that applies to any flow that is not steady, but **transient** is typically used for developing flows. When a rocket engine is fired up, for example, there are transient effects (the pressure builds up inside the rocket engine, the flow accelerates, etc.) until the engine settles down and operates steadily. The term **periodic** refers to the kind of unsteady flow in which the flow oscillates about a steady mean.

Many devices such as turbines, compressors, boilers, condensers, and heat exchangers operate for long periods of time under the same conditions, and they are classified as *steady-flow devices.* (Note that the flow field near the rotating blades of a turbomachine is of course unsteady, but we consider the overall flow field rather than the details at some localities when we classify devices.) During steady flow, the fluid properties can change from point to point within a device, but at any fixed point they remain constant. Therefore, the volume, the mass, and the total energy content of a steady-flow device or flow section remain constant in steady operation.

Steady-flow conditions can be closely approximated by devices that are intended for continuous operation such as turbines, pumps, boilers, condensers, and heat exchangers of power plants or refrigeration systems. Some cyclic devices, such as reciprocating engines or compressors, do not satisfy the steady-flow conditions since the flow at the inlets and the exits is pulsating and not steady. However, the fluid properties vary with time in a periodic manner, and the flow through these devices can still be analyzed as a steady-flow process by using time-averaged values for the properties.

Some fascinating visualizations of fluid flow are provided in the book *An Album of Fluid Motion* by Milton Van Dyke (1982). A nice illustration of an unsteady-flow field is shown in Fig. 10–9, taken from Van Dyke's book. Figure 10–9a is an instantaneous snapshot from a high-speed motion picture; it reveals large, alternating, swirling, turbulent eddies that are shed into the periodically oscillating wake from the blunt base of the object. The eddies produce shock waves that move upstream alternately over the top and bottom surfaces of the airfoil in an unsteady fashion. Figure 10–9b shows the *same* flow field, but the film is exposed for a longer time so that the image is time averaged over 12 cycles. The resulting time-averaged flow field appears "steady" since the details of the unsteady oscillations have been lost in the long exposure.

One of the most important jobs of an engineer is to determine whether it is sufficient to study only the time-averaged "steady" flow features of a problem, or whether a more detailed study of the unsteady features is required. If the engineer were interested only in the overall properties of the

**FIGURE 10–8**

In this schlieren image of a girl in a swimming suit, the rise of lighter, warmer air adjacent to her body indicates that humans and warm-blooded animals are surrounded by thermal plumes of rising warm air.

*G. S. Settles, Gas Dynamics Lab, Penn State University. Used by permission.*

(a)

(b)

**FIGURE 10–9**

Oscillating wake of a blunt-based airfoil at Mach number 0.6. Photo (a) is an instantaneous image, while photo (b) is a long-exposure (time-averaged) image.

*(a) Dyment, A., Flodrops, J. P. & Gryson, P. 1982 in Flow Visualization II, W. Merzkirch, ed., 331–336. Washington: Hemisphere. Used by permission of Arthur Dyment.*

*(b) Dyment, A. & Gryson, P. 1978 in Inst. Mèc. Fluides Lille, No. 78-5. Used by permission of Arthur Dyment.*

flow field (such as the time-averaged drag coefficient, the mean velocity, and pressure fields), a time-averaged description like that of Fig. 10–9b, time-averaged experimental measurements, or an analytical or numerical calculation of the time-averaged flow field would be sufficient. However, if the engineer were interested in details about the unsteady-flow field, such as flow-induced vibrations, unsteady pressure fluctuations, or the sound waves emitted from the turbulent eddies or the shock waves, a time-averaged description of the flow field would be insufficient.

Most of the analytical and computational examples provided in this textbook deal with steady or time-averaged flows, although we occasionally point out some relevant unsteady-flow features as well when appropriate.

## One-, Two-, and Three-Dimensional Flows

A flow field is best characterized by its velocity distribution, and thus a flow is said to be one-, two-, or three-dimensional if the flow velocity varies in one, two, or three primary dimensions, respectively. A typical fluid flow involves a three-dimensional geometry, and the velocity may vary in all three dimensions, rendering the flow three-dimensional [$\vec{V}(x, y, z)$ in rectangular or $\vec{V}(r, \theta, z)$ in cylindrical coordinates]. However, the variation of velocity in certain directions can be small relative to the variation in other directions and can be ignored with negligible error. In such cases, the flow can be modeled conveniently as being one- or two-dimensional, which is easier to analyze.

Consider steady flow of a fluid entering from a large tank into a circular pipe. The fluid velocity everywhere on the pipe surface is zero because of the no-slip condition, and the flow is two-dimensional in the entrance region of the pipe since the velocity changes in both the $r$- and $z$-directions, but not in the $\theta$-direction. The velocity profile develops fully and remains unchanged after some distance from the inlet (about 10 pipe diameters in turbulent flow, and less in laminar pipe flow, as in Fig. 10–10), and the flow in this region is said to be *fully developed*. The fully developed flow in a circular pipe is *one-dimensional* since the velocity varies in the radial $r$-direction but not in the angular $\theta$- or axial $z$-directions, as shown in Fig. 10–10. That is, the velocity profile is the same at any axial $z$-location, and it is symmetric about the axis of the pipe.

Note that the dimensionality of the flow also depends on the choice of coordinate system and its orientation. The pipe flow discussed, for example, is one-dimensional in cylindrical coordinates, but two-dimensional in Cartesian coordinates—illustrating the importance of choosing the most appropriate coordinate system. Also note that even in this simple flow, the velocity cannot be uniform across the cross section of the pipe because of the no-slip condition. However, at a well-rounded entrance to the pipe, the velocity

**FIGURE 10–10**

The development of the velocity profile in a circular pipe. $V = V(r, z)$ and thus the flow is two-dimensional in the entrance region, and becomes one-dimensional downstream when the velocity profile fully develops and remains unchanged in the flow direction, $V = V(r)$.

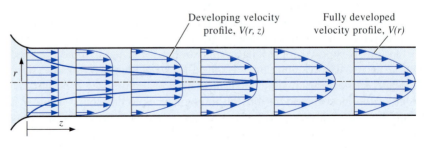

Developing velocity profile, V(r, z)          Fully developed velocity profile, V(r)

profile may be approximated as being nearly uniform across the pipe, since the velocity is nearly constant at all radii except very close to the pipe wall.

A flow may be approximated as *two-dimensional* when the aspect ratio is large and the flow does not change appreciably along the longer dimension. For example, the flow of air over a car antenna can be considered two-dimensional except near its ends since the antenna's length is much greater than its diameter, and the airflow hitting the antenna is fairly uniform (Fig. 10–11).

**FIGURE 10–11**

Flow over a car antenna is approximately two-dimensional except near the top and bottom of the antenna.

---

### EXAMPLE 10–1    Axisymmetric Flow over a Bullet

Consider a bullet piercing through calm air during a short time interval in which the bullet's speed is nearly constant. Determine if the time-averaged airflow over the bullet during its flight is one-, two-, or three-dimensional (Fig. 10–12).

**SOLUTION**   It is to be determined whether airflow over a bullet is one-, two-, or three-dimensional.
**Assumptions**   There are no significant winds and the bullet is not spinning.
**Analysis**   The bullet possesses an axis of symmetry and is therefore an axisymmetric body. The airflow upstream of the bullet is parallel to this axis, and we expect the time-averaged airflow to be rotationally symmetric about the axis—such flows are said to be axisymmetric. The velocity in this case varies with axial distance $z$ and radial distance $r$, but not with angle $\theta$. Therefore, the time-averaged airflow over the bullet is **two-dimensional**.
**Discussion**   While the time-averaged airflow is axisymmetric, the *instantaneous* airflow is not, as illustrated in Fig. 10–9. In Cartesian coordinates, the flow would be three-dimensional.

**FIGURE 10–12**

Axisymmetric flow over a bullet.

## 10–3 · VAPOR PRESSURE AND CAVITATION

It is well-established that temperature and pressure are dependent properties for pure substances during phase-change processes, and there is one-to-one correspondence between temperature and pressure. At a given pressure, the temperature at which a pure substance changes phase is called the **saturation temperature** $T_{sat}$. Likewise, at a given temperature, the pressure at which a pure substance changes phase is called the **saturation pressure** $P_{sat}$. At an absolute pressure of 1 standard atmosphere (1 atm or 101.325 kPa), for example, the saturation temperature of water is 100°C. Conversely, at a temperature of 100°C, the saturation pressure of water is 1 atm.

The **vapor pressure** $P_v$ of a pure substance is defined as *the pressure exerted by its vapor in phase equilibrium with its liquid at a given temperature* (Fig. 10–13) $P_v$ is a property of the pure substance, and turns out to be identical to the saturation pressure $P_{sat}$ of the liquid ($P_v = P_{sat}$). We must be careful not to confuse vapor pressure with *partial pressure*. **Partial pressure** is defined as *the pressure of a gas or vapor in a mixture with other gases*. For example, atmospheric air is a mixture of dry air and water vapor, and atmospheric pressure is the sum of the partial pressure of dry air and the partial pressure of water vapor. The partial pressure of water vapor constitutes a small fraction (usually under 3 percent) of the atmospheric pressure since air is mostly nitrogen and oxygen. The partial pressure of a vapor must be less than or equal to the vapor pressure

Water molecules—vapor phase

Water molecules—liquid phase

**FIGURE 10–13**

The vapor pressure (saturation pressure) of a pure substance (e.g., water) is the pressure exerted by its vapor molecules when the system is in phase equilibrium with its liquid molecules at a given temperature.

### TABLE 10–1

Saturation (or vapor) pressure of water at various temperatures

| Temperature $T$, °C | Saturation Pressure $P_{sat}$, kPa |
|---|---|
| −10 | 0.260 |
| −5 | 0.403 |
| 0 | 0.611 |
| 5 | 0.872 |
| 10 | 1.23 |
| 15 | 1.71 |
| 20 | 2.34 |
| 25 | 3.17 |
| 30 | 4.25 |
| 40 | 7.38 |
| 50 | 12.35 |
| 100 | 101.3 (1 atm) |
| 150 | 475.8 |
| 200 | 1554 |
| 250 | 3973 |
| 300 | 8581 |

### FIGURE 10–14

Cavitation damage on a 16-mm by 23-mm aluminum sample tested at 60 m/s for 2.5 hours. The sample was located at the cavity collapse region downstream of a cavity generator specifically designed to produce high damage potential.

*Photo by David Stinebring, ARL/Pennsylvania State University.*
*Used by permission.*

if there is no liquid present. However, when both vapor and liquid are present and the system is in phase equilibrium, the partial pressure of the vapor must equal the vapor pressure, and the system is said to be *saturated*. The rate of evaporation from open water bodies such as lakes is controlled by the difference between the vapor pressure and the partial pressure. For example, the vapor pressure of water at 20°C is 2.34 kPa. Therefore, a bucket of water at 20°C left in a room with dry air at 1 atm will continue evaporating until one of two things happens: the water evaporates away (there is not enough water to establish phase equilibrium in the room), or the evaporation stops when the partial pressure of the water vapor in the room rises to 2.34 kPa at which point phase equilibrium is established.

For phase-change processes between the liquid and vapor phases of a pure substance, the saturation pressure and the vapor pressure are equivalent since the vapor is pure. Note that the pressure value would be the same whether it is measured in the vapor or liquid phase (provided that it is measured at a location close to the liquid–vapor interface to avoid the hydrostatic effects). Vapor pressure increases with temperature. Thus, a substance at higher pressure boils at higher temperature. For example, water boils at 134°C in a pressure cooker operating at 3 atm absolute pressure, but it boils at 93°C in an ordinary pan at a 2000-m elevation, where the atmospheric pressure is 0.8 atm. The saturation (or vapor) pressures are given in Appendices 1 and 2 for various substances. An abridged table for water is given in Table 10–1 for easy reference.

The reason for our interest in vapor pressure is the possibility of the liquid pressure in liquid-flow systems dropping below the vapor pressure at some locations, and the resulting unplanned vaporization. For example, water at 10°C may vaporize and form bubbles at locations (such as the tip regions of impellers or suction sides of pumps) where the pressure drops below 1.23 kPa. The vapor bubbles (called **cavitation bubbles** since they form "cavities" in the liquid) collapse as they are swept away from the low-pressure regions, generating highly destructive, extremely high-pressure waves. This phenomenon, which is a common cause for drop in performance and even the erosion of impeller blades, is called **cavitation**, and it is an important consideration in the design of hydraulic turbines and pumps (Fig. 10–14).

Cavitation must be avoided (or at least minimized) in most flow systems since it reduces performance, generates annoying vibrations and noise, and causes damage to equipment. We note that some flow systems use cavitation to their *advantage*, e.g., high-speed "supercavitating" torpedoes. The pressure spikes resulting from the large number of bubbles collapsing near a solid surface over a long period of time may cause erosion, surface pitting, fatigue failure, and the eventual destruction of the components or machinery. The presence of cavitation in a flow system can be sensed by its characteristic tumbling sound.

### EXAMPLE 10–2  Minimum Pressure to Avoid Cavitation

In a water distribution system, the temperature of water is observed to be as high as 30°C. Determine the minimum pressure allowed in the system to avoid cavitation.

**SOLUTION** The minimum pressure in a water distribution system to avoid cavitation is to be determined.

**Properties** The vapor pressure of water at 30°C is 4.25 kPa.

**Analysis** To avoid cavitation, the pressure anywhere in the flow should not be allowed to drop below the vapor (or saturation) pressure at the given temperature. That is,

$$P_{min} = P_{sat@30°C} = \mathbf{4.25\ kPa}$$

Therefore, the pressure should be maintained above 4.25 kPa everywhere in the flow.

**Discussion** Note that the vapor pressure increases with increasing temperature, and thus the risk of cavitation is greater at higher fluid temperatures.

# 10–4 ▪ VISCOSITY

When two solid bodies in contact move relative to each other, a friction force develops at the contact surface in the direction opposite to motion. To move a table on the floor, for example, we have to apply a force to the table in the horizontal direction large enough to overcome the friction force. The magnitude of the force needed to move the table depends on the *friction coefficient* between the table legs and the floor.

The situation is similar when a fluid moves relative to a solid or when two fluids move relative to each other. We move with relative ease in air, but not so in water. Moving in oil would be even more difficult, as can be observed by the slower downward motion of a glass ball dropped in a tube filled with oil. It appears that there is a property that represents the internal resistance of a fluid to motion or the "fluidity," and that property is the **viscosity**. The force a flowing fluid exerts on a body in the flow direction is called the **drag force**, and the magnitude of this force depends, in part, on viscosity (Fig. 10–15).

To obtain a relation for viscosity, consider a fluid layer between two very large parallel plates (or equivalently, two parallel plates immersed in a large body of a fluid) separated by a distance $\ell$ (Fig. 10–16). Now a constant parallel force $F$ is applied to the upper plate while the lower plate is held fixed. After the initial transients, it is observed that the upper plate moves continuously under the influence of this force at a constant speed $V$. The fluid in contact with the upper plate sticks to the plate surface and moves with it at the same speed, and the shear stress $\tau$ acting on this fluid layer is

$$\tau = \frac{F}{A} \tag{10–1}$$

where $A$ is the contact area between the plate and the fluid. Note that the fluid layer deforms continuously under the influence of shear stress.

The fluid in contact with the lower plate assumes the velocity of that plate, which is zero (because of the no-slip condition—see Section 10–1). In steady laminar flow, the fluid velocity between the plates varies linearly between 0 and $V$, and thus the *velocity profile* and the *velocity gradient* are

$$u(y) = \frac{y}{\ell} V \quad \text{and} \quad \frac{du}{dy} = \frac{V}{\ell} \tag{10–2}$$

where $y$ is the vertical distance from the lower plate.

During a differential time interval $dt$, the sides of fluid particles along a vertical line $MN$ rotate through a differential angle $d\beta$ while the upper plate

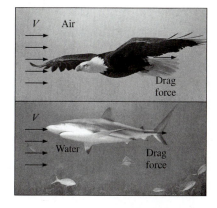

**FIGURE 10–15**
A fluid moving relative to a body exerts a drag force on the body, partly because of friction caused by viscosity.
© *Getty RF*

**FIGURE 10–16**
The behavior of a fluid in laminar flow between two parallel plates when the upper plate moves with a constant velocity.

moves a differential distance $da = V\,dt$. The angular displacement or deformation (or shear strain) can be expressed as

$$d\beta \approx \tan d\beta = \frac{da}{\ell} = \frac{V\,dt}{\ell} = \frac{du}{dy}\,dt \qquad (10\text{–}3)$$

Rearranging, the rate of deformation under the influence of shear stress $\tau$ becomes

$$\frac{d\beta}{dt} = \frac{du}{dy} \qquad (10\text{–}4)$$

Thus we conclude that the rate of deformation of a fluid element is equivalent to the velocity gradient $du/dy$. Furthermore, it can be verified experimentally that for most fluids the rate of deformation (and thus the velocity gradient) is directly proportional to the shear stress $\tau$,

$$\tau \propto \frac{d\beta}{dt} \qquad \text{or} \qquad \tau \propto \frac{du}{dy} \qquad (10\text{–}5)$$

Fluids for which the rate of deformation is linearly proportional to the shear stress are called **Newtonian fluids** after Sir Isaac Newton, who expressed it first in 1687. Most common fluids such as water, air, gasoline, and oils are Newtonian fluids. Blood and liquid plastics are examples of non-Newtonian fluids.

In one-dimensional shear flow of Newtonian fluids, shear stress can be expressed by the linear relationship

*Shear stress*: $$\tau = \mu\frac{du}{dy} \qquad (\text{N/m}^2) \qquad (10\text{–}6)$$

where the constant of proportionality $\mu$ is called the **coefficient of viscosity** or the **dynamic** (or **absolute**) **viscosity** of the fluid, whose unit is kg/m·s, or equivalently, N·s/m$^2$ (or Pa·s where Pa is the pressure unit pascal). A common viscosity unit is **poise**, which is equivalent to 0.1 Pa·s (or *centipoise*, which is one-hundredth of a poise). The viscosity of water at 20°C is 1.002 centipoise, and thus the unit centipoise serves as a useful reference. A plot of shear stress versus the rate of deformation (velocity gradient) for a Newtonian fluid is a straight line whose slope is the viscosity of the fluid, as shown in Fig. 10–17. Note that viscosity is independent of the rate of deformation for Newtonian fluids. Since the rate of deformation is proportional to the strain rate, Fig. 10–17 reveals that viscosity is actually a coefficient in a stress–strain relationship.

The **shear force** acting on a Newtonian fluid layer (or, by Newton's third law, the force acting on the plate) is

*Shear force*: $$F = \tau A = \mu A\frac{du}{dy} \qquad (\text{N}) \qquad (10\text{–}7)$$

where again $A$ is the contact area between the plate and the fluid. Then the force $F$ required to move the upper plate in Fig. 10–16 at a constant speed of $V$ while the lower plate remains stationary is

$$F = \mu A\frac{V}{\ell} \qquad (\text{N}) \qquad (10\text{–}8)$$

**FIGURE 10–17**

The rate of deformation (velocity gradient) of a Newtonian fluid is proportional to shear stress, and the constant of proportionality is the viscosity.

This relation can alternately be used to calculate $\mu$ when the force $F$ is measured. Therefore, the experimental setup just described can be used to measure the viscosity of fluids. Note that under identical conditions, the force $F$ would be very different for different fluids.

For non-Newtonian fluids, the relationship between shear stress and rate of deformation is not linear, as shown in Fig. 10–18. The slope of the curve on the $\tau$ versus $du/dy$ chart is referred to as the *apparent viscosity* of the fluid. Fluids for which the apparent viscosity increases with the rate of deformation (such as solutions with suspended starch or sand) are referred to as *dilatant* or *shear thickening fluids*, and those that exhibit the opposite behavior (the fluid becoming less viscous as it is sheared harder, such as some paints, polymer solutions, and fluids with suspended particles) are referred to as *pseudoplastic* or *shear thinning fluids*. Some materials such as toothpaste can resist a finite shear stress and thus behave as a solid, but deform continuously when the shear stress exceeds the yield stress and behave as a fluid. Such materials are referred to as Bingham plastics after Eugene C. Bingham (1878–1945), who did pioneering work on fluid viscosity for the U.S. National Bureau of Standards in the early twentieth century.

In fluid mechanics and heat transfer, the ratio of dynamic viscosity to density appears frequently. For convenience, this ratio is given the name **kinematic viscosity** $\nu$ and is expressed as $\nu = \mu/\rho$. Two common units of kinematic viscosity are $m^2/s$ and **stoke** ($1$ stoke $= 1$ $cm^2/s = 0.0001$ $m^2/s$).

In general, the viscosity of a fluid depends on both temperature and pressure, although the dependence on pressure is rather weak. For *liquids*, both the dynamic and kinematic viscosities are practically independent of pressure, and any small variation with pressure is usually disregarded, except at extremely high pressures. For *gases*, this is also the case for dynamic viscosity (at low to moderate pressures), but not for kinematic viscosity since the density of a gas is proportional to its pressure (Fig. 10–19).

The viscosity of a fluid is a measure of its "resistance to deformation." Viscosity is due to the internal frictional force that develops between different layers of fluids as they are forced to move relative to each other.

The viscosity of a fluid is directly related to the pumping power needed to transport a fluid in a pipe or to move a body (such as a car in air or a submarine in the sea) through a fluid. Viscosity is caused by the cohesive forces between the molecules in liquids and by the molecular collisions in gases, and it varies greatly with temperature. The viscosity of liquids decreases with temperature, whereas the viscosity of gases increases with temperature (Fig. 10–20). This is because in a liquid the molecules possess more energy at higher temperatures, and they can oppose the large cohesive intermolecular forces more strongly. As a result, the energized liquid molecules can move more freely.

In a gas, on the other hand, the intermolecular forces are negligible, and the gas molecules at high temperatures move randomly at higher velocities. This results in more molecular collisions per unit volume per unit time and therefore in greater resistance to flow. The kinetic theory of gases predicts the viscosity of gases to be proportional to the square root of temperature.

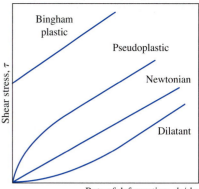

**FIGURE 10–18**

Variation of shear stress with the rate of deformation for Newtonian and non-Newtonian fluids (the slope of a curve at a point is the apparent viscosity of the fluid at that point).

Air at 20°C and 1 atm:
$\mu = 1.83 \times 10^{-5}$ kg/m · s
$\nu = 1.52 \times 10^{-5}$ m$^2$/s

Air at 20°C and 4 atm:
$\mu = 1.83 \times 10^{-5}$ kg/m · s
$\nu = 0.380 \times 10^{-5}$ m$^2$/s

**FIGURE 10–19**

Dynamic viscosity, in general, does not depend on pressure, but kinematic viscosity does.

Viscosity

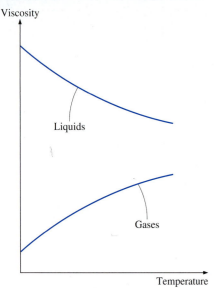

**FIGURE 10–20**

The viscosity of liquids decreases and the viscosity of gases increases with temperature.

**TABLE 10–2**

Dynamic viscosity of some fluids at 1 atm and 20°C (unless otherwise stated)

| Fluid | Dynamic Viscosity $\mu$, kg/m·s |
|---|---|
| Glycerin: | |
| −20°C | 134.0 |
| 0°C | 10.5 |
| 20°C | 1.52 |
| 40°C | 0.31 |
| Engine oil: | |
| SAE 10W | 0.10 |
| SAE 10W30 | 0.17 |
| SAE 30 | 0.29 |
| SAE 50 | 0.86 |
| Mercury | 0.0015 |
| Ethyl alcohol | 0.0012 |
| Water: | |
| 0°C | 0.0018 |
| 20°C | 0.0010 |
| 100°C (liquid) | 0.00028 |
| 100°C (vapor) | 0.000012 |
| Blood, 37°C | 0.00040 |
| Gasoline | 0.00029 |
| Ammonia | 0.00015 |
| Air | 0.000018 |
| Hydrogen, 0°C | 0.0000088 |

That is, $\mu_{gas} \propto \sqrt{T}$. This prediction is confirmed by practical observations, but deviations for different gases need to be accounted for by incorporating some correction factors. The viscosity of gases is expressed as a function of temperature by the Sutherland correlation (from The U.S. Standard Atmosphere) as

*Gases*:
$$\mu = \frac{aT^{1/2}}{1 + b/T} \tag{10–9}$$

where $T$ is absolute temperature and $a$ and $b$ are experimentally determined constants. Note that measuring viscosity at two different temperatures is sufficient to determine these constants. For air at atmospheric conditions, the values of these constants are $a = 1.458 \times 10^{-6}$ kg/(m·s·K$^{1/2}$) and $b = 110.4$ K. The viscosity of gases is independent of pressure at low to moderate pressures (from a few percent of 1 atm to several atm). But viscosity increases at high pressures due to the increase in density.

For liquids, the viscosity is approximated as

*Liquids*:
$$\mu = a10^{b/(T-c)} \tag{10–10}$$

where again $T$ is absolute temperature and $a$, $b$, and $c$ are experimentally determined constants. For water, using the values $a = 2.414 \times 10^{-5}$ N·s/m$^2$, $b = 247.8$ K, and $c = 140$ K results in less than 2.5 percent error in viscosity in the temperature range of 0°C to 370°C (Touloukian et al., 1975).

The viscosities of some fluids at room temperature are listed in Table 10–2. They are plotted against temperature in Fig. 10–21. Note that the viscosities of different fluids differ by several orders of magnitude. Also note that it is more difficult to move an object in a higher-viscosity fluid such as engine oil than it is in a lower-viscosity fluid such as water. Liquids, in general, are much more viscous than gases.

Consider a fluid layer of thickness $\ell$ within a small gap between two concentric cylinders, such as the thin layer of oil in a journal bearing. The gap between the cylinders can be modeled as two parallel flat plates separated by the fluid. Noting that torque is $T = FR$ (force times the moment arm, which is the radius $R$ of the inner cylinder in this case), the tangential velocity is $V = \omega R$ (angular velocity times the radius), and taking the wetted surface area of the inner cylinder to be $A = 2\pi RL$ by disregarding the shear stress acting on the two ends of the inner cylinder, torque can be expressed as

$$T = FR = \mu \frac{2\pi R^3 \omega L}{\ell} = \mu \frac{4\pi^2 R^3 \dot{n} L}{\ell} \tag{10–11}$$

where $L$ is the length of the cylinder and $\dot{n}$ is the number of revolutions per unit time, which is usually expressed in rpm (revolutions per minute). Note that the angular distance traveled during one rotation is $2\pi$ rad, and thus the relation between the angular velocity in rad/min and the rpm is $\omega = 2\pi\dot{n}$. Equation 10–11 can be used to calculate the viscosity of a fluid by measuring torque at a specified angular velocity. Therefore, two concentric cylinders can be used as a *viscometer*, a device that measures viscosity.

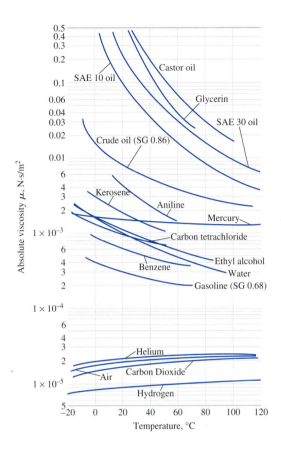

**FIGURE 10–21**

The variation of dynamic (absolute) viscosity of common fluids with temperature at 1 atm (1 N·s/m² = 1 kg/m·s = 0.020886 lbf·s/ft²).

*F. M. White, Fluid Mechanics 4e. Copyright © 1999 The McGraw-Hill Companies, Inc. Used by permission.*

**EXAMPLE 10–3    Determining the Viscosity of a Fluid**

The viscosity of a fluid is to be measured by a viscometer constructed of two 40-cm-long concentric cylinders (Fig. 10–22). The outer diameter of the inner cylinder is 12 cm, and the gap between the two cylinders is 0.15 cm. The inner cylinder is rotated at 300 rpm, and the torque is measured to be 1.8 N·m. Determine the viscosity of the fluid.

**SOLUTION**   The torque and the rpm of a double cylinder viscometer are given. The viscosity of the fluid is to be determined.
*Assumptions*   **1** The inner cylinder is completely submerged in the fluid. **2** The viscous effects on the two ends of the inner cylinder are negligible.
*Analysis*   The velocity profile is linear only when the curvature effects are negligible, and the profile can be approximated as being linear in this case since $\ell/R = 0.025 \ll 1$. Solving Eq. 10–11 for viscosity and substituting the given values, the viscosity of the fluid is determined to be

$$\mu = \frac{T\ell}{4\pi^2 R^3 \dot{n} L} = \frac{(1.8 \text{ N·m})(0.0015 \text{ m})}{4\pi^2 (0.06 \text{ m})^3 (300/60 \text{ 1/s})(0.4 \text{ m})} = \textbf{0.158 N·s/m}^2$$

*Discussion*   Viscosity is a strong function of temperature, and a viscosity value without a corresponding temperature is of little usefulness. Therefore, the temperature of the fluid should have also been measured during this experiment, and reported with this calculation.

Stationary cylinder

$\ell$

$R$

$\dot{n} = 300$ rpm

Shaft

Fluid

**FIGURE 10–22**

Schematic for Example 10–3.

(a)

(b)

**FIGURE 10–23**

Some consequences
of surface tension.

(a) © AEE Fotosearch RF
(b) © Dennis Drenner/Visuals Unlimited.

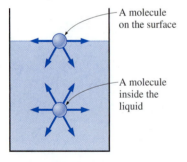

A molecule
on the surface

A molecule
inside the
liquid

**FIGURE 10–24**

Attractive forces acting on a liquid
molecule at the surface and deep
inside the liquid.

# 10–5 ■ SURFACE TENSION AND CAPILLARY EFFECT

It is often observed that a drop of blood forms a hump on a horizontal glass; a drop of mercury forms a near-perfect sphere and can be rolled just like a steel ball over a smooth surface; water droplets from rain or dew hang from branches or leaves of trees; a liquid fuel injected into an engine forms a mist of spherical droplets; water dripping from a leaky faucet falls as nearly spherical droplets; a soap bubble released into the air forms a nearly spherical shape; and water beads up into small drops on flower petals (Fig. 10–23a).

In these and other observances, liquid droplets behave like small balloons filled with the liquid, and the surface of the liquid acts like a stretched elastic membrane under tension. The pulling force that causes this tension acts parallel to the surface and is due to the attractive forces between the molecules of the liquid. The magnitude of this force per unit length is called **surface tension** or *coefficient of surface tension* $\sigma_s$ and is usually expressed in the unit N/m (or lbf/ft in English units). This effect is also called *surface energy* (per unit area) and is expressed in the equivalent unit of N·m/m$^2$ or J/m$^2$. In this case, $\sigma_s$ represents the stretching work that needs to be done to increase the surface area of the liquid by a unit amount.

To visualize how surface tension arises, we present a microscopic view in Fig. 10–24 by considering two liquid molecules, one at the surface and one deep within the liquid body. The attractive forces applied on the interior molecule by the surrounding molecules balance each other because of symmetry. But the attractive forces acting on the surface molecule are not symmetric, and the attractive forces applied by the gas molecules above are usually very small. Therefore, there is a net attractive force acting on the molecule at the surface of the liquid, which tends to pull the molecules on the surface toward the interior of the liquid. This force is balanced by the repulsive forces from the molecules below the surface that are trying to be compressed. The result is that the liquid minimizes its surface area. This is the reason for the tendency of liquid droplets to attain a spherical shape, which has the minimum surface area for a given volume.

You also may have observed, with amusement, that some insects can land on water or even walk on water (Fig. 10–23b) and that small steel needles can float on water. These phenomena are made possible by surface tension which balances the weights of these objects.

To understand the surface tension effect better, consider a liquid film (such as the film of a soap bubble) suspended on a U-shaped wire frame with a movable side (Fig. 10–25). Normally, the liquid film tends to pull the movable wire inward in order to minimize its surface area. A force $F$ needs to be applied on the movable wire in the opposite direction to balance this pulling effect. Both sides of the thin film are surfaces exposed to air, and thus the length along which the surface tension acts in this case is $2b$. Then a force balance on the movable wire gives $F = 2b\sigma_s$, and thus the surface tension can be expressed as

$$\sigma_s = \frac{F}{2b} \tag{10–12}$$

Note that for $b = 0.5$ m, the measured force $F$ (in N) is simply the surface tension in N/m. An apparatus of this kind with sufficient precision can be used to measure the surface tension of various liquids.

In the U-shaped wire frame apparatus, the movable wire is pulled to stretch the film and increase its surface area. When the movable wire is pulled a distance $\Delta x$, the surface area increases by $\Delta A = 2b\,\Delta x$, and the work $W$ done during this stretching process is

$$W = \text{Force} \times \text{Distance} = F\,\Delta x = 2b\sigma_s\,\Delta x = \sigma_s\,\Delta A$$

where we have assumed that the force remains constant over the small distance. This result can also be interpreted as *the surface energy of the film is increased by an amount $\sigma_s\,\Delta A$ during this stretching process*, which is consistent with the alternative interpretation of $\sigma_s$ as surface energy per unit area. This is similar to a rubber band having more potential (elastic) energy after it is stretched further. In the case of liquid film, the work is used to move liquid molecules from the interior parts to the surface against the attraction forces of other molecules. Therefore, surface tension also can be defined as *the work done per unit increase in the surface area of the liquid*.

The surface tension varies greatly from substance to substance, and with temperature for a given substance, as shown in Table 10–3. At 20°C, for example, the surface tension is 0.073 N/m for water and 0.440 N/m for mercury surrounded by atmospheric air. The surface tension of mercury is large enough that mercury droplets form nearly spherical balls that can be rolled like a solid ball on a smooth surface. The surface tension of a liquid, in general, decreases with temperature and becomes zero at the critical point (and thus there is no distinct liquid–vapor interface at temperatures above the critical point). The effect of pressure on surface tension is usually negligible.

The surface tension of a substance can be changed considerably by *impurities*. Therefore, certain chemicals, called *surfactants*, can be added to a liquid to decrease its surface tension. For example, soaps and detergents lower the surface tension of water and enable it to penetrate the small openings between fibers for more effective washing. But this also means that devices whose operation depends on surface tension (such as heat pipes) can be destroyed by the presence of impurities due to poor workmanship.

We speak of surface tension for liquids only at liquid–liquid or liquid–gas interfaces. Therefore, it is imperative that the adjacent liquid or gas be specified when specifying surface tension. Surface tension determines the size of the liquid droplets that form, and so a droplet that keeps growing by the addition of more mass breaks down when the surface tension can no longer hold it together. This is like a balloon that bursts while being inflated when the pressure inside rises above the strength of the balloon material.

A curved interface indicates a pressure difference (or "pressure jump") across the interface with pressure being higher on the concave side. Consider, for example, a droplet of liquid in air, an air (or other gas) bubble in water, or a soap bubble in air. The excess pressure $\Delta P$ above atmospheric pressure can be determined by considering a free-body diagram of half the droplet or bubble (Fig. 10–26). Noting that surface tension acts along the circumference and the pressure acts on the area, horizontal force balances for the droplet or air bubble and the soap bubble give

*Droplet or air bubble:* $(2\pi R)\sigma_s = (\pi R^2)\Delta P_{\text{droplet}} \rightarrow \Delta P_{\text{droplet}} = P_i - P_o = \dfrac{2\sigma_s}{R}$  (10–13)

*Soap bubble:* $2(2\pi R)\sigma_s = (\pi R^2)\Delta P_{\text{bubble}} \rightarrow \Delta P_{\text{bubble}} = P_i - P_o = \dfrac{4\sigma_s}{R}$  (10–14)

**FIGURE 10–25**

Stretching a liquid film with a U-shaped wire, and the forces acting on the movable wire of length $b$.

**TABLE 10–3**

Surface tension of some fluids in air at 1 atm and 20°C (unless otherwise stated)

| Fluid | Surface Tension $\sigma_s$, N/m* |
|---|---|
| †Water: | |
| 0°C | 0.076 |
| 20°C | 0.073 |
| 100°C | 0.059 |
| 300°C | 0.014 |
| Glycerin | 0.063 |
| SAE 30 oil | 0.035 |
| Mercury | 0.440 |
| Ethyl alcohol | 0.023 |
| Blood, 37°C | 0.058 |
| Gasoline | 0.022 |
| Ammonia | 0.021 |
| Soap solution | 0.025 |
| Kerosene | 0.028 |

\* Multiply by 0.06852 to convert to lbf/ft.
† See Appendices for more precise data for water.

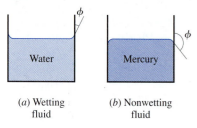

**FIGURE 10–26**
The free-body diagram of half of a droplet or air bubble and half of a soap bubble.

(a) Wetting fluid  (b) Nonwetting fluid

**FIGURE 10–27**
The contact angle for wetting and nonwetting fluids.

**FIGURE 10–28**
The meniscus of colored water in a 4-mm-inner-diameter glass tube. Note that the edge of the meniscus meets the wall of the capillary tube at a very small contact angle.

*Photo by Gabrielle Trembley, Pennsylvania State University. Used by permission.*

where $P_i$ and $P_o$ are the pressures inside and outside the droplet or bubble, respectively. When the droplet or bubble is in the atmosphere, $P_o$ is simply atmospheric pressure. The extra factor of 2 in the force balance for the soap bubble is due to the existence of a soap film with *two* surfaces (inner and outer surfaces) and thus two circumferences in the cross section.

The excess pressure in a droplet of liquid in a gas (or a bubble of gas in a liquid) can also be determined by considering a differential increase in the radius of the droplet due to the addition of a differential amount of mass and interpreting the surface tension as the increase in the surface energy per unit area. Then the increase in the surface energy of the droplet during this differential expansion process becomes

$$\delta W_{surface} = \sigma_s \, dA = \sigma_s \, d(4\pi R^2) = 8\pi R\sigma_s \, dR$$

The expansion work done during this differential process is determined by multiplying the force by distance to obtain

$$\delta W_{expansion} = \text{Force} \times \text{Distance} = F \, dR = (\Delta PA) \, dR = 4\pi R^2 \, \Delta P \, dR$$

Equating the two expressions above gives $\Delta P_{droplet} = 2\sigma_s / R$, which is the same relation obtained before and given in Eq. 10–13. Note that the excess pressure in a droplet or bubble is inversely proportional to the radius.

## Capillary Effect

Another interesting consequence of surface tension is the **capillary effect**, which is the rise or fall of a liquid in a small-diameter tube inserted into the liquid. Such narrow tubes or confined flow channels are called **capillaries**. The rise of kerosene through a cotton wick inserted into the reservoir of a kerosene lamp is due to this effect. The capillary effect is also partially responsible for the rise of water to the top of tall trees. The curved free surface of a liquid in a capillary tube is called the **meniscus**.

It is commonly observed that water in a glass container curves up slightly at the edges where it touches the glass surface; but the opposite occurs for mercury: it curves down at the edges (Fig. 10–27). This effect is usually expressed by saying that water *wets* the glass (by sticking to it) while mercury does not. The strength of the capillary effect is quantified by the **contact** (or *wetting*) **angle** $\phi$, defined as *the angle that the tangent to the liquid surface makes with the solid surface at the point of contact*. The surface tension force acts along this tangent line toward the solid surface. A liquid is said to wet the surface when $\phi < 90°$ and not to wet the surface when $\phi > 90°$. In atmospheric air, the contact angle of water (and most other organic liquids) with glass is nearly zero, $\phi \approx 0°$ (Fig. 10–28). Therefore, the surface tension force acts upward on water in a glass tube along the circumference, tending to pull the water up. As a result, water rises in the tube until the weight of the liquid in the tube above the liquid level of the reservoir balances the surface tension force. The contact angle is 130° for mercury–glass and 26° for kerosene–glass in air. Note that the contact angle, in general, is different in different environments (such as another gas or liquid in place of air).

The phenomenon of the capillary effect can be explained microscopically by considering *cohesive forces* (the forces between like molecules, such as water and water) and *adhesive forces* (the forces between unlike molecules, such as water and glass). The liquid molecules at the solid–liquid interface are subjected to both cohesive forces by other liquid molecules and adhesive forces by the molecules of the solid. The relative magnitudes of these forces determine whether a liquid wets a solid surface or not. Obviously, the water molecules are more strongly attracted to the glass molecules than they are to other water molecules, and thus water tends to rise along the glass surface. The opposite occurs for mercury, which causes the liquid surface near the glass wall to be suppressed (Fig. 10–29).

The magnitude of the capillary rise in a circular tube can be determined from a force balance on the cylindrical liquid column of height $h$ in the tube (Fig. 10–30). The bottom of the liquid column is at the same level as the free surface of the reservoir, and thus the pressure there must be atmospheric pressure. This balances the atmospheric pressure acting at the top surface of the liquid column, and thus these two effects cancel each other. The weight of the liquid column is approximately

$$W = mg = \rho \mathcal{V}g = \rho g(\pi R^2 h)$$

Equating the vertical component of the surface tension force to the weight gives

$$W = F_{surface} \rightarrow \rho g(\pi R^2 h) = 2\pi R \sigma_s \cos \phi$$

Solving for $h$ gives the capillary rise to be

*Capillary rise*:     $$h = \frac{2\sigma_s}{\rho g R} \cos \phi \quad (R = \text{constant})$$     **(10–15)**

This relation is also valid for nonwetting liquids (such as mercury in glass) and gives the capillary drop. In this case $\phi > 90°$ and thus $\cos \phi < 0$, which makes $h$ negative. Therefore, a negative value of capillary rise corresponds to a capillary drop (Fig. 10–29).

Note that the capillary rise is inversely proportional to the radius of the tube. Therefore, the thinner the tube is, the greater the rise (or fall) of the liquid in the tube. In practice, the capillary effect for water is usually negligible in tubes whose diameter is greater than 1 cm. When pressure measurements are made using manometers and barometers, it is important to use sufficiently large tubes to minimize the capillary effect. The capillary rise is also inversely proportional to the density of the liquid, as expected. Therefore, in general, lighter liquids experience greater capillary rises. Finally, it should be kept in mind that Eq. 10–15 is derived for constant-diameter tubes and should not be used for tubes of variable cross section.

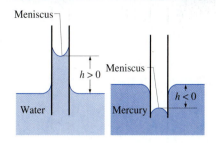

**FIGURE 10–29**

The capillary rise of water and the capillary fall of mercury in a small-diameter glass tube.

**FIGURE 10–30**

The forces acting on a liquid column that has risen in a tube due to the capillary effect.

---

**EXAMPLE 10–4    The Capillary Rise of Water in a Tube**

A 0.6-mm-diameter glass tube is inserted into water at 20°C in a cup. Determine the capillary rise of water in the tube (Fig. 10–31).

**SOLUTION**  The rise of water in a slender tube as a result of the capillary effect is to be determined.
**Assumptions**  1 There are no impurities in the water and no contamination on the surfaces of the glass tube. 2 The experiment is conducted in atmospheric air.
**Properties**  The surface tension of water at 20°C is 0.073 N/m (Table 10–3). The contact angle of water with glass is approximately 0° (from preceding text). We take the density of liquid water to be 1000 kg/m³.

**FIGURE 10–31**

Schematic for Example 10–4.

*Analysis*  The capillary rise is determined directly from Eq. 10–15 by substituting the given values, yielding

$$h = \frac{2\sigma_s}{\rho g R} \cos\phi = \frac{2(0.073 \text{ N/m})}{(1000 \text{ kg/m}^3)(9.81 \text{ m/s}^2)(0.3 \times 10^{-3}\text{m})} (\cos 0°)\left(\frac{1\text{kg}\cdot\text{m/s}^2}{1 \text{ N}}\right)$$

$$= 0.050 \text{ m} = \textbf{5.0 cm}$$

Therefore, water rises in the tube 5 cm above the liquid level in the cup.
*Discussion*  Note that if the tube diameter were 1 cm, the capillary rise would be 0.3 mm, which is hardly noticeable to the eye. Actually, the capillary rise in a large-diameter tube occurs only at the rim. The center does not rise at all. Therefore, the capillary effect can be ignored for large-diameter tubes.

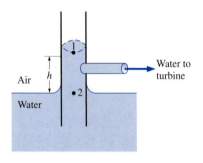

**FIGURE 10–32**

Schematic for Example 10–5.

**EXAMPLE 10–5**     **Using Capillary Rise to Generate Power in a Hydraulic Turbine**

Reconsider Example 10–4. Realizing that water rises by 5 cm under the influence of surface tension without requiring any energy input from an external source, a person conceives the idea that power can be generated by drilling a hole in the tube just below the water level and feeding the water spilling out of the tube into a turbine (Fig. 10–32). The person takes this idea even further by suggesting that a series of tube banks can be used for this purpose and cascading can be incorporated to achieve practically feasible flow rates and elevation differences. Determine if this idea has any merit.

**SOLUTION**  Water that rises in tubes under the influence of the capillary effect is to be used to generate power by feeding it into a turbine. The validity of this suggestion is to be evaluated.

*Analysis*  The proposed system may appear like a stroke of genius, since the commonly used hydroelectric power plants generate electric power by simply capturing the potential energy of elevated water, and the capillary rise provides the mechanism to raise the water to any desired height without requiring any energy input.

When viewed from a thermodynamic point of view, the proposed system immediately can be labeled as a perpetual motion machine (PMM) since it continuously generates electric power without requiring any energy input. That is, the proposed system creates energy, which is a clear violation of the first law of thermodynamics or the conservation of energy principle, and it does not warrant any further consideration. But the fundamental principle of conservation of energy did not stop many from dreaming about being the first to prove nature wrong, and to come up with a trick to permanently solve the world's energy problems. Therefore, the impossibility of the proposed system should be demonstrated.

As you may recall from your physics courses (also to be discussed in the next chapter), the pressure in a static fluid varies in the vertical direction only and increases with increasing depth linearly. Then the pressure difference across the 5-cm high water column in the tube becomes

$$\Delta P_{\text{water column in tube}} = P_2 - P_1 = \rho_{\text{water}}gh$$

$$= (1000 \text{ kg/m}^2)(9.81 \text{ m/s}^2)(0.05 \text{ m})\left(\frac{1 \text{ kN}}{1000 \text{ kg·m/s}^2}\right)$$

$$= 0.49 \text{ kN/m}^2 \ (\approx 0.005 \text{ atm})$$

That is, the pressure at the top of the water column in the tube is 0.005 atm *less* than the pressure at the bottom. Noting that the pressure at the bottom of the water column is atmospheric pressure (since it is at the same horizontal line as the water surface in the cup) the pressure anywhere in the tube is below atmospheric pressure with the difference reaching 0.005 atm at the top. Therefore, if a hole is drilled in the tube, air will leak into the tube rather than water leaking out.

**Discussion** The water column in the tube is motionless, and thus, there cannot be any unbalanced force acting on it (zero net force). The force due to the pressure difference across the meniscus between the atmospheric air and the water at the top of water column is balanced by the surface tension. If this surface-tension force were to disappear, the water in the tube would drop down under the influence of atmospheric pressure to the level of the free surface in the tube.

## SUMMARY

A fluid in direct contact with a solid surface sticks to the surface and there is no slip. This is known as the *no-slip condition*, and it is due to the viscosity of the fluid. At a given temperature, the pressure at which a pure substance changes phase is called the *saturation pressure*. For phase-change processes between the liquid and vapor phases of a pure substance, the saturation pressure is commonly called the *vapor pressure $P_v$*.

At a given temperature, the pressure at which a pure substance changes phase is called the *saturation pressure*. For phase-change processes between the liquid and vapor phases of a pure substance, the saturation pressure is commonly called the *vapor pressure $P_v$*. Vapor bubbles that form in the low-pressure regions in a liquid (a phenomenon called *cavitation*) collapse as they are swept away from the low-pressure regions, generating highly destructive, extremely high-pressure waves.

The *viscosity* of a fluid is a measure of its resistance to deformation. The tangential force per unit area is called *shear stress* and is expressed for simple shear flow between plates (one-dimensional flow) as

$$\tau = \mu \frac{du}{dy}$$

where $\mu$ is the coefficient of viscosity or the *dynamic (or absolute) viscosity* of the fluid, $u$ is the velocity component in the flow direction, and $y$ is the direction normal to the flow direction. Fluids that obey this linear relationship are called *Newtonian fluids*. The ratio of dynamic viscosity to density is called the *kinematic viscosity $v$*.

The pulling effect on the liquid molecules at an interface caused by the attractive forces of molecules per unit length is called *surface tension $\sigma_s$*. The excess pressure $\Delta P$ inside a spherical droplet or soap bubble, respectively, is given by

$$\Delta P_{\text{droplet}} = P_i - P_o = \frac{2\sigma_s}{R} \quad \text{and} \quad \Delta P_{\text{soap bubble}} = P_i - P_o = \frac{4\sigma_s}{R}$$

where $P_i$ and $P_o$ are the pressures inside and outside the droplet or soap bubble. The rise or fall of a liquid in a small-diameter tube inserted into the liquid due to surface tension is called the *capillary effect*. The capillary rise or drop is given by

$$h = \frac{2\sigma_s}{\rho gR} \cos \phi$$

where $\phi$ is the *contact angle*. The capillary rise is inversely proportional to the radius of the tube; for water, it is negligible for tubes whose diameter is larger than about 1 cm.

## REFERENCES AND SUGGESTED READING

1. E. C. Bingham. "An Investigation of the Laws of Plastic Flow," *U.S. Bureau of Standards Bulletin*, 13, pp. 309–353, 1916.

2. Y. A. Çengel and M. A. Boles. *Thermodynamics: An Engineering Approach*, 7th ed. New York: McGraw-Hill, 2011.

3. Y. A. Çengel and John M. Cimbala. *Fluid Mechanics: Fundamentals and Applications*. 2nd ed. New York: McGraw-Hill, 2010.

4. D. C. Giancoli. *Physics*, 3rd ed. Upper Saddle River, NJ: Prentice Hall, 1991.

5. G. M. Homsy, H. Aref, K. S. Breuer, S. Hochgreb, J. R. Koseff, B. R. Munson, K. G. Powell, C. R. Robertson, and S. T. Thoroddsen. *Multi-Media Fluid Mechanics* (CD). Cambridge: Cambridge University Press, 2000.

6. Y. S. Touloukian, S. C. Saxena, and P. Hestermans. *Thermophysical Properties of Matter, The TPRC Data Series*, Vol. 11, *Viscosity*. New York: Plenum, 1975.

7. L. Trefethen. "Surface Tension in Fluid Mechanics." In *Illustrated Experiments in Fluid Mechanics*. Cambridge, MA: MIT Press, 1972.

8. M. Van Dyke. *An Album of Fluid Motion*. Stanford, CA: Parabolic Press, 1982.

9. C. L. Yaws, X. Lin, and L. Bu. "Calculate Viscosities for 355 Compounds. An Equation Can Be Used to Calculate Liquid Viscosity as a Function of Temperature," *Chemical Engineering*, 101, no. 4, pp. 119–1128, April 1994.

10. C. L. Yaws. *Handbook of Viscosity*. 3 Vols. Houston, TX: Gulf Publishing, 1994.

## PROBLEMS*

### No-Slip Condition and Classification of Fluid Flows

**10–1C**  Define internal, external, and open-channel flows.

**10–2C**  Define incompressible flow and incompressible fluid. Must the flow of a compressible fluid necessarily be treated as compressible?

**10–3C**  Consider the flow of air over the wings of an aircraft. Is this flow internal or external? How about the flow of gases through a jet engine?

**10–4C**  What is the no-slip condition? What causes it?

**10–5C**  What is forced flow? How does it differ from natural flow? Is flow caused by winds forced or natural flow?

**10–6C**  What is a boundary layer? What causes a boundary layer to develop?

### Vapor Pressure and Cavitation

**10–7C**  What is vapor pressure? How is it related to saturation pressure?

**10–8C**  Does water boil at higher temperatures at higher pressures? Explain.

**10–9C**  If the pressure of a substance is increased during a boiling process, will the temperature also increase or will it remain constant? Why?

**10–10C**  What is cavitation? What causes it?

**10–11**  In a piping system, the water temperature remains under 35°C. Determine the minimum pressure allowed in the system to avoid cavitation.

**10–12**  The analysis of a propeller that operates in water at 20°C shows that the pressure at the tips of the propeller drops to 2 kPa at high speeds. Determine if there is a danger of cavitation for this propeller.

**10–13E**  The analysis of a propeller that operates in water at 70°F shows that the pressure at the tips of the propeller drops to 0.1 psia at high speeds. Determine if there is a danger of cavitation for this propeller.

### Viscosity

**10–14C**  What is viscosity? What is the cause of it in liquids and in gases? Do liquids or gases have higher dynamic viscosities?

**10–15C**  What is a Newtonian fluid? Is water a Newtonian fluid?

**10–16C**  Consider two identical small glass balls dropped into two identical containers, one filled with water and the other with oil. Which ball will reach the bottom of the container first? Why?

---

* Problems designated by a "C" are concept questions, and students are encouraged to answer them all. Problems designated by an "E" are in English units, and the SI users can ignore them. Problems with the 🕮 icon are solved using EES, and complete solutions together with parametric studies are included on the enclosed DVD. Problems with the 🖥 icon are comprehensive in nature and are intended to be solved with a computer, preferably using the EES software that accompanies this text.

**10–17C** How does the dynamic viscosity of (a) liquids and (b) gases vary with temperature?

**10–18C** How does the kinematic viscosity of (a) liquids and (b) gases vary with temperature?

**10–19** For flow over a plate, the variation of velocity with vertical distance y from the plate is given as $u(y) = ay - by^2$ where $a$ and $b$ are constants. Obtain a relation for the wall shear stress in terms of $a$, $b$, and $\mu$.

**10–20** A 50-cm $\times$ 30-cm $\times$ 20-cm block weighing 150 N is to be moved at a constant velocity of 0.8 m/s on an inclined surface with a friction coefficient of 0.27. (a) Determine the force $F$ that needs to be applied in the horizontal direction. (b) If a 0.4-mm-thick oil film with a dynamic viscosity of 0.012 Pa·s is applied between the block and inclined surface, determine the percent reduction in the required force.

**FIGURE P10–20**

**10–21** Consider the flow of a fluid with viscosity $\mu$ through a circular pipe. The velocity profile in the pipe is given as $u(r) = u_{max}(1 - r^n/R^n)$, where $u_{max}$ is the maximum flow velocity, which occurs at the centerline; $r$ is the radial distance from the centerline; and $u(r)$ is the flow velocity at any position $r$. Develop a relation for the drag force exerted on the pipe wall by the fluid in the flow direction per unit length of the pipe.

**FIGURE P10–21**

**10–22** A thin 40-cm $\times$ 40-cm flat plate is pulled at 2 m/s horizontally through a 3.6-mm-thick oil layer sandwiched between two plates, one stationary and the other moving at a constant velocity of 0.3 m/s, as shown in Fig. P10–22. The dynamic viscosity of oil is 0.027 Pa·s. Assuming the velocity in each oil layer to vary linearly, (a) plot the velocity profile and find the location where the oil velocity is zero and (b) determine the force that needs to be applied on the plate to maintain this motion.

**FIGURE P10–22**

**10–23** A frustum-shaped 'body is rotating at a constant angular speed of 200 rad/s in a container filled with SAE 10W oil at 20°C ($\mu = 0.1$ Pa·s), as shown in Fig. P10–23. If the thickness of the oil film on all sides is 1.2 mm, determine the power required to maintain this motion. Also determine the reduction in the required power input when the oil temperature rises to 80°C ($\mu = 0.0078$ Pa·s).

**FIGURE P10–23**

**10–24** The clutch system shown in Fig. P10–24 is used to transmit torque through a 2-mm-thick oil film with $\mu = 0.38$ N·s/m$^2$ between two identical 30-cm-diameter disks. When the driving shaft rotates at a speed of 1450 rpm, the driven shaft is observed to rotate at 1398 rpm. Assuming a linear velocity profile for the oil film, determine the transmitted torque.

**FIGURE P10–24**

**10–25** Reconsider Prob. 10–24. Using EES (or other) software, investigate the effect of oil film thickness on the torque transmitted. Let the film thickness vary from 0.1 mm to 10 mm. Plot your results, and state your conclusions.

**10–26** The viscosity of some fluids changes when a strong electric field is applied on them. This phenomenon is known as the electrorheological (ER) effect, and fluids that exhibit such behavior are known as ER fluids. The Bingham plastic model for shear stress, which is expressed as $\tau = \tau_y + \mu(du/dy)$ is widely used to describe ER fluid behavior because of its simplicity. One of the most promising applications of ER fluids is the ER clutch. A typical multidisk ER clutch consists of several equally spaced steel disks of inner radius $R_1$ and outer radius $R_2$, $N$ of them attached to the input shaft. The gap $h$ between the parallel disks is filled with a viscous fluid. (*a*) Find a relationship for the torque generated by the clutch when the output shaft is stationary and (*b*) calculate the torque for an ER clutch with $N = 11$ for $R_1 = 50$ mm, $R_2 = 200$ mm, and $\dot{n} = 2400$ rpm if the fluid is SAE 10 with $\mu = 0.1$ Pa·s, $\tau_y = 2.5$ kPa, and $h = 1.2$ mm. *Answer:* (*b*) 2060 N·m

**FIGURE P10–26**

**10–27** The viscosity of some fluids, called magnetorheological (MR) fluids, changes when a magnetic field is applied. Such fluids involve micron-sized magnetizable particles suspended in an appropriate carrier liquid, and are suitable for use in controllable hydraulic clutches. See Fig. P10–26. The MR fluids can have much higher viscosities than the ER fluids, and they often exhibit shear-thinning behavior in which the viscosity of the fluid decreases as the applied shear force increases. This behavior is also known as pseudoplastic behavior, and can be successfully represented by Herschel–Bulkley constitutive model expressed as $\tau = \tau_y + K(du/dy)^m$. Here $\tau$ is the shear stress applied, $\tau_y$ is the yield stress, $K$ is the consistency index, and $m$ is the power index. For a Herschel–Bulkley fluid with $\tau_y = 900$ Pa, $K = 58$ Pa·s$^m$, and $m = 0.82$, (*a*) find a relationship for the torque transmitted by an MR clutch for $N$ plates attached to the input shaft when the input shaft is rotating at an angular speed of $\omega$ while the output shaft is stationary and (*b*) calculate the torque transmitted by such a clutch with $N = 11$ plates for $R_1 = 50$ mm, $R_2 = 200$ mm, $\dot{n} = 3000$ rpm, and $h = 1.5$ mm.

**10–28** The viscosity of a fluid is to be measured by a viscometer constructed of two 75-cm-long concentric cylinders. The outer diameter of the inner cylinder is 15 cm, and the gap between the two cylinders is 1 mm. The inner cylinder is rotated at 300 rpm, and the torque is measured to be 0.8 N·m. Determine the viscosity of the fluid.

**FIGURE P10–28**

**10–29E** The viscosity of a fluid is to be measured by a viscometer constructed of two 3-ft-long concentric cylinders. The inner diameter of the outer cylinder is 6 in, and the gap between the two cylinders is 0.05 in. The outer cylinder is rotated at 250 rpm, and the torque is measured to be 1.2 lbf·ft. Determine the viscosity of the fluid. *Answer:* 0.000648 lb·s/ft$^2$

**10–30** In regions far from the entrance, fluid flow through a circular pipe is one-dimensional, and the velocity profile for laminar flow is given by $u(r) = u_{max}(1 - r^2/R^2)$, where $R$ is the radius of the pipe, $r$ is the radial distance from the center of the pipe, and $u_{max}$ is the maximum flow velocity, which occurs at the center. Obtain (*a*) a relation for the drag force applied by the fluid on a section of the pipe of length $L$ and (*b*) the value of the drag force for water flow at 20°C with $R = 0.08$ m, $L = 30$ m, $u_{max} = 3$ m/s, and $\mu = 0.0010$ kg/m·s.

**FIGURE P10–30**

**10–31** Repeat Prob. 10–30 for $u_{max} = 7$ m/s. *Answer:* (*b*) 2.64 N

## Surface Tension and Capillary Effect

**10–32C** What is surface tension? What is it caused by? Why is the surface tension also called surface energy?

**10–33C** Consider a soap bubble. Is the pressure inside the bubble higher or lower than the pressure outside?

**10–34C** What is the capillary effect? What is it caused by? How is it affected by the contact angle?

**10–35C** A small-diameter tube is inserted into a liquid whose contact angle is 110°. Will the level of liquid in the tube be higher or lower than the level of the rest of the liquid? Explain.

**10–36C** Is the capillary rise greater in small- or large-diameter tubes?

**10–37** Consider a 0.2-mm diameter air bubble in a liquid. Determine the pressure difference between the inside and outside of the air bubble if the surface tension at the air-liquid interface is (a) 0.08 N/m and (b) 0.12 N/m.

**10–38E** A 0.03-in-diameter glass tube is inserted into kerosene at 68°F. The contact angle of kerosene with a glass surface is 26°. Determine the capillary rise of kerosene in the tube. *Answer: 0.65 in*

←0.03 in

*h*

Kerosene

**FIGURE P10–38**

**10–39** A 1.2-mm-diameter tube is inserted into an unknown liquid whose density is 960 kg/m³, and it is observed that the liquid rises 5 mm in the tube, making a contact angle of 15°. Determine the surface tension of the liquid.

**10–40**  Determine the gage pressure inside a soap bubble of diameter (a) 0.2 cm and (b) 5 cm at 20°C.

**10–41** Nutrients dissolved in water are carried to upper parts of plants by tiny tubes partly because of the capillary effect. Determine how high the water solution will rise in a tree in a 0.002-mm-diameter tube as a result of the capillary effect. Treat the solution as water at 20°C with a contact angle of 15°. *Answer: 14.4 m*

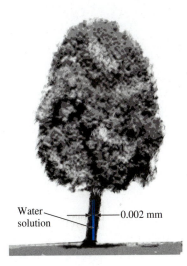

Water
solution

←0.002 mm

**FIGURE P10–41**

**10–42** The surface tension of a liquid is to be measured using a liquid film suspended on a U-shaped wire frame with an 8-cm-long movable side. If the force needed to move

the wire is 0.024 N, determine the surface tension of this liquid in air.

**10–43** Contrary to what you might expect, a solid steel ball can float on water due to the surface tension effect. Determine the maximum diameter of a steel ball that would float on water at 20°C. What would your answer be for an aluminum ball? Take the densities of steel and aluminum balls to be 7800 kg/m³ and 2700 kg/m³, respectively.

### Review problems

**10–44E** The pressure on the suction side of pumps is typically low, and the surfaces on that side of the pump are susceptible to cavitation, especially at high fluid temperatures. If the minimum pressure on the suction side of a water pump is 0.95 psia absolute, determine the maximum water temperature to avoid the danger of cavitation.

**10–45** A closed tank is partially filled with water at 80°C. If the air above the water is completely evacuated, determine the absolute pressure in the evacuated space. Assume the temperature to remain constant.

**10–46** [EES] The variation of the dynamic viscosity of water with absolute temperature is given as

| T, K | μ, Pa·s |
|---|---|
| 273.15 | $1.787 \times 10^{-3}$ |
| 278.15 | $1.519 \times 10^{-3}$ |
| 283.15 | $1.307 \times 10^{-3}$ |
| 293.15 | $1.002 \times 10^{-3}$ |
| 303.15 | $7.975 \times 10^{-4}$ |
| 313.15 | $6.529 \times 10^{-4}$ |
| 333.15 | $4.665 \times 10^{-4}$ |
| 353.15 | $3.547 \times 10^{-4}$ |
| 373.15 | $2.828 \times 10^{-4}$ |

Using tabulated data, develop a relation for viscosity in the form of $\mu = \mu(T) = A + BT + CT^2 + DT^3 + ET^4$. Using the relation developed, predict the dynamic viscosity of water at 50°C at which the reported value is $5.468 \times 10^{-4}$ Pa·s. Compare your result with the results of Andrade's equation, which is given in the form of $\mu = D \cdot e^{B/T}$, where $D$ and $B$ are constants whose values are to be determined using the viscosity data given.

**10–47** Consider laminar flow of a Newtonian fluid of viscosity $\mu$ between two parallel plates. The flow is one-dimensional, and the velocity profile is given as $u(y) = 4u_{\max}[y/h - (y/h)^2]$, where $y$ is the vertical coordinate from the bottom surface, $h$ is the distance between the two plates, and $u_{\max}$ is the maximum flow velocity that occurs at midplane. Develop a relation for the drag force exerted on both plates by the fluid in the flow direction per unit area of the plates.

$$u(y) = 4u_{max}[y/h - (y/h)^2]$$

**FIGURE P10–47**

**10–48** Some non-Newtonian fluids behave as a Bingham plastic for which shear stress can be expressed as $\tau = \tau_y + \mu(du/dr)$. For laminar flow of a Bingham plastic in a horizontal pipe of radius $R$, the velocity profile is given as $u(r) = (\Delta P/4\mu L)(r^2 - R^2) + (\tau_y/\mu)(r - R)$, where $\Delta P/L$ is the constant pressure drop along the pipe per unit length, $\mu$ is the dynamic viscosity, $r$ is the radial distance from the centerline, and $\tau_y$ is the yield stress of Bingham plastic. Determine (a) the shear stress at the pipe wall and (b) the drag force acting on a pipe section of length $L$.

**10–49E** A 0.018-in-diameter glass tube is inserted into mercury, which makes a contact angle of 140° with glass. Determine the capillary drop of mercury in the tube at 68°F. *Answer:* 0.874 in

**10–50** Derive a relation for the capillary rise of a liquid between two large parallel plates a distance $t$ apart inserted into the liquid vertically. Take the contact angle to be $\phi$.

**10–51** Consider a 40-cm-long journal bearing that is lubricated with oil whose viscosity is 0.1 kg/m·s at 20°C at the beginning of operation and 0.008 kg/m·s at the anticipated steady operating temperature of 80°C. The diameter of the shaft is 8 cm, and the average gap between the shaft and the journal is 0.08 cm. Determine the torque needed to overcome the bearing friction initially and during steady operation when the shaft is rotated at 800 rpm.

**10–52** The diameter of one arm of a U-tube is 5 mm while the other arm is large. If the U-tube contains some water, and both surface are exposed to atmospheric pressure, determine the difference between the water levels in the two arms.

**10–53** Some rocks or bricks contain small air pockets in them and have a spongy structure. Assuming the air spaces form columns of an average diameter of 0.004 mm, determine how high water can rise in such a material. Take the surface tension of the air–water interface in that material to be 0.085 N/m.

### Design and Essay Problems

**10–54** Design an experiment to measure the viscosity of liquids using a vertical funnel with a cylindrical reservoir of height $h$ and a narrow flow section of diameter $D$ and length $L$. Making appropriate assumptions, obtain a relation for viscosity in terms of easily measurable quantities such as density and volume flow rate.

**10–55** Write an essay on the rise of the fluid to the top of the trees by capillary and other effects.

**10–56** Write an essay on the oils used in car engines in different seasons and their viscosities.

**10–57** Even though steel is about 7 to 8 times denser than water, a steel paper clip or razor blade can be made to float on water! Explain and discuss. Predict what would happen if you mix some soap with the water.

**FIGURE P10–57**

*Photo by John M. Cimbala.*

# FLUID STATICS

This chapter deals with forces applied by fluids at rest. The fluid property responsible for those forces is *pressure*, which is a normal force exerted by a fluid per unit area. We start this chapter with a discussion of the *hydrostatic forces* applied on submerged bodies with plane or curved surfaces. We then consider the *buoyant force* applied by fluids on submerged or floating bodies, and discuss the *stability* of such bodies. This chapter makes extensive use of force balances for bodies in static equilibrium, and it will be helpful if the relevant topics from statics are first reviewed.

**Objectives**

The objectives of this chapter are to:

■ Calculate the forces exerted by a fluid at rest on plane or curved submerged surfaces.

■ Analyze the stability of floating and submerged bodies.

# 11–1 ▪ INTRODUCTION TO FLUID STATICS

**Fluid statics** deals with problems associated with fluids at rest. The fluid can be either gaseous or liquid. Fluid statics is generally referred to as *hydrostatics* when the fluid is a liquid and as *aerostatics* when the fluid is a gas. In fluid statics, there is no relative motion between adjacent fluid layers, and thus there are no shear (tangential) stresses in the fluid trying to deform it. The only stress we deal with in fluid statics is the *normal stress,* which is the pressure, and the variation of pressure is due only to the weight of the fluid. Therefore, the topic of fluid statics has significance only in gravity fields, and the force relations developed naturally involve the gravitational acceleration *g*. The force exerted on a surface by a fluid at rest is normal to the surface at the point of contact since there is no relative motion between the fluid and the solid surface, and thus there are no shear forces acting parallel to the surface.

Fluid statics is used to determine the forces acting on floating or submerged bodies and the forces developed by devices like hydraulic presses and car jacks. The design of many engineering systems such as water dams and liquid storage tanks requires the determination of the forces acting on their surfaces using fluid statics. The complete description of the resultant hydrostatic force acting on a submerged surface requires the determination of the magnitude, the direction, and the line of action of the force. In the following two sections, we consider the forces acting on both plane and curved surfaces of submerged bodies due to pressure.

# 11–2 ▪ HYDROSTATIC FORCES ON SUBMERGED PLANE SURFACES

A plate (such as a gate valve in a dam, the wall of a liquid storage tank, or the hull of a ship at rest) is subjected to fluid pressure distributed over its surface when exposed to a liquid (Fig. 11–1). On a *plane* surface, the hydrostatic forces form a system of parallel forces, and we often need to determine the *magnitude* of the force and its *point of application,* which is called the **center of pressure**. In most cases, the other side of the plate is open to the atmosphere (such as the dry side of a gate), and thus atmospheric pressure acts on both sides of the plate, yielding a zero resultant. In such cases, it is convenient to subtract atmospheric pressure and work with the gage pressure only (Fig. 11–2). For example, $P_{gage} = \rho g h$ at the bottom of the lake.

Consider the top surface of a flat plate of arbitrary shape completely submerged in a liquid, as shown in Fig. 11–3 together with its normal view. The plane of this surface (normal to the page) intersects the horizontal free surface at angle $\theta$, and we take the line of intersection to be the *x*-axis (out of the page). The absolute pressure above the liquid is $P_0$, which is the local atmospheric pressure $P_{atm}$ if the liquid is open to the atmosphere (but $P_0$ may be different than $P_{atm}$ if the space above the liquid is evacuated or pressurized). Then the absolute pressure at any point on the plate is

$$P = P_0 + \rho g h = P_0 + \rho g y \sin \theta \tag{11–1}$$

where *h* is the vertical distance of the point from the free surface and *y* is the distance of the point from the *x*-axis (from point *O* in Fig. 11–3). The resultant

**FIGURE 11–1**

Hoover Dam.

*Courtesy United States Department of the Interior, Bureau of Reclamation-Lower Colorado Region.*

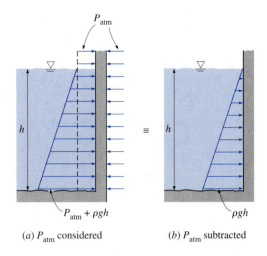

FIGURE 11–2

When analyzing hydrostatic forces on submerged surfaces, the atmospheric pressure can be subtracted for simplicity when it acts on both sides of the structure.

hydrostatic force $F_R$ acting on the surface is determined by integrating the force $P\,dA$ acting on a differential area $dA$ over the entire surface area,

$$F_R = \int_A P\,dA = \int_A (P_0 + \rho g y \sin\theta)\,dA = P_0 A + \rho g \sin\theta \int_A y\,dA \quad \text{(11–2)}$$

But the *first moment of area* $\int_A y\,dA$ is related to the $y$-coordinate of the centroid (or center) of the surface by

$$y_C = \frac{1}{A} \int_A y\,dA \quad \text{(11–3)}$$

Substituting,

$$F_R = (P_0 + \rho g y_C \sin\theta)A = (P_0 + \rho g h_C)A = P_C A = P_{avg} A \quad \text{(11–4)}$$

where $P_C = P_0 + \rho g h_C$ is the pressure at the centroid of the surface, which is equivalent to the *average* pressure $P_{avg}$ on the surface, and $h_C = y_C \sin\theta$

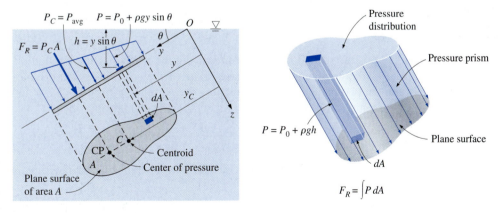

**FIGURE 11–3**

Hydrostatic force on an inclined plane surface completely submerged in a liquid.

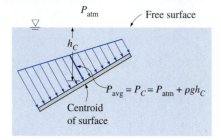

**FIGURE 11–4**

The pressure at the centroid of a plane surface is equivalent to the *average* pressure on the surface.

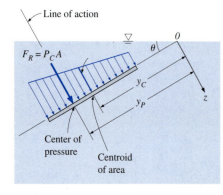

**FIGURE 11–5**

The resultant force acting on a plane surface is equal to the product of the pressure at the centroid of the surface and the surface area, and its line of action passes through the center of pressure.

is the *vertical distance* of the centroid from the free surface of the liquid (Fig. 11–4). Thus we conclude that:

> The magnitude of the resultant force acting on a plane surface of a completely submerged plate in a homogeneous (constant density) fluid is equal to the product of the pressure $P_C$ at the centroid of the surface and the area $A$ of the surface (Fig. 11–5).

The pressure $P_0$ is usually atmospheric pressure, which can be ignored in most force calculations since it acts on both sides of the plate. When this is not the case, a practical way of accounting for the contribution of $P_0$ to the resultant force is simply to add an equivalent depth $h_{equiv} = P_0/\rho g$ to $h_C$; that is, to assume the presence of an additional liquid layer of thickness $h_{equiv}$ on top of the liquid with absolute vacuum above.

Next we need to determine the line of action of the resultant force $F_R$. Two parallel force systems are equivalent if they have the same magnitude and the same moment about any point. The line of action of the resultant hydrostatic force, in general, does not pass through the centroid of the surface—it lies underneath where the pressure is higher. The point of intersection of the line of action of the resultant force and the surface is the **center of pressure**. The vertical location of the line of action is determined by equating the moment of the resultant force to the moment of the distributed pressure force about the $x$-axis:

$$y_P F_R = \int_A yP \, dA = \int_A y(P_0 + \rho gy \sin \theta) \, dA = P_0 \int_A y \, dA + \rho g \sin \theta \int_A y^2 \, dA$$

or

$$y_P F_R = P_0 y_C A + \rho g \sin \theta \, I_{xx, O} \tag{11–5}$$

where $y_P$ is the distance of the center of pressure from the $x$-axis (point $O$ in Fig. 11–5) and $I_{xx, O} = \int_A y^2 \, dA$ is the *second moment of area* (also called the *area moment of inertia*) about the $x$-axis. The second moments of area are widely available for common shapes in engineering handbooks, but they are usually given about the axes passing through the centroid of the area. Fortunately, the second moments of area about two parallel axes are related to each other by the *parallel axis theorem*, which in this case is expressed as

$$I_{xx, O} = I_{xx, C} + y_C^2 A \tag{11–6}$$

where $I_{xx, C}$ is the second moment of area about the $x$-axis passing through the centroid of the area and $y_C$ (the $y$-coordinate of the centroid) is the distance between the two parallel axes. Substituting the $F_R$ relation from Eq. 11–4 and the $I_{xx, O}$ relation from Eq. 11–6 into Eq. 11–5 and solving for $y_P$ yields

$$y_P = y_C + \frac{I_{xx, C}}{[y_C + P_0/(\rho g \sin \theta)]A} \tag{11–7a}$$

For $P_0 = 0$, which is usually the case when the atmospheric pressure is ignored, it simplifies to

$$y_P = y_C + \frac{I_{xx, C}}{y_C A} \tag{11–7b}$$

Knowing $y_P$, the vertical distance of the center of pressure from the free surface is determined from $h_P = y_P \sin \theta$.

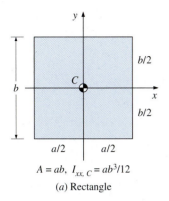

$A = ab, \; I_{xx, C} = ab^3/12$

(a) Rectangle

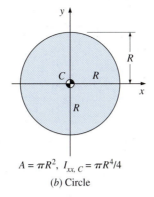

$A = \pi R^2, \; I_{xx, C} = \pi R^4/4$

(b) Circle

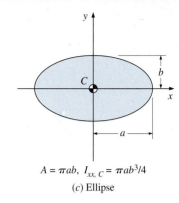

$A = \pi ab, \; I_{xx, C} = \pi ab^3/4$

(c) Ellipse

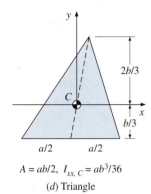

$A = ab/2, \; I_{xx, C} = ab^3/36$

(d) Triangle

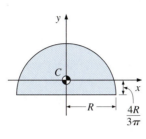

$A = \pi R^2/2, \; I_{xx, C} = 0.109757R^4$

(e) Semicircle

$A = \pi ab/2, \; I_{xx, C} = 0.109757ab^3$

(f) Semiellipse

**FIGURE 11–6**

The centroid and the centroidal moments of inertia for some common geometries.

The $I_{xx, C}$ values for some common areas are given in Fig. 11–6. For areas that possess symmetry about the $y$-axis, the center of pressure lies on the $y$-axis directly below the centroid. The location of the center of pressure in such cases is simply the point on the surface of the vertical plane of symmetry at a distance $h_P$ from the free surface.

Pressure acts normal to the surface, and the hydrostatic forces acting on a flat plate of any shape form a volume whose base is the plate area and whose length is the linearly varying pressure, as shown in Fig. 11–7. This virtual **pressure prism** has an interesting physical interpretation: its *volume* is equal to the *magnitude* of the resultant hydrostatic force acting on the plate since $F_R = \int P \, dA$, and the line of action of this force passes through the *centroid* of this homogeneous prism. The projection of the centroid on the plate is the *pressure center*. Therefore, with the concept of pressure prism, the problem of describing the resultant hydrostatic force on a plane surface reduces to finding the volume and the two coordinates of the centroid of this pressure prism.

## Special Case: Submerged Rectangular Plate

Consider a completely submerged rectangular flat plate of height $b$ and width $a$ tilted at an angle $\theta$ from the horizontal and whose top edge is horizontal and is at a distance $s$ from the free surface along the plane of the plate, as shown in Fig. 11–8a. The resultant hydrostatic force on the upper

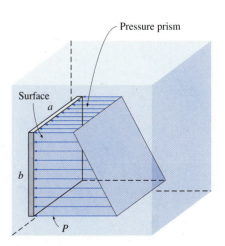

**FIGURE 11–7**

The hydrostatic forces acting on a plane surface form a pressure prism whose base (left face) is the surface and whose length is the pressure.

$$F_R = [P_0 + \rho g(s + b/2) \sin \theta]ab$$

(a) Tilted plate

$$F_R = [P_0 + \rho g(s + b/2)]ab$$

(b) Vertical plate

$$F_R = (P_0 + \rho gh)ab$$

(c) Horizontal plate

**FIGURE 11–8**

Hydrostatic force acting on the top surface of a submerged rectangular plate for tilted, vertical, and horizontal cases.

surface is equal to the average pressure, which is the pressure at the midpoint of the surface, times the surface area $A$. That is,

*Tilted rectangular plate*: $\quad F_R = P_C A = [P_0 + \rho g(s + b/2) \sin \theta]ab$ $\quad$ **(11–8)**

The force acts at a vertical distance of $h_P = y_P \sin \theta$ from the free surface directly beneath the centroid of the plate where, from Eq. 11–7a,

$$y_P = s + \frac{b}{2} + \frac{ab^3/12}{[s + b/2 + P_0/(\rho g \sin \theta)]ab}$$

$$= s + \frac{b}{2} + \frac{b^2}{12[s + b/2 + P_0/(\rho g \sin \theta)]} \quad \textbf{(11–9)}$$

When the upper edge of the plate is at the free surface and thus $s = 0$, Eq. 11–8 reduces to

*Tilted rectangular plate ($s = 0$)*: $\quad F_R = [P_0 + \rho g(b \sin \theta)/2]ab$ $\quad$ **(11–10)**

For a completely submerged *vertical* plate ($\theta = 90°$) whose top edge is horizontal, the hydrostatic force can be obtained by setting $\sin \theta = 1$ (Fig. 11–8b)

*Vertical rectangular plate*: $\quad F_R = [P_0 + \rho g(s + b/2)]ab$ $\quad$ **(11–11)**

*Vertical rectangular plate ($s = 0$)*: $\quad F_R = (P_0 + \rho/2)ab$ $\quad$ **(11–12)**

When the effect of $P_0$ is ignored since it acts on both sides of the plate, the hydrostatic force on a vertical rectangular surface of height $b$ whose top edge is horizontal and at the free surface is $F_R = \rho gab^2/2$ acting at a distance of $2b/3$ from the free surface directly beneath the centroid of the plate.

The pressure distribution on a submerged *horizontal* surface is uniform, and its magnitude is $P = P_0 + \rho gh$, where $h$ is the distance of the surface from the free surface. Therefore, the hydrostatic force acting on a horizontal rectangular surface is

*Horizontal rectangular plate*: $\quad F_R = (P_0 + \rho gh)ab$ $\quad$ **(11–13)**

and it acts through the midpoint of the plate (Fig. 11–8c).

## EXAMPLE 11–1   Hydrostatic Force Acting on the Door of a Submerged Car

A heavy car plunges into a lake during an accident and lands at the bottom of the lake on its wheels (Fig. 11–9). The door is 1.2 m high and 1 m wide, and the top edge of the door is 8 m below the free surface of the water. Determine the hydrostatic force on the door and the location of the pressure center, and discuss if the driver can open the door.

**Solution**   A car is submerged in water. The hydrostatic force on the door is to be determined, and the likelihood of the driver opening the door is to be assessed.

**Assumptions**   **1** The bottom surface of the lake is horizontal. **2** The passenger cabin is well-sealed so that no water leaks inside. **3** The door can be approximated as a vertical rectangular plate. **4** The pressure in the passenger cabin remains at atmospheric value since there is no water leaking in, and thus no compression of the air inside. Therefore, atmospheric pressure cancels out in the calculations since it acts on both sides of the door. **5** The weight of the car is larger than the buoyant force acting on it.

**Properties**   We take the density of lake water to be 1000 kg/m³ throughout.

**Analysis**   The average (gage) pressure on the door is the pressure value at the centroid (midpoint) of the door and is determined to be

$$P_{avg} = P_C = \rho g h_C = \rho g(s + b/2)$$

$$= (1000 \text{ kg/m}^3)(9.81 \text{ m/s}^2)(8 + 1.2/2 \text{ m})\left(\frac{1 \text{ kN}}{1000 \text{ kg·m/s}^2}\right)$$

$$= \mathbf{84.4 \text{ kN/m}^2}$$

Then the resultant hydrostatic force on the door becomes

$$F_R = P_{avg}A = (84.4 \text{ kN/m}^2)(1 \text{ m} \times 1.2 \text{ m}) = \mathbf{101.3 \text{ kN}}$$

The pressure center is directly under the midpoint of the door, and its distance from the surface of the lake is determined from Eq. 11–9 by setting $P_0 = 0$, yielding

$$y_P = s + \frac{b}{2} + \frac{b^2}{12(s + b/2)} = 8 + \frac{1.2}{2} + \frac{1.2^2}{12(8 + 1.2/2)} = \mathbf{8.61 \text{ m}}$$

**Discussion**   A strong person can lift 100 kg, whose weight is 981 N or about 1 kN. Also, the person can apply the force at a point farthest from the hinges (1 m farther) for maximum effect and generate a moment of 1 kN·m. The resultant hydrostatic force acts under the midpoint of the door, and thus a distance of 0.5 m from the hinges. This generates a moment of 50.6 kN·m, which is about 50 times the moment the driver can possibly generate. Therefore, it is impossible for the driver to open the door of the car. The driver's best bet is to let some water in (by rolling the window down a little, for example) and to keep his or her head close to the ceiling. The driver should be able to open the door shortly before the car is filled with water since at that point the pressures on both sides of the door are nearly the same and opening the door in water is almost as easy as opening it in air.

**FIGURE 11–9**

Schematic for Example 11–1.

# 11–3 · HYDROSTATIC FORCES ON SUBMERGED CURVED SURFACES

For a submerged curved surface, the determination of the resultant hydrostatic force is more involved since it typically requires integration of the pressure forces that change direction along the curved surface. The concept of the pressure prism in this case is not much help either because of the complicated shapes involved.

The easiest way to determine the resultant hydrostatic force $F_R$ acting on a two-dimensional curved surface is to determine the horizontal and vertical components $F_H$ and $F_V$ separately. This is done by considering the free-body diagram of the liquid block enclosed by the curved surface and the two plane surfaces (one horizontal and one vertical) passing through the two ends of the curved surface, as shown in Fig. 11–10. Note that the vertical surface of the liquid block considered is simply the projection of the curved surface on a *vertical plane,* and the horizontal surface is the projection of the curved surface on a *horizontal plane.* The resultant force acting on the curved solid surface is then equal and opposite to the force acting on the curved liquid surface (Newton's third law).

The force acting on the imaginary horizontal or vertical plane surface and its line of action can be determined as discussed in Section 11–2. The weight of the enclosed liquid block of volume $V$ is simply $W = \rho g V$, and it acts downward through the centroid of this volume. Noting that the fluid block is in static equilibrium, the force balances in the horizontal and vertical directions give

*Horizontal force component on curved surface:*     $F_H = F_x$     **(11–14)**

*Vertical force component on curved surface:*     $F_V = F_y \pm W$     **(11–15)**

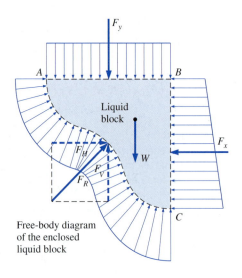

**FIGURE 11–10**

Determination of the hydrostatic force acting on a submerged curved surface.

where the summation $F_y \pm W$ is a vector addition (i.e., add magnitudes if both act in the same direction and subtract if they act in opposite directions). Thus, we conclude that

1. The horizontal component of the hydrostatic force acting on a curved surface is equal (in both magnitude and the line of action) to the hydrostatic force acting on the vertical projection of the curved surface.
2. The vertical component of the hydrostatic force acting on a curved surface is equal to the hydrostatic force acting on the horizontal projection of the curved surface, plus (minus, if acting in the opposite direction) the weight of the fluid block.

The magnitude of the resultant hydrostatic force acting on the curved surface is $F_R = \sqrt{F_H^2 + F_V^2}$, and the tangent of the angle it makes with the horizontal is $\tan \alpha = F_V/F_H$. The exact location of the line of action of the resultant force (e.g., its distance from one of the end points of the curved surface) can be determined by taking a moment about an appropriate point. These discussions are valid for all curved surfaces regardless of whether they are above or below the liquid. Note that in the case of a *curved surface above a liquid,* the weight of the liquid is *subtracted* from the vertical component of the hydrostatic force since they act in opposite directions (Fig. 11–11).

When the curved surface is a *circular arc* (full circle or any part of it), the resultant hydrostatic force acting on the surface always passes through the center of the circle. This is because the pressure forces are normal to the surface, and all lines normal to the surface of a circle pass through the center of the circle. Thus, the pressure forces form a concurrent force system at the center, which can be reduced to a single equivalent force at that point (Fig. 11–12).

Finally, hydrostatic forces acting on a plane or curved surface submerged in a **multilayered fluid** of different densities can be determined by considering different parts of surfaces in different fluids as different surfaces, finding the force on each part, and then adding them using vector addition. For a plane surface, it can be expressed as (Fig. 11–13)

*Plane surface in a multilayered fluid:* $\qquad F_R = \sum F_{R,i} = \sum P_{C,i} A_i \qquad$ **(11–16)**

where $P_{C,i} = P_0 + \rho_i g h_{C,i}$ is the pressure at the centroid of the portion of the surface in fluid $i$ and $A_i$ is the area of the plate in that fluid. The line of action of this equivalent force can be determined from the requirement that the moment of the equivalent force about any point is equal to the sum of the moments of the individual forces about the same point.

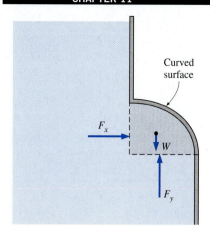

**FIGURE 11–11**

When a curved surface is above the liquid, the weight of the liquid and the vertical component of the hydrostatic force act in the opposite directions.

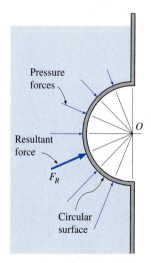

**FIGURE 11–12**

The hydrostatic force acting on a circular surface always passes through the center of the circle since the pressure forces are normal to the surface and they all pass through the center.

---

**EXAMPLE 11–2** **A Gravity-Controlled Cylindrical Gate**

A long solid cylinder of radius 0.8 m hinged at point $A$ is used as an automatic gate, as shown in Fig. 11–14 on page 457. When the water level reaches 5 m, the gate opens by turning about the hinge at point $A$. Determine (*a*) the hydrostatic force acting on the cylinder and its line of action when the gate opens and (*b*) the weight of the cylinder per m length of the cylinder.

**Solution** The height of a water reservoir is controlled by a cylindrical gate hinged to the reservoir. The hydrostatic force on the cylinder and the weight of the cylinder per m length are to be determined.

**FIGURE 11–13**

The hydrostatic force on a surface submerged in a multilayered fluid can be determined by considering parts of the surface in different fluids as different surfaces.

**Assumptions** **1** Friction at the hinge is negligible. **2** Atmospheric pressure acts on both sides of the gate, and thus it cancels out.

**Properties** We take the density of water to be 1000 kg/m³ throughout.

**Analysis** (*a*) We consider the free-body diagram of the liquid block enclosed by the circular surface of the cylinder and its vertical and horizontal projections. The hydrostatic forces acting on the vertical and horizontal plane surfaces as well as the weight of the liquid block are determined as

Horizontal force on vertical surface:

$$F_H = F_x = P_{avg}A = \rho gh_C A = \rho g(s + R/2)A$$

$$= (1000 \text{ kg/m}^3)(9.81 \text{ m/s}^2)(4.2 + 0.8/2 \text{ m})(0.8 \text{ m} \times 1 \text{ m})\left(\frac{1 \text{ kN}}{1000 \text{ kg·m/s}^2}\right)$$

$$= 36.1 \text{ kN}$$

Vertical force on horizontal surface (upward):

$$F_y = P_{avg}A = \rho gh_C A = \rho gh_{bottom}A$$

$$= (1000 \text{ kg/m}^3)(9.81 \text{ m/s}^2)(5 \text{ m})(0.8 \text{ m} \times 1 \text{ m})\left(\frac{1 \text{ kN}}{1000 \text{ kg·m/s}^2}\right)$$

$$= 39.2 \text{ kN}$$

Weight (downward) of fluid block for one m width into the page:

$$W = mg = \rho g V = \rho g(R^2 - \pi R^2/4)(1 \text{ m})$$

$$= (1000 \text{ kg/m}^3)(9.81 \text{ m/s}^2)(0.8 \text{ m})^2(1 - \pi/4)(1 \text{ m})\left(\frac{1 \text{ kN}}{1000 \text{ kg·m/s}^2}\right)$$

$$= 1.3 \text{ kN}$$

Therefore, the net upward vertical force is

$$F_V = F_y - W = 39.2 - 1.3 = 37.9 \text{ kN}$$

Then the magnitude and direction of the hydrostatic force acting on the cylindrical surface become

$$F_R = \sqrt{F_H^2 + F_V^2} = \sqrt{36.1^2 + 37.9^2} = \mathbf{52.3 \text{ kN}}$$

$$\tan \theta = F_V/F_H = 37.9/36.1 = 1.05 \rightarrow \theta = 46.4°$$

Therefore, the magnitude of the hydrostatic force acting on the cylinder is 52.3 kN per m length of the cylinder, and its line of action passes through the center of the cylinder making an angle 46.4° with the horizontal.

(*b*) When the water level is 5 m high, the gate is about to open and thus the reaction force at the bottom of the cylinder is zero. Then the forces other than those at the hinge acting on the cylinder are its weight, acting through the center, and the hydrostatic force exerted by water. Taking a moment about point *A* at the location of the hinge and equating it to zero gives

$$F_R R \sin \theta - W_{cyl} R = 0 \rightarrow W_{cyl} = F_R \sin \theta = (52.3 \text{ kN}) \sin 46.4° = \mathbf{37.9 \text{ kN}}$$

**Discussion** The weight of the cylinder per m length is determined to be 37.9 kN. It can be shown that this corresponds to a mass of 3863 kg per m length and to a density of 1921 kg/m³ for the material of the cylinder.

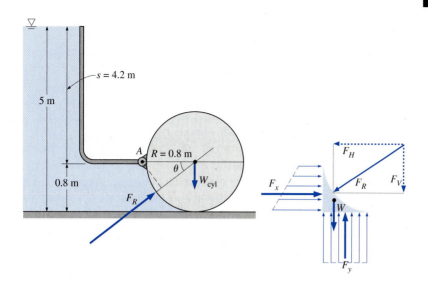

**FIGURE 11–14**

Schematic for Example 11–2 and
the free-body diagram of the liquid
underneath the cylinder.

## 11–4 · BUOYANCY AND STABILITY

It is a common experience that an object feels lighter and weighs less in a liquid than it does in air. This can be demonstrated easily by weighing a heavy object in water by a waterproof spring scale. Also, objects made of wood or other light materials float on water. These and other observations suggest that a fluid exerts an upward force on a body immersed in it. This force that tends to lift the body is called the **buoyant force** and is denoted by $F_B$.

The buoyant force is caused by the increase of pressure with depth in a fluid. Consider, for example, a flat plate of thickness $h$ submerged in a liquid of density $\rho_f$ parallel to the free surface, as shown in Fig. 11–15. The area of the top (and also bottom) surface of the plate is $A$, and its distance to the free surface is $s$. The gage pressures at the top and bottom surfaces of the plate are $\rho_f gs$ and $\rho_f g(s + h)$, respectively. Then the hydrostatic force $F_{top} = \rho_f gsA$ acts downward on the top surface, and the larger force $F_{bottom} = \rho_f g(s + h)A$ acts upward on the bottom surface of the plate. The difference between these two forces is a net upward force, which is the *buoyant force,*

$$F_B = F_{bottom} - F_{top} = \rho_f g(s + h)A - \rho_f gsA = \rho_f ghA = \rho_f g\mathcal{V} \quad \text{(11–17)}$$

where $\mathcal{V} = hA$ is the volume of the plate. But the relation $\rho_f g\mathcal{V}$ is simply the weight of the liquid whose volume is equal to the volume of the plate. Thus, we conclude that *the buoyant force acting on the plate is equal to the weight of the liquid displaced by the plate.* For a fluid with constant density, the buoyant force is independent of the distance of the body from the free surface. It is also independent of the density of the solid body.

The relation in Eq. 11–17 is developed for a simple geometry, but it is valid for any body regardless of its shape. This can be shown mathematically by a force balance, or simply by this argument: Consider an arbitrarily shaped solid body submerged in a fluid at rest and compare it to a body of fluid of the same shape indicated by dotted lines at the same vertical location

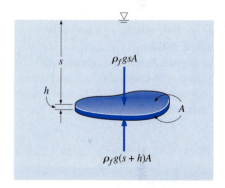

**FIGURE 11–15**

A flat plate of uniform thickness $h$ submerged in a liquid parallel to the free surface.

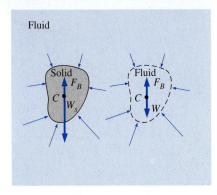

**FIGURE 11–16**

The buoyant forces acting on a solid body submerged in a fluid and on a fluid body of the same shape at the same depth are identical. The buoyant force $F_B$ acts upward through the centroid $C$ of the displaced volume and is equal in magnitude to the weight $W$ of the displaced fluid, but is opposite in direction. For a solid of uniform density, its weight $W_s$ also acts through the centroid, but its magnitude is not necessarily equal to that of the fluid it displaces. (Here $W_s > W$ and thus $W_s > F_B$; this solid body would sink.)

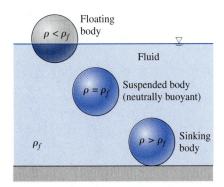

**FIGURE 11–17**

A solid body dropped into a fluid will sink, float, or remain at rest at any point in the fluid, depending on its average density relative to the density of the fluid.

(Fig. 11–16). The buoyant forces acting on these two bodies are the same since the pressure distributions, which depend only on elevation, are the same at the boundaries of both. The imaginary fluid body is in static equilibrium, and thus the net force and net moment acting on it are zero. Therefore, the upward buoyant force must be equal to the weight of the imaginary fluid body whose volume is equal to the volume of the solid body. Furthermore, the weight and the buoyant force must have the same line of action to have a zero moment. This is known as **Archimedes' principle**, after the Greek mathematician Archimedes (287–212 BC), and is expressed as

> The buoyant force acting on a body of uniform density immersed in a fluid is equal to the weight of the fluid displaced by the body, and it acts upward through the centroid of the displaced volume.

For *floating* bodies, the weight of the entire body must be equal to the buoyant force, which is the weight of the fluid whose volume is equal to the volume of the *submerged portion* of the floating body. That is,

$$F_B = W \rightarrow \rho_f g V_{sub} = \rho_{avg, body} g V_{total} \rightarrow \frac{V_{sub}}{V_{total}} = \frac{\rho_{avg, body}}{\rho_f} \quad (11\text{–}18)$$

Therefore, the submerged volume fraction of a floating body is equal to the ratio of the average density of the body to the density of the fluid. Note that when the density ratio is equal to or greater than one, the floating body becomes completely submerged.

It follows from these discussions that a body immersed in a fluid (1) remains at rest at any location in the fluid where its average density is equal to the density of the fluid, (2) sinks to the bottom when its average density is greater than the density of the fluid, and (3) rises to the surface of the fluid and floats when the average density of the body is less than the density of the fluid (Fig. 11–17).

The buoyant force is proportional to the density of the fluid, and thus we might think that the buoyant force exerted by gases such as air is negligible. This is certainly the case in general, but there are significant exceptions. For example, the volume of a person is about 0.1 m³, and taking the density of air to be 1.2 kg/m³, the buoyant force exerted by air on the person is

$$F_B = \rho_f g V = (1.2 \text{ kg/m}^3)(9.81 \text{ m/s}^2)(0.1 \text{ m}^3) \cong 1.2 \text{ N}$$

The weight of an 80-kg person is $80 \times 9.81 = 788$ N. Therefore, ignoring the buoyancy in this case results in an error in weight of just 0.15 percent, which is negligible. But the buoyancy effects in gases dominate some important natural phenomena such as the rise of warm air in a cooler environment and thus the onset of natural convection currents, the rise of hot-air or helium balloons, and air movements in the atmosphere. A helium balloon, for example, rises as a result of the buoyancy effect until it reaches an altitude where the density of air (which decreases with altitude) equals the density of helium in the balloon—assuming the balloon does not burst by then, and ignoring the weight of the balloon's skin. Hot air balloons (Fig. 11–18) work by similar principles.

Archimedes' principle is also used in geology by considering the continents to be floating on a sea of magma.

### EXAMPLE 11–3     Measuring Specific Gravity by a Hydrometer

If you have a seawater aquarium, you have probably used a small cylindrical glass tube with a lead-weight at its bottom to measure the salinity of the water by simply watching how deep the tube sinks. Such a device that floats in a vertical position and is used to measure the specific gravity of a liquid is called a *hydrometer* (Fig. 11–19). The top part of the hydrometer extends above the liquid surface, and the divisions on it allow one to read the specific gravity directly. The hydrometer is calibrated such that in pure water it reads exactly 1.0 at the air–water interface. (a) Obtain a relation for the specific gravity of a liquid as a function of distance $\Delta z$ from the mark corresponding to pure water and (b) determine the mass of lead that must be poured into a 1-cm-diameter, 20-cm-long hydrometer if it is to float halfway (the 10-cm mark) in pure water.

**Solution**   The specific gravity of a liquid is to be measured by a hydrometer. A relation between specific gravity and the vertical distance from the reference level is to be obtained, and the amount of lead that needs to be added into the tube for a certain hydrometer is to be determined.

**Assumptions**   **1** The weight of the glass tube is negligible relative to the weight of the lead added. **2** The curvature of the tube bottom is disregarded.

**Properties**   We take the density of pure water to be 1000 kg/m³.

**Analysis**   (a) Noting that the hydrometer is in static equilibrium, the buoyant force $F_B$ exerted by the liquid must always be equal to the weight $W$ of the hydrometer. In pure water (subscript $w$), we let the vertical distance between the bottom of the hydrometer and the free surface *of* water be $z_0$. Setting $F_{B, w} = W$ in this case gives

$$W_{\text{hydro}} = F_{B, w} = \rho_w g V_{\text{sub}} = \rho_w g A z_0 \tag{1}$$

where $A$ is the cross-sectional area of the tube, and $\rho_w$ is the density of pure water.

In a fluid lighter than water ($\rho_f < \rho_w$), the hydrometer will sink deeper, and the liquid level will be a distance of $\Delta z$ above $z_0$. Again setting $F_B = W$ gives

$$W_{\text{hydro}} = F_{B, f} = \rho_f g V_{\text{sub}} = \rho_f g A(z_0 + \Delta z) \tag{2}$$

This relation is also valid for fluids heavier than water by taking $\Delta z$ to be a negative quantity. Setting Eqs. (1) and (2) here equal to each other since the weight of the hydrometer is constant and rearranging gives

$$\rho_w g A z_0 = \rho_f g A(z_0 + \Delta z) \quad \rightarrow \quad \text{SG}_f = \frac{\rho_f}{\rho_w} = \frac{z_0}{z_0 + \Delta z}$$

which is the relation between the specific gravity of the fluid and $\Delta z$. Note that $z_0$ is constant for a given hydrometer and $\Delta z$ is negative for fluids heavier than pure water.

(b) Disregarding the weight of the glass tube, the amount of lead that needs to be added to the tube is determined from the requirement that the weight of the lead be equal to the buoyant force. When the hydrometer is floating with half of it submerged in water, the buoyant force acting on it is

$$F_B = \rho_w g V_{\text{sub}}$$

**FIGURE 11–18**

The altitude of a hot air balloon is controlled by the temperature difference between the air inside and outside the balloon, since warm air is less dense than cold air. When the balloon is neither rising nor falling, the upward buoyant force exactly balances the downward weight.

© Getty RF.

**FIGURE 11–19**

Schematic for Example 11–3.

Equating $F_B$ to the weight of lead gives

$$W = mg = \rho_w g V_{sub}$$

Solving for $m$ and substituting, the mass of lead is determined to be

$$m = \rho_w V_{sub} = \rho_w(\pi R^2 h_{sub}) = (1000 \text{ kg/m}^3)[\pi(0.005 \text{ m})^2(0.1 \text{ m})] = \mathbf{0.00785 \text{ kg}}$$

**Discussion** Note that if the hydrometer were required to sink only 5 cm in water, the required mass of lead would be one-half of this amount. Also, the assumption that the weight of the glass tube is negligible is questionable since the mass of lead is only 7.85 g.

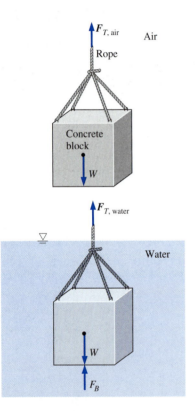

**FIGURE 11–20**
Schematic for Example 11–4.

**EXAMPLE 11–4**     **Weight Loss of an Object in Seawater**

A crane is used to lower weights into the sea (density = 1025 kg/m³) for an underwater construction project (Fig. 11–20). Determine the tension in the rope of the crane due to a rectangular 0.4-m × 0.4-m × 3-m concrete block (density = 2300 kg/m³) when it is (a) suspended in the air and (b) completely immersed in water.

**Solution** A concrete block is lowered into the sea. The tension in the rope is to be determined before and after the block is in water.
**Assumptions** 1 The buoyant force in air is negligible. 2 The weight of the ropes is negligible.
**Properties** The densities are given to be 1025 kg/m³ for seawater and 2300 kg/m³ for concrete.
**Analysis** (a) Consider a free-body diagram of the concrete block. The forces acting on the concrete block in air are its weight and the upward pull action (tension) by the rope. These two forces must balance each other, and thus the tension in the rope must be equal to the weight of the block:

$$V = (0.4 \text{ m})(0.4 \text{ m})(3 \text{ m}) = 0.48 \text{ m}^3$$

$$F_{T, \text{air}} = W = \rho_{\text{concrete}} g V$$

$$= (2300 \text{ kg/m}^3)(9.81 \text{ m/s}^2)(0.48 \text{ m}^3)\left(\frac{1 \text{ kN}}{1000 \text{ kg·m/s}^2}\right) = \mathbf{10.8 \text{ kN}}$$

(b) When the block is immersed in water, there is the additional force of buoyancy acting upward. The force balance in this case gives

$$F_B = \rho_f g V = (1025 \text{ kg/m}^3)(9.81 \text{ m/s}^2)(0.48 \text{ m}^3)\left(\frac{1 \text{ kN}}{1000 \text{ kg·m/s}^2}\right) = 4.8 \text{ kN}$$

$$F_{T, \text{water}} = W - F_B = 10.8 - 4.8 = \mathbf{6.0 \text{ kN}}$$

**Discussion** Note that the weight of the concrete block, and thus the tension of the rope, decreases by (10.8 − 6.0)/10.8 = 55 percent in water.

# Stability of Immersed and Floating Bodies

An important application of the buoyancy concept is the assessment of the stability of immersed and floating bodies with no external attachments. This topic is of great importance in the design of ships and submarines (Fig. 11–21). Here we provide some general qualitative discussions on vertical and rotational stability.

We use the classic "ball on the floor" analogy to explain the fundamental concepts of stability and instability. Three balls at rest on the floor are shown in Fig. 11–22. Case (*a*) is **stable** since any small disturbance (someone moves the ball to the right or left) generates a restoring force (due to gravity) that returns it to its initial position. Case (*b*) is **neutrally stable** because if someone moves the ball to the right or left, it would stay put at its new location. It has no tendency to move back to its original location, nor does it continue to move away. Case (*c*) is a situation in which the ball may be at rest at the moment, but any disturbance, even an infinitesimal one, causes the ball to roll off the hill—it does not return to its original position; rather it *diverges* from it. This situation is **unstable**. What about a case where the ball is on an *inclined* floor? It is not appropriate to discuss stability for this case since the ball is not in a state of equilibrium. In other words, it cannot be at rest and would roll down the hill even without any disturbance.

For an immersed or floating body in static equilibrium, the weight and the buoyant force acting on the body balance each other, and such bodies are inherently stable in the *vertical direction*. If an immersed neutrally buoyant body is raised or lowered to a different depth in an incompressible fluid, the body will remain in equilibrium at that location. If a floating body is raised or lowered somewhat by a vertical force, the body will return to its original position as soon as the external effect is removed. Therefore, a floating body possesses vertical stability, while an immersed neutrally buoyant body is neutrally stable since it does not return to its original position after a disturbance.

The *rotational stability* of an *immersed body* depends on the relative locations of the *center of gravity G* of the body and the *center of buoyancy B*, which is the centroid of the displaced volume. An immersed body is *stable* if the body is bottom-heavy and thus point *G* is directly below point *B* (Fig. 11–23*a*). A rotational disturbance of the body in such cases produces a *restoring moment* to return the body to its original stable position. Thus, a stable design for a submarine calls for the engines and the cabins for the crew to be located at the lower half in order to shift the weight to the bottom as much as possible. Hot-air or helium balloons (which can be viewed as being immersed in air) are also stable since the heavy cage that carries the load is at the bottom. An immersed body whose center of gravity *G* is directly above point *B* is *unstable,* and any disturbance will cause this body to turn upside down (Fig 11–23*c*). A body for which *G* and *B* coincide is *neutrally stable* (Fig 11–23*b*). This is the case for bodies whose density is constant throughout. For such bodies, there is no tendency to overturn or right themselves.

What about a case where the center of gravity is not vertically aligned with the center of buoyancy, as in Fig. 11–24? It is not appropriate to discuss stability for this case since the body is not in a state of equilibrium. In other words, it cannot be at rest and would rotate toward its stable state even without any disturbance. The restoring moment in the case shown in Fig. 11–24 is counterclockwise and causes the body to rotate counterclockwise

**FIGURE 11–21**

For floating bodies such as ships, stability is an important consideration for safety.
© *Vol. 96/Corbis RF.*

(*a*) Stable

(*b*) Neutrally stable

(*c*) Unstable

**FIGURE 11–22**

Stability is easily understood by analyzing a ball on the floor.

## FIGURE 11–23

An immersed neutrally buoyant body is (*a*) stable if the center of gravity *G* is directly below the center of buoyancy *B* of the body, (*b*) neutrally stable if *G* and *B* are coincident, and (*c*) unstable if *G* is directly above *B*.

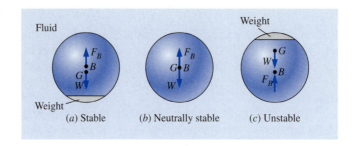

(*a*) Stable    (*b*) Neutrally stable    (*c*) Unstable

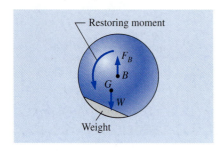

## FIGURE 11–24

When the center of gravity *G* of an immersed neutrally buoyant body is not vertically aligned with the center of buoyancy *B* of the body, it is not in an equilibrium state and would rotate to its stable state, even without any disturbance.

so as to align point *G* vertically with point *B*. Note that there may be some oscillation, but eventually the body settles down at its stable equilibrium state [case (*a*) of Fig. 11–23]. The initial stability of the body of Fig. 11–24 is analogous to that of the ball on an inclined floor. Can you predict what would happen if the weight in the body of Fig. 11–24 were on the opposite side of the body?

The rotational stability criteria are similar for *floating bodies*. Again, if the floating body is bottom-heavy and thus the center of gravity *G* is directly below the center of buoyancy *B*, the body is always stable. But unlike immersed bodies, a floating body may still be stable when *G* is directly above *B* (Fig. 11–25). This is because the centroid of the displaced volume shifts to the side to a point *B'* during a rotational disturbance while the center of gravity *G* of the body remains unchanged. If point *B'* is sufficiently far, these two forces create a restoring moment and return the body to the original position. A measure of stability for floating bodies is the **metacentric height** *GM*, which is the distance between the center of gravity *G* and the metacenter *M*—the intersection point of the lines of action of the buoyant force through the body before and after rotation. The metacenter may be considered to be a fixed point for most hull shapes for small rolling angles up to about 20°. A floating body is stable if point *M* is above point *G*, and thus *GM* is positive, and unstable if point *M* is below point *G*, and thus *GM* is negative. In the latter case, the weight and the buoyant force acting on the tilted body generate an overturning moment instead of a restoring moment, causing the body to capsize. The length of the metacentric height

## FIGURE 11–25

A floating body is *stable* if the body is (*a*) bottom-heavy and thus the center of gravity *G* is below the centroid *B* of the body, or (*b*) if the metacenter *M* is above point *G*. However, the body is (*c*) *unstable* if point *M* is below point *G*.

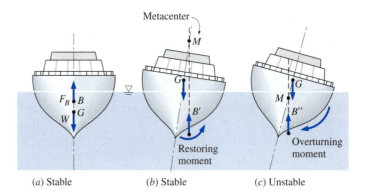

(*a*) Stable    (*b*) Stable    (*c*) Unstable

*GM* above *G* is a measure of the stability: the larger it is, the more stable is the floating body.

As already discussed, a boat can tilt to some maximum angle without capsizing, but beyond that angle it overturns (and sinks). We make a final analogy between the stability of floating objects and the stability of a ball rolling along the floor. Namely, imagine the ball in a trough between two hills (Fig. 11–26). The ball returns to its stable equilibrium position after being perturbed—up to a limit. If the perturbation amplitude is too great, the ball rolls down the opposite side of the hill and does not return to its equilibrium position. This situation is described as stable up to some limiting level of disturbance, but unstable beyond.

**FIGURE 11–26**

A ball in a trough between two hills is stable for small disturbances, but unstable for large disturbances.

## SUMMARY

*Fluid statics* deals with problems associated with fluids at rest, and it is called *hydrostatics* when the fluid is a liquid. The magnitude of the resultant force acting on a plane surface of a completely submerged plate in a homogeneous fluid is equal to the product of the pressure $P_C$ at the centroid of the surface and the area $A$ of the surface and is expressed as

$$F_R = (P_0 + \rho g h_C)A = P_C A = P_{avg}A$$

where $h_C = y_C \sin \theta$ is the *vertical distance* of the centroid from the free surface of the liquid. The pressure $P_0$ at the surface of the liquid is usually atmospheric pressure, which cancels out in most cases since it acts on both sides of the plate. The point of intersection of the line of action of the resultant force and the surface is the *center of pressure*. The vertical location of the line of action of the resultant force is given by

$$y_P = y_C + \frac{I_{xx, C}}{[y_C + P_0/(\rho g \sin \theta)]A}$$

where $I_{xx, C}$ is the second moment of area about the *x*-axis passing through the centroid of the area.

A fluid of density $P_f$ exerts an upward force on a body immersed in it. This force is called the *buoyant force* and is expressed as

$$F_B = \rho_f g V$$

where $V$ is the volume of the body. This is known as *Archimedes' principle* and is expressed as: the buoyant force acting on a body immersed in a fluid is equal to the weight of the fluid displaced by the body; it acts upward through the centroid of the displaced volume. In a fluid with constant density, the buoyant force is independent of the distance of the body from the free surface. For *floating* bodies, the submerged volume fraction of the body is equal to the ratio of the average density of the body to the density of the fluid.

## REFERENCES AND SUGGESTED READING

**1.** F. P. Beer, E. R. Johnston, Jr., E. R. Eisenberg, and G. H. Staab. *Vector Mechanics for Engineers, Statics*, 7th ed. New York: McGraw-Hill, 2004.

**2.** D. C. Giancoli. *Physics*, 3rd ed. Upper Saddle River, NJ: Prentice Hall, 1991.

# PROBLEMS*

## Fluid Statics: Hydrostatic Forces on Plane and Curved Surfaces

**11–1C**  Define the resultant hydrostatic force acting on a submerged surface, and the center of pressure.

**11–2C**  Someone claims that she can determine the magnitude of the hydrostatic force acting on a plane surface submerged in water regardless of its shape and orientation if she knew the vertical distance of the centroid of the surface from the free surface and the area of the surface. Is this a valid claim? Explain.

**11–3C**  A submerged horizontal flat plate is suspended in water by a string attached at the centroid of its upper surface. Now the plate is rotated 45° about an axis that passes through its centroid. Discuss the change on the hydrostatic force acting on the top surface of this plate as a result of this rotation. Assume the plate remains submerged at all times.

**11–4C**  You may have noticed that dams are much thicker at the bottom. Explain why dams are built that way.

**11–5C**  Consider a submerged curved surface. Explain how you would determine the horizontal component of the hydrostatic force acting on this surface.

**11–6C**  Consider a submerged curved surface. Explain how you would determine the vertical component of the hydrostatic force acting on this surface.

**11–7C**  Consider a circular surface subjected to hydrostatic forces by a constant density liquid. If the magnitudes of the horizontal and vertical components of the resultant hydrostatic force are determined, explain how you would find the line of action of this force.

**11–8**  Consider a heavy car submerged in water in a lake with a flat bottom. The driver's side door of the car is 1.1 m high and 0.9 m wide, and the top edge of the door is 10 m below the water surface. Determine the net force acting on the door (normal to its surface) and the location of the pressure center if (a) the car is well-sealed and it contains air at atmospheric pressure and (b) the car is filled with water.

**11–9E**  A long, solid cylinder of radius 2 ft hinged at point A is used as an automatic gate, as shown in Fig. P11–9E. When the water level reaches 15 ft, the cylindrical gate opens by turning about the hinge at point A. Determine (a) the hydrostatic

force acting on the cylinder and its line of action when the gate opens and (b) the weight of the cylinder per ft length of the cylinder.

**FIGURE P11–9E**

**11–10**  Consider a 4-m-long, 4-m-wide, and 1.5-m-high aboveground swimming pool that is filled with water to the rim. (a) Determine the hydrostatic force on each wall and the distance of the line of action of this force from the ground. (b) If the height of the walls of the pool is doubled and the pool is filled, will the hydrostatic force on each wall double or quadruple? Why?  *Answer: (a) 44.1 kN*

**11–11E**  Consider a 200-ft-high, 1200-ft-wide dam filled to capacity. Determine (a) the hydrostatic force on the dam and (b) the force per unit area of the dam near the top and near the bottom.

**11–12**  A room in the lower level of a cruise ship has a 30-cm-diameter circular window. If the midpoint of the window is 4 m below the water surface, determine the hydrostatic force acting on the window, and the pressure center. Take the specific gravity of seawater to be 1.025.  *Answers: 2840 N, 4.001 m*

**FIGURE P11–12**

**11–13**  The water side of the wall of a 100-m-long dam is a quarter circle with a radius of 10 m. Determine the hydrostatic force on the dam and its line of action when the dam is filled to the rim.

**11–14** A 6-m-high, 5-m-wide rectangular plate blocks the end of a 5-m-deep freshwater channel, as shown in Fig. P11–14. The plate is hinged about a horizontal axis along its upper edge through a point A and is restrained from opening by a fixed ridge at point B. Determine the force exerted on the plate by the ridge.

**FIGURE P11–14**

**11–15** Reconsider Prob. 11–14. Using EES (or other) software, investigate the effect of water depth on the force exerted on the plate by the ridge. Let the water depth vary from 0 to 5 m in increments of 0.5 m. Tabulate and plot your results.

**11–16E** The flow of water from a reservoir is controlled by a 5-ft-wide L-shaped gate hinged at point A, as shown in Fig. P11–16E. If it is desired that the gate open when the water height is 12 ft, determine the mass of the required weight W.   *Answer:* 30,900 lbm

**FIGURE P11–16E**

**11–17E** Repeat Prob. 11–16E for a water height of 8 ft.

**11–18** A water trough of semicircular cross section of radius 0.7 m consists of two symmetric parts hinged to each other at the bottom, as shown in Fig. P11–18. The two parts are held together by a cable and turnbuckle placed every 3 m along the length of the trough. Calculate the tension in each cable when the trough is filled to the rim.

**FIGURE P11–18**

**11–19** The two sides of a V-shaped water trough are hinged to each other at the bottom where they meet, as shown in Fig. P11–19, making an angle of 45° with the ground from both sides. Each side is 0.75 m wide, and the two parts are held together by a cable and turnbuckle placed every 6 m along the length of the trough. Calculate the tension in each cable when the trough is filled to the rim.   *Answer:* 5510 N

**FIGURE P11–19**

**11–20** Repeat Prob. 11–19 for the case of a partially filled trough with a water height of 0.4 m directly above the hinge.

**11–21** A retaining wall against a mud slide is to be constructed by placing 1.2-m-high and 0.25-m-wide rectangular concrete blocks ($\rho = 2700$ kg/m$^3$) side by side, as shown in Fig. P11–21. The friction coefficient between the ground and the concrete blocks is $f = 0.3$, and the density of the mud is about 1800 kg/m$^3$. There is concern that the concrete blocks may slide or tip over the lower left edge as the mud level rises. Determine the mud height at which (a) the blocks will overcome friction and start sliding and (b) the blocks will tip over.

**FIGURE P11–21**

**11–22** Repeat Prob. 11–21 for 0.4-m-wide concrete blocks.

**11–23** A 4-m-long quarter-circular gate of radius 3 m and of negligible weight is hinged about its upper edge $A$, as shown in Fig. P11–23. The gate controls the flow of water over the ledge at $B$, where the gate is pressed by a spring. Determine the minimum spring force required to keep the gate closed when the water level rises to $A$ at the upper edge of the gate.

**FIGURE P11–23**

**11–24** Repeat Prob. 11–23 for a radius of 4 m for the gate. *Answer:* 314 kN

### Buoyancy

**11–25C** What is buoyant force? What causes it? What is the magnitude of the buoyant force acting on a submerged body whose volume is $V$? What are the direction and the line of action of the buoyant force?

**11–26C** Consider two identical spherical balls submerged in water at different depths. Will the buoyant forces acting on these two balls be the same or different? Explain.

**11–27C** Consider two 5-cm-diameter spherical balls—one made of aluminum, the other of iron—submerged in water. Will the buoyant forces acting on these two balls be the same or different? Explain.

**11–28C** Consider a 3-kg copper cube and a 3-kg copper ball submerged in a liquid. Will the buoyant forces acting on these two bodies be the same or different? Explain.

**11–29C** Discuss the stability of ($a$) a submerged and ($b$) a floating body whose center of gravity is above the center of buoyancy.

**11–30** The density of a liquid is to be determined by an old 1-cm-diameter cylindrical hydrometer whose division marks are completely wiped out. The hydrometer is first dropped in water, and the water level is marked. The hydrometer is then dropped into the other liquid, and it is observed that the mark for water has risen 0.6 cm above the liquid–air interface (Fig. P11–30). If the height of the original water mark is 13.6 cm, determine the density of the liquid.

**FIGURE P11–30**

**11–31E** A crane is used to lower weights into a lake for an underwater construction project. Determine the tension in the rope of the crane due to a 3-ft-diameter spherical steel block (density = 494 lbm/ft$^3$) when it is ($a$) suspended in the air and ($b$) completely immersed in water.

**11–32** The volume and the average density of an irregularly shaped body are to be determined by using a spring scale. The body weighs 7200 N in air and 4790 N in water. Determine the volume and the density of the body. State your assumptions.

**11–33** Consider a large cubic ice block floating in seawater. The specific gravities of ice and seawater are 0.92 and 1.025, respectively. If a 15-cm-high portion of the ice block extends above the surface of the water, determine the height of the ice block below the surface. *Answer:* 1.31 m

**FIGURE P11–33**

**11–34** It is estimated that 90 percent of an iceberg's volume is below the surface, while only 10 percent is visible above the surface. For seawater with a density of 1025 kg/m$^3$, estimate the density of the iceberg.

**FIGURE P11-34**

**11-35** A 170-kg granite rock ($\rho = 2700$ kg/m$^3$) is dropped into a lake. A man dives in and tries to lift the rock. Determine how much force the man needs to apply to lift it from the bottom of the lake. Do you think he can do it?

**11-36** It is said that Archimedes discovered his principle during a bath while thinking about how he could determine if King Hiero's crown was actually made of pure gold. While in the bathtub, he conceived the idea that he could determine the average density of an irregularly shaped object by weighing it in air and also in water. If the crown weighed 3.20 kgf (= 31.4 N) in air and 2.95 kgf (= 28.9 N) in water, determine if the crown is made of pure gold. The density of gold is 19,300 kg/m$^3$. Discuss how you can solve this problem without weighing the crown in water but by using an ordinary bucket with no calibration for volume. You may weigh anything in air.

**11-37** One of the common procedures in fitness programs is to determine the fat-to-muscle ratio of the body. This is based on the principle that the muscle tissue is denser than the fat tissue, and, thus, the higher the average density of the body, the higher is the fraction of muscle tissue. The average density of the body can be determined by weighing the person in air and also while submerged in water in a tank. Treating all tissues and bones (other than fat) as muscle with an equivalent density of $\rho_{muscle}$, obtain a relation for the volume fraction of body fat $x_{fat}$. *Answer:* $x_{fat} = (\rho_{muscle} - \rho_{avg})/(\rho_{muscle} - \rho_{fat})$.

**11-38** The hull of a boat has a volume of 180 m$^3$, and the total mass of the boat when empty is 8560 kg. Determine how much load this boat can carry without sinking (*a*) in a lake and (*b*) in seawater with a specific gravity of 1.03.

### Review Problems

**11-39** An air-conditioning system requires a 20-m-long section of 15-cm-diameter ductwork to be laid underwater. Determine the upward force the water will exert on the duct. Take the densities of air and water to be 1.3 kg/m$^3$ and 1000 kg/m$^3$, respectively.

**11-40** The density of a floating body can be determined by tying weights to the body until both the body and the weights are completely submerged, and then weighing them separately in air. Consider a wood log that weighs 1540 N in air. If it takes 34 kg of lead ($\rho = 11,300$ kg/m$^3$) to completely sink the log and the lead in water, determine the average density of the log. *Answer:* 835 kg/m$^3$

**11-41** The 280-kg, 6-m-wide rectangular gate shown in Fig. P11-41 is hinged at *B* and leans against the floor at *A* making an angle of 45° with the horizontal. The gate is to be opened from its lower edge by applying a normal force at its center. Determine the minimum force *F* required to open the water gate. *Answer:* 626 kN

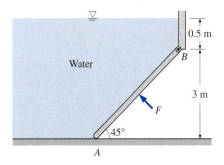

**FIGURE P11-41**

**11–42** Repeat Prob. 11–41 for a water height of 1.2 m above the hinge at *B*.

**11–43** A 3-m-high, 6-m-wide rectangular gate is hinged at the top edge at *A* and is restrained by a fixed ridge at *B*. Determine the hydrostatic force exerted on the gate by the 5-m-high water and the location of the pressure center.

**FIGURE P11–43**

**11–44** Repeat Prob. 11–43 for a total water height of 2 m.

**11–45E** A semicircular 40-ft-diameter tunnel is to be built under a 150-ft-deep, 800-ft-long lake, as shown in Fig. P11–45E. Determine the total hydrostatic force acting on the roof of the tunnel.

**FIGURE P11–45E**

**11–46** A 50-ton, 6-m-diameter hemispherical dome on a level surface is filled with water, as shown in Fig. P11–46. Someone claims that he can lift this dome by making use of Pascal's law by attaching a long tube to the top and filling it with water. Determine the required height of water in the tube to lift the dome. Disregard the weight of the tube and the water in it.   *Answer: 0.77 m*

**FIGURE P11–46**

**11–47** The water in a 25-m-deep reservoir is kept inside by a 150-m-wide wall whose cross section is an equilateral triangle, as shown in Fig. P11–47. Determine (*a*) the total force (hydrostatic + atmospheric) acting on the inner surface of the wall and its line of action and (*b*) the magnitude of the horizontal component of this force. Take $P_{atm} = 100$ kPa.

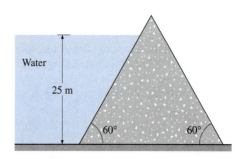

**FIGURE P11–47**

**11–48** An elastic air balloon having a diameter of 30 cm is attached to the base of a container partially filled with water at +4°C, as shown in Fig. P11–48. If the pressure of air above water is gradually increased from 100 kPa to 1.6 MPa, will the force on the cable change? If so, what is the percent change in the force? Assume the pressure on the free surface and the diameter of the balloon are related by $P = CD^n$, where $C$ is a constant and $n = -2$. The weight of the balloon and the air in it is negligible.   *Answer: 98.4 percent*

$P_1 = 100$ kPa

20 cm

$D_1 = 30$ cm

50 cm

Water

50 cm

**FIGURE P11–48**

**11–49** Reconsider Prob. 11–48. Using EES (or other) software, investigate the effect of air pressure above water on the cable force. Let this pressure vary from 0.5 MPa to 15 MPa. Plot the cable force versus the air pressure.

**11–50** The average density of icebergs is about 917 kg/m³. (*a*) Determine the percentage of the total volume of an iceberg submerged in seawater of density 1042 kg/m³. (*b*) Although icebergs are mostly submerged, they are observed to turn over. Explain how this can happen. (*Hint*: Consider the temperatures of icebergs and seawater.)

### Design and Essay Problems

**11–51** Shoes are to be designed to enable people of up to 80 kg to walk on freshwater or seawater. The shoes are to be made of blown plastic in the shape of a sphere, a (American) football, or a loaf of French bread. Determine the equivalent diameter of each shoe and comment on the proposed shapes from the stability point of view. What is your assessment of the marketability of these shoes?

**11–52** The volume of a rock is to be determined without using any volume measurement devices. Explain how you would do this with a waterproof spring scale.

# BERNOULLI AND ENERGY EQUATIONS

This chapter deals with two equations commonly used in fluid mechanics: Bernoulli and energy equations. The *Bernoulli equation* is concerned with the conservation of kinetic, potential, and flow energies of a fluid stream and their conversion to each other in regions of flow where net viscous forces are negligible and where other restrictive conditions apply. The *energy equation* is a statement of the conservation of energy principle. In fluid mechanics, it is found convenient to separate *mechanical energy* from *thermal energy* and to consider the conversion of mechanical energy to thermal energy as a result of frictional effects as *mechanical energy loss*. Then the energy equation becomes the *mechanical energy balance*.

In this chapter we derive the Bernoulli equation by applying Newton's second law to a fluid element along a streamline and demonstrate its use in a variety of applications. We continue with the development of the energy equation in a form suitable for use in fluid mechanics and introduce the concept of *head loss*. Finally, we apply the energy equation to various engineering systems.

## Objectives
The objectives of this chapter are to:

■ Understand the use and limitations of the Bernoulli equation, and apply it to solve a variety of fluid flow problems.

■ Work with the energy equation expressed in terms of heads, and use it to determine turbine power output and pumping power requirements.

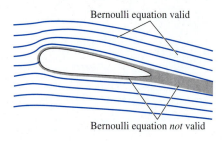

Bernoulli equation valid

Bernoulli equation *not* valid

**FIGURE 12–1**

The *Bernoulli equation* is an *approximate* equation that is valid only in *inviscid regions of flow* where net viscous forces are negligibly small compared to inertial, gravitational, or pressure forces. Such regions occur outside of *boundary layers* and *wakes*.

# 12–1 ▪ THE BERNOULLI EQUATION

The **Bernoulli equation** is *an approximate relation between pressure, velocity, and elevation,* and is valid in *regions of steady, incompressible flow where net frictional forces are negligible* (Fig. 12–1). Despite its simplicity, it has proven to be a very powerful tool in fluid mechanics. In this section, we derive the Bernoulli equation by applying the *conservation of linear momentum principle*, and we demonstrate both its usefulness and its limitations.

The key approximation in the derivation of the Bernoulli equation is that *viscous effects are negligibly small compared to inertial, gravitational, and pressure effects.* Since all fluids have viscosity (there is no such thing as an "inviscid fluid"), this approximation cannot be valid for an entire flow field of practical interest. In other words, we cannot apply the Bernoulli equation *everywhere* in a flow, no matter how small the fluid's viscosity. However, it turns out that the approximation *is* reasonable in certain *regions* of many practical flows. We refer to such regions as *inviscid regions of flow*, and we stress that they are *not* regions where the fluid itself is inviscid or frictionless, but rather they are regions where net viscous or frictional forces are negligibly small compared to other forces acting on fluid particles.

Care must be exercised when applying the Bernoulli equation since it is an approximation that applies only to inviscid regions of flow. In general, frictional effects are always important very close to solid walls (*boundary layers*) and directly downstream of bodies (*wakes*). Thus, the Bernoulli approximation is typically useful in flow regions outside of boundary layers and wakes, where the fluid motion is governed by the combined effects of pressure and gravity forces.

## Acceleration of a Fluid Particle

The motion of a particle and the path it follows are described by the *velocity vector* as a function of time and space coordinates and the initial position of the particle. When the flow is *steady* (no change with time at a specified location), all particles that pass through the same point follow the same path (which is the *streamline*), and the velocity vectors remain tangent to the path at every point.

Often it is convenient to describe the motion of a particle in terms of its distance $s$ along a streamline together with the radius of curvature along the streamline. The speed of the particle is related to the distance by $V = ds/dt$, which may vary along the streamline. In two-dimensional flow, the acceleration can be decomposed into two components: *streamwise acceleration* $a_s$ along the streamline and *normal acceleration* $a_n$ in the direction normal to the streamline, which is given as $a_n = V^2/R$. Note that streamwise acceleration is due to a change in speed along a streamline, and normal acceleration is due to a change in direction. For particles that move along a *straight path*, $a_n = 0$ since the radius of curvature is infinity and thus there is no change in direction. The Bernoulli equation results from a force balance along a streamline.

One may be tempted to think that acceleration is zero in steady flow since acceleration is the rate of change of velocity with time, and in steady flow there is no change with time. Well, a garden hose nozzle tells us that this understanding is not correct. Even in steady flow and thus constant mass flow rate, water accelerates through the nozzle (Fig. 12–2). *Steady* simply means *no change with time at a specified location,* but the value of a quantity may change from one location to another. In the case of a nozzle, the velocity of water remains constant at a specified point, but it changes from the inlet to the exit (water accelerates along the nozzle).

Mathematically, this can be expressed as follows: We take the velocity $V$ of a fluid particle to be a function of $s$ and $t$. Taking the total differential of $V(s, t)$ and dividing both sides by $dt$ yield

$$dV = \frac{\partial V}{\partial s}\, ds + \frac{\partial V}{\partial t}\, dt \quad \text{and} \quad \frac{dV}{dt} = \frac{\partial V}{\partial s}\frac{ds}{dt} + \frac{\partial V}{\partial t} \tag{12–1}$$

In steady flow $\partial V/\partial t = 0$ and thus $V = V(s)$, and the acceleration in the $s$-direction becomes

$$a_s = \frac{dV}{dt} = \frac{\partial V}{\partial s}\frac{ds}{dt} = \frac{\partial V}{\partial s}V = V\frac{dV}{ds} \tag{12–2}$$

where $V = ds/dt$ if we are following a fluid particle as it moves along a streamline. Therefore, acceleration in steady flow is due to the change of velocity with position.

**FIGURE 12–2**
During steady flow, a fluid may not accelerate in time at a fixed point, but it may accelerate in space.

## Derivation of the Bernoulli Equation

Consider the motion of a fluid particle in a flow field in steady flow. Applying Newton's second law (which is referred to as the *linear momentum equation* in fluid mechanics) in the $s$-direction on a particle moving along a streamline gives

$$\sum F_s = ma_s \tag{12–3}$$

In regions of flow where net frictional forces are negligible, there is no pump or turbine, and there is no heat transfer along the streamline, the significant forces acting in the $s$-direction are the pressure (acting on both sides) and the component of the weight of the particle in the $s$-direction (Fig. 12–3). Therefore, Eq. 12–3 becomes

$$P\, dA - (P + dP)\, dA - W\sin\theta = mV\frac{dV}{ds} \tag{12–4}$$

where $\theta$ is the angle between the normal of the streamline and the vertical $z$-axis at that point, $m = \rho V = \rho\, dA\, ds$ is the mass, $W = mg = \rho g\, dA\, ds$ is the weight of the fluid particle, and $\sin\theta = dz/ds$. Substituting,

$$-dP\, dA - \rho g\, dA\, ds\frac{dz}{ds} = \rho\, dA\, ds\, V\frac{dV}{ds} \tag{12–5}$$

Canceling $dA$ from each term and simplifying,

$$-dP - \rho g\, dz = \rho V\, dV \tag{12–6}$$

Noting that $V\, dV = \frac{1}{2}d(V^2)$ and dividing each term by $\rho$ gives

$$\frac{dP}{\rho} + \frac{1}{2}d(V^2) + g\, dz = 0 \tag{12–7}$$

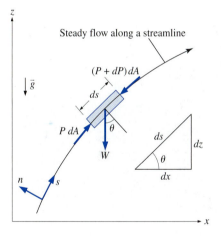

**FIGURE 12–3**
The forces acting on a fluid particle along a streamline.

(Steady flow along a streamline)

General:

$$\int \frac{dP}{\rho} + \frac{V^2}{2} + gz = \text{constant}$$

Incompressible flow ($\rho$ = constant):

$$\frac{P}{\rho} + \frac{V^2}{2} + gz = \text{constant}$$

**FIGURE 12–4**
The incompressible Bernoulli equation is derived assuming incompressible flow, and thus it should not be used for flows with significant compressibility effects.

Integrating,

*Steady flow:*

$$\int \frac{dP}{\rho} + \frac{V^2}{2} + gz = \text{constant (along a streamline)} \qquad (12\text{–}8)$$

since the last two terms are exact differentials. In the case of incompressible flow, the first term also becomes an exact differential, and integration gives

*Steady, incompressible flow:*

$$\frac{P}{\rho} + \frac{V^2}{2} + gz = \text{constant (along a streamline)} \qquad (12\text{–}9)$$

This is the famous **Bernoulli equation** (Fig. 12–4), which is commonly used in fluid mechanics for steady, incompressible flow along a streamline in inviscid regions of flow. The Bernoulli equation was first stated in words by the Swiss mathematician Daniel Bernoulli (1700–1782) in a text written in 1738 when he was working in St. Petersburg, Russia. It was later derived in equation form by his associate Leonhard Euler (1707–1783) in 1755.

The value of the constant in Eq. 12–9 can be evaluated at any point on the streamline where the pressure, density, velocity, and elevation are known. The Bernoulli equation can also be written between any two points on the same streamline as

*Steady, incompressible flow:*

$$\frac{P_1}{\rho} + \frac{V_1^2}{2} + gz_1 = \frac{P_2}{\rho} + \frac{V_2^2}{2} + gz_2 \qquad (12\text{–}10)$$

We recognize $V^2/2$ as *kinetic energy,* $gz$ as *potential energy,* and $P/\rho$ as *flow energy,* all per unit mass. Therefore, the Bernoulli equation can be viewed as an expression of *mechanical energy balance* and can be stated as follows (Fig. 12–5):

> The sum of the kinetic, potential, and flow energies of a fluid particle is constant along a streamline during steady flow when compressibility and frictional effects are negligible.

The kinetic, potential, and flow energies are the mechanical forms of energy, and the Bernoulli equation can be viewed as the "conservation of mechanical energy principle." This is equivalent to the general conservation of energy principle for systems that do not involve any conversion of mechanical energy and thermal energy to each other, and thus the mechanical energy and thermal energy are conserved separately. The Bernoulli equation states that during steady, incompressible flow with negligible friction, the various forms of mechanical energy are converted to each other, but their sum remains constant. In other words, there is no dissipation of mechanical energy during such flows since there is no friction that converts mechanical energy to sensible thermal (internal) energy.

Recall that energy is transferred to a system as work when a force is applied to the system through a distance. In the light of Newton's second law of motion, the Bernoulli equation can also be viewed as: *The work done by the pressure and gravity forces on the fluid particle is equal to the increase in the kinetic energy of the particle.*

The Bernoulli equation is obtained from Newton's second law for a fluid particle moving along a streamline. It can also be obtained from the *first law of thermodynamics* applied to a steady-flow system, as shown in Section 12–2.

**FIGURE 12–5**
The Bernoulli equation states that the sum of the kinetic, potential, and flow energies (all per unit mass) of a fluid particle is constant along a streamline during steady flow.

Despite the highly restrictive approximations used in its derivation, the Bernoulli equation is commonly used in practice since a variety of practical fluid flow problems can be analyzed to reasonable accuracy with it. This is because many flows of practical engineering interest are steady (or at least steady in the mean), compressibility effects are relatively small, and net frictional forces are negligible in some regions of interest in the flow.

## Force Balance across Streamlines

It is left as an exercise to show that a force balance in the direction $n$ normal to the streamline yields the following relation applicable *across* the streamlines for steady, incompressible flow:

$$\frac{P}{\rho} + \int \frac{V^2}{R}\, dn + gz = \text{constant} \qquad \text{(across streamlines)} \qquad \textbf{(12–11)}$$

where $R$ is the local radius of curvature of the streamline. For flow along curved streamlines (Fig 12–6a), the pressure *decreases* towards the center of curvature, and fluid particles experience a corresponding centripetal force and centripetal acceleration due to this pressure gradient.

For flow along a straight line, $R \rightarrow \infty$ and Eq. 12–11 reduces to $P/\rho + gz = $ constant or $P = -\rho gz + $ constant, which is an expression for the variation of hydrostatic pressure with vertical distance for a stationary fluid body. Therefore, the variation of pressure with elevation in steady, incompressible flow along a straight line in an inviscid region of flow is the same as that in the stationary fluid (Fig. 12–6b).

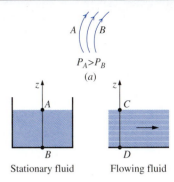

$$P_B - P_A = P_D - P_C$$
(b)

**FIGURE 12–6**
Pressure decreases towards the center of curvature when streamlines are curved (*a*), but the variation of pressure with elevation in steady, incompressible flow along a straight line (*b*) is the same as that in stationary fluid.

## Unsteady, Compressible Flow

Similarly, using both terms in the acceleration expression (Eq. 12–3), it can be shown that the Bernoulli equation for *unsteady, compressible flow* is

$$\textit{Unsteady, compressible flow:} \qquad \int \frac{dP}{\rho} + \int \frac{\partial V}{\partial t}\, ds + \frac{V^2}{2} + gz = \text{constant} \qquad \textbf{(12–12)}$$

## Static, Dynamic, and Stagnation Pressures

The Bernoulli equation states that the sum of the flow, kinetic, and potential energies of a fluid particle along a streamline is constant. Therefore, the kinetic and potential energies of the fluid can be converted to flow energy (and vice versa) during flow, causing the pressure to change. This phenomenon can be made more visible by multiplying the Bernoulli equation by the density $\rho$,

$$P + \rho \frac{V^2}{2} + \rho gz = \text{constant (along a streamline)} \qquad \textbf{(12–13)}$$

Each term in this equation has pressure units, and thus each term represents some kind of pressure:

- $P$ is the **static pressure** (it does not incorporate any dynamic effects); it represents the actual thermodynamic pressure of the fluid. This is the same as the pressure used in thermodynamics and property tables.
- $\rho V^2/2$ is the **dynamic pressure**; it represents the pressure rise when the fluid in motion is brought to a stop isentropically.

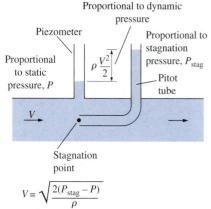

**FIGURE 12–7**
The static, dynamic, and stagnation pressures measured using piezometer tubes.

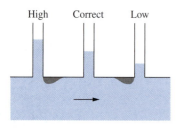

**FIGURE 12–8**
Close-up of a Pitot-static probe, showing the stagnation pressure hole and two of the five static circumferential pressure holes.

*Photo by Po-Ya Abel Chuang. Used by permission.*

**FIGURE 12–9**
Careless drilling of the static pressure tap may result in an erroneous reading of the static pressure head.

- $\rho gz$ is the **hydrostatic pressure** term, which is not pressure in a real sense since its value depends on the reference level selected; it accounts for the elevation effects, i.e., fluid weight on pressure. (Be careful of the sign—unlike hydrostatic pressure $\rho gh$ which *increases* with fluid depth $h$, the hydrostatic pressure term $\rho gz$ *decreases* with fluid depth.)

The sum of the static, dynamic, and hydrostatic pressures is called the **total pressure**. Therefore, the Bernoulli equation states that *the total pressure along a streamline is constant*.

The sum of the static and dynamic pressures is called the **stagnation pressure**, and it is expressed as

$$P_{stag} = P + \rho \frac{V^2}{2} \qquad \text{(kPa)} \qquad (12\text{–}14)$$

The stagnation pressure represents the pressure at a point where the fluid is brought to a complete stop isentropically. The static, dynamic, and stagnation pressures are shown in Fig. 12–7. When static and stagnation pressures are measured at a specified location, the fluid velocity at that location is calculated from

$$V = \sqrt{\frac{2(P_{stag} - P)}{\rho}} \qquad (12\text{–}15)$$

Equation 12–15 is useful in the measurement of flow velocity when a combination of a static pressure tap and a Pitot tube is used, as illustrated in Fig. 12–7. A **static pressure tap** is simply a small hole drilled into a wall such that the plane of the hole is parallel to the flow direction. It measures the static pressure. A **Pitot tube** is a small tube with its open end aligned *into* the flow so as to sense the full impact pressure of the flowing fluid. It measures the stagnation pressure. In situations in which the static and stagnation pressure of a flowing *liquid* are greater than atmospheric pressure, a vertical transparent tube called a **piezometer tube** (or simply a **piezometer**) can be attached to the pressure tap and to the Pitot tube, as sketched in Fig. 12–8. The liquid rises in the piezometer tube to a column height (*head*) that is proportional to the pressure being measured. If the pressures to be measured are below atmospheric, or if measuring pressures in *gases*, piezometer tubes do not work. However, the static pressure tap and Pitot tube can still be used, but they must be connected to some other kind of pressure measurement device such as a U-tube manometer or a pressure transducer (Chap. 11). Sometimes it is convenient to integrate static pressure holes on a Pitot probe. The result is a **Pitot-static probe** (also called a **Pitot-Darcy probe**), as shown in Fig. 12–9 and discussed in more detail in Chap. 14. A Pitot-static probe connected to a pressure transducer or a manometer measures the dynamic pressure (and thus infers the fluid velocity) directly.

When the static pressure is measured by drilling a hole in the tube wall, care must be exercised to ensure that the opening of the hole is flush with the wall surface, with no extrusions before or after the hole (Fig. 12–9). Otherwise the reading would incorporate some dynamic effects, and thus it would be in error.

When a stationary body is immersed in a flowing stream, the fluid is brought to a stop at the nose of the body (the **stagnation point**). The flow streamline that extends from far upstream to the stagnation point is called the **stagnation streamline** (Fig. 12–10). For a two-dimensional flow in the *xy*-plane, the stagnation point is actually a *line* parallel to the *z*-axis, and the stagnation streamline is actually a *surface* that separates fluid that flows *over* the body from fluid that flows *under* the body. In an incompressible flow, the fluid decelerates nearly isentropically from its free-stream velocity to zero at the stagnation point, and the pressure at the stagnation point is thus the stagnation pressure.

## Limitations on the Use of the Bernoulli Equation

The Bernoulli equation (Eq. 12–9) is one of the most frequently used and *misused* equations in fluid mechanics. Its versatility, simplicity, and ease of use make it a very valuable tool for use in analysis, but the same attributes also make it very tempting to misuse. Therefore, it is important to understand the restrictions on its applicability and observe the limitations on its use, as explained here:

1. **Steady flow**   The first limitation on the Bernoulli equation is that it is applicable to *steady flow*. Therefore, it should not be used during the transient start-up and shut-down periods, or during periods of change in the flow conditions. Note that there is an unsteady form of the Bernoulli equation (Eq. 12–12), discussion of which is beyond the scope of the present text (see Panton, 1996).

2. **Negligible viscous effects**   Every flow involves some friction, no matter how small, and *frictional effects* may or may not be negligible. The situation is complicated even more by the amount of error that can be tolerated. In general, frictional effects are negligible for short flow sections with large cross sections, especially at low flow velocities. Frictional effects are usually significant in long and narrow flow passages, in the wake region downstream of an object, and in *diverging flow sections* such as diffusers because of the increased possibility of the fluid separating from the walls in such geometries. Frictional effects are also significant near solid surfaces, and thus the Bernoulli equation is usually applicable along a streamline in the core region of the flow, but not along a streamline close to the surface (Fig. 12–11).

   A component that disturbs the streamlined structure of flow and thus causes considerable mixing and backflow such as a sharp entrance of a tube or a partially closed valve in a flow section can make the Bernoulli equation inapplicable.

3. **No shaft work**   The Bernoulli equation was derived from a force balance on a particle moving along a streamline. Therefore, the Bernoulli equation is not applicable in a flow section that involves a pump, turbine, fan, or any other machine or impeller since such devices disrupt the streamlines and carry out energy interactions with the fluid particles. When the flow section considered involves any of these devices, the energy equation should be used instead to account for the shaft work input or output. However, the Bernoulli equation can still be applied to a flow section prior to or past a machine (assuming, of course, that the other restrictions on its use are satisfied). In such cases, the Bernoulli constant changes from upstream to downstream of the device.

**FIGURE 12–10**
Streaklines produced by colored fluid introduced upstream of an airfoil; since the flow is steady, the streaklines are the same as streamlines and pathlines. The stagnation streamline is marked.
*Courtesy ONERA. Photograph by Werlé.*

**FIGURE 12–11**
Frictional effects, heat transfer, and components that disturb the streamlined structure of flow make the Bernoulli equation invalid. It should *not* be used in any of the flows shown here.

$$\frac{P_1}{\rho} + \frac{V_1^2}{2} + gz_1 = \frac{P_2}{\rho} + \frac{V_2^2}{2} + gz_2$$

**FIGURE 12–12**

When the flow is irrotational, the Bernoulli equation becomes applicable between any two points along the flow (not just on the same streamline).

**4. Incompressible flow** One of the approximations used in the derivation of the Bernoulli equation is that $\rho =$ constant and thus the flow is incompressible. This condition is satisfied by liquids and also by gases at Mach numbers less than about 0.3 since compressibility effects and thus density variations of gases are negligible at such relatively low velocities. Note that there is a compressible form of the Bernoulli equation (Eqs. 12–8 and 12–12).

**5. Negligible heat transfer** The density of a gas is inversely proportional to temperature, and thus the Bernoulli equation should not be used for flow sections that involve significant temperature change such as heating or cooling sections.

**6. Flow along a streamline** Strictly speaking, the Bernoulli equation $P/\rho + V^2/2 + gz = C$ is applicable along a streamline, and the value of the constant $C$ is generally different for different streamlines. However, when a region of the flow is *irrotational* and there is no *vorticity* in the flow field, the value of the constant $C$ remains the same for all streamlines, and the Bernoulli equation becomes applicable *across* streamlines as well (Fig. 12–12). Therefore, we do not need to be concerned about the streamlines when the flow is irrotational, and we can apply the Bernoulli equation between any two points in the irrotational region of the flow.

We derived the Bernoulli equation by considering two-dimensional flow in the *xz*-plane for simplicity, but the equation is valid for general three-dimensional flow as well, as long as it is applied along the same streamline. We should always keep in mind the approximations used in the derivation of the Bernoulli equation and make sure that they are valid before applying it.

# Hydraulic Grade Line (HGL) and Energy Grade Line (EGL)

It is often convenient to represent the level of mechanical energy graphically using *heights* to facilitate visualization of the various terms of the Bernoulli equation. This is done by dividing each term of the Bernoulli equation by $g$ to give

$$\frac{P}{\rho g} + \frac{V^2}{2g} + z = H = \text{constant} \qquad \text{(along a streamline)} \qquad \text{(12–16)}$$

**FIGURE 12–13**

An alternative form of the Bernoulli equation is expressed in terms of heads as: *The sum of the pressure, velocity, and elevation heads is constant along a streamline.*

Each term in this equation has the dimension of length and represents some kind of "head" of a flowing fluid as follows:

- $P/\rho g$ is the **pressure head**; it represents the height of a fluid column that produces the static pressure $P$.
- $V^2/2g$ is the **velocity head**; it represents the elevation needed for a fluid to reach the velocity $V$ during frictionless free fall.
- $z$ is the **elevation head**; it represents the potential energy of the fluid.

Also, $H$ is the **total head** for the flow. Therefore, the Bernoulli equation is expressed in terms of heads as: *The sum of the pressure, velocity, and elevation heads along a streamline is constant during steady flow when compressibility and frictional effects are negligible* (Fig. 12–13).

**FIGURE 12–14**
The *hydraulic grade line* (HGL) and
the *energy grade line* (EGL) for free
discharge from a reservoir through a
horizontal pipe with a diffuser.

If a piezometer (which measures static pressure) is tapped into a pressurized pipe, as shown in Fig. 12–14, the liquid would rise to a height of $P/\rho g$ above the pipe center. The *hydraulic grade line* (HGL) is obtained by doing this at several locations along the pipe and drawing a curve through the liquid levels in the piezometers. The vertical distance above the pipe center is a measure of pressure within the pipe. Similarly, if a Pitot tube (measures static + dynamic pressure) is tapped into a pipe, the liquid would rise to a height of $P/\rho g + V^2/2g$ above the pipe center, or a distance of $V^2/2g$ above the HGL. The *energy grade line* (EGL) is obtained by doing this at several locations along the pipe and drawing a curve through the liquid levels in the Pitot tubes.

Noting that the fluid also has elevation head $z$ (unless the reference level is taken to be the centerline of the pipe), the HGL and EGL are defined as follows: The line that represents the sum of the static pressure and the elevation heads, $P/\rho g + z$, is called the **hydraulic grade line**. The line that represents the total head of the fluid, $P/\rho g + V^2/2g + z$, is called the **energy grade line**. The difference between the heights of EGL and HGL is equal to the dynamic head, $V^2/2g$. We note the following about the HGL and EGL:

- For *stationary bodies* such as reservoirs or lakes, the EGL and HGL coincide with the free surface of the liquid. The elevation of the free surface $z$ in such cases represents both the EGL and the HGL since the velocity is zero and the static (gage) pressure is zero.

- The EGL is always a distance $V^2/2g$ above the HGL. These two curves approach each other as the velocity decreases, and they diverge as the velocity increases. The height of the HGL decreases as the velocity increases, and vice versa.

- In an *idealized Bernoulli-type flow,* EGL is horizontal and its height remains constant. This would also be the case for HGL when the flow velocity is constant (Fig. 12–15).

- For *open-channel flow,* the HGL coincides with the free surface of the liquid, and the EGL is a distance $V^2/2g$ above the free surface.

- At a *pipe exit,* the pressure head is zero (atmospheric pressure) and thus the HGL coincides with the pipe outlet (location 3 on Fig. 12–14).

- The *mechanical energy loss* due to frictional effects (conversion to thermal energy) causes the EGL and HGL to slope downward in the direction of flow. The slope is a measure of the head loss in the pipe.

**FIGURE 12–15**
In an idealized Bernoulli-type flow,
EGL is horizontal and its height
remains constant. But this is not
the case for HGL when the flow
velocity varies along the flow.

**FIGURE 12–16**

A *steep jump* occurs in EGL and HGL whenever mechanical energy is added to the fluid by a pump, and a *steep drop* occurs whenever mechanical energy is removed from the fluid by a turbine.

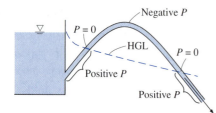

**FIGURE 12–17**

The gage pressure of a fluid is zero at locations where the HGL *intersects* the fluid, and the gage pressure is negative (vacuum) in a flow section that lies above the HGL.

A component that generates significant frictional effects such as a valve causes a sudden drop in both EGL and HGL at that location.

• A *steep jump* occurs in EGL and HGL whenever mechanical energy is added to the fluid (by a pump, for example). Likewise, a *steep drop* occurs in EGL and HGL whenever mechanical energy is removed from the fluid (by a turbine, for example), as shown in Fig. 12–16.

• The gage pressure of a fluid is zero at locations where the HGL *intersects* the fluid. The pressure in a flow section that lies above the HGL is negative, and the pressure in a section that lies below the HGL is positive (Fig. 12–17). Therefore, an accurate drawing of a piping system overlaid with the HGL can be used to determine the regions where the pressure in the pipe is negative (below atmospheric pressure).

The last remark enables us to avoid situations in which the pressure drops below the vapor pressure of the liquid (which may cause *cavitation*). Proper consideration is necessary in the placement of a liquid pump to ensure that the suction side pressure does not fall too low, especially at elevated temperatures where vapor pressure is higher than it is at low temperatures.

Now we examine Fig. 12–14 more closely. At point 0 (at the liquid surface), EGL and HGL are even with the liquid surface since there is no flow there. HGL decreases rapidly as the liquid accelerates into the pipe; however, EGL decreases very slowly through the well-rounded pipe inlet. EGL declines continually along the flow direction due to friction and other irreversible losses in the flow. EGL cannot increase in the flow direction unless energy is supplied to the fluid. HGL can rise or fall in the flow direction, but can never exceed EGL. HGL rises in the diffuser section as the velocity decreases, and the static pressure recovers somewhat; the total pressure does *not* recover, however, and EGL decreases through the diffuser. The difference between EGL and HGL is $V_1^2/2g$ at point 1, and $V_2^2/2g$ at point 2. Since $V_1 > V_2$, the difference between the two grade lines is larger at point 1 than at point 2. The downward slope of both grade lines is larger for the smaller diameter section of pipe since the frictional head loss is greater. Finally, HGL decays to the liquid surface at the outlet since the pressure there is atmospheric. However, EGL is still higher than HGL by the amount $V_2^2/2g$ since $V_3 = V_2$ at the outlet.

## Applications of the Bernoulli Equation

So far, we have discussed the fundamental aspects of the Bernoulli equation. Now, we demonstrate its use in a wide range of applications through examples.

---

**EXAMPLE 12–1    Spraying Water into the Air**

Water is flowing from a garden hose (Fig. 12–18). A child places his thumb to cover most of the hose outlet, causing a thin jet of high-speed water to emerge. The pressure in the hose just upstream of his thumb is 400 kPa. If the hose is held upward, what is the maximum height that the jet could achieve?

**Solution**  Water from a hose attached to the water main is sprayed into the air. The maximum height the water jet can rise is to be determined.

**Assumptions** 1 The flow exiting into the air is steady, incompressible, and irrotational (so that the Bernoulli equation is applicable). 2 The surface tension effects are negligible. 3 The friction between the water and air is negligible. 4 The irreversibilities that occur at the outlet of the hose due to abrupt contraction are not taken into account.

**Properties** We take the density of water to be 1000 kg/m³.

**Analysis** This problem involves the conversion of flow, kinetic, and potential energies to each other without involving any pumps, turbines, and wasteful components with large frictional losses, and thus it is suitable for the use of the Bernoulli equation. The water height will be maximum under the stated assumptions. The velocity inside the hose is relatively low ($V_1^2 \ll V_j^2$, and thus $V_1 \cong 0$ compared to $V_j$) and we take the elevation just below the hose outlet as the reference level ($z_1 = 0$). At the top of the water trajectory $V_2 = 0$, and atmospheric pressure pertains. Then the Bernoulli equation along a streamline from 1 to 2 simplifies to

$$\frac{P_1}{\rho g} + \frac{\overset{\cong 0}{V_1^2}}{2g} + \overset{0}{z_1} = \frac{P_2}{\rho g} + \frac{\overset{0}{V_2^2}}{2g} + z_2 \quad \rightarrow \quad \frac{P_1}{\rho g} = \frac{P_{atm}}{\rho g} + z_2$$

Solving for $z_2$ and substituting,

$$z_2 = \frac{P_1 - P_{atm}}{\rho g} = \frac{P_{1,\,gage}}{\rho g} = \frac{400\text{ kPa}}{(1000\text{ kg/m}^3)(9.81\text{ m/s}^2)}\left(\frac{1000\text{ N/m}^2}{1\text{ kPa}}\right)\left(\frac{1\text{ kg}\cdot\text{m/s}^2}{1\text{ N}}\right)$$

$$= \mathbf{40.8\text{ m}}$$

Therefore, the water jet can rise as high as 40.8 m into the sky in this case.

**Discussion** The result obtained by the Bernoulli equation represents the upper limit and should be interpreted accordingly. It tells us that the water cannot possibly rise more than 40.8 m, and, in all likelihood, the rise will be much less than 40.8 m due to irreversible losses that we neglected.

**FIGURE 12–18**
Schematic for Example 12–1. Inset shows a magnified view of the hose outlet region.

---

### EXAMPLE 12–2    Water Discharge from a Large Tank

A large tank open to the atmosphere is filled with water to a height of 5 m from the outlet tap (Fig. 12–19). A tap near the bottom of the tank is now opened, and water flows out from the smooth and rounded outlet. Determine the maximum water velocity at the outlet.

**Solution** A tap near the bottom of a tank is opened. The maximum exit velocity of water from the tank is to be determined.

**Assumptions** 1 The flow is incompressible and irrotational (except very close to the walls). 2 The water drains slowly enough that the flow can be approximated as steady (actually quasi-steady when the tank begins to drain). 3 Irreversible losses in the tap region are neglected.

**Analysis** This problem involves the conversion of flow, kinetic, and potential energies to each other without involving any pumps, turbines, and wasteful components with large frictional losses, and thus it is suitable for the use of the Bernoulli equation. We take point 1 to be at the free surface of water so that $P_1 = P_{atm}$ (open to the atmosphere), $V_1^2 \ll V_2^2$ and thus $V_1 \cong 0$ compared to $V_2$ (the tank is very large relative to the outlet), $z_1 = 5$ m and $z_2 = 0$ (we take the reference level at the center of the outlet). Also, $P_2 = P_{atm}$ (water discharges into the atmosphere). For flow along a streamline from 1 to 2, the Bernoulli equation simplifies to

**FIGURE 12–19**
Schematic for Example 12–2.

$$\frac{P_1}{\rho g} + \frac{V_1^{2\nearrow \approx 0}}{2g} + z_1 = \frac{P_2}{\rho g} + \frac{V_2^2}{2g} + z_2^{\nearrow 0} \quad \rightarrow \quad z_1 = \frac{V_2^2}{2g}$$

Solving for $V_2$ and substituting,

$$V_2 = \sqrt{2gz_1} = \sqrt{2(9.81 \text{ m/s}^2)(5 \text{ m})} = \textbf{9.9 m/s}$$

The relation $V = \sqrt{2gz}$ is called the **Torricelli equation**.

Therefore, the water leaves the tank with an initial maximum velocity of 9.9 m/s. This is the same velocity that would manifest if a solid were dropped a distance of 5 m in the absence of air friction drag. (What would the velocity be if the tap were at the bottom of the tank instead of on the side?)

**Discussion** If the orifice were sharp-edged instead of rounded, then the flow would be disturbed, and the average exit velocity would be less than 9.9 m/s. Care must be exercised when attempting to apply the Bernoulli equation to situations where abrupt expansions or contractions occur since the friction and flow disturbance in such cases may not be negligible. From conversion of mass, $(V_1/V_2)^2 = (D_2/D_1)^4$. So, for example, if $D_2/D_1 = 0.1$, then $(V_1/V_2)^2 = 0.0001$, and our approximation that $V_1^2 \ll V_2^2$ is justified.

**FIGURE 12–20**
Schematic for Example 12–3.

**EXAMPLE 12–3**    **Siphoning Out Gasoline from a Fuel Tank**

During a trip to the beach ($P_{atm}$ = 1 atm = 101.3 kPa), a car runs out of gasoline, and it becomes necessary to siphon gas out of the car of a Good Samaritan (Fig. 12–20). The siphon is a small-diameter hose, and to start the siphon it is necessary to insert one siphon end in the full gas tank, fill the hose with gasoline via suction, and then place the other end in a gas can below the level of the gas tank. The difference in pressure between point 1 (at the free surface of the gasoline in the tank) and point 2 (at the outlet of the tube) causes the liquid to flow from the higher to the lower elevation. Point 2 is located 0.75 m below point 1 in this case, and point 3 is located 2 m above point 1. The siphon diameter is 5 mm, and frictional losses in the siphon are to be disregarded. Determine (a) the minimum time to withdraw 4 L of gasoline from the tank to the can and (b) the pressure at point 3. The density of gasoline is 750 kg/m³.

**Solution** Gasoline is to be siphoned from a tank. The minimum time it takes to withdraw 4 L of gasoline and the pressure at the highest point in the system are to be determined.

**Assumptions** **1** The flow is steady and incompressible. **2** Even though the Bernoulli equation is not valid through the pipe because of frictional losses, we employ the Bernoulli equation anyway in order to obtain a *best-case estimate*. **3** The change in the gasoline surface level inside the tank is negligible compared to elevations $z_1$ and $z_2$ during the siphoning period.

**Properties** The density of gasoline is given to be 750 kg/m³.

**Analysis** (a) We take point 1 to be at the free surface of gasoline in the tank so that $P_1 = P_{atm}$ (open to the atmosphere), $V_1 \cong 0$ (the tank is large relative to the tube diameter), and $z_2 = 0$ (point 2 is taken as the reference level). Also, $P_2 = P_{atm}$ (gasoline discharges into the atmosphere). Then the Bernoulli equation simplifies to

$$\frac{P_1}{\rho g} + \frac{V_1^{2\nearrow \approx 0}}{2g} + z_1 = \frac{P_2}{\rho g} + \frac{V_2^2}{2g} + z_2^{\nearrow 0} \quad \rightarrow \quad z_1 = \frac{V_2^2}{2g}$$

Solving for $V_2$ and substituting,

$$V_2 = \sqrt{2gz_1} = \sqrt{2(9.81 \text{ m/s}^2)(0.75 \text{ m})} = 3.84 \text{ m/s}$$

The cross-sectional area of the tube and the flow rate of gasoline are

$$A = \pi D^2/4 = \pi(5 \times 10^{-3} \text{ m})^2/4 = 1.96 \times 10^{-5} \text{ m}^2$$

$$\dot{V} = V_2 A = (3.84 \text{ m/s})(1.96 \times 10^{-5} \text{ m}^2) = 7.53 \times 10^{-5} \text{ m}^3/\text{s} = 0.0753 \text{ L/s}$$

Then the time needed to siphon 4 L of gasoline becomes

$$\Delta t = \frac{V}{\dot{V}} = \frac{4 \text{ L}}{0.0753 \text{ L/s}} = \textbf{53.1 s}$$

(*b*) The pressure at point 3 is determined by writing the Bernoulli equation along a streamline between points 3 and 2. Noting that $V_2 = V_3$ (conservation of mass), $z_2 = 0$, and $P_2 = P_{\text{atm}}$,

$$\frac{P_2}{\rho g} + \frac{\cancel{V_2^2}}{\cancel{2g}} + \cancel{z_2}^{\,0} = \frac{P_3}{\rho g} + \frac{\cancel{V_3^2}}{\cancel{2g}} + z_3 \quad \rightarrow \quad \frac{P_{\text{atm}}}{\rho g} = \frac{P_3}{\rho g} + z_3$$

Solving for $P_3$ and substituting,

$$P_3 = P_{\text{atm}} - \rho g z_3$$

$$= 101.3 \text{ kPa} - (750 \text{ kg/m}^3)(9.81 \text{ m/s}^2)(2.75 \text{ m})\left(\frac{1 \text{ N}}{1 \text{ kg·m/s}^2}\right)\left(\frac{1 \text{ kPa}}{1000 \text{ N/m}^2}\right)$$

$$= \textbf{81.1 kPa}$$

**Discussion** The siphoning time is determined by neglecting frictional effects, and thus this is the *minimum time* required. In reality, the time will be longer than 53.1 s because of friction between the gasoline and the tube surface, along with other irreversible losses, as discussed in Chap. 14. Also, the pressure at point 3 is below the atmospheric pressure. If the elevation difference between points 1 and 3 is too high, the pressure at point 3 may drop below the vapor pressure of gasoline at the gasoline temperature, and some gasoline may evaporate (cavitate). The vapor then may form a pocket at the top and halt the flow of gasoline.

---

**EXAMPLE 12–4    Velocity Measurement by a Pitot Tube**

A piezometer and a Pitot tube are tapped into a horizontal water pipe, as shown in Fig. 12–21, to measure static and stagnation (static + dynamic) pressures. For the indicated water column heights, determine the velocity at the center of the pipe.

**Solution**  The static and stagnation pressures in a horizontal pipe are measured. The velocity at the center of the pipe is to be determined.
**Assumptions**  **1** The flow is steady and incompressible. **2** Points 1 and 2 are close enough together that the irreversible energy loss between these two points is negligible, and thus we can use the Bernoulli equation.
**Analysis**  We take points 1 and 2 along the streamline at the centerline of the pipe, with point 1 directly under the piezometer and point 2 at the tip

**FIGURE 12–21**
Schematic for Example 12–4.

of the Pitot tube. This is a steady flow with straight and parallel streamlines, and the gage pressures at points 1 and 2 can be expressed as

$$P_1 = \rho g(h_1 + h_2)$$
$$P_2 = \rho g(h_1 + h_2 + h_3)$$

Noting that $z_1 = z_2$, and point 2 is a stagnation point and thus $V_2 = 0$, the application of the Bernoulli equation between points 1 and 2 gives

$$\frac{P_1}{\rho g} + \frac{V_1^2}{2g} + \cancel{z_1} = \frac{P_2}{\rho g} + \frac{\cancel{V_2^2}^{0}}{2g} + \cancel{z_2} \quad \rightarrow \quad \frac{V_1^2}{2g} = \frac{P_2 - P_1}{\rho g}$$

Substituting the $P_1$ and $P_2$ expressions gives

$$\frac{V_1^2}{2g} = \frac{P_2 - P_1}{\rho g} = \frac{\rho g(h_1 + h_2 + h_3) - \rho g(h_1 + h_2)}{\rho g} = h_3$$

Solving for $V_1$ and substituting,

$$V_1 = \sqrt{2gh_3} = \sqrt{2(9.81 \text{ m/s}^2)(0.12 \text{ m})} = \mathbf{1.53 \text{ m/s}}$$

**Discussion**  Note that to determine the flow velocity, all we need is to measure the height of the excess fluid column in the Pitot tube compared to that in the piezometer tube.

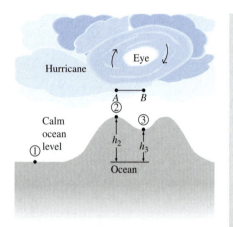

**FIGURE 12–22**
Schematic for Example 12–5. The vertical scale is greatly exaggerated.

**EXAMPLE 12–5**    **The Rise of the Ocean Due to a Hurricane**

A hurricane is a tropical storm formed over the ocean by low atmospheric pressures. As a hurricane approaches land, inordinate ocean swells (very high tides) accompany the hurricane. A Class-5 hurricane features winds in excess of 155 mph, although the wind velocity at the center "eye" is very low.

Figure 12–22 depicts a hurricane hovering over the ocean swell below. The atmospheric pressure 200 mi from the eye is 30.0 in Hg (at point 1, generally normal for the ocean) and the winds are calm. The atmospheric pressure at the eye of the storm is 22.0 in Hg. Estimate the ocean swell at (a) the eye of the hurricane at point 3 and (b) point 2, where the wind velocity is 155 mph. Take the density of seawater and mercury to be 64 lbm/ft³ and 848 lbm/ft³, respectively, and the density of air at normal sea-level temperature and pressure to be 0.076 lbm/ft³.

**Solution**  A hurricane is moving over the ocean. The amount of ocean swell at the eye and at active regions of the hurricane are to be determined.
**Assumptions**  **1** The airflow within the hurricane is steady, incompressible, and irrotational (so that the Bernoulli equation is applicable). (This is certainly a very questionable assumption for a highly turbulent flow, but it is justified in the discussion.) **2** The effect of water sucked into the air is negligible.
**Properties**  The densities of air at normal conditions, seawater, and mercury are given to be 0.076 lbm/ft³, 64.0 lbm/ft³, and 848 lbm/ft³, respectively.
**Analysis**  (a) Reduced atmospheric pressure over the water causes the water to rise. Thus, decreased pressure at point 2 relative to point 1 causes the ocean water to rise at point 2. The same is true at point 3, where the storm air

velocity is negligible. The pressure difference given in terms of the mercury column height is expressed in terms of the seawater column height by

$$\Delta P = (\rho g h)_{\text{Hg}} = (\rho g h)_{\text{sw}} \rightarrow h_{\text{sw}} = \frac{\rho_{\text{Hg}}}{\rho_{\text{sw}}} h_{\text{Hg}}$$

Then the pressure difference between points 1 and 3 in terms of the seawater column height becomes

$$h_3 = \frac{\rho_{\text{Hg}}}{\rho_{\text{sw}}} h_{\text{Hg}} = \left(\frac{848 \text{ lbm/ft}^3}{64.0 \text{ lbm/ft}^3}\right)[(30 - 22) \text{ in Hg}]\left(\frac{1 \text{ ft}}{12 \text{ in}}\right) = \textbf{8.83 ft}$$

which is equivalent to the storm surge at the *eye of the hurricane* (Fig. 12–23) since the wind velocity there is negligible and there are no dynamic effects.

(*b*) To determine the additional rise of ocean water at point 2 due to the high winds at that point, we write the Bernoulli equation between points *A* and *B*, which are on top of points 2 and 3, respectively. Noting that $V_B \cong 0$ (the eye region of the hurricane is relatively calm) and $z_A = z_B$ (both points are on the same horizontal line), the Bernoulli equation simplifies to

$$\frac{P_A}{\rho g} + \frac{V_A^2}{2g} + \cancel{z_A} = \frac{P_B}{\rho g} + \cancel{\frac{V_B^2}{2g}}^{\,0} + \cancel{z_B} \quad \rightarrow \quad \frac{P_B - P_A}{\rho g} = \frac{V_A^2}{2g}$$

Substituting,

$$\frac{P_B - P_A}{\rho g} = \frac{V_A^2}{2g} = \frac{(155 \text{ mph})^2}{2(32.2 \text{ ft/s}^2)}\left(\frac{1.4667 \text{ ft/s}}{1 \text{ mph}}\right)^2 = 803 \text{ ft}$$

where $\rho$ is the density of air in the hurricane. Noting that the density of an ideal gas at constant temperature is proportional to absolute pressure and the density of air at the normal atmospheric pressure of 14.7 psia $\cong$ 30 in Hg is 0.076 lbm/ft³, the density of air in the hurricane is

$$\rho_{\text{air}} = \frac{P_{\text{air}}}{P_{\text{atm air}}} \rho_{\text{atm air}} = \left(\frac{22 \text{ in Hg}}{30 \text{ in Hg}}\right)(0.076 \text{ lbm/ft}^3) = 0.056 \text{ lbm/ft}^3$$

Using the relation developed above in part (*a*), the seawater column height equivalent to 803 ft of air column height is determined to be

$$h_{\text{dynamic}} = \frac{\rho_{\text{air}}}{\rho_{\text{sw}}} h_{\text{air}} = \left(\frac{0.056 \text{ lbm/ft}^3}{64 \text{ lbm/ft}^3}\right)(803 \text{ ft}) = 0.70 \text{ ft}$$

Therefore, the pressure at point 2 is 0.70 ft seawater column lower than the pressure at point 3 due to the high wind velocities, causing the ocean to rise an additional 0.70 ft. Then the total storm surge at point 2 becomes

$$h_2 = h_3 + h_{\text{dynamic}} = 8.83 + 0.70 = \textbf{9.53 ft}$$

*Discussion* This problem involves highly turbulent flow and the intense breakdown of the streamlines, and thus the applicability of the Bernoulli equation in part (*b*) is questionable. Furthermore, the flow in the eye of the storm is *not* irrotational, and the Bernoulli equation constant changes across

**FIGURE 12–23**
The eye of hurricane Linda (1997 in the Pacific Ocean near Baja California) is clearly visible in this satellite photo.
© *PunchStock RF.*

streamlines. The Bernoulli analysis can be thought of as the limiting, ideal case, and shows that the rise of seawater due to high-velocity winds cannot be more than 0.70 ft.

The wind power of hurricanes is not the only cause of damage to coastal areas. Ocean flooding and erosion from excessive tides is just as serious, as are high waves generated by the storm turbulence and energy.

### EXAMPLE 12–6    Bernoulli Equation for Compressible Flow

Derive the Bernoulli equation when the compressibility effects are not negligible for an ideal gas undergoing (a) an isothermal process and (b) an isentropic process.

**Solution**   The Bernoulli equation for compressible flow is to be obtained for an ideal gas for isothermal and isentropic processes.

**Assumptions**   1 The flow is steady and frictional effects are negligible. 2 The fluid is an ideal gas, so the relation $P = \rho RT$ is applicable. 3 The specific heats are constant so that $P/\rho^k$ = constant during an isentropic process.

**Analysis**   (a) When the compressibility effects are significant and the flow cannot be assumed to be incompressible, the Bernoulli equation is given by Eq. 12–8 as

$$\int \frac{dP}{\rho} + \frac{V^2}{2} + gz = \text{constant} \qquad \text{(along a streamline)} \qquad \textbf{(1)}$$

The compressibility effects can be properly accounted for by performing the integration $\int dP/\rho$ in Eq. 1. But this requires a relation between $P$ and $\rho$ for the process. For the *isothermal* expansion or compression of an ideal gas, the integral in Eq. 1 is performed easily by noting that $T$ = constant and substituting $\rho = P/RT$,

$$\int \frac{dP}{\rho} = \int \frac{dP}{P/RT} = RT \ln P$$

Substituting into Eq. 1 gives the desired relation,

*Isothermal process:* $\qquad \boldsymbol{RT \ln P + \dfrac{V^2}{2} + gz = \text{constant}} \qquad \textbf{(2)}$

(b) A more practical case of compressible flow is the *isentropic flow of ideal gases* through equipment that involves high-speed fluid flow such as nozzles, diffusers, and the passages between turbine blades (Fig. 12–24). Isentropic (i.e., reversible and adiabatic) flow is closely approximated by these devices, and it is characterized by the relation $P/\rho^k = C$ = constant, where $k$ is the specific heat ratio of the gas. Solving for $\rho$ from $P/\rho^k = C$ gives $\rho = C^{-1/k}P^{1/k}$. Performing the integration,

$$\int \frac{dP}{\rho} = \int C^{1/k}P^{-1/k}\, dP = C^{1/k}\frac{P^{-1/k+1}}{-1/k+1} = \frac{P^{1/k}}{\rho}\frac{P^{-1/k+1}}{-1/k+1} = \left(\frac{k}{k-1}\right)\frac{P}{\rho} \qquad \textbf{(3)}$$

Substituting, the Bernoulli equation for steady, isentropic, compressible flow of an ideal gas becomes

*Isentropic flow:* $\qquad \boldsymbol{\left(\dfrac{k}{k-1}\right)\dfrac{P}{\rho} + \dfrac{V^2}{2} + gz = \text{constant}} \qquad \textbf{(4a)}$

**FIGURE 12–24**

Compressible flow of a gas through turbine blades is often modeled as isentropic, and the compressible form of the Bernoulli equation is a reasonable approximation.

© Corbis RF.

or

$$\left(\frac{k}{k-1}\right)\frac{P_1}{\rho_1} + \frac{V_1^2}{2} + gz_1 = \left(\frac{k}{k-1}\right)\frac{P_2}{\rho_2} + \frac{V_2^2}{2} + gz_2 \qquad \text{(4b)}$$

A common practical situation involves the acceleration of a gas from rest (stagnation conditions at state 1) with negligible change in elevation. In that case we have $z_1 = z_2$ and $V_1 = 0$. Noting that $\rho = P/RT$ for ideal gases, $P/\rho^k =$ constant for isentropic flow, and the Mach number is defined as $\mathrm{Ma} = V/c$ where $c = \sqrt{kRT}$ is the local speed of sound for ideal gases, Eq. 4b simplifies to

$$\frac{P_1}{P_2} = \left[1 + \left(\frac{k-1}{2}\right)\mathrm{Ma}_2^2\right]^{k/(k-1)} \qquad \text{(4c)}$$

where state 1 is the stagnation state and state 2 is any state along the flow.

*Discussion*   It can be shown that the results obtained using the compressible and incompressible equations deviate no more than 2 percent when the Mach number is less than 0.3. Therefore, the flow of an ideal gas can be considered to be incompressible when $\mathrm{Ma} \leqslant 0.3$. For atmospheric air at normal conditions, this corresponds to a flow speed of about 100 m/s or 360 km/h.

## 12–2 · GENERAL ENERGY EQUATION

One of the most fundamental laws in nature is the **first law of thermodynamics**, also known as the **conservation of energy principle**, which provides a sound basis for studying the relationships among the various forms of energy and energy interactions. It states that *energy can be neither created nor destroyed during a process; it can only change forms.* Therefore, every bit of energy must be accounted for during a process.

A rock falling off a cliff, for example, picks up speed as a result of its potential energy being converted to kinetic energy (Fig. 12–25). Experimental data show that the decrease in potential energy equals the increase in kinetic energy when the air resistance is negligible, thus confirming the conservation of energy principle. The conservation of energy principle also forms the backbone of the diet industry: a person who has a greater energy input (food) than energy output (exercise) will gain weight (store energy in the form of fat), and a person who has a smaller energy input than output will lose weight. The change in the energy content of a system is equal to the difference between the energy input and the energy output, and the conservation of energy principle for any system can be expressed simply as $E_{\text{in}} - E_{\text{out}} = \Delta E$.

The transfer of any quantity (such as mass, momentum, and energy) is recognized *at the boundary* as the quantity *crosses the boundary*. A quantity is said to *enter* a system (or control volume) if it crosses the boundary from the outside to the inside, and to *exit* the system if it moves in the reverse direction. A quantity that moves from one location to another within a system is not considered as a transferred quantity in an analysis since it does not enter or exit the system. Therefore, it is important to specify the system and thus clearly identify its boundaries before an engineering analysis is performed.

**FIGURE 12–25**
Energy cannot be created or destroyed during a process; it can only change forms.

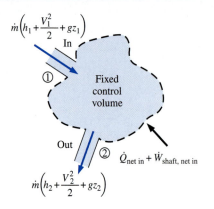

$$\dot{m}\left(h_1 + \frac{V_1^2}{2} + gz_1\right)$$

In

① Fixed control volume

Out

② $\dot{Q}_{\text{net in}} + \dot{W}_{\text{shaft, net in}}$

$$\dot{m}\left(h_2 + \frac{V_2^2}{2} + gz_2\right)$$

**FIGURE 12–26**
A control volume with only one inlet and one outlet and energy interactions.

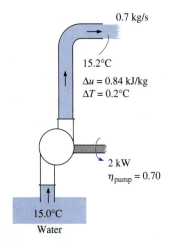

0.7 kg/s

15.2°C

$\Delta u = 0.84$ kJ/kg
$\Delta T = 0.2$°C

2 kW
$\eta_{\text{pump}} = 0.70$

15.0°C
Water

**FIGURE 12–27**
The lost mechanical energy in a fluid flow system results in an increase in the internal energy of the fluid and thus in a rise of fluid temperature.

where we used the definition of enthalpy $h = u + Pv = u + P/\rho$. The last two equations are fairly general expressions of conservation of energy, but their use is still limited to fixed control volumes, uniform flow at inlets and outlets, and negligible work due to viscous forces and other effects. Also, the subscript "net in" stands for "net input," and thus any heat or work transfer is positive if *to* the system and negative if *from* the system.

## 12–3 ■ ENERGY ANALYSIS OF STEADY FLOWS

For steady flows, the time rate of change of the energy content of the control volume is zero, and the energy equation can be expressed as

$$\dot{Q}_{\text{net in}} + \dot{W}_{\text{shaft, net in}} = \sum_{\text{out}} \dot{m}\left(h + \frac{V^2}{2} + gz\right) - \sum_{\text{in}} \dot{m}\left(h + \frac{V^2}{2} + gz\right) \quad \textbf{(12–17)}$$

It states that *during steady flow the net rate of energy transfer to a control volume by heat and work transfers is equal to the difference between the rates of outgoing and incoming energy flows by mass flow.*

Many practical problems involve just one inlet and one outlet (Fig. 12–26). The mass flow rate for such **single-stream devices** is the same at the inlet and outlet, and Eq. 12–17 reduces to

$$\dot{Q}_{\text{net in}} + \dot{W}_{\text{shaft, net in}} = \dot{m}\left(h_2 - h_1 + \frac{V_2^2 - V_1^2}{2} + g(z_2 - z_1)\right) \quad \textbf{(12–18)}$$

where subscripts 1 and 2 refer to the inlet and outlet, respectively. The steady-flow energy equation on a unit-mass basis is obtained by dividing Eq. 12–18 by the mass flow rate $\dot{m}$,

$$q_{\text{net in}} + w_{\text{shaft, net in}} = h_2 - h_1 + \frac{V_2^2 - V_1^2}{2} + g(z_2 - z_1) \quad \textbf{(12–19)}$$

where $q_{\text{net in}} = \dot{Q}_{\text{net in}}/\dot{m}$ is the net heat transfer to the fluid per unit mass and $w_{\text{shaft, net in}} = \dot{W}_{\text{shaft, net in}}/\dot{m}$ is the net shaft work input to the fluid per unit mass. Using the definition of enthalpy $h = u + P/\rho$ and rearranging, the steady-flow energy equation can also be expressed as

$$w_{\text{shaft, net in}} + \frac{P_1}{\rho_1} + \frac{V_1^2}{2} + gz_1 = \frac{P_2}{\rho_2} + \frac{V_2^2}{2} + gz_2 + (u_2 - u_1 - q_{\text{net in}}) \quad \textbf{(12–20)}$$

where $u$ is the *internal energy*, $P/\rho$ is the *flow energy*, $V^2/2$ is the *kinetic energy*, and $gz$ is the *potential energy* of the fluid, all per unit mass. These relations are valid for both compressible and incompressible flows.

The left side of Eq. 12–20 represents the mechanical energy input, while the first three terms on the right side represent the mechanical energy output. If the flow is ideal with no irreversibilities such as friction, the total mechanical energy must be conserved, and the term in parentheses ($u_2 - u_1 - q_{\text{net in}}$) must equal zero. That is,

*Ideal flow* (no mechanical energy loss): $\qquad q_{\text{net in}} = u_2 - u_1 \qquad$ **(12–21)**

Any increase in $u_2 - u_1$ above $q_{\text{net in}}$ is due to the irreversible conversion of mechanical energy to thermal energy, and thus $u_2 - u_1 - q_{\text{net in}}$ represents the mechanical energy loss per unit mass (Fig. 12–27). That is,

*Real flow* (with mechanical energy loss): $\qquad e_{\text{mech, loss}} = u_2 - u_1 - q_{\text{net in}} \qquad$ **(12–22)**

For single-phase fluids (a gas or a liquid), $u_2 - u_1 = c_v(T_2 - T_1)$ where $c_v$ is the constant-volume specific heat.

The steady-flow energy equation on a unit-mass basis can be written conveniently as a **mechanical energy** balance,

$$e_{mech, in} = e_{mech, out} + e_{mech, loss} \qquad (12\text{--}23)$$

or

$$w_{shaft, net in} + \frac{P_1}{\rho_1} + \frac{V_1^2}{2} + gz_1 = \frac{P_2}{\rho_2} + \frac{V_2^2}{2} + gz_2 + e_{mech, loss} \qquad (12\text{--}24)$$

Noting that $w_{shaft, net in} = w_{pump} - w_{turbine}$, the mechanical energy balance can be written more explicitly as

$$\frac{P_1}{\rho_1} + \frac{V_1^2}{2} + gz_1 + w_{pump} = \frac{P_2}{\rho_2} + \frac{V_2^2}{2} + gz_2 + w_{turbine} + e_{mech, loss} \qquad (12\text{--}25)$$

where $w_{pump}$ is the mechanical work input (due to the presence of a pump, fan, compressor, etc.) and $w_{turbine}$ is the mechanical work output (due to a turbine). When the flow is incompressible, either absolute or gage pressure can be used for $P$ since $P_{atm}/\rho$ would appear on both sides and would cancel out.

Multiplying Eq. 12–25 by the mass flow rate $\dot{m}$ gives

$$\dot{m}\left(\frac{P_1}{\rho_1} + \frac{V_1^2}{2} + gz_1\right) + \dot{W}_{pump} = \dot{m}\left(\frac{P_2}{\rho_2} + \frac{V_2^2}{2} + gz_2\right) + \dot{W}_{turbine} + \dot{E}_{mech, loss} \qquad (12\text{--}26)$$

where $\dot{W}_{pump}$ is the shaft power input through the pump's shaft, $\dot{W}_{turbine}$ is the shaft power output through the turbine's shaft, and $\dot{E}_{mech, loss}$ is the *total* mechanical power loss, which consists of pump and turbine losses as well as the frictional losses in the piping network. That is,

$$\dot{E}_{mech, loss} = \dot{E}_{mech loss, pump} + \dot{E}_{mech loss, turbine} + \dot{E}_{mech loss, piping}$$

By convention, irreversible pump and turbine losses are treated separately from irreversible losses due to other components of the piping system (Fig. 12–28). Thus, the energy equation is expressed in its most common form in terms of *heads* by dividing each term in Eq. 12–26 by $\dot{m}g$. The result is

$$\frac{P_1}{\rho_1 g} + \frac{V_1^2}{2g} + z_1 + h_{pump, u} = \frac{P_2}{\rho_2 g} + \frac{V_2^2}{2g} + z_2 + h_{turbine, e} + h_L \qquad (12\text{--}27)$$

where

- $h_{pump, u} = \dfrac{w_{pump, u}}{g} = \dfrac{\dot{W}_{pump, u}}{\dot{m}g} = \dfrac{\eta_{pump}\dot{W}_{pump}}{\dot{m}g}$ is the *useful head delivered*

  *to the fluid by the pump.* Because of irreversible losses in the pump, $h_{pump, u}$ is *less* than $\dot{W}_{pump}/\dot{m}g$ by the factor $\eta_{pump}$.

- $h_{turbine, e} = \dfrac{w_{turbine, e}}{g} = \dfrac{\dot{W}_{turbine, e}}{\dot{m}g} = \dfrac{\dot{W}_{turbine}}{\eta_{turbine}\dot{m}g}$ is the *extracted head removed*

  *from the fluid by the turbine.* Because of irreversible losses in the turbine, $h_{turbine, e}$ is *greater* than $\dot{W}_{turbine}/\dot{m}g$ by the factor $\eta_{turbine}$.

- $h_L = \dfrac{e_{mech loss, piping}}{g} = \dfrac{\dot{E}_{mech loss, piping}}{\dot{m}g}$ is the *irreversible head loss* between

  1 and 2 due to all components of the piping system other than the pump or turbine.

**FIGURE 12–28**
A typical power plant has numerous pipes, elbows, valves, pumps, and turbines, all of which have irreversible losses.

© *PunchStock RF.*

**FIGURE 12–29**

Mechanical energy flow chart for a fluid flow system that involves a pump and a turbine. Vertical dimensions show each energy term expressed as an equivalent column height of fluid, i.e., *head*, corresponding to each term of Eq. 12–27.

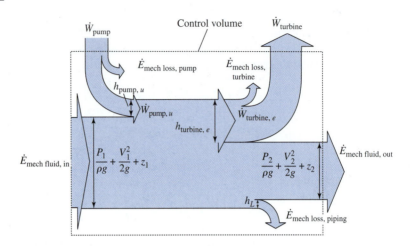

Note that the head loss $h_L$ represents the frictional losses associated with fluid flow in piping, and it does not include the losses that occur within the pump or turbine due to the inefficiencies of these devices—these losses are taken into account by $\eta_{pump}$ and $\eta_{turbine}$. Equation 12–27 is illustrated schematically in Fig. 12–29.

The *pump head* is zero if the piping system does not involve a pump, a fan, or a compressor, and the *turbine head* is zero if the system does not involve a turbine.

## Special Case: Incompressible Flow with No Mechanical Work Devices and Negligible Friction

When piping losses are negligible, there is negligible dissipation of mechanical energy into thermal energy, and thus $h_L = e_{mech\ loss,\ piping}/g \cong 0$. Also, $h_{pump,\ u} = h_{turbine,\ e} = 0$ when there are no mechanical work devices such as fans, pumps, or turbines. Then Eq. 12–27 reduces to

$$\frac{P_1}{\rho g} + \frac{V_1^2}{2g} + z_1 = \frac{P_2}{\rho g} + \frac{V_2^2}{2g} + z_2 \qquad \text{or} \qquad \frac{P}{\rho g} + \frac{V^2}{2g} + z = \text{constant} \qquad \textbf{(12–28)}$$

which is the **Bernoulli equation** derived earlier using Newton's second law of motion. Thus, the Bernoulli equation can be thought of as a degenerate form of the energy equation.

## Kinetic Energy Correction Factor, $\alpha$

The average flow velocity $V_{avg}$ was defined such that the relation $\rho V_{avg} A$ gives the actual mass flow rate. Therefore, there is no such thing as a correction factor for mass flow rate. However, as Gaspard Coriolis (1792–1843) showed, the kinetic energy of a fluid stream obtained from $V^2/2$ is not the same as the actual kinetic energy of the fluid stream since the square of a sum is not equal to the sum of the squares of its components (Fig. 12–30). This error can be corrected by replacing the kinetic energy terms $V^2/2$ in the energy equation by $\alpha V_{avg}^2/2$, where $\alpha$ is the **kinetic energy correction factor**. By using equations for the variation of velocity with the radial distance, it can be shown that the correction factor is 2.0 for fully developed laminar pipe flow, and it ranges between 1.04 and 1.11 for fully developed turbulent flow in a round pipe.

$$\dot{m} = \rho V_{avg} A, \qquad \rho = \text{constant}$$

$$\dot{KE}_{act} = \int ke\, \delta \dot{m} = \int_A \frac{1}{2}\,[V(r)]^2\,[\rho V(r)\, dA]$$

$$= \frac{1}{2}\,\rho \int_A [V(r)]^3\, dA$$

$$\dot{KE}_{avg} = \frac{1}{2}\,\dot{m} V_{avg}^2 = \frac{1}{2}\,\rho A V_{avg}^3$$

$$\alpha = \frac{\dot{KE}_{act}}{\dot{KE}_{avg}} = \frac{1}{A} \int_A \left(\frac{V(r)}{V_{avg}}\right)^3 dA$$

**FIGURE 12–30**

The determination of the *kinetic energy correction factor* using the actual velocity distribution $V(r)$ and the average velocity $V_{avg}$ at a cross section.

The kinetic energy correction factors are often ignored (i.e., $\alpha$ is set equal to 1) in an elementary analysis since (1) most flows encountered in practice are turbulent, for which the correction factor is near unity, and (2) the kinetic energy terms are often small relative to the other terms in the energy equation, and multiplying them by a factor less than 2.0 does not make much difference. When the velocity and thus the kinetic energy are high, the flow turns turbulent, and a unity correction factor is more appropriate. However, you should keep in mind that you may encounter some situations for which these factors *are* significant, especially when the flow is laminar. Therefore, we recommend that you always include the kinetic energy correction factor when analyzing fluid flow problems. When the kinetic energy correction factors are included, the energy equations for *steady incompressible flow* (Eqs. 12–26 and 12–27) become

$$\dot{m}\left(\frac{P_1}{\rho} + \alpha_1 \frac{V_1^2}{2} + gz_1\right) + \dot{W}_{\text{pump}} = \dot{m}\left(\frac{P_2}{\rho} + \alpha_2 \frac{V_2^2}{2} + gz_2\right) + \dot{W}_{\text{turbine}} + \dot{E}_{\text{mech, loss}} \quad \textbf{(12–29)}$$

$$\frac{P_1}{\rho g} + \alpha_1 \frac{V_1^2}{2g} + z_1 + h_{\text{pump, }u} = \frac{P_2}{\rho g} + \alpha_2 \frac{V_2^2}{2g} + z_2 + h_{\text{turbine, }e} + h_L \quad \textbf{(12–30)}$$

If the flow at an inlet or outlet is fully developed turbulent pipe flow, we recommend using $\alpha = 1.05$ as a reasonable estimate of the correction factor. This leads to a more conservative estimate of head loss, and it does not take much additional effort to include $\alpha$ in the equations.

---

### EXAMPLE 12–7    Pumping Power and Frictional Heating in a Pump

The pump of a water distribution system is powered by a 15-kW electric motor whose efficiency is 90 percent (Fig. 12–31). The water flow rate through the pump is 50 L/s. The diameters of the inlet and outlet pipes are the same, and the elevation difference across the pump is negligible. If the absolute pressures at the inlet and outlet of the pump are measured to be 100 kPa and 300 kPa, respectively, determine (*a*) the mechanical efficiency of the pump and (*b*) the temperature rise of water as it flows through the pump due to mechanical inefficiencies.

**Solution**  The pressures across a pump are measured. The mechanical efficiency of the pump and the temperature rise of water are to be determined.
**Assumptions**  **1** The flow is steady and incompressible. **2** The pump is driven by an external motor so that the heat generated by the motor is dissipated to the atmosphere. **3** The elevation difference between the inlet and outlet of the pump is negligible, $z_1 \cong z_2$. **4** The inlet and outlet diameters are the same and thus the average inlet and outlet velocities are equal, $V_1 = V_2$. **5** The kinetic energy correction factors are equal, $\alpha_1 = \alpha_2$.
**Properties**  We take the density of water to be 1 kg/L = 1000 kg/m³ and its specific heat to be 4.18 kJ/kg · °C.
**Analysis**  (*a*) The mass flow rate of water through the pump is

$$\dot{m} = \rho \dot{V} = (1 \text{ kg/L})(50 \text{ L/s}) = 50 \text{ kg/s}$$

The motor draws 15 kW of power and is 90 percent efficient. Thus the mechanical (shaft) power it delivers to the pump is

$$\dot{W}_{\text{pump, shaft}} = \eta_{\text{motor}}\dot{W}_{\text{electric}} = (0.90)(15 \text{ kW}) = 13.5 \text{ kW}$$

Water
50 L/s

300 kPa

②

$\eta_{\text{motor}} = 90\%$

Motor
15 kW

$\dot{W}_{\text{pump}}$

100 kPa
①

**FIGURE 12–31**
Schematic for Example 12–7.

To determine the mechanical efficiency of the pump, we need to know the increase in the mechanical energy of the fluid as it flows through the pump, which is

$$\Delta \dot{E}_{\text{mech, fluid}} = \dot{E}_{\text{mech, out}} - \dot{E}_{\text{mech, in}} = \dot{m}\left(\frac{P_2}{\rho} + \alpha_2 \frac{V_2^2}{2} + gz_2\right) - \dot{m}\left(\frac{P_1}{\rho} + \alpha_1 \frac{V_1^2}{2} + gz_1\right)$$

Simplifying it for this case and substituting the given values,

$$\Delta \dot{E}_{\text{mech, fluid}} = \dot{m}\left(\frac{P_2 - P_1}{\rho}\right) = (50 \text{ kg/s})\left(\frac{(300 - 100) \text{ kPa}}{1000 \text{ kg/m}^3}\right)\left(\frac{1 \text{ kJ}}{1 \text{ kP}\cdot\text{m}^3}\right) = 10.0 \text{ kW}$$

Then the mechanical efficiency of the pump becomes

$$\eta_{\text{pump}} = \frac{\dot{W}_{\text{pump, }u}}{\dot{W}_{\text{pump, shaft}}} = \frac{\Delta \dot{E}_{\text{mech, fluid}}}{\dot{W}_{\text{pump, shaft}}} = \frac{10.0 \text{ kW}}{13.5 \text{ kW}} = \textbf{0.741} \quad \text{or} \quad \textbf{74.1\%}$$

(b) Of the 13.5-kW mechanical power supplied by the pump, only 10.0 kW is imparted to the fluid as mechanical energy. The remaining 3.5 kW is converted to thermal energy due to frictional effects, and this "lost" mechanical energy manifests itself as a heating effect in the fluid,

$$\dot{E}_{\text{mech, loss}} = \dot{W}_{\text{pump, shaft}} - \Delta \dot{E}_{\text{mech, fluid}} = 13.5 - 10.0 = 3.5 \text{ kW}$$

The temperature rise of water due to this mechanical inefficiency is determined from the thermal energy balance, $\dot{E}_{\text{mech, loss}} = \dot{m}(u_2 - u_1) = \dot{m}c\Delta T$. Solving for $\Delta T$,

$$\Delta T = \frac{\dot{E}_{\text{mech, loss}}}{\dot{m}c} = \frac{3.5 \text{ kW}}{(50 \text{ kg/s})(4.18 \text{ kJ/ kg}\cdot°\text{C})} = \textbf{0.017}°\textbf{C}$$

Therefore, the water experiences a temperature rise of 0.017°C which is very small, due to mechanical inefficiency, as it flows through the pump.

**Discussion** In an actual application, the temperature rise of water would probably be less since part of the heat generated would be transferred to the casing of the pump and from the casing to the surrounding air. If the entire pump and motor were submerged in water, then the 1.5 kW dissipated due to motor inefficiency would also be transferred to the surrounding water as heat.

**FIGURE 12–32**
Schematic for Example 12–8.

**EXAMPLE 12–8    Hydroelectric Power Generation from a Dam**

In a hydroelectric power plant, 100 m³/s of water flows from an elevation of 120 m to a turbine, where electric power is generated (Fig. 12–32). The total irreversible head loss in the piping system from point 1 to point 2 (excluding the turbine unit) is determined to be 35 m. If the overall efficiency of the turbine–generator is 80 percent, estimate the electric power output.

**Solution** The available head, flow rate, head loss, and efficiency of a hydroelectric turbine are given. The electric power output is to be determined.
**Assumptions** 1 The flow is steady and incompressible. 2 Water levels at the reservoir and the discharge site remain constant.
**Properties** We take the density of water to be 1000 kg/m³.
**Analysis** The mass flow rate of water through the turbine is

$$\dot{m} = \rho\dot{V} = (1000 \text{ kg/m}^3)(100 \text{ m}^3/\text{s}) = 10^5 \text{ kg/s}$$

We take point 2 as the reference level, and thus $z_2 = 0$. Also, both points 1 and 2 are open to the atmosphere ($P_1 = P_2 = P_{atm}$) and the flow velocities are negligible at both points ($V_1 = V_2 = 0$). Then the energy equation for steady, incompressible flow reduces to

$$\cancel{\frac{P_1}{\rho g}} + \alpha_1 \cancel{\frac{V_1^2}{2g}} + z_1 + \cancel{h_{pump,\,u}}^{\;0} = \cancel{\frac{P_2}{\rho g}} + \alpha_2 \cancel{\frac{V_2^2}{2g}} + \cancel{z_2}^{\;0} + h_{turbine,\,e} + h_L$$

or

$$h_{turbine,\,e} = z_1 - h_L$$

Substituting, the extracted turbine head and the corresponding turbine power are

$$h_{turbine,\,e} = z_1 - h_L = 120 - 35 = 85\ \text{m}$$

$$\dot{W}_{turbine,\,e} = \dot{m}gh_{turbine,\,e} = (10^5\ \text{kg/s})(9.81\ \text{m/s}^2)(85\ \text{m})\left(\frac{1\ \text{kJ/kg}}{1000\ \text{m}^2/\text{s}^2}\right) = 83{,}400\ \text{kW}$$

Therefore, a perfect turbine–generator would generate 83,400 kW of electricity from this resource. The electric power generated by the actual unit is

$$\dot{W}_{electric} = \eta_{turbine-gen}\dot{W}_{turbine,\,e} = (0.80)(83.4\ \text{MW}) = \textbf{66.7 MW}$$

***Discussion*** Note that the power generation would increase by almost 1 MW for each percentage point improvement in the efficiency of the turbine–generator unit.

---

### EXAMPLE 12–9  Fan Selection for Air Cooling of a Computer

A fan is to be selected to cool a computer case whose dimensions are 12 cm $\times$ 40 cm $\times$ 40 cm (Fig. 12–33). Half of the volume in the case is expected to be filled with components and the other half to be air space. A 5-cm-diameter hole is available at the back of the case for the installation of the fan that is to replace the air in the void spaces of the case once every second. Small low-power fan–motor combined units are available in the market and their efficiency is estimated to be 30 percent. Determine (*a*) the wattage of the fan–motor unit to be purchased and (*b*) the pressure difference across the fan. Take the air density to be 1.20 kg/m$^3$.

**Solution**  A fan is to cool a computer case by completely replacing the air inside once every second. The power of the fan and the pressure difference across it are to be determined.
***Assumptions***  **1** The flow is steady and incompressible. **2** Losses other than those due to the inefficiency of the fan–motor unit are negligible. **3** The flow at the outlet is fairly uniform except near the center (due to the wake of the fan motor), and the kinetic energy correction factor at the outlet is 1.10.
***Properties***  The density of air is given to be 1.20 kg/m$^3$.
***Analysis***  (*a*) Noting that half of the volume of the case is occupied by the components, the air volume in the computer case is

$$V = (\text{Void fraction})(\text{Total case volume})$$

$$= 0.5(12\ \text{cm} \times 40\ \text{cm} \times 40\ \text{cm}) = 9600\ \text{cm}^3$$

**FIGURE 12–33**
Schematic for Example 12–9.

Therefore, the volume and mass flow rates of air through the case are

$$\dot{V} = \frac{V}{\Delta t} = \frac{9600 \text{ cm}^3}{1 \text{ s}} = 9600 \text{ cm}^3/\text{s} = 9.6 \times 10^{-3} \text{ m}^3/\text{s}$$

$$\dot{m} = \rho \dot{V} = (1.20 \text{ kg/m}^3)(9.6 \times 10^{-3} \text{ m}^3/\text{s}) = 0.0115 \text{ kg/s}$$

The cross-sectional area of the opening in the case and the average air velocity through the outlet are

$$A = \frac{\pi D^2}{4} = \frac{\pi (0.05 \text{ m})^2}{4} = 1.96 \times 10^{-3} \text{ m}^2$$

$$V = \frac{\dot{V}}{A} = \frac{9.6 \times 10^{-3} \text{ m}^3/\text{s}}{1.96 \times 10^{-3} \text{ m}^2} = 4.90 \text{ m/s}$$

We draw the control volume around the fan such that both the inlet and the outlet are at atmospheric pressure ($P_1 = P_2 = P_{\text{atm}}$), as shown in Fig. 12–33, where the inlet section 1 is large and far from the fan so that the flow velocity at the inlet section is negligible ($V_1 \cong 0$). Noting that $z_1 = z_2$ and frictional losses in the flow are disregarded, the mechanical losses consist of fan losses only and the energy equation (Eq. 12–29) simplifies to

$$\dot{m}\left(\frac{\cancel{P_1}^{\,0}}{\rho} + \alpha_1 \frac{\cancel{V_1^2}^{\,0}}{2} + \cancel{gz_1}\right) + \dot{W}_{\text{fan}} = \dot{m}\left(\frac{\cancel{P_2}^{\,0}}{\rho} + \alpha_2 \frac{V_2^2}{2} + \cancel{gz_2}\right) + \cancel{\dot{W}_{\text{turbine}}}^{\,0} + \dot{E}_{\text{mech loss, fan}}$$

Solving for $\dot{W}_{\text{fan}} - \dot{E}_{\text{mech loss, fan}} = \dot{W}_{\text{fan, }u}$ and substituting,

$$\dot{W}_{\text{fan, }u} = \dot{m}\alpha_2 \frac{V_2^2}{2} = (0.0115 \text{ kg/s})(1.10)\frac{(4.90 \text{ m/s})^2}{2}\left(\frac{1 \text{ N}}{1 \text{ kg·m/s}^2}\right) = 0.152 \text{ W}$$

Then the required electric power input to the fan is determined to be

$$\dot{W}_{\text{elect}} = \frac{\dot{W}_{\text{fan, }u}}{\eta_{\text{fan–motor}}} = \frac{0.152 \text{ W}}{0.3} = \mathbf{0.506 \text{ W}}$$

Therefore, a fan–motor rated at about a half watt is adequate for this job (Fig. 12–34). (b) To determine the pressure difference across the fan unit, we take points 3 and 4 to be on the two sides of the fan on a horizontal line. This time $z_3 = z_4$ again and $V_3 = V_4$ since the fan is a narrow cross section, and the energy equation reduces to

$$\dot{m}\frac{P_3}{\rho} + \dot{W}_{\text{fan}} = \dot{m}\frac{P_4}{\rho} + \dot{E}_{\text{mech loss, fan}} \quad \rightarrow \quad \dot{W}_{\text{fan, }u} = \dot{m}\frac{P_4 - P_3}{\rho}$$

Solving for $P_4 - P_3$ and substituting,

$$P_4 - P_3 = \frac{\rho \dot{W}_{\text{fan, }u}}{\dot{m}} = \frac{(1.2 \text{ kg/m}^3)(0.152 \text{ W})}{0.0115 \text{ kg/s}}\left(\frac{1 \text{ Pa·m}^3}{1 \text{ Ws}}\right) = \mathbf{15.8 \text{ Pa}}$$

Therefore, the pressure rise across the fan is 15.8 Pa.

**Discussion** The efficiency of the fan–motor unit is given to be 30 percent, which means 30 percent of the electric power $\dot{W}_{\text{electric}}$ consumed by the unit is converted to useful mechanical energy while the rest (70 percent) is "lost" and converted to thermal energy. Also, a more powerful fan is required in an actual system to overcome frictional losses inside the computer case. Note that if we had ignored the kinetic energy correction factor at the outlet, the required electrical power and pressure rise would have been 10 percent lower in this case (0.460 W and 14.4 Pa, respectively).

**FIGURE 12–34**

The cooling fans used in computers and computer power supplies are typically small and consume only a few watts of electrical power.

*© Getty RF.*

## SUMMARY

This chapter deals with the Bernoulli and energy equations and their applications.

The *Bernoulli equation* is a relation between pressure, velocity, and elevation in steady, incompressible flow, and is expressed along a streamline and in regions where net viscous forces are negligible as

$$\frac{P}{\rho} + \frac{V^2}{2} + gz = \text{constant}$$

The Bernoulli equation can also be considered as an expression of mechanical energy balance, stated as: *The sum of the kinetic, potential, and flow energies of a fluid particle is constant along a streamline during steady flow when the compressibility and frictional effects are negligible.* Multiplying the Bernoulli equation by density gives

$$P + \rho \frac{V^2}{2} + \rho gz = \text{constant}$$

where $P$ is the *static pressure*, which represents the actual pressure of the fluid; $\rho V^2/2$ is the *dynamic pressure*, which represents the pressure rise when the fluid in motion is brought to a stop; and $\rho gz$ is the *hydrostatic pressure*, which accounts for the effects of fluid weight on pressure. The sum of the static, dynamic, and hydrostatic pressures is called the *total pressure*. The Bernoulli equation states that *the total pressure along a streamline is constant*. The sum of the static and dynamic pressures is called the *stagnation pressure*, which represents the pressure at a point where the fluid is brought to a complete stop in a frictionless manner. The Bernoulli equation can also be represented in terms of "heads" by dividing each term by $g$,

$$\frac{P}{\rho g} + \frac{V^2}{2g} + z = H = \text{constant}$$

where $P/\rho g$ is the *pressure head*, which represents the height of a fluid column that produces the static pressure $P$; $V^2/2g$ is the *velocity head*, which represents the elevation needed for a fluid to reach the velocity $V$ during frictionless free fall; and $z$ is the *elevation head*, which represents the potential energy of the fluid. Also, $H$ is the *total head* for the flow. The curve that represents the sum of the static pressure and the elevation heads, $P/\rho g + z$, is called the *hydraulic grade line* (HGL), and the curve that represents the total head of the fluid, $P/\rho g + V^2/2g + z$, is called the *energy grade line* (EGL).

The *energy equation* for steady, incompressible flow is expressed as

$$\frac{P_1}{\rho g} + \alpha_1 \frac{V_1^2}{2g} + z_1 + h_{\text{pump, }u} = \frac{P_2}{\rho g} + \alpha_2 \frac{V_2^2}{2g} + z_2 + h_{\text{turbine, }e} + h_L$$

where

$$h_{\text{pump, }u} = \frac{w_{\text{pump, }u}}{g} = \frac{\dot{W}_{\text{pump, }u}}{\dot{m}g} = \frac{\eta_{\text{pump}}\dot{W}_{\text{pump}}}{\dot{m}g}$$

$$h_{\text{turbine, }e} = \frac{w_{\text{turbine, }e}}{g} = \frac{\dot{W}_{\text{turbine, }e}}{\dot{m}g} = \frac{\dot{W}_{\text{turbine}}}{\eta_{\text{turbine}}\dot{m}g}$$

$$h_L = \frac{e_{\text{mech loss, piping}}}{g} = \frac{\dot{E}_{\text{mech loss, piping}}}{\dot{m}g}$$

$$e_{\text{mech, loss}} = u_2 - u_1 - q_{\text{net in}}$$

The Bernoulli and energy equations are two of the most fundamental relations in fluid mechanics.

## REFERENCES AND SUGGESTED READING

**1.** R. C. Dorf, ed. in chief. *The Engineering Handbook.* Boca Raton, FL: CRC Press, 1995.

**2.** R. L. Panton. *Incompressible Flow*, 2nd ed. New York: Wiley, 1996.

**3.** M. Van Dyke. *An Album of Fluid Motion.* Stanford, CA: The Parabolic Press, 1982.

## PROBLEMS*

### Bernoulli Equation

**12–1C**  What is streamwise acceleration? How does it differ from normal acceleration? Can a fluid particle accelerate in steady flow?

**12–2C**  Express the Bernoulli equation in three different ways using (*a*) energies, (*b*) pressures, and (*c*) heads.

**12–3C**  What are the three major assumptions used in the derivation of the Bernoulli equation?

**12–4C**  Define static, dynamic, and hydrostatic pressure. Under what conditions is their sum constant for a flow stream?

**12–5C**  What is stagnation pressure? Explain how it can be measured.

**12–6C**  Define pressure head, velocity head, and elevation head for a fluid stream and express them for a fluid stream whose pressure is $P$, velocity is $V$, and elevation is $z$.

**12–7C**  What is the hydraulic grade line? How does it differ from the energy grade line? Under what conditions do both lines coincide with the free surface of a liquid?

**12–8C**  How is the location of the hydraulic grade line determined for open-channel flow? How is it determined at the outlet of a pipe discharging to the atmosphere?

**12–9C**  In a certain application, a siphon must go over a high wall. Can water or oil with a specific gravity of 0.8 go over a higher wall? Why?

**12–10C**  Explain how and why a siphon works. Someone proposes siphoning cold water over a 7-m-high wall. Is this feasible? Explain.

**12–11C**  A glass manometer with oil as the working fluid is connected to an air duct as shown in Fig. P12–11C. Will the oil levels in the manometer be as in Fig. P12–11C*a* or *b*? Explain. What would your response be if the flow direction is reversed?

* Problems designated by a "C" are concept questions, and students are encouraged to answer them all. Problems designated by an "E" are in English units, and the SI users can ignore them. Problems with the ⬤ icon are solved using EES, and complete solutions together with parametric studies are included on the text specific web site. Problems with the ⬛ icon are comprehensive in nature and are intended to be solved with a computer, preferably using the EES software that accompanies this text.

**FIGURE P12–11C**

**12–12C**  The velocity of a fluid flowing in a pipe is to be measured by two different Pitot-type mercury manometers shown in Fig. P12–12C. Would you expect both manometers to predict the same velocity for flowing water? If not, which would be more accurate? Explain. What would your response be if air were flowing in the pipe instead of water?

**FIGURE P12–12C**

**12–13C**  The water level of a tank on a building roof is 20 m above the ground. A hose leads from the tank bottom to the ground. The end of the hose has a nozzle, which is pointed straight up. What is the maximum height to which the water could rise? What factors would reduce this height?

**12–14C**  A student siphons water over a 8.5-m-high wall at sea level. She then climbs to the summit of Mount Shasta (elevation 4390 m, $P_{atm} = 58.5$ kPa) and attempts the same experiment. Comment on her prospects for success.

**12–15**  In a hydroelectric power plant, water enters the turbine nozzles at 800 kPa absolute with a low velocity. If the nozzle outlets are exposed to atmospheric pressure of 100 kPa, determine the maximum velocity to which water can be accelerated by the nozzles before striking the turbine blades.

**12–16**  A Pitot-static probe is used to measure the speed of an aircraft flying at 3000 m. If the differential pressure reading is 3 kPa, determine the speed of the aircraft.

**12–17** While traveling on a dirt road, the bottom of a car hits a sharp rock and a small hole develops at the bottom of its gas tank. If the height of the gasoline in the tank is 40 cm, determine the initial velocity of the gasoline at the hole. Discuss how the velocity will change with time and how the flow will be affected if the lid of the tank is closed tightly. *Answer:* 2.80 m/s

**12–18E** The drinking water needs of an office are met by large water bottles. One end of a 0.25-in-diameter plastic hose is inserted into the bottle placed on a high stand, while the other end with an on/off valve is maintained 2 ft below the bottom of the bottle. If the water level in the bottle is 1.5 ft when it is full, determine how long it will take at the minimum to fill an 8-oz glass (= 0.00835 ft$^3$) (a) when the bottle is first opened and (b) when the bottle is almost empty. Neglect frictional losses.

1.5 ft

Water

2 ft

**FIGURE P12–18E**

**12–19** A piezometer and a Pitot tube are tapped into a 4-cm-diameter horizontal water pipe, and the height of the water columns are measured to be 26 cm in the piezometer and 35 cm in the Pitot tube (both measured from the top surface of the pipe). Determine the velocity at the center of the pipe.

**12–20** The diameter of a cylindrical water tank is $D_o$ and its height is $H$. The tank is filled with water, which is open to the atmosphere. An orifice of diameter $D$ with a smooth entrance (i.e., negligible losses) is open at the bottom. Develop a relation for the time required for the tank (a) to empty halfway and (b) to empty completely.

**12–21E** A siphon pumps water from a large reservoir to a lower tank that is initially empty. The tank also has a rounded orifice 20 ft below the reservoir surface where the water leaves the tank. Both the siphon and the orifice diameters are 2 in. Ignoring frictional losses, determine to what height the water will rise in the tank at equilibrium.

**12–22** Water enters a tank of diameter $D_T$ steadily at a mass flow rate of $\dot{m}_{in}$. An orifice at the bottom with diameter $D_o$ allows water to escape. The orifice has a rounded entrance, so the frictional losses are negligible. If the tank is initially empty, (a) determine the maximum height that the water will reach in the tank and (b) obtain a relation for water height $z$ as a function of time.

$\dot{m}_{in}$

$D_T$

$z$

$D_o$

**FIGURE P12–22**

**12–23E** Water flows through a horizontal pipe at a rate of 1 gal/s. The pipe consists of two sections of diameters 4 in and 2 in with a smooth reducing section. The pressure difference between the two pipe sections is measured by a mercury manometer. Neglecting frictional effects, determine the differential height of mercury between the two pipe sections. *Answer:* 0.52 in

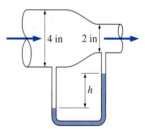

4 in    2 in

h

**FIGURE P12–23E**

**12–24** An airplane is flying at an altitude of 12,000 m. Determine the gage pressure at the stagnation point on the nose of the plane if the speed of the plane is 300 km/h. How would you solve this problem if the speed were 1050 km/h? Explain.

**12–25** The air velocity in the duct of a heating system is to be measured by a Pitot-static probe inserted into the duct parallel to the flow. If the differential height between the water columns connected to the two outlets of the probe is 2.4 cm, determine (a) the flow velocity and (b) the pressure rise at the tip of the probe. The air temperature and pressure in the duct are 45°C and 98 kPa, respectively.

**12–26** The water in a 8-m-diameter, 3-m-high aboveground swimming pool is to be emptied by unplugging a 3-cm-diameter, 25-m-long horizontal pipe attached to the bottom of the pool. Determine the maximum discharge rate of water through the pipe. Also, explain why the actual flow rate will be less.

**12–27** Reconsider Prob. 12–26. Determine how long it will take to empty the swimming pool completely.    *Answer: 15.4 h*

**12–28** ⎰EES⎱ Reconsider Prob. 12–27. Using EES (or other) software, investigate the effect of the discharge pipe diameter on the time required to empty the pool completely. Let the diameter vary from 1 to 10 cm in increments of 1 cm. Tabulate and plot the results.

**12–29** Air at 110 kPa and 50°C flows upward through a 6-cm-diameter inclined duct at a rate of 45 L/s. The duct diameter is then reduced to 4 cm through a reducer. The pressure change across the reducer is measured by a water manometer. The elevation difference between the two points on the pipe where the two arms of the manometer are attached is 0.20 m. Determine the differential height between the fluid levels of the two arms of the manometer.

**FIGURE P12–29**

**12–30E** Air is flowing through a venturi meter whose diameter is 2.6 in at the entrance part (location 1) and 1.8 in at the throat (location 2). The gage pressure is measured to be 12.2 psia at the entrance and 11.8 psia at the throat. Neglecting frictional effects, show that the volume flow rate can be expressed as

$$\dot{V} = A_2 \sqrt{\frac{2(P_1 - P_2)}{\rho(1 - A_2^2/A_1^2)}}$$

and determine the flow rate of air. Take the air density to be 0.075 lbm/ft³.

**FIGURE P12–30E**

**12–31** The water pressure in the mains of a city at a particular location is 350 kPa gage. Determine if this main can serve water to neighborhoods that are 50 m above this location.

**12–32** A pressurized tank of water has a 10-cm-diameter orifice at the bottom, where water discharges to the atmosphere. The water level is 2.5 m above the outlet. The tank air pressure above the water level is 250 kPa (absolute) while the atmospheric pressure is 100 kPa. Neglecting frictional effects, determine the initial discharge rate of water from the tank.    *Answer: 0.147 m³/s*

**FIGURE P12–32**

**12–33** ⎰EES⎱ Reconsider Prob. 12–32. Using EES (or other) software, investigate the effect of water height in the tank on the discharge velocity. Let the water height vary from 0 to 5 m in increments of 0.5 m. Tabulate and plot the results.

**12–34** A handheld bicycle pump can be used as an atomizer to generate a fine mist of paint or pesticide by forcing air at a high velocity through a small hole and placing a short tube between the liquid reservoir and the high-speed air jet. The pressure across a subsonic jet exposed to the atmosphere is nearly atmospheric, and the surface of the liquid in the reservoir is also open to atmospheric pressure. In light of this, explain how the liquid is sucked up the tube. *Hint:* Read Sec. 12–1 carefully.

**FIGURE P12–34**

**12–35**  The water level in a tank is 15 m above the ground. A hose is connected to the bottom of the tank, and the nozzle at the end of the hose is pointed straight up. The tank cover is airtight, and the air pressure above the water surface is 3 atm gage. The system is at sea level. Determine the maximum height to which the water stream could rise.  *Answer:* 46.0 m

**FIGURE P12–35**

**12–36**  A Pitot-static probe connected to a water manometer is used to measure the velocity of air. If the deflection (the vertical distance between the fluid levels in the two arms) is 7.3 cm, determine the air velocity. Take the density of air to be 1.25 kg/m$^3$.

**FIGURE P12–36**

**12–37E**  The air velocity in a duct is measured by a Pitot-static probe connected to a differential pressure gage. If the air is at 13.4 psia absolute and 70°F and the reading of the differential pressure gage is 0.15 psi, determine the air velocity. *Answer:* 143 ft/s

**12–38**  In cold climates, water pipes may freeze and burst if proper precautions are not taken. In such an occurrence, the exposed part of a pipe on the ground ruptures, and water shoots up to 42 m. Estimate the gage pressure of water in the pipe. State your assumptions and discuss if the actual pressure is more or less than the value you predicted.

## Energy Equation

**12–39C**  What is irreversible head loss? How is it related to the mechanical energy loss?

**12–40C**  What is useful pump head? How is it related to the power input to the pump?

**12–41C**  What is the kinetic energy correction factor? Is it significant?

**12–42C**  The water level in a tank is 20 m above the ground. A hose is connected to the bottom of the tank, and the nozzle at the end of the hose is pointed straight up. The water stream from the nozzle is observed to rise 25 m above the ground. Explain what may cause the water from the hose to rise above the tank level.

**12–43C**  A person is filling a knee-high bucket with water using a garden hose and holding it such that water discharges from the hose at the level of his waist. Someone suggests that the bucket will fill faster if the hose is lowered such that water discharges from the hose at the knee level. Do you agree with this suggestion? Explain. Disregard any frictional effects.

**12–44**  A 3-m-high tank filled with water has a discharge valve near the bottom and another near the top. (*a*) If these two valves are opened, will there be any difference between the discharge velocities of the two water streams? (*b*) If a hose whose discharge end is left open on the ground is first connected to the lower valve and then to the higher valve, will there be any difference between the discharge rates of water for the two cases? Disregard any frictional effects.

**12–45E**  In a hydroelectric power plant, water flows from an elevation of 240 ft to a turbine, where electric power is generated. For an overall turbine–generator efficiency of 83 percent, determine the minimum flow rate required to generate 100 kW of electricity.  *Answer:* 370 lbm/s

**12–46E**  Reconsider Prob. 12–45E. Determine the flow rate of water if the irreversible head loss of the piping system between the free surfaces of the source and the sink is 36 ft.

**12–47**  A fan is to be selected to ventilate a bathroom whose dimensions are 2 m $\times$ 3 m $\times$ 3 m. The air velocity is not to exceed 8 m/s to minimize vibration and noise. The combined efficiency of the fan–motor unit to be used can be taken to be 50 percent. If the fan is to replace the entire volume of air in 10 min, determine (*a*) the wattage of the fan–motor unit to be purchased, (*b*) the diameter of the fan casing, and (*c*) the pressure difference across the fan. Take the air density to be 1.25 kg/m$^3$ and disregard the effect of the kinetic energy correction factors.

**FIGURE P12–47**

**12–48** Water is being pumped from a large lake to a reservoir 25 m above at a rate of 25 L/s by a 10-kW (shaft) pump. If the irreversible head loss of the piping system is 5 m, determine the mechanical efficiency of the pump. *Answer:* 73.6 percent

**12–49** ![EES] Reconsider Prob. 12–48. Using EES (or other) software, investigate the effect of irreversible head loss on the mechanical efficiency of the pump. Let the head loss vary from 0 to 15 m in increments of 1 m. Plot the results, and discuss them.

**12–50** A 7-hp (shaft) pump is used to raise water to a 15-m higher elevation. If the mechanical efficiency of the pump is 82 percent, determine the maximum volume flow rate of water.

**12–51** Water flows at a rate of 0.035 m³/s in a horizontal pipe whose diameter is reduced from 15 cm to 8 cm by a reducer. If the pressure at the centerline is measured to be 480 kPa and 445 kPa before and after the reducer, respectively, determine the irreversible head loss in the reducer. Take the kinetic energy correction factors to be 1.05. *Answer:* 1.18 m

**12–52** The water level in a tank is 20 m above the ground. A hose is connected to the bottom of the tank, and the nozzle at the end of the hose is pointed straight up. The tank is at sea level, and the water surface is open to the atmosphere. In the line leading from the tank to the nozzle is a pump, which increases the pressure of water. If the water jet rises to a height of 27 m from the ground, determine the minimum pressure rise supplied by the pump to the water line.

**FIGURE P12–52**

**12–53** A hydraulic turbine has 85 m of head available at a flow rate of 0.25 m³/s, and its overall turbine–generator efficiency is 78 percent. Determine the electric power output of this turbine.

**12–54** An oil pump is drawing 25 kW of electric power while pumping oil with $\rho = 860$ kg/m³ at a rate of 0.1 m³/s. The inlet and outlet diameters of the pipe are 8 cm and 12 cm, respectively. If the pressure rise of oil in the pump is measured to be 250 kPa and the motor efficiency is 90 percent, determine the mechanical efficiency of the pump. Take the kinetic energy correction factor to be 1.05.

**FIGURE P12–54**

**12–55** Water flows at a rate of 20 L/s through a horizontal pipe whose diameter is constant at 3 cm. The pressure drop across a valve in the pipe is measured to be 2 kPa, as shown in Fig P12–55. Determine the irreversible head loss of the valve, and the useful pumping power needed to overcome the resulting pressure drop. *Answers:* 0.204 m, 40 W

**FIGURE P12–55**

**12–56E** The water level in a tank is 66 ft above the ground. A hose is connected to the bottom of the tank at the ground level and the nozzle at the end of the hose is pointed straight up. The tank cover is airtight, but the pressure over the water surface is unknown. Determine the minimum tank air pressure (gage) that will cause a water stream from the nozzle to rise 90 ft from the ground.

**12–57** A large tank is initially filled with water 5 m above the center of a sharp-edged 10-cm-diameter orifice. The tank water surface is open to the atmosphere, and the orifice drains to the atmosphere. If the total irreversible head loss in the system is 0.3 m, determine the initial discharge velocity of water from the tank. Take the kinetic energy correction factor at the orifice to be 1.2.

**12–58** Water enters a hydraulic turbine through a 30-cm-diameter pipe at a rate of 0.6 m³/s and exits through a 25-cm-diameter pipe. The pressure drop in the turbine is measured by a mercury manometer to be 1.2 m. For a combined turbine–generator efficiency of 83 percent, determine the net electric power output. Disregard the effect of the kinetic energy correction factors.

**FIGURE P12–58**

**12–59** The velocity profile for turbulent flow in a circular pipe is approximated as $u(r) = u_{max}(1 - r/R)^{1/n}$, where $n = 7$. Determine the kinetic energy correction factor for this flow.
*Answer:* 1.06

**12–60** Water is pumped from a lower reservoir to a higher reservoir by a pump that provides 20 kW of useful mechanical power to the water. The free surface of the upper reservoir is 45 m higher than the surface of the lower reservoir. If the flow rate of water is measured to be 0.03 m³/s, determine the irreversible head loss of the system and the lost mechanical power during this process.

**FIGURE P12–60**

**12–61** Water in a partially filled large tank is to be supplied to the roof top, which is 8 m above the water level in the tank, through a 2.5-cm-internal-diameter pipe by maintaining a constant air pressure of 300 kPa (gage) in the tank. If the head loss in the piping is 2 m of water, determine the discharge rate of the supply of water to the roof top.

**12–62** Underground water is to be pumped by a 78 percent efficient 5-kW submerged pump to a pool whose free surface is 30 m above the underground water level. The diameter of the pipe is 7 cm on the intake side and 5 cm on the discharge side. Determine (a) the maximum flow rate of water and (b) the pressure difference across the pump. Assume the elevation difference between the pump inlet and the outlet and the effect of the kinetic energy correction factors to be negligible.

**FIGURE P12–62**

**12–63** Reconsider Prob. 12–62. Determine the flow rate of water and the pressure difference across the pump if the irreversible head loss of the piping system is 4 m.

**12–64** A fireboat is to fight fires at coastal areas by drawing seawater with a density of 1030 kg/m³ through a 20-cm-diameter pipe at a rate of 0.1 m³/s and discharging it through a hose nozzle with an exit diameter of 5 cm. The total irreversible head loss of the system is 3 m, and the position of the nozzle is 3 m above sea level. For a pump efficiency of 70 percent, determine the required shaft power input to the pump and the water discharge velocity. *Answers: 199 kW, 50.9 m/s*

**FIGURE P12–64**

## Review Problems

**12–65** Air flows through a pipe at a rate of 170 L/s. The pipe consists of two sections of diameters 18 cm and 10 cm with a smooth reducing section that connects them. The pressure difference between the two pipe sections is measured by a water manometer. Neglecting frictional effects, determine the differential height of water between the two pipe sections. Take the air density to be 1.20 kg/m³. *Answer: 2.60 cm*

**FIGURE P12–65**

**12–66** Air at 100 kPa and 25°C flows in a horizontal duct of variable cross section. The water column in the manometer that measures the difference between two sections has a vertical displacement of 8 cm. If the velocity in the first section is low and the friction is negligible, determine the velocity at the second section. Also, if the manometer reading has a possible error of ±2 mm, conduct an error analysis to estimate the range of validity for the velocity found.

**12–67** A very large tank contains air at 102 kPa at a location where the atmospheric air is at 100 kPa and 20°C. Now a 2-cm-diameter tap is opened. Determine the maximum flow rate of air through the hole. What would your response be if air is discharged through a 2-m-long, 4-cm-diameter tube with a 2-cm-diameter nozzle? Would you solve the problem the same way if the pressure in the storage tank were 300 kPa?

**FIGURE P12–67**

**12–68** Water is flowing through a Venturi meter whose diameter is 7 cm at the entrance part and 4 cm at the throat. The pressure is measured to be 380 kPa at the entrance and 150 kPa at the throat. Neglecting frictional effects, determine the flow rate of water. *Answer: 0.0285 m³/s*

**12–69** Water flows at a rate of 0.025 m³/s in a horizontal pipe whose diameter increases from 6 to 11 cm by an enlargement section. If the head loss across the enlargement section is 0.45 m and the kinetic energy correction factor at both the inlet and the outlet is 1.05, determine the pressure change.

**12–70** A 3-m-high large tank is initially filled with water. The tank water surface is open to the atmosphere, and a sharp-edged 10-cm-diameter orifice at the bottom drains to the atmosphere through a horizontal 80-m-long pipe. If the total irreversible head loss of the system is determined to be 1.5 m, determine the initial velocity of the water from the tank. Disregard the effect of the kinetic energy correction factors. *Answer: 5.42 m/s*

**FIGURE P12–70**

**12–71** Reconsider Prob. 12–70. Using EES (or other) software, investigate the effect of the tank height on the initial discharge velocity of water from the completely filled tank. Let the tank height vary from 2 to 15 m in increments of 1 m, and assume the irreversible head loss to remain constant. Tabulate and plot the results.

**12–72** Reconsider Prob. 12–70. In order to drain the tank faster, a pump is installed near the tank exit. Determine the pump head input necessary to establish an average water velocity of 6.5 m/s when the tank is full.

**12–73E** The water level in a tank is 120 ft above the ground. A hose is connected to the bottom of the tank, and the nozzle at the end of the hose is pointed straight up. The tank is at sea level, and the water surface is open to the atmosphere. In the line leading from the tank to the nozzle is a pump, which increases the water pressure by 10 psia. Determine the maximum height to which the water stream could rise.

**12–74** A wind tunnel draws atmospheric air at 20°C and 101.3 kPa by a large fan located near the exit of the tunnel. If the air velocity in the tunnel is 80 m/s, determine the pressure in the tunnel.

## Design and Essay Problems

**12–75** Computer-aided designs, the use of better materials, and better manufacturing techniques have resulted in a tremendous increase in the efficiency of pumps, turbines, and electric motors. Contact one or more pump, turbine, and motor manufacturers and obtain information about the efficiency of their products. In general, how does efficiency vary with rated power of these devices?

**12–76** Using a handheld bicycle pump to generate an air jet, a soda can as the water reservoir, and a straw as the tube, design and build an atomizer. Study the effects of various parameters such as the tube length, the diameter of the exit hole, and the pumping speed on performance.

**12–77** Using a flexible drinking straw and a ruler, explain how you would measure the water flow velocity in a river.

**12–78** The power generated by a wind turbine is proportional to the cube of the wind velocity. Inspired by the acceleration of a fluid in a nozzle, someone proposes to install a reducer casing to capture the wind energy from a larger area and accelerate it before the wind strikes the turbine blades, as shown in Fig. P12–78. Evaluate if the proposed modification should be given a consideration in the design of new wind turbines.

**FIGURE P12–74**

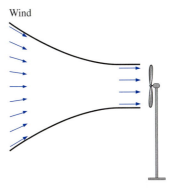

**FIGURE P12–78**

# MOMENTUM ANALYSIS OF FLOW SYSTEMS

When dealing with engineering problems, it is desirable to obtain fast and accurate solutions at minimal cost. Most engineering problems, including those associated with fluid flow, can be analyzed using one of three basic approaches: differential, experimental, and control volume. In *differential approaches*, the problem is formulated accurately using differential quantities, but the solution of the resulting differential equations is difficult, usually requiring the use of numerical methods with extensive computer codes. *Experimental approaches* complemented with dimensional analysis are highly accurate, but they are typically time consuming and expensive. The *finite control volume approach* described in this chapter is remarkably fast and simple and usually gives answers that are sufficiently accurate for most engineering purposes. Therefore, despite the approximations involved, the basic finite control volume analysis performed with paper and pencil has always been an indispensable tool for engineers.

In Chap. 12, the control volume energy analysis of fluid flow systems was presented. In this chapter, we present the finite control volume momentum analysis of fluid flow problems. First we give an overview of Newton's laws and the conservation relations for linear and angular momentum. Then using the Reynolds transport theorem, we develop the linear momentum and angular momentum equations for control volumes and use them to determine the forces and torques associated with fluid flow.

## Objectives

The objectives of this chapter are to:

- Identify the various kinds of forces and moments acting on a control volume.

- Use control volume analysis to determine the forces associated with fluid flow.

- Use control volume analysis to determine the moments caused by fluid flow and the torque transmitted.

# 13–1 ■ NEWTON'S LAWS

Newton's laws are relations between motions of bodies and the forces acting on them. Newton's first law states that *a body at rest remains at rest, and a body in motion remains in motion at the same velocity in a straight path when the net force acting on it is zero.* Therefore, a body tends to preserve its state of inertia. Newton's second law states that *the acceleration of a body is proportional to the net force acting on it and is inversely proportional to its mass.* Newton's third law states that *when a body exerts a force on a second body, the second body exerts an equal and opposite force on the first.* Therefore, the direction of an exposed reaction force depends on the body taken as the system.

For a rigid body of mass $m$, Newton's second law is expressed as

*Newton's second law:*
$$\vec{F} = m\vec{a} = m\frac{d\vec{V}}{dt} = \frac{d(m\vec{V})}{dt} \tag{13–1}$$

where $\vec{F}$ is the net force acting on the body and $\vec{a}$ is the acceleration of the body under the influence of $\vec{F}$.

The product of the mass and the velocity of a body is called the *linear momentum* or just the *momentum* of the body. The momentum of a rigid body of mass $m$ moving with velocity $\vec{V}$ is $m\vec{V}$ (Fig. 13–1). Then Newton's second law expressed in Eq. 13–1 can also be stated as *the rate of change of the momentum of a body is equal to the net force acting on the body* (Fig. 13–2). This statement is more in line with Newton's original statement of the second law, and it is more appropriate for use in fluid mechanics when studying the forces generated as a result of velocity changes of fluid streams. Therefore, in fluid mechanics, Newton's second law is usually referred to as the *linear momentum equation.*

The momentum of a system remains constant only when the net force acting on it is zero, and thus the momentum of such a system is conserved. This is known as the *conservation of momentum principle.* This principle has proven to be a very useful tool when analyzing collisions such as those between balls; between balls and rackets, bats, or clubs; and between atoms or subatomic particles; and explosions such as those that occur in rockets, missiles, and guns. In fluid mechanics, however, the net force acting on a system is typically *not* zero, and we prefer to work with the linear momentum equation rather than the conservation of momentum principle.

Note that force, acceleration, velocity, and momentum are vector quantities, and as such they have direction as well as magnitude. Also, momentum is a constant multiple of velocity, and thus the direction of momentum is the direction of velocity as shown in Fig 13–1. Any vector equation can be written in scalar form for a specified direction using magnitudes, e.g., $F_x = ma_x = d(mV_x)/dt$ in the $x$-direction.

The counterpart of Newton's second law for rotating rigid bodies is expressed as $\vec{M} = I\vec{\alpha}$, where $\vec{M}$ is the net moment or torque applied on the body, $I$ is the moment of inertia of the body about the axis of rotation, and $\vec{\alpha}$ is the angular acceleration. It can also be expressed in terms of the rate of change of angular momentum $d\vec{H}/dt$ as

*Angular momentum equation:*
$$\vec{M} = I\vec{\alpha} = I\frac{d\vec{\omega}}{dt} = \frac{d(I\vec{\omega})}{dt} = \frac{d\vec{H}}{dt} \tag{13–2}$$

**FIGURE 13–1**

Linear momentum is the product of mass and velocity, and its direction is the direction of velocity.

**FIGURE 13–2**

Newton's second law is also expressed as *the rate of change of the momentum of a body is equal to the net force acting on it.*

where $\vec{\omega}$ is the angular velocity. For a rigid body rotating about a fixed $x$-axis, the angular momentum equation is written in scalar form as

*Angular momentum about x-axis:*
$$M_x = I_x \frac{d\omega_x}{dt} = \frac{dH_x}{dt} \qquad \text{(13–3)}$$

The angular momentum equation can be stated as *the rate of change of the angular momentum of a body is equal to the net torque acting on it* (Fig. 13–3).

The total angular momentum of a rotating body remains constant when the net torque acting on it is zero, and thus the angular momentum of such systems is conserved. This is known as the *conservation of angular momentum principle* and is expressed as $I\omega$ = constant. Many interesting phenomena such as ice skaters spinning faster when they bring their arms close to their bodies and divers rotating faster when they curl after the jump can be explained easily with the help of the conservation of angular momentum principle (in both cases, the moment of inertia $I$ is decreased and thus the angular velocity $\omega$ is increased as the outer parts of the body are brought closer to the axis of rotation).

**FIGURE 13–3**

The rate of change of the angular momentum of a body is equal to the net torque acting on it.

## 13–2 ▪ CHOOSING A CONTROL VOLUME

We now briefly discuss how to *wisely* select a control volume. A control volume can be selected as any arbitrary region in space through which fluid flows, and its bounding control surface can be fixed, moving, and even deforming during flow. The application of a basic conservation law is a systematic procedure for bookkeeping or accounting of the quantity under consideration, and thus it is extremely important that the boundaries of the control volume are well defined during an analysis. Also, the flow rate of any quantity into or out of a control volume depends on the flow velocity *relative to the control surface*, and thus it is essential to know if the control volume remains at rest during flow or if it moves.

Many flow systems involve stationary hardware firmly fixed to a stationary surface, and such systems are best analyzed using *fixed* control volumes. When determining the reaction force acting on a tripod holding the nozzle of a hose, for example, a natural choice for the control volume is one that passes perpendicularly through the nozzle exit flow and through the bottom of the tripod legs (Fig. 13–4a). This is a fixed control volume, and the water velocity relative to a fixed point on the ground is the same as the water velocity relative to the nozzle exit plane.

When analyzing flow systems that are moving or deforming, it is usually more convenient to allow the control volume to *move* or *deform*. When determining the thrust developed by the jet engine of an airplane cruising at constant velocity, for example, a wise choice of control volume is one that encloses the airplane and cuts through the nozzle exit plane (Fig. 13–4b). The control volume in this case moves with velocity $\vec{V}_{CV}$, which is identical to the cruising velocity of the airplane relative to a fixed point on earth. When determining the flow rate of exhaust gases leaving the nozzle, the proper velocity to use is the velocity of the exhaust gases relative to the nozzle exit plane, that is, the *relative velocity* $\vec{V}_r$. Since the entire control volume moves at velocity $\vec{V}_{CV}$, the relative velocity becomes $\vec{V}_r = \vec{V} - \vec{V}_{CV}$, where $\vec{V}$ is the *absolute velocity* of the exhaust gases, i.e., the velocity relative to a

Fixed control volume

*(a)*

Moving control volume

*(b)*

Deforming control volume

*(c)*

**FIGURE 13–4**

Examples of (a) fixed, (b) moving, and (c) deforming control volumes.

fixed point on earth. Note that $\vec{V}_r$ is the fluid velocity expressed relative to a coordinate system moving *with* the control volume. Also, this is a vector equation, and velocities in opposite directions have opposite signs. For example, if the airplane is cruising at 500 km/h to the left, and the velocity of the exhaust gases is 800 km/h to the right relative to the ground, the velocity of the exhaust gases relative to the nozzle exit is

$$\vec{V}_r = \vec{V} - \vec{V}_{CV} = 800\,\vec{i} - (-500\,\vec{i}) = 1300\,\vec{i} \text{ km/h}$$

That is, the exhaust gases leave the nozzle at 1300 km/h to the right relative to the nozzle exit (in the direction opposite to that of the airplane); this is the velocity that should be used when evaluating the outflow of exhaust gases through the control surface (Fig. 13–4b). Note that the exhaust gases would appear motionless to an observer on the ground if the relative velocity were equal in magnitude to the airplane velocity.

When analyzing the purging of exhaust gases from a reciprocating internal combustion engine, a wise choice for the control volume is one that comprises the space between the top of the piston and the cylinder head (Fig. 13–4c). This is a *deforming* control volume, since part of the control surface moves relative to other parts. The relative velocity for an inlet or outlet on the deforming part of a control surface (there are no such inlets or outlets in Fig. 13–4c) is then given by $\vec{V}_r = \vec{V} - \vec{V}_{CS}$ where $\vec{V}$ is the absolute fluid velocity and $\vec{V}_{CS}$ is the control surface velocity, both relative to a fixed point outside the control volume. Note that $\vec{V}_{CS} = \vec{V}_{CV}$ for moving but nondeforming control volumes, and $\vec{V}_{CS} = \vec{V}_{CV} = 0$ for fixed ones.

## 13–3 ■ FORCES ACTING ON A CONTROL VOLUME

The forces acting on a control volume consist of **body forces** that act throughout the entire body of the control volume (such as gravity, electric, and magnetic forces) and **surface forces** that act on the control surface (such as pressure and viscous forces and reaction forces at points of contact). Only external forces are considered in the analysis. Internal forces (such as the pressure force between a fluid and the inner surfaces of the flow section) are not considered in a control volume analysis unless they are exposed by passing the control surface through that area.

In control volume analysis, the sum of all forces acting on the control volume at a particular instant in time is represented by $\sum \vec{F}$ and is expressed as

*Total force acting on control volume:*  $$\sum \vec{F} = \sum \vec{F}_{body} + \sum \vec{F}_{surface} \qquad \textbf{(13–4)}$$

Body forces act on each volumetric portion of the control volume. The body force acting on a differential element of fluid of volume $dV$ within the control volume is shown in Fig. 13–5, and we must perform a volume integral to account for the net body force on the entire control volume. *Surface forces* act on each portion of the control surface. A differential surface element of area $dA$ and unit outward normal $\vec{n}$ on the control surface is shown in Fig. 13–5, along with the surface force acting on it. We must perform an area integral to obtain the net surface force acting on the entire control surface. As sketched, the surface force may act in a direction independent of that of the outward normal vector.

The most common body force is that of **gravity**, which exerts a downward force on every differential element of the control volume. While other body forces, such as electric and magnetic forces, may be important in some analyses, we consider only gravitational forces here.

The differential body force $d\vec{F}_{body} = d\vec{F}_{gravity}$ acting on the small fluid element shown in Fig. 13–6 is simply its weight,

*Gravitational force acting on a fluid element:*     $$d\vec{F}_{gravity} = \rho\vec{g}\,dV \qquad \textbf{(13–5)}$$

where $\rho$ is the average density of the element and $\vec{g}$ is the gravitational vector. In Cartesian coordinates we adopt the convention that $\vec{g}$ acts in the negative $z$-direction, as in Fig. 13–6, so that

*Gravitational vector in Cartesian coordinates:*     $$\vec{g} = -g\vec{k} \qquad \textbf{(13–6)}$$

Note that the coordinate axes in Fig. 13–6 are oriented so that the gravity vector acts *downward* in the $-z$-direction. On earth at sea level, the gravitational constant $g$ is equal to 9.807 m/s². Since gravity is the only body force being considered, integration of Eq. 13–5 yields

*Total body force acting on control volume:*     $$\sum\vec{F}_{body} = \int_{CV}\rho\vec{g}\,d = m_{CV}\vec{g} \qquad \textbf{(13–7)}$$

Surface forces are not as simple to analyze since they consist of both *normal* and *tangential* components. **Normal stresses** are composed of pressure (which always acts inwardly normal) and viscous stresses. **Shear stresses** are composed entirely of viscous stresses.

A careful selection of the control volume enables us to write the total force acting on the control volume, $\sum\vec{F}$, as the sum of more readily available quantities like weight, pressure, and reaction forces. We recommend the following for control volume analysis:

*Total force:*     $$\underbrace{\sum\vec{F}}_{\text{total force}} = \underbrace{\sum\vec{F}_{gravity}}_{\text{body force}} + \underbrace{\sum\vec{F}_{pressure} + \sum\vec{F}_{viscous} + \sum\vec{F}_{other}}_{\text{surface forces}} \qquad \textbf{(13–8)}$$

The first term on the right-hand side of Eq. 13–8 is the body force *weight*, since gravity is the only body force we are considering. The other three terms combine to form the net surface force; they are pressure forces, viscous forces, and "other" forces acting on the control surface. $\sum\vec{F}_{other}$ is composed of reaction forces required to turn the flow; forces at bolts, cables, struts, or walls through which the control surface cuts; etc.

All these surface forces arise as the control volume is isolated from its surroundings for analysis, and the effect of any detached object is accounted for by a force at that location. This is similar to drawing a free-body diagram in your statics and dynamics classes. We should choose the control volume such that forces that are not of interest remain internal, and thus they do not complicate the analysis. A well-chosen control volume exposes only the forces that are to be determined (such as reaction forces) and a minimum number of other forces.

A common simplification in the application of Newton's laws of motion is to subtract the *atmospheric pressure* and work with gage pressures. This is because atmospheric pressure acts in all directions, and its effect cancels out in every direction (Fig. 13–7). This means we can also ignore the pressure

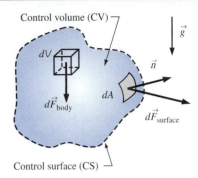

**FIGURE 13–5**

The total force acting on a control volume is composed of body forces and surface forces; body force is shown on a differential volume element, and surface force is shown on a differential surface element.

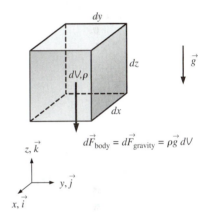

**FIGURE 13–6**

The gravitational force acting on a differential volume element of fluid is equal to its weight; the axes are oriented so that the gravity vector acts *downward* in the negative $z$-direction.

With atmospheric pressure considered       With atmospheric pressure cancelled out

**FIGURE 13–7**

Atmospheric pressure acts in all directions, and thus it can be ignored when performing force balances since its effect cancels out in every direction.

**FIGURE 13–8**

Cross section through a faucet assembly, illustrating the importance of choosing a control volume wisely; CV B is much easier to work with than CV A.

**FIGURE 13–9**

Two methods of analyzing the spraying of deodorant from a spray can: (*a*) We follow the fluid as it moves and deforms. This is the *system approach*—no mass crosses the boundary, and the total mass of the system remains fixed. (*b*) We consider a fixed interior volume of the can. This is the *control volume approach*—mass crosses the boundary.

forces at outlet sections where the fluid is discharged at subsonic velocities to the atmosphere since the discharge pressures in such cases are very near atmospheric pressure.

As an example of how to wisely choose a control volume, consider control volume analysis of water flowing steadily through a faucet with a partially closed gate valve spigot (Fig. 13–8). It is desired to calculate the net force on the flange to ensure that the flange bolts are strong enough. There are many possible choices for the control volume. Some engineers restrict their control volumes to the fluid itself, as indicated by CV A (the colored control volume) in Fig 13–8. With this control volume, there are pressure forces that vary along the control surface, there are viscous forces along the pipe wall and at locations inside the valve, and there is a body force, namely, the weight of the water in the control volume. Fortunately, to calculate the net force on the flange, we do *not* need to integrate the pressure and viscous stresses all along the control surface. Instead, we can lump the unknown pressure and viscous forces together into one reaction force, representing the net force of the walls on the water. This force, plus the weight of the faucet and the water, is equal to the net force on the flange. (We must be very careful with our signs, of course.)

When choosing a control volume, you are not limited to the fluid alone. Often it is more convenient to slice the control surface *through* solid objects such as walls, struts, or bolts as illustrated by CV B (the gray control volume) in Fig. 13–8. A control volume may even surround an entire object, like the one shown here. Control volume B is a wise choice because we are not concerned with any details of the flow or even the geometry inside the control volume. For the case of CV B, we assign a net reaction force acting at the portions of the control surface that slice through the flange bolts. Then, the only other things we need to know are the gage pressure of the water at the flange (the inlet to the control volume) and the weights of the water and the faucet assembly. The pressure everywhere else along the control surface is atmospheric (zero gage pressure) and cancels out.

## 13–4 ▪ THE REYNOLDS TRANSPORT THEOREM

In thermodynamics and solid mechanics we often work with a *system* (also called a *closed system*), defined as a *quantity of matter of fixed identity*. In fluid dynamics, it is more common to work with a *control volume* (also called an *open system*), defined as a *region in space chosen for study*. The size and shape of a system may change during a process, but no mass crosses its boundaries. A control volume, on the other hand, allows mass to flow in or out across its boundaries, which are called the **control surface**. A control volume may also move and deform during a process, but many real-world applications involve fixed, nondeformable control volumes.

Figure 13–9 illustrates both a system and a control volume for the case of deodorant being sprayed from a spray can. When analyzing the spraying process, a natural choice for our analysis is either the moving, deforming fluid (a system) or the volume bounded by the inner surfaces of the can (a control volume). These two choices are identical before the deodorant is sprayed. When some contents of the can are discharged, the system approach considers the discharged mass as part of the system and tracks it (a difficult job indeed); thus the mass of the system remains constant. Conceptually, this is equivalent to attaching a flat balloon to the nozzle of the can and letting the spray inflate the balloon. The inner surface of the balloon now becomes part of the boundary of the system. The control volume approach, however, is not concerned at all with the deodorant that has escaped the can (other than its properties at the exit), and thus the mass of the control volume decreases during this process while its volume remains constant. Therefore, the system approach treats the spraying process as an expansion of the system's volume, whereas the control volume approach considers it as a fluid discharge through the control surface of the fixed control volume.

Most principles of fluid mechanics are adopted from solid mechanics, where the physical laws dealing with the time rates of change of extensive properties are expressed for systems. In fluid mechanics, it is usually more convenient to work with control volumes, and thus there is a need to relate the changes in a control volume to the changes in a system. The relationship between the time rates of change of an extensive property for a system and for a control volume is expressed by the **Reynolds transport theorem (RTT)**, which provides the link between the system and control volume approaches (Fig. 13–10). RTT is named after the English engineer, Osborne Reynolds (1842–1912), who did much to advance its application in fluid mechanics.

The general form of the Reynolds transport theorem can be derived by considering a system with an arbitrary shape and arbitrary interactions, but the derivation is rather involved. To help you grasp the fundamental meaning of the theorem, we derive it first in a straightforward manner using a simple geometry and then generalize the results.

Consider flow from left to right through a diverging (expanding) portion of a flow field as sketched in Fig. 13–11. The upper and lower bounds of the fluid under consideration are *streamlines* of the flow, and we assume uniform flow through any cross section between these two streamlines. We choose the control volume to be fixed between sections (1) and (2) of the flow field. Both (1) and (2) are normal to the direction of flow. At some initial time $t$, the system coincides with the control volume, and thus the system and control volume are identical (the blue-shaded region in Fig. 13–11). During time interval $\Delta t$, the system moves in the flow direction at uniform speeds $V_1$ at section (1) and $V_2$ at section (2). The system at this later time is indicated by the hatched region. The region uncovered by the system during this motion is designated as section I (part of the CV), and the new region covered by the system is designated as section II (not part of the CV). Therefore, at time $t + \Delta t$, the system consists of the same fluid, but it occupies the region CV − I + II. The control volume is fixed in space, and thus it remains as the shaded region marked CV at all times.

Let $B$ represent any **extensive property** (such as mass, energy, or momentum), and let $b = B/m$ represent the corresponding **intensive property**. Noting that extensive properties are additive, the extensive property $B$ of the system at times $t$ and $t + \Delta t$ is expressed as

$$B_{\text{sys},t} = B_{\text{CV},t} \quad \text{(the system and CV coincide at time } t\text{)}$$

$$B_{\text{sys},t+\Delta t} = B_{\text{CV},t+\Delta t} - B_{\text{I},t+\Delta t} + B_{\text{II},t+\Delta t}$$

Subtracting the first equation from the second one and dividing by $\Delta t$ gives

$$\frac{B_{\text{sys},t+\Delta t} - B_{\text{sys},t}}{\Delta t} = \frac{B_{\text{CV},t+\Delta t} - B_{\text{CV},t}}{\Delta t} - \frac{B_{\text{I},t+\Delta t}}{\Delta t} + \frac{B_{\text{II},t+\Delta t}}{\Delta t}$$

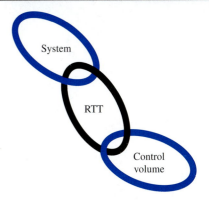

**FIGURE 13–10**

The *Reynolds transport theorem* (RTT) provides a link between the system approach and the control volume approach.

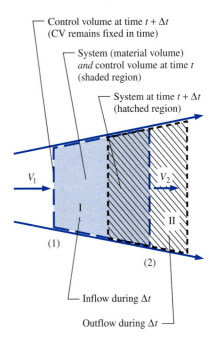

Control volume at time $t + \Delta t$ (CV remains fixed in time)

System (material volume) *and* control volume at time $t$ (shaded region)

System at time $t + \Delta t$ (hatched region)

Inflow during $\Delta t$

Outflow during $\Delta t$

At time $t$: Sys = CV
At time $t + \Delta t$: Sys = CV − I + II

**FIGURE 13–11**

A moving *system* (hatched region) and a fixed *control volume* (shaded region) in a diverging portion of a flow field at times $t$ and $t + \Delta t$. The upper and lower bounds are streamlines of the flow.

Taking the limit as $\Delta t \rightarrow 0$, and using the definition of derivative, we get

$$\frac{dB_{sys}}{dt} = \frac{dB_{CV}}{dt} - \dot{B}_{in} + \dot{B}_{out} \qquad (13\text{–}9)$$

or

$$\frac{dB_{sys}}{dt} = \frac{dB_{CV}}{dt} - b_1\rho_1 V_1 A_1 + b_2\rho_2 V_2 A_2$$

since

$$B_{I,\,t+\Delta t} = b_1 m_{I,\,t+\Delta t} = b_1\rho_1 V_{I,\,t+\Delta t} = b_1\rho_1 V_1\,\Delta t\,A_1$$
$$B_{II,\,t+\Delta t} = b_2 m_{II,\,t+\Delta t} = b_2\rho_2 V_{II,\,t+\Delta t} = b_2\rho_2 V_2\,\Delta t\,A_2$$

and

$$\dot{B}_{in} = \dot{B}_I = \lim_{\Delta t \to 0}\frac{B_{I,\,t+\Delta t}}{\Delta t} = \lim_{\Delta t \to 0}\frac{b_1\rho_1 V_1\,\Delta t\,A_1}{\Delta t} = b_1\rho_1 V_1 A_1$$
$$\dot{B}_{out} = \dot{B}_{II} = \lim_{\Delta t \to 0}\frac{B_{II,\,t+\Delta t}}{\Delta t} = \lim_{\Delta t \to 0}\frac{b_2\rho_2 V_2\,\Delta t\,A_2}{\Delta t} = b_2\rho_2 V_2 A_2$$

where $A_1$ and $A_2$ are the cross-sectional areas at locations 1 and 2. Equation 13–9 states that *the time rate of change of the property B of the system is equal to the time rate of change of B of the control volume plus the net flux of B out of the control volume by mass crossing the control surface.* This is the desired relation since it relates the change of a property of a system to the change of that property for a control volume. Note that Eq. 13–9 applies at any instant in time, where it is assumed that the system and the control volume occupy the same space at that particular instant in time.

The influx $\dot{B}_{in}$ and outflux $\dot{B}_{out}$ of the property $B$ in this case are easy to determine since there is only one inlet and one outlet, and the velocities are approximately normal to the surfaces at sections (1) and (2). In general, however, we may have several inlet and outlet ports, and the velocity may not be normal to the control surface at the point of entry. Also, the velocity may not be uniform. To generalize the process, we consider a differential surface area $dA$ on the control surface and denote its **unit outer normal** by $\vec{n}$. The flow rate of property $b$ through $dA$ is $\rho b \vec{V} \cdot \vec{n}\, dA$ since the dot product $\vec{V} \cdot \vec{n}$ gives the normal component of the velocity. Then the net rate of outflow through the entire control surface is determined by integration to be (Fig. 13–12)

$$\dot{B}_{net} = \dot{B}_{out} - \dot{B}_{in} = \int_{CS}\rho b \vec{V}\cdot\vec{n}\, dA \quad \text{(inflow if negative)} \qquad (13\text{–}10)$$

An important aspect of this relation is that it automatically subtracts the inflow from the outflow, as explained next. The dot product of the velocity vector at a point on the control surface and the outer normal at that point is $\vec{V}\cdot\vec{n} = |\vec{V}||\vec{n}|\cos\theta = |\vec{V}|\cos\theta$, where $\theta$ is the angle between the velocity vector and the outer normal, as shown in Fig. 13–13. For $\theta < 90°$, $\cos\theta > 0$ and thus $\vec{V}\cdot\vec{n} > 0$ for outflow of mass from the control volume, and for $\theta > 90°$, $\cos\theta < 0$ and thus $\vec{V}\cdot\vec{n} < 0$ for inflow of mass into the control volume. Therefore, the differential quantity $\rho b \vec{V}\cdot\vec{n}\, dA$ is positive for mass flowing out of the control volume, and negative for mass flowing into the control volume, and its integral over the entire control surface gives the rate of net outflow of the property $B$ by mass.

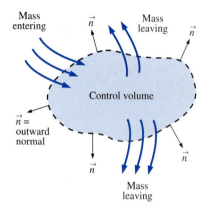

$$\dot{B}_{net} = \dot{B}_{out} - \dot{B}_{in} = \int_{CS}\rho b\vec{V}\cdot\vec{n}\,dA$$

**FIGURE 13–12**

The integral of $b\rho\vec{V}\cdot\vec{n}\, dA$ over the control surface gives the net amount of the property $B$ flowing out of the control volume (into the control volume if it is negative) per unit time.

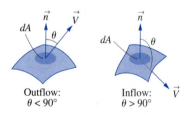

$$\vec{V}\cdot\vec{n} = |\vec{V}||\vec{n}|\cos\theta = V\cos\theta$$
If $\theta < 90°$, then $\cos\theta > 0$ (outflow).
If $\theta > 90°$, then $\cos\theta < 0$ (inflow).
If $\theta = 90°$, then $\cos\theta = 0$ (no flow).

**FIGURE 13–13**

Outflow and inflow of mass across the differential area of a control surface.

The properties within the control volume may vary with position, in general. In such a case, the total amount of property $B$ within the control volume must be determined by integration:

$$B_{CV} = \int_{CV} \rho b \, dV \qquad (13\text{--}11)$$

The term $dB_{CV}/dt$ in Eq. 13–9 is thus equal to $\dfrac{d}{dt} \displaystyle\int_{CV} \rho b \, dV$, and represents the time rate of change of the property $B$ content of the control volume. A positive value for $dB_{CV}/dt$ indicates an increase in the $B$ content, and a negative value indicates a decrease. Substituting Eqs. 13–10 and 13–11 into Eq. 13–9 yields the Reynolds transport theorem, also known as the *system-to-control-volume transformation* for a fixed control volume:

*RTT, fixed CV:* 
$$\frac{dB_{sys}}{dt} = \frac{d}{dt} \int_{CV} \rho b \, dV + \int_{CS} \rho b \vec{V} \cdot \vec{n} \, dA \qquad (13\text{--}12)$$

Since the control volume is not moving or deforming with time, the time derivative on the right-hand side can be moved inside the integral, since the domain of integration does not change with time. (In other words, it is irrelevant whether we differentiate or integrate first.) But the time derivative in that case must be expressed as a *partial* derivative ($\partial/\partial t$) since density and the quantity $b$ may depend not only on time, but also on the position within the control volume. Thus, an alternate form of the Reynolds transport theorem for a fixed control volume is

*Alternate RTT, fixed CV:* 
$$\frac{dB_{sys}}{dt} = \int_{CV} \frac{\partial}{\partial t} (\rho b) \, dV + \int_{CS} \rho b \vec{V} \cdot \vec{n} \, dA \qquad (13\text{--}13)$$

It turns out that Eq. 13–13 is also valid for the most general case of a moving and/or deforming control volume, provided that velocity vector $\vec{V}$ is an *absolute* velocity (as viewed from a fixed reference frame).

Next we consider yet *another* alternative form of the RTT. Equation 13–12 was derived for a *fixed* control volume. However, many practical systems such as turbine and propeller blades involve nonfixed control volumes. Fortunately, Eq. 13–12 is also valid for *moving* and/or *deforming* control volumes provided that the absolute fluid velocity $\vec{V}$ in the last term is replaced by the **relative velocity** $\vec{V}_r$,

*Relative velocity:* 
$$\vec{V}_r = \vec{V} - \vec{V}_{CS} \qquad (13\text{--}14)$$

where $\vec{V}_{CS}$ is the local velocity of the control surfsace (Fig. 13–14). The most general form of the Reynolds transport theorem is thus

*RTT, nonfixed CV:* 
$$\frac{dB_{sys}}{dt} = \frac{d}{dt} \int_{CV} \rho b \, dV + \int_{CS} \rho b \vec{V}_r \cdot \vec{n} \, dA \qquad (13\text{--}15)$$

Note that for a control volume that moves and/or deforms with time, the time derivative is applied *after* integration in Eq. 13–15. As a simple example of a moving control volume, consider a toy car moving at a constant absolute velocity $\vec{V}_{car} = 10$ km/h to the right. A high-speed jet of water (absolute velocity $= \vec{V}_{jet} = 25$ km/h to the right) strikes the back of the

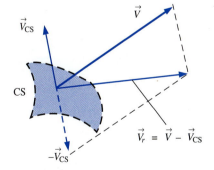

**FIGURE 13–14**

*Relative velocity* crossing a control surface is found by vector addition of the absolute velocity of the fluid and the negative of the local velocity of the control surface.

Absolute reference frame:

Relative reference frame:

**FIGURE 13–15**
Reynolds transport theorem applied to a control volume moving at constant velocity.

**FIGURE 13–16**
An example control volume in which there is one well-defined inlet (1) and two well-defined outlets (2 and 3). In such cases, the control surface integral in the RTT can be more conveniently written in terms of the average values of fluid properties crossing each inlet and outlet.

car and propels it (Fig. 13–15). If we draw a control volume around the car, the relative velocity is $\vec{V}_r = 25 - 10 = 15$ km/h to the right. This represents the velocity at which an observer moving with the control volume (moving with the car) would observe the fluid crossing the control surface. In other words, $\vec{V}_r$ is the fluid velocity expressed relative to a coordinate system moving *with* the control volume.

Finally, by application of the Leibniz theorem, it can be shown that the Reynolds transport theorem for a general moving and/or deforming control volume (Eq. 13–15) is equivalent to the form given by Eq. 13–13, which is repeated here:

*Alternate RTT, nonfixed CV:* $\quad \dfrac{dB_{sys}}{dt} = \displaystyle\int_{CV} \dfrac{\partial}{\partial t}(\rho b)\, dV + \int_{CS} \rho b \vec{V} \cdot \vec{n}\, dA \quad$ **(13–16)**

In contrast to Eq. 13–15, the velocity vector $\vec{V}$ in Eq. 13–16 must be taken as the *absolute* velocity (as viewed from a fixed reference frame) in order to apply to a nonfixed control volume.

During steady flow, the amount of the property $B$ within the control volume remains constant in time, and thus the time derivative in Eq. 13–15 becomes zero. Then the Reynolds transport theorem reduces to

*RTT, steady flow:* $\quad \dfrac{dB_{sys}}{dt} = \displaystyle\int_{CS} \rho b \vec{V}_r \cdot \vec{n}\, dA \quad$ **(13–17)**

Note that unlike the control volume, the property $B$ content of the system may still change with time during a steady process. But in this case the change must be equal to the net property transported by mass across the control surface (an advective rather than an unsteady effect).

In most practical engineering applications of the RTT, fluid crosses the boundary of the control volume at a finite number of well-defined inlets and outlets (Fig. 13–16). In such cases, it is convenient to cut the control surface directly across each inlet and outlet and replace the surface integral in Eq. 13–15 with approximate algebraic expressions at each inlet and outlet based on the *average* values of fluid properties crossing the boundary. We define $\rho_{avg}$, $b_{avg}$, and $V_{r,avg}$ as the average values of $\rho$, $b$, and $V_r$, respectively, across an inlet or outlet of cross-sectional area $A$ [e.g., $b_{avg} = \dfrac{1}{A}\displaystyle\int_{A} b\, dA$]. The surface integrals in the RTT (Eq. 13–15), when applied over an inlet or outlet of cross-sectional area $A$, are then *approximated* by pulling property $b$ out of the surface integral and replacing it with its average. This yields

$$\int_{A} \rho b \vec{V}_r \cdot \vec{n}\, dA \cong b_{avg} \int_{A} \rho \vec{V}_r \cdot \vec{n}\, dA = b_{avg}\dot{m}_r$$

where $\dot{m}_r$ is the mass flow rate through the inlet or outlet relative to the (moving) control surface. The approximation in this equation is exact when property $b$ is uniform over cross-sectional area $A$. Equation 13–15 thus becomes

$$\dfrac{dB_{sys}}{dt} = \dfrac{d}{dt}\int_{CV} \rho b\, dV + \underbrace{\sum_{out} \dot{m}_r b_{avg}}_{\text{for each outlet}} - \underbrace{\sum_{in} \dot{m}_r b_{avg}}_{\text{for each inlet}} \quad \textbf{(13–18)}$$

In some applications, we may wish to rewrite Eq. 13–18 in terms of volume (rather than mass) flow rate. In such cases, we make a further approximation that $\dot{m}_r \approx \rho_{avg}\dot{V}_r = \rho_{avg}V_{r,avg}A$. This approximation is exact when fluid density $\rho$ is uniform over $A$. Equation 13–18 then reduces to

*Approximate RTT for well-defined inlets and outlets*:

$$\frac{dB_{sys}}{dt} = \frac{d}{dt}\int_{CV}\rho b\, dV + \sum_{out}\underbrace{\rho_{avg}b_{avg}V_{r,\,avg}A}_{\text{for each outlet}} - \sum_{in}\underbrace{\rho_{avg}b_{avg}V_{r,avg}A}_{\text{for each inlet}} \quad \textbf{(13–19)}$$

Note that these approximations simplify the analysis greatly but may not always be accurate, especially in cases where the velocity distribution across the inlet or outlet is not very uniform (e.g., pipe flows; Fig. 13–16). In particular, the control surface integral of Eq. 13–16 becomes *nonlinear* when property $b$ contains a velocity term (e.g., when applying RTT to the linear momentum equation, $b = \vec{V}$), and the approximation of Eq. 13–19 leads to errors. Fortunately we can eliminate the errors by including *correction factors* in Eq. 13–19.

Equations 13–18 and 13–19 apply to fixed *or* moving control volumes, but as discussed previously, the *relative velocity* must be used for the case of a nonfixed control volume. In Eq. 13–18 for example, the mass flow rate $\dot{m}_r$ is relative to the (moving) control surface, hence the $r$ subscript.

## An Application: Conservation of Mass

The general conservation of mass relation for a control volume can also be derived using the Reynolds transport theorem by taking the property $B$ to be the mass, $m$. Then we have $b = 1$ since dividing the mass by mass to get the property per unit mass gives uity. Also, the mass of a system is constant, and thus its time derivative is zero. That is, $dm_{sys}/dt = 0$. Then the Reynolds transport equation in this case reduces to (Fig. 13–17)

*General conservation of mass:* $\quad \dfrac{d}{dt}\int_{CV}\rho b V + \int_{CV}\rho(\vec{V}\cdot\vec{n})\, dA = 0 \quad$ **(13–20)**

It states that *the time rate of change of mass within the control volume plus the net mass flow rate through the control surface is equal to zero.*

This demonstrates that the Reynolds transport theorem is a very powerful tool, and we can use it with confidence.

Splitting the surface integral in Eq. 13–20 in to two parts—one for the outgoing flow streams (positive) and one for the incoming streams (negative)—the general conservation of mass relation can also be expressed as

$$\frac{d}{dt}\int_{CV}\rho d\,V + \sum_{out}\int_{A}\rho^{\circ}V_n\, dA - \sum_{in}\int_{A}\rho^{\circ}V_n\, dA = 0 \quad \textbf{(13–21)}$$

Using the definition of mass flow rate, it can also be expressed as

$$\frac{d}{dt}\int_{CV}\rho dV = \sum_{out}\dot{m} - \sum_{in}\dot{m} \quad \text{or} \quad \frac{dm_{CV}}{dt} = \sum_{out}\dot{m} - \sum_{in}\dot{m} \quad \textbf{(13–22)}$$

where the summation signs are used to emphasize that *all* the inlets and outlets are to be considered.

## 13–5 · THE LINEAR MOMENTUM EQUATION

Newton's second law for a system of mass $m$ subjected to net force $\Sigma\vec{F}$ is expressed as

$$\sum\vec{F} = m\vec{a} = m\frac{d\vec{V}}{dt} = \frac{d}{dt}(m\vec{V}) \quad \textbf{(13–23)}$$

$$\frac{dB_{sys}}{dt}\; \frac{d}{dt}\int_{CV}\rho b dV + \int_{CS}\rho b(\vec{V}\cdot\vec{n})\, dA$$

$$B = m \qquad b = 1 \qquad b = 1$$
$$\downarrow \qquad\qquad \downarrow \qquad\qquad \downarrow$$

$$\frac{dm_{sys}}{dt}\; \frac{d}{dt}\int_{CV}\rho dV + \int_{CS}\rho(\vec{V}\cdot\vec{n})\, dA$$

**FIGURE 13–17**

The conservation of mass equation is obtained by replacing $B$ in the Reynolds transport theorem by mass $m$, and $b$ by 1 ($m$ per unit mass = $m/m = 1$)

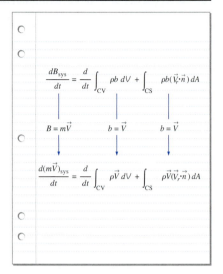

**FIGURE 13–18**

The linear momentum equation is obtained by replacing $B$ in the Reynolds transport theorem by the momentum $m\vec{V}$, and $b$ by the momentum per unit mass $\vec{V}$.

where $m\vec{V}$ is the **linear momentum** of the system. Noting that both the density and velocity may change from point to point within the system, Newton's second law can be expressed more generally as

$$\sum \vec{F} = \frac{d}{dt} \int_{sys} \rho \vec{V} \, dV \tag{13-24}$$

where $\rho \vec{V} \, dV$ is the momentum of a differential element $dV$, which has mass $\delta m = \rho dV$. Therefore, Newton's second law can be stated as *the sum of all external forces acting on a system is equal to the time rate of change of linear momentum of the system.* This statement is valid for a coordinate system that is at rest or moves with a constant velocity, called an *inertial coordinate system* or *inertial reference frame*. Accelerating systems such as aircraft during takeoff are best analyzed using noninertial (or accelerating) coordinate systems fixed to the aircraft. Note that Eq. 13–24 is a vector relation, and thus the quantities $\vec{F}$ and $\vec{V}$ have direction as well as magnitude.

Equation 13–24 is for a given mass of a solid or fluid and is of limited use in fluid mechanics since most flow systems are analyzed using control volumes. The *Reynolds transport theorem* developed in Section 13–4 provides the necessary tools to shift from the system formulation to the control volume formulation. Setting $b = \vec{V}$ and thus $B = m\vec{V}$, the Reynolds transport theorem is expressed for linear momentum as (Fig. 13–18)

$$\frac{d(m\vec{V})_{sys}}{dt} = \frac{d}{dt} \int_{CV} \rho \vec{V} \, dV + \int_{CS} \rho \vec{V} \, (\vec{V}_r \cdot \vec{n}) \, dA \tag{13-25}$$

The left-hand side of this equation is, from Eq. 13–23, equal to $\sum \vec{F}$. Substituting, the general form of the linear momentum equation that applies to fixed, moving, or deforming control volumes is

*General:* $$\sum \vec{F} = \frac{d}{dt} \int_{CV} \rho \vec{V} \, dV + \int_{CS} \rho \vec{V} (\vec{V}_r \cdot \vec{n}) \, dA \tag{13-26}$$

which is stated in words as

$$\begin{pmatrix} \text{The sum of all} \\ \text{external forces} \\ \text{acting on a CV} \end{pmatrix} = \begin{pmatrix} \text{The time rate of change} \\ \text{of the linear momentum} \\ \text{of the contents of the CV} \end{pmatrix} + \begin{pmatrix} \text{The net flow rate of} \\ \text{linear momentum out of the} \\ \text{control surface by mass flow} \end{pmatrix}$$

Here $\vec{V}_r = \vec{V} - \vec{V}_{CS}$ is the fluid velocity relative to the control surface (for use in mass flow rate calculations at all locations where the fluid crosses the control surface), and $\vec{V}$ is the fluid velocity as viewed from an inertial reference frame. The product $\rho(\vec{V}_r \cdot \vec{n}) \, dA$ represents the mass flow rate through area element $dA$ into or out of the control volume.

For a fixed control volume (no motion or deformation of control volume), $\vec{V}_r = \vec{V}$ and the linear momentum equation becomes

*Fixed CV:* $$\sum \vec{F} = \frac{d}{dt} \int_{CV} \rho \vec{V} \, dV + \int_{CS} \rho \vec{V} (\vec{V} \cdot \vec{n}) \, dA \tag{13-27}$$

Note that the momentum equation is a *vector equation*, and thus each term should be treated as a vector. Also, the components of this equation can be resolved along orthogonal coordinates (such as $x$, $y$, and $z$ in the Cartesian coordinate system) for convenience. The sum of forces $\sum \vec{F}$ in most cases consists of weights, pressure forces, and reaction forces (Fig. 13–19). The momentum equation is commonly used to calculate the forces (usually on support systems or connectors) induced by the flow.

## Special Cases

Most momentum problems considered in this text are steady. During *steady flow*, the amount of momentum within the control volume remains constant, and thus the time rate of change of linear momentum of the contents of the control volume (the second term of Eq. 13–26) is zero. Thus,

*Steady flow*:
$$\sum \vec{F} = \int_{CS} \rho \vec{V} (\vec{V}_r \cdot \vec{n}) \, dA \qquad \textbf{(13–28)}$$

For a case in which a non-deforming control volume moves at constant velocity (an inertial reference frame), the *first* $\vec{V}$ in Eq. 13-28 may *also* be taken relative to the moving control surface.

While Eq. 13–27 is exact for fixed control volumes, it is not always convenient when solving practical engineering problems because of the integrals. Instead, as we did for conservation of mass, we would like to rewrite Eq. 13–27 in terms of average velocities and mass flow rates through inlets and outlets. In other words, our desire is to rewrite the equation in *algebraic* rather than *integral* form. In many practical applications, fluid crosses the boundaries of the control volume at one or more inlets and one or more outlets, and carries with it some momentum into or out of the control volume. For simplicity, we always draw our control surface such that it slices normal to the inflow or outflow velocity at each such inlet or outlet (Fig. 13–20).

The mass flow rate $\dot{m}$ into or out of the control volume across an inlet or outlet at which $\rho$ is nearly constant is

*Mass flow rate across an inlet or outlet*:
$$\dot{m} = \int_{A_c} \rho(\vec{V} \cdot \vec{n}) \, dA_c = \rho V_{avg} A_c \qquad \textbf{(13–29)}$$

Comparing Eq. 13–29 to Eq. 13–27, we notice an extra velocity in the control surface integral of Eq. 13–27. If $\vec{V}$ were uniform ($\vec{V} = \vec{V}_{avg}$) across the inlet or outlet, we could simply take it outside the integral. Then we could write the rate of inflow or outflow of momentum through the inlet or outlet in simple algebraic form,

*Momentum flow rate across a uniform inlet or outlet*:
$$\int_{A_c} \rho \vec{V}(\vec{V} \cdot \vec{n}) \, dA_c = \rho V_{avg} A_c \vec{V}_{avg} = \dot{m} \vec{V}_{avg} \qquad \textbf{(13–30)}$$

The uniform flow approximation is reasonable at some inlets and outlets, e.g., the well-rounded entrance to a pipe, the flow at the entrance to a wind tunnel test section, and a slice through a water jet moving at nearly uniform speed through air (Fig. 13–21). At each such inlet or outlet, Eq. 13–30 can be applied directly.

## Momentum-Flux Correction Factor, β

Unfortunately, the velocity across most inlets and outlets of practical engineering interest is *not* uniform. Nevertheless, it turns out that we can still convert the control surface integral of Eq. 13–27 into algebraic form, but a

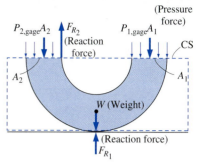

An 180° elbow supported by the ground

**FIGURE 13–19**

In most flow systems, the sum of forces $\sum \vec{F}$ consists of weights, pressure forces, and reaction forces. Gage pressures are used here since atmospheric pressure cancels out on all sides of the control surface.

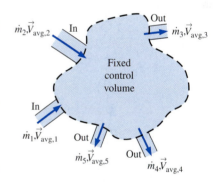

**FIGURE 13–20**

In a typical engineering problem, the control volume may contain many inlets and outlets; at each inlet or outlet we define the mass flow rate $\dot{m}$ and the average velocity $\vec{V}_{avg}$.

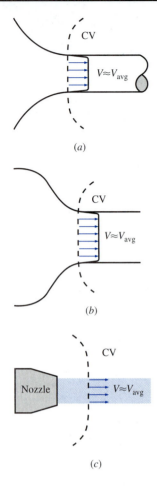

**FIGURE 13–21**

Examples of inlets or outlets in which the uniform flow approximation is reasonable: (a) the well-rounded entrance to a pipe, (b) the entrance to a wind tunnel test section, and (c) a slice through a free water jet in air.

dimensionless correction factor $\beta$, called the **momentum-flux correction factor**, is required, as first shown by the French scientist Joseph Boussinesq (1842–1929). The algebraic form of Eq. 13–27 for a fixed control volume is then written as

$$\sum \vec{F} = \frac{d}{dt} \int_{CV} \rho \vec{V} \, dV + \sum_{out} \beta \dot{m} \vec{V}_{avg} - \sum_{in} \beta \dot{m} \vec{V}_{avg} \tag{13–31}$$

where a unique value of momentum-flux correction factor is applied to each inlet and outlet in the control surface. Note that $\beta = 1$ *for the case of uniform flow* over an inlet or outlet, as in Fig. 13–24. For the general case, we define $\beta$ such that the integral form of the momentum flux into or out of the control surface at an inlet or outlet of cross-sectional area $A_c$ can be expressed in terms of mass flow rate $\dot{m}$ through the inlet or outlet and average velocity $\vec{V}_{avg}$ through the inlet or outlet,

*Momentum flux across an inlet or outlet:*  $\displaystyle \int_{A_c} \rho \vec{V} (\vec{V} \cdot \vec{n}) \, dA_c = \beta \dot{m} \vec{V}_{avg}$  (13–32)

For the case in which density is uniform over the inlet or outlet and $\vec{V}$ is in the same direction as $\vec{V}_{avg}$ over the inlet or outlet, we solve Eq. 13–32 for $\beta$,

$$\beta = \frac{\displaystyle \int_{A_c} \rho V (\vec{V} \cdot \vec{n}) \, dA_c}{\dot{m} V_{avg}} = \frac{\displaystyle \int_{A_c} \rho V (\vec{V} \cdot \vec{n}) \, dA_c}{\rho V_{avg} A_c V_{avg}} \tag{13–33}$$

where we have substituted $\rho V_{avg} A_c$ for $\dot{m}$ in the denominator. The densities cancel and since $V_{avg}$ is constant, it can be brought inside the integral. Furthermore, if the control surface slices normal to the inlet or outlet area, $(\vec{V} \cdot \vec{n}) \, dA_c = V \, dA_c$. Then, Eq. 13–33 simplifies to

*Momentum-flux correction factor:*  $\displaystyle \beta = \frac{1}{A_c} \int_{A_c} \left( \frac{V}{V_{avg}} \right)^2 dA_c$  (13–34)

It may be shown that $\beta$ is always greater than or equal to unity.

---

**EXAMPLE 13–1**  **Momentum-Flux Correction Factor for Laminar Pipe Flow**

Consider laminar flow through a very long straight section of round pipe. It is shown in Chap. 14 that the velocity profile through a cross-sectional area of the pipe is parabolic (Fig. 13–22), with the axial velocity component given by

$$V = 2V_{avg} \left( 1 - \frac{r^2}{R^2} \right) \tag{1}$$

where $R$ is the radius of the inner wall of the pipe and $V_{avg}$ is the average velocity. Calculate the momentum-flux correction factor through a cross section of the pipe for the case in which the pipe flow represents an outlet of the control volume, as sketched in Fig. 13–22.

**Solution** For a given velocity distribution we are to calculate the momentum-flux correction factor.

**Assumptions** **1** The flow is incompressible and steady. **2** The control volume slices through the pipe normal to the pipe axis, as sketched in Fig. 13–22.

**Analysis** We substitute the given velocity profile for $V$ in Eq. 13–34 and integrate, noting that $dA_c = 2\pi r\, dr$,

$$\beta = \frac{1}{A_c}\int_{A_c}\left(\frac{V}{V_{avg}}\right)^2 dA_c = \frac{4}{\pi R^2}\int_0^R\left(1-\frac{r^2}{R^2}\right)^2 2\pi r\, dr \tag{2}$$

Defining a new integration variable $y = 1 - r^2/R^2$ and thus $dy = -2r\, dr/R^2$ (also, $y = 1$ at $r = 0$, and $y = 0$ at $r = R$) and performing the integration, the momentum-flux correction factor for fully developed laminar flow becomes

*Laminar flow:* $$\beta = -4\int_1^0 y^2\, dy = -4\left[\frac{y^3}{3}\right]_1^0 = \frac{4}{3} \tag{3}$$

**Discussion** We have calculated $\beta$ for an outlet, but the same result would have been obtained if we had considered the cross section of the pipe as an *inlet* to the control volume.

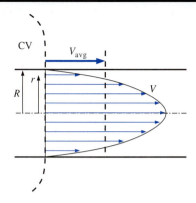

**FIGURE 13–22**

Velocity profile over a cross section of a pipe in which the flow is fully developed and laminar.

From Example 13–1 we see that $\beta$ is not very close to unity for fully developed laminar pipe flow, and ignoring $\beta$ could potentially lead to significant error. If we were to perform the same kind of integration as in Example 13–1 but for fully developed *turbulent* rather than laminar pipe flow, we would find that $\beta$ ranges from about 1.01 to 1.04. Since these values are so close to unity, many practicing engineers completely disregard the momentum-flux correction factor. While the neglect of $\beta$ in turbulent flow calculations may have an insignificant effect on the final results, it is wise to keep it in our equations. Doing so not only improves the accuracy of our calculations, but reminds us to include the momentum-flux correction factor when solving laminar flow control volume problems.

> For turbulent flow $\beta$ may have an insignificant effect at inlets and outlets, but for laminar flow $\beta$ may be important and should not be neglected. It is wise to include $\beta$ in all momentum control volume problems.

## Steady Flow

If the flow is also *steady*, the time derivative term in Eq. 13–31 vanishes and we are left with

*Steady linear momentum equation:* $$\sum \vec{F} = \sum_{out} \beta \dot{m}\vec{V} - \sum_{in} \beta \dot{m}\vec{V} \tag{13–35}$$

where we have dropped the subscript "avg" from average velocity. Equation 13–35 states that *the net force acting on the control volume during steady flow is equal to the difference between the rates of outgoing and incoming momentum flows.* This statement is illustrated in Fig. 13–23. It can also be expressed for any direction, since Eq. 13–35 is a vector equation.

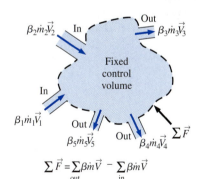

$$\sum \vec{F} = \sum_{out} \beta \dot{m}\vec{V} - \sum_{in} \beta \dot{m}\vec{V}$$

**FIGURE 13–23**

The net force acting on the control volume during steady flow is equal to the difference between the outgoing and the incoming momentum fluxes.

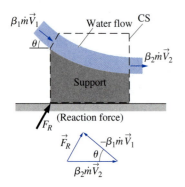

**FIGURE 13–24**

A control volume with only one inlet and one outlet.

Note: $\vec{V}_2 \neq \vec{V}_1$ even if $|\vec{V}_2| = |\vec{V}_1|$

**FIGURE 13–25**

The determination by vector addition of the reaction force on the support caused by a change of direction of water.

## Steady Flow with One Inlet and One Outlet

Many practical engineering problems involve just one inlet and one outlet (Fig. 13–24). The mass flow rate for such **single-stream systems** remains constant, and Eq. 13–35 reduces to

*One inlet and one outlet:*   $$\sum \vec{F} = \dot{m}\,(\beta_2\vec{V}_2 - \beta_1\vec{V}_1) \qquad (13\text{–}36)$$

where we have adopted the usual convention that subscript 1 implies the inlet and subscript 2 the outlet, and $\vec{V}_1$ and $\vec{V}_2$ denote the *average* velocities across the inlet and outlet, respectively.

We emphasize again that all the preceding relations are *vector* equations, and thus all the additions and subtractions are *vector* additions and subtractions. Recall that subtracting a vector is equivalent to adding it after reversing its direction (Fig. 13–25). When writing the momentum equation for a specified coordinate direction (such as the *x*-axis), we use the projections of the vectors on that axis. For example, Eq. 13–36 is written along the *x*-coordinate as

*Along x-coordinate:*   $$\sum F_x = \dot{m}\,(\beta_2 V_{2,\,x} - \beta_1 V_{1,\,x}) \qquad (13\text{–}37)$$

where $\sum F_x$ is the vector sum of the *x*-components of the forces, and $V_{2,\,x}$ and $V_{1,\,x}$ are the *x*-components of the outlet and inlet velocities of the fluid stream, respectively. The force or velocity components in the positive *x*-direction are positive quantities, and those in the negative *x*-direction are negative quantities. Also, it is good practice to take the direction of unknown forces in the positive directions (unless the problem is very straightforward). A negative value obtained for an unknown force indicates that the assumed direction is wrong and should be reversed.

## Flow with No External Forces

An interesting situation arises when there are no external forces (such as weight, pressure, and reaction forces) acting on the body in the direction of motion—a common situation for space vehicles and satellites. For a control volume with multiple inlets and outlets, Eq. 13–31 reduces in this case to

*No external forces:*   $$0 = \frac{d(m\vec{V})_{\text{CV}}}{dt} + \sum_{\text{out}} \beta \dot{m}\vec{V} - \sum_{\text{in}} \beta \dot{m}\vec{V} \qquad (13\text{–}38)$$

This is an expression of the conservation of momentum principle, which is stated in words as *in the absence of external forces, the rate of change of the momentum of a control volume is equal to the difference between the rates of incoming and outgoing momentum flow rates.*

When the mass $m$ of the control volume remains nearly constant, the first term of Eq. 13–38 becomes simply mass times acceleration, since

$$\frac{d(m\vec{V})_{\text{CV}}}{dt} = m_{\text{CV}}\frac{d\vec{V}_{\text{CV}}}{dt} = (m\vec{a})_{\text{CV}} = m_{\text{CV}}\vec{a}$$

Therefore, the control volume in this case can be treated as a solid body (a fixed-mass system) with a net thrusting force (or just **thrust**) of

*Thrust:*   $$\vec{F}_{\text{thrust}} = \vec{F}_{\text{body}} = m_{\text{body}}\vec{a} = \sum_{\text{in}} \beta \dot{m}\vec{V} - \sum_{\text{out}} \beta \dot{m}\vec{V} \qquad (13\text{–}39)$$

acting on the body. In Eq 13–39, fluid velocities are relative to an inertial reference frame—that is, a coordinate system that is fixed in space or is moving uniformly at constant velocity on a straight path. When analyzing the motion of bodies moving at constant velocity on a straight path, it is convenient to choose an inertial reference frame that moves with the body at the same velocity on the same path. In this case the velocities of fluid streams relative to the inertial reference frame are identical to the velocities relative to the moving body, which are much easier to apply. This approach offers great convenience when evaluating the linear acceleration of space vehicles when a rocket is fired (Fig. 13–26).

Recall that thrust is a mechanical force typically generated through the reaction of an accelerating fluid. In the jet engine of an aircraft, for example, hot exhaust gases are accelerated by the action of expansion and outflow of gases through the back of the engine, and a thrusting force is produced by a reaction in the opposite direction. The generation of thrust is based on Newton's third law of motion, which states that *for every action at a point there is an equal and opposite reaction*. In the case of a jet engine, if the engine exerts a force on exhaust gases, then the exhaust gases exert an equal force on the engine in the opposite direction. That is, the pushing force exerted on the departing gases by the engine is equal to the thrusting force the departing gases exert on the remaining mass of the aircraft in the opposite direction $\vec{F}_{thrust} = -\vec{F}_{push}$. On the free-body diagram of an aircraft, the effect of outgoing exhaust gases is accounted for by the insertion of a force in the opposite direction of motion of the exhaust gases.

**FIGURE 13–26**

The thrust needed to lift the space shuttle is generated by the rocket engines as a result of momentum change of the fuel as it is accelerated from about zero to an exit speed of about 2000 m/s after combustion.

*Courtest of NASA.*

---

### EXAMPLE 13–2    The Force to Hold a Deflector Elbow in Place

A reducing elbow is used to deflect water flow at a rate of 14 kg/s in a horizontal pipe upward 30° while accelerating it (Fig. 13–27). The elbow discharges water into the atmosphere. The cross-sectional area of the elbow is 113 cm² at the inlet and 7 cm² at the outlet. The elevation difference between the centers of the outlet and the inlet is 30 cm. The weight of the elbow and the water in it is considered to be negligible. Determine (a) the gage pressure at the center of the inlet of the elbow and (b) the anchoring force needed to hold the elbow in place.

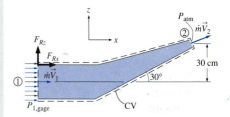

**FIGURE 13–27**

Schematic for Example 13–2.

**Solution**  A reducing elbow deflects water upward and discharges it to the atmosphere. The pressure at the inlet of the elbow and the force needed to hold the elbow in place are to be determined.

*Assumptions*  **1** The flow is steady, and the frictional effects are negligible. **2** The weight of the elbow and the water in it is negligible. **3** The water is discharged to the atmosphere, and thus the gage pressure at the outlet is zero. **4** The flow is turbulent and fully developed at both the inlet and outlet of the control volume, and we take the momentum-flux correction factor to be $\beta = 1.03$ (as a conservative estimate) at both the inlet and the outlet.

*Properties*  We take the density of water to be 1000 kg/m³.

*Analysis*  (a) We take the elbow as the control volume and designate the inlet by 1 and the outlet by 2. We also take the x- and z-coordinates as shown. The continuity equation for this one-inlet, one-outlet, steady-flow system is $\dot{m}_1 = \dot{m}_2 = \dot{m} = 14$ kg/s. Noting that $\dot{m} = \rho A V$, the inlet and outlet velocities of water are

$$V_1 = \frac{\dot{m}}{\rho A_1} = \frac{14 \text{ kg/s}}{(1000 \text{ kg/m}^3)(0.0113 \text{ m}^2)} = 1.24 \text{ m/s}$$

$$V_2 = \frac{\dot{m}}{\rho A_2} = \frac{14 \text{ kg/s}}{(1000 \text{ kg/m}^3)(7 \times 10^{-4} \text{ m}^2)} = 20.0 \text{ m/s}$$

We use the Bernoulli equation (Chap. 12) as a first approximation to calculate the pressure. In Chap. 14 we will learn how to account for frictional losses along the walls. Taking the center of the inlet cross section as the reference level ($z_1 = 0$) and noting that $P_2 = P_{atm}$, the Bernoulli equation for a streamline going through the center of the elbow is expressed as

$$\frac{P_1}{\rho g} + \frac{V_1^2}{2g} + z_1 = \frac{P_2}{\rho g} + \frac{V_2^2}{2g} + z_2$$

$$P_1 - P_2 = \rho g \left( \frac{V_2^2 - V_1^2}{2g} + z_2 - z_1 \right)$$

$$P_1 - P_{atm} = (1000 \text{ kg/m}^3)(9.81 \text{ m/s}^2)$$

$$\times \left( \frac{(20 \text{ m/s})^2 - (1.24 \text{ m/s})^2}{2(9.81 \text{ m/s}^2)} + 0.3 - 0 \right) \left( \frac{1 \text{ kN}}{1000 \text{ kg} \cdot \text{m/s}^2} \right)$$

$$P_{1, \text{gage}} = 202.2 \text{ kN/m}^2 = \textbf{202.2 kPa} \quad \text{(gage)}$$

(b) The momentum equation for steady flow is

$$\sum \vec{F} = \sum_{out} \beta \dot{m} \vec{V} - \sum_{in} \beta \dot{m} \vec{V}$$

We let the x- and z-components of the anchoring force of the elbow be $F_{Rx}$ and $F_{Rz}$, and assume them to be in the positive direction. We also use gage pressure since the atmospheric pressure acts on the entire control surface. Then the momentum equations along the x- and z-axes become

$$F_{Rx} + P_{1, \text{gage}} A_1 = \beta \dot{m} V_2 \cos \theta - \beta \dot{m} V_1$$

$$F_{Rz} = \beta \dot{m} V_2 \sin \theta$$

where we have set $\beta = \beta_1 = \beta_2$. Solving for $F_{Rx}$ and $F_{Rz}$, and substituting the given values,

$$F_{Rx} = \beta \dot{m} (V_2 \cos \theta - V_1) - P_{1, \text{gage}} A_1$$

$$= 1.03(14 \text{ kg/s}) \left[ (20 \cos 30° - 1.24) \text{ m/s} \right] \left( \frac{1 \text{ N}}{1 \text{ kg} \cdot \text{m/s}^2} \right)$$

$$- (202,200 \text{ N/m}^2)(0.0113 \text{ m}^2)$$

$$= 232 - 2285 = \textbf{2053 N}$$

$$F_{Rz} = \beta \dot{m} V_2 \sin \theta = 1.1032 (14 \text{ kg/s})(20 \sin 30° \text{ m/s}) \left( \frac{1 \text{ N}}{1 \text{ kg} \cdot \text{m/s}^2} \right) = \textbf{144 N}$$

The negative result for $F_{Rx}$ indicates that the assumed direction is wrong, and it should be reversed. Therefore, $F_{Rx}$ acts in the negative x-direction.

*Discussion* There is a nonzero pressure distribution along the inside walls of the elbow, but since the control volume is outside the elbow, these pressures do not appear in our analysis. The weight of the elbow and the water in it could be added to the vertical force for better accuracy. The actual value of $P_{1, \text{gage}}$ will be higher than that calculated here because of frictional and other irreversible losses in the elbow.

## EXAMPLE 13–3    The Force to Hold a Reversing Elbow in Place

The deflector elbow in Example 13–2 is replaced by a reversing elbow such that the fluid makes a 180° U-turn before it is discharged, as shown in Fig. 13–28. The elevation difference between the centers of the inlet and the exit sections is still 0.3 m. Determine the anchoring force needed to hold the elbow in place.

**Solution**  The inlet and the outlet velocities and the pressure at the inlet of the elbow remain the same, but the vertical component of the anchoring force at the connection of the elbow to the pipe is zero in this case ($F_{Rz} = 0$) since there is no other force or momentum flux in the vertical direction (we are neglecting the weight of the elbow and the water). The horizontal component of the anchoring force is determined from the momentum equation written in the x-direction. Noting that the outlet velocity is negative since it is in the negative x-direction, we have

$$F_{Rx} + P_{1,\,gage} A_1 = \beta_2 \dot{m}(-V_2) - \beta_1 \dot{m} V_1 = -\beta \dot{m}(V_2 + V_1)$$

Solving for $F_{Rx}$ and substituting the known values,

$$F_{Rx} = -\beta \dot{m}(V_2 + V_1) - P_{1,\,gage} A_1$$

$$= -(1.03)(14 \text{ kg/s})[(20 + 1.24) \text{ m/s}]\left(\frac{1 \text{ N}}{1 \text{ kg·m/s}^2}\right) - (202{,}200 \text{ N/m}^2)(0.0113 \text{ m}^2)$$

Therefore, the horizontal force on the flange is 2591 N acting in the negative x-direction (the elbow is trying to separate from the pipe). This force is equivalent to the weight of about 260 kg mass, and thus the connectors (such as bolts) used must be strong enough to withstand this force.

**Discussion**  The reaction force in the x-direction is larger than that of Example 13–2 since the walls turn the water over a much greater angle. If the reversing elbow is replaced by a straight nozzle (like one used by firefighters) such that water is discharged in the positive x-direction, the momentum equation in the x-direction becomes

$$F_{Rx} + P_{1,\,gage} A_1 = \beta \dot{m} V_2 - \beta \dot{m} V_1 \quad \rightarrow \quad F_{Rx} = \beta \dot{m}(V_2 - V_1) - P_{1,\,gage} A_1$$

since both $V_1$ and $V_2$ are in the positive x-direction. This shows the importance of using the correct sign (positive if in the positive direction and negative if in the opposite direction) for velocities and forces.

**FIGURE 13–28**

Schematic for Example 13–3.

## EXAMPLE 13–4    Water Jet Striking a Stationary Plate

Water is accelerated by a nozzle to an average speed of 20 m/s, and strikes a stationary vertical plate at a rate of 10 kg/s with a normal velocity of 20 m/s (Fig. 13–29). After the strike, the water stream splatters off in all directions in the plane of the plate. Determine the force needed to prevent the plate from moving horizontally due to the water stream.

**Solution**  A water jet strikes a vertical stationary plate normally. The force needed to hold the plate in place is to be determined.

**Assumptions**  **1** The flow of water at the nozzle outlet is steady. **2** The water splatters in directions normal to the approach direction of the water jet.

**FIGURE 13–29**

Schematic for Example 13–4.

**3** The water jet is exposed to the atmosphere, and thus the pressure of the water jet and the splattered water leaving the control volume is atmospheric pressure, which is disregarded since it acts on the entire system. **4** The vertical forces and momentum fluxes are not considered since they have no effect on the horizontal reaction force. **5** The effect of the momentum-flux correction factor is negligible, and thus $\beta \cong 1$ at the inlet.

*Analysis* We draw the control volume for this problem such that it contains the entire plate and cuts through the water jet and the support bar normally. The momentum equation for steady one-dimensional flow is given as

$$\sum \vec{F} = \sum_{out} \beta \dot{m} \vec{V} - \sum_{in} \beta \dot{m} \vec{V} \tag{1}$$

Writing Eq. 1 for this problem along the $x$-direction (without forgetting the negative sign for forces and velocities in the negative $x$-direction) and noting that $V_{1,x} = V_1$ and $V_{2,x} = 0$ gives

$$-F_R = 0 - \beta \dot{m} V_1$$

Substituting the given values,

$$F_R = \beta \dot{m} V_1 = (1)(10 \text{ kg/s})(20 \text{ m/s})\left(\frac{1 \text{ N}}{1 \text{ kg·m/s}^2}\right) = \textbf{200 N}$$

Therefore, the support must apply a 200-N horizontal force (equivalent to the weight of about a 20-kg mass) in the negative $x$-direction (the opposite direction of the water jet) to hold the plate in place.

*Discussion* The plate absorbs the full brunt of the momentum of the water jet since the $x$-direction momentum at the outlet of the control volume is zero. If the control volume were drawn instead along the interface between the water and the plate, there would be additional (unknown) pressure forces in the analysis. By cutting the control volume through the support, we avoid having to deal with this additional complexity. This is an example of a "wise" choice of control volume.

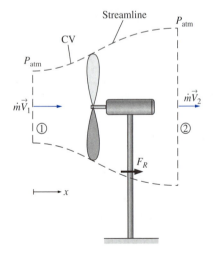

**FIGURE 13–30**

Schematic for Example 13–5.

### EXAMPLE 13–5    Power Generation and Wind Loading of a Wind Turbine

A wind generator with a 30-ft-diameter blade span has a cut-in wind speed (minimum speed for power generation) of 7 mph, at which velocity the turbine generates 0.4 kW of electric power (Fig. 13–30). Determine (a) the efficiency of the wind turbine–generator unit and (b) the horizontal force exerted by the wind on the supporting mast of the wind turbine. What is the effect of doubling the wind velocity to 14 mph on power generation and the force exerted? Assume the efficiency remains the same, and take the density of air to be 0.076 lbm/ft$^3$.

*Solution* The power generation and loading of a wind turbine are to be analyzed. The efficiency and the force exerted on the mast are to be determined, and the effects of doubling the wind velocity are to be investigated.

*Assumptions* **1** The wind flow is steady and incompressible. **2** The efficiency of the turbine–generator is independent of wind speed. **3** The frictional effects are negligible, and thus none of the incoming kinetic energy is converted to thermal energy. **4** The average velocity of air through the wind turbine is the same as the wind velocity. **5** The wind flow is nearly uniform upstream and downstream of the wind turbine and thus the momentum-flux correction factor is $\beta = \beta_1 = \beta_2 \cong 1$.

**Properties** The density of air is given to be 0.076 lbm/ft³.

**Analysis** Kinetic energy is a mechanical form of energy, and thus it can be converted to work entirely. Therefore, the power potential of the wind is proportional to its kinetic energy, which is $V^2/2$ per unit mass, and thus the maximum power is $\dot{m}V^2/2$ for a given mass flow rate:

$$V_1 = (7\text{ mph})\left(\frac{1.4667\text{ ft/s}}{1\text{ mph}}\right) = 10.27\text{ ft/s}$$

$$\dot{m} = \rho_1 V_1 A_1 = \rho_1 V_1 \frac{\pi D^2}{4} = (0.076\text{ lbm/ft}^3)(10.27\text{ ft/s})\frac{\pi(30\text{ ft})^2}{4} = 551.7\text{ lbm/s}$$

$$\dot{W}_{max} = \dot{m}\text{ke}_1 = \dot{m}\frac{V_1^2}{2}$$

$$= (551.7\text{ lbm/s})\frac{(10.27\text{ ft/s})^2}{2}\left(\frac{1\text{ lbf}}{32.2\text{ lbm·ft/s}^2}\right)\left(\frac{1\text{ kW}}{737.56\text{ lbf·ft/s}}\right)$$

$$= 1.225\text{ kW}$$

Therefore, the available power to the wind turbine is 1.225 kW at the wind velocity of 7 mph. Then the turbine–generator efficiency becomes

$$\eta_{\text{wind turbine}} = \frac{\dot{W}_{act}}{\dot{W}_{max}} = \frac{0.4\text{ kW}}{1.225\text{ kW}} = \textbf{0.327} \qquad (\text{or } \textbf{32.7\%})$$

(*b*) The frictional effects are assumed to be negligible, and thus the portion of incoming kinetic energy not converted to electric power leaves the wind turbine as outgoing kinetic energy. Noting that the mass flow rate remains constant, the exit velocity is determined to be

$$\dot{m}\text{ke}_2 = \dot{m}\text{ke}_1(1 - \eta_{\text{wind turbine}}) \quad \rightarrow \quad \dot{m}\frac{V_2^2}{2} = \dot{m}\frac{V_1^2}{2}(1 - \eta_{\text{wind turbine}}) \qquad (1)$$

or

$$V_2 = V_1\sqrt{1 - \eta_{\text{wind turbine}}} = (10.27\text{ ft/s})\sqrt{1 - 0.327} = 8.43\text{ ft/s}$$

We draw a control volume around the wind turbine such that the wind is normal to the control surface at the inlet and the outlet and the entire control surface is at atmospheric pressure (Fig. 13–30). The momentum equation for steady one-dimensional flow is given as

$$\sum\vec{F} = \sum_{\text{out}}\beta\dot{m}\vec{V} - \sum_{\text{in}}\beta\dot{m}\vec{V} \qquad (2)$$

Writing Eq. 2 along the *x*-direction and noting that $\beta = 1$, $V_{1,x} = V_1$, and $V_{2,x} = V_2$ give

$$F_R = \dot{m}V_2 - \dot{m}V_1 = \dot{m}(V_2 - V_1) \qquad (3)$$

Substituting the known values into Eq. 3 gives

$$F_R = \dot{m}(V_2 - V_1) = (551.7\text{ lbm/s})(8.43 - 10.27\text{ ft/s})\left(\frac{1\text{ lbf}}{32.2\text{ lbm·ft/s}^2}\right)$$

$$= -31.5\text{ lbf}$$

The negative sign indicates that the reaction force acts in the negative $x$-direction, as expected. Then the force exerted by the wind on the mast becomes $F_{mast} = -F_R = $ **31.5 lbf**.

The power generated is proportional to $V^3$ since the mass flow rate is proportional to $V$ and the kinetic energy to $V^2$. Therefore, doubling the wind velocity to 14 mph will increase the power generation by a factor of $2^3 = 8$ to $0.4 \times 8 = 3.2$ kW. The force exerted by the wind on the support mast is proportional to $V^2$. Therefore, doubling the wind velocity to 14 mph will increase the wind force by a factor of $2^2 = 4$ to $31.5 \times 4 = 126$ lbf.

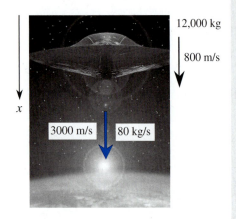

12,000 kg

800 m/s

$x$

3000 m/s    80 kg/s

**FIGURE 13–31**

Schematic for Example 13–6.

### EXAMPLE 13–6    Deceleration of a Spacecraft

A spacecraft with a mass of 12,000 kg is dropping vertically towards a planet at a constant speed of 800 m/s (Fig. 13–31). To slow down the spacecraft, a solid-fuel rocket at the bottom is fired, and combustion gases leave the rocket at a constant rate of 80 kg/s and at a velocity of 3000 m/s relative to the spacecraft in the direction of motion of the spacecraft for a period of 5 s. Disregarding the small change in the mass of the spacecraft, determine (a) the deceleration of the spacecraft during this period, (b) the change of velocity of the spacecraft, and (c) the thrust exerted on the spacecraft.

**Solution**  The rocket of a spacecraft is fired in the direction of motion. The deceleration, the velocity change, and the thrust are to be determined.

**Assumptions**  1 The flow of combustion gases is steady and one-dimensional during the firing period, but the flight of the spacecraft is unsteady. 2 There are no external forces acting on the spacecraft, and the effect of pressure force at the nozzle outlet is negligible. 3 The mass of discharged fuel is negligible relative to the mass of the spacecraft, and thus, the spacecraft may be treated as a solid body with a constant mass. 4 The nozzle is well designed such that the effect of the momentum-flux correction factor is negligible, and thus, $\beta \cong 1$.

**Analysis**  (a) For convenience, we choose an inertial reference frame that moves with the spacecraft at the same initial velocity. Then the velocities of the fluid stream relative to an inertial reference frame become simply the velocities relative to the spacecraft. We take the direction of motion of the spacecraft as the positive direction along the $x$-axis. There are no external forces acting on the spacecraft, and its mass is essentially constant. Therefore, the spacecraft can be treated as a solid body with constant mass, and the momentum equation in this case is, from Eq. 13–39,

$$\vec{F}_{thrust} = m_{spacecraft}\,\vec{a}_{spacecraft} = \sum_{in} \beta \dot{m}\vec{V} - \sum_{out} \beta \dot{m}\vec{V}$$

where the fluid stream velocities relative to the inertial reference frame in this case are identical to the velocities relative to the spacecraft. Noting that the motion is on a straight line and the discharged gases move in the positive $x$-direction, we write the momentum equation using magnitudes as

$$m_{spacecraft}a_{spacecraft} = m_{spacecraft}\frac{dV_{spacecraft}}{dt} = -\dot{m}_{gas}V_{gas}$$

Noting that gases leave in the positive $x$ direction and substituting, the acceleration of the spacecraft during the first 5 seconds is determined to be

$$a_{\text{spacecraft}} = \frac{dV_{\text{spacecraft}}}{dt} = -\frac{\dot{m}_{\text{gas}}}{m_{\text{spacecraft}}}V_{\text{gas}} = -\frac{80 \text{ kg/s}}{12{,}000 \text{ kg}}(+3000 \text{ m/s}) = -20 \text{ m/s}^2$$

The negative value confirms that the spacecraft is decelerating in the positive $x$ direction at a rate of 20 m/s².

(b) Knowing the deceleration, which is constant, the velocity change of the spacecraft during the first 5 seconds is determined from the definition of acceleration to be

$$dV_{\text{spacecraft}} = a_{\text{spacecraft}}dt \rightarrow \Delta V_{\text{spacecraft}} = a_{\text{spacecraft}}\Delta t = (-20 \text{ m/s}^2)(5 \text{ s})$$
$$= -100 \text{ m/s}$$

(c) The thrusting force exerted on the space aircraft is, from Eq. 13–39,

$$F_{\text{thrust}} = 0 - \dot{m}_{\text{gas}}V_{\text{gas}} = 0 - (80 \text{ kg/s})(+3000 \text{ m/s})\left(\frac{1 \text{ kN}}{1000 \text{ kg·m/s}^2}\right) = -240 \text{ kN}$$

The negative sign indicates that the trusting force due to firing of the rocket acts on the aircraft in the negative $x$-direction.

**Discussion** Note that if this fired rocket were attached somewhere on a test stand, it would exert a force of 240 kN (equivalent to the weight of about 24 tons of mass) to its support in the opposite direction of the discharged gases.

## SUMMARY

This chapter deals mainly with the conservation of momentum for finite control volumes. The forces acting on the control volume consist of *body forces* that act throughout the entire body of the control volume (such as gravity, electric, and magnetic forces) and *surface forces* that act on the control surface (such as the pressure forces and reaction forces at points of contact). The sum of all forces acting on the control volume at a particular instant in time is represented by $\Sigma \vec{F}$ and is expressed as

$$\underbrace{\Sigma \vec{F}}_{\text{total force}} = \underbrace{\Sigma \vec{F}_{\text{gravity}}}_{\text{body force}} + \underbrace{\Sigma \vec{F}_{\text{pressure}} + \Sigma \vec{F}_{\text{viscous}} + \Sigma \vec{F}_{\text{other}}}_{\text{surface forces}}$$

Newton's second law can be stated as *the sum of all external forces acting on a system is equal to the time rate of change of linear momentum of the system*. Setting $b = \vec{V}$ and thus $B = m\vec{V}$ in the Reynolds transport theorem and utilizing Newton's second law gives the *linear momentum equation* for a control volume as

$$\Sigma \vec{F} = \frac{d}{dt}\int_{\text{CV}} \rho\vec{V}\, dV + \int_{\text{CS}} \rho\vec{V}(\vec{V_r}\cdot\vec{n})\, dA$$

which reduces to the following special cases:

*Steady flow:* 
$$\Sigma \vec{F} = \int_{\text{CS}} \rho\vec{V}(\vec{V_r}\cdot\vec{n})\, dA$$

*Unsteady flow (algebraic form):*

$$\Sigma \vec{F} = \frac{d}{dt}\int_{\text{CV}} \rho\vec{V}\, dV + \sum_{\text{out}} \beta\dot{m}\vec{V} - \sum_{\text{in}} \beta\dot{m}\vec{V}$$

*Steady flow (algebraic form):* 
$$\Sigma \vec{F} = \sum_{\text{out}} \beta\dot{m}\vec{V} - \sum_{\text{in}} \beta\dot{m}\vec{V}$$

*No external forces:* 
$$0 = \frac{d(m\vec{V})_{\text{CV}}}{dt} + \sum_{\text{out}} \beta\dot{m}\vec{V} - \sum_{\text{in}} \beta\dot{m}\vec{V}$$

where $\beta$ is the momentum-flux correction factor. A control volume whose mass $m$ remains constant can be treated as a solid body (a fixed-mass system) with a net *thrusting force* (also called simply the *thrust*) of

$$\vec{F}_{\text{thrust}} = m_{\text{CV}}\vec{a} = \sum_{\text{in}} \beta\dot{m}\vec{V} - \sum_{\text{out}} \beta\dot{m}\vec{V}$$

acting on the body.

Newton's second law can also be stated as *the rate of change of angular momentum of a system is equal to the net torque acting on the system*. Setting $b = \vec{r} \times \vec{V}$ and thus $B = \vec{H}$ in the general Reynolds transport theorem gives the *angular momentum equation* as

$$\sum \vec{M} = \frac{d}{dt} \int_{CV} (\vec{r} \times \vec{V}) \rho \, dV + \int_{CS} (\vec{r} \times \vec{V}) \rho (\vec{V_r} \cdot \vec{n}) \, dA$$

which reduces to the following special cases:

*Steady flow:* 
$$\sum \vec{M} = \int_{CS} (\vec{r} \times \vec{V}) \rho (\vec{V_r} \cdot \vec{n}) \, dA$$

*Unsteady flow (algebraic form):*

$$\sum \vec{M} = \frac{d}{dt} \int_{CV} (\vec{r} \times \vec{V}) \rho \, dV + \sum_{out} \vec{r} \times \dot{m} \vec{V} - \sum_{in} \vec{r} \times \dot{m} \vec{V}$$

*Steady and uniform flow:*

$$\sum \vec{M} = \sum_{out} \vec{r} \times \dot{m} \vec{V} - \sum_{in} \vec{r} \times \dot{m} \vec{V}$$

*Scalar form for one direction:*

$$\sum M = \sum_{out} r \dot{m} V - \sum_{in} r \dot{m} V$$

*No external moments:*

$$0 = \frac{d\vec{H}_{CV}}{dt} + \sum_{out} \vec{r} \times \dot{m} \vec{V} - \sum_{in} \vec{r} \times \dot{m} \vec{V}$$

## REFERENCES AND SUGGESTED READING

**1.** P. K. Kundu and I. M. Cohen. *Fluid Mechanics,* ed. 4. San Diego, CA: Academic Press, 2008.

**2.** Terry Wright, *Fluid Machinery: Performance, Analysis, and Design*, Boca Raton, FL: CRC Press, 1999.

## PROBLEMS*

### Newton's Laws

**13–1C**  Is momentum a vector? If so, in what direction does it point?

**13–2C**  Express the conservation of momentum principle. What can you say about the momentum of a body if the net force acting on it is zero?

**13–3C**  Express Newton's first, second, and third laws.

---

* Problems designated by a "C" are concept questions, and students are encouraged to answer them all. Problems designated by an "E" are in English units, and the SI users can ignore them. Problems with the 🖱 icon are solved using EES, and complete solutions together with parametric studies are included on the text specific web site. Problems with the 🖳 icon are comprehensive in nature and are intended to be solved with a computer, preferably using the EES software that accompanies this text.

**13–4C**  Express Newton's second law of motion for rotating bodies. What can you say about the angular velocity and angular momentum of a rotating nonrigid body of constant mass if the net torque acting on it is zero?

### Linear Momentum Equation

**13–5C**  Describe body forces and surface forces, and explain how the net force acting on a control volume is determined. Is fluid weight a body force or surface force? How about pressure?

**13–6C**  How do surface forces arise in the momentum analysis of a control volume? How can we minimize the number of surface forces exposed during analysis?

**13–7C**  Explain the importance of the Reynolds transport theorem in fluid mechanics, and describe how the linear momentum equation is obtained from it.

**13–8C** What is the importance of the momentum-flux correction factor in the momentum analysis of flow systems? For which type(s) of flow is it significant and must it be considered in analysis: laminar flow, turbulent flow, or jet flow?

**13–9C** Write the momentum equation for steady one-dimensional flow for the case of no external forces and explain the physical significance of its terms.

**13–10C** In the application of the momentum equation, explain why we can usually disregard the atmospheric pressure and work with gage pressures only.

**13–11C** A rocket in space (no friction or resistance to motion) can expel gases relative to itself at some high velocity $V$. Is $V$ the upper limit to the rocket's ultimate velocity?

**13–12C** Describe in terms of momentum and airflow how a helicopter is able to hover.

**FIGURE P13–12C**

**13–13C** Does it take more, equal, or less power for a helicopter to hover at the top of a high mountain than it does at sea level? Explain.

**13–14C** In a given location, would a helicopter require more energy in summer or winter to achieve a specified performance? Explain.

**13–15C** A horizontal water jet from a nozzle of constant exit cross section impinges normally on a stationary vertical flat plate. A certain force $F$ is required to hold the plate against the water stream. If the water velocity is doubled, will the necessary holding force also be doubled? Explain.

**13–16C** Two firefighters are fighting a fire with identical water hoses and nozzles, except that one is holding the hose straight so that the water leaves the nozzle in the same direction it comes, while the other holds it backward so that the water makes a U-turn before being discharged. Which firefighter will experience a greater reaction force?

**13–17C** A constant-velocity horizontal water jet from a stationary nozzle impinges normally on a vertical flat plate that rides on a nearly frictionless track. As the water jet hits the plate, it begins to move due to the water force. Will the accel-

eration of the plate remain constant or change? Explain.

**FIGURE P13–17C**

**13–18C** A horizontal water jet of constant velocity $V$ from a stationary nozzle impinges normally on a vertical flat plate that rides on a nearly frictionless track. As the water jet hits the plate, it begins to move due to the water force. What is the highest velocity the plate can attain? Explain.

**13–19** A horizontal water jet of constant velocity $V$ impinges normally on a vertical flat plate and splashes off the sides in the vertical plane. The plate is moving toward the oncoming water jet with velocity $\frac{1}{2}V$. If a force $F$ is required to maintain the plate stationary, how much force is required to move the plate toward the water jet?

**FIGURE P13–19**

**13–20** A 90° elbow in a horizontal pipe is used to direct water flow upward at a rate of 30 kg/s. The diameter of the entire elbow is 10 cm. The elbow discharges water into the atmosphere, and thus the pressure at the exit is the local atmospheric pressure. The elevation difference between the centers of the exit and the inlet of the elbow is 40 cm. The weight of the elbow and the water in it is considered to be negligible. Determine (a) the gage pressure at the center of the inlet of the elbow and (b) the anchoring force needed to hold the elbow in place. Take the momentum-flux correction factor to be 1.03 at both the inlet and the outlet.

**FIGURE P13–20**

**13–21** Repeat Prob. 13–20 for the case of another (identical) elbow attached to the existing elbow so that the fluid makes a U-turn. *Answers: (a)* 7.85 kPa, *(b)* 298 N

**13–22E** A horizontal water jet impinges against a vertical flat plate at 25 ft/s and splashes off the sides in the vertical plane. If a horizontal force of 350 lbf is required to hold the plate against the water stream, determine the volume flow rate of the water.

**13–23** A reducing elbow in a horizontal pipe is used to deflect water flow by an angle $\theta = 45°$ from the flow direction while accelerating it. The elbow discharges water into the atmosphere. The cross-sectional area of the elbow is 150 cm$^2$ at the inlet and 25 cm$^2$ at the exit. The elevation difference between the centers of the exit and the inlet is 40 cm. The mass of the elbow and the water in it is 50 kg. Determine the anchoring force needed to hold the elbow in place. Take the momentum-flux correction factor to be 1.03 at both the inlet and outlet.

**FIGURE P13–23**

**13–24** Repeat Prob. 13–23 for the case of $\theta = 110°$.

**13–25** Water accelerated by a nozzle to 20 m/s strikes the vertical back surface of a cart moving horizontally at a constant velocity of 5 m/s in the flow direction. The mass flow rate of water through the stationary nozzle is 30 kg/s. After the strike, the water stream splatters off in all directions in the plane of the back surface. (*a*) Determine the force that needs to be applied by the brakes of the cart to prevent it from accelerating. (*b*) If this force were used to generate power instead of wasting it on the brakes, determine the maximum amount of power that could ideally be generated. *Answers: (a)* −338 N, *(b)* 1.69 kW

**FIGURE P13–25**

**13–26** Reconsider Prob. 13–25. If the mass of the cart is 400 kg and the brakes fail, determine the acceleration of the

cart when the water first strikes it. Assume the mass of water that wets the back surface is negligible.

**13–27E** A 100-ft$^3$/s water jet is moving in the positive *x*-direction at 18 ft/s. The stream hits a stationary splitter, such that half of the flow is diverted upward at 45° and the other half is directed downward, and both streams have a final speed of 18 ft/s. Disregarding gravitational effects, determine the *x*- and *z*-components of the force required to hold the splitter in place against the water force.

**FIGURE P13–27E**

**13–28E**  Reconsider Prob. 13–27E. Using EES (or other) software, investigate the effect of the splitter angle on the force exerted on the splitter in the incoming flow direction. Let the half splitter angle vary from 0° to 180° in increments of 10°. Tabulate and plot your results, and draw some conclusions.

**13–29** A horizontal 5-cm-diameter water jet with a velocity of 18 m/s impinges normally upon a vertical plate of mass 1000 kg. The plate rides on a nearly frictionless track and is initially stationary. When the jet strikes the plate, the plate begins to move in the direction of the jet. The water always splatters in the plane of the retreating plate. Determine (*a*) the acceleration of the plate when the jet first strikes it (time = 0), (*b*) the time it takes for the plate to reach a velocity of 9 m/s, and (*c*) the plate velocity 20 s after the jet first strikes the plate. For simplicity, assume the velocity of the jet is increased as the cart moves such that the impulse force exerted by the water jet on the plate remains constant.

**13–30E** A 4-in-diameter horizontal water jet having a velocity of 120 ft/s strikes a curved plate, which deflects the water 180° at the same speed. Ignoring the frictional effects, determine the force required to hold the plate against the water stream.

**FIGURE P13–30E**

**13–31E** A 3-in-diameter horizontal jet of water, with velocity 140 ft/s, strikes a bent plate, which deflects the water by 135° from its original direction. How much force is required to hold the plate against the water stream and what is its direction? Disregard frictional and gravitational effects.

**13–32** Firefighters are holding a nozzle at the end of a hose while trying to extinguish a fire. If the nozzle exit diameter is 6 cm and the water flow rate is 5 m³/min, determine (a) the average water exit velocity and (b) the horizontal resistance force required of the firefighters to hold the nozzle. *Answers:* (a) 29.5 m/s, (b) 2457 N

5 m³/min

**FIGURE P13–32**

**13–33** A 5-cm-diameter horizontal jet of water with a velocity of 40 m/s relative to the ground strikes a flat plate that is moving in the same direction as the jet at a velocity of 10 m/s. The water splatters in all directions in the plane of the plate. How much force does the water stream exert on the plate?

**13–34** Reconsider Prob. 13–33. Using EES (or other) software, investigate the effect of the plate velocity on the force exerted on the plate. Let the plate velocity vary from 0 to 30 m/s, in increments of 3 m/s. Tabulate and plot your results.

**13–35E** A fan with 24-in-diameter blades moves 2000 cfm (cubic feet per minute) of air at 70°F at sea level. Determine (a) the force required to hold the fan and (b) the minimum power input required for the fan. Choose a control volume sufficiently large to contain the fan, with the inlet sufficiently far upstream so that the gage pressure at the inlet is nearly zero. Assume air approaches the fan through a large area with negligible velocity and air exits the fan with a uniform velocity at atmospheric pressure through an imaginary cylinder whose diameter is the fan blade diameter. *Answers:* (a) 0.820 lbf, (b) 5.91 W

**13–36** An unloaded helicopter of mass 10,000 kg hovers at sea level while it is being loaded. In the unloaded hover mode, the blades rotate at 400 rpm. The horizontal blades above the helicopter cause a 18-m-diameter air mass to move downward at an average velocity proportional to the overhead blade rotational velocity (rpm). A load of 14,000 kg is loaded onto the helicopter, and the helicopter slowly rises. Determine (a) the volumetric airflow rate downdraft that the helicopter generates during unloaded hover and the required

power input and (b) the rpm of the helicopter blades to hover with the 14,000-kg load and the required power input. Take the density of atmospheric air to be 1.18 kg/m³. Assume air approaches the blades from the top through a large area with negligible velocity and air is forced by the blades to move down with a uniform velocity through an imaginary cylinder whose base is the blade span area.

18 m

Load
14,000 kg

**FIGURE P13–36**

**13–37** Reconsider the helicopter in Prob. 13–36, except that it is hovering on top of a 2800-m-high mountain where the air density is 0.928 kg/m³. Noting that the unloaded helicopter blades must rotate at 400 rpm to hover at sea level, determine the blade rotational velocity to hover at the higher altitude. Also determine the percent increase in the required power input to hover at 3000-m altitude relative to that at sea level. *Answers:* 451 rpm, 13 percent

**13–38** A sluice gate, which controls flow rate in a channel by simply raising or lowering a vertical plate, is commonly used in irrigation systems. A force is exerted on the gate due to the difference between the water heights $y_1$ and $y_2$ and the flow velocities $V_1$ and $V_2$ upstream and downstream from the gate, respectively. Take the width of the sluice gate (into the page) to be $w$. Wall shear stresses along the channel walls may be ignored, and for simplicity, we assume steady, uniform flow at locations 1 and 2. Develop a relationship for the force $F_R$ acting on the sluice gate as a function of depths $y_1$ and $y_2$, mass flow rate $\dot{m}$, gravitational constant $g$, gate width $w$, and water density $\rho$.

$V_1$

$y_1$

Sluice gate

$y_2$

$V_2$

**FIGURE P13–38**

**13–39** Water enters a centrifugal pump axially at atmospheric pressure at a rate of 0.12 m³/s and at a velocity of 7 m/s, and leaves in the normal direction along the pump casing, as shown in Fig. P13–39. Determine the force acting on the shaft (which is also the force acting on the bearing of the shaft) in the axial direction.

Blade

Impeller shroud

0.12 m³/s

$\dot{n}$

Shaft

**FIGURE P13–39**

**13–40** An incompressible fluid of density $\rho$ and viscosity $\mu$ flows through a curved duct that turns the flow 180°. The duct cross-sectional area remains constant. The average velocity, momentum flux correction factor, and gage pressure are known at the inlet (1) and outlet (2), as in Fig. P13–40. (*a*) Write an expression for the horizontal force $F_x$ of the fluid on the walls of the duct in terms of the given variables. (*b*) Verify your expression by plugging in the following values: $\rho = 998.2$ kg/m³, $\mu = 1.003 \times 10^{-3}$ kg/m·s, $A_1 = A_2 = 0.025$ m², $\beta_1 = 1.01$, $\beta_2 = 1.03$, $V_1 = 10$ m/s, $P_{1,\text{gage}} = 78.47$ kPa, and $P_{2,\text{gage}} = 65.23$ kPa. *Answer:* (*b*) $F_x = 8680$ N to the right

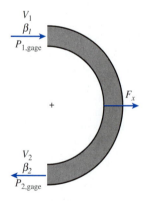

$V_1$
$\beta_1$
$P_{1,\text{gage}}$

$F_x$

$V_2$
$\beta_2$
$P_{2,\text{gage}}$

**FIGURE P13–40**

**13–41** Consider the curved duct of Prob. 13–40, except allow the cross-sectional area to vary along the duct ($A_1 \neq A_2$). (*a*) Write an expression for the horizontal force $F_x$ of the fluid on the walls of the duct in terms of the given variables. (*b*) Verify your expression by plugging in the following values: $\rho = 998.2$ kg/m³, $A_1 = 0.025$ m², $A_2 = 0.015$ m², $\beta_1 = 1.02$, $\beta_2 = 1.04$, $V_1 = 20$ m/s, $P_{1,\text{gage}} = 88.34$ kPa, and $P_{2,\text{gage}} = 67.48$ kPa. *Answer:* (*b*) $F_x = 30,700$ N to the right

**13–42** Water flowing in a horizontal 25-cm-diameter pipe at 8 m/s and 300 kPa gage enters a 90° bend reducing section, which connects to a 15-cm-diameter vertical pipe. The inlet of the bend is 50 cm above the exit. Neglecting any frictional and gravitational effects, determine the net resultant force exerted on the reducer by the water. Take the momentum-flux correction factor to be 1.04.

**13–43** Commercially available large wind turbines have blade span diameters as large as 100 m and generate over 3 MW of electric power at peak design conditions. Consider a wind turbine with a 90-m blade span subjected to 25-km/h steady winds. If the combined turbine–generator efficiency of the wind turbine is 32 percent, determine (*a*) the power generated by the turbine and (*b*) the horizontal force exerted by the wind on the supporting mast of the turbine. Take the density of air to be 1.25 kg/m³, and disregard frictional effects.

25 km/h

90 m

**FIGURE P13–43**

### Review Problems

**13–44** Water enters vertically and steadily at a rate of 15 L/s into the sprinkler shown in Fig. P13–44. Both water jets have a diameter of 1.2 cm. Disregarding any frictional effects, determine (*a*) the rotational speed of the sprinkler in rpm and (*b*) the torque required to prevent the sprinkler from rotating.

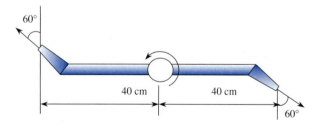

60°

40 cm

40 cm

60°

**FIGURE P13–44**

**13–45** Repeat Prob. 13–44 for the case of unequal arms—the left one being 60 cm and the right one 20 cm from the axis of rotation.

**13–46** A 60-kg ice skater is standing on ice with ice skates (negligible friction). She is holding a flexible hose (essentially weightless) that directs a 2-cm-diameter stream of water horizontally parallel to her skates. The water velocity at the hose outlet is 10 m/s relative to the skater. If she is initially standing still, determine (a) the velocity of the skater and the distance she travels in 5 s and (b) how long it will take to move 5 m and the velocity at that moment. *Answers:* (*a*) 2.62 m/s, 6.54 m, (*b*) 4.4 s, 2.3 m/s

Ice skater

10 m/s

$D = 2$ cm

**FIGURE P13–46**

**13–47** Show that the force exerted by a liquid jet on a stationary nozzle as it leaves with a velocity $V$ is proportional to $V^2$ or, alternatively, to $\dot{m}^2$. Assume the jet stream is perpendicular to the incoming liquid flow line.

**13–48** A tripod holding a nozzle, which directs a 5-cm-diameter stream of water from a hose, is shown in Fig. P13–48. The nozzle mass is 10 kg when filled with water. The tripod is rated to provide 1800 N of holding force. A firefighter was standing 60 cm behind the nozzle and was hit by the nozzle when the tripod suddenly failed and released the nozzle. You have been hired as an accident reconstructionist and, after testing the tripod, have determined that as water flow rate increased, it did collapse at 1800 N. In your final report you must state the water velocity and the flow rate consistent with the failure and the nozzle velocity when it hit the firefighter. For simplicity, ignore pressure and momentum effects in the upstream portion of the hose. *Answers:* 30.3 m/s, 0.0595 m3/s, 14.7 m/s

Nozzle

$D = 5$ cm

Tripod

**FIGURE P13–48**

**13–49** Consider an airplane with a jet engine attached to the tail section that expels combustion gases at a rate of 18 kg/s with a velocity of $V = 300$ m/s relative to the plane. During landing, a thrust reverser (which serves as a brake for the aircraft and facilitates landing on a short runway) is lowered in the path of the exhaust jet, which deflects the exhaust from rearward to 150°. Determine (a) the thrust (forward force) that the engine produces prior to the insertion of the thrust reverser and (b) the braking force produced after the thrust reverser is deployed.

Thrust reverser

150°

300 m/s    Thrust reverser

**FIGURE P13–49**

**13–50** Reconsider Prob. 13–49. Using EES (or other) software, investigate the effect of thrust reverser angle on the braking force exerted on the airplane. Let the reverser angle vary from 0° (no reversing) to 180° (full reversing) in increments of 10°. Tabulate and plot your results and draw conclusions.

**13–51E** A spacecraft cruising in space at a constant velocity of 1500 ft/s has a mass of 18,000 lbm. To slow down the spacecraft, a solid fuel rocket is fired, and the combustion gases leave the rocket at a constant rate of 150 lbm/s at a velocity of 5000 ft/s in the same direction as the spacecraft for a period of 5 s. Assuming the mass of the spacecraft remains constant, determine (a) the deceleration of the spacecraft during this 5-s period, (b) the change of velocity of the spacecraft during this time period, and (c) the thrust exerted on the spacecraft.

**13–52** A 6-cm-diameter horizontal water jet having a velocity of 25 m/s strikes a vertical stationary flat plate. The water splatters in all directions in the plane of the plate. How much force is required to hold the plate against the water stream? *Answer:* 1770 N

**13–53** A 5-cm-diameter horizontal jet of water, with velocity 30 m/s, strikes the tip of a horizontal cone, which deflects the water by 45° from its original direction. How much force is required to hold the cone against the water stream?

**13–54** Water is flowing into and discharging from a pipe U-section as shown in Fig. P13–54. At flange (1), the total absolute pressure is 200 kPa, and 42 kg/s flows into the pipe. At flange (2), the total pressure is 150 kPa. At location (3), 10 kg/s of water discharges to the atmosphere, which is at 100 kPa. Determine the total $x$- and $z$-forces at the two flanges connecting the pipe. Discuss the significance of gravity force for this problem. Take the momentum-flux correction factor to be 1.03 throughout the pipes.

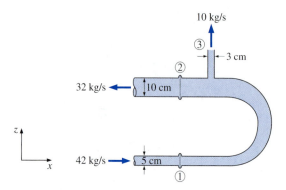

**FIGURE P13–54**

**13–55** Indiana Jones needs to ascend a 10-m-high building. There is a large hose filled with pressurized water hanging down from the building top. He builds a square platform and mounts four 4-cm-diameter nozzles pointing down at each corner. By connecting hose branches, a water jet with 15-m/s velocity can be produced from each nozzle. Jones, the platform, and the nozzles have a combined mass of 150 kg. Determine (*a*) the minimum water jet velocity needed to raise the system, (*b*) how long it takes for the system to rise 10 m when the water jet velocity is 18 m/s and the velocity of the platform at that moment, and (*c*) how much higher will the momentum raise Jones if he shuts off the water at the moment the platform reaches 10 m above the ground. How much time does he have to jump from the platform to the roof? *Answers:* (*a*) 17.1 m/s, (*b*) 4.37 s, 4.57 m/s, (*c*) 1.07 m, 0.933 s

**FIGURE P13–55**

**13–56E** An engineering student considers using a fan as a levitation demonstration. She plans to face the box-enclosed fan so the air blast is directed face down through a 3-ft-diameter blade span area. The system weighs 5 lbf, and the student will secure the system from rotating. By increasing the power to the fan, she plans to increase the blade rpm and air exit velocity until the exhaust provides sufficient upward force to cause the box fan to hover in the air. Determine (*a*) the air exit velocity to produce 5 lbf, (*b*) the volumetric flow rate needed, and (*c*) the minimum mechanical power that must be supplied to the airstream. Take the air density to be 0.078 lbm/ft$^3$.

**FIGURE P13–56E**

**13–57** An orbiting satellite has a mass of 5000 kg and is traveling at a constant velocity of $V_0$. To alter its orbit, an attached rocket discharges 100 kg of gases from the reaction of solid fuel at a speed of 3000 m/s relative to the satellite in a direction opposite $V_0$. The fuel discharge rate is constant for 2 s. Determine (a) the thrust exerted on the satellite, (b) the acceleration of the satellite during this 2-s period, and (c) the change of velocity of the satellite during this time period.

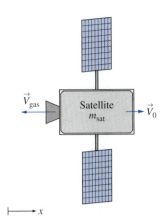

**FIGURE P13–57**

**13–58** A 7-cm diameter vertical water jet is injected upwards by a nozzle at a speed of 15 m/s. Determine the maximum weight of a flat plate that can be supported by this water jet at a height of 2 m from the nozzle.

**13–59** A soldier jumps from a plane and opens his parachute when his velocity reaches the terminal velocity $V_T$. The para-chute slows him down to his landing velocity of $V_F$. After the parachute is deployed, the air resistance is proportional to the velocity squared (i.e., $F = kV^2$). The soldier, his parachute, and his gear have a total mass of $m$. Show that $k = mg/V_F^2$ and develop a relation for the soldier's velocity after he opens the parachute at time $t = 0$.

$$Answer: \quad V = V_F \frac{V_T + V_F + (V_T - V_F)e^{-2gt/V_F}}{V_T + V_F - (V_T - V_F)e^{-2gt/V_F}}$$

**13–60** Nearly frictionless vertical guide rails maintain a plate of mass $m_p$ in a horizontal position, such that it can slide freely in the vertical direction. A nozzle directs a water stream of area $A$ against the plate underside. The water jet splatters in the plate plane, applying an upward force against the plate. The water flow rate $\dot{m}$ (kg/s) can be controlled. Assume that distances are short, so the velocity of the rising jet can be considered constant with height. (a) Determine the minimum mass flow rate $\dot{m}_{min}$ necessary to just levitate the plate and obtain a relation for the steady-state velocity of the upward moving plate for $\dot{m} > \dot{m}_{min}$. (b) At time $t = 0$, the plate is at rest, and the water jet with $\dot{m} > \dot{m}_{min}$ is suddenly turned on. Apply a force balance to the plate and obtain the integral that relates velocity to time (do not solve).

**FIGURE P13–59**

**FIGURE P13–60**

**Design and Essay Problem**

**13–61** Visit a fire station and obtain information about flow rates through hoses and discharge diameters. Using this information, calculate the impulse force to which the firefighters are subjected.

# INTERNAL FLOW

luid flow is classified as *external* or *internal*, depending on whether the
fluid is forced to flow over a surface or in a conduit. Internal and external flows exhibit very different characteristics. In this chapter we consider *internal flow* where the conduit is completely filled with the fluid, and
the flow is driven primarily by a pressure difference. This should not be
confused with *open-channel flow* where the conduit is partially filled by the
fluid and thus the flow is partially bounded by solid surfaces, as in an irrigation ditch, and the flow is driven by gravity alone.

We start this chapter with a general physical description of internal flow
through pipes and ducts including the *entrance region* and the *fully developed* region. We continue with a discussion of the dimensionless *Reynolds
number* and its physical significance. We then introduce the *pressure drop*
correlations associated with pipe flow for both laminar and turbulent flows.
Finally, we discuss minor losses and determine the pressure drop and pumping power requirements for real-world piping systems.

## Objectives

The objectives of this chapter are to:

- Have a deeper understanding of
  laminar and turbulent flow in
  pipes and the analysis of fully
  developed flow.

- Calculate the major and minor
  losses associated with pipe
  flow in piping networks and
  determine the pumping power
  requirements.

## FIGURE 14–1

Circular pipes can withstand large pressure differences between the inside and the outside without undergoing any significant distortion, but noncircular pipes cannot.

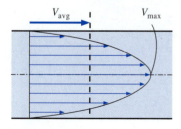

## FIGURE 14–2

Average velocity $V_{avg}$ is defined as the average speed through a cross section. For fully developed laminar pipe flow, $V_{avg}$ is half of the maximum velocity.

# 14–1 · INTRODUCTION

Liquid or gas flow through *pipes* or *ducts* is commonly used in heating and cooling applications and fluid distribution networks. The fluid in such applications is usually forced to flow by a fan or pump through a flow section. We pay particular attention to *friction,* which is directly related to the *pressure drop* and *head loss* during flow through pipes and ducts. The pressure drop is then used to determine the pumping power requirement. A typical piping system involves pipes of different diameters connected to each other by various fittings or elbows to route the fluid, valves to control the flow rate, and pumps to pressurize the fluid.

The terms *pipe, duct,* and *conduit* are usually used interchangeably for flow sections. In general, flow sections of circular cross section are referred to as *pipes* (especially when the fluid is a liquid), and flow sections of noncircular cross section as *ducts* (especially when the fluid is a gas). Small-diameter pipes are usually referred to as *tubes.* Given this uncertainty, we will use more descriptive phrases (such as *a circular pipe* or *a rectangular duct*) whenever necessary to avoid any misunderstandings.

You have probably noticed that most fluids, especially liquids, are transported in *circular pipes.* This is because pipes with a circular cross section can withstand large pressure differences between the inside and the outside without undergoing significant distortion. *Noncircular pipes* are usually used in applications such as the heating and cooling systems of buildings where the pressure difference is relatively small, the manufacturing and installation costs are lower, and the available space is limited for ductwork (Fig. 14–1).

Although the theory of fluid flow is reasonably well understood, theoretical solutions are obtained only for a few simple cases such as fully developed laminar flow in a circular pipe. Therefore, we must rely on experimental results and empirical relations for most fluid flow problems rather than closed-form analytical solutions. Noting that the experimental results are obtained under carefully controlled laboratory conditions and that no two systems are exactly alike, we must not be so naive as to view the results obtained as "exact." An error of 10 percent (or more) in friction factors calculated using the relations in this chapter is the "norm" rather than the "exception."

The fluid velocity in a pipe changes from *zero* at the wall because of the no-slip condition to a maximum at the pipe center. In fluid flow, it is convenient to work with an *average* velocity $V_{avg}$, which remains constant in incompressible flow when the cross-sectional area of the pipe is constant (Fig. 14–2). The average velocity in heating and cooling applications may change somewhat because of changes in density with temperature. But, in practice, we evaluate the fluid properties at some average temperature and treat them as constants. The convenience of working with constant properties usually more than justifies the slight loss in accuracy.

Also, the friction between the fluid particles in a pipe does cause a slight rise in fluid temperature as a result of the mechanical energy being converted to sensible thermal energy. But this temperature rise due to *frictional heating* is usually too small to warrant any consideration in calculations and thus is disregarded. For example, in the absence of any heat transfer, no noticeable difference can be detected between the inlet and outlet temperatures of water flowing in a pipe. The primary consequence of friction in fluid flow is pressure drop, and thus any significant temperature change in the fluid is due to heat transfer.

The value of the average velocity $V_{avg}$ at some streamwise cross-section is determined from the requirement that the *conservation of mass* principle be satisfied (Fig. 14–2). That is,

$$\dot{m} = \rho V_{avg} A_c = \int_{A_c} \rho u(r) \, dA_c \qquad (14\text{–}1)$$

where $\dot{m}$ is the mass flow rate, $\rho$ is the density, $A_c$ is the cross-sectional area, and $u(r)$ is the velocity profile. Then the average velocity for incompressible flow in a circular pipe of radius $R$ is expressed as

$$V_{avg} = \frac{\int_{A_c} \rho u(r) \, dA_c}{\rho A_c} = \frac{\int_0^R \rho u(r) 2\pi r \, dr}{\rho \pi R^2} = \frac{2}{R^2} \int_0^R u(r) r \, dr \qquad (14\text{–}2)$$

Therefore, when we know the flow rate or the velocity profile, the average velocity can be determined easily.

## 14–2 · LAMINAR AND TURBULENT FLOWS

If you have been around smokers, you probably noticed that the cigarette smoke rises in a smooth plume for the first few centimeters and then starts fluctuating randomly in all directions as it continues its rise. Other plumes behave similarly (Fig. 14–3). Likewise, a careful inspection of flow in a pipe reveals that the fluid flow is streamlined at low velocities but turns chaotic as the velocity is increased above a critical value, as shown in Fig. 14–4. The flow regime in the first case is said to be **laminar**, characterized by *smooth streamlines* and *highly ordered motion,* and **turbulent** in the second case, where it is characterized by *velocity fluctuations* and *highly disordered motion.* The **transition** from laminar to turbulent flow does not occur suddenly; rather, it occurs over some region in which the flow fluctuates between laminar and turbulent flows before it becomes fully turbulent. Most flows encountered in practice are turbulent. Laminar flow is encountered when highly viscous fluids such as oils flow in small pipes or narrow passages.

We can verify the existence of these laminar, transitional, and turbulent flow regimes by injecting some dye streaks into the flow in a glass pipe, as the British engineer Osborne Reynolds (1842–1912) did over a century ago. We observe that the dye streak forms a *straight and smooth line* at low velocities when the flow is laminar (we may see some blurring because of molecular diffusion), has *bursts of fluctuations* in the transitional regime, and *zigzags rapidly and disorderly* when the flow becomes fully turbulent. These zigzags and the dispersion of the dye are indicative of the fluctuations in the main flow and the rapid mixing of fluid particles from adjacent layers.

The *intense mixing* of the fluid in turbulent flow as a result of rapid fluctuations enhances momentum transfer between fluid particles, which increases the friction force on the pipe wall and thus the required pumping power. The friction factor reaches a maximum when the flow becomes fully turbulent.

**FIGURE 14–3**

Laminar and turbulent flow regimes of a candle smoke plume.

(a) Laminar flow

(b) Turbulent flow

**FIGURE 14–4**

The behavior of colored fluid injected into the flow in (a) laminar and (b) turbulent flow in a pipe.

**FIGURE 14–5**

The Reynolds number can be viewed as the ratio of inertial forces to viscous forces acting on a fluid element.

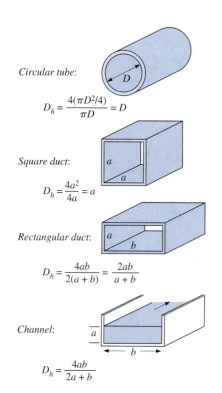

**FIGURE 14–6**

The hydraulic diameter $D_h = 4A_c/p$ is defined such that it reduces to ordinary diameter for circular tubes. When there is a free surface, such as in open-channel flow, the wetted perimeter includes only the walls in contact with the fluid.

# Reynolds Number

The transition from laminar to turbulent flow depends on the *geometry, surface roughness, flow velocity, surface temperature,* and *type of fluid,* among other things. After exhaustive experiments in the 1880s, Osborne Reynolds discovered that the flow regime depends mainly on the ratio of *inertial forces* to *viscous forces* in the fluid. This ratio is called the **Reynolds number** and is expressed for internal flow in a circular pipe as (Fig. 14–5)

$$\text{Re} = \frac{\text{Inertial forces}}{\text{Viscous forces}} = \frac{V_{avg}D}{\nu} = \frac{\rho V_{avg}D}{\mu} \qquad (14\text{–}3)$$

where $V_{avg}$ = average flow velocity (m/s), $D$ = characteristic length of the geometry (diameter in this case, in m), and $\nu = \mu/\rho$ = kinematic viscosity of the fluid (m²/s). Note that the Reynolds number is a *dimensionless* quantity. Also, kinematic viscosity has the unit m²/s, and can be viewed as *viscous diffusivity* or *diffusivity for momentum.*

At large Reynolds numbers, the inertial forces, which are proportional to the fluid density and the square of the fluid velocity, are large relative to the viscous forces, and thus the viscous forces cannot prevent the random and rapid fluctuations of the fluid. At *small* or *moderate* Reynolds numbers, however, the viscous forces are large enough to suppress these fluctuations and to keep the fluid "in line." Thus the flow is *turbulent* in the first case and *laminar* in the second.

The Reynolds number at which the flow becomes turbulent is called the **critical Reynolds number**, $\text{Re}_{cr}$. The value of the critical Reynolds number is different for different geometries and flow conditions. For internal flow in a circular pipe, the generally accepted value of the critical Reynolds number is $\text{Re}_{cr} = 2300$.

For flow through noncircular pipes, the Reynolds number is based on the **hydraulic diameter** $D_h$ defined as (Fig. 14–6)

$$\text{Hydraulic diameter:} \qquad D_h = \frac{4A_c}{p} \qquad (14\text{–}4)$$

where $A_c$ is the cross-sectional area of the pipe and $p$ is its wetted perimeter. The hydraulic diameter is defined such that it reduces to ordinary diameter $D$ for circular pipes,

$$\text{Circular pipes:} \qquad D_h = \frac{4A_c}{p} = \frac{4(\pi D^2/4)}{\pi D} = D$$

It certainly is desirable to have precise values of Reynolds numbers for laminar, transitional, and turbulent flows, but this is not the case in practice. It turns out that the transition from laminar to turbulent flow also depends on the degree of disturbance of the flow by *surface roughness, pipe vibrations,* and *fluctuations in the upstream flow.* Under most practical conditions, the flow in a circular pipe is laminar for $\text{Re} \lesssim 2300$, turbulent for $\text{Re} \gtrsim 4000$, and transitional in between. That is,

$$\text{Re} \lesssim 2300 \qquad \text{laminar flow}$$
$$2300 \lesssim \text{Re} \lesssim 4000 \qquad \text{transitional flow}$$
$$\text{Re} \gtrsim 4000 \qquad \text{turbulent flow}$$

In transitional flow, the flow switches between laminar and turbulent in a disorderly fashion (Fig. 14–7). It should be kept in mind that laminar flow can be maintained at much higher Reynolds numbers in very smooth pipes by avoiding flow disturbances and pipe vibrations. In such carefully controlled laboratory experiments, laminar flow has been maintained at Reynolds numbers of up to 100,000.

## 14–3 · THE ENTRANCE REGION

Consider a fluid entering a circular pipe at a uniform velocity. Because of the no-slip condition, the fluid particles in the layer in contact with the wall of the pipe come to a complete stop. This layer also causes the fluid particles in the adjacent layers to slow down gradually as a result of friction. To make up for this velocity reduction, the velocity of the fluid at the midsection of the pipe has to increase to keep the mass flow rate through the pipe constant. As a result, a velocity gradient develops along the pipe.

The region of the flow in which the effects of the viscous shearing forces caused by fluid viscosity are felt is called the **velocity boundary layer** or just the **boundary layer**. The hypothetical boundary surface divides the flow in a pipe into two regions: the **boundary layer region**, in which the viscous effects and the velocity changes are significant, and the **irrotational (core) flow region**, in which the frictional effects are negligible and the velocity remains essentially constant in the radial direction.

The thickness of this boundary layer increases in the flow direction until the boundary layer reaches the pipe center and thus fills the entire pipe, as shown in Fig. 14–8, and the velocity becomes fully developed a little farther downstream. The region from the pipe inlet to the point at which the velocity profile is fully developed is called the **hydrodynamic entrance region**, and the length of this region is called the **hydrodynamic entry length** $L_h$. Flow in the entrance region is called *hydrodynamically developing flow* since this is the region where the velocity profile develops. The region beyond the entrance region in which the velocity profile is fully developed and remains unchanged is called the **hydrodynamically fully developed region**. The flow is said to be **fully developed** when the normalized temperature profile remains unchanged as well. Hydrodynamically fully developed flow is equivalent to fully developed flow when the fluid in the pipe is not heated or cooled since the fluid temperature in this case remains

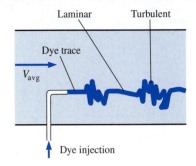

**FIGURE 14–7**

In the transitional flow region of $2300 \leq \text{Re} \leq 4000$, the flow switches between laminar and turbulent somewhat randomly.

**FIGURE 14–8**

The development of the velocity boundary layer in a pipe. (The developed average velocity profile is parabolic in laminar flow, as shown, but much flatter or fuller in turbulent flow.)

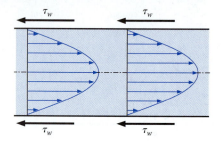

**FIGURE 14–9**

In the fully developed flow region of a pipe, the velocity profile does not change downstream, and thus the wall shear stress remains constant as well.

essentially constant throughout. The velocity profile in the fully developed region is *parabolic* in laminar flow and much *flatter* (or *fuller*) in turbulent flow due to eddy motion and more vigorous mixing in the radial direction. The time-averaged velocity profile remains unchanged when the flow is fully developed, and thus

$$\text{Hydrodynamically fully developed:} \quad \frac{\partial u(r, x)}{\partial x} = 0 \quad \rightarrow \quad u = u(r) \tag{14–5}$$

The shear stress at the pipe wall $\tau_w$ is related to the slope of the velocity profile at the surface. Noting that the velocity profile remains unchanged in the hydrodynamically fully developed region, the wall shear stress also remains constant in that region (Fig. 14–9).

Consider fluid flow in the hydrodynamic entrance region of a pipe. The wall shear stress is the *highest* at the pipe inlet where the thickness of the boundary layer is smallest, and decreases gradually to the fully developed value, as shown in Fig. 14–10. Therefore, the pressure drop is *higher* in the entrance regions of a pipe, and the effect of the entrance region is always to *increase* the average friction factor for the entire pipe. This increase may be significant for short pipes but is negligible for long ones.

## Entry Lengths

The hydrodynamic entry length is usually taken to be the distance from the pipe entrance to where the wall shear stress (and thus the friction factor) reaches within about 2 percent of the fully developed value. In *laminar flow,* the nondimensional hydrodynamic entry length is given approximately as [see Kays and Crawford (1993) and Shah and Bhatti (1987)]

$$\frac{L_{h,\, \text{laminar}}}{D} \cong 0.05 \text{Re} \tag{14–6}$$

**FIGURE 14–10**

The variation of wall shear stress in the flow direction for flow in a pipe from the entrance region into the fully developed region.

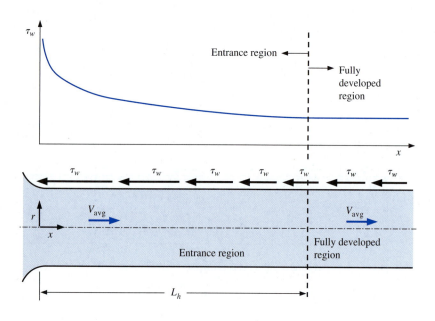

For Re = 20, the hydrodynamic entry length is about the size of the diameter, but increases linearly with velocity. In the limiting laminar case of Re = 2300, the hydrodynamic entry length is 115D.

In *turbulent flow*, the intense mixing during random fluctuations usually overshadows the effects of molecular diffusion. The nondimensional hydrodynamic entry length for turbulent flow is approximated as [see Bhatti and Shah (1987) and Zhi-qing (1982)]

$$\frac{L_{h,\, turbulent}}{D} = 1.359 Re^{1/4} \qquad (14\text{–}7)$$

The entry length is much shorter in turbulent flow, as expected, and its dependence on the Reynolds number is weaker. In many pipe flows of practical engineering interest, the entrance effects become insignificant beyond a pipe length of about 10 diameters, and the nondimensional hydrodynamic entry length is approximated as

$$\frac{L_{h,\, turbulent}}{D} \approx 10 \qquad (14\text{–}8)$$

Precise correlations for calculating the frictional head losses in entrance regions are available in the literature. However, the pipes used in practice are usually several times the length of the entrance region, and thus the flow through the pipes is often assumed to be fully developed for the entire length of the pipe. This simplistic approach gives *reasonable* results for long pipes but sometimes poor results for short ones since it underpredicts the wall shear stress and thus the friction factor.

## 14–4 · LAMINAR FLOW IN PIPES

We mentioned in Section 14–2 that flow in pipes is laminar for Re ≲ 2300, and that the flow is fully developed if the pipe is sufficiently long (relative to the entry length) so that the entrance effects are negligible. In this section, we consider the steady, laminar, incompressible flow of fluid with constant properties in the fully developed region of a straight circular pipe. We obtain the momentum equation by applying a momentum balance to a differential volume element, and obtain the velocity profile by solving it. Then we use it to obtain a relation for the friction factor. An important aspect of the analysis here is that it is one of the few available for viscous flow.

In fully developed laminar flow, each fluid particle moves at a constant axial velocity along a streamline and the velocity profile $u(r)$ remains unchanged in the flow direction. There is no motion in the radial direction, and thus the velocity component in the direction normal to the pipe axis is everywhere zero. There is no acceleration since the flow is steady and fully developed.

Now consider a ring-shaped differential volume element of radius $r$, thickness $dr$, and length $dx$ oriented coaxially with the pipe, as shown in Fig. 14–11. The volume element involves only pressure and viscous effects and thus the pressure and shear forces must balance each other. The pressure force acting on a submerged plane surface is the product of the pressure at the centroid of the surface and the surface area. A force balance on the volume element in the flow direction gives

$$(2\pi r\, dr\, P)_x - (2\pi r\, dr\, P)_{x+dx} + (2\pi r\, dx\, \tau)_r - (2\pi r\, dx\, \tau)_{r+dr} = 0 \qquad (14\text{–}9)$$

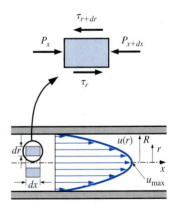

**FIGURE 14–11**

Free-body diagram of a ring-shaped differential fluid element of radius $r$, thickness $dr$, and length $dx$ oriented coaxially with a horizontal pipe in fully developed laminar flow. (The size of the fluid element is greatly exaggerated for clarity.)

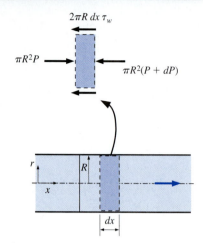

**Force balance:**
$$\pi R^2 P - \pi R^2 (P + dP) - 2\pi R\, dx\, \tau_w = 0$$

**Simplifying:**
$$\frac{dP}{dx} = -\frac{2\tau_w}{R}$$

**FIGURE 14–12**

Free-body diagram of a fluid disk element of radius $R$ and length $dx$ in fully developed laminar flow in a horizontal pipe.

which indicates that in fully developed flow in a horizontal pipe, the viscous and pressure forces balance each other. Dividing by $2\pi dr dx$ and rearranging,

$$r\frac{P_{x+dx} - P_x}{dx} + \frac{(r\tau)_{r+dr} - (r\tau)_r}{dr} = 0 \qquad (14\text{–}10)$$

Taking the limit as $dr$, $dx \rightarrow 0$ gives

$$r\frac{dP}{dx} + \frac{d(r\tau)}{dr} = 0 \qquad (14\text{–}11)$$

Substituting $\tau = -\mu(du/dr)$, dividing by $r$, and taking $\mu = $ constant gives the desired equation,

$$\frac{\mu}{r}\frac{d}{dr}\left(r\frac{du}{dr}\right) = \frac{dP}{dx} \qquad (14\text{–}12)$$

The quantity $du/dr$ is negative in pipe flow, and the negative sign is included to obtain positive values for $\tau$. (Or, $du/dr = -du/dy$ if we define $y = R - r$.) The left side of Eq. 14–12 is a function of $r$, and the right side is a function of $x$. The equality must hold for any value of $r$ and $x$, and an equality of the form $f(r) = g(x)$ can be satisfied only if both $f(r)$ and $g(x)$ are equal to the same constant. Thus we conclude that $dP/dx = $ constant. This is verified by writing a force balance on a volume element of radius $R$ and thickness $dx$ (a slice of the pipe as in Fig. 14–12), which gives

$$\frac{dP}{dx} = -\frac{2\tau_w}{R} \qquad (14\text{–}13)$$

Here $\tau_w$ is constant since the viscosity and the velocity profile are constants in the fully developed region. Therefore, $dP/dx = $ constant.

Equation 14–12 is solved by rearranging and integrating it twice to give

$$u(r) = \frac{r^2}{4\mu}\left(\frac{dP}{dx}\right) + C_1 \ln r + C_2 \qquad (14\text{–}14)$$

The velocity profile $u(r)$ is obtained by applying the boundary conditions $\partial u/\partial r = 0$ at $r = 0$ (because of symmetry about the centerline) and $u = 0$ at $r = R$ (the no-slip condition at the pipe wall),

$$u(r) = -\frac{R^2}{4\mu}\left(\frac{dP}{dx}\right)\left(1 - \frac{r^2}{R^2}\right) \qquad (14\text{–}15)$$

Therefore, the velocity profile in fully developed laminar flow in a pipe is *parabolic* with a maximum at the centerline and a minimum (zero) at the pipe wall. Also, the axial velocity $u$ is positive for any $r$, and thus the axial pressure gradient $dP/dx$ must be negative (i.e., pressure must decrease in the flow direction because of viscous effects—it takes pressure to push the fluid through the pipe).

The average velocity is determined from its definition by substituting Eq. 14–15 into Eq. 14–2, and performing the integration, yielding

$$V_{avg} = \frac{2}{R^2}\int_0^R u(r)r\,dr = \frac{-2}{R^2}\int_0^R \frac{R^2}{4\mu}\left(\frac{dP}{dx}\right)\left(1 - \frac{r^2}{R^2}\right)r\,dr = -\frac{R^2}{8\mu}\left(\frac{dP}{dx}\right) \qquad (14\text{–}16)$$

Combining the last two equations, the velocity profile is rewritten as

$$u(r) = 2V_{avg}\left(1 - \frac{r^2}{R^2}\right) \qquad (14\text{–}17)$$

This is a convenient form for the velocity profile since $V_{avg}$ can be determined easily from the flow rate information.

The maximum velocity occurs at the centerline and is determined from Eq. 14–17 by substituting $r = 0$,

$$u_{max} = 2V_{avg} \tag{14–18}$$

Therefore, *the average velocity in fully developed laminar pipe flow is one-half of the maximum velocity.*

## Pressure Drop and Head Loss

A quantity of interest in the analysis of pipe flow is the *pressure drop* $\Delta P$ since it is directly related to the power requirements of the fan or pump to maintain flow. We note that $dP/dx = $ constant, and integrating from $x = x_1$ where the pressure is $P_1$ to $x = x_1 + L$ where the pressure is $P_2$ gives

$$\frac{dP}{dx} = \frac{P_2 - P_1}{L} \tag{14–19}$$

Substituting Eq. 14–19 into the $V_{avg}$ expression in Eq. 14–16, the pressure drop is expressed as

*Laminar flow:* $\qquad \Delta P = P_1 - P_2 = \dfrac{8\mu L V_{avg}}{R^2} = \dfrac{32\mu L V_{avg}}{D^2} \tag{14–20}$

The symbol $\Delta$ is typically used to indicate the difference between the final and initial values, like $\Delta y = y_2 - y_1$. But in fluid flow, $\Delta P$ is used to designate pressure drop, and thus it is $P_1 - P_2$. A pressure drop due to viscous effects represents an irreversible pressure loss, and it is sometimes called **pressure loss** $\Delta P_L$ to emphasize that it is a *loss* (just like the head loss $h_L$, which as we shall see is proportional to $\Delta P_L$).

Note from Eq. 14–20 that the pressure drop is proportional to the viscosity $\mu$ of the fluid, and $\Delta P$ would be zero if there were no friction. Therefore, the drop of pressure from $P_1$ to $P_2$ in this case is due entirely to viscous effects, and Eq. 14–20 represents the pressure loss $\Delta P_L$ when a fluid of viscosity $\mu$ flows through a pipe of constant diameter $D$ and length $L$ at average velocity $V_{avg}$.

In practice, it is convenient to express the pressure loss for all types of fully developed internal flows (laminar or turbulent flows, circular or noncircular pipes, smooth or rough surfaces, horizontal or inclined pipes) as (Fig. 14–13)

*Pressure loss:* $\qquad \Delta P_L = f\dfrac{L}{D}\dfrac{\rho V_{avg}^2}{2} \tag{14–21}$

where $\rho V_{avg}^2/2$ is the *dynamic pressure* and $f$ is the **Darcy friction factor**,

$$f = \frac{8\tau_w}{\rho V_{avg}^2} \tag{14–22}$$

It is also called the **Darcy–Weisbach friction factor**, named after the Frenchman Henry Darcy (1803–1858) and the German Julius Weisbach (1806–1871), the two engineers who provided the greatest contribution in its development. It should not be confused with the *friction coefficient* $C_f$

**FIGURE 14–13**

The relation for pressure loss (and head loss) is one of the most general relations in fluid mechanics, and it is valid for laminar or turbulent flows, circular or noncircular pipes, and pipes with smooth or rough surfaces.

[also called the *Fanning friction factor*, named after the American engineer John Fanning (1837–1911)], which is defined as $C_f = 2\tau_w/(\rho V_{avg}^2) = f/4$.

Setting Eqs. 14–20 and 14–21 equal to each other and solving for $f$ gives the friction factor for fully developed laminar flow in a circular pipe,

*Circular pipe, laminar:*
$$f = \frac{64\mu}{\rho D V_{avg}} = \frac{64}{Re} \qquad (14\text{–}23)$$

This equation shows that *in laminar flow, the friction factor is a function of the Reynolds number only and is independent of the roughness of the pipe surface.*

In the analysis of piping systems, pressure losses are commonly expressed in terms of the *equivalent fluid column height*, called the **head loss** $h_L$. Noting from fluid statics that $\Delta P = \rho g h$ and thus a pressure difference of $\Delta P$ corresponds to a fluid height of $h = \Delta P/\rho g$, the *pipe head loss* is obtained by dividing $\Delta P_L$ by $\rho g$ to give

*Head loss:*
$$h_L = \frac{\Delta P_L}{\rho g} = f \frac{L}{D} \frac{V_{avg}^2}{2g} \qquad (14\text{–}24)$$

The head loss $h_L$ represents *the additional height that the fluid needs to be raised by a pump in order to overcome the frictional losses in the pipe.* The head loss is caused by viscosity, and it is directly related to the wall shear stress. Equations 14–21 and 14–24 are valid for both laminar and turbulent flows in both circular and noncircular pipes, but Eq. 14–23 is valid only for fully developed laminar flow in circular pipes.

Once the pressure loss (or head loss) is known, the required pumping power *to overcome the pressure loss* is determined from

$$\dot{W}_{pump, L} = \dot{V}\Delta P_L = \dot{V}\rho g h_L = \dot{m} g h_L \qquad (14\text{–}25)$$

where $\dot{V}$ is the volume flow rate and $\dot{m}$ is the mass flow rate.

The average velocity for laminar flow in a horizontal pipe is, from Eq. 14–20,

*Horizontal pipe:*
$$V_{avg} = \frac{(P_1 - P_2)R^2}{8\mu L} = \frac{(P_1 - P_2)D^2}{32\mu L} = \frac{\Delta P D^2}{32\mu L} \qquad (14\text{–}26)$$

Then the volume flow rate for laminar flow through a horizontal pipe of diameter $D$ and length $L$ becomes

$$\dot{V} = V_{avg}A_c = \frac{(P_1 - P_2)R^2}{8\mu L}\pi R^2 = \frac{(P_1 - P_2)\pi D^4}{128\mu L} = \frac{\Delta P \pi D^4}{128\mu L} \qquad (14\text{–}27)$$

This equation is known as **Poiseuille's law**, and this flow is called *Hagen–Poiseuille flow* in honor of the works of G. Hagen (1797–1884) and J. Poiseuille (1799–1869) on the subject. Note from Eq. 14–27 that *for a specified flow rate, the pressure drop and thus the required pumping power is proportional to the length of the pipe and the viscosity of the fluid, but it is inversely proportional to the fourth power of the radius (or diameter) of the pipe.* Therefore, the pumping power requirement for a laminar-flow piping system can be reduced by a factor of 16 by doubling the pipe diameter (Fig. 14–14). Of course the benefits of the reduction in the energy costs must be weighed against the increased cost of constructiosn due to using a larger-diameter pipe.

The pressure drop $\Delta P$ equals the pressure loss $\Delta P_L$ in the case of a horizontal pipe, but this is not the case for inclined pipes or pipes with variable cross-sectional area. This can be demonstrated by writing the energy

**FIGURE 14–14**

The pumping power requirement for a laminar-flow piping system can be reduced by a factor of 16 by doubling the pipe diameter.

equation for steady, incompressible one-dimensional flow in terms of heads as (see Chap. 12)

$$\frac{P_1}{\rho g} + \alpha_1 \frac{V_1^2}{2g} + z_1 + h_{\text{pump}, u} = \frac{P_2}{\rho g} + \alpha_2 \frac{V_2^2}{2g} + z_2 + h_{\text{turbine}, e} + h_L \quad \text{(14–28)}$$

where $h_{\text{pump}, u}$ is the useful pump head delivered to the fluid, $h_{\text{turbine}, e}$ is the turbine head extracted from the fluid, $h_L$ is the irreversible head loss between sections 1 and 2, $V_1$ and $V_2$ are the average velocities at sections 1 and 2, respectively, and $\alpha_1$ and $\alpha_2$ are the *kinetic energy correction factors* at sections 1 and 2 (it can be shown that $\alpha = 2$ for fully developed laminar flow and about 1.05 for fully developed turbulent flow). Equation 14–28 can be rearranged as

$$P_1 - P_2 = \rho\left(\alpha_2 V_2^2 - \alpha_1 V_1^2\right)/2 + \rho g[(z_2 - z_1) + h_{\text{turbine}, e} - h_{\text{pump}, u} + h_L] \quad \text{(14–29)}$$

Therefore, the pressure drop $\Delta P = P_1 - P_2$ and pressure loss $\Delta P_L = \rho g h_L$ for a given flow section are equivalent if (1) the flow section is horizontal so that there are no hydrostatic or gravity effects ($z_1 = z_2$), (2) the flow section does not involve any work devices such as a pump or a turbine since they change the fluid pressure ($h_{\text{pump}, u} = h_{\text{turbine}, e} = 0$), (3) the cross-sectional area of the flow section is constant and thus the average flow velocity is constant ($V_1 = V_2$), and (4) the velocity profiles at sections 1 and 2 are the same shape ($\alpha_1 = \alpha_2$).

## Effect of Gravity on Velocity and Flow Rate in Laminar Flow

Gravity has no affect on flow in horizontal pipes, but it has a significant effect on both the velocity and the flow rate in uphill or downhill pipes. Relations for inclined pipes can be obtained in a similar manner from a force balance in the direction of flow. The only additional force in this case is the component of the fluid weight in the flow direction, whose magnitude is

$$W_x = W \sin\theta = \rho g V_{\text{element}} \sin\theta = \rho g (2\pi r \, dr \, dx) \sin\theta \quad \text{(14–30)}$$

where $\theta$ is the angle between the horizontal and the flow direction (Fig. 14–15). The force balance in Eq. 14–9 now becomes

$$(2\pi r \, dr \, P)_x - (2\pi r \, dr \, P)_{x+dx} + (2\pi r \, dx \, \tau)_r$$
$$- (2\pi r \, dx \, \tau)_{r+dr} - \rho g (2\pi r \, dr \, dx) \sin\theta = 0 \quad \text{(14–31)}$$

which results in the differential equation

$$\frac{\mu}{r} \frac{d}{dr}\left(r \frac{du}{dr}\right) = \frac{dP}{dx} + \rho g \sin\theta \quad \text{(14–32)}$$

Following the same solution procedure as previously, the velocity profile is

$$u(r) = -\frac{R^2}{4\mu}\left(\frac{dP}{dx} + \rho g \sin\theta\right)\left(1 - \frac{r^2}{R^2}\right) \quad \text{(14–33)}$$

From Eq. 14–33, the *average velocity* and the *volume flow rate* relations for laminar flow through inclined pipes are, respectively,

$$V_{\text{avg}} = \frac{(\Delta P - \rho g L \sin\theta)D^2}{32\mu L} \quad \text{and} \quad \dot{V} = \frac{(\Delta P - \rho g L \sin\theta)\pi D^4}{128\mu L} \quad \text{(14–34)}$$

which are identical to the corresponding relations for horizontal pipes, except that $\Delta P$ is replaced by $\Delta P - \rho g L \sin\theta$. Therefore, the results already obtained for horizontal pipes can also be used for inclined pipes provided

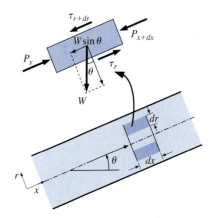

**FIGURE 14–15**

Free-body diagram of a ring-shaped differential fluid element of radius $r$, thickness $dr$, and length $dx$ oriented coaxially with an inclined pipe in fully developed laminar flow.

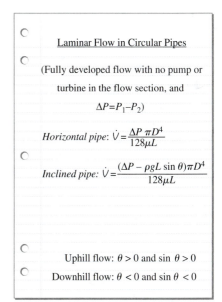

Laminar Flow in Circular Pipes

(Fully developed flow with no pump or turbine in the flow section, and $\Delta P = P_1 - P_2$)

Horizontal pipe: $\dot{V} = \dfrac{\Delta P \, \pi D^4}{128\mu L}$

Inclined pipe: $\dot{V} = \dfrac{(\Delta P - \rho g L \sin\theta)\pi D^4}{128\mu L}$

Uphill flow: $\theta > 0$ and $\sin\theta > 0$

Downhill flow: $\theta < 0$ and $\sin\theta < 0$

**FIGURE 14–16**

The relations developed for fully developed laminar flow through horizontal pipes can also be used for inclined pipes by replacing $\Delta P$ with $\Delta P - \rho g L \sin\theta$.

that $\Delta P$ is replaced by $\Delta P - \rho g L \sin \theta$ (Fig. 14–16). Note that $\theta > 0$ and thus $\sin \theta > 0$ for uphill flow, and $\theta < 0$ and thus $\sin \theta < 0$ for downhill flow.

In inclined pipes, the combined effect of pressure difference and gravity drives the flow. Gravity helps downhill flow but opposes uphill flow. Therefore, much greater pressure differences need to be applied to maintain a specified flow rate in uphill flow although this becomes important only for liquids, because the density of gases is generally low. In the special case of *no flow* ($\dot{V} = 0$), we have $\Delta P = \rho g L \sin \theta$, which is what we would obtain from fluid statics (Chap. 11).

## Laminar Flow in Noncircular Pipes

The friction factor $f$ relations are given in Table 14–1 for *fully developed laminar flow* in pipes of various cross sections. The Reynolds number for flow in these pipes is based on the hydraulic diameter $D_h = 4A_c/p$, where $A_c$ is the cross-sectional area of the pipe and $p$ is its wetted perimeter.

**TABLE 14–1**

Friction factor for fully developed *laminar flow* in pipes of various cross sections ($D_h = 4A_c/p$ and Re $= V_{avg} D_h/\nu$)

| Tube Geometry | a/b or θ° | Friction Factor f |
|---|---|---|
| Circle | — | 64.00/Re |
| | | |
| Rectangle | a/b | |
| | 1 | 56.92/Re |
| | 2 | 62.20/Re |
| | 3 | 68.36/Re |
| | 4 | 72.92/Re |
| | 6 | 78.80/Re |
| | 8 | 82.32/Re |
| | ∞ | 96.00/Re |
| Ellipse | a/b | |
| | 1 | 64.00/Re |
| | 2 | 67.28/Re |
| | 4 | 72.96/Re |
| | 8 | 76.60/Re |
| | 16 | 78.16/Re |
| Isosceles triangle | θ | |
| | 10° | 50.80/Re |
| | 30° | 52.28/Re |
| | 60° | 53.32/Re |
| | 90° | 52.60/Re |
| | 120° | 50.96/Re |

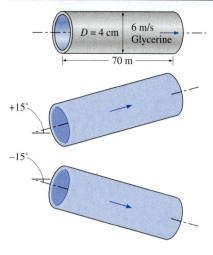

**FIGURE 14–17**
Schematic for Example 14–1.

*EXAMPLE 14–1*    **Laminar Flow in Horizontal and Inclined Pipes**

Consider the fully developed flow of glycerin at 40°C through a 70-m-long, 4-cm-diameter, horizontal, circular pipe (Fig. 14–17). If the flow velocity at the centerline is measured to be 6 m/s, determine the velocity profile and the pressure difference across this 70-m-long section of the pipe, and the useful pumping power required to maintain this flow. For the same useful pumping power input, determine the percent increase of the flow rate if the pipe is inclined 15° downward and the percent decrease if it is inclined 15° upward. The pump is located outside this pipe section.

**Solution**   The centerline velocity in a horizontal pipe in fully developed flow is measured. The velocity profile, the pressure difference across the pipe, and the pumping power required are to be determined. The effects of downward and upward tilting of the pipe on the flow rate is to be investigated.
**Assumptions**   **1** The flow is steady, laminar, incompressible, and fully developed. **2** There are no pumps or turbines in the flow section. **3** There are no valves, elbows, or other devices that may cause local losses.
**Properties**   The density and dynamic viscosity of glycerin at 40°C are $\rho = 1252$ kg/m$^3$ and $\mu = 0.3073$ kg/m·s, respectively.
**Analysis**   The velocity profile in fully developed laminar flow in a circular pipe is expressed as

$$u(r) = u_{max}\left(1 - \frac{r^2}{R^2}\right)$$

Substituting, the velocity profile is determined to be

$$u(r) = (6 \text{ m/s})\left(1 - \frac{r^2}{(0.02 \text{ m})^2}\right) = 6(1 - 2500r^2)$$

where $u$ is in m/s and $r$ is in m. The average velocity, the flow rate, and the Reynolds number are

$$V = V_{avg} = \frac{u_{max}}{2} = \frac{6 \text{ m/s}}{2} = 3 \text{ m/s}$$

$$\dot{V} = V_{avg} A_c = V(\pi D^2/4) = (3 \text{ m/s})[\pi (0.04 \text{ m})^2/4] = 3.77 \times 10^{-3} \text{ m}^3/\text{s}$$

$$\text{Re} = \frac{\rho V D}{\mu} = \frac{(1252 \text{ kg/m}^3)(3 \text{ m/s})(0.04 \text{ m})}{0.3073 \text{ kg/m·s}} = 488.9$$

which is less than 2300. Therefore, the flow is indeed laminar. Then the friction factor and the head loss become

$$f = \frac{64}{\text{Re}} = \frac{64}{488.9} = 0.1309$$

$$h_L = f\frac{LV^2}{D\,2g} = 0.1309 \frac{(70 \text{ m})}{(0.04 \text{ m})} \frac{(3 \text{ m/s})^2}{2(9.81 \text{ m/s}^2)} = 105.1\text{m}$$

The energy balance for steady, incompressible one-dimensional flow is given by Eq. 14–28 as

$$\frac{P_1}{\rho g} + \alpha_1 \frac{V_1^2}{2g} + z_1 + h_{pump, u} = \frac{P_2}{\rho g} + \alpha_2 \frac{V_2^2}{2g} + z_2 + h_{turbine, e} + h_L$$

For fully developed flow in a constant diameter pipe with no pumps or turbines, it reduces to

$$\Delta P = P_1 - P_2 = \rho g(z_2 - z_1 + h_L)$$

Then the pressure difference and the required useful pumping power for the horizontal case become

$$\Delta P = \rho g(z_2 - z_1 + h_L)$$

$$= (1252 \text{ kg/m}^3)(9.81 \text{ m/s}^2)(0 + 105.1 \text{ m})\left(\frac{1 \text{ kPa}}{1000 \text{ kg/m}\cdot\text{s}^2}\right)$$

$$= 1291 \text{ kPa}$$

$$\dot{W}_{\text{pump, u}} = \dot{V}\Delta P = (3.77 \times 10^3 \text{ m}^3/\text{s})(1291 \text{ kPa})\left(\frac{1 \text{ kW}}{\text{kPa}\cdot\text{m}^3/\text{s}}\right) = 4.87 \text{ kW}$$

The elevation difference and the pressure difference for a pipe inclined upwards 15° is

$$\Delta z = z_2 - z_1 = L\sin 15° = (70 \text{ m})\sin 15° = 18.1 \text{ m}$$

$$\Delta P_{\text{upward}} = (1252 \text{ kg/m}^3)(9.81 \text{ m/s}^2)(18.1 \text{ m} + 105.1 \text{ m})\left(\frac{1 \text{ kPa}}{1000 \text{ kg/m}\cdot\text{s}^2}\right)$$

$$= 1366 \text{ kPa}$$

Then the flow rate through the upward inclined pipe becomes

$$\dot{V}_{\text{upward}} = \frac{\dot{W}_{\text{pump, u}}}{\Delta P_{\text{upward}}} = \frac{4.87 \text{ kW}}{1366 \text{ kPa}}\left(\frac{1 \text{ kPa}\cdot\text{m}^3/\text{s}}{1 \text{ kW}}\right) = 3.57 \times 10^{-3} \text{ m}^3/\text{s}$$

which is a decrease of **5.6 percent** in flow rate. It can be shown similarly that when the pipe is inclined 15° downward from the horizontal, the flow rate will increase by **5.6 percent**.

**Discussion** Note that the flow is driven by the combined effect of pumping power and gravity. As expected, gravity opposes uphill flow, enhances downhill flow, and has no effect on horizontal flow. Downhill flow can occur even in the absence of a pressure difference applied by a pump. For the case of $P_1 = P_2$ (i.e., no applied pressure difference), the pressure throughout the entire pipe would remain constant, and the fluid would flow through the pipe under the influence of gravity at a rate that depends on the angle of inclination, reaching its maximum value when the pipe is vertical. When solving pipe flow problems, it is always a good idea to calculate the Reynolds number to verify the flow regime—laminar or turbulent.

**EXAMPLE 14–2** **Pressure Drop and Head Loss in a Pipe**

Water at 40°F ($\rho = 62.42 \text{ lbm/ft}^3$ and $\mu = 1.038 \times 10^{-3} \text{ lbm/ft}\cdot\text{s}$) is flowing steadily through a 0.12-in- (= 0.010 ft) diameter 30-ft-long horizontal pipe at an average velocity of 3.0 ft/s (Fig. 14–18). Determine (a) the head loss, (b) the pressure drop, and (c) the pumping power requirement to overcome this pressure drop.

**FIGURE 14–18**

Schematic for Example 14–2.

**Solution** The average flow velocity in a pipe is given. The head loss, the pressure drop, and the pumping power are to be determined.

**Assumptions** 1 The flow is steady and incompressible. 2 The entrance effects are negligible, and thus the flow is fully developed. 3 The pipe involves no components such as bends, valves, and connectors.

**Properties** The density and dynamic viscosity of water are given to be $\rho = 62.42$ lbm/ft$^3$ and $\mu = 1.038 \times 10^{-3}$ lbm/ft·s, respectively.

**Analysis** (a) First we need to determine the flow regime. The Reynolds number is

$$\text{Re} = \frac{\rho V_{\text{avg}} D}{\mu} = \frac{(62.42 \text{ lbm/ft}^3)(3 \text{ ft/s})(0.01 \text{ ft})}{1.038 \times 10^{-3} \text{ lbm/ft·s}} = 1803$$

which is less than 2300. Therefore, the flow is laminar. Then the friction factor and the head loss become

$$f = \frac{64}{\text{Re}} = \frac{64}{1803} = 0.0355$$

$$h_L = f \frac{L}{D} \frac{V_{\text{avg}}^2}{2g} = 0.0355 \frac{30 \text{ ft}}{0.01 \text{ ft}} \frac{(3 \text{ ft/s})^2}{2(32.2 \text{ ft/s}^2)} = \textbf{14.9 ft}$$

(b) Noting that the pipe is horizontal and its diameter is constant, the pressure drop in the pipe is due entirely to the frictional losses and is equivalent to the pressure loss,

$$\Delta P = \Delta P_L = f \frac{L}{D} \frac{\rho V_{\text{avg}}^2}{2} = 0.0355 \frac{30 \text{ ft}}{0.01 \text{ ft}} \frac{(62.42 \text{ lbm/ft}^3)(3 \text{ ft/s})^2}{2} \left( \frac{1 \text{ lbf}}{32.2 \text{ lbm·ft/s}^2} \right)$$

$$= \textbf{929 lbf/ft}^2 = \textbf{6.45 psi}$$

(c) The volume flow rate and the pumping power requirements are

$$\dot{V} = V_{\text{avg}} A_c = V_{\text{avg}}(\pi D^2/4) = (3 \text{ ft/s})[\pi(0.01 \text{ ft})^2/4] = 0.000236 \text{ ft}^3/\text{s}$$

$$\dot{W}_{\text{pump}} = \dot{V} \Delta P = (0.000236 \text{ ft}^3/\text{s})(929 \text{ lbf/ft}^2) \left( \frac{1 \text{ W}}{0.737 \text{ lbf·ft/s}} \right) = \textbf{0.30 W}$$

Therefore, power input in the amount of 0.30 W is needed to overcome the frictional losses in the flow due to viscosity.

**Discussion** The pressure rise provided by a pump is often listed by a pump manufacturer in units of head. Thus, the pump in this flow needs to provide 14.9 ft of water head in order to overcome the irreversible head loss.

# 14–5 · TURBULENT FLOW IN PIPES

Most flows encountered in engineering practice are turbulent, and thus it is important to understand how turbulence affects wall shear stress. However, turbulent flow is a complex mechanism dominated by fluctuations, and despite tremendous amounts of work done in this area by researchers, turbulent flow still is not fully understood. Therefore, we must rely on experiments and the empirical or semi-empirical correlations developed for various situations.

(a) Before turbulence

(b) After turbulence

**FIGURE 14–19**

The intense mixing in turbulent flow brings fluid particles at different momentums into close contact and thus enhances momentum transfer.

Laminar flow

Turbulent flow

**FIGURE 14–20**

The velocity profile in fully developed pipe flow is parabolic in laminar flow, but much fuller in turbulent flow. Note that $u(r)$ in the turbulent case is the *time-averaged* velocity component in the axial direction (the overbar on $u$ has been dropped for simplicity).

Turbulent flow is characterized by disorderly and rapid fluctuations of swirling regions of fluid, called **eddies**, throughout the flow. These fluctuations provide an additional mechanism for momentum and energy transfer. In laminar flow, fluid particles flow in an orderly manner along pathlines, and momentum and energy are transferred across streamlines by molecular diffusion. In turbulent flow, the swirling eddies transport mass, momentum, and energy to other regions of flow much more rapidly than molecular diffusion, greatly enhancing mass, momentum, and heat transfer. As a result, turbulent flow is associated with much higher values of friction, heat transfer, and mass transfer coefficients (Fig. 14–19).

## Turbulent Velocity Profile

Unlike laminar flow, the expressions for the velocity profile in a turbulent flow are based on both analysis and measurements, and thus they are semi-empirical in nature with constants determined from experimental data. Consider fully developed turbulent flow in a pipe, and let $u$ denote the time-averaged velocity in the axial direction (and thus drop the overbar from $\bar{u}$ for simplicity).

Typical velocity profiles for fully developed laminar and turbulent flows are given in Fig. 14–20. Note that the velocity profile is parabolic in laminar flow but is much fuller in turbulent flow, with a sharp drop near the pipe wall. Turbulent flow along a wall can be considered to consist of four regions, characterized by the distance from the wall (Fig. 14–20). The very thin layer next to the wall where viscous effects are dominant is the **viscous** (or **laminar** or **linear** or **wall**) **sublayer**. The velocity profile in this layer is very nearly *linear*, and the flow is streamlined. Next to the viscous sublayer is the **buffer layer**, in which turbulent effects are becoming significant, but the flow is still dominated by viscous effects. Above the buffer layer is the **overlap** (or **transition**) **layer**, also called the **inertial sublayer**, in which the turbulent effects are much more significant, but still not dominant. Above that is the **outer** (or **turbulent**) **layer** in the remaining part of the flow in which turbulent effects dominate over molecular diffusion (viscous) effects.

Flow characteristics are quite different in different regions, and thus it is difficult to come up with an analytic relation for the velocity profile for the entire flow as we did for laminar flow. The best approach in the turbulent case turns out to be to identify the key variables and functional forms using dimensional analysis, and then to use experimental data to determine the numerical values of any constants. Several equations have been generated in this manner to approximate the velocity profile shape in fully developed trubulent pipe fow. Discussion of these equations is beyond the scope of the present text.

## The Moody Chart and the Colebrook Equation

The friction factor in fully developed turbulent pipe flow depends on the Reynolds number and the **relative roughness** $\varepsilon/D$, which is the ratio of the mean height of roughness of the pipe to the pipe diameter. The functional form of this dependence cannot be obtained from a theoretical analysis, and all available results are obtained from painstaking experiments using artificially roughened surfaces (usually by gluing sand grains of a known size on the inner surfaces of the pipes). Most such experiments were conducted by Prandtl's student J. Nikuradse in 1933, followed by the works of others. The friction factor was calculated from measurements of the flow rate and the pressure drop.

The experimental results are presented in tabular, graphical, and functional forms obtained by curve-fitting experimental data. In 1939, Cyril F. Colebrook (1910–1997) combined the available data for transition and turbulent flow in smooth as well as rough pipes into the following implicit relation known as the **Colebrook equation**:

$$\frac{1}{\sqrt{f}} = -2.0 \log\left(\frac{\varepsilon/D}{3.7} + \frac{2.51}{Re\sqrt{f}}\right) \qquad \text{(turbulent flow)} \qquad \text{(14–35)}$$

We note that the logarithm in Eq. 14–35 is a base 10 rather than a natural logarithm. In 1942, the American engineer Hunter Rouse (1906–1996) verified Colebrook's equation and produced a graphical plot of $f$ as a function of Re and the product $Re\sqrt{f}$. He also presented the laminar flow relation and a table of commercial pipe roughness. Two years later, Lewis F. Moody (1880–1953) redrew Rouse's diagram into the form commonly used today. The now famous **Moody chart** is given in the appendix as Fig. A–27. It presents the Darcy friction factor for pipe flow as a function of the Reynolds number and $\varepsilon/D$ over a wide range. It is probably one of the most widely accepted and used charts in engineering. Although it is developed for circular pipes, it can also be used for noncircular pipes by replacing the diameter with the hydraulic diameter.

Commercially available pipes differ from those used in the experiments in that the roughness of pipes in the market is not uniform and it is difficult to give a precise description of it. Equivalent roughness values for some commercial pipes are given in Table 14–2 as well as on the Moody chart. But it should be kept in mind that these values are for new pipes, and the relative roughness of pipes may increase with use as a result of corrosion, scale buildup, and precipitation. As a result, the friction factor may increase by a factor of 5 to 10. Actual operating conditions must be considered in the design of piping systems. Also, the Moody chart and its equivalent Colebrook equation involve several uncertainties (the roughness size, experimental error, curve fitting of data, etc.), and thus the results obtained should not be treated as "exact." They are is usually considered to be accurate to ±15 percent over the entire range in the figure.

The Colebrook equation is implicit in $f$, and thus the determination of the friction factor requires iteration. An approximate explicit relation for $f$ was given by S. E. Haaland in 1983 as

$$\frac{1}{\sqrt{f}} \cong -1.8 \log\left[\frac{6.9}{Re} + \left(\frac{\varepsilon/D}{3.7}\right)^{1.11}\right] \qquad \text{(14–36)}$$

**TABLE 14–2**

Equivalent roughness values for new commercial pipes*

| Material | Roughness, $\varepsilon$ | |
|---|---|---|
| | ft | mm |
| Glass, plastic | 0 (smooth) | |
| Concrete | 0.003–0.03 | 0.9–9 |
| Wood stave | 0.0016 | 0.5 |
| Rubber, smoothed | 0.000033 | 0.01 |
| Copper or brass tubing | 0.000005 | 0.0015 |
| Cast iron | 0.00085 | 0.26 |
| Galvanized iron | 0.0005 | 0.15 |
| Wrought iron | 0.00015 | 0.046 |
| Stainless steel | 0.000007 | 0.002 |
| Commercial steel | 0.00015 | 0.045 |

* The uncertainty in these values can be as much as ±60 percent.

| Relative Roughness, $\varepsilon/D$ | Friction Factor, $f$ |
|---|---|
| 0.0* | 0.0119 |
| 0.00001 | 0.0119 |
| 0.0001 | 0.0134 |
| 0.0005 | 0.0172 |
| 0.001 | 0.0199 |
| 0.005 | 0.0305 |
| 0.01 | 0.0380 |
| 0.05 | 0.0716 |

* Smooth surface. All values are for Re = $10^6$ and are calculated from the Colebrook equation.

**FIGURE 14–21**

The friction factor is minimum for a smooth pipe and increases with roughness.

The results obtained from this relation are within 2 percent of those obtained from the Colebrook equation. If more accurate results are desired, Eq. 14–36 can be used as a good *first guess* in a Newton iteration when using a programmable calculator or a spreadsheet to solve for $f$ with Eq. 14–35.

We make the following observations from the Moody chart:

- For laminar flow, the friction factor decreases with increasing Reynolds number, and it is independent of surface roughness.

- The friction factor is a minimum for a smooth pipe (but still not zero because of the no-slip condition) and increases with roughness (Fig. 14–21). The Colebrook equation in this case ($\varepsilon = 0$) reduces to the **Prandtl equation** expressed as $1/\sqrt{f} = 2.0 \log(\mathrm{Re}\sqrt{f}) - 0.8$.

- The transition region from the laminar to turbulent regime ($2300 < \mathrm{Re} < 4000$) is indicated by the shaded area in the Moody chart (Figs. 14–22 and A–27). The flow in this region may be laminar or turbulent, depending on flow disturbances, or it may alternate between laminar and turbulent, and thus the friction factor may also alternate between the values for laminar and turbulent flow. The data in this range are the least reliable. At small relative roughnesses, the friction factor increases in the transition region and approaches the value for smooth pipes.

- At very large Reynolds numbers (to the right of the dashed line on the Moody chart) the friction factor curves corresponding to specified relative roughness curves are nearly horizontal, and thus the friction factors are independent of the Reynolds number (Fig. 14–22). The flow in that region is called *fully rough turbulent flow* or just *fully rough flow* because the thickness of the viscous sublayer decreases with increasing Reynolds

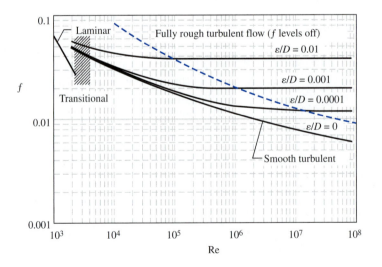

**FIGURE 14–22**

At very large Reynolds numbers, the friction factor curves on the Moody chart are nearly horizontal, and thus the friction factors are independent of the Reynolds number. See Fig. A–27 for a full-page moody chart.

number, and it becomes so thin that it is negligibly small compared to the surface roughness height. The viscous effects in this case are produced in the main flow primarily by the protruding roughness elements, and the contribution of the viscous sublayer is negligible. The Colebrook equation in the *fully rough* zone (Re → ∞) reduces to the **von Kármán equation** expressed as $1/\sqrt{f} = -2.0 \log[(\varepsilon/D)/3.7]$, which is explicit in $f$. Some authors call this zone *completely* (or *fully*) *turbulent flow,* but this is misleading since the flow to the left of the dashed blue line in Fig. 14–22 is also fully turbulent.

In calculations, we should make sure that we use the actual internal diameter of the pipe, which may be different than the nominal diameter. For example, the internal diameter of a steel pipe whose nominal diameter is 1 in is 1.049 in (Table 14–3).

## Types of Fluid Flow Problems

In the design and analysis of piping systems that involve the use of the Moody chart (or the Colebrook equation), we usually encounter three types of problems (the fluid and the roughness of the pipe are assumed to be specified in all cases) (Fig. 14–23):

1. Determining the **pressure drop** (or head loss) when the pipe length and diameter are given for a specified flow rate (or velocity)
2. Determining the **flow rate** when the pipe length and diameter are given for a specified pressure drop (or head loss)
3. Determining the **pipe diameter** when the pipe length and flow rate are given for a specified pressure drop (or head loss)

Problems of the *first type* are straightforward and can be solved directly by using the Moody chart. Problems of the *second type* and *third type* are commonly encountered in engineering design (in the selection of pipe diameter, for example, that minimizes the sum of the construction and pumping costs), but the use of the Moody chart with such problems requires an iterative approach—an equation solver (such as EES) is recommended.

In problems of the *second type*, the diameter is given but the flow rate is unknown. A good guess for the friction factor in that case is obtained from the completely turbulent flow region for the given roughness. This is true for large Reynolds numbers, which is often the case in practice. Once the flow rate is obtained, the friction factor is corrected using the Moody chart or the Colebrook equation, and the process is repeated until the solution converges. (Typically only a few iterations are required for convergence to three or four digits of precision.)

In problems of the *third type*, the diameter is not known and thus the Reynolds number and the relative roughness cannot be calculated. Therefore, we start calculations by assuming a pipe diameter. The pressure drop calculated for the assumed diameter is then compared to the specified pressure drop, and calculations are repeated with another pipe diameter in an iterative fashion until convergence.

**TABLE 14–3**

Standard sizes for Schedule 40 steel pipes

| Nominal Size, in | Actual Inside Diameter, in |
|---|---|
| $\frac{1}{8}$ | 0.269 |
| $\frac{1}{4}$ | 0.364 |
| $\frac{3}{8}$ | 0.493 |
| $\frac{1}{2}$ | 0.622 |
| $\frac{3}{4}$ | 0.824 |
| 1 | 1.049 |
| $1\frac{1}{2}$ | 1.610 |
| 2 | 2.067 |
| $2\frac{1}{2}$ | 2.469 |
| 3 | 3.068 |
| 5 | 5.047 |
| 10 | 10.02 |

| Problem type | Given | Find |
|---|---|---|
| 1 | $L, D, \dot{V}$ | $\Delta P$ (or $h_L$) |
| 2 | $L, D, \Delta P$ | $\dot{V}$ |
| 3 | $L, \Delta P, \dot{V}$ | $D$ |

**FIGURE 14–23**

The three types of problems encountered in pipe flow.

To avoid tedious iterations in head loss, flow rate, and diameter calculations, Swamee and Jain (1976) proposed the following explicit relations that are accurate to within 2 percent of the Moody chart:

$$h_L = 1.07 \frac{\dot{V}^2 L}{g D^5}\left\{\ln\left[\frac{\varepsilon}{3.7D} + 4.62\left(\frac{\nu D}{\dot{V}}\right)^{0.9}\right]\right\}^{-2} \qquad \begin{array}{l} 10^{-6} < \varepsilon/D < 10^{-2} \\ 3000 < \mathrm{Re} < 3 \times 10^8 \end{array} \qquad (14\text{–}37)$$

$$\dot{V} = -0.965\left(\frac{g D^5 h_L}{L}\right)^{0.5}\ln\left[\frac{\varepsilon}{3.7D} + \left(\frac{3.17 V^2 L}{g D^3 h_L}\right)^{0.5}\right] \qquad \mathrm{Re} > 2000 \qquad (14\text{–}38)$$

$$D = 0.66\left[\varepsilon^{1.25}\left(\frac{L\dot{V}^2}{g h_L}\right)^{4.75} + \nu \dot{V}^{9.4}\left(\frac{L}{g h_L}\right)^{5.2}\right]^{0.04} \qquad \begin{array}{l} 10^{-6} < \varepsilon/D < 10^{-2} \\ 5000 < \mathrm{Re} < 3 \times 10^8 \end{array} \qquad (14\text{–}39)$$

Note that all quantities are dimensional and the units simplify to the desired unit (for example, to m or ft in the last relation) when consistent units are used. Noting that the Moody chart is accurate to within 15 percent of experimental data, we should have no reservation in using these approximate relations in the design of piping systems.

**FIGURE 14–24**

Schematic for Example 14–3.

**EXAMPLE 14–3    Determining the Head Loss in a Water Pipe**

Water at 60°F ($\rho$ = 62.36 lbm/ft³ and $\mu$ = 7.536 × 10⁻⁴ lbm/ft·s) is flowing steadily in a 2-in-diameter horizontal pipe made of stainless steel at a rate of 0.2 ft³/s (Fig. 14–24). Determine the pressure drop, the head loss, and the required pumping power input for flow over a 200-ft-long section of the pipe.

**Solution** The flow rate through a specified water pipe is given. The pressure drop, the head loss, and the pumping power requirements are to be determined.

**Assumptions** **1** The flow is steady and incompressible. **2** The entrance effects are negligible, and thus the flow is fully developed. **3** The pipe involves no components such as bends, valves, and connectors. **4** The piping section involves no work devices such as a pump or a turbine.

**Properties** The density and dynamic viscosity of water are given to be $\rho$ = 62.36 lbm/ft³ and $\mu$ = 7.536 × 10⁻⁴ lbm/ft·s, respectively.

**Analysis** We recognize this as a problem of the first type, since flow rate, pipe length, and pipe diameter are known. First we calculate the average velocity and the Reynolds number to determine the flow regime:

$$V = \frac{\dot{V}}{A_c} = \frac{\dot{V}}{\pi D^2/4} = \frac{0.2\ \text{ft}^3/\text{s}}{\pi (2/12\ \text{ft})^2/4} = 9.17\ \text{ft/s}$$

$$\mathrm{Re} = \frac{\rho V D}{\mu} = \frac{(62.36\ \text{lbm/ft}^3)(9.17\ \text{ft/s})(2/12\ \text{ft})}{7.536 \times 10^{-4}\ \text{lbm/ft·s}} = 126{,}400$$

Since Re is greater than 4000, the flow is turbulent. The relative roughness of the pipe is estimated using Table 14–2

$$\varepsilon/D = \frac{0.000007\ \text{ft}}{2/12\ \text{ft}} = 0.000042$$

The friction factor corresponding to this relative roughness and Reynolds number is determined from the Moody chart. To avoid any reading error, we determine $f$ from the Colebrook equation on which the Moody chart is based:

$$\frac{1}{\sqrt{f}} = -2.0 \log\left(\frac{\varepsilon/D}{3.7} + \frac{2.51}{\mathrm{Re}\sqrt{f}}\right) \rightarrow \frac{1}{\sqrt{f}} = -2.0 \log\left(\frac{0.000042}{3.7} + \frac{2.51}{126{,}400\sqrt{f}}\right)$$

Using an equation solver or an iterative scheme, the friction factor is determined to be $f = 0.0174$. Then the pressure drop (which is equivalent to pressure loss in this case), head loss, and the required power input become

$$\Delta P = \Delta P_L = f\frac{L}{D}\frac{\rho V^2}{2} = 0.0174\,\frac{200\ \text{ft}}{2/12\ \text{ft}}\,\frac{(62.36\ \text{lbm/ft}^3)(9.17\ \text{ft/s})^2}{2}\left(\frac{1\ \text{lbf}}{32.2\ \text{lbm·ft/s}^2}\right)$$

$$= \mathbf{1700\ lbf/ft^2} = \mathbf{11.8\ psi}$$

$$h_L = \frac{\Delta P_L}{\rho g} = f\frac{L}{D}\frac{V^2}{2g} = 0.0174\,\frac{200\ \text{ft}}{2/12\ \text{ft}}\,\frac{(9.17\ \text{ft/s})^2}{2(32.2\ \text{ft/s}^2)} = \mathbf{27.3\ ft}$$

$$\dot{W}_{\text{pump}} = \dot{V}\,\Delta P = (0.2\ \text{ft}^3/\text{s})(1700\ \text{lbf/ft}^2)\left(\frac{1\ \text{W}}{0.737\ \text{lbf·ft/s}}\right) = \mathbf{461\ W}$$

Therefore, power input in the amount of 461 W is needed to overcome the frictional losses in the pipe.

*Discussion* It is common practice to write our final answers to three significant digits, even though we know that the results are accurate to at most two significant digits because of inherent inaccuracies in the Colebrook equation, as discussed previously. The friction factor could also be determined easily from the explicit Haaland relation (Eq. 14–36). It would give $f = 0.0172$, which is sufficiently close to 0.0174. Also, the friction factor corresponding to $\varepsilon = 0$ in this case is 0.0171, which indicates that this stainless-steel pipe can be approximated as smooth with negligible error.

---

**EXAMPLE 14–4    Determining the Diameter of an Air Duct**

Heated air at 1 atm and 35°C is to be transported in a 150-m-long circular plastic duct at a rate of 0.35 m³/s (Fig. 14–25). If the head loss in the pipe is not to exceed 20 m, determine the minimum diameter of the duct.

**Solution** The flow rate and the head loss in an air duct are given. The diameter of the duct is to be determined.

*Assumptions* **1** The flow is steady and incompressible. **2** The entrance effects are negligible, and thus the flow is fully developed. **3** The duct involves no components such as bends, valves, and connectors. **4** Air is an ideal gas. **5** The duct is smooth since it is made of plastic. **6** The flow is turbulent (to be verified).

*Properties* The density, dynamic viscosity, and kinematic viscosity of air at 35°C are $\rho = 1.145$ kg/m³, $\mu = 1.895 \times 10^{-5}$ kg/m·s, and $\nu = 1.655 \times 10^{-5}$ m²/s.

**FIGURE 14–25**

Schematic for Example 14–4

*Analysis* This is a problem of the third type since it involves the determination of diameter for specified flow rate and head loss. We can solve this problem by three different approaches: (1) an iterative approach by assuming a pipe diameter, calculating the head loss, comparing the result to the specified head loss, and repeating calculations until the calculated head loss matches the specified value; (2) writing all the relevant equations (leaving the diameter as an unknown) and solving them simultaneously using an equation solver; and (3) using the third Swamee–Jain formula. We will demonstrate the use of the last two approaches.

The average velocity, the Reynolds number, the friction factor, and the head loss relations are expressed as ($D$ is in m, $V$ is in m/s, and Re and $f$ are dimensionless)

$$V = \frac{\dot{V}}{A_c} = \frac{\dot{V}}{\pi D^2/4} = \frac{0.35 \text{ m}^3/\text{s}}{\pi D^2/4}$$

$$\text{Re} = \frac{VD}{\nu} = \frac{VD}{1.655 \times 10^{-5} \text{ m}^2/\text{s}}$$

$$\frac{1}{\sqrt{f}} = -2.0 \log\left(\frac{\varepsilon/D}{3.7} + \frac{2.51}{\text{Re}\sqrt{f}}\right) = -2.0 \log\left(\frac{2.51}{\text{Re}\sqrt{f}}\right)$$

$$h_L = f\frac{L}{D}\frac{V^2}{2g} \quad \rightarrow \quad 20 \text{ m} = f\frac{150 \text{ m}}{D}\frac{V^2}{2(9.81 \text{ m/s}^2)}$$

The roughness is approximately zero for a plastic pipe (Table 14–2). Therefore, this is a set of four equations and four unknowns, and solving them with an equation solver such as EES gives

$$D = \mathbf{0.267 \text{ m}}, \quad f = 0.0180, \quad V = 6.24 \text{ m/s}, \quad \text{and} \quad \text{Re} = 100{,}800$$

Therefore, the diameter of the duct should be more than 26.7 cm if the head loss is not to exceed 20 m. Note that Re > 4000, and thus the turbulent flow assumption is verified.

The diameter can also be determined directly from the third Swamee–Jain formula to be

$$D = 0.66\left[\varepsilon^{1.25}\left(\frac{L\dot{V}^2}{gh_L}\right)^{4.75} + \nu\dot{V}^{9.4}\left(\frac{L}{gh_L}\right)^{5.2}\right]^{0.04}$$

$$= 0.66\left[0 + (1.655 \times 10^{-5} \text{ m}^2/\text{s})(0.35 \text{ m}^3/\text{s})^{9.4}\left(\frac{150 \text{ m}}{(9.81 \text{ m/s}^2)(20 \text{ m})}\right)^{5.2}\right]^{0.04}$$

$$= \mathbf{0.271 \text{ m}}$$

*Discussion* Note that the difference between the two results is less than 2 percent. Therefore, the simple Swamee–Jain relation can be used with confidence. Finally, the first (iterative) approach requires an initial guess for $D$. If we use the Swamee–Jain result as our initial guess, the diameter converges to $D = 0.267$ m in short order.

---

**EXAMPLE 14–5**    **Determining the Flow Rate of Air in a Duct**

Reconsider Example 14–4. Now the duct length is doubled while its diameter is maintained constant. If the total head loss is to remain constant, determine the drop in the flow rate through the duct.

**Solution**  The diameter and the head loss in an air duct are given. The drop in the flow rate is to be determined.

**Analysis**  This is a problem of the second type since it involves the determination of the flow rate for a specified pipe diameter and head loss. The solution involves an iterative approach since the flow rate (and thus the flow velocity) is not known.

The average velocity, Reynolds number, friction factor, and the head loss relations are expressed as ($D$ is in m, $V$ is in m/s, and Re and $f$ are dimensionless)

$$V = \frac{\dot{V}}{A_c} = \frac{\dot{V}}{\pi D^2/4} \quad \rightarrow \quad V = \frac{\dot{V}}{\pi (0.267 \text{ m})^2/4}$$

$$\text{Re} = \frac{VD}{\nu} \quad \rightarrow \quad \text{Re} = \frac{V(0.267 \text{ m})}{1.655 \times 10^{-5} \text{ m}^2/\text{s}}$$

$$\frac{1}{\sqrt{f}} = -2.0 \log\left(\frac{\varepsilon/D}{3.7} + \frac{2.51}{\text{Re}\sqrt{f}}\right) \quad \rightarrow \quad \frac{1}{\sqrt{f}} = -2.0 \log\left(\frac{2.51}{\text{Re}\sqrt{f}}\right)$$

$$h_L = f\frac{L}{D}\frac{V^2}{2g} \quad \rightarrow \quad 20 \text{ m} = f\frac{300 \text{ m}}{0.267 \text{ m}}\frac{V^2}{2(9.81 \text{ m/s}^2)}$$

This is a set of four equations in four unknowns and solving them with an equation solver such as EES gives

$$\dot{V} = 0.24 \text{ m}^3/\text{s}, \quad f = 0.0195, \quad V = 4.23 \text{ m/s}, \quad \text{and} \quad \text{Re} = 68,300$$

Then the drop in the flow rate becomes

$$\dot{V}_{\text{drop}} = \dot{V}_{\text{old}} - \dot{V}_{\text{new}} = 0.35 - 0.24 = \textbf{0.11 m}^3\textbf{/s} \quad \text{(a drop of 31 percent)}$$

Therefore, for a specified head loss (or available head or fan pumping power), the flow rate drops by about 31 percent from 0.35 to 0.24 m³/s when the duct length doubles.

**Alternative Solution**  If a computer is not available (as in an exam situation), another option is to set up a *manual iteration loop*. We have found that the best convergence is usually realized by first guessing the friction factor $f$, and then solving for the velocity $V$. The equation for $V$ as a function of $f$ is

*Average velocity through the pipe*:
$$V = \sqrt{\frac{2gh_L}{f L/D}}$$

Once $V$ is calculated, the Reynolds number can be calculated, from which a *corrected* friction factor is obtained from the Moody chart or the Colebrook equation. We repeat the calculations with the corrected value of $f$ until convergence. We guess $f = 0.04$ for illustration:

| Iteration | $f$ (guess) | $V$, m/s | Re | Corrected $f$ |
|---|---|---|---|---|
| 1 | 0.04 | 2.955 | $4.724 \times 10^4$ | 0.0212 |
| 2 | 0.0212 | 4.059 | $6.489 \times 10^4$ | 0.01973 |
| 3 | 0.01973 | 4.207 | $6.727 \times 10^4$ | 0.01957 |
| 4 | 0.01957 | 4.224 | $6.754 \times 10^4$ | 0.01956 |
| 5 | 0.01956 | 4.225 | $6.756 \times 10^4$ | 0.01956 |

Notice that the iteration has converged to three digits in only three iterations and to four digits in only four iterations. The final results are identical to those obtained with EES, yet do not require a computer.

*Discussion* The new flow rate can also be determined directly from the second Swamee–Jain formula to be

$$\dot{V} = -0.965\left(\frac{gD^5 h_L}{L}\right)^{0.5} \ln\left[\frac{\varepsilon}{3.7D} + \left(\frac{3.17\nu^2 L}{gD^3 h_L}\right)^{0.5}\right]$$

$$= -0.965\left(\frac{(9.81 \text{ m/s}^2)(0.267 \text{ m})^5(20 \text{ m})}{300 \text{ m}}\right)^{0.5}$$

$$\times \ln\left[0 + \left(\frac{3.17(1.655 \times 10^{-5} \text{ m}^2/\text{s})^2(300 \text{ m})}{(9.81 \text{ m/s}^2)(0.267 \text{ m})^3(20 \text{ m})}\right)^{0.5}\right]$$

$$= 0.24 \text{ m}^3/\text{s}$$

Note that the result from the Swamee–Jain relation is the same (to two significant digits) as that obtained with the Colebrook equation using EES or using our manual iteration technique. Therefore, the simple Swamee–Jain relation can be used with confidence.

## 14–6 · MINOR LOSSES

The fluid in a typical piping system passes through various fittings, valves, bends, elbows, tees, inlets, exits, expansions, and contractions in addition to the straight sections of piping. These components interrupt the smooth flow of the fluid and cause additional losses because of the flow separation and mixing they induce. In a typical system with long pipes, these losses are minor compared to the head loss in the straight sections (the **major losses**) and are called **minor losses**. Although this is generally true, in some cases the minor losses may be greater than the major losses. This is the case, for example, in systems with several turns and valves in a short distance. The head loss introduced by a completely open valve, for example, may be negligible. But a partially closed valve may cause the largest head loss in the system, as evidenced by the drop in the flow rate. Flow through valves and fittings is very complex, and a theoretical analysis is generally not plausible. Therefore, minor losses are determined experimentally, usually by the manufacturers of the components.

Minor losses are usually expressed in terms of the **loss coefficient** $K_L$ (also called the **resistance coefficient**), defined as (Fig. 14–26)

Pipe section with valve:

Pipe section without valve:

$$\Delta P_L = (P_1 - P_2)_{\text{valve}} - (P_1 - P_2)_{\text{pipe}}$$

### FIGURE 14–26

For a constant-diameter section of a pipe with a minor loss component, the loss coefficient of the component (such as the gate valve shown) is determined by measuring the additional pressure loss it causes and dividing it by the dynamic pressure in the pipe.

*Loss coefficient:* $$K_L = \frac{h_L}{V^2/(2g)}$$ (14–40)

where $h_L$ is the *additional* irreversible head loss in the piping system caused by insertion of the component, and is defined as $h_L = \Delta P_L/\rho g$. For example, imagine replacing the valve in Fig. 14–26 with a section of constant diameter pipe from location 1 to location 2. $\Delta P_L$ is defined as the pressure drop from 1 to 2 for the case *with* the valve, $(P_1 - P_2)_{\text{valve}}$, *minus* the pressure drop that would occur in the imaginary straight pipe section from 1 to 2 *without* the valve, $(P_1 - P_2)_{\text{pipe}}$ at the same flow rate. While the majority of the irreversible head loss occurs locally near the valve, some of it occurs downstream of the valve due to induced swirling turbulent eddies that are produced in the valve and continue downstream. These eddies "waste" mechanical energy because they are ultimately dissipated into heat while the flow in the downstream section of pipe eventually returns to fully developed conditions. When measuring minor losses in some minor loss components, such as *elbows,* for example, location 2 must be considerably far downstream (tens of pipe diameters) in order to fully account for the additional irreversible losses due to these decaying eddies.

When the pipe diameter downstream of the component *changes,* determination of the minor loss is even more complicated. In all cases, however, it is based on the *additional* irreversible loss of mechanical energy that would otherwise not exist if the minor loss component were not there. For simplicity, you may think of the minor loss as occurring *locally* across the minor loss component, but keep in mind that the component influences the flow for several pipe diameters downstream. By the way, this is the reason why most flow meter manufacturers recommend installing their flow meter at least 10 to 20 pipe diameters downstream of any elbows or valves—this allows the swirling turbulent eddies generated by the elbow or valve to largely disappear and the velocity profile to become fully developed before entering the flow meter. (Most flow meters are calibrated with a fully developed velocity profile at the flow meter inlet, and yield the best accuracy when such conditions also exist in the actual application.)

When the inlet diameter equals the outlet diameter, the loss coefficient of a component can also be determined by measuring the pressure loss across the component and dividing it by the dynamic pressure, $K_L = \Delta P_L/(\frac{1}{2}\rho V^2)$. When the loss coefficient for a component is available, the head loss for that component is determined from

*Minor loss*:
$$h_L = K_L \frac{V^2}{2g} \tag{14–41}$$

The loss coefficient, in general, depends on the geometry of the component and the Reynolds number, just like the friction factor. However, it is usually assumed to be independent of the Reynolds number. This is a reasonable approximation since most flows in practice have large Reynolds numbers and the loss coefficients (including the friction factor) tend to be independent of the Reynolds number at large Reynolds numbers.

Minor losses are also expressed in terms of the **equivalent length** $L_{equiv}$, defined as (Fig. 14–27)

*Equivalent length*:
$$h_L = K_L \frac{V^2}{2g} = f \frac{L_{equiv}}{D} \frac{V^2}{2g} \rightarrow L_{equiv} = \frac{D}{f} K_L \tag{14–42}$$

where $f$ is the friction factor and $D$ is the diameter of the pipe that contains the component. The head loss caused by the component is equivalent to the head loss caused by a section of the pipe whose length is $L_{equiv}$. Therefore, the contribution of a component to the head loss is accounted for by simply adding $L_{equiv}$ to the total pipe length.

Both approaches are used in practice, but the use of loss coefficients is more common. Therefore, we also use that approach in this book. Once all the loss coefficients are available, the total head loss in a piping system is determined from

*Total head loss (general)*:
$$h_{L, \text{total}} = h_{L, \text{major}} + h_{L, \text{minor}}$$

$$= \sum_i f_i \frac{L_i}{D_i} \frac{V_i^2}{2g} + \sum_j K_{L,j} \frac{V_j^2}{2g} \tag{14–43}$$

$\Delta P = P_1 - P_2 = P_3 - P_4$

**FIGURE 14–27**

The head loss caused by a component (such as the angle valve shown) is equivalent to the head loss caused by a section of the pipe whose length is the equivalent length.

where *i* represents each pipe section with constant diameter and *j* represents each component that causes a minor loss. If the entire piping system being analyzed has a constant diameter, Eq. 14–43 reduces to

*Total head loss (D = constant):*  $$h_{L,\,\text{total}} = \left( f\frac{L}{D} + \sum K_L \right)\frac{V^2}{2g} \qquad (14\text{–}44)$$

where $V$ is the average flow velocity through the entire system (note that $V$ = constant since $D$ = constant).

Representative loss coefficients $K_L$ are given in Table 14–4 for inlets, exits, bends, sudden and gradual area changes, and valves. There is considerable uncertainty in these values since the loss coefficients, in general, vary with the pipe diameter, the surface roughness, the Reynolds number, and the details of the design. The loss coefficients of two seemingly identical valves by two different manufacturers, for example, can differ by a factor of 2 or more. Therefore, the particular manufacturer's data should be consulted in the final design of piping systems rather than relying on the representative values in handbooks.

The head loss at the inlet of a pipe is a strong function of geometry. It is almost negligible for well-rounded inlets ($K_L$ = 0.03 for $r/D$ > 0.2), but increases to about 0.50 for sharp-edged inlets (Fig. 14–28). That is, a sharp-edged inlet causes half of the velocity head to be lost as the fluid enters the pipe. This is because the fluid cannot make sharp 90° turns easily, especially at high velocities. As a result, the flow separates at the corners, and the flow is constricted into the **vena contracta** region formed in the midsection of the pipe (Fig. 14–29 on page 564). Therefore, a sharp-edged inlet acts like a flow constriction. The velocity increases in the vena contracta region (and the pressure decreases) because of the reduced effective flow area and then decreases as the flow fills the entire cross section of the pipe. There would be negligible loss if the pressure were increased in accordance with Bernoulli's equation (the velocity head would simply be converted into pressure head). However, this deceleration process is far from ideal and the viscous dissipation caused by intense mixing and the turbulent eddies converts part of the kinetic energy into frictional heating, as evidenced by a slight rise in fluid temperature. The end result is a drop in velocity without much pressure recovery, and the inlet loss is a measure of this irreversible pressure drop.

Even slight rounding of the edges can result in significant reduction of $K_L$, as shown in Fig. 14–30 on page 565. The loss coefficient rises sharply (to about $K_L$ = 0.8) when the pipe protrudes into the reservoir since some fluid near the edge in this case is forced to make a 180° turn.

The loss coefficient for a submerged pipe exit is often listed in handbooks as $K_L$ = 1. More precisely, however, $K_L$ is equal to the kinetic energy correction factor $\alpha$ at the exit of the pipe. Although $\alpha$ is indeed close to 1 for fully developed *turbulent* pipe flow, it is equal to 2 for fully developed *laminar* pipe flow. To avoid possible errors when analyzing laminar pipe flow, then, it is best to always set $K_L = \alpha$ at a submerged pipe exit. At any such exit, whether laminar or turbulent, the fluid leaving the pipe loses *all* of its kinetic energy as it mixes with the reservoir fluid and eventually comes to rest through the irreversible action of viscosity. This is true regardless of the shape of the exit (Table 14–4, and Fig. 14–31 on page 565). Therefore, there is no need to round off the sharp edges of pipe exits.

Sharp-edged inlet
$K_L$ = 0.50      Vena contracta

Recirculating flow

Well-rounded inlet
$K_L$ = 0.03

$D$

• $r$

**FIGURE 14–28**

The head loss at the inlet of a pipe is almost negligible for well-rounded inlets ($K_L$ = 0.03 for $r/D$ > 0.2) but increases to about 0.50 for sharp-edged inlets.

## TABLE 14–4

Loss coefficients $K_L$ of various pipe components for turbulent flow (for use in the relation $h_L = K_L V^2/(2g)$, where $V$ is the average velocity in the pipe that contains the component)[*]

**Pipe Inlet**

Reentrant: $K_L = 0.80$
($t \ll D$ and $l \approx 0.1D$)

Sharp-edged: $K_L = 0.50$

Well-rounded ($r/D > 0.2$): $K_L = 0.03$
Slightly rounded ($r/D = 0.1$): $K_L = 0.12$
(see Fig. 14–36)

**Pipe Exit**

Reentrant: $K_L = \alpha$

Sharp-edged: $K_L = \alpha$

Rounded: $K_L = \alpha$

Note: The kinetic energy correction factor is $\alpha = 2$ for fully developed laminar flow, and $\alpha \approx 1.05$ for fully developed turbulent flow.

**Sudden Expansion and Contraction** (based on the velocity in the smaller-diameter pipe)

Sudden expansion: $K_L = \alpha\left(1 - \dfrac{d^2}{D^2}\right)^2$

Sudden contraction: See chart.

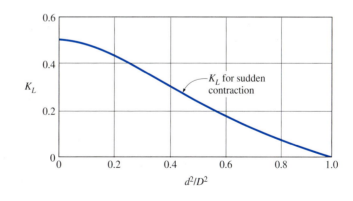

**Gradual Expansion and Contraction** (based on the velocity in the smaller-diameter pipe)

Expansion (for $\theta = 20°$):
$K_L = 0.30$ for $d/D = 0.2$
$K_L = 0.25$ for $d/D = 0.4$
$K_L = 0.15$ for $d/D = 0.6$
$K_L = 0.10$ for $d/D = 0.8$

Contraction:
$K_L = 0.02$ for $\theta = 30°$
$K_L = 0.04$ for $\theta = 45°$
$K_L = 0.07$ for $\theta = 60°$

## TABLE 14–4 (CONCLUDED)

*Bends and Branches*

| | | | |
|---|---|---|---|
| 90° *smooth bend:*<br>Flanged: $K_L = 0.3$<br>Threaded: $K_L = 0.9$ | 90° *miter bend*<br>(without vanes): $K_L = 1.1$ | 90° *miter bend*<br>(with vanes): $K_L = 0.2$ | 45° *threaded elbow:*<br>$K_L = 0.4$ |

| | | | |
|---|---|---|---|
| 180° *return bend:*<br>Flanged: $K_L = 0.2$<br>Threaded: $K_L = 1.5$ | *Tee* (branch flow):<br>Flanged: $K_L = 1.0$<br>Threaded: $K_L = 2.0$ | *Tee* (line flow):<br>Flanged: $K_L = 0.2$<br>Threaded: $K_L = 0.9$ | *Threaded union:*<br>$K_L = 0.08$ |

*Valves*

| | |
|---|---|
| *Globe valve,* fully open: $K_L = 10$ | *Gate valve,* fully open: $K_L = 0.2$ |
| *Angle valve,* fully open: $K_L = 5$ | $\frac{1}{4}$ closed: $K_L = 0.3$ |
| *Ball valve,* fully open: $K_L = 0.05$ | $\frac{1}{2}$ closed: $K_L = 2.1$ |
| *Swing check valve:* $K_L = 2$ | $\frac{3}{4}$ closed: $K_L = 17$ |

* These are representative values for loss coefficients. Actual values strongly depend on the design and manufacture of the components and may differ from the given values considerably (especially for valves). Actual manufacturer's data should be used in the final design.

**FIGURE 14–29**

Graphical representation of flow contraction and the associated head loss at a sharp-edged pipe inlet.

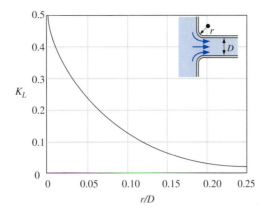

**FIGURE 14–30**

The effect of rounding of a pipe inlet on the loss coefficient.

*From ASHRAE* Handbook of Fundamentals.

Piping systems often involve *sudden* or *gradual* expansion or contraction sections to accommodate changes in flow rates or properties such as density and velocity. The losses are usually much greater in the case of *sudden* expansion and contraction (or wide-angle expansion) because of flow separation. By combining the equations of mass, momentum, and energy balance, the loss coefficient for the case of a **sudden expansion** is approximated as

$$K_L = \alpha \left( 1 - \frac{A_{small}}{A_{large}} \right)^2 \qquad \text{(sudden expansion)} \qquad \textbf{(14–45)}$$

where $A_{small}$ and $A_{large}$ are the cross-sectional areas of the small and large pipes, respectively. Note that $K_L = 0$ when there is no area change ($A_{small} = A_{large}$) and $K_L = \alpha$ when a pipe discharges into a reservoir ($A_{large} \gg A_{small}$). No such relation exists for a sudden contraction, and the $K_L$ values in that case must be read from a chart or table (e.g., Table 14–4). The losses due to expansions and contractions can be reduced significantly by installing conical gradual area changers (nozzles and diffusers) between the small and large pipes. The $K_L$ values for representative cases of gradual expansion and contraction are given in Table 14–4. Note that in head loss calculations, the velocity in the *small pipe* is to be used as the reference velocity in Eq. 14–41. Losses during expansion are usually much higher than the losses during contraction because of flow separation.

Piping systems also involve changes in direction without a change in diameter, and such flow sections are called *bends* or *elbows*. The losses in these devices are due to flow separation (just like a car being thrown off the road when it enters a turn too fast) on the inner side and the swirling secondary flows that result. The losses during changes of direction can be minimized by making the turn "easy" on the fluid by using circular arcs (like 90° elbows) instead of sharp turns (like miter bends) (Fig. 14–32). But the use of sharp turns (and thus suffering a penalty in loss coefficient) may be necessary when the turning space is limited. In such cases, the losses can be minimized by utilizing properly placed guide vanes to help the flow turn in an orderly manner without being thrown off the course. The loss coefficients for some elbows and miter bends as well as tees are given in Table 14–4. These coefficients do not include the frictional losses along the pipe bend. Such losses should be calculated as in straight pipes (using the length of the centerline as the pipe length) and added to other losses.

**FIGURE 14–31**

All the kinetic energy of the flow is "lost" (turned into thermal energy) through friction as the jet decelerates and mixes with ambient fluid downstream of a submerged outlet.

Flanged elbow
$K_L = 0.3$

Sharp turn
$K_L = 1.1$

**FIGURE 14–32**

The losses during changes of direction can be minimized by making the turn "easy" on the fluid by using circular arcs instead of sharp turns.

*(a)*

A globe valve

$V_1$ ⟶

⟶ $V_2$

Constriction

$V_2 = V_1$
$V_{constriction} > V_1$

*(b)*

**FIGURE 14–33**

(*a*) The large head loss in a partially closed globe valve is due to irreversible deceleration, flow separation, and mixing of high-velocity fluid coming from the narrow valve passage. (*b*) The head loss through a fully-open ball valve, on the other hand, is quite small.

*Photo by John M. Cimbala.*

Water
7 m/s
150 kPa

6 cm      9 cm

**FIGURE 14–34**

Schematic for Example 14–6.

*Valves* are commonly used in piping systems to control flow rates by simply altering the head loss until the desired flow rate is achieved. For valves it is desirable to have a very low loss coefficient when they are fully open, such as with a *ball valve,* so that they cause minimal head loss during full-load operation (Fig. 14–33b). Several different valve designs, each with its own advantages and disadvantages, are in common use today. The *gate valve* slides up and down like a gate, the *globe valve* (Fig. 14–33a) closes a hole placed in the valve, the *angle valve* is a globe valve with a 90° turn, and the *check valve* allows the fluid to flow only in one direction like a diode in an electric circuit. Table 14–4 lists the representative loss coefficients of the popular designs. Note that the loss coefficient increases drastically as a valve is closed. Also, the deviation in the loss coefficients for different manufacturers is greatest for valves because of their complex geometries.

### EXAMPLE 14–6    Head Loss and Pressure Rise during Gradual Expansion

A 6-cm-diameter horizontal water pipe expands gradually to a 9-cm-diameter pipe (Fig. 14–34). The walls of the expansion section are angled 10° from the axis. The average velocity and pressure of water before the expansion section are 7 m/s and 150 kPa, respectively. Determine the head loss in the expansion section and the pressure in the larger-diameter pipe.

**Solution** A horizontal water pipe expands gradually into a larger-diameter pipe. The head loss and pressure after the expansion are to be determined.

**Assumptions** 1 The flow is steady and incompressible. 2 The flow at sections 1 and 2 is fully developed and turbulent with $\alpha_1 = \alpha_2 \cong 1.06$.

**Properties** We take the density of water to be $\rho = 1000$ kg/m³. The loss coefficient for a gradual expansion of total included angle $\theta = 20°$ and diameter ratio $d/D = 6/9$ is $K_L = 0.133$ (by interpolation using Table 14–4).

**Analysis** Noting that the density of water remains constant, the downstream velocity of water is determined from conservation of mass to be

$$\dot{m}_1 = \dot{m}_2 \rightarrow \rho V_1 A_1 = \rho V_2 A_2 \rightarrow V_2 = \frac{A_1}{A_2} V_1 = \frac{D_1^2}{D_2^2} V_1$$

$$V_2 = \frac{(0.06 \text{ m})^2}{(0.09 \text{ m})^2} (7 \text{ m/s}) = 3.11 \text{ m/s}$$

Then the irreversible head loss in the expansion section becomes

$$h_L = K_L \frac{V_1^2}{2g} = (0.133) \frac{(7 \text{ m/s})^2}{2(9.81 \text{ m/s}^2)} = \mathbf{0.333 \text{ m}}$$

Noting that $z_1 = z_2$ and there are no pumps or turbines involved, the energy equation for the expansion section is expressed in terms of heads as

$$\frac{P_1}{\rho g} + \alpha_1 \frac{V_1^2}{2g} + z_1 + h_{pump, u}^{\,0} = \frac{P_2}{\rho g} + \alpha_2 \frac{V_2^2}{2g} + z_2 + h_{turbine, e}^{\,0} + h_L$$

or

$$\frac{P_1}{\rho g} + \alpha_1 \frac{V_1^2}{2g} = \frac{P_2}{\rho g} + \alpha_2 \frac{V_2^2}{2g} + h_L$$

Solving for $P_2$ and substituting,

$$P_2 = P_1 + \rho\left\{\frac{\alpha_1 V_1^2 - \alpha_2 V_2^2}{2} - gh_L\right\} = (150 \text{ kPa}) + (1000 \text{ kg/m}^3)$$

$$\times \left\{\frac{1.06(7 \text{ m/s})^2 - 1.06(3.11 \text{ m/s})^2}{2} - (9.81 \text{ m/s}^2)(0.333 \text{ m})\right\}$$

$$\times \left(\frac{1 \text{ kN}}{1000 \text{ kg·m/s}^2}\right)\left(\frac{1 \text{ kPa}}{1 \text{ kN/m}^2}\right)$$

$$= \mathbf{168 \text{ kPa}}$$

Therefore, despite the head (and pressure) loss, the pressure *increases* from 150 to 168 kPa after the expansion. This is due to the conversion of dynamic pressure to static pressure when the average flow velocity is decreased in the larger pipe.

**Discussion** It is common knowledge that higher pressure upstream is necessary to cause flow, and it may come as a surprise to you that the downstream pressure has *increased* after the expansion, despite the loss. This is because the flow is driven by the sum of the three heads that comprise the total head (namely, the pressure head, velocity head, and elevation head). During flow expansion, the higher velocity head upstream is converted to pressure head downstream, and this increase outweighs the nonrecoverable head loss. Also, you may be tempted to solve this problem using the Bernoulli equation. Such a solution would ignore the head (and the associated pressure) loss and result in an incorrect higher pressure for the fluid downstream.

# 14–7 ▪ PIPING NETWORKS AND PUMP SELECTION

## Series and Parallel Pipes

Most piping systems encountered in practice such as the water distribution systems in cities or commercial or residential establishments involve numerous parallel and series connections as well as several sources (supply of fluid into the system) and loads (discharges of fluid from the system) (Fig. 14–35). A piping project may involve the design of a new system or the expansion of an existing system. The engineering objective in such projects is to design a piping system that will reliably deliver the specified flow rates at specified pressures at minimum total (initial plus operating and maintenance) cost. Once the layout of the system is prepared, the determination of the pipe diameters and the pressures throughout the system, while remaining within the budget constraints, typically requires solving the system repeatedly until the optimal solution is reached. Computer modeling and analysis of such systems make this tedious task a simple chore.

Piping systems typically involve several pipes connected to each other in series and/or in parallel, as shown in Figs. 14–36 and 14–37. When the pipes are connected **in series**, the flow rate through the entire system remains constant regardless of the diameters of the individual pipes in the system. This is a natural consequence of the conservation of mass principle for steady incompressible flow. The total head loss in this case is equal to the sum of the head losses in individual pipes in the system, including the

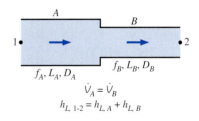

**FIGURE 14–35**

A piping network in an industrial facility.

*Courtesy UMDE Engineering, Contracting, and Trading. Used by permission.*

$$\dot{V}_A = \dot{V}_B$$
$$h_{L,\,1\text{-}2} = h_{L,\,A} + h_{L,\,B}$$

**FIGURE 14–36**

For pipes *in series*, the flow rate is the same in each pipe, and the total head loss is the sum of the head losses in individual pipes.

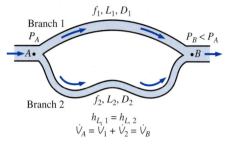

**FIGURE 14–37**

For pipes *in parallel*, the head loss is the same in each pipe, and the total flow rate is the sum of the flow rates in individual pipes.

minor losses. The expansion or contraction losses at connections are considered to belong to the smaller-diameter pipe since the expansion and contraction loss coefficients are defined on the basis of the average velocity in the smaller-diameter pipe.

For a pipe that branches out into two (or more) **parallel pipes** and then rejoins at a junction downstream, the total flow rate is the sum of the flow rates in the individual pipes. The pressure drop (or head loss) in each individual pipe connected in parallel must be the same since $\Delta P = P_A - P_B$ and the junction pressures $P_A$ and $P_B$ are the same for all the individual pipes. For a system of two parallel pipes 1 and 2 between junctions $A$ and $B$ with negligible minor losses, this is expressed as

$$h_{L,1} = h_{L,2} \quad \rightarrow \quad f_1 \frac{L_1}{D_1} \frac{V_1^2}{2g} = f_2 \frac{L_2}{D_2} \frac{V_2^2}{2g}$$

Then the ratio of the average velocities and the flow rates in the two parallel pipes become

$$\frac{V_1}{V_2} = \left( \frac{f_2 L_2 D_1}{f_1 L_1 D_2} \right)^{1/2} \quad \text{and} \quad \frac{\dot{V}_1}{\dot{V}_2} = \frac{A_{c,1} V_1}{A_{c,2} V_2} = \frac{D_1^2}{D_2^2} \left( \frac{f_2 L_2 D_1}{f_1 L_1 D_2} \right)^{1/2}$$

Therefore, the relative flow rates in parallel pipes are established from the requirement that the head loss in each pipe be the same. This result can be extended to any number of pipes connected in parallel. The result is also valid for pipes for which the minor losses are significant if the equivalent lengths for components that contribute to minor losses are added to the pipe length. Note that the flow rate in one of the parallel branches is proportional to its diameter to the power 5/2 and is inversely proportional to the square root of its length and friction factor.

The analysis of piping networks, no matter how complex they are, is based on two simple principles:

1. *Conservation of mass throughout the system must be satisfied.* This is done by requiring the total flow into a junction to be equal to the total flow out of the junction for all junctions in the system. Also, the flow rate must remain constant in pipes connected in series regardless of the changes in diameters.

2. *Pressure drop (and thus head loss) between two junctions must be the same for all paths between the two junctions.* This is because pressure is a point function and it cannot have two values at a specified point. In practice this rule is used by requiring that the algebraic sum of head losses in a loop (for all loops) be equal to zero. (A head loss is taken to be positive for flow in the clockwise direction and negative for flow in the counterclockwise direction.)

Therefore, the analysis of piping networks is very similar to the analysis of electric circuits (Kirchhoff's laws), with flow rate corresponding to electric current and pressure corresponding to electric potential. However, the situation is much more complex here since, unlike the electrical resistance, the "flow resistance" is a highly nonlinear function. Therefore, the analysis of piping networks requires the simultaneous solution of a system of nonlinear equations, which requires software such as EES, Mathcad, Matlab, etc., or commercially available software designed specifically for such applications.

# Piping Systems with Pumps and Turbines

When a piping system involves a pump and/or turbine, the steady-flow energy equation on a unit-mass basis is expressed as (see Section 12–2)

$$\frac{P_1}{\rho} + \alpha_1 \frac{V_1^2}{2} + gz_1 + w_{pump,\,u} = \frac{P_2}{\rho} + \alpha_2 \frac{V_2^2}{2} + gz_2 + w_{turbine,\,e} + gh_L \quad \textbf{(14–46)}$$

or in terms of heads as

$$\frac{P_1}{\rho g} + \alpha_1 \frac{V_1^2}{2g} + z_1 + h_{pump,\,u} = \frac{P_2}{\rho g} + \alpha_2 \frac{V_2^2}{2g} + z_2 + h_{turbine,\,e} + h_L \quad \textbf{(14–47)}$$

where $h_{pump,\,u} = w_{pump,\,u}/g$ is the useful pump head delivered to the fluid, $h_{turbine,\,e} = w_{turbine,\,e}/g$ is the turbine head extracted from the fluid, $\alpha$ is the kinetic energy correction factor whose value is about 1.05 for most (turbulent) flows encountered in practice, and $h_L$ is the total head loss in the piping (including the minor losses if they are significant) between points 1 and 2. The pump head is zero if the piping system does not involve a pump or a fan, the turbine head is zero if the system does not involve a turbine, and both are zero if the system does not involve any mechanical work-producing or work-consuming devices.

Many practical piping systems involve a pump to move a fluid from one reservoir to another. Taking points 1 and 2 to be at the *free surfaces* of the reservoirs (Fig. 14–38), the energy equation is solved for the required useful pump head, yielding

$$h_{pump,\,u} = (z_2 - z_1) + h_L \quad \textbf{(14–48)}$$

since the velocities at free surfaces are negligible for large reservoirs and the pressures are at atmospheric pressure. Therefore, the useful pump head is equal to the elevation difference between the two reservoirs plus the head loss. If the head loss is negligible compared to $z_2 - z_1$, the useful pump head is equal to the elevation difference between the two reservoirs. In the case of $z_1 > z_2$ (the first reservoir being at a higher elevation than the second one) with no pump, the flow is driven by gravity at a flow rate that causes a head loss equal to the elevation difference. A similar argument can be given for the turbine head for a hydroelectric power plant by replacing $h_{pump,\,u}$ in Eq. 14–48 by $-h_{turbine,\,e}$.

Once the useful pump head is known, the *mechanical power that needs to be delivered by the pump to the fluid* and the *electric power consumed by the motor of the pump* for a specified flow rate are determined from

$$\dot{W}_{pump,\,shaft} = \frac{\rho \dot{V} g h_{pump,\,u}}{\eta_{pump}} \quad \text{and} \quad \dot{W}_{elect} = \frac{\rho \dot{V} g h_{pump,\,u}}{\eta_{pump-motor}} \quad \textbf{(14–49)}$$

where $\eta_{pump-motor}$ is the *efficiency of the pump–motor combination*, which is the product of the pump and the motor efficiencies (Fig. 14–39). The pump–motor efficiency is defined as the ratio of the net mechanical energy delivered to the fluid by the pump to the electric energy consumed by the motor of the pump, and it typically ranges between 50 and 85 percent.

The head loss of a piping system increases (usually quadratically) with the flow rate. A plot of required useful pump head $h_{pump,\,u}$ as a function of flow rate is called the **system** (or **demand**) **curve**. The head produced by a pump is not a constant either. Both the pump head and the pump efficiency vary

**FIGURE 14–38**

When a pump moves a fluid from one reservoir to another, the useful pump head requirement is equal to the elevation difference between the two reservoirs plus the head loss.

$$h_{pump,\,u} = (z_2 - z_1) + h_L$$
$$\dot{W}_{pump,\,u} = \rho \dot{V} g h_{pump,\,u}$$

$\eta_{motor} = 0.90$

Liquid out

Liquid in

Pump
$\eta_{pump} = 0.70$

$$\eta_{pump-motor} = \eta_{pump}\eta_{motor}$$
$$= 0.70 \times 0.90 = 0.63$$

**FIGURE 14–39**

The efficiency of the pump–motor combination is the product of the pump and the motor efficiencies.

*Photo by Yunus Çengel.*

**FIGURE 14–40**

Characteristic pump curves for centrifugal pumps, the system curve for a piping system, and the operating point.

with the flow rate, and pump manufacturers supply this variation in tabular or graphical form, as shown in Fig. 14–40. These experimentally determined $h_{pump,u}$ and $\eta_{pump,u}$ versus $\dot{V}$ curves are called **characteristic** (or **supply** or **performance**) **curves**. Note that the flow rate of a pump increases as the required head decreases. The intersection point of the pump head curve with the vertical axis typically represents the *maximum head* (called the **shutoff head**) the pump can provide, while the intersection point with the horizontal axis indicates the *maximum flow rate* (called the **free delivery**) that the pump can supply.

The *efficiency* of a pump is highest at a certain combination of head and flow rate. Therefore, a pump that can supply the required head and flow rate is not necessarily a good choice for a piping system unless the efficiency of the pump at those conditions is sufficiently high. The pump installed in a piping system will operate at the point where the *system curve* and the *characteristic curve* intersect. This point of intersection is called the **operating point**, as shown in Fig. 14–40. The useful head produced by the pump at this point matches the head requirements of the system at that flow rate. Also, the efficiency of the pump during operation is the value corresponding to that flow rate.

---

### EXAMPLE 14–7    Pumping Water through Two Parallel Pipes

Water at 20°C is to be pumped from a reservoir ($z_A = 5$ m) to another reservoir at a higher elevation ($z_B = 13$ m) through two 36-m-long pipes connected in parallel, as shown in Fig. 14–41. The pipes are made of commercial steel, and the diameters of the two pipes are 4 and 8 cm. Water is to be pumped by a 70 percent efficient motor–pump combination that draws 8 kW of electric power during operation. The minor losses and the head loss in pipes that connect the parallel pipes to the two reservoirs are considered to be negligible. Determine the total flow rate between the reservoirs and the flow rate through each of the parallel pipes.

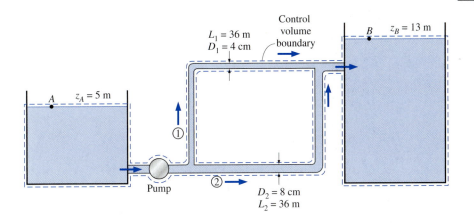

**FIGURE 14–41**

The piping system discussed in Example 14–7.

**Solution** The pumping power input to a piping system with two parallel pipes is given. The flow rates are to be determined.

**Assumptions** 1 The flow is steady (since the reservoirs are large) and incompressible. 2 The entrance effects are negligible, and thus the flow is fully developed. 3 The elevations of the reservoirs remain constant. 4 The minor losses and the head loss in pipes other than the parallel pipes are negligible. 5 Flows through both pipes are turbulent (to be verified).

**Properties** The density and dynamic viscosity of water at 20°C are $\rho = 998$ kg/m$^3$ and $\mu = 1.002 \times 10^{-3}$ kg/m·s. The roughness of commercial steel pipe is $\varepsilon = 0.000045$ m (Table 14–2).

**Analysis** This problem cannot be solved directly since the velocities (or flow rates) in the pipes are not known. Therefore, we would normally use a trial-and-error approach here. However, equation solvers such as EES are widely available, and thus, we simply set up the equations to be solved by an equation solver. The useful head supplied by the pump to the fluid is determined from

$$\dot{W}_{elect} = \frac{\rho \dot{V} g h_{pump,\,u}}{\eta_{pump-motor}} \quad \rightarrow \quad 8000\text{ W} = \frac{(998 \text{ kg/m}^3)\dot{V}(9.81 \text{ m/s}^2)h_{pump,\,u}}{0.70} \quad (1)$$

We choose points $A$ and $B$ at the free surfaces of the two reservoirs. Noting that the fluid at both points is open to the atmosphere (and thus $P_A = P_B = P_{atm}$) and that the fluid velocities at both points are nearly zero ($V_A \approx V_B \approx 0$) since the reservoirs are large, the energy equation for a control volume between these two points simplifies to

$$\frac{P_A}{\rho g} + \alpha_A \frac{V_A^2}{2g}\overset{0}{\nearrow} + z_A + h_{pump,\,u} = \frac{P_B}{\rho g} + \alpha_B \frac{V_B^2}{2g}\overset{0}{\nearrow} + z_B + h_L$$

or

$$h_{pump,\,u} = (z_B - z_A) + h_L$$

or

$$h_{pump,\,u} = (13 \text{ m} - 5 \text{ m}) + h_L \quad (2)$$

where

$$h_L = h_{L,1} = h_{L,2} \quad (3)(4)$$

We designate the 4-cm-diameter pipe by 1 and the 8-cm-diameter pipe by 2. Equations for the average velocity, the Reynolds number, the friction factor, and the head loss in each pipe are

$$V_1 = \frac{\dot{V}_1}{A_{c,1}} = \frac{\dot{V}_1}{\pi D_1^2/4} \quad \rightarrow \quad V_1 = \frac{\dot{V}_1}{\pi (0.04 \text{ m})^2/4} \tag{5}$$

$$V_2 = \frac{\dot{V}_2}{A_{c,2}} = \frac{\dot{V}_2}{\pi D_2^2/4} \quad \rightarrow \quad V_2 = \frac{\dot{V}_2}{\pi (0.08 \text{ m})^2/4} \tag{6}$$

$$\text{Re}_1 = \frac{\rho V_1 D_1}{\mu} \quad \rightarrow \quad \text{Re}_1 = \frac{(998 \text{ kg/m}^3) V_1 (0.04 \text{ m})}{1.002 \times 10^{-3} \text{ kg/m·s}} \tag{7}$$

$$\text{Re}_2 = \frac{\rho V_2 D_2}{\mu} \quad \rightarrow \quad \text{Re}_2 = \frac{(998 \text{ kg/m}^3) V_2 (0.08 \text{ m})}{1.002 \times 10^{-3} \text{ kg/m·s}} \tag{8}$$

$$\frac{1}{\sqrt{f_1}} = -2.0 \log\left(\frac{\varepsilon/D_1}{3.7} + \frac{2.51}{\text{Re}_1 \sqrt{f_1}}\right)$$

$$\rightarrow \quad \frac{1}{\sqrt{f_1}} = -2.0 \log\left(\frac{0.000045}{3.7 \times 0.04} + \frac{2.51}{\text{Re}_1 \sqrt{f_1}}\right) \tag{9}$$

$$\frac{1}{\sqrt{f_2}} = -2.0 \log\left(\frac{\varepsilon/D_2}{3.7} + \frac{2.51}{\text{Re}_2 \sqrt{f_2}}\right)$$

$$\rightarrow \quad \frac{1}{\sqrt{f_2}} = -2.0 \log\left(\frac{0.000045}{3.7 \times 0.08} + \frac{2.51}{\text{Re}_2 \sqrt{f_2}}\right) \tag{10}$$

$$h_{L,1} = f_1 \frac{L_1}{D_1} \frac{V_1^2}{2g} \quad \rightarrow \quad h_{L,1} = f_1 \frac{36 \text{ m}}{0.04 \text{ m}} \frac{V_1^2}{2(9.81 \text{ m/s}^2)} \tag{11}$$

$$h_{L,2} = f_2 \frac{L_2}{D_2} \frac{V_2^2}{2g} \quad \rightarrow \quad h_{L,2} = f_2 \frac{36 \text{ m}}{0.08 \text{ m}} \frac{V_2^2}{2(9.81 \text{ m/s}^2)} \tag{12}$$

$$\dot{V} = \dot{V}_1 + \dot{V}_2 \tag{13}$$

This is a system of 13 equations in 13 unknowns, and their simultaneous solution by an equation solver gives

$$\dot{V} = \mathbf{0.0300 \text{ m}^3/s}, \quad \dot{V}_1 = \mathbf{0.00415 \text{ m}^3/s}, \quad \dot{V}_2 = \mathbf{0.0259 \text{ m}^3/s}$$

$$V_1 = 3.30 \text{ m/s}, \quad V_2 = 5.15 \text{ m/s}, \quad h_L = h_{L,1} = h_{L,2} = 11.1 \text{ m}, \quad h_{\text{pump}} = 19.1 \text{ m}$$

$$\text{Re}_1 = 131{,}600, \quad \text{Re}_2 = 410{,}000, \quad f_1 = 0.0221, \quad f_2 = 0.0182$$

Note that Re > 4000 for both pipes, and thus the assumption of turbulent flow is verified.

**Discussion** The two parallel pipes have the same length and roughness, but the diameter of the first pipe is half the diameter of the second one. Yet only 14 percent of the water flows through the first pipe. This shows the strong dependence of the flow rate on diameter. Also, it can be shown that if the free surfaces of the two reservoirs were at the same elevation (and thus $z_A = z_B$), the flow rate would increase by 20 percent from 0.0300 to 0.0361 m³/s. Alternately, if the reservoirs were as given but the irreversible head losses were negligible, the flow rate would become 0.0715 m³/s (an increase of 138 percent).

## EXAMPLE 14–8    Gravity-Driven Water Flow in a Pipe

Water at 10°C flows from a large reservoir to a smaller one through a 5-cm-diameter cast iron piping system, as shown in Fig. 14–42. Determine the elevation $z_1$ for a flow rate of 6 L/s.

**Solution**  The flow rate through a piping system connecting two reservoirs is given. The elevation of the source is to be determined.

**Assumptions**  **1** The flow is steady and incompressible. **2** The elevations of the reservoirs remain constant. **3** There are no pumps or turbines in the line.

**Properties**  The density and dynamic viscosity of water at 10°C are $\rho = 999.7$ kg/m³ and $\mu = 1.307 \times 10^{-3}$ kg/m·s. The roughness of cast iron pipe is $\varepsilon = 0.00026$ m (Table 14–2).

**Analysis**  The piping system involves 89 m of piping, a sharp-edged entrance ($K_L = 0.5$), two standard flanged elbows ($K_L = 0.3$ each), a fully open gate valve ($K_L = 0.2$), and a submerged exit ($K_L = 1.06$). We choose points 1 and 2 at the free surfaces of the two reservoirs. Noting that the fluid at both points is open to the atmosphere (and thus $P_1 = P_2 = P_{atm}$) and that the fluid velocities at both points are nearly zero ($V_1 \approx V_2 \approx 0$), the energy equation for a control volume between these two points simplifies to

$$\cancel{\frac{P_1}{\rho g}} + \alpha_1 \cancel{\frac{V_1^2}{2g}}^{\,0} + z_1 = \cancel{\frac{P_2}{\rho g}} + \alpha_2 \cancel{\frac{V_2^2}{2g}}^{\,0} + z_2 + h_L \quad \rightarrow \quad z_1 = z_2 + h_L$$

where

$$h_L = h_{L,\,total} = h_{L,\,major} + h_{L,\,minor} = \left( f \frac{L}{D} + \sum K_L \right) \frac{V^2}{2g}$$

since the diameter of the piping system is constant. The average velocity in the pipe and the Reynolds number are

$$V = \frac{\dot{V}}{A_c} = \frac{\dot{V}}{\pi D^2/4} = \frac{0.006 \text{ m}^3/\text{s}}{\pi (0.05 \text{ m})^2/4} = 3.06 \text{ m/s}$$

$$Re = \frac{\rho V D}{\mu} = \frac{(999.7 \text{ kg/m}^3)(3.06 \text{ m/s})(0.05 \text{ m})}{1.307 \times 10^{-3} \text{ kg/m·s}} = 117,000$$

The flow is turbulent since Re > 4000. Noting that $\varepsilon/D = 0.00026/0.05 = 0.0052$, the friction factor is determined from the Colebrook equation (or the Moody chart),

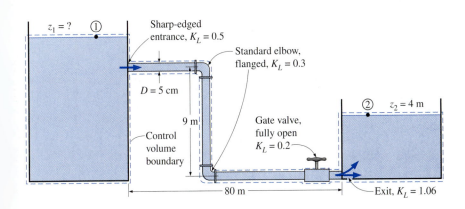

FIGURE 14–42

The piping system discussed in Example 14–8.

$$\frac{1}{\sqrt{f}} = -2.0 \log\left(\frac{\varepsilon/D}{3.7} + \frac{2.51}{\text{Re}\sqrt{f}}\right) \quad \rightarrow \quad \frac{1}{\sqrt{f}} = -2.0 \log\left(\frac{0.0052}{3.7} + \frac{2.51}{117{,}000\sqrt{f}}\right)$$

It gives $f = 0.0315$. The sum of the loss coefficients is

$$\sum K_L = K_{L,\,\text{entrance}} + 2K_{L,\,\text{elbow}} + K_{L,\,\text{valve}} + K_{L,\,\text{exit}}$$

$$= 0.5 + 2 \times 0.3 + 0.2 + 1.06 = 2.36$$

Then the total head loss and the elevation of the source become

$$h_L = \left(f\frac{L}{D} + \sum K_L\right)\frac{V^2}{2g} = \left(0.0315\frac{89\text{ m}}{0.05\text{ m}} + 2.36\right)\frac{(3.06\text{ m/s})^2}{2(9.81\text{ m/s}^2)} = 27.9\text{ m}$$

$$z_1 = z_2 + h_L = 4 + 27.9 = \mathbf{31.9\ m}$$

Therefore, the free surface of the first reservoir must be 31.9 m above the ground level to ensure water flow between the two reservoirs at the specified rate.

**Discussion** Note that $fL/D = 56.1$ in this case, which is about 24 times the total minor loss coefficient. Therefore, ignoring the sources of minor losses in this case would result in about 4 percent error. It can be shown that at the same flow rate, the total head loss would be 35.9 m (instead of 27.9 m) if the valve were three-fourths closed, and it would drop to 24.8 m if the pipe between the two reservoirs were straight at the ground level (thus eliminating the elbows and the vertical section of the pipe). The head loss could be reduced further (from 24.8 to 24.6 m) by rounding the entrance. The head loss can be reduced significantly (from 27.9 to 16.0 m) by replacing the cast iron pipes by smooth pipes such as those made of plastic.

# SUMMARY

In *internal flow*, a pipe is completely filled with a fluid. *Laminar flow* is characterized by smooth streamlines and highly ordered motion, and *turbulent flow* is characterized by unsteady disorderly velocity fluctuations and highly disordered motion. The *Reynolds number* is defined as

$$\text{Re} = \frac{\text{Inertial forces}}{\text{Viscous forces}} = \frac{V_{\text{avg}}D}{\nu} = \frac{\rho V_{\text{avg}}D}{\mu}$$

Under most practical conditions, the flow in a pipe is laminar at $\text{Re} < 2300$, turbulent at $\text{Re} > 4000$, and transitional in between.

The region of the flow in which the effects of the viscous shearing forces are felt is called the *velocity boundary layer*. The region from the pipe inlet to the point at which the flow

becomes fully developed is called the *hydrodynamic entrance region*, and the length of this region is called the *hydrodynamic entry length* $L_h$. It is given by

$$\frac{L_{h,\,\text{laminar}}}{D} \cong 0.05\text{ Re} \quad \text{and} \quad \frac{L_{h,\,\text{turbulent}}}{D} \cong 10$$

The friction coefficient in the fully developed flow region is constant. The *maximum* and *average* velocities in fully developed laminar flow in a circular pipe are

$$u_{\text{max}} = 2V_{\text{avg}} \quad \text{and} \quad V_{\text{avg}} = \frac{\Delta P D^2}{32\mu L}$$

The *volume flow rate* and the *pressure drop* for laminar flow in a horizontal pipe are

$$\dot{V} = V_{avg} A_c = \frac{\Delta P \pi D^4}{128 \mu L} \quad \text{and} \quad \Delta P = \frac{32 \mu L V_{avg}}{D^2}$$

The *pressure loss* and *head loss* for all types of internal flows (laminar or turbulent, in circular or noncircular pipes, smooth or rough surfaces) are expressed as

$$\Delta P_L = f \frac{L}{D} \frac{\rho V^2}{2} \quad \text{and} \quad h_L = \frac{\Delta P_L}{\rho g} = f \frac{L}{D} \frac{V^2}{2g}$$

where $\rho V^2/2$ is the *dynamic pressure* and the dimensionless quantity $f$ is the *friction factor*. For fully developed laminar flow in a round pipe, the friction factor is $f = 64/\text{Re}$.

For noncircular pipes, the diameter in the previous relations is replaced by the *hydraulic diameter* defined as $D_h = 4A_c/p$, where $A_c$ is the cross-sectional area of the pipe and $p$ is its wetted perimeter.

In fully developed turbulent flow, the friction factor depends on the Reynolds number and the *relative roughness* $\varepsilon/D$. The friction factor in turbulent flow is given by the *Colebrook equation*, expressed as

$$\frac{1}{\sqrt{f}} = -2.0 \log\left( \frac{\varepsilon/D}{3.7} + \frac{2.51}{\text{Re}\sqrt{f}} \right)$$

The plot of this formula is known as the *Moody chart*. The design and analysis of piping systems involve the determination of the head loss, flow rate, or the pipe diameter. Tedious iterations in these calculations can be avoided by the approximate *Swamee–Jain formulas* expressed as

$$h_L = 1.07 \frac{\dot{V}^2 L}{g D^5} \left\{ \ln \left[ \frac{\varepsilon}{3.7D} + 4.62 \left( \frac{\nu D}{\dot{V}} \right)^{0.9} \right] \right\}^{-2}$$

$$10^{-6} < \varepsilon/D < 10^{-2}$$
$$3000 < \text{Re} < 3 \times 10^8$$

$$\dot{V} = -0.965 \left( \frac{g D^5 h_L}{L} \right)^{0.5} \ln\left[ \frac{\varepsilon}{3.7D} + \left( \frac{3.17 \nu^2 L}{g D^3 h_L} \right)^{0.5} \right]$$

$$\text{Re} > 2000$$

$$D = 0.66 \left[ \varepsilon^{1.25} \left( \frac{L \dot{V}^2}{g h_L} \right)^{4.75} + \nu \dot{V}^{9.4} \left( \frac{L}{g h_L} \right)^{5.2} \right]^{0.04}$$

$$10^{-6} < \varepsilon/D < 10^{-2}$$
$$5000 < \text{Re} < 3 \times 10^8$$

The losses that occur in piping components such as fittings, valves, bends, elbows, tees, inlets, exits, expansions, and contractions are called *minor losses*. The minor losses are usually expressed in terms of the *loss coefficient* $K_L$. The head loss for a component is determined from

$$h_L = K_L \frac{V^2}{2g}$$

When all the loss coefficients are available, the total head loss in a piping system is

$$h_{L, total} = h_{L, major} + h_{L, minor} = \sum_i f_i \frac{L_i}{D_i} \frac{V_i^2}{2g} + \sum_j K_{L,j} \frac{V_j^2}{2g}$$

If the entire piping system is of constant diameter, the total head loss reduces to

$$h_{L, total} = \left( f \frac{L}{D} + \sum K_L \right) \frac{V^2}{2g}$$

The analysis of a piping system is based on two simple principles: (1) The conservation of mass throughout the system must be satisfied and (2) the pressure drop between two points must be the same for all paths between the two points. When the pipes are connected *in series*, the flow rate through the entire system remains constant regardless of the diameters of the individual pipes. For a pipe that branches out into two (or more) *parallel pipes* and then rejoins at a junction downstream, the total flow rate is the sum of the flow rates in the individual pipes but the head loss in each branch is the same.

When a piping system involves a pump and/or turbine, the steady-flow energy equation is expressed as

$$\frac{P_1}{\rho g} + \alpha_1 \frac{V_1^2}{2g} + z_1 + h_{pump,\, u}$$

$$= \frac{P_2}{\rho g} + \alpha_2 \frac{V_2^2}{2g} + z_2 + h_{turbine,\, e} + h_L$$

When the useful pump head $h_{pump,\, u}$ is known, the mechanical power that needs to be supplied by the pump to the fluid and the electric power consumed by the motor of the pump for a specified flow rate are

$$\dot{W}_{pump,\, shaft} = \frac{\rho \dot{V} g h_{pump,\, u}}{\eta_{pump}} \quad \text{and} \quad \dot{W}_{elect} = \frac{\rho \dot{V} g h_{pump,\, u}}{\eta_{pump-motor}}$$

where $\eta_{pump-motor}$ is the *efficiency of the pump–motor combination,* which is the product of the pump and the motor efficiencies.

The plot of the head loss versus the flow rate $\dot{V}$ is called the *system curve*. The head produced by a pump is not a constant, and the curves of $h_{pump,\, u}$ and $\eta_{pump}$ versus $\dot{V}$ are called the *characteristic curves*. A pump installed in a piping system operates at the *operating point*, which is the point of intersection of the system curve and the characteristic curve.

## REFERENCES AND SUGGESTED READING

1. M. S. Bhatti and R. K. Shah. "Turbulent and Transition Flow Convective Heat Transfer in Ducts." In *Handbook of Single-Phase Convective Heat Transfer*, ed. S. Kakaç, R. K. Shah, and W. Aung. New York: Wiley Interscience, 1987.

2. C. F. Colebrook. "Turbulent Flow in Pipes, with Particular Reference to the Transition between the Smooth and Rough Pipe Laws," *Journal of the Institute of Civil Engineers London.* 11 (1939), pp. 133–156.

3. S. E. Haaland. "Simple and Explicit Formulas for the Friction Factor in Turbulent Pipe Flow," *Journal of Fluids Engineering*, March 1983, pp. 89–90.

4. I. E. Idelchik. *Handbook of Hydraulic Resistance*, 3rd ed. Boca Raton, FL: CRC Press, 1993.

5. W. M. Kays and M. E. Crawford. *Convective Heat and Mass Transfer*, 3rd ed. New York: McGraw-Hill, 1993.

6. L. F. Moody. "Friction Factors for Pipe Flows," *Transactions of the ASME* 66 (1944), pp. 671–684.

7. O. Reynolds. "On the Experimental Investigation of the Circumstances Which Determine Whether the Motion of Water Shall Be Direct or Sinuous, and the Law of Resistance in Parallel Channels." *Philosophical Transactions of the Royal Society of London*, 174 (1883), pp. 935–982.

8. H. Schlichting. *Boundary Layer Theory*, 7th ed. New York: McGraw-Hill, 1979.

9. R. K. Shah and M. S. Bhatti. "Laminar Convective Heat Transfer in Ducts." In *Handbook of Single-Phase Convective Heat Transfer*, ed. S. Kakaç, R. K. Shah, and W. Aung. New York: Wiley Interscience, 1987.

10. P. K. Swamee and A. K. Jain. "Explicit Equations for Pipe-Flow Problems," *Journal of the Hydraulics Division.* ASCE 102, no. HY5 (May 1976), pp. 657–664.

11. W. Zhi-qing. "Study on Correction Coefficients of Laminar and Turbulent Entrance Region Effects in Round Pipes," *Applied Mathematical Mechanics*, 3 (1982), p. 433.

## PROBLEMS*

### Laminar and Turbulent Flow

**14–1C** What is hydraulic diameter? How is it defined? What is it equal to for a circular pipe of diameter $D$?

**14–2C** How is the hydrodynamic entry length defined for flow in a pipe? Is the entry length longer in laminar or turbulent flow?

**14–3C** Why are liquids usually transported in circular pipes?

**14–4C** What is the physical significance of the Reynolds number? How is it defined for (*a*) flow in a circular pipe of inner diameter $D$ and (*b*) flow in a rectangular duct of cross section $a \times b$?

**FIGURE P14–4C**

**14–5C** Consider a person walking first in air and then in water at the same speed. For which motion will the Reynolds number be higher?

**14–6C** Show that the Reynolds number for flow in a circular pipe of diameter $D$ can be expressed as $\text{Re} = 4\dot{m}/(\pi D\mu)$.

**14–7C** Which fluid at room temperature requires a larger pump to flow at a specified velocity in a given pipe: water or engine oil? Why?

**14–8C** Consider the flow of air and water in pipes of the same diameter, at the same temperature, and at the same mean velocity. Which flow is more likely to be turbulent? Why?

**14–9C** Consider laminar flow in a circular pipe. Is the wall shear stress $\tau_w$ higher near the inlet of the pipe or near the exit? Why? What would your response be if the flow were turbulent?

**14–10C** How does surface roughness affect the pressure drop in a pipe if the flow is turbulent? What would your response be if the flow were laminar?

**14–11E** Shown here is a cool picture of water being released at 300,000 gallons per second in the spring of 2008. This was part of a revitalization effort for the ecosystem of the Grand Canyon and the Colorado River. Estimate the Reynolds number of the pipe flow. Is it laminar or turbulent? (*Hint:* For a length scale, approximate the height of the man directly above the pipe to be 6 ft.)

**FIGURE P14–11E**

*Photo courtesy of Don Becker, U.S. Geological Survey.*

## Fully Developed Flow in Pipes

**14–12C** Someone claims that the shear stress at the center of a circular pipe during fully developed laminar flow is zero. Do you agree with this claim? Explain.

**14–13C** Someone claims that in fully developed turbulent flow in a pipe, the shear stress is a maximum at the pipe wall. Do you agree with this claim? Explain.

**14–14C** How does the wall shear stress $\tau_w$ vary along the flow direction in the fully developed region in (*a*) laminar flow and (*b*) turbulent flow?

**14–15C** In the fully developed region of flow in a circular pipe, does the velocity profile change in the flow direction?

**14–16C** How is the friction factor for flow in a pipe related to the pressure loss? How is the pressure loss related to the pumping power requirement for a given mass flow rate?

**14–17C** Discuss whether fully developed pipe flow is one-, two-, or three-dimensional.

**14–18C** Consider fully developed flow in a circular pipe with negligible entrance effects. If the length of the pipe is doubled, the head loss will (*a*) double, (*b*) more than double, (*c*) less than double, (*d*) reduce by half, or (*e*) remain constant.

**14–19C** Someone claims that the volume flow rate in a circular pipe with laminar flow can be determined by measuring the velocity at the centerline in the fully developed region, multiplying it by the cross-sectional area, and dividing the result by 2. Do you agree? Explain.

**14–20C** Someone claims that the average velocity in a circular pipe in fully developed laminar flow can be determined by simply measuring the velocity at $R/2$ (midway between the wall surface and the centerline). Do you agree? Explain.

**14–21C** Consider fully developed laminar flow in a circular pipe. If the diameter of the pipe is reduced by half while the flow rate and the pipe length are held constant, the head loss will (*a*) double, (*b*) triple, (*c*) quadruple, (*d*) increase by a factor of 8, or (*e*) increase by a factor of 16.

**14–22C** What is the physical mechanism that causes the friction factor to be higher in turbulent flow?

**14–23C** What is turbulent viscosity? What causes it?

**14–24C** Consider fully developed laminar flow in a circular pipe. If the viscosity of the fluid is reduced by half by heating while the flow rate is held constant, how does the head loss change?

**14–25C** How is head loss related to pressure loss? For a given fluid, explain how you would convert head loss to pressure loss.

**14–26C** Consider laminar flow of air in a circular pipe with perfectly smooth surfaces. Do you think the friction factor for this flow is zero? Explain.

**14–27C** Explain why the friction factor is independent of the Reynolds number at very large Reynolds numbers.

**14–28** Water at 10°C ($\rho = 999.7$ kg/m³ and $\mu = 1.307 \times 10^{-3}$ kg/m·s) is flowing steadily in a 0.20-cm-diameter, 15-m-long pipe at an average velocity of 1.2 m/s. Determine (*a*) the pressure drop, (*b*) the head loss, and (*c*) the pumping power requirement to overcome this pressure drop. *Answers:* (*a*) 188 kPa, (*b*) 19.2 m, (*c*) 0.71 W

**14–29** Water at 15°C ($\rho = 999.1$ kg/m³ and $\mu = 1.138 \times 10^{-3}$ kg/m·s) is flowing steadily in a 30-m-long and 5-cm-diameter horizontal pipe made of stainless steel at a rate of 9 L/s. Determine (*a*) the pressure drop, (*b*) the head loss, and (*c*) the pumping power requirement to overcome this pressure drop.

**FIGURE P14–29**

**14–30E** Heated air at 1 atm and 100°F is to be transported in a 400-ft-long circular plastic duct at a rate of 12 ft³/s. If the head loss in the pipe is not to exceed 50 ft, determine the minimum diameter of the duct.

**14–31** In fully developed laminar flow in a circular pipe, the velocity at $R/2$ (midway between the wall surface and the centerline) is measured to be 8 m/s. Determine the velocity at the center of the pipe. *Answer:* 10.7 m/s

**14–32** The velocity profile in fully developed laminar flow in a circular pipe of inner radius $R = 2$ cm, in m/s, is given by $u(r) = 4(1 - r^2/R^2)$. Determine the average and maximum velocities in the pipe and the volume flow rate.

**FIGURE P14–32**

**14–33** Repeat Prob. 14–32 for a pipe of inner radius 7 cm.

**14–34** Consider an air solar collector that is 1 m wide and 5 m long and has a constant spacing of 3 cm between the glass cover and the collector plate. Air flows at an average temperature of 45°C at a rate of 0.15 m³/s through the 1-m-wide edge of the collector along the 5-m-long passageway. Disregarding the entrance and roughness effects and the 90° bend, determine the pressure drop in the collector. *Answer:* 32.3 Pa

**FIGURE P14–34**

**14–35** Consider laminar flow of a fluid through a square channel with smooth surfaces. Now the average velocity of the fluid is doubled. Determine the change in the head loss of the fluid. Assume the flow regime remains unchanged.

**14–36** Repeat Prob. 14–35 for turbulent flow in smooth pipes for which the friction factor is given as $f = 0.184\text{Re}^{-0.2}$. What would your answer be for fully turbulent flow in a rough pipe?

**14–37** Air enters a 10-m-long section of a rectangular duct of cross section 15 cm × 20 cm made of commercial steel at 1 atm and 35°C at an average velocity of 7 m/s. Disregarding the entrance effects, determine the fan power needed to overcome the pressure losses in this section of the duct. *Answer:* 7.00 W

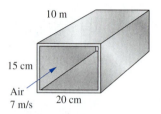

**FIGURE P14–37**

**14–38E** Water at 60°F passes through 0.75-in-internal-diameter copper tubes at a rate of 1.2 lbm/s. Determine the pumping power per ft of pipe length required to maintain this flow at the specified rate.

**14–39** Oil with $\rho = 876$ kg/m³ and $\mu = 0.24$ kg/m·s is flowing through a 1.5-cm-diameter pipe that discharges into the atmosphere at 88 kPa. The absolute pressure 15 m before the exit is measured to be 135 kPa. Determine the flow rate of oil through the pipe if the pipe is (*a*) horizontal, (*b*) inclined 8° upward from the horizontal, and (*c*) inclined 8° downward from the horizontal.

**FIGURE P14–39**

**14–40** Glycerin at 40°C with $\rho = 1252$ kg/m³ and $\mu = 0.27$ kg/m·s is flowing through a 2-cm-diameter, 25-m-long pipe that discharges into the atmosphere at 100 kPa. The flow rate through the pipe is 0.048 L/s. (*a*) Determine the absolute pressure 25 m before the pipe exit. (*b*) At what angle $\theta$ must the pipe be inclined downward from the horizontal for the pressure in the entire pipe to be atmospheric pressure and the flow rate to be maintained the same?

**14–41E** Oil at 80°F ($\rho = 56.8$ lbm/ft³ and $\mu = 0.0278$ lbm/ft·s) is flowing steadily in a 0.5-in-diameter, 120-ft-long pipe. During the flow, the pressure at the pipe inlet and exit is measured to be 120 psi and 14 psi, respectively. Determine the flow rate of oil through the pipe assuming the pipe is (a) horizontal, (b) inclined 20° upward, and (c) inclined 20° downward.

**14–42** Oil with a density of 850 kg/m³ and kinematic viscosity of 0.00062 m²/s is being discharged by a 8-mm-diameter, 40-m-long horizontal pipe from a storage tank open to the atmosphere. The height of the liquid level above the center of the pipe is 4 m. Disregarding the minor losses, determine the flow rate of oil through the pipe.

**FIGURE P14–42**

**14–43** In an air heating system, heated air at 40°C and 105 kPa absolute is distributed through a 0.2 m × 0.3 m rectangular duct made of commercial steel at a rate of 0.5 m³/s. Determine the pressure drop and head loss through a 40-m-long section of the duct. *Answers: 124 Pa, 10.8 m*

**14–44** Glycerin at 40°C with $\rho = 1252$ kg/m³ and $\mu = 0.27$ kg/m·s is flowing through a 4-cm-diameter horizontal smooth pipe with an average velocity of 3.5 m/s. Determine the pressure drop per 10 m of the pipe.

**14–45**  Reconsider Prob. 14–44. Using EES (or other) software, investigate the effect of the pipe diameter on the pressure drop for the same constant flow rate. Let the pipe diameter vary from 1 to 10 cm in increments of 1 cm. Tabulate and plot the results, and draw conclusions.

**14–46E** Air at 1 atm and 60°F is flowing through a 1 ft × 1 ft square duct made of commercial steel at a rate of 1600 cfm. Determine the pressure drop and head loss per ft of the duct.

**FIGURE P14–46E**

**14–47** Liquid ammonia at −20°C is flowing through a 30-m-long section of a 5-mm-diameter copper tube at a rate of 0.15 kg/s. Determine the pressure drop, the head loss, and the pumping power required to overcome the frictional losses in the tube. *Answers: 4790 kPa, 734 m, 1.08 kW*

**Minor Losses**

**14–48C** What is minor loss in pipe flow? How is the minor loss coefficient $K_L$ defined?

**14–49C** Define equivalent length for minor loss in pipe flow. How is it related to the minor loss coefficient?

**14–50C** The effect of rounding of a pipe inlet on the loss coefficient is (a) negligible, (b) somewhat significant, or (c) very significant.

**14–51C** Which has a greater minor loss coefficient during pipe flow: gradual expansion or gradual contraction? Why?

**14–52C** A piping system involves sharp turns, and thus large minor head losses. One way of reducing the head loss is to replace the sharp turns by circular elbows. What is another way?

**14–53C** During a retrofitting project of a fluid flow system to reduce the pumping power, it is proposed to install vanes into the miter elbows or to replace the sharp turns in 90° miter elbows by smooth curved bends. Which approach will result in a greater reduction in pumping power requirements?

**14–54** Water is to be withdrawn from a 5-m-high water reservoir by drilling a 1.5-cm-diameter hole at the bottom surface. Disregarding the effect of the kinetic energy correction factor, determine the flow rate of water through the hole if (a) the entrance of the hole is well-rounded and (b) the entrance is sharp-edged.

**14–55** A horizontal pipe has an abrupt expansion from $D_1 = 8$ cm to $D_2 = 16$ cm. The water velocity in the smaller section is 10 m/s and the flow is turbulent. The pressure in the smaller section is $P_1 = 410$ kPa. Taking the kinetic energy correction factor to be 1.06 at both the inlet and the outlet, determine the downstream pressure $P_2$, and estimate the error that would have occurred if Bernoulli's equation had been used. *Answers: 432 kPa, 25.4 kPa*

**FIGURE P14–55**

**14–56** Consider flow from a water reservoir through a circular hole of diameter $D$ at the side wall at a vertical distance $H$ from the free surface. The flow rate through an actual hole with a sharp-edged entrance ($K_L = 0.5$) is considerably less than the flow rate calculated assuming "frictionless" flow and thus zero loss for the hole. Disregarding the effect of the kinetic energy correction factor, obtain a relation for the "equivalent diameter" of the sharp-edged hole for use in frictionless flow relations.

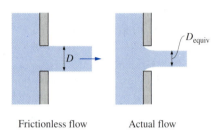

Frictionless flow          Actual flow

**FIGURE P14–56**

**14–57** Repeat Prob. 14–56 for a slightly rounded entrance ($K_L = 0.12$).

## Piping Systems and Pump Selection

**14–58C** A person filling a bucket with water using a garden hose suddenly remembers that attaching a nozzle to the hose increases the discharge velocity of water and wonders if this increased velocity would decrease the filling time of the bucket. What would happen to the filling time if a nozzle were attached to the hose: increase it, decrease it, or have no effect? Why?

**14–59C** Consider two identical 2-m-high open tanks filled with water on top of a 1-m-high table. The discharge valve of one of the tanks is connected to a hose whose other end is left open on the ground while the other tank does not have a hose connected to its discharge valve. Now the discharge valves of both tanks are opened. Disregarding any frictional loses in the hose, which tank do you think empties completely first? Why?

**14–60C** A piping system involves two pipes of different diameters (but of identical length, material, and roughness) connected in series. How would you compare the (*a*) flow rates and (*b*) pressure drops in these two pipes?

**14–61C** A piping system involves two pipes of identical diameters but of different lengths connected in parallel. How would you compare the pressure drops in these two pipes?

**14–62C** Water is pumped from a large lower reservoir to a higher reservoir. Someone claims that if the head loss is negligible, the required pump head is equal to the elevation difference between the free surfaces of the two reservoirs. Do you agree?

**14–63C** A piping system equipped with a pump is operating steadily. Explain how the operating point (the flow rate and the head loss) is established.

**14–64C** For a piping system, define the system curve, the characteristic curve, and the operating point on a head versus flow rate chart.

**14–65** The water needs of a small farm are to be met by pumping water from a well that can supply water continuously at a rate of 4 L/s. The water level in the well is 20 m below the ground level, and water is to be pumped to a large tank on a hill, which is 58 m above the ground level of the well, using 5-cm internal diameter plastic pipes. The required length of piping is measured to be 420 m, and the total minor loss coefficient due to the use of elbows, vanes, etc. is estimated to be 12. Taking the efficiency of the pump to be 75 percent, determine the rated power of the pump that needs to be purchased, in kW. The density and viscosity of water at anticipated operation conditions are taken to be 1000 kg/m³ and 0.00131 kg/m·s, respectively. Is it wise to purchase a suitable pump that meets the total power requirements, or is it necessary to also pay particular attention to the large elevation head in this case? Explain.  *Answer:* 6 kW

**14–66E** Water at 70°F flows by gravity from a large reservoir at a high elevation to a smaller one through a 90-ft-long, 2-in-diameter cast iron piping system that includes four standard flanged elbows, a well-rounded entrance, a sharp-edged exit, and a fully open gate valve. Taking the free surface of the lower reservoir as the reference level, determine the elevation $z_1$ of the higher reservoir for a flow rate of 10 ft³/min.  *Answer:* 17.9 ft

**14–67** A 2.4-m-diameter tank is initially filled with water 4 m above the center of a sharp-edged 10-cm-diameter orifice. The tank water surface is open to the atmosphere, and the orifice drains to the atmosphere. Neglecting the effect of the kinetic energy correction factor, calculate (*a*) the initial velocity from the tank and (*b*) the time required to empty the tank. Does the loss coefficient of the orifice cause a significant increase in the draining time of the tank?

**FIGURE P14–67**

**14–68** A 3-m-diameter tank is initially filled with water 2 m above the center of a sharp-edged 10-cm-diameter orifice. The tank water surface is open to the atmosphere, and the orifice drains to the atmosphere through a 100-m-long pipe. The friction coefficient of the pipe is taken to be 0.015 and the effect of the kinetic energy correction factor can be neglected. Determine (a) the initial velocity from the tank and (b) the time required to empty the tank.

**14–69** Reconsider Prob. 14–68. In order to drain the tank faster, a pump is installed near the tank exit as in Fig. P14–69. Determine how much pump power input is necessary to establish an average water velocity of 4 m/s when the tank is full at $z = 2$ m. Also, assuming the discharge velocity to remain constant, estimate the time required to drain the tank.

Someone suggests that it makes no difference whether the pump is located at the beginning or at the end of the pipe, and that the performance will be the same in either case, but another person argues that placing the pump near the end of the pipe may cause cavitation. The water temperature is 30°C, so the water vapor pressure is $P_v = 4.246$ kPa $= 0.43$ m-H$_2$O, and the system is located at sea level. Investigate if there is the possibility of cavitation and if we should be concerned about the location of the pump.

**FIGURE P14–69**

**14–70** Water to a residential area is transported at a rate of 3 m³/s via 90-cm-internal-diameter concrete pipes with a surface roughness of 3 mm and a total length of 1500 m. In order to reduce pumping power requirements, it is proposed to line the interior surfaces of the concrete pipe with 2-cm-thick petroleum-based lining that has a surface roughness thickness of 0.04 mm. There is a concern that the reduction of pipe diameter to 86 cm and the increase in average velocity may offset any gains. Taking $\rho = 1000$ kg/m³ and $\nu = 1 \times 10^{-6}$ m²/s for water, determine the percent increase or decrease in the pumping power requirements due to pipe frictional losses as a result of lining the concrete pipes.

**14–71** Water is transported by gravity through a 12-cm-diameter 800-m-long plastic pipe with an elevation gradient of 0.01 (i.e., an elevation drop of 1 m per 100 m of pipe length). Taking $\rho = 1000$ kg/m³ and $\nu = 1 \times 10^{-6}$ m²/s for

water, determine the flow rate of water through the pipe. If the pipe were horizontal, what would the power requirements be to maintain the same flow rate?

**14–72** Water at 15°C is drained from a large reservoir using two horizontal plastic pipes connected in series. The first pipe is 20 m long and has a 10-cm diameter, while the second pipe is 35 m long and has a 4-cm diameter. The water level in the reservoir is 18 m above the centerline of the pipe. The pipe entrance is sharp-edged, and the contraction between the two pipes is sudden. Neglecting the effect of the kinetic energy correction factor, determine the discharge rate of water from the reservoir.

**FIGURE P14–72**

**14–73E** A farmer is to pump water at 70°F from a river to a water storage tank nearby using a 125-ft-long, 5-in-diameter plastic pipe with three flanged 90° smooth bends. The water velocity near the river surface is 6 ft/s, and the pipe inlet is placed in the river normal to the flow direction of water to take advantage of the dynamic pressure. The elevation difference between the river and the free surface of the tank is 12 ft. For a flow rate of 1.5 ft³/s and an overall pump efficiency of 70 percent, determine the required electric power input to the pump.

**14–74E** Reconsider Prob. 14–73E. Using EES (or other) software, investigate the effect of the pipe diameter on the required electric power input to the pump. Let the pipe diameter vary from 1 to 10 in, in increments of 1 in. Tabulate and plot the results, and draw conclusions.

**14–75** A water tank filled with solar-heated water at 40°C is to be used for showers in a field using gravity-driven flow. The system includes 25 m of 1.5-cm-diameter galvanized iron piping with four miter bends (90°) without vanes and a wide-open globe valve. If water is to flow at a rate of 0.8 L/s through the shower head, determine how high the water level in the tank must be from the exit level of the shower. Disregard the losses at the entrance and at the shower head, and neglect the effect of the kinetic energy correction factor.

**14–76** Two water reservoirs A and B are connected to each other through a 40-m-long, 2-cm-diameter cast iron pipe with a sharp-edged entrance. The pipe also involves a swing check valve and a fully open gate valve. The water level in both reservoirs is the same, but reservoir $A$ is pressurized by compressed air while reservoir $B$ is open to the atmosphere at 88 kPa. If the initial flow rate through the pipe is 1.2 L/s, determine the absolute air pressure on top of reservoir $A$. Take the water temperature to be 10°C. *Answer: 733 kPa*

**FIGURE P14–76**

14–77  A vented tanker is to be filled with fuel oil with $\rho = 920$ kg/m³ and $\mu = 0.045$ kg/m·s from an underground reservoir using a 25-m-long, 4-cm-diameter plastic hose with a slightly rounded entrance and two 90° smooth bends. The elevation difference between the oil level in the reservoir and the top of the tanker where the hose is discharged is 5 m. The capacity of the tanker is 18 m³ and the filling time is 30 min. Taking the kinetic energy correction factor at the hose discharge to be 1.05 and assuming an overall pump efficiency of 82 percent, determine the required power input to the pump.

**FIGURE P14–77**

14–78  Two pipes of identical length and material are connected in parallel. The diameter of pipe $A$ is twice the diameter of pipe $B$. Assuming the friction factor to be the same in both cases and disregarding minor losses, determine the ratio of the flow rates in the two pipes.

14–79  A certain part of cast iron piping of a water distribution system involves a parallel section. Both parallel pipes have a diameter of 30 cm, and the flow is fully turbulent. One of the branches (pipe $A$) is 1000 m long while the other branch (pipe $B$) is 3000 m long. If the flow rate through pipe $A$ is 0.4 m³/s, determine the flow rate through pipe $B$. Disregard minor losses and assume the water temperature to be

15°C. Show that the flow is fully rough, and thus the friction factor is independent of Reynolds number. *Answer: 0.231 m³/s*

**FIGURE P14–79**

14–80  Repeat Prob. 14–79 assuming pipe $A$ has a halfway-closed gate valve ($K_L = 2.1$) while pipe $B$ has a fully open globe valve ($K_L = 10$), and the other minor losses are negligible.

14–81  Gasoline ($\rho = 680$ kg/m³ and $\nu = 4.29 \times 10^{-7}$ m²/s) is transported at a rate of 400 L/s for a distance of 2 km. The surface roughness of the piping is 0.03 mm. If the head loss due to pipe friction is not to exceed 8 m, determine the minimum diameter of the pipe.

14–82  In large buildings, hot water in a water tank is circulated through a loop so that the user doesn't have to wait for all the water in long piping to drain before hot water starts coming out. A certain recirculating loop involves 40-m-long, 1.2-cm-diameter cast iron pipes with six 90° threaded smooth bends and two fully open gate valves. If the average flow velocity through the loop is 2 m/s, determine the required power input for the recirculating pump. Take the average water temperature to be 60°C and the efficiency of the pump to be 70 percent. *Answer: 0.111 kW*

14–83  Reconsider Prob. 14–82. Using EES (or other) software, investigate the effect of the average flow velocity on the power input to the recirculating pump. Let the velocity vary from 0 to 3 m/s in increments of 0.3 m/s. Tabulate and plot the results.

14–84  Repeat Prob. 14–82 for plastic (smooth) pipes.

14–85  Water at 20°C is to be pumped from a reservoir ($z_A = 2$ m) to another reservoir at a higher elevation ($z_B = 9$ m) through two 25-m-long plastic pipes connected in parallel. The diameters of the two pipes are 3 cm and 5 cm. Water is to be pumped by a 68 percent efficient motor–pump unit that draws 7 kW of electric power during operation. The minor losses and the head loss in the pipes that connect the parallel pipes to the two reservoirs are considered to be negligible. Determine the total flow rate between the reservoirs and the flow rates through each of the parallel pipes.

FIGURE P14–85

FIGURE P14–87

## Review Problems

**14–86** Shell-and-tube heat exchangers with hundreds of tubes housed in a shell are commonly used in practice for heat transfer between two fluids. Such a heat exchanger used in an active solar hot-water system transfers heat from a water-antifreeze solution flowing through the shell and the solar collector to fresh water flowing through the tubes at an average temperature of 60°C at a rate of 15 L/s. The heat exchanger contains 80 brass tubes 1 cm in inner diameter and 1.5 m in length. Disregarding inlet, exit, and header losses, determine the pressure drop across a single tube and the pumping power required by the tube-side fluid of the heat exchanger.

After operating a long time, 1-mm-thick scale builds up on the inner surfaces with an equivalent roughness of 0.4 mm. For the same pumping power input, determine the percent reduction in the flow rate of water through the tubes.

FIGURE P14–86

**14–87** The compressed air requirements of a manufacturing facility are met by a 150-hp compressor that draws in air from the outside through an 12-m-long, 18-cm-diameter duct made of thin galvanized iron sheets. The compressor takes in air at a rate of 0.27 m³/s at the outdoor conditions of 15°C and 95 kPa. Disregarding any minor losses, determine the useful power used by the compressor to overcome the frictional losses in this duct. *Answer: 24.7 W*

**14–88** A house built on a riverside is to be cooled in summer by utilizing the cool water of the river. A 15-m-long section of a circular stainless-steel duct of 20-cm diameter passes through the water. Air flows through the underwater section of the duct at 3 m/s at an average temperature of 15°C. For an overall fan efficiency of 62 percent, determine the fan power needed to overcome the flow resistance in this section of the duct.

FIGURE P14–88

**14–89** The velocity profile in fully developed laminar flow in a circular pipe, in m/s, is given by $u(r) = 6(1 - 100r^2)$, where $r$ is the radial distance from the centerline of the pipe in m. Determine (a) the radius of the pipe, (b) the average velocity through the pipe, and (c) the maximum velocity in the pipe.

**14–90E** The velocity profile in a fully developed laminar flow of water at 40°F in a 115-ft-long horizontal circular pipe, in ft/s, is given by $u(r) = 0.8(1 - 625r^2)$, where $r$ is the radial distance from the centerline of the pipe in ft. Determine (a) the volume flow rate of water through the pipe,

(b) the pressure drop across the pipe, and (c) the useful pumping power required to overcome this pressure drop.

**14–91E** Repeat Prob. 14–90E assuming the pipe is inclined 12° from the horizontal and the flow is uphill.

**14–92** A highly viscous liquid discharges from a large container through a small-diameter tube in laminar flow. Disregarding entrance effects and velocity heads, obtain a relation for the variation of fluid depth in the tank with time.

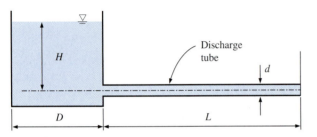

**FIGURE P14–92**

**14–93** A student is to determine the kinematic viscosity of an oil using the system shown in Prob. 14–92. The initial fluid height in the tank is $H = 40$ cm, the tube diameter is $d = 6$ mm, the tube length is $L = 0.65$ m, and the tank diameter is $D = 0.63$ m. The student observes that it takes 1750 s for the fluid level in the tank to drop to 36 cm. Find the fluid viscosity.

**14–94** Water is to be withdrawn from a 7-m-high water reservoir by drilling a well-rounded 4-cm-diameter hole with negligible loss near the bottom and attaching a horizontal 90° bend of negligible length. Taking the kinetic energy correction factor to be 1.05, determine the flow rate of water through the bend if (a) the bend is a flanged smooth bend and (b) the bend is a miter bend without vanes. *Answers:* (a) 12.7 L/s, (b) 10.0 L/s

**FIGURE P14–94**

**14–95** In a geothermal district heating system, 10,000 kg/s of hot water must be delivered a distance of 10 km in a horizontal pipe. The minor losses are negligible, and the only significant energy loss arises from pipe friction. The friction factor is taken to be 0.015. Specifying a larger-diameter pipe would reduce water velocity, velocity head, pipe friction, and thus power consumption. But a larger pipe would also cost more money initially to purchase and install. Otherwise stated, there is an optimum pipe diameter that will minimize the sum of pipe cost and future electric power cost.

Assume the system will run 24 h/day, every day, for 30 years. During this time the cost of electricity remains constant at \$0.06/kWh. Assume system performance stays constant over the decades (this may not be true, especially if highly mineralized water is passed through the pipeline—scale may form). The pump has an overall efficiency of 80 percent. The cost to purchase, install, and insulate a 10-km pipe depends on the diameter $D$ and is given by *Cost* $= \$10^6 D^2$, where $D$ is in m. Assuming zero inflation and interest rate for simplicity and zero salvage value and zero maintenance cost, determine the optimum pipe diameter.

**14–96** Two pipes of identical diameter and material are connected in parallel. The length of pipe $A$ is three times the length of pipe $B$. Assuming the flow is fully turbulent in both pipes and thus the friction factor is independent of the Reynolds number and disregarding minor losses, determine the ratio of the flow rates in the two pipes. *Answer:* 0.577

**14–97** A pipeline that transports oil at 40°C at a rate of 3 m³/s branches out into two parallel pipes made of commercial steel that reconnect downstream. Pipe $A$ is 500 m long and has a diameter of 30 cm while pipe $B$ is 800 m long and has a diameter of 45 cm. The minor losses are considered to be negligible. Determine the flow rate through each of the parallel pipes.

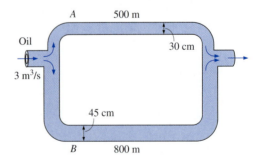

**FIGURE P14–97**

**14–98** Repeat Prob. 14–97 for hot-water flow of a district heating system at 100°C.

**14–99E** A water fountain is to be installed at a remote location by attaching a cast iron pipe directly to a water main through which water is flowing at 70°F and 60 psig. The entrance to the pipe is sharp-edged, and the 70-ft-long piping system involves three 90° miter bends without vanes, a fully open gate valve, and an angle valve with a loss coefficient of 5 when fully open. If the system is to provide water at a rate of 15 gal/min and the elevation difference between the pipe and the fountain is negligible, determine the minimum diameter of the piping system. *Answer: 0.71 in*

FIGURE P14–103E

**FIGURE P14–99E**

**14–100E** Repeat Prob. 14–99E for plastic (smooth) pipes.

**14–101** In a hydroelectric power plant, water at 20°C is supplied to the turbine at a rate of 1.2 m³/s through a 200-m-long, 0.35-m-diameter cast iron pipe. The elevation difference between the free surface of the reservoir and the turbine discharge is 115 m, and the combined turbine–generator efficiency is 80 percent. Disregarding the minor losses because of the large length-to-diameter ratio, determine the electric power output of this plant.

**14–102** In Prob. 14–101, the pipe diameter is tripled in order to reduce the pipe losses. Determine the percent increase in the net power output as a result of this modification.

**14–103E** The drinking water needs of an office are met by large water bottles. One end of a 0.35-in-diameter, 6-ft-long plastic hose is inserted into the bottle placed on a high stand, while the other end with an on/off valve is maintained 3 ft below the bottom of the bottle. If the water level in the bottle is 1 ft when it is full, determine how long it would take to fill an 8-oz glass (= 0.00835 ft³) (*a*) when the bottle is first opened and (*b*) when the bottle is almost empty. Take the total minor loss coefficient, including the on/off valve, to be 2.8 when it is fully open. Assume the water temperature to be the same as the room temperature of 70°F. *Answers: (a) 2.4 s, (b) 2.8 s*

**14–104E** Reconsider Prob. 14–103E. Using EES (or other) software, investigate the effect of the hose diameter on the time required to fill a glass when the bottle is full. Let the diameter vary from 0.2 to 2 in, in increments of 0.2 in. Tabulate and plot the results.

**14–105E** Reconsider Prob. 14–103E. The office worker who set up the siphoning system purchased a 12-ft-long reel of the plastic tube and wanted to use the whole thing to avoid cutting it in pieces, thinking that it is the elevation difference that makes siphoning work, and the length of the tube is not important. So he used the entire 12-ft-long tube. Assuming the turns or constrictions in the tube are not significant (being very optimistic) and the same elevation is maintained, determine the time it takes to fill a glass of water for both cases (bottle nearly full and bottle nearly empty).

**14–106** A circular water pipe has an abrupt expansion from diameter $D_1$ = 15 cm to $D_2$ = 20 cm. The pressure and the average water velocity in the smaller pipe are $P_1$ = 120 kPa and 10 m/s, respectively, and the flow is turbulent. By applying the continuity, momentum, and energy equations and disregarding the effects of the kinetic energy and momentum-flux correction factors, show that the loss coefficient for sudden expansion is $K_L = (1 - D_1^2/D_2^2)^2$, and calculate $K_L$ and $P_2$ for the given case.

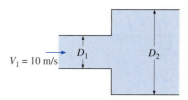

**FIGURE P14–106**

**14–107** The water at 20°C in a 10-m-diameter, 2-m-high aboveground swimming pool is to be emptied by unplugging a 5-cm-diameter, 25-m-long horizontal plastic pipe attached to the bottom of the pool. Determine the initial rate of discharge of water through the pipe and the time (hours) it would take to empty the swimming pool completely assuming the entrance to the pipe is well-rounded with negligible loss. Take the friction factor of the pipe to be 0.022. Using the initial discharge velocity, check if this is a reasonable value for the friction factor. *Answers:* 3.55 L/s, 24.6 h

**FIGURE P14–107**

**14–108** Reconsider Prob. 14–107. Using EES (or other) software, investigate the effect of the discharge pipe diameter on the time required to empty the pool completely. Let the diameter vary from 1 to 10 cm, in increments of 1 cm. Tabulate and plot the results.

**14–109** Repeat Prob. 14–107 for a sharp-edged entrance to the pipe with $K_L = 0.5$. Is this "minor loss" truly "minor" or not?

## Design and Essay Problems

**14–110** A pump is to be selected for a waterfall in a garden. The water collects in a pond at the bottom, and the elevation difference between the free surface of the pond and the location where the water is discharged is 3 m. The flow rate of water is to be at least 8 L/s. Select an appropriate motor–pump unit for this job and identify three manufacturers with product model numbers and prices. Make a selection and explain why you selected that particular product. Also estimate the cost of annual power consumption of this unit assuming continuous operation.

**14–111** During a camping trip you notice that water is discharged from a high reservoir to a stream in the valley through a 30-cm-diameter plastic pipe. The elevation difference between the free surface of the reservoir and the stream is 70 m. You conceive the idea of generating power from this water. Design a power plant that will produce the most power from this resource. Also, investigate the effect of power generation on the discharge rate of water. What discharge rate maximizes the power production?

# EXTERNAL FLOW: DRAG AND LIFT

I n this chapter we consider *external flow*—flow over bodies that are immersed in a fluid, with emphasis on the resulting lift and drag forces. In external flow, the viscous effects are confined to a portion of the flow field such as the boundary layers and wakes, which are surrounded by an outer flow region that involves small velocity and temperature gradients.

When a fluid moves over a solid body, it exerts pressure forces normal to the surface and shear forces parallel to the surface of the body. We are usually interested in the *resultant* of the pressure and shear forces acting on the body rather than the details of the distributions of these forces along the entire surface of the body. The component of the resultant pressure and shear forces that acts in the flow direction is called the *drag force* (or just *drag*), and the component that acts normal to the flow direction is called the *lift force* (or just *lift*).

We start this chapter with a discussion of drag and lift, and explore the concepts of pressure drag, friction drag, and flow separation. We continue with the drag coefficients of various two- and three-dimensional geometries encountered in practice and determine the drag force using experimentally determined drag coefficients. We then examine the development of the velocity boundary layer during parallel flow over a flat surface, and develop relations for the skin friction and drag coefficients for flow over flat plates, cylinders, and spheres. Finally, we discuss the lift developed by airfoils and the factors that affect the lift characteristics of bodies.

## Objectives

The objectives of this chapter are to:

■ Have an intuitive understanding of the various physical phenomena associated with external flow such as drag, friction and pressure drag, drag reduction, and lift.

■ Calculate the drag force associated with flow over common geometries.

■ Understand the effects of flow regime on the drag coefficients associated with flow over cylinders and spheres.

■ Understand the fundamentals of flow over airfoils, and calculate the drag and lift forces acting on airfoils.

## 15–1 ▪ INTRODUCTION

Fluid flow over solid bodies frequently occurs in practice, and it is responsible for numerous physical phenomena such as the *drag force* acting on automobiles, power lines, trees, and underwater pipelines; the *lift* developed by airplane wings; *upward draft* of rain, snow, hail, and dust particles in high winds; the transportation of red blood cells by blood flow; the entrainment and disbursement of liquid droplets by sprays; the vibration and noise generated by bodies moving in a fluid; and the power generated by wind turbines (Fig. 15–1). Therefore, developing a good understanding of external flow is important in the design of many engineering systems such as aircraft, automobiles, buildings, ships, submarines, and all kinds of turbines. Late-model cars, for example, have been designed with particular emphasis on aerodynamics. This has resulted in significant reductions in fuel consumption and noise, and considerable improvement in handling.

Sometimes a fluid moves over a stationary body (such as the wind blowing over a building), and other times a body moves through a quiescent fluid (such as a car moving through air). These two seemingly different processes are equivalent to each other; what matters is the relative motion between the fluid and the body. Such motions are conveniently analyzed by fixing the coordinate system on the body and are referred to as **flow over bodies** or **external flow**. The aerodynamic aspects of different airplane wing designs, for example, are studied conveniently in a lab by placing the wings in a wind tunnel and blowing air over them by large fans. Also, a flow can be classified as being steady or unsteady, depending on the reference frame selected. Flow around an airplane, for example, is always unsteady with respect to the ground, but it is steady with respect to a frame of reference moving with the airplane at cruise conditions.

The flow fields and geometries for most external flow problems are too complicated to be solved analytically, and thus we have to rely on correlations based on experimental data. The availability of high-speed computers has made it possible to conduct a series of "numerical experiments" quickly by solving the governing equations numerically and to resort to the expensive and time-consuming testing and experimentation only in the final stages of design. Such testing is done in wind tunnels. H. F. Phillips (1845–1912) built the first wind tunnel in 1894 and measured lift and drag. In this chapter we mostly rely on relations developed experimentally.

The velocity of the fluid approaching a body is called the **free-stream velocity** and is denoted by V. It is also denoted by $u_\infty$ or $U_\infty$ when the flow is aligned with the x-axis since u is used to denote the x-component of velocity. The fluid velocity ranges from zero at the body surface (the no-slip condition) to the free-stream value away from the body surface, and the subscript "infinity" serves as a reminder that this is the value at a distance where the presence of the body is not felt. The free-stream velocity may vary with location and time (e.g., the wind blowing past a building). But in the design and analysis, the free-stream velocity is usually assumed to be *uniform* and *steady* for convenience, and this is what we do in this chapter.

The shape of a body has a profound influence on the flow over the body and the velocity field. The flow over a body is said to be **two-dimensional** when the body is very long and of constant cross section and the flow is

(a)  (b)  (c)

(d)  (e)

(f)  (g)

**FIGURE 15–1**

Flow over bodies is commonly encountered in practice.

*(a) © Corbis RF; (b) The McGraw-Hill Companies, Inc./ John Flournoy, photographer; (c) © It/age Fotostock RF; (d) © Corbis RF; (e) © StockTrek/Superstock RF; (f) © Corbis RF; (g) © Alamy RF.*

Long cylinder (2-D)

(a)

Bullet (axisymmetric)

(b)

Car (3-D)

(c)

**FIGURE 15–2**

Two-dimensional, axisymmetric, and three-dimensional flows.

*(a) Photo by John M. Cimbala; (b) © CorbisRF; (c) Chevy Volt, courtesy of G.M. Corporation.*

normal to the body. The wind blowing over a long pipe perpendicular to its axis is an example of two-dimensional flow. Note that the velocity component in the axial direction is zero in this case, and thus the velocity is two-dimensional.

The two-dimensional idealization is appropriate when the body is sufficiently long so that the end effects are negligible and the approach flow is uniform. Another simplification occurs when the body possesses rotational symmetry about an axis in the flow direction. The flow in this case is also two-dimensional and is said to be **axisymmetric**. A bullet piercing through air is an example of axisymmetric flow. The velocity in this case varies with the axial distance $x$ and the radial distance $r$. Flow over a body that cannot be modeled as two-dimensional or axisymmetric, such as flow over a car, is **three-dimensional** (Fig. 15–2).

Flow over bodies can also be classified as **incompressible flows** (e.g., flows over automobiles, submarines, and buildings) and **compressible flows** (e.g., flows over high-speed aircraft, rockets, and missiles). Compressibility effects are negligible at low velocities (flows with Ma ≲ 0.3), and such flows can be treated as incompressible with little loss in accuracy. Compressible flows that involve partially immersed bodies with a free surface (such as a ship cruising in water) are beyond the scope of this introductory text.

Bodies subjected to fluid flow are classified as being streamlined or bluff, depending on their overall shape. A body is said to be **streamlined** if a conscious effort is made to align its shape with the anticipated streamlines in the flow. Streamlined bodies such as race cars and airplanes appear to be contoured and sleek. Otherwise, a body (such as a building) tends to block the flow and is said to be **bluff** or blunt. Usually it is much easier to force a streamlined body through a fluid, and thus streamlining has been of great importance in the design of vehicles and airplanes (Fig. 15–3).

## 15–2 ▪ DRAG AND LIFT

It is a common experience that a body meets some resistance when it is forced to move through a fluid, especially a liquid. As you may have noticed, it is very difficult to walk in water because of the much greater resistance it offers to motion compared to air. Also, you may have seen high winds knocking down trees, power lines, and even trailers and felt the strong "push" the wind exerts on your body (Fig. 15–4). You experience the same feeling when you extend your arm out of the window of a moving car. A fluid may exert forces and moments on a body in and about various directions. The force a flowing fluid exerts on a body in the flow direction is called **drag**. The drag force can be measured directly by simply attaching the body subjected to fluid flow to a calibrated spring and measuring the displacement in the flow direction (just like measuring weight with a spring scale). More sophisticated drag-measuring devices, called drag balances, use flexible beams fitted with strain gages to measure the drag electronically.

Drag is usually an undesirable effect, like friction, and we do our best to minimize it. Reduction of drag is closely associated with the reduction of fuel consumption in automobiles, submarines, and aircraft; improved safety and durability of structures subjected to high winds; and reduction of noise and vibration. But in some cases drag produces a beneficial effect and we try to maximize it. Friction, for example, is a "life saver" in the brakes of automobiles. Likewise, it is the drag that makes it possible for people to parachute, for pollens to fly to distant locations, and for us all to enjoy the waves of the oceans and the relaxing movements of the leaves of trees.

A stationary fluid exerts only normal pressure forces on the surface of a body immersed in it. A moving fluid, however, also exerts tangential shear forces on the surface because of the no-slip condition caused by viscous effects. Both of these forces, in general, have components in the direction of flow, and thus the drag force is due to the combined effects of pressure and wall shear forces in the flow direction. The components of the pressure and wall shear forces in the direction *normal* to the flow tend to move the body in that direction, and their sum is called **lift**.

For two-dimensional flows, the resultant of the pressure and shear forces can be split into two components: one in the direction of flow, which is the drag force, and another in the direction normal to flow, which is the lift, as shown in Fig. 15–5. For three-dimensional flows, there is also a side force component in the direction normal to the page that tends to move the body in that direction.

The fluid forces may also generate moments and cause the body to rotate. The moment about the flow direction is called the *rolling moment,* the moment about the lift direction is called the *yawing moment,* and the moment about the side force direction is called the *pitching moment.* For bodies that possess symmetry about the lift–drag plane such as cars, airplanes, and ships, the time-averaged side force, yawing moment, and rolling moment are zero when the wind and wave forces are aligned with the body. What remain for such bodies are the drag and lift forces and the pitching moment. For axisymmetric bodies aligned with the flow, such as a bullet, the only time-averaged force exerted by the fluid on the body is the drag force.

The pressure and shear forces acting on a differential area $dA$ on the surface are $P \, dA$ and $\tau_w \, dA$, respectively. The differential drag force and the lift force acting on $dA$ in two-dimensional flow are (Fig. 15–5)

$$dF_D = -P \, dA \cos \theta + \tau_w \, dA \sin \theta \qquad (15\text{–}1)$$

and

$$dF_L = -P \, dA \sin \theta - \tau_w \, dA \cos \theta \qquad (15\text{–}2)$$

where $\theta$ is the angle that the outer normal of $dA$ makes with the positive flow direction. The total drag and lift forces acting on the body are determined by integrating Eqs. 15–1 and 15–2 over the entire surface of the body,

*Drag force:* $\qquad F_D = \displaystyle\int_A dF_D = \int_A (-P \cos \theta + \tau_w \sin \theta) \, dA \qquad (15\text{–}3)$

and

*Lift force:* $\qquad F_L = \displaystyle\int_A dF_L = -\int_A (P \sin \theta + \tau_w \cos \theta) \, dA \qquad (15\text{–}4)$

These are the equations used to predict the net drag and lift forces on bodies when the flow is simulated on a computer. However, when we perform experimental analyses, Eqs. 15–3 and 15–4 are not practical since the detailed distributions of pressure and shear forces are difficult to obtain by

**FIGURE 15–3**

It is much easier to force a streamlined body than a blunt body through a fluid.

**FIGURE 15–4**

High winds knock down trees, power lines, and even people as a result of the drag force.

$$F_D = F_R \cos \phi$$
$$F_L = F_R \sin \phi$$

**FIGURE 15–5**

The pressure and viscous forces acting on a two-dimensional body and the resultant lift and drag forces.

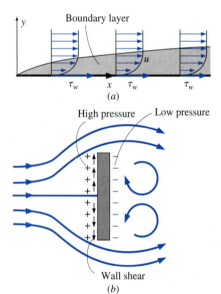

**FIGURE 15–6**

(a) Drag force acting on a flat plate parallel to the flow depends on wall shear only. (b) Drag force acting on a flat plate normal to the flow depends on the pressure only and is independent of the wall shear, which acts normal to the free-stream flow.

measurements. Fortunately, this information is often not needed. Usually all we need to know is the resultant drag force and lift acting on the entire body, which can be measured directly and easily in a wind tunnel.

Equations 15–1 and 15–2 show that both the skin friction (wall shear) and pressure, in general, contribute to the drag and the lift. In the special case of a thin *flat plate* aligned parallel to the flow direction, the drag force depends on the wall shear only and is independent of pressure since $\theta = 90°$. When the flat plate is placed normal to the flow direction, however, the drag force depends on the pressure only and is independent of wall shear since the shear stress in this case acts in the direction normal to flow and $\theta = 0°$ (Fig. 15–6). If the flat plate is tilted at an angle relative to the flow direction, then the drag force depends on both the pressure and the shear stress.

The wings of airplanes are shaped and positioned specifically to generate lift with minimal drag. This is done by maintaining an angle of attack during cruising, as shown in Fig. 15–7. Both lift and drag are strong functions of the angle of attack, as we discuss later in this chapter. The pressure difference between the top and bottom surfaces of the wing generates an upward force that tends to lift the wing and thus the airplane to which it is connected. For slender bodies such as wings, the shear force acts nearly parallel to the flow direction, and thus its contribution to the lift is small. The drag force for such slender bodies is mostly due to shear forces (the skin friction).

The drag and lift forces depend on the density $\rho$ of the fluid, the upstream velocity $V$, and the size, shape, and orientation of the body, among other things, and it is not practical to list these forces for a variety of situations. Instead, it is more convenient to work with appropriate dimensionless numbers that represent the drag and lift characteristics of the body. These numbers are the **drag coefficient** $C_D$, and the **lift coefficient** $C_L$, and they are defined as

*Drag coefficient:*
$$C_D = \frac{F_D}{\frac{1}{2}\rho V^2 A} \tag{15–5}$$

*Lift coefficient:*
$$C_L = \frac{F_L}{\frac{1}{2}\rho V^2 A} \tag{15–6}$$

where $A$ is ordinarily the **frontal area** (the area projected on a plane normal to the direction of flow) of the body. In other words, $A$ is the area seen by a person looking at the body from the direction of the approaching fluid. The frontal area of a cylinder of diameter $D$ and length $L$, for example, is $A = LD$. In lift and drag calculations of some thin bodies, such as airfoils, $A$ is taken to be the **planform area**, which is the area seen by a person looking at the body from above in a direction normal to the body. The drag and lift coefficients are primarily functions of the shape of the body, but in some cases they also depend on the Reynolds number and the surface roughness. The term $\frac{1}{2}\rho V^2$ in Eqs. 15–5 and 15–6 is the **dynamic pressure**.

The local drag and lift coefficients vary along the surface as a result of the changes in the velocity boundary layer in the flow direction. We are usually interested in the drag and lift forces for the *entire* surface, which can be determined using the *average* drag and lift coefficients. Therefore, we present correlations for both local (identified with the subscript $x$) and average drag and lift coefficients. When relations for local drag and lift coefficients for a surface of length $L$ are available, the *average* drag and lift coefficients for the entire surface are determined by integration from

$$C_D = \frac{1}{L} \int_0^L C_{D,x} \, dx \tag{15-7}$$

and

$$C_L = \frac{1}{L} \int_0^L C_{L,x} \, dx \tag{15-8}$$

The forces acting on a falling body are usually the drag force, the buoyant force, and the weight of the body. When a body is dropped into the atmosphere or a lake, it first accelerates under the influence of its weight. The motion of the body is resisted by the drag force, which acts in the direction opposite to motion. As the velocity of the body increases, so does the drag force. This continues until all the forces balance each other and the net force acting on the body (and thus its acceleration) is zero. Then the velocity of the body remains constant during the rest of its fall if the properties of the fluid in the path of the body remain essentially constant. This is the maximum velocity a falling body can attain and is called the **terminal velocity** (Fig. 15–8).

**FIGURE 15–7**

Airplane wings are shaped and positioned to generate sufficient lift during flight while keeping drag at a minimum. Pressures above and below atmospheric pressure are indicated by plus and minus signs, respectively.

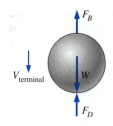

$$F_D = W - F_B$$
(No acceleration)

**FIGURE 15–8**

During a free fall, a body reaches its *terminal velocity* when the drag force equals the weight of the body minus the buoyant force.

---

**EXAMPLE 15–1    Measuring the Drag Coefficient of a Car**

The drag coefficient of a car at the design conditions of 1 atm, 70°F, and 60 mi/h is to be determined experimentally in a large wind tunnel in a full-scale test (Fig. 15–9). The frontal area of the car is 22.26 ft². If the force acting on the car in the flow direction is measured to be 68 lbf, determine the drag coefficient of this car.

**Solution**  The drag force acting on a car is measured in a wind tunnel. The drag coefficient of the car at test conditions is to be determined.
**Assumptions**  **1** The flow of air is steady and incompressible. **2** The cross section of the tunnel is large enough to simulate free flow over the car. **3** The bottom of the tunnel is also moving at the speed of air to approximate actual driving conditions or this effect is negligible.
**Properties**  The density of air at 1 atm and 70°F is $\rho = 0.07489$ lbm/ft³.
**Analysis**  The drag force acting on a body and the drag coefficient are given by

$$F_D = C_D A \frac{\rho V^2}{2} \quad \text{and} \quad C_D = \frac{2F_D}{\rho A V^2}$$

where $A$ is the frontal area. Substituting and noting that 1 mi/h = 1.467 ft/s, the drag coefficient of the car is determined to be

$$C_D = \frac{2 \times (68 \text{ lbf})}{(0.07489 \text{ lbm/ft}^3)(22.26 \text{ ft}^2)(60 \times 1.467 \text{ ft/s})^2} \left( \frac{32.2 \text{ lb} \cdot \text{ft/s}^2}{1 \text{ lbf}} \right) = \mathbf{0.34}$$

**Discussion**  Note that the drag coefficient depends on the design conditions, and its value may be different at different conditions such as the Reynolds number. Therefore, the published drag coefficients of different vehicles can be compared meaningfully only if they are determined under dynamically similar conditions or if Reynolds number independence is demonstrated. This shows the importance of developing standard testing procedures.

**FIGURE 15–9**

Schematic for Example 15–1.

**FIGURE 15–10**

Drag is due entirely to *friction drag* for a flat plate parallel to the flow; it is due entirely to pressure drag for a flat plate normal to the flow; and it is due to *both* (but mostly *pressure drag*) for a cylinder normal to the flow. The total drag coefficient $C_D$ is lowest for a parallel flat plate, highest for a vertical flat plate, and in between (but close to that of a vertical flat plate) for a cylinder.

*From G. M. Homsy, et al. (2000).*

# 15–3 ▪ FRICTION AND PRESSURE DRAG

As mentioned in Section 15–2, the drag force is the net force exerted by a fluid on a body in the direction of flow due to the combined effects of wall shear and pressure forces. It is often instructive to separate the two effects, and study them separately.

The part of drag that is due directly to wall shear stress $\tau_w$ is called the **skin friction drag** (or just *friction drag $F_{D,\text{friction}}$*) since it is caused by frictional effects, and the part that is due directly to pressure $P$ is called the **pressure drag** (also called the *form drag* because of its strong dependence on the form or shape of the body). The friction and pressure drag coefficients are defined as

$$C_{D,\text{friction}} = \frac{F_{D,\text{friction}}}{\frac{1}{2}\rho V^2 A} \quad \text{and} \quad C_{D,\text{pressure}} = \frac{F_{D,\text{pressure}}}{\frac{1}{2}\rho V^2 A} \tag{15–9}$$

When the friction and pressure drag coefficients (based on the same area $A$) or forces are available, the total drag coefficient or drag force is determined by simply adding them,

$$C_D = C_{D,\text{friction}} + C_{D,\text{pressure}} \quad \text{and} \quad F_D = F_{D,\text{friction}} + F_{D,\text{pressure}} \tag{15–10}$$

The *friction drag* is the component of the wall shear force in the direction of flow, and thus it depends on the orientation of the body as well as the magnitude of the wall shear stress $\tau_w$. The friction drag is *zero* for a flat surface normal to the flow, and *maximum* for a flat surface parallel to the flow since the friction drag in this case equals the total shear force on the surface. Therefore, for parallel flow over a flat surface, the drag coefficient is equal to the *friction drag coefficient*, or simply the *friction coefficient*. Friction drag is a strong function of viscosity, and increases with increasing viscosity.

The Reynolds number is inversely proportional to the viscosity of the fluid. Therefore, the contribution of friction drag to total drag for blunt bodies is less at higher Reynolds numbers and may be negligible at very high Reynolds numbers. The drag in such cases is mostly due to pressure drag. At low Reynolds numbers, most drag is due to friction drag. This is especially the case for highly streamlined bodies such as airfoils. The friction drag is also proportional to the surface area. Therefore, bodies with a larger surface area experience a larger friction drag. Large commercial airplanes, for example, reduce their total surface area and thus their drag by retracting their wing extensions when they reach cruising altitudes to save fuel. The friction drag coefficient is independent of *surface roughness* in laminar flow, but is a strong function of surface roughness in turbulent flow due to surface roughness elements protruding further into the boundary layer. The *friction drag coefficient* is analogous to the *friction factor* in pipe flow discussed in Chap. 14, and its value depends on the flow regime.

The pressure drag is proportional to the frontal area and to the *difference* between the pressures acting on the front and back of the immersed body. Therefore, the pressure drag is usually dominant for blunt bodies, small for streamlined bodies such as airfoils, and zero for thin flat plates parallel to the flow (Fig. 15–10). The pressure drag becomes most significant when the velocity of the fluid is too high for the fluid to be able to follow the curvature of the body, and thus the fluid *separates* from the body at some point and creates a very low pressure region in the back. The pressure drag in this case is due to the large pressure difference between the front and back sides of the body.

## Reducing Drag by Streamlining

The first thought that comes to mind to reduce drag is to streamline a body in order to reduce flow separation and thus to reduce pressure drag. Even car salespeople are quick to point out the low drag coefficients of their cars, owing to streamlining. But streamlining has opposite effects on pressure and friction drag forces. It decreases pressure drag by delaying boundary layer separation and thus reducing the pressure difference between the front and back of the body and increases the friction drag by increasing the surface area. The end result depends on which effect dominates. Therefore, any optimization study to reduce the drag of a body must consider both effects and must attempt to minimize the *sum* of the two, as shown in Fig. 15–11. The minimum total drag occurs at $D/L = 0.25$ for the case shown in Fig. 15–11. For the case of a circular cylinder with the same thickness as the streamlined shape of Fig. 15–11, the drag coefficient would be about five times as much. Therefore, it is possible to reduce the drag of a cylindrical component to nearly one-fifth by the use of proper fairings.

The effect of streamlining on the drag coefficient is described best by considering long elliptical cylinders with different aspect (or length-to-thickness) ratios $L/D$, where $L$ is the length in the flow direction and $D$ is the thickness, as shown in Fig. 15–12. Note that the drag coefficient decreases drastically as the ellipse becomes slimmer. For the special case of $L/D = 1$ (a circular cylinder), the drag coefficient is $C_D \cong 1$ at this Reynolds number. As the aspect ratio is decreased and the cylinder resembles a flat plate, the drag coefficient increases to 1.9, the value for a flat plate normal to flow. Note that the curve becomes nearly flat for aspect ratios greater than about 4. Therefore, for a given diameter $D$, elliptical shapes with an aspect ratio of about $L/D \cong 4$ usually offer a good compromise between the total drag coefficient and length $L$. The reduction in the

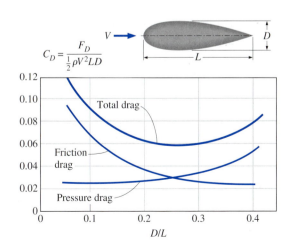

**FIGURE 15–11**

The variation of friction, pressure, and total drag coefficients of a two-dimensional streamlined strut with thickness-to-chord length ratio for Re = $4 \times 10^4$. Note that $C_D$ for airfoils and other thin bodies is based on *planform* area rather than frontal area.

*From Abbott and von Doenhoff (1959).*

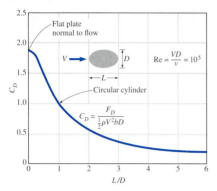

**FIGURE 15–12**

The variation of the drag coefficient of a long elliptical cylinder with aspect ratio. Here $C_D$ is based on the frontal area $bD$ where $b$ is the width of the body.

*From Blevins (1984).*

**FIGURE 15–13**

Flow separation in a waterfall.

**FIGURE 15–14**

Flow separation over a backward-facing step along a wall.

drag coefficient at high aspect ratios is primarily due to the boundary layer staying attached to the surface longer and the resulting pressure recovery. The pressure drag on an elliptical cylinder with an aspect ratio of 4 is negligible (less than 2 percent of total drag at this Reynolds number).

As the aspect ratio of an elliptical cylinder is increased by flattening it (i.e., decreasing $D$ while holding $L$ constant), the drag coefficient starts increasing and tends to infinity as $L/D \rightarrow \infty$ (i.e., as the ellipse resembles a flat plate parallel to flow). This is due to the frontal area, which appears in the denominator in the definition of $C_D$, approaching zero. It does not mean that the drag force increases drastically (actually, the drag force decreases) as the body becomes flat. This shows that the frontal area is inappropriate for use in the drag force relations for slim bodies such as thin airfoils and flat plates. In such cases, the drag coefficient is defined on the basis of the *planform area*, which is simply the surface area of one side (top or bottom) of a flat plate parallel to the flow. This is quite appropriate since for slim bodies the drag is almost entirely due to friction drag, which is proportional to the surface area.

Streamlining has the added benefit of *reducing vibration and noise*. Streamlining should be considered only for bluff bodies that are subjected to high-velocity fluid flow (and thus high Reynolds numbers) for which flow separation is a real possibility. It is not necessary for bodies that typically involve low Reynolds number flows (e.g., creeping flows in which Re < 1) since the drag in those cases is almost entirely due to friction drag, and streamlining would only increase the surface area and thus the total drag. Therefore, careless streamlining may actually increase drag instead of decreasing it.

## Flow Separation

When driving on country roads, it is a common safety measure to slow down at sharp turns in order to avoid being thrown off the road. Many drivers have learned the hard way that a car refuses to comply when forced to turn curves at excessive speeds. We can view this phenomenon as "the separation of cars" from roads. This phenomenon is also observed when fast vehicles jump off hills. At low velocities, the wheels of the vehicle always remain in contact with the road surface. But at high velocities, the vehicle is too fast to follow the curvature of the road and takes off at the hill, losing contact with the road.

A fluid acts much the same way when forced to flow over a curved surface at high velocities. A fluid follows the front portion of the curved surface with no problem, but it has difficulty remaining attached to the surface on the back side. At sufficiently high velocities, the fluid stream detaches itself from the surface of the body. This is called **flow separation** (Fig. 15–13). Flow can separate from a surface even if it is fully submerged in a liquid or immersed in a gas (Fig. 15–14). The location of the separation point depends on several factors such as the Reynolds number, the surface roughness, and the level of fluctuations in the free stream, and it is usually difficult to predict exactly where separation will occur unless there are sharp corners or abrupt changes in the shape of the solid surface.

When a fluid separates from a body, it forms a separated region between the body and the fluid stream. This low-pressure region behind the body where recirculating and backflows occur is called the **separated region**. The larger the separated region, the larger the pressure drag. The effects of flow separation are felt far downstream in the form of reduced velocity (relative to the upstream velocity). The region of flow trailing the body where the effects of the body on velocity are felt is called the **wake** (Fig. 15–15). The separated region comes to an end when the two flow streams reattach. Therefore, the separated region is an enclosed volume, whereas the wake keeps growing behind the body until the fluid in the wake region regains its velocity and the velocity profile becomes nearly flat again. Viscous and rotational effects are the most significant in the boundary layer, the separated region, and the wake.

The occurrence of separation is not limited to bluff bodies. Complete separation over the entire back surface may also occur on a streamlined body such as an airplane wing at a sufficiently large **angle of attack** (larger than about 15° for most airfoils), which is the angle the incoming fluid stream makes with the **chord** (the line that connects the nose and the trailing edge) of the wing. Flow separation on the top surface of a wing reduces lift drastically and may cause the airplane to **stall**. Stalling has been blamed for many airplane accidents and loss of efficiencies in turbomachinery (Fig. 15–16).

Note that drag and lift are strongly dependent on the shape of the body, and any effect that causes the shape to change has a profound effect on the drag and lift. For example, snow accumulation and ice formation on airplane wings may change the shape of the wings sufficiently to cause significant loss of lift. This phenomenon has caused many airplanes to lose altitude and crash and many others to abort takeoff. Therefore, it has become a routine safety measure to check for ice or snow buildup on critical components of airplanes before takeoff in bad weather. This is especially important for airplanes that have waited a long time on the runway before takeoff because of heavy traffic.

An important consequence of flow separation is the formation and shedding of circulating fluid structures, called **vortices**, in the wake region. The periodic generation of these vortices downstream is referred to as **vortex shedding**. This phenomenon usually occurs during normal flow over long cylinders or spheres for Re ≳ 90. The vibrations generated by vortices near the body may cause the body to resonate to dangerous levels if the frequency of the vortices is close to the natural frequency of the body—a situation that must be avoided in the design of equipment that is subjected to high-velocity fluid flow such as the wings of airplanes and suspended bridges subjected to steady high winds.

# 15–4 ▪ DRAG COEFFICIENTS OF COMMON GEOMETRIES

The concept of drag has important consequences in daily life, and the drag behavior of various natural and human-made bodies is characterized by their drag coefficients measured under typical operating conditions. Although drag is caused by two different effects (friction and pressure), it is usually difficult to determine them separately. Besides, in most cases, we are interested in the *total* drag rather than the individual drag components, and thus usually the *total* drag coefficient is reported. The determination of

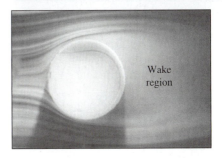

**FIGURE 15–15**

Flow separation and the wake region for flow over a tennis ball.

*Courtesy of NASA and Cislunar Aerospace, Inc.*

(a) 5°

(b) 15°

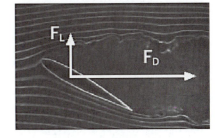

(c) 30°

**FIGURE 15–16**

At large angles of attack (usually larger than 15°), flow may separate completely from the top surface of an airfoil, reducing lift drastically and causing the airfoil to stall.

*From G. M. Homsy, et al. (2000).*

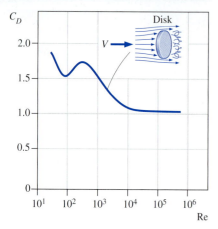

**FIGURE 15–17**

The drag coefficients for most geometries (but not all) remain essentially constant at Reynolds numbers above about $10^4$.

drag coefficients has been the topic of numerous studies (mostly experimental), and there is a huge amount of drag coefficient data in the literature for just about any geometry of practical interest.

The drag coefficient, in general, depends on the *Reynolds number*, especially for Reynolds numbers below about $10^4$. At higher Reynolds numbers, the drag coefficients for most geometries remain essentially constant (Fig. 15–17). This is due to the flow at high Reynolds numbers becoming fully turbulent. However, this is not the case for rounded bodies such as circular cylinders and spheres, as we discuss later in this section. The reported drag coefficients are usually applicable only to flows at high Reynolds numbers.

The drag coefficient exhibits different behavior in the low (creeping), moderate (laminar), and high (turbulent) regions of the Reynolds number. The inertia effects are negligible in low Reynolds number flows (Re $\lesssim 1$), called *creeping flows* and the fluid wraps around the body smoothly. The drag coefficient in this case is inversely proportional to the Reynolds number, and for a sphere it is determined to be

*Sphere*:
$$C_D = \frac{24}{\text{Re}} \quad (\text{Re} \lesssim 1) \tag{15–11}$$

Then the drag force acting on a spherical object at low Reynolds numbers becomes

$$F_D = C_D A \frac{\rho V^2}{2} = \frac{24}{\text{Re}} A \frac{\rho V^2}{2} = \frac{24}{\rho VD/\mu} \frac{\pi D^2}{4} \frac{\rho V^2}{2} = 3\pi \mu VD \tag{15–12}$$

which is known as **Stokes law**, after British mathematician and physicist G. G. Stokes (1819–1903). This relation shows that at very low Reynolds numbers, the drag force acting on spherical objects is proportional to the diameter, the velocity, and the viscosity of the fluid. This relation is often applicable to dust particles in the air and suspended solid particles in water.

The drag coefficients for low Reynolds number flows past some other geometries are given in Fig. 15–18. Note that at low Reynolds numbers, the shape of the body does not have a major influence on the drag coefficient.

**FIGURE 15–18**

Drag coefficients $C_D$ at low Reynolds numbers (Re $\lesssim 1$ where Re $= VD/\nu$ and $A = \pi D^2/4$).

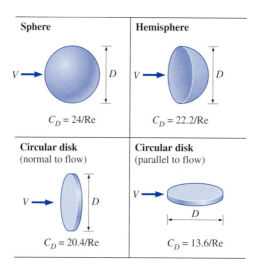

The drag coefficients for various two- and three-dimensional bodies are given in Tables 15–1 and 15–2 for large Reynolds numbers. We make several observations from these tables about the drag coefficient at high Reynolds numbers. First of all, the *orientation* of the body relative to the direction of flow has a major influence on the drag coefficient. For example, the drag coefficient for flow over a hemisphere is 0.4 when the spherical side faces the flow, but it increases threefold to 1.2 when the flat side faces the flow (Fig. 15–19 on page 601).

For blunt bodies with sharp corners, such as flow over a rectangular block or a flat plate normal to the flow, separation occurs at the edges of the front

**TABLE 15–1**

Drag coefficients $C_D$ of various two-dimensional bodies for Re $> 10^4$ based on the frontal area $A = bD$, where $b$ is the length in direction normal to the page (for use in the drag force relation $F_D = C_D A \rho V^2/2$ where $V$ is the upstream velocity)

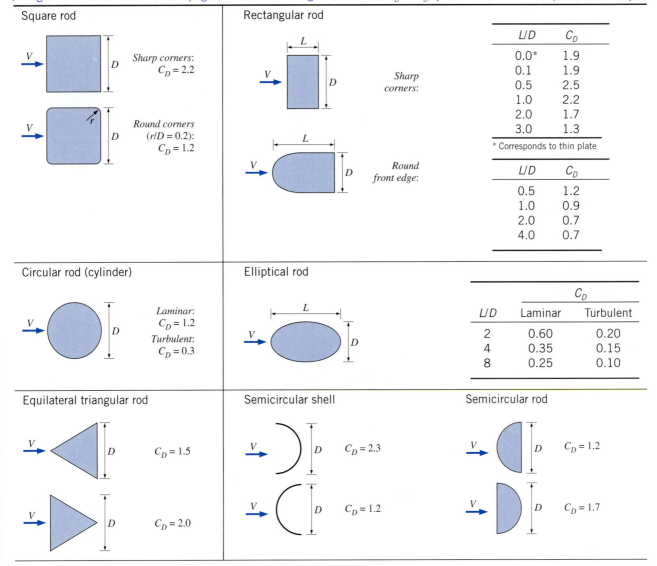

Square rod

Sharp corners:
$C_D = 2.2$

Round corners
$(r/D = 0.2)$:
$C_D = 1.2$

Rectangular rod

Sharp corners:

| $L/D$ | $C_D$ |
|-------|-------|
| 0.0*  | 1.9   |
| 0.1   | 1.9   |
| 0.5   | 2.5   |
| 1.0   | 2.2   |
| 2.0   | 1.7   |
| 3.0   | 1.3   |

* Corresponds to thin plate

Round
front edge:

| $L/D$ | $C_D$ |
|-------|-------|
| 0.5   | 1.2   |
| 1.0   | 0.9   |
| 2.0   | 0.7   |
| 4.0   | 0.7   |

Circular rod (cylinder)

Laminar:
$C_D = 1.2$
Turbulent:
$C_D = 0.3$

Elliptical rod

|       | $C_D$ | |
|-------|---------|-----------|
| $L/D$ | Laminar | Turbulent |
| 2     | 0.60    | 0.20      |
| 4     | 0.35    | 0.15      |
| 8     | 0.25    | 0.10      |

Equilateral triangular rod

$C_D = 1.5$

$C_D = 2.0$

Semicircular shell

$C_D = 2.3$

$C_D = 1.2$

Semicircular rod

$C_D = 1.2$

$C_D = 1.7$

**TABLE 15–2**

Representative drag coefficients $C_D$ for various three-dimensional bodies based on the frontal area for $Re > 10^4$ unless stated otherwise (for use in the drag force relation $F_D = C_D A \rho V^2/2$ where $V$ is the upstream velocity)

| Cube, $A = D^2$ | Thin circular disk, $A = \pi D^2/4$ | Cone (for $\theta = 30°$), $A = \pi D^2/4$ |
|---|---|---|

 $C_D = 1.05$

 $C_D = 1.1$

 $C_D = 0.5$

---

**Sphere**, $A = \pi D^2/4$

*Laminar:*
$Re \lesssim 2 \times 10^5$
$C_D = 0.5$
*Turbulent:*
$Re \gtrsim 2 \times 10^6$
$C_D = 0.2$

See Fig. 15–36 for $C_D$ vs. Re for smooth and rough spheres.

**Ellipsoid**, $A = \pi D^2/4$

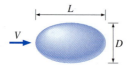

| | $C_D$ | |
|---|---|---|
| $L/D$ | Laminar | Turbulent |
| | $Re \lesssim 2 \times 10^5$ | $Re \gtrsim 2 \times 10^6$ |
| 0.75 | 0.5 | 0.2 |
| 1 | 0.5 | 0.2 |
| 2 | 0.3 | 0.1 |
| 4 | 0.3 | 0.1 |
| 8 | 0.2 | 0.1 |

---

**Hemisphere**, $A = \pi D^2/4$

 $C_D = 0.4$

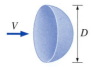 $C_D = 1.2$

**Finite cylinder, vertical**, $A = LD$

| $L/D$ | $C_D$ |
|---|---|
| 1 | 0.6 |
| 2 | 0.7 |
| 5 | 0.8 |
| 10 | 0.9 |
| 40 | 1.0 |
| $\infty$ | 1.2 |

Values are for laminar flow ($Re \lesssim 2 \times 10^5$)

**Finite cylinder, horizontal**, $A = \pi D^2/4$

| $L/D$ | $C_D$ |
|---|---|
| 0.5 | 1.1 |
| 1 | 0.9 |
| 2 | 0.9 |
| 4 | 0.9 |
| 8 | 1.0 |

---

**Streamlined body**, $A = \pi D^2/4$

 $C_D = 0$

**Rectangular plate**, $A = LD$

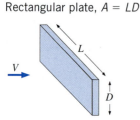

$C_D = 1.10 + 0.02\,(L/D + D/L)$
for $1/30 < (L/D) < 30$

**Parachute**, $A = \pi D^2/4$

$C_D = 1.3$

**Tree**, $A =$ frontal area

$A =$ frontal area

| $V$, m/s | $C_D$ |
|---|---|
| 10 | 0.4–1.2 |
| 20 | 0.3–1.0 |
| 30 | 0.2–0.7 |

*(continues)*

**TABLE 15–2 (Continued)**

| Person (average) | Bikes | |
|---|---|---|

Standing: $C_D A = 9 \text{ ft}^2 = 0.84 \text{ m}^2$

Upright:
$A = 5.5 \text{ ft}^2 = 0.51 \text{ m}^2$
$C_D = 1.1$

Racing:
$A = 3.9 \text{ ft}^2 = 0.36 \text{ m}^2$
$C_D = 0.9$

$C_D = 0.5$  $C_D = 0.9$

Drafting:
$A = 3.9 \text{ ft}^2 = 0.36 \text{ m}^2$
$C_D = 0.50$

With fairing:
$A = 5.0 \text{ ft}^2 = 0.46 \text{ m}^2$
$C_D = 0.12$

| Semitrailer, $A$ = frontal area | Automotive, $A$ = frontal area | High-rise buildings, $A$ = frontal area |
|---|---|---|

Without fairing:
$C_D = 0.96$

With fairing:
$C_D = 0.76$

Minivan:
$C_D = 0.4$

Passenger car
or sports car:
$C_D = 0.3$

$C_D \approx 1.0 \text{ to } 1.4$

and back surfaces, with no significant change in the character of flow. Therefore, the drag coefficient of such bodies is nearly independent of the Reynolds number. Note that the drag coefficient of a long rectangular rod is reduced almost by half from 2.2 to 1.2 by rounding the corners.

## Biological Systems and Drag

The concept of drag also has important consequences for biological systems. For example, the bodies of *fish*, especially the ones that swim fast for long distances (such as dolphins), are highly streamlined to minimize drag (the drag coefficient of dolphins based on the wetted skin area is about 0.0035, comparable to the value for a flat plate in turbulent flow). So it is no surprise that we build submarines that mimic large fish. Tropical fish with fascinating beauty and elegance, on the other hand, swim short distances only. Obviously grace, not high speed and drag, was the primary consideration in their design. Birds teach us a lesson on drag reduction by extending their beak forward and folding their feet backward during flight (Fig. 15–20). Airplanes, which look somewhat like big birds, retract their wheels after takeoff in order to reduce drag and thus fuel consumption.

The flexible structure of plants enables them to reduce drag at high winds by changing their shapes. Large flat leaves, for example, curl into a low-drag conical shape at high wind speeds, while tree branches cluster to reduce drag. Flexible trunks bend under the influence of the wind to reduce drag, and the bending moment is lowered by reducing frontal area.

If you watch the Olympic games, you have probably observed many instances of conscious effort by the competitors to reduce drag. Some examples:

A hemisphere at two different orientations for Re > $10^4$

$C_D = 0.4$

$C_D = 1.2$

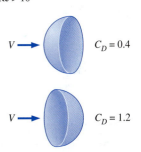

**FIGURE 15–19**
The drag coefficient of a body may change drastically by changing the body's orientation (and thus shape) relative to the direction of flow.

**FIGURE 15–20**

Birds teach us a lesson on drag reduction by extending their beak forward and folding their feet backward during flight.

*©PunchStock RF.*

**FIGURE 15–21**

This sleek-looking Toyota Prius has a drag coefficient of 0.26—one of the lowest for a passenger car.

*Courtesy Toyota.*

Flow @ Re = 10$^4$ | Potential Flow

Flow @ Re = 10$^4$ | Potential Flow

**FIGURE 15–22**

Streamlines around an aerodynamically designed modern car closely resemble the streamlines around the car in the ideal potential flow (assumes negligible friction), except near the rear end, resulting in a low drag coefficient.

*From G. M. Homsy, et al. (2000).*

During 100-m running, the runners hold their fingers together and straight and move their hands parallel to the direction of motion to reduce the drag on their hands. Swimmers with long hair cover their head with a tight and smooth cover to reduce head drag. They also wear well-fitting one-piece swimming suits. Horse and bicycle riders lean forward as much as they can to reduce drag (by reducing both the drag coefficient and frontal area). Speed skiers do the same thing.

## Drag Coefficients of Vehicles

The term *drag coefficient* is commonly used in various areas of daily life. Car manufacturers try to attract consumers by pointing out the *low drag coefficients* of their cars (Fig. 15–21). The drag coefficients of vehicles range from about 1.0 for large semitrailers to 0.4 for minivans and 0.3 for passenger cars. In general, the more blunt the vehicle, the higher the drag coefficient. Installing a fairing reduces the drag coefficient of tractor-trailer rigs by about 20 percent by making the frontal surface more streamlined. As a rule of thumb, the percentage of fuel savings due to reduced drag is about half the percentage of drag reduction at highway speeds.

When the effect of the road on air motion is disregarded, the ideal shape of a *vehicle* is the basic *teardrop*, with a drag coefficient of about 0.1 for the turbulent flow case. But this shape needs to be modified to accommodate several necessary external components such as wheels, mirrors, axles, and door handles. Also, the vehicle must be high enough for comfort and there must be a minimum clearance from the road. Furthermore, a vehicle cannot be too long to fit in garages and parking spaces. Controlling the material and manufacturing costs requires minimizing or eliminating any "dead" volume that cannot be utilized. The result is a shape that resembles more a box than a teardrop, and this was the shape of early cars with a drag coefficient of about 0.8 in the 1920s. This wasn't a problem in those days since the velocities were low, fuel was cheap, and drag was not a major design consideration.

The average drag coefficients of cars dropped to about 0.70 in the 1940s, to 0.55 in the 1970s, to 0.45 in the 1980s, and to 0.30 in the 1990s as a result of improved manufacturing techniques for metal forming and paying more attention to the shape of the car and streamlining (Fig. 15–22). The drag coefficient for well-built racing cars is about 0.2, but this is achieved after making the comfort of drivers a secondary consideration. Noting that the theoretical lower limit of $C_D$ is about 0.1 and the value for racing cars is 0.2, it appears that there is only a little room for further improvement in the drag coefficient of passenger cars from the current value of about 0.3. The drag coefficient of a Mazda 3, for example, is 0.29. For trucks and buses, the drag coefficient can be reduced further by optimizing the front and rear contours (by rounding, for example) to the extent it is practical while keeping the overall length of the vehicle the same.

When traveling as a group, a sneaky way of reducing drag is **drafting**, a phenomenon well known by bicycle riders and car racers. It involves approaching a moving body from behind and being *drafted* into the low-pressure region in the rear of the body. The drag coefficient of a racing bicyclist, for example, is reduced from 0.9 to 0.5 (Table 15–2) by drafting, as also shown in Fig. 15–23.

We also can help reduce the overall drag of a vehicle and thus fuel consumption by being more conscientious drivers. For example, drag force is proportional to the square of velocity. Therefore, driving over the speed limit on the highways not only increases the chances of getting speeding tickets or getting into an accident, but it also increases the amount of fuel consumption per mile. Therefore, driving at moderate speeds is safe and economical. Also, anything that extends from the car, even an arm, increases the drag coefficient. Driving with the windows rolled down also increases the drag and fuel consumption. At highway speeds, a driver can often save fuel in hot weather by running the air conditioner instead of driving with the windows rolled down. For many low-drag automobiles, the turbulence and additional drag generated by open windows consume more fuel than does the air conditioner, but this is not the case for high-drag vehicles.

**FIGURE 15–23**

The drag coefficients of bodies following other moving bodies closely is reduced considerably due to drafting (i.e., entering into the low pressure region created by the body in front).

© *Getty RF.*

## Superposition

The shapes of many bodies encountered in practice are not simple. But such bodies can be treated conveniently in drag force calculations by considering them to be composed of two or more simple bodies. A satellite dish mounted on a roof with a cylindrical bar, for example, can be considered to be a combination of a hemispherical body and a cylinder. Then the drag coefficient of the body can be determined approximately by using **superposition**. Such a simplistic approach does not account for the effects of components on each other, and thus the results obtained should be interpreted accordingly.

---

**EXAMPLE 15–2**    **Effect of Frontal Area on Fuel Efficiency of a Car**

Two common methods of improving fuel efficiency of a vehicle are to reduce the drag coefficient and the frontal area of the vehicle. Consider a car (Fig. 15–24) whose width ($W$) and height ($H$) are 1.85 m and 1.70 m, respectively, with a drag coefficient of 0.30. Determine the amount of fuel and money saved per year as a result of reducing the car height to 1.55 m while keeping its width the same. Assume the car is driven 18,000 km a year at an average speed of 95 km/h. Take the density and price of gasoline to be 0.74 kg/L and $0.95/L, respectively. Also take the density of air to be 1.20 kg/m³, the heating value of gasoline to be 44,000 kJ/kg, and the overall efficiency of the car's drive train to be 30 percent.

**SOLUTION**   The frontal area of a car is reduced by redesigning it. The resulting fuel and money savings per year are to be determined.
*Assumptions*   **1** The car is driven 18,000 km a year at an average speed of 95 km/h. **2** The effect of reduction of the frontal area on the drag coefficient is negligible.
*Properties*   The densities of air and gasoline are given to be 1.20 kg/m³ and 0.74 kg/L, respectively. The heating value of gasoline is given to be 44,000 kJ/kg.
*Analysis*   The drag force acting on a body is

$$F_D = C_D A \frac{\rho V^2}{2}$$

**FIGURE 15–24**

Schematic for Example 15–2.

where $A$ is the frontal area of the body. The drag force acting on the car before redesigning is

$$F_D = 0.3(1.85 \times 1.70 \text{ m}^2)\frac{(1.20 \text{ kg/m}^3)(95 \text{ km/h})^2}{2}\left(\frac{1 \text{ m/s}}{3.6 \text{ km/h}}\right)^2\left(\frac{1 \text{ N}}{1 \text{ kg·m/s}^2}\right)$$

$$= 394 \text{ N}$$

Noting that work is force times distance, the amount of work done to overcome this drag force and the required energy input for a distance of 18,000 km are

$$W_{\text{drag}} = F_D L = (394 \text{ N})(18,000 \text{ km/year})\left(\frac{1000 \text{ m}}{1 \text{ km}}\right)\left(\frac{1 \text{ kJ}}{1000 \text{ N·m}}\right)$$

$$= 7.092 \times 10^6 \text{ kJ/year}$$

$$E_{\text{in}} = \frac{W_{\text{drag}}}{\eta_{\text{car}}} = \frac{7.092 \times 10^6 \text{ kJ/year}}{0.30} = 2.364 \times 10^7 \text{ kJ/year}$$

The amount and the cost of the fuel that supplies this much energy are

$$\text{Amount of fuel} = \frac{m_{\text{fuel}}}{\rho_{\text{fuel}}} = \frac{E_{\text{in}}/HV}{\rho_{\text{fuel}}} = \frac{(2.364 \times 10^7 \text{ kJ/year})/(44,000 \text{ kJ/kg})}{0.74 \text{ kg/L}}$$

$$= 726 \text{ L/year}$$

$$\text{Cost} = (\text{Amount of fuel})(\text{Unit cost}) = (726 \text{ L/year})(\$0.95/\text{L}) = \$690/\text{year}$$

That is, the car uses about 730 liters of gasoline at a total cost of about $690 per year to overcome the drag.

The drag force and the work done to overcome it are directly proportional to the frontal area. Then the percent reduction in the fuel consumption due to reducing the frontal area is equal to the percent reduction in the frontal area:

$$\text{Reduction ratio} = \frac{A - A_{\text{new}}}{A} = \frac{H - H_{\text{new}}}{H} = \frac{1.70 - 1.55}{1.70} = 0.0882$$

$$\text{Amount reduction} = (\text{Reduction ratio})(\text{Amount})$$

$$\text{Fuel reduction} = 0.0882(726 \text{ L/year}) = \textbf{64 L/year}$$

$$\text{Cost reduction} = (\text{Reduction ratio})(\text{Cost}) = 0.0882(\$690/\text{year}) = \textbf{\$61/year}$$

Therefore, reducing the car's height reduces the fuel consumption due to drag by nearly 9 percent.

**Discussion** Answers are given to 2 significant digits. This example demonstrates that significant reductions in drag and fuel consumption can be achieved by reducing the frontal area of a vehicle as well as its drag coefficient.

Example 15–2 is indicative of the tremendous amount of effort put into redesigning various parts of cars such as the window moldings, the door handles, the windshield, and the front and rear ends in order to reduce aerodynamic drag. For a car moving on a level road at constant speed, the power developed by the engine is used to overcome rolling resistance, friction between moving components, aerodynamic drag, and driving the auxiliary equipment. The aerodynamic drag is negligible at low speeds, but becomes significant at speeds above about 30 mi/h. Reduction of the frontal area of the cars (to the dislike of tall drivers) has also contributed greatly to the reduction of drag and fuel consumption.

# 15–5 ▪ PARALLEL FLOW OVER FLAT PLATES

Consider the flow of a fluid over a *flat plate,* as shown in Fig. 15–25. Surfaces that are slightly contoured (such as turbine blades) also can be approximated as flat plates with reasonable accuracy. The *x*-coordinate is measured along the plate surface from the *leading edge* of the plate in the direction of the flow, and *y* is measured from the surface in the normal direction. The fluid approaches the plate in the *x*-direction with a uniform velocity *V*, which is equivalent to the velocity over the plate away from the surface.

For the sake of discussion, we consider the fluid to consist of adjacent layers piled on top of each other. The velocity of the particles in the first fluid layer adjacent to the plate is zero because of the no-slip condition. This motionless layer slows down the particles of the neighboring fluid layer as a result of friction between the particles of these two adjoining fluid layers at different velocities. This fluid layer then slows down the molecules of the next layer, and so on. Thus, the presence of the plate is felt up to some normal distance $\delta$ from the plate beyond which the free-stream velocity remains virtually unchanged. As a result, the *x*-component of the fluid velocity, *u*, varies from 0 at $y = 0$ to nearly *V* (typically 0.99*V*) at $y = \delta$ (Fig. 15–26).

The region of the flow above the plate bounded by $\delta$ in which the effects of the viscous shearing forces caused by fluid viscosity are felt is called the **velocity boundary layer**. The *boundary layer thickness* $\delta$ is typically defined as the distance *y* from the surface at which $u = 0.99V$.

The hypothetical line of $u = 0.99V$ divides the flow over a plate into two regions: the **boundary layer region**, in which the viscous effects and the velocity changes are significant, and the **irrotational flow region**, in which the frictional effects are negligible and the velocity remains essentially constant.

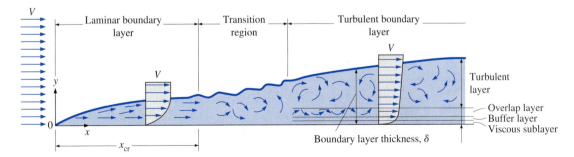

**FIGURE 15–25**

The development of the boundary layer for flow over a flat plate, and the different flow regimes. Not to scale.

**FIGURE 15–26**

The development of a boundary layer on a surface is due to the no-slip condition and friction.

**FIGURE 15–27**

For parallel flow over a flat plate, the pressure drag is zero, and thus the drag coefficient is equal to the friction coefficient and the drag force is equal to the friction force.

For parallel flow over a flat plate, the pressure drag is zero, and thus the drag coefficient is equal to the *friction drag coefficient,* or simply the *friction coefficient* (Fig. 15–27). That is,

*Flat plate*:
$$C_D = C_{D,\,\text{friction}} = C_f \tag{15–13}$$

Once the average friction coefficient $C_f$ is available, the drag (or friction) force over the surface is determined from

*Friction force on a flat plate*:
$$F_D = F_f = \tfrac{1}{2} C_f A \rho V^2 \tag{15–14}$$

where $A$ is the surface area of the plate exposed to fluid flow. When both sides of a thin plate are subjected to flow, $A$ becomes the total area of the top and bottom surfaces. Note that both the average friction coefficient $C_f$ and the local friction coefficient $C_{f,\,x}$, in general, vary with location along the surface.

Typical average velocity profiles in laminar and turbulent flow are sketched in Fig. 15–25. Note that the velocity profile in turbulent flow is much fuller than that in laminar flow, with a sharp drop near the surface. The turbulent boundary layer can be considered to consist of four regions, characterized by the distance from the wall. The very thin layer next to the wall where viscous effects are dominant is the **viscous sublayer**. The velocity profile in this layer is very nearly *linear*, and the flow is nearly parallel. Next to the viscous sublayer is the **buffer layer**, in which turbulent effects are becoming significant, but the flow is still dominated by viscous effects. Above the buffer layer is the **overlap layer**, in which the turbulent effects are much more significant, but still not dominant. Above that is the **turbulent** (or **outer**) **layer** in which turbulent effects dominate over viscous effects. Note that the turbulent boundary layer profile on a flat plate closely resembles the boundary layer profile in fully developed turbulent pipe flow (Chap. 14).

The transition from laminar to turbulent flow depends on the *surface geometry, surface roughness, upstream velocity, surface temperature,* and the *type of fluid,* among other things, and is best characterized by the Reynolds number. The Reynolds number at a distance $x$ from the leading edge of a flat plate is expressed as

$$\text{Re}_x = \frac{\rho V x}{\mu} = \frac{V x}{\nu} \tag{15–15}$$

where $V$ is the upstream velocity and $x$ is the characteristic length of the geometry, which, for a flat plate, is the length of the plate in the flow direction. Note that unlike pipe flow, the Reynolds number varies for a flat plate along the flow, reaching $\text{Re}_L = VL/\nu$ at the end of the plate. For any point on a flat plate, the characteristic length is the distance $x$ of the point from the leading edge in the flow direction.

For flow over a smooth flat plate, transition from laminar to turbulent begins at about $\text{Re} \cong 1 \times 10^5$, but does not become fully turbulent before the Reynolds number reaches much higher values, typically around $3 \times 10^6$. In engineering analysis, a generally accepted value for the critical Reynolds number is

$$\text{Re}_{x,\,\text{cr}} = \frac{\rho V x_{\text{cr}}}{\mu} = 5 \times 10^5$$

The actual value of the engineering critical Reynolds number for a flat plate may vary somewhat from about $10^5$ to $3 \times 10^6$ depending on the surface roughness, the turbulence level, and the variation of pressure along the surface.

# Friction Coefficient

The friction coefficient for laminar flow over a flat plate can be determined theoretically by solving the conservation of mass and linear momentum equations numerically. For turbulent flow, however, it must be determined experimentally and expressed by empirical correlations.

The local friction coefficient *varies* along the surface of the flat plate as a result of the changes in the velocity boundary layer in the flow direction. We are usually interested in the drag force on the *entire* surface, which can be determined using the *average* friction coefficient. But sometimes we are also interested in the drag force at a certain location, and in such cases, we need to know the *local* value of the friction coefficient. With this in mind, we present correlations for both local (identified with the subscript $x$) and average friction coefficients over a flat plate for *laminar, turbulent,* and *combined laminar and turbulent* flow conditions. Once the local values are available, the *average* friction coefficient for the entire plate is determined by integration as

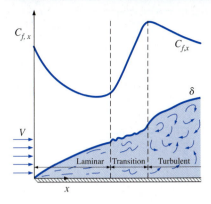

$$C_f = \frac{1}{L} \int_0^L C_{f,x} \, dx \qquad (15\text{–}16)$$

Based on analysis, the boundary layer thickness and the local friction coefficient at location $x$ for laminar flow over a flat plate are determined to be

Laminar: $\quad \delta = \dfrac{4.91x}{\text{Re}_x^{1/2}} \quad$ and $\quad C_{f,x} = \dfrac{0.664}{\text{Re}_x^{1/2}}, \quad \text{Re}_x \lesssim 5 \times 10^5 \qquad (15\text{–}17)$

The corresponding relations for turbulent flow are

Turbulent: $\quad \delta = \dfrac{0.38x}{\text{Re}_x^{1/5}} \quad$ and $\quad C_{f,x} = \dfrac{0.059}{\text{Re}_x^{1/5}}, \quad 5 \times 10^5 \lesssim \text{Re}_x \lesssim 10^7 \qquad (15\text{–}18)$

where $x$ is the distance from the leading edge of the plate and $\text{Re}_x = Vx/\nu$ is the Reynolds number at location $x$. Note that $C_{f,x}$ is proportional to $1/\text{Re}_x^{1/2}$ and thus to $x^{-1/2}$ for laminar flow and it is proportional to $x^{-1/5}$ for turbulent flow. In either case, $C_{f,x}$ is infinite at the leading edge ($x = 0$), and therefore Eqs. 15–17 and 15–18 are not valid close to the leading edge. The variation of the boundary layer thickness $\delta$ and the friction coefficient $C_{f,x}$ along a flat plate is sketched in Fig. 15–28. The local friction coefficients are higher in turbulent flow than they are in laminar flow because of the intense mixing that occurs in the turbulent boundary layer. Note that $C_{f,x}$ reaches its highest values when the flow becomes fully turbulent, and then decreases by a factor of $x^{-1/5}$ in the flow direction, as shown in the figure.

The *average* friction coefficient over the entire plate is determined by substituting Eqs. 15–17 and 15–18 into Eq. 15–16 and performing the integrations (Fig. 15–29). We get

Laminar: $\quad C_f = \dfrac{1.33}{\text{Re}_L^{1/2}} \quad \text{Re}_L \lesssim 5 \times 10^5 \qquad (15\text{–}19)$

Turbulent: $\quad C_f = \dfrac{0.074}{\text{Re}_L^{1/5}} \quad 5 \times 10^5 \lesssim \text{Re}_L \lesssim 10^7 \qquad (15\text{–}20)$

| Relative Roughness, $\varepsilon/L$ | Friction Coefficient, $C_f$ |
|---|---|
| 0.0* | 0.0029 |
| $1 \times 10^{-5}$ | 0.0032 |
| $1 \times 10^{-4}$ | 0.0049 |
| $1 \times 10^{-3}$ | 0.0084 |

* Smooth surface for Re = $10^7$. Others calculated from Eq. 15–23 for fully rough flow.

**FIGURE 15–30**

For turbulent flow, surface roughness may cause the friction coefficient to increase severalfold.

The first of these relations gives the average friction coefficient for the entire plate when the flow is *laminar* over the *entire* plate. The second relation gives the average friction coefficient for the entire plate only when the flow is *turbulent* over the *entire* plate, or when the laminar flow region of the plate is negligibly small relative to the turbulent flow region (that is, $x_{cr} \ll L$ where the length of the plate $x_{cr}$ over which the flow is laminar is determined from $Re_{cr} = 5 \times 10^5 = Vx_{cr}/\nu$).

In some cases, a flat plate is sufficiently long for the flow to become turbulent, but not long enough to disregard the laminar flow region. In such cases, the *average* friction coefficient over the entire plate is determined by performing the integration in Eq. 15–16 over two parts: the laminar region $0 \le x \le x_{cr}$ and the turbulent region $x_{cr} < x \le L$ as

$$C_f = \frac{1}{L}\left( \int_0^{x_{cr}} C_{f,x,\text{laminar}}\ dx + \int_{x_{cr}}^{L} C_{f,x,\text{turbulent}}\ dx \right) \tag{15–21}$$

Note that we included the transition region with the turbulent region. Again taking the critical Reynolds number to be $Re_{cr} = 5 \times 10^5$ and performing these integrations after substituting the indicated expressions, the *average* friction coefficient over the *entire* plate is determined to be

$$C_f = \frac{0.074}{Re_L^{1/5}} - \frac{1742}{Re_L} \qquad 5 \times 10^5 \lesssim Re_L \lesssim 10^7 \tag{15–22}$$

The constants in this relation would be different for different critical Reynolds numbers. Also, the surfaces are assumed to be *smooth,* and the free stream to be of very low turbulence intensity. For laminar flow, the friction coefficient depends on only the Reynolds number, and the surface roughness has no effect. For turbulent flow, however, surface roughness causes the friction coefficient to increase severalfold, to the point that in the fully rough turbulent regime the friction coefficient is a function of surface roughness alone and is independent of the Reynolds number (Fig. 15–30). This is analogous to flow in pipes.

A curve fit of experimental data for the average friction coefficient in this regime is given by Schlichting (1979) as

*Fully rough turbulent regime:* $\qquad C_f = \left( 1.89 - 1.62 \log \frac{\varepsilon}{L} \right)^{-2.5} \tag{15–23}$

where $\varepsilon$ is the surface roughness and $L$ is the length of the plate in the flow direction. In the absence of a better one, this relation can be used for turbulent flow on rough surfaces for $Re > 10^6$, especially when $\varepsilon/L > 10^{-4}$.

Friction coefficients $C_f$ for parallel flow over smooth and rough flat plates are plotted in Fig. 15–31 for both laminar and turbulent flows. Note that $C_f$ increases severalfold with roughness in turbulent flow. Also note that $C_f$ is independent of the Reynolds number in the fully rough region. This chart is the flat-plate analog of the Moody chart for pipe flows.

---

**EXAMPLE 15–3**     **Flow of Hot Oil over a Flat Plate**

Engine oil at 40°C flows over a 5-m-long flat plate with a free-stream velocity of 2 m/s (Fig. 15–32). Determine the drag force acting on the top side of the plate per unit width.

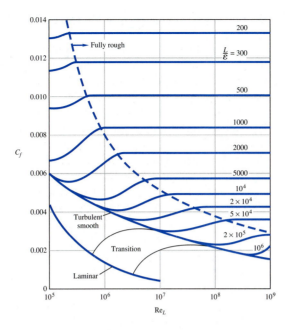

**FIGURE 15–31**

Friction coefficient for parallel flow over smooth and rough flat plates.

*From White (2003).*

$V = 2$ m/s

Oil

$A$

$L = 5$ m

**FIGURE 15–32**

Schematic for Example 15–3.

**SOLUTION**  Engine oil flows over a flat plate. The drag force per unit width of the plate is to be determined.

**Assumptions**  **1** The flow is steady and incompressible. **2** The critical Reynolds number is $Re_{cr} = 5 \times 10^5$.

**Properties**  The density and kinematic viscosity of engine oil at 40°C are $\rho = 876$ kg/m³ and $\nu = 2.485 \times 10^{-4}$ m²/s.

**Analysis**  Noting that $L = 5$ m, the Reynolds number at the end of the plate is

$$Re_L = \frac{VL}{\nu} = \frac{(2 \text{ m/s})(5 \text{ m})}{2.485 \times 10^{-4} \text{ m}^2/\text{s}} = 4.024 \times 10^4$$

which is less than the critical Reynolds number. Thus we have *laminar flow* over the entire plate, and the average friction coefficient is (Fig. 15–29)

$$C_f = 1.328 Re_L^{-0.5} = 1.328 \times (4.024 \times 10^4)^{-0.5} = 0.00662$$

Noting that the pressure drag is zero and thus $C_D = C_f$ for parallel flow over a flat plate, the drag force acting on the plate per unit width becomes

$$F_D = C_f A \frac{\rho V^2}{2} = 0.00662(5 \times 1 \text{ m}^2) \frac{(876 \text{ kg/m}^3)(2 \text{ m/s})^2}{2} \left( \frac{1 \text{ N}}{1 \text{ kg·m/s}^2} \right) = \textbf{58.0 N}$$

The total drag force acting on the entire plate can be determined by multiplying the value just obtained by the width of the plate.

**Discussion**  The force per unit width corresponds to the weight of a mass of about 6 kg. Therefore, a person who applies an equal and opposite force to the plate to keep it from moving will feel like he or she is using as much force as is necessary to hold a 6-kg mass from dropping.

## 15–6 ▪ FLOW OVER CYLINDERS AND SPHERES

Flow over cylinders and spheres is frequently encountered in practice. For example, the tubes in a shell-and-tube heat exchanger involve both *internal flow* through the tubes and *external flow* over the tubes, and both flows must be considered in the analysis of the heat exchanger. Also, many sports such as soccer, tennis, and golf involve flow over spherical balls.

The characteristic length for a circular cylinder or sphere is taken to be the *external diameter D*. Thus, the Reynolds number is defined as $Re = VD/\nu$ where $V$ is the uniform velocity of the fluid as it approaches the cylinder or sphere. The critical Reynolds number for flow across a circular cylinder or sphere is about $Re_{cr} \cong 2 \times 10^5$. That is, the boundary layer remains laminar for about $Re \lesssim 2 \times 10^5$, is transitional for $2 \times 10^5 \lesssim Re \lesssim 2 \times 10^6$, and becomes fully turbulent for $Re \gtrsim 2 \times 10^6$.

Cross-flow over a cylinder exhibits complex flow patterns, as shown in Fig. 15–33. The fluid approaching the cylinder branches out and encircles the cylinder, forming a boundary layer that wraps around the cylinder. The fluid particles on the midplane strike the cylinder at the stagnation point, bringing the fluid to a complete stop and thus raising the pressure at that point. The pressure decreases in the flow direction while the fluid velocity increases.

At very low upstream velocities ($Re \lesssim 1$), the fluid completely wraps around the cylinder and the two arms of the fluid meet on the rear side of the cylinder in an orderly manner. Thus, the fluid follows the curvature of the cylinder. At higher velocities, the fluid still hugs the cylinder on the frontal side, but it is too fast to remain attached to the surface as it approaches the top (or bottom) of the cylinder. As a result, the boundary layer detaches from the surface, forming a separation region behind the cylinder. Flow in the wake region is characterized by periodic vortex formation and pressures much lower than the stagnation point pressure.

The nature of the flow across a cylinder or sphere strongly affects the total drag coefficient $C_D$. Both the *friction drag* and the *pressure drag* can be significant. The high pressure in the vicinity of the stagnation point and the low pressure on the opposite side in the wake produce a net force on the body in the direction of flow. The drag force is primarily due to friction drag at low Reynolds numbers ($Re \lesssim 10$) and to pressure drag at high Reynolds numbers ($Re \gtrsim 5000$). Both effects are significant at intermediate Reynolds numbers.

**FIGURE 15–33**

Laminar boundary layer separation with a turbulent wake; flow over a circular cylinder at Re = 2000.

*Courtesy ONERA, photograph by Werlé.*

The average drag coefficients $C_D$ for cross-flow over a smooth single circular cylinder and a sphere are given in Fig. 15–34. The curves exhibit different behaviors in different ranges of Reynolds numbers:

- For Re ≲ 1, we have creeping flow, and the drag coefficient decreases with increasing Reynolds number. For a sphere, it is $C_D = 24/\text{Re}$. There is no flow separation in this regime.

- At about Re ≅ 10, separation starts occurring on the rear of the body with vortex shedding starting at about Re ≅ 90. The region of separation increases with increasing Reynolds number up to about Re ≅ $10^3$. At this point, the drag is mostly (about 95 percent) due to pressure drag. The drag coefficient continues to decrease with increasing Reynolds number in this range of $10 ≲ \text{Re} ≲ 10^3$. (A decrease in the drag coefficient does not necessarily indicate a decrease in drag. The drag force is proportional to the square of the velocity, and the increase in velocity at higher Reynolds numbers usually more than offsets the decrease in the drag coefficient.)

- In the moderate range of $10^3 ≲ \text{Re} ≲ 10^5$, the drag coefficient remains relatively constant. This behavior is characteristic of bluff bodies. The flow in the boundary layer is laminar in this range, but the flow in the separated region past the cylinder or sphere is highly turbulent with a wide turbulent wake.

- There is a sudden drop in the drag coefficient somewhere in the range of $10^5 ≲ \text{Re} ≲ 10^6$ (usually, at about $2 \times 10^5$). This large reduction in $C_D$ is due to the flow in the boundary layer becoming *turbulent*, which moves the separation point further on the rear of the body, reducing the size of the wake and thus the magnitude of the pressure drag. This is in contrast to streamlined bodies, which experience an increase in the drag coefficient (mostly due to friction drag) when the boundary layer becomes turbulent.

- There is a "transitional" regime for $2 \times 10^5 ≲ \text{Re} ≲ 2 \times 10^6$, in which $C_D$ dips to a minimum value and then slowly rises to its final turbulent value.

Flow separation occurs at about $\theta \cong 80°$ (measured from the front stagnation point of a cylinder) when the boundary layer is *laminar* and at about $\theta \cong 140°$ when it is *turbulent* (Fig. 15–35). The delay of separation in turbulent flow is caused by the rapid fluctuations of the fluid in the transverse

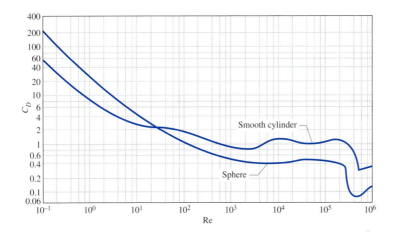

**FIGURE 15–34**

Average drag coefficient for cross-flow over a smooth circular cylinder and a smooth sphere.

*From H. Schlichting,* Boundary Layer Theory *7e. Copyright © 1979 The McGraw-Hill Companies, Inc. Used by permission.*

(a)

(b)

## FIGURE 15–35

Flow visualization of flow over
(a) a smooth sphere at Re = 15,000,
and (b) a sphere at Re = 30,000 with
a trip wire. The delay of boundary
layer separation is clearly seen by
comparing the two photographs.

*Courtesy ONERA, photograph by Werlé.*

direction, which enables the turbulent boundary layer to travel farther along the surface before separation occurs, resulting in a narrower wake and a smaller pressure drag. Keep in mind that turbulent flow has a fuller velocity profile as compared to the laminar case, and thus it requires a stronger adverse pressure gradient to overcome the additional momentum close to the wall. In the range of Reynolds numbers where the flow changes from laminar to turbulent, even the drag force $F_D$ decreases as the velocity (and thus the Reynolds number) increases. This results in a sudden decrease in drag of a flying body (sometimes called the *drag crisis*) and instabilities in flight.

## Effect of Surface Roughness

We mentioned earlier that *surface roughness,* in general, increases the drag coefficient in turbulent flow. This is especially the case for streamlined bodies. For blunt bodies such as a circular cylinder or sphere, however, an increase in the surface roughness may actually *decrease* the drag coefficient, as shown in Fig. 15–36 for a sphere. This is done by tripping the boundary layer into turbulence at a lower Reynolds number, and thus delaying flow separation, causing the fluid to close in behind the body, narrowing the wake, and reducing pressure drag considerably. This results in a much smaller drag coefficient and thus drag force for a rough-surfaced cylinder or sphere in a certain range of Reynolds number compared to a smooth one of identical size at the same velocity. At Re = $2 \times 10^5$, for example, $C_D \cong 0.1$ for a rough sphere with $\varepsilon/D = 0.0015$, whereas $C_D \cong 0.5$ for a smooth one. Therefore, the drag coefficient in this case is reduced by a factor of 5 by simply roughening the surface. Note, however, that at Re = $10^6$, $C_D \cong 0.4$ for a very rough sphere while $C_D \cong 0.1$ for the smooth one. Obviously, roughening the sphere in this case increases the drag by a factor of 4 (Fig. 15–37).

The preceding discussion shows that roughening the surface can be used to great advantage in reducing drag, but it can also backfire on us if we are not careful—specifically, if we do not operate in the right range of the Reynolds number. With this consideration, golf balls are intentionally roughened to induce *turbulence* at a lower Reynolds number to take advantage of the sharp

## FIGURE 15–36

The effect of surface roughness on the drag coefficient of a sphere.

*From Blevins (1984).*

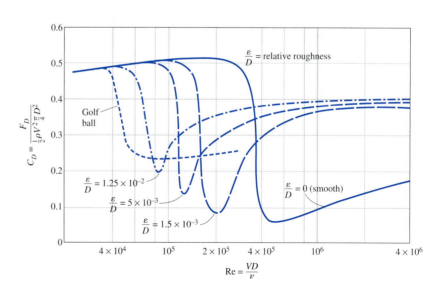

*drop* in the drag coefficient at the onset of turbulence in the boundary layer (the typical velocity range of golf balls is 15 to 150 m/s, and the Reynolds number is less than $4 \times 10^5$). The critical Reynolds number of dimpled golf balls is about $4 \times 10^4$. The occurrence of turbulent flow at this Reynolds number reduces the drag coefficient of a golf ball by about half, as shown in Fig. 15–36. For a given hit, this means a longer distance for the ball. Experienced golfers also give the ball a spin during the hit, which helps the rough ball develop a lift and thus travel higher and farther. A similar argument can be given for a tennis ball. For a table tennis ball, however, the speeds are slower and the ball is smaller—it never reaches the turbulent range. Therefore, the surfaces of table tennis balls are smooth.

Once the drag coefficient is available, the drag force acting on a body in cross-flow is determined from Eq. 15–5 where $A$ is the *frontal area* ($A = LD$ for a cylinder of length $L$ and $A = \pi D^2/4$ for a sphere). It should be kept in mind that free-stream turbulence and disturbances by other bodies in the flow (such as flow over tube bundles) may affect the drag coefficient significantly.

| | $C_D$ | |
| Re | Smooth Surface | Rough Surface, $\varepsilon/D = 0.0015$ |
| --- | --- | --- |
| $2 \times 10^5$ | 0.5 | 0.1 |
| $10^6$ | 0.1 | 0.4 |

**FIGURE 15–37**

Surface roughness may increase or decrease the drag coefficient of a spherical object, depending on the value of the Reynolds number.

---

### EXAMPLE 15–4    Drag Force Acting on a Pipe in a River

A 2.2-cm-outer-diameter pipe is to span across a river at a 30-m-wide section while being completely immersed in water (Fig. 15–38). The average flow velocity of water is 4 m/s and the water temperature is 15°C. Determine the drag force exerted on the pipe by the river.

**SOLUTION** A pipe is submerged in a river. The drag force that acts on the pipe is to be determined.
**Assumptions** **1** The outer surface of the pipe is smooth so that Fig. 15–34 can be used to determine the drag coefficient. **2** Water flow in the river is steady. **3** The direction of water flow is normal to the pipe. **4** Turbulence in river flow is not considered.
**Properties** The density and dynamic viscosity of water at 15°C are $\rho = 999.1$ kg/m³ and $\mu = 1.138 \times 10^{-3}$ kg/m·s.
**Analysis** Noting that $D = 0.022$ m, the Reynolds number is

$$\text{Re} = \frac{VD}{\nu} = \frac{\rho VD}{\mu} = \frac{(999.1 \text{ kg/m}^3)(4 \text{ m/s})(0.022 \text{ m})}{1.138 \times 10^{-3} \text{ kg/m·s}} = 7.73 \times 10^4$$

The drag coefficient corresponding to this value is, from Fig. 15–34, $C_D = 1.0$. Also, the frontal area for flow past a cylinder is $A = LD$. Then the drag force acting on the pipe becomes

$$F_D = C_D A \frac{\rho V^2}{2} = 1.0(30 \times 0.022 \text{ m}^2)\frac{(999.1 \text{ kg/m}^3)(4 \text{ m/s})^2}{2}\left(\frac{1 \text{ N}}{1 \text{ kg·m/s}^2}\right)$$

$$= 5275 \text{ N} \cong \textbf{5300 N}$$

**Discussion** Note that this force is equivalent to the weight of a mass over 500 kg. Therefore, the drag force the river exerts on the pipe is equivalent to hanging a total of over 500 kg in mass on the pipe supported at its ends 30 m apart. The necessary precautions should be taken if the pipe cannot support this force. If the river were to flow at a faster speed or if turbulent fluctuations in the river were more significant, the drag force would be even larger. *Unsteady* forces on the pipe might then be significant.

River

Pipe

30 m

**FIGURE 15–38**

Schematic for Example 15–4.

# 15–7 · LIFT

Lift was defined earlier as the component of the net force (due to viscous and pressure forces) that is perpendicular to the flow direction, and the lift coefficient was expressed in Eq. 15–6 as

$$C_L = \frac{F_L}{\frac{1}{2}\rho V^2 A}$$

where $A$ in this case is normally the *planform area,* which is the area that would be seen by a person looking at the body from above in a direction normal to the body, and $V$ is the upstream velocity of the fluid (or, equivalently, the velocity of a flying body in a quiescent fluid). For an airfoil of width (or span) $b$ and chord length $c$ (the length between the leading and trailing edges), the planform area is $A = bc$. The distance between the two ends of a wing or airfoil is called the **wingspan** or just the **span**. For an aircraft, the wingspan is taken to be the total distance between the tips of the two wings, which includes the width of the fuselage between the wings (Fig. 15–39). The average lift per unit planform area $F_L/A$ is called the **wing loading**, which is simply the ratio of the weight of the aircraft to the planform area of the wings (since lift equals weight when flying at constant altitude).

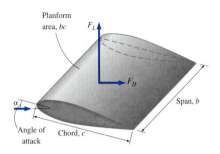

**FIGURE 15–39**

Definition of various terms associated with an airfoil.

Airplane flight is based on lift, and thus developing a better understanding of lift as well as improving the lift characteristics of bodies have been the focus of numerous studies. Our emphasis in this section is on devices such as *airfoils* that are specifically designed to generate lift while keeping the drag at a minimum. But it should be kept in mind that some devices such as *spoilers* and *inverted airfoils* on racing cars are designed for the opposite purpose of avoiding lift or even generating negative lift to improve traction and control (some early race cars actually "took off" at high speeds as a result of the lift produced, which alerted the engineers to come up with ways to reduce lift in their design).

For devices that are intended to generate lift such as airfoils, the contribution of *viscous effects* to lift is usually negligible since the bodies are streamlined, and wall shear is parallel to the surfaces of such devices and thus nearly normal to the direction of lift (Fig. 15–40). Therefore, lift in practice can be approximated as due entirely to the pressure distribution on the surfaces of the body, and thus the shape of the body has the primary influence on lift. Then the primary consideration in the design of airfoils is minimizing the average pressure at the upper surface while maximizing it at the lower surface. The Bernoulli equation can be used as a guide in identifying the high- and low-pressure regions: *Pressure is low at locations where the flow velocity is high, and pressure is high at locations where the flow velocity is low.* Also, at moderate angles of attack, lift is practically independent of the surface roughness since roughness affects the wall shear, not the pressure. The contribution of shear to lift is significant only for very small (lightweight) bodies that fly at low velocities (and thus low Reynolds numbers).

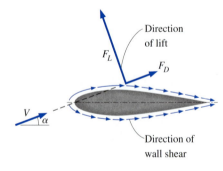

**FIGURE 15–40**

For airfoils, the contribution of viscous effects to lift is usually negligible since wall shear is parallel to the surfaces and thus nearly normal to the direction of lift.

Noting that the contribution of viscous effects to lift is negligible, we should be able to determine the lift acting on an airfoil by simply integrating the pressure distribution around the airfoil. The pressure changes in the flow direction along the surface, but it remains essentially constant through the boundary layer in a direction normal to the surface. Therefore, it seems reasonable to ignore the very thin boundary layer on the airfoil and calculate the pressure distribution around the airfoil from the relatively simple potential flow theory (zero vorticity, irrotational flow) for which net viscous forces are zero for flow past an airfoil.

The flow fields obtained from such calculations are sketched in Fig. 15–41 for both symmetrical and nonsymmetrical airfoils by ignoring the thin boundary layer. At zero angle of attack, the lift produced by the symmetrical airfoil is zero, as expected because of symmetry, and the stagnation points are at the leading and trailing edges. For the nonsymmetrical airfoil, which is at a small angle of attack, the front stagnation point has moved down below the leading edge, and the rear stagnation point has moved up to the upper surface close to the trailing edge. To our surprise, the lift produced is calculated again to be zero—a clear contradiction of experimental observations and measurements. Obviously, the theory needs to be modified to bring it in line with the observed phenomenon.

The source of inconsistency is the rear stagnation point being at the upper surface instead of the trailing edge. This requires the lower side fluid to make a nearly U-turn and flow around the sharp trailing edge toward the stagnation point while remaining attached to the surface, which is a physical impossibility since the observed phenomenon is the separation of flow at sharp turns (imagine a car attempting to make this turn at high speed). Therefore, the lower side fluid separates smoothly off the trailing edge, and the upper side fluid responds by pushing the rear stagnation point downstream. In fact, the stagnation point at the upper surface moves all the way to the trailing edge. This way the two flow streams from the top and the bottom sides of the airfoil meet at the trailing edge, yielding a smooth flow downstream parallel to the sharp trailing edge. Lift is generated because the flow velocity at the top surface is higher, and thus the pressure on that surface is lower due to the Bernoulli effect.

The potential flow theory and the observed phenomenon can be reconciled as follows: Flow starts out as predicted by theory, with no lift, but the lower fluid stream separates at the trailing edge when the velocity reaches a certain value. This forces the separated upper fluid stream to close in at the trailing edge, initiating clockwise circulation around the airfoil. This clockwise circulation increases the velocity of the upper stream while decreasing that of the lower stream, causing lift. A **starting vortex** of opposite sign (counterclockwise circulation) is then shed downstream (Fig. 15–42), and smooth streamlined flow is established over the airfoil. When the potential flow theory is modified by the addition of an appropriate amount of circulation to move the stagnation point down to the trailing edge, excellent agreement is obtained between theory and experiment for both the flow field and the lift.

It is desirable for airfoils to generate the most lift while producing the least drag. Therefore, a measure of performance for airfoils is the **lift-to-drag ratio**, which is equivalent to the ratio of the lift-to-drag coefficients $C_L/C_D$. This information is provided either by plotting $C_L$ versus $C_D$ for different values of the angle of attack (a lift–drag polar) or by plotting the ratio $C_L/C_D$ versus the angle of attack. The latter is done for a particular airfoil design in Fig. 15–43. Note that the $C_L/C_D$ ratio increases with the angle of

(*a*) Irrotational flow past a symmetrical airfoil (zero lift)

(*b*) Irrotational flow past a nonsymmetrical airfoil (zero lift)

(*c*) Actual flow past a nonsymmetrical airfoil (positive lift)

**FIGURE 15–41**

Irrotational and actual flow past symmetrical and nonsymmetrical two-dimensional airfoils.

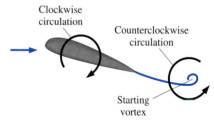

**FIGURE 15–42**

Shortly after a sudden increase in angle of attack, a counterclockwise starting vortex is shed from the airfoil, while clockwise circulation appears around the airfoil, causing lift to be generated.

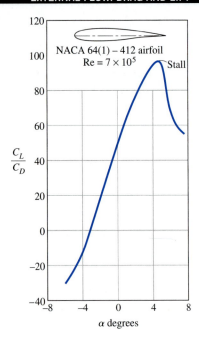

**FIGURE 15–43**

The variation of the lift-to-drag ratio with angle of attack for a two-dimensional airfoil.

*From Abbott, von Doenhoff, and Stivers (1945).*

attack until the airfoil stalls, and the value of the lift-to-drag ratio can be of the order of 100 for a two-dimensional airfoil.

One obvious way to change the lift and drag characteristics of an airfoil is to change the angle of attack. On an airplane, for example, the entire plane is pitched up to increase lift, since the wings are fixed relative to the fuselage. Another approach is to change the shape of the airfoil by the use of movable *leading edge* and *trailing edge flaps,* as is commonly done in modern large aircraft (Fig. 15–44). The flaps are used to alter the shape of the wings during takeoff and landing to maximize lift and to enable the aircraft to land or take off at low speeds. The increase in drag during this takeoff and landing is not much of a concern because of the relatively short time periods involved. Once at cruising altitude, the flaps are retracted, and the wing is returned to its "normal" shape with minimal drag coefficient and adequate lift coefficient to minimize fuel consumption while cruising at a constant altitude. Note that even a small lift coefficient can generate a large lift force during normal operation because of the large cruising velocities of aircraft and the proportionality of lift to the square of flow velocity.

The effects of flaps on the lift and drag coefficients are shown in Fig. 15–45 for an airfoil. Note that the maximum lift coefficient increases from about 1.5 for the airfoil with no flaps to 3.5 for the double-slotted flap case. But also note that the maximum drag coefficient increases from about 0.06 for the airfoil with no flaps to about 0.3 for the double-slotted flap case. This is a five-fold increase in the drag coefficient, and the engines must work much harder to provide the necessary thrust to overcome this drag. The angle of attack of the flaps can be increased to maximize the lift coefficient. Also, the flaps extend the chord length, and thus enlarge the wing area $A$. The Boeing 727 uses a triple-slotted flap at the trailing edge and a slot at the leading edge.

The minimum flight velocity is determined from the requirement that the total weight $W$ of the aircraft be equal to lift and $C_L = C_{L, \text{max}}$. That is,

$$W = F_L = \tfrac{1}{2} C_{L, \text{max}} \rho V_{\text{min}}^2 A \quad \rightarrow \quad V_{\text{min}} = \sqrt{\frac{2W}{\rho C_{L, \text{max}} A}} \qquad (15\text{–}24)$$

For a given weight, the landing or takeoff speed can be minimized by maximizing the product of the lift coefficient and the wing area, $C_{L, \text{max}} A$. One way of doing that is to use flaps, as already discussed. Another way is to control the boundary layer, which can be accomplished simply by leaving flow sections (slots) between the flaps, as shown in Fig. 15–46. Slots are used to prevent the separation of the boundary layer from the upper surface of the wings and the flaps. This is done by allowing air to move from the high-pressure region under the wing into the low-pressure region at the top surface. Note that the lift coefficient reaches its maximum value $C_L = C_{L, \text{max}}$,

**FIGURE 15–44**

The lift and drag characteristics of an airfoil during takeoff and landing are changed by changing the shape of the airfoil by the use of movable flaps.

*Photos by Yunus Çengel.*

*(a)* Flaps extended (takeoff)

*(b)* Flaps retracted (cruising)

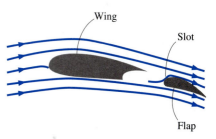

**FIGURE 15–45**

Effect of flaps on the lift and drag coefficients of an airfoil.

*From Abbott and von Doenhoff, for NACA 23012 (1959).*

and thus the flight velocity reaches its minimum, at stall conditions, which is a region of unstable operation and must be avoided. The Federal Aviation Administration (FAA) does not allow operation below 1.2 times the stall speed for safety.

Another thing we notice from this equation is that the minimum velocity for takeoff or landing is inversely proportional to the square root of density. Noting that air density decreases with altitude (by about 15 percent at 1500 m), longer runways are required at airports at higher altitudes such as Denver to accommodate higher minimum takeoff and landing velocities. The situation becomes even more critical on hot summer days since the density of air is inversely proportional to temperature.

The development of efficient (low-drag) airfoils was the subject of intense experimental investigations in the 1930s. These airfoils were standardized by the National Advisory Committee for Aeronautics (NACA, which is now NASA), and extensive lists of data on lift coefficients were reported. The variation of the lift coefficient $C_L$ with angle of attack for two 2-D (infinite span) airfoils (NACA 0012 and NACA 2412) is given in Fig. 15–47. We make the following observations from this figure:

- The lift coefficient increases almost linearly with angle of attack $\alpha$, reaches a maximum at about $\alpha = 16°$, and then starts to decrease sharply. This decrease of lift with further increase in the angle of attack is called *stall,* and it is caused by flow separation and the formation of a wide wake region over the top surface of the airfoil. Stall is highly undesirable since it also increases drag.

- At zero angle of attack ($\alpha = 0°$), the lift coefficient is zero for symmetrical airfoils but nonzero for nonsymmetrical ones with greater curvature at the top surface. Therefore, planes with symmetrical wing sections must fly with their wings at higher angles of attack in order to produce the same lift.

- The lift coefficient is increased by severalfold by adjusting the angle of attack (from 0.25 at $\alpha = 0°$ for the nonsymmetrical airfoil to 1.25 at $\alpha = 10°$).

- The drag coefficient also increases with angle of attack, often exponentially (Fig. 15–48). Therefore, large angles of attack should be used sparingly for short periods of time for fuel efficiency.

**FIGURE 15–46**

A flapped airfoil with a slot to prevent the separation of the boundary layer from the upper surface and to increase the lift coefficient.

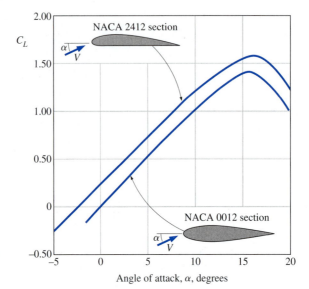

**FIGURE 15–47**

The variation of the lift coefficient with angle of attack for a symmetrical and a nonsymmetrical airfoil.

*From Abbott (1945, 1959).*

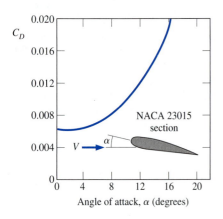

**FIGURE 15–48**

The variation of the drag coefficient of an airfoil with angle of attack.

*From Abbott and von Doenhoff (1959).*

## Finite-Span Wings and Induced Drag

For airplane wings and other airfoils of finite span, the end effects at the tips become important because of the fluid leakage between the lower and upper surfaces. The pressure difference between the lower surface (high-pressure region) and the upper surface (low-pressure region) drives the fluid at the tips upward while the fluid is swept toward the back because of the relative motion between the fluid and the wing. This results in a swirling motion that spirals along the flow, called the **tip vortex**, at the tips of both wings. Vortices are also formed along the airfoil between the tips of the wings. These distributed vortices collect toward the edges after being shed from the trailing edges of the wings and combine with the tip vortices to form two streaks of powerful **trailing vortices** along the tips of the wings (Fig. 15–49). Trailing vortices generated by large aircraft persist for a long time for long distances (over 10 km) before they gradually disappear due to viscous dissipation. Such vortices and the accompanying downdraft are strong enough to cause a small aircraft to lose control and flip over if it flies through the wake of a larger aircraft. Therefore, following a large aircraft closely (within 10 km) poses a real danger for smaller aircraft. This issue is the controlling factor that governs the spacing of aircraft at takeoff, which limits the flight capacity at airports. In nature, this effect is used to advantage by birds that migrate in V-formation by utilizing the updraft generated by the bird in front. It has been determined that the birds in a typical flock can fly to their destination in V-formation with one-third less energy. Military jets also occasionally fly in V-formation for the same reason (Fig. 15–50).

Tip vortices that interact with the free stream impose forces on the wing tips in all directions, including the flow direction. The component of the force in the flow direction adds to drag and is called **induced drag**. The total drag of a wing is then the sum of the induced drag (3-D effects) and the drag of the airfoil section (2-D effects).

The ratio of the square of the average span of an airfoil to the planform area is called the **aspect ratio**. For an airfoil with a rectangular planform of chord $c$ and span $b$, it is expressed as

$$\text{AR} = \frac{b^2}{A} = \frac{b^2}{bc} = \frac{b}{c} \qquad (15\text{–}25)$$

Therefore, the aspect ratio is a measure of how (relatively) narrow an airfoil is in the flow direction. The lift coefficient of wings, in general, increases while the drag coefficient decreases with increasing aspect ratio. This is because a long narrow wing (large aspect ratio) has a shorter tip length and thus smaller tip losses and smaller induced drag than a short and wide wing of the same planform area. Therefore, bodies with large aspect ratios fly more efficiently, but they are less maneuverable because of their larger moment of inertia (owing to the greater distance from the center). Bodies with smaller aspect ratios maneuver better since the wings are closer to the central part. So it is no surprise that *fighter planes* (and fighter birds like falcons) have short and wide wings while *large commercial planes* (and soaring birds like albatrosses) have long and narrow wings.

The end effects can be minimized by attaching **endplates** or **winglets** at the tips of the wings perpendicular to the top surface. The endplates function by blocking some of the leakage around the wing tips, which results in a considerable reduction in the strength of the tip vortices and the induced drag. Wing tip feathers on birds fan out for the same purpose (Fig. 15–51).

(a)

(b)

(c)

**FIGURE 15–49**

Trailing vortices visualized in various ways: (*a*) Smoke streaklines in a wind tunnel show vortex cores leaving the trailing edge of a rectangular wing; (*b*) Four contrails initially formed by condensation of water vapor in the low pressure region behind the jet engines eventually merge into the two counter-rotating trailing vortices that persist very far downstream; (*c*) A crop duster flies through smoky air which swirls around in one of the tip vortices from the aircraft's wing.

*(a) Courtesy of the Parabolic Press, Stanford, California; (b) © Getty RF; (c) NASA Langley Research Center*

---

**EXAMPLE 15–5    Lift and Drag of a Commercial Airplane**

A commercial airplane has a total mass of 70,000 kg and a wing planform area of 150 $m^2$ (Fig. 15–52, on page 621). The plane has a cruising speed of 558 km/h and a cruising altitude of 12,000 m, where the air density is 0.312 kg/$m^3$. The plane has double-slotted flaps for use during takeoff and landing, but it cruises with all flaps retracted. Assuming the lift and the drag characteristics of the wings can be approximated by NACA 23012 (Fig. 15–45), determine (*a*) the minimum safe speed for takeoff and landing with and without extending the flaps, (*b*) the angle of attack to cruise steadily at the cruising altitude, and (*c*) the power that needs to be supplied to provide enough thrust to overcome wing drag.

**SOLUTION**  The cruising conditions of a passenger plane and its wing characteristics are given. The minimum safe landing and takeoff speeds, the angle of attack during cruising, and the power required are to be determined.
*Assumptions*  **1** The drag and lift produced by parts of the plane other than the wings, such as the fuselage are not considered. **2** The wings are assumed to be two-dimensional airfoil sections, and the tip effects of the wings are not considered. **3** The lift and the drag characteristics of the wings are approximated by NACA 23012 so that Fig. 15–45 is applicable. **4** The average density of air on the ground is 1.20 kg/$m^3$.
*Properties*  The density of air is 1.20 kg/$m^3$ on the ground and 0.312 kg/$m^3$ at cruising altitude. The maximum lift coefficient $C_{L, max}$ of the wing is 3.48 and 1.52 with and without flaps, respectively (Fig. 15–45).

**FIGURE 15–50**

(*a*) Geese flying in their characteristic V-formation to save energy.
(*b*) Military jets imitating nature.

*(a) (b) © Corbis RF.*

(*a*) A bearded vulture with its wing feathers fanned out during flight.

(*b*) Winglets are used on this sailplane to reduce induced drag.

**FIGURE 15–51**

Induced drag is reduced by (*a*) wing tip feathers on bird wings and (*b*) endplates or other disruptions on airplane wings.

*(a) © Vol. 44/PhotoDisc. (b) Courtesy of Schempp-Hirth. Used by permission.*

*Analysis* (*a*) The weight and cruising speed of the airplane are

$$W = mg = (70{,}000 \text{ kg})(9.81 \text{ m/s}^2)\left(\frac{1 \text{ N}}{1 \text{ kg·m/s}^2}\right) = 686{,}700 \text{ N}$$

$$V = (558 \text{ km/h})\left(\frac{1 \text{ m/s}}{3.6 \text{ km/h}}\right) = 155 \text{ m/s}$$

The minimum velocities corresponding to the stall conditions without and with flaps, respectively, are obtained from Eq. 15–24,

$$V_{\min 1} = \sqrt{\frac{2W}{\rho C_{L, \max 1} A}} = \sqrt{\frac{2(686{,}700 \text{ N})}{(1.2 \text{ kg/m}^3)(1.52)(150 \text{ m}^2)}\left(\frac{1 \text{ kg·m/s}^2}{1 \text{ N}}\right)} = 70.9 \text{ m/s}$$

$$V_{\min 2} = \sqrt{\frac{2W}{\rho C_{L, \max 2} A}} = \sqrt{\frac{2(686{,}700 \text{ N})}{(1.2 \text{ kg/m}^3)(3.48)(150 \text{ m}^2)}\left(\frac{1 \text{ kg·m/s}^2}{1 \text{ N}}\right)} = 46.8 \text{ m/s}$$

Then the "safe" minimum velocities to avoid the stall region are obtained by multiplying the values above by 1.2:

*Without flaps:* $V_{\min 1, \text{safe}} = 1.2 V_{\min 1} = 1.2(70.9 \text{ m/s}) = 85.1 \text{ m/s} = \textbf{306 km/h}$

*With flaps:* $V_{\min 2, \text{safe}} = 1.2 V_{\min 2} = 1.2(46.8 \text{ m/s}) = 56.2 \text{ m/s} = \textbf{202 km/h}$

since 1 m/s = 3.6 km/h. Note that the use of flaps allows the plane to take off and land at considerably lower velocities, and thus on a shorter runway.

(*b*) When an aircraft is cruising steadily at a constant altitude, the lift must be equal to the weight of the aircraft, $F_L = W$. Then the lift coefficient is

$$C_L = \frac{F_L}{\frac{1}{2}\rho V^2 A} = \frac{686{,}700 \text{ N}}{\frac{1}{2}(0.312 \text{ kg/m}^3)(155 \text{ m/s})^2(150 \text{ m}^2)}\left(\frac{1 \text{ kg·m/s}^2}{1 \text{ N}}\right) = 1.22$$

For the case with no flaps, the angle of attack corresponding to this value of $C_L$ is determined from Fig. 15–45 to be $\alpha \cong 10°$.

(*c*) When the aircraft is cruising steadily at a constant altitude, the net force acting on the aircraft is zero, and thus thrust provided by the engines must be equal to the drag force. The drag coefficient corresponding to the cruising lift coefficient of 1.22 is determined from Fig. 15–45 to be $C_D \cong 0.03$ for the case with no flaps. Then the drag force acting on the wings becomes

$$F_D = C_D A \frac{\rho V^2}{2} = (0.03)(150 \text{ m}^2)\frac{(0.312 \text{ kg/m}^3)(155 \text{ m/s})^2}{2}\left(\frac{1 \text{ kN}}{1000 \text{ kg·m/s}^2}\right)$$

$$= 16.9 \text{ kN}$$

Noting that power is force times velocity (distance per unit time), the power required to overcome this drag is equal to the thrust times the cruising velocity:

$$\text{Power} = \text{Thrust} \times \text{Velocity} = F_D V = (16.9 \text{ kN})(155 \text{ m/s})\left(\frac{1 \text{ kW}}{1 \text{ kN·m/s}}\right)$$

$$= \textbf{2620 kW}$$

Therefore, the engines must supply 2620 kW of power to overcome the drag on the wings during cruising. For a propulsion efficiency of 30 percent (i.e., 30 percent of the energy of the fuel is utilized to propel the aircraft), the plane requires energy input at a rate of 8730 kJ/s.

**Discussion** The power determined is the power to overcome the drag that acts on the wings only and does not include the drag that acts on the remaining parts of the aircraft (the fuselage, the tail, etc.). Therefore, the total power required during cruising will be much greater. Also, it does not consider induced drag, which can be dominant during takeoff when the angle of attack is high (Fig. 15–45 is for a 2-D airfoil, and does not include 3-D effects).

558 km/h
70,000 kg
150 m², double-flapped
12,000 m

**FIGURE 15–52**
Schematic for Example 15–5.

No discussion on lift and drag would be complete without mentioning the contributions of Wilbur (1867–1912) and Orville (1871–1948) Wright. The Wright Brothers are truly the most impressive engineering team of all time. Self-taught, they were well informed of the contemporary theory and practice in aeronautics. They both corresponded with other leaders in the field and published in technical journals. While they cannot be credited with developing the concepts of lift and drag, they used them to achieve the first powered, manned, heavier-than-air, controlled flight (Fig. 15–53). They succeeded, while so many before them failed, because they evaluated and designed parts separately. Before the Wrights, experimenters were building and testing whole airplanes. While intuitively appealing, the approach did not allow the determination of how to make the craft better. When a flight lasts only a moment, you can only guess at the weakness in the design. Thus, a new craft did not necessarily perform any better than its predecessor. Testing was simply one belly flop followed by another. The Wrights changed all that. They studied each part using scale and full-size models in wind tunnels and in the field. Well before the first powered flyer was assembled, they knew the area required for their best wing shape to support a plane carrying a man and the engine horsepower required to provide adequate thrust with their improved impeller. The Wright Brothers not only showed the world how to fly, they showed engineers how to use the equations presented here to design even better aircraft.

**FIGURE 15–53**
The Wright Brothers take flight at Kitty Hawk.
*National Air and Space Museum/Smithsonian Institution.*

## SUMMARY

In this chapter, we study flow of fluids over immersed bodies with emphasis on the resulting lift and drag forces. A fluid may exert forces and moments on a body in and about various directions. The force a flowing fluid exerts on a body in the flow direction is called *drag* while that in the direction normal to the flow is called *lift*. The part of drag that is due directly to wall shear stress $\tau_w$ is called the *skin friction drag* since it is caused by frictional effects, and the part that is due directly to pressure $P$ is called the *pressure drag* or *form drag* because of its strong dependence on the form or shape of the body.

The *drag coefficient* $C_D$ and the *lift coefficient* $C_L$ are dimensionless numbers that represent the drag and the lift characteristics of a body and are defined as

$$C_D = \frac{F_D}{\frac{1}{2}\rho V^2 A} \quad \text{and} \quad C_L = \frac{F_L}{\frac{1}{2}\rho V^2 A}$$

where $A$ is usually the *frontal area* (the area projected on a plane normal to the direction of flow) of the body. For plates and airfoils, $A$ is taken to be the *planform area*, which is the area that would be seen by a person looking at the body from directly above. The drag coefficient, in general, depends on the *Reynolds number*, especially for Reynolds numbers below $10^4$. At higher Reynolds numbers, the drag coefficients for many geometries remain essentially constant.

A body is said to be *streamlined* if a conscious effort is made to align its shape with the anticipated streamlines in the flow in order to reduce drag. Otherwise, a body (such as a building) tends to block the flow and is said to be *bluff*. At sufficiently high velocities, the fluid stream detaches itself from the surface of the body. This is called *flow separation*. When a fluid stream separates from the body, it forms a *separated region* between the body and the fluid stream. Separation may also occur on a streamlined body such as an airplane wing at a sufficiently large *angle of attack*, which is the angle the incoming fluid stream makes with the *chord* (the line that connects the nose and the end) of the body. Flow separation on the top surface of a wing reduces lift drastically and may cause the airplane to *stall*.

The region of flow above a surface in which the effects of the viscous shearing forces caused by fluid viscosity are felt is called the *velocity boundary layer* or just the *boundary layer*. The *thickness* of the boundary layer, $\delta$, is defined as the distance from the surface at which the velocity is $0.99V$. The hypothetical line of velocity $0.99V$ divides the flow over a plate into two regions: the *boundary layer region*, in which the viscous effects and the velocity changes are significant, and the *irrotational outer flow region*, in which the frictional effects are negligible and the velocity remains essentially constant.

For external flow, the Reynolds number is expressed as

$$\text{Re}_L = \frac{\rho VL}{\mu} = \frac{VL}{\nu}$$

where $V$ is the upstream velocity and $L$ is the characteristic length of the geometry, which is the length of the plate in the flow direction for a flat plate and the diameter $D$ for a cylinder or sphere. The *average* friction coefficients over an entire flat plate are

*Laminar flow*: $\qquad C_f = \dfrac{1.33}{\text{Re}_L^{1/2}} \qquad \text{Re}_L \lesssim 5 \times 10^5$

*Turbulent flow*: $\qquad C_f = \dfrac{0.074}{\text{Re}_L^{1/5}} \quad 5 \times 10^5 \lesssim \text{Re}_L \lesssim 10^7$

If the flow is approximated as laminar up to the engineering critical number of $\text{Re}_{cr} = 5 \times 10^5$, and then turbulent beyond, the average friction coefficient over the entire flat plate becomes

$$C_f = \frac{0.074}{\text{Re}_L^{1/5}} - \frac{1742}{\text{Re}_L} \quad 5 \times 10^5 \lesssim \text{Re}_L \lesssim 10^7$$

A curve fit of experimental data for the average friction coefficient in the fully rough turbulent regime is

*Rough surface*: $\qquad C_f = \left(1.89 - 1.62 \log \dfrac{\varepsilon}{L}\right)^{-2.5}$

where $\varepsilon$ is the surface roughness and $L$ is the length of the plate in the flow direction. In the absence of a better one, this relation can be used for turbulent flow on rough surfaces for $\text{Re} > 10^6$, especially when $\varepsilon/L > 10^{-4}$.

Surface roughness, in general, increases the drag coefficient in turbulent flow. For bluff bodies such as a circular cylinder or sphere, however, an increase in the surface roughness may *decrease* the drag coefficient. This is done by tripping the flow into turbulence at a lower Reynolds number, and thus causing the fluid to close in behind the body, narrowing the wake and reducing pressure drag considerably.

It is desirable for airfoils to generate the most lift while producing the least drag. Therefore, a measure of performance for airfoils is the *lift-to-drag ratio*, $C_L/C_D$.

The minimum safe flight velocity of an aircraft is determined from

$$V_{\min} = \sqrt{\frac{2W}{\rho C_{L,\max} A}}$$

For a given weight, the landing or takeoff speed can be minimized by maximizing the product of the lift coefficient and the wing area, $C_{L,\max} A$.

For airplane wings and other airfoils of finite span, the pressure difference between the lower and the upper surfaces drives the fluid at the tips upward. This results in swirling eddies, called *tip vortices*. Tip vortices that interact with the free stream impose forces on the wing tips in all directions, including the flow direction. The component of the force in the flow direction adds to drag and is called *induced drag*. The total drag of a wing is then the sum of the induced drag (3-D effects) and the drag of the airfoil section (2-D effects).

## REFERENCES AND SUGGESTED READING

**1.** I. H. Abbott. "The Drag of Two Streamline Bodies as Affected by Protuberances and Appendages," *NACA Report* 451, 1932.

**2.** I. H. Abbott and A. E. von Doenhoff. *Theory of Wing Sections, Including a Summary of Airfoil Data.* New York: Dover, 1959.

**3.** I. H. Abbott, A. E. von Doenhoff, and L. S. Stivers. "Summary of Airfoil Data," *NACA Report* 824, Langley Field, VA, 1945.

**4.** J. D. Anderson. *Fundamentals of Aerodynamics,* 2nd ed. New York: McGraw-Hill, 1991.

5. R. D. Blevins. *Applied Fluid Dynamics Handbook*. New York: Van Nostrand Reinhold, 1984.

6. S. W. Churchill and M. Bernstein. "A Correlating Equation for Forced Convection from Gases and Liquids to a Circular Cylinder in Cross Flow," *Journal of Heat Transfer* 99, pp. 300–306, 1977.

7. J. Happel and H. Brenner. *Low Reynolds Number Hydrodynamics with Special Applications to Particulate Media*. Norwell, MA: Kluwer Academic Publishers, 1983.

8. S. F. Hoerner. *Fluid-Dynamic Drag*. [Published by the author.] Library of Congress No. 64, 1966.

9. J. D. Holmes. *Wind Loading of Structures*. London: Spon Press (Taylor and Francis), 2001.

10. G. M. Homsy, H. Aref, K. S. Breuer, S. Hochgreb, J. R. Koseff, B. R. Munson, K. G. Powell, C. R. Roberston, S. T. Thoroddsen. *Multi-Media Fluid Mechanics* (CD). Cambridge University Press, 2000.

11. W. H. Hucho. *Aerodynamics of Road Vehicles*. London: Butterworth-Heinemann, 1987.

12. H. Schlichting. *Boundary Layer Theory,* 7th ed. New York: McGraw-Hill, 1979.

13. M. Van Dyke. *An Album of Fluid Motion*. Stanford, CA: The Parabolic Press, 1982.

14. J. Vogel. *Life in Moving Fluids,* 2nd ed. Boston: Willard Grand Press, 1994.

15. F. M. White. *Fluid Mechanics,* 5th ed. New York: McGraw-Hill, 2003.

## PROBLEMS*

### Drag, Lift, and Drag Coefficients

**15–1C**  Which car is more likely to be more fuel-efficient: one with sharp corners or one that is contoured to resemble an ellipse? Why?

**15–2C**  Consider laminar flow over a flat plate. How does the local friction coefficient change with position?

**15–3C**  Explain when an external flow is two-dimensional, three-dimensional, and axisymmetric. What type of flow is the flow of air over a car?

**15–4C**  What is the difference between the upstream velocity and the free-stream velocity? For what types of flow are these two velocities equal to each other?

**15–5C**  What is the difference between streamlined and bluff bodies? Is a tennis ball a streamlined or bluff body?

**15–6C**  Name some applications in which a large drag is desired.

**15–7C**  What is drag? What causes it? Why do we usually try to minimize it?

**15–8C**  What is lift? What causes it? Does wall shear contribute to the lift?

**15–9C**  During flow over a given body, the drag force, the upstream velocity, and the fluid density are measured. Explain how you would determine the drag coefficient. What area would you use in the calculations?

**15–10C**  Define the frontal area of a body subjected to external flow. When is it appropriate to use the frontal area in drag and lift calculations?

**15–11C**  Define the planform area of a body subjected to external flow. When is it appropriate to use the planform area in drag and lift calculations?

**15–12C**  What is terminal velocity? How is it determined?

**15–13C**  What is the difference between skin friction drag and pressure drag? Which is usually more significant for slender bodies such as airfoils?

**15–14C**  What is the effect of surface roughness on the friction drag coefficient in laminar and turbulent flows?

**15–15C**  In general, how does the drag coefficient vary with the Reynolds number at (*a*) low and moderate Reynolds numbers and (*b*) at high Reynolds numbers (Re $> 10^4$)?

**15–16C**  What is the effect of streamlining on (*a*) friction drag and (*b*) pressure drag? Does the total drag acting on a body necessarily decrease as a result of streamlining? Explain.

**15–17C**  What is flow separation? What causes it? What is the effect of flow separation on the drag coefficient?

**15–18C**  What is drafting? How does it affect the drag coefficient of the drafted body?

**15–19**  A car is moving at a constant velocity of 80 km/h. Determine the upstream velocity to be used in fluid flow analysis if (*a*) the air is calm, (*b*) wind is blowing against the

---

* Problems designated by a "C" are concept questions, and students are encouraged to answer them all. Problems designated by an "E" are in English units, and the SI users can ignore them. Problems with the ✿ icon are solved using EES, and complete solutions together with parametric studies are included on the text specific web site. Problems with the 🖥 icon are comprehensive in nature and are intended to be solved with a computer, preferably using the EES software that accompanies this text.

direction of motion of the car at 30 km/h, and (*c*) wind is blowing in the same direction of motion of the car at 50 km/h.

**15–20** The resultant of the pressure and wall shear forces acting on a body is measured to be 580 N, making 35° with the direction of flow. Determine the drag and the lift forces acting on the body.

**FIGURE P15–20**

**15–21** During a high Reynolds number experiment, the total drag force acting on a spherical body of diameter $D = 12$ cm subjected to airflow at 1 atm and 5°C is measured to be 5.2 N. The pressure drag acting on the body is calculated by integrating the pressure distribution (measured by the use of pressure sensors throughout the surface) to be 4.9 N. Determine the friction drag coefficient of the sphere. *Answer:* 0.0115

**15–22** The drag coefficient of a car at the design conditions of 1 atm, 25°C, and 90 km/h is to be determined experimentally in a large wind tunnel in a full-scale test. The height and width of the car are 1.40 m and 1.65 m, respectively. If the horizontal force acting on the car is measured to be 280 N, determine the total drag coefficient of this car. *Answer:* 0.33

**15–23E** To reduce the drag coefficient and thus to improve the fuel efficiency, the frontal area of a car is to be reduced. Determine the amount of fuel and money saved per year as a result of reducing the frontal area from 18 to 15 ft². Assume the car is driven 12,000 mi a year at an average speed of 55 mi/h. Take the density and price of gasoline to be 50 lbm/ft³ and $3.10/gal, respectively; the density of air to be 0.075 lbm/ft³, the heating value of gasoline to be 20,000 Btu/lbm; and the overall efficiency of the engine to be 30 percent.

**15–24E** Reconsider Prob. 15–23E. Using EES (or other) software, investigate the effect of frontal area on the annual fuel consumption of the car. Let the frontal area vary from 10 to 30 ft² in increments of 2 ft². Tabulate and plot the results.

**15–25** A circular sign has a diameter of 50 cm and is subjected to normal winds up to 150 km/h at 10°C and 100 kPa. Determine the drag force acting on the sign. Also determine the bending moment at the bottom of its pole whose height from the ground to the bottom of the sign is 1.5 m. Disregard the drag on the pole.

**FIGURE P15–25**

**15–26E** Wind loading is a primary consideration in the design of the supporting mechanisms of billboards, as evidenced by many billboards being knocked down during high winds. Determine the wind force acting on an 8-ft-high, 20-ft-wide billboard due to 75-mi/h winds in the normal direction when the atmospheric conditions are 14.3 psia and 40°F. *Answer:* 4420 lbf

**15–27** Advertisement signs are commonly carried by taxicabs for additional income, but they also increase the fuel cost. Consider a sign that consists of a 0.30-m-high, 0.9-m-wide, and 0.9-m-long rectangular block mounted on top of a taxicab such that the sign has a frontal area of 0.3 m by 0.9 m from all four sides. Determine the increase in the annual fuel cost of this taxicab due to this sign. Assume the taxicab is driven 60,000 km a year at an average speed of 50 km/h and the overall efficiency of the engine is 28 percent. Take the density, unit price, and heating value of gasoline to be 0.72 kg/L, $1.10/L, and 42,000 kJ/kg, respectively, and the density of air to be 1.25 kg/m³.

**FIGURE P15–27**

**15–28E** At highway speeds, about half of the power generated by the car's engine is used to overcome aerodynamic drag, and thus the fuel consumption is nearly proportional to the drag force on a level road. Determine the percentage increase in fuel consumption of a car per unit time when a person who normally drives at 55 mi/h now starts driving at 75 mi/h.

**15–29E** Bill gets a job delivering pizzas. The pizza company makes him mount a sign on the roof of his car. The frontal area of the sign is $A = 0.612$ ft², and he estimates the drag coefficient to be $C_D = 0.94$ at nearly all air speeds. Estimate how much additional money it costs Bill per year in fuel to drive with the sign on his roof compared to without the sign. Use the following additional information: He drives about 10,000 miles per year at an average speed of 45 mph. The overall car efficiency is 0.332, $\rho_{fuel} = 50.2$ lbm/ft³, and the heating value of the fuel is $1.53 \times 10^7$ ft·lbf/lbm.

The fuel costs $3.50 per gallon. Use standard air properties. Be careful with unit conversions.

**15–30** During major windstorms, high vehicles such as RVs and semis may be thrown off the road and boxcars off their tracks, especially when they are empty and in open areas. Consider a 5000-kg semi that is 9 m long, 2.5 m high, and 2 m wide. The distance between the bottom of the truck and the road is 0.75 m. Now the truck is exposed to winds from its side surface. Determine the wind velocity that will tip the truck over to its side. Take the air density to be 1.1 kg/m³ and assume the weight to be uniformly distributed.

**FIGURE P15–30**

**15–31** An 80-kg bicyclist is riding her 15-kg bicycle downhill on a road with a slope of 12° without pedaling or braking. The bicyclist has a frontal area of 0.45 m² and a drag coefficient of 1.1 in the upright position, and a frontal area of 0.4 m² and a drag coefficient of 0.9 in the racing position. Disregarding the rolling resistance and friction at the bearings, determine the terminal velocity of the bicyclist for both positions. Take the air density to be 1.25 kg/m³. *Answers: 90 km/h, 106 km/h*

**15–32** A wind turbine with two or four hollow hemispherical cups connected to a pivot is commonly used to measure wind speed. Consider a wind turbine with four 8-cm-diameter cups with a center-to-center distance of 40 cm, as shown in

**FIGURE P15–32**

Fig. P15–32. The pivot is stuck as a result of some malfunction, and the cups stop rotating. For a wind speed of 15 m/s and air density of 1.25 kg/m³, determine the maximum torque this turbine applies on the pivot.

**15–33** Reconsider Prob. 15–32. Using EES (or other) software, investigate the effect of wind speed on the torque applied on the pivot. Let the wind speed vary from 0 to 50 m/s in increments of 5 m/s. Tabulate and plot the results.

**15–34E** A 5-ft-diameter spherical tank completely submerged in freshwater is being towed by a ship at 12 ft/s. Assuming turbulent flow, determine the required towing power.

**15–35** During steady motion of a vehicle on a level road, the power delivered to the wheels is used to overcome aerodynamic drag and rolling resistance (the product of the rolling resistance coefficient and the weight of the vehicle), assuming the friction at the bearings of the wheels is negligible. Consider a car that has a total mass of 950 kg, a drag coefficient of 0.32, a frontal area of 1.8 m², and a rolling resistance coefficient of 0.04. The maximum power the engine can deliver to the wheels is 80 kW. Determine (a) the speed at which the rolling resistance is equal to the aerodynamic drag force and (b) the maximum speed of this car. Take the air density to be 1.20 kg/m³.

**15–36** Reconsider Prob. 15–35. Using EES (or other) software, investigate the effect of car speed on the required power to overcome (a) rolling resistance, (b) the aerodynamic drag, and (c) their combined effect. Let the car speed vary from 0 to 150 km/h in increments of 15 km/h. Tabulate and plot the results.

**15–37** A submarine can be treated as an ellipsoid with a diameter of 5 m and a length of 25 m. Determine the power required for this submarine to cruise horizontally and steadily at 65 km/h in seawater whose density is 1025 kg/m³. Also determine the power required to tow this submarine in air whose density is 1.30 kg/m³. Assume the flow is turbulent in both cases.

**FIGURE P15–37**

**15–38** An 0.90-m-diameter, 1.1-m-high garbage can is found in the morning tipped over due to high winds during the night. Assuming the average density of the garbage inside to be 150 kg/m³ and taking the air density to be 1.25 kg/m³, estimate the wind velocity during the night when the can was tipped over. Take the drag coefficient of the can to be 0.7. *Answer: 159 km/h*

**15–39** To reduce the drag coefficient and thus to improve the fuel efficiency of cars, the design of side rearview mirrors has changed drastically in recent decades from a simple circular plate to a streamlined shape. Determine the amount of fuel and money saved per year as a result of replacing a 13-cm-diameter flat mirror by one with a hemispherical back, as shown in the figure. Assume the car is driven 24,000 km a year at an average speed of 95 km/h. Take the density and price of gasoline to be 0.75 kg/L and $0.90/L, respectively; the heating value of gasoline to be 44,000 kJ/kg; and the overall efficiency of the engine to be 30 percent.

Flat mirror
95 km/h
$D = 13$ cm

Rounded mirror
95 km/h
$D = 13$ cm

**FIGURE P15–39**

**15–40** A 5-m-diameter hot air balloon that has a total mass of 230 kg is standing still in air on a windless day. The balloon is suddenly subjected to 40 km/h winds. Determine the initial acceleration of the balloon in the horizontal direction.

**15–41** A 6-mm-diameter plastic sphere whose density is 1150 kg/m³ is dropped into water at 20°C. Determine the terminal velocity of the sphere in water.

### Flow over Flat Plates

**15–42C** What does the friction coefficient represent in flow over a flat plate? How is it related to the drag force acting on the plate?

**15–43C** What fluid property is responsible for the development of the velocity boundary layer? What is the effect of the velocity on the thickness of the boundary layer?

**15–44C** How is the average friction coefficient determined in flow over a flat plate?

**15–45E** Light oil at 75°F flows over a 22-ft-long flat plate with a free-stream velocity of 6 ft/s. Determine the total drag force per unit width of the plate.

**15–46** The local atmospheric pressure in Denver, Colorado (elevation 1610 m) is 83.4 kPa. Air at this pressure and at 25°C flows with a velocity of 6 m/s over a 2.5-m × 8-m flat plate. Determine the drag force acting on the top surface of the plate if the air flows parallel to the (*a*) 8-m-long side and (*b*) the 2.5-m-long side.

**15–47** During a winter day, wind at 55 km/h, 5°C, and 1 atm is blowing parallel to a 4-m-high and 10-m-long wall of a house. Approximating the wall surfaces as smooth, determine the friction drag acting on the wall. What would your answer be if the wind velocity has doubled? How realistic is it to treat the flow over side wall surfaces as flow over a flat plate? *Answers:* 16 N, 58 N

Air
5°C
55 km/h

4 m

10 m

**FIGURE P15–47**

**15–48E** Air at 70°F flows over a 10-ft-long flat plate at 25 ft/s. Determine the local friction coefficient at intervals of 1 ft and plot the results against the distance from the leading edge.

**15–49** Consider laminar flow of a fluid over a flat plate. Now the free-stream velocity of the fluid is doubled. Determine the change in the drag force on the plate. Assume the flow to remain laminar. *Answer:* A 2.83-fold increase

**15–50E** Consider a refrigeration truck traveling at 70 mi/h at a location where the air is at 1 atm and 80°F. The refrigerated compartment of the truck can be considered to be a 9-ft-wide, 8-ft-high, and 20-ft-long rectangular box. Assuming the airflow over the entire outer surface to be turbulent and attached (no flow separation), determine the drag force acting on the top and side surfaces and the power required to overcome this drag.

Air, 80°F
$V = 70$ mi/h

20 ft

8 ft

Refrigeration truck

**FIGURE P15–50E**

**15–51E** Reconsider Prob. 15–50E. Using EES (or other) software, investigate the effect of truck speed on the total drag force acting on the top and side surfaces, and the power required to overcome it. Let the truck speed vary from 0 to 100 mi/h in increments of 10 mi/h. Tabulate and plot the results.

**15–52** Air at 25°C and 1 atm is flowing over a long flat plate with a velocity of 8 m/s. Determine the distance from the leading edge of the plate where the flow becomes turbulent, and the thickness of the boundary layer at that location.

**15–53** Repeat Prob. 15–52 for water.

**15–54** The top surface of the passenger car of a train moving at a velocity of 70 km/h is 3.2 m wide and 8 m long. If the outdoor air is at 1 atm and 25°C, determine the drag force acting on the top surface of the car.

Air 70 km/h
25°C

**FIGURE P15–54**

**15–55** The weight of a thin flat plate 50 cm × 50 cm in size is balanced by a counterweight that has a mass of 2 kg, as shown in Fig. P15–55. Now a fan is turned on, and air at 1 atm and 25°C flows downward over both surfaces of the plate (front and back in the sketch) with a free-stream velocity of 10 m/s. Determine the mass of the counterweight that needs to be added in order to balance the plate in this case.

Air
25°C, 10 m/s

Plate    50 cm

50 cm

**FIGURE P15–55**

## Flow across Cylinders and Spheres

**15–56C** In flow over cylinders, why does the drag coefficient suddenly drop when the boundary layer becomes turbulent? Isn't turbulence supposed to increase the drag coefficient instead of decreasing it?

**15–57C** In flow over bluff bodies such as a cylinder, how does the pressure drag differ from the friction drag?

**15–58C** Why is flow separation in flow over cylinders delayed when the boundary layer is turbulent?

**15–59E** A 1.2-in-outer-diameter pipe is to span across a river at a 140-ft-wide section while being completely immersed in water. The average flow velocity of the water is 10 ft/s, and its temperature is 70°F. Determine the drag force exerted on the pipe by the river. *Answer:* 1490 lbf

**15–60** A long 8-cm-diameter steam pipe passes through some area open to the wind. Determine the drag force acting on the pipe per unit of its length when the air is at 1 atm and 5°C and the wind is blowing across the pipe at a speed of 50 km/h.

**15–61** Consider 0.8-cm-diameter hail that is falling freely in atmospheric air at 1 atm and 5°C. Determine the terminal velocity of the hail. Take the density of hail to be 910 kg/m$^3$.

**15–62** A 0.1-mm-diameter dust particle whose density is 2.1 g/cm$^3$ is observed to be suspended in the air at 1 atm and 25°C at a fixed point. Estimate the updraft velocity of air motion at that location. Assume Stokes law to be applicable. Is this a valid assumption? *Answer:* 0.62 m/s

**15–63** Dust particles of diameter 0.06 mm and density 1.6 g/cm$^3$ are unsettled during high winds and rise to a height of 350 m by the time things calm down. Estimate how long it takes for the dust particles to fall back to the ground in still air at 1 atm and 15°C, and their velocity. Disregard the initial transient period during which the dust particles accelerate to their terminal velocity, and assume Stokes law to be applicable.

**15–64** A 2-m-long, 0.2-m-diameter cylindrical pine log (density = 513 kg/m$^3$) is suspended by a crane in the horizontal position. The log is subjected to normal winds of 40 km/h at 5°C and 88 kPa. Disregarding the weight of the cable and its drag, determine the angle $\theta$ the cable will make with the horizontal and the tension on the cable.

40 km/h

0.2 m

2 m

$\theta$

**FIGURE P15–64**

**15–65E** A person extends his uncovered arms into the windy air outside at 1 atm and 60°F and 20 mi/h in order to feel nature closely. Treating the arm as a 2-ft-long and 3-in-diameter cylinder, determine the combined drag force on both arms. *Answer:* 1.02 lbf

Air
60°F, 20 mi/h

**FIGURE P15–65E**

**15–66** A 6-mm-diameter electrical transmission line is exposed to windy air. Determine the drag force exerted on a 160-m-long section of the wire during a windy day when the air is at 1 atm and 15°C and the wind is blowing across the transmission line at 65 km/h.

## Lift

**15–67C** Air is flowing past a nonsymmetrical airfoil at zero angle of attack. Is the (*a*) lift and (*b*) drag acting on the airfoil zero or nonzero?

**15–68C** Why is the contribution of viscous effects to lift usually negligible for airfoils?

**15–69C** Air is flowing past a symmetrical airfoil at an angle of attack of 5°. Is the (*a*) lift and (*b*) drag acting on the airfoil zero or nonzero?

**15–70C** What is stall? What causes an airfoil to stall? Why are commercial aircraft not allowed to fly at conditions near stall?

**15–71C** Both the lift and the drag of an airfoil increase with an increase in the angle of attack. In general, which increases at a higher rate, the lift or the drag?

**15–72C** Why are flaps used at the leading and trailing edges of the wings of large aircraft during takeoff and landing? Can an aircraft take off or land without them?

**15–73C** How do flaps affect the lift and the drag of wings?

**15–74C** What is the effect of wing tip vortices (the air circulation from the lower part of the wings to the upper part) on the drag and the lift?

**15–75C** What is induced drag on wings? Can induced drag be minimized by using long and narrow wings or short and wide wings?

**15–76C** Explain why endplates or winglets are added to some airplane wings.

**15–77** Consider an aircraft that takes off at 190 km/h when it is fully loaded. If the weight of the aircraft is increased by 20 percent as a result of overloading, determine the speed at which the overloaded aircraft will take off. *Answer:* 208 km/h

**15–78** Consider an airplane whose takeoff speed is 220 km/h and that takes 15 s to take off at sea level. For an airport at an elevation of 1600 m (such as Denver), determine (*a*) the takeoff speed, (*b*) the takeoff time, and (*c*) the additional runway length required for this airplane. Assume constant acceleration for both cases.

220 km/h

**FIGURE P15–78**

**15–79E** An airplane is consuming fuel at a rate of 7 gal/min when cruising at a constant altitude of 10,000 ft at constant speed. Assuming the drag coefficient and the engine efficiency to remain the same, determine the rate of fuel consumption at an altitude of 30,000 ft at the same speed.

**15–80** A jumbo jet airplane has a mass of about 400,000 kg when fully loaded with over 400 passengers and takes off at a speed of 250 km/h. Determine the takeoff speed when the airplane has 100 empty seats. Assume each passenger with luggage is 140 kg and the wing and flap settings are maintained the same. *Answer:* 246 km/h

**15–81** Reconsider Prob. 15–80. Using EES (or other) software, investigate the effect of passenger count on the takeoff speed of the aircraft. Let the number of passengers vary from 0 to 500 in increments of 50. Tabulate and plot the results.

**15–82** A small aircraft has a wing area of 28 m², a lift coefficient of 0.45 at takeoff settings, and a total mass of 2500 kg. Determine (*a*) the takeoff speed of this aircraft at sea level at standard atmospheric conditions, (*b*) the wing loading, and (*c*) the required power to maintain a constant cruising speed of 300 km/h for a cruising drag coefficient of 0.035.

**15–83** A small airplane has a total mass of 1800 kg and a wing area of 42 m². Determine the lift and drag coefficients of this airplane while cruising at an altitude of 4000 m at a constant speed of 280 km/h and generating 190 kW of power.

**15–84** The NACA 64(1)–412 airfoil has a lift-to-drag ratio of 50 at 0° angle of attack, as shown in Fig. 15–43. At what angle of attack does this ratio increase to 80?

**15–85** Consider a light plane that has a total weight of 15,000 N and a wing area of 46 m$^2$ and whose wings resemble the NACA 23012 airfoil with no flaps. Using data from Fig. 15–45, determine the takeoff speed at an angle of attack of 5° at sea level. Also determine the stall speed. *Answers:* 110 km/h, 67.4 km/h

**15–86** An airplane has a mass of 50,000 kg, a wing area of 300 m$^2$, a maximum lift coefficient of 3.2, and a cruising drag coefficient of 0.03 at an altitude of 12,000 m. Determine (*a*) the takeoff speed at sea level, assuming it is 20 percent over the stall speed, and (*b*) the thrust that the engines must deliver for a cruising speed of 700 km/h.

## Review Problems

**15–87E** The passenger compartment of a minivan traveling at 60 mi/h in ambient air at 1 atm and 80°F is modeled as a 3.2-ft-high, 6-ft-wide, and 11-ft-long rectangular box. The airflow over the exterior surfaces is assumed to be turbulent because of the intense vibrations involved. Determine the drag force acting on the top and the two side surfaces of the van and the power required to overcome it.

**FIGURE P15–87E**

**15–88** A 1.2-m-external-diameter spherical tank is located outdoors at 1 atm and 25°C and is subjected to winds at 48 km/h. Determine the drag force exerted on it by the wind. *Answer:* 16.7 N

**15–89** A 2-m-high, 4-m-wide rectangular advertisement panel is attached to a 4-m-wide, 0.15-m-high rectangular concrete block (density = 2300 kg/m$^3$) by two 5-cm-diameter, 4-m-high (exposed part) poles, as shown in Fig. P15–89. If the sign is to withstand 150 km/h winds from any direction, determine (*a*) the maximum drag force on the panel, (*b*) the drag force acting on the poles, and (*c*) the minimum length *L* of the concrete block for the panel to resist the winds. Take the density of air to be 1.30 kg/m$^3$.

**FIGURE P15–89**

**15–90** A plastic boat whose bottom surface can be approximated as a 1.5-m-wide, 2-m-long flat surface is to move through water at 15°C at speeds up to 45 km/h. Determine the friction drag exerted on the boat by the water and the power needed to overcome it.

**FIGURE P15–90**

**15–91** Reconsider Prob. 15–90. Using EES (or other) software, investigate the effect of boat speed on the drag force acting on the bottom surface of the boat, and the power needed to overcome it. Let the boat speed vary from 0 to 100 km/h in increments of 10 km/h. Tabulate and plot the results.

**15–92** The cylindrical chimney of a factory has an external diameter of 1.6 m and is 35 m high. Determine the bending moment at the base of the chimney when winds at 110 km/h are blowing across it. Take the atmospheric conditions to be 20°C and 1 atm.

**15–93** Consider a blimp that can be approximated as a 3-m diameter, 8-m long ellipsoid and is connected to the ground. On a windless day, the rope tension due to the net buoyancy effect is measured to be 120 N. Determine the rope tension when there are 50 km/h winds blowing along the blimp (parallel to the blimp axis).

**FIGURE P15–93**

**FIGURE P15–96**

**15–94E** A commercial airplane has a total mass of 150,000 lbm and a wing planform area of 1800 ft². The plane has a cruising speed of 550 mi/h and a cruising altitude of 38,000 ft where the air density is 0.0208 lbm/ft³. The plane has double-slotted flaps for use during takeoff and landing, but it cruises with all flaps retracted. Assuming the lift and drag characteristics of the wings can be approximated by NACA 23012, determine (a) the minimum safe speed for takeoff and landing with and without extending the flaps, (b) the angle of attack to cruise steadily at the cruising altitude, and (c) the power that needs to be supplied to provide enough thrust to overcome drag. Take the air density on the ground to be 0.075 lbm/ft³.

**15–95** A 9-cm-diameter smooth sports ball has a velocity of 36 km/h during a typical hit. Determine the percent increase in the drag coefficient if the ball is given a spin of 3500 rpm in air at 1 atm and 25°C.

**15–96** A paratrooper and his 8-m-diameter parachute weigh 950 N. Taking the average air density to be 1.2 kg/m³, determine the terminal velocity of the paratrooper. *Answer: 4.9 m/s*

**15–97** It is proposed to meet the water needs of a recreational vehicle (RV) by installing a 2-m-long, 0.5-m-diameter cylindrical tank on top of the vehicle. Determine the additional power requirement of the RV at a speed of 95 km/h when the tank is installed such that its circular surfaces face (a) the front and back (as sketched) and (b) the sides of the RV. Assume atmospheric conditions are 87 kPa and 20°C. *Answers: (a) 1.67 kW, (b) 7.55 kW*

**FIGURE P15–97**

**15–98**  A 17,000-kg tractor-trailer rig has a frontal area of 9.2 m², a drag coefficient of 0.96, a rolling resistance coefficient of 0.05 (multiplying the weight of a vehicle by the rolling resistance coefficient gives the rolling resistance), a bearing friction resistance of 350 N, and a maximum speed of 110 km/h on a level road during steady cruising in calm weather with an air density of 1.25 kg/m³. Now a fairing is installed to the front of the rig to suppress separation and to streamline the flow to the top surface, and the drag coefficient is reduced to 0.76. Determine the maximum speed of the rig with the fairing.   *Answer:* 133 km/h

**15–99**  Stokes law can be used to determine the viscosity of a fluid by dropping a spherical object in it and measuring the terminal velocity of the object in that fluid. This can be done by plotting the distance traveled against time and observing when the curve becomes linear. During such an experiment a 3-mm-diameter glass ball ($\rho = 2500$ kg/m³) is dropped into a fluid whose density is 875 kg/m³, and the terminal velocity is measured to be 0.16 m/s. Disregarding the wall effects, determine the viscosity of the fluid.

Glass ball      0.16 m/s

**FIGURE P15–99**

**15–100**  During an experiment, three aluminum balls ($\rho_s = 2600$ kg/m³) having diameters 2, 4, and 10 mm, respectively, are dropped into a tank filled with glycerin at 22°C ($\rho_f = 1274$ kg/m³ and $\mu = 1$ kg/m·s). The terminal settling velocities of the balls are measured to be 3.2, 12.8, and 60.4 mm/s, respectively. Compare these values with the velocities predicted by Stokes law for drag force $F_D = 3\pi\mu DV$, which is valid for very low Reynolds numbers (Rc ≪ 1). Determine the error involved for each case and assess the accuracy of Stokes law.

**15–101**  Repeat Prob. 15–100 by considering the more general form of Stokes law expressed as $F_D = 3\pi\mu DV + (9\pi/16)\rho V^2 D^2$ where $\rho$ is the fluid density.

**15–102**  Engine oil at 40°C is flowing over a long flat plate with a velocity of 6 m/s. Determine the distance $x_{cr}$ from the leading edge of the plate where the flow becomes turbulent, and calculate and plot the thickness of the boundary layer over a length of $2x_{cr}$.

**15–103E**  Janie likes to drive with a tennis ball on her car antenna. The ball diameter is $D = 2.62$ in and its equivalent roughness factor is $\varepsilon/D = 1.5 \times 10^{-3}$. Her friends tell her she is wasting gas because of the additional drag on the ball. Estimate how much money (in dollars) she wastes per year by driving around with this tennis ball on her antenna. Use the following additional information: She drives mostly on the highway, about 12,000 miles per year at an average speed of 55 mph. The overall car efficiency is 0.308, $\rho_{fuel} = 50.2$ lbm/ft³, and the heating value of the fuel is $1.47 \times 10^7$ ft·lbf/lbm. The fuel costs $4.00 per gallon. Use standard air properties. Be careful with unit conversions. Should Janie remove the tennis ball?

**FIGURE P15–103E**

*Photo by John M. Cimbala.*

## Design and Essay Problems

**15–104**  Write a report on the flaps used at the leading and trailing edges of the wings of large commercial aircraft. Discuss how the flaps affect the drag and lift coefficients during takeoff and landing.

**15–105**  Large commercial airplanes cruise at high altitudes (up to about 40,000 ft) to save fuel. Discuss how flying at high altitudes reduces drag and saves fuel. Also discuss why small planes fly at relatively low altitudes.

**15–106**  Many drivers turn off their air conditioners and roll down the car windows in hopes of saving fuel. But it is claimed that this apparent "free cooling" actually increases the fuel consumption of some cars. Investigate this matter and write a report on which practice saves gasoline under what conditions.

# HEAT TRANSFER

# MECHANISMS OF HEAT TRANSFER

The science of thermodynamics deals with the *amount* of heat transfer as a system undergoes a process from one equilibrium state to another, and makes no reference to *how long* the process will take. But in engineering, we are often interested in the *rate* of heat transfer, which is the topic of the science of *heat transfer*.

We start this chapter with an overview of the three basic mechanisms of heat transfer, which are conduction, convection, and radiation, and discuss thermal conductivity. *Conduction* is the transfer of energy from the more energetic particles of a substance to the adjacent, less energetic ones as a result of interactions between the particles. *Convection* is the mode of heat transfer between a solid surface and the adjacent liquid or gas that is in motion, and it involves the combined effects of conduction and fluid motion. *Radiation* is the energy emitted by matter in the form of electromagnetic waves (or photons) as a result of the changes in the electronic configurations of the atoms or molecules. We close this chapter with a discussion of simultaneous heat transfer.

## Objectives

The objectives of this chapter are to:

- Understand the basic mechanisms of heat transfer, which are conduction, convection, and radiation, and Fourier's law of heat conduction, Newton's law of cooling, and the Stefan–Boltzmann law of radiation.

- Identify the mechanisms of heat transfer that occur simultaneously in practice.

- Develop an awareness of the cost associated with heat losses.

- Solve various heat transfer problems encountered in practice.

## 16–1 ▪ INTRODUCTION

In Chap. 3, we defined *heat* as the form of energy that can be transferred from one system to another as a result of temperature difference. A thermodynamic analysis is concerned with the *amount* of heat transfer as a system undergoes a process from one equilibrium state to another. The science that deals with the determination of the *rates* of such energy transfers is the *heat transfer*. The transfer of energy as heat is always from the higher-temperature medium to the lower-temperature one, and heat transfer stops when the two mediums reach the same temperature.

Heat can be transferred in three different modes: *conduction, convection,* and *radiation*. All modes of heat transfer require the existence of a temperature difference, and all modes are from the high-temperature medium to a lower-temperature one. Below we give a brief description of each mode. A detailed study of these modes is given in later chapters of this text.

## 16–2 ▪ CONDUCTION

**Conduction** is the transfer of energy from the more energetic particles of a substance to the adjacent less energetic ones as a result of interactions between the particles. Conduction can take place in solids, liquids, or gases. In gases and liquids, conduction is due to the *collisions* and *diffusion* of the molecules during their random motion. In solids, it is due to the combination of *vibrations* of the molecules in a lattice and the energy transport by *free electrons*. A cold canned drink in a warm room, for example, eventually warms up to the room temperature as a result of heat transfer from the room to the drink through the aluminum can by conduction.

The *rate* of heat conduction through a medium depends on the *geometry* of the medium, its *thickness*, and the *material* of the medium, as well as the *temperature difference* across the medium. We know that wrapping a hot water tank with glass wool (an insulating material) reduces the rate of heat loss from the tank. The thicker the insulation, the smaller the heat loss. We also know that a hot water tank loses heat at a higher rate when the temperature of the room housing the tank is lowered. Furthermore, the larger the tank, the larger the surface area and thus the rate of heat loss.

Consider steady heat conduction through a large plane wall of thickness $\Delta x = L$ and area $A$, as shown in Fig. 16–1. The temperature difference across the wall is $\Delta T = T_2 - T_1$. Experiments have shown that the rate of heat transfer $\dot{Q}$ through the wall is *doubled* when the temperature difference $\Delta T$ across the wall or the area $A$ normal to the direction of heat transfer is doubled, but is *halved* when the wall thickness $L$ is doubled. Thus we conclude that *the rate of heat conduction through a plane layer is proportional to the temperature difference across the layer and the heat transfer area, but is inversely proportional to the thickness of the layer.* That is,

$$\text{Rate of heat conduction} \propto \frac{(\text{Area})(\text{Temperature difference})}{\text{Thickness}}$$

**FIGURE 16–1**

Heat conduction through a large plane wall of thickness $\Delta x$ and area $A$.

or,

$$\dot{Q}_{cond} = kA \frac{T_1 - T_2}{\Delta x} = -kA \frac{\Delta T}{\Delta x} \qquad (W) \qquad (16\text{-}1)$$

where the constant of proportionality $k$ is the **thermal conductivity** of the material, which is a *measure of the ability of a material to conduct heat* (Fig. 16–2). In the limiting case of $\Delta x \rightarrow 0$, the equation above reduces to the differential form

$$\dot{Q}_{cond} = -kA \frac{dT}{dx} \qquad (W) \qquad (16\text{-}2)$$

which is called **Fourier's law of heat conduction** after J. Fourier who expressed it first in his heat transfer text in 1822. Here $dT/dx$ is the **temperature gradient**, which is the slope of the temperature curve on a $T$-$x$ diagram (the rate of change of $T$ with $x$), at location $x$. The relation above indicates that the rate of heat conduction in a given direction is proportional to the temperature gradient in that direction. Heat is conducted in the direction of decreasing temperature, and the temperature gradient becomes negative when temperature decreases with increasing $x$. The *negative sign* in Eq. 16–2 ensures that heat transfer in the positive $x$ direction is a positive quantity.

The heat transfer area $A$ is always *normal* to the direction of heat transfer. For heat loss through a 5-m-long, 3-m-high, and 25-cm-thick wall, for example, the heat transfer area is $A = 15 \text{ m}^2$. Note that the thickness of the wall has no effect on $A$ (Fig. 16–3).

(a) Copper ($k$ = 401 W/m·K)

(b) Silicon ($k$ = 148 W/m·K)

**FIGURE 16–2**

The rate of heat conduction through a solid is directly proportional to its thermal conductivity.

**FIGURE 16–3**

In heat conduction analysis, $A$ represents the area *normal* to the direction of heat transfer.

---

**EXAMPLE 16–1    The Cost of Heat Loss through a Roof**

The roof of an electrically heated home is 6 m long, 8 m wide, and 0.25 m thick, and is made of a flat layer of concrete whose thermal conductivity is $k$ = 0.8 W/m·K (Fig. 16–4). The temperatures of the inner and the outer surfaces of the roof one night are measured to be 15°C and 4°C, respectively, for a period of 10 hours. Determine (a) the rate of heat loss through the roof that night and (b) the cost of that heat loss to the home owner if the cost of electricity is $0.08/kWh.

**SOLUTION**   The inner and outer surfaces of the flat concrete roof of an electrically heated home are maintained at specified temperatures during a night. The heat loss through the roof and its cost that night are to be determined.

**Assumptions**   **1** Steady operating conditions exist during the entire night since the surface temperatures of the roof remain constant at the specified values. **2** Constant properties can be used for the roof.

**Properties**   The thermal conductivity of the roof is given to be $k$ = 0.8 W/m·K.

**Analysis**   (a) Noting that heat transfer through the roof is by conduction and the area of the roof is $A = 6 \text{ m} \times 8 \text{ m} = 48 \text{ m}^2$, the steady rate of heat transfer through the roof is

$$\dot{Q} = kA \frac{T_1 - T_2}{L} = (0.8 \text{ K})(48 \text{ m}^2) \frac{(15 - 4)°C}{0.25 \text{ m}} = \mathbf{1690 \text{ W} = 1.69 \text{ kW}}$$

**FIGURE 16–4**

Schematic for Example 16–1.

**TABLE 16–1**

The thermal conductivities of some materials at room temperature

| Material | $k$, W/m·K* |
|---|---|
| Diamond | 2300 |
| Silver | 429 |
| Copper | 401 |
| Gold | 317 |
| Aluminum | 237 |
| Iron | 80.2 |
| Mercury (l) | 8.54 |
| Glass | 0.78 |
| Brick | 0.72 |
| Water (l) | 0.607 |
| Human skin | 0.37 |
| Wood (oak) | 0.17 |
| Helium (g) | 0.152 |
| Soft rubber | 0.13 |
| Glass fiber | 0.043 |
| Air (g) | 0.026 |
| Urethane, rigid foam | 0.026 |

*Multiply by 0.5778 to convert to Btu/h·ft·°F.

(*b*) The amount of heat lost through the roof during a 10-hour period and its cost is

$$Q = \dot{Q}\,\Delta t = (1.69\text{ kW})(10\text{ h}) = 16.9\text{ kWh}$$

$$\text{Cost} = (\text{Amount of energy})(\text{Unit cost of energy})$$

$$= (16.9\text{ kWh})(\$0.08/\text{kWh}) = \mathbf{\$1.35}$$

**Discussion** The cost to the home owner of the heat loss through the roof that night was $1.35. The total heating bill of the house will be much larger since the heat losses through the walls are not considered in these calculations.

## Thermal Conductivity

We have seen that different materials store heat differently, and we have defined the property specific heat $c_p$ as a measure of a material's ability to store thermal energy. For example, $c_p = 4.18$ kJ/kg·°C for water and $c_p = 0.45$ kJ/kg·°C for iron at room temperature, which indicates that water can store almost 10 times the energy that iron can per unit mass. Likewise, the thermal conductivity $k$ is a measure of a material's ability to conduct heat. For example, $k = 0.607$ W/m·K for water and $k = 80.2$ W/m·K for iron at room temperature, which indicates that iron conducts heat more than 100 times faster than water can. Thus we say that water is a poor heat conductor relative to iron, although water is an excellent medium to store thermal energy.

Equation 16–1 for the rate of conduction heat transfer under steady conditions can also be viewed as the defining equation for thermal conductivity. Thus the **thermal conductivity** of a material can be defined as *the rate of heat transfer through a unit thickness of the material per unit area per unit temperature difference.* The thermal conductivity of a material is a measure of the ability of the material to conduct heat. A high value for thermal conductivity indicates that the material is a good heat conductor, and a low value indicates that the material is a poor heat conductor or *insulator.* The thermal conductivities of some common materials at room temperature are given in Table 16–1. The thermal conductivity of pure copper at room temperature is $k = 401$ W/m·K, which indicates that a 1-m-thick copper wall will conduct heat at a rate of 401 W per m² area per K temperature difference across the wall. Note that materials such as copper and silver that are good electric conductors are also good heat conductors, and have high values of thermal conductivity. Materials such as rubber, wood, and Styrofoam are poor conductors of heat and have low conductivity values.

A layer of material of known thickness and area can be heated from one side by an electric resistance heater of known output. If the outer surfaces of the heater are well insulated, all the heat generated by the resistance heater will be transferred through the material whose conductivity is to be determined. Then measuring the two surface temperatures of the material when steady heat transfer is reached and substituting them into Eq. 16–1 together with other known quantities give the thermal conductivity (Fig. 16–5).

$$k = \frac{L}{A(T_1 - T_2)}\dot{Q}$$

**FIGURE 16–5**

A simple experimental setup to determine the thermal conductivity of a material.

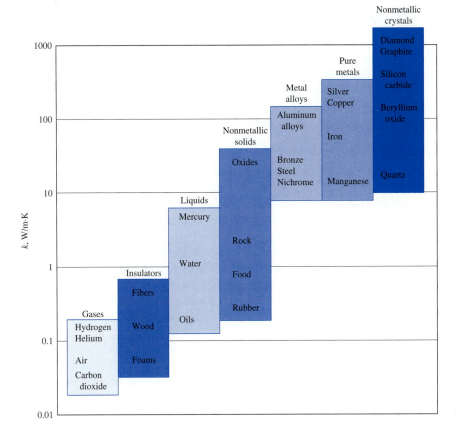

**FIGURE 16–6**

The range of thermal conductivity of various materials at room temperature.

The thermal conductivities of materials vary over a wide range, as shown in Fig. 16–6. The thermal conductivities of gases such as air vary by a factor of $10^4$ from those of pure metals such as copper. Note that pure crystals and metals have the highest thermal conductivities, and gases and insulating materials the lowest.

Temperature is a measure of the kinetic energies of the particles such as the molecules or atoms of a substance. In a liquid or gas, the kinetic energy of the molecules is due to their random translational motion as well as their vibrational and rotational motions. When two molecules possessing different kinetic energies collide, part of the kinetic energy of the more energetic (higher-temperature) molecule is transferred to the less energetic (lower-temperature) molecule, much the same as when two elastic balls of the same mass at different velocities collide, part of the kinetic energy of the faster ball is transferred to the slower one. The higher the temperature, the faster the molecules move and the higher the number of such collisions, and the better the heat transfer.

The *Kinetic theory* of gases predicts and the experiments confirm that the thermal conductivity of gases is proportional to the *square root of the thermodynamic temperature T*, and inversely proportional to the *square root of the molar mass M*. Therefore, for a particular gas (fixed *M*), the thermal

**FIGURE 16–7**

The mechanisms of heat conduction in different phases of a substance.

**TABLE 16–2**

The thermal conductivity of an alloy is usually much lower than the thermal conductivity of either metal of which it is composed

| Pure metal or alloy | $k$, W/m·K, at 300 K |
|---|---|
| Copper | 401 |
| Nickel | 91 |
| *Constantan* (55% Cu, 45% Ni) | 23 |
| Copper | 401 |
| Aluminum | 237 |
| *Commercial bronze* (90% Cu, 10% Al) | 52 |

conductivity increases with increasing temperature and at a fixed temperature the thermal conductivity decreases with increasing $M$. For example, at a fixed temperature of 1000 K, the thermal conductivity of helium ($M = 4$) is 0.343 W/m·K and that of air ($M = 29$) is 0.0667 W/m·K, which is much lower than that of helium.

The thermal conductivities of *gases* at 1 atm pressure are listed in Table A–23. However, they can also be used at pressures other than 1 atm, since the thermal conductivity of gases is *independent of pressure* in a wide range of pressures encountered in practice.

The mechanism of heat conduction in a *liquid* is complicated by the fact that the molecules are more closely spaced, and they exert a stronger intermolecular force field. The thermal conductivities of liquids usually lie between those of solids and gases. The thermal conductivity of a substance is normally highest in the solid phase and lowest in the gas phase. The thermal conductivity of liquids is generally insensitive to pressure except near the thermodynamic critical point. Unlike gases, the thermal conductivities of most liquids decrease with increasing temperature, with water being a notable exception. Like gases, the conductivity of liquids decreases with increasing molar mass. Liquid metals such as mercury and sodium have high thermal conductivities and are very suitable for use in applications where a high heat transfer rate to a liquid is desired, as in nuclear power plants.

In *solids,* heat conduction is due to two effects: the *lattice vibrational waves* induced by the vibrational motions of the molecules positioned at relatively fixed positions in a periodic manner called a lattice, and the energy transported via the *free flow of electrons* in the solid (Fig. 16–7). The thermal conductivity of a solid is obtained by adding the lattice and electronic components. The relatively high thermal conductivities of pure metals are primarily due to the electronic component. The lattice component of thermal conductivity strongly depends on the way the molecules are arranged. For example, diamond, which is a highly ordered crystalline solid, has the highest known thermal conductivity at room temperature.

Unlike metals, which are good electrical and heat conductors, *crystalline solids* such as diamond and semiconductors such as silicon are good heat conductors but poor electrical conductors. As a result, such materials find widespread use in the electronics industry. Despite their higher price, diamond heat sinks are used in the cooling of sensitive electronic components because of the excellent thermal conductivity of diamond. Silicon oils and gaskets are commonly used in the packaging of electronic components because they provide both good thermal contact and good electrical insulation.

Pure metals have high thermal conductivities, and one would think that *metal alloys* should also have high conductivities. One would expect an alloy made of two metals of thermal conductivities $k_1$ and $k_2$ to have a conductivity $k$ between $k_1$ and $k_2$. But this turns out not to be the case. The thermal conductivity of an alloy of two metals is usually much lower than that of either metal, as shown in Table 16–2. Even small amounts in a pure metal of "foreign" molecules that are good conductors themselves seriously disrupt the transfer of heat in that metal. For example, the thermal conductivity of steel containing just 1 percent of chrome is 62 W/m·K, while the thermal conductivities of iron and chromium are 83 and 95 W/m·K, respectively.

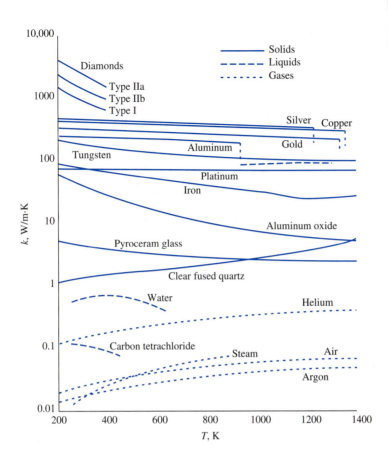

**TABLE 16–3**

Thermal conductivities of materials vary with temperature

| | k, W/m·K | |
| --- | --- | --- |
| T, K | Copper | Aluminum |
| 100 | 482 | 302 |
| 200 | 413 | 237 |
| 300 | 401 | 237 |
| 400 | 393 | 240 |
| 600 | 379 | 231 |
| 800 | 366 | 218 |

**FIGURE 16–8**

The variation of the thermal conductivity of various solids, liquids, and gases with temperature.

The thermal conductivities of materials vary with temperature (Table 16–3). The variation of thermal conductivity over certain temperature ranges is negligible for some materials, but significant for others, as shown in Fig. 16–8. The thermal conductivities of certain solids exhibit dramatic increases at temperatures near absolute zero, when these solids become *superconductors*. For example, the conductivity of copper reaches a maximum value of about 20,000 W/m·K at 20 K, which is about 50 times the conductivity at room temperature. The thermal conductivities and other thermal properties of various materials are given in Tables A–15 to A–25.

Values of thermal conductivity given in Tables A–15 to A–25 are appropriate when the physical dimensions of the material under consideration are relatively large. In some of the emerging areas of technology, such as microelectronics, the physical dimensions are typically in the order of micro or nano meters. For these applications the small physical dimensions most likely will influence the value of thermal conductivity in the solid and liquid states. In these situations, as the physical dimensions decrease, the average net distance traveled by the energy carriers typically decreases and this reduces the value of thermal conductivity.

The temperature dependence of thermal conductivity causes considerable complexity in conduction analysis. Therefore, it is common practice to evaluate the thermal conductivity k at the *average temperature* and treat it as a *constant* in calculations.

**TABLE 16–4**

The thermal diffusivities of some materials at room temperature

| Material | $\alpha$, m²/s* |
|---|---|
| Silver | $149 \times 10^{-6}$ |
| Gold | $127 \times 10^{-6}$ |
| Copper | $113 \times 10^{-6}$ |
| Aluminum | $97.5 \times 10^{-6}$ |
| Iron | $22.8 \times 10^{-6}$ |
| Mercury (l) | $4.7 \times 10^{-6}$ |
| Marble | $1.2 \times 10^{-6}$ |
| Ice | $1.2 \times 10^{-6}$ |
| Concrete | $0.75 \times 10^{-6}$ |
| Brick | $0.52 \times 10^{-6}$ |
| Heavy soil (dry) | $0.52 \times 10^{-6}$ |
| Glass | $0.34 \times 10^{-6}$ |
| Glass wool | $0.23 \times 10^{-6}$ |
| Water (l) | $0.14 \times 10^{-6}$ |
| Beef | $0.14 \times 10^{-6}$ |
| Wood (oak) | $0.13 \times 10^{-6}$ |

*Multiply by 10.76 to convert to ft²/s.

In heat transfer analysis, a material is normally assumed to be *isotropic;* that is, to have uniform properties in all directions. This assumption is realistic for most materials, except those that exhibit different structural characteristics in different directions, such as laminated composite materials and wood. The thermal conductivity of wood across the grain, for example, is different than that parallel to the grain.

## Thermal Diffusivity

The product $\rho c_p$, which is frequently encountered in heat transfer analysis, is called the **heat capacity** of a material. Both the specific heat $c_p$ and the heat capacity $\rho c_p$ represent the heat storage capability of a material. But $c_p$ expresses it *per unit mass* whereas $\rho c_p$ expresses it *per unit volume,* as can be noticed from their units J/kg·K and J/m³·K, respectively.

Another material property that appears in the transient heat conduction analysis is the **thermal diffusivity**, which represents how fast heat diffuses through a material and is defined as

$$\alpha = \frac{\text{Heat conduction}}{\text{Heat storage}} = \frac{k}{\rho c_p} \quad \text{(m}^2\text{/s)} \quad \text{(16–3)}$$

Note that the thermal conductivity $k$ represents how well a material conducts heat, and the heat capacity $\rho c_p$ represents how much energy a material stores per unit volume. Therefore, the thermal diffusivity of a material can be viewed as the ratio of the *heat conducted* through the material to the *heat stored* per unit volume. A material that has a high thermal conductivity or a low heat capacity will obviously have a large thermal diffusivity. The larger the thermal diffusivity, the faster the propagation of heat into the medium. A small value of thermal diffusivity means that heat is mostly absorbed by the material and a small amount of heat is conducted further.

The thermal diffusivities of some common materials at 20°C are given in Table 16–4. Note that the thermal diffusivity ranges from $\alpha = 0.14 \times 10^{-6}$ m²/s for water to $149 \times 10^{-6}$ m²/s for silver, which is a difference of more than a thousand times. Also note that the thermal diffusivities of beef and water are the same. This is not surprising, since meat as well as fresh vegetables and fruits are mostly water, and thus they possess the thermal properties of water.

**FIGURE 16–9**

Apparatus to measure the thermal conductivity of a material using two identical samples and a thin resistance heater (Example 16–2).

### EXAMPLE 16–2    Measuring the Thermal Conductivity of a Material

A common way of measuring the thermal conductivity of a material is to sandwich an electric thermofoil heater between two identical samples of the material, as shown in Fig. 16–9. The thickness of the resistance heater, including its cover, which is made of thin silicon rubber, is usually less than 0.5 mm. A circulating fluid such as tap water keeps the exposed ends of the samples at constant temperature. The lateral surfaces of the samples are well insulated to ensure that heat transfer through the samples is one-dimensional. Two thermocouples are embedded into each sample some distance $L$ apart, and a differential thermometer reads the temperature drop $\Delta T$ across this distance along each sample. When steady operating conditions are reached, the total rate of heat transfer through both samples becomes equal to the electric power drawn by the heater.

In a certain experiment, cylindrical samples of diameter 5 cm and length 10 cm are used. The two thermocouples in each sample are placed 3 cm apart. After initial transients, the electric heater is observed to draw 0.4 A at 110 V, and both differential thermometers read a temperature difference of 15°C. Determine the thermal conductivity of the sample.

**SOLUTION**  The thermal conductivity of a material is to be determined by ensuring one-dimensional heat conduction, and by measuring temperatures when steady operating conditions are reached.

*Assumptions*  **1** Steady operating conditions exist since the temperature readings do not change with time. **2** Heat losses through the lateral surfaces of the apparatus are negligible since those surfaces are well insulated, and thus the entire heat generated by the heater is conducted through the samples. **3** The apparatus possesses thermal symmetry.

*Analysis*  The electrical power consumed by the resistance heater and converted to heat is

$$\dot{W}_e = \mathbf{V}I = (110 \text{ V})(0.4 \text{ A}) = 44 \text{ W}$$

The rate of heat flow through each sample is

$$\dot{Q} = \tfrac{1}{2}\dot{W}_e = \tfrac{1}{2} \times (44 \text{ W}) = 22 \text{ W}$$

since only half of the heat generated flows through each sample because of symmetry. Reading the same temperature difference across the same distance in each sample also confirms that the apparatus possesses thermal symmetry. The heat transfer area is the area normal to the direction of heat transfer, which is the cross-sectional area of the cylinder in this case:

$$A = \tfrac{1}{4}\pi D^2 = \tfrac{1}{4}\pi(0.05 \text{ m})^2 = 0.001963 \text{ m}^2$$

Noting that the temperature drops by 15°C within 3 cm in the direction of heat flow, the thermal conductivity of the sample is determined to be

$$\dot{Q} = kA \frac{\Delta T}{L} \quad \rightarrow \quad k = \frac{\dot{Q} L}{A \, \Delta T} = \frac{(22 \text{ W})(0.03 \text{ m})}{(0.001963 \text{ m}^2)(15°C)} = \mathbf{22.4 \text{ W/m·K}}$$

*Discussion*  Perhaps you are wondering if we really need to use two samples in the apparatus, since the measurements on the second sample do not give any additional information. It seems like we can replace the second sample by insulation. Indeed, we do not need the second sample; however, it enables us to verify the temperature measurements on the first sample and provides thermal symmetry, which reduces experimental error.

**EXAMPLE 16–3** **Conversion between SI and English Units**

An engineer who is working on the heat transfer analysis of a brick building in English units needs the thermal conductivity of brick. But the only value he can find from his handbooks is 0.72 W/m·°C, which is in SI units. To make matters worse, the engineer does not have a direct conversion factor between the two unit systems for thermal conductivity. Can you help him out?

**SOLUTION** The situation this engineer is facing is not unique, and most engineers often find themselves in a similar position. A person must be very careful during unit conversion not to fall into some common pitfalls and to avoid some costly mistakes. Although unit conversion is a simple process, it requires utmost care and careful reasoning.

The conversion factors for W and m are straightforward and are given in conversion tables to be

$$1 \text{ W} = 3.41214 \text{ Btu/h}$$
$$1 \text{ m} = 3.2808 \text{ ft}$$

But the conversion of °C into °F is not so simple, and it can be a source of error if one is not careful. Perhaps the first thought that comes to mind is to replace °C by (°F − 32)/1.8 since $T(°C) = [T(°F) − 32]/1.8$. But this will be wrong since the °C in the unit W/m·°C represents *per °C change in temperature*. Noting that 1°C change in temperature corresponds to 1.8°F, the proper conversion factor to be used is

$$1°C = 1.8°F$$

Substituting, we get

$$1 \text{ W/m·°C} = \frac{3.41214 \text{ Btu/h}}{(3.2808 \text{ ft})(1.8°F)} = 0.5778 \text{ Btu/h·ft·°F}$$

which is the desired conversion factor. Therefore, the thermal conductivity of the brick in English units is

$$k_{\text{brick}} = 0.72 \text{ W/m·°C}$$
$$= 0.72 \times (0.5778 \text{ Btu/h·ft·°F})$$
$$= \mathbf{0.42 \text{ Btu/h·ft·°F}}$$

***Discussion*** Note that the thermal conductivity value of a material in English units is about half that in SI units (Fig. 16–10). Also note that we rounded the result to two significant digits (the same number in the original value) since expressing the result in more significant digits (such as 0.4160 instead of 0.42) would falsely imply a more accurate value than the original one.

$k = 0.72$ W/m·°C
$= 0.42$ Btu/h·ft·°F

**FIGURE 16–10**

The thermal conductivity value in English units is obtained by multiplying the value in SI units by 0.5778.

## 16–3 · CONVECTION

**Convection** is the mode of energy transfer between a solid surface and the adjacent liquid or gas that is in motion, and it involves the combined effects of *conduction* and *fluid motion*. The faster the fluid motion, the greater the convection heat transfer. In the absence of any bulk fluid motion, heat transfer between a solid surface and the adjacent fluid is by pure conduction. The presence of bulk motion of the fluid enhances the heat transfer between the solid surface and the fluid, but it also complicates the determination of heat transfer rates.

Consider the cooling of a hot block by blowing cool air over its top surface (Fig. 16–11). Heat is first transferred to the air layer adjacent to the block by conduction. This heat is then carried away from the surface by convection, that is, by the combined effects of conduction within the air that is due to random motion of air molecules and the bulk or macroscopic motion of the air that removes the heated air near the surface and replaces it by the cooler air.

Convection is called **forced convection** if the fluid is forced to flow over the surface by external means such as a fan, pump, or the wind. In contrast, convection is called **natural** (or **free**) **convection** if the fluid motion is caused by buoyancy forces that are induced by density differences due to the variation of temperature in the fluid (Fig. 16–12). For example, in the absence of a fan, heat transfer from the surface of the hot block in Fig. 16–11 is by natural convection since any motion in the air in this case is due to the rise of the warmer (and thus lighter) air near the surface and the fall of the cooler (and thus heavier) air to fill its place. Heat transfer between the block and the surrounding air is by conduction if the temperature difference between the air and the block is not large enough to overcome the resistance of air to movement and thus to initiate natural convection currents.

Heat transfer processes that involve *change of phase* of a fluid are also considered to be convection because of the fluid motion induced during the process, such as the rise of the vapor bubbles during boiling or the fall of the liquid droplets during condensation.

Despite the complexity of convection, the rate of *convection heat transfer* is observed to be proportional to the temperature difference, and is conveniently expressed by **Newton's law of cooling**

$$\dot{Q}_{conv} = hA_s (T_s - T_\infty) \qquad \text{(W)} \qquad \text{(16–4)}$$

where $h$ is the *convection heat transfer coefficient* in W/m²·K or Btu/h·ft²·°F, $A_s$ is the surface area through which convection heat transfer takes place, $T_s$ is the surface temperature, and $T_\infty$ is the temperature of the fluid sufficiently far from the surface. Note that at the surface, the fluid temperature equals the surface temperature of the solid.

The convection heat transfer coefficient $h$ is not a property of the fluid. It is an experimentally determined parameter whose value depends on all the variables influencing convection such as the surface geometry, the nature of fluid motion, the properties of the fluid, and the bulk fluid velocity. Typical values of $h$ are given in Table 16–5.

Some people do not consider convection to be a fundamental mechanism of heat transfer since it is essentially heat conduction in the presence of fluid motion. But we still need to give this combined phenomenon a name, unless we are willing to keep referring to it as "conduction with fluid motion." Thus, it is practical to recognize convection as a separate heat transfer mechanism despite the valid arguments to the contrary.

### EXAMPLE 16–4    Measuring Convection Heat Transfer Coefficient

A 2-m-long, 0.3-cm-diameter electrical wire extends across a room at 15°C, as shown in Fig. 16–13. Heat is generated in the wire as a result of resistance heating, and the surface temperature of the wire is measured to be 152°C in

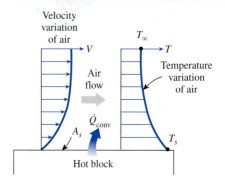

**FIGURE 16–11**

Heat transfer from a hot surface to air by convection.

**FIGURE 16–12**

The cooling of a boiled egg by forced and natural convection.

**TABLE 16–5**

Typical values of convection heat transfer coefficient

| Type of convection | $h$, W/m²·K* |
|---|---|
| Free convection of gases | 2–25 |
| Free convection of liquids | 10–1000 |
| Forced convection of gases | 25–250 |
| Forced convection of liquids | 50–20,000 |
| Boiling and condensation | 2500–100,000 |

*Multiply by 0.176 to convert to Btu/h·ft²·°F.

**FIGURE 16–13**

Schematic for Example 16–4.

steady operation. Also, the voltage drop and electric current through the wire are measured to be 60 V and 1.5 A, respectively. Disregarding any heat transfer by radiation, determine the convection heat transfer coefficient for heat transfer between the outer surface of the wire and the air in the room.

**SOLUTION** The convection heat transfer coefficient for heat transfer from an electrically heated wire to air is to be determined by measuring temperatures when steady operating conditions are reached and the electric power consumed.

**Assumptions** **1** Steady operating conditions exist since the temperature readings do not change with time. **2** Radiation heat transfer is negligible.

**Analysis** When steady operating conditions are reached, the rate of heat loss from the wire equals the rate of heat generation in the wire as a result of resistance heating. That is,

$$\dot{Q} = \dot{E}_{generated} = \mathbf{V}I = (60 \text{ V})(1.5 \text{ A}) = 90 \text{ W}$$

The surface area of the wire is

$$A_s = \pi DL = \pi (0.003 \text{ m})(2 \text{ m}) = 0.01885 \text{ m}^2$$

Newton's law of cooling for convection heat transfer is expressed as

$$\dot{Q}_{conv} = hA_s (T_s - T_\infty)$$

Disregarding any heat transfer by radiation and thus assuming all the heat loss from the wire to occur by convection, the convection heat transfer coefficient is determined to be

$$h = \frac{\dot{Q}_{conv}}{A_s(T_s - T_\infty)} = \frac{90 \text{ W}}{(0.01885 \text{ m}^2)(152 - 15)°C} = \mathbf{34.9 \ W/m^2·K}$$

**Discussion** Note that the simple setup described above can be used to determine the average heat transfer coefficients from a variety of surfaces in air. Also, heat transfer by radiation can be eliminated by keeping the surrounding surfaces at the temperature of the wire.

# 16–4 · RADIATION

**Radiation** is the energy emitted by matter in the form of *electromagnetic waves* (or *photons*) as a result of the changes in the electronic configurations of the atoms or molecules. Unlike conduction and convection, the transfer of heat by radiation does not require the presence of an *intervening medium*. In fact, heat transfer by radiation is fastest (at the speed of light) and it suffers no attenuation in a vacuum. This is how the energy of the sun reaches the earth.

In heat transfer studies we are interested in *thermal radiation*, which is the form of radiation emitted by bodies because of their temperature. It differs from other forms of electromagnetic radiation such as x-rays, gamma rays, microwaves, radio waves, and television waves that are not related to temperature. All bodies at a temperature above absolute zero emit thermal radiation.

Radiation is a *volumetric phenomenon*, and all solids, liquids, and gases emit, absorb, or transmit radiation to varying degrees. However, radiation is usually considered to be a *surface phenomenon* for solids that are opaque to thermal radiation such as metals, wood, and rocks since the radiation emitted by the interior regions of such material can never reach the surface, and the radiation incident on such bodies is usually absorbed within a few microns from the surface.

The maximum rate of radiation that can be emitted from a surface at a thermodynamic temperature $T_s$ (in K or R) is given by the **Stefan–Boltzmann law** as

$$\dot{Q}_{emit,\,max} = \sigma A_s T_s^4 \qquad (\text{W}) \qquad (16\text{–}5)$$

where $\sigma = 5.670 \times 10^{-8}$ W/m²·K⁴ or $0.1714 \times 10^{-8}$ Btu/h·ft²·R⁴ is the *Stefan–Boltzmann constant*. The idealized surface that emits radiation at this maximum rate is called a **blackbody**, and the radiation emitted by a blackbody is called **blackbody radiation** (Fig. 16–14). The radiation emitted by all real surfaces is less than the radiation emitted by a blackbody at the same temperature, and is expressed as

$$\dot{Q}_{emit} = \varepsilon \sigma A_s T_s^4 \qquad (\text{W}) \qquad (16\text{–}6)$$

where $\varepsilon$ is the **emissivity** of the surface. The property emissivity, whose value is in the range $0 \le \varepsilon \le 1$, is a measure of how closely a surface approximates a blackbody for which $\varepsilon = 1$. The emissivities of some surfaces are given in Table 16–6.

Another important radiation property of a surface is its **absorptivity** $\alpha$, which is the fraction of the radiation energy incident on a surface that is absorbed by the surface. Like emissivity, its value is in the range $0 \le \alpha \le 1$. A blackbody absorbs the entire radiation incident on it. That is, a blackbody is a perfect absorber ($\alpha = 1$) as it is a perfect emitter.

In general, both $\varepsilon$ and $\alpha$ of a surface depend on the temperature and the wavelength of the radiation. **Kirchhoff's law** of radiation states that the emissivity and the absorptivity of a surface at a given temperature and wavelength are equal. In many practical applications, the surface temperature and the temperature of the source of incident radiation are of the same order of magnitude, and the average absorptivity of a surface is taken to be equal to its average emissivity. The rate at which a surface absorbs radiation is determined from (Fig. 16–15)

$$\dot{Q}_{absorbed} = \alpha \dot{Q}_{incident} \qquad (\text{W}) \qquad (16\text{–}7)$$

where $\dot{Q}_{incident}$ is the rate at which radiation is incident on the surface and $\alpha$ is the absorptivity of the surface. For opaque (nontransparent) surfaces, the portion of incident radiation not absorbed by the surface is reflected back.

The difference between the rates of radiation emitted by the surface and the radiation absorbed is the *net* radiation heat transfer. If the rate of radiation absorption is greater than the rate of radiation emission, the surface is said to be *gaining* energy by radiation. Otherwise, the surface is said to be *losing* energy by radiation. In general, the determination of the net rate of heat transfer by radiation between two surfaces is a complicated matter since it depends on the properties of the surfaces, their orientation relative to each other, and the interaction of the medium between the surfaces with radiation.

**FIGURE 16–14**

Blackbody radiation represents the *maximum amount of radiation that can be emitted from a surface at a specified temperature.*

**TABLE 16–6**

Emissivities of some materials at 300 K

| Material | Emissivity |
| --- | --- |
| Aluminum foil | 0.07 |
| Anodized aluminum | 0.82 |
| Polished copper | 0.03 |
| Polished gold | 0.03 |
| Polished silver | 0.02 |
| Polished stainless steel | 0.17 |
| Black paint | 0.98 |
| White paint | 0.90 |
| White paper | 0.92–0.97 |
| Asphalt pavement | 0.85–0.93 |
| Red brick | 0.93–0.96 |
| Human skin | 0.95 |
| Wood | 0.82–0.92 |
| Soil | 0.93–0.96 |
| Water | 0.96 |
| Vegetation | 0.92–0.96 |

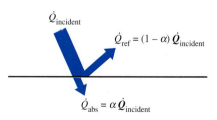

**FIGURE 16–15**

The absorption of radiation incident on an opaque surface of absorptivity $\alpha$.

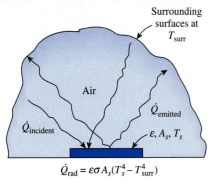

$$\dot{Q}_{rad} = \varepsilon\sigma A_s(T_s^4 - T_{surr}^4)$$

**FIGURE 16–16**

Radiation heat transfer between a surface and the surfaces surrounding it.

When a surface of emissivity $\varepsilon$ and surface area $A_s$ at a *thermodynamic temperature* $T_s$ is *completely enclosed* by a much larger (or black) surface at thermodynamic temperature $T_{surr}$ separated by a gas (such as air) that does not intervene with radiation, the net rate of radiation heat transfer between these two surfaces is given by (Fig. 16–16)

$$\dot{Q}_{rad} = \varepsilon\sigma A_s\,(T_s^4 - T_{surr}^4) \qquad (W) \qquad \text{(16–8)}$$

In this special case, the emissivity and the surface area of the surrounding surface do not have any effect on the net radiation heat transfer.

Radiation heat transfer to or from a surface surrounded by a gas such as air occurs *parallel* to conduction (or convection, if there is bulk gas motion) between the surface and the gas. Thus the total heat transfer is determined by *adding* the contributions of both heat transfer mechanisms. For simplicity and convenience, this is often done by defining a **combined heat transfer coefficient** $h_{combined}$ that includes the effects of both convection and radiation. Then the *total* heat transfer rate to or from a surface by convection and radiation is expressed as

$$\dot{Q}_{total} = \dot{Q}_{conv} + \dot{Q}_{rad} = h_{conv}A_s\,(T_s - T_{surr}) + \varepsilon\sigma A_s\,(T_s^4 - T_{surr}^4)$$

$$\dot{Q}_{total} = h_{combined}A_s\,(T_s - T_\infty) \qquad (W) \qquad \text{(16–9)}$$

$$h_{combined} = h_{conv} + h_{rad} = h_{conv} + \varepsilon\sigma\,(T_s + T_{surr})(T_s^2 + T_{surr}^2)$$

Note that the combined heat transfer coefficient is essentially a convection heat transfer coefficient modified to include the effects of radiation.

Radiation is usually significant relative to conduction or natural convection, but negligible relative to forced convection. Thus radiation in forced convection applications is usually disregarded, especially when the surfaces involved have low emissivities and low to moderate temperatures.

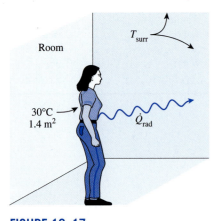

**FIGURE 16–17**

Schematic for Example 16–5.

**EXAMPLE 16–5**   **Radiation Effect on Thermal Comfort**

It is a common experience to feel "chilly" in winter and "warm" in summer in our homes even when the thermostat setting is kept the same. This is due to the so called "radiation effect" resulting from radiation heat exchange between our bodies and the surrounding surfaces of the walls and the ceiling.

Consider a person standing in a room maintained at 22°C at all times. The inner surfaces of the walls, floors, and the ceiling of the house are observed to be at an average temperature of 10°C in winter and 25°C in summer. Determine the rate of radiation heat transfer between this person and the surrounding surfaces if the exposed surface area and the average outer surface temperature of the person are 1.4 m² and 30°C, respectively (Fig. 16–17).

**SOLUTION**   The rates of radiation heat transfer between a person and the surrounding surfaces at specified temperatures are to be determined in summer and winter.

**Assumptions**   **1** Steady operating conditions exist. **2** Heat transfer by convection is not considered. **3** The person is completely surrounded by the interior surfaces of the room. **4** The surrounding surfaces are at a uniform temperature.

**Properties**   The emissivity of a person is $\varepsilon = 0.95$ (Table 16–6).

**Analysis** The net rates of radiation heat transfer from the body to the surrounding walls, ceiling, and floor in winter and summer are

$$\dot{Q}_{rad,\,winter} = \varepsilon\sigma A_s\,(T_s^4 - T_{surr,\,winter}^4)$$
$$= (0.95)(5.67 \times 10^{-8}\ W/m^2{\cdot}K^4)(1.4\ m^2)$$
$$\times\ [(30 + 273)^4 - (10 + 273)^4]\ K^4$$
$$= \mathbf{152\ W}$$

and

$$\dot{Q}_{rad,\,summer} = \varepsilon\sigma A_s\,(T_s^4 - T_{surr,\,summer}^4)$$
$$= (0.95)(5.67 \times 10^{-8}\ W/m^2{\cdot}K^4)(1.4\ m^2)$$
$$\times\ [(30 + 273)^4 - (25 + 273)^4]\ K^4$$
$$= \mathbf{40.9\ W}$$

**Discussion** Note that we must use *thermodynamic (i.e., absolute) temperatures* in radiation calculations. Also note that the rate of heat loss from the person by radiation is almost four times as large in winter than it is in summer, which explains the "chill" we feel in winter even if the thermostat setting is kept the same.

## 16–5 ▪ SIMULTANEOUS HEAT TRANSFER MECHANISMS

We mentioned that there are three mechanisms of heat transfer, but not all three can exist simultaneously in a medium. For example, heat transfer is only by conduction in *opaque solids,* but by conduction and radiation in *semitransparent solids.* Thus, a solid may involve conduction and radiation but not convection. However, a solid may involve heat transfer by convection and/or radiation on its surfaces exposed to a fluid or other surfaces. For example, the outer surfaces of a cold piece of rock will warm up in a warmer environment as a result of heat gain by convection (from the air) and radiation (from the sun or the warmer surrounding surfaces). But the inner parts of the rock will warm up as this heat is transferred to the inner region of the rock by conduction.

Heat transfer is by conduction and possibly by radiation in a *still fluid* (no bulk fluid motion) and by convection and radiation in a *flowing fluid.* In the absence of radiation, heat transfer through a fluid is either by conduction or convection, depending on the presence of any bulk fluid motion. Convection can be viewed as combined conduction and fluid motion, and conduction in a fluid can be viewed as a special case of convection in the absence of any fluid motion (Fig. 16–18).

Thus, when we deal with heat transfer through a *fluid,* we have either *conduction* or *convection,* but not both. Also, gases are practically transparent to radiation, except that some gases are known to absorb radiation strongly at certain wavelengths. Ozone, for example, strongly absorbs ultraviolet radiation. But in most cases, a gas between two solid surfaces does not interfere with radiation and acts effectively as a vacuum. Liquids, on the other hand, are usually strong absorbers of radiation.

Finally, heat transfer through a *vacuum* is by radiation only since conduction or convection requires the presence of a material medium.

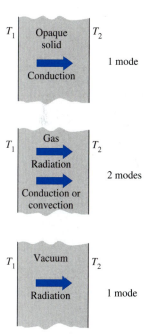

**FIGURE 16–18**

Although there are three mechanisms of heat transfer, a medium may involve only two of them simultaneously.

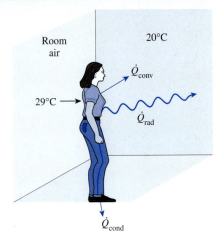

**FIGURE 16–19**

Heat transfer from the person
described in Example 16–6.

**EXAMPLE 16–6**    **Heat Loss from a Person**

Consider a person standing in a breezy room at 20°C. Determine the total rate of heat transfer from this person if the exposed surface area and the average outer surface temperature of the person are 1.6 m² and 29°C, respectively, and the convection heat transfer coefficient is 6 W/m²·K (Fig. 16–19).

**SOLUTION**    The total rate of heat transfer from a person by both convection and radiation to the surrounding air and surfaces at specified temperatures is to be determined.

**Assumptions**   **1** Steady operating conditions exist. **2** The person is completely surrounded by the interior surfaces of the room. **3** The surrounding surfaces are at the same temperature as the air in the room. **4** Heat conduction to the floor through the feet is negligible.

**Properties**   The emissivity of a person is $\varepsilon = 0.95$ (Table 16–6).

**Analysis**   The heat transfer between the person and the air in the room is by convection (instead of conduction) since it is conceivable that the air in the vicinity of the skin or clothing warms up and rises as a result of heat transfer from the body, initiating natural convection currents. It appears that the experimentally determined value for the rate of convection heat transfer in this case is 6 W per unit surface area (m²) per unit temperature difference (in K or °C) between the person and the air away from the person. Thus, the rate of convection heat transfer from the person to the air in the room is

$$\dot{Q}_{conv} = hA_s \, (T_s - T_\infty)$$
$$= (6 \text{ W/m}^2\text{·K})(1.6 \text{ m}^2)(29 - 20)\text{°C}$$
$$= 86.4 \text{ W}$$

The person also loses heat by radiation to the surrounding wall surfaces. We take the temperature of the surfaces of the walls, ceiling, and floor to be equal to the air temperature in this case for simplicity, but we recognize that this does not need to be the case. These surfaces may be at a higher or lower temperature than the average temperature of the room air, depending on the outdoor conditions and the structure of the walls. Considering that air does not intervene with radiation and the person is completely enclosed by the surrounding surfaces, the net rate of radiation heat transfer from the person to the surrounding walls, ceiling, and floor is

$$\dot{Q}_{rad} = \varepsilon\sigma A_s \, (T_s^4 - T_{surr}^4)$$
$$= (0.95)(5.67 \times 10^{-8} \text{ W/m}^2\text{·K}^4)(1.6 \text{ m}^2)$$
$$\times \, [(29 + 273)^4 - (20 + 273)^4] \text{ K}^4$$
$$= 81.7 \text{ W}$$

Note that we must use *thermodynamic* temperatures in radiation calculations. Also note that we used the emissivity value for the skin and clothing at room temperature since the emissivity is not expected to change significantly at a slightly higher temperature.

Then the rate of total heat transfer from the body is determined by adding these two quantities:

$$\dot{Q}_{total} = \dot{Q}_{conv} + \dot{Q}_{rad} = (86.4 + 81.7) \text{ W} \cong \textbf{168 W}$$

*Discussion*  The heat transfer would be much higher if the person were not dressed since the exposed surface temperature would be higher. Thus, an important function of the clothes is to serve as a barrier against heat transfer.

In these calculations, heat transfer through the feet to the floor by conduction, which is usually very small, is neglected. Heat transfer from the skin by perspiration, which is the dominant mode of heat transfer in hot environments, is not considered here.

Also, the units W/m²·°C and W/m²·K for heat transfer coefficient are equivalent, and can be interchanged.

---

### EXAMPLE 16–7    Heat Transfer between Two Isothermal Plates

Consider steady heat transfer between two large parallel plates at constant temperatures of $T_1 = 300$ K and $T_2 = 200$ K that are $L = 1$ cm apart, as shown in Fig. 16–20. Assuming the surfaces to be black (emissivity $\varepsilon = 1$), determine the rate of heat transfer between the plates per unit surface area assuming the gap between the plates is (*a*) filled with atmospheric air, (*b*) evacuated, (*c*) filled with urethane insulation, and (*d*) filled with superinsulation that has an apparent thermal conductivity of 0.00002 W/m·K.

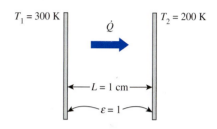

**FIGURE 16–20**

Schematic for Example 16–7.

**SOLUTION**  The total rate of heat transfer between two large parallel plates at specified temperatures is to be determined for four different cases.

*Assumptions*  **1** Steady operating conditions exist. **2** There are no natural convection currents in the air between the plates. **3** The surfaces are black and thus $\varepsilon = 1$.

*Properties*  The thermal conductivity at the average temperature of 250 K is $k = 0.0219$ W/m·K for air (Table A–22), 0.026 W/m·K for urethane insulation, and 0.00002 W/m·K for the superinsulation.

*Analysis*  (*a*) The rates of conduction and radiation heat transfer between the plates through the air layer are

$$\dot{Q}_{cond} = kA\frac{T_1 - T_2}{L} = (0.0219 \text{ W/m·K})(1 \text{ m}^2)\frac{(300 - 200)\text{K}}{0.01 \text{ m}} = 219 \text{ W}$$

and

$$\dot{Q}_{rad} = \varepsilon\sigma A(T_1^4 - T_2^4)$$
$$= (1)(5.67 \times 10^{-8} \text{ W/m}^2\text{·K}^4)(1 \text{ m}^2)[(300 \text{ K})^4 - (200 \text{ K})^4] = 369 \text{ W}$$

Therefore,

$$\dot{Q}_{total} = \dot{Q}_{cond} + \dot{Q}_{rad} = 219 + 369 = \mathbf{588 \ W}$$

The heat transfer rate in reality will be higher because of the natural convection currents that are likely to occur in the air space between the plates.

(*b*) When the air space between the plates is evacuated, there will be no conduction or convection, and the only heat transfer between the plates will be by radiation. Therefore,

$$\dot{Q}_{total} = \dot{Q}_{rad} = \mathbf{369 \ W}$$

(*c*) An opaque solid material placed between two plates blocks direct radiation heat transfer between the plates. Also, the thermal conductivity of an insulating material accounts for the radiation heat transfer that may be occurring

300 K      200 K     300 K      200 K     300 K      200 K     300 K      200 K

$\dot{Q} = 588$ W        $\dot{Q} = 369$ W        $\dot{Q} = 260$ W        $\dot{Q} = 0.2$ W

← 1 cm →        ← 1 cm →        ← 1 cm →        ← 1 cm →

(a) Air space       (b) Vacuum       (c) Insulation       (d) Superinsulation

**FIGURE 16–21**

Different ways of reducing heat transfer between two isothermal plates, and their effectiveness.

through the voids in the insulating material. The rate of heat transfer through the urethane insulation is

$$\dot{Q}_{total} = \dot{Q}_{cond} = kA \frac{T_1 - T_2}{L} = (0.026 \text{ W/m·K})(1 \text{ m}^2)\frac{(300 - 200)\text{K}}{0.01 \text{ m}} = \textbf{260 W}$$

Note that heat transfer through the urethane material is less than the heat transfer through the air determined in (a), although the thermal conductivity of the insulation is higher than that of air. This is because the insulation blocks the radiation whereas air transmits it.

(d) The layers of the superinsulation prevent any direct radiation heat transfer between the plates. However, radiation heat transfer between the sheets of superinsulation does occur, and the apparent thermal conductivity of the superinsulation accounts for this effect. Therefore,

$$\dot{Q}_{total} = kA \frac{T_1 - T_2}{L} = (0.00002 \text{ W/m·K})(1 \text{ m}^2)\frac{(300 - 200)\text{K}}{0.01 \text{ m}} = \textbf{0.2 W}$$

which is $\frac{1}{1845}$ of the heat transfer through the vacuum. The results of this example are summarized in Fig. 16–21 to put them into perspective.

**Discussion** This example demonstrates the effectiveness of superinsulations and explains why they are the insulation of choice in critical applications despite their high cost.

**FIGURE 16–22**

A chicken being cooked in a microwave oven (Example 16–8).

**EXAMPLE 16–8**    **Heat Transfer in Conventional and Microwave Ovens**

The fast and efficient cooking of microwave ovens made them one of the essential appliances in modern kitchens (Fig. 16–22). Discuss the heat transfer mechanisms associated with the cooking of a chicken in microwave and conventional ovens, and explain why cooking in a microwave oven is more efficient.

**SOLUTION** Food is cooked in a microwave oven by absorbing the electromagnetic radiation energy generated by the microwave tube, called the magnetron.

The radiation emitted by the magnetron is not thermal radiation, since its emission is not due to the temperature of the magnetron; rather, it is due to the conversion of electrical energy into electromagnetic radiation at a specified wavelength. The wavelength of the microwave radiation is such that it is *reflected* by metal surfaces; *transmitted* by the cookware made of glass, ceramic, or plastic; and *absorbed* and converted to internal energy by food (especially the water, sugar, and fat) molecules.

In a microwave oven, the *radiation* that strikes the chicken is absorbed by the skin of the chicken and the outer parts. As a result, the temperature of the chicken at and near the skin rises. Heat is then *conducted* toward the inner parts of the chicken from its outer parts. Of course, some of the heat absorbed by the outer surface of the chicken is lost to the air in the oven by *convection*.

In a conventional oven, the air in the oven is first heated to the desired temperature by the electric or gas heating element. This preheating may take several minutes. The heat is then transferred from the air to the skin of the chicken by *natural convection* in older ovens or by *forced convection* in the newer convection ovens that utilize a fan. The air motion in convection ovens increases the convection heat transfer coefficient and thus decreases the cooking time. Heat is then *conducted* toward the inner parts of the chicken from its outer parts as in microwave ovens.

Microwave ovens replace the slow convection heat transfer process in conventional ovens by the instantaneous radiation heat transfer. As a result, microwave ovens transfer energy to the food at full capacity the moment they are turned on, and thus they cook faster while consuming less energy.

## EXAMPLE 16–9    Heating of a Plate by Solar Energy

A thin metal plate is insulated on the back and exposed to solar radiation at the front surface (Fig. 16–23). The exposed surface of the plate has an absorptivity of 0.6 for solar radiation. If solar radiation is incident on the plate at a rate of 700 W/m$^2$ and the surrounding air temperature is 25°C, determine the surface temperature of the plate when the heat loss by convection and radiation equals the solar energy absorbed by the plate. Assume the combined convection and radiation heat transfer coefficient to be 50 W/m$^2$·K.

**SOLUTION**   The back side of the thin metal plate is insulated and the front side is exposed to solar radiation. The surface temperature of the plate is to be determined when it stabilizes.
*Assumptions*   **1** Steady operating conditions exist. **2** Heat transfer through the insulated side of the plate is negligible. **3** The heat transfer coefficient remains constant.
*Properties*   The solar absorptivity of the plate is given to be $\alpha = 0.6$.
*Analysis*   The absorptivity of the plate is 0.6, and thus 60 percent of the solar radiation incident on the plate is absorbed continuously. As a result, the temperature of the plate rises, and the temperature difference between the plate and the surroundings increases. This increasing temperature difference causes the rate of heat loss from the plate to the surroundings to increase. At some

**FIGURE 16–23**
Schematic for Example 16–9.

point, the rate of heat loss from the plate equals the rate of solar energy absorbed, and the temperature of the plate no longer changes. The temperature of the plate when steady operation is established is determined from

$$\dot{E}_{\text{gained}} = \dot{E}_{\text{lost}} \qquad \text{or} \qquad \alpha A_s \, \dot{q}_{\text{incident, solar}} = h_{\text{combined}} A_s (T_s - T_\infty)$$

Solving for $T_s$ and substituting, the plate surface temperature is determined to be

$$T_s = T_\infty + \alpha \frac{\dot{q}_{\text{incident, solar}}}{h_{\text{combined}}} = 25°C + \frac{0.6 \times (700 \text{ W/m}^2)}{50 \text{ W/m}^2 \cdot \text{K}} = \mathbf{33.4°C}$$

**Discussion** Note that the heat losses prevent the plate temperature from rising above 33.4°C. Also, the combined heat transfer coefficient accounts for the effects of both convection and radiation, and thus it is very convenient to use in heat transfer calculations when its value is known with reasonable accuracy.

## SUMMARY

Heat can be transferred in three different modes: conduction, convection, and radiation. *Conduction* is the transfer of heat from the more energetic particles of a substance to the adjacent less energetic ones as a result of interactions between the particles, and is expressed by *Fourier's law of heat conduction* as

$$\dot{Q}_{\text{cond}} = -kA \frac{dT}{dx}$$

where $k$ is the *thermal conductivity* of the material, $A$ is the *area* normal to the direction of heat transfer, and $dT/dx$ is the *temperature gradient*. The magnitude of the rate of heat conduction across a plane layer of thickness $L$ is given by

$$\dot{Q}_{\text{cond}} = kA \frac{\Delta T}{L}$$

where $\Delta T$ is the temperature difference across the layer.

*Convection* is the mode of heat transfer between a solid surface and the adjacent liquid or gas that is in motion, and involves the combined effects of conduction and fluid motion. The rate of convection heat transfer is expressed by *Newton's law of cooling* as

$$\dot{Q}_{\text{convection}} = hA_s (T_s - T_\infty)$$

where $h$ is the *convection heat transfer coefficient* in W/m²·K or Btu/h·ft²·R, $A_s$ is the *surface area* through which

convection heat transfer takes place, $T_s$ is the *surface temperature*, and $T_\infty$ is the *temperature of the fluid* sufficiently far from the surface.

*Radiation* is the energy emitted by matter in the form of electromagnetic waves (or photons) as a result of the changes in the electronic configurations of the atoms or molecules. The maximum rate of radiation that can be emitted from a surface at a thermodynamic temperature $T_s$ is given by the *Stefan–Boltzmann law* as $\dot{Q}_{\text{emit, max}} = \sigma A_s T_s^4$, where $\sigma = 5.67 \times 10^{-8}$ W/m²·K⁴ or $0.1714 \times 10^{-8}$ Btu/h·ft²·R⁴ is the *Stefan–Boltzmann constant*.

When a surface of emissivity $\varepsilon$ and surface area $A_s$ at a temperature $T_s$ is completely enclosed by a much larger (or black) surface at a temperature $T_{\text{surr}}$ separated by a gas (such as air) that does not intervene with radiation, the net rate of radiation heat transfer between these two surfaces is given by

$$\dot{Q}_{\text{rad}} = \varepsilon \sigma A_s (T_s^4 - T_{\text{surr}}^4)$$

In this case, the emissivity and the surface area of the surrounding surface do not have any effect on the net radiation heat transfer.

The rate at which a surface absorbs radiation is determined from $\dot{Q}_{\text{absorbed}} = \alpha \dot{Q}_{\text{incident}}$ where $\dot{Q}_{\text{incident}}$ is the rate at which radiation is incident on the surface and $\alpha$ is the absorptivity of the surface.

## REFERENCES AND SUGGESTED READING

1. Y. A. Çengel and A. J. Ghajar. *Heat and Mass Transfer: Fundamentals and Applications*. 4th ed. New York: McGraw-Hill, 2011.

2. Y. A. Çengel and M. A. Boles. *Thermodynamics: An Engineering Approach*. 7th ed. New York: McGraw-Hill, 2011.

3. Robert J. Ribando. *Heat Transfer Tools*. New York: McGraw-Hill, 2002.

## PROBLEMS*

### Heat Transfer Mechanisms

**16–1C** Which is a better heat conductor, diamond or silver?

**16–2C** Define thermal conductivity and explain its significance in heat transfer.

**16–3C** What are the mechanisms of heat transfer? How are they distinguished from each other?

**16–4C** What is the physical mechanism of heat conduction in a solid, a liquid, and a gas?

**16–5C** Consider heat transfer through a windowless wall of a house on a winter day. Discuss the parameters that affect the rate of heat conduction through the wall.

**16–6C** Write down the expressions for the physical laws that govern each mode of heat transfer, and identify the variables involved in each relation.

**16–7C** How does heat conduction differ from convection?

**16–8C** Does any of the energy of the sun reach the earth by conduction or convection?

**16–9C** How does forced convection differ from natural convection?

**16–10C** Define emissivity and absorptivity. What is Kirchhoff's law of radiation?

**16–11C** What is a blackbody? How do real bodies differ from blackbodies?

**16–12C** Judging from its unit W/m·K, can we define thermal conductivity of a material as the rate of heat transfer through the material per unit thickness per unit temperature difference? Explain.

**16–13C** Consider heat loss through the two walls of a house on a winter night. The walls are identical, except that one of them has a tightly fit glass window. Through which wall will the house lose more heat? Explain.

**16–14C** Consider two houses that are identical, except that the walls are built using bricks in one house, and wood in the other. If the walls of the brick house are twice as thick, which house do you think will be more energy efficient?

**16–15C** Consider two walls of a house that are identical except that one is made of 10-cm-thick wood, while the other is made of 25-cm-thick brick. Through which wall will the house lose more heat in winter?

**16–16C** How do the thermal conductivity of gases and liquids vary with temperature?

**16–17C** Why is the thermal conductivity of superinsulation orders of magnitude lower than the thermal conductivity of ordinary insulation?

**16–18C** Why do we characterize the heat conduction ability of insulators in terms of their apparent thermal conductivity instead of the ordinary thermal conductivity?

**16–19C** Consider an alloy of two metals whose thermal conductivities are $k_1$ and $k_2$. Will the thermal conductivity of the alloy be less than $k_1$, greater than $k_2$, or between $k_1$ and $k_2$?

**16–20** Two surfaces of a 2-cm-thick plate are maintained at 0°C and 80°C, respectively. If it is determined that heat is transferred through the plate at a rate of 500 W/m², determine its thermal conductivity.

**16–21** Four power transistors, each dissipating 12 W, are mounted on a thin vertical aluminum plate 22 cm × 22 cm in size. The heat generated by the transistors is to be dissipated by both surfaces of the plate to the surrounding air at 25°C, which is blown over the plate by a fan. The entire plate can be assumed to be nearly isothermal, and the exposed surface area of the transistor can be taken to be equal to its base area. If the

---

\* Problems designated by a "C" are concept questions, and students are encouraged to answer them all. Problems designated by an "E" are in English units, and the SI users can ignore them. Problems with the ⚛ icon are solved using EES, and complete solutions together with parametric studies are included on the enclosed DVD. Problems with the 🖳 icon are comprehensive in nature and are intended to be solved with a computer, preferably using the EES software that accompanies this text.

average convection heat transfer coefficient is 25 W/m²·K, determine the temperature of the aluminum plate. Disregard any radiation effects.

**16–22** In a power plant, pipes transporting superheated vapor are very common. Superheated vapor is flowing at a rate of 0.3 kg/s inside a pipe with 5 cm in diameter and 10 m in length. The pipe is located in a power plant at 20°C, and has a uniform pipe surface temperature of 100°C. If the temperature drop between the inlet and exit of the pipe is 30°C, and the specific heat of the vapor is 2190 J/kg·K, determine the heat transfer coefficient as a result of convection between the pipe surface and the surrounding.

Air, 20°C
$T_s = 100°C$
$D = 5$ cm

Superheated
vapor
0.3 kg/s

$L = 10$ m
$T_{in} - T_{out} = 30°C$

**FIGURE P16–22**

**16–23** An electric current of 5 A passing through a resistor has a measured voltage of 6 V across the resistor. The resistor is cylindrical with a diameter of 2.5 cm and length of 15 cm. The resistor has a uniform temperature of 90°C and the room air temperature is 20°C. Assuming that heat transfer by radiation is negligible, determine the heat transfer coefficient by convection. *Answer:* 33.6 W/m²·K

**16–24** The inner and outer surfaces of a 4-m × 7-m brick wall of thickness 30 cm and thermal conductivity 0.69 W/m·K are maintained at temperatures of 26°C and 8°C, respectively. Determine the rate of heat transfer through the wall, in W.

Brick
wall

26°C          8°C

←30 cm→

**FIGURE P16–24**

**16–25** The inner and outer surfaces of a 0.5-cm thick 2-m × 2-m window glass in winter are 10°C and 3°C, respectively. If the thermal conductivity of the glass is 0.78 W/m·K, determine the amount of heat loss through the glass over a period of 5 h. What would your answer be if the glass were 1 cm thick? *Answers:* 78.6 MJ, 39.3 MJ

**16–26** [EES] Reconsider Prob. 16–25. Using EES (or other) software, plot the amount of heat loss through the glass as a function of the window glass thickness in the range of 0.1 cm to 1.0 cm. Discuss the results.

**16–27** An aluminum pan whose thermal conductivity is 237 W/m·K has a flat bottom with diameter 15 cm and thickness 0.4 cm. Heat is transferred steadily to boiling water in the pan through its bottom at a rate of 1400 W. If the inner surface of the bottom of the pan is at 105°C, determine the temperature of the outer surface of the bottom of the pan.

105°C

0.4 cm

1400 W

**FIGURE P16–27**

**16–28E** The north wall of an electrically heated home is 20 ft long, 10 ft high, and 1 ft thick, and is made of brick whose thermal conductivity is $k = 0.42$ Btu/h·ft·°F. On a certain winter night, the temperatures of the inner and the outer surfaces of the wall are measured to be at about 62°F and 25°F, respectively, for a period of 8 h. Determine (*a*) the rate of heat loss through the wall that night and (*b*) the cost of that heat loss to the home owner if the cost of electricity is $0.07/kWh.

**16–29** In a certain experiment, cylindrical samples of diameter 4 cm and length 7 cm are used (see Fig. 16–9). The two thermocouples in each sample are placed 3 cm apart. After initial transients, the electric heater is observed to draw 0.6 A at 110 V, and both differential thermometers read a temperature difference of 8°C. Determine the thermal conductivity of the sample. *Answer:* 98.5 W/m·K

**16–30** One way of measuring the thermal conductivity of a material is to sandwich an electric thermofoil heater between two identical rectangular samples of the material and to heavily insulate the four outer edges, as shown in the figure. Thermocouples attached to the inner and outer surfaces of the samples record the temperatures.

During an experiment, two 0.5-cm thick samples 10 cm × 10 cm in size are used. When steady operation is reached, the heater is observed to draw 25 W of electric power, and the temperature of each sample is observed to drop from 82°C at the inner surface to 74°C at the outer surface. Determine the thermal conductivity of the material at the average temperature.

**Samples** ← Insulation
**Wattmeter**
← Insulation
**Source**
0.5 cm
**Resistance heater**

**FIGURE P16–30**

**16–31** Repeat Prob. 16–30 for an electric power consumption of 20 W.

**16–32** A heat flux meter attached to the inner surface of a 3-cm-thick refrigerator door indicates a heat flux of 32 W/m² through the door. Also, the temperatures of the inner and the outer surfaces of the door are measured to be 7°C and 15°C, respectively. Determine the average thermal conductivity of the refrigerator door. *Answer: 0.120 W/m·K*

**16–33** Consider a person standing in a room maintained at 20°C at all times. The inner surfaces of the walls, floors, and ceiling of the house are observed to be at an average temperature of 12°C in winter and 23°C in summer. Determine the rates of radiation heat transfer between this person and the surrounding surfaces in both summer and winter if the exposed surface area, emissivity, and the average outer surface temperature of the person are 1.6 m², 0.95, and 32°C, respectively.

**16–34** Reconsider Prob. 16–33. Using EES (or other) software, plot the rate of radiation heat transfer in winter as a function of the temperature of the inner surface of the room in the range of 8°C to 18°C. Discuss the results.

**16–35** For heat transfer purposes, a standing man can be modeled as a 30-cm-diameter, 170-cm-long vertical cylinder with both the top and bottom surfaces insulated and with the side surface at an average temperature of 34°C. For a convection heat transfer coefficient of 8 W/m²·K, determine the rate of heat loss from this man by convection in an environment at 18°C. *Answer: 205 W*

**16–36** Hot air at 80°C is blown over a 2-m × 4-m flat surface at 30°C. If the average convection heat transfer coefficient is 55 W/m²·K, determine the rate of heat transfer from the air to the plate, in kW. *Answer: 22 kW*

**16–37** Reconsider Prob. 16–36. Using EES (or other) software, plot the rate of heat transfer as a function of the heat transfer coefficient in the range of 20 W/m²·K to 100 W/m²·K. Discuss the results.

**16–38** The outer surface of a spacecraft in space has an emissivity of 0.8 and a solar absorptivity of 0.3. If solar radiation is incident on the spacecraft at a rate of 950 W/m², determine the surface temperature of the spacecraft when the radiation emitted equals the solar energy absorbed.

**16–39** The heat generated in the circuitry on the surface of a silicon chip ($k$ = 130 W/m·K) is conducted to the ceramic substrate to which it is attached. The chip is 6 mm × 6 mm in size and 0.5 mm thick and dissipates 5 W of power. Disregarding any heat transfer through the 0.5-mm high side surfaces, determine the temperature difference between the front and back surfaces of the chip in steady operation.

Silicon chip
0.5 mm
5 W
6 mm
6 mm
Ceramic substrate

**FIGURE P16–39**

**16–40** A 40-cm-long, 800-W electric resistance heating element with diameter 0.5 cm and surface temperature 120°C is immersed in 75 kg of water initially at 20°C. Determine how long it will take for this heater to raise the water temperature to 80°C. Also, determine the convection heat transfer coefficients at the beginning and at the end of the heating process.

**16–41** A 5-cm-external-diameter, 10-m-long hot-water pipe at 80°C is losing heat to the surrounding air at 5°C by natural convection with a heat transfer coefficient of 25 W/m²·K. Determine the rate of heat loss from the pipe by natural convection. *Answer: 2945 W*

**16–42** A hollow spherical iron container with outer diameter 20 cm and thickness 0.2 cm is filled with iced water at 0°C.

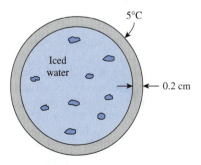

5°C
Iced water
0.2 cm

**FIGURE P16–42**

If the outer surface temperature is 5°C, determine the approximate rate of heat loss from the sphere, in kW, and the rate at which ice melts in the container. The heat of fusion of water is 333.7 kJ/kg.

**16–43** Reconsider Prob. 16–42. Using EES (or other) software, plot the rate at which ice melts as a function of the container thickness in the range of 0.1 cm to 1.0 cm. Discuss the results.

**16–44E** The inner and outer glasses of a 4-ft × 4-ft double-pane window are at 60°F and 48°F, respectively. If the 0.25-in. space between the two glasses is filled with still air, determine the rate of heat transfer through the window.
*Answer:* 131 Btu/h

**16–45** A transistor with a height of 0.4 cm and a diameter of 0.6 cm is mounted on a circuit board. The transistor is cooled by air flowing over it with an average heat transfer coefficient of 30 W/m²·K. If the air temperature is 55°C and the transistor case temperature is not to exceed 70°C, determine the amount of power this transistor can dissipate safely. Disregard any heat transfer from the transistor base.

Air
55°C

Power
transistor
$T_s \leq 70°C$

0.6 cm

$\leftarrow$ 0.4 cm $\rightarrow$

**FIGURE P16–45**

**16–46** Reconsider Prob. 16–45. Using EES (or other) software, plot the amount of power the transistor can dissipate safely as a function of the maximum case temperature in the range of 60°C to 90°C. Discuss the results.

**16–47E** A 300-ft-long section of a steam pipe whose outer diameter is 4 in passes through an open space at 50°F. The average temperature of the outer surface of the pipe is measured to be 280°F, and the average heat transfer coefficient on that surface is determined to be 6 Btu/h·ft²·°F. Determine (a) the rate of heat loss from the steam pipe and (b) the annual cost of this energy loss if steam is generated in a natural gas furnace having an efficiency of 86 percent, and the price of natural gas is $1.10/therm (1 therm = 100,000 Btu).
*Answers:* (a) 433,500 Btu/h, (b) $48,576/yr

**16–48** The boiling temperature of nitrogen at atmospheric pressure at sea level (1 atm) is −196°C. Therefore, nitrogen is commonly used in low temperature scientific studies since the temperature of liquid nitrogen in a tank open to the atmosphere remains constant at −196°C until the liquid nitrogen in the tank is depleted. Any heat transfer to the tank results in the evaporation of some liquid nitrogen, which has a heat of vaporization of 198 kJ/kg and a density of 810 kg/m³ at 1 atm.

Consider a 4-m-diameter spherical tank initially filled with liquid nitrogen at 1 atm and −196°C. The tank is exposed to 20°C ambient air with a heat transfer coefficient of 25 W/m²·K. The temperature of the thin-shelled spherical tank is observed to be almost the same as the temperature of the nitrogen inside. Disregarding any radiation heat exchange, determine the rate of evaporation of the liquid nitrogen in the tank as a result of the heat transfer from the ambient air.

$N_2$ vapor

$T_{air} = 20°C$

1 atm
Liquid $N_2$
−196°C

**FIGURE P16–48**

**16–49** Repeat Prob. 16–48 for liquid oxygen, which has a boiling temperature of −183°C, a heat of vaporization of 213 kJ/kg, and a density of 1140 kg/m³ at 1 atm pressure.

**16–50** Reconsider Prob. 16–48. Using EES (or other) software, plot the rate of evaporation of liquid nitrogen as a function of the ambient air temperature in the range of 0°C to 40°C. Discuss the results.

**16–51** Consider a person whose exposed surface area is 1.7 m², emissivity is 0.5, and surface temperature is 32°C. Determine the rate of heat loss from that person by radiation in a large room having walls at a temperature of (a) 300 K and (b) 280 K. *Answers:* (a) 26.7 W, (b) 121 W

**16–52** A 0.3-cm-thick, 12-cm-high, and 18-cm-long circuit board houses 80 closely spaced logic chips on one side, each dissipating 0.06 W. The board is impregnated with copper fillings and has an effective thermal conductivity of 16 W/m·K. All the heat generated in the chips is conducted across the circuit board and is dissipated from the back side of the board to the ambient air. Determine the temperature difference between the two sides of the circuit board.    *Answer:* 0.042°C

**16–53**   Consider a sealed 20-cm-high electronic box whose base dimensions are 50 cm × 50 cm placed in a vacuum chamber. The emissivity of the outer surface of the box is 0.95. If the electronic components in the box dissipate a total of 120 W of power and the outer surface temperature of the box is not to exceed 55°C, determine the temperature at which the surrounding surfaces must be kept if this box is to be cooled by radiation alone. Assume the heat transfer from the bottom surface of the box to the stand to be negligible.

**FIGURE P16–53**

**16–54E**   Using the conversion factors between W and Btu/h, m and ft, and K and R, express the Stefan–Boltzmann constant $\sigma = 5.67 \times 10^{-8}$ W/m²·K⁴ in the English unit Btu/h·ft²·R⁴.

**16–55E**   An engineer who is working on the heat transfer analysis of a house in English units needs the convection heat transfer coefficient on the outer surface of the house. But the only value he can find from his handbooks is 22 W/m²·K, which is in SI units. The engineer does not have a direct conversion factor between the two unit systems for the convection heat transfer coefficient. Using the conversion factors between W and Btu/h, m and ft, and °C and °F, express the given convection heat transfer coefficient in Btu/h·ft²·°F.    *Answer:* 3.87 Btu/h·ft²·°F

**16–56**   Water at 0°C releases 333.7 kJ/kg of heat as it freezes to ice ($\rho = 920$ kg/m³) at 0°C. An aircraft flying under icing conditions maintains a heat transfer coefficient of 150 W/m²·K between the air and wing surfaces. What temperature must the wings be maintained at to prevent ice from forming on them during icing conditions at a rate of 1 mm/min or less?

**16–57**   A 2.1-m-long, 0.2-cm-diameter electrical wire extends across a room that is maintained at 20°C. Heat is generated in the wire as a result of resistance heating, and the surface temperature of the wire is measured to be 180°C in steady operation. Also, the voltage drop and electric current through the wire are measured to be 110 V and 3 A, respectively.

Disregarding any heat transfer by radiation, determine the convection heat transfer coefficient for heat transfer between the outer surface of the wire and the air in the room.
*Answer:* 156 W/m²·K

**FIGURE P16–57**

**16–58**   Reconsider Prob. 16–57. Using EES (or other) software, plot the convection heat transfer coefficient as a function of the wire surface temperature in the range of 100°C to 300°C. Discuss the results.

### Simultaneous Heat Transfer Mechanisms

**16–59C**   Can all three modes of heat transfer occur simultaneously (in parallel) in a medium?

**16–60C**   Can a medium involve (*a*) conduction and convection, (*b*) conduction and radiation, or (*c*) convection and radiation simultaneously? Give examples for the "yes" answers.

**16–61C**   The deep human body temperature of a healthy person remains constant at 37°C while the temperature and the humidity of the environment change with time. Discuss the heat transfer mechanisms between the human body and the environment both in summer and winter, and explain how a person can keep cooler in summer and warmer in winter.

**16–62C**   We often turn the fan on in summer to help us cool. Explain how a fan makes us feel cooler in the summer. Also explain why some people use ceiling fans also in winter.

**16–63**   Consider a person standing in a room at 18°C. Determine the total rate of heat transfer from this person if the exposed surface area and the skin temperature of the person are 1.7 m² and 32°C, respectively, and the convection heat transfer coefficient is 5 W/m²·K. Take the emissivity of the skin and the clothes to be 0.9, and assume the temperature of the inner surfaces of the room to be the same as the air temperature. *Answer:* 248 W

**16–64**   Consider steady heat transfer between two large parallel plates at constant temperatures of $T_1 = 290$ K and $T_2 = 150$ K that are $L = 2$ cm apart. Assuming the surfaces to be black (emissivity $\varepsilon = 1$), determine the rate of heat transfer between the plates per unit surface area assuming the gap between the plates is (*a*) filled with atmospheric air, (*b*) evacuated, (*c*) filled with fiberglass insulation, and (*d*) filled with superinsulation having an apparent thermal conductivity of 0.00015 W/m·K.

**16–65**  The inner and outer surfaces of a 25-cm-thick wall in summer are at 27°C and 44°C, respectively. The outer surface of the wall exchanges heat by radiation with surrounding surfaces at 40°C, and convection with ambient air also at 40°C with a convection heat transfer coefficient of 8 W/m²·K. Solar radiation is incident on the surface at a rate of 150 W/m². If both the emissivity and the solar absorptivity of the outer surface are 0.8, determine the effective thermal conductivity of the wall.

27°C

150 W/m²

44°C

$\alpha_s = \varepsilon = 0.8$
air, 40°C
$h$

$\dot{Q}_{rad}$

**FIGURE P16–65**

**16–66E**  A 2-in-diameter spherical ball whose surface is maintained at a temperature of 170°F is suspended in the middle of a room at 70°F. If the convection heat transfer coefficient is 15 Btu/h·ft²·°F and the emissivity of the surface is 0.8, determine the total rate of heat transfer from the ball.

**16–67**  An 800-W iron is left on the iron board with its base exposed to the air at 20°C. The convection heat transfer coefficient between the base surface and the surrounding air is 35 W/m²·K. If the base has an emissivity of 0.6 and a surface area of 0.02 m², determine the temperature of the base of the iron.  *Answer: 601°C*

20°C

Iron
800 W

**FIGURE P16–67**

**16–68**  A 3-m-internal-diameter spherical tank made of 1-cm-thick stainless steel is used to store iced water at 0°C. The tank is located outdoors at 25°C. Assuming the entire steel tank to be at 0°C and thus the thermal resistance of the tank to be negligible, determine (*a*) the rate of heat transfer to the iced water in the tank and (*b*) the amount of ice at 0°C that melts during a 24-hour period. The heat of fusion of water at atmospheric pressure is $h_{if}$ = 333.7 kJ/kg. The emissivity of the outer surface of the tank is 0.75, and the convection heat transfer coefficient on the outer surface can be taken to be 30 W/m²·K. Assume the average surrounding surface temperature for radiation exchange to be 15°C. *Answers: (a) 23.1 kW, (b) 5980 kg*

**16–69**  The roof of a house consists of a 22-cm-thick concrete slab ($k$ = 2 W/m·K) that is 15 m wide and 20 m long. The emissivity of the outer surface of the roof is 0.9, and the convection heat transfer coefficient on that surface is estimated to be 15 W/m²·K. The inner surface of the roof is maintained at 15°C. On a clear winter night, the ambient air is reported to be at 10°C while the night sky temperature for radiation heat transfer is 255 K. Considering both radiation and convection heat transfer, determine the outer surface temperature and the rate of heat transfer through the roof.

If the house is heated by a furnace burning natural gas with an efficiency of 85 percent, and the unit cost of natural gas is $1.20/therm (1 therm = 105,500 kJ of energy content), determine the money lost through the roof that night during a 14-hour period.

**16–70E**  Consider a flat-plate solar collector placed horizontally on the flat roof of a house. The collector is 5 ft wide and 15 ft long, and the average temperature of the exposed surface of the collector is 100°F. The emissivity of the exposed surface of the collector is 0.9. Determine the rate of heat loss from the collector by convection and radiation during a calm day when the ambient air temperature is 70°F and the effective sky temperature for radiation exchange is 50°F. Take the convection heat transfer coefficient on the exposed surface to be 2.5 Btu/h·ft²·°F.

$T_{sky}$ = 50°F

70°F

Solar collector

**FIGURE P16–70E**

## Review Problems

**16–71**  A soldering iron has a cylindrical tip of 2.5 mm in diameter and 20 mm in length. With age and usage, the tip has oxidized and has an emissivity of 0.80. Assuming that the average convection heat transfer coefficient over the soldering iron tip is 25 W/m$^2$·K, and the surrounding air temperature is 20°C, determine the power required to maintain the tip at 400°C.

Air, 20°C

$T_s = 400$°C

$D = 2.5$ mm

$L = 20$ mm

**FIGURE P16–71**

**16–72**  It is well known that wind makes the cold air feel much colder as a result of the *wind chill effect* that is due to the increase in the convection heat transfer coefficient with increasing air velocity. The wind chill effect is usually expressed in terms of the *wind chill temperature* (WCT), which is the apparent temperature felt by exposed skin. For outdoor air temperature of 0°C, for example, the wind chill temperature is −5°C at 20 km/h winds and −9°C at 60 km/h winds. That is, a person exposed to 0°C windy air at 20 km/h will feel as cold as a person exposed to −5°C calm air (air motion under 5 km/h).

For heat transfer purposes, a standing man can be modeled as a 30-cm-diameter, 170-cm-long vertical cylinder with both the top and bottom surfaces insulated and with the side surface at an average temperature of 34°C. For a convection heat transfer coefficient of 15 W/m$^2$·K, determine the rate of heat loss from this man by convection in still air at 20°C. What would your answer be if the convection heat transfer coefficient is increased to 30 W/m$^2$·K as a result of winds? What is the wind chill temperature in this case? *Answers:* 336 W, 672 W, 6°C

**16–73**  A thin metal plate is insulated on the back and exposed to solar radiation on the front surface. The exposed surface of the plate has an absorptivity of 0.7 for solar radiation. If solar radiation is incident on the plate at a rate of 550 W/m$^2$ and the surrounding air temperature is 10°C, determine the surface temperature of the plate when the heat loss by convection equals the solar energy absorbed by the plate. Take the convection heat transfer coefficient to be 25 W/m$^2$·K, and disregard any heat loss by radiation.

550 W/m$^2$

$\alpha = 0.7$      10°C

**FIGURE P16–73**

**16–74**  Consider a 3-m × 3-m × 3-m cubical furnace whose top and side surfaces closely approximate black surfaces at a temperature of 1200 K. The base surface has an emissivity of $\varepsilon = 0.4$, and is maintained at 800 K. Determine the net rate of radiation heat transfer to the base surface from the top and side surfaces.  *Answer:* 340 kW

**16–75**  Engine valves ($c_p = 440$ J/kg·K and $\rho = 7840$ kg/m$^3$) are to be heated from 40°C to 800°C in 5 min in the heat treatment section of a valve manufacturing facility. The valves have a cylindrical stem with a diameter of 8 mm and a length of 10 cm. The valve head and the stem may be assumed to be of equal surface area, with a total mass of 0.0788 kg. For a single valve, determine (*a*) the amount of heat transfer, (*b*) the average rate of heat transfer, (*c*) the average heat flux, and (*d*) the number of valves that can be heat treated per day if the heating section can hold 25 valves and it is used 10 h per day.

**16–76**  Consider a flat-plate solar collector placed on the roof of a house. The temperatures at the inner and outer surfaces of the glass cover are measured to be 33°C and 31°C, respectively. The glass cover has a surface area of 2.5 m$^2$, a thickness of 0.6 cm, and a thermal conductivity of 0.7 W/m·K. Heat is lost from the outer surface of the cover by convection and radiation with a convection heat transfer coefficient of 10 W/m$^2$·K and an ambient temperature of 15°C. Determine the fraction of heat lost from the glass cover by radiation.

**16–77**  The rate of heat loss through a unit surface area of a window per unit temperature difference between the indoors and the outdoors is called the *U-factor*. The value of the *U*-factor ranges from about 1.25 W/m$^2$·K (or 0.22 Btu/h·ft$^2$·°F) for low-*e* coated, argon-filled, quadruple-pane windows to 6.25 W/m$^2$·K (or 1.1 Btu/h·ft$^2$·°F) for a single-pane window

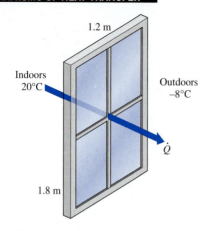

**FIGURE P16–77**

with aluminum frames. Determine the range for the rate of heat loss through a 1.2-m × 1.8-m window of a house that is maintained at 20°C when the outdoor air temperature is −8°C.

**16–78** Reconsider Prob. 16–77. Using EES (or other) software, plot the rate of heat loss through the window as a function of the U-factor. Discuss the results.

**16–79** An electric heater with the total surface area of 0.25 m² and emissivity 0.75 is in a room where the air has a temperature of 20°C and the walls are at 10°C. When the heater consumes 500 W of electric power, its surface has a steady temperature of 120°C. Determine the temperature of the heater surface when it consumes 700 W. Solve the problem (a) assuming negligible radiation and (b) taking radiation into consideration. Based on your results, comment on the assumption made in part (a).

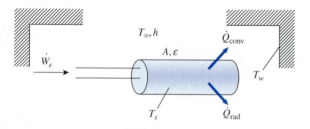

**FIGURE P16–79**

## Design and Essay Problems

**16–80** Write an essay on how microwave ovens work, and explain how they cook much faster than conventional ovens. Discuss whether conventional electric or microwave ovens consume more electricity for the same task.

**16–81** Using information from the utility bills for the coldest month last year, estimate the average rate of heat loss from your house for that month. In your analysis, consider the contribution of the internal heat sources such as people, lights, and appliances. Identify the primary sources of heat loss from your house and propose ways of improving the energy efficiency of your house.

**16–82** Conduct this experiment to determine the combined heat transfer coefficient between an incandescent lightbulb and the surrounding air and surfaces using a 60-W lightbulb. You will need a thermometer, which can be purchased in a hardware store, and a metal glue. You will also need a piece of string and a ruler to calculate the surface area of the lightbulb. First, measure the air temperature in the room, and then glue the tip of the thermocouple wire of the thermometer to the glass of the lightbulb. Turn the light on and wait until the temperature reading stabilizes. The temperature reading will give the surface temperature of the lightbulb. Assuming 10 percent of the rated power of the bulb is converted to light and is transmitted by the glass, calculate the heat transfer coefficient from Newton's law of cooling.

**16–83** It is well-known that at the same outdoor air temperature a person is cooled at a faster rate under windy conditions than under calm conditions due to the higher convection heat transfer coefficients associated with windy air. The phrase *wind chill* is used to relate the rate of heat loss from people under windy conditions to an equivalent air temperature for calm conditions (considered to be a wind or walking speed of 3 mph or 5 km/h). The hypothetical *wind chill temperature* (WCT), called the *wind chill temperature index* (WCTI), is an equivalent air temperature equal to the air temperature needed to produce the same cooling effect under calm conditions. A 2003 report on wind chill temperature by the U.S. National Weather Service gives the WCTI in metric units as

$$\text{WCTI (°C)} = 13.12 + 0.6215T - 11.37V^{0.16} + 0.3965TV^{0.16}$$

where $T$ is the air temperature in °C and $V$ the wind speed in km/h at 10 m elevation. Show that this relation can be expressed in English units as

$$\text{WCTI (°F)} = 35.74 + 0.6215T - 35.75V^{0.16} + 0.4275TV^{0.16}$$

where $T$ is the air temperature in °F and $V$ the wind speed in mph at 33 ft elevation. Also, prepare a table for WCTI for air temperatures ranging from 10 to −60°C and wind speeds ranging from 10 to 80 km/h. Comment on the magnitude of the cooling effect of the wind and the danger of frostbite.

# STEADY HEAT CONDUCTION

I n heat transfer analysis, we are often interested in the rate of heat transfer through a medium under steady conditions and surface temperatures. Such problems can be solved easily without involving any differential equations by the introduction of the *thermal resistance concept* in an analogous manner to electrical circuit problems. In this case, the thermal resistance corresponds to electrical resistance, temperature difference corresponds to voltage, and the heat transfer rate corresponds to electric current.

We start this chapter with *one-dimensional steady heat conduction* in a plane wall, a cylinder, and a sphere, and develop relations for *thermal resistances* in these geometries. We also develop thermal resistance relations for convection and radiation conditions at the boundaries. We apply this concept to heat conduction problems in *multilayer* plane walls, cylinders, and spheres and generalize it to systems that involve heat transfer in two or three dimensions. We also discuss the *thermal contact resistance* and the *overall heat transfer coefficient* and develop relations for the critical radius of insulation for a cylinder and a sphere. Finally, we discuss steady heat transfer from *finned surfaces*.

## Objectives

The objectives of this chapter are to:

- Understand the concept of thermal resistance and its limitations, and develop thermal resistance networks for practical heat conduction problems.

- Solve steady conduction problems that involve multilayer rectangular, cylindrical, or spherical geometries.

- Develop an intuitive understanding of thermal contact resistance, and circumstances under which it may be significant.

- Identify applications in which insulation may actually increase heat transfer.

- Analyze finned surfaces, and assess how efficiently and effectively fins enhance heat transfer.

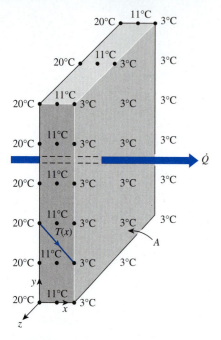

**FIGURE 17–1**
Heat transfer through a wall is one-dimensional when the temperature of the wall varies in one direction only.

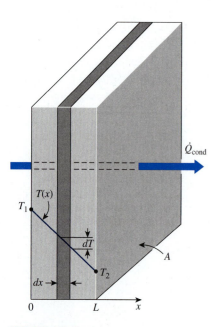

**FIGURE 17–2**
Under steady conditions, the temperature distribution in a plane wall is a straight line.

# 17–1 ▪ STEADY HEAT CONDUCTION IN PLANE WALLS

Consider steady heat conduction through the walls of a house during a winter day. We know that heat is continuously lost to the outdoors through the wall. We intuitively feel that heat transfer through the wall is in the *normal direction* to the wall surface, and no significant heat transfer takes place in the wall in other directions (Fig. 17–1).

Recall that heat transfer in a certain direction is driven by the *temperature gradient* in that direction. There is no heat transfer in a direction in which there is no change in temperature. Temperature measurements at several locations on the inner or outer wall surface will confirm that a wall surface is nearly *isothermal*. That is, the temperatures at the top and bottom of a wall surface as well as at the right and left ends are almost the same. Therefore, there is no heat transfer through the wall from the top to the bottom, or from left to right, but there is considerable temperature difference between the inner and the outer surfaces of the wall, and thus significant heat transfer in the direction from the inner surface to the outer one.

The small thickness of the wall causes the temperature gradient in that direction to be large. Further, if the air temperatures in and outside the house remain constant, then heat transfer through the wall of a house can be modeled as *steady* and *one-dimensional*. The temperature of the wall in this case depends on one direction only (say the $x$-direction) and can be expressed as $T(x)$.

Noting that heat transfer is the only energy interaction involved in this case and there is no heat generation, the *energy balance* for the wall can be expressed as

$$\begin{pmatrix} \text{Rate of} \\ \text{heat transfer} \\ \text{into the wall} \end{pmatrix} - \begin{pmatrix} \text{Rate of} \\ \text{heat transfer} \\ \text{out of the wall} \end{pmatrix} = \begin{pmatrix} \text{Rate of change} \\ \text{of the energy} \\ \text{of the wall} \end{pmatrix}$$

or

$$\dot{Q}_{in} - \dot{Q}_{out} = \frac{dE_{wall}}{dt} \qquad (17\text{–}1)$$

But $dE_{wall}/dt = 0$ for *steady* operation, since there is no change in the temperature of the wall with time at any point. Therefore, the rate of heat transfer into the wall must be equal to the rate of heat transfer out of it. In other words, *the rate of heat transfer through the wall must be constant*, $\dot{Q}_{cond, wall}$ = constant.

Consider a plane wall of thickness $L$ and average thermal conductivity $k$. The two surfaces of the wall are maintained at constant temperatures of $T_1$ and $T_2$. For one-dimensional steady heat conduction through the wall, we have $T(x)$. Then Fourier's law of heat conduction for the wall can be expressed as

$$\dot{Q}_{cond, wall} = -kA\frac{dT}{dx} \qquad \text{(W)} \qquad (17\text{–}2)$$

where the rate of conduction heat transfer $\dot{Q}_{cond, wall}$ and the wall area $A$ are constant. Thus $dT/dx$ = constant, which means that *the temperature through the wall varies linearly with x*. That is, the temperature distribution in the wall under steady conditions is a *straight line* (Fig. 17–2).

Separating the variables in the preceding equation and integrating from $x = 0$, where $T(0) = T_1$, to $x = L$, where $T(L) = T_2$, we get

$$\int_{x=0}^{L} \dot{Q}_{\text{cond, wall}}\, dx = -\int_{T=T_1}^{T_2} kA\, dT$$

Performing the integrations and rearranging gives

$$\dot{Q}_{\text{cond, wall}} = kA \frac{T_1 - T_2}{L} \qquad \text{(W)} \qquad\qquad (17\text{–}3)$$

which is identical to Eq. 16–1. Again, *the rate of heat conduction through a plane wall is proportional to the average thermal conductivity, the wall area, and the temperature difference, but is inversely proportional to the wall thickness.* Also, once the rate of heat conduction is available, the temperature $T(x)$ at any location $x$ can be determined by replacing $T_2$ in Eq. 17–3 by $T$, and $L$ by $x$.

## Thermal Resistance Concept

Equation 17–3 for heat conduction through a plane wall can be rearranged as

$$\dot{Q}_{\text{cond, wall}} = \frac{T_1 - T_2}{R_{\text{wall}}} \qquad \text{(W)} \qquad\qquad (17\text{–}4)$$

where

$$R_{\text{wall}} = \frac{L}{kA} \qquad \text{(K/W)} \qquad\qquad (17\text{–}5)$$

is the *thermal resistance* of the wall against heat conduction or simply the **conduction resistance** of the wall. Note that the thermal resistance of a medium depends on the *geometry* and the *thermal properties* of the medium. Note that thermal resistance can also be expressed as $R_{\text{wall}} = \Delta T/\dot{Q}_{\text{cond, wall}}$, which is the ratio of the driving potential $\Delta T$ to the corresponding heat transfer rate $\dot{Q}_{\text{cond, wall}}$.

This equation for heat transfer is analogous to the relation for *electric current flow I*, expressed as

$$I = \frac{\mathbf{V}_1 - \mathbf{V}_2}{R_e} \qquad\qquad (17\text{–}6)$$

where $R_e = L/\sigma_e A$ is the *electric resistance* and $\mathbf{V}_1 - \mathbf{V}_2$ is the *voltage difference* across the resistance ($\sigma_e$ is the electrical conductivity). Thus, the *rate of heat transfer* through a layer corresponds to the *electric current*, the *thermal resistance* corresponds to *electrical resistance*, and the *temperature difference* corresponds to *voltage difference* across the layer (Fig. 17–3).

Consider convection heat transfer from a solid surface of area $A_s$ and temperature $T_s$ to a fluid whose temperature sufficiently far from the surface is $T_\infty$, with a convection heat transfer coefficient $h$. Newton's law of cooling for convection heat transfer rate $\dot{Q}_{\text{conv}} = hA_s(T_s - T_\infty)$ can be rearranged as

$$\dot{Q}_{\text{conv}} = \frac{T_s - T_\infty}{R_{\text{conv}}} \qquad \text{(W)} \qquad\qquad (17\text{–}7)$$

$$\dot{Q} = \frac{T_1 - T_2}{R}$$

(a) Heat flow

$$I = \frac{\mathbf{V}_1 - \mathbf{V}_2}{R_e}$$

(b) Electric current flow

**FIGURE 17–3**

Analogy between thermal and electrical resistance concepts.

**FIGURE 17–4**
Schematic for convection resistance at a surface.

where

$$R_{conv} = \frac{1}{hA_s} \qquad \text{(K/W)} \qquad (17\text{–}8)$$

is the *thermal resistance* of the surface against heat convection, or simply the **convection resistance** of the surface (Fig. 17–4). Note that when the convection heat transfer coefficient is very large ($h \rightarrow \infty$), the convection resistance becomes *zero* and $T_s \approx T_\infty$. That is, the surface offers *no resistance to convection,* and thus it does not slow down the heat transfer process. This situation is approached in practice at surfaces where boiling and condensation occur. Also note that the surface does not have to be a plane surface. Equation 17–8 for convection resistance is valid for surfaces of any shape, provided that the assumption of $h$ = constant and uniform is reasonable.

When the wall is surrounded by a gas, the *radiation effects,* which we have ignored so far, can be significant and may need to be considered. The rate of radiation heat transfer between a surface of emissivity $\varepsilon$ and area $A_s$ at temperature $T_s$ and the surrounding surfaces at some average temperature $T_{surr}$ can be expressed as

$$\dot{Q}_{rad} = \varepsilon \sigma A_s (T_s^4 - T_{surr}^4) = h_{rad} A_s (T_s - T_{surr}) = \frac{T_s - T_{surr}}{R_{rad}} \qquad \text{(W)} \qquad (17\text{–}9)$$

where

$$R_{rad} = \frac{1}{h_{rad} A_s} \qquad \text{(K/W)} \qquad (17\text{–}10)$$

is the *thermal resistance* of a surface against radiation, or the **radiation resistance**, and

$$h_{rad} = \frac{\dot{Q}_{rad}}{A_s(T_s - T_{surr})} = \varepsilon \sigma (T_s^2 + T_{surr}^2)(T_s + T_{surr}) \qquad \text{(W/m}^2\cdot\text{K)} \qquad (17\text{–}11)$$

is the **radiation heat transfer coefficient**. Note that both $T_s$ and $T_{surr}$ *must* be in K in the evaluation of $h_{rad}$. The definition of the radiation heat transfer coefficient enables us to express radiation conveniently in an analogous manner to convection in terms of a temperature difference. But $h_{rad}$ depends strongly on temperature while $h_{conv}$ usually does not.

A surface exposed to the surrounding air involves convection and radiation simultaneously, and the total heat transfer at the surface is determined by adding (or subtracting, if in the opposite direction) the radiation and convection components. The convection and radiation resistances are parallel to each other, as shown in Fig. 17–5, and may cause some complication in the thermal resistance network. When $T_{surr} \approx T_\infty$, the radiation effect can properly be accounted for by replacing $h$ in the convection resistance relation by

$$h_{combined} = h_{conv} + h_{rad} \qquad \text{(W/m}^2\cdot\text{K)} \qquad (17\text{–}12)$$

where $h_{combined}$ is the **combined heat transfer coefficient** discussed in Chapter 16. This way all complications associated with radiation are avoided.

**FIGURE 17–5**
Schematic for convection and radiation resistances at a surface.

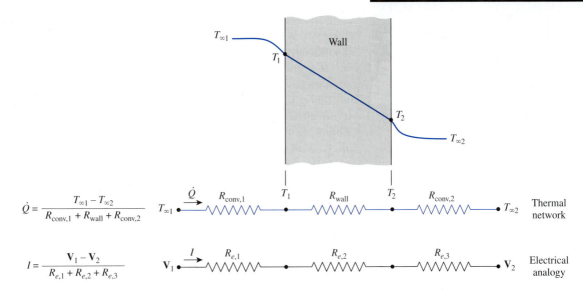

$$\dot{Q} = \frac{T_{\infty 1} - T_{\infty 2}}{R_{\mathrm{conv},1} + R_{\mathrm{wall}} + R_{\mathrm{conv},2}}$$

$$I = \frac{V_1 - V_2}{R_{e,1} + R_{e,2} + R_{e,3}}$$

**FIGURE 17–6**

The thermal resistance network for heat transfer through a plane wall subjected to convection on both sides, and the electrical analogy.

## Thermal Resistance Network

Now consider steady one-dimensional heat transfer through a plane wall of thickness $L$, area $A$, and thermal conductivity $k$ that is exposed to convection on both sides to fluids at temperatures $T_{\infty 1}$ and $T_{\infty 2}$ with heat transfer coefficients $h_1$ and $h_2$, respectively, as shown in Fig. 17–6. Assuming $T_{\infty 2} < T_{\infty 1}$, the variation of temperature will be as shown in the figure. Note that the temperature varies linearly in the wall, and asymptotically approaches $T_{\infty 1}$ and $T_{\infty 2}$ in the fluids as we move away from the wall.

Under steady conditions we have

$$\begin{pmatrix} \text{Rate of} \\ \textit{heat convection} \\ \text{into the wall} \end{pmatrix} = \begin{pmatrix} \text{Rate of} \\ \textit{heat conduction} \\ \text{through the wall} \end{pmatrix} = \begin{pmatrix} \text{Rate of} \\ \textit{heat convection} \\ \text{from the wall} \end{pmatrix}$$

or

$$\dot{Q} = h_1 A(T_{\infty 1} - T_1) = kA\frac{T_1 - T_2}{L} = h_2 A(T_2 - T_{\infty 2}) \qquad \textbf{(17–13)}$$

which can be rearranged as

$$\dot{Q} = \frac{T_{\infty 1} - T_1}{1/h_1 A} = \frac{T_1 - T_2}{L/kA} = \frac{T_2 - T_{\infty 2}}{1/h_2 A}$$

$$= \frac{T_{\infty 1} - T_1}{R_{\mathrm{conv},\,1}} = \frac{T_1 - T_2}{R_{\mathrm{wall}}} = \frac{T_2 - T_{\infty 2}}{R_{\mathrm{conv},\,2}} \qquad \textbf{(17–14)}$$

If

$$\frac{a_1}{b_1} = \frac{a_2}{b_2} = \cdots = \frac{a_n}{b_n} = c$$

then

$$\frac{a_1 + a_2 + \cdots + a_n}{b_1 + b_2 + \cdots + b_n} = c$$

For example,

$$\frac{1}{4} = \frac{2}{8} = \frac{5}{20} = 0.25$$

and

$$\frac{1 + 2 + 5}{4 + 8 + 20} = 0.25$$

**FIGURE 17–7**
A useful mathematical identity.

Once the rate of heat transfer is calculated, Eq. 17–14 can also be used to determine the intermediate temperatures $T_1$ or $T_2$. Adding the numerators and denominators yields (Fig. 17–7)

$$\dot{Q} = \frac{T_{\infty 1} - T_{\infty 2}}{R_{\text{total}}} \quad \text{(W)} \qquad (17\text{–}15)$$

where

$$R_{\text{total}} = R_{\text{conv, 1}} + R_{\text{wall}} + R_{\text{conv, 2}} = \frac{1}{h_1 A} + \frac{L}{kA} + \frac{1}{h_2 A} \quad \text{(K/W)} \qquad (17\text{–}16)$$

Note that the heat transfer area $A$ is constant for a plane wall, and the rate of heat transfer through a wall separating two media is equal to the overall temperature difference $(T_{\infty 1} - T_{\infty 2})$ divided by the total thermal resistance between the media. Also note that the thermal resistances are in *series,* and the equivalent thermal resistance is determined by simply *adding* the individual resistances, just like the electrical resistances connected in series. Thus, the electrical analogy still applies. We summarize this as *the rate of steady heat transfer between two surfaces is equal to the temperature difference divided by the total thermal resistance between those two surfaces.*

Another observation that can be made from Eq. 17–15 is that the ratio of the temperature drop to the thermal resistance across any layer is constant, and thus the temperature drop across any layer is proportional to the thermal resistance of the layer. The larger the resistance, the larger the temperature drop. In fact, the equation $\dot{Q} = \Delta T / R$ can be rearranged as

$$\Delta T = \dot{Q} R \quad \text{(°C)} \qquad (17\text{–}17)$$

which indicates that the *temperature drop* across any layer is equal to the *rate of heat transfer* times the *thermal resistance* across that layer (Fig. 17–8). You may recall that this is also true for voltage drop across an electrical resistance when the electric current is constant.

It is sometimes convenient to express heat transfer through a medium in an analogous manner to Newton's law of cooling as

$$\dot{Q} = U A \, \Delta T \quad \text{(W)} \qquad (17\text{–}18)$$

where $U$ is the **overall heat transfer coefficient** with the unit W/m²·K. The overall heat transfer coefficient is usually used in heat transfer calculations associated with heat exchangers (Chapter 22). It is also used in heat transfer calculations through windows, commonly referred to as U-factor. A comparison of Eqs. 17–15 and 17–18 reveals that

$$UA = \frac{1}{R_{\text{total}}} \quad \text{(W/K)} \qquad (17\text{–}19)$$

**FIGURE 17–8**
The temperature drop across a layer is proportional to its thermal resistance.

Therefore, for a unit area, the overall heat transfer coefficient is equal to the inverse of the total thermal resistance.

Note that we do not need to know the surface temperatures of the wall in order to evaluate the rate of steady heat transfer through it. All we need to know is the convection heat transfer coefficients and the fluid temperatures on both sides of the wall. The *surface temperature* of the wall can be determined as described above using the thermal resistance concept, but by taking the surface at which the temperature is to be determined as one of the terminal surfaces. For example, once $\dot{Q}$ is evaluated, the surface temperature $T_1$ can be determined from

$$\dot{Q} = \frac{T_{\infty 1} - T_1}{R_{\text{conv, 1}}} = \frac{T_{\infty 1} - T_1}{1/h_1 A} \qquad \text{(17–20)}$$

## Multilayer Plane Walls

In practice we often encounter plane walls that consist of several layers of different materials. The thermal resistance concept can still be used to determine the rate of steady heat transfer through such *composite* walls. As you may have already guessed, this is done by simply noting that the conduction resistance of each wall is $L/kA$ connected in series, and using the electrical analogy. That is, by dividing the *temperature difference* between two surfaces at known temperatures by the *total thermal resistance* between them.

Consider a plane wall that consists of two layers (such as a brick wall with a layer of insulation). The rate of steady heat transfer through this two-layer composite wall can be expressed as (Fig. 17–9)

$$\dot{Q} = \frac{T_{\infty 1} - T_{\infty 2}}{R_{\text{total}}} \qquad \text{(17–21)}$$

where $R_{\text{total}}$ is the *total thermal resistance*, expressed as

$$R_{\text{total}} = R_{\text{conv, 1}} + R_{\text{wall, 1}} + R_{\text{wall, 2}} + R_{\text{conv, 2}}$$

$$= \frac{1}{h_1 A} + \frac{L_1}{k_1 A} + \frac{L_2}{k_2 A} + \frac{1}{h_2 A} \qquad \text{(17–22)}$$

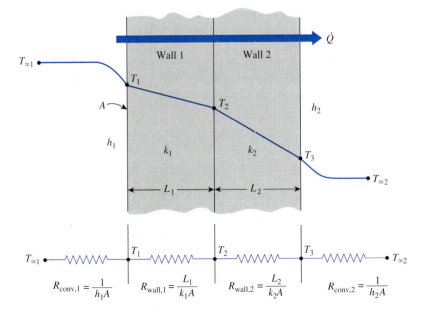

**FIGURE 17–9**
The thermal resistance network for heat transfer through a two-layer plane wall subjected to convection on both sides.

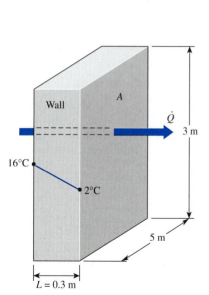

To find $T_1$: $\dot{Q} = \dfrac{T_{\infty 1} - T_1}{R_{conv,1}}$

To find $T_2$: $\dot{Q} = \dfrac{T_{\infty 1} - T_2}{R_{conv,1} + R_{wall,1}}$

To find $T_3$: $\dot{Q} = \dfrac{T_3 - T_{\infty 2}}{R_{conv,2}}$

**FIGURE 17–10**
The evaluation of the surface and interface temperatures when $T_{\infty 1}$ and $T_{\infty 2}$ are given and $\dot{Q}$ is calculated.

**FIGURE 17–11**
Schematic for Example 17–1.

The subscripts 1 and 2 in the $R_{wall}$ relations above indicate the first and the second layers, respectively. We could also obtain this result by following the approach already used for the single-layer case by noting that the rate of steady heat transfer $\dot{Q}$ through a multilayer medium is constant, and thus it must be the same through each layer. Note from the thermal resistance network that the resistances are *in series*, and thus the *total thermal resistance* is simply the *arithmetic sum* of the individual thermal resistances in the path of heat transfer.

This result for the *two-layer* case is analogous to the *single-layer* case, except that an *additional resistance* is added for the *additional layer*. This result can be extended to plane walls that consist of *three* or *more layers* by adding an *additional resistance* for each *additional layer*.

Once $\dot{Q}$ is *known*, an unknown surface temperature $T_j$ at any surface or interface $j$ can be determined from

$$\dot{Q} = \frac{T_i - T_j}{R_{total,\,i-j}} \qquad (17\text{–}23)$$

where $T_i$ is a *known* temperature at location $i$ and $R_{total,\,i-j}$ is the total thermal resistance between locations $i$ and $j$. For example, when the fluid temperatures $T_{\infty 1}$ and $T_{\infty 2}$ for the two-layer case shown in Fig. 17–9 are available and $\dot{Q}$ is calculated from Eq. 17–21, the interface temperature $T_2$ between the two walls can be determined from (Fig. 17–10)

$$\dot{Q} = \frac{T_{\infty 1} - T_2}{R_{conv,\,1} + R_{wall,\,1}} = \frac{T_{\infty 1} - T_2}{\dfrac{1}{h_1 A} + \dfrac{L_1}{k_1 A}} \qquad (17\text{–}24)$$

The temperature drop across a layer is easily determined from Eq. 17–17 by multiplying $\dot{Q}$ by the thermal resistance of that layer.

The thermal resistance concept is widely used in practice because it is intuitively easy to understand and it has proven to be a powerful tool in the solution of a wide range of heat transfer problems. But its use is limited to systems through which the rate of heat transfer $\dot{Q}$ remains *constant*; that is, to systems involving *steady* heat transfer with *no heat generation* (such as resistance heating or chemical reactions) within the medium.

**EXAMPLE 17–1    Heat Loss through a Wall**

Consider a 3-m-high, 5-m-wide, and 0.3-m-thick wall whose thermal conductivity is $k = 0.9$ W/m·K (Fig. 17–11). On a certain day, the temperatures of the inner and the outer surfaces of the wall are measured to be 16°C and 2°C, respectively. Determine the rate of heat loss through the wall on that day.

**SOLUTION** The two surfaces of a wall are maintained at specified temperatures. The rate of heat loss through the wall is to be determined.

*Assumptions* **1** Heat transfer through the wall is steady since the surface temperatures remain constant at the specified values. **2** Heat transfer is one-dimensional since any significant temperature gradients exist in the direction from the indoors to the outdoors. **3** Thermal conductivity is constant.

*Properties* The thermal conductivity is given to be $k = 0.9$ W/m·K.

*Analysis* Noting that heat transfer through the wall is by conduction and the area of the wall is $A = 3 \text{ m} \times 5 \text{ m} = 15 \text{ m}^2$, the steady rate of heat transfer through the wall can be determined from Eq. 17–3 to be

$$\dot{Q} = kA\frac{T_1 - T_2}{L} = (0.9 \text{ W/m·°C})(15 \text{ m}^2)\frac{(16 - 2)°C}{0.3 \text{ m}} = \textbf{630 W}$$

We could also determine the steady rate of heat transfer through the wall by making use of the thermal resistance concept from

$$\dot{Q} = \frac{\Delta T_{\text{wall}}}{R_{\text{wall}}}$$

where

$$R_{\text{wall}} = \frac{L}{kA} = \frac{0.3 \text{ m}}{(0.9 \text{ W/m·°C})(15 \text{ m}^2)} = 0.02222°C/W$$

Substituting, we get

$$\dot{Q} = \frac{(16 - 2)°C}{0.02222°C/W} = 630 \text{ W}$$

*Discussion* This is the same result obtained earlier. Note that heat conduction through a plane wall with specified surface temperatures can be determined directly and easily without utilizing the thermal resistance concept. However, the thermal resistance concept serves as a valuable tool in more complex heat transfer problems, as you will see in the following examples. Also, the units W/m·°C and W/m·K for thermal conductivity are equivalent, and thus interchangeable. This is also the case for °C and K for temperature differences.

---

*EXAMPLE 17–2*     **Heat Loss through a Single-Pane Window**

Consider a 0.8-m-high and 1.5-m-wide glass window with a thickness of 8 mm and a thermal conductivity of $k = 0.78$ W/m·K. Determine the steady rate of heat transfer through this glass window and the temperature of its inner surface for a day during which the room is maintained at 20°C while the temperature of the outdoors is −10°C. Take the heat transfer coefficients on the inner and outer surfaces of the window to be $h_1 = 10$ W/m²·K and $h_2 = 40$ W/m²·K, which includes the effects of radiation.

**SOLUTION** Heat loss through a window glass is considered. The rate of heat transfer through the window and the inner surface temperature are to be determined.

**FIGURE 17–12**
Schematic for Example 17–2.

*Assumptions* **1** Heat transfer through the window is steady since the surface temperatures remain constant at the specified values. **2** Heat transfer through the wall is one-dimensional since any significant temperature gradients exist in the direction from the indoors to the outdoors. **3** Thermal conductivity is constant.

*Properties* The thermal conductivity is given to be $k = 0.78$ W/m·K.

*Analysis* This problem involves conduction through the glass window and convection at its surfaces, and can best be handled by making use of the thermal resistance concept and drawing the thermal resistance network, as shown in Fig. 17–12. Noting that the area of the window is $A = 0.8$ m $\times$ 1.5 m $=$ 1.2 m², the individual resistances are evaluated from their definitions to be

$$R_i = R_{\text{conv, 1}} = \frac{1}{h_1 A} = \frac{1}{(10 \text{ W/m}^2\cdot\text{K})(1.2 \text{ m}^2)} = 0.08333°\text{C/W}$$

$$R_{\text{glass}} = \frac{L}{kA} = \frac{0.008 \text{ m}}{(0.78 \text{ W/m·K})(1.2 \text{ m}^2)} = 0.00855°\text{C/W}$$

$$R_o = R_{\text{conv, 2}} = \frac{1}{h_2 A} = \frac{1}{(40 \text{ W/m}^2\cdot\text{K})(1.2 \text{ m}^2)} = 0.02083°\text{C/W}$$

Noting that all three resistances are in series, the total resistance is

$$R_{\text{total}} = R_{\text{conv, 1}} + R_{\text{glass}} + R_{\text{conv, 2}} = 0.08333 + 0.00855 + 0.02083$$
$$= 0.1127°\text{C/W}$$

Then the steady rate of heat transfer through the window becomes

$$\dot{Q} = \frac{T_{\infty 1} - T_{\infty 2}}{R_{\text{total}}} = \frac{[20 - (-10)]°\text{C}}{0.1127°\text{C/W}} = \textbf{266 W}$$

Knowing the rate of heat transfer, the inner surface temperature of the window glass can be determined from

$$\dot{Q} = \frac{T_{\infty 1} - T_1}{R_{\text{conv, 1}}} \quad \longrightarrow \quad T_1 = T_{\infty 1} - \dot{Q}R_{\text{conv, 1}}$$
$$= 20°\text{C} - (266 \text{ W})(0.08333°\text{C/W})$$
$$= \textbf{-2.2°C}$$

*Discussion* Note that the inner surface temperature of the window glass is $-2.2°$C even though the temperature of the air in the room is maintained at 20°C. Such low surface temperatures are highly undesirable since they cause the formation of fog or even frost on the inner surfaces of the glass when the humidity in the room is high.

**EXAMPLE 17–3**    **Heat Loss through Double-Pane Windows**

Consider a 0.8-m-high and 1.5-m-wide double-pane window consisting of two 4-mm-thick layers of glass ($k = 0.78$ W/m·K) separated by a 10-mm-wide stagnant air space ($k = 0.026$ W/m·K). Determine the steady rate of heat transfer through this double-pane window and the temperature of its inner surface for a day during which the room is maintained at 20°C while the temperature of the outdoors is −10°C. Take the convection heat transfer coefficients on the inner and outer surfaces of the window to be $h_1 = 10$ W/m²·K and $h_2 = 40$ W/m²·K, which includes the effects of radiation.

**SOLUTION** A double-pane window is considered. The rate of heat transfer through the window and the inner surface temperature are to be determined.

*Analysis* This example problem is identical to the previous one except that the single 8-mm-thick window glass is replaced by two 4-mm-thick glasses that enclose a 10-mm-wide stagnant air space. Therefore, the thermal resistance network of this problem involves two additional conduction resistances corresponding to the two additional layers, as shown in Fig. 17–13. Noting that the area of the window is again $A = 0.8$ m × 1.5 m = 1.2 m², the individual resistances are evaluated from their definitions to be

$$R_i = R_{\text{conv, 1}} = \frac{1}{h_1 A} = \frac{1}{(10 \text{ W/m}^2\text{·K})(1.2 \text{ m}^2)} = 0.08333°\text{C/W}$$

$$R_1 = R_3 = R_{\text{glass}} = \frac{L_1}{k_1 A} = \frac{0.004 \text{ m}}{(0.78 \text{ W/m·K})(1.2 \text{ m}^2)} = 0.00427°\text{C/W}$$

$$R_2 = R_{\text{air}} = \frac{L_2}{k_2 A} = \frac{0.01 \text{ m}}{(0.026 \text{ W/m·K})(1.2 \text{ m}^2)} = 0.3205°\text{C/W}$$

$$R_o = R_{\text{conv, 2}} = \frac{1}{h_2 A} = \frac{1}{(40 \text{ W/m}^2\text{·K})(1.2 \text{ m}^2)} = 0.02083°\text{C/W}$$

Noting that all three resistances are in series, the total resistance is

$$R_{\text{total}} = R_{\text{conv, 1}} + R_{\text{glass, 1}} + R_{\text{air}} + R_{\text{glass, 2}} + R_{\text{conv, 2}}$$
$$= 0.08333 + 0.00427 + 0.3205 + 0.00427 + 0.02083$$
$$= 0.4332°\text{C/W}$$

Then the steady rate of heat transfer through the window becomes

$$\dot{Q} = \frac{T_{\infty 1} - T_{\infty 2}}{R_{\text{total}}} = \frac{[20 - (-10)]°\text{C}}{0.4332°\text{C/W}} = \mathbf{69.2 \text{ W}}$$

which is about one-fourth of the result obtained in the previous example. This explains the popularity of the double- and even triple-pane windows in cold climates. The drastic reduction in the heat transfer rate in this case is due to the large thermal resistance of the air layer between the glasses.

The inner surface temperature of the window in this case will be

$$T_1 = T_{\infty 1} - \dot{Q} R_{\text{conv, 1}} = 20°\text{C} - (69.2 \text{ W})(0.08333°\text{C/W}) = \mathbf{14.2°\text{C}}$$

which is considerably higher than the −2.2°C obtained in the previous example. Therefore, a double-pane window will rarely get fogged. A double-pane window will also reduce the heat gain in summer, and thus reduce the air-conditioning costs.

**FIGURE 17–13**
Schematic for Example 17–3.

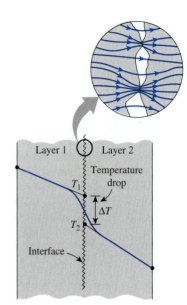

**FIGURE 17–14**

Temperature distribution and heat flow lines along two solid plates pressed against each other for the case of perfect and imperfect contact.

(a) Ideal (perfect) thermal contact

(b) Actual (imperfect) thermal contact

**FIGURE 17–15**

A typical experimental setup for the determination of thermal contact resistance.

*From Song et al.*

## 17–2 ▪ THERMAL CONTACT RESISTANCE

In the analysis of heat conduction through multilayer solids, we assumed "perfect contact" at the interface of two layers, and thus no temperature drop at the interface. This would be the case when the surfaces are perfectly smooth and they produce a perfect contact at each point. In reality, however, even flat surfaces that appear smooth to the eye turn out to be rather rough when examined under a microscope, as shown in Fig. 17–14, with numerous peaks and valleys. That is, a surface is *microscopically rough* no matter how smooth it appears to be.

When two such surfaces are pressed against each other, the peaks form good material contact but the valleys form voids filled with air in most cases. As a result, an interface contains numerous *air gaps* of varying sizes that act as *insulation* because of the low thermal conductivity of air. Thus, an interface offers some resistance to heat transfer, and this resistance for a unit interface area is called the **thermal contact resistance**, $R_c$. The value of $R_c$ is determined experimentally using a setup like the one shown in Fig. 17–15, and as expected, there is considerable scatter of data because of the difficulty in characterizing the surfaces.

Consider heat transfer through two metal rods of cross-sectional area $A$ that are pressed against each other. Heat transfer through the interface of these two rods is the sum of the heat transfers through the *solid contact spots* (solid-to-solid conduction) and the *gaps* (conduction and/or radiation across the gaps) in the noncontact areas (which is a major contributor to heat transfer) and can be expressed as

$$\dot{Q} = \dot{Q}_{\text{contact}} + \dot{Q}_{\text{gap}} \tag{17–25}$$

It can also be expressed in an analogous manner to Newton's law of cooling as

$$\dot{Q} = h_c A \, \Delta T_{\text{interface}} \tag{17–26}$$

where $A$ is the apparent interface area (which is the same as the cross-sectional area of the rods) and $\Delta T_{\text{interface}}$ is the effective temperature difference at the interface. The quantity $h_c$, which corresponds to the convection heat transfer coefficient, is called the **thermal contact conductance** and is expressed as

$$h_c = \frac{\dot{Q}/A}{\Delta T_{\text{interface}}} \qquad (\text{W/m}^2\cdot\text{K}) \qquad (17\text{--}27)$$

It is related to thermal contact resistance by

$$R_c = \frac{1}{h_c} = \frac{\Delta T_{\text{interface}}}{\dot{Q}/A} \qquad (\text{m}^2\cdot\text{K/W}) \qquad (17\text{--}28)$$

That is, thermal contact resistance is the inverse of thermal contact conductance. Usually, thermal contact conductance is reported in the literature, but the concept of thermal contact resistance serves as a better vehicle for explaining the effect of interface on heat transfer. Note that $R_c$ represents thermal contact resistance for a *unit area*. The thermal resistance for the entire interface is obtained by dividing $R_c$ by the apparent interface area $A$.

The thermal contact resistance can be determined from Eq. 17–28 by measuring the temperature drop at the interface and dividing it by the heat flux under steady conditions. The value of thermal contact resistance depends on the *surface roughness* and the *material properties* as well as the *temperature* and *pressure* at the interface and the *type of fluid* trapped at the interface. The situation becomes more complex when plates are fastened by bolts, screws, or rivets since the interface pressure in this case is nonuniform. The thermal contact resistance in that case also depends on the plate thickness, the bolt radius, and the size of the contact zone. Thermal contact resistance is observed to *decrease* with *decreasing surface roughness* and *increasing interface pressure*, as expected. Most experimentally determined values of the thermal contact resistance fall between 0.000005 and 0.0005 m²·K/W (the corresponding range of thermal contact conductance is 2000 to 200,000 W/m²·K).

When we analyze heat transfer in a medium consisting of two or more layers, the first thing we need to know is whether the thermal contact resistance is *significant* or not. We can answer this question by comparing the magnitudes of the thermal resistances of the layers with typical values of thermal contact resistance. For example, the thermal resistance of a 1-cm-thick layer of an insulating material for a unit surface area is

$$R_{c, \text{insulation}} = \frac{L}{k} = \frac{0.01 \text{ m}}{0.04 \text{ W/m·K}} = 0.25 \text{ m}^2\cdot\text{K/W}$$

whereas for a 1-cm-thick layer of copper, it is

$$R_{c, \text{copper}} = \frac{L}{k} = \frac{0.01 \text{ m}}{386 \text{ W/m·K}} = 0.000026 \text{ m}^2\cdot\text{K/W}$$

Comparing the values above with typical values of thermal contact resistance, we conclude that thermal contact resistance is significant and can even dominate the heat transfer for good heat conductors such as metals, but can be

**TABLE 17–1**

Thermal contact conductance for aluminum plates with different fluids at the interface for a surface roughness of 10 μm and interface pressure of 1 atm (from Fried, 1969).

| Fluid at the interface | Contact conductance, $h_c$, W/m²·K |
|---|---|
| Air | 3640 |
| Helium | 9520 |
| Hydrogen | 13,900 |
| Silicone oil | 19,000 |
| Glycerin | 37,700 |

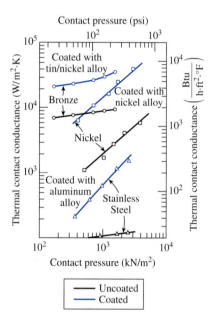

**FIGURE 17–16**

Effect of metallic coatings on thermal contact conductance

*From Peterson, 1987.*

disregarded for poor heat conductors such as insulations. This is not surprising since insulating materials consist mostly of air space just like the inter-face itself.

The thermal contact resistance can be minimized by applying a thermally conducting liquid called a *thermal grease* such as silicon oil on the surfaces before they are pressed against each other. This is commonly done when attaching electronic components such as power transistors to heat sinks. The thermal contact resistance can also be reduced by replacing the air at the interface by a *better conducting gas* such as helium or hydrogen, as shown in Table 17–1.

Another way to minimize the contact resistance is to insert a *soft metallic foil* such as tin, silver, copper, nickel, or aluminum between the two surfaces. Experimental studies show that the thermal contact resistance can be reduced by a factor of up to 7 by a metallic foil at the interface. For maximum effectiveness, the foils must be very thin. The effect of metallic coatings on thermal contact conductance is shown in Fig. 17–16 for various metal surfaces.

There is considerable uncertainty in the contact conductance data reported in the literature, and care should be exercised when using them. In Table 17–2 some experimental results are given for the contact conductance between similar and dissimilar metal surfaces for use in preliminary design calculations. Note that the *thermal contact conductance* is *highest* (and thus the contact resistance is lowest) for *soft metals* with *smooth surfaces* at *high pressure*.

---

**EXAMPLE 17–4**    **Equivalent Thickness for Contact Resistance**

The thermal contact conductance at the interface of two 1-cm-thick aluminum plates is measured to be 11,000 W/m²·K. Determine the thickness of the aluminum plate whose thermal resistance is equal to the thermal resistance of the interface between the plates (Fig. 17–17).

**SOLUTION**    The thickness of the aluminum plate whose thermal resistance is equal to the thermal contact resistance is to be determined.

**Properties**    The thermal conductivity of aluminum at room temperature is $k = 237$ W/m·K (Table A–24).

**Analysis**    Noting that thermal contact resistance is the inverse of thermal contact conductance, the thermal contact resistance is

$$R_c = \frac{1}{h_c} = \frac{1}{11,000 \text{ W/m}^2\cdot\text{K}} = 0.909 \times 10^{-4} \text{ m}^2\cdot\text{K/W}$$

For a unit surface area, the thermal resistance of a flat plate is defined as

$$R = \frac{L}{k}$$

where $L$ is the thickness of the plate and $k$ is the thermal conductivity. Setting $R = R_c$, the equivalent thickness is determined from the relation above to be

$$L = kR_c = (237 \text{ W/m}\cdot\text{K})(0.909 \times 10^{-4} \text{ m}^2\cdot\text{K/W}) = 0.0215 \text{ m} = \textbf{2.15 cm}$$

## TABLE 17–2

Thermal contact conductance of some metal surfaces in air (from various sources)

| Material | Surface condition | Roughness, μm | Temperature, °C | Pressure, MPa | $h_c$,* W/m²·K |
|----------|-------------------|---------------|------------------|---------------|----------------|
| **Identical Metal Pairs** | | | | | |
| 416 Stainless steel | Ground | 2.54 | 90–200 | 0.17–2.5 | 3800 |
| 304 Stainless steel | Ground | 1.14 | 20 | 4–7 | 1900 |
| Aluminum | Ground | 2.54 | 150 | 1.2–2.5 | 11,400 |
| Copper | Ground | 1.27 | 20 | 1.2–20 | 143,000 |
| Copper | Milled | 3.81 | 20 | 1–5 | 55,500 |
| Copper (vacuum) | Milled | 0.25 | 30 | 0.17–7 | 11,400 |
| **Dissimilar Metal Pairs** | | | | | |
| Stainless steel– Aluminum | | 20–30 | 20 | 10 / 20 | 2900 / 3600 |
| Stainless steel– Aluminum | | 1.0–2.0 | 20 | 10 / 20 | 16,400 / 20,800 |
| Steel Ct-30– Aluminum | Ground | 1.4–2.0 | 20 | 10 / 15–35 | 50,000 / 59,000 |
| Steel Ct-30– Aluminum | Milled | 4.5–7.2 | 20 | 10 / 30 | 4800 / 8300 |
| Aluminum-Copper | Ground | 1.17–1.4 | 20 | 5 / 15 | 42,000 / 56,000 |
| Aluminum-Copper | Milled | 4.4–4.5 | 20 | 10 / 20–35 | 12,000 / 22,000 |

*Divide the given values by 5.678 to convert to Btu/h·ft²·°F.

**Discussion** Note that the interface between the two plates offers as much resistance to heat transfer as a 2.15-cm-thick aluminum plate. It is interesting that the thermal contact resistance in this case is greater than the sum of the thermal resistances of both plates.

**FIGURE 17–17**
Schematic for Example 17–4.

## EXAMPLE 17–5    Contact Resistance of Transistors

Four identical power transistors with aluminum casing are attached on one side of a 1-cm-thick 20-cm × 20-cm square copper plate ($k$ = 386 W/m·K) by screws that exert an average pressure of 6 MPa (Fig. 17–18). The base area of each transistor is 8 cm², and each transistor is placed at the center of a 10-cm × 10-cm quarter section of the plate. The interface roughness is estimated to be about 1.5 μm. All transistors are covered by a thick Plexiglas layer, which is a poor conductor of heat, and thus all the heat generated at the junction of the transistor must be dissipated to the ambient at 20°C through the back surface of the copper plate. The combined convection/radiation heat transfer coefficient at the back surface can be taken to be 25 W/m²·K. If the case temperature of

1 cm

20 cm

20°C

Copper plate

Plexiglas cover

70°C

**FIGURE 17–18**
Schematic for Example 17–5.

the transistor is not to exceed 70°C, determine the maximum power each transistor can dissipate safely, and the temperature jump at the case-plate interface.

**SOLUTION** Four identical power transistors are attached on a copper plate. For a maximum case temperature of 70°C, the maximum power dissipation and the temperature jump at the interface are to be determined.

*Assumptions* **1** Steady operating conditions exist. **2** Heat transfer can be approximated as being one-dimensional, although it is recognized that heat conduction in some parts of the plate will be two-dimensional since the plate area is much larger than the base area of the transistor. But the large thermal conductivity of copper will minimize this effect. **3** All the heat generated at the junction is dissipated through the back surface of the plate since the transistors are covered by a thick Plexiglas layer. **4** Thermal conductivities are constant.

*Properties* The thermal conductivity of copper is given to be $k = 386$ W/m·K. The contact conductance is obtained from Table 17–2 to be $h_c = 42,000$ W/m²·K, which corresponds to copper-aluminum interface for the case of 1.17–1.4 μm roughness and 5 MPa pressure, which is sufficiently close to what we have.

*Analysis* The contact area between the case and the plate is given to be 8 cm², and the plate area for each transistor is 100 cm². The thermal resistance network of this problem consists of three resistances in series (interface, plate, and convection), which are determined to be

$$R_{\text{interface}} = \frac{1}{h_c A_c} = \frac{1}{(42,000 \text{ W/m}^2 \cdot \text{K})(8 \times 10^{-4} \text{ m}^2)} = 0.030°\text{C/W}$$

$$R_{\text{plate}} = \frac{L}{kA} = \frac{0.01 \text{ m}}{(386 \text{ W/m} \cdot \text{K})(0.01 \text{ m}^2)} = 0.0026°\text{C/W}$$

$$R_{\text{conv}} = \frac{1}{h_o A} = \frac{1}{(25 \text{ W/m}^2 \cdot \text{K})(0.01 \text{ m}^2)} = 4.0°\text{C/W}$$

The total thermal resistance is then

$$R_{\text{total}} = R_{\text{interface}} + R_{\text{plate}} + R_{\text{ambient}} = 0.030 + 0.0026 + 4.0 = 4.0326°\text{C/W}$$

Note that the thermal resistance of a copper plate is very small and can be ignored altogether. Then the rate of heat transfer is determined to be

$$\dot{Q} = \frac{\Delta T}{R_{\text{total}}} = \frac{(70 - 20)°\text{C}}{4.0326°\text{C/W}} = \mathbf{12.4 \text{ W}}$$

Therefore, the power transistor should not be operated at power levels greater than 12.4 W if the case temperature is not to exceed 70°C.

The temperature jump at the interface is determined from

$$\Delta T_{\text{interface}} = \dot{Q} R_{\text{interface}} = (12.4 \text{ W})(0.030°\text{C/W}) = \mathbf{0.37°C}$$

which is not very large. Therefore, even if we eliminate the thermal contact resistance at the interface completely, we lower the operating temperature of the transistor in this case by less than 0.4°C.

# 17–3 ▪ GENERALIZED THERMAL RESISTANCE NETWORKS

The *thermal resistance* concept or the *electrical analogy* can also be used to solve steady heat transfer problems that involve parallel layers or combined series-parallel arrangements. Although such problems are often two- or even three-dimensional, approximate solutions can be obtained by assuming one-dimensional heat transfer and using the thermal resistance network.

Consider the composite wall shown in Fig. 17–19, which consists of two parallel layers. The thermal resistance network, which consists of two parallel resistances, can be represented as shown in the figure. Noting that the total heat transfer is the sum of the heat transfers through each layer, we have

$$\dot{Q} = \dot{Q}_1 + \dot{Q}_2 = \frac{T_1 - T_2}{R_1} + \frac{T_1 - T_2}{R_2} = (T_1 - T_2)\left(\frac{1}{R_1} + \frac{1}{R_2}\right) \qquad \text{(17–29)}$$

Utilizing electrical analogy, we get

$$\dot{Q} = \frac{T_1 - T_2}{R_{total}} \qquad \text{(17–30)}$$

where

$$\frac{1}{R_{total}} = \frac{1}{R_1} + \frac{1}{R_2} \quad \longrightarrow \quad R_{total} = \frac{R_1 R_2}{R_1 + R_2} \qquad \text{(17–31)}$$

since the resistances are in parallel.

Now consider the combined series-parallel arrangement shown in Fig. 17–20. The total rate of heat transfer through this composite system can again be expressed as

$$\dot{Q} = \frac{T_1 - T_\infty}{R_{total}} \qquad \text{(17–32)}$$

where

$$R_{total} = R_{12} + R_3 + R_{conv} = \frac{R_1 R_2}{R_1 + R_2} + R_3 + R_{conv} \qquad \text{(17–33)}$$

and

$$R_1 = \frac{L_1}{k_1 A_1} \qquad R_2 = \frac{L_2}{k_2 A_2} \qquad R_3 = \frac{L_3}{k_3 A_3} \qquad R_{conv} = \frac{1}{h A_3} \qquad \text{(17–34)}$$

Once the individual thermal resistances are evaluated, the total resistance and the total rate of heat transfer can easily be determined from the relations above.

The result obtained is somewhat approximate, since the surfaces of the third layer are probably not isothermal, and heat transfer between the first two layers is likely to occur.

Two assumptions commonly used in solving complex multidimensional heat transfer problems by treating them as one-dimensional (say, in the

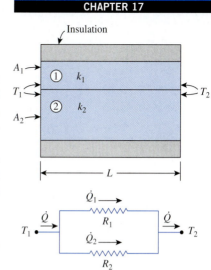

**FIGURE 17–19**
Thermal resistance network for two parallel layers.

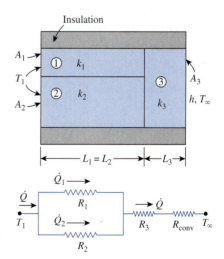

**FIGURE 17–20**
Thermal resistance network for combined series-parallel arrangement.

*x*-direction) using the thermal resistance network are (1) any plane wall normal to the *x*-axis is *isothermal* (i.e., to assume the temperature to vary in the *x*-direction only) and (2) any plane parallel to the *x*-axis is *adiabatic* (i.e., to assume heat transfer to occur in the *x*-direction only). These two assumptions result in different resistance networks, and thus different (but usually close) values for the total thermal resistance and thus heat transfer. The actual result lies between these two values. In geometries in which heat transfer occurs predominantly in one direction, either approach gives satisfactory results.

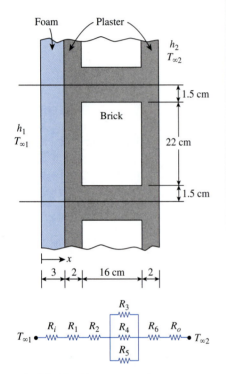

**FIGURE 17–21**
Schematic for Example 17–6.

### EXAMPLE 17–6    Heat Loss through a Composite Wall

A 3-m-high and 5-m-wide wall consists of long 16-cm × 22-cm cross section horizontal bricks ($k = 0.72$ W/m·K) separated by 3-cm-thick plaster layers ($k = 0.22$ W/m·K). There are also 2-cm-thick plaster layers on each side of the brick and a 3-cm-thick rigid foam ($k = 0.026$ W/m·K) on the inner side of the wall, as shown in Fig. 17–21. The indoor and the outdoor temperatures are 20°C and −10°C, respectively, and the convection heat transfer coefficients on the inner and the outer sides are $h_1 = 10$ W/m²·K and $h_2 = 25$ W/m²·K, respectively. Assuming one-dimensional heat transfer and disregarding radiation, determine the rate of heat transfer through the wall.

**SOLUTION**   The composition of a composite wall is given. The rate of heat transfer through the wall is to be determined.

*Assumptions*   **1** Heat transfer is steady since there is no indication of change with time. **2** Heat transfer can be approximated as being one-dimensional since it is predominantly in the *x*-direction. **3** Thermal conductivities are constant. **4** Heat transfer by radiation is negligible.

*Properties*   The thermal conductivities are given to be $k = 0.72$ W/m·K for bricks, $k = 0.22$ W/m·K for plaster layers, and $k = 0.026$ W/m·K for the rigid foam.

*Analysis*   There is a pattern in the construction of this wall that repeats itself every 25-cm distance in the vertical direction. There is no variation in the horizontal direction. Therefore, we consider a 1-m-deep and 0.25-m-high portion of the wall, since it is representative of the entire wall.

Assuming any cross section of the wall normal to the *x*-direction to be *isothermal*, the thermal resistance network for the representative section of the wall becomes as shown in Fig. 17–21. The individual resistances are evaluated as:

$$R_i = R_{\text{conv, 1}} = \frac{1}{h_1 A} = \frac{1}{(10 \text{ W/m}^2\text{·K})(0.25 \times 1 \text{ m}^2)} = 0.40°\text{C/W}$$

$$R_1 = R_{\text{foam}} = \frac{L}{kA} = \frac{0.03 \text{ m}}{(0.026 \text{ W/m·K})(0.25 \times 1 \text{ m}^2)} = 4.62°\text{C/W}$$

$$R_2 = R_6 = R_{\text{plaster, side}} = \frac{L}{kA} = \frac{0.02 \text{ m}}{(0.22 \text{ W/m·K})(0.25 \times 1 \text{ m}^2)}$$
$$= 0.36°\text{C/W}$$

$$R_3 = R_5 = R_{\text{plaster, center}} = \frac{L}{kA} = \frac{0.16 \text{ m}}{(0.22 \text{ W/m·K})(0.015 \times 1 \text{ m}^2)}$$
$$= 48.48°\text{C/W}$$

$$R_4 = R_{brick} = \frac{L}{kA} = \frac{0.16 \text{ m}}{(0.72 \text{ W/m·K})(0.22 \times 1 \text{ m}^2)} = 1.01°C/W$$

$$R_o = R_{conv, 2} = \frac{1}{h_2 A} = \frac{1}{(25 \text{ W/m}^2\text{·K})(0.25 \times 1 \text{ m}^2)} = 0.16°C/W$$

The three resistances $R_3$, $R_4$, and $R_5$ in the middle are parallel, and their equivalent resistance is determined from

$$\frac{1}{R_{mid}} = \frac{1}{R_3} + \frac{1}{R_4} + \frac{1}{R_5} = \frac{1}{48.48} + \frac{1}{1.01} + \frac{1}{48.48} = 1.03 \text{ W/°C}$$

which gives

$$R_{mid} = 0.97°C/W$$

Now all the resistances are in series, and the total resistance is

$$\begin{aligned}R_{total} &= R_i + R_1 + R_2 + R_{mid} + R_6 + R_o \\ &= 0.40 + 4.62 + 0.36 + 0.97 + 0.36 + 0.16 \\ &= 6.87°C/W\end{aligned}$$

Then the steady rate of heat transfer through the wall becomes

$$\dot{Q} = \frac{T_{\infty 1} - T_{\infty 2}}{R_{total}} = \frac{[20 - (-10)]°C}{6.87°C/W} = 4.37 \text{ W} \qquad \text{(per 0.25 m}^2 \text{ surface area)}$$

or $4.37/0.25 = 17.5$ W per $m^2$ area. The total area of the wall is $A = 3 \text{ m} \times 5 \text{ m} = 15 \text{ m}^2$. Then the rate of heat transfer through the entire wall becomes

$$\dot{Q}_{total} = (17.5 \text{ W/m}^2)(15 \text{ m}^2) = \textbf{263 W}$$

Of course, this result is approximate, since we assumed the temperature within the wall to vary in one direction only and ignored any temperature change (and thus heat transfer) in the other two directions.

***Discussion*** In the above solution, we assumed the temperature at any cross section of the wall normal to the $x$-direction to be *isothermal*. We could also solve this problem by going to the other extreme and assuming the surfaces parallel to the $x$-direction to be *adiabatic*. The thermal resistance network in this case will be as shown in Fig. 17–22. By following the approach outlined above, the total thermal resistance in this case is determined to be $R_{total} = 6.97°C/W$, which is very close to the value $6.85°C/W$ obtained before. Thus either approach gives roughly the same result in this case. This example demonstrates that either approach can be used in practice to obtain satisfactory results.

**FIGURE 17–22**
Alternative thermal resistance network for Example 17–6 for the case of surfaces parallel to the primary direction of heat transfer being adiabatic.

**FIGURE 17–23**

Heat is lost from a hot-water pipe to the air outside in the radial direction, and thus heat transfer from a long pipe is one-dimensional.

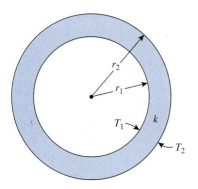

**FIGURE 17–24**

A long cylindrical pipe (or spherical shell) with specified inner and outer surface temperatures $T_1$ and $T_2$.

# 17–4 ■ HEAT CONDUCTION IN CYLINDERS AND SPHERES

Consider steady heat conduction through a hot-water pipe. Heat is continuously lost to the outdoors through the wall of the pipe, and we intuitively feel that heat transfer through the pipe is in the normal direction to the pipe surface and no significant heat transfer takes place in the pipe in other directions (Fig. 17–23). The wall of the pipe, whose thickness is rather small, separates two fluids at different temperatures, and thus the temperature gradient in the radial direction is relatively large. Further, if the fluid temperatures inside and outside the pipe remain constant, then heat transfer through the pipe is *steady*. Thus heat transfer through the pipe can be modeled as *steady* and *one-dimensional*. The temperature of the pipe in this case depends on one direction only (the radial *r*-direction) and can be expressed as $T = T(r)$. The temperature is independent of the azimuthal angle or the axial distance. This situation is approximated in practice in long cylindrical pipes and spherical containers.

In *steady* operation, there is no change in the temperature of the pipe with time at any point. Therefore, the rate of heat transfer into the pipe must be equal to the rate of heat transfer out of it. In other words, heat transfer through the pipe must be constant, $\dot{Q}_{\text{cond, cyl}} = \text{constant}$.

Consider a long cylindrical layer (such as a circular pipe) of inner radius $r_1$, outer radius $r_2$, length $L$, and average thermal conductivity $k$ (Fig. 17–24). The two surfaces of the cylindrical layer are maintained at constant temperatures $T_1$ and $T_2$. There is no heat generation in the layer and the thermal conductivity is constant. For one-dimensional heat conduction through the cylindrical layer, we have $T(r)$. Then Fourier's law of heat conduction for heat transfer through the cylindrical layer can be expressed as

$$\dot{Q}_{\text{cond, cyl}} = -kA \frac{dT}{dr} \quad \text{(W)} \tag{17–35}$$

where $A = 2\pi rL$ is the heat transfer area at location $r$. Note that $A$ depends on $r$, and thus it *varies* in the direction of heat transfer. Separating the variables in the above equation and integrating from $r = r_1$, where $T(r_1) = T_1$, to $r = r_2$, where $T(r_2) = T_2$, gives

$$\int_{r=r_1}^{r_2} \frac{\dot{Q}_{\text{cond, cyl}}}{A} \, dr = -\int_{T=T_1}^{T_2} k \, dT \tag{17–36}$$

Substituting $A = 2\pi rL$ and performing the integrations give

$$\dot{Q}_{\text{cond, cyl}} = 2\pi Lk \frac{T_1 - T_2}{\ln(r_2/r_1)} \quad \text{(W)} \tag{17–37}$$

since $\dot{Q}_{\text{cond, cyl}} = \text{constant}$. This equation can be rearranged as

$$\dot{Q}_{\text{cond, cyl}} = \frac{T_1 - T_2}{R_{\text{cyl}}} \quad \text{(W)} \tag{17–38}$$

where

$$R_{cyl} = \frac{\ln(r_2/r_1)}{2\pi L k} = \frac{\ln(\text{Outer radius/Inner radius})}{2\pi \times \text{Length} \times \text{Thermal conductivity}} \qquad \textbf{(17–39)}$$

is the *thermal resistance* of the cylindrical layer against heat conduction, or simply the *conduction resistance* of the cylinder layer.

We can repeat the analysis for a *spherical layer* by taking $A = 4\pi r^2$ and performing the integrations in Eq. 17–36. The result can be expressed as

$$\dot{Q}_{cond, sph} = \frac{T_1 - T_2}{R_{sph}} \qquad \textbf{(17–40)}$$

where

$$R_{sph} = \frac{r_2 - r_1}{4\pi r_1 r_2 k} = \frac{\text{Outer radius} - \text{Inner radius}}{4\pi(\text{Outer radius})(\text{Inner radius})(\text{Thermal conductivity})} \qquad \textbf{(17–41)}$$

is the *thermal resistance* of the spherical layer against heat conduction, or simply the *conduction resistance* of the spherical layer.

Now consider steady one-dimensional heat transfer through a cylindrical or spherical layer that is exposed to convection on both sides to fluids at temperatures $T_{\infty 1}$ and $T_{\infty 2}$ with heat transfer coefficients $h_1$ and $h_2$, respectively, as shown in Fig. 17–25. The thermal resistance network in this case consists of one conduction and two convection resistances in series, just like the one for the plane wall, and the rate of heat transfer under steady conditions can be expressed as

$$\dot{Q} = \frac{T_{\infty 1} - T_{\infty 2}}{R_{total}} \qquad \textbf{(17–42)}$$

where

$$R_{total} = R_{conv, 1} + R_{cyl} + R_{conv, 2}$$
$$= \frac{1}{(2\pi r_1 L)h_1} + \frac{\ln(r_2/r_1)}{2\pi L k} + \frac{1}{(2\pi r_2 L)h_2} \qquad \textbf{(17–43)}$$

for a *cylindrical* layer, and

$$R_{total} = R_{conv, 1} + R_{sph} + R_{conv, 2}$$
$$= \frac{1}{(4\pi r_1^2)h_1} + \frac{r_2 - r_1}{4\pi r_1 r_2 k} + \frac{1}{(4\pi r_2^2)h_2} \qquad \textbf{(17–44)}$$

*for a spherical* layer. Note that $A$ in the convection resistance relation $R_{conv} = 1/hA$ is the *surface area at which convection occurs*. It is equal to $A = 2\pi r L$ for a cylindrical surface and $A = 4\pi r^2$ for a spherical surface of radius $r$. Also note that the thermal resistances are in series, and thus the total thermal resistance is determined by simply adding the individual resistances, just like the electrical resistances connected in series.

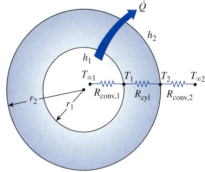

$$R_{total} = R_{conv,1} + R_{cyl} + R_{conv,2}$$

**FIGURE 17–25**

The thermal resistance network for a cylindrical (or spherical) shell subjected to convection from both the inner and the outer sides.

## Multilayered Cylinders and Spheres

Steady heat transfer through multilayered cylindrical or spherical shells can be handled just like multilayered plane walls discussed earlier by simply adding an *additional resistance* in series for each *additional layer.* For example, the steady heat transfer rate through the three-layered composite cylinder of length $L$ shown in Fig. 17–26 with convection on both sides can be expressed as

$$\dot{Q} = \frac{T_{\infty 1} - T_{\infty 2}}{R_{total}} \tag{17–45}$$

where $R_{total}$ is the *total thermal resistance,* expressed as

$$R_{total} = R_{conv,\,1} + R_{cyl,\,1} + R_{cyl,\,2} + R_{cyl,\,3} + R_{conv,\,2}$$
$$= \frac{1}{h_1 A_1} + \frac{\ln(r_2/r_1)}{2\pi L k_1} + \frac{\ln(r_3/r_2)}{2\pi L k_2} + \frac{\ln(r_4/r_3)}{2\pi L k_3} + \frac{1}{h_2 A_4} \tag{17–46}$$

where $A_1 = 2\pi r_1 L$ and $A_4 = 2\pi r_4 L$. Equation 17–46 can also be used for a three-layered spherical shell by replacing the thermal resistances of cylindrical layers by the corresponding spherical ones. Again, note from the thermal resistance network that the resistances are in series, and thus the total thermal resistance is simply the *arithmetic sum* of the individual thermal resistances in the path of heat flow.

Once $\dot{Q}$ is known, we can determine any intermediate temperature $T_j$ by applying the relation $\dot{Q} = (T_i - T_j)/R_{total,\,i-j}$ across any layer or layers such that $T_i$ is a *known* temperature at location $i$ and $R_{total,\,i-j}$ is the total thermal resistance between locations $i$ and $j$ (Fig. 17–27). For example, once $\dot{Q}$ has been calculated, the interface temperature $T_2$ between the first and second cylindrical layers can be determined from

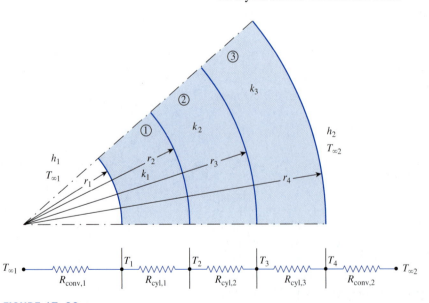

**FIGURE 17–26**
The thermal resistance network for heat transfer through a three-layered composite cylinder subjected to convection on both sides.

$$\dot{Q} = \frac{T_{\infty 1} - T_2}{R_{\text{conv}, 1} + R_{\text{cyl}, 1}} = \frac{T_{\infty 1} - T_2}{\dfrac{1}{h_1(2\pi r_1 L)} + \dfrac{\ln(r_2/r_1)}{2\pi L k_1}}$$ (17–47)

We could also calculate $T_2$ from

$$\dot{Q} = \frac{T_2 - T_{\infty 2}}{R_2 + R_3 + R_{\text{conv}, 2}} = \frac{T_2 - T_{\infty 2}}{\dfrac{\ln(r_3/r_2)}{2\pi L k_2} + \dfrac{\ln(r_4/r_3)}{2\pi L k_3} + \dfrac{1}{h_o(2\pi r_4 L)}}$$ (17–48)

Although both relations give the same result, we prefer the first one since it involves fewer terms and thus less work.

The thermal resistance concept can also be used for *other geometries*, provided that the proper conduction resistances and the proper surface areas in convection resistances are used.

$$\dot{Q} = \frac{T_{\infty 1} - T_1}{R_{\text{conv}, 1}}$$

$$= \frac{T_{\infty 1} - T_2}{R_{\text{conv}, 1} + R_1}$$

$$= \frac{T_1 - T_3}{R_1 + R_2}$$

$$= \frac{T_2 - T_3}{R_2}$$

$$= \frac{T_2 - T_{\infty 2}}{R_2 + R_{\text{conv}, 2}}$$

$$= \cdots$$

**FIGURE 17–27**
The ratio $\Delta T/R$ across any layer is equal to $\dot{Q}$, which remains constant in one-dimensional steady conduction.

---

## EXAMPLE 17–7    Heat Transfer to a Spherical Container

A 3-m internal diameter spherical tank made of 2-cm-thick stainless steel ($k = 15$ W/m·K) is used to store iced water at $T_{\infty 1} = 0°C$. The tank is located in a room whose temperature is $T_{\infty 2} = 22°C$. The walls of the room are also at 22°C. The outer surface of the tank is black and heat transfer between the outer surface of the tank and the surroundings is by natural convection and radiation. The convection heat transfer coefficients at the inner and the outer surfaces of the tank are $h_1 = 80$ W/m²·K and $h_2 = 10$ W/m²·K, respectively. Determine (a) the rate of heat transfer to the iced water in the tank and (b) the amount of ice at 0°C that melts during a 24-h period.

**SOLUTION**   A spherical container filled with iced water is subjected to convection and radiation heat transfer at its outer surface. The rate of heat transfer and the amount of ice that melts per day are to be determined.

*Assumptions*   **1** Heat transfer is steady since the specified thermal conditions at the boundaries do not change with time. **2** Heat transfer is one-dimensional since there is thermal symmetry about the midpoint. **3** Thermal conductivity is constant.

*Properties*   The thermal conductivity of steel is given to be $k = 15$ W/m·K. The heat of fusion of water at atmospheric pressure is $h_{if} = 333.7$ kJ/kg. The outer surface of the tank is black and thus its emissivity is $\varepsilon = 1$.

*Analysis*   (a) The thermal resistance network for this problem is given in Fig. 17–28. Noting that the inner diameter of the tank is $D_1 = 3$ m and the outer diameter is $D_2 = 3.04$ m, the inner and the outer surface areas of the tank are

$$A_1 = \pi D_1^2 = \pi(3 \text{ m})^2 = 28.3 \text{ m}^2$$
$$A_2 = \pi D_2^2 = \pi(3.04 \text{ m})^2 = 29.0 \text{ m}^2$$

Also, the radiation heat transfer coefficient is given by

$$h_{\text{rad}} = \varepsilon\sigma(T_2^2 + T_{\infty 2}^2)(T_2 + T_{\infty 2})$$

But we do not know the outer surface temperature $T_2$ of the tank, and thus we cannot calculate $h_{\text{rad}}$. Therefore, we need to assume a $T_2$ value now and check

**FIGURE 17–28**
Schematic for Example 17–7.

the accuracy of this assumption later. We will repeat the calculations if necessary using a revised value for $T_2$.

We note that $T_2$ must be between 0°C and 22°C, but it must be closer to 0°C, since the heat transfer coefficient inside the tank is much larger. Taking $T_2 = 5°C = 278$ K, the radiation heat transfer coefficient is determined to be

$$h_{rad} = (1)(5.67 \times 10^{-8} \text{ W/m}^2 \cdot \text{K}^4)[(295 \text{ K})^2 + (278 \text{ K})^2][(295 + 278) \text{ K}]$$
$$= 5.34 \text{ W/m}^2 \cdot \text{K} = 5.34 \text{ W/m}^2 \cdot °\text{C}$$

Then the individual thermal resistances become

$$R_i = R_{conv, 1} = \frac{1}{h_1 A_1} = \frac{1}{(80 \text{ W/m}^2 \cdot \text{K})(28.3 \text{ m}^2)} = 0.000442°\text{C/W}$$

$$R_1 = R_{sphere} = \frac{r_2 - r_1}{4\pi k r_1 r_2} = \frac{(1.52 - 1.50) \text{ m}}{4\pi (15 \text{ W/m} \cdot \text{K})(1.52 \text{ m})(1.50 \text{ m})}$$
$$= 0.000047°\text{C/W}$$

$$R_o = R_{conv, 2} = \frac{1}{h_2 A_2} = \frac{1}{(10 \text{ W/m}^2 \cdot \text{K})(29.0 \text{ m}^2)} = 0.00345°\text{C/W}$$

$$R_{rad} = \frac{1}{h_{rad} A_2} = \frac{1}{(5.34 \text{ W/m}^2 \cdot \text{K})(29.0 \text{ m}^2)} = 0.00646°\text{C/W}$$

The two parallel resistances $R_o$ and $R_{rad}$ can be replaced by an equivalent resistance $R_{equiv}$ determined from

$$\frac{1}{R_{equiv}} = \frac{1}{R_o} + \frac{1}{R_{rad}} = \frac{1}{0.00345} + \frac{1}{0.00646} = 444.7 \text{ W/°C}$$

which gives

$$R_{equiv} = 0.00225°\text{C/W}$$

Now all the resistances are in series, and the total resistance is

$$R_{total} = R_i + R_1 + R_{equiv} = 0.000442 + 0.000047 + 0.00225 = 0.00274°\text{C/W}$$

Then the steady rate of heat transfer to the iced water becomes

$$\dot{Q} = \frac{T_{\infty 2} - T_{\infty 1}}{R_{total}} = \frac{(22 - 0)°\text{C}}{0.00274°\text{C/W}} = \textbf{8029 W} \qquad (\text{or } \dot{Q} = 8.029 \text{ kJ/s})$$

To check the validity of our original assumption, we now determine the outer surface temperature from

$$\dot{Q} = \frac{T_{\infty 2} - T_2}{R_{equiv}} \longrightarrow T_2 = T_{\infty 2} - \dot{Q} R_{equiv}$$
$$= 22°\text{C} - (8029 \text{ W})(0.00225°\text{C/W}) = 4°\text{C}$$

which is sufficiently close to the 5°C assumed in the determination of the radiation heat transfer coefficient. Therefore, there is no need to repeat the calculations using 4°C for $T_2$.

(*b*) The total amount of heat transfer during a 24-h period is

$$Q = \dot{Q}\,\Delta t = (8.029 \text{ kJ/s})(24 \times 3600 \text{ s}) = 693{,}700 \text{ kJ}$$

Noting that it takes 333.7 kJ of energy to melt 1 kg of ice at 0°C, the amount of ice that will melt during a 24-h period is

$$m_{ice} = \frac{Q}{h_{if}} = \frac{693{,}700 \text{ kJ}}{333.7 \text{ kJ/kg}} = \textbf{2079 kg}$$

Therefore, about 2 metric tons of ice will melt in the tank every day.

*Discussion*   An easier way to deal with combined convection and radiation at a surface when the surrounding medium and surfaces are at the same temperature is to add the radiation and convection heat transfer coefficients and to treat the result as the convection heat transfer coefficient. That is, to take $h = 10 + 5.34 = 15.34 \text{ W/m}^2\cdot\text{K}$ in this case. This way, we can ignore radiation since its contribution is accounted for in the convection heat transfer coefficient. The convection resistance of the outer surface in this case would be

$$R_{combined} = \frac{1}{h_{combined}A_2} = \frac{1}{(15.34 \text{ W/m}^2\cdot\text{K})(29.0 \text{ m}^2)} = 0.00225°\text{C/W}$$

which is identical to the value obtained for equivalent resistance for the parallel convection and the radiation resistances.

---

### EXAMPLE 17–8     Heat Loss through an Insulated Steam Pipe

Steam at $T_{\infty 1} = 320°\text{C}$ flows in a cast iron pipe ($k = 80 \text{ W/m}\cdot\text{K}$) whose inner and outer diameters are $D_1 = 5$ cm and $D_2 = 5.5$ cm, respectively. The pipe is covered with 3-cm-thick glass wool insulation with $k = 0.05 \text{ W/m}\cdot\text{K}$. Heat is lost to the surroundings at $T_{\infty 2} = 5°\text{C}$ by natural convection and radiation, with a combined heat transfer coefficient of $h_2 = 18 \text{ W/m}^2\cdot\text{K}$. Taking the heat transfer coefficient inside the pipe to be $h_1 = 60 \text{ W/m}^2\cdot\text{K}$, determine the rate of heat loss from the steam per unit length of the pipe. Also determine the temperature drops across the pipe shell and the insulation.

**SOLUTION**   A steam pipe covered with glass wool insulation is subjected to convection on its surfaces. The rate of heat transfer per unit length and the temperature drops across the pipe and the insulation are to be determined.

*Assumptions*   **1** Heat transfer is steady since there is no indication of any change with time. **2** Heat transfer is one-dimensional since there is thermal symmetry about the centerline and no variation in the axial direction. **3** Thermal conductivities are constant. **4** The thermal contact resistance at the interface is negligible.

*Properties*   The thermal conductivities are given to be $k = 80 \text{ W/m}\cdot\text{K}$ for cast iron and $k = 0.05 \text{ W/m}\cdot\text{K}$ for glass wool insulation.

**FIGURE 17–29**
Schematic for Example 17–8.

*Analysis* The thermal resistance network for this problem involves four resistances in series and is given in Fig. 17–29. Taking $L = 1$ m, the areas of the surfaces exposed to convection are determined to be

$$A_1 = 2\pi r_1 L = 2\pi(0.025 \text{ m})(1 \text{ m}) = 0.157 \text{ m}^2$$
$$A_3 = 2\pi r_3 L = 2\pi(0.0575 \text{ m})(1 \text{ m}) = 0.361 \text{ m}^2$$

Then the individual thermal resistances become

$$R_i = R_{\text{conv, 1}} = \frac{1}{h_1 A_1} = \frac{1}{(60 \text{ W/m}^2\cdot\text{K})(0.157 \text{ m}^2)} = 0.106°\text{C/W}$$

$$R_1 = R_{\text{pipe}} = \frac{\ln(r_2/r_1)}{2\pi k_1 L} = \frac{\ln(2.75/2.5)}{2\pi(80 \text{ W/m}\cdot\text{K})(1 \text{ m})} = 0.0002°\text{C/W}$$

$$R_2 = R_{\text{insulation}} = \frac{\ln(r_3/r_2)}{2\pi k_2 L} = \frac{\ln(5.75/2.75)}{2\pi(0.05 \text{ W/m}\cdot\text{K})(1 \text{ m})} = 2.35°\text{C/W}$$

$$R_o = R_{\text{conv, 2}} = \frac{1}{h_2 A_3} = \frac{1}{(18 \text{ W/m}^2\cdot\text{K})(0.361 \text{ m}^2)} = 0.154°\text{C/W}$$

Noting that all resistances are in series, the total resistance is determined to be

$$R_{\text{total}} = R_i + R_1 + R_2 + R_o = 0.106 + 0.0002 + 2.35 + 0.154 = 2.61°\text{C/W}$$

Then the steady rate of heat loss from the steam becomes

$$\dot{Q} = \frac{T_{\infty 1} - T_{\infty 2}}{R_{\text{total}}} = \frac{(320 - 5)°\text{C}}{2.61°\text{C/W}} = \textbf{121 W} \qquad \text{(per m pipe length)}$$

The heat loss for a given pipe length can be determined by multiplying the above quantity by the pipe length $L$.

The temperature drops across the pipe and the insulation are determined from Eq. 17–17 to be

$$\Delta T_{\text{pipe}} = \dot{Q} R_{\text{pipe}} = (121 \text{ W})(0.0002°\text{C/W}) = \textbf{0.02°C}$$
$$\Delta T_{\text{insulation}} = \dot{Q} R_{\text{insulation}} = (121 \text{ W})(2.35°\text{C/W}) = \textbf{284°C}$$

That is, the temperatures between the inner and the outer surfaces of the pipe differ by 0.02°C, whereas the temperatures between the inner and the outer surfaces of the insulation differ by 284°C.

*Discussion* Note that the thermal resistance of the pipe is too small relative to the other resistances and can be neglected without causing any significant error. Also note that the temperature drop across the pipe is practically zero, and thus the pipe can be assumed to be isothermal. The resistance to heat flow in insulated pipes is primarily due to insulation.

# 17–5 · CRITICAL RADIUS OF INSULATION

We know that adding more insulation to a wall or to the attic always decreases heat transfer. The thicker the insulation, the lower the heat transfer rate. This is expected, since the heat transfer area $A$ is constant, and adding insulation always increases the thermal resistance of the wall without increasing the convection resistance.

Adding insulation to a cylindrical pipe or a spherical shell, however, is a different matter. The additional insulation increases the conduction resistance of the insulation layer but decreases the convection resistance of the surface because of the increase in the outer surface area for convection. The heat transfer from the pipe may increase or decrease, depending on which effect dominates.

Consider a cylindrical pipe of outer radius $r_1$ whose outer surface temperature $T_1$ is maintained constant (Fig. 17–30). The pipe is now insulated with a material whose thermal conductivity is $k$ and outer radius is $r_2$. Heat is lost from the pipe to the surrounding medium at temperature $T_\infty$, with a convection heat transfer coefficient $h$. The rate of heat transfer from the insulated pipe to the surrounding air can be expressed as (Fig. 17–31)

$$\dot{Q} = \frac{T_1 - T_\infty}{R_{ins} + R_{conv}} = \frac{T_1 - T_\infty}{\dfrac{\ln(r_2/r_1)}{2\pi Lk} + \dfrac{1}{h(2\pi r_2 L)}} \qquad \textbf{(17–49)}$$

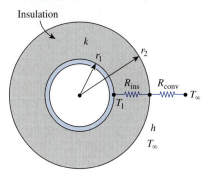

**FIGURE 17–30**
An insulated cylindrical pipe exposed to convection from the outer surface and the thermal resistance network associated with it.

variation of $\dot{Q}$ with the outer radius of the insulation $r_2$ is plotted in Fig. 17–31. The value of $r_2$ at which $\dot{Q}$ reaches a maximum is determined from the requirement that $d\dot{Q}/dr_2 = 0$ (zero slope). Performing the differentiation and solving for $r_2$ yields the **critical radius of insulation** for a cylindrical body to be

$$r_{cr,\,cylinder} = \frac{k}{h} \qquad \text{(m)} \qquad \textbf{(17–50)}$$

Note that the critical radius of insulation depends on the thermal conductivity of the insulation $k$ and the external convection heat transfer coefficient $h$. The rate of heat transfer from the cylinder increases with the addition of insulation for $r_2 < r_{cr}$, reaches a maximum when $r_2 = r_{cr}$, and starts to decrease for $r_2 > r_{cr}$. Thus, insulating the pipe may actually increase the rate of heat transfer from the pipe instead of decreasing it when $r_2 < r_{cr}$.

The important question to answer at this point is whether we need to be concerned about the critical radius of insulation when insulating hot-water pipes or even hot-water tanks. Should we always check and make sure that the outer radius of insulation sufficiently exceeds the critical radius before we install any insulation? Probably not, as explained here.

The value of the critical radius $r_{cr}$ is the largest when $k$ is large and $h$ is small. Noting that the lowest value of $h$ encountered in practice is about 5 W/m²·K for the case of natural convection of gases, and that the thermal conductivity of common insulating materials is about 0.05 W/m²·K, the largest value of the critical radius we are likely to encounter is

$$r_{cr,\,max} = \frac{k_{max,\,insulation}}{h_{min}} \approx \frac{0.05 \text{ W/m·K}}{5 \text{ W/m}^2\text{·K}} = 0.01 \text{ m} = 1 \text{ cm}$$

This value would be even smaller when the radiation effects are considered. The critical radius would be much less in forced convection, often less than 1 mm, because of much larger $h$ values associated with forced convection. Therefore, we can insulate hot-water or steam pipes freely without worrying about the possibility of increasing the heat transfer by insulating the pipes.

The radius of electric wires may be smaller than the critical radius. Therefore, the plastic electrical insulation may actually *enhance* the heat transfer

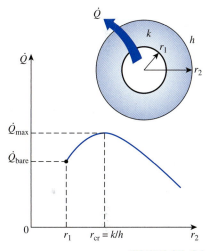

**FIGURE 17–31**
The variation of heat transfer rate with the outer radius of the insulation $r_2$ when $r_1 < r_{cr}$.

from electric wires and thus keep their steady operating temperatures at lower and thus safer levels.

The discussions above can be repeated for a sphere, and it can be shown in a similar manner that the critical radius of insulation for a spherical shell is

$$r_{\text{cr, sphere}} = \frac{2k}{h} \tag{17–51}$$

where $k$ is the thermal conductivity of the insulation and $h$ is the convection heat transfer coefficient on the outer surface.

---

**EXAMPLE 17–9    Heat Loss from an Insulated Electric Wire**

A 3-mm-diameter and 5-m-long electric wire is tightly wrapped with a 2-mm-thick plastic cover whose thermal conductivity is $k = 0.15$ W/m·K. Electrical measurements indicate that a current of 10 A passes through the wire and there is a voltage drop of 8 V along the wire. If the insulated wire is exposed to a medium at $T_\infty = 30°C$ with a heat transfer coefficient of $h = 12$ W/m²·K, determine the temperature at the interface of the wire and the plastic cover in steady operation. Also determine whether doubling the thickness of the plastic cover will increase or decrease this interface temperature.

**SOLUTION**    An electric wire is tightly wrapped with a plastic cover. The interface temperature and the effect of doubling the thickness of the plastic cover on the interface temperature are to be determined.

**Assumptions**    **1** Heat transfer is steady since there is no indication of any change with time. **2** Heat transfer is one-dimensional since there is thermal symmetry about the centerline and no variation in the axial direction. **3** Thermal conductivities are constant. **4** The thermal contact resistance at the interface is negligible. **5** Heat transfer coefficient incorporates the radiation effects, if any.

**Properties**    The thermal conductivity of plastic is given to be $k = 0.15$ W/m·K.

**Analysis**    Heat is generated in the wire and its temperature rises as a result of resistance heating. We assume heat is generated uniformly throughout the wire and is transferred to the surrounding medium in the radial direction. In steady operation, the rate of heat transfer becomes equal to the heat generated within the wire, which is determined to be

$$\dot{Q} = \dot{W}_e = \mathbf{V}I = (8 \text{ V})(10 \text{ A}) = 80 \text{ W}$$

The thermal resistance network for this problem involves a conduction resistance for the plastic cover and a convection resistance for the outer surface in series, as shown in Fig. 17–32. The values of these two resistances are

$$A_2 = (2\pi r_2)L = 2\pi(0.0035 \text{ m})(5 \text{ m}) = 0.110 \text{ m}^2$$

$$R_{\text{conv}} = \frac{1}{hA_2} = \frac{1}{(12 \text{ W/m}^2\text{·K})(0.110 \text{ m}^2)} = 0.76°C/\text{W}$$

$$R_{\text{plastic}} = \frac{\ln(r_2/r_1)}{2\pi kL} = \frac{\ln(3.5/1.5)}{2\pi(0.15 \text{ W/m·K})(5 \text{ m})} = 0.18°C/\text{W}$$

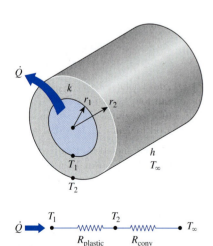

**FIGURE 17–32**
Schematic for Example 17–9.

and therefore

$$R_{\text{total}} = R_{\text{plastic}} + R_{\text{conv}} = 0.76 + 0.18 = 0.94°C/W$$

Then the interface temperature can be determined from

$$\dot{Q} = \frac{T_1 - T_\infty}{R_{\text{total}}} \longrightarrow \begin{aligned} T_1 &= T_\infty + \dot{Q}R_{\text{total}} \\ &= 30°C + (80\ \text{W})(0.94°C/W) = \mathbf{105°C} \end{aligned}$$

Note that we did not involve the electrical wire directly in the thermal resistance network, since the wire involves heat generation.

To answer the second part of the question, we need to know the critical radius of insulation of the plastic cover. It is determined from Eq. 17–50 to be

$$r_{\text{cr}} = \frac{k}{h} = \frac{0.15\ \text{W/m·K}}{12\ \text{W/m}^2\text{·K}} = 0.0125\ \text{m} = 12.5\ \text{mm}$$

which is larger than the radius of the plastic cover. Therefore, increasing the thickness of the plastic cover will *enhance* heat transfer until the outer radius of the cover reaches 12.5 mm. As a result, the rate of heat transfer $\dot{Q}$ will *increase* when the interface temperature $T_1$ is held constant, or $T_1$ will *decrease* when $\dot{Q}$ is held constant, which is the case here.

**Discussion** It can be shown by repeating the calculations above for a 4-mm-thick plastic cover that the interface temperature drops to 90.6°C when the thickness of the plastic cover is doubled. It can also be shown in a similar manner that the interface reaches a minimum temperature of 83°C when the outer radius of the plastic cover equals the critical radius.

# 17–6 ▪ HEAT TRANSFER FROM FINNED SURFACES

The rate of heat transfer from a surface at a temperature $T_s$ to the surrounding medium at $T_\infty$ is given by Newton's law of cooling as

$$\dot{Q}_{\text{conv}} = hA_s(T_s - T_\infty)$$

where $A_s$ is the heat transfer surface area and $h$ is the convection heat transfer coefficient. When the temperatures $T_s$ and $T_\infty$ are fixed by design considerations, as is often the case, there are *two ways* to increase the rate of heat transfer: to increase the *convection heat transfer coefficient h* or to increase the *surface area $A_s$*. Increasing $h$ may require the installation of a pump or fan, or replacing the existing one with a larger one, but this approach may or may not be practical. Besides, it may not be adequate. The alternative is to increase the surface area by attaching to the surface *extended surfaces* called *fins* made of highly conductive materials such as aluminum. Finned surfaces are manufactured by extruding, welding, or wrapping a thin metal sheet on a surface. Fins enhance heat transfer from a surface by exposing a larger surface area to convection and radiation.

An interesting application of fins from about 150 million years ago, the Jurassic era, is shown in Fig. 17–33. The dinosaur stegosaurus lived during

**FIGURE 17–33**
Presumed cooling fins on dinosaur stegosaurus.
*© Alamy RF.*

### FIGURE 17–34

The thin plate fins of a car radiator greatly increase the rate of heat transfer to the air.
*© Yunus Çengel, photo by James Kleiser;*
*© The McGraw-Hill Companies,*
*Inc./Christopher Kerrigan, Photographer.*

### FIGURE 17–35

Some innovative fin designs.

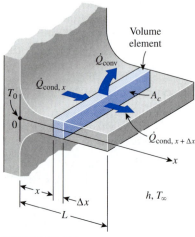

### FIGURE 17–36

Volume element of a fin at location $x$ having a length of $\Delta x$, cross-sectional area of $A_c$, and perimeter of $p$.

this era and it had two rows of big (and bizarre) bony plates down its back. For a long time, scientists thought that the plates were some kind of armor to protect the vegetarian from predators. We now know that a lot of blood flowed through the plates, and they may have acted like a car radiator. The heart pumped blood through the plates, and the plates acted like cooling fins to cool the blood down.

Finned surfaces are commonly used in practice to enhance heat transfer, and they often increase the rate of heat transfer from a surface severalfold. The car radiator shown in Fig. 17–34 is an example of a finned surface. The closely packed thin metal sheets attached to the hot-water tubes increase the surface area for convection and thus the rate of convection heat transfer from the tubes to the air many times. There are a variety of innovative fin designs available in the market, and they seem to be limited only by imagination (Fig. 17–35).

In the analysis of fins, we consider *steady* operation with *no heat generation* in the fin, and we assume the thermal conductivity $k$ of the material to remain constant. We also assume the convection heat transfer coefficient $h$ to be *constant* and *uniform* over the entire surface of the fin for convenience in the analysis. We recognize that the convection heat transfer coefficient $h$, in general, varies along the fin as well as its circumference, and its value at a point is a strong function of the *fluid motion* at that point. The value of $h$ is usually much lower at the *fin base* than it is at the *fin tip* because the fluid is surrounded by solid surfaces near the base, which seriously disrupt its motion to the point of "suffocating" it, while the fluid near the fin tip has little contact with a solid surface and thus encounters little resistance to flow. Therefore, adding too many fins on a surface may actually decrease the overall heat transfer when the decrease in $h$ offsets any gain resulting from the increase in the surface area.

## Fin Equation

Consider a volume element of a fin at location $x$ having a length of $\Delta x$, cross-sectional area of $A_c$, and a perimeter of $p$, as shown in Fig. 17–36. Under steady conditions, the energy balance on this volume element can be expressed as

$$
\begin{pmatrix} \text{Rate of } heat \\ conduction \text{ into} \\ \text{the element at } x \end{pmatrix} = \begin{pmatrix} \text{Rate of } heat \\ conduction \text{ from the} \\ \text{element at } x + \Delta x \end{pmatrix} + \begin{pmatrix} \text{Rate of } heat \\ convection \text{ from} \\ \text{the element} \end{pmatrix}
$$

or

$$\dot{Q}_{cond,\,x} = \dot{Q}_{cond,\,x + \Delta x} + \dot{Q}_{conv}$$

where

$$\dot{Q}_{conv} = h(p\,\Delta x)(T - T_\infty)$$

Substituting and dividing by $\Delta x$, we obtain

$$\frac{\dot{Q}_{cond,\,x+\Delta x} - \dot{Q}_{cond,\,x}}{\Delta x} + hp(T - T_\infty) = 0 \qquad \textbf{(17–52)}$$

Taking the limit as $\Delta x \to 0$ gives

$$\frac{d\dot{Q}_{cond}}{dx} + hp(T - T_\infty) = 0 \qquad \textbf{(17–53)}$$

From Fourier's law of heat conduction we have

$$\dot{Q}_{cond} = -kA_c \frac{dT}{dx} \qquad \textbf{(17–54)}$$

where $A_c$ is the cross-sectional area of the fin at location $x$. Substitution of this relation into Eq. 17–53 gives the differential equation governing heat transfer in fins,

$$\frac{d}{dx}\left(kA_c \frac{dT}{dx}\right) - hp(T - T_\infty) = 0 \qquad \textbf{(17–55)}$$

In general, the cross-sectional area $A_c$ and the perimeter $p$ of a fin vary with $x$, which makes this differential equation difficult to solve. In the special case of *constant cross section* and *constant thermal conductivity*, the differential equation 17–55 reduces to

$$\frac{d^2T}{dx^2} - \frac{hp}{kA_c}(T - T_\infty) = 0 \qquad \text{or} \qquad \frac{d^2\theta}{dx^2} - m^2\theta = 0 \qquad \textbf{(17–56)}$$

where

$$m^2 = \frac{hp}{kA_c} \qquad \textbf{(17–57)}$$

and $\theta = T - T_\infty$ is the *temperature excess*. At the fin base we have $\theta_b = T_b - T_\infty$.

Equation 17–56 is a linear, homogeneous, second-order differential equation with constant coefficients. A fundamental theory of differential equations states that such an equation has two linearly independent solution functions, and its general solution is the linear combination of those two solution functions. A careful examination of the differential equation reveals that subtracting a constant multiple of the solution function $\theta$ from its second derivative yields zero. Thus we conclude that the function $\theta$ and its second derivative must be *constant multiples* of each other. The only functions whose derivatives are constant multiples of the functions themselves are the *exponential functions* (or a linear combination of exponential functions such as sine and cosine hyperbolic functions). Therefore, the solution functions of the differential equation above are the exponential functions $e^{-mx}$ or $e^{mx}$ or constant multiples of them. This can be verified by direct substitution. For example, the second derivative of $e^{-mx}$ is $m^2 e^{-mx}$, and its substitution into Eq. 17–56

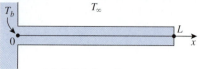

1. Infinitely long fin
2. Negligible heat loss (adiabatic tip)
3. Specified temperature
4. Convection

**FIGURE 17–37**
Boundary conditions at the fin base and the fin tip.

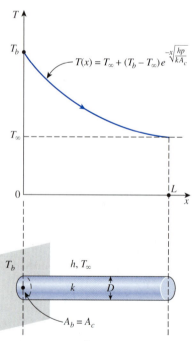

$(p = \pi D, A_c = \pi D^2/4$ for a cylindrical fin)

**FIGURE 17–38**
A long circular fin of uniform cross section and the variation of temperature along it.

yields zero. Therefore, the general solution of the differential equation Eq. 17–56 is

$$\theta(x) = C_1 e^{mx} + C_2 e^{-mx} \tag{17-58}$$

where $C_1$ and $C_2$ are arbitrary constants whose values are to be determined from the boundary conditions at the base and at the tip of the fin. Note that we need only two conditions to determine $C_1$ and $C_2$ uniquely.

The temperature of the plate to which the fins are attached is normally known in advance. Therefore, at the fin base we have a *specified temperature* boundary condition, expressed as

*Boundary condition at fin base:* $\quad \theta(0) = \theta_b = T_b - T_\infty \tag{17-59}$

At the fin tip we have several possibilities, including infinitely long fins, negligible heat loss (idealized as an adiabatic tip), specified temperature, and convection (Fig. 17–37). Next, we consider each case separately.

## 1 Infinitely Long Fin ($T_{\text{fin tip}} = T_\infty$)

For a sufficiently long fin of *uniform* cross section ($A_c$ = constant), the temperature of the fin at the fin tip approaches the environment temperature $T_\infty$ and thus $\theta$ approaches zero. That is,

*Boundary condition at fin tip:* $\quad \theta(L) = T(L) - T_\infty = 0 \quad$ as $\quad L \rightarrow \infty$

This condition is satisfied by the function $e^{-mx}$, but not by the other prospective solution function $e^{mx}$ since it tends to infinity as $x$ gets larger. Therefore, the general solution in this case will consist of a constant multiple of $e^{-mx}$. The value of the constant multiple is determined from the requirement that at the fin base where $x = 0$ the value of $\theta$ is $\theta_b$. Noting that $e^{-mx} = e^0 = 1$, the proper value of the constant is $\theta_b$, and the solution function we are looking for is $\theta(x) = \theta_b e^{-mx}$. This function satisfies the differential equation as well as the requirements that the solution reduce to $\theta_b$ at the fin base and approach zero at the fin tip for large $x$. Noting that $\theta = T - T_\infty$ and $m = \sqrt{hp/kA_c}$, the variation of temperature along the fin in this case can be expressed as

*Very long fin:* $\qquad \dfrac{T(x) - T_\infty}{T_b - T_\infty} = e^{-mx} = e^{-x\sqrt{hp/kA_c}} \tag{17-60}$

Note that the temperature along the fin in this case decreases *exponentially* from $T_b$ to $T_\infty$, as shown in Fig. 17–38. The steady rate of *heat transfer* from the entire fin can be determined from Fourier's law of heat conduction

*Very long fin:* $\qquad \dot{Q}_{\text{long fin}} = -kA_c \dfrac{dT}{dx}\bigg|_{x=0} = \sqrt{hpkA_c}\,(T_b - T_\infty) \tag{17-61}$

where $p$ is the perimeter, $A_c$ is the cross-sectional area of the fin, and $x$ is the distance from the fin base. Alternatively, the rate of heat transfer from the fin could also be determined by considering heat transfer from a differential volume element of the fin and integrating it over the entire surface of the fin:

$$\dot{Q}_{\text{fin}} = \int_{A_{\text{fin}}} h[T(x) - T_\infty]\,dA_{\text{fin}} = \int_{A_{\text{fin}}} h\theta(x)\,dA_{\text{fin}} \tag{17-62}$$

The two approaches described are equivalent and give the same result since, under steady conditions, the heat transfer from the exposed surfaces of the fin is equal to the heat transfer to the fin at the base (Fig. 17–39).

## 2 Negligible Heat Loss from the Fin Tip (Adiabatic fin tip, $\dot{Q}_{\text{fin tip}} = 0$)

Fins are not likely to be so long that their temperature approaches the surrounding temperature at the tip. A more realistic situation is for heat transfer from the fin tip to be negligible since the heat transfer from the fin is proportional to its surface area, and the surface area of the fin tip is usually a negligible fraction of the total fin area. Then the fin tip can be assumed to be adiabatic, and the condition at the fin tip can be expressed as

*Boundary condition at fin tip:* $\qquad \dfrac{d\theta}{dx}\bigg|_{x=L} = 0 \qquad$ **(17–63)**

The condition at the fin base remains the same as expressed in Eq. 17–59. The application of the boundary conditions given by Eqs. (17–59) and (17–63) on the general solution (Eq. 17–58) requires that $\theta(0) = \theta_b = C_1 + C_2$ and $mC_1e^{mL} - mC_2e^{-mL} = 0$, respectively. Solving these two equations simultaneously for $C_1$ and $C_2$ yields $C_1 = \theta_b/(1 + e^{2mL})$ and $C_2 = \theta_b/(1 + e^{-2mL})$. Substituting the relations for $C_1$ and $C_2$ into Eq. 17–58 and using the definition of the hyperbolic cosine function $\cosh x = (e^x + e^{-x})/2$ gives the desired relation for the temperature distribution:

*Adiabatic fin tip:* $\qquad \dfrac{T(x) - T_\infty}{T_b - T_\infty} = \dfrac{\cosh m(L - x)}{\cosh mL} \qquad$ **(17–64)**

*The rate of heat transfer* from the fin can be determined again from Fourier's law of heat conduction:

*Adiabatic fin tip:* $\quad \dot{Q}_{\text{adiabatic tip}} = -kA_c \dfrac{dT}{dx}\bigg|_{x=0} = \sqrt{hpkA_c}\,(T_b - T_\infty)\tanh mL \quad$ **(17–65)**

where the equation for the hyperbolic tangent function is

$$\tanh x = \sinh x/\cosh x = (e^x - e^{-x})/(e^x + e^{-x}).$$

Note that the heat transfer relations for the very long fin and the fin with negligible heat loss at the tip differ by the factor $\tanh mL$, which approaches 1 as $L$ becomes very large.

## 3 Specified Temperature ($T_{\text{fin,tip}} = T_L$)

In this case the temperature at the end of the fin (the fin tip) is fixed at a specified temperature $T_L$. This case could be considered as a generalization of the case of *Infinitely Long Fin* where the fin tip temperature was fixed at $T_\infty$. The condition at the fin tip for this case is

*Boundary condition at fin tip:* $\qquad \theta(L) = \theta_L = T_L - T_\infty \qquad$ **(17–66)**

The fin base boundary condition remains the same as given in Eq. 17–59. Applying the boundary conditions given by Eqs. 17–59 and 17–66 on the general solution (Eq. 17–58) gives, after some lengthy algebra and using the definition

**FIGURE 17–39**
Under steady conditions, heat transfer from the exposed surfaces of the fin is equal to heat conduction to the fin at the base.

of the hyperbolic sine function, $\sinh x = (e^x - e^{-x})/2$, the desired temperature distribution:

*Specified fin tip temperature:*

$$\frac{T(x) - T_\infty}{T_b - T_\infty} = \frac{[(T_L - T_\infty)/(T_b - T_\infty)]\sinh\ mx + \sinh m(L - x)}{\sinh mL} \tag{17–67}$$

Using the Fourier's law of heat conduction, the *rate of heat transfer* from the fin is

*Specified fin tip temperature:*

$$\dot{Q}_{\text{specified temp.}} = -kA_c \frac{dT}{dx}\bigg|_{x=0} \tag{17–68}$$

$$= \sqrt{h\ p\ k\ A_c}(T_b - T_\infty)\frac{\cosh mL - [(T_L - T_\infty)/(T_b - T_\infty)]}{\sinh mL}$$

Note that Eqs. 17–67 and 17–68 reduce to Eqs. 17–60 and 17–61 for the case of *infinitely long fin* ($L \to \infty$).

## 4   Convection from Fin Tip

The fin tips, in practice, are exposed to the surroundings, and thus the proper boundary condition for the fin tip is convection that may also include the effects of radiation. Consider the case of convection only at the tip. The condition at the fin tip can be obtained from an energy balance at the fin tip ($\dot{Q}_{\text{cond}} = \dot{Q}_{\text{conv}}$). That is,

*Boundary condition at fin tip:* $\qquad -kA_c \frac{dT}{dx}\bigg|_{x=L} = hA_c[T(L) - T_\infty]$ **(17–69)**

The boundary condition at the fin base is Eq. 17–59, which is the same as the three previous cases. Substituting the two boundary conditions given by Eqs. 17–59 and 17–69 in the general solution (Eq. 17–58), it may be shown, after some lengthy manipulation that the temperature distribution is

*Convection from fin tip:* $\quad \dfrac{T(x) - T_\infty}{T_b - T_\infty} = \dfrac{\cosh m(L - x)\ +\ (h/mk)\sinh m(L - x)}{\cosh mL + (h/mk)\sinh mL}$ **(17–70)**

The *rate of heat transfer* from the fin can be found by substituting the temperature gradient at the base of the fin, obtained from Eq. 17–70, into the Fourier's law of heat conduction. The result is

*Convection from fin tip:*

$$\dot{Q}_{\text{convection}} = -kA_c \frac{dT}{dx}\bigg|_{x=0}$$

$$= \sqrt{hpkA_c}\ (T_b - T_\infty)\frac{\sinh mL + (h/mk)\cosh mL}{\cosh mL + (h/mk)\sinh mL} \tag{17–71}$$

The solution to the general fin equation for the case of *convection from fin tip* is rather complex. An approximate, yet practical and accurate, way of accounting for the loss from the fin tip is to replace the *fin length L* in the relation for the *insulated tip* case by a **corrected fin length** (defined as (Fig. 17–40)

*Corrected fin length:* $\qquad\qquad L_c = L + \dfrac{A_c}{p}$ **(17–72)**

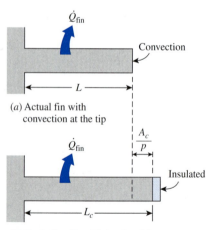

(a) Actual fin with convection at the tip

(b) Equivalent fin with insulated tip

**FIGURE 17–40**
Corrected fin length $L_c$ is defined such that heat transfer from a fin of length $L_c$ with insulated tip is equal to heat transfer from the actual fin of length $L$ with convection at the fin tip.

where $A_c$ is the cross-sectional area and $p$ is the perimeter of the fin at the tip. Multiplying the relation above by the perimeter gives $A_{\text{corrected}} = A_{\text{fin (lateral)}} + A_{\text{tip}}$, which indicates that the fin area determined using the corrected length is equivalent to the sum of the lateral fin area plus the fin tip area.

The corrected length approximation gives very good results when the variation of temperature near the fin tip is small (which is the case when $mL \geq 1$) and the heat transfer coefficient at the fin tip is about the same as that at the lateral surface of the fin. Therefore, *fins subjected to convection at their tips can be treated as fins with insulated tips by replacing the actual fin length by the corrected length in Eqs. 17–64 and 17–65.*

Using the proper relations for $A_c$ and $p$, the corrected lengths for rectangular and cylindrical fins are easily determined to be

$$L_{c,\text{ rectangular fin}} = L + \frac{t}{2} \quad \text{and} \quad L_{c,\text{ cylindrical fin}} = L + \frac{D}{4}$$

where $t$ is the thickness of the rectangular fins and $D$ is the diameter of the cylindrical fins.

## Fin Efficiency

Consider the surface of a *plane wall* at temperature $T_b$ exposed to a medium at temperature $T_\infty$. Heat is lost from the surface to the surrounding medium by convection with a heat transfer coefficient of $h$. Disregarding radiation or accounting for its contribution in the convection coefficient $h$, heat transfer from a surface area $A_s$ is expressed as $\dot{Q} = hA_s(T_s - T_\infty)$.

Now let us consider a fin of constant cross-sectional area $A_c = A_b$ and length $L$ that is attached to the surface with a perfect contact (Fig. 17–41). This time heat is transferrd from the surface to the fin *by conduction* and from the fin to the surrounding medium *by convection* with the same heat transfer coefficient $h$. The temperature of the fin is $T_b$ at the fin base and gradually decreases toward the fin tip. Convection from the fin surface causes the temperature at any cross section to drop somewhat from the midsection toward the outer surfaces. However, the cross-sectional area of the fins is usually very small, and thus the temperature at any cross section can be considered to be uniform. Also, the fin tip can be assumed for convenience and simplicity to be adiabatic by using the corrected length for the fin instead of the actual length.

In the limiting case of *zero thermal resistance* or *infinite thermal conductivity* ($k \rightarrow \infty$), the temperature of the fin is uniform at the base value of $T_b$. The heat transfer from the fin is *maximum* in this case and can be expressed as

$$\dot{Q}_{\text{fin, max}} = hA_{\text{fin}}(T_b - T_\infty) \tag{17–73}$$

In reality, however, the temperature of the fin drops along the fin, and thus the heat transfer from the fin is less because of the decreasing temperature difference $T(x) - T_\infty$ toward the fin tip, as shown in Fig. 17–42. To account for the effect of this decrease in temperature on heat transfer, we define a **fin efficiency** as

$$\eta_{\text{fin}} = \frac{\dot{Q}_{\text{fin}}}{\dot{Q}_{\text{fin, max}}} = \frac{\text{Actual heat transfer rate from the fin}}{\substack{\text{Ideal heat transfer rate from the fin} \\ \text{if the entire fin were at base temperature}}} \tag{17–74}$$

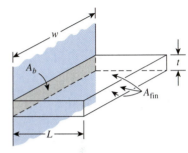

(a) Surface without fins

(b) Surface with a fin

$A_{\text{fin}} = 2 \times w \times L + w \times t$
$\cong 2 \times w \times L$

**FIGURE 17–41**
Fins enhance heat transfer from a surface by enhancing surface area.

(a) Ideal

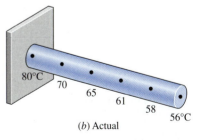

(b) Actual

**FIGURE 17–42**
Ideal and actual temperature distribution along a fin.

## TABLE 17–3

Efficiency and surface areas of common fin configurations

**Straight rectangular fins**

$m = \sqrt{2h/kt}$

$L_c = L + t/2$

$A_{fin} = 2wL_c$

$$\eta_{fin} = \frac{\tanh mL_c}{mL_c}$$

**Straight triangular fins**

$m = \sqrt{2h/kt}$

$A_{fin} = 2w\sqrt{L^2 + (t/2)^2}$

$$\eta_{fin} = \frac{1}{mL} \frac{I_1(2mL)}{I_0(2mL)}$$

$y = (t/2)(1 - x/L)$

**Straight parabolic fins**

$m = \sqrt{2h/kt}$

$A_{fin} = wL[C_1 + (L/t)\ln(t/L + C_1)]$

$C_1 = \sqrt{1 + (t/L)^2}$

$$\eta_{fin} = \frac{2}{1 + \sqrt{(2mL)^2 + 1}}$$

$y = (t/2)(1 - x/L)^2$

**Circular fins of rectangular profile**

$m = \sqrt{2h/kt}$

$r_{2c} = r_2 + t/2$

$A_{fin} = 2\pi(r_{2c}^2 - r_1^2)$

$$\eta_{fin} = C_2 \frac{K_1(mr_1)I_1(mr_{2c}) - I_1(mr_1)K_1(mr_{2c})}{I_0(mr_1)K_1(mr_{2c}) + K_0(mr_1)I_1(mr_{2c})}$$

$$C_2 = \frac{2r_1/m}{r_{2c}^2 - r_1^2}$$

**Pin fins of rectangular profile**

$m = \sqrt{4h/kD}$

$L_c = L + D/4$

$A_{fin} = \pi D L_c$

$$\eta_{fin} = \frac{\tanh mL_c}{mL_c}$$

**Pin fins of triangular profile**

$m = \sqrt{4h/kD}$

$A_{fin} = \frac{\pi D}{2}\sqrt{L^2 + (D/2)^2}$

$$\eta_{fin} = \frac{2}{mL} \frac{I_2(2mL)}{I_1(2mL)}$$

$$I_2(x) = I_0(x) - (2/x)I_1(x) \text{ where } x = 2mL$$

$y = (D/2)(1 - x/L)$

**Pin fins of parabolic profile**

$m = \sqrt{4h/kD}$

$A_{fin} = \frac{\pi L^3}{8D}\left[C_3 C_4 - \frac{L}{2D}\ln(2DC_4/L + C_3)\right]$

$C_3 = 1 + 2(D/L)^2$

$C_4 = \sqrt{1 + (D/L)^2}$

$$\eta_{fin} = \frac{2}{1 + \sqrt{(2mL/3)^2 + 1}}$$

$y = (D/2)(1 - x/L)^2$

**Pin fins of parabolic profile (blunt tip)**

$m = \sqrt{4h/kD}$

$A_{fin} = \frac{\pi D^4}{96L^2}\{[16(L/D)^2 + 1]^{3/2} - 1\}$

$$\eta_{fin} = \frac{3}{2mL} \frac{I_1(4mL/3)}{I_0(4mL/3)}$$

$y = (D/2)(1 - x/L)^{1/2}$

or

$$\dot{Q}_{\text{fin}} = \eta_{\text{fin}} \dot{Q}_{\text{fin, max}} = \eta_{\text{fin}} h A_{\text{fin}} (T_b - T_\infty) \qquad (17\text{--}75)$$

where $A_{\text{fin}}$ is the total surface area of the fin. This relation enables us to determine the heat transfer from a fin when its efficiency is known. For the cases of constant cross section of *very long fins* and *fins with adiabatic tips*, the fin efficiency can be expressed as

$$\eta_{\text{long fin}} = \frac{\dot{Q}_{\text{fin}}}{\dot{Q}_{\text{fin, max}}} = \frac{\sqrt{hpkA_c}\,(T_b - T_\infty)}{hA_{\text{fin}}(T_b - T_\infty)} = \frac{1}{L}\sqrt{\frac{kA_c}{hp}} = \frac{1}{mL} \qquad (17\text{--}76)$$

and

$$\eta_{\text{adiabatic tip}} = \frac{\dot{Q}_{\text{fin}}}{\dot{Q}_{\text{fin, max}}} = \frac{\sqrt{hpkA_c}\,(T_b - T_\infty)\tanh mL}{hA_{\text{fin}}(T_b - T_\infty)} = \frac{\tanh mL}{mL} \qquad (17\text{--}77)$$

since $A_{\text{fin}} = pL$ for fins with constant cross section. Equation 17–77 can also be used for fins subjected to convection provided that the fin length $L$ is replaced by the corrected length $L_c$.

Table 17–3 provides fin efficiency relations for fins with uniform and non-uniform cross section. For fins with non-uniform profile, Eq. 17–56 is no longer valid and the general form of the differential equation governing heat transfer in fins of arbitrary shape, Eq. 17–55, must be used. For these cases the solution is no longer in the form of simple exponential or hyperbolic functions. The mathematical functions $I$ and $K$ that appear in some of these relations are the *modified Bessel functions,* and their values are given in Table 17–4. Efficiencies are plotted in Fig. 17–43 for fins on a *plain surface* and in Fig. 17–44 for *circular fins* of constant thickness. For most fins of constant thickness encountered in practice, the fin thickness $t$ is too small relative to the fin length $L$, and thus the fin tip area is negligible.

Note that fins with triangular and parabolic profiles contain less material and are more efficient than the ones with rectangular profiles, and thus are more suitable for applications requiring minimum weight such as space applications.

An important consideration in the design of finned surfaces is the selection of the proper *fin length L.* Normally the *longer* the fin, the *larger* the heat transfer area and thus the *higher* the rate of heat transfer from the fin. But also the larger the fin, the bigger the mass, the higher the price, and the larger the fluid friction. Therefore, increasing the length of the fin beyond a certain value cannot be justified unless the added benefits outweigh the added cost. Also, the fin efficiency decreases with increasing fin length because of the decrease in fin temperature with length. Fin lengths that cause the fin efficiency to drop below 60 percent usually cannot be justified economically and should be avoided. The efficiency of most fins used in practice is above 90 percent.

## TABLE 17–4

Modified Bessel functions of the first and second kinds*

| $x$ | $e^{-x}I_0(x)$ | $e^{-x}I_1(x)$ | $e^x K_0(x)$ | $e^x K_1(x)$ |
|---|---|---|---|---|
| 0.0 | 1.0000 | 0.0000 | — | — |
| 0.2 | 0.8269 | 0.0823 | 2.1408 | 5.8334 |
| 0.4 | 0.6974 | 0.1368 | 1.6627 | 3.2587 |
| 0.6 | 0.5993 | 0.1722 | 1.4167 | 2.3739 |
| 0.8 | 0.5241 | 0.1945 | 1.2582 | 1.9179 |
| 1.0 | 0.4658 | 0.2079 | 1.1445 | 1.6362 |
| 1.2 | 0.4198 | 0.2153 | 1.0575 | 1.4429 |
| 1.4 | 0.3831 | 0.2185 | 0.9881 | 1.3011 |
| 1.6 | 0.3533 | 0.2190 | 0.9309 | 1.1919 |
| 1.8 | 0.3289 | 0.2177 | 0.8828 | 1.1048 |
| 2.0 | 0.3085 | 0.2153 | 0.8416 | 1.0335 |
| 2.2 | 0.2913 | 0.2121 | 0.8057 | 0.9738 |
| 2.4 | 0.2766 | 0.2085 | 0.7740 | 0.9229 |
| 2.6 | 0.2639 | 0.2047 | 0.7459 | 0.8790 |
| 2.8 | 0.2528 | 0.2007 | 0.7206 | 0.8405 |
| 3.0 | 0.2430 | 0.1968 | 0.6978 | 0.8066 |
| 3.2 | 0.2343 | 0.1930 | 0.6770 | 0.7763 |
| 3.4 | 0.2264 | 0.1892 | 0.6580 | 0.7491 |
| 3.6 | 0.2193 | 0.1856 | 0.6405 | 0.7245 |
| 3.8 | 0.2129 | 0.1821 | 0.6243 | 0.7021 |
| 4.0 | 0.2070 | 0.1788 | 0.6093 | 0.6816 |
| 4.2 | 0.2016 | 0.1755 | 0.5953 | 0.6627 |
| 4.4 | 0.1966 | 0.1725 | 0.5823 | 0.6454 |
| 4.6 | 0.1919 | 0.1695 | 0.5701 | 0.6292 |
| 4.8 | 0.1876 | 0.1667 | 0.5586 | 0.6143 |
| 5.0 | 0.1835 | 0.1640 | 0.5478 | 0.6003 |
| 5.2 | 0.1797 | 0.1614 | 0.5376 | 0.5872 |
| 5.4 | 0.1762 | 0.1589 | 0.5280 | 0.5749 |
| 5.6 | 0.1728 | 0.1565 | 0.5188 | 0.5634 |
| 5.8 | 0.1697 | 0.1542 | 0.5101 | 0.5525 |
| 6.0 | 0.1667 | 0.1521 | 0.5019 | 0.5422 |
| 6.5 | 0.1598 | 0.1469 | 0.4828 | 0.5187 |
| 7.0 | 0.1537 | 0.1423 | 0.4658 | 0.4981 |
| 7.5 | 0.1483 | 0.1380 | 0.4505 | 0.4797 |
| 8.0 | 0.1434 | 0.1341 | 0.4366 | 0.4631 |
| 8.5 | 0.1390 | 0.1305 | 0.4239 | 0.4482 |
| 9.0 | 0.1350 | 0.1272 | 0.4123 | 0.4346 |
| 9.5 | 0.1313 | 0.1241 | 0.4016 | 0.4222 |
| 10.0 | 0.1278 | 0.1213 | 0.3916 | 0.4108 |

*Evaluated from EES using the mathematical functions Bessel_I(x) and Bessel_K(x)

## Fin Effectiveness

Fins are used to *enhance* heat transfer, and the use of fins on a surface cannot be recommended unless the enhancement in heat transfer justifies the added cost and complexity associated with the fins. In fact, there is no assurance that

**FIGURE 17–43**
Efficiency of straight fins of rectangular, triangular, and parabolic profiles.

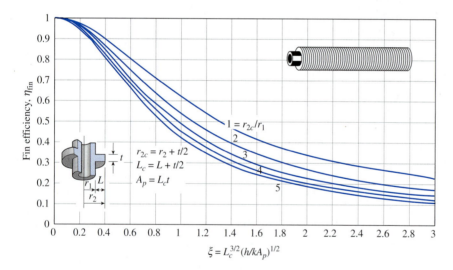

**FIGURE 17–44**
Efficiency of annular fins of constant thickness $t$.

adding fins on a surface will *enhance* heat transfer. The performance of the fins is judged on the basis of the enhancement in heat transfer relative to the no-fin case. The performance of fins is expressed in terms of the *fin effectiveness* $\varepsilon_{\text{fin}}$ defined as (Fig. 17–45)

$$\varepsilon_{\text{fin}} = \frac{\dot{Q}_{\text{fin}}}{\dot{Q}_{\text{no fin}}} = \frac{\dot{Q}_{\text{fin}}}{hA_b\,(T_b - T_\infty)} = \frac{\text{Heat transfer rate from the fin of } \textit{base area } A_b}{\text{Heat transfer rate from the surface of } \textit{area } A_b} \qquad (17\text{–}78)$$

Here, $A_b$ is the cross-sectional area of the fin at the base and $\dot{Q}_{\text{no fin}}$ represents the rate of heat transfer from this area if no fins are attached to the surface. An effectiveness of $\varepsilon_{\text{fin}} = 1$ indicates that the addition of fins to the surface does not affect heat transfer at all. That is, heat conducted to the fin through the base area $A_b$ is equal to the heat transferred from the same area $A_b$ to the surrounding medium. An effectiveness of $\varepsilon_{\text{fin}} < 1$ indicates that the fin actually acts as *insulation*, slowing down the heat transfer from the surface. This situation can occur when fins made of low thermal conductivity materials are used. An effectiveness of $\varepsilon_{\text{fin}} > 1$ indicates that fins are *enhancing* heat transfer from the surface, as they should. However, the use of fins cannot be justified unless $\varepsilon_{\text{fin}}$ is sufficiently larger than 1. Finned surfaces are designed on the basis of *maximizing* effectiveness for a specified cost or *minimizing* cost for a desired effectiveness.

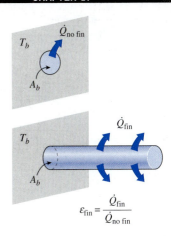

$$\varepsilon_{\text{fin}} = \frac{\dot{Q}_{\text{fin}}}{\dot{Q}_{\text{no fin}}}$$

**FIGURE 17–45**
The effectiveness of a fin.

Note that both the fin efficiency and fin effectiveness are related to the performance of the fin, but they are different quantities. However, they are related to each other by

$$\varepsilon_{\text{fin}} = \frac{\dot{Q}_{\text{fin}}}{\dot{Q}_{\text{no fin}}} = \frac{\dot{Q}_{\text{fin}}}{hA_b\,(T_b - T_\infty)} = \frac{\eta_{\text{fin}}\,hA_{\text{fin}}\,(T_b - T_\infty)}{hA_b\,(T_b - T_\infty)} = \frac{A_{\text{fin}}}{A_b}\,\eta_{\text{fin}} \qquad \text{(17–79)}$$

Therefore, the fin effectiveness can be determined easily when the fin efficiency is known, or vice versa.

The rate of heat transfer from a sufficiently *long* fin of *uniform* cross section under steady conditions is given by Eq. 17–61. Substituting this relation into Eq. 17–78, the effectiveness of such a long fin is determined to be

$$\varepsilon_{\text{long fin}} = \frac{\dot{Q}_{\text{fin}}}{\dot{Q}_{\text{no fin}}} = \frac{\sqrt{hpkA_c}\,(T_b - T_\infty)}{hA_b\,(T_b - T_\infty)} = \sqrt{\frac{kp}{hA_c}} \qquad \text{(17–80)}$$

*since* $A_c = A_b$ in this case. We can draw several important conclusions from the fin effectiveness relation above for consideration in the design and selection of the fins:

- The *thermal conductivity $k$* of the fin material should be as high as possible. Thus it is no coincidence that fins are made from metals, with copper, aluminum, and iron being the most common ones. Perhaps the most widely used fins are made of aluminum because of its low cost and weight and its resistance to corrosion.

- The ratio of the *perimeter* to the *cross-sectional area* of the fin $p/A_c$ should be as high as possible. This criterion is satisfied by *thin* plate fins and *slender* pin fins.

- The use of fins is *most effective* in applications involving a *low convection heat transfer coefficient*. Thus, the use of fins is more easily justified when the medium is a *gas* instead of a *liquid* and the heat transfer is by *natural convection* instead of by forced convection. Therefore, it is no coincidence that in liquid-to-gas heat exchangers such as the car radiator, fins are placed on the *gas* side.

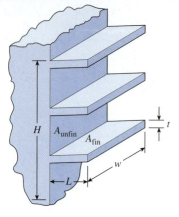

$A_{\text{no fin}} = w \times H$
$A_{\text{unfin}} = w \times H - 3 \times (t \times w)$
$A_{\text{fin}} = 2 \times L \times w + t \times w$
$\qquad \cong 2 \times L \times w \text{ (one fin)}$

**FIGURE 17–46**
Various surface areas associated with a rectangular surface with three fins.

When determining the rate of heat transfer from a finned surface, we must consider the *unfinned portion* of the surface as well as the *fins*. Therefore, the rate of heat transfer for a surface containing $n$ fins can be expressed as

$$
\begin{aligned}
\dot{Q}_{\text{total, fin}} &= \dot{Q}_{\text{unfin}} + \dot{Q}_{\text{fin}} \\
&= hA_{\text{unfin}} (T_b - T_\infty) + \eta_{\text{fin}} hA_{\text{fin}} (T_b - T_\infty) \\
&= h(A_{\text{unfin}} + \eta_{\text{fin}} A_{\text{fin}})(T_b - T_\infty)
\end{aligned}
\qquad \textbf{(17–81)}
$$

We can also define an **overall effectiveness** for a finned surface as the ratio of the total heat transfer from the finned surface to the heat transfer from the same surface if there were no fins,

$$
\varepsilon_{\text{fin, overall}} = \frac{\dot{Q}_{\text{total, fin}}}{\dot{Q}_{\text{total, no fin}}} = \frac{h(A_{\text{unfin}} + \eta_{\text{fin}} A_{\text{fin}})(T_b - T_\infty)}{hA_{\text{no fin}} (T_b - T_\infty)} = \frac{A_{\text{unfin}} + \eta_{\text{fin}} A_{\text{fin}}}{A_{\text{no fin}}}
\qquad \textbf{(17–82)}
$$

where $A_{\text{no fin}}$ is the area of the surface when there are no fins, $A_{\text{fin}}$ is the total surface area of all the fins on the surface, and $A_{\text{unfin}}$ is the area of the unfinned portion of the surface (Fig. 17–46). Note that the overall fin effectiveness depends on the fin density (number of fins per unit length) as well as the effectiveness of the individual fins. The overall effectiveness is a better measure of the performance of a finned surface than the effectiveness of the individual fins.

## Proper Length of a Fin

An important step in the design of a fin is the determination of the appropriate length of the fin once the fin material and the fin cross section are specified. You may be tempted to think that the longer the fin, the larger the surface area and thus the higher the rate of heat transfer. Therefore, for maximum heat transfer, the fin should be infinitely long. However, the temperature drops along the fin exponentially and reaches the environment temperature at some length. The part of the fin beyond this length does not contribute to heat transfer since it is at the temperature of the environment, as shown in Fig. 17–47. Therefore, designing such an "extra long" fin is out of the question since it results in material waste, excessive weight, and increased size and thus increased cost with no benefit in return (in fact, such a long fin will hurt performance since it will suppress fluid motion and thus reduce the convection heat transfer coefficient). Fins that are so long that the temperature approaches the environment temperature cannot be recommended either since the little increase in heat transfer at the tip region cannot justify the disproportionate increase in the weight and cost.

To get a sense of the proper length of a fin, we compare heat transfer from a fin of finite length to heat transfer from an infinitely long fin under the same conditions. The ratio of these two heat transfers is

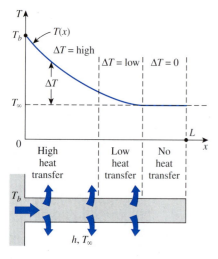

**FIGURE 17–47**
Because of the gradual temperature drop along the fin, the region near the fin tip makes little or no contribution to heat transfer.

*Heat transfer ratio:*
$$
\frac{\dot{Q}_{\text{fin}}}{\dot{Q}_{\text{long fin}}} = \frac{\sqrt{hpkA_c} (T_b - T_\infty) \tanh mL}{\sqrt{hpkA_c} (T_b - T_\infty)} = \tanh mL
\qquad \textbf{(17–83)}
$$

Using a hand calculator, the values of $\tanh mL$ are evaluated for some values of $mL$ and the results are given in Table 17–5. We observe from the table that heat transfer from a fin increases with $mL$ almost linearly at first, but the curve reaches a plateau later and reaches a value for the infinitely long fin at about

$mL = 5$. Therefore, a fin whose length is $L = L = \frac{1}{5}m$ can be considered to be an infinitely long fin. We also observe that reducing the fin length by half in that case (from $mL = 5$ to $mL = 2.5$) causes a drop of just 1 percent in heat transfer. We certainly would not hesitate sacrificing 1 percent in heat transfer performance in return for 50 percent reduction in the size and possibly the cost of the fin. In practice, a fin length that corresponds to about $mL = 1$ will transfer 76.2 percent of the heat that can be transferred by an infinitely long fin, and thus it should offer a good compromise between heat transfer performance and the fin size.

A common approximation used in the analysis of fins is to assume the fin temperature to vary in one direction only (along the fin length) and the temperature variation along other directions is negligible. Perhaps you are wondering if this one-dimensional approximation is a reasonable one. This is certainly the case for fins made of thin metal sheets such as the fins on a car radiator, but we wouldn't be so sure for fins made of thick materials. Studies have shown that the error involved in one-dimensional fin analysis is negligible (less than about 1 percent) when

$$\frac{h\delta}{k} < 0.2$$

where $\delta$ is the characteristic thickness of the fin, which is taken to be the plate thickness $t$ for rectangular fins and the diameter $D$ for cylindrical ones.

Specially designed finned surfaces called *heat sinks,* which are commonly used in the cooling of electronic equipment, involve one-of-a-kind complex geometries, as shown in Table 17–6. The heat transfer performance of heat sinks is usually expressed in terms of their *thermal resistances R* in °C/W, which is defined as

$$\dot{Q}_{fin} = \frac{T_b - T_\infty}{R} = hA_{fin} \, \eta_{fin} \, (T_b - T_\infty) \qquad \text{(17–84)}$$

A small value of thermal resistance indicates a small temperature drop across the heat sink, and thus a high fin efficiency.

### TABLE 17–5

The variation of heat transfer from a fin relative to that from an infinitely long fin

| $mL$ | $\dfrac{\dot{Q}_{fin}}{\dot{Q}_{long\,fin}} = \tanh mL$ |
|---|---|
| 0.1 | 0.100 |
| 0.2 | 0.197 |
| 0.5 | 0.462 |
| 1.0 | 0.762 |
| 1.5 | 0.905 |
| 2.0 | 0.964 |
| 2.5 | 0.987 |
| 3.0 | 0.995 |
| 4.0 | 0.999 |
| 5.0 | 1.000 |

---

### EXAMPLE 17–10    Maximum Power Dissipation of a Transistor

Power transistors that are commonly used in electronic devices consume large amounts of electric power. The failure rate of electronic components increases almost exponentially with operating temperature. As a rule of thumb, the failure rate of electronic components is halved for each 10°C reduction in the junction operating temperature. Therefore, the operating temperature of electronic components is kept below a safe level to minimize the risk of failure.

The sensitive electronic circuitry of a power transistor at the junction is protected by its case, which is a rigid metal enclosure. Heat transfer characteristics of a power transistor are usually specified by the manufacturer in terms of the case-to-ambient thermal resistance, which accounts for both the natural convection and radiation heat transfers.

The case-to-ambient thermal resistance of a power transistor that has a maximum power rating of 10 W is given to be 20°C/W. If the case temperature of

(Continued on page 705)

### TABLE 17–6

Combined natural convection and radiation thermal resistance of various heat sinks used in the cooling of electronic devices between the heat sink and the surroundings. All fins are made of aluminum 6063T-5, are black anodized, and are 76 mm (3 in) long.

**HS 5030**

$R = 0.9°C/W$ (vertical)
$R = 1.2°C/W$ (horizontal)

Dimensions: 76 mm × 105 mm × 44 mm
Surface area: 677 cm$^2$

**HS 6065**

$R = 5°C/W$

Dimensions: 76 mm × 38 mm × 24 mm
Surface area: 387 cm$^2$

**HS 6071**

$R = 1.4°C/W$ (vertical)
$R = 1.8°C/W$ (horizontal)

Dimensions: 76 mm × 92 mm × 26 mm
Surface area: 968 cm$^2$

**HS 6105**

$R = 1.8°C/W$ (vertical)
$R = 2.1°C/W$ (horizontal)

Dimensions: 76 mm × 127 mm × 91 mm
Surface area: 677 cm$^2$

**HS 6115**

$R = 1.1°C/W$ (vertical)
$R = 1.3°C/W$ (horizontal)

Dimensions: 76 mm × 102 mm × 25 mm
Surface area: 929 cm$^2$

**HS 7030**

$R = 2.9°C/W$ (vertical)
$R = 3.1°C/W$ (horizontal)

Dimensions: 76 mm × 97 mm × 19 mm
Surface area: 290 cm$^2$

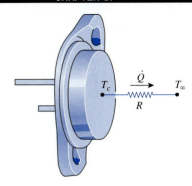

the transistor is not to exceed 85°C, determine the power at which this transistor can be operated safely in an environment at 25°C.

**SOLUTION** The maximum power rating of a transistor whose case temperature is not to exceed 85°C is to be determined.

*Assumptions* **1** Steady operating conditions exist. **2** The transistor case is isothermal at 85°C.

*Properties* The case-to-ambient thermal resistance is given to be 20°C/W.

*Analysis* The power transistor and the thermal resistance network associated with it are shown in Fig. 17–48. We notice from the thermal resistance network that there is a single resistance of 20°C/W between the case at $T_c = 85$°C and the ambient at $T_\infty = 25$°C, and thus the rate of heat transfer is

$$\dot{Q} = \left(\frac{\Delta T}{R}\right)_{\text{case-ambient}} = \frac{T_c - T_\infty}{R_{\text{case-ambient}}} = \frac{(85 - 25)°C}{20°C/W} = \textbf{17 W}$$

Therefore, this power transistor should not be operated at power levels above 3 W if its case temperature is not to exceed 85°C.

*Discussion* This transistor can be used at higher power levels by attaching it to a heat sink (which lowers the thermal resistance by increasing the heat transfer surface area, as discussed in the next example) or by using a fan (which lowers the thermal resistance by increasing the convection heat transfer coefficient).

**FIGURE 17–48**
Schematic for Example 17–10.

---

**EXAMPLE 17–11      Selecting a Heat Sink for a Transistor**

A 60-W power transistor is to be cooled by attaching it to one of the commercially available heat sinks shown in Table 17–6. Select a heat sink that will allow the case temperature of the transistor not to exceed 90°C in the ambient air at 30°C.

**SOLUTION** A commercially available heat sink from Table 17–6 is to be selected to keep the case temperature of a transistor below 90°C.

*Assumptions* **1** Steady operating conditions exist. **2** The transistor case is isothermal at 90°C. **3** The contact resistance between the transistor and the heat sink is negligible.

*Analysis* The rate of heat transfer from a 60-W transistor at full power is $\dot{Q} = 60$ W. The thermal resistance between the transistor attached to the heat sink and the ambient air for the specified temperature difference is determined to be

$$\dot{Q} = \frac{\Delta T}{R} \quad \longrightarrow \quad R = \frac{\Delta T}{\dot{Q}} = \frac{(90 - 30)°C}{60\ W} = 1.0°C/W$$

Therefore, the thermal resistance of the heat sink should be below 1.0°C/W. An examination of Table 17–6 reveals that the HS 5030, whose thermal resistance is 0.9°C/W in the vertical position, is the only heat sink that will meet this requirement.

**FIGURE 17–49**
Schematic for Example 17–12.

**EXAMPLE 17–12    Heat Transfer from Fins of Variable Cross-Section**

Aluminum pin fins of parabolic profile with blunt tips are attached on a plane wall with surface temperature of 200°C (Fig. 17–49). Each fin has a length of 20 mm and a base diameter of 5 mm. The fins are exposed to an ambient air condition of 25°C and the convection heat transfer coefficient is 50 W/m²·K. If the thermal conductivity of the fins is 240 W/m²·K, determine the efficiency, heat transfer rate, and effectiveness of each fin.

**SOLUTION**   The efficiency, heat transfer rate, and effectiveness of a pin fin of parabolic profile with blunt tips are to be determined.

**Assumptions**   **1** Heat conduction is steady and one-dimensional. **2** Thermal properties are constant. **3** Heat transfer by radiation is negligible.

**Properties**   The thermal conductivity of the fin is given as 240 W/m²·K.

**Analysis**   From Table 17–3, for pin fins of parabolic profile (blunt tip), we have

$$mL = \sqrt{\frac{4h}{kD}}L = \sqrt{\frac{4(50 \text{ W/m}^2\cdot\text{K})}{(240 \text{ W/m}\cdot\text{K})(0.005\text{ m})}}\,(0.020 \text{ m}) = 0.2582$$

$$A_{\text{fin}} = \frac{\pi D^4}{96L^2}\left\{\left[16\left(\frac{L}{D}\right)^2+1\right]^{3/2}-1\right\} = \frac{\pi(0.005\text{ m})^4}{96(0.020\text{ m})^2}\left\{\left[16\left(\frac{0.020\text{ m}}{0.005\text{ m}}\right)^2+1\right]^{3/2}-1\right\}$$

$$= 2.106 \times 10^{-4}\text{ m}^2$$

$$\eta_{\text{fin}} = \frac{3}{2mL}\frac{I_1(4mL/3)}{I_0(4mL/3)} = \frac{3}{2(0.2582)}\frac{I_1[4(0.2582)/3]}{I_0[4(0.2582)/3]} = 5.8095\frac{I_1[0.3443]}{I_0[0.3443]}$$

The values of the Bessel functions corresponding to $x = 0.3443$ are determined from Table 17–4 to be $I_0 = 1.0350$ and $I_1 = 0.1716$. Substituting, the fin efficiency is determined to be

$$\eta_{\text{fin}} = 5.8095\frac{0.1716}{1.0350} = \mathbf{0.9632}$$

The heat transfer rate for a single fin is

$$\dot{Q}_{\text{fin}} = \eta_{\text{fin}}\, hA_{\text{fin}}\,(T_b - T_\infty)$$

$$= (0.9632)\,(50 \text{ W/m}^2\cdot\text{K})(2.106 \times 10^{-4}\text{ m}^2)(200 - 25)°\text{C} = \mathbf{1.77\ W}$$

The fin effectiveness is

$$\varepsilon_{\text{fin}} = \frac{\dot{Q}_{\text{fin}}}{hA_b(T_b - T_\infty)} = \frac{\dot{Q}_{\text{fin}}}{h(\pi D^2/4)\,(T_b - T_\infty)}$$

$$= \frac{1.77\text{ W}}{(50\text{W/m}^2\cdot\text{K})\,[\pi(0.005\text{ m})^2/4]\,(200 - 25)°\text{C}}$$

$$= \mathbf{10.3}$$

That is, over a 10-fold increase in heat transfer is achieved by using a pin fin in this case.

*Discussion* The fin efficiency can be determined more accurately by avoiding the interpolation error by using an equation solver with built-in mathematical functions such as EES. Copying the line

eta_fin = 3/(2*0.2582)*Bessel_I 1(4*0.2582/3)/Bessel_I0 (4*0.2582/3)

on a blank EES screen and hitting the 'solve' button gives the fin efficiency to be $\eta_{\text{fin}} = 0.9855$, which is about 2 percent higher than the result obtained above using the tables.

## SUMMARY

One-dimensional heat transfer through a simple or composite body exposed to convection from both sides to mediums at temperatures $T_{\infty 1}$ and $T_{\infty 2}$ can be expressed as

$$\dot{Q} = \frac{T_{\infty 1} - T_{\infty 2}}{R_{\text{total}}}$$

where $R_{\text{total}}$ is the total thermal resistance between the two mediums. For a plane wall exposed to convection on both sides, the total resistance is expressed as

$$R_{\text{total}} = R_{\text{conv, 1}} + R_{\text{wall}} + R_{\text{conv, 2}} = \frac{1}{h_1 A} + \frac{L}{kA} + \frac{1}{h_2 A}$$

This relation can be extended to plane walls that consist of two or more layers by adding an additional resistance for each additional layer. The elementary thermal resistance relations can be expressed as follows:

*Conduction resistance (plane wall):* $\quad R_{\text{wall}} = \dfrac{L}{kA}$

*Conduction resistance (cylinder):* $\quad R_{\text{cyl}} = \dfrac{\ln(r_2/r_1)}{2\pi Lk}$

*Conduction resistance (sphere):* $\quad R_{\text{sph}} = \dfrac{r_2 - r_1}{4\pi r_1 r_2 k}$

*Convection resistance:* $\quad R_{\text{conv}} = \dfrac{1}{hA}$

*Interface resistance:* $\quad R_{\text{interface}} = \dfrac{1}{h_c A} = \dfrac{R_c}{A}$

*Radiation resistance:* $\quad R_{\text{rad}} = \dfrac{1}{h_{\text{rad}} A}$

where $h_c$ is the thermal contact conductance, $R_c$ is the thermal contact resistance, and the radiation heat transfer coefficient is defined as

$$h_{\text{rad}} = \varepsilon\sigma(T_s^2 + T_{\text{surr}}^2)(T_s + T_{\text{surr}})$$

Once the rate of heat transfer is available, the *temperature drop* across any layer can be determined from

$$\Delta T = \dot{Q}R$$

The thermal resistance concept can also be used to solve steady heat transfer problems involving parallel layers or combined series-parallel arrangements.

Adding insulation to a cylindrical pipe or a spherical shell increases the rate of heat transfer if the outer radius of the insulation is less than the *critical radius of insulation*, defined as

$$r_{\text{cr, cylinder}} = \frac{k_{\text{ins}}}{h}$$

$$r_{\text{cr, sphere}} = \frac{2k_{\text{ins}}}{h}$$

The effectiveness of an insulation is often given in terms of its *R-value*, the thermal resistance of the material for a unit surface area, expressed as

$$R\text{-value} = \frac{L}{k} \qquad \text{(flat insulation)}$$

where $L$ is the thickness and $k$ is the thermal conductivity of the material.

Finned surfaces are commonly used in practice to enhance heat transfer. Fins enhance heat transfer from a surface by exposing a larger surface area to convection. The temperature distribution along the fin are given by

*Very long fin:* $\quad \dfrac{T(x) - T_{\infty}}{T_b - T_{\infty}} = e^{-x\sqrt{hp/kA_c}}$

*Adiabatic fin tip:* $\quad \dfrac{T(x) - T_{\infty}}{T_b - T_{\infty}} = \dfrac{\cosh m(L - x)}{\cosh mL}$

*Specified temperature at fin tip:*

$$\frac{T(x) - T_\infty}{T_b - T_\infty} = \frac{[(T_L - T_\infty)/(T_b - T_\infty)] \sinh mx + \sinh m(L - x)}{\sinh mL}$$

*Convection from fin tip:*

$$\frac{T(x) - T_\infty}{T_b - T_\infty} = \frac{\cosh m(L - x) + (h/mk) \sinh m(L - x)}{\cosh mL + (h/mk) \sinh mL}$$

where $m = \sqrt{hp/kA_c}$, $p$ is the perimeter, and $A_c$ is the cross-sectional area of the fin. The rates of heat transfer for these cases are given to be

*Very long fin:*

$$\dot{Q}_{\text{long fin}} = -kA_c \frac{dT}{dx}\Big|_{x=0} = \sqrt{hpkA_c}\,(T_b - T_\infty)$$

*Adiabatic fin tip:*

$$\dot{Q}_{\text{adiabatic tip}} = -kA_c \frac{dT}{dx}\Big|_{x=0} = \sqrt{hpkA_c}\,(T_b - T_\infty) \tanh mL$$

*Specified temperature at fin tip:*

$$\dot{Q}_{\text{specified temp.}} = \sqrt{hpkA_c}\,(T_b - T_\infty) \frac{\cosh mL - [(T_L - T_\infty)/(T_b - T_\infty)]}{\sinh mL}$$

*Convection from the fin tip:*

$$\dot{Q}_{\text{convection}} = \sqrt{hpkA_c}\,(T_b - T_\infty) \frac{\sinh mL + (h/mk) \cosh mL}{\cosh mL + (h/mk) \sinh mL}$$

Fins exposed to convection at their tips can be treated as fins with adiabatic tips by using the corrected length $L_c = L + A_c/p$ instead of the actual fin length.

The temperature of a fin drops along the fin, and thus the heat transfer from the fin is less because of the decreasing temperature difference toward the fin tip. To account for the effect of this decrease in temperature on heat transfer, we define *fin efficiency* as

$$\eta_{\text{fin}} = \frac{\dot{Q}_{\text{fin}}}{\dot{Q}_{\text{fin, max}}} = \frac{\text{Actual heat transfer rate from the fin}}{\begin{array}{c}\text{Ideal heat transfer rate from the fin if}\\ \text{the entire fin were at base temperature}\end{array}}$$

When the fin efficiency is available, the rate of heat transfer from a fin can be determined from

$$\dot{Q}_{\text{fin}} = \eta_{\text{fin}} \dot{Q}_{\text{fin, max}} = \eta_{\text{fin}} hA_{\text{fin}}\,(T_b - T_\infty)$$

The performance of the fins is judged on the basis of the enhancement in heat transfer relative to the no-fin case and is expressed in terms of the *fin effectiveness* $\varepsilon_{\text{fin}}$, defined as

$$\varepsilon_{\text{fin}} = \frac{\dot{Q}_{\text{fin}}}{\dot{Q}_{\text{no fin}}} = \frac{\dot{Q}_{\text{fin}}}{hA_b\,(T_b - T_\infty)} = \frac{\begin{array}{c}\text{Heat transfer rate from}\\ \text{the fin of } \textit{base area } A_b\end{array}}{\begin{array}{c}\text{Heat transfer rate from}\\ \text{the surface of } \textit{area } A_b\end{array}}$$

Here, $A_b$ is the cross-sectional area of the fin at the base and $\dot{Q}_{\text{no fin}}$ represents the rate of heat transfer from this area if no fins are attached to the surface. The *overall effectiveness* for a finned surface is defined as the ratio of the total heat transfer from the finned surface to the heat transfer from the same surface if there were no fins,

$$\varepsilon_{\text{fin, overall}} = \frac{\dot{Q}_{\text{total, fin}}}{\dot{Q}_{\text{total, no fin}}} = \frac{h(A_{\text{unfin}} + \eta_{\text{fin}} A_{\text{fin}})(T_b - T_\infty)}{hA_{\text{no fin}}\,(T_b - T_\infty)}$$

Fin efficiency and fin effectiveness are related to each other by

$$\varepsilon_{\text{fin}} = \frac{A_{\text{fin}}}{A_b} \eta_{\text{fin}}$$

## REFERENCES AND SUGGESTED READINGS

**1.** American Society of Heating, Refrigeration, and Air Conditioning Engineers. *Handbook of Fundamentals.* Atlanta: ASHRAE, 1993.

**2.** R. V. Andrews. "Solving Conductive Heat Transfer Problems with Electrical-Analogue Shape Factors." *Chemical Engineering Progress* 5 (1955), p. 67.

**3.** L. S. Fletcher. "Recent Developments in Contact Conductance Heat Transfer." *Journal of Heat Transfer* 110, no. 4B (1988), pp. 1059–79.

**4.** E. Fried. "Thermal Conduction Contribution to Heat Transfer at Contacts." *Thermal Conductivity,* vol. 2, ed. R. P. Tye. London: Academic Press, 1969.

**5.** K. A. Gardner. "Efficiency of Extended Surfaces." *Trans. ASME* 67 (1945), pp. 621–31. Reprinted by permission of ASME International.

**6.** D. Q. Kern and A. D. Kraus. *Extended Surface Heat Transfer.* New York: McGraw-Hill, 1972.

**7.** G. P. Peterson. "Thermal Contact Resistance in Waste Heat Recovery Systems." *Proceedings of the 18th ASME/ETCE Hydrocarbon Processing Symposium.* Dallas, TX, 1987, pp. 45–51. Reprinted by permission of ASME International.

**8.** S. Song, M. M. Yovanovich, and F. O. Goodman. "Thermal Gap Conductance of Conforming Surfaces in Contact." *Journal of Heat Transfer* 115 (1993), p. 533.

# PROBLEMS*

## Steady Heat Conduction in Plane Walls

**17–1C** Consider heat conduction through a wall of thickness $L$ and area $A$. Under what conditions will the temperature distributions in the wall be a straight line?

**17–2C** Consider heat conduction through a plane wall. Does the energy content of the wall change during steady heat conduction? How about during transient conduction? Explain.

**17–3C** Consider steady heat transfer through the wall of a room in winter. The convection heat transfer coefficient at the outer surface of the wall is three times that of the inner surface as a result of the winds. On which surface of the wall do you think the temperature will be closer to the surrounding air temperature? Explain.

**17–4C** Consider one-dimensional heat conduction through a cylindrical rod of diameter $D$ and length $L$. What is the heat transfer area of the rod if (*a*) the lateral surfaces of the rod are insulated and (*b*) the top and bottom surfaces of the rod are insulated?

**17–5C** What does the thermal resistance of a medium represent?

**17–6C** How is the combined heat transfer coefficient defined? What convenience does it offer in heat transfer calculations?

**17–7C** Can we define the convection resistance for a unit surface area as the inverse of the convection heat transfer coefficient?

**17–8C** Why are the convection and the radiation resistances at a surface in parallel instead of being in series?

**17–9C** Consider a surface of area $A$ at which the convection and radiation heat transfer coefficients are $h_{conv}$ and $h_{rad}$, respectively. Explain how you would determine (*a*) the single equivalent heat transfer coefficient, and (*b*) the equivalent thermal resistance. Assume the medium and the surrounding surfaces are at the same temperature.

**17–10C** How does the thermal resistance network associated with a single-layer plane wall differ from the one associated with a five-layer composite wall?

**17–11C** Consider steady one-dimensional heat transfer through a multilayer medium. If the rate of heat transfer $\dot{Q}$ is known, explain how you would determine the temperature drop across each layer.

**17–12C** Consider steady one-dimensional heat transfer through a plane wall exposed to convection from both sides to environments at known temperatures $T_{\infty 1}$ and $T_{\infty 2}$ with known heat transfer coefficients $h_1$ and $h_2$. Once the rate of heat transfer $\dot{Q}$ has been evaluated, explain how you would determine the temperature of each surface.

**17–13C** Someone comments that a microwave oven can be viewed as a conventional oven with zero convection resistance at the surface of the food. Is this an accurate statement?

**17–14C** Consider a window glass consisting of two 4-mm-thick glass sheets pressed tightly against each other. Compare the heat transfer rate through this window with that of one consisting of a single 8-mm-thick glass sheet under identical conditions.

**17–15C** Consider two cold canned drinks, one wrapped in a blanket and the other placed on a table in the same room. Which drink will warm up faster?

**17–16** Consider a 3-m-high, 6-m-wide, and 0.25-m-thick brick wall whose thermal conductivity is $k = 0.8$ W/m·K. On a certain day, the temperatures of the inner and the outer surfaces of the wall are measured to be 14°C and 5°C, respectively. Determine the rate of heat loss through the wall on that day.

**17–17** Consider a 1.5-m-high and 2.4-m-wide glass window whose thickness is 6 mm and thermal conductivity is $k = 0.78$ W/m·K. Determine the steady rate of heat transfer through this glass window and the temperature of its inner surface for a day during which the room is maintained at 24°C while the temperature of the outdoors is −5°C. Take the convection heat transfer coefficients on the inner and outer surfaces of the window to be $h_1 = 10$ W/m²·K and $h_2 = 25$ W/m²·K, and disregard any heat transfer by radiation.

**17–18** Consider a 1.5-m-high and 2.4-m-wide double-pane window consisting of two 3-mm-thick layers of glass ($k = 0.78$ W/m·K) separated by a 12-mm-wide stagnant air space ($k = 0.026$ W/m·K). Determine the steady rate of heat transfer through this double-pane window and the temperature of its inner surface for a day during which the room is maintained at 21°C while the temperature of the outdoors is −5°C. Take the convection heat transfer coefficients on the inner and outer surfaces of the window to be $h_1 = 10$ W/m²·K and $h_2 = 25$ W/m²·K, and disregard any heat transfer by radiation. *Answers:* 154 W, 16.7°C

---

*Problems designated by a "C" are concept questions, and students are encouraged to answer them all. Problems designated by an "E" are in English units, and the SI users can ignore them. Problems with the icon ✏ are solved using EES, and complete solutions together with parametric studies are included on the enclosed DVD. Problems with the icon 🖳 are comprehensive in nature, and are intended to be solved with a computer, preferably using the EES software that accompanies this text.

Glass

3  12  3 mm

Frame

**FIGURE P17–18**

**17–19**  Repeat Prob. 17–18, assuming the space between the two glass layers is evacuated.

**17–20**  [EES] Reconsider Prob. 17–18. Using EES (or other) software, plot the rate of heat transfer through the window as a function of the width of air space in the range of 2 mm to 20 mm, assuming pure conduction through the air. Discuss the results.

**17–21E**  Consider an electrically heated brick house ($k$ = 0.40 Btu/h·ft·°F) whose walls are 9 ft high and 1 ft thick. Two of the walls of the house are 50 ft long and the others are 35 ft long. The house is maintained at 70°F at all times while the temperature of the outdoors varies. On a certain day, the temperature of the inner surface of the walls is measured to be at 55°F while the average temperature of the outer surface is observed to remain at 45°F during the day for 10 h and at 35°F at night for 14 h. Determine the amount of heat lost from the house that day. Also determine the cost of that heat loss to the home owner for an electricity price of $0.09/kWh.

$T_{in}$ = 70°F

9 ft

50 ft

35 ft

**FIGURE P17–21E**

**17–22**  A cylindrical resistor element on a circuit board dissipates 0.15 W of power in an environment at 35°C. The resistor is 1.2 cm long, and has a diameter of 0.3 cm. Assuming heat to be transferred uniformly from all surfaces, determine (a) the amount of heat this resistor dissipates during a 24-h period; (b) the heat flux on the surface of the resistor, in W/m²; and (c) the surface temperature of the resistor for a combined convection and radiation heat transfer coefficient of 9 W/m²·K.

**17–23**  To defog the rear window of an automobile, a very thin transparent heating element is attached to the inner surface of the window. A uniform heat flux of 1300 W/m² is provided to the heating element for defogging a rear window with thickness of 5 mm. The interior temperature of the automobile is 22°C and the convection heat transfer coefficient is 15 W/m²·K. The outside ambient temperature is −5°C and the convection heat transfer coefficient is 100 W/m²·K. If the thermal conductivity of the window is 1.2 W/m·K, determine the inner surface temperature of the window.

Heating element
$\dot{q}_h$ = 1300 W/m²

Inside air, 22°C
h = 15 W/m²·K

Rear window
k = 1.2 W/m·K

Outside air, −5°C
h = 100 W/m²·K

$T_1$   L = 5 mm

**FIGURE P17–23**

**17–24**  Consider a power transistor that dissipates 0.15 W of power in an environment at 30°C. The transistor is 0.4 cm long and has a diameter of 0.5 cm. Assuming heat to be transferred uniformly from all surfaces, determine (a) the amount of heat this resistor dissipates during a 24-h period, in kWh; (b) the heat flux on the surface of the transistor, in W/m²; and (c) the surface temperature of the resistor for a combined convection and radiation heat transfer coefficient of 18 W/m²·K.

30°C

Power
transistor
0.15 W

0.5 cm

0.4 cm

**FIGURE P17–24**

**17–25** A 12-cm × 18-cm circuit board houses on its surface 100 closely spaced logic chips, each dissipating 0.06 W in an environment at 40°C. The heat transfer from the back surface of the board is negligible. If the heat transfer coefficient on the surface of the board is 10 W/m²·K, determine (a) the heat flux on the surface of the circuit board, in W/m²; (b) the surface temperature of the chips; and (c) the thermal resistance between the surface of the circuit board and the cooling medium, in °C/W.

**17–26** Consider a person standing in a room at 20°C with an exposed surface area of 1.7 m². The deep body temperature of the human body is 37°C, and the thermal conductivity of the human tissue near the skin is about 0.3 W/m·K. The body is losing heat at a rate of 150 W by natural convection and radiation to the surroundings. Taking the body temperature 0.5 cm beneath the skin to be 37°C, determine the skin temperature of the person. *Answer: 35.5°C*

**17–27** A 1.0 m × 1.5 m double-pane window consists of two 4-mm-thick layers of glass ($k = 0.78$ W/m·K) that are the separated by a 5-mm air gap ($k_{air} = 0.025$ W/m·K). The heat flow through the air gap is assumed to be by conduction. The inside and outside air temperatures are 20°C and −20°C, respectively, and the inside and outside heat transfer coefficients are 40 and 20 W/m²·K. Determine (a) the daily rate of heat loss through the window in steady operation and (b) the temperature difference across the largest thermal resistence.

**17–28E** A wall is constructed of two layers of 0.6-in-thick sheetrock ($k = 0.10$ Btu/h·ft·°F), which is a plasterboard made of two layers of heavy paper separated by a layer of gypsum, placed 7 in apart. The space between the sheetrocks is filled with fiberglass insulation ($k = 0.020$ Btu/h·ft·°F). Determine (a) the thermal resistance of the wall and (b) its R-value of insulation in English units.

**FIGURE P17–28E**

**17–29** The roof of a house consists of a 15-cm-thick concrete slab ($k = 2$ W/m·K) that is 15 m wide and 20 m long. The convection heat transfer coefficients on the inner and outer

surfaces of the roof are 5 and 12 W/m²·K, respectively. On a clear winter night, the ambient air is reported to be at 10°C, while the night sky temperature is 100 K. The house and the interior surfaces of the wall are maintained at a constant temperature of 20°C. The emissivity of both surfaces of the concrete roof is 0.9. Considering both radiation and convection heat transfers, determine the rate of heat transfer through the roof, and the inner surface temperature of the roof.

If the house is heated by a furnace burning natural gas with an efficiency of 80 percent, and the price of natural gas is $1.20/therm (1 therm = 105,500 kJ of energy content), determine the money lost through the roof that night during a 14-h period.

**FIGURE P17–29**

**17–30** A 2-m × 1.5-m section of wall of an industrial furnace burning natural gas is not insulated, and the temperature at the outer surface of this section is measured to be 110°C. The temperature of the furnace room is 32°C, and the combined convection and radiation heat transfer coefficient at the surface of the outer furnace is 10 W/m²·K. It is proposed to insulate this section of the furnace wall with glass wool insulation ($k = 0.038$ W/m·K) in order to reduce the heat loss by 90 percent. Assuming the outer surface temperature of the metal section still remains at about 110°C, determine the thickness of the insulation that needs to be used.

The furnace operates continuously and has an efficiency of 78 percent. The price of the natural gas is $1.10/therm (1 therm = 105,500 kJ of energy content). If the installation of the insulation will cost $250 for materials and labor, determine how long it will take for the insulation to pay for itself from the energy it saves.

**17–31** Repeat Prob. 17–30 for expanded perlite insulation assuming conductivity is $k = 0.052$ W/m·K.

**17–32** Reconsider Prob. 17–30. Using EES (or other) software, investigate the effect of thermal conductivity on the required insulation thickness. Plot the thickness of insulation as a function of the thermal conductivity of the insulation in the range of 0.02 W/m·K to 0.08 W/m·K, and discuss the results.

**17–33** The wall of a refrigerator is constructed of fiberglass insulation ($k = 0.035$ W/m·K) sandwiched between two layers of 1-mm-thick sheet metal ($k = 15.1$ W/m·K). The refrigerated space is maintained at 2°C, and the average heat transfer coefficients at the inner and outer surfaces of the wall are 4 W/m²·K and 9 W/m²·K, respectively. The kitchen temperature averages 24°C. It is observed that condensation occurs on the outer surfaces of the refrigerator when the temperature of the outer surface drops to 20°C. Determine the minimum thickness of fiberglass insulation that needs to be used in the wall in order to avoid condensation on the outer surfaces.

Sheet metal

Kitchen air 24°C

Refrigerated space 2°C

Insulation

1 mm — L — 1 mm

**FIGURE P17–33**

**17–34** [EES] Reconsider Prob. 17–33. Using EES (or other) software, investigate the effects of the thermal conductivities of the insulation material and the sheet metal on the thickness of the insulation. Let the thermal conductivity vary from 0.02 W/m·K to 0.08 W/m·K for insulation and 10 W/m·K to 400 W/m·K for sheet metal. Plot the thickness of the insulation as the functions of the thermal conductivities of the insulation and the sheet metal, and discuss the results.

**17–35** Heat is to be conducted along a circuit board that has a copper layer on one side. The circuit board is 15 cm long and 15 cm wide, and the thicknesses of the copper and epoxy layers are 0.1 mm and 1.2 mm, respectively. Disregarding heat transfer from side surfaces, determine the percentages of heat conduction along the copper ($k = 386$ W/m·K) and epoxy ($k = 0.26$ W/m·K) layers. Also determine the effective thermal conductivity of the board.
*Answers:* 0.8 percent, 99.2 percent, and 29.9 W/m·K

**17–36E** A 0.05-in-thick copper plate ($k = 223$ Btu/h·ft·°F) is sandwiched between two 0.15-in-thick epoxy boards ($k = 0.15$ Btu/h·ft·°F) that are 7 in × 9 in in size. Determine the effective thermal conductivity of the board along its 9-in-long side. What fraction of the heat conducted along that side is conducted through copper?

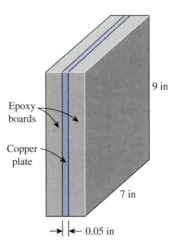

Epoxy boards

Copper plate

9 in

7 in

0.05 in

**FIGURE P17–36E**

**17–37** To defrost ice accumulated on the outer surface of an automobile windshield, warm air is blown over the inner surface of the windshield. Consider an automobile windshield with thickness of 5 mm and thermal conductivity of 1.4 W/m·K. The outside ambient temperature is −10°C and the convection heat transfer coefficient is 200 W/m²·K, while the ambient temperature inside the automobile is 25°C. Determine the value of the convection heat transfer coefficient for the warm air blowing over the inner surface of the windshield necessary to cause the accumulated ice to begin melting.

L = 5 mm

Outside air, −10°C
$h_o = 200$ W/m²·K

Windshield
$k = 1.4$ W/m·K

Inside air, 25°C

$T_1 = 0$°C

**FIGURE P17–37**

**17–38** An aluminum plate of 25 mm thick ($k = 235$ W/m·K) is attached on a copper plate with thickness of 10 mm. The copper plate is heated electrically to dissipate a uniform heat flux of 5300 W/m². The upper surface of the aluminum plate is exposed to convection heat transfer in a condition such that the convection heat transfer coefficient is 67 W/m²·K and the surrounding room temperature is 20°C. Other surfaces of the two attached plates are insulated such that heat only dissipates through the upper surface of the aluminum plate. If the surface of the copper plate that is attached to the aluminum plate has a temperature of 100°C, determine the thermal contact conductance of the aluminum/copper interface. *Answer:* 16 kW/m²·K

Air, 20°C
$h = 67$ W/m²·K

Aluminum plate
$k = 235$ W/m·K

$L = 25$ mm

$\dot{q}_{elec} = 5300$ W/m²

$T_1 = 100°C$

**FIGURE P17–38**

## Thermal Contact Resistance

**17–39C**  What is thermal contact resistance? How is it related to thermal contact conductance?

**17–40C**  Will the thermal contact resistance be greater for smooth or rough plain surfaces?

**17–41C**  A wall consists of two layers of insulation pressed against each other. Do we need to be concerned about the thermal contact resistance at the interface in a heat transfer analysis or can we just ignore it?

**17–42C**  A plate consists of two thin metal layers pressed against each other. Do we need to be concerned about the thermal contact resistance at the interface in a heat transfer analysis or can we just ignore it?

**17–43C**  Consider two surfaces pressed against each other. Now the air at the interface is evacuated. Will the thermal contact resistance at the interface increase or decrease as a result?

**17–44C**  Explain how the thermal contact resistance can be minimized.

**17–45**  The thermal contact conductance at the interface of two 1-cm-thick copper plates is measured to be 14,000 W/m²·K. Determine the thickness of the copper plate whose thermal resistance is equal to the thermal resistance of the interface between the plates.

**17–46**  A 1-mm-thick copper plate ($k = 386$ W/m·K) is sandwiched between two 7-mm-thick epoxy boards ($k = 0.26$ W/m·K) that are 15 cm × 20 cm in size. If the thermal contact conductance on both sides of the copper plate is estimated to be 6000 W/m·K, determine the error involved in the total thermal resistance of the plate if the thermal contact conductances are ignored.

Copper

Epoxy

Epoxy

$h_c$

$\dot{Q}$

7 mm   7 mm

**FIGURE P17–46**

**17–47**  Two 5-cm-diameter, 15-cm-long aluminum bars ($k = 176$ W/m·K) with ground surfaces are pressed against each other with a pressure of 20 atm. The bars are enclosed in an insulation sleeve and, thus, heat transfer from the lateral surfaces is negligible. If the top and bottom surfaces of the two-bar system are maintained at temperatures of 150°C and 20°C, respectively, determine (*a*) the rate of heat transfer along the cylinders under steady conditions and (*b*) the temperature drop at the interface.  *Answers:* (*a*) 142.4 W, (*b*) 6.4°C

## Generalized Thermal Resistance Networks

**17–48C**  What are the two approaches used in the development of the thermal resistance network for two-dimensional problems?

**17–49C**  The thermal resistance networks can also be used approximately for multidimensional problems. For what kind of multidimensional problems will the thermal resistance approach give adequate results?

**17–50C**  When plotting the thermal resistance network associated with a heat transfer problem, explain when two resistances are in series and when they are in parallel.

**17–51**  A typical section of a building wall is shown in Fig. P17–51. This section extends in and out of the page and is

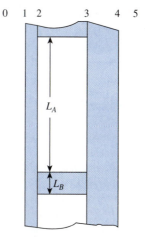

0  1  2      3    4  5

$L_A$

$L_B$

**FIGURE P17–51**

repeated in the vertical direction. The wall support members are made of steel ($k = 50$ W/m·K). The support members are 8 cm ($t_{23}$) × 0.5 cm ($L_B$). The remainder of the inner wall space is filled with insulation ($k = 0.03$ W/m·K) and measures 8 cm ($t_{23}$) × 60 cm ($L_B$). The inner wall is made of gypsum board ($k = 0.5$ W/m·K) that is 1 cm thick ($t_{12}$) and the outer wall is made of brick ($k = 1.0$ W/m·K) that is 10 cm thick ($t_{34}$). What is the average heat flux through this wall when $T_1 = 20°C$ and $T_4 = 35°C$?

**17–52** A 4-m-high and 6-m-wide wall consists of a long 15-cm × 25-cm cross section of horizontal bricks ($k = 0.72$ W/m·K) separated by 3-cm-thick plaster layers ($k = 0.22$ W/m·K). There are also 2-cm-thick plaster layers on each side of the wall, and a 2-cm-thick rigid foam ($k = 0.026$ W/m·K) on the inner side of the wall. The indoor and the outdoor temperatures are 22°C and −4°C, and the convection heat transfer coefficients on the inner and the outer sides are $h_1 = 10$ W/m²·K and $h_2 = 20$ W/m²·K, respectively. Assuming one-dimensional heat transfer and disregarding radiation, determine the rate of heat transfer through the wall.

**FIGURE P17–52**

**17–53** ![EES] Reconsider Prob. 17–52. Using EES (or other) software, plot the rate of heat transfer through the wall as a function of the thickness of the rigid foam in the range of 1 cm to 10 cm. Discuss the results.

**17–54** A 12-m-long and 5-m-high wall is constructed of two layers of 1-cm-thick sheetrock ($k = 0.17$ W/m·K) spaced 16 cm by wood studs ($k = 0.11$ W/m·K) whose cross section is 16 cm × 5 cm. The studs are placed vertically 60 cm apart, and the space between them is filled with fiberglass insulation ($k = 0.034$ W/m·K). The house is maintained at 20°C and the

ambient temperature outside is −9°C. Taking the heat transfer coefficients at the inner and outer surfaces of the house to be 8.3 and 34 W/m²·K, respectively, determine (a) the thermal resistance of the wall considering a representative section of it and (b) the rate of heat transfer through the wall.

**17–55** A 10-cm-thick wall is to be constructed with 2.5-m-long wood studs ($k = 0.11$ W/m·K) that have a cross section of 10 cm × 10 cm. At some point the builder ran out of those studs and started using pairs of 2.5-m-long wood studs that have a cross section of 5 cm × 10 cm nailed to eachother instead. The manganese steel nails ($k = 50$ W/m·K) are 10 cm long and have a diameter of 0.4 cm. A total of 50 nails are used to connect the two studs, which are mounted to the wall such that the nails cross the wall. The temperature difference between the inner and outer surfaces of the wall is 8°C. Assuming the thermal contact resistance between the two layers to be negligible, determine the rate of heat transfer (a) through a solid stud and (b) through a stud pair of equal length and width nailed to each other. (c) Also determine the effective conductivity of the nailed stud pair.

**17–56E** A 10-in-thick, 30-ft-long, and 10-ft-high wall is to be constructed using 9-in-long solid bricks ($k = 0.40$ Btu/h·ft·°F) of cross section 7 in × 7 in, or identical size bricks with nine square air holes ($k = 0.015$ Btu/h·ft·°F) that are 9 in long and have a cross section of 1.5 in × 1.5 in. There is a 0.5-in-thick plaster layer ($k = 0.10$ Btu/h·ft·°F) between two adjacent bricks on all four sides and on both sides of the wall. The house is maintained at 80°F and the ambient temperature outside is 35°F. Taking the heat transfer coefficients at the inner and outer surfaces of the wall to be 1.5 and 6 Btu/h·ft²·°F, respectively, determine the rate of heat transfer through the wall constructed of (a) solid bricks and (b) bricks with air holes.

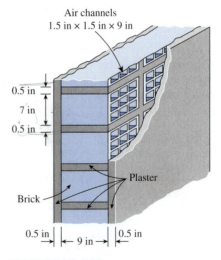

**FIGURE P17–56E**

**17–57** Consider a 5-m-high, 8-m-long, and 0.22-m-thick wall whose representative cross section is given in the figure. The thermal conductivities of various materials used, in W/m·K, are $k_A = k_F = 2$, $k_B = 8$, $k_C = 20$, $k_D = 15$, and $k_E = 35$. The left and right surfaces of the wall are maintained at uniform temperatures of 300°C and 100°C, respectively. Assuming heat transfer through the wall to be one-dimensional, determine (a) the rate of heat transfer through the wall; (b) the temperature at the point where the sections B, D, and E meet; and (c) the temperature drop across the section F. Disregard any contact resistances at the interfaces.

Multilayered ski jacket

**FIGURE P17–59**

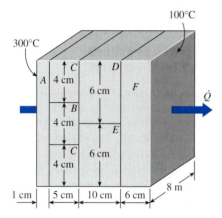

**FIGURE P17–57**

**17–58** Repeat Prob. 17–57 assuming that the thermal contact resistance at the interfaces D-F and E-F is 0.00012 m²·°C/W.

**17–59** Clothing made of several thin layers of fabric with trapped air in between, often called ski clothing, is commonly used in cold climates because it is light, fashionable, and a very effective thermal insulator. So it is no surprise that such clothing has largely replaced thick and heavy old-fashioned coats.

Consider a jacket made of five layers of 0.15-mm-thick synthetic fabric ($k = 0.13$ W/m·K) with 1.5-mm-thick air space ($k = 0.026$ W/m·K) between the layers. Assuming the inner surface temperature of the jacket to be 25°C and the surface area to be 1.25 m², determine the rate of heat loss through the jacket when the temperature of the outdoors is 0°C and the heat transfer coefficient at the outer surface is 25 W/m²·K.

What would your response be if the jacket is made of a single layer of 0.75-mm-thick synthetic fabric? What should be the thickness of a wool fabric ($k = 0.035$ W/m·K) if the person is to achieve the same level of thermal comfort wearing a thick wool coat instead of a five-layer ski jacket?

**17–60** Repeat Prob. 17–59 assuming the layers of the jacket are made of cotton fabric ($k = 0.06$ W/m·K).

**17–61** In an experiment to measure convection heat transfer coefficients, a very thin metal foil of very low emissivity (e.g., highly polished copper) is attached on the surface of a slab of material with very low thermal conductivity. The other surface of the metal foil is exposed to convection heat transfer by flowing fluid over the foil surface. This setup diminishes heat conduction through the slab and radiation on the metal foil surface, while heat convection plays the prominent role. The slab on which the metal foil is attached to has a thickness of 25 mm and a thermal conductivity of 0.023 W/m·K. In a condition where the surrounding room temperature is 20°C, the metal foil is heated electrically with a uniform heat flux of 5000 W/m². If the bottom surface of the slab is 20°C and the metal foil has an emissivity of 0.02, determine.

(a) The convection heat transfer coefficient if air is flowing over the metal foil and the surface temperature of the foil is 150°C.

(b) The convection heat transfer coefficient if water is flowing over the metal foil and the surface temperature of the foil is 30°C.

**FIGURE P17–61**

**17–62E** Consider a 10-in $\times$ 12-in epoxy glass laminate ($k = 0.10$ Btu/h·ft·°F) whose thickness is 0.05 in. In order to reduce the thermal resistance across its thickness, cylindrical copper fillings ($k = 223$ Btu/h·ft·°F) of 0.02 in diameter are to be planted throughout the board, with a center-to-center distance of 0.06 in. Determine the new value of the thermal resistance of the epoxy board for heat conduction across its thickness as a result of this modification. *Answer:* 0.000256 h·°F/Btu

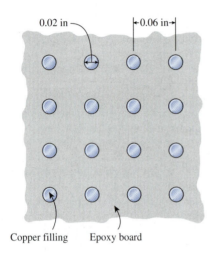

0.02 in — ⊢0.06 in→

Copper filling    Epoxy board

**FIGURE P17–62E**

## Heat Conduction in Cylinders and Spheres

**17–63C** What is an infinitely long cylinder? When is it proper to treat an actual cylinder as being infinitely long, and when is it not?

**17–64C** Can the thermal resistance concept be used for a solid cylinder or sphere in steady operation? Explain.

**17–65C** Consider a short cylinder whose top and bottom surfaces are insulated. The cylinder is initially at a uniform temperature $T_i$ and is subjected to convection from its side surface to a medium at temperature $T_\infty$, with a heat transfer coefficient of $h$. Is the heat transfer in this short cylinder one- or two-dimensional? Explain.

**17–66** An 8-m-internal-diameter spherical tank made of 1.5-cm-thick stainless steel ($k = 15$ W/m·K) is used to store iced water at 0°C. The tank is located in a room whose temperature is 25°C. The walls of the room are also at 25°C. The outer surface of the tank is black (emissivity $\varepsilon = 1$), and heat transfer between the outer surface of the tank and the surroundings is by natural convection and radiation. The convection heat transfer coefficients at the inner and the outer

surfaces of the tank are 80 W/m²·K and 10 W/m²·K, respectively. Determine (a) the rate of heat transfer to the iced water in the tank and (b) the amount of ice at 0°C that melts during a 24-h period. The heat of fusion of water at atmospheric pressure is $h_{if} = 333.7$ kJ/kg.

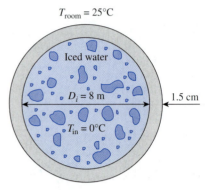

$T_{room} = 25°C$

Iced water

$D_i = 8$ m     1.5 cm

$T_{in} = 0°C$

**FIGURE P17–66**

**17–67** Steam at 280°C flows in a stainless steel pipe ($k = 15$ W/m·K) whose inner and outer diameters are 5 cm and 5.5 cm, respectively. The pipe is covered with 3-cm-thick glass wool insulation ($k = 0.038$ W/m·K). Heat is lost to the surroundings at 5°C by natural convection and radiation, with a combined natural convection and radiation heat transfer coefficient of 22 W/m²·K. Taking the heat transfer coefficient inside the pipe to be 80 W/m²·K, determine the rate of heat loss from the steam per unit length of the pipe. Also determine the temperature drops across the pipe shell and the insulation.

**17–68** Reconsider Prob. 17–67. Using EES (or other) software, investigate the effect of the thickness of the insulation on the rate of heat loss from the steam and the temperature drop across the insulation layer. Let the insulation thickness vary from 1 cm to 10 cm. Plot the rate of heat loss and the temperature drop as a function of insulation thickness, and discuss the results.

**17–69** A 50-m-long section of a steam pipe whose outer diameter is 10 cm passes through an open space at 15°C. The average temperature of the outer surface of the pipe is measured to be 150°C. If the combined heat transfer coefficient on the outer surface of the pipe is 20 W/m²·K, determine (a) the rate of heat loss from the steam pipe; (b) the annual cost of this energy lost if steam is generated in a natural gas furnace that has an efficiency of 75 percent and the price of natural gas is \$0.52/therm (1 therm = 105,500 kJ); and (c) the thickness of fiberglass insulation ($k = 0.035$ W/m·K) needed in order to save 90 percent of the heat lost. Assume the pipe temperature to remain constant at 150°C.

$T_{air} = 15°C$

150°C

Steam

50 m

Fiberglass insulation

**FIGURE P17–69**

**17–70**  Consider a 1.5-m-high electric hot-water heater that has a diameter of 40 cm and maintains the hot water at 60°C. The tank is located in a small room whose average temperature is 27°C, and the heat transfer coefficients on the inner and outer surfaces of the heater are 50 and 12 W/m²·K, respectively. The tank is placed in another 46-cm-diameter sheet metal tank of negligible thickness, and the space between the two tanks is filled with foam insulation ($k = 0.03$ W/m·K). The thermal resistances of the water tank and the outer thin sheet metal shell are very small and can be neglected. The price of electricity is $0.08/kWh, and the home owner pays $280 a year for water heating. Determine the fraction of the hot-water energy cost of this household that is due to the heat loss from the tank.

Hot-water tank insulation kits consisting of 3-cm-thick fiberglass insulation ($k = 0.035$ W/m·K) large enough to wrap the entire tank are available in the market for about $30. If such an insulation is installed on this water tank by the home owner himself, how long will it take for this additional insulation to pay for itself?  *Answers: 15.2 percent, 21 months*

3 cm

40 cm

27°C

$T_w = 60°C$

Foam insulation

1.5 m

Water heater

**FIGURE P17–70**

**17–71**  Reconsider Prob. 17–70. Using EES (or other) software, plot the fraction of energy cost of hot water due to the heat loss from the tank as a function of the hot-water temperature in the range of 40°C to 90°C. Discuss the results.

**17–72**  Chilled water enters a thin-shelled 4-cm-diameter, 200-m-long pipe at 7°C at a rate of 0.98 kg/s and leaves at 8°C. The pipe is exposed to ambient air at 30°C with a heat transfer coefficient of 9 W/m²·K. If the pipe is to be insulated with glass wool insulation ($k = 0.05$ W/m·K) in order to decrease the temperature rise of water to 0.25°C, determine the required thickness of the insulation.

**17–73**  Superheated steam at an average temperature 200°C is transported through a steel pipe ($k = 50$ W/m·K, $D_o = 8.0$ cm, $D_i = 6.0$ cm, and $L = 20.0$ m). The pipe is insulated with a 4-cm thick layer of gypsum plaster ($k = 0.5$ W/m·K). The insulated pipe is placed horizontally inside a warehouse where the average air temperature is 10°C. The steam and the air heat transfer coefficients are estimated to be 800 and 200 W/m²·K, respectively. Calculate (*a*) the daily rate of heat transfer from the superheated steam, and (*b*) the temperature on the outside surface of the gypsum plaster insulation.

**17–74E**  Steam at 450°F is flowing through a steel pipe ($k = 8.7$ Btu/h·ft·°F) whose inner and outer diameters are 3.5 in and 4.0 in, respectively, in an environment at 55°F. The pipe is insulated with 2-in-thick fiberglass insulation ($k = 0.020$ Btu/h·ft·°F). If the heat transfer coefficients on the inside and the outside of the pipe are 30 and 5 Btu/h·ft²·°F, respectively, determine the rate of heat loss from the steam per foot length of the pipe. What is the error involved in neglecting the thermal resistance of the steel pipe in calculations?

**17–75**  Hot water at an average temperature of 90°C is flowing through a 15-m section of a cast iron pipe ($k = 52$ W/m·K) whose inner and outer diameters are 4 cm and 4.6 cm, respectively. The outer surface of the pipe, whose emissivity is 0.7, is exposed to the cold air at 10°C in the basement, with a heat transfer coefficient of 15 W/m²·K. The heat transfer coefficient at the inner surface of the pipe is 120 W/m²·K. Taking the walls of the basement to be at 10°C also, determine the rate of heat loss from the hot water. Also, determine the average velocity of the water in the pipe if the temperature of the water drops by 3°C as it passes through the basement.

Steel pipe

Steam 450°F

Insulation

**FIGURE P17–75**

**17–76** Repeat Prob. 17–75 for a pipe made of copper ($k = 386$ W/m·K) instead of cast iron.

**17–77E** Steam exiting the turbine of a steam power plant at 100°F is to be condensed in a large condenser by cooling water flowing through copper pipes ($k = 223$ Btu/h·ft·°F) of inner diameter 0.4 in and outer diameter 0.6 in at an average temperature of 70°F. The heat of vaporization of water at 100°F is 1037 Btu/lbm. The heat transfer coefficients are 2400 Btu/h·ft²·°F on the steam side and 35 Btu/h·ft²·°F on the water side. Determine the length of the tube required to condense steam at a rate of 250 lbm/h. *Answer:* 2380 ft

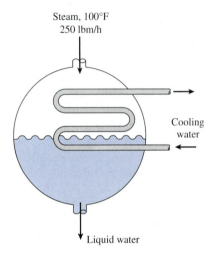

Steam, 100°F
250 lbm/h

Cooling
water

Liquid water

**FIGURE P17–77E**

**17–78E** Repeat Prob. 17–77E, assuming that a 0.01-in-thick layer of mineral deposit ($k = 0.5$ Btu/h·ft·°F) has formed on the inner surface of the pipe.

**17–79E** *EES* Reconsider Prob. 17–77E. Using EES (or other) software, investigate the effects of the thermal conductivity of the pipe material and the outer diameter of the pipe on the length of the tube required. Let the thermal conductivity vary from 10 Btu/h·ft·°F to 400 Btu/h·ft·°F and the outer diameter from 0.5 in to 1.0 in. Plot the length of the tube as functions of pipe conductivity and the outer pipe diameter, and discuss the results.

**17–80** A 2.2-mm-diameter and 14-m-long electric wire is tightly wrapped with a 1-mm-thick plastic cover whose thermal conductivity is $k = 0.15$ W/m·K. Electrical measurements indicate that a current of 13 A passes through the wire and there is a voltage drop of 8 V along the wire. If the insulated wire is exposed to a medium at $T_\infty = 30$°C with a heat transfer coefficient of $h = 24$ W/m²·K, determine the temperature at the interface of the wire and the plastic cover in steady operation. Also determine if doubling the thickness of the plastic cover will increase or decrease this interface temperature.

Electrical wire

Insulation

$T_\infty = 30$°C

14 m

**FIGURE P17–80**

**17–81** In a pharmaceutical plant, a copper pipe ($k_c = 400$ W/m·K) with inner diameter of 20 mm and wall thickness of 2.5 mm is used for carrying liquid oxygen to a storage tank. The liquid oxygen flowing in the pipe has an average temperature of −200°C and a convection heat transfer coefficient of 120 W/m²·K. The condition surrounding the pipe has an ambient air temperature of 20°C and a combined heat transfer coefficient of 20 W/m²·K. If the dew point is 10°C, determine the thickness of the insulation ($k_i = 0.05$ W/m·K) around the copper pipe to avoid condensation on the outer surface. Assume thermal contact resistance is negligible.

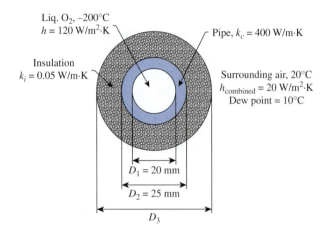

Liq. O₂, −200°C
$h = 120$ W/m²·K

Pipe, $k_c = 400$ W/m·K

Insulation
$k_i = 0.05$ W/m·K

Surrounding air, 20°C
$h_{combined} = 20$ W/m²·K
Dew point = 10°C

$D_1 = 20$ mm

$D_2 = 25$ mm

$D_3$

**FIGURE P17–81**

## Critical Radius of Insulation

**17–82C** What is the critical radius of insulation? How is it defined for a cylindrical layer?

**17–83C** Consider an insulated pipe exposed to the atmosphere. Will the critical radius of insulation be greater on calm days or on windy days? Why?

**17–84C** A pipe is insulated to reduce the heat loss from it. However, measurements indicate that the rate of heat loss has increased instead of decreasing. Can the measurements be right?

**17–85C** Consider a pipe at a constant temperature whose radius is greater than the critical radius of insulation. Someone claims that the rate of heat loss from the pipe has increased when some insulation is added to the pipe. Is this claim valid?

**17–86C** A pipe is insulated such that the outer radius of the insulation is less than the critical radius. Now the insulation is taken off. Will the rate of heat transfer from the pipe increase or decrease for the same pipe surface temperature?

**17–87E** A 0.083-in-diameter electrical wire at 90°F is covered by 0.02-in-thick plastic insulation ($k = 0.075$ Btu/h·ft·°F). The wire is exposed to a medium at 50°F, with a combined convection and radiation heat transfer coefficient of 2.5 Btu/h·ft²·°F. Determine if the plastic insulation on the wire will increase or decrease heat transfer from the wire. *Answer: It helps*

**17–88E** Repeat Prob. 17–87E, assuming a thermal contact resistance of 0.001 h·ft²·°F/Btu at the interface of the wire and the insulation.

**17–89** A 4-mm-diameter spherical ball at 50°C is covered by a 1-mm-thick plastic insulation ($k = 0.13$ W/m·K). The ball is exposed to a medium at 15°C, with a combined convection and radiation heat transfer coefficient of 20 W/m²·K. Determine if the plastic insulation on the ball will help or hurt heat transfer from the ball.

Plastic insulation

4 mm | 1 mm

**FIGURE P17–89**

**17–90** Reconsider Prob. 17–89. Using EES (or other) software, plot the rate of heat transfer from the ball as a function of the plastic insulation thickness in the range of 0.5 mm to 20 mm. Discuss the results.

## Heat Transfer from Finned Surfaces

**17–91C** Hot air is to be cooled as it is forced to flow through the tubes exposed to atmospheric air. Fins are to be added in order to enhance heat transfer. Would you recommend attaching the fins inside or outside the tubes? Why? When would you recommend attaching fins both inside and outside the tubes?

**17–92C** What is the reason for the widespread use of fins on surfaces?

**17–93C** What is the difference between the fin effectiveness and the fin efficiency?

**17–94C** The fins attached to a surface are determined to have an effectiveness of 0.9. Do you think the rate of heat transfer from the surface has increased or decreased as a result of the addition of these fins?

**17–95C** Explain how the fins enhance heat transfer from a surface. Also, explain how the addition of fins may actually decrease heat transfer from a surface.

**17–96C** How does the overall effectiveness of a finned surface differ from the effectiveness of a single fin?

**17–97C** Hot water is to be cooled as it flows through the tubes exposed to atmospheric air. Fins are to be attached in order to enhance heat transfer. Would you recommend attaching the fins inside or outside the tubes? Why?

**17–98C** Consider two finned surfaces that are identical except that the fins on the first surface are formed by casting or extrusion, whereas they are attached to the second surface afterwards by welding or tight fitting. For which case do you think the fins will provide greater enhancement in heat transfer? Explain.

**17–99C** The heat transfer surface area of a fin is equal to the sum of all surfaces of the fin exposed to the surrounding medium, including the surface area of the fin tip. Under what conditions can we neglect heat transfer from the fin tip?

**17–100C** Does the (*a*) efficiency and (*b*) effectiveness of a fin increase or decrease as the fin length is increased?

**17–101C** Two pin fins are identical, except that the diameter of one of them is twice the diameter of the other. For which fin is the (*a*) fin effectiveness and (*b*) fin efficiency higher? Explain.

**17–102C** Two plate fins of constant rectangular cross section are identical, except that the thickness of one of them is twice the thickness of the other. For which fin is the (*a*) fin effectiveness and (*b*) fin efficiency higher? Explain.

**17–103C** Two finned surfaces are identical, except that the convection heat transfer coefficient of one of them is twice that of the other. For which finned surface is the (*a*) fin effectiveness and (*b*) fin efficiency higher? Explain.

**17–104** The case-to-ambient thermal resistance of a power transistor that has a maximum power rating of 15 W is given to be 25°C/W. If the case temperature of the transistor is not to exceed 80°C, determine the power at which this transistor can be operated safely in an environment at 35°C.

**17–105** A 4-mm-diameter and 10-cm-long aluminum fin ($k = 237$ W/m·K) is attached to a surface. If the heat transfer coefficient is 12 W/m²·K, determine the percent error in the rate of heat transfer from the fin when the infinitely long fin assumption is used instead of the adiabatic fin tip assumption.

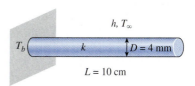

$h$, $T_\infty$

$T_b$ | $k$ | $D = 4$ mm

$L = 10$ cm

**FIGURE P17–105**

**17–106** Consider a very long rectangular fin attached to a flat surface such that the temperature at the end of the fin is essentially that of the surrounding air, i.e. 20°C. Its width is 5.0 cm; thickness is 1.0 mm; thermal conductivity is 200 W/m·K; and base temperature is 40°C. The heat transfer coefficient is 20 W/m²·K. Estimate the fin temperature at a distance of 5.0 cm from the base and the rate of heat loss from the entire fin.

**17–107** A DC motor delivers mechanical power to a rotating stainless steel shaft ($k = 15.1$ W/m·K) with a length of 25 cm and a diameter of 25 mm. In a surrounding with ambient air temperature of 20°C and convection heat transfer coefficient of 25 W/m²·K, the surface area of the motor housing that is exposed to the ambient air is 0.075 m². The motor uses 300 W of electrical power and delivers 55% of it as mechanical power to rotate the stainless steel shaft. If the tip of the stainless steel shaft has a temperature of 22°C, determine the surface temperature of the motor housing. Assume the base temperature of the shaft is equal to the surface temperature of the motor housing. *Answer: 87.7°C.*

**FIGURE P17–108**

**FIGURE P17–107**

**17–108** Two 4-m-long and 0.4-cm-thick cast iron ($k = 52$ W/m·K) steam pipes of outer diameter 10 cm are connected to each other through two 1-cm-thick flanges of outer diameter 18 cm. The steam flows inside the pipe at an average temperature of 200°C with a heat transfer coefficient of 180 W/m²·K. The outer surface of the pipe is exposed to an ambient at 12°C, with a heat transfer coefficient of 25 W/m²·K. (*a*) Disregarding the flanges, determine the average outer surface temperature of the pipe. (*b*) Using this temperature for the base of the flange and treating the flanges as the fins, determine the fin efficiency and the rate of heat transfer from the flanges. (*c*) What length of pipe is the flange section equivalent to for heat transfer purposes?

**17–109** A 40-W power transistor is to be cooled by attaching it to one of the commercially available heat sinks shown in Table 17–6. Select a heat sink that will allow the case temperature of the transistor not to exceed 90°C in the ambient air at 20°C.

**FIGURE P17–109**

**17–110** A 25-W power transistor is to be cooled by attaching it to one of the commercially available heat sinks shown in Table 17–6. Select a heat sink that will allow the case temperature of the transistor not to exceed 55°C in the ambient air at 18°C.

**17–111** A plane wall with surface temperature of 350°C is attached with straight rectangular fins ($k = 235$ W/m·K). The fins are exposed to an ambient air condition of 25°C and the convection heat transfer coefficient is 154 W/m²·K. Each fin has a length of 50 mm, a base of 5 mm thick and a width of 100 mm. Determine the efficiency, heat transfer rate, and effectiveness of each fin, using (*a*) Table 17–3 and (*b*) Figure 17–43.

**FIGURE P17–111**

**FIGURE P17–113E**

**17–112** Steam in a heating system flows through tubes whose outer diameter is 5 cm and whose walls are maintained at a temperature of 130°C. Circular aluminum alloy 2024-T6 fins ($k = 186$ W/m·K) of outer diameter 6 cm and constant thickness 1 mm are attached to the tube. The space between the fins is 3 mm, and thus there are 250 fins per meter length of the tube. Heat is transferred to the surrounding air at $T_\infty = 25°C$, with a heat transfer coefficient of 40 W/m²·K. Determine the increase in heat transfer from the tube per meter of its length as a result of adding fins. *Answer:* 1788 W

**FIGURE P17–112**

**17–113E** Consider a stainless steel spoon ($k = 8.7$ Btu/h·ft·°F) partially immersed in boiling water at 200°F in a kitchen at 75°F. The handle of the spoon has a cross section of 0.08 in × 0.5 in, and extends 7 in in the air from the free surface of the water. If the heat transfer coefficient at the exposed surfaces of the spoon handle is 3 Btu/h·ft²·°F, determine the temperature difference across the exposed surface of the spoon handle. State your assumptions. *Answer:* 124.6°F

**17–114E** Repeat Prob. 17–113E for a silver spoon ($k = 247$ Btu/h·ft·°F).

**17–115E** Reconsider Prob. 17–113E. Using EES (or other) software, investigate the effects of the thermal conductivity of the spoon material and the length of its extension in the air on the temperature difference across the exposed surface of the spoon handle. Let the thermal conductivity vary from 5 Btu/h·ft·°F to 225 Btu/h·ft·°F and the length from 5 in to 12 in. Plot the temperature difference as the functions of thermal conductivity and length, and discuss the results.

**17–116** A 0.4-cm-thick, 12-cm-high, and 18-cm-long circuit board houses 80 closely spaced logic chips on one side, each dissipating 0.04 W. The board is impregnated with copper fillings and has an effective thermal conductivity of 30 W/m·K. All the heat generated in the chips is conducted across the circuit board and is dissipated from the back side of the board to a medium at 40°C, with a heat transfer coefficient of 52 W/m²·K. (a) Determine the temperatures on the two sides of the circuit board. (b) Now a 0.2-cm-thick, 12-cm-high, and 18-cm-long aluminum plate ($k = 237$ W/m·K) with 864 2-cm-long aluminum pin fins of diameter 0.25 cm is attached to the back side of the circuit board with a 0.02-cm-thick epoxy adhesive ($k = 1.8$ W/m·K). Determine the new temperatures on the two sides of the circuit board.

**17–117** Repeat Prob. 17–116 using a copper plate with copper fins ($k = 386$ W/m·K) instead of aluminum ones.

**17–118** A hot surface at 100°C is to be cooled by attaching 3-cm-long, 0.25-cm-diameter aluminum pin fins ($k = 237$ W/m·K) to it, with a center-to-center distance of 0.6 cm. The temperature of the surrounding medium is 30°C, and the heat transfer coefficient on the surfaces is 35 W/m²·K. Determine the rate of heat transfer from the surface for a 1-m × 1-m section of the plate. Also determine the overall effectiveness of the fins.

**FIGURE P17–118**

**17–119** Repeat Prob. 17–118 using copper fins ($k = 386$ W/m·K) instead of aluminum ones.

**17–120** ⌨ Reconsider Prob. 17–118. Using EES (or other) software, investigate the effect of the center-to-center distance of the fins on the rate of heat transfer from the surface and the overall effectiveness of the fins. Let the center-to-center distance vary from 0.4 cm to 2.0 cm. Plot the rate of heat transfer and the overall effectiveness as a function of the center-to-center distance, and discuss the results.

**17–121** Circular cooling fins of diameter $D = 1$ mm and length $L = 30$ mm, made of copper ($k = 380$ W/m·K), are used to enhance heat transfer from a surface that is maintained at temperature $T_{s1} = 132$°C. Each rod has one end attached to this surface ($x = 0$), while the opposite end ($x = L$) is joined to a second surface, which is maintained at $T_{s2} = 0$°C. The air flowing between the surfaces and the rods is also at $T_\infty = 0$°C, and the convection coefficient is $h = 100$ W/m²·K.

(a) Express the function $\theta(x) = T(x) - T_\infty$ along a fin, and calculate the temperature at $x = L/2$.

**FIGURE P17–121**

(b) Determine the rate of heat transferred from the hot surface through each fin and the fin effectiveness. Is the use of fins justified? Why?

(c) What is the total rate of heat transfer from a 10-cm by 10-cm section of the wall, which has 625 uniformly distributed fins? Assume the same convection coefficient for the fin and for the unfinned wall surface.

## Review Problems

**17–122** Steam in a heating system flows through tubes whose outer diameter is 3 cm and whose walls are maintained at a temperature of 120°C. Circular aluminum alloy fins ($k = 180$ W/m·K) of outer diameter 6 cm and constant thickness $t = 2$ mm are attached to the tube, as shown in Fig. P17–122. The space between the fins is 3 mm, and thus there are 200 fins per meter length of the tube. Heat is transferred to the surrounding air at 25°C, with a combined heat transfer coefficient of 60 W/m²·K. Determine the increase in heat transfer from the tube per meter of its length as a result of adding fins.

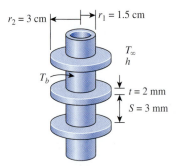

**FIGURE P17–122**

**17–123E** Steam is produced in the copper tubes ($k = 223$ Btu/h·ft·°F) of a heat exchanger at a temperature of 280°F by another fluid condensing on the outside surfaces of the tubes at 350°F. The inner and outer diameters of the tube are 1 in and 1.3 in, respectively. When the heat exchanger was new, the rate of heat transfer per foot length of the tube was $2 \times 10^4$ Btu/h. Determine the rate of heat transfer per foot length of the tube when a 0.01-in-thick layer of limestone ($k = 1.7$ Btu/h·ft·°F) has formed on the inner surface of the tube after extended use.

**17–124E** Repeat Prob. 17–123E, assuming that a 0.01-in-thick limestone layer has formed on both the inner and outer surfaces of the tube.

**17–125** Hot water is flowing at an average velocity of 1.5 m/s through a cast iron pipe ($k = 52$ W/m·K) whose inner and outer diameters are 3 cm and 3.5 cm, respectively. The pipe passes through a 15-m-long section of a basement whose temperature is 15°C. If the temperature of the water drops from 70°C to 67°C as it passes through the basement and the heat transfer coefficient on the inner surface of the pipe is

400 W/m²·K, determine the combined convection and radiation heat transfer coefficient at the outer surface of the pipe. *Answer: 272.5 W/m²·K*

**17–126** Newly formed concrete pipes are usually cured first overnight by steam in a curing kiln maintained at a temperature of 45°C before the pipes are cured for several days outside. The heat and moisture to the kiln is provided by steam flowing in a pipe whose outer diameter is 12 cm. During a plant inspection, it was noticed that the pipe passes through a 8-m section that is completely exposed to the ambient air before it reaches the kiln. The temperature measurements indicate that the average temperature of the outer surface of the steam pipe is 90°C when the ambient temperature is 8°C. The combined convection and radiation heat transfer coefficient at the outer surface of the pipe is estimated to be 35 W/m²·K. Determine the amount of heat lost from the steam during a 10-h curing process that night.

Steam is supplied by a gas-fired steam generator that has an efficiency of 85 percent, and the plant pays $1.20/therm of natural gas (1 therm = 105,500 kJ). If the pipe is insulated and 90 percent of the heat loss is saved as a result, determine the amount of money this facility will save a year as a result of insulating the steam pipes. Assume that the concrete pipes are cured 110 nights a year. State your assumptions.

**FIGURE P17–126**

**17–127** Consider an 18-cm × 18-cm multilayer circuit board dissipating 27 W of heat. The board consists of four layers of 0.2-mm-thick copper ($k = 386$ W/m·K) and three layers of 1.5-mm-thick epoxy glass ($k = 0.26$ W/m·K) sandwiched together, as shown in the figure. The circuit board is attached to a heat sink from both ends, and the temperature of the board at those ends is 35°C. Heat is considered to be uniformly generated in the epoxy layers of the board at a rate of 0.5 W per 1-cm × 18-cm epoxy laminate strip (or 1.5 W per 1-cm × 18-cm strip of the board). Considering only a portion of the board because of symmetry, determine the magnitude and location of the maximum temperature that occurs in the board. Assume heat transfer from the top and bottom faces of the board to be negligible.

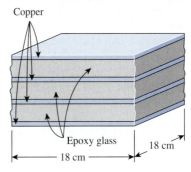

**FIGURE P17–127**

**17–128E** The surface temperature of a 3-in-diameter baked potato is observed to drop from 300°F to 200°F in 5 min in an environment at 70°F. Determine the average heat transfer coefficient between the potato and its surroundings. Using this heat transfer coefficient and the same surface temperature, determine how long it will take for the potato to experience the same temperature drop if it is wrapped completely in a 0.12-in-thick towel ($k = 0.035$ Btu/h·ft·°F). You may use the properties of water for potato.

**17–129E** Repeat Prob. 17–128E assuming there is a 0.02-in-thick air space ($k = 0.015$ Btu/h·ft·°F) between the potato and the towel.

**17–130** A 6-m-wide, 2.8-m-high wall is constructed of one layer of common brick ($k = 0.72$ W/m·K) of thickness 20 cm, one inside layer of light-weight plaster ($k = 0.36$ W/m·K) of thickness 1 cm, and one outside layer of cement based covering ($k = 1.40$ W/m·K) of thickness 2 cm. The inner surface of the wall is maintained at 23°C while the outer surface is exposed to outdoors at 8°C with a combined convection and radiation heat transfer coefficient of 17 W/m²·K. Determine the rate of heat transfer through the wall and temperature drops across the plaster, brick, covering, and surface-ambient air.

**17–131** Reconsider Prob. 17–130. It is desired to insulate the wall in order to decrease the heat loss by 90 percent. For the same inner surface temperature, determine the thickness of insulation and the outer surface temperature if the wall is insulated with (*a*) polyurethane foam ($k = 0.025$ W/m·K) and (*b*) glass fiber ($k = 0.036$ W/m·K).

**17–132** A 0.2-cm-thick, 10-cm-high, and 15-cm-long circuit board houses electronic components on one side that dissipate a total of 15 W of heat uniformly. The board is impregnated with conducting metal fillings and has an effective thermal conductivity of 12 W/m·K. All the heat generated in the components is conducted across the circuit board and is dissipated from the back side of the board to a medium at 37°C, with a heat transfer coefficient of 45 W/m²·K. (*a*) Determine the surface temperatures on the two sides of the circuit board. (*b*) Now a 0.1-cm-thick, 10-cm-high, and 15-cm-long aluminum plate ($k = 237$ W/m·K) with 20 0.2-cm-thick, 2-cm-long, and 15-cm-wide aluminum fins of rectangular profile are attached to the back side of the circuit board with a 0.03-cm-thick epoxy adhesive ($k$ = 1.8 W/m·K). Determine the new temperatures on the two sides of the circuit board.

**FIGURE P17–132**

**17–133** Repeat Prob. 17–132 using a copper plate with copper fins ($k$ = 386 W/m·K) instead of aluminum ones.

**17–134** Consider two identical people each generating 60 W of metabolic heat steadily while doing sedentary work, and dissipating it by convection and perspiration. The first person is wearing clothes made of 1-mm-thick leather ($k$ = 0.159 W/m·K) that covers half of the body while the second one is wearing clothes made of 1-mm-thick synthetic fabric ($k$ = 0.13 W/m·K) that covers the body completely. The ambient air is at 30°C, the heat transfer coefficient at the outer surface is 15 W/m²·K, and the inner surface temperature of the clothes can be taken to be 32°C. Treating the body of each person as a 25-cm-diameter, 1.7-m-long cylinder, determine the fractions of heat lost from each person by perspiration.

**17–135** A 4-m-high and 6-m-long wall is constructed of two large 0.8-cm-thick steel plates ($k$ = 15 W/m·K) separated by 1-cm-thick and 22-cm wide steel bars placed 99 cm apart. The remaining space between the steel plates is filled with fiberglass insulation ($k$ = 0.035 W/m·K). If the temperature difference between the inner and the outer surfaces of the walls is 22°C, determine the rate of heat transfer through the wall. Can we ignore the steel bars between the plates in heat transfer analysis since they occupy only 1 percent of the heat transfer surface area?

**FIGURE P17–135**

**17–136** Cold conditioned air at 12°C is flowing inside a 1.5-cm-thick square aluminum ($k$ = 237 W/m·K) duct of inner cross section 22 cm × 22 cm at a mass flow rate of 0.8 kg/s. The duct is exposed to air at 33°C with a combined convection-radiation heat transfer coefficient of 13 W/m²·K. The convection heat transfer coefficient at the inner surface is 75 W/m²·K. If the air temperature in the duct should not increase by more than 1°C determine the maximum length of the duct.

**17–137** When analyzing heat transfer through windows, it is important to consider the frame as well as the glass area. Consider a 2-m-wide, 1.5-m-high wood-framed window with 85 percent of the area covered by 3-mm-thick single-pane glass ($k$ = 0.7 W/m·K). The frame is 5 cm thick, and is made of pine wood ($k$ = 0.12 W/m·K). The heat transfer coefficient is 7 W/m²·K inside and 13 W/m²·K outside. The room is maintained at 24°C, and the outdoor temperature is 40°C. Determine the percent error involved in heat transfer when the window is assumed to consist of glass only.

**17–138** Steam at 260°C is flowing inside a steel pipe ($k$ = 61 W/m·K) whose inner and outer diameters are 10 cm and 12 cm, respectively, in an environment at 20°C. The heat transfer coefficients inside and outside the pipe are 120 W/m²·K and 14 W/m²·K, respectively. Determine (a) the thickness of the insulation ($k$ = 0.038 W/m·K) needed to reduce the heat loss by 95 percent and (b) the thickness of the insulation needed to reduce the exposed surface temperature of insulated pipe to 40°C for safety reasons.

**17–139** When the transportation of natural gas in a pipeline is not feasible for economic or other reasons, it is first liquefied at about $-160°C$, and then transported in specially insulated tanks placed in marine ships. Consider a 4-m-diameter spherical tank that is filled with liquefied natural gas (LNG) at $-160°C$. The tank is exposed to ambient air at $24°C$ with a heat transfer coefficient of 22 W/m²·K. The tank is thin-shelled and its temperature can be taken to be the same as the LNG temperature. The tank is insulated with 5-cm-thick super insulation that has an effective thermal conductivity of 0.00008 W/m·K. Taking the density and the specific heat of LNG to be 425 kg/m³ and 3.475 kJ/kg·°C, respectively, estimate how long it will take for the LNG temperature to rise to $-150°C$.

**17–140**   A 15-cm × 20-cm hot surface at 85°C is to be cooled by attaching 4-cm-long aluminum ($k = 237$ W/m·K) fins of 2-mm × 2-mm square cross section. The temperature of surrounding medium is 25°C and the heat transfer coefficient on the surfaces can be taken to be 20 W/m²·K. If it is desired to triple the rate of heat transfer from the bare hot surface, determine the number of fins that needs to be attached.

**17–141**   Reconsider Prob. 17–140. Using EES (or other) software, plot the number of fins as a function of the increase in the heat loss by fins relative to no fin case (i.e., overall effectiveness of the fins) in the range of 1.5 to 5. Discuss the results. Is it realistic to assume the heat transfer coefficient to remain constant?

**17–142**   An agitated vessel is used for heating 500 kg/min of an aqueous solution at 15°C by saturated steam condensing in the jacket outside the vessel. The vessel can hold 6200 kg of the aqueous solution. It is fabricated from 15-mm-thick sheet of 1.0 percent carbon steel ($k = 43$ W/m·K), and it provides a heat transfer area of 12.0 m². The heat transfer coefficient due to agitation is 5.5 kW/m²·K, while the steam condensation at 115°C in the jacket gives a heat transfer coefficient of 10.0 kW/m²·K. All properties of the aqueous solution are comparable to those of pure water. Calculate the temperature of the outlet stream in steady operation.

**17–143**   A typical section of a building wall is shown in Fig. P17–143. This section extends in and out of the page and is repeated in the vertical direction. The wall support members are made of steel ($k = 50$ W/m·K). The support members are 8 cm ($t_{23}$) × 0.5 cm ($L_B$). The remainder of the inner wall space is filled with insulation ($k = 0.03$ W/m·K) and measures 8 cm ($t_{23}$) × 60 cm ($L_B$). The inner wall is made of gypsum board ($k = 0.5$ W/m·K) that is 1 cm thick ($t_{12}$) and the outer wall is made of brick ($k = 1.0$ W/m·K) that is 10 cm thick ($t_{34}$). What is the temperature on the interior brick surface, 3, when $T_1 = 20$ °C and $T_4 = 35°C$?

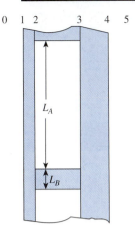

**FIGURE P17–143**

**17–144**   A total of 10 rectangular aluminum fins ($k = 203$ W/m·K) are placed on the outside flat surface of an electronic device. Each fin is 100 mm wide, 20 mm high and 4 mm thick. The fins are located parallel to each other at a center-to-center distance of 8 mm. The temperature at the outside surface of the electronic device is 72°C. The air is at 20°C, and the heat transfer coefficient is 80 W/m²·K. Determine (a) the rate of heat loss from the electronic device to the surrounding air and (b) the fin effectiveness.

**17–145**   One wall of a refrigerated warehouse is 10.0-m-high and 5.0-m-wide. The wall is made of three layers: 1.0-cm-thick aluminum ($k = 200$ W/m·K), 8.0-cm-thick fibreglass ($k = 0.038$ W/m·K), and 3.0-cm thick gypsum board ($k = 0.48$ W/m·K). The warehouse inside and outside temperatures are $-10°C$ and 20°C, respectively, and the average value of both inside and outside heat transfer coefficients is 40 W/m²·K.

(a) Calculate the rate of heat transfer across the warehouse wall in steady operation.

(b) Suppose that 400 metal bolts ($k = 43$ W/m·K), each 2.0 cm in diameter and 12.0 cm long, are used to fasten (i.e., hold together) the three wall layers. Calculate the rate of heat transfer for the "bolted" wall.

(c) What is the percent change in the rate of heat transfer across the wall due to metal bolts?

**17–146**   A 12-cm-long bar with a square cross-section, as shown in Fig. P17–146, consists of a 1-cm-thick copper layer ($k = 380$ W/m·K) and a 1-cm-thick epoxy composite layer ($k = 0.4$ W/m·K). Calculate the rate of heat transfer under a thermal driving force of 50°C, when the direction of steady one-dimensional heat transfer is (a) from front to back (i.e., along its length), (b) from left to right, and (c) from top to bottom.

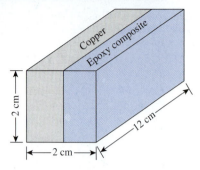

**FIGURE P17–146**

## Design and Essay Problems

**17–147** The temperature in deep space is close to absolute zero, which presents thermal challenges for the astronauts who do space walks. Propose a design for the clothing of the astronauts that will be most suitable for the thermal environment in space. Defend the selections in your design.

**17–148** In the design of electronic components, it is very desirable to attach the electronic circuitry to a substrate material that is a very good thermal conductor but also a very effective electrical insulator. If the high cost is not a major concern, what material would you propose for the substrate?

**17–149** Using cylindrical samples of the same material, devise an experiment to determine the thermal contact resistance.

Cylindrical samples are available at any length, and the thermal conductivity of the material is known.

**17–150** A house with 200-m² floor space is to be heated with geothermal water flowing through pipes laid in the ground under the floor. The walls of the house are 4 m high, and there are 10 single-paned windows in the house that are 1.2 m wide and 1.8 m high. The house has $R$-19 (in h·ft²·°F/Btu) insulation in the walls and $R$-30 on the ceiling. The floor temperature is not to exceed 40°C. Hot geothermal water is available at 90°C, and the inner and outer diameter of the pipes to be used are 2.4 cm and 3.0 cm. Design such a heating system for this house in your area.

**17–151** Using a timer (or watch) and a thermometer, conduct this experiment to determine the rate of heat gain of your refrigerator. First, make sure that the door of the refrigerator is not opened for at least a few hours to make sure that steady operating conditions are established. Start the timer when the refrigerator stops running and measure the time $\Delta t_1$ it stays off before it kicks in. Then measure the time $\Delta t_2$ it stays on. Noting that the heat removed during $\Delta t_2$ is equal to the heat gain of the refrigerator during $\Delta t_1 + \Delta t_2$ and using the power consumed by the refrigerator when it is running, determine the average rate of heat gain for your refrigerator, in watts. Take the COP (coefficient of performance) of your refrigerator to be 1.3 if it is not available.

Now, clean the condenser coils of the refrigerator and remove any obstacles on the way of airflow through the coils. By replacing these measurements, determine the improvement in the COP of the refrigerator.

# TRANSIENT HEAT CONDUCTION

The temperature of a body, in general, varies with time as well as position. In rectangular coordinates, this variation is expressed as $T(x, y, z, t)$, where $(x, y, z)$ indicate variation in the $x$-, $y$-, and $z$-directions, and $t$ indicates variation with time. In the preceding chapter, we considered heat conduction under *steady* conditions, for which the temperature of a body at any point does not change with time. This certainly simplified the analysis, especially when the temperature varied in one direction only, and we were able to obtain analytical solutions. In this chapter, we consider the variation of temperature with *time* as well as *position* in one- and multidimensional systems.

We start this chapter with the analysis of *lumped systems* in which the temperature of a body varies with time but remains uniform throughout at any time. Then we consider the variation of temperature with time as well as position for one-dimensional heat conduction problems such as those associated with a large plane wall, a long cylinder, a sphere, and a semi-infinite medium using *transient temperature charts* and analytical solutions. Finally, we consider transient heat conduction in multidimensional systems by utilizing the *product solution*.

## Objectives

The objectives of this chapter are to:

- Assess when the spatial variation of temperature is negligible, and temperature varies nearly uniformly with time, making the simplified lumped system analysis applicable.

- Obtain analytical solutions for transient one-dimensional conduction problems in rectangular, cylindrical, and spherical geometries using the method of separation of variables, and understand why a one-term solution is usually a reasonable approximation.

- Solve the transient conduction problem in large mediums using the similarity variable, and predict the variation of temperature with time and distance from the exposed surface.

- Construct solutions for multidimensional transient conduction problems using the product solution approach.

(a) Copper ball

(b) Roast beef

**FIGURE 18–1**
A small copper ball can be modeled as a lumped system, but a roast beef cannot.

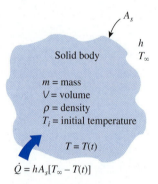

**FIGURE 18–2**
The geometry and parameters involved in the lumped system analysis.

# 18–1 · LUMPED SYSTEM ANALYSIS

In heat transfer analysis, some bodies are observed to behave like a "lump" whose interior temperature remains essentially uniform at any times during a heat transfer process. The temperature of such bodies can be taken to be a function of time only, $T(t)$. Heat transfer analysis that utilizes this idealization is known as **lumped system analysis**, which provides great simplification in certain classes of heat transfer problems without much sacrifice from accuracy.

Consider a small hot copper ball coming out of an oven (Fig. 18–1a). Measurements indicate that the temperature of the copper ball changes with time, but it does not change much with position at any given time. Thus the temperature of the ball remains nearly uniform at all times, and we can talk about the temperature of the ball with no reference to a specific location.

Now let us go to the other extreme and consider a large roast in an oven (Fig. 18–1b). If you have done any roasting, you must have noticed that the temperature distribution within the roast is not even close to being uniform. You can easily verify this by taking the roast out before it is completely done and cutting it in half. You will see that the outer parts of the roast are well done while the center part is barely warm. Thus, lumped system analysis is not applicable in this case. Before presenting a criterion about applicability of lumped system analysis, we develop the formulation associated with it.

Consider a body of arbitrary shape of mass $m$, volume $V$, surface area $A_s$, density $\rho$, and specific heat $c_p$ initially at a uniform temperature $T_i$ (Fig. 18–2). At time $t = 0$, the body is placed into a medium at temperature $T_\infty$, and heat transfer takes place between the body and its environment, with a heat transfer coefficient $h$. For the sake of discussion, we assume that $T_\infty > T_i$, but the analysis is equally valid for the opposite case. We assume lumped system analysis to be applicable, so that the temperature remains uniform within the body at all times and changes with time only, $T = T(t)$.

During a differential time interval $dt$, the temperature of the body rises by a differential amount $dT$. An energy balance of the solid for the time interval $dt$ can be expressed as

$$\left( \begin{array}{c} \text{Heat transfer into the body} \\ \text{during } dt \end{array} \right) = \left( \begin{array}{c} \text{The increase in the} \\ \text{energy of the body} \\ \text{during } dt \end{array} \right)$$

or

$$hA_s(T_\infty - T)\, dt = mc_p\, dT \tag{18–1}$$

Noting that $m = \rho V$ and $dT = d(T - T_\infty)$ since $T_\infty = $ constant, Eq. 18–1 can be rearranged as

$$\frac{d(T - T_\infty)}{T - T_\infty} = -\frac{hA_s}{\rho V c_p}\, dt \tag{18–2}$$

Integrating from $t = 0$, at which $T = T_i$, to any time $t$, at which $T = T(t)$, gives

$$\ln \frac{T(t) - T_\infty}{T_i - T_\infty} = -\frac{hA_s}{\rho V c_p}\, t \tag{18–3}$$

Taking the exponential of both sides and rearranging, we obtain

$$\frac{T(t) - T_\infty}{T_i - T_\infty} = e^{-bt} \tag{18-4}$$

where

$$b = \frac{hA_s}{\rho V c_p} \quad (1/s) \tag{18-5}$$

is a positive quantity whose dimension is (time)$^{-1}$. The reciprocal of $b$ has time unit (usually s), and is called the **time constant**. Equation 18–4 is plotted in Fig. 18–3 for different values of $b$. There are two observations that can be made from this figure and the relation above:

1. Equation 18–4 enables us to determine the temperature $T(t)$ of a body at time $t$, or alternatively, the time $t$ required for the temperature to reach a specified value $T(t)$.
2. The temperature of a body approaches the ambient temperature $T_\infty$ exponentially. The temperature of the body changes rapidly at the beginning, but rather slowly later on. A large value of $b$ indicates that the body approaches the environment temperature in a short time. The larger the value of the exponent $b$, the higher the rate of decay in temperature. Note that $b$ is proportional to the surface area, but inversely proportional to the mass and the specific heat of the body. This is not surprising since it takes longer to heat or cool a larger mass, especially when it has a large specific heat.

Once the temperature $T(t)$ at time $t$ is available from Eq. 18–4, the *rate* of convection heat transfer between the body and its environment at that time can be determined from Newton's law of cooling as

$$\dot{Q}(t) = hA_s[T(t) - T_\infty] \quad (W) \tag{18-6}$$

The *total amount* of heat transfer between the body and the surrounding medium over the time interval $t = 0$ to $t$ is simply the change in the energy content of the body:

$$Q = mc_p[T(t) - T_i] \quad (kJ) \tag{18-7}$$

The amount of heat transfer reaches its upper limit when the body reaches the surrounding temperature $T_\infty$. Therefore, the *maximum* heat transfer between the body and its surroundings is (Fig. 18–4)

$$Q_{max} = mc_p(T_\infty - T_i) \quad (kJ) \tag{18-8}$$

We could also obtain this equation by substituting the $T(t)$ relation from Eq. 18–4 into the $\dot{Q}(t)$ relation in Eq. 18–6 and integrating it from $t = 0$ to $t \to \infty$.

# Criteria for Lumped System Analysis

The lumped system analysis certainly provides great convenience in heat transfer analysis, and naturally we would like to know when it is appropriate

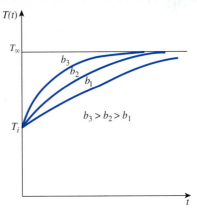

**FIGURE 18–3**
The temperature of a lumped system approaches the environment temperature as time gets larger.

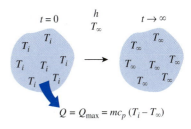

$$Q = Q_{max} = mc_p(T_i - T_\infty)$$

**FIGURE 18–4**
Heat transfer to or from a body reaches its maximum value when the body reaches the environment temperature.

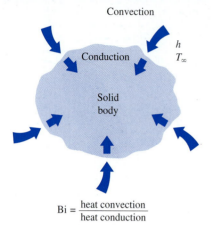

$$Bi = \frac{\text{heat convection}}{\text{heat conduction}}$$

**FIGURE 18–5**
The Biot number can be viewed as the ratio of the convection at the surface to conduction within the body.

to use it. The first step in establishing a criterion for the applicability of the lumped system analysis is to define a **characteristic length** as

$$L_c = \frac{V}{A_s}$$

and a dimensionless **Biot number** Bi as

$$Bi = \frac{hL_c}{k} \tag{18–9}$$

The characteristic length $L_c$ to be used in the Biot number for simple geometries in which heat transfer is one-dimensional, such as a large plane wall of thickness $2L$, a long cylinder of radius $r_o$, and a sphere of radius $r_o$, becomes $L$ (half thickness), $r_o/2$, and $r_o/3$, respectively. Equation 18–9 can also be expressed as (Fig. 18–5).

$$Bi = \frac{h}{k/L_c} \frac{\Delta T}{\Delta T} = \frac{\text{Convection at the surface of the body}}{\text{Conduction within the body}}$$

or

$$Bi = \frac{L_c/k}{1/h} = \frac{\text{Conduction resistance within the body}}{\text{Convection resistance at the surface of the body}}$$

When a solid body is being heated by the hotter fluid surrounding it (such as a potato being baked in an oven), heat is first *convected* to the body and subsequently *conducted* within the body. The Biot number is the *ratio* of the internal resistance of a body to *heat conduction* to its external resistance to *heat convection*. Therefore, a small Biot number represents small resistance to heat conduction, and thus small temperature gradients within the body.

Lumped system analysis assumes a *uniform* temperature distribution throughout the body, which is the case only when the thermal resistance of the body to heat conduction (the *conduction resistance*) is zero. Thus, lumped system analysis is *exact* when Bi = 0 and *approximate* when Bi > 0. Of course, the smaller the Bi number, the more accurate the lumped system analysis. Then the question we must answer is, how much accuracy are we willing to sacrifice for the convenience of the lumped system analysis?

Before answering this question, we should mention that a 15 percent uncertainty in the convection heat transfer coefficient $h$ in most cases is considered "normal" and "expected." Assuming $h$ to be *constant* and *uniform* is also an approximation of questionable validity, especially for irregular geometries. Therefore, in the absence of sufficient experimental data for the specific geometry under consideration, we cannot claim our results to be better than ±15 percent, even when Bi = 0. This being the case, introducing another source of uncertainty in the problem will not have much effect on the overall uncertainty, provided that it is minor. It is generally accepted that lumped system analysis is *applicable* if

$$Bi \leq 0.1$$

When this criterion is satisfied, the temperatures within the body relative to the surroundings (i.e., $T - T_\infty$) remain within 5 percent of each other even for well-rounded geometries such as a spherical ball. Thus, when Bi < 0.1, the variation of temperature with location within the body is slight and can reasonably be approximated as being uniform.

The first step in the application of lumped system analysis is the calculation of the *Biot number*, and the assessment of the applicability of this approach. One may still wish to use lumped system analysis even when the criterion Bi < 0.1 is not satisfied, if high accuracy is not a major concern.

Note that the Biot number is the ratio of the *convection* at the surface to *conduction* within the body, and this number should be as small as possible for lumped system analysis to be applicable. Therefore, *small bodies* with *high thermal conductivity* are good candidates for lumped system analysis, especially when they are in a medium that is a poor conductor of heat (such as air or another gas) and motionless. Thus, the hot small copper ball placed in quiescent air, discussed earlier, is most likely to satisfy the criterion for lumped system analysis (Fig. 18–6).

$$L_c = \frac{V}{A_s} = \frac{\frac{1}{6}\pi D^3}{\pi D^2} = \frac{1}{6}D = 0.02 \text{ m}$$

$$Bi = \frac{hL_c}{k} = \frac{15 \times 0.02}{401} = 0.00075 < 0.1$$

**FIGURE 18–6**
Small bodies with high thermal conductivities and low convection coefficients are most likely to satisfy the criterion for lumped system analysis.

## Some Remarks on Heat Transfer in Lumped Systems

To understand the heat transfer mechanism during the heating or cooling of a solid by the fluid surrounding it, and the criterion for lumped system analysis, consider this analogy (Fig. 18–7). People from the mainland are to go *by boat* to an island whose entire shore is a harbor, and from the harbor to their destinations on the island *by bus*. The overcrowding of people at the harbor depends on the boat traffic to the island and the ground transportation system on the island. If there is an excellent ground transportation system with plenty of buses, there will be no overcrowding at the harbor, especially when the boat traffic is light. But when the opposite is true, there will be a huge overcrowding at the harbor, creating a large difference between the populations at the harbor and inland. The chance of overcrowding is much lower in a small island with plenty of fast buses.

In heat transfer, a poor ground transportation system corresponds to poor heat conduction in a body, and overcrowding at the harbor to the accumulation of thermal energy and the subsequent rise in temperature near the surface of the body relative to its inner parts. Lumped system analysis is obviously not applicable when there is overcrowding at the surface. Of course, we have disregarded radiation in this analogy and thus the air traffic to the island. Like passengers at the harbor, heat changes *vehicles* at the surface from *convection* to *conduction*. Noting that a surface has zero thickness and thus cannot store any energy, heat reaching the surface of a body by convection must continue its journey within the body by conduction.

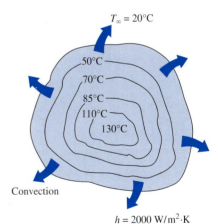

**FIGURE 18–7**
Analogy between heat transfer to a solid and passenger traffic to an island.

Consider heat transfer from a hot body to its cooler surroundings. Heat is transferred from the body to the surrounding fluid as a result of a temperature difference. But this energy comes from the region near the surface, and thus the temperature of the body near the surface will drop. This creates a *temperature gradient* between the inner and outer regions of the body and initiates heat transfer by conduction from the interior of the body toward the outer surface.

When the convection heat transfer coefficient $h$ and thus the rate of convection from the body are high, the temperature of the body near the surface drops quickly (Fig. 18–8). This creates a larger temperature difference between the inner and outer regions unless the body is able to transfer heat from the inner to the outer regions just as fast. Thus, the magnitude of the maximum temperature difference within the body depends strongly on the ability of a

**FIGURE 18–8**
When the convection coefficient $h$ is high and $k$ is low, large temperature differences occur between the inner and outer regions of a large solid.

body to conduct heat toward its surface relative to the ability of the surrounding medium to convect heat away from the surface. The Biot number is a measure of the relative magnitudes of these two competing effects.

Recall that heat conduction in a specified direction $n$ per unit surface area is expressed as $\dot{q} = -k \partial T / \partial n$, where $\partial T / \partial n$ is the temperature gradient and $k$ is the thermal conductivity of the solid. Thus, the temperature distribution in the body will be *uniform* only when its thermal conductivity is *infinite*, and no such material is known to exist. Therefore, temperature gradients and thus temperature differences must exist within the body, no matter how small, in order for heat conduction to take place. Of course, the temperature gradient and the thermal conductivity are inversely proportional for a given heat flux. Therefore, the larger the thermal conductivity, the smaller the temperature gradient.

**FIGURE 18–9**
Schematic for Example 18–1.

---

### EXAMPLE 18–1    Temperature Measurement by Thermocouples

The temperature of a gas stream is to be measured by a thermocouple whose junction can be approximated as a 1-mm-diameter sphere, as shown in Fig. 18–9. The properties of the junction are $k = 35$ W/m·K, $\rho = 8500$ kg/m³, and $c_p = 320$ J/kg·K, and the convection heat transfer coefficient between the junction and the gas is $h = 210$ W/m²·K. Determine how long it will take for the thermocouple to read 99 percent of the initial temperature difference.

**SOLUTION** The temperature of a gas stream is to be measured by a thermocouple. The time it takes to register 99 percent of the initial $\Delta T$ is to be determined.

**Assumptions** **1** The junction is spherical in shape with a diameter of $D = 0.001$ m. **2** The thermal properties of the junction and the heat transfer coefficient are constant. **3** Radiation effects are negligible.

**Properties** The properties of the junction are given in the problem statement.

**Analysis** The characteristic length of the junction is

$$L_c = \frac{V}{A_s} = \frac{\frac{1}{6}\pi D^3}{\pi D^2} = \frac{1}{6}D = \frac{1}{6}(0.001 \text{ m}) = 1.67 \times 10^{-4} \text{ m}$$

Then the Biot number becomes

$$\text{Bi} = \frac{hL_c}{k} = \frac{(210 \text{ W/m}^2\text{·K})(1.67 \times 10^{-4} \text{ m})}{35 \text{ W/m·K}} = 0.001 < 0.1$$

Therefore, lumped system analysis is applicable, and the error involved in this approximation is negligible.

In order to read 99 percent of the initial temperature difference $T_i - T_\infty$ between the junction and the gas, we must have

$$\frac{T(t) - T_\infty}{T_i - T_\infty} = 0.01$$

For example, when $T_i = 0$°C and $T_\infty = 100$°C, a thermocouple is considered to have read 99 percent of this applied temperature difference when its reading indicates $T(t) = 99$°C.

The value of the exponent $b$ is

$$b = \frac{hA_s}{\rho c_p V} = \frac{h}{\rho c_p L_c} = \frac{210 \text{ W/m}^2 \cdot \text{K}}{(8500 \text{ kg/m}^3)(320 \text{ J/kg} \cdot \text{K})(1.67 \times 10^{-4} \text{ m})} = 0.462 \text{ s}^{-1}$$

We now substitute these values into Eq. 18–4 and obtain

$$\frac{T(t) - T_\infty}{T_i - T_\infty} = e^{-bt} \longrightarrow 0.01 = e^{-(0.462 \text{ s}^{-1})t}$$

which yields

$$t = \mathbf{10 \text{ s}}$$

Therefore, we must wait at least 10 s for the temperature of the thermocouple junction to approach within 99 percent of the initial junction-gas temperature difference.

*Discussion* Note that conduction through the wires and radiation exchange with the surrounding surfaces affect the result, and should be considered in a more refined analysis.

---

**EXAMPLE 18–2**   **Quench Hardening of Steel Rods**

In a quench hardening process, steel rods ($\rho = 7832$ kg/m³, $c_p = 434$ J/kg·K, and $k = 63.9$ W/m·K) are heated in a furnace to 850°C and then cooled in a water bath to an average temperature of 95°C (Fig. 18–10). The water bath has a uniform temperature of 40°C and convection heat transfer coefficient of 450 W/m²·K. If the steel rods have a diameter of 50 mm and a length of 2 m, determine (a) the time required to cool a steel rod from 850°C to 95°C in the water bath, and (b) the total amount of heat transferred to water during the quenching of a single rod.

**SOLUTION**   In a quench hardening process, the time required to cool a steel rod in a water bath and the total amount of heat transferred are to be determined.

*Assumptions*   **1** Thermal properties of steel rod are constant. **2** Convection heat transfer coefficient is uniform. **3** Heat transfer by radiation is negligible.

*Properties*   The properties of the steel rod are given as $\rho = 7832$ kg/m³, $c_p = 434$ J/kg·K, and $k = 63.9$ W/m·K.

*Analysis*   (a) For a cylindrical rod, the characteristic length and the Biot number are

$$L_c = \frac{V}{A_s} = \frac{(\pi D^2/4)L}{\pi D L} = \frac{D}{4} = \frac{0.050 \text{ m}}{4} = 0.0125 \text{ m}$$

$$Bi = \frac{hL_c}{k} = \frac{(450 \text{ W/m}^2 \cdot \text{K})(0.0125 \text{ m})}{63.9 \text{ W/m} \cdot \text{K}} = 0.088 < 0.1$$

Steel rod

Water, 40°C
$h = 450$ W/m²·K

**FIGURE 18–10**
Schematic for Example 18–2

Since $Bi < 0.1$, the lumped system analysis is applicable. Then the time required to cool a steel rod in the water bath from 850°C to 95°C is

$$b = \frac{hA_s}{\rho c_p V} = \frac{h}{\tilde{n} c_p L_c} = \frac{450 \text{ W/m}^2 \cdot \text{K}}{(7832 \text{ kg/m}^3)(434 \text{ J/kg} \cdot \text{K})(0.0125 \text{ m})} = 0.01059 \text{ s}^{-1}$$

$$\frac{T(t) - T_\infty}{T_i - T_\infty} = e^{-bt} \rightarrow t = -\frac{1}{b} \ln \left[ \frac{T(t) - T_\infty}{T_i - T_\infty} \right] = -\frac{1}{0.01059 \text{ s}^{-1}} \ln \left[ \frac{95 - 40}{850 - 40} \right] = \textbf{254 s}$$

(b) The total amount of heat transferred during the cooling process of the steel rod in the water bath is

$$Q = m c_p [T_i - T(t)] = \rho V c_p [T_i - T(t)] = \frac{\pi D^2 L \rho c_p}{4} [T_i - T(t)]$$

$$= \frac{\pi (0.050 \text{ m})^2 (2 \text{ m})(7832 \text{ kg/m}^3)(434 \text{ J/kg} \cdot \text{K})}{4} (850 - 95) \text{ K}$$

$$= 1.01 \times 10^7 \text{ J} = \textbf{10.1 MJ}$$

**Discussion** For the temperature of the water bath to remain constant, the water body should be very large or the water should be cooled externally.

## 18–2 ▪ TRANSIENT HEAT CONDUCTION IN LARGE PLANE WALLS, LONG CYLINDERS, AND SPHERES WITH SPATIAL EFFECTS

In Section 18–1, we considered bodies in which the variation of temperature within the body is negligible; that is, bodies that remain nearly *isothermal* during a process. Relatively small bodies of highly conductive materials approximate this behavior. In general, however, the temperature within a body changes from point to point as well as with time. In this section, we consider the variation of temperature with *time* and *position* in one-dimensional problems such as those associated with a large plane wall, a long cylinder, and a sphere.

Consider a plane wall of thickness $2L$, a long cylinder of radius $r_o$, and a sphere of radius $r_o$ initially at a *uniform temperature* $T_i$, as shown in Fig. 18–12. At time $t = 0$, each geometry is placed in a large medium that is at a constant temperature $T_\infty$ and kept in that medium for $t > 0$. Heat transfer takes place between these bodies and their environments by convection with a *uniform* and *constant* heat transfer coefficient $h$. Note that all three cases possess geometric and thermal symmetry: the plane wall is symmetric about its *center plane* ($x = 0$), the cylinder is symmetric about its *centerline* ($r = 0$), and the sphere is symmetric about its *center point* ($r = 0$). We neglect *radiation* heat transfer between these bodies and their surrounding surfaces, or incorporate the radiation effect into the convection heat transfer coefficient $h$.

The variation of the temperature profile with *time* in the plane wall is illustrated in Fig. 18–12. When the wall is first exposed to the surrounding medium at $T_\infty < T_i$ at $t = 0$, the entire wall is at its initial temperature $T_i$. But the wall temperature at and near the surfaces starts to drop as a result of heat transfer from the wall to the surrounding medium. This creates a

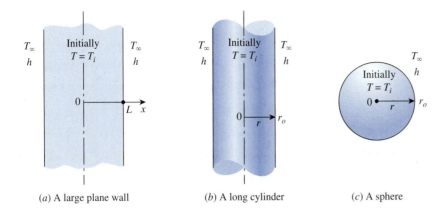

(a) A large plane wall    (b) A long cylinder    (c) A sphere

**FIGURE 18–11**
Schematic of the simple
geometries in which heat
transfer is one-dimensional.

*temperature gradient* in the wall and initiates heat conduction from the inner parts of the wall toward its outer surfaces. Note that the temperature at the center of the wall remains at $T_i$ until $t = t_2$, and that the temperature profile within the wall remains symmetric at all times about the center plane. The temperature profile gets flatter and flatter as time passes as a result of heat transfer, and eventually becomes uniform at $T = T_\infty$. That is, the wall reaches *thermal equilibrium* with its surroundings. At that point, heat transfer stops since there is no longer a temperature difference. Similar discussions can be given for the long cylinder or sphere.

## Nondimensionalized One-Dimensional Transient Conduction Problem

The formulation of heat conduction problems for the determination of the one-dimensional transient temperature distribution in a plane wall, a cylinder, or a sphere results in a partial differential equation whose solution typically involves infinite series and transcendental equations, which are inconvenient to use. But the analytical solution provides valuable insight to the physical problem, and thus it is important to go through the steps involved. Below we demonstrate the solution procedure for the case of plane wall.

Consider a plane wall of thickness $2L$ initially at a uniform temperature of $T_i$, as shown in Fig. 18–11a. At time $t = 0$, the wall is immersed in a fluid at temperature $T_\infty$ and is subjected to convection heat transfer from both sides with a convection coefficient of $h$. The height and the width of the wall are large relative to its thickness, and thus heat conduction in the wall can be approximated to be one-dimensional. Also, there is thermal symmetry about the midplane passing through $x = 0$, and thus the temperature distribution must be symmetrical about the midplane. Therefore, the value of temperature at any $-x$ value in $-L \leq x \leq 0$ at any time $t$ must be equal to the value at $+x$ in $0 \leq x \leq L$ at the same time. This means we can formulate and solve the heat conduction problem in the positive half domain $0 \leq x \leq L$, and then apply the solution to the other half.

Under the conditions of constant thermophysical properties, no heat generation, thermal symmetry about the midplane, uniform initial temperature, and constant convection coefficient, the one-dimensional transient heat conduction

**FIGURE 18–12**
Transient temperature profiles in a
plane wall exposed to convection
from its surfaces for $T_i > T_\infty$.

problem in the half-domain $0 \leq x \leq L$ of the plane wall can be expressed as (see Chapter 2 of Çengel and Ghajar, Ref. 3)

Differential equation:
$$\frac{\partial^2 T}{\partial x^2} = \frac{1}{\alpha} \frac{\partial T}{\partial t} \qquad (18-10a)$$

Boundary conditions:
$$\frac{\partial T(0, t)}{\partial x} = 0 \quad \text{and} \quad -k \frac{\partial T(L, t)}{\partial x} = h[T(L, t) - T_\infty] \quad (18-10b)$$

Initial condition:
$$T(x, 0) = T_i \qquad (18-10c)$$

where the property $\alpha = k/\rho c_p$ is the thermal diffusivity of the material.

We now attempt to nondimensionalize the problem by defining a dimensionless space variable $X = x/L$ and dimensionless temperature $\theta(x, t) = [T(x, t) - T_\infty]/[T_i - T_\infty]$. These are convenient choices since both $X$ and $\theta$ vary between 0 and 1. However, there is no clear guidance for the proper form of the dimensionless time variable and the $h/k$ ratio, so we will let the analysis indicate them. We note that

$$\frac{\partial \theta}{\partial X} = \frac{\partial \theta}{\partial (x/L)} = \frac{L}{T_i - T_\infty} \frac{\partial T}{\partial x}, \quad \frac{\partial^2 \theta}{\partial X^2} = \frac{L^2}{T_i - T_\infty} \frac{\partial^2 T}{\partial x^2} \quad \text{and} \quad \frac{\partial \theta}{\partial t} = \frac{1}{T_i - T_\infty} \frac{\partial T}{\partial t}$$

Substituting into Eqs. 18–10a and 18–10b and rearranging give

$$\frac{\partial^2 \theta}{\partial X^2} = \frac{L^2}{\alpha} \frac{\partial \theta}{\partial t} \quad \text{and} \quad \frac{\partial \theta(1, t)}{\partial X} = \frac{hL}{k} \theta(1, t) \qquad (18-11)$$

Therefore, the proper form of the dimensionless time is $\tau = \alpha t/L^2$, which is called the **Fourier number** Fo, and we recognize Bi $= k/hL$ as the Biot number defined in Section 18–1. Then the formulation of the one-dimensional transient heat conduction problem in a plane wall can be expressed in nondimensional form as

Dimensionless differential equation:
$$\frac{\partial^2 \theta}{\partial X^2} = \frac{\partial \theta}{\partial \tau} \qquad (18-12a)$$

Dimensionless BC's:
$$\frac{\partial \theta(0, \tau)}{\partial X} = 0 \quad \text{and} \quad \frac{\partial \theta(1, \tau)}{\partial X} = -\text{Bi}\theta(1, \tau)$$

$$(18-12b)$$

Dimensionless initial condition:
$$\theta(X, 0) = 1 \qquad (18-12c)$$

where

$$\theta(X, \tau) = \frac{T(x, t) - T_\infty}{T_i - T_\infty} \qquad \text{\textit{Dimensionless temperature}}$$

$$X = \frac{x}{L} \qquad \text{\textit{Dimensionless distance from the center}}$$

$$\text{Bi} = \frac{hL}{k} \qquad \text{\textit{Dimensionless heat transfer coefficient}} \text{ (Biot number)}$$

$$\tau = \frac{\alpha t}{L^2} = \text{Fo} \qquad \text{\textit{Dimensionless time}} \text{ (Fourier number)}$$

The heat conduction equation in cylindrical or spherical coordinates can be nondimensionalized in a similar way. Note that nondimensionalization reduces the number of independent variables and parameters from 8 to 3— from $x$, $L$, $t$, $k$, $\alpha$, $h$, $T_i$, and $T_\infty$ to $X$, Bi, and Fo (Fig. 18–13). That is,

$$\theta = f(X, \text{Bi}, \text{Fo}) \qquad (18\text{–}13)$$

This makes it very practical to conduct parametric studies and avoid results in graphical form. Equation 18–13 is the generalized version of Eq. 18–4 for the lumped system analysis (no space variables). This can be shown by using the definitions of $\theta$, $\alpha$, $L_c$, Bi, and Fo in Eq. 18–4. The final result is

$$\theta = \frac{T(t) - T_\infty}{T_i - T_\infty} = e^{-bt} = e^{-\frac{hA_s t}{\rho V c_p}} = e^{-\text{BiFo}}$$

or $\theta = f(\text{Fo}, \text{Bi})$ which is the special case of Eq. 18–13 with no space variable.

The non-dimensionalized partial differential equation given in Eqs. 18–12 together with its boundary and initial conditions can be solved using several analytical and numerical techniques, including the Laplace or other transform method of separation of variables, the finite difference method, and the finite-element method. Using the method of separation of variables (see Chatper 4 of Çengel and Ghajar, Ref. 3, for mathematical details), solutions for all three geometries are obtained in terms of infinite series, and are summarized in Table 18–1. The solution for the plane wall is also applicable for a plane wall of thickness $L$ whose left surface at $x = 0$ is insulated and the right surface at $x = L$ is subjected to convection since this is precisely the mathematical problem we solved.

The analytical solutions of transient conduction problems typically involve infinite series, and thus the evaluation of an infinite number of terms to determine the temperature at a specified location and time. This may look intimidating at first, but there is no need to worry. As demonstrated in Fig. 18–14, the terms in the summation decline rapidly as $n$ and thus $\lambda_n$ increases because of the exponential decay function $e^{-\lambda_n^2 \tau}$. This is especially the case when the dimensionless time $\tau$ is large. Therefore, the evaluation of the first few terms of the infinite series (in this case just the first term) is usually adequate to determine the dimensionless temperature $\theta$.

(a) Original heat conduction problem:

$$\frac{\partial^2 T}{\partial x^2} = \frac{1}{\alpha} \frac{\partial T}{\partial x}, \quad T(x, 0) = T_i$$

$$\frac{\partial T(0, t)}{\partial x} = 0, \quad -k \frac{\partial T(L, t)}{\partial x} = h[T(L, t) - T_\infty]$$

$$T = F(x, L, t, k, \alpha, h, T_i, T_\infty)$$

(b) Nondimensionalized problem:

$$\frac{\partial^2 \theta}{\partial X^2} = \frac{\partial \theta}{\partial \tau}, \quad \theta(X, 0) = 1$$

$$\frac{\partial \theta(0, \tau)}{\partial X} = 0, \quad \frac{\partial \theta(1, \tau)}{\partial X} = -\text{Bi}\,\theta(1, \tau)$$

$$\theta = f(X, \text{Bi}, \tau)$$

**FIGURE 18–13**

Nondimensionalization reduces the number of independent variables in one-dimensional transient conduction problems from 8 to 3, offering great convenience in the presentation of results.

## TABLE 18–1

Summary of the solutions for one-dimensional transient conduction in a plane wall of thickness $2L$, a cylinder of radius $r_o$ and a sphere of radius $r_o$ subjected to convention from all surfaces.*

| Geometry | Solution | $\lambda_n$'s are the roots of |
|---|---|---|
| Plane wall | $\theta = \displaystyle\sum_{n=1}^{\infty} \frac{4 \sin \lambda_n}{2\lambda_n + \sin(2\lambda_n)} e^{-\lambda_n^2 \tau} \cos(\lambda_n x/L)$ | $\lambda_n \tan \lambda_n = \text{Bi}$ |
| Cylinder | $\theta = \displaystyle\sum_{n=1}^{\infty} \frac{2}{\lambda_n} \frac{J_1(\lambda_n)}{J_0^2(\lambda_n) + J_1^2(\lambda_n)} e^{-\lambda_n^2 \tau} J_0(\lambda_n r/r_o)$ | $\lambda_n \dfrac{J_1(\lambda_n)}{J_0(\lambda_n)} = \text{Bi}$ |
| Sphere | $\theta = \displaystyle\sum_{n=1}^{\infty} \frac{4(\sin \lambda_n - \lambda_n \cos \lambda_n)}{2\lambda_n - \sin(2\lambda_n)} e^{-\lambda_n^2 \tau} \frac{\sin(\lambda_n x/L)}{\lambda_n x/L}$ | $1 - \lambda_n \cot \lambda_n = \text{Bi}$ |

*Here $\theta = (T - T_\infty)/(T_i - T_\infty)$ is the dimensionless temperature, Bi = $hL/k$ or $hr_o/k$ is the Biot number, Fo = $\tau = \alpha t/L^2$ or $\alpha t/r_o^2$ is the Fourier number, and $J_0$ and $J_1$ are the Bessel functions of the first kind whose values are given in Table 4–3.

$$\theta_n = A_n e^{-\lambda_n^2 \tau} \cos(\lambda_n X)$$

$$A_n = \frac{4 \sin \lambda_n}{2\lambda_n + \sin(2\lambda_n)}$$

$$\lambda_n \tan \lambda_n = Bi$$

For Bi = 5, X = 1, and $t$ = 0.2:

| $n$ | $\lambda_n$ | $A_n$ | $\theta_n$ |
|---|---|---|---|
| 1 | 1.3138 | 1.2402 | 0.22321 |
| 2 | 4.0336 | −0.3442 | 0.00835 |
| 3 | 6.9096 | 0.1588 | 0.00001 |
| 4 | 9.8928 | −0.876 | 0.00000 |

**FIGURE 18–14**

The term in the series solution of transient conduction problems decline rapidly as $n$ and thus $\lambda_n$ increases because of the exponential decay function with the exponent $-\lambda_n \tau$.

## Approximate Analytical and Graphical Solutions

The analytical solution obtained above for one-dimensional transient heat conduction in a plane wall involves infinite series and implicit equations, which are difficult to evaluate. Therefore, there is clear motivation to simplify the analytical solutions and to present the solutions in *tabular* or *graphical* form using simple relations.

The dimensionless quantities defined above for a plane wall can also be used for a *cylinder* or *sphere* by replacing the space variable $x$ by $r$ and the half-thickness $L$ by the outer radius $r_o$. Note that the characteristic length in the definition of the Biot number is taken to be the *half-thickness* $L$ for the plane wall, and the *radius* $r_o$ for the long cylinder and sphere instead of $V/A$ used in lumped system analysis.

We mentioned earlier that the terms in the series solutions in Table 18–1 converge rapidly with increasing time, and for $\tau > 0.2$, keeping the first term and neglecting all the remaining terms in the series results in an error under 2 percent. We are usually interested in the solution for times with $\tau > 0.2$, and thus it is very convenient to express the solution using this **one-term approximation**, given as

Plane wall: $\quad \theta_{\text{wall}} = \dfrac{T(x, t) - T_\infty}{T_i - T_\infty} = A_1 e^{-\lambda_1^2 \tau} \cos(\lambda_1 x/L), \quad \tau > 0.2 \quad$ **(18–14)**

Cylinder: $\quad \theta_{\text{cyl}} = \dfrac{T(r, t) - T_\infty}{T_i - T_\infty} = A_1 e^{-\lambda_1^2 \tau} J_0(\lambda_1 r/r_o), \quad \tau > 0.2 \quad$ **(18–15)**

Sphere: $\quad \theta_{\text{sph}} = \dfrac{T(r, t) - T_\infty}{T_i - T_\infty} = A_1 e^{-\lambda_1^2 \tau} \dfrac{\sin(\lambda_1 r/r_o)}{\lambda_1 r/r_o}, \quad \tau > 0.2 \quad$ **(18–16)**

where the constants $A_1$ and $\lambda_1$ are functions of the Bi number only, and their values are listed in Table 18–2 against the Bi number for all three geometries. The function $J_0$ is the zeroth-order Bessel function of the first kind, whose value can be determined from Table 18–3. Noting that $\cos(0) = J_0(0) = 1$ and the limit of $(\sin x)/x$ is also 1, these relations simplify to the next ones at the center of a plane wall, cylinder, or sphere:

Center of plane wall ($x$ = 0): $\quad \theta_{0,\text{wall}} = \dfrac{T_0 - T_\infty}{T_i - T_\infty} = A_1 e^{-\lambda_1^2 \tau} \quad$ **(18–17)**

Center of cylinder ($r$ = 0): $\quad \theta_{0,\text{cyl}} = \dfrac{T_0 - T_\infty}{T_i - T_\infty} = A_1 e^{-\lambda_1^2 \tau} \quad$ **(18–18)**

Center of sphere ($r$ = 0): $\quad \theta_{0,\text{sph}} = \dfrac{T_0 - T_\infty}{T_i - T_\infty} = A_1 e^{-\lambda_1^2 \tau} \quad$ **(18–19)**

Comparing the two sets of equations above, we notice that the dimensionless temperatures anywhere in a plane wall, cylinder, and sphere are related to the center temperature by

$$\frac{\theta_{\text{wall}}}{\theta_{0,\text{wall}}} = \cos\left(\frac{\lambda_1 x}{L}\right), \quad \frac{\theta_{\text{cyl}}}{\theta_{0,\text{cyl}}} = J_0\left(\frac{\lambda_1 r}{r_o}\right), \quad \text{and} \quad \frac{\theta_{\text{sph}}}{\theta_{0,\text{sph}}} = \frac{\sin(\lambda_1 r/r_o)}{\lambda_1 r/r_o} \quad \textbf{(18–20)}$$

which shows that time dependence of dimensionless temperature within a given geometry is the same throughout. That is, if the dimensionless center

## TABLE 18–2

Coefficients used in the one-term approximate solution of transient one-dimensional heat conduction in plane walls, cylinders, and spheres (Bi $= hL/k$ for a plane wall of thickness $2L$, and Bi $= hr_o/k$ for a cylinder or sphere of radius $r_o$)

| | Plane Wall | | Cylinder | | Sphere | |
|---|---|---|---|---|---|---|
| **Bi** | $\lambda_1$ | $A_1$ | $\lambda_1$ | $A_1$ | $\lambda_1$ | $A_1$ |
| 0.01 | 0.0998 | 1.0017 | 0.1412 | 1.0025 | 0.1730 | 1.0030 |
| 0.02 | 0.1410 | 1.0033 | 0.1995 | 1.0050 | 0.2445 | 1.0060 |
| 0.04 | 0.1987 | 1.0066 | 0.2814 | 1.0099 | 0.3450 | 1.0120 |
| 0.06 | 0.2425 | 1.0098 | 0.3438 | 1.0148 | 0.4217 | 1.0179 |
| 0.08 | 0.2791 | 1.0130 | 0.3960 | 1.0197 | 0.4860 | 1.0239 |
| 0.1 | 0.3111 | 1.0161 | 0.4417 | 1.0246 | 0.5423 | 1.0298 |
| 0.2 | 0.4328 | 1.0311 | 0.6170 | 1.0483 | 0.7593 | 1.0592 |
| 0.3 | 0.5218 | 1.0450 | 0.7465 | 1.0712 | 0.9208 | 1.0880 |
| 0.4 | 0.5932 | 1.0580 | 0.8516 | 1.0931 | 1.0528 | 1.1164 |
| 0.5 | 0.6533 | 1.0701 | 0.9408 | 1.1143 | 1.1656 | 1.1441 |
| 0.6 | 0.7051 | 1.0814 | 1.0184 | 1.1345 | 1.2644 | 1.1713 |
| 0.7 | 0.7506 | 1.0918 | 1.0873 | 1.1539 | 1.3525 | 1.1978 |
| 0.8 | 0.7910 | 1.1016 | 1.1490 | 1.1724 | 1.4320 | 1.2236 |
| 0.9 | 0.8274 | 1.1107 | 1.2048 | 1.1902 | 1.5044 | 1.2488 |
| 1.0 | 0.8603 | 1.1191 | 1.2558 | 1.2071 | 1.5708 | 1.2732 |
| 2.0 | 1.0769 | 1.1785 | 1.5995 | 1.3384 | 2.0288 | 1.4793 |
| 3.0 | 1.1925 | 1.2102 | 1.7887 | 1.4191 | 2.2889 | 1.6227 |
| 4.0 | 1.2646 | 1.2287 | 1.9081 | 1.4698 | 2.4556 | 1.7202 |
| 5.0 | 1.3138 | 1.2403 | 1.9898 | 1.5029 | 2.5704 | 1.7870 |
| 6.0 | 1.3496 | 1.2479 | 2.0490 | 1.5253 | 2.6537 | 1.8338 |
| 7.0 | 1.3766 | 1.2532 | 2.0937 | 1.5411 | 2.7165 | 1.8673 |
| 8.0 | 1.3978 | 1.2570 | 2.1286 | 1.5526 | 2.7654 | 1.8920 |
| 9.0 | 1.4149 | 1.2598 | 2.1566 | 1.5611 | 2.8044 | 1.9106 |
| 10.0 | 1.4289 | 1.2620 | 2.1795 | 1.5677 | 2.8363 | 1.9249 |
| 20.0 | 1.4961 | 1.2699 | 2.2880 | 1.5919 | 2.9857 | 1.9781 |
| 30.0 | 1.5202 | 1.2717 | 2.3261 | 1.5973 | 3.0372 | 1.9898 |
| 40.0 | 1.5325 | 1.2723 | 2.3455 | 1.5993 | 3.0632 | 1.9942 |
| 50.0 | 1.5400 | 1.2727 | 2.3572 | 1.6002 | 3.0788 | 1.9962 |
| 100.0 | 1.5552 | 1.2731 | 2.3809 | 1.6015 | 3.1102 | 1.9990 |
| $\infty$ | 1.5708 | 1.2732 | 2.4048 | 1.6021 | 3.1416 | 2.0000 |

## TABLE 18–3

The zeroth- and first-order Bessel functions of the first kind

| $\eta$ | $J_0(\eta)$ | $J_1(\eta)$ |
|---|---|---|
| 0.0 | 1.0000 | 0.0000 |
| 0.1 | 0.9975 | 0.0499 |
| 0.2 | 0.9900 | 0.0995 |
| 0.3 | 0.9776 | 0.1483 |
| 0.4 | 0.9604 | 0.1960 |
| 0.5 | 0.9385 | 0.2423 |
| 0.6 | 0.9120 | 0.2867 |
| 0.7 | 0.8812 | 0.3290 |
| 0.8 | 0.8463 | 0.3688 |
| 0.9 | 0.8075 | 0.4059 |
| 1.0 | 0.7652 | 0.4400 |
| 1.1 | 0.7196 | 0.4709 |
| 1.2 | 0.6711 | 0.4983 |
| 1.3 | 0.6201 | 0.5220 |
| 1.4 | 0.5669 | 0.5419 |
| 1.5 | 0.5118 | 0.5579 |
| 1.6 | 0.4554 | 0.5699 |
| 1.7 | 0.3980 | 0.5778 |
| 1.8 | 0.3400 | 0.5815 |
| 1.9 | 0.2818 | 0.5812 |
| 2.0 | 0.2239 | 0.5767 |
| 2.1 | 0.1666 | 0.5683 |
| 2.2 | 0.1104 | 0.5560 |
| 2.3 | 0.0555 | 0.5399 |
| 2.4 | 0.0025 | 0.5202 |
| 2.6 | −0.0968 | −0.4708 |
| 2.8 | −0.1850 | −0.4097 |
| 3.0 | −0.2601 | −0.3391 |
| 3.2 | −0.3202 | −0.2613 |

temperature $\theta_0$ drops by 20 percent at a specified time, so does the dimensionless temperature $\theta_0$ anywhere else in the medium at the same time.

Once the Bi number is known, these relations can be used to determine the temperature anywhere in the medium. The determination of the constants $A_1$ and $\lambda_1$ usually requires interpolation. For those who prefer reading charts to interpolating, these relations are plotted and the one-term approximation solutions are presented in graphical form, known as the *transient temperature charts*. Note that the charts are sometimes difficult to read, and they are subject to reading errors. Therefore, the relations above should be preferred to the charts.

The transient temperature charts in Figs. 18–15 on the next page, 18–16 on page 741, and 18–17 on page 742 for a large plane wall, long cylinder, and sphere were presented by M. P. Heisler in 1947 and are called **Heisler charts**.

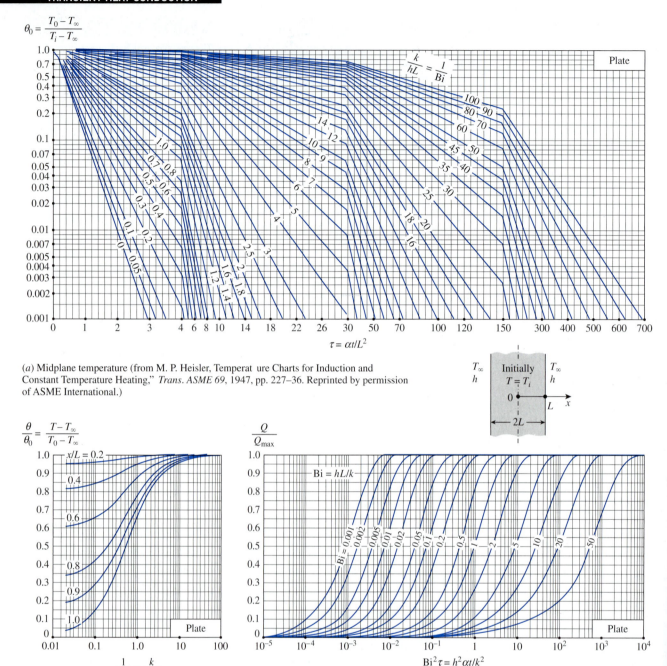

(a) Midplane temperature (from M. P. Heisler, Temperat ure Charts for Induction and Constant Temperature Heating," *Trans. ASME 69*, 1947, pp. 227–36. Reprinted by permission of ASME International.)

(b) Temperature distribution (from M. P. Heisler, "Temperature Charts for Induction and Constant Temperature Heating," *Trans. ASME 69*, 1947, pp. 227–36. Reprinted by permission of ASME International.)

(c) Heat transfer (from H. Gröber et al.)

## FIGURE 18–15

Transient temperature and heat transfer charts for a plane wall of thickness $2L$ initially at a uniform temperature $T_i$ subjected to convection from both sides to an environment at temperature $T_\infty$ with a convection coefficient of $h$.

$$\theta_0 = \frac{T_0 - T_\infty}{T_i - T_\infty}$$

$$\tau = \alpha t/r_o^2$$

(a) Centerline temperature (from M. P. Heisler, "Temperature Charts for Induction and Constant Temperature Heating," *Trans. ASME 69*, 1947, pp. 227–36. Reprinted by permission of ASME International.)

$$\frac{\theta}{\theta_0} = \frac{T - T_\infty}{T_0 - T_\infty}$$

$$\frac{1}{\text{Bi}} = \frac{k}{hr_o}$$

$$\text{Bi}^2\tau = h^2\alpha t/k^2$$

(b) Temperature distribution (from M. P. Heisler, "Temperature Charts for Induction and Constant Temperature Heating," *Trans. ASME 69*, 1947, pp. 227–36. Reprinted by permission of ASME International.)

(c) Heat transfer (from H. Gröber et al.)

**FIGURE 18–16**

Transient temperature and heat transfer charts for a long cylinder of radius $r_o$ initially at a uniform temperature $T_i$ subjected to convection from all sides to an environment at temperature $T_\infty$ with a convection coefficient of $h$.

$$\theta_0 = \frac{T_0 - T_\infty}{T_i - T_\infty}$$

(a) Midpoint temperature (from M. P. Heisler, "Temperature Charts for Induction and Constant Temperature Heating," *Trans. ASME 69*, 1947, pp. 227–36. Reprinted by permission of ASME International.)

$$\frac{\theta}{\theta_0} = \frac{T - T_\infty}{T_0 - T_\infty}$$

$$\frac{Q}{Q_{max}}$$

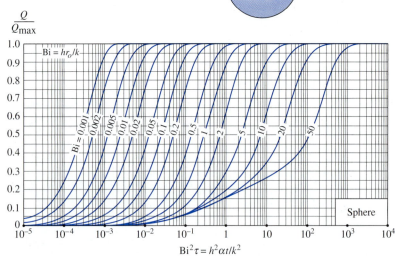

(b) Temperature distribution (from M. P. Heisler, "Temperature Charts for Induction and Constant Temperature Heating," *Trans. ASME 69*, 1947, pp. 227–36. Reprinted by permission of ASME International.)

(c) Heat transfer (from H. Gröber et al.)

### FIGURE 18–17

Transient temperature and heat transfer charts for a sphere of radius $r_o$ initially at a uniform temperature $T_i$ subjected to convection from all sides to an environment at temperature $T_\infty$ with a convection coefficient of $h$.

They were supplemented in 1961 with transient heat transfer charts by H. Gröber. There are *three* charts associated with each geometry: the first chart is to determine the temperature $T_0$ at the *center* of the geometry at a given time $t$. The second chart is to determine the temperature at *other locations* at the same time in terms of $T_0$. The third chart is to determine the total amount of *heat transfer* up to the time $t$. These plots are valid for $\tau > 0.2$.

Note that the case $1/\text{Bi} = k/hL = 0$ corresponds to $h \to \infty$, which corresponds to the case of *specified surface temperature* $T_\infty$. That is, the case in which the surfaces of the body are suddenly brought to the temperature $T_\infty$ at $t = 0$ and kept at $T_\infty$ at all times can be handled by setting $h$ to infinity (Fig. 18–18).

The temperature of the body changes from the initial temperature $T_i$ to the temperature of the surroundings $T_\infty$ at the end of the transient heat conduction process. Thus, the *maximum* amount of heat that a body can gain (or lose if $T_i > T_\infty$) is simply the *change* in the *energy content* of the body. That is,

$$Q_{\text{max}} = mc_p(T_\infty - T_i) = \rho V c_p(T_\infty - T_i) \qquad \text{(kJ)} \qquad \text{(18–21)}$$

where $m$ is the mass, $V$ is the volume, $\rho$ is the density, and $c_p$ is the specific heat of the body. Thus, $Q_{\text{max}}$ represents the amount of heat transfer for $t \to \infty$. The amount of heat transfer $Q$ at a finite time $t$ is obviously less than this maximum, and it can be expressed as the sum of the internal energy changes throughout the entire geometry as

$$Q = \int_V \rho c_p[T(x, t) - T_i]dV \qquad \text{(18–22)}$$

where $T(x, t)$ is the temperature distribution in the medium at time $t$. Assuming constant properties, the ratio of $Q/Q_{\text{max}}$ becomes

$$\frac{Q}{Q_{\text{max}}} = \frac{\int_V \rho c_p[T(x, t) - T_i]dV}{\rho c_p(T_\infty - T_i)V} = \frac{1}{V}\int_V(1 - \theta)dV \qquad \text{(18–23)}$$

Using the appropriate nondimensional temperature relations based on the one-term approximation for the plane wall, cylinder, and sphere, and performing the indicated integrations, we obtain the following relations for the fraction of heat transfer in those geometries:

*Plane wall:* 
$$\left(\frac{Q}{Q_{\text{max}}}\right)_{\text{wall}} = 1 - \theta_{0,\text{ wall}}\frac{\sin \lambda_1}{\lambda_1} \qquad \text{(18–24)}$$

*Cylinder:* 
$$\left(\frac{Q}{Q_{\text{max}}}\right)_{\text{cyl}} = 1 - 2\theta_{0,\text{ cyl}}\frac{J_1(\lambda_1)}{\lambda_1} \qquad \text{(18–25)}$$

*Sphere:* 
$$\left(\frac{Q}{Q_{\text{max}}}\right)_{\text{sph}} = 1 - 3\theta_{0,\text{ sph}}\frac{\sin \lambda_1 - \lambda_1 \cos \lambda_1}{\lambda_1^3} \qquad \text{(18–26)}$$

These $Q/Q_{\text{max}}$ ratio relations based on the one-term approximation are also plotted in Figures 18–15c, 18–16c, and 18–17c, against the variables Bi and $h^2\alpha t/k^2$ for the large plane wall, long cylinder, and sphere, respectively. Note that once the *fraction* of heat transfer $Q/Q_{\text{max}}$ has been determined from these charts or equations for the given $t$, the actual amount of heat transfer by that time can be evaluated by multiplying this fraction by $Q_{\text{max}}$. A *negative* sign for $Q_{\text{max}}$ indicates that the body is *rejecting* heat (Fig. 18–19).

*(a)* Finite convection coefficient

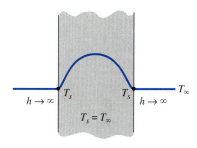

*(b)* Infinite convection coefficient

**FIGURE 18–18**
The specified surface temperature corresponds to the case of convection to an environment at $T_\infty$ with a convection coefficient $h$ that is *infinite*.

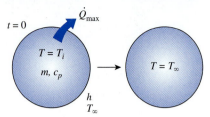

*(a)* Maximum heat transfer ($t \rightarrow \infty$)

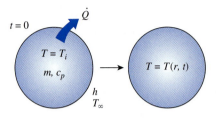

$$\left.\begin{array}{l} \text{Bi} = \ldots \\ \dfrac{h^2 \alpha t}{k^2} = \text{Bi}^2 \tau = \ldots \end{array}\right\} \quad \dfrac{Q}{Q_{\text{max}}} = \ldots$$

(Gröber chart)

*(b)* Actual heat transfer for time $t$

### FIGURE 18–19

The fraction of total heat transfer $Q/Q_{\text{max}}$ up to a specified time $t$ is determined using the Gröber charts.

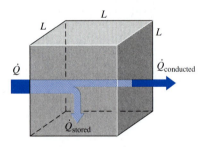

Fourier number: $\tau = \dfrac{\alpha t}{L^2} = \dfrac{\dot{Q}_{\text{conducted}}}{\dot{Q}_{\text{stored}}}$

### FIGURE 18–20

Fourier number at time $t$ can be viewed as the ratio of the rate of heat conducted to the rate of heat stored at that time.

The use of the Heisler/Gröber charts and the one-term solutions already discussed is limited to the conditions specified at the beginning of this section: the body is initially at a *uniform* temperature, the temperature of the medium surrounding the body and the convection heat transfer coefficient are *constant* and *uniform,* and there is no *heat generation* in the body.

We discussed the physical significance of the *Biot number* earlier and indicated that it is a measure of the relative magnitudes of the two heat transfer mechanisms: *convection* at the surface and *conduction* through the solid. A *small* value of Bi indicates that the inner resistance of the body to heat conduction is *small* relative to the resistance to convection between the surface and the fluid. As a result, the temperature distribution within the solid becomes fairly uniform, and lumped system analysis becomes applicable. Recall that when Bi < 0.1, the error in assuming the temperature within the body to be *uniform* is negligible.

To understand the physical significance of the *Fourier number* $\tau$ (or Fo), we express it as (Fig. 18–20)

$$\tau = \frac{\alpha t}{L^2} = \frac{kL^2(1/L)}{\rho c_p L^3/t} \frac{\Delta T}{\Delta T} = \frac{\begin{array}{c}\text{The rate at which heat is } conducted \\ \text{across a body of thickness } L \text{ and} \\ \text{normal area } L^2 \text{ (and thus volume } L^3)\end{array}}{\begin{array}{c}\text{The rate at which heat is } stored \\ \text{in a body of volume } L^3\end{array}} \quad \textbf{(18–27)}$$

Therefore, the Fourier number is a measure of *heat conducted* through a body relative to *heat stored.* Thus, a large value of the Fourier number indicates faster propagation of heat through a body.

Perhaps you are wondering about what constitutes an infinitely large plate or an infinitely long cylinder. After all, nothing in this world is infinite. A plate whose thickness is small relative to the other dimensions can be modeled as an infinitely large plate, except very near the outer edges. But the edge effects on large bodies are usually negligible, and thus a large plane wall such as the wall of a house can be modeled as an infinitely large wall for heat transfer purposes. Similarly, a long cylinder whose diameter is small relative to its length can be analyzed as an infinitely long cylinder. The use of the transient temperature charts and the one-term solutions is illustrated in Examples 18–3, 18–4, and 18–5.

### EXAMPLE 18–3    Boiling Eggs

An ordinary egg can be approximated as a 5-cm-diameter sphere (Fig. 18–21). The egg is initially at a uniform temperature of 5°C and is dropped into boiling water at 95°C. Taking the convection heat transfer coefficient to be $h = 1200 \ W/m^2 \cdot K$, determine how long it will take for the center of the egg to reach 70°C.

**SOLUTION**   An egg is cooked in boiling water. The cooking time of the egg is to be determined.

*Assumptions*  **1** The egg is spherical in shape with a radius of $r_0 = 2.5$ cm. **2** Heat conduction in the egg is one-dimensional because of thermal symmetry about the midpoint. **3** The thermal properties of the egg and the heat transfer coefficient are constant. **4** The Fourier number is $\tau > 0.2$ so that the one-term approximate solutions are applicable.

**FIGURE 4–21**
Schematic for Example 18–3.

**Properties** The water content of eggs is about 74 percent, and thus the thermal conductivity and diffusivity of eggs can be approximated by those of water at the average temperature of $(5 + 70)/2 = 37.5°C$; $k = 0.627$ W/m·K and $\alpha = k/\rho c_p = 0.151 \times 10^{-6}$ m²/s (Table A–15).

**Analysis** Egg white begins to thicken at 63°C and turns solid at 65°C. The yolk begins to thicken at 65°C and sets at 70°C. The whole egg sets at temperatures above 70°C. Therefore, the egg in this case will qualify as hard boiled. The temperature within the egg varies with radial distance as well as time, and the temperature at a specified location at a given time can be determined from the Heisler charts or the one-term solutions. Here we use the latter to demonstrate their use. The Biot number for this problem is

$$\text{Bi} = \frac{hr_o}{k} = \frac{(1200 \text{ W/m}^2\text{·K})(0.025 \text{ m})}{0.627 \text{ W/m·K}} = 47.8$$

which is much greater than 0.1, and thus the lumped system analysis is not applicable. The coefficients $\lambda_1$ and $A_1$ for a sphere corresponding to this Bi are, from Table 18–2,

$$\lambda_1 = 3.0754, \qquad A_1 = 1.9958$$

Substituting these and other values into Eq. 18–28 and solving for $\tau$ gives

$$\frac{T_0 - T_\infty}{T_i - T_\infty} = A_1 e^{-\lambda_1^2 \tau} \longrightarrow \frac{70 - 95}{5 - 95} = 1.9958 e^{-(3.0754)^2 \tau} \longrightarrow \tau = 0.209$$

which is greater than 0.2, and thus the one-term solution is applicable with an error of less than 2 percent. Then the cooking time is determined from the definition of the Fourier number to be

$$t = \frac{\tau r_o^2}{\alpha} = \frac{(0.209)(0.025 \text{ m})^2}{0.151 \times 10^{-6} \text{ m}^2/\text{s}} = 865 \text{ s} \approx \textbf{14.4 min}$$

Therefore, it will take about 15 min for the center of the egg to be heated from 5°C to 70°C.

**Discussion** Note that the Biot number in lumped system analysis was defined differently as $\text{Bi} = hL_c/k = h(r_o/3)/k$. However, either definition can be used in determining the applicability of the lumped system analysis unless $\text{Bi} \approx 0.1$.

Also note that the cooking time depends on many parameters such as the size of the egg, its temperature before cooking, the boiling temperature of water (and thus altitude), the heat transfer coefficient (and thus the level of bubble motion during boiling). Therefore, there is a considerable amount of science or a good amount of experience behind boiling eggs to the correct amount of doneness.

---

## EXAMPLE 18–4    Heating of Brass Plates in an Oven

In a production facility, large brass plates of 4-cm thickness that are initially at a uniform temperature of 20°C are heated by passing them through an oven that is maintained at 500°C (Fig. 18–22). The plates remain in the oven for a period of 7 min. Taking the combined convection and radiation heat transfer coefficient to be $h = 120$ W/m²·K, determine the surface temperature of the plates when they come out of the oven.

**FIGURE 18–22**
Schematic for Example 18–4.

**SOLUTION** Large brass plates are heated in an oven. The surface temperature of the plates leaving the oven is to be determined.

*Assumptions* **1** Heat conduction in the plate is one-dimensional since the plate is large relative to its thickness and there is thermal symmetry about the center plane. **2** The thermal properties of the plate and the heat transfer coefficient are constant. **3** The Fourier number is $\tau > 0.2$ so that the one-term approximate solutions are applicable.

*Properties* The properties of brass at room temperature are $k = 110$ W/m·K, $\rho = 8530$ kg/m$^3$, $c_p = 380$ J/kg·K, and $\alpha = 33.9 \times 10^{-6}$ m$^2$/s (Table A–24). More accurate results are obtained by using properties at average temperature.

*Analysis* The temperature at a specified location at a given time can be determined from the Heisler charts or one-term solutions. Here we use the charts to demonstrate their use. Noting that the half-thickness of the plate is $L = 0.02$ m, from Fig. 18–15 we have

$$\left.\begin{array}{l} \dfrac{1}{\text{Bi}} = \dfrac{k}{hL} = \dfrac{110 \text{ W/m·K}}{(120 \text{ W/m}^2\text{·K})(0.02 \text{ m})} = 45.8 \\[3mm] \tau = \dfrac{\alpha t}{L^2} = \dfrac{(33.9 \times 10^{-6} \text{ m}^2\text{/s})(7 \times 60 \text{ s})}{(0.02 \text{ m})^2} = 35.6 \end{array}\right\} \dfrac{T_0 - T_\infty}{T_i - T_\infty} = 0.46$$

Also,

$$\left.\begin{array}{l} \dfrac{1}{\text{Bi}} = \dfrac{k}{hL} = 45.8 \\[3mm] \dfrac{x}{L} = \dfrac{L}{L} = 1 \end{array}\right\} \dfrac{T - T_\infty}{T_0 - T_\infty} = 0.99$$

Therefore,

$$\frac{T - T_\infty}{T_i - T_\infty} = \frac{T - T_\infty}{T_0 - T_\infty}\frac{T_0 - T_\infty}{T_i - T_\infty} = 0.46 \times 0.99 = 0.455$$

and

$$T = T_\infty + 0.455(T_i - T_\infty) = 500 + 0.455(20 - 500) = \textbf{282°C}$$

Therefore, the surface temperature of the plates will be 282°C when they leave the oven.

*Discussion* We notice that the Biot number in this case is Bi = 1/45.8 = 0.022, which is much less than 0.1. Therefore, we expect the lumped system analysis to be applicable. This is also evident from $(T - T_\infty)/(T_0 - T_\infty) = 0.99$, which indicates that the temperatures at the center and the surface of the plate relative to the surrounding temperature are within 1 percent of each other. Noting that the error involved in reading the Heisler charts is typically a few percent, the lumped system analysis in this case may yield just as accurate results with less effort.

The heat transfer surface area of the plate is 2A, where A is the face area of the plate (the plate transfers heat through both of its surfaces), and the volume of the plate is $V = (2L)A$, where L is the half-thickness of the plate. The exponent b used in the lumped system analysis is

$$b = \frac{hA_s}{\rho c_p V} = \frac{h(2A)}{\rho c_p (2LA)} = \frac{h}{\rho c_p L}$$

$$= \frac{120 \text{ W/m}^2\text{·K}}{(8530 \text{ kg/m}^3)(380 \text{ J/kg·K})(0.02 \text{ m})} = 0.00185 \text{ s}^{-1}$$

Then the temperature of the plate at $t = 7$ min $= 420$ s is determined from

$$\frac{T(t) - T_\infty}{T_i - T_\infty} = e^{-bt} \quad \longrightarrow \quad \frac{T(t) - 500}{20 - 500} = e^{-(0.00185 \text{ s}^{-1})(420 \text{ s})}$$

It yields

$$T(t) = 279°C$$

which is practically identical to the result obtained above using the Heisler charts. Therefore, we can use lumped system analysis with confidence when the Biot number is sufficiently small.

---

**EXAMPLE 18–5**    **Cooling of a Long Stainless Steel Cylindrical Shaft**

A long 20-cm-diameter cylindrical shaft made of stainless steel 304 comes out of an oven at a uniform temperature of 600°C (Fig. 18–23). The shaft is then allowed to cool slowly in an environment chamber at 200°C with an average heat transfer coefficient of $h = 80$ W/m²·K. Determine the temperature at the center of the shaft 45 min after the start of the cooling process. Also, determine the heat transfer per unit length of the shaft during this time period.

**SOLUTION**   A long cylindrical shaft is allowed to cool slowly. The center temperature and the heat transfer per unit length are to be determined.

*Assumptions* **1** Heat conduction in the shaft is one-dimensional since it is long and it has thermal symmetry about the centerline. **2** The thermal properties of the shaft and the heat transfer coefficient are constant. **3** The Fourier number is $\tau > 0.2$ so that the one-term approximate solutions are applicable.

*Properties* The properties of stainless steel 304 at room temperature are $k = 14.9$ W/m·K, $\rho = 7900$ kg/m³, $c_p = 477$ J/kg·K, and $\alpha = 3.95 \times 10^{-6}$ m²/s (Table A–24). More accurate results can be obtained by using properties at average temperature.

*Analysis* The temperature within the shaft may vary with the radial distance $r$ as well as time, and the temperature at a specified location at a given time can be determined from the Heisler charts. Noting that the radius of the shaft is $r_o = 0.1$ m, from Fig. 18–16 we have

$$\left. \begin{aligned} \frac{1}{\text{Bi}} &= \frac{k}{hr_o} = \frac{14.9 \text{ W/m·K}}{(80 \text{ W/m}^2\text{·K})(0.1 \text{ m})} = 1.86 \\[2mm] \tau &= \frac{\alpha t}{r_o^2} = \frac{(3.95 \times 10^{-6} \text{ m}^2\text{/s})(45 \times 60 \text{ s})}{(0.1 \text{ m})^2} = 1.07 \end{aligned} \right\} \quad \frac{T_0 - T_\infty}{T_i - T_\infty} = 0.40$$

and

$$T_0 = T_\infty + 0.4(T_i - T_\infty) = 200 + 0.4(600 - 200) = \textbf{360°C}$$

Therefore, the center temperature of the shaft drops from 600°C to 360°C in 45 min.

$T_\infty = 200°C$
$h = 80$ W/m²·K

Stainless steel shaft

$T_i = 600°C$    $D = 20$ cm

**FIGURE 18–23**
Schematic for Example 18–5.

To determine the actual heat transfer, we first need to calculate the maximum heat that can be transferred from the cylinder, which is the sensible energy of the cylinder relative to its environment. Taking $L = 1$ m,

$$m = \rho V = \rho \pi r_o^2 L = (7900 \text{ kg/m}^3)\pi(0.1 \text{ m})^2(1 \text{ m}) = 248.2 \text{ kg}$$
$$Q_{max} = mc_p(T_\infty - T_i) = (248.2 \text{ kg})(0.477 \text{ kJ/kg·K})(600 - 200)°\text{C}$$
$$= 47{,}350 \text{ kJ}$$

The dimensionless heat transfer ratio is determined from Fig. 18–16c for a long cylinder to be

$$\left.\begin{array}{l} \text{Bi} = \dfrac{1}{1/\text{Bi}} = \dfrac{1}{1.86} = 0.537 \\[2mm] \dfrac{h^2\alpha t}{k^2} = \text{Bi}^2\tau = (0.537)^2(1.07) = 0.309 \end{array}\right\} \quad \dfrac{Q}{Q_{max}} = 0.62$$

Therefore,

$$Q = 0.62Q_{max} = 0.62 \times (47{,}350 \text{ kJ}) = \textbf{29{,}360 kJ}$$

which is the total heat transfer from the shaft during the first 45 min of the cooling.

**Alternative solution**   We could also solve this problem using the one-term solution relation instead of the transient charts. First we find the Biot number

$$\text{Bi} = \frac{hr_o}{k} = \frac{(80 \text{ W/m}^2\text{·K})(0.1 \text{ m})}{14.9 \text{ W/m·K}} = 0.537$$

The coefficients $\lambda_1$ and $A_1$ for a cylinder corresponding to this Bi are determined from Table 18–2 to be

$$\lambda_1 = 0.970, \qquad A_1 = 1.122$$

Substituting these values into Eq. 18–18 gives

$$\theta_0 = \frac{T_0 - T_\infty}{T_i - T_\infty} = A_1 e^{-\lambda_1^2 \tau} = 1.122e^{-(0.970)^2(1.07)} = 0.41$$

and thus

$$T_0 = T_\infty + 0.41(T_i - T_\infty) = 200 + 0.41(600 - 200) = \textbf{364°C}$$

The value of $J_1(\lambda_1)$ for $\lambda_1 = 0.970$ is determined from Table 18–3 to be 0.430. Then the fractional heat transfer is determined from Eq. 18–25 to be

$$\frac{Q}{Q_{max}} = 1 - 2\theta_0 \frac{J_1(\lambda_1)}{\lambda_1} = 1 - 2 \times 0.41 \frac{0.430}{0.970} = 0.636$$

and thus

$$Q = 0.636Q_{max} = 0.636 \times (47{,}350 \text{ kJ}) = \textbf{30{,}120 kJ}$$

**Discussion**   The slight difference between the two results is due to the reading error of the charts.

# 18–3 ▪ TRANSIENT HEAT CONDUCTION IN SEMI-INFINITE SOLIDS

A semi-infinite solid is an idealized body that has a *single plane surface* and extends to infinity in all directions, as shown in Figure 18–24. This idealized body is used to indicate that the temperature change in the part of the body in which we are interested (the region close to the surface) is due to the thermal conditions on a single surface. The earth, for example, can be considered to be a semi-infinite medium in determining the variation of temperature near its surface. Also, a thick wall can be modeled as a semi-infinite medium if all we are interested in is the variation of temperature in the region near one of the surfaces, and the other surface is too far to have any impact on the region of interest during the time of observation. The temperature in the core region of the wall remains unchanged in this case.

For short periods of time, most bodies can be modeled as semi-infinite solids since heat does not have sufficient time to penetrate deep into the body, and the thickness of the body does not enter into the heat transfer analysis. A steel piece of any shape, for example, can be treated as a semi-infinite solid when it is quenched rapidly to harden its surface. A body whose surface is heated by a laser pulse can be treated the same way.

Consider a semi-infinite solid with constant thermophysical properties, no internal heat generation, uniform thermal conditions on its exposed surface, and initially a uniform temperature of $T_i$ throughout. Heat transfer in this case occurs only in the direction normal to the surface (the $x$ direction), and thus it is one-dimensional. Differential equations are independent of the boundary or initial conditions, and thus Eq. 18–10a for one-dimensional transient conduction in Cartesian coordinates applies. The depth of the solid is large ($x \to \infty$) compared to the depth that heat can penetrate, and these phenomena can be expressed mathematically as a boundary condition as $T(x \to \infty, t) = T_i$.

Heat conduction in a semi-infinite solid is governed by the thermal conditions imposed on the exposed surface, and thus the solution depends strongly on the boundary condition at $x = 0$. Below we present a detailed analytical solution for the case of constant temperature $T_s$ on the surface, and give the results for other more complicated boundary conditions. When the surface temperature is changed to $T_s$ at $t = 0$ and held constant at that value at all times, the formulation of the problem can be expressed as

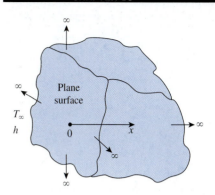

**FIGURE 18–24**
Schematic of a semi-infinite body.

*Differential equation:*
$$\frac{\partial^2 T}{\partial x^2} = \frac{1}{\alpha}\frac{\partial T}{\partial t} \qquad (18\text{–}28a)$$

*Boundary conditions:* $\quad T(0, t) = T_s \quad$ and $\quad T(x \to \infty, t) = T_i \qquad (18\text{–}28b)$

*Initial condition:* $\qquad\qquad T(x, 0) = T_i \qquad (18\text{–}28c)$

The separation of variables technique does not work in this case since the medium is infinite. But another clever approach that converts the partial differential equation into an ordinary differential equation by combining the two independent variables $x$ and $t$ into a single variable $\eta$, called the **similarity variable**, works well. For transient conduction in a semi-infinite medium, it is defined as

Similarity variable: $\qquad\qquad \eta = \dfrac{x}{\sqrt{4\alpha t}} \qquad (18\text{–}29)$

$$\frac{\partial^2 T}{\partial x^2} = \frac{1}{\alpha}\frac{\partial T}{\partial t} \quad \text{and} \quad \eta = \frac{x}{\sqrt{4\alpha t}}$$

$$\frac{\partial T}{\partial t} = \frac{dT}{d\eta}\frac{\partial \eta}{\partial t} = \frac{-x}{2t\sqrt{4\alpha t}}\frac{dT}{d\eta}$$

$$\frac{\partial T}{\partial x} = \frac{dT}{d\eta}\frac{\partial \eta}{\partial x} = \frac{1}{\sqrt{4\alpha t}}\frac{dT}{d\eta}$$

$$\frac{\partial^2 T}{\partial x^2} = \frac{d}{d\eta}\left(\frac{\partial T}{\partial x}\right)\frac{\partial \eta}{\partial x} = \frac{1}{4\alpha t}\frac{d^2 T}{d\eta^2}$$

**FIGURE 18–25**
Transformation of variables in the derivatives of the heat conduction equation by the use of chain rule.

Assuming $T = T(\eta)$ (to be verified) and using the chain rule, all derivatives in the heat conduction equation can be transformed into the new variable, as shown in Fig. 18–25. Noting that $\eta = 0$ at $x = 0$ and $\eta \to \infty$ as $x \to \infty$ (and also at $t = 0$) and substituting into Eqs. 18–28 give, after simplification,

$$\frac{d^2 T}{d\eta^2} = -2\eta\frac{dT}{d\eta} \tag{18–30a}$$

$$T(0) = T_s \quad \text{and} \quad T(\eta \to \infty) = T_i \tag{18–30b}$$

Note that the second boundary condition and the initial condition result in the same boundary condition. Both the transformed equation and the boundary conditions depend on $\eta$ only and are independent of $x$ and $t$. Therefore, transformation is successful, and $\eta$ is indeed a similarity variable.

To solve the 2nd order ordinary differential equation in Eqs. 18–30, we define a new variable $w$ as $w = dT/d\eta$. This reduces Eq. 18–30a into a first order differential equation than can be solved by separating variables,

$$\frac{dw}{d\eta} = -2\eta w \quad \to \quad \frac{dw}{w} = -2\eta d\eta \quad \to \quad \ln w = -\eta^2 + C_0 \quad \to \quad w = C_1 e^{-\eta^2}$$

where $C_1 = \ln C_0$. Back substituting $w = dT/d\eta$ and integrating again,

$$T = C_1 \int_0^\eta e^{-u^2}\, du + C_2 \tag{18–31}$$

where $u$ is a dummy integration variable. The boundary condition at $\eta = 0$ gives $C_2 = T_s$, and the one for $\eta \to \infty$ gives

$$T_i = C_1 \int_0^\infty e^{-u^2}\, du + C_2 = C_1\frac{\sqrt{\pi}}{2} + T_s \quad \to \quad C_1 = \frac{2(T_i - T_s)}{\sqrt{\pi}} \tag{18–32}$$

Substituting the $C_1$ and $C_2$ expressions into Eq. 18–31 and rearranging, the variation of temperature becomes

$$\frac{T - T_s}{T_i - T_s} = \frac{2}{\sqrt{\pi}}\int_0^\eta e^{-u^2}\, du = \text{erf}(\eta) = 1 - \text{erfc}(\eta) \tag{18–33}$$

where the mathematical functions

$$\text{erf}(\eta) = \frac{2}{\sqrt{\pi}}\int_0^\eta e^{-u^2}\, du \quad \text{and} \quad \text{erfc}(\eta) = 1 - \text{erf}(\eta) = 1 - \frac{2}{\sqrt{\pi}}\int_0^\eta e^{-u^2}\, du \tag{18–34}$$

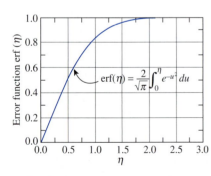

**FIGURE 18–26**
Error function is a standard mathematical function, just like the sinus and tangent functions, whose value varies between 0 and 1.

are called the **error function** and the **complementary error function**, respectively, of argument $\eta$ (Fig. 18–26). Despite its simple appearance, the integral in the definition of the error function cannot be performed analytically. Therefore, the function $\text{erfc}(\eta)$ is evaluated numerically for different values of $\eta$, and the results are listed in Table 18–4.

## TABLE 18–4

The complementary error function

| $\eta$ | erfc $(\eta)$ | $\eta$ | erfc $(\eta)$ | $\eta$ | erfc $(\eta)$ | $\eta$ | erfc $(\eta)$ | $\eta$ | erfc $(\eta)$ | $\eta$ | erfc $(\eta)$ |
|------|---------|------|---------|------|---------|------|---------|------|---------|------|---------|
| 0.00 | 1.00000 | 0.38 | 0.5910 | 0.76 | 0.2825 | 1.14 | 0.1069 | 1.52 | 0.03159 | 1.90 | 0.00721 |
| 0.02 | 0.9774 | 0.40 | 0.5716 | 0.78 | 0.2700 | 1.16 | 0.10090 | 1.54 | 0.02941 | 1.92 | 0.00662 |
| 0.04 | 0.9549 | 0.42 | 0.5525 | 0.80 | 0.2579 | 1.18 | 0.09516 | 1.56 | 0.02737 | 1.94 | 0.00608 |
| 0.06 | 0.9324 | 0.44 | 0.5338 | 0.82 | 0.2462 | 1.20 | 0.08969 | 1.58 | 0.02545 | 1.96 | 0.00557 |
| 0.08 | 0.9099 | 0.46 | 0.5153 | 0.84 | 0.2349 | 1.22 | 0.08447 | 1.60 | 0.02365 | 1.98 | 0.00511 |
| 0.10 | 0.8875 | 0.48 | 0.4973 | 0.86 | 0.2239 | 1.24 | 0.07950 | 1.62 | 0.02196 | 2.00 | 0.00468 |
| 0.12 | 0.8652 | 0.50 | 0.4795 | 0.88 | 0.2133 | 1.26 | 0.07476 | 1.64 | 0.02038 | 2.10 | 0.00298 |
| 0.14 | 0.8431 | 0.52 | 0.4621 | 0.90 | 0.2031 | 1.28 | 0.07027 | 1.66 | 0.01890 | 2.20 | 0.00186 |
| 0.16 | 0.8210 | 0.54 | 0.4451 | 0.92 | 0.1932 | 1.30 | 0.06599 | 1.68 | 0.01751 | 2.30 | 0.00114 |
| 0.18 | 0.7991 | 0.56 | 0.4284 | 0.94 | 0.1837 | 1.32 | 0.06194 | 1.70 | 0.01612 | 2.40 | 0.00069 |
| 0.20 | 0.7773 | 0.58 | 0.4121 | 0.96 | 0.1746 | 1.34 | 0.05809 | 1.72 | 0.01500 | 2.50 | 0.00041 |
| 0.22 | 0.7557 | 0.60 | 0.3961 | 0.98 | 0.1658 | 1.36 | 0.05444 | 1.74 | 0.01387 | 2.60 | 0.00024 |
| 0.24 | 0.7343 | 0.62 | 0.3806 | 1.00 | 0.1573 | 1.38 | 0.05098 | 1.76 | 0.01281 | 2.70 | 0.00013 |
| 0.26 | 0.7131 | 0.64 | 0.3654 | 1.02 | 0.1492 | 1.40 | 0.04772 | 1.78 | 0.01183 | 2.80 | 0.00008 |
| 0.28 | 0.6921 | 0.66 | 0.3506 | 1.04 | 0.1413 | 1.42 | 0.04462 | 1.80 | 0.01091 | 2.90 | 0.00004 |
| 0.30 | 0.6714 | 0.68 | 0.3362 | 1.06 | 0.1339 | 1.44 | 0.04170 | 1.82 | 0.01006 | 3.00 | 0.00002 |
| 0.32 | 0.6509 | 0.70 | 0.3222 | 1.08 | 0.1267 | 1.46 | 0.03895 | 1.84 | 0.00926 | 3.20 | 0.00001 |
| 0.34 | 0.6306 | 0.72 | 0.3086 | 1.10 | 0.1198 | 1.48 | 0.03635 | 1.86 | 0.00853 | 3.40 | 0.00000 |
| 0.36 | 0.6107 | 0.74 | 0.2953 | 1.12 | 0.1132 | 1.50 | 0.03390 | 1.88 | 0.00784 | 3.60 | 0.00000 |

Knowing the temperature distribution, the heat flux at the surface can be determined from the Fourier's law to be

$$\dot{q}_s = -k\frac{\partial T}{\partial x}\bigg|_{x=0} = -k\frac{dT}{d\eta}\frac{\partial \eta}{\partial x}\bigg|_{\eta=0} = -kC_1 e^{-\eta^2}\frac{1}{\sqrt{4\alpha t}}\bigg|_{\eta=0} = \frac{k(T_s - T_i)}{\sqrt{\pi \alpha t}} \quad \text{(18–35)}$$

The solutions in Eqs. 18–33 and 18–34 correspond to the case when the temperature of the exposed surface of the medium is suddenly raised (or lowered) to $T_s$ at $t = 0$ and is maintained at that value at all times. The specified surface temperature case is closely approximated in practice when condensation or boiling takes place on the surface. Using a similar approach or the Laplace transform technique, analytical solutions can be obtained for other boundary conditions on the surface, with the following results.

**Case 1: Specified Surface Temperature, $T_s$ = constant (Fig. 18–27).**

$$\frac{T(x, t) - T_i}{T_s - T_i} = \text{erfc}\left(\frac{x}{2\sqrt{\alpha t}}\right) \quad \text{and} \quad \dot{q}_s(t) = \frac{k(T_s - T_i)}{\sqrt{\pi \alpha t}} \quad \text{(18–36)}$$

**Case 2: Specified Surface Heat Flux, $\dot{q}_s$ = constant.**

$$T(x, t) - T_i = \frac{\dot{q}_s}{k}\left[\sqrt{\frac{4\alpha t}{\pi}}\exp\left(-\frac{x^2}{4\alpha t}\right) - x\,\text{erfc}\left(\frac{x}{2\sqrt{\alpha t}}\right)\right] \quad \text{(18–37)}$$

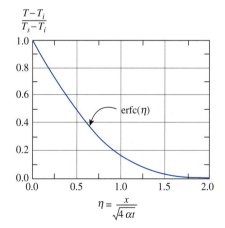

**FIGURE 18–27**

Dimensionless temperature distribution for transient conduction in a semi-infinite solid whose surface is maintained at a constant temperature $T_s$.

## Case 3: Convection on the Surface, $\dot{q}_s(t) = h[T_\infty - T(0, t)]$.

$$\frac{T(x, t) - T_i}{T_\infty - T_i} = \text{erfc}\left(\frac{x}{2\sqrt{\alpha t}}\right) - \exp\left(\frac{hx}{k} + \frac{h^2\alpha t}{k^2}\right)\text{erfc}\left(\frac{x}{2\sqrt{\alpha t}} + \frac{h\sqrt{\alpha t}}{k}\right) \quad \text{(18–38)}$$

### Case 4: Energy Pulse at Surface, $e_s$ = constant.

Energy in the amount of $e_s$ per unit surface area (in J/m$^2$) is supplied to the semi-infinite body instantaneously at time $t = 0$ (by a laser pulse, for example), and the entire energy is assumed to enter the body, with no heat loss from the surface.

$$T(x, t) - T_i = \frac{e_s}{k\sqrt{\pi t/\alpha}} \exp\left(-\frac{x^2}{4\alpha t}\right) \quad \text{(18–39)}$$

Note that Cases 1 and 3 are closely related. In Case 1, the surface $x = 0$ is brought to a temperature $T_s$ at time $t = 0$, and kept at that value at all times. In Case 3, the surface is exposed to convection by a fluid at a constant temperature $T_\infty$ with a heat transfer coefficient $h$.

The solutions for all four cases are plotted in Fig. 18–28 for a representative case using a large cast iron block initially at 0°C throughout. In Case 1, the surface temperature remains constant at the specified value of $T_s$, and temperature increases gradually within the medium as heat penetrates deeper into the solid. Note that during initial periods only a thin slice near the surface is affected by heat transfer. Also, the temperature gradient at the surface and thus the rate of heat transfer into the solid decreases with time. In Case 2, heat is continually supplied to the solid, and thus the temperature within the solid, including the surface, increases with time. This is also the case with convection (Case 3), except that the surrounding fluid temperature $T_\infty$ is the highest temperature that the solid body can rise to. In Case 4, the surface is subjected to an instant burst of heat supply at time $t = 0$, such as heating by a laser pulse, and then the surface is covered with insulation. The result is an instant rise in surface temperature, followed by a temperature drop as heat is conducted deeper into the solid. Note that the temperature profile is always normal to the surface at all times. (Why?)

The variation of temperature with position and time in a semi-infinite solid subjected to convection heat transfer is plotted in Fig. 18–28 for the nondimensionalized temperature against the dimensionless similarity variable $\eta = x/\sqrt{4\alpha t}$ for various values of the parameter $h\sqrt{\alpha t}/k$. Although the graphical solution given in Fig. 18–29 on page 754 is simply a plot of the exact analytical solution, it is subject to reading errors, and thus is of limited accuracy compared to the analytical solution. Also, the values on the vertical axis of Fig. 18–30 on page 754 correspond to $x = 0$, and thus represent the surface temperature. The curve $h\sqrt{\alpha t}/k = \infty$ corresponds to $h \to \infty$, which corresponds to the case of *specified temperature* $T_\infty$ at the surface at $x = 0$. That is, the case in which the surface of the semi-infinite body is suddenly brought to temperature $T_\infty$ at $t = 0$ and kept at $T_\infty$ at all times can be handled by setting $h$ to infinity. For a *finite* heat transfer coefficient $h$, the surface temperature approaches the fluid temperature $T_\infty$ as the time $t$ approaches infinity.

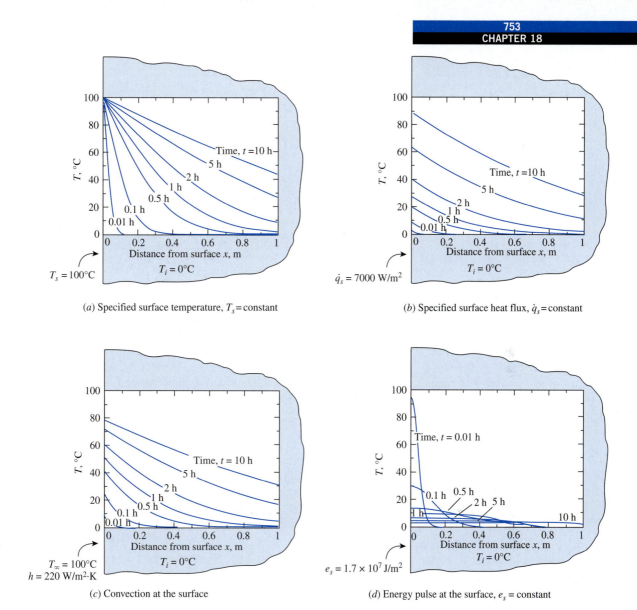

(a) Specified surface temperature, $T_s$ = constant

(b) Specified surface heat flux, $\dot{q}_s$ = constant

(c) Convection at the surface

(d) Energy pulse at the surface, $e_s$ = constant

**FIGURE 18–28**

Variations of temperature with position and time in a large cast iron block ($\alpha = 2.31 \times 10^{-5}$ m²/s, $k = 80.2$ W/m·K) initially at 0 °C under different thermal conditions on the surface.

## Contact of Two Semi-Infinite Solids

When two large bodies $A$ and $B$, initially at uniform temperatures $T_{A,i}$ and $T_{B,i}$ are brought into contact, they instantly achieve temperature equality at the contact surface (temperature equality is achieved over the entire surface if the contact resistance is negligible). If the two bodies are of the same material with constant properties, thermal symmetry requires the contact surface temperature to be the arithmetic average, $T_s = (T_{A,i} + T_{B,i})/2$ and to remain constant at that value at all times.

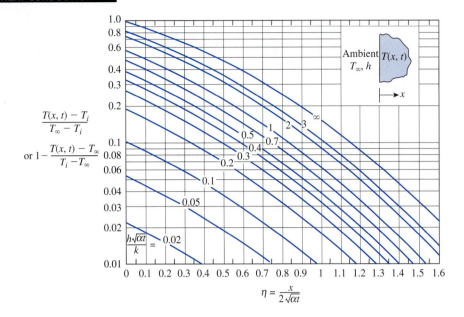

**FIGURE 18–29**
Variation of temperature with position and time in a semi-infinite solid initially at temperature $T_i$ subjected to convection to an environment at $T_\infty$ with a convection heat transfer coefficient of $h$ (plotted using EES).

If the bodies are of different materials, they still achieve a temperature equality, but the surface temperature $T_s$ in this case will be different than the arithmetic average. Noting that both bodies can be treated as semi-infinite solids with the same specified surface temperature, the energy balance on the contact surface gives, from Eq. 18–36,

$$\dot{q}_{s,A} = \dot{q}_{s,B} \;\rightarrow\; -\frac{k_A(T_s - T_{A,i})}{\sqrt{\pi\alpha_A t}} = \frac{k_B(T_s - T_{B,i})}{\sqrt{\pi\alpha_B t}} \;\rightarrow\; \frac{T_{A,i} - T_s}{T_s - T_{B,i}} = \sqrt{\frac{(k\rho c_p)_B}{(k\rho c_p)_A}}$$

Then $T_s$ is determined to be (Fig. 18–30)

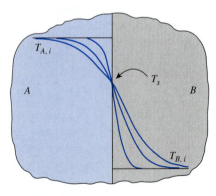

**FIGURE 18–30**
Contact of two semi-infinite solids of different initial temperatures.

$$T_s = \frac{\sqrt{(k\rho c_p)_A}\,T_{A,i} + \sqrt{(k\rho c_p)_B}\,T_{B,i}}{\sqrt{(k\rho c_p)_A} + \sqrt{(k\rho c_p)_B}} \tag{18–40}$$

Therefore, the interface temperature of two bodies brought into contact is dominated by the body with the larger $k\rho c_p$. This also explains why a metal at room temperature feels colder than wood at the same temperature. At room temperature, the $\sqrt{k\rho c_p}$ value is 24 kJ/m²·K for aluminum, 0.38 kJ/m²·K for wood, and 1.1 kJ/m²·K for the human flesh. Using Eq. 18–40, it can be shown that when a person with a skin temperature of 35°C touches an aluminum block and then a wood block both at 15°C, the contact surface temperature will be 15.9°C in the case of aluminum and 30°C in the case of wood.

## EXAMPLE 18–6    Minimum Burial Depth of Water Pipes to Avoid Freezing

In areas where the air temperature remains below 0°C for prolonged periods of time, the freezing of water in underground pipes is a major concern. Fortunately, the soil remains relatively warm during those periods, and it takes weeks for the subfreezing temperatures to reach the water mains in the ground. Thus, the soil effectively serves as an insulation to protect the water from subfreezing temperatures in winter.

The ground at a particular location is covered with snow pack at $-10°C$ for a continuous period of three months, and the average soil properties at that location are $k = 0.4$ W/m·K and $\alpha = 0.15 \times 10^{-6}$ m²/s (Fig. 18–31). Assuming an initial uniform temperature of 15°C for the ground, determine the minimum burial depth to prevent the water pipes from freezing.

$T_s = -10°C$

Soil

$x$

Water pipe

$T_i = 15°C$

**FIGURE 18–31**
Schematic for Example 18–6.

**SOLUTION** The water pipes are buried in the ground to prevent freezing. The minimum burial depth at a particular location is to be determined.

**Assumptions**  **1** The temperature in the soil is affected by the thermal conditions at one surface only, and thus the soil can be considered to be a semi-infinite medium. **2** The thermal properties of the soil are constant.

**Properties**  The properties of the soil are as given in the problem statement.

**Analysis**  The temperature of the soil surrounding the pipes will be 0°C after three months in the case of minimum burial depth. Therefore, from Fig. 18–29, we have

$$\left. \begin{array}{l} \dfrac{h\sqrt{\alpha t}}{k} = \infty \quad (\text{since } h \to \infty) \\[2mm] \dfrac{T(x, t) - T_i}{T_\infty - T_i} = \dfrac{0 - 15}{-10 - 15} = 0.6 \end{array} \right\} \quad \eta = \dfrac{x}{2\sqrt{\alpha t}} = 0.36$$

We note that

$$t = (90 \text{ days})(24 \text{ h/day})(3600 \text{ s/h}) = 7.78 \times 10^6 \text{ s}$$

and thus

$$x = 2\eta\sqrt{\alpha t} = 2 \times 0.36\sqrt{(0.15 \times 10^{-6} \text{ m}^2/\text{s})(7.78 \times 10^6 \text{ s})} = \textbf{0.78 m}$$

Therefore, the water pipes must be buried to a depth of at least 78 cm to avoid freezing under the specified harsh winter conditions.

**ALTERNATIVE SOLUTION**  The solution of this problem could also be determined from Eq. 18–36:

$$\dfrac{T(x, t) - T_i}{T_s - T_i} = \text{erfc}\left(\dfrac{x}{2\sqrt{\alpha t}}\right) \longrightarrow \dfrac{0 - 15}{-10 - 15} = \text{erfc}\left(\dfrac{x}{2\sqrt{\alpha t}}\right) = 0.60$$

The argument that corresponds to this value of the complementary error function is determined from Table 18–4 to be $\eta = 0.37$. Therefore,

$$x = 2\eta\sqrt{\alpha t} = 2 \times 0.37\sqrt{(0.15 \times 10^{-6} \text{ m}^2/\text{s})(7.78 \times 10^6 \text{ s})} = \textbf{0.80 m}$$

Again, the slight difference is due to the reading error of the chart.

$\dot{q}_s = 1250 \ \text{W/m}^2$

Wood block

$T_i = 20°C$

**FIGURE 18–32**
Schematic for Example 18–7.

**FIGURE 18–33**
Variation of temperature within the wood and aluminum blocks at $t = 20$ min.

*EXAMPLE 18–7*    **Surface Temperature Rise of Heated Blocks**

A thick black-painted wood block at 20°C is subjected to constant solar heat flux of 1250 W/m² (Fig. 18–32). Determine the exposed surface temperature of the block after 20 minutes. What would your answer be if the block were made of aluminum?

**SOLUTION**  A wood block is subjected to solar heat flux. The surface temperature of the block is to be determined, and to be compared to the value for an aluminum block.

*Assumptions*  **1** All incident solar radiation is absorbed by the block. **2** Heat loss from the block is disregarded (and thus the result obtained is the maximum temperature). **3** The block is sufficiently thick to be treated as a semi-infinite solid, and the properties of the block are constant.

*Properties*  Thermal conductivity and diffusivity values at room temperature are $k = 1.26$ W/m·K and $\alpha = 1.1 \times 10^{-5}$ m²/s for wood, and $k = 237$ W/m·K and $\alpha = 9.71 \times 10^{-5}$ m²/s for aluminum.

*Analysis*  This is a transient conduction problem in a semi-infinite medium subjected to constant surface heat flux, and the surface temperature can be expressed from Eq. 18–37 as

$$T_s = T(0, t) = T_i + \frac{\dot{q}_s}{k}\sqrt{\frac{4\alpha t}{\pi}}$$

Substituting the given values, the surface temperatures for both the wood and aluminum blocks are determined to be

$$T_{s, \text{wood}} = 20°C + \frac{1250 \ \text{W/m}^2}{1.26 \ \text{W/m·K}}\sqrt{\frac{4(1.1 \times 10^{-5} \ \text{m}^2/\text{s})(20 \times 60 \ \text{s})}{\pi}} = \textbf{149°C}$$

$$T_{s, \text{Al}} = 20°C + \frac{1250 \ \text{W/m}^2}{237 \ \text{W/m·K}}\sqrt{\frac{4(9.71 \times 10^{-5} \ \text{m}^2/\text{s})(20 \times 60 \ \text{s})}{\pi}} = \textbf{22.0°C}$$

Note that thermal energy supplied to the wood accumulates near the surface because of the low conductivity and diffusivity of wood, causing the surface temperature to rise to high values. Metals, on the other hand, conduct the heat they receive to inner parts of the block because of their high conductivity and diffusivity, resulting in minimal temperature rise at the surface. In reality, both temperatures will be lower because of heat losses.

*Discussion*  The temperature profiles for both the wood and aluminum blocks at $t = 20$ min are evaluated and plotted in Fig. 18–33 using EES. At a depth of $x = 0.41$ m, the temperature in both blocks is 20.6°C. At a depth of 0.5 m, the temperatures become 20.1°C for wood and 20.4°C for aluminum block, which confirms that heat penetrates faster and further in metals compared to nonmetals.

# 18–4 ▪ TRANSIENT HEAT CONDUCTION IN MULTIDIMENSIONAL SYSTEMS

The transient temperature charts and analytical solutions presented earlier can be used to determine the temperature distribution and heat transfer in *one-dimensional* heat conduction problems associated with a large plane wall, a long cylinder, a sphere, and a semi-infinite medium. Using a superposition approach called the **product solution**, these charts and solutions can also be used to construct solutions for the *two-dimensional* transient heat conduction problems encountered in geometries such as a short cylinder, a long rectangular bar, or a semi-infinite cylinder or plate, and even *three-dimensional* problems associated with geometries such as a rectangular prism or a semi-infinite rectangular bar, provided that *all* surfaces of the solid are subjected to convection to the *same* fluid at temperature $T_\infty$, with the *same* heat transfer coefficient $h$, and the body involves no heat generation (Fig. 18–34). The solution in such multidimensional geometries can be expressed as the *product* of the solutions for the one-dimensional geometries whose intersection is the multidimensional geometry.

Consider a *short cylinder* of height $a$ and radius $r_o$ initially at a uniform temperature $T_i$. There is no heat generation in the cylinder. At time $t = 0$, the cylinder is subjected to convection from all surfaces to a medium at temperature $T_\infty$ with a heat transfer coefficient $h$. The temperature within the cylinder will change with $x$ as well as $r$ and time $t$ since heat transfer occurs from the top and bottom of the cylinder as well as its side surfaces. That is, $T = T(r, x, t)$ and thus this is a two-dimensional transient heat conduction problem. When the properties are assumed to be constant, it can be shown that the solution of this two-dimensional problem can be expressed as

$$\left( \frac{T(r, x, t) - T_\infty}{T_i - T_\infty} \right)_{\substack{\text{short} \\ \text{cylinder}}} = \left( \frac{T(x, t) - T_\infty}{T_i - T_\infty} \right)_{\substack{\text{plane} \\ \text{wall}}} \left( \frac{T(r, t) - T_\infty}{T_i - T_\infty} \right)_{\substack{\text{infinite} \\ \text{cylinder}}} \quad \textbf{(18–41)}$$

That is, the solution for the two-dimensional short cylinder of height $a$ and radius $r_o$ is equal to the *product* of the nondimensionalized solutions for the one-dimensional plane wall of thickness $a$ and the long cylinder of radius $r_o$, which are the two geometries whose intersection is the short cylinder, as shown in Fig. 18–35. We generalize this as follows: *the solution for a multidimensional geometry is the product of the solutions of the one-dimensional geometries whose intersection is the multidimensional body.*

For convenience, the one-dimensional solutions are denoted by

$$\theta_{\text{wall}}(x, t) = \left( \frac{T(x, t) - T_\infty}{T_i - T_\infty} \right)_{\substack{\text{plane} \\ \text{wall}}}$$

$$\theta_{\text{cyl}}(r, t) = \left( \frac{T(r, t) - T_\infty}{T_i - T_\infty} \right)_{\substack{\text{infinite} \\ \text{cylinder}}}$$

$$\theta_{\text{semi-inf}}(x, t) = \left( \frac{T(x, t) - T_\infty}{T_i - T_\infty} \right)_{\substack{\text{semi infinite} \\ \text{solid}}} \quad \textbf{(18–42)}$$

For example, the solution for a long solid bar whose cross section is an $a \times b$ rectangle is the intersection of the two infinite plane walls of thicknesses $a$ and $b$, as shown in Fig. 18–36, and thus the transient temperature distribution for this rectangular bar can be expressed as

$$\left( \frac{T(x, y, t) - T_\infty}{T_i - T_\infty} \right)_{\substack{\text{rectangular} \\ \text{bar}}} = \theta_{\text{wall}}(x, t)\theta_{\text{wall}}(y, t) \quad \textbf{(18–43)}$$

The proper forms of the product solutions for some other geometries are given

(a) Long cylinder

(b) Short cylinder (two-dimensional)

**FIGURE 184–34**
The temperature in a short cylinder exposed to convection from all surfaces varies in both the radial and axial directions, and thus heat is transferred in both directions.

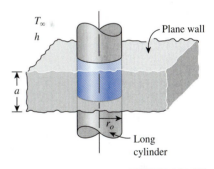

**FIGURE 18–35**
A short cylinder of radius $r_o$ and height $a$ is the *intersection* of a long cylinder of radius $r_o$ and a plane wall of thickness $a$.

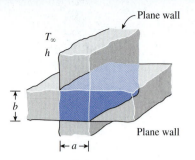

Plane wall

$T_\infty$
$h$

$b$

$a$

Plane wall

**FIGURE 18–36**
A long solid bar of rectangular profile $a \times b$ is the *intersection* of two plane walls of thicknesses $a$ and $b$.

in Table 18–5. It is important to note that the $x$-coordinate is measured from the *surface* in a semi-infinite solid, and from the *midplane* in a plane wall. The radial distance $r$ is always measured from the centerline.

Note that the solution of a *two-dimensional* problem involves the product of *two* one-dimensional solutions, whereas the solution of a *three-dimensional* problem involves the product of *three* one-dimensional solutions.

A modified form of the product solution can also be used to determine the total transient heat transfer to or from a multidimensional geometry by using the one-dimensional values, as shown by L. S. Langston in 1982. The transient heat transfer for a two-dimensional geometry formed by the intersection of two one-dimensional geometries 1 and 2 is

$$\left(\frac{Q}{Q_{max}}\right)_{total,\,2D} = \left(\frac{Q}{Q_{max}}\right)_1 + \left(\frac{Q}{Q_{max}}\right)_2 \left[1 - \left(\frac{Q}{Q_{max}}\right)_1\right] \tag{18–44}$$

Transient heat transfer for a three-dimensional body formed by the intersection of three one-dimensional bodies 1, 2, and 3 is given by

$$\left(\frac{Q}{Q_{max}}\right)_{total,\,3D} = \left(\frac{Q}{Q_{max}}\right)_1 + \left(\frac{Q}{Q_{max}}\right)_2 \left[1 - \left(\frac{Q}{Q_{max}}\right)_1\right]$$

$$+ \left(\frac{Q}{Q_{max}}\right)_3 \left[1 - \left(\frac{Q}{Q_{max}}\right)_1\right]\left[1 - \left(\frac{Q}{Q_{max}}\right)_2\right] \tag{18–45}$$

The use of the product solution in transient two- and three-dimensional heat conduction problems is illustrated in the following examples.

---

### EXAMPLE 18–8    Cooling of a Short Brass Cylinder

A short brass cylinder of diameter $D = 10$ cm and height $H = 12$ cm is initially at a uniform temperature $T_i = 120°C$. The cylinder is now placed in atmospheric air at 25°C, where heat transfer takes place by convection, with a heat transfer coefficient of $h = 60$ W/m²·K. Calculate the temperature at (*a*) the center of the cylinder and (*b*) the center of the top surface of the cylinder 15 min after the start of the cooling.

**SOLUTION**  A short cylinder is allowed to cool in atmospheric air. The temperatures at the centers of the cylinder and the top surface are to be determined.
**Assumptions**  **1** Heat conduction in the short cylinder is two-dimensional, and thus the temperature varies in both the axial $x$- and the radial $r$-directions. **2** The thermal properties of the cylinder and the heat transfer coefficient are constant. **3** The Fourier number is $\tau > 0.2$ so that the one-term approximate solutions are applicable.
**Properties**  The properties of brass at room temperature are $k = 110$ W/m·K and $\alpha = 33.9 \times 10^{-6}$ m²/s (Table A–24). More accurate results can be obtained by using properties at average temperature.
**Analysis**  (*a*) This short cylinder can physically be formed by the intersection of a long cylinder of radius $r_o = 5$ cm and a plane wall of thickness $2L = 12$ cm,

(*Continued on page 760*)

## TABLE 18–5

Multidimensional solutions expressed as products of one-dimensional solutions for bodies that are initially at a uniform temperature $T_i$ and exposed to convection from all surfaces to a medium at $T_\infty$

$\theta(r, t) = \theta_{cyl}(r, t)$
**Infinite cylinder**

$\theta(x, r, t) = \theta_{cyl}(r, t)\, \theta_{semi\text{-}inf}(x, t)$
**Semi-infinite cylinder**

$\theta(x, r, t) = \theta_{cyl}(r, t)\, \theta_{wall}(x, t)$
**Short cylinder**

$\theta(x, t) = \theta_{semi\text{-}inf}(x, t)$
**Semi-infinite medium**

$\theta(x, y, t) = \theta_{semi\text{-}inf}(x, t)\, \theta_{semi\text{-}inf}(y, t)$
**Quarter-infinite medium**

$\theta(x, y, z, t) =$
$\theta_{semi\text{-}inf}(x, t)\, \theta_{semi\text{-}inf}(y, t)\, \theta_{semi\text{-}inf}(z, t)$
**Corner region of a large medium**

$\theta(x, t) = \theta_{wall}(x, t)$
**Infinite plate (or plane wall)**

$\theta(x, y, t) = \theta_{wall}(x, t)\, \theta_{semi\text{-}inf}(y, t)$
**Semi-infinite plate**

$\theta(x, y, z, t) =$
$\theta_{wall}(x, t)\, \theta_{semi\text{-}inf}(y, t)\, \theta_{semi\text{-}inf}(z, t)$
**Quarter-infinite plate**

$\theta(x, y, t) = \theta_{wall}(x, t)\, \theta_{wall}(y, t)$
**Infinite rectangular bar**

$\theta(x, y, z, t) =$
$\theta_{wall}(x, t)\, \theta_{wall}(y, t)\, \theta_{semi\text{-}inf}(z, t)$
**Semi-infinite rectangular bar**

$\theta(x, y, z, t) =$
$\theta_{wall}(x, t)\, \theta_{wall}(y, t)\, \theta_{wall}(z, t)$
**Rectangular parallelepiped**

$T_\infty = 25°C$

$h = 60\ \text{W/m}^2\cdot\text{K}$

**FIGURE 18–37**
Schematic for Example 18–8.

as shown in Fig. 18–37. The dimensionless temperature at the center of the plane wall is determined from Fig. 18–15a to be

$$\left.\begin{array}{l}\tau = \dfrac{\alpha t}{L^2} = \dfrac{(3.39 \times 10^{-5}\ \text{m}^2/\text{s})(900\ \text{s})}{(0.06\ \text{m})^2} = 8.48 \\[12pt] \dfrac{1}{\text{Bi}} = \dfrac{k}{hL} = \dfrac{110\ \text{W/m}\cdot\text{K}}{(60\ \text{W/m}^2\cdot\text{K})(0.06\ \text{m})} = 30.6\end{array}\right\} \theta_{\text{wall}}(0, t) = \dfrac{T(0, t) - T_\infty}{T_i - T_\infty} = 0.8$$

Similarly, at the center of the cylinder, we have

$$\left.\begin{array}{l}\tau = \dfrac{\alpha t}{r_o^2} = \dfrac{(3.39 \times 10^{-5}\ \text{m}^2/\text{s})(900\ \text{s})}{(0.05\ \text{m})^2} = 12.2 \\[12pt] \dfrac{1}{\text{Bi}} = \dfrac{k}{hr_o} = \dfrac{110\ \text{W/m}\cdot\text{K}}{(60\ \text{W/m}^2\cdot\text{K})(0.05\ \text{m})} = 36.7\end{array}\right\} \theta_{\text{cyl}}(0, t) = \dfrac{T(0, t) - T_\infty}{T_i - T_\infty} = 0.5$$

Therefore,

$$\left(\frac{T(0, 0, t) - T_\infty}{T_i - T_\infty}\right)_{\substack{\text{short}\\\text{cylinder}}} = \theta_{\text{wall}}(0, t) \times \theta_{\text{cyl}}(0, t) = 0.8 \times 0.5 = 0.4$$

and

$$T(0, 0, t) = T_\infty + 0.4(T_i - T_\infty) = 25 + 0.4(120 - 25) = \mathbf{63°C}$$

This is the temperature at the center of the short cylinder, which is also the center of both the long cylinder and the plate.

(b) The center of the top surface of the cylinder is still at the center of the long cylinder ($r = 0$), but at the outer surface of the plane wall ($x = L$). Therefore, we first need to find the surface temperature of the wall. Noting that $x = L = 0.06$ m,

$$\left.\begin{array}{l}\dfrac{x}{L} = \dfrac{0.06\ \text{m}}{0.06\ \text{m}} = 1 \\[12pt] \dfrac{1}{\text{Bi}} = \dfrac{k}{hL} = \dfrac{110\ \text{W/m}\cdot\text{K}}{(60\ \text{W/m}^2\cdot\text{K})(0.06\ \text{m})} = 30.6\end{array}\right\} \dfrac{T(L, t) - T_\infty}{T_0 - T_\infty} = 0.98$$

Then

$$\theta_{\text{wall}}(L, t) = \frac{T(L, t) - T_\infty}{T_i - T_\infty} = \left(\frac{T(L, t) - T_\infty}{T_0 - T_\infty}\right)\left(\frac{T_0 - T_\infty}{T_i - T_\infty}\right) = 0.98 \times 0.8 = 0.784$$

Therefore,

$$\left(\frac{T(L, 0, t) - T_\infty}{T_i - T_\infty}\right)_{\substack{\text{short}\\\text{cylinder}}} = \theta_{\text{wall}}(L, t)\theta_{\text{cyl}}(0, t) = 0.784 \times 0.5 = 0.392$$

and

$$T(L, 0, t) = T_\infty + 0.392(T_i - T_\infty) = 25 + 0.392(120 - 25) = \mathbf{62.2°C}$$

which is the temperature at the center of the top surface of the cylinder.

## EXAMPLE 18–9    Heat Transfer from a Short Cylinder

Determine the total heat transfer from the short brass cylinder ($\rho = 8530$ kg/m³, $c_p = 0.380$ kJ/kg·K) discussed in Example 18–8.

**SOLUTION**   We first determine the maximum heat that can be transferred from the cylinder, which is the sensible energy content of the cylinder relative to its environment:

$$m = \rho V = \rho \pi r_o^2 H = (8530 \text{ kg/m}^3)\pi(0.05 \text{ m})^2(0.12 \text{ m}) = 8.04 \text{ kg}$$

$$Q_{max} = mc_p(T_i - T_\infty) = (8.04 \text{ kg})(0.380 \text{ kJ/kg·K})(120 - 25)°\text{C} = 290.2 \text{ kJ}$$

Then we determine the dimensionless heat transfer ratios for both geometries. For the plane wall, it is determined from Fig. 18–15c to be

$$\left. \begin{array}{l} \text{Bi} = \dfrac{1}{1/\text{Bi}} = \dfrac{1}{30.6} = 0.0327 \\[2ex] \dfrac{h^2 \alpha t}{k^2} = \text{Bi}^2 \tau = (0.0327)^2(8.48) = 0.0091 \end{array} \right\} \left(\dfrac{Q}{Q_{max}}\right)_{\substack{\text{plane} \\ \text{wall}}} = 0.23$$

Similarly, for the cylinder, we have

$$\left. \begin{array}{l} \text{Bi} = \dfrac{1}{1/\text{Bi}} = \dfrac{1}{36.7} = 0.0272 \\[2ex] \dfrac{h^2 \alpha t}{k^2} = \text{Bi}^2 \tau = (0.0272)^2(12.2) = 0.0090 \end{array} \right\} \left(\dfrac{Q}{Q_{max}}\right)_{\substack{\text{infinite} \\ \text{cylinder}}} = 0.47$$

Then the heat transfer ratio for the short cylinder is, from Eq. 18–44,

$$\left(\dfrac{Q}{Q_{max}}\right)_{\text{short cyl}} = \left(\dfrac{Q}{Q_{max}}\right)_1 + \left(\dfrac{Q}{Q_{max}}\right)_2 \left[1 - \left(\dfrac{Q}{Q_{max}}\right)_1\right]$$

$$= 0.23 + 0.47(1 - 0.23) = 0.592$$

Therefore, the total heat transfer from the cylinder during the first 15 min of cooling is

$$Q = 0.592 Q_{max} = 0.592 \times (290.2 \text{ kJ}) = \textbf{172 kJ}$$

## SUMMARY

In this chapter, we considered the variation of temperature with time as well as position in one- or multidimensional systems. We first considered the *lumped systems* in which the temperature varies with time but remains uniform throughout the system at any time. The temperature of a lumped body of arbitrary shape of mass $m$, volume $V$, surface area $A_s$, density $\rho$, and specific heat $c_p$ initially at a uniform temperature $T_i$ that is exposed to convection at time $t = 0$ in a medium at temperature $T_\infty$ with a heat transfer coefficient $h$ is expressed as

$$\dfrac{T(t) - T_\infty}{T_i - T_\infty} = e^{-bt}$$

where

$$b = \dfrac{hA_s}{\rho c_p V} = \dfrac{h}{\rho c_p L_c}$$

is a positive quantity whose dimension is $(\text{time})^{-1}$. This relation can be used to determine the temperature $T(t)$ of a body at time $t$ or, alternatively, the time $t$ required for the temperature to reach a specified value $T(t)$. Once the temperature $T(t)$ at time $t$ is available, the *rate* of convection heat transfer between the body and its environment at that time can be determined from Newton's law of cooling as

$$\dot{Q}(t) = hA_s[T(t) - T_\infty]$$

The *total amount* of heat transfer between the body and the surrounding medium over the time interval $t = 0$ to $t$ is simply the change in the energy content of the body,

$$Q = mc_p[T(t) - T_i]$$

The *maximum* heat transfer between the body and its surroundings is

$$Q_{\max} = mc_p(T_\infty - T_i)$$

The error involved in lumped system analysis is negligible when

$$\text{Bi} = \frac{hL_c}{k} < 0.1$$

where Bi is the *Biot number* and $L_c = V/A_s$ is the *characteristic length.*

When the lumped system analysis is not applicable, the variation of temperature with position as well as time can be determined using the *transient temperature charts* given in Figs. 18–16, 18–17, 18–18, and 18–30 for a large plane wall, a long cylinder, a sphere, and a semi-infinite medium, respectively. These charts are applicable for one-dimensional heat transfer in those geometries. Therefore, their use is limited to situations in which the body is initially at a uniform temperature, all surfaces are subjected to the same thermal conditions, and the body does not involve any heat generation. These charts can also be used to determine the total heat transfer from the body up to a specified time $t$.

Using the *one-term approximation,* the solutions of one-dimensional transient heat conduction problems are expressed analytically as

*Plane wall:* $\quad \theta_{\text{wall}} = \dfrac{T(x, t) - T_\infty}{T_i - T_\infty} = A_1 e^{-\lambda_1^2 \tau} \cos(\lambda_1 x/L)$

*Cylinder:* $\quad \theta_{\text{cyl}} = \dfrac{T(r, t) - T_\infty}{T_i - T_\infty} = A_1 e^{-\lambda_1^2 \tau} J_0(\lambda_1 r/r_o)$

*Sphere:* $\quad \theta_{\text{sph}} = \dfrac{T(r, t) - T_\infty}{T_i - T_\infty} = A_1 e^{-\lambda_1^2 \tau} \dfrac{\sin(\lambda_1 r/r_o)}{\lambda_1 r/r_o}$

where the constants $A_1$ and $\lambda_1$ are functions of the Bi number only, and their values are listed in Table 18–2 against the Bi

number for all three geometries. The error involved in one-term solutions is less than 2 percent when $\tau > 0.2$.

Using the one-term solutions, the fractional heat transfers in different geometries are expressed as

*Plane wall:* $\quad \left(\dfrac{Q}{Q_{\max}}\right)_{\text{wall}} = 1 - \theta_{0,\text{wall}} \dfrac{\sin \lambda_1}{\lambda_1}$

*Cylinder:* $\quad \left(\dfrac{Q}{Q_{\max}}\right)_{\text{cyl}} = 1 - 2\theta_{0,\text{cyl}} \dfrac{J_1(\lambda_1)}{\lambda_1}$

*Sphere:* $\quad \left(\dfrac{Q}{Q_{\max}}\right)_{\text{sph}} = 1 - 3\theta_{0,\text{sph}} \dfrac{\sin \lambda_1 - \lambda_1 \cos \lambda_1}{\lambda_1^3}$

The solutions of transient heat conduction in a semi-infinite solid with constant properties under various boundary conditions at the surface are given as follows:

*Specified Surface Temperature, $T_s = $ constant:*

$$\frac{T(x, t) - T_i}{T_s - T_i} = \text{erfc}\left(\frac{x}{2\sqrt{\alpha t}}\right) \quad \text{and} \quad \dot{q}_s(t) = \frac{k(T_s - T_i)}{\sqrt{\pi \alpha t}}$$

*Specified Surface Heat Flux, $\dot{q}_s = $ constant:*

$$T(x, t) - T_i = \frac{\dot{q}_s}{k}\left[\sqrt{\frac{4\alpha t}{\pi}} \exp\left(-\frac{x^2}{4\alpha t}\right) - x\,\text{erfc}\left(\frac{x}{2\sqrt{\alpha t}}\right)\right]$$

*Convection on the Surface, $\dot{q}_s(t) = h[T_\infty - T(0, t)]$:*

$$\frac{T(x, t) - T_i}{T_\infty - T_i} = \text{erfc}\left(\frac{x}{2\sqrt{\alpha t}}\right) - \exp\left(\frac{hx}{k} + \frac{h^2 \alpha t}{k^2}\right)$$

$$\times\,\text{erfc}\left(\frac{x}{2\sqrt{\alpha t}} + \frac{h\sqrt{\alpha t}}{k}\right)$$

*Energy Pulse at Surface, $e_s = $ constant:*

$$T(x, t) - T_i = \frac{e_s}{k\sqrt{\pi t/\alpha}} \exp\left(-\frac{x^2}{4\alpha t}\right)$$

where $\text{erfc}(\eta)$ is the *complementary error function* of argument $\eta$.

Using a superposition principle called the *product solution* these charts can also be used to construct solutions for the *two-dimensional* transient heat conduction problems encountered in geometries such as a short cylinder, a long rectangular bar, or a semi-infinite cylinder or plate, and even *three-dimensional* problems associated with geometries such as a rectangular prism or a semi-infinite rectangular bar, provided that all surfaces of the solid are subjected to convection to the same fluid at temperature $T_\infty$, with the same convection heat transfer coefficient $h$, and the body involves no heat generation. The solution in such multidimensional geometries can be expressed as the product of the solutions for the one-dimensional geometries whose intersection is the multidimensional geometry.

The total heat transfer to or from a multidimensional geometry can also be determined by using the one-dimensional values. The transient heat transfer for a two-dimensional geometry formed by the intersection of two one-dimensional geometries 1 and 2 is

$$\left(\frac{Q}{Q_{max}}\right)_{total, 2D} = \left(\frac{Q}{Q_{max}}\right)_1 + \left(\frac{Q}{Q_{max}}\right)_2 \left[1 - \left(\frac{Q}{Q_{max}}\right)_1\right]$$

Transient heat transfer for a three-dimensional body formed by the intersection of three one-dimensional bodies 1, 2, and 3 is given by

$$\left(\frac{Q}{Q_{max}}\right)_{total, 3D} = \left(\frac{Q}{Q_{max}}\right)_1 + \left(\frac{Q}{Q_{max}}\right)_2 \left[1 - \left(\frac{Q}{Q_{max}}\right)_1\right]$$
$$+ \left(\frac{Q}{Q_{max}}\right)_3 \left[1 - \left(\frac{Q}{Q_{max}}\right)_1\right]\left[1 - \left(\frac{Q}{Q_{max}}\right)_2\right]$$

## REFERENCES AND SUGGESTED READINGS

1. ASHRAE. *Handbook of Fundamentals.* SI version. Atlanta, GA: American Society of Heating, Refrigerating, and Air-Conditioning Engineers, Inc., 1993.

2. H. S. Carslaw and J. C. Jaeger. *Conduction of Heat in Solids.* 2nd ed. London: Oxford University Press, 1959.

3. Y. A. Çengel and A. J. Ghajar. *Heat and Mass Transfer: Fundamentals and Applications.* 4th ed. New York: McGraw-Hill, 2011.

4. H. Gröber, S. Erk, and U. Grigull. *Fundamentals of Heat Transfer.* New York: McGraw-Hill, 1961.

5. M. P. Heisler. "Temperature Charts for Induction and Constant Temperature Heating." *ASME Transactions* 69 (1947), pp. 227–36.

6. L. S. Langston. "Heat Transfer from Multidimensional Objects Using One-Dimensional Solutions for Heat Loss." *International Journal of Heat and Mass Transfer* 25 (1982), pp. 149–50.

7. P. J. Schneider. *Conduction Heat Transfer.* Reading, MA: Addison-Wesley, 1955.

## PROBLEMS*

### Lumped System Analysis

**18–1C** What is lumped system analysis? When is it applicable?

**18–2C** For which solid is the lumped system analysis more likely to be applicable: an actual apple or a golden apple of the same size? Why?

---

*Problems designated by a "C" are concept questions, and students are encouraged to answer them all. Problems designated by an "E" are in English units, and the SI users can ignore them. Problems with the icon 🌸 are solved using EES, and complete solutions together with parametric studies are included on the enclosed DVD. Problems with the icon 🖳 are comprehensive in nature, and are intended to be solved with a computer, preferably using the EES software that accompanies this text.

**18–3C** For which kind of bodies made of the same material is the lumped system analysis more likely to be applicable: slender ones or well-rounded ones of the same volume? Why?

**18–4C** Consider heat transfer between two identical hot solid bodies and the air surrounding them. The first solid is being cooled by a fan while the second one is allowed to cool naturally. For which solid is the lumped system analysis more likely to be applicable? Why?

**18–5C** Consider heat transfer between two identical hot solid bodies and their environments. The first solid is dropped in a large container filled with water, while the second one is allowed to cool naturally in the air. For which solid is the lumped system analysis more likely to be applicable? Why?

**18–6C** Consider a hot baked potato on a plate. The temperature of the potato is observed to drop by 4°C during the first minute. Will the temperature drop during the second minute be less than, equal to, or more than 4°C? Why?

Cool air

Hot baked potato

**FIGURE P18–6C**

**18–7C** Consider a potato being baked in an oven that is maintained at a constant temperature. The temperature of the potato is observed to rise by 5°C during the first minute. Will the temperature rise during the second minute be less than, equal to, or more than 5°C? Why?

**18–8C** What is the physical significance of the Biot number? Is the Biot number more likely to be larger for highly conducting solids or poorly conducting ones?

**18–9C** Consider two identical 4-kg pieces of roast beef. The first piece is baked as a whole, while the second is baked after being cut into two equal pieces in the same oven. Will there be any difference between the cooking times of the whole and cut roasts? Why?

**18–10C** Consider a sphere and a cylinder of equal volume made of copper. Both the sphere and the cylinder are initially at the same temperature and are exposed to convection in the same environment. Which do you think will cool faster, the cylinder or the sphere? Why?

**18–11C** In what medium is the lumped system analysis more likely to be applicable: in water or in air? Why?

**18–12** To warm up some milk for a baby, a mother pours milk into a thin-walled glass whose diameter is 6 cm. The height of the milk in the glass is 7 cm. She then places the glass into a large pan filled with hot water at 70°C. The milk is stirred constantly, so that its temperature is uniform at all times. If the heat transfer coefficient between the water and the glass is 120 W/m²·K, determine how long it will take for the milk to warm up from 3°C to 38°C. Take the properties of the milk to be the same as those of water. Can the milk in this case be treated as a lumped system? Why? *Answer:* 4.50 min

**18–13** Repeat Prob. 18–12 for the case of water also being stirred, so that the heat transfer coefficient is doubled to 240 W/m²·K.

**18–14** Obtain a relation for the time required for a lumped system to reach the average temperature $\frac{1}{2}(T_i + T_\infty)$, where $T_i$ is the initial temperature and $T_\infty$ is the temperature of the environment.

**18–15** The temperature of a gas stream is to be measured by a thermocouple whose junction can be approximated as a 1.2-mm-diameter sphere. The properties of the junction are $k = 35$ W/m·K, $\rho = 8500$ kg/m³, and $c_p = 320$ J/kg·K, and the heat transfer coefficient between the junction and the gas is $h = 110$ W/m²·K. Determine how long it will take for the thermocouple to read 99 percent of the initial temperature difference. *Answer:* 22.8s

**18–16E** In a manufacturing facility, 2-in-diameter brass balls ($k = 64.1$ Btu/h·ft·°F, $\rho = 532$ lbm/ft³, and $c_p = 0.092$ Btu/lbm·°F) initially at 250°F are quenched in a water bath at 120°F for a period of 2 min at a rate of 120 balls per minute. If the convection heat transfer coefficient is 42 Btu/h·ft²·°F, determine (a) the temperature of the balls after quenching and (b) the rate at which heat needs to be removed from the water in order to keep its temperature constant at 120°F.

250°F   Brass balls

120°F

Water bath

**FIGURE P18–16E**

**18–17E** Repeat Prob. 18–16E for aluminum balls.

**18–18** Obtain relations for the characteristic lengths of a large plane wall of thickness $2L$, a very long cylinder of radius $r_o$, and a sphere of radius $r_o$.

**18–19** A long copper rod of diameter 2.0 cm is initially at a uniform temperature of 100°C. It is now exposed to an air stream at 20°C with a heat transfer coefficient of 200 W/m²·K. How long would it take for the copper road to cool to an average temperature of 25°C?

**18–20** Consider a sphere of diameter 5 cm, a cube of side length 5 cm, and a rectangular prism of dimension 4 cm × 5 cm × 6 cm, all initially at 0°C and all made of silver ($k = 429$ W/m·K, $\rho = 10,500$ kg/m³, $c_p = 0.235$ kJ/kg·K). Now all three of these geometries are exposed to ambient air at 33°C on all of their surfaces with a heat transfer coefficient of 12 W/m²·K. Determine how long it will take for the temperature of each geometry to rise to 25°C.

**18–21** Consider a 800-W iron whose base plate is made of 0.5-cm-thick aluminum alloy 20218–T6 ($\rho = 2770$ kg/m³, $c_p = 875$ J/kg·K, $\alpha = 7.3 \times 10^{-5}$ m²/s). The base plate has a surface area of 0.03 m². Initially, the iron is in thermal equilibrium with the ambient air at 22°C. Taking the heat transfer coefficient at the surface of the base plate to be 12 W/m²·K and assuming 85 percent of the heat generated in the resistance wires is transferred to the plate, determine how long it will take for the plate temperature to reach 140°C. Is it realistic to assume the plate temperature to be uniform at all times?

Air
22°C

800 W
iron

**FIGURE P18–21**

**18–22** Reconsider Prob. 18–21. Using EES (or other) software, investigate the effects of the heat transfer coefficient and the final plate temperature on the time it will take for the plate to reach this temperature. Let the heat transfer coefficient vary from 5 W/m²·K to 25 W/m²·K and the temperature from 30°C to 200°C. Plot the time as functions of the heat transfer coefficient and the temperature, and discuss the results.

**18–23** A person is found dead at 5PM in a room whose temperature is 20°C. The temperature of the body is measured to be 25°C when found, and the heat transfer coefficient is estimated to be 8 W/m²·K. Modeling the body as a 30-cm-diameter, 1.70-m-long cylinder and using the lumped system analysis as a rough approximation, estimate the time of death of that person.

**18–24** Springs in a suspension system of automobiles are made of steel rods heated and wound into coils while ductile. Consider steel rods ($\rho = 7832$ kg/m³, $c_p = 434$ J/kg·K, and $k = 63.9$ W/m·K) with diameter of 2.5 cm and length of 1.27 m. The steel rods are heated in an oven with a uniform convection heat transfer coefficient of 20 W/m²·K. The steel rods were heated from an initial temperature of 20°C to the desired temperature of 450°C before being wound into coils. Determine the ambient temperature in the oven, if the steel rods were to be heated to the desired temperature within 10 minutes. *Answer: 1016°C*

**18–25** Steel rods ($\rho = 7832$ kg/m³, $c_p = 434$ J/kg·K, and $k = 63.9$ W/m·K) are heated in a furnace to 850°C and then quenched in a water bath at 50°C for a period of 40 seconds as part of a hardening process. The convection heat transfer coefficient is 650 W/m²·K. If the steel rods have diameter of 40 mm and length of 2 m, determine their average temperature when they are taken out of the water bath.

**18–26** A brick of 203 × 102 × 57 mm in dimension is being burned in a kiln to 1100°C, and then allowed to cool in a room with ambient air temperature of 30°C and convection heat transfer coefficient of 5 W/m²·K. If the brick has properties of $\rho = 1920$ kg/m³, $c_p = 790$ J/kg·K, and $k = 0.90$ W/m·K, determine the time required to cool the brick to a temperature difference of 5°C from the ambient air temperature. *Answer: 7 h*

**18–27** Carbon steel balls ($\rho = 7833$ kg/m³, $k = 54$ W/m·K, $c_p = 0.465$ kJ/kg·°C, and $\alpha = 1.474 \times 10^{-6}$ m²/s) 8 mm in diameter are annealed by heating them first to 900°C in a furnace and then allowing them to cool slowly to 100°C in ambient air at 35°C. If the average heat transfer coefficient is 75 W/m²·K, determine how long the annealing process will take. If 2500 balls are to be annealed per hour, determine the total rate of heat transfer from the balls to the ambient air.

Furnace

Air, 35°C

900°C

Steel ball 100°C

**FIGURE P18–27**

**18–28** Reconsider Prob. 18–27. Using EES (or other) software, investigate the effect of the initial temperature of the balls on the annealing time and the total rate of heat transfer. Let the temperature vary from 500°C to 1000°C. Plot the time and the total rate of heat transfer as a function of the initial temperature, and discuss the results.

**18–29** An electronic device dissipating 18 W has a mass of 20 g, a specific heat of 850 J/kg·K, and a surface area of 4 cm². The device is lightly used, and it is on for 5 min and then off for several hours, during which it cools to the ambient temperature of 25°C. Taking the heat transfer coefficient to be 12 W/m²·K, determine the temperature of the device at the end of the 5-min operating period. What would your answer be if the device were attached to an aluminum heat sink having a mass of 200 g and a surface area of 80 cm²? Assume the device and the heat sink to be nearly isothermal.

## Transient Heat Conduction in Large Plane Walls, Long Cylinders, and Spheres with Spatial Effects

**18–30C** An egg is to be cooked to a certain level of doneness by being dropped into boiling water. Can the cooking time be shortened by turning up the heat and bringing water to a more rapid boiling?

**18–31C** What is an infinitely long cylinder? When is it proper to treat an actual cylinder as being infinitely long, and when is it not? For example, is it proper to use this model when finding the temperatures near the bottom or top surfaces of a cylinder? Explain.

**18–32C** Can the transient temperature charts in Fig. 18–15 for a plane wall exposed to convection on both sides be used for a plane wall with one side exposed to convection while the other side is insulated? Explain.

**18–33C** Why are the transient temperature charts prepared using nondimensionalized quantities such as the Biot and Fourier numbers instead of the actual variables such as thermal conductivity and time?

**18–34C** What is the physical significance of the Fourier number? Will the Fourier number for a specified heat transfer problem double when the time is doubled?

**18–35C** How can we use the transient temperature charts when the surface temperature of the geometry is specified instead of the temperature of the surrounding medium and the convection heat transfer coefficient?

**18–36C** A body at an initial temperature of $T_i$ is brought into a medium at a constant temperature of $T_\infty$. How can you determine the maximum possible amount of heat transfer between the body and the surrounding medium?

**18–37C** The Biot number during a heat transfer process between a sphere and its surroundings is determined to be 0.02. Would you use lumped system analysis or the transient temperature charts when determining the midpoint temperature of the sphere? Why?

**18–38** A long cylindrical wood log ($k = 0.17$ W/m·K and $\alpha = 1.28 \times 10^{-7}$ m²/s) is 10 cm in diameter and is initially at a uniform temperature of 25°C. It is exposed to hot gases at 600°C in a fireplace with a heat transfer coefficient of 13.6 W/m²·K on the surface. If the ignition temperature of the wood is 420°C, determine how long it will be before the log ignites.

**18–39** An experiment is to be conducted to determine heat transfer coefficient on the surfaces of tomatoes that are placed in cold water at 7°C. The tomatoes ($k = 0.59$ W/m·K, $\alpha = 0.141 \times 10^{-6}$ m²/s, $\rho = 999$ kg/m³, $c_p = 3.99$ kJ/kg·K) with an initial uniform temperature of 30°C are spherical in shape with a diameter of 8 cm. After a period of 2 hours, the temperatures at the center and the surface of the tomatoes are measured to be 10.0°C and 7.1°C, respectively. Using analytical one-term approximation method (not the Heisler charts), determine the heat transfer coefficient and the amount of heat transfer during this period if there are eight such tomatoes in water.

**18–40** In an annealing process, a 50-mm-thick stainless steel plate ($\rho = 8238$ kg/m³, $c_p = 468$ J/kg·K, $k = 13.4$ W/m·K, and $\alpha = 3.48 \times 10^{-6}$ m²/s) was reheated in a furnace from an initial uniform temperature of 230°C. The ambient temperature inside the furnace is at a uniform temperature of 1000°C and has a convection heat transfer coefficient

of 215 W/m²·K. If the entire stainless steel plate is to be heated to at least 600°C, determine the time that the plate should be heated in the furnace using (a) Table 18–2 and (b) the Heisler chart (Figure 18–15).

**18–41** An ordinary egg can be approximated as a 5.5-cm-diameter sphere whose properties are roughly $k = 0.6$ W/m·K and $\alpha = 0.14 \times 10^{-6}$ m²/s. The egg is initially at a uniform temperature of 4°C and is dropped into boiling water at 97°C. Taking the convection heat transfer coefficient to be $h = 1400$ W/m²·K, determine how long it will take for the center of the egg to reach 70°C.

**FIGURE P18–41**

**18–42**  Reconsider Prob. 18–41. Using EES (or other) software, investigate the effect of the final center temperature of the egg on the time it will take for the center to reach this temperature. Let the temperature vary from 50°C to 95°C. Plot the time versus the temperature, and discuss the results.

**18–43** In a production facility, 3-cm-thick large brass plates ($k = 110$ W/m·K, $\rho = 8530$ kg/m³, $c_p = 380$ J/kg·K, and $\alpha = 33.9 \times 10^{-6}$ m²/s) that are initially at a uniform temperature of 25°C are heated by passing them through an oven maintained at 700°C. The plates remain in the oven for a period of 10 min. Taking the convection heat transfer coefficient to be $h = 80$ W/m²·K, determine the surface temperature of the plates when they come out of the oven.

**FIGURE P18–43**

**18–44** Reconsider Prob. 18–43. Using EES (or other) software, investigate the effects of the temperature of the oven and the heating time on the final surface temperature of the plates. Let the oven temperature vary from 500°C to 900°C and the time from 2 min to 30 min. Plot the surface temperature as the functions of the oven temperature and the time, and discuss the results.

**18–45** A long 35-cm-diameter cylindrical shaft made of stainless steel 304 ($k$ = 14.9 W/m·K, $\rho$ = 7900 kg/m³, $c_p$ = 477 J/kg·K, and $\alpha$ = 3.95 × 10⁻⁶ m²/s) comes out of an oven at a uniform temperature of 500°C. The shaft is then allowed to cool slowly in a chamber at 150°C with an average convection heat transfer coefficient of $h$ = 60 W/m²·K. Determine the temperature at the center of the shaft 20 min after the start of the cooling process. Also, determine the heat transfer per unit length of the shaft during this time period. *Answers: 486°C, 22,270 kJ*

**18–46** Reconsider Prob. 18–45. Using EES (or other) software, investigate the effect of the cooling time on the final center temperature of the shaft and the amount of heat transfer. Let the time vary from 5 min to 60 min. Plot the center temperature and the heat transfer as a function of the time, and discuss the results.

**18–47E** Long cylindrical AISI stainless steel rods ($k$ = 7.74 Btu/h·ft·°F and $\alpha$ = 0.135 ft²/h) of 4-in-diameter are heat treated by drawing them at a velocity of 7 ft/min through a 21-ft-long oven maintained at 1700°F. The heat transfer coefficient in the oven is 20 Btu/h·ft²·°F. If the rods enter the oven at 70°F, determine their centerline temperature when they leave.

Oven
1700°F

7 ft/min

|← 21 ft →|

Stainless steel
70°F

**FIGURE P18–47E**

**18–48** In a meat processing plant, 2-cm-thick steaks ($k$ = 0.45 W/m·K and $\alpha$ = 0.91 × 10⁻⁷ m²/s) that are initially at 25°C are to be cooled by passing them through a refrigeration room at −11°C. The heat transfer coefficient on both sides of the steaks is 9 W/m²·K. If both surfaces of the steaks are to be cooled to 2°C, determine how long the steaks should be kept in the refrigeration room.

**18–49** A student calculates that the total heat transfer from a spherical copper ball of diameter 18 cm initially at 200°C and its environment at a constant temperature of 25°C during the first 20 min of cooling is 3150 kJ. Is this result reasonable? Why?

**18–50** In *Betty Crocker's Cookbook,* it is stated that it takes 2 h 45 min to roast a 3.2-kg rib initially at 4.5°C "rare" in an oven maintained at 163°C. It is recommended that a meat thermometer be used to monitor the cooking, and the rib is considered rare done when the thermometer inserted into the center of the thickest part of the meat registers 60°C. The rib can be treated as a homogeneous spherical object with the properties $\rho$ = 1200 kg/m³, $c_p$ = 4.1 kJ/kg·K, $k$ = 0.45 W/m·K, and $\alpha$ = 0.91 × 10⁻⁷ m²/s. Determine (*a*) the heat transfer coefficient at the surface of the rib; (*b*) the temperature of the outer surface of the rib when it is done; and (*c*) the amount of heat transferred to the rib. (*d*) Using the values obtained, predict how long it will take to roast this rib to "medium" level, which occurs when the innermost temperature of the rib reaches 71°C. Compare your result to the listed value of 3 h 20 min.

If the roast rib is to be set on the counter for about 15 min before it is sliced, it is recommended that the rib be taken out of the oven when the thermometer registers about 4°C below the indicated value because the rib will continue cooking even after it is taken out of the oven. Do you agree with this recommendation? *Answers: (a) 156.9 W/m²·K, (b) 159.5°C, (c) 1629 kJ, (d) 3.0 h*

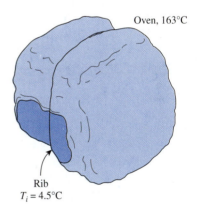

Oven, 163°C

Rib
$T_i$ = 4.5°C

**FIGURE P18–50**

**18–51** Repeat Prob. 18–50 for a roast rib that is to be "well-done" instead of "rare." A rib is considered to be well-done when its center temperature reaches 77°C, and the roasting in this case takes about 4 h 15 min.

**18–52** Citrus fruits are very susceptible to cold weather, and extended exposure to subfreezing temperatures can destroy them. Consider an 8-cm-diameter orange that is initially at 15°C. A cold front moves in one night, and the ambient temperature suddenly drops to −6°C, with a heat transfer coefficient of 15 W/m²·K. Using the properties of water for the orange and assuming the ambient conditions to remain constant for 4 h before the cold front moves out, determine if any part of the orange will freeze that night.

**18–53E**  In a chicken processing plant, whole chickens averaging 5 lbm each and initially at 65°F are to be cooled in the racks of a large refrigerator that is maintained at 5°F. The entire chicken is to be cooled below 45°F, but the temperature of the chicken is not to drop below 35°F at any point during refrigeration. The convection heat transfer coefficient and thus the rate of heat transfer from the chicken can be controlled by varying the speed of a circulating fan inside. Determine the heat transfer-coefficient that will enable us to meet both temperature constraints while keeping the refrigeration time to a minimum. The chicken can be treated as a homogeneous spherical object having the properties $\rho = 74.9$ lbm/ft$^3$, $c_p = 0.98$ Btu/lbm·°F, $k = 0.26$ Btu/h·ft·°F, and $\alpha = 0.0035$ ft$^2$/h.

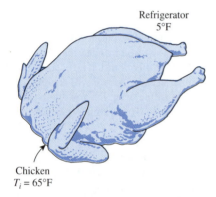

Refrigerator
5°F

Chicken
$T_i = 65°F$

**FIGURE P18–53E**

**18–54**  A person puts a few apples into the freezer at $-15°C$ to cool them quickly for guests who are about to arrive. Initially, the apples are at a uniform temperature of 25°C, and the heat transfer coefficient on the surfaces is 8 W/m$^2$·K. Treating the apples as 9-cm-diameter spheres and taking their properties to be $\rho = 840$ kg/m$^3$, $c_p = 3.81$ kJ/kg·K, $k = 0.418$ W/m·K, and $\alpha = 1.3 \times 10^{-7}$ m$^2$/s, determine the center and surface temperatures of the apples in 1 h. Also, determine the amount of heat transfer from each apple.

**18–55**  Reconsider Prob. 18–54. Using EES (or other) software, investigate the effect of the initial temperature of the apples on the final center and surface temperatures and the amount of heat transfer. Let the initial temperature vary from 2°C to 30°C. Plot the center temperature, the surface temperature, and the amount of heat transfer as a function of the initial temperature, and discuss the results.

**18–56**  A 9-cm-diameter potato ($\rho = 1100$ kg/m$^3$, $c_p = 3900$ J/kg·K, $k = 0.6$ W/m·K, and $\alpha = 1.4 \times 10^{-7}$ m$^2$/s) that is initially at a uniform temperature of 25°C is baked in an oven at 170°C until a temperature sensor inserted to the center of the potato indicates a reading of 70°C. The potato is

then taken out of the oven and wrapped in thick towels so that almost no heat is lost from the baked potato. Assuming the heat transfer coefficient in the oven to be 40 W/m$^2$·K, determine (a) how long the potato is baked in the oven and (b) the final equilibrium temperature of the potato after it is wrapped.

**18–57**  White potatoes ($k = 0.50$ W/m·K and $\alpha = 0.13 \times 10^{-6}$ m$^2$/s) that are initially at a uniform temperature of 20°C and have an average diameter of 6 cm are to be cooled by refrigerated air at 2°C flowing at a velocity of 3 m/s. The average heat transfer coefficient between the potatoes and the air is experimentally determined to be 19 W/m$^2$·K. Determine how long it will take for the center temperature of the potatoes to drop to 6°C. Also, determine if any part of the potatoes will experience chilling injury during this process.

Air
2°C
3 m/s

Potato
$T_i = 20°C$

**FIGURE P18–57**

**18–58**  Layers of 23-cm-thick meat slabs ($k = 0.47$ W/m·K and $\alpha = 0.13 \times 10^{-6}$ m$^2$/s) initially at a uniform temperature of 7°C are to be frozen by refrigerated air at $-30°C$ flowing at a velocity of 1.4 m/s. The average heat transfer coefficient between the meat and the air is 20 W/m$^2$·K. Assuming the size of the meat slabs to be large relative to their thickness, determine how long it will take for the center temperature of the slabs to drop to $-18°C$. Also, determine the surface temperature of the meat slab at that time.

**18–59E**  Layers of 6-in-thick meat slabs ($k = 0.26$ Btu/h·ft·°F and $\alpha = 1.4 \times 10^{-6}$ ft$^2$/s) initially at a uniform temperature of 50°F are cooled by refrigerated air at 23°F to a temperature of 36°F at their center in 12 h. Estimate the average heat transfer coefficient during this cooling process.  *Answer:* 1.5 Btu/h·ft$^2$·°F

**18–60E**  Oranges of 2.5-in-diameter ($k = 0.26$ Btu/h·ft·°F and $\alpha = 1.4 \times 10^{-6}$ ft$^2$/s) initially at a uniform temperature of 78°F are to be cooled by refrigerated air at 25°F flowing at a velocity of 1 ft/s. The average heat transfer coefficient between the oranges and the air is experimentally determined to be 4.6 Btu/h·ft$^2$·°F. Determine how long it will take for the center temperature of the oranges to drop to 40°F. Also, determine if any part of the oranges will freeze during this process.

**18–61**  A heated 6-mm-thick Pyroceram plate ($\rho = 2600$ kg/m$^3$, $c_p = 808$ J/kg·K, $k = 3.98$ W/m·K, and $\alpha = 1.89 \times 10^{-6}$ m$^2$/s) is being cooled in a room with air temperature of 25°C and convection heat transfer coefficient of

13.3 W/m²·K. The heated Pyroceram plate had an initial temperature of 500°C, and is allowed to cool for 286 seconds. If the mass of the Pyroceram plate is 10 kg, determine the heat transfer from the Pyroceram plate during the cooling process using (a) Table 18–2 and (b) Figure 18–15.

**18–62** A 10-cm thick aluminum plate ($\rho = 2702$ kg/m³, $c_p = 903$ J/kg·K, $k = 237$ W/m·K, and $\alpha = 97.1 \times 10^{-6}$ m²/s) is being heated in liquid with temperature of 500°C. The aluminum plate has a uniform initial temperature of 25°C. If the surface temperature of the aluminum plate is approximately the liquid temperature, determine the temperature at the center plane of the aluminum plate after 15 seconds of heating.

Aluminum plate
$T_s = 500°C$

Liquid, 500°C

**FIGURE P18–62**

**18–63** A long iron rod ($\rho = 7870$ kg/m³, $c_p = 447$ J/kg·K, $k = 80.2$ W/m·K, and $\alpha = 23.1 \times 10^{-6}$ m²/s) with diameter of 25 mm is initially heated to a uniform temperature of 700°C. The iron rod is then quenched in a large water bath that is maintained at constant temperature of 50°C and convection heat transfer coefficient of 128 W/m²·K. Determine the time required for the iron rod surface temperature to cool to 200°C. *Answer: 248 s.*

Iron rod

Water, 50°C
$h = 128$ W/m²·K

**FIGURE P18–63**

**18–64** A long Pyroceram rod ($\rho = 2600$ kg/m³, $c_p = 808$ J/kg·K, $k = 3.98$ W/m·K, and $\alpha = 1.89 \times 10^{-6}$ m²/s) with diameter of 10 mm has an initial uniform temperature of 1000°C. The Pyroceram rod is allowed to cool in ambient temperature of 25°C and convection heat transfer coefficient of 80 W/m²·K. If the Pyroceram rod is allowed to

cool for 3 minutes, determine the temperature at the center of the rod using (a) Table 18–2 and (b) the Heisler chart (Figure 18–16).

**18–65** A 2-cm-diameter plastic rod has a thermocouple inserted to measure temperature at the center of the rod. The plastic rod ($\rho = 1190$ kg/m³, $c_p = 1465$ J/kg·K, and $k = 0.19$ W/m·K) was initially heated to a uniform temperature of 70°C, and allowed to be cooled in ambient air temperature of 25°C. After 1388 s of cooling, the thermocouple measured the temperature at the center of the rod to be 30°C. Determine the convection heat transfer coefficient for this process.

Air, 25°C

Plastic rod

Thermocouple
wires

**FIGURE P18–65**

**18–66** Steel rods, 2 m in length and 60 mm in diameter, are being drawn through an oven that maintains a temperature of 800°C and convection heat transfer coefficient of 128 W/m²·K. The steel rods ($\rho = 7832$ kg/m³, $c_p = 434$ J/kg·K, $k = 63.9$ W/m·K, and $\alpha = 18.8 \times 10^{-6}$ m²/s) were initially in uniform temperature of 30°C. Using (a) Table 18–2 and (b) Figure 18–16, determine the amount of heat is transferred to the steel rod after 133 s of heating. *Answer: 4140 kJ*

### Transient Heat Conduction in Semi-Infinite Solids

**18–67C** What is a semi-infinite medium? Give examples of solid bodies that can be treated as semi-infinite mediums for heat transfer purposes.

**18–68C** Under what conditions can a plane wall be treated as a semi-infinite medium?

**18–69C** Consider a hot semi-infinite solid at an initial temperature of $T_i$ that is exposed to convection to a cooler medium at a constant temperature of $T_\infty$, with a heat transfer coefficient of $h$. Explain how you can determine the total amount of heat transfer from the solid up to a specified time $t_o$.

**18–70** A thick wood slab ($k = 0.17$ W/m·K and $\alpha = 1.28 \times 10^{-7}$ m²/s) that is initially at a uniform temperature of 25°C is exposed to hot gases at 550°C for a period of 5 min. The heat transfer coefficient between the gases and the wood slab is 35 W/m²·K. If the ignition temperature of the wood is 450°C, determine if the wood will ignite.

**18–71** The soil temperature in the upper layers of the earth varies with the variations in the atmospheric conditions. Before a cold front moves in, the earth at a location is initially at a uniform temperature of 15°C. Then the area is subjected to a temperature of −8°C and high winds that resulted in a convection heat transfer coefficient of 40 W/m²·K on the earth's surface for a period of 10 h. Taking the properties of the soil at that location to be $k = 0.9$ W/m·K and $\alpha = 1.6 \times 10^{-5}$ m$^2$/s, determine the soil temperature at distances 0, 10, 20, and 50 cm from the earth's surface at the end of this 10-h period.

**FIGURE P18–71**

**18–72** Reconsider Prob. 18–71. Using EES (or other) software, plot the soil temperature as a function of the distance from the earth's surface as the distance varies from 0 m to 1m, and discuss the results.

**18–73** A thick aluminum block initially at 20°C is subjected to constant heat flux of 4000 W/m$^2$ by an electric resistance heater whose top surface is insulated. Determine how much the surface temperature of the block will rise after 30 minutes.

**18–74** A bare-footed person whose feet are at 32°C steps on a large aluminum block at 20°C. Treating both the feet and the aluminum block as semi-infinite solids, determine the contact surface temperature. What would your answer be if the person stepped on a wood block instead? At room temperature, the $\sqrt{k\rho c_p}$ value is 218 kJ/m$^2$·°C for aluminum, 0.38 kJ/m$^2$·°C for wood, and 1.1 kJ/m$^2$·°C for human flesh.

**18–75E** The walls of a furnace are made of 1.5-ft-thick concrete ($k = 0.64$ Btu/h·ft·°F and $\alpha = 0.023$ ft$^2$/h). Initially, the furnace and the surrounding air are in thermal equilibrium at 70°F. The furnace is then fired, and the inner surfaces of the furnace are subjected to hot gases at 1800°F with a very large heat transfer coefficient. Determine how long it will take for the temperature of the outer surface of the furnace walls to rise to 70.1°F. *Answer:* 3.0 h

**18–76** In areas where the air temperature remains below 0°C for prolonged periods of time, the freezing of water in underground pipes is a major concern. Fortunately, the soil remains relatively warm during those periods, and it takes weeks for the subfreezing temperatures to reach the water mains in the ground. Thus, the soil effectively serves as an insulation to protect the water from the freezing atmospheric temperatures in winter.

The ground at a particular location is covered with snow pack at −8°C for a continuous period of 60 days, and the average soil properties at that location are $k = 0.35$ W/m·K

and $\alpha = 0.15 \times 10^{-6}$ m$^2$/s. Assuming an initial uniform temperature of 8°C for the ground, determine the minimum burial depth to prevent the water pipes from freezing.

**18–77** A highway made of asphalt is initially at a uniform temperature of 55°C. Suddenly the highway surface temperature is reduced to 25°C by rain. Determine the temperature at the depth of 3 cm from the highway surface and the heat flux transferred from the highway after 60 minutes. Assume the highway surface temperature is maintained at 25°C. *Answers:* 53.6°C, 98 W/m$^2$

**18–78** In a vacuum chamber, a thick slab is placed under an array of laser diodes with an output constant pulse. A thermocouple is inserted inside the slab at 25 mm from the surface and the slab has an initial uniform temperature of 20°C. The known properties of the slab are $k = 63.9$ W/m·K and $\alpha = 18.8 \times 10^{-6}$ m$^2$/s. If the thermocouple measured a temperature of 130°C after 30 s the slab surface has been exposed to the laser pulse, determine (a) the amount of energy per unit surface area directed on the slab surface and (b) the thermocouple reading after 60 s has elapsed.

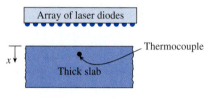

**FIGURE P18–78**

### Transient Heat Conduction in Multidimensional Systems

**18–79C** What is the product solution method? How is it used to determine the transient temperature distribution in a two-dimensional system?

**18–80C** How is the product solution used to determine the variation of temperature with time and position in three-dimensional systems?

**18–81C** A short cylinder initially at a uniform temperature $T_i$ is subjected to convection from all of its surfaces to a medium at temperature $T_\infty$. Explain how you can determine the temperature of the midpoint of the cylinder at a specified time $t$.

**18–82C** Consider a short cylinder whose top and bottom surfaces are insulated. The cylinder is initially at a uniform temperature $T_i$ and is subjected to convection from its side surface to a medium at temperature $T_\infty$ with a heat transfer coefficient of $h$. Is the heat transfer in this short cylinder one- or two-dimensional? Explain.

**18–83** Consider a cubic block whose sides are 5 cm long and a cylindrical block whose height and diameter are also 5 cm. Both blocks are initially at 20°C and are made of granite ($k = 2.5$ W/m·K and $\alpha = 1.15 \times 10^{-6}$ m²/s). Now both blocks are exposed to hot gases at 500°C in a furnace on all of their surfaces with a heat transfer coefficient of 40 W/m²·K. Determine the center temperature of each geometry after 10, 20, and 60 min.

5 cm
5 cm
5 cm
$T_i = 20°C$
5 cm
5 cm
$T_i = 20°C$

Hot gases, 500°C

**FIGURE P18–83**

**18–84** Repeat Prob. 18–83 with the heat transfer coefficient at the top and the bottom surfaces of each block being doubled to 80 W/m²·K.

**18–85** A short brass cylinder ($\rho = 8530$ kg/m³, $c_p = 0.389$ kJ/kg·K, $k = 110$ W/m·K, and $\alpha = 3.39 \times 10^{-5}$ m²/s) of diameter 4 cm and height 20 cm is initially at a uniform temperature of 150°C. The cylinder is now placed in atmospheric air at 20°C, where heat transfer takes place by convection with a heat transfer coefficient of 40 W/m²·K. Calculate (a) the center temperature of the cylinder; (b) the center temperature of the top surface of the cylinder; and (c) the total heat transfer from the cylinder 15 min after the start of the cooling.

Ambient
air
20°C

Brass
cylinder

20 cm

4 cm

150°C

**FIGURE P18–85**

**18–86** Reconsider Prob. 18–85. Using EES (or other) software, investigate the effect of the cooling time on the center temperature of the cylinder, the center temperature of the top surface of the cylinder, and the total heat transfer. Let the time vary from 5 min to 60 min. Plot the center temperature of the cylinder, the center temperature of the top surface, and the total heat transfer as a function of the time, and discuss the results.

**18–87** A semi-infinite aluminum cylinder ($k = 237$ W/m·K, $\alpha = 9.71 \times 10^{-5}$ m²/s) of diameter $D = 15$ cm is initially at a uniform temperature of $T_i = 115°C$. The cylinder is now placed in water at 10°C, where heat transfer takes place by convection with a heat transfer coefficient of $h = 140$ W/m²·K. Determine the temperature at the center of the cylinder 5 cm from the end surface 8 min after the start of cooling.

**18–88E** A hot dog can be considered to be a cylinder 5 in long and 0.8 in in diameter whose properties are $\rho = 61.2$ lbm/ft³, $c_p = 0.93$ Btu/lbm·°F, $k = 0.44$ Btu/h·ft·°F, and $\alpha = 0.0077$ ft²/h. A hot dog initially at 40°F is dropped into boiling water at 212°F. If the heat transfer coefficient at the surface of the hot dog is estimated to be 120 Btu/h·ft²·°F, determine the center temperature of the hot dog after 5, 10, and 15 min by treating the hot dog as (a) a finite cylinder and (b) an infinitely long cylinder.

**18–89** A 2-cm-high cylindrical ice block ($k = 2.22$ W/m·K and $\alpha = 0.124 \times 10^{-7}$ m²/s) is placed on a table on its base of diameter 2 cm in a room at 24°C. The heat transfer coefficient on the exposed surfaces of the ice block is 13 W/m²·K, and heat transfer from the base of the ice block to the table is negligible. If the ice block is not to start melting at any point for at least 3 h, determine what the initial temperature of the ice block should be.

**18–90** A 30-cm-long cylindrical aluminum block ($\rho = 2702$ kg/m³, $c_p = 0.896$ kJ/kg·K, $k = 236$ W/m·K, and $\alpha = 9.75 \times 10^{-5}$ m²/s), 15 cm in diameter, is initially at a uniform temperature of 20°C. The block is to be heated in a furnace at 1200°C until its center temperature rises to 300°C. If the heat transfer coefficient on all surfaces of the block is 80 W/m²·K, determine how long the block should be kept in the furnace. Also, determine the amount of heat transfer from the aluminum block if it is allowed to cool in the room until its temperature drops to 20°C throughout.

**18–91** Repeat Prob. 18–90 for the case where the aluminum block is inserted into the furnace on a low-conductivity material so that the heat transfer to or from the bottom surface of the block is negligible.

**18–92** Reconsider Prob. 18–90. Using EES (or other) software, investigate the effect of the final center temperature of the block on the heating time and the amount of heat transfer. Let the final center temperature vary from 50°C to 1000°C. Plot the time and the heat transfer as a function of the final center temperature, and discuss the results.

## Review Problems

**18–93** Spherical glass beads coming out of a kiln are allowed to cool in a room temperature of 30°C. A glass bead with a diameter of 10 mm and an initial temperature of 400°C is allowed to cool for 3 minutes. If the convection heat transfer coefficient is 28 W/m²·K, determine the temperature at the center of the glass bead using (*a*) Table 18–2 and (*b*) the Heisler chart (Figure 18–17). The glass bead has properties of $\rho = 2800$ kg/m³, $c_p = 750$ J/kg·K, and $k = 0.7$ W/m·K.

**18–94** A large iron slab ($\rho = 7870$ kg/m³, $c_p = 447$ J/kg·K, and $k = 80.2$ W/m·K) was initially heated to a uniform temperature of 150°C and then placed on concrete floor ($\rho = 1600$ kg/m³, $c_p = 840$ J/kg·K, and $k = 0.79$ W/m·K). The concrete floor was initially at a uniform temperature of 30°C. Determine (*a*) The surface temperature between the iron slab and concrete floor and (*b*) The temperature of the concrete floor at the depth of 25 mm, if the surface temperature remains constant after 15 minutes.

**FIGURE P18–94**

**18–95** In desert climates, rainfall is not a common occurrence since the rain droplets formed in the upper layer of the atmosphere often evaporate before they reach the ground. Consider a raindrop that is initially at a temperature of 5°C and has a diameter of 5 mm. Determine how long it will take for the diameter of the raindrop to reduce to 3 mm as it falls through ambient air at 18°C with a heat transfer coefficient of 400 W/m²·K. The water temperature can be assumed to remain constant and uniform at 5°C at all times.

**18–96** The water main in the cities must be placed at sufficient depth below the earth's surface to avoid freezing during extended periods of subfreezing temperatures. Determine the minimum depth at which the water main must be placed at a location where the soil is initially at 15°C and the earth's surface temperature under the worst conditions is expected to remain at −10°C for a period of 75 days. Take the properties of soil at that location to be $k = 0.7$ W/m·K and $\alpha = 1.4 \times 10^{-5}$ m²/s. *Answer: 7.05 m*

**18–97** A hot dog can be considered to be a 12-cm-long cylinder whose diameter is 2 cm and whose properties are $\rho = 980$ kg/m³, $c_p = 3.9$ kJ/kg·K, $k = 0.76$ W/m·K, and $\alpha = 2 \times 10^{-7}$ m²/s. A hot dog initially at 5°C is dropped into boiling water at 100°C. The heat transfer coefficient at the surface of the hot dog is estimated to be 600 W/m²·K. If the hot dog is

considered cooked when its center temperature reaches 80°C, determine how long it will take to cook it in the boiling water.

**FIGURE P18–97**

**18–98** A long roll of 2-m-wide and 0.5-cm-thick 1-Mn manganese steel plate coming off a furnace at 820°C is to be quenched in an oil bath ($c_p = 2.0$ kJ/kg·K) at 185°C. The metal sheet is moving at a steady velocity of 20 m/min, and the oil bath is 9 m long. Taking the convection heat transfer coefficient on both sides of the plate to be 860 W/m²·K, determine the temperature of the sheet metal when it leaves the oil bath. Also, determine the required rate of heat removal from the oil to keep its temperature constant at 45°C.

**FIGURE P18–98**

**18–99E** In *Betty Crocker's Cookbook*, it is stated that it takes 5 h to roast a 14-lb stuffed turkey initially at 40°F in an oven maintained at 325°F. It is recommended that a meat thermometer be used to monitor the cooking, and the turkey is considered done when the thermometer inserted deep into the thickest part of the breast or thigh without touching the bone registers 185°F. The turkey can be treated as a homogeneous spherical object with the properties $\rho = 75$ lbm/ft³, $c_p = 0.98$ Btu/lbm·°F, $k = 0.26$ Btu/h·ft·°F, and $\alpha = 0.0035$ ft²/h. Assuming the tip of the thermometer is at one-third radial distance from the center of the turkey, determine (*a*) the average heat transfer coefficient at the surface of the turkey; (*b*) the temperature of the skin of the turkey when it is done; and (*c*) the total amount of heat transferred to the turkey in the oven. Will the reading of the thermometer be more or less than 185°F 5 min after the turkey is taken out of the oven?

**18–100** During a fire, the trunks of some dry oak trees ($k = 0.17$ W/m·K and $\alpha = 1.28 \times 10^{-7}$ m²/s) that are initially at a uniform temperature of 30°C are exposed to hot gases at 600°C for a period of 5 h, with a heat transfer coefficient of 65 W/m²·K on the surface. The ignition temperature of the trees is 410°C. Treating the trunks of the trees as long cylindrical rods of diameter 20 cm, determine if these dry trees will ignite as the fire sweeps through them.

**FIGURE P18–100**

**18–101** Consider two 2-cm-thick large steel plates ($k = 43$ W/m·K and $\alpha = 1.17 \times 10^{-5}$ m²/s) that were put on top of each other while wet and left outside during a cold winter night at $-10°C$. The next day, a worker needs one of the plates, but the plates are stuck together because the freezing of the water between the two plates has bonded them together. In an effort to melt the ice between the plates and separate them, the worker takes a large hair dryer and blows hot air at 80°C all over the exposed surface of the plate on the top. The convection heat transfer coefficient at the top surface is estimated to be 40 W/m²·K. Determine how long the worker must keep blowing hot air before the two plates separate. *Answer:* 217 s

**18–102E** Consider a plate of thickness 1 in, a long cylinder of diameter 1 in, and a sphere of diameter 1 in, all initially at 400°F and all made of bronze ($k = 15.0$ Btu/h·ft·°F and $\alpha = 0.333$ ft²/h). Now all three of these geometries are exposed to cool air at 75°F on all of their surfaces, with a heat transfer coefficient of 7 Btu/h·ft²·°F. Determine the center temperature of each geometry after 10 and 20 min. Explain why the center temperature of the sphere is always the lowest.

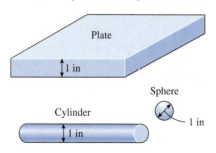

**FIGURE P18–102E**

**18–103E** Repeat Prob. 18–102E for cast iron geometries ($k = 29$ Btu/h·ft·°F and $\alpha = 0.61$ ft²/h).

**18–104E** Reconsider Prob. 18–102E. Using EES (or other) software, plot the center temperature of each geometry as a function of the cooling time as the time varies from 5 min to 60 min, and discuss the results.

**18–105** Engine valves ($k = 48$ W/m·K, $c_p = 440$ J/kg·K, and $\rho = 7840$ kg/m³) are heated to 800°C in the heat treatment section of a valve manufacturing facility. The valves are then quenched in a large oil bath at an average temperature of 50°C. The heat transfer coefficient in the oil bath is 800 W/m²·K. The valves have a cylindrical stem with a diameter of 8 mm and a length of 10 cm. The valve head and the stem may be assumed to be of equal surface area, and the volume of the valve head can be taken to be 80 percent of the volume of stem. Determine how long it will take for the valve temperature to drop to (a) 400°C, (b) 200°C, and (c) 51°C, and (d) the maximum heat transfer from a single valve.

**18–106** Consider the engine block of a car made of cast iron ($k = 52$ W/m·K and $\alpha = 1.7 \times 10^{-5}$ m²/s). The engine can be considered to be a rectangular block whose sides are 80 cm, 40 cm, and 40 cm. The engine is at a temperature of 150°C when it is turned off. The engine is then exposed to atmospheric air at 17°C with a heat transfer coefficient of 6 W/m²·K. Determine (a) the center temperature of the top surface whose sides are 80 cm and 40 cm and (b) the corner temperature after 45 min of cooling.

**18–107** 8-cm-thick large food slabs tightly wrapped by thin paper are to be cooled in a refrigeration room maintained at 0°C. The heat transfer coefficient on the box surfaces is 25 W/m²·K and the boxes are to be kept in the refrigeration room for a period of 6 h. If the initial temperature of the boxes is 30°C determine the center temperature of the boxes if the boxes contain (a) margarine ($k = 0.233$ W/m·K and $\alpha = 0.11 \times 10^{-6}$ m²/s), (b) white cake ($k = 0.082$ W/m·K and $\alpha = 0.10 \times 10^{-6}$ m²/s), and (c) chocolate cake ($k = 0.106$ W/m·K and $\alpha = 0.12 \times 10^{-6}$ m²/s).

**18–108** Long aluminum wires of diameter 3 mm ($\rho = 2702$ kg/m³, $c_p = 0.896$ kJ/kg·K, $k = 236$ W/m·K, and $\alpha = 9.75 \times 10^{-5}$ m²/s) are extruded at a temperature of 350°C and exposed to atmospheric air at 30°C with a heat transfer coefficient of 35 W/m²·K. (a) Determine how long it will take for the wire temperature to drop to 50°C. (b) If the wire is extruded at a velocity of 10 m/min, determine how far the

**FIGURE P18–108**

wire travels after extrusion by the time its temperature drops to 50°C. What change in the cooling process would you propose to shorten this distance? (c) Assuming the aluminum wire leaves the extrusion room at 50°C, determine the rate of heat transfer from the wire to the extrusion room. *Answers: (a) 144 s, (b) 24 m, (c) 856 W*

**18–109** Repeat Prob. 18–108 for a copper wire ($\rho = 8950$ kg/m³, $c_p = 0.383$ kJ/kg·K, $k = 386$ W/m·K, and $\alpha = 1.13 \times 10^{-4}$ m²/s).

**18–110** A 40-cm-thick brick wall ($k = 0.72$ W/m·K, and $\alpha = 1.6 \times 10^{-6}$ m²/s) is heated to an average temperature of 18°C by the heating system and the solar radiation incident on it during the day. During the night, the outer surface of the wall is exposed to cold air at $-3$°C with an average heat transfer coefficient of 20 W/m²·K, determine the wall temperatures at distances 15, 30, and 40 cm from the outer surface for a period of 2 h.

**18–111** Large steel plates 1.0-cm in thickness are quenched from 600°C to 100°C by submerging them in an oil reservoir held at 30°C. The average heat transfer coefficient for both faces of steel plates is 400 W/m²·K. Average steel properties are $k = 45$ W/m·K, $\rho = 7800$ kg/m³, and $c_p = 470$ J/kg·K. Calculate the quench time for steel plates.

**18–112** Aluminium wires, 4 mm in diameter, are produced by extrusion. The wires leave the extruder at an average temperature of 350°C and at a linear rate of 10 m/min. Before leaving the extrusion room, the wires are cooled to an average temperature of 50°C by transferring heat to the surrounding air at 25°C with a heat transfer coefficient of 50 W/m²·K. Calculate the necessary length of the wire cooling section in the extrusion room.

**18–113** Two metal rods are being heated in an oven with uniform ambient temperature of 1000°C and convection heat transfer coefficient of 25 W/m²·K. Rod A is made of aluminum ($\rho = 2702$ kg/m³, $c_p = 903$ J/kg·K, and $k = 237$ W/m·K) and rod B is made of stainless steel ($\rho = 8238$ kg/m³, $c_p = 468$ J/kg·K, and $k = 13.4$ W/m·K). Both rods have diameter of 25 mm and length of 1 m. If the initial temperature of both rods is 15°C, determine the average temperatures of both rods after 5 minutes elapsed time.

**18–114** In an event of a volcano eruption, lava at 1200°C is found flowing on the ground. The ground was initially at 15°C and the lava flow has a convection heat transfer coefficient of 3500 W/m²·K. Determine the ground surface (a) temperature and (b) heat flux after 2 s of lava flow. Assume the ground has properties of dry soil. *Answers: (a) 983°C, (b) 759.5 kW/m²*

**18–115** A large heated steel block ($\rho = 7832$ kg/m³, $c_p = 434$ J/kg·K, $k = 63.9$ W/m·K, and $\alpha = 18.8 \times 10^{-6}$ m²/s) is allowed to cool in a room at 25°C. The steel block has an initial temperature of 450°C and the convection heat transfer coefficient is 25 W/m²·K. Assuming that the steel block can be treated as a quarter-infinite medium, determine the temperature at the edge of the steel block after 10 minutes of cooling.

## Design and Essay Problems

**18–116** Conduct the following experiment at home to determine the combined convection and radiation heat transfer coefficient at the surface of an apple exposed to the room air. You will need two thermometers and a clock.

First, weigh the apple and measure its diameter. You can measure its volume by placing it in a large measuring cup halfway filled with water, and measuring the change in volume when it is completely immersed in the water. Refrigerate the apple overnight so that it is at a uniform temperature in the morning and measure the air temperature in the kitchen. Then take the apple out and stick one of the thermometers to its middle and the other just under the skin. Record both temperatures every 5 min for an hour. Using these two temperatures, calculate the heat transfer coefficient for each interval and take their average. The result is the combined convection and radiation heat transfer coefficient for this heat transfer process. Using your experimental data, also calculate the thermal conductivity and thermal diffusivity of the apple and compare them to the values given above.

**18–117** Repeat Prob. 18–116 using a banana instead of an apple. The thermal properties of bananas are practically the same as those of apples.

**18–118** Conduct the following experiment to determine the time constant for a can of soda and then predict the temperature of the soda at different times. Leave the soda in the refrigerator overnight. Measure the air temperature in the kitchen and the temperature of the soda while it is still in the refrigerator by taping the sensor of the thermometer to the outer surface of the can. Then take the soda out and measure its temperature again in 5 min. Using these values, calculate the exponent b. Using this b-value, predict the temperatures of the soda in 10, 15, 20, 30, and 60 min and compare the results with the actual temperature measurements. Do you think the lumped system analysis is valid in this case?

# FORCED CONVECTION

So far, we have considered *conduction*, which is the mechanism of heat transfer through a solid or a quiescent fluid. We now consider *convection*, which is the mechanism of heat transfer through a fluid in the presence of bulk fluid motion.

Convection is classified as *natural* (or *free*) and *forced convection*, depending on how the fluid motion is initiated. In forced convection, the fluid is forced to flow over a surface or in a pipe by external means such as a pump or a fan. In natural convection, any fluid motion is caused by natural means such as the buoyancy effect, which manifests itself as the rise of warmer fluid and the fall of the cooler fluid. Convection is also classified as *external* and *internal*, depending on whether the fluid is forced to flow over a surface or in a pipe.

We start this chapter with a general physical description of the *convection* mechanism and *thermal boundary* layer. We continue with the discussion of the dimensionless *Prandtl* and *Nusselt numbers*, and their physical significance. We then present empirical relations for the *heat transfer coefficients* for flow over various geometries such as a flat plate, cylinder, and sphere, for both laminar and turbulent flow conditions. Finally, we discuss the characteristics of flow inside tubes and present the heat transfer correlations associated with it. The relevant concepts from Chaps. 14 and 15 should be reviewed before this chapter is studied.

## Objectives

The objectives of this chapter are to:

- Understand the physical mechanism of convection and its classification.

- Visualize the development of thermal boundary layer during flow over surfaces.

- Gain a working knowledge of the dimensionless Prandtl and Nusselt numbers.

- Develop an understanding of the mechanism of heat transfer in turbulent flow.

- Evaluate the heat transfer associated with flow over a flat plate for both laminar and turbulent flow, and flow over cylinders and spheres.

- Have a visual understanding of different flow regions in internal flow, and calculate hydrodynamic and thermal entry lengths.

- Analyze heating and cooling of a fluid flowing in a tube under constant surface temperature and constant surface heat flux conditions, and work with the logarithmic mean temperature difference.

- Determine the Nusselt number in fully developed turbulent flow using empirical relations, and calculate the heat transfer rate.

(a) Forced convection

(b) Free convection

(c) Conduction

**FIGURE 19–1**

Heat transfer from a hot surface to the surrounding fluid by convection and conduction.

**FIGURE 19–2**

Heat transfer through a fluid sandwiched between two parallel plates.

# 19–1 ▪ PHYSICAL MECHANISM OF CONVECTION

We mentioned in Chap. 16 that there are three basic mechanisms of heat transfer: conduction, convection, and radiation. Conduction and convection are similar in that both mechanisms require the presence of a material medium. But they are different in that convection requires the presence of fluid motion.

Heat transfer through a solid is always by conduction, since the molecules of a solid remain at relatively fixed positions. Heat transfer through a liquid or gas, however, can be by conduction or convection, depending on the presence of any bulk fluid motion. Heat transfer through a fluid is by convection in the presence of bulk fluid motion and by conduction in the absence of it. Therefore, conduction in a fluid can be viewed as the limiting case of convection, corresponding to the case of quiescent fluid (Fig. 19–1).

Convection heat transfer is complicated by the fact that it involves fluid motion as well as heat conduction. The fluid motion enhances heat transfer, since it brings warmer and cooler chunks of fluid into contact, initiating higher rates of conduction at a greater number of sites in a fluid. Therefore, the rate of heat transfer through a fluid is much higher by convection than it is by conduction. In fact, the higher the fluid velocity, the higher the rate of heat transfer.

To clarify this point further, consider steady heat transfer through a fluid contained between two parallel plates maintained at different temperatures, as shown in Figure 19–2. The temperatures of the fluid and the plate are the same at the points of contact because of the continuity of temperature. Assuming no fluid motion, the energy of the hotter fluid molecules near the hot plate is transferred to the adjacent cooler fluid molecules. This energy is then transferred to the next layer of the cooler fluid molecules. This energy is then transferred to the next layer of the cooler fluid, and so on, until it is finally transferred to the other plate. This is what happens during conduction through a fluid. Now let us use a syringe to draw some fluid near the hot plate and inject it next to the cold plate repeatedly. You can imagine that this will speed up the heat transfer process considerably, since some energy is carried to the other side as a result of fluid motion.

Consider the cooling of a hot block with a fan blowing air over its top surface. We know that heat is transferred from the hot block to the surrounding cooler air, and the block eventually cools. We also know that the block cools faster if the fan is switched to a higher speed. Replacing air by water enhances the convection heat transfer even more.

Experience shows that convection heat transfer strongly depends on the fluid properties *dynamic viscosity* $\mu$, *thermal conductivity k*, *density* $\rho$, and *specific heat* $c_p$, as well as the *fluid velocity V*. It also depends on the *geometry* and the *roughness* of the solid surface, in addition to the *type of fluid flow* (such as being streamlined or turbulent). Thus, we expect the convection heat transfer relations to be rather complex because of the dependence of convection on so many variables. This is not surprising, since convection is the most complex mechanism of heat transfer.

Despite the complexity of convection, the rate of convection heat transfer is observed to be proportional to the temperature difference and is conveniently expressed by **Newton's law of cooling** as

$$\dot{q}_{\text{conv}} = h(T_s - T_\infty) \qquad (\text{W/m}^2) \qquad (19\text{–}1)$$

or

$$\dot{Q}_{conv} = hA_s(T_s - T_\infty) \qquad (W) \qquad (19\text{–}2)$$

where

$h$ = convection heat transfer coefficient, W/m²·°C

$A_s$ = heat transfer surface area, m²

$T_s$ = temperature of the surface, °C

$T_\infty$ = temperature of the fluid sufficiently far from the surface, °C

Judging from its units, the **convection heat transfer coefficient** $h$ can be defined as *the rate of heat transfer between a solid surface and a fluid per unit surface area per unit temperature difference.*

You should not be deceived by the simple appearance of this relation, because the convection heat transfer coefficient $h$ depends on the several of the mentioned variables, and thus is difficult to determine.

Fluid flow is often confined by solid surfaces, and it is important to understand how the presence of solid surfaces affects fluid flow. Consider the flow of a fluid in a stationary pipe or over a solid surface that is nonporous (i.e., impermeable to the fluid). All experimental observations indicate that a fluid in motion comes to a complete stop at the surface and assumes a zero velocity relative to the surface. That is, a fluid in direct contact with a solid "sticks" to the surface due to viscous effects, and there is no slip. This is known as the **no-slip condition**.

The photo in Fig. 19–3 obtained from a video clip clearly shows the evolution of a velocity gradient as a result of the fluid sticking to the surface of a blunt nose. The layer that sticks to the surface slows the adjacent fluid layer because of viscous forces between the fluid layers, which slows the next layer, and so on. Therefore, the no-slip condition is responsible for the development of the velocity profile. The flow region adjacent to the wall in which the viscous effects (and thus the velocity gradients) are significant is called the **boundary layer**. The fluid property responsible for the no-slip condition and the development of the boundary layer is *viscosity*.

A fluid layer adjacent to a moving surface has the same velocity as the surface. A consequence of the no-slip condition is that all velocity profiles must have zero values with respect to the surface at the points of contact between a fluid and a solid surface (Fig. 19–4). Another consequence of the no-slip condition is the *surface drag*, which is the force a fluid exerts on a surface in the flow direction.

An implication of the no-slip condition is that heat transfer from the solid surface to the fluid layer adjacent to the surface is by *pure conduction*, since the fluid layer is motionless, and can be expressed as

$$\dot{q}_{conv} = \dot{q}_{cond} = -k_{fluid} \left.\frac{\partial T}{\partial y}\right|_{y=0} \qquad (W/m^2) \qquad (19\text{–}3)$$

where $T$ represents the temperature distribution in the fluid and $(\partial T/\partial y)_{y=0}$ is the *temperature gradient* at the surface. Heat is then *convected away* from the surface as a result of fluid motion. Note that convection heat transfer from a solid surface to a fluid is merely the conduction heat transfer from

**FIGURE 19–3**

The development of a velocity profile due to the no-slip condition as a fluid flows over a blunt nose.

*"Hunter Rouse: Laminar and Turbulent Flow Film."*
*Copyright IIHR-Hydroscience & Engineering,*
*The University of Iowa. Used by permission.*

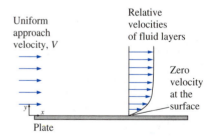

**FIGURE 19–4**

A fluid flowing over a stationary surface comes to a complete stop at the surface because of the no-slip condition.

the solid surface to the fluid layer adjacent to the surface. Therefore, we can equate Eqs. 19–1 and 19–3 for the heat flux to obtain

$$h = \frac{-k_{\text{fluid}}(\partial T/\partial y)_{y=0}}{T_s - T_\infty} \qquad (\text{W/m}^2\cdot{}^\circ\text{C}) \qquad (19\text{–}4)$$

for the determination of the *convection heat transfer coefficient* when the temperature distribution within the fluid is known.

The convection heat transfer coefficient, in general, varies along the flow (or *x*-) direction. The *average* or *mean* convection heat transfer coefficient for a surface in such cases is determined by properly averaging the *local* convection heat transfer coefficients over the entire surface area $A_s$ or length $L$ as:

$$h = \frac{1}{A_s}\int_{A_s} h_{\text{local}}dA_s \quad \text{and} \quad h = \frac{1}{L}\int_0^L h_x dx \qquad (19\text{–}5)$$

## Nusselt Number

In convection studies, it is common practice to nondimensionalize the governing equations and combine the variables, which group together into *dimensionless numbers* in order to reduce the number of total variables. It is also common practice to nondimensionalize the heat transfer coefficient $h$ with the Nusselt number, defined as

$$\text{Nu} = \frac{hL_c}{k} \qquad (19\text{–}6)$$

where $k$ is the thermal conductivity of the fluid and $L_c$ is the *characteristic length*. The Nusselt number is named after Wilhelm Nusselt, who made significant contributions to convective heat transfer in the first half of the twentieth century, and it is viewed as the *dimensionless convection heat transfer coefficient*.

To understand the physical significance of the Nusselt number, consider a fluid layer of thickness $L$ and temperature difference $\Delta T = T_2 - T_1$, as shown in Fig. 19–5. Heat transfer through the fluid layer is by *convection* when the fluid involves some motion and by *conduction* when the fluid layer is motionless. Heat flux (the rate of heat transfer per unit surface area) in either case is

$$\dot{q}_{\text{conv}} = h\Delta T \qquad (19\text{–}7)$$

and

$$\dot{q}_{\text{cond}} = k\frac{\Delta T}{L} \qquad (19\text{–}8)$$

Taking their ratio gives

$$\frac{\dot{q}_{\text{conv}}}{\dot{q}_{\text{cond}}} = \frac{h\Delta T}{k\Delta T/L} = \frac{hL}{k} = \text{Nu} \qquad (19\text{–}9)$$

which is the Nusselt number. Therefore, the Nusselt number represents the enhancement of heat transfer through a fluid layer as a result of convection relative to conduction across the same fluid layer. The larger the Nusselt number, the more effective the convection. A Nusselt number of Nu = 1 for a fluid layer represents heat transfer across the layer by pure conduction.

$\Delta T = T_2 - T_1$

**FIGURE 19–5**

Heat transfer through a fluid layer of thickness $L$ and temperature difference $\Delta T$.

We use forced convection in daily life more often than you might think (Fig. 19–6). We resort to forced convection whenever we want to increase the rate of heat transfer from a hot object. For example, we turn on the fan on hot summer days to help our body cool more effectively. The higher the fan speed, the better we feel. We *stir* our soup and *blow* on a hot slice of pizza to make them cool faster. The air on windy winter days feels much colder than it actually is. The simplest solution to heating problems in electronics packaging is to use a large enough fan.

Blowing on food

**FIGURE 19–6**
We resort to forced convection whenever we need to increase the rate of heat transfer.

## 19–2 ▪ THERMAL BOUNDARY LAYER

We have seen in Chap. 15 that a velocity boundary layer develops when a fluid flows over a surface as a result of the fluid layer adjacent to the surface assuming the surface velocity (i.e., zero velocity relative to the surface). Also, we defined the velocity boundary layer as the region in which the fluid velocity varies from zero to $0.99V$. Likewise, a *thermal boundary layer* develops when a fluid at a specified temperature flows over a surface that is at a different temperature, as shown in Fig. 19–7.

Consider the flow of a fluid at a uniform temperature of $T_\infty$ over an isothermal flat plate at temperature $T_s$. The fluid particles in the layer adjacent to the surface reach thermal equilibrium with the plate and assume the surface temperature $T_s$. These fluid particles then exchange energy with the particles in the adjoining-fluid layer, and so on. As a result, a temperature profile develops in the flow field that ranges from $T_s$ at the surface to $T_\infty$ sufficiently far from the surface. The flow region over the surface in which the temperature variation in the direction normal to the surface is significant is the **thermal boundary layer**. The *thickness* of the thermal boundary layer $\delta_t$ at any location along the surface is defined as *the distance from the surface at which the temperature difference $T - T_s$ equals $0.99(T_\infty - T_s)$.* Note that for the special case of $T_s = 0$, we have $T = 0.99T_\infty$ at the outer edge of the thermal boundary layer, which is analogous to $u = 0.99V$ for the velocity boundary layer.

The thickness of the thermal boundary layer increases in the flow direction, since the effects of heat transfer are felt at greater distances from the surface further down stream.

The convection heat transfer rate anywhere along the surface is directly related to the temperature gradient at that location. Therefore, the shape of the temperature profile in the thermal boundary layer dictates the convection heat transfer between a solid surface and the fluid flowing over it. In flow over a heated (or cooled) surface, both velocity and thermal boundary layers develop simultaneously. Noting that the fluid velocity has a strong influence on the temperature profile, the development of the velocity boundary layer relative to the thermal boundary layer will have a strong effect on the convection heat transfer.

**FIGURE 19–7**
Thermal boundary layer on a flat plate (the fluid is hotter than the plate surface).

## Prandtl Number

The relative thickness of the velocity and the thermal boundary layers is best described by the *dimensionless* parameter **Prandtl number**, defined as

$$Pr = \frac{\text{Molecular diffusivity of momentum}}{\text{Molecular diffusivity of heat}} = \frac{\nu}{\alpha} = \frac{\mu c_p}{k} \qquad \text{(19–10)}$$

It is named after Ludwig Prandtl, who introduced the concept of boundary layer in 1904 and made significant contributions to boundary layer theory. The Prandtl numbers of fluids range from less than 0.01 for liquid metals to more than 100,000 for heavy oils (Table 19–1). Note that the Prandtl number is in the order of 10 for water.

The Prandtl numbers of gases are about 1, which indicates that both momentum and heat dissipate through the fluid at about the same rate. Heat diffuses very quickly in liquid metals ($Pr \ll 1$) and very slowly in oils ($Pr \gg 1$) relative to momentum. Consequently the thermal boundary layer is much thicker for liquid metals and much thinner for oils relative to the velocity boundary layer.

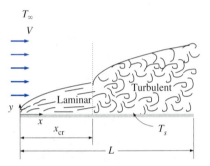

**FIGURE 19–8**

Laminar and turbulent regions of the boundary layer during flow over a flat plate.

## 19–3 ■ PARALLEL FLOW OVER FLAT PLATES

Consider the parallel flow of a fluid over a flat plate of length $L$ in the flow direction, as shown in Fig. 19–8. The $x$-coordinate is measured along the plate surface from the leading edge in the direction of the flow. The fluid approaches the plate in the $x$-direction with a uniform velocity $V$ and temperature $T_\infty$. The flow in the velocity boundary layers starts out as laminar, but if the plate is sufficiently long, the flow becomes turbulent at a distance $x_{cr}$ from the leading edge where the Reynolds number reaches its critical value for transition.

The transition from laminar to turbulent flow depends on the *surface geometry, surface roughness, upstream velocity, surface temperature,* and the *type of fluid,* among other things, and is best characterized by the Reynolds number. The Reynolds number at a distance $x$ from the leading edge of a flat plate is expressed as

$$Re_x = \frac{\rho V x}{\mu} = \frac{V x}{v} \tag{19–11}$$

Note that the value of the Reynolds number varies for a flat plate along the flow, reaching $Re_L = VL/v$ at the end of the plate.

For flow over a flat plate, transition from laminar to turbulent begins at about $Re \cong 1 \times 10^5$, but does not become fully turbulent before the Reynolds number reaches much higher values, typically around $3 \times 10^6$. In engineering analysis, a generally accepted value for the critical Reynold number is

$$Re_{cr} = \frac{\rho V x_{cr}}{\mu} = 5 \times 10^5 \tag{19–12}$$

The actual value of the engineering critical Reynolds number for a flat plate may vary somewhat from $10^5$ to $3 \times 10^6$, depending on the surface roughness, the turbulence level, and the variation of pressure along the surface.

The local Nusselt number at a location $x$ for laminar flow over a flat plate may be obtained by solving the differential energy equation to be

*Laminar:*
$$Nu_x = \frac{h_x x}{k} = 0.332 \, Re_x^{0.5} \, Pr^{1/3} \qquad Pr > 0.6 \tag{19–13}$$

From experiments, the corresponding relation for turbulent flow is

*Turbulent:* $\quad \text{Nu}_x = \dfrac{h_x x}{k} = 0.0296\, \text{Re}_x^{0.8}\, \text{Pr}^{1/3} \quad \begin{array}{l} 0.6 \le \text{Pr} \le 60 \\ 5 \times 10^5 \le \text{Re}_x \le 10^7 \end{array}$ $\quad$ **(19–14)**

Note that $h_x$ is proportional to $\text{Re}_x^{0.5}$ and thus to $x^{-0.5}$ for laminar flow. Therefore, $h_x$ is *infinite* at the leading edge ($x = 0$) and decreases by a factor of $x^{-0.5}$ in the flow direction. The variation of the boundary layer thickness $\delta$ and the friction and heat transfer coefficients along an isothermal flat plate are shown in Fig. 19–9. The local friction and heat transfer coefficients are higher in turbulent flow than they are in laminar flow. Also, $h_x$ reaches its highest values when the flow becomes fully turbulent, and then decreases by a factor of $x^{-0.2}$ in the flow direction, as shown in the figure.

The *average* Nusselt number over the entire plate is determined by substituting the relations above into Eq. 19–5 and performing the integrations. We get

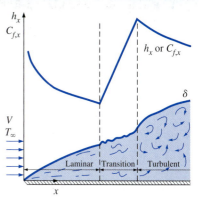

**FIGURE 19–9**

The variation of the local friction and heat transfer coefficients for flow over a flat plate.

*Laminar:* $\quad \text{Nu} = \dfrac{hL}{k} = 0.664\, \text{Re}_L^{0.5}\, \text{Pr}^{1/3} \quad \text{Re}_L < 5 \times 10^5$ $\quad$ **(19–15)**

*Turbulent:* $\quad \text{Nu} = \dfrac{hL}{k} = 0.037\, \text{Re}_L^{0.8}\, \text{Pr}^{1/3} \quad \begin{array}{l} 0.6 \le \text{Pr} \le 60 \\ 5 \times 10^5 \le \text{Re}_L \le 10^7 \end{array}$ $\quad$ **(19–16)**

The first relation gives the average heat transfer coefficient for the entire plate when the flow is *laminar* over the *entire* plate. The second relation gives the average heat transfer coefficient for the entire plate only when the flow is *turbulent* over the *entire* plate, or when the laminar flow region of the plate is too small relative to the turbulent flow region.

In some cases, a flat plate is sufficiently long for the flow to become turbulent, but not long enough to disregard the laminar flow region. In such cases, the *average* heat transfer coefficient over the entire plate is determined by performing the integration in Eq. 19–5 over two parts as

$$h = \frac{1}{L}\left( \int_0^{x_{cr}} h_{x,\,\text{laminar}}\, dx + \int_{x_{cr}}^{L} h_{x,\,\text{turbulent}}\, dx \right) \quad \textbf{(19–17)}$$

Again taking the critical Reynolds number to be $\text{Re}_{cr} = 5 \times 10^5$ and performing the integrations in Eq. 19–17 after substituting the indicated expressions, the *average* Nusselt number over the *entire* plate is determined to be (Fig. 19–10)

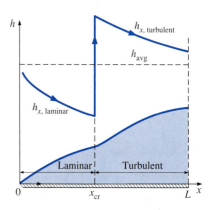

**FIGURE 19–10**

Graphical representation of the average heat transfer coefficient for a flat plate with combined laminar and turbulent flow.

$$\text{Nu} = \frac{hL}{k} = (0.037\, \text{Re}_L^{0.8} - 871)\text{Pr}^{1/3} \quad \begin{array}{l} 0.6 \le \text{Pr} \le 60 \\ 5 \times 10^5 \le \text{Re}_L \le 10^7 \end{array} \quad \textbf{(19–18)}$$

The constants in this relation will be different for different critical Reynolds numbers.

*Liquid metals* such as mercury have high thermal conductivities, and are commonly used in applications that require high heat transfer rates. However, they have very small Prandtl numbers, and thus the thermal boundary layer develops much faster than the velocity boundary layer. Then we can assume the velocity in the thermal boundary layer to be constant at the free stream value and solve the energy equation. It gives

$$\text{Nu}_x = 0.565(\text{Re}_x\, \text{Pr})^{1/2} \quad \text{Pr} < 0.05 \quad \textbf{(19–19)}$$

It is desirable to have a single correlation that applies to *all fluids*, including liquid metals. By curve-fitting existing data, Churchill and Ozoe (1973) proposed the following relation which is applicable for *all Prandtl numbers* and is claimed to be accurate to $\pm 1\%$,

$$\text{Nu}_x = \frac{h_x x}{k} = \frac{0.3387 \, \text{Pr}^{1/3} \, \text{Re}_x^{1/2}}{[1 + (0.0468/\text{Pr})^{2/3}]^{1/4}} \qquad (19\text{--}20)$$

These relations have been obtained for the case of *isothermal* surfaces but could also be used approximately for the case of nonisothermal surfaces by assuming the surface temperature to be constant at some average value. Also, the surfaces are assumed to be *smooth,* and the free stream to be *turbulent free.* The effect of variable properties can be accounted for by evaluating all properties at the film temperature.

## Flat Plate with Unheated Starting Length

So far we have limited our consideration to situations for which the entire plate is heated from the leading edge. But many practical applications involve surfaces with an unheated starting section of length $\xi$, shown in Fig. 19–11, and thus there is no heat transfer for $0 < x < \xi$. In such cases, the velocity boundary layer starts to develop at the leading edge ($x = 0$), but the thermal boundary layer starts to develop where heating starts ($x = \xi$).

Consider a flat plate whose heated section is maintained at a constant temperature ($T = T_s$ constant for $x > \xi$). Using integral solution methods (see Kays and Crawford, 1994), the local Nusselt numbers for both laminar and turbulent flows are determined to be

*Laminar:* $$\text{Nu}_x = \frac{\text{Nu}_{x \, (\text{for } \xi=0)}}{[1 - (\xi/x)^{3/4}]^{1/3}} = \frac{0.332 \, \text{Re}_x^{0.5} \, \text{Pr}^{1/3}}{[1 - (\xi/x)^{3/4}]^{1/3}} \qquad (19\text{--}21)$$

*Turbulent:* $$\text{Nu}_x = \frac{\text{Nu}_{x \, (\text{for } \xi=0)}}{[1 - (\xi/x)^{9/10}]^{1/9}} = \frac{0.0296 \, \text{Re}_x^{0.8} \, \text{Pr}^{1/3}}{[1 - (\xi/x)^{9/10}]^{1/9}} \qquad (19\text{--}22)$$

for $x > \xi$. Note that for $\xi = 0$, these $\text{Nu}_x$ relations reduce to $\text{Nu}_{x \, (\text{for } \xi=0)}$, which is the Nusselt number relation for a flat plate without an unheated starting length. Therefore, the terms in brackets in the denominator serve as correction factors for plates with unheated starting lengths.

The determination of the average Nusselt number for the heated section of a plate requires the integration of the local Nusselt number relations above, which cannot be done analytically. Therefore, integrations must be done numerically. The results of numerical integrations have been correlated for the average convection coefficients (Thomas, 1977) as

*Laminar:* $$h = \frac{2[1 - (\xi/x)^{3/4}]}{1 - \xi/L} \, h_{x=L} \qquad (19\text{--}23)$$

*Turbulent:* $$h = \frac{5[1 - (\xi/x)^{9/10}]}{4(1 - \xi/L)} \, h_{x=L} \qquad (19\text{--}24)$$

The first relation gives the average convection coefficient for the entire heated section of the plate when the flow is laminar over the entire plate.

$T_\infty$
$V$

Thermal boundary layer

Velocity boundary layer

$T_s$

$\xi$

$x$

**FIGURE 19–11**
Flow over a flat plate with an unheated starting length.

Note that for $\xi = 0$ it reduces to $h_L = 2h_{x = L}$, as expected. The second relation gives the average convection coefficient for the case of turbulent flow over the entire plate or when the laminar flow region is small relative to the turbulent region.

## Uniform Heat Flux

When a flat plate is subjected to *uniform heat flux* instead of uniform temperature, the local Nusselt number is given by

*Laminar*:
$$\text{Nu}_x = 0.453\,\text{Re}_x^{0.5}\,\text{Pr}^{1/3} \tag{19–25}$$

*Turbulent*:
$$\text{Nu}_x = 0.0308\,\text{Re}_x^{0.8}\,\text{Pr}^{1/3} \tag{19–26}$$

These relations give values that are 36 percent higher for laminar flow and 4 percent higher for turbulent flow relative to the isothermal plate case. When the plate involves an unheated starting length, the relations developed for the uniform surface temperature case can still be used provided that Eqs. 19–25 and 19–26 are used for $\text{Nu}_{x(\text{for }\xi=0)}$ in Eqs. 19–21 and 19–22, respectively.

When heat flux $\dot{q}_s$ is prescribed, the rate of heat transfer to or from the plate and the surface temperature at a distance $x$ are determined from

$$\dot{Q} = \dot{q}_s A_s \tag{19–27}$$

and

$$\dot{q}_s = h_x[T_s(x) - T_\infty] \longrightarrow T_s(x) = T_\infty + \frac{\dot{q}_s}{h_x} \tag{19–28}$$

where $A_s$ is the heat transfer surface area.

---

### EXAMPLE 19–1    Defrosting Ice from an Automobile Windshield

To defrost ice accumulated on the outer surface of an automobile windshield, warm air is blown over the inner surface of the windshield. Consider an automobile windshield ($k_w = 1.4$ W/m·K) with an overall height of 0.5 m and thickness of 5 mm (Fig. 19–12). The outside air (1 atm) ambient temperature is −20°C and the average airflow velocity over the outer windshield surface is 80 km/h, while the ambient temperature inside the automobile is 25°C. Determine the value of the convection heat transfer coefficient, for the warm air blowing over the inner surface of the windshield, necessary to cause the accumulated ice to begin melting. Assume the windshield surface can be treated as a flat plate surface.

**SOLUTION**   Warm air blowing over the inner surface of a windshield is used for defrosting. The required inner convection heat transfer coefficient to cause the accumulated ice to begin melting is to be determined.

**Assumptions**   **1** Steady operating conditions exist. **2** Heat transfer through the windshield is one-dimensional. **3** Thermal properties are constant. **4** Heat transfer by radiation is negligible. **5** The outside air pressure is 1 atm. **6** The critical Reynolds number is $\text{Re}_{cr} = 5 \times 10^5$.

**FIGURE 19–12**
Schematic for Example 19–1.

**Properties** The properties of air at the film temperature of $T_f = (-20°C + 0°C)/2 = -10°C$ are $k = 0.02288$ W/m·K, $v = 1.252 \times 10^{-5}$ m²/s, and Pr = 0.7387 (from Table A–22).

**Analysis** On the outer surface of the windshield, the Reynolds number at $L = 0.5$ m is

$$\text{Re}_L = \frac{VL}{v} = \frac{(80/3.60 \, \text{m/s})(0.5 \, \text{m})}{1.252 \times 10^{-5} \, \text{m}^2/\text{s}} = 8.875 \times 10^5$$

Since $5 \times 10^5 < \text{Re}_L < 10^7$ the flow is a combined laminar and turbulent flow. Using the proper relation for Nusselt number, the average heat transfer coefficient on the outer surface of the windshield is

$$\text{Nu}_o = \frac{h_o L}{k} = (0.037 \text{Re}_L^{0.8} - 871)\text{Pr}^{1/3} = [0.037(8.875 \times 10^5)^{0.8} - 871](0.7387)^{1/3} = 1131$$

$$h_o = 1131 \frac{k}{L} = 1131 \frac{0.02288 \, \text{W/m·K}}{0.5 \, \text{m}} = 51.75 \, \text{W/m}^2\text{·K}$$

From energy balance, the heat transfer through the windshield thickness can be written as

$$\frac{T_{\infty,o} - T_{s,o}}{1/h_o} = \frac{T_{s,o} - T_{\infty,i}}{t/k_w + 1/h_i}$$

For the ice to begin melting, the outer surface temperature of the windshield ($T_{s,o}$) should be at least 0°C. Then the convection heat transfer coefficient for the warm air blowing over the inner surface of the windshield must be

$$h_i = \left( \frac{1}{h_o} \frac{T_{s,o} - T_{\infty,i}}{T_{\infty,o} - T_{s,o}} - \frac{t}{k_w} \right)^{-1}$$

$$= \left[ \frac{(0 - 25)°C}{(-20-0)°C} \left( \frac{1}{51.75 \, \text{W/m}^2\text{·K}} \right) - \frac{0.005 \, \text{m}}{1.4 \, \text{W/m·K}} \right]^{-1}$$

$$= \textbf{48.6 W/m}^2\textbf{·K}$$

**Discussion** In practical situations the ambient temperature and the convective heat transfer coefficient outside the automobile vary with weather conditions and the automobile speed. Therefore, the convection heat transfer coefficient of the warm air necessary to melt the ice should be varied as well. This is done by adjusting the warm air flow rate and temperature.

**FIGURE 19–13**
Schematic for Example 19–2.

**EXAMPLE 19–2    Cooling of Plastic Sheets by Forced Air**

The forming section of a plastics plant puts out a continuous sheet of plastic that is 4 ft wide and 0.04 in thick at a velocity of 30 ft/min. The temperature of the plastic sheet is 200°F when it is exposed to the surrounding air, and a 2-ft-long section of the plastic sheet is subjected to air flow at 80°F at a velocity of 10 ft/s on both sides along its surfaces normal to the direction of motion of the sheet, as shown in Fig. 19–13. Determine (a) the rate of heat transfer from the plastic sheet to air by forced convection and radiation and (b) the temperature of the plastic sheet at the end of the cooling section. Take the density, specific heat, and emissivity of the plastic sheet to be $\rho = 75$ lbm/ft³, $c_p = 0.4$ Btu/lbm·°F, and $\varepsilon = 0.9$.

**SOLUTION** Plastic sheets are cooled as they leave the forming section of a plastics plant. The rate of heat loss from the plastic sheet by convection and radiation and the exit temperature of the plastic sheet are to be determined.

*Assumptions* **1** Steady operating conditions exist. **2** The critical Reynolds number is $Re_{cr} = 5 \times 10^5$. **3** Air is an ideal gas. **4** The local atmospheric pressure is 1 atm. **5** The surrounding surfaces are at the temperature of the room air.

*Properties* The properties of the plastic sheet are given in the problem statement. The properties of air at the film temperature of $T_f = (T_s + T_\infty)/2 = (200 + 80)/2 = 140°F$ and 1 atm pressure are (Table A–22E)

$$k = 0.01623 \text{ Btu/h·ft·°F} \qquad Pr = 0.7202$$
$$\nu = 0.204 \times 10^{-3} \text{ ft}^2\text{/s}$$

*Analysis* (*a*) We expect the temperature of the plastic sheet to drop somewhat as it flows through the 2-ft-long cooling section, but at this point we do not know the magnitude of that drop. Therefore, we assume the plastic sheet to be isothermal at 200°F to get started. We will repeat the calculations if necessary to account for the temperature drop of the plastic sheet.

Noting that $L = 4$ ft, the Reynolds number at the end of the air flow across the plastic sheet is

$$Re_L = \frac{VL}{\nu} = \frac{(10 \text{ ft/s})(4 \text{ ft})}{0.204 \times 10^{-3} \text{ ft}^2\text{/s}} = 1.961 \times 10^5$$

which is less than the critical Reynolds number. Thus, we have *laminar flow* over the entire sheet, and the Nusselt number is determined from the laminar flow relations for a flat plate to be

$$Nu = \frac{hL}{k} = 0.664 \, Re_L^{0.5} \, Pr^{1/3} = 0.664 \times (1.961 \times 10^5)^{0.5} \times (0.7202)^{1/3} = 263.6$$

Then,

$$h = \frac{k}{L} Nu = \frac{0.01623 \text{ Btu/h·ft·°F}}{4 \text{ ft}} (263.6) = 1.07 \text{ Btu/h·ft}^2\text{·°F}$$
$$A_s = (2 \text{ ft})(4 \text{ ft})(2 \text{ sides}) = 16 \text{ ft}^2$$

and

$$\dot{Q}_{conv} = hA_s(T_s - T_\infty)$$
$$= (1.07 \text{ Btu/h·ft}^2\text{·°F})(16 \text{ ft}^2)(200 - 80)°F$$
$$= 2054 \text{ Btu/h}$$
$$\dot{Q}_{rad} = \varepsilon\sigma A_s(T_s^4 - T_{surr}^4)$$
$$= (0.9)(0.1714 \times 10^{-8} \text{ Btu/h·ft}^2\text{·R}^4)(16 \text{ ft}^2)[(660 \text{ R})^4 - (540 \text{ R})^4]$$
$$= 2585 \text{ Btu/h}$$

Therefore, the rate of cooling of the plastic sheet by combined convection and radiation is

$$\dot{Q}_{total} = \dot{Q}_{conv} + \dot{Q}_{rad} = 2054 + 2585 = \textbf{4639 Btu/h}$$

(b) To find the temperature of the plastic sheet at the end of the cooling section, we need to know the mass of the plastic rolling out per unit time (or the mass flow rate), which is determined from

$$\dot{m} = r A_c V_{\text{plastic}} = (75 \text{ lbm/ft}^3)\left(\frac{4 \times 0.04}{12} \text{ ft}^2\right)\left(\frac{30}{60} \text{ ft/s}\right) = 0.5 \text{ lbm/s}$$

Then, an energy balance on the cooled section of the plastic sheet yields

$$\dot{Q} = \dot{m} c_p (T_2 - T_1) \rightarrow T_2 = T_1 + \frac{\dot{Q}}{\dot{m} c_p}$$

Noting that $\dot{Q}$ is a negative quantity (heat loss) for the plastic sheet and substituting, the temperature of the plastic sheet as it leaves the cooling section is determined to be

$$T_2 = 200°\text{F} + \frac{-4639 \text{ Btu/h}}{(0.5 \text{ lbm/s})(0.4 \text{ Btu/lbm}\cdot°\text{F})}\left(\frac{1 \text{ h}}{3600 \text{ s}}\right) = \mathbf{193.6°F}$$

**DISCUSSION** The average temperature of the plastic sheet drops by about 6.4°F as it passes through the cooling section. The calculations now can be repeated by taking the average temperature of the plastic sheet to be 196.8°F instead of 200°F for better accuracy, but the change in the results will be insignificant because of the small change in temperature.

**FIGURE 19–14**

Variation of the local heat transfer coefficient along the circumference of a circular cylinder in cross flow of air.

*From Giedt, 1949.*

## 19–4 ■ FLOW ACROSS CYLINDERS AND SPHERES

Flows across cylinders and spheres, in general, involve *flow separation*, which is difficult to handle analytically. Therefore, such flows must be studied experimentally or numerically. Indeed, flow across cylinders and spheres has been studied experimentally by numerous investigators, and several empirical correlations have been developed for the heat transfer coefficient.

The complicated flow pattern across a cylinder greatly influences heat transfer. The variation of the local Nusselt number $Nu_\theta$ around the periphery of a cylinder subjected to cross flow of air is given in Fig. 19–14. Note that, for all cases, the value of $Nu_\theta$ starts out relatively high at the stagnation point ($\theta = 0°$) but decreases with increasing $\theta$ as a result of the thickening of the laminar boundary layer. On the two curves at the bottom corresponding to Re = 70,800 and 101,300, $Nu_\theta$ reaches a minimum at $\theta \approx 80°$, which is the separation point in laminar flow. Then $Nu_\theta$ increases with increasing $\theta$ as a result of the intense mixing in the separated flow region (the wake). The curves at the top corresponding to Re = 140,000 to 219,000 differ from the first two curves in that they have *two* minima for $Nu_\theta$. The sharp increase in $Nu_\theta$ at about $\theta \approx 90°$ is due to the transition from laminar to turbulent flow. The later decrease in $Nu_\theta$ is again due to the thickening of the boundary layer. $Nu_\theta$ reaches its second minimum at about $\theta \approx 140°$, which is the flow separation point in turbulent flow, and increases with $\theta$ as a result of the intense mixing in the turbulent wake region.

The discussions above on the local heat transfer coefficients are insightful; however, they are of limited value in heat transfer calculations since the calculation of heat transfer requires the *average* heat transfer coefficient over the entire surface. Of the several such relations available in the literature for the average Nusselt number for cross flow over a cylinder, we present the one proposed by Churchill and Bernstein (1977):

$$Nu_{cyl} = \frac{hD}{k} = 0.3 + \frac{0.62\,Re^{1/2}\,Pr^{1/3}}{[1 + (0.4/Pr)^{2/3}]^{1/4}}\left[1 + \left(\frac{Re}{282{,}000}\right)^{5/8}\right]^{4/5} \tag{19-29}$$

This relation is quite comprehensive in that it correlates available data well for $RePr > 0.2$. The fluid properties are evaluated at the *film temperature* $T_f = \frac{1}{2}(T_\infty + T_s)$, which is the average of the free-stream and surface temperatures.

For flow over a *sphere*, Whitaker (1972) recommends the following comprehensive correlation:

$$Nu_{sph} = \frac{hD}{k} = 2 + [0.4\,Re^{1/2} + 0.06\,Re^{2/3}]\,Pr^{0.4}\left(\frac{\mu_\infty}{\mu_s}\right)^{1/4} \tag{19-30}$$

which is valid for $3.5 \le Re \le 80{,}000$ and $0.7 \le Pr \le 380$. The fluid properties in this case are evaluated at the free-stream temperature $T_\infty$, except for $\mu_s$, which is evaluated at the surface temperature $T_s$. Although the two relations above are considered to be quite accurate, the results obtained from them can be off by as much as 30 percent.

The average Nusselt number for flow across cylinders can be expressed compactly as

$$Nu_{cyl} = \frac{hD}{k} = C\,Re^m\,Pr^n \tag{19-31}$$

where $n = \frac{1}{3}$ and the experimentally determined constants $C$ and $m$ are given in Table 19–2 for circular as well as various noncircular cylinders. The characteristic length $D$ for use in the calculation of the Reynolds and the Nusselt numbers for different geometries is as indicated on the figure. All fluid properties are evaluated at the film temperature.

The relations for cylinders above are for *single* cylinders or cylinders oriented such that the flow over them is not affected by the presence of others. Also, they are applicable to *smooth* surfaces. *Surface roughness* and the *free-stream turbulence* may affect the drag and heat transfer coefficients significantly. Eq. 19–31 provides a simpler alternative to Eq. 19–29 for flow over cylinders. However, Eq. 19–29 is more accurate, and thus should be preferred in calculations whenever possible.

## TABLE 19–2

Empirical correlations for the average Nusselt number for forced convection over circular and noncircular cylinders in cross flow (from Zukauskas, 1972 and Jakob, 1949)

| Cross-section of the cylinder | Fluid | Range of Re | Nusselt number |
|---|---|---|---|
| Circle | Gas or liquid | 0.4–4<br>4–40<br>40–4000<br>4000–40,000<br>40,000–400,000 | $Nu = 0.989Re^{0.330} Pr^{1/3}$<br>$Nu = 0.911Re^{0.385} Pr^{1/3}$<br>$Nu = 0.683Re^{0.466} Pr^{1/3}$<br>$Nu = 0.193Re^{0.618} Pr^{1/3}$<br>$Nu = 0.027Re^{0.805} Pr^{1/3}$ |
| Square | Gas | 5000–100,000 | $Nu = 0.102Re^{0.675} Pr^{1/3}$ |
| Square (tilted 45°) | Gas | 5000–100,000 | $Nu = 0.246Re^{0.588} Pr^{1/3}$ |
| Hexagon | Gas | 5000–100,000 | $Nu = 0.153Re^{0.638} Pr^{1/3}$ |
| Hexagon (tilted 45°) | Gas | 5000–19,500<br>19,500–100,000 | $Nu = 0.160Re^{0.638} Pr^{1/3}$<br>$Nu = 0.0385Re^{0.782} Pr^{1/3}$ |
| Vertical plate | Gas | 4000–15,000 | $Nu = 0.228Re^{0.731} Pr^{1/3}$ |
| Ellipse | Gas | 2500–15,000 | $Nu = 0.248Re^{0.612} Pr^{1/3}$ |

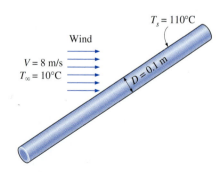

$T_s = 110°C$

Wind

$V = 8$ m/s
$T_\infty = 10°C$

$D = 0.1$ m

**FIGURE 19–15**
Schematic for Example 19–3.

## EXAMPLE 19–3  Heat Loss from a Steam Pipe in Windy Air

A long 10-cm-diameter steam pipe whose external surface temperature is 110°C passes through some open area that is not protected against the winds (Fig. 19–15). Determine the rate of heat loss from the pipe per unit of its length when the air is at 1 atm pressure and 10°C and the wind is blowing across the pipe at a velocity of 8 m/s.

**Solution** A steam pipe is exposed to windy air. The rate of heat loss from the steam is to be determined.

**Assumptions** 1 Steady operating conditions exist. 2 Radiation effects are negligible. 3 Air is an ideal gas.

**Properties** The properties of air at the average film temperature of $T_f = (T_s + T_\infty)/2 = (110 + 10)/2 = 60°C$ and 1 atm pressure are (Table A–22)

$$k = 0.02808 \text{ W/m·°C} \qquad \text{Pr} = 0.7202$$
$$\nu = 1.896 \times 10^{-5} \text{ m}^2/\text{s}$$

**Analysis** The Reynolds number is

$$\text{Re} = \frac{VD}{\nu} = \frac{(8 \text{ m/s})(0.1 \text{ m})}{1.896 \times 10^{-5} \text{ m}^2/\text{s}} = 4.219 \times 10^4$$

The Nusselt number can be determined from

$$\text{Nu} = \frac{hD}{k} = 0.3 + \frac{0.62 \, \text{Re}^{1/2} \, \text{Pr}^{1/3}}{[1 + (0.4/\text{Pr})^{2/3}]^{1/4}} \left[ 1 + \left( \frac{\text{Re}}{282,000} \right)^{5/8} \right]^{4/5}$$

$$= 0.3 + \frac{0.62(4.219 \times 10^4)^{1/2} (0.7202)^{1/3}}{[1 + (0.4/0.7202)^{2/3}]^{1/4}} \left[ 1 + \left( \frac{4.219 \times 10^4}{282,000} \right)^{5/8} \right]^{4/5}$$

$$= 124$$

and

$$h = \frac{k}{D} \text{Nu} = \frac{0.02808 \text{ W/m·°C}}{0.1 \text{ m}} (124) = 34.8 \text{ W/m}^2\text{·°C}$$

Then the rate of heat transfer from the pipe per unit of its length becomes

$$A_s = pL = \pi DL = \pi(0.1 \text{ m})(1 \text{ m}) = 0.314 \text{ m}^2$$
$$\dot{Q} = hA_s(T_s - T_\infty) = (34.8 \text{ W/m}^2\text{·C})(0.314 \text{ m}^2)(110 - 10)°C = \mathbf{1093 \text{ W}}$$

The rate of heat loss from the entire pipe can be obtained by multiplying the value above by the length of the pipe in m.

**Discussion** The simpler Nusselt number relation in Table 19–2 in this case would give Nu = 128, which is 3 percent higher than the value obtained above using Eq. 19–29.

---

### EXAMPLE 19–4     Cooling of a Steel Ball by Forced Air

A 25-cm-diameter stainless steel ball ($\rho = 8055 \text{ kg/m}^3$, $c_p = 480 \text{ J/kg·°C}$) is removed from the oven at a uniform temperature of 300°C (Fig. 19–16). The ball is then subjected to the flow of air at 1 atm pressure and 25°C with a velocity of 3 m/s. The surface temperature of the ball eventually drops to 200°C. Determine the average convection heat transfer coefficient during this cooling process and estimate how long the process will take.

**Solution** A hot stainless steel ball is cooled by forced air. The average convection heat transfer coefficient and the cooling time are to be determined.

**FIGURE 19–16**
Schematic for Example 19–4.

**Assumptions** **1** Steady operating conditions exist. **2** Radiation effects are negligible. **3** Air is an ideal gas. **4** The outer surface temperature of the ball is uniform at all times. **5** The surface temperature of the ball during cooling is changing. Therefore, the convection heat transfer coefficient between the ball and the air will also change. To avoid this complexity, we take the surface temperature of the ball to be constant at the average temperature of $(300 + 200)/2 = 250°C$ in the evaluation of the heat transfer coefficient and use the value obtained for the entire cooling process.

**Properties** The dynamic viscosity of air at the average surface temperature is $\mu_s = \mu_{@ 250°C} = 2.76 \times 10^{-5}$ kg/m·s. The properties of air at the free-stream temperature of 25°C and 1 atm are (Table A–22)

$$k = 0.02551 \text{ W/m·°C} \qquad \nu = 1.562 \times 10^{-5} \text{ m}^2\text{/s}$$
$$\mu = 1.849 \times 10^{-5} \text{ kg/m·s} \qquad \text{Pr} = 0.7296$$

**Analysis** The Reynolds number is determined from

$$\text{Re} = \frac{VD}{\nu} = \frac{(3 \text{ m/s})(0.25 \text{ m})}{1.562 \times 10^{-5} \text{ m}^2\text{/s}} = 4.802 \times 10^4$$

The Nusselt number is

$$\text{Nu} = \frac{hD}{k} = 2 + [0.4 \, \text{Re}^{1/2} + 0.06 \, \text{Re}^{2/3}] \, \text{Pr}^{0.4} \left(\frac{\mu_\infty}{\mu_s}\right)^{1/4}$$

$$= 2 + [0.4(4.802 \times 10^4)^{1/2} + 0.06(4.802 \times 10^4)^{2/3}](0.7296)^{0.4}$$
$$\times \left(\frac{1.849 \times 10^{-5}}{2.76 \times 10^{-5}}\right)^{1/4}$$

$$= 135$$

Then the average convection heat transfer coefficient becomes

$$h = \frac{k}{D} \text{Nu} = \frac{0.02551 \text{ W/m·°C}}{0.25 \text{ m}} (135) = \mathbf{13.8 \text{ W/m}^2\text{·°C}}$$

In order to estimate the time of cooling of the ball from 300°C to 200°C, we determine the *average* rate of heat transfer from Newton's law of cooling by using the *average* surface temperature. That is,

$$A_s = \pi D^2 = \pi (0.25 \text{ m})^2 = 0.1963 \text{ m}^2$$
$$\dot{Q}_{avg} = hA_s(T_{s, avg} - T_\infty) = (13.8 \text{ W/m}^2\text{·°C})(0.1963 \text{ m}^2)(250 - 25)°C = 610 \text{ W}$$

Next we determine the *total* heat transferred from the ball, which is simply the change in the energy of the ball as it cools from 300°C to 200°C:

$$m = \rho V = \rho \tfrac{1}{6} \pi D^3 = (8055 \text{ kg/m}^3) \tfrac{1}{6} \pi (0.25 \text{ m})^3 = 65.9 \text{ kg}$$
$$Q_{total} = mc_p(T_2 - T_1) = (65.9 \text{ kg})(480 \text{ J/kg·°C})(300 - 200)°C = 3{,}163{,}000 \text{ J}$$

In this calculation, we assumed that the entire ball is at 200°C, which is not necessarily true. The inner region of the ball will probably be at a higher temperature than its surface. With this assumption, the time of cooling is determined to be

$$\Delta t \approx \frac{Q}{\dot{Q}_{avg}} = \frac{3{,}163{,}000 \text{ J}}{610 \text{ J/s}} = 5185 \text{ s} = \mathbf{1 \text{ h } 26 \text{ min}}$$

*Discussion* The time of cooling could also be determined more accurately using the transient temperature charts or relations introduced in Chapter 18. But the simplifying assumptions we made above can be justified if all we need is a ballpark value. It will be naive to expect the time of cooling to be exactly 1 h 26 min, but, using our engineering judgment, it is realistic to expect the time of cooling to be somewhere between one and two hours.

# 19–5 · GENERAL CONSIDERATIONS FOR PIPE FLOW

Liquid or gas flow through pipes or ducts is commonly used in practice in heating and cooling applications. The fluid in such applications is forced to flow by a fan or pump through a conduit that is sufficiently long to accomplish the desired heat transfer.

The general aspects of flow in pipes were considered in Chap. 14. The Reynolds number for flow through a pipe of inside diameter $D$ was defined as

$$\text{Re} = \frac{\rho V_{\text{avg}} D}{\mu} = \frac{V_{\text{avg}} D}{\nu} \tag{19–32}$$

where $V_{\text{avg}}$ is the average velocity and $\nu = \mu/\rho$ is the kinematic viscosity of the fluid. Transition from laminar to turbulent flow depends on the Reynolds number as well as the degree of disturbance of the flow by *surface roughness*, *pipe vibrations*, and the *fluctuations in the flow*. Under most practical conditions, the flow in a pipe is laminar for $\text{Re} < 2300$, fully turbulent for $\text{Re} > 10,000$, and transitional in between. But it should be kept in mind that in many cases the flow becomes fully turbulent for $\text{Re} > 4000$. When designing piping networks and determining pumping power, a conservative approach is taken and flows with $\text{Re} > 4000$ are assumed to be turbulent.

When a fluid is heated or cooled as it flows through a tube, the temperature of the fluid at any cross section changes from $T_s$ at the surface of the wall to some maximum (or minimum in the case of heating) at the tube center. In fluid flow it is convenient to work with an **average** or **mean temperature** $T_m$, which remains constant at a cross section. Unlike the mean velocity, the mean temperature $T_m$ *changes* in the flow direction whenever the fluid is heated or cooled.

The value of the mean temperature $T_m$ is determined from the requirement that the *conservation of energy* principle be satisfied. That is, the energy transported by the fluid through a cross section in actual flow must be equal to the energy that would be transported through the same cross section if the fluid were at a constant temperature $T_m$. This can be expressed mathematically as (Fig. 19–17)

$$\dot{E}_{\text{fluid}} = \dot{m} c_p T_m = \int_{\dot{m}} c_p T(r) \delta \dot{m} = \int_{A_c} \rho c_p T(r) u(r) V dA_c \tag{19–33}$$

where $c_p$ is the specific heat of the fluid. Note that the product $\dot{m} c_p T_m$ at any cross section along the tube represents the *energy flow* with the fluid at that cross section. Then the mean temperature of a fluid with constant density and specific heat flowing in a circular pipe of radius $R$ can be expressed as

$$T_m = \frac{\int_{\dot{m}} c_p T(r) \delta \dot{m}}{\dot{m} c_p} = \frac{\int_0^R c_p T(r) \rho u(r) 2\pi r dr}{\rho V_{\text{avg}} (\pi R^2) c_p} = \frac{2}{V_{\text{avg}} R^2} \int_0^R T(r) u(r) \, r dr \tag{19–34}$$

(a) Actual

(b) Idealized

**FIGURE 19–17**

Actual and idealized temperature profiles for flow in a tube (the rate at which energy is transported with the fluid is the same for both cases).

Note that the mean temperature $T_m$ of a fluid changes during heating or cooling. Also, the fluid properties in internal flow are usually evaluated at the *bulk mean fluid temperature*, which is the arithmetic average of the mean temperatures at the inlet and the exit. That is,

$$T_b = (T_{m,i} + T_{m,e})/2 \qquad (19\text{–}35)$$

## Thermal Entrance Region

The development of the velocity boundary layer was discussed in Chap. 14 (Fig. 19–18). Now consider a fluid at a uniform temperature entering a circular tube whose surface is maintained at a different temperature. This time, the fluid particles in the layer in contact with the surface of the tube assume the surface temperature. This initiates convection heat transfer in the tube and the development of a *thermal boundary layer* along the tube. The thickness of this boundary layer also increases in the flow direction until the boundary layer reaches the tube center and thus fills the entire tube, as shown in Fig. 19–19.

The region of flow over which the thermal boundary layer develops and reaches the tube center is called the **thermal entrance region**, and the length of this region is called the **thermal entry length** $L_t$. Flow in the thermal entrance region is called *thermally developing flow* since this is the region where the temperature profile develops. The region beyond the thermal entrance region in which the dimensionless temperature profile expressed as $(T_s - T)/(T_s - T_m)$ remains unchanged is called the **thermally fully developed region**. The region in which the flow is both hydrodynamically and thermally developed and thus both the velocity and dimensionless temperature profiles remain unchanged is called *fully developed flow*. That is,

**FIGURE 19–18**

The development of the velocity boundary layer in a tube. (The developed average velocity profile is parabolic in laminar flow, as shown, but somewhat flatter or fuller in turbulent flow.)

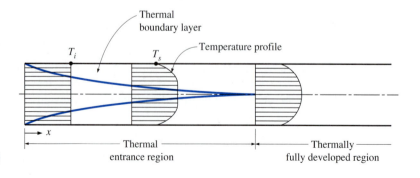

**FIGURE 19–19**

The development of the thermal boundary layer in a tube. (The fluid in the tube is being cooled.)

*Hydrodynamically fully developed*:

$$\frac{\partial u(r, x)}{\partial x} = 0 \longrightarrow u = u(r) \qquad \text{(19-36)}$$

*Thermally fully developed*:

$$\frac{\partial}{\partial x}\left[\frac{T_s(x) - T(r, x)}{T_s(x) - T_m(x)}\right] = 0 \qquad \text{(19-37)}$$

The shear stress at the tube wall $\tau_w$ is related to the slope of the velocity profile at the surface. Noting that the velocity profile remains unchanged in the hydrodynamically fully developed region, the wall shear stress also remains constant in that region. A similar argument can be given for the heat transfer coefficient in the thermally fully developed region.

In a thermally fully developed region, the derivative of $(T_s - T)/(T_s - T_m)$ with respect to $x$ is zero by definition, and thus $(T_s - T)/(T_s - T_m)$ is independent of $x$. Then the derivative of $(T_s - T)/(T_s - T_m)$ with respect $r$ must also be independent of $x$. That is,

$$\frac{\partial}{\partial r}\left(\frac{T_s - T}{T_s - T_m}\right)\bigg|_{r=R} = \frac{-(\partial T/\partial r)|_{r=R}}{T_s - T_m} \neq f(x) \qquad \text{(19-38)}$$

Surface heat flux can be expressed as

$$\dot{q}_s = h_x(T_s - T_m) = k\frac{\partial T}{\partial r}\bigg|_{r=R} \longrightarrow h_x = \frac{k(\partial T/\partial r)|_{r=R}}{T_s - T_m} \qquad \text{(19-39)}$$

which, from Eq. 19–37, is independent of $x$. Thus we conclude that *in the thermally fully developed region of a tube, the local convection coefficient is constant* (does not vary with $x$). Therefore, both *the friction (which is related to wall shear stress) and convection coefficients remain constant in the fully developed region of a tube.*

Note that the *temperature profile* in the thermally fully developed region may vary with $x$ in the flow direction. That is, unlike the velocity profile, the temperature profile can be different at different cross sections of the tube in the developed region, and it usually is, because the fluid is changing temperature with length. However, the dimensionless temperature profile defined previously remains unchanged in the thermally developed region when the temperature or heat flux at the tube surface remains constant.

During laminar flow in a tube, the magnitude of the dimensionless Prandtl number Pr is a measure of the relative growth of the velocity and thermal boundary layers. For fluids with Pr $\approx$ 1, such as gases, the two boundary layers essentially coincide with each other. For fluids with Pr $\gg$ 1, such as oils, the velocity boundary layer outgrows the thermal boundary layer. As a result, the hydrodynamic entry length is smaller than the thermal entry length. The opposite is true for fluids with Pr $\ll$ 1 such as liquid metals.

Consider a fluid that is being heated (or cooled) in a tube as it flows through it. The wall shear stress and the heat transfer coefficient are *highest* at the tube inlet where the thickness of the boundary layers is smallest, and decrease gradually to the fully developed values, as shown in Fig. 19–20. Therefore, the pressure drop and heat flux are *higher* in the entrance regions of a tube, and the effect of the entrance region is always to *increase* the average friction factor and heat transfer coefficient for the entire tube. This enhancement can be significant for short tubes but negligible for long ones.

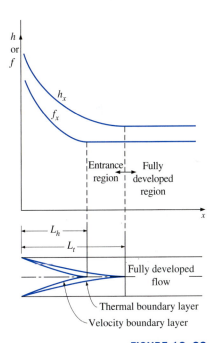

**FIGURE 19–20**

Variation of the friction factor and the convection heat transfer coefficient in the flow direction for flow in a tube (Pr > 1).

# Entry Lengths

The hydrodynamic entry length is usually taken to be the distance from the tube entrance where the wall shear stress (and thus the friction factor) reaches within about 2 percent of the fully developed value. In *laminar flow,* the hydrodynamic and thermal entry lengths are given approximately as (see Kays and Crawford, 1993 and Shah and Bhatti, 1987)

$$L_{h, \text{laminar}} \approx 0.05 \, \text{Re} \, D \tag{19–40}$$

$$L_{t, \text{laminar}} \approx 0.05 \, \text{Re} \, \text{Pr} \, D = \text{Pr} \, L_{h, \text{laminar}} \tag{19–41}$$

For Re $= 20$, the hydrodynamic entry length is about the size of the diameter, but increases linearly with velocity. In the limiting case of Re $= 2300$, the hydrodynamic entry length is $115D$.

In *turbulent flow,* the intense mixing during random fluctuations usually overshadows the effects of molecular diffusion, and therefore the hydrodynamic and thermal entry lengths are of about the same size and independent of the Prandtl number. The entry length is much shorter in turbulent flow, as expected, and its dependence on the Reynolds number is weaker. In many tube flows of practical interest, the entrance effects become insignificant beyond a tube length of 10 diameters, and the hydrodynamic and thermal entry lengths are approximately taken to be

$$L_{h, \text{turbulent}} \approx L_{t, \text{turbulent}} \approx 10D \tag{19–42}$$

The variation of local Nusselt number along a tube in turbulent flow for both uniform surface temperature and uniform surface heat flux is given in Fig. 19–21 for the range of Reynolds numbers encountered in heat transfer equipment. We make these important observations from this figure:

- The Nusselt numbers and thus the convection heat transfer coefficients are much higher in the entrance region.

- The Nusselt number reaches a constant value at a distance of less than 10 diameters, and thus the flow can be assumed to be fully developed for $x > 10D$.

- The Nusselt numbers for the uniform surface temperature and uniform surface heat flux conditions are identical in the fully developed regions, and nearly identical in the entrance regions. Therefore, Nusselt number is insensitive to the type of thermal boundary condition, and the turbulent flow correlations can be used for either type of boundary condition.

Precise correlations for the friction and heat transfer coefficients for the entrance regions are available in the literature. However, the tubes used in practice in forced convection are usually several times the length of either entrance region, and thus the flow through the tubes is often assumed to be fully developed for the entire length of the tube. This simplistic approach gives *reasonable* results for the rate of heat transfer for long tubes and *conservative* results for short ones.

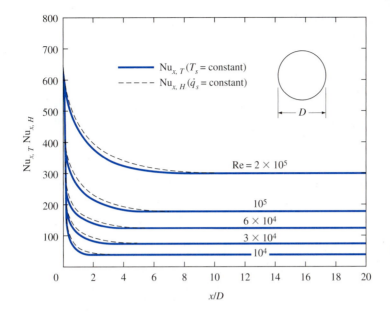

**FIGURE 19–21**

Variation of local Nusselt number along a tube in turbulent flow for both uniform surface temperature and uniform surface heat flux.

*Deissler, 1953.*

# 19–6 · GENERAL THERMAL ANALYSIS

In the absence of any work interactions (such as electric resistance heating), the conservation of energy equation for the steady flow of a fluid in a tube can be expressed as (Fig. 19–22)

$$\dot{Q} = \dot{m} c_p (T_e - T_i) \qquad \text{(W)} \qquad (19\text{–}43)$$

where $T_i$ and $T_e$ are the mean fluid temperatures at the inlet and exit of the tube, respectively, and $\dot{Q}$ is the rate of heat transfer to or from the fluid. Note that the temperature of a fluid flowing in a tube remains constant in the absence of any energy interactions through the wall of the tube.

The thermal conditions at the surface can usually be approximated with reasonable accuracy to be *constant surface temperature* ($T_s$ = constant) or *constant surface heat flux* ($\dot{q}_s$ = constant). For example, the constant surface temperature condition is realized when a phase change process such as boiling or condensation occurs at the outer surface of a tube. The constant surface heat flux condition is realized when the tube is subjected to radiation or electric resistance heating uniformly from all directions.

Surface heat flux is expressed as

$$\dot{q}_s = h_x (T_s - T_m) \qquad \text{(W/m}^2) \qquad (19\text{–}44)$$

where $h_x$ is the *local* heat transfer coefficient and $T_s$ and $T_m$ are the surface and the mean fluid temperatures at that location. Note that the mean fluid temperature $T_m$ of a fluid flowing in a tube must change during heating or cooling. Therefore, when $h_x = h$ = constant, the surface temperature $T_s$

Energy balance:
$$\dot{Q} = \dot{m} c_p (T_e - T_i)$$

**FIGURE 19–22**

The heat transfer to a fluid flowing in a tube is equal to the increase in the energy of the fluid.

must change when $\dot{q}_s$ = constant, and the surface heat flux $\dot{q}_s$ must change when $T_s$ = constant. Thus we may have either $T_s$ = constant or $\dot{q}_s$ = constant at the surface of a tube, but not both. Next we consider convection heat transfer for these two common cases.

## Constant Surface Heat Flux ($\dot{q}_s$ = constant)

In the case of $\dot{q}_s$ = constant, the rate of heat transfer can also be expressed as

$$\dot{Q} = \dot{q}_s A_s = \dot{m} c_p (T_e - T_i) \qquad \text{(W)} \qquad \text{(19–45)}$$

Then the mean fluid temperature at the tube exit becomes

$$T_e = T_i + \frac{\dot{q}_s A_s}{\dot{m} c_p} \qquad \text{(19–46)}$$

Note that the mean fluid temperature increases *linearly* in the flow direction in the case of constant surface heat flux, since the surface area increases linearly in the flow direction ($A_s$ is equal to the perimeter, which is constant, times the tube length).

The surface temperature in the case of constant surface heat flux $\dot{q}_s$ can be determined from

$$\dot{q}_s = h(T_s - T_m) \qquad \longrightarrow \qquad T_s = T_m + \frac{\dot{q}_s}{h} \qquad \text{(19–47)}$$

In the fully developed region, the surface temperature $T_s$ will also increase linearly in the flow direction since $h$ is constant and thus $T_s - T_m$ = constant (Fig. 19–23). Of course this is true when the fluid properties remain constant during flow.

The slope of the mean fluid temperature $T_m$ on a $T$-$x$ diagram can be determined by applying the steady-flow energy balance to a tube slice of thickness $dx$ shown in Fig. 19–24. It gives

$$\dot{m} c_p dT_m = \dot{q}_s (p\,dx) \qquad \longrightarrow \qquad \frac{dT_m}{dx} = \frac{\dot{q}_s p}{\dot{m} c_p} = \text{constant} \qquad \text{(19–48)}$$

where $p$ is the perimeter of the tube.

Noting that both $\dot{q}_s$ and $h$ are constants, the differentiation of Eq. 19–47 with respect to $x$ gives

$$\frac{dT_m}{dx} = \frac{dT_s}{dx} \qquad \text{(19–49)}$$

Also, the requirement that the dimensionless temperature profile remains unchanged in the fully developed region gives

$$\frac{\partial}{\partial x}\left(\frac{T_s - T}{T_s - T_m}\right) = 0 \quad \longrightarrow \quad \frac{1}{T_s - T_m}\left(\frac{\partial T_s}{\partial x} - \frac{\partial T}{\partial x}\right) = 0 \quad \longrightarrow \quad \frac{\partial T}{\partial x} = \frac{dT_s}{dx} \quad \text{(19–50)}$$

since $T_s - T_m$ = constant. Combining Eqs. 19–48, 19–49, and 19–50 gives

$$\frac{\partial T}{\partial x} = \frac{dT_s}{dx} = \frac{dT_m}{dx} = \frac{\dot{q}_s p}{\dot{m} c_p} = \text{constant} \qquad \text{(19–51)}$$

**FIGURE 19–23**
Variation of the *tube surface* and the *mean fluid* temperatures along the tube for the case of constant surface heat flux.

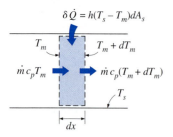

**FIGURE 19–24**
Energy interactions for a differential control volume in a tube.

Then we conclude that *in fully developed flow in a tube subjected to constant surface heat flux, the temperature gradient is independent of x and thus the shape of the temperature profile does not change along the tube* (Fig. 19–25).

For a circular tube, $p = 2\pi R$ and $\dot{m} = \rho V_{avg} A_c = \rho V_{avg}(\pi R^2)$, and Eq. 19–51 becomes

*Circular tube:*
$$\frac{\partial T}{\partial x} = \frac{dT_s}{dx} = \frac{dT_m}{dx} = \frac{2\dot{q}_s}{\rho V_{avg} c_p R} = \text{constant} \qquad \textbf{(19–52)}$$

where $V_{avg}$ is the mean velocity of the fluid.

## Constant Surface Temperature ($T_s = $ constant)

From Newton's law of cooling, the rate of heat transfer to or from a fluid flowing in a tube can be expressed as

$$\dot{Q} = hA_s\Delta T_{avg} = hA_s(T_s - T_m)_{avg} \qquad \text{(W)} \qquad \textbf{(19–53)}$$

where $h$ is the average convection heat transfer coefficient, $A_s$ is the heat transfer surface area (it is equal to $\pi DL$ for a circular pipe of length $L$), and $\Delta T_{avg}$ is some appropriate *average* temperature difference between the fluid and the surface. Below we discuss two suitable ways of expressing $\Delta T_{avg}$.

In the constant surface temperature ($T_s = $ constant) case, $\Delta T_{avg}$ can be expressed *approximately* by the **arithmetic mean temperature difference** $\Delta T_{am}$ as

$$\Delta T_{avg} \approx \Delta T_{am} = \frac{\Delta T_i + \Delta T_e}{2} = \frac{(T_s - T_i) + (T_s - T_e)}{2} = T_s - \frac{T_i + T_e}{2}$$

$$= T_s - T_b \qquad \textbf{(19–54)}$$

where $T_b = (T_i + T_e)/2$ is the *bulk mean fluid temperature*, which is the *arithmetic average* of the mean fluid temperatures at the inlet and the exit of the tube.

Note that the *arithmetic mean temperature difference* $\Delta T_{am}$ is simply the *average* of the *temperature differences* between the surface and the fluid at the inlet and the exit of the tube. Inherent in this definition is the assumption that the mean fluid temperature varies linearly along the tube, which is hardly ever the case when $T_s = $ constant. This simple approximation often gives acceptable results, but not always. Therefore, we need a better way to evaluate $\Delta T_{avg}$.

Consider the heating of a fluid in a tube of constant cross section whose inner surface is maintained at a constant temperature of $T_s$. We know that the mean temperature of the fluid $T_m$ increases in the flow direction as a result of heat transfer. The energy balance on a differential control volume shown in Fig. 19–24 gives

$$\dot{m}c_p dT_m = h(T_s - T_m)dA_s \qquad \textbf{(19–55)}$$

That is, the increase in the energy of the fluid (represented by an increase in its mean temperature by $dT_m$) is equal to the heat transferred to the fluid from the tube surface by convection. Noting that the differential surface area is

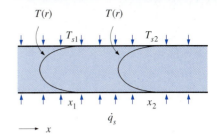

**FIGURE 19–25**
The shape of the temperature profile remains unchanged in the fully developed region of a tube subjected to constant surface heat flux.

$dA_s = pdx$, where $p$ is the perimeter of the tube, and that $dT_m = -d(T_s - T_m)$, since $T_s$ is constant, the relation above can be rearranged as

$$\frac{d(T_s - T_m)}{T_s - T_m} = -\frac{hp}{\dot{m}c_p}\, dx \qquad (19\text{-}56)$$

Integrating from $x = 0$ (tube inlet where $T_m = T_i$) to $x = L$ (tube exit where $T_m = T_e$) gives

$$\ln\frac{T_s - T_e}{T_s - T_i} = -\frac{hA_s}{\dot{m}c_p} \qquad (19\text{-}57)$$

where $A_s = pL$ is the surface area of the tube and $h$ is the constant *average* convection heat transfer coefficient. Taking the exponential of both sides and solving for $T_e$ gives the following relation which is very useful for the determination of the *mean fluid temperature at the tube exit:*

$$T_e = T_s - (T_s - T_i)\exp(-hA_s/\dot{m}c_p) \qquad (19\text{-}58)$$

This relation can also be used to determine the mean fluid temperature $T_m(x)$ at any $x$ by replacing $A_s = pL$ by $px$.

Note that the temperature difference between the fluid and the surface *decays exponentially* in the flow direction, and the rate of decay depends on the magnitude of the exponent $hA_s/\dot{m}c_p$, as shown in Fig. 19–26. This dimensionless parameter is called the *number of transfer units,* denoted by NTU, and is a measure of the effectiveness of the heat transfer systems. For NTU > 5, the exit temperature of the fluid becomes almost equal to the surface temperature, $T_e \approx T_s$ (Fig. 19–27). Noting that the fluid temperature can approach the surface temperature but cannot cross it, an NTU of about 5 indicates that the limit is reached for heat transfer, and the heat transfer does not increase no matter how much we extend the length of the tube. A small value of NTU, on the other hand, indicates more opportunities for heat transfer, and the heat transfer continues to increase as the tube length is increased. A large NTU and thus a large heat transfer surface area (which means a large tube) may be desirable from a heat transfer point of view, but it may be unacceptable from an economic point of view. The selection of heat transfer equipment usually reflects a compromise between heat transfer performance and cost.

Solving Eq. 19–57 for $\dot{m}c_p$ gives

$$\dot{m}c_p = -\frac{hA_s}{\ln[(T_s - T_e)/(T_s - T_i)]} \qquad (19\text{-}59)$$

Substituting this into Eq. 19–43, we obtain

$$\dot{Q} = hA_s\Delta T_{\ln} \qquad (19\text{-}60)$$

where

$$\Delta T_{\ln} = \frac{T_i - T_e}{\ln[(T_s - T_e)/(T_s - T_i)]} = \frac{\Delta T_e - \Delta T_i}{\ln(\Delta T_e/\Delta T_i)} \qquad (19\text{-}61)$$

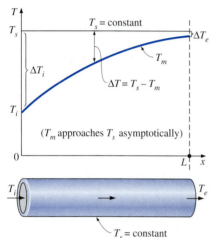

**FIGURE 19–26**

The variation of the *mean fluid temperature* along the tube for the case of constant temperature.

| NTU $= hA_s/\dot{m}c_p$ | $T_e$, °C |
|---|---|
| 0.01 | 20.8 |
| 0.05 | 23.9 |
| 0.10 | 27.6 |
| 0.50 | 51.5 |
| 1.00 | 70.6 |
| 5.00 | 99.5 |
| 10.00 | 100.0 |

**FIGURE 19–27**

An NTU greater than 5 indicates that the fluid flowing in a tube will reach the surface temperature at the exit regardless of the inlet temperature.

is the **logarithmic mean temperature difference**. Note that $\Delta T_i = T_s - T_i$ and $\Delta T_e = T_s - T_e$ are the temperature differences between the surface and the fluid at the inlet and the exit of the tube, respectively. This $\Delta T_{\ln}$ relation appears to be prone to misuse, but it is practically fail-safe, since using $T_i$ in place of $T_e$ and vice versa in the numerator and/or the denominator will, at most, affect the sign, not the magnitude. Also, it can be used for both heating ($T_s > T_i$ and $T_e$) and cooling ($T_s < T_i$ and $T_e$) of a fluid in a tube.

The logarithmic mean temperature difference $\Delta T_{\ln}$ is obtained by tracing the actual temperature profile of the fluid along the tube, and is an *exact* representation of the *average temperature difference* between the fluid and the surface. It truly reflects the exponential decay of the local temperature difference. When $\Delta T_e$ differs from $\Delta T_i$ by no more than 40 percent, the error in using the arithmetic mean temperature difference is less than 1 percent. But the error increases to undesirable levels when $\Delta T_e$ differs from $\Delta T_i$ by greater amounts. Therefore, we should always use the logarithmic mean temperature difference when determining the convection heat transfer in a tube whose surface is maintained at a constant temperature $T_s$.

---

**EXAMPLE 19–5   Heating of Water in a Tube by Steam**

Water enters a 2.5-cm-internal-diameter thin copper tube of a heat exchanger at 15°C at a rate of 0.3 kg/s, and is heated by steam condensing outside at 120°C. If the average heat transfer coefficient is 800 W/m²·°C, determine the length of the tube required in order to heat the water to 115°C (Fig. 19–28).

**Solution**  Water is heated by steam in a circular tube. The tube length required to heat the water to a specified temperature is to be determined.
**Assumptions**  **1** Steady operating conditions exist. **2** Fluid properties are constant. **3** The convection heat transfer coefficient is constant. **4** The conduction resistance of copper tube is negligible so that the inner surface temperature of the tube is equal to the condensation temperature of steam.
**Properties**  The specific heat of water at the bulk mean temperature of (15 + 115)/2 = 65°C is 4187 J/kg·°C. The heat of condensation of steam at 120°C is 2203 kJ/kg.
**Analysis**  Knowing the inlet and exit temperatures of water, the rate of heat transfer is determined to be

$$\dot{Q} = \dot{m}c_p(T_e - T_i) = (0.3 \text{ kg/s})(4.187 \text{ kJ/kg·°C})(115°C - 15°C)$$
$$= 125.6 \text{ kW}$$

The logarithmic mean temperature difference is

$$\Delta T_e = T_s - T_e = 120°C - 115°C = 5°C$$
$$\Delta T_i = T_s - T_i = 120°C - 15°C = 105°C$$
$$\Delta T_{\ln} = \frac{\Delta T_e - \Delta T_i}{\ln(\Delta T_e/\Delta T_i)} = \frac{5 - 105}{\ln(5/105)} = 32.85°C$$

The heat transfer surface area is

$$\dot{Q} = hA_s\Delta T_{\ln} \longrightarrow A_s = \frac{\dot{Q}}{h\Delta T_{\ln}} = \frac{125.6 \text{ kW}}{(0.8 \text{ kW/m}^2\cdot°C)(32.85°C)} = 4.78 \text{ m}^2$$

Steam
$T_s = 120°C$

Water
15°C
0.3 kg/s

115°C

$D = 2.5$ cm

**FIGURE 19–28**
Schematic for Example 19–5.

Then the required tube length becomes

$$A_s = \pi DL \quad \longrightarrow \quad L = \frac{A_s}{\pi D} = \frac{4.78 \text{ m}^2}{\pi(0.025 \text{ m})} = 61 \text{ m}$$

**Discussion** The bulk mean temperature of water during this heating process is 65°C, and thus the *arithmetic* mean temperature difference is $\Delta T_{am} = 120 - 65 = 55$°C. Using $\Delta T_{am}$ instead of $\Delta T_{ln}$ would give $L = 36$ m, which is grossly in error. This shows the importance of using the logarithmic mean temperature in calculations.

## 19–7 · LAMINAR FLOW IN TUBES

Consider steady laminar flow of a fluid in a circular tube of radius $R$. The fluid properties $\rho$, $k$, and $c_p$ are constant, and the work done by viscous forces is negligible. The fluid flows along the $x$-axis with velocity $u$. The flow is fully developed so that $u$ is independent of $x$ and thus $u = u(r)$. Noting that energy is transferred by mass in the $x$-direction, and by conduction in the $r$-direction (heat conduction in the $x$-direction is assumed to be negligible), the steady-flow energy balance for a cylindrical shell element of thickness $dr$ and length $dx$ can be expressed as (Fig. 19–29)

$$\dot{m}c_p T_x - \dot{m}c_p T_{x+dx} + \dot{Q}_r - \dot{Q}_{r+dr} = 0 \qquad (19\text{–}62)$$

where $\dot{m} = \rho u A_c = \rho u(2\pi r dr)$. Substituting and dividing by $2\pi r dr dx$ gives, after rearranging,

$$\rho c_p u \frac{T_{x+dx} - T_x}{dx} = -\frac{1}{2\pi r dx} \frac{\dot{Q}_{r+dr} - \dot{Q}_r}{dr} \qquad (19\text{–}63)$$

or

$$u \frac{\partial T}{\partial x} = -\frac{1}{2\rho c_p \pi r dx} \frac{\partial \dot{Q}}{\partial r} \qquad (19\text{–}64)$$

But

$$\frac{\partial \dot{Q}}{\partial r} = \frac{\partial}{\partial r}\left(-k2\pi r dx \frac{\partial T}{\partial r}\right) = -2\pi k dx \frac{\partial}{\partial r}\left(r \frac{\partial T}{\partial r}\right) \qquad (19\text{–}65)$$

Substituting and using $\alpha = k/\rho c_p$ gives

$$u \frac{\partial T}{\partial x} = \frac{\alpha}{r} \frac{\partial}{\partial r}\left(r \frac{\partial T}{\partial r}\right) \qquad (19\text{–}66)$$

which states that *the rate of net energy transfer to the control volume by mass flow is equal to the net rate of heat conduction in the radial direction.*

**FIGURE 19–29**
The differential volume element used in the derivation of energy balance relation.

# Constant Surface Heat Flux

For fully developed flow in a circular tube subjected to constant surface heat flux, we have, from Eq. 19–52,

$$\frac{\partial T}{\partial x} = \frac{dT_s}{dx} = \frac{dT_m}{dx} = \frac{2\dot{q}_s}{\rho V_{avg} c_p R} = \text{constant} \qquad (19\text{--}67)$$

If heat conduction in the $x$-direction were considered in the derivation of Eq. 19–66, it would give an additional term $\alpha \partial^2 T/\partial x^2$, which would be equal to zero since $\partial T/\partial x = \text{constant}$ and thus $T = T(r)$. Therefore, the assumption that there is no axial heat conduction is satisfied exactly in this case.

Substituting Eq. 19–67 and the relation for velocity profile from Chap. 14 into Eq. 19–66 gives

$$\frac{4\dot{q}_s}{kR}\left(1 - \frac{r^2}{R^2}\right) = \frac{1}{r}\frac{d}{dr}\left(r\frac{dT}{dr}\right) \qquad (19\text{--}68)$$

which is a second-order ordinary differential equation. Its general solution is obtained by separating the variables and integrating twice to be

$$T = \frac{\dot{q}_s}{kR}\left(r^2 - \frac{r^4}{4R^2}\right) + C_1 \ln r + C_2 \qquad (19\text{--}69)$$

The desired solution to the problem is obtained by applying the boundary conditions $\partial T/\partial r = 0$ at $r = 0$ (because of symmetry) and $T = T_s$ at $r = R$. We get

$$T = T_s - \frac{\dot{q}_s R}{k}\left(\frac{3}{4} - \frac{r^2}{R^2} + \frac{r^4}{4R^4}\right) \qquad (19\text{--}70)$$

The bulk mean temperature $T_m$ is determined by substituting the velocity and temperature profile relation 19–70 into Eq. 19–34 and performing the integration. It gives

$$T_m = T_s - \frac{11}{24}\frac{\dot{q}_s R}{k} \qquad (19\text{--}71)$$

Combining this relation with $\dot{q}_s = h(T_s - T_m)$ gives

$$h = \frac{24}{11}\frac{k}{R} = \frac{48}{11}\frac{k}{D} = 4.36\frac{k}{D} \qquad (19\text{--}72)$$

or

*Circular tube, laminar ($\dot{q}_s = \text{constant}$):* $\qquad \text{Nu} = \frac{hD}{k} = 4.36 \qquad (19\text{--}73)$

Therefore, for fully developed laminar flow in a circular tube subjected to constant surface heat flux, the Nusselt number is a constant. There is no dependence on the Reynolds or the Prandtl numbers.

$$T_s = \text{constant}$$

$$f = \frac{64}{\text{Re}}$$

$u(r)$

$D$

$$\text{Nu} = 3.66$$

Fully developed
laminar flow

**FIGURE 19–30**

In laminar flow in a tube with
constant surface temperature,
both the *friction factor* and
the *heat transfer coefficient*
remain constant in the fully
developed region.

## Constant Surface Temperature

A similar analysis can be performed for fully developed laminar flow in a circular tube for the case of constant surface temperature $T_s$. The solution procedure in this case is more complex as it requires iterations, but the Nusselt number relation obtained is equally simple (Fig. 19–30):

$$\textit{Circular tube, laminar } (T_s = \text{constant}): \qquad \text{Nu} = \frac{hD}{k} = 3.66 \qquad \text{(19–74)}$$

The thermal conductivity $k$ for use in the Nu relations above should be evaluated at the bulk mean fluid temperature, which is the arithmetic average of the mean fluid temperatures at the inlet and the exit of the tube. For laminar flow, the effect of *surface roughness* on the friction factor and the heat transfer coefficient is negligible.

## Laminar Flow in Noncircular Tubes

Nusselt number relations are given in Table 19–3 for *fully developed laminar flow* in tubes of various cross sections. The Reynolds and Nusselt numbers

**TABLE 19–3**

Nusselt number and friction factor for fully developed laminar flow in tubes of various cross sections ($D_h = 4A_c/p$, $\text{Re} = V_{avg}D_h/\nu$, and $\text{Nu} = hD_h/k$)

| Tube Geometry | $a/b$ or $\theta°$ | Nusselt Number $T_s = $ Const. | Nusselt Number $\dot{q}_s = $ Const. |
|---|---|---|---|
| Circle | — | 3.66 | 4.36 |
| Rectangle | $a/b$ | | |
| | 1 | 2.98 | 3.61 |
| | 2 | 3.39 | 4.12 |
| | 3 | 3.96 | 4.79 |
| | 4 | 4.44 | 5.33 |
| | 6 | 5.14 | 6.05 |
| | 8 | 5.60 | 6.49 |
| | ∞ | 7.54 | 8.24 |
| Ellipse | $a/b$ | | |
| | 1 | 3.66 | 4.36 |
| | 2 | 3.74 | 4.56 |
| | 4 | 3.79 | 4.88 |
| | 8 | 3.72 | 5.09 |
| | 16 | 3.65 | 5.18 |
| Isosceles Triangle | $\theta$ | | |
| | 10° | 1.61 | 2.45 |
| | 30° | 2.26 | 2.91 |
| | 60° | 2.47 | 3.11 |
| | 90° | 2.34 | 2.98 |
| | 120° | 2.00 | 2.68 |

for flow in these tubes are based on the hydraulic diameter $D_h = 4A_c/p$, where $A_c$ is the cross-sectional area of the tube and $p$ is its perimeter. Once the Nusselt number is available, the convection heat transfer coefficient is determined from $h = k\text{Nu}/D_h$.

## Developing Laminar Flow in the Entrance Region

For a circular tube of length $L$ subjected to constant surface temperature, the average Nusselt number for the *thermal entrance region* can be determined from (Edwards et al., 1979)

*Entry region, laminar:*
$$\text{Nu} = 3.66 + \frac{0.065 \, (D/L) \, \text{Re Pr}}{1 + 0.04[(D/L)\text{Re Pr}]^{2/3}} \qquad \textbf{(19–75)}$$

Note that the average Nusselt number is larger at the entrance region, as expected, and it approaches asymptotically to the fully developed value of 3.66 as $L \to \infty$. This relation assumes that the flow is hydrodynamically developed when the fluid enters the heating section, but it can also be used approximately for flow developing hydrodynamically.

When the difference between the surface and the fluid temperatures is large, it may be necessary to account for the variation of viscosity with temperature.

The average Nusselt number for developing laminar flow in a circular tube in that case can be determined from (Sieder and Tate, 1936)

$$\text{Nu} = 1.86 \left( \frac{\text{Re Pr } D}{L} \right)^{1/3} \left( \frac{\mu_b}{\mu_s} \right)^{0.14} \qquad \textbf{(19–76)}$$

All properties are evaluated at the bulk mean fluid temperature, except for $\mu_s$, which is evaluated at the surface temperature.

The average Nusselt number for the thermal entrance region of flow between *isothermal parallel plates* of length $L$ is expressed as (Edwards et al., 1979)

*Entry region, laminar:*
$$\text{Nu} = 7.54 + \frac{0.03 \, (D_h/L) \, \text{Re Pr}}{1 + 0.016[(D_h/L) \, \text{Re Pr}]^{2/3}} \qquad \textbf{(19–77)}$$

where $D_h$ is the hydraulic diameter, which is twice the spacing of the plates. This relation can be used for $\text{Re} \le 2800$.

---

### EXAMPLE 19–6    Flow of Oil in a Pipeline through a Lake

Consider the flow of oil at 20°C in a 30-cm-diameter pipeline at an average velocity of 2 m/s (Fig. 19–31). A 200-m-long section of the horizontal pipeline passes through icy waters of a lake at 0°C. Measurements indicate that the surface temperature of the pipe is very nearly 0°C. Disregarding the thermal resistance of the pipe material, determine (a) the temperature of the oil when the pipe leaves the lake, (b) the rate of heat transfer from the oil, and (c) the pumping power required to overcome the pressure losses and to maintain the flow of the oil in the pipe.

**FIGURE 19–31**
Schematic for Example 19–6.

**Solution** Oil flows in a pipeline that passes through icy waters of a lake at 0°C. The exit temperature of the oil, the rate of heat loss, and the pumping power needed to overcome pressure losses are to be determined.

**Assumptions** 1 Steady operating conditions exist. 2 The surface temperature of the pipe is very nearly 0°C. 3 The thermal resistance of the pipe is negligible. 4 The inner surfaces of the pipeline are smooth. 5 The flow is hydrodynamically developed when the pipeline reaches the lake.

**Properties** We do not know the exit temperature of the oil, and thus we cannot determine the bulk mean temperature, which is the temperature at which the properties of oil are to be evaluated. The mean temperature of the oil at the inlet is 20°C, and we expect this temperature to drop somewhat as a result of heat loss to the icy waters of the lake. We evaluate the properties of the oil at the inlet temperature, but we will repeat the calculations, if necessary, using properties at the evaluated bulk mean temperature. At 20°C we read (Table A-19)

$$\rho = 888.1 \text{ kg/m}^3 \qquad \nu = 9.429 \times 10^{-4} \text{ m}^2/\text{s}$$

$$k = 0.145 \text{ W/m·°C} \qquad c_p = 1880 \text{ J/kg·°C} \qquad \text{Pr} = 10{,}863$$

**Analysis** (a) The Reynolds number is

$$\text{Re} = \frac{V_{avg} D}{n} = \frac{(2 \text{ m/s})(0.3 \text{ m})}{9.429 \times 10^{-4} \text{ m}^2/\text{s}} = 636$$

which is less than the critical Reynolds number of 2300. Therefore, the flow is laminar, and the thermal entry length in this case is roughly

$$L_t \approx 0.05 \text{ Re Pr } D = 0.05 \times 636 \times 10{,}863 \times (0.3 \text{ m}) \approx 103{,}600 \text{ m}$$

which is much greater than the total length of the pipe. This is typical of fluids with high Prandtl numbers. Therefore, we assume thermally developing flow and determine the Nusselt number from

$$\text{Nu} = \frac{hD}{k} = 3.66 + \frac{0.065 \, (D/L) \text{ Re Pr}}{1 + 0.04 \left[ (D/L) \text{ Re Pr} \right]^{2/3}}$$

$$= 3.66 + \frac{0.065(0.3/200) \times 636 \times 10{,}863}{1 + 0.04[(0.3/200) \times 636 \times 10{,}863]^{2/3}}$$

$$= 33.7$$

Note that this Nusselt number is considerably higher than the fully developed value of 3.66. Then,

$$h = \frac{k}{D} \text{ Nu} = \frac{0.145 \text{ W/m·°C}}{0.3 \text{ m}} (33.7) = 16.3 \text{ W/m}^2\text{·°C}$$

Also,

$$A_s = \pi DL = \pi(0.3 \text{ m})(200 \text{ m}) = 188.5 \text{ m}^2$$

$$\dot{m} = \rho A_c V_{avg} = (888.1 \text{ kg/m}^3)[\tfrac{1}{4}\pi(0.3 \text{ m})^2](2 \text{ m/s}) = 125.6 \text{ kg/s}$$

Next we determine the exit temperature of oil,

$$T_e = T_s - (T_s - T_i) \exp(-hA_s/\dot{m}c_p)$$

$$= 0°C - [(0 - 20)°C] \exp\left[-\frac{(16.3 \text{ W/m}^2\cdot°C)(188.5 \text{ m}^2)}{(125.6 \text{ kg/s})(1881 \text{ J/kg}\cdot°C)}\right]$$

$$= \mathbf{19.74°C}$$

Thus, the mean temperature of oil drops by a mere 0.26°C as it crosses the lake. This makes the bulk mean oil temperature 19.87°C, which is practically identical to the inlet temperature of 20°C. Therefore, we do not need to re-evaluate the properties.

(b) The logarithmic mean temperature difference and the rate of heat loss from the oil are

$$\Delta T_{\text{ln}} = \frac{T_i - T_e}{\ln\dfrac{T_s - T_e}{T_s - T_i}} = \frac{20 - 19.74}{\ln\dfrac{0 - 19.74}{0 - 20}} = -19.87°C$$

$$\dot{Q} = hA_s \Delta T_{\text{ln}} = (16.3 \text{ W/m}^2\cdot°C)(188.5 \text{ m}^2)(-19.87°C) = \mathbf{-6.11 \times 10^4 \text{ W}}$$

Therefore, the oil will lose heat at a rate of 61.1 kW as it flows through the pipe in the icy waters of the lake. Note that $\Delta T_{\text{ln}}$ is identical to the arithmetic mean temperature in this case, since $\Delta T_i \approx \Delta T_e$.

(c) The laminar flow of oil is hydrodynamically developed. Therefore, the friction factor can be determined from (see Chap. 14)

$$f = \frac{64}{\text{Re}} = \frac{64}{636} = 0.1006$$

Then the pressure drop in the pipe and the required pumping power become

$$\Delta P = f\frac{L}{D}\frac{\rho V_{\text{avg}}^2}{2} = 0.1006\frac{200 \text{ m}}{0.3 \text{ m}}\frac{(888.1 \text{ kg/m}^3)(2 \text{ m/s})^2}{2} = 1.19 \times 10^5 \text{ N/m}^2$$

$$\dot{W}_{\text{pump}} = \frac{\dot{m}\Delta P}{r} = \frac{(125.6 \text{ kg/s})(1.19 \times 10^5 \text{ N/m}^2)}{888.1 \text{ kg/m}^3} = \mathbf{16.8 \text{ kW}}$$

**Discussion** We need a 16.8-kW pump just to overcome the friction in the pipe as the oil flows in the 200-m-long pipe through the lake. If the pipe is long, a more optimal design to reduce pump work would be to increase the pipe diameter $D$, since pump work is inversely proportional to $D^5$. Optimal pipe sizing requires an economic life-cycle cost analysis.

# 19–8 · TURBULENT FLOW IN TUBES

We mentioned in Chap. 14 that flow in smooth tubes is usually fully turbulent for Re > 10,000. Turbulent flow is commonly utilized in practice because of the higher heat transfer coefficients associated with it. Most correlations for the friction and heat transfer coefficients in turbulent flow are based on experimental studies because of the difficulty in dealing with turbulent flow theoretically.

For *smooth* tubes, the friction factor in turbulent flow can be determined from the explicit *first Petukhov equation* (Petukhov, 1970) given as

*Smooth tubes:* $\qquad f = (0.790 \ln \text{Re} - 1.64)^{-2} \qquad 3000 < \text{Re} < 5 \times 10^6$ **(19–78)**

The Nusselt number in turbulent flow is related to the friction factor through the *Chilton–Colburn analogy* expressed as

$$Nu = 0.125 f \, RePr^{1/3} \qquad \text{(19–79)}$$

Once the friction factor is available, this equation can be used conveniently to evaluate the Nusselt number for both smooth and rough tubes.

For fully developed turbulent flow in *smooth tubes,* a simple relation for the Nusselt number can be obtained by substituting the simple power law relation $f = 0.184 \, Re^{-0.2}$ for the friction factor into Eq. 19–79. It gives

$$Nu = 0.023 \, Re^{0.8} \, Pr^{1/3} \qquad \left( \begin{array}{c} 0.7 \leq Pr \leq 160 \\ Re > 10{,}000 \end{array} \right) \qquad \text{(19–80)}$$

which is known as the *Colburn equation.* The accuracy of this equation can be improved by modifying it as

$$Nu = 0.023 \, Re^{0.8} \, Pr^{n} \qquad \text{(19–81)}$$

where $n = 0.4$ for *heating* and 0.3 for *cooling* of the fluid flowing through the tube. This equation is known as the *Dittus–Boelter equation* (Dittus and Boelter, 1930) and it is preferred to the Colburn equation.

The proceeding equations can be used when the temperature difference between the fluid and wall surface is not large by evaluating all fluid properties at the *bulk mean fluid temperature* $T_b = (T_i + T_e)/2$. When the variation in properties is large due to a large temperature difference, the following equation due to Sieder and Tate (1936) can be used:

$$Nu = 0.027 \, Re^{0.8} Pr^{1/3} \left( \frac{\mu}{\mu_s} \right)^{0.14} \qquad \left( \begin{array}{c} 0.7 \leq Pr \leq 17{,}600 \\ Re \geq 10{,}000 \end{array} \right) \qquad \text{(19–82)}$$

Here all properties are evaluated at $T_b$ except $\mu_s$, which is evaluated at $T_s$.

The Nusselt number relations above are fairly simple, but they may give errors as large as 25 percent. This error can be reduced considerably to less than 10 percent by using more complex but accurate relations such as the *second Petukhov equation* expressed as

$$Nu = \frac{(f/8) \, Re \, Pr}{1.07 + 12.7 \, (f/8)^{0.5} \, (Pr^{2/3} - 1)} \qquad \left( \begin{array}{c} 0.5 \leq Pr \leq 2000 \\ 10^4 < Re < 5 \times 10^6 \end{array} \right) \qquad \text{(19–83)}$$

The accuracy of this relation at lower Reynolds numbers is improved by modifying it as (Gnielinski, 1976)

$$Nu = \frac{(f/8) \, (Re - 1000) \, Pr}{1 + 12.7 \, (f/8)^{0.5} \, (Pr^{2/3} - 1)} \qquad \left( \begin{array}{c} 0.5 \leq Pr \leq 2000 \\ 3 \times 10^3 < Re < 5 \times 10^6 \end{array} \right) \qquad \text{(19–84)}$$

where the friction factor $f$ can be determined from an appropriate relation such as the first Petukhov equation. Gnielinski's equation should be preferred in calculations. Again properties should be evaluated at the bulk mean fluid temperature.

The relations above are not very sensitive to the *thermal conditions* at the tube surfaces and can be used for both $T_s$ = constant and $\dot{q}_s$ = constant cases. Despite their simplicity, the correlations already presented give sufficiently accurate results for most engineering purposes. They can also be used to obtain rough estimates of the friction factor and the heat transfer coefficients in the transition region.

The relations given so far do not apply to liquid metals because of their very low Prandtl numbers. For liquid metals ($0.004 < Pr < 0.01$), the following relations are recommended by Sleicher and Rouse (1975) for $10^4 < Re < 10^6$:

*Liquid metals, $T_s$ = constant:*     $Nu = 4.8 + 0.0156\ Re^{0.85}\ Pr_s^{0.93}$     **(19–85)**

*Liquid metals, $\dot{q}_s$ = constant:*     $Nu = 6.3 + 0.0167\ Re^{0.85}\ Pr_s^{0.93}$     **(19–86)**

where the subscript $s$ indicates that the Prandtl number is to be evaluated at the surface temperature.

In turbulent flow, wall roughness increases the heat transfer coefficient $h$ by a factor of 2 or more (Dipprey and Sabersky, 1963). The convection heat transfer coefficient for rough tubes can be calculated approximately from the Nusselt number relations such as Eq. 19–84 by using the friction factor determined from the Moody chart or the Colebrook equation. However, this approach is not very accurate since there is no further increase in $h$ with $f$ for $f > 4f_{smooth}$ (Norris, 1970) and correlations developed specifically for rough tubes should be used when more accuracy is desired.

## Developing Turbulent Flow in the Entrance Region

The entry lengths for turbulent flow are typically short, often just 10 tube diameters long, and thus the Nusselt number determined for fully developed turbulent flow can be used approximately for the entire tube. This simple approach gives reasonable results for pressure drop and heat transfer for long tubes and conservative results for short ones. Correlations for the friction and heat transfer coefficients for the entrance regions are available in the literature for better accuracy.

## Turbulent Flow in Noncircular Tubes

The velocity and temperature profiles in turbulent flow are nearly straight lines in the core region, and any significant velocity and temperature gradients occur in the viscous sublayer (Fig. 19–32). Despite the small thickness of viscous sublayer (usually much less than 1 percent of the pipe diameter), the characteristics of the flow in this layer are very important since they set the stage for flow in the rest of the pipe. Therefore, pressure drop and heat transfer characteristics of turbulent flow in tubes are dominated by the very thin viscous sublayer next to the wall surface, and the shape of the core region is not of much significance. Consequently, the turbulent flow relations given above for circular tubes can also be used for noncircular tubes with reasonable accuracy by replacing the diameter $D$ in the evaluation of the Reynolds number by the hydraulic diameter $D_h = 4A_c/p$.

**TABLE 19–4**

Equivalent roughness values for new commercial pipes*

| Material | Roughness, $\varepsilon$ | |
| --- | --- | --- |
| | ft | mm |
| Glass, plastic | 0 (smooth) | |
| Concrete | 0.003–0.03 | 0.9–9 |
| Wood stave | 0.0016 | 0.5 |
| Rubber, smoothed | 0.000033 | 0.01 |
| Copper or brass tubing | 0.000005 | 0.0015 |
| Cast iron | 0.00085 | 0.26 |
| Galvanized iron | 0.0005 | 0.15 |
| Wrought iron | 0.00015 | 0.046 |
| Stainless steel | 0.000007 | 0.002 |
| Commercial steel | 0.00015 | 0.045 |

*The uncertainty in these values can be as much as ±60 percent.

**FIGURE 19–32**
In turbulent flow, the velocity profile is nearly a straight line in the core region, and any significant velocity gradients occur in the viscous sublayer.

**FIGURE 19–33**
A double-tube heat exchanger that consists of two concentric tubes.

**TABLE 19–5**

Nusselt number for fully developed laminar flow in an annulus with one surface isothermal and the other adiabatic (Kays and Perkins, 1972)

| $D_i/D_o$ | $Nu_i$ | $Nu_o$ |
|-----------|--------|--------|
| 0         | —      | 3.66   |
| 0.05      | 17.46  | 4.06   |
| 0.10      | 11.56  | 4.11   |
| 0.25      | 7.37   | 4.23   |
| 0.50      | 5.74   | 4.43   |
| 1.00      | 4.86   | 4.86   |

(a) Finned surface

(b) Roughened surface

**FIGURE 19–34**
Tube surfaces are often *roughened, corrugated,* or *finned* in order to *enhance* convection heat transfer.

# Flow through Tube Annulus

Some simple heat transfer equipments consist of two concentric tubes, and are properly called *double-tube heat exchangers* (Fig. 19–33). In such devices, one fluid flows through the tube while the other flows through the annular space. The governing differential equations for both flows are identical. Therefore, steady laminar flow through an annulus can be studied analytically by using suitable boundary conditions.

Consider a concentric annulus of inner diameter $D_i$ and outer diameter $D_o$. The hydraulic diameter of annulus is

$$D_h = \frac{4A_c}{p} = \frac{4\pi(D_o^2 - D_i^2)/4}{\pi(D_o + D_i)} = D_o - D_i$$

Annular flow is associated with two Nusselt numbers—$Nu_i$ on the inner tube surface and $Nu_o$ on the outer tube surface—since it may involve heat transfer on both surfaces. The Nusselt numbers for fully developed laminar flow with one surface isothermal and the other adiabatic are given in Table 19–5. When Nusselt numbers are known, the convection coefficients for the inner and the outer surfaces are determined from

$$Nu_i = \frac{h_i D_h}{k} \quad \text{and} \quad Nu_o = \frac{h_o D_h}{k} \tag{19–87}$$

For fully developed turbulent flow, the inner and outer convection coefficients are approximately equal to each other, and the tube annulus can be treated as a noncircular duct with a hydraulic diameter of $D_h = D_o - D_i$. The Nusselt number in this case can be determined from a suitable turbulent flow relation such as the Gnielinski equation. To improve the accuracy of Nusselt numbers obtained from these relations for annular flow, Petukhov and Roizen (1964) recommend multiplying them by the following correction factors when one of the tube walls is adiabatic and heat transfer is through the other wall:

$$F_i = 0.86 \left(\frac{D_i}{D_o}\right)^{-0.16} \quad \text{(outer wall adiabatic)} \tag{19–88}$$

$$F_o = 0.86 \left(\frac{D_i}{D_o}\right)^{-0.16} \quad \text{(inner wall adiabatic)} \tag{19–89}$$

# Heat Transfer Enhancement

Tubes with rough surfaces have much higher heat transfer coefficients than tubes with smooth surfaces. Therefore, tube surfaces are often intentionally *roughened, corrugated,* or *finned* in order to *enhance* the convection heat transfer coefficient and thus the convection heat transfer rate (Fig. 19–34). Heat transfer in turbulent flow in a tube has been increased by as much as 400 percent by roughening the surface. Roughening the surface, of course, also increases the friction factor and thus the power requirement for the pump or the fan.

The convection heat transfer coefficient can also be increased by inducing pulsating flow by pulse generators, by inducing swirl by inserting a twisted tape into the tube, or by inducing secondary flows by coiling the tube.

## EXAMPLE 19–7    Heating of Water by Resistance Heaters in a Tube

Water is to be heated from 15°C to 65°C as it flows through a 3-cm-internal-diameter 5-m-long tube (Fig. 19–35). The tube is equipped with an electric resistance heater that provides uniform heating throughout the surface of the tube. The outer surface of the heater is well insulated, so that in steady operation all the heat generated in the heater is transferred to the water in the tube. If the system is to provide hot water at a rate of 10 L/min, determine the power rating of the resistance heater. Also, estimate the inner surface temperature of the tube at the exit.

**Solution**   Water is to be heated in a tube equipped with an electric resistance heater on its surface. The power rating of the heater and the inner surface temperature at the exit are to be determined.

**Assumptions**   **1** Steady flow conditions exist. **2** The surface heat flux is uniform. **3** The inner surfaces of the tube are smooth.

**Properties**   The properties of water at the bulk mean temperature of $T_b =$ $(T_i + T_e)/2 = (15 + 65)/2 = 40°C$ are (Table A–15)

$$\rho = 992.1 \text{ kg/m}^3 \qquad\qquad c_p = 4179 \text{ J/kg·°C}$$
$$k = 0.631 \text{ W/m·°C} \qquad\qquad \text{Pr} = 4.32$$
$$\nu = \mu/\rho = 0.658 \times 10^{-6} \text{ m}^2/\text{s}$$

**Analysis**   The cross sectional and heat transfer surface areas are

$$A_c = \tfrac{1}{4}\pi D^2 = \tfrac{1}{4}\pi(0.03 \text{ m})^2 = 7.069 \times 10^{-4} \text{ m}^2$$
$$A_s = \pi DL = \pi(0.03 \text{ m})(5 \text{ m}) = 0.471 \text{ m}^2$$

The volume flow rate of water is given as $\dot{V} = 10$ L/min $= 0.01$ m³/min. Then the mass flow rate becomes

$$\dot{m} = \rho\dot{V} = (992.1 \text{ kg/m}^3)(0.01 \text{ m}^3/\text{min}) = 9.921 \text{ kg/min} = 0.1654 \text{ kg/s}$$

To heat the water at this mass flow rate from 15°C to 65°C, heat must be supplied to the water at a rate of

$$\dot{Q} = \dot{m}c_p(T_e - T_i)$$
$$= (0.1654 \text{ kg/s})(4.179 \text{ kJ/kg·°C})(65 - 15)°C$$
$$= 34.6 \text{ kJ/s} = 34.6 \text{ kW}$$

All of this energy must come from the resistance heater. Therefore, the power rating of the heater must be **34.6 kW**.

The surface temperature $T_s$ of the tube at any location can be determined from

$$\dot{q}_s = h(T_s - T_m) \qquad T_s = T_m + \frac{\dot{q}_s}{h}$$

where $h$ is the heat transfer coefficient and $T_m$ is the mean temperature of the fluid at that location. The surface heat flux is constant in this case, and its value can be determined from

$$\dot{q}_s = \frac{\dot{Q}}{A_s} = \frac{34.6 \text{ kW}}{0.471 \text{ m}^2} = 73.46 \text{ kW/m}^2$$

$\dot{q}_s = $ constant

15°C    Water    $D = 3$ cm    65°C

5 m

**FIGURE 19–35**
Schematic for Example 19–7.

To determine the heat transfer coefficient, we first need to find the mean velocity of water and the Reynolds number:

$$V_{avg} = \frac{\dot{V}}{A_c} = \frac{0.010 \text{ m}^3/\text{min}}{7.069 \times 10^{-4} \text{ m}^2} = 14.15 \text{ m/min} = 0.236 \text{ m/s}$$

$$\text{Re} = \frac{V_{avg} D}{\nu} = \frac{(0.236 \text{ m/s})(0.03 \text{ m})}{0.658 \times 10^{-6} \text{ m}^2/\text{s}} = 10,760$$

which is greater than 10,000. Therefore, the flow is turbulent and the entry length is roughly

$$L_h \approx L_t \approx 10D = 10 \times 0.03 = 0.3 \text{ m}$$

which is much shorter than the total length of the tube. Therefore, we can assume fully developed turbulent flow in the entire tube and determine the Nusselt number from

$$\text{Nu} = \frac{hD}{k} = 0.023 \text{ Re}^{0.8} \text{ Pr}^{0.4} = 0.023(10,760)^{0.8} (4.32)^{0.4} = 69.4$$

Then,

$$h = \frac{k}{D} \text{Nu} = \frac{0.631 \text{ W/m} \cdot °\text{C}}{0.03 \text{ m}} (69.4) = 1460 \text{ W/m}^2 \cdot °\text{C}$$

and the surface temperature of the pipe at the exit becomes

$$T_s = T_m + \frac{\dot{q}_s}{h} = 65°\text{C} + \frac{73,460 \text{ W/m}^2}{1460 \text{ W/m}^2 \cdot °\text{C}} = \mathbf{115°C}$$

**Discussion**  Note that the inner surface temperature of the tube will be 50°C higher than the mean water temperature at the tube exit. This temperature difference of 50°C between the water and the surface will remain constant throughout the fully developed flow region.

**FIGURE 19–36**
Schematic for Example 19–8.

### EXAMPLE 19–8    Heat Loss from the Ducts of a Heating System

Hot air at atmospheric pressure and 80°C enters an 8-m-long uninsulated square duct of cross section 0.2 m × 0.2 m that passes through the attic of a house at a rate of 0.15 m³/s (Fig. 19–36). The duct is observed to be nearly isothermal at 60°C. Determine the exit temperature of the air and the rate of heat loss from the duct to the attic space.

**Solution**  Heat loss from uninsulated square ducts of a heating system in the attic is considered. The exit temperature and the rate of heat loss are to be determined.
**Assumptions**  1 Steady operating conditions exist. 2 The inner surfaces of the duct are smooth. 3 Air is an ideal gas.

*Properties* We do not know the exit temperature of the air in the duct, and thus we cannot determine the bulk mean temperature of air, which is the temperature at which the properties are to be determined. The temperature of air at the inlet is 80°C and we expect this temperature to drop somewhat as a result of heat loss through the duct whose surface is at 60°C. At 80°C and 1 atm we read (Table A–22)

$$\rho = 0.9994 \text{ kg/m}^3 \qquad c_p = 1008 \text{ J/kg·°C}$$
$$k = 0.02953 \text{ W/m·°C} \qquad \text{Pr} = 0.7154$$
$$\nu = 2.097 \times 10^{-5} \text{ m}^2\text{/s}$$

*Analysis* The characteristic length (which is the hydraulic diameter), the mean velocity, and the Reynolds number in this case are

$$D_h = \frac{4A_c}{p} = \frac{4a^2}{4a} = a = 0.2 \text{ m}$$

$$V_{avg} = \frac{\dot{V}}{A_c} = \frac{0.15 \text{ m}^3\text{/s}}{(0.2 \text{ m})^2} = 3.75 \text{ m/s}$$

$$\text{Re} = \frac{V_{avg} D_h}{n} = \frac{(3.75 \text{ m/s})(0.2 \text{ m})}{2.097 \times 10^{-5} \text{ m}^2\text{/s}} = 35{,}765$$

which is greater than 10,000. Therefore, the flow is turbulent and the entry lengths in this case are roughly

$$L_h \approx L_t \approx 10D = 10 \times 0.2 \text{ m} = 2 \text{ m}$$

which is much shorter than the total length of the duct. Therefore, we can assume fully developed turbulent flow in the entire duct and determine the Nusselt number from

$$\text{Nu} = \frac{hD_h}{k} = 0.023 \text{ Re}^{0.8} \text{ Pr}^{0.3} = 0.023(35{,}765)^{0.8}(0.7154)^{0.3} = 91.4$$

Then,

$$h = \frac{k}{D_h} \text{Nu} = \frac{0.02953 \text{ W/m·°C}}{0.2 \text{ m}}(91.4) = 13.5 \text{ W/m}^2\text{·°C}$$

$$A_s = 4aL = 4 \times (0.2 \text{ m})(8 \text{ m}) = 6.4 \text{ m}^2$$

$$\dot{m} = \rho\dot{V} = (0.9994 \text{ kg/m}^3)(0.15 \text{ m}^3\text{/s}) = 0.150 \text{ kg/s}$$

Next, we determine the exit temperature of air from

$$T_e = T_s - (T_s - T_i) \exp(-hA_s/\dot{m}c_p)$$

$$= 60°C - [(60 - 80)°C] \exp\left[-\frac{(13.5 \text{ W/m}^2\text{·°C})(6.4 \text{ m}^2)}{(0.150 \text{ kg/s})(1008 \text{ J/kg·°C})}\right]$$

$$= \mathbf{71.3°C}$$

Then the logarithmic mean temperature difference and the rate of heat loss from the air become

$$\Delta T_{\ln} = \frac{T_i - T_e}{\ln \dfrac{T_s - T_e}{T_s - T_i}} = \frac{80 - 71.3}{\ln \dfrac{60 - 71.3}{60 - 80}} = -15.2°C$$

$$\dot{Q} = hA_s\,\Delta T_{\ln} = (13.5 \text{ W/m}^2\cdot°C)(6.4 \text{ m}^2)(-15.2°C) = \mathbf{-1313 \text{ W}}$$

Therefore, air will lose heat at a rate of 1313 W as it flows through the duct in the attic.

**Discussion** The average fluid temperature is (80 + 71.3)/2 = 75.7°C, which is sufficiently close to 80°C at which we evaluated the properties of air. Therefore, it is not necessary to re-evaluate the properties at this temperature and to repeat the calculations.

## SUMMARY

Convection is the mode of heat transfer that involves conduction as well as bulk fluid motion. The rate of convection heat transfer in external flow is expressed by *Newton's law of cooling* as

$$\dot{Q} = hA_s(T_s - T_\infty)$$

where $T_s$ is the surface temperature and $T_\infty$ is the free-stream temperature. The heat transfer coefficient $h$ is usually expressed in the dimensionless form as the *Nusselt number* as $Nu = hL_c/k$ where $L_c$ is the *characteristic length*. The characteristic length for noncircular tubes is the *hydraulic diameter* $D_h$ defined as $D_h = 4A_c/p$ where $A_c$ is the cross-sectional area of the tube and $p$ is its perimeter. The value of the critical Reynolds number is about $5 \times 10^5$ for flow over a flat plate, $2 \times 10^5$ for flow over cylinders and spheres, and 2300 for flow inside tubes.

The average Nusselt number relations for flow over a flat plate are:

*Laminar:* $Nu = \dfrac{hL}{k} = 0.664\, Re_L^{0.5}\, Pr^{1/3}$, $\qquad Re_L < 5 \times 10^5$

*Turbulent:*

$$Nu = \frac{hL}{k} = 0.037\, Re_L^{0.8}\, Pr^{1/3}, \quad \begin{array}{l} 0.6 \le Pr \le 60 \\ 5 \times 10^5 \le Re_L \le 10^7 \end{array}$$

*Combined:*
$$Nu = \frac{hL}{k} = (0.037\, Re_L^{0.8} - 871)\, Pr^{1/3}, \quad \begin{array}{l} 0.6 \le Pr \le 60 \\ 5 \times 10^5 \le Re_L \le 10^7 \end{array}$$

For isothermal surfaces with an unheated starting section of length $\xi$, the local Nusselt number and the average convection coefficient relations are

*Laminar:* $\quad Nu_x = \dfrac{Nu_{x\,(\text{for } \xi=0)}}{\left[1 - (\xi/x)^{3/4}\right]^{1/3}} = \dfrac{0.332\, Re_x^{0.5}\, Pr^{1/3}}{\left[1 - (\xi/x)^{3/4}\right]^{1/3}}$

*Turbulent:* $\quad Nu_x = \dfrac{Nu_{x\,(\text{for } \xi=0)}}{\left[1 - (\xi/x)^{9/10}\right]^{1/9}} = \dfrac{0.0296\, Re_x^{0.8}\, Pr^{1/3}}{\left[1 - (\xi/x)^{9/10}\right]^{1/9}}$

*Laminar:* $\quad h = \dfrac{2\left[1 - (\xi/x)^{3/4}\right]}{1 - \xi/L}\, h_{x=L}$

*Turbulent:* $\quad h = \dfrac{5\left[1 - (\xi/x)^{9/10}\right]}{(1 - \xi/L)}\, h_{x=L}$

These relations are for the case of *isothermal* surfaces. When a flat plate is subjected to *uniform heat flux,* the local Nusselt number is given by

*Laminar:* $\qquad Nu_x = 0.453\, Re_x^{0.5} Pr^{1/3}$

*Turbulent:* $\qquad Nu_x = 0.0308\, Re_x^{0.8}\, Pr^{1/3}$

The average Nusselt numbers for cross flow over a *cylinder* and *sphere* are

$$Nu_{cyl} = \frac{hD}{k} = 0.3 + \frac{0.62\, Re^{1/2}\, Pr^{1/3}}{\left[1 + (0.4/Pr)^{2/3}\right]^{1/4}}\left[1 + \left(\frac{Re}{282{,}000}\right)^{5/8}\right]^{4/5}$$

which is valid for $Re\,Pr > 0.2$, and

$$Nu_{sph} = \frac{hD}{k} = 2 + \left[0.4\, Re^{1/2} + 0.06\, Re^{2/3}\right]Pr^{0.4}\left(\frac{\mu_\infty}{\mu_s}\right)^{1/4}$$

which is valid for $3.5 \le Re \le 80{,}000$ and $0.7 \le Pr \le 380$. The fluid properties are evaluated at the film temperature $T_f = (T_\infty + T_s)/2$ in the case of a cylinder, and at the free-stream temperature $T_\infty$ (except for $\mu_s$, which is evaluated at the surface temperature $T_s$) in the case of a sphere.

The Reynolds number for internal flow and the hydraulic diameter are defined as

$$Re = \frac{\rho V_{avg} D}{\mu} = \frac{V_{avg} D}{\nu} \quad \text{and} \quad D_h = \frac{4A_c}{p}$$

The flow in a tube is laminar for Re < 2300, turbulent for about Re > 10,000, and transitional in between.

The length of the region from the tube inlet to the point at which the boundary layer merges at the centerline is the *hydrodynamic entry length* $L_h$. The region beyond the entrance region in which the velocity profile is fully developed is the *hydrodynamically fully developed region*. The length of the region of flow over which the thermal boundary layer develops and reaches the tube center is the *thermal entry length* $L_t$. The region in which the flow is both hydrodynamically and thermally developed is the *fully developed flow region*. The entry lengths are given by

$$L_{h, \text{laminar}} \approx 0.05 \, \text{Re} \, D$$

$$L_{t, \text{laminar}} \approx 0.05 \, \text{Re} \, \text{Pr} \, D = \text{Pr} \, L_{h, \text{laminar}}$$

$$L_{h, \text{turbulent}} \approx L_{t, \text{turbulent}} = 10D$$

For $\dot{q}_s$ = constant, the rate of heat transfer is expressed as

$$\dot{Q} = \dot{q}_s A_s = \dot{m} c_p (T_e - T_i)$$

For $T_s$ = constant, we have

$$\dot{Q} = h A_s \Delta T_{\text{ln}} = \dot{m} c_p (T_e - T_i)$$

$$T_e = T_s - (T_s - T_i) \exp(-h A_s / \dot{m} c_p)$$

$$\Delta T_{\text{ln}} = \frac{T_i - T_e}{\ln[(T_s - T_e)/(T_s - T_i)]} = \frac{\Delta T_e - \Delta T_i}{\ln(\Delta T_e / \Delta T_i)}$$

For fully developed laminar flow in a circular pipe, we have

*Circular tube, laminar* ($\dot{q}_s$ = constant):  $\text{Nu} = \dfrac{hD}{k} = 4.36$

*Circular tube, laminar* ($T_s$ = constant):  $\text{Nu} = \dfrac{hD}{k} = 3.66$

For *developing laminar flow* in the entrance region with constant surface temperature, we have

*Circular tube*:  $\text{Nu} = 3.66 + \dfrac{0.065 \, (D/L) \, \text{Re} \, \text{Pr}}{1 + 0.04[(D/L) \, \text{Re} \, \text{Pr}]^{2/3}}$

*Circular tube*:  $\text{Nu} = 1.86 \left( \dfrac{\text{Re} \, \text{Pr} \, D}{L} \right)^{1/3} \left( \dfrac{\mu_b}{\mu_s} \right)^{0.14}$

*Parallel plates*:  $\text{Nu} = 7.54 + \dfrac{0.03 \, (D_h/L) \, \text{Re} \, \text{Pr}}{1 + 0.016[(D_h/L) \, \text{Re} \, \text{Pr}]^{2/3}}$

For *fully developed turbulent flow with smooth surfaces*, we have

$$f = (0.790 \ln \text{Re} - 1.64)^{-2} \qquad 10^4 < \text{Re} < 10^6$$

$$\text{Nu} = 0.125 f \, \text{Re} \, \text{Pr}^{1/3}$$

$$\text{Nu} = 0.023 \, \text{Re}^{0.8} \, \text{Pr}^{1/3} \qquad \left( \begin{array}{l} 0.7 \leq \text{Pr} \leq 160 \\ \text{Re} > 10{,}000 \end{array} \right)$$

$\text{Nu} = 0.023 \, \text{Re}^{0.8} \, \text{Pr}^n$ with $n = 0.4$ for *heating* and 0.3 for *cooling* of fluid

$$\text{Nu} = \dfrac{(f/8)(\text{Re} - 1000) \, \text{Pr}}{1 + 12.7 \, (f/8)^{0.5} \, (\text{Pr}^{2/3} - 1)} \quad \left( \begin{array}{l} 0.5 \leq \text{Pr} \leq 2000 \\ 3 \times 10^3 < \text{Re} < 5 \times 10^6 \end{array} \right)$$

The fluid properties are evaluated at the *bulk mean fluid temperature* $T_b = (T_i + T_e)/2$. For liquid metal flow in the range of $10^4 < \text{Re} < 10^6$ we have:

$T_s$ = constant:  $\text{Nu} = 4.8 + 0.0156 \, \text{Re}^{0.85} \, \text{Pr}_s^{0.93}$

$\dot{q}_s$ = constant:  $\text{Nu} = 6.3 + 0.0167 \, \text{Re}^{0.85} \, \text{Pr}_s^{0.93}$

## REFERENCES AND SUGGESTED READING

**1.** Y. A. Çengel and J. M. Cimbala. *Fluid Mechanics: Fundamentals and Applications.* 2nd ed. New York: McGraw-Hill, 2010.

**2.** S. W. Churchill and M. Bernstein. "A Correlating Equation for Forced Convection from Gases and Liquids to a Circular Cylinder in Cross Flow." *Journal of Heat Transfer* 99 (1977), pp. 300–306.

**3.** S. W. Churchill and H. Ozoe. "Correlations for Laminar Forced Convection in Flow over an Isothermal Flat Plate and in Developing and Fully Developed Flow in an Isothermal Tube." *Journal of Heat Transfer* 95 (Feb. 1973), pp. 78–84.

**4.** A. P. Colburn. *Transactions of the AIChE* 26 (1933), p. 174.

**5.** C. F. Colebrook. "Turbulent flow in Pipes, with Particular Reference to the Transition between the Smooth and Rough Pipe Laws." *Journal of the Institute of Civil Engineers London.* 11 (1939), pp. 133–156.

**6.** R. G. Deissler. "Analysis of Turbulent Heat Transfer and Flow in the Entrance Regions of Smooth Passages." 1953. Referred to in *Handbook of Single-Phase Convective Heat Transfer,* ed. S. Kakaç, R. K. Shah, and W. Aung. New York: Wiley Interscience, 1987.

7. D. F. Dipprey and D. H. Sabersky. "Heat and Momentum Transfer in Smooth and Rough Tubes at Various Prandtl Numbers." *International Journal of Heat Mass Transfer* 6 (1963), pp. 329–353.

8. F. W. Dittus and L. M. K. Boelter. *University of California Publications on Engineering* 2 (1930), p. 433.

9. D. K. Edwards, V. E. Denny, and A. F. Mills. *Transfer Processes.* 2nd ed. Washington, D.C.: Hemisphere, 1979.

10. W. H. Giedt. "Investigation of Variation of Point Unit-Heat Transfer Coefficient around a Cylinder Normal to an Air Stream." *Transactions of the ASME* 71 (1949), pp. 375–381.

11. V. Gnielinski. "New Equations for Heat and Mass Transfer in Turbulent Pipe and Channel Flow." *International Chemical Engineering* 16 (1976), pp. 359–368.

12. S. E. Haaland. "Simple and Explicit Formulas for the Friction Factor in Turbulent Pipe Flow." *Journal of Fluids Engineering* (March 1983), pp. 89–90.

13. M. Jakob. *Heat Transfer.* Vol. 1. New York: John Wiley & Sons, 1949.

14. S. Kakaç, R. K. Shah, and W. Aung, eds. *Handbook of Single-Phase Convective Heat Transfer.* New York: Wiley Interscience, 1987.

15. W. M. Kays and H. C. Perkins. Chapter 7. In *Handbook of Heat Transfer,* ed. W. M. Rohsenow and J. P. Hartnett. New York: McGraw-Hill, 1972.

16. W. M. Kays and M. E. Crawford. *Convective Heat and Mass Transfer.* 3rd ed. New York: McGraw-Hill, 1993.

17. M. Molki and E. M. Sparrow. "An Empirical Correlation for the Average Heat Transfer Coefficient in Circular Tubes." *Journal of Heat Transfer* 108 (1986), pp. 482–484.

18. L. F. Moody. "Friction Factors for Pipe Flows." *Transactions of the ASME* 66 (1944), pp. 671–684.

19. R. H. Norris. "Some Simple Approximate Heat Transfer Correlations for Turbulent Flow in Ducts with Rough Surfaces." In *Augmentation of Convective Heat Transfer,* ed. A. E. Bergles and R. L. Webb. New York: ASME, 1970.

20. B. S. Petukhov and L. I. Roizen. "Generalized Relationships for Heat Transfer in a Turbulent Flow of a Gas in Tubes of Annular Section." *High Temperature* (USSR) 2 (1964), pp. 65–68.

21. B. S. Petukhov. "Heat Transfer and Friction in Turbulent Pipe Flow with Variable Physical Properties." In *Advances in Heat Transfer,* ed. T. F. Irvine and J. P. Hartnett, Vol. 6. New York: Academic Press, 1970.

22. H. Schlichting. *Boundary Layer Theory,* 7th ed. New York, McGraw-Hill, 1979.

23. R. K. Shah and M. S. Bhatti. "Laminar Convective Heat Transfer in Ducts." In *Handbook of Single-Phase Convective Heat Transfer,* ed. S. Kakaç, R. K. Shah, and W. Aung. New York: Wiley Interscience, 1987.

24. E. N. Sieder and G. E. Tate. "Heat Transfer and Pressure Drop of Liquids in Tubes." *Industrial Engineering Chemistry* 28 (1936), pp. 1429–1435.

25. C. A. Sleicher and M. W. Rouse. "A Convenient Correlation for Heat Transfer to Constant and Variable Property Fluids in Turbulent Pipe Flow." *International Journal of Heat Mass Transfer* 18 (1975), pp. 1429–1435.

26. W. C. Thomas. "Note on the Heat Transfer Equation for Forced Convection Flow over a Flat Plate with an Unheated Starting Length." *Mechanical Engineering News,* 9, no.1 (1977), p. 361.

27. S. Whitaker. "Forced Convection Heat Transfer Correlations for Flow in Pipes, Past Flat Plates, Single Cylinders, and for Flow in Packed Beds and Tube Bundles." *AIChE Journal* 18 (1972), pp. 361–371.

# PROBLEMS*

## Physical Mechanism of Convection

**19–1C** Define incompressible flow and incompressible fluid. Must the flow of a compressible fluid necessarily be treated as compressible?

**19–2C** What is forced convection? How does it differ from natural convection? Is convection caused by winds forced or natural convection?

**19–3C** What is external forced convection? How does it differ from internal forced convection? Can a heat transfer system involve both internal and external convection at the same time? Give an example.

**19–4C** In which mode of heat transfer is the convection heat transfer coefficient usually higher, natural convection or forced convection? Why?

**19–5C** Consider a hot baked potato. Will the potato cool faster or slower when we blow the warm air coming from our lungs on it instead of letting it cool naturally in the cooler air in the room? Explain.

**19–6C** What is the physical significance of the Nusselt number? How is it defined?

**19–7C** When is heat transfer through a fluid conduction and when is it convection? For what case is the rate of heat transfer higher? How does the convection heat transfer coefficient differ from the thermal conductivity of a fluid?

**19–8** During air cooling of potatoes, the heat transfer coefficient for combined convection, radiation, and evaporation is determined experimentally to be as shown:

| Air Velocity, m/s | Heat Transfer Coefficient, W/m²·K |
|---|---|
| 0.66 | 14.0 |
| 1.00 | 19.1 |
| 1.36 | 20.2 |
| 1.73 | 24.4 |

Consider an 8-cm-diameter potato initially at 20°C with a thermal conductivity of 0.49 W/m·K. Potatoes are cooled by refrigerated air at 5°C at a velocity of 1 m/s. Determine the initial rate of heat transfer from a potato, and the initial value of the temperature gradient in the potato at the surface.
*Answers: 5.8 W, −585°C/m*

**19–9** An average man has a body surface area of 1.8 m² and a skin temperature of 33°C. The convection heat transfer coefficient for a clothed person walking in still air is expressed as $h = 8.6V^{0.53}$ for $0.5 < V < 2$ m/s, where $V$ is the walking velocity in m/s. Assuming the average surface temperature of the clothed person to be 30°C, determine the rate of heat loss from an average man walking in still air at 7°C by convection at a walking velocity of (a) 0.5 m/s, (b) 1.0 m/s, (c) 1.5 m/s, and (d) 2.0 m/s.

**19–10** The upper surface of a 50-cm-thick solid plate ($k = 237$ W/m·K) is being cooled by water with temperature of 20°C. The upper and lower surfaces of the solid plate maintained at constant temperatures of 60°C and 120°C, respectively. Determine the water convection heat transfer coefficient and the water temperature gradient at the upper plate surface.

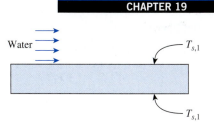

**FIGURE P19–10**

**19–11** Consider airflow over a plate surface maintained at a temperature of 220°C. The temperature profile of the airflow is given as

$$T(y) = T_\infty - (T_\infty - T_s)\exp\left(-\frac{V}{\alpha_{\text{fluid}}}y\right)$$

The airflow at 1 atm has a free stream velocity and temperature of 0.08 m/s and 20°C, respectively. Determine the heat flux on the plate surface and the convection heat transfer coefficient of the airflow. *Answers: $1.45 \times 10^4$ W/m², 72.6 W/m²·K*

**19–12** The convection heat transfer coefficient for a clothed person standing in moving air is expressed as $h = 14.8V^{0.69}$ for $0.15 < V < 1.5$ m/s, where $V$ is the air velocity. For a person with a body surface area of 1.7 m² and an average surface temperature of 29°C, determine the rate of heat loss from the person in windy air at 10°C by convection for air velocities of (a) 0.5 m/s, (b) 1.0 m/s, and (c) 1.5 m/s.

**19–13** During air cooling of oranges, grapefruit, and tangelos, the heat transfer coefficient for combined convection, radiation, and evaporation for air velocities of $0.11 < V < 0.33$ m/s is determined experimentally and is expressed as $h = 5.05\,k_{\text{air}}\text{Re}^{1/3}/D$, where the diameter $D$ is the characteristic length. Oranges are cooled by refrigerated air at 3°C and 1 atm at a velocity of 0.3 m/s. Determine (a) the initial rate of heat transfer from a 7-cm-diameter orange initially at 15°C with a thermal conductivity of 0.70 W/m·K, (b) the value of the initial temperature gradient inside the orange at the surface, and (c) the value of the Nusselt number.

Air
3°C
1 atm

Orange

**FIGURE P19–13**

## Flow over Flat Plates

**19–14C** What does the friction coefficient represent in flow over a flat plate? How is it related to the drag force acting on the plate?

**19–15** In an experiment, the local heat transfer over a flat plate was correlated in the form of local Nusselt number as expressed by the following correlation

$$Nu_x = 0.035 Re_x^{0.8} Pr^{1/3}$$

Determine the ratio of the average convection heat transfer coefficient ($h$) over the entire plate length to the local convection heat transfer coefficient ($h_x$) at $x = L$. *Answer: 1.25*

**19–16** A 5-m long strip of sheet metal is being transported on a conveyor at a velocity of 5 m/s, while the coating on the upper surface is being cured by infrared lamps. The coating on the upper surface of the metal strip has an absorptivity of 0.6 and an emissivity of 0.7, while the surrounding ambient air temperature is 25°C. Radiation heat transfer occurs only on the upper surface, while convection heat transfer occurs on both upper and lower surfaces of the sheet metal. If the infrared lamps supply a constant heat flux of 5000 W/m², determine the surface temperature of the sheet metal. Evaluate the properties of air at 80°C. *Answer: 138°C*

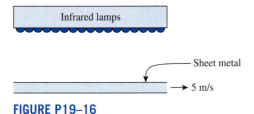

**FIGURE P19–16**

**19–17** The local atmospheric pressure in Denver, Colorado (elevation 1610 m), is 83.4 kPa. Air at this pressure and at 30°C flows with a velocity of 6 m/s over a 2.5-m × 8-m flat plate whose temperature is 120°C. Determine the rate of heat transfer from the plate if the air flows parallel to the (a) 8-m-long side and (b) the 2.5-m side.

**19–18** During a cold winter day, wind at 42 km/h is blowing parallel to a 6-m-high and 10-m-long wall of a house. If the air outside is at 5°C and the surface temperature of the wall is 12°C, determine the rate of heat loss from that wall by convection. What would your answer be if the wind velocity was doubled? *Answers: 10.8 kW, 19.4 kW*

**FIGURE P19–18**

**19–19** Reconsider Prob. 19–18. Using EES (or other) software, investigate the effects of wind velocity and outside air temperature on the rate of heat loss from the wall by convection. Let the wind velocity vary from 10 km/h to 80 km/h and the outside air temperature from 0°C to 10°C. Plot the rate of heat loss as a function of the wind velocity and of the outside temperature, and discuss the results.

**19–20** Water at 43.3°C flows over a large plate at a velocity of 30.0 cm/s. The plate is 1.0 m long (in the flow direction), and its surface is maintained at a uniform temperature of 10.0°C. Calculate the steady rate of heat transfer per unit width of the plate.

**19–21E** Warm air is blown over the inner surface of an automobile windshield to defrost ice accumulated on the outer surface of the windshield. Consider an automobile windshield ($k_w = 0.8$ Btu/h·ft·R) with an overall height of 20 in. and thickness of 0.2 in. The outside air (1 atm) ambient temperature is 8°F and the average airflow velocity over the outer windshield surface is 50 mph, while the ambient temperature inside the automobile is 77°F. Determine the value of the convection heat transfer coefficient, for the warm air blowing over the inner surface of the windshield, necessary to cause the accumulated ice to begin melting. Assume the windshield surface can be treated as a flat plate surface. *Answer: 5.36 Btu/h·ft²·R*

**19–22** Hot carbon dioxide exhaust gas at 1 atm is being cooled by flat plates. The gas at 220°C flows in parallel over the upper and lower surfaces of a 1.5 m-long flat plate at a velocity of 3 m/s. If the flat plate surface temperature is maintained at 80°C, determine (a) the local convection heat transfer coefficient at 1 m from the leading edge, (b) the average convection heat transfer coefficient over the entire plate, and (c) the total heat flux transfer to the plate.

**19–23** Hot engine oil at 150°C is flowing in parallel over a flat plate at a velocity of 2 m/s. Surface temperature of the 0.5-m long flat plate is constant at 50°C. Determine (a) the local convection heat transfer coefficient at 0.2 m from the leading edge and the average convection heat transfer coefficient, and (b) repeat part (a) using the Churchill and Ozoe (1973) relation.

**19–24** Mercury at 25°C flows over a 3-m-long and 2-m-wide flat plate maintained at 75°C with a velocity of 0.8 m/s. Determine the rate of heat transfer from the entire plate.

Mercury
$V = 0.8$ m/s
$T_\infty = 25°C$
$T_s = 75°C$

$L$

**FIGURE P19–24**

**19–25** Parallel plates form a solar collector that covers a roof, as shown in the figure. The plates are maintained at 15°C, while ambient air at 10°C flows over the roof with $V = 4$ m/s. Determine the rate of convective heat loss from (a) the first plate and (b) the third plate.

1 m

4 m

$V, T_\infty$

**FIGURE P19–25**

**19–26** Consider a hot automotive engine, which can be approximated as a 0.5-m-high, 0.40-m-wide, and 0.8-m-long rectangular block. The bottom surface of the block is at a temperature of 100°C and has an emissivity of 0.95. The ambient air is at 20°C, and the road surface is at 25°C. Determine the rate of heat transfer from the bottom surface of the engine block by convection and radiation as the car travels at a velocity of 80 km/h. Assume the flow to be turbulent over the entire surface because of the constant agitation of the engine block.

**19–27** The top surface of the passenger car of a train moving at a velocity of 95 km/h is 2.8 m wide and 8 m long. The top surface is absorbing solar radiation at a rate of 380 W/m²,

and the temperature of the ambient air is 30°C. Assuming the roof of the car to be perfectly insulated and the radiation heat exchange with the surroundings to be small relative to convection, determine the equilibrium temperature of the top surface of the car. *Answer:* 37.5°C

380 W/m²

Air
30°C

$V = 95$ km/h

**FIGURE P19–27**

**19–28** ![EES] Reconsider Prob. 19–27. Using EES (or other) software, investigate the effects of the train velocity and the rate of absorption of solar radiation on the equilibrium temperature of the top surface of the car. Let the train velocity vary from 10 km/h to 120 km/h and the rate of solar absorption from 100 W/m² to 500 W/m². Plot the equilibrium temperature as functions of train velocity and solar radiation absorption rate, and discuss the results.

**19–29** A 15-cm × 15-cm circuit board dissipating 20 W of power uniformly is cooled by air, which approaches the circuit board at 20°C with a velocity of 6 m/s. Disregarding any heat transfer from the back surface of the board, determine the surface temperature of the electronic components (a) at the leading edge and (b) at the end of the board. Assume the flow to be turbulent since the electronic components are expected to act as turbulators.

**19–30E** Consider a refrigeration truck traveling at 70 mph at a location where the air temperature is 80°F. The refrigerated compartment of the truck can be considered to be a 9-ft-wide, 7-ft-high, and 20-ft-long rectangular box. The refrigeration system of the truck can provide 3 tons of refrigeration (i.e., it can remove heat at a rate of 600 Btu/min). The outer surface of the truck is coated with a low-emissivity material, and thus radiation heat transfer is very small. Determine the average temperature of the outer surface of the refrigeration compartment of the truck if the refrigeration system is observed to be operating at half the capacity. Assume the air flow over the entire outer surface to be turbulent and the heat transfer coefficient at the front and rear surfaces to be equal to that on side surfaces.

**FIGURE P19–30E**

**FIGURE P19–35**

**19–31** Liquid mercury at 250°C is flowing in parallel over a flat plate at a velocity of 0.3 m/s. Surface temperature of the 0.1-m-long flat plate is constant at 50°C. Determine (a) the local convection heat transfer coefficient at 5 cm from the leading edge and (b) the average convection heat transfer coefficient over the entire plate. *Answers:* (a) 5343 W/m²·K, (b) 7555 W/m²·K

**19–32** Liquid mercury at 250°C is flowing with a velocity of 0.3 m/s in parallel over a 0.1-m long flat plate where there is an unheated starting length of 5 cm. The heated section of the flat plate is maintained at a constant temperature of 50°C. Determine (a) the local convection heat transfer coefficient at the trailing edge, (b) the average convection heat transfer coefficient for the heated section, and (c) the rate of heat transfer per unit width for the heated section.

**19–33E** Air at 60°F flows over a 10-ft-long flat plate at 7 ft/s. Determine the local heat transfer coefficient at intervals of 1 ft, and plot the results against the distance from the leading edge.

**19–34E** Reconsider Prob. 19–33E. Using EES (or other) software, evaluate the local heat transfer coefficient along the plate at intervals of 0.1 ft, and plot them against the distance from the leading edge.

**19–35** An array of power transistors, dissipating 5 W of power each, are to be cooled by mounting them on a 25-cm × 25-cm square aluminum plate and blowing air at 35°C over the plate with a fan at a velocity of 4 m/s. The average temperature of the plate is not to exceed 65°C. Assuming the heat transfer from the back side of the plate to be negligible and disregarding radiation, determine the number of transistors that can be placed on this plate.

**19–36** Repeat Prob. 19–35 for a location at an elevation of 1610 m where the atmospheric pressure is 83.4 kPa. *Answer:* 5

**19–37** Air at 25°C and 1 atm is flowing over a long flat plate with a velocity of 8 m/s. Determine the distance from the leading edge of the plate where the flow becomes turbulent, and the thickness of the boundary layer at that location.

**19–38** Repeat Prob. 19–37 for water.

**19–39** Solar radiation is incident on the glass cover of a solar collector at a rate of 700 W/m². The glass transmits 88 percent of the incident radiation and has an emissivity of 0.90. The entire hot water needs of a family in summer can be met by two collectors 1.2 m high and 1 m wide. The two collectors are attached to each other on one side so that they appear like a single collector 1.2 m × 2 m in size. The temperature of the glass cover is measured to be 35°C on a day when the surrounding air temperature is 25°C and the wind is blowing at 30 km/h. The effective sky temperature for radiation exchange between the glass cover and the open sky is −40°C. Water enters the tubes attached to the absorber plate at a rate of 1 kg/min. Assuming the back surface of the absorber plate to be

**FIGURE P19–39**

heavily insulated and the only heat loss to occur through the glass cover, determine (a) the total rate of heat loss from the collector, (b) the collector efficiency, which is the ratio of the amount of heat transferred to the water to the solar energy incident on the collector, and (c) the temperature rise of water as it flows through the collector.

**19–40**   A 15-mm × 15-mm silicon chip is mounted such that the edges are flush in a substrate. The chip dissipates 1.4 W of power uniformly, while air at 20°C (1 atm) with a velocity of 25 m/s is used to cool the upper surface of the chip. If the substrate provides an unheated starting length of 15 mm, determine the surface temperature at the trailing edge of the chip. Evaluate the air properties at 50°C.

**19–41**   Air at 1 atm and 20°C is flowing over the top surface of a 0.5-m long thin flat plate. The air stream velocity is 50 m/s and the plate is maintained at a constant surface temperature of 180°C. Determine (a) the average convection heat transfer coefficient, and (b) repeat part (a) using the modified Reynolds analogy. *Answers:* (a) 89.5 W/m²·K, (b) 89.5 W/m²·K

Air, $T_\infty$

**FIGURE P19–41**

## Flow across Cylinders and Spheres

**19–42C**   Consider laminar flow of air across a hot circular cylinder. At what point on the cylinder will the heat transfer be highest? What would your answer be if the flow were turbulent?

**19–43**   A heated long cylindrical rod is placed in a cross flow of air at 20°C (1 atm) with velocity of 10 m/s. The rod has a diameter of 5 mm and its surface has an emissivity of 0.95. If the surrounding temperature is 20°C and the heat flux dissipated from the rod is 16000 W/m², determine the surface temperature of the rod. Evaluate the air properties at 70°C.

**19–44**   A stainless steel ball ($\rho$ = 8055 kg/m³, $c_p$ = 480 J/kg·K) of diameter $D$ = 15 cm is removed from the oven at a uniform temperature of 350°C. The ball is then subjected to the flow of air at 1 atm pressure and 30°C with a velocity of 6 m/s. The surface temperature of the ball eventually drops to 250°C. Determine the average convection heat transfer coefficient during this cooling process and estimate how long this process has taken.

**19–45**   Reconsider Prob. 19–44. Using EES (or other) software, investigate the effect of air velocity on the average convection heat transfer coefficient and the cooling time. Let the air velocity vary from 1 m/s to 10 m/s. Plot the heat transfer coefficient and the cooling time as a function of air velocity, and discuss the results.

**19–46E**   A person extends his uncovered arms into the windy air outside at 54°F and 30 mph in order to feel nature closely. Initially, the skin temperature of the arm is 84°F. Treating the arm as a 2-ft-long and 3-in-diameter cylinder, determine the rate of heat loss from the arm.

Air
54°F, 30 mph

84°F

**FIGURE P19–46E**

**19–47E**   Reconsider Prob. 19–46E. Using EES (or other) software, investigate the effects of air temperature and wind velocity on the rate of heat loss from the arm. Let the air temperature vary from 20°F to 80°F and the wind velocity from 10 mph to 40 mph. Plot the rate of heat loss as a function of air temperature and of wind velocity, and discuss the results.

**19–48**   An average person generates heat at a rate of 84 W while resting. Assuming one-quarter of this heat is lost from the head and disregarding radiation, determine the average surface temperature of the head when it is not covered and is subjected to winds at 10°C and 25 km/h. The head can be approximated as a 30-cm-diameter sphere. *Answer:* 13.2°C

**19–49**   A long 12-cm-diameter steam pipe whose external surface temperature is 90°C passes through some open area that is not protected against the winds. Determine the rate of heat loss from the pipe per unit of its length when the air is at 1 atm pressure and 7°C and the wind is blowing across the pipe at a velocity of 65 km/h.

**19–50**   In a geothermal power plant, the used geothermal water at 80°C enters a 15-cm-diameter and 400-m-long uninsulated pipe at a rate of 8.5 kg/s and leaves at 70°C before being reinjected back to the ground. Windy air at 15°C flows normal to the pipe. Disregarding radiation, determine the average wind velocity in km/h.

**FIGURE P19–50**

**19–51** A 5-mm-diameter electrical transmission line carries an electric current of 50 A and has a resistance of 0.002 ohm per meter length. Determine the surface temperature of the wire during a windy day when the air temperature is 10°C and the wind is blowing across the transmission line at 50 km/h.

**FIGURE P19–51**

**19–52** Reconsider Prob. 19–51. Using EES (or other) software, investigate the effect of the wind velocity on the surface temperature of the wire. Let the wind velocity vary from 10 km/h to 80 km/h. Plot the surface temperature as a function of wind velocity, and discuss the results.

**19–53** A heating system is to be designed to keep the wings of an aircraft cruising at a velocity of 900 km/h above freezing temperatures during flight at 12,200-m altitude where the standard atmospheric conditions are −55.4°C and 18.8 kPa. Approximating the wing as a cylinder of elliptical cross section whose minor axis is 50 cm and disregarding radiation, determine the average convection heat transfer coefficient on the wing surface and the average rate of heat transfer per unit surface area.

**19–54** A long aluminum wire of diameter 3 mm is extruded at a temperature of 280°C. The wire is

**FIGURE P19–54**

subjected to cross air flow at 20°C at a velocity of 6 m/s. Determine the rate of heat transfer from the wire to the air per meter length when it is first exposed to the air.

**19–55** An incandescent lightbulb is an inexpensive but highly inefficient device that converts electrical energy into light. It converts about 10 percent of the electrical energy it consumes into light while converting the remaining 90 percent into heat. (A fluorescent lightbulb will give the same amount of light while consuming only one-fourth of the electrical energy, and it will last 10 times longer than an incandescent lightbulb.) The glass bulb of the lamp heats up very quickly as a result of absorbing all that heat and dissipating it to the surroundings by convection and radiation.

Consider a 10-cm-diameter 100-W lightbulb cooled by a fan that blows air at 30°C to the bulb at a velocity of 2 m/s. The surrounding surfaces are also at 30°C, and the emissivity of the glass is 0.9. Assuming 10 percent of the energy passes through the glass bulb as light with negligible absorption and the rest of the energy is absorbed and dissipated by the bulb itself, determine the equilibrium temperature of the glass bulb.

**FIGURE P19–55**

**19–56E** A 12-ft-long, 1.5-kW electrical resistance wire is made of 0.1-in-diameter stainless steel ($k = 8.7$ Btu/h·ft·°F). The resistance wire operates in an environment at 85°F. Determine the surface temperature of the wire if it is cooled by a fan blowing air at a velocity of 20 ft/s.

**FIGURE P19–56E**

**19–57** The components of an electronic system are located in a 1.5-m-long horizontal duct whose cross section is 20 cm × 20 cm. The components in the duct are not allowed to come into direct contact with cooling air, and thus are cooled by air at 30°C flowing over the duct with a velocity of 200 m/min. If the surface temperature of the duct is not to exceed 65°C, determine the total power rating of the electronic devices that can be mounted into the duct. *Answer: 640 W*

**FIGURE P19–57**

**19–58** Repeat Prob. 19–57 for a location at 3000-m altitude where the atmospheric pressure is 70.12 kPa.

**19–59** A 0.4-W cylindrical electronic component with diameter 0.3 cm and length 1.8 cm and mounted on a circuit board is cooled by air flowing across it at a velocity of 240 m/min. If the air temperature is 35°C, determine the surface temperature of the component.

**19–60** Consider a 50-cm-diameter and 95-cm-long hot water tank. The tank is placed on the roof of a house. The water inside the tank is heated to 80°C by a flat-plate solar collector during the day. The tank is then exposed to windy air at 18°C with an average velocity of 40 km/h during the night. Estimate the temperature of the tank after a 45-min period. Assume the tank surface to be at the same temperature as the water inside, and the heat transfer coefficient on the top and bottom surfaces to be the same as that on the side surface.

**19–61** Reconsider Prob. 19–60. Using EES (or other) software, plot the temperature of the tank as a function of the cooling time as the time varies from 30 min to 5 h, and discuss the results.

**19–62** A 2.2-m-diameter spherical tank of negligible thickness contains iced water at 0°C. Air at 25°C flows over the tank with a velocity of 6 m/s. Determine the rate of heat transfer to the tank and the rate at which ice melts. The heat of fusion of water at 0°C is 333.7 kJ/kg.

**19–63** During a plant visit, it was noticed that a 12-m-long section of a 12-cm-diameter steam pipe is completely exposed to the ambient air. The temperature measurements indicate that the average temperature of the outer surface of the steam pipe is 75°C when the ambient temperature is 5°C. There are also light winds in the area at 25 km/h. The emissivity of the outer surface of the pipe is 0.8, and the average temperature of the surfaces surrounding the pipe, including the sky, is estimated to be 0°C. Determine the amount of heat lost from the steam during a 10-h-long work day.

Steam is supplied by a gas-fired steam generator that has an efficiency of 80 percent, and the plant pays $1.05/therm of natural gas. If the pipe is insulated and 90 percent of the heat loss is saved, determine the amount of money this facility will save a year as a result of insulating the steam pipes. Assume the plant operates every day of the year for 10 h. State your assumptions.

**FIGURE P19–63**

**19–64** Reconsider Prob. 19–63. There seems to be some uncertainty about the average temperature of the surfaces surrounding the pipe used in radiation calculations, and you are asked to determine if it makes any significant difference in overall heat transfer. Repeat the calculations for average surrounding and surface temperatures of −20°C and 25°C, respectively, and determine the change in the values obtained.

**19–65** A 10-cm-diameter, 30-cm-high cylindrical bottle contains cold water at 3°C. The bottle is placed in windy air at 27°C. The water temperature is measured to be 11°C after 45 min of cooling. Disregarding radiation effects and heat transfer from the top and bottom surfaces, estimate the average wind velocity.

**19–66** A 0.2 m × 0.2 m street sign surface has an absorptivity of 0.6 and an emissivity of 0.7, while the street sign is subjected to a cross flow wind at 20°C with a velocity of 1 m/s. Solar radiation is incident on the street sign at a rate of 1100 W/m², and the surrounding temperature is 20°C. Determine the surface temperature of the street sign. Evaluate the air properties at 30°C.

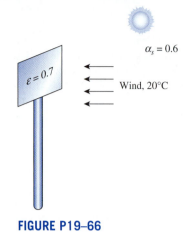

$\alpha_s = 0.6$

$\varepsilon = 0.7$

Wind, 20°C

**FIGURE P19–66**

## Flow in Tubes

**19–67C** What is the physical significance of the number of transfer units NTU $= hA_s/\dot{m}c_p$? What do small and large NTU values tell about a heat transfer system?

**19–68C** Consider the flow of mercury (a liquid metal) in a tube. How will the hydrodynamic and thermal entry lengths compare if the flow is laminar? How would they compare if the flow were turbulent?

**19–69C** What do the average velocity $V_{avg}$ and the mean temperature $T_m$ represent in flow through circular tubes of constant diameter?

**19–70C** What does the logarithmic mean temperature difference represent for flow in a tube whose surface temperature is constant? Why do we use the logarithmic mean temperature instead of the arithmetic mean temperature?

**19–71C** How is the thermal entry length defined for flow in a tube? In what region is the flow in a tube fully developed?

**19–72C** Consider laminar forced convection in a circular tube. Will the heat flux be higher near the inlet of the tube or near the exit? Why?

**19–73C** Consider turbulent forced convection in a circular tube. Will the heat flux be higher near the inlet of the tube or near the exit? Why?

**19–74C** Consider fluid flow in a tube whose surface temperature remains constant. What is the appropriate temperature difference for use in Newton's law of cooling with an average heat transfer coefficient?

**19–75** Air enters an 18-cm-diameter 12-m-long underwater duct at 50°C and 1 atm at a mean velocity of 7 m/s, and is cooled by the water outside. If the average heat transfer coefficient is 65 W/m²·K and the tube temperature is nearly equal to the water temperature of 10°C, determine the exit temperature of air and the rate of heat transfer.

**19–76** Cooling water available at 10°C is used to condense steam at 30°C in the condenser of a power plant at a rate of 0.15 kg/s by circulating the cooling water through a bank of 5-m-long 1.2-cm-internal-diameter thin copper tubes. Water enters the tubes at a mean velocity of 4 m/s and leaves at a temperature of 24°C. The tubes are nearly isothermal at 30°C. Determine the average heat transfer coefficient between the water, the tubes, and the number of tubes needed to achieve the indicated heat transfer rate in the condenser.

**19–77** Repeat Prob. 19–76 for steam condensing at a rate of 0.60 kg/s.

**19–78** Combustion gases passing through a 5-cm-internal-diameter circular tube are used to vaporize waste water at atmospheric pressure. Hot gases enter the tube at 115 kPa and 250°C at a mean velocity of 5 m/s, and leave at 150°C. If the average heat transfer coefficient is 120 W/m²·K and the inner surface temperature of the tube is 110°C, determine (a) the tube length and (b) the rate of evaporation of water.

**19–79** Repeat Prob. 19–78 for a heat transfer coefficient of 40 W/m²·K.

## Laminar and Turbulent Flow

**19–80** Determine the convection heat transfer coefficient for the flow of (a) air and (b) water at a velocity of 5 m/s in an 8-cm-diameter and 10-m-long tube when the tube is subjected to uniform heat flux from all surfaces. Use fluid properties at 25°C.

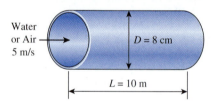

Water or Air
5 m/s

$D = 8$ cm

$L = 10$ m

**FIGURE P19–80**

**19–81** An 8-m long, uninsulated square duct of cross section 0.2 m × 0.2 m and relative roughness $10^{-3}$ passes through the attic space of a house. Hot air enters the duct at 1 atm and 80°C at a volume flow rate of 0.15 m³/s. The duct surface is nearly isothermal at 60°C. Determine the rate of heat loss from the duct to the attic space and the pressure difference between the inlet and outlet sections of the duct.

**19–82** A 12-m long and 12-mm inner-diameter pipe made of commercial steel is used to heat a liquid in an industrial process. The liquid enters the pipe with $T_i = 25°C$, $V = 0.8$ m/s. A uniform heat flux is maintained by an electric resistance heater wrapped around the outer surface of the pipe, so that the fluid exits at 75°C. Assuming fully developed flow and taking the average fluid properties to be $\rho = 1000$ kg/m³, $c_p = 4000$ J/kg·K, $\mu = 2 \times 10^{-3}$ kg/m·s, $k = 0.48$ W/m·K, and Pr = 10, determine:
(a) The required surface heat flux $\dot{q}_s$, produced by the heater
(b) The surface temperature at the exit, $T_s$

**19–83** Determine the hydrodynamic and thermal entry lengths for water, engine oil, and liquid mercury flowing through a 2.5-cm-diameter smooth tube with mass flow rate of 0.01 kg/s and temperature of 100°C.

**19–84** Water is to be heated from 10°C to 80°C as it flows through a 2-cm-internal-diameter, 13-m-long tube. The tube is equipped with an electric resistance heater, which provides uniform heating throughout the surface of the tube. The outer surface of the heater is well insulated, so that in steady operation all the heat generated in the heater is transferred to the water in the tube. If the system is to provide hot water at a rate of 5 L/min, determine the power rating of the resistance heater. Also, estimate the inner surface temperature of the pipe at the exit.

**19–85** Hot air at atmospheric pressure and 75°C enters a 10-m-long uninsulated square duct of cross section 0.15 m × 0.15 m that passes through the attic of a house at a rate of 0.2 m³/s. The duct is observed to be nearly isothermal at 70°C. Determine the exit temperature of the air and the rate of heat loss from the duct to the air space in the attic. *Answers: 72.2°C, 582 W*

**FIGURE P19–85**

**19–86** [EES] Reconsider Prob. 19–85. Using EES (or other) software, investigate the effect of the volume flow rate of air on the exit temperature of air and the rate of heat loss. Let the flow rate vary from 0.05 m³/s to 0.15 m³/s. Plot the exit temperature and the rate of heat loss as a function of flow rate, and discuss the results.

**19–87** Consider a fluid with a Prandtl number of 7 flowing through a smooth circular tube. Using the Colburn, Petukhov, and Gnielinski equations, determine the Nusselt numbers for Reynolds numbers at 3500, 10⁴, and 5 × 10⁵. Compare and discuss the results.

**19–88** A concentric annulus tube has inner and outer diameters of 25 mm and 100 mm, respectively. Liquid water flows at a mass flow rate of 0.05 kg/s through the annulus with the inlet and outlet mean temperatures of 20°C and 80°C, respectively. The inner tube wall is maintained with a constant surface temperature of 120°C, while the outer tube surface is insulated. Determine the length of the concentric annulus tube. *Answer: 38.5 m*

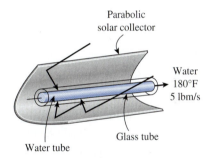

**FIGURE P19–88**

**19–89E** The hot water needs of a household are to be met by heating water at 55°F to 180°F by a parabolic solar collector at a rate of 5 lbm/s. Water flows through a 1.25-in-diameter thin aluminum tube whose outer surface is blackanodized in order to maximize its solar absorption ability. The centerline of the tube coincides with the focal line of the collector, and a glass sleeve is placed outside the tube to minimize the heat losses. If solar energy is transferred to water at a net rate of 350 Btu/h per ft length of the tube, determine the required length of the parabolic collector to meet the hot water requirements of this house. Also, determine the surface temperature of the tube at the exit.

**FIGURE P19–89E**

**19–90E** Determine the average velocity, hydrodynamic and thermal entry lengths for water, engine oil, and liquid mercury flowing through a standard 2-in Schedule 40 pipe with mass flow rate of 0.1 lbm/s and temperature of 100°F.

**19–91** In a manufacturing plant that produces cosmetic products, glycerin is being heated by flowing through a 25-mm-diameter and 10-m long tube. With a mass flow rate of 0.5 kg/s, the flow of glycerin enters the tube at 25°C. The tube surface is maintained at a constant surface temperature of 140°C. Determine the outlet mean temperature and the total rate of heat transfer for the tube. Evaluate the properties for glycerin at 30°C. *Answers: 35.7°C, 13.1 kW*

**19–92** Liquid glycerin is flowing through a 25-mm-diameter and 10-m-long tube. The liquid glycerin enters the tube at 20°C with a mass flow rate of 0.5 kg/s. If the outlet mean temperature is 40°C and the tube surface temperature is constant, determine the surface temperature of the tube.

**19–93** A 15-cm × 20-cm printed circuit board whose components are not allowed to come into direct contact with air for reliability reasons is to be cooled by passing cool air through a 20-cm-long channel of rectangular cross section 0.2 cm × 14 cm drilled into the board. The heat generated by the electronic components is conducted across the thin layer of the board to the channel, where it is removed by air that enters the channel at 15°C. The heat flux at the top surface of the channel can be considered to be uniform, and heat transfer through other surfaces is negligible. If the velocity of the air at the inlet of the channel is not to exceed 4 m/s and the surface temperature of the channel is to remain under 50°C, determine the maximum total power of the electronic components that can safely be mounted on this circuit board.

**FIGURE P19–93**

Air
15°C

Air channel
0.2 cm × 14 cm

Electronic
components

**19–94** Repeat Prob. 19–93 by replacing air with helium, which has six times the thermal conductivity of air.

**19–95** Reconsider Prob. 19–93. Using EES (or other) software, investigate the effects of air velocity at the inlet of the channel and the maximum surface temperature on the maximum total power dissipation of electronic components. Let the air velocity vary from 1 m/s to 20 m/s and the surface temperature from 30°C to 90°C. Plot the power dissipation as functions of air velocity and surface temperature, and discuss the results.

**19–96** Hot air at 60°C leaving the furnace of a house enters a 12-m-long section of a sheet metal duct of rectangular cross section 20 cm × 20 cm at an average velocity of 4 m/s. The thermal resistance of the duct is negligible, and the outer surface of the duct, whose emissivity is 0.3, is exposed to the cold air at 10°C in the basement, with a convection heat transfer coefficient of 10 W/m²·K. Taking the walls of the basement to be at 10°C also, determine (*a*) the temperature at which the hot air will leave the basement and (*b*) the rate of heat loss from the hot air in the duct to the basement.

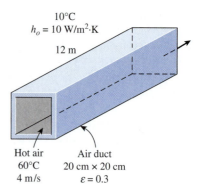

10°C
$h_o = 10$ W/m²·K

12 m

Hot air
60°C
4 m/s

Air duct
20 cm × 20 cm
$\varepsilon = 0.3$

**FIGURE P19–96**

**19–97** Reconsider Prob. 19–96. Using EES (or other) software, investigate the effect of air velocity and the surface emissivity on the exit temperature of air and the rate of heat loss. Let the air velocity vary from 1 m/s to 10 m/s and the emissivity from 0.1 to 1.0. Plot the exit temperature and the rate of heat loss functions of the air velocity and emissivity, and discuss the results.

**19–98** Liquid water flows at a mass flow rate of 0.7 kg/s through a concentric annulus tube with the inlet and outlet mean temperatures of 20°C and 80°C, respectively. The concentric annulus tube has inner and outer diameters of 10 mm and 100 mm, respectively. The inner tube wall is maintained with a constant surface temperature of 120°C, while the outer tube surface is insulated. Determine the length of the concentric annulus tube.

$T_s$
Insulation
Water
$D_o$
$D_i$

**FIGURE P19–98**

**19–99** Consider a 10-m-long smooth rectangular tube, with $a = 50$ mm and $b = 25$ mm, that is maintained at a constant surface temperature. Liquid water enters the tube at 20°C with a mass flow rate of 0.01 kg/s. Determine the tube surface temperature necessary to heat the water to the desired outlet temperature of 80°C.

$b$

$a$

**FIGURE P19–99**

**19–100** The components of an electronic system dissipating 220 W are located in a 1-m-long horizontal duct whose cross section is 16 cm × 16 cm. The components in the duct are cooled by forced air, which enters at 27°C at a rate of 0.65 m³/min. Assuming 85 percent of the heat generated inside is transferred to air flowing through the duct and the remaining 15 percent is lost through the outer surfaces of the duct, determine (a) the exit temperature of air and (b) the highest component surface temperature in the duct.

**19–101** Repeat Prob. 19–100 for a circular horizontal duct of 15-cm diameter.

**19–102** A computer cooled by a fan contains eight printed circuit boards (PCBs), each dissipating 12 W of power. The height of the PCBs is 12 cm and the length is 15 cm. The clearance between the tips of the components on the PCB and the back surface of the adjacent PCB is 0.3 cm. The cooling air is supplied by a 10-W fan mounted at the inlet. If the temperature rise of air as it flows through the case of the computer is not to exceed 10°C, determine (a) the flow rate of the air that the fan needs to deliver, (b) the fraction of the temperature rise of air that is due to the heat generated by the fan and its motor, and (c) the highest allowable inlet air temperature if the surface temperature of the components is not to exceed 70°C anywhere in the system. Use air properties at 25°C.

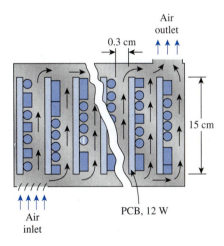

**FIGURE P19–102**

**19–103** Consider a 25-mm diameter and 15-m long smooth tube that is used for heating fluids. The wall is heated electrically to provide a constant surface heat flux along the entire tube. Fluids enter the tube at 50°C and exit at 150°C. If the mass flow rate is maintained at 0.01 kg/s, determine the convection heat transfer coefficients at the tube outlet for water, engine oil, and liquid mercury.

**19–104** Consider a 25-mm diameter and 15-m long smooth tube that is maintained at a constant surface temperature. Fluids enter the tube at 50°C with a mass flow rate of 0.01 kg/s. Determine the tube surface temperatures necessary to heat water, engine oil, and liquid mercury to the desired outlet temperature of 150°C. *Answers:* 157°C, 203°C, 150°C

**19–105** Air at 20°C (1 atm) enters into a 5-mm diameter and 10-cm long circular tube at an average velocity of 5 m/s. The tube wall is maintained at a constant surface temperature of 160°C. Determine the convection heat transfer coefficient and the outlet mean temperature. Evaluate the air properties at 50°C.

**19–106** Air (1 atm) enters into a 5-cm diameter circular tube at 20°C with an average velocity of 5 m/s. The tube wall is maintained at a constant surface temperature of 160°C, and the outlet mean temperature is 80°C. Estimate the length of the tube.

### Review Problems

**19–107** A 20 mm × 20 mm silicon chip is mounted such that the edges are flush in a substrate. The substrate provides an unheated starting length of 20 mm that acts as turbulator. Airflow at 25°C (1 atm) with a velocity of 25 m/s is used to cool the upper surface of the chip. If the maximum surface temperature of the chip cannot exceed 75°C, determine the maximum allowable power dissipation on the chip surface.

**19–108** In the effort to increase the removal of heat from a hot surface at 120°C, a cylindrical pin fin ($k_f = 237$ W/m·K) with diameter of 5 mm is attached to the hot surface. Air at 20°C (1 atm) is flowing across the pin fin with a velocity of 10 m/s. Determine the maximum possible rate of heat transfer from the pin fin. Evaluate the air properties at 70°C. *Answer: 10.4 W*

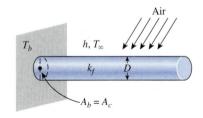

**FIGURE P19–108**

**19–109E** The passenger compartment of a minivan traveling at 70 mph can be modeled as a 3.2-ft-high, 6-ft-wide, and 12-ft-long rectangular box whose walls have an insulating value of R-3 (i.e., a wall thickness-to-thermal conductivity ratio of 3 h·ft²·°F/Btu). The interior of a minivan is maintained at an average temperature of 70°F during a trip at night while the outside air temperature is 90°F.

The average heat transfer coefficient on the interior surfaces of the van is 1.2 Btu/h·ft²·°F. The air flow over the exterior surfaces can be assumed to be turbulent because of the intense vibrations involved, and the heat transfer coefficient on the front and back surfaces can be taken to be equal to that on the top surface. Disregarding any heat gain or loss by radiation, determine the rate of heat transfer from the ambient air to the van.

**FIGURE P19–109E**

**19–110** Consider a house that is maintained at a constant temperature of 22°C. One of the walls of the house has three single-pane glass windows that are 1.5 m high and 1.8 m long. The glass ($k = 0.78$ W/m.K) is 0.5 cm thick, and the heat transfer coefficient on the inner surface of the glass is 8 W/m²·K. Now winds at 35 km/h start to blow parallel to the surface of this wall. If the air temperature outside is −2°C, determine the rate of heat loss through the windows of this wall. Assume radiation heat transfer to be negligible.

**19–111** Consider a person who is trying to keep cool on a hot summer day by turning a fan on and exposing his body to air flow. The air temperature is 32°C, and the fan is blowing air at a velocity of 5 m/s. The surrounding surfaces are at 40°C, and the emissivity of the person can be taken to be 0.9. If the person is doing light work and generating sensible heat at a rate of 90 W, determine the average temperature of the outer surface (skin or clothing) of the person. The average human body can be treated as a 30-cm-diameter cylinder with an exposed surface area of 1.7 m². *Answer: 36.2°C*

**19–112** A 0.5-cm-thick, 12-cm-high, and 18-cm-long circuit board houses 80 closely spaced logic chips on one side, each dissipating 0.06 W. The board is impregnated with copper fillings and has an effective thermal conductivity of 16 W/m·K. All the heat generated in the chips is conducted across the circuit board and is dissipated from the back side of the board to the ambient air at 30°C, which is forced to flow over the surface by a fan at a free-stream velocity of 300 m/min. Determine the temperatures on the two sides of the circuit board.

**19–113** A 3-m-internal-diameter spherical tank made of 1-cm-thick stainless steel ($k = 15$ W/m·K) is used to store iced water at 0°C. The tank is located outdoors at 30°C

and is subjected to winds at 25 km/h. Assuming the entire steel tank to be at 0°C and thus its thermal resistance to be negligible, determine (a) the rate of heat transfer to the iced water in the tank and (b) the amount of ice at 0°C that melts during a 24-h period. The heat of fusion of water at atmospheric pressure is $h_{if} = 333.7$ kJ/kg. Disregard any heat transfer by radiation.

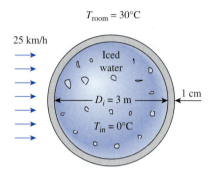

**FIGURE P19–113**

**19–114** Repeat Prob. 19–113, assuming the inner surface of the tank to be at 0°C but by taking the thermal resistance of the tank and heat transfer by radiation into consideration. Assume the average surrounding surface temperature for radiation exchange to be 25°C and the outer surface of the tank to have an emissivity of 0.75. *Answers: (a) 10,530 W, (b) 2727 kg*

**19–115E** A transistor with a height of 0.32 in and a diameter of 0.22 in is mounted on a circuit board. The transistor is cooled by air flowing over it at a velocity of 600 ft/min. If the air temperature is 120°F and the transistor case temperature is not to exceed 180°F, determine the amount of power this transistor can dissipate safely.

**FIGURE P19–115E**

**19–116** The roof of a house consists of a 15-cm-thick concrete slab ($k = 2$ W/m·K) that is 15 m wide and 20 m long. The convection heat transfer coefficient on the inner surface of the roof is 5 W/m²·K. On a clear winter night, the ambient air is reported to be at 10°C, while the night sky temperature is 100 K. The house and the interior surfaces of the wall are maintained at a constant temperature of 20°C. The emissivity of both surfaces of the concrete roof is 0.9. Considering both radiation and convection heat transfer, determine the rate of heat transfer through the roof when wind at 60 km/h is blowing over the roof.

If the house is heated by a furnace burning natural gas with an efficiency of 85 percent, and the price of natural gas is $1.20/therm, determine the money lost through the roof that night during a 14-h period. *Answers: 28 kW, $18.9*

**FIGURE P19–116**

**19–117** Steam at 250°C flows in a stainless steel pipe ($k = 15$ W/m·K) whose inner and outer diameters are 4 cm and 4.6 cm, respectively. The pipe is covered with 3.5-cm-thick glass wool insulation ($k = 0.038$ W/m·K) whose outer surface has an emissivity of 0.3. Heat is lost to the surrounding air and surfaces at 3°C by convection and radiation. Taking the heat transfer coefficient inside the pipe to be 80 W/m²·K, determine the rate of heat loss from the steam per unit length of the pipe when air is flowing across the pipe at 4 m/s.

**19–118** Consider a house that is maintained at 24°C at all times. The walls of the house have R-3.38 insulation in SI units (i.e., an $L/k$ value or a thermal resistance of 3.38 m²·°C/W). During a cold winter night, the outside air temperature is 6°C and wind at 50 km/h is blowing parallel to a 4-m-high and 8-m-long wall of the house. If the heat transfer coefficient on the interior surface of the wall is 12 W/m²·K, determine the rate of heat loss from that wall of the house. Draw the thermal resistance network and disregard radiation heat transfer. *Answer: 165 W*

**19–119** An automotive engine can be approximated as a 0.4-m-high, 0.60-m-wide, and 0.7-m-long rectangular block. The bottom surface of the block is at a temperature of 75°C and has an emissivity of 0.92. The ambient air is at 5°C, and the road surface is at 10°C. Determine the rate of heat transfer from the bottom surface of the engine block by convection and radiation as the car travels at a velocity of 60 km/h. Assume the flow to be turbulent over the entire surface because of the constant agitation of the engine block. How will the heat transfer be affected when a 2-mm-thick gunk ($k = 3$ W/m·K) has formed at the bottom surface as a result of the dirt and oil collected at that surface over time? Assume the metal temperature under the gunk still to be 75°C.

**FIGURE P19–119**

**19–120** The boiling temperature of nitrogen at atmospheric pressure at sea level (1 atm pressure) is −196°C. Therefore, nitrogen is commonly used in low-temperature scientific studies, since the temperature of liquid nitrogen in a tank open to the atmosphere will remain constant at −196°C until it is depleted. Any heat transfer to the tank will result in the evaporation of some liquid nitrogen, which has a heat of vaporization of 198 kJ/kg and a density of 810 kg/m³ at 1 atm.

**FIGURE P19–120**

Consider a 4-m-diameter spherical tank that is initially filled with liquid nitrogen at 1 atm and $-196°C$. The tank is exposed to 20°C ambient air and 55 km/h winds. The temperature of the thin-shelled spherical tank is observed to be almost the same as the temperature of the nitrogen inside. Disregarding any radiation heat exchange, determine the rate of evaporation of the liquid nitrogen in the tank as a result of heat transfer from the ambient air if the tank is (a) not insulated, (b) insulated with 5-cm-thick fiberglass insulation ($k = 0.035$ W/m·K), and (c) insulated with 2-cm-thick superinsulation that has an effective thermal conductivity of 0.00005 W/m·K.

**19–121**  Repeat Prob. 19–120 for liquid oxygen, which has a boiling temperature of $-183°C$, a heat of vaporization of 213 kJ/kg, and a density of 1140 kg/m³ at 1 atm pressure.

**19–122**  Four power transistors, each dissipating 10 W, are mounted on a thin vertical aluminum plate ($k = 237$ W/m·K) 22 cm × 22 cm in size. The heat generated by the transistors is to be dissipated by both surfaces of the plate to the surrounding air at 20°C, which is blown over the plate by a fan at a velocity of 5 m/s. The entire plate can be assumed to be nearly isothermal, and the exposed surface area of the transistor can be taken to be equal to its base area. Determine the temperature of the aluminum plate.

**19–123E**  It is well known that cold air feels much colder in windy weather than what the thermometer reading indicates because of the "chilling effect" of the wind. This effect is due to the increase in the convection heat transfer coefficient with increasing air velocities. The *equivalent temperature* in °F is given by (1993 *ASHRAE Handbook of Fundamentals,* Atlanta, GA, p. 8.15)

$$T_{equiv} = 91.4 - (91.4 - T_{ambient})(0.475 - 0.0203V + 0.304\sqrt{V})$$

**FIGURE P19–123E**

where $V$ is the wind velocity in mph and $T_{ambient}$ is the ambient air temperature in °F in calm air, which is taken to be air with light winds at speeds up to 4 mph. The constant 91.4°F in the above equation is the mean skin temperature of a resting person in a comfortable environment. Windy air at a temperature $T_{ambient}$ and velocity $V$ will feel as cold as calm air at a temperature $T_{equiv}$. The equation above is valid for winds up to 43 mph. Winds at higher velocities produce little additional chilling effect. Determine the equivalent temperature of an environment at 10°F at wind speeds of 10, 20, 30, and 40 mph. Exposed flesh can freeze within one minute at a temperature below $-25°F$ in calm weather. Does a person need to be concerned about this possibility in any of the cases above?

**19–124E**  [EES]  Reconsider Prob. 19–123E. Using EES (or other) software, plot the equivalent wind chill temperatures in °F as a function of wind velocity in the range of 5 mph to 50 mph for ambient temperatures of 20°F, 40°F and 60°F. Discuss the results.

**19–125**  To cool a storehouse in the summer without using a conventional air-conditioning system, the owner decided to hire an engineer to design an alternative system that would make use of the water in the nearby lake. The engineer decided to flow air through a thin smooth 10-cm diameter copper tube that is submerged in the nearby lake. The water in the lake is typically maintained at a constant temperature of 15°C and a convection heat transfer coefficient of 1000 W/m²·K. If air (1 atm) enters the copper tube at a mean temperature of 30°C with an average velocity of 2.5 m/s, determine the necessary copper tube length so that the outlet mean temperature of the air is 20°C.  *Answer: 6.74 m*

**FIGURE P19–125**

**19–126**  Liquid mercury is flowing at 0.6 kg/s through a 5-cm diameter tube with inlet and outlet mean temperatures of 100°C and 200°C, respectively. The tube surface temperature is maintained constant at 250°C. Determine the tube length using (a) the appropriate Nusselt number relation for liquid metals and (b) the Dittus-Boelter equation. (c) Compare the results of (a) and (b).

**19–127** In the effort to find the best way to cool a smooth thin-walled copper tube, an engineer decided to flow air either through the tube or across the outer tube surface. The tube has a diameter of 5 cm, and the surface temperature is maintained constant. Determine (a) the convection heat transfer coefficient when air is flowing through its inside at 25 m/s with bulk mean temperature of 50°C and (b) the convection heat transfer coefficient when air is flowing across its outer surface at 25 m/s with film temperature of 50°C.

**19–128** A silicon chip is cooled by passing water through microchannels etched in the back of the chip, as shown in Fig. P19–128. The channels are covered with a silicon cap. Consider a 12-mm × 12-mm square chip in which $N = 60$ rectangular microchannels, each of width $W = 50$ μm and height $H = 200$ μm have been etched. Water enters the microchannels at a temperature $T_i = 290$ K, and a total flow rate of 0.005 kg/s. The chip and cap are maintained at a uniform temperature of 350 K. Assuming that the flow in the channels is fully developed, all the heat generated by the circuits on the top surface of the chip is transferred to the water, and using circular tube correlations, determine:
(a) The water outlet temperature, $T_e$
(b) The chip power dissipation, $\dot{W}_e$

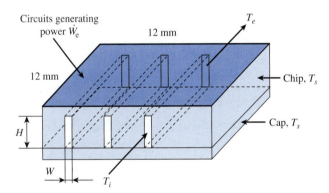

**FIGURE P19–128**

**19–129** Water is heated at a rate of 10 kg/s from a temperature of 15°C to 35°C by passing it through five identical tubes, each 5.0 cm in diameter, whose surface temperature is 60.0°C. Estimate (a) the steady rate of heat transfer and (b) the length of tubes necessary to accomplish this task.

**19–130** Repeat Prob. 19–129 for a flow rate of 20 kg/s.

**19–131** Water at 1800 kg/h and 10°C enters a 10-mm diameter smooth tube whose wall temperature is maintained at 49°C. Calculate (a) the tube length necessary to heat the water to 40°C, and (b) the water outlet temperature if the tube length is doubled. Assume average water properties to be the same as in (a).

**19–132** Liquid water enters a 10-m-long smooth rectangular tube with $a = 50$ mm and $b = 25$ mm. The surface temperature is maintained constant, and water enters the tube at 20°C with a mass flow rate of 0.25 kg/s. Determine the tube surface temperature necessary to heat the water to the desired outlet temperature of 80°C.

**FIGURE P19–132**

**19–133E** Air is flowing through a smooth thin-walled 4-in diameter copper tube that is submerged in water. The water maintains a constant temperature of 60°F and a convection heat transfer coefficient of 176 Btu/h·ft²·R. If air (1 atm) enters the copper tube at a mean temperature of 90°F with an average velocity of 8 ft/s, determine the necessary copper tube length so that the outlet mean temperature of the air is 70°F.
*Answer:* 22.9 ft

**FIGURE P19–133E**

**19–134** Hot water at 90°C enters a 15-m section of a cast iron pipe ($k = 52$ W/m·K) whose inner and outer diameters are 4 and 4.6 cm, respectively, at an average velocity of 1.2 m/s. The outer surface of the pipe, whose emissivity is 0.7, is exposed to the cold air at 10°C in a basement, with a convection heat transfer coefficient of 12 W/m²·K. Taking the walls of the basement to be at 10°C also, determine (a) the rate of heat loss from the water and (b) the temperature at which the water leaves the basement.

**FIGURE P19–134**

**19–135**   Repeat Prob. 19–134 for a pipe made of copper ($k$ = 386 W/m·K) instead of cast iron.

**19–136**   Hot exhaust gases leaving a stationary diesel engine at 450°C enter a 15-cm-diameter pipe at an average velocity of 7.2 m/s. The surface temperature of the pipe is 180°C. Determine the pipe length if the exhaust gases are to leave the pipe at 250°C after transferring heat to water in a heat recovery unit. Use properties of air for exhaust gases.

**19–137**   Geothermal steam at 165°C condenses in the shell side of a heat exchanger over the tubes through which water flows. Water enters the 4-cm-diameter, 14-m-long tubes at 20°C at a rate of 0.8 kg/s. Determine the exit temperature of water and the rate of condensation of geothermal steam.

**19–138**   Cold air at 5°C enters a 12-cm-diameter 20-m-long isothermal pipe at a velocity of 2.5 m/s and leaves at 19°C. Estimate the surface temperature of the pipe.

**19–139**   Oil at 15°C is to be heated by saturated steam at 1 atm in a double-pipe heat exchanger to a temperature of 25°C. The inner and outer diameters of the annular space are 3 cm and 5 cm, respectively, and oil enters at with a mean velocity of 1.2 m/s. The inner tube may be assumed to be isothermal at 100°C, and the outer tube is well insulated. Assuming fully developed flow for oil, determine the tube length required to heat the oil to the indicated temperature. In reality, will you need a shorter or longer tube? Explain.

**19–140**   100 kg/s of a crude oil is heated from 20°C to 40°C through the tube side of a multitube heat exchanger. The crude oil flow is divided evenly among all 100 tubes in the tube bundle. The ID of each tube is 10 mm, and the inside tube-wall temperature is maintained at 100°C. Average properties of the crude oil are $\rho$ = 950 kg/m³, $c_p$ = 1.9 kJ/kg·K, $k$ = 0.25 W/m·K, $\mu$ = 12 mPa·s, and $\mu_w$ = 4 mPa·s. Estimate the rate of heat transfer and the tube length.

**19–141**   A fluid ($\rho$ = 1000 kg/m³, $\mu$ = 1.4 × 10⁻³kg/m·s, $c_p$ = 4.2 kJ/kg·K, and $k$ = 0.58 W/m·K) flows with an average velocity of 0.3 m/s through a 14-m long tube with inside diameter of 0.01 m. Heat is uniformly added to the entire tube at the rate of 1500 W/m². Determine (*a*) the value of convection heat transfer coefficient at the exit, (*b*) the value of $T_s - T_m$, and (*c*) the value of $T_e - T_i$.

**19–142**   Air (1 atm) entered into a 5-mm diameter circular tube at an average velocity of 5 m/s. The tube wall is maintained at a constant surface temperature. Determine the convection heat transfer coefficient for (*a*) a 10-cm long tube and (*b*) a 50-cm long tube. Evaluate the air properties at 50°C.
*Answers:* (*a*) 36.5 W/m²·K, (*b*) 20.0 W/m²·K

## Design and Essay Problems

**19–143**   Conduct this experiment to determine the heat loss coefficient of your house or apartment in W/°C or Btu/h·°F. First make sure that the conditions in the house are steady and the house is at the set temperature of the thermostat. Use an outdoor thermometer to monitor outdoor temperature. One evening, using a watch or timer, determine how long the heater was on during a 3-h period and the average outdoor temperature during that period. Then using the heat output rating of your heater, determine the amount of heat supplied. Also, estimate the amount of heat generation in the house during that period by noting the number of people, the total wattage of lights that were on, and the heat generated by the appliances and equipment. Using that information, calculate the average rate of heat loss from the house and the heat loss coefficient.

**19–144**   Electronic boxes such as computers are commonly cooled by a fan. Write an essay on forced air cooling of electronic boxes and on the selection of the fan for electronic devices.

**19–145**   Design a heat exchanger to pasteurize milk by steam in a dairy plant. Milk is to flow through a bank of 1.2-cm internal diameter tubes while steam condenses outside the tubes at 1 atm. Milk is to enter the tubes at 4°C, and it is to be heated to 72°C at a rate of 15 L/s. Making reasonable assumptions, you are to specify the tube length and the number of tubes, and the pump for the heat exchanger.

**19–146**   A desktop computer is to be cooled by a fan. The electronic components of the computer consume 80 W of power under full-load conditions. The computer is to operate in environments at temperatures up to 50°C and at elevations up to 3000 m where the atmospheric pressure is 70.12 kPa. The exit temperature of air is not to exceed 60°C to meet the reliability requirements. Also, the average velocity of air is not to exceed 120 m/min at the exit of the computer case, where the fan is installed to keep the noise level down. Specify the flow rate of the fan that needs to be installed and the diameter of the casing of the fan.

# NATURAL CONVECTION

I n Chapter 19, we considered heat transfer by *forced convection,* where a fluid was *forced* to move over a surface or in a tube by external means such as a pump or a fan. In this chapter, we consider *natural convection,* alternatively called free convection, where any fluid motion occurs by natural means such as buoyancy. The fluid motion in forced convection is quite *noticeable,* since a fan or a pump can transfer enough momentum to the fluid to move it in a certain direction. The fluid motion in natural convection, however, is often not noticeable because of the low velocities involved.

The convection heat transfer coefficient is a strong function of *velocity:* the higher the velocity, the higher the convection heat transfer coefficient. The fluid velocities associated with natural convection are low, typically less than 1 m/s. Therefore, the heat transfer coefficients encountered in natural convection are usually much lower than those encountered in forced convection. Yet several types of heat transfer equipment are designed to operate under natural convection conditions instead of forced convection, because natural convection does not require the use of a fluid mover.

We start this chapter with a discussion of the physical mechanism of *natural convection* and the *Grashof number.* We then present the correlations to evaluate heat transfer by natural convection for various geometries, including enclosures.

## Objectives

The objectives of this chapter are to:

- Understand the physical mechanism of natural convection.

- Derive the governing equations of natural convection, and obtain the dimensionless Grashof number by nondimensionalizing them.

- Evaluate the Nusselt number for natural convection associated with vertical, horizontal, and inclined plates as well as cylinders and spheres.

- Analyze natural convection inside enclosures such as double-pane windows.

**FIGURE 20–1**
The cooling of a boiled egg in a cooler environment by natural convection.

**FIGURE 20–2**
The warming up of a cold drink in a warmer environment by natural convection.

# 20–1 · PHYSICAL MECHANISM OF NATURAL CONVECTION

Many familiar heat transfer applications involve natural convection as the primary mechanism of heat transfer. Some examples are cooling of electronic equipment such as power transistors, TVs, and DVDs; heat transfer from electric baseboard heaters or steam radiators; heat transfer from the refrigeration coils and power transmission lines; and heat transfer from the bodies of animals and human beings. Natural convection in gases is usually accompanied by radiation of comparable magnitude except for low-emissivity surfaces.

We know that a hot boiled egg (or a hot baked potato) on a plate eventually cools to the surrounding air temperature (Fig. 20–1). The egg is cooled by transferring heat by convection to the air and by radiation to the surrounding surfaces. Disregarding heat transfer by radiation, the physical mechanism of cooling a hot egg (or any hot object) in a cooler environment can be explained as follows:

As soon as the hot egg is exposed to cooler air, the temperature of the outer surface of the egg shell drops somewhat, and the temperature of the air adjacent to the shell rises as a result of heat conduction from the shell to the air. Consequently, the egg is surrounded by a thin layer of warmer air, and heat is then transferred from this warmer layer to the outer layers of air. The cooling process in this case is rather slow since the egg would always be blanketed by warm air, and it has no direct contact with the cooler air farther away. We may not notice any air motion in the vicinity of the egg, but careful measurements would indicate otherwise.

The temperature of the air adjacent to the egg is higher and thus its density is lower, since at constant pressure the density of a gas is inversely proportional to its temperature. Thus, we have a situation in which some low-density or "light" gas is surrounded by a high-density or "heavy" gas, and the natural laws dictate that the *light gas rise*. This is no different than the oil in a vinegar-and-oil salad dressing rising to the top (since $\rho_{oil} < \rho_{vinegar}$). This phenomenon is characterized incorrectly by the phrase "heat rises," which is understood to mean *heated air rises*. The space vacated by the warmer air in the vicinity of the egg is replaced by the cooler air nearby, and the presence of cooler air in the vicinity of the egg speeds up the cooling process. The rise of warmer air and the flow of cooler air into its place continues until the egg is cooled to the temperature of the surrounding air. The motion that results from the continual replacement of the heated air in the vicinity of the egg by the cooler air nearby is called a **natural convection current**, and the heat transfer that is enhanced as a result of this natural convection current is called **natural convection heat transfer**. Note that in the absence of natural convection currents, heat transfer from the egg to the air surrounding it would be by conduction only, and the rate of heat transfer from the egg would be much lower.

Natural convection is just as effective in the heating of cold surfaces in a warmer environment as it is in the cooling of hot surfaces in a cooler environment, as shown in Fig. 20–2. Note that the direction of fluid motion is reversed in this case.

In a gravitational field, there is a net force that pushes upward a light fluid placed in a heavier fluid. The upward force exerted by a fluid on a body completely or partially immersed in it is called the **buoyancy force**. The magnitude of the buoyancy force is equal to the weight of the *fluid displaced* by the body. That is,

$$F_{buoyancy} = \rho_{fluid}\, g V_{body} \qquad (20\text{--}1)$$

where $\rho_{fluid}$ is the average density of the *fluid* (not the body), $g$ is the gravitational acceleration, and $V_{body}$ is the volume of the portion of the body immersed in the fluid (for bodies completely immersed in the fluid, it is the total volume of the body). In the absence of other forces, the net vertical force acting on a body is the difference between the weight of the body and the buoyancy force. That is,

$$
\begin{aligned}
F_{net} &= W - F_{buoyancy} \\
&= \rho_{body}\, g V_{body} - \rho_{fluid}\, g V_{body} \\
&= (\rho_{body} - \rho_{fluid})\, g V_{body} \qquad (20\text{--}2)
\end{aligned}
$$

Note that this force is proportional to the difference in the densities of the fluid and the body immersed in it. Thus, a body immersed in a fluid will experience a "weight loss" in an amount equal to the weight of the fluid it displaces. This is known as *Archimedes' principle.*

To have a better understanding of the buoyancy effect, consider an egg dropped into water. If the average density of the egg is greater than the density of water (a sign of freshness), the egg settles at the bottom of the container. Otherwise, it rises to the top. When the density of the egg equals the density of water, the egg settles somewhere in the water while remaining completely immersed, acting like a "weightless object" in space. This occurs when the upward buoyancy force acting on the egg equals the weight of the egg, which acts downward.

The *buoyancy effect* has far-reaching implications in life. For one thing, without buoyancy, heat transfer between a hot (or cold) surface and the fluid surrounding it would be by *conduction* instead of by *natural convection.* The natural convection currents encountered in the oceans, lakes, and the atmosphere owe their existence to buoyancy. Also, light boats as well as heavy warships made of steel float on water because of buoyancy (Fig. 20–3). Ships are designed on the basis of the principle that the entire weight of a ship and its contents is equal to the weight of the water that the submerged volume of the ship can contain. The "chimney effect" that induces the upward flow of hot combustion gases through a chimney is also due to the buoyancy effect, and the upward force acting on the gases in the chimney is proportional to the difference between the densities of the hot gases in the chimney and the cooler air outside. Note that there is *no noticable gravity* in space, and thus there can be no natural convection heat transfer in a spacecraft, even if the spacecraft is filled with atmospheric air.

In heat transfer studies, the primary variable is *temperature,* and it is desirable to express the net buoyancy force (Eq. 20–2) in terms of temperature differences. But this requires expressing the density difference in terms of a temperature difference, which requires a knowledge of a property that represents the *variation of the density of a fluid with temperature at constant pressure. The property that provides that information is the* **volume expansion coefficient** $\beta$, defined as (Fig. 20–4)

$$\beta = \frac{1}{V}\left(\frac{\partial V}{\partial T}\right)_P = -\frac{1}{\rho}\left(\frac{\partial \rho}{\partial T}\right)_P \qquad (1/K) \qquad (20\text{--}3)$$

**FIGURE 20–3**
It is the buoyancy force that keeps the ships afloat in water ($W = F_{buoyancy}$ for floating objects).

(a) A substance with a large $\beta$

(b) A substance with a small $\beta$

**FIGURE 20–4**
The coefficient of volume expansion is a measure of the change in volume of a substance with temperature at constant pressure.

In natural convection studies, the condition of the fluid sufficiently far from the hot or cold surface is indicated by the subscript "infinity" to serve as a reminder that this is the value at a distance where the presence of the surface is not felt. In such cases, the volume expansion coefficient can be expressed approximately by replacing differential quantities by differences as

$$\beta \approx -\frac{1}{\rho}\frac{\Delta\rho}{\Delta T} = -\frac{1}{\rho}\frac{\rho_\infty - \rho}{T_\infty - T} \qquad \text{(at constant } P) \qquad \text{(20–4)}$$

or

$$\rho_\infty - \rho = \rho\beta(T - T_\infty) \text{ (at constant } P) \qquad \text{(20–5)}$$

where $\rho_\infty$ is the density and $T_\infty$ is the temperature of the quiescent fluid away from the surface.

We can show easily that the volume expansion coefficient $\beta$ of an *ideal gas* ($P = \rho RT$) at a temperature $T$ is equivalent to the inverse of the temperature:

$$\beta_{\text{ideal gas}} = \frac{1}{T} \qquad (1/K) \qquad \text{(20–6)}$$

where $T$ is the *thermodynamic* temperature. Note that a large value of $\beta$ for a fluid means a large change in density with temperature, and that the product $\beta\Delta T$ represents the fraction of volume change of a fluid that corresponds to a temperature change $\Delta T$ at constant pressure. Also note that the buoyancy force is proportional to the *density difference,* which is proportional to the *temperature difference* at constant pressure. Therefore, the larger the temperature difference between the fluid adjacent to a hot (or cold) surface and the fluid away from it, the *larger* the buoyancy force and the *stronger* the natural convection currents, and thus the *higher* the heat transfer rate.

The magnitude of the natural convection heat transfer between a surface and a fluid is directly related to the *flow rate* of the fluid. The higher the flow rate, the higher the heat transfer rate. In fact, it is the very high flow rates that increase the heat transfer coefficient by orders of magnitude when forced convection is used. In natural convection, no blowers are used, and therefore the flow rate cannot be controlled externally. The flow rate in this case is established by the dynamic balance of *buoyancy* and *friction.*

As we have discussed earlier, the buoyancy force is caused by the density difference between the heated (or cooled) fluid adjacent to the surface and the fluid surrounding it, and is proportional to this density difference and the volume occupied by the warmer fluid. It is also well known that whenever two bodies in contact (solid–solid, solid–fluid, or fluid–fluid) move relative to each other, a *friction force* develops at the contact surface in the direction opposite to that of the motion. This opposing force slows down the fluid and thus reduces the flow rate of the fluid. Under steady conditions, the airflow rate driven by buoyancy is established at the point where these two effects *balance* each other. The friction force increases as more and more solid surfaces are introduced, seriously disrupting the fluid flow and heat transfer. For that reason, heat sinks with closely spaced fins are not suitable for natural convection cooling.

Most heat transfer correlations in natural convection are based on experimental measurements. The instrument often used in natural convection experiments is the *Mach–Zehnder interferometer,* which gives a plot of isotherms in the fluid in the vicinity of a surface. The operation principle of interferometers is based on the fact that at low pressure, the lines of constant temperature for a gas correspond to the lines of constant density, and that the index of refraction of a gas is a function of its density. Therefore, the degree of refraction of light at some point in a gas is a measure of the temperature gradient at that point. An interferometer produces a map of interference fringes, which can be interpreted as lines of *constant temperature* as shown in Fig. 20–5. The smooth and parallel lines in (*a*) indicate that the flow is *laminar,* whereas the eddies and irregularities in (*b*) indicate that the flow is *turbulent.* Note that the lines are closest near the surface, indicating a *higher temperature gradient.*

## 20–2 ▪ EQUATION OF MOTION AND THE GRASHOF NUMBER

In this section we derive the equation of motion that governs the natural convection flow in laminar boundary layer. The conservation of mass and energy equations for forced convection are also applicable for natural convection, but the momentum equation needs to be modified to incorporate buoyancy.

Consider a vertical hot flat plate immersed in a quiescent fluid body. We assume the natural convection flow to be steady, laminar, and two-dimensional, and the fluid to be Newtonian with constant properties, including density, with one exception: the density difference $\rho - \rho_\infty$ is to be considered since it is this density difference between the inside and the outside of the boundary layer that gives rise to buoyancy force and sustains flow. (This is known as the *Boussinesq approximation.*) We take the upward direction along the plate to be *x,* and the direction normal to surface to be *y,* as shown in Fig. 20–6. Therefore, gravity acts in the $-x$-direction. Noting that the flow is steady and two-dimensional, the *x*- and *y*-components of velocity within boundary layer are $u = u(x, y)$ and $v = v(x, y)$, respectively.

The velocity and temperature profiles for natural convection over a vertical hot plate are also shown in Fig. 20–6. Note that as in forced convection, the thickness of the boundary layer increases in the flow direction. Unlike forced convection, however, the fluid velocity is *zero* at the outer edge of the velocity boundary layer as well as at the surface of the plate. This is expected since the fluid beyond the boundary layer is motionless. Thus, the fluid velocity increases with distance from the surface, reaches a maximum, and gradually decreases to zero at a distance sufficiently far from the surface. At the surface, the fluid temperature is equal to the plate temperature, and gradually decreases to the temperature of the surrounding fluid at a distance sufficiently far from the surface, as shown in the figure. In the case of *cold surfaces,* the shape of the velocity and temperature profiles remains the same but their direction is reversed.

Consider a differential volume element of height *dx,* length *dy,* and unit depth in the *z*-direction (normal to the paper) for analysis. The forces acting on this volume element are shown in Fig. 20–7. Newton's second law of motion for this volume element can be expressed as

$$\delta m \cdot a_x = F_x \qquad (20\text{–}7)$$

(*a*) Laminar flow    (*b*) Turbulent flow

**FIGURE 20–5**
Isotherms in natural convection over a hot plate in air.

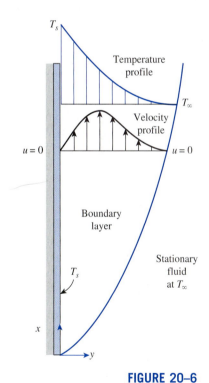

**FIGURE 20–6**
Typical velocity and temperature profiles for natural convection flow over a hot vertical plate at temperature $T_s$ inserted in a fluid at temperature $T_\infty$.

**FIGURE 20–7**
Forces acting on a differential volume element in the natural convection boundary layer over a vertical flat plate.

where $\delta m = \rho(dx \cdot dy \cdot 1)$ is the mass of the fluid within the volume element. The acceleration in the $x$-direction is obtained by taking the total differential of $u(x, y)$, which is $du = (\partial u/\partial x)dx + (\partial u/\partial y)dy$, and dividing it by $dt$. We get

$$a_x = \frac{du}{dt} = \frac{\partial u}{\partial x}\frac{dx}{dt} + \frac{\partial u}{\partial y}\frac{dy}{dt} = u\frac{\partial u}{\partial x} + v\frac{\partial u}{\partial y} \tag{20-8}$$

The forces acting on the differential volume element in the vertical direction are the pressure forces acting on the top and bottom surfaces, the shear stresses acting on the side surfaces (the normal stresses acting on the top and bottom are small and are disregarded), and the force of gravity acting on the entire volume element. Then the net surface force acting in the $x$-direction becomes

$$F_x = \left(\frac{\partial \tau}{\partial y} dy\right)(dx \cdot 1) - \left(\frac{\partial P}{\partial x} dx\right)(dy \cdot 1) - \rho g(dx \cdot dy \cdot 1)$$

$$= \left(\mu \frac{\partial^2 u}{\partial y^2} - \frac{\partial P}{\partial x} - \rho g\right)(dx \cdot dy \cdot 1) \tag{20-9}$$

since $\tau = \mu(\partial u/\partial y)$. Substituting Eqs. 20–8 and 20–9 into Eq. 20–7 and dividing by $\rho \cdot dx \cdot dy \cdot 1$ gives Newton's second-law of motion in the $x$-direction (the $x$-momentum equation) as

$$\rho\left(u\frac{\partial u}{\partial x} + v\frac{\partial u}{\partial y}\right) = \mu\frac{\partial^2 u}{\partial y^2} - \frac{\partial P}{\partial x} - \rho g \tag{20-10}$$

The $x$-momentum equation in the quiescent fluid outside the boundary layer can be obtained from the relation above as a special case by setting $u = 0$. It gives

$$\frac{\partial P_\infty}{\partial x} = -\rho_\infty g \tag{20-11}$$

which is simply the relation for the variation of hydrostatic pressure in a quiescent fluid with height, as expected. Also, noting that $v \ll u$ in the boundary layer and thus $\partial v/\partial x \approx \partial v/\partial y \approx 0$, and that there are no body forces (including gravity) in the $y$-direction, the force balance in that direction gives $\partial P/\partial y = 0$. That is, the variation of pressure in the direction normal to the surface is negligible, and for a given $x$ the pressure in the boundary layer is equal to the pressure in the quiescent fluid. Therefore, $P = P(x) = P_\infty(x)$ and $\partial P/\partial x = \partial P_\infty/\partial x = -\rho_\infty g$. Substituting into Eq. 20–10,

$$\rho\left(u\frac{\partial u}{\partial x} + v\frac{\partial u}{\partial y}\right) = \mu\frac{\partial^2 u}{\partial y^2} + (\rho_\infty - \rho)g \tag{20-12}$$

The last term represents the net upward force per unit volume of the fluid (the difference between the buoyant force and the fluid weight). This is the force that initiates and sustains convection currents.

From Eq. 20–5, we have $\rho_\infty - \rho = \rho\beta(T - T_\infty)$. Substituting it into the last equation and dividing both sides by $\rho$ gives the desired form of the $x$-momentum equation,

$$u\frac{\partial u}{\partial x} + v\frac{\partial u}{\partial y} = v\frac{\partial^2 u}{\partial y^2} + g\beta(T - T_\infty) \tag{20-13}$$

This is the equation that governs the fluid motion in the boundary layer due to the effect of buoyancy. Note that the momentum equation involves the temperature, and thus the momentum and energy equations must be solved simultaneously.

The complete set of conservation equations, continuity, momentum (Eq. 20–13), and energy that govern natural convection flow over vertical isothermal plates are:

Continuity:
$$\frac{\partial u}{\partial x} + \frac{\partial v}{\partial y} = 0$$

Momentum:
$$u\frac{\partial u}{\partial x} + v\frac{\partial u}{\partial y} = \nu\frac{\partial^2 u}{\partial y^2} + g\beta(T - T_\infty) \tag{20–13}$$

Energy:
$$u\frac{\partial T}{\partial x} + v\frac{\partial T}{\partial y} = \alpha\frac{\partial^2 T}{\partial y^2}$$

with the following boundary conditions (see Fig. 20–6):

At $y = 0$:  $u(x, 0) = 0$, $v(x, 0) = 0$,  $T(x, 0) = T_s$

At $y \to \infty$:  $u(x, \infty) \to 0$, $v(x, \infty) \to 0$,  $T(x, \infty) \to T_\infty$

## The Grashof Number

The governing equations of natural convection and the boundary conditions can be nondimensionalized by dividing all dependent and independent variables by suitable constant quantities: all lengths by a characteristic length $L_c$, all velocities by an arbitrary reference velocity $V$ (which, from the definition of Reynolds number, is taken to be $V = \text{Re}_L\, \nu/L_c$), and temperature by a suitable temperature difference (which is taken to be $T_s - T_\infty$) as

$$x^* = \frac{x}{L_c} \qquad y^* = \frac{y}{L_c} \qquad u^* = \frac{u}{V} \qquad v^* = \frac{v}{V} \quad \text{and} \quad T^* = \frac{T - T_\infty}{T_s - T_\infty}$$

where asterisks are used to denote nondimensional variables. The nondimensionalized forms of continuity and energy equations are still applicable here. However, the nondimensionalized momentum equation due to buoyancy effects will be different. Nondimesionalizing the momentum equation (Eq. 20–13) with the above parameters and simplifying gives

$$u^*\frac{\partial u^*}{\partial x^*} + v^*\frac{\partial u^*}{\partial y^*} = \left[\frac{g\beta(T_s - T_x)L_c^3}{\nu^2}\right]\frac{T^*}{\text{Re}_L^2} + \frac{1}{\text{Re}_L}\frac{\partial^2 u^*}{\partial y^{*2}} \tag{20–14}$$

The dimensionless parameter in the brackets represents the natural convection effects, and is called the **Grashof number** $\text{Gr}_L$

$$\text{Gr}_L = \frac{g\beta(T_s - T_\infty)L_c^3}{\nu^2} \tag{20–15}$$

where

**FIGURE 20–8**
The Grashof number Gr is a measure of the relative magnitudes of the *buoyancy force* and the opposing *viscous force* acting on the fluid.

*(a) Natural convection ($Gr_L / Re_L^2 \gg 1$)*

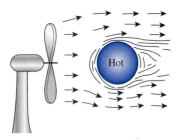

*(b) Forced convection ($Gr_L / Re_L^2 \ll 1$)*

*(c) Mixed convection ($Gr_L / Re_L^2 \approx 1$)*

**FIGURE 20–9**
The relative importance of convection heat transfer regimes for flow near a hot sphere.

$g$ = gravitational acceleration, m/s²

$\beta$ = coefficient of volume expansion, 1/K ($\beta = 1/T$ for ideal gases)

$T_s$ = temperature of the surface, °C

$T_\infty$ = temperature of the fluid sufficiently far from the surface, °C

$L_c$ = characteristic length of the geometry, m

$\nu$ = kinematic viscosity of the fluid, m²/s

We mentioned in the preceding chapters that the flow regime in forced convection is governed by the dimensionless *Reynolds number*, which represents the ratio of inertial forces to viscous forces acting on the fluid. The flow regime in natural convection is governed by the dimensionless *Grashof number*, which represents the ratio of the *buoyancy force* to the *viscous force* acting on the fluid (Fig. 20–8).

The role played by the Reynolds number in forced convection is played by the Grashof number in natural convection. As such, the Grashof number provides the main criterion in determining whether the fluid flow is laminar or turbulent in natural convection. For vertical plates, for example, the critical Grashof number is observed to be about $10^9$. Therefore, the flow regime on a vertical plate becomes *turbulent* at Grashof numbers greater than $10^9$.

When a surface is subjected to external flow, the problem involves both natural and forced convection. The relative importance of each mode of heat transfer is determined by the value of the coefficient $Gr_L/Re_L^2$, which appears in Eq. 20–14 (Fig. 20–9). If $Gr_L/Re_L^2 \gg 1$, inertia forces are negligible and natural convection effects dominate. Conversely, if $Gr_L/Re_L^2 \ll 1$, buoyancy forces are negligible and forced convection must be considered. For the case, that $Gr_L/Re_L^2 \approx 1$, both inertia and buoyancy forces are equally present and both natural and forced convection effects must be considered. In this case, the flow is referred to as **mixed convection**.

## 20–3 · NATURAL CONVECTION OVER SURFACES

Natural convection heat transfer on a surface depends on the geometry of the surface as well as its orientation. It also depends on the variation of temperature on the surface and the thermophysical properties of the fluid involved.

Although we understand the mechanism of natural convection well, the complexities of fluid motion make it very difficult to obtain simple analytical relations for heat transfer by solving the governing equations of motion and energy. Some analytical solutions exist for natural convection, but such solutions lack generality since they are obtained for simple geometries under some simplifying assumptions. Therefore, with the exception of some simple cases, heat transfer relations in natural convection are based on experimental studies. Of the numerous such correlations of varying complexity and claimed accuracy available in the literature for any given geometry, we present here the ones that are best known and widely used.

The simple empirical correlations for the average *Nusselt number* Nu in natural convection are of the form (Fig. 20–10)

$$\text{Nu} = \frac{hL_c}{k} = C\,(\text{Gr}_L\,\text{Pr})^n = C\,\text{Ra}_L^n \qquad \textbf{(20-16)}$$

where $\text{Re}_L$ is the **Rayleigh number**, which is the product of the Grashof number, which describes the relationship between buoyancy and viscosity within the fluid, and the Prandtl number, which describes the relationship between momentum diffusivity and thermal diffusivity. Hence the Rayleigh number itself may also be viewed as the ratio of buoyancy forces and (the products of) thermal and momentum diffusitivities.

$$\text{Ra}_L = \text{Gr}_L\,\text{Pr} = \frac{g\beta(T_s - T_\infty)L_c^3}{\nu^2}\,\text{Pr} = \frac{g\beta(T_s - T_x)L_c^3}{\nu\alpha} \qquad \textbf{(20-17)}$$

The values of the constants $C$ and $n$ depend on the *geometry* of the surface and the *flow regime,* which is characterized by the range of the Rayleigh number. The value of $n$ is usually $\frac{1}{4}$ for laminar flow and $\frac{1}{3}$ for turbulent flow. The value of the constant $C$ is normally less than 1.

Simple relations for the average Nusselt number for various geometries are given in Table 20–1, together with sketches of the geometries. Also given in this table are the characteristic lengths of the geometries and the ranges of Rayleigh number in which the relation is applicable. All fluid properties are to be evaluated at the film temperature $T_f = \frac{1}{2}\,(T_s + T_\infty)$.

When the average Nusselt number and thus the average convection coefficient is known, the rate of heat transfer by natural convection from a solid surface at a uniform temperature $T_s$ to the surrounding fluid is expressed by Newton's law of cooling as

$$\dot{Q}_{\text{conv}} = hA_s(T_s - T_\infty) \qquad (\text{W}) \qquad \textbf{(20-18)}$$

where $A_s$ is the heat transfer surface area and $h$ is the average heat transfer coefficient on the surface.

## Vertical Plates ($T_s$ = constant)

For a vertical flat plate, the characteristic length is the plate height $L$. In Table 20–1 we give three relations for the average Nusselt number for an isothermal vertical plate. The first two relations are very simple. Despite its complexity, we suggest using the third one (Eq. 20–21) recommended by Churchill and Chu (1975) since it is applicable over the entire range of Rayleigh number. This relation is most accurate in the range of $10^{-1} < \text{Ra}_L < 10^9$.

## Vertical Plates ($\dot{q}_s$ = constant)

In the case of constant surface heat flux, the rate of heat transfer is known (it is simply $\dot{Q} = \dot{q}_s A_s$), but the surface temperature $T_s$ is not. In fact, $T_s$ increases with height along the plate. It turns out that the Nusselt number relations for the constant surface temperature and constant surface

**FIGURE 20–10**
Natural convection heat transfer correlations are usually expressed in terms of the Rayleigh number raised to a constant $n$ multiplied by another constant $C$, both of which are determined experimentally.

**TABLE 20–1**

Empirical correlations for the average Nusselt number for natural convection over surfaces

| Geometry | Characteristic length $L_c$ | Range of Ra | Nu | |
|---|---|---|---|---|
| Vertical plate | $L$ | $10^4$–$10^9$ <br> $10^{10}$–$10^{13}$ <br><br> Entire range | $Nu = 0.59Ra_L^{1/4}$ <br> $Nu = 0.1Ra_L^{1/3}$ <br><br> $Nu = \left\{ 0.825 + \dfrac{0.387Ra_L^{1/6}}{[1 + (0.492/Pr)^{9/16}]^{8/27}} \right\}^2$ <br><br> (complex but more accurate) | (20–19) <br> (20–20) <br><br> (20–21) |
| Inclined plate | $L$ | | Use vertical plate equations for the upper surface of a cold plate and the lower surface of a hot plate <br><br> Replace $g$ by $g\cos\theta$    for    $0 < \theta < 60°$ | |
| Horizontal plate <br> (Surface area $A$ and perimeter $p$) <br> (a) Upper surface of a hot plate <br> (or lower surface of a cold plate) <br><br> Hot surface    $T_s$ <br> <br><br> (b) Lower surface of a hot plate <br> (or upper surface of a cold plate) <br> <br> $T_s$ <br> Hot surface | $A_s/p$ | $10^4$–$10^7$ <br> $10^7$–$10^{11}$ <br><br><br><br><br><br> $10^5$–$10^{11}$ | $Nu = 0.54Ra_L^{1/4}$ <br> $Nu = 0.15Ra_L^{1/3}$ <br><br><br><br><br><br> $Nu = 0.27Ra_L^{1/4}$ | (20–22) <br> (20–23) <br><br><br><br><br><br> (20–24) |
| Vertical cylinder | $L$ | | A vertical cylinder can be treated as a vertical plate when <br><br> $D \geq \dfrac{35L}{Gr_L^{1/4}}$ | |
| Horizontal cylinder | $D$ | $Ra_D \leq 10^{12}$ | $Nu = \left\{ 0.6 + \dfrac{0.387Ra_D^{1/6}}{[1 + (0.559/Pr)^{9/16}]^{8/27}} \right\}^2$ | (20–25) |
| Sphere | $D$ | $Ra_D \leq 10^{11}$ <br> $(Pr \geq 0.7)$ | $Nu = 2 + \dfrac{0.589Ra_D^{1/4}}{[1 + (0.469/Pr)^{9/16}]^{4/9}}$ | (20–26) |

heat flux cases are nearly identical [Churchill and Chu (1975)]. Therefore, the relations for isothermal plates can also be used for plates subjected to constant heat flux, provided that the plate midpoint temperature $T_{L/2}$ is used for $T_s$ in the evaluation of the film temperature, Rayleigh number, and the Nusselt number. Noting that $h = \dot{q}_s/(T_{L/2} - T_\infty)$, the average Nusselt number in this case can be expressed as

$$\text{Nu} = \frac{hL}{k} = \frac{\dot{q}_s L}{k(T_{L/2} - T_\infty)} \qquad \textbf{(20–27)}$$

The midpoint temperature $T_{L/2}$ is determined by iteration so that the Nusselt numbers determined from Eqs. 20–21 and 20–27 match.

## Vertical Cylinders

An outer surface of a vertical cylinder can be treated as a vertical plate when the diameter of the cylinder is sufficiently large so that the curvature effects are negligible. This condition is satisfied if

$$D \geq \frac{35L}{\text{Gr}_L^{1/4}} \qquad \textbf{(20–28)}$$

When this criteria is met, the relations for vertical plates can also be used for vertical cylinders. Nusselt number relations for slender cylinders that do not meet this criteria are available in the literature [e.g., Cebeci (1974)].

## Inclined Plates

Consider an inclined hot plate that makes an angle $\theta$ from the vertical, as shown in Fig. 20–11, in a cooler environment. The net force $F = g(\rho_\infty - \rho)$ (the difference between the buoyancy and gravity) acting on a unit volume of the fluid in the boundary layer is always in the vertical direction. In the case of inclined plate, this force can be resolved into two components: $F_x = F \cos \theta$ parallel to the plate that drives the flow along the plate, and $F_y = F \sin \theta$ normal to the plate. Noting that the force that drives the motion is reduced, we expect the convection currents to be weaker, and the rate of heat transfer to be lower relative to the vertical plate case.

The experiments confirm what we suspect for the lower surface of a hot plate, but the opposite is observed on the upper surface. The reason for this curious behavior for the upper surface is that the force component $F_y$ initiates upward motion in addition to the parallel motion along the plate, and thus the boundary layer breaks up and forms plumes, as shown in the figure. As a result, the thickness of the boundary layer and thus the resistance to heat transfer decreases, and the rate of heat transfer increases relative to the vertical orientation.

In the case of a cold plate in a warmer environment, the opposite occurs as expected: The boundary layer on the upper surface remains intact with

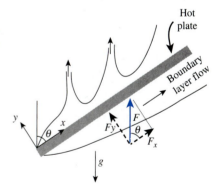

**FIGURE 20–11**
Natural convection flows on the upper and lower surfaces of an inclined hot plate.

weaker boundary layer flow and thus lower rate of heat transfer, and the boundary layer on the lower surface breaks apart (the colder fluid falls down) and thus enhances heat transfer.

When the boundary layer remains intact (the lower surface of a hot plate or the upper surface of a cold plate), the Nusselt number can be determined from the vertical plate relations provided that $g$ in the Rayleigh number relation is replaced by $g \cos\theta$ for $\theta < 60°$. Nusselt number relations for the other two surfaces (the upper surface of a hot plate or the lower surface of a cold plate) are available in the literature [e.g., Fujii and Imura (1972)].

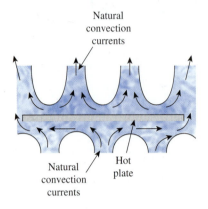

Natural convection
currents

Natural
convection
currents

Hot
plate

**FIGURE 20–12**
Natural convection flows on the upper and lower surfaces of a horizontal hot plate.

## Horizontal Plates

The rate of heat transfer to or from a horizontal surface depends on whether the surface is facing upward or downward. For a hot surface in a cooler environment, the net force acts upward, forcing the heated fluid to rise. If the hot surface is facing upward, the heated fluid rises freely, inducing strong natural convection currents and thus effective heat transfer, as shown in Fig. 20–12. But if the hot surface is facing downward, the plate blocks the heated fluid that tends to rise (except near the edges), impeding heat transfer. The opposite is true for a cold plate in a warmer environment since the net force (weight minus buoyancy force) in this case acts downward, and the cooled fluid near the plate tends to descend.

The average Nusselt number for horizontal surfaces can be determined from the simple power-law relations given in Table 20–1. The characteristic length for horizontal surfaces is calculated from

$$L_c = \frac{A_s}{p} \tag{20–29}$$

where $A_s$ is the surface area and $p$ is the perimeter. Note that $L_c = a/4$ for a horizontal square surface of length $a$, and $D/4$ for a horizontal circular surface of diameter $D$.

Boundary
layer flow

Hot
cylinder

**FIGURE 20–13**
Natural convection flow over a horizontal hot cylinder.

## Horizontal Cylinders and Spheres

The boundary layer over a hot horizontal cylinder starts to develop at the bottom, increasing in thickness along the circumference, and forming a rising plume at the top, as shown in Fig. 20–13. Therefore, the local Nusselt number is highest at the bottom, and lowest at the top of the cylinder when the boundary layer flow remains laminar. The opposite is true in the case of a cold horizontal cylinder in a warmer medium, and the boundary layer in this case starts to develop at the top of the cylinder and ending with a descending plume at the bottom.

The average Nusselt number over the entire surface can be determined from Eq. 20–25 [Churchill and Chu (1975)] for an isothermal horizontal cylinder, and from Eq. 20–26 for an isothermal sphere [Churchill (1983)], both given in Table 20–1.

**FIGURE 20–14**
Schematic for Example 20–1.

## EXAMPLE 20–1  Heat Loss from Hot-Water Pipes

A 6-m-long section of an 8-cm-diameter horizontal hot-water pipe shown in Fig. 20–14 passes through a large room whose temperature is 20°C. If the outer surface temperature of the pipe is 70°C, determine the rate of heat loss from the pipe by natural convection.

**SOLUTION**  A horizontal hot-water pipe passes through a large room. The rate of heat loss from the pipe by natural convection is to be determined.

**Assumptions**  **1** Steady operating conditions exist. **2** Air is an ideal gas. **3** The local atmospheric pressure is 1 atm.

**Properties**  The properties of air at the film temperature of $T_f = (T_s + T_\infty)/2 = (70 + 20)/2 = 45°C$ and 1 atm are (Table A–22)

$$k = 0.02699 \text{ W/m·K} \qquad Pr = 0.7241$$

$$\nu = 1.750 \times 10^{-5} \text{ m}^2\text{/s} \qquad \beta = \frac{1}{T_f} = \frac{1}{318 \text{ K}}$$

**Analysis**  The characteristic length in this case is the outer diameter of the pipe, $L_c = D = 0.08$ m. Then the Rayleigh number becomes

$$Ra_D = \frac{g\beta(T_s - T_\infty)D^3}{\nu^2} Pr$$

$$= \frac{(9.81 \text{ m/s}^2)[1/(318 \text{ K})](70 - 20 \text{ K})(0.08 \text{ m})^3}{(1.750 \times 10^{-5} \text{ m}^2\text{/s})^2}(0.7241) = 1.867 \times 10^6$$

$$= 1.867 \times 10^6$$

The natural convection Nusselt number in this case can be determined from Eq. 20–25 to be

$$Nu = \left\{ 0.6 + \frac{0.387 Ra_D^{1/6}}{[1 + (0.559/Pr)^{9/16}]^{8/27}} \right\}^2 = \left\{ 0.6 + \frac{0.387(1.867 \times 10^6)^{1/6}}{[1 + (0.559/0.7241)^{9/16}]^{8/27}} \right\}^2$$

$$= 17.39$$

Then,

$$h = \frac{k}{D} Nu = \frac{0.02699 \text{ W/m·K}}{0.08 \text{ m}}(17.39) = 5.867 \text{ W/m·K}$$

$$A_s = \pi DL = \pi(0.08 \text{ m})(6 \text{ m}) = 1.508 \text{ m}^2$$

and

$$\dot{Q} = hA_s(T_s - T_\infty) = (5.867 \text{ W/m}^2\text{·K})(1.508 \text{ m}^2)(70 - 20)°C = \textbf{442 W}$$

Therefore, the pipe loses heat to the air in the room at a rate of 442 W by natural convection.

**Discussion**  The pipe loses heat to the surroundings by radiation as well as by natural convection. Assuming the outer surface of the pipe to be black (emissivity $\varepsilon = 1$) and the inner surfaces of the walls of the room to be at room temperature, the radiation heat transfer is determined to be (Fig. 20–15)

$$\dot{Q}_{rad} = \varepsilon A_s \sigma(T_s^4 - T_{surr}^4)$$

$$= (1)(1.508 \text{ m}^2)(5.67 \times 10^{-8} \text{ W/m}^2\text{·K}^4)[(70 + 273 \text{ K})^4 - (20 + 273 \text{ K})^4]$$

$$= 553 \text{ W}$$

**FIGURE 20–15**
Radiation heat transfer is usually comparable to natural convection in magnitude and should be considered in heat transfer analysis.

(a) Vertical

(b) Hot surface facing up

(c) Hot surface facing down

**FIGURE 20–16**
Schematic for Example 20–2.

which is larger than natural convection. The emissivity of a real surface is less than 1, and thus the radiation heat transfer for a real surface will be less. But radiation will still be significant for most systems cooled by natural convection. Therefore, a radiation analysis should normally accompany a natural convection analysis unless the emissivity of the surface is low.

**EXAMPLE 20–2    Cooling of a Plate in Different Orientations**

Consider a 0.6-m × 0.6-m thin square plate in a room at 30°C. One side of the plate is maintained at a temperature of 90°C, while the other side is insulated, as shown in Fig. 20–16. Determine the rate of heat transfer from the plate by natural convection if the plate is (a) vertical, (b) horizontal with hot surface facing up, and (c) horizontal with hot surface facing down.

**SOLUTION**  A hot plate with an insulated back is considered. The rate of heat loss by natural convection is to be determined for different orientations.
**Assumptions**  1 Steady operating conditions exist. 2 Air is an ideal gas. 3 The local atmospheric pressure is 1 atm.
**Properties**  The properties of air at the film temperature of $T_f = (T_s + T_\infty)/2 = (90 + 30)/2 = 60°C$ and 1 atm are (Table A–22)

$$k = 0.02808 \text{ W/m·K} \qquad \text{Pr} = 0.7202$$

$$\nu = 1.896 \times 10^{-5} \text{ m}^2/\text{s} \qquad \beta = \frac{1}{T_f} = \frac{1}{333 \text{ K}}$$

**Analysis**  (a) *Vertical.* The characteristic length in this case is the height of the plate, which is $L = 0.6$ m. The Rayleigh number is

$$\text{Ra}_L = \frac{g\beta(T_s - T_\infty)L^3}{\nu^2} \text{Pr}$$

$$= \frac{(9.81 \text{ m/s}^2)[1/(333 \text{ K})](90 - 30 \text{ K})(0.6 \text{ m})^3}{(1.896 \times 10^{-5} \text{ m}^2/\text{s})^2}(0.7202) = 7.649 \times 10^8$$

Then the natural convection Nusselt number can be determined from Eq. 20–21 to be

$$\text{Nu} = \left\{ 0.825 + \frac{0.387\text{Ra}_L^{1/6}}{[1 + (0.492/\text{Pr})^{9/16}]^{8/27}} \right\}^2$$

$$= \left\{ 0.825 + \frac{0.387(7.649 \times 10^8)^{1/6}}{[1 + (0.492/0.7202)^{9/16}]^{8/27}} \right\}^2 = 113.3$$

Note that the simpler relation Eq. 20–19 would give $\text{Nu} = 0.59 \text{ Ra}_L^{1/4} = 98.12$, which is 13 percent lower. Then,

$$h = \frac{k}{L} \text{Nu} = \frac{0.02808 \text{ W/m·K}}{0.6 \text{ m}}(113.3) = 5.302 \text{ W/m}^2\cdot\text{K}$$

$$A_s = L^2 = (0.6 \text{ m})^2 = 0.36 \text{ m}^2$$

and

$$\dot{Q} = hA_s(T_s - T_\infty) = (5.302 \text{ W/m}^2\cdot\text{K})(0.36 \text{ m}^2)(90 - 30)°C = \textbf{115 W}$$

(b) *Horizontal with hot surface facing up.* The characteristic length and the Rayleigh number in this case are

$$L_c = \frac{A_s}{p} = \frac{L^2}{4L} = \frac{L}{4} = \frac{0.6 \text{ m}}{4} = 0.15 \text{ m}$$

$$\text{Ra}_L = \frac{g\beta(T_s - T_\infty)L_c^3}{\nu^2} \text{Pr}$$

$$= \frac{(9.81 \text{ m/s}^2)[1/(333 \text{ K})](90 - 30 \text{ K})(0.15 \text{ m})^3}{(1.896 \times 10^{-5} \text{ m}^2/\text{s})^2} (0.7202) = 1.195 \times 10^7$$

The natural convection Nusselt number can be determined from Eq. 20–22 to be

$$\text{Nu} = 0.54\text{Ra}_L^{1/4} = 0.54(1.195 \times 10^7)^{1/4} = 31.75$$

Then,

$$h = \frac{k}{L_c} \text{Nu} = \frac{0.02808 \text{ W/m·K}}{0.15 \text{ m}} (31.75) = 5.944 \text{ W/m}^2\text{·K}$$

and

$$\dot{Q} = hA_s(T_s - T_\infty) = (5.944 \text{ W/m}^2\text{·K})(0.36 \text{ m}^2)(90 - 30)°\text{C} = \mathbf{128 \text{ W}}$$

(c) *Horizontal with hot surface facing down.* The characteristic length and the Rayleigh number in this case are the same as those determined in (b). But the natural convection Nusselt number is to be determined from Eq. 20–24,

$$\text{Nu} = 0.27\text{Ra}_L^{1/4} = 0.27(1.195 \times 10^7)^{1/4} = 15.87$$

Then,

$$h = \frac{k}{L_c} \text{Nu} = \frac{0.02808 \text{ W/m·K}}{0.15 \text{ m}} (15.87) = 2.971 \text{ W/m}^2\text{·K}$$

and

$$\dot{Q} = hA_s(T_s - T_\infty) = (2.971 \text{ W/m}^2\text{·K})(0.36 \text{ m}^2)(90 - 30)°\text{C} = \mathbf{64.2 \text{ W}}$$

Note that the natural convection heat transfer is the lowest in the case of the hot surface facing down. This is not surprising, since the hot air is "trapped" under the plate in this case and cannot get away from the plate easily. As a result, the cooler air in the vicinity of the plate will have difficulty reaching the plate, which results in a reduced rate of heat transfer.

*Discussion*   The plate will lose heat to the surroundings by radiation as well as by natural convection. Assuming the surface of the plate to be black (emissivity $\varepsilon = 1$) and the inner surfaces of the walls of the room to be at room temperature, the radiation heat transfer in this case is determined to be

$$\dot{Q}_{\text{rad}} = \varepsilon A_s \sigma(T_s^4 - T_{\text{surr}}^4)$$

$$= (1)(0.36 \text{ m}^2)(5.67 \times 10^{-8} \text{ W/m}^2\text{·K}^4)[(90 + 273 \text{ K})^4 - (30 + 273 \text{ K})^4]$$

$$= 182 \text{ W}$$

which is larger than that for natural convection heat transfer for each case. Therefore, radiation can be significant and needs to be considered in surfaces cooled by natural convection.

**FIGURE 20–17**
Convective currents in a vertical rectangular enclosure.

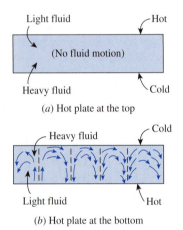

**FIGURE 20–18**
Convective currents in a horizontal enclosure with (a) hot plate at the top and (b) hot plate at the bottom.

# 20–4 · NATURAL CONVECTION INSIDE ENCLOSURES

A considerable portion of heat loss from a typical residence occurs through the windows. We certainly would insulate the windows, if we could, in order to conserve energy. The problem is finding an insulating material that is transparent. An examination of the thermal conductivities of the insulating materials reveals that *air is a better insulator* than most common insulating materials. Besides, it is transparent. Therefore, it makes sense to insulate the windows with a layer of air. Of course, we need to use another sheet of glass to trap the air. The result is an *enclosure*, which is known as a *double-pane window* in this case. Other examples of enclosures include wall cavities, solar collectors, and cryogenic chambers involving concentric cylinders or spheres.

Enclosures are frequently encountered in practice, and heat transfer through them is of practical interest. Heat transfer in enclosed spaces is complicated by the fact that the fluid in the enclosure, in general, does not remain stationary. In a vertical enclosure, the fluid adjacent to the hotter surface rises and the fluid adjacent to the cooler one falls, setting off a rotationary motion within the enclosure that enhances heat transfer through the enclosure. Typical flow patterns in vertical and horizontal rectangular enclosures are shown in Figs. 20–17 and 20–18.

The characteristics of heat transfer through a horizontal enclosure depend on whether the hotter plate is at the top or at the bottom, as shown in Fig. 20–18. When the hotter plate is at the *top,* no convection currents develop in the enclosure, since the lighter fluid is always on top of the heavier fluid. Heat transfer in this case is by *pure conduction,* and we have Nu = 1. When the hotter plate is at the *bottom,* the heavier fluid will be on top of the lighter fluid, and there will be a tendency for the lighter fluid to topple the heavier fluid and rise to the top, where it comes in contact with the cooler plate and cools down. Until that happens, however, heat transfer is still by *pure conduction* and Nu = 1. When $\text{Ra}_L > 1708$, the buoyant force overcomes the fluid resistance and initiates natural convection currents, which are observed to be in the form of hexagonal cells called *Bénard cells*. For $\text{Ra}_L < 3 \times 10^5$, the cells break down and the fluid motion becomes turbulent.

The Rayleigh number for an enclosure is determined from

$$\text{Ra}_L = \frac{g\beta(T_1 - T_2)L_c^3}{\nu^2}\,\text{Pr} \tag{20–30}$$

where the characteristic length $L_c$ is the distance between the hot and cold surfaces, and $T_1$ and $T_2$ are the temperatures of the hot and cold surfaces, respectively. All fluid properties are to be evaluated at the average fluid temperature $T_{avg} = (T_1 + T_2)/2$.

## Effective Thermal Conductivity

When the Nusselt number is known, the rate of heat transfer through the enclosure can be determined from

$$\dot{Q} = hA_s(T_1 - T_2) = k\text{Nu}A_s\frac{T_1 - T_2}{L_c} \tag{20–31}$$

since $h = k\mathrm{Nu}/L$. The rate of steady heat conduction across a layer of thickness $L_c$, area $A_s$, and thermal conductivity $k$ is expressed as

$$\dot{Q}_{\mathrm{cond}} = kA_s \frac{T_1 - T_2}{L_c} \qquad \text{(20–32)}$$

where $T_1$ and $T_2$ are the temperatures on the two sides of the layer. A comparison of this relation with Eq. 20–31 reveals that the convection heat transfer in an enclosure is analogous to heat conduction across the fluid layer in the enclosure provided that the thermal conductivity $k$ is replaced by $k\mathrm{Nu}$. That is, *the fluid in an enclosure behaves like a fluid whose thermal conductivity is $k\mathrm{Nu}$ as a result of convection currents*. Therefore, the quantity $k\mathrm{Nu}$ is called the **effective thermal conductivity** of the enclosure. That is,

$$k_{\mathrm{eff}} = k\mathrm{Nu} \qquad \text{(20–33)}$$

Note that for the special case of $\mathrm{Nu} = 1$, the effective thermal conductivity of the enclosure becomes equal to the conductivity of the fluid. This is expected since this case corresponds to pure conduction (Fig. 20–19).

Natural convection heat transfer in enclosed spaces has been the subject of many experimental and numerical studies, and numerous correlations for the Nusselt number exist. Simple power-law type relations in the form of $\mathrm{Nu} = C\mathrm{Ra}_L^n$, where $C$ and $n$ are constants, are sufficiently accurate, but they are usually applicable to a narrow range of Prandtl and Rayleigh numbers and aspect ratios. The relations that are more comprehensive are naturally more complex. Next we present some widely used relations for various types of enclosures.

## Horizontal Rectangular Enclosures

We need no Nusselt number relations for the case of the hotter plate being at the top, since there are no convection currents in this case and heat transfer is downward by conduction ($\mathrm{Nu} = 1$). When the hotter plate is at the bottom, however, significant convection currents set in for $\mathrm{Ra}_L > 1708$, and the rate of heat transfer increases (Fig. 20–20).

For horizontal enclosures that contain air, Jakob (1949) recommends the following simple correlations

$$\mathrm{Nu} = 0.195\mathrm{Ra}_L^{1/4} \qquad 10^4 < \mathrm{Ra}_L < 4 \times 10^5 \qquad \text{(20–34)}$$
$$\mathrm{Nu} = 0.068\mathrm{Ra}_L^{1/3} \qquad 4 \times 10^5 < \mathrm{Ra}_L < 10^7 \qquad \text{(20–35)}$$

These relations can also be used for other gases with $0.5 < \mathrm{Pr} < 2$. Using water, silicone oil, and mercury in their experiments, Globe and Dropkin (1959) obtained this correlation for horizontal enclosures heated from below,

$$\mathrm{Nu} = 0.069\mathrm{Ra}_L^{1/3} \mathrm{Pr}^{0.074} \qquad 3 \times 10^5 < \mathrm{Ra}_L < 7 \times 10^9 \qquad \text{(20–36)}$$

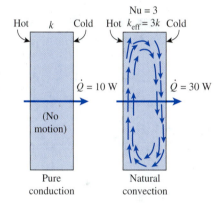

**FIGURE 20–19**
A Nusselt number of 3 for an enclosure indicates that heat transfer through the enclosure by *natural convection* is three times that by *pure conduction*.

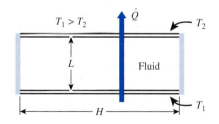

**FIGURE 20–20**
A horizontal rectangular enclosure with isothermal surfaces.

Based on experiments with air, Hollands et al. (1976) recommend this correlation for horizontal enclosures,

$$\text{Nu} = 1 + 1.44\left[1 - \frac{1708}{\text{Ra}_L}\right]^+ + \left[\frac{\text{Ra}_L^{1/3}}{18} - 1\right]^+ \qquad \text{Ra}_L < 10^8 \qquad \textbf{(20–37)}$$

The notation [ ]⁺ indicates that if the quantity in the bracket is negative, it should be set equal to zero. This relation also correlates data well for liquids with moderate Prandtl numbers for $\text{Ra}_L < 10^5$, and thus it can also be used for water.

## Inclined Rectangular Enclosures

Air spaces between two inclined parallel plates are commonly encountered in flat-plate solar collectors (between the glass cover and the absorber plate) and the double-pane skylights on inclined roofs. Heat transfer through an inclined enclosure depends on the **aspect ratio** *H/L* as well as the tilt angle $\theta$ from the horizontal (Fig. 20–21).

For large aspect ratios ($H/L \geq 12$), this equation (Hollands et al., 1976) correlates experimental data extremely well for tilt angles up to 70°,

$$\text{Nu} = 1 + 1.44\left[1 - \frac{1708}{\text{Ra}_L \cos \theta}\right]^+ \left(1 - \frac{1708(\sin 1.8\theta)^{1.6}}{\text{Ra}_L \cos \theta}\right) + \left[\frac{(\text{Ra}_L \cos \theta)^{1/3}}{18} - 1\right]^+$$

$$\textbf{(20–38)}$$

for $\text{Ra}_L < 10^5$, $0 < \theta < 70°$, and $H/L \geq 12$. Again, any quantity in [ ]⁺ should be set equal to zero if it is negative. This is to ensure that Nu = 1 for $\text{Ra}_L \cos \theta < 1708$. Note that this relation reduces to Eq. 20–37 for horizontal enclosures for $\theta = 0°$, as expected.

For enclosures with smaller aspect ratios ($H/L < 12$), the next correlation can be used provided that the tilt angle is less than the critical value $\theta_{cr}$ listed in Table 20–2 [Catton (1978)]

$$\text{Nu} = \text{Nu}_{\theta=0°}\left(\frac{\text{Nu}_{\theta=90°}}{\text{Nu}_{\theta=0°}}\right)^{\theta/\theta_{cr}} (\sin \theta_{cr})^{\theta/(4\theta_{cr})} \qquad 0° < \theta < \theta_{cr} \qquad \textbf{(20–39)}$$

For tilt angles greater than the critical value ($\theta_{cr} < \theta < 90°$), the Nusselt number can be obtained by multiplying the Nusselt number for a vertical enclosure by $(\sin \theta)^{1/4}$ [Ayyaswamy and Catton (1973)],

$$\text{Nu} = \text{Nu}_{\theta=90°}(\sin \theta)^{1/4} \qquad \theta_{cr} < \theta < 90°, \text{ any } H/L \qquad \textbf{(20–40)}$$

For enclosures tilted more than 90°, the recommended relation is [Arnold et al. (1974)]

$$\text{Nu} = 1 + (\text{Nu}_{\theta=90°} - 1)\sin \theta \qquad 90° < \theta < 180°, \text{ any } H/L \qquad \textbf{(20–41)}$$

More recent but more complex correlations are also available in the literature [e.g., ElSherbiny et al. (1982)].

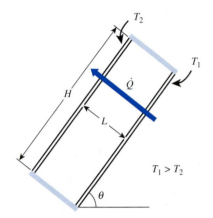

**FIGURE 20–21**
An inclined rectangular enclosure with isothermal surfaces.

**TABLE 20–2**

Critical angles for inclined rectangular enclosures

| Aspect ratio, $H/L$ | Critical angle, $\theta_{cr}$ |
|---|---|
| 1 | 25° |
| 3 | 53° |
| 6 | 60° |
| 12 | 67° |
| >12 | 70° |

## Vertical Rectangular Enclosures

For vertical enclosures (Fig. 20–22), Catton (1978) recommends these two correlations due to Berkovsky and Polevikov (1977),

$$Nu = 0.18 \left( \frac{Pr}{0.2 + Pr} Ra_L \right)^{0.29} \qquad \begin{array}{l} 1 < H/L < 2 \\ \text{any Prandtl number} \\ Ra_L \, Pr/(0.2 + Pr) > 10^3 \end{array} \qquad \textbf{(20–42)}$$

$$Nu = 0.22 \left( \frac{Pr}{0.2 + Pr} Ra_L \right)^{0.28} \left( \frac{H}{L} \right)^{-1/4} \qquad \begin{array}{l} 2 < H/L < 10 \\ \text{any Prandtl number} \\ Ra_L < 10^{10} \end{array} \qquad \textbf{(20–43)}$$

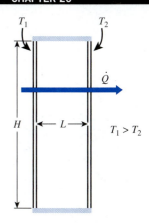

**FIGURE 20–22**
A vertical rectangular enclosure with isothermal surfaces.

For vertical enclosures with larger aspect ratios, the following correlations can be used [MacGregor and Emery (1969)]

$$Nu = 0.42 Ra_L^{1/4} \, Pr^{0.012} \left( \frac{H}{L} \right)^{-0.3} \qquad \begin{array}{l} 10 < H/L < 40 \\ 1 < Pr < 2 \times 10^4 \\ 10^4 < Ra_L < 10^7 \end{array} \qquad \textbf{(20–44)}$$

$$Nu = 0.046 Ra_L^{1/3} \qquad \begin{array}{l} 1 < H/L < 40 \\ 1 < Pr < 20 \\ 10^6 < Ra_L < 10^9 \end{array} \qquad \textbf{(20–45)}$$

Again, all fluid properties are to be evaluated at the average temperature $(T_1 + T_2)/2$.

## Concentric Cylinders

Consider two long concentric horizontal cylinders maintained at uniform but different temperatures of $T_i$ and $T_o$, as shown in Fig. 20–23. The diameters of the inner and outer cylinders are $D_i$ and $D_o$, respectively, and the characteristic length is the spacing between the cylinders, $L_c = (D_o - D_i)/2$. The rate of heat transfer through the annular space between the cylinders by natural convection per unit length is expressed as

$$\dot{Q} = \frac{2\pi k_{eff}}{\ln(D_o/D_i)} (T_i - T_o) \qquad (\text{W/m}) \qquad \textbf{(20–46)}$$

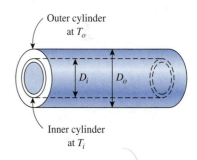

**FIGURE 20–23**
Two concentric horizontal isothermal cylinders.

The recommended relation for effective thermal conductivity is [Raithby and Hollands (1975)]

$$\frac{k_{eff}}{k} = 0.386 \left( \frac{Pr}{0.861 + Pr} \right)^{1/4} (F_{cyl} Ra_L)^{1/4} \qquad \textbf{(20–47)}$$

where the geometric factor for concentric cylinders $F_{cyl}$ is

$$F_{cyl} = \frac{[\ln(D_o/D_i)]^4}{L_c^3 (D_i^{-3/5} + D_o^{-3/5})^5} \qquad \textbf{(20–48)}$$

The $k_{eff}$ relation in Eq. 20–47 is applicable for $0.70 \leq Pr \leq 6000$ and $10^2 \leq F_{cyl}Ra_L \leq 10^7$. For $F_{cyl}Ra_L < 100$, natural convection currents are negligible and thus $k_{eff} = k$. Note that $k_{eff}$ cannot be less than $k$, and thus we should set $k_{eff} = k$ if $k_{eff}/k < 1$. The fluid properties are evaluated at the average temperature of $(T_i + T_o)/2$.

## Concentric Spheres

For concentric isothermal spheres, the rate of heat transfer through the gap between the spheres by natural convection is expressed as (Fig. 20–24)

$$\dot{Q} = k_{eff} \frac{\pi D_i D_o}{L_c} (T_i - T_o) \qquad (W) \qquad \text{(20–49)}$$

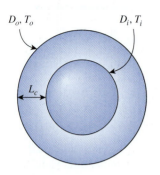

**FIGURE 20–24**
Two concentric isothermal spheres.

where $L_c = (D_o - D_i)/2$ is the characteristic length. The recommended relation for effective thermal conductivity is [Raithby and Hollands (1975)]

$$\frac{k_{eff}}{k} = 0.74 \left( \frac{Pr}{0.861 + Pr} \right)^{1/4} (F_{sph}Ra_L)^{1/4} \qquad \text{(20–50)}$$

where the geometric factor for concentric spheres $F_{sph}$ is

$$F_{sph} = \frac{L_c}{(D_i D_o)^4 (D_i^{-7/5} + D_o^{-7/5})^5} \qquad \text{(20–51)}$$

The $k_{eff}$ relation in Eq. 20–50 is applicable for $0.70 \leq Pr \leq 4200$ and $10^2 \leq F_{sph}Ra_L \leq 10^4$. If $k_{eff}/k < 1$, we should set $k_{eff} = k$.

## Combined Natural Convection and Radiation

Gases are nearly transparent to radiation, and thus heat transfer through a gas layer is by simultaneous convection (or conduction, if the gas is quiescent) and radiation. Natural convection heat transfer coefficients are typically very low compared to those for forced convection. Therefore, radiation is usually disregarded in forced convection problems, but it must be considered in natural convection problems that involve a gas. This is especially the case for surfaces with high emissivities. For example, about half of the heat transfer through the air space of a double-pane window is by radiation. The total rate of heat transfer is determined by adding the convection and radiation components,

$$\dot{Q}_{total} = \dot{Q}_{conv} + \dot{Q}_{rad} \qquad \text{(20–52)}$$

Radiation heat transfer from a surface at temperature $T_s$ surrounded by surfaces at a temperature $T_{surr}$ (both in K) is determined from

$$\dot{Q}_{rad} = \varepsilon \sigma A_s (T_s^4 - T_{surr}^4) \qquad (W) \qquad \text{(20–53)}$$

where $\varepsilon$ is the emissivity of the surface, $A_s$ is the surface area, and $\sigma = 5.67 \times 10^{-8}$ W/m²·K⁴ is the Stefan–Boltzmann constant.

When the end effects are negligible, radiation heat transfer between two large parallel plates at temperatures $T_1$ and $T_2$ is expressed as (see Chapter 21 for details)

$$\dot{Q}_{rad} = \frac{\sigma A_s(T_1^4 - T_2^4)}{1/\varepsilon_1 + 1/\varepsilon_2 - 1} = \varepsilon_{effective}\, \sigma A_s(T_1^4 - T_2^4) \qquad \text{(W)} \qquad \textbf{(20–54)}$$

where $\varepsilon_1$ and $\varepsilon_2$ are the emissivities of the plates, and $\varepsilon_{effective}$ is the *effective emissivity* defined as

$$\varepsilon_{effective} = \frac{1}{1/\varepsilon_1 + 1/\varepsilon_2 - 1} \qquad \textbf{(20–55)}$$

The emissivity of an ordinary glass surface, for example, is 0.84. Therefore, the effective emissivity of two parallel glass surfaces facing each other is 0.72. Radiation heat transfer between concentric cylinders and spheres is discussed in Chapter 21.

Note that in some cases the temperature of the surrounding medium may be below the surface temperature ($T_\infty < T_s$), while the temperature of the surrounding surfaces is above the surface temperature ($T_{surr} < T_s$). In such cases, convection and radiation heat transfers are subtracted from each other instead of being added since they are in opposite directions. Also, for a metal surface, the radiation effect can be reduced to negligible levels by polishing the surface and thus lowering the surface emissivity to a value near zero.

---

### EXAMPLE 20–3  Heat Loss through a Double-Pane Window

The vertical 0.8-m-high, 2-m-wide double-pane window shown in Fig. 20–25 consists of two sheets of glass separated by a 2-cm air gap at atmospheric pressure. If the glass surface temperatures across the air gap are measured to be 12°C and 2°C, determine the rate of heat transfer through the window.

**SOLUTION** Two glasses of a double-pane window are maintained at specified temperatures. The rate of heat transfer through the window is to be determined.

**Assumptions** 1 Steady operating conditions exist. 2 Air is an ideal gas. 3 Radiation heat transfer is not considered.

**Properties** The properties of air at the average temperature of $T_{avg} = (T_1 + T_2)/2 = (12 + 2)/2 = 7°C$ and 1 atm pressure are (Table A–22)

$$k = 0.02416 \text{ W/m·K} \qquad Pr = 0.7344$$

$$\nu = 1.400 \times 10^{-5} \text{ m}^2/\text{s} \qquad \beta = \frac{1}{T_{avg}} = \frac{1}{280 \text{ K}}$$

**Analysis** We have a rectangular enclosure filled with air. The characteristic length in this case is the distance between the two glasses, $L_c = L = 0.02$ m. Then the Rayleigh number becomes

$$Ra_L = \frac{g\beta(T_1 - T_2)L_c^3}{\nu^2} Pr$$

$$= \frac{(9.81 \text{ m/s}^2)[1/(280 \text{ K})](12 - 2 \text{ K})(0.02 \text{ m})^3}{(1.400 \times 10^{-5} \text{ m}^2/\text{s})^2}(0.7344) = 1.050 \times 10^4$$

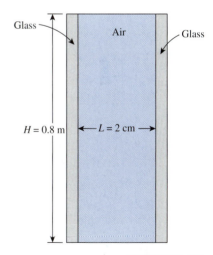

**FIGURE 20–25**
Schematic for Example 20–3.

Glass — Air — Glass

$H = 0.8$ m    $\leftarrow L = 2 \text{ cm} \rightarrow$

The aspect ratio of the geometry is $H/L = 0.8/0.02 = 40$. Then the Nusselt number in this case can be determined from Eq. 20–44 to be

$$\text{Nu} = 0.42\text{Ra}_L^{1/4}\,\text{Pr}^{0.012}\left(\frac{H}{L}\right)^{-0.3}$$

$$= 0.42(1.050 \times 10^4)^{1/4}(0.7344)^{0.012}\left(\frac{0.8}{0.02}\right)^{-0.3} = 1.40$$

Then,

$$A_s = H \times W = (0.8\text{ m})(2\text{ m}) = 1.6\text{ m}^2$$

and

$$\dot{Q} = hA_s(T_1 - T_2) = k\text{Nu}A_s\frac{T_1 - T_2}{L}$$

$$= (0.02416\text{ W/m·K})(1.40)(1.6\text{ m}^2)\frac{(12 - 2)°\text{C}}{0.02\text{ m}} = \mathbf{27.1\ W}$$

Therefore, heat is lost through the window at a rate of 27.1 W.

**Discussion** Recall that a Nusselt number of Nu = 1 for an enclosure corresponds to pure conduction heat transfer through the enclosure. The air in the enclosure in this case remains still, and no natural convection currents occur in the enclosure. The Nusselt number in our case is 1.40, which indicates that heat transfer through the enclosure is 1.40 times that by pure conduction. The increase in heat transfer is due to the natural convection currents that develop in the enclosure.

---

### EXAMPLE 20–4    Heat Transfer through a Spherical Enclosure

The two concentric spheres of diameters $D_i = 20$ cm and $D_o = 30$ cm shown in Fig. 20–26 are separated by air at 1 atm pressure. The surface temperatures of the two spheres enclosing the air are $T_i = 320$ K and $T_o = 280$ K, respectively. Determine the rate of heat transfer from the inner sphere to the outer sphere by natural convection.

**SOLUTION** Two surfaces of a spherical enclosure are maintained at specified temperatures. The rate of heat transfer through the enclosure is to be determined.

**Assumptions** 1 Steady operating conditions exist. 2 Air is an ideal gas. 3 Radiation heat transfer is not considered.

**Properties** The properties of air at the average temperature of $T_{avg} = (T_i + T_o)/2 = (320 + 280)/2 = 300$ K = 27°C and 1 atm pressure are (Table A–22)

$$k = 0.02566\text{ W/m·K} \qquad \text{Pr} = 0.7290$$

$$\nu = 1.580 \times 10^{-5}\text{ m}^2\text{/s} \qquad \beta = \frac{1}{T_{avg}} = \frac{1}{300\text{ K}}$$

**Analysis** We have a spherical enclosure filled with air. The characteristic length in this case is the distance between the two spheres,

$$L_c = (D_o - D_i)/2 = (0.3 - 0.2)/2 = 0.05\text{ m}$$

The Rayleigh number is

$$\text{Ra}_L = \frac{g\beta(T_i - T_o)L_c^3}{\nu^2}\,\text{Pr}$$

$$= \frac{(9.81\text{ m/s}^2)[1/(300\text{ K})](320 - 280)(0.05\text{ m})^3}{(1.58 \times 10^{-5}\text{ m}^2\text{/s})^2}(0.729)$$

$$= 4.775 \times 10^5$$

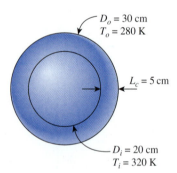

$D_o = 30$ cm
$T_o = 280$ K

$L_c = 5$ cm

$D_i = 20$ cm
$T_i = 320$ K

**FIGURE 20–26**
Schematic for Example 20–4.

The effective thermal conductivity is

$$F_{sph} = \frac{L_c}{(D_i D_o)^4 (D_i^{-7/5} + D_o^{-7/5})^5}$$

$$= \frac{0.05 \text{ m}}{[(0.2 \text{ m})(0.3 \text{ m})]^4[(0.2 \text{ m} 2^{-7/5} + (0.3 \text{ m})^{-7/5}]^5} = 0.005229$$

$$k_{eff} = 0.74k \left(\frac{Pr}{0.861 + Pr}\right)^{1/4} (F_{sph} Ra_L)^{1/4}$$

$$= 0.74(0.02566 \text{ W/m·K})\left(\frac{0.729}{0.861 + 0.729}\right)^{1/4} (0.005229 \times 4.775 \times 10^5)^{1/4}$$

$$= 0.1105 \text{ W/m·K}$$

Then the rate of heat transfer between the spheres becomes

$$\dot{Q} = k_{eff} \frac{\pi D_i D_o}{L_c}(T_i - T_o)$$

$$= (0.1105 \text{ W/m·K})\frac{\pi(0.2 \text{ m})(0.3 \text{ m})}{0.05 \text{ m}}(320 - 280) \text{ K} = \textbf{16.7 W}$$

Therefore, heat will be lost from the inner sphere to the outer one at a rate of 16.7 W.

**Discussion** Note that the air in the spherical enclosure acts like a stationary fluid whose thermal conductivity is $k_{eff}/k$ = 0.1105/0.02566 = 4.3 times that of air as a result of natural convection currents. Also, radiation heat transfer between spheres is usually significant, and should be considered in a complete analysis.

---

## EXAMPLE 20–5    Heating Water in a Tube by Solar Energy

A solar collector consists of a horizontal aluminum tube having an outer diameter of 2 in enclosed in a concentric thin glass tube of 4-in-diameter (Fig. 20–27). Water is heated as it flows through the tube, and the annular space between the aluminum and the glass tubes is filled with air at 1 atm pressure. The pump circulating the water fails during a clear day, and the water temperature in the tube starts rising. The aluminum tube absorbs solar radiation at a rate of 30 Btu/h per foot length, and the temperature of the ambient air outside is 70°F. Disregarding any heat loss by radiation, determine the temperature of the aluminum tube when steady operation is established (i.e., when the rate of heat loss from the tube equals the amount of solar energy gained by the tube).

**SOLUTION** The circulating pump of a solar collector that consists of a horizontal tube and its glass cover fails. The equilibrium temperature of the tube is to be determined.

**Assumptions** 1 Steady operating conditions exist. 2 The tube and its cover are isothermal. 3 Air is an ideal gas. 4 Heat loss by radiation is negligible.

**Properties** The properties of air should be evaluated at the average temperature. But we do not know the exit temperature of the air in the duct, and thus we cannot determine the bulk fluid and glass cover temperatures at this point, and we cannot evaluate the average temperatures. Therefore, we assume the glass temperature to be 110°F, and use properties at an anticipated average temperature of (70 + 110)/2 = 90°F (Table A–22E),

$$k = 0.01505 \text{ Btu/h·ft·°F} \qquad Pr = 0.7275$$

$$\nu = 1.753 \times 10^{-4} \text{ ft}^2/\text{s} \qquad \beta = \frac{1}{T_{avg}} = \frac{1}{550 \text{ R}}$$

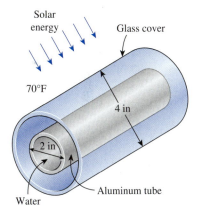

**FIGURE 20–27**
Schematic for Example 20–5.

*Analysis* We have a horizontal cylindrical enclosure filled with air at 1 atm pressure. The problem involves heat transfer from the aluminum tube to the glass cover and from the outer surface of the glass cover to the surrounding ambient air. When steady operation is reached, these two heat transfer rates must equal the rate of heat gain. That is,

$$\dot{Q}_{\text{tube-glass}} = \dot{Q}_{\text{glass-ambient}} = \dot{Q}_{\text{solar gain}} = 30 \text{ Btu/h} \qquad \text{(per foot of tube)}$$

The heat transfer surface area of the glass cover is

$$A_o = A_{\text{glass}} = (\pi D_o L) = \pi (4/12 \text{ ft})(1 \text{ ft}) = 1.047 \text{ ft}^2 \qquad \text{(per foot of tube)}$$

To determine the Rayleigh number, we need to know the surface temperature of the glass, which is not available. Therefore, it is clear that the solution will require a trial-and-error approach. Assuming the glass cover temperature to be 110°F, the Rayleigh number, the Nusselt number, the convection heat transfer coefficient, and the rate of natural convection heat transfer from the glass cover to the ambient air are determined to be

$$\text{Ra}_{D_o} = \frac{g\beta(T_s - T_\infty)D_o^3}{\nu^2} \text{Pr}$$

$$= \frac{(32.2 \text{ ft/s}^2)[1/(550 \text{ R})](110 - 70 \text{ R})(4/12 \text{ ft})^3}{(1.753 \times 10^{-4} \text{ ft}^2/\text{s})^2}(0.7275) = 2.054 \times 10^6$$

$$\text{Nu} = \left\{ 0.6 + \frac{0.387 \text{ Ra}_D^{1/6}}{[1 + (0.559/\text{Pr})^{9/16}]^{8/27}} \right\}^2 = \left\{ 0.6 + \frac{0.387(2.054 \times 10^6)^{1/6}}{[1 + (0.559/0.7275)^{9/16}]^{8/27}} \right\}^2$$

$$= 17.89$$

$$h_o = \frac{k}{D_o} \text{Nu} = \frac{0.01505 \text{ Btu/h} \cdot \text{ft} \cdot °\text{F}}{4/12 \text{ ft}}(17.89) = 0.8077 \text{ Btu/h} \cdot \text{ft}^2 \cdot °\text{F}$$

$$\dot{Q}_o = h_o A_o (T_o - T_\infty) = (0.8077 \text{ Btu/h} \cdot \text{ft}^2 \cdot °\text{F})(1.047 \text{ ft}^2)(110 - 70)°\text{F}$$
$$= 33.8 \text{ Btu/h}$$

which is more than 30 Btu/h. Therefore, the assumed temperature of 110°F for the glass cover is high. Repeating the calculations with lower temperatures, the glass cover temperature corresponding to 30 Btu/h is determined to be 106°F.

The temperature of the aluminum tube is determined in a similar manner using the natural convection relations for two horizontal concentric cylinders. The characteristic length in this case is the distance between the two cylinders, which is

$$L_c = (D_o - D_i)/2 = (4 - 2)/2 = 1 \text{ in} = 1/12 \text{ ft}$$

We start the calculations by assuming the tube temperature to be 200°F, and thus an average temperature of $(106 + 200)/2 = 153°\text{F} = 613 \text{ R}$. Using air properties at this temperature gives

$$\text{Ra}_L = \frac{g\beta(T_i - T_o)L_c^3}{\nu^2} \text{Pr}$$

$$= \frac{(32.2 \text{ ft/s}^2)[1/613 \text{ R}](200 - 106 \text{ R})(1/12 \text{ ft})^3}{(2.117 \times 10^{-4} \text{ ft}^2/\text{s})^2}(0.7184) = 4.580 \times 10^4$$

The effective thermal conductivity is

$$F_{cyl} = \frac{[\ln(D_o/D_i)]^4}{L_c^3(D_i^{-3/5} + D_o^{-3/5})^5}$$

$$= \frac{[\ln(4/2)]^4}{(1/12 \text{ ft})^3[(2/12 \text{ ft})^{-3/5} + (4/12 \text{ ft})^{-3/5}]^5} = 0.1466$$

$$k_{eff} = 0.386k\left(\frac{Pr}{0.861 + Pr}\right)^{1/4}(F_{cyl}Ra_L)^{1/4}$$

$$= 0.386(0.01653 \text{ Btu/h·ft·°F})\left(\frac{0.7184}{0.861 + 0.7184}\right)^{1/4}$$

$$\times (0.1466 \times 4.580 \times 10^4)^{1/4}$$

$$= 0.04743 \text{ Btu/h·ft·°F}$$

Then the rate of heat transfer between the cylinders becomes

$$\dot{Q} = \frac{2\pi k_{eff}}{\ln(D_o/D_i)}(T_i - T_o)$$

$$= \frac{2\pi(0.04743 \text{ Btu/h·ft·°F})}{\ln(4/2)}(200 - 106)°F = 40.4 \text{ Btu/h}$$

which is more than 30 Btu/h. Therefore, the assumed temperature of 200°F for the tube is high. By trying other values, the tube temperature corresponding to 30 Btu/h is determined to be **180°F**. Therefore, the tube will reach an equilibrium temperature of 180°F when the pump fails.

**Discussion** Note that we have not considered heat loss by radiation in the calculations, and thus the tube temperature determined is probably too high.

## SUMMARY

In this chapter, we have considered natural convection heat transfer where any fluid motion occurs by natural means such as buoyancy. The volume expansion coefficient of a substance represents the variation of the density of that substance with temperature at constant pressure, and for an ideal gas, it is expressed as $\beta = 1/T$, where $T$ is the absolute temperature in K or R.

The flow regime in natural convection is governed by a dimensionless number called the *Grashof number,* which represents the ratio of the buoyancy force to the viscous force acting on the fluid and is expressed as

$$Gr_L = \frac{g\beta(T_s - T_\infty)L_c^3}{\nu^2}$$

where $L_c$ is the *characteristic length,* which is the height $L$ for a vertical plate and the diameter $D$ for a horizontal cylinder. The correlations for the Nusselt number $Nu = hL_c/k$ in natural convection are expressed in terms of the *Rayleigh number* defined as

$$Ra_L = Gr_L Pr = \frac{g\beta(T_s - T_\infty)L_c^3}{\nu^2}Pr$$

Nusselt number relations for various surfaces are given in Table 20–1. All fluid properties are evaluated at the film temperature of $T_f = \frac{1}{2}(T_s + T_\infty)$. The outer surface of a vertical cylinder can be treated as a vertical plate when the curvature effects are negligible. The characteristic length for a horizontal surface is $L_c = A_s/p$, where $A_s$ is the surface area and $p$ is the perimeter.

The average Nusselt number for vertical isothermal *parallel plates* of spacing $S$ and height $L$ is given as

$$Nu = \frac{hS}{k} = \left[\frac{576}{(Ra_S S/L)^2} + \frac{2.873}{(Ra_S S/L)^{0.5}}\right]^{-0.5}$$

In a *horizontal rectangular enclosure* with the hotter plate at the top, heat transfer is by pure conduction and Nu = 1. When the hotter plate is at the bottom, the Nusselt number is

$$Nu = 1 + 1.44\left[1 - \frac{1708}{Ra_L}\right]^+ + \left[\frac{Ra_L^{1/3}}{18} - 1\right]^+ \qquad Ra_L < 10^8$$

The notation [ ]$^+$ indicates that if the quantity in the bracket is negative, it should be set equal to zero. For *vertical horizontal enclosures,* the Nusselt number can be determined from

$$Nu = 0.18\left(\frac{Pr}{0.2 + Pr}Ra_L\right)^{0.29} \qquad \begin{matrix} 1 < H/L < 2 \\ \text{any Prandtl number} \\ Ra_L\,Pr/(0.2 + Pr) > 10^3 \end{matrix}$$

$$Nu = 0.22\left(\frac{Pr}{0.2 + Pr}Ra_L\right)^{0.28}\left(\frac{H}{L}\right)^{-1/4} \qquad \begin{matrix} 2 < H/L < 10 \\ \text{any Prandtl number} \\ Ra_L < 10^{10} \end{matrix}$$

For aspect ratios greater than 10, Eqs. 20–44 and 20–45 should be used. For inclined enclosures, Eqs. 20–38 through 20–41 should be used.

For *concentric horizontal cylinders,* the rate of heat transfer through the annular space between the cylinders by natural convection per unit length is

$$\dot{Q} = \frac{2\pi k_{eff}}{\ln(D_o/D_i)}(T_i - T_o)$$

where

$$\frac{k_{eff}}{k} = 0.386\left(\frac{Pr}{0.861 + Pr}\right)^{1/4}(F_{cyl}Ra_L)^{1/4}$$

and

$$F_{cyl} = \frac{[\ln(D_o/D_i)]^4}{L_c^3(D_i^{-3/5} + D_o^{-3/5})^5}$$

For a *spherical enclosure,* the rate of heat transfer through the space between the spheres by natural convection is expressed as

$$\dot{Q} = k_{eff}\frac{\pi D_i D_o}{L_c}(T_i - T_o)$$

where

$$\frac{k_{eff}}{k} = 0.74\left(\frac{Pr}{0.861 + Pr}\right)^{1/4}(F_{sph}Ra_L)^{1/4}$$

$$L_c = (D_o - D_i)/2$$

$$F_{sph} = \frac{L_c}{(D_iD_o)^4(D_i^{-7/5} + D_o^{-7/5})^5}$$

The quantity $k$Nu is called the *effective thermal conductivity* of the enclosure, since a fluid in an enclosure behaves like a quiescent fluid whose thermal conductivity is $k$Nu as a result of convection currents. The fluid properties are evaluated at the average temperature of $(T_i + T_o)/2$.

## REFERENCES AND SUGGESTED READINGS

1. J. N. Arnold, I. Catton, and D. K. Edwards. "Experimental Investigation of Natural Convection in Inclined Rectangular Region of Differing Aspects Ratios." ASME Paper No. 75-HT-62, 1975.

2. L. S. Asgeirsson and A. J. Ghajar. "Prediction of Thermal Conductivity and Viscosity for Some Fluids in the Near-Critical Region." *Chemical Engineering Communications,* Vol. 43, Nos. 1–3 (1986), pp. 165–184.

3. P. S. Ayyaswamy and I. Catton. "The Boundary–Layer Regime for Natural Convection in a Differently Heated Tilted Rectangular Cavity." *Journal of Heat Transfer* 95 (1973), p. 543.

4. B. M. Berkovsky and V. K. Polevikov. "Numerical Study of Problems on High-Intensive Free Convection." In *Heat Transfer and Turbulent Buoyant Convection,* eds. D. B. Spalding and N. Afgan, pp. 443–445. Washington, DC: Hemisphere, 1977.

5. I. Catton. "Natural Convection in Enclosures." *Proceedings of Sixth International Heat Transfer Conference.* Toronto: Canada, 1978, Vol. 6, pp. 13–31.

6. T. Cebeci. "Laminar Free Convection Heat Transfer from the Outer Surface of a Vertical Slender Circular Cylinder." *Proceedings of Fifth International Heat Transfer Conference* paper NCI.4, 1974 pp. 15–19.

7. S. W. Churchill. "A Comprehensive Correlating Equation for Laminar Assisting Forced and Free Convection." *AIChE Journal* 23 (1977), pp. 10–16.

8. S. W. Churchill. "Free Convection around Immersed Bodies." In *Heat Exchanger Design Handbook,* ed. E. U. Schlünder, Section 2.5.7. New York: Hemisphere, 1983.

9. S. W. Churchill. "Combined Free and Forced Convection around Immersed Bodies." In *Heat Exchanger Design Handbook,* Section 2.5.9. New York: Hemisphere Publishing, 1986.

10. S. W. Churchill and H. H. S. Chu. "Correlating Equations for Laminar and Turbulent Free Convection from a Horizontal Cylinder." *International Journal of Heat Mass Transfer* 18 (1975), p. 1049.

11. S. W. Churchill and H. H. S. Chu. "Correlating Equations for Laminar and Turbulent Free Convection from a Vertical Plate." *International Journal of Heat Mass Transfer* 18 (1975), p. 1323.

12. E. R. G. Eckert and E. Soehngen. "Studies on Heat Transfer in Laminar Free Convection with Zehnder–Mach Interferometer." USAF Technical Report 5747, December 1948.

13. E. R. G. Eckert and E. Soehngen. "Interferometric Studies on the Stability and Transition to Turbulence of a Free Convection Boundary Layer." *Proceedings of General Discussion, Heat Transfer ASME-IME,* London, 1951.

14. S. M. ElSherbiny, G. D. Raithby, and K. G. T. Hollands. "Heat Transfer by Natural Convection Across Vertical and Inclined Air Layers. *Journal of Heat Transfer* 104 (1982), pp. 96–102.

15. T. Fujii and H. Imura. "Natural Convection Heat Transfer from a Plate with Arbitrary Inclination." *International Journal of Heat Mass Transfer* 15 (1972), p. 755.

16. A. J. Ghajar and A. Asadi, "Improved Forced Convective Heat Transfer Correlations for Liquids in the Near-Critical Region." *AIAA Journal*, Vol. 24, No. 12 (1986), pp. 2030–2037.

17. S. Ghorbani-Tari and A. J. Ghajar. "Improved Free Convective Heat Transfer Correlations in the Near-Critical Region." *AIAA Journal*, Vol. 23, No. 10 (1985), pp. 1647–1649.

18. K. G. T. Hollands, T. E. Unny, G. D. Raithby, and L. Konicek. "Free Convective Heat Transfer Across Inclined Air Layers." *Journal of Heat Transfer* 98 (1976), pp. 182–193.

19. M. Jakob. *Heat Transfer.* New York: Wiley, 1949.

20. W. M. Kays, M. E. Crawford, and B. Weigand. *Convective Heat and Mass Transfer*. 4th ed. New York: McGraw-Hill, 2005.

21. R. K. MacGregor and A. P. Emery. "Free Convection Through Vertical Plane Layers: Moderate and High Prandtl Number Fluids." *Journal of Heat Transfer* 91 (1969), p. 391.

22. T. S. Najjar and A. J. Ghajar. "Prediction of Density and Constant Pressure Specific Heat for Several Fluids in the Near-Critical Region." AIAA paper No. 83-1476, *AIAA 18th Thermophysics Conference*, Montreal, Canada, June 1-3 (1983).

23. G. D. Raithby and K. G. T. Hollands. "A General Method of Obtaining Approximate Solutions to Laminar and Turbulent Free Convection Problems." In *Advances in Heat Transfer,* ed. F. Irvine and J. P. Hartnett, Vol. II, pp. 265–315. New York: Academic Press, 1975.

24. E. M. Sparrow and J. L. Gregg. "Laminar Free Convection from a Vertical Flat Plate." *Transactions of the ASME* 78 (1956), p. 438.

25. E. M. Sparrow and J. L. Gregg. "Laminar Free Convection Heat Transfer from the Outer Surface of a Vertical Circular Cylinder." *ASME* 78 (1956), p. 1823.

26. L. M. Tam and A. J. Ghajar. "Transitional Heat Transfer in Plain Horizontal Tubes." *Heat Transfer Engineering*, Vol. 27, No. 5 (2006), pp. 23–38.

## PROBLEMS*

### Physical Mechanism of Natural Convection

**20–1C** What is natural convection? How does it differ from forced convection? What force causes natural convection currents?

**20–2C** In which mode of heat transfer is the convection heat transfer coefficient usually higher, natural convection or forced convection? Why?

*Problems designated by a "C" are concept questions, and students are encouraged to answer them all. Problems designated by an "E" are in English units, and the SI users can ignore them. Problems with the icon 🐕 are solved using EES, and complete solutions together with parametric studies are included on the enclosed DVD. Problems with the icon 🖥 are comprehensive in nature, and are intended to be solved with a computer, preferably using the EES software that accompanies this text.

**20–3C** What is buoyancy force? Compare the relative magnitudes of the buoyancy force acting on a body immersed in these mediums: (*a*) air, (*b*) water, (*c*) mercury, and (*d*) an evacuated chamber.

**20–4C** Consider a hot boiled egg in a spacecraft that is filled with air at atmospheric pressure and temperature at all times. Will the egg cool faster or slower when the spacecraft is in space instead of on the ground? Explain.

**20–5C** When will the hull of a ship sink in water deeper: when the ship is sailing in fresh water or in seawater? Why?

**20–6C** Consider two fluids, one with a large coefficient of volume expansion and the other with a small one. In what fluid will a hot surface initiate stronger natural convection currents? Why? Assume the viscosity of the fluids to be the same.

**20–7C** Physically, what does the Grashof number represent? How does the Grashof number differ from the Reynolds number?

**20–8** Show that the volume expansion coefficient of an ideal gas is $\beta = 1/T$, where $T$ is the absolute temperature.

**20–9** Using its definition and the values listed in Table A–15, determine the volume expression coefficient of saturated liquid water at 70°C. Compare the result with the value tabulated in Table A–15. *Answer: 5.83 × 10⁻⁴ K⁻¹*

**20–10** The density of liquid water can be correlated as $\rho(T) = 1000 - 0.0736T - 0.00355T^2$ where $\rho$ and $T$ are in kg/m and °C, respectively. Determine the volume expansion coefficient at 70°C. Compare the result with the value tabulated in Table A–15.

**20–11** A 10 cm × 10 cm plate has a constant surface temperature of 150°C. Determine the Grashof number if the chip is placed in the following fluids: air (1 atm, 30°C), liquid water (30°C), engine oil (10°C). Discuss how the Grashof number affects the natural convection flow.

**20–12** A long 2-in.-diameter rod with surface temperature of 200°F is submerged in a bath of fluid. Determine the Grashof and Rayleigh numbers if the fluid is (*a*) liquid water at 40°F, (*b*) liquid ammonia at 40°F, (*c*) engine oil at 50°F, and (*d*) air at 40°F (1 atm).

## Natural Convection over Surfaces

**20–13C** Consider laminar natural convection from a vertical hot-plate. Will the heat flux be higher at the top or at the bottom of the plate? Why?

**20–14C** Under what conditions can the outer surface of a vertical cylinder be treated as a vertical plate in natural convection calculations?

**20–15C** Will a hot horizontal plate whose back side is insulated cool faster or slower when its hot surface is facing down instead of up?

**20–16** An aluminum soda can of 150-mm in length and 60-mm in diameter is placed horizontally inside a refrigerator compartment that maintains a temperature of 4°C. If the surface temperature of the can is 36°C, estimate heat transfer rate from the can. Neglect the heat transfer from the ends of the can. *Answer: 5.08 W*

**20–17** A can of engine oil with a length of 150 mm and a diameter of 100 mm is placed vertically in the trunk of a car. In a hot summer day, the temperature in the trunk is 43°C. If the surface temperature of the can is 17°C, determine heat transfer rate from the can surface. Neglect the heat transfer from the ends of the can.

**20–18** Consider a thin 24-cm-long and 20-cm-wide horizontal plate suspended in air at 20°C. The plate is equipped with electric resistance heating elements with a rating of 20 W. Now the heater is turned on and the plate temperature rises. Determine the temperature of the plate when steady operating conditions are reached. The plate has an emissivity of 0.90 and the surrounding surfaces are at 17°C.

Air
$T_\infty = 20°C$

L = 24 cm

Resistance wire

**FIGURE P20–18**

**20–19** Flue gases from an incinerator are released to atmosphere using a stack that is 0.6 m in diameter and 10.0 m high. The outer surface of the stack is at 40°C and the surrounding air is at 10°C. Determine the rate of heat transfer from the stack assuming (*a*) there is no wind and (*b*) the stack is exposed to 20 km/h winds.

**20–20** Thermal energy generated by the electrical resistance of a 5-mm-diameter and 4-m-long bare cable is dissipated to the surrounding air at 20°C. The voltage drop and the electric current across the cable in steady operation are measured to be 60 V and 1.5 A, respectively. Disregarding radiation, estimate the surface temperature of the cable.

**20–21** A 12-m-long section of a 5-cm-diameter horizontal hot-water pipe passes through a large room whose temperature is 27°C. If the temperature and the emissivity of the outer surface of the pipe are 73°C and 0.8, respectively, determine the rate of heat loss from the pipe by (*a*) natural convection and (*b*) radiation.

**20–22E** An average person generates heat at a rate of 240 Btu/h while resting in a room at 70°F. Assuming one-quarter of this heat is lost from the head and taking the emissivity of the skin to be 0.9, determine the average surface temperature of the head when it is not covered. The head can be approximated as a 12-in-diameter sphere, and the interior surfaces of the room can be assumed to be at the room temperature.

**20–23** An incandescent lightbulb is an inexpensive but highly inefficient device that converts electrical energy into light. It converts about 5 percent of the electrical energy it consumes into light while converting the remaining 95 percent into heat. The glass bulb of the lamp heats up very quickly as a result of absorbing all that heat and dissipating it to the surroundings by

25°C

60 W

$\varepsilon = 0.9$

Light, 3 W

**FIGURE P20–23**

convection and radiation. Consider an 8-cm-diameter 60-W lightbulb in a room at 25°C. The emissivity of the glass is 0.9. Assuming that 5 percent of the energy passes through the glass bulb as light with negligible absorption and the rest of the energy is absorbed and dissipated by the bulb itself by natural convection and radiation, determine the equilibrium temperature of the glass bulb. Assume the interior surfaces of the room to be at room temperature. *Answer:* 175°C

**20–24**  A 450-W cylindrical resistance heater is 0.75 m long and 0.5 cm in diameter. The resistance wire is placed horizontally in a fluid at 20°C. Determine the outer surface temperature of the resistance wire in steady operation if the fluid is (*a*) air and (*b*) water. Ignore any heat transfer by radiation. Use properties at 500°C for air and 40°C for water.

**20–25**  In a plant that manufactures canned aerosol paints, the cans are temperature-tested in water baths at 60°C before they are shipped to ensure that they withstand temperatures up to 55°C during transportation and shelving. The cans, moving on a conveyor, enter the open hot water bath, which is 0.5 m deep, 1 m wide, and 3.5 m long, and move slowly in the hot water toward the other end. Some of the cans fail the test and explode in the water bath. The water container is made of sheet metal, and the entire container is at about the same temperature as the hot water. The emissivity of the outer surface of the container is 0.7. If the temperature of the surrounding air and surfaces is 20°C, determine the rate of heat loss from the four side surfaces of the container (disregard the top surface, which is open).

The water is heated electrically by resistance heaters, and the cost of electricity is $0.085/kWh. If the plant operates 24 h a day 365 days a year and thus 8760 h a year, determine the annual cost of the heat losses from the container for this facility.

**FIGURE P20–25**

**20–26**  Reconsider Prob. 20–25. In order to reduce the heating cost of the hot water, it is proposed to insulate the side and bottom surfaces of the container with 5-cm-thick fiberglass insulation ($k = 0.035$ W/m·K) and to wrap the insulation with aluminum foil ($\varepsilon = 0.1$) in order to minimize the heat loss by radiation. An estimate is obtained from a local insulation contractor, who proposes to do the insulation job for $350, including materials and labor. Would you support this proposal? How long will it take for the insulation to pay for itself from the energy it saves?

**20–27**  Consider a 15-cm × 20-cm printed circuit board (PCB) that has electronic components on one side. The board is placed in a room at 20°C. The heat loss from the back surface of the board is negligible. If the circuit board is dissipating 8 W of power in steady operation, determine the average temperature of the hot surface of the board, assuming the board is (*a*) vertical; (*b*) horizontal with hot surface facing up; and (*c*) horizontal with hot surface facing down. Take the emissivity of the surface of the board to be 0.8 and assume the surrounding surfaces to be at the same temperature as the air in the room. *Answers:* (*a*) 46.6°C, (*b*) 42.6°C, (*c*) 50.3°C

**FIGURE P20–27**

**20–28**  Reconsider Prob. 20–27. Using EES (or other) software, investigate the effects of the room temperature and the emissivity of the board on the temperature of the hot surface of the board for different orientations of the board. Let the room temperature vary from 5°C to 35°C and the emissivity from 0.1 to 1.0. Plot the hot surface temperature for different orientations of the board as the functions of the room temperature and the emissivity, and discuss the results.

**20–29**  Consider a vertical plate with length $L$, placed in quiescent air. If the film temperature is 20°C and the average Nusselt number in natural convection is of the form $Nu = CRa_L^n$, show that the average heat transfer coefficient can be expressed as

$$h = 1.51(\Delta T/L)^{1/4} \qquad 10^4 < Ra_L < 10^9$$
$$h = 1.19\Delta T^{1/3} \qquad 10^{10} < Ra_L < 10^{13}$$

**20–30**  A circular grill of diameter 0.25 m has an emissivity of 0.8. If the surface temperature is maintained at 150°C, determine the required electrical power when the room air and surroundings are at 30°C.

**20–31**  A manufacturer makes absorber plates that are 1.2 m × 0.8 m in size for use in solar collectors. The back side of the plate is heavily insulated, while its front surface is coated with black chrome, which has an absorptivity of 0.87 for solar radiation and an emissivity of 0.09. Consider such a plate placed horizontally outdoors in

**FIGURE P20–31**

**FIGURE P20–34**

calm air at 25°C. Solar radiation is incident on the plate at a rate of 600 W/m². Taking the effective sky temperature to be 10°C, determine the equilibrium temperature of the absorber plate. What would your answer be if the absorber plate is made of ordinary aluminum plate that has a solar absorptivity of 0.28 and an emissivity of 0.07?

**20–32**  Repeat Prob. 20–31 for an aluminum plate painted flat black (solar absorptivity 0.98 and emissivity 0.98) and also for a plate painted white (solar absorptivity 0.26 and emissivity 0.90).

**20–33**  A room is to be heated by a coal-burning stove, which is a cylindrical cavity with an outer diameter of 50 cm and a height of 120 cm. The rate of heat loss from the room is estimated to be 1.5 kW when the air temperature in the room is maintained constant at 24°C. The emissivity of the stove surface is 0.85, and the average temperature of the surrounding wall surfaces is 14°C. Determine the surface temperature of the stove. Neglect the heat transfer from the bottom surface and take the heat transfer coefficient at the top surface to be the same as that on the side surface.

The heating value of the coal is 30,000 kJ/kg, and the combustion efficiency is 65 percent. Determine the amount of coal burned a day if the stove operates 14 h a day.

**20–34**  Thick fluids such as asphalt and waxes and the pipes in which they flow are often heated in order to reduce the viscosity of the fluids and thus to reduce the pumping costs. Consider the flow of such a fluid through a 100-m-long pipe of outer diameter 30 cm in calm ambient air at 0°C. The pipe is heated electrically, and a thermostat keeps the outer surface temperature of the pipe constant at 25°C. The emissivity of the outer surface of the pipe is 0.8, and the effective sky temperature is −30°C. Determine the power rating of the electric resistance heater, in kW, that needs to be used. Also, determine the cost of electricity associated with heating the pipe during a 15-h period under the above conditions if the price of electricity is $0.09/kWh. *Answers:* 29.1 kW, $39.3

**20–35**  Reconsider Prob. 20–34. To reduce the heating cost of the pipe, it is proposed to insulate it with sufficiently thick fiberglass insulation ($k$ = 0.035 W/m·K) wrapped with aluminum foil ($\varepsilon$ = 0.1) to cut down the heat losses by 85 percent. Assuming the pipe temperature to remain constant at 25°C, determine the thickness of the insulation that needs to be used. How much money will the insulation save during this 15-h period?  *Answers:* 1.3 cm, $33.4

**20–36**  Consider a 1.2-m-high and 2-m-wide glass window with a thickness of 6 mm, thermal conductivity $k$ = 0.78 W/m·K, and emissivity $\varepsilon$ = 0.9. The room and the walls that face the window are maintained at 25°C, and the average temperature of the inner surface of the window is measured to be 5°C. If the temperature of the outdoors is −5°C, determine (*a*) the convection heat transfer coefficient on the inner surface of the window, (*b*) the rate of total heat transfer through the window, and (*c*) the combined natural convection and radiation heat transfer coefficient on the outer surface of the window. Is it reasonable to neglect the thermal resistance of the glass in this case?

**FIGURE P20–36**

**20–37** A 3-mm-diameter and 12-m-long electric wire is tightly wrapped with a 1.5-mm-thick plastic cover whose thermal conductivity and emissivity are $k = 0.20$ W/m·K and $\varepsilon = 0.9$. Electrical measurements indicate that a current of 10 A passes through the wire and there is a voltage drop of 7 V along the wire. If the insulated wire is exposed to calm atmospheric air at $T_\infty = 30°C$, determine the temperature at the interface of the wire and the plastic cover in steady operation. Take the surrounding surfaces to be at about the same temperature as the air.

**20–38** During a visit to a plastic sheeting plant, it was observed that a 45-m-long section of a 2-in nominal (6.03-cm-outer-diameter) steam pipe extended from one end of the plant to the other with no insulation on it. The temperature measurements at several locations revealed that the average temperature of the exposed surfaces of the steam pipe was 170°C, while the temperature of the surrounding air was 20°C. The outer surface of the pipe appeared to be oxidized, and its emissivity can be taken to be 0.7. Taking the temperature of the surrounding surfaces to be 20°C also, determine the rate of heat loss from the steam pipe.

Steam is generated in a gas furnace that has an efficiency of 84 percent, and the plant pays $1.10 per therm (1 therm = 105,500 kJ) of natural gas. The plant operates 24 h a day 365 days a year, and thus 8760 h a year. Determine the annual cost of the heat losses from the steam pipe for this facility.

170°C
$\varepsilon = 0.7$
20°C
6.03 cm
45 m
Steam

**FIGURE P20–38**

**20–39** Reconsider Prob. 20–38. Using EES (or other) software, investigate the effect of the surface temperature of the steam pipe on the rate of heat loss from the pipe and the annual cost of this heat loss. Let the surface temperature vary from 100°C to 200°C. Plot the rate of heat loss and the annual cost as a function of the surface temperature, and discuss the results.

**20–40** Reconsider Prob. 20–38. In order to reduce heat losses, it is proposed to insulate the steam pipe with 5-cm-thick fiberglass insulation ($k = 0.038$ W/m·K) and to wrap it with aluminum foil ($\varepsilon = 0.1$) in order to minimize the radiation losses. Also, an estimate is obtained from a local insulation contractor, who proposed to do the insulation job for

$750, including materials and labor. Would you support this proposal? How long will it take for the insulation to pay for itself from the energy it saves? Assume the temperature of the steam pipe to remain constant at 170°C.

**20–41** A 50-cm × 50-cm circuit board that contains 121 square chips on one side is to be cooled by combined natural convection and radiation by mounting it on a vertical surface in a room at 25°C. Each chip dissipates 0.18 W of power, and the emissivity of the chip surfaces is 0.7. Assuming the heat transfer from the back side of the circuit board to be negligible, and the temperature of the surrounding surfaces to be the same as the air temperature of the room, determine the surface temperature of the chips. *Answer: 36.2°C*

**20–42** Repeat Prob. 20–41 assuming the circuit board to be positioned horizontally with (*a*) chips facing up and (*b*) chips facing down.

**20–43** The side surfaces of a 3-m-high cubic industrial furnace burning natural gas are not insulated, and the temperature at the outer surface of this section is measured to be 110°C. The temperature of the furnace room, including its surfaces, is 30°C, and the emissivity of the outer surface of the furnace is 0.7. It is proposed that this section of the furnace wall be insulated with glass wool insulation ($k = 0.038$ W/m·K) wrapped by a reflective sheet ($\varepsilon = 0.2$) in order to reduce the heat loss by 90 percent. Assuming the outer surface temperature of the metal section still remains at about 110°C, determine the thickness of the insulation that needs to be used.

The furnace operates continuously throughout the year and has an efficiency of 78 percent. The price of the natural gas is $1.10/therm (1 therm = 105,500 kJ of energy content). If the installation of the insulation will cost $550 for materials and labor, determine how long it will take for the insulation to pay for itself from the energy it saves.

Hot gases
30°C
3 m
Furnace
110°C
$\varepsilon = 0.7$
3 m
3 m

**FIGURE P20–43**

**20–44** A 1.5-m-diameter, 4-m-long cylindrical propane tank is initially filled with liquid propane, whose density is 581 kg/m³. The tank is exposed to the ambient air at 25°C in calm weather. The outer surface of the tank is polished so that the radiation heat transfer is negligible. Now a crack develops at the top of the tank, and the pressure inside drops to 1 atm while the temperature drops to −42°C, which is the boiling temperature of propane at 1 atm. The heat of vaporization of propane at 1 atm is 425 kJ/kg. The propane is slowly vaporized as a result of the heat transfer from the ambient air into the tank, and the propane vapor escapes the tank at −42°C through the crack. Assuming the propane tank to be at about the same temperature as the propane inside at all times, determine how long it will take for the tank to empty if it is not insulated.

**FIGURE P20–44**

**20–45E** Consider a 2-ft × 2-ft thin square plate in a room at 75°F. One side of the plate is maintained at a temperature of 130°F, while the other side is insulated. Determine the rate of heat transfer from the plate by natural convection if the plate is (a) vertical; (b) horizontal with hot surface facing up; and (c) horizontal with hot surface facing down.

**20–46E** Reconsider Prob. 20–45E. Using EES (or other) software, plot the rate of natural convection heat transfer for different orientations of the plate as a function of the plate temperature as the temperature varies from 80°F to 180°F, and discuss the results.

**20–47** A 28-cm-high, 18-cm-long, and 18-cm-wide rectangular container suspended in a room at 24°C is initially filled with cold water at 2°C. The surface temperature of the container is observed to be nearly the same as the water temperature inside. The emissivity of the container surface is 0.6, and the temperature of the surrounding surfaces is about the same as the air temperature. Determine the water temperature in the container after 3 h, and the average rate of heat transfer to the water. Assume the heat transfer coefficient on the top and bottom surfaces to be the same as that on the side surfaces.

**20–48** Reconsider Prob. 20–47. Using EES (or other) software, plot the water temperature in the container as a function of the heating time as the time varies from 30 min to 10 h, and discuss the results.

**20–49** The following experiment is conducted to determine the natural convection heat transfer coefficient for a horizontal cylinder that is 80 cm long and 2 cm in diameter. A 80-cm-long resistance heater is placed along the centerline of the cylinder, and the surfaces of the cylinder are polished to minimize the radiation effect. The two circular side surfaces of the cylinder are well insulated. The resistance heater is turned on, and the power dissipation is maintained constant at 60 W. If the average surface temperature of the cylinder is measured to be 120°C in the 20°C room air when steady operation is reached, determine the natural convection heat transfer coefficient. If the emissivity of the outer surface of the cylinder is 0.1 and a 5 percent error is acceptable, do you think we used to do any correction for the radiation effect? Assume the surrounding surfaces to be at 20°C also.

**FIGURE P20–49**

**20–50** The water in a 40-L tank is to be heated from 15°C to 45°C by a 6-cm-diameter spherical heater whose surface temperature is maintained at 85°C. Determine how long the heater should be kept on.

**20–51** Consider a cylinder with a length of 15 cm and a diameter of 10 cm. The cylinder has a surface temperature of 43°C, while the room air temperature is 17°C. Determine whether placing the cylinder horizontally or vertically would achieve higher heat transfer rate.

### Natural Convection inside Enclosures

**20–52C** The upper and lower compartments of a well-insulated container are separated by two parallel sheets of glass with an air space between them. One of the compartments is to be filled with a hot fluid and the other with a cold fluid. If it is desired that heat transfer between the two compartments be minimal, would you recommend putting the hot fluid into the upper or the lower compartment of the container? Why?

**20–53C** Someone claims that the air space in a double-pane window enhances the heat transfer from a house because of the natural convection currents that occur in the air space and recommends that the double-pane window be replaced by a single sheet of glass whose thickness is equal to the sum of the thicknesses of the two glasses of the double-pane window to save energy. Do you agree with this claim?

**20–54C** Consider a double-pane window consisting of two glass sheets separated by a 1-cm-wide air space. Someone suggests inserting a thin vinyl sheet in the middle of the two glasses to form two 0.5-cm-wide compartments in the window in order to reduce natural convection heat transfer through the window. From a heat transfer point of view, would you be in favor of this idea to reduce heat losses through the window?

**20–55C** What does the effective conductivity of an enclosure represent? How is the ratio of the effective conductivity to thermal conductivity related to the Nusselt number?

**20–56** Consider a 3-m high rectangular enclosure consisting of two surfaces separated by a 0.1-m air gap at 1 atm. If the surface temperatures across the air gap are 30°C and –10°C, determine the ratio of the heat transfer rate for the horizontal orientation (with hotter surface at the bottom) to that for vertical orientation. *Answer*: 1.66

**FIGURE P20–56**

**20–57** Two concentric spheres with radii of 5 cm and 10 cm are having the surface temperatures maintained at 100°C and 200°C, respectively. The enclosure between the two concentric spherical surfaces is filled with nitrogen gas at 1 atm. Determine the rate of heat transfer through the enclosure.

**FIGURE P20–57**

**20–58** Show that the thermal resistance of a rectangular enclosure can be expressed as $R = L_c/(Ak\ \mathrm{Nu})$, where $k$ is the thermal conductivity of the fluid in the enclosure.

**20–59** A vertical 1.5-m-high and 3.0-m-wide enclosure consists of two surfaces separated by a 0.4-m air gap at atmospheric pressure. If the surface temperatures across the air gap are measured to be 280 K and 336 K and the surface emissivities to be 0.15 and 0.90, determine the fraction of heat transferred through the enclosure by radiation. *Answer*: 0.30

**20–60** A vertical 1.5-m-high, 2.8-m-wide double-pane window consists of two layers of glass separated by a 2.0-cm air gap at atmospheric pressure. The room temperature is 26°C while the inner glass temperature is 18°C. Disregarding radiation heat transfer, determine the temperature of the outer glass layer and the rate of heat loss through the window by natural convection.

**20–61** Two concentric spheres of diameters 15 cm and 25 cm are separated by air at 1 atm pressure. The surface temperatures of the two spheres enclosing the air are $T_1 = 350$ K and $T_2 = 275$ K, respectively. Determine the rate of heat transfer from the inner sphere to the outer sphere by natural convection.

**20–62** Reconsider Prob. 20–61. Using EES (or other) software, plot the rate of natural convection heat transfer as a function of the hot surface temperature of the sphere as the temperature varies from 250 K to 450 K, and discuss the results.

**20–63** Flat-plate solar collectors are often tilted up toward the sun in order to intercept a greater amount of direct solar radiation. The tilt angle from the horizontal also affects the rate of heat loss from the collector. Consider a 1.5-m-high and 3-m-wide solar collector that is tilted at an angle $\theta$ from the horizontal. The back side of the absorber is heavily insulated. The absorber plate and the glass cover, which are spaced 2.5 cm from each other, are maintained at temperatures of 80°C and 40°C, respectively. Determine the rate of heat loss from the absorber plate by natural convection for $\theta = 0°$, 30°, and 90°.

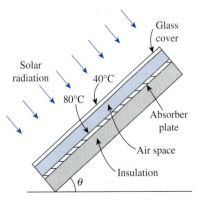

**FIGURE P20–63**

**20–64** A simple solar collector is built by placing a 5-cm-diameter clear plastic tube around a garden hose whose outer diameter is 1.6 cm. The hose is painted black to maximize solar absorption, and some plastic rings are used to keep the spacing between the hose and the clear plastic cover constant. During a clear day, the temperature of the hose is measured to be 65°C, while the ambient air temperature is 26°C. Determine the rate of heat loss from the water in the hose per meter of its length by natural convection. Also, discuss how the performance of this solar collector can be improved.
*Answer:* 8.2 W

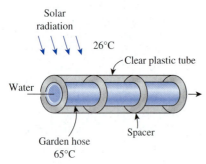

Solar radiation

26°C

Clear plastic tube

Water

Garden hose 65°C

Spacer

**FIGURE P20–64**

**20–65** Reconsider Prob. 20–64. Using EES (or other) software, plot the rate of heat loss from the water by natural convection as a function of the ambient air temperature as the temperature varies from 4°C to 40°C, and discuss the results.

**20–66E** A vertical 5-ft-high and 6-ft-wide double-pane window consists of two sheets of glass separated by a 0.8-in air gap at atmospheric pressure. If the glass surface temperatures across the air gap are measured to be 65°F and 40°F, determine the rate of heat transfer through the window by (a) natural convection and (b) radiation. Also, determine the R-value of insulation of this window such that multiplying the

65°F

40°F

Glass

0.8 in

5 ft

Frame

**FIGURE P20–66E**

inverse of the R-value by the surface area and the temperature difference gives the total rate of heat transfer through the window. The effective emissivity for use in radiation calculations between two large parallel glass plates can be taken to be 0.82.

**20–67E** Reconsider Prob. 20–66E. Using EES (or other) software, investigate the effect of the air gap thickness on the rates of heat transfer by natural convection and radiation, and the R-value of insulation. Let the air gap thickness vary from 0.2 in to 2.0 in. Plot the rates of heat transfer by natural convection and radiation, and the R-value of insulation as a function of the air gap thickness, and discuss the results.

**20–68** Consider two concentric horizontal cylinders of diameters 55 cm and 65 cm, and length 125 cm. The surfaces of the inner and outer cylinders are maintained at 54°C and 106°C, respectively. Determine the rate of heat transfer between the cylinders by natural convection if the annular space is filled with (a) water and (b) air.

## Review Problems

**20–69** Consider an $L \times L$ horizontal plate that is placed in quiescent air with the hot surface facing up. If the film temperature is 20°C and the average Nusselt number in natural convection is of the form $Nu = CRa_L^n$, show that the average heat transfer coefficient can be expressed as

$$h = 1.95(\Delta T/L)^{1/4} \qquad 10^4 < Ra_L < 10^7$$
$$h = 1.79\Delta T^{1/3} \qquad 10^7 < Ra_L < 10^{11}$$

**20–70** A plate (0.5 m × 0.5 m) is inclined at an angle of 30°. The top surface of the plate is well insulated. Estimate the rate of heat loss from the plate when the bottom surface is maintained at 60°C and the surrounding atmospheric quiescent air is at 0°C. *Answer:* 81.5 W

**20–71** A 12-cm-diameter and 15-m-long cylinder with a surface temperature of 10°C is placed horizontally in air at 40°C. Calculate the steady rate of heat transfer for the cases of (a) free-stream air velocity of 10 m/s due to normal winds and (b) no winds and thus a free stream velocity of zero.

**20–72** A spherical vessel, with 30.0-cm outside diameter, is used as a reactor for a slow endothermic reaction. The vessel is completely submerged in a large water-filled tank, held at a constant temperature of 30°C. The outside surface temperature of the vessel is 20°C. Calculate the rate of heat transfer in steady operation for the following cases: (a) the water in the tank is still, (b) the water in the tank is still (as in a part a), however, the buoyancy force caused by the difference in water density is assumed to be negligible, and (c) the water in the tank is circulated at an average velocity of 20 cm/s.

**20–73E** Hot water is flowing at an average velocity of 4 ft/s through a cast iron pipe (k = 30 Btu/h·ft·°F) whose inner and outer diameters are 1.0 in and 1.2 in, respectively. The pipe passes through a 50-ft-long section of a basement whose temperature is 60°F. The emissivity of the outer surface of the pipe is 0.5, and the walls of the basement are also at about 60°F. If the inlet temperature of the water is 150°F and the heat transfer coefficient on the inner surface of the pipe is 30 Btu/h·ft²·°F, determine the temperature drop of water as it passes through the basement.

**20–74** Consider a flat-plate solar collector placed horizontally on the flat roof of a house. The collector is 1.5 m wide and 4.5 m long, and the average temperature of the exposed surface of the collector is 42°C. Determine the rate of heat loss from the collector by natural convection during a calm day when the ambient air temperature is 8°C. Also, determine the heat loss by radiation by taking the emissivity of the collector surface to be 0.85 and the effective sky temperature to be −15°C. *Answers:* 1314 W, 1762 W

**20–75** An electronic box that consumes 200 W of power is cooled by a fan blowing air into the box enclosure. The dimensions of the electronic box are 15 cm × 50 cm × 50 cm, and all surfaces of the box are exposed to the ambient except the base surface. Temperature measurements indicate that the box is at an average temperature of 32°C when the ambient temperature and the temperature of the surrounding walls are 25°C. If the emissivity of the outer surface of the box is 0.75, determine the fraction of the heat lost from the outer surfaces of the electronic box.

25°C
32°C
$\varepsilon = 0.75$
15 cm
50 cm
50 cm

**FIGURE P20–75**

**20–76** A 6-m-internal-diameter spherical tank made of 1.5-cm-thick stainless steel ($k = 15$ W/m·K) is used to store iced water at 0°C in a room at 20°C. The walls of the room are also at 20°C. The outer surface of the tank is black (emissivity $\varepsilon = 1$), and heat transfer between the outer surface of the tank and the surroundings is by natural convection and radiation. Assuming the entire steel tank to be at 0°C and thus the thermal resistance of the tank to be negligible, determine (a) the rate of heat transfer to the iced water in the tank and (b) the amount of ice at 0°C that melts during a 24-h period. The heat of fusion of water is 333.7 kJ/kg. *Answers:* (a) 15.4 kW, (b) 3988 kg

**20–77** Consider a 1.2-m-high and 2-m-wide double-pane window consisting of two 3-mm-thick layers of glass ($k = 0.78$ W/m · K) separated by a 2.5-cm-wide air space. Determine the steady rate of heat transfer through this window and the temperature of its inner surface for a day during which the room is maintained at 20°C while the temperature of the outdoors is 0°C. Take the heat transfer coefficients on the inner and outer surfaces of the window to be $h_1 = 10$ W/m² · K and $h_2 = 25$ W/m²·K and disregard any heat transfer by radiation.

**20–78** An electric resistance space heater is designed such that it resembles a rectangular box 50 cm high, 80 cm long, and 15 cm wide filled with 45 kg of oil. The heater is to be placed against a wall, and thus heat transfer from its back surface is negligible. The surface temperature of the heater is not to exceed 75°C in a room at 25°C for safety considerations. Disregarding heat transfer from the bottom and top surfaces of the heater in anticipation that the top surface will be used as a shelf, determine the power rating of the heater in W. Take the emissivity of the outer surface of the heater to be 0.8 and the average temperature of the ceiling and wall surfaces to be the same as the room air temperature.

Also, determine how long it will take for the heater to reach steady operation when it is first turned on (i.e., for the oil temperature to rise from 25°C to 75°C). State your assumptions in the calculations.

Wall
15 cm
80 cm
$T_s = 75°C$
$\varepsilon = 0.8$
50 cm
Oil
Electric heater
Heating element

**FIGURE P20–78**

**20–79** During a plant visit, it was observed that a 2-m high and 1-m-wide section of the vertical front section of a natural gas furnace wall was too hot to touch. The temperature measurements on the surface revealed that the average temperature of the exposed hot surface was 110°C, while the temperature of the surrounding air was 25°C. The surface appeared to be oxidized, and its emissivity can be taken to be 0.6. Taking the temperature of the surrounding surfaces to be 25°C also, determine the rate of heat loss from this furnace.

The furnace has an efficiency of 79 percent, and the plant pays $1.20 per therm of natural gas. If the plant operates 10 h a day, 310 days a year, and thus 3100 h a year, determine the annual cost of the heat loss from this vertical hot surface on the front section of the furnace wall.

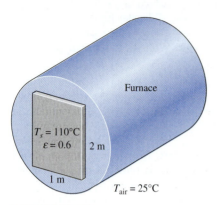

Furnace

$T_s = 110°C$
$\varepsilon = 0.6$   2 m

1 m

$T_{air} = 25°C$

**FIGURE P20–79**

**20–80** A solar collector consists of a horizontal copper tube of outer diameter 5 cm enclosed in a concentric thin glass tube of 9 cm diameter. Water is heated as it flows through the tube, and the annular space between the copper and glass tube is filled with air at 1 atm pressure. During a clear day, the temperatures of the tube surface and the glass cover are measured to be 60°C and 32°C, respectively. Determine the rate of heat loss from the collector by natural convection per meter length of the tube.   *Answer:* 17.4 W

9 cm

5 cm

Glass cover

**FIGURE P20–80**

**20–81** A solar collector consists of a horizontal aluminum tube of outer diameter 5 cm enclosed in a concentric thin glass tube of 7 cm diameter. Water is heated as it flows through the aluminum tube, and the annular space between the aluminum and glass tubes is filled with air at 1 atm pressure. The pump circulating the water fails during a clear day, and the water temperature in the tube starts rising. The aluminum tube absorbs solar radiation at a rate of 20 W per meter length, and the temperature of the ambient air outside is 30°C. Approximating the surfaces of the tube and the glass cover as being black (emissivity $\varepsilon = 1$) in radiation calculations and taking the effective sky temperature to be 20°C, determine the temperature of the aluminum tube when equilibrium is established (i.e., when the net heat loss from the tube by convection and radiation equals the amount of solar energy absorbed by the tube).

**20–82E** The components of an electronic system dissipating 150 W are located in a 5-ft-long horizontal duct whose cross section is 6 in × 6 in. The components in the duct are cooled by forced air, which enters at 85°F at a rate of 22 cfm and leaves at 100°F. The surfaces of the sheet metal duct are not painted, and thus radiation heat transfer from the outer surfaces is negligible. If the ambient air temperature is 80°F, determine (*a*) the heat transfer from the outer surfaces of the duct to the ambient air by natural convection and (*b*) the average temperature of the duct.

Natural
convection

80°F            100°F

150 W

85°F
22 cfm         5 ft

**FIGURE P20–82E**

**20–83E** Repeat Prob. 20–82E for a circular horizontal duct of diameter 4 in.

**20–84E** Repeat Prob. 20–82E assuming the fan fails and thus the entire heat generated inside the duct must be rejected to the ambient air by natural convection through the outer surfaces of the duct.

**20–85** Consider a cold aluminum canned drink that is initially at a uniform temperature of 5°C. The can is 12.5 cm high and has a diameter of 6 cm. The emissivity of the outer surface of the can is 0.6. Disregarding any heat transfer from the bottom surface of the can, determine how long it will take for the average temperature of the drink to rise to 7°C if the surrounding air and surfaces are at 25°C.   *Answer:* 12.1 min

**20–86** A group of 25 power transistors, dissipating 1.5 W each, are to be cooled by attaching them to a black-anodized square aluminum plate and mounting the plate on the wall of a room at 30°C. The emissivity of the transistor and the plate surfaces is 0.9. Assuming the heat transfer from the back side of the plate to be negligible and the temperature of the surrounding surfaces to be the same as the air temperature of the room, determine the size of the plate if the average surface temperature of the plate is not to exceed 50°C. *Answer:* 43 cm × 43 cm

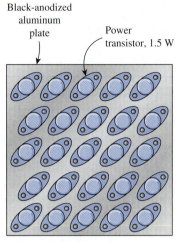

**FIGURE P20–86**

**20–87** Repeat Prob. 20–86 assuming the plate to be positioned horizontally with (a) transistors facing up and (b) transistors facing down.

**20–88E** A 0.1-W small cylindrical resistor mounted on a lower part of a vertical circuit board is 0.3 in long and has a diameter of 0.2 in. The view of the resistor is largely blocked by another circuit board facing it, and the heat transfer through the connecting wires is negligible. The air is free to flow through the large parallel flow passages between the boards as a result of natural convection currents. If the air temperature at the vicinity of the resistor is 120°F, determine the approximate surface temperature of the resistor. *Answer*: 211°F

**FIGURE P20–88E**

**20–89** An ice chest whose outer dimensions are 30 cm × 40 cm × 40 cm is made of 2-cm-thick Styrofoam ($k = 0.033$ W/m·K). Initially, the chest is filled with 35 kg of ice at 0°C, and the inner surface temperature of the ice chest can be taken to be 0°C at all times. The heat of fusion of water at 0°C is 337.7 kJ/kg, and the surrounding ambient air is at 20°C. Disregarding any heat transfer from the 40 cm × 40 cm base of the ice chest, determine how long it will take for the ice in the chest to melt completely if the ice chest is subjected to (a) calm air and (b) winds at 50 km/h. Assume the heat transfer coefficient on the front, back, and top surfaces to be the same as that on the side surfaces.

**20–90** Solar radiation is incident on the glass cover of a solar collector at a rate of 650 W/m². The glass transmits 88 percent of the incident radiation and has an emissivity of 0.90. The hot water needs of a family in summer can be met completely by a collector 1.5 m high and 2 m wide, and tilted 40° from the horizontal. The temperature of the glass cover is measured to be 40°C on a calm day when the surrounding air temperature is 20°C. The effective sky temperature for radiation exchange between the glass cover and the open sky is −40°C. Water enters the tubes attached to the absorber plate at a rate of 1 kg/min. Assuming the back surface of the absorber plate to be heavily insulated and the only heat loss occurs through the glass cover, determine (a) the total rate of heat loss from the collector; (b) the collector efficiency, which is the ratio of the amount of heat transferred to the water to the solar energy incident on the collector; and (c) the temperature rise of water as it flows through the collector.

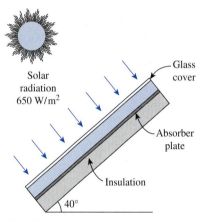

**FIGURE P20–90**

**20–91** Two concentric spheres with radii of 5 cm and 10 cm are having the surface temperatures maintained at 200°C and 100°C, respectively. The enclosure between the two concentric spherical surfaces is filled with nitrogen gas at 1 atm. If both concentric sphere surfaces are black ($\varepsilon = 1$), determine the rate of heat transfer on the inner surface. *Answer*: 18.5 W

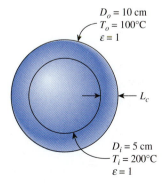

**FIGURE P20–91**

**20–92** Exhaust gases from a manufacturing plant are being discharged through a 10-m tall exhaust stack with outer diameter of 1 m. The exhaust gases are discharged at a rate of 0.125 kg/s, while temperature drop between inlet and exit of the exhaust stack is 30°C, and the constant pressure-specific heat of the exhaust gases is 1600 J/kg·K. On a particular calm day, the surrounding quiescent air temperature is 33°C. Solar radiation is incident on the exhaust stack outer surface at a rate of 500 W/m², and both the emissivity and solar absorptivity of the outer surface are 0.9. Determine the exhaust stack outer surface temperature. Assume the film temperature is 60°C.

Exhaust
gases

$\alpha_s = \varepsilon = 0.9$

Exhaust
stack    Air, 33°C

$L = 10$ m

$T_{in} - T_{out} = 30°C$

Manufacturing
plant

**FIGURE P20–92**

### Design and Essay Problems

**20–93** Contact a manufacturer of aluminum heat sinks and obtain their product catalog for cooling electronic components by natural convection and radiation. Write an essay on how to select a suitable heat sink for an electronic component when its maximum power dissipation and maximum allowable surface temperature are specified.

**20–94** The top surfaces of practically all flat-plate solar collectors are covered with glass in order to reduce the heat losses from the absorber plate underneath. Although the glass cover reflects or absorbs about 15 percent of the incident solar radiation, it saves much more from the potential heat losses from the absorber plate, and thus it is considered to be an essential part of a well-designed solar collector. Inspired by the energy efficiency of double-pane windows, someone proposes to use double glazing on solar collectors instead of a single glass. Investigate if this is a good idea for the town in which you live. Use local weather data and base your conclusion on heat transfer analysis and economic considerations.

# RADIATION HEAT TRANSFER

So far, we have considered the conduction and convection modes of heat transfer, which are related to the nature of the media involved and the presence of fluid motion, among other things. We now turn our attention to the third mechanism of heat transfer: *radiation*, which is characteristically different from the other two.

We start this chapter with a discussion of *electromagnetic waves* and the *electromagnetic spectrum*, with particular emphasis on *thermal radiation*. Then we introduce the idealized *blackbody, blackbody radiation, and blackbody radiation function*, together with the *Stefan–Boltzmann law, Planck's law*, and *Wien's displacement law*. Various radiation fluxes such as emissive power, irradiation, and *radiosity* are defined. This is followed by a discussion of radiative properties of materials such as *emissivity, absorptivity, reflectivity*, and *transmissivity* and their dependence on wavelength and temperature. The *greenhouse effect* is presented as an example of the consequences of the wavelength dependence of radiation properties.

We continue with a discussion of view factors and the rules associated with them. View factor expressions and charts for some common configurations are given, and the crossed-strings method is presented. Finally, we discuss radiation heat transfer, first between black surfaces and then between nonblack surfaces using the radiation network approach.

## Objectives

The objectives of this chapter are to:

- Classify electromagnetic radiation, and identify thermal radiation.

- Understand the idealized blackbody, and calculate the total and spectral blackbody emissive power.

- Calculate the fraction of radiation emitted in a specified wavelength band using the blackbody radiation functions.

- Develop a clear understanding of the properties emissivity, absorptivity, reflectivity, and transmissivity on spectral and total basis.

- Apply Kirchhoff's law to determine the absorptivity of a surface when its emissivity is known.

- Define view factor, and understand its importance in radiation heat transfer calculations.

- Calculate radiation heat transfer between black surfaces.

- Obtain relations for net rate of radiation heat transfer between the surfaces of a two-zone enclosure, including two large parallel plates, two long concentric cylinders, and two concentric spheres.

- Understand radiation heat transfer in three-surface enclosures.

**FIGURE 21–1**

A hot object in a vacuum chamber loses heat by radiation only.

**FIGURE 21–2**

Unlike conduction and convection, heat transfer by radiation can occur between two bodies, even when they are separated by a medium colder than both.

# 21–1 · INTRODUCTION

Consider a hot object that is suspended in an evacuated chamber whose walls are at room temperature (Fig. 21–1). The hot object will eventually cool down and reach thermal equilibrium with its surroundings. That is, it will lose heat until its temperature reaches the temperature of the walls of the chamber. Heat transfer between the object and the chamber could not have taken place by conduction or convection, because these two mechanisms cannot occur in a vacuum. Therefore, heat transfer must have occurred through another mechanism that involves the emission of the internal energy of the object. This mechanism is *radiation*.

Radiation differs from the other two heat transfer mechanisms in that it does not require the presence of a material medium to take place. In fact, energy transfer by radiation is fastest (at the speed of light) and it suffers no attenuation in a *vacuum*. Also, radiation transfer occurs in solids as well as liquids and gases. In most practical applications, all three modes of heat transfer occur concurrently at varying degrees. But heat transfer through an evacuated space can occur only by radiation. For example, the energy of the sun reaches the earth by radiation.

You will recall that heat transfer by conduction or convection takes place in the direction of decreasing temperature; that is, from a high-temperature medium to a lower-temperature one. It is interesting that radiation heat transfer can occur between two bodies separated by a medium colder than both bodies (Fig. 21–2). For example, solar radiation reaches the surface of the earth after passing through cold air layers at high altitudes. Also, the radiation-absorbing surfaces inside a greenhouse reach high temperatures even when its plastic or glass cover remains relatively cool.

The theoretical foundation of radiation was established in 1864 by physicist James Clerk Maxwell, who postulated that accelerated charges or changing electric currents give rise to electric and magnetic fields. These rapidly moving fields are called **electromagnetic waves** or **electromagnetic radiation**, and they represent the energy emitted by matter as a result of the changes in the electronic configurations of the atoms or molecules. In 1887, Heinrich Hertz experimentally demonstrated the existence of such waves. Electromagnetic waves transport energy just like other waves, and all electromagnetic waves travel at the *speed of light* in a vacuum, which is $c_0 = 2.9979 \times 10^8$ m/s. Electromagnetic waves are characterized by their *frequency* $\nu$ or *wavelength* $\lambda$. These two properties in a medium are related by

$$\lambda = \frac{c}{\nu} \qquad (21-1)$$

where $c$ is the speed of propagation of a wave in that medium. The speed of propagation in a medium is related to the speed of light in a vacuum by $c = c_0/n$, where $n$ is the *index of refraction* of that medium. The refractive index is essentially unity for air and most gases, about 1.5 for glass, and 1.33 for water. The commonly used unit of wavelength is the *micrometer* ($\mu$m) or micron, where 1 $\mu$m $= 10^{-6}$ m. Unlike the wavelength and the speed of propagation, the frequency of an electromagnetic wave depends only on the source and is independent of the medium through which the wave travels. The *frequency* (the number of oscillations per second) of an electromagnetic wave can range from less than a million Hz to a septillion Hz or higher, depending on the source. Note from Eq. 21–1 that the wavelength and the frequency of electromagnetic radiation are inversely proportional.

It has proven useful to view electromagnetic radiation as the propagation of a collection of discrete packets of energy called **photons** or **quanta**, as proposed by Max Planck in 1900 in conjunction with his *quantum theory*. In this view, each photon of frequency $\nu$ is considered to have an energy of

$$e = h\nu = \frac{hc}{\lambda} \qquad (21\text{--}2)$$

where $h = 6.626069 \times 10^{-34}$ J·s is *Planck's constant*. Note from the second part of Eq. 21–2 that the energy of a photon is inversely proportional to its wavelength. Therefore, shorter-wavelength radiation possesses larger photon energies. It is no wonder that we try to avoid very-short-wavelength radiation such as gamma rays and X-rays since they are highly destructive.

## 21–2 ▪ THERMAL RADIATION

Although all electromagnetic waves have the same general features, waves of different wavelength differ significantly in their behavior. The electromagnetic radiation encountered in practice covers a wide range of wavelengths, varying from less than $10^{-10}$ μm for cosmic rays to more than $10^{10}$ μm for electrical power waves. The **electromagnetic spectrum** also includes gamma rays, X-rays, ultraviolet radiation, visible light, infrared radiation, thermal radiation, microwaves, and radio waves, as shown in Fig. 21–3.

Different types of electromagnetic radiation are produced through various mechanisms. For example, *gamma rays* are produced by nuclear reactions, *X-rays* by the bombardment of metals with high-energy electrons, *microwaves* by special types of electron tubes such as klystrons and magnetrons, and *radio waves* by the excitation of some crystals or by the flow of alternating current through electric conductors.

The short-wavelength gamma rays and X-rays are primarily of concern to nuclear engineers, while the long-wavelength microwaves and radio waves are of concern to electrical engineers. The type of electromagnetic radiation that is pertinent to heat transfer is the **thermal radiation** emitted as a result of energy transitions of molecules, atoms, and electrons of a substance. Temperature is a measure of the strength of these activities at the microscopic level, and the rate of thermal radiation emission increases with increasing temperature. Thermal radiation is continuously emitted by all matter whose temperature is above absolute zero. That is, everything around us such as walls, furniture, and our friends constantly emits (and absorbs) radiation (Fig. 21–4). Thermal radiation is also defined as the portion of the electromagnetic spectrum that extends from about 0.1 to 100 μm, since the radiation emitted by bodies due to their temperature falls almost entirely into this wavelength range. Thus, thermal radiation includes the entire visible and infrared (IR) radiation as well as a portion of the ultraviolet (UV) radiation.

What we call **light** is simply the *visible* portion of the electromagnetic spectrum that lies between 0.40 and 0.76 μm. Light is characteristically no different than other electromagnetic radiation, except that it happens to trigger the

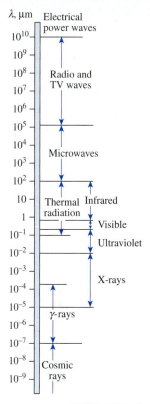

**FIGURE 21–3**
The electromagnetic wave spectrum.

**FIGURE 21–4**
Everything around us constantly emits thermal radiation.

## TABLE 21–1

The wavelength ranges of different colors

| Color | Wavelength band |
|-------|-----------------|
| Violet | 0.40–0.44 μm |
| Blue | 0.44–0.49 μm |
| Green | 0.49–0.54 μm |
| Yellow | 0.54–0.60 μm |
| Orange | 0.60–0.67 μm |
| Red | 0.63–0.76 μm |

**FIGURE 21–5**

Food is heated or cooked in a microwave oven by absorbing the electromagnetic radiation energy generated by the magnetron of the oven.

sensation of seeing in the human eye. Light, or the visible spectrum, consists of narrow bands of color from violet (0.40–0.44 μm) to red (0.63–0.76 μm), as shown in Table 21–1.

A body that emits some radiation in the visible range is called a light source. The sun is obviously our primary light source. The electromagnetic radiation emitted by the sun is known as **solar radiation**, and nearly all of it falls into the wavelength band 0.3–3 μm. Almost *half* of solar radiation is light (i.e., it falls into the visible range), with the remaining being ultraviolet and infrared.

The radiation emitted by bodies at room temperature falls into the **infrared** region of the spectrum, which extends from 0.76 to 100 μm. Bodies start emitting noticeable visible radiation at temperatures above 800 K. The tungsten filament of a lightbulb must be heated to temperatures above 2000 K before it can emit any significant amount of radiation in the visible range.

The **ultraviolet** radiation includes the low-wavelength end of the thermal radiation spectrum and lies between the wavelengths 0.01 and 0.40 μm. Ultraviolet rays are to be avoided since they can kill microorganisms and cause serious damage to humans and other living beings. About 21 percent of solar radiation is in the ultraviolet range, and it would be devastating if it were to reach the surface of the earth. Fortunately, the ozone ($O_3$) layer in the atmosphere acts as a protective blanket and absorbs most of this ultraviolet radiation. The ultraviolet rays that remain in sunlight are still sufficient to cause serious sunburns to sun worshippers, and prolonged exposure to direct sunlight is the leading cause of skin cancer, which can be lethal. Recent discoveries of "holes" in the ozone layer have prompted the international community to ban the use of ozone-destroying chemicals such as the refrigerant Freon-21 in order to save the earth. Ultraviolet radiation is also produced artificially in fluorescent lamps for use in medicine as a bacteria killer and in tanning parlors as an artificial tanner.

Microwave ovens utilize electromagnetic radiation in the **microwave** region of the spectrum generated by microwave tubes called *magnetrons*. Microwaves in the range of $10^2$–$10^5$ μm are very suitable for use in cooking since they are *reflected* by metals, *transmitted* by glass and plastics, and *absorbed* by food (especially water) molecules. Thus, the electric energy converted to radiation in a microwave oven eventually becomes part of the internal energy of the food. The fast and efficient cooking of microwave ovens has made them one of the essential appliances in modern kitchens (Fig. 21–5).

Radars and cordless telephones also use electromagnetic radiation in the microwave region. The wavelength of the electromagnetic waves used in radio and TV broadcasting usually ranges between 1 and 1000 m in the **radio wave** region of the spectrum.

In heat transfer studies, we are interested in the energy emitted by bodies because of their temperature only. Therefore, we limit our consideration to *thermal radiation,* which we simply call *radiation*. The relations developed below are restricted to thermal radiation only and may not be applicable to other forms of electromagnetic radiation.

The electrons, atoms, and molecules of all solids, liquids, and gases above absolute zero temperature are constantly in motion, and thus radiation is constantly emitted, as well as being absorbed or transmitted throughout the entire volume of matter. That is, radiation is a **volumetric phenomenon**. However, for opaque (nontransparent) solids such as metals, wood, and rocks, radiation is considered to be a **surface phenomenon**, since the radiation emitted by the interior regions can never reach the surface, and the radiation incident on such bodies is usually absorbed within a few microns from the surface (Fig. 21–6). Note that the radiation characteristics of surfaces can be changed completely by applying thin layers of coatings on them.

# 21–3 · BLACKBODY RADIATION

A body at a thermodynamic (or absolute) temperature above zero emits radiation in all directions over a wide range of wavelengths. The amount of radiation energy emitted from a surface at a given wavelength depends on the material of the body and the condition of its surface as well as the surface temperature. Therefore, different bodies may emit different amounts of radiation per unit surface area, even when they are at the same temperature. Thus, it is natural to be curious about the *maximum* amount of radiation that can be emitted by a surface at a given temperature. Satisfying this curiosity requires the definition of an idealized body, called a *blackbody*, to serve as a standard against which the radiative properties of real surfaces may be compared.

A **blackbody** is defined as *a perfect emitter and absorber of radiation.* At a specified temperature and wavelength, no surface can emit more energy than a blackbody. A blackbody absorbs *all* incident radiation, regardless of wavelength and direction. Also, a blackbody emits radiation energy uniformly in all directions per unit area normal to direction of emission (Fig. 21–7). That is, a blackbody is a *diffuse* emitter. The term *diffuse* means "independent of direction."

The radiation energy emitted by a blackbody per unit time and per unit surface area was determined experimentally by Joseph Stefan in 1879 and expressed as

$$E_b(T) = \sigma T^4 \quad (\text{W/m}^2) \tag{21–3}$$

where $\sigma = 5.670 \times 10^{-8}$ W/m$^2$·K$^4$ is the *Stefan–Boltzmann constant* and $T$ is the absolute temperature of the surface in K. This relation was theoretically verified in 1884 by Ludwig Boltzmann. Equation 21–3 is known as the **Stefan–Boltzmann law** and $E_b$ is called the **blackbody emissive power**. Note that the emission of thermal radiation is proportional to the *fourth power* of the absolute temperature.

Although a blackbody would appear *black* to the eye, a distinction should be made between the idealized blackbody and an ordinary black surface. Any surface that absorbs light (the visible portion of radiation) would appear black to the eye, and a surface that reflects it completely would appear white. Considering that visible radiation occupies a very narrow band of the spectrum from 0.4 to 0.76 μm, we cannot make any judgments about the blackness of a surface on the basis of visual observations. For example, snow and white paint reflect light and thus appear white. But they are essentially black for infrared radiation since they strongly absorb long-wavelength radiation. Surfaces coated with lampblack paint approach idealized blackbody behavior.

Another type of body that closely resembles a blackbody is a *large cavity with a small opening,* as shown in Fig. 21–8. Radiation coming in through the opening of area $A$ undergoes multiple reflections, and thus it has several

**FIGURE 21–6**
Radiation in opaque solids is considered a surface phenomenon since the radiation emitted only by the molecules at the surface can escape the solid.

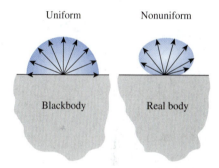

**FIGURE 21–7**
A blackbody is said to be a *diffuse* emitter since it emits radiation energy uniformly in all directions.

**FIGURE 21–8**
A large isothermal cavity at temperature $T$ with a small opening of area $A$ closely resembles a blackbody of surface area $A$ at the same temperature.

chances to be absorbed by the interior surfaces of the cavity before any part of it can possibly escape. Also, if the surface of the cavity is isothermal at temperature $T$, the radiation emitted by the interior surfaces streams through the opening after undergoing multiple reflections, and thus it has a diffuse nature. Therefore, the cavity acts as a perfect absorber and perfect emitter, and the opening will resembles a blackbody of surface area $A$ at temperature $T$, regardless of the actual radiative properties of the cavity.

The Stefan–Boltzmann law in Eq. 21–3 gives the *total* blackbody emissive power $E_b$, which is the sum of the radiation emitted over all wavelengths. Sometimes we need to know the **spectral blackbody emissive power**, which is *the amount of radiation energy emitted by a blackbody at a thermodynamic temperature T per unit time, per unit surface area, and per unit wavelength about the wavelength λ*. For example, we are more interested in the amount of radiation an incandescent lightbulb emits in the visible wavelength spectrum than we are in the total amount emitted.

The relation for the spectral blackbody emissive power $E_{b\lambda}$ was developed by Max Planck in 1901 in conjunction with his famous quantum theory. This relation is known as **Planck's law** and is expressed as

$$E_{b\lambda}(\lambda, T) = \frac{C_1}{\lambda^5[\exp{(C_2/\lambda T)} - 1]} \quad \text{(W/m}^2\cdot\mu\text{m)} \quad \text{(21–4)}$$

where

$$C_1 = 2\pi h c_0^2 = 3.74177 \times 10^8 \text{ W}\cdot\mu\text{m}^4/\text{m}^2$$
$$C_2 = h c_0/k = 1.43878 \times 10^4 \text{ }\mu\text{m}\cdot\text{K}$$

Also, $T$ is the absolute temperature of the surface, $\lambda$ is the wavelength of the radiation emitted, and $k = 1.38065 \times 10^{-23}$ J/K is *Boltzmann's constant*. This relation is valid for a surface in a *vacuum* or a *gas*. For other mediums, it needs to be modified by replacing $C_1$ by $C_1/n^2$, where $n$ is the index of refraction of the medium. Note that the term *spectral* indicates dependence on wavelength.

The variation of the spectral blackbody emissive power with wavelength is plotted in Fig. 21–9 for selected temperatures. Several observations can be made from this figure:

1. The emitted radiation is a continuous function of *wavelength*. At any specified temperature, it increases with wavelength, reaches a peak, and then decreases with increasing wavelength.
2. At any wavelength, the amount of emitted radiation *increases* with increasing temperature.
3. As temperature increases, the curves shift to the left to the shorter-wavelength region. Consequently, a larger fraction of the radiation is emitted at *shorter wavelengths* at higher temperatures.
4. The radiation emitted by the *sun*, which is considered to be a blackbody at 5780 K (or roughly at 5800 K), reaches its peak in the visible region of the spectrum. Therefore, the sun is in tune with our eyes. On the other hand, surfaces at $T \leq 800$ K emit almost entirely in the infrared region and thus are not visible to the eye unless they reflect light coming from other sources.

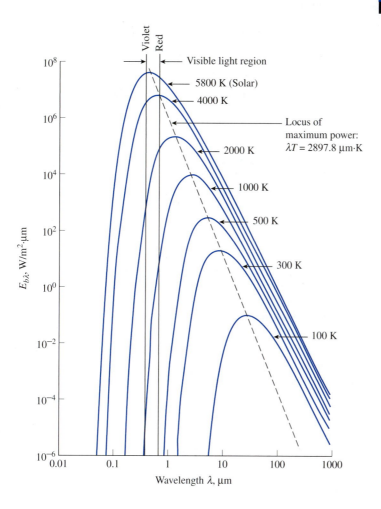

Graph of blackbody emissive power $E_{b\lambda}$ in W/m²·μm versus wavelength λ in μm, showing curves for 5800 K (Solar), 4000 K, 2000 K, 1000 K, 500 K, 300 K, and 100 K, with a visible light region marked between Violet and Red, and a dashed line for the Locus of maximum power: $\lambda T = 2897.8\ \mu m\cdot K$.

**FIGURE 21–9**
The variation of the blackbody emissive power with wavelength for several temperatures.

As the temperature increases, the peak of the curve in Fig. 21–9 shifts toward shorter wavelengths. The wavelength at which the peak occurs for a specified temperature is given by **Wien's displacement law** as

$$(\lambda T)_{max\ power} = 2897.8\ \mu m\cdot K \qquad (21\text{-}5)$$

This relation was originally developed by Willy Wien in 1894 using classical thermodynamics, but it can also be obtained by differentiating Eq. 21–4 with respect to $\lambda$ while holding $T$ constant and setting the result equal to zero. A plot of Wien's displacement law, which is the locus of the peaks of the radiation emission curves, is also given in Fig. 21–9.

The peak of the solar radiation, for example, occurs at $\lambda = 2897.8/5780 = 0.50\ \mu m$, which is near the middle of the visible range. The peak of the radiation emitted by a surface at room temperature ($T = 298$ K) occurs at 9.72 $\mu m$, which is well into the infrared region of the spectrum.

An electrical resistance heater starts radiating heat soon after it is plugged in, and we can feel the emitted radiation energy by holding our hands against the heater. But this radiation is entirely in the infrared region and thus cannot be sensed by our eyes. The heater would appear dull red when its temperature reaches about 1000 K, since it starts emitting a detectable amount (about 1 W/m²·μm) of visible red radiation at that temperature. As the temperature rises even more, the heater appears bright red and is said to be *red hot*. When the temperature reaches about 1500 K, the heater emits enough radiation in the entire visible range of the spectrum to appear almost *white* to the eye, and it is called *white hot*.

Although it cannot be sensed directly by the human eye, infrared radiation can be detected by infrared cameras, which transmit the information to microprocessors to display visual images of objects at night. *Rattlesnakes* can sense the infrared radiation or the "body heat" coming off warm-blooded animals, and thus they can see at night without using any instruments. Similarly, honeybees are sensitive to ultraviolet radiation. A surface that reflects all of the light appears *white*, while a surface that absorbs all of the light incident on it appears black. (Then how do we see a black surface?)

It should be clear from this discussion that the color of an object is not due to emission, which is primarily in the infrared region, unless the surface temperature of the object exceeds about 1000 K. Instead, the color of a surface depends on the absorption and reflection characteristics of the surface and is due to selective absorption and reflection of the incident visible radiation coming from a light source such as the sun or an incandescent lightbulb. A piece of clothing containing a pigment that reflects red while absorbing the remaining parts of the incident light appears "red" to the eye (Fig. 21–10). Leaves appear "green" because their cells contain the pigment chlorophyll, which strongly reflects green while absorbing other colors.

It is left as an exercise to show that integration of the *spectral* blackbody emissive power $E_{b\lambda}$ over the entire wavelength spectrum gives the *total* blackbody emissive power $E_b$:

$$E_b(T) = \int_0^\infty E_{b\lambda}(\lambda, T)\, d\lambda = \sigma T^4 \qquad (\text{W/m}^2) \qquad (21\text{–}6)$$

Thus, we obtained the Stefan–Boltzmann law (Eq. 21–3) by integrating Planck's law (Eq. 21–4) over all wavelengths. Note that on an $E_{b\lambda}$–$\lambda$ chart, $E_{b\lambda}$ corresponds to any value on the curve, whereas $E_b$ corresponds to the area under the entire curve for a specified temperature (Fig. 21–11). Also, the term *total* means "integrated over all wavelengths."

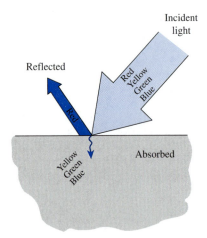

**FIGURE 21–10**
A surface that reflects red while absorbing the remaining parts of the incident light appears red to the eye.

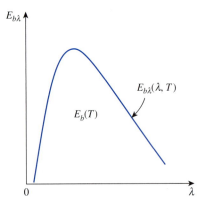

**FIGURE 21–11**
On an $E_{b\lambda}$–$\lambda$ chart, the area under a curve for a given temperature represents the total radiation energy emitted by a blackbody at that temperature.

---

**EXAMPLE 21–1    Radiation Emission from a Black Ball**

Consider a 20-cm-diameter spherical ball at 800 K suspended in air as shown in Fig. 21–12. Assuming the ball closely approximates a blackbody, determine (*a*) the total blackbody emissive power, (*b*) the total amount of radiation emitted by the ball in 5 min, and (*c*) the spectral blackbody emissive power at a wavelength of 3 μm.

**SOLUTION**  An isothermal sphere is suspended in air. The total blackbody emissive power, the total radiation emitted in 5 min, and the spectral blackbody emissive power at 3 μm are to be determined.
*Assumptions*  The ball behaves as a blackbody.

**FIGURE 21–12**
The spherical ball considered in
Example 21–1.

**Analysis** (a) The total blackbody emissive power is determined from the Stefan–Boltzmann law to be

$$E_b = \sigma T^4 = (5.67 \times 10^{-8} \text{ W/m}^2 \cdot \text{K}^4)(800 \text{ K})^4 = \textbf{23.2 kW/m}^2$$

That is, the ball emits 23.2 kJ of energy in the form of electromagnetic radiation per second per m² of the surface area of the ball.

(b) The total amount of radiation energy emitted from the entire ball in 5 min is determined by multiplying the blackbody emissive power obtained above by the total surface area of the ball and the given time interval:

$$A_s = \pi D^2 = \pi(0.2 \text{ m})^2 = 0.1257 \text{ m}^2$$

$$\Delta t = (5 \text{ min})\left(\frac{60 \text{ s}}{1 \text{ min}}\right) = 300 \text{ s}$$

$$Q_{\text{rad}} = E_b A_s \,\Delta t = (23.2 \text{ kW/m}^2)(0.1257 \text{ m}^2)(300 \text{ s})\left(\frac{1 \text{ kJ}}{1 \text{ kW·s}}\right)$$

$$= \textbf{875 kJ}$$

That is, the ball loses 875 kJ of its internal energy in the form of electromagnetic waves to the surroundings in 5 min, which is enough energy to heat 20 kg of water from 0°C to 100°C. Note that the surface temperature of the ball cannot remain constant at 800 K unless there is an equal amount of energy flow to the surface from the surroundings or from the interior regions of the ball through some mechanisms such as chemical or nuclear reactions.
(c) The spectral blackbody emissive power at a wavelength of 3 μm is determined from Planck's distribution law to be

$$E_{b\lambda} = \frac{C_1}{\lambda^5\left[\exp\left(\dfrac{C_2}{\lambda T}\right) - 1\right]} = \frac{3.74177 \times 10^8 \text{ W·}\mu\text{m}^4/\text{m}^2}{13 \;\mu\text{m}2^5\left[\exp\left(\dfrac{1.43878 \times 10_4 \;\mu\text{m·K}}{13 \;\mu\text{m}2 \; 1800 \text{ K}2}\right) - 1\right]}$$

$$= \textbf{3846 W/m}^2\textbf{·}\mu\textbf{m}$$

The Stefan–Boltzmann law $E_b(T) = \sigma T^4$ gives the *total* radiation emitted by a blackbody at all wavelengths from $\lambda = 0$ to $\lambda = \infty$. But we are often interested in the amount of radiation emitted over *some wavelength band*. For example, an incandescent lightbulb is judged on the basis of the radiation it emits in the visible range rather than the radiation it emits at all wavelengths.

The radiation energy emitted by a blackbody per unit area over a wavelength band from $\lambda = 0$ to $\lambda$ is determined from (Fig. 21–13)

$$E_{b,\,0-\lambda}(T) = \int_0^\lambda E_{b\lambda}(\lambda, T)\, d\lambda \qquad (\text{W/m}^2) \qquad (21\text{--}7)$$

It looks like we can determine $E_{b,\,0-\lambda}$ by substituting the $E_{b\lambda}$ relation from Eq. 21–4 and performing this integration. But it turns out that this integration does not have a simple closed-form solution, and performing a numerical

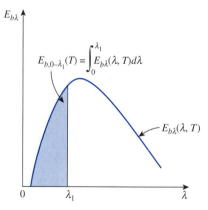

**FIGURE 21–13**
On an $E_{b\lambda}$–$\lambda$ chart, the area under the curve to the left of the $\lambda = \lambda_1$ line represents the radiation energy emitted by a blackbody in the wavelength range $0$–$\lambda_1$ for the given temperature.

integration each time we need a value of $E_{b,0-\lambda}$ is not practical. Therefore, we define a dimensionless quantity $f_\lambda$ called the **blackbody radiation function** as

$$f_\lambda(T) = \frac{\displaystyle\int_0^\lambda E_{b\lambda}(\lambda, T)\, d\lambda}{\sigma T^4} \qquad (21\text{--}8)$$

The function $f_\lambda$ represents *the fraction of radiation emitted from a blackbody at temperature T in the wavelength band from $\lambda = 0$ to $\lambda$.* In reference to Fig. 21–13, $f_\lambda$ can be interpreted as the ratio of the shaded region (radiation energy emitted by the blackbody in the wavelength range $0 - \lambda_1$) over the total area under the curve (total radiation energy emitted by the blackbody in the wavelength range $0 - \infty$). The values of $f_\lambda$ are listed in Table 21–2 as a function of $\lambda T$, where $\lambda$ is in $\mu$m and $T$ is in K.

**TABLE 21–2**

Blackbody radiation functions $f_\lambda$

| $\lambda T$, $\mu$m·K | $f_\lambda$ | $\lambda T$, $\mu$m·K | $f_\lambda$ |
|---|---|---|---|
| 200 | 0.000000 | 6200 | 0.754140 |
| 400 | 0.000000 | 6400 | 0.769234 |
| 600 | 0.000000 | 6600 | 0.783199 |
| 800 | 0.000016 | 6800 | 0.796129 |
| 1000 | 0.000321 | 7000 | 0.808109 |
| 1200 | 0.002134 | 7200 | 0.819217 |
| 1400 | 0.007790 | 7400 | 0.829527 |
| 1600 | 0.019718 | 7600 | 0.839102 |
| 1800 | 0.039341 | 7800 | 0.848005 |
| 2000 | 0.066728 | 8000 | 0.856288 |
| 2200 | 0.100888 | 8500 | 0.874608 |
| 2400 | 0.140256 | 9000 | 0.890029 |
| 2600 | 0.183120 | 9500 | 0.903085 |
| 2800 | 0.227897 | 10,000 | 0.914199 |
| 3000 | 0.273232 | 10,500 | 0.923710 |
| 3200 | 0.318102 | 11,000 | 0.931890 |
| 3400 | 0.361735 | 11,500 | 0.939959 |
| 3600 | 0.403607 | 12,000 | 0.945098 |
| 3800 | 0.443382 | 13,000 | 0.955139 |
| 4000 | 0.480877 | 14,000 | 0.962898 |
| 4200 | 0.516014 | 15,000 | 0.969981 |
| 4400 | 0.548796 | 16,000 | 0.973814 |
| 4600 | 0.579280 | 18,000 | 0.980860 |
| 4800 | 0.607559 | 20,000 | 0.985602 |
| 5000 | 0.633747 | 25,000 | 0.992215 |
| 5200 | 0.658970 | 30,000 | 0.995340 |
| 5400 | 0.680360 | 40,000 | 0.997967 |
| 5600 | 0.701046 | 50,000 | 0.998953 |
| 5800 | 0.720158 | 75,000 | 0.999713 |
| 6000 | 0.737818 | 100,000 | 0.999905 |

The fraction of radiation energy emitted by a blackbody at temperature $T$ over a finite wavelength band from $\lambda = \lambda_1$ to $\lambda = \lambda_2$ is determined from (Fig. 21–14)

$$f_{\lambda_1-\lambda_2}(T) = \frac{\int_0^{\lambda_2} E_{b\lambda}(\lambda, T)d\lambda - \int_0^{\lambda_1} E_{b\lambda}(\lambda, T)d\lambda}{\sigma T^4} = f_{\lambda_2}(T) - f_{\lambda_1}(T) \quad (21\text{–}9)$$

where $f_{\lambda_1}(T)$ and $f_{\lambda_2}(T)$ are blackbody radiation functions corresponding to $\lambda_1 T$ and $\lambda_2 T$, respectively.

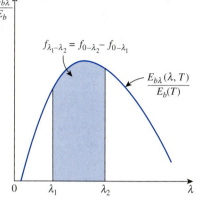

**FIGURE 21–14**
Graphical representation of the fraction of radiation emitted in the wavelength band from $\lambda_1$ to $\lambda_2$.

---

### EXAMPLE 21–2  Light Emitted by the Sun and by a Lightbulb

Charge-coupled device (CCD) image sensors, that are common in modern digital cameras, respond differently to light sources with different spectral distributions. Daylight and incandescent light may be approximated as a blackbody at the effective surface temperatures of 5800 K and 2800 K, respectively. Determine the fraction of radiation emitted within the visible spectrum wavelengths, from 0.40 μm (violet) to 0.76 μm (red), for each of the lighting sources.

**SOLUTION**  For specified blackbody temperatures, the fraction of visible radiation emitted by the sun and the filament of an incandescent lightbulb are to be determined.

**Assumptions**  The sun and the incandescent light filament behave as black bodies.

**Analysis**  The visible range of the electromagnetic spectrum extends from $\lambda_1 = 0.40$ μm to $\lambda_2 = 0.76$ μm. For the sun at $T = 5800$ K, the blackbody radiation functions corresponding to $\lambda_1 T$ and $\lambda_2 T$ are determined from Table 21-2 to be (Fig. 21–15)

$$\lambda_1 T = (0.40 \ \mu\text{m})(5800 \text{K}) = 2320 \ \mu\text{m}\cdot\text{K} \rightarrow f_{\lambda_1, \text{daylight}} = 0.124509$$
$$\lambda_2 T = (0.76 \ \mu\text{m})(5800 \text{K}) = 4408 \ \mu\text{m}\cdot\text{K} \rightarrow f_{\lambda_2, \text{daylight}} = 0.550015$$

Then the fraction of visible radiation emitted by the sun becomes

$$f_{\lambda_1-\lambda_2, \text{daylight}} = 0.550015 - 0.124509 = \mathbf{0.426} \text{ or } \mathbf{42.6 \ percent}$$

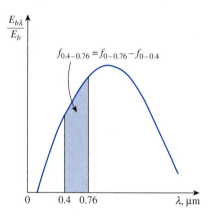

**FIGURE 21–15**
Graphical representation of the fraction of radiation emitted in the visible range in Example 21–2.

For incandescent lightbulb at $T = 2800$ K, the blackbody radiation functions corresponding to $\lambda_1 T$ and $\lambda_2 T$ are determined from Table 21–2 to be

$$\lambda_1 T = (0.40 \ \mu\text{m})(2800 \text{K}) = 1120 \ \mu\text{m}\cdot\text{K} \rightarrow f_{\lambda_1, \text{incandescent}} = 0.001409$$
$$\lambda_2 T = (0.76 \ \mu\text{m})(5800 \text{K}) = 2128 \ \mu\text{m}\cdot\text{K} \rightarrow f_{\lambda_2, \text{incandescent}} = 0.088590$$

Then the fraction of radiation the lightbulb emits in the visible range becomes

$$f_{\lambda_1-\lambda_2, \text{incandescent}} = 0.088590 - 0.001409 = \mathbf{0.087} \text{ or } \mathbf{8.7 \ percent}$$

**Discussion**  Note that almost half of the radiation emitted by the sun is in the visible range, and thus the sun is a very efficient light source. But less than 10 percent of the radiation emitted by the incandescent lightbulb is in the form of visible light, and thus incandescent lightbulbs are inefficient as light sources. Consequently, they are being replaced by the highly efficient fluorescent and LED light sources.

## 21–4 ▪ RADIATIVE PROPERTIES

Most materials encountered in practice, such as metals, wood, and bricks, are *opaque* to thermal radiation, and radiation is considered to be a *surface phenomenon* for such materials. That is, thermal radiation is emitted or absorbed within the first few microns of the surface, and thus we speak of radiative properties of *surfaces* for opaque materials.

Some other materials, such as glass and water, allow visible radiation to penetrate to considerable depths before any significant absorption takes place. Radiation through such *semitransparent* materials obviously cannot be considered to be a surface phenomenon since the entire volume of the material interacts with radiation. On the other hand, both glass and water are practically opaque to infrared radiation. Therefore, materials can exhibit different behavior at different wavelengths, and the dependence on wavelength is an important consideration in the study of radiative properties such as emissivity, absorptivity, reflectivity, and transmissivity of materials.

In the preceding section, we defined a *blackbody* as a perfect emitter and absorber of radiation and said that no body can emit more radiation than a blackbody at the same temperature. Therefore, a blackbody can serve as a convenient *reference* in describing the emission and absorption characteristics of real surfaces.

## Emissivity

The **emissivity** of a surface represents *the ratio of the radiation emitted by the surface at a given temperature to the radiation emitted by a blackbody at the same temperature*. The emissivity of a surface is denoted by $\varepsilon$, and it varies between zero and one, $0 \leq \varepsilon \leq 1$. Emissivity is a measure of how closely a surface approximates a blackbody, for which $\varepsilon = 1$.

The emissivity of a real surface is not a constant. Rather, it varies with the *temperature* of the surface as well as the *wavelength* and the *direction* of the emitted radiation. Therefore, different emissivities can be defined for a surface, depending on the effects considered. For example, the emissivity of a surface at a specified wavelength is called *spectral emissivity* and is denoted by $\varepsilon_\lambda$. Likewise, the emissivity in a specified direction is called *directional emissivity*, denoted by $\varepsilon_\theta$, where $\theta$ is the angle between the direction of radiation and the normal of the surface. The most elemental emissivity of a surface at a given temperature is the *spectral directional emissivity* $\varepsilon_{\lambda,\theta}$, which is the emissivity at a specified wavelength in a specified direction. Note that blackbody radiation is independent of direction.

In practice, it is usually more convenient to work with radiation properties averaged over all directions, called *hemispherical properties*. Noting that the integral of the rate of radiation energy emitted at a specified wavelength per unit surface area over the entire hemisphere is *spectral emissive power*, the **spectral hemispherical emissivity** can be expressed as

$$\varepsilon_\lambda(\lambda, T) = \frac{E_\lambda(\lambda, T)}{E_{b\lambda}(\lambda, T)} \tag{21–10}$$

Note that the emissivity of a surface at a given wavelength can be different at different temperatures since the spectral distribution of emitted radiation (and thus the amount of radiation emitted at a given wavelength) changes with temperature.

The **total hemispherical emissivity** is defined in terms of the radiation energy emitted over all wavelengths in all directions as

$$\varepsilon(T) = \frac{E(T)}{E_b(T)} \qquad (21\text{--}11)$$

Therefore, the total hemispherical emissivity (or simply the "average emissivity") of a surface at a given temperature represents the ratio of the total radiation energy emitted by the surface to the radiation emitted by a blackbody of the same surface area at the same temperature.

Noting that $E = \int_0^\infty E_\lambda d\lambda$ and $E_\lambda(\lambda, T) = \varepsilon_\lambda(\lambda, T)E_{b\lambda}(\lambda, T)$, the total hemispherical emissivity can also be expressed as

$$\varepsilon(T) = \frac{E(T)}{E_b(T)} = \frac{\int_0^\infty \varepsilon_\lambda(\lambda, T)E_{b\lambda}(\lambda, T)d\lambda}{\sigma T^4} \qquad (21\text{--}12)$$

since $E_b(T) = \sigma T^4$. To perform this integration, we need to know the variation of spectral emissivity with wavelength at the specified temperature. The integrand is usually a complicated function, and the integration has to be performed numerically. However, the integration can be performed quite easily by dividing the spectrum into a sufficient number of *wavelength bands* and assuming the emissivity to remain constant over each band; that is, by expressing the function $\varepsilon_\lambda(\lambda, T)$ as a step function. This simplification offers great convenience for little sacrifice of accuracy, since it allows us to transform the integration into a summation in terms of blackbody emission functions.

As an example, consider the emissivity function plotted in Fig. 21–16. It seems like this function can be approximated reasonably well by a step function of the form

$$\varepsilon_\lambda = \begin{cases} \varepsilon_1 = \text{constant}, & 0 \le \lambda < \lambda_1 \\ \varepsilon_2 = \text{constant}, & \lambda_1 \le \lambda < \lambda_2 \\ \varepsilon_3 = \text{constant}, & \lambda_2 \le \lambda < \infty \end{cases} \qquad (21\text{--}13)$$

Then the average emissivity can be determined from Eq. 21–12 by breaking the integral into three parts and utilizing the definition of the blackbody radiation function as

$$\varepsilon(T) = \frac{\varepsilon_1 \int_0^{\lambda_1} E_{b\lambda} d\lambda}{E_b} + \frac{\varepsilon_2 \int_{\lambda_1}^{\lambda_2} E_{b\lambda} d\lambda}{E_b} + \frac{\varepsilon_3 \int_{\lambda_2}^{\infty} E_{b\lambda} d\lambda}{E_b} \qquad (21\text{--}14)$$

$$= \varepsilon_1 f_{0-\lambda_1}(T) + \varepsilon_2 f_{\lambda_1-\lambda_2}(T) + \varepsilon_3 f_{\lambda_2-\infty}(T)$$

Radiation is a complex phenomenon as it is, and the consideration of the wavelength and direction dependence of properties, assuming sufficient data exist, makes it even more complicated. Therefore, the *gray* and *diffuse* approximations are often utilized in radiation calculations. A surface is said to be *diffuse* if its properties are *independent of direction*, and *gray* if its properties are *independent of wavelength*. Therefore, the emissivity of a gray, diffuse surface is simply the total hemispherical emissivity of that surface because of independence of direction and wavelength (Fig. 21–17).

**FIGURE 21–16**
Approximating the actual variation of emissivity with wavelength by a step function.

Real surface:
  $\varepsilon_\theta \ne$ constant
  $\varepsilon_\lambda \ne$ constant

Diffuse surface:
  $\varepsilon_\theta =$ constant

Gray surface:
  $\varepsilon_\lambda =$ constant

Diffuse, gray surface:
  $\varepsilon = \varepsilon_\lambda = \varepsilon_\theta =$ constant

**FIGURE 21–17**
The effect of diffuse and gray approximations on the emissivity of a surface.

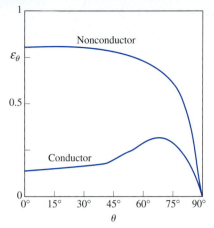

**FIGURE 21–18**

Typical variations of emissivity with direction for electrical conductors and nonconductors.

A few comments about the validity of the diffuse approximation are in order. Although real surfaces do not emit radiation in a perfectly diffuse manner as a blackbody does, they often come close. The variation of emissivity with direction for both electrical conductors and nonconductors is given in Fig. 21–18. Here $\theta$ is the angle measured from the normal of the surface, and thus $\theta = 0$ for radiation emitted in a direction normal to the surface. Note that $\varepsilon_\theta$ remains nearly constant for about $\theta < 40°$ for conductors such as metals and for $\theta < 70°$ for nonconductors such as plastics. Therefore, the directional emissivity of a surface in the normal direction is representative of the hemispherical emissivity of the surface. In radiation analysis, it is common practice to assume the surfaces to be diffuse emitters with an emissivity equal to the value in the normal ($\theta = 0$) direction.

The effect of the gray approximation on emissivity and emissive power of a real surface is illustrated in Fig. 21–19. Note that the radiation emission from a real surface, in general, differs from the Planck distribution, and the emission curve may have several peaks and valleys. A gray surface should emit as much radiation as the real surface it represents at the same temperature. Therefore, the areas under the emission curves of the real and gray surfaces must be equal.

The emissivities of common materials are listed in Table A–26 in the appendix, and the variation of emissivity with wavelength and temperature is illustrated in Fig. 21–20. Typical ranges of emissivity of various materials are given in Fig. 21–21. Note that metals generally have low emissivities, as low as 0.02 for polished surfaces, and nonmetals such as ceramics and organic materials have high ones. The emissivity of metals increases with temperature. Also, oxidation causes significant increases in the emissivity of metals. Heavily oxidized metals can have emissivities comparable to those of nonmetals.

Care should be exercised in the use and interpretation of radiation property data reported in the literature, since the properties strongly depend on the surface conditions such as oxidation, roughness, type of finish, and cleanliness. Consequently, there is considerable discrepancy and uncertainty in the reported values. This uncertainty is largely due to the difficulty in characterizing and describing the surface conditions precisely.

**FIGURE 21–19**

Comparison of the emissivity (*a*) and emissive power (*b*) of a real surface with those of a gray surface and a blackbody at the same temperature.

*(a)*        *(b)*

(a)

(b)

**FIGURE 21–20**

The variation of normal emissivity with (a) wavelength and (b) temperature for various materials.

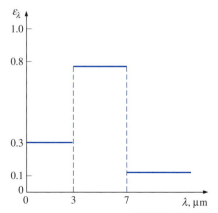

**FIGURE 21–21**

Typical ranges of emissivity for various materials.

### EXAMPLE 21–3    Emissivity of a Surface and Emissive Power

The spectral emissivity function of an opaque surface at 800 K is approximated as (Fig. 21–22)

$$\varepsilon_\lambda = \begin{cases} \varepsilon_1 = 0.3, & 0 \le \lambda < 3 \ \mu m \\ \varepsilon_2 = 0.8, & 3 \ \mu m \le \lambda < 7 \ \mu m \\ \varepsilon_3 = 0.1, & 7 \ \mu m \le \lambda < \infty \end{cases}$$

Determine the average emissivity of the surface and its emissive power.

**SOLUTION** The variation of emissivity of a surface at a specified temperature with wavelength is given. The average emissivity of the surface and its emissive power are to be determined.

**Analysis** The variation of the emissivity with wavelength is given as a step function. Therefore, the average emissivity of the surface can be determined from Eq. 21–12 by breaking the integral into three parts,

$$\varepsilon(T) = \frac{\varepsilon_1 \int_0^{\lambda_1} E_{b\lambda} \, d\lambda}{\sigma T^4} + \frac{\varepsilon_2 \int_{\lambda_1}^{\lambda_2} E_{b\lambda} \, d\lambda}{\sigma T^4} + \frac{\varepsilon_3 \int_{\lambda_2}^{\infty} E_{b\lambda} \, d\lambda}{\sigma T^4}$$

$$= \varepsilon_1 f_{0-\lambda_1}(T) + \varepsilon_2 f_{\lambda_1 - \lambda_2}(T) + \varepsilon_3 f_{\lambda_2 - \infty}(T)$$

$$= \varepsilon_1 f_{\lambda_1} + \varepsilon_2 (f_{\lambda_2} - f_{\lambda_1}) + \varepsilon_3 (1 - f_{\lambda_2})$$

where $f_{\lambda_1}$ and $f_{\lambda_2}$ are blackbody radiation functions and are determined from Table 21–2 to be

$$\lambda_1 T = (3 \ \mu m)(800 \ K) = 2400 \ \mu m \cdot K \quad \rightarrow \quad f_{\lambda_1} = 0.140256$$

$$\lambda_2 T = (7 \ \mu m)(800 \ K) = 5600 \ \mu m \cdot K \quad \rightarrow \quad f_{\lambda_2} = 0.701046$$

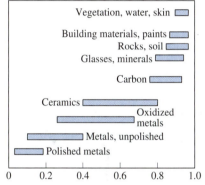

**FIGURE 21–22**

The spectral emissivity of the surface considered in Example 21–3.

Note that $f_{0-\lambda_1} = f_{\lambda_1} - f_0 = f_{\lambda_1}$ since $f_0 = 0$, and $f_{\lambda_2-\infty} = f_\infty - f_{\lambda_2} = 1 - f_{\lambda_2}$ since $f_\infty = 1$. Substituting,

$$\varepsilon = 0.3 \times 0.140256 + 0.8(0.701046 - 0.140256) + 0.1(1 - 0.701046)$$
$$= \mathbf{0.521}$$

That is, the surface will emit as much radiation energy at 800 K as a gray surface having a constant emissivity of $\varepsilon = 0.521$. The emissive power of the surface is

$$E = \varepsilon \sigma T^4 = 0.521(5.67 \times 10^{-8} \text{ W/m}^2 \cdot \text{K}^4)(800 \text{ K})^4 = \mathbf{12{,}100 \text{ W/m}^2}$$

**Discussion** Note that the surface emits 12.1 kJ of radiation energy per second per m$^2$ area of the surface.

## Absorptivity, Reflectivity, and Transmissivity

Everything around us constantly emits radiation, and the emissivity represents the emission characteristics of those bodies. This means that every body, including our own, is constantly bombarded by radiation coming from all directions over a range of wavelengths. Recall that radiation flux *incident on a surface* is called **irradiation** and is denoted by $G$.

When radiation strikes a surface, part of it is absorbed, part of it is reflected, and the remaining part, if any, is transmitted, as illustrated in Fig. 21–23. *The fraction of irradiation absorbed by the surface* is called the **absorptivity** $\alpha$, *the fraction reflected by the surface* is called the **reflectivity** $\rho$, and *the fraction transmitted* is called the **transmissivity** $\tau$. That is,

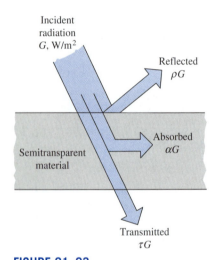

Incident radiation $G$, W/m$^2$

Reflected $\rho G$

Absorbed $\alpha G$

Semitransparent material

Transmitted $\tau G$

**FIGURE 21–23**
The absorption, reflection, and transmission of incident radiation by a semitransparent material.

*Absorptivity*: $$\alpha = \frac{\text{Absorbed radiation}}{\text{Incident radiation}} = \frac{G_{\text{abs}}}{G}, \qquad 0 \le \alpha \le 1 \qquad \text{(21–15)}$$

*Reflectivity*: $$\rho = \frac{\text{Reflected radiation}}{\text{Incident radiation}} = \frac{G_{\text{ref}}}{G}, \qquad 0 \le \rho \le 1 \qquad \text{(21–16)}$$

*Transmissivity*: $$\tau = \frac{\text{Transmitted radiation}}{\text{Incident radiation}} = \frac{G_{\text{tr}}}{G}, \qquad 0 \le \tau \le 1 \qquad \text{(21–17)}$$

where $G$ is the radiation flux incident on the surface, and $G_{\text{abs}}$, $G_{\text{ref}}$, and $G_{\text{tr}}$ are the absorbed, reflected, and transmitted portions of it, respectively. The first law of thermodynamics requires that the sum of the absorbed, reflected, and transmitted radiation be equal to the incident radiation. That is,

$$G_{\text{abs}} + G_{\text{ref}} + G_{\text{tr}} = G \qquad \text{(21–18)}$$

Dividing each term of this relation by $G$ yields

$$\alpha + \rho + \tau = 1 \qquad \text{(21–19)}$$

For opaque surfaces, $\tau = 0$, and thus

$$\alpha + \rho = 1 \qquad \text{(21–20)}$$

This is an important property relation since it enables us to determine both the absorptivity and reflectivity of an opaque surface by measuring either of these properties.

These definitions are for *total hemispherical* properties, since $G$ represents the radiation flux incident on the surface from all directions over the hemispherical space and over all wavelengths. Thus, $\alpha$, $\rho$, and $\tau$ are the *average* properties of a medium for all directions and all wavelengths. However, like emissivity, these properties can also be defined for a specific wavelength and/or direction. For example,

$$\alpha_\lambda(\lambda) = \frac{G_{\lambda,\,abs}(\lambda)}{G_\lambda(\lambda)} \quad \text{and} \quad \rho_\lambda(\lambda) = \frac{G_{\lambda,\,ref}(\lambda)}{G_\lambda(\lambda)} \tag{21–21}$$

where $G_\lambda$ is the spectral irradiation (in W/m²·μm) incident on the surface, and $G_{\lambda,\,abs}$ and $G_{\lambda,\,ref}$ are the reflected and absorbed portions of it, respectively.

Similar quantities can be defined for the transmissivity of semitransparent materials. For example, the **spectral hemispherical transmissivity** of a medium can be expressed as

$$\tau_\lambda(\lambda) = \frac{G_{\lambda,\,tr}(\lambda)}{G_\lambda(\lambda)} \tag{21–22}$$

The average absorptivity, reflectivity, and transmissivity of a surface can also be defined in terms of their spectral counterparts as

$$\alpha = \frac{\int_0^\infty \alpha_\lambda G_\lambda \, d\lambda}{\int_0^\infty G_\lambda \, d\lambda}, \quad \rho = \frac{\int_0^\infty \rho_\lambda G_\lambda \, d\lambda}{\int_0^\infty G_\lambda \, d\lambda}, \quad \tau = \frac{\int_0^\infty \tau_\lambda G_\lambda \, d\lambda}{\int_0^\infty G_\lambda \, d\lambda} \tag{21–23}$$

The reflectivity differs somewhat from the other properties in that it is *bidirectional* in nature. That is, the value of the reflectivity of a surface depends not only on the direction of the incident radiation but also the direction of reflection. Therefore, the reflected rays of a radiation beam incident on a real surface in a specified direction forms an irregular shape, as shown in Fig. 21–24. Such detailed reflectivity data do not exist for most surfaces, and even if they did, they would be of little value in radiation calculations since this would usually add more complication to the analysis.

In practice, for simplicity, surfaces are assumed to reflect in a perfectly *specular* or *diffuse* manner. In **specular** (or *mirrorlike*) **reflection**, *the angle of reflection equals the angle of incidence of the radiation beam*. In **diffuse reflection**, *radiation is reflected equally in all directions*, as shown in Fig. 21–24. Reflection from smooth and polished surfaces approximates specular reflection, whereas reflection from rough surfaces approximates diffuse reflection. In radiation analysis, smoothness is defined relative to wavelength. A surface is said to be *smooth* if the height of the surface roughness is much smaller than the wavelength of the incident radiation.

Unlike emissivity, the absorptivity of a material is practically independent of surface temperature. However, the absorptivity depends strongly on the temperature of the source at which the incident radiation is originating. This is also evident from Fig. 21–25, which shows the absorptivities of various materials at room temperature as functions of the temperature of the radiation source. For example, the absorptivity of the concrete roof of a house is about 0.6 for solar radiation (source temperature: 5780 K) and 0.9 for radiation originating from the surrounding trees and buildings (source temperature: 300 K), as illustrated in Fig. 21–26.

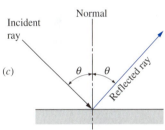

**FIGURE 21–24**
Different types of reflection from a surface: (*a*) actual or irregular, (*b*) diffuse, and (*c*) specular or mirrorlike.

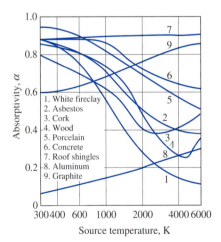

**FIGURE 21–25**
Variation of absorptivity with the temperature of the source of irradiation for various common materials at room temperature.

**FIGURE 21–26**
The absorptivity of a material may be quite different for radiation originating from sources at different temperatures.

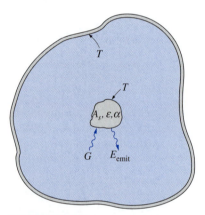

**FIGURE 21–27**
The small body contained in a large isothermal enclosure used in the development of Kirchhoff's law.

Notice that the absorptivity of aluminum increases with the source temperature, a characteristic for metals, and the absorptivity of electric nonconductors, in general, decreases with temperature. This decrease is most pronounced for surfaces that appear white to the eye. For example, the absorptivity of a white painted surface is low for solar radiation, but it is rather high for infrared radiation.

## Kirchhoff's Law

Consider a small body of surface area $A_s$, emissivity $\varepsilon$, and absorptivity $\alpha$ at temperature $T$ contained in a large isothermal enclosure at the same temperature, as shown in Fig. 21–27. Recall that a large isothermal enclosure forms a blackbody cavity regardless of the radiative properties of the enclosure surface, and the body in the enclosure is too small to interfere with the blackbody nature of the cavity. Therefore, the radiation incident on any part of the surface of the small body is equal to the radiation emitted by a blackbody at temperature $T$. That is, $G = E_b(T) = \sigma T^4$, and the radiation absorbed by the small body per unit of its surface area is

$$G_{abs} = \alpha G = \alpha \sigma T^4$$

The radiation emitted by the small body is

$$E_{emit} = \varepsilon \sigma T^4$$

Considering that the small body is in thermal equilibrium with the enclosure, the net rate of heat transfer to the body must be zero. Therefore, the radiation emitted by the body must be equal to the radiation absorbed by it:

$$A_s \varepsilon \sigma T^4 = A_s \alpha \sigma T^4 \tag{21–24}$$

Thus, we conclude that

$$\varepsilon(T) = \alpha(T) \tag{21–25}$$

That is, *the total hemispherical emissivity of a surface at temperature T is equal to its total hemispherical absorptivity for radiation coming from a blackbody at the same temperature.* This relation, which greatly simplifies the radiation analysis, was first developed by Gustav Kirchhoff in 1860 and is now called **Kirchhoff's law**. Note that this relation is derived under the condition that the surface temperature is equal to the temperature of the source of irradiation, and the reader is cautioned against using it when considerable difference (more than a few hundred degrees) exists between the surface temperature and the temperature of the source of irradiation.

The derivation above can also be repeated for radiation at a specified wavelength to obtain the *spectral* form of Kirchhoff's law:

$$\varepsilon_\lambda(T) = \alpha_\lambda(T) \tag{21–26}$$

This relation is valid when the irradiation or the emitted radiation is independent of direction. The form of Kirchhoff's law that involves no restrictions is the *spectral directional* form expressed as $\varepsilon_{\lambda, \theta}(T) = \alpha_{\lambda, \theta}(T)$. That is, the emissivity of a surface at a specified wavelength, direction, and temperature is always equal to its absorptivity at the same wavelength, direction, and temperature.

It is very tempting to use Kirchhoff's law in radiation analysis since the relation $\varepsilon = \alpha$ together with $\rho = 1 - \alpha$ enables us to determine all three properties of an opaque surface from a knowledge of only *one* property. Although Eq. 21–25 gives acceptable results in most cases, in practice, care should be exercised when there is considerable difference between the surface temperature and the temperature of the source of incident radiation.

## The Greenhouse Effect

You have probably noticed that when you leave your car under direct sunlight on a sunny day, the interior of the car gets much warmer than the air outside, and you may have wondered why the car acts like a *heat trap*. The answer lies in the spectral transmissivity curve of the *glass*, which resembles an inverted U, as shown in Fig. 21–28. We observe from this figure that glass at thicknesses encountered in practice transmits over 90 percent of radiation in the visible range and is practically opaque (nontransparent) to radiation in the longer-wavelength infrared regions of the electromagnetic spectrum (roughly $\lambda > 3 \ \mu m$). Therefore, glass has a transparent window in the wavelength range $0.3 \ \mu m < \lambda < 3 \ \mu m$ in which over 90 percent of solar radiation is emitted. On the other hand, the entire radiation emitted by surfaces at room temperature falls in the infrared region. Consequently, glass allows the solar radiation to enter but does not allow the infrared radiation from the interior surfaces to escape. This causes a rise in the interior temperature as a result of the energy buildup in the car. This heating effect, which is due to the nongray characteristic of glass (or clear plastics), is known as the **greenhouse effect**, since it is utilized extensively in greenhouses (Fig. 21–29) and also in flat plate solar collectors.

The greenhouse effect is also experienced on a larger scale on earth. The surface of the earth, which warms up during the day as a result of the absorption of solar energy, cools down at night by radiating its energy into deep space as infrared radiation. The combustion gases such as $CO_2$ and water vapor in the atmosphere transmit the bulk of the solar radiation but absorb the infrared radiation emitted by the surface of the earth. Thus, there is concern that the energy trapped on earth will eventually cause global warming and thus drastic changes in weather patterns.

In *humid* places such as coastal areas, there is not a large change between the daytime and nighttime temperatures, because the humidity acts as a barrier on the path of the infrared radiation coming from the earth, and thus slows down the cooling process at night. In areas with clear skies such as deserts, there is a large swing between the daytime and nighttime temperatures because of the absence of such barriers for infrared radiation.

## 21–5 · THE VIEW FACTOR

Radiation heat transfer between surfaces depends on the *orientation* of the surfaces relative to each other as well as their radiation properties and temperatures, as illustrated in Fig. 21–30. For example, a camper can make the most use of a campfire on a cold night by standing as close to the fire as possible and by blocking as much of the radiation coming from the fire by turning his or her front to the fire instead of the side. Likewise, a person can maximize the amount of solar radiation incident on him or her and take a sunbath by lying down on his or her back instead of standing.

**FIGURE 21–28**
The spectral transmissivity of low-iron glass at room temperature for different thicknesses.

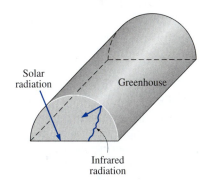

**FIGURE 21–29**
A greenhouse traps energy by allowing the solar radiation to come in but not allowing the infrared radiation to go out.

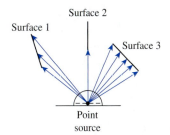

**FIGURE 21–30**
Radiation heat exchange between surfaces depends on the *orientation* of the surfaces relative to each other, and this dependence on orientation is accounted for by the *view factor*.

To account for the effects of orientation on radiation heat transfer between two surfaces, we define a new parameter called the *view factor*, which is a purely geometric quantity and is independent of the surface properties and temperature. It is also called the *shape factor, configuration factor*, and *angle factor*. The view factor based on the assumption that the surfaces are diffuse emitters and diffuse reflectors is called the *diffuse view factor*, and the view factor based on the assumption that the surfaces are diffuse emitters but specular reflectors is called the *specular view factor*. In this book, we consider radiation exchange between diffuse surfaces only, and thus the term *view factor* simply means *diffuse view factor*.

The view factor from a surface $i$ to a surface $j$ is denoted by $F_{i \rightarrow j}$ or just $F_{ij}$, and is defined as

$F_{ij}$ = *the fraction of the radiation leaving surface i that strikes surface j directly*

The notation $F_{i \rightarrow j}$ is instructive for beginners, since it emphasizes that the view factor is for radiation that travels from surface $i$ to surface $j$. However, this notation becomes rather awkward when it has to be used many times in a problem. In such cases, it is convenient to replace it by its *shorthand* version $F_{ij}$.

The view factor $F_{12}$ represents the fraction of radiation leaving surface 1 that strikes surface 2 directly, and $F_{21}$ represents the fraction of radiation leaving surface 2 that strikes surface 1 directly. Note that the radiation that strikes a surface does not need to be absorbed by that surface. Also, radiation that strikes a surface after being reflected by other surfaces is not considered in the evaluation of view factors. For the special case of $j = i$, we have

$F_{i \rightarrow i}$ = *the fraction of radiation leaving surface i that strikes itself directly*

Noting that in the absence of strong electromagnetic fields radiation beams travel in straight paths, the view factor from a surface to itself is zero unless the surface "sees" itself. Therefore, $F_{i \rightarrow i} = 0$ for *plane* or *convex* surfaces and $F_{i \rightarrow i} \neq 0$ for concave surfaces, as illustrated in Fig. 21–31.

The value of the view factor ranges between *zero* and *one*. The limiting case $F_{i \rightarrow j} = 0$ indicates that the two surfaces do not have a direct view of each other, and thus radiation leaving surface $i$ cannot strike surface $j$ directly. The other limiting case $F_{i \rightarrow j} = 1$ indicates that surface $j$ completely surrounds surface $i$, so that the entire radiation leaving surface $i$ is intercepted by surface $j$. For example, in a geometry consisting of two concentric spheres, the entire radiation leaving the surface of the smaller sphere (surface 1) strikes the larger sphere (surface 2), and thus $F_{1 \rightarrow 2} = 1$, as illustrated in Fig. 21–32.

The view factor has proven to be very useful in radiation analysis because it allows us to express the *fraction of radiation* leaving a surface that strikes another surface in terms of the orientation of these two surfaces relative to each other. The underlying assumption in this process is that the radiation a surface receives from a source is directly proportional to the angle the surface subtends when viewed from the source. This would be the case only if the radiation coming off the source is *uniform* in all directions throughout its surface and the medium between the surfaces does not *absorb, emit*, or *scatter* radiation. That is, it is the case when the surfaces are *isothermal* and *diffuse* emitters and reflectors and the surfaces are separated by a *nonparticipating* medium such as a vacuum or air.

$F_{1 \rightarrow 1} = 0$

(a) Plane surface

$F_{2 \rightarrow 2} = 0$

(b) Convex surface

$F_{3 \rightarrow 3} \neq 0$

(c) Concave surface

**FIGURE 21–31**
The view factor from a surface to itself is *zero* for *plane* or *convex* surfaces and *nonzero* for *concave* surfaces.

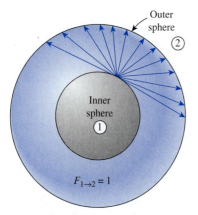

Outer sphere

②

Inner sphere

①

$F_{1 \rightarrow 2} = 1$

**FIGURE 21–32**
In a geometry that consists of two concentric spheres, the view factor $F_{1 \rightarrow 2} = 1$ since the entire radiation leaving the surface of the smaller sphere is intercepted by the larger sphere.

The view factor $F_{1 \to 2}$ between two surfaces $A_1$ and $A_2$ can be determined in a systematic manner first by expressing the view factor between two differential areas $dA_1$ and $dA_2$ in terms of the spatial variables and then by performing the necessary integrations. However, this approach is not practical, since, even for simple geometries, the resulting integrations are usually very complex and difficult to perform.

View factors for hundreds of common geometries are evaluated and the results are given in analytical, graphical, and tabular form in several publications. View factors for selected geometries are given in Tables 21–3 and 21–4 in *analytical* form and in Figs. 21–33 to 21–36 in *graphical* form. The view factors in Table 21–3 are for three-dimensional geometries. The view factors in Table 21–4, on the other hand, are for geometries that are *infinitely long* in the direction perpendicular to the plane of the paper and are therefore two-dimensional.

## TABLE 21–3

View factor expressions for some common geometries of finite size (3-D)

| Geometry | Relation |
|---|---|
| Aligned parallel rectangles | $\overline{X} = X/L$ and $\overline{Y} = Y/L$ <br><br> $F_{i \to j} = \dfrac{2}{\pi \overline{X}\,\overline{Y}} \left\{ \ln\left[ \dfrac{(1+\overline{X}^2)(1+\overline{Y}^2)}{1+\overline{X}^2+\overline{Y}^2} \right]^{1/2} + \overline{X}(1+\overline{Y}^2)^{1/2} \tan^{-1} \dfrac{\overline{X}}{(1+\overline{Y}^2)^{1/2}} \right.$ <br><br> $\left. + \overline{Y}(1+\overline{X}^2)^{1/2} \tan^{-1} \dfrac{\overline{Y}}{(1+\overline{X}^2)^{1/2}} - \overline{X} \tan^{-1} \overline{X} - \overline{Y} \tan^{-1} \overline{Y} \right\}$ |
| Coaxial parallel disks | $R_i = r_i/L$ and $R_j = r_j/L$ <br><br> $S = 1 + \dfrac{1+R_j^2}{R_i^2}$ <br><br> $F_{i \to j} = \dfrac{1}{2} \left\{ S - \left[ S^2 - 4\left(\dfrac{r_j}{r_i}\right)^2 \right]^{1/2} \right\}$ <br><br> For $r_i = r_j = r$ and $R = r/L$: $\quad F_{i \to j} = F_{j \to i} = 1 + \dfrac{1 - \sqrt{4R^2 + 1}}{2R^2}$ |
| Perpendicular rectangles with a common edge | $H = Z/X$ and $W = Y/X$ <br><br> $F_{i \to j} = \dfrac{1}{\pi W} \left( W \tan^{-1} \dfrac{1}{W} + H \tan^{-1} \dfrac{1}{H} - (H^2 + W^2)^{1/2} \tan^{-1} \dfrac{1}{(H^2+W^2)^{1/2}} \right.$ <br><br> $+ \dfrac{1}{4} \ln \left\{ \dfrac{(1+W^2)(1+H^2)}{1+W^2+H^2} \left[ \dfrac{W^2(1+W^2+H^2)}{(1+W^2)(W^2+H^2)} \right]^{W^2} \right.$ <br><br> $\left. \left. \times \left[ \dfrac{H^2(1+H^2+W^2)}{(1+H^2)(H^2+W^2)} \right]^{H^2} \right\} \right)$ |

**TABLE 21–4**

View factor expressions for some infinitely long (2-D) geometries

| Geometry | Relation |
|---|---|
| Parallel plates with midlines connected by perpendicular line | $W_i = w_i/L$ and $W_j = w_j/L$ <br><br> $F_{i \to j} = \dfrac{[(W_i + W_j)^2 + 4]^{1/2} - (W_j - W_i)^2 + 4]^{1/2}}{2W_i}$ |
| Inclined plates of equal width and with a common edge | $F_{i \to j} = 1 - \sin \dfrac{1}{2}\alpha$ |
| Perpendicular plates with a common edge | $F_{i \to j} = \dfrac{1}{2}\left\{ 1 + \dfrac{w_j}{w_i} - \left[ 1 + \left(\dfrac{w_j}{w_i}\right)^2 \right]^{1/2} \right\}$ |
| Three-sided enclosure | $F_{i \to j} = \dfrac{w_i + w_j - w_k}{2w_i}$ |
| Infinite plane and row of cylinders | $F_{i \to j} = 1 - \left[ 1 - \left(\dfrac{D}{s}\right)^2 \right]^{1/2}$ <br><br> $+ \dfrac{D}{s} \tan^{-1}\left(\dfrac{s^2 - D^2}{D^2}\right)^{1/2}$ |

# View Factor Relations

Radiation analysis on an enclosure consisting of $N$ surfaces requires the evaluation of $N^2$ view factors, and this evaluation process is probably the most time-consuming part of a radiation analysis. However, it is neither practical nor necessary to evaluate all of the view factors directly. Once a sufficient number of view factors are available, the rest of them can be determined by utilizing some fundamental relations for view factors, as discussed next.

**FIGURE 21–33**
View factor between two aligned parallel rectangles of equal size.

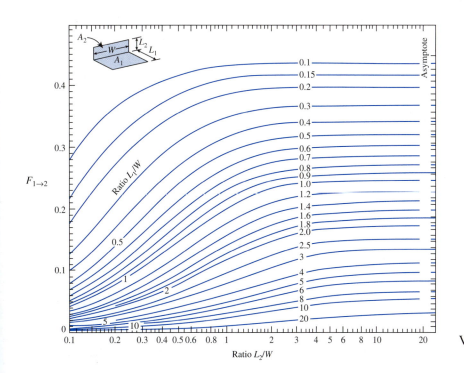

**FIGURE 21–34**
View factor between two perpendicular rectangles with a common edge.

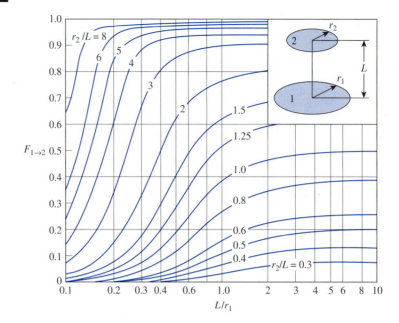

**FIGURE 21–35**
View factor between two coaxial parallel disks.

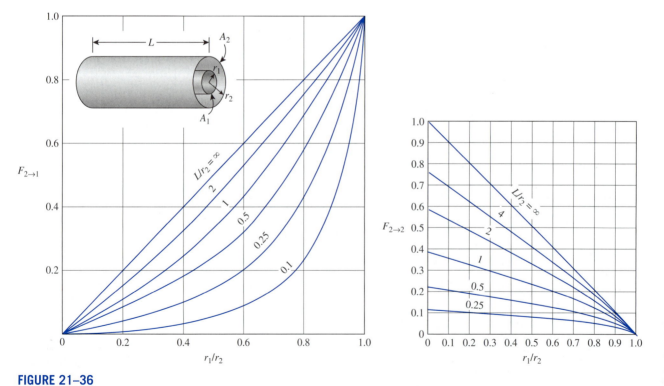

**FIGURE 21–36**
View factors for two concentric cylinders of finite length: (*a*) outer cylinder to inner cylinder; (*b*) outer cylinder to itself.

# 1  The Reciprocity Relation

The view factors $F_{i \rightarrow j}$ and $F_{j \rightarrow i}$ are *not* equal to each other unless the areas of the two surfaces are. That is,

$$F_{j \rightarrow i} = F_{i \rightarrow j} \qquad \text{when} \qquad A_i = A_j$$
$$F_{j \rightarrow i} \neq F_{i \rightarrow j} \qquad \text{when} \qquad A_i \neq A_j$$

We have shown earlier that the pair of view factors $F_{i \rightarrow j}$ and $F_{j \rightarrow i}$ are related to each other by

$$A_i F_{i \rightarrow j} = A_j F_{j \rightarrow i} \tag{21-27}$$

This relation is referred to as the **reciprocity relation** or the **reciprocity rule**, and it enables us to determine the counterpart of a view factor from a knowledge of the view factor itself and the areas of the two surfaces. When determining the pair of view factors $F_{i \rightarrow j}$ and $F_{j \rightarrow i}$, it makes sense to evaluate first the easier one directly and then the more difficult one by applying the reciprocity relation.

# 2  The Summation Rule

The radiation analysis of a surface normally requires the consideration of the radiation coming in or going out in all directions. Therefore, most radiation problems encountered in practice involve enclosed spaces. When formulating a radiation problem, we usually form an *enclosure* consisting of the surfaces interacting radiatively. Even openings are treated as imaginary surfaces with radiation properties equivalent to those of the opening.

The conservation of energy principle requires that the entire radiation leaving any surface $i$ of an enclosure be intercepted by the surfaces of the enclosure. Therefore, *the sum of the view factors from surface $i$ of an enclosure to all surfaces of the enclosure, including to itself, must equal unity.* This is known as the **summation rule** for an enclosure and is expressed as (Fig. 21–37)

$$\sum_{j=1}^{N} F_{i \rightarrow j} = 1 \tag{21-28}$$

where $N$ is the number of surfaces of the enclosure. For example, applying the summation rule to surface 1 of a three-surface enclosure yields

$$\sum_{j=1}^{3} F_{1 \rightarrow j} = F_{1 \rightarrow 1} + F_{1 \rightarrow 2} + F_{1 \rightarrow 3} = 1$$

The summation rule can be applied to each surface of an enclosure by varying $i$ from 1 to $N$. Therefore, the summation rule applied to each of the $N$ surfaces of an enclosure gives $N$ relations for the determination of the view factors. Also, the reciprocity rule gives $\frac{1}{2}N(N - 1)$ additional relations. Then the total number of view factors that need to be evaluated directly for an $N$-surface enclosure becomes

$$N^2 - [N + \tfrac{1}{2}N(N - 1)] = \tfrac{1}{2}N(N - 1)$$

For example, for a six-surface enclosure, we need to determine only $\frac{1}{2} \times 6$ $(6 - 1) = 15$ of the $6^2 = 36$ view factors directly. The remaining 21 view factors can be determined from the 21 equations that are obtained by applying the reciprocity and the summation rules.

Surface $i$

**FIGURE 21–37**
Radiation leaving any surface $i$ of an enclosure must be intercepted completely by the surfaces of the enclosure. Therefore, the sum of the view factors from surface $i$ to each one of the surfaces of the enclosure must be unity.

**FIGURE 21–38**

The geometry considered in Example 21–4.

*EXAMPLE 21–4* **View Factors Associated with Two Concentric Spheres**

Determine the view factors associated with an enclosure formed by two concentric spheres, shown in Fig. 21–38.

**SOLUTION** The view factors associated with two concentric spheres are to be determined.

*Assumptions* The surfaces are diffuse emitters and reflectors.

*Analysis* The outer surface of the smaller sphere (surface 1) and inner surface of the larger sphere (surface 2) form a two-surface enclosure. Therefore, $N = 2$ and this enclosure involves $N^2 = 2^2 = 4$ view factors, which are $F_{11}$, $F_{12}$, $F_{21}$, and $F_{22}$. In this two-surface enclosure, we need to determine only

$$\tfrac{1}{2}N(N - 1) = \tfrac{1}{2} \times 2(2 - 1) = 1$$

view factor directly. The remaining three view factors can be determined by the application of the summation and reciprocity rules. But it turns out that we can determine not only one but *two* view factors directly in this case by a simple *inspection:*

$$F_{11} = 0, \quad \text{since no radiation leaving surface 1 strikes itself}$$
$$F_{12} = 1, \quad \text{since all radiation leaving surface 1 strikes surface 2}$$

Actually it would be sufficient to determine only one of these view factors by inspection, since we could always determine the other one from the summation rule applied to surface 1 as $F_{11} + F_{12} = 1$.

The view factor $F_{21}$ is determined by applying the reciprocity relation to surfaces 1 and 2:

$$A_1 F_{12} = A_2 F_{21}$$

which yields

$$F_{21} = \frac{A_1}{A_2} F_{12} = \frac{4\pi r_1^2}{4\pi r_2^2} \times 1 = \left(\frac{r_1}{r_2}\right)^2$$

Finally, the view factor $F_{22}$ is determined by applying the summation rule to surface 2:

$$F_{21} + F_{22} = 1$$

and thus

$$F_{22} = 1 - F_{21} = 1 - \left(\frac{r_1}{r_2}\right)^2$$

*Discussion* Note that when the outer sphere is much larger than the inner sphere ($r_2 \gg r_1$), $F_{22}$ approaches one. This is expected, since the fraction of radiation leaving the outer sphere that is intercepted by the inner sphere will be negligible in that case. Also note that the two spheres considered above do not need to be concentric. However, the radiation analysis will be most accurate for the case of concentric spheres, since the radiation is most likely to be uniform on the surfaces in that case.

# 3 The Superposition Rule

Sometimes the view factor associated with a given geometry is not available in standard tables and charts. In such cases, it is desirable to express the given geometry as the sum or difference of some geometries with known view factors, and then to apply the **superposition rule**, which can be expressed as *the view factor from a surface i to a surface j is equal to the sum of the view factors from surface i to the parts of surface j*. Note that the reverse of this is not true. That is, the view factor from a surface *j* to a surface *i* is *not* equal to the sum of the view factors from the parts of surface *j* to surface *i*.

Consider the geometry in Fig. 21–39, which is infinitely long in the direction perpendicular to the plane of the paper. The radiation that leaves surface 1 and strikes the combined surfaces 2 and 3 is equal to the sum of the radiation that strikes surfaces 2 and 3. Therefore, the view factor from surface 1 to the combined surfaces of 2 and 3 is

$$F_{1 \to (2,3)} = F_{1 \to 2} + F_{1 \to 3} \tag{21-29}$$

Suppose we need to find the view factor $F_{1 \to 3}$. A quick check of the view factor expressions and charts in this section reveals that such a view factor cannot be evaluated directly. However, the view factor $F_{1 \to 3}$ can be determined from Eq. 21–29 after determining both $F_{1 \to 2}$ and $F_{1 \to (2,3)}$ from the chart in Table 21–4. Therefore, it may be possible to determine some difficult view factors with relative ease by expressing one or both of the areas as the sum or differences of areas and then applying the superposition rule.

To obtain a relation for the view factor $F_{(2,3) \to 1}$, we multiply Eq. 21–29 by $A_1$,

$$A_1 F_{1 \to (2,3)} = A_1 F_{1 \to 2} + A_1 F_{1 \to 3}$$

and apply the reciprocity relation to each term to get

$$(A_2 + A_3) F_{(2,3) \to 1} = A_2 F_{2 \to 1} + A_3 F_{3 \to 1}$$

or

$$F_{(2,3) \to 1} = \frac{A_2 F_{2 \to 1} + A_3 F_{3 \to 1}}{A_2 + A_3} \tag{21-30}$$

Areas that are expressed as the sum of more than two parts can be handled in a similar manner.

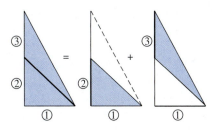

$$F_{1 \to (2,3)} = F_{1 \to 2} + F_{1 \to 3}$$

**FIGURE 21–39**

The view factor from a surface to a composite surface is equal to the sum of the view factors from the surface to the parts of the composite surface.

---

**EXAMPLE 21–5**  **Fraction of Radiation Leaving through an Opening**

Determine the fraction of the radiation leaving the base of the cylindrical enclosure shown in Fig. 21–40 that escapes through a coaxial ring opening at its top surface. The radius and the length of the enclosure are $r_1 = 10$ cm and $L = 10$ cm, while the inner and outer radii of the ring are $r_2 = 5$ cm and $r_3 = 8$ cm, respectively.

**FIGURE 21–40**

The cylindrical enclosure considered in Example 21–5.

**SOLUTION** The fraction of radiation leaving the base of a cylindrical enclosure through a coaxial ring opening at its top surface is to be determined.

**Assumptions** The base surface is a diffuse emitter and reflector.

**Analysis** We are asked to determine the fraction of the radiation leaving the base of the enclosure that escapes through an opening at the top surface. Actually, what we are asked to determine is simply the *view factor* $F_{1 \rightarrow ring}$ from the base of the enclosure to the ring-shaped surface at the top.

We do not have an analytical expression or chart for view factors between a circular area and a coaxial ring, and so we cannot determine $F_{1 \rightarrow ring}$ directly. However, we do have a chart for view factors between two coaxial parallel disks, and we can always express a ring in terms of disks.

Let the base surface of radius $r_1 = 10$ cm be surface 1, the circular area of $r_2 = 5$ cm at the top be surface 2, and the circular area of $r_3 = 8$ cm be surface 3. Using the superposition rule, the view factor from surface 1 to surface 3 can be expressed as

$$F_{1 \rightarrow 3} = F_{1 \rightarrow 2} + F_{1 \rightarrow ring}$$

since surface 3 is the sum of surface 2 and the ring area. The view factors $F_{1 \rightarrow 2}$ and $F_{1 \rightarrow 3}$ are determined from the chart in Fig. 21–35.

$$\frac{L}{r_1} = \frac{10 \text{ cm}}{10 \text{ cm}} = 1 \quad \text{and} \quad \frac{r_2}{L} = \frac{5 \text{ cm}}{10 \text{ cm}} = 0.5 \xrightarrow{\text{(Fig. 21-35)}} = 0.11$$

$$\frac{L}{r_1} = \frac{10 \text{ cm}}{10 \text{ cm}} = 1 \quad \text{and} \quad \frac{r_3}{L} = \frac{8 \text{ cm}}{10 \text{ cm}} = 0.8 \xrightarrow{\text{(Fig. 21-35)}} = 0.28$$

Therefore,

$$F_{1 \rightarrow ring} = F_{1 \rightarrow 3} - F_{1 \rightarrow 2} = 0.28 - 0.11 = \mathbf{0.17}$$

which is the desired result. Note that $F_{1 \rightarrow 2}$ and $F_{1 \rightarrow 3}$ represent the fractions of radiation leaving the base that strike the circular surfaces 2 and 3, respectively, and their difference gives the fraction that strikes the ring area.

## 4   The Symmetry Rule

The determination of the view factors in a problem can be simplified further if the geometry involved possesses some sort of symmetry. Therefore, it is good practice to check for the presence of any *symmetry* in a problem before attempting to determine the view factors directly. The presence of symmetry can be determined *by inspection*, keeping the definition of the view factor in mind. Identical surfaces that are oriented in an identical manner with respect to another surface will intercept identical amounts of radiation leaving that surface. Therefore, the **symmetry rule** can be expressed as *two (or more) surfaces that possess symmetry about a third surface will have identical view factors from that surface* (Fig. 21–41).

The symmetry rule can also be expressed as *if the surfaces j and k are symmetric about the surface i then $F_{i \rightarrow j} = F_{i \rightarrow k}$.* Using the reciprocity rule, we can show that the relation $F_{j \rightarrow i} = F_{k \rightarrow i}$ is also true in this case.

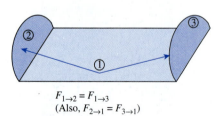

$$F_{1 \rightarrow 2} = F_{1 \rightarrow 3}$$
$$(\text{Also, } F_{2 \rightarrow 1} = F_{3 \rightarrow 1})$$

**FIGURE 21–41**

Two surfaces that are symmetric about a third surface will have the same view factor from the third surface.

**FIGURE 21–42**
The pyramid considered
in Example 21–6.

*EXAMPLE 21–6*     **View Factors Associated with a Tetragon**

Determine the view factors from the base of the pyramid shown in Fig. 21–42 to each of its four side surfaces. The base of the pyramid is a square, and its side surfaces are isosceles triangles.

**SOLUTION**    The view factors from the base of a pyramid to each of its four side surfaces for the case of a square base are to be determined.
*Assumptions*   The surfaces are diffuse emitters and reflectors.
*Analysis*   The base of the pyramid (surface 1) and its four side surfaces (surfaces 2, 3, 4, and 5) form a five-surface enclosure. The first thing we notice about this enclosure is its symmetry. The four side surfaces are symmetric about the base surface. Then, from the *symmetry rule,* we have

$$F_{12} = F_{13} = F_{14} = F_{15}$$

Also, the *summation rule* applied to surface 1 yields

$$\sum_{j=1}^{5} F_{1j} = F_{11} + F_{12} + F_{13} + F_{14} + F_{15} = 1$$

However, $F_{11} = 0$, since the base is a *flat* surface. Then the two relations above yield

$$F_{12} = F_{13} = F_{14} = F_{15} = \mathbf{0.25}$$

*Discussion*   Note that each of the four side surfaces of the pyramid receive one-fourth of the entire radiation leaving the base surface, as expected. Also note that the presence of symmetry greatly simplified the determination of the view factors.

*EXAMPLE 21–7*     **View Factors Associated with a Triangular Duct**

Determine the view factor from any one side to any other side of the infinitely long triangular duct whose cross section is given in Fig. 21–43.

**SOLUTION**    The view factors associated with an infinitely long triangular duct are to be determined.
*Assumptions*   The surfaces are diffuse emitters and reflectors.
*Analysis*   The widths of the sides of the triangular cross section of the duct are $L_1$, $L_2$, and $L_3$, and the surface areas corresponding to them are $A_1$, $A_2$, and $A_3$, respectively. Since the duct is infinitely long, the fraction of radiation leaving any surface that escapes through the ends of the duct is negligible. Therefore, the infinitely long duct can be considered to be a three-surface enclosure, $N = 3$.

This enclosure involves $N^2 = 3^2 = 9$ view factors, and we need to determine

$$\tfrac{1}{2}N(N - 1) = \tfrac{1}{2} \times 3(3 - 1) = 3$$

of these view factors directly. Fortunately, we can determine all three of them by inspection to be

$$F_{11} = F_{22} = F_{33} = 0$$

since all three surfaces are flat. The remaining six view factors can be determined by the application of the summation and reciprocity rules.

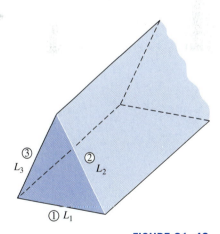

**FIGURE 21–43**
The infinitely long triangular duct considered in Example 21–7.

Applying the summation rule to each of the three surfaces gives

$$F_{11} + F_{12} + F_{13} = 1$$
$$F_{21} + F_{22} + F_{23} = 1$$
$$F_{31} + F_{32} + F_{33} = 1$$

Noting that $F_{11} = F_{22} = F_{33} = 0$ and multiplying the first equation by $A_1$, the second by $A_2$, and the third by $A_3$ gives

$$A_1 F_{12} + A_1 F_{13} = A_1$$
$$A_2 F_{21} + A_2 F_{23} = A_2$$
$$A_3 F_{31} + A_3 F_{32} = A_3$$

Finally, applying the three reciprocity relations $A_1 F_{12} = A_2 F_{21}$, $A_1 F_{13} = A_3 F_{31}$, and $A_2 F_{23} = A_3 F_{32}$ gives

$$A_1 F_{12} + A_1 F_{13} = A_1$$
$$A_1 F_{12} + A_2 F_{23} = A_2$$
$$A_1 F_{13} + A_2 F_{23} = A_3$$

This is a set of three algebraic equations with three unknowns, which can be solved to obtain

$$F_{12} = \frac{A_1 + A_2 - A_3}{2A_1} = \frac{L_1 + L_2 - L_3}{2L_1}$$

$$F_{13} = \frac{A_1 + A_3 - A_2}{2A_1} = \frac{L_1 + L_3 - L_2}{2L_1} \qquad (21\text{--}31)$$

$$F_{23} = \frac{A_2 + A_3 - A_1}{2A_2} = \frac{L_2 + L_3 - L_1}{2L_2}$$

**Discussion** Note that we have replaced the areas of the side surfaces by their corresponding widths for simplicity, since $A = Ls$ and the length $s$ can be factored out and canceled. We can generalize this result as *the view factor from a surface of a very long triangular duct to another surface is equal to the sum of the widths of these two surfaces minus the width of the third surface, divided by twice the width of the first surface.*

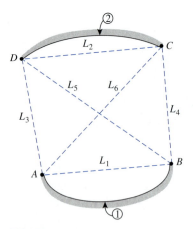

**FIGURE 21–44**

Determination of the view factor $F_{1 \rightarrow 2}$ by the application of the crossed-strings method.

# View Factors between Infinitely Long Surfaces: The Crossed-Strings Method

Many problems encountered in practice involve geometries of constant cross section such as channels and ducts that are *very long* in one direction relative to the other directions. Such geometries can conveniently be considered to be *two-dimensional*, since any radiation interaction through their end surfaces is negligible. These geometries can subsequently be modeled as being *infinitely long*, and the view factor between their surfaces can be determined by the amazingly simple *crossed-strings method* developed by H. C. Hottel in the 1950s. The surfaces of the geometry do not need to be flat; they can be convex, concave, or any irregular shape.

To demonstrate this method, consider the geometry shown in Fig. 21–44, and let us try to find the view factor $F_{1 \rightarrow 2}$ between surfaces 1 and 2. The first thing we do is identify the endpoints of the surfaces (the points $A$, $B$, $C$, and $D$) and connect them to each other with tightly stretched strings, which are indicated by dashed lines. Hottel has shown that the view factor $F_{1 \rightarrow 2}$ can be expressed in terms of the lengths of these stretched strings, which are straight lines, as

$$F_{1 \rightarrow 2} = \frac{(L_5 + L_6) - (L_3 + L_4)}{2L_1} \qquad \text{(21–32)}$$

Note that $L_5 + L_6$ is the sum of the lengths of the *crossed strings,* and $L_3 + L_4$ is the sum of the lengths of the *uncrossed strings* attached to the endpoints. Therefore, Hottel's crossed-strings method can be expressed verbally as

$$F_{i \rightarrow j} = \frac{\Sigma(\text{Crossed strings}) - \Sigma(\text{Uncrossed strings})}{2 \times (\text{String on surface } i)} \qquad \text{(21–33)}$$

The crossed-strings method is applicable even when the two surfaces considered share a common edge, as in a triangle. In such cases, the common edge can be treated as an imaginary string of zero length. The method can also be applied to surfaces that are partially blocked by other surfaces by allowing the strings to bend around the blocking surfaces.

---

### EXAMPLE 21–8    The Crossed-Strings Method for View Factors

Two infinitely long parallel plates of widths $a = 12$ cm and $b = 5$ cm are located a distance $c = 6$ cm apart, as shown in Fig. 21–45. (*a*) Determine the view factor $F_{1 \rightarrow 2}$ from surface 1 to surface 2 by using the crossed-strings method. (*b*) Derive the crossed-strings formula by forming triangles on the given geometry and using Eq. 21–31 for view factors between the sides of triangles.

**SOLUTION**   The view factors between two infinitely long parallel plates are to be determined using the crossed-strings method, and the formula for the view factor is to be derived.

**Assumptions**   The surfaces are diffuse emitters and reflectors.

**Analysis**   (*a*) First we label the endpoints of both surfaces and draw straight dashed lines between the endpoints, as shown in Fig. 21–45. Then we identify the crossed and uncrossed strings and apply the crossed-strings method (Eq. 21–33) to determine the view factor $F_{1 \rightarrow 2}$:

$$F_{1 \rightarrow 2} = \frac{\Sigma(\text{Crossed strings}) - \Sigma(\text{Uncrossed strings})}{2 \times (\text{String on surface 1})} = \frac{(L_5 + L_6) - (L_3 + L_4)}{2L_1}$$

where

$$\begin{aligned}
L_1 &= a = 12 \text{ cm} & L_4 &= \sqrt{7^2 + 6^2} = 9.22 \text{ cm} \\
L_2 &= b = 5 \text{ cm} & L_5 &= \sqrt{5^2 + 6^2} = 7.81 \text{ cm} \\
L_3 &= c = 6 \text{ cm} & L_6 &= \sqrt{12^2 + 6^2} = 13.42 \text{ cm}
\end{aligned}$$

Substituting,

$$F_{1 \rightarrow 2} = \frac{[(7.81 + 13.42) - (6 + 9.22)] \text{ cm}}{2 \times 12 \text{ cm}} = \mathbf{0.250}$$

(*b*) The geometry is infinitely long in the direction perpendicular to the plane of the paper, and thus the two plates (surfaces 1 and 2) and the two openings

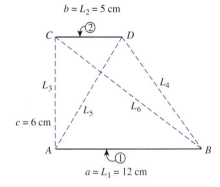

**FIGURE 21–45**

The two infinitely long parallel plates considered in Example 21–8.

(imaginary surfaces 3 and 4) form a four-surface enclosure. Then applying the summation rule to surface 1 yields

$$F_{11} + F_{12} + F_{13} + F_{14} = 1$$

But $F_{11} = 0$ since it is a flat surface. Therefore,

$$F_{12} = 1 - F_{13} - F_{14}$$

where the view factors $F_{13}$ and $F_{14}$ can be determined by considering the triangles *ABC* and *ABD*, respectively, and applying Eq. 21–31 for view factors between the sides of triangles. We obtain

$$F_{13} = \frac{L_1 + L_3 - L_6}{2L_1}, \qquad F_{14} = \frac{L_1 + L_4 - L_5}{2L_1}$$

Substituting,

$$F_{12} = 1 - \frac{L_1 + L_3 - L_6}{2L_1} - \frac{L_1 + L_4 - L_5}{2L_1}$$

$$= \frac{(L_5 + L_6) - (L_3 + L_4)}{2L_1}$$

which is the desired result. This is also a miniproof of the crossed-strings method for the case of two infinitely long plain parallel surfaces.

# 21–6 ▪ RADIATION HEAT TRANSFER: BLACK SURFACES

So far, we have considered the nature of radiation, the radiation properties of materials, and the view factors, and we are now in a position to consider the rate of heat transfer between surfaces by radiation. The analysis of radiation exchange between surfaces, in general, is complicated because of reflection: a radiation beam leaving a surface may be reflected several times, with partial reflection occurring at each surface, before it is completely absorbed. The analysis is simplified greatly when the surfaces involved can be approximated as blackbodies because of the absence of reflection. In this section, we consider radiation exchange between *black surfaces* only; we extend the analysis to reflecting surfaces in the next section.

Consider two black surfaces of arbitrary shape maintained at uniform temperatures $T_1$ and $T_2$, as shown in Fig. 21–46. Recognizing that radiation leaves a black surface at a rate of $E_b = \sigma T^4$ per unit surface area and that the view factor $F_{1 \rightarrow 2}$ represents the fraction of radiation leaving surface 1 that strikes surface 2, the *net* rate of radiation heat transfer from surface 1 to surface 2 can be expressed as

$$\dot{Q}_{1 \rightarrow 2} = \begin{pmatrix} \text{Radiation leaving} \\ \text{the entire surface 1} \\ \text{that strikes surface 2} \end{pmatrix} - \begin{pmatrix} \text{Radiation leaving} \\ \text{the entire surface 2} \\ \text{that strikes surface 1} \end{pmatrix}$$

$$= A_1 E_{b1} F_{1 \rightarrow 2} - A_2 E_{b2} F_{2 \rightarrow 1} \quad \text{(W)} \qquad \text{(21–34)}$$

Applying the reciprocity relation $A_1 F_{1 \rightarrow 2} = A_2 F_{2 \rightarrow 1}$ yields

$$\dot{Q}_{1 \rightarrow 2} = A_1 F_{1 \rightarrow 2} \, \sigma (T_1^4 - T_2^4) \quad \text{(W)} \qquad \text{(21–35)}$$

**FIGURE 21–46**
Two general black surfaces maintained at uniform temperatures $T_1$ and $T_2$.

which is the desired relation. A negative value for $\dot{Q}_{1 \to 2}$ indicates that net radiation heat transfer is from surface 2 to surface 1.

Now consider an *enclosure* consisting of *N black* surfaces maintained at specified temperatures. The *net* radiation heat transfer *from* any surface *i* of this enclosure is determined by adding up the net radiation heat transfers from surface *i* to each of the surfaces of the enclosure:

$$\dot{Q}_i = \sum_{j=1}^{N} \dot{Q}_{i \to j} = \sum_{j=1}^{N} A_i F_{i \to j} \sigma (T_i^4 - T_j^4) \qquad (\text{W}) \qquad \textbf{(21–36)}$$

Again a negative value for $\dot{Q}$ indicates that net radiation heat transfer is *to* surface *i* (i.e., surface *i gains* radiation energy instead of losing). Also, the net heat transfer from a surface to itself is zero, regardless of the shape of the surface.

---

### EXAMPLE 21–9    Radiation Heat Transfer in a Black Furnace

Consider the 5-m × 5-m × 5-m cubical furnace shown in Fig. 21–47, whose surfaces closely approximate black surfaces. The base, top, and side surfaces of the furnace are maintained at uniform temperatures of 800 K, 1500 K, and 500 K, respectively. Determine (*a*) the net rate of radiation heat transfer between the base and the side surfaces, (*b*) the net rate of radiation heat transfer between the base and the top surface, and (*c*) the net radiation heat transfer from the base surface.

**SOLUTION**   The surfaces of a cubical furnace are black and are maintained at uniform temperatures. The net rate of radiation heat transfer between the base and side surfaces, between the base and the top surface, and from the base surface are to be determined.

*Assumptions*   The surfaces are black and isothermal.

*Analysis*   (*a*) The geometry involves six surfaces, and thus we may be tempted at first to treat the furnace as a six-surface enclosure. However, the four side surfaces possess the same properties, and thus we can treat them as a single side surface in radiation analysis. We consider the base surface to be surface 1, the top surface to be surface 2, and the side surfaces to be surface 3. Then the problem reduces to determining $\dot{Q}_{1 \to 3}$, $\dot{Q}_{1 \to 2}$, and $\dot{Q}_1$.

The net rate of radiation heat transfer $\dot{Q}_{1 \to 3}$ from surface 1 to surface 3 can be determined from Eq. 21–35, since both surfaces involved are black, by replacing the subscript 2 by 3:

$$\dot{Q}_{1 \to 3} = A_1 F_{1 \to 3} \sigma (T_1^4 - T_3^4)$$

But first we need to evaluate the view factor $F_{1 \to 3}$. After checking the view factor charts and tables, we realize that we cannot determine this view factor directly. However, we can determine the view factor $F_{1 \to 2}$ from Fig. 21–33 to be $F_{1 \to 2} = 0.2$, and we know that $F_{1 \to 1} = 0$ since surface 1 is a plane. Then applying the summation rule to surface 1 yields

$$F_{1 \to 1} + F_{1 \to 2} + F_{1 \to 3} = 1$$

or

$$F_{1 \to 3} = 1 - F_{1 \to 1} - F_{1 \to 2} = 1 - 0 - 0.2 = 0.8$$

Substituting,

$$\dot{Q}_{1 \to 3} = (25 \text{ m}^2)(0.8)(5.67 \times 10^{-8} \text{ W/m}^2 \cdot \text{K}^4)[(800 \text{ K})^4 - (500 \text{ K})^4]$$
$$= \textbf{394 kW}$$

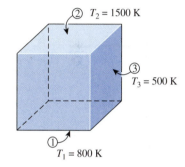

**FIGURE 21–47**

The cubical furnace of black surfaces considered in Example 21–9.

(b) The net rate of radiation heat transfer $\dot{Q}_{1 \to 2}$ from surface 1 to surface 2 is determined in a similar manner from Eq. 21–35 to be

$$\dot{Q}_{1 \to 2} = A_1 F_{1 \to 2} \sigma (T_1^4 - T_2^4)$$

$$= (25 \text{ m}^2)(0.2)(5.67 \times 10^{-8} \text{ W/m}^2 \cdot \text{K}^4)[(800 \text{ K})^4 - (1500 \text{ K})^4]$$

$$= -1319 \text{ kW}$$

The negative sign indicates that net radiation heat transfer is from surface 2 to surface 1.

(c) The net radiation heat transfer from the base surface $\dot{Q}_1$ is determined from Eq. 21–36 by replacing the subscript $i$ by 1 and taking $N = 3$:

$$\dot{Q}_1 = \sum_{j=1}^{3} \dot{Q}_{1 \to j} = \dot{Q}_{1 \to 1} + \dot{Q}_{1 \to 2} + \dot{Q}_{1 \to 3}$$

$$= 0 + (-1319 \text{ kW}) + (394 \text{ kW})$$

$$= -925 \text{ kW}$$

Again the negative sign indicates that net radiation heat transfer is *to* surface 1. That is, the base of the furnace is gaining net radiation at a rate of 925 kW.

## 21–7 · RADIATION HEAT TRANSFER: DIFFUSE, GRAY SURFACES

The analysis of radiation transfer in enclosures consisting of black surfaces is relatively easy, as we have seen, but most enclosures encountered in practice involve nonblack surfaces, which allow multiple reflections to occur. Radiation analysis of such enclosures becomes very complicated unless some simplifying assumptions are made.

To make a simple radiation analysis possible, it is common to assume the surfaces of an enclosure to be *opaque, diffuse,* and *gray.* That is, the surfaces are nontransparent, they are diffuse emitters and diffuse reflectors, and their radiation properties are independent of wavelength. Also, each surface of the enclosure is *isothermal,* and both the incoming and outgoing radiation are *uniform* over each surface. But first we introduce the concept of radiosity.

### Radiosity

Surfaces emit radiation as well as reflect it, and thus the radiation leaving a surface consists of emitted and reflected parts. The calculation of radiation heat transfer between surfaces involves the *total* radiation energy streaming away from a surface, with no regard for its origin. The *total radiation energy leaving a surface per unit time and per unit area* is the **radiosity** and is denoted by $J$ (Fig. 21–48).

For a surface $i$ that is *gray* and *opaque* ($\varepsilon_i = \alpha_i$ and $\alpha_i + \pi_i = 1$), the radiosity can be expressed as

$$J_i = \left( \begin{array}{c} \text{Radiation emitted} \\ \text{by surface } i \end{array} \right) + \left( \begin{array}{c} \text{Radiation reflected} \\ \text{by surface } i \end{array} \right)$$

$$= \varepsilon_i E_{bi} + \rho_i G_i$$

$$= \varepsilon_i E_{bi} + (1 - \varepsilon_i) G_i \qquad (\text{W/m}^2) \qquad (21\text{–}37)$$

**FIGURE 21–48**
Radiosity represents the sum of the radiation energy emitted and reflected by a surface.

where $E_{bi} = \sigma T_i^4$ is the blackbody emissive power of surface $i$ and $G_i$ is irradiation (i.e., the radiation energy incident on surface $i$ per unit time per unit area).

For a surface that can be approximated as a *blackbody* ($\varepsilon_i = 1$), the radiosity relation reduces to

$$J_i = E_{bi} = \sigma T_i^4 \qquad \text{(blackbody)} \tag{21-38}$$

That is, *the radiosity of a blackbody is equal to its emissive power.* This is expected, since a blackbody does not reflect any radiation, and thus radiation coming from a blackbody is due to emission only.

## Net Radiation Heat Transfer to or from a Surface

During a radiation interaction, a surface *loses* energy by emitting radiation and *gains* energy by absorbing radiation emitted by other surfaces. A surface experiences a net gain or a net loss of energy, depending on which quantity is larger. The *net* rate of radiation heat transfer from a surface $i$ of surface area $A_i$ is denoted by $\dot{Q}_i$ and is expressed as

$$\dot{Q}_i = \left( \begin{array}{c} \text{Radiation leaving} \\ \text{entire surface } i \end{array} \right) - \left( \begin{array}{c} \text{Radiation incident} \\ \text{on entire surface } i \end{array} \right)$$

$$= A_i(J_i - G_i) \qquad \text{(W)} \tag{21-39}$$

Solving for $G_i$ from Eq. 21–37 and substituting into Eq. 21–39 yields

$$\dot{Q}_i = A_i \left( J_i - \frac{J_i - \varepsilon_i E_{bi}}{1 - \varepsilon_i} \right) = \frac{A_i \varepsilon_i}{1 - \varepsilon_i} (E_{bi} - J_i) \qquad \text{(W)} \tag{21-40}$$

In an electrical analogy to Ohm's law, this equation can be rearranged as

$$\dot{Q}_i = \frac{E_{bi} - J_i}{R_i} \qquad \text{(W)} \tag{21-41}$$

where

$$R_i = \frac{1 - \varepsilon_i}{A_i \varepsilon_i} \tag{21-42}$$

is the **surface resistance** to radiation. The quantity $E_{bi} - J_i$ corresponds to a *potential difference* and the net rate of radiation heat transfer corresponds to *current* in the electrical analogy, as illustrated in Fig. 21–49.

The direction of the net radiation heat transfer depends on the relative magnitudes of $J_i$ (the radiosity) and $E_{bi}$ (the emissive power of a blackbody at the temperature of the surface). It is *from* the surface if $E_{bi} > J_i$ and *to* the surface if $J_i > E_{bi}$. A negative value for $\dot{Q}_i$ indicates that heat transfer is *to* the surface. All of this radiation energy gained must be removed from the other side of the surface through some mechanism if the surface temperature is to remain constant.

The surface resistance to radiation for a *blackbody* is *zero* since $\varepsilon_i = 1$ and $J_i = E_{bi}$. The net rate of radiation heat transfer in this case is determined directly from Eq. 21–39.

Some surfaces encountered in numerous practical heat transfer applications are modeled as being *adiabatic* since their back sides are well insulated and the net heat transfer through them is zero. When the convection effects on the front (heat transfer) side of such a surface is negligible and steady-state

**FIGURE 21–49**

Electrical analogy of surface resistance to radiation.

conditions are reached, the surface must lose as much radiation energy as it gains, and thus $\dot{Q}_i = 0$. In such cases, the surface is said to *reradiate* all the radiation energy it receives, and such a surface is called a **reradiating surface**. Setting $\dot{Q}_i = 0$ in Eq. 21–41 yields

$$J_i = E_{bi} = \sigma T_i^4 \qquad (\text{W/m}^2) \qquad \text{(21–43)}$$

Therefore, the *temperature* of a reradiating surface under steady conditions can easily be determined from the equation above once its radiosity is known. Note that the temperature of a reradiating surface is *independent of its emissivity*. In radiation analysis, the surface resistance of a reradiating surface is disregarded since there is no net heat transfer through it. (This is like the fact that there is no need to consider a resistance in an electrical network if no current is flowing through it.)

## Net Radiation Heat Transfer between Any Two Surfaces

Consider two diffuse, gray, and opaque surfaces of arbitrary shape maintained at uniform temperatures, as shown in Fig. 21–50. Recognizing that the radiosity $J$ represents the rate of radiation leaving a surface per unit surface area and that the view factor $F_{i \rightarrow j}$ represents the fraction of radiation leaving surface $i$ that strikes surface $j$, the *net* rate of radiation heat transfer from surface $i$ to surface $j$ can be expressed as

$$\dot{Q}_{i \rightarrow j} = \begin{pmatrix} \text{Radiation leaving} \\ \text{the entire surface } i \\ \text{that strikes surface } j \end{pmatrix} - \begin{pmatrix} \text{Radiation leaving} \\ \text{the entire surface } j \\ \text{that strikes surface } i \end{pmatrix} \qquad \text{(21–44)}$$

$$= A_i J_i F_{i \rightarrow j} - A_j J_j F_{j \rightarrow i} \qquad (\text{W})$$

Applying the reciprocity relation $A_i F_{i \rightarrow j} = A_j F_{j \rightarrow i}$ yields

$$\dot{Q}_{i \rightarrow j} = A_i F_{i \rightarrow j} (J_i - J_j) \qquad (\text{W}) \qquad \text{(21–45)}$$

Again in analogy to Ohm's law, this equation can be rearranged as

$$\dot{Q}_{i \rightarrow j} = \frac{J_i - J_j}{R_{i \rightarrow j}} \qquad (\text{W}) \qquad \text{(21–46)}$$

where

$$R_{i \rightarrow j} = \frac{1}{A_i F_{i \rightarrow j}} \qquad \text{(21–47)}$$

is the **space resistance** to radiation. Again the quantity $J_i - J_j$ corresponds to a *potential difference,* and the net rate of heat transfer between two surfaces corresponds to *current* in the electrical analogy, as illustrated in Fig. 21–50.

The direction of the net radiation heat transfer between two surfaces depends on the relative magnitudes of $J_i$ and $J_j$. A positive value for $\dot{Q}_{i \rightarrow j}$ indicates that net heat transfer is *from* surface $i$ *to* surface $j$. A negative value indicates the opposite.

In an $N$-surface enclosure, the conservation of energy principle requires that the net heat transfer from surface $i$ be equal to the sum of the net heat transfers from surface $i$ to each of the $N$ surfaces of the enclosure. That is,

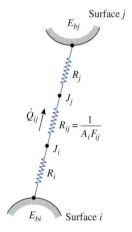

**FIGURE 21–50**
Electrical analogy of space resistance to radiation.

$$\dot{Q}_i = \sum_{j=1}^{N} \dot{Q}_{i \to j} = \sum_{j=1}^{N} A_i F_{i \to j} (J_i - J_j) = \sum_{j=1}^{N} \frac{J_i - J_j}{R_{i \to j}} \quad \text{(W)} \qquad \textbf{(21–48)}$$

The network representation of net radiation heat transfer from surface $i$ to the remaining surfaces of an $N$-surface enclosure is given in Fig. 21–51. Note that $\dot{Q}_{i \to i}$ (the net rate of heat transfer from a surface to itself) is zero regardless of the shape of the surface. Combining Eqs. 21–41 and 21–48 gives

$$\frac{E_{bi} - J_i}{R_i} = \sum_{j=1}^{N} \frac{J_i - J_j}{R_{i \to j}} \quad \text{(W)} \qquad \textbf{(21–49)}$$

which has the electrical analogy interpretation that *the net radiation flow from a surface through its surface resistance is equal to the sum of the radiation flows from that surface to all other surfaces through the corresponding space resistances.*

**FIGURE 21–51**
Network representation of net radiation heat transfer from surface $i$ to the remaining surfaces of an $N$-surface enclosure.

## Methods of Solving Radiation Problems

In the radiation analysis of an enclosure, either the temperature or the net rate of heat transfer must be given for each of the surfaces to obtain a unique solution for the unknown surface temperatures and heat transfer rates. There are two methods commonly used to solve radiation problems. In the first method, Eqs. 21–48 (for surfaces with specified heat transfer rates) and 21–49 (for surfaces with specified temperatures) are simplified and rearranged as

*Surfaces with specified net heat transfer rate $\dot{Q}$*
$$\dot{Q}_i = A_i \sum_{j=1}^{N} F_{i \to j}(J_i - J_j) \qquad \textbf{(21–50)}$$

*Surfaces with specified temperature $T_i$*
$$\sigma T_i^4 = J_i + \frac{1 - \varepsilon_i}{\varepsilon_i} \sum_{j=1}^{N} F_{i \to j}(J_i - J_j) \qquad \textbf{(21–51)}$$

Note that $\dot{Q}_i = 0$ for insulated (or reradiating) surfaces, and $\sigma T_i^4 = J_i$ for black surfaces since $\varepsilon_i = 1$ in that case. Also, the term corresponding to $j = i$ drops out from either relation since $J_i - J_j = J_i - J_i = 0$ in that case.

The equations above give $N$ linear algebraic equations for the determination of the $N$ unknown radiosities for an $N$-surface enclosure. Once the radiosities $J_1, J_2, \ldots, J_N$ are available, the unknown heat transfer rates can be determined from Eq. 21–50 while the unknown surface temperatures can be determined from Eq. 21–51. The temperatures of insulated or reradiating surfaces can be determined from $\sigma T_i^4 = J_i$. A positive value for $\dot{Q}_i$ indicates net radiation heat transfer *from* surface $i$ to other surfaces in the enclosure while a negative value indicates net radiation heat transfer *to* the surface.

The systematic approach described above for solving radiation heat transfer problems is very suitable for use with today's popular equation solvers such as EES, Mathcad, and Matlab, especially when there are a large number of surfaces, and is known as the **direct method** (formerly, the *matrix method,* since it resulted in matrices and the solution required a knowledge of linear algebra). The second method described below, called the **network method**, is based on the electrical network analogy.

The network method was first introduced by A. K. Oppenheim in the 1950s and found widespread acceptance because of its simplicity and emphasis on the physics of the problem. The application of the method is straightforward: draw a surface resistance associated with each surface of an enclosure and connect them with space resistances. Then solve the radiation problem by treating it as an electrical network problem where the radiation heat transfer replaces the current and radiosity replaces the potential.

The network method is not practical for enclosures with more than three or four surfaces, however, because of the increased complexity of the network. Next we apply the method to solve radiation problems in two- and three-surface enclosures.

## Radiation Heat Transfer in Two-Surface Enclosures

Consider an enclosure consisting of two opaque surfaces at specified temperatures $T_1$ and $T_2$, as shown in Fig. 21–52, and try to determine the net rate of radiation heat transfer between the two surfaces with the network method. Surfaces 1 and 2 have emissivities $\varepsilon_1$ and $\varepsilon_2$ and surface areas $A_1$ and $A_2$ and are maintained at uniform temperatures $T_1$ and $T_2$, respectively. There are only two surfaces in the enclosure, and thus we can write

$$\dot{Q}_{12} = \dot{Q}_1 = -\dot{Q}_2$$

That is, the net rate of radiation heat transfer from surface 1 to surface 2 must equal the net rate of radiation heat transfer *from* surface 1 and the net rate of radiation heat transfer *to* surface 2.

The radiation network of this two-surface enclosure consists of two surface resistances and one space resistance, as shown in Fig. 21–52. In an electrical network, the electric current flowing through these resistances connected in series would be determined by dividing the potential difference between points $A$ and $B$ by the total resistance between the same two points. The net rate of radiation transfer is determined in the same manner and is expressed as

$$\dot{Q}_{12} = \frac{E_{b1} - E_{b2}}{R_1 + R_{12} + R_2} = \dot{Q}_1 = -\dot{Q}_2$$

or

$$\dot{Q}_{12} = \frac{\sigma(T_1^4 - T_2^4)}{\dfrac{1 - \varepsilon_1}{A_1\varepsilon_1} + \dfrac{1}{A_1 F_{12}} + \dfrac{1 - \varepsilon_2}{A_2\varepsilon_2}} \quad \text{(W)} \qquad \text{(21-52)}$$

This important result is applicable to any two gray, diffuse, and opaque surfaces that form an enclosure. The view factor $F_{12}$ depends on the geometry and must be determined first. Simplified forms of Eq. 21–52 for some familiar arrangements that form a two-surface enclosure are given in Table 21–5. Note that $F_{12} = 1$ for all of these special cases.

$$R_1 = \frac{1 - \varepsilon_1}{A_1\varepsilon_1} \qquad R_{12} = \frac{1}{A_1 F_{12}} \qquad R_2 = \frac{1 - \varepsilon_2}{A_2\varepsilon_2}$$

**FIGURE 21–52**
Schematic of a two-surface enclosure and the radiation network associated with it.

**FIGURE 21–53**
The two parallel plates considered in Example 21–10.

### EXAMPLE 21–10  Radiation Heat Transfer between Parallel Plates

Two very large parallel plates are maintained at uniform temperatures $T_1 =$ 800 K and $T_2 = 500$ K and have emissivities $\varepsilon_1 = 0.2$ and $\varepsilon_2 = 0.7$, respectively, as shown in Fig. 21–53. Determine the net rate of radiation heat transfer between the two surfaces per unit surface area of the plates.

**SOLUTION**  Two large parallel plates are maintained at uniform temperatures. The net rate of radiation heat transfer between the plates is to be determined.

**TABLE 21–5**

Radiation heat transfer relations for some familiar two-surface arrangments.

| | | | |
|---|---|---|---|
| Small object in a large cavity <br> 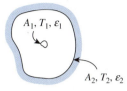 | $\dfrac{A_1}{A_2} \approx 0$ <br><br> $F_{12} = 1$ | $\dot{Q}_{12} = A_1 \sigma \varepsilon_1 (T_1^4 - T_2^4)$ | **(21–53)** |
| Infinitely large parallel plates <br><br> $\overline{\quad A_1, T_1, \varepsilon_1 \quad}$ <br><br> $\overline{\quad A_2, T_2, \varepsilon_2 \quad}$ | $A_1 = A_2 = A$ <br> $F_{12} = 1$ | $\dot{Q}_{12} = \dfrac{A \sigma (T_1^4 - T_2^4)}{\dfrac{1}{\varepsilon_1} + \dfrac{1}{\varepsilon_2} - 1}$ | **(21–54)** |
| Infinitely long concentric cylinders <br>  | $\dfrac{A_1}{A_2} = \dfrac{r_1}{r_2}$ <br><br> $F_{12} = 1$ | $\dot{Q}_{12} = \dfrac{A_1 \sigma (T_1^4 - T_2^4)}{\dfrac{1}{\varepsilon_1} + \dfrac{1 - \varepsilon_2}{\varepsilon_2}\left(\dfrac{r_1}{r_2}\right)}$ | **(21–55)** |
| Concentric spheres <br>  | $\dfrac{A_1}{A_2} = \left(\dfrac{r_1}{r_2}\right)^2$ <br><br> $F_{12} = 1$ | $\dot{Q}_{12} = \dfrac{A_1 \sigma (T_1^4 - T_2^4)}{\dfrac{1}{\varepsilon_2} + \dfrac{1 - \varepsilon_2}{\varepsilon_2}\left(\dfrac{r_1}{r_2}\right)^2}$ | **(21–56)** |

*Assumptions*  Both surfaces are opaque, diffuse, and gray.

*Analysis*  The net rate of radiation heat transfer between the two plates per unit area is readily determined from Eq. 21–54 to be

$$\dot{q}_{12} = \frac{\dot{Q}_{12}}{A} = \frac{\sigma(T_1^4 - T_2^4)}{\dfrac{1}{\varepsilon_1} + \dfrac{1}{\varepsilon_2} - 1} = \frac{(5.67 \times 10^{-8}\ \text{W/m}^2\cdot\text{K}^4)\left[(800\ \text{K})^4 - (500\ \text{K})^4\right]}{\dfrac{1}{0.2} + \dfrac{1}{0.7} - 1}$$

$$= \mathbf{3625\ W/m^2}$$

*Discussion*  Note that heat at a net rate of 3625 W is transferred from plate 1 to plate 2 by radiation per unit surface area of either plate.

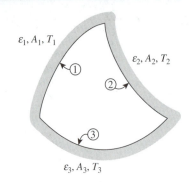

$\varepsilon_1, A_1, T_1$

$\varepsilon_2, A_2, T_2$

$\varepsilon_3, A_3, T_3$

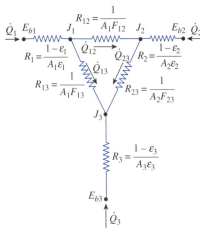

## FIGURE 21–54

Schematic of a three-surface enclosure and the radiation network associated with it.

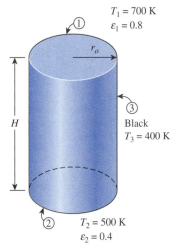

$T_1 = 700$ K
$\varepsilon_1 = 0.8$

$r_o$

$H$

Black
$T_3 = 400$ K

$T_2 = 500$ K
$\varepsilon_2 = 0.4$

## FIGURE 21–55

The cylindrical furnace considered in Example 21–11.

# Radiation Heat Transfer in Three-Surface Enclosures

We now consider an enclosure consisting of three opaque, diffuse, and gray surfaces, as shown in Fig. 21–54. Surfaces 1, 2, and 3 have surface areas $A_1$, $A_2$, and $A_3$; emissivities $\varepsilon_1$, $\varepsilon_2$, and $\varepsilon_3$; and uniform temperatures $T_1$, $T_2$, and $T_3$, respectively. The radiation network of this geometry is constructed by following the standard procedure: draw a surface resistance associated with each of the three surfaces and connect these surface resistances with space resistances, as shown in the figure. Relations for the surface and space resistances are given by Eqs. 21–42 and 21–47. The three endpoint potentials $E_{b1}$, $E_{b2}$, and $E_{b3}$ are considered known, since the surface temperatures are specified. Then all we need to find are the radiosities $J_1$, $J_2$, and $J_3$. The three equations for the determination of these three unknowns are obtained from the requirement that *the algebraic sum of the currents (net radiation heat transfer) at each node must equal zero.* That is,

$$\frac{E_{b1} - J_1}{R_1} + \frac{J_2 - J_1}{R_{12}} + \frac{J_3 - J_1}{R_{13}} = 0$$

$$\frac{J_1 - J_2}{R_{12}} + \frac{E_{b2} - J_2}{R_2} + \frac{J_3 - J_2}{R_{23}} = 0 \qquad (21\text{–}57)$$

$$\frac{J_1 - J_3}{R_{13}} + \frac{J_2 - J_3}{R_{23}} + \frac{E_{b3} - J_3}{R_3} = 0$$

Once the radiosities $J_1$, $J_2$, and $J_3$ are available, the net rate of radiation heat transfers at each surface can be determined from Eq. 21–58.

The set of equations above simplify further if one or more surfaces are "special" in some way. For example, $J_i = E_{bi} = \sigma T_i^4$ for a *black* or *reradiating* surface. Also, $\dot{Q}_i = 0$ for a reradiating surface. Finally, when the net rate of radiation heat transfer $\dot{Q}_i$ is specified at surface $i$ instead of the temperature, the term $(E_{bi} - J_i)/R_i$ should be replaced by the specified $\dot{Q}_i$.

---

**EXAMPLE 21–11**    **Radiation Heat Transfer in a Cylindrical Furnace**

Consider a cylindrical furnace with $r_o = H = 1$ m, as shown in Fig. 21–55. The top (surface 1) and the base (surface 2) of the furnace have emissivities $\varepsilon_1 = 0.8$ and $\varepsilon_2 = 0.4$, respectively, and are maintained at uniform temperatures $T_1 = 700$ K and $T_2 = 500$ K. The side surface closely approximates a blackbody and is maintained at a temperature of $T_3 = 400$ K. Determine the net rate of radiation heat transfer at each surface during steady operation and explain how these surfaces can be maintained at specified temperatures.

**SOLUTION**    The surfaces of a cylindrical furnace are maintained at uniform temperatures. The net rate of radiation heat transfer at each surface during steady operation is to be determined.

*Assumptions*    1 Steady operating conditions exist. 2 The surfaces are opaque, diffuse, and gray. 3 Convection heat transfer is not considered.

*Analysis*  We will solve this problem systematically using the direct method to demonstrate its use. The cylindrical furnace can be considered to be a three-surface enclosure with surface areas of

$$A_1 = A_2 = \pi r_o^2 = \pi(1 \text{ m})^2 = 3.14 \text{ m}^2$$
$$A_3 = 2\pi r_o H = 2\pi(1 \text{ m})(1 \text{ m}) = 6.28 \text{ m}^2$$

The view factor from the base to the top surface is, from Fig. 21–35, $F_{12} = 0.38$. Then the view factor from the base to the side surface is determined by applying the summation rule to be

$$F_{11} + F_{12} + F_{13} = 1 \rightarrow F_{13} = 1 - F_{11} - F_{12} = 1 - 0 - 0.38 = 0.62$$

since the base surface is flat and thus $F_{11} = 0$. Noting that the top and bottom surfaces are symmetric about the side surface, $F_{21} = F_{12} = 0.38$ and $F_{23} = F_{13} = 0.62$. The view factor $F_{31}$ is determined from the reciprocity relation,

$$A_1 F_{13} = A_3 F_{31} \rightarrow F_{31} = F_{13}(A_1/A_3) = (0.62)(0.314/0.628) = 0.31$$

Also, $F_{32} = F_{31} = 0.31$ because of symmetry. Now that all the view factors are available, we apply Eq. 21–51 to each surface to determine the radiosities:

*Top surface (i = 1):*  $\sigma T_1^4 = J_1 + \dfrac{1 - \varepsilon_1}{\varepsilon_1}[F_{12}(J_1 - J_2) + F_{13}(J_1 - J_3)]$

*Bottom surface (i = 2):* $\sigma T_2^4 = J_2 + \dfrac{1 - \varepsilon_2}{\varepsilon_2}[F_{21}(J_2 - J_1) + F_{23}(J_2 - J_3)]$

*Side surface (i = 3):*  $\sigma T_3^4 = J_3 + 0$ (since surface 3 is black and thus $\varepsilon_3 = 1$)

Substituting the known quantities,

$$(5.67 \times 10^{-8} \text{ W/m}^2\cdot\text{K}^4)(700 \text{ K})^4 = J_1 + \frac{1 - 0.8}{0.8}[0.38(J_1 - J_2) + 0.62(J_1 - J_3)]$$

$$(5.67 \times 10^{-8} \text{ W/m}^2\cdot\text{K}^4)(500 \text{ K})^4 = J_2 + \frac{1 - 0.4}{0.4}[0.38(J_2 - J_1) + 0.62(J_2 - J_3)]$$

$$(5.67 \times 10^{-8} \text{ W/m}^2\cdot\text{K}^4)(400 \text{ K})^4 = J_3$$

Solving these equations for $J_1$, $J_2$, and $J_3$ gives

$$J_1 = 11{,}418 \text{ W/m}^2, J_2 = 4562 \text{ W/m}^2, \quad \text{and} \quad J_3 = 1452 \text{ W/m}^2$$

Then the net rates of radiation heat transfer at the three surfaces are determined from Eq. 21–50 to be

$$\dot{Q}_1 = A_1[F_{1\rightarrow2}(J_1 - J_2) + F_{1\rightarrow3}(J_1 - J_3)]$$
$$= (3.14 \text{ m}^2)[0.38(11{,}418 - 4562) + 0.62(11{,}418 - 1452)] \text{ W/m}^2$$
$$= \textbf{27.6 kW}$$

$$\dot{Q}_2 = A_2[F_{2\rightarrow1}(J_2 - J_1) + F_{2\rightarrow3}(J_2 - J_3)]$$
$$= (3.14 \text{ m}^2)[0.38(4562 - 11{,}418) + 0.62(4562 - 1452)] \text{ W/m}^2$$
$$= \textbf{-2.13 kW}$$

$$\dot{Q}_3 = A_3[F_{3 \to 1}(J_3 - J_1) + F_{3 \to 2}(J_3 - J_2)]$$
$$= (6.28 \text{ m}^2)[0.31(1452 - 11{,}418) + 0.31(1452 - 4562)] \text{ W/m}^2$$
$$= -25.5 \text{ kW}$$

Note that the direction of net radiation heat transfer is *from* the top surface *to* the base and side surfaces, and the algebraic sum of these three quantities must be equal to zero. That is,

$$\dot{Q}_1 + \dot{Q}_2 + \dot{Q}_3 = 27.6 + (-2.13) + (-25.5) \cong 0$$

**Discussion**  To maintain the surfaces at the specified temperatures, we must supply heat to the top surface continuously at a rate of 27.6 kW while removing 2.13 kW from the base and 25.5 kW from the side surfaces.

The direct method presented here is straightforward, and it does not require the evaluation of radiation resistances. Also, it can be applied to enclosures with any number of surfaces in the same manner.

## EXAMPLE 21–12  Radiation Heat Transfer in a Triangular Furnace

A furnace is shaped like a long equilateral triangular duct, as shown in Fig. 21–56. The width of each side is 1 m. The base surface has an emissivity of 0.7 and is maintained at a uniform temperature of 600 K. The heated left-side surface closely approximates a blackbody at 1000 K. The right-side surface is well insulated. Determine the rate at which heat must be supplied to the heated side externally per unit length of the duct in order to maintain these operating conditions.

**SOLUTION**  Two of the surfaces of a long equilateral triangular furnace are maintained at uniform temperatures while the third surface is insulated. The external rate of heat transfer to the heated side per unit length of the duct during steady operation is to be determined.

**Assumptions**  **1** Steady operating conditions exist. **2** The surfaces are opaque, diffuse, and gray. **3** Convection heat transfer is not considered.

**Analysis**  The furnace can be considered to be a three-surface enclosure with a radiation network as shown in the figure, since the duct is very long and thus the end effects are negligible. We observe that the view factor from any surface to any other surface in the enclosure is 0.5 because of symmetry. Surface 3 is a reradiating surface since the net rate of heat transfer at that surface is zero. Then we must have $\dot{Q}_1 = -\dot{Q}_2$, since the entire heat lost by surface 1 must be gained by surface 2. The radiation network in this case is a simple series–parallel connection, and we can determine $\dot{Q}_1$ directly from

$$\dot{Q}_1 = \frac{E_{b1} - E_{b2}}{R_1 + \left(\dfrac{1}{R_{12}} + \dfrac{1}{R_{13} + R_{23}}\right)^{-1}} = \frac{E_{b1} - E_{b2}}{\dfrac{1 - \varepsilon_1}{A_1 \varepsilon_1} + \left(A_1 A_{12} + \dfrac{1}{1/F_1 A_{13} + 1/A_2 F_{23}}\right)^{-1}}$$

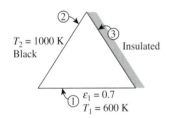

$T_2 = 1000$ K
Black

Insulated

$\varepsilon_1 = 0.7$
$T_1 = 600$ K

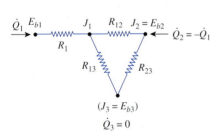

$\dot{Q}_1 \, E_{b1}$  $J_1$  $R_{12}$  $J_2 = E_{b2}$  $\dot{Q}_2 = -\dot{Q}_1$

$R_1$

$R_{13}$  $R_{23}$

$(J_3 = E_{b3})$

$\dot{Q}_3 = 0$

**FIGURE 21–56**
The triangular furnace considered in Example 21–12.

where

$$A_1 = A_2 = A_3 = wL = 1 \text{ m} \times 1 \text{ m} = 1 \text{ m}^2 \qquad \text{(per unit length of the duct)}$$

$$F_{12} = F_{13} = F_{23} = 0.5 \qquad \text{(symmetry)}$$

$$E_{b1} = \sigma T_1^4 = (5.67 \times 10^{-8} \text{ W/m}^2\text{·K}^4)(600 \text{ K})^4 = 7348 \text{ W/m}^2$$

$$E_{b2} = \sigma T_2^4 = (5.67 \times 10^{-8} \text{ W/m}^2\text{·K}^4)(1000 \text{ K})^4 = 56{,}700 \text{ W/m}^2$$

Substituting,

$$\dot{Q}_1 = \frac{(56{,}700 - 7348) \text{ W/m}^2}{\dfrac{1 - 0.7}{0.7 \times 1 \text{ m}^2} + \left[ (0.5 \times 1 \text{ m}^2) + \dfrac{1}{1/(0.5 \times 1 \text{ m}^2) + 1/(0.5 \times 1 \text{ m}^2)} \right]^{-1}}$$

$$= 28.0 \text{ kW}$$

Therefore, heat at a rate of 28 kW must be supplied to the heated surface per unit length of the duct to maintain steady operation in the furnace.

## SUMMARY

Radiation propagates in the form of electromagnetic waves. The *frequency $v$* and *wavelength $\lambda$* of electromagnetic waves in a medium are related by $\lambda = c/v$, where $c$ is the speed of propagation in that medium. All matter continuously emits *thermal radiation* as a result of vibrational and rotational motions of molecules, atoms, and electrons of a substance.

A *blackbody* is defined as a perfect emitter and absorber of radiation. At a specified temperature and wavelength, no surface can emit more energy than a blackbody. A blackbody absorbs *all* incident radiation, regardless of wavelength and direction. The radiation energy emitted by a blackbody per unit time and per unit surface area is called the *blackbody emissive power $E_b$* and is expressed by the *Stefan–Boltzmann law* as

$$E_b(T) = \sigma T^4$$

where $\sigma = 5.670 \times 10^{-8} \text{ W/m}^2 \cdot \text{K}^4$ is the *Stefan–Boltzmann constant* and $T$ is the absolute temperature of the surface in K. At any specified temperature, the spectral blackbody emissive power $E_{b\lambda}$ increases with wavelength, reaches a peak, and then decreases with increasing wavelength. The wavelength at which the peak occurs for a specified temperature is given by *Wien's displacement law* as

$$(\lambda T)_{\text{max power}} = 2897.8 \text{ μm·K}$$

The *blackbody radiation function $f_\lambda$* represents the fraction of radiation emitted by a blackbody at temperature $T$ in the wavelength band from $\lambda = 0$ to $\lambda$. The fraction of radiation energy emitted by a blackbody at temperature $T$ over a finite wavelength band from $\lambda = \lambda_1$ to $\lambda = \lambda_2$ is determined from

$$f_{\lambda_1 - \lambda_2}(T) = f_{\lambda_2}(T) - f_{\lambda_1}(T)$$

where $f_{\lambda_1}(T)$ and $f_{\lambda_2}(T)$ are the blackbody radiation functions corresponding to $\lambda_1 T$ and $\lambda_2 T$, respectively.

The *emissivity* of a surface represents the ratio of the radiation emitted by the surface at a given temperature to the radiation emitted by a blackbody at the same temperature. Different emissivities are defined as

*Spectral hemispherical emissivity:*

$$\varepsilon_\lambda(\lambda, T) = \frac{E_\lambda(\lambda, T)}{E_{b\lambda}(\lambda, T)}$$

*Total hemispherical emissivity:*

$$\varepsilon(T) = \frac{E(T)}{E_b(T)} = \frac{\displaystyle\int_0^\infty E_\lambda(\lambda, T) E_{b\lambda}(\lambda, T)\, d\lambda}{\sigma T^4}$$

Emissivity can also be expressed as a step function by dividing the spectrum into a sufficient number of *wavelength bands* of constant emissivity as, for example,

$$\varepsilon(T) = \varepsilon_1 f_{0-\lambda_1}(T) + \varepsilon_2 f_{\lambda_1 - \lambda_2}(T) + \varepsilon_3 f_{\lambda_2 - \infty}(T)$$

The *total hemispherical emissivity* $\varepsilon$ of a surface is the average emissivity over all directions and wavelengths.

When radiation strikes a surface, part of it is absorbed, part of it is reflected, and the remaining part, if any, is transmitted. The fraction of incident radiation (irradiation $G$) absorbed by the surface is called the *absorptivity*, the fraction reflected by the surface is called the *reflectivity*, and the fraction transmitted is called the *transmissivity*. Various absorptivities, reflectivities, and transmissivities for a medium are expressed as

$$\alpha_\lambda(\lambda) = \frac{G_{\lambda,\,abs}(\lambda)}{G_\lambda(\lambda)}, \quad \rho_\lambda(\lambda) = \frac{G_{\lambda,\,ref}(\lambda)}{G_\lambda(\lambda)}, \quad \text{and} \quad \tau_\lambda(\lambda) = \frac{G_{\lambda,\,tr}(\lambda)}{G_\lambda(\lambda)}$$

$$\alpha = \frac{G_{abs}}{G}, \quad \rho = \frac{G_{ref}}{G}, \quad \text{and} \quad \tau = \frac{G_{tr}}{G}$$

The consideration of wavelength and direction dependence of properties makes radiation calculations very complicated. Therefore, the *gray* and *diffuse* approximations are commonly utilized in radiation calculations. A surface is said to be *diffuse* if its properties are independent of direction and *gray* if its properties are independent of wavelength.

The sum of the absorbed, reflected, and transmitted fractions of radiation energy must be equal to unity,

$$\alpha + \rho + \tau = 1$$

For *opaque* surfaces, $\tau = 0$, and thus

$$\alpha + \rho = 1$$

Surfaces are usually assumed to reflect in a perfectly *specular* or *diffuse* manner for simplicity. In *specular* (or *mirrorlike*) reflection, the angle of reflection equals the angle of incidence of the radiation beam. In *diffuse reflection,* radiation is reflected equally in all directions. Reflection from smooth and polished surfaces approximates specular reflection, whereas reflection from rough surfaces approximates diffuse reflection. *Kirchhoff's law* of radiation is expressed as

$$\varepsilon_{\lambda,\,\theta}(T) = \alpha_{\lambda,\,\theta}(T), \quad \varepsilon_\lambda(T) = \alpha_\lambda(T), \quad \text{and} \quad \varepsilon(T) = \alpha(T)$$

Radiaton heat transfer between surfaces depends on the orientation of the surfaces relative to each other. In a radiation analysis, this effect is accounted for by the geometric parameter *view factor*. The *view factor* from a surface $i$ to a surface $j$ is denoted by $F_{i \to j}$ or $F_{ij}$, and is defined as the fraction of the radiation leaving surface $i$ that strikes surface $j$ directly. The view factor $F_{i \to i}$ represents the fraction of the radiation leaving a surface $i$ that strikes itself directly; $F_{i \to i} = 0$ for *plane or convex surfaces* and $F_{i \to i} \neq 0$ for *concave surfaces.*

For view factors, the *reciprocity rule* is expressed as

$$A_i F_{i \to j} = A_j F_{j \to i}$$

The sum of the view factors from surface $i$ of an enclosure to all surfaces of the enclosure, including to itself, must equal unity. This is known as the *summation rule* for an enclosure. The *superposition rule* is expressed as the view factor from a surface $i$ to a surface $j$ is equal to the sum of the view factors from surface $i$ to the parts of surface $j$. The symmetry rule is expressed as if the surfaces $j$ and $k$ are symmetric about the surface $i$ then $F_{i \to j} = F_{i \to k}$.

The rate of net radiation heat transfer between two *black* surfaces is determined from

$$\dot{Q}_{1 \to 2} = A_1 F_{1 \to 2} \sigma(T_1^4 - T_2^4)$$

The *net* radiation heat transfer from any surface $i$ of a *black* enclosure is determined by adding up the net radiation heat transfers from surface $i$ to each of the surfaces of the enclosure:

$$\dot{Q}_i = \sum_{j=1}^{N} \dot{Q}_{i \to j} = \sum_{j=1}^{N} A_i F_{i \to j} \sigma(T_i^4 - T_j^4)$$

The total radiation energy leaving a surface per unit time and per unit area is called the *radiosity* and is denoted by $J$. The *net* rate of radiation heat transfer from a surface $i$ of surface area $A_i$ is expressed as

$$\dot{Q}_i = \frac{E_{bi} - J_i}{R_i} \quad \text{where} \quad R_i = \frac{1 - \varepsilon_i}{A_i \varepsilon_i}$$

is the *surface resistance* to radiation. The *net* rate of radiation heat transfer from surface $i$ to surface $j$ can be expressed as

$$\dot{Q}_{i \to j} = \frac{J_i - J_j}{R_{i \to j}} \quad \text{where} \quad R_{i \to j} = \frac{1}{A_i F_{i \to j}}$$

is the *space resistance* to radiation. The *network method* is applied to radiation enclosure problems by drawing a surface resistance associated with each surface of an enclosure and connecting them with space resistances. Then the problem is solved by treating it as an electrical network problem where the radiation heat transfer replaces the current and the radiosity replaces the potential. The *direct method* is based on the following two equations:

*Surfaces with specified, net heat transfer rate $\dot{Q}_i$*

$$\dot{Q}_i = A_i \sum_{j=1}^{N} F_{i \to j}(J_i - J_j)$$

*Surfaces with specified temperature $T_i$*

$$\sigma T_i^4 = J_i + \frac{1 - \varepsilon_i}{\varepsilon_i} \sum_{j=1}^{N} F_{i \to j}(J_i - J_j)$$

The first and the second groups of equations give $N$ linear algebraic equations for the determination of the $N$ unknown radiosities for an $N$-surface enclosure. Once the radiosities $J_1, J_2, \ldots, J_N$ are available, the unknown surface temperatures and heat transfer rates can be determined from the equations just shown.

The net rate of radiation transfer between any two gray, diffuse, opaque surfaces that form an enclosure is given by

$$\dot{Q}_{12} = \frac{\sigma(T_1^4 - T_2^4)}{\dfrac{1 - \varepsilon_1}{A_1 \varepsilon_1} + \dfrac{1}{A_1 F_{12}} + \dfrac{1 - \varepsilon_2}{A_2 \varepsilon_2}}$$

## REFERENCES AND SUGGESTED READINGS

1. American Society of Heating, Refrigeration, and Air Conditioning Engineers, *Handbook of Fundamentals*, Atlanta, ASHRAE, 1993.

2. A. G. H. Dietz. "Diathermanous Materials and Properties of Surfaces." In *Space Heating with Solar Energy,* ed. R. W. Hamilton. Cambridge, MA: MIT Press, 1954.

3. D. C. Hamilton and W. R. Morgan. "Radiation Interchange Configuration Factors." National Advisory Committee for Aeronautics, Technical Note 2836, 1952.

4. H. C. Hottel. "Radiant Heat Transmission." In *Heat Transmission*. 3rd ed., ed. W. H. McAdams. New York: McGraw-Hill, 1954.

5. J. R. Howell. *A Catalog of Radiation Configuration Factors.* New York: McGraw-Hill, 1982.

6. M. F. Modest. *Radiative Heat Transfer.* New York: McGraw-Hill, 1993.

7. A. K. Oppenheim. "Radiation Analysis by the Network Method." *Transactions of the ASME* 78 (1956), pp. 725–735.

8. M. Planck. *The Theory of Heat Radiation.* New York: Dover, 1959.

9. W. Sieber. *Zeitschrift für Technische Physics* 22 (1941), pp. 130–135.

10. R. Siegel and J. R. Howell. *Thermal Radiation Heat Transfer.* 3rd ed. Washington, DC: Hemisphere, 1992.

11. Y. S. Touloukian and D. P. DeWitt. "Nonmetallic Solids." In *Thermal Radiative Properties*. Vol. 8. New York: IFI/Plenum, 1970.

12. Y. S. Touloukian and D. P. DeWitt. "Metallic Elements and Alloys." In *Thermal Radiative Properties,* Vol. 7. New York: IFI/Plenum, 1970.

## PROBLEMS*

### Electromagnetic and Thermal Radiation

**21–1C**  By what properties is an electromagnetic wave characterized? How are these properties related to each other?

**21–2C**  What is visible light? How does it differ from the other forms of electromagnetic radiation?

**21–3C**  How does microwave cooking differ from conventional cooking?

**21–4C**  What is thermal radiation? How does it differ from the other forms of electromagnetic radiation?

*Problems designated by a "C" are concept questions, and students are encouraged to answer them all. Problems designated by an "E" are in English units, and the SI users can ignore them. Problems with the icon 🔧 are solved using EES, and complete solutions together with parametric studies are included on the enclosed DVD. Problems with the icon 🖥 are comprehensive in nature, and are intended to be solved with a computer, preferably using the EES software that accompanies this text.

**21–5C**  What is the cause of color? Why do some objects appear blue to the eye while others appear red? Is the color of a surface at room temperature related to the radiation it emits?

**21–6C**  Why is radiation usually treated as a surface phenomenon?

**21–7C**  How do ultraviolet and infrared radiation differ? Do you think your body emits any radiation in the ultraviolet range? Explain.

**21–8**  A cordless telephone is designed to operate at a frequency of $8.5 \times 10^8$ Hz. Determine the wavelength of these telephone waves.

**21–9**  Electricity is generated and transmitted in power lines at a frequency of 50 Hz (1 Hz = 1 cycle per second). Determine the wavelength of the electromagnetic waves generated by the passage of electricity in power lines.

**21–10**  A microwave oven is designed to operate at a frequency of $2.2 \times 10^9$ Hz. Determine the wavelength of these microwaves and the energy of each microwave.

**21–11**  A radio station is broadcasting radio waves at a wavelength of 150 m. Determine the frequency of these waves.
*Answer:* $2 \times 10^6$ Hz

**21–12**  The speed of light in vacuum is given to be $3.0 \times 10^8$ m/s. Determine the speed of light in air ($n = 1$), in water ($n = 1.33$), and in glass ($n = 1.5$).

## Blackbody Radiation

**21–13C**  What is a blackbody? Does a blackbody actually exist?

**21–14C**  Why did we define the blackbody radiation function? What does it represent? For what is it used?

**21–15C**  Define the total and spectral blackbody emissive powers. How are they related to each other? How do they differ?

**21–16**  The temperature of the filament of an incandescent lightbulb is 2500 K. Assuming the filament to be a blackbody, determine the fraction of the radiant energy emitted by the filament that falls in the visible range. Also, determine the wavelength at which the emission of radiation from the filament peaks.

**21–17E**  Consider a black surface at 2060°F. Determine the radiation energy emitted by the black surface per unit area (in Btu/h·ft²) for $\lambda \geq 4$ $\mu$m.

**21–18**  Consider a surface at a uniform temperature of 1000 K. Determine the maximum rate of thermal radiation that can be emitted by this surface, in W/m².

**21–19**  An incandescent lightbulb is desired to emit at least 15 percent of its energy at wavelengths shorter than 0.8 $\mu$m. Determine the minimum temperature to which the filament of the lightbulb must be heated.

**21–20E**  The sun can be treated as a blackbody at an effective surface temperature of 10,400 R. Determine the rate at which infrared radiation energy ($\lambda = 0.76 - 100$ $\mu$m) is emitted by the sun, in Btu/h·ft².

**21–21**  [EES] The sun can be treated as a blackbody at 5780 K. Using EES (or other) software, calculate and plot the spectral blackbody emissive power $E_{b\lambda}$ of the sun versus wavelength in the range of 0.01 $\mu$m to 1000 $\mu$m. Discuss the results.

**21–22**  The temperature of the filament of an incandescent lightbulb is 2800 K. Treating the filament as a blackbody, determine the fraction of the radiant energy emitted by the filament that falls in the visible range. Also, determine the wavelength at which the emission of radiation from the filament peaks.

**21–23**  [EES] Reconsider Prob. 21–22. Using EES (or other) software, investigate the effect of temperature on the fraction of radiation emitted in the visible range. Let the surface temperature vary from 1000 K to 4000 K, and plot fraction of radiation emitted in the visible range versus the surface temperature.

**21–24**  It is desired that the radiation energy emitted by a light source reach a maximum in the blue range ($\lambda = 0.47$ $\mu$m).

Determine the temperature of this light source and the fraction of radiation it emits in the visible range ($\lambda = 0.40$–0.76 $\mu$m).

**21–25**  A 3-mm-thick glass window transmits 90 percent of the radiation between $\lambda = 0.3$ and 3.0 $\mu$m and is essentially opaque for radiation at other wavelengths. Determine the rate of radiation transmitted through a 3-m $\times$ 3-m glass window from blackbody sources at (*a*) 5800 K and (*b*) 1000 K.
*Answers:* (*a*) 491,300 kW, (*b*) 125.5 kW

**21–26**  Daylight and candlelight may be approximated as a blackbody at the effective surface temperatures of 5800 K and 1800 K, respectively. Determine the radiation energy (in W/m²) emitted by both lighting sources (daylight and candlelight) within the visible light region (0.40 to 0.76 $\mu$m).
*Answers:* $2.73 \times 10^7$ W/m², 4090 W/m²

**21–27**  A flame from a match may be approximated as a blackbody at the effective surface temperature of 1700 K, while moonlight may be approximated as a blackbody at the effective surface temperature of 4000 K, respectively. Determine the peak spectral blackbody emissive power for both lighting sources (match flame and moonlight).

## Radiation Properties

**21–28C**  What is a graybody? How does it differ from a blackbody? What is a diffuse gray surface?

**21–29C**  Define the properties emissivity and absorptivity. When are these two properties equal to each other?

**21–30C**  What is the greenhouse effect? Why is it a matter of great concern among atmospheric scientists?

**21–31**  Consider an opaque horizontal plate that is well insulated on the edges and the lower surface. The plate is maintained at 500 K, has a total hemispherical absorptivity of 0.51 and the following spectral emissivity function:

$$\varepsilon_\lambda = \begin{cases} \varepsilon_1 = 0.4, & 0 \leq \lambda < 4 \ \mu\text{m} \\ \varepsilon_2 = 0.8, & 4 \ \mu\text{m} \leq \lambda < \infty \end{cases}$$

If the plate is subjected to an irradiation of 5600 W/m², determine (*a*) the total hemispherical emissivity and (*b*) the radiosity of the plate surface.

**21–32**  The spectral emissivity function of an opaque surface at 1000 K is approximated as

$$\varepsilon_\lambda = \begin{cases} \varepsilon_1 = 0.4, & 0 \leq \lambda < 3 \ \mu\text{m} \\ \varepsilon_2 = 0.7, & 3 \ \mu\text{m} \leq \lambda < 6 \ \mu\text{m} \\ \varepsilon_3 = 0.3, & 6 \ \mu\text{m} \leq \lambda < \infty \end{cases}$$

Determine the average emissivity of the surface and the rate of radiation emission from the surface, in W/m².
*Answers:* 0.513, 29.1 kW/m²

**21–33E**  A 5-in-diameter spherical ball is known to emit radiation at a rate of 480 Btu/h when its surface temperature is 950 R. Determine the average emissivity of the ball at this temperature.

**21–34** A furnace that has a 40-cm × 40-cm glass window can be considered to be a blackbody at 1200 K. If the transmissivity of the glass is 0.7 for radiation at wavelengths less than 3 μm and zero for radiation at wavelengths greater than 3 μm, determine the fraction and the rate of radiation coming from the furnace and transmitted through the window.

**21–35** The emissivity of a tungsten filament can be approximated to be 0.5 for radiation at wavelengths less than 1 μm and 0.15 for radiation at greater than 1 μm. Determine the average emissivity of the filament at (a) 1500 K and (b) 2500 K. Also, determine the absorptivity and reflectivity of the filament at both temperatures.

**21–36** The variations of the spectral emissivity of two surfaces are as given in Fig. P21–36. Determine the average emissivity of each surface at $T = 3000$ K. Also, determine the average absorptivity and reflectivity of each surface for radiation coming from a source at 3000 K. Which surface is more suitable to serve as a solar absorber?

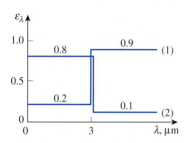

**FIGURE P21–36**

**21–37** The emissivity of a surface coated with aluminum oxide can be approximated to be 0.15 for radiation at wavelengths less than 5 μm and 0.9 for radiation at wavelengths greater than 5 μm. Determine the average emissivity of this surface at (a) 5800 K and (b) 300 K. What can you say about the absorptivity of this surface for radiation coming from sources at 5800 K and 300 K?  *Answers: (a) 0.154, (b) 0.89*

**21–38** The variation of the spectral absorptivity of a surface is as given in Fig. P21–38. Determine the average absorptivity and reflectivity of the surface for radiation that originates from a source at $T = 2500$ K. Also, determine the average emissivity of this surface at 3000 K.

**FIGURE P21–38**

**21–39** The reflectivity of aluminum coated with lead sulfate is 0.35 for radiation at wavelengths less than 3 μm and 0.95 for radiation greater than 3 μm. Determine the average reflectivity of this surface for solar radiation ($T \approx 5800$ K) and radiation coming from surfaces at room temperature ($T \approx 300$ K). Also, determine the emissivity and absorptivity of this surface at both temperatures. Do you think this material is suitable for use in solar collectors?

**21–40** The variation of the spectral transmissivity of a 0.6-cm-thick glass window is as given in Fig. P21–40. Determine the average transmissivity of this window for solar radiation ($T \approx 5800$ K) and radiation coming from surfaces at room temperature ($T \approx 300$ K). Also, determine the amount of solar radiation transmitted through the window for incident solar radiation of 650 W/m².  *Answers: 0.870, 0.00016, 566 W/m²*

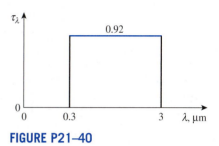

**FIGURE P21–40**

**21–41** An opaque horizontal plate is well insulated on the edges and the lower surface. The irradiation on the plate is 3000 W/m², of which 500 W/m² is reflected. The plate has a uniform temperature of 700 K and has an emissive power of 5000 W/m². Determine the total emissivity and absorptivity of the plate.

**FIGURE P21–41**

**21–42** A horizontal plate is experiencing uniform irradiation on the both upper and lower surfaces. The ambient air temperature surrounding the plate is 290 K with a convection heat

**FIGURE P21–42**

transfer coefficient of 30 W/m²·K. Both upper and lower surfaces of the plate have a radiosity of 4000 W/m², and the plate temperature is maintained uniformly at 390 K. If the plate is not opaque and has an absorptivity of 0.527, determine the irradiation and emissivity of the plate. *Answers: 7000 W/m², 0.525*

## The View Factor

**21–43C** What does the view factor represent? When is the view factor from a surface to itself not zero?

**21–44C** What is the crossed-strings method? For what kind of geometries is the crossed-strings method applicable?

**21–45C** What are the summation rule and the superposition rule for view factors?

**21–46C** How can you determine the view factor $F_{12}$ when the view factor $F_{21}$ and the surface areas are available?

**21–47** Consider an enclosure consisting of five surfaces. How many view factors does this geometry involve? How many of these view factors can be determined by the application of the reciprocity and summation rules?

**21–48** Consider a cylindrical enclosure with $A_1$, $A_2$, and $A_3$ representing the internal base, top, and side surfaces, respectively. Using the length to diameter ratio, $K = L/D$, determine (*a*) the expression for the view factor between the base and the side surface $F_{13}$ in terms of $K$ and (*b*) the value of the view factor $F_{13}$ for $L = D$. *Answers: (a)* $F_{13} = 2K\sqrt{(K^2 + 1)} - 2K^2$, (*b*) 0.828

**FIGURE P21–48**

**21–49** Two infinitely long parallel plates of width $w$ are located at $w$ distance apart, as shown in the figure. Using the Hottel's crossed-strings method, determine the view factor $F_{12}$.

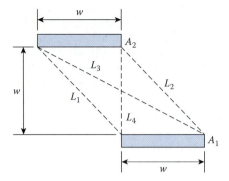

**FIGURE P21–49**

**21–50** Determine the view factors $F_{13}$ and $F_{23}$ between the rectangular surfaces shown in Fig. P21–50.

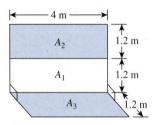

**FIGURE P21–50**

**21–51** Consider a cylindrical enclosure whose height is twice the diameter of its base. Determine the view factor from the side surface of this cylindrical enclosure to its base surface.

**21–52** Consider a hemispherical furnace with a flat circular base of diameter $D$. Determine the view factor from the dome of this furnace to its base. *Answer: 0.5*

**21–53** Three infinitely long parallel cylinders of diameter $D$ are located a distance $s$ apart from each other. Determine the view factor between the cylinder in the middle and the surroundings.

**FIGURE P21–53**

**21–54** Determine the view factors from the base of a cube to each of the other five surfaces.

**21–55** Consider a conical enclosure of height $h$ and base diameter $D$. Determine the view factor from the conical side surface to a hole of diameter $d$ located at the center of the base.

**FIGURE P21–55**

**21–56** Determine the four view factors associated with an enclosure formed by two very long concentric cylinders of radii $r_1$ and $r_2$. Neglect the end effects.

**21–57** Two infinitely long parallel cylinders of diameter $D$ are located a distance $s$ apart from each other. Determine the view factor $F_{12}$ between these two cylinders.

**21–58** Determine the view factors from the very long grooves shown in Fig. P21–58 to the surroundings without using any view factor tables or charts. Neglect end effects.

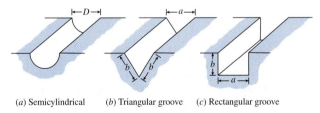

(a) Semicylindrical    (b) Triangular groove    (c) Rectangular groove

**FIGURE P21–58**

**21–59** A circular cone of diameter $D$ is positioned on a common axis with a circular disk, also of diameter, $D$, at a distance $L$, as shown in the figure. With a hypothetical area ($A_3$) corresponding to the opening of the cone, determine the values of $F_{11}$ and $F_{12}$ for $L = D = 100$ mm.

**FIGURE P21–59**

**21–60** Consider a cylindrical surface and a disk oriented coaxially as shown in the figure. The cylinder has a diameter $D$ and a length $L$. The disk of diameter $D$ is placed coaxially with the cylinder at a distance $L$. If $L = 2D$, determine the view factor $F_{12}$ between the cylindrical surface (1) and the disk (2) facing it.

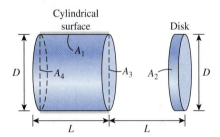

**FIGURE P21–60**

**21–61** Repeat Prob. 21–60 with $L = D$.

## Radiation Heat Transfer between Surfaces

**21–62C** What are the two methods used in radiation analysis? How do these two methods differ?

**21–63C** How does radiosity for a surface differ from the emitted energy? For what kind of surfaces are these two quantities identical?

**21–64C** What are the radiation surface and space resistances? How are they expressed? For what kind of surfaces is the radiation surface resistance zero?

**21–65C** Why is the radiation analysis of enclosures that consist of black surfaces relatively easy? How is the rate of radiation heat transfer between two surfaces expressed in this case?

**21–66C** What is a reradiating surface? What simplifications does a reradiating surface offer in the radiation analysis?

**21–67** Two aligned parallel rectangles with dimensions 6 m × 8 m are spaced apart by a distance of 2 m. If the two parallel rectangles are experiencing radiation heat transfer as black surfaces, determine the percentage of change in radiation heat transfer rate when the rectangles are moved 8 m apart.

*Answer: 71.6 percent*

**FIGURE P21–67**

**21–68** A solid sphere of 1 m diameter at 600 K is kept in an evacuated, long, equilateral triangular enclosure whose sides are 2 m long. The emissivity of the sphere is 0.45 and the temperature of the enclosure is 420 K. If heat is generated uniformly within the sphere at a rate of 3100 W, determine (a) the view factor from the enclosure to the sphere and (b) the emissivity of the enclosure.

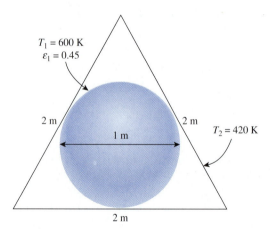

**FIGURE P21–68**

**21–69E** Consider a 10-ft × 10-ft × 10-ft cubical furnace whose top and side surfaces closely approximate black surfaces and whose base surface has an emissivity $\varepsilon = 0.4$. The base, top, and side surfaces of the furnace are maintained at uniform temperatures of 800 R, 1600 R, and 2400 R, respectively. Determine the net rate of radiation heat transfer between (a) the base and the side surfaces and (b) the base and the top surfaces. Also, determine the net rate of radiation heat transfer to the base surface.

**21–70E** Reconsider Prob. 21–69E. Using EES (or other) software, investigate the effect of base surface emissivity on the net rates of radiation heat transfer between the base and the side surfaces, between the base and top surfaces, and to the base surface. Let the emissivity vary from 0.1 to 0.9. Plot the rates of heat transfer as a function of emissivity, and discuss the results.

**21–71** Two very large parallel plates are maintained at uniform temperatures of $T_1 = 600$ K and $T_2 = 400$ K and have emissivities $\varepsilon_1 = 0.5$ and $\varepsilon_2 = 0.9$, respectively. Determine the net rate of radiation heat transfer between the two surfaces per unit area of the plates.

**21–72** Reconsider Prob. 21–71. Using EES (or other) software, investigate the effects of the temperature and the emissivity of the hot plate on the net rate of radiation heat transfer between the plates. Let the temperature vary from 500 K to 1000 K and the emissivity from 0.1 to 0.9. Plot the net rate of radiation heat transfer as functions of temperature and emissivity, and discuss the results.

**21–73** Two infinitely long parallel plates of width $w$ are located at $w$ distance apart, as shown in the figure. The two plates behave as black surfaces, where surface $A_1$ has a temperature of 700 K and surface $A_2$ has a temperature of 300 K. Determine the radiation heat flux between the two surfaces.

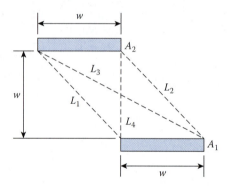

**FIGURE P21–73**

**21–74** A dryer is shaped like a long semi-cylindrical duct of diameter 1.5 m. The base of the dryer is occupied with water soaked materials to be dried. The base is maintained at

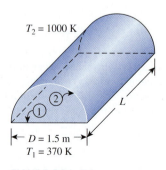

**FIGURE P21–74**

a temperature of 370 K, while the dome of the dryer is maintained at 1000 K. If both surfaces behave as blackbody, determine the drying rate per unit length experienced by the wet materials. *Answer*: 0.0370 kg/s·m

**21–75**   A furnace is of cylindrical shape with $R = H = 3$ m. The base, top, and side surfaces of the furnace are all black and are maintained at uniform temperatures of 500, 700, and 1400 K, respectively. Determine the net rate of radiation heat transfer to or from the top surface during steady operation.

**FIGURE P21–75**

**21–76E**   A 9-ft-high room with a base area of 12 ft × 12 ft is to be heated by electric resistance heaters placed on the ceiling, which is maintained at a uniform temperature of 90°F at all times. The floor of the room is at 65°F and has an emissivity of 0.8. The side surfaces are well insulated. Treating the ceiling as a blackbody, determine the rate of heat loss from the room through the floor.

**21–77**   Two very long concentric cylinders of diameters $D_1 = 0.35$ m and $D_2 = 0.5$ m are maintained at uniform temperatures of $T_1 = 950$ K and $T_2 = 500$ K and have emissivities $\varepsilon_1 = 1$ and $\varepsilon_2 = 0.55$, respectively. Determine the net rate of radiation heat transfer between the two cylinders per unit length of the cylinders.

**21–78**   This question deals with steady–state radiation heat transfer between a sphere ($r_1 = 30$ cm) and a circular disk ($r_2 = 120$ cm), which are separated by a center-to-center distance $h = 60$ cm. When the normal to the center of disk passes through the center of the sphere, the radiation view factor is given by

$$F_{12} = 0.5 \left\{ 1 - \left[ 1 + \left( \frac{r_2}{h} \right)^2 \right]^{-0.5} \right\}$$

Surface temperatures of the sphere and the disk are 600°C and 200°C, respectively; and their emissivities are 0.9 and 0.5, respectively.
   (a) Calculate the view factors $F_{12}$ and $F_{21}$.
   (b) Calculate the net rate of radiation heat exchange between the sphere and the disk.
   (c) For the given radii and temperatures of the sphere and the disk, the following four possible modifications could increase the net rate of radiation heat exchange: paint each of the two surfaces to alter their emissivities, adjust the distance between them, and provide an (refractory) enclosure. Calculate the net rate of radiation heat exchange between the two bodies if the *best values* are selected for each of the above modifications.

**FIGURE P21–78**

**21–79**   Two parallel disks of diameter $D = 0.6$ m separated by $L = 0.4$ m are located directly on top of each other. Both disks are black and are maintained at a temperature of 450 K. The back sides of the disks are insulated, and the environment that the disks are in can be considered to be a blackbody at 300 K. Determine the net rate of radiation heat transfer from the disks to the environment.   *Answer*: 781 W

**21–80E**   Two black parallel rectangles with dimensions 3 ft × 5 ft are spaced apart by a distance of 1 ft. The two parallel rectangles are experiencing radiation heat transfer as black surfaces, where the top rectangle receives a total of 180,000 Btu/h radiation heat transfer rate from the bottom rectangle. If the top rectangle has a uniform temperature of 60°F, determine the temperature of the bottom rectangle.

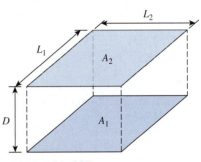

**FIGURE P21–80E**

**21–81**   Two coaxial parallel disks of equal diameter 1 m are originally placed at a distance of 1 m apart. If both disks behave as black surfaces, determine the new distance between the disks such that there is a 75% reduction in radiation heat transfer rate from the original distance of 1 m.

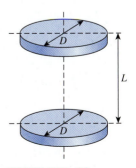

**FIGURE P21–81**

**21–82** A furnace is shaped like a long equilateral-triangular duct where the width of each side is 2 m. Heat is supplied from the base surface, whose emissivity is $\varepsilon_1 = 0.8$, at a rate of 800 W/m² while the side surfaces, whose emissivities are 0.4, are maintained at 600 K. Neglecting the end effects, determine the temperature of the base surface. Can you treat this geometry as a two-surface enclosure?

**21–83** Reconsider Prob. 21–82. Using EES (or other) software, investigate the effects of the rate of the heat transfer at the base surface and the temperature of the side surfaces on the temperature of the base surface. Let the rate of heat transfer vary from 500 W/m² to 1000 W/m² and the temperature from 300 K to 700 K. Plot the temperature of the base surface as functions of the rate of heat transfer and the temperature of the side surfaces, and discuss the results.

**21–84** Consider a 4-m × 4-m × 4-m cubical furnace whose floor and ceiling are black and whose side surfaces are reradiating. The floor and the ceiling of the furnace are maintained at temperatures of 550 K and 1100 K, respectively. Determine the net rate of radiation heat transfer between the floor and the ceiling of the furnace.

**21–85** Two concentric spheres of diameters $D_1 = 0.3$ m and $D_2 = 0.6$ m are maintained at uniform temperatures $T_1 = 800$ K and $T_2 = 500$ K and have emissivities $\varepsilon_1 = 0.5$ and $\varepsilon_2 = 0.7$, respectively. Determine the net rate of radiation heat transfer between the two spheres. Also, determine the convection heat transfer coefficient at the outer surface if both the surrounding medium and the surrounding surfaces are at 30°C. Assume the emissivity of the outer surface is 0.35.

**21–86** A spherical tank of diameter $D = 2$ m that is filled with liquid nitrogen at 100 K is kept in an evacuated cubic enclosure whose sides are 3 m long. The emissivities of the spherical tank and the enclosure are $\varepsilon_1 = 0.1$ and $\varepsilon_2 = 0.8$, respectively. If the temperature of the cubic enclosure is measured to be 240 K, determine the net rate of radiation heat transfer to the liquid nitrogen. *Answer: 228 W*

**FIGURE P21–86**

**21–87** Repeat Prob. 21–86 by replacing the cubic enclosure by a spherical enclosure whose diameter is 3 m.

**21–88** Reconsider Prob. 21–86. Using EES (or other) software, investigate the effects of the side length and the emissivity of the cubic enclosure, and the emissivity of the spherical tank on the net rate of radiation heat transfer. Let the side length vary from 2 m to 12 m and both emissivities from 0.1 to 0.9. Plot the net rate of radiation heat transfer as functions of side length and emissivities, and discuss the results.

**21–89** Consider a circular grill whose diameter is 0.3 m. The bottom of the grill is covered with hot coal bricks at 950 K, while the wire mesh on top of the grill is covered with steaks initially at 5°C. The distance between the coal bricks and the steaks is 0.20 m. Treating both the steaks and the coal bricks as blackbodies, determine the initial rate of radiation heat transfer from the coal bricks to the steaks. Also, determine the initial rate of radiation heat transfer to the steaks if the side opening of the grill is covered by aluminum foil, which can be approximated as a reradiating surface. *Answers: 928 W, 2085 W*

**FIGURE P21–89**

**21–90** Consider a hemispherical furnace of diameter $D = 5$ m with a flat base. The dome of the furnace is black, and the base has an emissivity of 0.7. The base and the dome of the furnace are maintained at uniform temperatures of 400 and 1000 K, respectively. Determine the net rate of radiation heat transfer from the dome to the base surface during steady operation. *Answer: 759 kW*

**FIGURE P21–90**

**21–91** Consider two rectangular surfaces perpendicular to each other with a common edge which is 1.6 m long. The horizontal surface is 0.8 m wide and the vertical surface is 1.2 m high. The horizontal surface has an emissivity of 0.75 and is maintained at 450 K. The vertical surface is black and is maintained at 700 K. The back sides of the surfaces are insulated. The surrounding surfaces are at 290 K, and can be considered to have an emissivity of 0.85. Determine the net rate of radiation heat transfers between the two surfaces, and between the horizontal surface and the surroundings.

**FIGURE P21–91**

**21–92** Two long parallel 20-cm-diameter cylinders are located 30 cm apart from each other. Both cylinders are black, and are maintained at temperatures 425 K and 275 K. The surroundings can be treated as a blackbody at 300 K. For a 1-m-long section of the cylinders, determine the rates of radiation heat transfer between the cylinders and between the hot cylinder and the surroundings.

**FIGURE P21–92**

**21–93** Consider a long semicylindrical duct of diameter 1.0 m. Heat is supplied from the base surface, which is black, at a rate of 1200 W/m$^2$, while the side surface with an emissivity of 0.4 is maintained at 650 K. Neglecting the end effects, determine the temperature of the base surface.

**21–94** Consider a 30-cm-diameter hemispherical enclosure. The dome is maintained at 600 K and heat is supplied from the dome at a rate of 65 W while the base surface with an emissivity of 0.55 is maintained at 400 K. Determine the emissivity of the dome.

**21–95** Consider a hot cylindrical surface and a disk oriented as shown in the figure. Both the cylindrical surface and the disk behave as blackbody. The cylinder has a diameter $D$ and a length $L$ of 0.2 m. The disk of diameter $D = 0.2$ m is placed coaxially with the cylinder at a distance of $L = 0.2$ m apart. The cylindrical surface and the disk are maintained at uniform temperatures of 1000 K and 300 K, respectively. Determine the rate of heat transfer by radiation from cylindrical surface to the disk.

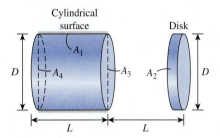

**FIGURE P21–95**

**21–96** Two parallel black disks are positioned coaxially with a distance of 0.25 m apart in a surrounding with a constant temperature of 300 K. The lower disk is 0.2 m is diameter and the upper disk is 0.4 m in diameter. If the lower disk is heated electrically at 100 W to maintain a uniform temperature of 500 K, determine the temperature of the upper disk.

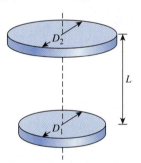

**FIGURE P21–96**

**21–97** Two-phase gas-liquid oxygen is stored in a spherical tank of 1-m diameter, where it is maintained at its normal boiling point. The spherical tank is enclosed by a 1.6-m diameter concentric spherical surface at 273 K. Both spherical surfaces have an emissivity of 0.01, and the gap between the inner sphere and outer sphere is vacuumed. Assuming that the spherical tank surface has the same temperature as the oxygen, determined the heat transfer rate at the spherical tank surface.

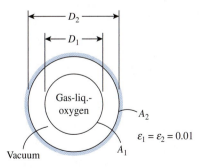

**FIGURE P21–97**

## Review Problems

**21–98**  Daylight and incandescent light may be approximated as a blackbody at the effective surface temperatures of 5800 K and 2800 K, respectively. Determine the wavelength at maximum emission of radiation for each of the lighting sources.

**21–99**  A horizontal opaque flat plate is well insulated on the edges and the lower surface. The top surface has an area of 5 m$^2$, and it experiences uniform irradiation at a rate of 5000 W. The plate absorbs 4000 W of the irradiation, and the surface is losing heat at a rate of 500 W by convection. If the plate maintains a uniform temperature of 350 K, determine the absorptivity, reflectivity, and emissivity of the plate.

**FIGURE P21–99**

**21–100**  The spectral emissivity of an opaque surface at 1200 K is approximated as

$$\varepsilon_1 = 0 \qquad \text{for} \qquad \lambda < 2 \ \mu m$$
$$\varepsilon_2 = 0.85 \qquad \text{for} \qquad 2 \le \lambda \le 6 \ \mu m$$
$$\varepsilon_3 = 0 \qquad \text{for} \qquad \lambda > 6 \ \mu m$$

Determine the total emissivity and the emissive flux of the surface.

**21–101**  The spectral transmissivity of a glass cover used in a solar collector is given as

$$\tau_1 = 0 \qquad \text{for} \qquad \lambda < 0.3 \ \mu m$$
$$\tau_2 = 0.9 \qquad \text{for} \qquad 0.3 < \lambda < 3 \ \mu m$$
$$\tau_3 = 0 \qquad \text{for} \qquad \lambda > 3 \ \mu m$$

Solar radiation is incident at a rate of 950 W/m$^2$, and the absorber plate, which can be considered to be black, is maintained at 340 K by the cooling water. Determine (a) the solar flux incident on the absorber plate, (b) the transmissivity of the glass cover for radiation emitted by the absorber plate, and (c) the rate of heat transfer to the cooling water if the glass cover temperature is also 340 K.

**21–102**  The spectral absorptivity of an opaque surface is as shown on the graph. Determine the absorptivity of the surface for radiation emitted by a source at (a) 1000 K and (b) 3000 K.

**FIGURE P21–102**

**21–103**  The surface in Prob. 21–102 receives solar radiation at a rate of 470 W/m$^2$. Determine the solar absorptivity of the surface and the rate of absorption of solar radiation.

**21–104**  The spectral transmissivity of a 3-mm-thick regular glass can be expressed as

$$\tau_1 = 0 \qquad \text{for} \qquad \lambda < 0.35 \ \mu m$$
$$\tau_2 = 0.85 \qquad \text{for} \qquad 0.35 < \lambda < 2.5 \ \mu m$$
$$\tau_3 = 0 \qquad \text{for} \qquad \lambda > 2.5 \ \mu m$$

Determine the transmissivity of this glass for solar radiation. What is the transmissivity of this glass for light?

**21–105**  A 1-m-diameter spherical cavity is maintained at a uniform temperature of 600 K. Now a 5-mm-diameter hole is drilled. Determine the maximum rate of radiation energy streaming through the hole. What would your answer be if the diameter of the cavity were 3 m?

**21–106**  The human skin is "selective" when it comes to the absorption of the solar radiation that strikes it perpendicularly. The skin absorbs only 50 percent of the incident radiation with wavelengths between $\lambda_1 = 0.517 \ \mu m$ and $\lambda_2 = 1.552 \ \mu m$. The radiation with wavelengths shorter than $\lambda_1$ and longer than $\lambda_2$ is fully absorbed. The solar surface may be modeled as a blackbody with effective surface temperature of 5800 K. Calculate the fraction of the incident solar radiation that is absorbed by the human skin.

**21–107**  Consider a cylindrical enclosure with $A_1$, $A_2$ and $A_3$ representing the internal base, top, and, side surfaces, respectively. Using the length to diameter ratio, $K = L/D$, determine (a) the expression for the view factor from the side surface to itself $F_{33}$ in terms of $K$ and (b) the value of the view factor $F_{33}$ for $L = D$.

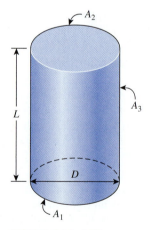

**FIGURE P21–107**

**21–108**  A long cylindrical black surface fuel rod of diameter 25 mm is shielded by a surface concentric to the rod. The shield has diameter of 50 mm and a constant surface temperature of 320 K. Both inner and outer surfaces of the shield have an emissivity of 0.05, and the gap between the fuel rod and the shield is a vacuum. If the fuel rod is generating 0.5 MW/m³ of heat per unit length, determine the surface temperature of the fuel rod.

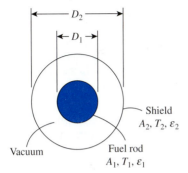

**FIGURE P21–108**

**21–109** A dryer is shaped like a long semi-cylindrical duct of diameter 1.5 m. The base of the dryer is occupied with water soaked materials to be dried, and maintained at a temperature of 370 K and emissivity of 0.5. The dome of the dryer is maintained at 1000 K with emissivity of 0.8. Determine the drying rate per unit length experienced by the wet materials.

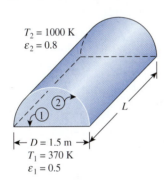

**FIGURE P21–109**

**21–110** Two concentric spheres of diameters $D_1 = 15$ cm and $D_2 = 25$ cm are separated by air at 1 atm pressure. The surface temperatures of the two spheres enclosing the air are $T_1 = 350$ K and $T_2 = 275$ K, respectively, and their emissivities are 0.75. Determine the rate of heat transfer from the inner sphere to the outer sphere by (a) natural convection and (b) radiation.

**21–111** Consider the two parallel coaxial disks of diameters $a$ and $b$, shown in Fig. P21–111. For this geometry, the view factor from the smaller disk to the larger disk can be calculated from

$$F_{ij} = 0.5\left(\frac{B}{A}\right)^2 \left\{ C - \left[ C^2 - 4\left(\frac{A}{B}\right)^2 \right]^{0.5} \right\}$$

where, $A = a/2L$, $B = b/2L$, and $C = 1 + [(1 + A^2)/B^2]$. The diameter, emissivity and temperature are 20 cm, 0.60, and 600°C,

respectively, for disk $a$, and 40 cm, 0.80 and 200°C for disk $b$. The distance between the two disks is $L = 10$ cm.

(a) Calculate $F_{ab}$ and $F_{ba}$.
(b) Calculate the net rate of radiation heat exchange between disks $a$ and $b$ in steady operation.
(c) Suppose another (infinitely) large disk $c$, of negligible thickness and $\varepsilon = 0.7$, is inserted between disks $a$ and $b$ such that it is parallel and equidistant to both disks. Calculate the net rate of radiation heat exchange between disks $a$ and $c$ and disks $c$ and $b$ in steady operation.

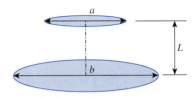

**FIGURE P21–111**

**21–112** A large number of long tubes, each of diameter $D$, are placed parallel to each other and at a center-to-center distance of $s$. Since all of the tubes are geometrically similar and at the same temperature, these could be treated collectively as one surface ($A_j$) for radiation heat transfer calculations. As shown in Fig. P21–112, the tube-bank ($A_j$) is placed opposite a large flat wall ($A_i$) such that the tube-bank is parallel to the wall. The radiation view factor, $F_{ij}$, for this arrangement is given by

$$F_{ij} = 1 - \left[ 1 - \left(\frac{D}{s}\right)^2 \right]^{0.5} + \frac{D}{s}\left\{ \tan^{-1}\left[ \left(\frac{s}{D}\right)^2 - 1 \right]^{0.5} \right\}$$

(a) Calculate the view factors $F_{ij}$ and $F_{ji}$ for $s = 3.0$ cm and $D = 1.5$ cm.
(b) Calculate the net rate of radiation heat transfer between the wall and the tube-bank per unit area of the wall when $T_i = 900°C$, $T_j = 60°C$, $\varepsilon_i = 0.8$, and $\varepsilon_j = 0.9$.
(c) A fluid flows through the tubes at an average temperature of 40°C, resulting in a heat transfer coefficient of 2.0 kW/m²·K. Assuming $T_i = 900°C$, $\varepsilon_i = 0.8$ and $\varepsilon_j = 0.9$ (as above) and neglecting the tube wall thickness and convection from the outer surface, calculate the temperature of the tube surface in steady operation.

**FIGURE P21–112**

**21–113** 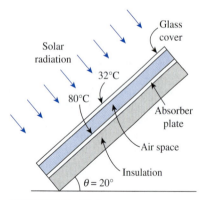 A simple solar collector is built by placing a 6-cm-diameter clear plastic tube around a garden hose whose outer diameter is 2 cm. The hose is painted black to maximize solar absorption, and some plastic rings are used to keep the spacing between the hose and the clear plastic cover constant. The emissivities of the hose surface and the glass cover are 0.9, and the effective sky temperature is estimated to be 15°C. The temperature of the plastic tube is measured to be 40°C, while the ambient air temperature is 25°C. Determine the rate of heat loss from the water in the hose by natural convection and radiation per meter of its length under steady conditions.
*Answers: 12.7 W, 26.1 W*

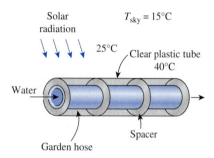

**FIGURE P21–113**

**21–114** Consider a 1.5-m-high and 3-m-wide solar collector that is tilted at an angle 20° from the horizontal. The distance between the glass cover and the absorber plate is 3 cm, and the back side of the absorber is heavily insulated. The absorber plate and the glass cover are maintained at temperatures of 80°C and 32°C, respectively. The emissivity of the glass surface is 0.9 and that of the absorber plate is 0.8. Determine the rate of heat loss from the absorber plate by natural convection and radiation. *Answers: 750 W, 1289 W*

**FIGURE P21–114**

**21–115E** 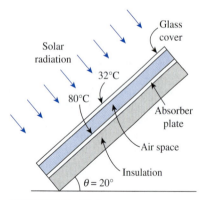 A solar collector consists of a horizontal aluminum tube having an outer diameter of 2.5 in enclosed in a concentric thin glass tube of diameter 5 in. Water is heated as it flows through the tube, and the annular space between

the aluminum and the glass tube is filled with air at 0.5 atm pressure. The pump circulating the water fails during a clear day, and the water temperature in the tube starts rising. The aluminum tube absorbs solar radiation at a rate of 40 Btu/h per foot length, and the temperature of the ambient air outside is 75°F. The emissivities of the tube and the glass cover are 0.9. Taking the effective sky temperature to be 60°F, determine the temperature of the aluminum tube when thermal equilibrium is established (i.e., when the rate of heat loss from the tube equals the amount of solar energy gained by the tube).

**21–116** A vertical 2-m-high and 5-m-wide double-pane window consists of two sheets of glass separated by a 3-cm-thick air gap. In order to reduce heat transfer through the window, the air space between the two glasses is partially evacuated to 0.3 atm pressure. The emissivities of the glass surfaces are 0.9. Taking the glass surface temperatures across the air gap to be 15°C and 5°C, determine the rate of heat transfer through the window by natural convection and radiation.

**FIGURE P21–116**

**21–117** A 2-m-internal-diameter double-walled spherical tank is used to store iced water at 0°C. Each wall is 0.5 cm thick, and the 1.5-cm-thick air space between the two walls of the tank is evacuated in order to minimize heat transfer. The surfaces surrounding the evacuated space are polished so that each surface has an emissivity of 0.15. The temperature of the outer wall of the tank is measured to be 20°C. Assuming the inner wall of the steel tank to be at 0°C, determine (a) the rate of heat transfer to the iced water in the tank and (b) the amount of ice at 0°C that melts during a 24-h period.

**FIGURE P21–117**

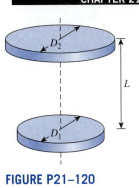

**FIGURE P21–120**

**21–118** A solar collector consists of a horizontal copper tube of outer diameter 5 cm enclosed in a concentric thin glass tube of diameter 12 cm. Water is heated as it flows through the tube, and the annular space between the copper and the glass tubes is filled with air at 1 atm pressure. The emissivities of the tube surface and the glass cover are 0.85 and 0.9, respectively. During a clear day, the temperatures of the tube surface and the glass cover are measured to be 60°C and 40°C, respectively. Determine the rate of heat loss from the collector by natural convection and radiation per meter length of the tube.

**21–119** Two parallel concentric disks, 20 cm and 40 cm in diameter, are separated by a distance of 10 cm. The smaller disk ($\varepsilon = 0.80$) is at a temperature of 300°C. The larger disk ($\varepsilon = 0.60$) is at a temperature of 800°C.
  (a) Calculate the radiation view factors.
  (b) Determine the rate of radiation heat exchange between the two disks.
  (c) Suppose that the space between the two disks is completely surrounded by a reflective surface. Estimate the rate of radiation heat exchange between the two disks.

**21–120** Two parallel back disks are positioned coaxially with a distance of 0.25 m apart. The lower disk is 0.2 m in diameter and the upper disk is 0.4 m in diameter. If the lower disk is heated electrically at 20 W to maintain a uniform temperature of 500 K, determine the temperature of the upper disk.
*Answer*: 423 K

## Design and Essay Problems

**21–121** Write an essay on the radiation properties of selective surfaces used on the absorber plates of solar collectors. Find out about the various kinds of such surfaces, and discuss the performance and cost of each type. Recommend a selective surface that optimizes cost and performance.

**21–122** According to an Atomic Energy Commission report, a hydrogen bomb can be approximated as a large fireball at a temperature of 7200 K. You are to assess the impact if such a bomb exploded 5 km above a city. Assume the diameter of the fireball to be 1 km, and the blast to last 15 s. Investigate the level of radiation energy people, plants, and houses will be exposed to, and how adversely they will be affected by the blast.

**21–123** Radiation shields are commonly used in the design of superinsulations for use in space and cryogenics applications. Write an essay on superinsulations and how they are used in different applications.

**21–124** Thermal comfort in a house is strongly affected by the so-called radiation effect, which is due to radiation heat transfer between the person and surrounding surface. A person feels much colder in the morning, for example, because of the lower surface temperature of the walls at that time, although the thermostat setting of the house is fixed. Write an essay on the radiation effect, how it affect human comfort, and how it is accounted for in heating and air-conditioning applications.

# HEAT EXCHANGERS

**H**eat exchangers are devices that facilitate the *exchange of heat* between two fluids that are at different temperatures while keeping them from mixing with each other. Heat exchangers are commonly used in practice in a wide range of applications, from heating and air-conditioning systems in a household, to chemical processing and power production in large plants. Heat exchangers differ from mixing chambers in that they do not allow the two fluids involved to mix.

Heat transfer in a heat exchanger usually involves *convection* in each fluid and *conduction* through the wall separating the two fluids. In the analysis of heat exchangers, it is convenient to work with an *overall heat transfer coefficient U* that accounts for the contribution of all these effects on heat transfer. The rate of heat transfer between the two fluids at a location in a heat exchanger depends on the magnitude of the temperature difference at that location, which varies along the heat exchanger.

Heat exchangers are manufactured in a variety of types, and thus we start this chapter with the *classification* of heat exchangers. We then discuss the determination of the overall heat transfer coefficient in heat exchangers, and the *log mean temperature difference* (LMTD) for some configurations. We then introduce the *correction factor F* to account for the deviation of the mean temperature difference from the LMTD in complex configurations. Finally, we discuss the effectiveness–NTU method, which enables us to analyze heat exchangers when the outlet temperatures of the fluids are not known.

## Objectives

The objectives of this chapter are to:

■ Recognize numerous types of heat exchangers, and classify them.

■ Develop an awareness of fouling on surfaces, and determine the overall heat transfer coefficient for a heat exchanger.

■ Perform a general energy analysis on heat exchangers.

■ Obtain a relation for the logarithmic mean temperature difference for use in the LMTD method, and modify it for different types of heat exchangers using the correction factor.

■ Develop relations for effectiveness, and analyze heat exchangers when outlet temperatures are not known using the effectiveness–NTU method.

## 22–1 ■ TYPES OF HEAT EXCHANGERS

Different heat transfer applications require different types of hardware and different configurations of heat transfer equipment. The attempt to match the heat transfer hardware to the heat transfer requirements within the specified constraints has resulted in numerous types of innovative heat exchanger designs.

The simplest type of heat exchanger consists of two concentric pipes of different diameters, as shown in Fig. 22–1, called the **double-pipe** heat exchanger. One fluid in a double-pipe heat exchanger flows through the smaller pipe while the other fluid flows through the annular space between the two pipes. Two types of flow arrangement are possible in a double-pipe heat exchanger: in **parallel flow**, both the hot and cold fluids enter the heat exchanger at the same end and move in the *same* direction. In **counter flow**, on the other hand, the hot and cold fluids enter the heat exchanger at opposite ends and flow in *opposite* directions.

Another type of heat exchanger, which is specifically designed to realize a large heat transfer surface area per unit volume, is the **compact** heat exchanger. The ratio of the heat transfer surface area of a heat exchanger to its volume is called the *area density* $\beta$. A heat exchanger with $\beta > 700$ m$^2$/m$^3$ (or 200 ft$^2$/ft$^3$) is classified as being compact. Examples of compact heat exchangers are car radiators ($\beta \approx 1000$ m$^2$/m$^3$), glass-ceramic gas turbine heat exchangers ($\beta \approx 6000$ m$^2$/m$^3$), the regenerator of a Stirling engine ($\beta \approx 15{,}000$ m$^2$/m$^3$), and the human lung ($\beta \approx 20{,}000$ m$^2$/m$^3$). The flow passages in these compact heat exchangers are usually small and the flow can be considered to be laminar.

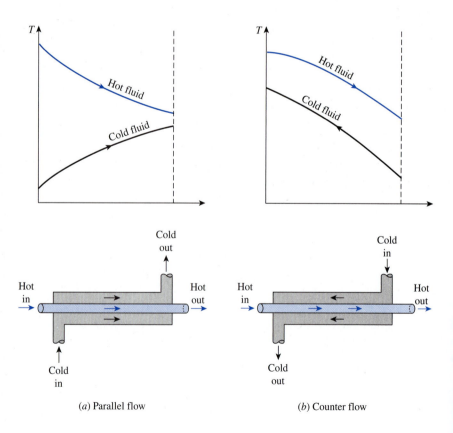

**FIGURE 22–1**
Different flow regimes and associated temperature profiles in a double-pipe heat exchanger.

(*a*) Parallel flow

(*b*) Counter flow

Compact heat exchangers enable us to achieve high heat transfer rates between two fluids in a small volume, and they are commonly used in applications with strict limitations on the weight and volume of heat exchangers (Fig. 22–2).

The large surface area in compact heat exchangers is obtained by attaching closely spaced *thin plate* or *corrugated fins* to the walls separating the two fluids. Compact heat exchangers are commonly used in gas-to-gas and gas-to-liquid (or liquid-to-gas) heat exchangers to counteract the low heat transfer coefficient associated with gas flow with increased surface area. In a car radiator, which is a water-to-air compact heat exchanger, for example, it is no surprise that fins are attached to the air side of the tube surface.

In compact heat exchangers, the two fluids usually move *perpendicular* to each other, and such flow configuration is called **cross-flow**. The cross-flow is further classified as *unmixed* and *mixed flow,* depending on the flow configuration, as shown in Fig. 22–3. In (*a*) the cross-flow is said to be *unmixed* since the plate fins force the fluid to flow through a particular interfin spacing and prevent it from moving in the transverse direction (i.e., parallel to the tubes). The cross-flow in (*b*) is said to be *mixed* since the fluid now is free to move in the transverse direction. Both fluids are unmixed in a car radiator. The presence of mixing in the fluid can have a significant effect on the heat transfer characteristics of the heat exchanger.

Perhaps the most common type of heat exchanger in industrial applications is the **shell-and-tube** heat exchanger, shown in Fig. 22–4. Shell-and-tube heat exchangers contain a large number of tubes (sometimes several hundred) packed in a shell with their axes parallel to that of the shell. Heat transfer takes place as one fluid flows inside the tubes while the other fluid flows outside the tubes through the shell. *Baffles* are commonly placed in the shell to force the shell-side fluid to flow across the shell to enhance heat transfer and to maintain uniform spacing between the tubes. Despite their widespread use, shell-and-tube heat exchangers are not suitable for use in automotive and aircraft applications because of their relatively large size and weight. Note that the tubes in a shell-and-tube heat exchanger open to some large flow areas called *headers* at both ends of the shell, where the tube-side fluid accumulates before entering the tubes and after leaving them.

Shell-and-tube heat exchangers are further classified according to the number of shell and tube passes involved. Heat exchangers in which all the tubes make one U-turn in the shell, for example, are called *one-shell-pass and*

**FIGURE 22–2**
A gas-to-liquid compact heat exchanger for a residential air-conditioning system.
*© Yunus Cengel.*

Cross-flow (*unmixed*)
Tube flow (*unmixed*)
(*a*) Both fluids unmixed

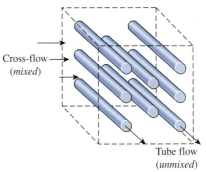

Cross-flow (*mixed*)
Tube flow (*unmixed*)
(*b*) One fluid mixed, one fluid unmixed

**FIGURE 22–3**
Different flow configurations in cross-flow heat exchangers.

Tube outlet  Shell inlet  Baffles  Front-end header  Rear-end header  Tubes  Shell  Shell outlet  Tube inlet

**FIGURE 22–4**
The schematic of a shell-and-tube heat exchanger (one-shell pass and one-tube pass).

(a) One-shell pass and two-tube passes

(b) Two-shell passes and four-tube passes

**FIGURE 22–5**
Multipass flow arrangements in shell-and-tube heat exchangers.

*two-tube-passes* heat exchangers. Likewise, a heat exchanger that involves two passes in the shell and four passes in the tubes is called a *two-shell-passes and four-tube-passes* heat exchanger (Fig. 22–5).

An innovative type of heat exchanger that has found widespread use is the **plate and frame** (or just plate) heat exchanger, which consists of a series of plates with corrugated flat flow passages (Fig. 22–6). The hot and cold fluids flow in alternate passages, and thus each cold fluid stream is surrounded by two hot fluid streams, resulting in very effective heat transfer. Also, plate heat exchangers can grow with increasing demand for heat transfer by simply mounting more plates. They are well suited for liquid-to-liquid heat exchange applications, provided that the hot and cold fluid streams are at about the same pressure.

Another type of heat exchanger that involves the alternate passage of the hot and cold fluid streams through the same flow area is the **regenerative** heat exchanger. The *static*-type regenerative heat exchanger is basically a porous mass that has a large heat storage capacity, such as a ceramic wire mesh. Hot and cold fluids flow through this porous mass alternatively. Heat is transferred from the hot fluid to the matrix of the regenerator during the flow of the hot fluid, and from the matrix to the cold fluid during the flow of the cold fluid. Thus, the matrix serves as a temporary heat storage medium.

The *dynamic*-type regenerator involves a rotating drum and continuous flow of the hot and cold fluid through different portions of the drum so that any portion of the drum passes periodically through the hot stream, storing heat, and then through the cold stream, rejecting this stored heat. Again the drum serves as the medium to transport the heat from the hot to the cold fluid stream.

Heat exchangers are often given specific names to reflect the specific application for which they are used. For example, a *condenser* is a heat exchanger in which one of the fluids is cooled and condenses as it flows through the heat

**FIGURE 22–6**
A plate-and-frame liquid-to-liquid heat exchanger.
*Courtesy of Tranter PHE, Inc.*

Nozzles attached to end frames allow for entrance and exit of fluids.

Plates supported by an upper guide bar are held in a frame which is bolted together.

Portholes and gaskets allow fluids to flow in alternate channels.

Special gaskets on end plates prevent fluids from contacting the frames.

A gasket mounted on each plate seals the channel between it and the next plate.

A Plate
B Plate
A Plate

A and B plates are arranged alternately.

Tightening bolt.

The lower rectangular guide bar assures absolute plate alignment preventing lateral movement.

exchanger. A *boiler* is another heat exchanger in which one of the fluids absorbs heat and vaporizes. A *space radiator* is a heat exchanger that transfers heat from the hot fluid to the surrounding space by radiation.

## 22–2 ▪ THE OVERALL HEAT TRANSFER COEFFICIENT

A heat exchanger typically involves two flowing fluids separated by a solid wall. Heat is first transferred from the hot fluid to the wall by *convection,* through the wall by *conduction,* and from the wall to the cold fluid again by *convection.* Any radiation effects are usually included in the convection heat transfer coefficients.

The thermal resistance network associated with this heat transfer process involves two convection and one conduction resistances, as shown in Fig. 22–7. The analysis is very similar to the material presented in Chapter 17. Here the subscripts $i$ and $o$ represent the inner and outer surfaces of the inner tube. For a double-pipe heat exchanger, the *thermal resistance* of the tube wall is (also see Eq. 17–39)

$$R_{wall} = \frac{\ln (D_o/D_i)}{2\pi kL} \qquad (22–1)$$

where $k$ is the thermal conductivity of the wall material and $L$ is the length of the tube. Then the *total thermal resistance* becomes (also see Eq. 17–43)

$$R = R_{total} = R_i + R_{wall} + R_o = \frac{1}{h_i A_i} + \frac{\ln (D_o/D_i)}{2\pi kL} + \frac{1}{h_o A_o} \qquad (22–2)$$

The $A_i$ is the area of the inner surface of the wall that separates the two fluids, and $A_o$ is the area of the outer surface of the wall. In other words, $A_i$ and $A_o$ are surface areas of the separating wall wetted by the inner and the outer fluids, respectively. When one fluid flows inside a circular tube and the other outside of it, we have $A_i = \pi D_i L$ and $A_o = \pi D_o L$ (Fig. 22–8).

In the analysis of heat exchangers, it is convenient to combine all the thermal resistances in the path of heat flow from the hot fluid to the cold one into a single resistance $R$, and to express the rate of heat transfer between the two fluids as

$$\dot{Q} = \frac{\Delta T}{R} = UA_s \Delta T = U_i A_i \Delta T = U_o A_o \Delta T \qquad (22–3)$$

where $A_s$ is the surface area and $U$ is the **overall heat transfer coefficient**, whose unit is $W/m^2 \cdot K$, which is identical to the unit of the ordinary convection coefficient $h$. Canceling $\Delta T$, Eq. 22–3 reduces to

$$\frac{1}{UA_s} = \frac{1}{U_i A_i} = \frac{1}{U_o A_o} = R = \frac{1}{h_i A_i} + R_{wall} + \frac{1}{h_o A_o} \qquad (22–4)$$

Perhaps you are wondering why we have two overall heat transfer coefficients $U_i$ and $U_o$ for a heat exchanger. The reason is that every heat exchanger has two heat transfer surface areas $A_i$ and $A_o$, which, in general, are not equal to each other.

Note that $U_i A_i = U_o A_o$, but $U_i \neq U_o$ unless $A_i = A_o$. Therefore, the overall heat transfer coefficient $U$ of a heat exchanger is meaningless unless the area on which it is based is specified. This is especially the case when one side of the tube wall is finned and the other side is not, since the surface area of the finned side is several times that of the unfinned side.

**FIGURE 22–7**
Thermal resistance network associated with heat transfer in a double-pipe heat exchanger.

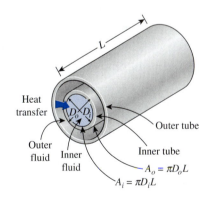

**FIGURE 22–8**
The two heat transfer surface areas associated with a double-pipe heat exchanger (for thin tubes, $D_i \approx D_o$ and thus $A_i \approx A_o$).

When the wall thickness of the tube is small and the thermal conductivity of the tube material is high, as is usually the case, the thermal resistance of the tube is negligible ($R_{wall} \approx 0$) and the inner and outer surfaces of the tube are almost identical ($A_i \approx A_o \approx A_s$). Then Eq. 22–4 for the overall heat transfer coefficient simplifies to

$$\frac{1}{U} \approx \frac{1}{h_i} + \frac{1}{h_o} \tag{22–5}$$

where $U \approx U_i \approx U_o$. The individual convection heat transfer coefficients inside and outside the tube, $h_i$ and $h_o$, are determined using the convection relations discussed in Chap. 19 for internal and external forced convection flows, respectively.

The overall heat transfer coefficient $U$ in Eq. 22–5 is dominated by the *smaller* convection coefficient, since the inverse of a large number is small. When one of the convection coefficients is *much smaller* than the other (say, $h_i \ll h_o$), we have $1/h_i \gg 1/h_o$, and thus $U \approx h_i$. Therefore, the smaller heat transfer coefficient creates a *bottleneck* on the path of heat transfer and seriously impedes heat transfer. This situation arises frequently when one of the fluids is a gas and the other is a liquid. In such cases, fins are commonly used on the gas side to enhance the product $UA$ and thus the heat transfer on that side.

Representative values of the overall heat transfer coefficient $U$ are given in Table 22–1. Note that the overall heat transfer coefficient ranges from about 10 W/m²·K for gas-to-gas heat exchangers to about 10,000 W/m²·K for heat exchangers that involve phase changes. This is not surprising, since gases have very low thermal conductivities, and phase-change processes involve very high heat transfer coefficients.

## TABLE 22–1

Representative values of the overall heat transfer coefficients in heat exchangers

| Type of heat exchanger | $U$, W/m²·K* |
|---|---|
| Water-to-water | 850–1700 |
| Water-to-oil | 100–350 |
| Water-to-gasoline or kerosene | 300–1000 |
| Feedwater heaters | 1000–8500 |
| Steam-to-light fuel oil | 200–400 |
| Steam-to-heavy fuel oil | 50–200 |
| Steam condenser | 1000–6000 |
| Freon condenser (water cooled) | 300–1000 |
| Ammonia condenser (water cooled) | 800–1400 |
| Alcohol condensers (water cooled) | 250–700 |
| Gas-to-gas | 10–40 |
| Water-to-air in finned tubes (water in tubes) | 30–60[†] |
|  | 400–850[†] |
| Steam-to-air in finned tubes (steam in tubes) | 30–300[†] |
|  | 400–4000[‡] |

*Multiply the listed values by 0.176 to convert them to Btu/h·ft²·°F.
[†]Based on air-side surface area.
[‡]Based on water- or steam-side surface area.

When the tube is *finned* on one side to enhance heat transfer, the total heat transfer surface area on the finned side becomes

$$A_s = A_{\text{total}} = A_{\text{fin}} + A_{\text{unfinned}} \tag{22–6}$$

where $A_{\text{fin}}$ is the surface area of the fins and $A_{\text{unfinned}}$ is the area of the unfinned portion of the tube surface. For short fins of high thermal conductivity, we can use this total area in the convection resistance relation $R_{\text{conv}} = 1/hA_s$ since the fins in this case will be very nearly isothermal. Otherwise, we should determine the effective surface area from

$$A_s = A_{\text{unfinned}} + \eta_{\text{fin}} A_{\text{fin}} \tag{22–7}$$

where $\eta_{\text{fin}}$ is the fin efficiency which can be obtained from the relationships discussed in Chapter 17. This way, the temperature drop along the fins is accounted for. Note that $\eta_{\text{fin}} = 1$ for isothermal fins, and thus Eq. 22–7 reduces to Eq. 22–6 in that case. Note that for the finned surface, the relevant area ($A_i$ or $A_o$) in Eq. 22–4 should be calculated from Eq. 22–7.

## Fouling Factor

The performance of heat exchangers usually deteriorates with time as a result of accumulation of *deposits* on heat transfer surfaces. The layer of deposits represents *additional resistance* to heat transfer and causes the rate of heat transfer in a heat exchanger to decrease. The net effect of these accumulations on heat transfer is represented by a **fouling factor** $R_f$, which is a measure of the *thermal resistance* introduced by fouling.

The most common type of fouling is the *precipitation* of solid deposits in a fluid on the heat transfer surfaces. You can observe this type of fouling even in your house. If you check the inner surfaces of your teapot after prolonged use, you will probably notice a layer of calcium-based deposits on the surfaces at which boiling occurs. This is especially the case in areas where the water is hard. The scales of such deposits come off by scratching, and the surfaces can be cleaned of such deposits by chemical treatment. Now imagine those mineral deposits forming on the inner surfaces of fine tubes in a heat exchanger (Fig. 22–9) and the detrimental effect it may have on the flow passage area and the heat transfer. To avoid this potential problem, water in power and process plants is extensively treated and its solid contents are removed before it is allowed to circulate through the system. The solid ash particles in the flue gases accumulating on the surfaces of air preheaters create similar problems.

Another form of fouling, which is common in the chemical process industry, is *corrosion* and other *chemical fouling*. In this case, the surfaces are fouled by the accumulation of the products of chemical reactions on the surfaces. This form of fouling can be avoided by coating metal pipes with glass or using plastic pipes instead of metal ones. Heat exchangers may also be fouled by the growth of algae in warm fluids. This type of fouling is called *biological fouling* and can be prevented by chemical treatment.

In applications where it is likely to occur, fouling should be considered in the design and selection of heat exchangers. In such applications, it may be necessary to select a larger and thus more expensive heat exchanger to ensure that it meets the design heat transfer requirements even after fouling occurs.

**FIGURE 22–9**
Precipitation fouling of ash particles on superheater tubes.

*From Steam: Its Generation, and Use, Babcock and Wilcox Co., 1978. Reprinted by permission.*

## TABLE 22–2

Representative fouling factors (thermal resistance due to folling for a unit surface area)

| Fluid | $R_f$, m²·K/W |
|---|---|
| Distilled water, sea-water, river water, boiler feedwater: | |
| Below 50°C | 0.0001 |
| Above 50°C | 0.0002 |
| Fuel oil | 0.0009 |
| Steam (oil-free) | 0.0001 |
| Refrigerants (liquid) | 0.0002 |
| Refrigerants (vapor) | 0.0004 |
| Alcohol vapors | 0.0001 |
| Air | 0.0004 |

(*Source:* Tubular Exchange Manufacturers Association.)

The periodic cleaning of heat exchangers and the resulting down time are additional penalties associated with fouling.

The fouling factor is obviously zero for a new heat exchanger and increases with time as the solid deposits build up on the heat exchanger surface. The fouling factor depends on the *operating temperature* and the *velocity* of the fluids, as well as the length of service. Fouling increases with *increasing temperature* and *decreasing velocity*.

The overall heat transfer coefficient relation Eq. 22–4 or 22–5, for an unfinned double-pipe heat exchanger is valid for clean surfaces and needs to be modified to account for the effects of fouling on both the inner and the outer surfaces of the tube. For an unfinned double-pipe heat exchanger, it can be expressed as

$$\frac{1}{UA_s} = \frac{1}{U_i A_i} = \frac{1}{U_o A_o} = R = \frac{1}{h_i A_i} + \frac{R_{f,i}}{A_i} + \frac{\ln(D_o/D_i)}{2\pi k L} + \frac{R_{f,o}}{A_o} + \frac{1}{h_o A_o} \quad (22\text{–}8)$$

where $R_{f,i}$ and $R_{f,o}$ are the fouling factors at those surfaces.

Representative values of fouling factors are given in Table 22–2. More comprehensive tables of fouling factors are available in handbooks. As you would expect, considerable uncertainty exists in these values, and they should be used as a guide in the selection and evaluation of heat exchangers to account for the effects of anticipated fouling on heat transfer. Note that most fouling factors in the table are of the order of $10^{-4}$ m²·K/W, which is equivalent to the thermal resistance of a 0.2-mm-thick limestone layer ($k = 2.9$ W/m·K) per unit surface area. Therefore, in the absence of specific data, we can assume the surfaces to be coated with 0.2 mm of limestone as a starting point to account for the effects of fouling.

---

### EXAMPLE 22–1 Overall Heat Transfer Coefficient of a Heat Exchanger

Hot oil is to be cooled in a double-tube counter-flow heat exchanger. The copper inner tubes have a diameter of 2 cm and negligible thickness. The inner diameter of the outer tube (the shell) is 3 cm. Water flows through the tube at a rate of 0.5 kg/s, and the oil through the shell at a rate of 0.8 kg/s. Taking the average temperatures of the water and the oil to be 45°C and 80°C, respectively, determine the overall heat transfer coefficient of this heat exchanger.

**SOLUTION** Hot oil is cooled by water in a double-tube counter-flow heat exchanger. The overall heat transfer coefficient is to be determined.

*Assumptions* **1** The thermal resistance of the inner tube is negligible since the tube material is highly conductive and its thickness is negligible. **2** Both the oil and water flow are fully developed. **3** Properties of the oil and water are constant.

*Properties* The properties of water at 45°C are (Table A–15)

$$\rho = 990.1 \text{ kg/m}^3 \qquad \text{Pr} = 3.91$$

$$k = 0.637 \text{ W/m·K} \qquad \nu = \mu/\rho = 0.602 \times 10^{-6} \text{ m}^2/\text{s}$$

The properties of oil at 80°C are (Table A–19)

$$\rho = 852 \text{ kg/m}^3 \qquad \text{Pr} = 499.3$$

$$k = 0.138 \text{ W/m·K} \qquad \nu = 3.794 \times 10^{-5} \text{ m}^2/\text{s}$$

*Analysis* The schematic of the heat exchanger is given in Fig. 22–10. The overall heat transfer coefficient $U$ can be determined from Eq. 22–5:

$$\frac{1}{U} \approx \frac{1}{h_i} + \frac{1}{h_o}$$

where $h_i$ and $h_o$ are the convection heat transfer coefficients inside and outside the tube, respectively, which are to be determined using the forced convection relations.

The hydraulic diameter for a circular tube is the diameter of the tube itself, $D_h = D = 0.02$ m. The average velocity of water in the tube and the Reynolds number are

$$V = \frac{\dot{m}}{\rho A_c} = \frac{\dot{m}}{\rho\left(\frac{1}{4}\pi D^2\right)} = \frac{0.5 \text{ kg/s}}{(990.1 \text{ kg/m}^3)\left[\frac{1}{4}\pi (0.02 \text{ m})^2\right]} = 1.61 \text{ m/s}$$

and

$$\text{Re} = \frac{VD}{\nu} = \frac{(1.61 \text{ m/s})(0.02 \text{ m})}{0.602 \times 10^{-6} \text{ m}^2/\text{s}} = 53{,}490$$

which is greater than 10,000. Therefore, the flow of water is turbulent. Assuming the flow to be fully developed, the Nusselt number can be determined from

$$\text{Nu} = \frac{hD}{k} = 0.023 \, \text{Re}^{0.8}\text{Pr}^{0.4} = 0.023(53{,}490)^{0.8}(3.91)^{0.4} = 240.6$$

Then,

$$h = \frac{k}{D} \, \text{Nu} = \frac{0.637 \text{ W/m·K}}{0.02 \text{ m}} (240.6) = 7663 \text{ W/m}^2\text{·K}$$

Now we repeat the analysis above for oil. The properties of oil at 80°C are

$$\rho = 852 \text{ kg/m}^3 \qquad \nu = 37.5 \times 10^{-6} \text{ m}^2/\text{s}$$

$$k = 0.138 \text{ W/m·K} \qquad \text{Pr} = 490$$

The hydraulic diameter for the annular space is

$$D_h = D_o - D_i = 0.03 - 0.02 = 0.01 \text{ m}$$

The average velocity and the Reynolds number in this case are

$$V = \frac{\dot{m}}{\rho A_c} = \frac{\dot{m}}{\rho\left[\frac{1}{4}\pi(D_o^2 - D_i^2)\right]} = \frac{0.8 \text{ kg/s}}{(852 \text{ kg/m}^3)\left[\frac{1}{4}\pi(0.03^2 - 0.02^2)\right]\text{m}^2} = 2.39 \text{ m/s}$$

and

$$\text{Re} = \frac{VD}{\nu} = \frac{(2.39 \text{ m/s})(0.01 \text{ m})}{3.794 \times 10^{-5} \text{ m}^2/\text{s}} = 630$$

which is less than 2300. Therefore, the flow of oil is laminar. Assuming fully developed flow, the Nusselt number on the tube side of the annular space

**FIGURE 22–10**
Schematic for Example 22–1.

## TABLE 22-3

Nusselt number for fully developed laminar flow in a circular annulus with one surface insulated and the other isothermal (Kays and Perkins, 1972)

| $D_i/D_o$ | $Nu_i$ | $Nu_o$ |
|-----------|--------|--------|
| 0.00 | — | 3.66 |
| 0.05 | 17.46 | 4.06 |
| 0.10 | 11.56 | 4.11 |
| 0.25 | 7.37 | 4.23 |
| 0.50 | 5.74 | 4.43 |
| 1.00 | 4.86 | 4.86 |

$Nu_i$ corresponding to $D_i/D_o = 0.02/0.03 = 0.667$ can be determined from Table 22-3 by interpolation to be

$$Nu = 5.45$$

and

$$h_o = \frac{k}{D_h} Nu = \frac{0.138 \text{ W/m·K}}{0.01 \text{ m}} (5.45) = 75.2 \text{ W/m}^2\text{·K}$$

Then the overall heat transfer coefficient for this heat exchanger becomes

$$U = \frac{1}{\dfrac{1}{h_i} + \dfrac{1}{h_o}} = \frac{1}{\dfrac{1}{7663 \text{ W/m}^2\text{·K}} + \dfrac{1}{75.2 \text{ W/m}^2\text{·K}}} = \textbf{74.5 W/m}^2\textbf{·K}$$

**Discussion**  Note that $U \approx h_o$ in this case, since $h_i \gg h_o$. This confirms our earlier statement that the overall heat transfer coefficient in a heat exchanger is dominated by the smaller heat transfer coefficient when the difference between the two values is large.

To improve the overall heat transfer coefficient and thus the heat transfer in this heat exchanger, we must use some enhancement techniques on the oil side, such as a finned surface.

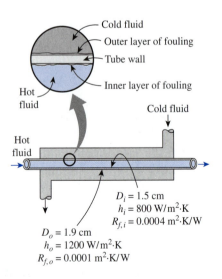

**FIGURE 22-11**
Schematic for Example 22-2.

Cold fluid
Outer layer of fouling
Tube wall
Inner layer of fouling
Hot fluid
Cold fluid
Hot fluid

$D_i = 1.5$ cm
$h_i = 800$ W/m²·K
$R_{f,i} = 0.0004$ m²·K/W

$D_o = 1.9$ cm
$h_o = 1200$ W/m²·K
$R_{f,o} = 0.0001$ m²·K/W

### EXAMPLE 22-2    Effect of Fouling on the Overall Heat Transfer Coefficient

A double-pipe (shell-and-tube) heat exchanger is constructed of a stainless steel ($k = 15.1$ W/m·K) inner tube of inner diameter $D_i = 1.5$ cm and outer diameter $D_o = 1.9$ cm and an outer shell of inner diameter 3.2 cm. The convection heat transfer coefficient is given to be $h_i = 800$ W/m²·K on the inner surface of the tube and $h_o = 1200$ W/m²·K on the outer surface. For a fouling factor of $R_{f,i} = 0.0004$ m²·K/W on the tube side and $R_{f,o} = 0.0001$ m²·K/W on the shell side, determine (a) the thermal resistance of the heat exchanger per unit length and (b) the overall heat transfer coefficients, $U_i$ and $U_o$ based on the inner and outer surface areas of the tube, respectively.

**SOLUTION**  The heat transfer coefficients and the fouling factors on the tube and shell sides of a heat exchanger are given. The thermal resistance and the overall heat transfer coefficients based on the inner and outer areas are to be determined.

**Assumptions**  The heat transfer coefficients and the fouling factors are constant and uniform.

**Analysis**  (a) The schematic of the heat exchanger is given in Fig. 22-11. The thermal resistance for an unfinned shell-and-tube heat exchanger with fouling on both heat transfer surfaces is given by Eq. 22-8 as

$$R = \frac{1}{UA_s} = \frac{1}{U_i A_i} = \frac{1}{U_o A_o} = \frac{1}{h_i A_i} + \frac{R_{f,i}}{A_i} + \frac{\ln(D_o/D_i)}{2\pi kL} + \frac{R_{f,o}}{A_o} + \frac{1}{h_o A_o}$$

where

$$A_i = \pi D_i L = \pi(0.015 \text{ m})(1 \text{ m}) = 0.0471 \text{ m}^2$$

$$A_o = \pi D_o L = \pi(0.019 \text{ m})(1 \text{ m}) = 0.0597 \text{ m}^2$$

Substituting, the total thermal resistance is determined to be

$$R = \frac{1}{(800 \text{ W/m}^2\cdot\text{K})(0.0471 \text{ m}^2)} + \frac{0.0004 \text{ m}^2\cdot\text{K/W}}{0.0471 \text{ m}^2}$$

$$+ \frac{\ln(0.019/0.015)}{2\pi(15.1 \text{ W/m}\cdot\text{K})(1 \text{ m})}$$

$$+ \frac{0.0001 \text{ m}^2\cdot\text{K/W}}{0.0597 \text{ m}^2} + \frac{1}{(1200 \text{ W/m}^2\cdot\text{K})(0.0597 \text{ m}^2)}$$

$$= (0.02654 + 0.00849 + 0.0025 + 0.00168 + 0.01396)\text{K/W}$$

$$= \mathbf{0.0532°C/W}$$

Note that about 19 percent of the total thermal resistance in this case is due to fouling and about 5 percent of it is due to the steel tube separating the two fluids. The rest (76 percent) is due to the convection resistances.

(*b*) Knowing the total thermal resistance and the heat transfer surface areas, the overall heat transfer coefficients based on the inner and outer surfaces of the tube are

$$U_i = \frac{1}{RA_i} = \frac{1}{(0.0532 \text{ K/W})(0.0471 \text{ m}^2)} = \mathbf{399 \text{ W/m}^2\cdot\text{K}}$$

and

$$U_o = \frac{1}{RA_o} = \frac{1}{(0.0532 \text{ K/W})(0.0597 \text{ m}^2)} = \mathbf{315 \text{ W/m}^2\cdot\text{K}}$$

**Discussion** Note that the two overall heat transfer coefficients differ significantly (by 27 percent) in this case because of the considerable difference between the heat transfer surface areas on the inner and the outer sides of the tube. For tubes of negligible thickness, the difference between the two overall heat transfer coefficients would be negligible.

# 22–3 · ANALYSIS OF HEAT EXCHANGERS

Heat exchangers are commonly used in practice, and an engineer often finds himself or herself in a position to *select a heat exchanger* that will achieve a *specified temperature change* in a fluid stream of known mass flow rate, or to *predict the outlet temperatures* of the hot and cold fluid streams in a *specified heat exchanger.*

In upcoming sections, we discuss the two methods used in the analysis of heat exchangers. Of these, the *log mean temperature difference* (or LMTD) method is best suited for the first task and the *effectiveness–NTU* (the number of transfer units) method for the second task. But first we present some general considerations.

Heat exchangers usually operate for long periods of time with no change in their operating conditions. Therefore, they can be modeled as *steady-flow* devices. As such, the mass flow rate of each fluid remains constant, and the fluid properties such as temperature and velocity at any inlet or outlet remain the same. Also, the fluid streams experience little or no change in their velocities and elevations, and thus the kinetic and potential energy changes are negligible. The specific heat of a fluid, in general, changes with temperature. But, in a specified temperature range, it can be treated as a constant at some average value with little loss in accuracy. Axial heat conduction along the tube is usually insignificant and can be considered negligible. Finally, the outer surface of the heat exchanger is assumed to be *perfectly insulated,* so that there is no heat loss to the surrounding medium, and any heat transfer occurs between the two fluids only.

The idealizations stated above are closely approximated in practice, and they greatly simplify the analysis of a heat exchanger with little sacrifice from accuracy. Therefore, they are commonly used. Under these assumptions, the *first law of thermodynamics* requires that the rate of heat transfer from the hot fluid be equal to the rate of heat transfer to the cold one. That is,

$$\dot{Q} = \dot{m} c_{pc}(T_{c,\,out} - T_{c,\,in}) \tag{22–9}$$

and

$$\dot{Q} = \dot{m}_h c_{ph}(T_{h,\,in} - T_{h,\,out}) \tag{22–10}$$

where the subscripts $c$ and $h$ stand for *cold* and *hot* fluids, respectively, and

$$\dot{m}_c,\ \dot{m}_h = \text{mass flow rates}$$

$$c_{pc},\ c_{ph} = \text{specific heats}$$

$$T_{c,\,out},\ T_{h,\,out} = \text{outlet temperatures}$$

$$T_{c,\,in},\ T_{h,\,in} = \text{inlet temperatures}$$

Note that the heat transfer rate $\dot{Q}$ is taken to be a positive quantity, and its direction is understood to be from the hot fluid to the cold one in accordance with the second law of thermodynamics.

In heat exchanger analysis, it is often convenient to combine the product of the mass flow rate and the specific heat of a fluid into a single quantity. This quantity is called the **heat capacity rate** and is defined for the hot and cold fluid streams as

$$C_h = \dot{m}_h c_{ph} \quad \text{and} \quad C_c = \dot{m}_c c_{pc} \tag{22–11}$$

The heat capacity rate of a fluid stream represents the rate of heat transfer needed to change the temperature of the fluid stream by 1°C as it flows through a heat exchanger. Note that in a heat exchanger, the fluid with a *large* heat capacity rate experiences a *small* temperature change, and the fluid with a *small* heat capacity rate experiences a *large* temperature change. Therefore, *doubling* the mass flow rate of a fluid while leaving everything else unchanged will *halve* the temperature change of that fluid.

With the definition of the heat capacity rate above, Eqs. 22–9 and 22–10 can also be expressed as

$$\dot{Q} = C_c(T_{c,\,out} - T_{c,\,in}) \tag{22–12}$$

and

$$\dot{Q} = C_h(T_{h,\,in} - T_{h,\,out}) \tag{22-13}$$

That is, the heat transfer rate in a heat exchanger is equal to the heat capacity rate of either fluid multiplied by the temperature change of that fluid. Note that *the only time the temperature rise of a cold fluid is equal to the temperature drop of the hot fluid is when the heat capacity rates of the two fluids are equal to each other* (Fig. 22–12).

Two special types of heat exchangers commonly used in practice are *condensers* and *boilers*. One of the fluids in a condenser or a boiler undergoes a phase-change process, and the rate of heat transfer is expressed as

$$\dot{Q} = \dot{m}h_{fg} \tag{22-14}$$

where $\dot{m}$ is the rate of evaporation or condensation of the fluid and $h_{fg}$ is the enthalpy of vaporization of the fluid at the specified temperature or pressure.

An ordinary fluid absorbs or releases a large amount of heat essentially at constant temperature during a phase-change process, as shown in Fig. 22–13. The heat capacity rate of a fluid during a phase-change process must approach infinity since the temperature change is practically zero. That is, $C = \dot{m}c_p \to \infty$ when $\Delta T \to 0$, so that the heat transfer rate $\dot{Q} = \dot{m}c_p \, \Delta T$ is a finite quantity. Therefore, in heat exchanger analysis, a condensing or boiling fluid is conveniently modeled as a fluid whose heat capacity rate is *infinity*.

The rate of heat transfer in a heat exchanger can also be expressed in an analogous manner to Newton's law of cooling as

$$\dot{Q} = UA_s \, \Delta T_m \tag{22-15}$$

where $U$ is the overall heat transfer coefficient, $A_s$ is the heat transfer surface area, and $\Delta T_m$ is an appropriate mean temperature difference between the two fluids. Here the surface area $A_s$ can be determined precisely using the dimensions of the heat exchanger. However, the overall heat transfer coefficient $U$ and the temperature difference $\Delta T$ between the hot and cold fluids, in general, may vary along the heat exchanger.

The average value of the overall heat transfer coefficient can be determined as described in the preceding section by using the average convection coefficients for each fluid. It turns out that the appropriate form of the average temperature difference between the two fluids is *logarithmic* in nature, and its determination is presented in Section 22–4. It should be noted that the average temperature difference $\Delta T_m$ is dependent on the heat exchanger flow arrangement and its type of construction.

## 22–4 ▪ THE LOG MEAN TEMPERATURE DIFFERENCE METHOD

Earlier, we mentioned that the temperature difference between the hot and cold fluids varies along the heat exchanger, and it is convenient to have a *mean temperature difference* $\Delta T_m$ for use in the relation $\dot{Q} = UA_s \, \Delta T_m$.

In order to develop a relation for the equivalent average temperature difference between the two fluids, consider the *parallel-flow double-pipe* heat exchanger

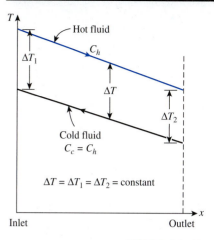

**FIGURE 22–12**
Two fluid streams that have the same capacity rates experience the same temperature change in a well-insulated heat exchanger.

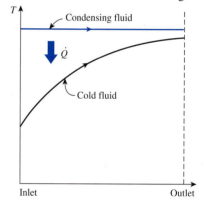

(a) Condenser ($C_h \to \infty$)

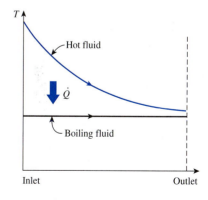

(b) Boiler ($C_c \to \infty$)

**FIGURE 22–13**
Variation of fluid temperatures in a heat exchanger when one of the fluids condenses or boils.

**FIGURE 22–14**

Variation of the fluid temperatures in a parallel-flow double-pipe heat exchanger.

shown in Fig. 22–14. Note that the temperature difference $\Delta T$ between the hot and cold fluids is large at the inlet of the heat exchanger but decreases exponentially toward the outlet. As you would expect, the temperature of the hot fluid decreases and the temperature of the cold fluid increases along the heat exchanger, but the temperature of the cold fluid can never exceed that of the hot fluid no matter how long the heat exchanger is.

Assuming the outer surface of the heat exchanger to be well insulated so that any heat transfer occurs between the two fluids, and disregarding any changes in kinetic and potential energy, an energy balance on each fluid in a differential section of the heat exchanger can be expressed as

$$\delta \dot{Q} = -\dot{m}_h c_{ph}\, dT_h \tag{22–16}$$

and

$$\delta \dot{Q} = \dot{m}_c c_{pc}\, dT_c \tag{22–17}$$

That is, the rate of heat loss from the hot fluid at any section of a heat exchanger is equal to the rate of heat gain by the cold fluid in that section. The temperature change of the hot fluid is a *negative* quantity, and so a *negative sign* is added to Eq. 22–16 to make the heat transfer rate $\dot{Q}$ a positive quantity. Solving the equations above for $dT_h$ and $dT_c$ gives

$$dT_h = -\frac{\delta \dot{Q}}{\dot{m}_h c_{ph}} \tag{22–18}$$

and

$$dT_c = \frac{\delta \dot{Q}}{\dot{m}_c c_{pc}} \tag{22–19}$$

Taking their difference, we get

$$dT_h - dT_c = d(T_h - T_c) = -\delta \dot{Q}\left(\frac{1}{\dot{m}_h c_{ph}} + \frac{1}{\dot{m}_c c_{pc}}\right) \tag{22–20}$$

The rate of heat transfer in the differential section of the heat exchanger can also be expressed as

$$\delta \dot{Q} = U(T_h - T_c)\, dA_s \tag{22–21}$$

Substituting this equation into Eq. 22–20 and rearranging give

$$\frac{d(T_h - T_c)}{T_h - T_c} = -U\, dA_s \left(\frac{1}{\dot{m}_h c_{ph}} + \frac{1}{\dot{m}_c c_{pc}}\right) \tag{22–22}$$

Integrating from the inlet of the heat exchanger to its outlet, we obtain

$$\ln \frac{T_{h,\,out} - T_{c,\,out}}{T_{h,\,in} - T_{c,\,in}} = -UA_s \left(\frac{1}{\dot{m}_h c_{ph}} + \frac{1}{\dot{m}_c c_{pc}}\right) \tag{22–23}$$

Finally, solving Eqs. 22–9 and 22–10 for $\dot{m}_c c_{pc}$ and $\dot{m}_h c_{ph}$ and substituting into Eq. 22–23 give, after some rearrangement,

$$\dot{Q} = UA_s\, \Delta T_{lm} \tag{22–24}$$

where

$$\Delta T_{lm} = \frac{\Delta T_1 - \Delta T_2}{\ln(\Delta T_1/\Delta T_2)} \tag{22-25}$$

is the **log mean temperature difference**, which is the suitable form of the average temperature difference for use in the analysis of heat exchangers. Here $\Delta T_1$ and $\Delta T_2$ represent the temperature difference between the two fluids at the two ends (inlet and outlet) of the heat exchanger. It makes no difference which end of the heat exchanger is designated as the inlet or the outlet (Fig. 22–15). It should be noted that Eqs. 22–24 and 22–25 are good for any heat exchanger provided that the end point temperature differences are defined properly. For multipass and cross-flow heat exchangers, as will be shown later, the log mean temperature difference should be corrected through a correction factor.

The temperature difference between the two fluids decreases from $\Delta T_1$ at the inlet to $\Delta T_2$ at the outlet. Thus, it is tempting to use the arithmetic mean temperature $\Delta T_{am} = \frac{1}{2}(\Delta T_1 + \Delta T_2)$ as the average temperature difference. The logarithmic mean temperature difference $\Delta T_{lm}$ is obtained by tracing the actual temperature profile of the fluids along the heat exchanger and is an *exact* representation of the *average temperature difference* between the hot and cold fluids. It truly reflects the exponential decay of the local temperature difference.

Note that $\Delta T_{lm}$ is always less than $\Delta T_{am}$. Therefore, using $\Delta T_{am}$ in calculations instead of $\Delta T_{lm}$ will overestimate the rate of heat transfer in a heat exchanger between the two fluids. When $\Delta T_1$ differs from $\Delta T_2$ by no more than 40 percent, the error in using the arithmetic mean temperature difference is less than 1 percent. But the error increases to undesirable levels when $\Delta T_1$ differs from $\Delta T_2$ by greater amounts. Therefore, we should always use the *logarithmic mean temperature difference* when determining the rate of heat transfer in a heat exchanger.

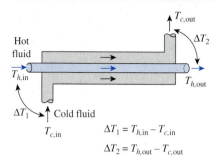

$$\Delta T_1 = T_{h,in} - T_{c,in}$$
$$\Delta T_2 = T_{h,out} - T_{c,out}$$

(a) Parallel-flow heat exchangers

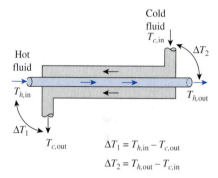

$$\Delta T_1 = T_{h,in} - T_{c,out}$$
$$\Delta T_2 = T_{h,out} - T_{c,in}$$

(b) Counter-flow heat exchangers

**FIGURE 22–15**
The $\Delta T_1$ and $\Delta T_2$ expressions in parallel-flow and counter-flow heat exchangers.

## Counter-Flow Heat Exchangers

The variation of temperatures of hot and cold fluids in a counter-flow heat exchanger is given in Fig. 22–16. Note that the hot and cold fluids enter the heat exchanger from opposite ends, and the outlet temperature of the *cold fluid* in this case may exceed the outlet temperature of the *hot fluid*. In the limiting case, the cold fluid will be heated to the inlet temperature of the hot fluid. However, the outlet temperature of the cold fluid can *never* exceed the inlet temperature of the hot fluid, since this would be a violation of the second law of thermodynamics.

The relation already given for the log mean temperature difference is developed using a parallel-flow heat exchanger, but we can show by repeating the analysis for a counter-flow heat exchanger that is also applicable to counter-flow heat exchangers. But this time, $\Delta T_1$ and $\Delta T_2$ are expressed as shown in Fig. 22–15.

For specified inlet and outlet temperatures, the log mean temperature difference for a counter-flow heat exchanger is always greater than that for a parallel-flow heat exchanger. That is, $\Delta T_{lm, CF} > \Delta T_{lm, PF}$, and thus a smaller surface area (and thus a smaller heat exchanger) is needed to achieve a specified

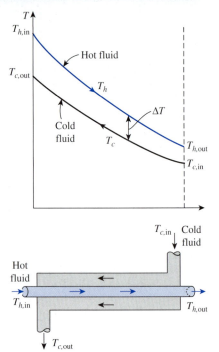

**FIGURE 22–16**
The variation of the fluid temperatures in a counter-flow double-pipe heat exchanger.

Heat transfer rate:
$$\dot{Q} = UA_s F \Delta T_{lm,CF}$$

where
$$\Delta T_{lm,CF} = \frac{\Delta T_1 - \Delta T_2}{\ln(\Delta T_1/\Delta T_2)}$$

$$\Delta T_1 = T_{h,in} - T_{c,out}$$
$$\Delta T_2 = T_{h,out} - T_{c,in}$$

and    $F = ...$ (Fig. 22–18)

**FIGURE 22–17**
The determination of the heat transfer rate for cross-flow and multipass shell-and-tube heat exchangers using the correction factor.

heat transfer rate in a counter-flow heat exchanger, assuming the same value of the overall heat transfer coefficient. Therefore, it is common practice to use counter-flow arrangements in heat exchangers.

In a counter-flow heat exchanger, the temperature difference between the hot and the cold fluids remains constant along the heat exchanger when the *heat capacity rates* of the two fluids are *equal* (that is, $\Delta T$ = constant when $C_h = C_c$ or $\dot{m}_h c_{ph} = \dot{m}_c c_{pc}$). Then we have $\Delta T_1 = \Delta T_2$, and the log mean temperature difference relation gives $\Delta T_{lm} = \frac{0}{0}$, which is indeterminate. It can be shown by the application of l'Hôpital's rule that in this case we have $\Delta T_{lm} = \Delta T_1 = \Delta T_2$, as expected.

A *condenser* or a *boiler* can be considered to be either a parallel- or counter-flow heat exchanger since both approaches give the same result.

## Multipass and Cross-Flow Heat Exchangers: Use of a Correction Factor

The log mean temperature difference $\Delta T_{lm}$ relation developed earlier is limited to parallel-flow and counter-flow heat exchangers only. Similar relations are also developed for *cross-flow* and *multipass shell-and-tube* heat exchangers, but the resulting expressions are too complicated because of the complex flow conditions.

In such cases, it is convenient to relate the equivalent temperature difference to the log mean temperature difference relation for the counter-flow case as

$$\Delta T_{lm} = F \Delta T_{lm, CF} \qquad (22\text{–}26)$$

where $F$ is the **correction factor**, which depends on the *geometry* of the heat exchanger and the inlet and outlet temperatures of the hot and cold fluid streams. The $\Delta T_{lm, CF}$ is the log mean temperature difference for the case of a *counter-flow* heat exchanger with the same inlet and outlet temperatures and is determined from Eq. 22–25 by taking $\Delta T_1 = T_{h, in} - T_{c, out}$ and $\Delta T_2 = T_{h, out} - T_{c, in}$ (Fig. 22–17).

The correction factor is less than unity for a cross-flow and multipass shell-and-tube heat exchanger. That is, $F \leq 1$. The limiting value of $F = 1$ corresponds to the counter-flow heat exchanger. Thus, the correction factor $F$ for a heat exchanger is *a measure of deviation of the $\Delta T_{lm}$ from the corresponding values for the counter-flow case.*

The correction factor $F$ for common cross-flow and shell-and-tube heat exchanger configurations is given in Fig. 22–18 versus two temperature ratios $P$ and $R$ defined as

$$P = \frac{t_2 - t_1}{T_1 - t_1} \qquad (22\text{–}27)$$

and

$$R = \frac{T_1 - T_2}{t_2 - t_1} = \frac{(\dot{m}c_p)_{tube\,side}}{(\dot{m}c_p)_{shell\,side}} \qquad (22\text{–}28)$$

where the subscripts 1 and 2 represent the *inlet* and *outlet*, respectively. Note that for a shell-and-tube heat exchanger, $T$ and $t$ represent the *shell-* and *tube-side* temperatures, respectively, as shown in the correction factor charts.

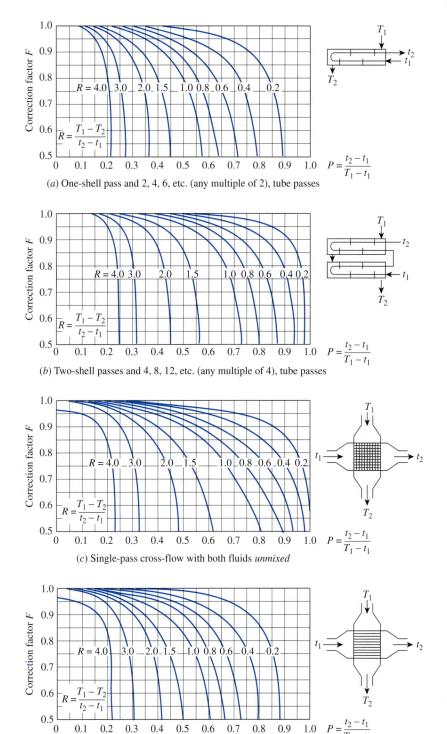

(a) One-shell pass and 2, 4, 6, etc. (any multiple of 2), tube passes

(b) Two-shell passes and 4, 8, 12, etc. (any multiple of 4), tube passes

(c) Single-pass cross-flow with both fluids *unmixed*

(d) Single-pass cross-flow with one fluid *mixed* and the other *unmixed*

**FIGURE 22–18**

Correction factor $F$ charts for common shell-and-tube and cross-flow heat exchangers.

*From Bowman, Mueller, and Nagle, 1940.*

It makes no difference whether the hot or the cold fluid flows through the shell or the tube. The determination of the correction factor $F$ requires the availability of the inlet and the outlet temperatures for both the cold and hot fluids.

Note that the value of $P$ ranges from 0 to 1. The value of $R$, on the other hand, ranges from 0 to infinity, with $R = 0$ corresponding to the phase-change (condensation or boiling) on the shell-side and $R \to \infty$ to phase-change on the tube side. The correction factor is $F = 1$ for both of these limiting cases. Therefore, the correction factor for a condenser or boiler is $F = 1$, regardless of the configuration of the heat exchanger.

---

### EXAMPLE 22–3    The Condensation of Steam in a Condenser

Steam in the condenser of a power plant is to be condensed at a temperature of 30°C with cooling water from a nearby lake, which enters the tubes of the condenser at 14°C and leaves at 22°C. The surface area of the tubes is 45 m², and the overall heat transfer coefficient is 2100 W/m²·K. Determine the mass flow rate of the cooling water needed and the rate of condensation of the steam in the condenser.

**SOLUTION**   Steam is condensed by cooling water in the condenser of a power plant. The mass flow rate of the cooling water and the rate of condensation are to be determined.

**Assumptions**   **1** Steady operating conditions exist. **2** The heat exchanger is well insulated so that heat loss to the surroundings is negligible. **3** Changes in the kinetic and potential energies of fluid streams are negligible. **4** There is no fouling. **5** Fluid properties are constant.

**Properties**   The heat of vaporization of water at 30°C is $h_{fg}$ = 2431 kJ/kg and the specific heat of cold water at the average temperature of 18°C is $c_p$ = 4184 J/kg·K (Table A–15).

**Analysis**   The schematic of the condenser is given in Fig. 22–19. The condenser can be treated as a counter-flow heat exchanger since the temperature of one of the fluids (the steam) remains constant.

The temperature difference between the steam and the cooling water at the two ends of the condenser is

$$\Delta T_1 = T_{h,\,in} - T_{c,\,out} = (30 - 22)°C = 8°C$$

$$\Delta T_2 = T_{h,\,out} - T_{c,\,in} = (30 - 14)°C = 16°C$$

That is, the temperature difference between the two fluids varies from 8°C at one end to 16°C at the other. The proper average temperature difference between the two fluids is the *log mean temperature difference* (not the arithmetic), which is determined from

$$\Delta T_{lm} = \frac{\Delta T_1 - \Delta T_2}{\ln(\Delta T_1/\Delta T_2)} = \frac{8 - 16}{\ln(8/16)} = 11.5°C$$

This is a little less than the arithmetic mean temperature difference of $\frac{1}{2}(8 + 16) = 12°C$. Then the heat transfer rate in the condenser is determined from

$$\dot{Q} = UA_s\,\Delta T_{lm} = (2100\ \text{W/m}^2\text{·K})(45\ \text{m}^2)(11.5°C) = 1.087 \times 10^6\ \text{W} = 1087\ \text{kW}$$

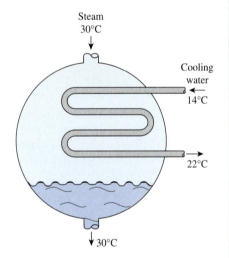

Steam
30°C

Cooling water
14°C

22°C

30°C

**FIGURE 22–19**
Schematic for Example 22–3.

Therefore, steam will lose heat at a rate of 1087 kW as it flows through the condenser, and the cooling water will gain practically all of it, since the condenser is well insulated.

The mass flow rate of the cooling water and the rate of the condensation of the steam are determined from $\dot{Q} = [\dot{m}c_p(T_{out} - T_{in})]_{cooling\ water} = (\dot{m}h_{fg})_{steam}$ to be

$$\dot{m}_{cooling\ water} = \frac{\dot{Q}}{c_p\,(T_{out} - T_{in})} = \frac{1087\ kJ/s}{(4.184\ kJ/kg\cdot K)(22 - 14)^\circ C} = \textbf{32.5 kg/s}$$

and

$$\dot{m}_{steam} = \frac{\dot{Q}}{h_{fg}} = \frac{1087\ kJ/s}{2431\ kJ/kg} = \textbf{0.45 kg/s}$$

Therefore, we need to circulate about 72 kg of cooling water for each 1 kg of steam condensing to remove the heat released during the condensation process.

## EXAMPLE 22–4   Heating Water in a Counter-Flow Heat Exchanger

A counter-flow double-pipe heat exchanger is to heat water from 20°C to 80°C at a rate of 1.2 kg/s. The heating is to be accomplished by geothermal water available at 160°C at a mass flow rate of 2 kg/s. The inner tube is thin-walled and has a diameter of 1.5 cm. If the overall heat transfer coefficient of the heat exchanger is 640 W/m²·K, determine the length of the heat exchanger required to achieve the desired heating.

**SOLUTION**   Water is heated in a counter-flow double-pipe heat exchanger by geothermal water. The required length of the heat exchanger is to be determined.

**Assumptions**   **1** Steady operating conditions exist. **2** The heat exchanger is well insulated so that heat loss to the surroundings is negligible. **3** Changes in the kinetic and potential energies of fluid streams are negligible. **4** There is no fouling. **5** Fluid properties are constant.

**Properties**   We take the specific heats of water and geothermal fluid to be 4.18 and 4.31 kJ/kg·K, respectively.

**Analysis**   The schematic of the heat exchanger is given in Fig. 22–20. The rate of heat transfer in the heat exchanger can be determined from

$$\dot{Q} = [\dot{m}c_p(T_{out} - T_{in})]_{water} = (1.2\ kg/s)(4.18\ kJ/kg\cdot K)(80 - 20)^\circ C = 301\ kW$$

Noting that all of this heat is supplied by the geothermal water, the outlet temperature of the geothermal water is determined to be

$$\dot{Q} = [\dot{m}c_p(T_{in} - T_{out})]_{geothermal} \longrightarrow T_{out} = T_{in} - \frac{\dot{Q}}{\dot{m}c_p}$$

$$= 160^\circ C - \frac{301\ kW}{(2\ kg/s)(4.31\ kJ/kg\cdot K)}$$

$$= 125^\circ C$$

**FIGURE 22–20**
Schematic for Example 22–4.

Knowing the inlet and outlet temperatures of both fluids, the logarithmic mean temperature difference for this counter-flow heat exchanger becomes

$$\Delta T_1 = T_{h,\,in} - T_{c,\,out} = (160 - 80)°C = 80°C$$

$$\Delta T_2 = T_{h,\,out} - T_{c,\,in} = (125 - 20)°C = 105°C$$

and

$$\Delta T_{lm} = \frac{\Delta T_1 - \Delta T_2}{\ln(\Delta T_1/\Delta T_2)} = \frac{80 - 105}{\ln(80/105)} = 91.9°C$$

Then the surface area of the heat exchanger is determined to be

$$\dot{Q} = UA_s\,\Delta T_{lm} \longrightarrow A_s = \frac{\dot{Q}}{U\,\Delta T_{lm}} = \frac{301,000 \text{ W}}{(640 \text{ W/m}^2\cdot\text{K})(91.9°C)} = 5.12 \text{ m}^2$$

To provide this much heat transfer surface area, the length of the tube must be

$$A_s = \pi DL \longrightarrow L = \frac{A_s}{\pi D} = \frac{5.12 \text{ m}^2}{\pi(0.015 \text{ m})} = \mathbf{109 \text{ m}}$$

**Discussion** The inner tube of this counter-flow heat exchanger (and thus the heat exchanger itself) needs to be over 100 m long to achieve the desired heat transfer, which is impractical. In cases like this, we need to use a plate heat exchanger or a multipass shell-and-tube heat exchanger with multiple passes of tube bundles.

Cold
glycerin
20°C

40°C
Hot
water

80°C

50°C

**FIGURE 22–21**
Schematic for Example 22–5.

**EXAMPLE 22–5   Heating of Glycerin in a Multipass Heat Exchanger**

A 2-shell passes and 4-tube passes heat exchanger is used to heat glycerin from 20°C to 50°C by hot water, which enters the thin-walled 2-cm-diameter tubes at 80°C and leaves at 40°C (Fig. 22–21). The total length of the tubes in the heat exchanger is 60 m. The convection heat transfer coefficient is 25 W/m²·K on the glycerin (shell) side and 160 W/m²·K on the water (tube) side. Determine the rate of heat transfer in the heat exchanger (a) before any fouling and (b) after fouling with a fouling factor of 0.0006 m²·K/W occurs on the outer surfaces of the tubes.

**SOLUTION** Glycerin is heated in a 2-shell passes and 4-tube passes heat exchanger by hot water. The rate of heat transfer for the cases of fouling and no fouling are to be determined.

**Assumptions** 1 Steady operating conditions exist. 2 The heat exchanger is well insulated so that heat loss to the surroundings is negligible. 3 Changes in the kinetic and potential energies of fluid streams are negligible. 4 Heat transfer coefficients and fouling factors are constant and uniform. 5 The thermal resistance of the inner tube is negligible since the tube is thin-walled and highly conductive.

*Analysis* The tubes are said to be thin-walled, and thus it is reasonable to assume the inner and outer surface areas of the tubes to be equal. Then the heat transfer surface area becomes

$$A_s = \pi D L = \pi(0.02 \text{ m})(60 \text{ m}) = 3.77 \text{ m}^2$$

The rate of heat transfer in this heat exchanger can be determined from

$$\dot{Q} = U A_s F \, \Delta T_{\text{lm, CF}}$$

where $F$ is the correction factor and $\Delta T_{\text{lm, CF}}$ is the log mean temperature difference for the counter-flow arrangement. These two quantities are determined from

$$\Delta T_1 = T_{h,\text{ in}} - T_{c,\text{ out}} = (80 - 50)°\text{C} = 30°\text{C}$$

$$\Delta T_2 = T_{h,\text{ out}} - T_{c,\text{ in}} = (40 - 20)°\text{C} = 20°\text{C}$$

$$\Delta T_{\text{lm, CF}} = \frac{\Delta T_1 - \Delta T_2}{\ln (\Delta T_1/\Delta T_2)} = \frac{30 - 20}{\ln (30/20)} = 24.7°\text{C}$$

and

$$\left.\begin{array}{l} P = \dfrac{t_2 - t_1}{T_1 - t_1} = \dfrac{40 - 80}{20 - 80} = 0.67 \\[2mm] R = \dfrac{T_1 - T_2}{t_2 - t_1} = \dfrac{20 - 50}{40 - 80} = 0.75 \end{array}\right\} \quad F = 0.91 \qquad \text{(Fig. 22–18}b\text{)}$$

(*a*) In the case of no fouling, the overall heat transfer coefficient $U$ is

$$U = \frac{1}{\dfrac{1}{h_i} + \dfrac{1}{h_o}} = \frac{1}{\dfrac{1}{160 \text{ W/m}^2\cdot\text{K}} + \dfrac{1}{25 \text{ W/m}^2\cdot\text{K}}} = 21.6 \text{ W/m}^2\cdot\text{K}$$

Then the rate of heat transfer becomes

$$\dot{Q} = U A_s F \, \Delta T_{\text{lm, CF}} = (21.6 \text{ W/m}^2\cdot\text{K})(3.77 \text{ m}^2)(0.91)(24.7°\text{C}) = \mathbf{1830 \text{ W}}$$

(*b*) When there is fouling on one of the surfaces, we have

$$U = \frac{1}{\dfrac{1}{h_i} + \dfrac{1}{h_o} + R_f} = \frac{1}{\dfrac{1}{160 \text{ W/m}^2\cdot\text{K}} + \dfrac{1}{25 \text{ W/m}^2\cdot\text{K}} + 0.0006 \text{ m}^2\cdot\text{K/ W}}$$

$$= 21.3 \text{ W/m}^2\cdot\text{K}$$

and

$$\dot{Q} = U A_s F \, \Delta T_{\text{lm, CF}} = (21.3 \text{ W/m}^2\cdot\text{K})(3.77 \text{ m}^2)(0.91)(24.7°\text{C}) = \mathbf{1805 \text{ W}}$$

*Discussion* Note that the rate of heat transfer decreases as a result of fouling, as expected. The decrease is not dramatic, however, because of the relatively low convection heat transfer coefficients involved.

90°C

Air flow
(*unmixed*)
20°C

40°C

65°C
Water flow
(*unmixed*)

**FIGURE 22–22**
Schematic for Example 22–6.

### EXAMPLE 22–6    Cooling of Water in an Automotive Radiator

A test is conducted to determine the overall heat transfer coefficient in an automotive radiator that is a compact cross-flow water-to-air heat exchanger with both fluids (air and water) unmixed (Fig. 22–22). The radiator has 40 tubes of internal diameter 0.5 cm and length 65 cm in a closely spaced plate-finned matrix. Hot water enters the tubes at 90°C at a rate of 0.6 kg/s and leaves at 65°C. Air flows across the radiator through the interfin spaces and is heated from 20°C to 40°C. Determine the overall heat transfer coefficient $U_i$ of this radiator based on the inner surface area of the tubes.

**SOLUTION**    During an experiment involving an automotive radiator, the inlet and exit temperatures of water and air and the mass flow rate of water are measured. The overall heat transfer coefficient based on the inner surface area is to be determined.

**Assumptions**    1 Steady operating conditions exist. 2 Changes in the kinetic and potential energies of fluid streams are negligible. 3 Fluid properties are constant.

**Properties**    The specific heat of water at the average temperature of (90 + 65)/2 = 77.5°C is 4.195 kJ/kg·K (Table A–15).

**Analysis**    The rate of heat transfer in this radiator from the hot water to the air is determined from an energy balance on water flow,

$$\dot{Q} = [\dot{m}c_p(T_{in} - T_{out})]_{water} = (0.6 \text{ kg/s})(4.195 \text{ kJ/kg·K})(90 - 65)°C$$

$$= 62.93 \text{ kW}$$

The tube-side heat transfer area is the total surface area of the tubes, and is determined from

$$A_i = n\pi D_i L = (40)\pi(0.005 \text{ m})(0.65 \text{ m}) = 0.408 \text{ m}^2$$

Knowing the rate of heat transfer and the surface area, the overall heat transfer coefficient can be determined from

$$\dot{Q} = U_i A_i F \, \Delta T_{lm, \, CF} \longrightarrow U_i = \frac{\dot{Q}}{A_i F \, \Delta T_{lm, \, CF}}$$

where $F$ is the correction factor and $\Delta T_{lm, \, CF}$ is the log mean temperature difference for the counter-flow arrangement. These two quantities are found to be

$$\Delta T_1 = T_{h, \, in} - T_{c, \, out} = (90 - 40)°C = 50°C$$

$$\Delta T_2 = T_{h, \, out} - T_{c, \, in} = (65 - 20)°C = 45°C$$

$$\Delta T_{lm, \, CF} = \frac{\Delta T_1 - \Delta T_2}{\ln (\Delta T_1/\Delta T_2)} = \frac{50 - 45}{\ln (50/45)} = 47.5°C$$

and

$$\left. \begin{aligned} P &= \frac{t_2 - t_1}{T_1 - t_1} = \frac{65 - 90}{20 - 90} = 0.36 \\ R &= \frac{T_1 - T_2}{t_2 - t_1} = \frac{20 - 40}{65 - 90} = 0.80 \end{aligned} \right\} \quad F = 0.97 \qquad \text{(Fig. 22–18c)}$$

Substituting, the overall heat transfer coefficient $U_i$ is determined to be

$$U_i = \frac{\dot{Q}}{A_i F \, \Delta T_{lm,\,CF}} = \frac{62{,}930 \text{ W}}{(0.408 \text{ m}^2)(0.97)(47.5°C)} = 3347 \text{ W/m}^2 \cdot \text{K}$$

**Discussion**  Note that the overall heat transfer coefficient on the air side will be much lower because of the large surface area involved on that side.

## 22–5 ▪ THE EFFECTIVENESS–NTU METHOD

The log mean temperature difference (LMTD) method discussed in Section 22–4 is easy to use in heat exchanger analysis when the inlet and the outlet temperatures of the hot and cold fluids are known or can be determined from an energy balance. Once $\Delta T_{lm}$, the mass flow rates, and the overall heat transfer coefficient are available, the heat transfer surface area of the heat exchanger can be determined from

$$\dot{Q} = UA_s \, \Delta T_{lm}$$

Therefore, the LMTD method is very suitable for determining the *size* of a heat exchanger to realize prescribed outlet temperatures when the mass flow rates and the inlet and outlet temperatures of the hot and cold fluids are specified.

With the LMTD method, the task is to *select* a heat exchanger that will meet the prescribed heat transfer requirements. The procedure to be followed by the selection process is:

1. Select the type of heat exchanger suitable for the application.
2. Determine any unknown inlet or outlet temperature and the heat transfer rate using an energy balance.
3. Calculate the log mean temperature difference $\Delta T_{lm}$ and the correction factor $F$, if necessary.
4. Obtain (select or calculate) the value of the overall heat transfer coefficient $U$.
5. Calculate the heat transfer surface area $A_s$.

The task is completed by selecting a heat exchanger that has a heat transfer surface area equal to or larger than $A_s$.

A second kind of problem encountered in heat exchanger analysis is the determination of the *heat transfer rate* and the *outlet temperatures* of the hot and cold fluids for prescribed fluid mass flow rates and inlet temperatures when the *type* and *size* of the heat exchanger are specified. The heat transfer surface area of the heat exchanger in this case is known, but the *outlet temperatures* are not. Here the task is to determine the heat transfer performance of a specified heat exchanger or to determine if a heat exchanger available in storage will do the job.

The LMTD method could still be used for this alternative problem, but the procedure would require tedious iterations, and thus it is not practical. In an attempt to eliminate the iterations from the solution of such problems, Kays and London came up with a method in 1955 called the **effectiveness–NTU method**, which greatly simplified heat exchanger analysis.

This method is based on a dimensionless parameter called the **heat transfer effectiveness ε**, defined as

$$\varepsilon = \frac{\dot{Q}}{Q_{max}} = \frac{\text{Actual heat transfer rate}}{\text{Maximum possible heat transfer rate}} \quad \text{(22–29)}$$

The *actual* heat transfer rate in a heat exchanger can be determined from an energy balance on the hot or cold fluids and can be expressed as

$$\dot{Q} = C_c(T_{c,\,out} - T_{c,\,in}) = C_h(T_{h,\,in} - T_{h,\,out}) \quad \text{(22–30)}$$

where $C_c = \dot{m}_c c_{pc}$ and $C_h = \dot{m}_c c_{ph}$ are the heat capacity rates of the cold and hot fluids, respectively.

To determine the maximum possible heat transfer rate in a heat exchanger, we first recognize that the *maximum temperature difference* in a heat exchanger is the difference between the *inlet* temperatures of the hot and cold fluids. That is,

$$\Delta T_{max} = T_{h,\,in} - T_{c,\,in} \quad \text{(22–31)}$$

The heat transfer in a heat exchanger will reach its maximum value when (1) the cold fluid is heated to the inlet temperature of the hot fluid or (2) the hot fluid is cooled to the inlet temperature of the cold fluid. These two limiting conditions will not be reached simultaneously unless the heat capacity rates of the hot and cold fluids are identical (i.e., $C_c = C_h$). When $C_c \neq C_h$, which is usually the case, the fluid with the *smaller* heat capacity rate will experience a larger temperature change, and thus it will be the first to experience the maximum temperature, at which point the heat transfer will come to a halt. Therefore, the maximum possible heat transfer rate in a heat exchanger is (Fig. 22–23)

$$\dot{Q}_{max} = C_{min}(T_{h,\,in} - T_{c,\,in}) \quad \text{(22–32)}$$

where $C_{min}$ is the smaller of $C_h$ and $C_c$. This is further clarified by Example 22–7.

20°C
25 kg/s
4.18 kJ/kg·K
Cold water

Hot oil
130°C
40 kg/s
2.3 kJ/kg·K

$C_c = \dot{m}_c c_{pc} = 104.5$ kW/K

$C_h = \dot{m}_c c_{ph} = 92$ kW/K

$C_{min} = 92$ kW/K

$\Delta T_{max} = T_{h,in} - T_{c,in} = 110°C$

$\dot{Q}_{max} = C_{min} \Delta T_{max} = 10{,}120$ kW

**FIGURE 22–23**
The determination of the maximum rate of heat transfer in a heat exchanger.

---

**EXAMPLE 22–7**    **Upper Limit for Heat Transfer in a Heat Exchanger**

Cold water enters a counter-flow heat exchanger at 10°C at a rate of 8 kg/s, where it is heated by a hot-water stream that enters the heat exchanger at 70°C at a rate of 2 kg/s. Assuming the specific heat of water to remain constant at $c_p = 4.18$ kJ/kg·K, determine the maximum heat transfer rate and the outlet temperatures of the cold- and the hot-water streams for this limiting case.

**SOLUTION**   Cold- and hot-water streams enter a heat exchanger at specified temperatures and flow rates. The maximum rate of heat transfer in the heat exchanger and the outlet temperatures are to be determined.

*Assumptions*   1 Steady operating conditions exist. 2 The heat exchanger is well insulated so that heat loss to the surroundings is negligible. 3 Changes in the kinetic and potential energies of fluid streams are negligible. **4** Fluid properties are constant.

**Properties** The specific heat of water is given to be $c_p = 4.18$ kJ/kg·K.

**Analysis** A schematic of the heat exchanger is given in Fig. 22–24. The heat capacity rates of the hot and cold fluids are

$$C_h = \dot{m}_h c_{ph} = (2 \text{ kg/s})(4.18 \text{ kJ/kg·K}) = 8.36 \text{ kW/K}$$

and

$$C_c = \dot{m}_c c_{pc} = (8 \text{ kg/s})(4.18 \text{ kJ/kg·K}) = 33.4 \text{ kW/K}$$

Therefore,

$$C_{min} = C_h = 8.36 \text{ kW/K}$$

which is the smaller of the two heat capacity rates. Then the maximum heat transfer rate is determined from Eq. 22–32 to be

$$\dot{Q}_{max} = C_{min}(T_{h, in} - T_{c, in})$$

$$= (8.36 \text{ kW/K})(70 - 10)°C$$

$$= \mathbf{502 \text{ kW}}$$

That is, the maximum possible heat transfer rate in this heat exchanger is 502 kW. This value would be approached in a counter-flow heat exchanger with a *very large* heat transfer surface area.

The maximum temperature difference in this heat exchanger is $\Delta T_{max} = T_{h, in} - T_{c, in} = (70 - 10)°C = 60°C$. Therefore, the hot water cannot be cooled by more than 60°C (to 10°C) in this heat exchanger, and the cold water cannot be heated by more than 60°C (to 70°C), no matter what we do. The outlet temperatures of the cold and the hot streams in this limiting case are determined to be

$$\dot{Q} = C_c(T_{c, out} - T_{c, in}) \longrightarrow T_{c, out} = T_{c, in} + \frac{\dot{Q}}{C_c} = 10°C + \frac{502 \text{ kW}}{33.4 \text{ kW/K}} = \mathbf{25°C}$$

$$\dot{Q} = C_h(T_{h, in} - T_{h, out}) \longrightarrow T_{h, out} = T_{h, in} - \frac{\dot{Q}}{C_h} = 70°C - \frac{502 \text{ kW}}{8.38 \text{ kW/K}} = \mathbf{10°C}$$

**Discussion** Note that the hot water is cooled to the limit of 10°C (the inlet temperature of the cold-water stream), but the cold water is heated to 25°C only when maximum heat transfer occurs in the heat exchanger. This is not surprising, since the mass flow rate of the hot water is only one-fourth that of the cold water, and, as a result, the temperature of the cold water increases by 0.25°C for each 1°C drop in the temperature of the hot water.

You may be tempted to think that the cold water should be heated to 70°C in the limiting case of maximum heat transfer. But this will require the temperature of the hot water to drop to −170°C (below 10°C), which is impossible. Therefore, heat transfer in a heat exchanger reaches its maximum value when the fluid with the smaller heat capacity rate (or the smaller mass flow rate when both fluids have the same specific heat value) experiences the maximum

**FIGURE 22–24**
Schematic for Example 22–7.

$$\dot{Q} = \dot{m}_h c_{ph} \Delta T_h$$
$$= \dot{m}_c c_{pc} \Delta T_c$$

If $\quad \dot{m}_c c_{pc} = \dot{m}_h c_{ph}$

then $\quad \Delta T_h = \Delta T_c$

**FIGURE 22–25**
The temperature rise of the cold fluid in a heat exchanger will be equal to the temperature drop of the hot fluid when the heat capacity rates of the hot and cold fluids are identical.

temperature change. This example explains why we use $C_{min}$ in the evaluation of $\dot{Q}_{max}$ instead of $C_{max}$.

We can show that the hot water will leave at the inlet temperature of the cold water and vice versa in the limiting case of maximum heat transfer when the mass flow rates of the hot- and cold-water streams are identical (Fig. 22–25). We can also show that the outlet temperature of the cold water will reach the 70°C limit when the mass flow rate of the hot water is greater than that of the cold water.

The determination of $\dot{Q}_{max}$ requires the availability of the *inlet temperature* of the hot and cold fluids and their *mass flow rates*, which are usually specified. Then, once the effectiveness of the heat exchanger is known, the actual heat transfer rate $\dot{Q}$ can be determined from

$$\dot{Q} = \varepsilon \dot{Q}_{max} = \varepsilon C_{min}(T_{h, in} - T_{c, in}) \tag{22–33}$$

where

if $C_c = C_{min}$:  $\quad \varepsilon = \dfrac{\dot{Q}}{\dot{Q}_{max}} = \dfrac{C_c(T_{c, out} - T_{c, in})}{C_c(T_{h, in} - T_{c, in})} = \dfrac{T_{c, out} - T_{c, in}}{T_{h, in} - T_{c, in}}$

if $C_h = C_{min}$:  $\quad \varepsilon = \dfrac{\dot{Q}}{\dot{Q}_{max}} = \dfrac{C_h(T_{h, in} - T_{h, out})}{C_h(T_{h, in} - T_{c, in})} = \dfrac{T_{h, out} - T_{h, out}}{T_{h, in} - T_{c, in}}$

Therefore, the effectiveness of a heat exchanger enables us to determine the heat transfer rate without knowing the *outlet temperatures* of the fluids.

The effectiveness of a heat exchanger depends on the *geometry* of the heat exchanger as well as the *flow arrangement.* Therefore, different types of heat exchangers have different effectiveness relations. Below we illustrate the development of the effectiveness $\varepsilon$ relation for the double-pipe *parallel-flow* heat exchanger.

Equation 22–23 developed in Section 22–4 for a parallel-flow heat exchanger can be rearranged as

$$\ln \frac{T_{h, out} - T_{c, out}}{T_{h, in} - T_{c, in}} = - \frac{UA_s}{C_c} \left(1 + \frac{C_c}{C_h}\right) \tag{22–34}$$

Also, solving Eq. 22–30 for $T_{h, out}$ gives

$$T_{h, out} = T_{h, in} - \frac{C_c}{C_h}(T_{c, out} - T_{c, in}) \tag{22–35}$$

Substituting this relation into Eq. 22–34 after adding and subtracting $T_{c, in}$ gives

$$\ln \frac{T_{h, in} - T_{c, in} + T_{c, in} - T_{c, out} - \dfrac{C_c}{C_h}(T_{c, out} - T_{c, in})}{T_{h, in} - T_{c, in}} = - \frac{UA_s}{C_c}\left(1 + \frac{C_c}{C_h}\right)$$

which simplifies to

$$\ln\left[1 - \left(1 + \frac{C_c}{C_h}\right)\frac{T_{c,\,out} - T_{c,\,in}}{T_{h,\,in} - T_{c,\,in}}\right] = -\frac{UA_s}{C_c}\left(1 + \frac{C_c}{C_h}\right) \qquad (22\text{–}36)$$

We now manipulate the definition of effectiveness to obtain

$$\varepsilon = \frac{\dot{Q}}{\dot{Q}_{max}} = \frac{C_c(T_{c,\,out} - T_{c,\,in})}{C_{min}(T_{h,\,in} - T_{c,\,in})} \longrightarrow \frac{T_{c,\,out} - T_{c,\,in}}{T_{h,\,in} - T_{c,\,in}} = \varepsilon\frac{C_{min}}{C_c}$$

Substituting this result into Eq. 22–36 and solving for $\varepsilon$ gives the following relation for the effectiveness of a *parallel-flow* heat exchanger:

$$\varepsilon_{\text{parallel flow}} = \frac{1 - \exp\left[-\dfrac{UA_s}{C_c}\left(1 + \dfrac{C_c}{C_h}\right)\right]}{\left(1 + \dfrac{C_c}{C_h}\right)\dfrac{C_{min}}{C_c}} \qquad (22\text{–}37)$$

Taking either $C_c$ or $C_h$ to be $C_{min}$ (both approaches give the same result), the relation above can be expressed more conveniently as

$$\varepsilon_{\text{parallel flow}} = \frac{1 - \exp\left[-\dfrac{UA_s}{C_{min}}\left(1 + \dfrac{C_{min}}{C_{max}}\right)\right]}{1 + \dfrac{C_{min}}{C_{max}}} \qquad (22\text{–}38)$$

Again $C_{min}$ is the *smaller* heat capacity ratio and $C_{max}$ is the larger one, and it makes no difference whether $C_{min}$ belongs to the hot or cold fluid.

Effectiveness relations of the heat exchangers typically involve the *dimensionless* group $UA_s/C_{min}$. This quantity is called the **number of transfer units NTU** and is expressed as

$$\text{NTU} = \frac{UA_s}{C_{min}} = \frac{UA_s}{(\dot{m}c_p)_{min}} \qquad (22\text{–}39)$$

where $U$ is the overall heat transfer coefficient and $A_s$ is the heat transfer surface area of the heat exchanger. Note that NTU is proportional to $A_s$. Therefore, for specified values of $U$ and $C_{min}$, the value of NTU *is a measure of the heat transfer surface area $A_s$*. Thus, the larger the NTU, the larger the heat exchanger.

In heat exchanger analysis, it is also convenient to define another dimensionless quantity called the **capacity ratio $c$** as

$$c = \frac{C_{min}}{C_{max}} \qquad (22\text{–}40)$$

It can be shown that the effectiveness of a heat exchanger is a function of the number of transfer units NTU and the capacity ratio $c$. That is,

$$\varepsilon = \text{function}\,(UA_s/C_{min}, C_{min}/C_{max}) = \text{function}\,(\text{NTU}, c)$$

**TABLE 22–4**

Effectiveness relations for heat exchangers: $NTU = UA_s/C_{min}$ and $c = C_{min}/C_{max} = (\dot{m}c_p)_{min}/(\dot{m}c_p)_{max}$

| Heat exchanger type | Effectiveness relation |
|---|---|
| 1 *Double pipe:* | |
|    Parallel-flow | $\varepsilon = \dfrac{1 - \exp[-NTU(1 + c)]}{1 + c}$ |
|    Counter-flow | $\varepsilon = \dfrac{1 - \exp[-NTU(1 - c)]}{1 - c\exp[-NTU(1 - c)]}$ |
| 2 *Shell-and-tube:* <br>   One-shell pass <br>   2, 4,...tube <br>   passes | $\varepsilon = 2\left\{1 + c + \sqrt{1 + c^2}\,\dfrac{1 + \exp[-NTU\sqrt{1 + c^2}]}{1 - \exp[-NTU\sqrt{1 + c^2}]}\right\}^{-1}$ |
| 3 *Cross-flow* <br>   *(single-pass)* <br>   Both fluids unmixed | $\varepsilon = 1 - \exp\left\{\dfrac{NTU^{0.22}}{c}[\exp(-c\,NTU^{0.78}) - 1]\right\}$ |
|   $C_{max}$ mixed, <br>   $C_{min}$ unmixed | $\varepsilon = \dfrac{1}{c}(1 - \exp\{-c[1 - \exp(-NTU)]\})$ |
|   $C_{min}$ mixed, <br>   $C_{max}$ unmixed | $\varepsilon = 1 - \exp\left\{-\dfrac{1}{c}[1 - \exp(-c\,NTU)]\right\}$ |
| 4 *All heat exchangers* <br>   with $c = 0$ | $\varepsilon = 1 - \exp(-NTU)$ |

From W. M. Kays and A. L. London. *Compact Heat Exchangers, 3/e.* McGraw-Hill, 1984. Reprinted by permission of William M. Kays.

Effectiveness relations have been developed for a large number of heat exchangers, and the results are given in Table 22–4. The effectivenesses of some common types of heat exchangers are also plotted in Fig. 22–26.

More extensive effectiveness charts and relations are available in the literature. The dashed lines in Fig. 22–26*f* are for the case of $C_{min}$ unmixed and $C_{max}$ mixed and the solid lines are for the opposite case. The analytic relations for the effectiveness give more accurate results than the charts, since reading errors in charts are unavoidable, and the relations are very suitable for computerized analysis of heat exchangers.

We make these observations from the effectiveness relations and charts already given:

1. The value of the effectiveness ranges from 0 to 1. It increases rapidly with NTU for small values (up to about NTU = 1.5) but rather slowly for larger values. Therefore, the use of a heat exchanger with a large NTU (usually larger than 3) and thus a large size cannot be justified economically, since a large increase in NTU in this case corresponds to a small increase in effectiveness. Thus, a heat exchanger with a very high effectiveness may be desirable from a heat transfer point of view but undesirable from an economical point of view.

2. For a given NTU and capacity ratio $c = C_{min}/C_{max}$, the *counter-flow* heat exchanger has the *highest* effectiveness, followed closely by the cross-flow heat exchangers with both fluids unmixed. As you might

(a) Parallel-flow

(b) Counter-flow

(c) One-shell pass and 2, 4, 6, ... tube passes

(d) Two-shell passes and 4, 8, 12, ... tube passes

(e) Cross-flow with both fluids unmixed

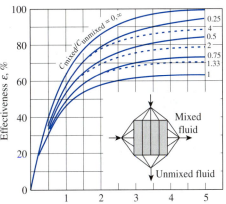

(f) Cross-flow with one fluid mixed and the other unmixed

**FIGURE 22–26**
Effectiveness for heat exchangers.
*From Kays and London, 1984.*

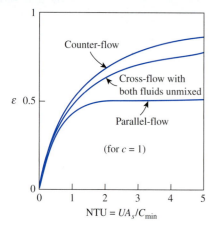

**FIGURE 22–27**

For a specified NTU and capacity ratio $c$, the counter-flow heat exchanger has the highest effectiveness and the parallel-flow the lowest.

**FIGURE 22–28**

The effectiveness relation reduces to $\varepsilon = \varepsilon_{max} = 1 - \exp(-NTU)$ for all heat exchangers when the capacity ratio $c = 0$.

expect, the lowest effectiveness values are encountered in parallel-flow heat exchangers (Fig. 22–27).

3. The effectiveness of a heat exchanger is independent of the capacity ratio $c$ for NTU values of less than about 0.3.

4. The value of the capacity ratio $c$ ranges between 0 and 1. For a given NTU, the effectiveness becomes a *maximum* for $c = 0$ and a *minimum* for $c = 1$. The case $c = C_{min}/C_{max} \rightarrow 0$ corresponds to $C_{max} \rightarrow \infty$, which is realized during a phase-change process in a *condenser* or *boiler*. All effectiveness relations in this case reduce to

$$\varepsilon = \varepsilon_{max} = 1 - \exp(-NTU) \tag{22–41}$$

regardless of the type of heat exchanger (Fig. 22–28). Note that the temperature of the condensing or boiling fluid remains constant in this case. The effectiveness is the *lowest* in the other limiting case of $c = C_{min}/C_{max} = 1$, which is realized when the heat capacity rates of the two fluids are equal.

Once the quantities $c = C_{min}/C_{max}$ and NTU $= UA_s/C_{min}$ have been evaluated, the effectiveness $\varepsilon$ can be determined from either the charts or the effectiveness relation for the specified type of heat exchanger. Then the rate of heat transfer $\dot{Q}$ and the outlet temperatures $T_{h,\,out}$ and $T_{c,\,out}$ can be determined from Eqs. 22–33 and 22–30, respectively. Note that the analysis of heat exchangers with unknown outlet temperatures is a straightforward matter with the effectiveness–NTU method but requires rather tedious iterations with the LMTD method.

We mentioned earlier that when all the inlet and outlet temperatures are specified, the *size* of the heat exchanger can easily be determined using the LMTD method. Alternatively, it can also be determined from the effectiveness–NTU method by first evaluating the effectiveness $\varepsilon$ from its definition (Eq. 22–29) and then the NTU from the appropriate NTU relation in Table 22–5.

**TABLE 22–5**

NTU relations for heat exchangers: NTU $= UA_s/C_{min}$ and $c = C_{min}/C_{max} = (\dot{m}c_p)_{min}/(\dot{m}c_p)_{max}$

| Heat exchanger type | NTU relation |
|---|---|
| 1 *Double-pipe:* | |
|   Parallel-flow | $NTU = -\dfrac{\ln\,[1 - \varepsilon(1 + c)]}{1 + c}$ |
|   Counter-flow | $NTU = \dfrac{1}{c - 1}\,\ln\left(\dfrac{\varepsilon - 1}{\varepsilon\,c - 1}\right)$ |
| 2 *Shell and tube:* | |
|   One-shell pass<br>  2, 4,...tube passes | $NTU = -\dfrac{1}{\sqrt{1 + c^2}}\,\ln\left(\dfrac{2/\varepsilon - 1 - c - \sqrt{1 + c^2}}{2/\varepsilon - 1 - c + \sqrt{1 + c^2}}\right)$ |
| 3 *Cross-flow (single-pass):*<br>  $C_{max}$ mixed,<br>  $C_{min}$ unmixed | $NTU = -\ln\left[1 + \dfrac{\ln\,(1 - \varepsilon c)}{c}\right]$ |
|   $C_{min}$ mixed,<br>  $C_{max}$ unmixed | $NTU = -\dfrac{\ln\,[c\,\ln\,(1 - \varepsilon) + 1]}{c}$ |
| 4 *All heat exchangers*<br>  *with $c = 0$* | $NTU = -\ln(1 - \varepsilon)$ |

From W. M. Kays and A. L. London. *Compact Heat Exchangers, 3/e.* McGraw-Hill, 1984. Reprinted by permission of William M. Kays.

Note that the relations in Table 22–5 are equivalent to those in Table 22–4. Both sets of relations are given for convenience. The relations in Table 22–4 give the effectiveness directly when NTU is known, and the relations in Table 22–5 give the NTU directly when the effectiveness $\varepsilon$ is known.

---

### EXAMPLE 22–8   The Effectiveness-NTU Method

A counter-flow double-pipe heat exchanger is to heat water from 20°C to 80°C at a rate of 1.2 kg/s (Fig. 22–29). The heating is to be accomplished by geothermal water available at 160°C at a mass flow rate of 2 kg/s. The inner tube is thin-walled and has a diameter of 1.5 cm. The overall heat transfer coefficient of the heat exchanger is 640 W/m²·K. Using the effectiveness-NTU method determine the length of the heat exchanger required to achieve the desired heating.

**FIGURE 22–29**
Schematic for Example 22–8.

**SOLUTION**   Water is heated in a counter-flow double-pipe heat exchanger by geothermal water. Using the $\varepsilon$-NTU method, the required length of the heat exchanger is to be determined.
*Analysis*   In the effectiveness-NTU method, we first determine the heat capacity rates of the hot and cold fluids and identify the smaller one.

$$C_h = \dot{m}_h c_{ph} = (2 \text{ kg/s}) (4.31 \text{ kJ/kg·K}) = 8.62 \text{ kW/K}$$
$$C_c = \dot{m}_c c_{pc} = (1.2 \text{ kg/s}) (4.18 \text{ kJ/kg·K}) = 5.02 \text{ kW/K}$$

Therefore,

$$C_{min} = C_c = 5.02 \text{ kW/K}$$

and

$$c = C_{min}/C_{max} = 5.02/8.62 = 0.582$$

Then the maximum heat transfer rate is determined from Eq. 22–32 to be

$$\dot{Q}_{max} = C_{min}(T_{h, in} - T_{c, in})$$
$$= (5.02 \text{ kW/K}) (160 - 20)°C$$
$$= 702.8 \text{ kW}$$

That is, the maximum possible heat transfer rate in this heat exchanger is 702.8 kW. The actual rate of heat transfer is

$$\dot{Q} = [\dot{m}c_p(T_{out} - T_{in}2]_{water} = 11.2 \text{ kg/s2 14.18 kJ/kg·K2 180} - 20)°C$$

$$= 301.0 \text{ kW}$$

Thus, the effectiveness of the heat exchanger is

$$\varepsilon = \frac{\dot{Q}}{\dot{Q}_{max}} = \frac{301.0 \text{ kW}}{702.8 \text{ kW}} = 0.428$$

Knowing the effectiveness, the NTU of this counter-flow heat exchanger can be determined from Fig. 22–26b or the appropriate relation from Table 22–5. We choose the latter approach for greater accuracy:

$$\text{NTU} = \frac{1}{c - 1} \ln\left(\frac{\varepsilon - 1}{\varepsilon c - 1}\right) = \frac{1}{0.582 - 1} \ln\left(\frac{0.428 - 1}{0.428 \times 0.582 - 1}\right) = 0.651$$

Then the heat transfer surface area becomes

$$NTU = \frac{UA_s}{C_{min}} \longrightarrow A_s = \frac{NTU\, C_{min}}{U} = \frac{(0.651)\,(5020\ \text{W/K})}{640\ \text{W/m}^2\cdot\text{K}} = 5.11\ \text{m}^2$$

To provide this much heat transfer surface area, the length of the tube must be

$$A_s = \pi DL \longrightarrow L = \frac{A_s}{\pi D} = \frac{5.11\ \text{m}^2}{\pi(0.015\ \text{m})} = \textbf{1.08 m}$$

**Discussion** This problem was solved in Example 22–4 using the LMTD method. Note that we obtained practically the same result in a systematic and straightforward manner using the effectiveness–NTU method.

---

**EXAMPLE 22–9    Cooling Hot Oil by Water in a Multipass Heat Exchanger**

Hot oil is to be cooled by water in a 1-shell-pass and 8-tube-passes heat exchanger. The tubes are thin-walled and are made of copper with an internal diameter of 1.4 cm. The length of each tube pass in the heat exchanger is 5 m, and the overall heat transfer coefficient is 310 W/m²·K. Water flows through the tubes at a rate of 0.2 kg/s, and the oil through the shell at a rate of 0.3 kg/s. The water and the oil enter at temperatures of 20°C and 150°C, respectively. Determine the rate of heat transfer in the heat exchanger and the outlet temperatures of the water and the oil.

**SOLUTION** Hot oil is to be cooled by water in a heat exchanger. The mass flow rates and the inlet temperatures are given. The rate of heat transfer and the outlet temperatures are to be determined.

**Assumptions** **1** Steady operating conditions exist. **2** The heat exchanger is well insulated so that heat loss to the surroundings is negligible. **3** The thickness of the tube is negligible since it is thin-walled. **4** Changes in the kinetic and potential energies of fluid streams are negligible. **5** The overall heat transfer coefficient is constant and uniform.

**Properties** We take the specific heats of water and oil to be 4.18 and 2.13 kJ/kg·°C, respectively.

**Analysis** The schematic of the heat exchanger is given in Fig. 22–30. The outlet temperatures are not specified, and they cannot be determined from an energy balance. The use of the LMTD method in this case will involve tedious iterations, and thus the ε–NTU method is indicated. The first step in the ε–NTU method is to determine the heat capacity rates of the hot and cold fluids and identify the smaller one:

$$C_h = \dot{m}_h c_{ph} = (0.3\ \text{kg/s})(2.13\ \text{kJ/kg}\cdot°\text{C}) = 0.639\ \text{kW/K}$$

$$C_c = \dot{m}_c c_{pc} = (0.2\ \text{kg/s})(4.18\ \text{kJ/kg}\cdot°\text{C}) = 0.836\ \text{kW/K}$$

Therefore,

$$C_{min} = C_h = 0.639\ \text{kW/K} \quad \text{and} \quad c = \frac{C_{min}}{C_{max}} = \frac{0.639}{0.836} = 0.764$$

Oil
150°C  0.3 kg/s

2°C
Water
0.2 kg/s

**FIGURE 22–30**
Schematic for Example 22–9.

Then the maximum heat transfer rate is determined from Eq. 22–32 to be

$$\dot{Q}_{max} = C_{min}(T_{h,\,in} - T_{c,\,in}) = (0.639 \text{ kW/K})(150 - 20)°C = 83.1 \text{ kW}$$

That is, the maximum possible heat transfer rate in this heat exchanger is 83.1 kW. The heat transfer surface area is

$$A_s = n(\pi DL) = 8\pi(0.014 \text{ m})(5 \text{ m}) = 1.76 \text{ m}^2$$

Then the NTU of this heat exchanger becomes

$$\text{NTU} = \frac{UA_s}{C_{min}} = \frac{(310 \text{ W/m}^2 \cdot \text{K})(1.76 \text{ m}^2)}{639 \text{ W/K}} = 0.854$$

The effectiveness of this heat exchanger corresponding to $c = 0.764$ and NTU = 0.854 is determined from Fig. 22–26c to be

$$\varepsilon = 0.47$$

We could also determine the effectiveness from the third relation in Table 22–4 more accurately but with more labor. Then the actual rate of heat transfer becomes

$$\dot{Q} = \varepsilon \dot{Q}_{max} = (0.47)(83.1 \text{ kW}) = \textbf{39.1 kW}$$

Finally, the outlet temperatures of the cold and the hot fluid streams are determined to be

$$\dot{Q} = C_c(T_{c,\,out} - T_{c,\,in}) \longrightarrow T_{c,\,out} = T_{c,\,in} + \frac{\dot{Q}}{C_c}$$

$$= 20°C + \frac{39.1 \text{ kW}}{0.836 \text{ kW/K}} = \textbf{66.8°C}$$

$$\dot{Q} = C_h(T_{h,\,in} - T_{h,\,out}) \longrightarrow T_{h,\,out} = T_{h,\,in} - \frac{\dot{Q}}{C_h}$$

$$= 150°C - \frac{39.1 \text{ kW}}{0.639 \text{ kW/K}} = \textbf{88.8°C}$$

Therefore, the temperature of the cooling water will rise from 20°C to 66.8°C as it cools the hot oil from 150°C to 88.8°C in this heat exchanger.

## SUMMARY

Heat exchangers are devices that allow the exchange of heat between two fluids without allowing them to mix with each other. Heat exchangers are manufactured in a variety of types, the simplest being the *double-pipe* heat exchanger. In a *parallel-flow* type, both the hot and cold fluids enter the heat exchanger at the same end and move in the same direction, whereas in a *counter-flow* type, the hot and cold fluids enter the heat exchanger at opposite ends and flow in opposite directions. In *compact* heat exchangers, the two fluids move perpendicular to each other, and such a flow configuration is called *cross-flow*. Other common types of heat exchangers in industrial applications are the *plate* and the *shell-and-tube* heat exchangers.

Heat transfer in a heat exchanger usually involves convection in each fluid and conduction through the wall separating the two fluids. In the analysis of heat exchangers, it is convenient to work with an *overall heat transfer coefficient U* or a *total thermal resistance R*, expressed as

$$\frac{1}{UA_s} = \frac{1}{U_iA_i} = \frac{1}{U_oA_o} = R = \frac{1}{h_iA_i} + R_{wall} + \frac{1}{h_oA_o}$$

where the subscripts $i$ and $o$ stand for the inner and outer surfaces of the wall that separates the two fluids, respectively. When the wall thickness of the tube is small and the thermal conductivity of the tube material is high, the relation simplifies to

$$\frac{1}{U} \approx \frac{1}{h_i} + \frac{1}{h_o}$$

where $U \approx U_i \approx U_o$. The effects of fouling on both the inner and the outer surfaces of the tubes of a heat exchanger can be accounted for by

$$\frac{1}{UA_s} = \frac{1}{U_i A_i} = \frac{1}{U_o A_o} = R$$

$$= \frac{1}{h_i A_i} + \frac{R_{f,i}}{A_i} + \frac{\ln (D_o/D_i)}{2\pi kL} + \frac{R_{f,o}}{A_o} + \frac{1}{h_o A_o}$$

where $A_i = \pi D_i L$ and $A_o = \pi D_o L$ are the areas of the inner and outer surfaces and $R_{f,i}$ and $R_{f,o}$ are the fouling factors at those surfaces.

In a well-insulated heat exchanger, the rate of heat transfer from the hot fluid is equal to the rate of heat transfer to the cold one. That is,

$$\dot{Q} = \dot{m}_c c_{pc}(T_{c,\,out} - T_{c,\,in}) = C_c(T_{c,\,out} - T_{c,\,in})$$

and

$$\dot{Q} = \dot{m}_h c_{ph}(T_{h,\,in} - T_{h,\,out}) = C_h(T_{h,\,in} - T_{h,\,out})$$

where the subscripts $c$ and $h$ stand for the cold and hot fluids, respectively, and the product of the mass flow rate and the specific heat of a fluid $\dot{m}c_p$ is called the *heat capacity rate*.

Of the two methods used in the analysis of heat exchangers, the *log mean temperature difference* (or LMTD) method is best suited for determining the size of a heat exchanger when all the inlet and the outlet temperatures are known. The *effectiveness–NTU* method is best suited to predict the outlet temperatures of the hot and cold fluid streams in a specified heat exchanger. In the LMTD method, the rate of heat transfer is determined from

$$\dot{Q} = UA_s \Delta T_{lm}$$

where

$$\Delta T_{lm} = \frac{\Delta T_1 - \Delta T_2}{\ln (\Delta T_1/\Delta T_2)}$$

is the *log mean temperature difference,* which is the suitable form of the average temperature difference for use in the analysis of heat exchangers. Here $\Delta T_1$ and $\Delta T_2$ represent the temperature differences between the two fluids at the two ends (inlet and outlet) of the heat exchanger. For cross-flow and multipass shell-and-tube heat exchangers, the logarithmic mean temperature difference is related to the counter-flow one $\Delta T_{lm,\,CF}$ as

$$\Delta T_{lm} = F \Delta T_{lm,\,CF}$$

where $F$ is the *correction factor,* which depends on the geometry of the heat exchanger and the inlet and outlet temperatures of the hot and cold fluid streams.

The *effectiveness* of a heat exchanger is defined as

$$\varepsilon = \frac{\dot{Q}}{Q_{max}} = \frac{\text{Actual heat transfer rate}}{\text{Maximum possible heat transfer rate}}$$

where

$$\dot{Q}_{max} = C_{min}(T_{h,\,in} - T_{c,\,in})$$

and $C_{min}$ is the smaller of $C_h = \dot{m}_h c_{ph}$ and $C_c = \dot{m}_c c_{pc}$. The effectiveness of heat exchangers can be determined from effectiveness relations or charts.

## REFERENCES AND SUGGESTED READINGS

1. N. Afgan and E. U. Schlunder. *Heat Exchanger: Design and Theory Sourcebook.* Washington, DC: McGraw-Hill/Scripta, 1974.

2. R. A. Bowman, A. C. Mueller, and W. M. Nagle. "Mean Temperature Difference in Design." *Trans. ASME* 62, 1940, p. 283. Reprined with permission of ASME International.

3. A. P. Fraas. *Heat Exchanger Design.* 2d ed. New York: John Wiley & Sons, 1989.

4. K. A. Gardner. "Variable Heat Transfer Rate Correction in Multipass Exchangers, Shell Side Film Controlling." *Transactions of the ASME* 67 (1945), pp. 31–38.

5. W. M. Kays and A. L. London. *Compact Heat Exchangers.* 3rd ed. New York: McGraw-Hill, 1984.

6. W. M. Kays and H. C. Perkins. In *Handbook of Heat Transfer,* ed. W. M. Rohsenow and J. P. Hartnett. New York: McGraw-Hill, 1972, Chap. 7.

7. A. C. Mueller. "Heat Exchangers." In *Handbook of Heat Transfer,* ed. W. M. Rohsenow and J. P. Hartnett. New York: McGraw-Hill, 1972, Chap. 18.

8. M. N. Özişik. *Heat Transfer—A Basic Approach.* New York: McGraw-Hill, 1985.

9. E. U. Schlunder. *Heat Exchanger Design Handbook.* Washington, DC: Hemisphere, 1982.

**10.** *Standards of Tubular Exchanger Manufacturers Association.* New York: Tubular Exchanger Manufacturers Association, 9th ed, 2007.

**11.** R. A. Stevens, J. Fernandes, and J. R. Woolf. "Mean Temperature Difference in One, Two, and Three Pass Crossflow Heat Exchangers." *Transactions of the ASME* 79 (1957), pp. 287–297.

**12.** J. Taborek, G. F. Hewitt, and N. Afgan. *Heat Exchangers: Theory and Practice.* New York: Hemisphere, 1983.

**13.** G. Walker. *Industrial Heat Exchangers.* Washington, DC: Hemisphere, 1982.

## PROBLEMS*

### Types of Heat Exchangers

**22–1C**  Classify heat exchangers according to flow type and explain the characteristics of each type.

**22–2C**  When is a heat exchanger classified as being compact? Do you think a double-pipe heat exchanger can be classified as a compact heat exchanger?

**22–3C**  What is a regenerative heat exchanger? How does a static type of regenerative heat exchanger differ from a dynamic type?

**22–4C**  What is the role of the baffles in a shell-and-tube exchanger? How does the presence of baffles affect the heat transfer and the pumping power requirements? Explain.

**22–5C**  How does a cross-flow heat exchanger differ from a counter-flow one? What is the difference between mixed and unmixed fluids in cross-flow?

### The Overall Heat Transfer Coefficient

**22–6C**  What are the heat transfer mechanisms involved during heat transfer in a liquid-to-liquid heat exchanger from the hot to the cold fluid?

**22–7C**  Under what conditions is the thermal resistance of the tube in a heat exchanger negligible?

**22–8C**  Consider a double-pipe parallel-flow heat exchanger of length $L$. The inner and outer diameters of the inner tube are $D_1$ and $D_2$, respectively, and the inner diameter of the outer tube is $D_3$. Explain how you would determine the two heat transfer surface areas $A_i$ and $A_o$. When is it reasonable to assume $A_i \approx A_o \approx A_s$?

**22–9C**  How is the thermal resistance due to fouling in a heat exchanger accounted for? How do the fluid velocity and temperature affect fouling?

**22–10C**  In a thin-walled double-pipe heat exchanger, when is the approximation $U = h_i$ a reasonable one? Here $U$ is the overall heat transfer coefficient and $h_i$ is the convection heat transfer coefficient inside the tube.

**22–11C**  What are the common causes of fouling in a heat exchanger? How does fouling affect heat transfer and pressure drop?

**22–12C**  Under what conditions can the overall heat transfer coefficient of a heat exchanger be determined from $U = (1/h_i + 1/h_o)^{-1}$?

**22–13E**  Water at an average temperature of 180°F and an average velocity of 4 ft/s flows through a thin-walled $\frac{3}{4}$-in-diameter tube. The water is cooled by air that flows across the tube with a velocity of 12 ft/s at an average temperature of 80°F. Determine the overall heat transfer coefficient.

**22–14**  Water at an average temperature of 110°C and an average velocity of 3.5 m/s flows through a 7-m-long stainless steel tube ($k = 14.2$ W/m·K) in a boiler. The inner and outer diameters of the tube are $D_i = 1.0$ cm and $D_o = 1.4$ cm, respectively. If the convection heat transfer coefficient at the outer surface of the tube where boiling is taking place is $h_o = 7200$ W/m²·K, determine the overall heat transfer coefficient $U_i$ of this boiler based on the inner surface area of the tube.

**22–15**  Repeat Prob. 22–14, assuming a fouling factor $R_{f, i} = 0.0005$ m²·K/W on the inner surface of the tube.

**22–16**  [EES] Reconsider Prob. 22–15. Using EES (or other) software, plot the overall heat transfer coefficient based on the inner surface as a function of fouling factor as it varies from 0.0001 m²·K/W to 0.0008 m²·K/W, and discuss the results.

**22–17E**  A counter-flow heat exchanger is stated to have an overall heat transfer coefficient, based on outside tube area of 50 Btu/h·ft²·°F when operating at design and clean conditions. After a period of use, scale built-up in the heat exchanger gives a fouling factor of 0.002 h·ft²·°F/Btu. Determine (*a*) the overall heat transfer coefficient of the heat exchanger and (*b*) the percentage change in the overall heat transfer coefficient due to the scale built-up.

---

*Problems designated by a "C" are concept questions, and students are encouraged to answer them all. Problems designated by an "E" are in English units, and the SI users can ignore them. Problems with the icon 🌐 are solved using EES, and complete solutions together with parametric studies are included on the enclosed DVD. Problems with the icon 💻 are comprehensive in nature, and are intended to be solved with a computer, preferably using the EES software that accompanies this text.

**22–18E** The tube in a heat exchanger has a 2-in. inner diameter and a 3-in. outer diameter. The thermal conductivity of the tube material is 0.5 Btu/h·ft·°F, while the inner surface heat transfer coefficient is 50 Btu/h·ft²·°F and the outer surface heat transfer coefficient is 10 Btu/h·ft²·°F. Determine the overall heat transfer coefficients based on the outer and inner surfaces. *Answers:* 4.32 Btu/h·ft²·°F, 6.48 Btu/h·ft²·°F

**22–19** A long thin-walled double-pipe heat exchanger with tube and shell diameters of 1.0 cm and 2.5 cm, respectively, is used to condense refrigerant-134a by water at 20°C. The refrigerant flows through the tube, with a convection heat transfer coefficient of $h_i = 4100$ W/m²·K. Water flows through the shell at a rate of 0.3 kg/s. Determine the overall heat transfer coefficient of this heat exchanger. *Answer:* 1856 W/m²·K

**22–20** Repeat Prob. 22–19 by assuming a 2-mm-thick layer of limestone ($k = 1.3$ W/m·K) forms on the outer surface of the inner tube.

**22–21** Reconsider Prob. 22–20. Using EES (or other) software, plot the overall heat transfer coefficient as a function of the limestone thickness as it varies from 1 mm to 3 mm, and discuss the results.

**22–22** A jacketted-agitated vessel, fitted with a turbine agitator, is used for heating a water stream from 10°C to 54°C. The average heat transfer coefficient for water at the vessel's inner-wall can be estimated from $Nu = 0.76Re^{2/3}Pr^{1/3}$. Saturated steam at 100°C condenses in the jacket, for which the average heat transfer coefficient in kW/m²·K is: $h_o = 13.1(T_g - T_w)^{-0.25}$. The vessel dimensions are: $D_t = 0.6$ m, $H = 0.6$ m and $D_a = 0.2$ m. The agitator speed is 60 rpm. Calculate the mass rate of water that can be heated in this agitated vessel steadily.

## Analysis of Heat Exchangers

**22–23C** What are the common approximations made in the analysis of heat exchangers?

**22–24C** Under what conditions will the temperature rise of the cold fluid in a heat exchanger be equal to the temperature drop of the hot fluid?

**22–25C** What is the heat capacity rate? What can you say about the temperature changes of the hot and cold fluids in a heat exchanger if both fluids have the same capacity rate? What does a heat capacity of infinity for a fluid in a heat exchanger mean?

**22–26C** Consider a condenser in which steam at a specified temperature is condensed by rejecting heat to the cooling water. If the heat transfer rate in the condenser and the temperature rise of the cooling water is known, explain how the rate of condensation of the steam and the mass flow rate of the cooling water can be determined. Also, explain how the total thermal resistance $R$ of this condenser can be evaluated in this case.

**22–27C** Under what conditions is the heat transfer relation

$$\dot{Q} = \dot{m}_c c_{pc}(T_{c,\,out} - T_{c,\,in}) = \dot{m}_h c_{ph}(T_{h,\,in} - T_{h,\,out})$$

valid for a heat exchanger?

## The Log Mean Temperature Difference Method

**22–28C** In the heat transfer relation $\dot{Q} = UA_s\,\Delta T_{lm}$ for a heat exchanger, what is $\Delta T_{lm}$ called? How is it calculated for a parallel-flow and counter-flow heat exchanger?

**22–29C** How does the log mean temperature difference for a heat exchanger differ from the arithmetic mean temperature difference? For specified inlet and outlet temperatures, which one of these two quantities is larger?

**22–30C** The temperature difference between the hot and cold fluids in a heat exchanger is given to be $\Delta T_1$ at one end and $\Delta T_2$ at the other end. Can the logarithmic temperature difference $\Delta T_{lm}$ of this heat exchanger be greater than both $\Delta T_1$ and $\Delta T_2$? Explain.

**22–31C** Can the outlet temperature of the cold fluid in a heat exchanger be higher than the outlet temperature of the hot fluid in a parallel-flow heat exchanger? How about in a counter-flow heat exchanger? Explain.

**22–32C** Explain how the LMTD method can be used to determine the heat transfer surface area of a multipass shell-and-tube heat exchanger when all the necessary information, including the outlet temperatures, is given.

**22–33C** In the heat transfer relation $\dot{Q} = UA_sF\Delta T_{lm}$ for a heat exchanger, what is the quantity $F$ called? What does it represent? Can $F$ be greater than one?

**22–34C** When the outlet temperatures of the fluids in a heat exchanger are not known, is it still practical to use the LMTD method? Explain.

**22–35C** For specified inlet and outlet temperatures, for what kind of heat exchanger will the $\Delta T_{lm}$ be greatest: double-pipe parallel-flow, double-pipe counter-flow, cross-flow, or multipass shell-and-tube heat exchanger?

**22–36** A counter-flow heat exchanger is stated to have an overall heat transfer coefficient of 284 W/m²·K when operating at design and clean conditions. Hot fluid enters the tube side at 93°C and exits at 71°C, while cold fluid enters the shell side at 27°C and exits at 38°C. After a period of use, built-up scale in the heat exchanger gives a fouling factor of 0.0004 m²·K/W. If the surface area is 93 m², determine (*a*) the rate of heat transfer in the heat exchanger and (*b*) the mass flow rates of both hot and cold fluids. Assume both hot and cold fluids have a specific heat of 4.2 kJ/kg·K.

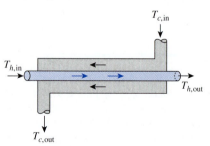

**FIGURE P22–36**

**22–37E** A single-pass cross-flow heat exchanger is used to cool jacket water ($c_p = 1.0$ Btu/lbm·°F) of a diesel engine from 190°F to 140°F, using air ($c_p = 0.245$ Btu/lbm·°F) with inlet temperature of 90°F. Both air flow and water flow are unmixed. If the water and air mass flow rates are 92,000 lbm/h and 400,000 lbm/h, respectively, determine the log mean temperature difference for this heat exchanger.   *Answer: 47.4°F*

Air flow (*unmixed*)

Water flow (*unmixed*)

**FIGURE P22–37E**

**22–38** Ethylene glycol is heated from 25°C to 40°C at a rate of 1.5 kg/s in a horizontal copper tube ($k = 386$ W/m·K) with an inner diameter of 2.0 cm and an outer diameter of 2.5 cm. A saturated vapor ($T_g = 110$°C) condenses on the outside-tube surface with the heat transfer coefficient (in kW/m²·K) given by $9.2/(T_g - T_w)^{0.25}$, where $T_w$ is the average outside-tube wall temperature. What tube length must be used? Take the properties of ethylene glycol to be $\rho = 1109$ kg/m³, $c_p = 2428$ kJ/kg·K, $k = 0.253$ W/m·K, $\mu = 0.01545$ kg/m·s, and Pr = 148.5.

**22–39** A test is conducted to determine the overall heat transfer coefficient in a shell-and-tube oil-to-water heat exchanger that has 24 tubes of internal diameter 1.2 cm and length 2 m in a single shell. Cold water ($c_p = 4180$ J/kg·K) enters the tubes at 20°C at a rate of 3 kg/s and leaves at 55°C. Oil ($c_p = 2150$ J/kg·K) flows through the shell and is cooled from 120°C to 45°C. Determine the overall heat transfer coefficient $U_i$ of this heat exchanger based on the inner surface area of the tubes.   *Answer: 8.31 kW/m²·K*

**22–40** A stream of hydrocarbon ($c_p = 2.2$ kJ/kg·K) is cooled at a rate of 720 kg/h from 150°C to 40°C in the tube side of a double-pipe counter-flow heat exchanger. Water ($c_p = 4.18$ kJ/kg·K) enters the heat exchanger at 10°C at a rate of 540 kg/h. The outside diameter of the inner tube is 2.5 cm, and its length is 6.0 m. Calculate the overall heat transfer coefficient.

**22–41** A shell-and-tube heat exchanger is used for heating 14 kg/s of oil ($c_p = 2.0$ kJ/kg·K) from 20°C to 46°C. The heat exchanger has 1-shell pass and 6-tube passes. Water enters the shell side at 80°C and leaves at 60°C. The overall heat transfer coefficient is estimated to be 1000 W/m²·K. Calculate the rate of heat transfer and the heat transfer area.

**22–42** A double-pipe parallel-flow heat exchanger is to heat water ($c_p = 4180$ J/kg·K) from 25°C to 60°C at a rate of 0.2 kg/s. The heating is to be accomplished by geothermal water ($c_p = 4310$ J/kg·K) available at 140°C at a mass flow rate of 0.3 kg/s. The inner tube is thin-walled and has a diameter of 0.8 cm. If the overall heat transfer coefficient of the heat exchanger is 550 W/m²·K, determine the length of the tube required to achieve the desired heating.

**22–43** Reconsider Prob. 22–42. Using EES (or other) software, investigate the effects of temperature and mass flow rate of geothermal water on the length of the tube. Let the temperature vary from 100°C to 200°C, and the mass flow rate from 0.1 kg/s to 0.5 kg/s. Plot the length of the tube as functions of temperature and mass flow rate, and discuss the results.

**22–44E** A 1-shell-pass and 8-tube-passes heat exchanger is used to heat glycerin ($c_p = 0.60$ Btu/lbm·°F) from 80°F to 140°F by hot water ($c_p = 1.0$ Btu/lbm·°F) that enters the thin-walled 0.5-in-diameter tubes at 175°F and leaves at 120°F. The total length of the tubes in the heat exchanger is 400 ft. The convection heat transfer coefficient is 4 Btu/h·ft²·°F on the glycerin (shell) side and 50 Btu/h·ft²·°F on the water (tube) side. Determine the rate of heat transfer in the heat exchanger (a) before any fouling occurs and (b) after fouling with a fouling factor of 0.002 h·ft²·°F/Btu on the outer surfaces of the tubes.

**22–45** A double-pipe parallel-flow heat exchanger is used to heat cold tap water with hot water. Hot water ($c_p = 4.25$ kJ/kg·K) enters the tube at 85°C at a rate of 1.4 kg/s and leaves at 50°C. The heat exchanger is not well insulated, and it is estimated that 3 percent of the heat given up by the hot fluid is lost from the heat exchanger. If the overall heat transfer coefficient and the surface area of the heat exchanger are 1150 W/m²·K and 4 m², respectively, determine the rate of heat transfer to the cold water and the log mean temperature difference for this heat exchanger.

Hot water 85°C

Cold water

50°C

**FIGURE P22–45**

**22–46** A thin-walled double-pipe counter-flow heat exchanger is to be used to cool oil ($c_p = 2200$ J/kg·K) from 150°C to 50°C at a rate of 2.5 kg/s by water ($c_p = 4180$ J/kg·K) that enters at 22°C at a rate of 1.5 kg/s. The diameter of the tube is 2.5 cm, and its length is 6 m. Determine the overall heat transfer coefficient of this heat exchanger.

**22–47**  Reconsider Prob. 22–46. Using EES (or other) software, investigate the effects of oil exit temperature and water inlet temperature on the overall heat transfer coefficient of the heat exchanger. Let the oil exit temperature vary from 30°C to 70°C and the water inlet temperature from 5°C to 25°C. Plot the overall heat transfer coefficient as functions of the two temperatures, and discuss the results.

**22–48** A shell-and-tube heat exchanger with 2-shell passes and 12-tube passes is used to heat water ($c_p$ = 4180 J/kg·K) with ethylene glycol ($c_p$ = 2680 J/kg·K). Water enters the tubes at 22°C at a rate of 0.8 kg/s and leaves at 70°C. Ethylene glycol enters the shell at 110°C and leaves at 60°C. If the overall heat transfer coefficient based on the tube side is 280 W/m²·K, determine the rate of heat transfer and the heat transfer surface area on the tube side.

**22–49** Reconsider Prob. 22–48. Using EES (or other) software, investigate the effect of the mass flow rate of water on the rate of heat transfer and the tube-side surface area. Let the mass flow rate vary from 0.4 kg/s to 2.2 kg/s. Plot the rate of heat transfer and the surface area as a function of the mass flow rate, and discuss the results.

**22–50** Glycerin ($c_p$ = 2400 J/kg·K) at 20°C and 0.5 kg/s is to be heated by ethylene glycol ($c_p$ = 2500 J/kg·K) at 70°C in a thin-walled double-pipe parallel-flow heat exchanger. The temperature difference between the two fluids is 15°C at the outlet of the heat exchanger. If the overall heat transfer coefficient is 240 W/m²·K and the heat transfer surface area is 3.2 m², determine (a) the rate of heat transfer, (b) the outlet temperature of the glycerin, and (c) the mass flow rate of the ethylene glycol.

**22–51** Air ($c_p$ = 1005 J/kg·K) is to be preheated by hot exhaust gases in a cross-flow heat exchanger before it enters the furnace. Air enters the heat exchanger at 95 kPa and 20°C at a rate of 0.8 m³/s. The combustion gases ($c_p$ = 1100 J/kg·K) enter at 180°C at a rate of 1.1 kg/s and leave at 95°C. The product of the overall heat transfer coefficient and the heat transfer surface area is $UA_s$ = 1200 W/K. Assuming both fluids to be unmixed, determine the rate of heat transfer and the outlet temperature of the air.

Air
95 kPa
20°C
0.8 m³/s

Exhaust gases
1.1 kg/s
95°C

**FIGURE P22–51**

**22–52** A shell-and-tube heat exchanger with 2-shell passes and 12-tube passes is used to heat water ($c_p$ = 4180 J/kg·K) in the tubes from 20°C to 70°C at a rate of 4.5 kg/s. Heat is supplied by hot oil ($c_p$ = 2300 J/kg·K) that enters the shell side at 170°C at a rate of 10 kg/s. For a tube-side overall heat transfer coefficient of 350 W/m²·K, determine the heat transfer surface area on the tube side. *Answer:* 25.7 m²

**22–53** Repeat Prob. 22–52 for a mass flow rate of 3 kg/s for water.

**22–54** A shell-and-tube heat exchanger with 2-shell passes and 8-tube passes is used to heat ethyl alcohol ($c_p$ = 2670 J/kg·K) in the tubes from 25°C to 70°C at a rate of 2.1 kg/s. The heating is to be done by water ($c_p$ = 4190 J/kg·K) that enters the shell side at 95°C and leaves at 45°C. If the overall heat transfer coefficient is 950 W/m²·K, determine the heat transfer surface area of the heat exchanger.

Water
95°C

70°C

Ethyl
alcohol
25°C
2.1 kg/s

45°C

(8-tube passes)

**FIGURE P22–54**

**22–55** Consider a water-to-water double-pipe heat exchanger whose flow arrangement is not known. The temperature measurements indicate that the cold water enters at 20°C and leaves at 50°C, while the hot water enters at 80°C and leaves at 45°C. Do you think this is a parallel-flow or counter-flow heat exchanger? Explain.

**22–56** Cold water ($c_p$ = 4180 J/kg·K) leading to a shower enters a thin-walled double-pipe counter-flow heat exchanger at 15°C at a rate of 1.25 kg/s and is heated to 60°C by hot water ($c_p$ = 4190 J/kg·K) that enters at 100°C at a rate of 4 kg/s. If the overall heat transfer coefficient is 880 W/m²·K, determine the rate of heat transfer and the heat transfer surface area of the heat exchanger.

**22–57E** Steam is to be condensed on the shell side of a 1-shell-pass and 8-tube-passes condenser, with 60 tubes in each pass at 90°F ($h_{fg}$ = 1043 Btu/lbm). Cooling water ($c_p$ = 1.0 Btu/lbm·°F) enters the tubes at 55°F and leaves at 70°F. The tubes are thin walled and have a diameter of 3/4 in and length of 5 ft per pass. If the overall heat transfer coefficient is 600 Btu/h·ft²·°F, determine (a) the rate of heat transfer, (b) the rate of condensation of steam, and (c) the mass flow rate of cold water.

Steam
90°F
70°F
55°F
Water
90°F

**FIGURE P22–57E**

Fresh
water
10°C

Dyeing
water
80°C

$T_{h,\text{out}}$

$T_{c,\text{out}}$

**FIGURE P22–61**

**22–58E** Reconsider Prob. 22–57E. Using EES (or other) software, investigate the effect of the condensing steam temperature on the rate of heat transfer, the rate of condensation of steam, and the mass flow rate of cold water. Let the steam temperature vary from 80°F to 120°F. Plot the rate of heat transfer, the condensation rate of steam, and the mass flow rate of cold water as a function of steam temperature, and discuss the results.

**22–59** Hot exhaust gases of a stationary diesel engine are to be used to generate steam in an evaporator. Exhaust gases ($c_p$ = 1051 J/kg·K) enter the heat exchanger at 550°C at a rate of 0.25 kg/s while water enters as saturated liquid and evaporates at 200°C ($h_{fg}$ = 1941 kJ/kg). The heat transfer surface area of the heat exchanger based on water side is 0.5 m² and overall heat transfer coefficient is 1780 W/m²·K. Determine the rate of heat transfer, the exit temperature of exhaust gases, and the rate of evaporation of water.

**22–60** Reconsider Prob. 22–59. Using EES (or other) software, investigate the effect of the exhaust gas inlet temperature on the rate of heat transfer, the exit temperature of exhaust gases, and the rate of evaporation of water. Let the temperature of exhaust gases vary from 300°C to 600°C. Plot the rate of heat transfer, the exit temperature of exhaust gases, and the rate of evaporation of water as a function of the temperature of the exhaust gases, and discuss the results.

**22–61** In a textile manufacturing plant, the waste dyeing water ($c_p$ = 4295 J/kg·K) at 80°C is to be used to preheat fresh water ($c_p$ = 4180 J/kg·K) at 10°C at the same flow rate in a double-pipe counter-flow heat exchanger. The heat transfer surface area of the heat exchanger is 1.65 m² and the overall heat transfer coefficient is 625 W/m²·K. If the rate of heat transfer in the heat exchanger is 35 kW, determine the outlet temperature and the mass flow rate of each fluid stream.

**22–62** A heat exchanger contains 400 tubes with inner diameter of 23 mm and outer diameter of 25 mm. The length of each tube is 3.7 m. The corrected log mean temperature difference is 23°C, while the inner surface convection heat transfer coefficient is 3410 W/m²·K and the outer surface convection heat transfer coefficient is 6820 W/m²·K. If the thermal resistance of the tubes is negligible, determine the heat transfer rate. *Answer:* 5750 kW

**22–63E** A single pass heat exchanger is to be designed to heat 100,000 lbm of water in an hour from 60°F to 100°F by condensation of water vapor at 230°F on the shell side. Each tube has an inner diameter of 1.2 in. and a wall thickness of 0.12 in. The inner surface convection heat transfer coefficient is 480 Btu/h·ft²·°F and the outer surface convection heat transfer coefficient is 2000 Btu/h·ft²·°F. If the inlet velocity of water ($c_{pc}$ = 1 Btu/lbm·°F and $\rho$ = 62.3 lbm/ft³) on the tube side is 4 ft/s, determine the required number of tubes and length of tubes. Assume the thermal resistance of the tubes is negligible.

## The Effectiveness–NTU Method

**22–64C** What does the effectiveness of a heat exchanger represent? Can effectiveness be greater than one? On what factors does the effectiveness of a heat exchanger depend?

**22–65C** For a specified fluid pair, inlet temperatures, and mass flow rates, what kind of heat exchanger will have the highest effectiveness: double-pipe parallel-flow, double-pipe counter-flow, cross-flow, or multipass shell-and-tube heat exchanger?

**22–66C** Explain how you can evaluate the outlet temperatures of the cold and hot fluids in a heat exchanger after its effectiveness is determined.

**22–67C** Can the temperature of the cold fluid rise above the inlet temperature of the hot fluid at any location in a heat exchanger? Explain.

**22–68C** Consider a heat exchanger in which both fluids have the same specific heats but different mass flow rates. Which fluid will experience a larger temperature change: the one with the lower or higher mass flow rate?

**22–69C** Explain how the maximum possible heat transfer rate $\dot{Q}_{max}$ in a heat exchanger can be determined when the mass flow rates, specific heats, and the inlet temperatures of the two fluids are specified. Does the value of $\dot{Q}_{max}$ depend on the type of heat exchanger?

**22–70C** Under what conditions can a counter-flow heat exchanger have an effectiveness of one? What would your answer be for a parallel-flow heat exchanger?

**22–71C** Consider a double-pipe counter-flow heat exchanger. In order to enhance heat transfer, the length of the heat exchanger is now doubled. Do you think its effectiveness will also double?

**22–72C** Consider two double-pipe counter-flow heat exchangers that are identical except that one is twice as long as the other one. Which heat exchanger is more likely to have a higher effectiveness?

**22–73C** How is the NTU of a heat exchanger defined? What does it represent? Is a heat exchanger with a very large NTU (say, 10) necessarily a good one to buy?

**22–74C** Consider a heat exchanger that has an NTU of 4. Someone proposes to double the size of the heat exchanger and thus double the NTU to 8 in order to increase the effectiveness of the heat exchanger and thus save energy. Would you support this proposal?

**22–75C** Consider a heat exchanger that has an NTU of 0.1. Someone proposes to triple the size of the heat exchanger and thus triple the NTU to 0.3 in order to increase the effectiveness of the heat exchanger and thus save energy. Would you support this proposal?

**22–76E** A 1-shell and 2-tube type heat exchanger has an overall heat transfer coefficient of 300 Btu/h·ft²·°F. The shell side fluid has a heat capacity rate of 20,000 Btu/h·°F, while the tube side fluid has a heat capacity rate of 40,000 Btu/h·°F. The inlet temperatures on the shell side and tube side are 200°F and 90°F, respectively. If the total heat transfer area is 100 ft², determine (a) the heat transfer effectiveness and (b) the actual heat transfer rate in the heat exchanger.

**22–77** A cross-flow heat exchanger with both fluids unmixed has an overall heat transfer coefficient of 200 W/m²·K, and a heat transfer surface area of 400 m². The hot fluid has a heat capacity of 40,000 W/K, while the cold fluid has a heat capacity of 80,000 W/K. If the inlet temperatures of both hot and cold fluids are 80°C and 20°C, respectively, determine the exit temperature of the cold fluid. *Answer: 42.2°C*

**22–78** In a 1-shell and 2-tube heat exchanger, cold water with inlet temperature of 20°C is heated by hot water supplied at the inlet at 80°C. The cold and hot water flow rates are 5000 kg/h and 10,000 kg/h, respectively. If the shell-and-tube heat exchanger has a $UA_s$ value of 11,600 W/K, determine the cold water and hot water outlet temperatures. Assume $c_{pc}$ = 4178 J/kg·K and $c_{ph}$ = 4188 J/kg·K.

**22–79** The radiator in an automobile is a cross-flow heat exchanger ($UA_s$ = 10 kW/K) that uses air ($c_p$ = 1.00 kJ/kg·K) to cool the engine-coolant fluid ($c_p$ = 4.00 kJ/kg·K). The engine fan draws 22°C air through this radiator at a rate of 8 kg/s while the coolant pump circulates the engine coolant at a rate of 5 kg/s. The coolant enters this radiator at 80°C. Under these conditions, the effectiveness of the radiator is 0.4. Determine (a) the outlet temperature of the air and (b) the rate of heat transfer between the two fluids.

**22–80** Water ($c_p$ = 4180 J/kg·K) enters the 2.5-cm-internal-diameter tube of a double-pipe counter-flow heat exchanger at 20°C at a rate of 2.2 kg/s. Water is heated by steam condensing at 120°C ($h_{fg}$ = 2203 kJ/kg) in the shell. If the overall heat transfer coefficient of the heat exchanger is 700 W/m²·K, determine the length of the tube required in order to heat the water to 80°C using (a) the LMTD method and (b) the $\varepsilon$–NTU method.

**22–81** Ethanol is vaporized at 78°C ($h_{fg}$ = 846 kJ/kg) in a double-pipe parallel-flow heat exchanger at a rate of 0.04 kg/s by hot oil ($c_p$ = 2200 J/kg·K) that enters at 115°C. If the heat transfer surface area and the overall heat transfer coefficients are 6.2 m² and 320 W/m²·K, respectively, determine the outlet temperature and the mass flow rate of oil using (a) the LMTD method and (b) the $\varepsilon$–NTU method.

**FIGURE P22–81**

**22–82** Air ($c_p$ = 1005 J/kg·K) enters a cross-flow heat exchanger at 20°C at a rate of 3 kg/s, where it is heated by a hot water stream ($c_p$ = 4190 J/kg·K) that enters the heat exchanger at 70°C at a rate of 1 kg/s. Determine the maximum heat transfer rate and the outlet temperatures of both fluids for that case.

**22–83** Hot oil ($c_p$ = 2200 J/kg·K) is to be cooled by water ($c_p$ = 4180 J/kg·K) in a 2-shell-passes and 12-tube-passes heat exchanger. The tubes are thin-walled and are made of copper with a diameter of 1.8 cm. The length of each tube pass in the heat exchanger is 3 m, and the overall heat transfer coefficient is 340 W/m²·K. Water flows through the tubes at a total rate of 0.1 kg/s, and the oil through the shell at a rate of 0.2 kg/s. The water and the oil enter at temperatures 18°C and 160°C, respectively. Determine the rate of heat transfer in the heat exchanger and the outlet temperatures of the water and the oil. *Answers: 36.2 kW, 104.6°C, 77.7°C*

Oil
160°C
0.2 kg/s

Water
18°C →
0.1 kg/s

(12-tube passes)

**FIGURE P22–83**

**22–84** Consider an oil-to-oil double-pipe heat exchanger whose flow arrangement is not known. The temperature measurements indicate that the cold oil enters at 15°C and leaves at 55°C, while the hot oil enters at 80°C and leaves at 40°C. Do you think this is a parallel-flow or counter-flow heat exchanger? Why? Assuming the mass flow rates of both fluids to be identical, determine the effectiveness of this heat exchanger.

**22–85E** Hot water enters a double-pipe counter-flow water-to-oil heat exchanger at 190°F and leaves at 100°F. Oil enters at 70°F and leaves at 130°F. Determine which fluid has the smaller heat capacity rate and calculate the effectiveness of this heat exchanger.

**22–86** Saturated water vapor at 100°C condenses in a 1-shell and 2-tube heat exchanger with a surface area of 0.5 m² and an overall heat transfer coefficient of 2000 W/m²·K. Cold water ($c_{pc}$ = 4179 J/kg·K) flowing at 0.5 kg/s enters the tube side at 15°C, determine (a) the heat transfer effectiveness, (b) the outlet temperature of the cold water, and (c) the heat transfer rate for the heat exchanger.

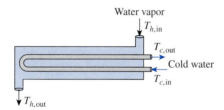

Water vapor

$T_{h,\text{in}}$

$T_{c,\text{out}}$

Cold water

$T_{c,\text{in}}$

$T_{h,\text{out}}$

**FIGURE P22–86**

**22–87** Hot water ($c_{ph}$ = 4188 J/kg·K) with mass flow rate of 2.5 kg/s at 100°C enters a thin-walled concentric tube counterflow heat exchanger with a surface area of 23 m² and an overall heat transfer coefficient of 1000 W/m²·K. Cold water ($c_{pc}$ = 4178 J/kg·K) with mass flow rate of 5 kg/s enters the heat exchanger at 20°C, determine (a) the heat transfer rate for the heat exchanger and (b) the outlet temperatures of the cold and hot fluids. After a period of operation, the overall heat transfer coefficient is reduced to 500 W/m²·K, determine (c) the fouling factor that caused the reduction in the overall heat transfer coefficient.

**22–88** A cross-flow air-to-water heat exchanger with an effectiveness of 0.65 is used to heat water ($c_p$ = 4180 J/kg·K) with hot air ($c_p$ = 1010 J/kg·K). Water enters the heat exchanger at 20°C at a rate of 4 kg/s, while air enters at 100°C at a rate of 9 kg/s. If the overall heat transfer coefficient based on the water side is 260 W/m²·K, determine the heat transfer surface area of the heat exchanger on the water side. Assume both fluids are unmixed. *Answer: 52.4 m²*

**22–89** Cold water ($c_p$ = 4.18 kJ/kg·K) enters a cross-flow heat exchanger at 14°C at a rate of 0.35 kg/s where it is heated by hot air ($c_p$ = 1.0 kJ/kg·K) that enters the heat exchanger at 65°C at a rate of 0.8 kg/s and leaves at 25°C. Determine the maximum outlet temperature of the cold water and the effectiveness of this heat exchanger.

14°C
0.35 kg/s

Air
65°C
0.8 kg/s

**FIGURE P22–89**

**22–90** Water from a lake is used as the cooling agent in a power plant. To achieve condensation of 3.2 kg/s of steam exiting the turbine, a shell-and-tube heat exchanger is used, which has a single shell and 220 thin-walled, 25-mm-diameter tubes, each tube making two passes. Steam flows through the shell, while cooling water flows through the tubes. Steam enters as saturated vapor at 60°C and leaves as saturated liquid. Cooling water at 20°C is available at a rate of 200 kg/s. The convection coefficient at the outer surface of the tubes is 8500 W/m²·K. Determine (a) the temperature of the cooling water leaving the condenser and (b) the required tube length per pass. (Use the following average properties for water: $c_p$ = 4180 J/kg·K, $\mu$ = 8 × 10⁻⁴ N·s/m², k = 0.6 W/m·K, Pr = 6).

**22–91** Water ($c_p$ = 4180 J/kg·K) is to be heated by solar-heated hot air ($c_p$ = 1010 J/kg·K) in a double-pipe counterflow heat exchanger. Air enters the heat exchanger at 90°C at a rate of 0.3 kg/s, while water enters at 22°C at a rate of 0.1 kg/s. The overall heat transfer coefficient based on the inner side of the tube is given to be 80 W/m²·K. The length of the tube is 12 m and the internal diameter of the tube is 1.2 cm. Determine the outlet temperatures of the water and the air.

**22–92**  Reconsider Prob. 22–91. Using EES (or other) software, investigate the effects of the mass flow rate of water and the tube length on the outlet temperatures of water and air. Let the mass flow rate vary from 0.05 kg/s to 1.0 kg/s and the tube length from 5 m to 25 m. Plot the outlet temperatures of the water and the air as functions of the mass flow rate and the tube length, and discuss the results.

**22–93E** A thin-walled double-pipe, counter-flow heat exchanger is to be used to cool oil ($c_p = 0.525$ Btu/lbm·°F) from 300°F to 105°F at a rate of 5 lbm/s by water ($c_p = 1.0$ Btu/lbm·°F) that enters at 70°F at a rate of 3 lbm/s. The diameter of the tube is 5 in and its length is 200 ft. Determine the overall heat transfer coefficient of this heat exchanger using (a) the LMTD method and (b) the $\varepsilon - NTU$ method.

**22–94** A shell-and-tube heat exchanger with 2-shell passes and 8-tube passes is used to heat ethyl alcohol ($c_p = 2670$ J/kg·K) in the tubes from 25°C to 70°C at a rate of 2.1 kg/s. The heating is to be done by water ($c_p = 4190$ J/kg·K) that enters the shell at 95°C and leaves at 60°C. If the overall heat transfer coefficient is 800 W/m²·K, determine the heat transfer surface area of the heat exchanger using (a) the LMTD method and (b) the $\varepsilon - NTU$ method. *Answer: (a) 11.4 m²*

**22–95** Steam is to be condensed on the shell side of a 1-shell-pass and 8-tube-passes condenser, with 50 tubes in each pass, at 30°C ($h_{fg} = 2431$ kJ/kg). Cooling water ($c_p = 4180$ J/kg·K) enters the tubes at 18°C at a rate of 2200 kg/h. The tubes are thin-walled, and have a diameter of 1.5 cm and length of 2 m per pass. If the overall heat transfer coefficient is 3000 W/m²·K, determine (a) the rate of heat transfer and (b) the rate of condensation of steam.

Steam
30°C

18°C

Water
2200 kg/h

30°C

**FIGURE P22–95**

**22–96**  Reconsider Prob. 22–95. Using EES (or other) software, investigate the effects of the condensing steam temperature and the tube diameter on the rate of heat transfer and the rate of condensation of steam. Let the steam temperature vary from 20°C to 70°C and the tube diameter from 1.0 cm to 2.0 cm. Plot the rate of heat transfer and the rate of condensation as functions of steam temperature and tube diameter, and discuss the results.

**22–97** Cold water ($c_p = 4180$ J/kg·K) leading to a shower enters a thin-walled double-pipe counter-flow heat exchanger at 15°C at a rate of 0.25 kg/s and is heated to 45°C by hot water ($c_p = 4190$ J/kg·K) that enters at 100°C at a rate of 3 kg/s. If the overall heat transfer coefficient is 950 W/m²·K, determine the rate of heat transfer and the heat transfer surface area of the heat exchanger using the $\varepsilon - NTU$ method. *Answers: 31.35 kW, 0.482 m²*

Cold water
15°C
0.25 kg/s

Hot
water
100°C
3 kg/s

45°C

**FIGURE P22–97**

**22–98** Reconsider Prob. 22–97. Using EES (or other) software, investigate the effects of the inlet temperature of hot water and the heat transfer coefficient on the rate of heat transfer and the surface area. Let the inlet temperature vary from 60°C to 120°C and the overall heat transfer coefficient from 750 W/m²·K to 1250 W/m²·K. Plot the rate of heat transfer and surface area as functions of the inlet temperature and the heat transfer coefficient, and discuss the results.

**22–99E** Oil is being cooled from 180°F to 120°F in a 1-shell and 2-tube heat exchanger with an overall heat transfer coefficient of 40 Btu/h·ft²·°F. Water ($c_{pc} = 1.0$ Btu/lbm·°F) enters at 80°F and exits at 100°F with a mass flow rate of 20,000 lbm/h, determine (a) the NTU value and (b) the surface area of the heat exchanger. *Answers: (a) 1.11, (b) 185 ft²*

Oil flow
$T_{h,in}$

$T_{c,out}$

Water flow

$T_{c,in}$

$T_{h,out}$

**FIGURE P22–99E**

**22–100** Oil in an engine is being cooled by air in a cross-flow heat exchanger, where both fluids are unmixed. Oil ($c_{ph} = 2047$ J/kg·K) flowing with a flow rate of 0.026 kg/s enters the heat exchanger at 75°C, while air ($c_{pc} = 1007$ J/kg·K) enters at 30°C with a flow rate of 0.21 kg/s. The overall heat transfer coefficient of the heat exchanger is 53 W/m²·K and the total surface area is 1 m². Determine (a) the heat transfer effectiveness and (b) the outlet temperature of the oil.

**FIGURE P22–100**

## Review Problems

**22–101** A cross-flow heat exchanger with both fluids unmixed has an overall heat transfer coefficient of 200 W/m²·K, and a heat transfer surface area of 400 m². The hot fluid has a heat capacity of 40,000 W/K, while the cold fluid has a heat capacity of 80,000 W/K. If the inlet temperatures of both hot and cold fluids are 80°C and 20°C, respectively, determine (a) the exit temperature of the hot fluid and (b) the rate of heat transfer in the heat exchanger.

**22–102** A single-pass cross-flow heat exchanger uses hot air (mixed) to heat water (unmixed), flowing with a mass flow rate of 3 kg/s, from 30°C to 80°C. The hot air enters and exits the heat exchanger at 220°C and 100°C, respectively. If the overall heat transfer coefficient is 200 W/m²·K, determine the required surface area.

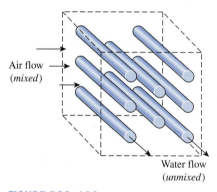

**FIGURE P22–102**

**22–103** The mass flow rate, specific heat, and inlet temperature of the tube-side stream in a double-pipe, parallel-flow heat exchanger are 3200 kg/h, 2.0 kJ/kg·K, and 120°C, respectively. The mass flow rate, specific heat, and inlet temperature of the other stream are 2000 kg/h, 4.2 kJ/kg·K, and 20°C, respectively. The heat transfer area and overall heat transfer coefficient are 0.50 m² and 2.0 kW/m²·K, respectively. Find the outlet temperatures of both streams in steady operation using (a) the LMTD method and (b) the effectiveness–NTU method.

**22–104** A shell-and-tube heat exchanger with 1-shell pass and 14-tube passes is used to heat water in the tubes with geothermal steam condensing at 120°C ($h_{fg}$ = 2203 kJ/kg) on the shell side. The tubes are thin-walled and have a diameter of 2.4 cm and length of 3.2 m per pass. Water ($c_p$ = 4180 J/kg·K) enters the tubes at 18°C at a rate of 6.2 kg/s. If the temperature difference between the two fluids at the exit is 46°C, determine (a) the rate of heat transfer, (b) the rate of condensation of steam, and (c) the overall heat transfer coefficient.

**FIGURE P22–104**

**22–105** Geothermal water ($c_p$ = 4250 J/kg·K) at 75°C is to be used to heat fresh water ($c_p$ = 4180 J/kg·K) at 17°C at a rate of 1.2 kg/s in a double-pipe counter-flow heat exchanger. The heat transfer surface area is 25 m², the overall heat transfer coefficient is 480 W/m²·K, and the mass flow rate of geothermal water is larger than that of fresh water. If the effectiveness of the heat exchanger is desired to be 0.823, determine the mass flow rate of geothermal water and the outlet temperatures of both fluids.

**22–106** Hot water at 60°C is cooled to 36°C through the tube side of a 1–shell pass and 2-tube passes heat exchanger. The coolant is also a water stream, for which the inlet and outlet temperatures are 7°C and 31°C, respectively. The overall heat transfer coefficient and the heat transfer area are 950 W/m²·K and 15 m², respectively. Calculate the mass flow rates of hot and cold water streams in steady operation.

**22–107** Cold water ($c_p$ = 4180 J/kg·K) enters the tubes of a heat exchanger with 2-shell passes and 20-tube passes at 15°C

**FIGURE P22–107**

at a rate of 4 kg/s, while hot oil ($c_p$ = 2200 J/kg·K) enters the shell at 130°C at the same mass flow rate and leaves at 60°C. If the overall heat transfer coefficient based on the outer surface of the tube is 220 W/m²·K, determine (a) the rate of heat transfer and (b) the heat transfer surface area on the outer side of the tube. *Answers: (a) 616 kW, (b) 48.0 m²*

**22–108E** Water ($c_p$ = 1.0 Btu/lbm·°F) is to be heated by solar-heated hot air ($c_p$ = 0.24 Btu/lbm·°F) in a double-pipe counterflow heat exchanger. Air enters the heat exchanger at 190°F at a rate of 0.7 lbm/s and leaves at 135°F. Water enters at 70°F at a rate of 0.35 lbm/s. The overall heat transfer coefficient based on the inner side of the tube is given to be 20 Btu/h·ft²·°F. Determine the length of the tube required for a tube internal diameter of 0.5 in.

**22–109** By taking the limit as $\Delta T_2 \rightarrow \Delta T_1$, show that when $\Delta T_1 = \Delta T_2$ for a heat exchanger, the $\Delta T_{lm}$ relation reduces to $\Delta T_{lm} = \Delta T_1 = \Delta T_2$.

**22–110** The condenser of a room air conditioner is designed to reject heat at a rate of 22,500 kJ/h from refrigerant-134a as the refrigerant is condensed at a temperature of 40°C. Air ($c_p$ = 1005 J/kg·K) flows across the finned condenser coils, entering at 25°C and leaving at 32°C. If the overall heat transfer coefficient based on the refrigerant side is 150 W/m²·K, determine the heat transfer area on the refrigerant side. *Answer: 3.74 m²*

R-134a
40°C

Air
25°C

32°C

40°C

**FIGURE P22–110**

**22–111** Air ($c_p$ = 1005 J/kg·K) is to be preheated by hot exhaust gases in a cross-flow heat exchanger before it enters the furnace. Air enters the heat exchanger at 95 kPa and 20°C at a rate of 0.4 m³/s. The combustion gases ($c_p$ = 1100 J/kg·K) enter at 180°C at a rate of 0.65 kg/s and leave at 95°C. The product of the overall heat transfer coefficient and the heat transfer surface area is $UA_s$ = 1620 W/K. Assuming both fluids to be unmixed, determine the rate of heat transfer.

**22–112** In a chemical plant, a certain chemical is heated by hot water supplied by a natural gas furnace. The hot water ($c_p$ = 4180 J/kg·K) is then discharged at 60°C at a rate of 8 kg/min. The plant operates 8 h a day, 5 days a week, 52 weeks a year. The furnace has an efficiency of 78 percent, and the cost of the natural gas is $1.00 per therm (1 therm = 105,500 kJ). The average temperature of the cold water entering the furnace throughout the year is 14°C. In order to save energy, it is proposed to install a water-to-water heat exchanger to preheat the incoming cold water by the drained hot water. Assuming that the heat exchanger will recover 72 percent of the available heat in the hot water, determine the heat transfer rating of the heat exchanger that needs to be purchased and suggest a suitable type. Also, determine the amount of money this heat exchanger will save the company per year from natural gas savings.

**22–113** A shell-and-tube heat exchanger is used for cooling 47 kg/s of a process stream flowing through the tubes from 160°C to 100°C. This heat exchanger has a total of 100 identical tubes, each with an inside diameter of 2.5 cm and negligible wall thickness. The average properties of the process stream are: $\rho$ = 950 kg/m³, $k$ = 0.50 W/m·K, $c_p$ = 3.5 kJ/kg·K and $\mu$ = 2.0 mPa·s. The coolant stream is water ($c_p$ = 4.18 kJ/kg·K) at a flow rate of 66 kg/s and an inlet temperature of 10°C, which yields an average shell-side heat transfer coefficient of 4.0 kW/m²·K. Calculate the tube length if the heat exchanger has (a) a 1-shell pass and a 1-tube pass and (b) a 1-shell pass and 4-tube passes.

**22–114** A 2-shell passes and 4-tube passes heat exchanger is used for heating a hydrocarbon stream ($c_p$ = 2.0 kJ/kg·K) steadily from 20°C to 50°C. A water stream enters the shell-side at 80°C and leaves at 40°C. There are 160 thin-walled tubes, each with a diameter of 2.0 cm and length of 1.5 m. The tube-side and shell-side heat transfer coefficients are 1.6 and 2.5 kW/m²·K, respectively. (a) Calculate the rate of heat transfer and the mass rates of water and hydrocarbon streams. (b) With usage, the outlet hydrocarbon-stream temperature was found to decrease by 5°C due to the deposition of solids on the tube surface. Estimate the magnitude of fouling factor.

**22–115** Air at 18°C ($c_p$ = 1006 J/kg·K) is to be heated to 58°C by hot oil at 80°C ($c_p$ = 2150 J/kg·K) in a cross-flow heat exchanger with air mixed and oil unmixed. The product of heat transfer surface area and the overall heat transfer coefficient is 750 W/K and the mass flow rate of air is twice that of oil. Determine (a) the effectiveness of the heat exchanger, (b) the mass flow rate of air, and (c) the rate of heat transfer.

**22–116** During an experiment, a plate heat exchanger that is used to transfer heat from a hot-water stream to a cold-water stream is tested, and the following measurements are taken:

| | Hot Water Stream | Cold Water Stream |
|---|---|---|
| Inlet temperature, °C | 38.9 | 14.3 |
| Outlet temperature, °C | 27.0 | 19.8 |
| Volume flow rate, L/min | 2.5 | 4.5 |

The heat transfer area is calculated to be 0.056 m².
(a) Calculate the rate of heat transfer to the cold water.
(b) Calculate the overall heat transfer coefficient.

(c) Determine if the heat exchanger is truly adiabatic. If not, determine the fraction of heat loss and calculate the heat transfer efficiency.

(d) Determine the effectiveness and the NTU values of the heat exchanger.

Also, discuss if the measured values are reasonable.

**22–117** A shell-and-tube heat exchanger with 2-shell passes and 4-tube passes is used for cooling oil ($c_p = 2.0$ kJ/kg·K) from 125°C to 55°C. The coolant is water, which enters the shell side at 25°C and leaves at 46°C. The overall heat transfer coefficient is 900 W/m²·K. For an oil flow rate of 10 kg/s, calculate the cooling water flow rate and the heat transfer area.

**22–118** Consider a water-to-water counter-flow heat exchanger with these specifications. Hot water enters at 90°C while cold water enters at 20°C. The exit temperature of hot water is 15°C greater than that of cold water, and the mass flow rate of hot water is 50 percent greater than that of cold water. The product of heat transfer surface area and the overall heat transfer coefficient is 2200 W/K. Taking the specific heat of both cold and hot water to be $c_p = 4180$ J/kg·K, determine (a) the outlet temperature of the cold water, (b) the effectiveness of the heat exchanger, (c) the mass flow rate of the cold water, and (d) the heat transfer rate.

**FIGURE P22–118**

**22–119** A single-pass cross-flow heat exchanger with both fluids unmixed has water entering at 16°C and exiting at 33°C, while oil ($c_p = 1.93$ kJ/kg·K and $\rho = 870$ kg/m³) flowing at 0.19 m³/min enters at 38°C and exits at 29°C. If the surface area of the heat exchanger is 20 m², determine (a) the NTU value and (b) the value of the overall heat transfer coefficient.
*Answers: (a) 2.39, (b) 336 W/m²·K*

**FIGURE P22–119**

**22–120** Saturated water vapor at 100°C condenses in the shell side of a 1-shell and 2-tube heat exchanger with a surface area of 0.5 m² and an overall heat transfer coefficient of 2000 W/m²·K. Cold water ($c_{pc} = 4179$ J/kg·K) flowing at 0.5 kg/s enters the tube side at 15°C, determine the outlet temperature of the cold water and the heat transfer rate for the heat exchanger.

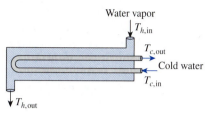

**FIGURE P22–120**

**22–121** Oil in an engine is being cooled by air in a cross-flow heat exchanger, where both fluids are unmixed. Oil ($c_{ph} = 2047$ J/kg·K) flowing with a flow rate of 0.026 kg/s enters the tube side at 75°C, while air ($c_{pc} = 1007$ J/kg·K) enters the shell side at 30°C with a flow rate of 0.21 kg/s. The overall heat transfer coefficient of the heat exchanger is 53 W/m²·K and the total surface area is 1 m². If the correction factor is $F = 0.96$, determine the outlet temperatures of the oil and air.

**FIGURE P22–121**

### Design and Essay Problems

**22–122** Water flows through a shower head steadily at a rate of 8 kg/min. The water is heated in an electric water heater from 15°C to 45°C. In an attempt to conserve energy, it is proposed to pass the drained warm water at a temperature of 38°C through a heat exchanger to preheat the incoming cold water. Design a heat exchanger that is suitable for this task, and discuss the potential savings in energy and money for your area.

**22–123** Design a hydrocooling unit that can cool fruits and vegetables from 30°C to 5°C at a rate of 20,000 kg/h under the following conditions:

The unit will be of flood type that will cool the products as they are conveyed into the channel filled with water. The products will be dropped into the channel filled with water at one end and picked up at the other end. The channel can be as wide as 3 m and as high as 90 cm. The water is to be circulated and cooled by the evaporator section of a refrigeration system. The refrigerant temperature inside the coils is to be −2°C, and the water temperature is not to drop below 1°C and not to exceed 6°C.

Assuming reasonable values for the average product density, specific heat, and porosity (the fraction of air volume in a box), recommend reasonable values for the quantities related to the thermal aspects of the hydrocooler, including (a) how long the fruits and vegetables need to remain in the channel, (b) the length of the channel, (c) the water velocity through the channel, (d) the velocity of the conveyor and thus the fruits and vegetables through the channel, (e) the refrigeration capacity of the refrigeration system, and (f) the type of heat exchanger for the evaporator and the surface area on the water side.

**22–124**  A company owns a refrigeration system whose refrigeration capacity is 200 tons (1 ton of refrigeration = 211 kJ/min), and you are to design a forced-air cooling system for fruits whose diameters do not exceed 7 cm under the following conditions:

The fruits are to be cooled from 28°C to an average temperature of 8°C. The air temperature is to remain above −2°C and below 10°C at all times, and the velocity of air approaching the fruits must remain under 2 m/s. The cooling section can be as wide as 3.5 m and as high as 2 m.

Assuming reasonable values for the average fruit density, specific heat, and porosity (the fraction of air volume in a box),

recommend reasonable values for the quantities related to the thermal aspects of the forced-air cooling, including (a) how long the fruits need to remain in the cooling section, (b) the length of the cooling section, (c) the air velocity approaching the cooling section, (d) the product cooling capacity of the system, in kg·fruit/h, (e) the volume flow rate of air, and (f) the type of heat exchanger for the evaporator and the surface area on the air side.

**22–125**  A counterflow double-pipe heat exchanger with $A_s = 9.0$ m$^2$ is used for cooling a liquid stream ($c_p = 3.15$ kJ/kg·K) at a rate of 10.0 kg/s with an inlet temperature of 90°C. The coolant ($c_p = 4.2$ kJ/kg·K) enters the heat exchanger at a rate of 8.0 kg/s with an inlet temperature of 10°C. The plant data gave the following equation for the overall heat transfer coefficient in W/m$^2$·K: $U = 600/(1/\dot{m}_c^{0.8} + 2/\dot{m}_h^{0.8})$, where $\dot{m}_c$ and $\dot{m}_h$ are the cold-and hot-stream flow rates in kg/s, respectively. (a) Calculate the rate of heat transfer and the outlet stream temperatures for this unit. (b) The existing unit is to be replaced. A vendor is offering a very attractive discount on two identical heat exchangers that are presently stocked in its warehouse, each with $A_s = 5$ m$^2$. Because the tube diameters in the existing and new units are the same, the above heat transfer coefficient equation is expected to be valid for the new units as well. The vendor is proposing that the two new units could be operated in parallel, such that each unit would process exactly one-half the flow rate of each of the hot and cold streams in a counterflow manner; hence, they together would meet (or exceed) the present plant heat duty. Give your recommendation, with supporting calculations, on this replacement proposal.

# PROPERTY TABLES AND CHARTS (SI UNITS)

## TABLE A–1

Molar mass, gas constant, and critical-point properties

| Substance | Formula | Molar mass, $M$ kg/kmol | Gas constant, $R$ kJ/kg·K* | Critical-point properties | | |
|---|---|---|---|---|---|---|
| | | | | Temperature, K | Pressure, MPa | Volume, $m^3$/kmol |
| Air | — | 28.97 | 0.2870 | 132.5 | 3.77 | 0.0883 |
| Ammonia | $NH_3$ | 17.03 | 0.4882 | 405.5 | 11.28 | 0.0724 |
| Argon | Ar | 39.948 | 0.2081 | 151 | 4.86 | 0.0749 |
| Benzene | $C_6H_6$ | 78.115 | 0.1064 | 562 | 4.92 | 0.2603 |
| Bromine | $Br_2$ | 159.808 | 0.0520 | 584 | 10.34 | 0.1355 |
| n-Butane | $C_4H_{10}$ | 58.124 | 0.1430 | 425.2 | 3.80 | 0.2547 |
| Carbon dioxide | $CO_2$ | 44.01 | 0.1889 | 304.2 | 7.39 | 0.0943 |
| Carbon monoxide | CO | 28.011 | 0.2968 | 133 | 3.50 | 0.0930 |
| Carbon tetrachloride | $CCl_4$ | 153.82 | 0.05405 | 556.4 | 4.56 | 0.2759 |
| Chlorine | $Cl_2$ | 70.906 | 0.1173 | 417 | 7.71 | 0.1242 |
| Chloroform | $CHCl_3$ | 119.38 | 0.06964 | 536.6 | 5.47 | 0.2403 |
| Dichlorodifluoromethane (R-12) | $CCl_2F_2$ | 120.91 | 0.06876 | 384.7 | 4.01 | 0.2179 |
| Dichlorofluoromethane (R-21) | $CHCl_2F$ | 102.92 | 0.08078 | 451.7 | 5.17 | 0.1973 |
| Ethane | $C_2H_6$ | 30.070 | 0.2765 | 305.5 | 4.48 | 0.1480 |
| Ethyl alcohol | $C_2H_5OH$ | 46.07 | 0.1805 | 516 | 6.38 | 0.1673 |
| Ethylene | $C_2H_4$ | 28.054 | 0.2964 | 282.4 | 5.12 | 0.1242 |
| Helium | He | 4.003 | 2.0769 | 5.3 | 0.23 | 0.0578 |
| n-Hexane | $C_6H_{14}$ | 86.179 | 0.09647 | 507.9 | 3.03 | 0.3677 |
| Hydrogen (normal) | $H_2$ | 2.016 | 4.1240 | 33.3 | 1.30 | 0.0649 |
| Krypton | Kr | 83.80 | 0.09921 | 209.4 | 5.50 | 0.0924 |
| Methane | $CH_4$ | 16.043 | 0.5182 | 191.1 | 4.64 | 0.0993 |
| Methyl alcohol | $CH_3OH$ | 32.042 | 0.2595 | 513.2 | 7.95 | 0.1180 |
| Methyl chloride | $CH_3Cl$ | 50.488 | 0.1647 | 416.3 | 6.68 | 0.1430 |
| Neon | Ne | 20.183 | 0.4119 | 44.5 | 2.73 | 0.0417 |
| Nitrogen | $N_2$ | 28.013 | 0.2968 | 126.2 | 3.39 | 0.0899 |
| Nitrous oxide | $N_2O$ | 44.013 | 0.1889 | 309.7 | 7.27 | 0.0961 |
| Oxygen | $O_2$ | 31.999 | 0.2598 | 154.8 | 5.08 | 0.0780 |
| Propane | $C_3H_8$ | 44.097 | 0.1885 | 370 | 4.26 | 0.1998 |
| Propylene | $C_3H_6$ | 42.081 | 0.1976 | 365 | 4.62 | 0.1810 |
| Sulfur dioxide | $SO_2$ | 64.063 | 0.1298 | 430.7 | 7.88 | 0.1217 |
| Tetrafluoroethane (R-134a) | $CF_3CH_2F$ | 102.03 | 0.08149 | 374.2 | 4.059 | 0.1993 |
| Trichlorofluoromethane (R-11) | $CCl_3F$ | 137.37 | 0.06052 | 471.2 | 4.38 | 0.2478 |
| Water | $H_2O$ | 18.015 | 0.4615 | 647.1 | 22.06 | 0.0560 |
| Xenon | Xe | 131.30 | 0.06332 | 289.8 | 5.88 | 0.1186 |

*The unit kJ/kg·K is equivalent to kPa·$m^3$/kg·K. The gas constant is calculated from $R = R_u/M$, where $R_u$ = 8.31447 kJ/kmol·K and $M$ is the molar mass.

Source: K. A. Kobe and R. E. Lynn, Jr., Chemical Review 52 (1953), pp. 117–236; and ASHRAE, Handbook of Fundamentals (Atlanta, GA: American Society of Heating, Refrigerating and Air-Conditioning Engineers, Inc., 1993), pp. 16.4 and 36.1.

## TABLE A–2

Ideal-gas specific heats of various common gases

(a) At 300 K

| Gas | Formula | Gas constant, $R$ kJ/kg·K | $c_p$ kJ/kg·K | $c_v$ kJ/kg·K | $k$ |
|---|---|---|---|---|---|
| Air | — | 0.2870 | 1.005 | 0.718 | 1.400 |
| Argon | Ar | 0.2081 | 0.5203 | 0.3122 | 1.667 |
| Butane | $C_4H_{10}$ | 0.1433 | 1.7164 | 1.5734 | 1.091 |
| Carbon dioxide | $CO_2$ | 0.1889 | 0.846 | 0.657 | 1.289 |
| Carbon monoxide | CO | 0.2968 | 1.040 | 0.744 | 1.400 |
| Ethane | $C_2H_6$ | 0.2765 | 1.7662 | 1.4897 | 1.186 |
| Ethylene | $C_2H_4$ | 0.2964 | 1.5482 | 1.2518 | 1.237 |
| Helium | He | 2.0769 | 5.1926 | 3.1156 | 1.667 |
| Hydrogen | $H_2$ | 4.1240 | 14.307 | 10.183 | 1.405 |
| Methane | $CH_4$ | 0.5182 | 2.2537 | 1.7354 | 1.299 |
| Neon | Ne | 0.4119 | 1.0299 | 0.6179 | 1.667 |
| Nitrogen | $N_2$ | 0.2968 | 1.039 | 0.743 | 1.400 |
| Octane | $C_8H_{18}$ | 0.0729 | 1.7113 | 1.6385 | 1.044 |
| Oxygen | $O_2$ | 0.2598 | 0.918 | 0.658 | 1.395 |
| Propane | $C_3H_8$ | 0.1885 | 1.6794 | 1.4909 | 1.126 |
| Steam | $H_2O$ | 0.4615 | 1.8723 | 1.4108 | 1.327 |

*Note:* The unit kJ/kg·K is equivalent to kJ/kg·°C.

*Source: Chemical and Process Thermodynamics* 3/E by Kyle, B. G., © 2000. Adapted by permission of Pearson Education, Inc., Upper Saddle River, NJ.

## TABLE A–2

Ideal-gas specific heats of various common gases (*Continued*)

(*b*) At various temperatures

| Temperature, K | $c_p$ kJ/kg·K | $c_v$ kJ/kg·K | k | $c_p$ kJ/kg·K | $c_v$ kJ/kg·K | k | $c_p$ kJ/kg·K | $c_v$ kJ/kg·K | k |
|---|---|---|---|---|---|---|---|---|---|
| | Air | | | Carbon dioxide, $CO_2$ | | | Carbon monoxide, CO | | |
| 250 | 1.003 | 0.716 | 1.401 | 0.791 | 0.602 | 1.314 | 1.039 | 0.743 | 1.400 |
| 300 | 1.005 | 0.718 | 1.400 | 0.846 | 0.657 | 1.288 | 1.040 | 0.744 | 1.399 |
| 350 | 1.008 | 0.721 | 1.398 | 0.895 | 0.706 | 1.268 | 1.043 | 0.746 | 1.398 |
| 400 | 1.013 | 0.726 | 1.395 | 0.939 | 0.750 | 1.252 | 1.047 | 0.751 | 1.395 |
| 450 | 1.020 | 0.733 | 1.391 | 0.978 | 0.790 | 1.239 | 1.054 | 0.757 | 1.392 |
| 500 | 1.029 | 0.742 | 1.387 | 1.014 | 0.825 | 1.229 | 1.063 | 0.767 | 1.387 |
| 550 | 1.040 | 0.753 | 1.381 | 1.046 | 0.857 | 1.220 | 1.075 | 0.778 | 1.382 |
| 600 | 1.051 | 0.764 | 1.376 | 1.075 | 0.886 | 1.213 | 1.087 | 0.790 | 1.376 |
| 650 | 1.063 | 0.776 | 1.370 | 1.102 | 0.913 | 1.207 | 1.100 | 0.803 | 1.370 |
| 700 | 1.075 | 0.788 | 1.364 | 1.126 | 0.937 | 1.202 | 1.113 | 0.816 | 1.364 |
| 750 | 1.087 | 0.800 | 1.359 | 1.148 | 0.959 | 1.197 | 1.126 | 0.829 | 1.358 |
| 800 | 1.099 | 0.812 | 1.354 | 1.169 | 0.980 | 1.193 | 1.139 | 0.842 | 1.353 |
| 900 | 1.121 | 0.834 | 1.344 | 1.204 | 1.015 | 1.186 | 1.163 | 0.866 | 1.343 |
| 1000 | 1.142 | 0.855 | 1.336 | 1.234 | 1.045 | 1.181 | 1.185 | 0.888 | 1.335 |
| | Hydrogen, $H_2$ | | | Nitrogen, $N_2$ | | | Oxygen, $O_2$ | | |
| 250 | 14.051 | 9.927 | 1.416 | 1.039 | 0.742 | 1.400 | 0.913 | 0.653 | 1.398 |
| 300 | 14.307 | 10.183 | 1.405 | 1.039 | 0.743 | 1.400 | 0.918 | 0.658 | 1.395 |
| 350 | 14.427 | 10.302 | 1.400 | 1.041 | 0.744 | 1.399 | 0.928 | 0.668 | 1.389 |
| 400 | 14.476 | 10.352 | 1.398 | 1.044 | 0.747 | 1.397 | 0.941 | 0.681 | 1.382 |
| 450 | 14.501 | 10.377 | 1.398 | 1.049 | 0.752 | 1.395 | 0.956 | 0.696 | 1.373 |
| 500 | 14.513 | 10.389 | 1.397 | 1.056 | 0.759 | 1.391 | 0.972 | 0.712 | 1.365 |
| 550 | 14.530 | 10.405 | 1.396 | 1.065 | 0.768 | 1.387 | 0.988 | 0.728 | 1.358 |
| 600 | 14.546 | 10.422 | 1.396 | 1.075 | 0.778 | 1.382 | 1.003 | 0.743 | 1.350 |
| 650 | 14.571 | 10.447 | 1.395 | 1.086 | 0.789 | 1.376 | 1.017 | 0.758 | 1.343 |
| 700 | 14.604 | 10.480 | 1.394 | 1.098 | 0.801 | 1.371 | 1.031 | 0.771 | 1.337 |
| 750 | 14.645 | 10.521 | 1.392 | 1.110 | 0.813 | 1.365 | 1.043 | 0.783 | 1.332 |
| 800 | 14.695 | 10.570 | 1.390 | 1.121 | 0.825 | 1.360 | 1.054 | 0.794 | 1.327 |
| 900 | 14.822 | 10.698 | 1.385 | 1.145 | 0.849 | 1.349 | 1.074 | 0.814 | 1.319 |
| 1000 | 14.983 | 10.859 | 1.380 | 1.167 | 0.870 | 1.341 | 1.090 | 0.830 | 1.313 |

*Source:* Kenneth Wark, *Thermodynamics,* 4th ed. (New York: McGraw-Hill, 1983), p. 783, Table A–4M. Originally published in *Tables of Thermal Properties of Gases,* NBS Circular 564, 1955.

## TABLE A–2

Ideal-gas specific heats of various common gases (*Concluded*)

(*c*) As a function of temperature

$$\bar{c}_p = a + bT + cT^2 + dT^3$$

$$(T \text{ in K}, c_p \text{ in kJ/kmol·K})$$

| Substance | Formula | $a$ | $b$ | $c$ | $d$ | Temperature range, K | % error Max. | % error Avg. |
|---|---|---|---|---|---|---|---|---|
| Nitrogen | $N_2$ | 28.90 | $-0.1571 \times 10^{-2}$ | $0.8081 \times 10^{-5}$ | $-2.873 \times 10^{-9}$ | 273–1800 | 0.59 | 0.34 |
| Oxygen | $O_2$ | 25.48 | $1.520 \times 10^{-2}$ | $-0.7155 \times 10^{-5}$ | $1.312 \times 10^{-9}$ | 273–1800 | 1.19 | 0.28 |
| Air | — | 28.11 | $0.1967 \times 10^{-2}$ | $0.4802 \times 10^{-5}$ | $-1.966 \times 10^{-9}$ | 273–1800 | 0.72 | 0.33 |
| Hydrogen | $H_2$ | 29.11 | $-0.1916 \times 10^{-2}$ | $0.4003 \times 10^{-5}$ | $-0.8704 \times 10^{-9}$ | 273–1800 | 1.01 | 0.26 |
| Carbon monoxide | CO | 28.16 | $0.1675 \times 10^{-2}$ | $0.5372 \times 10^{-5}$ | $-2.222 \times 10^{-9}$ | 273–1800 | 0.89 | 0.37 |
| Carbon dioxide | $CO_2$ | 22.26 | $5.981 \times 10^{-2}$ | $-3.501 \times 10^{-5}$ | $7.469 \times 10^{-9}$ | 273–1800 | 0.67 | 0.22 |
| Water vapor | $H_2O$ | 32.24 | $0.1923 \times 10^{-2}$ | $1.055 \times 10^{-5}$ | $-3.595 \times 10^{-9}$ | 273–1800 | 0.53 | 0.24 |
| Nitric oxide | NO | 29.34 | $-0.09395 \times 10^{-2}$ | $0.9747 \times 10^{-5}$ | $-4.187 \times 10^{-9}$ | 273–1500 | 0.97 | 0.36 |
| Nitrous oxide | $N_2O$ | 24.11 | $5.8632 \times 10^{-2}$ | $-3.562 \times 10^{-5}$ | $10.58 \times 10^{-9}$ | 273–1500 | 0.59 | 0.26 |
| Nitrogen dioxide | $NO_2$ | 22.9 | $5.715 \times 10^{-2}$ | $-3.52 \times 10^{-5}$ | $7.87 \times 10^{-9}$ | 273–1500 | 0.46 | 0.18 |
| Ammonia | $NH_3$ | 27.568 | $2.5630 \times 10^{-2}$ | $0.99072 \times 10^{-5}$ | $-6.6909 \times 10^{-9}$ | 273–1500 | 0.91 | 0.36 |
| Sulfur | $S_2$ | 27.21 | $2.218 \times 10^{-2}$ | $-1.628 \times 10^{-5}$ | $3.986 \times 10^{-9}$ | 273–1800 | 0.99 | 0.38 |
| Sulfur dioxide | $SO_2$ | 25.78 | $5.795 \times 10^{-2}$ | $-3.812 \times 10^{-5}$ | $8.612 \times 10^{-9}$ | 273–1800 | 0.45 | 0.24 |
| Sulfur trioxide | $SO_3$ | 16.40 | $14.58 \times 10^{-2}$ | $-11.20 \times 10^{-5}$ | $32.42 \times 10^{-9}$ | 273–1300 | 0.29 | 0.13 |
| Acetylene | $C_2H_2$ | 21.8 | $9.2143 \times 10^{-2}$ | $-6.527 \times 10^{-5}$ | $18.21 \times 10^{-9}$ | 273–1500 | 1.46 | 0.59 |
| Benzene | $C_6H_6$ | -36.22 | $48.475 \times 10^{-2}$ | $-31.57 \times 10^{-5}$ | $77.62 \times 10^{-9}$ | 273–1500 | 0.34 | 0.20 |
| Methanol | $CH_4O$ | 19.0 | $9.152 \times 10^{-2}$ | $-1.22 \times 10^{-5}$ | $-8.039 \times 10^{-9}$ | 273–1000 | 0.18 | 0.08 |
| Ethanol | $C_2H_6O$ | 19.9 | $20.96 \times 10^{-2}$ | $-10.38 \times 10^{-5}$ | $20.05 \times 10^{-9}$ | 273–1500 | 0.40 | 0.22 |
| Hydrogen chloride | HCl | 30.33 | $-0.7620 \times 10^{-2}$ | $1.327 \times 10^{-5}$ | $-4.338 \times 10^{-9}$ | 273–1500 | 0.22 | 0.08 |
| Methane | $CH_4$ | 19.89 | $5.024 \times 10^{-2}$ | $1.269 \times 10^{-5}$ | $-11.01 \times 10^{-9}$ | 273–1500 | 1.33 | 0.57 |
| Ethane | $C_2H_6$ | 6.900 | $17.27 \times 10^{-2}$ | $-6.406 \times 10^{-5}$ | $7.285 \times 10^{-9}$ | 273–1500 | 0.83 | 0.28 |
| Propane | $C_3H_8$ | -4.04 | $30.48 \times 10^{-2}$ | $-15.72 \times 10^{-5}$ | $31.74 \times 10^{-9}$ | 273–1500 | 0.40 | 0.12 |
| *n*-Butane | $C_4H_{10}$ | 3.96 | $37.15 \times 10^{-2}$ | $-18.34 \times 10^{-5}$ | $35.00 \times 10^{-9}$ | 273–1500 | 0.54 | 0.24 |
| *i*-Butane | $C_4H_{10}$ | -7.913 | $41.60 \times 10^{-2}$ | $-23.01 \times 10^{-5}$ | $49.91 \times 10^{-9}$ | 273–1500 | 0.25 | 0.13 |
| *n*-Pentane | $C_5H_{12}$ | 6.774 | $45.43 \times 10^{-2}$ | $-22.46 \times 10^{-5}$ | $42.29 \times 10^{-9}$ | 273–1500 | 0.56 | 0.21 |
| *n*-Hexane | $C_6H_{14}$ | 6.938 | $55.22 \times 10^{-2}$ | $-28.65 \times 10^{-5}$ | $57.69 \times 10^{-9}$ | 273–1500 | 0.72 | 0.20 |
| Ethylene | $C_2H_4$ | 3.95 | $15.64 \times 10^{-2}$ | $-8.344 \times 10^{-5}$ | $17.67 \times 10^{-9}$ | 273–1500 | 0.54 | 0.13 |
| Propylene | $C_3H_6$ | 3.15 | $23.83 \times 10^{-2}$ | $-12.18 \times 10^{-5}$ | $24.62 \times 10^{-9}$ | 273–1500 | 0.73 | 0.17 |

*Source:* B. G. Kyle, *Chemical and Process Thermodynamics* (Englewood Cliffs, NJ: Prentice-Hall, 1984). Used with permission.

## TABLE A–3

Properties of common liquids, solids, and foods

(a) Liquids

| Substance | Boiling data at 1 atm | | Freezing data | | Liquid properties | | |
|---|---|---|---|---|---|---|---|
| | Normal boiling point, °C | Latent heat of vaporization $h_{fg}$, kJ/kg | Freezing point, °C | Latent heat of fusion $h_{if}$, kJ/kg | Temperature, °C | Density $\rho$, kg/m³ | Specific heat $c_p$, kJ/kg·K |
| Ammonia | −33.3 | 1357 | −77.7 | 322.4 | −33.3 | 682 | 4.43 |
| | | | | | −20 | 665 | 4.52 |
| | | | | | 0 | 639 | 4.60 |
| | | | | | 25 | 602 | 4.80 |
| Argon | −185.9 | 161.6 | −189.3 | 28 | −185.6 | 1394 | 1.14 |
| Benzene | 80.2 | 394 | 5.5 | 126 | 20 | 879 | 1.72 |
| Brine (20% sodium chloride by mass) | 103.9 | — | −17.4 | — | 20 | 1150 | 3.11 |
| n-Butane | −0.5 | 385.2 | −138.5 | 80.3 | −0.5 | 601 | 2.31 |
| Carbon dioxide | −78.4* | 230.5 (at 0°C) | −56.6 | | 0 | 298 | 0.59 |
| Ethanol | 78.2 | 838.3 | −114.2 | 109 | 25 | 783 | 2.46 |
| Ethyl alcohol | 78.6 | 855 | −156 | 108 | 20 | 789 | 2.84 |
| Ethylene glycol | 198.1 | 800.1 | −10.8 | 181.1 | 20 | 1109 | 2.84 |
| Glycerine | 179.9 | 974 | 18.9 | 200.6 | 20 | 1261 | 2.32 |
| Helium | −268.9 | 22.8 | — | — | −268.9 | 146.2 | 22.8 |
| Hydrogen | −252.8 | 445.7 | −259.2 | 59.5 | −252.8 | 70.7 | 10.0 |
| Isobutane | −11.7 | 367.1 | −160 | 105.7 | −11.7 | 593.8 | 2.28 |
| Kerosene | 204–293 | 251 | −24.9 | — | 20 | 820 | 2.00 |
| Mercury | 356.7 | 294.7 | −38.9 | 11.4 | 25 | 13,560 | 0.139 |
| Methane | −161.5 | 510.4 | −182.2 | 58.4 | −161.5 | 423 | 3.49 |
| | | | | | −100 | 301 | 5.79 |
| Methanol | 64.5 | 1100 | −97.7 | 99.2 | 25 | 787 | 2.55 |
| Nitrogen | −195.8 | 198.6 | −210 | 25.3 | −195.8 | 809 | 2.06 |
| | | | | | −160 | 596 | 2.97 |
| Octane | 124.8 | 306.3 | −57.5 | 180.7 | 20 | 703 | 2.10 |
| Oil (light) | | | | | 25 | 910 | 1.80 |
| Oxygen | −183 | 212.7 | −218.8 | 13.7 | −183 | 1141 | 1.71 |
| Petroleum | — | 230–384 | | | 20 | 640 | 2.0 |
| Propane | −42.1 | 427.8 | −187.7 | 80.0 | −42.1 | 581 | 2.25 |
| | | | | | 0 | 529 | 2.53 |
| | | | | | 50 | 449 | 3.13 |
| Refrigerant-134a | −26.1 | 217.0 | −96.6 | — | −50 | 1443 | 1.23 |
| | | | | | −26.1 | 1374 | 1.27 |
| | | | | | 0 | 1295 | 1.34 |
| | | | | | 25 | 1207 | 1.43 |
| Water | 100 | 2257 | 0.0 | 333.7 | 0 | 1000 | 4.22 |
| | | | | | 25 | 997 | 4.18 |
| | | | | | 50 | 988 | 4.18 |
| | | | | | 75 | 975 | 4.19 |
| | | | | | 100 | 958 | 4.22 |

* Sublimation temperature. (At pressures below the triple-point pressure of 518 kPa, carbon dioxide exists as a solid or gas. Also, the freezing-point temperature of carbon dioxide is the triple-point temperature of −56.5°C.)

## TABLE A–3

Properties of common liquids, solids, and foods (*Concluded*)

### (*b*) Solids (values are for room temperature unless indicated otherwise)

| Substance | Density, $\rho$ kg/m$^3$ | Specific heat, $c_p$ kJ/kg·K | Substance | Density, $\rho$ kg/m$^3$ | Specific heat, $c_p$ kJ/kg·K |
|---|---|---|---|---|---|
| Metals | | | Nonmetals | | |
| Aluminum | | | Asphalt | 2110 | 0.920 |
| 200 K | | 0.797 | Brick, common | 1922 | 0.79 |
| 250 K | | 0.859 | Brick, fireclay (500°C) | 2300 | 0.960 |
| 300 K | 2,700 | 0.902 | Concrete | 2300 | 0.653 |
| 350 K | | 0.929 | Clay | 1000 | 0.920 |
| 400 K | | 0.949 | Diamond | 2420 | 0.616 |
| 450 K | | 0.973 | Glass, window | 2700 | 0.800 |
| 500 K | | 0.997 | Glass, pyrex | 2230 | 0.840 |
| Bronze (76% Cu, 2% Zn, 2% Al) | 8,280 | 0.400 | Graphite | 2500 | 0.711 |
| | | | Granite | 2700 | 1.017 |
| Brass, yellow (65% Cu, 35% Zn) | 8,310 | 0.400 | Gypsum or plaster board | 800 | 1.09 |
| Copper | | | Ice | | |
| −173°C | | 0.254 | 200 K | | 1.56 |
| −100°C | | 0.342 | 220 K | | 1.71 |
| −50°C | | 0.367 | 240 K | | 1.86 |
| 0°C | | 0.381 | 260 K | | 2.01 |
| 27°C | 8,900 | 0.386 | 273 K | 921 | 2.11 |
| 100°C | | 0.393 | Limestone | 1650 | 0.909 |
| 200°C | | 0.403 | Marble | 2600 | 0.880 |
| Iron | 7,840 | 0.45 | Plywood (Douglas Fir) | 545 | 1.21 |
| Lead | 11,310 | 0.128 | Rubber (soft) | 1100 | 1.840 |
| Magnesium | 1,730 | 1.000 | Rubber (hard) | 1150 | 2.009 |
| Nickel | 8,890 | 0.440 | Sand | 1520 | 0.800 |
| Silver | 10,470 | 0.235 | Stone | 1500 | 0.800 |
| Steel, mild | 7,830 | 0.500 | Woods, hard (maple, oak, etc.) | 721 | 1.26 |
| Tungsten | 19,400 | 0.130 | Woods, soft (fir, pine, etc.) | 513 | 1.38 |

### (*c*) Foods

| Food | Water content, % (mass) | Freezing point, °C | Specific heat, kJ/kg·K Above freezing | Specific heat, kJ/kg·K Below freezing | Latent heat of fusion, kJ/kg | Food | Water content, % (mass) | Freezing point, °C | Specific heat, kJ/kg·K Above freezing | Specific heat, kJ/kg·K Below freezing | Latent heat of fusion, kJ/kg |
|---|---|---|---|---|---|---|---|---|---|---|---|
| Apples | 84 | −1.1 | 3.65 | 1.90 | 281 | Lettuce | 95 | −0.2 | 4.02 | 2.04 | 317 |
| Bananas | 75 | −0.8 | 3.35 | 1.78 | 251 | Milk, whole | 88 | −0.6 | 3.79 | 1.95 | 294 |
| Beef round | 67 | — | 3.08 | 1.68 | 224 | Oranges | 87 | −0.8 | 3.75 | 1.94 | 291 |
| Broccoli | 90 | −0.6 | 3.86 | 1.97 | 301 | Potatoes | 78 | −0.6 | 3.45 | 1.82 | 261 |
| Butter | 16 | — | — | 1.04 | 53 | Salmon fish | 64 | −2.2 | 2.98 | 1.65 | 214 |
| Cheese, swiss | 39 | −10.0 | 2.15 | 1.33 | 130 | Shrimp | 83 | −2.2 | 3.62 | 1.89 | 277 |
| Cherries | 80 | −1.8 | 3.52 | 1.85 | 267 | Spinach | 93 | −0.3 | 3.96 | 2.01 | 311 |
| Chicken | 74 | −2.8 | 3.32 | 1.77 | 247 | Strawberries | 90 | −0.8 | 3.86 | 1.97 | 301 |
| Corn, sweet | 74 | −0.6 | 3.32 | 1.77 | 247 | Tomatoes, ripe | 94 | −0.5 | 3.99 | 2.02 | 314 |
| Eggs, whole | 74 | −0.6 | 3.32 | 1.77 | 247 | Turkey | 64 | — | 2.98 | 1.65 | 214 |
| Ice cream | 63 | −5.6 | 2.95 | 1.63 | 210 | Watermelon | 93 | −0.4 | 3.96 | 2.01 | 311 |

*Source:* Values are obtained from various handbooks and other sources or are calculated. Water content and freezing-point data of foods are from *ASHRAE, Handbook of Fundamentals,* SI version (Atlanta, GA: American Society of Heating, Refrigerating and Air-Conditioning Engineers, Inc., 1993), Chapter 30, Table 1. Freezing point is the temperature at which freezing starts for fruits and vegetables, and the average freezing temperature for other foods.

## TABLE A–4

Saturated water—Temperature table

| Temp., $T\,°C$ | Sat. press., $P_{sat}$ kPa | Specific volume, m³/kg | | Internal energy, kJ/kg | | | Enthalpy, kJ/kg | | | Entropy, kJ/kg·K | | |
|---|---|---|---|---|---|---|---|---|---|---|---|---|
| | | Sat. liquid, $v_f$ | Sat. vapor, $v_g$ | Sat. liquid, $u_f$ | Evap., $u_{fg}$ | Sat. vapor, $u_g$ | Sat. liquid, $h_f$ | Evap., $h_{fg}$ | Sat. vapor, $h_g$ | Sat. liquid, $s_f$ | Evap., $s_{fg}$ | Sat. vapor, $s_g$ |
| 0.01 | 0.6117 | 0.001000 | 206.00 | 0.000 | 2374.9 | 2374.9 | 0.001 | 2500.9 | 2500.9 | 0.0000 | 9.1556 | 9.1556 |
| 5 | 0.8725 | 0.001000 | 147.03 | 21.019 | 2360.8 | 2381.8 | 21.020 | 2489.1 | 2510.1 | 0.0763 | 8.9487 | 9.0249 |
| 10 | 1.2281 | 0.001000 | 106.32 | 42.020 | 2346.6 | 2388.7 | 42.022 | 2477.2 | 2519.2 | 0.1511 | 8.7488 | 8.8999 |
| 15 | 1.7057 | 0.001001 | 77.885 | 62.980 | 2332.5 | 2395.5 | 62.982 | 2465.4 | 2528.3 | 0.2245 | 8.5559 | 8.7803 |
| 20 | 2.3392 | 0.001002 | 57.762 | 83.913 | 2318.4 | 2402.3 | 83.915 | 2453.5 | 2537.4 | 0.2965 | 8.3696 | 8.6661 |
| 25 | 3.1698 | 0.001003 | 43.340 | 104.83 | 2304.3 | 2409.1 | 104.83 | 2441.7 | 2546.5 | 0.3672 | 8.1895 | 8.5567 |
| 30 | 4.2469 | 0.001004 | 32.879 | 125.73 | 2290.2 | 2415.9 | 125.74 | 2429.8 | 2555.6 | 0.4368 | 8.0152 | 8.4520 |
| 35 | 5.6291 | 0.001006 | 25.205 | 146.63 | 2276.0 | 2422.7 | 146.64 | 2417.9 | 2564.6 | 0.5051 | 7.8466 | 8.3517 |
| 40 | 7.3851 | 0.001008 | 19.515 | 167.53 | 2261.9 | 2429.4 | 167.53 | 2406.0 | 2573.5 | 0.5724 | 7.6832 | 8.2556 |
| 45 | 9.5953 | 0.001010 | 15.251 | 188.43 | 2247.7 | 2436.1 | 188.44 | 2394.0 | 2582.4 | 0.6386 | 7.5247 | 8.1633 |
| 50 | 12.352 | 0.001012 | 12.026 | 209.33 | 2233.4 | 2442.7 | 209.34 | 2382.0 | 2591.3 | 0.7038 | 7.3710 | 8.0748 |
| 55 | 15.763 | 0.001015 | 9.5639 | 230.24 | 2219.1 | 2449.3 | 230.26 | 2369.8 | 2600.1 | 0.7680 | 7.2218 | 7.9898 |
| 60 | 19.947 | 0.001017 | 7.6670 | 251.16 | 2204.7 | 2455.9 | 251.18 | 2357.7 | 2608.8 | 0.8313 | 7.0769 | 7.9082 |
| 65 | 25.043 | 0.001020 | 6.1935 | 272.09 | 2190.3 | 2462.4 | 272.12 | 2345.4 | 2617.5 | 0.8937 | 6.9360 | 7.8296 |
| 70 | 31.202 | 0.001023 | 5.0396 | 293.04 | 2175.8 | 2468.9 | 293.07 | 2333.0 | 2626.1 | 0.9551 | 6.7989 | 7.7540 |
| 75 | 38.597 | 0.001026 | 4.1291 | 313.99 | 2161.3 | 2475.3 | 314.03 | 2320.6 | 2634.6 | 1.0158 | 6.6655 | 7.6812 |
| 80 | 47.416 | 0.001029 | 3.4053 | 334.97 | 2146.6 | 2481.6 | 335.02 | 2308.0 | 2643.0 | 1.0756 | 6.5355 | 7.6111 |
| 85 | 57.868 | 0.001032 | 2.8261 | 355.96 | 2131.9 | 2487.8 | 356.02 | 2295.3 | 2651.4 | 1.1346 | 6.4089 | 7.5435 |
| 90 | 70.183 | 0.001036 | 2.3593 | 376.97 | 2117.0 | 2494.0 | 377.04 | 2282.5 | 2659.6 | 1.1929 | 6.2853 | 7.4782 |
| 95 | 84.609 | 0.001040 | 1.9808 | 398.00 | 2102.0 | 2500.1 | 398.09 | 2269.6 | 2667.6 | 1.2504 | 6.1647 | 7.4151 |
| 100 | 101.42 | 0.001043 | 1.6720 | 419.06 | 2087.0 | 2506.0 | 419.17 | 2256.4 | 2675.6 | 1.3072 | 6.0470 | 7.3542 |
| 105 | 120.90 | 0.001047 | 1.4186 | 440.15 | 2071.8 | 2511.9 | 440.28 | 2243.1 | 2683.4 | 1.3634 | 5.9319 | 7.2952 |
| 110 | 143.38 | 0.001052 | 1.2094 | 461.27 | 2056.4 | 2517.7 | 461.42 | 2229.7 | 2691.1 | 1.4188 | 5.8193 | 7.2382 |
| 115 | 169.18 | 0.001056 | 1.0360 | 482.42 | 2040.9 | 2523.3 | 482.59 | 2216.0 | 2698.6 | 1.4737 | 5.7092 | 7.1829 |
| 120 | 198.67 | 0.001060 | 0.89133 | 503.60 | 2025.3 | 2528.9 | 503.81 | 2202.1 | 2706.0 | 1.5279 | 5.6013 | 7.1292 |
| 125 | 232.23 | 0.001065 | 0.77012 | 524.83 | 2009.5 | 2534.3 | 525.07 | 2188.1 | 2713.1 | 1.5816 | 5.4956 | 7.0771 |
| 130 | 270.28 | 0.001070 | 0.66808 | 546.10 | 1993.4 | 2539.5 | 546.38 | 2173.7 | 2720.1 | 1.6346 | 5.3919 | 7.0265 |
| 135 | 313.22 | 0.001075 | 0.58179 | 567.41 | 1977.3 | 2544.7 | 567.75 | 2159.1 | 2726.9 | 1.6872 | 5.2901 | 6.9773 |
| 140 | 361.53 | 0.001080 | 0.50850 | 588.77 | 1960.9 | 2549.6 | 589.16 | 2144.3 | 2733.5 | 1.7392 | 5.1901 | 6.9294 |
| 145 | 415.68 | 0.001085 | 0.44600 | 610.19 | 1944.2 | 2554.4 | 610.64 | 2129.2 | 2739.8 | 1.7908 | 5.0919 | 6.8827 |
| 150 | 476.16 | 0.001091 | 0.39248 | 631.66 | 1927.4 | 2559.1 | 632.18 | 2113.8 | 2745.9 | 1.8418 | 4.9953 | 6.8371 |
| 155 | 543.49 | 0.001096 | 0.34648 | 653.19 | 1910.3 | 2563.5 | 653.79 | 2098.0 | 2751.8 | 1.8924 | 4.9002 | 6.7927 |
| 160 | 618.23 | 0.001102 | 0.30680 | 674.79 | 1893.0 | 2567.8 | 675.47 | 2082.0 | 2757.5 | 1.9426 | 4.8066 | 6.7492 |
| 165 | 700.93 | 0.001108 | 0.27244 | 696.46 | 1875.4 | 2571.9 | 697.24 | 2065.6 | 2762.8 | 1.9923 | 4.7143 | 6.7067 |
| 170 | 792.18 | 0.001114 | 0.24260 | 718.20 | 1857.5 | 2575.7 | 719.08 | 2048.8 | 2767.9 | 2.0417 | 4.6233 | 6.6650 |
| 175 | 892.60 | 0.001121 | 0.21659 | 740.02 | 1839.4 | 2579.4 | 741.02 | 2031.7 | 2772.7 | 2.0906 | 4.5335 | 6.6242 |
| 180 | 1002.8 | 0.001127 | 0.19384 | 761.92 | 1820.9 | 2582.8 | 763.05 | 2014.2 | 2777.2 | 2.1392 | 4.4448 | 6.5841 |
| 185 | 1123.5 | 0.001134 | 0.17390 | 783.91 | 1802.1 | 2586.0 | 785.19 | 1996.2 | 2781.4 | 2.1875 | 4.3572 | 6.5447 |
| 190 | 1255.2 | 0.001141 | 0.15636 | 806.00 | 1783.0 | 2589.0 | 807.43 | 1977.9 | 2785.3 | 2.2355 | 4.2705 | 6.5059 |
| 195 | 1398.8 | 0.001149 | 0.14089 | 828.18 | 1763.6 | 2591.7 | 829.78 | 1959.0 | 2788.8 | 2.2831 | 4.1847 | 6.4678 |
| 200 | 1554.9 | 0.001157 | 0.12721 | 850.46 | 1743.7 | 2594.2 | 852.26 | 1939.8 | 2792.0 | 2.3305 | 4.0997 | 6.4302 |

# TABLE A–4

Saturated water—Temperature table (*Concluded*)

| Temp., $T\,°C$ | Sat. press., $P_{sat}$ kPa | Specific volume, m³/kg | | Internal energy, kJ/kg | | | Enthalpy, kJ/kg | | | Entropy, kJ/kg·K | | |
|---|---|---|---|---|---|---|---|---|---|---|---|---|
| | | Sat. liquid, $v_f$ | Sat. vapor, $v_g$ | Sat. liquid, $u_f$ | Evap., $u_{fg}$ | Sat. vapor, $u_g$ | Sat. liquid, $h_f$ | Evap., $h_{fg}$ | Sat. vapor, $h_g$ | Sat. liquid, $s_f$ | Evap., $s_{fg}$ | Sat. vapor, $s_g$ |
| 205 | 1724.3 | 0.001164 | 0.11508 | 872.86 | 1723.5 | 2596.4 | 874.87 | 1920.0 | 2794.8 | 2.3776 | 4.0154 | 6.3930 |
| 210 | 1907.7 | 0.001173 | 0.10429 | 895.38 | 1702.9 | 2598.3 | 897.61 | 1899.7 | 2797.3 | 2.4245 | 3.9318 | 6.3563 |
| 215 | 2105.9 | 0.001181 | 0.094680 | 918.02 | 1681.9 | 2599.9 | 920.50 | 1878.8 | 2799.3 | 2.4712 | 3.8489 | 6.3200 |
| 220 | 2319.6 | 0.001190 | 0.086094 | 940.79 | 1660.5 | 2601.3 | 943.55 | 1857.4 | 2801.0 | 2.5176 | 3.7664 | 6.2840 |
| 225 | 2549.7 | 0.001199 | 0.078405 | 963.70 | 1638.6 | 2602.3 | 966.76 | 1835.4 | 2802.2 | 2.5639 | 3.6844 | 6.2483 |
| 230 | 2797.1 | 0.001209 | 0.071505 | 986.76 | 1616.1 | 2602.9 | 990.14 | 1812.8 | 2802.9 | 2.6100 | 3.6028 | 6.2128 |
| 235 | 3062.6 | 0.001219 | 0.065300 | 1010.0 | 1593.2 | 2603.2 | 1013.7 | 1789.5 | 2803.2 | 2.6560 | 3.5216 | 6.1775 |
| 240 | 3347.0 | 0.001229 | 0.059707 | 1033.4 | 1569.8 | 2603.1 | 1037.5 | 1765.5 | 2803.0 | 2.7018 | 3.4405 | 6.1424 |
| 245 | 3651.2 | 0.001240 | 0.054656 | 1056.9 | 1545.7 | 2602.7 | 1061.5 | 1740.8 | 2802.2 | 2.7476 | 3.3596 | 6.1072 |
| 250 | 3976.2 | 0.001252 | 0.050085 | 1080.7 | 1521.1 | 2601.8 | 1085.7 | 1715.3 | 2801.0 | 2.7933 | 3.2788 | 6.0721 |
| 255 | 4322.9 | 0.001263 | 0.045941 | 1104.7 | 1495.8 | 2600.5 | 1110.1 | 1689.0 | 2799.1 | 2.8390 | 3.1979 | 6.0369 |
| 260 | 4692.3 | 0.001276 | 0.042175 | 1128.8 | 1469.9 | 2598.7 | 1134.8 | 1661.8 | 2796.6 | 2.8847 | 3.1169 | 6.0017 |
| 265 | 5085.3 | 0.001289 | 0.038748 | 1153.3 | 1443.2 | 2596.5 | 1159.8 | 1633.7 | 2793.5 | 2.9304 | 3.0358 | 5.9662 |
| 270 | 5503.0 | 0.001303 | 0.035622 | 1177.9 | 1415.7 | 2593.7 | 1185.1 | 1604.6 | 2789.7 | 2.9762 | 2.9542 | 5.9305 |
| 275 | 5946.4 | 0.001317 | 0.032767 | 1202.9 | 1387.4 | 2590.3 | 1210.7 | 1574.5 | 2785.2 | 3.0221 | 2.8723 | 5.8944 |
| 280 | 6416.6 | 0.001333 | 0.030153 | 1228.2 | 1358.2 | 2586.4 | 1236.7 | 1543.2 | 2779.9 | 3.0681 | 2.7898 | 5.8579 |
| 285 | 6914.6 | 0.001349 | 0.027756 | 1253.7 | 1328.1 | 2581.8 | 1263.1 | 1510.7 | 2773.7 | 3.1144 | 2.7066 | 5.8210 |
| 290 | 7441.8 | 0.001366 | 0.025554 | 1279.7 | 1296.9 | 2576.5 | 1289.8 | 1476.9 | 2766.7 | 3.1608 | 2.6225 | 5.7834 |
| 295 | 7999.0 | 0.001384 | 0.023528 | 1306.0 | 1264.5 | 2570.5 | 1317.1 | 1441.6 | 2758.7 | 3.2076 | 2.5374 | 5.7450 |
| 300 | 8587.9 | 0.001404 | 0.021659 | 1332.7 | 1230.9 | 2563.6 | 1344.8 | 1404.8 | 2749.6 | 3.2548 | 2.4511 | 5.7059 |
| 305 | 9209.4 | 0.001425 | 0.019932 | 1360.0 | 1195.9 | 2555.8 | 1373.1 | 1366.3 | 2739.4 | 3.3024 | 2.3633 | 5.6657 |
| 310 | 9865.0 | 0.001447 | 0.018333 | 1387.7 | 1159.3 | 2547.1 | 1402.0 | 1325.9 | 2727.9 | 3.3506 | 2.2737 | 5.6243 |
| 315 | 10,556 | 0.001472 | 0.016849 | 1416.1 | 1121.1 | 2537.2 | 1431.6 | 1283.4 | 2715.0 | 3.3994 | 2.1821 | 5.5816 |
| 320 | 11,284 | 0.001499 | 0.015470 | 1445.1 | 1080.9 | 2526.0 | 1462.0 | 1238.5 | 2700.6 | 3.4491 | 2.0881 | 5.5372 |
| 325 | 12,051 | 0.001528 | 0.014183 | 1475.0 | 1038.5 | 2513.4 | 1493.4 | 1191.0 | 2684.3 | 3.4998 | 1.9911 | 5.4908 |
| 330 | 12,858 | 0.001560 | 0.012979 | 1505.7 | 993.5 | 2499.2 | 1525.8 | 1140.3 | 2666.0 | 3.5516 | 1.8906 | 5.4422 |
| 335 | 13,707 | 0.001597 | 0.011848 | 1537.5 | 945.5 | 2483.0 | 1559.4 | 1086.0 | 2645.4 | 3.6050 | 1.7857 | 5.3907 |
| 340 | 14,601 | 0.001638 | 0.010783 | 1570.7 | 893.8 | 2464.5 | 1594.6 | 1027.4 | 2622.0 | 3.6602 | 1.6756 | 5.3358 |
| 345 | 15,541 | 0.001685 | 0.009772 | 1605.5 | 837.7 | 2443.2 | 1631.7 | 963.4 | 2595.1 | 3.7179 | 1.5585 | 5.2765 |
| 350 | 16,529 | 0.001741 | 0.008806 | 1642.4 | 775.9 | 2418.3 | 1671.2 | 892.7 | 2563.9 | 3.7788 | 1.4326 | 5.2114 |
| 355 | 17,570 | 0.001808 | 0.007872 | 1682.2 | 706.4 | 2388.6 | 1714.0 | 812.9 | 2526.9 | 3.8442 | 1.2942 | 5.1384 |
| 360 | 18,666 | 0.001895 | 0.006950 | 1726.2 | 625.7 | 2351.9 | 1761.5 | 720.1 | 2481.6 | 3.9165 | 1.1373 | 5.0537 |
| 365 | 19,822 | 0.002015 | 0.006009 | 1777.2 | 526.4 | 2303.6 | 1817.2 | 605.5 | 2422.7 | 4.0004 | 0.9489 | 4.9493 |
| 370 | 21,044 | 0.002217 | 0.004953 | 1844.5 | 385.6 | 2230.1 | 1891.2 | 443.1 | 2334.3 | 4.1119 | 0.6890 | 4.8009 |
| 373.95 | 22,064 | 0.003106 | 0.003106 | 2015.7 | 0 | 2015.7 | 2084.3 | 0 | 2084.3 | 4.4070 | 0 | 4.4070 |

*Source:* Tables A–4 through A–8 are generated using the Engineering Equation Solver (EES) software developed by S. A. Klein and F. L. Alvarado. The routine used in calculations is the highly accurate Steam_IAPWS, which incorporates the 1995 Formulation for the Thermodynamic Properties of Ordinary Water Substance for General and Scientific Use, issued by The International Association for the Properties of Water and Steam (IAPWS). This formulation replaces the 1984 formulation of Haar, Gallagher, and Kell (NBS/NRC Steam Tables, Hemisphere Publishing Co., 1984), which is also available in EES as the routine STEAM. The new formulation is based on the correlations of Saul and Wagner (J. Phys. Chem. Ref. Data, 16, 893, 1987) with modifications to adjust to the International Temperature Scale of 1990. The modifications are described by Wagner and Pruss (J. Phys. Chem. Ref. Data, 22, 783, 1993). The properties of ice are based on Hyland and Wexler, "Formulations for the Thermodynamic Properties of the Saturated Phases of $H_2O$ from 173.15 K to 473.15 K," *ASHRAE Trans.*, Part 2A, Paper 2793, 1983.

## TABLE A–5

Saturated water—Pressure table

| Press., P kPa | Sat. temp., $T_{sat}$ °C | Specific volume, m³/kg | | Internal energy, kJ/kg | | | Enthalpy, kJ/kg | | | Entropy, kJ/kg·K | | |
|---|---|---|---|---|---|---|---|---|---|---|---|---|
| | | Sat. liquid, $v_f$ | Sat. vapor, $v_g$ | Sat. liquid, $u_f$ | Evap., $u_{fg}$ | Sat. vapor, $u_g$ | Sat. liquid, $h_f$ | Evap., $h_{fg}$ | Sat. vapor, $h_g$ | Sat. liquid, $s_f$ | Evap., $s_{fg}$ | Sat. vapor, $s_g$ |
| 1.0 | 6.97 | 0.001000 | 129.19 | 29.302 | 2355.2 | 2384.5 | 29.303 | 2484.4 | 2513.7 | 0.1059 | 8.8690 | 8.9749 |
| 1.5 | 13.02 | 0.001001 | 87.964 | 54.686 | 2338.1 | 2392.8 | 54.688 | 2470.1 | 2524.7 | 0.1956 | 8.6314 | 8.8270 |
| 2.0 | 17.50 | 0.001001 | 66.990 | 73.431 | 2325.5 | 2398.9 | 73.433 | 2459.5 | 2532.9 | 0.2606 | 8.4621 | 8.7227 |
| 2.5 | 21.08 | 0.001002 | 54.242 | 88.422 | 2315.4 | 2403.8 | 88.424 | 2451.0 | 2539.4 | 0.3118 | 8.3302 | 8.6421 |
| 3.0 | 24.08 | 0.001003 | 45.654 | 100.98 | 2306.9 | 2407.9 | 100.98 | 2443.9 | 2544.8 | 0.3543 | 8.2222 | 8.5765 |
| 4.0 | 28.96 | 0.001004 | 34.791 | 121.39 | 2293.1 | 2414.5 | 121.39 | 2432.3 | 2553.7 | 0.4224 | 8.0510 | 8.4734 |
| 5.0 | 32.87 | 0.001005 | 28.185 | 137.75 | 2282.1 | 2419.8 | 137.75 | 2423.0 | 2560.7 | 0.4762 | 7.9176 | 8.3938 |
| 7.5 | 40.29 | 0.001008 | 19.233 | 168.74 | 2261.1 | 2429.8 | 168.75 | 2405.3 | 2574.0 | 0.5763 | 7.6738 | 8.2501 |
| 10 | 45.81 | 0.001010 | 14.670 | 191.79 | 2245.4 | 2437.2 | 191.81 | 2392.1 | 2583.9 | 0.6492 | 7.4996 | 8.1488 |
| 15 | 53.97 | 0.001014 | 10.020 | 225.93 | 2222.1 | 2448.0 | 225.94 | 2372.3 | 2598.3 | 0.7549 | 7.2522 | 8.0071 |
| 20 | 60.06 | 0.001017 | 7.6481 | 251.40 | 2204.6 | 2456.0 | 251.42 | 2357.5 | 2608.9 | 0.8320 | 7.0752 | 7.9073 |
| 25 | 64.96 | 0.001020 | 6.2034 | 271.93 | 2190.4 | 2462.4 | 271.96 | 2345.5 | 2617.5 | 0.8932 | 6.9370 | 7.8302 |
| 30 | 69.09 | 0.001022 | 5.2287 | 289.24 | 2178.5 | 2467.7 | 289.27 | 2335.3 | 2624.6 | 0.9441 | 6.8234 | 7.7675 |
| 40 | 75.86 | 0.001026 | 3.9933 | 317.58 | 2158.8 | 2476.3 | 317.62 | 2318.4 | 2636.1 | 1.0261 | 6.6430 | 7.6691 |
| 50 | 81.32 | 0.001030 | 3.2403 | 340.49 | 2142.7 | 2483.2 | 340.54 | 2304.7 | 2645.2 | 1.0912 | 6.5019 | 7.5931 |
| 75 | 91.76 | 0.001037 | 2.2172 | 384.36 | 2111.8 | 2496.1 | 384.44 | 2278.0 | 2662.4 | 1.2132 | 6.2426 | 7.4558 |
| 100 | 99.61 | 0.001043 | 1.6941 | 417.40 | 2088.2 | 2505.6 | 417.51 | 2257.5 | 2675.0 | 1.3028 | 6.0562 | 7.3589 |
| 101.325 | 99.97 | 0.001043 | 1.6734 | 418.95 | 2087.0 | 2506.0 | 419.06 | 2256.5 | 2675.6 | 1.3069 | 6.0476 | 7.3545 |
| 125 | 105.97 | 0.001048 | 1.3750 | 444.23 | 2068.8 | 2513.0 | 444.36 | 2240.6 | 2684.9 | 1.3741 | 5.9100 | 7.2841 |
| 150 | 111.35 | 0.001053 | 1.1594 | 466.97 | 2052.3 | 2519.2 | 467.13 | 2226.0 | 2693.1 | 1.4337 | 5.7894 | 7.2231 |
| 175 | 116.04 | 0.001057 | 1.0037 | 486.82 | 2037.7 | 2524.5 | 487.01 | 2213.1 | 2700.2 | 1.4850 | 5.6865 | 7.1716 |
| 200 | 120.21 | 0.001061 | 0.88578 | 504.50 | 2024.6 | 2529.1 | 504.71 | 2201.6 | 2706.3 | 1.5302 | 5.5968 | 7.1270 |
| 225 | 123.97 | 0.001064 | 0.79329 | 520.47 | 2012.7 | 2533.2 | 520.71 | 2191.0 | 2711.7 | 1.5706 | 5.5171 | 7.0877 |
| 250 | 127.41 | 0.001067 | 0.71873 | 535.08 | 2001.8 | 2536.8 | 535.35 | 2181.2 | 2716.5 | 1.6072 | 5.4453 | 7.0525 |
| 275 | 130.58 | 0.001070 | 0.65732 | 548.57 | 1991.6 | 2540.1 | 548.86 | 2172.0 | 2720.9 | 1.6408 | 5.3800 | 7.0207 |
| 300 | 133.52 | 0.001073 | 0.60582 | 561.11 | 1982.1 | 2543.2 | 561.43 | 2163.5 | 2724.9 | 1.6717 | 5.3200 | 6.9917 |
| 325 | 136.27 | 0.001076 | 0.56199 | 572.84 | 1973.1 | 2545.9 | 573.19 | 2155.4 | 2728.6 | 1.7005 | 5.2645 | 6.9650 |
| 350 | 138.86 | 0.001079 | 0.52422 | 583.89 | 1964.6 | 2548.5 | 584.26 | 2147.7 | 2732.0 | 1.7274 | 5.2128 | 6.9402 |
| 375 | 141.30 | 0.001081 | 0.49133 | 594.32 | 1956.6 | 2550.9 | 594.73 | 2140.4 | 2735.1 | 1.7526 | 5.1645 | 6.9171 |
| 400 | 143.61 | 0.001084 | 0.46242 | 604.22 | 1948.9 | 2553.1 | 604.66 | 2133.4 | 2738.1 | 1.7765 | 5.1191 | 6.8955 |
| 450 | 147.90 | 0.001088 | 0.41392 | 622.65 | 1934.5 | 2557.1 | 623.14 | 2120.3 | 2743.4 | 1.8205 | 5.0356 | 6.8561 |
| 500 | 151.83 | 0.001093 | 0.37483 | 639.54 | 1921.2 | 2560.7 | 640.09 | 2108.0 | 2748.1 | 1.8604 | 4.9603 | 6.8207 |
| 550 | 155.46 | 0.001097 | 0.34261 | 655.16 | 1908.8 | 2563.9 | 655.77 | 2096.6 | 2752.4 | 1.8970 | 4.8916 | 6.7886 |
| 600 | 158.83 | 0.001101 | 0.31560 | 669.72 | 1897.1 | 2566.8 | 670.38 | 2085.8 | 2756.2 | 1.9308 | 4.8285 | 6.7593 |
| 650 | 161.98 | 0.001104 | 0.29260 | 683.37 | 1886.1 | 2569.4 | 684.08 | 2075.5 | 2759.6 | 1.9623 | 4.7699 | 6.7322 |
| 700 | 164.95 | 0.001108 | 0.27278 | 696.23 | 1875.6 | 2571.8 | 697.00 | 2065.8 | 2762.8 | 1.9918 | 4.7153 | 6.7071 |
| 750 | 167.75 | 0.001111 | 0.25552 | 708.40 | 1865.6 | 2574.0 | 709.24 | 2056.4 | 2765.7 | 2.0195 | 4.6642 | 6.6837 |

## TABLE A–5

Saturated water—Pressure table (*Concluded*)

| Press., $P$ kPa | Sat. temp., $T_{sat}$ °C | Specific volume, m³/kg | | Internal energy, kJ/kg | | | Enthalpy, kJ/kg | | | Entropy, kJ/kg·K | | |
|---|---|---|---|---|---|---|---|---|---|---|---|---|
| | | Sat. liquid, $v_f$ | Sat. vapor, $v_g$ | Sat. liquid, $u_f$ | Evap., $u_{fg}$ | Sat. vapor, $u_g$ | Sat. liquid, $h_f$ | Evap., $h_{fg}$ | Sat. vapor, $h_g$ | Sat. liquid, $s_f$ | Evap., $s_{fg}$ | Sat. vapor, $s_g$ |
| 800 | 170.41 | 0.001115 | 0.24035 | 719.97 | 1856.1 | 2576.0 | 720.87 | 2047.5 | 2768.3 | 2.0457 | 4.6160 | 6.6616 |
| 850 | 172.94 | 0.001118 | 0.22690 | 731.00 | 1846.9 | 2577.9 | 731.95 | 2038.8 | 2770.8 | 2.0705 | 4.5705 | 6.6409 |
| 900 | 175.35 | 0.001121 | 0.21489 | 741.55 | 1838.1 | 2579.6 | 742.56 | 2030.5 | 2773.0 | 2.0941 | 4.5273 | 6.6213 |
| 950 | 177.66 | 0.001124 | 0.20411 | 751.67 | 1829.6 | 2581.3 | 752.74 | 2022.4 | 2775.2 | 2.1166 | 4.4862 | 6.6027 |
| 1000 | 179.88 | 0.001127 | 0.19436 | 761.39 | 1821.4 | 2582.8 | 762.51 | 2014.6 | 2777.1 | 2.1381 | 4.4470 | 6.5850 |
| 1100 | 184.06 | 0.001133 | 0.17745 | 779.78 | 1805.7 | 2585.5 | 781.03 | 1999.6 | 2780.7 | 2.1785 | 4.3735 | 6.5520 |
| 1200 | 187.96 | 0.001138 | 0.16326 | 796.96 | 1790.9 | 2587.8 | 798.33 | 1985.4 | 2783.8 | 2.2159 | 4.3058 | 6.5217 |
| 1300 | 191.60 | 0.001144 | 0.15119 | 813.10 | 1776.8 | 2589.9 | 814.59 | 1971.9 | 2786.5 | 2.2508 | 4.2428 | 6.4936 |
| 1400 | 195.04 | 0.001149 | 0.14078 | 828.35 | 1763.4 | 2591.8 | 829.96 | 1958.9 | 2788.9 | 2.2835 | 4.1840 | 6.4675 |
| 1500 | 198.29 | 0.001154 | 0.13171 | 842.82 | 1750.6 | 2593.4 | 844.55 | 1946.4 | 2791.0 | 2.3143 | 4.1287 | 6.4430 |
| 1750 | 205.72 | 0.001166 | 0.11344 | 876.12 | 1720.6 | 2596.7 | 878.16 | 1917.1 | 2795.2 | 2.3844 | 4.0033 | 6.3877 |
| 2000 | 212.38 | 0.001177 | 0.099587 | 906.12 | 1693.0 | 2599.1 | 908.47 | 1889.8 | 2798.3 | 2.4467 | 3.8923 | 6.3390 |
| 2250 | 218.41 | 0.001187 | 0.088717 | 933.54 | 1667.3 | 2600.9 | 936.21 | 1864.3 | 2800.5 | 2.5029 | 3.7926 | 6.2954 |
| 2500 | 223.95 | 0.001197 | 0.079952 | 958.87 | 1643.2 | 2602.1 | 961.87 | 1840.1 | 2801.9 | 2.5542 | 3.7016 | 6.2558 |
| 3000 | 233.85 | 0.001217 | 0.066667 | 1004.6 | 1598.5 | 2603.2 | 1008.3 | 1794.9 | 2803.2 | 2.6454 | 3.5402 | 6.1856 |
| 3500 | 242.56 | 0.001235 | 0.057061 | 1045.4 | 1557.6 | 2603.0 | 1049.7 | 1753.0 | 2802.7 | 2.7253 | 3.3991 | 6.1244 |
| 4000 | 250.35 | 0.001252 | 0.049779 | 1082.4 | 1519.3 | 2601.7 | 1087.4 | 1713.5 | 2800.8 | 2.7966 | 3.2731 | 6.0696 |
| 5000 | 263.94 | 0.001286 | 0.039448 | 1148.1 | 1448.9 | 2597.0 | 1154.5 | 1639.7 | 2794.2 | 2.9207 | 3.0530 | 5.9737 |
| 6000 | 275.59 | 0.001319 | 0.032449 | 1205.8 | 1384.1 | 2589.9 | 1213.8 | 1570.9 | 2784.6 | 3.0275 | 2.8627 | 5.8902 |
| 7000 | 285.83 | 0.001352 | 0.027378 | 1258.0 | 1323.0 | 2581.0 | 1267.5 | 1505.2 | 2772.6 | 3.1220 | 2.6927 | 5.8148 |
| 8000 | 295.01 | 0.001384 | 0.023525 | 1306.0 | 1264.5 | 2570.5 | 1317.1 | 1441.6 | 2758.7 | 3.2077 | 2.5373 | 5.7450 |
| 9000 | 303.35 | 0.001418 | 0.020489 | 1350.9 | 1207.6 | 2558.5 | 1363.7 | 1379.3 | 2742.9 | 3.2866 | 2.3925 | 5.6791 |
| 10,000 | 311.00 | 0.001452 | 0.018028 | 1393.3 | 1151.8 | 2545.2 | 1407.8 | 1317.6 | 2725.5 | 3.3603 | 2.2556 | 5.6159 |
| 11,000 | 318.08 | 0.001488 | 0.015988 | 1433.9 | 1096.6 | 2530.4 | 1450.2 | 1256.1 | 2706.3 | 3.4299 | 2.1245 | 5.5544 |
| 12,000 | 324.68 | 0.001526 | 0.014264 | 1473.0 | 1041.3 | 2514.3 | 1491.3 | 1194.1 | 2685.4 | 3.4964 | 1.9975 | 5.4939 |
| 13,000 | 330.85 | 0.001566 | 0.012781 | 1511.0 | 985.5 | 2496.6 | 1531.4 | 1131.3 | 2662.7 | 3.5606 | 1.8730 | 5.4336 |
| 14,000 | 336.67 | 0.001610 | 0.011487 | 1548.4 | 928.7 | 2477.1 | 1571.0 | 1067.0 | 2637.9 | 3.6232 | 1.7497 | 5.3728 |
| 15,000 | 342.16 | 0.001657 | 0.010341 | 1585.5 | 870.3 | 2455.7 | 1610.3 | 1000.5 | 2610.8 | 3.6848 | 1.6261 | 5.3108 |
| 16,000 | 347.36 | 0.001710 | 0.009312 | 1622.6 | 809.4 | 2432.0 | 1649.9 | 931.1 | 2581.0 | 3.7461 | 1.5005 | 5.2466 |
| 17,000 | 352.29 | 0.001770 | 0.008374 | 1660.2 | 745.1 | 2405.4 | 1690.3 | 857.4 | 2547.7 | 3.8082 | 1.3709 | 5.1791 |
| 18,000 | 356.99 | 0.001840 | 0.007504 | 1699.1 | 675.9 | 2375.0 | 1732.2 | 777.8 | 2510.0 | 3.8720 | 1.2343 | 5.1064 |
| 19,000 | 361.47 | 0.001926 | 0.006677 | 1740.3 | 598.9 | 2339.2 | 1776.8 | 689.2 | 2466.0 | 3.9396 | 1.0860 | 5.0256 |
| 20,000 | 365.75 | 0.002038 | 0.005862 | 1785.8 | 509.0 | 2294.8 | 1826.6 | 585.5 | 2412.1 | 4.0146 | 0.9164 | 4.9310 |
| 21,000 | 369.83 | 0.002207 | 0.004994 | 1841.6 | 391.9 | 2233.5 | 1888.0 | 450.4 | 2338.4 | 4.1071 | 0.7005 | 4.8076 |
| 22,000 | 373.71 | 0.002703 | 0.003644 | 1951.7 | 140.8 | 2092.4 | 2011.1 | 161.5 | 2172.6 | 4.2942 | 0.2496 | 4.5439 |
| 22,064 | 373.95 | 0.003106 | 0.003106 | 2015.7 | 0 | 2015.7 | 2084.3 | 0 | 2084.3 | 4.4070 | 0 | 4.4070 |

## TABLE A–6

Superheated water

| T °C | v m³/kg | u kJ/kg | h kJ/kg | s kJ/kg·K | v m³/kg | u kJ/kg | h kJ/kg | s kJ/kg·K | v m³/kg | u kJ/kg | h kJ/kg | s kJ/kg·K |
|---|---|---|---|---|---|---|---|---|---|---|---|---|
| | P = 0.01 MPa (45.81°C)* | | | | P = 0.05 MPa (81.32°C) | | | | P = 0.10 MPa (99.61°C) | | | |
| Sat.† | 14.670 | 2437.2 | 2583.9 | 8.1488 | 3.2403 | 2483.2 | 2645.2 | 7.5931 | 1.6941 | 2505.6 | 2675.0 | 7.3589 |
| 50 | 14.867 | 2443.3 | 2592.0 | 8.1741 | | | | | | | | |
| 100 | 17.196 | 2515.5 | 2687.5 | 8.4489 | 3.4187 | 2511.5 | 2682.4 | 7.6953 | 1.6959 | 2506.2 | 2675.8 | 7.3611 |
| 150 | 19.513 | 2587.9 | 2783.0 | 8.6893 | 3.8897 | 2585.7 | 2780.2 | 7.9413 | 1.9367 | 2582.9 | 2776.6 | 7.6148 |
| 200 | 21.826 | 2661.4 | 2879.6 | 8.9049 | 4.3562 | 2660.0 | 2877.8 | 8.1592 | 2.1724 | 2658.2 | 2875.5 | 7.8356 |
| 250 | 24.136 | 2736.1 | 2977.5 | 9.1015 | 4.8206 | 2735.1 | 2976.2 | 8.3568 | 2.4062 | 2733.9 | 2974.5 | 8.0346 |
| 300 | 26.446 | 2812.3 | 3076.7 | 9.2827 | 5.2841 | 2811.6 | 3075.8 | 8.5387 | 2.6389 | 2810.7 | 3074.5 | 8.2172 |
| 400 | 31.063 | 2969.3 | 3280.0 | 9.6094 | 6.2094 | 2968.9 | 3279.3 | 8.8659 | 3.1027 | 2968.3 | 3278.6 | 8.5452 |
| 500 | 35.680 | 3132.9 | 3489.7 | 9.8998 | 7.1338 | 3132.6 | 3489.3 | 9.1566 | 3.5655 | 3132.2 | 3488.7 | 8.8362 |
| 600 | 40.296 | 3303.3 | 3706.3 | 10.1631 | 8.0577 | 3303.1 | 3706.0 | 9.4201 | 4.0279 | 3302.8 | 3705.6 | 9.0999 |
| 700 | 44.911 | 3480.8 | 3929.9 | 10.4056 | 8.9813 | 3480.6 | 3929.7 | 9.6626 | 4.4900 | 3480.4 | 3929.4 | 9.3424 |
| 800 | 49.527 | 3665.4 | 4160.6 | 10.6312 | 9.9047 | 3665.2 | 4160.4 | 9.8883 | 4.9519 | 3665.0 | 4160.2 | 9.5682 |
| 900 | 54.143 | 3856.9 | 4398.3 | 10.8429 | 10.8280 | 3856.8 | 4398.2 | 10.1000 | 5.4137 | 3856.7 | 4398.0 | 9.7800 |
| 1000 | 58.758 | 4055.3 | 4642.8 | 11.0429 | 11.7513 | 4055.2 | 4642.7 | 10.3000 | 5.8755 | 4055.0 | 4642.6 | 9.9800 |
| 1100 | 63.373 | 4260.0 | 4893.8 | 11.2326 | 12.6745 | 4259.9 | 4893.7 | 10.4897 | 6.3372 | 4259.8 | 4893.6 | 10.1698 |
| 1200 | 67.989 | 4470.9 | 5150.8 | 11.4132 | 13.5977 | 4470.8 | 5150.7 | 10.6704 | 6.7988 | 4470.7 | 5150.6 | 10.3504 |
| 1300 | 72.604 | 4687.4 | 5413.4 | 11.5857 | 14.5209 | 4687.3 | 5413.3 | 10.8429 | 7.2605 | 4687.2 | 5413.3 | 10.5229 |
| | P = 0.20 MPa (120.21°C) | | | | P = 0.30 MPa (133.52°C) | | | | P = 0.40 MPa (143.61°C) | | | |
| Sat. | 0.88578 | 2529.1 | 2706.3 | 7.1270 | 0.60582 | 2543.2 | 2724.9 | 6.9917 | 0.46242 | 2553.1 | 2738.1 | 6.8955 |
| 150 | 0.95986 | 2577.1 | 2769.1 | 7.2810 | 0.63402 | 2571.0 | 2761.2 | 7.0792 | 0.47088 | 2564.4 | 2752.8 | 6.9306 |
| 200 | 1.08049 | 2654.6 | 2870.7 | 7.5081 | 0.71643 | 2651.0 | 2865.9 | 7.3132 | 0.53434 | 2647.2 | 2860.9 | 7.1723 |
| 250 | 1.19890 | 2731.4 | 2971.2 | 7.7100 | 0.79645 | 2728.9 | 2967.9 | 7.5180 | 0.59520 | 2726.4 | 2964.5 | 7.3804 |
| 300 | 1.31623 | 2808.8 | 3072.1 | 7.8941 | 0.87535 | 2807.0 | 3069.6 | 7.7037 | 0.65489 | 2805.1 | 3067.1 | 7.5677 |
| 400 | 1.54934 | 2967.2 | 3277.0 | 8.2236 | 1.03155 | 2966.0 | 3275.5 | 8.0347 | 0.77265 | 2964.9 | 3273.9 | 7.9003 |
| 500 | 1.78142 | 3131.4 | 3487.7 | 8.5153 | 1.18672 | 3130.6 | 3486.6 | 8.3271 | 0.88936 | 3129.8 | 3485.5 | 8.1933 |
| 600 | 2.01302 | 3302.2 | 3704.8 | 8.7793 | 1.34139 | 3301.6 | 3704.0 | 8.5915 | 1.00558 | 3301.0 | 3703.3 | 8.4580 |
| 700 | 2.24434 | 3479.9 | 3928.8 | 9.0221 | 1.49580 | 3479.5 | 3928.2 | 8.8345 | 1.12152 | 3479.0 | 3927.6 | 8.7012 |
| 800 | 2.47550 | 3664.7 | 4159.8 | 9.2479 | 1.65004 | 3664.3 | 4159.3 | 9.0605 | 1.23730 | 3663.9 | 4158.9 | 8.9274 |
| 900 | 2.70656 | 3856.3 | 4397.7 | 9.4598 | 1.80417 | 3856.0 | 4397.3 | 9.2725 | 1.35298 | 3855.7 | 4396.9 | 9.1394 |
| 1000 | 2.93755 | 4054.8 | 4642.3 | 9.6599 | 1.95824 | 4054.5 | 4642.0 | 9.4726 | 1.46859 | 4054.3 | 4641.7 | 9.3396 |
| 1100 | 3.16848 | 4259.6 | 4893.3 | 9.8497 | 2.11226 | 4259.4 | 4893.1 | 9.6624 | 1.58414 | 4259.2 | 4892.9 | 9.5295 |
| 1200 | 3.39938 | 4470.5 | 5150.4 | 10.0304 | 2.26624 | 4470.3 | 5150.2 | 9.8431 | 1.69966 | 4470.2 | 5150.0 | 9.7102 |
| 1300 | 3.63026 | 4687.1 | 5413.1 | 10.2029 | 2.42019 | 4686.9 | 5413.0 | 10.0157 | 1.81516 | 4686.7 | 5412.8 | 9.8828 |
| | P = 0.50 MPa (151.83°C) | | | | P = 0.60 MPa (158.83°C) | | | | P = 0.80 MPa (170.41°C) | | | |
| Sat. | 0.37483 | 2560.7 | 2748.1 | 6.8207 | 0.31560 | 2566.8 | 2756.2 | 6.7593 | 0.24035 | 2576.0 | 2768.3 | 6.6616 |
| 200 | 0.42503 | 2643.3 | 2855.8 | 7.0610 | 0.35212 | 2639.4 | 2850.6 | 6.9683 | 0.26088 | 2631.1 | 2839.8 | 6.8177 |
| 250 | 0.47443 | 2723.8 | 2961.0 | 7.2725 | 0.39390 | 2721.2 | 2957.6 | 7.1833 | 0.29321 | 2715.9 | 2950.4 | 7.0402 |
| 300 | 0.52261 | 2803.3 | 3064.6 | 7.4614 | 0.43442 | 2801.4 | 3062.0 | 7.3740 | 0.32416 | 2797.5 | 3056.9 | 7.2345 |
| 350 | 0.57015 | 2883.0 | 3168.1 | 7.6346 | 0.47428 | 2881.6 | 3166.1 | 7.5481 | 0.35442 | 2878.6 | 3162.2 | 7.4107 |
| 400 | 0.61731 | 2963.7 | 3272.4 | 7.7956 | 0.51374 | 2962.5 | 3270.8 | 7.7097 | 0.38429 | 2960.2 | 3267.7 | 7.5735 |
| 500 | 0.71095 | 3129.0 | 3484.5 | 8.0893 | 0.59200 | 3128.2 | 3483.4 | 8.0041 | 0.44332 | 3126.6 | 3481.3 | 7.8692 |
| 600 | 0.80409 | 3300.4 | 3702.5 | 8.3544 | 0.66976 | 3299.8 | 3701.7 | 8.2695 | 0.50186 | 3298.7 | 3700.1 | 8.1354 |
| 700 | 0.89696 | 3478.6 | 3927.0 | 8.5978 | 0.74725 | 3478.1 | 3926.4 | 8.5132 | 0.56011 | 3477.2 | 3925.3 | 8.3794 |
| 800 | 0.98966 | 3663.6 | 4158.4 | 8.8240 | 0.82457 | 3663.2 | 4157.9 | 8.7395 | 0.61820 | 3662.5 | 4157.0 | 8.6061 |
| 900 | 1.08227 | 3855.4 | 4396.6 | 9.0362 | 0.90179 | 3855.1 | 4396.2 | 8.9518 | 0.67619 | 3854.5 | 4395.5 | 8.8185 |
| 1000 | 1.17480 | 4054.0 | 4641.4 | 9.2364 | 0.97893 | 4053.8 | 4641.1 | 9.1521 | 0.73411 | 4053.3 | 4640.5 | 9.0189 |
| 1100 | 1.26728 | 4259.0 | 4892.6 | 9.4263 | 1.05603 | 4258.8 | 4892.4 | 9.3420 | 0.79197 | 4258.3 | 4891.9 | 9.2090 |
| 1200 | 1.35972 | 4470.0 | 5149.8 | 9.6071 | 1.13309 | 4469.8 | 5149.6 | 9.5229 | 0.84980 | 4469.4 | 5149.3 | 9.3898 |
| 1300 | 1.45214 | 4686.6 | 5412.6 | 9.7797 | 1.21012 | 4686.4 | 5412.5 | 9.6955 | 0.90761 | 4686.1 | 5412.2 | 9.5625 |

*The temperature in parentheses is the saturation temperature at the specified pressure.

† Properties of saturated vapor at the specified pressure.

## TABLE A–6

Superheated water (*Continued*)

| T °C | v m³/kg | u kJ/kg | h kJ/kg | s kJ/kg·K | v m³/kg | u kJ/kg | h kJ/kg | s kJ/kg·K | v m³/kg | u kJ/kg | h kJ/kg | s kJ/kg·K |
|---|---|---|---|---|---|---|---|---|---|---|---|---|
| | P = 1.00 MPa (179.88°C) | | | | P = 1.20 MPa (187.96°C) | | | | P = 1.40 MPa (195.04°C) | | | |
| Sat. | 0.19437 | 2582.8 | 2777.1 | 6.5850 | 0.16326 | 2587.8 | 2783.8 | 6.5217 | 0.14078 | 2591.8 | 2788.9 | 6.4675 |
| 200 | 0.20602 | 2622.3 | 2828.3 | 6.6956 | 0.16934 | 2612.9 | 2816.1 | 6.5909 | 0.14303 | 2602.7 | 2803.0 | 6.4975 |
| 250 | 0.23275 | 2710.4 | 2943.1 | 6.9265 | 0.19241 | 2704.7 | 2935.6 | 6.8313 | 0.16356 | 2698.9 | 2927.9 | 6.7488 |
| 300 | 0.25799 | 2793.7 | 3051.6 | 7.1246 | 0.21386 | 2789.7 | 3046.3 | 7.0335 | 0.18233 | 2785.7 | 3040.9 | 6.9553 |
| 350 | 0.28250 | 2875.7 | 3158.2 | 7.3029 | 0.23455 | 2872.7 | 3154.2 | 7.2139 | 0.20029 | 2869.7 | 3150.1 | 7.1379 |
| 400 | 0.30661 | 2957.9 | 3264.5 | 7.4670 | 0.25482 | 2955.5 | 3261.3 | 7.3793 | 0.21782 | 2953.1 | 3258.1 | 7.3046 |
| 500 | 0.35411 | 3125.0 | 3479.1 | 7.7642 | 0.29464 | 3123.4 | 3477.0 | 7.6779 | 0.25216 | 3121.8 | 3474.8 | 7.6047 |
| 600 | 0.40111 | 3297.5 | 3698.6 | 8.0311 | 0.33395 | 3296.3 | 3697.0 | 7.9456 | 0.28597 | 3295.1 | 3695.5 | 7.8730 |
| 700 | 0.44783 | 3476.3 | 3924.1 | 8.2755 | 0.37297 | 3475.3 | 3922.9 | 8.1904 | 0.31951 | 3474.4 | 3921.7 | 8.1183 |
| 800 | 0.49438 | 3661.7 | 4156.1 | 8.5024 | 0.41184 | 3661.0 | 4155.2 | 8.4176 | 0.35288 | 3660.3 | 4154.3 | 8.3458 |
| 900 | 0.54083 | 3853.9 | 4394.8 | 8.7150 | 0.45059 | 3853.3 | 4394.0 | 8.6303 | 0.38614 | 3852.7 | 4393.3 | 8.5587 |
| 1000 | 0.58721 | 4052.7 | 4640.0 | 8.9155 | 0.48928 | 4052.2 | 4639.4 | 8.8310 | 0.41933 | 4051.7 | 4638.8 | 8.7595 |
| 1100 | 0.63354 | 4257.9 | 4891.4 | 9.1057 | 0.52792 | 4257.5 | 4891.0 | 9.0212 | 0.45247 | 4257.0 | 4890.5 | 8.9497 |
| 1200 | 0.67983 | 4469.0 | 5148.9 | 9.2866 | 0.56652 | 4468.7 | 5148.5 | 9.2022 | 0.48558 | 4468.3 | 5148.1 | 9.1308 |
| 1300 | 0.72610 | 4685.8 | 5411.9 | 9.4593 | 0.60509 | 4685.5 | 5411.6 | 9.3750 | 0.51866 | 4685.1 | 5411.3 | 9.3036 |
| | P = 1.60 MPa (201.37°C) | | | | P = 1.80 MPa (207.11°C) | | | | P = 2.00 MPa (212.38°C) | | | |
| Sat. | 0.12374 | 2594.8 | 2792.8 | 6.4200 | 0.11037 | 2597.3 | 2795.9 | 6.3775 | 0.09959 | 2599.1 | 2798.3 | 6.3390 |
| 225 | 0.13293 | 2645.1 | 2857.8 | 6.5537 | 0.11678 | 2637.0 | 2847.2 | 6.4825 | 0.10381 | 2628.5 | 2836.1 | 6.4160 |
| 250 | 0.14190 | 2692.9 | 2919.9 | 6.6753 | 0.12502 | 2686.7 | 2911.7 | 6.6088 | 0.11150 | 2680.3 | 2903.3 | 6.5475 |
| 300 | 0.15866 | 2781.6 | 3035.4 | 6.8864 | 0.14025 | 2777.4 | 3029.9 | 6.8246 | 0.12551 | 2773.2 | 3024.2 | 6.7684 |
| 350 | 0.17459 | 2866.6 | 3146.0 | 7.0713 | 0.15460 | 2863.6 | 3141.9 | 7.0120 | 0.13860 | 2860.5 | 3137.7 | 6.9583 |
| 400 | 0.19007 | 2950.8 | 3254.9 | 7.2394 | 0.16849 | 2948.3 | 3251.6 | 7.1814 | 0.15122 | 2945.9 | 3248.4 | 7.1292 |
| 500 | 0.22029 | 3120.1 | 3472.6 | 7.5410 | 0.19551 | 3118.5 | 3470.4 | 7.4845 | 0.17568 | 3116.9 | 3468.3 | 7.4337 |
| 600 | 0.24999 | 3293.9 | 3693.9 | 7.8101 | 0.22200 | 3292.7 | 3692.3 | 7.7543 | 0.19962 | 3291.5 | 3690.7 | 7.7043 |
| 700 | 0.27941 | 3473.5 | 3920.5 | 8.0558 | 0.24822 | 3472.6 | 3919.4 | 8.0005 | 0.22326 | 3471.7 | 3918.2 | 7.9509 |
| 800 | 0.30865 | 3659.5 | 4153.4 | 8.2834 | 0.27426 | 3658.8 | 4152.4 | 8.2284 | 0.24674 | 3658.0 | 4151.5 | 8.1791 |
| 900 | 0.33780 | 3852.1 | 4392.6 | 8.4965 | 0.30020 | 3851.5 | 4391.9 | 8.4417 | 0.27012 | 3850.9 | 4391.1 | 8.3925 |
| 1000 | 0.36687 | 4051.2 | 4638.2 | 8.6974 | 0.32606 | 4050.7 | 4637.6 | 8.6427 | 0.29342 | 4050.2 | 4637.1 | 8.5936 |
| 1100 | 0.39589 | 4256.6 | 4890.0 | 8.8878 | 0.35188 | 4256.2 | 4889.6 | 8.8331 | 0.31667 | 4255.7 | 4889.1 | 8.7842 |
| 1200 | 0.42488 | 4467.9 | 5147.7 | 9.0689 | 0.37766 | 4467.6 | 5147.3 | 9.0143 | 0.33989 | 4467.2 | 5147.0 | 8.9654 |
| 1300 | 0.45383 | 4684.8 | 5410.9 | 9.2418 | 0.40341 | 4684.5 | 5410.6 | 9.1872 | 0.36308 | 4684.2 | 5410.3 | 9.1384 |
| | P = 2.50 MPa (223.95°C) | | | | P = 3.00 MPa (233.85°C) | | | | P = 3.50 MPa (242.56°C) | | | |
| Sat. | 0.07995 | 2602.1 | 2801.9 | 6.2558 | 0.06667 | 2603.2 | 2803.2 | 6.1856 | 0.05706 | 2603.0 | 2802.7 | 6.1244 |
| 225 | 0.08026 | 2604.8 | 2805.5 | 6.2629 | | | | | | | | |
| 250 | 0.08705 | 2663.3 | 2880.9 | 6.4107 | 0.07063 | 2644.7 | 2856.5 | 6.2893 | 0.05876 | 2624.0 | 2829.7 | 6.1764 |
| 300 | 0.09894 | 2762.2 | 3009.6 | 6.6459 | 0.08118 | 2750.8 | 2994.3 | 6.5412 | 0.06845 | 2738.8 | 2978.4 | 6.4484 |
| 350 | 0.10979 | 2852.5 | 3127.0 | 6.8424 | 0.09056 | 2844.4 | 3116.1 | 6.7450 | 0.07680 | 2836.0 | 3104.9 | 6.6601 |
| 400 | 0.12012 | 2939.8 | 3240.1 | 7.0170 | 0.09938 | 2933.6 | 3231.7 | 6.9235 | 0.08456 | 2927.2 | 3223.2 | 6.8428 |
| 450 | 0.13015 | 3026.2 | 3351.6 | 7.1768 | 0.10789 | 3021.2 | 3344.9 | 7.0856 | 0.09198 | 3016.1 | 3338.1 | 7.0074 |
| 500 | 0.13999 | 3112.8 | 3462.8 | 7.3254 | 0.11620 | 3108.6 | 3457.2 | 7.2359 | 0.09919 | 3104.5 | 3451.7 | 7.1593 |
| 600 | 0.15931 | 3288.5 | 3686.8 | 7.5979 | 0.13245 | 3285.5 | 3682.8 | 7.5103 | 0.11325 | 3282.5 | 3678.9 | 7.4357 |
| 700 | 0.17835 | 3469.3 | 3915.2 | 7.8455 | 0.14841 | 3467.0 | 3912.2 | 7.7590 | 0.12702 | 3464.7 | 3909.3 | 7.6855 |
| 800 | 0.19722 | 3656.2 | 4149.2 | 8.0744 | 0.16420 | 3654.3 | 4146.9 | 7.9885 | 0.14061 | 3652.5 | 4144.6 | 7.9156 |
| 900 | 0.21597 | 3849.4 | 4389.3 | 8.2882 | 0.17988 | 3847.9 | 4387.5 | 8.2028 | 0.15410 | 3846.4 | 4385.7 | 8.1304 |
| 1000 | 0.23466 | 4049.0 | 4635.6 | 8.4897 | 0.19549 | 4047.7 | 4634.2 | 8.4045 | 0.16751 | 4046.4 | 4632.7 | 8.3324 |
| 1100 | 0.25330 | 4254.7 | 4887.9 | 8.6804 | 0.21105 | 4253.6 | 4886.7 | 8.5955 | 0.18087 | 4252.5 | 4885.6 | 8.5236 |
| 1200 | 0.27190 | 4466.3 | 5146.0 | 8.8618 | 0.22658 | 4465.3 | 5145.1 | 8.7771 | 0.19420 | 4464.4 | 5144.1 | 8.7053 |
| 1300 | 0.29048 | 4683.4 | 5409.5 | 9.0349 | 0.24207 | 4682.6 | 5408.8 | 8.9502 | 0.20750 | 4681.8 | 5408.0 | 8.8786 |

## TABLE A–6

Superheated water (*Continued*)

| T °C | v m³/kg | u kJ/kg | h kJ/kg | s kJ/kg·K | v m³/kg | u kJ/kg | h kJ/kg | s kJ/kg·K | v m³/kg | u kJ/kg | h kJ/kg | s kJ/kg·K |
|---|---|---|---|---|---|---|---|---|---|---|---|---|
| | | P = 4.0 MPa (250.35°C) | | | | P = 4.5 MPa (257.44°C) | | | | P = 5.0 MPa (263.94°C) | | |
| Sat. | 0.04978 | 2601.7 | 2800.8 | 6.0696 | 0.04406 | 2599.7 | 2798.0 | 6.0198 | 0.03945 | 2597.0 | 2794.2 | 5.9737 |
| 275 | 0.05461 | 2668.9 | 2887.3 | 6.2312 | 0.04733 | 2651.4 | 2864.4 | 6.1429 | 0.04144 | 2632.3 | 2839.5 | 6.0571 |
| 300 | 0.05887 | 2726.2 | 2961.7 | 6.3639 | 0.05138 | 2713.0 | 2944.2 | 6.2854 | 0.04535 | 2699.0 | 2925.7 | 6.2111 |
| 350 | 0.06647 | 2827.4 | 3093.3 | 6.5843 | 0.05842 | 2818.6 | 3081.5 | 6.5153 | 0.05197 | 2809.5 | 3069.3 | 6.4516 |
| 400 | 0.07343 | 2920.8 | 3214.5 | 6.7714 | 0.06477 | 2914.2 | 3205.7 | 6.7071 | 0.05784 | 2907.5 | 3196.7 | 6.6483 |
| 450 | 0.08004 | 3011.0 | 3331.2 | 6.9386 | 0.07076 | 3005.8 | 3324.2 | 6.8770 | 0.06332 | 3000.6 | 3317.2 | 6.8210 |
| 500 | 0.08644 | 3100.3 | 3446.0 | 7.0922 | 0.07652 | 3096.0 | 3440.4 | 7.0323 | 0.06858 | 3091.8 | 3434.7 | 6.9781 |
| 600 | 0.09886 | 3279.4 | 3674.9 | 7.3706 | 0.08766 | 3276.4 | 3670.9 | 7.3127 | 0.07870 | 3273.3 | 3666.9 | 7.2605 |
| 700 | 0.11098 | 3462.4 | 3906.3 | 7.6214 | 0.09850 | 3460.0 | 3903.3 | 7.5647 | 0.08852 | 3457.7 | 3900.3 | 7.5136 |
| 800 | 0.12292 | 3650.6 | 4142.3 | 7.8523 | 0.10916 | 3648.8 | 4140.0 | 7.7962 | 0.09816 | 3646.9 | 4137.7 | 7.7458 |
| 900 | 0.13476 | 3844.8 | 4383.9 | 8.0675 | 0.11972 | 3843.3 | 4382.1 | 8.0118 | 0.10769 | 3841.8 | 4380.2 | 7.9619 |
| 1000 | 0.14653 | 4045.1 | 4631.2 | 8.2698 | 0.13020 | 4043.9 | 4629.8 | 8.2144 | 0.11715 | 4042.6 | 4628.3 | 8.1648 |
| 1100 | 0.15824 | 4251.4 | 4884.4 | 8.4612 | 0.14064 | 4250.4 | 4883.2 | 8.4060 | 0.12655 | 4249.3 | 4882.1 | 8.3566 |
| 1200 | 0.16992 | 4463.5 | 5143.2 | 8.6430 | 0.15103 | 4462.6 | 5142.2 | 8.5880 | 0.13592 | 4461.6 | 5141.3 | 8.5388 |
| 1300 | 0.18157 | 4680.9 | 5407.2 | 8.8164 | 0.16140 | 4680.1 | 5406.5 | 8.7616 | 0.14527 | 4679.3 | 5405.7 | 8.7124 |
| | | P = 6.0 MPa (275.59°C) | | | | P = 7.0 MPa (285.83°C) | | | | P = 8.0 MPa (295.01°C) | | |
| Sat. | 0.03245 | 2589.9 | 2784.6 | 5.8902 | 0.027378 | 2581.0 | 2772.6 | 5.8148 | 0.023525 | 2570.5 | 2758.7 | 5.7450 |
| 300 | 0.03619 | 2668.4 | 2885.6 | 6.0703 | 0.029492 | 2633.5 | 2839.9 | 5.9337 | 0.024279 | 2592.3 | 2786.5 | 5.7937 |
| 350 | 0.04225 | 2790.4 | 3043.9 | 6.3357 | 0.035262 | 2770.1 | 3016.9 | 6.2305 | 0.029975 | 2748.3 | 2988.1 | 6.1321 |
| 400 | 0.04742 | 2893.7 | 3178.3 | 6.5432 | 0.039958 | 2879.5 | 3159.2 | 6.4502 | 0.034344 | 2864.6 | 3139.4 | 6.3658 |
| 450 | 0.05217 | 2989.9 | 3302.9 | 6.7219 | 0.044187 | 2979.0 | 3288.3 | 6.6353 | 0.038194 | 2967.8 | 3273.3 | 6.5579 |
| 500 | 0.05667 | 3083.1 | 3423.1 | 6.8826 | 0.048157 | 3074.3 | 3411.4 | 6.8000 | 0.041767 | 3065.4 | 3399.5 | 6.7266 |
| 550 | 0.06102 | 3175.2 | 3541.3 | 7.0308 | 0.051966 | 3167.9 | 3531.6 | 6.9507 | 0.045172 | 3160.5 | 3521.8 | 6.8800 |
| 600 | 0.06527 | 3267.2 | 3658.8 | 7.1693 | 0.055665 | 3261.0 | 3650.6 | 7.0910 | 0.048463 | 3254.7 | 3642.4 | 7.0221 |
| 700 | 0.07355 | 3453.0 | 3894.3 | 7.4247 | 0.062850 | 3448.3 | 3888.3 | 7.3487 | 0.054829 | 3443.6 | 3882.2 | 7.2822 |
| 800 | 0.08165 | 3643.2 | 4133.1 | 7.6582 | 0.069856 | 3639.5 | 4128.5 | 7.5836 | 0.061011 | 3635.7 | 4123.8 | 7.5185 |
| 900 | 0.08964 | 3838.8 | 4376.6 | 7.8751 | 0.076750 | 3835.7 | 4373.0 | 7.8014 | 0.067082 | 3832.7 | 4369.3 | 7.7372 |
| 1000 | 0.09756 | 4040.1 | 4625.4 | 8.0786 | 0.083571 | 4037.5 | 4622.5 | 8.0055 | 0.073079 | 4035.0 | 4619.6 | 7.9419 |
| 1100 | 0.10543 | 4247.1 | 4879.7 | 8.2709 | 0.090341 | 4245.0 | 4877.4 | 8.1982 | 0.079025 | 4242.8 | 4875.0 | 8.1350 |
| 1200 | 0.11326 | 4459.8 | 5139.4 | 8.4534 | 0.097075 | 4457.9 | 5137.4 | 8.3810 | 0.084934 | 4456.1 | 5135.5 | 8.3181 |
| 1300 | 0.12107 | 4677.7 | 5404.1 | 8.6273 | 0.103781 | 4676.1 | 5402.6 | 8.5551 | 0.090817 | 4674.5 | 5401.0 | 8.4925 |
| | | P = 9.0 MPa (303.35°C) | | | | P = 10.0 MPa (311.00°C) | | | | P = 12.5 MPa (327.81°C) | | |
| Sat. | 0.020489 | 2558.5 | 2742.9 | 5.6791 | 0.018028 | 2545.2 | 2725.5 | 5.6159 | 0.013496 | 2505.6 | 2674.3 | 5.4638 |
| 325 | 0.023284 | 2647.6 | 2857.1 | 5.8738 | 0.019877 | 2611.6 | 2810.3 | 5.7596 | | | | |
| 350 | 0.025816 | 2725.0 | 2957.3 | 6.0380 | 0.022440 | 2699.6 | 2924.0 | 5.9460 | 0.016138 | 2624.9 | 2826.6 | 5.7130 |
| 400 | 0.029960 | 2849.2 | 3118.8 | 6.2876 | 0.026436 | 2833.1 | 3097.5 | 6.2141 | 0.020030 | 2789.6 | 3040.0 | 6.0433 |
| 450 | 0.033524 | 2956.3 | 3258.0 | 6.4872 | 0.029782 | 2944.5 | 3242.4 | 6.4219 | 0.023019 | 2913.7 | 3201.5 | 6.2749 |
| 500 | 0.036793 | 3056.3 | 3387.4 | 6.6603 | 0.032811 | 3047.0 | 3375.1 | 6.5995 | 0.025630 | 3023.2 | 3343.6 | 6.4651 |
| 550 | 0.039885 | 3153.0 | 3512.0 | 6.8164 | 0.035655 | 3145.4 | 3502.0 | 6.7585 | 0.028033 | 3126.1 | 3476.5 | 6.6317 |
| 600 | 0.042861 | 3248.4 | 3634.1 | 6.9605 | 0.038378 | 3242.0 | 3625.8 | 6.9045 | 0.030306 | 3225.8 | 3604.6 | 6.7828 |
| 650 | 0.045755 | 3343.4 | 3755.2 | 7.0954 | 0.041018 | 3338.0 | 3748.1 | 7.0408 | 0.032491 | 3324.1 | 3730.2 | 6.9227 |
| 700 | 0.048589 | 3438.8 | 3876.1 | 7.2229 | 0.043597 | 3434.0 | 3870.0 | 7.1693 | 0.034612 | 3422.0 | 3854.6 | 7.0540 |
| 800 | 0.054132 | 3632.0 | 4119.2 | 7.4606 | 0.048629 | 3628.2 | 4114.5 | 7.4085 | 0.038724 | 3618.8 | 4102.8 | 7.2967 |
| 900 | 0.059562 | 3829.6 | 4365.7 | 7.6802 | 0.053547 | 3826.5 | 4362.0 | 7.6290 | 0.042720 | 3818.9 | 4352.9 | 7.5195 |
| 1000 | 0.064919 | 4032.4 | 4616.7 | 7.8855 | 0.058391 | 4029.9 | 4613.8 | 7.8349 | 0.046641 | 4023.5 | 4606.5 | 7.7269 |
| 1100 | 0.070224 | 4240.7 | 4872.7 | 8.0791 | 0.063183 | 4238.5 | 4870.3 | 8.0289 | 0.050510 | 4233.1 | 4864.5 | 7.9220 |
| 1200 | 0.075492 | 4454.2 | 5133.6 | 8.2625 | 0.067938 | 4452.4 | 5131.7 | 8.2126 | 0.054342 | 4447.7 | 5127.0 | 8.1065 |
| 1300 | 0.080733 | 4672.9 | 5399.5 | 8.4371 | 0.072667 | 4671.3 | 5398.0 | 8.3874 | 0.058147 | 4667.3 | 5394.1 | 8.2819 |

## TABLE A–6

Superheated water (*Concluded*)

| T °C | v m³/kg | u kJ/kg | h kJ/kg | s kJ/kg·K | v m³/kg | u kJ/kg | h kJ/kg | s kJ/kg·K | v m³/kg | u kJ/kg | h kJ/kg | s kJ/kg·K |
|---|---|---|---|---|---|---|---|---|---|---|---|---|
| | *P* = 15.0 MPa (342.16°C) | | | | *P* = 17.5 MPa (354.67°C) | | | | *P* = 20.0 MPa (365.75°C) | | | |
| Sat. | 0.010341 | 2455.7 | 2610.8 | 5.3108 | 0.007932 | 2390.7 | 2529.5 | 5.1435 | 0.005862 | 2294.8 | 2412.1 | 4.9310 |
| 350 | 0.011481 | 2520.9 | 2693.1 | 5.4438 | | | | | | | | |
| 400 | 0.015671 | 2740.6 | 2975.7 | 5.8819 | 0.012463 | 2684.3 | 2902.4 | 5.7211 | 0.009950 | 2617.9 | 2816.9 | 5.5526 |
| 450 | 0.018477 | 2880.8 | 3157.9 | 6.1434 | 0.015204 | 2845.4 | 3111.4 | 6.0212 | 0.012721 | 2807.3 | 3061.7 | 5.9043 |
| 500 | 0.020828 | 2998.4 | 3310.8 | 6.3480 | 0.017385 | 2972.4 | 3276.7 | 6.2424 | 0.014793 | 2945.3 | 3241.2 | 6.1446 |
| 550 | 0.022945 | 3106.2 | 3450.4 | 6.5230 | 0.019305 | 3085.8 | 3423.6 | 6.4266 | 0.016571 | 3064.7 | 3396.2 | 6.3390 |
| 600 | 0.024921 | 3209.3 | 3583.1 | 6.6796 | 0.021073 | 3192.5 | 3561.3 | 6.5890 | 0.018185 | 3175.3 | 3539.0 | 6.5075 |
| 650 | 0.026804 | 3310.1 | 3712.1 | 6.8233 | 0.022742 | 3295.8 | 3693.8 | 6.7366 | 0.019695 | 3281.4 | 3675.3 | 6.6593 |
| 700 | 0.028621 | 3409.8 | 3839.1 | 6.9573 | 0.024342 | 3397.5 | 3823.5 | 6.8735 | 0.021134 | 3385.1 | 3807.8 | 6.7991 |
| 800 | 0.032121 | 3609.3 | 4091.1 | 7.2037 | 0.027405 | 3599.7 | 4079.3 | 7.1237 | 0.023870 | 3590.1 | 4067.5 | 7.0531 |
| 900 | 0.035503 | 3811.2 | 4343.7 | 7.4288 | 0.030348 | 3803.5 | 4334.6 | 7.3511 | 0.026484 | 3795.7 | 4325.4 | 7.2829 |
| 1000 | 0.038808 | 4017.1 | 4599.2 | 7.6378 | 0.033215 | 4010.7 | 4592.0 | 7.5616 | 0.029020 | 4004.3 | 4584.7 | 7.4950 |
| 1100 | 0.042062 | 4227.7 | 4858.6 | 7.8339 | 0.036029 | 4222.3 | 4852.8 | 7.7588 | 0.031504 | 4216.9 | 4847.0 | 7.6933 |
| 1200 | 0.045279 | 4443.1 | 5122.3 | 8.0192 | 0.038806 | 4438.5 | 5117.6 | 7.9449 | 0.033952 | 4433.8 | 5112.9 | 7.8802 |
| 1300 | 0.048469 | 4663.3 | 5390.3 | 8.1952 | 0.041556 | 4659.2 | 5386.5 | 8.1215 | 0.036371 | 4655.2 | 5382.7 | 8.0574 |
| | *P* = 25.0 MPa | | | | *P* = 30.0 MPa | | | | *P* = 35.0 MPa | | | |
| 375 | 0.001978 | 1799.9 | 1849.4 | 4.0345 | 0.001792 | 1738.1 | 1791.9 | 3.9313 | 0.001701 | 1702.8 | 1762.4 | 3.8724 |
| 400 | 0.006005 | 2428.5 | 2578.7 | 5.1400 | 0.002798 | 2068.9 | 2152.8 | 4.4758 | 0.002105 | 1914.9 | 1988.6 | 4.2144 |
| 425 | 0.007886 | 2607.8 | 2805.0 | 5.4708 | 0.005299 | 2452.9 | 2611.8 | 5.1473 | 0.003434 | 2253.3 | 2373.5 | 4.7751 |
| 450 | 0.009176 | 2721.2 | 2950.6 | 5.6759 | 0.006737 | 2618.9 | 2821.0 | 5.4422 | 0.004957 | 2497.5 | 2671.0 | 5.1946 |
| 500 | 0.011143 | 2887.3 | 3165.9 | 5.9643 | 0.008691 | 2824.0 | 3084.8 | 5.7956 | 0.006933 | 2755.3 | 2997.9 | 5.6331 |
| 550 | 0.012736 | 3020.8 | 3339.2 | 6.1816 | 0.010175 | 2974.5 | 3279.7 | 6.0403 | 0.008348 | 2925.8 | 3218.0 | 5.9093 |
| 600 | 0.014140 | 3140.0 | 3493.5 | 6.3637 | 0.011445 | 3103.4 | 3446.8 | 6.2373 | 0.009523 | 3065.6 | 3399.0 | 6.1229 |
| 650 | 0.015430 | 3251.9 | 3637.7 | 6.5243 | 0.012590 | 3221.7 | 3599.4 | 6.4074 | 0.010565 | 3190.9 | 3560.7 | 6.3030 |
| 700 | 0.016643 | 3359.9 | 3776.0 | 6.6702 | 0.013654 | 3334.3 | 3743.9 | 6.5599 | 0.011523 | 3308.3 | 3711.6 | 6.4623 |
| 800 | 0.018922 | 3570.7 | 4043.8 | 6.9322 | 0.015628 | 3551.2 | 4020.0 | 6.8301 | 0.013278 | 3531.6 | 3996.3 | 6.7409 |
| 900 | 0.021075 | 3780.2 | 4307.1 | 7.1668 | 0.017473 | 3764.6 | 4288.8 | 7.0695 | 0.014904 | 3749.0 | 4270.6 | 6.9853 |
| 1000 | 0.023150 | 3991.5 | 4570.2 | 7.3821 | 0.019240 | 3978.6 | 4555.8 | 7.2880 | 0.016450 | 3965.8 | 4541.5 | 7.2069 |
| 1100 | 0.025172 | 4206.1 | 4835.4 | 7.5825 | 0.020954 | 4195.2 | 4823.9 | 7.4906 | 0.017942 | 4184.4 | 4812.4 | 7.4118 |
| 1200 | 0.027157 | 4424.6 | 5103.5 | 7.7710 | 0.022630 | 4415.3 | 5094.2 | 7.6807 | 0.019398 | 4406.1 | 5085.0 | 7.6034 |
| 1300 | 0.029115 | 4647.2 | 5375.1 | 7.9494 | 0.024279 | 4639.2 | 5367.6 | 7.8602 | 0.020827 | 4631.2 | 5360.2 | 7.7841 |
| | *P* = 40.0 MPa | | | | *P* = 50.0 MPa | | | | *P* = 60.0 MPa | | | |
| 375 | 0.001641 | 1677.0 | 1742.6 | 3.8290 | 0.001560 | 1638.6 | 1716.6 | 3.7642 | 0.001503 | 1609.7 | 1699.9 | 3.7149 |
| 400 | 0.001911 | 1855.0 | 1931.4 | 4.1145 | 0.001731 | 1787.8 | 1874.4 | 4.0029 | 0.001633 | 1745.2 | 1843.2 | 3.9317 |
| 425 | 0.002538 | 2097.5 | 2199.0 | 4.5044 | 0.002009 | 1960.3 | 2060.7 | 4.2746 | 0.001816 | 1892.9 | 2001.8 | 4.1630 |
| 450 | 0.003692 | 2364.2 | 2511.8 | 4.9449 | 0.002487 | 2160.3 | 2284.7 | 4.5896 | 0.002086 | 2055.1 | 2180.2 | 4.4140 |
| 500 | 0.005623 | 2681.6 | 2906.5 | 5.4744 | 0.003890 | 2528.1 | 2722.6 | 5.1762 | 0.002952 | 2393.2 | 2570.3 | 4.9356 |
| 550 | 0.006985 | 2875.1 | 3154.4 | 5.7857 | 0.005118 | 2769.5 | 3025.4 | 5.5563 | 0.003955 | 2664.6 | 2901.9 | 5.3517 |
| 600 | 0.008089 | 3026.8 | 3350.4 | 6.0170 | 0.006108 | 2947.1 | 3252.6 | 5.8245 | 0.004833 | 2866.8 | 3156.8 | 5.6527 |
| 650 | 0.009053 | 3159.5 | 3521.6 | 6.2078 | 0.006957 | 3095.6 | 3443.5 | 6.0373 | 0.005591 | 3031.3 | 3366.8 | 5.8867 |
| 700 | 0.009930 | 3282.0 | 3679.2 | 6.3740 | 0.007717 | 3228.7 | 3614.6 | 6.2179 | 0.006265 | 3175.4 | 3551.3 | 6.0814 |
| 800 | 0.011521 | 3511.8 | 3972.6 | 6.6613 | 0.009073 | 3472.2 | 3925.8 | 6.5225 | 0.007456 | 3432.6 | 3880.0 | 6.4033 |
| 900 | 0.012980 | 3733.3 | 4252.5 | 6.9107 | 0.010296 | 3702.0 | 4216.8 | 6.7819 | 0.008519 | 3670.9 | 4182.1 | 6.6725 |
| 1000 | 0.014360 | 3952.9 | 4527.3 | 7.1355 | 0.011441 | 3927.4 | 4499.4 | 7.0131 | 0.009504 | 3902.0 | 4472.2 | 6.9099 |
| 1100 | 0.015686 | 4173.7 | 4801.1 | 7.3425 | 0.012534 | 4152.2 | 4778.9 | 7.2244 | 0.010439 | 4130.9 | 4757.3 | 7.1255 |
| 1200 | 0.016976 | 4396.9 | 5075.9 | 7.5357 | 0.013590 | 4378.6 | 5058.1 | 7.4207 | 0.011339 | 4360.5 | 5040.8 | 7.3248 |
| 1300 | 0.018239 | 4623.3 | 5352.8 | 7.7175 | 0.014620 | 4607.5 | 5338.5 | 7.6048 | 0.012213 | 4591.8 | 5324.5 | 7.5111 |

## TABLE A–7

Compressed liquid water

| T °C | v m³/kg | u kJ/kg | h kJ/kg | s kJ/kg·K | v m³/kg | u kJ/kg | h kJ/kg | s kJ/kg·K | v m³/kg | u kJ/kg | h kJ/kg | s kJ/kg·K |
|---|---|---|---|---|---|---|---|---|---|---|---|---|
| | P = 5 MPa (263.94°C) | | | | P = 10 MPa (311.00°C) | | | | P = 15 MPa (342.16°C) | | | |
| Sat. | 0.0012862 | 1148.1 | 1154.5 | 2.9207 | 0.0014522 | 1393.3 | 1407.9 | 3.3603 | 0.0016572 | 1585.5 | 1610.3 | 3.6848 |
| 0 | 0.0009977 | 0.04 | 5.03 | 0.0001 | 0.0009952 | 0.12 | 10.07 | 0.0003 | 0.0009928 | 0.18 | 15.07 | 0.0004 |
| 20 | 0.0009996 | 83.61 | 88.61 | 0.2954 | 0.0009973 | 83.31 | 93.28 | 0.2943 | 0.0009951 | 83.01 | 97.93 | 0.2932 |
| 40 | 0.0010057 | 166.92 | 171.95 | 0.5705 | 0.0010035 | 166.33 | 176.37 | 0.5685 | 0.0010013 | 165.75 | 180.77 | 0.5666 |
| 60 | 0.0010149 | 250.29 | 255.36 | 0.8287 | 0.0010127 | 249.43 | 259.55 | 0.8260 | 0.0010105 | 248.58 | 263.74 | 0.8234 |
| 80 | 0.0010267 | 333.82 | 338.96 | 1.0723 | 0.0010244 | 332.69 | 342.94 | 1.0691 | 0.0010221 | 331.59 | 346.92 | 1.0659 |
| 100 | 0.0010410 | 417.65 | 422.85 | 1.3034 | 0.0010385 | 416.23 | 426.62 | 1.2996 | 0.0010361 | 414.85 | 430.39 | 1.2958 |
| 120 | 0.0010576 | 501.91 | 507.19 | 1.5236 | 0.0010549 | 500.18 | 510.73 | 1.5191 | 0.0010522 | 498.50 | 514.28 | 1.5148 |
| 140 | 0.0010769 | 586.80 | 592.18 | 1.7344 | 0.0010738 | 584.72 | 595.45 | 1.7293 | 0.0010708 | 582.69 | 598.75 | 1.7243 |
| 160 | 0.0010988 | 672.55 | 678.04 | 1.9374 | 0.0010954 | 670.06 | 681.01 | 1.9316 | 0.0010920 | 667.63 | 684.01 | 1.9259 |
| 180 | 0.0011240 | 759.47 | 765.09 | 2.1338 | 0.0011200 | 756.48 | 767.68 | 2.1271 | 0.0011160 | 753.58 | 770.32 | 2.1206 |
| 200 | 0.0011531 | 847.92 | 853.68 | 2.3251 | 0.0011482 | 844.32 | 855.80 | 2.3174 | 0.0011435 | 840.84 | 858.00 | 2.3100 |
| 220 | 0.0011868 | 938.39 | 944.32 | 2.5127 | 0.0011809 | 934.01 | 945.82 | 2.5037 | 0.0011752 | 929.81 | 947.43 | 2.4951 |
| 240 | 0.0012268 | 1031.6 | 1037.7 | 2.6983 | 0.0012192 | 1026.2 | 1038.3 | 2.6876 | 0.0012121 | 1021.0 | 1039.2 | 2.6774 |
| 260 | 0.0012755 | 1128.5 | 1134.9 | 2.8841 | 0.0012653 | 1121.6 | 1134.3 | 2.8710 | 0.0012560 | 1115.1 | 1134.0 | 2.8586 |
| 280 | | | | | 0.0013226 | 1221.8 | 1235.0 | 3.0565 | 0.0013096 | 1213.4 | 1233.0 | 3.0410 |
| 300 | | | | | 0.0013980 | 1329.4 | 1343.3 | 3.2488 | 0.0013783 | 1317.6 | 1338.3 | 3.2279 |
| 320 | | | | | | | | | 0.0014733 | 1431.9 | 1454.0 | 3.4263 |
| 340 | | | | | | | | | 0.0016311 | 1567.9 | 1592.4 | 3.6555 |
| | P = 20 MPa (365.75°C) | | | | P = 30 MPa | | | | P = 50 MPa | | | |
| Sat. | 0.0020378 | 1785.8 | 1826.6 | 4.0146 | | | | | | | | |
| 0 | 0.0009904 | 0.23 | 20.03 | 0.0005 | 0.0009857 | 0.29 | 29.86 | 0.0003 | 0.0009767 | 0.29 | 49.13 | −0.0010 |
| 20 | 0.0009929 | 82.71 | 102.57 | 0.2921 | 0.0009886 | 82.11 | 111.77 | 0.2897 | 0.0009805 | 80.93 | 129.95 | 0.2845 |
| 40 | 0.0009992 | 165.17 | 185.16 | 0.5646 | 0.0009951 | 164.05 | 193.90 | 0.5607 | 0.0009872 | 161.90 | 211.25 | 0.5528 |
| 60 | 0.0010084 | 247.75 | 267.92 | 0.8208 | 0.0010042 | 246.14 | 276.26 | 0.8156 | 0.0009962 | 243.08 | 292.88 | 0.8055 |
| 80 | 0.0010199 | 330.50 | 350.90 | 1.0627 | 0.0010155 | 328.40 | 358.86 | 1.0564 | 0.0010072 | 324.42 | 374.78 | 1.0442 |
| 100 | 0.0010337 | 413.50 | 434.17 | 1.2920 | 0.0010290 | 410.87 | 441.74 | 1.2847 | 0.0010201 | 405.94 | 456.94 | 1.2705 |
| 120 | 0.0010496 | 496.85 | 517.84 | 1.5105 | 0.0010445 | 493.66 | 525.00 | 1.5020 | 0.0010349 | 487.69 | 539.43 | 1.4859 |
| 140 | 0.0010679 | 580.71 | 602.07 | 1.7194 | 0.0010623 | 576.90 | 608.76 | 1.7098 | 0.0010517 | 569.77 | 622.36 | 1.6916 |
| 160 | 0.0010886 | 665.28 | 687.05 | 1.9203 | 0.0010823 | 660.74 | 693.21 | 1.9094 | 0.0010704 | 652.33 | 705.85 | 1.8889 |
| 180 | 0.0011122 | 750.78 | 773.02 | 2.1143 | 0.0011049 | 745.40 | 778.55 | 2.1020 | 0.0010914 | 735.49 | 790.06 | 2.0790 |
| 200 | 0.0011390 | 837.49 | 860.27 | 2.3027 | 0.0011304 | 831.11 | 865.02 | 2.2888 | 0.0011149 | 819.45 | 875.19 | 2.2628 |
| 220 | 0.0011697 | 925.77 | 949.16 | 2.4867 | 0.0011595 | 918.15 | 952.93 | 2.4707 | 0.0011412 | 904.39 | 961.45 | 2.4414 |
| 240 | 0.0012053 | 1016.1 | 1040.2 | 2.6676 | 0.0011927 | 1006.9 | 1042.7 | 2.6491 | 0.0011708 | 990.55 | 1049.1 | 2.6156 |
| 260 | 0.0012472 | 1109.0 | 1134.0 | 2.8469 | 0.0012314 | 1097.8 | 1134.7 | 2.8250 | 0.0012044 | 1078.2 | 1138.4 | 2.7864 |
| 280 | 0.0012978 | 1205.6 | 1231.5 | 3.0265 | 0.0012770 | 1191.5 | 1229.8 | 3.0001 | 0.0012430 | 1167.7 | 1229.9 | 2.9547 |
| 300 | 0.0013611 | 1307.2 | 1334.4 | 3.2091 | 0.0013322 | 1288.9 | 1328.9 | 3.1761 | 0.0012879 | 1259.6 | 1324.0 | 3.1218 |
| 320 | 0.0014450 | 1416.6 | 1445.5 | 3.3996 | 0.0014014 | 1391.7 | 1433.7 | 3.3558 | 0.0013409 | 1354.3 | 1421.4 | 3.2888 |
| 340 | 0.0015693 | 1540.2 | 1571.6 | 3.6086 | 0.0014932 | 1502.4 | 1547.1 | 3.5438 | 0.0014049 | 1452.9 | 1523.1 | 3.4575 |
| 360 | 0.0018248 | 1703.6 | 1740.1 | 3.8787 | 0.0016276 | 1626.8 | 1675.6 | 3.7499 | 0.0014848 | 1556.5 | 1630.7 | 3.6301 |
| 380 | | | | | 0.0018729 | 1782.0 | 1838.2 | 4.0026 | 0.0015884 | 1667.1 | 1746.5 | 3.8102 |

## TABLE A–8

Saturated ice–water vapor

| Temp., $T\,°C$ | Sat. press., $P_{sat}$ kPa | Specific volume, m³/kg | | Internal energy, kJ/kg | | | Enthalpy, kJ/kg | | | Entropy, kJ/kg·K | | |
|---|---|---|---|---|---|---|---|---|---|---|---|---|
| | | Sat. ice, $v_i$ | Sat. vapor, $v_g$ | Sat. ice, $u_i$ | Subl., $u_{ig}$ | Sat. vapor, $u_g$ | Sat. ice, $h_i$ | Subl., $h_{ig}$ | Sat. vapor, $h_g$ | Sat. ice, $s_i$ | Subl., $s_{ig}$ | Sat. vapor, $s_g$ |
| 0.01 | 0.61169 | 0.001091 | 205.99 | −333.40 | 2707.9 | 2374.5 | −333.40 | 2833.9 | 2500.5 | −1.2202 | 10.374 | 9.154 |
| 0 | 0.61115 | 0.001091 | 206.17 | −333.43 | 2707.9 | 2374.5 | −333.43 | 2833.9 | 2500.5 | −1.2204 | 10.375 | 9.154 |
| −2 | 0.51772 | 0.001091 | 241.62 | −337.63 | 2709.4 | 2371.8 | −337.63 | 2834.5 | 2496.8 | −1.2358 | 10.453 | 9.218 |
| −4 | 0.43748 | 0.001090 | 283.84 | −341.80 | 2710.8 | 2369.0 | −341.80 | 2835.0 | 2493.2 | −1.2513 | 10.533 | 9.282 |
| −6 | 0.36873 | 0.001090 | 334.27 | −345.94 | 2712.2 | 2366.2 | −345.93 | 2835.4 | 2489.5 | −1.2667 | 10.613 | 9.347 |
| −8 | 0.30998 | 0.001090 | 394.66 | −350.04 | 2713.5 | 2363.5 | −350.04 | 2835.8 | 2485.8 | −1.2821 | 10.695 | 9.413 |
| −10 | 0.25990 | 0.001089 | 467.17 | −354.12 | 2714.8 | 2360.7 | −354.12 | 2836.2 | 2482.1 | −1.2976 | 10.778 | 9.480 |
| −12 | 0.21732 | 0.001089 | 554.47 | −358.17 | 2716.1 | 2357.9 | −358.17 | 2836.6 | 2478.4 | −1.3130 | 10.862 | 9.549 |
| −14 | 0.18121 | 0.001088 | 659.88 | −362.18 | 2717.3 | 2355.2 | −362.18 | 2836.9 | 2474.7 | −1.3284 | 10.947 | 9.618 |
| −16 | 0.15068 | 0.001088 | 787.51 | −366.17 | 2718.6 | 2352.4 | −366.17 | 2837.2 | 2471.0 | −1.3439 | 11.033 | 9.689 |
| −18 | 0.12492 | 0.001088 | 942.51 | −370.13 | 2719.7 | 2349.6 | −370.13 | 2837.5 | 2467.3 | −1.3593 | 11.121 | 9.761 |
| −20 | 0.10326 | 0.001087 | 1131.3 | −374.06 | 2720.9 | 2346.8 | −374.06 | 2837.7 | 2463.6 | −1.3748 | 11.209 | 9.835 |
| −22 | 0.08510 | 0.001087 | 1362.0 | −377.95 | 2722.0 | 2344.1 | −377.95 | 2837.9 | 2459.9 | −1.3903 | 11.300 | 9.909 |
| −24 | 0.06991 | 0.001087 | 1644.7 | −381.82 | 2723.1 | 2341.3 | −381.82 | 2838.1 | 2456.2 | −1.4057 | 11.391 | 9.985 |
| −26 | 0.05725 | 0.001087 | 1992.2 | −385.66 | 2724.2 | 2338.5 | −385.66 | 2838.2 | 2452.5 | −1.4212 | 11.484 | 10.063 |
| −28 | 0.04673 | 0.001086 | 2421.0 | −389.47 | 2725.2 | 2335.7 | −389.47 | 2838.3 | 2448.8 | −1.4367 | 11.578 | 10.141 |
| −30 | 0.03802 | 0.001086 | 2951.7 | −393.25 | 2726.2 | 2332.9 | −393.25 | 2838.4 | 2445.1 | −1.4521 | 11.673 | 10.221 |
| −32 | 0.03082 | 0.001086 | 3610.9 | −397.00 | 2727.2 | 2330.2 | −397.00 | 2838.4 | 2441.4 | −1.4676 | 11.770 | 10.303 |
| −34 | 0.02490 | 0.001085 | 4432.4 | −400.72 | 2728.1 | 2327.4 | −400.72 | 2838.5 | 2437.7 | −1.4831 | 11.869 | 10.386 |
| −36 | 0.02004 | 0.001085 | 5460.1 | −404.40 | 2729.0 | 2324.6 | −404.40 | 2838.4 | 2434.0 | −1.4986 | 11.969 | 10.470 |
| −38 | 0.01608 | 0.001085 | 6750.5 | −408.07 | 2729.9 | 2321.8 | −408.07 | 2838.4 | 2430.3 | −1.5141 | 12.071 | 10.557 |
| −40 | 0.01285 | 0.001084 | 8376.7 | −411.70 | 2730.7 | 2319.0 | −411.70 | 2838.3 | 2426.6 | −1.5296 | 12.174 | 10.644 |

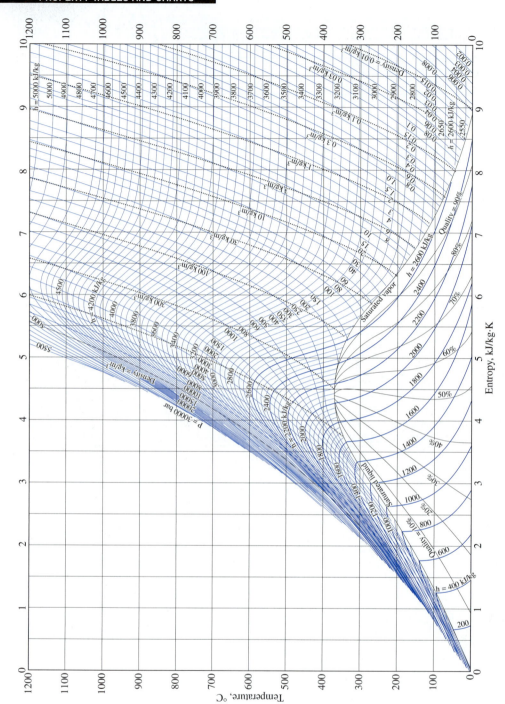

**FIGURE A–9**

*T-s* diagram for water.

*Copyright © 1984. From NBS/NRC Steam Tables/1 by Lester Haar, John S. Gallagher, and George S. Kell. Reproduced by permission of Routledge/Taylor & Francis Books, Inc.*

**FIGURE A–10**

Mollier diagram for water.

## TABLE A–11

Saturated refrigerant-134a—Temperature table

| Temp., $T\,°C$ | Sat. press., $P_{sat}$ kPa | Specific volume, m³/kg | | Internal energy, kJ/kg | | | Enthalpy, kJ/kg | | | Entropy, kJ/kg·K | | |
|---|---|---|---|---|---|---|---|---|---|---|---|---|
| | | Sat. liquid, $v_f$ | Sat. vapor, $v_g$ | Sat. liquid, $u_f$ | Evap., $u_{fg}$ | Sat. vapor, $u_g$ | Sat. liquid, $h_f$ | Evap., $h_{fg}$ | Sat. vapor, $h_g$ | Sat. liquid, $s_f$ | Evap., $s_{fg}$ | Sat. vapor, $s_g$ |
| −40 | 51.25 | 0.0007054 | 0.36081 | −0.036 | 207.40 | 207.37 | 0.000 | 225.86 | 225.86 | 0.00000 | 0.96866 | 0.96866 |
| −38 | 56.86 | 0.0007083 | 0.32732 | 2.475 | 206.04 | 208.51 | 2.515 | 224.61 | 227.12 | 0.01072 | 0.95511 | 0.96584 |
| −36 | 62.95 | 0.0007112 | 0.29751 | 4.992 | 204.67 | 209.66 | 5.037 | 223.35 | 228.39 | 0.02138 | 0.94176 | 0.96315 |
| −34 | 69.56 | 0.0007142 | 0.27090 | 7.517 | 203.29 | 210.81 | 7.566 | 222.09 | 229.65 | 0.03199 | 0.92859 | 0.96058 |
| −32 | 76.71 | 0.0007172 | 0.24711 | 10.05 | 201.91 | 211.96 | 10.10 | 220.81 | 230.91 | 0.04253 | 0.91560 | 0.95813 |
| −30 | 84.43 | 0.0007203 | 0.22580 | 12.59 | 200.52 | 213.11 | 12.65 | 219.52 | 232.17 | 0.05301 | 0.90278 | 0.95579 |
| −28 | 92.76 | 0.0007234 | 0.20666 | 15.13 | 199.12 | 214.25 | 15.20 | 218.22 | 233.43 | 0.06344 | 0.89012 | 0.95356 |
| −26 | 101.73 | 0.0007265 | 0.18946 | 17.69 | 197.72 | 215.40 | 17.76 | 216.92 | 234.68 | 0.07382 | 0.87762 | 0.95144 |
| −24 | 111.37 | 0.0007297 | 0.17395 | 20.25 | 196.30 | 216.55 | 20.33 | 215.59 | 235.92 | 0.08414 | 0.86527 | 0.94941 |
| −22 | 121.72 | 0.0007329 | 0.15995 | 22.82 | 194.88 | 217.70 | 22.91 | 214.26 | s237.17 | 0.09441 | 0.85307 | 0.94748 |
| −20 | 132.82 | 0.0007362 | 0.14729 | 25.39 | 193.45 | 218.84 | 25.49 | 212.91 | 238.41 | 0.10463 | 0.84101 | 0.94564 |
| −18 | 144.69 | 0.0007396 | 0.13583 | 27.98 | 192.01 | 219.98 | 28.09 | 211.55 | 239.64 | 0.11481 | 0.82908 | 0.94389 |
| −16 | 157.38 | 0.0007430 | 0.12542 | 30.57 | 190.56 | 221.13 | 30.69 | 210.18 | 240.87 | 0.12493 | 0.81729 | 0.94222 |
| −14 | 170.93 | 0.0007464 | 0.11597 | 33.17 | 189.09 | 222.27 | 33.30 | 208.79 | 242.09 | 0.13501 | 0.80561 | 0.94063 |
| −12 | 185.37 | 0.0007499 | 0.10736 | 35.78 | 187.62 | 223.40 | 35.92 | 207.38 | 243.30 | 0.14504 | 0.79406 | 0.93911 |
| −10 | 200.74 | 0.0007535 | 0.099516 | 38.40 | 186.14 | 224.54 | 38.55 | 205.96 | 244.51 | 0.15504 | 0.78263 | 0.93766 |
| −8 | 217.08 | 0.0007571 | 0.092352 | 41.03 | 184.64 | 225.67 | 41.19 | 204.52 | 245.72 | 0.16498 | 0.77130 | 0.93629 |
| −6 | 234.44 | 0.0007608 | 0.085802 | 43.66 | 183.13 | 226.80 | 43.84 | 203.07 | 246.91 | 0.17489 | 0.76008 | 0.93497 |
| −4 | 252.85 | 0.0007646 | 0.079804 | 46.31 | 181.61 | 227.92 | 46.50 | 201.60 | 248.10 | 0.18476 | 0.74896 | 0.93372 |
| −2 | 272.36 | 0.0007684 | 0.074304 | 48.96 | 180.08 | 229.04 | 49.17 | 200.11 | 249.28 | 0.19459 | 0.73794 | 0.93253 |
| 0 | 293.01 | 0.0007723 | 0.069255 | 51.63 | 178.53 | 230.16 | 51.86 | 198.60 | 250.45 | 0.20439 | 0.72701 | 0.93139 |
| 2 | 314.84 | 0.0007763 | 0.064612 | 54.30 | 176.97 | 231.27 | 54.55 | 197.07 | 251.61 | 0.21415 | 0.71616 | 0.93031 |
| 4 | 337.90 | 0.0007804 | 0.060338 | 56.99 | 175.39 | 232.38 | 57.25 | 195.51 | 252.77 | 0.22387 | 0.70540 | 0.92927 |
| 6 | 362.23 | 0.0007845 | 0.056398 | 59.68 | 173.80 | 233.48 | 59.97 | 193.94 | 253.91 | 0.23356 | 0.69471 | 0.92828 |
| 8 | 387.88 | 0.0007887 | 0.052762 | 62.39 | 172.19 | 234.58 | 62.69 | 192.35 | 255.04 | 0.24323 | 0.68410 | 0.92733 |
| 10 | 414.89 | 0.0007930 | 0.049403 | 65.10 | 170.56 | 235.67 | 65.43 | 190.73 | 256.16 | 0.25286 | 0.67356 | 0.92641 |
| 12 | 443.31 | 0.0007975 | 0.046295 | 67.83 | 168.92 | 236.75 | 68.18 | 189.09 | 257.27 | 0.26246 | 0.66308 | 0.92554 |
| 14 | 473.19 | 0.0008020 | 0.043417 | 70.57 | 167.26 | 237.83 | 70.95 | 187.42 | 258.37 | 0.27204 | 0.65266 | 0.92470 |
| 16 | 504.58 | 0.0008066 | 0.040748 | 73.32 | 165.58 | 238.90 | 73.73 | 185.73 | 259.46 | 0.28159 | 0.64230 | 0.92389 |
| 18 | 537.52 | 0.0008113 | 0.038271 | 76.08 | 163.88 | 239.96 | 76.52 | 184.01 | 260.53 | 0.29112 | 0.63198 | 0.92310 |

## TABLE A–11

Saturated refrigerant-134a—Temperature table (*Concluded*)

| Temp., $T$ °C | Sat. press., $P_{sat}$ kPa | Specific volume, m³/kg | | Internal energy, kJ/kg | | | Enthalpy, kJ/kg | | | Entropy, kJ/kg·K | | |
|---|---|---|---|---|---|---|---|---|---|---|---|---|
| | | Sat. liquid, $v_f$ | Sat. vapor, $v_g$ | Sat. liquid, $u_f$ | Evap., $u_{fg}$ | Sat. vapor, $u_g$ | Sat. liquid, $h_f$ | Evap., $h_{fg}$ | Sat. vapor, $h_g$ | Sat. liquid, $s_f$ | Evap., $s_{fg}$ | Sat. vapor, $s_g$ |
| 20 | 572.07 | 0.0008161 | 0.035969 | 78.86 | 162.16 | 241.02 | 79.32 | 182.27 | 261.59 | 0.30063 | 0.62172 | 0.92234 |
| 22 | 608.27 | 0.0008210 | 0.033828 | 81.64 | 160.42 | 242.06 | 82.14 | 180.49 | 262.64 | 0.31011 | 0.61149 | 0.92160 |
| 24 | 646.18 | 0.0008261 | 0.031834 | 84.44 | 158.65 | 243.10 | 84.98 | 178.69 | 263.67 | 0.31958 | 0.60130 | 0.92088 |
| 26 | 685.84 | 0.0008313 | 0.029976 | 87.26 | 156.87 | 244.12 | 87.83 | 176.85 | 264.68 | 0.32903 | 0.59115 | 0.92018 |
| 28 | 727.31 | 0.0008366 | 0.028242 | 90.09 | 155.05 | 245.14 | 90.69 | 174.99 | 265.68 | 0.33846 | 0.58102 | 0.91948 |
| 30 | 770.64 | 0.0008421 | 0.026622 | 92.93 | 153.22 | 246.14 | 93.58 | 173.08 | 266.66 | 0.34789 | 0.57091 | 0.91879 |
| 32 | 815.89 | 0.0008478 | 0.025108 | 95.79 | 151.35 | 247.14 | 96.48 | 171.14 | 267.62 | 0.35730 | 0.56082 | 0.91811 |
| 34 | 863.11 | 0.0008536 | 0.023691 | 98.66 | 149.46 | 248.12 | 99.40 | 169.17 | 268.57 | 0.36670 | 0.55074 | 0.91743 |
| 36 | 912.35 | 0.0008595 | 0.022364 | 101.55 | 147.54 | 249.08 | 102.33 | 167.16 | 269.49 | 0.37609 | 0.54066 | 0.91675 |
| 38 | 963.68 | 0.0008657 | 0.021119 | 104.45 | 145.58 | 250.04 | 105.29 | 165.10 | 270.39 | 0.38548 | 0.53058 | 0.91606 |
| 40 | 1017.1 | 0.0008720 | 0.019952 | 107.38 | 143.60 | 250.97 | 108.26 | 163.00 | 271.27 | 0.39486 | 0.52049 | 0.91536 |
| 42 | 1072.8 | 0.0008786 | 0.018855 | 110.32 | 141.58 | 251.89 | 111.26 | 160.86 | 272.12 | 0.40425 | 0.51039 | 0.91464 |
| 44 | 1130.7 | 0.0008854 | 0.017824 | 113.28 | 139.52 | 252.80 | 114.28 | 158.67 | 272.95 | 0.41363 | 0.50027 | 0.91391 |
| 46 | 1191.0 | 0.0008924 | 0.016853 | 116.26 | 137.42 | 253.68 | 117.32 | 156.43 | 273.75 | 0.42302 | 0.49012 | 0.91315 |
| 48 | 1253.6 | 0.0008996 | 0.015939 | 119.26 | 135.29 | 254.55 | 120.39 | 154.14 | 274.53 | 0.43242 | 0.47993 | 0.91236 |
| 52 | 1386.2 | 0.0009150 | 0.014265 | 125.33 | 130.88 | 256.21 | 126.59 | 149.39 | 275.98 | 0.45126 | 0.45941 | 0.91067 |
| 56 | 1529.1 | 0.0009317 | 0.012771 | 131.49 | 126.28 | 257.77 | 132.91 | 144.38 | 277.30 | 0.47018 | 0.43863 | 0.90880 |
| 60 | 1682.8 | 0.0009498 | 0.011434 | 137.76 | 121.46 | 259.22 | 139.36 | 139.10 | 278.46 | 0.48920 | 0.41749 | 0.90669 |
| 65 | 1891.0 | 0.0009750 | 0.009950 | 145.77 | 115.05 | 260.82 | 147.62 | 132.02 | 279.64 | 0.51320 | 0.39039 | 0.90359 |
| 70 | 2118.2 | 0.0010037 | 0.008642 | 154.01 | 108.14 | 262.15 | 156.13 | 124.32 | 280.46 | 0.53755 | 0.36227 | 0.89982 |
| 75 | 2365.8 | 0.0010372 | 0.007480 | 162.53 | 100.60 | 263.13 | 164.98 | 115.85 | 280.82 | 0.56241 | 0.33272 | 0.89512 |
| 80 | 2635.3 | 0.0010772 | 0.006436 | 171.40 | 92.23 | 263.63 | 174.24 | 106.35 | 280.59 | 0.58800 | 0.30111 | 0.88912 |
| 85 | 2928.2 | 0.0011270 | 0.005486 | 180.77 | 82.67 | 263.44 | 184.07 | 95.44 | 279.51 | 0.61473 | 0.26644 | 0.88117 |
| 90 | 3246.9 | 0.0011932 | 0.004599 | 190.89 | 71.29 | 262.18 | 194.76 | 82.35 | 277.11 | 0.64336 | 0.22674 | 0.87010 |
| 95 | 3594.1 | 0.0012933 | 0.003726 | 202.40 | 56.47 | 258.87 | 207.05 | 65.21 | 272.26 | 0.67578 | 0.17711 | 0.85289 |
| 100 | 3975.1 | 0.0015269 | 0.002630 | 218.72 | 29.19 | 247.91 | 224.79 | 33.58 | 258.37 | 0.72217 | 0.08999 | 0.81215 |

*Source:* Tables A–11 through A–13 are generated using the Engineering Equation Solver (EES) software developed by S. A. Klein and F. L. Alvarado. The routine used in calculations is the R134a, which is based on the fundamental equation of state developed by R. Tillner-Roth and H.D. Baehr, "An International Standard Formulation for the Thermodynamic Properties of 1,1,1,2-Tetrafluoroethane (HFC-134a) for temperatures from 170 K to 455 K and Pressures up to 70 MPa," *J. Phys. Chem, Ref. Data*, Vol. 23, No. 5, 1994. The enthalpy and entropy values of saturated liquid are set to zero at −40°C (and −40°F).

## TABLE A–12

Saturated refrigerant-134a—Pressure table

| Press., P kPa | Sat. temp., $T_{sat}$ °C | Specific volume, m³/kg | | Internal energy, kJ/kg | | | Enthalpy, kJ/kg | | | Entropy, kJ/kg·K | | |
|---|---|---|---|---|---|---|---|---|---|---|---|---|
| | | Sat. liquid, $v_f$ | Sat. vapor, $v_g$ | Sat. liquid, $u_f$ | Evap., $u_{fg}$ | Sat. vapor, $u_g$ | Sat. liquid, $h_f$ | Evap., $h_{fg}$ | Sat. vapor, $h_g$ | Sat. liquid, $s_f$ | Evap., $s_{fg}$ | Sat. vapor, $s_g$ |
| 60 | −36.95 | 0.0007098 | 0.31121 | 3.798 | 205.32 | 209.12 | 3.841 | 223.95 | 227.79 | 0.01634 | 0.94807 | 0.96441 |
| 70 | −33.87 | 0.0007144 | 0.26929 | 7.680 | 203.20 | 210.88 | 7.730 | 222.00 | 229.73 | 0.03267 | 0.92775 | 0.96042 |
| 80 | −31.13 | 0.0007185 | 0.23753 | 11.15 | 201.30 | 212.46 | 11.21 | 220.25 | 231.46 | 0.04711 | 0.90999 | 0.95710 |
| 90 | −28.65 | 0.0007223 | 0.21263 | 14.31 | 199.57 | 213.88 | 14.37 | 218.65 | 233.02 | 0.06008 | 0.89419 | 0.95427 |
| 100 | −26.37 | 0.0007259 | 0.19254 | 17.21 | 197.98 | 215.19 | 17.28 | 217.16 | 234.44 | 0.07188 | 0.87995 | 0.95183 |
| 120 | −22.32 | 0.0007324 | 0.16212 | 22.40 | 195.11 | 217.51 | 22.49 | 214.48 | 236.97 | 0.09275 | 0.85503 | 0.94779 |
| 140 | −18.77 | 0.0007383 | 0.14014 | 26.98 | 192.57 | 219.54 | 27.08 | 212.08 | 239.16 | 0.11087 | 0.83368 | 0.94456 |
| 160 | −15.60 | 0.0007437 | 0.12348 | 31.09 | 190.27 | 221.35 | 31.21 | 209.90 | 241.11 | 0.12693 | 0.81496 | 0.94190 |
| 180 | −12.73 | 0.0007487 | 0.11041 | 34.83 | 188.16 | 222.99 | 34.97 | 207.90 | 242.86 | 0.14139 | 0.79826 | 0.93965 |
| 200 | −10.09 | 0.0007533 | 0.099867 | 38.28 | 186.21 | 224.48 | 38.43 | 206.03 | 244.46 | 0.15457 | 0.78316 | 0.93773 |
| 240 | −5.38 | 0.0007620 | 0.083897 | 44.48 | 182.67 | 227.14 | 44.66 | 202.62 | 247.28 | 0.17794 | 0.75664 | 0.93458 |
| 280 | −1.25 | 0.0007699 | 0.072352 | 49.97 | 179.50 | 229.46 | 50.18 | 199.54 | 249.72 | 0.19829 | 0.73381 | 0.93210 |
| 320 | 2.46 | 0.0007772 | 0.063604 | 54.92 | 176.61 | 231.52 | 55.16 | 196.71 | 251.88 | 0.21637 | 0.71369 | 0.93006 |
| 360 | 5.82 | 0.0007841 | 0.056738 | 59.44 | 173.94 | 233.38 | 59.72 | 194.08 | 253.81 | 0.23270 | 0.69566 | 0.92836 |
| 400 | 8.91 | 0.0007907 | 0.051201 | 63.62 | 171.45 | 235.07 | 63.94 | 191.62 | 255.55 | 0.24761 | 0.67929 | 0.92691 |
| 450 | 12.46 | 0.0007985 | 0.045619 | 68.45 | 168.54 | 237.00 | 68.81 | 188.71 | 257.53 | 0.26465 | 0.66069 | 0.92535 |
| 500 | 15.71 | 0.0008059 | 0.041118 | 72.93 | 165.82 | 238.75 | 73.33 | 185.98 | 259.30 | 0.28023 | 0.64377 | 0.92400 |
| 550 | 18.73 | 0.0008130 | 0.037408 | 77.10 | 163.25 | 240.35 | 77.54 | 183.38 | 260.92 | 0.29461 | 0.62821 | 0.92282 |
| 600 | 21.55 | 0.0008199 | 0.034295 | 81.02 | 160.81 | 241.83 | 81.51 | 180.90 | 262.40 | 0.30799 | 0.61378 | 0.92177 |
| 650 | 24.20 | 0.0008266 | 0.031646 | 84.72 | 158.48 | 243.20 | 85.26 | 178.51 | 263.77 | 0.32051 | 0.60030 | 0.92081 |
| 700 | 26.69 | 0.0008331 | 0.029361 | 88.24 | 156.24 | 244.48 | 88.82 | 176.21 | 265.03 | 0.33230 | 0.58763 | 0.91994 |
| 750 | 29.06 | 0.0008395 | 0.027371 | 91.59 | 154.08 | 245.67 | 92.22 | 173.98 | 266.20 | 0.34345 | 0.57567 | 0.91912 |
| 800 | 31.31 | 0.0008458 | 0.025621 | 94.79 | 152.00 | 246.79 | 95.47 | 171.82 | 267.29 | 0.35404 | 0.56431 | 0.91835 |
| 850 | 33.45 | 0.0008520 | 0.024069 | 97.87 | 149.98 | 247.85 | 98.60 | 169.71 | 268.31 | 0.36413 | 0.55349 | 0.91762 |
| 900 | 35.51 | 0.0008580 | 0.022683 | 100.83 | 148.01 | 248.85 | 101.61 | 167.66 | 269.26 | 0.37377 | 0.54315 | 0.91692 |
| 950 | 37.48 | 0.0008641 | 0.021438 | 103.69 | 146.10 | 249.79 | 104.51 | 165.64 | 270.15 | 0.38301 | 0.53323 | 0.91624 |
| 1000 | 39.37 | 0.0008700 | 0.020313 | 106.45 | 144.23 | 250.68 | 107.32 | 163.67 | 270.99 | 0.39189 | 0.52368 | 0.91558 |
| 1200 | 46.29 | 0.0008934 | 0.016715 | 116.70 | 137.11 | 253.81 | 117.77 | 156.10 | 273.87 | 0.42441 | 0.48863 | 0.91303 |
| 1400 | 52.40 | 0.0009166 | 0.014107 | 125.94 | 130.43 | 256.37 | 127.22 | 148.90 | 276.12 | 0.45315 | 0.45734 | 0.91050 |
| 1600 | 57.88 | 0.0009400 | 0.012123 | 134.43 | 124.04 | 258.47 | 135.93 | 141.93 | 277.86 | 0.47911 | 0.42873 | 0.90784 |
| 1800 | 62.87 | 0.0009639 | 0.010559 | 142.33 | 117.83 | 260.17 | 144.07 | 135.11 | 279.17 | 0.50294 | 0.40204 | 0.90498 |
| 2000 | 67.45 | 0.0009886 | 0.009288 | 149.78 | 111.73 | 261.51 | 151.76 | 128.33 | 280.09 | 0.52509 | 0.37675 | 0.90184 |
| 2500 | 77.54 | 0.0010566 | 0.006936 | 166.99 | 96.47 | 263.45 | 169.63 | 111.16 | 280.79 | 0.57531 | 0.31695 | 0.89226 |
| 3000 | 86.16 | 0.0011406 | 0.005275 | 183.04 | 80.22 | 263.26 | 186.46 | 92.63 | 279.09 | 0.62118 | 0.25776 | 0.87894 |

## TABLE A–13

Superheated refrigerant-134a

| T °C | v m³/kg | u kJ/kg | h kJ/kg | s kJ/kg·K | v m³/kg | u kJ/kg | h kJ/kg | s kJ/kg·K | v m³/kg | u kJ/kg | h kJ/kg | s kJ/kg·K |
|------|---------|---------|---------|-----------|---------|---------|---------|-----------|---------|---------|---------|-----------|
| | $P = 0.06$ MPa ($T_{sat} = -36.95$°C) | | | | $P = 0.10$ MPa ($T_{sat} = -26.37$°C) | | | | $P = 0.14$ MPa ($T_{sat} = -18.77$°C) | | | |
| Sat. | 0.31121 | 209.12 | 227.79 | 0.9644 | 0.19254 | 215.19 | 234.44 | 0.9518 | 0.14014 | 219.54 | 239.16 | 0.9446 |
| −20 | 0.33608 | 220.60 | 240.76 | 1.0174 | 0.19841 | 219.66 | 239.50 | 0.9721 | | | | |
| −10 | 0.35048 | 227.55 | 248.58 | 1.0477 | 0.20743 | 226.75 | 247.49 | 1.0030 | 0.14605 | 225.91 | 246.36 | 0.9724 |
| 0 | 0.36476 | 234.66 | 256.54 | 1.0774 | 0.21630 | 233.95 | 255.58 | 1.0332 | 0.15263 | 233.23 | 254.60 | 1.0031 |
| 10 | 0.37893 | 241.92 | 264.66 | 1.1066 | 0.22506 | 241.30 | 263.81 | 1.0628 | 0.15908 | 240.66 | 262.93 | 1.0331 |
| 20 | 0.39302 | 249.35 | 272.94 | 1.1353 | 0.23373 | 248.79 | 272.17 | 1.0918 | 0.16544 | 248.22 | 271.38 | 1.0624 |
| 30 | 0.40705 | 256.95 | 281.37 | 1.1636 | 0.24233 | 256.44 | 280.68 | 1.1203 | 0.17172 | 255.93 | 279.97 | 1.0912 |
| 40 | 0.42102 | 264.71 | 289.97 | 1.1915 | 0.25088 | 264.25 | 289.34 | 1.1484 | 0.17794 | 263.79 | 288.70 | 1.1195 |
| 50 | 0.43495 | 272.64 | 298.74 | 1.2191 | 0.25937 | 272.22 | 298.16 | 1.1762 | 0.18412 | 271.79 | 297.57 | 1.1474 |
| 60 | 0.44883 | 280.73 | 307.66 | 1.2463 | 0.26783 | 280.35 | 307.13 | 1.2035 | 0.19025 | 279.96 | 306.59 | 1.1749 |
| 70 | 0.46269 | 288.99 | 316.75 | 1.2732 | 0.27626 | 288.64 | 316.26 | 1.2305 | 0.19635 | 288.28 | 315.77 | 1.2020 |
| 80 | 0.47651 | 297.41 | 326.00 | 1.2997 | 0.28465 | 297.08 | 325.55 | 1.2572 | 0.20242 | 296.75 | 325.09 | 1.2288 |
| 90 | 0.49032 | 306.00 | 335.42 | 1.3260 | 0.29303 | 305.69 | 334.99 | 1.2836 | 0.20847 | 305.38 | 334.57 | 1.2553 |
| 100 | 0.50410 | 314.74 | 344.99 | 1.3520 | 0.30138 | 314.46 | 344.60 | 1.3096 | 0.21449 | 314.17 | 344.20 | 1.2814 |
| | $P = 0.18$ MPa ($T_{sat} = -12.73$°C) | | | | $P = 0.20$ MPa ($T_{sat} = -10.09$°C) | | | | $P = 0.24$ MPa ($T_{sat} = -5.38$°C) | | | |
| Sat. | 0.11041 | 222.99 | 242.86 | 0.9397 | 0.09987 | 224.48 | 244.46 | 0.9377 | 0.08390 | 227.14 | 247.28 | 0.9346 |
| −10 | 0.11189 | 225.02 | 245.16 | 0.9484 | 0.09991 | 224.55 | 244.54 | 0.9380 | | | | |
| 0 | 0.11722 | 232.48 | 253.58 | 0.9798 | 0.10481 | 232.09 | 253.05 | 0.9698 | 0.08617 | 231.29 | 251.97 | 0.9519 |
| 10 | 0.12240 | 240.00 | 262.04 | 1.0102 | 0.10955 | 239.67 | 261.58 | 1.0004 | 0.09026 | 238.98 | 260.65 | 0.9831 |
| 20 | 0.12748 | 247.64 | 270.59 | 1.0399 | 0.11418 | 247.35 | 270.18 | 1.0303 | 0.09423 | 246.74 | 269.36 | 1.0134 |
| 30 | 0.13248 | 255.41 | 279.25 | 1.0690 | 0.11874 | 255.14 | 278.89 | 1.0595 | 0.09812 | 254.61 | 278.16 | 1.0429 |
| 40 | 0.13741 | 263.31 | 288.05 | 1.0975 | 0.12322 | 263.08 | 287.72 | 1.0882 | 0.10193 | 262.59 | 287.06 | 1.0718 |
| 50 | 0.14230 | 271.36 | 296.98 | 1.1256 | 0.12766 | 271.15 | 296.68 | 1.1163 | 0.10570 | 270.71 | 296.08 | 1.1001 |
| 60 | 0.14715 | 279.56 | 306.05 | 1.1532 | 0.13206 | 279.37 | 305.78 | 1.1441 | 0.10942 | 278.97 | 305.23 | 1.1280 |
| 70 | 0.15196 | 287.91 | 315.27 | 1.1805 | 0.13641 | 287.73 | 315.01 | 1.1714 | 0.11310 | 287.36 | 314.51 | 1.1554 |
| 80 | 0.15673 | 296.42 | 324.63 | 1.2074 | 0.14074 | 296.25 | 324.40 | 1.1983 | 0.11675 | 295.91 | 323.93 | 1.1825 |
| 90 | 0.16149 | 305.07 | 334.14 | 1.2339 | 0.14504 | 304.92 | 333.93 | 1.2249 | 0.12038 | 304.60 | 333.49 | 1.2092 |
| 100 | 0.16622 | 313.88 | 343.80 | 1.2602 | 0.14933 | 313.74 | 343.60 | 1.2512 | 0.12398 | 313.44 | 343.20 | 1.2356 |
| | $P = 0.28$ MPa ($T_{sat} = -1.25$°C) | | | | $P = 0.32$ MPa ($T_{sat} = 2.46$°C) | | | | $P = 0.40$ MPa ($T_{sat} = 8.91$°C) | | | |
| Sat. | 0.07235 | 229.46 | 249.72 | 0.9321 | 0.06360 | 231.52 | 251.88 | 0.9301 | 0.051201 | 235.07 | 255.55 | 0.9269 |
| 0 | 0.07282 | 230.44 | 250.83 | 0.9362 | | | | | | | | |
| 10 | 0.07646 | 238.27 | 259.68 | 0.9680 | 0.06609 | 237.54 | 258.69 | 0.9544 | 0.051506 | 235.97 | 256.58 | 0.9305 |
| 20 | 0.07997 | 246.13 | 268.52 | 0.9987 | 0.06925 | 245.50 | 267.66 | 0.9856 | 0.054213 | 244.18 | 265.86 | 0.9628 |
| 30 | 0.08338 | 254.06 | 277.41 | 1.0285 | 0.07231 | 253.50 | 276.65 | 1.0157 | 0.056796 | 252.36 | 275.07 | 0.9937 |
| 40 | 0.08672 | 262.10 | 286.38 | 1.0576 | 0.07530 | 261.60 | 285.70 | 1.0451 | 0.059292 | 260.58 | 284.30 | 1.0236 |
| 50 | 0.09000 | 270.27 | 295.47 | 1.0862 | 0.07823 | 269.82 | 294.85 | 1.0739 | 0.061724 | 268.90 | 293.59 | 1.0528 |
| 60 | 0.09324 | 278.56 | 304.67 | 1.1142 | 0.08111 | 278.15 | 304.11 | 1.1021 | 0.064104 | 277.32 | 302.96 | 1.0814 |
| 70 | 0.09644 | 286.99 | 314.00 | 1.1418 | 0.08395 | 286.62 | 313.48 | 1.1298 | 0.066443 | 285.86 | 312.44 | 1.1094 |
| 80 | 0.09961 | 295.57 | 323.46 | 1.1690 | 0.08675 | 295.22 | 322.98 | 1.1571 | 0.068747 | 294.53 | 322.02 | 1.1369 |
| 90 | 0.10275 | 304.29 | 333.06 | 1.1958 | 0.08953 | 303.97 | 332.62 | 1.1840 | 0.071023 | 303.32 | 331.73 | 1.1640 |
| 100 | 0.10587 | 313.15 | 342.80 | 1.2222 | 0.09229 | 312.86 | 342.39 | 1.2105 | 0.073274 | 312.26 | 341.57 | 1.1907 |
| 110 | 0.10897 | 322.16 | 352.68 | 1.2483 | 0.09503 | 321.89 | 352.30 | 1.2367 | 0.075504 | 321.33 | 351.53 | 1.2171 |
| 120 | 0.11205 | 331.32 | 362.70 | 1.2742 | 0.09775 | 331.07 | 362.35 | 1.2626 | 0.077717 | 330.55 | 361.63 | 1.2431 |
| 130 | 0.11512 | 340.63 | 372.87 | 1.2997 | 0.10045 | 340.39 | 372.54 | 1.2882 | 0.079913 | 339.90 | 371.87 | 1.2688 |
| 140 | 0.11818 | 350.09 | 383.18 | 1.3250 | 0.10314 | 349.86 | 382.87 | 1.3135 | 0.082096 | 349.41 | 382.24 | 1.2942 |

## TABLE A–13

Superheated refrigerant-134a (*Continued*)

| T °C | v m³/kg | u kJ/kg | h kJ/kg | s kJ/kg·K | v m³/kg | u kJ/kg | h kJ/kg | s kJ/kg·K | v m³/kg | u kJ/kg | h kJ/kg | s kJ/kg·K |
|---|---|---|---|---|---|---|---|---|---|---|---|---|
| | $P = 0.50$ MPa ($T_{sat} = 15.71$°C) | | | | $P = 0.60$ MPa ($T_{sat} = 21.55$°C) | | | | $P = 0.70$ MPa ($T_{sat} = 26.69$°C) | | | |
| Sat. | 0.041118 | 238.75 | 259.30 | 0.9240 | 0.034295 | 241.83 | 262.40 | 0.9218 | 0.029361 | 244.48 | 265.03 | 0.9199 |
| 20 | 0.042115 | 242.40 | 263.46 | 0.9383 | | | | | | | | |
| 30 | 0.044338 | 250.84 | 273.01 | 0.9703 | 0.035984 | 249.22 | 270.81 | 0.9499 | 0.029966 | 247.48 | 268.45 | 0.9313 |
| 40 | 0.046456 | 259.26 | 282.48 | 1.0011 | 0.037865 | 257.86 | 280.58 | 0.9816 | 0.031696 | 256.39 | 278.57 | 0.9641 |
| 50 | 0.048499 | 267.72 | 291.96 | 1.0309 | 0.039659 | 266.48 | 290.28 | 1.0121 | 0.033322 | 265.20 | 288.53 | 0.9954 |
| 60 | 0.050485 | 276.25 | 301.50 | 1.0599 | 0.041389 | 275.15 | 299.98 | 1.0417 | 0.034875 | 274.01 | 298.42 | 1.0256 |
| 70 | 0.052427 | 284.89 | 311.10 | 1.0883 | 0.043069 | 283.89 | 309.73 | 1.0705 | 0.036373 | 282.87 | 308.33 | 1.0549 |
| 80 | 0.054331 | 293.64 | 320.80 | 1.1162 | 0.044710 | 292.73 | 319.55 | 1.0987 | 0.037829 | 291.80 | 318.28 | 1.0835 |
| 90 | 0.056205 | 302.51 | 330.61 | 1.1436 | 0.046318 | 301.67 | 329.46 | 1.1264 | 0.039250 | 300.82 | 328.29 | 1.1114 |
| 100 | 0.058053 | 311.50 | 340.53 | 1.1705 | 0.047900 | 310.73 | 339.47 | 1.1536 | 0.040642 | 309.95 | 338.40 | 1.1389 |
| 110 | 0.059880 | 320.63 | 350.57 | 1.1971 | 0.049458 | 319.91 | 349.59 | 1.1803 | 0.042010 | 319.19 | 348.60 | 1.1658 |
| 120 | 0.061687 | 329.89 | 360.73 | 1.2233 | 0.050997 | 329.23 | 359.82 | 1.2067 | 0.043358 | 328.55 | 358.90 | 1.1924 |
| 130 | 0.063479 | 339.29 | 371.03 | 1.2491 | 0.052519 | 338.67 | 370.18 | 1.2327 | 0.044688 | 338.04 | 369.32 | 1.2186 |
| 140 | 0.065256 | 348.83 | 381.46 | 1.2747 | 0.054027 | 348.25 | 380.66 | 1.2584 | 0.046004 | 347.66 | 379.86 | 1.2444 |
| 150 | 0.067021 | 358.51 | 392.02 | 1.2999 | 0.055522 | 357.96 | 391.27 | 1.2838 | 0.047306 | 357.41 | 390.52 | 1.2699 |
| 160 | 0.068775 | 368.33 | 402.72 | 1.3249 | 0.057006 | 367.81 | 402.01 | 1.3088 | 0.048597 | 367.29 | 401.31 | 1.2951 |
| | $P = 0.80$ MPa ($T_{sat} = 31.31$°C) | | | | $P = 0.90$ MPa ($T_{sat} = 35.51$°C) | | | | $P = 1.00$ MPa ($T_{sat} = 39.37$°C) | | | |
| Sat. | 0.025621 | 246.79 | 267.29 | 0.9183 | 0.022683 | 248.85 | 269.26 | 0.9169 | 0.020313 | 250.68 | 270.99 | 0.9156 |
| 40 | 0.027035 | 254.82 | 276.45 | 0.9480 | 0.023375 | 253.13 | 274.17 | 0.9327 | 0.020406 | 251.30 | 271.71 | 0.9179 |
| 50 | 0.028547 | 263.86 | 286.69 | 0.9802 | 0.024809 | 262.44 | 284.77 | 0.9660 | 0.021796 | 260.94 | 282.74 | 0.9525 |
| 60 | 0.029973 | 272.83 | 296.81 | 1.0110 | 0.026146 | 271.60 | 295.13 | 0.9976 | 0.023068 | 270.32 | 293.38 | 0.9850 |
| 70 | 0.031340 | 281.81 | 306.88 | 1.0408 | 0.027413 | 280.72 | 305.39 | 1.0280 | 0.024261 | 279.59 | 303.85 | 1.0160 |
| 80 | 0.032659 | 290.84 | 316.97 | 1.0698 | 0.028630 | 289.86 | 315.63 | 1.0574 | 0.025398 | 288.86 | 314.25 | 1.0458 |
| 90 | 0.033941 | 299.95 | 327.10 | 1.0981 | 0.029806 | 299.06 | 325.89 | 1.0860 | 0.026492 | 298.15 | 324.64 | 1.0748 |
| 100 | 0.035193 | 309.15 | 337.30 | 1.1258 | 0.030951 | 308.34 | 336.19 | 1.1140 | 0.027552 | 307.51 | 335.06 | 1.1031 |
| 110 | 0.036420 | 318.45 | 347.59 | 1.1530 | 0.032068 | 317.70 | 346.56 | 1.1414 | 0.028584 | 316.94 | 345.53 | 1.1308 |
| 120 | 0.037625 | 327.87 | 357.97 | 1.1798 | 0.033164 | 327.18 | 357.02 | 1.1684 | 0.029592 | 326.47 | 356.06 | 1.1580 |
| 130 | 0.038813 | 337.40 | 368.45 | 1.2061 | 0.034241 | 336.76 | 367.58 | 1.1949 | 0.030581 | 336.11 | 366.69 | 1.1846 |
| 140 | 0.039985 | 347.06 | 379.05 | 1.2321 | 0.035302 | 346.46 | 378.23 | 1.2210 | 0.031554 | 345.85 | 377.40 | 1.2109 |
| 150 | 0.041143 | 356.85 | 389.76 | 1.2577 | 0.036349 | 356.28 | 389.00 | 1.2467 | 0.032512 | 355.71 | 388.22 | 1.2368 |
| 160 | 0.042290 | 366.76 | 400.59 | 1.2830 | 0.037384 | 366.23 | 399.88 | 1.2721 | 0.033457 | 365.70 | 399.15 | 1.2623 |
| 170 | 0.043427 | 376.81 | 411.55 | 1.3080 | 0.038408 | 376.31 | 410.88 | 1.2972 | 0.034392 | 375.81 | 410.20 | 1.2875 |
| 180 | 0.044554 | 386.99 | 422.64 | 1.3327 | 0.039423 | 386.52 | 422.00 | 1.3221 | 0.035317 | 386.04 | 421.36 | 1.3124 |
| | $P = 1.20$ MPa ($T_{sat} = 46.29$°C) | | | | $P = 1.40$ MPa ($T_{sat} = 52.40$°C) | | | | $P = 1.60$ MPa ($T_{sat} = 57.88$°C) | | | |
| Sat. | 0.016715 | 253.81 | 273.87 | 0.9130 | 0.014107 | 256.37 | 276.12 | 0.9105 | 0.012123 | 258.47 | 277.86 | 0.9078 |
| 50 | 0.017201 | 257.63 | 278.27 | 0.9267 | | | | | | | | |
| 60 | 0.018404 | 267.56 | 289.64 | 0.9614 | 0.015005 | 264.46 | 285.47 | 0.9389 | 0.012372 | 260.89 | 280.69 | 0.9163 |
| 70 | 0.019502 | 277.21 | 300.61 | 0.9938 | 0.016060 | 274.62 | 297.10 | 0.9733 | 0.013430 | 271.76 | 293.25 | 0.9535 |
| 80 | 0.020529 | 286.75 | 311.39 | 1.0248 | 0.017023 | 284.51 | 308.34 | 1.0056 | 0.014362 | 282.09 | 305.07 | 0.9875 |
| 90 | 0.021506 | 296.26 | 322.07 | 1.0546 | 0.017923 | 294.28 | 319.37 | 1.0364 | 0.015215 | 292.17 | 316.52 | 1.0194 |
| 100 | 0.022442 | 305.80 | 332.73 | 1.0836 | 0.018778 | 304.01 | 330.30 | 1.0661 | 0.016014 | 302.14 | 327.76 | 1.0500 |
| 110 | 0.023348 | 315.38 | 343.40 | 1.1118 | 0.019597 | 313.76 | 341.19 | 1.0949 | 0.016773 | 312.07 | 338.91 | 1.0795 |
| 120 | 0.024228 | 325.03 | 354.11 | 1.1394 | 0.020388 | 323.55 | 352.09 | 1.1230 | 0.017500 | 322.02 | 350.02 | 1.1081 |
| 130 | 0.025086 | 334.77 | 364.88 | 1.1664 | 0.021155 | 333.41 | 363.02 | 1.1504 | 0.018201 | 332.00 | 361.12 | 1.1360 |
| 140 | 0.025927 | 344.61 | 375.72 | 1.1930 | 0.021904 | 343.34 | 374.01 | 1.1773 | 0.018882 | 342.05 | 372.26 | 1.1632 |
| 150 | 0.026753 | 354.56 | 386.66 | 1.2192 | 0.022636 | 353.37 | 385.07 | 1.2038 | 0.019545 | 352.17 | 383.44 | 1.1900 |
| 160 | 0.027566 | 364.61 | 397.69 | 1.2449 | 0.023355 | 363.51 | 396.20 | 1.2298 | 0.020194 | 362.38 | 394.69 | 1.2163 |
| 170 | 0.028367 | 374.78 | 408.82 | 1.2703 | 0.024061 | 373.75 | 407.43 | 1.2554 | 0.020830 | 372.69 | 406.02 | 1.2421 |
| 180 | 0.029158 | 385.08 | 420.07 | 1.2954 | 0.024757 | 384.10 | 418.76 | 1.2807 | 0.021456 | 383.11 | 417.44 | 1.2676 |

**FIGURE A–14**

*P-h* diagram for refrigerant-134a.

*Note:* The reference point used for the chart is different than that used in the R-134a tables. Therefore, problems should be solved using all property data either from the tables or from the chart, but not from both.

*Reprinted by permission of American Society of Heating, Refrigerating, and Air-Conditioning Engineers, Inc., Atlanta, GA.*

## TABLE A–15

### Properties of saturated water

| Temp. $T$, °C | Saturation Pressure $P_{sat}$, kPa | Density $\rho$, kg/m³ Liquid | Density $\rho$, kg/m³ Vapor | Enthalpy of Vaporization $h_{fg}$, kJ/kg | Specific Heat $c_p$, J/kg·K Liquid | Specific Heat $c_p$, J/kg·K Vapor | Thermal Conductivity $k$, W/m·K Liquid | Thermal Conductivity $k$, W/m·K Vapor | Dynamic Viscosity $\mu$, kg/m·s Liquid | Dynamic Viscosity $\mu$, kg/m·s Vapor | Prandtl Number Pr Liquid | Prandtl Number Pr Vapor | Volume Expansion Coefficient $\beta$, 1/K Liquid |
|---|---|---|---|---|---|---|---|---|---|---|---|---|---|
| 0.01 | 0.6113 | 999.8 | 0.0048 | 2501 | 4217 | 1854 | 0.561 | 0.0171 | $1.792 \times 10^{-3}$ | $0.922 \times 10^{-5}$ | 13.5 | 1.00 | $-0.068 \times 10^{-3}$ |
| 5 | 0.8721 | 999.9 | 0.0068 | 2490 | 4205 | 1857 | 0.571 | 0.0173 | $1.519 \times 10^{-3}$ | $0.934 \times 10^{-5}$ | 11.2 | 1.00 | $0.015 \times 10^{-3}$ |
| 10 | 1.2276 | 999.7 | 0.0094 | 2478 | 4194 | 1862 | 0.580 | 0.0176 | $1.307 \times 10^{-3}$ | $0.946 \times 10^{-5}$ | 9.45 | 1.00 | $0.733 \times 10^{-3}$ |
| 15 | 1.7051 | 999.1 | 0.0128 | 2466 | 4185 | 1863 | 0.589 | 0.0179 | $1.138 \times 10^{-3}$ | $0.959 \times 10^{-5}$ | 8.09 | 1.00 | $0.138 \times 10^{-3}$ |
| 20 | 2.339 | 998.0 | 0.0173 | 2454 | 4182 | 1867 | 0.598 | 0.0182 | $1.002 \times 10^{-3}$ | $0.973 \times 10^{-5}$ | 7.01 | 1.00 | $0.195 \times 10^{-3}$ |
| 25 | 3.169 | 997.0 | 0.0231 | 2442 | 4180 | 1870 | 0.607 | 0.0186 | $0.891 \times 10^{-3}$ | $0.987 \times 10^{-5}$ | 6.14 | 1.00 | $0.247 \times 10^{-3}$ |
| 30 | 4.246 | 996.0 | 0.0304 | 2431 | 4178 | 1875 | 0.615 | 0.0189 | $0.798 \times 10^{-3}$ | $1.001 \times 10^{-5}$ | 5.42 | 1.00 | $0.294 \times 10^{-3}$ |
| 35 | 5.628 | 994.0 | 0.0397 | 2419 | 4178 | 1880 | 0.623 | 0.0192 | $0.720 \times 10^{-3}$ | $1.016 \times 10^{-5}$ | 4.83 | 1.00 | $0.337 \times 10^{-3}$ |
| 40 | 7.384 | 992.1 | 0.0512 | 2407 | 4179 | 1885 | 0.631 | 0.0196 | $0.653 \times 10^{-3}$ | $1.031 \times 10^{-5}$ | 4.32 | 1.00 | $0.377 \times 10^{-3}$ |
| 45 | 9.593 | 990.1 | 0.0655 | 2395 | 4180 | 1892 | 0.637 | 0.0200 | $0.596 \times 10^{-3}$ | $1.046 \times 10^{-5}$ | 3.91 | 1.00 | $0.415 \times 10^{-3}$ |
| 50 | 12.35 | 988.1 | 0.0831 | 2383 | 4181 | 1900 | 0.644 | 0.0204 | $0.547 \times 10^{-3}$ | $1.062 \times 10^{-5}$ | 3.55 | 1.00 | $0.451 \times 10^{-3}$ |
| 55 | 15.76 | 985.2 | 0.1045 | 2371 | 4183 | 1908 | 0.649 | 0.0208 | $0.504 \times 10^{-3}$ | $1.077 \times 10^{-5}$ | 3.25 | 1.00 | $0.484 \times 10^{-3}$ |
| 60 | 19.94 | 983.3 | 0.1304 | 2359 | 4185 | 1916 | 0.654 | 0.0212 | $0.467 \times 10^{-3}$ | $1.093 \times 10^{-5}$ | 2.99 | 1.00 | $0.517 \times 10^{-3}$ |
| 65 | 25.03 | 980.4 | 0.1614 | 2346 | 4187 | 1926 | 0.659 | 0.0216 | $0.433 \times 10^{-3}$ | $1.110 \times 10^{-5}$ | 2.75 | 1.00 | $0.548 \times 10^{-3}$ |
| 70 | 31.19 | 977.5 | 0.1983 | 2334 | 4190 | 1936 | 0.663 | 0.0221 | $0.404 \times 10^{-3}$ | $1.126 \times 10^{-5}$ | 2.55 | 1.00 | $0.578 \times 10^{-3}$ |
| 75 | 38.58 | 974.7 | 0.2421 | 2321 | 4193 | 1948 | 0.667 | 0.0225 | $0.378 \times 10^{-3}$ | $1.142 \times 10^{-5}$ | 2.38 | 1.00 | $0.607 \times 10^{-3}$ |
| 80 | 47.39 | 971.8 | 0.2935 | 2309 | 4197 | 1962 | 0.670 | 0.0230 | $0.355 \times 10^{-3}$ | $1.159 \times 10^{-5}$ | 2.22 | 1.00 | $0.653 \times 10^{-3}$ |
| 85 | 57.83 | 968.1 | 0.3536 | 2296 | 4201 | 1977 | 0.673 | 0.0235 | $0.333 \times 10^{-3}$ | $1.176 \times 10^{-5}$ | 2.08 | 1.00 | $0.670 \times 10^{-3}$ |
| 90 | 70.14 | 965.3 | 0.4235 | 2283 | 4206 | 1993 | 0.675 | 0.0240 | $0.315 \times 10^{-3}$ | $1.193 \times 10^{-5}$ | 1.96 | 1.00 | $0.702 \times 10^{-3}$ |
| 95 | 84.55 | 961.5 | 0.5045 | 2270 | 4212 | 2010 | 0.677 | 0.0246 | $0.297 \times 10^{-3}$ | $1.210 \times 10^{-5}$ | 1.85 | 1.00 | $0.716 \times 10^{-3}$ |
| 100 | 101.33 | 957.9 | 0.5978 | 2257 | 4217 | 2029 | 0.679 | 0.0251 | $0.282 \times 10^{-3}$ | $1.227 \times 10^{-5}$ | 1.75 | 1.00 | $0.750 \times 10^{-3}$ |
| 110 | 143.27 | 950.6 | 0.8263 | 2230 | 4229 | 2071 | 0.682 | 0.0262 | $0.255 \times 10^{-3}$ | $1.261 \times 10^{-5}$ | 1.58 | 1.00 | $0.798 \times 10^{-3}$ |
| 120 | 198.53 | 943.4 | 1.121 | 2203 | 4244 | 2120 | 0.683 | 0.0275 | $0.232 \times 10^{-3}$ | $1.296 \times 10^{-5}$ | 1.44 | 1.00 | $0.858 \times 10^{-3}$ |
| 130 | 270.1 | 934.6 | 1.496 | 2174 | 4263 | 2177 | 0.684 | 0.0288 | $0.213 \times 10^{-3}$ | $1.330 \times 10^{-5}$ | 1.33 | 1.01 | $0.913 \times 10^{-3}$ |
| 140 | 361.3 | 921.7 | 1.965 | 2145 | 4286 | 2244 | 0.683 | 0.0301 | $0.197 \times 10^{-3}$ | $1.365 \times 10^{-5}$ | 1.24 | 1.02 | $0.970 \times 10^{-3}$ |
| 150 | 475.8 | 916.6 | 2.546 | 2114 | 4311 | 2314 | 0.682 | 0.0316 | $0.183 \times 10^{-3}$ | $1.399 \times 10^{-5}$ | 1.16 | 1.02 | $1.025 \times 10^{-3}$ |
| 160 | 617.8 | 907.4 | 3.256 | 2083 | 4340 | 2420 | 0.680 | 0.0331 | $0.170 \times 10^{-3}$ | $1.434 \times 10^{-5}$ | 1.09 | 1.05 | $1.145 \times 10^{-3}$ |
| 170 | 791.7 | 897.7 | 4.119 | 2050 | 4370 | 2490 | 0.677 | 0.0347 | $0.160 \times 10^{-3}$ | $1.468 \times 10^{-5}$ | 1.03 | 1.05 | $1.178 \times 10^{-3}$ |
| 180 | 1,002.1 | 887.3 | 5.153 | 2015 | 4410 | 2590 | 0.673 | 0.0364 | $0.150 \times 10^{-3}$ | $1.502 \times 10^{-5}$ | 0.983 | 1.07 | $1.210 \times 10^{-3}$ |
| 190 | 1,254.4 | 876.4 | 6.388 | 1979 | 4460 | 2710 | 0.669 | 0.0382 | $0.142 \times 10^{-3}$ | $1.537 \times 10^{-5}$ | 0.947 | 1.09 | $1.280 \times 10^{-3}$ |
| 200 | 1,553.8 | 864.3 | 7.852 | 1941 | 4500 | 2840 | 0.663 | 0.0401 | $0.134 \times 10^{-3}$ | $1.571 \times 10^{-5}$ | 0.910 | 1.11 | $1.350 \times 10^{-3}$ |
| 220 | 2,318 | 840.3 | 11.60 | 1859 | 4610 | 3110 | 0.650 | 0.0442 | $0.122 \times 10^{-3}$ | $1.641 \times 10^{-5}$ | 0.865 | 1.15 | $1.520 \times 10^{-3}$ |
| 240 | 3,344 | 813.7 | 16.73 | 1767 | 4760 | 3520 | 0.632 | 0.0487 | $0.111 \times 10^{-3}$ | $1.712 \times 10^{-5}$ | 0.836 | 1.24 | $1.720 \times 10^{-3}$ |
| 260 | 4,688 | 783.7 | 23.69 | 1663 | 4970 | 4070 | 0.609 | 0.0540 | $0.102 \times 10^{-3}$ | $1.788 \times 10^{-5}$ | 0.832 | 1.35 | $2.000 \times 10^{-3}$ |
| 280 | 6,412 | 750.8 | 33.15 | 1544 | 5280 | 4835 | 0.581 | 0.0605 | $0.094 \times 10^{-3}$ | $1.870 \times 10^{-5}$ | 0.854 | 1.49 | $2.380 \times 10^{-3}$ |
| 300 | 8,581 | 713.8 | 46.15 | 1405 | 5750 | 5980 | 0.548 | 0.0695 | $0.086 \times 10^{-3}$ | $1.965 \times 10^{-5}$ | 0.902 | 1.69 | $2.950 \times 10^{-3}$ |
| 320 | 11,274 | 667.1 | 64.57 | 1239 | 6540 | 7900 | 0.509 | 0.0836 | $0.078 \times 10^{-3}$ | $2.084 \times 10^{-5}$ | 1.00 | 1.97 | |
| 340 | 14,586 | 610.5 | 92.62 | 1028 | 8240 | 11,870 | 0.469 | 0.110 | $0.070 \times 10^{-3}$ | $2.255 \times 10^{-5}$ | 1.23 | 2.43 | |
| 360 | 18,651 | 528.3 | 144.0 | 720 | 14,690 | 25,800 | 0.427 | 0.178 | $0.060 \times 10^{-3}$ | $2.571 \times 10^{-5}$ | 2.06 | 3.73 | |
| 374.14 | 22,090 | 317.0 | 317.0 | 0 | — | — | — | — | $0.043 \times 10^{-3}$ | $4.313 \times 10^{-5}$ | | | |

*Note 1:* Kinematic viscosity $\nu$ and thermal diffusivity $\alpha$ can be calculated from their definitions, $\nu = \mu/\rho$ and $\alpha = k/\rho c_p = \nu/Pr$. The temperatures 0.01°C, 100°C, and 374.14°C are the triple-, boiling-, and critical-point temperatures of water, respectively. The properties listed above (except the vapor density) can be used at any pressure with negligible error except at temperatures near the critical-point value.

*Note 2:* The unit kJ/kg·°C for specific heat is equivalent to kJ/kg·K, and the unit W/m·°C for thermal conductivity is equivalent to W/m·K.

*Source:* Viscosity and thermal conductivity data are from J. V. Sengers and J. T. R. Watson, *Journal of Physical and Chemical Reference Data* 15 (1986), pp. 1291–1322. Other data are obtained from various sources or calculated.

## TABLE A–16

Properties of saturated refrigerant-134a

| Temp. T, °C | Saturation Pressure P, kPa | Density ρ, kg/m³ Liquid | Density ρ, kg/m³ Vapor | Enthalpy of Vaporization $h_{fg}$, kJ/kg | Specific Heat $c_p$, J/kg·K Liquid | Specific Heat $c_p$, J/kg·K Vapor | Thermal Conductivity k, W/m·K Liquid | Thermal Conductivity k, W/m·K Vapor | Dynamic Viscosity μ, kg/m·s Liquid | Dynamic Viscosity μ, kg/m·s Vapor | Prandtl Number Pr Liquid | Prandtl Number Pr Vapor | Volume Expansion Coefficient β, 1/K Liquid | Surface Tension, N/m Liquid |
|---|---|---|---|---|---|---|---|---|---|---|---|---|---|---|
| −40 | 51.2 | 1418 | 2.773 | 225.9 | 1254 | 748.6 | 0.1101 | 0.00811 | $4.878 \times 10^{-4}$ | $2.550 \times 10^{-6}$ | 5.558 | 0.235 | 0.00205 | 0.01760 |
| −35 | 66.2 | 1403 | 3.524 | 222.7 | 1264 | 764.1 | 0.1084 | 0.00862 | $4.509 \times 10^{-4}$ | $3.003 \times 10^{-6}$ | 5.257 | 0.266 | 0.00209 | 0.01682 |
| −30 | 84.4 | 1389 | 4.429 | 219.5 | 1273 | 780.2 | 0.1066 | 0.00913 | $4.178 \times 10^{-4}$ | $3.504 \times 10^{-6}$ | 4.992 | 0.299 | 0.00215 | 0.01604 |
| −25 | 106.5 | 1374 | 5.509 | 216.3 | 1283 | 797.2 | 0.1047 | 0.00963 | $3.882 \times 10^{-4}$ | $4.054 \times 10^{-6}$ | 4.757 | 0.335 | 0.00220 | 0.01527 |
| −20 | 132.8 | 1359 | 6.787 | 213.0 | 1294 | 814.9 | 0.1028 | 0.01013 | $3.614 \times 10^{-4}$ | $4.651 \times 10^{-6}$ | 4.548 | 0.374 | 0.00227 | 0.01451 |
| −15 | 164.0 | 1343 | 8.288 | 209.5 | 1306 | 833.5 | 0.1009 | 0.01063 | $3.371 \times 10^{-4}$ | $5.295 \times 10^{-6}$ | 4.363 | 0.415 | 0.00233 | 0.01376 |
| −10 | 200.7 | 1327 | 10.04 | 206.0 | 1318 | 853.1 | 0.0989 | 0.01112 | $3.150 \times 10^{-4}$ | $5.982 \times 10^{-6}$ | 4.198 | 0.459 | 0.00241 | 0.01302 |
| −5 | 243.5 | 1311 | 12.07 | 202.4 | 1330 | 873.8 | 0.0968 | 0.01161 | $2.947 \times 10^{-4}$ | $6.709 \times 10^{-6}$ | 4.051 | 0.505 | 0.00249 | 0.01229 |
| 0 | 293.0 | 1295 | 14.42 | 198.7 | 1344 | 895.6 | 0.0947 | 0.01210 | $2.761 \times 10^{-4}$ | $7.471 \times 10^{-6}$ | 3.919 | 0.553 | 0.00258 | 0.01156 |
| 5 | 349.9 | 1278 | 17.12 | 194.8 | 1358 | 918.7 | 0.0925 | 0.01259 | $2.589 \times 10^{-4}$ | $8.264 \times 10^{-6}$ | 3.802 | 0.603 | 0.00269 | 0.01084 |
| 10 | 414.9 | 1261 | 20.22 | 190.8 | 1374 | 943.2 | 0.0903 | 0.01308 | $2.430 \times 10^{-4}$ | $9.081 \times 10^{-6}$ | 3.697 | 0.655 | 0.00280 | 0.01014 |
| 15 | 488.7 | 1244 | 23.75 | 186.6 | 1390 | 969.4 | 0.0880 | 0.01357 | $2.281 \times 10^{-4}$ | $9.915 \times 10^{-6}$ | 3.604 | 0.708 | 0.00293 | 0.00944 |
| 20 | 572.1 | 1226 | 27.77 | 182.3 | 1408 | 997.6 | 0.0856 | 0.01406 | $2.142 \times 10^{-4}$ | $1.075 \times 10^{-5}$ | 3.521 | 0.763 | 0.00307 | 0.00876 |
| 25 | 665.8 | 1207 | 32.34 | 177.8 | 1427 | 1028 | 0.0833 | 0.01456 | $2.012 \times 10^{-4}$ | $1.160 \times 10^{-5}$ | 3.448 | 0.819 | 0.00324 | 0.00808 |
| 30 | 770.6 | 1188 | 37.53 | 173.1 | 1448 | 1061 | 0.0808 | 0.01507 | $1.888 \times 10^{-4}$ | $1.244 \times 10^{-5}$ | 3.383 | 0.877 | 0.00342 | 0.00742 |
| 35 | 887.5 | 1168 | 43.41 | 168.2 | 1471 | 1098 | 0.0783 | 0.01558 | $1.772 \times 10^{-4}$ | $1.327 \times 10^{-5}$ | 3.328 | 0.935 | 0.00364 | 0.00677 |
| 40 | 1017.1 | 1147 | 50.08 | 163.0 | 1498 | 1138 | 0.0757 | 0.01610 | $1.660 \times 10^{-4}$ | $1.408 \times 10^{-5}$ | 3.285 | 0.995 | 0.00390 | 0.00613 |
| 45 | 1160.5 | 1125 | 57.66 | 157.6 | 1529 | 1184 | 0.0731 | 0.01664 | $1.554 \times 10^{-4}$ | $1.486 \times 10^{-5}$ | 3.253 | 1.058 | 0.00420 | 0.00550 |
| 50 | 1318.6 | 1102 | 66.27 | 151.8 | 1566 | 1237 | 0.0704 | 0.01720 | $1.453 \times 10^{-4}$ | $1.562 \times 10^{-5}$ | 3.231 | 1.123 | 0.00455 | 0.00489 |
| 55 | 1492.3 | 1078 | 76.11 | 145.7 | 1608 | 1298 | 0.0676 | 0.01777 | $1.355 \times 10^{-4}$ | $1.634 \times 10^{-5}$ | 3.223 | 1.193 | 0.00500 | 0.00429 |
| 60 | 1682.8 | 1053 | 87.38 | 139.1 | 1659 | 1372 | 0.0647 | 0.01838 | $1.260 \times 10^{-4}$ | $1.704 \times 10^{-5}$ | 3.229 | 1.272 | 0.00554 | 0.00372 |
| 65 | 1891.0 | 1026 | 100.4 | 132.1 | 1722 | 1462 | 0.0618 | 0.01902 | $1.167 \times 10^{-4}$ | $1.771 \times 10^{-5}$ | 3.255 | 1.362 | 0.00624 | 0.00315 |
| 70 | 2118.2 | 996.2 | 115.6 | 124.4 | 1801 | 1577 | 0.0587 | 0.01972 | $1.077 \times 10^{-4}$ | $1.839 \times 10^{-5}$ | 3.307 | 1.471 | 0.00716 | 0.00261 |
| 75 | 2365.8 | 964 | 133.6 | 115.9 | 1907 | 1731 | 0.0555 | 0.02048 | $9.891 \times 10^{-5}$ | $1.908 \times 10^{-5}$ | 3.400 | 1.612 | 0.00843 | 0.00209 |
| 80 | 2635.2 | 928.2 | 155.3 | 106.4 | 2056 | 1948 | 0.0521 | 0.02133 | $9.011 \times 10^{-5}$ | $1.982 \times 10^{-5}$ | 3.558 | 1.810 | 0.01031 | 0.00160 |
| 85 | 2928.2 | 887.1 | 182.3 | 95.4 | 2287 | 2281 | 0.0484 | 0.02233 | $8.124 \times 10^{-5}$ | $2.071 \times 10^{-5}$ | 3.837 | 2.116 | 0.01336 | 0.00114 |
| 90 | 3246.9 | 837.7 | 217.8 | 82.2 | 2701 | 2865 | 0.0444 | 0.02357 | $7.203 \times 10^{-5}$ | $2.187 \times 10^{-5}$ | 4.385 | 2.658 | 0.01911 | 0.00071 |
| 95 | 3594.1 | 772.5 | 269.3 | 64.9 | 3675 | 4144 | 0.0396 | 0.02544 | $6.190 \times 10^{-5}$ | $2.370 \times 10^{-5}$ | 5.746 | 3.862 | 0.03343 | 0.00033 |
| 100 | 3975.1 | 651.7 | 376.3 | 33.9 | 7959 | 8785 | 0.0322 | 0.02989 | $4.765 \times 10^{-5}$ | $2.833 \times 10^{-5}$ | 11.77 | 8.326 | 0.10047 | 0.00004 |

*Note 1*: Kinematic viscosity $\nu$ and thermal diffusivity $\alpha$ can be calculated from their definitions, $\nu = \mu/\rho$ and $\alpha = k/\rho c_p = \nu/Pr$. The properties listed here (except the vapor density) can be used at any pressures with negligible error except at temperatures near the critical-point value.

*Note 2*: The unit kJ/kg·°C for specific heat is equivalent to kJ/kg·K, and the unit W/m·°C for thermal conductivity is equivalent to W/m·K.

*Source*: Data generated from the EES software developed by S. A. Klein and F. L. Alvarado. Original sources: R. Tillner-Roth and H. D. Baehr, "An International Standard Formulation for the Thermodynamic Properties of 1,1,1,2-Tetrafluoroethane (HFC-134a) for Temperatures from 170 K to 455 K and Pressures up to 70 MPa," *J. Phys. Chem, Ref. Data*, Vol. 23, No. 5, 1994; M.J. Assael, N. K. Dalaouti, A. A. Griva, and J. H. Dymond, "Viscosity and Thermal Conductivity of Halogenated Methane and Ethane Refrigerants," *IJR*, Vol. 22, pp. 525–535, 1999; NIST REFPROP 6 program (M. O. McLinden, S. A. Klein, E. W. Lemmon, and A. P. Peskin, Physical and Chemical Properties Division, National Institute of Standards and Technology, Boulder, CO 80303, 1995).

## TABLE A–17

Properties of saturated ammonia

| Temp. $T$, °C | Saturation Pressure $P$, kPa | Density $\rho$, kg/m³ Liquid | Density $\rho$, kg/m³ Vapor | Enthalpy of Vaporization $h_{fg}$, kJ/kg | Specific Heat $c_p$, J/kg·K Liquid | Specific Heat $c_p$, J/kg·K Vapor | Thermal Conductivity $k$, W/m·K Liquid | Thermal Conductivity $k$, W/m·K Vapor | Dynamic Viscosity $\mu$, kg/m·s Liquid | Dynamic Viscosity $\mu$, kg/m·s Vapor | Prandtl Number Pr Liquid | Prandtl Number Pr Vapor | Volume Expansion Coefficient $\beta$, I/K Liquid | Surface Tension, N/m |
|---|---|---|---|---|---|---|---|---|---|---|---|---|---|---|
| −40 | 71.66 | 690.2 | 0.6435 | 1389 | 4414 | 2242 | — | 0.01792 | $2.926 \times 10^{-4}$ | $7.957 \times 10^{-6}$ | — | 0.9955 | 0.00176 | 0.03565 |
| −30 | 119.4 | 677.8 | 1.037 | 1360 | 4465 | 2322 | — | 0.01898 | $2.630 \times 10^{-4}$ | $8.311 \times 10^{-6}$ | — | 1.017 | 0.00185 | 0.03341 |
| −25 | 151.5 | 671.5 | 1.296 | 1345 | 4489 | 2369 | 0.5968 | 0.01957 | $2.492 \times 10^{-4}$ | $8.490 \times 10^{-6}$ | 1.875 | 1.028 | 0.00190 | 0.03229 |
| −20 | 190.1 | 665.1 | 1.603 | 1329 | 4514 | 2420 | 0.5853 | 0.02015 | $2.361 \times 10^{-4}$ | $8.669 \times 10^{-6}$ | 1.821 | 1.041 | 0.00194 | 0.03118 |
| −15 | 236.2 | 658.6 | 1.966 | 1313 | 4538 | 2476 | 0.5737 | 0.02075 | $2.236 \times 10^{-4}$ | $8.851 \times 10^{-6}$ | 1.769 | 1.056 | 0.00199 | 0.03007 |
| −10 | 290.8 | 652.1 | 2.391 | 1297 | 4564 | 2536 | 0.5621 | 0.02138 | $2.117 \times 10^{-4}$ | $9.034 \times 10^{-6}$ | 1.718 | 1.072 | 0.00205 | 0.02896 |
| −5 | 354.9 | 645.4 | 2.886 | 1280 | 4589 | 2601 | 0.5505 | 0.02203 | $2.003 \times 10^{-4}$ | $9.218 \times 10^{-6}$ | 1.670 | 1.089 | 0.00210 | 0.02786 |
| 0 | 429.6 | 638.6 | 3.458 | 1262 | 4617 | 2672 | 0.5390 | 0.02270 | $1.896 \times 10^{-4}$ | $9.405 \times 10^{-6}$ | 1.624 | 1.107 | 0.00216 | 0.02676 |
| 5 | 516 | 631.7 | 4.116 | 1244 | 4645 | 2749 | 0.5274 | 0.02341 | $1.794 \times 10^{-4}$ | $9.593 \times 10^{-6}$ | 1.580 | 1.126 | 0.00223 | 0.02566 |
| 10 | 615.3 | 624.6 | 4.870 | 1226 | 4676 | 2831 | 0.5158 | 0.02415 | $1.697 \times 10^{-4}$ | $9.784 \times 10^{-6}$ | 1.539 | 1.147 | 0.00230 | 0.02457 |
| 15 | 728.8 | 617.5 | 5.729 | 1206 | 4709 | 2920 | 0.5042 | 0.02492 | $1.606 \times 10^{-4}$ | $9.978 \times 10^{-6}$ | 1.500 | 1.169 | 0.00237 | 0.02348 |
| 20 | 857.8 | 610.2 | 6.705 | 1186 | 4745 | 3016 | 0.4927 | 0.02573 | $1.519 \times 10^{-4}$ | $1.017 \times 10^{-5}$ | 1.463 | 1.193 | 0.00245 | 0.02240 |
| 25 | 1003 | 602.8 | 7.809 | 1166 | 4784 | 3120 | 0.4811 | 0.02658 | $1.438 \times 10^{-4}$ | $1.037 \times 10^{-5}$ | 1.430 | 1.218 | 0.00254 | 0.02132 |
| 30 | 1167 | 595.2 | 9.055 | 1144 | 4828 | 3232 | 0.4695 | 0.02748 | $1.361 \times 10^{-4}$ | $1.057 \times 10^{-5}$ | 1.399 | 1.244 | 0.00264 | 0.02024 |
| 35 | 1351 | 587.4 | 10.46 | 1122 | 4877 | 3354 | 0.4579 | 0.02843 | $1.288 \times 10^{-4}$ | $1.078 \times 10^{-5}$ | 1.372 | 1.272 | 0.00275 | 0.01917 |
| 40 | 1555 | 579.4 | 12.03 | 1099 | 4932 | 3486 | 0.4464 | 0.02943 | $1.219 \times 10^{-4}$ | $1.099 \times 10^{-5}$ | 1.347 | 1.303 | 0.00287 | 0.01810 |
| 45 | 1782 | 571.3 | 13.8 | 1075 | 4993 | 3631 | 0.4348 | 0.03049 | $1.155 \times 10^{-4}$ | $1.121 \times 10^{-5}$ | 1.327 | 1.335 | 0.00301 | 0.01704 |
| 50 | 2033 | 562.9 | 15.78 | 1051 | 5063 | 3790 | 0.4232 | 0.03162 | $1.094 \times 10^{-4}$ | $1.143 \times 10^{-5}$ | 1.310 | 1.371 | 0.00316 | 0.01598 |
| 55 | 2310 | 554.2 | 18.00 | 1025 | 5143 | 3967 | 0.4116 | 0.03283 | $1.037 \times 10^{-4}$ | $1.166 \times 10^{-5}$ | 1.297 | 1.409 | 0.00334 | 0.01493 |
| 60 | 2614 | 545.2 | 20.48 | 997.4 | 5234 | 4163 | 0.4001 | 0.03412 | $9.846 \times 10^{-5}$ | $1.189 \times 10^{-5}$ | 1.288 | 1.452 | 0.00354 | 0.01389 |
| 65 | 2948 | 536.0 | 23.26 | 968.9 | 5340 | 4384 | 0.3885 | 0.03550 | $9.347 \times 10^{-5}$ | $1.213 \times 10^{-5}$ | 1.285 | 1.499 | 0.00377 | 0.01285 |
| 70 | 3312 | 526.3 | 26.39 | 939.0 | 5463 | 4634 | 0.3769 | 0.03700 | $8.879 \times 10^{-5}$ | $1.238 \times 10^{-5}$ | 1.287 | 1.551 | 0.00404 | 0.01181 |
| 75 | 3709 | 516.2 | 29.90 | 907.5 | 5608 | 4923 | 0.3653 | 0.03862 | $8.440 \times 10^{-5}$ | $1.264 \times 10^{-5}$ | 1.296 | 1.612 | 0.00436 | 0.01079 |
| 80 | 4141 | 505.7 | 33.87 | 874.1 | 5780 | 5260 | 0.3538 | 0.04038 | $8.030 \times 10^{-5}$ | $1.292 \times 10^{-5}$ | 1.312 | 1.683 | 0.00474 | 0.00977 |
| 85 | 4609 | 494.5 | 38.36 | 838.6 | 5988 | 5659 | 0.3422 | 0.04232 | $7.646 \times 10^{-5}$ | $1.322 \times 10^{-5}$ | 1.338 | 1.768 | 0.00521 | 0.00876 |
| 90 | 5116 | 482.8 | 43.48 | 800.6 | 6242 | 6142 | 0.3306 | 0.04447 | $7.284 \times 10^{-5}$ | $1.354 \times 10^{-5}$ | 1.375 | 1.871 | 0.00579 | 0.00776 |
| 95 | 5665 | 470.2 | 49.35 | 759.8 | 6561 | 6740 | 0.3190 | 0.04687 | $6.946 \times 10^{-5}$ | $1.389 \times 10^{-5}$ | 1.429 | 1.999 | 0.00652 | 0.00677 |
| 100 | 6257 | 456.6 | 56.15 | 715.5 | 6972 | 7503 | 0.3075 | 0.04958 | $6.628 \times 10^{-5}$ | $1.429 \times 10^{-5}$ | 1.503 | 2.163 | 0.00749 | 0.00579 |

*Note 1*: Kinematic viscosity $\nu$ and thermal diffusivity $\alpha$ can be calculated from their definitions, $\nu = \mu/\rho$ and $\alpha = k/\rho c_p = \nu/\text{Pr}$. The properties listed here (except the vapor density) can be used at any pressures with negligible error except at temperatures near the critical-point value.

*Note 2*: The unit kJ/kg·°C for specific heat is equivalent to kJ/kg·K, and the unit W/m·°C for thermal conductivity is equivalent to W/m·K.

*Source*: Data generated from the EES software developed by S. A. Klein and F. L. Alvarado. Original sources: Tillner-Roth, Harms-Watzenberg, and Baehr, "Eine neue Fundamentalgleichung fur Ammoniak," DKV-Tagungsbericht 20:167–181, 1993; Liley and Desai, "Thermophysical Properties of Refrigerants," ASHRAE, 1993, ISBN 1-1883413-10-9.

## TABLE A–18

### Properties of saturated propane

| Temp. $T$, °C | Saturation Pressure $P$, kPa | Density $\rho$, kg/m³ | | Enthalpy of Vaporization $h_{fg}$, kJ/kg | Specific Heat $c_p$, J/kg·K | | Thermal Conductivity $k$, W/m·K | | Dynamic Viscosity $\mu$, kg/m·s | | Prandtl Number Pr | | Volume Expansion Coefficient $\beta$, 1/K | Surface Tension, N/m |
|---|---|---|---|---|---|---|---|---|---|---|---|---|---|---|
| | | Liquid | Vapor | | Liquid | Vapor | Liquid | Vapor | Liquid | Vapor | Liquid | Vapor | Liquid | |
| −120 | 0.4053 | 664.7 | 0.01408 | 498.3 | 2003 | 1115 | 0.1802 | 0.00589 | $6.136 \times 10^{-4}$ | $4.372 \times 10^{-6}$ | 6.820 | 0.827 | 0.00153 | 0.02630 |
| −110 | 1.157 | 654.5 | 0.03776 | 489.3 | 2021 | 1148 | 0.1738 | 0.00645 | $5.054 \times 10^{-4}$ | $4.625 \times 10^{-6}$ | 5.878 | 0.822 | 0.00157 | 0.02486 |
| −100 | 2.881 | 644.2 | 0.08872 | 480.4 | 2044 | 1183 | 0.1672 | 0.00705 | $4.252 \times 10^{-4}$ | $4.881 \times 10^{-6}$ | 5.195 | 0.819 | 0.00161 | 0.02344 |
| −90 | 6.406 | 633.8 | 0.1870 | 471.5 | 2070 | 1221 | 0.1606 | 0.00769 | $3.635 \times 10^{-4}$ | $5.143 \times 10^{-6}$ | 4.686 | 0.817 | 0.00166 | 0.02202 |
| −80 | 12.97 | 623.2 | 0.3602 | 462.4 | 2100 | 1263 | 0.1539 | 0.00836 | $3.149 \times 10^{-4}$ | $5.409 \times 10^{-6}$ | 4.297 | 0.817 | 0.00171 | 0.02062 |
| −70 | 24.26 | 612.5 | 0.6439 | 453.1 | 2134 | 1308 | 0.1472 | 0.00908 | $2.755 \times 10^{-4}$ | $5.680 \times 10^{-6}$ | 3.994 | 0.818 | 0.00177 | 0.01923 |
| −60 | 42.46 | 601.5 | 1.081 | 443.5 | 2173 | 1358 | 0.1407 | 0.00985 | $2.430 \times 10^{-4}$ | $5.956 \times 10^{-6}$ | 3.755 | 0.821 | 0.00184 | 0.01785 |
| −50 | 70.24 | 590.3 | 1.724 | 433.6 | 2217 | 1412 | 0.1343 | 0.01067 | $2.158 \times 10^{-4}$ | $6.239 \times 10^{-6}$ | 3.563 | 0.825 | 0.00192 | 0.01649 |
| −40 | 110.7 | 578.8 | 2.629 | 423.1 | 2258 | 1471 | 0.1281 | 0.01155 | $1.926 \times 10^{-4}$ | $6.529 \times 10^{-6}$ | 3.395 | 0.831 | 0.00201 | 0.01515 |
| −30 | 167.3 | 567.0 | 3.864 | 412.1 | 2310 | 1535 | 0.1221 | 0.01250 | $1.726 \times 10^{-4}$ | $6.827 \times 10^{-6}$ | 3.266 | 0.839 | 0.00213 | 0.01382 |
| −20 | 243.8 | 554.7 | 5.503 | 400.3 | 2368 | 1605 | 0.1163 | 0.01351 | $1.551 \times 10^{-4}$ | $7.136 \times 10^{-6}$ | 3.158 | 0.848 | 0.00226 | 0.01251 |
| −10 | 344.4 | 542.0 | 7.635 | 387.8 | 2433 | 1682 | 0.1107 | 0.01459 | $1.397 \times 10^{-4}$ | $7.457 \times 10^{-6}$ | 3.069 | 0.860 | 0.00242 | 0.01122 |
| 0 | 473.3 | 528.7 | 10.36 | 374.2 | 2507 | 1768 | 0.1054 | 0.01576 | $1.259 \times 10^{-4}$ | $7.794 \times 10^{-6}$ | 2.996 | 0.875 | 0.00262 | 0.00996 |
| 5 | 549.8 | 521.8 | 11.99 | 367.0 | 2547 | 1814 | 0.1028 | 0.01637 | $1.195 \times 10^{-4}$ | $7.970 \times 10^{-6}$ | 2.964 | 0.883 | 0.00273 | 0.00934 |
| 10 | 635.1 | 514.7 | 13.81 | 359.5 | 2590 | 1864 | 0.1002 | 0.01701 | $1.135 \times 10^{-4}$ | $8.151 \times 10^{-6}$ | 2.935 | 0.893 | 0.00286 | 0.00872 |
| 15 | 729.8 | 507.5 | 15.85 | 351.7 | 2637 | 1917 | 0.0977 | 0.01767 | $1.077 \times 10^{-4}$ | $8.339 \times 10^{-6}$ | 2.909 | 0.905 | 0.00301 | 0.00811 |
| 20 | 834.4 | 500.0 | 18.13 | 343.4 | 2688 | 1974 | 0.0952 | 0.01836 | $1.022 \times 10^{-4}$ | $8.534 \times 10^{-6}$ | 2.886 | 0.918 | 0.00318 | 0.00751 |
| 25 | 949.7 | 492.2 | 20.68 | 334.8 | 2742 | 2036 | 0.0928 | 0.01908 | $9.702 \times 10^{-5}$ | $8.738 \times 10^{-6}$ | 2.866 | 0.933 | 0.00337 | 0.00691 |
| 30 | 1076 | 484.2 | 23.53 | 325.8 | 2802 | 2104 | 0.0904 | 0.01982 | $9.197 \times 10^{-5}$ | $8.952 \times 10^{-6}$ | 2.850 | 0.950 | 0.00358 | 0.00633 |
| 35 | 1215 | 475.8 | 26.72 | 316.2 | 2869 | 2179 | 0.0881 | 0.02061 | $8.710 \times 10^{-5}$ | $9.178 \times 10^{-6}$ | 2.837 | 0.971 | 0.00384 | 0.00575 |
| 40 | 1366 | 467.1 | 30.29 | 306.1 | 2943 | 2264 | 0.0857 | 0.02142 | $8.240 \times 10^{-5}$ | $9.417 \times 10^{-6}$ | 2.828 | 0.995 | 0.00413 | 0.00518 |
| 45 | 1530 | 458.0 | 34.29 | 295.3 | 3026 | 2361 | 0.0834 | 0.02228 | $7.785 \times 10^{-5}$ | $9.674 \times 10^{-6}$ | 2.824 | 1.025 | 0.00448 | 0.00463 |
| 50 | 1708 | 448.5 | 38.79 | 283.9 | 3122 | 2473 | 0.0811 | 0.02319 | $7.343 \times 10^{-5}$ | $9.950 \times 10^{-5}$ | 2.826 | 1.061 | 0.00491 | 0.00408 |
| 60 | 2110 | 427.5 | 49.66 | 258.4 | 3283 | 2769 | 0.0765 | 0.02517 | $6.487 \times 10^{-5}$ | $1.058 \times 10^{-5}$ | 2.784 | 1.164 | 0.00609 | 0.00303 |
| 70 | 2580 | 403.2 | 64.02 | 228.0 | 3595 | 3241 | 0.0717 | 0.02746 | $5.649 \times 10^{-5}$ | $1.138 \times 10^{-5}$ | 2.834 | 1.343 | 0.00811 | 0.00204 |
| 80 | 3127 | 373.0 | 84.28 | 189.7 | 4501 | 4173 | 0.0663 | 0.03029 | $4.790 \times 10^{-5}$ | $1.249 \times 10^{-5}$ | 3.251 | 1.722 | 0.01248 | 0.00114 |
| 90 | 3769 | 329.1 | 118.6 | 133.2 | 6977 | 7239 | 0.0595 | 0.03441 | $3.807 \times 10^{-5}$ | $1.448 \times 10^{-5}$ | 4.465 | 3.047 | 0.02847 | 0.00037 |

*Note 1*: Kinematic viscosity $\nu$ and thermal diffusivity $\alpha$ can be calculated from their definitions, $\nu = \mu/\rho$ and $\alpha = k/\mu c_p = \nu/\text{Pr}$. The properties listed here (except the vapor density) can be used at any pressures with negligible error except at temperatures near the critical-point value.

*Note 2*: The unit kJ/kg·°C for specific heat is equivalent to kJ/kg·K, and the unit W/m·°C for thermal conductivity is equivalent to W/m·K.

*Source*: Data generated from the EES software developed by S. A. Klein and F. L. Alvarado. Original sources: Reiner Tillner-Roth, "Fundamental Equations of State," Shaker, Verlag, Aachan, 1998; B. A. Younglove and J. F. Ely, "Thermophysical Properties of Fluids. II Methane, Ethane, Propane, Isobutane, and Normal Butane," *J. Phys. Chem. Ref. Data*, Vol. 16, No. 4, 1987; G.R. Somayajulu, "A Generalized Equation for Surface Tension from the Triple-Point to the Critical-Point," *International Journal of Thermophysics*, Vol. 9, No. 4, 1988.

## TABLE A–19

Properties of liquids

| Temp. $T$, °C | Density $\rho$, kg/m³ | Specific Heat $c_p$, J/kg·K | Thermal Conductivity $k$, W/m·K | Thermal Diffusivity $\alpha$, m²/s | Dynamic Viscosity $\mu$, kg/m·s | Kinematic Viscosity $\nu$, m²/s | Prandtl Number Pr | Volume Expansion Coeff. $\beta$, 1/K |
|---|---|---|---|---|---|---|---|---|
| | | | | Methane [$CH_4$] | | | | |
| −160 | 420.2 | 3492 | 0.1863 | $1.270 \times 10^{-7}$ | $1.133 \times 10^{-4}$ | $2.699 \times 10^{-7}$ | 2.126 | 0.00352 |
| −150 | 405.0 | 3580 | 0.1703 | $1.174 \times 10^{-7}$ | $9.169 \times 10^{-5}$ | $2.264 \times 10^{-7}$ | 1.927 | 0.00391 |
| −140 | 388.8 | 3700 | 0.1550 | $1.077 \times 10^{-7}$ | $7.551 \times 10^{-5}$ | $1.942 \times 10^{-7}$ | 1.803 | 0.00444 |
| −130 | 371.1 | 3875 | 0.1402 | $9.749 \times 10^{-8}$ | $6.288 \times 10^{-5}$ | $1.694 \times 10^{-7}$ | 1.738 | 0.00520 |
| −120 | 351.4 | 4146 | 0.1258 | $8.634 \times 10^{-8}$ | $5.257 \times 10^{-5}$ | $1.496 \times 10^{-7}$ | 1.732 | 0.00637 |
| −110 | 328.8 | 4611 | 0.1115 | $7.356 \times 10^{-8}$ | $4.377 \times 10^{-5}$ | $1.331 \times 10^{-7}$ | 1.810 | 0.00841 |
| −100 | 301.0 | 5578 | 0.0967 | $5.761 \times 10^{-8}$ | $3.577 \times 10^{-5}$ | $1.188 \times 10^{-7}$ | 2.063 | 0.01282 |
| −90 | 261.7 | 8902 | 0.0797 | $3.423 \times 10^{-8}$ | $2.761 \times 10^{-5}$ | $1.055 \times 10^{-7}$ | 3.082 | 0.02922 |
| | | | | Methanol [$CH_3(OH)$] | | | | |
| 20 | 788.4 | 2515 | 0.1987 | $1.002 \times 10^{-7}$ | $5.857 \times 10^{-4}$ | $7.429 \times 10^{-7}$ | 7.414 | 0.00118 |
| 30 | 779.1 | 2577 | 0.1980 | $9.862 \times 10^{-8}$ | $5.088 \times 10^{-4}$ | $6.531 \times 10^{-7}$ | 6.622 | 0.00120 |
| 40 | 769.6 | 2644 | 0.1972 | $9.690 \times 10^{-8}$ | $4.460 \times 10^{-4}$ | $5.795 \times 10^{-7}$ | 5.980 | 0.00123 |
| 50 | 760.1 | 2718 | 0.1965 | $9.509 \times 10^{-8}$ | $3.942 \times 10^{-4}$ | $5.185 \times 10^{-7}$ | 5.453 | 0.00127 |
| 60 | 750.4 | 2798 | 0.1957 | $9.320 \times 10^{-8}$ | $3.510 \times 10^{-4}$ | $4.677 \times 10^{-7}$ | 5.018 | 0.00132 |
| 70 | 740.4 | 2885 | 0.1950 | $9.128 \times 10^{-8}$ | $3.146 \times 10^{-4}$ | $4.250 \times 10^{-7}$ | 4.655 | 0.00137 |
| | | | | Isobutane (R600a) | | | | |
| −100 | 683.8 | 1881 | 0.1383 | $1.075 \times 10^{-7}$ | $9.305 \times 10^{-4}$ | $1.360 \times 10^{-6}$ | 12.65 | 0.00142 |
| −75 | 659.3 | 1970 | 0.1357 | $1.044 \times 10^{-7}$ | $5.624 \times 10^{-4}$ | $8.531 \times 10^{-7}$ | 8.167 | 0.00150 |
| −50 | 634.3 | 2069 | 0.1283 | $9.773 \times 10^{-8}$ | $3.769 \times 10^{-4}$ | $5.942 \times 10^{-7}$ | 6.079 | 0.00161 |
| −25 | 608.2 | 2180 | 0.1181 | $8.906 \times 10^{-8}$ | $2.688 \times 10^{-4}$ | $4.420 \times 10^{-7}$ | 4.963 | 0.00177 |
| 0 | 580.6 | 2306 | 0.1068 | $7.974 \times 10^{-8}$ | $1.993 \times 10^{-4}$ | $3.432 \times 10^{-7}$ | 4.304 | 0.00199 |
| 25 | 550.7 | 2455 | 0.0956 | $7.069 \times 10^{-8}$ | $1.510 \times 10^{-4}$ | $2.743 \times 10^{-7}$ | 3.880 | 0.00232 |
| 50 | 517.3 | 2640 | 0.0851 | $6.233 \times 10^{-8}$ | $1.155 \times 10^{-4}$ | $2.233 \times 10^{-7}$ | 3.582 | 0.00286 |
| 75 | 478.5 | 2896 | 0.0757 | $5.460 \times 10^{-8}$ | $8.785 \times 10^{-5}$ | $1.836 \times 10^{-7}$ | 3.363 | 0.00385 |
| 100 | 429.6 | 3361 | 0.0669 | $4.634 \times 10^{-8}$ | $6.483 \times 10^{-5}$ | $1.509 \times 10^{-7}$ | 3.256 | 0.00628 |
| | | | | Glycerin | | | | |
| 0 | 1276 | 2262 | 0.2820 | $9.773 \times 10^{-8}$ | 10.49 | $8.219 \times 10^{-3}$ | 84,101 | |
| 5 | 1273 | 2288 | 0.2835 | $9.732 \times 10^{-8}$ | 6.730 | $5.287 \times 10^{-3}$ | 54,327 | |
| 10 | 1270 | 2320 | 0.2846 | $9.662 \times 10^{-8}$ | 4.241 | $3.339 \times 10^{-3}$ | 34,561 | |
| 15 | 1267 | 2354 | 0.2856 | $9.576 \times 10^{-8}$ | 2.496 | $1.970 \times 10^{-3}$ | 20,570 | |
| 20 | 1264 | 2386 | 0.2860 | $9.484 \times 10^{-8}$ | 1.519 | $1.201 \times 10^{-3}$ | 12,671 | |
| 25 | 1261 | 2416 | 0.2860 | $9.388 \times 10^{-8}$ | 0.9934 | $7.878 \times 10^{-4}$ | 8,392 | |
| 30 | 1258 | 2447 | 0.2860 | $9.291 \times 10^{-8}$ | 0.6582 | $5.232 \times 10^{-4}$ | 5,631 | |
| 35 | 1255 | 2478 | 0.2860 | $9.195 \times 10^{-8}$ | 0.4347 | $3.464 \times 10^{-4}$ | 3,767 | |
| 40 | 1252 | 2513 | 0.2863 | $9.101 \times 10^{-8}$ | 0.3073 | $2.455 \times 10^{-4}$ | 2,697 | |
| | | | | Engine Oil (unused) | | | | |
| 0 | 899.0 | 1797 | 0.1469 | $9.097 \times 10^{-8}$ | 3.814 | $4.242 \times 10^{-3}$ | 46,636 | 0.00070 |
| 20 | 888.1 | 1881 | 0.1450 | $8.680 \times 10^{-8}$ | 0.8374 | $9.429 \times 10^{-4}$ | 10,863 | 0.00070 |
| 40 | 876.0 | 1964 | 0.1444 | $8.391 \times 10^{-8}$ | 0.2177 | $2.485 \times 10^{-4}$ | 2,962 | 0.00070 |
| 60 | 863.9 | 2048 | 0.1404 | $7.934 \times 10^{-8}$ | 0.07399 | $8.565 \times 10^{-5}$ | 1,080 | 0.00070 |
| 80 | 852.0 | 2132 | 0.1380 | $7.599 \times 10^{-8}$ | 0.03232 | $3.794 \times 10^{-5}$ | 499.3 | 0.00070 |
| 100 | 840.0 | 2220 | 0.1367 | $7.330 \times 10^{-8}$ | 0.01718 | $2.046 \times 10^{-5}$ | 279.1 | 0.00070 |
| 120 | 828.9 | 2308 | 0.1347 | $7.042 \times 10^{-8}$ | 0.01029 | $1.241 \times 10^{-5}$ | 176.3 | 0.00070 |
| 140 | 816.8 | 2395 | 0.1330 | $6.798 \times 10^{-8}$ | 0.006558 | $8.029 \times 10^{-6}$ | 118.1 | 0.00070 |
| 150 | 810.3 | 2441 | 0.1327 | $6.708 \times 10^{-8}$ | 0.005344 | $6.595 \times 10^{-6}$ | 98.31 | 0.00070 |

*Source*: Data generated from the EES software developed by S. A. Klein and F. L. Alvarado. Originally based on various sources.

## TABLE A–20

Properties of liquid metals

| Temp. $T$, °C | Density $\rho$, kg/m$^3$ | Specific Heat $c_p$, J/kg·K | Thermal Conductivity $k$, W/m·K | Thermal Diffusivity $\alpha$, m$^2$/s | Dynamic Viscosity $\mu$, kg/m·s | Kinematic Viscosity $\nu$, m$^2$/s | Prandtl Number Pr | Volume Expansion Coeff. $\beta$, 1/K |
|---|---|---|---|---|---|---|---|---|
| | | | | *Mercury (Hg) Melting Point: −39°C* | | | | |
| 0 | 13595 | 140.4 | 8.18200 | $4.287 \times 10^{-6}$ | $1.687 \times 10^{-3}$ | $1.241 \times 10^{-7}$ | 0.0289 | $1.810 \times 10^{-4}$ |
| 25 | 13534 | 139.4 | 8.51533 | $4.514 \times 10^{-6}$ | $1.534 \times 10^{-3}$ | $1.133 \times 10^{-7}$ | 0.0251 | $1.810 \times 10^{-4}$ |
| 50 | 13473 | 138.6 | 8.83632 | $4.734 \times 10^{-6}$ | $1.423 \times 10^{-3}$ | $1.056 \times 10^{-7}$ | 0.0223 | $1.810 \times 10^{-4}$ |
| 75 | 13412 | 137.8 | 9.15632 | $4.956 \times 10^{-6}$ | $1.316 \times 10^{-3}$ | $9.819 \times 10^{-8}$ | 0.0198 | $1.810 \times 10^{-4}$ |
| 100 | 13351 | 137.1 | 9.46706 | $5.170 \times 10^{-6}$ | $1.245 \times 10^{-3}$ | $9.326 \times 10^{-8}$ | 0.0180 | $1.810 \times 10^{-4}$ |
| 150 | 13231 | 136.1 | 10.07780 | $5.595 \times 10^{-6}$ | $1.126 \times 10^{-3}$ | $8.514 \times 10^{-8}$ | 0.0152 | $1.810 \times 10^{-4}$ |
| 200 | 13112 | 135.5 | 10.65465 | $5.996 \times 10^{-6}$ | $1.043 \times 10^{-3}$ | $7.959 \times 10^{-8}$ | 0.0133 | $1.815 \times 10^{-4}$ |
| 250 | 12993 | 135.3 | 11.18150 | $6.363 \times 10^{-6}$ | $9.820 \times 10^{-4}$ | $7.558 \times 10^{-8}$ | 0.0119 | $1.829 \times 10^{-4}$ |
| 300 | 12873 | 135.3 | 11.68150 | $6.705 \times 10^{-6}$ | $9.336 \times 10^{-4}$ | $7.252 \times 10^{-8}$ | 0.0108 | $1.854 \times 10^{-4}$ |
| | | | | *Bismuth (Bi) Melting Point: 271°C* | | | | |
| 350 | 9969 | 146.0 | 16.28 | $1.118 \times 10^{-5}$ | $1.540 \times 10^{-3}$ | $1.545 \times 10^{-7}$ | 0.01381 | |
| 400 | 9908 | 148.2 | 16.10 | $1.096 \times 10^{-5}$ | $1.422 \times 10^{-3}$ | $1.436 \times 10^{-7}$ | 0.01310 | |
| 500 | 9785 | 152.8 | 15.74 | $1.052 \times 10^{-5}$ | $1.188 \times 10^{-3}$ | $1.215 \times 10^{-7}$ | 0.01154 | |
| 600 | 9663 | 157.3 | 15.60 | $1.026 \times 10^{-5}$ | $1.013 \times 10^{-3}$ | $1.048 \times 10^{-7}$ | 0.01022 | |
| 700 | 9540 | 161.8 | 15.60 | $1.010 \times 10^{-5}$ | $8.736 \times 10^{-4}$ | $9.157 \times 10^{-8}$ | 0.00906 | |
| | | | | *Lead (Pb) Melting Point: 327°C* | | | | |
| 400 | 10506 | 158 | 15.97 | $9.623 \times 10^{-6}$ | $2.277 \times 10^{-3}$ | $2.167 \times 10^{-7}$ | 0.02252 | |
| 450 | 10449 | 156 | 15.74 | $9.649 \times 10^{-6}$ | $2.065 \times 10^{-3}$ | $1.976 \times 10^{-7}$ | 0.02048 | |
| 500 | 10390 | 155 | 15.54 | $9.651 \times 10^{-6}$ | $1.884 \times 10^{-3}$ | $1.814 \times 10^{-7}$ | 0.01879 | |
| 550 | 10329 | 155 | 15.39 | $9.610 \times 10^{-6}$ | $1.758 \times 10^{-3}$ | $1.702 \times 10^{-7}$ | 0.01771 | |
| 600 | 10267 | 155 | 15.23 | $9.568 \times 10^{-6}$ | $1.632 \times 10^{-3}$ | $1.589 \times 10^{-7}$ | 0.01661 | |
| 650 | 10206 | 155 | 15.07 | $9.526 \times 10^{-6}$ | $1.505 \times 10^{-3}$ | $1.475 \times 10^{-7}$ | 0.01549 | |
| 700 | 10145 | 155 | 14.91 | $9.483 \times 10^{-6}$ | $1.379 \times 10^{-3}$ | $1.360 \times 10^{-7}$ | 0.01434 | |
| | | | | *Sodium (Na) Melting Point: 98°C* | | | | |
| 100 | 927.3 | 1378 | 85.84 | $6.718 \times 10^{-5}$ | $6.892 \times 10^{-4}$ | $7.432 \times 10^{-7}$ | 0.01106 | |
| 200 | 902.5 | 1349 | 80.84 | $6.639 \times 10^{-5}$ | $5.385 \times 10^{-4}$ | $5.967 \times 10^{-7}$ | 0.008987 | |
| 300 | 877.8 | 1320 | 75.84 | $6.544 \times 10^{-5}$ | $3.878 \times 10^{-4}$ | $4.418 \times 10^{-7}$ | 0.006751 | |
| 400 | 853.0 | 1296 | 71.20 | $6.437 \times 10^{-5}$ | $2.720 \times 10^{-4}$ | $3.188 \times 10^{-7}$ | 0.004953 | |
| 500 | 828.5 | 1284 | 67.41 | $6.335 \times 10^{-5}$ | $2.411 \times 10^{-4}$ | $2.909 \times 10^{-7}$ | 0.004593 | |
| 600 | 804.0 | 1272 | 63.63 | $6.220 \times 10^{-5}$ | $2.101 \times 10^{-4}$ | $2.614 \times 10^{-7}$ | 0.004202 | |
| | | | | *Potassium (K) Melting Point: 64°C* | | | | |
| 200 | 795.2 | 790.8 | 43.99 | $6.995 \times 10^{-5}$ | $3.350 \times 10^{-4}$ | $4.213 \times 10^{-7}$ | 0.006023 | |
| 300 | 771.6 | 772.8 | 42.01 | $7.045 \times 10^{-5}$ | $2.667 \times 10^{-4}$ | $3.456 \times 10^{-7}$ | 0.004906 | |
| 400 | 748.0 | 754.8 | 40.03 | $7.090 \times 10^{-5}$ | $1.984 \times 10^{4}$ | $2.652 \times 10^{-7}$ | 0.00374 | |
| 500 | 723.9 | 750.0 | 37.81 | $6.964 \times 10^{-5}$ | $1.668 \times 10^{-4}$ | $2.304 \times 10^{-7}$ | 0.003309 | |
| 600 | 699.6 | 750.0 | 35.50 | $6.765 \times 10^{-5}$ | $1.487 \times 10^{-4}$ | $2.126 \times 10^{-7}$ | 0.003143 | |
| | | | | *Sodium-Potassium (%22Na-%78K) Melting Point: −11°C* | | | | |
| 100 | 847.3 | 944.4 | 25.64 | $3.205 \times 10^{-5}$ | $5.707 \times 10^{-4}$ | $6.736 \times 10^{-7}$ | 0.02102 | |
| 200 | 823.2 | 922.5 | 26.27 | $3.459 \times 10^{-5}$ | $4.587 \times 10^{-4}$ | $5.572 \times 10^{-7}$ | 0.01611 | |
| 300 | 799.1 | 900.6 | 26.89 | $3.736 \times 10^{-5}$ | $3.467 \times 10^{-4}$ | $4.339 \times 10^{-7}$ | 0.01161 | |
| 400 | 775.0 | 879.0 | 27.50 | $4.037 \times 10^{-5}$ | $2.357 \times 10^{-4}$ | $3.041 \times 10^{-7}$ | 0.00753 | |
| 500 | 751.5 | 880.1 | 27.89 | $4.217 \times 10^{-5}$ | $2.108 \times 10^{-4}$ | $2.805 \times 10^{-7}$ | 0.00665 | |
| 600 | 728.0 | 881.2 | 28.28 | $4.408 \times 10^{-5}$ | $1.859 \times 10^{-4}$ | $2.553 \times 10^{-7}$ | 0.00579 | |

*Source*: Data generated from the EES software developed by S. A. Klein and F. L. Alvarado. Originally based on various sources.

## TABLE A–21

Ideal-gas properties of air

| T K | h kJ/kg | $P_r$ | u kJ/kg | $v_r$ | $s°$ kJ/kg·K | T K | h kJ/kg | $P_r$ | u kJ/kg | $v_r$ | $s°$ kJ/kg·K |
|---|---|---|---|---|---|---|---|---|---|---|---|
| 200 | 199.97 | 0.3363 | 142.56 | 1707.0 | 1.29559 | 580 | 586.04 | 14.38 | 419.55 | 115.7 | 2.37348 |
| 210 | 209.97 | 0.3987 | 149.69 | 1512.0 | 1.34444 | 590 | 596.52 | 15.31 | 427.15 | 110.6 | 2.39140 |
| 220 | 219.97 | 0.4690 | 156.82 | 1346.0 | 1.39105 | 600 | 607.02 | 16.28 | 434.78 | 105.8 | 2.40902 |
| 230 | 230.02 | 0.5477 | 164.00 | 1205.0 | 1.43557 | 610 | 617.53 | 17.30 | 442.42 | 101.2 | 2.42644 |
| 240 | 240.02 | 0.6355 | 171.13 | 1084.0 | 1.47824 | 620 | 628.07 | 18.36 | 450.09 | 96.92 | 2.44356 |
| 250 | 250.05 | 0.7329 | 178.28 | 979.0 | 1.51917 | 630 | 638.63 | 19.84 | 457.78 | 92.84 | 2.46048 |
| 260 | 260.09 | 0.8405 | 185.45 | 887.8 | 1.55848 | 640 | 649.22 | 20.64 | 465.50 | 88.99 | 2.47716 |
| 270 | 270.11 | 0.9590 | 192.60 | 808.0 | 1.59634 | 650 | 659.84 | 21.86 | 473.25 | 85.34 | 2.49364 |
| 280 | 280.13 | 1.0889 | 199.75 | 738.0 | 1.63279 | 660 | 670.47 | 23.13 | 481.01 | 81.89 | 2.50985 |
| 285 | 285.14 | 1.1584 | 203.33 | 706.1 | 1.65055 | 670 | 681.14 | 24.46 | 488.81 | 78.61 | 2.52589 |
| 290 | 290.16 | 1.2311 | 206.91 | 676.1 | 1.66802 | 680 | 691.82 | 25.85 | 496.62 | 75.50 | 2.54175 |
| 295 | 295.17 | 1.3068 | 210.49 | 647.9 | 1.68515 | 690 | 702.52 | 27.29 | 504.45 | 72.56 | 2.55731 |
| 298 | 298.18 | 1.3543 | 212.64 | 631.9 | 1.69528 | 700 | 713.27 | 28.80 | 512.33 | 69.76 | 2.57277 |
| 300 | 300.19 | 1.3860 | 214.07 | 621.2 | 1.70203 | 710 | 724.04 | 30.38 | 520.23 | 67.07 | 2.58810 |
| 305 | 305.22 | 1.4686 | 217.67 | 596.0 | 1.71865 | 720 | 734.82 | 32.02 | 528.14 | 64.53 | 2.60319 |
| 310 | 310.24 | 1.5546 | 221.25 | 572.3 | 1.73498 | 730 | 745.62 | 33.72 | 536.07 | 62.13 | 2.61803 |
| 315 | 315.27 | 1.6442 | 224.85 | 549.8 | 1.75106 | 740 | 756.44 | 35.50 | 544.02 | 59.82 | 2.63280 |
| 320 | 320.29 | 1.7375 | 228.42 | 528.6 | 1.76690 | 750 | 767.29 | 37.35 | 551.99 | 57.63 | 2.64737 |
| 325 | 325.31 | 1.8345 | 232.02 | 508.4 | 1.78249 | 760 | 778.18 | 39.27 | 560.01 | 55.54 | 2.66176 |
| 330 | 330.34 | 1.9352 | 235.61 | 489.4 | 1.79783 | 780 | 800.03 | 43.35 | 576.12 | 51.64 | 2.69013 |
| 340 | 340.42 | 2.149 | 242.82 | 454.1 | 1.82790 | 800 | 821.95 | 47.75 | 592.30 | 48.08 | 2.71787 |
| 350 | 350.49 | 2.379 | 250.02 | 422.2 | 1.85708 | 820 | 843.98 | 52.59 | 608.59 | 44.84 | 2.74504 |
| 360 | 360.58 | 2.626 | 257.24 | 393.4 | 1.88543 | 840 | 866.08 | 57.60 | 624.95 | 41.85 | 2.77170 |
| 370 | 370.67 | 2.892 | 264.46 | 367.2 | 1.91313 | 860 | 888.27 | 63.09 | 641.40 | 39.12 | 2.79783 |
| 380 | 380.77 | 3.176 | 271.69 | 343.4 | 1.94001 | 880 | 910.56 | 68.98 | 657.95 | 36.61 | 2.82344 |
| 390 | 390.88 | 3.481 | 278.93 | 321.5 | 1.96633 | 900 | 932.93 | 75.29 | 674.58 | 34.31 | 2.84856 |
| 400 | 400.98 | 3.806 | 286.16 | 301.6 | 1.99194 | 920 | 955.38 | 82.05 | 691.28 | 32.18 | 2.87324 |
| 410 | 411.12 | 4.153 | 293.43 | 283.3 | 2.01699 | 940 | 977.92 | 89.28 | 708.08 | 30.22 | 2.89748 |
| 420 | 421.26 | 4.522 | 300.69 | 266.6 | 2.04142 | 960 | 1000.55 | 97.00 | 725.02 | 28.40 | 2.92128 |
| 430 | 431.43 | 4.915 | 307.99 | 251.1 | 2.06533 | 980 | 1023.25 | 105.2 | 741.98 | 26.73 | 2.94468 |
| 440 | 441.61 | 5.332 | 315.30 | 236.8 | 2.08870 | 1000 | 1046.04 | 114.0 | 758.94 | 25.17 | 2.96770 |
| 450 | 451.80 | 5.775 | 322.62 | 223.6 | 2.11161 | 1020 | 1068.89 | 123.4 | 776.10 | 23.72 | 2.99034 |
| 460 | 462.02 | 6.245 | 329.97 | 211.4 | 2.13407 | 1040 | 1091.85 | 133.3 | 793.36 | 23.29 | 3.01260 |
| 470 | 472.24 | 6.742 | 337.32 | 200.1 | 2.15604 | 1060 | 1114.86 | 143.9 | 810.62 | 21.14 | 3.03449 |
| 480 | 482.49 | 7.268 | 344.70 | 189.5 | 2.17760 | 1080 | 1137.89 | 155.2 | 827.88 | 19.98 | 3.05608 |
| 490 | 492.74 | 7.824 | 352.08 | 179.7 | 2.19876 | 1100 | 1161.07 | 167.1 | 845.33 | 18.896 | 3.07732 |
| 500 | 503.02 | 8.411 | 359.49 | 170.6 | 2.21952 | 1120 | 1184.28 | 179.7 | 862.79 | 17.886 | 3.09825 |
| 510 | 513.32 | 9.031 | 366.92 | 162.1 | 2.23993 | 1140 | 1207.57 | 193.1 | 880.35 | 16.946 | 3.11883 |
| 520 | 523.63 | 9.684 | 374.36 | 154.1 | 2.25997 | 1160 | 1230.92 | 207.2 | 897.91 | 16.064 | 3.13916 |
| 530 | 533.98 | 10.37 | 381.84 | 146.7 | 2.27967 | 1180 | 1254.34 | 222.2 | 915.57 | 15.241 | 3.15916 |
| 540 | 544.35 | 11.10 | 389.34 | 139.7 | 2.29906 | 1200 | 1277.79 | 238.0 | 933.33 | 14.470 | 3.17888 |
| 550 | 555.74 | 11.86 | 396.86 | 133.1 | 2.31809 | 1220 | 1301.31 | 254.7 | 951.09 | 13.747 | 3.19834 |
| 560 | 565.17 | 12.66 | 404.42 | 127.0 | 2.33685 | 1240 | 1324.93 | 272.3 | 968.95 | 13.069 | 3.21751 |
| 570 | 575.59 | 13.50 | 411.97 | 121.2 | 2.35531 | | | | | | |

## TABLE A–21

Ideal-gas properties of air (*Concluded*)

| T K | h kJ/kg | $P_r$ | u kJ/kg | $v_r$ | s° kJ/kg·K | T K | h kJ/kg | $P_r$ | u kJ/kg | $v_r$ | s° kJ/kg·K |
|---|---|---|---|---|---|---|---|---|---|---|---|
| 1260 | 1348.55 | 290.8 | 986.90 | 12.435 | 3.23638 | 1600 | 1757.57 | 791.2 | 1298.30 | 5.804 | 3.52364 |
| 1280 | 1372.24 | 310.4 | 1004.76 | 11.835 | 3.25510 | 1620 | 1782.00 | 834.1 | 1316.96 | 5.574 | 3.53879 |
| 1300 | 1395.97 | 330.9 | 1022.82 | 11.275 | 3.27345 | 1640 | 1806.46 | 878.9 | 1335.72 | 5.355 | 3.55381 |
| 1320 | 1419.76 | 352.5 | 1040.88 | 10.747 | 3.29160 | 1660 | 1830.96 | 925.6 | 1354.48 | 5.147 | 3.56867 |
| 1340 | 1443.60 | 375.3 | 1058.94 | 10.247 | 3.30959 | 1680 | 1855.50 | 974.2 | 1373.24 | 4.949 | 3.58335 |
| 1360 | 1467.49 | 399.1 | 1077.10 | 9.780 | 3.32724 | 1700 | 1880.1 | 1025 | 1392.7 | 4.761 | 3.5979 |
| 1380 | 1491.44 | 424.2 | 1095.26 | 9.337 | 3.34474 | 1750 | 1941.6 | 1161 | 1439.8 | 4.328 | 3.6336 |
| 1400 | 1515.42 | 450.5 | 1113.52 | 8.919 | 3.36200 | 1800 | 2003.3 | 1310 | 1487.2 | 3.994 | 3.6684 |
| 1420 | 1539.44 | 478.0 | 1131.77 | 8.526 | 3.37901 | 1850 | 2065.3 | 1475 | 1534.9 | 3.601 | 3.7023 |
| 1440 | 1563.51 | 506.9 | 1150.13 | 8.153 | 3.39586 | 1900 | 2127.4 | 1655 | 1582.6 | 3.295 | 3.7354 |
| 1460 | 1587.63 | 537.1 | 1168.49 | 7.801 | 3.41247 | 1950 | 2189.7 | 1852 | 1630.6 | 3.022 | 3.7677 |
| 1480 | 1611.79 | 568.8 | 1186.95 | 7.468 | 3.42892 | 2000 | 2252.1 | 2068 | 1678.7 | 2.776 | 3.7994 |
| 1500 | 1635.97 | 601.9 | 1205.41 | 7.152 | 3.44516 | 2050 | 2314.6 | 2303 | 1726.8 | 2.555 | 3.8303 |
| 1520 | 1660.23 | 636.5 | 1223.87 | 6.854 | 3.46120 | 2100 | 2377.7 | 2559 | 1775.3 | 2.356 | 3.8605 |
| 1540 | 1684.51 | 672.8 | 1242.43 | 6.569 | 3.47712 | 2150 | 2440.3 | 2837 | 1823.8 | 2.175 | 3.8901 |
| 1560 | 1708.82 | 710.5 | 1260.99 | 6.301 | 3.49276 | 2200 | 2503.2 | 3138 | 1872.4 | 2.012 | 3.9191 |
| 1580 | 1733.17 | 750.0 | 1279.65 | 6.046 | 3.50829 | 2250 | 2566.4 | 3464 | 1921.3 | 1.864 | 3.9474 |

*Note:* The properties $P_r$ (relative pressure) and $v_r$ (relative specific volume) are dimensionless quantities used in the analysis of isentropic processes, and should not be confused with the properties pressure and specific volume.

*Source:* Kenneth Wark, *Thermodynamics,* 4th ed. (New York: McGraw-Hill, 1983), pp. 785–86, table A–5. Originally published in J. H. Keenan and J. Kaye, *Gas Tables* (New York: John Wiley & Sons, 1948).

## TABLE A–22

Properties of air at 1 atm pressure

| Temp. T, °C | Density $\rho$, kg/m$^3$ | Specific Heat $c_p$, J/kg·K | Thermal Conductivity $k$, W/m·K | Thermal Diffusivity $\alpha$, m$^2$/s | Dynamic Viscosity $\mu$, kg/m·s | Kinematic Viscosity $\nu$, m$^2$/s | Prandtl Number Pr |
|---|---|---|---|---|---|---|---|
| −150 | 2.866 | 983 | 0.01171 | $4.158 \times 10^{-6}$ | $8.636 \times 10^{-6}$ | $3.013 \times 10^{-6}$ | 0.7246 |
| −100 | 2.038 | 966 | 0.01582 | $8.036 \times 10^{-6}$ | $1.189 \times 10^{-5}$ | $5.837 \times 10^{-6}$ | 0.7263 |
| −50 | 1.582 | 999 | 0.01979 | $1.252 \times 10^{-5}$ | $1.474 \times 10^{-5}$ | $9.319 \times 10^{-6}$ | 0.7440 |
| −40 | 1.514 | 1002 | 0.02057 | $1.356 \times 10^{-5}$ | $1.527 \times 10^{-5}$ | $1.008 \times 10^{-5}$ | 0.7436 |
| −30 | 1.451 | 1004 | 0.02134 | $1.465 \times 10^{-5}$ | $1.579 \times 10^{-5}$ | $1.087 \times 10^{-5}$ | 0.7425 |
| −20 | 1.394 | 1005 | 0.02211 | $1.578 \times 10^{-5}$ | $1.630 \times 10^{-5}$ | $1.169 \times 10^{-5}$ | 0.7408 |
| −10 | 1.341 | 1006 | 0.02288 | $1.696 \times 10^{-5}$ | $1.680 \times 10^{-5}$ | $1.252 \times 10^{-5}$ | 0.7387 |
| 0 | 1.292 | 1006 | 0.02364 | $1.818 \times 10^{-5}$ | $1.729 \times 10^{-5}$ | $1.338 \times 10^{-5}$ | 0.7362 |
| 5 | 1.269 | 1006 | 0.02401 | $1.880 \times 10^{-5}$ | $1.754 \times 10^{-5}$ | $1.382 \times 10^{-5}$ | 0.7350 |
| 10 | 1.246 | 1006 | 0.02439 | $1.944 \times 10^{-5}$ | $1.778 \times 10^{-5}$ | $1.426 \times 10^{-5}$ | 0.7336 |
| 15 | 1.225 | 1007 | 0.02476 | $2.009 \times 10^{-5}$ | $1.802 \times 10^{-5}$ | $1.470 \times 10^{-5}$ | 0.7323 |
| 20 | 1.204 | 1007 | 0.02514 | $2.074 \times 10^{-5}$ | $1.825 \times 10^{-5}$ | $1.516 \times 10^{-5}$ | 0.7309 |
| 25 | 1.184 | 1007 | 0.02551 | $2.141 \times 10^{-5}$ | $1.849 \times 10^{-5}$ | $1.562 \times 10^{-5}$ | 0.7296 |
| 30 | 1.164 | 1007 | 0.02588 | $2.208 \times 10^{-5}$ | $1.872 \times 10^{-5}$ | $1.608 \times 10^{-5}$ | 0.7282 |
| 35 | 1.145 | 1007 | 0.02625 | $2.277 \times 10^{-5}$ | $1.895 \times 10^{-5}$ | $1.655 \times 10^{-5}$ | 0.7268 |
| 40 | 1.127 | 1007 | 0.02662 | $2.346 \times 10^{-5}$ | $1.918 \times 10^{-5}$ | $1.702 \times 10^{-5}$ | 0.7255 |
| 45 | 1.109 | 1007 | 0.02699 | $2.416 \times 10^{-5}$ | $1.941 \times 10^{-5}$ | $1.750 \times 10^{-5}$ | 0.7241 |
| 50 | 1.092 | 1007 | 0.02735 | $2.487 \times 10^{-5}$ | $1.963 \times 10^{-5}$ | $1.798 \times 10^{-5}$ | 0.7228 |
| 60 | 1.059 | 1007 | 0.02808 | $2.632 \times 10^{-5}$ | $2.008 \times 10^{-5}$ | $1.896 \times 10^{-5}$ | 0.7202 |
| 70 | 1.028 | 1007 | 0.02881 | $2.780 \times 10^{-5}$ | $2.052 \times 10^{-5}$ | $1.995 \times 10^{-5}$ | 0.7177 |
| 80 | 0.9994 | 1008 | 0.02953 | $2.931 \times 10^{-5}$ | $2.096 \times 10^{-5}$ | $2.097 \times 10^{-5}$ | 0.7154 |
| 90 | 0.9718 | 1008 | 0.03024 | $3.086 \times 10^{-5}$ | $2.139 \times 10^{-5}$ | $2.201 \times 10^{-5}$ | 0.7132 |
| 100 | 0.9458 | 1009 | 0.03095 | $3.243 \times 10^{-5}$ | $2.181 \times 10^{-5}$ | $2.306 \times 10^{-5}$ | 0.7111 |
| 120 | 0.8977 | 1011 | 0.03235 | $3.565 \times 10^{-5}$ | $2.264 \times 10^{-5}$ | $2.522 \times 10^{-5}$ | 0.7073 |
| 140 | 0.8542 | 1013 | 0.03374 | $3.898 \times 10^{-5}$ | $2.345 \times 10^{-5}$ | $2.745 \times 10^{-5}$ | 0.7041 |
| 160 | 0.8148 | 1016 | 0.03511 | $4.241 \times 10^{-5}$ | $2.420 \times 10^{-5}$ | $2.975 \times 10^{-5}$ | 0.7014 |
| 180 | 0.7788 | 1019 | 0.03646 | $4.593 \times 10^{-5}$ | $2.504 \times 10^{-5}$ | $3.212 \times 10^{-5}$ | 0.6992 |
| 200 | 0.7459 | 1023 | 0.03779 | $4.954 \times 10^{-5}$ | $2.577 \times 10^{-5}$ | $3.455 \times 10^{-5}$ | 0.6974 |
| 250 | 0.6746 | 1033 | 0.04104 | $5.890 \times 10^{-5}$ | $2.760 \times 10^{-5}$ | $4.091 \times 10^{-5}$ | 0.6946 |
| 300 | 0.6158 | 1044 | 0.04418 | $6.871 \times 10^{-5}$ | $2.934 \times 10^{-5}$ | $4.765 \times 10^{-5}$ | 0.6935 |
| 350 | 0.5664 | 1056 | 0.04721 | $7.892 \times 10^{-5}$ | $3.101 \times 10^{-5}$ | $5.475 \times 10^{-5}$ | 0.6937 |
| 400 | 0.5243 | 1069 | 0.05015 | $8.951 \times 10^{-5}$ | $3.261 \times 10^{-5}$ | $6.219 \times 10^{-5}$ | 0.6948 |
| 450 | 0.4880 | 1081 | 0.05298 | $1.004 \times 10^{-4}$ | $3.415 \times 10^{-5}$ | $6.997 \times 10^{-5}$ | 0.6965 |
| 500 | 0.4565 | 1093 | 0.05572 | $1.117 \times 10^{-4}$ | $3.563 \times 10^{-5}$ | $7.806 \times 10^{-5}$ | 0.6986 |
| 600 | 0.4042 | 1115 | 0.06093 | $1.352 \times 10^{-4}$ | $3.846 \times 10^{-5}$ | $9.515 \times 10^{-5}$ | 0.7037 |
| 700 | 0.3627 | 1135 | 0.06581 | $1.598 \times 10^{-4}$ | $4.111 \times 10^{-5}$ | $1.133 \times 10^{-4}$ | 0.7092 |
| 800 | 0.3289 | 1153 | 0.07037 | $1.855 \times 10^{-4}$ | $4.362 \times 10^{-5}$ | $1.326 \times 10^{-4}$ | 0.7149 |
| 900 | 0.3008 | 1169 | 0.07465 | $2.122 \times 10^{-4}$ | $4.600 \times 10^{-5}$ | $1.529 \times 10^{-4}$ | 0.7206 |
| 1000 | 0.2772 | 1184 | 0.07868 | $2.398 \times 10^{-4}$ | $4.826 \times 10^{-5}$ | $1.741 \times 10^{-4}$ | 0.7260 |
| 1500 | 0.1990 | 1234 | 0.09599 | $3.908 \times 10^{-4}$ | $5.817 \times 10^{-5}$ | $2.922 \times 10^{-4}$ | 0.7478 |
| 2000 | 0.1553 | 1264 | 0.11113 | $5.664 \times 10^{-4}$ | $6.630 \times 10^{-5}$ | $4.270 \times 10^{-4}$ | 0.7539 |

*Note*: For ideal gases, the properties $c_p$, $k$, $\mu$, and Pr are independent of pressure. The properties $\rho$, $\nu$, and $\alpha$ at a pressure $P$ (in atm) other than 1 atm are determined by multiplying the values of $\rho$ at the given temperature by $P$ and by dividing $\nu$ and $\alpha$ by $P$.

*Source:* Data generated from the EES software developed by S. A. Klein and F. L. Alvarado. Original sources: Keenan, Chao, Keyes, Gas Tables, Wiley, 1984; and Thermophysical Properties of Matter. Vol. 3: Thermal Conductivity, Y. S. Touloukian, P. E. Liley, S. C. Saxena, Vol. 11: Viscosity, Y. S. Touloukian, S. C. Saxena, and P. Hestermans, IFI/Plenun, NY, 1970, ISBN 0-306067020-8.

## TABLE A–23

Properties of gases at 1 atm pressure

| Temp. $T$, °C | Density $\rho$, kg/m³ | Specific Heat $c_p$, J/kg·K | Thermal Conductivity $k$, W/m·K | Thermal Diffusivity $\alpha$, m²/s | Dynamic Viscosity $\mu$, kg/m·s | Kinematic Viscosity $\nu$, m²/s | Prandtl Number Pr |
|---|---|---|---|---|---|---|---|
| | | | Carbon Dioxide, $CO_2$ | | | | |
| −50 | 2.4035 | 746 | 0.01051 | $5.860 \times 10^{-6}$ | $1.129 \times 10^{-5}$ | $4.699 \times 10^{-6}$ | 0.8019 |
| 0 | 1.9635 | 811 | 0.01456 | $9.141 \times 10^{-6}$ | $1.375 \times 10^{-5}$ | $7.003 \times 10^{-6}$ | 0.7661 |
| 50 | 1.6597 | 866.6 | 0.01858 | $1.291 \times 10^{-5}$ | $1.612 \times 10^{-5}$ | $9.714 \times 10^{-6}$ | 0.7520 |
| 100 | 1.4373 | 914.8 | 0.02257 | $1.716 \times 10^{-5}$ | $1.841 \times 10^{-5}$ | $1.281 \times 10^{-5}$ | 0.7464 |
| 150 | 1.2675 | 957.4 | 0.02652 | $2.186 \times 10^{-5}$ | $2.063 \times 10^{-5}$ | $1.627 \times 10^{-5}$ | 0.7445 |
| 200 | 1.1336 | 995.2 | 0.03044 | $2.698 \times 10^{-5}$ | $2.276 \times 10^{-5}$ | $2.008 \times 10^{-5}$ | 0.7442 |
| 300 | 0.9358 | 1060 | 0.03814 | $3.847 \times 10^{-5}$ | $2.682 \times 10^{-5}$ | $2.866 \times 10^{-5}$ | 0.7450 |
| 400 | 0.7968 | 1112 | 0.04565 | $5.151 \times 10^{-5}$ | $3.061 \times 10^{-5}$ | $3.842 \times 10^{-5}$ | 0.7458 |
| 500 | 0.6937 | 1156 | 0.05293 | $6.600 \times 10^{-5}$ | $3.416 \times 10^{-5}$ | $4.924 \times 10^{-5}$ | 0.7460 |
| 1000 | 0.4213 | 1292 | 0.08491 | $1.560 \times 10^{-4}$ | $4.898 \times 10^{-5}$ | $1.162 \times 10^{-4}$ | 0.7455 |
| 1500 | 0.3025 | 1356 | 0.10688 | $2.606 \times 10^{-4}$ | $6.106 \times 10^{-5}$ | $2.019 \times 10^{-4}$ | 0.7745 |
| 2000 | 0.2359 | 1387 | 0.11522 | $3.521 \times 10^{-4}$ | $7.322 \times 10^{-5}$ | $3.103 \times 10^{-4}$ | 0.8815 |
| | | | Carbon Monoxide, CO | | | | |
| −50 | 1.5297 | 1081 | 0.01901 | $1.149 \times 10^{-5}$ | $1.378 \times 10^{-5}$ | $9.012 \times 10^{-6}$ | 0.7840 |
| 0 | 1.2497 | 1048 | 0.02278 | $1.739 \times 10^{-5}$ | $1.629 \times 10^{-5}$ | $1.303 \times 10^{-5}$ | 0.7499 |
| 50 | 1.0563 | 1039 | 0.02641 | $2.407 \times 10^{-5}$ | $1.863 \times 10^{-5}$ | $1.764 \times 10^{-5}$ | 0.7328 |
| 100 | 0.9148 | 1041 | 0.02992 | $3.142 \times 10^{-5}$ | $2.080 \times 10^{-5}$ | $2.274 \times 10^{-5}$ | 0.7239 |
| 150 | 0.8067 | 1049 | 0.03330 | $3.936 \times 10^{-5}$ | $2.283 \times 10^{-5}$ | $2.830 \times 10^{-5}$ | 0.7191 |
| 200 | 0.7214 | 1060 | 0.03656 | $4.782 \times 10^{-5}$ | $2.472 \times 10^{-5}$ | $3.426 \times 10^{-5}$ | 0.7164 |
| 300 | 0.5956 | 1085 | 0.04277 | $6.619 \times 10^{-5}$ | $2.812 \times 10^{-5}$ | $4.722 \times 10^{-5}$ | 0.7134 |
| 400 | 0.5071 | 1111 | 0.04860 | $8.628 \times 10^{-5}$ | $3.111 \times 10^{-5}$ | $6.136 \times 10^{-5}$ | 0.7111 |
| 500 | 0.4415 | 1135 | 0.05412 | $1.079 \times 10^{-4}$ | $3.379 \times 10^{-5}$ | $7.653 \times 10^{-5}$ | 0.7087 |
| 1000 | 0.2681 | 1226 | 0.07894 | $2.401 \times 10^{-4}$ | $4.557 \times 10^{-5}$ | $1.700 \times 10^{-4}$ | 0.7080 |
| 1500 | 0.1925 | 1279 | 0.10458 | $4.246 \times 10^{-4}$ | $6.321 \times 10^{-5}$ | $3.284 \times 10^{-4}$ | 0.7733 |
| 2000 | 0.1502 | 1309 | 0.13833 | $7.034 \times 10^{-4}$ | $9.826 \times 10^{-5}$ | $6.543 \times 10^{-4}$ | 0.9302 |
| | | | Methane, $CH_4$ | | | | |
| −50 | 0.8761 | 2243 | 0.02367 | $1.204 \times 10^{-5}$ | $8.564 \times 10^{-6}$ | $9.774 \times 10^{-6}$ | 0.8116 |
| 0 | 0.7158 | 2217 | 0.03042 | $1.917 \times 10^{-5}$ | $1.028 \times 10^{-5}$ | $1.436 \times 10^{-5}$ | 0.7494 |
| 50 | 0.6050 | 2302 | 0.03766 | $2.704 \times 10^{-5}$ | $1.191 \times 10^{-5}$ | $1.969 \times 10^{-5}$ | 0.7282 |
| 100 | 0.5240 | 2443 | 0.04534 | $3.543 \times 10^{-5}$ | $1.345 \times 10^{-5}$ | $2.567 \times 10^{-5}$ | 0.7247 |
| 150 | 0.4620 | 2611 | 0.05344 | $4.431 \times 10^{-5}$ | $1.491 \times 10^{-5}$ | $3.227 \times 10^{-5}$ | 0.7284 |
| 200 | 0.4132 | 2791 | 0.06194 | $5.370 \times 10^{-5}$ | $1.630 \times 10^{-5}$ | $3.944 \times 10^{-5}$ | 0.7344 |
| 300 | 0.3411 | 3158 | 0.07996 | $7.422 \times 10^{-5}$ | $1.886 \times 10^{-5}$ | $5.529 \times 10^{-5}$ | 0.7450 |
| 400 | 0.2904 | 3510 | 0.09918 | $9.727 \times 10^{-5}$ | $2.119 \times 10^{-5}$ | $7.297 \times 10^{-5}$ | 0.7501 |
| 500 | 0.2529 | 3836 | 0.11933 | $1.230 \times 10^{-4}$ | $2.334 \times 10^{-5}$ | $9.228 \times 10^{-5}$ | 0.7502 |
| 1000 | 0.1536 | 5042 | 0.22562 | $2.914 \times 10^{-4}$ | $3.281 \times 10^{-5}$ | $2.136 \times 10^{-4}$ | 0.7331 |
| 1500 | 0.1103 | 5701 | 0.31857 | $5.068 \times 10^{-4}$ | $4.434 \times 10^{-5}$ | $4.022 \times 10^{-4}$ | 0.7936 |
| 2000 | 0.0860 | 6001 | 0.36750 | $7.120 \times 10^{-4}$ | $6.360 \times 10^{-5}$ | $7.395 \times 10^{-4}$ | 1.0386 |
| | | | Hydrogen, $H_2$ | | | | |
| −50 | 0.11010 | 12635 | 0.1404 | $1.009 \times 10^{-4}$ | $7.293 \times 10^{-6}$ | $6.624 \times 10^{-5}$ | 0.6562 |
| 0 | 0.08995 | 13920 | 0.1652 | $1.319 \times 10^{-4}$ | $8.391 \times 10^{-6}$ | $9.329 \times 10^{-5}$ | 0.7071 |
| 50 | 0.07603 | 14349 | 0.1881 | $1.724 \times 10^{-4}$ | $9.427 \times 10^{-6}$ | $1.240 \times 10^{-4}$ | 0.7191 |
| 100 | 0.06584 | 14473 | 0.2095 | $2.199 \times 10^{-4}$ | $1.041 \times 10^{-5}$ | $1.582 \times 10^{-4}$ | 0.7196 |
| 150 | 0.05806 | 14492 | 0.2296 | $2.729 \times 10^{-4}$ | $1.136 \times 10^{-5}$ | $1.957 \times 10^{-4}$ | 0.7174 |
| 200 | 0.05193 | 14482 | 0.2486 | $3.306 \times 10^{-4}$ | $1.228 \times 10^{-5}$ | $2.365 \times 10^{-4}$ | 0.7155 |

## TABLE A–23

Properties of gases at 1 atm pressure (*Concluded*)

| Temp. T, °C | Density $\rho$, kg/m³ | Specific Heat $c_p$, J/kg·K | Thermal Conductivity $k$, W/m·K | Thermal Diffusivity $\alpha$, m²/s | Dynamic Viscosity $\mu$, kg/m·s | Kinematic Viscosity $\nu$, m²/s | Prandtl Number Pr |
|---|---|---|---|---|---|---|---|
| 300 | 0.04287 | 14481 | 0.2843 | $4.580 \times 10^{-4}$ | $1.403 \times 10^{-5}$ | $3.274 \times 10^{-4}$ | 0.7149 |
| 400 | 0.03650 | 14540 | 0.3180 | $5.992 \times 10^{-4}$ | $1.570 \times 10^{-5}$ | $4.302 \times 10^{-4}$ | 0.7179 |
| 500 | 0.03178 | 14653 | 0.3509 | $7.535 \times 10^{-4}$ | $1.730 \times 10^{-5}$ | $5.443 \times 10^{-4}$ | 0.7224 |
| 1000 | 0.01930 | 15577 | 0.5206 | $1.732 \times 10^{-3}$ | $2.455 \times 10^{-5}$ | $1.272 \times 10^{-3}$ | 0.7345 |
| 1500 | 0.01386 | 16553 | 0.6581 | $2.869 \times 10^{-3}$ | $3.099 \times 10^{-5}$ | $2.237 \times 10^{-3}$ | 0.7795 |
| 2000 | 0.01081 | 17400 | 0.5480 | $2.914 \times 10^{-3}$ | $3.690 \times 10^{-5}$ | $3.414 \times 10^{-3}$ | 1.1717 |

Nitrogen, $N_2$

| Temp. T, °C | Density $\rho$, kg/m³ | Specific Heat $c_p$, J/kg·K | Thermal Conductivity $k$, W/m·K | Thermal Diffusivity $\alpha$, m²/s | Dynamic Viscosity $\mu$, kg/m·s | Kinematic Viscosity $\nu$, m²/s | Prandtl Number Pr |
|---|---|---|---|---|---|---|---|
| −50 | 1.5299 | 957.3 | 0.02001 | $1.366 \times 10^{-5}$ | $1.390 \times 10^{-5}$ | $9.091 \times 10^{-6}$ | 0.6655 |
| 0 | 1.2498 | 1035 | 0.02384 | $1.843 \times 10^{-5}$ | $1.640 \times 10^{-5}$ | $1.312 \times 10^{-5}$ | 0.7121 |
| 50 | 1.0564 | 1042 | 0.02746 | $2.494 \times 10^{-5}$ | $1.874 \times 10^{-5}$ | $1.774 \times 10^{-5}$ | 0.7114 |
| 100 | 0.9149 | 1041 | 0.03090 | $3.244 \times 10^{-5}$ | $2.094 \times 10^{-5}$ | $2.289 \times 10^{-5}$ | 0.7056 |
| 150 | 0.8068 | 1043 | 0.03416 | $4.058 \times 10^{-5}$ | $2.300 \times 10^{-5}$ | $2.851 \times 10^{-5}$ | 0.7025 |
| 200 | 0.7215 | 1050 | 0.03727 | $4.921 \times 10^{-5}$ | $2.494 \times 10^{-5}$ | $3.457 \times 10^{-5}$ | 0.7025 |
| 300 | 0.5956 | 1070 | 0.04309 | $6.758 \times 10^{-5}$ | $2.849 \times 10^{-5}$ | $4.783 \times 10^{-5}$ | 0.7078 |
| 400 | 0.5072 | 1095 | 0.04848 | $8.727 \times 10^{-5}$ | $3.166 \times 10^{-5}$ | $6.242 \times 10^{-5}$ | 0.7153 |
| 500 | 0.4416 | 1120 | 0.05358 | $1.083 \times 10^{-4}$ | $3.451 \times 10^{-5}$ | $7.816 \times 10^{-5}$ | 0.7215 |
| 1000 | 0.2681 | 1213 | 0.07938 | $2.440 \times 10^{-4}$ | $4.594 \times 10^{-5}$ | $1.713 \times 10^{-4}$ | 0.7022 |
| 1500 | 0.1925 | 1266 | 0.11793 | $4.839 \times 10^{-4}$ | $5.562 \times 10^{-5}$ | $2.889 \times 10^{-4}$ | 0.5969 |
| 2000 | 0.1502 | 1297 | 0.18590 | $9.543 \times 10^{-4}$ | $6.426 \times 10^{-5}$ | $4.278 \times 10^{-4}$ | 0.4483 |

Oxygen, $O_2$

| Temp. T, °C | Density $\rho$, kg/m³ | Specific Heat $c_p$, J/kg·K | Thermal Conductivity $k$, W/m·K | Thermal Diffusivity $\alpha$, m²/s | Dynamic Viscosity $\mu$, kg/m·s | Kinematic Viscosity $\nu$, m²/s | Prandtl Number Pr |
|---|---|---|---|---|---|---|---|
| −50 | 1.7475 | 984.4 | 0.02067 | $1.201 \times 10^{-5}$ | $1.616 \times 10^{-5}$ | $9.246 \times 10^{-6}$ | 0.7694 |
| 0 | 1.4277 | 928.7 | 0.02472 | $1.865 \times 10^{-5}$ | $1.916 \times 10^{-5}$ | $1.342 \times 10^{-5}$ | 0.7198 |
| 50 | 1.2068 | 921.7 | 0.02867 | $2.577 \times 10^{-5}$ | $2.194 \times 10^{-5}$ | $1.818 \times 10^{-5}$ | 0.7053 |
| 100 | 1.0451 | 931.8 | 0.03254 | $3.342 \times 10^{-5}$ | $2.451 \times 10^{-5}$ | $2.346 \times 10^{-5}$ | 0.7019 |
| 150 | 0.9216 | 947.6 | 0.03637 | $4.164 \times 10^{-5}$ | $2.694 \times 10^{-5}$ | $2.923 \times 10^{-5}$ | 0.7019 |
| 200 | 0.8242 | 964.7 | 0.04014 | $5.048 \times 10^{-5}$ | $2.923 \times 10^{-5}$ | $3.546 \times 10^{-5}$ | 0.7025 |
| 300 | 0.6804 | 997.1 | 0.04751 | $7.003 \times 10^{-5}$ | $3.350 \times 10^{-5}$ | $4.923 \times 10^{-5}$ | 0.7030 |
| 400 | 0.5793 | 1025 | 0.05463 | $9.204 \times 10^{-5}$ | $3.744 \times 10^{-5}$ | $6.463 \times 10^{-5}$ | 0.7023 |
| 500 | 0.5044 | 1048 | 0.06148 | $1.163 \times 10^{-4}$ | $4.114 \times 10^{-5}$ | $8.156 \times 10^{-5}$ | 0.7010 |
| 1000 | 0.3063 | 1121 | 0.09198 | $2.678 \times 10^{-4}$ | $5.732 \times 10^{-5}$ | $1.871 \times 10^{-4}$ | 0.6986 |
| 1500 | 0.2199 | 1165 | 0.11901 | $4.643 \times 10^{-4}$ | $7.133 \times 10^{-5}$ | $3.243 \times 10^{-4}$ | 0.6985 |
| 2000 | 0.1716 | 1201 | 0.14705 | $7.139 \times 10^{-4}$ | $8.417 \times 10^{-5}$ | $4.907 \times 10^{-4}$ | 0.6873 |

Water Vapor, $H_2O$

| Temp. T, °C | Density $\rho$, kg/m³ | Specific Heat $c_p$, J/kg·K | Thermal Conductivity $k$, W/m·K | Thermal Diffusivity $\alpha$, m²/s | Dynamic Viscosity $\mu$, kg/m·s | Kinematic Viscosity $\nu$, m²/s | Prandtl Number Pr |
|---|---|---|---|---|---|---|---|
| −50 | 0.9839 | 1892 | 0.01353 | $7.271 \times 10^{-6}$ | $7.187 \times 10^{-6}$ | $7.305 \times 10^{-6}$ | 1.0047 |
| 0 | 0.8038 | 1874 | 0.01673 | $1.110 \times 10^{-5}$ | $8.956 \times 10^{-6}$ | $1.114 \times 10^{-5}$ | 1.0033 |
| 50 | 0.6794 | 1874 | 0.02032 | $1.596 \times 10^{-5}$ | $1.078 \times 10^{-5}$ | $1.587 \times 10^{-5}$ | 0.9944 |
| 100 | 0.5884 | 1887 | 0.02429 | $2.187 \times 10^{-5}$ | $1.265 \times 10^{-5}$ | $2.150 \times 10^{-5}$ | 0.9830 |
| 150 | 0.5189 | 1908 | 0.02861 | $2.890 \times 10^{-5}$ | $1.456 \times 10^{-5}$ | $2.806 \times 10^{-5}$ | 0.9712 |
| 200 | 0.4640 | 1935 | 0.03326 | $3.705 \times 10^{-5}$ | $1.650 \times 10^{-5}$ | $3.556 \times 10^{-5}$ | 0.9599 |
| 300 | 0.3831 | 1997 | 0.04345 | $5.680 \times 10^{-5}$ | $2.045 \times 10^{-5}$ | $5.340 \times 10^{-5}$ | 0.9401 |
| 400 | 0.3262 | 2066 | 0.05467 | $8.114 \times 10^{-5}$ | $2.446 \times 10^{-5}$ | $7.498 \times 10^{-5}$ | 0.9240 |
| 500 | 0.2840 | 2137 | 0.06677 | $1.100 \times 10^{-4}$ | $2.847 \times 10^{-5}$ | $1.002 \times 10^{-4}$ | 0.9108 |
| 1000 | 0.1725 | 2471 | 0.13623 | $3.196 \times 10^{-4}$ | $4.762 \times 10^{-5}$ | $2.761 \times 10^{-4}$ | 0.8639 |
| 1500 | 0.1238 | 2736 | 0.21301 | $6.288 \times 10^{-4}$ | $6.411 \times 10^{-5}$ | $5.177 \times 10^{-4}$ | 0.8233 |
| 2000 | 0.0966 | 2928 | 0.29183 | $1.032 \times 10^{-3}$ | $7.808 \times 10^{-5}$ | $8.084 \times 10^{-4}$ | 0.7833 |

*Note*: For ideal gases, the properties $c_p$, $k$, $\mu$, and Pr are independent of pressure. The properties $\rho$, $\nu$, and $\alpha$ at a pressure $P$ (in atm) other than 1 atm are determined by multiplying the values of $\rho$ at the given temperature by $P$ and by dividing $\nu$ and $\alpha$ by $P$.

*Source*: Data generated from the EES software developed by S. A. Klein and F. L. Alvarado. Originally based on various sources.

## TABLE A–24

Properties of solid metals

| Composition | Melting Point, K | Properties at 300 K | | | | Properties at Various Temperatures (K), $k(W/m\cdot K)/c_p(J/kg\cdot K)$ | | | | | |
| | | $\rho$ kg/m³ | $c_p$ J/kg·K | $k$ W/m·K | $\alpha \times 10^6$ m²/s | 100 | 200 | 400 | 600 | 800 | 1000 |
|---|---|---|---|---|---|---|---|---|---|---|---|
| **Aluminum:** | | | | | | | | | | | |
| Pure | 933 | 2702 | 903 | 237 | 97.1 | 302 | 237 | 240 | 231 | 218 | |
| | | | | | | 482 | 798 | 949 | 1033 | 1146 | |
| Alloy 2024-T6 (4.5% Cu, 1.5% Mg, 0.6% Mn) | 775 | 2770 | 875 | 177 | 73.0 | 65 | 163 | 186 | 186 | | |
| | | | | | | 473 | 787 | 925 | 1042 | | |
| Alloy 195, Cast (4.5% Cu) | | 2790 | 883 | 168 | 68.2 | | | 174 | 185 | | |
| Beryllium | 1550 | 1850 | 1825 | 200 | 59.2 | 990 | 301 | 161 | 126 | 106 | 90.8 |
| | | | | | | 203 | 1114 | 2191 | 2604 | 2823 | 3018 |
| Bismuth | 545 | 9780 | 122 | 7.86 | 6.59 | 16.5 | 9.69 | 7.04 | | | |
| | | | | | | 112 | 120 | 127 | | | |
| Boron | 2573 | 2500 | 1107 | 27.0 | 9.76 | 190 | 55.5 | 16.8 | 10.6 | 9.60 | 9.85 |
| | | | | | | 128 | 600 | 1463 | 1892 | 2160 | 2338 |
| Cadmium | 594 | 8650 | 231 | 96.8 | 48.4 | 203 | 99.3 | 94.7 | | | |
| | | | | | | 198 | 222 | 242 | | | |
| Chromium | 2118 | 7160 | 449 | 93.7 | 29.1 | 159 | 111 | 90.9 | 80.7 | 71.3 | 65.4 |
| | | | | | | 192 | 384 | 484 | 542 | 581 | 616 |
| Cobalt | 1769 | 8862 | 421 | 99.2 | 26.6 | 167 | 122 | 85.4 | 67.4 | 58.2 | 52.1 |
| | | | | | | 236 | 379 | 450 | 503 | 550 | 628 |
| **Copper:** | | | | | | | | | | | |
| Pure | 1358 | 8933 | 385 | 401 | 117 | 482 | 413 | 393 | 379 | 366 | 352 |
| | | | | | | 252 | 356 | 397 | 417 | 433 | 451 |
| Commercial bronze (90% Cu, 10% Al) | 1293 | 8800 | 420 | 52 | 14 | | 42 | 52 | 59 | | |
| | | | | | | | 785 | 160 | 545 | | |
| Phosphor gear bronze (89% Cu, 11% Sn) | 1104 | 8780 | 355 | 54 | 17 | | 41 | 65 | 74 | | |
| | | | | | | | — | — | — | | |
| Cartridge brass (70% Cu, 30% Zn) | 1188 | 8530 | 380 | 110 | 33.9 | 75 | 95 | 137 | 149 | | |
| | | | | | | | 360 | 395 | 425 | | |
| Constantan (55% Cu, 45% Ni) | 1493 | 8920 | 384 | 23 | 6.71 | 17 | 19 | | | | |
| | | | | | | 237 | 362 | | | | |
| Germanium | 1211 | 5360 | 322 | 59.9 | 34.7 | 232 | 96.8 | 43.2 | 27.3 | 19.8 | 17.4 |
| | | | | | | 190 | 290 | 337 | 348 | 357 | 375 |
| Gold | 1336 | 19,300 | 129 | 317 | 127 | 327 | 323 | 311 | 298 | 284 | 270 |
| | | | | | | 109 | 124 | 131 | 135 | 140 | 145 |
| Iridium | 2720 | 22,500 | 130 | 147 | 50.3 | 172 | 153 | 144 | 138 | 132 | 126 |
| | | | | | | 90 | 122 | 133 | 138 | 144 | 153 |
| **Iron:** | | | | | | | | | | | |
| Pure | 1810 | 7870 | 447 | 80.2 | 23.1 | 134 | 94.0 | 69.5 | 54.7 | 43.3 | 32.8 |
| | | | | | | 216 | 384 | 490 | 574 | 680 | 975 |
| Armco (99.75% pure) | | 7870 | 447 | 72.7 | 20.7 | 95.6 | 80.6 | 65.7 | 53.1 | 42.2 | 32.3 |
| | | | | | | 215 | 384 | 490 | 574 | 680 | 975 |
| **Carbon steels:** | | | | | | | | | | | |
| Plain carbon (Mn ≤ 1% Si ≤ 0.1%) | | 7854 | 434 | 60.5 | 17.7 | | | 56.7 | 48.0 | 39.2 | 30.0 |
| | | | | | | | | 487 | 559 | 685 | 1169 |
| AISI 1010 | | 7832 | 434 | 63.9 | 18.8 | | | 58.7 | 48.8 | 39.2 | 31.3 |
| | | | | | | | 487 | | 559 | 685 | 1168 |
| Carbon–silicon (Mn ≤ 1% 0.1% < Si ≤ 0.6%) | | 7817 | 446 | 51.9 | 14.9 | | | 49.8 | 44.0 | 37.4 | 29.3 |
| | | | | | | | | 501 | 582 | 699 | 971 |

## TABLE A–24

Properties of solid metals (Continued)

| Composition | Melting Point, K | Properties at 300 K | | | | Properties at Various Temperatures (K), $k$(W/m·K)/$c_p$(J/kg·K) | | | | | |
| --- | --- | --- | --- | --- | --- | --- | --- | --- | --- | --- | --- |
| | | $\rho$ kg/m³ | $c_p$ J/kg·K | $k$ W/m·K | $\alpha \times 10^6$ m²/s | 100 | 200 | 400 | 600 | 800 | 1000 |
| Carbon–manganese–silicon | | 8131 | 434 | 41.0 | 11.6 | | | 42.2 | 39.7 | 35.0 | 27.6 |
| (1% < Mn < 1.65% | | | | | | | | 487 | 559 | 685 | 1090 |
| 0.1% < Si < 0.6%) | | | | | | | | | | | |
| Chromium (low) steels: | | | | | | | | | | | |
| $\frac{1}{2}$ Cr–$\frac{1}{4}$ Mo–Si (0.18% C, | | 7822 | 444 | 37.7 | 10.9 | | | 38.2 | 36.7 | 33.3 | 26.9 |
| 0.65% Cr, 0.23% Mo, | | | | | | | | | | | |
| 0.6% Si) | | | | | | | | 492 | 575 | 688 | 969 |
| 1 Cr–$\frac{1}{2}$ Mo | | 7858 | 442 | 42.3 | 12.2 | | | 42.0 | 39.1 | 34.5 | 27.4 |
| (0.16% C, 1% Cr, | | | | | | | | | | | |
| 0.54% Mo, | | | | | | | | | | | |
| 0.39% Si) | | | | | | | | 492 | 575 | 688 | 969 |
| 1 Cr–V | | 7836 | 443 | 48.9 | 14.1 | | | 46.8 | 42.1 | 36.3 | 28.2 |
| (0.2% C, 1.02% Cr, | | | | | | | | | | | |
| 0.15% V) | | | | | | | | 492 | 575 | 688 | 969 |
| Stainless steels: | | | | | | | | | | | |
| AISI 302 | | 8055 | 480 | 15.1 | 3.91 | | | 17.3 | 20.0 | 22.8 | 25.4 |
| | | | | | | | | 512 | 559 | 585 | 606 |
| AISI 304 | 1670 | 7900 | 477 | 14.9 | 3.95 | 9.2 | 12.6 | 16.6 | 19.8 | 22.6 | 25.4 |
| | | | | | | 272 | 402 | 515 | 557 | 582 | 611 |
| AISI 316 | | 8238 | 468 | 13.4 | 3.48 | | | 15.2 | 18.3 | 21.3 | 24.2 |
| | | | | | | | | 504 | 550 | 576 | 602 |
| AISI 347 | | 7978 | 480 | 14.2 | 3.71 | | | 15.8 | 18.9 | 21.9 | 24.7 |
| | | | | | | | | 513 | 559 | 585 | 606 |
| Lead | 601 | 11,340 | 129 | 35.3 | 24.1 | 39.7 | 36.7 | 34.0 | 31.4 | | |
| | | | | | | 118 | 125 | 132 | 142 | | |
| Magnesium | 923 | 1740 | 1024 | 156 | 87.6 | 169 | 159 | 153 | 149 | 146 | |
| | | | | | | 649 | 934 | 1074 | 1170 | 1267 | |
| Molybdenum | 2894 | 10,240 | 251 | 138 | 53.7 | 179 | 143 | 134 | 126 | 118 | 112 |
| | | | | | | 141 | 224 | 261 | 275 | 285 | 295 |
| Nickel: | | | | | | | | | | | |
| Pure | 1728 | 8900 | 444 | 90.7 | 23.0 | 164 | 107 | 80.2 | 65.6 | 67.6 | 71.8 |
| | | | | | | 232 | 383 | 485 | 592 | 530 | 562 |
| Nichrome | 1672 | 8400 | 420 | 12 | 3.4 | | | 14 | 16 | 21 | |
| (80% Ni, 20% Cr) | | | | | | | | 480 | 525 | 545 | |
| Inconel X-750 | 1665 | 8510 | 439 | 11.7 | 3.1 | 8.7 | 10.3 | 13.5 | 17.0 | 20.5 | 24.0 |
| (73% Ni, 15% Cr, | | | | | | | | | | | |
| 6.7% Fe) | | | | | | — | 372 | 473 | 510 | 546 | 626 |
| Niobium | 2741 | 8570 | 265 | 53.7 | 23.6 | 55.2 | 52.6 | 55.2 | 58.2 | 61.3 | 64.4 |
| | | | | | | 188 | 249 | 274 | 283 | 292 | 301 |
| Palladium | 1827 | 12,020 | 244 | 71.8 | 24.5 | 76.5 | 71.6 | 73.6 | 79.7 | 86.9 | 94.2 |
| | | | | | | 168 | 227 | 251 | 261 | 271 | 281 |
| Platinum: | | | | | | | | | | | |
| Pure | 2045 | 21,450 | 133 | 71.6 | 25.1 | 77.5 | 72.6 | 71.8 | 73.2 | 75.6 | 78.7 |
| | | | | | | 100 | 125 | 136 | 141 | 146 | 152 |
| Alloy 60Pt–40Rh | 1800 | 16,630 | 162 | 47 | 17.4 | | | 52 | 59 | 65 | 69 |
| (60% Pt, 40% Rh) | | | | | | | | — | — | — | — |
| Rhenium | 3453 | 21,100 | 136 | 47.9 | 16.7 | 58.9 | 51.0 | 46.1 | 44.2 | 44.1 | 44.6 |
| | | | | | | 97 | 127 | 139 | 145 | 151 | 156 |
| Rhodium | 2236 | 12,450 | 243 | 150 | 49.6 | 186 | 154 | 146 | 136 | 127 | 121 |
| | | | | | | 147 | 220 | 253 | 274 | 293 | 311 |

## TABLE A–24

Properties of solid metals (*Concluded*)

| Composition | Melting Point, K | Properties at 300 K | | | | Properties at Various Temperatures (K), $k(\text{W/m·K})/c_p(\text{J/kg·K})$ | | | | | |
|---|---|---|---|---|---|---|---|---|---|---|---|
| | | $\rho$ kg/m$^3$ | $c_p$ J/kg·K | $k$ W/m·K | $\alpha \times 10^6$ m$^2$/s | 100 | 200 | 400 | 600 | 800 | 1000 |
| Silicon | 1685 | 2330 | 712 | 148 | 89.2 | 884 | 264 | 98.9 | 61.9 | 42.4 | 31.2 |
| | | | | | | 259 | 556 | 790 | 867 | 913 | 946 |
| Silver | 1235 | 10,500 | 235 | 429 | 174 | 444 | 430 | 425 | 412 | 396 | 379 |
| | | | | | | 187 | 225 | 239 | 250 | 262 | 277 |
| Tantalum | 3269 | 16,600 | 140 | 57.5 | 24.7 | 59.2 | 57.5 | 57.8 | 58.6 | 59.4 | 60.2 |
| | | | | | | 110 | 133 | 144 | 146 | 149 | 152 |
| Thorium | 2023 | 11,700 | 118 | 54.0 | 39.1 | 59.8 | 54.6 | 54.5 | 55.8 | 56.9 | 56.9 |
| | | | | | | 99 | 112 | 124 | 134 | 145 | 156 |
| Tin | 505 | 7310 | 227 | 66.6 | 40.1 | 85.2 | 73.3 | 62.2 | | | |
| | | | | | | 188 | 215 | 243 | | | |
| Titanium | 1953 | 4500 | 522 | 21.9 | 9.32 | 30.5 | 24.5 | 20.4 | 19.4 | 19.7 | 20.7 |
| | | | | | | 300 | 465 | 551 | 591 | 633 | 675 |
| Tungsten | 3660 | 19,300 | 132 | 174 | 68.3 | 208 | 186 | 159 | 137 | 125 | 118 |
| | | | | | | 87 | 122 | 137 | 142 | 146 | 148 |
| Uranium | 1406 | 19,070 | 116 | 27.6 | 12.5 | 21.7 | 25.1 | 29.6 | 34.0 | 38.8 | 43.9 |
| | | | | | | 94 | 108 | 125 | 146 | 176 | 180 |
| Vanadium | 2192 | 6100 | 489 | 30.7 | 10.3 | 35.8 | 31.3 | 31.3 | 33.3 | 35.7 | 38.2 |
| | | | | | | 258 | 430 | 515 | 540 | 563 | 597 |
| Zinc | 693 | 7140 | 389 | 116 | 41.8 | 117 | 118 | 111 | 103 | | |
| | | | | | | 297 | 367 | 402 | 436 | | |
| Zirconium | 2125 | 6570 | 278 | 22.7 | 12.4 | 33.2 | 25.2 | 21.6 | 20.7 | 21.6 | 23.7 |
| | | | | | | 205 | 264 | 300 | 332 | 342 | 362 |

*Source:* From Frank P. Incropera and David P. DeWitt, *Fundamentals of Heat and Mass Transfer, 3rd ed.*, 1990. This material is used by permission of John Wiley & Sons, Inc.

## TABLE A–25

Properties of solid nonmetals

| Composition | Melting Point, K | Properties at 300 K | | | | Properties at Various Temperatures (K), $k$ (W/m·K)/$c_p$(J/kg·K) | | | | | |
|---|---|---|---|---|---|---|---|---|---|---|---|
| | | $\rho$ kg/m³ | $c_p$ J/kg·K | $k$ W/m·K | $\alpha \times 10^6$ m²/s | 100 | 200 | 400 | 600 | 800 | 1000 |
| Aluminum oxide, sapphire | 2323 | 3970 | 765 | 46 | 15.1 | 450 | 82 | 32.4 | 18.9 | 13.0 | 10.5 |
| | | | | | | — | — | 940 | 1110 | 1180 | 1225 |
| Aluminum oxide, polycrystalline | 2323 | 3970 | 765 | 36.0 | 11.9 | 133 | 55 | 26.4 | 15.8 | 10.4 | 7.85 |
| | | | | | | — | — | 940 | 1110 | 1180 | 1225 |
| Beryllium oxide | 2725 | 3000 | 1030 | 272 | 88.0 | | | 196 | 111 | 70 | 47 |
| | | | | | | | | 1350 | 1690 | 1865 | 1975 |
| Boron | 2573 | 2500 | 1105 | 27.6 | 9.99 | 190 | 52.5 | 18.7 | 11.3 | 8.1 | 6.3 |
| | | | | | | — | — | 1490 | 1880 | 2135 | 2350 |
| Boron fiber epoxy (30% vol.) composite | 590 | 2080 | | | | | | | | | |
|   $k$, ∥ to fibers | | | | 2.29 | | 2.10 | 2.23 | 2.28 | | | |
|   $k$, ⊥ to fibers | | | | 0.59 | | 0.37 | 0.49 | 0.60 | | | |
|   $c_p$ | | | 1122 | | | 364 | 757 | 1431 | | | |
| Carbon Amorphous | 1500 | 1950 | — | 1.60 | — | 0.67 | 1.18 | 1.89 | 21.9 | 2.37 | 2.53 |
| | | | | | | — | — | — | — | — | — |
| Diamond, type IIa insulator | — | 3500 | 509 | 2300 | | 10,000 | 4000 | 1540 | | | |
| | | | | | | 21 | 194 | 853 | | | |
| Graphite, pyrolytic | 2273 | 2210 | | | | | | | | | |
|   $k$, ∥ to layers | | | | 1950 | | 4970 | 3230 | 1390 | 892 | 667 | 534 |
|   $k$, ⊥ to layers | | | | 5.70 | | 16.8 | 9.23 | 4.09 | 2.68 | 2.01 | 1.60 |
|   $c_p$ | | | 709 | | | 136 | 411 | 992 | 1406 | 1650 | 1793 |
| Graphite fiber epoxy (25% vol.) composite | 450 | 1400 | | | | | | | | | |
|   $k$, heat flow ∥ to fibers | | | | 11.1 | | 5.7 | 8.7 | 13.0 | | | |
|   $k$, heat flow ⊥ to fibers | | | 0.87 | | 0.46 | 0.68 | 1.1 | | | | |
|   $c_p$ | | | 935 | | | 337 | 642 | 1216 | | | |
| Pyroceram, Corning 9606 | 1623 | 2600 | 808 | 3.98 | 1.89 | 5.25 | 4.78 | 3.64 | 3.28 | 3.08 | 2.96 |
| | | | | | | — | — | 908 | 1038 | 1122 | 1197 |
| Silicon carbide | 3100 | 3160 | 675 | 490 | 230 | | | — | — | — | 87 |
| | | | | | | | | 880 | 1050 | 1135 | 1195 |
| Silicon dioxide, crystalline (quartz) | 1883 | 2650 | | | | | | | | | |
|   $k$, ∥ to $c$-axis | | | | 10.4 | | 39 | 16.4 | 7.6 | 5.0 | 4.2 | |
|   $k$, ⊥ to $c$-axis | | | | 6.21 | | 20.8 | 9.5 | 4.70 | 3.4 | 3.1 | |
|   $c_p$ | | | 745 | | | — | — | 885 | 1075 | 1250 | |
| Silicon dioxide, polycrystalline (fused silica) | 1883 | 2220 | 745 | 1.38 | 0.834 | 0.69 | 1.14 | 1.51 | 1.75 | 2.17 | 2.87 |
| | | | | | | — | — | 905 | 1040 | 1105 | 1155 |
| Silicon nitride | 2173 | 2400 | 691 | 16.0 | 9.65 | — | — | 13.9 | 11.3 | 9.88 | 8.76 |
| | | | | | | | 578 | 778 | 937 | 1063 | 1155 |
| Sulfur | 392 | 2070 | 708 | 0.206 | 0.141 | 0.165 | 0.185 | | | | |
| | | | | | | 403 | 606 | | | | |
| Thorium dioxide | 3573 | 9110 | 235 | 13 | 6.1 | | | 10.2 | 6.6 | 4.7 | 3.68 |
| | | | | | | | | 255 | 274 | 285 | 295 |
| Titanium dioxide, polycrystalline | 2133 | 4157 | 710 | 8.4 | 2.8 | | | 7.01 | 5.02 | 8.94 | 3.46 |
| | | | | | | | | 805 | 880 | 910 | 930 |

## TABLE A–26

Emissivities of surfaces
(a) Metals

| Material | Temperature, K | Emissivity, $\varepsilon$ | Material | Temperature, K | Emissivity, $\varepsilon$ |
|---|---|---|---|---|---|
| Aluminum | | | Magnesium, polished | 300–500 | 0.07–0.13 |
| Polished | 300–900 | 0.04–0.06 | Mercury | 300–400 | 0.09–0.12 |
| Commercial sheet | 400 | 0.09 | Molybdenum | | |
| Heavily oxidized | 400–800 | 0.20–0.33 | Polished | 300–2000 | 0.05–0.21 |
| Anodized | 300 | 0.8 | Oxidized | 600–800 | 0.80–0.82 |
| Bismuth, bright | 350 | 0.34 | Nickel | | |
| Brass | | | Polished | 500–1200 | 0.07–0.17 |
| Highly polished | 500–650 | 0.03–0.04 | Oxidized | 450–1000 | 0.37–0.57 |
| Polished | 350 | 0.09 | Platinum, polished | 500–1500 | 0.06–0.18 |
| Dull plate | 300–600 | 0.22 | Silver, polished | 300–1000 | 0.02–0.07 |
| Oxidized | 450–800 | 0.6 | Stainless steel | | |
| Chromium, polished | 300–1400 | 0.08–0.40 | Polished | 300–1000 | 0.17–0.30 |
| Copper | | | Lightly oxidized | 600–1000 | 0.30–0.40 |
| Highly polished | 300 | 0.02 | Highly oxidized | 600–1000 | 0.70–0.80 |
| Polished | 300–500 | 0.04–0.05 | Steel | | |
| Commercial sheet | 300 | 0.15 | Polished sheet | 300–500 | 0.08–0.14 |
| Oxidized | 600–1000 | 0.5–0.8 | Commercial sheet | 500–1200 | 0.20–0.32 |
| Black oxidized | 300 | 0.78 | Heavily oxidized | 300 | 0.81 |
| Gold | | | Tin, polished | 300 | 0.05 |
| Highly polished | 300–1000 | 0.03–0.06 | Tungsten | | |
| Bright foil | 300 | 0.07 | Polished | 300–2500 | 0.03–0.29 |
| Iron | | | Filament | 3500 | 0.39 |
| Highly polished | 300–500 | 0.05–0.07 | Zinc | | |
| Case iron | 300 | 0.44 | Polished | 300–800 | 0.02–0.05 |
| Wrought iron | 300–500 | 0.28 | Oxidized | 300 | 0.25 |
| Rusted | 300 | 0.61 | | | |
| Oxidized | 500–900 | 0.64–0.78 | | | |
| Lead | | | | | |
| Polished | 300–500 | 0.06–0.08 | | | |
| Unoxidized, rough | 300 | 0.43 | | | |
| Oxidized | 300 | 0.63 | | | |

## TABLE A–26

Emissivities of surfaces (*Concluded*)
(*b*) Nonmetals

| Material | Temperature, K | Emissivity, $\varepsilon$ | Material | Temperature, K | Emissivity, $\varepsilon$ |
|---|---|---|---|---|---|
| Alumina | 800–1400 | 0.65–0.45 | Paper, white | 300 | 0.90 |
| Aluminum oxide | 600–1500 | 0.69–0.41 | Plaster, white | 300 | 0.93 |
| Asbestos | 300 | 0.96 | Porcelain, glazed | 300 | 0.92 |
| Asphalt pavement | 300 | 0.85–0.93 | Quartz, rough, fused | 300 | 0.93 |
| Brick | | | Rubber | | |
|   Common | 300 | 0.93–0.96 |   Hard | 300 | 0.93 |
|   Fireclay | 1200 | 0.75 |   Soft | 300 | 0.86 |
| Carbon filament | 2000 | 0.53 | Sand | 300 | 0.90 |
| Cloth | 300 | 0.75–0.90 | Silicon carbide | 600–1500 | 0.87–0.85 |
| Concrete | 300 | 0.88–0.94 | Skin, human | 300 | 0.95 |
| Glass | | | Snow | 273 | 0.80–0.90 |
|   Window | 300 | 0.90–0.95 | Soil, earth | 300 | 0.93–0.96 |
|   Pyrex | 300–1200 | 0.82–0.62 | Soot | 300–500 | 0.95 |
|   Pyroceram | 300–1500 | 0.85–0.57 | Teflon | 300–500 | 0.85–0.92 |
| Ice | 273 | 0.95–0.99 | Water, deep | 273–373 | 0.95–0.96 |
| Magnesium oxide | 400–800 | 0.69–0.55 | Wood | | |
| Masonry | 300 | 0.80 |   Beech | 300 | 0.94 |
| Paints | | |   Oak | 300 | 0.90 |
|   Aluminum | 300 | 0.40–0.50 | | | |
|   Black, lacquer, shiny | 300 | 0.88 | | | |
|   Oils, all colors | 300 | 0.92–0.96 | | | |
|   Red primer | 300 | 0.93 | | | |
|   White acrylic | 300 | 0.90 | | | |
|   White enamel | 300 | 0.90 | | | |

**FIGURE A–27**

The Moody chart for the friction factor for fully developed flow in circular pipes for use in the head loss relation $\Delta P_L = f \dfrac{L}{D} \dfrac{\rho V^2}{2}$. Friction factors in the turbulent flow are evaluated from the Colebrook equation $\dfrac{1}{\sqrt{f}} = -2\log_{10}\left(\dfrac{\varepsilon/D}{3.7} + \dfrac{2.51}{\mathrm{Re}\sqrt{f}}\right)$.

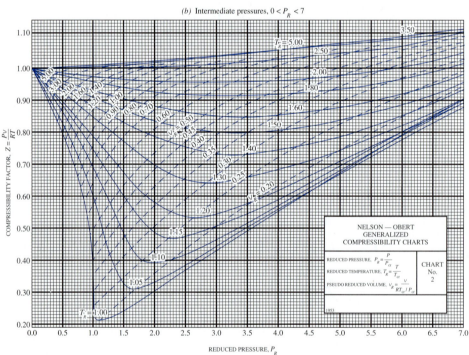

## FIGURE A–28

Nelson–Obert generalized compressibility chart.

*Used with permission of Dr. Edward E. Obert, University of Wisconsin.*

# PROPERTY TABLES AND CHARTS (ENGLISH UNITS)

## TABLE A–1E

Molar mass, gas constant, and critical-point properties

| Substance | Formula | Molar mass, $M$ lbm/lbmol | Gas constant, $R$ | | Critical-point properties | | |
|---|---|---|---|---|---|---|---|
| | | | Btu/ lbm·R* | psia·ft³/ lbm·R* | Temperature, R | Pressure, psia | Volume, ft³/lbmol |
| Air | — | 28.97 | 0.06855 | 0.3704 | 238.5 | 547 | 1.41 |
| Ammonia | $NH_3$ | 17.03 | 0.1166 | 0.6301 | 729.8 | 1636 | 1.16 |
| Argon | Ar | 39.948 | 0.04971 | 0.2686 | 272 | 705 | 1.20 |
| Benzene | $C_6H_6$ | 78.115 | 0.02542 | 0.1374 | 1012 | 714 | 4.17 |
| Bromine | $Br_2$ | 159.808 | 0.01243 | 0.06714 | 1052 | 1500 | 2.17 |
| *n*-Butane | $C_4H_{10}$ | 58.124 | 0.03417 | 0.1846 | 765.2 | 551 | 4.08 |
| Carbon dioxide | $CO_2$ | 44.01 | 0.04513 | 0.2438 | 547.5 | 1071 | 1.51 |
| Carbon monoxide | CO | 28.011 | 0.07090 | 0.3831 | 240 | 507 | 1.49 |
| Carbon tetrachloride | $CCl_4$ | 153.82 | 0.01291 | 0.06976 | 1001.5 | 661 | 4.42 |
| Chlorine | $Cl_2$ | 70.906 | 0.02801 | 0.1517 | 751 | 1120 | 1.99 |
| Chloroform | $CHCl_3$ | 119.38 | 0.01664 | 0.08988 | 965.8 | 794 | 3.85 |
| Dichlorodifluoromethane (R-12) | $CCl_2F_2$ | 120.91 | 0.01643 | 0.08874 | 692.4 | 582 | 3.49 |
| Dichlorofluoromethane (R-21) | $CHCl_2F$ | 102.92 | 0.01930 | 0.1043 | 813.0 | 749 | 3.16 |
| Ethane | $C_2H_6$ | 30.020 | 0.06616 | 0.3574 | 549.8 | 708 | 2.37 |
| Ethyl alcohol | $C_2H_5OH$ | 46.07 | 0.04311 | 0.2329 | 929.0 | 926 | 2.68 |
| Ethylene | $C_2H_4$ | 28.054 | 0.07079 | 0.3825 | 508.3 | 742 | 1.99 |
| Helium | He | 4.003 | 0.4961 | 2.6809 | 9.5 | 33.2 | 0.926 |
| *n*-Hexane | $C_6H_{14}$ | 86.178 | 0.02305 | 0.1245 | 914.2 | 439 | 5.89 |
| Hydrogen (normal) | $H_2$ | 2.016 | 0.9851 | 5.3224 | 59.9 | 188.1 | 1.04 |
| Krypton | Kr | 83.80 | 0.02370 | 0.1280 | 376.9 | 798 | 1.48 |
| Methane | $CH_4$ | 16.043 | 0.1238 | 0.6688 | 343.9 | 673 | 1.59 |
| Methyl alcohol | $CH_3OH$ | 32.042 | 0.06198 | 0.3349 | 923.7 | 1154 | 1.89 |
| Methyl chloride | $CH_3Cl$ | 50.488 | 0.03934 | 0.2125 | 749.3 | 968 | 2.29 |
| Neon | Ne | 20.183 | 0.09840 | 0.5316 | 80.1 | 395 | 0.668 |
| Nitrogen | $N_2$ | 28.013 | 0.07090 | 0.3830 | 227.1 | 492 | 1.44 |
| Nitrous oxide | $N_2O$ | 44.013 | 0.04512 | 0.2438 | 557.4 | 1054 | 1.54 |
| Oxygen | $O_2$ | 31.999 | 0.06206 | 0.3353 | 278.6 | 736 | 1.25 |
| Propane | $C_3H_8$ | 44.097 | 0.04504 | 0.2433 | 665.9 | 617 | 3.20 |
| Propylene | $C_3H_6$ | 42.081 | 0.04719 | 0.2550 | 656.9 | 670 | 2.90 |
| Sulfur dioxide | $SO_2$ | 64.063 | 0.03100 | 1.1675 | 775.2 | 1143 | 1.95 |
| Tetrafluoroethane (R-134a) | $CF_3CH_2F$ | 102.03 | 0.01946 | 0.1052 | 673.6 | 588.7 | 3.19 |
| Trichlorofluoromethane (R-11) | $CCl_3F$ | 137.37 | 0.01446 | 0.07811 | 848.1 | 635 | 3.97 |
| Water | $H_2O$ | 18.015 | 0.1102 | 0.5956 | 1164.8 | 3200 | 0.90 |
| Xenon | Xe | 131.30 | 0.01513 | 0.08172 | 521.55 | 852 | 1.90 |

*Calculated from $R = R_u/M$, where $R_u$ = 1.98588 Btu/lbmol·R = 10.7316 psia·ft³/lbmol·R and $M$ is the molar mass.

*Source:* K. A. Kobe and R. E. Lynn, Jr., *Chemical Review* 52 (1953), pp. 117–236, and ASHRAE, *Handbook of Fundamentals* (Atlanta, GA: American Society of Heating, Refrigerating, and Air-Conditioning Engineers, Inc., 1993), pp. 16.4 and 36.1.

## TABLE A–2E

Ideal-gas specific heats of various common gases
(a) At 80°F

| Gas | Formula | Gas constant, $R$ Btu/lbm·R | $c_p$ Btu/lbm·R | $c_v$ Btu/lbm·R | $k$ |
|---|---|---|---|---|---|
| Air | — | 0.06855 | 0.240 | 0.171 | 1.400 |
| Argon | Ar | 0.04971 | 0.1253 | 0.0756 | 1.667 |
| Butane | $C_4H_{10}$ | 0.03424 | 0.415 | 0.381 | 1.09 |
| Carbon dioxide | $CO_2$ | 0.04513 | 0.203 | 0.158 | 1.285 |
| Carbon monoxide | CO | 0.07090 | 0.249 | 0.178 | 1.399 |
| Ethane | $C_2H_6$ | 0.06616 | 0.427 | 0.361 | 1.183 |
| Ethylene | $C_2H_4$ | 0.07079 | 0.411 | 0.340 | 1.208 |
| Helium | He | 0.4961 | 1.25 | 0.753 | 1.667 |
| Hydrogen | $H_2$ | 0.9851 | 3.43 | 2.44 | 1.404 |
| Methane | $CH_4$ | 0.1238 | 0.532 | 0.403 | 1.32 |
| Neon | Ne | 0.09840 | 0.246 | 0.1477 | 1.667 |
| Nitrogen | $N_2$ | 0.07090 | 0.248 | 0.177 | 1.400 |
| Octane | $C_8H_{18}$ | 0.01742 | 0.409 | 0.392 | 1.044 |
| Oxygen | $O_2$ | 0.06206 | 0.219 | 0.157 | 1.395 |
| Propane | $C_3H_8$ | 0.04504 | 0.407 | 0.362 | 1.124 |
| Steam | $H_2O$ | 0.1102 | 0.445 | 0.335 | 1.329 |

*Source:* Gordon J. Van Wylen and Richard E. Sonntag, *Fundamentals of Classical Thermodynamics*, English/SI Version, 3rd ed. (New York: John Wiley & Sons, 1986), p. 687, Table A–8E.

## TABLE A–2E

Ideal-gas specific heats of various common gases (*Continued*)
(*b*) At various temperatures

| Temp., °F | $c_p$ Btu/lbm·R | $c_v$ Btu/lbm·R | $k$ | $c_p$ Btu/lbm·R | $c_v$ Btu/lbm·R | $k$ | $c_p$ Btu/lbm·R | $c_v$ Btu/lbm·R | $k$ |
|---|---|---|---|---|---|---|---|---|---|
| | Air | | | Carbon dioxide, $CO_2$ | | | Carbon monoxide, CO | | |
| 40 | 0.240 | 0.171 | 1.401 | 0.195 | 0.150 | 1.300 | 0.248 | 0.177 | 1.400 |
| 100 | 0.240 | 0.172 | 1.400 | 0.205 | 0.160 | 1.283 | 0.249 | 0.178 | 1.399 |
| 200 | 0.241 | 0.173 | 1.397 | 0.217 | 0.172 | 1.262 | 0.249 | 0.179 | 1.397 |
| 300 | 0.243 | 0.174 | 1.394 | 0.229 | 0.184 | 1.246 | 0.251 | 0.180 | 1.394 |
| 400 | 0.245 | 0.176 | 1.389 | 0.239 | 0.193 | 1.233 | 0.253 | 0.182 | 1.389 |
| 500 | 0.248 | 0.179 | 1.383 | 0.247 | 0.202 | 1.223 | 0.256 | 0.185 | 1.384 |
| 600 | 0.250 | 0.182 | 1.377 | 0.255 | 0.210 | 1.215 | 0.259 | 0.188 | 1.377 |
| 700 | 0.254 | 0.185 | 1.371 | 0.262 | 0.217 | 1.208 | 0.262 | 0.191 | 1.371 |
| 800 | 0.257 | 0.188 | 1.365 | 0.269 | 0.224 | 1.202 | 0.266 | 0.195 | 1.364 |
| 900 | 0.259 | 0.191 | 1.358 | 0.275 | 0.230 | 1.197 | 0.269 | 0.198 | 1.357 |
| 1000 | 0.263 | 0.195 | 1.353 | 0.280 | 0.235 | 1.192 | 0.273 | 0.202 | 1.351 |
| 1500 | 0.276 | 0.208 | 1.330 | 0.298 | 0.253 | 1.178 | 0.287 | 0.216 | 1.328 |
| 2000 | 0.286 | 0.217 | 1.312 | 0.312 | 0.267 | 1.169 | 0.297 | 0.226 | 1.314 |
| | Hydrogen, $H_2$ | | | Nitrogen, $N_2$ | | | Oxygen, $O_2$ | | |
| 40 | 3.397 | 2.412 | 1.409 | 0.248 | 0.177 | 1.400 | 0.219 | 0.156 | 1.397 |
| 100 | 3.426 | 2.441 | 1.404 | 0.248 | 0.178 | 1.399 | 0.220 | 0.158 | 1.394 |
| 200 | 3.451 | 2.466 | 1.399 | 0.249 | 0.178 | 1.398 | 0.223 | 0.161 | 1.387 |
| 300 | 3.461 | 2.476 | 1.398 | 0.250 | 0.179 | 1.396 | 0.226 | 0.164 | 1.378 |
| 400 | 3.466 | 2.480 | 1.397 | 0.251 | 0.180 | 1.393 | 0.230 | 0.168 | 1.368 |
| 500 | 3.469 | 2.484 | 1.397 | 0.254 | 0.183 | 1.388 | 0.235 | 0.173 | 1.360 |
| 600 | 3.473 | 2.488 | 1.396 | 0.256 | 0.185 | 1.383 | 0.239 | 0.177 | 1.352 |
| 700 | 3.477 | 2.492 | 1.395 | 0.260 | 0.189 | 1.377 | 0.242 | 0.181 | 1.344 |
| 800 | 3.494 | 2.509 | 1.393 | 0.262 | 0.191 | 1.371 | 0.246 | 0.184 | 1.337 |
| 900 | 3.502 | 2.519 | 1.392 | 0.265 | 0.194 | 1.364 | 0.249 | 0.187 | 1.331 |
| 1000 | 3.513 | 2.528 | 1.390 | 0.269 | 0.198 | 1.359 | 0.252 | 0.190 | 1.326 |
| 1500 | 3.618 | 2.633 | 1.374 | 0.283 | 0.212 | 1.334 | 0.263 | 0.201 | 1.309 |
| 2000 | 3.758 | 2.773 | 1.355 | 0.293 | 0.222 | 1.319 | 0.270 | 0.208 | 1.298 |

*Note:* The unit Btu/lbm·R is equivalent to Btu/lbm·°F.

*Source:* Kenneth Wark, *Thermodynamics*, 4th ed. (New York: McGraw-Hill, 1983), p. 830, Table A–4. Originally published in *Tables of Properties of Gases*, NBS Circular 564, 1955.

## TABLE A–2E

Ideal-gas specific heats of various common gases (*Concluded*)
(*c*) As a function of temperature

$$\bar{c}_p = a + bT + cT^2 + dT^3$$

(*T* in R, $c_p$ in Btu/lbmol·R)

| Substance | Formula | a | b | c | d | Temperature range, R | % error Max. | % error Avg. |
|---|---|---|---|---|---|---|---|---|
| Nitrogen | $N_2$ | 6.903 | $-0.02085 \times 10^{-2}$ | $0.05957 \times 10^{-5}$ | $-0.1176 \times 10^{-9}$ | 491-3240 | 0.59 | 0.34 |
| Oxygen | $O_2$ | 6.085 | $0.2017 \times 10^{-2}$ | $-0.05275 \times 10^{-5}$ | $0.05372 \times 10^{-9}$ | 491-3240 | 1.19 | 0.28 |
| Air | — | 6.713 | $0.02609 \times 10^{-2}$ | $0.03540 \times 10^{-5}$ | $-0.08052 \times 10^{-9}$ | 491-3240 | 0.72 | 0.33 |
| Hydrogen | $H_2$ | 6.952 | $-0.02542 \times 10^{-2}$ | $0.02952 \times 10^{-5}$ | $-0.03565 \times 10^{-9}$ | 491-3240 | 1.02 | 0.26 |
| Carbon monoxide | CO | 6.726 | $0.02222 \times 10^{-2}$ | $0.03960 \times 10^{-5}$ | $-0.09100 \times 10^{-9}$ | 491-3240 | 0.89 | 0.37 |
| Carbon dioxide | $CO_2$ | 5.316 | $0.79361 \times 10^{-2}$ | $-0.2581 \times 10^{-5}$ | $0.3059 \times 10^{-9}$ | 491-3240 | 0.67 | 0.22 |
| Water vapor | $H_2O$ | 7.700 | $0.02552 \times 10^{-2}$ | $0.07781 \times 10^{-5}$ | $-0.1472 \times 10^{-9}$ | 491-3240 | 0.53 | 0.24 |
| Nitric oxide | NO | 7.008 | $-0.01247 \times 10^{-2}$ | $0.07185 \times 10^{-5}$ | $-0.1715 \times 10^{-9}$ | 491-2700 | 0.97 | 0.36 |
| Nitrous oxide | $N_2O$ | 5.758 | $0.7780 \times 10^{-2}$ | $-0.2596 \times 10^{-5}$ | $0.4331 \times 10^{-9}$ | 491-2700 | 0.59 | 0.26 |
| Nitrogen dioxide | $NO_2$ | 5.48 | $0.7583 \times 10^{-2}$ | $-0.260 \times 10^{-5}$ | $0.322 \times 10^{-9}$ | 491-2700 | 0.46 | 0.18 |
| Ammonia | $NH_3$ | 6.5846 | $0.34028 \times 10^{-2}$ | $0.073034 \times 10^{-5}$ | $-0.27402 \times 10^{-9}$ | 491-2700 | 0.91 | 0.36 |
| Sulfur | $S_2$ | 6.499 | $0.2943 \times 10^{-2}$ | $-0.1200 \times 10^{-5}$ | $0.1632 \times 10^{-9}$ | 491-3240 | 0.99 | 0.38 |
| Sulfur dioxide | $SO_2$ | 6.157 | $0.7689 \times 10^{-2}$ | $-0.2810 \times 10^{-5}$ | $0.3527 \times 10^{-9}$ | 491-3240 | 0.45 | 0.24 |
| Sulfur trioxide | $SO_3$ | 3.918 | $1.935 \times 10^{-2}$ | $-0.8256 \times 10^{-5}$ | $1.328 \times 10^{-9}$ | 491-2340 | 0.29 | 0.13 |
| Acetylene | $C_2H_2$ | 5.21 | $1.2227 \times 10^{-2}$ | $-0.4812 \times 10^{-5}$ | $0.7457 \times 10^{-9}$ | 491-2700 | 1.46 | 0.59 |
| Benzene | $C_6H_6$ | $-8.650$ | $6.4322 \times 10^{-2}$ | $-2.327 \times 10^{-5}$ | $3.179 \times 10^{-9}$ | 491-2700 | 0.34 | 0.20 |
| Methanol | $CH_4O$ | 4.55 | $1.214 \times 10^{-2}$ | $-0.0898 \times 10^{-5}$ | $-0.329 \times 10^{-9}$ | 491-1800 | 0.18 | 0.08 |
| Ethanol | $C_2H_6O$ | 4.75 | $2.781 \times 10^{-2}$ | $-0.7651 \times 10^{-5}$ | $0.821 \times 10^{-9}$ | 491-2700 | 0.40 | 0.22 |
| Hydrogen chloride | HCl | 7.244 | $-0.1011 \times 10^{-2}$ | $0.09783 \times 10^{-5}$ | $-0.1776 \times 10^{-9}$ | 491-2740 | 0.22 | 0.08 |
| Methane | $CH_4$ | 4.750 | $0.6666 \times 10^{-2}$ | $0.09352 \times 10^{-5}$ | $-0.4510 \times 10^{-9}$ | 491-2740 | 1.33 | 0.57 |
| Ethane | $C_2H_6$ | 1.648 | $2.291 \times 10^{-2}$ | $-0.4722 \times 10^{-5}$ | $0.2984 \times 10^{-9}$ | 491-2740 | 0.83 | 0.28 |
| Propane | $C_3H_8$ | $-0.966$ | $4.044 \times 10^{-2}$ | $-1.159 \times 10^{-5}$ | $1.300 \times 10^{-9}$ | 491-2740 | 0.40 | 0.12 |
| *n*-Butane | $C_4H_{10}$ | 0.945 | $4.929 \times 10^{-2}$ | $-1.352 \times 10^{-5}$ | $1.433 \times 10^{-9}$ | 491-2740 | 0.54 | 0.24 |
| *i*-Butane | $C_4H_{10}$ | $-1.890$ | $5.520 \times 10^{-2}$ | $-1.696 \times 10^{-5}$ | $2.044 \times 10^{-9}$ | 491-2740 | 0.25 | 0.13 |
| *n*-Pentane | $C_5H_{12}$ | 1.618 | $6.028 \times 10^{-2}$ | $-1.656 \times 10^{-5}$ | $1.732 \times 10^{-9}$ | 491-2740 | 0.56 | 0.21 |
| *n*-Hexane | $C_6H_{14}$ | 1.657 | $7.328 \times 10^{-2}$ | $-2.112 \times 10^{-5}$ | $2.363 \times 10^{-9}$ | 491-2740 | 0.72 | 0.20 |
| Ethylene | $C_2H_4$ | 0.944 | $2.075 \times 10^{-2}$ | $-0.6151 \times 10^{-5}$ | $0.7326 \times 10^{-9}$ | 491-2740 | 0.54 | 0.13 |
| Propylene | $C_3H_6$ | 0.753 | $3.162 \times 10^{-2}$ | $-0.8981 \times 10^{-5}$ | $1.008 \times 10^{-9}$ | 491-2740 | 0.73 | 0.17 |

*Source: Chemical and Process Thermodynamics* 3/E by Kyle, B. G., © 2000. Adapted by permission of Pearson Education, Inc., Upper Saddle River, NJ.

## TABLE A–3E

Properties of common liquids, solids, and foods
(a) Liquids

| Substance | Boiling data at 1 atm | | Freezing data | | Liquid properties | | |
|---|---|---|---|---|---|---|---|
| | Normal boiling point, °F | Latent heat of vaporization, $h_{fg}$ Btu/lbm | Freezing point, °F | Latent heat of fusion, $h_{if}$ Btu/lbm | Temperature, °F | Density, $\rho$ lbm/ft³ | Specific heat, $c_p$ Btu/lbm·R |
| Ammonia | −27.9 | 24.54 | −107.9 | 138.6 | −27.9 | 42.6 | 1.06 |
| | | | | | 0 | 41.3 | 1.083 |
| | | | | | 40 | 39.5 | 1.103 |
| | | | | | 80 | 37.5 | 1.135 |
| Argon | −302.6 | 69.5 | −308.7 | 12.0 | −302.6 | 87.0 | 0.272 |
| Benzene | 176.4 | 169.4 | 41.9 | 54.2 | 68 | 54.9 | 0.411 |
| Brine (20% sodium chloride by mass) | 219.0 | — | 0.7 | — | 68 | 71.8 | 0.743 |
| n-Butane | 31.1 | 165.6 | −217.3 | 34.5 | 31.1 | 37.5 | 0.552 |
| Carbon dioxide | −109.2* | 99.6 (at 32°F) | −69.8 | — | 32 | 57.8 | 0.583 |
| Ethanol | 172.8 | 360.5 | −173.6 | 46.9 | 77 | 48.9 | 0.588 |
| Ethyl alcohol | 173.5 | 368 | −248.8 | 46.4 | 68 | 49.3 | 0.678 |
| Ethylene glycol | 388.6 | 344.0 | 12.6 | 77.9 | 68 | 69.2 | 0.678 |
| Glycerine | 355.8 | 419 | 66.0 | 86.3 | 68 | 78.7 | 0.554 |
| Helium | −452.1 | 9.80 | — | — | −452.1 | 9.13 | 5.45 |
| Hydrogen | −423.0 | 191.7 | −434.5 | 25.6 | −423.0 | 4.41 | 2.39 |
| Isobutane | 10.9 | 157.8 | −255.5 | 45.5 | 10.9 | 37.1 | 0.545 |
| Kerosene | 399–559 | 108 | −12.8 | — | 68 | 51.2 | 0.478 |
| Mercury | 674.1 | 126.7 | −38.0 | 4.90 | 77 | 847 | 0.033 |
| Methane | −258.7 | 219.6 | 296.0 | 25.1 | −258.7 | 26.4 | 0.834 |
| | | | | | −160 | 20.0 | 1.074 |
| Methanol | 148.1 | 473 | −143.9 | 42.7 | 77 | 49.1 | 0.609 |
| Nitrogen | −320.4 | 85.4 | −346.0 | 10.9 | −320.4 | 50.5 | 0.492 |
| | | | | | −260 | 38.2 | 0.643 |
| Octane | 256.6 | 131.7 | −71.5 | 77.9 | 68 | 43.9 | 0.502 |
| Oil (light) | — | — | | | 77 | 56.8 | 0.430 |
| Oxygen | −297.3 | 91.5 | −361.8 | 5.9 | −297.3 | 71.2 | 0.408 |
| Petroleum | — | 99–165 | | | 68 | 40.0 | 0.478 |
| Propane | −43.7 | 184.0 | −305.8 | 34.4 | −43.7 | 36.3 | 0.538 |
| | | | | | 32 | 33.0 | 0.604 |
| | | | | | 100 | 29.4 | 0.673 |
| Refrigerant-134a | −15.0 | 93.3 | −141.9 | — | −40 | 88.5 | 0.283 |
| | | | | | −15 | 86.0 | 0.294 |
| | | | | | 32 | 80.9 | 0.318 |
| | | | | | 90 | 73.6 | 0.348 |
| Water | 212 | 970.1 | 32 | 143.5 | 32 | 62.4 | 1.01 |
| | | | | | 90 | 62.1 | 1.00 |
| | | | | | 150 | 61.2 | 1.00 |
| | | | | | 212 | 59.8 | 1.01 |

*Sublimation temperature. (At pressures below the triple-point pressure of 75.1 psia, carbon dioxide exists as a solid or gas. Also, the freezing-point temperature of carbon dioxide is the triple-point temperature of −69.8°F.)

## TABLE A–3E

Properties of common liquids, solids, and foods (*Concluded*)
(*b*) Solids (values are for room temperature unless indicated otherwise)

| Substance | Density, $\rho$ lbm/ft$^3$ | Specific heat, $c_p$ Btu/lbm·R | Substance | Density, $\rho$ lbm/ft$^3$ | Specific heat, $c_p$ Btu/lbm·R |
|---|---|---|---|---|---|
| **Metals** | | | **Nonmetals** | | |
| Aluminum | | | Asphalt | 132 | 0.220 |
| −100°F | | 0.192 | Brick, common | 120 | 0.189 |
| 32°F | | 0.212 | Brick, fireclay (500°C) | 144 | 0.229 |
| 100°F | 170 | 0.218 | Concrete | 144 | 0.156 |
| 200°F | | 0.224 | Clay | 62.4 | 0.220 |
| 300°F | | 0.229 | Diamond | 151 | 0.147 |
| 400°F | | 0.235 | Glass, window | 169 | 0.191 |
| 500°F | | 0.240 | Glass, pyrex | 139 | 0.200 |
| Bronze (76% Cu, 2% Zn, 2% Al) | 517 | 0.0955 | Graphite | 156 | 0.170 |
| Brass, yellow (65% Cu, 35% Zn) | 519 | 0.0955 | Granite | 169 | 0.243 |
| | | | Gypsum or plaster board | 50 | 0.260 |
| Copper | | | Ice | | |
| −60°F | | 0.0862 | −50°F | | 0.424 |
| 0°F | | 0.0893 | 0°F | | 0.471 |
| 100°F | 555 | 0.0925 | 20°F | | 0.491 |
| 200°F | | 0.0938 | 32°F | 57.5 | 0.502 |
| 390°F | | 0.0963 | Limestone | 103 | 0.217 |
| Iron | 490 | 0.107 | Marble | 162 | 0.210 |
| Lead | 705 | 0.030 | Plywood (Douglas fir) | 34.0 | |
| Magnesium | 108 | 0.239 | Rubber (soft) | 68.7 | |
| Nickel | 555 | 0.105 | Rubber (hard) | 71.8 | |
| Silver | 655 | 0.056 | Sand | 94.9 | |
| Steel, mild | 489 | 0.119 | Stone | 93.6 | |
| Tungsten | 1211 | 0.031 | Woods, hard (maple, oak, etc.) | 45.0 | |
| | | | Woods, soft (fir, pine, etc.) | 32.0 | |

(*c*) Foods

| Food | Water content, % (mass) | Freezing point, °F | Specific heat, Btu/lbm·R Above freezing | Specific heat, Btu/lbm·R Below freezing | Latent heat of fusion, Btu/lbm | Food | Water content, % (mass) | Freezing point, °F | Specific heat, Btu/lbm·R Above freezing | Specific heat, Btu/lbm·R Below freezing | Latent heat of fusion, Btu/lbm |
|---|---|---|---|---|---|---|---|---|---|---|---|
| Apples | 84 | 30 | 0.873 | 0.453 | 121 | Lettuce | 95 | 32 | 0.961 | 0.487 | 136 |
| Bananas | 75 | 31 | 0.801 | 0.426 | 108 | Milk, whole | 88 | 31 | 0.905 | 0.465 | 126 |
| Beef round | 67 | — | 0.737 | 0.402 | 96 | Oranges | 87 | 31 | 0.897 | 0.462 | 125 |
| Broccoli | 90 | 31 | 0.921 | 0.471 | 129 | Potatoes | 78 | 31 | 0.825 | 0.435 | 112 |
| Butter | 16 | — | — | 0.249 | 23 | Salmon fish | 64 | 28 | 0.713 | 0.393 | 92 |
| Cheese, Swiss | 39 | 14 | 0.513 | 0.318 | 56 | Shrimp | 83 | 28 | 0.865 | 0.450 | 119 |
| Cherries | 80 | 29 | 0.841 | 0.441 | 115 | Spinach | 93 | 31 | 0.945 | 0.481 | 134 |
| Chicken | 74 | 27 | 0.793 | 0.423 | 106 | Strawberries | 90 | 31 | 0.921 | 0.471 | 129 |
| Corn, sweet | 74 | 31 | 0.793 | 0.423 | 106 | Tomatoes, ripe | 94 | 31 | 0.953 | 0.484 | 135 |
| Eggs, whole | 74 | 31 | 0.793 | 0.423 | 106 | Turkey | 64 | — | 0.713 | 0.393 | 92 |
| Ice cream | 63 | 22 | 0.705 | 0.390 | 90 | Watermelon | 93 | 31 | 0.945 | 0.481 | 134 |

*Source:* Values are obtained from various handbooks and other sources or are calculated. Water content and freezing-point data of foods are from ASHRAE, *Handbook of Fundamentals*, I-P version (Atlanta, GA: American Society of Heating, Refrigerating, and Air-Conditioning Engineers, Inc., 1993), Chap. 30, Table 1. Freezing point is the temperature at which freezing starts for fruits and vegetables, and the average freezing temperature for other foods.

## TABLE A–4E

Saturated water—Temperature table

| Temp., $T$ °F | Sat. press., $P_{sat}$ psia | Specific volume, ft³/lbm | | Internal energy, Btu/lbm | | | Enthalpy, Btu/lbm | | | Entropy, Btu/lbm·R | | |
|---|---|---|---|---|---|---|---|---|---|---|---|---|
| | | Sat. liquid, $v_f$ | Sat. vapor, $v_g$ | Sat. liquid, $u_f$ | Evap., $u_{fg}$ | Sat. vapor, $u_g$ | Sat. liquid, $h_f$ | Evap., $h_{fg}$ | Sat. vapor, $h_g$ | Sat. liquid, $s_f$ | Evap., $s_{fg}$ | Sat. vapor, $s_g$ |
| 32.018 | 0.08871 | 0.01602 | 3299.9 | 0.000 | 1021.0 | 1021.0 | 0.000 | 1075.2 | 1075.2 | 0.00000 | 2.18672 | 2.1867 |
| 35 | 0.09998 | 0.01602 | 2945.7 | 3.004 | 1019.0 | 1022.0 | 3.004 | 1073.5 | 1076.5 | 0.00609 | 2.17011 | 2.1762 |
| 40 | 0.12173 | 0.01602 | 2443.6 | 8.032 | 1015.6 | 1023.7 | 8.032 | 1070.7 | 1078.7 | 0.01620 | 2.14271 | 2.1589 |
| 45 | 0.14756 | 0.01602 | 2035.8 | 13.05 | 1012.2 | 1025.3 | 13.05 | 1067.8 | 1080.9 | 0.02620 | 2.11587 | 2.1421 |
| 50 | 0.17812 | 0.01602 | 1703.1 | 18.07 | 1008.9 | 1026.9 | 18.07 | 1065.0 | 1083.1 | 0.03609 | 2.08956 | 2.1256 |
| 55 | 0.21413 | 0.01603 | 1430.4 | 23.07 | 1005.5 | 1028.6 | 23.07 | 1062.2 | 1085.3 | 0.04586 | 2.06377 | 2.1096 |
| 60 | 0.25638 | 0.01604 | 1206.1 | 28.08 | 1002.1 | 1030.2 | 28.08 | 1059.4 | 1087.4 | 0.05554 | 2.03847 | 2.0940 |
| 65 | 0.30578 | 0.01604 | 1020.8 | 33.08 | 998.76 | 1031.8 | 33.08 | 1056.5 | 1089.6 | 0.06511 | 2.01366 | 2.0788 |
| 70 | 0.36334 | 0.01605 | 867.18 | 38.08 | 995.39 | 1033.5 | 38.08 | 1053.7 | 1091.8 | 0.07459 | 1.98931 | 2.0639 |
| 75 | 0.43016 | 0.01606 | 739.27 | 43.07 | 992.02 | 1035.1 | 43.07 | 1050.9 | 1093.9 | 0.08398 | 1.96541 | 2.0494 |
| 80 | 0.50745 | 0.01607 | 632.41 | 48.06 | 988.65 | 1036.7 | 48.07 | 1048.0 | 1096.1 | 0.09328 | 1.94196 | 2.0352 |
| 85 | 0.59659 | 0.01609 | 542.80 | 53.06 | 985.28 | 1038.3 | 53.06 | 1045.2 | 1098.3 | 0.10248 | 1.91892 | 2.0214 |
| 90 | 0.69904 | 0.01610 | 467.40 | 58.05 | 981.90 | 1040.0 | 58.05 | 1042.4 | 1100.4 | 0.11161 | 1.89630 | 2.0079 |
| 95 | 0.81643 | 0.01612 | 403.74 | 63.04 | 978.52 | 1041.6 | 63.04 | 1039.5 | 1102.6 | 0.12065 | 1.87408 | 1.9947 |
| 100 | 0.95052 | 0.01613 | 349.83 | 68.03 | 975.14 | 1043.2 | 68.03 | 1036.7 | 1104.7 | 0.12961 | 1.85225 | 1.9819 |
| 110 | 1.2767 | 0.01617 | 264.96 | 78.01 | 968.36 | 1046.4 | 78.02 | 1031.0 | 1109.0 | 0.14728 | 1.80970 | 1.9570 |
| 120 | 1.6951 | 0.01620 | 202.94 | 88.00 | 961.56 | 1049.6 | 88.00 | 1025.2 | 1113.2 | 0.16466 | 1.76856 | 1.9332 |
| 130 | 2.2260 | 0.01625 | 157.09 | 97.99 | 954.73 | 1052.7 | 97.99 | 1019.4 | 1117.4 | 0.18174 | 1.72877 | 1.9105 |
| 140 | 2.8931 | 0.01629 | 122.81 | 107.98 | 947.87 | 1055.9 | 107.99 | 1013.6 | 1121.6 | 0.19855 | 1.69024 | 1.8888 |
| 150 | 3.7234 | 0.01634 | 96.929 | 117.98 | 940.98 | 1059.0 | 117.99 | 1007.8 | 1125.7 | 0.21508 | 1.65291 | 1.8680 |
| 160 | 4.7474 | 0.01639 | 77.185 | 127.98 | 934.05 | 1062.0 | 128.00 | 1001.8 | 1129.8 | 0.23136 | 1.61670 | 1.8481 |
| 170 | 5.9999 | 0.01645 | 61.982 | 138.00 | 927.08 | 1065.1 | 138.02 | 995.88 | 1133.9 | 0.24739 | 1.58155 | 1.8289 |
| 180 | 7.5197 | 0.01651 | 50.172 | 148.02 | 920.06 | 1068.1 | 148.04 | 989.85 | 1137.9 | 0.26318 | 1.54741 | 1.8106 |
| 190 | 9.3497 | 0.01657 | 40.920 | 158.05 | 912.99 | 1071.0 | 158.08 | 983.76 | 1141.8 | 0.27874 | 1.51421 | 1.7930 |
| 200 | 11.538 | 0.01663 | 33.613 | 168.10 | 905.87 | 1074.0 | 168.13 | 977.60 | 1145.7 | 0.29409 | 1.48191 | 1.7760 |
| 210 | 14.136 | 0.01670 | 27.798 | 178.15 | 898.68 | 1076.8 | 178.20 | 971.35 | 1149.5 | 0.30922 | 1.45046 | 1.7597 |
| 212 | 14.709 | 0.01671 | 26.782 | 180.16 | 897.24 | 1077.4 | 180.21 | 970.09 | 1150.3 | 0.31222 | 1.44427 | 1.7565 |
| 220 | 17.201 | 0.01677 | 23.136 | 188.22 | 891.43 | 1079.6 | 188.28 | 965.02 | 1153.3 | 0.32414 | 1.41980 | 1.7439 |
| 230 | 20.795 | 0.01684 | 19.374 | 198.31 | 884.10 | 1082.4 | 198.37 | 958.59 | 1157.0 | 0.33887 | 1.38989 | 1.7288 |
| 240 | 24.985 | 0.01692 | 16.316 | 208.41 | 876.70 | 1085.1 | 208.49 | 952.06 | 1160.5 | 0.35342 | 1.36069 | 1.7141 |
| 250 | 29.844 | 0.01700 | 13.816 | 218.54 | 869.21 | 1087.7 | 218.63 | 945.41 | 1164.0 | 0.36779 | 1.33216 | 1.6999 |
| 260 | 35.447 | 0.01708 | 11.760 | 228.68 | 861.62 | 1090.3 | 228.79 | 938.65 | 1167.4 | 0.38198 | 1.30425 | 1.6862 |
| 270 | 41.877 | 0.01717 | 10.059 | 238.85 | 853.94 | 1092.8 | 238.98 | 931.76 | 1170.7 | 0.39601 | 1.27694 | 1.6730 |
| 280 | 49.222 | 0.01726 | 8.6439 | 249.04 | 846.16 | 1095.2 | 249.20 | 924.74 | 1173.9 | 0.40989 | 1.25018 | 1.6601 |
| 290 | 57.573 | 0.01735 | 7.4607 | 259.26 | 838.27 | 1097.5 | 259.45 | 917.57 | 1177.0 | 0.42361 | 1.22393 | 1.6475 |
| 300 | 67.028 | 0.01745 | 6.4663 | 269.51 | 830.25 | 1099.8 | 269.73 | 910.24 | 1180.0 | 0.43720 | 1.19818 | 1.6354 |
| 310 | 77.691 | 0.01755 | 5.6266 | 279.79 | 822.11 | 1101.9 | 280.05 | 902.75 | 1182.8 | 0.45065 | 1.17289 | 1.6235 |
| 320 | 89.667 | 0.01765 | 4.9144 | 290.11 | 813.84 | 1104.0 | 290.40 | 895.09 | 1185.5 | 0.46396 | 1.14802 | 1.6120 |
| 330 | 103.07 | 0.01776 | 4.3076 | 300.46 | 805.43 | 1105.9 | 300.80 | 887.25 | 1188.1 | 0.47716 | 1.12355 | 1.6007 |
| 340 | 118.02 | 0.01787 | 3.7885 | 310.85 | 796.87 | 1107.7 | 311.24 | 879.22 | 1190.5 | 0.49024 | 1.09945 | 1.5897 |
| 350 | 134.63 | 0.01799 | 3.3425 | 321.29 | 788.16 | 1109.4 | 321.73 | 870.98 | 1192.7 | 0.50321 | 1.07570 | 1.5789 |
| 360 | 153.03 | 0.01811 | 2.9580 | 331.76 | 779.28 | 1111.0 | 332.28 | 862.53 | 1194.8 | 0.51607 | 1.05227 | 1.5683 |
| 370 | 173.36 | 0.01823 | 2.6252 | 342.29 | 770.23 | 1112.5 | 342.88 | 853.86 | 1196.7 | 0.52884 | 1.02914 | 1.5580 |
| 380 | 195.74 | 0.01836 | 2.3361 | 352.87 | 761.00 | 1113.9 | 353.53 | 844.96 | 1198.5 | 0.54152 | 1.00628 | 1.5478 |
| 390 | 220.33 | 0.01850 | 2.0842 | 363.50 | 751.58 | 1115.1 | 364.25 | 835.81 | 1200.1 | 0.55411 | 0.98366 | 1.5378 |

## TABLE A–4E

Saturated water—Temperature table (*Concluded*)

| Temp., $T$ °F | Sat. press., $P_{sat}$ psia | Specific volume, ft³/lbm | | Internal energy, Btu/lbm | | | Enthalpy, Btu/lbm | | | Entropy, Btu/lbm·R | | |
|---|---|---|---|---|---|---|---|---|---|---|---|---|
| | | Sat. liquid, $v_f$ | Sat. vapor, $v_g$ | Sat. liquid, $u_f$ | Evap., $u_{fg}$ | Sat. vapor, $u_g$ | Sat. liquid, $h_f$ | Evap., $h_{fg}$ | Sat. vapor, $h_g$ | Sat. liquid, $s_f$ | Evap., $s_{fg}$ | Sat. vapor, $s_g$ |
| 400 | 247.26 | 0.01864 | 1.8639 | 374.19 | 741.97 | 1116.2 | 375.04 | 826.39 | 1201.4 | 0.56663 | 0.96127 | 1.5279 |
| 410 | 276.69 | 0.01878 | 1.6706 | 384.94 | 732.14 | 1117.1 | 385.90 | 816.71 | 1202.6 | 0.57907 | 0.93908 | 1.5182 |
| 420 | 308.76 | 0.01894 | 1.5006 | 395.76 | 722.08 | 1117.8 | 396.84 | 806.74 | 1203.6 | 0.59145 | 0.91707 | 1.5085 |
| 430 | 343.64 | 0.01910 | 1.3505 | 406.65 | 711.80 | 1118.4 | 407.86 | 796.46 | 1204.3 | 0.60377 | 0.89522 | 1.4990 |
| 440 | 381.49 | 0.01926 | 1.2178 | 417.61 | 701.26 | 1118.9 | 418.97 | 785.87 | 1204.8 | 0.61603 | 0.87349 | 1.4895 |
| 450 | 422.47 | 0.01944 | 1.0999 | 428.66 | 690.47 | 1119.1 | 430.18 | 774.94 | 1205.1 | 0.62826 | 0.85187 | 1.4801 |
| 460 | 466.75 | 0.01962 | 0.99510 | 439.79 | 679.39 | 1119.2 | 441.48 | 763.65 | 1205.1 | 0.64044 | 0.83033 | 1.4708 |
| 470 | 514.52 | 0.01981 | 0.90158 | 451.01 | 668.02 | 1119.0 | 452.90 | 751.98 | 1204.9 | 0.65260 | 0.80885 | 1.4615 |
| 480 | 565.96 | 0.02001 | 0.81794 | 462.34 | 656.34 | 1118.7 | 464.43 | 739.91 | 1204.3 | 0.66474 | 0.78739 | 1.4521 |
| 490 | 621.24 | 0.02022 | 0.74296 | 473.77 | 644.32 | 1118.1 | 476.09 | 727.40 | 1203.5 | 0.67686 | 0.76594 | 1.4428 |
| 500 | 680.56 | 0.02044 | 0.67558 | 485.32 | 631.94 | 1117.3 | 487.89 | 714.44 | 1202.3 | 0.68899 | 0.74445 | 1.4334 |
| 510 | 744.11 | 0.02067 | 0.61489 | 496.99 | 619.17 | 1116.2 | 499.84 | 700.99 | 1200.8 | 0.70112 | 0.72290 | 1.4240 |
| 520 | 812.11 | 0.02092 | 0.56009 | 508.80 | 605.99 | 1114.8 | 511.94 | 687.01 | 1199.0 | 0.71327 | 0.70126 | 1.4145 |
| 530 | 884.74 | 0.02118 | 0.51051 | 520.76 | 592.35 | 1113.1 | 524.23 | 672.47 | 1196.7 | 0.72546 | 0.67947 | 1.4049 |
| 540 | 962.24 | 0.02146 | 0.46553 | 532.88 | 578.23 | 1111.1 | 536.70 | 657.31 | 1194.0 | 0.73770 | 0.65751 | 1.3952 |
| 550 | 1044.8 | 0.02176 | 0.42465 | 545.18 | 563.58 | 1108.8 | 549.39 | 641.47 | 1190.9 | 0.75000 | 0.63532 | 1.3853 |
| 560 | 1132.7 | 0.02207 | 0.38740 | 557.68 | 548.33 | 1106.0 | 562.31 | 624.91 | 1187.2 | 0.76238 | 0.61284 | 1.3752 |
| 570 | 1226.2 | 0.02242 | 0.35339 | 570.40 | 532.45 | 1102.8 | 575.49 | 607.55 | 1183.0 | 0.77486 | 0.59003 | 1.3649 |
| 580 | 1325.5 | 0.02279 | 0.32225 | 583.37 | 515.84 | 1099.2 | 588.95 | 589.29 | 1178.2 | 0.78748 | 0.56679 | 1.3543 |
| 590 | 1430.8 | 0.02319 | 0.29367 | 596.61 | 498.43 | 1095.0 | 602.75 | 570.04 | 1172.8 | 0.80026 | 0.54306 | 1.3433 |
| 600 | 1542.5 | 0.02362 | 0.26737 | 610.18 | 480.10 | 1090.3 | 616.92 | 549.67 | 1166.6 | 0.81323 | 0.51871 | 1.3319 |
| 610 | 1660.9 | 0.02411 | 0.24309 | 624.11 | 460.73 | 1084.8 | 631.52 | 528.03 | 1159.5 | 0.82645 | 0.49363 | 1.3201 |
| 620 | 1786.2 | 0.02464 | 0.22061 | 638.47 | 440.14 | 1078.6 | 646.62 | 504.92 | 1151.5 | 0.83998 | 0.46765 | 1.3076 |
| 630 | 1918.9 | 0.02524 | 0.19972 | 653.35 | 418.12 | 1071.5 | 662.32 | 480.07 | 1142.4 | 0.85389 | 0.44056 | 1.2944 |
| 640 | 2059.3 | 0.02593 | 0.18019 | 668.86 | 394.36 | 1063.2 | 678.74 | 453.14 | 1131.9 | 0.86828 | 0.41206 | 1.2803 |
| 650 | 2207.8 | 0.02673 | 0.16184 | 685.16 | 368.44 | 1053.6 | 696.08 | 423.65 | 1119.7 | 0.88332 | 0.38177 | 1.2651 |
| 660 | 2364.9 | 0.02767 | 0.14444 | 702.48 | 339.74 | 1042.2 | 714.59 | 390.84 | 1105.4 | 0.89922 | 0.34906 | 1.2483 |
| 670 | 2531.2 | 0.02884 | 0.12774 | 721.23 | 307.22 | 1028.5 | 734.74 | 353.54 | 1088.3 | 0.91636 | 0.31296 | 1.2293 |
| 680 | 2707.3 | 0.03035 | 0.11134 | 742.11 | 269.00 | 1011.1 | 757.32 | 309.57 | 1066.9 | 0.93541 | 0.27163 | 1.2070 |
| 690 | 2894.1 | 0.03255 | 0.09451 | 766.81 | 220.77 | 987.6 | 784.24 | 253.96 | 1038.2 | 0.95797 | 0.22089 | 1.1789 |
| 700 | 3093.0 | 0.03670 | 0.07482 | 801.75 | 146.50 | 948.3 | 822.76 | 168.32 | 991.1 | 0.99023 | 0.14514 | 1.1354 |
| 705.10 | 3200.1 | 0.04975 | 0.04975 | 866.61 | 0 | 866.6 | 896.07 | 0 | 896.1 | 1.05257 | 0 | 1.0526 |

*Source:* Tables A–4E through A–8E are generated using the Engineering Equation Solver (EES) software developed by S. A. Klein and F. L. Alvarado. The routine used in calculations is the highly accurate Steam_IAPWS, which incorporates the 1995 Formulation for the Thermodynamic Properties of Ordinary Water Substance for General and Scientific Use, issued by The International Association for the Properties of Water and Steam (IAPWS). This formulation replaces the 1984 formulation of Haar, Gallagher, and Kell (NBS/NRC Steam Tables, Hemisphere Publishing Co., 1984), which is also available in EES as the routine STEAM. The new formulation is based on the correlations of Saul and Wagner (J. Phys. Chem. Ref. Data, 16, 893, 1987) with modifications to adjust to the International Temperature Scale of 1990. The modifications are described by Wagner and Pruss (J. Phys. Chem. Ref. Data, 22, 783, 1993). The properties of ice are based on Hyland and Wexler, "Formulations for the Thermodynamic Properties of the Saturated Phases of $H_2O$ from 173.15 K to 473.15 K," *ASHRAE Trans.*, Part 2A, Paper 2793, 1983.

## TABLE A–5E

Saturated water—Pressure table

| Press., $P$ psia | Sat. temp., $T_{sat}$ °F | Specific volume, ft³/lbm | | Internal energy, Btu/lbm | | | Enthalpy, Btu/lbm | | | Entropy, Btu/lbm·R | | |
|---|---|---|---|---|---|---|---|---|---|---|---|---|
| | | Sat. liquid, $v_f$ | Sat. vapor, $v_g$ | Sat. liquid, $u_f$ | Evap., $u_{fg}$ | Sat. vapor, $u_g$ | Sat. liquid, $h_f$ | Evap., $h_{fg}$ | Sat. vapor, $h_g$ | Sat. liquid, $s_f$ | Evap., $s_{fg}$ | Sat. vapor, $s_g$ |
| 1 | 101.69 | 0.01614 | 333.49 | 69.72 | 973.99 | 1043.7 | 69.72 | 1035.7 | 1105.4 | 0.13262 | 1.84495 | 1.9776 |
| 2 | 126.02 | 0.01623 | 173.71 | 94.02 | 957.45 | 1051.5 | 94.02 | 1021.7 | 1115.8 | 0.17499 | 1.74444 | 1.9194 |
| 3 | 141.41 | 0.01630 | 118.70 | 109.39 | 946.90 | 1056.3 | 109.40 | 1012.8 | 1122.2 | 0.20090 | 1.68489 | 1.8858 |
| 4 | 152.91 | 0.01636 | 90.629 | 120.89 | 938.97 | 1059.9 | 120.90 | 1006.0 | 1126.9 | 0.21985 | 1.64225 | 1.8621 |
| 5 | 162.18 | 0.01641 | 73.525 | 130.17 | 932.53 | 1062.7 | 130.18 | 1000.5 | 1130.7 | 0.23488 | 1.60894 | 1.8438 |
| 6 | 170.00 | 0.01645 | 61.982 | 138.00 | 927.08 | 1065.1 | 138.02 | 995.88 | 1133.9 | 0.24739 | 1.58155 | 1.8289 |
| 8 | 182.81 | 0.01652 | 47.347 | 150.83 | 918.08 | 1068.9 | 150.86 | 988.15 | 1139.0 | 0.26757 | 1.53800 | 1.8056 |
| 10 | 193.16 | 0.01659 | 38.425 | 161.22 | 910.75 | 1072.0 | 161.25 | 981.82 | 1143.1 | 0.28362 | 1.50391 | 1.7875 |
| 14.696 | 211.95 | 0.01671 | 26.805 | 180.12 | 897.27 | 1077.4 | 180.16 | 970.12 | 1150.3 | 0.31215 | 1.44441 | 1.7566 |
| 15 | 212.99 | 0.01672 | 26.297 | 181.16 | 896.52 | 1077.7 | 181.21 | 969.47 | 1150.7 | 0.31370 | 1.44441 | 1.7549 |
| 20 | 227.92 | 0.01683 | 20.093 | 196.21 | 885.63 | 1081.8 | 196.27 | 959.93 | 1156.2 | 0.33582 | 1.39606 | 1.7319 |
| 25 | 240.03 | 0.01692 | 16.307 | 208.45 | 876.67 | 1085.1 | 208.52 | 952.03 | 1160.6 | 0.35347 | 1.36060 | 1.7141 |
| 30 | 250.30 | 0.01700 | 13.749 | 218.84 | 868.98 | 1087.8 | 218.93 | 945.21 | 1164.1 | 0.36821 | 1.33132 | 1.6995 |
| 35 | 259.25 | 0.01708 | 11.901 | 227.92 | 862.19 | 1090.1 | 228.03 | 939.16 | 1167.2 | 0.38093 | 1.30632 | 1.6872 |
| 40 | 267.22 | 0.01715 | 10.501 | 236.02 | 856.09 | 1092.1 | 236.14 | 933.69 | 1169.8 | 0.39213 | 1.28448 | 1.6766 |
| 45 | 274.41 | 0.01721 | 9.4028 | 243.34 | 850.52 | 1093.9 | 243.49 | 928.68 | 1172.2 | 0.40216 | 1.26506 | 1.6672 |
| 50 | 280.99 | 0.01727 | 8.5175 | 250.05 | 845.39 | 1095.4 | 250.21 | 924.03 | 1174.2 | 0.41125 | 1.24756 | 1.6588 |
| 55 | 287.05 | 0.01732 | 7.7882 | 256.25 | 840.61 | 1096.9 | 256.42 | 919.70 | 1176.1 | 0.41958 | 1.23162 | 1.6512 |
| 60 | 292.69 | 0.01738 | 7.1766 | 262.01 | 836.13 | 1098.1 | 262.20 | 915.61 | 1177.8 | 0.42728 | 1.21697 | 1.6442 |
| 65 | 297.95 | 0.01743 | 6.6560 | 267.41 | 831.90 | 1099.3 | 267.62 | 911.75 | 1179.4 | 0.43443 | 1.20341 | 1.6378 |
| 70 | 302.91 | 0.01748 | 6.2075 | 272.50 | 827.90 | 1100.4 | 272.72 | 908.08 | 1180.8 | 0.44112 | 1.19078 | 1.6319 |
| 75 | 307.59 | 0.01752 | 5.8167 | 277.31 | 824.09 | 1101.4 | 277.55 | 904.58 | 1182.1 | 0.44741 | 1.17895 | 1.6264 |
| 80 | 312.02 | 0.01757 | 5.4733 | 281.87 | 820.45 | 1102.3 | 282.13 | 901.22 | 1183.4 | 0.45335 | 1.16783 | 1.6212 |
| 85 | 316.24 | 0.01761 | 5.1689 | 286.22 | 816.97 | 1103.2 | 286.50 | 898.00 | 1184.5 | 0.45897 | 1.15732 | 1.6163 |
| 90 | 320.26 | 0.01765 | 4.8972 | 290.38 | 813.62 | 1104.0 | 290.67 | 894.89 | 1185.6 | 0.46431 | 1.14737 | 1.6117 |
| 95 | 324.11 | 0.01770 | 4.6532 | 294.36 | 810.40 | 1104.8 | 294.67 | 891.89 | 1186.6 | 0.46941 | 1.13791 | 1.6073 |
| 100 | 327.81 | 0.01774 | 4.4327 | 298.19 | 807.29 | 1105.5 | 298.51 | 888.99 | 1187.5 | 0.47427 | 1.12888 | 1.6032 |
| 110 | 334.77 | 0.01781 | 4.0410 | 305.41 | 801.37 | 1106.8 | 305.78 | 883.44 | 1189.2 | 0.48341 | 1.11201 | 1.5954 |
| 120 | 341.25 | 0.01789 | 3.7289 | 312.16 | 795.79 | 1107.9 | 312.55 | 878.20 | 1190.8 | 0.49187 | 1.09646 | 1.5883 |
| 130 | 347.32 | 0.01796 | 3.4557 | 318.48 | 790.51 | 1109.0 | 318.92 | 873.21 | 1192.1 | 0.49974 | 1.08204 | 1.5818 |
| 140 | 353.03 | 0.01802 | 3.2202 | 324.45 | 785.49 | 1109.9 | 324.92 | 868.45 | 1193.4 | 0.50711 | 1.06858 | 1.5757 |
| 150 | 358.42 | 0.01809 | 3.0150 | 330.11 | 780.69 | 1110.8 | 330.61 | 863.88 | 1194.5 | 0.51405 | 1.05595 | 1.5700 |
| 160 | 363.54 | 0.01815 | 2.8347 | 335.49 | 776.10 | 1111.6 | 336.02 | 859.49 | 1195.5 | 0.52061 | 1.04405 | 1.5647 |
| 170 | 368.41 | 0.01821 | 2.6749 | 340.62 | 771.68 | 1112.3 | 341.19 | 855.25 | 1196.4 | 0.52682 | 1.03279 | 1.5596 |
| 180 | 373.07 | 0.01827 | 2.5322 | 345.53 | 767.42 | 1113.0 | 346.14 | 851.16 | 1197.3 | 0.53274 | 1.02210 | 1.5548 |
| 190 | 377.52 | 0.01833 | 2.4040 | 350.24 | 763.31 | 1113.6 | 350.89 | 847.19 | 1198.1 | 0.53839 | 1.01191 | 1.5503 |
| 200 | 381.80 | 0.01839 | 2.2882 | 354.78 | 759.32 | 1114.1 | 355.46 | 843.33 | 1198.8 | 0.54379 | 1.00219 | 1.5460 |
| 250 | 400.97 | 0.01865 | 1.8440 | 375.23 | 741.02 | 1116.3 | 376.09 | 825.47 | 1201.6 | 0.56784 | 0.95912 | 1.5270 |
| 300 | 417.35 | 0.01890 | 1.5435 | 392.89 | 724.77 | 1117.7 | 393.94 | 809.41 | 1203.3 | 0.58818 | 0.92289 | 1.5111 |
| 350 | 431.74 | 0.01912 | 1.3263 | 408.55 | 709.98 | 1118.5 | 409.79 | 794.65 | 1204.4 | 0.60590 | 0.89143 | 1.4973 |
| 400 | 444.62 | 0.01934 | 1.1617 | 422.70 | 696.31 | 1119.0 | 424.13 | 780.87 | 1205.0 | 0.62168 | 0.86350 | 1.4852 |
| 450 | 456.31 | 0.01955 | 1.0324 | 435.67 | 683.52 | 1119.2 | 437.30 | 767.86 | 1205.2 | 0.63595 | 0.83828 | 1.4742 |
| 500 | 467.04 | 0.01975 | 0.92819 | 447.68 | 671.42 | 1119.1 | 449.51 | 755.48 | 1205.0 | 0.64900 | 0.81521 | 1.4642 |
| 550 | 476.97 | 0.01995 | 0.84228 | 458.90 | 659.91 | 1118.8 | 460.93 | 743.60 | 1204.5 | 0.66107 | 0.79388 | 1.4550 |
| 600 | 486.24 | 0.02014 | 0.77020 | 469.46 | 648.88 | 1118.3 | 471.70 | 732.15 | 1203.9 | 0.67231 | 0.77400 | 1.4463 |

## TABLE A–5E

Saturated water—Pressure table (*Concluded*)

| Press., P psia | Sat. temp., $T_{sat}$ °F | Specific volume, ft³/lbm | | Internal energy, Btu/lbm | | | Enthalpy, Btu/lbm | | | Entropy, Btu/lbm·R | | |
|---|---|---|---|---|---|---|---|---|---|---|---|---|
| | | Sat. liquid, $v_f$ | Sat. vapor, $v_g$ | Sat. liquid, $u_f$ | Evap., $u_{fg}$ | Sat. vapor, $u_g$ | Sat. liquid, $h_f$ | Evap., $h_{fg}$ | Sat. vapor, $h_g$ | Sat. liquid, $s_f$ | Evap., $s_{fg}$ | Sat. vapor, $s_g$ |
| 700 | 503.13 | 0.02051 | 0.65589 | 488.96 | 627.98 | 1116.9 | 491.62 | 710.29 | 1201.9 | 0.69279 | 0.73771 | 1.4305 |
| 800 | 518.27 | 0.02087 | 0.56920 | 506.74 | 608.30 | 1115.0 | 509.83 | 689.48 | 1199.3 | 0.71117 | 0.70502 | 1.4162 |
| 900 | 532.02 | 0.02124 | 0.50107 | 523.19 | 589.54 | 1112.7 | 526.73 | 669.46 | 1196.2 | 0.72793 | 0.67505 | 1.4030 |
| 1000 | 544.65 | 0.02159 | 0.44604 | 538.58 | 571.49 | 1110.1 | 542.57 | 650.03 | 1192.6 | 0.74341 | 0.64722 | 1.3906 |
| 1200 | 567.26 | 0.02232 | 0.36241 | 566.89 | 536.87 | 1103.8 | 571.85 | 612.39 | 1184.2 | 0.77143 | 0.59632 | 1.3677 |
| 1400 | 587.14 | 0.02307 | 0.30161 | 592.79 | 503.50 | 1096.3 | 598.76 | 575.66 | 1174.4 | 0.79658 | 0.54991 | 1.3465 |
| 1600 | 604.93 | 0.02386 | 0.25516 | 616.99 | 470.69 | 1087.7 | 624.06 | 539.18 | 1163.2 | 0.81972 | 0.50645 | 1.3262 |
| 1800 | 621.07 | 0.02470 | 0.21831 | 640.03 | 437.86 | 1077.9 | 648.26 | 502.35 | 1150.6 | 0.84144 | 0.46482 | 1.3063 |
| 2000 | 635.85 | 0.02563 | 0.18815 | 662.33 | 404.46 | 1066.8 | 671.82 | 464.60 | 1136.4 | 0.86224 | 0.42409 | 1.2863 |
| 2500 | 668.17 | 0.02860 | 0.13076 | 717.67 | 313.53 | 1031.2 | 730.90 | 360.79 | 1091.7 | 0.91311 | 0.31988 | 1.2330 |
| 3000 | 695.41 | 0.03433 | 0.08460 | 783.39 | 186.41 | 969.8 | 802.45 | 214.32 | 1016.8 | 0.97321 | 0.18554 | 1.1587 |
| 3200.1 | 705.10 | 0.04975 | 0.04975 | 866.61 | 0 | 866.6 | 896.07 | 0 | 896.1 | 1.05257 | 0 | 1.0526 |

## TABLE A–6E

Superheated water

| T °F | v ft³/lbm | u Btu/lbm | h Btu/lbm | s Btu/lbm·R | v ft³/lbm | u Btu/lbm | h Btu/lbm | s Btu/lbm·R | v ft³/lbm | u Btu/lbm | h Btu/lbm | s Btu/lbm·R |
|---|---|---|---|---|---|---|---|---|---|---|---|---|
| | P = 1.0 psia (101.69°F)* | | | | P = 5.0 psia (162.18°F) | | | | P = 10 psia (193.16°F) | | | |
| Sat.† | 333.49 | 1043.7 | 1105.4 | 1.9776 | 73.525 | 1062.7 | 1130.7 | 1.8438 | 38.425 | 1072.0 | 1143.1 | 1.7875 |
| 200 | 392.53 | 1077.5 | 1150.1 | 2.0509 | 78.153 | 1076.2 | 1148.5 | 1.8716 | 38.849 | 1074.5 | 1146.4 | 1.7926 |
| 240 | 416.44 | 1091.2 | 1168.3 | 2.0777 | 83.009 | 1090.3 | 1167.1 | 1.8989 | 41.326 | 1089.1 | 1165.5 | 1.8207 |
| 280 | 440.33 | 1105.0 | 1186.5 | 2.1030 | 87.838 | 1104.3 | 1185.6 | 1.9246 | 43.774 | 1103.4 | 1184.4 | 1.8469 |
| 320 | 464.20 | 1118.9 | 1204.8 | 2.1271 | 92.650 | 1118.4 | 1204.1 | 1.9490 | 46.205 | 1117.6 | 1203.1 | 1.8716 |
| 360 | 488.07 | 1132.9 | 1223.3 | 2.1502 | 97.452 | 1132.5 | 1222.6 | 1.9722 | 48.624 | 1131.9 | 1221.8 | 1.8950 |
| 400 | 511.92 | 1147.1 | 1241.8 | 2.1722 | 102.25 | 1146.7 | 1241.3 | 1.9944 | 51.035 | 1146.2 | 1240.6 | 1.9174 |
| 440 | 535.77 | 1161.3 | 1260.4 | 2.1934 | 107.03 | 1160.9 | 1260.0 | 2.0156 | 53.441 | 1160.5 | 1259.4 | 1.9388 |
| 500 | 571.54 | 1182.8 | 1288.6 | 2.2237 | 114.21 | 1182.6 | 1288.2 | 2.0461 | 57.041 | 1182.2 | 1287.8 | 1.9693 |
| 600 | 631.14 | 1219.4 | 1336.2 | 2.2709 | 126.15 | 1219.2 | 1335.9 | 2.0933 | 63.029 | 1219.0 | 1335.6 | 2.0167 |
| 700 | 690.73 | 1256.8 | 1384.6 | 2.3146 | 138.09 | 1256.7 | 1384.4 | 2.1371 | 69.007 | 1256.5 | 1384.2 | 2.0605 |
| 800 | 750.31 | 1295.1 | 1433.9 | 2.3553 | 150.02 | 1294.9 | 1433.7 | 2.1778 | 74.980 | 1294.8 | 1433.5 | 2.1013 |
| 1000 | 869.47 | 1374.2 | 1535.1 | 2.4299 | 173.86 | 1374.2 | 1535.0 | 2.2524 | 86.913 | 1374.1 | 1534.9 | 2.1760 |
| 1200 | 988.62 | 1457.1 | 1640.0 | 2.4972 | 197.70 | 1457.0 | 1640.0 | 2.3198 | 98.840 | 1457.0 | 1639.9 | 2.2433 |
| 1400 | 1107.8 | 1543.7 | 1748.7 | 2.5590 | 221.54 | 1543.7 | 1748.7 | 2.3816 | 110.762 | 1543.6 | 1748.6 | 2.3052 |
| | P = 15 psia (212.99°F) | | | | P = 20 psia (227.92°F) | | | | P = 40 psia (267.22°F) | | | |
| Sat. | 26.297 | 1077.7 | 1150.7 | 1.7549 | 20.093 | 1081.8 | 1156.2 | 1.7319 | 10.501 | 1092.1 | 1169.8 | 1.6766 |
| 240 | 27.429 | 1087.8 | 1163.9 | 1.7742 | 20.478 | 1086.5 | 1162.3 | 1.7406 | | | | |
| 280 | 29.085 | 1102.4 | 1183.2 | 1.8010 | 21.739 | 1101.4 | 1181.9 | 1.7679 | 10.713 | 1097.3 | 1176.6 | 1.6858 |
| 320 | 30.722 | 1116.9 | 1202.2 | 1.8260 | 22.980 | 1116.1 | 1201.2 | 1.7933 | 11.363 | 1112.9 | 1197.1 | 1.7128 |
| 360 | 32.348 | 1131.3 | 1221.1 | 1.8496 | 24.209 | 1130.7 | 1220.2 | 1.8171 | 11.999 | 1128.1 | 1216.9 | 1.7376 |
| 400 | 33.965 | 1145.7 | 1239.9 | 1.8721 | 25.429 | 1145.1 | 1239.3 | 1.8398 | 12.625 | 1143.1 | 1236.5 | 1.7610 |
| 440 | 35.576 | 1160.1 | 1258.8 | 1.8936 | 26.644 | 1159.7 | 1258.3 | 1.8614 | 13.244 | 1157.9 | 1256.0 | 1.7831 |
| 500 | 37.986 | 1181.9 | 1287.3 | 1.9243 | 28.458 | 1181.6 | 1286.9 | 1.8922 | 14.165 | 1180.2 | 1285.0 | 1.8143 |
| 600 | 41.988 | 1218.7 | 1335.3 | 1.9718 | 31.467 | 1218.5 | 1334.9 | 1.9398 | 15.686 | 1217.5 | 1333.6 | 1.8625 |
| 700 | 45.981 | 1256.3 | 1383.9 | 2.0156 | 34.467 | 1256.1 | 1383.7 | 1.9837 | 17.197 | 1255.3 | 1382.6 | 1.9067 |
| 800 | 49.967 | 1294.6 | 1433.3 | 2.0565 | 37.461 | 1294.5 | 1433.1 | 2.0247 | 18.702 | 1293.9 | 1432.3 | 1.9478 |
| 1000 | 57.930 | 1374.0 | 1534.8 | 2.1312 | 43.438 | 1373.8 | 1534.6 | 2.0994 | 21.700 | 1373.4 | 1534.1 | 2.0227 |
| 1200 | 65.885 | 1456.9 | 1639.8 | 2.1986 | 49.407 | 1456.8 | 1639.7 | 2.1668 | 24.691 | 1456.5 | 1639.3 | 2.0902 |
| 1400 | 73.836 | 1543.6 | 1748.5 | 2.2604 | 55.373 | 1543.5 | 1748.4 | 2.2287 | 27.678 | 1543.3 | 1748.1 | 2.1522 |
| 1600 | 81.784 | 1634.0 | 1861.0 | 2.3178 | 61.335 | 1633.9 | 1860.9 | 2.2861 | 30.662 | 1633.7 | 1860.7 | 2.2096 |
| | P = 60 psia (292.69°F) | | | | P = 80 psia (312.02°F) | | | | P = 100 psia (327.81°F) | | | |
| Sat. | 7.1766 | 1098.1 | 1177.8 | 1.6442 | 5.4733 | 1102.3 | 1183.4 | 1.6212 | 4.4327 | 1105.5 | 1187.5 | 1.6032 |
| 320 | 7.4863 | 1109.6 | 1192.7 | 1.6636 | 5.5440 | 1105.9 | 1187.9 | 1.6271 | | | | |
| 360 | 7.9259 | 1125.5 | 1213.5 | 1.6897 | 5.8876 | 1122.7 | 1209.9 | 1.6545 | 4.6628 | 1119.8 | 1206.1 | 1.6263 |
| 400 | 8.3548 | 1140.9 | 1233.7 | 1.7138 | 6.2187 | 1138.7 | 1230.8 | 1.6794 | 4.9359 | 1136.4 | 1227.8 | 1.6521 |
| 440 | 8.7766 | 1156.1 | 1253.6 | 1.7364 | 6.5420 | 1154.3 | 1251.2 | 1.7026 | 5.2006 | 1152.4 | 1248.7 | 1.6759 |
| 500 | 9.4005 | 1178.8 | 1283.1 | 1.7682 | 7.0177 | 1177.3 | 1281.2 | 1.7350 | 5.5876 | 1175.9 | 1279.3 | 1.7088 |
| 600 | 10.4256 | 1216.5 | 1332.2 | 1.8168 | 7.7951 | 1215.4 | 1330.8 | 1.7841 | 6.2167 | 1214.4 | 1329.4 | 1.7586 |
| 700 | 11.4401 | 1254.5 | 1381.6 | 1.8613 | 8.5616 | 1253.8 | 1380.5 | 1.8289 | 6.8344 | 1253.0 | 1379.5 | 1.8037 |
| 800 | 12.4484 | 1293.3 | 1431.5 | 1.9026 | 9.3218 | 1292.6 | 1430.6 | 1.8704 | 7.4457 | 1292.0 | 1429.8 | 1.8453 |
| 1000 | 14.4543 | 1373.0 | 1533.5 | 1.9777 | 10.8313 | 1372.6 | 1532.9 | 1.9457 | 8.6575 | 1372.2 | 1532.4 | 1.9208 |
| 1200 | 16.4525 | 1456.2 | 1638.9 | 2.0454 | 12.3331 | 1455.9 | 1638.5 | 2.0135 | 9.8615 | 1455.6 | 1638.1 | 1.9887 |
| 1400 | 18.4464 | 1543.0 | 1747.8 | 2.1073 | 13.8306 | 1542.8 | 1747.5 | 2.0755 | 11.0612 | 1542.6 | 1747.2 | 2.0508 |
| 1600 | 20.438 | 1633.5 | 1860.5 | 2.1648 | 15.3257 | 1633.3 | 1860.2 | 2.1330 | 12.2584 | 1633.2 | 1860.0 | 2.1083 |
| 1800 | 22.428 | 1727.6 | 1976.6 | 2.2187 | 16.8192 | 1727.5 | 1976.5 | 2.1869 | 13.4541 | 1727.3 | 1976.3 | 2.1622 |
| 2000 | 24.417 | 1825.2 | 2096.3 | 2.2694 | 18.3117 | 1825.0 | 2096.1 | 2.2376 | 14.6487 | 1824.9 | 2096.0 | 2.2130 |

*The temperature in parentheses is the saturation temperature at the specified pressure.

† Properties of saturated vapor at the specified pressure.

## TABLE A–6E

Superheated water (*Continued*)

| T °F | v ft³/lbm | u Btu/lbm | h Btu/lbm | s Btu/ lbm·R | v ft³/lbm | u Btu/lbm | h Btu/lbm | s Btu/ lbm·R | v ft³/lbm | u Btu/lbm | h Btu/lbm | s Btu/ lbm·R |
|---|---|---|---|---|---|---|---|---|---|---|---|---|
| | *P* = 120 psia (341.25°F) | | | | *P* = 140 psia (353.03°F) | | | | *P* = 160 psia (363.54°F) | | | |
| Sat. | 3.7289 | 1107.9 | 1190.8 | 1.5883 | 3.2202 | 1109.9 | 1193.4 | 1.5757 | 2.8347 | 1111.6 | 1195.5 | 1.5647 |
| 360 | 3.8446 | 1116.7 | 1202.1 | 1.6023 | 3.2584 | 1113.4 | 1197.8 | 1.5811 | | | | |
| 400 | 4.0799 | 1134.0 | 1224.6 | 1.6292 | 3.4676 | 1131.5 | 1221.4 | 1.6092 | 3.0076 | 1129.0 | 1218.0 | 1.5914 |
| 450 | 4.3613 | 1154.5 | 1251.4 | 1.6594 | 3.7147 | 1152.6 | 1248.9 | 1.6403 | 3.2293 | 1150.7 | 1246.3 | 1.6234 |
| 500 | 4.6340 | 1174.4 | 1277.3 | 1.6872 | 3.9525 | 1172.9 | 1275.3 | 1.6686 | 3.4412 | 1171.4 | 1273.2 | 1.6522 |
| 550 | 4.9010 | 1193.9 | 1302.8 | 1.7131 | 4.1845 | 1192.7 | 1301.1 | 1.6948 | 3.6469 | 1191.4 | 1299.4 | 1.6788 |
| 600 | 5.1642 | 1213.4 | 1328.0 | 1.7375 | 4.4124 | 1212.3 | 1326.6 | 1.7195 | 3.8484 | 1211.3 | 1325.2 | 1.7037 |
| 700 | 5.6829 | 1252.2 | 1378.4 | 1.7829 | 4.8604 | 1251.4 | 1377.3 | 1.7652 | 4.2434 | 1250.6 | 1376.3 | 1.7498 |
| 800 | 6.1950 | 1291.4 | 1429.0 | 1.8247 | 5.3017 | 1290.8 | 1428.1 | 1.8072 | 4.6316 | 1290.2 | 1427.3 | 1.7920 |
| 1000 | 7.2083 | 1371.7 | 1531.8 | 1.9005 | 6.1732 | 1371.3 | 1531.3 | 1.8832 | 5.3968 | 1370.9 | 1530.7 | 1.8682 |
| 1200 | 8.2137 | 1455.3 | 1637.7 | 1.9684 | 7.0367 | 1455.0 | 1637.3 | 1.9512 | 6.1540 | 1454.7 | 1636.9 | 1.9363 |
| 1400 | 9.2149 | 1542.3 | 1746.9 | 2.0305 | 7.8961 | 1542.1 | 1746.6 | 2.0134 | 6.9070 | 1541.8 | 1746.3 | 1.9986 |
| 1600 | 10.2135 | 1633.0 | 1859.8 | 2.0881 | 8.7529 | 1632.8 | 1859.5 | 2.0711 | 7.6574 | 1632.6 | 1859.3 | 2.0563 |
| 1800 | 11.2106 | 1727.2 | 1976.1 | 2.1420 | 9.6082 | 1727.0 | 1975.9 | 2.1250 | 8.4063 | 1726.9 | 1975.7 | 2.1102 |
| 2000 | 12.2067 | 1824.8 | 2095.8 | 2.1928 | 10.4624 | 1824.6 | 2095.7 | 2.1758 | 9.1542 | 1824.5 | 2095.5 | 2.1610 |
| | *P* = 180 psia (373.07°F) | | | | *P* = 200 psia (381.80°F) | | | | *P* = 225 psia (391.80°F) | | | |
| Sat. | 2.5322 | 1113.0 | 1197.3 | 1.5548 | 2.2882 | 1114.1 | 1198.8 | 1.5460 | 2.0423 | 1115.3 | 1200.3 | 1.5360 |
| 400 | 2.6490 | 1126.3 | 1214.5 | 1.5752 | 2.3615 | 1123.5 | 1210.9 | 1.5602 | 2.0728 | 1119.7 | 1206.0 | 1.5427 |
| 450 | 2.8514 | 1148.7 | 1243.7 | 1.6082 | 2.5488 | 1146.7 | 1241.0 | 1.5943 | 2.2457 | 1144.1 | 1237.6 | 1.5783 |
| 500 | 3.0433 | 1169.8 | 1271.2 | 1.6376 | 2.7247 | 1168.2 | 1269.0 | 1.6243 | 2.4059 | 1166.2 | 1266.3 | 1.6091 |
| 550 | 3.2286 | 1190.2 | 1297.7 | 1.6646 | 2.8939 | 1188.9 | 1296.0 | 1.6516 | 2.5590 | 1187.2 | 1293.8 | 1.6370 |
| 600 | 3.4097 | 1210.2 | 1323.8 | 1.6897 | 3.0586 | 1209.1 | 1322.3 | 1.6771 | 2.7075 | 1207.7 | 1320.5 | 1.6628 |
| 700 | 3.7635 | 1249.8 | 1375.2 | 1.7361 | 3.3796 | 1249.0 | 1374.1 | 1.7238 | 2.9956 | 1248.0 | 1372.7 | 1.7099 |
| 800 | 4.1104 | 1289.5 | 1426.5 | 1.7785 | 3.6934 | 1288.9 | 1425.6 | 1.7664 | 3.2765 | 1288.1 | 1424.5 | 1.7528 |
| 900 | 4.4531 | 1329.7 | 1478.0 | 1.8179 | 4.0031 | 1329.2 | 1477.3 | 1.8059 | 3.5530 | 1328.5 | 1476.5 | 1.7925 |
| 1000 | 4.7929 | 1370.5 | 1530.1 | 1.8549 | 4.3099 | 1370.1 | 1529.6 | 1.8430 | 3.8268 | 1369.5 | 1528.9 | 1.8296 |
| 1200 | 5.4674 | 1454.3 | 1636.5 | 1.9231 | 4.9182 | 1454.0 | 1636.1 | 1.9113 | 4.3689 | 1453.6 | 1635.6 | 1.8981 |
| 1400 | 6.1377 | 1541.6 | 1746.0 | 1.9855 | 5.5222 | 1541.4 | 1745.7 | 1.9737 | 4.9068 | 1541.1 | 1745.4 | 1.9606 |
| 1600 | 6.8054 | 1632.4 | 1859.1 | 2.0432 | 6.1238 | 1632.2 | 1858.8 | 2.0315 | 5.4422 | 1632.0 | 1858.6 | 2.0184 |
| 1800 | 7.4716 | 1726.7 | 1975.6 | 2.0971 | 6.7238 | 1726.5 | 1975.4 | 2.0855 | 5.9760 | 1726.4 | 1975.2 | 2.0724 |
| 2000 | 8.1367 | 1824.4 | 2095.4 | 2.1479 | 7.3227 | 1824.3 | 2095.3 | 2.1363 | 6.5087 | 1824.1 | 2095.1 | 2.1232 |
| | *P* = 250 psia (400.97°F) | | | | *P* = 275 psia (409.45°F) | | | | *P* = 300 psia (417.35°F) | | | |
| Sat. | 1.8440 | 1116.3 | 1201.6 | 1.5270 | 1.6806 | 1117.0 | 1202.6 | 1.5187 | 1.5435 | 1117.7 | 1203.3 | 1.5111 |
| 450 | 2.0027 | 1141.3 | 1234.0 | 1.5636 | 1.8034 | 1138.5 | 1230.3 | 1.5499 | 1.6369 | 1135.6 | 1226.4 | 1.5369 |
| 500 | 2.1506 | 1164.1 | 1263.6 | 1.5953 | 1.9415 | 1162.0 | 1260.8 | 1.5825 | 1.7670 | 1159.8 | 1257.9 | 1.5706 |
| 550 | 2.2910 | 1185.6 | 1291.5 | 1.6237 | 2.0715 | 1183.9 | 1289.3 | 1.6115 | 1.8885 | 1182.1 | 1287.0 | 1.6001 |
| 600 | 2.4264 | 1206.3 | 1318.6 | 1.6499 | 2.1964 | 1204.9 | 1316.7 | 1.6380 | 2.0046 | 1203.5 | 1314.8 | 1.6270 |
| 650 | 2.5586 | 1226.8 | 1345.1 | 1.6743 | 2.3179 | 1225.6 | 1343.5 | 1.6627 | 2.1172 | 1224.4 | 1341.9 | 1.6520 |
| 700 | 2.6883 | 1247.0 | 1371.4 | 1.6974 | 2.4369 | 1246.0 | 1370.0 | 1.6860 | 2.2273 | 1244.9 | 1368.6 | 1.6755 |
| 800 | 2.9429 | 1287.3 | 1423.5 | 1.7406 | 2.6699 | 1286.5 | 1422.4 | 1.7294 | 2.4424 | 1285.7 | 1421.3 | 1.7192 |
| 900 | 3.1930 | 1327.9 | 1475.6 | 1.7804 | 2.8984 | 1327.3 | 1474.8 | 1.7694 | 2.6529 | 1326.6 | 1473.9 | 1.7593 |
| 1000 | 3.4403 | 1369.0 | 1528.2 | 1.8177 | 3.1241 | 1368.5 | 1527.4 | 1.8068 | 2.8605 | 1367.9 | 1526.7 | 1.7968 |
| 1200 | 3.9295 | 1453.3 | 1635.0 | 1.8863 | 3.5700 | 1452.9 | 1634.5 | 1.8755 | 3.2704 | 1452.5 | 1634.0 | 1.8657 |
| 1400 | 4.4144 | 1540.8 | 1745.0 | 1.9488 | 4.0116 | 1540.5 | 1744.6 | 1.9381 | 3.6759 | 1540.2 | 1744.2 | 1.9284 |
| 1600 | 4.8969 | 1631.7 | 1858.3 | 2.0066 | 4.4507 | 1631.5 | 1858.0 | 1.9960 | 4.0789 | 1631.3 | 1857.7 | 1.9863 |
| 1800 | 5.3777 | 1726.2 | 1974.9 | 2.0607 | 4.8882 | 1726.0 | 1974.7 | 2.0501 | 4.4803 | 1725.8 | 1974.5 | 2.0404 |
| 2000 | 5.8575 | 1823.9 | 2094.9 | 2.1116 | 5.3247 | 1823.8 | 2094.7 | 2.1010 | 4.8807 | 1823.6 | 2094.6 | 2.0913 |

## TABLE A–6E

Superheated water (*Continued*)

| T °F | v ft³/lbm | u Btu/lbm | h Btu/lbm | s Btu/ lbm·R | v ft³/lbm | u Btu/lbm | h Btu/lbm | s Btu/ lbm·R | v ft³/lbm | u Btu/lbm | h Btu/lbm | s Btu/ lbm·R |
|---|---|---|---|---|---|---|---|---|---|---|---|---|
| | | *P* = 350 psia (431.74°F) | | | | *P* = 400 psia (444.62°F) | | | | *P* = 450 psia (456.31°F) | | |
| Sat. | 1.3263 | 1118.5 | 1204.4 | 1.4973 | 1.1617 | 1119.0 | 1205.0 | 1.4852 | 1.0324 | 1119.2 | 1205.2 | 1.4742 |
| 450 | 1.3739 | 1129.3 | 1218.3 | 1.5128 | 1.1747 | 1122.5 | 1209.4 | 1.4901 | | | | |
| 500 | 1.4921 | 1155.2 | 1251.9 | 1.5487 | 1.2851 | 1150.4 | 1245.6 | 1.5288 | 1.1233 | 1145.4 | 1238.9 | 1.5103 |
| 550 | 1.6004 | 1178.6 | 1282.2 | 1.5795 | 1.3840 | 1174.9 | 1277.3 | 1.5610 | 1.2152 | 1171.1 | 1272.3 | 1.5441 |
| 600 | 1.7030 | 1200.6 | 1310.9 | 1.6073 | 1.4765 | 1197.6 | 1306.9 | 1.5897 | 1.3001 | 1194.6 | 1302.8 | 1.5737 |
| 650 | 1.8018 | 1221.9 | 1338.6 | 1.6328 | 1.5650 | 1219.4 | 1335.3 | 1.6158 | 1.3807 | 1216.9 | 1331.9 | 1.6005 |
| 700 | 1.8979 | 1242.8 | 1365.8 | 1.6567 | 1.6507 | 1240.7 | 1362.9 | 1.6401 | 1.4584 | 1238.5 | 1360.0 | 1.6253 |
| 800 | 2.0848 | 1284.1 | 1419.1 | 1.7009 | 1.8166 | 1282.5 | 1417.0 | 1.6849 | 1.6080 | 1280.8 | 1414.7 | 1.6706 |
| 900 | 2.2671 | 1325.3 | 1472.2 | 1.7414 | 1.9777 | 1324.0 | 1470.4 | 1.7257 | 1.7526 | 1322.7 | 1468.6 | 1.7117 |
| 1000 | 2.4464 | 1366.9 | 1525.3 | 1.7791 | 2.1358 | 1365.8 | 1523.9 | 1.7636 | 1.8942 | 1364.7 | 1522.4 | 1.7499 |
| 1200 | 2.7996 | 1451.7 | 1633.0 | 1.8483 | 2.4465 | 1450.9 | 1632.0 | 1.8331 | 2.1718 | 1450.1 | 1631.0 | 1.8196 |
| 1400 | 3.1484 | 1539.6 | 1743.5 | 1.9111 | 2.7527 | 1539.0 | 1742.7 | 1.8960 | 2.4450 | 1538.4 | 1742.0 | 1.8827 |
| 1600 | 3.4947 | 1630.8 | 1857.1 | 1.9691 | 3.0565 | 1630.3 | 1856.5 | 1.9541 | 2.7157 | 1629.8 | 1856.0 | 1.9409 |
| 1800 | 3.8394 | 1725.4 | 1974.0 | 2.0233 | 3.3586 | 1725.0 | 1973.6 | 2.0084 | 2.9847 | 1724.6 | 1973.2 | 1.9952 |
| 2000 | 4.1830 | 1823.3 | 2094.2 | 2.0742 | 3.6597 | 1823.0 | 2093.9 | 2.0594 | 3.2527 | 1822.6 | 2093.5 | 2.0462 |
| | | *P* = 500 psia (467.04°F) | | | | *P* = 600 psia (486.24°F) | | | | *P* = 700 psia (503.13°F) | | |
| Sat. | 0.92815 | 1119.1 | 1205.0 | 1.4642 | 0.77020 | 1118.3 | 1203.9 | 1.4463 | 0.65589 | 1116.9 | 1201.9 | 1.4305 |
| 500 | 0.99304 | 1140.1 | 1231.9 | 1.4928 | 0.79526 | 1128.2 | 1216.5 | 1.4596 | | | | |
| 550 | 1.07974 | 1167.1 | 1267.0 | 1.5284 | 0.87542 | 1158.7 | 1255.9 | 1.4996 | 0.72799 | 1149.5 | 1243.8 | 1.4730 |
| 600 | 1.15876 | 1191.4 | 1298.6 | 1.5590 | 0.94605 | 1184.9 | 1289.9 | 1.5325 | 0.79332 | 1177.9 | 1280.7 | 1.5087 |
| 650 | 1.23312 | 1214.3 | 1328.4 | 1.5865 | 1.01133 | 1209.0 | 1321.3 | 1.5614 | 0.85242 | 1203.4 | 1313.8 | 1.5393 |
| 700 | 1.30440 | 1236.4 | 1357.0 | 1.6117 | 1.07316 | 1231.9 | 1351.0 | 1.5877 | 0.90769 | 1227.2 | 1344.8 | 1.5666 |
| 800 | 1.44097 | 1279.2 | 1412.5 | 1.6576 | 1.19038 | 1275.8 | 1408.0 | 1.6348 | 1.01125 | 1272.4 | 1403.4 | 1.6150 |
| 900 | 1.57252 | 1321.4 | 1466.9 | 1.6992 | 1.30230 | 1318.7 | 1463.3 | 1.6771 | 1.10921 | 1316.0 | 1459.7 | 1.6581 |
| 1000 | 1.70094 | 1363.6 | 1521.0 | 1.7376 | 1.41097 | 1361.4 | 1518.1 | 1.7160 | 1.20381 | 1359.2 | 1515.2 | 1.6974 |
| 1100 | 1.82726 | 1406.2 | 1575.3 | 1.7735 | 1.51749 | 1404.4 | 1572.9 | 1.7522 | 1.29621 | 1402.5 | 1570.4 | 1.7341 |
| 1200 | 1.95211 | 1449.4 | 1630.0 | 1.8075 | 1.62252 | 1447.8 | 1627.9 | 1.7865 | 1.38709 | 1446.2 | 1625.9 | 1.7685 |
| 1400 | 2.1988 | 1537.8 | 1741.2 | 1.8708 | 1.82957 | 1536.6 | 1739.7 | 1.8501 | 1.56586 | 1535.4 | 1738.2 | 1.8324 |
| 1600 | 2.4430 | 1629.4 | 1855.4 | 1.9291 | 2.0340 | 1628.4 | 1854.2 | 1.9085 | 1.74192 | 1627.5 | 1853.1 | 1.8911 |
| 1800 | 2.6856 | 1724.2 | 1972.7 | 1.9834 | 2.2369 | 1723.4 | 1971.8 | 1.9630 | 1.91643 | 1722.7 | 1970.9 | 1.9457 |
| 2000 | 2.9271 | 1822.3 | 2093.1 | 2.0345 | 2.4387 | 1821.7 | 2092.4 | 2.0141 | 2.08987 | 1821.0 | 2091.7 | 1.9969 |
| | | *P* = 800 psia (518.27°F) | | | | *P* = 1000 psia (544.65°F) | | | | *P* = 1250 psia (572.45°F) | | |
| Sat. | 0.56920 | 1115.0 | 1199.3 | 1.4162 | 0.44604 | 1110.1 | 1192.6 | 1.3906 | 0.34549 | 1102.0 | 1181.9 | 1.3623 |
| 550 | 0.61586 | 1139.4 | 1230.5 | 1.4476 | 0.45375 | 1115.2 | 1199.2 | 1.3972 | | | | |
| 600 | 0.67799 | 1170.5 | 1270.9 | 1.4866 | 0.51431 | 1154.1 | 1249.3 | 1.4457 | 0.37894 | 1129.5 | 1217.2 | 1.3961 |
| 650 | 0.73279 | 1197.6 | 1306.0 | 1.5191 | 0.56411 | 1185.1 | 1289.5 | 1.4827 | 0.42703 | 1167.5 | 1266.3 | 1.4414 |
| 700 | 0.78330 | 1222.4 | 1338.4 | 1.5476 | 0.60844 | 1212.4 | 1325.0 | 1.5140 | 0.46735 | 1198.7 | 1306.8 | 1.4771 |
| 750 | 0.83102 | 1246.0 | 1369.1 | 1.5735 | 0.64944 | 1237.6 | 1357.8 | 1.5418 | 0.50344 | 1226.4 | 1342.9 | 1.5076 |
| 800 | 0.87678 | 1268.9 | 1398.7 | 1.5975 | 0.68821 | 1261.7 | 1389.0 | 1.5670 | 0.53687 | 1252.2 | 1376.4 | 1.5347 |
| 900 | 0.96434 | 1313.3 | 1456.0 | 1.6413 | 0.76136 | 1307.7 | 1448.6 | 1.6126 | 0.59876 | 1300.5 | 1439.0 | 1.5826 |
| 1000 | 1.04841 | 1357.0 | 1512.2 | 1.6812 | 0.83078 | 1352.5 | 1506.2 | 1.6535 | 0.65656 | 1346.7 | 1498.6 | 1.6249 |
| 1100 | 1.13024 | 1400.7 | 1568.0 | 1.7181 | 0.89783 | 1396.9 | 1563.1 | 1.6911 | 0.71184 | 1392.2 | 1556.8 | 1.6635 |
| 1200 | 1.21051 | 1444.6 | 1623.8 | 1.7528 | 0.96327 | 1441.4 | 1619.7 | 1.7263 | 0.76545 | 1437.4 | 1614.5 | 1.6993 |
| 1400 | 1.36797 | 1534.2 | 1736.7 | 1.8170 | 1.09101 | 1531.8 | 1733.7 | 1.7911 | 0.86944 | 1528.7 | 1729.8 | 1.7649 |
| 1600 | 1.52283 | 1626.5 | 1851.9 | 1.8759 | 1.21610 | 1624.6 | 1849.6 | 1.8504 | 0.97072 | 1622.2 | 1846.7 | 1.8246 |
| 1800 | 1.67606 | 1721.9 | 1970.0 | 1.9306 | 1.33956 | 1720.3 | 1968.2 | 1.9053 | 1.07036 | 1718.4 | 1966.0 | 1.8799 |
| 2000 | 1.82823 | 1820.4 | 2091.0 | 1.9819 | 1.46194 | 1819.1 | 2089.6 | 1.9568 | 1.16892 | 1817.5 | 2087.9 | 1.9315 |

## TABLE A–6E

Superheated water (*Concluded*)

| T °F | v ft³/lbm | u Btu/lbm | h Btu/lbm | s Btu/ lbm·R | v ft³/lbm | u Btu/lbm | h Btu/lbm | s Btu/ lbm·R | v ft³/lbm | u Btu/lbm | h Btu/lbm | s Btu/ lbm·R |
|---|---|---|---|---|---|---|---|---|---|---|---|---|
| | *P* = 1500 psia (596.26°F) | | | | *P* = 1750 psia (617.17°F) | | | | *P* = 2000 psia (635.85°F) | | | |
| Sat. | 0.27695 | 1092.1 | 1169.0 | 1.3362 | 0.22681 | 1080.5 | 1153.9 | 1.3112 | 0.18815 | 1066.8 | 1136.4 | 1.2863 |
| 600 | 0.28189 | 1097.2 | 1175.4 | 1.3423 | | | | | | | | |
| 650 | 0.33310 | 1147.2 | 1239.7 | 1.4016 | 0.26292 | 1122.8 | 1207.9 | 1.3607 | 0.20586 | 1091.4 | 1167.6 | 1.3146 |
| 700 | 0.37198 | 1183.6 | 1286.9 | 1.4433 | 0.30252 | 1166.8 | 1264.7 | 1.4108 | 0.24894 | 1147.6 | 1239.8 | 1.3783 |
| 750 | 0.40535 | 1214.4 | 1326.9 | 1.4771 | 0.33455 | 1201.5 | 1309.8 | 1.4489 | 0.28074 | 1187.4 | 1291.3 | 1.4218 |
| 800 | 0.43550 | 1242.2 | 1363.1 | 1.5064 | 0.36266 | 1231.7 | 1349.1 | 1.4807 | 0.30763 | 1220.5 | 1334.3 | 1.4567 |
| 850 | 0.46356 | 1268.2 | 1396.9 | 1.5328 | 0.38835 | 1259.3 | 1385.1 | 1.5088 | 0.33169 | 1250.0 | 1372.8 | 1.4867 |
| 900 | 0.49015 | 1293.1 | 1429.2 | 1.5569 | 0.41238 | 1285.4 | 1419.0 | 1.5341 | 0.35390 | 1277.5 | 1408.5 | 1.5134 |
| 1000 | 0.54031 | 1340.9 | 1490.8 | 1.6007 | 0.45719 | 1334.9 | 1482.9 | 1.5796 | 0.39479 | 1328.7 | 1474.9 | 1.5606 |
| 1100 | 0.58781 | 1387.3 | 1550.5 | 1.6402 | 0.49917 | 1382.4 | 1544.1 | 1.6201 | 0.43266 | 1377.5 | 1537.6 | 1.6021 |
| 1200 | 0.63355 | 1433.3 | 1609.2 | 1.6767 | 0.53932 | 1429.2 | 1603.9 | 1.6572 | 0.46864 | 1425.1 | 1598.5 | 1.6400 |
| 1400 | 0.72172 | 1525.7 | 1726.0 | 1.7432 | 0.61621 | 1522.6 | 1722.1 | 1.7245 | 0.53708 | 1519.5 | 1718.3 | 1.7081 |
| 1600 | 0.80714 | 1619.8 | 1843.8 | 1.8033 | 0.69031 | 1617.4 | 1840.9 | 1.7852 | 0.60269 | 1615.0 | 1838.0 | 1.7693 |
| 1800 | 0.89090 | 1716.4 | 1963.7 | 1.8589 | 0.76273 | 1714.5 | 1961.5 | 1.8410 | 0.66660 | 1712.5 | 1959.2 | 1.8255 |
| 2000 | 0.97358 | 1815.9 | 2086.1 | 1.9108 | 0.83406 | 1814.2 | 2084.3 | 1.8931 | 0.72942 | 1812.6 | 2082.6 | 1.8778 |
| | *P* = 2500 psia (668.17°F) | | | | *P* = 3000 psia (695.41°F) | | | | *P* = 3500 psia | | | |
| Sat. | 0.13076 | 1031.2 | 1091.7 | 1.2330 | 0.08460 | 969.8 | 1016.8 | 1.1587 | | | | |
| 650 | | | | | | | | | 0.02492 | 663.7 | 679.9 | 0.8632 |
| 700 | 0.16849 | 1098.4 | 1176.3 | 1.3072 | 0.09838 | 1005.3 | 1059.9 | 1.1960 | 0.03065 | 760.0 | 779.9 | 0.9511 |
| 750 | 0.20327 | 1154.9 | 1249.0 | 1.3686 | 0.14840 | 1114.1 | 1196.5 | 1.3118 | 0.10460 | 1057.6 | 1125.4 | 1.2434 |
| 800 | 0.22949 | 1195.9 | 1302.0 | 1.4116 | 0.17601 | 1167.5 | 1265.3 | 1.3676 | 0.13639 | 1134.3 | 1222.6 | 1.3224 |
| 850 | 0.25174 | 1230.1 | 1346.6 | 1.4463 | 0.19771 | 1208.2 | 1317.9 | 1.4086 | 0.15847 | 1183.8 | 1286.5 | 1.3721 |
| 900 | 0.27165 | 1260.7 | 1386.4 | 1.4761 | 0.21640 | 1242.8 | 1362.9 | 1.4423 | 0.17659 | 1223.4 | 1337.8 | 1.4106 |
| 950 | 0.29001 | 1289.1 | 1423.3 | 1.5028 | 0.23321 | 1273.9 | 1403.3 | 1.4716 | 0.19245 | 1257.8 | 1382.4 | 1.4428 |
| 1000 | 0.30726 | 1316.1 | 1458.2 | 1.5271 | 0.24876 | 1302.8 | 1440.9 | 1.4978 | 0.20687 | 1289.0 | 1423.0 | 1.4711 |
| 1100 | 0.33949 | 1367.3 | 1524.4 | 1.5710 | 0.27732 | 1356.8 | 1510.8 | 1.5441 | 0.23289 | 1346.1 | 1496.9 | 1.5201 |
| 1200 | 0.36966 | 1416.6 | 1587.6 | 1.6103 | 0.30367 | 1408.0 | 1576.6 | 1.5850 | 0.25654 | 1399.3 | 1565.4 | 1.5627 |
| 1400 | 0.42631 | 1513.3 | 1710.5 | 1.6802 | 0.35249 | 1507.0 | 1702.7 | 1.6567 | 0.29978 | 1500.7 | 1694.8 | 1.6364 |
| 1600 | 0.48004 | 1610.1 | 1832.2 | 1.7424 | 0.39830 | 1605.3 | 1826.4 | 1.7199 | 0.33994 | 1600.4 | 1820.5 | 1.7006 |
| 1800 | 0.53205 | 1708.6 | 1954.8 | 1.7991 | 0.44237 | 1704.7 | 1950.3 | 1.7773 | 0.37833 | 1700.8 | 1945.8 | 1.7586 |
| 2000 | 0.58295 | 1809.4 | 2079.1 | 1.8518 | 0.48532 | 1806.1 | 2075.6 | 1.8304 | 0.41561 | 1802.9 | 2072.1 | 1.8121 |
| | *P* = 4000 psia | | | | *P* = 5000 psia | | | | *P* = 6000 psia | | | |
| 650 | 0.02448 | 657.9 | 676.1 | 0.8577 | 0.02379 | 648.3 | 670.3 | 0.8485 | 0.02325 | 640.3 | 666.1 | 0.8408 |
| 700 | 0.02871 | 742.3 | 763.6 | 0.9347 | 0.02678 | 721.8 | 746.6 | 0.9156 | 0.02564 | 708.1 | 736.5 | 0.9028 |
| 750 | 0.06370 | 962.1 | 1009.2 | 1.1410 | 0.03373 | 821.8 | 853.0 | 1.0054 | 0.02981 | 788.7 | 821.8 | 0.9747 |
| 800 | 0.10520 | 1094.2 | 1172.1 | 1.2734 | 0.05937 | 986.9 | 1041.8 | 1.1581 | 0.03949 | 897.1 | 941.0 | 1.0711 |
| 850 | 0.12848 | 1156.7 | 1251.8 | 1.3355 | 0.08551 | 1092.4 | 1171.5 | 1.2593 | 0.05815 | 1018.6 | 1083.1 | 1.1819 |
| 900 | 0.14647 | 1202.5 | 1310.9 | 1.3799 | 0.10390 | 1155.9 | 1252.1 | 1.3198 | 0.07584 | 1103.5 | 1187.7 | 1.2603 |
| 950 | 0.16176 | 1240.7 | 1360.5 | 1.4157 | 0.11863 | 1203.9 | 1313.6 | 1.3643 | 0.09010 | 1163.7 | 1263.7 | 1.3153 |
| 1000 | 0.17538 | 1274.6 | 1404.4 | 1.4463 | 0.13128 | 1244.0 | 1365.5 | 1.4004 | 0.10208 | 1211.4 | 1324.7 | 1.3578 |
| 1100 | 0.19957 | 1335.1 | 1482.8 | 1.4983 | 0.15298 | 1312.2 | 1453.8 | 1.4590 | 0.12211 | 1288.4 | 1424.0 | 1.4237 |
| 1200 | 0.22121 | 1390.3 | 1554.1 | 1.5426 | 0.17185 | 1372.1 | 1531.1 | 1.5070 | 0.13911 | 1353.4 | 1507.8 | 1.4758 |
| 1300 | 0.24128 | 1443.0 | 1621.6 | 1.5821 | 0.18902 | 1427.8 | 1602.7 | 1.5490 | 0.15434 | 1412.5 | 1583.8 | 1.5203 |
| 1400 | 0.26028 | 1494.3 | 1687.0 | 1.6182 | 0.20508 | 1481.4 | 1671.1 | 1.5868 | 0.16841 | 1468.4 | 1655.4 | 1.5598 |
| 1600 | 0.29620 | 1595.5 | 1814.7 | 1.6835 | 0.23505 | 1585.6 | 1803.1 | 1.6542 | 0.19438 | 1575.7 | 1791.5 | 1.6294 |
| 1800 | 0.33033 | 1696.8 | 1941.4 | 1.7422 | 0.26320 | 1689.0 | 1932.5 | 1.7142 | 0.21853 | 1681.1 | 1923.7 | 1.6907 |
| 2000 | 0.36335 | 1799.7 | 2068.6 | 1.7961 | 0.29023 | 1793.2 | 2061.7 | 1.7689 | 0.24155 | 1786.7 | 2054.9 | 1.7463 |

## TABLE A–7E

Compressed liquid water

| $T$ °F | $v$ ft³/lbm | $u$ Btu/lbm | $h$ Btu/lbm | $s$ Btu/lbm·R | $v$ ft³/lbm | $u$ Btu/lbm | $h$ Btu/lbm | $s$ Btu/lbm·R | $v$ ft³/lbm | $u$ Btu/lbm | $h$ Btu/lbm | $s$ Btu/lbm·R |
|---|---|---|---|---|---|---|---|---|---|---|---|---|
| | P = 500 psia (467.04°F) | | | | P = 1000 psia (544.65°F) | | | | P = 1500 psia (596.26°F) | | | |
| Sat. | 0.019750 | 447.68 | 449.51 | 0.64900 | 0.021595 | 538.58 | 542.57 | 0.74341 | 0.023456 | 605.07 | 611.58 | 0.80836 |
| 32 | 0.015994 | 0.01 | 1.49 | 0.00001 | 0.015966 | 0.03 | 2.99 | 0.00005 | 0.015939 | 0.05 | 4.48 | 0.00008 |
| 50 | 0.015998 | 18.03 | 19.51 | 0.03601 | 0.015972 | 17.99 | 20.95 | 0.03593 | 0.015946 | 17.95 | 22.38 | 0.03584 |
| 100 | 0.016107 | 67.86 | 69.35 | 0.12930 | 0.016083 | 67.69 | 70.67 | 0.12899 | 0.016059 | 67.53 | 71.98 | 0.12869 |
| 150 | 0.016317 | 117.70 | 119.21 | 0.21462 | 0.016292 | 117.42 | 120.43 | 0.21416 | 0.016267 | 117.14 | 121.66 | 0.21369 |
| 200 | 0.016607 | 167.70 | 169.24 | 0.29349 | 0.016580 | 167.31 | 170.38 | 0.29289 | 0.016553 | 166.92 | 171.52 | 0.29229 |
| 250 | 0.016972 | 218.04 | 219.61 | 0.36708 | 0.016941 | 217.51 | 220.65 | 0.36634 | 0.016911 | 217.00 | 221.69 | 0.36560 |
| 300 | 0.017417 | 268.92 | 270.53 | 0.43641 | 0.017380 | 268.24 | 271.46 | 0.43551 | 0.017345 | 267.57 | 272.39 | 0.43463 |
| 350 | 0.017954 | 320.64 | 322.30 | 0.50240 | 0.017910 | 319.77 | 323.08 | 0.50132 | 0.017866 | 318.91 | 323.87 | 0.50025 |
| 400 | 0.018609 | 373.61 | 375.33 | 0.56595 | 0.018552 | 372.48 | 375.91 | 0.56463 | 0.018496 | 371.37 | 376.51 | 0.56333 |
| 450 | 0.019425 | 428.44 | 430.24 | 0.62802 | 0.019347 | 426.93 | 430.51 | 0.62635 | 0.019271 | 425.47 | 430.82 | 0.62472 |
| 500 | | | | | 0.020368 | 484.03 | 487.80 | 0.68764 | 0.020258 | 482.01 | 487.63 | 0.68550 |
| 550 | | | | | | | | | 0.021595 | 542.50 | 548.50 | 0.74731 |
| | P = 2000 psia (635.85°F) | | | | P = 3000 psia (695.41°F) | | | | P = 5000 psia | | | |
| Sat. | 0.025634 | 662.33 | 671.82 | 0.86224 | 0.034335 | 783.39 | 802.45 | 0.97321 | | | | |
| 32 | 0.015912 | 0.07 | 5.96 | 0.00010 | 0.015859 | 0.10 | 8.90 | 0.00011 | 0.015756 | 0.13 | 14.71 | 0.00002 |
| 50 | 0.015921 | 17.91 | 23.80 | 0.03574 | 0.015870 | 17.83 | 26.64 | 0.03554 | 0.015773 | 17.65 | 32.25 | 0.03505 |
| 100 | 0.016035 | 67.36 | 73.30 | 0.12838 | 0.015988 | 67.04 | 75.91 | 0.12776 | 0.015897 | 66.41 | 81.12 | 0.12652 |
| 200 | 0.016527 | 166.54 | 172.66 | 0.29170 | 0.016475 | 165.79 | 174.94 | 0.29053 | 0.016375 | 164.36 | 179.51 | 0.28824 |
| 300 | 0.017310 | 266.92 | 273.33 | 0.43376 | 0.017242 | 265.65 | 275.22 | 0.43204 | 0.017112 | 263.24 | 279.07 | 0.42874 |
| 400 | 0.018442 | 370.30 | 377.12 | 0.56205 | 0.018338 | 368.22 | 378.41 | 0.55959 | 0.018145 | 364.35 | 381.14 | 0.55492 |
| 450 | 0.019199 | 424.06 | 431.16 | 0.62314 | 0.019062 | 421.36 | 431.94 | 0.62010 | 0.018812 | 416.40 | 433.80 | 0.61445 |
| 500 | 0.020154 | 480.08 | 487.54 | 0.68346 | 0.019960 | 476.45 | 487.53 | 0.67958 | 0.019620 | 469.94 | 488.10 | 0.67254 |
| 560 | 0.021739 | 552.21 | 560.26 | 0.75692 | 0.021405 | 546.59 | 558.47 | 0.75126 | 0.020862 | 537.08 | 556.38 | 0.74154 |
| 600 | 0.023317 | 605.77 | 614.40 | 0.80898 | 0.022759 | 597.42 | 610.06 | 0.80086 | 0.021943 | 584.42 | 604.72 | 0.78803 |
| 640 | | | | | 0.024765 | 654.52 | 668.27 | 0.85476 | 0.023358 | 634.95 | 656.56 | 0.83603 |
| 680 | | | | | 0.028821 | 728.63 | 744.64 | 0.92288 | 0.025366 | 690.67 | 714.14 | 0.88745 |
| 700 | | | | | | | | | 0.026777 | 721.78 | 746.56 | 0.91564 |

## TABLE A–8E

Saturated ice—water vapor

| Temp., $T$ °F | Sat. press., $P_{sat}$ psia | Specific volume, ft³/lbm | | Internal energy, Btu/lbm | | | Enthalpy, Btu/lbm | | | Entropy, Btu/lbm·R | | |
|---|---|---|---|---|---|---|---|---|---|---|---|---|
| | | Sat. ice, $v_i$ | Sat. vapor, $v_g$ | Sat. ice, $u_i$ | Subl., $u_{ig}$ | Sat. vapor, $u_g$ | Sat. ice, $h_i$ | Subl., $h_{ig}$ | Sat. vapor, $h_g$ | Sat. ice, $s_i$ | Subl., $s_{ig}$ | Sat. vapor, $s_g$ |
| 32.018 | 0.08871 | 0.01747 | 3299.6 | −143.34 | 1164.2 | 1020.9 | −143.34 | 1218.3 | 1075.0 | −0.29146 | 2.4779 | 2.1864 |
| 32 | 0.08864 | 0.01747 | 3302.6 | −143.35 | 1164.2 | 1020.9 | −143.35 | 1218.4 | 1075.0 | −0.29148 | 2.4779 | 2.1865 |
| 30 | 0.08086 | 0.01747 | 3605.8 | −144.35 | 1164.6 | 1020.2 | −144.35 | 1218.5 | 1074.2 | −0.29353 | 2.4883 | 2.1948 |
| 25 | 0.06405 | 0.01746 | 4505.8 | −146.85 | 1165.4 | 1018.6 | −146.85 | 1218.8 | 1072.0 | −0.29865 | 2.5146 | 2.2160 |
| 20 | 0.05049 | 0.01746 | 5657.6 | −149.32 | 1166.2 | 1016.9 | −149.32 | 1219.1 | 1069.8 | −0.30377 | 2.5414 | 2.2376 |
| 15 | 0.03960 | 0.01745 | 7138.9 | −151.76 | 1167.0 | 1015.2 | −151.76 | 1219.3 | 1067.6 | −0.30889 | 2.5687 | 2.2598 |
| 10 | 0.03089 | 0.01744 | 9054.0 | −154.18 | 1167.8 | 1013.6 | −154.18 | 1219.5 | 1065.4 | −0.31401 | 2.5965 | 2.2825 |
| 5 | 0.02397 | 0.01743 | 11,543 | −156.57 | 1168.5 | 1011.9 | −156.57 | 1219.7 | 1063.1 | −0.31913 | 2.6248 | 2.3057 |
| 0 | 0.01850 | 0.01743 | 14,797 | −158.94 | 1169.2 | 1010.3 | −158.94 | 1219.9 | 1060.9 | −0.32426 | 2.6537 | 2.3295 |
| −5 | 0.01420 | 0.01742 | 19,075 | −161.28 | 1169.9 | 1008.6 | −161.28 | 1220.0 | 1058.7 | −0.32938 | 2.6832 | 2.3538 |
| −10 | 0.01083 | 0.01741 | 24,731 | −163.60 | 1170.6 | 1007.0 | −163.60 | 1220.1 | 1056.5 | −0.33451 | 2.7133 | 2.3788 |
| −15 | 0.00821 | 0.01740 | 32,257 | −165.90 | 1171.2 | 1005.3 | −165.90 | 1220.2 | 1054.3 | −0.33964 | 2.7440 | 2.4044 |
| −20 | 0.00619 | 0.01740 | 42,335 | −168.16 | 1171.8 | 1003.6 | −168.16 | 1220.3 | 1052.1 | −0.34478 | 2.7754 | 2.4306 |
| −25 | 0.00463 | 0.01739 | 55,917 | −170.41 | 1172.4 | 1002.0 | −170.41 | 1220.3 | 1049.9 | −0.34991 | 2.8074 | 2.4575 |
| −30 | 0.00344 | 0.01738 | 74,345 | −172.63 | 1173.0 | 1000.3 | −172.63 | 1220.3 | 1047.7 | −0.35505 | 2.8401 | 2.4850 |
| −35 | 0.00254 | 0.01738 | 99,526 | −174.83 | 1173.5 | 998.7 | −174.83 | 1220.3 | 1045.5 | −0.36019 | 2.8735 | 2.5133 |
| −40 | 0.00186 | 0.01737 | 134,182 | −177.00 | 1174.0 | 997.0 | −177.00 | 1220.3 | 1043.3 | −0.36534 | 2.9076 | 2.5423 |

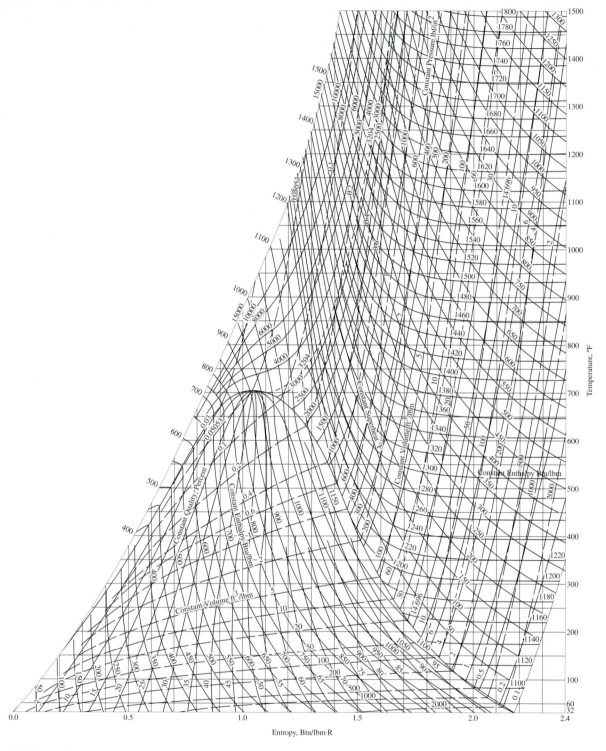

**FIGURE A–9E**

*T-s* diagram for water.

*Source: Joseph H. Keenan, Frederick G. Keyes, Philip G. Hill, and Joan G. Moore,* Steam Tables *(New York: John Wiley & Sons, 1969).*

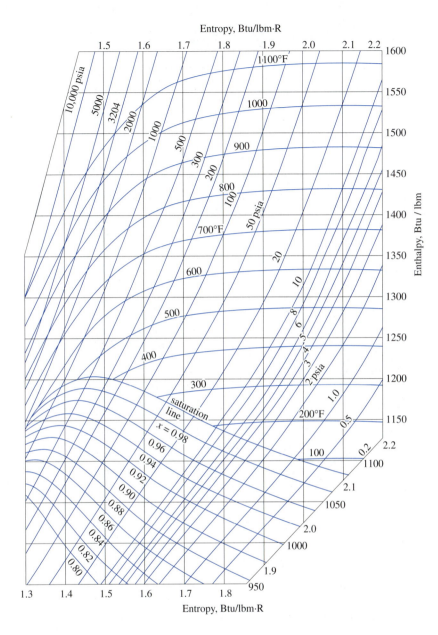

**FIGURE A–10E**

Mollier diagram for water.

*Source: Joseph H. Keenan, Frederick G. Keyes, Philip G. Hill, and Joan G. Moore,* Steam Tables *(New York: John Wiley & Sons, 1969).*

## TABLE A–11E

Saturated refrigerant-134a—Temperature table

| Temp., $T$ °F | Sat. press., $P_{sat}$ psia | Specific volume, ft³/lbm | | Internal energy, Btu/lbm | | | Enthalpy, Btu/lbm | | | | Entropy, Btu/lbm·R | | |
|---|---|---|---|---|---|---|---|---|---|---|---|---|---|
| | | Sat. liquid, $v_f$ | Sat. vapor, $v_g$ | Sat. liquid, $u_f$ | Evap., $u_{fg}$ | Sat. vapor, $u_g$ | Sat. liquid, $h_f$ | Evap., $h_{fg}$ | Sat. vapor, $h_g$ | Sat. liquid, $s_f$ | Evap., $s_{fg}$ | Sat. vapor, $s_g$ |
| −40 | 7.432 | 0.01130 | 5.7796 | −0.016 | 89.167 | 89.15 | 0.000 | 97.100 | 97.10 | 0.00000 | 0.23135 | 0.23135 |
| −35 | 8.581 | 0.01136 | 5.0509 | 1.484 | 88.352 | 89.84 | 1.502 | 96.354 | 97.86 | 0.00355 | 0.22687 | 0.23043 |
| −30 | 9.869 | 0.01143 | 4.4300 | 2.990 | 87.532 | 90.52 | 3.011 | 95.601 | 98.61 | 0.00708 | 0.22248 | 0.22956 |
| −25 | 11.306 | 0.01150 | 3.8988 | 4.502 | 86.706 | 91.21 | 4.526 | 94.839 | 99.36 | 0.01058 | 0.21817 | 0.22875 |
| −20 | 12.906 | 0.01156 | 3.4426 | 6.019 | 85.874 | 91.89 | 6.047 | 94.068 | 100.12 | 0.01405 | 0.21394 | 0.22798 |
| −15 | 14.680 | 0.01163 | 3.0494 | 7.543 | 85.036 | 92.58 | 7.574 | 93.288 | 100.86 | 0.01749 | 0.20978 | 0.22727 |
| −10 | 16.642 | 0.01171 | 2.7091 | 9.073 | 84.191 | 93.26 | 9.109 | 92.498 | 101.61 | 0.02092 | 0.20569 | 0.22660 |
| −5 | 18.806 | 0.01178 | 2.4137 | 10.609 | 83.339 | 93.95 | 10.650 | 91.698 | 102.35 | 0.02431 | 0.20166 | 0.22598 |
| 0 | 21.185 | 0.01185 | 2.1564 | 12.152 | 82.479 | 94.63 | 12.199 | 90.886 | 103.08 | 0.02769 | 0.19770 | 0.22539 |
| 5 | 23.793 | 0.01193 | 1.9316 | 13.702 | 81.610 | 95.31 | 13.755 | 90.062 | 103.82 | 0.03104 | 0.19380 | 0.22485 |
| 10 | 26.646 | 0.01201 | 1.7345 | 15.259 | 80.733 | 95.99 | 15.318 | 89.226 | 104.54 | 0.03438 | 0.18996 | 0.22434 |
| 15 | 29.759 | 0.01209 | 1.5612 | 16.823 | 79.846 | 96.67 | 16.889 | 88.377 | 105.27 | 0.03769 | 0.18617 | 0.22386 |
| 20 | 33.147 | 0.01217 | 1.4084 | 18.394 | 78.950 | 97.34 | 18.469 | 87.514 | 105.98 | 0.04098 | 0.18243 | 0.22341 |
| 25 | 36.826 | 0.01225 | 1.2732 | 19.973 | 78.043 | 98.02 | 20.056 | 86.636 | 106.69 | 0.04426 | 0.17874 | 0.22300 |
| 30 | 40.813 | 0.01234 | 1.1534 | 21.560 | 77.124 | 98.68 | 21.653 | 85.742 | 107.40 | 0.04752 | 0.17509 | 0.22260 |
| 35 | 45.124 | 0.01242 | 1.0470 | 23.154 | 76.195 | 99.35 | 23.258 | 84.833 | 108.09 | 0.05076 | 0.17148 | 0.22224 |
| 40 | 49.776 | 0.01251 | 0.95205 | 24.757 | 75.253 | 100.01 | 24.873 | 83.907 | 108.78 | 0.05398 | 0.16791 | 0.22189 |
| 45 | 54.787 | 0.01261 | 0.86727 | 26.369 | 74.298 | 100.67 | 26.497 | 82.963 | 109.46 | 0.05720 | 0.16437 | 0.22157 |
| 50 | 60.175 | 0.01270 | 0.79136 | 27.990 | 73.329 | 101.32 | 28.131 | 82.000 | 110.13 | 0.06039 | 0.16087 | 0.22127 |
| 55 | 65.957 | 0.01280 | 0.72323 | 29.619 | 72.346 | 101.97 | 29.775 | 81.017 | 110.79 | 0.06358 | 0.15740 | 0.22098 |
| 60 | 72.152 | 0.01290 | 0.66195 | 31.258 | 71.347 | 102.61 | 31.431 | 80.013 | 111.44 | 0.06675 | 0.15396 | 0.22070 |
| 65 | 78.780 | 0.01301 | 0.60671 | 32.908 | 70.333 | 103.24 | 33.097 | 78.988 | 112.09 | 0.06991 | 0.15053 | 0.22044 |
| 70 | 85.858 | 0.01312 | 0.55681 | 34.567 | 69.301 | 103.87 | 34.776 | 77.939 | 112.71 | 0.07306 | 0.14713 | 0.22019 |
| 75 | 93.408 | 0.01323 | 0.51165 | 36.237 | 68.251 | 104.49 | 36.466 | 76.866 | 113.33 | 0.07620 | 0.14375 | 0.21995 |
| 80 | 101.45 | 0.01334 | 0.47069 | 37.919 | 67.181 | 105.10 | 38.169 | 75.767 | 113.94 | 0.07934 | 0.14038 | 0.21972 |
| 85 | 110.00 | 0.01347 | 0.43348 | 39.612 | 66.091 | 105.70 | 39.886 | 74.641 | 114.53 | 0.08246 | 0.13703 | 0.21949 |
| 90 | 119.08 | 0.01359 | 0.39959 | 41.317 | 64.979 | 106.30 | 41.617 | 73.485 | 115.10 | 0.08559 | 0.13368 | 0.21926 |
| 95 | 128.72 | 0.01372 | 0.36869 | 43.036 | 63.844 | 106.88 | 43.363 | 72.299 | 115.66 | 0.08870 | 0.13033 | 0.21904 |
| 100 | 138.93 | 0.01386 | 0.34045 | 44.768 | 62.683 | 107.45 | 45.124 | 71.080 | 116.20 | 0.09182 | 0.12699 | 0.21881 |
| 105 | 149.73 | 0.01400 | 0.31460 | 46.514 | 61.496 | 108.01 | 46.902 | 69.825 | 116.73 | 0.09493 | 0.12365 | 0.21858 |
| 110 | 161.16 | 0.01415 | 0.29090 | 48.276 | 60.279 | 108.56 | 48.698 | 68.533 | 117.23 | 0.09804 | 0.12029 | 0.21834 |
| 115 | 173.23 | 0.01430 | 0.26913 | 50.054 | 59.031 | 109.08 | 50.512 | 67.200 | 117.71 | 0.10116 | 0.11693 | 0.21809 |
| 120 | 185.96 | 0.01446 | 0.24909 | 51.849 | 57.749 | 109.60 | 52.346 | 65.823 | 118.17 | 0.10428 | 0.11354 | 0.21782 |
| 130 | 213.53 | 0.01482 | 0.21356 | 55.495 | 55.071 | 110.57 | 56.080 | 62.924 | 119.00 | 0.11054 | 0.10670 | 0.21724 |
| 140 | 244.06 | 0.01521 | 0.18315 | 59.226 | 52.216 | 111.44 | 59.913 | 59.801 | 119.71 | 0.11684 | 0.09971 | 0.21655 |
| 150 | 277.79 | 0.01567 | 0.15692 | 63.059 | 49.144 | 112.20 | 63.864 | 56.405 | 120.27 | 0.12321 | 0.09251 | 0.21572 |
| 160 | 314.94 | 0.01619 | 0.13410 | 67.014 | 45.799 | 112.81 | 67.958 | 52.671 | 120.63 | 0.12970 | 0.08499 | 0.21469 |
| 170 | 355.80 | 0.01681 | 0.11405 | 71.126 | 42.097 | 113.22 | 72.233 | 48.499 | 120.73 | 0.13634 | 0.07701 | 0.21335 |
| 180 | 400.66 | 0.01759 | 0.09618 | 75.448 | 37.899 | 113.35 | 76.752 | 43.726 | 120.48 | 0.14323 | 0.06835 | 0.21158 |
| 190 | 449.90 | 0.01860 | 0.07990 | 80.082 | 32.950 | 113.03 | 81.631 | 38.053 | 119.68 | 0.15055 | 0.05857 | 0.20911 |
| 200 | 504.00 | 0.02009 | 0.06441 | 85.267 | 26.651 | 111.92 | 87.140 | 30.785 | 117.93 | 0.15867 | 0.04666 | 0.20533 |
| 210 | 563.76 | 0.02309 | 0.04722 | 91.986 | 16.498 | 108.48 | 94.395 | 19.015 | 113.41 | 0.16922 | 0.02839 | 0.19761 |

*Source:* Tables A–11E through A–13E are generated using the Engineering Equation Solver (EES) software developed by S. A. Klein and F. L. Alvarado. The routine used in calculations is the R134a, which is based on the fundamental equation of state developed by R. Tillner-Roth and H.D. Baehr, "An International Standard Formulation for the Thermodynamic Properties of 1,1,1,2-Tetrafluoroethane (HFC-134a) for Temperatures from 170 K to 455 K and Pressures up to 70 MPa," *J. Phys. Chem, Ref. Data*, Vol. 23, No. 5, 1994. The enthalpy and entropy values of saturated liquid are set to zero at −40°C (and −40°F).

## TABLE A–12E

Saturated refrigerant-134a—Pressure table

| Press., P psia | Sat. temp., T °F | Specific volume, ft³/lbm | | Internal energy, Btu/lbm | | | Enthalpy, Btu/lbm | | | Entropy, Btu/lbm·R | | |
|---|---|---|---|---|---|---|---|---|---|---|---|---|
| | | Sat. liquid, $v_f$ | Sat. vapor, $v_g$ | Sat. liquid, $u_f$ | Evap., $u_{fg}$ | Sat. vapor, $u_g$ | Sat. liquid, $h_f$ | Evap., $h_{fg}$ | Sat. vapor, $h_g$ | Sat. liquid, $s_f$ | Evap., $s_{fg}$ | Sat. vapor, $s_g$ |
| 5 | −53.09 | 0.01113 | 8.3785 | −3.918 | 91.280 | 87.36 | −3.907 | 99.022 | 95.11 | −0.00945 | 0.24353 | 0.23408 |
| 10 | −29.52 | 0.01144 | 4.3753 | 3.135 | 87.453 | 90.59 | 3.156 | 95.528 | 98.68 | 0.00742 | 0.22206 | 0.22948 |
| 15 | −14.15 | 0.01165 | 2.9880 | 7.803 | 84.893 | 92.70 | 7.835 | 93.155 | 100.99 | 0.01808 | 0.20908 | 0.22715 |
| 20 | −2.43 | 0.01182 | 2.2772 | 11.401 | 82.898 | 94.30 | 11.445 | 91.282 | 102.73 | 0.02605 | 0.19962 | 0.22567 |
| 25 | 7.17 | 0.01196 | 1.8429 | 14.377 | 81.231 | 95.61 | 14.432 | 89.701 | 104.13 | 0.03249 | 0.19213 | 0.22462 |
| 30 | 15.37 | 0.01209 | 1.5492 | 16.939 | 79.780 | 96.72 | 17.006 | 88.313 | 105.32 | 0.03793 | 0.18589 | 0.22383 |
| 35 | 22.57 | 0.01221 | 1.3369 | 19.205 | 78.485 | 97.69 | 19.284 | 87.064 | 106.35 | 0.04267 | 0.18053 | 0.22319 |
| 40 | 29.01 | 0.01232 | 1.1760 | 21.246 | 77.307 | 98.55 | 21.337 | 85.920 | 107.26 | 0.04688 | 0.17580 | 0.22268 |
| 45 | 34.86 | 0.01242 | 1.0497 | 23.110 | 76.221 | 99.33 | 23.214 | 84.858 | 108.07 | 0.05067 | 0.17158 | 0.22225 |
| 50 | 40.23 | 0.01252 | 0.94791 | 24.832 | 75.209 | 100.04 | 24.948 | 83.863 | 108.81 | 0.05413 | 0.16774 | 0.22188 |
| 55 | 45.20 | 0.01261 | 0.86400 | 26.435 | 74.258 | 100.69 | 26.564 | 82.924 | 109.49 | 0.05733 | 0.16423 | 0.22156 |
| 60 | 49.84 | 0.01270 | 0.79361 | 27.939 | 73.360 | 101.30 | 28.080 | 82.030 | 110.11 | 0.06029 | 0.16098 | 0.22127 |
| 65 | 54.20 | 0.01279 | 0.73370 | 29.357 | 72.505 | 101.86 | 29.510 | 81.176 | 110.69 | 0.06307 | 0.15796 | 0.22102 |
| 70 | 58.30 | 0.01287 | 0.68205 | 30.700 | 71.688 | 102.39 | 30.867 | 80.357 | 111.22 | 0.06567 | 0.15512 | 0.22080 |
| 75 | 62.19 | 0.01295 | 0.63706 | 31.979 | 70.905 | 102.88 | 32.159 | 79.567 | 111.73 | 0.06813 | 0.15245 | 0.22059 |
| 80 | 65.89 | 0.01303 | 0.59750 | 33.201 | 70.151 | 103.35 | 33.394 | 78.804 | 112.20 | 0.07047 | 0.14993 | 0.22040 |
| 85 | 69.41 | 0.01310 | 0.56244 | 34.371 | 69.424 | 103.79 | 34.577 | 78.064 | 112.64 | 0.07269 | 0.14753 | 0.22022 |
| 90 | 72.78 | 0.01318 | 0.53113 | 35.495 | 68.719 | 104.21 | 35.715 | 77.345 | 113.06 | 0.07481 | 0.14525 | 0.22006 |
| 95 | 76.02 | 0.01325 | 0.50301 | 36.578 | 68.035 | 104.61 | 36.811 | 76.645 | 113.46 | 0.07684 | 0.14307 | 0.21991 |
| 100 | 79.12 | 0.01332 | 0.47760 | 37.623 | 67.371 | 104.99 | 37.869 | 75.962 | 113.83 | 0.07879 | 0.14097 | 0.21976 |
| 110 | 85.00 | 0.01347 | 0.43347 | 39.612 | 66.091 | 105.70 | 39.886 | 74.641 | 114.53 | 0.08246 | 0.13703 | 0.21949 |
| 120 | 90.49 | 0.01360 | 0.39644 | 41.485 | 64.869 | 106.35 | 41.787 | 73.371 | 115.16 | 0.08589 | 0.13335 | 0.21924 |
| 130 | 95.64 | 0.01374 | 0.36491 | 43.258 | 63.696 | 106.95 | 43.589 | 72.144 | 115.73 | 0.08911 | 0.12990 | 0.21901 |
| 140 | 100.51 | 0.01387 | 0.33771 | 44.945 | 62.564 | 107.51 | 45.304 | 70.954 | 116.26 | 0.09214 | 0.12665 | 0.21879 |
| 150 | 105.12 | 0.01400 | 0.31401 | 46.556 | 61.467 | 108.02 | 46.945 | 69.795 | 116.74 | 0.09501 | 0.12357 | 0.21857 |
| 160 | 109.50 | 0.01413 | 0.29316 | 48.101 | 60.401 | 108.50 | 48.519 | 68.662 | 117.18 | 0.09774 | 0.12062 | 0.21836 |
| 170 | 113.69 | 0.01426 | 0.27466 | 49.586 | 59.362 | 108.95 | 50.035 | 67.553 | 117.59 | 0.10034 | 0.11781 | 0.21815 |
| 180 | 117.69 | 0.01439 | 0.25813 | 51.018 | 58.345 | 109.36 | 51.497 | 66.464 | 117.96 | 0.10284 | 0.11511 | 0.21795 |
| 190 | 121.53 | 0.01452 | 0.24327 | 52.402 | 57.349 | 109.75 | 52.912 | 65.392 | 118.30 | 0.10524 | 0.11250 | 0.21774 |
| 200 | 125.22 | 0.01464 | 0.22983 | 53.743 | 56.371 | 110.11 | 54.285 | 64.335 | 118.62 | 0.10754 | 0.10998 | 0.21753 |
| 220 | 132.21 | 0.01490 | 0.20645 | 56.310 | 54.458 | 110.77 | 56.917 | 62.256 | 119.17 | 0.11192 | 0.10517 | 0.21710 |
| 240 | 138.73 | 0.01516 | 0.18677 | 58.746 | 52.591 | 111.34 | 59.419 | 60.213 | 119.63 | 0.11603 | 0.10061 | 0.21665 |
| 260 | 144.85 | 0.01543 | 0.16996 | 61.071 | 50.757 | 111.83 | 61.813 | 58.192 | 120.00 | 0.11992 | 0.09625 | 0.21617 |
| 280 | 150.62 | 0.01570 | 0.15541 | 63.301 | 48.945 | 112.25 | 64.115 | 56.184 | 120.30 | 0.12362 | 0.09205 | 0.21567 |
| 300 | 156.09 | 0.01598 | 0.14266 | 65.452 | 47.143 | 112.60 | 66.339 | 54.176 | 120.52 | 0.12715 | 0.08797 | 0.21512 |
| 350 | 168.64 | 0.01672 | 0.11664 | 70.554 | 42.627 | 113.18 | 71.638 | 49.099 | 120.74 | 0.13542 | 0.07814 | 0.21356 |
| 400 | 179.86 | 0.01757 | 0.09642 | 75.385 | 37.963 | 113.35 | 76.686 | 43.798 | 120.48 | 0.14314 | 0.06848 | 0.21161 |
| 450 | 190.02 | 0.01860 | 0.07987 | 80.092 | 32.939 | 113.03 | 81.641 | 38.041 | 119.68 | 0.15056 | 0.05854 | 0.20911 |
| 500 | 199.29 | 0.01995 | 0.06551 | 84.871 | 27.168 | 112.04 | 86.718 | 31.382 | 118.10 | 0.15805 | 0.04762 | 0.20566 |

## TABLE A–13E

Superheated refrigerant-134a

| T °F | v ft³/lbm | u Btu/lbm | h Btu/lbm | s Btu/ lbm·R | v ft³/lbm | u Btu/lbm | h Btu/lbm | s Btu/ lbm·R | v ft³/lbm | u Btu/lbm | h Btu/lbm | s Btu/ lbm·R |
|---|---|---|---|---|---|---|---|---|---|---|---|---|
| | $P = 10$ psia ($T_{sat} = -29.52°F$) | | | | $P = 15$ psia ($T_{sat} = -14.15°F$) | | | | $P = 20$ psia ($T_{sat} = -2.43°F$) | | | |
| Sat. | 4.3753 | 90.59 | 98.68 | 0.22948 | 2.9880 | 92.70 | 100.99 | 0.22715 | 2.2772 | 94.30 | 102.73 | 0.22567 |
| −20 | 4.4856 | 92.13 | 100.43 | 0.23350 | | | | | | | | |
| 0 | 4.7135 | 95.41 | 104.14 | 0.24174 | 3.1001 | 95.08 | 103.68 | 0.23310 | 2.2922 | 94.72 | 103.20 | 0.22671 |
| 20 | 4.9380 | 98.77 | 107.91 | 0.24976 | 3.2551 | 98.48 | 107.52 | 0.24127 | 2.4130 | 98.19 | 107.12 | 0.23504 |
| 40 | 5.1600 | 102.20 | 111.75 | 0.25761 | 3.4074 | 101.95 | 111.41 | 0.24922 | 2.5306 | 101.70 | 111.07 | 0.24311 |
| 60 | 5.3802 | 105.72 | 115.67 | 0.26531 | 3.5577 | 105.50 | 115.38 | 0.25700 | 2.6461 | 105.28 | 115.07 | 0.25097 |
| 80 | 5.5989 | 109.32 | 119.68 | 0.27288 | 3.7064 | 109.13 | 119.42 | 0.26463 | 2.7600 | 108.93 | 119.15 | 0.25866 |
| 100 | 5.8165 | 113.01 | 123.78 | 0.28033 | 3.8540 | 112.84 | 123.54 | 0.27212 | 2.8726 | 112.66 | 123.29 | 0.26621 |
| 120 | 6.0331 | 116.79 | 127.96 | 0.28767 | 4.0006 | 116.63 | 127.74 | 0.27950 | 2.9842 | 116.47 | 127.52 | 0.27363 |
| 140 | 6.2490 | 120.66 | 132.22 | 0.29490 | 4.1464 | 120.51 | 132.02 | 0.28677 | 3.0950 | 120.37 | 131.82 | 0.28093 |
| 160 | 6.4642 | 124.61 | 136.57 | 0.30203 | 4.2915 | 124.48 | 136.39 | 0.29393 | 3.2051 | 124.35 | 136.21 | 0.28812 |
| 180 | 6.6789 | 128.65 | 141.01 | 0.30908 | 4.4361 | 128.53 | 140.84 | 0.30100 | 3.3146 | 128.41 | 140.67 | 0.29521 |
| 200 | 6.8930 | 132.77 | 145.53 | 0.31604 | 4.5802 | 132.66 | 145.37 | 0.30798 | 3.4237 | 132.55 | 145.22 | 0.30221 |
| 220 | 7.1068 | 136.98 | 150.13 | 0.32292 | 4.7239 | 136.88 | 149.99 | 0.31487 | 3.5324 | 136.78 | 149.85 | 0.30912 |
| | $P = 30$ psia ($T_{sat} = 15.37°F$) | | | | $P = 40$ psia ($T_{sat} = 29.01°F$) | | | | $P = 50$ psia ($T_{sat} = 40.23°F$) | | | |
| Sat. | 1.5492 | 96.72 | 105.32 | 0.22383 | 1.1760 | 98.55 | 107.26 | 0.22268 | 0.9479 | 100.04 | 108.81 | 0.22188 |
| 20 | 1.5691 | 97.56 | 106.27 | 0.22581 | | | | | | | | |
| 40 | 1.6528 | 101.17 | 110.35 | 0.23414 | 1.2126 | 100.61 | 109.58 | 0.22738 | | | | |
| 60 | 1.7338 | 104.82 | 114.45 | 0.24219 | 1.2768 | 104.34 | 113.79 | 0.23565 | 1.0019 | 103.84 | 113.11 | 0.23031 |
| 80 | 1.8130 | 108.53 | 118.59 | 0.25002 | 1.3389 | 108.11 | 118.02 | 0.24363 | 1.0540 | 107.68 | 117.43 | 0.23847 |
| 100 | 1.8908 | 112.30 | 122.80 | 0.25767 | 1.3995 | 111.93 | 122.29 | 0.25140 | 1.1043 | 111.55 | 121.77 | 0.24637 |
| 120 | 1.9675 | 116.15 | 127.07 | 0.26517 | 1.4588 | 115.82 | 126.62 | 0.25900 | 1.1534 | 115.48 | 126.16 | 0.25406 |
| 140 | 2.0434 | 120.08 | 131.42 | 0.27254 | 1.5173 | 119.78 | 131.01 | 0.26644 | 1.2015 | 119.47 | 130.59 | 0.26159 |
| 160 | 2.1185 | 124.08 | 135.84 | 0.27979 | 1.5750 | 123.81 | 135.47 | 0.27375 | 1.2488 | 123.53 | 135.09 | 0.26896 |
| 180 | 2.1931 | 128.16 | 140.34 | 0.28693 | 1.6321 | 127.91 | 140.00 | 0.28095 | 1.2955 | 127.66 | 139.65 | 0.27621 |
| 200 | 2.2671 | 132.32 | 144.91 | 0.29398 | 1.6887 | 132.10 | 144.60 | 0.28803 | 1.3416 | 131.87 | 144.28 | 0.28333 |
| 220 | 2.3408 | 136.57 | 149.56 | 0.30092 | 1.7449 | 136.36 | 149.27 | 0.29501 | 1.3873 | 136.15 | 148.98 | 0.29036 |
| 240 | 2.4141 | 140.89 | 154.29 | 0.30778 | 1.8007 | 140.70 | 154.03 | 0.30190 | 1.4326 | 140.50 | 153.76 | 0.29728 |
| 260 | 2.4871 | 145.30 | 159.10 | 0.31456 | 1.8562 | 145.12 | 158.86 | 0.30871 | 1.4776 | 144.93 | 158.60 | 0.30411 |
| 280 | 2.5598 | 149.78 | 163.99 | 0.32126 | 1.9114 | 149.61 | 163.76 | 0.31543 | 1.5223 | 149.44 | 163.53 | 0.31086 |
| | $P = 60$ psia ($T_{sat} = 49.84°F$) | | | | $P = 70$ psia ($T_{sat} = 58.30°F$) | | | | $P = 80$ psia ($T_{sat} = 65.89°F$) | | | |
| Sat. | 0.7936 | 101.30 | 110.11 | 0.22127 | 0.6821 | 102.39 | 111.22 | 0.22080 | 0.59750 | 103.35 | 112.20 | 0.22040 |
| 60 | 0.8179 | 103.31 | 112.39 | 0.22570 | 0.6857 | 102.73 | 111.62 | 0.22155 | | | | |
| 80 | 0.8636 | 107.23 | 116.82 | 0.23407 | 0.7271 | 106.76 | 116.18 | 0.23016 | 0.62430 | 106.26 | 115.51 | 0.22661 |
| 100 | 0.9072 | 111.16 | 121.24 | 0.24211 | 0.7662 | 110.76 | 120.68 | 0.23836 | 0.66009 | 110.34 | 120.11 | 0.23499 |
| 120 | 0.9495 | 115.14 | 125.68 | 0.24991 | 0.8037 | 114.78 | 125.19 | 0.24628 | 0.69415 | 114.42 | 124.69 | 0.24304 |
| 140 | 0.9908 | 119.16 | 130.16 | 0.25751 | 0.8401 | 118.85 | 129.73 | 0.25398 | 0.72698 | 118.52 | 129.29 | 0.25083 |
| 160 | 1.0312 | 123.25 | 134.70 | 0.26496 | 0.8756 | 122.97 | 134.31 | 0.26149 | 0.75888 | 122.68 | 133.91 | 0.25841 |
| 180 | 1.0709 | 127.41 | 139.30 | 0.27226 | 0.9105 | 127.15 | 138.94 | 0.26885 | 0.79003 | 126.89 | 138.58 | 0.26583 |
| 200 | 1.1101 | 131.63 | 143.96 | 0.27943 | 0.9447 | 131.40 | 143.63 | 0.27607 | 0.82059 | 131.16 | 143.31 | 0.27310 |
| 220 | 1.1489 | 135.93 | 148.69 | 0.28649 | 0.9785 | 135.71 | 148.39 | 0.28317 | 0.85065 | 135.49 | 148.09 | 0.28024 |
| 240 | 1.1872 | 140.30 | 153.48 | 0.29344 | 1.0118 | 140.10 | 153.21 | 0.29015 | 0.88030 | 139.90 | 152.93 | 0.28726 |
| 260 | 1.2252 | 144.75 | 158.35 | 0.30030 | 1.0449 | 144.56 | 158.10 | 0.29704 | 0.90961 | 144.37 | 157.84 | 0.29418 |
| 280 | 1.2629 | 149.27 | 163.29 | 0.30707 | 1.0776 | 149.10 | 163.06 | 0.30384 | 0.93861 | 148.92 | 162.82 | 0.30100 |
| 300 | 1.3004 | 153.87 | 168.31 | 0.31376 | 1.1101 | 153.71 | 168.09 | 0.31055 | 0.96737 | 153.54 | 167.86 | 0.30773 |
| 320 | 1.3377 | 158.54 | 173.39 | 0.32037 | 1.1424 | 158.39 | 173.19 | 0.31718 | 0.99590 | 158.24 | 172.98 | 0.31438 |

## TABLE A–13E

Superheated refrigerant-134a (*Concluded*)

| T °F | v ft³/lbm | u Btu/lbm | h Btu/lbm | s Btu/ lbm·R | v ft³/lbm | u Btu/lbm | h Btu/lbm | s Btu/ lbm·R | v ft³/lbm | u Btu/lbm | h Btu/lbm | s Btu/ lbm·R |
|---|---|---|---|---|---|---|---|---|---|---|---|---|
| | $P = 90$ psia ($T_{sat} = 72.78°F$) | | | | $P = 100$ psia ($T_{sat} = 79.12°F$) | | | | $P = 120$ psia ($T_{sat} = 90.49°F$) | | | |
| Sat. | 0.53113 | 104.21 | 113.06 | 0.22006 | 0.47760 | 104.99 | 113.83 | 0.21976 | 0.39644 | 106.35 | 115.16 | 0.21924 |
| 80 | 0.54388 | 105.74 | 114.80 | 0.22330 | 0.47906 | 105.18 | 114.05 | 0.22016 | | | | |
| 100 | 0.57729 | 109.91 | 119.52 | 0.23189 | 0.51076 | 109.45 | 118.90 | 0.22900 | 0.41013 | 108.48 | 117.59 | 0.22362 |
| 120 | 0.60874 | 114.04 | 124.18 | 0.24008 | 0.54022 | 113.66 | 123.65 | 0.23733 | 0.43692 | 112.84 | 122.54 | 0.23232 |
| 140 | 0.63885 | 118.19 | 128.83 | 0.24797 | 0.56821 | 117.86 | 128.37 | 0.24534 | 0.46190 | 117.15 | 127.41 | 0.24058 |
| 160 | 0.66796 | 122.38 | 133.51 | 0.25563 | 0.59513 | 122.08 | 133.09 | 0.25309 | 0.48563 | 121.46 | 132.25 | 0.24851 |
| 180 | 0.69629 | 126.62 | 138.22 | 0.26311 | 0.62122 | 126.35 | 137.85 | 0.26063 | 0.50844 | 125.79 | 137.09 | 0.25619 |
| 200 | 0.72399 | 130.92 | 142.97 | 0.27043 | 0.64667 | 130.67 | 142.64 | 0.26801 | 0.53054 | 130.17 | 141.95 | 0.26368 |
| 220 | 0.75119 | 135.27 | 147.78 | 0.27762 | 0.67158 | 135.05 | 147.47 | 0.27523 | 0.55206 | 134.59 | 146.85 | 0.27100 |
| 240 | 0.77796 | 139.69 | 152.65 | 0.28468 | 0.69605 | 139.49 | 152.37 | 0.28233 | 0.57312 | 139.07 | 151.80 | 0.27817 |
| 260 | 0.80437 | 144.19 | 157.58 | 0.29162 | 0.72016 | 143.99 | 157.32 | 0.28931 | 0.59379 | 143.61 | 156.79 | 0.28521 |
| 280 | 0.83048 | 148.75 | 162.58 | 0.29847 | 0.74396 | 148.57 | 162.34 | 0.29618 | 0.61413 | 148.21 | 161.85 | 0.29214 |
| 300 | 0.85633 | 153.38 | 167.64 | 0.30522 | 0.76749 | 153.21 | 167.42 | 0.30296 | 0.63420 | 152.88 | 166.96 | 0.29896 |
| 320 | 0.88195 | 158.08 | 172.77 | 0.31189 | 0.79079 | 157.93 | 172.56 | 0.30964 | 0.65402 | 157.62 | 172.14 | 0.30569 |
| | $P = 140$ psia ($T_{sat} = 100.50°F$) | | | | $P = 160$ psia ($T_{sat} = 109.50°F$) | | | | $P = 180$ psia ($T_{sat} = 117.69°F$) | | | |
| Sat. | 0.33771 | 107.51 | 116.26 | 0.21879 | 0.29316 | 108.50 | 117.18 | 0.21836 | 0.25813 | 109.36 | 117.96 | 0.21795 |
| 120 | 0.36243 | 111.96 | 121.35 | 0.22773 | 0.30578 | 111.01 | 120.06 | 0.22337 | 0.26083 | 109.94 | 118.63 | 0.21910 |
| 140 | 0.38551 | 116.41 | 126.40 | 0.23628 | 0.32774 | 115.62 | 125.32 | 0.23230 | 0.28231 | 114.77 | 124.17 | 0.22850 |
| 160 | 0.40711 | 120.81 | 131.36 | 0.24443 | 0.34790 | 120.13 | 130.43 | 0.24069 | 0.30154 | 119.42 | 129.46 | 0.23718 |
| 180 | 0.42766 | 125.22 | 136.30 | 0.25227 | 0.36686 | 124.62 | 135.49 | 0.24871 | 0.31936 | 124.00 | 134.64 | 0.24540 |
| 200 | 0.44743 | 129.65 | 141.24 | 0.25988 | 0.38494 | 129.12 | 140.52 | 0.25645 | 0.33619 | 128.57 | 139.77 | 0.25330 |
| 220 | 0.46657 | 134.12 | 146.21 | 0.26730 | 0.40234 | 133.64 | 145.55 | 0.26397 | 0.35228 | 133.15 | 144.88 | 0.26094 |
| 240 | 0.48522 | 138.64 | 151.21 | 0.27455 | 0.41921 | 138.20 | 150.62 | 0.27131 | 0.36779 | 137.76 | 150.01 | 0.26837 |
| 260 | 0.50345 | 143.21 | 156.26 | 0.28166 | 0.43564 | 142.81 | 155.71 | 0.27849 | 0.38284 | 142.40 | 155.16 | 0.27562 |
| 280 | 0.52134 | 147.85 | 161.35 | 0.28864 | 0.45171 | 147.48 | 160.85 | 0.28554 | 0.39751 | 147.10 | 160.34 | 0.28273 |
| 300 | 0.53895 | 152.54 | 166.50 | 0.29551 | 0.46748 | 152.20 | 166.04 | 0.29246 | 0.41186 | 151.85 | 165.57 | 0.28970 |
| 320 | 0.55630 | 157.30 | 171.71 | 0.30228 | 0.48299 | 156.98 | 171.28 | 0.29927 | 0.42594 | 156.66 | 170.85 | 0.29656 |
| 340 | 0.57345 | 162.13 | 176.98 | 0.30896 | 0.49828 | 161.83 | 176.58 | 0.30598 | 0.43980 | 161.53 | 176.18 | 0.30331 |
| 360 | 0.59041 | 167.02 | 182.32 | 0.31555 | 0.51338 | 166.74 | 181.94 | 0.31260 | 0.45347 | 166.46 | 181.56 | 0.30996 |
| | $P = 200$ psia ($T_{sat} = 125.22°F$) | | | | $P = 300$ psia ($T_{sat} = 156.09°F$) | | | | $P = 400$ psia ($T_{sat} = 179.86°F$) | | | |
| Sat. | 0.22983 | 110.11 | 118.62 | 0.21753 | 0.14266 | 112.60 | 120.52 | 0.21512 | 0.09642 | 113.35 | 120.48 | 0.21161 |
| 140 | 0.24541 | 113.85 | 122.93 | 0.22481 | | | | | | | | |
| 160 | 0.26412 | 118.66 | 128.44 | 0.23384 | 0.14656 | 113.82 | 121.95 | 0.21745 | | | | |
| 180 | 0.28115 | 123.35 | 133.76 | 0.24229 | 0.16355 | 119.52 | 128.60 | 0.22802 | 0.09658 | 113.41 | 120.56 | 0.21173 |
| 200 | 0.29704 | 128.00 | 138.99 | 0.25035 | 0.17776 | 124.78 | 134.65 | 0.23733 | 0.11440 | 120.52 | 128.99 | 0.22471 |
| 220 | 0.31212 | 132.64 | 144.19 | 0.25812 | 0.19044 | 129.85 | 140.42 | 0.24594 | 0.12746 | 126.44 | 135.88 | 0.23500 |
| 240 | 0.32658 | 137.30 | 149.38 | 0.26565 | 0.20211 | 134.83 | 146.05 | 0.25410 | 0.13853 | 131.95 | 142.20 | 0.24418 |
| 260 | 0.34054 | 141.99 | 154.59 | 0.27298 | 0.21306 | 139.77 | 151.59 | 0.26192 | 0.14844 | 137.26 | 148.25 | 0.25270 |
| 280 | 0.35410 | 146.72 | 159.82 | 0.28015 | 0.22347 | 144.70 | 157.11 | 0.26947 | 0.15756 | 142.48 | 154.14 | 0.26077 |
| 300 | 0.36733 | 151.50 | 165.09 | 0.28718 | 0.23346 | 149.65 | 162.61 | 0.27681 | 0.16611 | 147.65 | 159.94 | 0.26851 |
| 320 | 0.38029 | 156.33 | 170.40 | 0.29408 | 0.24310 | 154.63 | 168.12 | 0.28398 | 0.17423 | 152.80 | 165.70 | 0.27599 |
| 340 | 0.39300 | 161.22 | 175.77 | 0.30087 | 0.25246 | 159.64 | 173.66 | 0.29098 | 0.18201 | 157.97 | 171.44 | 0.28326 |
| 360 | 0.40552 | 166.17 | 181.18 | 0.30756 | 0.26159 | 164.70 | 179.22 | 0.29786 | 0.18951 | 163.15 | 177.18 | 0.29035 |

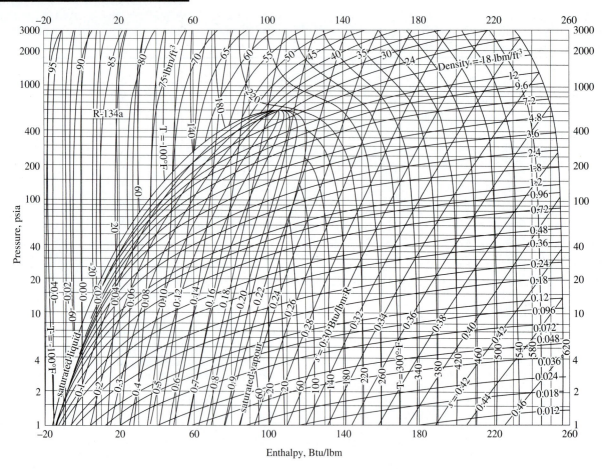

**FIGURE A–14E**

*P-h* diagram for refrigerant-134a.

*Reprinted by permission of American Society of Heating, Refrigerating, and Air-Conditioning Engineers, Inc., Atlanta, GA.*

## TABLE A–15E

Properties of saturated water

| Temp. $T$, °F | Saturation Pressure $P_{sat}$, psia | Density $\rho$, lbm/ft³ Liquid | Density $\rho$, lbm/ft³ Vapor | Enthalpy of Vaporization $h_{fg}$, Btu/lbm | Specific Heat $c_p$, Btu/lbm·R Liquid | Specific Heat $c_p$, Btu/lbm·R Vapor | Thermal Conductivity $k$, Btu/h·ft·R Liquid | Thermal Conductivity $k$, Btu/h·ft·R Vapor | Dynamic Viscosity $\mu$, lbm/ft·s Liquid | Dynamic Viscosity $\mu$, lbm/ft·s Vapor | Prandtl Number Pr Liquid | Prandtl Number Pr Vapor | Volume Expansion Coefficient $\beta$, 1/R Liquid |
|---|---|---|---|---|---|---|---|---|---|---|---|---|---|
| 32.02 | 0.0887 | 62.41 | 0.00030 | 1075 | 1.010 | 0.446 | 0.324 | 0.0099 | $1.204 \times 10^{-3}$ | $6.194 \times 10^{-6}$ | 13.5 | 1.00 | $-0.038 \times 10^{-3}$ |
| 40 | 0.1217 | 62.42 | 0.00034 | 1071 | 1.004 | 0.447 | 0.329 | 0.0100 | $1.308 \times 10^{-3}$ | $6.278 \times 10^{-6}$ | 11.4 | 1.01 | $A0.003 \times 10^{-3}$ |
| 50 | 0.1780 | 62.41 | 0.00059 | 1065 | 1.000 | 0.448 | 0.335 | 0.0102 | $8.781 \times 10^{-4}$ | $6.361 \times 10^{-6}$ | 9.44 | 1.01 | $0.047 \times 10^{-3}$ |
| 60 | 0.2563 | 62.36 | 0.00083 | 1060 | 0.999 | 0.449 | 0.341 | 0.0104 | $7.536 \times 10^{-4}$ | $6.444 \times 10^{-6}$ | 7.95 | 1.00 | $0.080 \times 10^{-3}$ |
| 70 | 0.3632 | 62.30 | 0.00115 | 1054 | 0.999 | 0.450 | 0.347 | 0.0106 | $6.556 \times 10^{-4}$ | $6.556 \times 10^{-6}$ | 6.79 | 1.00 | $0.115 \times 10^{-3}$ |
| 80 | 0.5073 | 62.22 | 0.00158 | 1048 | 0.999 | 0.451 | 0.352 | 0.0108 | $5.764 \times 10^{-4}$ | $6.667 \times 10^{-6}$ | 5.89 | 1.00 | $0.145 \times 10^{-3}$ |
| 90 | 0.6988 | 62.12 | 0.00214 | 1043 | 0.999 | 0.453 | 0.358 | 0.0110 | $5.117 \times 10^{-4}$ | $6.778 \times 10^{-6}$ | 5.14 | 1.00 | $0.174 \times 10^{-3}$ |
| 100 | 0.9503 | 62.00 | 0.00286 | 1037 | 0.999 | 0.454 | 0.363 | 0.0112 | $4.578 \times 10^{-4}$ | $6.889 \times 10^{-6}$ | 4.54 | 1.01 | $0.200 \times 10^{-3}$ |
| 110 | 1.2763 | 61.86 | 0.00377 | 1031 | 0.999 | 0.456 | 0.367 | 0.0115 | $4.128 \times 10^{-4}$ | $7.000 \times 10^{-6}$ | 4.05 | 1.00 | $0.224 \times 10^{-3}$ |
| 120 | 1.6945 | 61.71 | 0.00493 | 1026 | 0.999 | 0.458 | 0.371 | 0.0117 | $3.744 \times 10^{-4}$ | $7.111 \times 10^{-6}$ | 3.63 | 1.00 | $0246 \times 10^{-3}$ |
| 130 | 2.225 | 61.55 | 0.00636 | 1020 | 0.999 | 0.460 | 0.375 | 0.0120 | $3.417 \times 10^{-4}$ | $7.222 \times 10^{-6}$ | 3.28 | 1.00 | $0.267 \times 10^{-3}$ |
| 140 | 2.892 | 61.38 | 0.00814 | 1014 | 0.999 | 0.463 | 0.378 | 0.0122 | $3.136 \times 10^{-4}$ | $7.333 \times 10^{-6}$ | 2.98 | 1.00 | $0.287 \times 10^{-3}$ |
| 150 | 3.722 | 61.19 | 0.0103 | 1008 | 1.000 | 0.465 | 0.381 | 0.0125 | $2.889 \times 10^{-4}$ | $7.472 \times 10^{-6}$ | 2.73 | 1.00 | $0.306 \times 10^{-3}$ |
| 160 | 4.745 | 60.99 | 0.0129 | 1002 | 1.000 | 0.468 | 0.384 | 0.0128 | $2.675 \times 10^{-4}$ | $7.583 \times 10^{-6}$ | 2.51 | 1.00 | $0.325 \times 10^{-3}$ |
| 170 | 5.996 | 60.79 | 0.0161 | 996 | 1.001 | 0.472 | 0.386 | 0.0131 | $2.483 \times 10^{-4}$ | $7.722 \times 10^{-6}$ | 2.90 | 1.00 | $0.346 \times 10^{-3}$ |
| 180 | 7.515 | 60.57 | 0.0199 | 990 | 1.002 | 0.475 | 0.388 | 0.0134 | $2.317 \times 10^{-4}$ | $7.833 \times 10^{-6}$ | 2.15 | 1.00 | $0.367 \times 10^{-3}$ |
| 190 | 9.343 | 60.35 | 0.0244 | 984 | 1.004 | 0.479 | 0.390 | 0.0137 | $2.169 \times 10^{-4}$ | $7.972 \times 10^{-6}$ | 2.01 | 1.00 | $0.382 \times 10^{-3}$ |
| 200 | 11.53 | 60.12 | 0.0297 | 978 | 1.005 | 0.483 | 0.391 | 0.0141 | $2.036 \times 10^{-4}$ | $8.083 \times 10^{-6}$ | 1.88 | 1.00 | $0.395 \times 10^{-3}$ |
| 210 | 14.125 | 59.87 | 0.0359 | 972 | 1.007 | 0.487 | 0.392 | 0.0144 | $1.917 \times 10^{-4}$ | $8.222 \times 10^{-6}$ | 1.77 | 1.00 | $0.412 \times 10^{-3}$ |
| 212 | 14.698 | 59.82 | 0.0373 | 970 | 1.007 | 0.488 | 0.392 | 0.0145 | $1.894 \times 10^{-4}$ | $8.250 \times 10^{-6}$ | 1.75 | 1.00 | $0.417 \times 10^{-3}$ |
| 220 | 17.19 | 59.62 | 0.0432 | 965 | 1.009 | 0.492 | 0.393 | 0.0148 | $1808 \times 10^{-4}$ | $8.333 \times 10^{-6}$ | 1.67 | 1.00 | $0.429 \times 10^{-3}$ |
| 230 | 20.78 | 59.36 | 0.0516 | 959 | 1.011 | 0.497 | 0.394 | 0.0152 | $1.711 \times 10^{-4}$ | $8.472 \times 10^{-6}$ | 1.58 | 1.00 | $0.443 \times 10^{-3}$ |
| 240 | 24.97 | 59.09 | 0.0612 | 952 | 1.013 | 0.503 | 0.394 | 0.0156 | $1.625 \times 10^{-4}$ | $8.611 \times 10^{-6}$ | 1.50 | 1.00 | $0.462 \times 10^{-3}$ |
| 250 | 29.82 | 58.82 | 0.0723 | 946 | 1.015 | 0.509 | 0.395 | 0.0160 | $1.544 \times 10^{-4}$ | $8.611 \times 10^{-6}$ | 1.43 | 1.00 | $0.480 \times 10^{-3}$ |
| 260 | 35.42 | 58.53 | 0.0850 | 939 | 1.018 | 0.516 | 0.395 | 0.0164 | $1.472 \times 10^{-4}$ | $8.861 \times 10^{-6}$ | 1.37 | 1.00 | $0.497 \times 10^{-3}$ |
| 270 | 41.85 | 58.24 | 0.0993 | 932 | 1.020 | 0.523 | 0.395 | 0.0168 | $1.406 \times 10^{-4}$ | $9.000 \times 10^{-6}$ | 1.31 | 1.01 | $0.514 \times 10^{-3}$ |
| 280 | 49.18 | 57.94 | 0.1156 | 926 | 1.023 | 0.530 | 0.395 | 0.0172 | $1.344 \times 10^{-4}$ | $9.111 \times 10^{-6}$ | 1.25 | 1.01 | $0.532 \times 10^{-3}$ |
| 290 | 57–53 | 57.63 | 0.3390 | 918 | 1.026 | 0.538 | 0.395 | 0.0177 | $1.289 \times 10^{-4}$ | $9.250 \times 10^{-6}$ | 1.21 | 1.01 | $0.549 \times 10^{-3}$ |
| 300 | 66.98 | 57.31 | 0.1545 | 910 | 1.029 | 0.547 | 0.394 | 0.0182 | $1.236 \times 10^{-4}$ | $9.389 \times 10^{-6}$ | 1.16 | 1.02 | $0.566 \times 10^{-3}$ |
| 320 | 89.60 | 56.65 | 0.2033 | 895 | 1.036 | 0.567 | 0.393 | 0.0191 | $1.144 \times 10^{-4}$ | $9.639 \times 10^{-6}$ | 1.09 | 1.03 | $0.636 \times 10^{-3}$ |
| 340 | 117.93 | 55.95 | 0.2637 | 880 | 1.044 | 0.590 | 0.391 | 0.0202 | $1.063 \times 10^{-4}$ | $9.889 \times 10^{-6}$ | 1.02 | 1.04 | $0.656 \times 10^{-3}$ |
| 360 | 152.92 | 56.22 | 0.3377 | 863 | 1.054 | 0.617 | 0.389 | 0.0213 | $9.972 \times 10^{-5}$ | $1.013 \times 10^{-5}$ | 0.973 | 1.06 | $0.681 \times 10^{-3}$ |
| 380 | 195.60 | 54.46 | 0.4275 | 845 | 1.065 | 0.647 | 0.385 | 0.0224 | $9.361 \times 10^{-5}$ | $1.041 \times 10^{-5}$ | 0.932 | 1.08 | $0.720 \times 10^{-3}$ |
| 400 | 241.1 | 53.65 | 0.5359 | 827 | 1.078 | 0.683 | 0.382 | 0.0237 | $8.833 \times 10^{-5}$ | $1.066 \times 10^{-5}$ | 0.893 | 1.11 | $0.771 \times 10^{-3}$ |
| 450 | 422.1 | 51.46 | 0.9082 | 775 | 1.121 | 0.799 | 0.370 | 0.0271 | $7.722 \times 10^{-5}$ | $1.130 \times 10^{-5}$ | 0.842 | 1.20 | $0.912 \times 10^{-3}$ |
| 500 | 680.0 | 48.95 | 1.479 | 715 | 1.188 | 0.972 | 0.352 | 0.0312 | $6.833 \times 10^{-5}$ | $1.200 \times 10^{-5}$ | 0.830 | 1.35 | $1.111 \times 10^{-3}$ |
| 550 | 1046.7 | 45.96 | 4.268 | 641 | 1.298 | 1.247 | 0.329 | 0.0368 | $6.083 \times 10^{-5}$ | $1.280 \times 10^{-5}$ | 0.864 | 1.56 | $1.445 \times 10^{-3}$ |
| 600 | 1541 | 42.32 | 3.736 | 550 | 1.509 | 1.759 | 0.299 | 0.0461 | $5.389 \times 10^{-5}$ | $1.380 \times 10^{-5}$ | 0.979 | 1.90 | $1.883 \times 10^{-3}$ |
| 650 | 2210 | 37.31 | 6.152 | 422 | 2.086 | 3.103 | 0.267 | 0.0677 | $4.639 \times 10^{-5}$ | $1.542 \times 10^{-5}$ | 1.30 | 2.54 | |
| 700 | 3090 | 27.28 | 13.44 | 168 | 13.80 | 25.90 | 0.254 | 0.1964 | $3.417 \times 10^{-5}$ | $2.044 \times 10^{-5}$ | 6.68 | 9.71 | |
| 705.44 | 3204 | 19.79 | 19.79 | 0 | $\infty$ | $\infty$ | $\infty$ | $\infty$ | $2.897 \times 10^{-5}$ | $2.897 \times 10^{-5}$ | | | |

*Note 1:* Kinematic viscosity $\nu$ and thermal diffusivity $\alpha$ can be calculated from their definitions, $\nu = \mu/\rho$ and $\alpha = k/\rho c_p = \nu/Pr$. The temperatures 32.02°F, 212°F, and 705.44°F are the triple-, boiling-, and critical-point temperatures of water, respectively. All properties listed above (except the vapor density) can be used at any pressures with negligible error except at temperatures near the critical-point value.

*Note 2:* The unit Btu/lbm·°F for specific heat is equivalent to Btu/lbm·R, and the unit Btu/h·ft·°F for thermal conductivity is equivalent to Btu/h·ft·R.

*Source:* Viscosity and thermal conductivity data are from J. V. Sengers and J. T. T. Watson, *Journal of Physical and Chemical Reference Data* 15 (1986), pp. 1291–1322. Other data are obtained from various sources or calculated.

## TABLE A–16E

Properties of saturated refrigerant-134a

| Temp. $T$, °F | Saturation Pressure $P_{sat}$, psia | Density $\rho$, lbm/ft³ Liquid | Density $\rho$, lbm/ft³ Vapor | Enthalpy of Vaporization $h_{fg}$, Btu/lbm | Specific Heat $c_p$, Btu/lbm·R Liquid | Specific Heat $c_p$, Btu/lbm·R Vapor | Thermal Conductivity $k$, Btu/h·ft·R Liquid | Thermal Conductivity $k$, Btu/h·ft·R Vapor | Dynamic Viscosity $\mu$, lbm/ft·s Liquid | Dynamic Viscosity $\mu$, lbm/ft·s Vapor | Prandtl Number Pr Liquid | Prandtl Number Pr Vapor | Volume Expansion Coefficient $\beta$, 1/R Liquid | Surface Tension lbf/ft |
|---|---|---|---|---|---|---|---|---|---|---|---|---|---|---|
| −40 | 7.4 | 88.51 | 0.1731 | 97.1 | 0.2996 | 0.1788 | 0.0636 | 0.00466 | $3.278 \times 10^{-4}$ | $1.714 \times 10^{-6}$ | 5.558 | 0.237 | 0.00114 | 0.001206 |
| −30 | 9.9 | 87.5 | 0.2258 | 95.6 | 0.3021 | 0.1829 | 0.0626 | 0.00497 | $3.004 \times 10^{-4}$ | $2.053 \times 10^{-6}$ | 5.226 | 0.272 | 0.00117 | 0.001146 |
| −20 | 12.9 | 86.48 | 0.2905 | 94.1 | 0.3046 | 0.1872 | 0.0613 | 0.00529 | $2.762 \times 10^{-4}$ | $2.433 \times 10^{-6}$ | 4.937 | 0.310 | 0.00120 | 0.001087 |
| −10 | 16.6 | 85.44 | 0.3691 | 92.5 | 0.3074 | 0.1918 | 0.0602 | 0.00559 | $2.546 \times 10^{-4}$ | $2.856 \times 10^{-6}$ | 4.684 | 0.352 | 0.00124 | 0.001029 |
| 0 | 21.2 | 84.38 | 0.4635 | 90.9 | 0.3103 | 0.1966 | 0.0589 | 0.00589 | $2.345 \times 10^{-4}$ | $3.314 \times 10^{-6}$ | 4.463 | 0.398 | 0.00128 | 0.000972 |
| 10 | 26.6 | 83.31 | 0.5761 | 89.3 | 0.3134 | 0.2017 | 0.0576 | 0.00619 | $2.181 \times 10^{-4}$ | $3.811 \times 10^{-4}$ | 4.269 | 0.447 | 0.00132 | 0.000915 |
| 20 | 33.1 | 82.2 | 0.7094 | 87.5 | 0.3167 | 0.2070 | 0.0563 | 0.00648 | $2.024 \times 10^{-4}$ | $4.342 \times 10^{-6}$ | 4.098 | 0.500 | 0.00132 | 0.000859 |
| 30 | 40.8 | 81.08 | 0.866 | 85.8 | 0.3203 | 0.2127 | 0.0550 | 0.00676 | $1.883 \times 10^{-4}$ | $4.906 \times 10^{-6}$ | 3.947 | 0.555 | 0.00142 | 0.000803 |
| 40 | 49.8 | 79.92 | 1.049 | 83.9 | 0.3240 | 0.2188 | 0.0536 | 0.00704 | $1.752 \times 10^{-4}$ | $5.494 \times 10^{-6}$ | 3.814 | 0.614 | 0.00149 | 0.000749 |
| 50 | 60.2 | 78.73 | 1.262 | 82.0 | 0.3281 | 0.2253 | 0.0522 | 0.00732 | $1.633 \times 10^{-4}$ | $6.103 \times 10^{-6}$ | 3.697 | 0.677 | 0.00156 | 0.000695 |
| 60 | 72.2 | 77.51 | 1.509 | 80.0 | 0.3325 | 0.2323 | 0.0507 | 0.00758 | $1.522 \times 10^{-4}$ | $6.725 \times 10^{-6}$ | 3.594 | 0.742 | 0.00163 | 0.000642 |
| 70 | 85.9 | 76.25 | 1.794 | 78.0 | 0.3372 | 0.2398 | 0.0492 | 0.00785 | $1.420 \times 10^{-4}$ | $7.356 \times 10^{-6}$ | 3.504 | 0.810 | 0.00173 | 0.000590 |
| 80 | 101.4 | 74.94 | 2.122 | 75.8 | 0.3424 | 0.2481 | 0.0476 | 0.00810 | $1.324 \times 10^{-4}$ | $7.986 \times 10^{-6}$ | 3.425 | 0.880 | 0.00183 | 0.000538 |
| 90 | 119.1 | 73.59 | 2.5 | 73.5 | 0.3481 | 0.2572 | 0.0460 | 0.00835 | $1.234 \times 10^{-4}$ | $8.611 \times 10^{-6}$ | 3.357 | 0.955 | 0.00195 | 0.000488 |
| 100 | 138.9 | 72.17 | 2.935 | 71.1 | 0.3548 | 0.2674 | 0.0444 | 0.00860 | $1.149 \times 10^{-4}$ | $9.222 \times 10^{-6}$ | 3.303 | 1.032 | 0.00210 | 0.000439 |
| 110 | 161.2 | 70.69 | 3.435 | 68.5 | 0.3627 | 0.2790 | 0.0427 | 0.00884 | $1.068 \times 10^{-4}$ | $9.814 \times 10^{-6}$ | 3.262 | 1.115 | 0.00227 | 0.000391 |
| 120 | 186.0 | 69.13 | 4.012 | 65.8 | 0.3719 | 0.2925 | 0.0410 | 0.00908 | $9.911 \times 10^{-5}$ | $1.038 \times 10^{-5}$ | 3.235 | 1.204 | 0.00248 | 0.000344 |
| 130 | 213.5 | 67.48 | 4.679 | 62.9 | 0.3829 | 0.3083 | 0.0392 | 0.00931 | $9.175 \times 10^{-5}$ | $1.092 \times 10^{-5}$ | 3.223 | 1.303 | 0.00275 | 0.000299 |
| 140 | 244.1 | 65.72 | 5.455 | 59.8 | 0.3963 | 0.3276 | 0.0374 | 0.00954 | $8.464 \times 10^{-5}$ | $1.144 \times 10^{-5}$ | 3.229 | 1.416 | 0.00308 | 0.000255 |
| 150 | 277.8 | 63.83 | 6.367 | 56.4 | 0.4131 | 0.3520 | 0.0355 | 0.00976 | $7.778 \times 10^{-5}$ | $1.195 \times 10^{-5}$ | 3.259 | 1.551 | 0.00351 | 0.000212 |
| 160 | 314.9 | 61.76 | 7.45 | 52.7 | 0.4352 | 0.3839 | 0.0335 | 0.00998 | $7.108 \times 10^{-5}$ | $1.245 \times 10^{-5}$ | 3.324 | 1.725 | 0.00411 | 0.000171 |
| 170 | 355.8 | 59.47 | 8.762 | 48.5 | 0.4659 | 0.4286 | 0.0314 | 0.01020 | $6.450 \times 10^{-5}$ | $1.298 \times 10^{-5}$ | 3.443 | 1.963 | 0.00498 | 0.000132 |
| 180 | 400.7 | 56.85 | 10.4 | 43.7 | 0.5123 | 0.4960 | 0.0292 | 0.01041 | $5.792 \times 10^{-5}$ | $1.366 \times 10^{-5}$ | 3.661 | 2.327 | 0.00637 | 0.000095 |
| 190 | 449.9 | 53.75 | 12.53 | 38.0 | 0.5929 | 0.6112 | 0.0267 | 0.01063 | $5.119 \times 10^{-5}$ | $1.431 \times 10^{-5}$ | 4.090 | 2.964 | 0.00891 | 0.000061 |
| 200 | 504.0 | 49.75 | 15.57 | 30.7 | 0.7717 | 0.8544 | 0.0239 | 0.01085 | $4.397 \times 10^{-5}$ | $1.544 \times 10^{-5}$ | 5.119 | 4.376 | 0.01490 | 0.000031 |
| 210 | 563.8 | 43.19 | 21.18 | 18.9 | 1.4786 | 1.6683 | 0.0199 | 0.01110 | $3.483 \times 10^{-5}$ | $1.787 \times 10^{-5}$ | 9.311 | 9.669 | 0.04021 | 0.000006 |

*Note 1:* KInematic viscosity $\nu$ and thermal diffusivity $\alpha$ can be calculated from their definitions, $\nu = \mu/\rho$ and $\alpha = k/\rho c_p = \nu/Pr$. The properties listed here (except the vapor density) can be used at any pressures with negligible error except at temperatures near the critical-point value.

*Note 2:* The unit Btu/lbm . °F for specific heat is equivalent to Btu/lbm·R, and the unit Btu/h·ft·°F for thermal conductivity is equivalent to Btu/h·ft·R.

*Source:* Data generated from the EES software developed by S. A. Klein and F. L. Alvarado. Original sources: R. Tilner-Roth and H. D. Baehr, "An International Standard Formulation for the Thermodynamic Properties of 1,1,1,2-Tetrafluorethane (HFC-134a) for Temperatures from 170 K to 455 K and Pressures up to 70 Mpa," *J. Phys. Chem. Ref. Data*, Vol. 23, No.5, 1994: M. J. Assael, N. K. Dalaouti, A. A. Griva, and J. H. Dymond, "Viscosity and Thermal Conductivity of Halogenated Methane and Ethane Refrigerants," *IJR*, Vol. 22, pp. 525–535, 1999: NIST REPROP 6 program (M. O. McLinden, S. A. Klein, E. W. Lemmon, and A. P. Peskin, Physicial and Chemical Properties Division, National Institute of Standards and Technology, Boulder, CO 80303. 1995).

## TABLE A–17E

Properties of saturated ammonia

| Temp. $T$, °F | Saturation Pressure $P_{sat}$, psia | Density $\rho$, lbm/ft³ | | Enthalpy of Vaporization $h_{fg}$, Btu/lbm | Specific Heat $c_p$, Btu/lbm·R | | Thermal Conductivity $k$, Btu/h·ft·R | | Dynamic Viscosity $\mu$, lbm/ft·s | | Prandtl Number Pr | | Volume Expansion Coefficient $\beta$, 1/R | Surface Tension lbf/ft |
|---|---|---|---|---|---|---|---|---|---|---|---|---|---|---|
| | | Liquid | Vapor | | Liquid | Vapor | Liquid | Vapor | Liquid | Vapor | Liquid | Vapor | Liquid | |
| −40 | 10.4 | 43.08 | 0.0402 | 597.0 | 1.0542 | 0.5354 | – | 0.01026 | $1.966 \times 10^{-4}$ | $5.342 \times 10^{-6}$ | – | 1.003 | 0.00098 | 0.002443 |
| −30 | 13.9 | 42.66 | 0.0527 | 590.2 | 1.0610 | 0.5457 | – | 0.01057 | $1.853 \times 10^{-4}$ | $5.472 \times 10^{-6}$ | – | 1.017 | 0.00101 | 0.002357 |
| −20 | 18.3 | 42.33 | 0.0681 | 583.2 | 1.0677 | 0.5571 | 0.3501 | 0.01089 | $1.746 \times 10^{-4}$ | $5.600 \times 10^{-6}$ | 1.917 | 1.031 | 0.00103 | 0.002272 |
| −10 | 23.7 | 41.79 | 0.0869 | 575.9 | 1.0742 | 0.5698 | 0.3426 | 0.01121 | $1.645 \times 10^{-4}$ | $5.731 \times 10^{-6}$ | 1.856 | 1.048 | 0.00106 | 0.002187 |
| 0 | 30–4 | 41.34 | 0.1097 | 568.4 | 1.0807 | 0.5838 | 0.3352 | 0.01154 | $1.549 \times 10^{-4}$ | $5.861 \times 10^{-6}$ | 1.797 | 1.068 | 0.00109 | 0.002103 |
| 10 | 38.5 | 40.89 | 0.1370 | 560.7 | 1.0873 | 0.5992 | 0.3278 | 0.01187 | $1.458 \times 10^{-4}$ | $5.994 \times 10^{-6}$ | 1.740 | 1.089 | 0.00112 | 0.002018 |
| 20 | 48.2 | 40.43 | 0.1694 | 552.6 | 1.0941 | 0.6160 | 0.3203 | 0.01220 | $1.371 \times 10^{-4}$ | $6.125 \times 10^{-6}$ | 1.686 | 1.113 | 0.00116 | 0.001934 |
| 30 | 59.8 | 39.96 | 0.2075 | 544.4 | 1.1012 | 0.6344 | 0.3129 | 0.01254 | $1290 \times 10^{-4}$ | $6.256 \times 10^{-6}$ | 1.634 | 1.140 | 0.00119 | 0.001850 |
| 40 | 73.4 | 39.48 | 0.2521 | 535.8 | 1.1087 | 0.6544 | 0.3055 | 0.01288 | $1.213 \times 10^{-4}$ | $6.389 \times 10^{-6}$ | 1.585 | 1.168 | 0.00123 | 0.001767 |
| 50 | 89.2 | 38.99 | 0.3040 | 526.9 | 1.1168 | 0.6762 | 0.2980 | 0.01323 | $1.140 \times 10^{-4}$ | $6.522 \times 10^{-6}$ | 1.539 | 1.200 | 0.00128 | 0.001684 |
| 60 | 107.7 | 38.50 | 0.3641 | 517.7 | 1.1256 | 0.6999 | 0.2906 | 0.01358 | $1.072 \times 10^{-4}$ | $6.656 \times 10^{-6}$ | 1.495 | 1.234 | 0.00132 | 0.001601 |
| 70 | 128.9 | 37.99 | 0.4332 | 508.1 | 1.1353 | 0.7257 | 0.2832 | 0.01394 | $1.008 \times 10^{-4}$ | $6.786 \times 10^{-6}$ | 1.456 | 1.272 | 0.00137 | 0.001518 |
| 80 | 153.2 | 37.47 | 0.5124 | 498.2 | 1.1461 | 0.7539 | 0.2757 | 0.01431 | $9.486 \times 10^{-5}$ | $6.922 \times 10^{-6}$ | 1.419 | 1.313 | 0.00143 | 0.001436 |
| 90 | 180.8 | 36.94 | 0.6029 | 487.8 | 1.1582 | 0.7846 | 0.2683 | 0.01468 | $8.922 \times 10^{-5}$ | $7.056 \times 10^{-6}$ | 1.387 | 1.358 | 0.00149 | 0.001354 |
| 100 | 212.0 | 36.40 | 0.7060 | 477.0 | 1.1719 | 0.8183 | 0.2609 | 0.01505 | $8.397 \times 10^{-5}$ | $7.189 \times 10^{-6}$ | 1.358 | 1.407 | 0.00156 | 0.001273 |
| 110 | 247.2 | 35.83 | 0.8233 | 465.8 | 1.1875 | 0.8554 | 0.2535 | 0.01543 | $7.903 \times 10^{-5}$ | $7.325 \times 10^{-6}$ | 1.333 | 1.461 | 0.00164 | 0.001192 |
| 120 | 286.5 | 35.26 | 0.9564 | 454.1 | 1.2054 | 0.8965 | 0.2460 | 0.01582 | $7.444 \times 10^{-5}$ | $7.458 \times 10^{-6}$ | 1.313 | 1.522 | 0.00174 | 0.001111 |
| 130 | 330.4 | 34.66 | 1.1074 | 441.7 | 1.2261 | 0.9425 | 0.2386 | 0.01621 | $7.017 \times 10^{-5}$ | $7.594 \times 10^{-6}$ | 1.298 | 1.589 | 0.00184 | 0.001031 |
| 140 | 379.4 | 34.04 | 1.2786 | 428.8 | 1.2502 | 0.9943 | 0.2312 | 0.01661 | $6.617 \times 10^{-5}$ | $7.731 \times 10^{-6}$ | 1.288 | 1.666 | 0.00196 | 0.000951 |
| 150 | 433.2 | 33.39 | 1.4730 | 415.2 | 1.2785 | 1.0533 | 0.2237 | 0.01702 | $6.244 \times 10^{-5}$ | $7.867 \times 10^{-4}$ | 1.285 | 1.753 | 0.00211 | 0.000872 |
| 160 | 492.7 | 32.72 | 1.6940 | 400.8 | 1.3120 | 1.1214 | 0.2163 | 0.01744 | $5.900 \times 10^{-5}$ | $8.006 \times 10^{-6}$ | 1.288 | 1.853 | 0.00228 | 0.000794 |
| 170 | 558.2 | 32.01 | 1.9460 | 385.4 | 1.3523 | 1.2012 | 0.2089 | 0.01786 | $5.578 \times 10^{-5}$ | $8.142 \times 10^{-6}$ | 1.300 | 1.971 | 0.00249 | 0.000716 |
| 180 | 630.1 | 31.26 | 2.2346 | 369.1 | 1.4015 | 1.2965 | 0.2014 | 0.01829 | $5.278 \times 10^{-5}$ | $8.281 \times 10^{-6}$ | 1.322 | 2.113 | 0.00274 | 0.000638 |
| 190 | 708.5 | 30.47 | 2.5670 | 351.6 | 1.4624 | 1.4128 | 0.1940 | 0.01874 | $5.000 \times 10^{-5}$ | $8.419 \times 10^{-6}$ | 1.357 | 2.286 | 0.00306 | 0.000562 |
| 200 | 794.4 | 29.62 | 2.9527 | 332.7 | 1.5397 | 1.5586 | 0.1866 | 0.01919 | $4.742 \times 10^{-5}$ | $8.561 \times 10^{-6}$ | 1.409 | 2.503 | 0.00348 | 0.000486 |
| 210 | 887.9 | 28.70 | 3.4053 | 312.0 | 1.6411 | 1.7473 | 0.1791 | 0.01966 | $4500 \times 10^{-5}$ | $8.703 \times 10^{-6}$ | 1.484 | 2.784 | 0.00403 | 0.000411 |
| 220 | 989.5 | 27.69 | 3.9440 | 289.2 | 1.7798 | 2.0022 | 0.1717 | 0.02015 | $4.275 \times 10^{-5}$ | $8.844 \times 10^{-6}$ | 1.595 | 3.164 | 0.00480 | 0.000338 |
| 230 | 1099.0 | 25.57 | 4.5987 | 263.5 | 1.9824 | 2.3659 | 0.1643 | 0.02065 | $4.064 \times 10^{-5}$ | $8.989 \times 10^{-6}$ | 1.765 | 3.707 | 0.00594 | 0.000265 |
| 240 | 1219.4 | 25.28 | 5.4197 | 234.0 | 2.3100 | 2.9264 | 0.1568 | 0.02119 | $3.864 \times 10^{-5}$ | $9.136 \times 10^{-6}$ | 2.049 | 4.542 | 0.00784 | 0.000194 |

*Note 1:* Kinematic viscosity $\nu$ and thermal diffusivity $\alpha$ can be calculated from their definitions, $\nu = \mu/p$ and $\alpha = k/\rho c_p = \nu/\text{Pr}$. The properties listed here (except the vapor density) can be used at any pressures with negligible error except at temperatures near the critical-point value.

*Note 2:* The unit Btu/lbm·°F for specific heat is equivalent to Btu/lbm·R, and the unit Btu/h·ft·°F for thermal conductivity is equivalent to Btu/h·ft·R.

*Source:* Data generated from the EES software developed by S. A. Klein and F. L. Alvarado. Orginal sources: Tillner-Roth, Harms-Watzenterg, and Baehr, "Eine neue Fundamentalgleichung fur Ammoniak," *DKV-Tagungsbericht* 20: 167–181, 1993; Liley and Desai, "Thermophysical Properties of Refrigerants," *ASHRAE*, 1993, ISBN 1-1883413-10-9.

## TABLE A–18E

Properties of saturated propane

| Temp. $T$, °F | Saturation Pressure $P_{sat}$, psia | Density $\rho$, lbm/ft³ Liquid | Density $\rho$, lbm/ft³ Vapor | Enthalpy of Vaporization $h_{ig}$, Btu/lbm | Specific Heat $c_p$, Btu/lbm·R Liquid | Specific Heat $c_p$, Btu/lbm·R Vapor | Thermal Conductivity $k$, Btu/h·ft·R Liquid | Thermal Conductivity $k$, Btu/h·ft·R Vapor | Dynamic Viscosity $\mu$, lbm/ft·s Liquid | Dynamic Viscosity $\mu$, lbm/ft·s Vapor | Prandtl Number Pr Liquid | Prandtl Number Pr Vapor | Volume Expansion Coefficient $\beta$, 1/R Liquid | Surface Tension lbf/ft |
|---|---|---|---|---|---|---|---|---|---|---|---|---|---|---|
| −200 | 0.0201 | 42.06 | 0.0003 | 217.7 | 0.4750 | 0.2595 | 0.1073 | 0.00313 | $5.012 \times 10^{-4}$ | $2.789 \times 10^{-6}$ | 7.991 | 0.833 | 0.00083 | 0.001890 |
| −180 | 0.0752 | 41.36 | 0.0011 | 213.4 | 0.4793 | 0.2680 | 0.1033 | 0.00347 | $3.941 \times 10^{-4}$ | $2.975 \times 10^{-6}$ | 6.582 | 0.826 | 0.00086 | 0.001780 |
| −160 | 0.2307 | 40.65 | 0.0032 | 209.1 | 0.4845 | 0.2769 | 0.0992 | 0.00384 | $3.199 \times 10^{-4}$ | $3.164 \times 10^{-6}$ | 5.626 | 0.821 | 0.00088 | 0.001671 |
| −140 | 0.6037 | 39.93 | 0.0078 | 204.8 | 0.4907 | 0.2866 | 0.0949 | 0.00423 | $2.660 \times 10^{-4}$ | $3.358 \times 10^{-6}$ | 4.951 | 0.818 | 0.00091 | 0.001563 |
| −120 | 1.389 | 39.20 | 0.0170 | 200.5 | 0.4982 | 0.2971 | 0.0906 | 0.00465 | $2.252 \times 10^{-4}$ | $3.556 \times 10^{-6}$ | 4.457 | 0.817 | 0.00094 | 0.001455 |
| −100 | 2.878 | 38.46 | 0.0334 | 196.1 | 0.5069 | 0.3087 | 0.0863 | 0.00511 | $1.934 \times 10^{-4}$ | $3.756 \times 10^{-6}$ | 4.087 | 0.817 | 0.00097 | 0.001349 |
| −90 | 4.006 | 38.08 | 0.0453 | 193.9 | 0.5117 | 0.3150 | 0.0842 | 0.00534 | $1.799 \times 10^{-4}$ | $3.858 \times 10^{-6}$ | 3.936 | 0.819 | 0.00099 | 0.001297 |
| −80 | 5.467 | 37.70 | 0.0605 | 191.6 | 0.5169 | 0.3215 | 0.0821 | 0.00559 | $1.678 \times 10^{-4}$ | $3.961 \times 10^{-6}$ | 3.803 | 0.820 | 0.00101 | 0.001244 |
| −70 | 7.327 | 37.32 | 0.0793 | 189.3 | 0.5224 | 0.3284 | 0.0800 | 0.00585 | $1.569 \times 10^{-4}$ | $4.067 \times 10^{-6}$ | 3.686 | 0.822 | 0.00104 | 0.001192 |
| −60 | 9.657 | 36.93 | 0.1024 | 186.9 | 0.5283 | 0.3357 | 0.0780 | 0.00611 | $1.469 \times 10^{-4}$ | $4.172 \times 10^{-6}$ | 3.582 | 0.825 | 0.00106 | 0.001140 |
| −50 | 12.54 | 36.54 | 0.1305 | 184.4 | 0.5345 | 0.3433 | 0.0760 | 0.00639 | $1.378 \times 10^{-4}$ | $4.278 \times 10^{-6}$ | 3.490 | 0.828 | 0.00109 | 0.001089 |
| −40 | 16.05 | 36.13 | 0.1641 | 181.9 | 0.5392 | 0.3513 | 0.0740 | 0.00568 | $1.294 \times 10^{-4}$ | $4.386 \times 10^{-6}$ | 3.395 | 0.831 | 0.00112 | 0.001038 |
| −30 | 20.29 | 35.73 | 0.2041 | 179.3 | 0.5460 | 0.3596 | 0.0721 | 0.00697 | $1.217 \times 10^{-4}$ | $4.497 \times 10^{-6}$ | 3.320 | 0.835 | 0.00115 | 0.000987 |
| −20 | 25.34 | 35.31 | 0.2512 | 176.6 | 0.5531 | 0.3684 | 0.0702 | 0.00728 | $1.146 \times 10^{-4}$ | $4.611 \times 10^{-6}$ | 3.253 | 0.840 | 0.00119 | 0.000937 |
| −10 | 31.3 | 34.89 | 0.3063 | 173.8 | 0.5607 | 0.3776 | 0.0683 | 0.00761 | $1.079 \times 10^{-4}$ | $4.725 \times 10^{-6}$ | 3.192 | 0.845 | 0.00123 | 0.000887 |
| 0 | 38.28 | 34.46 | 0.3703 | 170.9 | 0.5689 | 0.3874 | 0.0665 | 0.00794 | $1.018 \times 10^{-4}$ | $4.842 \times 10^{-6}$ | 3.137 | 0.850 | 0.00127 | 0.000838 |
| 10 | 46.38 | 34.02 | 0.4441 | 167.9 | 0.5775 | 0.3976 | 0.0647 | 0.00829 | $9.606 \times 10^{-5}$ | $4.961 \times 10^{-6}$ | 3.088 | 0.857 | 0.00132 | 0.000789 |
| 20 | 55.7 | 33.56 | 0.5289 | 164.8 | 0.5867 | 0.4084 | 0.0629 | 0.00865 | $9.067 \times 10^{-5}$ | $5.086 \times 10^{-6}$ | 3.043 | 0.864 | 0.00138 | 0.000740 |
| 30 | 66.35 | 33.10 | 0.6259 | 161.6 | 0.5966 | 0.4199 | 0.0512 | 0.00903 | $8.561 \times 10^{-5}$ | $5.211 \times 10^{-6}$ | 3.003 | 0.873 | 0.00144 | 0.000692 |
| 40 | 78.45 | 32.62 | 0.7365 | 158.1 | 0.6072 | 0.4321 | 0.0595 | 0.00942 | $8.081 \times 10^{-5}$ | $5.342 \times 10^{-6}$ | 2.967 | 0.882 | 0.00151 | 0.000644 |
| 50 | 92.12 | 32.13 | 0.8621 | 154.6 | 0.6187 | 0.4452 | 0.0579 | 0.00983 | $7.631 \times 10^{-5}$ | $5.478 \times 10^{-6}$ | 2.935 | 0.893 | 0.00159 | 0.000597 |
| 60 | 107.5 | 31.63 | 1.0046 | 150.8 | 0.6311 | 0.4593 | 0.0563 | 0.01025 | $7.200 \times 10^{-5}$ | $5.617 \times 10^{-6}$ | 2.906 | 0.906 | 0.00168 | 0.000551 |
| 70 | 124.6 | 31.11 | 1.1659 | 146.8 | 0.6447 | 0.4746 | 0.0547 | 0.01070 | $6.794 \times 10^{-5}$ | $5.764 \times 10^{-6}$ | 2.881 | 0.921 | 0.00179 | 0.000505 |
| 80 | 143.7 | 30.56 | 1.3484 | 142.7 | 0.6596 | 0.4915 | 0.0532 | 0.01116 | $6.406 \times 10^{-5}$ | $5.919 \times 10^{-6}$ | 2.860 | 0.938 | 0.00191 | 0.000460 |
| 90 | 164.8 | 30.00 | 1.5549 | 138.2 | 0.6762 | 0.5103 | 0.0517 | 0.01165 | $6.033 \times 10^{-5}$ | $6.081 \times 10^{-6}$ | 2.843 | 0.959 | 0.00205 | 0.000416 |
| 100 | 188.1 | 29.41 | 1.7887 | 133.6 | 0.6947 | 0.5315 | 0.0501 | 0.01217 | $5.675 \times 10^{-5}$ | $6.256 \times 10^{-6}$ | 2.831 | 0.984 | 0.00222 | 0.000372 |
| 120 | 241.8 | 28.13 | 2.3562 | 123.2 | 0.7403 | 0.5844 | 0.0472 | 0.01328 | $5.000 \times 10^{-6}$ | $6.644 \times 10^{-6}$ | 2.825 | 1.052 | 0.00267 | 0.000288 |
| 140 | 306.1 | 26.69 | 3.1003 | 111.1 | 0.7841 | 0.6613 | 0.0442 | 0.01454 | $4.358 \times 10^{-5}$ | $7.111 \times 10^{-6}$ | 2.784 | 1.164 | 0.00338 | 0.000208 |
| 160 | 382.4 | 24.98 | 4.1145 | 96.4 | 0.8696 | 0.7911 | 0.0411 | 0.01603 | $3.733 \times 10^{-5}$ | $7.719 \times 10^{-6}$ | 2.845 | 1.371 | 0.00459 | 0.000133 |
| 180 | 472.9 | 22.79 | 5.6265 | 77.1 | 1.1436 | 1.0813 | 0.0376 | 0.01793 | $3.083 \times 10^{-5}$ | $8.617 \times 10^{-6}$ | 3.380 | 1.870 | 0.00791 | 0.000065 |

*Note 1:* Kinematic viscosity $\nu$ and thermal diffusivity $\alpha$ can be calculated from their definitions, $\nu = \mu/\rho$ and $\alpha = k/\rho c_p = \nu/Pr$. The properties listed here (except the vapor density) can be used at any pressures with negligible error at temperatures near the critical-point value.

*Note 2:* The unit Btu/lbm·°F for specific heat is equivalent to Btu/lbm·R, and the unit Btu/h·ft·°F for thermal conductivity is equivalent to Btu/h·ft·R.

*Source:* Data generated from the EES software developed by S. A. Klein and F. L. Alvarado. Original sources: Reiner Tillner-Roth, "Fundamental Equations of State," Shaker, Verlag, Aachan, 1998; B. A. Younglove and J. F. Ely. "Thermophysical Properties of Fluids. II Methane, Ethane, Propane, Isobutane, and Normal Butane," *J. Phys. Chem. Ref. Data*, Vol. 16, No. 4, 1987; G. R. Somayajulu, "A Generalized Equation for Surface Tension from the Triple-Point to the Critical-Point," *International Journal of Thermophysics*, Vol. 9, No. 4, 1988.

## TABLE A–19E

Properties of liquids

| Temp. $T$, °F | Density $\rho$, lbm/ft³ | Specific Heat $c_p$, Btu/lbm·R | Thermal Conductivity $k$, Btu/h·ft·R | Thermal Diffusivity $\alpha$, ft²/s | Dynamic Viscosity $\mu$, lbm/ft·s | Kinematic Viscosity $\nu$, ft²/s | Prandtl Number Pr | Volume Expansion Coeff. $\beta$, 1/R |
|---|---|---|---|---|---|---|---|---|
| \multicolumn{9}{c}{Methane (CH₄)} |

| Temp. $T$, °F | Density $\rho$, lbm/ft³ | Specific Heat $c_p$, Btu/lbm·R | Thermal Conductivity $k$, Btu/h·ft·R | Thermal Diffusivity $\alpha$, ft²/s | Dynamic Viscosity $\mu$, lbm/ft·s | Kinematic Viscosity $\nu$, ft²/s | Prandtl Number Pr | Volume Expansion Coeff. $\beta$, 1/R |
|---|---|---|---|---|---|---|---|---|
| \multicolumn{9}{c}{**Methane (CH₄)**} |
| −280 | 27.41 | 0.8152 | 0.1205 | $1.497 \times 10^{-6}$ | $1.057 \times 10^{-4}$ | $3.857 \times 10^{-6}$ | 2.575 | 0.00175 |
| −260 | 26.43 | 0.8301 | 0.1097 | $1.389 \times 10^{-6}$ | $8.014 \times 10^{-5}$ | $3.032 \times 10^{-6}$ | 2.183 | 0.00192 |
| −240 | 25.39 | 0.8523 | 0.0994 | $1.276 \times 10^{-6}$ | $6.303 \times 10^{-5}$ | $2.482 \times 10^{-6}$ | 1.945 | 0.00215 |
| −220 | 24.27 | 0.8838 | 0.0896 | $1.159 \times 10^{-6}$ | $5.075 \times 10^{-5}$ | $2.091 \times 10^{-6}$ | 1.803 | 0.00247 |
| −200 | 23.04 | 0.9314 | 0.0801 | $1.036 \times 10^{-6}$ | $4.142 \times 10^{-5}$ | $1.798 \times 10^{-6}$ | 1.734 | 0.00295 |
| −180 | 21.64 | 1.010 | 0.0709 | $9.008 \times 10^{-7}$ | $3.394 \times 10^{-5}$ | $1.568 \times 10^{-6}$ | 1.741 | 0.00374 |
| −160 | 19.99 | 1.158 | 0.0616 | $7.397 \times 10^{-7}$ | $2.758 \times 10^{-5}$ | $1.379 \times 10^{-6}$ | 1.865 | 0.00526 |
| −140 | 17.84 | 1.542 | 0.0518 | $5.234 \times 10^{-7}$ | $2.168 \times 10^{-5}$ | $1.215 \times 10^{-6}$ | 2.322 | 0.00943 |
| \multicolumn{9}{c}{**Methanol [CH₃(OH)]**} |
| 70 | 49.15 | 0.6024 | 0.1148 | $1.076 \times 10^{-6}$ | $3.872 \times 10^{-4}$ | $7.879 \times 10^{-6}$ | 7.317 | 0.000656 |
| 90 | 48.50 | 0.6189 | 0.1143 | $1.057 \times 10^{-6}$ | $3.317 \times 10^{-4}$ | $6.840 \times 10^{-6}$ | 6.468 | 0.000671 |
| 110 | 47.85 | 0.6373 | 0.1138 | $1.036 \times 10^{-6}$ | $2.872 \times 10^{-4}$ | $6.005 \times 10^{-6}$ | 5.793 | 0.000691 |
| 130 | 47.18 | 0.6576 | 0.1133 | $1.014 \times 10^{-6}$ | $2.513 \times 10^{-4}$ | $5.326 \times 10^{-6}$ | 5.250 | 0.000716 |
| 150 | 46.50 | 0.6796 | 0.1128 | $9.918 \times 10^{-7}$ | $2.218 \times 10^{-4}$ | $4.769 \times 10^{-6}$ | 4.808 | 0.000749 |
| 170 | 45.80 | 0.7035 | 0.1124 | $9.687 \times 10^{-7}$ | $1.973 \times 10^{-4}$ | $4.308 \times 10^{-6}$ | 4.447 | 0.000789 |
| \multicolumn{9}{c}{**Isobutane (R600a)**} |
| −150 | 42.75 | 0.4483 | 0.0799 | $1.157 \times 10^{-6}$ | $6.417 \times 10^{-4}$ | $1.500 \times 10^{-5}$ | 12.96 | 0.000785 |
| −100 | 41.06 | 0.4721 | 0.0782 | $1.120 \times 10^{-6}$ | $3.669 \times 10^{-4}$ | $8.939 \times 10^{-6}$ | 7.977 | 0.000836 |
| −50 | 39.31 | 0.4986 | 0.0731 | $1.036 \times 10^{-6}$ | $2.376 \times 10^{-4}$ | $6.043 \times 10^{-6}$ | 5.830 | 0.000908 |
| 0 | 37.48 | 0.5289 | 0.0664 | $9.299 \times 10^{-7}$ | $1.651 \times 10^{-4}$ | $4.406 \times 10^{-6}$ | 4.738 | 0.001012 |
| 50 | 35.52 | 0.5643 | 0.0591 | $8.187 \times 10^{-7}$ | $1.196 \times 10^{-4}$ | $3.368 \times 10^{-6}$ | 4.114 | 0.001169 |
| 100 | 33.35 | 0.6075 | 0.0521 | $7.139 \times 10^{-7}$ | $8.847 \times 10^{-5}$ | $2.653 \times 10^{-6}$ | 3.716 | 0.001421 |
| 150 | 30.84 | 0.6656 | 0.0457 | $6.188 \times 10^{-7}$ | $6.558 \times 10^{-5}$ | $2.127 \times 10^{-6}$ | 3.437 | 0.001883 |
| 200 | 27.73 | 0.7635 | 0.0400 | $5.249 \times 10^{-7}$ | $4.750 \times 10^{-5}$ | $1.713 \times 10^{-6}$ | 3.264 | 0.002970 |
| \multicolumn{9}{c}{**Glycerin**} |
| 32 | 79.65 | 0.5402 | 0.163 | $1.052 \times 10^{-6}$ | 7.047 | 0.08847 | 84101 | |
| 40 | 79.49 | 0.5458 | 0.1637 | $1.048 \times 10^{-6}$ | 4.803 | 0.06042 | 57655 | |
| 50 | 79.28 | 0.5541 | 0.1645 | $1.040 \times 10^{-6}$ | 2.850 | 0.03594 | 34561 | |
| 60 | 79.07 | 0.5632 | 0.1651 | $1.029 \times 10^{-6}$ | 1.547 | 0.01956 | 18995 | |
| 70 | 78.86 | 0.5715 | 0.1652 | $1.018 \times 10^{-6}$ | 0.9422 | 0.01195 | 11730 | |
| 80 | 78.66 | 0.5794 | 0.1652 | $1.007 \times 10^{-6}$ | 0.5497 | 0.00699 | 6941 | |
| 90 | 78.45 | 0.5878 | 0.1652 | $9.955 \times 10^{-7}$ | 0.3756 | 0.004787 | 4809 | |
| 100 | 78.24 | 0.5964 | 0.1653 | $9.841 \times 10^{-7}$ | 0.2277 | 0.00291 | 2957 | |
| \multicolumn{9}{c}{**Engine Oil (unused)**} |
| 32 | 56.12 | 0.4291 | 0.0849 | $9.792 \times 10^{-7}$ | 2.563 | $4.566 \times 10^{-2}$ | 46636 | 0.000389 |
| 50 | 55.79 | 0.4395 | 0.08338 | $9.448 \times 10^{-7}$ | 1.210 | $2.169 \times 10^{-2}$ | 22963 | 0.000389 |
| 75 | 55.3 | 0.4531 | 0.08378 | $9.288 \times 10^{-7}$ | 0.4286 | $7.751 \times 10^{-3}$ | 8345 | 0.000389 |
| 100 | 54.77 | 0.4669 | 0.08367 | $9.089 \times 10^{-7}$ | 0.1630 | $2.977 \times 10^{-3}$ | 3275 | 0.000389 |
| 125 | 54.24 | 0.4809 | 0.08207 | $8.740 \times 10^{-7}$ | $7.617 \times 10^{-2}$ | $1.404 \times 10^{-3}$ | 1607 | 0.000389 |
| 150 | 53.73 | 0.4946 | 0.08046 | $8.411 \times 10^{-7}$ | $3.833 \times 10^{-2}$ | $7.135 \times 10^{-4}$ | 848.3 | 0.000389 |
| 200 | 52.68 | 0.5231 | 0.07936 | $7.999 \times 10^{-7}$ | $1.405 \times 10^{-2}$ | $2.668 \times 10^{-4}$ | 333.6 | 0.000389 |
| 250 | 51.71 | 0.5523 | 0.07776 | $7.563 \times 10^{-7}$ | $6.744 \times 10^{-3}$ | $1.304 \times 10^{-4}$ | 172.5 | 0.000389 |
| 300 | 50.63 | 0.5818 | 0.07673 | $7.236 \times 10^{-7}$ | $3.661 \times 10^{-3}$ | $7.232 \times 10^{-5}$ | 99.94 | 0.000389 |

*Source:* Data generated from the EES software developed by S. A. Klein and F. L. Alvarado. Originally based on various sources.

## TABLE A–20E

Properties of liquid metals

| Temp. $T$, °F | Density $\rho$, lbm/ft$^3$ | Specific Heat $c_p$, Btu/lbm·R | Thermal Conductivity $k$, Btu/h·ft·R | Thermal Diffusivity $\alpha$, ft$^2$/s | Dynamic Viscosity $\mu$, lbm/ft·s | Kinematic Viscosity $\nu$, ft$^2$/s | Prandtl Number Pr | Volume Expansion Coeff. $\beta$, 1/R |
|---|---|---|---|---|---|---|---|---|
| *Mercury (Hg) Melting Point: −38°F* | | | | | | | | |
| 32 | 848.7 | 0.03353 | 4.727 | $4.614 \times 10^{-5}$ | $1.133 \times 10^{-3}$ | $1.335 \times 10^{-6}$ | 0.02895 | $1.005 \times 10^{-4}$ |
| 50 | 847.2 | 0.03344 | 4.805 | $4.712 \times 10^{-5}$ | $1.092 \times 10^{-3}$ | $1.289 \times 10^{-6}$ | 0.02737 | $1.005 \times 10^{-4}$ |
| 100 | 842.9 | 0.03319 | 5.015 | $4.980 \times 10^{-5}$ | $9.919 \times 10^{-4}$ | $1.176 \times 10^{-6}$ | 0.02363 | $1.005 \times 10^{-4}$ |
| 150 | 838.7 | 0.03298 | 5.221 | $5.244 \times 10^{-5}$ | $9.122 \times 10^{-4}$ | $1.087 \times 10^{-6}$ | 0.02074 | $1.005 \times 10^{-4}$ |
| 200 | 834.5 | 0.03279 | 5.422 | $5.504 \times 10^{-5}$ | $8.492 \times 10^{-4}$ | $1.017 \times 10^{-6}$ | 0.01849 | $1.005 \times 10^{-4}$ |
| 300 | 826.2 | 0.03252 | 5.815 | $6.013 \times 10^{-5}$ | $7.583 \times 10^{-4}$ | $9.180 \times 10^{-7}$ | 0.01527 | $1.005 \times 10^{-4}$ |
| 400 | 817.9 | 0.03236 | 6.184 | $6.491 \times 10^{-5}$ | $6.972 \times 10^{-4}$ | $8.524 \times 10^{-7}$ | 0.01313 | $1.008 \times 10^{-4}$ |
| 500 | 809.6 | 0.03230 | 6.518 | $6.924 \times 10^{-5}$ | $6.525 \times 10^{-4}$ | $8.061 \times 10^{-7}$ | 0.01164 | $1.018 \times 10^{-4}$ |
| 600 | 801.3 | 0.03235 | 6.839 | $7.329 \times 10^{-5}$ | $6.186 \times 10^{-4}$ | $7.719 \times 10^{-7}$ | 0.01053 | $1.035 \times 10^{-4}$ |
| *Bismuth (Bi) Melting Point: 520°F* | | | | | | | | |
| 700 | 620.7 | 0.03509 | 9.361 | $1.193 \times 10^{-4}$ | $1.001 \times 10^{-3}$ | $1.614 \times 10^{-6}$ | 0.01352 | |
| 800 | 616.5 | 0.03569 | 9.245 | $1.167 \times 10^{-4}$ | $9.142 \times 10^{-4}$ | $1.482 \times 10^{-6}$ | 0.01271 | |
| 900 | 612.2 | 0.0363 | 9.129 | $1.141 \times 10^{-4}$ | $8.267 \times 10^{-4}$ | $1.350 \times 10^{-6}$ | 0.01183 | |
| 1000 | 608.0 | 0.0369 | 9.014 | $1.116 \times 10^{-4}$ | $7.392 \times 10^{-4}$ | $1.215 \times 10^{-6}$ | 0.0109 | |
| 1100 | 603.7 | 0.0375 | 9.014 | $1.105 \times 10^{-4}$ | $6.872 \times 10^{-4}$ | $1.138 \times 10^{-6}$ | 0.01029 | |
| *Lead (Pb) Melting Point: 621°F* | | | | | | | | |
| 700 | 658 | 0.03797 | 9.302 | $1.034 \times 10^{-4}$ | $1.612 \times 10^{-3}$ | $2.450 \times 10^{-6}$ | 0.02369 | |
| 800 | 654 | 0.03750 | 9.157 | $1.037 \times 10^{-4}$ | $1.453 \times 10^{-3}$ | $2.223 \times 10^{-6}$ | 0.02143 | |
| 900 | 650 | 0.03702 | 9.013 | $1.040 \times 10^{-4}$ | $1.296 \times 10^{-3}$ | $1.994 \times 10^{-6}$ | 0.01917 | |
| 1000 | 645.7 | 0.03702 | 8.912 | $1.035 \times 10^{-4}$ | $1.202 \times 10^{-3}$ | $1.862 \times 10^{-6}$ | 0.01798 | |
| 1100 | 641.5 | 0.03702 | 8.810 | $1.030 \times 10^{-4}$ | $1.108 \times 10^{-3}$ | $1.727 \times 10^{-6}$ | 0.01676 | |
| 1200 | 637.2 | 0.03702 | 8.709 | $1.025 \times 10^{-4}$ | $1.013 \times 10^{-3}$ | $1.590 \times 10^{-6}$ | 0.01551 | |
| *Sodium (Na) Melting Point: 208°F* | | | | | | | | |
| 300 | 57.13 | 0.3258 | 48.19 | $7.192 \times 10^{-4}$ | $4.136 \times 10^{-4}$ | $7.239 \times 10^{-6}$ | 0.01007 | |
| 400 | 56.28 | 0.3219 | 46.58 | $7.142 \times 10^{-4}$ | $3.572 \times 10^{-4}$ | $6.350 \times 10^{-6}$ | 0.008891 | |
| 500 | 55.42 | 0.3181 | 44.98 | $7.087 \times 10^{-4}$ | $3.011 \times 10^{-4}$ | $5.433 \times 10^{-6}$ | 0.007667 | |
| 600 | 54.56 | 0.3143 | 43.37 | $7.026 \times 10^{-4}$ | $2.448 \times 10^{-4}$ | $4.488 \times 10^{-6}$ | 0.006387 | |
| 800 | 52.85 | 0.3089 | 40.55 | $6.901 \times 10^{-4}$ | $1.772 \times 10^{-4}$ | $3.354 \times 10^{-6}$ | 0.004860 | |
| 1000 | 51.14 | 0.3057 | 38.12 | $6.773 \times 10^{-4}$ | $1.541 \times 10^{-4}$ | $3.014 \times 10^{-6}$ | 0.004449 | |
| *Potassium (K) Melting Point: 147°F* | | | | | | | | |
| 300 | 50.40 | 0.1911 | 26.00 | $7.500 \times 10^{-4}$ | $2.486 \times 10^{-4}$ | $4.933 \times 10^{-6}$ | 0.006577 | |
| 400 | 49.58 | 0.1887 | 25.37 | $7.532 \times 10^{-4}$ | $2.231 \times 10^{-4}$ | $4.500 \times 10^{-6}$ | 0.005975 | |
| 500 | 48.76 | 0.1863 | 24.73 | $7.562 \times 10^{-4}$ | $1.976 \times 10^{-4}$ | $4.052 \times 10^{-6}$ | 0.005359 | |
| 600 | 47.94 | 0.1839 | 24.09 | $7.591 \times 10^{-4}$ | $1.721 \times 10^{-4}$ | $3.589 \times 10^{-6}$ | 0.004728 | |
| 800 | 46.31 | 0.1791 | 22.82 | $7.643 \times 10^{-4}$ | $1.210 \times 10^{-4}$ | $2.614 \times 10^{-6}$ | 0.003420 | |
| 1000 | 44.62 | 0.1791 | 21.34 | $7.417 \times 10^{-4}$ | $1.075 \times 10^{-4}$ | $2.409 \times 10^{-6}$ | 0.003248 | |
| *Sodium-Potassium (%22Na-%78K) Melting Point: 12°F* | | | | | | | | |
| 200 | 52.99 | 0.2259 | 14.79 | $3.432 \times 10^{-4}$ | $3.886 \times 10^{-4}$ | $7.331 \times 10^{-6}$ | 0.02136 | |
| 300 | 52.16 | 0.2230 | 14.99 | $3.580 \times 10^{-4}$ | $3.467 \times 10^{-4}$ | $6.647 \times 10^{-6}$ | 0.01857 | |
| 400 | 51.32 | 0.2201 | 15.19 | $3.735 \times 10^{-4}$ | $3.050 \times 10^{-4}$ | $5.940 \times 10^{-6}$ | 0.0159 | |
| 600 | 49.65 | 0.2143 | 15.59 | $4.070 \times 10^{-4}$ | $2.213 \times 10^{-4}$ | $4.456 \times 10^{-6}$ | 0.01095 | |
| 800 | 47.99 | 0.2100 | 15.95 | $4.396 \times 10^{-4}$ | $1.539 \times 10^{-4}$ | $3.207 \times 10^{-6}$ | 0.007296 | |
| 1000 | 46.36 | 0.2103 | 16.20 | $4.615 \times 10^{-4}$ | $1.353 \times 10^{-4}$ | $2.919 \times 10^{-6}$ | 0.006324 | |

*Source:* Data generated from the EES software developed by S. A. Klein and F. L. Alvarado. Originally based on various sources.

## TABLE A–21E

Ideal-gas properties of air

| T R | h Btu/lbm | $P_r$ | u Btu/lbm | $v_r$ | $s°$ Btu/lbm·R | T R | h Btu/lbm | $P_r$ | u Btu/lbm | $v_r$ | $s°$ Btu/lbm·R |
|---|---|---|---|---|---|---|---|---|---|---|---|
| 360 | 85.97 | 0.3363 | 61.29 | 396.6 | 0.50369 | 1600 | 395.74 | 71.13 | 286.06 | 8.263 | 0.87130 |
| 380 | 90.75 | 0.4061 | 64.70 | 346.6 | 0.51663 | 1650 | 409.13 | 80.89 | 296.03 | 7.556 | 0.87954 |
| 400 | 95.53 | 0.4858 | 68.11 | 305.0 | 0.52890 | 1700 | 422.59 | 90.95 | 306.06 | 6.924 | 0.88758 |
| 420 | 100.32 | 0.5760 | 71.52 | 270.1 | 0.54058 | 1750 | 436.12 | 101.98 | 316.16 | 6.357 | 0.89542 |
| 440 | 105.11 | 0.6776 | 74.93 | 240.6 | 0.55172 | 1800 | 449.71 | 114.0 | 326.32 | 5.847 | 0.90308 |
| 460 | 109.90 | 0.7913 | 78.36 | 215.33 | 0.56235 | 1850 | 463.37 | 127.2 | 336.55 | 5.388 | 0.91056 |
| 480 | 114.69 | 0.9182 | 81.77 | 193.65 | 0.57255 | 1900 | 477.09 | 141.5 | 346.85 | 4.974 | 0.91788 |
| 500 | 119.48 | 1.0590 | 85.20 | 174.90 | 0.58233 | 1950 | 490.88 | 157.1 | 357.20 | 4.598 | 0.92504 |
| 520 | 124.27 | 1.2147 | 88.62 | 158.58 | 0.59173 | 2000 | 504.71 | 174.0 | 367.61 | 4.258 | 0.93205 |
| 537 | 128.10 | 1.3593 | 91.53 | 146.34 | 0.59945 | 2050 | 518.71 | 192.3 | 378.08 | 3.949 | 0.93891 |
| 540 | 129.06 | 1.3860 | 92.04 | 144.32 | 0.60078 | 2100 | 532.55 | 212.1 | 388.60 | 3.667 | 0.94564 |
| 560 | 133.86 | 1.5742 | 95.47 | 131.78 | 0.60950 | 2150 | 546.54 | 223.5 | 399.17 | 3.410 | 0.95222 |
| 580 | 138.66 | 1.7800 | 98.90 | 120.70 | 0.61793 | 2200 | 560.59 | 256.6 | 409.78 | 3.176 | 0.95919 |
| 600 | 143.47 | 2.005 | 102.34 | 110.88 | 0.62607 | 2250 | 574.69 | 281.4 | 420.46 | 2.961 | 0.96501 |
| 620 | 148.28 | 2.249 | 105.78 | 102.12 | 0.63395 | 2300 | 588.82 | 308.1 | 431.16 | 2.765 | 0.97123 |
| 640 | 153.09 | 2.514 | 109.21 | 94.30 | 0.64159 | 2350 | 603.00 | 336.8 | 441.91 | 2.585 | 0.97732 |
| 660 | 157.92 | 2.801 | 112.67 | 87.27 | 0.64902 | 2400 | 617.22 | 367.6 | 452.70 | 2.419 | 0.98331 |
| 680 | 162.73 | 3.111 | 116.12 | 80.96 | 0.65621 | 2450 | 631.48 | 400.5 | 463.54 | 2.266 | 0.98919 |
| 700 | 167.56 | 3.446 | 119.58 | 75.25 | 0.66321 | 2500 | 645.78 | 435.7 | 474.40 | 2.125 | 0.99497 |
| 720 | 172.39 | 3.806 | 123.04 | 70.07 | 0.67002 | 2550 | 660.12 | 473.3 | 485.31 | 1.996 | 1.00064 |
| 740 | 177.23 | 4.193 | 126.51 | 65.38 | 0.67665 | 2600 | 674.49 | 513.5 | 496.26 | 1.876 | 1.00623 |
| 760 | 182.08 | 4.607 | 129.99 | 61.10 | 0.68312 | 2650 | 688.90 | 556.3 | 507.25 | 1.765 | 1.01172 |
| 780 | 186.94 | 5.051 | 133.47 | 57.20 | 0.68942 | 2700 | 703.35 | 601.9 | 518.26 | 1.662 | 1.01712 |
| 800 | 191.81 | 5.526 | 136.97 | 53.63 | 0.69558 | 2750 | 717.83 | 650.4 | 529.31 | 1.566 | 1.02244 |
| 820 | 196.69 | 6.033 | 140.47 | 50.35 | 0.70160 | 2800 | 732.33 | 702.0 | 540.40 | 1.478 | 1.02767 |
| 840 | 201.56 | 6.573 | 143.98 | 47.34 | 0.70747 | 2850 | 746.88 | 756.7 | 551.52 | 1.395 | 1.03282 |
| 860 | 206.46 | 7.149 | 147.50 | 44.57 | 0.71323 | 2900 | 761.45 | 814.8 | 562.66 | 1.318 | 1.03788 |
| 880 | 211.35 | 7.761 | 151.02 | 42.01 | 0.71886 | 2950 | 776.05 | 876.4 | 573.84 | 1.247 | 1.04288 |
| 900 | 216.26 | 8.411 | 154.57 | 39.64 | 0.72438 | 3000 | 790.68 | 941.4 | 585.04 | 1.180 | 1.04779 |
| 920 | 221.18 | 9.102 | 158.12 | 37.44 | 0.72979 | 3050 | 805.34 | 1011 | 596.28 | 1.118 | 1.05264 |
| 940 | 226.11 | 9.834 | 161.68 | 35.41 | 0.73509 | 3100 | 820.03 | 1083 | 607.53 | 1.060 | 1.05741 |
| 960 | 231.06 | 10.61 | 165.26 | 33.52 | 0.74030 | 3150 | 834.75 | 1161 | 618.82 | 1.006 | 1.06212 |
| 980 | 236.02 | 11.43 | 168.83 | 31.76 | 0.74540 | 3200 | 849.48 | 1242 | 630.12 | 0.955 | 1.06676 |
| 1000 | 240.98 | 12.30 | 172.43 | 30.12 | 0.75042 | 3250 | 864.24 | 1328 | 641.46 | 0.907 | 1.07134 |
| 1040 | 250.95 | 14.18 | 179.66 | 27.17 | 0.76019 | 3300 | 879.02 | 1418 | 652.81 | 0.8621 | 1.07585 |
| 1080 | 260.97 | 16.28 | 186.93 | 24.58 | 0.76964 | 3350 | 893.83 | 1513 | 664.20 | 0.8202 | 1.08031 |
| 1120 | 271.03 | 18.60 | 194.25 | 22.30 | 0.77880 | 3400 | 908.66 | 1613 | 675.60 | 0.7807 | 1.08470 |
| 1160 | 281.14 | 21.18 | 201.63 | 20.29 | 0.78767 | 3450 | 923.52 | 1719 | 687.04 | 0.7436 | 1.08904 |
| 1200 | 291.30 | 24.01 | 209.05 | 18.51 | 0.79628 | 3500 | 938.40 | 1829 | 698.48 | 0.7087 | 1.09332 |
| 1240 | 301.52 | 27.13 | 216.53 | 16.93 | 0.80466 | 3550 | 953.30 | 1946 | 709.95 | 0.6759 | 1.09755 |
| 1280 | 311.79 | 30.55 | 224.05 | 15.52 | 0.81280 | 3600 | 968.21 | 2068 | 721.44 | 0.6449 | 1.10172 |
| 1320 | 322.11 | 34.31 | 231.63 | 14.25 | 0.82075 | 3650 | 983.15 | 2196 | 732.95 | 0.6157 | 1.10584 |
| 1360 | 332.48 | 38.41 | 239.25 | 13.12 | 0.82848 | 3700 | 998.11 | 2330 | 744.48 | 0.5882 | 1.10991 |
| 1400 | 342.90 | 42.88 | 246.93 | 12.10 | 0.83604 | 3750 | 1013.1 | 2471 | 756.04 | 0.5621 | 1.11393 |
| 1440 | 353.37 | 47.75 | 254.66 | 11.17 | 0.84341 | 3800 | 1028.1 | 2618 | 767.60 | 0.5376 | 1.11791 |
| 1480 | 363.89 | 53.04 | 262.44 | 10.34 | 0.85062 | 3850 | 1043.1 | 2773 | 779.19 | 0.5143 | 1.12183 |
| 1520 | 374.47 | 58.78 | 270.26 | 9.578 | 0.85767 | 3900 | 1058.1 | 2934 | 790.80 | 0.4923 | 1.12571 |
| 1560 | 385.08 | 65.00 | 278.13 | 8.890 | 0.86456 | 3950 | 1073.2 | 3103 | 802.43 | 0.4715 | 1.12955 |

## TABLE A–21E

Ideal-gas properties of air (*Concluded*)

| T R | h Btu/lbm | $P_r$ | u Btu/lbm | $v_r$ | $s°$ Btu/lbm·R | T R | h Btu/lbm | $P_r$ | u Btu/lbm | $v_r$ | $s°$ Btu/lbm·R |
|------|--------|------|--------|--------|---------|------|--------|--------|--------|--------|---------|
| 4000 | 1088.3 | 3280 | 814.06 | 0.4518 | 1.13334 | 4600 | 1270.4 | 6089 | 955.04 | 0.2799 | 1.17575 |
| 4050 | 1103.4 | 3464 | 825.72 | 0.4331 | 1.13709 | 4700 | 1300.9 | 6701 | 978.73 | 0.2598 | 1.18232 |
| 4100 | 1118.5 | 3656 | 837.40 | 0.4154 | 1.14079 | 4800 | 1331.5 | 7362 | 1002.5 | 0.2415 | 1.18876 |
| 4150 | 1133.6 | 3858 | 849.09 | 0.3985 | 1.14446 | 4900 | 1362.2 | 8073 | 1026.3 | 0.2248 | 1.19508 |
| 4200 | 1148.7 | 4067 | 860.81 | 0.3826 | 1.14809 | 5000 | 1392.9 | 8837 | 1050.1 | 0.2096 | 1.20129 |
| 4300 | 1179.0 | 4513 | 884.28 | 0.3529 | 1.15522 | 5100 | 1423.6 | 9658 | 1074.0 | 0.1956 | 1.20738 |
| 4400 | 1209.4 | 4997 | 907.81 | 0.3262 | 1.16221 | 5200 | 1454.4 | 10,539 | 1098.0 | 0.1828 | 1.21336 |
| 4500 | 1239.9 | 5521 | 931.39 | 0.3019 | 1.16905 | 5300 | 1485.3 | 11,481 | 1122.0 | 0.1710 | 1.21923 |

*Note:* The properties $P_r$ (relative pressure) and $v_r$ (relative specific volume) are dimensionless quantities used in the analysis of isentropic processes, and should not be confused with the properties pressure and specific volume.

*Source:* Kenneth Wark, *Thermodynamics*, 4th ed. (New York: McGraw-Hill, 1983), pp. 832–33, Table A–5. Originally published in J. H. Keenan and J. Kaye, *Gas Tables* (New York: John Wiley & Sons, 1948).

## TABLE A–22E

Properties of air at 1 atm pressure

| Temp. $T$, °F | Density $\rho$, lbm/ft$^3$ | Specific Heat $c_p$, Btu/lbm·R | Thermal Conductivity $k$, Btu/h·ft·R | Thermal Diffusivity $\alpha$, ft$^2$/s | Dynamic Viscosity $\mu$, lbm/ft·s | Kinematic Viscosity $\nu$, ft$^2$/s | Prandtl Number Pr |
|---|---|---|---|---|---|---|---|
| −300 | 0.24844 | 0.5072 | 0.00508 | $1.119 \times 10^{-5}$ | $4.039 \times 10^{-6}$ | $1.625 \times 10^{-5}$ | 1.4501 |
| −200 | 0.15276 | 0.2247 | 0.00778 | $6.294 \times 10^{-5}$ | $6.772 \times 10^{-6}$ | $4.433 \times 10^{-5}$ | 0.7042 |
| −100 | 0.11029 | 0.2360 | 0.01037 | $1.106 \times 10^{-4}$ | $9.042 \times 10^{-6}$ | $8.197 \times 10^{-5}$ | 0.7404 |
| −50 | 0.09683 | 0.2389 | 0.01164 | $1.397 \times 10^{-4}$ | $1.006 \times 10^{-5}$ | $1.039 \times 10^{-4}$ | 0.7439 |
| 0 | 0.08630 | 0.2401 | 0.01288 | $1.726 \times 10^{-4}$ | $1.102 \times 10^{-5}$ | $1.278 \times 10^{-4}$ | 0.7403 |
| 10 | 0.08446 | 0.2402 | 0.01312 | $1.797 \times 10^{-4}$ | $1.121 \times 10^{-5}$ | $1.328 \times 10^{-4}$ | 0.7391 |
| 20 | 0.08270 | 0.2403 | 0.01336 | $1.868 \times 10^{-4}$ | $1.140 \times 10^{-5}$ | $1.379 \times 10^{-4}$ | 0.7378 |
| 30 | 0.08101 | 0.2403 | 0.01361 | $1.942 \times 10^{-4}$ | $1.158 \times 10^{-5}$ | $1.430 \times 10^{-4}$ | 0.7365 |
| 40 | 0.07939 | 0.2404 | 0.01385 | $2.016 \times 10^{-4}$ | $1.176 \times 10^{-5}$ | $1.482 \times 10^{-4}$ | 0.7350 |
| 50 | 0.07783 | 0.2404 | 0.01409 | $2.092 \times 10^{-4}$ | $1.194 \times 10^{-5}$ | $1.535 \times 10^{-4}$ | 0.7336 |
| 60 | 0.07633 | 0.2404 | 0.01433 | $2.169 \times 10^{-4}$ | $1.212 \times 10^{-5}$ | $1.588 \times 10^{-4}$ | 0.7321 |
| 70 | 0.07489 | 0.2404 | 0.01457 | $2.248 \times 10^{-4}$ | $1.230 \times 10^{-5}$ | $1.643 \times 10^{-4}$ | 0.7306 |
| 80 | 0.07350 | 0.2404 | 0.01481 | $2.328 \times 10^{-4}$ | $1.247 \times 10^{-5}$ | $1.697 \times 10^{-4}$ | 0.7290 |
| 90 | 0.07217 | 0.2404 | 0.01505 | $2.409 \times 10^{-4}$ | $1.265 \times 10^{-5}$ | $1.753 \times 10^{-4}$ | 0.7275 |
| 100 | 0.07088 | 0.2405 | 0.01529 | $2.491 \times 10^{-4}$ | $1.281 \times 10^{-5}$ | $1.809 \times 10^{-4}$ | 0.7260 |
| 110 | 0.06963 | 0.2405 | 0.01552 | $2.575 \times 10^{-4}$ | $1.299 \times 10^{-5}$ | $1.866 \times 10^{-4}$ | 0.7245 |
| 120 | 0.06843 | 0.2405 | 0.01576 | $2.660 \times 10^{-4}$ | $1.316 \times 10^{-5}$ | $1.923 \times 10^{-4}$ | 0.7230 |
| 130 | 0.06727 | 0.2405 | 0.01599 | $2.746 \times 10^{-4}$ | $1.332 \times 10^{-5}$ | $1.981 \times 10^{-4}$ | 0.7216 |
| 140 | 0.06615 | 0.2406 | 0.01623 | $2.833 \times 10^{-4}$ | $1.349 \times 10^{-5}$ | $2.040 \times 10^{-4}$ | 0.7202 |
| 150 | 0.06507 | 0.2406 | 0.01646 | $2.921 \times 10^{-4}$ | $1.365 \times 10^{-5}$ | $2.099 \times 10^{-4}$ | 0.7188 |
| 160 | 0.06402 | 0.2406 | 0.01669 | $3.010 \times 10^{-4}$ | $1.382 \times 10^{-5}$ | $2.159 \times 10^{-4}$ | 0.7174 |
| 170 | 0.06300 | 0.2407 | 0.01692 | $3.100 \times 10^{-4}$ | $1.398 \times 10^{-5}$ | $2.220 \times 10^{-4}$ | 0.7161 |
| 180 | 0.06201 | 0.2408 | 0.01715 | $3.191 \times 10^{-4}$ | $1.414 \times 10^{-5}$ | $2.281 \times 10^{-4}$ | 0.7148 |
| 190 | 0.06106 | 0.2408 | 0.01738 | $3.284 \times 10^{-4}$ | $1.430 \times 10^{-5}$ | $2.343 \times 10^{-4}$ | 0.7136 |
| 200 | 0.06013 | 0.2409 | 0.01761 | $3.377 \times 10^{-4}$ | $1.446 \times 10^{-5}$ | $2.406 \times 10^{-4}$ | 0.7124 |
| 250 | 0.05590 | 0.2415 | 0.01874 | $3.857 \times 10^{-4}$ | $1.524 \times 10^{-5}$ | $2.727 \times 10^{-4}$ | 0.7071 |
| 300 | 0.05222 | 0.2423 | 0.01985 | $4.358 \times 10^{-4}$ | $1.599 \times 10^{-5}$ | $3.063 \times 10^{-4}$ | 0.7028 |
| 350 | 0.04899 | 0.2433 | 0.02094 | $4.879 \times 10^{-4}$ | $1.672 \times 10^{-5}$ | $3.413 \times 10^{-4}$ | 0.6995 |
| 400 | 0.04614 | 0.2445 | 0.02200 | $5.419 \times 10^{-4}$ | $1.743 \times 10^{-5}$ | $3.777 \times 10^{-4}$ | 0.6971 |
| 450 | 0.04361 | 0.2458 | 0.02305 | $5.974 \times 10^{-4}$ | $1.812 \times 10^{-5}$ | $4.154 \times 10^{-4}$ | 0.6953 |
| 500 | 0.04134 | 0.2472 | 0.02408 | $6.546 \times 10^{-4}$ | $1.878 \times 10^{-5}$ | $4.544 \times 10^{-4}$ | 0.6942 |
| 600 | 0.03743 | 0.2503 | 0.02608 | $7.732 \times 10^{-4}$ | $2.007 \times 10^{-5}$ | $5.361 \times 10^{-4}$ | 0.6934 |
| 700 | 0.03421 | 0.2535 | 0.02800 | $8.970 \times 10^{-4}$ | $2.129 \times 10^{-5}$ | $6.225 \times 10^{-4}$ | 0.6940 |
| 800 | 0.03149 | 0.2568 | 0.02986 | $1.025 \times 10^{-3}$ | $2.247 \times 10^{-5}$ | $7.134 \times 10^{-4}$ | 0.6956 |
| 900 | 0.02917 | 0.2599 | 0.03164 | $1.158 \times 10^{-3}$ | $2.359 \times 10^{-5}$ | $8.087 \times 10^{-4}$ | 0.6978 |
| 1000 | 0.02718 | 0.2630 | 0.03336 | $1.296 \times 10^{-3}$ | $2.467 \times 10^{-5}$ | $9.080 \times 10^{-4}$ | 0.7004 |
| 1500 | 0.02024 | 0.2761 | 0.04106 | $2.041 \times 10^{-3}$ | $2.957 \times 10^{-5}$ | $1.460 \times 10^{-3}$ | 0.7158 |
| 2000 | 0.01613 | 0.2855 | 0.04752 | $2.867 \times 10^{-3}$ | $3.379 \times 10^{-5}$ | $2.095 \times 10^{-3}$ | 0.7308 |
| 2500 | 0.01340 | 0.2922 | 0.05309 | $3.765 \times 10^{-3}$ | $3.750 \times 10^{-5}$ | $2.798 \times 10^{-3}$ | 0.7432 |
| 3000 | 0.01147 | 0.2972 | 0.05811 | $4.737 \times 10^{-3}$ | $4.082 \times 10^{-5}$ | $3.560 \times 10^{-3}$ | 0.7516 |
| 3500 | 0.01002 | 0.3010 | 0.06293 | $5.797 \times 10^{-3}$ | $4.381 \times 10^{-5}$ | $4.373 \times 10^{-3}$ | 0.7543 |
| 4000 | 0.00889 | 0.3040 | 0.06789 | $6.975 \times 10^{-3}$ | $4.651 \times 10^{-5}$ | $5.229 \times 10^{-3}$ | 0.7497 |

*Note:* For ideal gases, the properties $c_p$, $k$, $\mu$, and Pr are independent of pressure. The properties $\rho$, $\nu$, and $\alpha$ at a pressure $P$ (in atm) other than 1 atm are determined by multiplying the values of $\rho$ at the given temperature by $P$ and by dividing $\nu$ and $\alpha$ by $P$.

*Source:* Data generated from the EES software developed by S. A. Klein and F. L. Alvarado. Original sources: Keenan, Chao, Keyes, Gas Tables, Wiley, 1984; and *Thermophysical Properties of Matter*, Vol. 3: *Thermal Conductivity*, Y. S. Touloukian, P. E. Liley, S. C. Saxena, Vol. 11: *Viscosity*, Y. S. Touloukian, S. C. Saxena, and P. Hestermans, IFI/Plenun, NY, 1970, ISBN 0-306067020-8.

## TABLE A–23E

Properties of gases at 1 atm pressure

| Temp. $T$, °F | Density $\rho$, lbm/ft$^3$ | Specific Heat $c_p$, Btu/lbm·R | Thermal Conductivity $k$, Btu/h·ft·R | Thermal Diffusivity $\alpha$, ft$^2$/s | Dynamic Viscosity $\mu$, lbm/ft·s | Kinematic Viscosity $\nu$, ft$^2$/s | Prandtl Number Pr |
|---|---|---|---|---|---|---|---|
| | | | Carbon Dioxide, $CO_2$ | | | | |
| −50 | 0.14712 | 0.1797 | 0.00628 | $6.600 \times 10^{-5}$ | $7.739 \times 10^{-6}$ | $5.261 \times 10^{-5}$ | 0.7970 |
| 0 | 0.13111 | 0.1885 | 0.00758 | $8.522 \times 10^{-5}$ | $8.661 \times 10^{-6}$ | $6.606 \times 10^{-5}$ | 0.7751 |
| 50 | 0.11825 | 0.1965 | 0.00888 | $1.061 \times 10^{-4}$ | $9.564 \times 10^{-6}$ | $8.086 \times 10^{-5}$ | 0.7621 |
| 100 | 0.10769 | 0.2039 | 0.01017 | $1.286 \times 10^{-4}$ | $1.045 \times 10^{-5}$ | $9.703 \times 10^{-5}$ | 0.7543 |
| 200 | 0.09136 | 0.2171 | 0.01273 | $1.784 \times 10^{-4}$ | $1.217 \times 10^{-5}$ | $1.332 \times 10^{-4}$ | 0.7469 |
| 300 | 0.07934 | 0.2284 | 0.01528 | $2.341 \times 10^{-4}$ | $1.382 \times 10^{-5}$ | $1.743 \times 10^{-4}$ | 0.7445 |
| 500 | 0.06280 | 0.2473 | 0.02027 | $3.626 \times 10^{-4}$ | $1.696 \times 10^{-5}$ | $2.700 \times 10^{-4}$ | 0.7446 |
| 1000 | 0.04129 | 0.2796 | 0.03213 | $7.733 \times 10^{-4}$ | $2.381 \times 10^{-5}$ | $5.767 \times 10^{-4}$ | 0.7458 |
| 1500 | 0.03075 | 0.2995 | 0.04281 | $1.290 \times 10^{-3}$ | $2.956 \times 10^{-5}$ | $9.610 \times 10^{-4}$ | 0.7445 |
| 2000 | 0.02450 | 0.3124 | 0.05193 | $1.885 \times 10^{-3}$ | $3.451 \times 10^{-5}$ | $1.408 \times 10^{-3}$ | 0.7474 |
| | | | Carbon Monoxide, CO | | | | |
| −50 | 0.09363 | 0.2571 | 0.01118 | $1.290 \times 10^{-4}$ | $9.419 \times 10^{-6}$ | $1.005 \times 10^{-4}$ | 0.7798 |
| 0 | 0.08345 | 0.2523 | 0.01240 | $1.636 \times 10^{-4}$ | $1.036 \times 10^{-5}$ | $1.242 \times 10^{-4}$ | 0.7593 |
| 50 | 0.07526 | 0.2496 | 0.01359 | $2.009 \times 10^{-4}$ | $1.127 \times 10^{-5}$ | $1.498 \times 10^{-4}$ | 0.7454 |
| 100 | 0.06854 | 0.2484 | 0.01476 | $2.408 \times 10^{-4}$ | $1.214 \times 10^{-5}$ | $1.772 \times 10^{-4}$ | 0.7359 |
| 200 | 0.05815 | 0.2485 | 0.01702 | $3.273 \times 10^{-4}$ | $1.379 \times 10^{-5}$ | $2.372 \times 10^{-4}$ | 0.7247 |
| 300 | 0.05049 | 0.2505 | 0.01920 | $4.217 \times 10^{-4}$ | $1.531 \times 10^{-5}$ | $3.032 \times 10^{-4}$ | 0.7191 |
| 500 | 0.03997 | 0.2567 | 0.02331 | $6.311 \times 10^{-4}$ | $1.802 \times 10^{-5}$ | $4.508 \times 10^{-4}$ | 0.7143 |
| 1000 | 0.02628 | 0.2732 | 0.03243 | $1.254 \times 10^{-3}$ | $2.334 \times 10^{-5}$ | $8.881 \times 10^{-4}$ | 0.7078 |
| 1500 | 0.01957 | 0.2862 | 0.04049 | $2.008 \times 10^{-3}$ | $2.766 \times 10^{-5}$ | $1.413 \times 10^{-3}$ | 0.7038 |
| 2000 | 0.01559 | 0.2958 | 0.04822 | $2.903 \times 10^{-3}$ | $3.231 \times 10^{-5}$ | $2.072 \times 10^{-3}$ | 0.7136 |
| | | | Methane, $CH_4$ | | | | |
| −50 | 0.05363 | 0.5335 | 0.01401 | $1.360 \times 10^{-4}$ | $5.861 \times 10^{-6}$ | $1.092 \times 10^{-4}$ | 0.8033 |
| 0 | 0.04779 | 0.5277 | 0.01616 | $1.780 \times 10^{-4}$ | $6.506 \times 10^{-6}$ | $1.361 \times 10^{-4}$ | 0.7649 |
| 50 | 0.04311 | 0.5320 | 0.01839 | $2.228 \times 10^{-4}$ | $7.133 \times 10^{-6}$ | $1.655 \times 10^{-4}$ | 0.7428 |
| 100 | 0.03925 | 0.5433 | 0.02071 | $2.698 \times 10^{-4}$ | $7.742 \times 10^{-6}$ | $1.972 \times 10^{-4}$ | 0.7311 |
| 200 | 0.03330 | 0.5784 | 0.02559 | $3.690 \times 10^{-4}$ | $8.906 \times 10^{-6}$ | $2.674 \times 10^{-4}$ | 0.7245 |
| 300 | 0.02892 | 0.6226 | 0.03077 | $4.748 \times 10^{-4}$ | $1.000 \times 10^{-5}$ | $3.457 \times 10^{-4}$ | 0.7283 |
| 500 | 0.02289 | 0.7194 | 0.04195 | $7.075 \times 10^{-4}$ | $1.200 \times 10^{-5}$ | $5.244 \times 10^{-4}$ | 0.7412 |
| 1000 | 0.01505 | 0.9438 | 0.07346 | $1.436 \times 10^{-3}$ | $1.620 \times 10^{-5}$ | $1.076 \times 10^{-3}$ | 0.7491 |
| 1500 | 0.01121 | 1.1162 | 0.10766 | $2.390 \times 10^{-3}$ | $1.974 \times 10^{-5}$ | $1.760 \times 10^{-3}$ | 0.7366 |
| 2000 | 0.00893 | 1.2419 | 0.14151 | $3.544 \times 10^{-3}$ | $2.327 \times 10^{-5}$ | $2.605 \times 10^{-3}$ | 0.7353 |
| | | | Hydrogen, $H_2$ | | | | |
| −50 | 0.00674 | 3.0603 | 0.08246 | $1.110 \times 10^{-3}$ | $4.969 \times 10^{-6}$ | $7.373 \times 10^{-4}$ | 0.6638 |
| 0 | 0.00601 | 3.2508 | 0.09049 | $1.287 \times 10^{-3}$ | $5.381 \times 10^{-6}$ | $8.960 \times 10^{-4}$ | 0.6960 |
| 50 | 0.00542 | 3.3553 | 0.09818 | $1.500 \times 10^{-3}$ | $5.781 \times 10^{-6}$ | $1.067 \times 10^{-3}$ | 0.7112 |
| 100 | 0.00493 | 3.4118 | 0.10555 | $1.742 \times 10^{-3}$ | $6.167 \times 10^{-6}$ | $1.250 \times 10^{-3}$ | 0.7177 |
| 200 | 0.00419 | 3.4549 | 0.11946 | $2.295 \times 10^{-3}$ | $6.911 \times 10^{-6}$ | $1.652 \times 10^{-3}$ | 0.7197 |
| 300 | 0.00363 | 3.4613 | 0.13241 | $2.924 \times 10^{-3}$ | $7.622 \times 10^{-6}$ | $2.098 \times 10^{-3}$ | 0.7174 |
| 500 | 0.00288 | 3.4572 | 0.15620 | $4.363 \times 10^{-3}$ | $8.967 \times 10^{-6}$ | $3.117 \times 10^{-3}$ | 0.7146 |
| 1000 | 0.00189 | 3.5127 | 0.20989 | $8.776 \times 10^{-3}$ | $1.201 \times 10^{-5}$ | $6.354 \times 10^{-3}$ | 0.7241 |
| 1500 | 0.00141 | 3.6317 | 0.26381 | $1.432 \times 10^{-2}$ | $1.477 \times 10^{-5}$ | $1.048 \times 10^{-2}$ | 0.7323 |
| 2000 | 0.00112 | 3.7656 | 0.31923 | $2.098 \times 10^{-2}$ | $1.734 \times 10^{-5}$ | $1.544 \times 10^{-2}$ | 0.7362 |
| | | | Nitrogen, $N_2$ | | | | |
| −50 | 0.09364 | 0.2320 | 0.01176 | $1.504 \times 10^{-4}$ | $9.500 \times 10^{-6}$ | $1.014 \times 10^{-4}$ | 0.6746 |
| 0 | 0.08346 | 0.2441 | 0.01300 | $1.773 \times 10^{-4}$ | $1.043 \times 10^{-5}$ | $1.251 \times 10^{-4}$ | 0.7056 |
| 50 | 0.07527 | 0.2480 | 0.01420 | $2.113 \times 10^{-4}$ | $1.134 \times 10^{-5}$ | $1.507 \times 10^{-4}$ | 0.7133 |

**TABLE A–23E**

Properties of gases at 1 atm pressure *(Concluded)*

| Temp. $T$, °F | Density $\rho$, lbm/ft$^3$ | Specific Heat $c_p$, Btu/lbm·R | Thermal Conductivity $k$, Btu/h·ft·R | Thermal Diffusivity $\alpha$, ft$^2$/s | Dynamic Viscosity $\mu$, lbm/ft·s | Kinematic Viscosity $\nu$, ft$^2$/s | Prandtl Number Pr |
|---|---|---|---|---|---|---|---|
| 100 | 0.06854 | 0.2489 | 0.01537 | $2.502 \times 10^{-4}$ | $1.221 \times 10^{-5}$ | $1.783 \times 10^{-4}$ | 0.7126 |
| 200 | 0.05815 | 0.2487 | 0.01760 | $3.379 \times 10^{-4}$ | $1.388 \times 10^{-5}$ | $2.387 \times 10^{-4}$ | 0.7062 |
| 300 | 0.05050 | 0.2492 | 0.01970 | $4.349 \times 10^{-4}$ | $1.543 \times 10^{-5}$ | $3.055 \times 10^{-4}$ | 0.7025 |
| 500 | 0.03997 | 0.2535 | 0.02359 | $6.466 \times 10^{-4}$ | $1.823 \times 10^{-5}$ | $4.559 \times 10^{-4}$ | 0.7051 |
| 1000 | 0.02628 | 0.2697 | 0.03204 | $1.255 \times 10^{-3}$ | $2.387 \times 10^{-5}$ | $9.083 \times 10^{-4}$ | 0.7232 |
| 1500 | 0.01958 | 0.2831 | 0.04002 | $2.006 \times 10^{-3}$ | $2.829 \times 10^{-5}$ | $1.445 \times 10^{-3}$ | 0.7202 |
| 2000 | 0.01560 | 0.2927 | 0.04918 | $2.992 \times 10^{-3}$ | $3.212 \times 10^{-5}$ | $2.059 \times 10^{-3}$ | 0.6882 |
| | | | Oxygen, $O_2$ | | | | |
| −50 | 0.10697 | 0.2331 | 0.01216 | $1.355 \times 10^{-4}$ | $1.104 \times 10^{-5}$ | $1.032 \times 10^{-4}$ | 0.7622 |
| 0 | 0.09533 | 0.2245 | 0.01346 | $1.747 \times 10^{-4}$ | $1.218 \times 10^{-5}$ | $1.277 \times 10^{-4}$ | 0.7312 |
| 50 | 0.08598 | 0.2209 | 0.01475 | $2.157 \times 10^{-4}$ | $1.326 \times 10^{-5}$ | $1.543 \times 10^{-4}$ | 0.7152 |
| 100 | 0.07830 | 0.2200 | 0.01601 | $2.582 \times 10^{-4}$ | $1.429 \times 10^{-5}$ | $1.826 \times 10^{-4}$ | 0.7072 |
| 200 | 0.06643 | 0.2221 | 0.01851 | $3.484 \times 10^{-4}$ | $1.625 \times 10^{-5}$ | $2.446 \times 10^{-4}$ | 0.7020 |
| 300 | 0.05768 | 0.2262 | 0.02096 | $4.463 \times 10^{-4}$ | $1.806 \times 10^{-5}$ | $3.132 \times 10^{-4}$ | 0.7018 |
| 500 | 0.04566 | 0.2352 | 0.02577 | $6.665 \times 10^{-4}$ | $2.139 \times 10^{-5}$ | $4.685 \times 10^{-4}$ | 0.7029 |
| 1000 | 0.03002 | 0.2520 | 0.03698 | $1.357 \times 10^{-3}$ | $2.855 \times 10^{-5}$ | $9.509 \times 10^{-4}$ | 0.7005 |
| 1500 | 0.02236 | 0.2626 | 0.04701 | $2.224 \times 10^{-3}$ | $3.474 \times 10^{-5}$ | $1.553 \times 10^{-3}$ | 0.6985 |
| 2000 | 0.01782 | 0.2701 | 0.05614 | $3.241 \times 10^{-3}$ | $4.035 \times 10^{-5}$ | $2.265 \times 10^{-3}$ | 0.6988 |
| | | | Water Vapor, $H_2O$ | | | | |
| −50 | 0.06022 | 0.4512 | 0.00797 | $8.153 \times 10^{-5}$ | $4.933 \times 10^{-6}$ | $8.192 \times 10^{-5}$ | 1.0050 |
| 0 | 0.05367 | 0.4484 | 0.00898 | $1.036 \times 10^{-4}$ | $5.592 \times 10^{-6}$ | $1.041 \times 10^{-4}$ | 1.0049 |
| 50 | 0.04841 | 0.4472 | 0.01006 | $1.291 \times 10^{-4}$ | $6.261 \times 10^{-6}$ | $1.293 \times 10^{-4}$ | 1.0018 |
| 100 | 0.04408 | 0.4473 | 0.01121 | $1.579 \times 10^{-4}$ | $6.942 \times 10^{-6}$ | $1.574 \times 10^{-4}$ | 0.9969 |
| 200 | 0.03740 | 0.4503 | 0.01372 | $2.263 \times 10^{-4}$ | $8.333 \times 10^{-6}$ | $2.228 \times 10^{-4}$ | 0.9845 |
| 300 | 0.03248 | 0.4557 | 0.01648 | $3.093 \times 10^{-4}$ | $9.756 \times 10^{-6}$ | $3.004 \times 10^{-4}$ | 0.9713 |
| 500 | 0.02571 | 0.4707 | 0.02267 | $5.204 \times 10^{-4}$ | $1.267 \times 10^{-5}$ | $4.931 \times 10^{-4}$ | 0.9475 |
| 1000 | 0.01690 | 0.5167 | 0.04134 | $1.314 \times 10^{-3}$ | $2.014 \times 10^{-5}$ | $1.191 \times 10^{-3}$ | 0.9063 |
| 1500 | 0.01259 | 0.5625 | 0.06315 | $2.477 \times 10^{-3}$ | $2.742 \times 10^{-5}$ | $2.178 \times 10^{-3}$ | 0.8793 |
| 2000 | 0.01003 | 0.6034 | 0.08681 | $3.984 \times 10^{-3}$ | $3.422 \times 10^{-5}$ | $3.411 \times 10^{-3}$ | 0.8563 |

*Note:* For ideal gases, the properties $c_p$, $k$, $\mu$, and Pr are independent of pressure. The properties $\rho$, $\nu$, and $\alpha$ at a pressure $P$ (in atm) other than 1 atm are determined by multiplying the values of $\rho$ at the given temperature by $P$ and by dividing $\nu$ and $\alpha$ by $P$.

*Source:* Data generated from the EES software developed by S. A. Klein and F. L. Alvarado. Originally based on various sources.

## TABLE A–24E

Properties of solid metals

| Composition | Melting Point, R | $\rho$ lbm/ft³ | $c_p$(Btu/ lbm·R) | k (Btu/ h·ft·R) | $\alpha \times 10^6$ ft²/s | Properties at Various Temperatures (R), k (Btu/h·ft·R)/$c_p$(Btu/lbm·R) | | | | | |
|---|---|---|---|---|---|---|---|---|---|---|---|
| | | | | | | 180 | 360 | 720 | 1080 | 1440 | 1800 |
| Aluminum | 1679 | 168 | 0.216 | 137 | 1045 | 174.5 | 137 | 138.6 | 133.4 | 126 | |
| Pure | | | | | | 0.115 | 0.191 | 0.226 | 0.246 | 0.273 | |
| Alloy 2024-T6 | 1395 | 173 | 0.209 | 102.3 | 785.8 | 37.6 | 94.2 | 107.5 | 107.5 | | |
| (4.5% Cu, 1.5% Mg, | | | | | | | | | | | |
| 0.6% Mn) | | | | | | | 0.113 | 0.188 | 0.22 | 0.249 | |
| Alloy 195, cast (4.5% Cu) | | 174.2 | 0.211 | 97 | 734 | | | 100.5 | 106.9 | | |
| Beryllium | 2790 | 115.5 | 0.436 | 115.6 | 637.2 | 572 | 174 | 93 | 72.8 | 61.3 | 52.5 |
| | | | | | | 0.048 | 0.266 | 0.523 | 0.621 | 0.624 | 0.72 |
| Bismuth | 981 | 610.5 | 0.029 | 4.6 | 71 | 9.5 | 5.6 | 4.06 | | | |
| | | | | | | 0.026 | 0.028 | 0.03 | | | |
| Boron | 4631 | 156 | 0.264 | 15.6 | 105 | 109.7 | 32.06 | 9.7 | 6.1 | 5.5 | 5.7 |
| | | | | | | 0.03 | 0.143 | 0.349 | 0.451 | 0.515 | 0.558 |
| Cadmium | 1069 | 540 | 0.055 | 55.6 | 521 | 117.3 | 57.4 | 54.7 | | | |
| | | | | | | 0.047 | 0.053 | 0.057 | | | |
| Chromium | 3812 | 447 | 0.107 | 54.1 | 313.2 | 91.9 | 64.1 | 52.5 | 46.6 | 41.2 | 37.8 |
| | | | | | | 0.045 | 0.091 | 0.115 | 0.129 | 0.138 | 0.147 |
| Cobalt | 3184 | 553.2 | 0.101 | 57.3 | 286.3 | 96.5 | 70.5 | 49.3 | 39 | 33.6 | 80.1 |
| | | | | | | 0.056 | 0.09 | 0.107 | 0.12 | 0.131 | 0.145 |
| Copper | 2445 | 559 | 0.092 | 231.7 | 1259.3 | 278.5 | 238.6 | 227.07 | 219 | 212 | 203.4 |
| Pure | | | | | | 0.06 | 0.085 | 0.094 | 0.01 | 0.103 | 0.107 |
| Commercial bronze | 2328 | 550 | 0.1 | 30 | 150.7 | 24.3 | 30 | 34 | | | |
| (90% Cu, 10% Al) | | | | | | | 0.187 | 0.109 | 0.130 | | |
| Phosphor gear bronze | 1987 | 548.1 | 0.084 | 31.2 | 183 | 23.7 | 37.6 | 42.8 | | | |
| (89% Cu, 11% Sn) | | | | | | — | — | — | | | |
| Cartridge brass | 2139 | 532.5 | 0.09 | 63.6 | 364.9 | 43.3 | 54.9 | 79.2 | 86.0 | | |
| (70% Cu, 30% Zn) | | | | | | | 0.09 | 0.09 | 0.101 | | |
| Constantan | 2687 | 557 | 0.092 | 13.3 | 72.3 | 9.8 | 1.1 | | | | |
| (55% Cu, 45% Ni) | | | | | | 0.06 | 0.09 | | | | |
| Germanium | 2180 | 334.6 | 0.08 | 34.6 | 373.5 | 134 | 56 | 25 | 15.7 | 11.4 | 10.05 |
| | | | | | | 0.045 | 0.069 | 0.08 | 0.083 | 0.085 | 0.089 |
| Gold | 2405 | 1205 | 0.03 | 183.2 | 1367 | 189 | 186.6 | 179.7 | 172.2 | 164.09 | 156 |
| | | | | | | 0.026 | 0.029 | 0.031 | 0.032 | 0.033 | 0.034 |
| Iridium | 4896 | 1404.6 | 0.031 | 85 | 541.4 | 99.4 | 88.4 | 83.2 | 79.7 | 76.3 | 72.8 |
| | | | | | | 0.021 | 0.029 | 0.031 | 0.032 | 0.034 | 0.036 |
| Iron: | 3258 | 491.3 | 0.106 | 46.4 | 248.6 | 77.4 | 54.3 | 40.2 | 31.6 | 25.01 | 19 |
| Pure | | | | | | 0.051 | 0.091 | 0.117 | 0.137 | 0.162 | 0.232 |
| Armco | | 491.3 | 0.106 | 42 | 222.8 | 55.2 | 46.6 | 38 | 30.7 | 24.4 | 18.7 |
| (99.75% pure) | | | | | | 0.051 | 0.091 | 0.117 | 0.137 | 0.162 | 0.233 |
| Carbon steels | | 490.3 | 0.103 | 35 | 190.6 | | | 32.8 | 27.7 | 22.7 | 17.4 |
| Plain carbon (Mn ≤ 1%, | | | | | | | | | | | |
| Si ≤ 0.1%) | | | | | | | 0.116 | 0.113 | 0.163 | 0.279 | |
| AISI 1010 | | 489 | 0.103 | 37 | 202.4 | | | 33.9 | 28.2 | 22.7 | 18 |
| | | | | | | | | 0.116 | 0.133 | 0.163 | 0.278 |
| Carbon-silicon | | 488 | 0.106 | 30 | 160.4 | | | 28.8 | 25.4 | 21.6 | 17 |
| (Mn ≤ 1%, | | | | | | | | | | | |
| 0.1% < Si ≤ 0.6%) | | | | | | | | 0.119 | 0.139 | 0.166 | 0.231 |
| Carbon-manganese-silicon | | 508 | 0.104 | 23.7 | 125 | | | 24.4 | 23 | 20.2 | 16 |
| (1% < Mn ≤ 1.65%, | | | | | | | | 0.116 | 0.133 | 0.163 | 0.260 |
| 0.1% < Si ≤ 0.6%) | | | | | | | | | | | |
| Chromium (low) steels: | | 488.3 | 0.106 | 21.8 | 117.4 | | | 22 | 21.2 | 19.3 | 15.6 |
| $\frac{1}{2}$ Cr–$\frac{1}{4}$ Mo–Si | | | | | | | | 0.117 | 0.137 | 0.164 | 0.23 |
| (0.18% C, 0.65% Cr, | | | | | | | | | | | |
| 0.23% Mo, 0.6% Si)1 | | | | | | | | | | | |
| 1 Cr–$\frac{1}{2}$ Mo (0.16% C, | | 490.6 | 0.106 | 24.5 | 131.3 | | | 24.3 | 22.6 | 20 | 15.8 |
| 1% Cr, 0.54% Mo, 0.39% Si) | | | | | | | | 0.117 | 0.137 | 0.164 | 0.231 |
| 1 Cr–V | | 489.2 | 0.106 | 28.3 | 151.8 | | | 27.0 | 24.3 | 21 | 16.3 |
| (0.2% C, 1.02% Cr, 0.15% V) | | | | | | | | 0.117 | 0.137 | 0.164 | 0.231 |

## TABLE A–24E

Properties of solid metals (Concluded)

| Composition | Melting Point, R | $\rho$ lbm/ft$^3$ | $c_p$(Btu/ lbm·R) | $k$ (Btu/ h·ft·R) | $\alpha \times 10^6$ ft$^2$/s | Properties at Various Temperatures (R), $k$ (Btu/h·ft·R)/$c_p$(Btu/lbm·R) | | | | | |
|---|---|---|---|---|---|---|---|---|---|---|---|
| | | Properties at 540 R | | | | 180 | 360 | 720 | 1080 | 1440 | 1800 |
| Stainless steels: | | 503 | 0.114 | 8.7 | 42 | | | 10 | 11.6 | 13.2 | 14.7 |
| AISI 302 | | | | | | | | 0.122 | 0.133 | 0.140 | 0.144 |
| AISI 304 | 3006 | 493.2 | 0.114 | 8.6 | 42.5 | 5.31 | 7.3 | 9.6 | 11.5 | 13 | 14.7 |
| | | | | | | 0.064 | 0.096 | 0.123 | 0.133 | 0.139 | 0.145 |
| AISI 316 | | 514.3 | 0.111 | 7.8 | 37.5 | | | 8.8 | 10.6 | 12.3 | 14 |
| | | | | | | 0.12 | 0.131 | 0.137 | 0.143 | | |
| AISI 347 | | 498 | 0.114 | 8.2 | 40 | | | 9.1 | 1.1 | 12.7 | 14.3 |
| | | | | | | | | 0.122 | 0.133 | 0.14 | 0.144 |
| Lead | 1082 | 708 | 0.03 | 20.4 | 259.4 | 23 | 21.2 | 19.7 | 18.1 | | |
| | | | | | | 0.028 | 0.029 | 0.031 | 0.034 | | |
| Magnesium | 1661 | 109 | 0.245 | 90.2 | 943 | 87.9 | 91.9 | 88.4 | 86.0 | 84.4 | |
| | | | | | | 0.155 | 0.223 | 0.256 | 0.279 | 0.302 | |
| Molybdenum | 5209 | 639.3 | 0.06 | 79.7 | 578 | 1034 | 82.6 | 77.4 | 72.8 | 68.2 | 64.7 |
| | | | | | | 0.038 | 0.053 | 0.062 | 0.065 | 0.068 | 0.070 |
| Nickel: | 3110 | 555.6 | 0.106 | 52.4 | 247.6 | 94.8 | 61.8 | 46.3 | 37.9 | 39 | 41.4 |
| Pure | | | | | | 0.055 | 0.091 | 0.115 | 0.141 | 0.126 | 0.134 |
| Nichrome | 3010 | 524.4 | 0.1 | 6.9 | 36.6 | | | 8.0 | 9.3 | 12.2 | |
| (80% Ni, 20% Cr) | | | | | | | 0.114 | 0.125 | 0.130 | | |
| Inconel X-750 | 2997 | 531.3 | 0.104 | 6.8 | 33.4 | 5 | 5.9 | 7.8 | 9.8 | 11.8 | 13.9 |
| (73% Ni, 15% Cr, 6.7% Fe) | | | | | | — | 0.088 | 0.112 | 0.121 | 0.13 | 0.149 |
| Niobium | 4934 | 535 | 0.063 | 31 | 254 | 31.9 | 30.4 | 32 | 33.6 | 35.4 | 32.2 |
| | | | | | | 0.044 | 0.059 | 0.065 | 0.067 | 0.069 | 0.071 |
| Palladium | 3289 | 750.4 | 0.058 | 41.5 | 263.7 | 44.2 | 41.4 | 42.5 | 46 | 50 | 54.4 |
| | | | | | | 0.04 | 0.054 | 0.059 | 0.062 | 0.064 | 0.067 |
| Platinum: | 3681 | 1339 | 0.031 | 41.4 | 270 | 44.7 | 42 | 41.5 | 42.3 | 43.7 | 45.5 |
| Pure | | | | | | 0.024 | 0.03 | 0.032 | 0.034 | 0.035 | 0.036 |
| Alloy 60Pt-40Rh | 3240 | 1038.2 | 0.038 | 27.2 | 187.3 | | | 30 | 34 | 37.5 | 40 |
| (60% Pt, 40% Rh) | | | | | | | | — | — | — | |
| Rhenium | 6215 | 1317.2 | 0.032 | 27.7 | 180 | 34 | 30 | 26.6 | 25.5 | 25.4 | 25.8 |
| | | | | | | 0.023 | 0.03 | 0.033 | 0.034 | 0.036 | 0.037 |
| Rhodium | 4025 | 777.2 | 0.058 | 86.7 | 534 | 107.5 | 89 | 84.3 | 78.5 | 73.4 | 70 |
| | | | | | | 0.035 | 0.052 | 0.06 | 0.065 | 0.069 | 0.074 |
| Silicon | 3033 | 145.5 | 0.17 | 85.5 | 960.2 | 510.8 | 152.5 | 57.2 | 35.8 | 24.4 | 18.0 |
| | | | | | | 0.061 | 0.132 | 0.189 | 0.207 | 0.218 | 0.226 |
| Silver | 2223 | 656 | 0.056 | 248 | 1873 | 257 | 248.4 | 245.5 | 238 | 228.8 | 219 |
| | | | | | | 0.044 | 0.053 | 0.057 | 0.059 | 0.062 | 0.066 |
| Tantalum | 5884 | 1036.3 | 0.033 | 33.2 | 266 | 34.2 | 33.2 | 33.4 | 34 | 34.3 | 34.8 |
| | | | | | | 0.026 | 0.031 | 0.034 | 0.035 | 0.036 | 0.036 |
| Thorium | 3641 | 730.4 | 0.028 | 31.2 | 420.9 | 34.6 | 31.5 | 31.4 | 32.2 | 32.9 | 32.9 |
| | | | | | | 0.024 | 0.027 | 0.029 | 0.032 | 0.035 | 0.037 |
| Tin | 909 | 456.3 | 0.054 | 38.5 | 431.6 | 49.2 | 42.4 | 35.9 | | | |
| | | | | | | 0.044 | 0.051 | 0.058 | | | |
| Titanium | 3515 | 281 | 0.013 | 12.7 | 100.3 | 17.6 | 14.2 | 11.8 | 11.2 | 11.4 | 12 |
| | | | | | | 0.071 | 0.111 | 0.131 | 0.141 | 0.151 | 0.161 |
| Tungsten | 6588 | 1204.9 | 0.031 | 100.5 | 735.2 | 120.2 | 107.5 | 92 | 79.2 | 72.2 | 68.2 |
| | | | | | | 0.020 | 0.029 | 0.032 | 0.033 | 0.034 | 0.035 |
| Uranium | 2531 | 1190.5 | 0.027 | 16 | 134.5 | 12.5 | 14.5 | 17.1 | 19.6 | 22.4 | 25.4 |
| | | | | | | 0.022 | 0.026 | 0.029 | 0.035 | 0.042 | 0.043 |
| Vanadium | 3946 | 381 | 0.117 | 17.7 | 110.9 | 20.7 | 18 | 18 | 19.3 | 20.6 | 22.0 |
| | | | | | | 0.061 | 0.102 | 0.123 | 0.128 | 0.134 | 0.142 |
| Zinc | 1247 | 445.7 | 0.093 | 67 | 450 | 67.6 | 68.2 | 64.1 | 59.5 | | |
| | | | | | | 0.07 | 0.087 | 0.096 | 0.104 | | |
| Zirconium | 3825 | 410.2 | 0.067 | 13.1 | 133.5 | 19.2 | 14.6 | 12.5 | 12 | 12.5 | 13.7 |
| | | | | | | 0.049 | 0.063 | 0.072 | 0.77 | 0.082 | 0.087 |

Source: Tables A–24E and A–25E are obtained from the respective tables in SI units in Appendix 1 using proper conversion factors.

## TABLE A–25E

Properties of solid nonmentals

| Composition | Melting Point, R | Properties at 540 R | | | | Properties at Various Temperatures (R), $k$ (Btu/h·ft·R)/$c_p$(Btu/lbm·R) | | | | | |
|---|---|---|---|---|---|---|---|---|---|---|---|
| | | $\rho$ lbm/ft³ | $c_p$(Btu/ lbm·R) | $k$ (Btu/ h·ft·R) | $\alpha \times 10^6$ ft²/s | 180 | 360 | 720 | 1080 | 1440 | 1800 |
| Aluminum oxide, sapphire | 4181 | 247.8 | 0.182 | 26.6 | 162.5 | 260 — | 47.4 — | 18.7 0.224 | 11 0.265 | 7.5 0.281 | 6 0.293 |
| Aluminum oxide polycrystalline | 4181 | 247.8 | 0.182 | 20.8 | 128 | 76.8 | 31.7 | 15.3 0.244 | 9.3 0.265 | 6 0.281 | 4.5 0.293 |
| Beryilium oxide | 4905 | 187.3 | 0.246 | 157.2 | 947.3 | | | 113.2 0.322 | 64.2 0.40 | 40.4 0.44 | 27.2 0.459 |
| Boron | 4631 | 156 | 0.264 | 16 | 107.5 | 109.8 | 30.3 | 10.8 0.355 | 6.5 0.445 | 4.6 0.509 | 3.6 0.561 |
| Boron fiber epoxy (30% vol.) composite | 1062 | 130 | | | | | | | | | |
| $k$, ‖ to fibers | | | | 1.3 | | 1.2 | 1.3 | 1.31 | | | |
| $k$, ⊥ to fibers | | | | 0.34 | | 0.21 | 0.28 | 0.34 | | | |
| $c_p$ | | | 0.268 | | | 0.086 | 0.18 | 0.34 | | | |
| Carbon Amorphous | 2700 | 121.7 | — | 0.92 | — | 0.38 | 0.68 | 1.09 | 1.26 | 1.36 | 1.46 |
| Diamond, type IIa insulator | — | 219 | 0.121 | 1329 | — | 5778 | 2311.2 0.005 | 889.8 0.046 | 0.203 | | |
| Graphite, pyrolytic | 4091 | 138 | | | | | | | | | |
| $k$, ‖ to layers | | | | 1126.7 | | 2871.6 | 1866.3 | 803.2 | 515.4 | 385.4 | 308.5 |
| $k$, ⊥ to layers | | | | 3.3 | | 9.7 | 5.3 | 2.4 | 1.5 | 1.16 | 0.92 |
| $c_p$ | | | 0.169 | | | 0.32 | 0.098 | 0.236 | 0.335 | 0.394 | 0.428 |
| Graphite fiber epoxy (25% vol.) composite | 810 | 87.4 | | | | | | | | | |
| $k$, heat flow ‖ to fibers | | | | 6.4 | | 3.3 | 5.0 | 7.5 | | | |
| $k$, heat flow ⊥ to fibers | | | | 0.5 | 5 | 0.4 | 0.63 | | | | |
| $c_p$ | | | 0.223 | | | 0.08 | 0.153 | 0.29 | | | |
| Pyroceram, Corning 9606 | 2921 | 162.3 | 0.193 | 2.3 | 20.3 | 3.0 | 2.3 | 2.1 | 1.9 | 1.7 | 1.7 |
| Silicon carbide, | 5580 | 197.3 | 0.161 | 283.1 | 2475.7 | | | — 0.210 | — 0.25 | — 0.27 | 50.3 0.285 |
| Silicon dioxide, crystalline (quartz) | 3389 | 165.4 | | | | | | | | | |
| $k$, ‖ to c-axis | | | | 6 | | 22.5 | 9.5 | 4.4 | 2.9 | 2.4 | |
| $k$, ⊥ to c-axis | | | | 3.6 | | 12.0 | 5.9 | 2.7 | 2 | 1.8 | |
| $c_p$ | | | 0.177 | | | | | 0.211 | 0.256 | 0.298 | |
| Silicon dioxide, polycrystalline (fused silica) | 3389 | 138.6 | 0.177 | 0.79 | 9 | 0.4 — | 0.65 — | 0.87 0.216 | 1.01 0.248 | 1.25 0.264 | 1.65 0.276 |
| Silicon nitride | 3911 | 150 | 0.165 | 9.2 | 104 | — — | — 0.138 | 8.0 0.185 | 6.5 0.223 | 5.7 0.253 | 5.0 0.275 |
| Sulfur | 706 | 130 | 0.169 | 0.1 | 1.51 | 0.095 0.962 | 0.1 0.144 | | | | |
| Thorium dioxide | 6431 | 568.7 | 0.561 | 7.5 | 65.7 | | | 5.9 0.609 | 3.8 0.654 | 2.7 0.680 | 2.12 0.704 |
| Titanium dioxide, polycrystalline | 3840 | 259.5 | 0.170 | 4.9 | 30.1 | | | 4.0 0.192 | 2.9 0.210 | 2.3 0.217 | 2 0.222 |

# INDEX

# NOMENCLATURE

| | | | | |
|---|---|---|---|---|
| $a$ | Acceleration, m/s$^2$ | | $G$ | Incident radiation, W/m$^2$ |
| $A_s$ | Surface area, m$^2$ | | Gr | Grashof number |
| $A_c$ | Cross-sectional area, m$^2$ | | $h$ | Convection heat transfer coefficient, W/m$^2$·°C |
| Bi | Biot number | | | |
| $C$ | Speed of sound, m/s | | $h$ | Specific enthalpy, $u + Pv$, kJ/kg |
| $c$ | Specific heat, kJ/kg·K | | $h_c$ | Thermal contact conductance, W/m$^2$·°C |
| $C_c, C_h$ | Heat capacity rate, W/°C | | $h_{fg}$ | Enthalpy of vaporization, kJ/kg |
| $C_D$ | Drag coefficient | | $h_{if}$ | Latent heat of fusion, kJ/kg |
| $C_f$ | Fanning friction factor or friction coefficient | | $h_L$ | Head loss, m |
| $C_L$ | Lift coefficient | | $H$ | Total enthalpy, $U + PV$, kJ |
| $c_p$ | Constant-pressure specific heat, kJ/kg·K | | $I$ | Electric current, A; Bessel function |
| $c_v$ | Constant-volume specific heat, kJ/kg·K | | $J$ | Radiosity, W/m$^2$; Bessel function |
| COP | Coefficient of performance | | $k$ | Specific heat ratio, $c_p/c_v$ |
| COP$_{HP}$ | Coefficient of performance of a heat pump | | $k$ | Thermal conductivity |
| COP$_R$ | Coefficient of performance of a refrigerator | | $k_{eff}$ | Effective thermal conductivity, W/m·°C |
| $d, D$ | Diameter, m | | $k_s$ | Spring constant |
| $D_h$ | Hydraulic diameter, m | | ke | Specific kinetic energy, $V^2/2$, kJ/kg |
| $e$ | Specific total energy, kJ/kg | | $K_L$ | Minor loss coefficient |
| erfc | Complementary error function | | KE | Total kinetic energy, $mV^2/2$, kJ |
| $E$ | Total energy, kJ | | $L$ | Length; half thickness of a plane wall |
| $E_b$ | Blackbody emissive flux, W/m$^2$ | | $L_c$ | Characteristic or corrected length |
| EER | Energy efficiency rating | | $L_h$ | Hydrodynamic entry length |
| $f$ | Darcy friction factor | | $L_t$ | Thermal entry length |
| $f_\lambda$ | Blackbody radiation function | | $m$ | Mass, kg |
| $F$ | Force, N | | $\dot{m}$ | Mass flow rate, kg/s |
| $F_D$ | Drag force, N | | $M$ | Molar mass, kg/kmol |
| $F_{ij}, F_{i \rightarrow j}$ | View factor | | $M$ | Moment of force, N·m |
| $F_L$ | Lift force, N | | Ma | Mach number |
| Fo | Fourier number | | $n$ | Polytropic exponent |
| $g$ | Gravitational acceleration, m/s$^2$ | | $N$ | Number of moles, kmol |

| | | | | |
|---|---|---|---|---|
| NTU | Number of transfer units | | T | Torque, N·m |
| Nu | Nusselt number | | $T_b$ | Bulk fluid temperature, °C |
| $p$ | Perimeter or wetted perimeter, m | | $T_{cr}$ | Critical temperature, K |
| pe | Specific potential energy, $gz$, kJ/kg | | $T_f$ | Film temperature, °C |
| $P$ | Pressure, kPa | | $T_H$ | Temperature of high-temperature body, K |
| $P_{cr}$ | Critical pressure, kPa | | $T_i$ | Initial temperature, °C |
| $P_r$ | Relative pressure | | $T_\infty$ | Free stream or surrounding medium temperature, °C |
| $P_R$ | Reduced pressure | | | |
| $P_v$ | Vapor pressure, kPa | | $T_L$ | Temperature of low-temperature body, K |
| PE | Total potential energy, $mgz$, kJ | | $T_R$ | Reduced temperature |
| Pr | Prandtl number | | $T_s$ | Surface temperature, °C or K |
| $q$ | Heat transfer per unit mass, kJ/kg | | $T_{sat}$ | Saturation temperature, °C |
| $\dot{q}$ | Heat flux, W/m$^2$ | | $u$ | Specific internal energy, kJ/kg |
| $Q$ | Total heat transfer, kJ | | $u,v$ | x- and y-components of velocity, m/s |
| $\dot{Q}$ | Heat transfer rate, kW | | $U$ | Total internal energy, kJ |
| $Q_H$ | Heat transfer with high-temperature body, kJ | | $U$ | Overall heat transfer coefficient, W/m$^2$·°C |
| $Q_L$ | Heat transfer with low-temperature body, kJ | | $v$ | Specific volume, m$^3$/kg |
| $r_{cr}$ | Critical radius of insulation, m | | $v_{cr}$ | Critical specific volume, m$^3$/kg |
| $R$ | Gas constant, kJ/kg·K | | $v_r$ | Relative specific volume |
| $R$ | Radius, m; Thermal resistance, °C/W | | $\vee$ | Total volume, m$^3$ |
| $R_c$ | Thermal contact resistance, m$^2$·°C/W | | $\dot{\vee}$ | Volume flow rate, m$^3$/s |
| $R_f$ | Fouling factor, m$^2$·°C/W | | $V$ | Velocity, m/s |
| $R_u$ | Universal gas constant, kJ/kmol·K | | $V_m$ | Mean velocity, m/s |
| R-value | R-value of insulation | | $V_\infty$ | Free-stream velocity, m/s |
| Ra | Rayleigh number | | $w$ | Work per unit mass, kJ/kg |
| Re | Reynolds number | | $W$ | Total work, kJ |
| $s$ | Specific entropy, kJ/kg·K | | $\dot{W}$ | Power, kW |
| $s_{gen}$ | Specific entropy generation, kJ/kg·K | | $W_{in}$ | Work input, kJ |
| $S$ | Total entropy, kJ/K; Conduction shape factor | | $W_{out}$ | Work output, kJ |
| $S_{gen}$ | Total entropy generation, kJ/K | | $x$ | Quality |
| $t$ | Time, s; Thickness, m | | $z$ | Elevation, m |
| $T$ | Temperature, °C or K | | $Z$ | Compressibility factor |